MW01517670

Contents

GEORGIAN COLLEGE LIBRARY

GFOB-BK
(REF)
#265.00

ASM Handbook®

Volume 22B
Metals Process Simulation

Prepared under the direction of the
ASM International Handbook Committee

D.U. Furrer and **S.L. Semiatin**, Volume Editors

Eileen DeGuire, Content Developer
Steve Lampman, Content Developer
Charles Moosbrugger, Content Developer
Ann Britton, Editorial Assistant
Madrid Tramble, Senior Production Coordinator
Patty Conti, Production Coordinator
Diane Whitelaw, Production Coordinator
Scott D. Henry, Senior Manager, Content Development
Bonnie R. Sanders, Manager of Production

Editorial Assistance
Elizabeth Marquard
Buz Riley

Library Commons
Georgian College
One Georgian Drive
Barrie, ON
L4M 3X9

**The Materials
Information Society**

Materials Park, Ohio 44073-0002
www.asminternational.org

Copyright © 2010
by
ASM International®
All rights reserved

No part of this book may be reproduced, stored in a retrieval system, or transmitted, in any form or by any means, electronic, mechanical, photocopying, recording, or otherwise, without the written permission of the copyright owner.

First printing, September 2010

This book is a collective effort involving hundreds of technical specialists. It brings together a wealth of information from worldwide sources to help scientists, engineers, and technicians solve current and long-range problems.

Great care is taken in the compilation and production of this Volume, but it should be made clear that NO WARRANTIES, EXPRESS OR IMPLIED, INCLUDING, WITHOUT LIMITATION, WARRANTIES OF MERCHANTABILITY OR FITNESS FOR A PARTICULAR PURPOSE, ARE GIVEN IN CONNECTION WITH THIS PUBLICATION. Although this information is believed to be accurate by ASM, ASM cannot guarantee that favorable results will be obtained from the use of this publication alone. This publication is intended for use by persons having technical skill, at their sole discretion and risk. Since the conditions of product or material use are outside of ASM's control, ASM assumes no liability or obligation in connection with any use of this information. No claim of any kind, whether as to products or information in this publication, and whether or not based on negligence, shall be greater in amount than the purchase price of this product or publication in respect of which damages are claimed. THE REMEDY HEREBY PROVIDED SHALL BE THE EXCLUSIVE AND SOLE REMEDY OF BUYER, AND IN NO EVENT SHALL EITHER PARTY BE LIABLE FOR SPECIAL, INDIRECT OR CONSEQUENTIAL DAMAGES WHETHER OR NOT CAUSED BY OR RESULTING FROM THE NEGLIGENCE OF SUCH PARTY. As with any material, evaluation of the material under end-use conditions prior to specification is essential. Therefore, specific testing under actual conditions is recommended.

Nothing contained in this book shall be construed as a grant of any right of manufacture, sale, use, or reproduction, in connection with any method, process, apparatus, product, composition, or system, whether or not covered by letters patent, copyright, or trademark, and nothing contained in this book shall be construed as a defense against any alleged infringement of letters patent, copyright, or trademark, or as a defense against liability for such infringement.

Comments, criticisms, and suggestions are invited, and should be forwarded to ASM International.

Library of Congress Cataloging-in-Publication Data

ASM International

ASM Handbook
Includes bibliographical references and indexes
Contents: v.1. Properties and selection—irons, steels, and high-performance alloys—v.2. Properties and selection—nonferrous alloys and special-purpose materials—[etc.]—v.22B. Metals Process Simulation

1. Metals—Handbooks, manuals, etc. 2. Metal-work—Handbooks, manuals, etc. I. ASM International. Handbook Committee. II. Metals Handbook.
TA459.M43 1990 620.1'6 90-115
SAN: 204-7586

ISBN-13: 978-1-61503-005-7
ISBN-10: 1-61503-005-0

ASM International®
Materials Park, OH 44073-0002
www.asminternational.org

Printed in the United States of America

Multiple copy reprints of individual articles are available from Technical Department, ASM International.

Foreword

Publication of Volume 22B, *Metals Process Simulation*, completes an ambitious undertaking begun in 2007 to compile an all-new, comprehensive reference resource on modeling as it applies to metals processing. The first part, Volume 22A, *Fundamentals of Modeling for Metals Processing*, was published in 2009.

Many of the sections in this Volume will be familiar to *ASM Handbook* users, as they have been covered extensively across the *ASM Handbook* series: phase diagrams, casting and solidification, forming, machining, powder metallurgy, joining, heat treatment, and design. This Volume interprets these subjects in the interdisciplinary context of modeling, simulation, and computational engineering.

The high cost of capital investment in manufacturing can be mitigated by modeling and simulating the options. The effects of processing on materials can be tested and understood through modeling. This Volume and its companion, Volume 22A, provide materials engineers and scientists with the information they need to understand the potential and advantages of modeling and simulation and to provide them with the tools they need to work with the modeling experts.

When the first *ASM Handbook* was published in 1923 by ASM International's predecessor, the American Society for Steel Treaters, the computational tools of choice were a slide rule, paper, pencil, and data tables—all conveniently sized to slip into a lab coat pocket. Today, computational tools are almost entirely software based, although some handheld electronics are also conveniently sized to slip into a lab coat pocket. Many of the basic concerns between then and now are the same: how to control properties during processing, how to minimize waste, how to maintain quality, and so on. Additional contemporary concerns include automated manufacturing, new alloys, new applications such as aerospace and medical devices, environmental responsibility, tracking, and so on.

ASM International is indebted to co-editors David Furrer and S. Lee Semiatin for their vision and leadership in bringing Volumes 22A and 22B to completion. The many authors and reviewers who worked on these Volumes shared that vision. Unlike the subjects about which they wrote, a technical article cannot be modeled or simulated; it must take tangible form as text and images, and this Volume is the direct result of the contributors' generosity in sharing their time and expertise.

That first *ASM Handbook* was published as a loose-leaf collection of data sheets assembled in a leather-bound binder. Today's *ASM Handbooks* are available online, in hardcover, or as DVDs. Times have changed, and ASM International continues to provide the quality information that materials science professionals need to chart the course of the future for their industries.

Frederick J. Lisy
President
ASM International

Stanley C. Theobald
Managing Director
ASM International

Preface

Computer-aided engineering and design have substantially changed the way new products are developed and defined. The pencil and drafting table have long since been replaced by the mouse and computer monitor. To date, much of this engineering transformation has been limited to geometric design, or the *form* and *fit* of a component. Efforts are now ongoing to develop computer-based tools to assess the *function* of components under the intended final application conditions (i.e., temperature, environment, stress, and time).

There have been substantial efforts over the past 25 years to develop and implement computer-based models to simulate manufacturing processes and the evolution of microstructure and accompanying mechanical properties within component materials. The rate of change within this area of engineering has continued to increase with increasing industrial application benefits from the use of such engineering tools, accompanied by the reduced cost and increased speed of computing systems required to perform increasingly complex simulations.

Volumes 22A and 22B of the *ASM Handbook* series summarize models that describe the behavior of metallic materials under processing conditions and describe the development and application of simulation methods for a wide range of materials and manufacturing processes. Such information allows the sharing of best practices among diverse scientific, engineering, and manufacturing disciplines. Background information on fundamental modeling methods detailed in Volume 22A provides the user with a solid foundation of the underlying physics that support many industrial simulation software packages. The present Volume provides an overview of a number of specific metals processing simulation tools applicable in the metals manufacturing industry for a wide range of engineering materials.

All simulation tools require a variety of inputs. For example, details regarding material and process boundary conditions are critical to the success of any computer-based simulation. Thus, this Handbook also provides information regarding material and process boundary conditions that are applicable to manufacturing methods. Additionally, this Volume provides guidance regarding how to develop and assess required thermophysical material data for materials that have not been previously characterized, so practitioners of simulation software packages can effectively generate required material and manufacturing process databases to enable successful predictions for metals processing methods.

The benefits provided by integrated computational materials engineering include reduced component development time, enhanced optimization of component design (design for performance, design for manufacturing, and design for cost), and increased right-the-first-time manufacturing. These benefits have led to an overwhelming pull for materials and manufacturing process simulation integration with early stages of component design.

D.U. Furrer, FASM
Rolls-Royce Corporation

S.L. Semiatin, FASM
Air Force Research Laboratory

Policy on Units of Measure

By a resolution of its Board of Trustees, ASM International has adopted the practice of publishing data in both metric and customary U.S. units of measure. In preparing this Handbook, the editors have attempted to present data in metric units based primarily on Système International d'Unités (SI), with secondary mention of the corresponding values in customary U.S. units. The decision to use SI as the primary system of units was based on the aforementioned resolution of the Board of Trustees and the widespread use of metric units throughout the world.

For the most part, numerical engineering data in the text and in tables are presented in SI-based units with the customary U.S. equivalents in parentheses (text) or adjoining columns (tables). For example, pressure, stress, and strength are shown both in SI units, which are pascals (Pa) with a suitable prefix, and in customary U.S. units, which are pounds per square inch (psi). To save space, large values of psi have been converted to kips per square inch (ksi), where 1 ksi = 1000 psi. The metric tonne (kg $\times$ 10^3) has sometimes been shown in megagrams (Mg). Some strictly scientific data are presented in SI units only.

To clarify some illustrations, only one set of units is presented on artwork. References in the accompanying text to data in the illustrations are presented in both SI-based and customary U.S. units. On graphs and charts, grids corresponding to SI-based units usually appear along the left and bottom edges. Where appropriate, corresponding customary U.S. units appear along the top and right edges.

Data pertaining to a specification published by a specification-writing group may be given in only the units used in that specification or in dual units, depending on the nature of the data. For example, the typical yield strength of steel sheet made to a specification written in customary U.S. units would be presented in dual units, but the sheet thickness specified in that specification might be presented only in inches.

Data obtained according to standardized test methods for which the standard recommends a particular system of units are presented in the units of that system. Wherever feasible, equivalent units are also presented. Some statistical data may also be presented in only the original units used in the analysis.

Conversions and rounding have been done in accordance with IEEE/ASTM SI-10, with attention given to the number of significant digits in the original data. For example, an annealing temperature of 1570 °F contains three significant digits. In this case, the equivalent temperature would be given as 855 °C; the exact conversion to 854.44 °C would not be appropriate. For an invariant physical phenomenon that occurs at a precise temperature (such as the melting of pure silver), it would be appropriate to report the temperature as 961.93 °C or 1763.5 °F. In some instances (especially in tables and data compilations), temperature values in °C and °F are alternatives rather than conversions.

The policy of units of measure in this Handbook contains several exceptions to strict conformance to IEEE/ASTM SI-10; in each instance the exception has been made in an effort to improve the clarity of the Handbook. The most notable exception is the use of g/cm^3 rather than kg/m^3 as the unit of measure for density (mass per unit volume).

SI practice requires that only one virgule (diagonal) appear in units formed by combination of several basic units. Therefore, all of the units preceding the virgule are in the numerator and all units following the virgule are in the denominator of the expression; no parentheses are required to prevent ambiguity.

Officers and Trustees of ASM International (2009–2010)

Frederick J. Lisy
President and Trustee
Orbital Research Incorporated
Mark F. Smith
Vice President and Trustee
Sandia National Laboratories
Paul L. Huber
Treasurer and Trustee
Seco/Warwick Corporation
Roger J. Fabian
Immediate Past President and Trustee
Bodycote Thermal Processing
Stanley C. Theobald
Managing Director and Secretary
ASM International

Mufit Akinc
Iowa State University
Riad I. Asfahani
United States Steel Corporation
Sunniva R. Collins
Swagelok
Robert J. Fulton
Hoeganaes Corporation (retired)
Richard Knight
Drexel University
Sunniva R. Collins
Swagelok
John J. Letcavits
AEP

Digby D. Macdonald
Penn State University
Charles A. Parker
Honeywell Aerospace
Jon D. Tirpak
ATI
[Student board representatives]
Joshua Holzhausen
Missouri University of Science and Technology
Kelsi Hurley
University of Washington
Natasha Rajan
University of Alberta

Members of the ASM Handbook Committee (2009–2010)

Kent L. Johnson
(Chair 2008–; Member 1999–)
Materials Engineering Inc.
Craig D. Clauser
(Vice Chair 2009–; Member 2005–)
Craig Clauser Engineering Consulting
Incorporated
Larry D. Hanke
(Immediate Past Chair; Member 1994–)
Materials Evaluation and Engineering Inc.
Viola L. Acoff (2005–)
University of Alabama
Lichun Leigh Chen (2002–)
Technical Materials Incorporated
Sarup K. Chopra (2007–)
Consultant
Craig V. Darragh (1989–)
The Timken Company (ret.)

Jon L. Dossett (2006–)
Consultant
Alan P. Druschitz (2009–)
University of Alabama-Birmingham
David U. Furrer (2006–)
Rolls-Royce Corporation
Jeffrey A. Hawk (1997–)
National Energy Technology Laboratory
William L. Mankins (1989–)
Metallurgical Services Inc.
Joseph W. Newkirk (2005–)
Missouri University of Science and Technology
Robert P. O'Shea, Jr. (2008–)
Baker Engineering and Risk Consultants
Cory J. Padfield (2006–)
American Axle & Manufacturing
Toby V. Padfield (2004–)
ZF Sachs Automotive of America

Cynthia A. Powell (2009–)
DoE National Energy Technology Lab
Elwin L. Rooy (2007–)
Elwin Rooy & Associates
Jeffrey S. Smith (2009–)
Material Processing Technology LLC
Kenneth B. Tator (1991–)
KTA-Tator Inc.
George F. Vander Voort (1997–)
Buehler Ltd.
Michael K. West (2008–)
South Dakota School of Mines
and Technology

Chairs of the ASM Handbook Committee

J.F. Harper
(1923–1926) (Member 1923–1926)
W.J. Merten
(1927–1930) (Member 1923–1933)
L.B. Case
(1931–1933) (Member 1927–1933)
C.H. Herty, Jr.
(1934–1936) (Member 1930–1936)
J.P. Gill
(1937) (Member 1934–1937)
R.L. Dowdell
(1938–1939) (Member 1935–1939)
G.V. Luerssen
(1943–1947) (Member 1942–1947)
J.B. Johnson
(1948–1951) (Member 1944–1951)
E.O. Dixon
(1952–1954) (Member 1947–1955)
N.E. Promisel
(1955–1961) (Member 1954–1963)
R.W.E. Leiter
(1962–1963) (Member 1955–1958, 1960–1964)

D.J. Wright
(1964–1965) (Member 1959–1967)
J.D. Graham
(1966–1968) (Member 1961–1970)
W.A. Stadtler
(1969–1972) (Member 1962–1972)
G.J. Shubat
(1973–1975) (Member 1966–1975)
R. Ward
(1976–1978) (Member 1972–1978)
G.N. Maniar
(1979–1980) (Member 1974–1980)
M.G.H. Wells
(1981) (Member 1976–1981)
J.L. McCall
(1982) (Member 1977–1982)
L.J. Korb
(1983) (Member 1978–1983)
T.D. Cooper
(1984–1986) (Member 1981–1986)
D.D. Huffman
(1986–1990) (Member 1982–2005)

D.L. Olson
(1990–1992) (Member 1982–1992)
R.J. Austin
(1992–1994) (Member 1984–1985)
W.L. Mankins
(1994–1997) (Member 1989–)
M.M. Gauthier
(1997–1998) (Member 1990–2000)
C.V. Darragh
(1999–2002) (Member 1989–)
Henry E. Fairman
(2002–2004) (Member 1993–2005)
Jeffrey A. Hawk
(2004–2006) (Member 1997–)
Larry D. Hanke
(2006–2008) (Member 1994–)
Kent L. Johnson
(2008–2010) (Member 1999–)

Authors and Contributors

John Agren
Royal Institute of Technology, Stockholm, Sweden

Seokyoung Ahn
The University of Texas-Pan American

Janet K. Allen
University of Oklahoma

Taylan Altan
The Ohio State University

Sudarsanam Suresh Babu
The Ohio State University

C. C. Bampton
Pratt & Whitney Rocketdyne

Jeff J. Bernath
Edison Welding Institute Incorporated

Bernard Billia
Aix-Marseille Université, France

Robert Brooks
National Physical Laboratory, UK

Dennis J. Buchanan
University of Dayton Research Institute

W.S. Cao
CompuTherm LLC

Y.A. Chang
University of Wisconsin

Anil Chaudhary
Applied Optimization Inc.

S.L. Chen
CompuTherm LLC

Suk Hwan Chung
Hyundai Steel Co, South Korea

Seong-Taek Chung
CetaTech, Inc.

Anders Engström
Thermo-Calc Software AB, Stockholm, Sweden

Hans J. Fecht
Ulm University, Germany

Chris Fischer
Scientific Forming Technologies Corporation

D. U. Furrer
Rolls-Royce Corporation

Ch.-A. Gandin
Centre de Mise en Forme des Matériaux, Sophia Antipolis, France

Randall M. German
San Diego State University

Somnath Ghosh
The Ohio State University

Robert Goetz
Rolls-Royce Corporation

Vassily Goloveshkin
Moscow State University of Instrument Engineering and Computer Sciences (MGUPI)

G. Gottstein
Institute of Physical Metallurgy and Metal Physics, RWTH Aachen University, Germany

Jianzheng Guo
ESI US R&D

Samuel Hallström
Thermo-Calc Software AB, Stockholm, Sweden

A. Jacot
Ecole Polytechnique Fédérale de Lausanne, Lausanne, Switzerland

JongTae Jinn
Scientific Forming Technologies Corporation

D. Kammer
Northwestern University

Kanchan M. Kelkar
Innovative Research Inc.

Pat Koch
Engineous Software

M. V. Kral
University of Canterbury, New Zealand

Matthew John M. Krane
Purdue University

Howard Kuhn
University of Pittsburgh

Young-Sam Kwon
CetaTech, Inc.

Peter D. Lee
Department of Materials, Imperial College, London, U.K

Guoji Li
Scientific Forming Technologies Corporation

Ming Li
Alcoa Technical Center

Kong Ma
Rolls-Royce Corporation

Paul Mason
Thermo-Calc Software Inc., Stockholm, Sweden

Ramesh S. Minisandram
ATI Allvac

Alec Mitchell
University of British Columbia

D. A. Molodov
Institute of Physical Metallurgy and Metal Physics, RWTH Aachen University, Germany;

Seong Jin Park
Mississippi State University

Suhas V. Patankar
Innovative Research Inc.

Ashish D. Patel
Carpenter Technologies

Michael Preuss
Manchester University, UK

Peter J. Quested
National Physical Laboratory, UK

A. D. Rollett
Carnegie Mellon University

Yiming Rong
Worcester Polytechnic Institute

D. J. Rowenhorst
US Naval Research Laboratory

Valery Rudnev
Inductoheat Incorporated

Victor Samarov
Synertech PM

Mark Samonds
ESI US R&D

N. Saunders
Thermotech / Sente Software Ltd., UK

S. L. Semiatin
Air Force Research Laboratory

L. S. Shvindlerman
Institute of Solid State Physics, Russian Academy of Sciences, Chernogolovka, Russia

Richard D. Sisson, Jr.
Worcester Polytechnic Institute

G. Spanos
US Naval Research Laboratory

Shesh K. Srivatsa
GE Aviation

Santosh Tiwari
Engineous Software

Juan J. Valencia
Concurrent Technologies Corporation

Alex Van der Velden
Engineous Software

P. W. Voorhees
Northwestern University

Ronald A. Wallis
Wyman Gordon Forgings

Gang Wang
Worcester Polytechnic Institute

Junsheng Wang
Department of Materials, Imperial College, London, UK

Philip J. Withers
Manchester University, UK

K.S. Wu
CompuTherm LLC

Wei-Tsu Wu
Scientific Forming Technologies Corporation

Junde Xu
Edison Welding Institute Incorporated

Jaebong Yang
Scientific Forming Technologies Corporation

Y. Yang
CompuTherm LLC

F. Zhang
CompuTherm LLC

Reviewers

Taylan Altan
The Ohio State University

Egbert Baake
Leibniz Universität Hannover

L. Battezzati
Università di Torino

Michel Bellet
Centre de Mise en Forme des Matériaux,
Sophia Antipolis, France

Hongbo Cao
General Electric Global Research Center

Qing Chen
Thermo-Calc Software AB, Stockholm,
Sweden

Jon Dantzig
University of Illinois at Urbana-Champaign

Uwe Diekmann
Metatech GmbH

Rollie Dutton
Air Force Research Laboratory

D.U. Furrer
Rolls-Royce Corporation

Martin E. Glicksman
University of Florida

Janez Grum
University Of Ljubljana

Jianzheng Guo
ESI US R&D

Larry Hanke
Materials Evaluation and Engineering Inc

Jeffrey Hawk
U.S. Department of Energy

Edmond Ilia
Metaldyne

Richard Johnson

Ursula Kattner
National Institute of Standards and
Technology

Leijun Li
Utah State University

Daan Maijer
University of British Columbia

William Mankins
Metallurgical Services Incorporated

David McDowell
Georgia Institute of Technology

Tugrul Ozel
Rutgers University

S.L. Semiatin
Air Force Research Laboratory

Brian Thomas
University of Illinois at Urbana-Champaign

Ray Walker
Keystone Synergistic Enterprises, Inc.

Michael West
South Dakota School of Mines and
Technology

John Wooten
CalRAM, Inc

Contents

Input Data for Simulations

ASM Handbook, Volume 22B, Metals Process Simulation
D.U. Furrer and S.L. Semiatin, editors

Copyright © 2010, ASM International®
All rights reserved.
www.asminternational.org

Introduction to Metals Process Simulation

D.U. Furrer, Rolls-Royce Corporation
S.L. Semiatin, Air Force Research Laboratory

TECHNOLOGY CHANGE and the adoption of new technologies by industry are largely dictated by economics. Tools and methods that enable reduced-cost design approaches, manufacturing methods, or maintenance costs for components or products are sought to enable greater competitiveness or product differentiation within the marketplace.

Technology change and advancement is currently occurring at a very high rate. The current rate of technology change is due in part to the development and application of computers and computational methods (Ref 1). Nearly all areas of engineering and industry are seeing rapid changes in technology, including data-analysis methods, automation, enhanced sensors and feedback systems, rapid prototyping and manufacturing methods, and computer-enhanced optimization of materials, processes, and designs.

Computational modeling and simulation methods for materials and processes have been growing rapidly. As an example, in the 1970s, the development of computer codes to simulate solidification was initiated; commercial codes such as AFS-Solids became available to industry shortly thereafter, followed by many other software packages, including ProCast, MagmaSoft, and SOLIDCast. During this time period, codes for deformation processes, such as ALPID, were being developed to enable industrial engineers to simulate metal flow and predict the occurrence of flow defects. A number of commercial codes are now available for use by industry, including DEFORM, FORGE, and Simufact.

Today (2010), commercial off-the-shelf codes are available for a wide range of manufacturing processes for use in material and process design and analysis. Nearly every known metallurgical mechanism has been modeled or is being studied to enable quantitative predictions, as detailed in Fundamentals of Modeling for Metals Processing, Volume 22A of the ASM Handbook. The simulation of crystallographic texture evolution during deformation processes is now possible as well with both research and commercial computer codes. The various models are leading to the simulation of microstructure and defect occurrence during real-world manufacturing and in-service applications.

As the sophistication of computational modeling and simulation of materials and processes increases, so do the technologies within design and manufacturing engineering to which these modeling and simulation efforts are being applied. With the ability to predict microstructure and mechanical properties or the potential for defect occurrence within components, designers and manufacturing engineers can assess virtual designs and manufacturing processes. This can lead to right-the-first-time design and manufacturing as well as unique component and process designs that would otherwise not be discovered. In fact, holistic engineering approaches are now increasingly possible through the linkage of design tools with material and process models and simulation tools. In addition, these simulation tools are enabling the prediction and control of location-specific microstructure and mechanical properties within components. These location-specific mechanical properties can, in some cases, be directly used to assess adequacy of designs for specific service applications.

The use of material and process modeling and simulation tools has also led to a greater understanding of materials and manufacturing processes. For example, simulation tools have enabled the assessment and quantification of sources of variation within materials and manufacturing processes. This has led to enhanced manufacturing methods, equipment, and controls to ensure materials and products meet design specifications. Further utilization of material and process simulation tools will result in an increased rate of technology change within alloy and manufacturing process designs, which is already being seen, as exemplified by the Defense Advanced Research Projects Agency (DARPA)-funded program known as Accelerated Insertion of Materials (AIM) (Ref 2).

Future challenges for material and process simulation technology include further linkage of these tools with design functions, increasing the range and fidelity of simulation tool capability and reducing cost, and the further increase in computation speed. The benefits of material and process simulation are clear, as it continues to drive development and application within a variety of legacy and emerging manufacturing sectors.

Metals Process Simulation

Simulation Applications

A wide range of engineered-materials industries are employing simulation methods. These industries are using advanced tools to simulate a range of metallurgical processes and properties within materials and components. Volume 22A of the ASM Handbook detailed models for a number of metallurgical phenomena that play a role in industrial processes for metals and alloys, such as recrystallization and grain growth; defect formation of various types, such as cavitation/porosity and strain localization; the development of chemical or microstructural gradients; and the effect of such metallurgical conditions on mechanical properties.

The metalworking industry has used simulation tools for a number of years. For example, forging process modeling had been used initially to assess bulk metal flow and to predict metal-flow defects, such as laps/folds during deformation. Various articles in this Volume (e.g., "Finite-Element Method Applications in Bulk Forming" and "Sheet Metal Forming Simulation") and its companion Volume (22A) provide information on the current state of metal working simulation capabilities, including the prediction of recrystallization, deformation texture, and cavitation.

Casting processes have been and continue to be simulated using a wide range of computational methods. A number of software companies have developed simulation tools for the prediction of solidification and defect formation. The capabilities and application of current simulation tools in this area are summarized in the articles "Modeling of Transport Phenomena during Solidification Processes," "Modeling of Porosity Formation during Solidification," "Computational Analysis of the Vacuum Arc Remelt (VAR) and Electroslag Remelt (ESR) Processes," "Simulation of Casting and Solidification Processes," "Modeling of Dendritic Grain Solidification," "Modeling of Laser-Additive Manufacturing Processes," and

"Formation of Microstructures, Grain Textures, and Defects during Solidification." Prediction of solidification shrinkage, porosity, grain structure, and solidification defects, such as freckles, can be readily obtained through current commercial simulation methods.

Powder metallurgy processing is used in many industries due to the economics of manufacturing near-net or net components by means of powder material compaction, sintering, and shaping. The articles "Modeling of Powder Metallurgy Processes," "Modeling and Simulation of Press and Sinter Powder Metallurgy," "Modeling of Hot Isostatic Pressing," and "Modeling and Simulation of Metal Powder Injection Molding" provide details on the simulation of these manufacturing processes through the use of various models and computational tools. The issue of shape change during consolidation is a critical aspect in these manufacturing processes and is reviewed within these articles.

Machining processes are used within the manufacturing sequence of nearly every engineered component produced today (2010). Simulation of machining processes is of growing interest due to its associated cost. Increased machining speed, reduced tool wear, and the elimination of distortion and machining-induced damage are key quality and economic drivers. The articles "Modeling and Simulation of Machining," "Modeling Sheet Shearing Processes for Process Design," and "Modeling of Residual Stress and Machining Distortion in Aerospace Components" provide insight into how simulation tools can provide guidance to engineering decisions relative to machining processes, cutting tools, and machining fixtures. The simulation of chip formation is leading to new machining methods and control schemes for improved component surface/dimensional quality and reduced cost.

Surface Engineering. Simulation tools are also being developed and used to support surface engineering applications. The articles "Simulation of Induction Heat Treating" and "Simulation of Diffusion in Surface and Interface Reactions" provide insight into the state of the art of these methods. Specific applications, such as local heat treatment and chemistry enhancement for local property optimization, are provided in these articles. Gradients in chemistry or microstructure are often the goal for predictions within metallurgical simulations of surface heat treatment, carburizing, nitriding, or diffusion-coating processes. Greater understanding of critical processing parameters and control requirements can be obtained through the simulation of these processes and conducting virtual processing experiments.

Heat treatment processes are also common in the manufacturing milieu for metallic components to develop the required final component microstructure and mechanical properties. The simulation of thermal processes provides detailed temperature and stress histories. The articles "Heating and Heat-Flow Simulation,"

"Modeling of Quenching, Residual-Stress Formation, and Quench Cracking," and "Simulation of Induction Heating Prior to Hot Working and Coating" provide details of the models that are used to simulate heating and cooling processes and provide details of the capabilities of current simulation methods. Simulation tools are also providing greater understanding of equipment and processes and are being used to guide the application of process controls to further enhance the repeatability of thermal processes. Prediction of thermally induced residual stresses is being accomplished with the simulation of thermal histories within manufactured components. Understanding of bulk residual stresses is having a profound impact on how components are machined and how components are being analyzed for service performance.

Joining processes for many applications can now be simulated. The articles "Introduction to Integrated Weld Modeling," "Simulation of Rotational Welding Operations," "Simulation of Joining Operation: Friction Stir Welding," and "Modeling of Diffusion Bonding" provide detailed analysis and examples of simulation tools and methods for various joining processes. The simulation of a joining process frequently uses software developed for other types of simulations but often with very different boundary conditions.

Alloy Development. The development of new alloys with unique properties that can fulfill the requirements for new applications is seeing significant benefit from the use of modeling and simulation tools. Historical trial-and-error methods of alloy design are extremely costly, time-consuming, and often result in less than optimal or robust results. The simulation of alloy phases and phase diagrams can thus be a critical first step in computational alloy design. The articles "Commercial Alloy Phase Diagrams and Their Industrial Applications" and "The Application of Thermodynamic and Material Property Modeling to Process Simulation" provide excellent examples of how thermodynamic modeling can be used to simulate and design new alloys. The predicted phase equilibria information can also be used to predict microstructure and be linked to microstructure-sensitive mechanical property models.

Within an integrated computational materials engineering approach to alloy design, organizations are working to identify the metallurgical mechanisms that control or could control the behavior of an alloy during a specified-use application. Using simulation tools to assess variations in alloy chemistry and microstructure, new alloys can be investigated computationally. Process simulation tools are also incorporated into the alloy design process to assess the ease or potential challenges in achieving the goal alloy chemistry and microstructure.

A number of organizations are using material and process modeling and simulation tools to guide alloy development through "alloy-by-

design" methods (Ref 3, 4) or a "Materials by Design" (Ref 5) approach. For instance, the recent development of alloys such as Ferrium C61 and S53 are examples of how modeling and simulation tools can enable rapid development, maturation, and implementation of new alloys.

Optimization. Lastly, simulations can be used to assess parameters for a unique or optimal final component process or design. In this regard, uncertainties in input information and the propagation of errors from one model to another must be assessed and managed. The articles entitled "Design Optimization Methodologies," "Stress-Relief Simulation," "Uncertainty Management in Materials Design and Analysis," and "Manufacturing Cost Estimating" provide insight into these issues.

Input Data and Boundary Conditions for Process Simulations

The implementation of modeling and simulation tools requires accurate descriptions of material properties and process boundary conditions. Inaccurate input data can easily lead to poor or even misleading predictions.

Input Data. The articles "Measurement and Interpretation of Flow Stress Data for the Simulation of Metal-Forming Processes," "Thermophysical Properties," "Thermophysical Properties of Liquids and Solidification Microstructure Characteristics: Benchmark Data Generated in Microgravity," "Measurement of Thermophysical Properties at High Temperatures for Liquid, Semisolid, and Solid Commercial Alloys," "Grain-Boundary Energy and Mobility," "Texture Measurement and Analysis," and "Three-Dimensional Microstructure Representation" in this Volume provide information on the generation and representation of accurate physical and mechanical properties. Standard methods have been established in many cases, although there are still challenges in developing the needed material property data over the entire range of processing conditions (time, temperature, pressure, environment, transient conditions, etc.) encountered within the manufacturing environment. Many current material property characterization methods and databases are for narrow use conditions and often do not apply to or include the range of conditions within other processes.

Boundary Conditions. In addition to material characteristics, process boundary conditions are critical to accurate simulation predictions. Each manufacturing process has unique boundary conditions that must be identified, understood, and characterized for the specific application being simulated. Moreover, boundary conditions can be equipment specific, meaning that one press or furnace may not give rise to the same boundary conditions as another press or furnace of a similar type used under the same nominal processing conditions. A number of the articles in this Volume touch upon the boundary-condition information needed and the criticality of its accuracy. Examples include the heating and cooling of metals; for example, the variation in heat-transfer coefficients during quenching processes and how to determine them are reviewed. Understanding of process boundary conditions and detailed assessment of these boundary conditions within simulation efforts can provide guidance for the level of control needed to establish robust and repeatable processes and equipment that produce results within the range required for specific applications.

The article "Solid Modeling" also provides information on how to represent the solid object that is being simulated and how to link a solid model with finite-element analysis software. Such geometry representation is a critical component in computational materials and process simulation.

Linking Academia and Industry

Modeling and simulation tools often have roots and ties to university or research laboratory efforts throughout the world. The science of material and process modeling is very challenging and often requires dedicated research to establish correct physics-based relationships, computational methods, and validation procedures. Industrial companies often do not have the capability to develop fundamental models to describe the physics of materials and manufacturing processes with which they work. However, industry has realized the importance of these models and subsequent simulation tools to the development of component designs and manufacturing methods and, as such, is incorporating modeling and simulation tools that have been initially conceived in basic research/academic environments. There is actually significant pull from industry for continued academic research in the area of material and process modeling and simulation. This is evident through the number of consortia that have been developed to link companies and universities to guide the development of new tools.

The increase in the development and application of modeling and simulation tools is bringing academia and industry closer together. While academia may focus on establishing new theories and models, industry is applying them in a practical, purposeful manner. The strengthened linkage between fundamental materials science and pragmatic engineering is also enhancing academic studies. Some universities have previously taught courses in materials processing, but many programs have had to eliminate industrial-scale or industrial-emulating processing labs due to shrinking budgets and space/safety issues. Nevertheless, students are still learning about manufacturing processes through simulation tools and applications pertinent to industrial-scale shop-floor processes.

Students are also learning to link component-design methods with material and process simulation tools to further develop skills in holistic

component and systems engineering. A recent survey has shown that many universities in the United States are using simulation software within their curriculum to aid in teaching general materials science and engineering principles (Ref 6). The application of simulation software in universities will develop capable scientists and engineers for the future who are more knowledgeable and equipped to develop and apply additional uses of modeling and simulation tools.

Computational Material and Process Modeling and Simulation Enablers and Challenges

Computational materials engineering has been growing and increasingly deployed throughout industry due to several critical enablers:

- Computational speed
- Computational materials engineering software/hardware supply chain
- Cost structure for virtual versus physical manufacturing and analysis

Computational Speed. The issue of computational speed is critical for industrial applications. Rapid simulation and analysis of materials and processes are required within industry, where design and manufacturing decisions require near-instant turnaround. To support design and manufacturing decision-making, simulation tools must provide real-time input. A rule of thumb within industry is that simulation run time is "acceptable" if it

runs overnight, with simulation times much shorter than this being the preference. In some cases, multiday simulations are still necessary to enable predictions to the required level of accuracy to be sufficiently useful for the intended purpose. These long computational times are a challenge for industrial applications.

Computational speeds have increased via three primary means: computer processor speed increases, increased efficiency of computational methods, and simplification of models to increase computation speed while still providing the required level of prediction accuracy and precision. Computer processor speeds have increased continually in recent years. Figure 1, for example, shows the rapid increase in computer processor speeds over the past two decades. It is significant to note that computer memory as well as computational speed is important in enabling rapid computer-based simulation. The amount of computer memory for computational processing has actually increased at a higher rate than processing speed, as shown in Fig. 1. Both of these factors have supported the further development and application of material and process simulation tools.

Computer codes and computational methods have been and continue to be developed to enable parallel processing. Dual- and quad-core personal computers (PCs) are providing increased computational speed through parallel processing. The PC clusters enable even further parallel-processing capability for software that can use large numbers of computer processors at the same time. Further efforts to establish and use distributed processing are continuing. Local, flexible networks with computing resources of various types and architectures are being linked

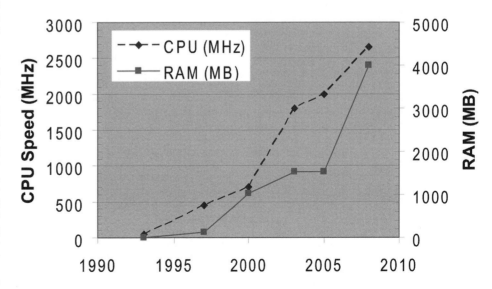

PC Processor Performance

Fig. 1 Historical comparison of approximate speed and memory capabilities for personal computers (PCs) during the last two decades. CPU, central processing unit; RAM, random access memory

together to enable multiple computational tasks simultaneously, such as large numbers of simulations within a design-of-experiments task or large-scale Monte Carlo simulations. Software tools are now available that can set up and support sharing of simulation workflows and to distribute their execution to available computer resources.

Computational Supply Chain. Efficient simulation codes are being written and maintained by code developers and software suppliers. This is a critical enabler and challenge for continued development and deployment of computational materials engineering tools. A computational engineering supply chain is needed to provide software solutions for various industries and to develop, enhance, and maintain these tools for the future. Many software companies are relatively small and depend on small niche markets. Survival of such companies and/or simulation tools is critical to the continued development and growth of computational materials engineering within industry. The added value obtained from a number of software packages is well known, such that continued pull from industry will sustain the growth of the software market. However, the market will also determine the number of sustainable codes for niche applications. It is critical that specialized codes for specific applications are flexible and enable linkage with other codes, thereby allowing seamless data passage for maximum benefit for component and manufacturing process design applications.

Cost Structure. The cost of computational materials engineering is a major challenge for this growing technology. Software suppliers are working to increase the value of their products through code enhancements and increased functionality. However, these efforts can lead to niche software tools and greater software maintenance and overhead cost that must be passed along to users. Computational speed is also critical, so high-end computers or computer clusters are sometimes required to run enhanced versions of some codes. The challenges of software, infrastructure, and personnel cost may lead to major barriers for the further utilization of these tools by small organizations.

The potential of material and process simulation is nearly limitless, but the challenges associated with tool enhancement, linkage with design tools and methods, and the cost relative to the value added by their application must be managed. Joint industrial sector efforts and industry/government consortia are potential approaches to contend with these challenges, while further enhancing and enabling the effective use and growth of this technology (Ref 7–9).

Benefits of Modeling and Simulation

The modeling and simulation of metallurgical processes and manufacturing methods is being used for a wide range of applications. Initial utilization has been for problem solving, in which simulation tools have been used to guide engineers to the root cause and potential corrective actions. High-visibility, high-cost challenges have led to the use of simulation tools for greater understanding of failures or suboptimal component or process capabilities.

Simulation tools have been and will continue to be used for trend analysis and providing direction for material or process changes. As further understanding and validation of models for specific metallurgical processes is attained, modeling and simulation tools can be developed to provide increasingly quantitative predictions to support design and component-capability analysis. There are many benefits from the use of computational material and process simulation during the early stages of component and system design. Simulation of materials and processes provides greater understanding of component designs and enables greater product definition. Furthermore, commercial products can exhibit variation in properties and performance. Through material and process simulation, these variations can often be understood and assigned a cause. If reduced variation is required, then simulation tools can provide a means to rapidly assess the impact of material and process parameter and tolerance changes.

Materials and process simulation also enables large numbers of virtual experiments to be conducted in a very short time period and with significantly less cost than physical trials. This ability to assess alternative material and process parameters allows engineers to gain increased understanding of materials and processes at an accelerated rate.

The knowledge captured by conventional empirical material and process engineering methods often resides with experienced engineers and scientists or is recorded in limited-distribution reports and documents. The development, utilization, and validation of new modeling and simulation tools is a means of capturing knowledge that can be effectively conveyed to subsequent generations of engineers and scientists. Simulation tools provide a means of examining specific material and process examples and providing quantitative predictions. Empirical knowledge tends to provide rules of thumb or information from historical examples that are similar to, but do not exactly match, current or future needs. In contrast, validated simulation tools are an enhanced means of capturing knowledge that can be readily reused for future specific examples.

Modeling and simulation tools within the materials industry have been used extensively for industrial problem-solving. Solving industrial problems is often time-sensitive and requires immediate action. Hence, funding for simulation capabilities to solve manufacturing issues has become available along with the development of new simulation tools themselves. As such, simulation tools initially used in a "reactive" mode to solve problems have now gained support for introduction into early design stages in a more proactive manner.

The evolution of material and process modeling and simulation tools within industry has been for direct-benefit applications, such as process development. These applications are so-called "hand-to-mouth"-type applications, in which simulations are conducted to guide engineering decisions on specific manufacturing methods. Examples of these types of process design applications are forging and casting in which manufacturing process engineers obtain information from simulations to establish shop-floor process parameters and tolerances. The benefit of the simulation effort for these applications is nearly immediately realized after initial manufacture of a component.

The next step in the industrial implementation of material and process modeling and simulation tools is to the early stages of component design. The benefits in this case are not realized until much later, when the component design is finalized, the exact processing method and source (often external for design organizations) have been defined, and components are manufactured and tested to validate acceptance to application requirements. The extended chain of engineering, manufacturing, and testing distances the benefits of modeling and simulation from the resultant benefits. A holistic approach to component and system design should therefore be adopted to capture and realize the benefits of early-design-stage simulation efforts. This approach is now being used within a range of industries for component design and life-cycle savings and capability enhancements (Ref 10–12).

Material and process simulation has the potential to substantially reduce the financial risk associated with new materials and processes. Companies that develop new alloys can reduce the risk of spending large amounts of funding and resources on developing, characterizing, and certifying an alloy that may have a critical issue limiting its widespread use. Similarly, manufacturing companies often must develop or specify new equipment for traditional or new manufacturing processes. Understanding of the level of control and range of flexibility of processing parameters can be critical for the success of a new piece of equipment. Installing a new piece of equipment that cannot control a critical parameter to the level required for final component property control could be a disaster for a company. Simulating manufacturing processes to support equipment design and equipment capability specifications can result in reduced capital expenditure risk, increase process performance when a new system is installed, and enable greater understanding of process windows.

Future Perspective for Computer-Based Modeling and Simulation

Many examples of material and process modeling and simulation pertain to solutions for specific examples. Exact simulation results for a single set of material and process inputs are very useful but do not by themselves provide guidance or understanding to the potential range of these predicted results in real-world materials and manufacturing. Characterizing variation in material and process inputs and using these sources of variation to provide understanding for the range of variation of final predictions in component properties is needed for future simulation efforts. Probabilistic methods for design must incorporate distributions in component properties. Further work is required to establish standardized methods to predict statistically relevant distributions of properties for engineering materials, processes, and component designs.

Certification of materials and processes will be another major area for the application and benefit of computational materials and process modeling and simulation. Efforts have already been accomplished in the area of materials certification (Ref 13). Material properties are now being predicted and applied for certification or confirmation of property compliance, which underlies needed manufacturing-process controls.

Component and system performance simulations incorporating location-specific material properties from prior process simulations are also being performed. For example, the automotive industry is using crash-simulation tools that take into account component forming methods and properties to provide greater fidelity of service performance predictions (Ref 14, 15). Further application of materials and process simulation to certification processes will result in reduced costs associated with physical tests and will greatly reduce the lead time for component and system certification and entry into the market.

The articles in this Handbook provide an introduction for both students and practicing engineers. The materials engineering tools have the potential for location-specific component design and optimization for enhanced component performance and reduced cost. The future of materials and process modeling and simulation appears to be past a critical point where benefits are becoming clear and industrial utilization and pull are greatly expanding.

REFERENCES

1. S.C. Glotzer et al., "International Assessment of Research and Development in Simulation-Based Engineering and Science," WTEC Panel Report, World Technology Evaluation Center, Inc., Baltimore, MD
2. Defense Advanced Research Projects Agency–Accelerated Insertion of Materials, (DARPA-AIM), http://www.darpa.mil/dso/thrusts/matdev/aim/index.html, accessed Jan 22, 2010
3. C. Rae, Alloys by Design: Modeling Next Generation Superalloys, *Mater. Sci. Technol.,* Vol 25 (No. 4), 2009, p 479–487
4. R. Reed, T. Tao, and N. Warnken, Alloys-by-Design: Application to Nickel-Based Single Crystal Superalloys, *Acta Mater.,* Vol 57 (No. 19), 2009, p 5898–5913
5. C.J. Kuehmann and G.B. Olson, Computational Materials Design and Engineering, *Mater. Sci. Technol.,* Vol 25 (No. 4), 2009, p 472–478
6. K. Thornton et al., Computational Materials Science and Engineering Education: A Survey of Trends and Needs, *JOM,* Oct 2009, p 12–17
7. M.E. Kinsella and D. Evans, Government and Industry Partnering: Technology Transition through Collaborative R&D; Metals Affordability Initiative: A Government-Industry Technical Program, *Defense AT&L,* March-April 2007, p 12–15
8. Center for Computational Materials Design, Pennslvania State University and Georgia Institute of Technology, www.ccmd.psu.edu, accessed Jan 22, 2010
9. Center for the Accelerated Maturation of Materials, The Ohio State University, http://www.camm.ohio-state.edu/index.html, accessed Jan 22, 2010
10. D.G. Backman, D.Y. Wei, D.D. Whitis, M.B. Buczek, P.M. Finnigan, and G. Gao, ICME at GE: Accelerating the Insertion of New Materials and Processes, *JOM,* Nov 2006, p 36–41
11. J. Allison, M. Li, C. Wolverton, and X. Su, Virtual Aluminum Castings: An Industrial Application of ICME, *JOM,* Vol 58 (No. 11), 2006, p 28–35
12. *Integrated Computational Material Engineering—A Transformational Discipline for Improved Competitiveness and National Security,* National Academies Press, 2008
13. Y.R. Im, J.H. Lee, H.J Kim, and J.K. Lee, The Implementation of a Structure-Property Prediction Model in Current Hot Strip Mills, *POSCO Tech. Rep.,* Vol. 10 (No. 1), 2006, p 21–26
14. S.-J. Hong, D.-C. Lee, J.-H. Jang, C.-S. Han, and K. Hedrick, Systematic Design Process for Frontal Crashworthiness of Aluminum-Intensive Electrical Vehicle Bodies, *J. Auto. Eng.,* Vol 220 (No. 12), 2006, p 1667–1678
15. K. Takashina, K. Ueda, and T. Ohtsuka, Influence of Work Hardening during Metal Forming on Crashworthiness Simulation, *Mitsubishi Motors Tech. Rev.,* No. 20, 2008, p 117–120

ASM Handbook, Volume 22B, *Metals Process Simulation*
D.U. Furrer and S.L. Semiatin, editors

Copyright © 2010, ASM International®
All rights reserved.
www.asminternational.org

Thermophysical Properties of Liquids and Solidification Microstructure Characteristics—Benchmark Data Generated in Microgravity

Hans J. Fecht, Ulm University, Germany
Bernard Billia, Aix-Marseille Université and CNRS, France

FOR A WIDE RANGE of new or better products, solidification processing of metallic materials from the melt is a step of uppermost importance in the industrial production chain. Examples of such advanced products include turbine blades for energy production in land-based power plants and for jet engines, low-emission energy-effective engines for cars, lightweight metallic foams for absorbing crashes, so-called supermetals (amorphous metallic alloys) as thin sheets for electronic components with ultimate strength, high-performance magnets, medical implants such as hip replacements, and fine metallic powders to catalyze chemical reactions, to name just a few. Furthermore, the production and fabrication of alloys, together with the casting and foundry industry, generate a considerable amount of wealth (Fig. 1a, c). The market is huge; the millions of tons of total castings produced worldwide are worth approximately 100 billion U.S. dollars per year. For example, Hydro (a division of Norsk Hydro ASA, Oslo, Norway) is providing more than 200,000 units of the V6 diesel engine block shown in Fig. 1 (a) to Mercedes-Benz (a division of Daimler AG, Stuttgart, Germany); the Engine Alliance (GE and Pratt and Whitney) has developed the new high-bypass GP7200 engine (Fig. 1c) for the Airbus A380. The continuation of wealth generation by the casting and foundry industry relies on the design and optimization of advanced materials processing, with increased efficiency at reduced cost.

Accordingly, to produce materials meeting specifications and performance requirements never seen in the past, which often implies breaking existing technology barriers, the solidification processing of metallic structural materials must be controlled with ever-increasing

precision, especially because it is expected that materials for tomorrow will be optimized in their design and more efficiently produced. Theoretical modeling and predictive quantitative numerical simulations of grain structure formation in solidification processes using sophisticated integrated software (Ref 1, 2) have become the manufacturers' tool-of-choice for optimizing processing routes and to improve casting quality and reproducibility, yield strength, and other properties (Fig. 1b, d).

There are two essential aspects for the continuous progress of integrated software serving materials design and processing at casthouses:

- The reliable determination of the thermophysical and related properties of metallic melts in order to understand the fundamentals of complex melts (e.g., multicomponent alloys, intermetallics, semiconductors, etc.)
- The reliable determination of the formation and selection mechanisms at microstructure scales in order to understand the fundamentals of casting and other solidification processes (foundry, welding, microelectronic soldering, etc.) and to foster the development of quantitative predictive numerical simulation

In practice, the improvement of the design and processing of advanced materials follows the progress of the sophisticated numerical simulations developed following the explosion in available computational resources. Close comparison with precisely controlled benchmark experiments is necessary for guidance and validation. Researchers and engineers are seeking simpler model experiments that will capture the essence of the thermophysical phenomena to feed numerical simulations with better fundamentals of solidification microstructure formation for

process modeling and uncorrupted physicochemical materials parameters. The goal is to generate reliable data for comparison of numerical simulation predictions of microstructure evolution and solidification processing.

Research conducted in the reduced-gravity environment of space is making significant contributions by rendering negligible one of the most critical parameters, gravity, which can interact in multiple ways in experiments on the ground. The basic understandings learned from these studies, freed from the churning of the alloy melt by gravity-driven fluid flow, are fostering the development of physically better and more accurate models for solidification microstructure development, the prediction of materials properties, and, eventually, may open ways to improve manufacturing on Earth.

Materials processing in space affords a way to remedy the lack of standards that modeling and numerical simulation teams need and is therefore a key area of microgravity research (Ref 3–8). Aside from the basics of materials processing, that is, grain structure, dendrite morphology, mushy zone characteristics, columnar-to-equiaxed transition conditions, micro- and macrosegregation of chemical species, and so on, a number of topics have been included in the research program: eutectic, peritectic, monotectic, and intermetallic alloy growth, and, more generally, multiphase multicomponent alloy solidification.

Casting and Solidification Processing from the Melt

Solidification of a molten metallic alloy is a thermodynamically nonequilibrium process. Without special attention, the solidification process installs a polycrystalline grain structure in

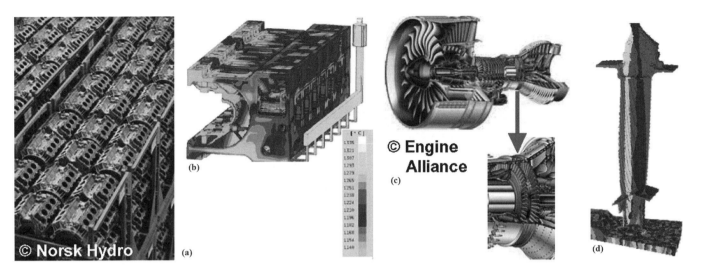

Fig. 1 Representative examples of high-tech castings and integrated-software contributions to processing control. (a) V6 diesel engine blocks cast for Mercedes cars by Hydro Aluminium. ©Norsk Hydro. Used with permission. (b) MAGMASOFT numerical simulation of the surface temperature distribution over an engine car block immediately after mold filling. For such a large piece of metal, the temperature varies considerably because the cooling rate is high close to the surface. Courtesy of H. Fecht. (c) Cutaway view of the Engine Alliance GP7200 engine for the Airbus A380, with a closeup of the turbine blades. ©The Engine Alliance. (d) THERCAST simulated three-dimensional grain structure in a turbine blade geometry produced by investment casting, with the selection with time of a few columnar grains visible on the outer surface. Courtesy of Ch.-A. Gandin

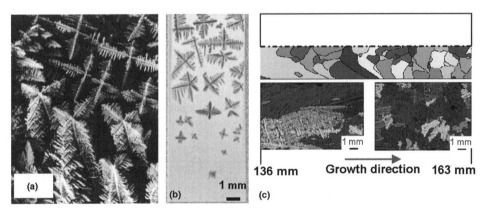

Fig. 2 (a) Columnar dendritic growth in a directionally solidified Co-Sm-Cu peritectic alloy showing primary and secondary arms. The view of the dendrite array is obtained by etching away the $Co_{17}Sm_2$ matrix from the primary cobalt dendrites. Courtesy of R. Glardon and W. Kurz, École Polytechnique Fédérale de Lausanne. (b) Equiaxed grains growing in the melt during cooling of an Al-4wt%Cu alloy. Observed by synchrotron x-ray radiography at the European Synchrotron Radiation Facility (Institut Matériaux Microélectronique Nanosciences de Provence, Univ. Paul Cézanne). Equiaxed crystals have dendritic branches that are not identical to one another, because the arms are not feeling neighbor interactions at the same time. (c) Input on the columnar-to-equiaxed transition in unrefined Al-7wt%Si from the microgravity experiment MACE A in the sounding rocket Maxus 7 and corresponding R2sol axisymmetric two-dimensional numerical simulation of grain structure (Ref 9) showing qualitative agreement. Courtesy of the Columnar-Equiaxed Transition in Solidification Processing (CETSOL) European Space Agency Microgravity Applications Promotion Program

the material (Fig. 1d). In each grain, the growth front reorganizes itself into a diversity of microstructures rather than evolving evenly in space and smoothly in time. The most prevalent solidification microstructures exhibit a dendritic morphology. Columnar dendrites (Fig. 2a) are required for aeroengine turbine blades capable of operating at ever higher temperatures with excellent creep properties and long service life. A dendritic equiaxed grain structure (Fig. 2b) is desirable for homogeneous, macroscopic behavior under mechanical stress, such as for car engine blocks. The columnar-to-equiaxed transition is thus a critical issue in casting

(Fig. 2c) (Ref 9). Besides the grain structure, the dendrites, which control the mechanical properties through their branch spacing and concomitant microsegregation of the chemical species, must be designed in accordance with new materials specifications.

Actually, the relevant length scales in casting are widespread over 10 orders of magnitude. At the nanometer scale, atomic attachment determines the growth kinetics, and the change in atomic arrangement between solid and melt determines the solid-liquid interfacial energy; crystalline defects such as dislocations also are observed. Macroscopic fluid flow driven by

gravity or imposed by an external stimulus (electromagnetic field, vibration, etc.) occurs in the melt at the meter scale of the cast product. The characteristic scales associated with the solidification microstructures are mesoscopic, that is, intermediate, ranging from dendrite tip/arm scale (1 to 100 μm) to the grain size (millimeter to centimeter). It follows that controlling the grain structure of the product and inner microstructure of the grains during the liquid-to-solid phase transition is paramount for the quality and reliability of castings, as well as for tailoring new advanced materials for specific technological applications.

To reliably determine the thermophysical and related properties of metallic melts and the formation and selection mechanisms at microstructure scales, a comprehensive strategy has been established that is based on benchmark experiments on technical and selected model alloy systems. Most of these simplified model experiments are carried out under the terrestrial conditions encountered in practice at industrial casthouses, where gravity-driven effects are unavoidable.

First, significant fluid flow in the melt drives segregation of the chemical species that can be characterized by x-ray radiography at the macroscopic sample scale for both thermal convection and solutal convection. Thermal convection effects are shown in Fig. 3. The horizontal bands in Fig. 3(a) show how the indium concentration stratifies in the liquid ahead of the growth front (Ref 10, 11). The indium concentration stratifications and the growth front predicted by the cellular automaton finite-element (CAFE) model are illustrated in Fig. 3(b) (Ref 12). Solutal convection during directional solidification is shown in Fig. 4 (Ref 13). The development of solute plumes in the dendritic

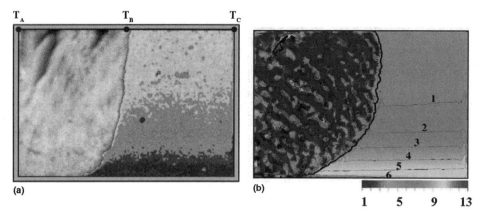

Fig. 3 Solidifying Ga-5wt%In alloy 165 min after beginning of cooling. (a) Indium stratification in the liquid (colored horizontal bands) and growth front morphology (field of view: 0.048 m width, 0.033 m height; T_A, T_B, and T_C indicate the positions of the thermocouples) (Ref 10, 11). (b) Map of indium concentration (wt%) predicted by the cellular automaton finite-element model (thick black line: growth front deduced from the cellular automaton model) (Ref 12)

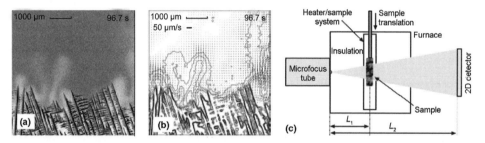

Fig. 4 Dendritic growth in directional solidification of Ga-30wt%In (Ref 13). (a) Snapshot of growing dendritic mush with solutal plumes in the melt showing solute segregation obtained by in situ x-ray radioscopic imaging. (b) Corresponding flow field (vector plot) calculated by optical flow approach at the given time offsets relative to the first image. Contour lines correspond to lines of constant brightness/solute concentration in (a). (c) Schematic drawing of the radiography setup based on the microfocus x-ray source that will be used for sounding-rocket experiment MASER 12 on aluminum-copper alloy. Field of view and magnification are controlled by L_1 and L_2

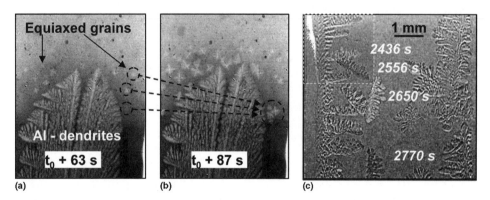

Fig. 5 (a, b) Sedimentation of growing equiaxed grains recorded by synchrotron x-ray radiography in columnar-to-equiaxed transition following a jump in pulling rate from 1.5 to 15 μm/s at t_0. Directional solidification of a refined Al-3.5wt%Ni alloy, $G = 30$ K/cm (Ref 14). (c) Sedimentation of a detached secondary arm (composite radiograph) during directional cooling down of Al-7wt%Si (Institut Matériaux Microélectronique Nanosciences de Provence, Univ. Paul Cézanne)

mush is shown in Fig. 4(a), and the corresponding flow fields are represented in Fig. 4(b). An x-ray radiography configuration is shown in Fig. 4(c).

Second, sedimentation of solid pieces (e.g., equiaxed grains, detached dendrite fragments, etc.) is observed (Fig. 5), the description of which involves evolution at thinner microstructure scale(s) and is still a largely pending issue in modeling (Ref 14). In solidification processing, for example, casting or cooling down, a characteristic sedimentation cone of fine equiaxed crystals generally is seen at the bottom part of the ingot. A longitudinal cross section (Fig. 6a) of an aluminum-base alloy casting shows the grain morphologies that develop during solidification (Ref 15). When the equiaxed grains are allowed to move during CAFE numerical simulation, the predicted grain structure (Fig. 6b) is in qualitative agreement with experiment (Ref 16). The simulation without grain movement is drastically different, showing a homogeneous structure of equiaxed grains (Fig. 6c), which is actually very similar to the grain structure obtained in microgravity processing (Fig. 6d) (Ref 17). Grain growth during cooling down at 1 g with strong settling effects yields a microstructure with fine equiaxed crystals at the bottom of the ingot (Fig. 6e).

Experiments in the limit of diffusion-controlled heat and mass transport in the low-gravity environment of space, and free from sedimentation, eliminate the intricate interactions of various gravity-driven factors (fluid flow, particle/grain sedimentation, bending under weight, etc.) that take place at the multiple length scales of solidification processing (Fig. 7) (Ref 18, 19). The dendrite length scales are the most critical for material strength and malleability/forming. These experiments provide timely benchmark data on material physicochemical properties and the solidification microstructure that are necessary to the clarification of critical pending issues and the sound advancement and validation of numerical predictions (Ref 20–22).

Reduced-gravity platforms have thus risen as unique research tools. The reduced-gravity facilities currently in use, which include drop towers, parabolic flights, sounding rockets, and satellites, are now complemented by the International Space Station (ISS), affording a variety of materials science facilities in a microgravity environment.

Materials Processing in Space

Low-Gravity Platforms and Facilities

Fresh insights into the fundamentals of metallic alloy solidification can be gained with the potential of engineering novel microstructures. Examples for high-tech systems are high-temperature superalloys (e.g., nickel base and titanium base); low-weight, high-strength cast-to-shape metals (aluminum, magnesium, steel); metallic glasses; and nanomaterials. The space environment allows levitated melts to be controlled effectively at temperatures up to 2200 K (3500 °F), which, in turn, enables critical liquid physicochemical parameters to be measured much more accurately than in an Earth-based laboratory. Cooperatively, these cornerstone research activities ultimately contribute to the improved validation of numerical models of solidification processing from the melt, as was mentioned earlier.

To perform these experiments, it is important to have access to extended periods of reduced

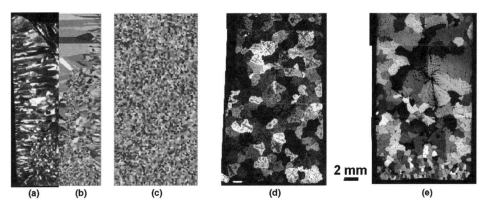

Fig. 6 (a) Longitudinal cross section of an aluminum-base alloy (height = 120mm) cast in a steel mold (Ref 15). Numerical simulation of grain structure in a cast Al-7wt%Si alloy (Ref 16) when grain movement is (b) included and (c) impeded. Equiaxed growth of refined Al-4wt%Cu alloy (Ref 17) in microgravity (d) with regular equiaxed grains and (e) at 1 g with a strong settling effect

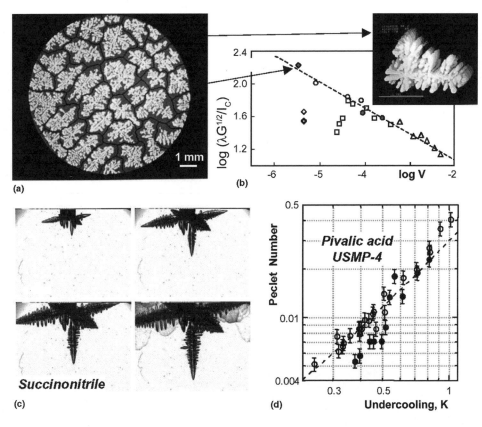

Fig. 7 (a) Postmortem observation of transverse cross section of a columnar dendritic array formed in microgravity at V = 4.2 µm/s and G = 30 K/cm in Al-26wt%Cu (D1-Spacelab mission, 1985), with huge dendrite spacing (1.5 mm) compared to that on Earth (450 µm). (b) Outputs: microgravity data (filled symbols) following the diffusion theory (dashed line) in the nondimensional diagram of dendrite spacing against growth velocity. The inset is a three-dimensional reconstruction of the morphology of an individual dendrite from a series of closely spaced cross sections and shows the coarsening of side arms (Ref 18, 19). In situ characterization of dendrite-free growth in an undercooled melt of transparent organic surrogates for metals (Ref 20–22). (c) Images of the formation of a succinonitrile dendrite. (d) Diagram showing a slowing down of dendrite growth velocity (Pe ~ $V^{1/2}$) in the absence of fluid flow (filled circles: microgravity data, Isothermal Dendritic Growth Experiment/U.S. Microgravity Payload (USMP)-4, 1997; open circles: 1 g data)

assembly and utilization and also the unfortunate *Columbia* accident, a series of microgravity experiments in materials research is now commencing onboard the ISS in a number of multiuser facilities afforded by major space agencies, such as the European Space Agency (ESA) in the Columbus module (Fig. 8). For the sake of completeness, it is valuable to briefly introduce some of the available tools of interest to the materials science community in the ISS era that are expected be used to 2020 and perhaps somewhat beyond.

In the field of materials research, both directional and isothermal solidification experiments will be performed in the Materials Science Laboratory (MSL) using dedicated furnace inserts, with the possibility of applying stimuli such as a rotating magnetic field to force fluid flow under microgravity conditions. The electromagnetic levitator (EML) will enable containerless melting and solidification of alloys and semiconductor samples, either under ultrahigh vacuum or high-purity gaseous atmospheres. Furthermore, the EML is equipped with highly advanced diagnostic tools that allow accurate measurements of thermophysical properties as well as direct observation of the experiment during flight by high-speed videography (see further discussion in the next section of this article).

The synergy of competences gained by teaming together scientists from academia and research-and-development engineers from industry on focused critical fields, which has been fostered by the space agencies involved in the ISS utilization (Ref 23), is proving to have a significant, positive return on the intellectual and financial investment. A striking example is the ESA-European Union integrated project IMPRESS (Intermetallic Materials Processing in Relation to Earth and Space Solidification) that comprises approximately 40 industrial and academic research groups from 15 countries in Europe, as shown in Fig. 9(a) (Ref 24). Among others, a new intermetallic aluminum-titanium alloy was developed for investment casting of lightweight, high-strength turbine blades for next-generation aeroengines (Fig. 9b), which are expected to contribute to reducing fuel consumption. Also, the first fractallike aggregates of catalytic nickel (Fig. 9c), for potential high-performance fuel cell applications, were produced in low gravity, where collapse under their own weight is suppressed.

Selected Highlights: Microgravity Research Inputs to Modeling and Numerical Simulation

Liberated from bothersome gravity-driven effects, new, major outcomes are expected from the ISS era that will build on pioneer and precursor experiments that have paved the way (e.g., Ref 3–8), in particular by making use of NASA's space shuttles or the Salyut and Mir space stations. Rather than an exhaustive

gravity, available in unmanned satellites and on retrievable platforms such as Foton (Russian science satellite) and the European Retrievable Carrier (European Space Agency science satellite), National Aeronautics and Space Administration (NASA) shuttles, and space stations (Mir, ISS). After many years of definition and preparation, and despite occasional slowdowns brought on by fluctuations in governmental space policies regarding support of ISS

compilation, representative success stories that have impacted science and applications are highlighted next.

Dendritic Growth Studies. Two striking examples of important low-gravity solidification experiments are shown in Fig. 7. Dendritic directional solidification of aluminum-copper alloys was performed in microgravity in the Gradient Heating Facility instrument of the French space agency, CNES, during the D1-Spacelab mission in 1985. The drastic reduction in primary dendrite spacing on Earth due to gravity-driven fluid flow was unambiguously demonstrated (Ref 18, 19).

The Isothermal Dendritic Growth Experiment (IDGE) provided a separate and quantitative test of the Ivantsov steady-state shape and selection theory of free dendrite growth under diffusion transport, which was not possible before (Ref 20–22). Hundreds of IDGE experiments were carried out aboard the Space Shuttle *Columbia* as part of NASA's U.S. Microgravity Payload (USMP)-2 to -4 Shuttle missions in 1994, 1996, and 1997. Dendrites were monitored in situ in pure succinonitrile and pivalic acid, which exhibit significantly different tip shapes and overall morphologies. Both steady-state tip speed and tip radii were

measured as functions of the initial supercooling from the global dendrite images provided by the IDGE flight instruments (35 mm films and then full gray-scale videos).

Electromagnetic Levitation. In recent years, there has been a great deal of work on preparatory experiments in precursor instruments of ISS facilities. For instance, using an electromagnetic levitator, surface oscillations of the liquid hot drop with a diameter of 8 mm (Fig. 10a) can be introduced by an electromagnetic pulse and the results captured by a high-speed, high-resolution video camera. The surface tension as a function of temperature in the range

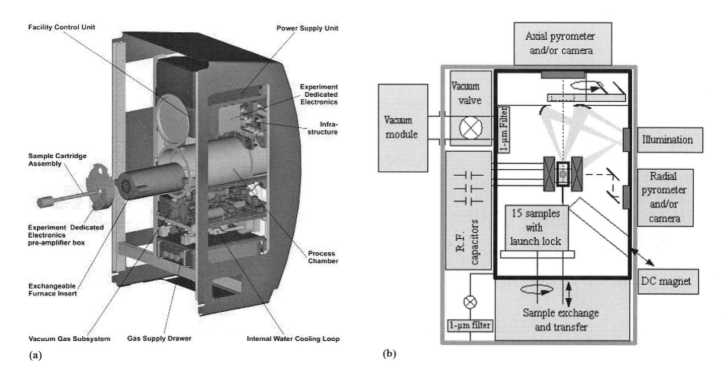

Fig. 8 Schematic presentation of (a) the Materials Science Laboratory and (b) the electromagnetic levitator reaching temperatures up to 2200 K (3500 °F). Courtesy of EADS Astrium, Germany

Fig. 9 Intermetallic Materials Processing in Relation to Earth and Space Solidification (IMPRESS) integrated project. (a) Geographical distribution of IMPRESS partners in Europe. Courtesy of European Space Agency (ESA). (b) Lightweight, high-strength TiAl turbine blades. Courtesy of ACCESS e.V., Germany. (c) Fractal-like aggregate of catalytic nickel. Courtesy of IFAM, Bremen, Germany, via ESA

of 1300 to 1750 K (1100 to 1480 °C, or 1880 to 2690 °F) for liquid Ni-75at.%Al processed under 10 s of low gravity in four parabolic flights is shown in Fig. 10(b).

It should be stressed that the EML is perfectly adequate for measuring thermophysical properties of deeply undercooled melts. Indeed, such undercoolings are required to freeze metallic alloys as amorphous materials with unique mechanical properties (Fig. 11a). Due to their excellent forming ability (Fig. 11b), metallic glasses have found diverse applications in common life, from cellular phones (Fig. 11c) to sporting goods (golf club heads, baseball bats, tennis rackets, etc.). (Because the material absorbs less energy than traditional materials, more energy is transferred to the ball and less to the player's hands, giving a "soft" feel.) Also, metallic glasses are used in high-tech applications such as microelectromechanical systems technology (Ref 26). It is worth noting that samples from California Institute of Technology processed on the space shuttle *Columbia*

(Microgravity Sciences Laboratory-1 mission, 1997) in the electromagnetic containerless metal-processing facility TEMPUS of the German Space Agency DLR yielded the first measurements of specific heat and thermal expansion of glass-forming metallic alloys and the highest temperature (2273 K, or 2000 °C, or 3630 °F) and largest undercooling (340 K, or 340 °C, or 610 °F) ever achieved in space (Ref 27).

Levitation of molten metals and other materials has been developed to solve a series of practical issues. First, high-temperature chemical reactivity between the liquid and container, if any, can be devastating (e.g., silicon in contact with graphite); sample contamination can lead to drastically erroneous phase transformation and thermophysical properties data. Second, heterogeneous nucleation of solid phases easily occurs at the container wall, making deep undercooling and growth of metastable phases unattainable. Conversely, without a container, undercoolings reaching several hundred degrees Kelvin below the melting point are now produced.

Microgravity has additional advantages over containerless levitation processing. Indeed, segregation and sedimentation effects become negligible, and uniform mixing is achieved; also, much larger samples can be levitated than is possible on Earth. However, levitation is still not straightforward in microgravity, where residual gravity and capillary forces at the fluid surfaces are still at work. (Everyone has seen astronauts on television catching free-floating water drops to drink.) Stable sample positioning, which must be ensured, for example, for heating with lasers or pyrometry measurements, is easier in microgravity and requires weaker electromagnetic fields.

For more than 20 years, space agencies have been developing levitators using electromagnetic forces as well as acoustic waves or electrostatic forces. Electromagnetic levitators (e.g., Fig. 8b) take advantage of well-known phenomena that follow from Maxwell's equations. High-frequency alternating and/or static magnetic fields are generated by passing a current through an assembly of coils shaped for positioning and heating. The applied electromagnetic field induces eddy currents in the metallic sample placed between the coils that, due to the Joule effect, heat and may even melt the sample. The coupling of these currents with the applied electromagnetic field is used to impose a lifting force on the sample, which is concomitantly undergoing fluid flow driven by the electromagnetic forces when molten.

Also in the EML, solidification of the metallic drop can be triggered and resolved in time with a high-speed camera (Fig. 12a) (Ref 28). This mimics the growth of a single equiaxed dendritic grain in a casting (Ref 29), which affords a case study for interactive feedback with numerical modeling. A comparison shows that the prediction by a three-dimensional CAFE model (Fig. 12b), which integrates dynamic memory allocation methods for the creation of the cellular automaton grid based on a finite-element mesh (Fig. 12c), is already in qualitative agreement with observation.

Solidification and Casting. A series of ground and sounding-rocket experiments were

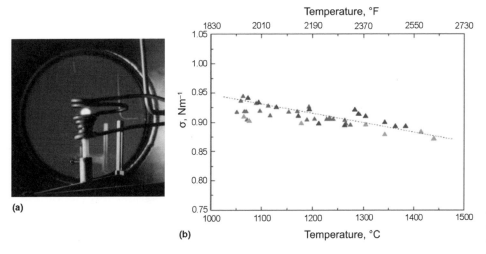

(b)

Fig. 10 (a) Video image of a fully spherical liquid sample of titanium in an electromagnetic levitator obtained in a parabolic flight. This allows measuring the surface tension and viscosity of liquid metallic alloys at elevated temperature with high accuracy. Courtesy of European Space Agency. (b) Surface tension of a drop of molten Ni-75at.% Al between 1300 and 1750 K (1100 and 1480 °C, or 1880 and 2690 °F). Adapted from R. Wunderlich, Ulm University. European Space Agency-German Space Agency (DLR) ThermoLab project

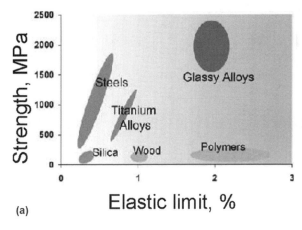

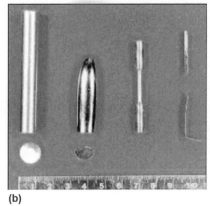

Fig. 11 (a) Elastic limit and strength of bulk metallic glasses (BMGs) compared to other materials. (b) Examples of shapes obtained for the BMG $Zr_{41}Ti_{14}Cu_{12.5}Ni_{10}Be_{22.5}$ (Ref 25). (c) Vertu cell phone with subframe and battery case made of thin BMG sheets with extremely high strength, hardness, and scratch resistance (Ref 26)

carried out within the Microstructure Formation in Castings of Technical Alloys under Diffusive and Magnetically Controlled Convective Conditions (MICAST), Columnar-Equiaxed Transition in Solidification Processing (CETSOL), and IMPRESS projects to prepare the utilization of the Low-Gradient Facility and the Solidification with Quenching Facility in the MSL. Unexpectedly, it was found that fluid flow, either driven by gravity or forced by a magnetic field, was enhancing dendrite arm coarsening despite the tip radius (Fig. 13a) not being affected. Benchmarking of the columnar-to-equiaxed transition (CET) has begun on Al-7wt%Si alloys in the Maxus 7 sounding rocket, and results suggest that the presence or absence of CET should be linked to the effectiveness of melt inoculation by dendrite arm fragmentation. For the sake of comparison, development of phase-field modeling (Ref 30–33) is pursued to reach quantitative numerical prediction from the scale of the dendrite tip (Fig. 13b) (Ref 34) to the scale of cooperative array growth (Fig. 13c).

In Situ and Real-Time Monitoring of Solidification Processing

The development of in situ, real-time diagnostics to clarify the dynamic phenomena acting and interacting in the formation of the solidification microstructure in processing was, and is still, a major challenge. This has been successfully accomplished using optical methods on transparent systems that freeze similar to metals, with the main focus on growth morphology (Fig. 7c, Fig. 14a–c). More recently, x-ray imaging of representative metallic systems (e.g., aluminum-silicon alloys are the most widely used alloys for aluminum cast parts), with pioneering work at synchrotrons such as the European Synchrotron Radiation Facility (Fig. 5, Fig.14d, e), has made possible the in situ monitoring of solidification of "real" materials of interest to industry, which are opaque to visible light (Ref 14, 35–43). Beyond the formation, selection, if any, and stability of the solidification microstructure,

unprecedented insight can be gained on chemical segregation and mechanical effects. This endeavor is resulting in new facilities for the ISS and sounding rockets.

Conclusion and Perspectives

Cast materials are common objects in everyday life (car and jet engines, metallic skeletons of buildings, dental and orthopedic implants, etc.). They are used primarily because of their attractive mechanical properties and the ability to sustain and/or transmit forces with negligible damage. The latest developments have been achieved through solidification processing from the melt on Earth and in space, to improve casting facilities. Examples are chill casting of complex parts into net shape products and gas atomization and powder production of catalytic compounds. To achieve these goals reliably, there has been a worldwide effort to develop sophisticated computer models, which, in all aspects of industrial materials processing, supports long-term sustainability and technology leadership on a global scale. Thus, most national space agencies have established or are establishing strong scientific programs to use space as a critical, complementary venue to expedite the development of new products through the fundamental advances acquired from parabolic flights, sounding rockets, and the ISS.

Since the Bronze Age, mankind has engineered materials to suit its purposes, and, acknowledging that today (2010) man is living "on the boundary between the Iron Age and the New Material Age" (Ref 44), basic research on solidification processing in microgravity can provide important insights to the engineering of materials on Earth. Interest in multicomponent, multiphase alloys, including metastable phases, and functional materials is expanding. For instance, eutectic ceramic oxides are promising in situ composites for high-temperature structural applications (Fig. 15a), but only if processing is carefully designed to tailor the solid-solid

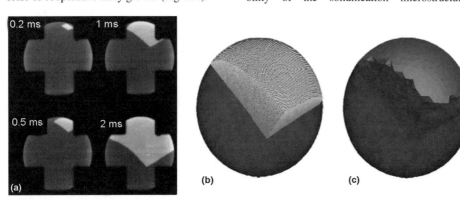

Fig. 12 (a) High-speed video images obtained in a parabolic flight during solidification at 70 K (126 °F) undercooling of an Fe-0.0046wt%C-0.636Mn sample in an electromagnetic levitator. The images show the growth of a dendritic grain (light gray) into the alloy melt (dark gray). Courtesy of D. Herlach et al., German Space Agency (DLR)-Köln. Corresponding numerical simulation using a three-dimensional cellular automaton finite-element model shows (b) the envelope of the growing grain and (c) the active (darker or red) chevron-shaped region through the midsection of the sphere and deactivated (bright or green area at top of sphere) elements at this computation time step. Courtesy of Ch.-A. Gandin, Centre for Material Forming (CEMEF), Chill Cooling for the Electro-Magnetic Levitator in Relation with Continuous Casting of Steel (CCEMLCC), European Space Agency-Microgravity Applications Promotion Project

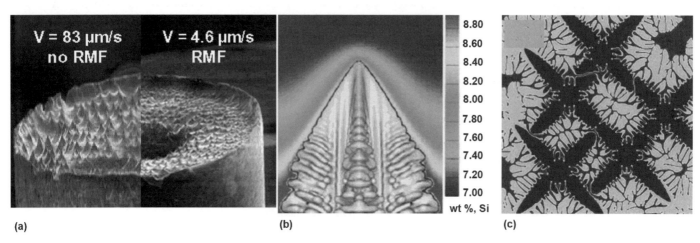

Fig. 13 (a) Decanted solid-liquid interfaces of Al-7Si-Mg samples directionally solidified without and with rotating magnetic field (RMF). (b) Three-dimensional phase-field simulation of AlSi7 dendrite with silicon solute field, revealing the dendrite tips (Ref 34). Courtesy of ACCESS e.V., Germany. (c) Two-dimensional phase-field simulation of nickel-aluminum solidification showing primary Ni_2Al_3 dendrites (dark-gray cross-hair morphology regions, or blue), $NiAl_3$ peritectic (bright areas, or green), and the remaining liquid (dark regions within the other two regions, or red/brown). Courtesy of A. Mullis, University of Leeds

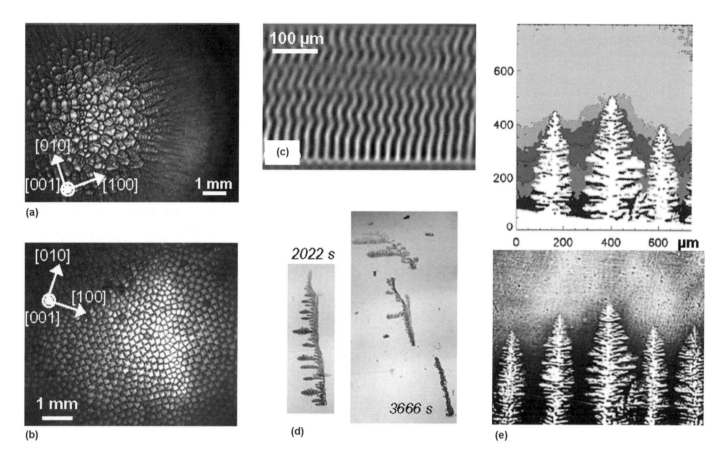

Fig. 14 In situ and real-time observation of the solidification microstructure in three-dimensional samples. Optical methods on transparent model systems. (a, b) [001]-dendrites viewed from above during directional solidification of dilute succinonitrile-camphor alloys in the Device for the Study of Critical Liquids and Crystallization (DECLIC)-Directional Solidification Insert of the French Space Agency, CNES (Ref 41, 42), showing that the localization of vertical dendrites at the center in the ground experiment (a) is effectively cured in microgravity experiments under diffusion transport (b), as recently observed in the first microgravity experiments carried out on the International Space Station, where laterally extended arrays of dendrites formed. Courtesy of Institut Matériaux Microélectronique Nanosciences de Provence (IM2NP), Univ. Paul Cézanne. (c) Zigzag instability of lamellae in directional solidification of CBr_4-C_2Cl_6 eutectic in Directional Solidification Laboratory (DIRSOL) model (Ref 43). Synchrotron x-ray imaging on metallic systems: (d) Disorienting on the dendrite stem (evidenced by the splitting of the dendrite image into pieces on topographs) during directional cooling of Al-7wt%Si. Courtesy of IM2NP, Univ. Paul Cézanne. (e) Solute field in the interdendritic liquid, varying from 33 wt% Cu (black) to 30 wt% Cu (light-gray region at top of image, or yellow online) in the melt ahead (dendrites are white), determined from the gray levels in the radiograph below (Ref 35, 37)

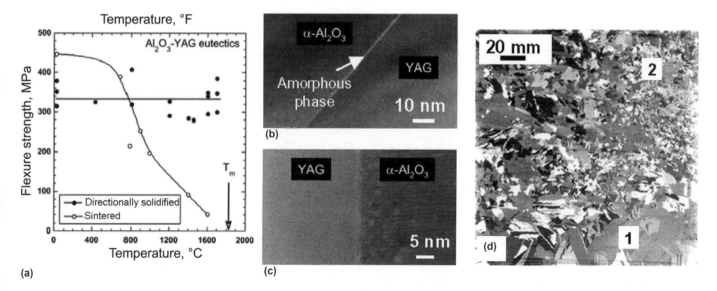

Fig. 15 Al_2O_3/yttrium-aluminum garnet (YAG) eutectic ceramics. (a) Dependence of high-temperature flexure strength on processing. Transmission electron microscopy revealing (b) the presence of an amorphous phase at the grain boundary in the sintered material and (c) a perfect interface in the directionally solidified material (Ref 45). (d) Grain structure in a polycrystalline wafer cut in directionally solidified silicon showing normal size grains (1) and small grains (2), or grits (Ref 46)

interfaces at the nanometer scale (Fig. 15b, c). The interfaces act as barriers to high-temperature dislocation motion, which increases resistance to plastic deformation (Ref 45). Polycrystalline silicon for photovoltaic solar cells is another example of a functional material produced by solidification processing. As the demand for clean-energy production increases, the need for efficient solar cells is expected to rise exponentially. To achieve solar cell efficiency, the research focus is on understanding and preventing the transition to small-grain, equiaxed-like growth, because grain boundaries impede the flow of electrons and favor detrimental electron-hole recombination, thereby reducing the power output (Fig. 15d) (Ref 46).

Using the reduced-gravity environment of space to generate benchmark data to input modeling and numerical simulation of materials processing and advance materials engineering is doubtless appealing. Yet, potential users must be aware that weightlessness is only approached in space experiments. Residual gravity (10^{-6} to 10^{-3} Earth gravity) is a vector field that varies in space and over a frequency spectrum in time and also depends on crew activity on manned spacecrafts. Thus, the efficiency of gravity reduction must be estimated case by case using an order-of-magnitude analysis (Ref 47; see also Chapter 14 in Ref 4) or direct numerical simulation. Modeling and simulation are all the more important because costs are huge for academic and research-and-development laboratories, and the return on own-money investment is not necessarily as rapid as expected in industry. Furthermore, in light of the hazards that have impacted ISS assembling and utilization funding in the last decade, it is certainly a prerequisite that continued governmental political and financial support through the space agencies persist to give confidence to individuals or companies willing to invest careers or funds in using such a unique and opportune research tool that has now reached maturity.

REFERENCES

1. S. Hans, I. Poitrault, and V. Vidal, Industrial Use of Simulation Software Dedicated to the Solidification of Ingots and Casting Parts, *Modeling of Casting, Welding and Advanced Solidification Processes—XI*, Ch.-A. Gandin and M. Bellet, Ed., TMS, The Minerals, Metals & Materials Society, Warrendale, PA, 2006, p 1053–1060
2. Ch.-A. Gandin and I. Steinbach, Direct Modeling of Structure Formation, *Casting*, Vol 15, *ASM Handbook*, S. Viswanathan, Ed., ASM International, 2008, p 435–444
3. G.E. Rindone, Ed., *Materials Processing in the Reduced Gravity Environment of Space*, Elsevier, New York, 1982
4. H.U. Walter, Ed., *Fluid Sciences and Materials Science in Space*, Springer-Verlag, Berlin, 1987
5. D.J. Jarvis and O. Minster, Metallurgy in Space, *Mater. Sci. Forum*, Vol 508, 2006, p 1–18
6. L.L. Regel, *Materials Processing in Space: Theory, Experiments, and Technology*, Halsted Press, New York, 1987
7. G. Seibert, B. Fitton, and B. Battrick, Ed., *A World Without Gravity*, ESA SP-1251, ESA, Noordwijk, 2001
8. M.E. Glicksman, Solidification Research in Microgravity, *Casting*, Vol 15, *ASM Handbook*, S. Viswanathan, Ed., ASM International, 2008, p 398–401
9. Ch.-A. Gandin, J. Blaizot, S. Mosbah, M. Bellet, G. Zimmermann, L. Sturz, D.J. Browne, S. McFadden, H. Jung, B. Billia, N. Mangelinck, H. Nguyen-Thi, Y. Fautrelle, and X. Wang, Modeling of Heat and Solute Interactions upon Grain Structure Solidification, *Mater. Sci. Forum*, Vol 649, 2010, p 189–198
10. H. Yin and J.N. Koster, In Situ Observation of Concentrational Stratification and Solid-Liquid Interface Morphology during Ga-5% in Alloy Melt Solidification, *J. Cryst. Growth*, Vol 205, 1999, p 590–606
11. H. Yin and J.N. Koster, Chemical Stratification and Solidification in a Differentially Heated Melt, *J. Alloy. Compd.*, Vol 352, 2003, p 197–209
12. G. Guillemot, Ch.-A. Gandin, and M. Bellet, Interaction between Single Grain Solidification and Macrosegregation: Application of a Cellular Automaton—Finite Element Model, *J. Cryst. Growth*, Vol 303, 2007, p 58–68
13. S. Boden, S. Eckert, B. Willers, and G. Gerbeth, X-Ray Radioscopic Visualization of the Solutal Convection during Solidification of a Ga-30 wt% in Alloy, *Metall. Mater. Trans. A*, Vol 39, 2008, p 613–623
14. H. Nguyen-Thi, G. Reinhart, N. Mangelinck-Noel, H. Jung, B. Billia, T. Schenk, J. Gastaldi, J. Härtwig, and J. Baruchel, In Situ and Real-Time Investigation of Columnar-to-Equiaxed Transition in Metallic Alloy, *Metall. Mater. Trans. A*, Vol 38, 2007, p 1458–1464
15. N.L. Cupini and M. Prates de Campos Filho, Aluminium Grain Refinement by Crystal Multiplication Mechanism Stimulated by Hexachloroethane Additions to the Mould Casting, *Solidification and Casting of Metals*, Metals Society, London, 1979, p 193–197
16. Ch.-A. Gandin, T. Jalanti, J.-L. Desbiolles, and M. Rappaz, Stochastic Modelling of Dendritic Grain Structures in the Presence of Convection-Sedimentation, *Solidification Microstructures*, CD-ROM in honor of Prof. W. Kurz 60th Birthday, EPFL, Lausanne, 1998
17. M.D. Dupouy, D. Camel, J.E. Mazille, and I. Hugon, Columnar to Equiaxed Transition in a Refined Al-Cu Alloy under Diffusive and Convective Transport Conditions, *Mater. Sci. Forum*, Vol 329–330, 2000, p 25–30
18. M.D. Dupouy, D. Camel, and J.J. Favier, Natural Convection in Directional Dendritic Solidification of Metallic Alloys—I. Macroscopic Effects, *Acta Metall.*, Vol 37, 1989, p 1143–1157
19. M.D. Dupouy, D. Camel, and J.J. Favier, Natural Convective Effects in Directional Dendritic Solidification of Binary Metallic Alloys: Dendritic Array Primary Spacing, *Acta Metall. Mater.*, Vol 40, 1992, p 1791–1801
20. M.E. Glicksman and A.O. Lupulescu, Dendritic Crystal Growth in Pure Materials, *J. Cryst. Growth*, Vol 264, 2004, p 541–549
21. M.B. Koss, J.C. Lacombe, L.A. Tennenhouse, M.E. Glicksman, and E.A. Winsa, Dendritic Growth Tip Velocities and Radii of Curvature in Microgravity, *Metall. Mater. Trans. A*, Vol 30, 1999, p 3177–3190
22. J.C. LaCombe, M.B. Koss, and M.E. Glicksman, Tip Velocities and Radii of Curvature of Pivalic Acid Dendrites under Convection-Free Conditions, *Metall. Mater. Trans. A*, Vol 38, 2007, p 116–126
23. A. Wilson, Ed., *Microgravity Applications Programme—Successful Teaming of Science and Industry*, ESA-SP 1290, European Space Agency, Noordwijk, The Netherlands, 2005
24. IMPRESS, http://www.spaceflight.esa.int/impress/
25. A. Peker, Ph.D. dissertation, California Institute of Technology
26. J. Schroers and N. Paton, Amorphous Metal Alloys Form Like Plastics, *Adv. Mater. Proc.*, Jan 2006, p 61–63
27. S.C. Glade, R. Busch, D.S. Lee, W.L. Johnson, R.K. Wunderlich, and H.J. Fecht, Thermodynamics of $Cu_{47}Ti_{34}Zr_{11}Ni_8$, $Zr_{52.5}Cu_{17.9}Ni_{14.6}Al_{10}Ti_5$ and $Zr_{57}Cu_{15.4}Ni_{12.6}Al_{10}Nb_5$ Bulk Metallic Glass Forming Alloys, *J. Appl. Phys.*, Vol 87, 2000, p 7242–7248
28. D.M. Herlach, R. Lengsdorf, P. Galenko, H. Hartmann, Ch.-A. Gandin, S. Mosbah, A. Garcia-Escorial, and H. Henein, Non-Equilibrium and Near-Equilibrium Solidification of Undercooled Melts of Ni- and Al-Based Alloys, *Adv. Eng. Mater.*, Vol 10, 2008, p 444–452
29. R. Heringer, Ch.-A. Gandin, G. Lesoult, and H. Henein, Atomized Droplet Solidification as an Equiaxed Growth Model, *Acta Mater.*, Vol 54, 2006, p 4427–4440
30. I. Steinbach, C. Beckermann, B. Kauerauf, Q. Li, and J. Guo, Three-Dimensional Modeling of Equiaxed Dendritic Growth on a Mesoscopic Scale, *Acta Mater.*, Vol 47, 1999, p 971–982
31. H.J. Diepers, D. Ma, and I. Steinbach, History Effects during the Selection of Primary Dendrite Spacing. Comparison of Phase-Field Simulations with Experimental Observations, *J. Cryst. Growth*, Vol 237–239, 2002, p 149–153
32. B. Böttger, J. Eiken, and I. Steinbach, Phase-Field Simulation of Equiaxed Solidification in Technical Alloys, *Acta Mater.*, Vol 54, 2006, p 2697–2704

33. I. Steinbach, Pattern Formation in Constrained Dendritic Growth with Solutal Buoyancy, *Acta Mater.,* Vol 57, 2009, p 2640–2645
34. G. Zimmermann, L. Sturz, and M. Walterfang, Directional Solidification of Al-7Si Based Alloys in a Rotating Magnetic Field, *Solidification Processing 2007,* H. Jones, Ed., University of Sheffield, Sheffield, U.K., 2007, p 395–399
35. R.H. Mathiesen, L. Arnberg, K. Ramsøskar, T. Weitkamp, C. Rau, and A. Snigirev, Time-Resolved X-Ray Imaging of Aluminum Alloy Solidification Processes, *Metall. Mater. Trans. B,* Vol 33, 2002, p 613–623
36. B. Billia, N. Bergeon, H. Nguyen-Thi, H. Jamgotchian, J. Gastaldi, and G. Grange, Cumulative Mechanical Moments and Microstructure Deformation Induced by Growth Shape in Columnar Solidification, *Phys. Rev. Lett.,* Vol 93, 2004, p 126105
37. R.H. Mathiesen and L. Arnberg, X-ray Radiography Observations of Columnar Dendritic Growth and Constitutional Undercooling in an Al-30wt%Cu Alloy, *Acta Mater.,* Vol 53, 2005, p 947–956

38. H. Yasuda, I. Ohnaka, K. Kawasaki, A. Sugiyama, T. Ohmichi, J. Iwane, and K. Umetani, Direct Observation of Stray Crystal Formation in Unidirectional Solidification of Sn-Bi Alloy by X-Ray Imaging, *J. Cryst. Growth,* Vol 262, 2004, p 645–652
39. B. Li, H.D. Brody, and A. Kazimirov, Synchrotron Microradiography of Temperature Gradient Zone Melting in Directional Solidification, *Metall. Mater. Trans. A,* Vol 37, 2006, p 1039–1044
40. G. Reinhart, A. Buffet, H. Nguyen-Thi, B. Billia, H. Jung, N. Mangelinck-Noël, N. Bergeon, T. Schenk, J. Härtwig, and J. Baruchel, In-Situ and Real-Time Analysis of the Formation of Strains and Microstructure Defects during Solidification of Al-3.5 Wt Pct Ni Alloys, *Metall. Mater. Trans. A,* Vol 39, 2008, p 865–874
41. C. Weiss, N. Bergeon, N. Mangelinck-Noël, and B. Billia, Effects of the Interface Curvature on Cellular and Dendritic Microstructures, *Mater. Sci. Eng. A,* Vol 413–414, 2005, p 296–301

42. C. Weiss, N. Bergeon, N. Mangelinck-Noël, and B. Billia, Cellular Pattern Dynamics on a Concave Interface in Three-Dimensional Alloy Solidification, *Phys. Rev. E,* Vol 79, 2009, p 011605
43. S. Akamatsu, S. Bottin-Rousseau, and G. Faivre, Experimental Evidence for a Zigzag Bifurcation in Bulk Lamellar Eutectic Growth, *Phys. Rev. Lett.,* Vol 93, 2004, p 175701
44. G.P. Thomson, Nobel Prize in Physics, 1937
45. J. Llorca, and V.M. Orera, Directionally Solidified Eutectic Ceramic Oxides, *Prog. Mater. Sci.,* Vol 51, 2006, p 711–809
46. N. Mangelinck-Noel and T. Duffar, Modelling of the Transition from a Planar Faceted Front to Equiaxed Growth: Application to Photovoltaic Polycrystalline Silicon, *J. Cryst. Growth,* Vol 311, 2008, p 20–25
47. D. Camel and J.J. Favier, Scaling Analysis of Convective Solute Transport and Segregation in Bridgman Crystal Growth from the Doped Melt, *J. Phys.(France),* Vol 47, 1986, p 1001–1014

ASM Handbook, Volume 22B, *Metals Process Simulation*
D.U. Furrer and S.L. Semiatin, editors

Copyright © 2010, ASM International®
All rights reserved.
www.asminternational.org

Thermophysical Properties

Juan J. Valencia, Concurrent Technologies Corporation
Peter N. Quested, National Physical Laboratory

ADVANCED COMPUTER SIMULATION TECHNOLOGY is a powerful tool used to understand the critical aspects of heat-transfer and fluid-transport phenomena and their relationships to metallurgical structures and defect formation in metal casting and solidification processes. Computational models are enabling the design and production of more economical and higher-quality castings. To produce accurate and reliable simulation of the complex solidification processes, accurate, self-consistent, and realistic thermophysical properties input data are necessary. Unfortunately, reliable data for many alloys of industrial interest are very limited.

Sand, ceramic, and metal molds are extensively used to cast most metals. During the solidification process, the predominant resistance to heat flow is within the mold/metal interface and the mold itself; thus, the primary interest is not the mold thermal history but rather the rate at which the heat is extracted from the solidifying metal. Therefore, heat transfer is the governing phenomenon in any casting process. Heat transfer is fundamentally described by the heat-transfer coefficient, the temperature gradient, the geometry of the system, and the thermophysical properties of both metal and mold material.

Table 1 shows the required thermophysical properties that must be available for input before reliable numerical simulations of a casting or solidification process can be performed, as well as their influence in the prediction of defects. Current commercial software requires that the thermal conductivity, specific heat capacity, latent heat, solidus and liquidus temperatures, and density must be known for heat-transfer operations. Viscosity, density, wetting angle, and surface tension of the molten alloy are required for fluid flow operations. In addition to the metal properties, mold materials properties are also needed to conduct an effective simulation.

Sources and Availability of Reliable Data

The thermophysical property data found in the literature for engineering and design calculations of casting processes must not be used indiscriminately without knowing their source and reliability. Prior to using a given set of data, it is very important to critically evaluate and analyze the available thermophysical property data, to give judgment on their reliability and accuracy.

There are several main sources of thermophysical property data that provide the most authoritative and comprehensive compilations of critically and systematically evaluated data that are presently available. The challenge of finding data is discussed in Ref 2. The data have been published and can be found in the following resources:

- The Center for Information and Numerical Data Analysis and Synthesis (CINDAS) generated and recommended reference values for diverse materials (Ref 3 to 12).
- Smithells' *Metals Reference Book* provides an extensive compilation of thermochemical data for metals, alloys, and compounds of metallurgical importance (Ref 13).
- *Summary of Thermal Properties for Casting Alloys and Mold Materials* by R.D. Pehlke and co-workers (Ref 14)
- The ASM International Materials Properties Database Committee publishes a comprehensive thermal properties database of most commercially available metals (Ref 15).
- *Recommended Values of Thermophysical Properties for Selected Commercial Alloys* by K.C. Mills includes the experimental determination, estimation, and validation of the thermophysical properties in the solid and liquid states (Ref 16).

Computer models based on first principles of thermodynamics and kinetics of phase transformations have been developed to calculate thermophysical properties for various materials in the solid and liquid states (Ref 17 to 22). However, their use is still limited due to the lack of thermodynamic data and accurate measurements of thermophysical properties for materials of industrial interest. Also, sensitivity studies (Ref 23) are necessary to truly evaluate the reliability of calculated thermophysical property data from these models in actual casting/solidification processes.

Limitations and Warning on the Use of Data

The thermophysical properties data presented in this article are provided to assist in the materials properties selection for the simulation of casting processes. Great effort has been exercised in the compilation and analysis of the data, and careful attention has been taken to faithfully duplicate the data and their sources found in the literature. The thermophysical properties data provided here shall bear the warning "not for design purposes." It is the full responsibility of the reader to further investigate the sources of information and follow all necessary engineering steps to make sure the validity and quality of the data meet the requirements of the intended application.

Methods to Determine Thermophysical Properties

Experimental determinations of reliable thermophysical properties are difficult. In the solid state, the properties recorded in the technical literature are often widely diverging, conflicting, and subject to large uncertainties. This problem is particularly acute for materials in the mushy and liquid state. Also, accurate, consistent, and reliable thermophysical property measurements are experimentally difficult. Convection effects in molten samples and their interactions and reactivity with their containers and environment often exacerbate the difficulties.

The measurements are difficult because of high temperatures and the reactivity of some alloys. The methods and strategies adopted to minimize these effects are presented and discussed in the article "Measurement of Thermophysical Properties at High Temperatures for Liquid, Semisolid, and Solid Commercial Alloys" in this Volume. Table 2 lists some common techniques used for the measurement of relevant thermophysical properties.

Numerous methods exist for the measurement of thermophysical properties of metallic materials and are cited in the literature (Ref 16, 25 to 45). However, only a few of them

Table 1 Thermophysical property data required for metal casting

Casting process component		Transport phenomena for casting	Thermophysical data required	Computer modeling for process, part design, and defect prediction
Furnace metal	S	Heat transfer • Conduction • Convection • Radiation	Heat-transfer coefficient • Metal/mold • Metal/core • Metal/chill • Mold/chill • Mold/environment Emissivity—Metal/mold/furnace wall Temperature-dependent parameters • Density • Heat capacity • Conductivity Latent heat of fusion Liquidus and solidus	Effective design for: • Riser • Chill • Insulation Solidification direction Solidification shrinkage Porosity Hot spots
Mold core chill	O L I D I F	Mass transfer (fluid flow)	Temperature dependent • Viscosity • Surface tension • Density	Effective design for: • Ingate • Runner • Vents Pouring parameters • Temperature • Pouring rate Mold filling time Cold shut Missruns
Insulation	I C A T	Microstructural evolution	Phase diagram Phase chemical composition Capillarity effect (Gibbs-Thompson coefficient) Nucleation and growth parameters Solid fraction vs, temperature Diffusivity • Solubility	Microsegregation Macrosegreation Grain size Grain orientation Phase morphology Mechanical properties
	I O N	Stress analysis	Temperature-dependent parameters • Coefficient of thermal expansion • Stress/strain	Casting design for: • Dimension and distortion Internal stresses Hot tears and hot cracks

Source: Ref 1

Table 2 Thermophysical and mechanical properties needed for casting process simulation and common measurement techniques

Thermophysical property	Measurement technique
Thermal conductivity	Comparative stationary (solid), indirect (liquid)
Heat capacity	Differential scanning calorimetry, pulse heating; drop calorimetry
Density	Archimedian balance, push-rod dilatometry; levitation
Thermal diffusivity	Laser flash
Heat of fusion	Differential scanning calorimetry, pulse heating
Transformation temperatures	Differential scanning calorimetry, thermal analysis
Fraction solid	Differential scanning calorimetry, thermal analysis
Electrical resistivity	Pulse heating, four-point probing
Hemispherical emissivity	Pulse heating
Viscosity	Levitation, viscometer
Surface tension	Levitation, sessile drop
Young's modulus, Poisson's ratio	Tensile test, sound speed measurements
Thermal expansion	Push-rod dilatometry
Yield strength	Tensile test

Source: Ref 24

have been standardized, and most of them are limited to the solid-state properties. The current ASTM International standards and selected CEN and ISO standards include:

• *Specific heat capacity:* Differential scanning calorimetry, ASTM E 1269, E 967, E 968, E 2253, E 793, and D 2766 (Ref 46 to 51). Ceramics, EN 821–3 drop and DSC (Ref 52)
• *Thermal expansion:* Dilatometry, ASTM E 228; interferometry, ASTM E 289; and thermomechanical analysis, ASTM E 831 (Ref 53 to 55). Ceramics, EN 821–1 and ISO 17562, both dilatometry (Ref 56, 57)
• *Thermal conductivity:* Modulated-temperature scanning calorimetry, ASTM E 1952; thermal diffusivity of solids by the laser flash method, ASTM E 1461; and the steady-state heat flow, ASTM C 518 (Ref 58 to 60). Ceramics, EN 821–2 and ISO 18755, both laser flash (Ref 61, 62)
• *Thermal emittance:* Radiometric techniques, ASTM E 307 and E 408 (Ref 63, 64)

Standardized methods to determine the thermal properties of liquid metals are practically nonexistent. However, details of the various methods that have been used to successfully measure thermophysical properties of liquid metallic alloys are discussed in the article "Measurement of Thermophysical Properties at High Temperatures for Liquid, Semisolid, and Solid Commercial Alloys" in this Volume. In this article, the available thermophysical properties for pure metals and some commercial alloys are presented.

Specific Heat Capacity and Enthalpy of Transformation

Specific heat capacity of a material is the amount of thermal energy needed to change the temperature of a unit mass (m) of a substance by one degree Kelvin. The specific heat capacity is an extensive property of matter that depends on the amount of species in the system and is sensitive to phase changes. Specific heat capacity can be defined for a constant volume (C_v) or for a constant pressure (C_p). The specific heat capacity in SI units is expressed in J/kg · K. Also, the specific heat term is often used interchangeably with heat capacity. While this is not precisely correct, it is not a cause of misunderstanding.

The total amount of thermal energy or enthalpy, ΔH, associated with the specific heat capacity and a temperature change (T_1 to T_2) is given by:

$$\Delta H = \int_{T_1}^{T_2} C_p dT \qquad (Eq\ 1)$$

A good approximation to calculate the specific heat capacity as a function of temperature is given by:

$$C_p = a + bT + cT^{-2} + \dots \qquad (Eq\ 2)$$

where C_p is the molar heat capacity; a, b, and c, are constants; and T is the temperature in degrees Kelvin. Table 3 shows the specific heat capacity of solids as a function of temperature, the specific heat capacity of liquids at the melting point (T_m), and the enthalpy of fusion for most common elements found in cast metals.

Changes in specific heat capacity and most thermophysical properties with changing temperature in liquid metals may be gradual and continuous, rather than showing the abrupt effects of phase transitions that take place in the mushy and solid state. Thus, a reasonable estimate of the specific heat capacity for a

Table 3   Specific heat capacity values and enthalpy of fusion for pure metals

Element	$C_p = a + bT + cT^{-2} + ...$ (a), J/K·mol	Temperature range, K	T_m (b), K	C_{PL_m}, J/g·K	ΔH_f, J/g	Reference
Al(s)	$4.94 + 2.96 \times 10^{-3}T$	298–932	933	1.18	397	16, 65
Al(l)	7.0	932–1273				
C(graphite)	$4.1 + 1.02 \times 10^{-3}T - 2.10 \times 10^{-5}T^{-2}$	298–2300	4073	...	...	65
Co(β)	$3.3 + 5.86 \times 10^{-3}T$	715–1400	1768	0.59	275	13, 16, 65
Co(γ)	9.60	1400–T_m				
Co(l)	9.65	T_m–1900				
Cr(s)	$5.84 + 2.35 \times 10^{-3}T - 0.88 \times 10^{-5}T^{-2}$	298–T_m	2130	0.78	401.95	13
Cr(l)	9.4	T_m				
Cu(s)	$5.41 + 1.4 \times 10^{-3}T$	298–T_m	1356	0.495	208.7	13, 16
Cu(l)	7.50	T_m–1600				
Fe(α, δ)	$8.873 + 1.474 \times 10^{-3}T - 56.92T^{-1/2}$	298–T_m	1809	0.762	247	13, 16, 66
Fe(γ)	$5.85 + 2.02 \times 10^{-3}T$	1187–1664				
Fe(l)	$9.74 + 0.4 \times 10^{-3}T$	T_m–2000				
Hf(s)	$5.61 + 1.82 \times 10^{-3}T$	298–1346	2500	...	134.85	13
Li(s)	$3.33 + 8.21 \times 10^{-3}T$	273–T_m	454	...	422.1	13, 65
Li(l)	$5.85 + 1.31 \times 10^{-3}T + 2.07 \times 10^{-5}T^{-2} - 467 \times 10^{-6}T^{-2}$	T_m–580				
Mg(s)	$5.33 + 2.45 \times 10^{-3}T - 0.103 \times 10^{-5}T^{-2}$	298–T_m	922	1.32	349	13, 16
Mg(l)	7.68	T_m–1100				
Mn(α)	$5.70 + 3.38 \times 10^{-3}T - 0.375 \times 10^{-5}T^{-2}$	298–1000	1517	0.838	267.5	13
Mn(β)	$8.33 + 0.66 \times 10^{-3}T$	1108–1317				
Mn(γ)	$6.03 + 3.56 \times 10^{-3}T - 0.443 \times 10^{-5}T^{-2}$	1374–1410				
Mn(δ)	11.10	1410–1450				
Mn(l)	11.0	T_m–T_{bp}(c)				
Mo(s)	$5.77 + 0.28 \times 10^{-3}T + 2.26 \times 10^{-6}T^2$	298–2500	2893	0.57	371	13
Nb(s)	$5.66 + 0.96 \times 10^{-3}T$	298–1900	2740	0.334	315.4	13
Ni(α)	$7.80 - 0.47 \times 10^{-3}T - 1.335 \times 10^{-5}T^{-2}$	298–630	1726	0.63	292.4	13, 16, 65
Ni(β)	$7.10 + 1.0 \times 10^{-3}T - 2.23 \times 10^{-5}T^{-2}$	630–T_m				
Ni(l)	9.20	T_m–2200				
Pb(s)	$5.63 + 2.33 \times 10^{-3}T$	298–T_m	600	0.142	23.2	13
Pb(l)	$7.75 - 0.74 \times 10^{-3}T$	T_m–1300				
Si(s)	$5.72 + 0.59 \times 10^{-3}T - 0.99 \times 10^{-5}T^{-2}$	298–1200	1685	0.968	1877	13, 16
Si(l)	6.498	T_m–1873				
Ta	$6.65 - 0.52 \times 10^{-3}T - 0.45 \times 10^{-5}T^{-2} + 0.47 \times 10^{-6}T^2$	298–2300	3288	...	136.5	13, 16
Ti(α)	$5.28 + 2.4 \times 10^{-3}T$	298–1155	1940	0.965	295	13, 16
Ti(β)	$4.74 + 1.90 \times 10^{-3}T$	1155–1350				
Ti(l)	11.042	T_m–2073				
V(s)	$4.90 + 2.58 \times 10^{-3}T + 0.2 \times 10^{-5}T^{-2}$	298–1900	2175	...	328.6	13
W(s)	$5.74 + 0.76 \times 10^{-3}T$	298–2000	3673	...	176.8	...
Y(α)	$5.72 + 1.805 \times 10^{-3}T + 0.08 \times 10^{-5}T^{-2}$	298–1758	1803	0.394	128.6	13
Y(β)	8.37	1758–T_m				
Y(l)	9.51	T_m–1950				
Zn(s)	$5.35 + 2.4 \times 10^{-3}T$	298–T_m	692.5	0.481	112	13, 16
Zn(l)	7.5	T_m–T_{bp}				
Zr(α)	$5.25 + 2.78 \times 10^{-3}T - 0.91 \times 10^{-5}T^{-2}$	298–1135	2125	...	...	13
Zr(β)	$5.55 + 1.11 \times 10^{-3}T$	1135–T_m	...	...	...	...

(a) C_p in SI units (J/kg · K) when multiplied by 4.184 J/cal and divided by the corresponding element atomic mass (kg/mol). (b) T_m = melting point. (c) T_{bp} = boiling point

liquid alloy (C_{pL}) can be calculated from the elemental heat capacities of the components in the alloy by using the commonly known Kopp-Neumann rule of mixtures (Ref 67, 68):

$$C_{pL} = \sum_{i=1}^{n} Y_i (C_p)_{L_i} \text{ in cal/(mol · K)} \qquad \text{(Eq 3)}$$

where Y_i is the atomic fraction of element i in the alloy. The small changes in the specific heat capacity of the liquid with temperature allows for a reasonable estimate ($+3\%$) using Eq 3. Equation 3 can be expressed in SI units by multiplying C_{pL} by 4.184 J/cal and dividing by the atomic mass of the element i in kg/mol.

Major changes in the specific heat capacity with heating or cooling rates are observed in the solid-liquid range (mushy region). Macro- and microsegregation, as well as the presence of eutectics, peritectics, and other phase transformations that occur during solidification, cannot be easily described by Eq 1. Instead, the behavior in the solid or mushy state is more complex, because the phase transformations are dependent on the heating and cooling rates and on the chemistry of the alloy system. The dynamic characteristics of a given casting process require the input of all liquid-to-solid changes to understand the behavior of the solidifying metal. Therefore, in casting processes, determination of the specific heat capacity must be conducted on cooling for the solidifying metal and on heating for the mold and core materials, because the latter absorb most of the superheat and latent heat of solidification.

Enthalpy of Melting, Solidus and Liquidus Temperatures

The enthalpy or latent heat of melting (ΔH_f) is the heat that is required during solid-to-liquid transformations, and the latent heat of solidification is the heat released during liquid-to-solid transformations. The latent heats of melting and solidification in a multicomponent alloy system occur over a temperature range. The temperature at which the alloy starts to melt is called the solidus temperature, and the temperature at which the melting is completed is called the liquidus temperature.

In actual melting and casting processes, equilibrium conditions do not exist, because melting and solidification processes are ruled by the rate of phase transformations and by the heat- and mass-transfer phenomena. On melting, high heating rates may displace the solidus and liquidus temperatures to higher values, while on cooling and prior to the nucleation of the solid phase, the molten alloy is usually undercooled. High undercooling generally decreases the liquidus and solidus temperatures. Also, the degree of undercooling directly affects the kinetics of the liquid-solid phase transformation and the type of second phases that evolve during solidification. A more detailed discussion of nonequilibrium structures can be found in Ref 69. In actual casting processes, depending on the process and degree of inhomogeneity and impurities in a cast material, the liquidus and solidus temperatures can differ by several or even tens of degrees from equilibrium. The latent heat, solidus, liquidus, and other phase transformation temperatures are determined using the same techniques as for heat capacity and are described elsewhere in the literature (Ref 25).

The enthalpy of fusion and the solidus and liquidus temperatures for various alloys of commercial interest are shown in Table 4. The heat capacity, along with the thermal conductivity and density, for some commonly used sand molds is shown in Table 12.

Coefficient of Thermal Expansion

The coefficient of linear thermal expansion (α) is a material property that indicates the extent to which the material expands or contracts with temperature changes. At a constant pressure, the true coefficient of volumetric thermal expansion (α_V, or commonly β) is defined by the changes that occur by a differential temperature change (dT). This is usually expressed by the relationship:

$$\alpha_v = \frac{1}{V}\left(\frac{\partial V}{\partial T}\right)_P \qquad \text{(Eq 4)}$$

where V is the volume at a temperature, T, at a constant pressure, P.

The corresponding definition for the linear coefficient of expansion can be represented by the relationship:

$$\alpha_l = \frac{1}{l}\left(\frac{\partial l}{\partial T}\right)_P \qquad \text{(Eq 5)}$$

Usually, the coefficient of thermal expansion is not measured directly but is calculated by the

Table 4 Specific heat capacity data, latent heat of fusion, and solidus and liquidus temperatures for some alloys of commercial interest

Material	Nominal composition, wt%	Specific heat capacity, J/g · K	Temperature range, K	C_p at 25 °C	C_p at T_L	ΔH_f, J/g	T_S, K	T_L, K	Reference
Aluminum alloys									
A319 (LM4; Al-5Si-3Cu)	Al-3Cu-5Si-1Zn-0.35Ni-0.4Mn-0.1Mg	$C_p = 0.7473 + 2 \times 10^{-4}T + 5 \times 10^{-7}T^2$	298–780	0.87	1.17	393	798	898	16
A356 (LM25; Al-7Si-0.5 Mg)	Al-7Si-0.5Fe-0.4Mn-0.2Cu-0.1Ni-0.2Ti	$C_p = 0.7284 + 5 \times 10^{-4}T - 8 \times 10^{-7}T^2$	298–840 Trans. T 653	0.88	1.16	425	840	887	16
2003	Al-4.5Cu	$C_p = 0.749 + 4.44 \times 10^{-4}T$ $C_p = 1.287 - 2.5 \times 10^{-4}T$	$T < 775$ 775–991	0.882	1.059	...	775	911	14
3004	Al-0.2Cu-1Mg-1Mn-0.43Fe-0.14Si-0.25Zn	$C_p = 0.7989 + 3 \times 10^{-4}T - 9 \times 10^{-8}T^2$	298–873	0.90	1.22	383	890	929	16
2024-T4	Al-4.4Cu-1.5Mg-0.6Mn-0.5Fe-0.5Si-0.25Zn-0.1Cr-0.15Ti	$C_p = 0.7688 + 3 \times 10^{-4}T - 2 \times 10^{-7}T^2$	298–811	0.87	1.14	297	811	905	16
6061-T6	Al-0.3Cu-1Mg-0.15Mn-0.7Fe-0.6Si-0.25Zn-0.04Cr-0.15Ti	$C_p = 0.7067 + 6 \times 10^{-4}T - 1 \times 10^{-7}T^2$	298–873	0.87	1.17	380	873	915	16
7075-T6	Al-1.6Cu-2.5Mg-0.3Mn-0.5Fe-0.4Si-5.6Zn-0.2Cr-0.2Ti	$C_p = 0.7148 + 5 \times 10^{-4}T + 4 \times 10^{-10}T^2$	298–805	0.86	1.13	358	805	901	16
Copper alloys									
Cu-Al (Al-bronze)	Cu-9.7Al-4.6Fe-0.64Mn-4.6Ni	$C_p = 0.353 + 3 \times 10^{-4}T - 1 \times 10^{-7}T^2$ $C_p = 0.582$	298–1313 1313–1773	0.442	0.582	240	1313	1350	16
Brass	70Cu-30Zn	$C_p = 0.355 + 1.36 \times 10^{-4}T$ $C_p = 1.32 + 6.75 \times 10^{-4}T$ $C_p = 0.49$	298–1188 1188–1228 $1228 \leq T$	0.396	0.49	164.8	1188	1228	14
Brass	60Cu-40Zn	$C_p = 0.354 + 1.11 \times 10^{-4}T$ $C_p = -0.689 + 1.0 \times 10^{-3}T$ $C_p = 0.489$	$T \leq 1173$ 1173–1178 $1178 \leq T$	0.387	0.489	160.1	1173	1178	14
Copper-nickel	70Cu-30Ni-Fe-Mn	$C_p = 0.37 + 1.13 \times 10^{-4}T$ $C_p = 0.348 + 1.28 \times 10^{-4}T$ $C_p = 0.543$	$T \leq 1443$ 1443–1513 $1513 \leq T$	0.404	0.543	...	1443	1513	14
Iron and steel alloys									
Carbon steel AISI 1008	Fe-0.08C-0.31Mn-0.08Si-0.45Cr-0.03P-0.05S	$C_p = 0.593 + 4.8 \times 10^{-5}T$	1273–1550	0.469	...	...	1768	1808	14
Carbon steel AISI 1026	Fe-0.23C-0.63Mn-0.11Si-0.07Ni-0.03P-0.03S	$C_p = +2.1 \times 10^{-4}T$	1379–1550	0.469	...	...	1768	1798	14
1% Cr	Fe-0.3C-0.69Mn-0.2Si-1.1Cr-0.7Ni-0.012Mo-0.039P-0.036S	$C_p = 0.436 + 1.22 \times 10^{-3}T$	1173–1683	0.477	0.856	251	1693	1793	14
304 stainless steel	Fe-0.08C-19Cr-0.3Cu-2Mn-9.5Ni	$C_p = 0.443 + 2 \times 10^{-4}T - 8 \times 10^{-10}T^2$	298–1727	0.49	0.80	290	1673	1727	14, 16
316 stainless steel	Fe-0.08C-17Cr-0.3Cu-2Mn-2.5Mo-12Ni-1Si	$C_p = 0.412 + 2 \times 10^{-4}T - 2 \times 10^{-8}T^2$	298–1658	0.45	0.79	260	1658	1723	14, 16
420 stainless steel	Fe-0.3C-13Cr-0.12Cu-0.5Mn-0.06Mo-0.5Ni-0.4Si	$C_p = 1.92 - 1.587 \times 10^{-2}T$ $C_p = 0.569$	1150–1173 $T > 1173$	0.477	...	304	1727	1783	14
10-Ni steel	Fe-0.07C-0.55Mn-0.061Si-0.55Cr-1.25Mo-0.12V-9.5Ni	$C_p = 0.66 - 6.0 \times 10^{-4}T - 8.0 \times 10^{-7}T^2$ $C_p = 0.62 - 0.68T$	450–900 1800–1880	0.74	0.80	212	1763	1783	70
HSLA-65	Fe-0.07C-0.06Mn-0.55Cr-1.25Mo-0.12V-9.5Ni	$C_p = 0.54 - 2.0 \times 10^{-4}T - 4.0 \times 10^{-7}T^2$ $C_p = -0.276 + 1.4 \times 10^{-3}T - 6.0 \times 10^{-7}T^2$	298–973 1140–1550	0.51	...	455	1795	1811	71
HSLA-100	Fe-0.06C-3.5Ni-1.0Mn-0.55Cr-1.25Cu-0.6Mo-0.4Si	$C_p = -0.64 + 5.6 \times 10^{-3}T - 9.0 \times 10^{-6}T^2 + 5.0 \times 10^{-9}T^3$ $C_p = -7.45 + 1.5 \times 10^{-2}T - 7.0 \times 10^{-6}T^2$	298–875 1100–1210	0.40	...	384	1783	1798	72
HY-100	Fe-0.20C-3.5Ni-0.35Mn-1.6Cr-0.25Cu-0.6Mo-0.4Si	$C_p = 0.62 - 5 \times 10^{-4}T + 6.0 \times 10^{-7}T^2$ $C_p = 0.33 - 8 \times 10^{-4}T + 7.0 \times 10^{-7}T^2$	300–925 1150–1500	0.48	...	260	1781	1798	73
Ductile iron	Fe-3.61C-2.91Si -0.08Cr-0.12Cu-0.65Mn-0.02Mo-0.13Ni≤0.002Mg	$C_p = 0.80$	1373	0.48	0.83	220	1413	1451	16
Gray cast iron	Fe-3.72C-1.89Si-0.95Cr-0.66Mn-0.59Mo-0.19Ni≤0.002Mg	$C_p = 0.66$	1353	0.49	0.95	240	1353	1463	16
Magnesium alloys									
AZ31B	Mg-3Al-1Zn-0.5Mn	$C_p = 1.88 - 5.22 \times 10^{-4}T$ $C_p = 0.979 + 4.73 \times 10^{-4}T$	$839 \leq T \leq 905$ $905 \leq T$	1.01	1.415	...	839	905	14
AZ91B	Mg-9Al-0.6Zn-0.2Mn	$C_p = 0.251 + 1.354 \times 10^{-3}T$ $C_p = 1.43$	$742 \leq T \leq 869$ $869 \leq T$	0.98	1.428	...	742	869	14
KIA	Mg-0.7Zr	$C_p = 0.873 + 0.463 \times 10^{-3}T$ $C_p = 1.43$	$T \leq 922$ $923 \leq T$	1.005	1.428	...	922	923	14
ZK51A	Mg-4.6Zn-0.7Zr	$C_p = 1.945 - 6.28 \times 10^{-4}T$ $C_p = 0.90 + 0.516 \times 10^{-3}T$	$822 \leq T \leq 914$ $914 \leq T$	1.022	1.371	...	822	914	14
HM11A	Mg-1.2Mn-1.2Th	$C_p = 0.65 + 8.75 \times 10^{-3}T$	$903 \leq T \leq 923$	0.946	1.411	...	905	923	14

(continued)

Table 4 (continued)

Material	Nominal composition, wt%	Specific heat capacity, J/g · K	Temperature range, K	C_p at 25 °C	C_p at T_L	ΔH_f, J/g	T_S, K	T_L, K	Reference
EZ33	Mg-3Ce-2Zn-0.6Zr	$C_p = 1.482$ $C_p = 1.21$ $C_p = 1.336$	$923 \leq T$ 818 $913 \leq T$	0.98	1.336	343	818	913	...
Nickel alloys									
Single-crystal CMSX-4	Ni-10Co-6.5Cr-6.5Ta-6.4W-3Re-1Ti-0.15Fe-0.04Si-0.006C	$C_p = 0.675$	$1653 \leq T$	0.397	0.636	240	1593	1653	16
Hastelloy C	Ni-16Mo-16Cr	$C_p = 0.281 + 0.283 \times 10^{-3}T$	$T \leq 1534$	0.283	...	...	1534	1578	14
Hastelloy X	Ni-22Cr-18.5Fe-9Mo-1.5Co-0.6W-0.5Si-0.1C	$C_p = 0.4384 + 1 \times 10^{-4}T + 5 \times 10^{-8}T^2$	1073–1533	0.439	0.677	276	1533	1628	16
Inconel 718	Ni-19Cr-16.7Fe-5.2Nb-3.1Mo-1Co-0.9Ti-0.35Mn-0.35Si-0.08C	$C_p = 0.65$	1443	0.435	0.72	210	1533	1609	16
Titanium alloys									
Ti-6Al-4V	Ti-5.5–6.7Al-3.5–4.5V-0.25(O$_2$ + N$_2$)-0.03Fe-0.0125H$_2$	$C_p = 0.4115 + 2 \times 10^{-4}T + 5 \times 10^{-10}T^2$ $C_p = 0.83$	1268–1923 $T \geq 1923$	0.546	0.83	286	...	1923	16
Zinc alloys									
Zn-Al	Zn-4.5Al-0.05Mg	$C_p = 0.50$ $C_p = 0.52 - 6 \times 10^{-5}$ $(T - 387 °C)$	630 Liquid $T \geq 660$	0.41	0.51	114	630	660	16

derivative of the equation that represents the expansion. Also, the instantaneous coefficient of linear thermal expansion is frequently defined as the fractional increase of length per unit rise in temperature. Further analyses and the theory of thermal expansion can be found in the literature (Ref 6).

The temperature dependence of the coefficient of thermal expansion for solids is very complex, but it has been shown that it varies inversely with the melting point (T_m) of the material and is expressed by:

$$\alpha = \gamma_G/100T_m \qquad \text{(Eq 6)}$$

where γ_G is the Grüneisen parameter and, for most solid materials, is close to 1. Semiempirical analyses (Ref 6) of the thermal expansion of crystalline materials, such as close-packed metals, have shown that the mean coefficient of linear thermal expansion (α_m) can also be related to their melting temperature by the relationship:

$$(L_m - L_0)/L_0 - T_m\alpha_m \approx 0.0222 \qquad \text{(Eq 7)}$$

where L_m is the length at T_m.

It has been found that most metals with melting points above 900 K linearly expand approximately two percent on heating from 298 K to T_m (Ref 6). Many metals exhibit a thermal expansion $\alpha_m \approx 10^{-4}$ K^{-1} in the liquid phase just above the liquidus temperature. Table 5 shows the linear expansion ($\Delta L/L_0$) and the coefficient of linear thermal expansion (α) for some pure metals at temperatures closer to their melting points (Ref 6). Unfortunately, few data at the solidus and/or liquidus are available for alloys of commercial interest. Data have been determined in the liquid by levitation, and recent developments with piston dilatometry enable measurements across the liquid/solid region. Using this technique, Blumm and Henderson (Ref 74) and Morrell and Quested (Ref 75) have performed measurements on nickel, aluminum, and cast irons, and typical data for alloys are included in Tables 15 to 18.

Density

Density (ρ) is defined as the mass per unit volume of a material. The reciprocal of the density ($1/\rho$) is the specific volume. Room temperature and liquid density for pure metals and some materials of commercial interest are available in the literature (Ref 13, 15, 16, 26, 76).

Accurate density values are highly desirable because it is a variable in the calculation of thermal conductivity, surface tension, and viscosity. Further, the evaluation of fluid flow phenomena in a solidifying metal is dominated by the changes of density. Density of liquid metals is also useful to calculate volume changes during melting, solidification, and alloying. As a general rule, the average density change of non-close-packed metals on fusion is approximately three percent while the average volume change of a close-packed metal does not exceed five percent (Ref 66). Bismuth, gallium, antimony, germanium, silicon, cerium, and plutonium are the exceptions to the general rule, because these elements contract on melting.

Density of solid alloys as a function of temperature can be calculated from thermal expansion data using the following relationship:

$$\rho_T = \rho_{RT}/(1 + \Delta L_{exp})^3 \qquad \text{(Eq 8)}$$

where ΔL_{exp} is the linear expansion ($\Delta L_{exp} = (L_T - L_0)/L_0 = \alpha T$) at a temperature, T. Alternatively, the density of a heterogeneous phase mixture (ρ_m) containing a number, n, of phases can be roughly estimated using the empirical relationship:

$$\rho_m = \frac{1}{\sum_{i=1}^{n}(X_p/\rho_p)} \qquad \text{(Eq 9)}$$

where X_p and ρ_p are the fraction and density of the phase, respectively, at a given temperature.

Table 5 Thermal expansion of selected pure metals at temperatures close to melting

Element	Melting point (T_m), K	LE%(a), $\Delta L/L_0$	CTE (α)(b), 10^{-6}/K	T(c), K
Ag	1233.7	2.11	28.4	1200
Al	933.5	1.764	37.4	900
Au	1336	1.757	22.1	1300
Be	1562	2.315	23.7	1500
Bi	544.6	0.307	12.4	525
Cd	594	1.028	40	590
Ce	1072	0.512	9.4	1000
Cr	2148	2.02	19	1900
Co	1766	1.5	17.7	1200
Cu	1356	2.095	25.8	1300
α-Fe	1185	1.37	16.8	1185
Fe (α-γ)	1185	0.993	23.3	1185
γ-Fe	1811	2.077	23.3	1650
Hf	2216	0.712	8.4	1300
La	1195	0.497	11.3	1000
Li	453.5	0.804	56	450
Mg	924	1.886	37.6	900
Mn	1525	6.604	...	1500
Mo	2880	2.15	16.5	2800
Nb	2741	1.788	10.1	2300
Ni	1727	2.06	20.3	1500
Pb	660.6	0.988	36.7	600
Pd	1825	1.302	16.9	1200
Pt	2042	1.837	14.9	1900
Re	3431	1.941	9.8	2800
Rh	2236	1.526	15.4	1600
Sb	904	0.588	11.7	800
Si	1683	...	3.8	...
Sn	505	0.516	27.2	500
Ta	3250	3.126	24.4	3200
α-Ti	1156	0.918	11.8	1156
Ti (α-β)	1156	0.868	11	1156
Ti-β	1958	1.411	13.5	1600
V	2185	2.16	17.2	2000
W	3650	2.263	11.6	3600
Zn	692	1.291	34	690
Zr	2123	1.139	11.3	1800

(a) LE% = percent of linear expansion. (b) CTE = coefficient of linear expansion. (c) T = temperature. Source: Ref 6

The density of pure liquid metals as a function of temperature can be reasonably estimated from the following empirical equation (Ref 26):

$$\rho_L = a - b(T - T_m) \qquad \text{(Eq 10)}$$

where ρ_L is in g/cm^3, a (the density at the liquidus temperature, ρ_m) and b are dimensionless constants, and T is the temperature above the melting point (T_m). Both temperatures are in degrees Kelvin.

Table 6 shows the density at the melting point and the values for the parameters a and b for estimation of the liquid density as a function of temperature for various elements. Table 6 also shows the viscosity and activation energy for viscous flow for some pure metals.

Surface Tension

Knowledge of the surface tension phenomena of metals is essential in the understanding of solidification during casting. In solidification phenomena, the transformation of liquid into solid requires the creation of curved solid/liquid interfaces that lead to capillarity, microscopic heat flow, and solute diffusion effects. The interplay of the heat flow and solute diffusion effects determines the solidification morphologies. Solute diffusion effects have the tendency to minimize the scale of the morphology, while the capillarity effects tend to maximize the scale. A compromise between these two tendencies has a profound effect on the crystal morphologies with respect to nucleation, interface instability, and dendritic and eutectic growth (Ref 80). The surface energy has an important role because of the creation of the solid/liquid interface area. Also, during solidification, a decrease in the equilibrium melting point produces a positive undercooling and is associated to the solid/liquid interface curvature, which is usually convex toward the liquid phase. This curvature effect is often called the Gibbs-Thompson effect ($\Gamma = \gamma_{s/l}/\Delta S_f$), which is a function of the solid/liquid interfacial energy ($\gamma_{s/l}$) and the volume entropy of fusion (ΔS_f).

The Gibbs-Thompson effect is of the order of 10^{-7} mK for most metals. This indicates that the effect of the surface energy becomes important only when a given morphology such as nuclei, interface perturbations, dendrites, and eutectic phases have a radius less than 10 μm (Ref 80). Unfortunately, data on the surface energy or surface tension (mN/m) for most metals and their alloys are very limited or not available. Surface tension of liquid metals can be measured using various contact and noncontact techniques (Ref 31 to 34, 81, 82).

Semitheoretical models have been developed to calculate the surface tension of pure metals. The rigid sphere model (Ref 83) assumed a structure of the liquid metal where the collision diameter may be estimated by the molar volume (V) at the melting temperature. Then, the surface tension can be expressed by:

$$\gamma = (3.6 T_m V^{-2/3}) \times 10^{-3} \text{J/m}^2 \qquad \text{(Eq 11)}$$

Alternatively, the surface tension has been correlated to the heat vaporization (ΔH_v) caused by the breaking of the interatomic bonds during

Table 6 Density at the melting point, dimensionless values for the parameters a and b, viscosity, and activation energy for viscous flow for selected elements

Element	Melting point, K	Density, g/cm^3 Measured at T_m (Ref 13)	$\rho_L = a - b\,(T - T_m)$ (Ref 26) a	$b \times 10^{-4}$	Viscosity (Ref 13, 77, 78) η at T_m, mN·s/m^2	η_o mN·s/m^2	E, kJ/mol
Ag	1233.7	9.346	9.329	10.51	3.88	0.4532	22.2
Al(a)	933.5	2.385	2.378	3.111	1.34	0.185	15.4
Au	1336	17.36	17.346	17.020	5.38	1.132	15.9
B	2448	2.08 at 2346 K	...	...	...	...	...
Be	1550	1.690	1.690	1.165	...	...	...
Bi	544	10.068	10.031	12.367	1.85	0.4458	6.45
Cd	593	8.020	7.997	12.205	2.28	0.3001	10.9
Ce	1060	6.685	6.689	2.270	2.88	...	...
Cr	2148	6.28	6.280	7.230	...	...	...
Co	1766	7.760	7.740	9.500	4.49	0.2550	44.4
Cu	1356	8.000	8.033	7.953	4.10	0.3009	30.5
Fe(a)	1811	7.015	7.035	9.26	5.85	0.191	51.5
Hf	2216	11.10	...	...	...	...	...
La	1203	5.955	5.950	2.370	2.45	...	...
Li	453.5	0.525	0.5150	1.201	0.55	0.1456	5.56
Mg	924	1.590	1.589	2.658	1.32	0.0245	30.5
Mn	1525	5.730	5.750	9.300	...	...	...
Mo	2880	9.35	...	...	...	...	...
Nb	2741	7.830	...	...	...	...	...
Ni	1727	7.905	7.890	9.910	4.60	0.1663	50.2
Pb	660.6	10.678	10.587	12.220	2.61	0.4636	8.61
Pd	1825	10.495	10.495	12.416	...	...	...
Pt	2042	19.00	18.909	28.826	...	...	...
Re	3431	18.80	...	...	...	...	...
Sb	904	6.483	6.077	6.486	1.48	0.0812	22
Si	1683	2.524	2.524	3.487	0.94	...	...
Sn	505	7.000	6.973	7.125	2.00	0.5382	5.44
Ta	3250	15.00	...	...	...	...	...
Ti	1958	4.110	4.140	2.260	5.20	...	...
V	2185	5.700	5.36	3.20	...	...	...
W	3650	17.60	...	...	...	...	...
Zn	692	6.575	6.552	9.502	3.85	0.4131	12.7
Zr	2123	5.800	...	...	8.0	...	...

(a) Density data from Ref 79

evaporation in the liquid state (Ref 84). Usually, metals with large atomic volume have low energies of vaporization:

$$\gamma = 1.8 \times 10^{-9} (\Delta H_v / V^{-2/3}) \qquad \text{(Eq 12)}$$

The surface tension for pure metals as a function of temperature (γ f(T)) can also be calculated using Eq 13 and Table 7:

$$\gamma f(T) = \gamma_i^0 + c(T - T_m) \qquad \text{(Eq 13)}$$

where γ^0 is the surface tension at the melting point, and c is $d\gamma/dT$, T_m, and $T > T_m$; T and T_m are in degrees Kelvin.

There are two definitive compendia of surface tension values by Keene: the first for pure metals (Ref 85) and the other for iron and its binary alloys (Ref 86). The review of elements has recently been updated (Ref 87). Variations on reported surface tension of pure elements and alloys are expected because of the experimental techniques and the strong effect of surface-active impurities such as soluble oxygen, sulfur, and tellurium in liquid metals on their surface tension. Therefore, the levels of impurities in solution should be carefully controlled and taken into account in the determination and assessment of the surface tension of metallic alloys. It should be noted that in practice it is difficult to control soluble oxygen levels.

Marangoni flows are those driven by surface tension gradients. In general, surface tension depends on both the temperature and chemical composition at the interface; consequently, Marangoni flows may be generated by gradients in either temperature or chemical concentration at an interface. Because the temperature coefficient for surface tension, $d\gamma/dT$, can change sign from a negative to positive as the impurity concentration increases, the direction of flow in shallow pools of liquid (Marangoni flow) with uneven temperature distributions can be reversed with different materials with different surface-active element concentrations. These effects are established to have implications in melt processes such as the weld quality and flow during melt refining. Fifty parts per million of either oxygen or sulfur cause a decrease of 25 percent in γ and a change from negative ($d\gamma/dT$) to positive ($d\gamma/dT$), which will reverse the direction of any Marangoni convection.

Brooks and Quested have reviewed both the surface tension (γ) and $d\gamma/dT$ for a variety of ferritic and austenitic steels as a function of sulfur content. In general, (γ) decreases with increasing sulfur levels, and $d\gamma/dT$ increases from negative to positive with increasing sulfur levels. For detailed compositions of the steels, the reader is referred to Ref 88.

Thermodynamic models to calculate the surface tensions of liquid alloys have been developed (Ref 44, 89 to 92). These models were based on earlier work on surface tension prediction of binary (Ref 91) and ternary (Ref 92) solutions and on the Buttler equation (Ref 93). To calculate the surface tension of a liquid metal using the Buttler equation, the surface tensions and surface areas of the pure constituent elements and the excess Gibbs energy of the liquid metal must be known. The excess Gibbs energy is the same as that used for calculating the phase diagram and thermodynamic properties. In this thermodynamic approach, the description of the alloy system must be established before any property of the multicomponent alloy can be calculated. Model parameters for several alloy systems have been determined (Ref 94 to 105). Chemical thermodynamics has been also applied to describe the surface tension of multicomponent diluted solutions (Ref 106). The equations developed using this approach have been demonstrated to be in excellent agreement with the measured values for the industrially important Fe-O-N and Fe-O-S alloy systems (Ref 107, 108). Table 8 shows the surface tension values for some alloys of industrial interest (Ref 16), including those for Fe-O, Fe-O-N, and Fe-O-S alloys.

Viscosity

The viscosity is the resistance of the fluid to flow when subjected to an external shear force. The shear stress (τ), or the force per unit area, causing a relative motion of two adjacent layers in a liquid is proportional to the velocity gradient (du/dy), which is normal to the direction of the applied force ($\tau = -\eta\, du/dy$), where the proportionality factor, η, is termed the viscosity. This concept is known as Newton's law of viscosity. Most liquid metals are believed to follow a Newtonian behavior. The unit of viscosity is called Poise (P) ($1P = 1$ dyne $\cdot$ s/cm^2 = 1 g/cm $\cdot$ s = 1 mPa $\cdot$ s). The parameter (η/ρ) is referred to as kinematic viscosity and has units (m^2/s), which are identical to the units for diffusion coefficients and thermal diffusivity.

Several methods exist to measure the viscosity of liquid metals (Ref 109 to 111). However, accurate and suitable methods to measure the viscosity of liquid metals and their alloys are restricted due to the relatively low viscosity, high liquidus temperatures, and chemical reactivity of melts. A review of the most common methods to measure the viscosity is given in the article "Measurement of Thermophysical Properties at High Temperatures for Liquid, Semisolid, and Solid Commercial Alloys" in this Volume.

The reciprocal of the viscosity is known as the fluidity. The kinematic viscosity is the ratio of the viscosity to density ($\nu = \eta/\rho$). This is an important parameter in fluid mechanics. The kinematic viscosity represents the transverse diffusion of momentum down a velocity gradient that is necessary to describe mold filling in a casting process. Table 9 shows the kinematic

viscosity of AISI 316, 321, 446, and 660 stainless steels (Ref 112).

The Arrhenius equation is the most common form of representing the temperature dependence (Ref 110) of viscosity:

$$\eta = A\,\exp(E_v/RT) \qquad \text{(Eq 14)}$$

where E_v is the activation energy for viscous flow, and R is the ideal gas constant (8.3144 J/K).

Andrade (Ref 113) derived a semiempirical relation to determine the viscosity of elemental liquid metals at their melting temperatures. Andrade's relationship is based on the quasi-crystalline theory that assumes that the atoms in the liquid at the melting point are vibrating in random directions and periods, just as in the solid state.

Modification to Andrade's equation, based on the characteristics of atomic vibration frequency at the melting point, gives (Ref 110):

$$\eta_m = 1.7 \times 10^{-7}\rho^{2/3}T_m^{1/2}M^{-1/6}(\text{Pa}\cdot\text{s}) \qquad \text{(Eq 15)}$$

Table 7 Surface tension for pure liquid metals

Element	$\gamma_l^0\, T_m$, mN/m	$-d\gamma/dT$, mN/m $\cdot$ K
Ag	903	0.16
Al	914	0.35
Au	1140	0.52
B	1070	...
Be	1390	0.29
Bi	378	0.07
Cd	570	0.26
Ce	740	0.33
Cr	1700	0.32
Co	1873	0.49
Cu	1285	0.13
Fe	1872	0.49
Hf	1630	0.21
La	720	0.32
Li	395	0.15
Mg	559	0.35
Mn	1090	0.2
Mo	2250	0.30
Nb	1900	0.24
Ni	1778	0.38
Pb	468	0.13
Pd	1500	0.22
Pt	1800	0.17
Re	2700	0.34
Sb	367	0.05
Si	865	0.13
Sn	544	0.07
Ta	2150	0.25
Ti	1650	0.26
V	1950	0.31
W	2500	0.29
Zn	782	0.17
Zr	1480	0.20

Source: Ref 13

where ρ is the density, M is the atomic mass, and T_m is the melting point in degrees Kelvin.

The temperature dependence of the viscosity for most pure liquid metals can be expressed by:

$$\eta = \eta_o\,\exp(2.65T^{1.27}/RT)\ (\text{mPa}\cdot\text{s}) \qquad \text{(Eq 16)}$$

At $T = T_m$ and $\eta = \eta_m$, then the value for η_o is determined as follows:

$$\eta_o = \eta_m/[\exp(2.65T^{1.27}/RT)] \qquad \text{(Eq 17)}$$

With the exception of silicon, manganese, chromium, hafnium, palladium, and vanadium, these equations allow a reasonable prediction of the viscosity as a function of temperature for most pure liquid metals. Further modifications of Andrade's equation have been developed in an attempt to estimate more accurately the viscosity of liquid metals (Ref 114 to 118).

An empirical relationship between viscosity and surface tension for pure liquid metals has been developed (Ref 118):

$$\eta_m = 2.81 \times 10^{-4}[(M\gamma_m)^{1/2}/V^{1/3}] \qquad \text{(Eq 18)}$$

Table 8 Surface tension for some industrial alloys

Material(a)	Surface tension(b), mN/m	Temperature, K
Cu-Al	1240	1350
(Al-bronze)	1215	1473
Single-crystal	1850	1653
CMSX-4	1850	1873
Hastelloy X	1880	1628
	1865	1773
Inconel 718	1882	1609
	1866	1773
Zn-Al	830	660
	807	773
Fe-O-N(c)	= $1927 - 1.977 \times 10^4$ (wt% O) $- 5273$(wt%N) $+ 1.512 \times 10^5$ (wt%O)^2 $- 7452$(wt %N)^2 $+ 6.591 \times 10^4$ (wt% O) (wt%N)	1873
Fe-O-S(d)	= $1902 - 5.749 \times 10^4$ (wt% O) $- 3.374 \times 10^4$ (wt%S) $+ 1.186 \times 10^6$ (wt%O)^2 $+ 4.695 \times 10^5$ (wt%S)^2 $- 1.124 \times 10^6$ (wt%O) (wt%S)	1873

(a) Chemistries given in Table 4. (b) Values depend on oxygen and sulfur contents. Source: Ref 16. (c) The thermodynamic calculation was made in relation to three oxygen concentration measurements: wt% = 0.002 to 0.003, 0.020 to 0.030, and 0.045 to 0.050. Source: Ref 106, 107. (d) The thermodynamic calculation was made in relation to the experimental data for the Fe-O-S system. Source: Ref 106, 108

Table 9 Kinematic viscosity of AISI 316, 321, 446, and 660 stainless steels

AISI steel grade	Nominal composition	T_s, K	T_l, K	Kinematic viscosity, $\nu \times 10^{-7}$ m^2/s(a)	ν at T_l
316	Fe-0.08C-16Cr-0.3Cu-2Mn-2.5Mo-12Ni-1Si	1658	1723	$\nu = 13.242 - 0.0041T$	6.18
321	Fe-0.08C-17Cr-0.3Cu-2Mn-12Ni-1Si	1644	1671	$\nu = 28.7 - 0.020T + 4 \times 10^{-6}T^2$	6.45
446	Fe-0.2C-25Cr-1.5Mn-1Si-0.25N max	1698	1783	$\nu = 1.40 - 0.013T + 5 \times 10^{-6}T^2$	5.88
660	Fe-0.08C-13.5-16Cr-0.25Cu-2Mn-24–27Ni-1Si-0.2Ti-0.3V	1643	1698	$\nu = 110.20 - 0.0909T + 2 \times 10^{-5}T^2$	13.51

(a) Estimated polynomial equations from kinematic viscosity data obtained during cooling of the liquid steel. Source: Ref 112

where M is the atomic mass, V is the atomic volume, and γ_m is the surface tension at the melting point.

Table 6 shows the viscosity of liquid metals at their melting temperature and their activation energy for viscous flow (Ref 77, 78), and Table 10 shows the viscosities of some metals and some alloys of commercial interest (Ref 16).

Electrical and Thermal Conductivity

Thermal and electrical conductivities are intrinsic properties of materials, and they reflect the relative ease or difficulty of energy transfer through the material. Metals are well known for their high electrical conductivity, which arises from the easy migration of electrons through the crystal lattice. The conductivity on melting for most metals decreases markedly due to the exceptional disorder of the liquid state. Generally, the electrical resistivity of some liquid metals just above their melting points is approximately 1.5 to 2.3 greater than that of the solids just below their melting temperature (Ref 31). Examples of exceptions are iron, cobalt, and nickel. Mott (Ref 120) derived an empirical equation to estimate the ratio of liquid/solid electrical conductivity ($\sigma_{e,l}/\sigma_{e,s}$) at the melting point of the pure metal. The Mott equation is expressed as follows:

$$\ln(\sigma_{sol}/\sigma_{liq}) = C(\Delta H_m/T_m) = C(\Delta S_m) \qquad (Eq\ 19)$$

where ΔH_m is the enthalpy of fusion in KJ/mol, T_m is the melting temperature in degrees Kelvin, ΔS_m is the entropy of fusion, and C is a constant. With the exception of a few metals (e.g., antimony, bismuth, gallium, mercury, and tin), the simple relationship proposed by Mott is in good agreement with experimental measurements.

Reviews of the electrical and thermal conductivity theories can be found in the literature (Ref 121 to 127). Of particular interest to high-temperature technology, such as casting processes, is the theoretical relation between the thermal (k) and electrical (σ) conductivities known as the Wiedman-Franz law (WFL) and the constant of proportionality known as the Lorentz number, L (Ref 128). The WFL relationship (Eq 20) appears to hold reasonably well for pure metals around the melting point, but large departures can occur at lower temperatures in the solid:

$$L = (k/\sigma T) = 2.445 \times 10^{-8}\ W \cdot \Omega/K^2 \qquad (Eq\ 20)$$

Convective flow that usually occurs in the liquid should not have any effect on the electrical conductivity. Therefore, it should be possible to estimate the thermal conductivity of liquid alloys from the electrical conductivity values. The electrical conductivity for molten binary alloy has been estimated using Eq 21 (Ref 31):

$$\sigma_{T1} = \sigma_1 x_1 + \sigma_2 x_2 + \sigma_3 x_3 + \ldots \qquad (Eq\ 21)$$

A negative departure of less than 10 percent from linearity has been observed for most alloys. Similarly, the temperature dependence of the electrical conductivity can be estimated using Eq 22:

$$\sigma_T = \sigma_{T1}\{1 + (d\sigma/dT)_{alloy}\} \qquad (Eq\ 22)$$

where:

$$d\sigma/dT = x_1(d\sigma_1/dT) + x_2(d\sigma_2/dT) + x_3(d\sigma_3/dT) + \ldots \qquad (Eq\ 23)$$

When σ_T is calculated at a given T, then the thermal conductivity is calculated using the WLF equation (Eq 20).

Excellent compilations of conductivity properties of pure elements can be found in the literature (Ref 13, 129, 130). Table 11 shows the electrical resistivity for some solid and liquid metals at the melting point. The resistivity data for the liquid, $\rho_{e,l}$, from the melting point to a given temperature in Table 10 can be calculated by the expression:

$$\rho_{e,l} = \alpha T + \beta \qquad (Eq\ 24)$$

The values for constants α and β for the various metals are given in Table 11. Note also that the electrical resistivity data given in Table 11 are for bulk metals and may not be applicable to thin films.

Mills and co-workers (Ref 130) combined Mott (Eq 19) and WFL (Eq 20) to establish a relationship between the thermal conductivity of the solid, k_{sm}, and liquid, k_{lm}, at the melting point:

$$\ln\left(\frac{k_{sm}}{k_{lm}}\right) = K\Delta S_m \qquad (Eq\ 25)$$

where K is a constant. This equation would be useful to estimate the thermal conductivity of a liquid alloy by determining the thermal conductivity (or electrical conductivity) at the melting point. The only limitation is the constant K, which would not have a uniform value for all metals and alloys (Ref 130).

Table 11 shows the thermal conductivity for pure metals in the solid and liquid state at the melting point as well as the estimation of the constant K based on the WFL (Ref 130). The experimental errors in the measurement of the thermal conductivity of the liquid can be larger than five percent for the thermal conductivity and $\pm 3\%$ for electrical conductivity (Ref 130). Calculating thermal conductivity from measured thermal diffusivity values may reduce some of the experimental errors.

Thermal diffusivity is the ability of a material to self-diffuse thermal energy. This is determined by combining the material ability to conduct heat and its specific heat capacity. The density is involved due to the given specific heat capacity in units of heat per unit mass, while conductivity relates to the volume of material. Thus, the thermal conductivity (k) and thermal diffusivity (α) that measure the heat flow within materials are

related by their specific heat capacity (C_p) and density (ρ) by the relationship:

$$k = \alpha C_p \rho \qquad (Eq\ 26)$$

Table 10 shows some of the data available in the literature for ferrous and nonferrous alloys (Ref 14).

In casting processes, the need for thermal conductivity data for mold materials becomes more crucial. Mold materials are usually bulk, porous, complex sand-polymer mixtures and ceramic materials, and their thermal conductivity is certainly different from the intrinsic thermal conductivity of the base material. Methods to estimate the thermal conductivity of these materials can be found in the literature (Ref 131). Table 12 shows the thermal conductivity for some mold materials (Ref 14).

Emissivity

Thermal radiation is an important heat-transfer phenomenon in casting processes, because heat losses during pouring of the molten metal and heat radiation from the mold contribute to the overall heat balance during solidification of the cast product. The thermal radiation is defined by Planck's law. The integrated form of Planck's equation gives the total emissive power of a body (e). This is known as the Stefan-Boltzmann equation and is represented by:

$$e = \varepsilon \sigma T^4 \qquad (Eq\ 27)$$

where ε is the emissivity, and σ is the Stefan-Boltzmann constant. The Stefan-Boltzmann constant is given by:

$$\sigma = \frac{2\pi^5\kappa^4}{15c^2h} = 5.67 \times 10^{-8}\ J/m^2 \cdot s \cdot K^4 \qquad (Eq\ 28)$$

where κ and h are the Boltzmann's and Planck's constants, respectively, and c is the speed of light.

Because emissivity data throughout the wavelength spectrum are not available for most metallic materials, the following empirical equation has been employed to represent the total emissivity as a function of temperature (Ref 132):

$$\varepsilon_t = K_1\sqrt{(\rho T)} - K_2\rho T \qquad (Eq\ 29)$$

where K_1 and K_2 are constants, $K_1 = 5.736$, $K_2 = 1.769$, and ρ is the electrical resistivity in $\Omega \cdot m$.

Equation 29 is in reasonable agreement with experimental data, and it shows an increase of ε_t with T. However, deviations for some materials at high temperatures can be expected because of the spectral emissivity that changes with the wavelength and direction of emission. Nevertheless, in most practical situations, including in casting processes, an average emissivity for all directions and wavelengths is used. The emissivity values of metals or other nonmetallic materials depend on the nature of the surfaces, such as the degree of oxidation, surface finish, and the grain size. Table 13 gives

Table 10   Density, thermal conductivity, and viscosity for some commercial alloys

Material(a)	273 K	Density, g/cm³ T in K	Thermal conductivity(b), W/m · K (temp. range in K)		Estimated viscosity(c), mPa · s
Aluminum alloys					
A319	2.75	$\rho_s = 2.753 - 22.3 \times 10^{-2}\,(T - 298)$ $\rho_l = 2.492 - 27.0 \times 10^{-2}\,(T - 894)$	$k_s = 76.64 + 0.2633T - 2 \times 10^{-4}T^2$ $k_l = 70$ $k_l = 71$	$(T: 298-773)$ $(T: 894)$ $(T: 1073)$	1.3 at 894 K 1.1 at 1073 K
LM25 (A356)(d)	2.68	$\rho_s = 2.68 - 21.2 \times 10^{-2}\,(T - 298)$ $\rho_l = 2.401 - 26.4 \times 10^{-2}\,(T - 887)$	$k_s = 149.7 + 0.0809T - 1 \times 10^{-4}T^2$ $k_l = 65.8$ $k_l = 70$	$(T: 298-840)$ $(T: 887)$ $(T: 1073)$	1.38 at 887 K 1.1 at 1073 K
2003(e,f)	...	$\rho_l = 2.43 - 3.2 \times 10^{-4}\,(T - 922)$	$k_s = 192.5$ $k_l = 818.7 - 0.808T$ $k_l = 86.3$	$(T: 573-775)$ $(T: 775-911)$ $(T: 1023)$	$\eta = 0.196\exp(15{,}206/RT)$
3004	2.72	$\rho_s = 2.72 - 23.4 \times 10^{-2}\,(T - 298)$ $\rho_l = 2.4 - 27.0 \times 10^{-2}\,(T - 929)$	$k_s = 124.7 + 0.56T + 1 \times 10^{-5}T^2$ $k_l = 61$	$(T: 298-890)$ $(T \geq 929)$	1.15 at 929 K 1.05 at 973 K
2024-T4	2.785	$\rho_s = 2.785 - 21.3 \times 10^{-2}\,(T - 298)$ $\rho_l = 2.50 - 28.0 \times 10^{-2}\,(T - 905)$	$k_s = 188$ $k_l = 85.5$	$(T: 473-573)$ $(T: 811-905)$	1.30 at 905 K 1.1 at 1073 K
6061-T6	2.705	$\rho_s = 2.705 - 20.1 \times 10^{-2}\,(T - 298)$ $\rho_l = 2.415 - 28.0 \times 10^{-2}\,(T - 915)$	$k_s = 7.62 + 0.995T - 17 \times 10^{-4}T^2 + 1 \times 10^{-6}T^3$ $k_s = 66.5$ $k_l = 90$	$(T: 298-773)$ $(T: 873)$ $(T: 915)$	1.15 at 915 K 1.0 at 1073 K
7075-T6	2.805	$\rho_s = 2.805 - 22.4 \times 10^{-2}\,(T - 273)$ $\rho_l = 2.50 - 28.0 \times 10^{-2}\,(T - 901)$	$k_s = 196$ $k_s = 193$ $k_l = 85$	$(T: 673-773)$ $(T: 805)$ $(T: 901)$	1.3 at 901 K 1.1 at 1073 K
Copper alloys					
Cu-Al (Al-bronze)	7.262	$\rho_s = 7.262 - 48.6 \times 10^{-2}\,(T - 298)$ $\rho_l = 6.425 - 65.0 \times 10^{-2}\,(T - 1350)$	$k_s = 7.925 + 0.1375T - 6 \times 10^{-5}T^2$ $k_s = 42$ $k_l = 27$	$(T: 373-773)$ $(T: 1313)$ $(T: 1373)$	6.3 at 1350 K 5.2 at 1473 K
Brass(e) 70Cu-30Zn	...	...	$k_s = 140.62 + 112.14 \times 10^{-4}T$ $k_{s/l} = 2430.3 - 191.61 \times 10^{-2}T$ $k_l = 45.43 + 26 \times 10^{-3}T$	$(T: 460-1188)$ $(T: 1188-1228)$ $(T \geq 1228)$	...
Brass(e) 60Cu-40Zn	...	...	$k_s = 182.95 + 366.1 \times 10^{-4}T$ $k_{s/l} = 16479.5 - 13.93\,T$ $k_l = 39.724 + 26 \times 10^{-3}T$	$(T: 620-1173)$ $(T: 1173-1178)$ $(T \geq 1178)$	...
Copper-nickel(e) 70Cu-30Ni-Fe-Mn	...	...	$k_s = 16.041 + 438.9 \times 10^{-4}T$ $k_{s/l} = 796.018 - 502.63 \times 10^{-3}T$ $k_l = -3.8 + 26 \times 10^{-3}T$	$(T \leq 1443)$ $(T: 1443-1513)$ $(T \geq 1513)$	...
Iron and steel alloys					
Steel AISI 1008(e)	7.86	$\rho_l = 7.0$ at 1823 K	$k_s = 13.58 + 11.3 \times 10^{-3}T$ $k_l = 280.72 - 14.0 \times 10^{-3}T$	$(T: 1122-1768)$ $(T \geq 1768)$	...
1% Cr steel(e) Fe-0.3C-0.69Mn-0.2Si-1.1Cr-0.7Ni	...	...	$k_s = 14.53 + 105.0 \times 10^{-4}T$ $k_{s/l} = 91.74 - 351.0 \times 10^{-4}T$ $k_l = 7.85 + 116.8 \times 10^{-4}T$	$(T: 1073-1693)$ $(T: 1693-1793)$ $(T \geq 1793)$	...
304 stainless steel	8.02	$\rho_s = 8.02 - 50.1 \times 10^{-2}\,(T - 298)$ $\rho_l = 6.90 - 80.0 \times 10^{-2}\,(T - T_m)$	$k_s = 10.33 + 15.4 \times 10^{-3}T - 7.0 \times 10^{-7}T^2$ $k_{s/l} = 355.93 - 196.8 \times 10^{-3}T$ $k_l = 6.6 + 12.14 \times 10^{-3}T$	$(T: 298-1633)$ $(T: 1644-1672)$ $(T \geq 1793)$	8.0 at 1727 K
316 stainless steel	7.95	$\rho_s = 7.95 - 50.1 \times 10^{-2}\,(T - 298)$ $\rho_l = 6.881 - 77.0 \times 10^{-2}\,(T - T_m)$	$k_s = 6.31 + 27.2 \times 10^{-3}T - 7.0 \times 10^{-6}T^2$ $k_{s/l} = 355.93 - 196.8 \times 10^{-3}T$ $k_l = 6.6 + 121.4 \times 10^{-4}T$	$(T: 298-1573)$ $(T: 1644-1672)$ $(T \geq 1672)$	8.0 at 1723 K
420 stainless steel(e)	7.7	$\rho_l = 7.0$ at 1823 K	$k_s = 20 + 61.5 \times 10^{-4}T$ $k_{s/l} = 133.4 - 594.9 \times 10^{-4}T$ $k_l = 6.5 + 116.8 \times 10^{-4}T$	$(T \leq 1727)$ $(T: 1727-1783)$ $(T \geq 1783)$	...
10-Ni Steel	7.88	$\rho = 7.96 - 2 \times 10^{-4}T - 7 \times 10^{-8}T^2$ $\rho = 8.29 - 5 \times 10^{-4}T$	$k_s = 29-33$ $k_s = 29 - 26$	$(298-973)$ $(1123-1683)$	...
HSLA-65	7.73	$\rho = 7.83 - 3 \times 10^{-4}T$ $\rho = 8.05 - 4 \times 10^{-4}T$	$k_s = 60.7 - 0.024 \times 10^{-2}T - 1.3 \times 10^{-5}T^2$ $k_s = 26.5$	$(298-973)$ $(1173-1283)$	...
HSLA-100	7.84	$\rho = 7.93 - 3 \times 10^{-4}T$ $\rho = 8.16 - 4.5 \times 10^{-4}T$	$k_s = 38.05 + 1.35 \times 10^{-2}T - 2 \times 10^{-5}T^2$ $k_s = -43.5 + 9.6 \times 10^{-2}T - 2.9 \times 10^{-5}T^2$	$(298-973)$ $(1083-1263)$ $(1190-1600)$	...
HY-100	7.76	$\rho = 7.89 - 2 \times 10^{-4}T$ $\rho = 8.23 - 4.5 \times 10^{-4}T$ $\rho = 8.23 - 4.5 \times 10^{-4}T$	$k_s = 36.4 + 2.15 \times 10^{-2}T - 2.7 \times 10^{-5}T^2$ $k_s = 26.8 + 1.5 \times 10^{-5}T + 3 \times 10^{-6}T^2$ $k_l = 25.2$ $k_l = -79.8 + 8 \times 10^{-2}T - 1.2 \times 10^{-5}T^2$	$(298-1050)$ $(1123-1723)$ $T_L = 1798$ K $(1798-1900)$	...
Ductile iron	7.3	$\rho_s = 7.06$ at 1373 K $\rho_l = 6.62$ at 1451 K $\rho_l = 6.586$ at 1573 C	$k_s = 31$ $k_s = 28$ $k_l = 29$	$(T: 1373)$ $(T: 1451)$ $(T: 1673)$	14.0 at 1451 K 11.5 at 1573 K 9.0 at 1673 K
Gray cast iron	7.2	$\rho_s = 6.992$ at 1273 K $\rho = 6.964$ at 1353 K $\rho_l = 7.495 - 77.0 \times 10^{-2}T$	$k_s = 29$ $k_s = 26$ $k_l = 28$	$(T: 1353)$ $(T: 1463)$ $(T: 1673)$	14.3 at 1463 K 14.0 at 1473 K 10.5 at 1573 K
Magnesium alloys					
AZ31B(e)	...	...	$k_s = 67.12 + 655.7 \times 10^{-4}T$ $k_{s/l} = 830.9 - 844.8 \times 10^{-3}T$ $k_l = 3.05 + 70.0 \times 10^{-3}T$	$(T: 499-839)$ $(T: 839-905)$ $(T \geq 905)$	...
AZ91B(e)	...	...	$k_s = 18.27 + 112.11 \times 10^{-3}T$ $k_{s/l} = 372 - 364.6 \times 10^{-3}T$ $k_l = -5.63 + 7.0 \times 10^{-2}T$	$(T \leq 742)$ $(T: 742-869)$ $(T \geq 869)$	...
KIA(e)	...	...	$k_s = 127.16 + 142.9 \times 10^{-4}T$ $k_l = 11.66 + 7.0 \times 10^{-2}T$	$(T: 520-922)$ $(T \geq 923)$	...

(a) Chemistry of the alloys is given in Table 4. (b) The polynomial equations that represent the thermal conductivity for most of the alloy systems shown in this table are estimated from the data recommended in Ref 16. Other data indicated as (e) have been obtained from Ref 14. For further analysis of the data it is recommended to consult the original source cited in the given literature. (c) Estimated values of viscosity may vary from ± 10 to $\pm 30\%$ (Ref 16). (d) Aluminum alloy with the British designation LM25 is equivalent to the Aluminum Association designation of A356. (e) Data from Ref 14. (f) Density and viscosity data from Ref 119. (g) Based on estimated C_p value from $K_S = \alpha \rho C_p$ (Ref 16)

(continued)

Table 10 (continued)

Material(a)	273 K	Density, g/cm³ T in K	Thermal conductivity(b), W/m·K (temp. range in K)		Estimated viscosity(c), mPa·s
ZK51A(e)	...	...	$k_s = 71.96 + 154.35 \times 10^{-3}T - 93.8 \times 10^{-6}T^2$	($T \leq 822$)	...
			$k_{s/l} = 688 - 672.2 \times 10^{-3}T$	(T: 822–914)	
			$k_l = 9.62 + 70.0 \times 10^{-3}T$	($T \geq 914$)	
HM11A(e)	...	...	$k_s = 73.35 + 133.43 \times 10^{-3}T - 73.9 \times 10^{-6}T^2$	($T \leq 903$)	...
			$k_{s/l} = 2887 - 305.0 \times 10^{-2}T$	(T: 903–923)	
			$k_l = 8.0 + 70.0 \times 10^{-3}T$	($T \geq 923$)	
EZ33	1.8	$\rho_s = 1.8 - 14.3 \times 10^{-2} (T - 298)$	$k_s = 156$	(T: 818)	1.5 at 913 K
		$\rho_l = 1.663 - 27.0 \times 10^{-2} (T - 913)$	$k_l = 91$	(T: 913)	1.4 at 973 K
Nickel alloys					
Single-crystal	8.7	$\rho_s = 8.7 - 45.8 \times 10^{-2} (T - 298)$	$k_s = 27.2$	(T: 1573)(g)	6.7 at 1653 K
CMSX-4		$\rho_l = 7.754 - 90.0 \times 10^{-2} (T - 1653)$	$k_l = 25.6$	(T: 1653–1673)	5.3 at 1773 K
Hastelloy X	8.24	$\rho_s = 8.24 - 38.1 \times 10^{-2} (T - 298)$	$k_s = 3.36 + 17.3 \times 10^{-3}T + 2.0 \times 10^{-6}T^2$	(T: 1073–1533)	7.5 at 1628 K
		$\rho_l = 7.42 - 83.0 \times 10^{-2} (T - 1628)$	$k_l = 29.0$	(T: 1428–1773)	5.5 at 1773 K
Inconel 718	8.19	$\rho_s = 8.19 - 39.2 \times 10^{-2} (T - 298)$	$k_s = 39.73 + 32.4 \times 10^{-3}T + 2.0 \times 10^{-5}T^2$	(T: 1173–1443)	7.2 at 1609 K
		$\rho_l = 7.40 - 88.0 \times 10^{-2} (T - 1609)$	$k_l = 29.6$	(T: 1609–1873)	5.31 at 1773 K
Titanium alloys					
Ti-6Al-4V	4.42	$\rho_s = 4.42 - 15.4 \times 10^{-2} (T - 298)$	$k_s = -0.797 + 18.2 \times 10^{-3}T - 2.0 \times 10^{-6}T^2$	(T: 1268–1923)	3.25 at 1923 K
		$\rho_l = 3.92 - 68.0 \times 10^{-2} (T - 1923)$	$k_l = 33.4$	(T: 1923)	2.66 at 2073 K
			$k_l = 34.6$	(T: 1973)	
Zinc alloys					
Zn-Al	6.7	$\rho_s = 6.7 - 60.3 \times 10^{-2} (T - 298)$	$k_s = 98.0$	(T: 630)	3.5 at 673 K
		$\rho_l = 6.142 - 97.7 \times 10^{-2} (T - 660)$	$k_l = 40.0$	(T: 660)	2.6 at 773 K

(a) Chemistry of the alloys is given in Table 4. (b) The polynomial equations that represent the thermal conductivity for most of the alloy systems shown in this table are estimated from the data recommended in Ref 16. Other data indicated as (e) have been obtained from Ref 14. For further analysis of the data, it is recommended to consult the original source cited in the given literature. (c) Estimated values of viscosity may vary from ±10 to $\pm30\%$ (Ref 16). (d) Aluminum alloy with the British designation LM25 is equivalent to the Aluminum Association designation of A356. (e) Data from Ref 14. (f) Density and viscosity data from Ref 119. (g) Based on estimated C_p value from $K_S = \alpha\rho C_p$ (Ref 16)

Table 11 Electrical resistivity (ρ_e) and thermal conductivity of solid and liquid metals

Element	T_m, K	Electrical resistivity (Ref 13, 31)						Thermal conductivity (Ref 130)			
		$\rho_{e,s}$, μΩ·cm	$\rho_{e,l}$, μΩ·cm	$\rho_{e,l}/\rho_{e,s}$	$\rho_{e,l} = \alpha T + \beta$		T range T_m to T, K	k_{sm}	k_{lm}	ΔS_m, J/mol·K	K
					α, μΩ·cm/K	β, μΩ·cm		W/m·K			
Ag	1233.7	8.2	17.2	2.09	0.0090	6.2	1473	362	175	9.15	0.0794
Al	931	10.9	24.2	2.20	0.0145	10.7	1473	211	91	10.71	0.0785
Au	1336	13.68	31.2	2.28	0.0140	12.5	1473	247	105	9.39	0.0911
B	2448	...	210.0	...	...	...	...	...	...	...	...
Be	1550	...	45.0	...	...	...	...	...	...	...	...
Bi	544	...	129.0	...	...	...	...	7.6	12	20.75	0.022
Cd	593	17.1	33.7	1.97	Not linear	...	...	37 ± 3	90	10.42	...
Ce	1060	...	126.8	...	...	...	...	21	22	2.99	...
Cr	2148	...	31.6	...	...	...	...	45	35	9.63	0.0261
Co	1766	97	102	1.05	0.0612	-6.0	1973	45	36	9.16	0.0243
Cu	1356	9.4	20.0	2.1	0.0102	6.2	1873	330	163	9.77	0.0722
Fe	1811	122	110	0.9	0.033	50	1973	34	33	7.62	0.0039
Hf	2216	...	218.0	...	...	...	...	39	...	10.9	...
La	1203	...	138	...	...	...	...	...	17	5.19	...
Li	453.5	...	240.0	...	...	...	...	71	43	6.6	0.076
Mg	924	15.4	27.4	1.78	0.005	22.9	1173	145	79	9.18	0.066
Mn	1525	66	40	0.61	No data	...	...	24	22	8.3	0.0102
Mo	2880	...	60.5	...	...	...	...	87	72	12.95	...
Nb	2741	...	105.0	...	...	...	...	78	66	10.90	...
Ni	1727	65.4	85.0	1.3	0.0127	63	1973	70	60	10.11	0.0152
Pb	660.6	49.0	95.0	1.94	0.0479	66.6	1273	30	15	7.95	0.0872
Pd	1825	...	...	...	...	...	99	87	9.15	...	...
Pt	2042	...	73.0	...	...	...	...	80	53	10.86	0.0392
Re	3431	...	145	...	...	...	...	65 ± 5	55	17.5	...
Sb	904	183	113.5	0.61	0.270	87.9	1273	17	25	22	...
Si	1683	...	75	...	...	...	...	25	56	29.8	...
Sn	505	22.8	48.0	2.10	0.0249	35.4	1473	59.5	27	13.9	0.0567
Ta	3250	...	118.0	...	...	...	...	70	58	11.1	0.0169
Ti	1958	...	172	...	...	...	...	31	31	7.28	0.0
V	2185	...	71.0	...	...	...	...	51	43.5	9.85	0.0161
W	3650	...	127	...	...	...	...	95	63	14.2	0.029
Zn	692	16.7	37.4	2.24	Not linear	...	...	90	50	10.6	0.055
Zr	2123	...	153	...	...	...	...	38	36.5	9.87	0.0041

some total normal emissivity for some pristine and oxidized metals as well as the normal spectral emissivity of some liquid metals near their melting point (Ref 133), while Table 14 gives the total emissivity for some alloys and refractory materials (Ref 134).

Additional emissivity data for various materials have been compiled and can be found in the literature (Ref 5, 79, 132, 135 to 137).

The data quoted in Tables 13 and 14 provide a guideline and must be used with discretion because, in practical applications, the values of emissivity may change considerably with oxidation and roughening of the surfaces. Therefore, it is important that the total emissivity should be determined for the actual surface conditions of the materials in question.

Table 12 Heat capacity and thermal conductivity of some mold materials

Mold material	Dry density, g/cm³	Specific heat capacity, kJ/Kg · K		Thermal conductivity, W/m⁻¹ · K	
Silica molding sands					
20–30 mesh	1.730	$C_p = 0.5472 - 1.147 \times 10^{-3}T - 5.401 \times 10^{-7}T$	$(T < 1033)$	$k = 0.604 - 0.767 \times 10^{-3}T + 0.795 \times 10^{-6}T^2$	$(T < 1033$
50–70 mesh	1.634	$C_p = 1.066 + 8.676 \times 10^{-5}T$	$(T > 1033)$	$k = 0.676 - 0.793 \times 10^{-3}T + 0.556 \times 10^{-6}T^2$	$(T > 1033)$
80–120 mesh	1.458	...	...	...	...
Silica sand (−22 mesh)					
+ 7% bentonite	1.520	$C_p = 0.4071(T - 273)^{0.154}$	$(T \leq 1600)$	$k = 0.946 - 0.903 \times 10^{-3}T + 0.564 \times 10^{-6}T^2$	$(T < 1500)$
+ 4% bentonite	1.60	...	...	$k = 1.26 - 0.169 \times 10^{-2}T + 0.105 \times 10^{-5}T^{-2}$	$(T < 1500)$
Olivine sand	1.83	$C_p = 0.3891(T - 273)^{0.162}$	$(T \leq 1300)$	$k = 0.713 + 0.349 \times 10^{-4}T$	
+ 4% bentonite	2.125	...	...	$k = 1.82 - 1.88 \times 10^{-2}T + 0.10 \times 10^{-5}T^{-2}$	$(T < 1300)$
Zircon sand	2.780	...	...	$k = 1.19 - 0.948 \times 10^{-3}T + 0.608 \times 10^{-6}T^2$	...
+ 4% bentonite	2.96	$C_p = 0.2519(T - 273)^{0.170}$	$(T \leq 1300)$	$k = 1.82 - 0.176 \times 10^{-2}T + 0.984 \times 10^{-6}T^2$	$(T < 1500)$
Chromite sand					
3.9% bentonite	2.75	$C_p = 0.318(T - 273)^{0.158}$	$(T \leq 1300)$	$k = 941 - 0.753 \times 10^{-3}T + 0.561 \times 10^{-6}T^2$	...
Graphite (chill foundry grade)	1.922	$C_p = -0.11511 + 2.8168 \times 10^{-3}T$	$(T < 505)$	$k = 135.99 - 8.378 \times 10^{-2}T$	$(T < 873)$
		$C_p = 0.6484 + 1.305 \times 10^{-3}T$	$(505–811)$	...	...
		$C_p = 1.3596 + 4.2797 \times 10^{-4}T$	$(T > 811)$	$k = 103.415 - 4.647 \times 10^{-2}T$	$(T > 873)$
Investment casting					
Zircon-30% alumina-20% silica	2.48–2.54	...	...	$k = 3.03 - 3.98 \times 10^{-4}T + 508\,T^{-1}$	$(T: 375–1825)$

Source: Ref 14

Table 13 Total normal and spectral emissivity of pristine and oxidized metals

Metal	Total normal emissivity				Normal spectral emissivity of the liquid near the melting point(a)	
	Pristine	Temp., K	Oxidized	Temp., K	Emissivity at λ = 632.8 nm	Temp., K
Ag	0.02–0.03	773	...	...	0.09	1300
Al	0.064	773	0.19	873	0.04	1000
Au	...	...	...	...	0.30	1400
Be	0.87	1473	...	...	...	...
Cr	0.11–0.14	773	0.14–0.34	873	...	...
Co	0.34–0.46	773	...	...	...	...
Cu: solid	0.02	773	0.24	1073	...	...
Cu: liquid	0.12	1473	...	...	0.13	1400
Hf	0.32	1873	...	...	...	...
Fe	0.24	1273	0.57	873	...	...
Mo	0.27	1873	0.84	673	...	...
Ni	0.14–0.22	1273	0.49–0.71	1073	0.40	1800
Nb	0.18	1873	0.74	1073	0.32	2750
Pd	0.15	1473	0.124	1273	0.38	1925
Pt	0.16	1473	...	...	0.38	2250
Rh	0.09	1673	...	...	...	...
Si	...	...	...	...	0.19	1761
Ta	0.18	1873	0.42	873	...	...
Ti(b)	0.47	1673	...	...	0.31	1946
W	0.17	1673	...	...	...	...
	0.18	1873	...	...	...	...
	0.23	2273	...	...	0.439	2500
Zr	...	...	...	...	0.35	2125
Alloys	...	...	...	...	...	
Cast iron	0.29	1873	0.78	873	...	

(a) Emissivities from laser polarimetry at a wavelength of λ = 632.8 nm. Source: Ref 133. (b) Spectral normal emissivity at 65 μm wavelength.
Source: Ref 132

Typical Thermophysical Properties Ranges of Some Cast Alloys

Tables 15 to 18 show the typical range of thermophysical properties for aluminum, magnesium, cast iron, and nickel alloys. Unfortunately, numerical values of those thermophysical properties in the mushy zone or even in the liquid state are not available. However, one could infer the expected trends as noted in the tables.

Summary

This article introduces the typical values for the most used types of thermophysical property data that are needed for processing and modeling of solidification and casting processes. The following conclusions can be drawn:

- Throughout this survey, values have been quoted. These should be used with discretion and taken as representative values for materials and not necessarily values for a particular system or process.
- The properties considered are specific heat capacity, enthalpy of melting, solidus and liquidus temperatures, coefficient of thermal expansion, density, surface tension, viscosity, electrical and thermal conductivity, and emissivity appropriate to metals, alloys, and molds.
- The prediction of properties has advanced in recent years, but some care is needed in validating the data.
- Before commissioning expensive work to measure properties, the user is advised to check the sensitivity of the required predictions to changes in the input thermophysical data. This will enable measurement efforts to concentrate on the critical data needed.
- Among properties not considered are:
  a. Chemical diffusion coefficients and partition coefficients, which can strongly affect the chemistry at the microstructural scale
  b. Mechanical properties of the metal close to the solidus temperature where it is very weak, with implications to the strength such as susceptibility to hot tearing and cracking
  c. Mechanical properties of molds at appropriate temperatures during casting

REFERENCES

1. J.J. Valencia, Symposium on Thermophysical Properties: Metalworking Industry Needs and Resource, Concurrent Technologies Corporation, Oct 22–23, 1996
2. A. Ludwig, P. Quested, and G. Neuer, How to Find Thermophysical Material Property Data for Casting Simulations, *Adv. Eng. Mater.*, Vol 3, 2001, p 11–14
3. Y.S. Touloukian, C.Y. Ho, et al., *Thermophysical Properties Research Center Data Book*, Purdue University, 1960–1966
4. Y.S. Touloukian, Ed., *Thermophysical Properties of High Temperature Solid Materials*, MacMillan Co., 1967
5. Y.S. Touloukian, J.K. Gerritsen, and N.Y. Moore, Ed., *Thermophysical Properties Literature Retrieval Guide, Basic Edition*, Plenum Press, 1967
6. Y.S. Touloukian and C.Y. Ho, Ed., *Thermophysical Properties of Matter—The TPRC Data Series*, IFA/Plenum Data Co., 1970–1979

Table 14 Total normal emissivity of various alloys and refractory materials

Material	ε	Temperature, K
Alloys		
Commercial aluminum	0.09	373
Oxidized	0.11–0.19	472–872
Cast iron: solid	0.60–0.70	1155–1261
Liquid	0.29	1873
Commercial copper: heated at 872 K	0.57	472–872
Liquid	0.16–0.13	1349–1550
Steel: plate rough	0.94–0.97	273–644
Steel: liquid	0.42–0.53	1772–1922
Inconel 718(a)	0.356	1650–1950
Refractory materials		
Alumina: 85–99.5Al_2O_3-0–12SiO_2-0–1Fe_2O_3	0.5–0.18	1283–1839
Alumina-silica: 58–80Al_2O_3-16–18SiO_2-0.4Fe_2O_3	0.61–0.43	1283–1839
Fireclay brick	0.75	1273
Carbon rough plate	0.77–0.72	373–773
Magnesite brick	0.38	1273
Quartz, fused	0.93	294
Zirconium silicate	0.92–0.80	510–772
	0.80–0.52	772–1105

Note: A linear interpolation of the emissivity values with temperature can be done when the emissivity values are separated by a dash. It should be noted that some materials (such as alumina) are semitransparent, and their measured emissivity depends on the thickness of the sample. Source: Ref 24. (a) Spectral normal emissivity at 633 nm wavelength. Source: Ref 138

Table 15 Range of thermophysical and mechanical properties of cast aluminum alloys

Property	At room temperature	At solidus temperature	In the mushy zone	In the liquid range
Thermal conductivity, W/K · m	100–180	150–210	Dropping	60–80
Heat capacity, J/kg · K	880–920	1100–1200	Almost constant	1100–1200
Density, kg/m³	2600–2800	2400–2600	Dropping	2200–2400
Viscosity, mm²/s			Sharply decreasing	0.4–0.5
Young's modulus, GPa	68–75	40–50	Dropping to zero	
Thermal expansion, μm/m · °C	20–24		25–30	Sharply increasing
Heat of fusion, kJ/kg			400–500	

Source: Ref 24

Table 16 Range of thermophysical and mechanical properties of cast magnesium alloys

Property	At room temperature	At solidus temperature	In the mushy zone	In the liquid range
Thermal conductivity, W/K · m	50–85	80–120	Dropping	50–70
Heat capacity, J/kg · K	1000–1050	1150–1250	Almost constant	1200–1350
Density, kg/m³	1750–1850	1650–1750	Dropping	1550–1650
Viscosity, μm²/s			Sharply decreasing	0.6–0.7
Young's modulus, GPa	42–47		30–35	Dropping to zero
Thermal expansion, μm/m · °C	24–26		30–34	Sharply increasing
Heat of fusion, kJ/kg			280–380	

Source: Ref 24

Table 17 Range of thermophysical and mechanical properties of cast iron

Property	At room temperature	At solidus temperature	In the mushy zone	In the liquid range
Thermal conductivity, W/K · m	30–50	25–30	Almost constant	25–35
Heat capacity, J/kg · K	460–700	850–1050	Dropping	800–950
Density, kg/m³	6900–7400	6750–7350	Shrinking/expanding	6700–7300
Viscosity, μm²/s	...	...	Sharply decreasing	0.5–0.8
Young's modulus, GPa	80–160	60–100	Dropping to zero	...
Thermal expansion, μm/m · °C	11–14	17–22	Sharply increasing	...
Heat of fusion, kJ/kg	...	...	200–250	...

Source: Ref 24

7. Y.S. Touloukian, J.K. Gerritsen, and W.H. Shafer, Ed., *Thermophysical Properties Literature Retrieval Guide, Supplement I (1964–1970)*, IFI/Plenum Data Co., 1973

8. Y.S. Touloukian and C.Y. Ho, Ed., *Thermophysical Properties of Selected Aerospace Materials, Part I: Thermal Radiative Properties; Part II: Thermophysical Properties of Seven Materials*, Purdue University, TEPIAC/CINDAS, Part I, 1976; Part II, 1977

9. J.K. Gerritsen, V. Ramdas, and T.M. Putnam, Ed., *Thermophysical Properties Literature Retrieval Guide, Supplement II (1971–1977)*, IFI/Plenum Data Co., 1979

10. J.F. Chaney, T.M. Putnam, C.R. Rodriguez, and M.H. Wu, Ed., *Thermophysical Properties Literature Retrieval Guide (1900–1980)*, IFI/Plenum Data Co., 1981

11. Y.S. Touloukian and C.Y. Ho, Ed., *McGraw-Hill/CINDAS Data Series on Materials Properties*, McGraw-Hill Book Co., 1981

12. *Thermophysical Properties of Materials: Computer-Readable Bibliographic Files, Computer Magnetic Tapes*, TEPIAC/CINDAS, 1981

13. C.J. Smithells, General Physical Properties, *Metals Reference Book*, 7th ed., E.A. Brandes and G.B. Brook, Ed., Butterworth-Heinemann, 1992, p 14–1

14. R.D. Pehlke, A. Jeyarajan, and H. Wada, *Summary of Thermal Properties for Casting Alloys and Mold Materials*, Grant DAR78–26171, The University of Michigan—National Science Foundation, Applied Research Division, Dec 1982

15. Thermal Properties of Metals, *ASM Ready Reference*, F. Cverna, Ed., ASM International, 2002

16. K.C. Mills, *Recommended Values of Thermophysical Properties for Selected Commercial Alloys*, National Physical Laboratory and ASM International, Woodhead Publishing Limited, Cambridge, England, 2002

17. J. Miettien, *Metall. Trans. A*, Vol 23, 1992, p 1155–1170

18. J. Miettien and S. Louhenkilpi, Calculation of Thermophysical Properties of Carbon and Low Alloy Steels for Modeling of Solidification Processes, *Metall. Trans. B*, Vol 25, 1994, p 909–916

19. Y.A. Chang, A Thermodynamic Approach to Obtain Materials Properties for Engineering Applications, *Proceedings of a Workshop on the Thermophysical Properties of Molten Materials*, Oct 20–23, 1992 (Cleveland, OH), NASA Lewis Research Center, p 177–201

20. K.C. Mills, A.P. Day, and P.N. Quested, Estimating the Thermophysical Properties of Commercial Alloys, *Proceedings of Nottingham Univ.-Osaka Univ. Joint Symposium*, Sept 1995 (Nottingham)

21. N. Saunders, A.P. Midwnik, and J.-Ph. Schillé, Modelling of the Thermophysical and Physical Properties for Solidification of Ni-Based Superalloys, *Proceedings of the 2003 International Symposium on Liquid Metal Processing and Casting*, P. Lee et al., Ed., p 253–260

22. J.A.J. Robinson, A.T. Dinsdale, and P. N. Quested, The Prediction of

Table 18 Range of thermophysical and mechanical properties of nickel-base superalloys

Property	At room temperature	At solidus temperature	In the mushy zone	In the liquid range
Thermal conductivity, W/K · m	8–14	28–32	Slightly dropping	25–35
Heat capacity, J/kg · K	420–470	620–700	Rising	700–800
Density, kg/m³	8200–8700	7500–8000	Dropping	7000–7500
Viscosity, μm²/s			Sharply decreasing	0.6–1.0
Young's modulus, GPa	180–220	100–120	Dropping to zero	
Thermal expansion, μm/m · °C	12–13		20–23	Sharply increasing
Heat of fusion, kJ/kg			200–220	

Source: Ref 24

Thermophysical Properties of Steels and Slags, *Second International Congress on the Science and Technology of Steelmaking,* University of Wales, Swansea, April 2001

23. X.L. Yang, P.D. Lee, R.F. Brooks, and R. Wunderlich, The Sensitivity of Investment Casting Simulations to the Accuracy of Thermophysical Property Values, *Superalloys 2004,* K.A. Green et al., Ed., TMS, 2004, p 951–958

24. E. Kaschnitz, A. Bührig-Polaczek, and G. Pottlacher, "Thermophysical Properties of Casting Materials," World Foundrymen Organization—International Commission 3.3, Casting Simulation—Background and Examples from Europe and USA, WFO, 2002

25. K.D. Maglic, A. Cezairliyan, and V.E. Peletsky, Ed., *Survey of Measurement Techniques,* Vol 1, *Compendium of Thermophysical Property Measurement Methods,* Plenum Press, 1984

26. A.F. Crawley, Density of Liquid Metals and Alloys, *Int. Metall. Rev.,* Review 180, Vol 19, 1974, p 32–48

27. E.E. Shpil'rain, K.A. Yakimovich, and A.G. Mozgovoi, Apparatus for Continuous Measurement of Temperature Dependence of Density of Molten Metals by the Method of a Suspended Pycnometer at High Temperatures and Pressures, *Recommended Measurement Techniques and Practices,* Vol 2, *Compendium of Thermophysical Property Measurement Methods,* Plenum Press, 1992, p 601–624

28. S.D. Mark and S.D. Emanuelson, A Thermal Expansion Apparatus with a Silicon Carbide Dilatometer for Temperatures to 1500 °C, *Ceram. Bull.,* Vol 37 (No. 4), 1958, p 193–196

29. L.D. Lucas, *Physicochemical Measurements in Metals Research, Part 2,* R.A. Rapp, Ed., Interscience, New York, 1970, p 219

30. P. Parlouer, Calorimetry and Dilatometry at Very High Temperatures. Recent Developments in Instrumentation and Applications, *Rev. Int. Hautes Temp. Réfract.,* Vol 28, 1992–1993, p 101–117

31. T. Iida and R.I.L. Guthrie, *The Physical Properties of Liquid Metals,* Clarendon Press, 1988, p 188

32. W.K. Rhim and A.J. Rulison, Measuring Surface Tension and Viscosity of a Levitated Drop, Report of National Aeronautics and Space Administration Contract NASA 7–918, 1996, p 1–6

33. I. Egry, G. Lohofer, P. Neuhaus, and S. Sauerland, Surface Tension Measurements of Liquid Metals Using Levitation, Microgravity, and Image Processing, *Int. J. Thermophys.,* Vol 13 (No. 1), 1992, p 65–74

34. Y. Bayazitoglu and G.F. Mitchell, Experiments in Acoustic Levitation: Surface Tension Measurements of Deformed Droplets, *J. Thermophys. Heat Transf.,* Vol 9 (No. 4), 1995, p 694–701

35. W.J. Parker, R.J. Jenkins, C.P. Butler, and G.L. Abbott, Flash Method of Determining Thermal Diffusivity, Heat Capacity and Thermal Conductivity, *J. Appl. Phys.,* Vol 32 (No. 9), 1961, p 1679–1684

36. R.D. Cowan, Pulse Method of Measuring Thermal Diffusivity at High Temperatures, *J. Appl. Phys.,* Vol 34 (No. 4), 1963, p 926–927

37. J.T. Schriempf, A Laser Flash Technique for Determining Thermal Diffusivity of Liquid Metals at Elevated Temperatures, *High Temp.—High Press.,* Vol 4, 1972, p 411–416

38. K.D. Maglic and R.E. Taylor, The Apparatus for Thermal Diffusivity Measurement by the Laser Pulse Method, *Compendium of Thermophysical Property Measurement Methods,* Vol 2, 1992, p 281–314

39. R.E. Taylor and K.D. Maglic, Pulse Method for Thermal Diffusivity Measurement, *Survey of Measurement Techniques,* Vol 1, *Compendium of Thermophysical Property Measurement Methods,* Plenum Press, 1984, p 305–336

40. Y. Qingzhao and W. Likun, "Laser Pulse Method of Determining Thermal Diffusivity, Heat Capacity, and Thermal Conductivity," FATPC, 1986, p 325–330

41. Y. Maeda, H. Sagara, R.P. Tye, M. Masuda, H. Ohta, and Y. Waseda, A High-Temperature System Based on the Laser Flash Method to Measure the Thermal Diffusivity of Melts, *Int. J. Thermophys.,* Vol 17 (No. 1), 1996, p 253–261

42. G. Höhne, W. Hemminger, and H.-J. Flammershein, *Differential Scanning Calorimetry. An Introduction to Practitioners,* Springer-Verlag, Berlin, 1996

43. J.D. James, J.A. Spittle, S.G.R. Brown, and R.W. Evans, A Review of the Measurement Techniques for the Thermal Expansion Coefficient of Metals and Alloys at Elevated temperatures, *Meas. Sci. Technol.,* Vol 12, 2001, p R1–R15

44. Y. Su, K.C. Mills, and A. Dinsdale, A Model to Calculate Surface Tension of Commercial Alloys, *J. Mater. Sci.,* Vol 40, 2005, p 2185–2190

45. K.C. Mills, Measurement and Estimation of Physical Properties of Metals at High Temperatures, *Fundamentals of Metallurgy,* S. Seetharaman, Ed., Woodhead Publishing Limited, Cambridge, England, 2005

46. "Standard Test Method for Determining Specific Heat Capacity by Differential Scanning Calorimetry," E 1269, *Annual Book of ASTM Standards,* ASTM, 2001

47. "Practice for Temperature Calibration of DSC and DTA," E 967, ASTM

48. "Standard Practice for Heat Flow Calibration of DSC," E 968, ASTM

49. "Standard Method for Enthalpy Measurement Validation of Differential Scanning Calorimeters," E 2253, ASTM

50. "Standard Test Method for Enthalpies of Fusion and Crystallisation by DSC," E 793, ASTM

51. "Standard Test Method for Specific Heat of Liquids and Solids," D 2766, *Annual Book of ASTM Standards,* ASTM, 1995

52. "Advanced Technical Ceramics. Monolithic Ceramics. Thermo-Physical Properties. Determination of Specific Heat Capacity," EN 821–3:2005

53. "Standard Test Method for Linear Thermal Expansion of Solid Materials with a Vitreous Silica Dilatometer," E 228, *Annual Book of ASTM Standards,* ASTM, 1995

54. "Standard Test Method for Linear Thermal Expansion of Rigid Solids with Interferometry," E 289, *Annual Book of ASTM Standards,* ASTM, 1999

55. "Standard Test Method for Linear Thermal Expansion of Solid Materials by Thermomechanical Analysis," E 831, *Annual Book of ASTM Standards,* ASTM, 2000

56. "Advanced Technical Ceramics. Monolithic Ceramics. Thermo-Physical Properties. Determination of Thermal Expansion," EN 821–1:1995

57. "Fine Ceramics (Advanced Ceramics, Advanced Technical Ceramics)—Test Method for Linear Thermal Expansion of Monolithic Ceramics by Push-Rod Technique," ISO 17562:2001

58. "Standard Test Method for Thermal Conductivity and Thermal Diffusivity by Modulated Temperature Differential Scanning Calorimetry," E 1952, *Annual Book of ASTM Standards,* ASTM, 2001

59. "Standard Test Method for Thermal Diffusivity of Solids by the Laser Flash Method," E 1461, *Annual Book of ASTM Standards,* ASTM, 2001

60. "Standard Test Method for Steady-State Thermal Transmission Properties by

Means of the Heat Flow Meter Apparatus," C 518, *Annual Book of ASTM Standards,* ASTM, 1998

61. "Advanced Technical Ceramics. Monolithic Ceramics. Thermo-Physical Properties. Determination of Thermal Diffusivity by the Laser Flash (or Heat Pulse) Method," EN 821–2:1997

62. "Fine Ceramics (Advanced Ceramics, Advanced Technical Ceramics)—Determination of Thermal Diffusivity of Monolithic Ceramics by Laser Flash Method," ISO 18755:2005

63. "Standard Test Method for Normal Spectral Emittance at Elevated Temperatures," E 307, *Annual Book of ASTM Standards,* ASTM, 1972, 2002

64. "Standard Test Method for Total Normal Spectral Normal Emittance of Surfaces Using Inspection Meter Techniques," E 408, *Annual Book of ASTM Standards,* ASTM, 1971, 2002

65. D.R. Gaskell, *Introduction to Metallurgical Thermodynamics,* McGraw-Hill, 1973, p 497–501

66. L. Darken and R. Gurry, *Physical Chemistry of Metals,* McGraw-Hill, 1953, p 125–126

67. F.E. Neumann, Untersuchung Über die Specifische Wärme der Mineralien, *Pogend. Annal.,* Vol XXIII (No. 1831), p 1–39

68. H. Kopp, *Proc. R. Soc. (London),* Vol 13, 1863–1864, p 229–239

69. K.A. Jackson, *Liquid Metals and Solidification,* American Society for Metals, 1958, p 174

70. J.J. Valencia et al., "Weld Solidification and Joint Strength Analyses of NS-110 Weldments," Navy Metalworking Center Report, TR No. 00-100, Nov 2001

71. J.J. Valencia and C. Papesch, "Thermophysical Properties Characterization of HSLA-65 Rolled Plate," NMC Internal Report, Feb 2006

72. J.J. Valencia and C. Papesch, "Thermophysical Properties Characterization of HSLA-100 Rolled Plate," NMC Internal Report, Aug 2005

73. J.J. Valencia and C. Papesch, "Thermophysical Properties Characterization of HSLA-100 Rolled Plate," NMC Internal Report, Sept 1998

74. J. Blumm and J.B. Henderson, *High Temp.—High Press.,* Vol 32, 2003–2004, p 109–113

75. R. Morrell and P. Quested, Evaluation of Piston Dilatometry for Studying the Melting Behaviour of Metals and Alloys, *High Temp.—High Press.,* Vol 35–36, 2003–2004, p 417–435

76. G. Lang, Density of Liquid Elements, *CRC Handbook of Chemistry and Physics,* 75th ed., D.R. Lide and H.P.R. Fredererikse, Ed., CRC Press, 1994–1995, p 4–126 to 4–134

77. J.R. Wilson, *Metall. Rev.,* Vol 10, 1965, p 381

78. L.J. Wittenberg, Viscosity of Liquid Metals, *Physicochemical Measurements in Metal Research,* Vol IV, Part 2, R.A. Rapp, Ed., Wiley Interscience, 1970, p 193

79. M.J. Assael et al., Reference Data for the Density and Viscosity of Liquid Aluminum and Liquid Iron, *J. Phys. Chem. Ref. Data,* Vol 35, 2006, p 285–300

80. W. Kurz and D.J. Fisher, *Fundamentals of Solidification,* Trans Tech Publications, Switzerland, 1984, p 15

81. S. Sauerland, R.F. Brooks, I. Egry, and K.C. Mills, *Proc. TMS Conf. Ann. Conf. on Containerless Processing,* 1993, p 65–69

82. K.C. Mills and R.F. Brooks, *Mater. Sci. Eng. A,* Vol 178, 1994, p 77–81

83. B.C. Allen, *Liquid Metals: Chemistry and Physics,* S.Z. Beer, Ed., Marcel Dekker, 1972, p 161–197

84. A.S. Skapski, *J. Chem. Phys.,* Vol 16, 1948, p 389

85. B.J. Keene, Review of Data for the Surface Tension of Pure Metals, *Int. Mater. Rev.,* Vol 38, 1993, p 157–192

86. B.J. Keene, Review of Data for Surface Tension of Iron and Its Binary Alloys, *Int. Mater. Rev.,* Vol 33, 1988, p 1–37

87. K.C. Mills and Y.C. Su, Review of Surface Tension Data for Metallic Elements and Alloys, Part 1: Pure Metals, *Int. Mater. Rev.,* Vol 51, 2006, p 329–351

88. R.F. Brooks and P.N. Quested, The Surface Tension of Steels, *J. Mater. Sci.,* Vol 40, 2005, p 2233–2238

89. S.-L. Chen, W. Oldfield, Y.A. Chang, and M.K. Thomas, *Metall. Mater. Trans. A,* Vol 25, 1994, p 1525–1533

90. F. Zhang, Y.A. Chang, and J.S. Chou, A Thermodynamic Approach to Estimate Titanium Thermophysical Properties, *Proceedings of 1997 International Symposium on Liquid Metal Processing and Casting,* A. Mitchel and P. Auburtin, Ed., American Vacuum Society, 1997, p 35–59

91. K.S. Yeum, R. Speiser, and D.R. Poirier, *Metall. Trans. B,* Vol 20, 1989, p 693–703

92. H.-K. Lee, J.P. Hajra, and Z. Frohberg, *Metalkd.,* Vol 83, 1992, p 638–643

93. J.A.V. Buttler, *Proc. R. Soc. (London) A,* Vol 135, 1932, p 348

94. J.-S. Chou et al., "Thermophysical and Solidification Properties of Titanium Alloys," Report TR 98–87, National Center for Excellence in Metalworking Technology, June 30, 1999

95. F. Zhang, S.-L. Chen, and Y.A. Chang, Modeling and Simulation in Metallurgical Engineering and Materials Science, *Proc. Int. Conf., MSMM'96,* Z.-S. Yu, Ed. (Beijing, China), 1996, p 191–196

96. J.L. Murray, *Phase Diagrams of Binary Titanium Alloys,* ASM International, 1987

97. J.L. Murray, *Bull. Alloy Phase Diagrams,* Vol 2, 1981, p 185–192

98. N. Saunders and V.G. Rivlin, *Mater. Sci. Technol.,* Vol 2 1986, p 521

99. L. Kaufman and H. Nesor, *Metall. Trans.,* Vol 5, 1974, p 1623–1629

100. U.R. Kattner and B.P. Burton, *Phase Diagrams of Binary Titanium Alloys,* ASM International, 1987

101. K. Frisk and P. Gustafson, *CALPHAD,* Vol 12, 1988, p 247–254

102. K.-J. Zeng, H. Marko, and L. Kaj, *CALPHAD,* Vol 17, 1993, p 101–107

103. J.-O. Anderson and N. Lange, *Metall. Trans. A,* Vol 19, 1988

104. A.D. Pelton, *J. Nucl. Mater.,* Vol 201 (No. 218–224), 1993, p 1385–1394

105. W. Huang, *Z. Metallkd.,* Vol 82, 1991, p 391–401

106. K. Mukai et al., Surface Tension of Liquid Alloys—A Thermodynamic Approach, *Metall. Mater. Trans. B,* Vol 38, 2008, p 561–569

107. J. Zhu and K. Mukai, *ISIJ Int.,* Vol 38, 1998, p 1039–1044

108. K. Ogino, K. Nogi, and C. Hosoi, *Tetsu-to-Hagané,* Vol 69, 1983, p 1989–1994

109. R.F. Brooks, A.P. Day, K.C. Mills, and P.N. Quested, *Int. J. Thermophys.,* Vol 18, 1997, p 471–480

110. I. Takamichi and R.I.L. Guthrie, *The Physical Properties of Liquid Metals,* Clarendon Press, 1988, p 148–198

111. R.F. Brooks, A.T. Dinsdale, and P.N. Quested, The Measurement of Viscosity of Alloys—A Review of Methods, Data and Models, *Meas. Sci. Technol.,* Vol 16, 2005, p 354–362

112. C.Y. Ho, Development of Computerized Numerical Databases on Thermophysical and Other Properties of Molten as well as Solid Materials and Data Evaluation and Validation for Generating Recommended Reliable Reference Data, *Proceedings of a Workshop on the Thermophysical Properties of Molten Materials,* Oct 20–23, 1992 (Cleveland, OH), NASA Lewis Research Center, p 51–68

113. E.N. da C. Andrade, The Theory of the Viscosity of Liquids: I, *Philos. Mag.,* Vol 17 1934, p 497, 698

114. M. Hirai, Estimation of Viscosity of Liquid Alloys, *ISIJ,* Vol 33, 2002, p 281–285

115. D.U. Sichen, J. Bygén, and S. Seetharaman, A Model for Estimation of Viscosities of Complex Metallic and Ionic Melts, *Metall. Trans. B,* Vol 25, 1991, p 519

116. E.T. Turkdogan, *Physical Chemistry of High Temperature Technology,* Academic Press, 1980, p 109

117. T. Chapman, *AlChE J.,* Vol 12, 1966, p 395

118. T. Iida et al., Accurate Predictions for the Viscosities of Several Liquid Transition Metals, Plus Barium and Strontium, *Metall. Mater. Trans. B,* Vol 37, 2006, p 403–412

119. Y. Plevachuk et al., Density, Viscosity, and Electrical Conductivity of Hypoeutectic Al-Cu Liquid Alloys, *Metall. Mater. Trans. A,* Vol 39, 2008, p 3040–3045

120. N.F. Mott, *Proc. R. Soc. (London) A,* Vol 146, 1934, p 465

121. T.E. Faber, *Introduction to the Theory of Liquid Metals,* Cambridge University Press, 1972

122. J.M. Ziman, *Adv. Phys.,* Vol 13, 1964, p 89

123. R. Evans, D.A. Greenwood, and P. Lloyd, *Phys. Lett. A,* Vol 35, 1971, p 57

124. R. Evans, B.L. Gyorffy, N. Szabo, and J. M. Ziman. *Proceedings of the Second International Conference on Liquid Metals,* S. Takeuchi, Ed. (Tokyo), Taylor and Francis, London, 1973

125. Y. Waseda, *The Structure of Non-Crystalline Materials: Liquids and Amorphous Solids,* McGraw-Hill, 1980

126. S. Takeuchi and H. Endo, *J. Jpn. Inst. Met.,* Vol 26, 1962, p 498

127. J.L. Tomlinson and B.D. Lichter, *Trans. Met. Soc. AIME,* Vol 245, 1969, p 2261

128. L. Lorentz, *Ann. Phys. Chem.,* Vol 147, 1982, p 429

129. G.T. Meaden, *Electrical Resistance of Metals,* Plenum, 1965

130. K.C. Mills, B.J. Monaghan, and B.J. Keene, Thermal Conductivities of Molten Metals, Part 1: Pure Metals, *Int. Mater. Rev.,* Vol 41, 1996, p 209–242

131. G.H. Geiger and D.R. Poirier, *Transport Phenomena in Metallurgy,* Addison-Wesley Publishing Co., 1973, p 575–587

132. C.J. Smithells, Radiating Properties of Metals, *Metals Reference Book,* 7th ed., E. A. Brandes and G.B. Brook, Ed., Butterworth-Heinemann, 1992, p 17–1 to 17–12

133. S. Krishnan and P.C. Nordine, Optical Properties and Emissivities of Liquid Metals and Alloys, *Proceedings of a Workshop on the Thermophysical Properties of Molten Materials,* Oct 20–23, 1992 (Cleveland, OH), NASA Lewis Research Center, p 143–160

134. J. Szekely and N.J. Themelis, Chap. 9, *Rate Phenomena in Process Metallurgy,* Wiley-Interscience, John Wiley and Sons, Inc., 1971, p 251–300

135. E.M. Sparrow and R.C. Cess, *Radiation Heat Transfer,* Brooks/Cole, Belmont, CA, 1966

136. H.C. Hottel and A.F. Sarofim, *Radiative Transfer,* McGraw-Hill, 1967

137. T.J. Love, *Radiative Heat Transfer,* Merrill, Columbus, OH, 1968

138. R. Weber, Containerless Measurements on Liquid at High Temperatures, *Proceedings of a Workshop on the Thermophysical Properties of Molten Materials,* Oct 20–23, 1992 (Cleveland, OH), NASA Lewis Research Center, p 87–98

ASM Handbook, Volume 22B, *Metals Process Simulation*
D.U. Furrer and S.L. Semiatin, editors

Copyright © 2010, ASM International®
All rights reserved.
www.asminternational.org

Measurement of Thermophysical Properties at High Temperatures for Liquid, Semisolid, and Solid Commercial Alloys

Peter Quested and Robert Brooks, National Physical Laboratory, United Kingdom

THE MEASUREMENT OF THERMOPHYSICAL PROPERTIES of metal alloys, especially at high temperatures, is difficult because of the reactivity of some alloys. A variety of strategies has been adopted to overcome this difficulty:

- *Perform experiments quickly to minimize reactions or suppress other effects:* One example is the use of transient methods such as the transient hot wire technique for the measurement of thermal conductivity (Ref 1) to avoid the onset of density-driven convection effects. Another is the use of microsecond discharge methods, where the sample is heated very rapidly, minimizing reactions of the sample with its surroundings (Ref 2).
- *Control the surroundings of the sample:* An example is the use of thin-walled sapphire inserts between the sample and the platinum crucible in the differential scanning calorimeter to avoid reactions, but thin enough to maintain the high heat transfer of the metal crucible (Ref 3).
- *Elimination of sample container by levitation:* This technique is performed most popularly by electromagnetic radiation, although other techniques are used (acoustic, aerodynamic, or gas jet, and electrostatic levitation). Reactions of the sample are limited to the surrounding atmosphere, and large undercoolings can be achieved (Ref 4).
- *Measurement of properties in microgravity:* This technique uses drop tubes and suborbital (parabolic flights and sounding rockets) and space flights. A number of advantages are claimed, such as the precise control of the surrounding atmosphere, the elimination of gravitational effects such as density-induced convection, and greater control of levitation experiments with lower

electromagnetic forces. Unfortunately, the experiments are expensive but provide reference values and validation of terrestrial measurements, such as establishing the magnitude of errors arising from density-driven convection in terrestrial experiments (Ref 5).

This article reviews the methods available for measurement of thermal and other physical properties for liquid, semisolid, and solid commercial alloys. This does not imply that other properties of commercial alloys, such as the strength of the mush and other factors, such as the heat transfer of the molding materials, are unimportant in some applications. This section is only envisaged as an elementary introduction; for more detailed discussion of the techniques, the reader is referred to several excellent texts (Ref 6–10). Also see the articles "Thermophysical Properties of Solids," "Thermophysical Properties of Liquids," and "Thermophysical Properties for Solidification Models" in this Volume.

Measurement Methods

Specific Heat, Enthalpy, and Transition Temperatures

The specific heat at constant pressure (C_p) and constant volume (C_v); enthalpy (H), which is the sum of the internal energy in a system plus its volume (V) times external pressure (p) ($H = E + pV$); and transition temperatures are among thermodynamic properties that can be determined. A number of measurement methods are available, depending on the temperature range of interest, the thermal property being measured, and the material.

Adiabatic Calorimetry

This is the classic method to measure specific heat, where the temperature rise (ΔT) of a well-insulated sample of mass, m, is provided with a known amount of electrical energy (I^2R, where I is the direct current, and R is the resistance of the heating element):

$$C_p = I^2R/(m\,\Delta T) \qquad \text{(Eq 1)}$$

The method is capable of very high accuracy, but there are difficulties in using it at high temperatures. See Ref 11 for details.

Modulated Calorimetry

The sample is heated by a modulated energy input (simple alternating current, or ac, or direct current, or dc, with an ac component), frequently using an electron beam or laser as the source of heat. The resultant temperature fluctuations are monitored, and the temperature fluctuations (θ) about the mean are measured. Knowing the mass of the sample, m, the power, P, and the frequency of the modulation of the heat source, ω:

$$C_p = P/(m \cdot \theta \cdot \omega) \qquad \text{(Eq 2)}$$

Details are given in Ref 12 and 13. Modern variants are noncontact modulation calorimetry, based on levitation and electromagnetic heating. Corrections are made for radiative heat loss and finite thermal conductivity of the specimens. These experiments are made under microgravity conditions because there is insufficient temperature and positioning control in an Earth-based laboratory under ultrahigh-vacuum conditions. The method is described in detail in Ref 14, and the application to a eutectic glass-forming

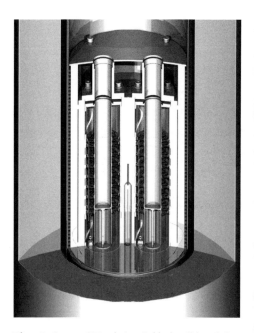

Fig. 1 Setaram C80 calorimetric block, a Calvet design calorimeter. Courtesy of Setaram Instrumentation

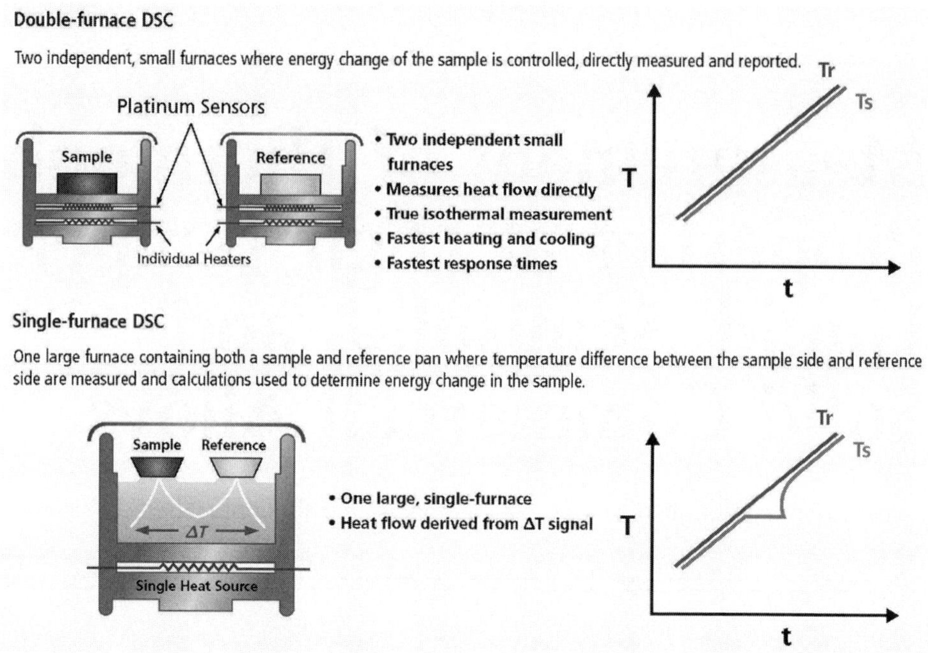

Double-furnace DSC

Two independent, small furnaces where energy change of the sample is controlled, directly measured and reported.

Platinum Sensors

Sample   Reference

Individual Heaters

- Two independent small furnaces
- Measures heat flow directly
- True isothermal measurement
- Fastest heating and cooling
- Fastest response times

Tr
Ts
T
t

Single-furnace DSC

One large furnace containing both a sample and reference pan where temperature difference between the sample side and reference side are measured and calculations used to determine energy change in the sample.

Sample   Reference

ΔT

Single Heat Source

- One large, single-furnace
- Heat flow derived from ΔT signal

Tr
Ts
T
t

Fig. 2 Schematic of single- and double-furnace-type differential scanning calorimeters (DSCs). © 2010, PerkinElmer, Inc. All rights reserved. Printed with permission

zirconium-base alloy is described in Ref 15 and 16. A recent example of the technique applied to measure the thermal conductivity, emissivity, and heat capacity of molten silicon uses noncontact modulated laser calorimetry in a dc magnetic field to suppress convection effects. By improving the quality of the temperature measurements, improved uncertainties of measurement are claimed (Ref 17).

Calvet Calorimetry

Two matched crucibles are placed in an isothermal block, and the temperature difference between each crucible and the block is monitored using thermopiles surrounding the sample. This provides measurement of the whole exchange of heat between the sample and its surroundings and greater sensitivity than a single thermocouple. Contrary to the differential scanning calorimeter plate system, the sensitivity of the Calvet calorimeter (Fig. 1) does not depend on the nature of the experimental vessel, the nature of the gas around the sample, or the sample size.

The system is used in one of two modes:

- *Scanning mode:* The temperature varies (scans) according to a programmed heating or cooling rate, and the thermopiles are monitored.
- *Isoperibol mode:* The temperature difference between the sample and the block is held constant; $T_{sample} - T_{block} =$ constant.

The instrument is calibrated by operation with a reference sample of known specific heat.

Commercial instruments are available, and the maximum temperature is 1000 °C (1830 °F), with a claimed accuracy in enthalpy measurement of $\pm 1\%$.

Differential Scanning Calorimetry and Differential Temperature Analysis

These two techniques are closely related, and they are commonly used methods for measuring transition temperatures, such as phase transformation temperatures or glass transition temperatures. Additionally, differential scanning calorimetry (DSC) instruments (Fig. 2) are used for measuring enthalpy and specific heat for commercial alloys (Ref 7, 18, 19). These techniques have the advantage of being relatively fast compared to adiabatic and drop calorimetry. Several commercial instruments are available from a variety of manufacturers.

In differential temperature analysis (DTA), the temperature difference between a sample relative to an inert reference material is monitored by a differential thermocouple as the two are heated or cooled at a constant rate (typically 10 °C/min). When a thermal transition occurs, the sample temperature will slow compared to the calibrant, there will be a deviation from the base line, and the onset temperature of this deviation is a measure of the transition temperature. Models of DTA, which achieve a maximum temperature capability of 2400 °C (4350 °F), are available.

There are two common designs of DSC instruments used for measuring specific heat:

- *Differential power scanning calorimeter or power compensation DSC:* The difference in power required to maintain the sample and reference at the same temperature is measured as a function of time. The maximum temperature is limited to approximately 730 °C (1350 °F), allowing the direct measurement of aluminum alloy transformations, including melting, and the properties of the solid for higher-melting-point alloys. A commercial instrument is available, and there is an International Standard describing its use (Ref 20). Uncertainty of measurement is approximately ± 1 to 2%.
- *Heat flux DSC with a disc-type measuring system:* Also known as a differential temperature scanning calorimeter (Fig. 3), this system records the temperature difference between the sample and reference cells. The maximum temperature attained by these instruments is frequently 1500 °C (2730 °F), but to allow measurements of iron alloys, a few systems can achieve 1600 °C (2910 °F). The uncertainty of measurement is of the order of ± 3 to 5%.

To measure specific heat, three runs are performed at the same cooling or heating rates:

- The first run measures the temperature difference between the sample and an empty pan. The S^{sample} is the recorded signal for mass, m^{sample}.
- The second run uses a calibrant, generally γ' alumina or sapphire, NIST SRM720, with an empty pan, providing the recorded signal S^{cal} for calibrant of mass m^{cal} and specific heat C_p^{cal}.

- The third run is with the two empty crucibles, which allows for a correction for slight asymmetric heat flows in the pans. S^{empty} is the recorded signal.

$$C_p^{sample} = \left(S^{sample} - S^{cal}\right) \cdot m^{cal} \\ \cdot C_p^{cal} / \left(S^{cal} - S^{empty}\right) \cdot m^{empt} \quad \text{(Eq 3)}$$

To aid heat transfer and avoid radiation transfer problems at high temperatures, since alumina is semitransparent to infrared radiation, platinum crucibles with lids mounted on a flux plate of platinum are employed. The crucibles may be protected from reaction with molten metals by a thin sapphire crucible insert (Ref 3).

Höhne et al. (Ref 7) and Richardson (Ref 18) have discussed some of the problems associated with these instruments. There is a temperature lag between the sensors and the sample temperature, because the sensors are not in contact with the material, and these authors describe correction schemes to allow for this thermal response. Difficulties are also associated during melt undercooling, and this makes it difficult to determine the liquidus temperature of the alloy on cooling. This, combined with the response time of the equipment, can lead to larger uncertainties in the determination of the enthalpy of fusion during cooling. Recently, Dong et al. (Ref 21, 22) presented a numerical model correcting for the effect of heat-transfer coefficients within a DSC. A guide to best practice with DTA and DSC instruments is provided for the interpretation of curves for analyzing melting and solidification behavior of metals (Ref 23).

Single-Pan Calorimetry

This technique can be traced to Smith (Ref 24), who devised a single-pan calorimeter (Fig. 4) for measuring the specific heat and transition temperatures in solid copper alloys. Although he was able to demonstrate its applicability, the advent of modern temperature controllers, which can maintain a better constant temperature difference, especially during a large change in heat evolution during a phase change, stimulated a renewed interest in these instruments. Calorimeters of this type were employed by Hayes et al. (Ref 25) for determining transition temperatures in the construction of phase diagrams of, for example, Au-Pb-Bi systems. Dong et al. (Ref 26, 27) have designed equipment for measuring the enthalpy and specific heat of aluminum casting alloys. The technique bears a close relationship to the cooling curve method, such as the classic work of Backerud et al. (Ref 28) for aluminum alloys.

The basis of the technique is to measure or control the heat flux across the wall of part of the calorimeter and to monitor the temperature of the sample as a function of time. There are two methods of employing it:

- By monitoring the heat flux at a constant rate of temperature change, thermal events may be monitored.
- By maintaining a fixed heat flux, thermal events will cause a response from the rate of temperature change.

The method is calibrated by running experiments with a standard specimen (in this case, copper) of known specific heat (cal) and an empty system (empty). The enthalpy and specific heats of the sample can be from the expression (Ref 27):

$$C_p^{sample} \cdot dT^{sample} = C_p^{cal} \frac{\Delta T^{Dsample} - \left(\frac{\Delta T^{Dempty} \cdot dT^{sample}}{dT^{empty}}\right)}{\left(\frac{\Delta T^{Dcal}}{dT^{cal}}\right) - \left(\frac{\Delta T^{Dempty}}{dT^{empty}}\right)} \quad \text{(Eq 4)}$$

The slope of the temperature rise with time is given by dT, and the temperature difference maintained between the inner and outer crucible for the respective runs is given by ΔT^D. (ΔT^D is the temperature difference between the inner and outer crucible, sample refers to the sample run, empty refers to the empty run, and cal refers to the calibrant run.)

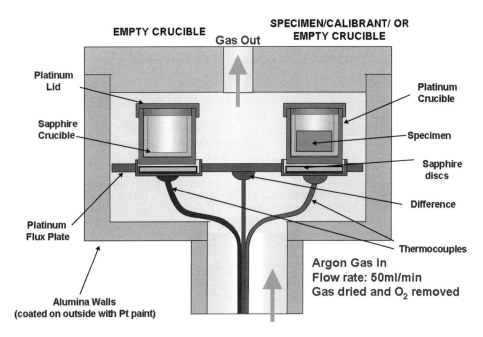

Fig. 3 Schematic of differential temperature calorimeter

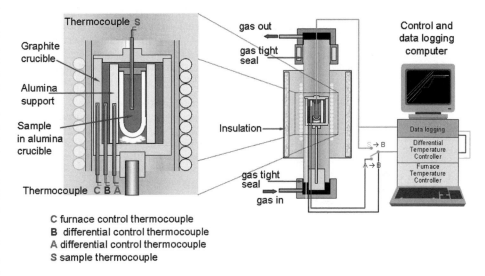

C furnace control thermocouple
B differential control thermocouple
A differential control thermocouple
S sample thermocouple

Fig. 4 Example design for a single-pan calorimeter

The major advantages compared to DSC are that there is only a small correction for the response of the specimen thermocouple (i.e., little smearing), and the temperature of transitions can be measured after undercooling. The technique has not been applied above 1000 °C (1830 °F).

Drop Calorimetry

Enthalpy may be measured using a drop calorimetry method (Ref 29, 30). A sample of mass m is held at a fixed temperature (T_s) and then dropped into a copper (or silver) receiving calorimeter, which is maintained at a constant temperature $T_{ambient}$ (preferably 25 °C, or 298 K). Braking devices are frequently used to reduce the velocity of the sample as it impacts with the calorimeter. The resulting temperature rise (ΔT) of the receiver is carefully measured, using a platinum resistance thermometer. From previous calibration of the temperature response of the receiver to a measured heat input (usually from an electrical heater measuring the power input), the thermal capacity ($m^{cal}C_p^{cal}$) of the calorimeter is determined. The enthalpy of the sample can be determined from:

$$H_T - H_{amb} = \left(m^{cal}C_p^{cal}\right)\Delta T \qquad \text{(Eq 5)}$$

Corrections for the difference between $T_{ambient}$ and 298 K and also the enthalpy of any sample container, which can significantly magnify the total errors of the enthalpy of the sample, are requirements.

Levitation methods can eliminate the need for a container and have been used for measuring the enthalpy of highly reactive systems, such as Ti-6Al-4V (Ref 31). The electromagnetic levitation technique is reviewed in more detail in Ref 32 to 34.

The problems with levitated drop calorimetry lie in uncertainties in temperature measurement arising from uncertainty in emissivity and the need to differentiate enthalpy/temperature curves to obtain C_p, which can be particularly difficult if there is a high-temperature phase transition (which is frequently the case for commercial materials).

The experimental uncertainties in enthalpy $H_T - H_{298}$ are estimated to be ±1 to 2%.

Microsecond Heating

This technique is also known as pulse heating or explosive wire. In this technique, an electrical current is discharged rapidly producing a rapid temperature rise (10^3 to 10^5 K s^{-1}) through a wire specimen. The temperature rise is measured by a pyrometer, and by knowing the electrical power applied (current, I, times voltage, V, over time, dt), the enthalpy, H, of the specimen can be determined according to:

$$H = (1/m)\,IV\,dt \qquad \text{(Eq 6)}$$

At faster rates, the discharge is so rapid that measurements can be made into the liquid, but

slow microstructural changes are masked. The minimum operating temperature is approximately 1000 °C (1830 °F). The technique was devised by Cezairliyan (Ref 35, 36), and more modern designs were developed by Pottlacher (Ref 37), who, for example, has shown good agreement between the enthalpy of fusion of the superalloy IN718 and measurements made by DSC.

Uncertainties in enthalpy are of the order of ±4% (Ref 38, 39) and ±8% for specific heat (Ref 39).

Thermal Conductivity/Thermal Diffusivity

The thermal diffusivity (α) and thermal conductivity (k) are related by:

$$k = C_p \cdot \rho \cdot \alpha \qquad \text{(Eq 7)}$$

where C_p is the specific heat, and ρ is the density.

Because thermal transport involves electron transport and the electrons are scattered by precipitates and boundaries in materials, thermal conductivity is dependent on the mechanical and thermal history of the material. This is very important in the solid state, and the condition of the material must be specified.

Also, for liquids it is very difficult to eliminate the effect of buoyancy on the measurement of thermal conductivity. Methods adopted are:

- Maintaining the upper surface at a slightly higher temperature than the lower surface
- Using transient methods so that convection is not established (Ref 1, 40)
- Minimizing temperature gradients in the liquid
- Performing experiments in microgravity to reduce density-driven convection. This showed that convection effects were largely suppressed for semiconductors (Ref 41) and metals (Ref 42) in microgravity experiments.
- Applying a magnetic field to counteract convection. Using this method, Nakamura et al. (Ref 43) showed that magnetic forces oppose convection with measurements using mercury.

These methods do not eliminate surface thermocapillary-driven flow (Marangoni forces), and Nagata et al. (Ref 42) placed a solid lid on the surface to counteract these effects.

The methods for measuring thermal conductivity may be divided into either steady-state or transient methods.

Steady-State Methods

Steady-state methods (sometimes known as classical methods) are normally limited to low temperatures because of the difficulties of eliminating convection in liquids at high temperatures. Heat losses around the specimen

must be eliminated, and guarding (the match of the axial gradient in the specimen to its surrounding environment) can be employed to minimize heat losses. The methods are frequently relative, comparing a material with unknown conductivity to a material of known conductivity. Some examples of steady-state methods are very briefly described.

Concentric Cylinder Method. The sample is placed in an annulus between two concentric cylinders, and a known heat flux (q) is applied. The temperature difference (ΔT) between temperature sensors (often thermocouples) in the two cylinders is determined when steady state is achieved. Knowing the radii of the two cylinders, r_1 and r_2, and their length, L, the thermal conductivity, k, is given by:

$$k = q(\ln(r_2/r_1)/2\pi L\,\Delta T \qquad \text{(Eq 8)}$$

Care is required in estimating end effects, providing good insulation, and, in the case of liquids, preventing convection.

Parallel Plate. The sample is placed between two semi-infinite plates, a known heat flux (q) is applied to the upper plate, and the temperature difference (ΔT) is monitored:

$$k = q\,L/A\,\Delta T \qquad \text{(Eq 9)}$$

where A is the area of the plate.

Axial and Radial Heat Flow Methods. A heat flux (q) is applied to one end of a long, thin bar, which provides one-dimensional heat transfer. The temperature gradient, $\Delta T/L$, is measured at steady state. To measure molten metals, the bar is vertical and heat is applied from the top to minimize convection:

$$k = (1/A)\{(q\,L/\Delta T) - k_c A_c\} \qquad \text{(Eq 10)}$$

where $k_c A_c$ refers to the properties of the measurement cell. More information about the methods is available in Ref 44 to 46.

Non-Steady-State and Transient Methods

Thermal Diffusivity by Laser Flash. This is a well-established technique for solids, first demonstrated by Parker (Ref 47) and reviewed in detail by Taylor (Ref 48, 49). Thermal conductivity can be derived from the diffusivity, knowing the density and heat capacity of the material, using the relation given in Eq 7.

Energy (from a laser or high-intensity xenon lamp) is fired at the front face of the sample (Fig. 5) (typically 12 mm in diameter with a thickness of 2 mm), and the rise in temperature on the rear face is measured as a function of time. The curve deviates from ideal by radiation losses, and there are several ways to correct for this effect. From the corrected plot, one of the simpler analyses to determine the thermal diffusivity (α, m^2/s) from the time to half the maximum temperature ($t_{0.5}$, s), provided thickness (L, m) of the sample is known, is:

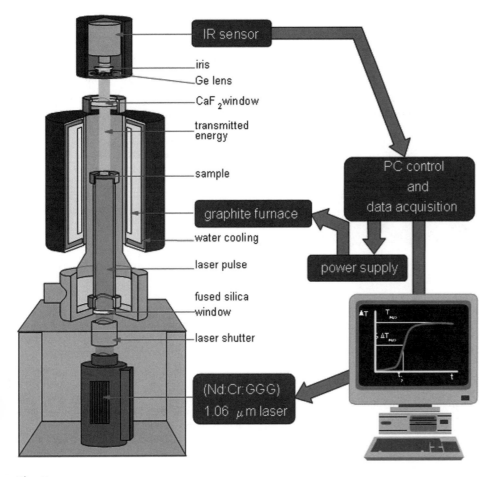

Fig. 5 Schematic laser flash apparatus for measuring thermal diffusivity

$$\alpha = 0.13888\, L^2/t_{0.5} \qquad \text{(Eq 11)}$$

To improve energy absorption, the front face is frequently coated with graphite, or roughening the solid surfaces can achieve a similar result.

Various workers have shown its application to molten liquids. Examples are by Henderson et al. (Ref 50), Monaghan (Ref 51), Seetharaman (Ref 52), and Kaschnitz (Ref 53). Significant difficulties arise from maintaining a known thickness and parallel sample faces in a liquid. There is also concern that density-driven convection effects could contribute to the result. Variants of this method exist, and for one commercial piece of equipment, the arrangement is vertical, with the laser energy incident on the bottom surface of the sample. The liquid is contained in a sapphire cassette, which is transparent to the wavelength of the laser. Tests with the solid show good agreement for the same sample with and without the cassette.

The thermal conductivity in the solid/liquid region is important and is difficult to measure because of the energy used in melting the sample. Two papers (Ref 54, 55) have dealt with this topic, presenting different approaches.

Uncertainties of measurement are of the order of ±5% in the liquid and less in the solid.

Line Source Method. Several researchers have attempted to use the line source (or hot wire) technique to measure the thermal conductivity of alloys. The technique relies on measurement of the temperature rise in an electrical conductor surrounded by a liquid when an electrical current is passed. The advantage is that the onset of convection in the liquid can be determined, so the inherent thermal conductivity of the liquid can be measured.

The rise in temperature (ΔT) in a conductor is monitored as the electrical current is applied. Plotting (ΔT) against ln(time) results in a curve with three regions:

- An initial transient, which is normally ignored
- A linear portion
- A nonlinear portion, indicating the onset of convection

The slope of the linear portion is proportional to the reciprocal of the thermal conductivity. Normally, the current is passed for no longer than 1 s.

For metals, the problem is that the conductor must be electrically insulated from the alloy. Hibiya and Nakamura (Ref 41) successfully insulated a thin metallic strip for measurement of indium-antimony at approximately 600 °C

(1110 °F), while Yamasue et al. (Ref 56) have developed the technique for molten silicon and germanium (up to 1400 °C, or 2550 °F) using a platinum or molybdenum wire insulated with silica. Recent developments by Wakeham and his group (Ref 57) using a modified insulated probe measuring molten lead give an estimate of uncertainty of measurement of the order of ±3%.

Electrical Conductivity

The review by Mills et al. (Ref 58) has confirmed that the thermal conductivities calculated from electrical resistivities by the Wiedemann-Franz-Lorenz rule for both solid and liquid elements near the melting point lie within experimental error for most metals. This finding is useful because, if confirmed for alloys, it would enable the easier measurement of electrical conductivity (resistivity) to derive the thermal conductivities of alloys. This would also allow the calculation of thermal conductivities of alloys in the solid/liquid region, because, in this region, heat-pulse methods of thermal conductivity are particularly difficult since they result in further melting of the sample from the energy input.

The four-wire probe is the simplest method to measure the electrical resistivity of molten alloys into the solid/liquid region (Ref 59). Reference 60 reviews a method based on a contactless, inductive resistivity technique including a microgravity facility. Microsecond, rapid heating techniques have been used to derive electrical conductivity values for the solid and liquid states of refractory metals and commercial alloys (Ref 61).

Density

The methods to determine linear thermal expansion coefficients of solids are reviewed in Ref 62 and 63. X-ray techniques and dilatometry are the principal methods used.

There is a wide variety of techniques for measuring the density of a molten alloy. Many are reviewed in Ref 6 and 64 and include classical techniques such as sessile drop, pendant drop, pycnometry, gamma ray attenuation, and capillary rise. The following brief review concentrates on other techniques that have been commonly used over the past ten years.

Density by Levitated Drop

In this method, a drop of metal of known mass is levitated and heated (Fig. 6, 7). Optical images of the drop are taken from three orthogonal directions and used to determine the drop volume and hence its density. This method is suitable for reactive materials at high temperatures. In recent years, there have been improvements in the imaging techniques and the methods used to analyze the geometry of the drops.

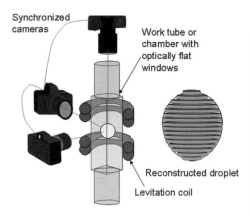

Fig. 6 Design of instrumentation for levitated drop density measurement showing camera placement

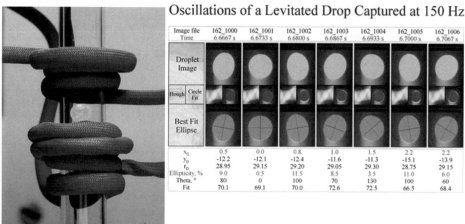

Fig. 7 Density determined by levitated drop oscillation measurement. Example images of a levitated drop captured using a high-speed camera. The images have been analyzed automatically and fitted to a generalized ellipse.

Details for electromagnetic levitation methods are given in Ref 65 and 66, for electrostatic levitation methods in Ref 67 and 68, and for aerodynamic levitation methods in Ref 69.

Although a simple technique in principle, there are a number of experimental difficulties, such as oscillations of the drop when electromagnetic levitation is used and inaccuracies in noncontact temperature measurement. Some of the other levitation techniques obviate the problem of the oscillation of the drop, but there can be uncertainties in the heating when a laser is used.

There is good agreement between this and conventional methods for a range of metals, such as iron, copper, and nickel. Some experiments have been performed in microgravity. Experimental uncertainty associated with the levitated drop is probably $\pm3\%$, although Rhim et al. (Ref 70) reported uncertainties of $\pm1\%$ when using very small drops.

Density by Displacement

The apparent change in mass of an inert probe of known dimensions, immersed in a liquid, provides a measurement of density. When stationary, the classic Archimedean method is used, but dynamic measurements may also be made as the probe is driven into the liquid at a constant rate. When the surface forces have been overcome, the density of the liquid is derived from the slope of the plot of mass against displacement (Ref 64).

Reactions between the furnace gases and the molten metal and between the probe and the crucible cause problems with this technique. Also, the presence of thick oxide films, for example, on aluminum, can cause measurement instability. Uncertainties of approximately $\pm2\%$ are probably achievable with these techniques.

Dilatometric Method

Dilatometric methods for measuring the density of the liquid and also the changes in density associated with melting have recently been proposed. The molten alloy is contained in a rigid cylindrical cell of a material of known expansion characteristics. Two protruding pistons are supplied so that the net changes in the volume of the specimen and the cylinder are measured, so the volume expansion and density of the alloy can be determined. It is particularly important that there is no leakage between the cylinder and the pistons.

This method has the advantage that the expansion through the solid/liquid during heating can be followed, although there are practical problems of feeding while following the cooling cycle. Preliminary results for copper, the aluminum-silicon casting alloy LM25, and the nickel-base alloy IN718 have been reported (Ref 71, 72).

Microsecond Technique

By monitoring the dimensions of a wire with a high-speed video camera as it is resistively heated, the density can be determined. Like the levitated drop, this method works well for reactive materials because there is no container, but the sample must be available as a wire. Work recently carried out on the nickel alloy IN718 shows good agreement with conventional methods (Ref 61).

Maximum Bubble Pressure

With this technique, the difference in the maximum pressures required to blow bubbles at the end of a capillary tube at two known depths in the liquid alloy is used to determine the density of the alloy. The main advantages of this technique are that a new surface is exposed for each bubble, so it is less susceptible to contamination; it will still operate when there is oxide on the surface of the melt; and there is no mass measurement. The disadvantages are that it can be difficult to detect the surface to measure the depth of immersion; the equilibrium bubble shape is easily distorted; capillary/sample reactions include blocking of the capillary by oxides; and the radius of the tube must be accurately measured (Ref 73).

Draining Crucible Method

In this method, the head and flow of a liquid metal draining through an orifice in a crucible are determined and the density derived by hydrodynamic analysis of the data (Ref 74).

Viscosity

There are many methods (Ref 6, 75, 76) to measure the viscosity of liquids, but those suitable for liquid metals are limited by their low viscosities (of the order of 1 mPa · s), their chemical reactivity, and generally high melting points. Proposed methods for measuring viscosity of metals include capillary, oscillating vessel, rotational bob or crucible, draining vessel, levitation using the damping of surface oscillations, and acoustic methods.

Capillary Method

The capillary rheometer is generally thought to be the best method for the measurement of the viscosity of liquids (Ref 6) and is based on the time for a finite volume of liquid to flow through a narrow bore tube under a given pressure. The relationship between viscosity, η, and efflux time, t, is given by the modified Poiseuille equation or the Hagen-Poiseuille equation:

$$\eta = (\pi r^4 \cdot \rho g h \cdot t)/(8V(l + \mathrm{n}r)) - (\mathrm{m}\rho V)/(8\pi(l + \mathrm{n}r)t) \qquad \text{(Eq 12)}$$

where r and l are the radius and length of the capillary, respectively; h is the effective height of the column of liquid; ρ is the liquid density; V is the volume discharged in time, t; and m and n are constants that can be determined experimentally. The value $\rho g h$ may be replaced by ΔP, the pressure drop along the capillary, and nr is called the end correction and corrects for surface tension effects as the liquid is expelled from the capillary. For liquid metals, with their relatively high densities, the second term, which corrects for kinetic energy, is particularly important.

This technique is often used as a relative, rather than absolute, method, because the experimental procedures are simple, and any errors incidental to the measurement of dimensions are thereby avoided. For a viscometer in which r, l, h, and V are fixed, Eq 12 reduces to:

$$\frac{\eta}{\rho} = C_1 t - \frac{C_2}{t} \qquad \text{(Eq 13)}$$

where the values of C_1 and C_2 are easily evaluated using viscosity standard reference samples but are constants equal to:

$$C_1 = \frac{\pi r^3 g h}{8V(1+nr)}, C_2 = \frac{mV}{8\pi(1+nr)} \qquad \text{(Eq 14)}$$

In determining the viscosities of metallic liquids by the capillary method, an especially fine and long-bore tube (in general, $r < 0.15$ to 0.2 mm, $l > 70$ to 80 mm) is needed to satisfy the condition of a low Reynolds number for ensuring laminar flow. This in turn requires a furnace with a similarly long and uniform hot zone. Blockage of the capillary by bubbles or oxide inclusions is a common problem, particularly with aluminum alloys. Materials problems often impose a temperature limit of approximately 1200 °C (2200 °F), but metals such as bismuth have been successfully measured (Ref 6).

Oscillating Vessel Viscometer

Most measurements of the viscosity of metals use some form of oscillating vessel viscometer (Fig. 8). A liquid contained in a vessel, normally a cylinder, is set in motion about a vertical axis, and the motion is damped by frictional energy absorption and dissipation within the liquid. The viscosity is determined from the decrement and time period of the motion. The main advantages of the method are that the time period and decrement are easily measured and the amount of liquid is relatively small, which allows stable temperature profiles to be attained.

For a right circular cylinder that is infinitely long and contains a fluid, the equation of motion of the damped cylinder is:

$$I_0(d^2\theta/dt^2) + L(d\theta/dt) + f\theta = 0 \qquad \text{(Eq 15)}$$

where I_o is the moment of inertia of an empty cup and suspension; t is the time; f is the force constant of the torsion wire; θ is the angle of displacement of any small segment of the fluid from its equilibrium position; and L is a function of the density and viscosity of the fluid, the internal radius of crucible and height of

liquid. Expressions for L are determined by solving the Navier-Stokes equations for the motion of the liquid within the vessel (neglecting nonlinear terms).

One of the major difficulties is relating the measured parameters to the viscosity through the second-order differential equation for the motion of an oscillating system, and there are a number of mathematical treatments that appear to yield different results (Ref 77–82) with the same experimental data. This problem was recognized by Iida (Ref 6) and is further discussed in Ref 75. The majority of measurements made by this method have used the analysis by Roscoe (Ref 79), mainly for reasons of simplicity of presentation of the working formulae of that reference.

There are several designs (examples may be found in Ref 83 to 85) of viscometers suitable for measuring the viscosity of metals at high temperatures. An example is the design described in Ref 83. The sample is contained within an alumina crucible, which is screwed into a molybdenum lid and suspension rod and suspended on a torsion wire. A rotary solenoid is used to impart oscillatory motion to the crucible, and an optical pointer with a diode array is used to measure the time constant and decrement of the system. The sample is heated by a two-zone furnace giving a maximum temperature capability of approximately 1650 °C (3000 °F). The authors claim an uncertainty of measurement of ±9% within a 95% confidence limit. A major contribution to the uncertainty is the extrapolation of the dimensions of the crucible and the height of liquid at high temperatures. Gruner et al. (Ref 86), using a statistical approach to estimate

experimental uncertainty, claim uncertainties in the range of 5 to10%.

Comparison of the oscillating viscometer with the capillary viscometer reveals some differences. Iida (Ref 6) suggests the end effect is inadequately weighted in the Roscoe treatment and should also include the effect of the liquid meniscus on the height of the liquid, suggesting that a correction factor, ξ, of $1 + 0.04$ be introduced into the (uncorrected) formula. Wetting of the crucible may also be important, and if the metal does not wet the crucible, it may slip during the oscillation and thus provide greater damping. The aspect ratio of the sample may be important in both of these assessments, and further work is required to justify modification of the (Roscoe) equation.

Rotating Cylinder

For the rotating cylinder technique, the torque on a cylinder rotated in a liquid is related to the viscosity of the fluid. Viscometers of this type consist of two concentric cylinders, that is, a bob and a crucible (Fig. 9). The viscosity is determined from measurements of the torque generated on the rotor arm of the rotating cylinder. When rotating the cylinder at a constant speed, the viscosity (η) can be obtained from:

$$\eta = \left(\frac{1}{r_1^2} - \frac{1}{r_1^2}\right)\frac{M}{8\pi^2 nh} \qquad \text{(Eq 16)}$$

where M is the torque, n is the number of revolutions per second, r_1 is the radius of the bob, r_0 is the radius of the crucible, and h is the height of the bob. The theory is applicable to infinitely

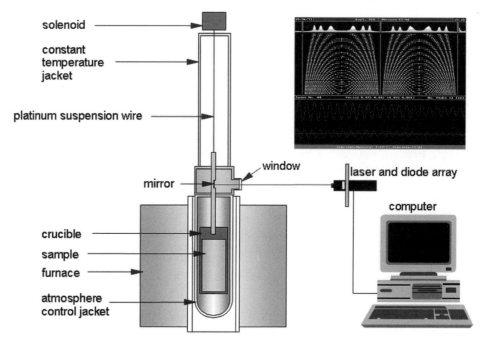

solenoid

constant temperature jacket

platinum suspension wire

window

laser and diode array

mirror

computer

crucible

sample

furnace

atmosphere control jacket

Fig. 8 Design of oscillation viscometer for metals

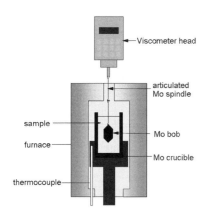

Fig. 9 Schematic of rotating bob viscometer for measuring viscosity of liquid metals

long cylinders, and it is normal to calibrate the system with reference materials using:

$$\eta = G \cdot S/n \qquad \text{(Eq 17)}$$

where S is the scale deflection, and G is the apparatus constant.

This is the most common method for the measurement of slag viscosity, and, for several practical reasons, it is more customary to rotate the bob rather than the crucible (Fig. 9). Probably the most important reason for its adoption is that the rotating bob viscometer is based on readily available and cheap commercial instruments. It is also easier to center the bob, but the viscosity range is less than for a rotating crucible instrument. It is critical that the bob rotates axisymetrically and concentrically within the sample. Any instability due to rotation speed or alignment will increase the apparent viscosity. Reference 87 shows a modern example of a rotating crucible method for high temperatures.

To obtain the necessary sensitivity to measure the low viscosity of liquid metals, the clearance between the stationary and rotating parts must be made very small, and it is difficult to maintain the system coaxially. In spite of the experimental difficulties, the rotating bob technique has been used to measure the viscosity of aluminum and its alloys (Ref 88). More recently, Bakhtiyarov and Overfelt (Ref 89) applied a rotational method to a low-melting-point alloy and paid particular attention to eliminating eccentricity and end effects.

Draining Vessel Method

It is common for comparative measurements of viscosity to be made for oils and slurries in industrial applications by use of a flow cup, where the time taken for a volume of sample to flow through a small orifice in the bottom of a cup is measured, and viscosity is derived from look-up tables. Roach and Henein (Ref 74) have derived equations to adapt this method for liquid metals and to provide values of viscosity, surface tension, and density. Values for

aluminum are lower than usually quoted, but the method is robust, and the experiments are simple to perform.

Oscillating (Levitated) Drop Method

The oscillating drop method is widely used for surface tension measurements of liquid, levitated samples. It is based on the fact that the frequencies of the surface oscillations of a liquid drop are related to the surface tension by Rayleigh's formula (Ref 90); for Earthbound levitation, his formula must be corrected as proposed by Cummings and Blackburn (Ref 91). For a viscous drop, these oscillations are damped due to the viscosity of the liquid. Therefore, it is, in principle, also possible to determine the viscosity from the damping. The damping constant, Γ, is given by:

$$\Gamma = \frac{20\pi}{3}\frac{R\eta}{m} \qquad \text{(Eq 18)}$$

where R and m are the radius and mass of the droplet, respectively; and η is the viscosity. To apply Eq 18, the following conditions must be satisfied:

- The liquid drop must be spherical.
- The oscillations must persist undisturbed for several cycles.
- There must be no additional damping mechanism present.

It is often thought that none of these conditions is met in terrestrial experiments; the sample is deformed, and there is an overlap of self-excited oscillations, making the decay of one single excitation hard to detect. Also, the electromagnetic fields necessary for earthbound levitation inevitably induce turbulent fluid flows inside the sample, which lead to additional damping, although Rhim (Ref 92) and Ishikawa (Ref 93) have published measurements of refractory metals made terrestrially using this method and electrostatic levitation.

Under microgravity conditions, the fields are much weaker, and a laminar fluid flow can be expected, at least for sufficiently viscous materials. In microgravity, an electromagnetic positioning field is used, with a separate heating coil. A pulse of power initiates the oscillations, which are monitored with a video camera. The temperature is recorded with a pyrometer. Image analysis is used to obtain the frequency and decay of the oscillations. Recently, Wunderlich (Ref 94) carried out measurements on parabolic and sounding rocket flights on the nickel-base superalloy CMSX-4, which showed reasonable agreement with terrestrial measurements of the same alloy using an oscillating cylinder method.

Damping of an Acoustic Wave

The viscosity of a liquid can be measured by measuring the damping of an acoustic wave.

There is one example for the measurement of aluminum (Ref 95) and some alloys using the hole theory to interpret the results.

Summary

A wide range of methods is available to measure the properties of liquid and solid alloys, each with their own strengths and weaknesses. Some methods require custom-made equipment, while for others, commercial equipment is readily available. Tables 1 to 4 provide a brief summary of the measurement methods discussed in this article. It is unlikely that, for a particular property, any one method is suitable for all materials.

The common challenges for these types of measurements are those of reactivity, with the formation of oxides, and temperature measurement. The article illustrates how these effects are minimized.

There are few methods that are successful in measuring properties in the important semisolid (mushy) region during solidification.

Before commissioning expensive work to measure properties, the modeler is advised to check the sensitivity of the required predictions to changes in the input thermophysical data. This will enable measurement efforts to concentrate on the critical data.

REFERENCES

1. J. Bilek, J.K. Atkinson, and W.A. Wakeham, Repeatability and Refinement of a Transient Hot-Wire Instrument for Measuring the Thermal Conductivity of High-Temperature Melts, *Int. J. Thermo.*, Vol 27, 2006, p 1626–1637
2. M. Boivineau and G. Pottlacher, Thermophysical Properties of Metals at Very High Temperatures Obtained by Dynamic Techniques: Recent Advances, *Int. J. Mat. Product Tech.*, Vol 26, 2006, p 217–246
3. L.A. Chapman, Application of High Temperature DSC Technique to Nickel Based Superalloy, *J. Mater. Sci.*, Vol 39, 2004, p 7229–7236
4. J. Brillo, G. Lohöfer, F. Schmidt-Hohagen, S. Schneider, and I. Egry, Thermophysical Property Measurements of Liquid Metals by Electromagnetic Levitation, *J. Mater. Prod. Technol.*, Vol 26, 2006, p 247–273
5. I. Egry, Noncontact Thermophysical Property Measurements of Metallic Melts under Microgravity, *High-Temperature Measurements of Materials*, H. Fukuyama and Y. Waseda, Ed., Springer, Berlin Heidelberg, 2009, p 131–147
6. T. Iida and R.I.L. Gutherie, *The Physical Properties of Liquid Metals,* Clarendon Press, Oxford, 1988
7. G. Höhne, W. Hemminger, and H.J. Flammersheim, *Differential Scanning Calorimetry,* Springer-Verlag, Berlin, 1996

Table 1 Selected methods for measuring the enthalpy, specific heat, and transition temperatures of alloys

Technique	Temperature range	Principle	Advantages	Disadvantages	Types of material
Differential power scanning calorimetry Power compensation differential scanning calorimetry	20–730 °C (70–1350 °F)	Measures the power difference required to maintain a reference sample and the specimen at the same temperature as the temperature is scanned. A secondary method	Relatively fast and can extract limited information about kinetics by varying cooling rates. A large background in their use from the polymer industry. Commercial instruments available	Requires baseline correction. Low-temperature capability. Preferably, the signal should be desmeared to allow for the response of the equipment during rapid changes.	Al and Mg alloys. High vapor pressure affects measurement.
Differential temperature scanning calorimetry Heat flux differential scanning calorimetry	20–1500 °C (70–2730 °F) (1650 °C, or 3000 °F)	Measures the temperature difference between a reference and the sample as the temperature is scanned. A secondary method	Relatively fast and can extract limited information about kinetics by varying cooling rates. Commercial instruments available	Preferably, the signal should be desmeared to allow for the response of the equipment during rapid changes. Stability of container and material (coat the crucible/alumina inserts). Importance of atmosphere control. Lack of temperature calibrants at higher temperatures	Cast iron, steels, Ni-base alloys
Single-pan calorimetry One of the modes of operation is the Smith calorimeter.	20–1000 °C (70–1830 °F)	Maintains a constant temperature difference between the sample and its enclosure and measures the temperature of the sample. Originally developed for measuring temperatures of phase boundaries	The apparatus is thermally symmetric, and desmearing of the signal is not required. Can extract limited information about kinetics by varying cooling rates	Limited temperature capability. No commercial instrument available	Al alloys
Drop calorimetry	500–1500 °C (930–2730 °F) (depends on container)	Heats a sample to a known temperature; drop into a calorimeter and measure the temperature rise to derive enthalpy	Commercial equipment	Liquid measurements limited by container. Cannot follow enthalpy changes with changes in microstructure, i.e., no information about variation of enthalpy in solid/liquid	Ni-base alloys, steels, Ti alloys
Levitated drop calorimetry	1000–2000 °C (1830–3630 °F)	Uses levitation to support the specimen	Levitated drop calorimetry can handle reactive materials because there is no container. Electromagnetic levitator is limited to electrical conductors.		
Modulation calorimetry	20–4000 °C (70–7230 °F)	Creates periodic oscillations of power that heat the sample and determine the oscillations of the sample temperature about its mean value. There are perceived advantages in using reduced gravity conditions.	Small samples	Temperature measurement is difficult because emissivity of liquid alloys is difficult to measure.	Ni-base alloys, refractory metals
Microsecond pulse calorimetry	1000–4000 °C (1830–7230 °F)	Direct electrical heating of a wire at microsecond rates. Power and temperature are monitored.	Handles reactive materials because there is no container; very fast	Must be able to manufacture a wire. Temperature measurement is difficult because emissivity of liquid alloys is difficult to measure. Cannot follow enthalpy changes with changes in microstructure, i.e., no information about variation of enthalpy in solid/liquid	Ni and Ti alloys, steels, refractory metals

8. *Measurement Techniques,* Vol 1, *Compendium of Thermophysical Property Measurement Methods,* K.D. Maglic, A. Cezairliyan, and V.E. Peletsky, Ed., Plenum Press, London and New York, 1984

9. *Recommended Measurement Techniques and Practices,* Vol 2, *Compendium of Thermophysical Property Measurement Methods,* K.D. Maglic, A. Cezairliyan, and V.E. Peletsky, Ed., Plenum Press, London and New York, 1992

10. *Physiochemical Measurements in Metals Research,* Part 2, *Techniques of Metal Research,* Vol IV, R.A. Rapp, Ed., Interscience Publishers, John Wiley, New York, 1970

11. D.N. Kagan, Adiabatic Calorimetry, Ch 12, *Measurement Techniques,* Vol 1, *Compendium of Thermophysical Property*

Measurement Methods, K.D. Maglic, A. Cezairliyan, and V.E. Peletsky, Ed., Plenum Press, London and New York, 1984, p 461–526

12. Y.A. Kraftmaker, Modulation Calorimetry, Ch 15, *Measurement Techniques,* Vol 1, *Compendium of Thermophysical Property Measurement Methods,* K.D. Maglic, A. Cezairliyan, and V.E. Peletsky, Ed., Plenum Press, London and New York, 1984, p 591–642

13. Y.A. Kraftmaker, Practical Modulation Calorimetry, Ch 14, *Recommended Measurement Techniques and Practices,* Vol 2, *Compendium of Thermophysical Property Measurement Methods,* K.D. Maglic, A. Cezairliyan, and V.E. Peletsky, Ed., Plenum Press, London and New York, 1992, p 409–436

14. R.K. Wunderlich and H.-J. Fecht, Modulated Electromagnetic Induction Calorimetry of Reactive Metallic Liquids, *Meas. Sci. Tech.,* Vol 16, 2005, p 402–416

15. R.K. Wunderlich and H.-J. Fecht, The Development of Containerless Modulation Calorimetry, *Int. J. Thermophys.,* Vol 17, 1996, p 1203

16. R.K. Wunderlich, D.S. Lees, W.I. Johnson, and H.-J. Fecht, Noncontact Modulation Calorimetry of Metallic Liquids in Low Earth Orbit, *Phys. Rev. B,* Vol 55, 1997, p 26–29

17. H. Kobatake, H. Fukuyama, T. Tsukada, and S. Awaji, Noncontact Modulated Laser Calorimetry in a DC Magnetic Field for Stable and Supercooled Liquid Silicon, *Meas. Sci. Tech.,* Vol 21, 2010

18. M.J. Richardson, Application of Differential Scanning Calorimetry to the Measurement

Table 2 Some methods for measuring the thermal conductivity and/or thermal diffusivity of alloys

Technique	Temperature range	Principle	Advantages	Disadvantages	Types of material
Pulse method for thermal diffusivity Normally laser flash technique	20–2000 °C (70–3630 °F)	A short-duration, high-energy pulse (laser or flash lamp) is absorbed at the front face of a thin, disk-shaped sample. The temperature rise of the rear surface is recorded, and the thermal diffusivity values are computed.	Applicable to a wide range of materials. Well-developed technique for solids. Several commercial models available	For liquids, the sample must be contained in a refractory holder. Reactions. Coat sample to absorb energy. Although the effects of convection are said to be minimized, there is no way of establishing if convection is occurring. To obtain thermal conductivity, specific heat data and density values are required.	Al, Mg, and Cu alloys, cast iron, steels, Ni-base alloys
Transient hot insulated conductor	20–1500 °C (70–2730 °F) (for metals)	The rise in temperature in an insulated conductor accompanies the input of electrical energy.	Separate convection and conduction effects	The compatibility of the coating limits the technique. No commercial equipment. Still in development	Sn, Si
Steady state	...	A temperature gradient is maintained between a known thickness of the sample and the heat flow monitored.	Methods are well developed for solids.	Difficult to control convection for metals	Sn, Pb
Electron bombardment heat input Variant of the temperature wave technique	730–3500 °C (1350–6330 °F)	The measurement of phase shift accompanying the passage of a temperature wave through a plane-parallel disc superimposed on a rising temperature	High-temperature capability. Small specimens	Complex experimental apparatus. High vacuum required. Arduous mathematical analysis. No commercial equipment available	Refractory metals
Electrical resistance by four-probe technique	20–1600 °C (70–2910 °F)	Measures the current and voltage with four probes	Experimentally easier than the direct measurement of thermal conductivity. No convection effects	Assumes that there is a relationship between electrical resistivity and thermal conductivity (such as the Wiedemann-Franz rule). Difficulties if alloy forms oxide	Cu, Cu-Sn, Fe
Electrical resistance by microsecond pulse technique	1000–4000 °C (1830–7230 °F)	Direct electrical heating of a wire at microsecond rates. Voltage and current monitored	No convection effects	Assumes that there is a relationship between electrical resistivity and thermal conductivity (such as the Wiedemann-Franz rule)	Ni and Ti alloys, steels, refractory metals

Table 3 Selected methods for measuring the viscosity of liquid alloys

Technique	Temperature range	Principle	Advantages	Disadvantages	Types of material
Capillary viscometer	Melting point to 1000 °C (1830 °F)	Time taken for a known volume of liquid to pass through a capillary of known length and radius	Applicable to a wide range of materials at low temperatures. Well-developed technique	Need accurate capillaries. Easily blocked by inclusions and oxides	Zn, Sn, Cu
Oscillating body viscometer (disk, cylinder, sphere, cup)	Melting point to max 2200 °C (4000 °F) (for metals)	The decay in the angular displacement (oscillation) and time period of a symmetrical body suspended on a torsional pendulum is measured. Liquid may be contained in or around the oscillating body.	The most common method for high-temperature liquid metals. Geometry can be modified to give high sensitivity. Although the theory is complicated, it is now well established (Roscoe equations). Recent reference data for Al and Fe to compare at high temperatures	Requires accurate dimensions of a contained sample. No readily available commercial equipment	Sn, Al, Cu, Ni, Fe alloys
Rotating bob Rotating cylinder	Melting point to 1600 °C (2910 °F)	The torque required to rotate a bob at a constant speed while immersed in a liquid is measured. In the rotating cylinder, the bob is stationary and the cylinder rotates. There are some theoretical advantages in rotating the cylinder, but it is often experimentally easier to rotate the bob.	Commercial apparatus available to provide measurements. Data analysis simplified by use of reference oils for calibration. Dynamic measurements possible. Variants employed for measurements in the solid/liquid region for metals	Use for low viscosity relies on accurate geometry of apparatus—correction of end effects and eccentricity of bob rotation. Some opinion that it is not reliable for metals	Al, low-melting-point metals, slags, oils, mixtures of solid/liquid metals
Oscillating spindle (Torsional crystal viscometer)	Melting point to 1600 °C (2910 °F)	A spindle (plate, sphere, cylinder, etc.) of known size is oscillated in the liquid, and measurements of resonant frequency and amplitude are compared with those of the plate in a gas or vacuum. Measures the kinematic viscosity	Plate is an adaptation for liquids.	Rarely used in liquid at high temperature	Iron
Levitation viscometer	1000–2000 °C (1830–3630 °F)	Measures decay of oscillations of a levitated droplet	Reduces risk of sample contamination. Fast measurements possible	Electromagnetic levitation works only in space. Some results reported with electrostatic levitation. Still in development	Liquid metals

of Specific Heat, Ch 17, *Measurement Techniques,* Vol 1, *Compendium of Thermophysical Property Measurement Methods,* K.D. Maglic, A. Cezairliyan, and V.E. Peletsky,

Ed., Plenum Press, London and New York, 1984, p 669–688
19. M.J. Richardson, The Application of Differential Scanning Calorimetry to the

Measurement of Specific Heat, Ch 18, *Recommended Measurement Techniques and Practices*, Vol 2, *Compendium of Thermophysical Property Measurement*

Table 4 Selected methods for measuring the density and surface tension of alloys

Technique	Temperature range	Principle	Advantages	Disadvantages	Types of material
Sessile drop	Liquidus to 1500 °C (2730 °F)	Measures the shape of a droplet sitting on a flat substrate, using image analysis—volume for density, curvature of surfaces for surface tension	Traditional technique, often used at low temperatures. Works best with fully nonwetting systems. Changes can be continually followed (temperature or composition). Principle can be used for solids	Edge definition may be difficult. Reactions with substrate affect wetting characteristics and may suppress surface tension. Will not work in presence of significant oxides. Must match sample to substrate material. Apparatus often requires vacuum.	Majority of metals with clean surface, low vapor pressure
Pendant drop	Melting point	Measures the shape of a suspended droplet as it melts on the end of a rod, using image analysis—volume for density, curvature for surface tension	Same as sessile drop but no reactions with substrate; no wetting problems. Fast if electron beam or induction heating used	Imaging the drop can be difficult. Only works at melting point. Determining size of droplet can be difficult. Measurement of mass is difficult for density. Oxides may distort shape.	Potentially all metals with clean surface
Drop weight	Melting point	Measures mass of a pendant drop as it detaches from a suspended rod	Used in conjunction with pendant drop. Can act as a check of surface tension and provides mass data for density. May be used on high-melting-point materials. Fast if electron beam or induction heating used	Only works at melting point. Small drop from breaking of surface often falls with big drop and produces error. Oxides can prevent measurement.	Potentially all metals with clean surface (better with high-melting-point, high-surface-tension materials)
Maximum bubble pressure	Liquidus to 1500 °C (2730 °F)	Measures pressure required to blow bubble at end of capillary at a known depth in a liquid	New surface for each bubble, so less susceptible to contamination. Will work with oxides on surface of sample. Does not require mass measurement	Difficult to detect surface to measure depth of immersion. Equilibrium bubble shape easily distorted, causing errors. Capillary/sample reactions. Measurement of capillary radius. Oxides on the surface can block the tube.	All materials
Hydrostatic probe Archimedean method Also detachment methods (surface tension) du Nouy ring Wilhelmy plate Maximum pull	Liquidus to 1000 °C (1830 °F) (potential to 1600 °C, or 2910 °F)	Measures apparent mass of a bob as it is immersed in a liquid (dynamic) or when immersed (Archimedean method density)	Wide range of materials. Can provide information about oxides. Better developed for density. Can be extremely accurate	Surface tension difficult to extract from data. Oxides may prevent repeat measurements.	All materials
Levitated (oscillating) drop	1000–2000 °C (1830–3630 °F)	Measures natural oscillation frequencies (surface tension) or volume (density) of levitated drop of known mass	Handles reactive materials because there is no container. Large temperature range of superheat and undercooling. Short run times to high temperatures	Large surface area maximizes evaporation and gas reactions. Oxides can distort shape, may prevent surface tension measurement. Temperature measurement is difficult because emissivity of liquid alloys is difficult to measure. Flat viewing port or tube required for density. Often requires separate measurement systems for density and surface tension	Cu and Ti alloys, Ni-base alloys, steels
Piston dilatometry	20–1500 °C (70–2730 °F)	Modification of standard dilatometer, with the sample contained in a cylinder with freely moving pistons to monitor change in length	Only method that gives density data in the solid/liquid region	Method relies on a good fit between piston and cylinder.	Al and Ni alloys
Microsecond technique Density only	1000–4000 °C (1830–7230 °F)	Direct electrical heating of a wire at microsecond rates. Changes in dimensions followed with high-speed camera	Handles reactive materials because there is no container	Need a wire. Care with temperature measurement	Ni and Ti alloys, steels, refractory metals
General notes Surface tension	The majority of methods rely on knowledge of the mass of the sample, which will vary if volatile elements are present. Surface tension is critically dependent on surface chemistry and solubility of surface-active elements in the sample. Oxides will therefore produce very different values from those of a clean surface. Difficulties with oxides are highlighted where the oxide will prevent measurement altogether.				

Methods, K.D. Maglic, A. Cezairliyan, and V.E. Peletsky, Ed., Plenum Press, London and New York, 1992, p 519–548

20. "Advanced Technical Ceramics—Monolithic Ceramics. Thermophysical Properties—Part 3: Determination of Specific Heat Capacity," BS EN 821-3:2005

21. H.B. Dong and R. Brooks, A Numerical Model for Heat Flux DSCs: Determining Heat Transfer Coefficients within a DSC, *Mater. Sci. Eng. A*, Vol 413–414, 2005, p 470–473

22. H.B. Dong and J.D. Hunt, A Numerical Model of Two-Pan Heat Flux DSC, *J. Therm. Anal. Cal.*, Vol 64, 2001, p 167–176

23. W.J. Boettinger, U.R. Kattner, K.-W. Moon, and J.H. Perepezko, "DTA and Heat Flux DSC Measurements of Alloy Melting and Freezing," NIST Recommended Practice Guide Special Publication 960-15, 2006

24. C.S. Smith, A Simple Method of Thermal Analysis Permitting Quantitative Measurements of Specific and Latent Heats, *Trans. Am. Inst. Met. Eng.*, Vol 137, 1939, p 236–245

25. F.H. Hayes, W.T. Chao, and J.A.J. Robinson, Phase Diagrams of (Gold+Binary Solder) Ternary Alloy Systems by Smith Thermal Analysis, *J. Therm. Anal.*, Vol 42, 1994, p 745–758

26. H.B. Dong, M.R.M. Shin, E.C. Kurum, J.D. Hunt, and H. Cama, A Study of Microsegregation in Al-Cu Using a Novel Single-Pan Calorimeter, *Metall. Mater. Trans. A*, Vol 34, 2003, p 441–447

27. H.B. Dong and J.D. Hunt, A Comparison of a Novel Scanning Calorimeter with a Conventional Two Pan Scanning Calorimeter, *High Temp.—High Press.*, Vol 32, 2000, p 311–319

28. L. Backerud, E. Krol, and J. Tamminen, Solidification Characteristics of Aluminum

Alloys, *Wrought Alloys*, Vol 1, *Skanaluminium*, Norway, 1986

29. D.A. Ditmars, Heat Capacity Calorimetry by the Method of Mixtures, Ch 13, *Measurement Techniques*, Vol 1, *Compendium of Thermophysical Property Measurement Methods*, K.D. Maglic, A. Cezairliyan, and V.E. Peletsky, Ed., Plenum Press, London and New York, 1984, p 527–553

30. D.A. Ditmars, Phase Change Calorimeter for Measuring the Relative Enthalpy in the Temperature Range 273.15 to 1200 K, Ch 15, *Recommended Measurement Techniques and Practices*, Vol 2, *Compendium of Thermophysical Property Measurement Methods*, K.D. Maglic, A. Cezairliyan, and V.E. Peletsky, Ed., Plenum Press, London and New York, 1992, p 437–456

31. P.N. Quested, K.C. Mills, R.F. Brooks, A.P. Day, R. Taylor, and H. Szealogowski, Physical Property Measurement for Simulation Modelling of Heat and Fluid Flow during Solidification, *Proc. 1977 International Symposium on Liquid Metal Processing and Casting*, Feb 16–19, 1977 (Santa Fe, NM), Vacuum Metallurgy Division, American Vacuum Society, p 1–17

32. V.Y. Chekhovskoi, Levitation Calorimetry, Ch 14, *Measurement Techniques*, Vol 1, *Compendium of Thermophysical Property Measurement Methods*, K.D. Maglic, A. Cezairliyan, and V.E. Peletsky, Ed., Plenum Press, London and New York, 1984, p 527–590

33. V.Y. Chekhovskoi, Apparatus for the Investigation of Thermodynamic Properties of Metals by Levitation Calorimetry, Ch 16, *Recommended Measurement Techniques and Practices*, Vol 2, *Compendium of Thermophysical Property Measurement Methods*, K.D. Maglic, A. Cezairliyan, and V.E. Peletsky, Ed., Plenum Press, London and New York, 1992, p 457–482

34. M.G. Frohberg, Thirty Years of Levitation—A Balance, *Thermochim. Acta*, Vol 337, 1999, p 7–17

35. A. Cezairliyan, Pulse Calorimetry, Ch 16, *Measurement Techniques*, Vol 1, *Compendium of Thermophysical Property Measurement Methods*, K.D. Maglic, A. Cezairliyan, and V.E. Peletsky, Ed., Plenum Press, London and New York, 1984, p 643–668

36. A. Cezairliyan, A Millisecond-Resolution Pulse Heating System for Specific Heat Measurement at High Temperatures, Ch 17, *Recommended Measurement Techniques and Practices*, Vol 2, *Compendium of Thermophysical Property Measurement Methods*, K.D. Maglic, A. Cezairliyan, and V.E. Peletsky, Ed., Plenum Press, London and New York, 1992, p 483–518

37. G. Pottlacher, H. Hosaeus E. Kaschnitz, and A. Seifter, Thermophysical Properties of Solid and Liquid Inconel 718 Alloy, *Scand. J. Met.*, Vol 31, 2002, p 161–168

38. B. Withan, C. Cagran, and G. Pottlacher, Combined DSC and Pulse Heating Measurements of the Electrical Resistivity and Enthalpy of Tungsten, Niobium and Titanium, *Int. J. Thermophys.*, Vol 26, 2005, p 1017–1039

39. R.S. Hixson and M.A. Winkler, Thermophysical Properties of Solid and Liquid Tungsten, *Int. J. Thermophys.*, Vol 11, 1990, p 709–718

Thermal Conductivity

40. E. Yamasue, M. Susa, H. Fukuyama, and K. Nagata, Nonstationary Hot Wire Method with Silica Coated Probe for Measuring the Thermal Conductivities of Molten Metals, *Met. Mater. Trans. A*, Vol 30, 1999, p 1071–1979

41. T. Hibiya and S. Nakamura, Thermophysical Property Measurement on Molten Semiconductors Using 10-S Microgravity in a Drop Shaft, *Int. J. Thermophys.*, Vol 17, 1996, p 1191–1201

42. K. Nagata, H. Fukuyama, K. Taguchi, H. Ishii, and M. Hayashi, Thermal Conductivity of Molten Al, Si and Ni Measured under Microgravity, *High Temp. Mater. Proc.*, Vol 22, 2003, p 267–273

43. S. Nakamura, T. Hibiya, T. Yokota, and F. Yamamoto, Thermal Conductivity Measurement of Mercury in a Magnetic Field, *J. Heat Mass. Transf.*, Vol 33, 1990, p 2609–2613

44. M.J. Laubitz, Axial Flow Methods of Measuring Thermal Conductivity, Ch 1, *Measurement Techniques*, Vol 1, *Compendium of Thermophysical Property Measurement Methods*, K.D. Maglic, A. Cezairliyan, and V.E. Peletsky, Ed., Plenum Press, London and New York, 1984, p 11–60

45. J.P. More, Analysis of Apparatus with Radial Symmetry for Steady-State Measurements of Thermal Conductivity, Ch 2, *Measurement Techniques*, Vol 1, *Compendium of Thermophysical Property Measurement Methods*, K.D. Maglic, A. Cezairliyan, and V.E. Peletsky, Ed., Plenum Press, London and New York, 1984, p 61–123

46. J.M. Corsan, Axial Heat Flow Methods of Thermal Conductivity Measurement for Good Conducting Materials, Ch 1, *Recommended Measurement Techniques and Practices*, Vol 2, *Compendium of Thermophysical Property Measurement Methods*, K.D. Maglic, A. Cezairliyan, and V.E. Peletsky, Ed., Plenum Press, London and New York, 1992, p 3–32

47. W.J. Parker, R.J. Jenkins, C.P. Butler, and G.L. Abbott, Flash Method of Thermal Diffusivity, Heat Capacity and Thermal Conductivity, *J. Appl. Phys.*, Vol 32, 1961, p 1679–1684

48. R.E. Taylor and K.D. Maglic, Pulse Method for Thermal Diffusivity Measurement, Ch 8, *Measurement Techniques*, Vol 1, *Compendium of Thermophysical Property Measurement Methods*, K.D. Maglic, A. Cezairliyan,

and V.E. Peletsky, Ed., Plenum Press, London and New York, 1984, p 305–336

49. K.D. Maglic and R.E. Taylor, The Apparatus for Thermal Diffusivity Measurement by the Laser Pulse Method, Ch 10, *Recommended Measurement Techniques and Practices*, Vol 2, *Compendium of Thermophysical Property Measurement Methods*, K.D. Maglic, A. Cezairliyan, and V.E. Peletsky, Ed., Plenum Press, London and New York, 1992, p 281–314

50. J.B. Henderson, R.E. Taylor, and H. Groot, Characterisation of Molten Metals Through Multiple Thermophysical Property Measurements, *High Temp.—High Press.*, Vol 25 1993, p 323–327

51. B.J. Monaghan and P.N. Quested, Thermal Diffusivity of Iron at High Temperature in Both the Liquid and Solid States, *ISIJ Int.*, Vol 41, 2001, p 1524–1528

52. R.A. Abas, M. Hayashi, and S. Seetharaman, Thermal Diffusivity Measurements of CMSX-4 Alloy by the Laser Flash Method, *Int. J. Thermophys.*, Vol 28, 2007, p 109–122

53. E. Kaschnitz and R. Ebner, Thermal Diffusivity of the Aluminium Alloy Al-17Si-4Cu (A390) in the Solid and Liquid States, *Int. J. Thermophys.*, Vol 28, 2007, p 711–722

54. H. Szelagowski, R. Taylor, J.D. Hunt, and P.N. Quested, A Numerical Analysis of the Laser Flash Technique Applied to a Semisolid Material, *Proceedings of the Fourth Decennial International Conference on Solidification Processing*, July 7–10, 1997 (Sheffield), p 151–154

55. K. Ravindran, S.G.R. Brown, and J.A. Spittle, Prediction of the Effective Thermal Conductivity of Three-Dimensional Regions by Finite Element Method, *Mater. Sci. Eng. A*, Vol 269, 1999, p 90–97

56. E. Yamasue, M. Susa, H. Fuyuyama, and K. Nagata, Thermal Conductivities of Silicon and Germanium in Solid and Liquid States Measured by Non-Stationary Hot Wire Method with Silica Coated Probe, *J. Cryst. Growth*, Vol 243, 2002, p 121–131

57. J. Bilek, J. Atkinson, and W. Wakeham, Measurements of the Thermal Conductivity of Molten Lead Using a New Transient Hot-Wire Sensor, *Int. J. Thermophys.*, Vol 28, 2007, p 496–505

58. K.C. Mills, B.J. Keene, and B.J. Monaghan, Thermal Conductivities of Molten Metals, Part 1: Pure Elements, *Int. Mater. Rev.*, Vol 41, 1996, p 209–242

59. B.J. Monaghan, A Four-Probe dc Method for Measuring the Electrical Resistivities of Molten Metals, *Int. J. Thermophys.*, Vol 20, 1999, p 677–690

60. G. Lohöfer, Electrical Resistivity Measurement of Liquid Metals, *Meas. Sci. Technol.*, Vol 16, 2005, p 417–425

61. H. Hosaeus, A. Seifter, G. Pottlacher, and E. Kaschnitz, Thermophysical Properties of Solid and Liquid Inconel 718 Alloy, *Scand. J. Metall.*, Vol 31, 2002, p 161–168

Density

62. R.K. Kirby, Methods of Measuring Thermal Expansion, *Recommended Measurement Techniques and Practices*, Vol 2, *Compendium of Thermophysical Property Measurement Methods*, K.D. Maglic, A. Cezairliyan, and V.E. Peletsky, Ed., Plenum Press, London and New York, 1992, p 549–567

63. J.D. James, J.A. Spittle, S.G.R. Brown, and R.W. Evans, A Review of Measurement Techniques for the Thermal Expansion Coefficient of Metals and Alloys at Elevated Temperatures, *Meas. Sci. Technol.*, Vol 12, 2001, p R1–R15

64. L.D. Lucas, Viscometry and Densitometry, Part B: Liquid Density Measurements, Ch 7B, *Physiochemical Measurements in Metals Research*, Part 2, *Techniques of Metal Research,* Vol IV, R.A. Rapp, Ed., Interscience Publishers, John Wiley, New York, 1970, p 219–292

65. J. Brillo and I. Egry, Density Determination of Liquid Copper, Nickel and Their Alloys, *Int. J. Thermophys.*, Vol 24, 2003, p 1155–1170

66. R.F. Brooks, A.P. Day, K.C. Mills, and P. N. Quested, Physical Property Measurements for the Mathematical Modelling of Fluid Flow in Solidification Processes, *Int. J. Thermophys.*, Vol 18, 1997, p 471–480

67. P.F. Paradis, T. Ishikawa, and S. Yoda, Electrostatic Levitation: Research and Development at JAXA: Past and Present Activities in Thermophysics, *Int. J. Thermophys.*, Vol 26, 2005, p 1031–1048

68. T. Ishikawa, P.F. Paradis, T. Itami, and S. Yoda, Non Contact Thermophysical Property Measurements of Refractory Metals Using an Electrostatic Levitator, *Meas. Sci. Technol.*, Vol 16, 2005, p 443–451

69. G. Wille, F. Millot, and J.C. Rifflet, Thermophysical Properties of Containerless Liquid Iron up to 2500 K, *Int. J. Thermophys.*, Vol 24, 2002, p 1197–1206

70. W.K. Rhim, S.K. Chung, A.J. Rulison, and R.E. Spjut, Measurements of Thermophysical Properties of Molten Silicon by a High-Temperature Electrostatic Levitator, *Int. J. Thermophys.*, Vol 18, 1997, p 459–470

71. J. Blumm and J.B. Henderson, Measurement of the Volumetric Expansion and Density Changes in the Solid and Molten Regions, *High Temp.—High Press.*, Vol 32, 2000, p 109–113

72. R. Morrell and P. Quested, Evaluation of Piston Dilatometry for Studying the Melting Behaviour of Metals and Alloys, *High Temp.—High Press.*, Vol 35/36, 2003/2004, p 417–435

73. C. Garcia-Cordonvilla, E. Louis, and A. Pamies, The Surface Tension of Liquid Pure Aluminium and Aluminium-Magnesium Alloy, *J. Mater. Sci.*, Vol 21, 1986, p 2287–2792

74. S.R. Roach and H. Henein, A New Method to Dynamically Measure the Surface Tension, Viscosity and Density of Melts, *Metall. Mater. Trans. B,* Vol 36, 2005, p 667-676

Viscosity

75. R.F. Brooks, A.T. Dinsdale, and P.N. Quested, The Measurement of Viscosity of Alloys—A Review of Methods, Data and Models, *Meas. Sci. Technol.*, Vol 16, 2005, p 354–362

76. L.J. Wittenberg, Viscometry and Densitometry, Part A: Viscosity of Liquid Metals, Ch 7A, *Physiochemical Measurements in Metals Research*, Part 2, *Techniques of Metal Research,* Vol IV, R.A. Rapp, Ed., Interscience Publishers, John Wiley, New York, 1970, p 193–217

77. A. Knappwost, A New Method of High Temperature Viscometry by the Method of Oscillating Hollow Bodies, *Z. Phys. Chem.*, Vol 200, 1952, p 81

78. Y.G. Shvidkovskiy, "Certain Problems Related to the Viscosity of Fused Metals," Translation of "Netotoryye Voprosy Vyazkosti Rasplavlennykh Metallow," Published by State Publishing House for Technical and Theoretical Literature (Moscow), 1955, NASA Technical Translation F-88, 1962

79. R. Roscoe, Viscosity Determination by the Oscillating Vessel Method 1: Theoretical Considerations, *Proc. Phys. Soc.*, Vol 72, 1958, p 576–584

80. J. Kestin and G.F. Newell, Theory of Oscillation Type Viscometers. The Oscillating Cup, *ZAMP*, Vol VIII, 1957, p 433

81. W. Brockner, K. Torklep, and H.A. Oye, Viscosity of Aluminium Chloride and Acidic Sodium Chloroaluminate Melts, *Ber. Bursenges Phys. Chem.*, Vol 83, 1979 p 1–11

82. D.A. Beckwith and G.F. Newell, Theory of Oscillation Type Viscometers. The Oscillating Cup. Part II, *ZAMP,* Vol VIII, 1957, p 450

83. R.F. Brooks, A.P. Day, R.J.L. Andon, L.A. Chapman, K.C. Mills, and P.N. Quested, Measurements of the Viscosities of Metals and Alloys with an Oscillating Viscometer, *High Temp.—High Press.*, Vol 33, 2001, p 73–82

84. M. Kehr, W. Hoyer, and I. Egry, A New High-Temperature Oscillating Cup Viscometer, *Int. J. Thermophys.*, Vol 28, 2007, p 1017–1025

85. Y. Sato, K. Sugisawa, D. Aoki, and T. Yamamura, Viscosities of Fe-Ni, Fe-Co, and Ni-Co Binary Melts, *Meas. Sci. Technol.*, Vol 16, 2005, p 363–371

86. S. Gruner and W. Hoyer, A Statistical Approach to Estimate the Experimental Uncertainty of Viscosity Data by Oscillating Cup Method, *J. Alloys Comps.*, Vol 480, 2009, p 629–633

87. K. Nakashima, T. Kawagoe, T. Ookado, and K. Mori, Viscosity of Binary Borate and Ternary Borosilicate, *Molten Slags, Fluxes and Salts '97 Conference,* Jan 5–8, 1997 (Sydney, Australia), Iron and Steel Society

88. W.R.D. Jones and W.L. Bartlett, The Viscosity of Aluminium and Binary Aluminium Alloys, *J. Inst. Met.*, Vol 81, 1952–1953, p 145–152

89. S.I. Bakhtiyarov and R.A. Overfelt, Measurement of Liquid Metal Viscosity by Rotational Technique, *Acta. Mater.*, Vol 47, 1999, p 4311–4319

90. Lord Rayleigh, On the Capillary Phenomena of Jets, *Proc. R. Soc. (London),* Ser a 29, 1879, p 71–97

91. D.L. Cummings and D.A. Blackburn, Oscillations of Magnetically Levitated Aspherical Drops, *J. Fluid Mech.*, Vol 221, 1991, p 395–416

92. W.K. Rhim, K. Ohsaka, P.F. Paradis, and R.E. Spjut, Noncontact Technique for Measuring Surface Tension and Viscosity of Molten Materials Using High Temperature Electrostatic Levitator, *Rev. Sci. Instr.*, Vol 70, 1999, p 2796–2801

93. T. Ishikawa, P.F. Paradis, T. Itami, and S. Yoda, Non Contact Thermophysical Property Measurements of Refractory Metals Using an Electrostatic Levitator, *Meas. Sci. Technol.*, Vol 16, 2005, p 443–451

94. K. Higuchi, H.J. Fecht, and R.K. Wunderlich, Surface Tension and Viscosity of the Ni-Based Superalloy CMSX-4 Measured by Oscillating Drop Method in Parabolic Flight Experiments, *Adv. Eng. Mater.*, Vol 9, 2007, p 349–354

95. V.F. Nozdrev, V.I. Strmousov, and V.V. Takuchev, Acoustic Study of the Viscosity of Liquid Aluminium Alloys, *Russ. J. Phys. Chem.*, Vol 53, 1979, p 677–679

ASM Handbook, Volume 22B, *Metals Process Simulation*
D.U. Furrer and S.L. Semiatin, editors

Copyright © 2010, ASM International®
All rights reserved.
www.asminternational.org

Measurement and Interpretation of Flow Stress Data for the Simulation of Metal-Forming Processes

S.L. Semiatin, Air Force Research Laboratory
T. Altan, The Ohio State University

THE YIELD STRESS of a metal under uniaxial conditions is often referred to as the flow stress. Metal starts to deform plastically when the applied stress (in uniaxial tension without necking or in uniaxial compression without bulging) reaches the value of the yield or flow stress. Metals that undergo flow hardening or softening exhibit an increasing or decreasing flow stress, respectively, with increasing strain. Furthermore, many metals show a small or large dependence of flow stress on both strain rate and temperature at cold or hot working temperatures, respectively. The quantification of flow stress constitutes one of the most important inputs to the simulation of a metal-forming process.

The flow stress of a metal may be quantified in terms of its dependence on strain, strain rate, and temperature. Such an approach yields a phenomenological description of flow behavior and is useful primarily for the specific material condition/microstructure and deformation regime in which actual measurements have been made. Alternatively, flow stress models can be based on so-called internal state variables such as dislocation density, grain size, phase fraction, strain rate, and temperature. In this case, the flow behavior can sometimes be extrapolated beyond the regime of measurements, provided the deformation mechanism is unchanged. Irrespective of whether flow stress is described phenomenologically or mechanistically, similar measurement techniques are used.

For a given microstructural condition, the flow stress, $\bar{\sigma}$, can be expressed as a function of the strain, $\bar{\varepsilon}$, the strain rate, $\dot{\bar{\varepsilon}}$, and the temperature, T:

$$\bar{\sigma} = f(T, \bar{\varepsilon}, \dot{\bar{\varepsilon}}) \qquad \text{(Eq 1)}$$

Under uniaxial stress conditions, the axial stress, σ, axial strain, ε, and axial strain rate, $\dot{\varepsilon}$, are equal to the effective stress, $\bar{\sigma}$, effective strain, $\bar{\varepsilon}$, and effective strain rate, $\dot{\bar{\varepsilon}}$. This is no longer true for multiaxial states of stress. Hence, effective quantities are used more often

to represent flow stress because of their general applicability. During hot forming of metals (at temperatures above approximately one-half of the melting point), the effect of strain on flow stress is often weak, and the influence of strain rate (i.e., rate of deformation) becomes increasingly important. Conversely, at room temperature (i.e., during cold forming), the effect of strain rate on flow stress is usually small. The degree of dependence of flow stress on temperature varies considerably among different materials. Therefore, temperature variations during the forming process can have different effects on load requirements and metal flow for different materials.

To be useful in the analysis of a forming process, the flow stress of a metal should be determined experimentally for the strain, strain rate, and temperature conditions that exist during the process and for the specific microstructural condition of the workpiece material. The most commonly used methods for determining flow stress are the tension, uniform compression, and torsion tests (Ref 1).

Tension Test

The tension test is commonly used to determine the mechanical (service) properties of metals. It is less frequently used to determine the large-strain flow stress of metals due the occurrence of necking at relatively small strains. Nevertheless, it does find application for the modeling of sheet-forming processes under ambient-temperature conditions (in which deformations can be moderate) and superplastic-forming operations at elevated temperatures, in which a large value of the strain-rate sensitivity exponent delays necking to large strains.

Two methods of representing flow stress data from the tension test are illustrated in Fig. 1 (Ref 2). In the classical engineering, or nominal, stress-strain diagram (Fig. 1a), the

engineering stress, S, is obtained by dividing the instantaneous tensile load, L, by the original cross-sectional area of the specimen, A_o. The stress S is then plotted against the engineering strain, $e = (l - l_o)/l_o$. During deformation, the specimen initially elongates in a uniform fashion. When the load reaches its maximum value, necking starts and the uniform uniaxial stress condition ceases to exist. Deformation is then concentrated in the neck region while the rest of the specimen undergoes very limited deformation.

Figure 1(b) illustrates the true stress-strain representation of the same tension-test data. In this case, before necking occurs, the following relationships are valid:

$$\begin{aligned} \bar{\sigma} &= \text{true stress (flow stress)} \\ &= \text{instantaneous load/instantaneous area} \\ &= L/A \\ &= S(1 + e) \end{aligned} \qquad \text{(Eq 2)}$$

and

$$\bar{\varepsilon} = \text{true strain} = \ln\left(\frac{l}{l_o}\right) = \ln\left(\frac{A_o}{A}\right) = \ln(1 + e) \qquad \text{(Eq 3)}$$

Prior to necking, the instantaneous load is given by $L = A\bar{\sigma}$. The criterion for necking can be formulated as the condition that L be maximum or that:

$$\frac{dL}{d\bar{\varepsilon}} = 0 \qquad \text{(Eq 4)}$$

Furthermore, prior to the attainment of maximum load, the uniform deformation conditions (Eq 2, 3) are valid (Ref 2), and the following useful relations can be derived:

$$A = A_o e^{-\bar{\varepsilon}}$$

and

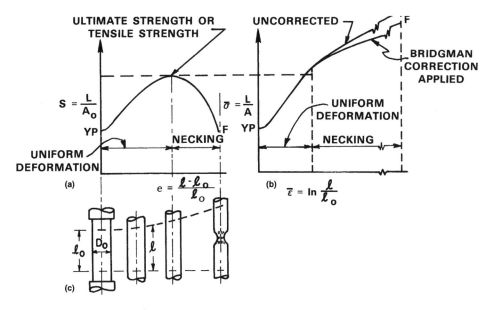

Fig. 1 Data from the uniaxial tension test. (a) Engineering stress-strain curve. (b) True stress-strain curve. (c) Schematic illustration of dimensional changes during the test. Source: Ref 2

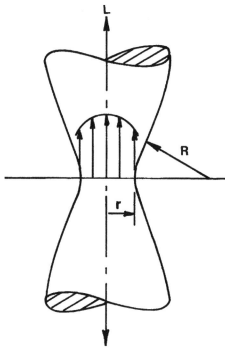

Fig. 3 Axial stress distribution at the symmetry plane of a necked portion of a tension specimen. Source: Ref 2, 3

$$L = A\bar{\sigma} = A_0\bar{\sigma}e^{-\bar{\varepsilon}} \quad \text{(Eq 5)}$$

Combining Eq 4 and 5 results in:

$$\frac{dL}{d\varepsilon} = 0 = A_0\left(\frac{d\bar{\sigma}}{d\bar{\varepsilon}}e^{-\bar{\varepsilon}} - \bar{\sigma}e^{-\bar{\varepsilon}}\right) \quad \text{(Eq 6)}$$

or

$$\frac{d\bar{\sigma}}{d\bar{\varepsilon}} = \bar{\sigma} \quad \text{(Eq 7)}$$

Very often, the flow stress curve (or simply flow curve) obtained at room temperature can be expressed in the form of a power law relation between stress and strain:

$$\bar{\sigma} = K\bar{\varepsilon}^n \quad \text{(Eq 8)}$$

in which K and n are material constants known as the strength coefficient and strain-hardening exponent, respectively. Combining Eq 7 and 8 results in:

$$\frac{d\bar{\sigma}}{d\bar{\varepsilon}} = Kn(\bar{\varepsilon})^{n-1} = \bar{\sigma} = K(\bar{\varepsilon})^n \quad \text{(Eq 9)}$$

or

$$\bar{\varepsilon} = n \quad \text{(Eq 10)}$$

The condition expressed by Eq 7 is shown schematically in Fig. 2. From this figure and Eq 10, it is evident that at low forming temperatures (for which Eq 8 is valid), a material with a large n has greater formability; that is, it sustains a larger amount of uniform deformation in tension than a material with a smaller n. It should be noted, however, that this statement is not true for materials and conditions for which the flow stress cannot be expressed by Eq 8. Such is the case at elevated (hot working) temperatures at which the material response is

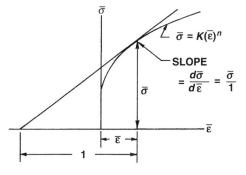

Fig. 2 Determination of the strain at the onset of necking during the tension test. Source: Ref 2

very rate sensitive but often exhibits limited strain hardening.

The determination of flow stress after the onset of necking (Fig. 1b) requires a correction, because a triaxial state of stress is induced (Fig. 3). Such a correction for a round bar specimen, derived by Bridgman (Ref 3), is given by:

$$\bar{\sigma} = \frac{L}{\pi r^2}\left[\left(1 + \frac{2R}{r}\right)\ln\left(1 + \frac{r}{2R}\right)\right]^{-1} \quad \text{(Eq 11)}$$

The quantities r and R are defined in Fig. 3. For the evaluation of Eq 11, the values of r and R must be measured continuously during the test. A similar expression was also derived by Bridgman to determine the flow stress during necking of sheet tension samples.

Uniaxial Compression Test

The compression test can be used to determine flow stress data for metals over a wide range of temperatures and strain rates. In this test, flat platens and a cylindrical sample are

heated and maintained at the same temperature so that die chilling, and its influence on metal flow, is prevented. To be applicable without corrections or errors, the cylindrical sample must be upset without any barreling; that is, the state of uniform stress in the sample must be maintained (Fig. 4). Barreling is prevented by using adequate lubrication. Teflon™ film, molybdenum sulfide, or machine oil is often used at room temperature. At hot working temperatures, graphite in oil is used for aluminum alloys, and melted glass is used for steel, titanium, and high-temperature alloys. To hold the lubricant, spiral grooves are often machined on both the flat surfaces of cylindrical test specimens (Fig. 5a). The load and displacement (or sample height) are measured during the test. From this information, the average pressure is calculated at each stage of deformation, that is, for increasing axial strain.

For frictionless, perfectly uniform compression, the average pressure-axial stress curve is equivalent to the flow curve. In this case, similar to the uniform elongation portion of the tension test, the following relationships are valid:

$$\bar{\varepsilon} = \ln\frac{h_0}{h} = \ln\frac{A}{A_0} \quad \text{(Eq 12)}$$

$$\bar{\sigma} = \frac{L}{A} \quad \text{(Eq 13)}$$

$$A = A_0e^{\bar{\varepsilon}} \quad \text{(Eq 14)}$$

$$\dot{\bar{\varepsilon}} = \frac{d\bar{\varepsilon}}{dt} = \frac{dh}{hdt} = \frac{v}{h} \quad \text{(Eq 15)}$$

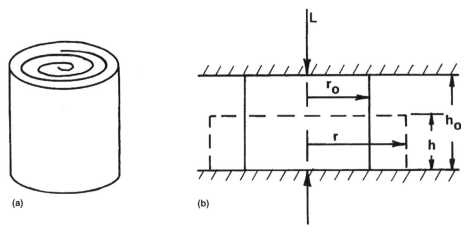

Fig. 4 Compression test specimen. (a) View of specimen showing lubricated shallow grooves on the ends. (b) Shape of the specimen before and after the test

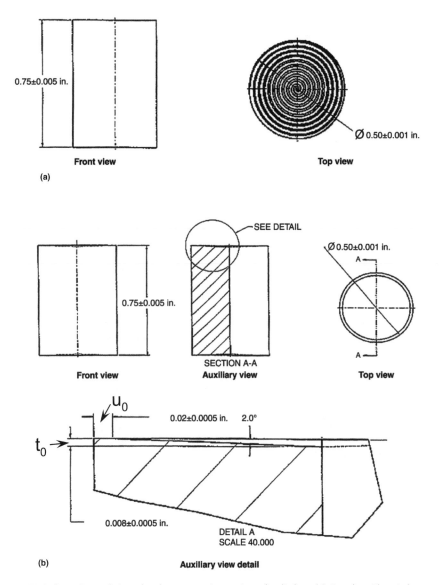

Fig. 5 Typical specimen designs for the compression testing of cylinders. (a) Sample with spiral grooves. (b) Rastegaev specimen. Source: Ref 4

in which v is instantaneous crosshead speed, h_o and h are the initial and instantaneous sample heights, respectively, and A_o and A are initial and instantaneous cross-sectional areas, respectively. Typically, the compression test can be conducted without barreling to ~50% height reduction ($\bar{\varepsilon} = 0.69$) or more. A typical load-displacement curve and the corresponding $\bar{\sigma}$-$\bar{\varepsilon}$ curve obtained for the uniform compression of annealed aluminum 1100 at room temperature is shown in Fig. 6.

At hot working temperatures (i.e., temperatures typically in excess of one-half of the absolute melting point), the flow stress of nearly all metals is very strain-rate dependent. Therefore, hot compression tests should be conducted using a test machine that provides a constant true strain rate, that is, a constant value of the ratio of the crosshead speed to the instantaneous sample height (Eq 15). For this purpose, programmable servohydraulic testing machines or cam plastometers are commonly used. Sometimes, a mechanical press is employed; however, an approximately constant strain rate is obtained for only the first half of the deformation when using such equipment. To maintain nearly isothermal and uniform compression conditions, hot compression tests are conducted in a furnace or using a preheated fixture such as that shown in Fig. 7. The dies are coated with an appropriate lubricant, for example, oil or graphite for temperatures to 425 °C (800 °F) and glass for temperatures to 1260 °C (2300 °F). The fixture/dies and the specimen are heated to the test temperature, soaked for a predetermined time (usually ~10 to 15 min), and then the test is initiated. Examples of tested hot compression samples are shown in Fig. 8. Examples of high-temperature $\bar{\sigma}$-$\bar{\varepsilon}$ data are given in Fig. 9.

Specimen Preparation

There are two machining techniques that can be used for preparing specimens for the cylinder-compression test, namely, the spiral-groove design (Fig. 5a) and the Rastegaev specimen (Fig. 5b). The spiral grooves and the recesses of the Rastegaev specimen serve the purpose of retaining the lubricant at the tool-workpiece interface during compression, thus minimizing barreling. It has been determined that Rastegaev specimens provide better lubrication and hold their form better during testing compared to the spiral-grooved specimens. The specifications for the specimens and the test conditions are as follows (Ref 4).

For a specimen with spiral grooves (Fig. 5a):

- Solid cylinder (diameter = 0.5 + 0.001 in., length = 0.75 + 0.005 in.)
- Ends should be flat and parallel within 0.0005 in./in.
- Surface should be free of grooves, nicks, and burrs.
- Spiral grooves machined at the flat ends of the specimen with approximately 0.01 in. depth.

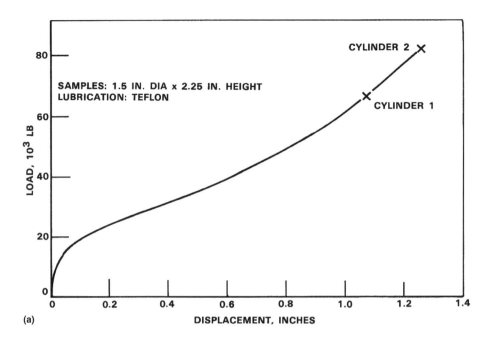

(a)

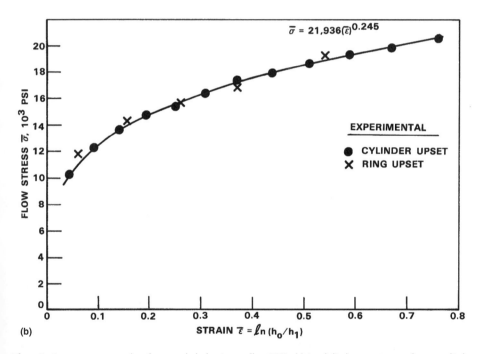

(b)

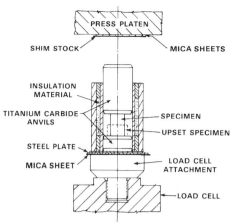

Fig. 7 Press setup and sample tooling design used for the hot compression of cylinders and rings

Fig. 6 Room-temperature data for annealed aluminum alloy 1100. (a) Load-displacement curve from a cylinder-compression test. (b) True stress-true strain (flow) curve results from both cylinder compression and ring compression. Source: Ref 5

For a Rastegaev specimen (Fig. 5b):

- Flat recesses at the ends should be filled with lubricant.
- Dimensions $t_0 = 0.008 \pm 0.0005$ in. and $u_o = 0.02 \pm 0.0005$ in. at the end faces have a significant effect on the lubrication conditions.
- Rastegaev specimen ensures good lubrication up to high strains of ~0.8 to 1; that is, the specimen remains cylindrical due to the radial pressure that the lubricant exerts on the ring.
- $t_0/u_o = 0.4$ for steels (optimum value at which the specimen retains cylindrical shape up to maximum strain before bulging occurs).

Parallelism of the Press (or Testing Machine) Dies

In a compression test, load is applied on the billet using flat dies. To ensure that a uniaxial state of stress exists during the experiment, the applied load should be exactly parallel to the axis of the cylindrical specimen. This calls for measurement of the parallelism of the platens of the testing machine or press. A commonly used technique for measuring parallelism involves compressing lead billets of the same height as the test samples; lead is used because it is soft and deforms easily at room temperature. The circumferential variation in billet height is an indication of the parallelism of the platens. Alternatively, for large-diameter dies, lead samples can be placed at different locations. The difference in the final height of the samples following compression can be used to correct for parallelism.

Errors in the Compression Test

Errors in the determination of flow stress by the compression test can be classified in three categories (Ref 4):

- Errors in the displacement readings, which result in errors in the calculated strain
- Errors in the load readings, which result in errors in the calculated stress
- Errors in the processing of the data due to barreling of the test specimens

The first and second type of errors may be reduced or eliminated by careful calibration of the transducers and data-acquisition equipment. However, barreling of the test specimens during compression cannot be entirely eliminated because there is always friction between the specimen and the tooling. The maximum error in determining flow stress via compression testing is thus usually that associated with friction. To correct the flow curve and to determine the percentage error in flow stress, finite-element method analysis is often used.

Average pressure (p_{av})-axial strain (ε) plots derived from measured load-stroke data (corrected for the test-machine compliance) and reduced assuming uniform deformation can also be corrected for friction effects using the following approximate relation (Ref 7):

$$\frac{\bar{\sigma}}{p_{av}} = \left(1 + \frac{m_s d_s}{(3\sqrt{3})h}\right)^{-1} \qquad \text{(Eq 16)}$$

in which m_s denotes the friction shear factor determined from a ring test (described next), and d_s and h represent the instantaneous sample diameter and height, respectively.

Gleeble systems (Dynamic Systems, Inc.) (Ref 8) can be used to conduct hot/warm compression or tension tests on various specimen geometries. The Gleeble 3500 system uses direct resistance heating capable of heating the specimen at a rate up to 10,000 °C/s. A high cooling rate of 10,000 °C/s can be achieved using an optional quench system. Temperature measurements are done using thermocouples or an infrared pyrometer. The Gleeble 3500 mechanical system has a complete integrated hydraulic servo system capable of exerting maximum tensile/compressive (static) forces of 10 tons. It also has Windows-based software for running the test and analyzing the data.

Ring Test

The ring test consists of compressing a flat ring-shaped specimen to a known reduction (Fig. 10). Changes in the external and internal diameters of the ring are very dependent on the friction at the tool/specimen interface (Ref 5). If the friction were equal to zero, the ring would deform in the same way as a solid disk, with each element flowing radially outward at a rate proportional to its distance from the center. With increasing deformation, the internal diameter of the ring is reduced if friction is large and is increased if friction is low. Thus, the change in the internal diameter represents a simple method for evaluating interface friction.

The ring test can be used to quantify friction in terms of either an interface friction shear factor:

$$m_s (= \sqrt{3}(\tau/\bar{\sigma}))$$

(in which τ denotes the interface shear stress) or a Coulomb coefficient of friction, μ. In either case, a numerical simulation of the ring test is conducted for the specific ring geometry, workpiece/die temperatures, and a range of friction factors/coefficients of friction to generate a series of so-called calibration curves describing the dependence of the percentage decrease in the ring inside diameter (ID) on height reduction. Corresponding measured values of the ID decrease (or increase) for several different height reductions are cross plotted on the set of calibration curves to determine the pertinent friction factor/coefficient of friction; the ID measurements are made at the internal bulge. A typical set of calibration curves for ring tests under isothermal conditions (die and workpiece at the same temperature) and various ring geometries (i.e., ratios of initial ring outside diameter, or OD: ring ID: thickness) are shown in Fig. 11.

Fig. 8 Left to right, specimen before compression and AISI 1018 steel, nickel alloy 718, and Ti-6Al-4V after hot deformation

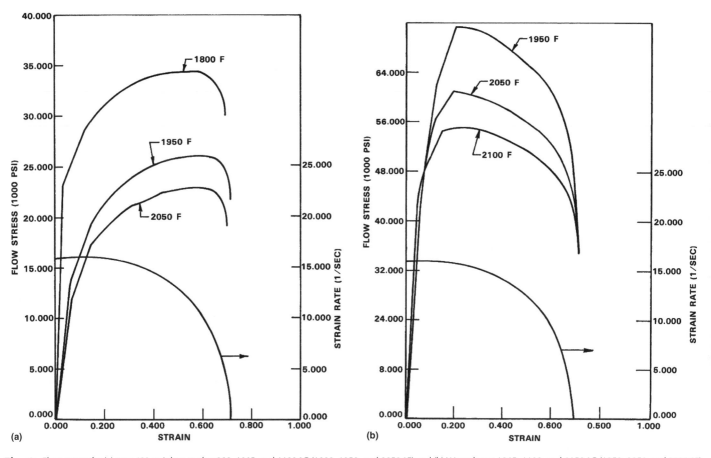

Fig. 9 Flow curves for (a) type 403 stainless steel at 980, 1065, and 1120 °C (1800, 1950, and 2050 °F) and (b) Waspaloy at 1065, 1120, and 1150 °C (1950, 2050, and 2100 °F). The tests were conducted in a mechanical press in which the strain rate was not constant. Source: Ref 6

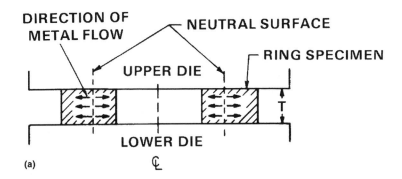

Fig. 10 The ring test. (a) Schematic of metal flow. (b) Example rings upset to various reductions in height

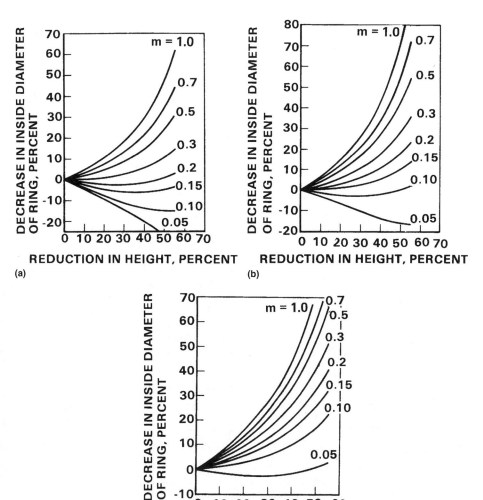

Fig. 11 Calibration curves for isothermal compression of rings having initial outside diameter: inside diameter: thickness ratios of (a) 6:3:2, (b) 6:3:1, or (c) 6:3:0.5. Source: Ref 7

Plane-Strain Compression Test

The plane-strain compression test (Fig. 12) was developed to establish stress-strain curves for the rolling process. According to Watts and Ford (Ref 9), the ratio of the width of the plate, b, to its thickness, h, should be greater than 6 (i.e., $b/h > 6$) to ensure plane-strain compression. The recommended value of b/h should be at least 10 (Ref 10). The ratio between breadth of the tool, a, and the plate thickness, h, should satisfy the inequality $2 < a/h < 4$ (Ref 11).

During the test, one starts with a tool whose breadth is twice the initial thickness of the strip. This tool pair is used to compress the specimen to half of its thickness. Then, the tool is exchanged with a second tool with half the breadth of the first tool, and compression goes on until the sheet is one-fourth its original thickness. A tool with one-half the breadth of the second can be inserted and so on. Thus, a plane-strain compression test can be carried out keeping a/h between the recommended limits (Ref 11).

The equivalent strain in the plane-strain compression test is calculated by using the following relation (Ref 11):

$$\bar{\varepsilon} = (2/\sqrt{3}) \ln(h_1/h) \qquad \text{(Eq 17)}$$

in which h_1 denotes original thickness of the specimen. The uniaxial flow stress (effective stress) is calculated using the expression (Ref 11):

$$\bar{\sigma} = (\sqrt{3}/2)(F/ab) \qquad \text{(Eq 18)}$$

The disadvantages of the plane-strain compression test are (Ref 12):

- The anvils must be kept exactly aligned under each other, because even a small lateral shift will decrease the area under load.
- Along the edges of the dies, there is a stress concentration that may cause crack initiation at a strain for which no cracks would occur under a uniaxial load.

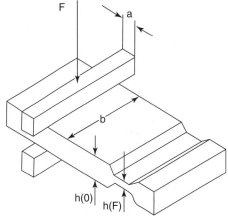

Fig. 12 Schematic illustration of the plane-strain compression test. Source: Ref 12

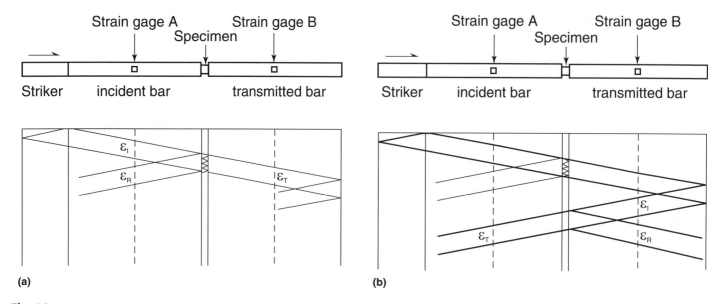

Fig. 13 Schematic diagram of specimen design and stress-wave propagation for (a) compressive and (b) tensile Hopkinson bar tests. Source: Ref 14

Torsion Test

Because complications associated with necking (tension test) and barreling (compression test) are avoided, the torsion test can be used to obtain $\bar{\sigma} - \bar{\varepsilon}$ data at higher strains, often in excess of $\bar{\varepsilon} = 2$. Therefore, it is used when $\bar{\sigma} - \bar{\varepsilon}$ must be known for bulk-forming operations such as extrusion, radial forging, or pilger rolling, in which very large strains are present.

In the torsion test, a hollow tube or solid bar is twisted at a constant rotational speed; the torque, M, and the number of rotations, θ (in radians), are measured.

For a tubular specimen (internal radius = r, wall thickness = t, and gage length = l), the average shear stress, τ, in the gage section is given by:

$$\tau = \frac{M}{2\pi r^2 t} \qquad \text{(Eq 19)}$$

The shear strain, γ, is:

$$\gamma = \frac{r\theta}{l} \qquad \text{(Eq 20)}$$

The corresponding shear strain rate, $\dot{\gamma}$, is:

$$\dot{\gamma} = \frac{r\dot{\theta}}{l} \qquad \text{(Eq 21)}$$

For a solid bar of radius R and gage length l, the shear stress is given by the following relation (Ref 13):

$$\tau = \frac{(3 + n^* + m^*)M}{2\pi R^3} \qquad \text{(Eq 22)}$$

Here, n^* and m^* denote the instantaneous slopes of $\log M$-versus-$\log\theta$ and $\log M$-versus-$\log\dot{\theta}$ plots, respectively. In most cases, $n^* \sim n$, the strain-hardening exponent, and $m^* \sim m$, the

strain-rate sensitivity exponent. The corresponding shear strain and shear strain rate are those pertaining to the outer surface of the specimen:

$$\gamma = \frac{R\theta}{l} \qquad \text{(Eq 23)}$$

$$\dot{\gamma} = \frac{R\dot{\theta}}{l} \qquad \text{(Eq 24)}$$

Assuming that the material can be considered to be isotropic, $\tau - \gamma$ results from the torsion test can be correlated to those from the uniform tension or compression tests using the following relations derived from the von Mises yield criterion:

$$\bar{\sigma} = \sqrt{3}\tau \qquad \text{(Eq 25)}$$

and

$$\bar{\varepsilon} = \gamma / \sqrt{3} \qquad \text{(Eq 26)}$$

Split-Hopkinson Bar Test

Forming processes such as hot or cold rolling that are carried out at high rates of deformation necessitate flow stress data at high strain rates. For this purpose, the split-Hopkinson pressure bar is used for compression tests (as well as tension or torsion tests) at high strain rates at room or elevated temperature.

A schematic illustration of the test apparatus is shown in Fig. 13. The apparatus contains a striker, an incident bar, and a transmitted bar. Figure 13(a) shows the general elastic wave propagation in a compressive test. In compression, when the striker bar impacts the incident bar, a compressive stress pulse is generated and travels through the incident bar until it hits the specimen. At the bar/specimen interface,

part of the incident stress pulse is reflected due to material (impedance) mismatch. The transmitted pulse emitted from the specimen reaches the free end of the transmitted bar and is reflected there as a tension pulse. The tensile stress pulse travels back through the transmitted bar and, upon reaching the specimen/transmitted bar interface, results in separation, thus ending the test. The stress, strain, and strain rate in the specimen are calculated in terms of strains recorded from the two strain gages, A and B.

In the tension version of the test (Fig. 13b), the specimen is attached to incident and transmitted bars. The compressive stress pulse generated in the incident bar travels along the specimen until it reaches the end of the transmitted bar. After reflection, the tensile stress pulse propagates through the specimen to the incident bar. Strains recorded by strain gages A and B are measured (Ref 14). Data collection and reduction are discussed in "Classic Split-Hopkinson Pressure Bar Testing" in *Mechanical Testing and Evaluation,* Volume 8 of *ASM Handbook,* 2000.

Indentation Tests

Indentation tests are attractive for determining flow-stress data under the following conditions (Ref 15):

- The sample size is limited because of the process technology involved or when the number of materials to be tested is large.
- Testing of coated components. A number of engineering components are coated with different materials to improve their durability. Using indentation tests, flow-stress behavior of coatings can be estimated by adjusting the indentation load.

At a given strain, the flow stress, $\bar{\sigma}$, and hardness value, H, are given by the relation (Ref 15):

$$H(\varepsilon) = C_f \bar{\sigma}(\bar{\varepsilon}) \qquad \text{(Eq 27)}$$

in which C_f is the constraint factor. The method for determining the flow stress-strain relationship from static or dynamic hardness tests using the constraint factor approach is given in Ref 15. Flow curves can also be obtained from hardness measurements by continuously measuring the force and depth of indentation. However, this requires an extremely high degree of measurement accuracy (Ref 15).

Effect of Deformation Heating on Flow Stress

The plastic work imposed during metalworking is dissipated by the formation of metallurgical defects (e.g., dislocations) and the generation of heat. The former usually accounts for 5 to 10% of the work, while deformation heating accounts for 90 to 95% of the work. Depending on the particular size of the workpiece, a greater or lesser amount of the deformation heat dissipates into the dies. For the sample sizes typically used to determine flow stress (e.g., ~12 mm diameter by 18 mm height cylinders for isothermal compression tests), a measurable fraction of the heat is retained in the workpiece for strain rates of the order of 0.01 s^{-1} or greater. For these strain rates, it becomes important to correct flow-stress data for the temperature rise associated with deformation heating.

The temperature increase, ΔT, can be estimated from the following relation:

$$\Delta T \sim \frac{0.90\eta \int \bar{\sigma} d\bar{\varepsilon}}{\rho_d c} \qquad \text{(Eq 28)}$$

Here, η denotes the adiabaticity factor (fraction of the deformation heat retained in the sample), c is the specific heat, and ρ_d is the density. The integral represents the area under the "uncorrected" true stress-strain curve from a strain of zero to the strain for which the temperature rise is to be calculated. The adiabaticity factor for hot compression testing of small samples is equal to ~0.5, ~0.9, or ~1 for strain rates of 0.1, 1, and 10 s^{-1} (Ref 16, 17).

The procedure for correcting flow curves for deformation heating comprises the following steps:

1. Calculate the temperature rise (Eq 28) for a number of specific strain levels for each of several flow curves measured at different nominal test temperatures and a given strain rate.
2. Construct plots of measured flow stress versus instantaneous temperature (equal to nominal test temperature + ΔT) for each of the given strains from step 1.
3. Determine the value of $d\bar{\sigma}/dT$ for a series of strains and nominal test temperatures from the stress-versus-temperature plots.

4. Using these values of $d\bar{\sigma}/dT$ and the calculated temperature rise, estimate what the isothermal flow stress would have been in the absence of deformation heating at a series of strains for each measured flow curve.

Fitting of Flow-Stress Data

Various analytical equations have been used to fit the flow-stress data obtained from tension, compression, and torsion tests. The specific form of the equation usually depends on the test temperature (i.e., cold working versus hot working temperatures) and, at hot working temperatures, on the strain rate. The definition of the temperature for cold versus hot working is not precise. However, the transition usually occurs at approximately one-half of the melting point on an absolute temperature scale. From a metallurgical standpoint, hot working is characterized by a steady-state flow stress beginning at modest strains (order of 0.25) due to dynamic recovery or by the occurrence of discontinuous dynamic recrystallization.

Cold Working Temperatures

A typical $\bar{\sigma}$-$\bar{\varepsilon}$ curve obtained at cold working temperatures is shown in Fig. 6(b). Strain hardening is pronounced, but the strain-rate dependence is usually minimal. The flow stress, $\bar{\sigma}$, increases with increasing $\bar{\varepsilon}$ and, for some materials, may eventually reach a saturation stress at very large true strains (usually $\bar{\varepsilon} > 1$). Because of the parabolic shape of the flow curve at cold working temperatures, true stress-strain data can often be fit by a power-law type of relation mentioned previously in Eq 8:

$$\bar{\sigma} = K\bar{\varepsilon}^n$$

in which K and n are material constants known as the strength coefficient and strain-hardening exponent, respectively. A log-log plot of experimental data can be used to determine whether Eq 8 provides a good fit. If so, the data fall on a straight line of slope equal to n, and the strength coefficient K is equal to the flow stress at a true strain of unity. Often at small strains, an experimentally determined curve may depart from linearity on the log-log plot. In this case, other values of n and K may be specified for different ranges of true strain. Typical values of n and K describing the flow-stress behavior of various metals at cold working temperatures are given in Tables 1 to 3.

Other analytical expressions have been used to fit the true stress-strain curves for metals at cold working temperatures. These include the following (in which a, b, and c are material/fitting constants that differ in each equation):

$$\text{Ludwik equation: } \bar{\sigma} = a + b(\bar{\varepsilon}) \qquad \text{(Eq 29)}$$

$$\text{Voce equation: } \bar{\sigma} = a + [b - a][1 - \exp(-c\bar{\varepsilon})] \qquad \text{(Eq 30)}$$

$$\text{Swift equation: } \bar{\sigma} = c(a + \bar{\varepsilon})^n \qquad \text{(Eq 31)}$$

The Ludwik equation approximates the stress-strain curves for annealed materials but tends to underestimate the stress at low strains (<0.2) and to overestimate the stress for high strains. For heavily prestrained materials, $c \sim 1$. The Voce and Swift equations tend to be used less frequently, partly because of their greater complexity.

Hot Working Temperatures

At hot working temperatures, $\bar{\sigma}$ increases with increasing $\bar{\varepsilon}$ and with decreasing temperature, T. Irrespective of strain rate, the flow curve generally exhibits a short strain-hardening transient followed by a peak stress. In materials whose principal dynamic restorative mechanism is dynamic recovery, subsequent deformation is characterized by a steady-state flow stress equal to the peak stress. In materials that undergo discontinuous dynamic recrystallization (characterized by the nucleation and growth of new, strain-free grains during deformation), flow softening occurs following the peak stress until a steady-state microstructure and flow stress are achieved. These phenomena are described more extensively in the subsequent section, "Metallurgical Considerations at Hot Working Temperatures."

In many engineering hot working applications, a simple power-law equation is used to describe the flow stress as a function of strain rate:

$$\bar{\sigma} = C\dot{\varepsilon}^m \qquad \text{(Eq 32)}$$

Here, C is a constant (or a function of strain sometimes needed to describe the low-strain, strain-hardening dependence), and m is the strain-rate sensitivity exponent. A log-log plot of $\bar{\sigma} - \dot{\varepsilon}$ data can be used to determine whether Eq 32 provides a good fit. If so, the data fall on a straight line of slope equal to m. Some typical C and m values for various metals are given in Tables 4 to 8 (Ref 18). It has also been found that the dependence of the m value on homologous temperature (the ratio of the test temperature to the melting point on the absolute temperature scale) is similar for many metallic materials (Fig. 14) (Ref 19).

Metallurgical Considerations at Hot Working Temperatures

The flow stress of a metallic material is closely coupled to its initial microstructure and how its microstructure evolves as a function of strain, strain rate, and temperature. Thus, the measured flow stress is actually the average macroscopic deformation resistance associated with a number of micromechanical processes, such as the glide (slip) and climb of dislocations, dislocation annihilation, slip transfer across grain boundaries, dynamic recrystallization, dynamic grain growth/coarsening, and so on. For this reason, a particular

Table 1 Summary of K and n values describing the flow stress-strain relation, $\bar{\sigma} = K(\bar{\varepsilon})^n$, for various steels

Steel	C	Mn	P	S	Si	N	Al	V	Ni	Cr	Mo	W	Material history(b)	°C	°F	Strain rate, 1/s	Strain range	K, 10³ psi	n
Armco iron	0.02	0.03	0.021	0.010	Tr	...	...	...	...	...	...	...	A	20	68	(c)	0.1–0.7	88.2	0.25
1006	0.06	0.29	0.02	0.042	Tr	0.004	...	...	...	...	...	...	A	20	68	(c)	0.1–0.7	89.6	0.31
1008	0.08	0.36	0.023	0.031	0.06	0.007	...	...	...	...	...	...	A	20	68	(c)	0.1–0.7	95.3	0.24
	0.07	0.28	...	...	0.27	...	...	...	...	...	...	...	A	20	68	(c)	0.1–0.7	95.3	0.17
1010	0.13	0.31	0.010	0.022	0.23	0.004	...	...	...	...	...	...	A	20	68	(c)	0.1–0.7	103.8	0.22
1015	0.15	0.40	0.01	0.016	Tr	...	...	...	...	...	...	...	F,A	0	32	30	0.2–0.7	91.4	0.116
1015	0.15	0.40	0.01	0.016	Tr	...	...	...	...	...	...	...	F,A	200	390	30	0.2–0.6	73.7	0.140
1015(d)	0.15	0.40	0.045	0.045	0.25	...	...	...	...	...	...	...	A	20	68	1.6	...	113.8	0.10
1015(d)	0.15	0.40	0.045	0.045	0.25	...	...	...	...	...	...	...	A	300	572	1.6	...	115.2	0.11
1020	0.22	0.44	0.017	0.043	Tr	0.005	...	...	...	...	...	...	A	20	68	(c)	0.1–0.7	108.1	0.20
1035	0.36	0.69	0.025	0.032	0.27	0.004	...	...	...	...	...	...	A	20	68	(c)	0.1–0.7	130.8	0.17
													A	20	68	1.6	...	139.4	0.11
													A	300	572	1.6	...	122.3	0.16
1045(d)	0.45	0.65	0.045	0.045	0.25	...	...	...	...	...	...	...	A	20	68	1.6	...	147.9	0.11
													A	20	68	1.5	...	137.9	0.14
													A	300	572	1.6	...	126.6	0.15
1050(e)	0.51	0.55	0.016	0.041	0.28	0.0062	0.03	...	...	...	...	...	A	20	68	(c)	0.1–0.7	140.8	0.16
1060													A	20	68	1.6	...	163.5	0.09
													A	20	68	1.5	...	157.8	0.12
2317(e)	0.19	0.55	0.057	0.023	0.26	0.016	...	...	...	...	...	...	A	20	68	(c)	0.2–1.0	111.2	0.170
5115	0.14	0.53	0.028	0.027	0.37	...	...	...	...	0.71	...	...	A	20	68	(c)	0.1–0.7	115.2	0.18
													A	20	68	1.6	...	123.7	0.09
													A	300	572	1.6	...	102.4	0.15
5120(e)	0.18	1.13	0.019	0.023	0.27	...	...	...	...	0.86	...	...	A	20	68	(c)	0.1–0.7	126.6	0.15
													A	20	68	1.6	...	116.6	0.09
													A	300	572	1.6	...	98.1	0.16
5140	0.41	0.67	0.04	0.019	0.35	...	...	...	...	1.07	...	...	A	20	68	(c)	0.1–0.7	125.1	0.15
													A	20	68	1.6	...	133.7	0.09
													A	300	572	1.6	...	112.3	0.12
D2 tool steel(e)	1.60	0.45	...	...	0.24	...	...	0.46	...	11.70	0.75	0.59	A	20	68	(c)	0.2–1.0	191.0	0.157
L6 tool steel	0.56	...	...	...	...	...	...	0.14	1.60	1.21	0.47	...	A	20	68	(c)	0.2–1.0	170.2	0.128
W1-1.0C special	1.05	0.21	...	...	0.16	...	...	...	...	...	...	...	A	20	68	(c)	0.2–1.0	135.6	0.179
302 SS	0.08	1.06	0.037	0.005	0.49	...	...	...	9.16	18.37	...	...	HR,A	0	32	10	0.25–0.7	186.7	0.295
													HR,A	200	390	30	0.25–0.7	120.8	0.278
													HR,A	400	750	30	0.25–0.7	92.7	0.279
302 SS	0.053	1.08	0.027	0.015	0.27	...	...	...	10.2	17.8	...	...	A	20	68	(c)	0.1–0.7	210.5	0.6
304 SS(e)	0.030	1.05	0.023	0.014	0.47	...	...	...	10.6	18.7	...	...	A	20	68	(c)	0.1–0.7	210.5	0.6
316 SS	0.055	0.92	0.030	0.008	0.49	...	...	...	12.9	18.1	2.05	...	A	20	68	(c)	0.1–0.7	182.0	0.59
410 SS	0.093	0.31	0.026	0.012	0.33	...	...	...	...	13.8	...	...	A	20	68	(c)	0.1–0.7	119.4	0.2
													A	20	68	1.6	...	137.9	0.09
431 SS	0.23	0.38	0.020	0.006	0.42	...	...	...	1.72	16.32	...	...	A	20	68	(c)	0.1–0.7	189.1	0.17

(a) Tr = trace. (b) A = annealed, F = forged, HR = hot rolled. (c) Low-speed testing machine; no specific rate given. (d) Composition given is nominal (analysis not given in original reference). (e) Approximate composition

Table 2 Summary of K and n values describing the flow stress-strain relation, $\bar{\sigma} = K(\bar{\varepsilon})^n$, for various aluminum alloys

Alloy	Al	Cu	Si	Fe	Mn	Mg	Zn	Ti	Cr	Pb	Material history(a)	°C	°F	Strain rate, 1/s	Strain range	K, 10³ psi	n
1100	99.0	0.10	0.15	0.50	0.01	0.01	...	...	...	...	CD,A	0	32	10	0.25–0.7	25.2	0.304
1100	bal	0.01	0.10	0.16	0.01	0.01	0.03	...	...	...	A	20	68	(b)	0.2–1.0	17.3	0.297
EC	99.5	0.01	0.092	0.23	0.026	0.033	0.01	...	...	...	A(c)	20	68	4	0.2–0.8	22.4	0.204
2017	bal	4.04	0.70	0.45	0.55	0.76	0.22	...	...	0.06	A	20	68	(b)	0.2–1.0	45.2	0.180
2024(d)	bal	4.48	0.60	0.46	0.87	1.12	0.20	...	...	0.056	A	20	68	(b)	0.2–1.0	56.1	0.154
5052	bal	0.068	0.10	0.19	0.04	2.74	0.01	0.003	...	...	A(e)	20	68	4	0.2–0.8	29.4	0.134
5052(d)	bal	0.09	0.13	0.16	0.23	2.50	0.05	...	...	...	A	20	68	(b)	0.2–1.0	55.6	0.189
5056	bal	0.036	0.15	0.22	0.04	4.83	0.01	...	0.14	...	A(e)	20	68	4	0.2–0.7	57.0	0.130
5083	bal	0.01	0.10	0.16	0.77	4.41	0.01	0.002	0.13	...	A	20	68	4	0.2–0.8	65.2	0.131
5454	bal	0.065	0.12	0.18	0.81	2.45	<0.01	0.002	...	...	A(e)	20	68	4	0.2–0.8	49.9	0.137
6062	bal	0.03	0.63	0.20	0.63	0.68	0.065	0.08	...	...	...	20	68	(b)	0.2–1.0	29.7	0.122

(a) CD = cold drawn, A = annealed. (b) Low-speed testing machine; no specific rate given. (c) Annealed for 4 h at 400 °C (752 °F). (d) Approximate composition. (e) Annealed for 4 h at 420 °C (788 °F)

set of flow-stress measurements is specific to the initial material and material condition for which it has been obtained.

Some of the important metallurgical factors that aid in the interpretation of flow curves at hot working temperatures are summarized in this section. These include the influence of dynamic recovery and dynamic recrystallization on flow response and the effect of microstructural features (e.g., grain size, crystallographic texture, and second phases) on plastic flow.

Conventional Metalworking Strain Rates

Deformation Mechanisms at Conventional Metalworking Strain Rates. The key mechanisms that control microstructure evolution and plastic flow during hot (and, to some extent, cold) working at conventional metalworking strain rates ($\dot{\varepsilon} \geq 0.1 \, \mathrm{s}^{-1}$) are dynamic recovery and dynamic recrystallization (Ref 20, 21). As the terms imply, dynamic recovery and recrystallization occur during hot working. As metals are worked, defects are generated in the crystal lattice. The most important defects are line defects known as dislocations. As deformation increases, the deformation resistance increases due to increasing dislocation content. However, the dislocation density does

Table 3 Summary of K and n values describing the flow stress-strain relation, $\bar{\sigma} = K(\bar{\varepsilon})^n$, for various copper alloys

Alloy(a)	Composition(b), %									Temperature		Material history(c)	Strain rate, 1/s	Strain range	K, 10^3 psi	n
	Cu	Si	Fe	Sb	Sn	Zn	S	Pb	Ni	°C	°F					
CDA110	99.94	...	0.0025	0.0003	...	...	0.0012	0.0012	0.001	18	64	HR,A	2.5	0.25–0.7	65.5	0.328
CDA110	...	...	...	...	...	...	...	...	...	20	68	F	(d)	0.2–1.0	54.0	0.275
CDA230	84.3	...	...	...	...	15.7	...	...	...	20	68	A	(d)	0.2–1.0	76.7	0.373
CDA260	70.8	...	...	...	...	29.2	...	...	...	20	68	A	(d)	0.2–1.0	98.1	0.412
CDA260	70.05	...	Tr	...	Tr	...	...	...	...	200	390	HR,A	...	0.25–0.7	71.7	0.414
CDA272	63.3	...	...	...	...	36.7	...	...	...	20	68	A	(d)	0.2–1.0	103.9	0.394
CDA377	58.6	...	Tr	...	...	39.6	...	1.7	...	20	68	A	(d)	0.2–1.0	115.3	0.334
CDA521(e)	91.0	...	...	...	9.0	...	...	...	...	20	68	F	(d)	0.2–1.0	130.8	0.486
CDA647	97.0	0.5	...	...	...	...	...	...	2.0	20	68	F	(d)	0.2–1.0	67.2	0.282
CDA757	65.1	...	...	...	...	22.4	...	<0.05	12.4	20	68	A	(d)	0.2–1.0	101.8	0.401
CDA794	61.7	...	Tr	...	...	20.6	...	Tr	17.5	20	68	A	(d)	0.2–1.0	107.0	0.336

(a) CDA = Copper Development Association. (b) Tr = trace. (c) HR = hot rolled, A = annealed, F = forged. (d) Low-speed testing machine; no specific rate given. (e) Approximate composition

Table 4 Summary of C (ksi) and m values describing the flow stress-strain rate relation, $\bar{\sigma} = C(\dot{\bar{\varepsilon}})^m$, for steels at various temperatures (C is in 10^3 psi)

Steel	Material history	Strain rate range, 1/s	Strain	C	m	C	m	C	m	C	m	C	m
			Test temperature, °C (°F):	600 (1110)		800 (1470)		1000 (1830)		1200 (2190)			
1015 0.15 C, trace Si, 0.40 Mn, 0.01 P, 0.016 S	Forged, annealed	0.2–30	0.2	36.8	0.112	...	...	...	...	...	...		
			0.25	...	...	19.9	0.105	17.0	0.045	7.2	0.137		
			0.4	40.6	0.131	...	...	...	...	...	...		
			0.5	...	...	21.5	0.104	18.8	0.058	6.8	0.169		
			0.6	40.0	0.121	...	...	...	...	...	...		
			0.7	39.5	0.114	21.1	0.109	18.3	0.068	5.7	0.181		
			Test temperature, °C (°F):	900 (1650)		1000 (1830)		1100 (2010)		1200 (2190)			
1016 0.15 C, 0.12 Si, 0.68 Mn, 0.034 S, 0.025 P	Hot rolled, annealed	1.5–100	0.10	16.6	0.092	13.4	0.100	9.9	0.124	7.5	0.143		
			0.30	22.7	0.082	18.2	0.085	13.3	0.115	9.4	0.153		
			0.50	23.7	0.087	18.2	0.105	12.7	0.146	8.5	0.191		
			0.70	23.1	0.099	16.1	0.147	11.9	0.166	7.5	0.218		
1016 0.15 C, 0.12 Si, 0.68 Mn, 0.034 S, 0.025 P	Hot rolled, annealed		0.05	11.8	0.133	10.7	0.124	9.0	0.117	6.4	0.150		
			0.1	16.5	0.099	13.7	0.099	9.7	0.130	7.1	0.157		
			0.2	20.8	0.082	16.5	0.090	12.1	0.119	9.1	0.140		
			0.3	22.8	0.085	18.2	0.088	13.4	0.109	9.5	0.148		
			0.4	23.0	0.084	18.2	0.098	12.9	0.126	9.1	0.164		
			0.5	23.9	0.084	18.1	0.109	12.5	0.141	8.2	0.189		
			0.6	23.3	0.097	16.9	0.127	12.1	0.156	7.8	0.205		
			0.7	22.8	0.104	17.1	0.127	12.4	0.151	8.1	0.196		
			Test temperature, °C (°F):	870 (1600)		980 (1800)		1090 (2000)		1205 (2200)		1180 (2150)	
1018	...	...	...	25.2	0.07	15.8	0.152	11.0	0.192	9.2	0.20	...	...
1025 0.25 C, 0.08 Si, 0.45 Mn, 0.012 P, 0.025 S	Forged, annealed	3.5–30	0.25	...	...	33.7	0.004	16.2	0.075	9.3	0.077	...	...
			0.50	...	...	41.4	−0.032	17.2	0.080	9.6	0.094	...	...
			0.70	...	...	41.6	−0.032	17.5	0.082	8.8	0.105	...	...
1043	Hot rolled, as received	0.1–100	0.3/0.5/0.7	...	...	...	...	...	...	...	...	10.8	0.21
			Test temperature, °C (°F):	900 (1650)		1000 (1830)		1100 (2010)		1200 (2190)			
1045(a) 0.46 C, 0.29 Si, 0.73 Mn, 0.018 P, 0.021 S, 0.08 Cr, 0.01 Mo, 0.04 Ni			0.05	25.4	0.080	15.1	0.089	11.2	0.100	8.0	0.175		
			0.10	28.9	0.082	18.8	0.103	13.5	0.125	9.4	0.168		
			0.20	33.3	0.086	22.8	0.108	15.4	0.128	10.5	0.167		
			0.30	35.4	0.083	24.6	0.110	15.8	0.162	10.8	0.180		
			0.40	35.4	0.105	24.7	0.134	15.5	0.173	10.8	0.188		
			Test temperature, °C (°F):	600 (1110)		800 (1470)		1000 (1830)		1200 (2190)			
1055 0.55 C, 0.24 Si, 0.73 Mn, 0.014 P, 0.016 S	Forged, annealed	3.5–30		...	...	29.4	0.087	14.9	0.126	7.4	0.145		
				...	...	32.5	0.076	13.3	0.191	7.4	0.178		
				...	...	32.7	0.066	11.5	0.237	6.4	0.229		
			Test temperature, °C (°F):	900 (1650)		1000 (1830)		1100 (2010)		1200 (2190)			
1060(a) 0.56 C, 0.26 Si, 0.28 Mn, 0.014 S, 0.013 P, 0.12 Cr, 0.09 Ni	Hot rolled, annealed	1.5–100	0.10	18.5	0.127	13.3	0.143	10.1	0.147	7.4	0.172		
			0.30	23.3	0.114	16.9	0.123	12.6	0.135	8.9	0.158		
			0.50	23.3	0.118	16.4	0.139	12.0	0.158	8.6	0.180		
			0.70	21.3	0.132	14.9	0.161	10.4	0.193	7.8	0.207		
1060(a) 0.56 C, 0.26 Si, 0.28 Mn, 0.014 S, 0.013 P, 0.12 Cr, 0.09 Ni			0.05	16.2	0.128	10.8	0.168	8.7	0.161	6.5	0.190		
			0.10	18.3	0.127	13.2	0.145	10.1	0.149	7.5	0.165		
			0.20	21.8	0.119	16.1	0.125	12.1	0.126	8.5	0.157		
			0.30	23.3	0.114	17.1	0.125	12.8	0.132	8.8	0.164		
			0.40	23.7	0.112	16.8	0.128	12.5	0.146	8.8	0.171		
			0.50	23.6	0.110	16.6	0.133	12.7	0.143	8.7	0.176		
			0.60	22.8	0.129	17.1	0.127	11.7	0.169	8.4	0.189		
			0.70	21.3	0.129	16.2	0.138	10.7	0.181	7.8	0.204		
1095(a) 100 C, 0.19 Si, 0.17 Mn, 0.027 S, 0.023 P, 0.10 Cr, 0.09 Ni	Hot rolled, annealed	1.5–100	0.10	18.3	0.146	13.9	0.143	9.8	0.159	7.1	0.184		
			0.30	21.9	0.133	16.6	0.132	11.7	0.147	8.0	0.183		
			0.50	21.8	0.130	15.7	0.151	10.6	0.176	7.3	0.209		
			0.70	21.0	0.128	13.6	0.179	9.7	0.191	6.5	0.232		

(continued)

(a) Approximate composition

Table 4 (continued)

Steel	Material history	Strain rate range, 1/s	Strain	C	m	C	m	C	m	C	m	C	m
		Test temperature, °C (°F):		930 (1705)		1000 (1830)		1060 (1940)		1135 (2075)		1200 (2190)	
1115	Hot rolled, as received	4.4–23.1	0.105	16.3	0.088	13.0	0.108	10.9	0.112	9.1	0.123	7.6	0.116
0.17 C, 0.153 Si, 0.62 Mn, 0.054 S, 0.032 P			0.223	19.4	0.084	15.6	0.100	12.9	0.107	10.5	0.129	8.6	0.122
			0.338	20.4	0.094	17.3	0.090	14.0	0.117	11.2	0.138	8.8	0.141
			0.512	20.9	0.099	18.0	0.093	14.4	0.127	11.0	0.159	8.3	0.173
			0.695	20.9	0.105	16.9	0.122	13.6	0.150	9.9	0.198	7.6	0.196
		Test temperature, °C (°F):		900 (1650)		1000 (1830)		1100 (2010)		1200 (2190)			
Alloy steel	...	...	0.05	16.6	0.102	12.2	0.125	9.4	0.150	7.4	0.161		
0.35 C, 0.27 Si, 1.49 Mn, 0.041 S, 0.037 P, 0.03 Cr, 0.11 Ni, 0.28 Mo			0.10	19.9	0.091	14.8	0.111	11.5	0.121	8.1	0.149		
			0.20	23.0	0.094	17.6	0.094	13.5	0.100	9.4	0.139		
			0.30	24.9	0.092	19.1	0.093	14.4	0.105	10.2	0.130		
			0.40	26.0	0.088	19.6	0.095	14.5	0.112	10.4	0.139		
			0.50	25.9	0.091	19.6	0.100	14.4	0.112	10.1	0.147		
			0.60	25.9	0.094	19.5	0.105	14.2	0.122	9.7	0.159		
			0.70	25.5	0.099	19.2	0.107	13.9	0.126	9.2	0.165		
4337(a)	Hot rolled, annealed	1.5–100	0.10	22.1	0.080	16.6	0.109	12.1	0.115	8.2	0.165		
0.35 C, 0.27 Si, 0.66 Mn, 0.023 S, 0.029 P, 0.59 Cr, 2.45 Ni, 0.59 Mo			0.30	28.1	0.077	20.8	0.098	15.0	0.111	10.7	0.138		
			0.50	29.2	0.075	21.8	0.096	15.7	0.112	11.3	0.133		
			0.70	28.1	0.080	21.3	0.102	15.5	0.122	11.3	0.135		
926(a)	Hot rolled, annealed	1.5–100	0.10	22.9	0.109	17.1	0.106	11.8	0.152	8.6	0.168		
0.61 C, 1.58 Si, 0.94 Mn, 0.038 S, 0.035 P, 0.12 Cr, 0.27 Ni, 0.06 Mo			0.30	28.2	0.101	20.4	0.106	14.3	0.140	10.1	0.162		
			0.50	27.8	0.104	20.0	0.120	13.8	0.154	9.1	0.193		
			0.70	25.8	0.112	18.2	0.146	11.8	0.179	7.5	0.235		
50100(a)	...	...	0.05	16.1	0.155	12.4	0.155	8.2	0.175	6.3	0.199		
1.00 C, 0.19 Si, 0.17 Mn, 0.027 S, 0.023 P, 0.10 Cr, 0.09 Ni			0.10	18.6	0.145	14.1	0.142	9.5	0.164	6.8	0.191		
			0.20	20.9	0.135	15.9	0.131	11.4	0.141	8.1	0.167		
			0.30	21.8	0.135	16.6	0.134	11.7	0.142	8.0	0.174		
			0.40	22.0	0.134	16.8	0.134	11.2	0.155	8.4	0.164		
			0.50	21.5	0.131	15.6	0.150	11.1	0.158	7.4	0.199		
			0.60	21.3	0.132	14.6	0.163	10.0	0.184	7.0	0.212		
			0.70	20.9	0.131	13.5	0.176	9.7	0.183	6.7	0.220		
52100	Hot rolled, annealed	1.5–100	0.10	20.9	0.123	14.3	0.146	9.5	0.169	6.7	0.203		
1.06 C, 0.22 Si, 0.46 Mn, 0.019 S, 0.031 P, 1.41 Cr, 0.17 Ni			0.30	25.5	0.107	17.7	0.127	12.0	0.143	8.3	0.171		
			0.50	25.9	0.107	17.7	0.129	12.3	0.143	8.3	0.178		
			0.70	23.3	0.131	16.8	0.134	12.0	0.148	7.7	0.192		
		Test temperature, °C (°F):		900 (1650)		1000 (1830)		1100 (2010)		1200 (2190)			
Mn-Si steel	...	...	0.05	19.2	0.117	14.8	0.119	9.7	0.172	7.5	0.181		
0.61 C, 1.58 Si, 0.94 Mn, 0.038 S, 0.035 P, 0.12 Cr, 0.27 Ni, 0.06 Mo			0.10	22.6	0.112	17.1	0.108	11.8	0.151	8.7	0.166		
			0.20	25.7	0.108	19.5	0.101	13.5	0.139	9.7	0.160		
			0.30	27.6	0.108	20.5	0.109	14.8	0.126	10.0	0.161		
			0.40	27.6	0.114	20.2	0.114	14.4	0.141	9.5	0.179		
			0.50	27.2	0.113	19.8	0.125	14.1	0.144	9.1	0.188		
			0.60	26.0	0.121	18.8	0.137	12.8	0.162	8.2	0.209		
			0.70	24.7	0.130	17.8	0.152	11.9	0.178	7.5	0.228		
Cr-Si steel	...	...	0.05	19.9	0.118	23.9	0.104	15.1	0.167	10.0	0.206		
0.47 C, 3.74 Si, 0.58 Mn, 8.20 Cr 20 Ni			0.10	19.9	0.136	25.6	0.120	16.8	0.162	11.1	0.189		
			0.20	19.9	0.143	27.6	0.121	18.5	0.153	11.9	0.184		
			0.30	19.9	0.144	28.4	0.119	19.1	0.148	12.1	0.182		
			0.40	19.3	0.150	28.2	0.125	18.9	0.150	12.1	0.178		
			0.50	18.5	0.155	26.6	0.132	18.5	0.155	11.8	0.182		
			0.60	17.5	0.160	25.2	0.142	17.5	0.160	11.5	0.182		
			0.70	16.1	0.163	23.3	0.158	16.1	0.162	10.7	0.199		
D3(a)	Hot rolled, annealed	1.5–100	0.10	39.2	0.087	29.0	0.108	21.0	0.123	14.6	0.121		
2.23 C, 0.43 Si, 0.37 Mn, 13.10 Cr, 0.33 Ni			0.30	43.7	0.087	30.4	0.114	21.0	0.139	13.9	0.130		
			0.50	39.7	0.101	27.1	0.125	18.4	0.155	12.2	0.124		
			0.70	33.3	0.131	22.5	0.145	15.3	0.168	10.7	0.108		
		Test temperature, °C (°F):		700 (1290)		820 (1510)		900 (1650)		1000 (1830)			
H-13	...	290–906	0.1	19.1	0.232	10.2	0.305	6.0	0.373	4.8	0.374		
0.39 C, 1.02 Si, 0.60 Mn, 0.016 P, 0.020 S, 5.29 Cr, 0.04 Ni, 1.35 Mo, 0.027 N, 0.83 V			0.2	30.1	0.179	13.7	0.275	8.2	0.341	9.0	0.295		
			0.3	31.0	0.179	15.1	0.265	10.8	0.305	11.6	0.267		
			0.4	25.9	0.204	12.3	0.295	12.5	0.287	11.8	0.269		
		Test temperature, °C (°F):		900 (1650)		1000 (1830)		1100 (2010)		1200 (2190)			
H-26(a)	Hot rolled, annealed	1.5–100	0.10	46.7	0.058	37.4	0.072	26.2	0.106	18.7	0.125		
0.80 C, 0.28 Si, 0.32 Mn, 4.30 Cr, 0.18 Ni, 0.55 Mo, 18.40 W, 1.54 V			0.30	49.6	0.075	38.1	0.087	26.0	0.121	18.3	0.140		
			0.50	44.6	0.096	33.7	0.102	23.6	0.131	16.2	0.151		
			0.70	39.1	0.115	27.9	0.124	20.1	0.149	13.8	0.162		
		Test temperature, °C (°F):		600 (1110)		800 (1470)		1000 (1830)		1200 (2190)			
301 SS(a)	Hot rolled, annealed	0.8–100	0.25	...	...	40.5	0.051	16.3	0.117	7.6	0.161		
0.08 C, 0.93 Si, 1.10 Mn, 0.009 P, 0.014 S, 16.99 Cr, 6.96 Ni, 0.31 Mo, 0.93 Al. 0.02 N, 0.063 Se			0.50	...	...	39.3	0.062	17.8	0.108	7.6	0.177		
			0.70	...	...	37.8	0.069	17.4	0.102	6.6	0.192		

(continued)

(a) Approximate composition

Table 4 (continued)

Steel	Material history	Strain rate range, 1/s	Strain	C	m	C	m	C	m	C	m	C	m
302 SS 0.07 C, 0.71 Si, 1.07 Mn, 0.03 P, 0.005 S, 18.34 Cr, 9.56 Ni	Hot rolled, annealed	310–460	0.25	26.5	0.147	25.1	0.129	11.0	0.206	4.6	0.281		
			0.40	31.3	0.153	30.0	0.121	13.5	0.188	4.7	0.284		
			0.60	17.5	0.270	45.4	0.063	16.8	0.161	4.1	0.310		
302 SS 0.08 C, 0.49 Si, 1.06 Mn, 0.037 P, 0.005 S, 18.37 Cr, 9.16 Ni	Hot rolled, annealed	0.2–30	0.25	52.2	0.031	36.6	0.042	23.1	0.040	12.8	0.082		
			0.40	58.9	0.022	40.4	0.032	24.7	0.050	13.6	0.083		
			0.60	63.2	0.020	41.9	0.030	24.9	0.053	13.5	0.091		
			0.70	64.0	0.023	42.0	0.031	24.7	0.052	13.4	0.096		
	Test temperature, °C (°F):			900 (1650)		1000 (1830)		1100 (2010)		1200 (2190)			
302 SS 0.07 C, 0.43 Si, 0.48 Mn, 18.60 Cr, 7.70 Ni	...	1.5–100	0.05	24.6	0.023	16.8	0.079	13.7	0.093	9.7	0.139		
			0.10	28.4	0.026	21.2	0.068	15.6	0.091	11.1	0.127		
			0.20	33.6	0.031	25.2	0.067	18.1	0.089	12.5	0.120		
			0.30	35.3	0.042	26.3	0.074	19.5	0.089	13.5	0.115		
			0.40	35.6	0.055	26.9	0.084	19.9	0.094	14.2	0.110		
			0.50	35.6	0.060	27.0	0.093	19.6	0.098	14.2	0.115		
			0.60	34.1	0.068	26.4	0.092	19.3	0.102	13.8	0.118		
			0.70	33.6	0.072	25.7	0.102	18.9	0.108	13.9	0.120		
	Test temperature, °C (°F):			600 (1110)		800 (1470)		1000 (1830)		1200 (2190)		900 (1650)	
309 SS 0.13 C, 0.42 Si, 1.30 Mn, 0.023 P, 0.008 S, 22.30 Cr, 12.99 Ni	Hot drawn, annealed	200–525	0.25	...	...	39.4	0.079	...	...	8.7	0.184	...	...
			0.40	...	...	45.1	0.074	...	...	9.6	0.178	...	...
			0.60	...	...	48.1	0.076	...	...	9.5	0.185	...	...
310 SS 0.12 C, 1.26 Si, 1.56 Mn, 0.01 P, 0.009 S, 25.49 Cr, 21.28 Ni	Hot drawn, annealed	310–460	0.25	50.3	0.080	32.3	0.127	27.5	0.101	12.0	0.154	...	...
			0.40	56.5	0.080	32.2	0.142	22.8	0.143	10.8	0.175	...	...
			0.60	61.8	0.067	21.9	0.212	9.7	0.284	4.5	0.326	...	...
316 SS 0.06 C, 0.52 Si, 1.40 Mn, 0.035 P, 0.005 S, 17.25 Cr, 12.23 Ni, 2.17 Mo	Hot drawn, annealed	310–460	0.25	13.5	0.263	22.2	0.149	6.4	0.317	8.0	0.204	...	...
			0.40	28.8	0.162	26.8	0.138	3.7	0.435	7.4	0.227	...	...
			0.60	39.3	0.128	30.1	0.133	6.1	0.365	6.5	0.254	...	...
403 SS 0.16 C, 0.37 Si, 0.44 Mn, 0.024 P, 0.007 S, 12.62 Cr	Hot rolled, annealed	0.8–100	0.25	...	...	26.3	0.079	15.4	0.125	7.3	0.157	...	...
			0.50	...	...	26.9	0.076	16.0	0.142	7.8	0.152	...	...
			0.70	...	...	24.6	0.090	15.3	0.158	7.5	0.155	...	...
SS 0.12 C, 0.12 Si, 0.29 Mn, 0.014 P, 0.016 S, 12.11 Cr, 0.50 Ni, 0.45 Mo	Hot rolled, annealed	0.8–100	0.25	...	...	28.7	0.082	17.2	0.082	11.9	0.079	...	...
			0.50	...	...	29.1	0.093	20.7	0.073	11.6	0.117	...	...
			0.70	...	...	28.7	0.096	22.5	0.067	11.2	0.131	...	...
SS 0.08 C, 0.45 Si, 0.43 Mn, 0.031 P, 0.005 S, 17.38 Cr, 0.31 Ni	Hot rolled, annealed	3.5–30	0.25	...	...	...	...	19.5	0.099	8.9	0.128	28.3	0.114
			0.50	...	...	...	...	22.3	0.097	9.5	0.145	34.9	0.105
			0.70	...	...	...	...	23.2	0.098	9.2	0.158	37.1	0.107
	Test temperature, °C (°F):			870 (1600)		925 (1700)		980 (1800)		1095 (2000)		1150 (2100)	
Maraging 300	...	...	...	43.4	0.077	36.4	0.095	30.6	0.113	21.5	0.145	18.0	0.165
	Test temperature, °C (°F):			1205 (2200)									
Maraging 300	...	...	...	12.8	0.185								

(a) Approximate composition

not increase without limit because of the occurrence of dynamic recovery and dynamic recrystallization.

In high-stacking-fault-energy (SFE) metals (e. g., aluminum and its alloys, iron in the ferrite phase field, titanium alloys in the beta phase field), dynamic recovery (DRV) predominates. During such processes, individual dislocations or pairs of dislocations are annihilated because of the ease of climb (and the subsequent annihilation of dislocations of opposite sign) and the formation of cells and subgrains that act as sinks for moving (mobile) dislocations. Because subgrains are formed and destroyed continuously during hot working, hot deformed metals often contain a collection of equiaxed subgrains (with low misorientations across their boundaries) contained within elongated primary grains (Ref 21, 22). Furthermore, the dynamic recovery process leads to low stresses at high temperatures, and thus, cavity nucleation and growth are retarded, and ductility

is high. The evolution of microstructure in high-SFE (and some low- SFE) materials worked at lower temperatures, such as those characteristic of cold working, is similar. At these temperatures, subgrains may also form and serve as sinks for dislocations. However, the subgrains are more stable. Thus, as more dislocations are absorbed into their boundaries, increasing misorientations are developed, eventually giving rise to an equiaxed structure of high-angle boundaries. Such a mechanism forms the basis for grain refinement in so-called severe plastic deformation processes. This mechanism of grain refinement is sometimes called continuous dynamic recrystallization because of the gradual nature of the formation of high-angle boundaries with increasing strain.

In low-SFE materials (e.g., iron and steel in the austenite phase field, copper, nickel), dynamic recovery occurs at a lower rate under hot working conditions, because mobile dislocations are dissociated, and therefore, climb is difficult. This leads

to somewhat higher densities of dislocations than in materials whose deformation is controlled by dynamic recovery. Furthermore, as the temperature is increased, the mobility of grain boundaries increases rapidly. Differences in dislocation density across the grain boundaries, coupled with high mobility, lead to the nucleation and growth of new, strain-free grains via a discontinuous dynamic recrystallization process (Ref 21, 23). At large strains, a fully recrystallized structure is obtained. However, even at this stage, recrystallized grains are being further strained and thus undergo additional cycles of dynamic recrystallization. Nevertheless, a steady state is reached in which the rate of dislocation input due to the imposed deformation is balanced by dislocation annihilation due to the nucleation and growth of new grains (as well as some dynamic recovery). Hence, although a nominally equiaxed grain structure is obtained at large strains, the distribution of stored energy is not uniform.

Table 5 Summary of C (ksi) and m values describing the flow stress-strain rate relation, $\bar{\sigma} = C(\dot{\bar{\varepsilon}})^m$, for aluminum alloys at various temperatures

Alloy	Material history	Strain rate range, 1/s	Strain	C	m	C	m	C	m	C	m	C	m
		Test temperature, °C (°F):		200 (390)		300 (570)		400 (750)		500 (930)		600 (1110)	
		0.4–311											
Super-pure 99.98 Al, 0.0017 Cu, 0.0026 Si, 0.0033 Fe, 0.006 Mn	Cold rolled, annealed 1/2 h at 1110°F		0.288	5.7	0.110	4.3	0.120	2.8	0.140	1.6	0.155	0.6	0.230
			2.88	8.7	0.050	4.9	0.095	2.8	0.125	1.6	0.175	0.6	0.215
		Test temperature, °C (°F):		240 (465)		360 (645)		480 (825)					
		0.25–63											
EC 0.01 Cu, 0.026 Mn, 0.033 Mg, 0.092 Si, 0.23 Fe, 0.01 Zn, 99.5 Al	Annealed 3 h at 750°F		0.20	10.9	0.066	5.9	0.141	3.4	0.168				
			0.40	12.3	0.069	6.3	0.146	3.3	0.169				
			0.60	13.1	0.067	6.4	0.147	3.2	0.173				
			0.80	13.8	0.064	6.7	0.135	3.4	0.161				
		Test temperature, °C (°F):		200 (390)		400 (750)		500 (930)					
		0.25–40											
1100 99.0 Al (min), 0.10 Cu, 0.15 Si, 0.50 Fe, 0.01 Mn, 0.01 Mg	Cold drawn, annealed		0.25	9.9	0.066	4.2	0.115	2.1	0.211				
			0.50	11.6	0.071	4.4	0.132	2.1	0.227				
			0.70	12.2	0.075	4.5	0.141	2.1	0.224				
		Test temperature, °C (°F):		150 (300)		250 (480)		350 (660)		450 (840)		550 (1020)	
		4–40											
1100(a) 0.10 Cu, 0.20 Si, 0.02 Mn, 0.46 Fe, 0.01 Zn, bal Al	Extruded, annealed 1 h at 750°F		0.105	11.4	0.022	9.1	0.026	6.3	0.055	3.9	0.100	2.2	0.130
			0.223	13.5	0.022	10.5	0.031	6.9	0.061	4.3	0.098	2.4	0.130
			0.338	15.0	0.021	11.4	0.035	7.2	0.073	4.5	0.104	2.5	0.141
			0.512	16.1	0.024	11.9	0.041	7.3	0.084	4.4	0.116	2.4	0.156
			0.695	17.0	0.026	12.3	0.041	7.4	0.088	4.3	0.130	2.4	0.155
		Test temperature, °C (°F):		200 (390)		400 (750)		500 (930)					
		0.2–30											
2017 94.95 Al, 3.50 Cu, 0.10 Si, 0.50 Fe, 0.50 Mn, 0.45 Mg	Cold drawn, annealed		0.250	34.5	0.014	14.8	0.110	5.8	0.126				
			0.500	32.2	−0.025	13.2	0.121	5.2	0.121				
			0.700	29.5	−0.038	12.5	0.128	5.1	0.119				
		Test temperature, °C (°F):		300 (570)		350 (660)		400 (750)		450 (840)		500 (930)	
		0.4–311											
2017(a) 0.89 Mg, 4.17 Cu, 0.61 Si, 0.41 Fe, 0.80 Mn, 0.052 Zn, 0.01 Pb, 92.9 Al	Solution treated 1 h at 950°F, water quenched, annealed 4 h at 750°F		0.115	10.8	0.695	9.1	0.100	7.5	0.110	6.2	0.145	5.1	0.155
			2.660	10.0	0.100	9.2	0.100	7.7	0.080	6.8	0.090	4.6	0.155
		Test temperature, °C (°F):		240 (465)		360 (645)		480 (825)					
		0.25–63											
5052 0.068 Cu, 0.04 Mn, 2.74 Mg, 0.10 Si, 0.19 Fe, 0.01 Zn, 0.003 Ti, bal Al	Annealed 3 h at 790°F		0.20	14.3	0.038	8.9	0.067	5.6	0.125				
			0.40	15.9	0.035	9.3	0.071	5.3	0.130				
			0.60	16.8	0.035	9.0	0.068	5.1	0.134				
			0.80	17.5	0.038	9.4	0.068	5.6	0.125				
5056 0.036 Cu, 0.04 Mn, 4.83 Mg, 0.15 Si, 0.22 Fe, 0.01 Zn, 0.14 Cr, bal Al	Annealed 3 h at 790°F	0.25–63	0.20	42.6	−0.032	20.9	0.138	11.7	0.200				
			0.40	44.0	−0.032	20.8	0.138	10.5	0.205				
			0.60	44.9	−0.031	19.9	0.143	10.3	0.202				
			0.70	45.6	−0.034	20.3	0.144	10.3	0.203				
5083 0.01 Cu, 0.77 Mn, 4.41 Mg, 0.10 Si, 0.16 Fe, 0.01 Zn, 0.13 Cr, 0.002 Ti, bal Al	Annealed 3 h at 790°F	0.25–63	0.20	43.6	−0.006	20.5	0.095	9.3	0.182				
			0.40	43.6	−0.001	19.7	0.108	8.3	0.208				
			0.60	41.9	0.003	18.8	0.111	8.5	0.201				
			0.80	40.2	0.002	19.1	0.105	9.7	0.161				
5454 0.065 Cu, 0.81 Mn, 2.45 Mg, 0.12 Si, 0.18 Fe, <0.01 Zn, 0.002 Ti, bal Al	Annealed 3 h at 790°F	0.25–63	0.20	33.6	−0.005	16.8	0.093	10.8	0.182				
			0.40	36.0	−0.009	16.3	0.104	10.7	0.188				
			0.60	36.9	−0.009	16.0	0.102	10.0	0.191				
			0.80	37.0	−0.009	16.2	0.097	10.2	0.183				
		Test temperature, °C (°F):		400 (750)		450 (840)		500 (930)		550 (1020)			
		0.4–311											
7075(a) 89.6 Al, 1.31 Cu, 2.21 Mg, 0.21 Si, 0.30 Fe, 0.34 Mn, 5.75 Zn, 0.01 Pb	Solution treated 1 h at 870°F, water quenched, aged at 285°F for 16 h		0.115	10.0	0.090	6.0	0.135	3.9	0.150	2.9	0.170		
			2.66	9.7	0.115	6.2	0.120	4.8	0.115	2.7	0.115		

(a) Approximate composition

The presence of second-phase particles may affect the evolution of microstructure during hot working of both high- and low-SFE materials. In high-SFE materials, particles may affect the homogeneity and magnitude of dislocation substructure that evolves. In low-SFE materials, particles may affect the evolution of substructure, serve as nucleation sites for dynamic recrystallization, as well as serve as obstacles to boundary migration during the recrystallization process.

Flow Curves at Conventional Metalworking Strain Rates. The stress-strain (flow) curves that are measured under conventional hot working conditions are a function of the predominant dynamic softening mechanism.

Flow Curves for Dynamic Recovery. As mentioned previously, the hot working response of high-SFE metals is controlled by dynamic recovery. In such cases, dislocation generation is offset by dislocation annihilation due to recovery processes. The flow curve thus exhibits an initial stage of strain hardening followed by a steady-state (constant) flow stress. Typical

Table 6 Summary of *C* (ksi) and *m* values describing the flow stress-strain rate relation, $\bar{\sigma} = C(\dot{\bar{\varepsilon}})^m$, for copper alloys at various temperatures

Alloy	Material history	Strain rate range, 1/s	Strain	C	m	C	m	C	m	C	m	C	m
			Test temperature, °C (°F):	300 (570)		450 (840)		600 (1110)		750 (1380)		900 (1650)	
Copper 0.018 P, 0.0010 Ni, 0.0003 Sn, 0.0002 Sb, 0.0005 Pb 0.0010 Fe, 0.0020 Mn <0.0005 Mg, <0.0005 As, <0.0001 Bi, 0.0014 S, less than 0.003 O$_2$, Se + Te not detected	Cold drawn, annealed 2 h at 1110°F	4–40	0.105	20.2	0.016	17.0	0.010	12.7	0.050	7.6	0.096	4.7	0.134
			0.223	26.5	0.018	22.5	0.004	16.8	0.043	9.7	0.097	6.3	0.110
			0.338	30.2	0.017	25.1	0.008	18.9	0.041	10.0	0.128	6.1	0.154
			0.512	32.2	0.025	26.6	0.014	19.4	0.056	8.5	0.186	5.5	0.195
			0.695	34.4	0.024	26.8	0.031	19.0	0.078	8.2	0.182	5.2	0.190
			Test temperature, °C (°F):	427 (800)									
OFHC copper	...	...		26.7	0.0413								
			Test temperature, °C (°F):	400 (750)		500 (930)		600 (1110)					
CDA 110 99.94 Cu, 0.0003 Sb, 0.0012 Pb 0.0012 S, 0.0025 Fe, 0.001 Ni	Hot rolled, annealed	0.25–40	0.25	23.0	0.046	12.9	0.136	6.6	0.160				
			0.50	27.4	0.049	13.7	0.150	6.9	0.168				
			0.70	28.8	0.057	13.3	0.165	6.8	0.176				
			Test temperature, °C (°F):	200 (390)		400 (750)		600 (1110)		800 (1470)			
CDA 220 90.06 Cu, 0.033 Fe, 0.004 Pb, 0.003 Sn, bal Zn	Extruded, cold drawn 30%; annealed 650 °C, 90 min	0.1–10	0.25	41.0	0.017	34.1	0.018	22.6	0.061	11.2	0.134		
			0.50	46.7	0.029	39.9	0.032	24.4	0.084	11.0	0.156		
			0.70	48.1	0.034	40.7	0.024	24.6	0.086	11.4	0.140		
CDA 260 70.05 Cu, trace Fe + Sn, bal Zn	Hot rolled, annealed	3.5–30	0.25	...	...	34.9	0.036	16.0	0.194	7.1	0.144		
			0.50	...	...	42.3	0.031	14.8	0.237	7.0	0.148		
			0.70	...	...	42.4	0.045	14.3	0.228	6.3	0.151		
CDA 280 60.44 Cu, 0.01 Pb, 0.02 Fe, trace Sn, bal Zn	Hot rolled, annealed	3.5–30	0.25	49.0	0.028	26.9	0.083	7.6	0.189	3.1	0.228		
			0.50	58.6	0.027	28.6	0.075	5.4	0.281	2.8	0.239		
			0.70	60.3	0.027	26.7	0.081	4.7	0.291	2.7	0.220		
CDA 365 59.78 Cu, 0.90 Pb, 0.02 Fe, trace Sn, bal Zn	Hot rolled, annealed	3.5–30	0.25	45.8	0.038	28.6	0.065	9.8	0.106	2.4	0.166		
			0.50	57.2	0.032	28.9	0.085	8.5	0.137	2.1	0.197		
			0.70	59.1	0.035	26.6	0.078	8.4	0.113	1.8	0.222		

flow curves for pure iron in the body-centered cubic phase field are shown in Fig. 15 (Ref 24). The magnitude of the steady-state flow stress, $\bar{\sigma}_{ss}$, decreases with increasing temperature, T, and decreasing strain rate, $\dot{\bar{\varepsilon}}$, typically according to a phenomenological (hyperbolic sine) relation, as follows (Ref 20, 21):

$$Z \equiv \dot{\bar{\varepsilon}} \exp(Q/RT) = C'[\sinh[(\alpha'\bar{\sigma}_{ss})]^{n_{drv}}$$ (Eq 33a)

in which Z denotes the Zener-Hollomon parameter, Q is an apparent activation energy, R is the gas constant, and C', α', and n_{drv} are constants. The constant n_{drv} is referred to as the stress exponent of the strain rate or simply the stress exponent. Equation 33(a) reduces to two simpler forms depending on whether deformation is imposed at high-temperature/low-strain-rate conditions (giving rise to low flow stresses) or at low-temperature/high-strain-rate conditions (giving rise to high flow stress):

Low stresses: $Z \equiv \dot{\bar{\varepsilon}} \exp(Q/RT) \sim \bar{\sigma}_{ss}^{n_{drv}}$ (Eq 33b)

High stresses: $Z \equiv \dot{\bar{\varepsilon}} \exp(Q/RT) \sim \exp(\beta\bar{\sigma}_{ss}^{n_{drv}})$ (Eq 33c)

The constants α', β, and n_{drv} are related by $\beta = \alpha'n_{drv}$. Rearrangement of Eq 33(b) yields an expression identical to Eq 32 in which the temperature dependence of the flow stress is incorporated into C, and the strain-rate sensitivity exponent, m, is equal to the inverse of the stress exponent, that is, $m = 1/n_{drv}$. The loss of the power-law dependence of flow stress on strain rate at high stresses, that is, Eq 33(c), is termed power-law breakdown.

The activation energy Q in Eq 33(a–c) can be determined from the slope of a plot of $\log_{10}\bar{\sigma}$ versus $1/T$ at fixed strain rate or $\log_{10}\dot{\bar{\varepsilon}}$ versus $1/T$ at fixed stress:

$$Q = \frac{2.3R}{m} \frac{\partial \log \bar{\sigma}}{\partial(1/T)}\Big|\dot{\varepsilon}$$ (Eq 34a)

or

$$Q = -2.3R \frac{\partial \log \dot{\bar{\varepsilon}}}{\partial(1/T)}\Big|\bar{\sigma}$$ (Eq 34b)

A more fundamental, mechanistic insight into the shape of the flow curve for cases involving dynamic recovery may be obtained by an analysis of the overall rate of change of (mobile) dislocation density, ρ, with strain, $\bar{\varepsilon}$, $d\rho/d\bar{\varepsilon}$:

$$d\rho/d\bar{\varepsilon} = d\rho/d\bar{\varepsilon}|_{storage} - d\rho/d\bar{\varepsilon}|_{recovery}$$ (Eq 35)

The specific functional form of the dislocation storage and annihilation terms in Eq 35 can be expressed as follows (Ref 25, 26):

$$d\rho/d\bar{\varepsilon} = U - \Omega\rho$$ (Eq 36)

In Eq 36, U denotes the rate of dislocation generation due to strain hardening, and Ω is a factor describing the rate of dynamic recovery;

the rate of recovery is also directly proportional to the instantaneous level of dislocation density, ρ. At hot working temperatures, U is independent of strain rate and temperature to a first order (Ref 26). Thus, the strain- rate and temperature dependence of the rate of dislocation multiplication is determined principally by $\Omega = \Omega(\dot{\bar{\varepsilon}}, T)$. An example of such a dependence for a low-carbon steel is shown in Fig. 16(a) (Ref 26).

To a first order, the flow stress, $\bar{\sigma}$, under working conditions is given by the following expression:

$$\bar{\sigma} = \alpha G\mathbf{b}\sqrt{\rho}$$ (Eq 37)

in which α denotes a constant whose magnitude is between 0.5 and 1.0, G is the shear modulus, and $\mathbf{b}$ is the length of the Burgers (slip) vector. Inspection of Eq 36 and 37 reveals that the strain-rate and temperature dependence of the overall rate of hardening in the flow curve is largely determined by Ω. The strain-hardening rate is frequently quantified in terms of plots of $d\bar{\sigma}/d\bar{\varepsilon}(\equiv \theta)$ as a function of $\bar{\sigma}$. Typical plots for low-carbon steel are shown in Fig. 16(b) (Ref 26).

Equation 36 reveals that a steady-state dislocation density, ρ_{ss}, is reached when $U = \Omega\rho$, or:

$$\rho_{ss} = U/\Omega$$ (Eq 38)

The steady-state flow stress, $\bar{\sigma}_{ss}$, is thus given by the following relation:

Table 7 Summary of *C* (ksi) and *m* values describing the flow stress-strain rate relation, $\bar{\sigma} = C(\dot{\bar{\varepsilon}})^m$, for titanium alloys at various temperatures

Alloy	Material history	Strain rate range, 1/s	Strain	C	m	C	m	C	m	C	m	C	m	C	m	C	m
			Test temperature, °C (°F):	20 (68)		200 (392)		400 (752)		600 (1112)		800 (1472)		900 (1652)		1000 (1832)	
Type 1 0.04 Fe, 0.02 C, 0.005 H₂, 0.01 N₂, 0.04 O₂ bal Ti	Annealed 15 min at 1200 °F in high vacuum	0.25–16.0	0.2	92.8	0.029	60.9	0.046	39.8	0.074	25.3	0.097	12.8	0.167	5.4	0.230	3.0	0.387
			0.4	113.7	0.029	73.3	0.056	48.8	0.061	29.6	0.115	14.6	0.181	5.5	0.248	3.6	0.289
			0.6	129.6	0.028	82.2	0.056	53.9	0.049	32.1	0.105	14.9	0.195	5.5	0.248	3.5	0.289
			0.8	142.5	0.027	87.7	0.058	56.3	0.042	32.7	0.099	15.4	0.180	5.9	0.186	3.2	0.264
			1.0	150.6	0.027	90.7	0.054	56.6	0.044	32.5	0.099	15.9	0.173	5.9	0.167	3.0	0.264
Type 2 0.15 Fe, 0.02 C, 0.005 H₂, 0.02 N₂, 0.12 O₂, bal Ti	Annealed 15 min at 1200 °F in high vacuum	0.25–16.0	0.2	143.3	0.021	92.7	0.043	54.5	0.051	33.6	0.092	17.5	0.167	6.9	0.135	4.2	0.220
			0.4	173.2	0.021	112.1	0.042	63.1	0.047	36.3	0.101	18.4	0.190	7.2	0.151	4.9	0.167
			0.6	193.8	0.024	125.3	0.045	65.6	0.047	36.9	0.104	18.4	0.190	7.8	0.138	4.5	0.167
			0.8	208.0	0.023	131.9	0.051	66.0	0.045	37.0	0.089	18.4	0.190	7.6	0.106	3.9	0.195
			1.0	216.8	0.023	134.8	0.056	65.3	0.045	36.9	0.092	18.6	0.190	6.8	0.097	3.7	0.167
			Test temperature, °C (°F):	600 (1110)		700 (1290)		800 (1470)		900 (1650)							
Unalloyed 0.03 Fe, 0.0084 N, 0.0025 H, bal Ti	Hot rolled, annealed 800 °C 90 min	0.1–10	0.25	23.4	0.062	14.3	0.115	8.2	0.236	1.8	0.324						
			0.50	27.9	0.066	17.8	0.111	10.0	0.242	2.1	0.326						
			0.70	30.1	0.065	20.0	0.098	12.2	0.185	2.5	0.316						
			Test temperature, °C (°F):	20 (68)		200 (392)		400 (752)		600 (1112)		800 (1472)		900 (1652)		1000 (1832)	
Ti-5Al-2.5 Sn 5.1 Al, 2.5 Sn, 0.06 Fe, 0.03 C, 0.01 H₂, 0.03 N₂, 0.1 O₂, bal Ti	Annealed 30 min at 1470 °F in high vacuum	0.25–16.0	0.1	173.6	0.046	125.6	0.028	97.6	0.028	...	...	...	...	...	...	...	...
			0.2	197.9	0.048	138.8	0.022	107.4	0.026	86.1	0.025	58.5	0.034	44.2	0.069	5.4	0.308
			0.3	215.6	0.046	147.4	0.021	112.5	0.027	92.8	0.020	...	...	...	...	...	...
			0.4	230.6	0.039	151.4	0.022	116.0	0.022	95.6	0.019	58.7	0.040	44.8	0.082	5.1	0.294
			0.5	...	...	...	...	...	...	96.7	0.021	...	...	...	...	...	...
			0.6	...	...	...	...	...	...	96.6	0.024	55.6	0.042	43.0	0.078	5.2	0.264
			0.8	...	...	...	...	...	...	...	...	50.2	0.033	39.1	0.073	5.2	0.264
			0.9	...	...	...	...	...	...	46.8	0.025	...	...	...	...	...	...
			1.0	...	...	...	...	...	...	...	...	...	...	35.2	0.056	5.3	0.280
Ti-6Al-4V 6.4 Al, 4.0 V, 0.14 Fe, 0.05 C, 0.01 H₂, 0.015 N₂, 0.1 O₂, bal Ti	Annealed 120 min at 1200 °F in high vacuum	0.25–16.0	0.1	203.3	0.017	143.8	0.026	119.4	0.025	...	...	...	...	...	...	...	...
			0.2	209.7	0.015	151.0	0.021	127.6	0.022	94.6	0.064	51.3	0.146	23.3	0.143	9.5	0.131
			0.3	206.0	0.015	152.0	0.017	126.2	0.017	91.2	0.073	...	...	...	...	...	...
			0.4	...	...	...	...	118.7	0.014	84.6	0.079	39.8	0.175	21.4	0.147	9.4	0.118
			0.5	...	...	...	...	...	...	77.9	0.080	...	...	...	...	...	...
			0.6	...	...	...	...	...	...	...	...	30.4	0.205	20.0	0.161	9.6	0.118
			0.8	...	...	...	...	...	...	...	...	26.6	0.199	19.5	0.172	9.3	0.154
			0.9	...	...	...	...	...	...	...	...	24.9	0.201	...	...	...	...
			1.0	...	...	...	...	...	...	...	...	...	...	20.3	0.146	8.9	0.192
			Test temperature, °C (°F):	843 (1550)		954 (1750)		982 (1800)									
Ti-6Al-4V	...	...		38.0	0.064	12.3	0.24	9.4	0.29								
			Test temperature, °C (°F):	20 (68)		200 (392)		400 (752)		600 (1112)		800 (1472)		900 (1652)		1000 (1832)	
Ti-13V-11Cr-3Al 3.6 Al, 14.1 V, 10.6 Cr, 0.27 Fe, 0.02 C, 0.014 H₂, 0.03 N₂, 0.11 O₂, bal Ti	Annealed 30 min at 1290 °F in high vacuum	0.25–16.0	0.1	173.1	0.041	...	...	...	...	...	...	...	...	...	...	...	...
			0.2	188.2	0.037	150.5	0.030	136.5	0.035	118.4	0.040	65.4	0.097	44.6	0.147	32.4	0.153
			0.3	202.3	0.034	...	...	...	...	...	...	...	...	...	...	...	...
			0.4	215.2	0.029	174.2	0.024	153.9	0.030	107.5	0.039	59.5	0.096	42.1	0.139	30.9	0.142
			0.5	226.3	0.026	181.1	0.023	...	...	...	...	...	...	...	...	...	...
			0.6	...	...	183.5	0.026	147.9	0.046	92.8	0.045	56.7	0.088	40.9	0.127	29.2	0.155
			0.7	...	...	181.4	0.029	...	...	...	...	...	...	...	...	...	...
			0.8	...	...	...	...	136.3	0.045	84.7	0.036	53.9	0.081	39.3	0.125	27.8	0.167
			0.9	...	...	...	...	...	...	...	...	52.9	0.080	...	...	...	...
			1.0	...	...	...	...	...	...	...	...	...	...	38.8	0.127	28.0	0.159

$$\bar{\sigma}_{ss} = \alpha G \mathbf{b} \sqrt{\rho_{ss}} = \alpha G \mathbf{b} \sqrt{U/\Omega} \qquad \text{(Eq 39)}$$

Plastic-flow formulations such as Eq 35 to 39 form the basis of so-called internal state variable relations of the flow stress. This specific case uses a single state variable, the mobile dislocation density, ρ.

There are a number of alternate approaches to the modeling of dynamic recovery under the broad framework of Eq 35. For example, Kocks (Ref 27) has shown that a linear dependence of strain-hardening rate ($d\bar{\sigma}/d\bar{\varepsilon}$) on stress ($\bar{\sigma}$) is consistent with the following relation for $d\rho/d\bar{\varepsilon}$:

$$d\rho/d\bar{\varepsilon} = (k_1\sqrt{\rho} - k_2\rho)/\mathbf{b} \qquad \text{(Eq 40)}$$

in which k_1 and k_2 are constants. Similarly, for a strain-hardening rate that varies linearly with $1/\bar{\sigma}$, Roberts (Ref 28) has shown that the following relation applies:

$$d\rho/d\bar{\varepsilon} = k_3 - k_4\sqrt{\rho} \qquad \text{(Eq 41)}$$

in which k_3 and k_4 are constants.

More information on internal-state variable models is contained in the article "Internal-State Variable Modeling of Plastic Flow" in *Fundamentals of Modeling for Metals Processing*, Volume 22A of the *ASM Handbook*, 2009.

Flow Curves for Discontinuous Dynamic Recrystallization. Flow curves for materials undergoing discontinuous dynamic recrystallization (DDRX) have shapes that are distinctively different from those that characterize materials which soften solely by dynamic recovery. Those for DDRX exhibit an initial strain-hardening transient, a peak stress, flow softening, and finally, a period of steady-state flow. Typical curves for two austenitic stainless steels are shown in Fig. 17 (Ref 29).

Dynamic recrystallization typically initiates at a strain of approximately five-sixths of the strain corresponding to the peak stress. Because of this behavior, the peak stress for a material that undergoes dynamic recrystallization is less than that which would be developed if the material softened solely by dynamic recovery (Fig. 18a). The strain at which DDRX initiates as well as the steady-state flow stress that would be developed in the absence of DDRX are readily determined from a plot of $d\bar{\sigma}/d\bar{\varepsilon}$ as a function of $\bar{\sigma}$ (Fig. 18b).

From a phenomenological standpoint, the strain at the peak stress, $\bar{\varepsilon}_p$, is usually found to depend on the initial grain size, d_o, and the Zener-Hollomon parameter, Z, in accordance with an expression of the following form (Ref 30 to 32):

$$\bar{\varepsilon}_p = C'' d_o^{n_3} Z^{n_4} \qquad \text{(Eq 42)}$$

in which C'', n_3, and n_4 are material-specific constants. The activation energy

Table 8 Summary of C (ksi) and m values describing the flow stress-strain rate relation, $\bar{\sigma} = C(\dot{\bar{\varepsilon}})^m$, for various materials

Alloy	Material history	Strain rate range, 1/s	Strain	C	m	C	m	C	m	C	m	C	m	C	m	C	m
		Test temperature, °C (°F):		22 (72)		110 (230)		170 (335)		215 (415)		260 (500)		300 (570)			
Lead 99.98 Pb, 0.003 Cu, 0.003 Fe, 0.002 Zn, 0.002 Ag	...	...	0.115	2.0	0.040	1.56	0.065	1.21	0.085	0.70	0.130	0.47	0.160	0.40	0.180		
			2.66	4.0	0.055	1.47	0.100	1.04	0.125	0.55	0.135	0.36	0.180	0.28	0.225		
		Test temperature, °C (°F):		200 (390) (13)		300 (570) (14)		400 (750) (13)		400 (930) (14)							
Magnesium 0.010 Al, 0.003 Zn, 0.008 Mn, 0.004 Si, 0.003 Cu, 0.0008 Ni, bal Mg	Extruded, cold drawn 15%, annealed 550 °C 90 min	0.1–10	0.25	19.1	0.069	9.8	0.215	4.1	0.263	1.7	0.337						
			0.50	17.2	0.093	8.4	0.211	4.0	0.234	1.7	0.302						
			0.70	15.5	0.094	8.3	0.152	4.3	0.215	2.1	0.210						
		Test temperature, °C (°F):		1080 (1975)		1166 (2030)											
U-700	...	...	...	26.6	0.21	22.1	0.21										
		Test temperature, °C (°F):		20 (68)		200 (392)		400 (752)		600 (1112)		800 (1472)		900 (1652)		1000 (1832)	
Zirconium 99.8 Zr, 0.009 Hf, 0.008 Al, 0.038 Fe, 0.0006 H₂, 0.0025 N₂, 0.0825 O₂, 0.0 Ni	Annealed 15 min at 1380 °F in high vacuum	0.25–16.0	0.2	117.4	0.031	74.0	0.052	40.2	0.050	23.8	0.069	16.8	0.069	6.8	0.227	4.6	0.301
			0.3	143.7	0.022	92.2	0.058	...	...	...	...	...	...	...	...	...	...
			0.4	159.5	0.017	105.1	0.046	54.4	0.085	29.4	0.09	18.2	0.116	7.1	0.252	4.0	0.387
			0.5	169.3	0.017	112.8	0.041	58.2	0.093	...	...	...	...	...	...	...	...
			0.6	...	...	118.5	0.042	60.2	0.095	31.3	0.089	18.8	0.118	7.2	0.264	4.0	0.387
			0.7	...	...	...	...	61.9	0.095	...	...	...	...	...	...	...	...
			0.8	...	...	...	...	...	...	32.0	0.081	19.4	0.101	6.9	0.252	4.1	0.403
			1.0	...	...	...	...	...	...	32.1	0.085	19.7	0.108	6.9	0.252	4.1	0.403
Zircaloy 2 98.35 Zr, 0.015 Hf, 1.4 Zn, 0.01 Al, 0.06 Fe, 0.045 Ni, 0.0006 H₂, 0.0023 N₂, 0.0765 O₂	Annealed 15 min at 1380 °F in high vacuum	0.25–16.0	0.1	96.8	0.031	65.9	0.046	...	...	...	...	...	...	...	...	...	...
			0.2	136.9	0.025	105.8	0.035	58.3	0.065	30.4	0.049	16.6	0.147	7.5	0.325	3.9	0.362
			0.3	178.5	0.034	131.4	0.035	67.9	0.056	...	...	...	...	...	...	...	...
			0.4	202.7	0.027	145.4	0.036	73.5	0.056	37.8	0.053	18.7	0.172	7.8	0.342	4.0	0.387
			0.5	...	...	154.2	0.034	77.3	0.057	...	...	...	...	...	...	...	...
			0.6	...	...	...	...	79.9	0.055	39.2	0.059	18.8	0.178	7.2	0.387	4.0	0.387
			0.8	...	...	...	...	...	...	40.4	0.057	18.8	0.178	7.9	0.342	4.8	0.333
			1.0	...	...	...	...	...	...	40.7	0.053	18.8	0.178	8.5	0.310	4.8	0.333
		Test temperature, °C (°F):		20 (68)		100 (212)		200 (392)		300 (572)		500 (932)		900 (1292)		900 (1652)	
Uranium 99.8 U, 0.0012 Mn, 0.0012 Ni, 0.00074 Cu, 0.00072 Cr, 0.0001 Co, 0.0047 H₂, 0.0041 N₂, 0.1760 O₂ (free of cadmium and boron)	Annealed 2 h at 1110 °F in high vacuum	0.25–16.0	0.2	151.0	0.043	113.0	0.042	77.4	0.034	45.9	0.044	31.9	0.051	16.0	0.081	4.5	0.069
			0.4	173.9	0.033	132.7	0.049	91.0	0.031	53.3	0.047	33.1	0.059	16.1	0.089	4.5	0.069
			0.6	184.9	0.023	143.1	0.047	98.1	0.032	56.0	0.056	33.4	0.054	16.1	0.089	4.5	0.069
			0.8	189.8	0.018	149.5	0.048	102.0	0.036	58.3	0.057	33.3	0.049	16.2	0.097	4.5	0.069
			1.0	...	...	...	...	...	...	59.0	0.056	32.5	0.055	16.4	0.097	4.5	0.069

specific to DDRX is used in the determination of Z.

The regime of steady-state flow in stress-strain curves for materials that undergo dynamic recrystallization may be smooth or exhibit an oscillatory behavior that dampens with increasing strain. Temperature-strain rate conditions for which the dynamically recrystallized grain size is less than one-half of the initial grain size show the former behavior. By contrast, those that give rise to grain-size coarsening or a reduction of less than one-half of the starting grain size exhibit the oscillatory behavior (Ref 33).

More information on the modeling of recrystallization is contained in the article "Models of Recrystallization" in *Fundamentals of Modeling for Metals Processing*, Volume 22A of the *ASM Handbook*, 2009.

Effect on Flow Stress

Effect of Microstructural Scale on Flow Stress at Hot Working Temperatures. The effect of microstructural scale (grain size, thickness of lamellae, etc.) on the flow stress at hot working temperatures varies from relatively weak to very strong. By and large, the influence is very small or negligible when the scale of the primary microstructural feature is of the order of 10 to 20 μm or greater. Below this size, the effect increases as the scale decreases. Two important examples include the plastic flow of materials with a lamellar (colony) or acicular microstructure and the superplastic flow of metals with a very fine equiaxed grain structure.

For materials with a colony or acicular microstructure, such as two-phase (alpha-beta) titanium alloys (Ref 34) and zirconium alloys (Ref 35), the thickness of the lamellae or lath-like features is typically less than or equal to a few micrometers. In these cases, dynamic recovery and the formation of subgrains are difficult. As such, flow curves exhibit a short strain-hardening transient, a peak stress, and then substantial flow softening over a wide range of strain rates (e.g., 0.001 to 10 s⁻¹) (Fig. 19a). The observed flow softening has been ascribed to slip transmission across interphase boundaries, dynamic spheroidization, and lamellar kinking, among other factors. For alpha-beta titanium alloys, such as Ti-6Al-4V, with a colony/acicular alpha microstructure, the peak stress, $\bar{\sigma}_p$, follows a Hall-Petch dependence on platelet thickness (Ref 34, 36):

$$\bar{\sigma}_p = M_T \left(\tau_o + k_s \, \lambda^{-1/2} \right) \tag{Eq 43}$$

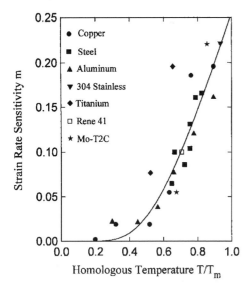

Fig. 14 Comparison of measurements (data points) of the strain-rate sensitivity exponent (*m*) as a function of the homologous temperature (fraction of the melting point) for various materials and an analytical model (solid line). Source: Ref 19

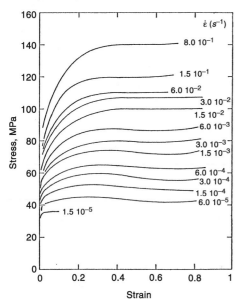

Fig. 15 Flow curves for Armco iron deformed under hot working conditions in the ferrite phase field. Source: Ref 24

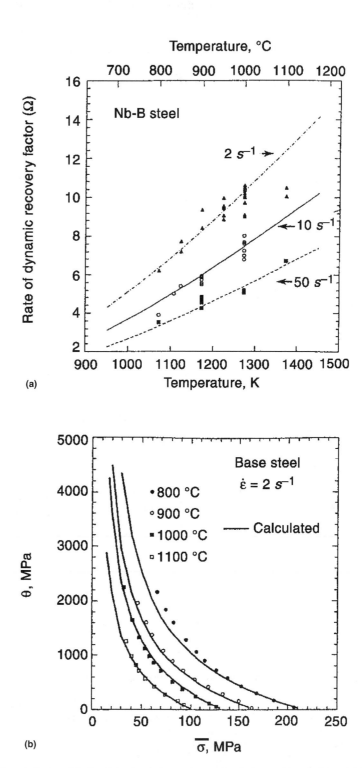

Fig. 16 Characterization of the flow behavior of a low-carbon steel under hot working conditions. (a) Ω as a function of temperature and strain rate. (b) Overall hardening rate, $\theta = d\bar{\sigma}/d\bar{\varepsilon}$, as a function of stress, $\bar{\sigma}$, at a strain rate of 2 s^{-1} and various temperatures. Source: Ref 26

in which M_T denotes the Taylor factor for the specific texture, τ_o is the friction (lattice) stress, k_s is the Hall-Petch constant ("reduced" by a factor equal to M_T), and λ is the platelet thickness. The slope of the lines on Hall-Petch plots (Fig. 19b) depend on strain rate. However, the overall magnitudes of the slope are comparable to that predicted by the classical Eshelby model (Ref 37). The loss of the Hall-Petch contribution to the strength (the term $M_T k_s \lambda^{-1/2}$ in Eq 43) has also been found to correlate to the level of flow softening observed in flow curves for Ti-6Al-4V with a colony/acicular microstructure (Ref 34).

Metals with a moderate-to-coarse equiaxed grain size tend to exhibit a rate sensitivity (*m*

value) that varies only slightly with strain rate in the conventional hot working regime ($0.01 \leq \dot{\bar{\varepsilon}} \leq 50$ s^{-1}). By contrast, the *m* values of metals with an ultrafine, equiaxed grain size ($d \leq 10$ μm) vary strongly with strain rate (Fig. 20). For such materials, a modest rate sensitivity ($m \sim 0.25$) is shown at both very low strain rates (the conventional creep regime, or

region I, in Fig. 20) and moderate-to-high strain rates (the so-called power-law creep regime, or region III, in Fig. 20). At intermediate rates (typically $0.0001 \leq \dot{\bar{\varepsilon}} \leq 0.005$ s^{-1}), or region II, *m* values are very high ($m \sim 0.4$ to 1), and superplastic behavior (tensile elongations of the order of 500% or more) is obtained. Under superplastic conditions, the majority of the

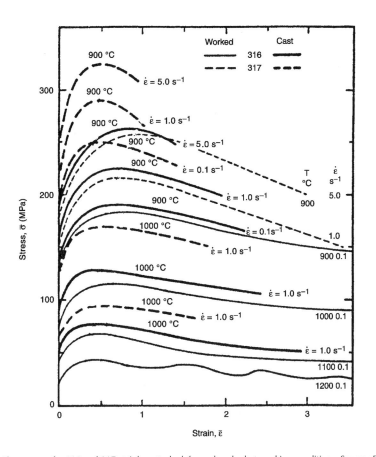

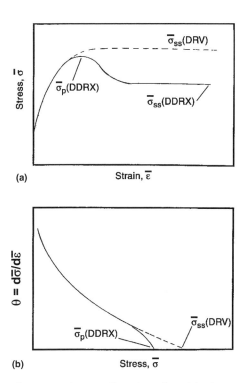

Fig. 17 Flow curves for 316 and 317 stainless steels deformed under hot working conditions. Source: Ref 29

Fig. 18 Schematic illustration of work-hardening behavior for a material undergoing dynamic recrystallization at hot working temperatures. (a) Stress-strain curve. (b) Corresponding plot of $d\bar{\sigma}/d\bar{\varepsilon}$ as a function of stress, $\bar{\sigma}$. DRV, dynamic recovery; DDRX, discontinuous dynamic recrystallization

deformation occurs by grain-boundary sliding (gbs) and grain rotation. The flow stress under superplastic conditions is controlled not by the grain-boundary sliding per se but by the kinetics of the micromechanical process by which stress concentrations developed at grain-boundary triple points (due to grain rotation) are relieved. These processes include climb-limited glide of dislocations in the vicinity of grain boundaries (i.e., mantle regions) or diffusional flow, either through the grains or along the grain boundaries. The former explanation (climb-limited glide of dislocations), first proposed by Gifkins (Ref 38) and later extended by Ghosh (Ref 39), appears to be the most successful explanation of superplasticity.

A generalized constitutive relation of the following form is often capable of describing the relationship between the flow stress, $\bar{\sigma}$, strain rate, $\dot{\bar{\varepsilon}}$, and grain size, d, of single-phase alloys during superplastic deformation (Ref 40, 41):

$$\dot{\bar{\varepsilon}} = \left(\frac{ADGb}{kT}\right)\left(\frac{\bar{\sigma}}{G}\right)^n\left(\frac{b}{d}\right)^p \qquad \text{(Eq 44)}$$

In Eq 44, A is a constant (usually of the order of 10), D is a diffusion parameter, k is Boltzmann's constant, T is absolute temperature, G is the shear modulus, b is the length of the Burgers vector, n is the stress exponent of the strain rate $(=1/m)$, and p is the grain size exponent of the strain rate. For superplastic flow

characterized by gbs accommodated by climb/glide of dislocations, $n \sim 2$ and $p \sim 2$. For gbs accommodated by diffusional flow, $n \sim 1$ and $p \sim 2$ or 3, depending on whether bulk (lattice) or boundary diffusion predominates, respectively.

The extension of the phenomenological relation between $\dot{\bar{\varepsilon}}, \bar{\sigma}, T$, and d expressed by Eq 44 to two- (or multi-) phase alloys is not obvious. This is because an ambiguity arises as to which phase the values of D, G, Ω, d, and b relate. For example, for fine, equiaxed two-phase titanium alloys, hard alpha phase particles are surrounded by much softer beta phase grains. In this instance, the alpha phase acts like the core (which deforms relatively little) and the beta phase like the mantle (which deforms to accommodate stress concentrations) in the Gifkins core-mantle model (Ref 38, 42). The alpha particle size is thus taken to be d, and all of the other quantities pertain to the beta phase (Ref 42). The applicability of this model for ultrafine Ti-6Al-4V deformed under superplastic conditions is shown in Fig. 21(a). The data in this figure have been plotted in accordance with Eq 44, rearranged to express AD as a function of $1/T$ and the measured/imposed values of $\bar{\sigma}, \dot{\bar{\varepsilon}}$, and so on. The plot also includes a line indicating the inverse temperature dependence of the diffusivity of vanadium solute in beta titanium. The similarity of the slope of this line and the trend line for the plastic flow measurements

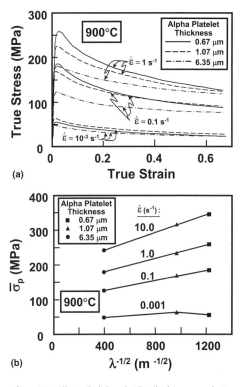

Fig. 19 Effect of alpha platelet thickness on plastic flow of Ti-6Al-4V (with a lamellar/acicular microstructure) at 900 °C. (a) Flow curves. (b) Hall-Petch plot for the peak flow stress, $\bar{\sigma}_p$. Source: Ref 34

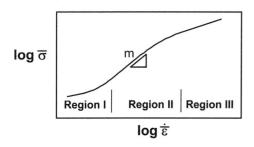

Fig. 20 Schematic illustration of the variation of flow stress with strain rate (on a log-log basis) for a fine-grained material that exhibits superplastic flow

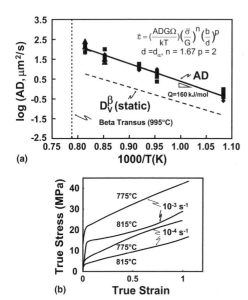

Fig. 21 Superplastic flow of ultrafine Ti-6Al-4V. (a) Plot illustrating applicability of the generalized constitutive relation (Eq 41). (b) Stress-strain data in the superplastic regime indicative of flow hardening due to dynamic coarsening. Source: Ref 42

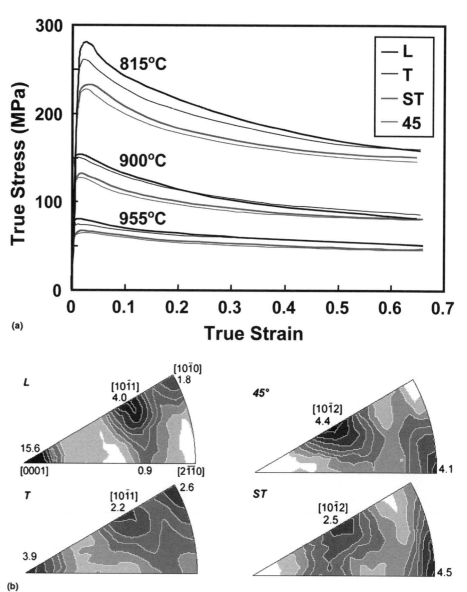

Fig. 22 Plastic flow behavior of textured plate of Ti-6Al-4V with a colony (lamellar) alpha microstructure. (a) Stress-strain curves of samples oriented along different directions in the plate. L, longitudinal; T, long transverse; ST, short transverse. (b) Corresponding inverse pole figure for each compression-test direction. Source: Ref 44

indicates an identical activation energy for the two processes. The fact that the plastic flow trend line lies above the diffusivity line by approximately one order of magnitude suggests that $A \sim 10$.

For materials that undergo grain growth or coarsening during superplastic flow, the value of d in Eq 44 increases with strain and must be taken into account when interpreting flow response. As an example, constant strain-rate flow curves for the superplastic deformation of Ti-6Al-4V (used in part to derive Fig. 21a) are shown in Fig. 21(b). The observed flow hardening is a result of the dynamic coarsening of the alpha particles (Ref 42, 43).

Effect of Crystallographic Texture on Flow Stress. Crystallographic texture, or the preferred orientation of the grains comprising a polycrystalline aggregate, can also have a major effect on the flow stress at hot (and cold) working temperatures. The influence is greatest for metals with low-symmetry crystal structures

and metals of any crystal structure having a very strong texture.

Single-phase alpha and alpha-beta titanium alloys can exhibit stress-strain curves that vary noticeably with test direction as a result of a strong texture of the hexagonal close-packed (hcp) alpha phase. For example, Fig. 22(a) shows stress-strain curves measured in compression on samples cut from the rolling or longitudinal (L), long transverse (T), 45°, and short transverse/thickness (ST) directions in a textured plate of Ti-6Al-4V with a colony alpha microstructure (Ref 44). Focusing on the peak stress, the plate was strongest along the rolling direction and weakest along the 45° direction in the plane. These trends correlated with the texture quantified in terms of inverse pole figures (Fig. 22b). Basal poles were preferentially

aligned with the "L" direction, thus forcing the activation of the strong $<c+a>$ slip system in the hcp alpha lamellae. Similarly, prism poles were preferentially aligned with the 45° direction, thereby favoring the activation of the complementary (softer) prism $<a>$ systems.

The presence of a strong crystallographic texture even in a metal with a high-symmetry crystal structure (e.g., cubic) can also lead to flow curves that exhibit a directionality with test direction. For example, cast ingots of face-centered cubic metals typically have strong 100 fiber textures associated with the development of coarse columnar grains during solidification. When tested parallel or perpendicular to the columnar grain/$<100>$ fiber direction, the plastic-flow response will be different, as shown in the results for a production-scale Waspaloy (Ni-Cr-Co-Mo,

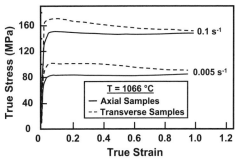

Fig. 23 Stress-strain curves from compression tests parallel (axial) or perpendicular (transverse) to the columnar grain/<100> fiber direction of a cast and homogenized Waspaloy ingot. Source: Ref 45

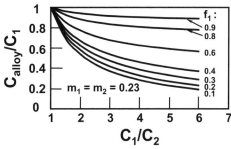

Fig. 24 Predictions from a self-consistent model of the dependence of the strength coefficient of a two-phase alloy (C_{alloy}) on the ratio of the strength coefficients of the two phases (C_1/C_2) and the volume fraction of the harder phase (f_1), assuming $m_1 = m_2 = 0.23$. Source: Ref 47

UNS N07001) ingot (Fig. 23) (Ref 45). Samples compressed transverse to the fiber-texture axis showed a peak stress following by flow softening, a behavior typical of a material undergoing discontinuous dynamic recrystallization. By contrast, the stress-strain curves from tests parallel to the fiber axis showed a lower peak stress followed by nearly steady-state flow. These differences can be explained in terms of the evolution of the Taylor factor (Ref 45). In the as-cast condition, the transverse samples had a higher Taylor factor than the axial samples and thus a higher peak stress. The Taylor factor after recrystallization of the transverse samples was similar to that initially, thus leading to the typical flow curve for material undergoing dynamic recrystallization. On the other hand, the Taylor factor of the axial samples increased during deformation due to recrystallization, thereby leading to an increment of texture hardening that counterbalanced the flow softening due to dynamic recrystallization.

Effect of Second Phases on Flow Stress. The flow stress of materials comprising two (or more) equiaxed phases (each of whose size is greater than or equal to ~10 μm) is usually a complex function of the specific volume fraction and individual flow stress of each constituent. In particular, the activation energy of the alloy (determined in accordance with Eq 34) is often found to be a function of the volume fraction and activation energy of each and thus exhibits a complex dependence on temperature.

Isostress, isostrain, and self-consistent modeling approaches have been used to understand the flow behavior of two-phase materials (Ref 46). The self-consistent analysis appears to be the best, as demonstrated by its application for alpha-beta titanium alloys (Ref 47). In brief, the analysis assumes that:

- The flow behavior of each of the two individual phases can be described by a power-law relation of the form of Eq 32.
- The m value is the same for both phases.
- The strength coefficients of the two phases, C_1 and C_2, are different.

The analysis yields the overall strength coefficient of the alloy (C_{alloy}) as a function of the

volume fraction of the harder phase, f_1, and the ratio C_1/C_2 and thus the alloy constitutive equation $\bar{\sigma} = C_{alloy}\dot{\bar{\varepsilon}}^m$. Parametric results for $m = 0.23$ are shown in Fig. 24 (Ref 47). Results for $m = 0.15$ and $m = 0.30$ (which typically span those commonly found during hot working) are similar.

The self-consistent analysis is also useful in quantifying the effect of temperature history on the flow stress of two-phase alloys (Ref 47, 48). This is especially important for conventional metalworking and high-strain-rate processes in which die chill or deformation heating, respectively, gives rise to large temperature transients. The nonequilibrium microstructure developed during such processes (which can be quite different from that in typical isothermal compression, tension, and torsion tests) is quantified by diffusion models and coupled with the self-consistent analysis to predict flow stress.

More information on the application of the self-consistent method of modeling plastic flow and texture evolution is contained in the article "Modeling and Simulation of Texture Evolution during the Thermomechanical Processing of Titanium Alloys" in *Fundamentals of Modeling for Metals Processing*, Volume 22A of the *ASM Handbook*, 2009.

ACKNOWLEDGMENTS

Portions of this article were excerpted from:

- M. Shirgaokar, Chapter 4, Flow Stress and Forgeability, *Cold and Hot Forging: Fundamentals and Applications*, T. Altan, G. Ngaile, and G. Shen, ASM International, Materials Park, OH, 2004
- S.L. Semiatin, Evolution of Microstructure during Hot Working, *Handbook of Workability and Process Design*, G.E. Dieter, H.A. Kuhn, and S.L. Semiatin, Ed., ASM International, Materials Park, OH, 2003, p 35

REFERENCES

1. T. Altan, G. Ngaile, and G. Shen, Chapter 4, *Cold and Hot Forging: Fundamentals and Applications*, ASM International, Materials Park, OH, 2004
2. E.G. Thomsen, C.T. Yang, and S. Kobayashi, *Mechanics of Plastic Deformation in Metal Processing*, MacMillan Company, New York, 1965
3. P.W. Bridgman, *Studies in Large Plastic Flow and Fracture*, McGraw-Hill, New York, 1952
4. C. Dahl, V. Vazquez, and T. Altan, "Determination of Flow Stress of 1524 Steel at Room Temperature Using the Compression Test," Report ERC/NSM-99-R-22, Engineering Research Center for Net Shape Manufacturing, The Ohio State University, Columbus, OH, 1999
5. C.H. Lee and T. Altan, Influence of Flow Stress and Friction upon Metal Flow in Upset Forging of Rings and Cylinders, *Trans. ASME, J. Eng. Ind.*, Vol 94, 1972, p 775
6. J.R. Douglas and T. Altan, Flow Stress Determination of Metals at Forging Rates and Temperatures, *Trans. ASME, J. Eng. Ind.*, Vol 97, 1975, p 66
7. T. Altan, S.-I. Oh, and H.L. Gegel, *Metals Forming: Fundamentals and Applications*, ASM International, Materials Park, OH, 1983
8. Gleeble systems, Dynamic Systems, Inc., Poestenkill, NY, http://www.gleeble.com
9. A.B. Watts and H. Ford, On the Basic Yield Stress Curve for a Metal, *Proc. Inst. Mech. Eng.*, Vol 169, 1955, p 1141
10. J. Vollmer, "Measurement of Flow Stress of Metallic Materials Mainly for High Strains and High Strain Rates," Thesis, TU Hannover, 1969 (in German)
11. G.E. Totten, K. Funatani, and L. Xie, *Handbook of Metallurgical Process Design*, CRC Press, 2004
12. K. Pohlandt, *Materials Testing for the Metal Forming Industry*, 1989
13. D.S. Fields and W.A. Backofen, Determination of Strain-Hardening Characteristics by Torsion Testing, *Proc. ASTM*, Vol 57, 1957, p 1259
14. O.S. Lee and S.M. Kim, Dynamic Material Property Characterization Using Split Hopkinson Pressure Bar (SHPB) Technique, *Nucl. Eng. Design*, Vol 226, 2003, p 119
15. G. Sundararajan and Y. Tirupataiah, The Hardness-Flow Stress Correlation in Metallic Materials, *Bull. Mater. Sci.*, Vol 17, 1994, p 747
16. S.I. Oh, S.L. Semiatin, and J.J. Jonas, An Analysis of the Isothermal Hot Compression Test, *Metall. Trans. A*, Vol 23, 1992, p 963
17. R.L. Goetz and S.L. Semiatin, The Adiabatic Correction Factor for Deformation Heating during the Uniaxial Compression Test, *J. Mater. Eng. Perf.*, Vol 10, 2001, p 710
18. T. Altan and F.W. Boulger, Flow Stress of Metals and Its Application in Metal Forming Analyses, *Trans. ASME, J. Eng. Ind.*, Vol 95, 1973, p 1009

19. F. Montheillet and J.J. Jonas, Temperature Dependence of the Rate Sensitivity and Its Effect on the Activation Energy for High-Temperature Flow, *Metall. Mater. Trans. A,* Vol 27, 1996, p 3346

20. F.J. Humphreys and M. Hatherly, *Recrystallization and Related Phenomena,* Elsevier, Oxford, U.K., 1995

21. J.J. Jonas and H.J. McQueen, Recovery and Recrystallization during High Temperature Deformation, *Treatise on Materials Science and Technology,* R.J. Arsenault, Ed., Academic Press, New York, 1975, p 394

22. H.J. McQueen and J.E. Hockett, Microstructures of Aluminum Compressed at Various Rates and Temperatures, *Metall. Trans.,* Vol 1, 1970, p 2997

23. R.D. Doherty, D.A. Hughes, F.J. Humphreys, J.J. Jonas, D. Juul Jensen, M.E. Kassner, W.E. King, T.R. McNelley, H.J. McQueen, and A.D. Rollett, Current Issues in Recrystallization: A Review, *Mater. Sci. Eng. A,* Vol 238, 1997, p 219

24. J.-P.A. Immarigeon and J.J. Jonas, The Deformation of Armco Iron and Silicon Steel in the Vicinity of the Curie Temperature, *Acta Metall.,* Vol 22, 1974, p 1235

25. A. Yoshie, H. Mirikawa, and Y. Onoe, Formulation of Static Recrystallization of Austenite in Hot Rolling Process of Steel Plate, *Trans. ISIJ,* Vol 27, 1987, p 425

26. A. Laasraoui and J.J. Jonas, Prediction of Steel Flow Stresses at High Temperatures and Strain Rates, *Metall. Trans. A,* Vol 22, 1991, p 1545

27. U.F. Kocks, Laws for Work Hardening and Low-Temperature Creep, *Trans. ASME, J. Eng. Mater. Technol.,* Vol 98, 1976, p 76

28. W. Roberts, Dynamic Changes That Occur during Hot Working and Their Significance Regarding Microstructural Development and Hot Workability, *Deformation, Processing, and Structure,* G. Krauss, Ed., American Society for Metals, Metals Park, OH, 1984, p 109

29. H.D. Ryan and H.J. McQueen, Comparative Hot Working Characteristics of 304, 316, and 317 Steels, Both Cast and Worked, *Inter. Conf. on New Developments in Stainless Steel Technology,* R.A. Lula, Ed., American Society for Metals, Metals Park, OH, 1985, p 293

30. C.M. Sellars, Modeling Microstructure Evolution, *Mater. Sci. Technol.,* Vol 6, 1990, p 1072

31. C. Devadas, I.V. Samarasekara, and E.B. Hawbolt, Thermal and Metallurgical State of Steel Strip during Hot Rolling, *Metall. Trans. A,* Vol 22, 1991, p 335

32. G. Shen, S.L. Semiatin, and R. Shivpuri, Modeling Microstructural Development during the Forging of Waspaloy, *Metall. Mater. Trans. A,* Vol 26, 1995, p 1795

33. T. Sakai and J.J. Jonas, Dynamic Recrystallization: Mechanical and Microstructural Considerations, *Acta Metall.,* Vol 32, 1984, p 189

34. S.L. Semiatin and T.R. Bieler, The Effect of Alpha Platelet Thickness on Plastic Flow during Hot Working of Ti-6Al-4V with a Transformed Microstructure, *Acta Mater.,* Vol 49, 2001, p 3565

35. D.J. Abson and J.J. Jonas, Hot Compression Behavior of Thermomechanically Processed Alpha Zirconium, *Metals Technol.,* Vol 4, 1977, p 462

36. R. Armstrong, I. Codd, R.M. Douthwaite, and N.J. Petch, The Plastic Deformation of Polycrystalline Aggregates, *Philos. Mag.,* Vol 7, 1962, p 45

37. J.D. Eshelby, The Distribution of Dislocations in an Elliptical Glide Zone, *Phys. Stat. Solidi,* Vol 3, 1963, p 2057

38. R.C. Gifkins, Grain-Boundary Sliding and Its Accommodation during Creep and Superplasticity, *Metall. Trans. A,* Vol 7, 1976, p 1225

39. A.K. Ghosh, *Metalworking: Bulk Forming,* Vol 14A, *ASM Handbook,* 10th ed., S.L. Semiatin, Ed., ASM International, Materials Park, OH, 2005, p 563

40. J.E. Bird, A.K. Mukherjee, and J.E. Dorn, *Quantitative Relation between Microstructure and Properties,* D.G. Brandon and A. Rosen, Ed., Israel Universities Press, Jerusalem, Israel, 1969, p 255

41. T.G. Langdon, Grain-Boundary Sliding Revisited: Developments in Sliding over Four Decades, *J. Mater. Sci.,* Vol 41, 2006, p 597

42. G.A. Sargent, A.P. Zane, P.N. Fagin, A.K. Ghosh, and S.L. Semiatin, Low-Temperature Coarsening and Plastic Flow Behavior of an Alpha/Beta Titanium Billet Material with an Ultrafine Microstructure, *Metall. Mater. Trans. A,* Vol 39, 2008, p 2949

43. S.L. Semiatin, M.W. Corbett, P.N. Fagin, G.A. Salishchev, and C.S. Lee, Dynamic-Coarsening Behavior of an Alpha/Beta Titanium Alloy, *Metall. Mater. Trans. A,* Vol 37, 2006, p 1125

44. S.L. Semiatin and T.R. Bieler, Effect of Texture and Slip Mode on the Anisotropy of Plastic Flow and Flow Softening during Hot Working of Ti-6Al-4V, *Metall. Mater. Trans. A,* Vol 32, 2001, p 1787

45. S.L. Semiatin, D.S. Weaver, P.N. Fagin, M.G. Glavicic, R.L. Goetz, N.D. Frey, R. C. Kramb, and M.M. Antony, Deformation and Recrystallization Behavior during Hot Working of a Coarse-Grain, Nickel-Base Superalloy Ingot Material, *Metall. Mater. Trans. A,* Vol 35, 2004, p 679

46. L. Briottet, J.J. Jonas, and F. Montheillet, A Mechanical Interpretation of the Activation Energy of High-Temperature Deformation in Two-Phase Materials, *Acta Mater.,* Vol 44, 1996, p 1665

47. S.L. Semiatin, F. Montheillet, G. Shen, and J.J. Jonas, Self-Consistent Modeling of the Flow Behavior of Wrought Alpha/Beta Titanium Alloys under Isothermal and Nonisothermal Hot-Working Conditions, *Metall. Mater. Trans. A,* Vol 33, 2002, p 2719

48. G. Shen, S.L. Semiatin, E. Kropp, and T.-Altan, A Technique to Compensate for Temperature-History Effects in the Simulation of Nonisothermal Forging Processes, *J. Mater. Proc. Technol.,* Vol 33, 1992, p 125

SELECTED REFERENCES

• T. Altan, S.L. Semiatin, and G.D. Lahoti, Determination of Flow Stress Data for Practical Metal Forming Analysis, *Ann. CIRP,* Vol 30 (No. 1), 1981, p 129

• G.D. Lahoti and T. Altan, Prediction of Temperature Distributions in Axisymmetric Compression and Torsion, *J. Eng. Mater. Technol.,* April 1975, p 113

ASM Handbook, Volume 22B, *Metals Process Simulation*
D.U. Furrer and S.L. Semiatin, editors

Copyright © 2010, ASM International®
All rights reserved.
www.asminternational.org

Grain-Boundary Energy and Mobility

G. Gottstein and D.A. Molodov, Institute of Physical Metallurgy and Metal Physics, RWTH Aachen University, Germany
L.S. Shvindlerman, Institute of Solid State Physics, Russian Academy of Sciences

GRAIN BOUNDARIES are interfaces between crystallites of the same phase but different crystallographic orientation, that is, different spatial orientation of the crystallographic unit cell. At the same time, they constitute the internal surfaces of the adjacent crystals. Due to the crystalline structure of the adjoining grains, the atomic arrangement in the boundary will depend on the misorientation across the boundary and the spatial orientation of the boundary (Ref 1). A flat grain-boundary element is geometrically characterized by (Fig. 1):

- The misorientation, which can be expressed as a rotation (three parameters, e.g., three Euler angles)
- The orientation of the boundary plane with respect to one of the two crystals (two parameters, e.g., unit vector of grain-boundary normal)
- A displacement of the two crystals (three parameters, i.e., the three-dimensional translation vector).

Additionally, because of the local atomic interactions, the atomic arrangement in the boundary will relax to assume a minimum energy configuration. Five of these parameters, that is, the misorientation and grain-boundary plane, can be controlled macroscopically. Crystal translation and local relaxations are adjustments on the atomic scale that are taken care of by the boundary itself. The classification of grain boundaries can therefore be confined to the five macroscopic parameters.

Because the misorientation consists of a rotation between the coordinate systems of the adjacent crystals, it can be expressed by a rotation axis, **a**, and angle, ω. The orientation of the boundary plane is unambiguously determined by the unit vector of the grain-boundary normal, **n**. The infinite number of possible combinations of crystal misorientation and boundary orientation can be subdivided into three groups (Fig. 2):

- Twist boundaries with the boundary plane perpendicular to the rotation axis (**a** ∥ **n**)
- Tilt boundaries with the boundary plane parallel to the rotation axis (**a**⊥**n**)

- Mixed boundaries that encompass all other boundaries but can be decomposed into twist and tilt components

There is an infinite number of planes parallel to a given direction; that is, there is an infinite number of tilt boundaries for a given misorientation. Among all tilt boundaries, those that stand out are where the boundary plane is a mirror plane of the two crystals. They are referred to as symmetrical tilt boundaries. Other tilt boundaries are termed asymmetrical tilt boundaries and characterized by an angular difference to the symmetric orientation (Ref 2). It is noted, however, that crystal symmetry renders the definition of grain-boundary character ambiguous. For instance, cubic crystal symmetry (e.g., a 90° rotation about a cube edge) generates 24 crystallographically equivalent orientation relationships. Therefore, a 60°<111> orientation relationship is equivalent to 70.5°<110> or 180°<112>, and so on. Hence, the common {111} boundary plane would be a twist boundary in the first notation but a tilt boundary in the second or third notation. This ambiguity can be removed by restricting the orientation relationship to the smallest angle of rotation.

Grain-Boundary Energy

Grain boundaries can be characterized as being low angle or high angle.

Low-Angle Boundaries

Read and Shockley (Ref 3) have shown that low-angle misorientations—typically rotation angles of less than 15°, regardless of the boundary plane—can be compensated by periodic dislocation arrangements (Fig. 3). A symmetrical low-angle tilt boundary is composed of a periodic arrangement of a single set of edge dislocations. Asymmetrical low-angle tilt boundaries require a second set of edge dislocations to accommodate the deviation from the symmetric boundary orientation. The simplest low-angle twist grain boundary is comprised of two periodic sets of mutually perpendicular screw dislocations, for instance, (100) twist

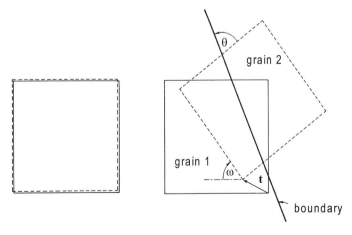

Fig. 1 Four parameters are needed to mathematically define a two-dimensional grain boundary: the orientation relationship as expressed by a rotation angle, ω; an angle, θ, that defines the spatial orientation of the grain boundary with respect to one of the grains; and the components t_1, t_2 of the translation vector **t** that characterizes the displacement of the two crystals with respect to each other. In three dimensions, the orientation relationship requires three parameters, such as the three Euler angles; the grain-boundary normal is given by its unit vector, that is, by two parameters; and the translation vector has three independent components.

boundaries. Other twist boundaries are typically comprised of hexagonal grids of three sets of screw dislocations. Low-angle mixed boundaries are composed of a network of mixed dislocations. For a low-angle boundary with a defined misorientation and boundary plane, the geometrically necessary dislocation content is given by Frank's rule (Ref 5):

$$\mathbf{B} = \Sigma\, c_i(\mathbf{Y})\mathbf{b}_i = 2\sin\frac{\omega}{2}(\mathbf{Y} \times \mathbf{a}) \qquad \text{(Eq 1)}$$

where $\mathbf{Y}$ is an arbitrary vector in the grain boundary; $\mathbf{B}$ is the closure failure, that is, the sum of the Burgers vectors, $\mathbf{b}_i$, of the dislocations intersected by $\mathbf{Y}$; and ω and $\mathbf{a}$ are the angle and axis, respectively, of misorientation. Frequently used symbols are listed in Table 1.

If the dislocation arrangement of a low-angle boundary is known, the grain-boundary energy per unit area, γ, can be calculated from the elastic strain energy of the structural (intrinsic) dislocations. For the energy per unit area of low-angle tilt boundaries, Read and Shockley obtained:

$$\gamma = \frac{\mu b(\cos\psi + \sin\psi)}{4\pi(1-\nu)}\cdot\omega[A - \ln\omega] \qquad \text{(Eq 2a)}$$

with

$$A = 1 + \ln\frac{b}{2\pi r_0} - \frac{\sin 2\psi}{2}$$
$$-\frac{\sin\psi\cdot\ln(\sin\psi) + \cos\psi\cdot\ln(\cos\psi)}{\sin\psi + \cos\psi} \qquad \text{(Eq 2b)}$$

That is, for the symmetrical tilt boundary ($\Psi = 0$):

$$\gamma = \frac{\mu b}{4\pi(1-\nu)}\cdot\omega\left[1 + \ln\frac{b}{2\pi r_0} - \ln\omega\right] \qquad \text{(Eq 2c)}$$

Here, μ is the shear modulus, ν is Poisson's ratio, r_0 is the radius of the dislocation core, and ψ is the deviation from the symmetrical boundary plane.

In the Read and Shockley model, the energy of the dislocation core is kept constant, regardless of the dislocation content. This certainly holds for low misorientation angles, ω, where the dislocation spacing, d, is large, because for symmetrical tilt boundaries:

$$\sin\frac{\omega}{2} = \frac{b}{2d} \qquad \text{(Eq 3a)}$$

or

$$\omega \approx \frac{b}{d} \qquad \text{(Eq 3b)}$$

For larger misorientations, the core structure of the dislocations may change and, correspondingly, the core energy and thus the grain-boundary energy. As a matter of fact, molecular statics simulations of grain-boundary structure demonstrate a significant deviation of the atomistically computed grain-boundary energy from the predictions of the Read-Shockley model for larger misorientation angles (Ref 6, 7). Because the energy of the boundary depends on the dislocation content and arrangement, it changes with misorientation axis, $\mathbf{a}$, angle, ω, and boundary plane, $\mathbf{n}$.

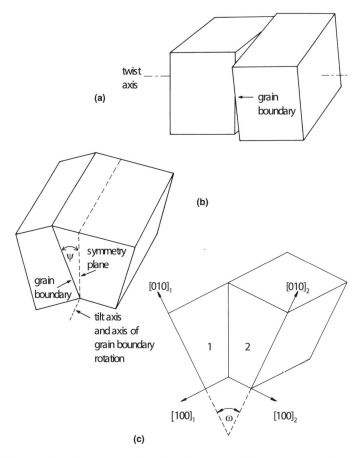

Fig. 2 Relative orientation of grain boundaries and rotation axes for different types of grain boundaries. (a) Twist boundary. (b) Asymmetrical tilt boundary. (c) Symmetrical tilt boundary

High-Angle Grain Boundaries

For misorientation angles larger than 15°, the Read-Shockley model breaks down, because the dislocation cores tend to overlap to form a contiguous boundary layer. Formally, the dislocation concept can also be extended to high-angle boundaries. In fact, a periodic arrangement of dislocations is equivalent to a so-called coincidence rotation, which generates a superlattice from the two rotated lattices called the coincidence site lattice (CSL) (Fig. 4). It is characterized by Σ, the volume of its unit cell normalized by the volume of the crystal unit cell. Deviations from the exact coincidence rotation can be compensated by secondary grain-boundary dislocations that have Burgers vectors of the so-called displacement shift complete lattice, which is composed of all difference vectors of the superimposed two lattices, that is, the coarsest grid that contains all lattice points (Ref 4). The introduction of secondary boundary dislocations does not change the size of the CSL but only the location of the coincidence sites. Therefore, the structure of a grain boundary can be formally understood as being composed of a primary dislocation arrangement that comprises the CSL and secondary dislocations that compensate the difference from the nearest CSL rotation. The CSL concept is based on the orientation relationship only; that is, it refers to the lattices of the adjoining crystals but disregards the grain-boundary plane. The density of coincidence sites may substantially change with the spatial orientation of the grain-boundary plane, which will affect grain-boundary properties. An example is the first-order twin boundary, which is described by a 60°<111> orientation relationship. In the common (111) plane (coherent twin boundary), all lattice sites are coincidence sites, whereas in planes deviating from the coherent boundary (incoherent twin boundaries), the density of coincidence sites is smaller, which is reflected by different properties, such as a larger grain-boundary energy.

The CSL construct, however, is only of geometrical nature and disregards atomic interactions to minimize the boundary energy. Computer simulations by molecular statics and molecular dynamics methods have been successfully used for a computation of grain-boundary structure and energy. The results show that the atomic relaxations destroy the coincidence sites but maintain the periodicity of the dislocation arrangements (Fig. 5) (Ref 8). The comparison of computed grain-boundary structure with high-resolution electron microscopy

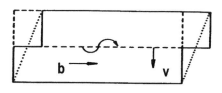

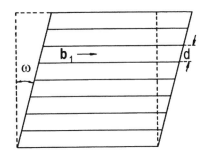

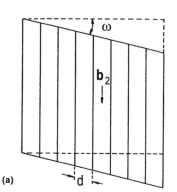

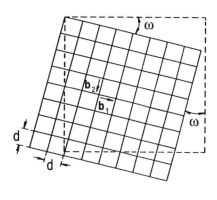

(a)

Table 1 List of symbols

Symbol	Meaning
a	Axis of rotation
ω	Angle of rotation
Y	Vector in boundary
B	Closure failure
b	Burgers vector
n	Grain-boundary normal vector
ψ	Angle of asymmetry
μ	Shear modulus
d	Dislocation spacing
α	Contact angle
Θ	Inclination of grain boundary
σ	Surface tension
D	Diameter
g	Gibbs free energy per unit volume
H	Enthalpy, magnetic field strength
k	Boltzmann constant, 1.38066×10^{-23} J/K
R	Radius of curvature
Φ	Interatomic potential
r_{ij}	Atomic distance
γ	Grain-boundary energy, grain-boundary surface tension
V	Grain-boundary velocity
P	Driving force for grain-boundary migration
m	Grain-boundary mobility
A	Reduced grain-boundary mobility (product $m \cdot \gamma$)
ρ	Dislocation density
χ	Magnetic susceptibility
τ	Shear stress

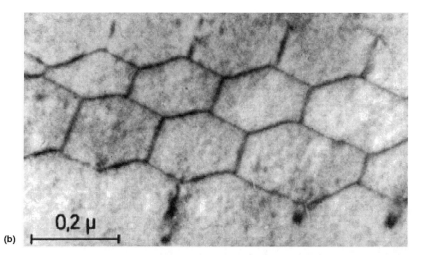

(b)

Fig. 3 Basic dislocation configuration of a low-angle twist boundary. (a) A single family of parallel screw dislocations results in a shear deformation, but two perpendicular families of dislocations result in a pure rotation. (b) Transmission electron microscopy image of a low-angle twist boundary in α-iron. The hexagonal dislocation configuration is composed of screw dislocations with three different Burgers vectors. Source: Ref 4

$$\frac{\gamma_1}{(1 + \varepsilon_2 - \varepsilon_3)\sin \alpha_1 + (\varepsilon_3 - \varepsilon_1)\cos \alpha_1} =$$
$$= \frac{\gamma_2}{(1 + \varepsilon_1 - \varepsilon_3)\sin \alpha_2 + (\varepsilon_1 - \varepsilon_3)\cos \alpha_2} = \quad \text{(Eq 5)}$$
$$= \frac{\gamma_3}{(1 + \varepsilon_1 - \varepsilon_2)\sin \alpha_3 + (\varepsilon_2 - \varepsilon_3)\cos \alpha_3}$$

The influence of these so-called torque terms is important in the vicinity of special misorientations or special grain-boundary planes, for example, twin relationships or coherent twin boundaries. For random grain boundaries and far-from-special misorientations, the values of $(\partial \gamma_i / \partial \Theta_i)$ are small enough, and Eq 4 gives a reasonable approximation.

The measurement of dihedral angles (Fig. 6a) to determine the grain-boundary surface tension was used in many experimental investigations (Ref 10 to 12). It is especially suitable for grown tricrystals, where two grain boundaries have the same misorientation and the surface tension of the third can be calculated, or when the observation of the change of the surface tension of the third grain boundary is the main purpose of the investigation (Fig. 7). Such experiments make it possible to compare the surface tensions of different grain boundaries or grain boundaries with different orientation in the same material. It was shown, in particular, that the minima of the misorientation dependence of the grain-boundary surface tension correlates with special low-Σ misorientations.

Another method of investigation of the relative grain-boundary surface tension was proposed by Wilson and Shewmon (Ref 14) and was applied in many experimental studies (Ref 15 to 17). If a small, single-crystal ball is placed on a single-crystal substrate of the same

images of such boundaries reveals good agreement. Therefore, most current information on grain-boundary structure and energy is derived from computer simulations.

Measurements of Grain-Boundary Energy

In a first-order approximation, the grain-boundary energy per unit area is equivalent to the grain-boundary surface tension. A brief summary of different schemes for measuring grain-boundary surface tension is given in Fig. 6. If there is a mechanical equilibrium at a grain-

boundary triple junction, and the grain-boundary surface tension, γ, does not depend on the orientation of the boundaries, the relation between the surface tension of different boundaries can be found from Young's theorem (Fig. 6a):

$$\frac{\gamma_1}{\sin \alpha_1} = \frac{\gamma_2}{\sin \alpha_2} = \frac{\gamma_3}{\sin \alpha_3} \quad \text{(Eq 4)}$$

where α_i are the dihedral angles opposite to γ_i. If the surface tensions of the grain boundaries depend on their orientation, the dependency $\varepsilon_i \equiv \partial \gamma_i / \partial \Theta_i$ also must be taken into account where Θ_i is the inclination of the boundary with regard to a reference plane (Ref 9):

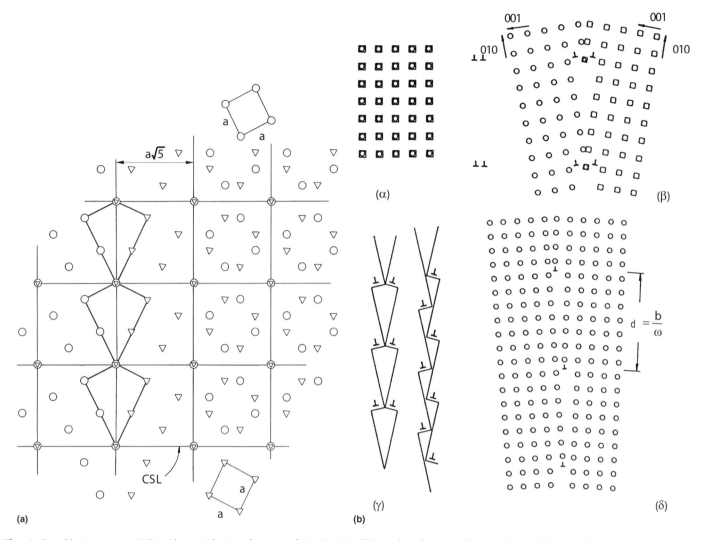

Fig. 4 Crystal lattice structure. (a) Coincidence site lattice and structure of a 36.9°<100> (Σ5) grain boundary in a cubic crystal lattice. Right side of figure: grain-boundary plane ∥ plane of the paper (twist boundary). Left side of figure: grain-boundary plane ⊥ plane of the paper (tilt boundary). (b) Relationship between the coincidence site lattice and the primary dislocation structure at a grain boundary. If two identical, interlocking lattices (α) are rotated symmetrically away from each other about an axis perpendicular to the plane of view (β), a coincidence site lattice forms at specific angles of rotation. The coincidence points are marked by overlapping circles and squares. The associated configuration of the resulting double dislocation is relaxed along the boundary (γ), and the structure of a symmetrical low-angle tilt boundary forms (δ)

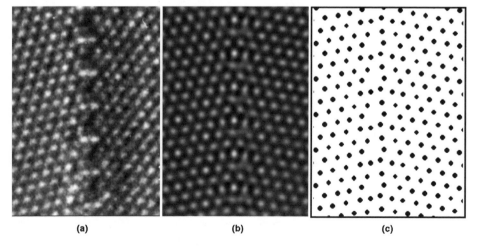

Fig. 5 Atomic structure of a Σ7 boundary in aluminum. (a) High-resolution electron microscopy (HREM) image. (b) Simulated HREM image using the relaxed grain-boundary structure. (c) Computed by molecular dynamics. Source: Ref 8

material at diffusion temperature, it will rotate during sintering to the substrate to decrease the surface tension of the newly formed grain boundary. The main condition for a proper conduct of the experiment is that the misorientation dependence of the grain-boundary surface tension must have singularities and sharp minima with a discontinuity of the derivative of the surface tension with respect to misorientation angle, ω (Fig. 6b). The driving force for the rotation of the balls is proportional to the derivative ($d\gamma/d\omega$).

To determine the absolute value of the grain-boundary surface tension, the energy of the reference surface must be known. For a known free surface tension, the grain-boundary surface tension can be determined from the equilibrium of the surface tensions at the root of a thermal groove (Fig. 6c). Unfortunately, there are only a few methods to determine the tension of a free surface of the crystal.

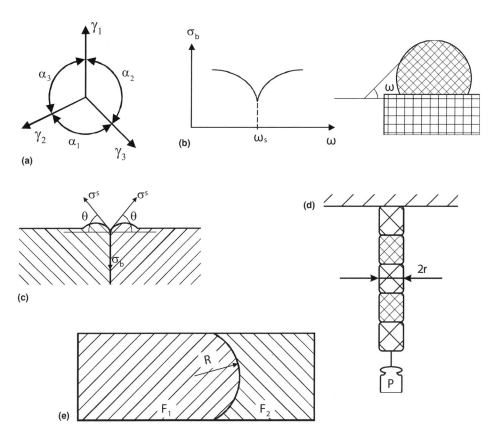

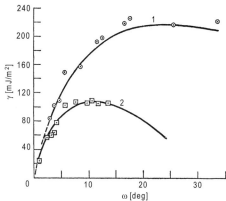

Fig. 7 Misorientation dependence of grain-boundary energy for <100> tilt grain boundaries in lead (curve 1) and tin (curve 2). The relative energy was measured according to the method in Fig. 6(a). The absolute energy was estimated by using the Read-Shockley equation. Source: Ref 13

Fig. 6 Methods of grain-boundary surface tension measurement. (a) Equilibrium angles at triple junction. (b) Rotating ball method: sintering of small, single-crystal balls to a single-crystal substrate. (c) Thermal groove method. (d) Zero-creep method. (e) Balance of grain-boundary surface tension and volume driving force

Table 2   Experimentally measured average grain-boundary energies

Metal	γ, mJ/m^2	T °C	γ/σ	γ_{ctw}/γ	γ_{nctw}/γ	$d\gamma/dT$, mJ/(m$^2 \cdot$ °C)
Ag	375	950	0.25	0.03	0.33	−0.10
Al	324	450	...	0.23	...	−0.12
Au	387	1000		0.039	...	−0.10
		1035	0.28			
Co	650	1354	...	...	...	...
Cr	920	1400		...	...	...
		1350	0.40			
Cu	625	925	0.35	0.035	0.80	−0.10
Fe (δ-phase)	468	1450	0.24	...	...	−0.25
Fe (γ-phase)	756	1350	0.36	...	...	−1.0
Fe-3wt%Si	617	1100	...	...	...	0.07
Fe-Cr-Ni stainless steel 304	835	1060	0.38	0.024	0.25	−0.49
Mo	575	2350	...	...	...	...
Nb	756	2250	0.36	...	...	...
Ni	866	1060	0.38	0.050	...	−0.2
Ni-20wt%Cr	756	1060	0.35	0.025	...	...
Pt	660	1300		...	...	−0.18
		1500	0.24			
Sn	164	223	0.24	...	...	...
W	1080	2000	0.36	...	...	...
Zn	340	300	...	...	...	...

γ, grain-boundary energy; σ, surface energy (solid-vapor case); γ_{ctw}, coherent twin grain-boundary energy; γ_{nctw}, noncoherent twin grain-boundary energy. Source: Ref 23

One of them is the zero-creep method (Ref 18) (Fig. 6d). Imagine a wire of small diameter D with perimeter $\pi \cdot D$ that is loaded with a weight, P, at diffusional creep temperature. The free energy of the wire can be defined as:

$$F = P_X \cdot \gamma \pi D L \qquad \text{(Eq 6a)}$$

where x is the coordinate of the weight P relative to the arbitrary point on the normal to the Earth's surface, L is the length of the wire, and γ is the surface tension of the free surface of the wire. The equilibrium ($\delta F = 0$) corresponds to such changes of x and L that compensate each other. Taking into account that $\delta x = -\delta L$:

$$\left(\frac{\delta F}{\delta x}\right)_{x=0} = P_0 - \gamma \pi D = 0 \qquad \text{(Eq 6b)}$$

and $\gamma = P_0/\pi D$, where P_0 is the zero-creep weight.

A more correct description takes into account the area of the grain boundaries and their shape (Ref 19 to 21). Evidently, the value of the free surface tension determined in this way is averaged over a large number of crystallographic planes.

The grain boundary is a degenerate kind of interface, because it separates two parts of the same phase. If an external tensorial field is applied to the system, the degeneracy will be removed, and, from a thermodynamical point of view, the grain boundary can be described as an interface. This provides a method to measure the absolute value of the grain-boundary surface tension. If an external field is applied to a system with a grain boundary, which is locked at the surface of a sample, a free energy differential of the adjacent grains will arise (per unit volume: g$_2$−g$_1$); due to this difference, the initially flat grain boundary will bend like a membrane under a pressure to assume a radius of curvature R. In this case the grain-boundary surface tension can be estimated as (Fig. 6e):

$$\gamma = R(g_2 - g_1) \qquad \text{(Eq 7)}$$

This method was applied to polycrystalline bismuth in a magnetic field (Ref 22).

Experimental Results. Several grain-boundary energy measurements have been determined. Average high-angle grain-boundary energies for various metals and alloys are listed in Table 2. They vary from 0.1 to 1 J/m^2, with an average of the order of 0.5 J/m^2. The dependency of grain-boundary energy on misorientation angle for a given axis of rotation shows an increase of grain-boundary energy with

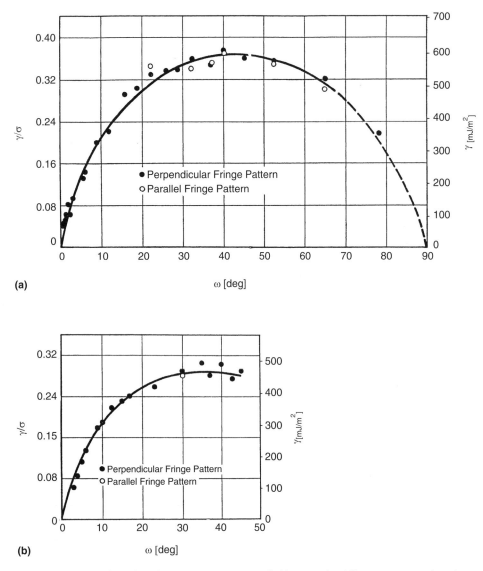

(a)

(b)

Fig. 8 Dependence of grain-boundary energy on misorientation for (a) <001> tilt and (b) <001> twist grain boundaries in copper at 1065 °C. The dihedral angles at thermal grooves were measured by interference microscopy; the absolute values for the boundary energy were determined by assuming a surface energy of 1.67 J/m². Source: Ref 24

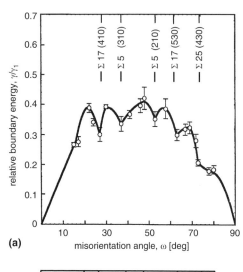

(a)

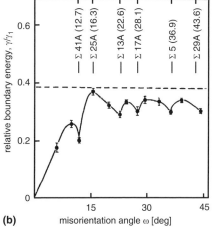

(b)

Fig. 9 Grain-boundary energy versus misorientation for (a) <001> symmetric tilt (Ref 25) and (b) <001> twist grain boundaries (Ref 26) in copper. $T = 1273$ K (100d °C). Grain boundaries with dispersed SiO$_2$ particles were observed by transmission electron microscopy. The boundary energies were determined with reference to isotropic Cu-SiO$_2$ interfacial energy by measuring the misorientation-dependent shape of the SiO$_2$ particles in the boundary.

misorientation angle in accordance with the predictions of the Read-Shockley theory (Fig. 7). Above a critical angle in the range of 15°, the Read-Shockley model breaks down, and the energy remains essentially constant, except for cusps at specific misorientation angles that relate to low-Σ coincidence boundaries (Fig. 8 to 11). For a given angle of misorientation, the grain-boundary energy will also change with rotation axis.

Experimental observations demonstrated that not all grain-boundary configurations in polycrystals occur with the same frequency and that the relative free energies of the different boundary configurations influence the population distribution (Ref 29).

Atomistic Simulations of Grain-Boundary Energy. In view of the experimental difficulties to determine grain-boundary energy, computer simulation tools have been used to calculate the grain-boundary energy. In fact, most currently available data on grain-boundary energy stem from computer simulations. There are essentially two atomistic approaches for calculating the grain-boundary energy: molecular statics and molecular dynamics simulations. Both methods consider an arrangement of atoms that comprises two adjacent grains with a grain boundary between them. The energy of the grain boundary is the difference of the energy of a perfect crystal and the energy of two crystals with a grain boundary. The energy results from the interaction of the atoms as described by the interatomic potential, Φ. In molecular statics simulations, the arrangement of the atoms in the boundary is changed step by step by displacement, removal, or addition of atoms, until the minimum energy configuration is found. In molecular dynamics simulations, Newton's equation of motion is solved

for each atom and degree of freedom, where the force is given by the gradient of the interatomic potential. The equilibrium structure is found for the condition that the force vanishes for each atom, and, from the energy of that configuration, the grain-boundary energy is calculated as for molecular statics computations.

It is evident that the quality of the computation sensitively depends on the interatomic potential used. Simple pair potentials, such as the most frequently used Lennard-Jones potential:

$$\Phi(r_{ij}) = \frac{A}{r_{ij}^6} - \frac{B}{r_{ij}^{12}} \qquad (Eq\ 8)$$

where r_{ij} is the radial distance of atoms i and j, and A and B are constants, may not correctly account for the electronic interactions. Embedded atom potentials (Ref 30), cluster potentials (Ref 31), or bond order potentials (Ref 32) offer

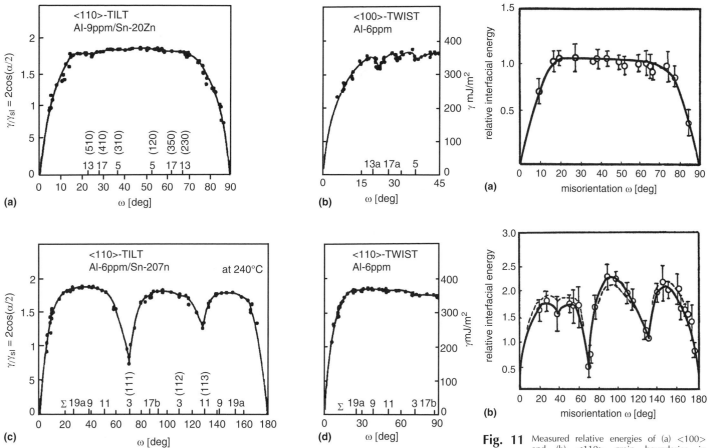

Fig. 10 Misorientation dependence of grain-boundary energy for (a) <100> tilt, (b) <100> twist, (c) Sn-20 Zn <110> tilt, and (d) <110> twist grain boundaries in aluminum. The dihedral angles, α, at the boundary grooves of aluminum wetted with tin and Sn-20%wtZn were measured by optical microscopy. The absolute values for the boundary energy were estimated assuming a solid-liquid interfacial energy of 200 mJ/m² that was estimated from the slope of the γ-ω plot according to the Read-Shockley equation. Source: Ref 27

Fig. 11 Measured relative energies of (a) <100> and (b) <110> grain boundaries in aluminum as a function of misorientation (between <001> directions). Relative energy has been measured from equilibrium angles at the triple junctions in tricrystals. Source: Ref 28

improved accuracy, but one must not forget that all potentials are approximations and do not predict all material constants with the same accuracy. The development of potentials from quantum-mechanical ab initio simulations is currently a promising new approach for improving the predictive power of atomistic simulations.

It is finally noted that the computed grain-boundary energies mostly pertain to absolutely pure materials. Real materials always contain a certain impurity level. Even a bulk material of 99.9999% purity still contains on the order of 10^{16} foreign atoms per cubic centimeter, most of which have a tendency to segregate to grain boundaries and decrease the grain-boundary energy. For grain-boundary energy computations, this may be less serious than for atomistic simulations of grain-boundary mobility.

Computed grain-boundary energies for various boundaries in copper and aluminum are listed in Table 3. The absolute values are of the same order as measurements, and very low values identify special boundaries with deep cusps, such as coherent twin boundaries.

The dependency of grain-boundary energy on misorientation angle confirms the general tendency described in the sections on low- and high-angle boundaries and also a dependence on rotation axis (Fig. 12, 13). Besides misorientation, inclination is also found to affect grain-boundary energy, not only for the first-order twin boundary (Σ3) (Fig. 14) but also for second-order twin boundaries (Σ9, Fig. 15).

An inclination dependence was also reported for intermetallic compounds (Fig. 16). Also, low-angle grain boundaries show a distinct dependence on grain-boundary inclination (Fig. 17).

Alloys. Grain boundaries constitute a perturbation of the perfect crystal structure, and thus, their introduction is associated with an increase of free energy. Consequently, the grain-boundary (free) energy in pure metals (or grain-boundary surface tension) is always positive. When alloying elements are added to a pure metal, the free energy decreases due to an increase of the entropy of mixing (Ref 40). Impurities tend to segregate to grain boundaries to harmonize their chemical potential. The adsorption of impurities at grain boundaries as

Table 3 Computed energies of symmetrical tilt grain boundaries in copper and aluminum

Embedded atom method was used

Grain boundary(a)	ω, degrees	γ, J/m²
Copper		
Σ 5 (210)[001]	53.13	0.952
Σ 5 (310)[001]	36.87	0.905
Σ 3 (1̄1̄1)[011]	109.47	0.022
Σ 3 (21̄1̄)[011]	70.53	0.594
Σ 9 (1̄22)[011]	141.06	0.834
Σ 9 (41̄1̄)[011]	38.94	0.662
Σ 11 (31̄1̄)[011]	50.48	0.301
Σ 11 (23̄3̄)[011]	129.52	0.703
Σ 7 (2̄31)[111]	38.21	0.867
Σ 7 (41̄5̄)[111]	141.79	0.775
Σ 13 (3̄4̄1)[111]	27.80	0.842
Σ 13 (52̄7̄)[111]	152.20	0.892
Aluminum		
Σ 5 (210)[001]	53.13	0.451
Σ 5 (310)[001]	36.87	0.466
Σ 9 (1̄22)[011]	141.06	0.440
Σ 11 (31̄1̄)[011]	50.48	0.146
Σ 11 (2̄31)[011]	38.21	0.470
Σ 13 (3̄4̄1)[111]	27.80	0.458

(a)The notation in terms of Σ y (hkl)[rst] refers to the value of Σ, y; the grain-boundary plane, (hkl); and the rotation axis [rst]. Source: Ref 33

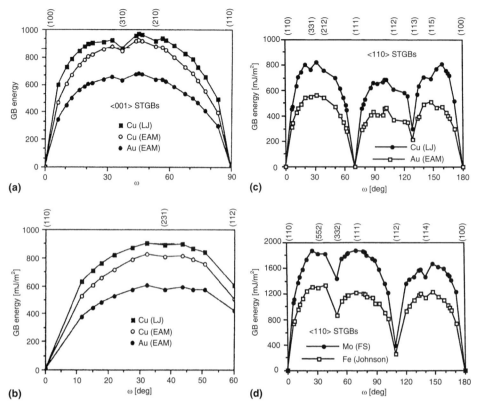

Fig. 12 Computed grain-boundary (GB) energy in copper and aluminum. Symmetrical tilt grain boundaries (STGBs) (a) <001>, (b) <111>, (c) <110>, (d) Symmetrical <110> tilt grain boundaries in molybdenum and iron, LJ, EAM, FS, and Johnson indicate the potential used. LJ, Lennard-Jones; EAM, embedded atom method; FS, Finnis-Sinclair; Johnson interatomic potential. Source: Ref 6, 7

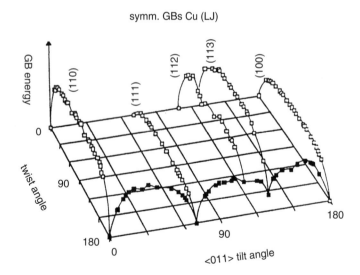

Fig. 13 Computed energies for various twist and <110> tilt grain boundaries (GBs) in copper (schematic). The energies of {110}, {111}, {112}, {113}, and {100} twist boundaries are represented by the curves running parallel to the twist angle axis. Note that at a twist angle of 180°, these boundaries are identical with symmetrical <110> tilt boundaries. LJ, Lennard-Jones. Source: Ref 6

a function of temperature is described by the adsorption isotherm. Overall, adsorption decreases with increasing temperature. The grain-boundary energy depends on the adsorption in such a way that the grain-boundary energy decreases with increasing segregation and thus

increasing alloy content (Fig. 18, 19). Principally, it may become zero or negative, which would have serious effects on grain-boundary-energy-driven processes such as grain growth (Ref 43). Grain-boundary adsorption not only affects grain-boundary energy but also grain-

boundary structure and mobility. The underlying thermodynamics are introduced in the section on grain-boundary mobility.

Grain-Boundary Mobility

Because in all relevant metallurgical phenomena, the driving force for grain-boundary migration is much smaller than the thermal energy, grain-boundary motion is a drift motion, and the velocity, V, is proportional to the acting driving force P. The ratio:

$$m = \frac{V}{P} \qquad \text{(Eq 9)}$$

is referred to as grain-boundary mobility (Ref 2).

Grain-boundary motion is a thermally activated process. Therefore, the mobility has an Arrhenius temperature dependence:

$$m = m_0 \exp\left(-\frac{H}{kT}\right) \qquad \text{(Eq 10a)}$$

where H is the activation enthalpy of grain-boundary migration.

The product:

$$A = m \cdot \gamma = A_0 \exp\left(-\frac{H}{kT}\right) \qquad \text{(Eq 10b)}$$

is referred to as the reduced mobility, where γ is the grain-boundary surface tension.

Equations (9) and (10) can be derived from simple rate theory by considering diffusive jumps of atoms through a boundary between crystals of different energy density. It provides the important information that the grain-boundary velocity is proportional to the driving force and that it has an Arrhenius-type temperature dependence. However, as evident from measurements and simulations of grain-boundary motion presented subsequently, the boundary migration process is much more complicated than assumed in this simple concept, so that absolute grain-boundary mobilities and activation parameters are likely to essentially differ from the predictions of this model.

It should be stressed that only the noncompensated exchange by the lattice points leads to grain-boundary displacement.

Measurement of Grain-Boundary Migration

During annealing of a polycrystal, grain growth occurs. The grain-boundary mobility can be extracted from the temporal change of the grain size during annealing. The major part of the data of grain-boundary mobility was derived just from such kind of experiments (see example in Fig. 20). However, measurements of grain-boundary mobility based on the change of the mean grain size with time of

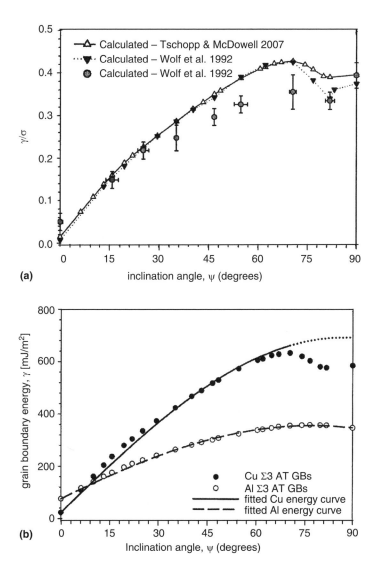

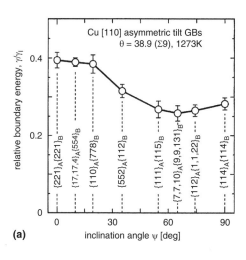

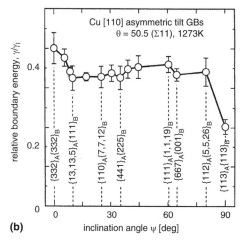

Fig. 15 Relative boundary energy versus the inclination angle, ψ. (a) Σ9 38.9° [110] asymmetric tilt grain boundaries (GBs) in copper at 1273 K (Ref 36). (b) Σ11 [110] asymmetric tilt grain boundaries in copper at 1273 K (Ref 37). Silica particle observations were used.

Fig. 14 Grain-boundary energy. (a) The energy, γ, of Σ3 <110> tilt grain boundaries in copper normalized by the surface energy, σ, as a function of inclination angle, ψ. The values calculated in Ref 34 are compared with both calculated values and measurements (Ref 35). (b) Computed energy of Σ3 <110> tilt grain boundaries as a function of inclination angle, ψ, for copper and aluminum (Ref 34)

polycrystals give only a mean value, averaged over many different types of grain boundaries. Specific data on grain-boundary mobility can only be obtained from the behavior of individual boundaries. This can be accomplished from bicrystal experiments.

Driving Forces. Table 4 gives comprehensive information on different types of driving forces used for bicrystal techniques. The driving force for grain-boundary migration, P, has the dimension of energy per unit volume, which is conceptually equivalent to a pressure—a force acting per unit area on a grain boundary. Driving forces can arise from various sources (Table 4) (Ref 2). In general, a driving force for grain-boundary migration occurs if the boundary displacement leads to a decrease of the total free energy of the system. A gradient of any intensive thermodynamic variable offers a source of such a driving force: a gradient of temperature, pressure, density of defects,

density of energy (for example, an energy of elastic deformation), content of impurities, a magnetic field strength, and so on. However, not all theoretically possible driving forces can be practically realized. To study grain-boundary motion, the following driving forces were most often used: curvature (capillary) driving force, magnetic energy, and mechanical stress.

Most frequently, the capillary driving force has been used in a variety of bicrystal geometries. Bicrystal techniques require substantial experimental efforts to manufacture bicrystals with precise orientations and misorientations as well as to prepare adequate specimens for investigations with the chosen technique. The different bicrystal arrangements designed to measure the grain-boundary velocity and, eventually, the grain-boundary mobility are given in Fig. 21 (Ref 2). A basic advantage of all techniques that use the capillary driving force is that the surface tension of a grain-boundary depends

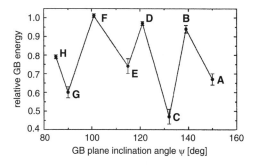

Fig. 16 Dependence of the measured relative grain-boundary (GB) energy of a 49° <332> tilt boundary in NiAl on the inclination angle. Measured using scanning probe microscopy of thermal grooving at 1400 °C. Although the value of inclination is not related directly to the grain-boundary crystallography, this dependence demonstrates qualitatively the effect of grain-boundary plane inclination on the grain-boundary energy for a grain boundary with fixed misorientational degrees of freedom. Source: Ref 38

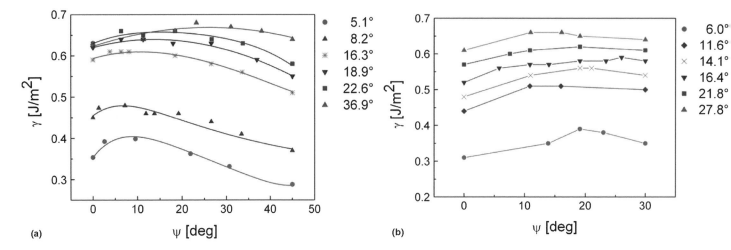

Fig. 17 Computed dependence of the grain-boundary energy on its boundary inclination for different (a) <100> tilt boundaries and (b) <111> tilt boundaries in aluminum at 0 K. Source: Ref 39

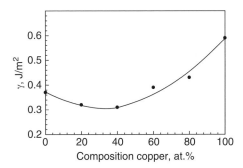

Fig. 18 Dependence of grain-boundary energy on composition in gold-copper alloys. From interferometric measurements of the boundary groove angles in specimens thermally etched at 850 °C. Source: Ref 41

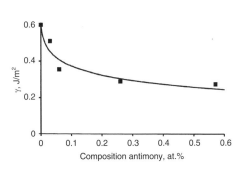

Fig. 19 Grain-boundary energy in copper-antimony as a function of antimony concentration. Source: Ref 42

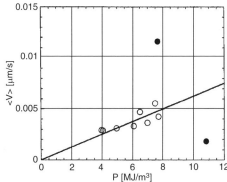

Fig. 20 Average interface velocity versus driving force for the migration of grain boundaries during recrystallization of cold-worked copper (99.96%). The boundary migration rates were estimated from stereological measurements. The instantaneous driving force for boundary migration was calculated from calorimetric measurements of the release of the stored energy of cold work. The average mobility of grain boundaries migrating during recrystallization of cold-worked copper at 121 °C was calculated to be 6.31×10^{-16} m^4/J · s. Source: Ref 44

$$P = \gamma/a \qquad \text{(Eq 11)}$$

In this technique, the driving force increases with progressing grain-boundary displacement. So, the driving force is very small in the beginning, and sources of pinning forces can manifest themselves to arrest the boundary.

Finally, the reversed-capillary technique (Fig. 21b) allows one to obtain relatively large driving forces, although unfortunately only in the beginning of the experiment. However, these are usually sufficient to force the boundary to break free from potential pinning centers (Ref 2, 48 to 52). The driving force for the reversed-capillary method is given by:

$$P = \gamma/R = \frac{\gamma}{a} f(\alpha) \qquad \text{(Eq 12)}$$

where R is the radius of boundary curvature. The major advantages of the reversed-capillary

technique are the relative ease of manufacturing and preparation of specimens and the possibility to change the driving force by varying the angle, α. The important disadvantage of the reversed-capillary technique is the lack of steady-state motion of the grain boundary. This disadvantage is avoided by the half-loop technique, where the driving force remains constant (Fig. 21c,d). The grain-boundary half-loop (Fig. 21d) or quarter-loop (Fig. 21c) moves as a whole, and its shape remains self-similar during migration (Ref 2). The average driving force on a half-loop and a quarter-loop is $P = \frac{2\gamma}{a}$ and $P = \frac{\gamma}{a}$, respectively (Ref 2). It is particularly emphasized that in all capillary techniques, except for the half-loop geometry, the moving grain boundary is exposed to the free lateral surface of the bicrystal.

When a bicrystal of a magnetically anisotropic solid with susceptibility $\chi \ll 1$ is subjected to a uniform magnetic field, a magnetic driving force, P_m, arises due to the difference of the magnetic energy density, Ω, in adjacent grains (Ref 53):

$$P_m = \Omega_1 - \Omega_2 = \frac{\mu_0 H^2}{2}(\chi_1 - \chi_2)$$
$$= \frac{1}{2}\mu_0 \Delta\chi H^2 (\cos^2 \varphi_1 - \cos^2 \varphi_2) \qquad \text{(Eq 13)}$$

where μ_0 is the magnetic constant; H is the magnetic field strength; χ_1 and χ_2 are the magnetic susceptibilities of the adjacent grains 1 and 2 along the field direction; $\Delta\chi$ is the difference of the susceptibilities parallel, $\chi_\parallel$, and perpendicular, $\chi_\perp$, to the principal (or c) axis of the crystal; and φ_1 and φ_2 are the angles between the c-axes in both neighboring grains and the magnetic field direction. In contrast to the curvature driving force, the magnetic one does not depend on the boundary properties, that is, on its energy and shape, but is determined only by the magnetic anisotropy of the material ($\Delta\chi$), the strength of the applied magnetic field, and its orientation with respect to

only slightly on temperature, and therefore, the driving force is practically constant over a wide temperature range.

The wedge bicrystal technique (Fig. 21a) was frequently used (Ref 45 to 47). There is a simple relation between driving force, P, and the macroscopic grain dimension, a (radius of curvature):

Table 4 Driving forces for grain-boundary migration

Source	Equation	Approximate value of parameters	Estimated driving force (P), MPa
Stored deformation energy	$P = \frac{1}{2}\rho\mu b^2$	ρ = dislocation density $\sim 10^{15}/m^2$ $\frac{\mu b^2}{2}$ = dislocation energy $\sim 10^{-8}$ J/m	10
Grain-boundary energy	$P = \frac{2\gamma}{R}$	γ = grain-boundary energy ~ 0.5 J/m^2 R = grain-boundary radius of curvature $\sim 10^{-4}$ m	10^{-2}
Surface energy	$P = \frac{2\Delta\sigma}{d}$	d = sample thickness $\sim 10^{-3}$ m $\Delta\sigma$ = surface energy difference of two neighboring grains ~ 0.1 J/m^2	2×10^{-4}
Chemical driving force	$P = R^* (T_1 - T_0)$ $c_0 \ln c_0$	R^* = gas constant = kN_A, N_A − Avogadro number c_0 = concentration = max solubility at T_0 T_1 ($<T_0$) annealing temperature (5% Ag in Cu at 300 °C)	6×10^2
Magnetic field	$P = \mu_0 H^2 \Delta_\chi \cdot$ $(\cos^2 \phi_1 - \cos^2 \phi_2)$	H = magnetic field strength; $\Delta\chi$ = difference of magnetic susceptibilities in adjacent grains along the field direction; φ = angle between c-axis and field direction For bismuth at 250 °C ($\Delta\chi \sim 0.23{\cdot}10^{-4}$), $H = 10^7$A/m; $\varphi_1 = 0°$; $\varphi_2 = 90°$	1.4×10^{-3}
Elastic energy	$P = \frac{\tau^2}{2}\left(\frac{1}{E_1} - \frac{1}{E_2}\right)$	τ = elastic stress ~ 10 MPa E_1, E_2 = elastic moduli of neighboring grains $\sim 10^5$ MPa	2.5×10^{-4}
Temperature gradient	$P = \frac{\Delta S \cdot 2\lambda \text{grad}T}{\Omega_a}$	ΔS = entropy difference between grain boundary and crystal (approx. equivalent to melting entropy) $\sim 8{\cdot}10^3$ J/ K·mol grad T = temperature gradient $\sim 10^4$ K/m 2λ = grain-boundary thickness $\sim 5{\cdot}10^{-10}$ m Ω_a = molar volume ~ 10 cm^3/mol	4×10^{-5}
Mechanical stress	$P = \tau\sin\omega$	τ = shear stress; ω = misorientation angle; for $\tau \sim 0.3$ MPa, $\omega \sim 12°$	6×10^{-2}

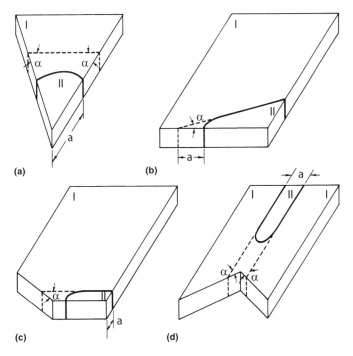

Fig. 21 Various boundary geometries in bicrystalline specimens for the study of grain-boundary migration. (a) Wedge technique. (b) Reversed-capillary technique. (c) Constant driving force technique (quarter-loop technique). (d) Constant driving force technique (half-loop technique). Source: Ref 2

the two grains. It moves the boundary from the grain with lower free energy toward the one with higher free energy and does not depend on the sign of the magnetic field (Fig. 22).

Low-angle symmetrical tilt boundaries are known to move under an applied mechanical stress (Ref 55 to 57). They move by collective glide of the dislocations that compose these boundaries (Ref 3). A shear stress acting on the boundary plane causes a force on each edge dislocation and thus results in a driving force for boundary migration (Ref 58):

$$p_s = \rho_{dis} \cdot F_{P-K} = \frac{2}{b}\sin\frac{\omega}{2} \cdot \tau b\cos\frac{\omega}{2} = \tau\sin\omega \quad \text{(Eq 14)}$$

where ρ_{dis} is the dislocations line length per unit area in the boundary, F_{P-K} is the Peach-Kohler force (per unit length) on a dislocation with Burgers vector, **b** (Ref 59), and τ is the shear stress.

According to the edge dislocation structure of such boundaries, their motion under a mechanical stress was confirmed to be coupled to a shear deformation, which is observed in

bicrystals with planar boundaries as a tangential translation of the adjoining grains (Fig. 23). Fukutomi et al. (Ref 61 to 63) also reported that high-angle boundaries with a low-Σ CSL orientation relationship in zinc and aluminum bicrystals can be moved by an applied shear stress, which causes a shape change of the bicrystal. A similar behavior was observed by Yoshida et al. (Ref 64) for a Σ11 boundary in cubic ZrO$_2$ bicrystals. Grain-boundary motion coupled to a lateral grain translation was also investigated by Cahn et al. (Ref 65, 66) in a computer simulation study of <100> tilt grain-boundary motion in copper. Shear-stress-induced motion of various low- and high-angle boundaries in aluminum bicrystals was also reported by Winning et al. (Ref 67, 68).

Techniques to Monitor Grain-Boundary Migration. There are two essentially different ways to determine the position of a grain boundary in crystalline materials and to measure the velocity of its motion: the continuous method and the discontinuous method.

The discontinuous method has been most commonly used. The location of the boundary is determined by its intersection with the crystal surface, and its change can be recorded after discrete time intervals. The position of a boundary can be revealed by the groove, which forms on sample cooling or by chemical etching of the crystal surface. The main advantage of this method is its simplicity; the major shortcoming is that the boundary cannot be observed during its motion, and consequently, it is necessary to average the measured boundary displacement over a large period of time between consecutive observations (Fig. 24) (Ref 2).

The continuous method requires determination of the boundary position at any moment of time and thus necessitates automation of the procedure to locate the boundary position. There are various techniques to distinguish different crystal orientations, such as reflection of polarized light (Ref 69), photoemission

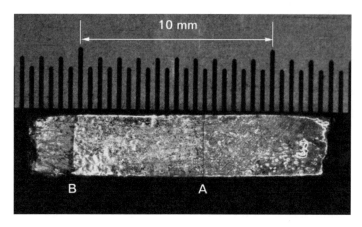

Fig. 22 Grain-boundary displacement in a bismuth bicrystal (from position A to position B) after annealing for 180 s at 252 °C in a magnetic field of 20.45 T. Source: Ref 54

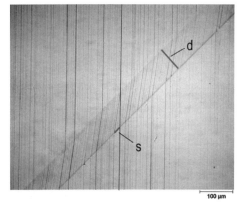

Fig. 23 Grain-boundary migration coupled to a shear deformation for a 17.8°<100> symmetrical tilt boundary after 68 min annealing at 375 °C under a tensile stress of 0.27 MPa. The coupling factor, β, is determined as the ratio of lateral grain translation (s) to normal boundary displacement (d). Source: Ref 60

Fig. 24 Traces of a moving boundary in a zinc bicrystal (reversed-capillary technique). Source: Ref 2

(Ref 70), x-ray topography (Ref 71, 72), x-ray diffraction (Ref 73 to 75), or backscattered electron intensity (Ref 76, 77). The most widely used techniques are the reflection of polarized light (Fig. 25), x-ray diffraction, or backscattered electron intensity in the scanning electron microscope (SEM).

The observation of grain-boundary motion by orientation contrast in the optical microscope is convenient and reliable and not only reveals the location of a moving grain boundary but also its shape at any moment (Fig. 25). The characteristic time resolution is better than

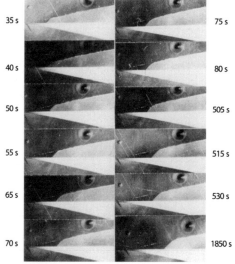

Fig. 25 Recorded grain-boundary migration in a zinc bicrystal by optical microscopy in polarized light (video frames). Source: Ref 2

10^{-2} s, the spatial resolution is a few micrometers, and the thickness of the sample is insignificant (Ref 2). Unfortunately, this technique is applicable only to optically anisotropic materials, that is, of material with lower-symmetry crystal structure, for instance, hexagonal crystals.

X-ray topography permits an image of a grain boundary to be obtained as the interface between next-neighbor grains. Synchrotron white beam x-ray topography (SWBXRT) offers excellent time resolution and thus can be used for in situ studies of grain growth, recrystallization, and grain-boundary motion (Ref 71, 72). SWBXRT allows a spatial resolution in the range of several micrometers, the time resolution is approximately 1 s, the field of view extends to 2 to 3 cm^2, and the sample thickness can range between 0.1 and 1 mm (Ref 72). However, the study of grain-boundary

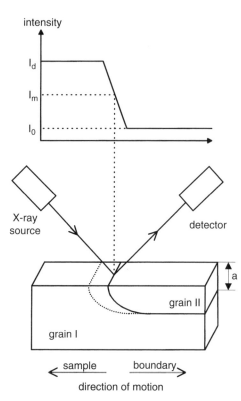

Fig. 26 Principle of operation of the interface continuous tracking device. The diffraction conditions generate an x-ray intensity gradient across the boundary. The specimen is moved to maintain a constant recorded intensity. Source: Ref 73 to 75

motion by SWBXRT is restricted to samples of high perfection; the dislocation density should not exceed 10^8 m^{-2}, which is very low.

A simple, precise, and very versatile technique to identify the position of a grain boundary uses x-ray diffraction (Ref 73 to 75). The principle of the method is illustrated in Fig. 26. The bicrystal is placed in a goniometer of the x-ray interface continuous tracking device (XICTD) in such a way that one grain (1) is in Bragg position, while the other is not. The maximum intensity I_d of the reflected x-ray beam is measured as long as the x-ray spot is located solely on the surface of crystal 1. If the spot illuminates the grain boundary, the intensity of the reflected beam should be intermediate between I_d and I_0, say I_m. When the boundary moves, the sample can be displaced accordingly so that the reflected x-ray intensity I_m remains constant during grain-boundary motion. Thus, the velocity of the moving grain boundary (generally speaking, interface) is equal to the speed of sample movement at any moment in the course of the experiment. This procedure does not interfere with the process of grain-boundary migration. The measurable velocity ranges from 1 to 1000 μm s^{-1} with a temporal resolution of approximately five measurements per second. The hot stage of the XICTD allows heating the sample to 1200 °C in nitrogen or an inert gas atmosphere to

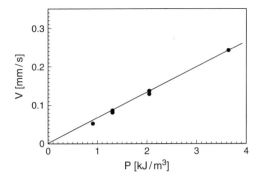

Fig. 27 Measured grain-boundary migration rate versus driving force of a flat boundary in a bicrystal of bismuth exposed to a magnetic field. Source: Ref 78

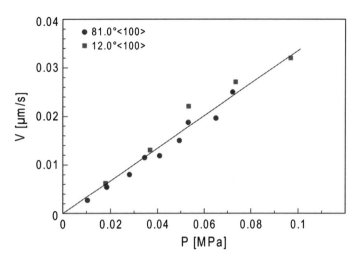

Fig. 29 Migration rate at 320 °C for low-angle 12.0° and 81.0°<100> tilt grain boundaries in aluminum as a function of a shear stress driving force. Source: Ref 58

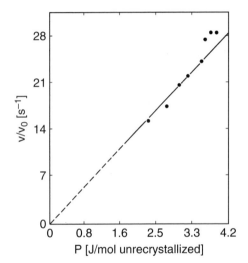

Fig. 28 Average growth rate during primary recrystallization at 125 °C (normalized by a constant v_0) as a function of driving force (p) for a polycrystalline aluminum alloy containing 17 ppm copper. Source: Ref 79

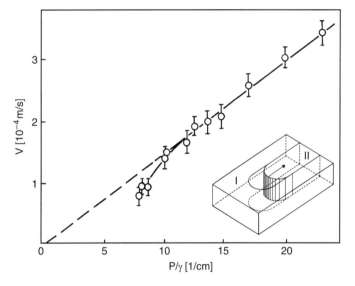

Fig. 30 Measured grain-boundary migration rate versus reduced driving force of U-shaped boundaries in aluminum bicrystals (half-loop technique). Source: Ref 80

suppress oxidation and grooving; the temperature is kept constant within ±0.3 °C. The accuracy of the velocity measurement of grain-boundary motion is better than 2%.

A recently developed elegant method is the investigation of crystal shape evolution by backscattered electrons in an SEM, because the contrast by backscattered electrons is very sensitive to orientation. The method also allows for orientation evaluation by electron backscattered diffraction (EBSD) with the same setup. By using a hot stage in the SEM, in situ studies can be conducted. However, caution must be exercised due to the use of high vacuum for operating the microscope, because this can aggravate the grooving problem. Also, photoemission microscopy can be used for continuous tracking of grain-boundary migration (Ref 70). However, this technique is restricted to metals with high melting points.

Dependence of Grain-Boundary Velocity on Driving Force. In accurate experiments Eq

9 has been unambiguously confirmed (Fig. 27 to 30). For small driving forces, sometimes a deviation to lower migration rates is observed (Fig. 30). The reason for such a deviation are drag effects (solute drag, groove dragging). A very large deviation from the linear dependence was observed in the works of Rath and Hu (Ref 45 to 48) (Fig. 31a). The observed deviation, however, can be explained by groove dragging (Fig. 31b) (Ref 80, 81).

Misorientation Dependence of Grain-Boundary Mobility. The long-known observation that pronounced crystallographic textures develop during annealing of deformed metals was the first indication that grain-boundary mobility may depend on misorientation. Correspondingly, the first findings on the orientation dependence of grain-boundary velocity were obtained in recrystallization experiments (Fig. 32) (Ref 82 to 84). Such dependencies

were confirmed subsequently by experiments on bicrystals, which not only permit the investigation of all orientation relationships but also the determination of the influence of various factors on grain-boundary mobility, for example, temperature, pressure, chemical nature and content of impurities, and so on. However, in the majority of investigations, grain-boundary migration was considered only for a very few specific boundaries (Table 5, Fig. 33) and provided only an incomplete characteristic of the migration capability of grain boundaries (Ref 45 to 50, 85 to 88). The misorientation dependence was studied extensively for lead, aluminum, and zinc.

In Fig. 34, the misorientation dependency of the mobility of low-angle boundaries and the activation energy of their migration is presented (Ref 89, 90) as determined from the growth rate of subgrains in deformed and recovered

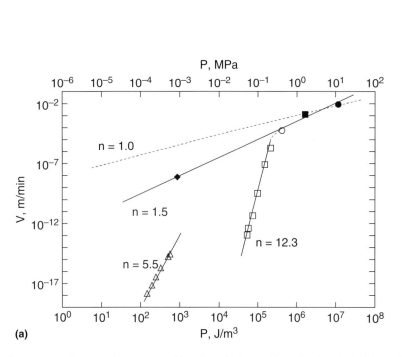

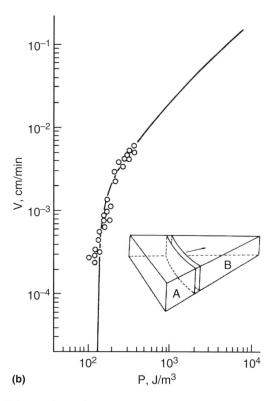

Fig. 31 Boundary migration rate versus driving force. (a) Reported by Rath and Hu (Ref 46) for wedge-shaped aluminum bicrystals. (b) Calculated $V(P)$ dependency (curve) in the presence of groove dragging (Ref 80, 81). Symbols are values reported by Rath and Hu (Ref 46)

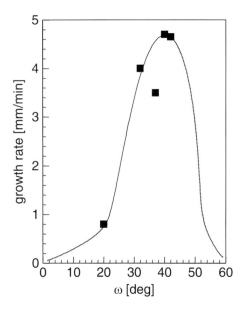

Fig. 32 Change of boundary migration rate with orientation difference about a common $<111>$ axis in aluminum. Source: Ref 82

Table 5 Activation energy for the migration of different $<001>$ tilt grain boundaries in copper

Misorientation angle, degree	Activation energy, kJ/mol
2	205.2
5	205.2
9	205.2
18	122.9
32	126.2

Source: Ref 85

dependence of specific boundaries but the kinetic coefficient of the growth rate of grains in a polycrystal.

The misorientation dependence of the velocity of single boundaries in zone-refined lead was studied under a constant driving force and in a range of temperatures (Ref 91). Well-pronounced extrema in Fig. 35 (maxima of grain-boundary velocity, minima of activation energy and preexponential factor, respectively) could be associated with low-Σ coincidence boundaries.

In aluminum, the misorientation dependency of grain-boundary mobility and its kinetic parameters (enthalpy of activation and pre-exponential factor) was investigated for $<100>$, $<111>$, and $<110>$ tilt grain boundaries with a total impurity amount of $(2-5) \times 10^{-4}$ at.% by the grain-boundary half-loop and quarter-loop technique (Fig. 36) (Ref 2, 92–94).

The misorientation dependency of the activation enthalpy of migration for $<10\overline{1}0>$ and

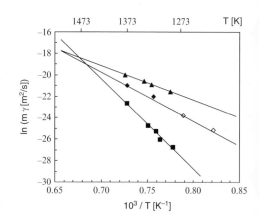

Fig. 33 Temperature dependence of the reduced mobility (product $m\gamma$) for symmetrical $<100>$ tilt boundaries with misorientation angles 13.8° (■), 34.3° (◆, ◇), and 37.6° (▲). Solid symbols: quarter-loop technique; open symbols: reversed capillary technique (Fig. 21b). Source: Ref 85

$<11\overline{2}0>$ tilt boundaries in zinc with a total impurity content of 5×10^{-4} at.% and 5×10^{-3} at.%, respectively, is shown in Fig. 37 (Ref 2, 95). The data were obtained by the reversed-capillary technique. Qualitatively, the behavior of the migration activation enthalpy in zinc is akin to face-centered cubic metals. Evidently, the dependency $H(\omega)$ reflects structural peculiarities of the grain boundaries.

In essence, all misorientation dependencies of grain-boundary mobility are nonmonotonous, and the minima of the activation enthalpy

aluminum polycrystals. The subgrains were reconstructed from an orientation map obtained by EBSD. The average misorientation of a particular subgrain to all next-neighbor subgrains was determined, and for a given initial microstructure, this parameter was averaged for all the subgrains in a map. In essence, these data do not reflect the misorientation

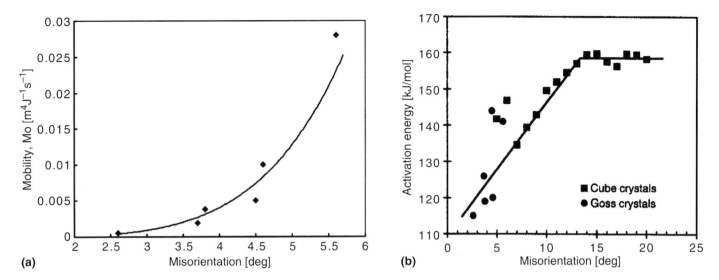

Fig. 34 Effect of average misorientation on the growth kinetics of subgrains in Al-0.05%Si. (a) Mobility and (b) activation energy versus misorientation angle (Ref 89, 90). It is noted that these data must not be confused with the mobility of individual boundaries with specific misorientation.

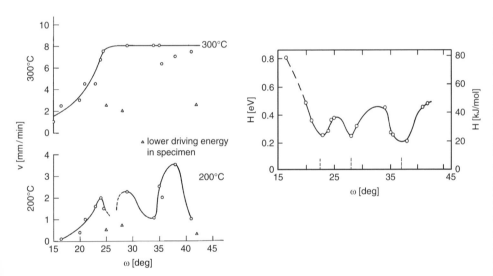

Fig. 35 Rate of boundary migration versus misorientation angle. (a) head at 300 °C (top) and 200 °C (bottom). (b) Measured activation energies versus misorientation angle for <100> tilt boundaries in zone-refined, lead. Source: Ref 91

correspond to special misorientations, namely low-Σ CSL misorientations. The magnitude of the oscillations is large: for the grain-boundary mobility, approximately 2 orders of magnitude; and for the activation enthalpy, a factor of 2, ~60 to 120 kJ/mol. These results reflect both the variance of grain-boundary structure with misorientation and the effect of misorientation on grain-boundary segregation (Ref 2).

Effect of Grain-Boundary Plane on Mobility. Grain-boundary mobility is known to depend not only on misorientation but also on the orientation of the grain-boundary plane. This is particularly evident for coherent twin boundaries, which are much less mobile than incoherent twin boundaries despite identical misorientation across the boundary. However, anisotropy of grain-boundary mobility can also

be observed for misorientations other than twin relationships, in particular, grain boundaries of a misorientation with <111> rotation axis. For such orientation relationships, tilt boundaries can move orders of magnitude faster than pure twist boundaries (Fig. 38) (Ref 84). Apparently, it is impossible to study the effect of grain-boundary orientation on its mobility by using grain-boundary curvature as a driving force. Such experiments provide only an average mobility of all involved boundary orientations. All pure tilt boundaries of the same misorientation exhibit essentially the same mobility, as evident from the conservation of shape of curved tilt boundaries during migration (Ref 96).

A more defined study of the dependence of grain-boundary mobility on the orientation of the grain-boundary plane requires experiments

where a grain boundary moves under a known driving force. Such an attempt was undertaken in Ref 78, where the motion of a flat grain boundary under the action of the magnetic field was studied (Fig. 39). The temperature dependence of the mobility (Eq 10a) of a symmetrical and an asymmetrical 90°<112> tilt grain boundary in bismuth revealed that the migration parameters (activation enthalpy, H, and mobility pre-exponential factor, m_0) for the symmetrical boundary ($H = 0.51$ eV, $m_0 = 0.67$ m^4/J·s) conspicuously differed from the migration parameters of the asymmetrical boundary ($H_\parallel = 3.38$ eV, $m_{0\parallel} = 2.04 \times 10^{24}$ m^4/J·s, and $H_\perp = 3.79$ eV, $m_{0\perp} = 1.10 \times 10^{28}$ m^4/J·s, where the symbols $\parallel$ and $\perp$ refer to the orientation of the c-axis with respect to the boundary plane normal). As a consequence, the symmetrical boundary has a much higher mobility than the asymmetrical boundary in the entire investigated temperature range up to the melting point of bismuth (Fig. 39), but particularly at low temperatures. Apparently, the inclination of tilt boundaries in bismuth has a very strong influence on boundary mobility (Ref 78).

Compensation Effect in Grain-Boundary Motion. It is textbook knowledge that the temperature dependence of the absolute rate, ρ, of a thermally activated process is governed by the Arrhenius relation:

$$\rho = \rho_0 \exp(-H/kT) \qquad \text{(Eq 15)}$$

There is experimental evidence that the pre-exponential factor, ρ_0, is strongly related to the activation enthalpy H; ρ_0 increases or decreases if H increases or decreases according to the relation:

$$H = \alpha \ln \rho_0 + \beta \qquad \text{(Eq 16)}$$

Here, α and β are constants, the meaning of which is dealt with subsequently. Equation 16 is referred to as the compensation effect, because it strongly moderates the effect of a variation of H on the value of the absolute reaction rate, ρ, such that above a so-called compensation temperature, $T_c = \alpha/k$, the process with the highest activation enthalpy has the highest reaction rate, whereas for $T < T_c$, the process with the lowest value of H proceeds fastest. At T_c, all reaction rates ρ of the considered group of thermally activated processes are the same; that is, the lines for the corresponding Arrhenius plots intersect at temperature T_c. Inserting Eq 16 into Eq 15 at $T = T_c$ yields:

$$\rho(T_c) = \exp\left(-\frac{\beta}{kT_c}\right) \qquad \text{(Eq 17)}$$

The thermodynamic fundamentals of the compensation effect are given in Ref 2 and 97. The compensation effect applies to many solid-state processes, in particular, grain-

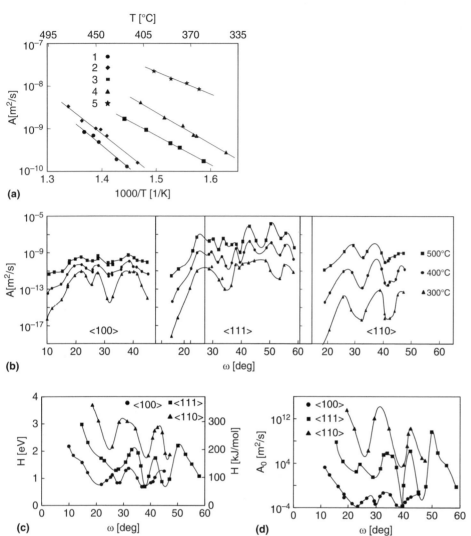

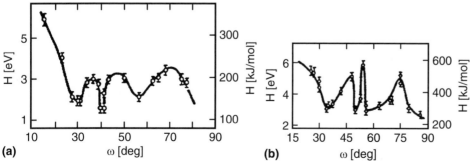

Fig. 36 Temperature and misorientation dependencies of reduced grain-boundary mobility for <100>, <111>, and <110> tilt grain boundaries in aluminum. (a) Temperature dependence, <111> tilt grain boundaries with various angles, ω, of misorientation: ●, 33°; ◆, 35.5°; ■, 28°; ▲, 36.5°; ★, 45°. (b) Misorientation dependence at 500, 400, and 300 °C, respectively. (c) Enthalpy of activation versus misorientation angle, ω, for motion of <100>, <111>, and <110> tilt boundaries in aluminum. (d) Pre-exponential factor of reduced grain-boundary mobility. Source: Ref 2

Fig. 37 Misorientation dependence of activation enthalpy of grain-boundary motion for (a) $<10\bar{1}0>$ and (b) $<11\bar{2}0>$ tilt boundaries in zinc. Total impurity content: 5×10^{-4} and 5×10^{-3} at.%, respectively. Source: Ref 2, 95

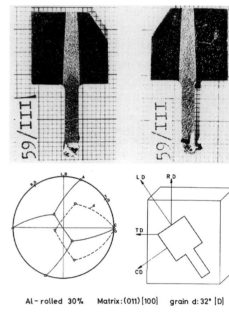

Al – rolled 30%   Matrix: (011) [100]   grain d: 32° [D]

Fig. 38 Anisotropic growth of a grain in rolled aluminum. Prior to annealing, the grain boundary was located at the top of the handle. The micrograph shows the front and back faces of the specimen. The long, straight grain boundaries are approximately perpendicular to the <111> rotation axis (twist boundaries). Source: Ref 84

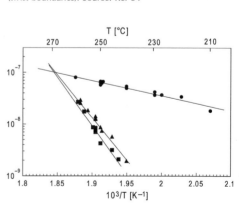

Fig. 39 Temperature dependence of mobility of 90°<112> symmetrical (●) and asymmetrical (▲, ■) boundaries in bismuth bicrystals. Trigonal axis in the growing grain parallel (▲) or perpendicular (■) to the growth direction. Source: Ref 78

boundary migration. Some examples of the compensation effect for various grain-boundary systems are given in Fig. 40 to 43.

Applications of the compensation effect to grain-boundary migration can be found in Ref 2. One of the applications is associated with the problem of the fastest grain boundary in aluminum, which is known to be in the vicinity of a 40°<111> misorientation. Numerous growth-selection experiments (Ref 104 to 106) provided strong experimental evidence that the maximum mobility was attained at misorientation angles above 40°, that is, close to 41° and not 38.2°, where a special grain boundary with Σ7 is situated. On the other hand, grain-boundary mobility measurements on bicrystals identified the Σ7 boundary as the fastest. The solution of this contradiction is connected with the compensation effect. The misorientation dependence of the activation enthalpy for migration of <111> tilt grain boundaries in the vicinity of the special grain boundary Σ7, 38.2°, as obtained from bicrystal experiments (Ref 107, 108), is given in Fig. 44(a). Obviously, the activation enthalpy is at maximum for a misorientation angle close to 41°. However, the misorientation dependence of the pre-exponential factor behaves the same way, that is, attains a maximum for 41° misorientation (Fig. 44b). In fact, the compensation effect

with a compensation temperature, T_c, of ~450 °C (Fig. 45) causes the grain boundaries with the lowest activation energy to be most mobile at $T < T_c$, whereas for $T > T_c$, the opposite is the case (Fig. 45b). Accordingly, which boundary moves fastest depends on the temperature range (relative to T_c), and this reconciles the contradiction between recrystallization and growth-selection experiments mentioned previously.

As a general rule, the compensation temperature divides the temperature range into two regimes, with different relations between the magnitude of reaction rate and energy of activation. When the experiments are conducted below T_c, the processes with low energies of activation prevail. By contrast, if the measurements are taken above T_c, the processes with high energies of activation dominate the kinetics (Fig. 46).

Computation of Grain-Boundary Mobility

The measurement of grain-boundary mobility with high accuracy is difficult and time-consuming. Correspondingly, atomistic simulations by molecular dynamics have been increasingly used to compute grain-boundary mobilities (Ref 109 to 118). The principle is

the same as for the computational derivation of grain-boundary energy. Newton's equation of motion is solved for each atom under the action of the interatomic potential (Ref 2). A motion of the boundary is enforced by an energy density differential across the boundary.

Mainly elastic driving forces or orientation-dependent interatomic potentials have been used to drive grain boundaries. In Ref 109, molecular-dynamics simulations were used to study migration of low- and high-angle [001] planar twist grain boundaries in copper. The temperature dependence of the grain-boundary mobility was determined over a wide misorientation range (Fig. 47). Additionally, grain-boundary self-diffusion was studied for all investigated [001] planar twist boundaries. A comparison of the determined activation energies shows that grain-boundary migration and self-diffusion are distinctly different processes.

It should be stressed in this context that the computed and measured mobilities are usually quite different. The computed boundary mobilities are typically very high, because the activation energy for migration is small compared to measured values (Fig. 48). This is commonly attributed to the segregation of impurities to grain boundaries in real materials, whereas totally pure materials are assumed in the simulations. Also, simulation results depend

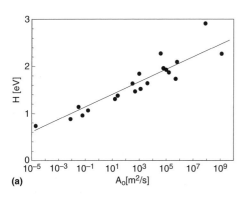

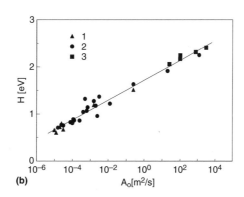

Fig. 40 Dependence of the migration activation enthalpy, H (measured in eV), on the (reduced) pre-exponential mobility factor, A_o, for (a) <111> and (b) <100> tilt grain boundaries in various grades of pure aluminum. ▲, Al 99.99995 at.%; ●, Al 99.9992 at.%; ■ Al 99.98 at.%. Refer to Eq 10b. Source: Ref 2

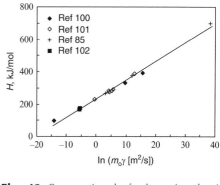

Fig. 42 Compensation plot for the motion of grain boundaries in Fe-3%Si. Source: Ref 85, 100 to 102

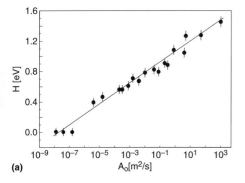

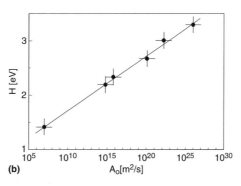

Fig. 41 Dependence of the migration activation enthalpy, H, on the (reduced) pre-exponential mobility factor, A_o, for (a) <11$\bar{2}$0> tilt grain boundaries in zinc (Ref 98) and (b) <001> tilt grain boundaries in tin (Ref 99)

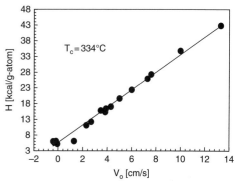

Fig. 43 Relationship between the migration activation enthalpy and pre-exponential mobility factor for migration of grain boundaries in lead as extracted from the results of Aust and Rutter (Ref 103). Source: Ref 96

strongly on the quality of the used interatomic potential and may be affected by the extreme or artificial conditions assumed for temperature and driving force that must be used to obtain results for the very small time interval covered by atomic simulations, typically some nanoseconds. Nevertheless, the computations can provide valuable information on the nature of boundary motion, its mechanisms, and its dependencies on crystallography and environmental conditions. Experimental results, however, as difficult as they are to generate, are thus far indispensible for the prediction of real materials behavior.

Effect of Impurities on Grain-Boundary Motion. It is well known that the purity of a material has a great influence on grain-boundary motion. Typically, foreign atoms reduce the rate of grain-boundary migration.

If there is an attractive interaction energy, U (energy gain), between boundary and impurity atoms, the solute atoms tend to segregate to the boundary. Due to thermal agitation, for $T > 0$, the concentration in the boundary, c_b, depends on temperature and will be different from the volume concentration, c_0, for example:

$$c_b = c_0 \exp\left(\frac{U}{kT}\right) \qquad \text{(Eq 18)}$$

Equation 18 describes the simplest (Henry) isotherm, which holds for an ideal solution at low impurity concentration. For high concentrations and strong solute-solute interactions, other, more complicated isotherms take effect (Ref 2). The influence of the impurity atoms on grain-boundary motion has been discussed (Ref 119 to 123) in terms of different

theoretical approaches to the problem, with the following principal results:

- Impurity atoms slow down grain-boundary motion due to the forced motion of grain-boundary impurities; the magnitude of

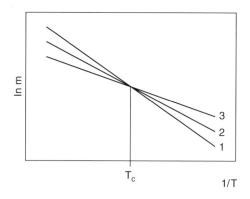

Fig. 46 Schematic sketch ($H_1 > H_2 > H_3$ and $m_{01} > m_{02} > m_{03}$) demonstrating the impact of the compensation effect on grain-boundary mobility, m (see Eq 10a) above and below the compensation temperature. Left of T_c where $T > T_c$: $m_1 > m_2 > m_3$; for $T < T_c$: $m_1 < m_2 < m_3$. Source: Ref 2

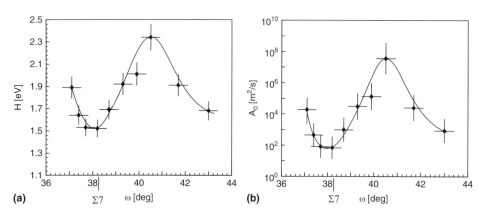

Fig. 44 Dependence on misorientation angle for enthalpy and pre-exponential factor, A_0. (a) Activation enthalpy, H. (b) Pre-exponential factor, A_0, for the migration of <111> tilt grain boundaries. Source: Ref 108

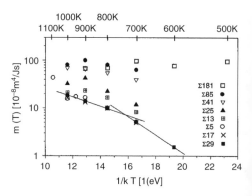

Fig. 47 Arrhenius plot of grain-boundary (GB) mobility for [001] twist grain boundaries in copper. For the $\Sigma 29$ GB (■), a high- and a low-temperature regime was found, as represented by the two different linear best data fits. Source: Ref 109

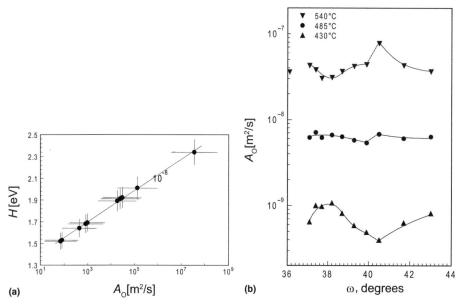

Fig. 45 (a) Relationship between the activation enthalpy and the pre-exponential factor for the motion of <111> tilt grain boundaries in aluminum. (b) Misorientation dependence of the grain-boundary reduced mobility at different temperatures. Source: Ref 108

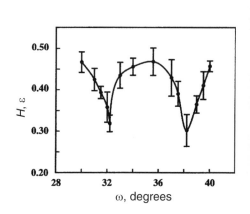

Fig. 48 Activation energy of grain-boundary migration ($\varepsilon = 0.57$ eV, Lennard-Jones potential for aluminum) as a function of misorientation angle from computer simulations of grain-boundary motion. Source: Ref 110

the effect depends on the diffusion characteristics of the impurities, their concentration, grain-boundary mobility, and the driving force of boundary migration.

- At large driving forces, high temperatures, or low impurity concentration, the grain boundary can detach from the impurity cloud (breakaway effect).

Figure 49 depicts the schematic dependency of grain-boundary velocity on driving force and elucidates the two major stages of grain-boundary motion in a system with impurities.

The effect of grain-boundary detachment from impurities was experimentally observed and studied on single <111> tilt grain boundaries in gold (Ref 124) and aluminum (Ref 125 to 127), on <10$\bar{1}$0> and <11$\bar{2}$0> tilt boundaries in zinc (Ref 128 to 130), and on <110> tilt boundaries in Fe-3%Si (Ref 102). Figures 50 and 51 depict the temperature and driving force dependency of grain-boundary velocity (mobility) in zinc and Fe-3%Si. A characteristic feature of the dependencies considered is a dramatic change of the grain-boundary velocity (more correctly, mobility) in a narrow temperature range. Above and below the breakaway region, the grain-boundary mobility shows a typical Arrhenius-type temperature dependence.

The temperature dependency of the mobility of 36.5°<111> tilt grain boundaries in specially doped high-purity aluminum samples was studied in Ref 131. The iron concentration, although rather low, was significantly higher than the concentration of any other solute element. A constant driving force was provided by the surface tension of a curved grain boundary (grain-boundary quarter-loop technique). The measurements were carried out by continuous boundary tracking. Breakaway was observed in the temperature range of 460 to 550 °C; the specific temperature depended on the driving force and impurity (iron) content. The experimental data correlated well with the predictions from the impurity drag theory. The energy of interaction between a grain boundary and iron atoms was found to be 0.134 + 0.02 eV (Ref 124).

Of special interest are the nature and number of adsorption sites in the grain boundary. The adsorption capacity of a special grain boundary is typically in the range of ~10^{18} cm^{-2}. This is distinctly lower than that measured by Auger spectroscopy on random grain boundaries in polycrystals, which proves a much higher adsorption capability of nonspecial grain boundaries.

Segregation Effect on Misorientation Dependence of Boundary Mobility. In spite of a rather large number of studies dedicated to the effect of impurities on grain-boundary motion, little is known on how solute atoms affect the motion of grain boundaries with different structure and segregation capacity. The structure dependence of segregation is most evident from mobility measurements on aluminum and zinc of various purity (Ref 2, 81, 92, 95, 132). The strong interaction of impurities

and grain-boundary structure is particularly obvious in <100> tilt boundaries in aluminum (Fig. 52) (Ref 92). For ultrapure and very impure material, the mobility of <100> tilt boundaries was found to be independent of rotation angle, irrespective of whether it involved a special or nonspecial boundary. For intermediate- (although high-) purity material, the mobility strongly depends on rotation angle, distinguishing special and nonspecial boundaries. A similar behavior of the misorientation dependence on impurity content was discovered for zinc of various purity (Fig. 53) (Ref 95).

Aust and Rutter (Ref 103,133) considered the impurity influence on the misorientation dependence of grain-boundary mobility as a

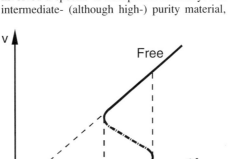

Fig. 49 Dependence of grain-boundary migration rate on driving force in the presence of impurity drag. In the interval denoted by T, the transition from the loaded to the free boundary and vice versa occurs discontinuously.

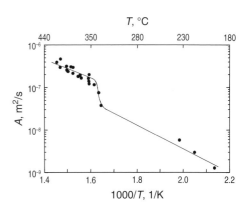

Fig. 50 Temperature dependence of the reduced mobility of 30° <10$\bar{1}$0> tilt grain boundaries in zinc. Source: Ref 129

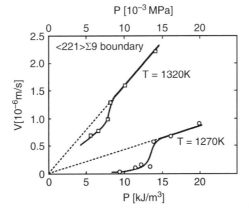

Fig. 51 Variation of boundary velocity, V, with driving force, P, for <221> tilt grain boundaries in iron-silicon. Source: Ref 102

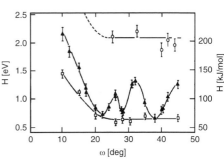

Fig. 52 Dependence of the activation enthalpy of migration for <100> tilt grain boundaries in aluminum of various purity □, 99.99995 at.%; ▲, 99.9992 at.%; ○, 99.98 at.%. Source: Ref 92

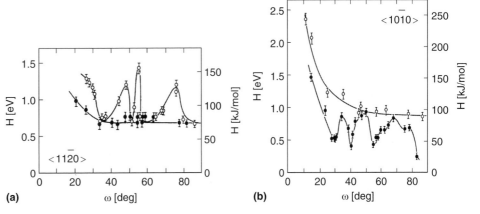

Fig. 53 Dependence of the activation enthalpy of migration on angle of rotation for (a) <11$\bar{2}$0> and (b) <10$\bar{1}$0> tilt grain boundaries in zinc of different impurity content ○, 99.995 at.%; ●, 99.9995 at.%. Source: Ref 2, 95

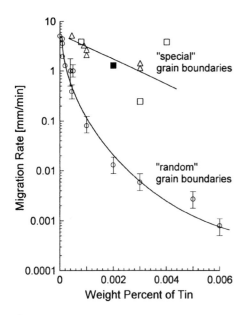

Fig. 54 Migration rate versus tin concentration in lead bicrystals according to Ref 133. Special and random boundaries behave differently.

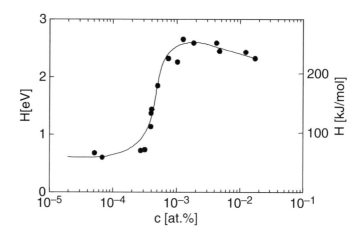

Fig. 55 Migration activation enthalpy for random tilt boundaries as a function of impurity content. Source: Ref 134

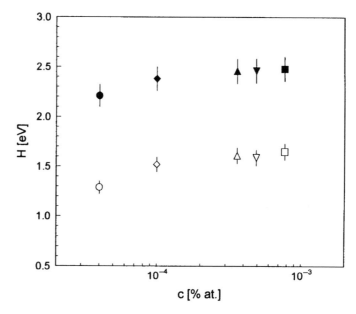

Fig. 56 Dependence of activation enthalpy for migration of 38.2° (open symbols) and 40.5° (filled symbols) <111> tilt grain boundaries on impurity concentration in differently pure aluminum. Source: Ref 108

segregation effect only; they postulated that strongly ordered boundaries, that is, low-Σ coincidence boundaries, segregate less and therefore move faster than random boundaries (Fig. 54). Current experimental observations and results of computer simulations suggest, however, that the observed segregation effect reflects the influence of impurities on both mobility and grain-boundary structure.

Dependence of Grain-Boundary Migration on Impurity Concentration. The results obtained from bicrystal experiments for random grain boundaries (Fig. 55) and <100> tilt boundaries in aluminum (Ref 134) agree with data of the activation enthalpy of grain growth in aluminum doped with copper, as measured by Gordon and Vandermeer (Ref 135).

The motion of <111> tilt boundaries with misorientation 38.2° (special misorientation $\Sigma 7$) and 40.5° in differently pure aluminum was studied in Ref 108. As evident from Fig. 56, even at the lowest impurity content, that is, in the material of highest purity, the activation enthalpy rises with increasing impurity concentration, contrary to predictions of the impurity drag theory. There are several reasons for the discrepancy between the theories of impurity drag and the experimental data. All theories are based on the assumption of a small concentration both in the bulk and in the grain boundary. However, the boundary impurity concentration may be high despite a small bulk impurity concentration. Furthermore, for high impurity concentrations in the boundary, it is necessary to take into account the mutual interaction of adsorbed atoms in the boundary. Also, as shown experimentally (Ref 136) and theoretically (Ref 1, 2), grain boundaries are inhomogeneous; that is, not every site in a grain boundary is equally favorable for impurity

segregation. Thus, the interaction between the adsorbed atoms should also be taken into account.

The strength of impurity drag varies with the chemical nature of the alloying element. For instance, iron in aluminum or niobium in steel generates a strong impurity drag, whereas silver is less effective in aluminum. Virtually all known experiments on bicrystals have demonstrated a reduction of the rate of grain-boundary migration by impurities. However, there are also some exceptions to the common rule. For instance, it was found that aluminum doped with minor amounts of gallium (10 ppm) experienced an increase grain-boundary mobility, that is, accelerated grain-boundary migration compared to pure aluminum (Fig. 57) (Ref 137). The reason for such behavior is not yet understood.

Drag Effects by Second-Phase Particles. The physical basis of this drag effect is the

attraction force between particles and a grain boundary, which is due to the reduction of grain-boundary energy upon contact of particle and grain boundary. The simplest and commonly considered attraction force is the well-known Zener force, which appears when a particle intersects the boundary and, in doing so, eliminates the intersected area of the boundary. For spherical particles with radius r and volume fraction f and a grain boundary with specific energy γ, the Zener force reads:

$$f_z = \frac{3}{2}\frac{\gamma f}{r} \qquad \text{(Eq 19)}$$

The Zener force, however, is not the only attraction force in a particle-boundary system. The physical sources and the description of other kinds of particle drag forces are given in (Ref 2). Traditionally, the dragging of a moving grain boundary by particles of a second phase is

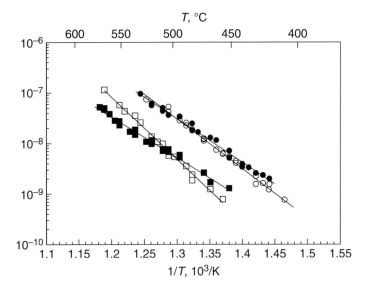

Fig. 57 Temperature dependence of mobility of 38.2° (●, ■) and 40.5 ° (○, □) <111> tilt grain boundaries in aluminum (□, ■) and aluminum plus 10 ppm gallium (○, ●). Source: Ref 137

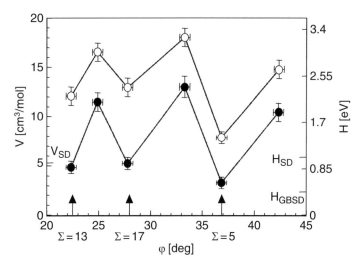

Fig. 58 Activation enthalpy, H (○), and activation volume, V^* (●), of <001> tilt grain-boundary (GB) migration in tin. SD subscript, self- diffusion. Source: Ref 144

considered in the approximation where the particles are immobile and act as stationary pinning centers for the boundaries (Ref 138 to 140). However, as considered in Ref 141, inclusions in solids are not necessarily immobile, and their mobility dramatically increases with decreasing particle size.

In the simple case of a single-size particle distribution, the velocity of the joint motion of grain boundary and particles reads (Ref 141):

$$V = \frac{P m_p(r_0)}{n_0} \qquad \text{(Eq 20)}$$

where P is the driving force of grain-boundary migration, $m_p(r_0)$ is the mobility of particles with the size (radius) r_0, and n_0 is the number of particles per unit area of the boundary. In this limit, the velocity of the grain boundary is determined by the mobility and density of the attached particles.

The collective movement of particles and grain boundary at subcritical driving forces and the detachment of particles at supercritical driving forces results in a bifurcation of the grain-boundary migration rate with increasing driving force (Ref 141).

A computer simulation study of grain growth in a system with mobile particles was carried out in Ref 142 and 143.

Effect of Pressure on Grain-Boundary Migration: Activation Volume. As mentioned at the begining of this section on grain-boundary mobility, the temperature dependence follows an Arrhenius dependence (Eq 10a). It is worth noting that the activation enthalpy depends on pressure:

$$H = E + pV^* \qquad \text{(Eq 21)}$$

where E is the activation energy, p is the hydrostatic pressure, and V^* is the activation volume.

The activation volume reflects the difference between the volume of the system in the activated and the ground state and can be obtained from measurements of the pressure dependence of grain-boundary mobility at constant temperature:

$$\left. \frac{\partial \ln m}{\partial p} \right|_T = -\frac{V^*}{kT} \qquad \text{(Eq 22)}$$

Special and nonspecial <001> tilt grain boundaries were studied in tin (Ref 144). Grain-boundary migration was measured at atmospheric pressure and at high hydrostatic pressures up to 16kbar. The orientation dependencies of the activation enthalpy and the activation volume of migration are depicted in Fig. 58. The activation enthalpy for migration of special grain boundaries was found to be 1.5 to 2 times larger than the energy of activation for bulk self-diffusion and almost an order of magnitude larger than for grain-boundary self-diffusion. The activation volume for special grain boundaries amounts to 0.6 to 0.96 V_{SD}^*, where V_{SD}^* is the activation volume for self-diffusion. However, for nonspecial (random) grain boundaries, the activation volume exceeds V_{SD}^* by a factor of 2 to 2.5. This is a first indication that the activation volume of grain-boundary migration can substantially exceed the diffusion activation volume.

The results of a comprehensive study of the temperature and pressure dependencies of <100>, <110>, and <111> tilt grain boundaries in bicrystals of pure aluminum are given in Fig. 59 (Ref 145). The activation volumes for <100> and <111> tilt boundaries are identical and independent of the angle of rotation, quite in contrast to the behavior of the activation enthalpy. The absolute value of approximately 12 cm³/mol corresponds to slightly more than one atomic volume in aluminum

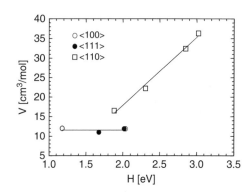

Fig. 59 Activation volume (V) versus activation enthalpy (H) of tilt grain-boundary migration in aluminum (tilt axis indicated). Source: Ref 145

and is therefore close to the activation volume for bulk self-diffusion.

By contrast, the activation volume for <110> tilt boundaries increases with increasing departure from the exact low-Σ coincidence misorientation, as is also the case for the activation energy. For a 30°<110> boundary, which deviates by ~8° from the Σ9 boundary, the activation volume amounts to 36cm³/mol, equivalent to more than three atomic volumes. Such values give evidence that more than a single atom is involved in the fundamental process of grain-boundary migration.

Motion of Connected Grain Boundaries in Polycrystals. Grain-boundary migration in a polycrystal requires the motion of connected grain boundaries. Traditionally, the evolution and properties of the granular assembly of a polycrystal are described entirely in terms of grain-boundary characteristics, while the junctions are tacitly assumed not to affect microstructure development and thus can be disregarded.

However, several recent theoretical and experimental studies provide evidence that the kinetics of triple junctions may be different from the kinetics of the adjoining grain boundaries. The microstructure evolution affected by triple junctions is distinctly different from the granular assembly in the course of normal grain growth. Also, the structure established under junction control is rather stable even under the conditions characteristic of grain growth governed by grain-boundary motion. This provides a means of controlling the grain microstructure evolution, in particular of ultrafine-grained and nanocrystalline materials.

There are two fundamental types of grain-boundary junctions in bulk polycrystalline materials: triple lines and quadruple points. A triple line forms where three grain boundaries meet, whereas a quadruple point is the geometrical location where four grains come into contact. There is a large variety of potential triple lines and quadruple points in polycrystalline materials, because their geometry is determined by the constituting grain boundaries, each of which has five degrees of freedom. Hence, a triple line is defined by 12 independent geometrical parameters; a quadruple point requires even 21 quantities for a unique geometrical characterization. Despite this large parameter space of potential triple lines and quadruple points, there are only very few configurations of boundaries that will cause a steady-state motion of the connected boundary systems. One of them is presented in Fig. 60(a).

The influence of triple junctions on grain-boundary migration can be expressed by the dimensionless criterion Λ (Ref 2, 146, 147):

$$\Lambda = \frac{m_{tj}a}{m} \qquad (Eq\ 23)$$

where m (m^4/Js) and m_{tj} (m^3/Js) are the mobility of the grain boundary and the mobility of the triple junction, respectively, and a(m) is the grain size (Fig. 60a). The dependency of the parameter Λ on the dihedral angle α at the triple junction for grains with a number of sides $n < 6$ (Fig. 60) and $n > 6$ makes it possible to measure the value of Λ experimentally and, as a result, the mobility of triple junctions for different grain-boundary systems (Ref 148, 149):

$$\Lambda = \frac{2\alpha}{2\cos\alpha - 1}, n < 6 \qquad (Eq\ 24)$$

$$\Lambda = -\frac{\ln\sin\alpha}{1 - 2\cos\alpha}, n > 6 \qquad (Eq\ 25)$$

Equations 24 and 25 constitute the basis of the experimental measurements of triple junction mobility. It follows from these equations that the strongest influence of triple junctions should be observed for small Λ; that is, the influence of grain-boundary junctions should be most pronounced for fine-grained and nanocrystalline materials. It was found experimentally that the mobility of triple junctions can be very low (Ref 147, 150 to 152).

The temperature dependence of the angle at the tip of a triple junction in zinc and the parameter Λ for individual junctions in zinc and aluminum tricrystals are presented in Fig. 60(b) and 61.

The temperature dependence of the mobility of a grain-boundary system with a triple junction is represented in Fig. 62 and 63. As a rule, the enthalpy of activation of triple junction motion is much higher than for grain-boundary migration. That is why grain-boundary motion at relatively low temperature is controlled by the motion of triple junctions.

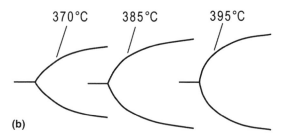

(a)

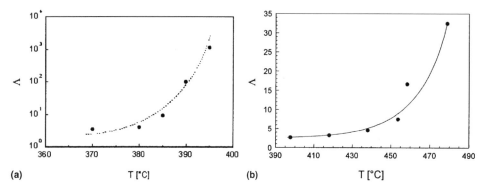

(b)

Fig. 60 Configuration of a grain-boundary triple junction that allows steady-state motion. (a) Tricrystal with triple junction (Ref 2, 146, 147). (b) Evolution of the shape of grain 3 in a zinc tricrystal with increasing temperature (Ref 147)

(a) T [°C]   (b) T [°C]

Fig. 61 Measured temperature dependence of the criterion Λ for (a) zinc and (b) aluminum tricrystals. Source: Ref 147, 152

REFERENCES

1. A.P. Sutton and R.W. Balluffi, *Interfaces of Crystalline Materials,* Clarendon Press, Oxford, 1995
2. G. Gottstein and L.S. Shvindlerman, *Grain Boundary Migration in Metals: Thermodynamics, Kinetics, Applications,* CRC Press, 1999
3. W.T. Read and W. Shockley, *Phys. Rev.,* Vol 78, 1950, p 275
4. W. Bollmann, *Crystal Defects and Crystalline Interfaces,* Springer-Verlag, Berlin, 1970
5. F.C. Frank, *Report on the Symposium on the Plastic Deformation of Crystalline Solids,* Carnegie Institute of Technology, Pittsburgh, 1950, p 150
6. D. Wolf, *Acta Metall. Mater.,* Vol 38, 1990, p 781
7. D. Wolf and K.L. Merkle, *Materials Interfaces,* D. Wolf and S. Yip, Eds., Chapman & Hall, New York, 1992
8. H. Hu, B. Schönfelder, and G. Gottstein, unpublished results
9. C. Herring, *Physics of Powder Metallurgy,* W.E. Kingston, Ed., McGraw-Hill, New York, 1951
10. R.L. Fullman, *J. Appl. Phys.,* Vol 22, 1951, p 456
11. J. Friedel, B. Cullity, and C. Crussard, *Acta Metall.,* Vol 1, 1953, p 79

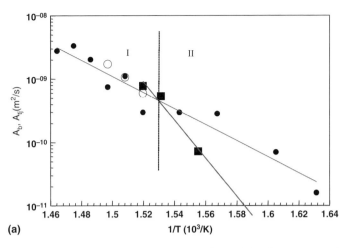
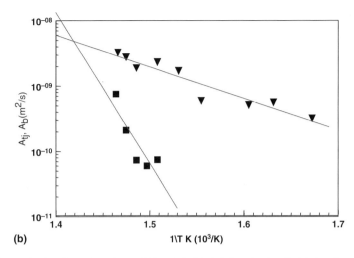

(a)

(b)

Fig. 62 Temperature dependence of grain-boundary mobility. (a) Temperature dependence of reduced grain-boundary mobility, A_b (solid circles), and reduced triple junction mobility, A_{tj} (solid squares), for the system with $<10\bar{1}0>$ and (b) $<11\bar{2}0>$ tilt boundaries (reduced grain-boundary mobility: A_b, solid triangles; reduced triple junction mobility: A_{tj}, solid squares). Source: Ref 153

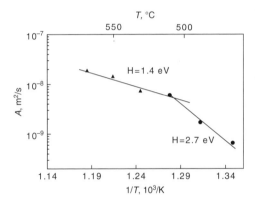

Fig. 63 Temperature dependence of triple junction (●) and grain-boundary mobility (▲) in tricrystal samples. Source: Ref 152

12. E.L. Maximova, L.S. Shvindlerman, and B.B. Straumal, *Acta Metall.,* Vol 36, 1988, p 1573

13. K.T. Aust and B. Chalmers, *Proc. Roy. Soc. A,* Vol 204, 1950, p 359

14. T.L. Wilson and P.G. Shewmon, *Trans. AIME,* Vol 236, 1966, p 48

15. G. Herrmann, H. Gleiter, and G. Bäro, *Acta Metall.,* Vol 24, 1976, p 353

16. T. Mori, H. Miura, T. Tokita, J. Haji, and M. Kato, *Phil. Mag. Letters,* Vol 58, 1988, p 11

17. S.W. Chan and R.W. Balluffi, *Acta Metall.,* Vol 33, 1985, p 1113

18. H. Udin, A.J. Shaler, and J. Wulff, *Trans. AIME,* Vol 185, 1949, p 186

19. H. Jones, *Scripta Mat.,* Vol 6, 1972, p 423

20. H. Jones, *Mater. Sci. Eng.,* Vol 4, 1969, p 106

21. L.E. Murr, G.I. Wong, and R.J. Honglew, *Acta Metall.,* Vol 21, 1973, p 595

22. W.W. Mullins, *Acta Metall.,* Vol 4, 1956, p 421

23. L.E. Murr, *Interfacial Phenomena in Metals and Alloys,* Addison-Wesley Publishing Company, Reading, 1975

24. N.A. Gjostein and F.N. Rhines, *Acta Metall.,* Vol 7, 1959, p 319

25. T. Mori, T. Ishii, M. Kajihara, and M. Kato, *Phil. Mag. Letters,* Vol 75, 1997, p 367

26. T. Mori, H. Miura, T. Tokita, J. Haji, and M. Kato, *Phil. Mag. Letters,* Vol 58, 1988, p 11

27. A. Otsuki and M. Mizuno, Grain Boundary Structure and Related Phenomena, *Proc. of JIMIS-4, Suppl. to Transactions of Jap. Inst. Met.,* 1986, p 789

28. G.C. Hasson and C. Goux, *Scripta Metall.,* Vol 5, 1971, p 889

29. D.M. Saylor, A. Morawiec and G.S. Rohrer, *J. Am. Ceram. Soc.,* Vol 85, 2002, p 3081

30. M.S. Daw and M.I. Baskes, *Phys. Rev. B,* Vol 29, 1984, p 6443–6453

31. P.N. Keating, *Phys. Rev.,* Vol 149, 1966, p 674

32. M. Albe, P. Erhard, and M. Müller, *Integral Materials Modeling,* G. Gottstein, Ed., Wiley-VCH, Weinheim, 2007, p 197

33. A. Suzuki and Y. Mishin, *Interface Sci.,* Vol 11, 2003, p 425

34. M.A. Tschopp and D.L. McDowell, *Phil. Mag.,* Vol 87, 2007, p 3147

35. U. Wolf, F. Ernst, T. Muschik, et al., *Phil. Mag. A,* Vol 66, 1992, p 991

36. N. Goukon and M. Kajihara, *Mater. Sci. Eng. A,* Vol 477, 2008, p 121

37. N. Goukon, T. Yamada, and M. Kajihara, *Acta Mater.,* Vol 48, 2000, p 2837

38. Y. Amouyal, E. Rabkin, and Y. Mishin, *Acta Mater.,* Vol 53, 2005, p 3795

39. D.M. Kirch, E. Jannot, L.A. Barrales-Mora, D.A. Molodov, and G. Gottstein, *Acta Mater.,* Vol 56, 2008, p 4998–5011

40. G. Gottstein, *Physical Foundations of Materials Science,* Springer-Verlag Berlin, 2004

41. J.E. Milliard, M. Cohen, and B.L. Averbach, *Acta Metall.,* Vol 8, 1960, p 26

42. M.C. Inman, D. McLean, and H.R. Tipler, *Proc. Roy. Soc. A,* Vol 273, 1963, p 538

43. R. Kirchheim, *Acta Mat.,* Vol 55, 2007, p 5129, 5139

44. R.A. Vandermeer, D. Juul-Jensen, and E. Woldt, *Metall. Mater. Trans. A,* Vol 28, 1997, p 749

45. B.B. Rath and H. Hu, *Met. Trans.,* Vol 245, 1969, p 1577

46. B.B. Rath and H. Hu, *The Nature and Behaviour of Grain Boundaries,* H. Hu, Ed., Plenum Press, New York-London, 1972

47. B.B. Rath and H. Hu, *Trans. TMS-AIME,* Vol 245, 1969, p 1243

48. B.B. Rath and H. Hu, *Trans. TMS-AIME,* Vol 236, 1966, p 1193

49. R.C. Sun and C.L. Bauer, *Acta Metall.,* Vol 18, 1970, p 635

50. R.C. Sun and C.L. Bauer, *Acta Metall.,* Vol 18, 1970, p 639

51. A.V. Antonov, Ch.V. Kopetskii, L.S. Shvindlerman, and Ya.M. Mukovskii, *Phys. Stat. Sol. (a),* Vol 9, 1972 p 45

52. M. Furtkamp, G. Gottstein, and L.S. Shvindlerman, *Interface Sci.,* Vol 6, 1998, p 279

53. W.W. Mullins, *Acta Metall.,* Vol 4, 1956, p 421–432

54. D.A. Molodov, G. Gottstein, F. Heringhaus, and L.S. Shvindlerman, *Scripta Mater.,* Vol 37, 1997, p 1207–1213

55. J. Washburn and E.R. Parker, *Trans. AIME,* Vol 194, 1952, p 1076–1078

56. C.H. Li, E.H. Edwards, J. Washburn, and E.R. Parker, *Acta Metall.,* Vol 1, 1953, p 223

57. D.W. Bainbridge, C.H. Li, and E.H. Edwards, *Acta Metall.,* Vol 2, 1954, p 322

58. D.A. Molodov, V.A. Ivanov, and G. Gottstein, *Acta Mater.,* Vol 55, 2007, p 1843

59. M. Peach and J.S. Koehler, *Phys. Rev.,* Vol 80, 1950, p 436

60. D.A. Molodov, T. Gorkaya, and G. Gottstein, *Mater Sci. Forum,* Vol 558–559, 2007, p 927

61. H. Fukutomi and R. Horiuchi, *Trans. Japan Inst. Metals*, Vol 22, 1981, p 633–642
62. H. Fukutomi and T. Kamijo, *Scripta Metall.*, Vol 19, 1985, p 195–197
63. H. Fukutomi, T. Iseki, T. Endo, and T. Kamijo, *Acta Metall. Mater.*, Vol 39, 1991, p 1445–1448
64. H. Yoshida, K. Yokoyama, N. Shibata, Y. Ikuhara, and T. Sakuma, *Acta Mater.*, Vol 52, 2004, p 2349–2357
65. A. Suzuki and Y. Mishin, *Mater. Sci. Forum*, Vol 502, 2005, p 157–162
66. J.W. Cahn, Y. Mishin, and A. Suzuki, *Acta Mat.*, Vol 54, 2006, p 4953–4975
67. M. Winning, G. Gottstein, and L.S. Shvindlerman, *Acta Mat.*, Vol 49, 2001, p 211–219
68. M. Winning, *Acta Mat.*, Vol 51, 2003, p 6465–6475
69. A.V. Antonov, Ch.V. Kopetskii, L.S. Shvindlerman, and V.G. Sursaeva, *Sov. Phys. - Dokl.*, Vol 18, 1974, p 736
70. P.F. Schmidt, Doctoral thesis, Universität Münster, 1977
71. J. Gastaldi and C. Jourdan, *Phys. Stat. Sol. (a)*, Vol 49, 1978, p 529
72. J. Gastaldi, C. Jourdan, and G. Grange, *Mater. Sci. Forum*, Vol 94–96, 1992 p 17
73. V.Yu Aristov, Ch.V. Kopetskii, and L.S. Shvindlerman, Author's Certificate of Invention, 642638 GOIN23/20 (in Russian), 1979
74. V.Yu Aristov, Ch.V. Kopetskii, D.A. Molodov, and L.S. Shvindlerman, *Sov. Phys. Solid State*, Vol 22, 1980, p 1900
75. U. Czubayko, D.A. Molodov, B.-C. Petersen, G. Gottstein, and L.S. Shvindlerman, *Meas. Sci. Technol.*, Vol 6, 1995, p 947
76. Y. Huang and F.J. Humphreys, *Proc. of the Third Int. Conf. on Grain Growth*, Carnegie Mellon University, Pittsburgh, PA, 1998
77. D. Mattissen, *In-situ Untersuchung des Einflusses der Tripelpunkte auf die Korngrenzebewegung in Aluminium*, Shaker Verlag, Aachen, 2004
78. D.A. Molodov, G. Gottstein, F. Heringhaus, and L.S. Shvindlerman, *Acta Mater.*, Vol 46, 1998, p 5627–5632
79. R. Vandermeer, *Trans. AIME*, Vol 233, 1965, p 266
80. V.Yu. Aristov, V.E. Fradkov, and L.S. Shvindlerman, *Phys. Met. Metall.*, Vol 45, 1979, p 83
81. G. Gottstein and L.S. Shvindlerman, *Scripta Metall. Mater.*, Vol 27, 1992, p 1521
82. B. Liebman, K. Lücke, and G. Masing, *Z. Metallk.*, Vol 47, 1956, p 57
83. M. Yoshida, B. Liebman, and K. Lücke, *Acta Metall.*, Vol 7, 1959, p 51
84. G. Gottstein, M.C. Murmann, G. Renner, C. Simpson, and K. Lücke, *Textures of Materials*, G. Gottstein and K. Lücke, Ed., Springer-Verlag, Berlin, 1978, p 521
85. R. Viswanathan and C.L. Bauer, *Acta Metall.*, Vol 21, 1975, p 1099
86. M. Furtkamp, G. Gottstein, D.A. Molodov, V.N. Semenov, and L.S. Shvindlerman, *Acta Mater.*, Vol 46, 1998, p 4103–4110
87. D.M. Deminczuk and K.T. Aust, *Acta Metall.*, Vol 21, 1975, p 1149
88. M. Furtkamp, P. Lejcek, and S. Tsurekawa, *Interface Sci.*, Vol 6, 1998, p 59–66
89. Y. Huang and F.J. Humpreys, *Acta Mater.*, Vol 48, 2000, p 2017
90. F.J. Humphreys and Y. Huang, *Proc. 21st Int. Risø Symposium on Materials Science*, N. Hansen et al., Ed., Risø, Denmark, 2000, p 71
91. K.T. Aust and J.W. Rutter, *Acta Metall.*, Vol 12, 1965, p 181
92. E.M. Fridman, Ch.V. Kopetskii, and L.S. Shvindlerman, *Z. Metallk.*, Vol 66, 1975, p 533
93. V.Yu. Aristov, Ch.V. Kopetskii, and L.S. Shvindlerman, in *Theoretical Fundamentals of Materials Science*, Moscow, Nauka, 84, 1981
94. V.Yu. Aristov, V.L. Mirochnik, and L.S. Shvindlerman, *Sov. Phys. - Solid State*, Vol 18, 1976, p 137
95. V.G. Sursaeva, A.V. Andreeva, Ch.V. Kopetskii, and L.S. Shvindlerman, *Phys. Met. Metall.*, Vol 41, 1976, p 98
96. D.A. Molodov, L.S. Shvindlerman, and G. Gottstein, *Zt. Metallk.*, Vol 94, 2003, p 1117–1126
97. G. Gottstein and L.S. Shvindlerman, *Interface Sci.*, Vol 6, 1998, p 267
98. Ch.V. Kopetskii, V.G. Sursaeva, and L.S. Shvindlerman, *Sov. Phys. Solid State*, Vol 21, 1979, p 238
99. D.A. Molodov, B.B. Straumal, and L.S. Shvindlerman, *Scripta Metall.*, Vol 18, 1984, p 207
100. P. Lejcek, V. Paidar, J. Adámek, and S. Kadecková, *Interface Sci.*, Vol 1, 1993, p 187–199
101. P. Lejcek and J. Adámek, *J. Physique IV*, Vol 5, 1995, p C3–107
102. S. Tsurekawa, T. Ueda, K. Ichikawa, H. Nakashima, Y. Yoshitomi, and H. Yoshinaga, *Mater. Sci. Forum*, Vol 204–206, 1996, p 221
103. K.T. Aust and J.W. Rutter, *Trans. AIME*, Vol 215, 1959, p 820
104. G. Ibe and K. Lücke, *Recrystallization, Grain Growth, and Textures*, American Society for Metals, Metals Park, Ohio, 1966, p 434
105. G. Ibe, W. Dietz, A.-C. Fraker and K. Lücke, *Z. Metallkde.*, Vol 6, 1970, p 498
106. G. Ibe and K. Lücke, *Texture*, Vol 1, 1972, p 87
107. D.A. Molodov, U. Czubayko, G. Gottstein, and L.S. Shvindlerman, *Scripta Metall. Mater.*, Vol 32, 1995, p 529
108. D.A. Molodov, U. Czubayko, G. Gottstein, and L.S. Shvindlerman, *Acta Mater.*, Vol 46, 1998, p 553
109. B. Schönfelder, G. Gottstein, and L.S. Shvindlerman, *Acta Mater.*, Vol 53, 2005, p 1597
110. M. Upmanyu, D.J. Srolovitz, L.S. Shvindlerman, and G. Gottstein, *Acta Mater.*, Vol 47, 1999, p 3901
111. M.I. Meldelev and D.J. Srolovitz, *Acta Mater.*, Vol 49, 2001, p 589
112. H. Zhang, M. Upmanyu, and D.J. Srolovitz, *Acta Mater.*, Vol 53, 2005, p 79
113. H. Zhang, M.I. Mendelev, and D.J. Srolovitz, *Scripta Mater.*, Vol 52, 2005, p 1193
114. L. Zhou, H. Zhang, and D.J. Srolovitz, *Acta Mater.*, Vol 53, 2005, p 5273
115. H. Zhang and D.J. Srolovitz, *Acta Mater.*, Vol 54, 2006, p 623
116. M. Upmanyu, D.J. Srolovitz, A.E. Lobkovsky, J.A. Warren, and W.C. Carter, *Acta Mater.*, Vol 54, 2006, p 1707
117. Y. Mishin, M. Asta, and J. Li, *Acta Mater.*, 2009
118. D.L. Olmsted, E.A. Holm, and S.M. Foiles, *Acta Mater.*, 2009
119. K. Lücke and K. Detert, *Acta Metall.*, Vol 5, 1957, p 628
120. J.W. Cahn, *Acta Metall.*, Vol 10, 1962, p 789
121. K. Lücke and H. Stüwe, *Acta Metall.*, Vol 19, 1971, p 1087
122. V.Yu. Aristov, V.E. Fradkov, and L.S. Shvindlerman, *Sov. Phys. Solid State*, Vol 22, 1980, p 1055
123. H. Westengen and N. Ryum, *Phil. Mag. A*, Vol 38, 1978, p 3279
124. W. Grünwald and F. Haessner, *Acta Metall.*, Vol 18, 1970, p 217
125. V.Yu. Aristov, Ch.V. Kopetskii, V.G. Sursaeva, and L.S. Shvindlerman, *Sov. Phys. Dokl.*, Vol 20, 1975, p 842
126. D.A. Molodov, Ch.V. Kopetskii, and L.S. Shvindlerman, *Sov. Phys. Solid State*, Vol 23, 1981, p 1718
127. V.Yu. Aristov, Ch.V. Kopetskii, D.A. Molodov, and L.S. Shvindlerman, *Sov. Phys. Solid State*, Vol 22, 1980, p 11
128. L.S. Shvindlerman, *Mat. Sci. Forum*, Vol 94–96, 1992, p 169
129. Ch.V. Kopetskii, V.G. Sursaeva, and L.S. Shvindlerman, *Scripta Mat.*, Vol 12, 1978, p 953
130. Ch.V. Kopetskii, V.G. Sursaeva, and L.S. Shvindlerman, *Sov. Phys. Dokl.*, Vol 23, 1978, p 137
131. V.Yu. Aristov, Ch.V. Kopetskii, D.A. Molodov, and L.S. Shvindlerman, *Sov. Phys. Solid State*, Vol 26, 1984, p 284
132. G. Gottstein, L.S. Shvindlerman, D.A. Molodov, and U. Czubayko, *Dynamics of Crystal Surfaces and Interfaces*, P.M. Duxburg and T.J. Pence, Ed., Plenum Press, New York, 1997, p 109
133. J.W. Rutter and K.T. Aust, *Trans. AIME*, Vol 215, 1959, p 119
134. E.M. Fridman, Ch.V. Kopetskii, L.S. Shvindlerman, and V.Yu. Aristov, *Zt. Metallk.*, Vol 64, 1973, p 458
135. P. Gordon and R.A. Vandermeer, *Trans. AIME*, Vol 24, 1962, p 917

136. D. Udler and D.N. Seidman, *Interface Sci.*, Vol 3, 1995, p 41
137. D.A. Molodov, U. Czubayko, G. Gottstein, L.S. Shvindlerman, W. Gust, and B.B. Straumal, *Phil. Mag. Lett.*, Vol 7, 1995, p 361
138. C.S. Smith, *Trans. AIME*, Vol 175, 1948, p 15
139. T. Gladman, *Proc. Roy Soc.*, Vol 294ff, 1966, p 298
140. P.A. Manohar, M. Ferry, and T. Chandra, *ISIJ Int.*, Vol 37, 1998, p 913
141. G. Gottstein and L.S. Shvindlerman, *Acta Metall. Mater.*, Vol 41, 1993, p 3267
142. V. Novikov, *Scripta Mater.*, Vol 55, 2006, p 243
143. V. Novikov, *IJMR*, Vol 98, 2007, p 18
144. D.A. Molodov, B.B. Straumal, and L.S. Shvindlerman, *Scripta Metall.*, Vol 18, 1984, p 207
145. D.A. Molodov, J. Swiderski, G. Gottstein, W. Lojkowski, and L.S. Shvindlerman, *Acta Metall. Mater.*, Vol 42, 1994, p 3397
146. A.V. Galina, V.E. Fradkov, and L.S. Shvindlerman, *Phys. Met. Metall.*, Vol 63, 1987, p 165
147. U. Czubayko, V.G. Sursaeva, G. Gottstein, and L.S. Shvindlerman, *Acta Mater.*, Vol 46, 1998, p 5863
148. G. Gottstein and L.S. Shvindlerman, *Scripta Metall.*, Vol 38, 1998, p 1541
149. G. Gottstein and L.S. Shvindlerman, *Acta Mat.*, Vol 50, 2002, p 703
150. D. Mattissen, D.A. Molodov, L.S. Shvindlerman, and G. Gottstein, *Acta Mater.*, Vol 53, 2005, p 2049
151. L.S. Shvindlerman and G. Gottstein, *J. Mat. Sci.*, Vol 40, 2005, p 819
152. S.G. Protasova, G. Gottstein, D.A. Molodov, V.G. Sursaeva, and L.S. Shvindlerman, *Acta Mater.*, Vol 49, 2001, p 2519
153. L.S. Shvindlerman, G. Gottstein, D.A. Molodov, and V.G. Sursaeva, *Proc. of the First Joint Int. Conf. Recrystallization and Grain Growth*, G. Gottstein and D.A. Molodov, Ed., 2001, p 177

Copyright © 2010, ASM International®
All rights reserved.
www.asminternational.org

Texture Measurement and Analysis

A.D. Rollett, Carnegie Mellon University

OTHER ARTICLES in this Handbook's first section, "Input Data for Simulations," describe methods for gathering input for simulations. This article deals with a central aspect of anisotropy modeling, namely that of texture measurement and analysis. Despite the fact that many laboratories have an electron backscatter diffraction (EBSD) system available, x-ray pole figures remain the most cost-effective way to quantify preferred crystallographic orientation in a given material. User facilities are also available for texture measurement with neutrons, which have exceptional penetrating power and are also very useful for measurement of elastic strains. For thin films and specimens that are not too thick, high-energy x-rays from synchrotrons or neutrons are also useful, albeit they are only available at specialized user facilities. Regardless of the data source, the subsequent calculation of the orientation distribution (OD) is a key step. As a point of historical interest, the relationship between a pole figure (PF) and the OD is known as the fundamental equation of texture analysis. The original solution to the problem in terms of series expansion, based on generalized spherical harmonics, enabled quantitative texture analysis to develop and replace what had previously been a qualitative topic.

Guide for Nonexperts

At this point, the reader may be thinking that none of the terms are familiar and that too many assumptions have already been made. Do not panic! Although texture and anisotropy is a specialized subfield of materials science with its own jargon, the basic ideas are readily accessible to a nonexpert. More to the point, the variation in properties with direction in a material can generally be well simulated if information about the texture is incorporated in the simulation. Now, texture expands to crystallographic preferred orientation, and so the experimental task consists of measuring those crystal orientations. The most straightforward method, conceptually, is to use an electron beam in a scanning electron microscope and to analyze a diffraction pattern from each grain. This provides an orientation, which immediately raises the next issue of why three numbers (e.g., three Euler angles) are needed to define an orientation. A reasonable analogy here is to recall that the attitude of an aircraft requires three things to be specified, that is, pitch, roll, and yaw. The three Euler angles commonly encountered in texture work can be used equally well for aircraft as for crystals. Given a list of orientations, the information must be transformed into a more digestible form known as an OD to standardize the presentation. (The significance of the term function in association with orientation distribution or misorientation distribution to make orientation distribution function or misorientation distribution function is that a series expansion is performed to fit coefficients of the generalized spherical harmonics to the data set. In other words, one assumes that the data can be approximated by a mathematical function.) Once again, the map analogy is useful because wherever a particular crystal orientation comprises a large fraction of the material, the intensity in the OD is high, which appears as a peak or hill in the graph of the OD. The axes of the OD are typically the three Euler angles used to characterize orientations. As with most three-dimensional distributions, slices or cross sections typically are used to make contoured graphs. The main point of the OD is that, given adequate input of data, it is a distribution that completely describes the material texture, and calculations of anisotropy can be based on it. Making use of the information contained in the OD may require fitting a set of individual orientations corresponding to the grains in, say, a finite-element mesh, or it may, at the continuum level, involve fitting a mathematical function to a yield surface derived from the OD. Extracting a list of (fitted) orientations is similar to sampling a distribution in statistics. Fitting a mathematical yield surface is admittedly a more complex operation.

Finally, it is necessary to say something about the pole figure, which is the traditional approach to texture measurement. Crystals have planes with high atomic densities that can be readily detected in diffraction experiments. Pole figure measurement is, accordingly, extending an ordinary diffraction experiment to measure the variation in diffracted intensity over all possible directions in a given sample. To summarize: given a set (typically three for cubic materials) of pole figures for several crystal directions (reflections), standardized and robust algorithms are available to obtain a complete OD for the material. Given an OD, all the standard texture-based analysis tools can be applied.

Pole Figure Measurement

X-Ray Diffraction for Pole Figure Measurement

The measurement of pole figures has been described in a number of texts, so only the basic information is provided here. Readers can find more detail in texts by Kocks et al. (Ref 1), Cullity (Ref 2), Randle and Engler (Ref 3), and so on. A sample of at least 10 mm on the side of a square is prepared to be optically flat, and the surface damage layer thickness must be substantially less than the x-ray penetration depth. The sample is placed in the diffractometer, which must have three types of motion in addition to the standard two (for θ and $2 - \theta$, where θ is the x-ray angle of incidence), namely tilting with respect to the specimen normal, turning (twisting) about the specimen normal, and translation in the plane of the specimen, as shown in Fig. 1. The tilt and twist motions enable a given diffraction peak to be measured over (almost) all directions in the material, and the same spherical angles are used to plot the measured intensities in a pole figure, where the tilt angle corresponds to co-latitude and the twist angle corresponds to the longitude. The translational motion (oscillation) allows a larger area of the specimen surface to be illuminated and therefore increases the number of grains included in the measurement. The choice of diffraction peaks or reflections to use is governed mainly by the separation between adjacent peaks. A pole figure measurement must be made with relatively large acceptance angles and therefore low resolution, which means that peaks that are less than approximately $2°$ apart tend to overlap at high tilt angles and should therefore not be used. For high-symmetry crystal types, such as the cubic metals, the first three reflections are adequate, for example, 111, 200, and 220 for face-cubic centered (fcc) materials. (For additional information about crystal structure and nomenclature, see the Appendix of *Alloy Phase Diagrams*, Volume 3 of *ASM Handbook*).

The last issue that must be mentioned is that, in recent years, pole figures have been measured in back-reflection mode (Ref 4), although transmission through a thin-enough specimen is also feasible, for example, for thin aluminum sheet. An inevitable consequence of using the back-reflection mode is that the beam spreads out on the surface of the specimen at high tilt angles (Fig. 2). This means that not all the diffracted x-rays enter the detector, and intensity is lost. One consequence is that the intensity measured from a perfectly uniformly textured (random) material is constant from zero tilt (which is the center of a pole figure) out to some limit and then decreases monotonically to zero at the edge. Correction of this defocusing is required for all pole figures measured in back-reflection. It is also one of the major sources of error in calculating ODs from pole figure data. Figure 3(a) shows typical (theoretical) defocusing curves for reflection geometry; the equation set for these curves is given by Kocks et al. (Ref 1, p 146), based on papers by Tenckhoff (Ref 5) and Gale and Griffiths (Ref 6). Figure 3(b) shows how films that are thin enough for significant intensity to be transmitted through them require additional corrections for the finite volume illuminated, which itself varies with the tilt angle.

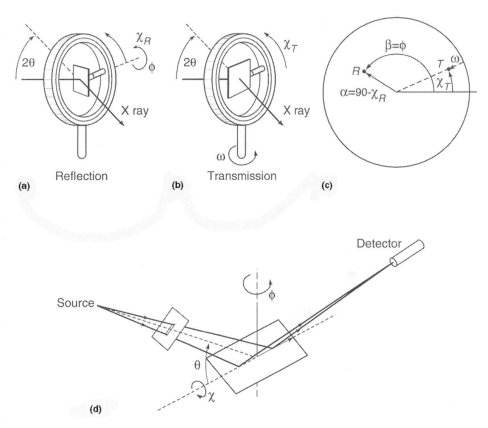

Fig. 1 Illustration of the diffraction geometries used for x-ray pole figure measurement. (a) Reflection geometry. (b) Transmission geometry. (c) Definition of a pole in reflection (R) by $\alpha = 90° - \chi$ and $\beta = \phi$, and in transmission (T) by $90° - \alpha = \omega$ and $\beta = \chi$. (d) In the Bragg-Brentano geometry, a divergent x-ray beam is focused on the detector. However, this no longer applies when the specimen is tilted, i.e., when $\chi \neq 0$. Reproduced by permission of Cambridge University Press. Source: Ref 1

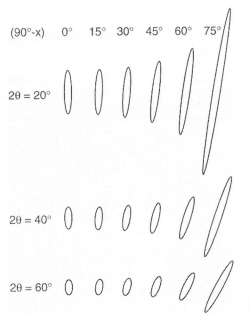

Fig. 2 Illustration of the way in which the illuminated spot on the specimen surface spreads out with increasing tilt angle, $\alpha = 90° - \chi$, and decreasing Bragg angle, 2θ. This spreading out causes a loss of intensity entering the detector and hence the need for a defocusing correction in reflection geometry. Reproduced by permission of Cambridge University Press. Source: Ref 1

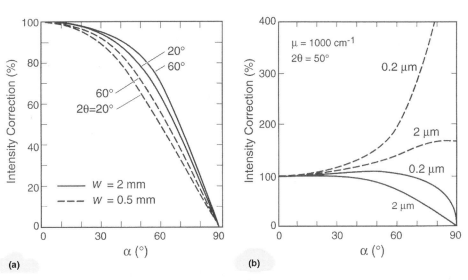

Fig. 3 Typical correction curves for reflection geometry for different values of the Bragg angle, 2θ, and as a function of the tilt angle, $\alpha = 90° - \chi$. The measured intensity is divided by the correction factor. (a) Theoretical curves (for an infinitely thick specimen) for both a wide slit (2 mm) and a narrow slit (0.5 mm), showing that more intensity is lost for the narrower slit, and therefore, a greater correction is required. (b) Theoretical correction curves for thin films with two different thicknesses, as indicated; the dashed lines show the correction from the finite volume of thin film irradiated, and the solid lines combine the volume correction with the defocusing correction. Reproduced by permission of Cambridge University Press. Source: Ref 1

Neutron Diffraction for Pole Figure Measurement

The measurement of pole figures with neutrons is more involved than for x-rays, because a source of neutrons requires either a nuclear reactor or an accelerator to drive a spallation source (Ref 7). Neutrons from a reactor are known as thermal neutrons, and their wavelengths cover a broad spectrum with a peak at approximately 1 Å. The fluxes are low, which means that counting times are long compared to x-rays. On the other hand, the penetration distances of neutrons are very large in most materials, which means that large samples with dimensions up to several centimeters can be measured. This means that texture measurement is truly a bulk measurement as opposed to the near-surface probe that x-rays or electrons represent. The large specimen capability also means that it is practicable to measure texture in coarse-grained materials, whereas a practical upper limit to grain size with x-rays is approximately 0.2 mm. Neutrons from a spallation source differ strongly from other sources in that they arrive as dispersed pulses, and the variation in neutron energy contained in the pulse makes it possible to base measurement on time of flight (Ref 8, 9). In effect, a wide range of wavelengths are dispersed in time such that a fixed detector position can be used to measure intensities for a wide range of reflections. In recent years, significant progress has been made in extending conventional analysis of neutron-scattering data to simultaneously obtain information on texture, crystal structure, elastic strains, and grain size (Ref 10). Thus, neutron diffraction proves to be useful (especially compared to x-rays) for large specimens, cases involving both elastic (residual) stresses and texture, and coarse-grained materials. The low absorption of neutrons by most materials also means that it is practical to surround samples with heating or cooling apparatuses and to perform diffraction experiments during thermal treatment (Ref 11). Despite these advantages, neutron diffraction can only be carried out at specialized facilities and thus remains a technique that is only applied to solve specific problems.

Stereographic Projection

To represent information that is a variation in a (normalized) intensity on the unit sphere, a method of projecting the data onto the (flat) page is needed. This is the same problem that geographers or cartographers have, and, of course, many different projections have been used over the years. It happens that only two are commonly encountered in materials work, one of which is the stereographic projection, also known as the equal-angle or Wulff projection (Fig. 4a). For technical details, the reader is referred to one of the standard texts (Ref 2, 3, 12).

Equal Area Projection

The other commonly encountered projection in materials science is the equal-area or Schmid projection (Fig. 4b). This preserves areas on the projection such that the same area on the original spherical surface occupies the same area on the projection plane. The geometry of the projection is not as important here as understanding the physical significance of a pole figure. Imagine a perfectly smooth sphere such as a ball bearing. The pole figure of a perfectly random material exhibits no variation in intensity with direction, which corresponds to the analogy of the smooth ball bearing. Imagine now the surface of a golf ball, which has a set of dimples in its surface. Now the surface height, compared to the minimum radius, varies with direction. Such a variation is not encountered in texture analysis but, if represented in a pole figure, would be revealed as sets of circular contours delineating each dimple. Another familiar shape is the nearly ellipsoidal shape of a lemon; translating this into a pole figure would yield circular contours whose spacing would be closest (for a constant interval between each contour) near each pole of the lemon, where the radius changes most rapidly.

Graphing Pole Figures

Given a set of orientations that represent a set of crystals or grains in a polycrystal, how can one set about plotting the associated pole figure? The answer is to realize that a pole figure represents the variation in intensity on the sphere that corresponds to a particular family, $\{hkl\}$, of diffracting planes, and so, graphing is very similar to the problem of mapping the globe, that is, cartography. A simple set of transformations of the measured intensity data is used to, in effect, flatten the spherical information onto the page. The two standard transformations are the stereographic and the equal-area projections, as described previously. Readers who are interested in developing their own plots of data (as opposed to the

commercially available software packages) may find helpful the open-source general mapping tools for cartographic plotting (Ref 13).

Reference Frame for Data

Implicit in any directional data, and too rarely discussed explicitly, is a reference frame. In a standard orthogonal X-Y-Z (right-handed) frame, the Z-axis is always perpendicular to the specimen surface. The position of the X-axis (and thus the Y-axis) is much less clear, however, but is critical to the correct interpretation of the data. It adds to the frequent state of confusion in this area that the mathematical convention in plotting data is to point the X-axis to the right on the page (horizontal) and the Y-axis vertically on the page. In plane-strain compression (rolling), one generally associates the X-axis with the extension (rolling) direction (RD) and the Z-axis with the compression (reduction, or normal) direction (ND) (Fig. 5). Unfortunately, it is also typical to point the RD vertically on the page in pole figure plots, which, visually speaking, aligns the RD with the Y-axis and is in conflict with the previously stated conventions! More seriously, if the actual measured data have the RD aligned with the Y-axis of the goniometer, then the calculated OD will have all the peaks and valleys in the distribution in the wrong locations, which means that no quantitative conclusions can be drawn. There is no simple answer to this problem, and the reader is urged to check their results by comparing ODs computed from measured pole figures against expected results for the particular material of interest.

Typical Textures from Deformation and Annealing

In this section, some information is provided on the textures that one can expect to find in deformed and annealed metals (Ref 14). There is similar information available for geological

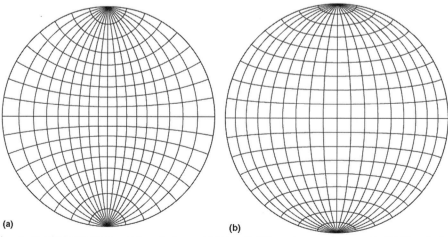

Fig. 4 Standard projections used for pole figures. (a) Grid with 10° spacing on a stereographic (Wulff) projection. (b) Grid with 10° spacing on an equal-area (Schmid) projection

materials in the appropriate literature (Chapter 4 in Ref 1). Table 1 provides a list, with Euler angles (in degrees) for both the Bunge and Kocks conventions (as explained subsequently in more detail), of texture components commonly found in rolled fcc metals. The last three (cube P, Q, and R) are commonly found after recrystallization. This list of texture components must be supplemented with the knowledge that orientations from deformation in fact concentrate along lines in orientation space, known as fibers. So, for example, there is generally a continuous distribution of intensity along a line between the brass, S, and copper components. Similarly in body-centered cubic metals, the two predominant fibers are the α fiber, with <110> parallel to the rolling direction, and the γ fiber, with <111> perpendicular to the rolling plane.

Orientation Distribution from Pole Figures

Before discussing the calculation of orientation distributions, recall that the OD is the basic description of texture. Because three Euler angles are needed to describe an orientation, the OD is necessarily a three-dimensional distribution, that is, the variation in intensity with respect to each of the three parameters. Any such distribution can be projected or collapsed to a two-dimensional or even a one-dimensional space. Pole figures (and inverse pole figures)

are, in fact, just projections of OD with some special consideration of geometry. The simplest such projection is to average an OD with respect to the third Euler angle, which, for any of the standard definitions used, yields a (001) pole figure. Other pole figures are, admittedly, more complicated to imagine, although the mathematical description is straightforward.

Given that Euler angles are so central to texture analysis, some basic information about them is provided. Figure 6 shows how each of the angles operates on a set of Cartesian axes to rotate (or, more properly, transform) from the sample frame to the crystal frame. Note that the Bunge Euler angles are illustrated, but (many) other conventions exist. It is always important to know which convention is used and to apply conversions as needed. Also, whereas the alignment of the Cartesian frame with the crystal is obvious for cubic, tetragonal, and orthorhombic crystals, hexagonal (and other lower-symmetry) crystals require more thought. One can align the Cartesian x-axis in hexagonals with either a $<2\bar{1}\bar{1}0>$ direction or a $<01\bar{1}0>$ direction. Accordingly, Table 2 gives interconversions between the more commonly found definitions of Euler angles.

The limits on the Euler angles depend on the symmetry of the crystals to be characterized. When there is no crystal symmetry (triclinic structure), the largest range applies (Eq 1), which covers all possible proper rotations:

$$0 \le \phi_1 < 2\pi, 0 \le \Phi \le \pi, 0 \le \phi_2 < 2\pi \qquad \text{(Eq 1)}$$

One can speak of an orientation space parameterized by the Euler angles, within which each point represents a unique orientation or texture component. As the crystal symmetry increases, the range of Euler angles required

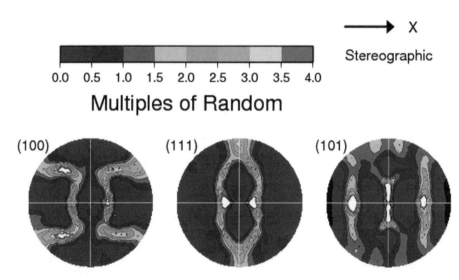

Fig. 5 Set of three pole figures showing a typical rolling texture in face-centered cubic metal. The rolling direction points to the right, and the normal direction points out of the plane of the figure.

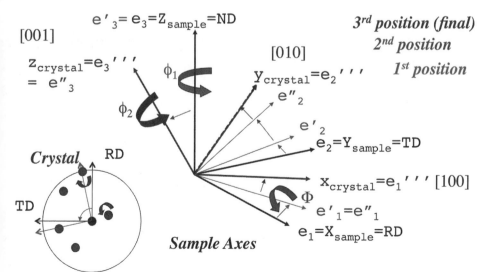

Fig. 6 Diagram showing how the three Euler angles relate a Cartesian frame associated with a crystal to a reference frame. The inset, lower left, shows a pole figure in standard position, with green points for the crystal axes in the second position and blue points for the third position. The sequence of angles corresponds to the Bunge Euler angles. In the case of Roe Euler angles, by contrast, the second Euler angle is about the local y-axis instead of the x-axis. TD, transverse direction; RD, rolling direction; ND, normal direction

Table 1 Rolling texture components in face-centered cubic metals

Name	Indices		Bunge (φ_1, Φ, φ_2)	Kocks (Ψ, Θ, φ)
Copper	{112}	$<11\bar{1}>$	90, 35, 45	0, 35, 45
S1	{124}	$<21\bar{1}>$	59, 29, 63	−31, 29, 27
S2	{123}	$<41\bar{2}>$	47, 37, 63	−43, 37, 27
S3	{123}	$<63\bar{4}>$	59, 37, 63	−31, 37, 27
Brass	{110}	$<1\bar{1}2>$	35, 45, 0	55, 45, 0
Taylor	{4, 4, 11}	$<11,11,\bar{8}>$	7, 71, 70	83, 27, 45
Goss	{110}	$<001>$	0, 45, 0	90, 45, 0
Cube	{001}	$<100>$	0, 0, 0	0, 0, 0
P	{011}	$<12\bar{2}>$	70, 45, 0	160, 45, 90
Q	{013}	$<2\bar{3}1>$	58, 18, 0	148, 18, 90
R	{124}	$<21\bar{1}>$	57, 29, 63	147, 29, 27

Table 2 Interconversions between different Euler angle conventions

Euler angle convention	1st	2nd	3rd	2nd angle about axis
Kocks	Ψ	Θ	φ	y
Bunge	$\varphi_1 - \pi/2$	Φ	$\pi/2 - \varphi_2$	x
Matthies	α	β	$\pi - \gamma$	y
Roe	Ψ	Θ	$\pi - \Phi$	y

decreases, because points in different parts of orientation space become related by symmetry operations. More precisely, a fundamental zone can be defined that is the subset of orientation space within which all possible orientations occur only once. Using group theory, it can be shown that the volume of the fundamental zone is equal to the volume of the entire space divided by the number of symmetry operators for the given crystal. Thus, cubic crystals require an orientation space volume that is only $\frac{1}{24}$ of the whole space. The effect of symmetry on the Euler angles is, unfortunately, not simple, especially when sample symmetry is also considered. The most commonly encountered case is the combination of cubic crystal symmetry with orthorhombic sample symmetry, for which the standard Euler range is described in Eq 2. What is confusing (but important to be aware of) is that this range includes three copies of the fundamental zone and is chosen because a true fundamental zone for cubics is an awkward shape in Euler space:

$$0 \leq \phi_1 < \pi/2, 0 \leq \Phi \leq \pi/2, 0 \leq \phi_2 < \pi/2 \quad \text{(Eq 2)}$$

The Euler angles are convenient for representation and visualizing orientation, but they cannot be used directly for calculations involving texture. Instead, one must convert them into the form of a rotation matrix, for example. The rather complicated set of trigonometric formulae shown in Table 3 comprise a matrix that can be used to transform tensor quantities (such as directions, stresses, elastic moduli) from the sample frame to the crystal frame:

This discussion sets the scene for what must be done to obtain an OD when the only data available are pole figures. In mathematical terms, each pole figure is drawn from the same underlying dataset (i.e., measured on the same material). Therefore, one has a set of simultaneous equations (Eq 3) to solve where the values in the OD are the unknowns (the vector $\mathbf{x}$ on the left side of the equation), the geometrical relationships between points in orientation space and pole figure space provide the equations (the $\mathbf{A}$ matrix on the left side of the equation), and the pole figure data provide the known inputs (the vector $\mathbf{b}$ on the right side of the equation). For a solution to be available, one must have enough pole figure data available. However, the number of independent figures needed depends on the inversion method used:

$$\mathbf{Ax} = \mathbf{b} \quad \text{(Eq 3)}$$

Although the simple equation implies that standard linear algebra methods can be applied, the size of the problem (on the order of 5000 unknowns for a 5° resolution in the OD) historically discouraged this until the recent efforts of Barton (Ref 15). The relationship that applies to pole figure data is as follows. P is the intensity in the pole figure associated with the crystal plane $\{hkl\}$, where α and β are the tilt and twist angles. The intensity in the OD, using Bunge Euler angles to parameterize the space, is represented by f, and Γ describes the path through orientation space that projects the OD onto the pole figure. Note that this fundamental equation of texture is an implicit equation, because one measures the pole figure intensities but would like to have the intensities of the OD. The aforementioned $\mathbf{A}$ is analogous to Γ in Eq 4, $\mathbf{x}$ is analogous to f, and $\mathbf{b}$ is analogous to P:

$$P_{(hkl)}(\alpha, \beta) = \frac{1}{2\pi} \int_0^{2\pi} f(\phi_1, \Phi, \phi_2) d\Gamma \quad \text{(Eq 4)}$$

The first solution procedure followed the series expansion approach after the realization that the essentially spherical data inherent in orientations could be fit with generalized spherical harmonics (Ref 16, 17). This permitted a compact and efficient computation and description of polycrystal textures along with calculation of properties. Elastic properties, in particular, are very efficiently computed based on the series expansion method. Two cautions are worth noting. One is that the approach is best suited to weak textures. As a texture approaches that of a single crystal, higher- and higher-order harmonics are required to describe it. In practical terms, however, the order is limited, with the consequence that pole figures of highly textured materials are commonly seen that have been recalculated from the coefficients with spurious intensity maxima. The other caution is that the method is most efficient for high crystal symmetries, because the number of independent harmonics rises as the crystal (and sample) symmetry decreases. Thus, three pole figures are sufficient for cubic crystal symmetry, whereas the desirable number for orthorhombic symmetry is five.

The commonly employed alternate method relies on direct numerical calculation of the OD based on a chosen discretization of the orientation space. As mentioned previously, pole figures are generally measured in reflection,

which means that they are incomplete; that is, they have unmeasured regions at the edges. As is the case for the series expansion method, iteration must be used in the discrete method. This method was impractical until fast personal computers became available after 1990. An advantage of the approach is that the texture to be calculated can be arbitrarily sharp up to the resolution limit imposed by the discretization of the orientation space.

Several software packages are available for calculating ODs from pole figure data. They also have tools for quantifying textures in polycrystals. Examples include BEARTEX (Ref 18), LaboTex (Ref 19), MulTex (Ref 20), MTEX (Ref 21), popLA (Ref 22), and TexTools (Ref 23).

Production of Discrete Orientation Lists

Finally, this section discusses how to obtain a weighted list from an OD that can be used in, say, finite-element simulations. At its most basic, this corresponds to sampling the distribution. In this case, using the series expansion approach is the easiest to understand, because the procedure has only two steps. The first step is to generate a set of Euler angles at random, that is, in such a fashion that the space is covered as uniformly as the number of points used will allow. The second step is to compute the value of the OD (by summing over the harmonics multiplied by their associated coefficients for that point) and use this as the weight associated with each sampled orientation.

If one has a discrete description of the OD (e.g., from the Williams-Imhot-Matthies-Vinel method, Ref 22), then an arbitrarily selected orientation will lie in between the gridpoints at which the OD is known. However, this just means that the second step of the two-step algorithm mentioned previously requires interpolation among some suitable set of nearest-neighbor points surrounding the orientation desired.

One caution is that if a finite list of orientations is generated that is, say, only a thousand entries, regenerating the texture in the form of an OD from that list is subject to error. There is no general rule that defines the minimum required number of orientations, but one should keep in mind that it increases with the size of the fundamental zone and decreases with increasing texture strength.

A more involved approach is required if orientations are to be assigned to a microstructure in the form of, say, a finite element mesh, where each grain is an aggregate of elements. The first key step is to generate a list that contains the volume of each grain. If the grain boundary character of the microstructure is to be fitted also, then the area of each grain boundary between a grain and its neighbors must also be determined. Based on this information, a list of orientations can be determined straightforwardly by using simulated annealing to minimize the difference between the generated OD

Table 3 Matrix formulae for transforming tensor quantities from the sample frame to the crystal frame

$$g(\phi_1, \Phi, \phi_2) = \begin{matrix} cos\phi_1 cos\phi_2 - sin\phi_1 sin\phi_2 cos\Phi & sin\phi_1 cos\phi_2 + cos\phi_1 sin\phi_2 cos\Phi & sin\phi_2 sin\Phi \\ -cos\phi_1 sin\phi_2 - sin\phi_1 cos\phi_2 cos\Phi & -sin\phi_1 sin\phi_2 + cos\phi_1 cos\phi_2 cos\Phi & cos\phi_2 sin\Phi \\ sin\phi_1 sin\Phi & -cos\phi_1 sin\Phi & cos\Phi \end{matrix}$$

Fig. 7 Example of an electron backscatter diffraction pattern. See text for details.

(together with the generated MDF, if applicable) and the measured OD.

Electron Backscatter Diffraction

The second major method for texture measurement that has reached maturity in the past decade or so is automated EBSD embedded in the scanning electron microscope (SEM). In brief, a specimen prepared such that it is flat and free from damage at the surface can be scanned for crystallographic orientation in a suitably equipped SEM (Fig. 7). Given a typical grain size in a polycrystalline material of between 2 and 200 μm, EBSD can map a large number of grains. A general rule for spatial resolution is to set the step size to be one tenth of the average grain size. Not only is the average texture (OD) of the material available directly, but local information is also available, such as grain size, grain shape, and the crystallographic character of the grain and phase boundaries. Especially when combined with other SEM-based techniques, such as energy-dispersive and wavelength-dispersive analysis of composition, phase identification is also possible. Readers may be tempted to think that the x-ray diffraction approach is old fashioned, but there are significant cautions to be applied, as discussed in more detail in the following sections. One example is that it is much easier to obtain a statistically valid measurement of the average texture with x-rays because, typically, a much larger surface area can be scanned in relation to the grain size. The EBSD method is outlined in this article; more detail is available in Ref 3 and 24.

EBSD Method

When the electron beam is left stationary at one location on the specimen surface with a 20° incidence angle between the beam and the surface, the backscattered electrons diffract, giving rise to sets of cones of high and low intensity. To image the pattern, a scintillation screen is placed near the specimen. In most configurations, the pattern appears as sets of nearly parallel lines, typically one pair per diffracting set of planes, as apparent in the example shown in Fig. 7. The scintillation screen converts the incident electrons to photons (light), which, in turn, are converted to a digital image with a charge-coupled device camera.

Although the eye and brain can readily index the pattern by recognizing specific crystallographic zones, automation in the computer is less simple because the characteristic parallel pairs lines cannot be easily identified in images. Accordingly, the diffraction patterns are transformed using the Radon or Hough mathematical transform. This procedure, in effect, converts the lines to high intensity peaks with a characteristic shape that can be located by traversing the image with a mask. Differences in the coordinates of the peaks represent angles between crystallographic planes in the crystal (or grain) that gave rise to the diffraction pattern. Because such interplanar angles are fixed by the lattice geometry and crystal symmetry, indexing the pattern is then reduced to matching the available set of interzonal angles to the measured and theoretical angles. When each peak has been identified, a full orientation can be calculated for the crystal, making appropriate allowance for the various geometrical relationships between the electron beam, the specimen, and the scintillation screen.

Limitations and Challenges Associated with EBSD

A significant challenge in using EBSD is specimen preparation. For a typical acceleration voltage in the range 15-25 kV in a SEM, the inelastic mean free path of electrons in most materials is only 10 nm. Therefore, any damage left over from specimen preparation means that the crystal lattice is disturbed, and the diffraction pattern is correspondingly smeared out and not indexable. Also, the surface must be inclined to the incident electron beam at 70° to maximize the intensity of the diffraction pattern, and deviations of a few degrees away from this setting significantly lower the intensity. Therefore, the surface must be flat to allow all of the scanned area to be indexed. The steep inclination also means that the spacing of points in the scan must compensate for foreshortening. The user should therefore be cautious and check for distortion of the EBSD image (e.g., grain shape).

Another potential challenge in acquiring EBSD scans is with materials with low electrical conductivity, such as most ceramics, because charging of the specimen can occur. Even mild charging can deflect the electron beam, thus displacing the measurement points from their intended locations and strongly distorting the image. The advent of environmental SEMs with higher pressure in the specimen chamber has substantially alleviated this problem.

All commercial EBSD systems permit samples to be scanned using either beam control or stage control. The former is faster, because stage control requires mechanical displacements of the stage under the beam. Although scanning via beam control is more efficient, the variation in the diffraction geometry for beams far from vertical means that the error in the orientation increases toward the edge of the image. If, however, EBSD is being used for texture measurement, then the scanned area must be large enough to include enough grains to obtain a statistically valid result (Ref 25, 26). Although the exact number depends on the crystal symmetry and required accuracy of result, including at least 1000 grains in the scan is generally necessary. This is important because it is all too common to be shown texture based on an EBSD scan that is too small (in relation to the grain size) to be representative. Note that misalignment of a specimen can lead to the same error in absolute texture measurements as noted for x-rays.

Applications of EBSD

The application of EBSD texture measurement is straightforward, with the caveats already mentioned concerning specimen preparation and the inclusion of a sufficient number of grains in the accumulated scanned area. The investigator may expect to see higher-than-expected noise in the OD plots if too few grains have been scanned. One crude test is to crop half (or some substantial fraction) of the area from the available data and reanalyze for texture (i.e., calculate the OD). If the maximum intensity has increased appreciably, then this is an indication that the number of grains scanned is insufficient. This is a variation on the bootstrap test in statistics (Ref 27).

Although beyond the scope of this article, it is worth mentioning the capability of EBSD to quantify the crystallographic character of grain boundaries, sometimes referred to as the grain-boundary character distribution. Because EBSD provides data on cross sections of materials, the full five degrees of freedom associated with grain-boundary character are not directly available, although Ref 28 contains a stereological approach to this problem. The lattice misorientation across boundaries (and phase boundaries) is directly accessible, however, so all EBSD software programs have the capability to calculate the misorientation distribution. This reveals, for example, whether a material contains a high or low fraction of high-angle boundaries or fractions of special boundaries that deviate markedly from that expected in a randomly oriented material. The word *special* in this context typically refers to whether or not the misorientation of a given boundary is close to that of one of the low-order coincident lattice site relationships. The reader is referred

Fig. 8 Example of an orientation map of a lightly deformed titanium showing a high density of deformation twins. Courtesy of Nathalie Bozzolo

to one of the standard texts for a full explanation of this concept (Ref 3, 29). Suffice it to say that certain special boundary types (such as the annealing twin boundary or pure twist $\Sigma 3$ in fcc metals) have higher-than-average resistance to corrosion, for example. This opens up the possibility of engineering high fractions of special boundaries, which is a practice known as grain-boundary engineering (Ref 30).

Any material that has been plastically deformed commonly contains orientation gradients within each grain (Ref 31, 32), which again are measurable in EBSD. In practice, such gradients reflect the presence of geometrically necessary dislocations (Ref 33). Figure 8 is an orientation map of a lightly deformed titanium that shows both orientation gradients within individual grains and a high density of deformation twins (Ref 34). There are many reasons to measure such gradients, but one practical metallurgical example is the characterization of recrystallization. It is useful to partition EBSD maps of deformed and annealed material into recrystallized and unrecrystallized regions to quantify the extent of recrystallization and the misorientation relationships of the process of microstructural evolution (Ref 35, 36).

There are a large and growing number of applications for EBSD, some of which have been reviewed recently (Ref 24). They include phase identification and strain analysis, among others.

Types or Classes of Materials

It is straightforward to write a description of texture characterization from a metals and ceramics perspective, because the main thrust to develop the topic came from the metallurgical community. It is worth noting, however, that Roe, one of the two individuals credited with the breakthrough development of the series expansion method for obtaining ODs from pole figure data, was mainly interested in polymers at the time. In semicrystalline polymers, texture can develop readily as a consequence of the processes used to form objects, such as rolling (calendering) and blowing, because the long-chain molecules are forced into alignment by the stretching

processes. There is a considerable body of literature on the subject, going back to Roe's original work with texture analysis (Ref 17). Naturally, biological materials that are largely polymeric in nature also can exhibit texture that is significant to their function. In biomaterials especially, the architecture of the materials (meaning microstructure at multiple length scales) is truly important. There are fascinating examples of texture investigations in the shells of the wide variety of creatures with external skeletons (Ref 37). Returning to ceramics, texture is of importance to many types of high-temperature superconductors and also to the materials used in piezoelectric sensors. Finally for metals, the effects of texture are found in the manufacturing and application of essentially all wrought products and thin films, although powder products are generally close to random.

Summary

This article provides an overview of the methods currently available for characterizing crystallographic preferred orientation, or texture, in polycrystalline materials. Enough detail is provided about the data acquisition and analysis methods to enable the reader to make use of such data. References are given to a variety of textbooks and journal papers, if the reader desires to learn more of the underlying mathematics and technical detail. Readers are also encouraged to explore the worldwide web, which, at the time of writing, has a growing amount of useful information about texture analysis.

REFERENCES

1. U.F. Kocks, C. Tome, and H.-R. Wenk, Ed., Texture and Anisotropy, Cambridge University Press, Cambridge, U.K., 1998
2. B.D. Cullity, *Elements of X-Ray Diffraction,* Addison-Wesley Publishing Co., Reading, MA, 1956
3. V. Randle and O. Engler, *Texture Analysis: Macrotexture, Microtexture and Orientation Mapping,* Gordon and Breach, Amsterdam, Holland, 2000
4. L. Schulz, A Direct Method of Determining Preferred Orientation of a Flat Reflection Sample Using a Geiger Counter X-Ray Spectrometer, *J. Appl. Phys.,* Vol 20, 1949
5. E. Tenckhoff, Defocussing for the Schulz Technique of Determining Preferred Orientation, *J. Appl. Phys.,* Vol 41, 1970, p 3944–3948
6. B. Gale and D. Griffiths, Influence of Instrumental Aberrations on the Schultz Technique for the Measurement of Pole Figures, *Br. J. Appl. Phys.,* Vol 11, 1960
7. H.-R. Wenk, Neutron Diffraction Texture Analysis, *Rev. Mineral. Geochem.,* Vol 63 (No. 1), 2006, p 399–426
8. K. Feldmann, M. Betzl, W. Kleinsteuber, and K. Walther, Neutron Time of Flight Texture Analysis, Part 1–2, *Textures Microstruct.,* Vol 14, 1991, p 59–64
9. K. Walther, J. Heinitz, K. Ullemeyer, M. Metzl, and H.R. Wenk, Time-of-Flight Texture Analysis of Limestone Standard—Dubna Results, Part 5, *J. Appl. Crystallogr.,* Vol 28, 1995, p 503–507
10. M. Morales, D. Chateigner, and L. Lutterotti, X-Ray Textural and Microstructural Characterisations by Using the Combined Analysis Approach for the Optical Optimisation of Micro- and Nano-Structured Thin Films, *Thin Solid Films,* Vol 517 (No. 23), 2009, p 6264–6270
11. D. Jensen and J. Kjems, Apparatus for Dynamical Texture Measurements by Neutron Diffraction Using a Position Sensitive Detector, *Textures Microstruct.,* Vol 5, 1983, p 239–251
12. H. Bunge, *Texture Analysis in Materials Science,* Butterworths, London, 1982
13. P. Wessel, *General Mapping Tools (GMT) Home Page,* 2010, accessed March 8, 2010
14. H.R. Wenk and P. Van Houtte, Texture and Anisotropy, *Rep. Prog. Phys.,* Vol 67 (No. 8), 2004, p 1367–1428
15. N. Barton, D. Boyce, and P. Dawson, Pole Figure Inversion Using Finite Elements Over Rodrigues Space, *Textures Microstruct.,* Vol 35, 2002, p 113–144
16. H.J. Bunge, Zur Darstellung Allgemeiner Texturen, *Z. Metallkd.,* Vol 56 (No. 12), 1965, p 872
17. R.J. Roe, Description of Crystallite Orientation in Poly Crystalline Materials, Part 3: General Solution to Pole Figure Inversion, *J. Appl. Phys.,* Vol 36 (No. 6), 1965, p 2024
18. H.R. Wenk, S. Matthies, J. Donovan, and D. Chateigner, Beartex: A Windows-Based Program System for Quantitative Texture Analysis, Part 2, *J. Appl. Crystallogr.,* Vol 31, 1998, p 262–269
19. K. Pawlik, J. Pospiech, and K. Lucke, The ODF Approximation from Pole Figures with the Aid of the ADC Method, Part 1–2, *Textures Microstruct.,* Vol 14, 1991, p 25–30
20. K. Helming, *MulTex,* 2002, accessed March 8, 2010
21. R. Hielscher, *mtex,* 2010, accessed March 8, 2010
22. J. Kallend, U. Kocks, A.D. Rollett, and H.-R. Wenk, Operational Texture Analysis, *Mater. Sci.Eng. A,* Vol 132, 1991, p 1–11
23. J. Szpunar, *ResMat,* 2003, accessed March 8, 2010
24. A. Schwartz, M. Kumar, and B. Adams, Ed., *Electron Backscatter Diffraction in Materials Science,* Kluwer, New York, NY, 2000
25. S. Matthies and F. Wagner, Using Sets of Individual Orientations for ODF Determination, *ICOTOM-12,* Vol 1, J.A. Szpunar, Ed., NRC Research Press, Montreal, Canada, 1999, p 40–45

26. N. Bozzolo, F. Gerspach, G. Sawina, and F. Wagner, Accuracy of Orientation Distribution Function Determination Based on EBSD Data—A Case Study of a Recrystallized Low Alloyed Zr Sheet, *J. Microsc.-Oxford*, Vol 227 (No. 3), 2007, p 275–283
27. *Bootstrap Test*, Wikipedia, Feb 2010, accessed March 8, 2010
28. D. Saylor, B. El-Dasher, B.L. Adams, and G.S. Rohrer, Measuring the Five Parameter Grain Boundary Distribution from Observations of Planar Sections, *Metall. Mater. Trans. A*, Vol 35, 2004, p 1981–1989
29. A. Morawiec, *Orientations and Rotations*, Springer, Berlin, 2003
30. G. Palumbo, E. Lehockey, and P. Lin, Applications for Grain Boundary Engineered Materials, *JOM*, Vol 50 (No. 2), 1998, p 40–43

31. S. Mishra, P. Pant, K. Narasimhan, A.D. Rollett, and I. Samajdar, On the Widths of Orientation Gradient Zones Adjacent to Grain Boundaries, *Scr. Mater.*, Vol 61 (No. 3), 2009, p 273–276
32. R.A. Lebensohn, R. Brenner, O. Castelnau, and A.D. Rollett, Orientation Image-Based Micromechanical Modelling of Subgrain Texture Evolution in Polycrystalline Copper, *Acta Mater.*, Vol 56 (No. 15), 2008, p 3914–3926
33. B. El-Dasher, B.L. Adams, and A.D. Rollett, Viewpoint: Experimental Recovery of Geometrically Necessary Dislocation Density in Polycrystals, *Scr. Mater.*, Vol 48, 2003, p 141–145
34. N. Bozzolo, L. Chan, and A.D. Rollett, Misorientations Induced by Deformation Twinning in Titanium, *J. Appl. Crystallogr.*, Vol 43 (No. 3), June 2010, p 596–602

35. M.H. Alvi, S.W. Cheong, J.P. Suni, H. Weiland, and A.D. Rollett, Cube Texture in Hot-Rolled Aluminum Alloy 1050 (AA1050)—Nucleation and Growth Behavior, *Acta Mater.*, Vol 56 (No. 13), 2008, p 3098–3108
36. T.A. Bennett, P. Kalu, and A.D. Rollett, Stored Energy Driven Abnormal Grain Growth in Fe-lSi, *COM-2006* (Montreal), METSOC, 2006, p 217–227
37. D. Raabe, P. Romano, C. Sachs, H. Fabritius, A. Al-Sawalmih, S. Vi, G. Servos, and H.G. Hartwig, Microstructure and Crystallographic Texture of the Chitin-Protein Network in the Biological Composite Material of the Exoskeleton of the Lobster *Homarus Americanus, Mater. Sci. Eng. A*, Vol 421 (No. 1–2), 2006, p 143–153

ASM Handbook, Volume 22B, *Metals Process Simulation*
D.U. Furrer and S.L. Semiatin, editors

Copyright © 2010, ASM International®
All rights reserved.
www.asminternational.org

Three-Dimensional Microstructure Representation

G. Spanos and D.J. Rowenhorst, U.S. Naval Research Laboratory
M.V. Kral, University of Canterbury, New Zealand
P.W. Voorhees and D. Kammer, Northwestern University

ACCURATE PREDICTIVE SIMULATIONS are critical for improving the precision and efficiency of the materials and process design cycle. In turn, modeling and simulation of materials processing and materials response require accurate and robust experimental microstructural data sets, to be used for both initial input into the simulations and to validate and enhance the models. However, it has become increasingly apparent that two-dimensional (2-D) microstructural characterization techniques are often inadequate for an accurate representation of all but the simplest structures in materials. Three-dimensional (3-D) microstructural characterization and analysis techniques have recently become much more common (Ref 1–4), and using the resultant 3-D experimental data as input for simulations, that is, employing image-based 3-D modeling (Ref 5, 6), has recently been shown to be quite useful for modeling both materials processing and materials response to external mechanical or thermal loads during service (Ref 4, 7–9). In general, three-dimensional analysis of microstructures is a rapidly developing field, and the reader is referred to a number of special journal issues that have appeared within the last few years that present a survey of the state-of-the-art in this field (Ref 1–3, 10).

This article reviews current characterization methods for producing 3-D microstructural data sets and describes how these experimental data sets are then used in realistic 3-D simulations of microstructural evolution during materials processing and materials response. Thus, the first section of this article reviews various 3-D characterization methods, while the second section describes how the 3-D experimental data are actually input and used in the simulations.

Three-Dimensional Characterization Methods

The article by Kral et al. (Ref 1) provides a detailed review of 3-D microscopy, from which much of this section is adapted, with additions where necessary to update certain advances in techniques and algorithms since its publication. Due to space limitations, this section focuses exclusively on experimental characterization methods involving serial-sectioning-based 3-D reconstructions centered about sequential planar milling (by either mechanical or ion beam methods); other sections of this article describe how these types of results can then be used in process and simulation modeling. Although there are a number of other very useful 3-D characterization techniques, such as x-ray tomography and 3-D atom probe tomography, it is beyond the scope of this article to cover them all. Instead, the reader is referred to a number of excellent review articles that discuss these and other types of 3-D characterization methods in some depth (Ref 1–3, 10).

A number of methodologies can be used to reveal the shape, distribution, and connectivity of 3-D features that lie buried within an opaque material. Serial sectioning techniques used to accomplish this task involve removing material from a bulk sample layer by layer, imaging each layer, and then reconstructing the resultant series of images using computer programs. Such methods include mechanical polishing, electrolytic dissolution, or focused ion beam ablation, all of which involve imaging periodically at a known increment in a continuous material-removal process. Other 3-D characterization methods, such as x-ray tomography, can nondestructively reveal the nature of the 3-D structure. It should be noted that any characteristic or attribute that can be associated with a particular location on a sample in 2-D, such as crystallographic orientation, chemical composition, or phase, can in principle be used to produce a 3-D image, if the material can be adequately sectioned and digitally reconstructed, or nondestructively imaged in 3-D.

Traditional 2-D metallographic methods provide single sections for observation. Quantitative metallography and stereological techniques can then be applied to estimate some parameters that describe 3-D features of polycrystalline or multiphase materials. While some average microstructural parameters can be accurately deduced in this manner, more precise descriptions of size and spatial distributions, shapes, and interconnectivities of complex microstructural features can only be obtained via true 3-D characterization methods (Ref 11). To cite a few representative examples, Lund and Voorhees (Ref 12) were able to observe for the first time the true 3-D morphologies, distributions, and interactions between γ' particles in a γ-γ' alloy, while 3-D analyses of proeutectoid cementite precipitates showed two different morphologies of Widmanstätten cementite (Ref 13) that were subsequently revealed to correspond to two known crystallographic orientation relationships between cementite and austenite in steel (Ref 14).

Continuous improvements to computer software and hardware now make the representation and dissemination of 3-D images, via the internet or digital media (CD ROMs, for example), a reality. Improvements in material-removal methods also offer a great opportunity for making 3-D analyses accessible to most materials scientists and engineers. Automated serial sectioning machines used in conjunction with optical or scanning electron microscopes, such as robotic polishers (Ref 15) or micromilling machines (Ref 16), have also been developed to reduce the time required for sectioning. Material can also be removed automatically using focused ion beam techniques (Ref 17–22). At much higher resolutions, the 3-D atom probe has been used to strip atoms from a specimen one at a time, while recording the atomic position and species in 3-D (Ref 23, 24). Three-dimensional reconstructions of microstructures can thus be made over a range of volumes from cubic millimeters down to the atomic level. This section presents several of the current serial-sectioning-based experimental techniques that can be used to generate 3-D images, including serial sectioning by mechanical material removal and by focused ion beam tomography methods.

Serial Sectioning by Mechanical Material-Removal Methods

The history of 3-D analysis of microstructures by serial sectioning spans at least from 1918 (Ref 25) with Forsman's effort to understand the 3-D structure of pearlite microstructures in steel. By projecting the images of each section onto cardboard layers of appropriate thickness, solid models of the cementite lamellae within the pearlite were constructed. In 1962, Hillert and Lange (Ref 26) produced a motion picture of serial sections to show the true 3-D structure of an entire pearlite colony. Eichen et al. (1964) studied the growth of Widmanstätten ferrite microstructures in steel by measuring the changing length of ferrite plates with increasing depth through serial sections (Ref 27). Hopkins and Kraft (1965) used a unique "cinephotomicrographic recording of the microstructure of a specimen undergoing controlled electrolytic dissolution" (Ref 28). Their results were subsequently represented by building a 3-D physical model of Plexiglas (Rohm & Haas Company) to show the eutectic fault structure in a copper-aluminum alloy. Hawbolt and Brown (1967) used serial sectioning to study the shapes of grain-boundary precipitates in a silver-aluminum alloy (Ref 29). Barrett and Yust (1967) showed the interconnectivity of voids in a sintered copper powder (Ref 30). Ziolkowski (1985) used a "mikrotom" to perform a serial sectioning study of grain-boundary precipitates in an α/β brass alloy (Ref 31). In most of the aforementioned cases, 3-D results were represented by hand-drawn sketches, graphical plots of length versus depth, or motion pictures. In 1983, R.T. DeHoff wrote:

> "In its current embryonic state of development, the use of serial sectioning analysis for all but the most rudimentary of measurements is prohibitively expensive and tedious" (Ref 11).

Relatively recent improvements in image processing and 3-D visualization capabilities in addition to the development of automatic sectioning devices have made 3-D reconstruction and visualization of serial sections much more practical. In 1991, Hull et al. (Ref 32) were among the first to use computer software to produce 3-D wire-frame drawings of microstructural features, in this case titanium prior-beta grain sizes and shapes. Brystrzycki and Przetakiewicz (1992) used a similar technique to study the sizes and shapes of annealing twins in a Ni-2%Mn alloy (Ref 33). A significant innovation was made in 1994 by two sets of independent researchers (Ref 34, 35). Mangan and Shiflet combined serial sectioning, scanning electron microscopy, manual electron backscatter diffraction (EBSD) analysis, and computerized 3-D reconstruction techniques in a high-manganese steel to produce digital 3-D reconstructions of pearlite colonies in which the crystallographic orientation of each crystal was also determined (Ref 34). That same year (1994), Wieland et al.

used serial sectioning and computerized 3-D reconstruction in a recrystallized aluminum-manganese alloy to simultaneously capture the crystallographic orientation and location of recrystallized grains using EBSD and imaging in a scanning electron microscope (SEM) (Ref 35). The significant step forward in these two studies (Ref 34, Ref 35) was that in both cases computer software and hardware were employed to produce smoothed, digital, 3-D reconstructions of the microstructure from the serial sections (which contained both spatial and crystallographic information).

Visual representations of 3-D data sets, such as microstructural data via optical or SEM images or crystallographic data via EBSD, obtained through serial sectioning, have continued to improve with advancements in computer visualization software. Among the earlier studies that took full advantage of advanced computer reconstructions of serial sectioning were investigations by Mangan et al. (Ref 36) and Wolsdorf et al. (Ref 16).

The effort required for serial sectioning and the subsequent 3-D analysis has always been a concern. Over recent years, steady improvements in automating material removal, digital image acquisition, and visualization of 3-D reconstructions using advanced computer software and hardware (Ref 4, 13, 15, 16, 36–39) have made 3-D analysis techniques more accessible to materials researchers. However, it is important to bear in mind that even with a relatively slow, completely manual serial sectioning technique, at 20 to 30 minutes per section on average and approximately 100 person-hours per 250 sections, the time expended in both image segmentation and in analyzing the volume data set will far exceed the time required to acquire the serial sections. (Image segmentation refers to segmenting the image into different

regions of interest, e.g., for the identification of individual grains.) In this regard, much progress is currently being made in the development and use of advanced image analysis software, segmentation algorithms, and data-mining techniques (Ref 4, 15, 39, 40). Three-dimensional analyses have thus resulted in many important new insights into microstructural evolution that not only have produced immediate rewards in understanding materials microstructures and resultant properties but have also led to new avenues of materials research.

Serial Sectioning by Mechanical Methods

Experimental Techniques. A generalized procedure for serial sectioning is described by the schematic diagram in Fig. 1; of course, different serial sectioning problems can require variations in this procedure. The various steps in this type of procedure are now considered individually.

Material Removal. One of the first steps in undertaking a serial sectioning project is to identify the microstructural feature (or set of microstructural features) for which some 3-D information is desired. When the subject of the study has been selected, the most important experimental option, that is, the material-removal technique, will probably have been decided by default, because each technique is only practical over a certain range of length scales and material-removal rates. The most common serial sectioning material-removal methods by mechanical means are metallographic polishing (which removes between approximately 0.1 and 5 μm per section) and micromilling (which removes between approximately 1.0 and 20 μm per section). Case studies of these techniques are presented later to illustrate the differences between various serial sectioning problems.

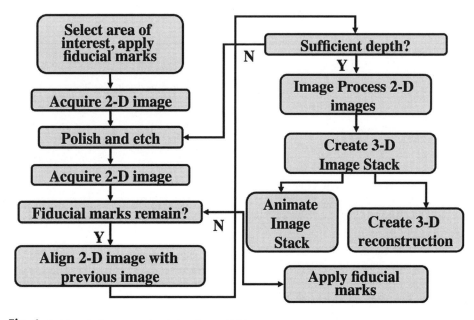

Fig. 1 Serial sectioning process description. Source: Ref 1

When the serial sectioning method and the serial section depth are determined (the latter is defined by the length scale of the features to be reconstructed in 3-D—typically, a minimum of ten sections per feature is desired), a number of other experimental parameters must be considered. These include the total number of sections required, whether an etchant is required, the use of fiducial marks for alignment of the individual 2-D images in the stack, the imaging method, image acquisition, and visualization and analysis software.

Etching. Some materials require etching to obtain sufficient contrast between phases subsequent to the removal of each layer. Potential problems include inconsistent etching between layers, boundaries that do not respond to the etchant due to their orientation, and etching artifacts such as pits due to overetching. It is essential that subsequent polishing steps remove all pits to prevent the pitting from continuing or even accelerating.

Imaging. Digital optical micrographs have been acquired via charge-coupled device video cameras with approximately 640 × 480 pixels as well as digital cameras with megapixel image size. There are a number of digital cameras now available specifically for optical microscopy (Olympus, Zeiss, Diagnostics Instruments, etc.). Scanning electron microscopes may also be used in conjunction with mechanical sectioning techniques, although with a considerable increase in the time required to remove and insert the sample into a vacuum chamber for each section. It should also be noted that the pixels in an SEM are often not square; unless there is a calibration or correction for the pixel aspect ratio, the sample must be placed into the SEM chamber in the same orientation for each section to make subsequent registration more convenient. Obviously, the resulting 3-D images will be distorted without a calibration or adjustment of pixel aspect ratio.

More recently, automated orientation mapping by EBSD techniques has been combined with sectioning by mechanical polishing, to provide imaging and crystallographic data from at least some of the sections used in 3-D reconstructions (Ref 4, 39). Because the time required to collect an EBSD orientation map from a given section is much greater than that needed to obtain a conventional SEM image or light optical micrograph, EBSD scans are sometimes taken only after some multiple of sections, and then correlated with optical micrographs taken at every section increment. For example, in a serial sectioning study in a titanium alloy, EBSD orientation maps were recorded every tenth section, guaranteeing crystallographic information was measured for all but the smallest grains in that study, while reducing the time required for data collection (Ref 4). The EBSD images were subsequently aligned with the optical micrographs (and thus to the final 3-D reconstruction) by a semiautomatic alignment routine. This study (Ref 4) is considered in more detail as example

2 later in this article. The EBSD orientation mapping in combination with serial sectioning by automated focused ion beam (FIB) milling allows for faster sectioning rates and EBSD map imaging of every section (Ref 18, Ref 20, Ref 41, Ref 42, Ref 43, Ref 44). This technique is discussed in detail in the subsequent section, "Focused Ion Beam Tomography," in this article.

Fiducial Marks. With manual or semiautomated polishing or grinding, fiducial marks such as microhardness indents have often been necessary to mark the area of interest, to align (or register) images in the *x-y* plane, and to track the actual material-removal rate. Prior to the commencement of actual image acquisition, the material-removal rate may be calibrated by measuring the change in an indent diagonal length when the indent diagonal-to-depth ratio is known. Figure 2 illustrates that fiducial

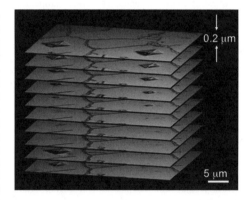

Fig. 2 Stack of serial sections showing fiducial hardness indents. Source: Ref 1

hardness indents must be renewed at relatively frequent intervals. More recently, FIB etching has been used to cut trenches into a side of the specimen that is perpendicular to the sectioning plane, before sectioning is started (Ref 4), as shown in Fig. 3. In this example, each FIB channel is approximately 10 µm wide by 10 µm deep, and when viewed from the sectioning plane ("top view"), the channels appear as individual notches (right side of Fig. 3). Vertical channels parallel to the sectioning direction (the horizontal trench in the left image in Fig. 3) are used for the translational and rotational alignments of the images. Diagonal channels (the angled trenches in the left image in Fig. 3) provide for measurement of the amount of material removed in each section. As pointed out (Ref 4), this fiducial marking system has several advantages over hardness indents, including:

- A pattern has to be applied only once.
- Long-range alignments are preserved (the last image can be aligned with the first).
- Depth calibrations can be used for removal of either small or large amounts of material. (Hardness indents are limited to the depth of the indent.)

Sectioning Increment and Number of Sections Required. In studies of grain shape and size in a single-phase polycrystalline material, Rhines et al. (Ref 45) recommended using a magnification sufficient to show several grains simultaneously and sectioning to a total depth of approximately twice the span of the largest grain in about 250 sections. In the general application of this rule of thumb, one must take into account the scale of the features being

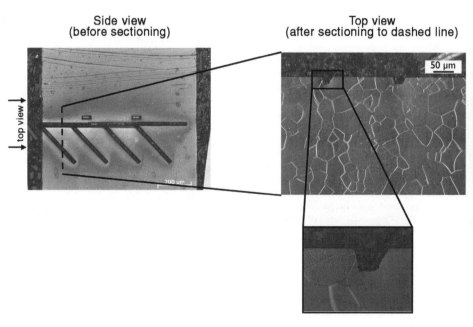

Fig. 3 Fiducial channels etched into the side of a titanium alloy (Ti-21S) specimen for serial sectioning experiments. Source: Ref 4

studied and the limitations of the material-removal technique being used. The resolution of the final 3-D representation in every serial sectioning project to date has been limited by the thickness of each serial section layer. This compromise is due to both the amount of time involved and the problems of storing and properly registering the images after they have been obtained. As mentioned earlier, the choice of serial section depth, which is defined by the length scale of the features to be reconstructed in 3-D, is typically on the order of a minimum of ten sections per critical feature to be reconstructed. Solutions for reducing the amount of time (or labor) required for sectioning have included microtoming (Ref 31) or micromilling (Ref 16). Regarding the computational challenge, consider that 250 images captured at low resolution (640 by 480 pixels) result in a 3-D image of 75 megabytes for gray-scale images (8 bits per pixel). In 1996, such a file size was approaching the practical limits for some computers. Now, typical gray-scale image file sizes are at least 1 megapixel. Furthermore, larger areas can be captured at a useful magnification by automontaging (Ref 46), and this quickly multiplies the overall size of the 3-D data set. However, with continuing advances in computing power, acquisition of such large data sets now presents much less of a problem.

Registration. The importance of proper alignment through the use of fiducial marks has been previously demonstrated (Ref 47). Alignment, or registry, of serial sections can be performed manually by overlaying fiducial marks in subsequent images, or in a more automated fashion, using common image-processing software (e.g., Adobe Photoshop, NIH Image, Interactive Data Language, etc.). Fiducial marks such as microhardness indents or FIB channels are also used to determine the distance between sections, as demonstrated in Fig. 2 and 3. If the sectioning depth is precisely known, it can improve efficiency to align automatically using microstructural features of individual cross sections either before or after an image is acquired (Ref 37, 48). However, this method assumes that the average displacement in the position of microstructural features is random; therefore, this technique must be applied carefully, because any morphological texture in the sample (such as columnar grains aligned in a direction that is not normal to the sectioning plane) will violate the assumption of random displacement, and this could cause real morphological textures to be removed in the final 3-D reconstruction (Ref 4).

Segmentation

As alluded to earlier, segmentation is the process of partitioning a digital image into specific regions or volumes of interest (for example, individual grains separated by grain boundaries) so that they can be easily visualized (Ref 49). It is one of the most difficult and time-intensive issues associated with serial sectioning and 3-D reconstruction using traditional imaging methods. In almost all cases, image-processing techniques must be applied to enhance the contrast of the microstructure. Additionally, because each microstructure is unique relative to imaging and segmentation, no single image-processing solution applies over a wide variety of materials systems. One method of segmentation can be accomplished by obtaining a histogram of pixel values that shows which pixel values are the most frequent, where peaks represent boundaries and/or phases in the histogram. Based on the range of pixel values, grain boundaries and/or interphase boundaries can be visualized by making the appropriate range of pixel values either opaque, transparent, or some specific color. Pixel values can be assigned a color (a red, green, blue, opacity quartet) by use of a transfer function. A transfer function maps a pixel value to a color and can be as simple as using a red-orange-yellow-green-blue color ramp. (In hue-saturation-value space, hue is varied from red to blue, i.e., from 0 to 255, with pure red being the highest pixel value and blue the lowest). The opacity can be assigned to elucidate features. For instance, in example 1 that follows, the austenite-cementite (matrix-precipitate) phase boundaries were assigned high opacity, and the matrix phase (austenite) was assigned low opacity to make it completely transparent. Although a few methodologies have been discussed here, segmentation is an active area of 3-D analysis, and there are many approaches that are used and are currently under development (Ref 4, 47).

Three-Dimensional Visualization and Analysis

There are a number of computer software packages currently available for 3-D reconstruction and analysis. Some programs are free (e.g., NIH ImageJ, etdips, IMOD, Paraview, etc.), while others are commercially available (e.g., Interactive Data Language, Advanced Visual Systems' AVS Express, Vaytek's VoxBlast, and MATLAB). Some researchers have chosen to develop their own software (Ref 36). The visualization and quantification of both 3-D experimentally determined data sets and 3-D simulations is essential for making critical comparisons between theory and experiment. Fortunately, the analysis techniques can almost always be applied equally well to experimental and simulation data.

There are two common models for analyzing 3-D microstructures. The most common method is to analyze the 3-D microstructures as a 3-D regular array of volume elements, or voxels. This is often described as a stack of 2-D images. Early efforts to quantify 3-D microstructures focused on performing 2-D analysis on each slice in the stack, then manually combining the 2-D data to form 3-D representations of the data (for more detail, see Ref 46 and 50). These methods have the advantage of using image-analysis programs that have already been developed for 2-D images, but at the cost of a large amount of manual labor to combine the 2-D data from each slice.

More recent applications have fully adapted common 2-D image-analysis routines to work on 3-D image arrays directly. The work by Rowenhorst et al. (Ref 51) measured the 3-D particle size distribution of tin particles in a liquid matrix of eutectic lead-tin by simply counting voxels assigned to each particle. Additionally, the researchers were able to determine the particle coordination by examining voxel nearest neighbors.

One of the most extensive characterizations of a 3-D microstructure was carried out by Groeber et al. (Ref 52, 53). Using FIB serial sectioning combined with EBSD, they considered the statistics of over 2700 nickel grains in a nickel-base superalloy. Again, the grain-size distribution was calculated by calculating the number of voxels in each grain. Additionally, the aspect ratio of each grain was determined by calculating the equivalent ellipsoid of revolution, which is calculated from the second moments of inertia of the object. This method also provides the orientation of the ellipsoid axis. The surface area of each grain was estimated by counting the number of voxel faces that are on the outer surface of each grain. The authors do point out that this is an estimate of the surface area, because the discrete nature of voxels cannot properly describe the smooth interface of the grains. (It should be noted that the same problem occurs in 2-D image analysis where the boundary length is estimated by the number of edge pixels.) Furthermore, the EBSD data collected allowed for the calculation of the orientation distribution function, which is a measure of the crystallographic texture, and the misorientation distribution function, which is a measure of the grain-boundary texture.

The significant advantage that 3-D techniques have over their 2-D analogs, as demonstrated by these examples, is that they provide a direct measurement of the properties of each individual region, not an estimation of the property derived from a random 2-D section through the material. A full accounting of image analysis is beyond the scope of this article, and the authors highly recommend *The Image Processing Handbook* by J.C. Russ (Ref 47) for further examples and techniques.

The second model for analyzing 3-D microstructural data is to create a surface mesh for each of the features of interest (e.g., each grain) wherein the exterior interface of the region of interest is described with a set of vertex points in space and a list of triangles that describe how to connect the vertex points to make a solid surface. One can translate volumetric data to such an isosurface using the fast-marching cubes algorithm (Ref 54). The fast-marching cubes algorithm determines if a threshold value is crossed for each voxel by comparing the voxel value with the values in the nearest-neighbor voxels. If the threshold value is not crossed, the algorithm moves to the next voxel.

If the threshold value is crossed, then the vertex points of the surface are determined by interpolating the intersection of the threshold with the edges of the voxel. These intersections then define a triangle (or quadrilateral in the case of four intersections with the voxel edges, which can easily be decomposed to two triangles) for that local patch of interface. Versions of the fast-marching cubes algorithm are contained within many visualization packages, including MatLab, Interactive Data Language, Amira, and Visualization Toolkit.

While the creation of a surface mesh is typically used as a step in visualization of 3-D structures, there are many analyses that can also benefit from this description as well. Jinnai et al. (Ref 55) used a surface mesh to calculate local surface curvatures in spinodal decomposition, and similar methods were used to examine curvatures during dendrite coarsening (Ref 56) and later in simulations by Mendoza et al. (Ref 57). The local surface normal is also easily calculated and is a critical parameter for understanding the microstructure in many systems. The investigation by Saylor et al. was one of the earliest studies to use a triangulated surface mesh to describe the grain-boundary geometry in MgO (Ref 58). Kammer and Voorhees (Ref 59) used a similar technique to examine the average interface texture during dendritic coarsening by examining the changes in the average normal direction. These techniques were similarly used by Rowenhorst et al. (Ref 39) to determine crystallographic facet planes of martensite crystals in a high-strength, low-alloy steel.

Example 1: Manual Serial Sectioning Using Mechanical Polishing. The purpose of this first study (published in 1999) was to characterize the 3-D morphology, distribution, and connectivity of proeutectoid cementite precipitates in a hypereutectoid steel (Fe-1.34%C-13.0%Mn alloy) (Ref 13). It is instructive to go through the details of this early study, to gain a sense of all of the individual steps that were undertaken at that time to produce the final 3-D reconstructions. Rapid solidification was used to refine the austenite grain size to approximately 25 μm to allow entire matrix austenite grains to be completely sectioned through using sectioning increments fine enough (≈0.2 μm) to allow enhanced resolution, facilitating 3-D reconstruction of entire austenite grains and groups of cementite precipitates within these grains. The splat-quenched specimens were austenitized for 30 s at 1100 °C in a deoxidized barium chloride salt bath, isothermally reacted at 650 °C in stirred, deoxidized lead baths for times ranging from 1 to 50 s to form the cementite precipitates, and finally quenched in room-temperature brine to halt the cementite transformation. Isothermal transformation at 650 °C for 50 s produced an adequate number of cementite precipitates of appropriate size and was selected for the 3-D analyses. A typical microstructure for this heat treatment is shown in Fig. 4. A 3 mm diameter specimen was mechanically punched, ground,

and mounted on a VCR Dimpler platen. Each layer was polished with a 0.06 μm silica slurry using a 7 mm wide, 25 mm diameter "flatting" tool covered with a Buehler Texmet cloth. Control of the sectioning depth was achieved through calibration of polishing load and time. The polished surface of each layer was lightly etched with 4% nital. Fiducial hardness indents were applied using a Buehler Micromet II hardness tester with a Vickers hardness indenter at a 10 g load. Optical microscopy was performed at a magnification of 1000× using an oil immersion objective lens, with digital acquisition of video images using PGT Imagist image analysis software on a Sun SPARCstation 5 computer. The images were "registered" with respect to each other in the plane of the image by aligning the hardness indents using Adobe Photoshop version 3.0 on a Power Macintosh 7200/120 personal computer. Thickness increments were calculated from the known indenter diagonal-to-depth ratio by measuring the change in diagonal length of the hardness indents after each polishing step. It was assumed that each successive polishing plane was parallel. Four overlapping areas encompassing a total area of approximately 150 by 180 μm were documented for each of 250 layers, with an average spacing of 0.17 ± 0.07 μm between layers. NIH Image software was used to store and view the resultant four "stacks" of TIFF images as video sequences that "step through" the microstructure slice by slice.

Each stack of TIFF images was saved to separate files using NIH Image's Stacks Macro, "Save Slices as files..." The separate TIFF images were then converted to ASCII Portable Pixel Map images and concatenated to a single file, resulting in a uniform volumetric data set. AVS 5.5 software on a Silicon Graphics Onyx workstation was used to read in these ASCII data sets, filter, and render them as isosurfaces and ray-traced volumetric images. (3-D reconstruction algorithms and isosurface versus volumetric images are discussed earlier.) Spot noise was removed using a "Median" image-processing filter with a 5×5 pixel-cross structuring element.

In this case, the austenite phase and the etched boundaries between the austenite and cementite phases were easily distinguished (segmented) by the different gray-scale value ranges of their pixels. The austenite matrix gray-scale values ranged from 0 to 100, and the etched cementite:austenite interphase boundaries were over 100, thus allowing a threshold to be set to automatically differentiate one phase from the other. After inverting the image, assigning translucency to the austenite phase, and assigning a large opacity value to the etched boundary between the cementite and austenite phases, ray-traced volumetric images were generated using a gray-scale, histogram-equalized color map along with Phong shading to best bring out the details of the surface. This gives the appearance of opaque cementite precipitates. The opacity is a linear

ramp from low to high pixel intensity. A single, uninterpolated ray-traced image required approximately 10 s of central processing unit time on an SGI Onyx 1 (R10K processor) to complete. A series of ray-traced images were generated to produce an animation of the 3-D microstructure as it is rotated about any axis.

Alternatively, isosurfaces were generated using the standard marching cubes algorithm (Ref 54). The exact isosurface value used is selected to be the average voxel intensity of the data set. A simple Gouraud lighting model was used to shade the isosurfaces along with employing depth cueing. Surface or volume-rendered models offer the capability of "real-time" manipulations on computer display devices that can take advantage of "stereo" images, giving the visual impression of 3-D objects in space.

The dimensions of individual precipitates were directly measured by digitally removing obstructing features (i.e., hardness indentations and overlapping precipitates) from a series of images and then projecting the resultant stack of images onto a projection plane using NIH Image version 1.61 software on a Power Macintosh 7200/120 personal computer. The length, width, and thickness of more than 200 individual precipitates were measured from appropriately rotated projected images with a calibrated scale. The length was taken as the maximum dimension of the precipitate when rotated into plan view. The width was taken as the largest dimension perpendicular to the length. The thickness was the maximum dimension measured with the precipitate rotated to an edge-on orientation. An error of approximately ± 0.4 μm was introduced by the apparent thickness of the etched boundaries of each precipitate and by the resolution of the image (approximately 5 pixels/μm).

Also, even with light etching, etch pits were often formed over the course of 250 sections, making subsequent image processing more difficult. Nevertheless, at least 20 entire matrix austenite grains and over 200 cementite

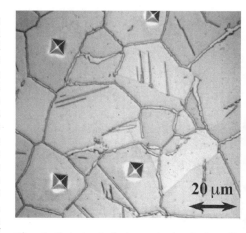

Fig. 4 Typical optical micrograph of an isothermally transformed Fe-13Mn-1.3%C alloy showing proeutectoid cementite precipitates. Source: Ref 13

precipitates within these austenite grains were entirely sectioned. A 3-D reconstruction showing a portion of a representative grain is shown in Fig. 5, and one precipitate selected from this grain is shown in Fig. 6. The ability to digitally remove individual precipitates for study enabled the measurement of each precipitate in three dimensions. Among other results, it was observed that all precipitates were either connected to an austenite grain boundary or another cementite precipitate. Also, in addition to grain-boundary precipitates, there appeared to be only two distinct types of Widmanstätten precipitates: those with relatively large length-to-width aspect ratios made up of several subunits (lathlike) and those with relatively small length-to-width aspect ratios (platelike). Thus, the Dubé

morphological classification system (Ref 60, 61) was simplified from nine types to only three for proeutectoid cementite (Ref 13).

Example 2: Manual Mechanical Polishing in Combination with Optical Microscopy and EBSD Orientation Mapping. This case study is from an investigation performed on a fully recrystallized β-grain microstructure in a β-stabilized titanium alloy known as Ti-21S (Ref 4, 62, 63), and most of the information in this section is summarized in the article by Spanos et al. (Ref 4). In that investigation (Ref 4, 62, 63), the alloy was heat treated for 15 min at 725 °C to form a thin film of α-phase at the β-phase grain boundaries, in order to provide very high contrast of the β grain boundaries. The serial sectioning was accomplished by mechanical polishing (using semiautomatic polishers), and the average depth between sections was approximately 1.5 µm. An 8 by 8 montage of optical micrographs was taken at each section, with a magnification of 500×. A fiducial marking system was employed in which a pattern of linear channels, each 10 µm wide by 10 µm deep, was etched directly into the side of the sample with a FIB, as shown in Fig. 3; this procedure only needed to be preformed once, prior to the start of sectioning. As mentioned previously in the subsection "Fiducial Marks," FIB channels parallel to the sectioning direction (e.g., the horizontal trench in Fig. 3) provided for translational and rotational alignments of the images between the different sections, while diagonal trenches were employed to measure the individual section depths (the angled trenches on the left side of Fig. 3). When viewed from the sectioning plane, the channels appear as individual notches (the top part of both micrographs on the right side of

Fig. 3). The EBSD orientation maps were taken every tenth section, to provide the crystallographic orientation of all but the smallest grains, while at the same time minimizing the time required to obtain the data. One EBSD map was thus typically obtained for the smallest grains sampled, while multiple EBSD orientation maps were obtained for the majority of grains in the final 3-D reconstruction. (In the latter case, all orientations obtained for a single grain were averaged over that grain.) Alternatively, typically ten sections or more were imaged per grain by optical microscopy, to enable higher-fidelity reconstruction of the individual grain features (grain shape, facets, etc.). The EBSD images were then aligned with the optical micrographs (and thus to the final 3-D reconstruction) by a semiautomatic procedure that matched the position of the center of the area of the grain in the optical micrographs of a section to the equivalent center of the same area of the grain in the EBSD map for that section, using approximately five to six grains per section. This alignment procedure thus allowed a crystallographic orientation to be assigned to each grain in the 3-D reconstruction (Ref 4).

A small portion of the entire 3-D reconstruction is provided in Fig. 7. This subset was produced from 95 serial sections and encompasses a material volume of 130 by 130 by 140 µm³. In Fig. 7(a), the colors correspond to the crystallographic direction that is parallel to the sectioning direction, relative to the body-centered cubic (bcc) lattice of each of the β-grains. The scripting Interactive Data Language was used to create this 3-D reconstruction, and within this subset of the data, 16 of the β-grains do not contact the outer edges of the reconstruction box and thus lie completely inside the analyzed

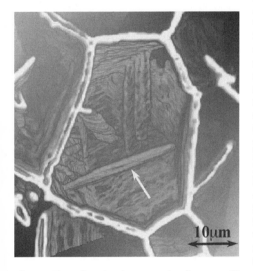

Fig. 5 Three-dimensional reconstruction of proeutectoid cementite precipitates in an isothermally transformed Fe-13Mn-1.3%C alloy. Source: Ref 13

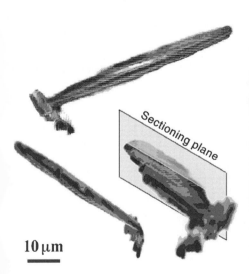

Fig. 6 Three perspective views of the Widmanstätten cementite precipitate indicated by an arrow in Fig. 5. Source: Ref 13

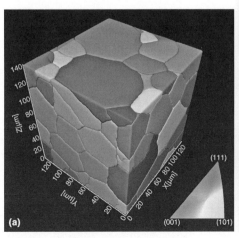

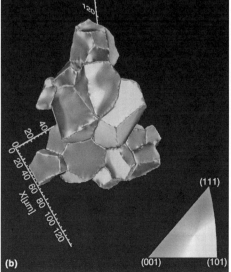

Fig. 7 (a) Reconstruction of β-grains in Ti-21S produced from 95 serial sections. Color indicates the crystallographic direction that is parallel with the sectioning direction (+z-direction). (b) Internal grains from the reconstruction. Color indicates the local crystallographic normal at that point on the interface. Source: Adapted from Ref 4

volume. These 16 grains are presented in Fig. 7(b), with the external grains removed and with a different crystallographic coloring scheme, in which the color at each location on the β grain boundaries now corresponds to the crystallographic direction that is normal to the grain-boundary interface, as opposed to normal to the sectioning direction (as in Fig. 7a). The crystallographic directions in Fig. 7(b) are relative to the bcc lattice of each individual β-grain enclosed by the corresponding grain boundaries. Thus, a grain-boundary region that is red corresponds to a {001}-type facet for the grain which that boundary encloses, while a {111} facet is represented by blue. This 3-D data set was then used as direct input into 3-D "image-based" finite-element simulations in which external loads were computationally applied to the reconstructed volume to provide a simulated microstructural response (Ref 4). (See the example presented in the subsection "Finite-Element Modeling" of the section "Simulations—Inputting and Using 3-D Data" in this article.)

Example 3: Micromilling. Alkemper and Voorhees (Ref 37) used a micromilling apparatus to obtain serial sections from a directionally solidified Al-15Cu (wt%) alloy. The experimental apparatus, shown in Fig. 8, allows one to take approximately 20 sections per hour with interlayer spacings of 1 to 20 μm. Because the sample remains attached to the micromiller at all times, the speed of the process is reduced dramatically over many other serial sectioning methods. The step height was measured using a profilometer and was shown to have a small systematic error from the set point of the Leitz Ultramiller. It has been shown that no translation between sections occurs in the x-direction, normal to the translation. The alignment of the cross sections in the y-direction was accomplished with a linear variable differential transformer (LVDT) and did not require the use of fiducial markers. The alignment procedure is illustrated in Fig. 9. Assuming that the original image is 1000 × 1000 pixels, the y-direction positions do not vary by more than 50 μm, there is a resolution of 1 μm per pixel, and the LVDT measures within 0.5 μm. A subset of images with which to form a stack can be created by deleting rows of pixels along the x-direction to produce a fully registered image without any intervention. In this case, the sections were 4.75 μm apart. The resulting 3-D microstructure is shown in Fig. 10. This work allowed a comparison of 3-D morphologies between samples of the Al-15wt%Cu alloy that had been allowed to coarsen for different times.

Example 4: Automated Serial Sectioning. Spowart (Ref 15) reported the development of a fully automated serial sectioning device called Robo-Met.3D, shown in Fig. 11. The original version of this machine was capable of acquiring ~20 cross sections per hour at 0.1 to 10 μm per section, with an accuracy in depth of approximately 0.03 μm per section. This device offers the additional capability of etching. An example application of this method is

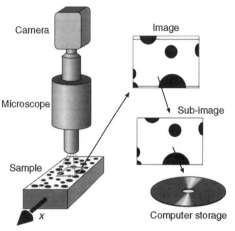

Fig. 9 Schematic illustration of the automated registration of images via a linear variable differential transformer on a micromilling device. Source: Ref 37

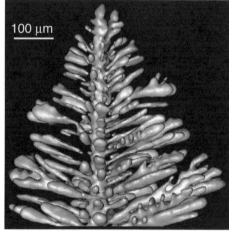

Fig. 10 Three-dimensional reconstruction of an aluminum dendrite in an aluminum-copper alloy, with the eutectic phase omitted for clarity. Source: Ref 37

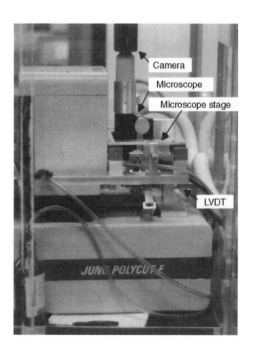

Fig. 8 Micromilling apparatus used for sectioning the aluminum-copper alloy. Source: Ref 37

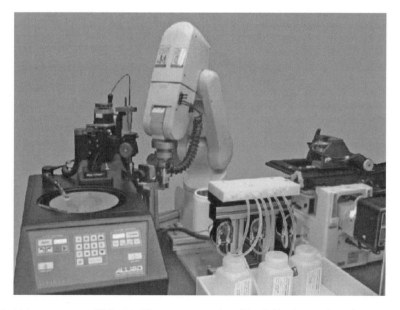

Fig. 11 Major components of Robo-Met.3D system: automatic polisher (left), robot arm (center), automatic etching station (lower center), and motorized inverted microscope (right). Source: Ref 15

presented in Fig. 12, which shows the 3-D microstructure of a sintered 67%Fe-33%Cu powder metallurgy alloy produced by Robo-Met.3D. The blue regions are the iron phase, and the orange regions are the copper phase. The 3-D analysis revealed the location and possible source of microporosity in this material.

Focused Ion Beam Tomography

The first portion of this section is adapted from Ref 1, while further studies are considered in the latter part of this section. The application of FIB instruments to microstructural tomography fills a void between methods that examine very small volumes with high lateral and depth resolution (such as 3-D atom probe tomography, field ion microscopy, and transmission electron microscopy tomography) and methods that examine very large volumes with much coarser resolution (such as mechanical serial sectioning coupled to optical microscopy, and x-ray tomography). The FIB instruments operate by rastering a focused beam of ions (typically gallium) over the surface of a sample. The interaction of the ion beam with the sample allows for direct imaging of the sample surface and also results in material removal through sputtering. Ion beam spot sizes are routinely of the order of tens of nanometers for FIB systems, and so, these instruments are capable of both high-precision material removal and high-resolution imaging. In particular, the micromachining capability of the FIB can be used to replace traditional mechanical polishing or cutting processes in a serial sectioning experiment, and FIB instruments can be used to examine volumes of material greater than 1000 μm^3 with high spatial resolution (<100 nm).

In using the FIB for sectioning, the FIB can be oriented in two different positions with respect to the sample surface. One position is to have the FIB orientation normal to the surface of interest—here, the FIB is used to mill the surface and collect structural or chemical data at the same time. This technique is termed image depth profiling (Ref 64) and is usually associated with surface science measurements using an ion beam (for example, 2-D elemental mapping using secondary ion mass spectrometry, or SIMS). Using image depth profiling, modern FIB instrumentation offers the potential for nanometer-level resolution in both the image plane and the sectioning depth. For image depth profiling, one of the most important considerations is the ability to controllably remove a planar section of material, and this can only be accomplished if the erosion rate of the selected area is both uniform and constant. Uniform erosion rates occur when the initial surface is flat and smooth and the microstructural constituents have similar sputtering rates. Because the local erosion rate is very sensitive to surface topology, chemical composition, and crystallographic orientation, this technique works best for single crystals of

uniform composition with a flat surface. As the ion beam is rastered over the surface, material is removed at a nominally fixed rate, and microstructural information can be collected via secondary electron imaging, ion imaging, and/or chemical analysis such as SIMS.

An alternate method for using the FIB for sectioning is to have the FIB orientation parallel to the surface of interest, that is, to use the FIB as a "nanoknife" to prepare a series of cross-sectional surfaces (Ref 17, 65, 66, 67). Unlike image depth profiling, this technique does not require that the microstructural constituents have the same milling rates in order to produce a series of planar surfaces. Note that the typical beam current can range from 100 to 1000 pA when preparing a cross section, because the user can determine whether to trade speed for cutting accuracy. For FIB instruments with a single ion beam, imaging the cross-sectioned surfaces requires tilting the sample so that the ion beam can be scanned across the newly prepared cross-sectional surface; after imaging, the sample is tilted back to prepare another cross section. Fiducial marks can be placed into the material surface to help with realignment and to ensure that the section slice thickness remains constant (Ref 17, 67).

Dual-beam FIB-SEM instruments—microscopes that have both an electron and ion column—eliminate the need to tilt between cutting and imaging operations by using the SEM column to image the cross-sectional surface (Fig. 13). Dual-beam instruments are relatively new and offer significant advantages for microstructural tomography applications, because the incorporation of an electron column potentially allows for additional characterization methods to be included into the 3-D characterization process (Ref 68). In addition, for some commercial instruments, the sectioning and imaging processes have been automated (e.g., FEI Company's Strata

DB 235 microscope using the "Auto(Slice and View)" software package); automation ensures that the sectioning slice thickness is very consistent and allows for the examination of much larger volumes of material than could be reasonably examined manually. The increasing availability of these systems to the materials science community will undoubtedly make FIB instruments a viable choice for microstructural tomography experiments.

Using FIB

Example 5: Nickel Superalloy (Ref 1). The nickel superalloy selected for this study has a nominal composition of 70Ni-20Cr-10Al (wt%), and experimental quantities of the alloy (400 g) were prepared by arc melting followed by thermomechanical processing. A thin slice approximately 1 mm thick was sectioned from the thermomechanically processed material, and from this slice a 3 mm disc was core drilled. The 3 mm disc was polished with SiC grinding paper and diamond lapping films and epoxied to a stub for insertion into a Strata DB 235 FIB-SEM manufactured by FEI Company. The DB 235 dual-beam microscope was used to section and image the superalloy microstructure. The DB 235 allows for various electron/ion detectors to be used with the ion beam. For these experiments, an Everhart-Thornley Thru-the-Lens Detector (TLD) was used to acquire the secondary electron signal. The secondary electron signal measured with the TLD provided reasonable contrast between the γ' precipitates and the γ matrix. As the ion beam was rastered over the surface, images of the microstructure were collected and saved, and these images were later reconstructed into the 3-D volume representation. The single slice

Fig. 12 Three-dimensional microstructure of a sintered 67%Fe-33%Cu powder metallurgy alloy produced by Robo-Met.3D. The blue regions are the iron phase, and the orange regions are the copper phase. The total volume of the data set is 1.7×10^6 μm^3, comprising 100 serial sections, 1.2 μm apart. Source: Ref 15

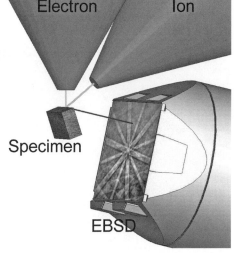

Fig. 13 Schematic illustration showing the geometry of the ion beam, the electron beam, the specimen, and the electron backscatter diffraction (EBSD) detector. The electron beam is vertical, the ion beam is at 52° to it, and the EBSD detector is at 12° below horizontal. Source: Ref 43

image shown in Fig. 14(a) was slightly processed (background has been leveled, histogram adjusted, and a median filter applied via Adobe Photoshop and Image Processing Toolkit).

Prior to sectioning and imaging, the FIB was used to prepare a smooth surface for imaging; a flat surface is necessary for producing uniform removal rates during milling. To produce an initially flat surface, a platinum cap was deposited at the cutting surface to minimize surface roughness (or curtaining) during preparation of the imaging surface. An initial surface trench was prepared using a 5000 pA beam, followed by 1000 and 300 pA beams to remove roughness produced during the previous cutting steps. After a smooth surface had been prepared, the sample was rotated so that the surface normal was parallel to the ion beam. The ion beam was simply rastered over the surface (using a 300 pA beam), and images were collected at a selected interval. This method allowed for the simultaneous milling/sectioning of the surface and the acquisition of microstructural information without moving the sample. In addition, the Model 235 Dual Beam is supplied with scripting software, which was used to automate the image-acquisition process, so that the sectioning and image collection took place without user intervention.

In this case, the time interval was approximately 3 min between 160 consecutive images, and thus, the experiment took approximately 8 h. The distance interval was 11 nm per section, and lateral resolution was 14.9 nm. For this material, the milling rates of γ and γ' in the <001> orientation were very similar, and the surface was uniformly milled away during the sectioning process. Note that because the TLD-S signal was noisy, the signal-to-noise ratio was improved by averaging images, and each individual image used in the volume reconstruction was an integrated set of 32 images. After the sequence of images was captured, commercially

available software (Interactive Data Language from Research Systems, Inc.) was used to process and render the stack of collected images into the 3-D volume reconstruction, as shown in Fig. 14(b).

As noted by Phaneuf, modern FIB systems can also be equipped with chemical and crystallographic analysis capabilities, such as energy-dispersive spectroscopy and EBSD, and thus allow for sectioning with submicron resolution while providing structural, chemical, and crystallographic analysis on each imaging plane (Ref 66). In fact, more recent studies have combined EBSD orientation mapping with dual-beam FIB milling to obtain 3-D reconstructions containing both spatial and crystallographic information. The remainder of this section briefly summarizes some of the techniques employed, and the types of results obtained, in these more recent studies.

As far as the authors are aware, Groeber, Uchic, and coworkers (Ref 41, 42) were the first to combine, in a semiautomated fashion, serial sectioning by FIB milling, EBSD orientation mapping, and computer-aided 3-D reconstruction, to produce 3-D data sets containing both morphological and crystallographic information. During the serial sectioning, the EBSD software paused between sections to allow the user to run the next EBSD scan. From the resultant data in a nickel-base superalloy, IN100 (UNS N13100), a number of microstructural attributes could be directly determined and analyzed in 3-D, including histograms of the distributions of grain diameter, number of grain faces, misorientation angle between grains, and number of grain neighbors. A follow-on, more thorough 3-D analysis of IN100 was subsequently published in which the combined serial sectioning and EBSD technique was fully automated to obtain 117 sections at a 250 nm interslice distance, in an unattended fashion

(Ref 20). Mulders and Day provided one of the first (if not the first) published renderings of 3-D EBSD maps and discussed the details of the dual-beam FIB-plus-EBSD methodology used to obtain such 3-D orientation maps (Ref 43). Their procedure was completely automated, allowing for controlled sectioning and EBSD mapping of a significant volume of material continuously without operator intervention over a period of many hours. Data were collected for three materials systems: aluminum, steel, and nickel (Ref 43). A schematic of the geometry of the dual-beam FIB-plus-EBSD setup (Fig. 13) that they employed is generally representative of the type of instrumentation typically used for automated sectioning and EBSD mapping for such 3-D reconstructions. Zaefferer (Ref 18) demonstrated the use of combining dual-beam FIB sectioning and EBSD mapping to study texture development and shear bands during recrystallization in cold-rolled Ni₃Al. Although the 2-D maps in the image stack were not combined to form a final 3-D reconstruction, this study demonstrated the utility of such techniques to elucidate the 3-D characteristics of shear bands and recrystallization microstructures (Ref 18).

In a later study of microstructural anisotropy and texture in a pipeline steel, Petrov et al. employed a dual-beam FIB and EBSD equipment to allow for repeated sectioning in an automatic mode and subsequent reconstruction of both the 3-D microstructure and the texture of the volume examined (Ref 44). Renderings of 3-D gray-scale maps based on both EBSD pattern quality and EBSD Euler angles were generated, and the 3-D data were used to correlate crystallographic orientation, grain shape, and anisotropy of Charpy impact toughness (Ref 44). Konrad et al. (Ref 21) also employed a 3-D EBSD-FIB technique, which they termed 3-D orientation microscopy, to study orientation gradients in Fe₃Al around second-phase Laves particles of (Fe,Al)₂Zr, because such orientation gradients can strongly affect the propensity for recrystallization in Fe₃Al. In particular, they used a dual-beam FIB to perform the milling and the EBSD orientation mapping and switched manually between milling and EBSD mapping modes for each section. Their results showed strong 3-D orientation gradients in the matrix material adjacent to the Laves particles, and the shapes and strengths of these texture gradients were quantified in 3-D. This type of 3-D experimental information could obviously serve as valuable input to both 3-D recrystallization models and 3-D finite-element models of the response of such orientation gradients to an externally applied stress field.

In two more recent articles, Uchic et al. (Ref 19, 22) reviewed the state-of-the-art of developments and applications of 3-D analyses using FIB tomography, including discussion of the combined use of FIB sectioning and EBSD for 3-D orientation mapping. In particular, they discussed the advantages of such techniques, such as the finer resolution and potential time-

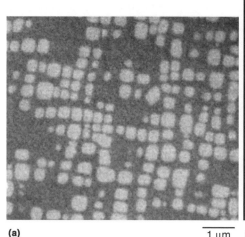
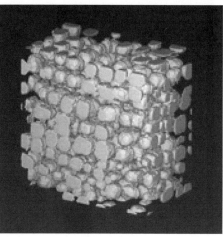

(a)   1 μm   **(b)**

Fig. 14 (a) Single two-dimensional secondary electron image of γ' precipitates in a γ matrix of a 70Ni-20Cr-10Al (wt %) alloy. (b) Three-dimensional (3-D) volume reconstruction of γ' precipitates in a 70Ni-20Cr-10Al (wt%) alloy. The 3-D image was generated from the stack of collected images (Fig. 14a) from a dual-beam focused ion beam-scanning electron microscope. Source: Ref 1

savings due to automation, and reviewed the use of FIB tomography in general as applied to a wide range of materials systems. Figure 15 (Ref 22) presents a typical example of the type for 3-D crystallographic reconstruction that can be obtained from such techniques that combine FIB milling and EBSD mapping.

A number of researchers (Ref 17, 18, 20, 21, 43, 44, 64–68) have thus used such FIB techniques to analyze the 3-D morphology, and 3-D crystallography, of a wide variety of materials systems and to investigate correlations between the 3-D microstructure and material behavior, including mechanical response and/or recrystallization. As is the case with experimental 3-D reconstructions developed with other techniques (mechanical serial sectioning, x-ray tomography, etc.), such 3-D data sets can serve as valuable input for predictive microstructural models, as considered in the next section.

Simulations—Inputting and Using 3-D Data

The development of robust 3-D experimental data sets has allowed for a relatively new type of microstructural modeling, namely 3-D image-based modeling (Ref 5, 6). Image-based modeling refers to a methodology where real, experimentally derived 3-D microstructures are used as the initial input into models of microstructural response and evolution, as opposed to the use of digitally created microstructures. The use of quantitative experimentally obtained 3-D results as input and validation for realistic 3-D image-based models is critical for accurate and efficient materials and process design methodologies. This section examines two different simulation methods that use experimental 3-D data as input: phase-field simulations and finite-element modeling (FEM). The usefulness of these methods is described and examples of image-based simulations are given.

Two recent review articles in the *MRS Bulletin on Three-Dimensional Materials Science* (Ref 10) covering 3-D image-based modeling in detail (Ref 4, 69) were written by the authors of this *ASM Handbook* article (along with other co-authors). A portion on image-based phase-field modeling is included in the article by Kammer and Voorhees (Ref 69), and a brief overview of 3-D image-based FEM modeling is included in the article by Spanos et al. (Ref 4). Most of the figures and much of the technical content from the following two subsections have thus been taken from appropriate sections of those two articles. For further details, the reader is referred directly to those two review articles (Ref 4, 69).

Image-based Modeling

Phase-Field Modeling. In the phase-field approach, the evolution of phase boundaries in material systems is modeled using a diffuse interface, in which the interface is taken to have a nonzero thickness. This is accomplished with the use of a field variable that changes smoothly across the interface from the bulk value of one phase to another. With the phase-field method, partial differential equations are employed that hold at all positions in space and therefore can be used to model the evolution of concentration, order parameter, stress, or magnetic field through the interface. This then eliminates the need for placing boundary conditions at the interface and for explicit tracking of the interface position. Instead, the position of the interface is inferred from the values of the field variable(s) (e.g., concentration) within the interface region. Using such diffuse-interface techniques provides for a number of advantages over the sharp interface approach, including the fact that it yields differential equations that are easy to solve numerically in three dimensions, and it handles topological singularities, such as the splitting or merging of phase interfaces during microstructural evolution.

Because the phase-field method can track the evolution of very complex grain and/or particle shapes, experimentally measured 3-D particle/grain interface morphologies can be used as the initial input into phase-field simulations. By importing the 3-D structure into a numerical model for the evolution of the microstructure, the microstructure can then be linked to its processing conditions. Because initial conditions can greatly affect the results of simulations, realistic initial conditions are frequently needed. Using an experimentally measured 3-D structure as an initial condition is a novel approach that avoids the need for making assumptions about the initial structure of a material. Combining experimentally obtained microstructures with phase-field simulations of microstructural evolution (i.e., image-based phase-field modeling) allows for examination of the validity of the models on which the simulations are based, provides new insights into physical processes controlling microstructural evolution, and enables significant improvements in the fidelity of such simulations, thereby

providing for improved processing-microstructure correlations. Some recent image-based phase-field simulation studies are now considered, in order to provide examples of the types of results provided by such simulations and to demonstrate their value in predictive microstructural evolution modeling (Ref 70, 71).

Coarsening of dendritic solid-liquid mixtures was recently examined (Ref 59, Ref 71). The morphology of the interface was characterized using interfacial shape distributions (ISDs), which are similar to the particle size distributions for systems with spherical particles. As an example of an ISD, Fig. 16 (Ref 59) shows the ISD that corresponds to the 3-D reconstruction of Fig. 17 (Ref 59). This experimental reconstruction corresponds to the microstructure of a lead-tin sample that had been coarsened for two days. The ISD is read with the help of Fig. 18 (Ref 59), a map of the different curvature regions, which allows for the determination of the various interfacial shapes. In Fig. 16 and 18, κ_1 and κ_2 are the two principal curvatures, and S_v is the surface area per unit volume. The color bar (top of Fig. 16) gives the probability of finding a patch of interface with a given pair of principal curvatures. The dominant platelike shapes of Fig. 17 give rise to the primary peak (red color, right end of bar) of the ISD (Fig. 16) in the saddle-shaped region (compare to Fig. 18, areas 2 and 3). The secondary peak (turquoise color) of the ISD in Fig. 16 is situated along the solid cylinder line ("S" in Fig. 18) as a result of the solid cylindrical-like shapes and the round edges of the platelike shapes in Fig. 17.

Very little is known about the processes that determine the evolution of the ISDs in systems with a dendritic structure. In the study by Mendoza et al. (Ref 71), 3-D microstructures were experimentally determined for different coarsening times and were then used to calculate ISDs similar to those presented in Fig. 16

Fig. 15 Three-dimensional reconstruction of IN100 created from electron backscatter diffraction data. The dimensions of the parallelepiped volume are 41 × 41 × 29 μm, and the cubic voxel dimension is 0.25 μm. Source: Ref 22

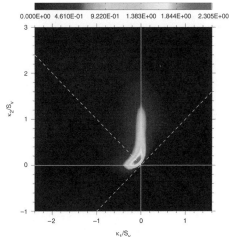

Fig. 16 Interfacial shape distribution of the two-day lead-tin coarsened sample. κ_1 and κ_2 are the two principal curvatures, and S_v is the surface area per unit volume. Source: Ref 59

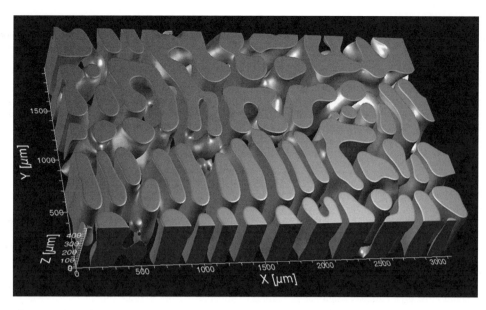

Fig. 17 Three-dimensional reconstruction (100 sections) of the two-day lead-tin coarsened sample. Source: Ref 72

and 18. The 3-D microstructure that was determined experimentally was input into the phase-field calculation and allowed to evolve for a short time. Thus, through the use of image-based phase-field modeling, it was possible to determine a flux of probability in the curvature space that governs the evolution of the ISD. It was found that this flux in probability, as determined by the phase-field calculations (and represented by the white arrows in Fig. 19), yielded a good description of the evolution of the ISD that was seen experimentally (represented by the colors in Fig. 19; the top bar ranges from violet on the left to red on the right). This comparison thus provides evidence for the validity of the phase-field approach and also indicates that a properly constructed flux in probability space can determine the evolution of the ISD.

In another example, Mendoza et al. used image-based phase-field modeling to examine discrete topological events, such as the breakup of tubes or the formation of holes in coarsened liquid-solid mixtures (Ref 70). Because the serial sectioning experiments are performed ex situ and are destructive to the samples, time-dependent microstructural evolution data cannot be directly obtained with such experimental techniques. Figure 20 thus shows an experimental 3-D microstructure for an aluminum-copper sample that had been coarsened in the liquid-solid state for 964 min, along with the interfacial velocity (represented by color) calculated from the phase-field simulations. In this figure, in which the solid phase has been made transparent, positive interfacial velocity points into the liquid and is represented by "warm" colors (a, b, c, and d). One interesting observation deduced from the interfacial velocity at point "c," for example, is that a topological singularity will lead to separation into two liquid regions. Another observation is that, based on its interfacial velocity, the liquid droplet at point "d" should shrink away. Mendoza et al. also showed that the predominant liquid tubes appearing in their structures are a result of the formation, growth, and coalescence of holes on the liquid walls of the structure, as demonstrated in Fig. 21, which shows a phase-field simulation of the evolution of a small region of the coarsened experimental structure. The images of the structure as a function of increasing simulation time demonstrate that a hole develops near the intersecting liquid walls (Fig. 21b), grows, and then separates the walls (Fig. 21c).

It should be mentioned that there are also a number of challenges associated with the phase-field approach. It is necessary to accurately resolve the diffuseness of the solid-liquid interface. Thus, several mesh points within the interfacial region are needed, and the interfacial thickness determines the maximum spacing between mesh points. In physical systems, the interfacial thicknesses are typically very small, on the order of 5 nm or less. On the other hand, the domain to be simulated is often

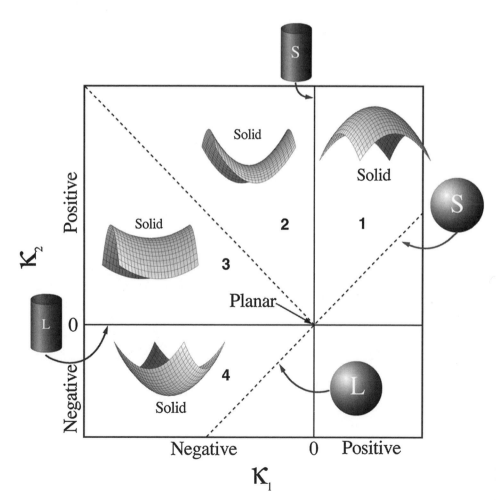

Fig. 18 Map of the local interfacial shapes for the interfacial shape distribution. The axes represent the two principal curvatures, κ_1 and κ_2. Source: Ref 59

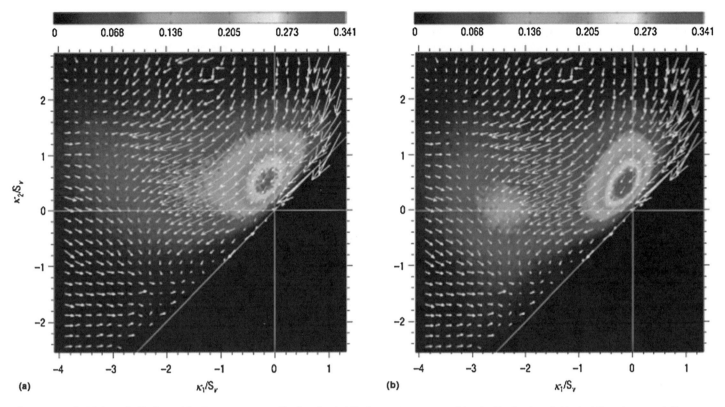

Fig. 19 Interfacial shape distribution and flux in curvature space. The flux of probability in curvature space, as determined by the phase-field calculations, is represented by arrows in (a) and (b). Flux arrows calculated from a 10 min coarsened aluminum-copper sample overlaid on an experimentally measured intensity plot of the interfacial shape distribution for (a) the 10 min coarsened sample and (b) a 90 min coarsened sample. The principal curvatures, κ_1 and κ_2, are scaled by the surface area per unit volume, S_v. Color scale indicates dimensionless probability density. Source: Ref 71

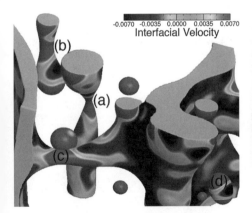

Fig. 20 Portion of the experimental three-dimensional microstructure for an aluminum-copper sample that has been coarsened for 964 min. The colors indicate the dimensionless interfacial velocity of the liquid front as computed from the three-dimensional phase-field simulations. The majority solid phase is transparent. The liquid that intersects the edges of the reconstruction box is capped with zero interfacial velocity.

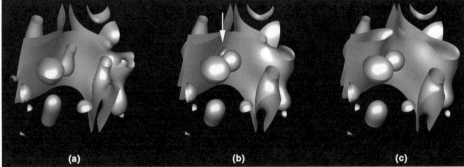

Fig. 21 Phase-field simulation snapshots showing the topological singularity of hole creation and growth within liquid walls. Only the interfaces appear. The arrow in (b) points at a hole that has been created in a wall. Source: Ref 70

considerably larger than the physical interfacial thickness, requiring an extremely large number of mesh points. For the model to be computationally manageable in such cases, some combination of two approaches can be taken:

- The simulated interfaces may be taken to be much thicker than the physical interfaces.
- Adaptive meshing procedures can be used to increase the number of mesh points near the interfaces.

In phase-field simulations where interfacial thicknesses much larger than the actual interfacial thickness are employed, matched asymptotic expansions of the phase-field equations are also typically used, so that the solutions of the phase-field equations will still capture the correct physical processes underlying the microstructural evolution (Ref 73–76). This approach is valid as long as the interfacial thickness employed in the simulations is smaller than any other length scale in the problem, such as the radius of curvature of the interfaces or the size of the domains bounded by the interfaces.

Finite-Element Modeling. Image-based 3-D FEM is critical for accurate and efficient materials and process design methodologies. In this regard, the use of 3-D microstructures in image-based modeling presents a number of advantages over the use of 2-D FEM simulations but also presents a new set of challenges. Specifically, although image-based 3-D simulations are very powerful in that they can be directly correlated with 3-D features in the real microstructure, which is important for determining critical structure-property relationships, computational power often limits such simulations to relatively small data sets. These simulations thus employ typically on the order of a few hundred grains or less.

Generating the finite-element mesh from experimentally obtained 3-D microstructures is a critical component of image-based FEM. Various approaches can be used to generate such meshes, and the most direct method involves assignment of voxels (3-D volume elements, equivalent to pixels in two dimensions) in the 3-D data set to eight-noded brick-shaped elements. Although an advantage of this method is the ease of generation of the mesh, there are shortcomings as well. In particular, this method creates a large number of elements, which makes the problem very computationally intensive. One way to reduce this effect is to sample the high-resolution data set such that only every third or fourth voxel is retained, thus reducing the number of elements in the mesh while retaining the entire volume (Ref 6, 77–79). However, using only a small number of voxels for each region of interest results in poor resolution of the shape at the object interfaces, resulting in stepped geometries at grain or interphase boundaries. Effects of mesh geometry in 3-D microstructural FEM simulations have been discussed in some detail (e.g., Ref 78, 79), and, in general, the ideal method of mesh generation depends on the specific problem being studied.

One important application of 3-D image-based FEM is to study the onset of yielding in real 3-D microstructures (Ref 77, 78). In such studies, 3-D microstructural features such as grain boundaries and triple junctions are often critical to the initiation of yielding and must therefore be represented as accurately as possible in the model. In such studies, important correlations between the 3-D microstructure and the onset of plasticity can often be determined by modeling the material response in the elastic (low-strain) regime, therefore requiring significantly fewer increments in strain or time in the simulations, ultimately reducing the computational cost, and thus allowing for the use of higher-density meshes with more elements.

The example that follows is from research on a β-stabilized titanium alloy that integrates 3-D experimental data with 3-D simulations (Ref 4, 62, 63). In this work, a 3-D reconstruction of beta grain morphology and crystallography in a single-phase, polycrystalline titanium alloy (example shown earlier) was combined with image-based 3-D FEM simulations. The

acquired 3-D data set was used as direct input into image-based 3-D finite-element simulations in which external loads were computationally applied to the reconstructed volume of this β-stabilized titanium alloy, to provide a simulated microstructural response.

The 3-D microstructure shown in Fig. 7(a) (the reconstruction of β-grains in alloy Ti-21S) was used as an initial input for the image-based FEM to simulate mechanical response to applied strains. For this simulation, only a portion of the entire 3-D data set was used to create the FEM mesh. Additionally, the FEM mesh was created by sampling every third voxel in the x- and y-directions and every second voxel in the z-direction from the experimentally obtained high-resolution data subset. Further details about the type of FEM elements, the elasticity/plasticity models and elastic constants, and the ABAQUS software employed can be found elsewhere (Ref 4, 62, 63).

To demonstrate the utility of these types of image-based 3-D FEM analyses, simulation results are presented from two studies (Ref 62, 63). Figure 22(a) is a contour plot of the response of the β-titanium microstructure shown in Fig. 7(a), for an applied uniaxial strain of 0.7% (Ref 4, 62, 63). This figure shows the von Mises stress scalar, which is a measure of the magnitude of stress, at a given location in the microstructure, allowing for visualization of the high local stress regions, or "hot spots." Microstructural features such as grain boundaries and grain triple junctions were identified as areas that produce high local stresses, which often serve as points of initiation of deformation and/or failure. To better visualize the effect of the grain interfaces, the interiors of the grains have been removed, revealing the grain-boundary network, with the mechanical response plotted as color on

the grain boundaries (Fig. 22b) (Ref 4, 63). In this way, both the grains and the grain boundaries corresponding to high local stresses during the initiation of yielding were identified (Ref 63). Combining experimentally-based 3-D reconstructions with 3-D FEM simulation of mechanical response in this way enables identification of critical microstructural and crystallographic features (microstructural hot spots). By using this information, it is possible to reveal the complex structure-property relationships, which can then be used to improve the efficiency and/or accuracy of the materials and process design cycle.

There are currently significant challenges remaining in the field of 3-D image-based FEM. One obstacle is that there is a lack of fundamental physical parameters for many of the important materials systems. For example, the mechanical anisotropies for both elasticity and plasticity in many alloy systems are not known. Many times, these parameters are estimated from other simplified simulations or extrapolated from similar alloy systems. Further experimental investigation into these parameters is needed.

Additionally, one must define what constitutes a representative volume for the simulations. When capturing the onset of plastic yielding, for instance, one requires roughly hundreds of grains to duplicate the bulk elastic response. However, in the case of critical failure flaws, thousands of grains and/or precipitates may be needed to obtain a representative volume. Such large simulations can quickly become so computationally intensive that they are unfeasible. On the other hand, as finite-element meshing techniques improve and computation speed and memory capabilities expand, simulations of such large volumes will become more common (Ref 4, 63).

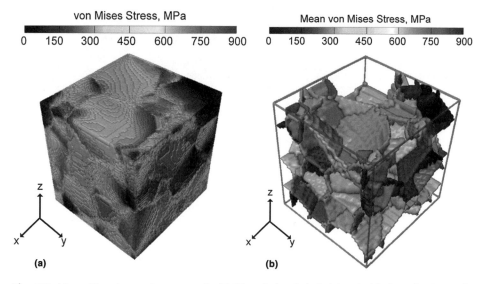

Fig. 22 (a) von Mises stress contour as a result of 0.7% applied strain (uniaxial tension) in the x-direction on the three-dimensional β-titanium microstructure shown in Fig. 7(a). (b) Mean von Mises stress as a result of 0.7% applied strain in the x-direction for each grain boundary in the β-titanium reconstruction. Source: Ref 4

ACKNOWLEDGMENT

Portions of this article have been adapted from "Three-Dimensional Microscopy" by M.V. Kral, G.E. Ice, M.K. Miller, M.D. Uchic, and R.O. Rosenberg in *Metallography and Microstructures,* Volume 9, *ASM Handbook,* 2004 (Ref 1).This includes portions of the sections "Three-Dimensional Characterization Methods" and "Focused Ion Beam Tomography."

REFERENCES

1. M.V. Kral, G.E. Ice, M.K. Miller, M.D. Uchic, and R.O. Rosenberg, Three-Dimensional Microscopy, *Metallography and Microstructures,* Vol 9, *ASM Handbook,* G.F. VanderVoort, Ed., ASM International, 2004
2. G. Spanos, Viewpoint Set on 3D Characterization and Analysis of Materials, *Scr. Mater.,* Vol 55, 2006, p 3–86
3. M.D. Uchic, *JOM, Special Issue on 3-D Characterization: Methods and Applications,* Vol 58, 2006, p 24–52
4. G. Spanos, D.J. Rowenhorst, A.C. Lewis, and A.B. Geltmacher, *MRS Bull.,* Vol 33, 2008, p 597–602
5. R.K. Everett, A.B. Geltmacher, and K.E. Simmonds, 3D Image-Based Modeling of Void Interactions in HY100 Steel, *Plasticity, Damage, and Fracture at Macro, Micro, and Nano Scales,* p 699–701
6. A.C. Lewis and A.B. Geltmacher, *Scr. Mater.,* Vol 55, 2006, p 81–86
7. N. Chawla and R.S. Sidhu, *J. Mater. Sci.— Mater. Electron.,* Vol 18, 2007, p 175–189
8. A.C. Lewis, J.F. Bingert, D.J. Rowenhorst, A. Gupta, A.B. Geltmacher, and G. Spanos, *Mater. Sci. Eng. A,* Vol 418, 2006, p 11–18
9. E. Nakamachi, N.N. Tam, and H. Morimoto, *Int. J. Plast.,* Vol 23, 2007, p 450–489
10. K. Thorton and H.F. Poulsen, *MRS Bull.— Special Issue on Three-Dimensional Materials Science,* Vol 33, 2008, p 587–629
11. R.T. DeHoff, *J. Microsc.,* Vol 131, 1983, p 259–263
12. A.C. Lund and P.W. Voorhees, *Acta Mater.,* Vol 50, 2002, p 2585–2598
13. M.V. Kral and G. Spanos, *Acta Metall.,* Vol 47, 1999, p 711–724
14. M.A. Mangan, M.V. Kral, and G. Spanos, *Acta Mater.,* Vol 47, 1999, p 4263–4274
15. J.E. Spowart, *Scr. Mater.,* Vol 55, 2006, p 5–10
16. T.L. Wolfsdorf, W.H. Bender, and P.W. Voorhees, *Acta Mater.,* Vol 45, 1997, p 2279–2295
17. B.J. Inkson, M. Mulvihill, and G. Möbus, *Scr. Mater.,* Vol 45, 2001, p 753–758
18. S. Zaefferer, *Mater. Sci. Forum,* Vol 495–497, 2005, p 3–12
19. M.D. Uchic, M.A. Groeber, D.M. Dimiduk, and J.P. Simmonds, *Scr. Mater.,* Vol 55, 2006, p 23–28

20. M. Groeber, B. Haley, M. Uchic, D.M. Dimiduk, and S. Ghosh, *Mater. Charact.,* Vol 57, 2006, p 259–273
21. J. Konrad, S. Zaefferer, and D. Raabe, *Acta Mater.,* Vol 54, 2006, p 1369–1380
22. M.D. Uchic, L. Holzer, B.J. Inkson, E.L. Principe, and P. Munroe, *MRS Bull.,* Vol 32, 2007, p 408–416
23. A. Cerezo, M.G. Hetherington, J.M. Hyde, M.K. Miller, G.D.W. Smith, and J.S. Underkoffler, *Surf. Sci.,* Vol 266, 1991, p 471–480
24. D.N. Seidman, C.K. Sudbrack, and K.E. Yoon, *JOM,* Vol 58, 2006, p 34–39
25. O. Forsman, *Jernkontorets Ann.,* Vol 102, 1918, p 1–30
26. M. Hillert, The Formation of Pearlite, *The Decomposition of Austenite by Diffusional Processes,* V.F. Zackay and H.I. Aaronson, Ed., Interscience, New York, 1962, p 197–237
27. E. Eichen, H.I. Aaronson, G.M. Pound, and R. Trivedi, *Acta Metall.,* Vol 12, 1964, p 1298–1301
28. R.H. Hopkins and R.W. Kraft, *Trans. AIME,* Vol 233, 1965, p 1526–1532
29. E.B. Hawbolt and L.C. Brown, *Trans. AIME,* Vol 239, 1967, p 1916–1924
30. L.K. Barrett and C.S. Yust, *Trans. AIME,* Vol 239, 1967, p 1172–1180
31. P.M. Ziolkowski, "The Three Dimensional Shapes of Grain Boundary Precipitates in α/β Brass," Ph.D. thesis, Michigan Technological University, 1985
32. D.A. Hull, D. McCammond, D.W. Hoeppner, and W.G. Hellier, *Mater. Charact.,* Vol 26, 1991, p 63–71
33. J. Brystrzycki and W. Przetakiewicz, *Scr. Metall. Mater.,* Vol 27, 1992, p 893–896
34. M.A. Mangan and G.J. Shiflet, Three-Dimensional Reconstruction of Pearlite Colonies, *Solid-Solid Phase Transformations Proceedings,* W.C. Johnson, J.M. Howe, D.E. Laughlin, and W.A. Soffa, Ed., TMS, Warrendale, PA, 1994, p 547–552
35. H. Wieland, T.N. Rouns, and J. Liu, *Z. Metallkd.,* Vol 85, 1994, p 592–597
36. M.A. Mangan, P.D. Lauren, and G.J. Shiflet, *J. Microsc.,* Vol 188, 1997, p 36–41
37. J. Alkemper and P.W. Voorhees, *J. Microsc.,* Vol 201, 2000, p 1–8
38. A. Tewari and A.M. Gokhale, *Mater. Charact.,* Vol 46, 2001, p 329–335
39. D.J. Rowenhorst, A. Gupta, C.R. Feng, and G. Spanos, *Scr. Mater.,* Vol 55, 2006, p 11–16
40. A.C. Lewis, C. Suh, M. Stukowski, A.B. Geltmacher, G. Spanos, and K. Rajan, *JOM,* Vol 58, 2006, p 52–56
41. M. Groeber, B. Haley, M. Uchic, and S. Ghosh, Microstructural Characterization Using 3-D Orientation Data Collected by an Automated FIB-EBSD System, *AIP Conference Proceedings—Materials Processing and Design: Modeling, Simulation,*

and Applications, NUMIFORM 2004, p 1712–1718
42. M.D. Uchic, M. Groeber, R. Wheeler, F. Scheltens, and D.M. Dimiduk, *Microsc. Microanal.,* Vol 10, 2004, p 1136–1137
43. J.J.L. Mulders and A.P. Day, *Mater. Sci. Forum,* Vol 495–497, 2005, p 237–242
44. R. Petrov, O.L. Garcìa, J.J.L. Mulders, A.C.C. Reis, J.-H. Bae, L. Kestens, and Y. Houbaert, *Mater. Sci. Forum,* Vol 550, 2007, p 625–630
45. F.N. Rhines, K.R. Craig, and D.A. Rousse, *Metall. Trans. A,* Vol 7, 1976, p 1729–1734
46. A. Tewari, A.M. Gokhale, and R.M. German, *Acta Mater.,* Vol 47, 1999, p 3721–3734
47. J.C. Russ, *The Image Processing Handbook,* CRC Press, 1998
48. H. Schaeben, M. Apel, T. Frank, M. Iwanowski, and S. Zaefferer, *Mater. Sci. Forum,* Vol 495–497, 2005, p 185–190
49. L.G. Shapiro and G.C. Stockman, *Computer Vision,* Prentice-Hall, New Jersey, 2001
50. C. Zhang, A. Suzuki, T. Ishimaru, and M. Enomoto, *Metall. Mater. Trans. A,* Vol 35, 2004
51. D.J. Rowenhorst, J. Kuang, K. Thornton, and P.W. Voorhees, *Acta Mater.,* 2006, p 2027–2039
52. M. Groeber, S. Ghosh, M.D. Uchic, and D.M. Dimiduk, *Acta Mater.,* Vol 56, 2008, p 1257–1273
53. M. Groeber, S. Ghosh, M.D. Uchic, and D.M. Dimiduk, *Acta Mater.,* Vol 56, 2008, p 1274–1287
54. W.E. Lorensen and H.E. Cline, *ACM SIGGRAPH Comput. Graph.,* Vol 21, 1987, p 163–169
55. H. Jinnai, T. Koga, Y. Nishikawa, T. Hashimoto, and S.T. Hyde, *Phys. Rev. Lett.,* Vol 78, 1997, p 2248–2251
56. J. Alkemper and P.W. Voorhees, *Acta Mater.,* Vol 49, 2001, p 897–902
57. R. Mendoza, J. Alkemper, and P.W. Voorhees, *Metall. Mater. Trans. A,* Vol 34, 2003, p 481–489
58. D.M. Saylor, A. Morawiec, and G.S. Rohrer, *Acta Mater.,* Vol 51, 2003, p 3663–3674
59. D. Kammer and P.W. Voorhees, *Acta Mater.,* Vol 54, 2006, p 1549–1558
60. C.A. Dubé, H.I. Aaronson, and R.F. Mehl, *Rev. Métall.,* Vol 55, 1958, p 201–210
61. H.I. Aaronson, Proeutectoid Ferrite and Cementite Reactions, *The Decomposition of Austenite by Diffusional Processes,* V.F. Zackay and H.I. Aaronson, Ed., Interscience, NY, 1962, p 387–546
62. A.B. Geltmacher, A.C. Lewis, D.J. Rowenhorst, M.A. Qidwai, and G. Spanos, Three-Dimensional Characterization and Mesoscale Mechanical Modeling of a Beta Titanium Alloy, *Proceedings of the 11th World Conference on Titanium: Ti-2007 Science*

and Technology, June 3, 2007 (Kyoto, Japan), p 475–478

63. M.A. Qidwai, A.B. Geltmacher, A.C. Lewis, D.J. Rowenhorst, and G. Spanos, *"High-Fidelity Reconstruction and Computational Modeling of Metallic Microstructures,"* Paper IMECE2007-42007, Proceedings of the 2007 ASME International Mechanical Engineering Congress and Exposition (IMECE), Nov 11, 2007 (Seattle, WA)

64. A.J. Patkin and G.H. Morrison, *Anal. Chem.,* Vol 54, 1982, p 2–5

65. D.N. Dunn and R. Hull, *Appl. Phys. Lett.,* Vol 75, 1999, p 3414–3416

66. M.W. Phaneuf and J. Li, FIB Techniques for Analysis of Metallurgical Specimens, *Microscopy & Microanalysis 2000,* Aug 13–17, 2000 (Philadelphia, PA), p 524–525

67. B.J. Inkson, T. Steer, G. Möbus, and T. Wagner, *J. Microsc.,* Vol 201, 2001, p 256–269

68. T. Sakamoto, Z. Cheng, M. Takahashi, M. Owari, and Y. Nihei, *Jpn. J. Appl. Phys.,* Vol 37 (Part 1), 1998, p 2051–2056

69. D. Kammer and P.W. Voorhees, *MRS Bull.,* Vol 33, 2008, p 603–610

70. R. Mendoza, K. Thornton, I. Savin, and P.W. Voorhees, *Acta Mater.,* Vol 54, 2006, p 743–750

71. R. Mendoza, I. Savin, K. Thornton, and P.W. Voorhees, *Nature Mater.,* Vol 3, 2004, p 385–388

72. D. Kammer, Ph.D. thesis, Northwestern University, Evanston, IL, 2006

73. P.W. Bates, P.C. Fife, R.A. Gardner, and C. Jones, *Physica D,* Vol 104, 1997, p 1–31

74. G. Caginalp and W. Xie, *Phys. Rev. E,* Vol 48, 1993, p 1897–1909

75. A. Karma and W.J. Rappel, *Phys. Rev. E,* Vol 53, 1996, p R3017–R3020

76. A.A. Wheeler, W.J. Boettinger, and G.B. McFadden, *Phys. Rev. A,* Vol 45, 1992, p 7424–7439

77. A.C. Lewis, K.A. Jordan, and A.B. Geltmacher, *Metall. Mater. Trans. A,* Vol 39, 2008, p 1109–1117

78. G. Cailletaud, S. Forest, D. Jeulin, F. Feyel, I. Galliet, V. Mounoury, and S. Quilici, *Comput. Mater. Sci.,* Vol 27, 2003, p 351–374

79. O. Diard, S. Leclercq, G. Rousselier, and G. Cailletaud, *Int. J. Plast.,* Vol 21, 2005, p 691–722

Simulation of Phase Diagrams and Transformations

ASM Handbook, Volume 22B, *Metals Process Simulation*
D.U. Furrer and S.L. Semiatin, editors

Copyright © 2010, ASM International®
All rights reserved.
www.asminternational.org

Commercial Alloy Phase Diagrams and Their Industrial Applications

F. Zhang, Y. Yang, W.S. Cao, S.L. Chen, and K.S. Wu, CompuTherm
Y.A. Chang, University of Wisconsin

THE INDUSTRIAL APPLICATION OF PHASE DIAGRAMS is the focus of this article. The first question one may ask is: What is a phase diagram and how is it useful for industrial application? As defined in "Phase Equilibria and Phase Diagram Modeling" in *Fundamentals of Modeling for Metals Processing*, Volume 22A, *ASM Handbook*, 2009, a phase diagram is a graphical representation of the phase equilibria of materials in terms of temperature, composition, and pressure. It is normal to think of binary phase diagrams as plots of temperature versus composition at a constant pressure of 1 bar. In reality, many other and equally useful representations exist, as discussed in the previously mentioned article. At a first glance, phase diagrams of a material do not seem to be directly useful to materials scientists/engineers whose ultimate goal is either to improve the performance of an existing material or to develop a new material with desirable mechanical properties, such as strength, toughness, and resistance to creep, in addition to being oxidation resistant. However, it is known that the mechanical behavior of a material is governed by its microstructure, which is developed due to the interplay of thermodynamics and kinetics; the former determines the phases to be present in the microstructure, and the latter determines the morphological features of these phases. A phase diagram thus plays an important role in understanding the microstructure that is developed and the ultimate performance of a material. In fact, phase diagrams are widely used by industry as road maps in materials design and development.

Phase diagrams have been traditionally determined purely by experimentation, which is costly and time-consuming. While the experimental approach is feasible for the determination of binary and simple ternary phase diagrams, it is less efficient for the complicated ternaries and becomes practically impossible for higher-order systems over a wide range of compositions and temperatures. On the other hand, commercial alloys are multicomponents in nature; a more efficient approach is therefore needed in the determination of multicomponent phase diagrams. In recent years, a phenomenological approach, or the CALPHAD approach (Ref 1), has been widely used for the study of phase equilibria of multicomponent systems. A comprehensive guide to the development and application of this approach has been presented by Saunders and Miodownik (Ref 2) and has been discussed in detail in "Phase Equilibria and Phase Diagram Modeling" in *Fundamentals of Modeling for Metals Processing*, Volume 22A, *ASM Handbook*, 2009. In this article, the focus is on the industrial applications of this approach.

The essence of this approach is to obtain self-consistent thermodynamic descriptions of the lower-order systems—binaries and ternaries—in terms of known thermodynamic data measured experimentally and/or calculated theoretically as well as the measured phase equilibria. The advantage of this method is that the separately measured phase diagrams and thermodynamic properties can be represented by the same "thermodynamic description" of a materials system in question. More importantly, on the basis of the known descriptions of the constituent lower-order systems, a reliable description of a higher-order system can be obtained by way of an extrapolation method (Ref 3). This description enables the calculation of phase diagrams of the multicomponent systems that are experimentally unavailable. Development of a thermodynamic description of a multicomponent system has been discussed elsewhere (Ref 4) and is not repeated here. It is worthy to note that the term *thermodynamic database* or simply *database* is usually used in the industrial community instead of *thermodynamic description*, particularly for multicomponent systems.

In addition to having a reliable thermodynamic database, robust computer software is another essential for the successful application of the CALPHAD approach. Because this topic has been covered in "Phase Equilibria and Phase Diagram Modeling" in *Fundamentals of Modeling for Metals Processing*, Volume 22A, *ASM Handbook*, 2009, the readers are referred to that article for details. In brief, given the initial conditions of alloy chemistry and temperature, one and only one stable equilibrium state exists. However, as pointed out by Chang et al. (Ref 5), it is not uncommon that the calculated equilibria are metastable if the software lacks the capability of searching for the global equilibria automatically. This problem was conquered due to the development of the second-generation software, such as Pandat.

The first focus of this discussion is on the industrial applications of phase diagram calculation, and a number of examples are presented. The integration of phase diagram calculation with kinetic and microstructural evolution models are then presented for the study of solidification process and precipitation kinetics. Finally, the achievements of the state-of-the-art CALPHAD approach, as well as the strength and limitation of this approach, are discussed. Note that the terms *phase diagram calculation*, *CALPHAD calculation*, *thermodynamic calculation*, and *thermodynamic modeling* are used in this article with no distinction. As stated earlier, robust computer software and reliable thermodynamic databases are the prerequisites for the successful application of the CALPHAD calculations. Several software packages, such as ThermoCalc, FactSage, JMatPro, and Pandat, have been developed in the past few decades to meet the needs of materials scientists/engineers working in various research fields. Each of these software packages has its own focuses, uniqueness, and strengths and may be used for various applications. In this article, Pandat software (Ref 6) and thermodynamic databases for aluminum-, iron-, nickel-, titanium-, and zirconium alloys developed at CompuTherm are used in the calculations.

Industrial Applications

Although industrial processes rarely reach an equilibrium state, knowledge of phase stabilities, phase transformation temperatures, phase amounts, and phase compositions in an equilibrium state is critically needed in the determination of processing parameters. In this section, four examples are presented to show how multicomponent phase diagram calculation can be readily useful for industrial applications.

Multicomponent Phase Diagram Calculation Examples

In example 1, phase diagram calculation is used to predict bulk metallic glass formability. In example 2, calculation is made for nickel alloys, focusing on the major concerns in the development of nickel-base superalloys, such as γ' solvus temperature, matrix (γ)/precipitate (γ') misfit, and the formation of deleterious topologically closed-packed phases. In example 3, phase diagram calculation is applied to commercial titanium alloys. The β-transus (temperature at which α starts to precipitate from β) and β-approach curve (fraction of β phase as a function of temperature) are calculated for a variety of titanium alloys and compared with experimental data. The effect of interstitial elements, such as oxygen, carbon, nitrogen, and hydrogen, on the β-transus temperatures is also discussed. In example 4, phase diagram calculation is used to predict dilation of iron alloys during phase transformations from austenite to ferrite and from austenite to martensite.

Example 1: Prediction of Bulk Metallic Glass Formability. Bulk metallic glass (BMG) materials exhibit unique properties, such as high strength, wear and corrosion resistance, castability, and fracture toughness. These properties make them extremely attractive as materials having great potential for practical applications in the areas of sports and luxury goods, electronics, medical, defense, and aerospace. Regardless of their importance, BMGs are mainly developed by experimental trial-and-error approaches, involving, in many cases, the production of hundreds to thousands of different alloy compositions (Ref 7). As pointed out by Gottschall (Ref 8), there remains an urgent need to either formulate a theoretical model or develop a computational approach for predicting families of alloy compositions with a greater tendency for glass formation. In recent years, Chang and his colleagues (Ref 9–11) have used multicomponent phase diagram calculation to identify the alloy compositions that have high BMG formability. The theoretical basis of their approach is that the reduced glass transition temperature (T_r), which is defined as the ratio of the glass formation temperature and the liquidus temperature, should be on the order of 0.6 or higher (Ref 12). According to this argument, BMGs are often found to form near deep eutectic invariant reactions where high T_r can be achieved. In addition, due to the low liquidus temperature at a deep eutectic region, the liquid shows exceptionally high viscosity, which further suppresses nucleation by inhibiting mass transport. It has been found that the formation

of BMG usually requires three or more elements, even though they do form in some binary systems. This implies that, in principle, if phase diagrams of multicomponent systems are known as functions of temperature and composition, potential alloy compositions that favor BMG formation can readily be identified. Unfortunately, accurate liquidus projections and phase diagrams of multicomponent systems, which are required to guide the exploration of BMGs, are rarely available, and this is true even for many ternary systems. Multicomponent phase diagram calculation thus becomes a desirable approach for efficiently scanning the vast number of possible compositions that may yield BMGs. The methodology is to develop a thermodynamic database for the system to be studied, from which the lowering liquidus valley can be calculated. Key experiments will then be arranged to validate the BMG formability for alloy compositions being identified by the calculation. Chang and his colleagues have explored the use of this approach to identify the BMGs for several systems (Ref 9–11).

To demonstrate the capability of this approach, two examples are given here. The first one is for the identification of BMG formers in the Zr-Ti-Cu-Ni system, and the second is for the effect of titanium on the glass formability of the Zr-Cu-Ni-Al alloys. In the first example (Ref 9), a thermodynamic database for the Zr-Ti-Cu-Ni system was developed. The liquidus projection was then calculated for this quaternary using Pandat software (Ref 6), and 18 five-phase invariant equilibria were identified that show low-lying liquidus temperatures. The compositions of the liquid phase at these invariant temperatures, shown as solid circles in Fig. 1, are found to be located in

two areas in the composition diagram, which is in excellent agreement with those identified experimentally (Ref 13). Should calculations be done first, a tremendous amount of experimental work can be saved. In the second example, the effect of titanium on the glass-forming ability of Zr-Cu-Ni-Al quaternary alloys was studied first by thermodynamic calculation and then validated by experimental approach (Ref 10). The quaternary $Zr_{56.28}Cu_{31.3}Ni_{4.0}Al_{8.5}$ alloy was first found to be a BMG-forming alloy based on the calculated low-lying liquidus surface of the quaternary Zr-Cu-Ni-Al system. By calculating the liquidus projection of the Zr-Cu-Ni-Al-Ti system, the liquidus temperature was found to decrease rapidly from the quaternary alloy $Zr_{56.28}Cu_{31.3}Ni_{4.0}Al_{8.5}$ (at.%) due to the addition of titanium. Further calculation of several isopleths reveals that the liquidus temperature decreased the fastest when zirconium was replaced by titanium and reached the minimum at 4.9 at.% Ti, as shown in Fig. 2. Alloy ingots of nominal compositions $Zr_{56.2-c}Ti_cCu_{31.3}Ni_{4.0}Al_{8.5}$ (c = 0 ~ 10.0 at.%) with different amounts of titanium were then prepared for experimental study of the glass-forming ability (GFA) of these alloys.

As shown in Fig. 3, the critical diameters of the amorphous rods formed increased from 6 mm for the base quaternary alloy to 14 mm for the quinary alloy with 4.9 at.% Ti and then decreased again. It is worth pointing out that 14 mm has reached the diameter limit of the casting mold used in the experiment. As indicated in Ref 10, an amorphous rod with larger diameter may be obtained if a larger casting mold was used. It is not difficult to imagine how many alloys must be studied in a five-component system to identify the alloys with high GFA if a

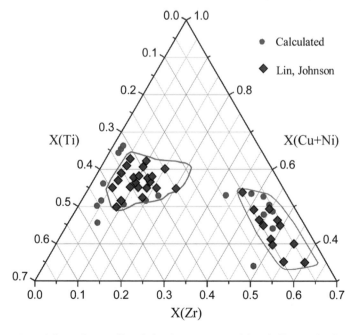

Fig. 1 Comparison of thermodynamically calculated compositions of liquid alloys at the five-phase invariant equilibria with those identified experimentally as bulk metallic glass formers by Lin and Johnson for the Zr-Cu-Ni-Ti system. Source: Ref 13

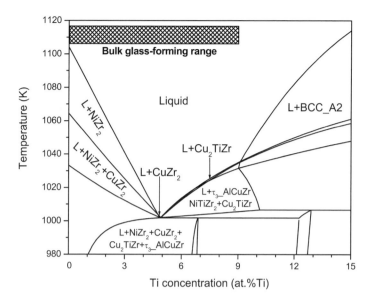

Fig. 2 Calculated isopleth of the quinary Al-Cu-Ni-Ti-Zr system expressed in terms of temperature as a function of titanium concentration (at.%). The composition of titanium at the origin is 0; corresponding to $Zr_{56.2}Cu_{31.3}Ni_{4.0}Al_{8.5}$, the compositions of copper, nickel, and aluminum are fixed at 31.3, 4.0, and 8.5 at.%, respectively. The shaded area denotes the experimentally observed bulk glass-forming range

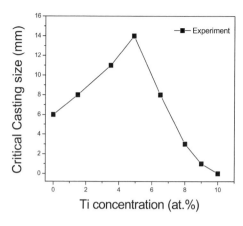

Fig. 3 Critical diameters of the cast glassy rods as a function of the titanium concentration in atomic percent

purely trial-and-error experimental method is used. These two examples thus demonstrate the great power of thermodynamic calculation in quickly locating the alloy chemistries with a high potential of GFA and guiding the experimental study. By using the same approach, Chang et al. successfully developed copper-rich BMGs by adding a small amount of yttrium into the Cu-Zr-Ti system, and as-cast amorphous rods as big as 10 mm in diameter were obtained (Ref 14).

Example 2: Design and Development of Nickel-Base Superalloys. Nickel-base superalloys are a unique class of metallic materials that possess an exceptional combination of high-temperature strength, toughness, and resistance to degradation under harsh operating environment. These materials are widely used in aircraft and power-generation turbines, rocket engines, and other challenging environments, including nuclear power and chemical processing plants (Ref 15). To improve the performance of nickel-base superalloys, alloy composition and processing conditions must be carefully optimized to promote the formation of desired phases and microstructure and to avoid the formation of deleterious phases. Multicomponent phase diagram calculation therefore plays an important role in nickel-base superalloy design. With this approach, not only equilibrium phases but also phase amounts and phase transformation temperatures can be predicted, given alloy chemistry. As an example, the liquidus, solidus, and γ' solvus for a number of René N6 alloys in the specification range are calculated and compared with those determined by experiments. As shown in Fig. 4, the liquidus varies in the temperature range of 1660 to 1700 K, the solidus of 1615 to 1675 K, and the γ' solvus of 1525 to 1585 K. Obviously,

even for the same group of alloys whose compositions are all within the specification range of René N6, their phase transformation temperatures can differ significantly, depending strongly on the alloy chemistry. It is important that these temperatures can be calculated, because they are critical parameters needed for the determination of processing conditions.

Another example is shown in Fig. 5, which compares the calculated and experimental determined volume percent of γ' phase for a number of nickel-base superalloys containing aluminum, chromium, molybdenum, titanium, and tungsten. It is known that the two major phases of nickel-base superalloys are γ and γ', and the presence of a high volume fraction of γ' is the key to strengthening. This figure clearly shows that a high volume fraction of γ' can be obtained through the optimization of alloy composition. Both calculations are carried out using PanNi thermodynamic database (Ref 17), and the experimental data are from Ritzert et al. (Ref 16) for the phase transformation temperatures and Dreshfield and Wallace (Ref 18) for the γ' phase fraction. Good agreement is obtained in both figures.

Because the matrix phase (γ) and the precipitate phase (γ') of nickel-base superalloys are crystallographically coherent, the precipitate-matrix misfit results in internal stresses that influence the shape of the precipitates and thus the final mechanical properties. The precipitate-matrix misfit, δ, can be calculated as (Ref 15):

$$\delta = \frac{a_{\gamma'} - a_{\gamma}}{0.5 \cdot (a_{\gamma'} + a_{\gamma})} \qquad \text{(Eq 1)}$$

where $a_{\gamma'}$ and a_{γ} are the lattice parameters of the γ' and γ phases, respectively. Obviously, the

precipitate-matrix misfit is determined by the lattice parameters of the γ' and γ phases, which depend on the compositions of these two phases. Elemental partitioning is therefore an important alloy-design consideration, because the compositions of the constituent phases will directly impact both the mechanical and environmental characteristics of the alloy (Ref 15). With the computational tool for multicomponent phase diagram calculation, the phase composition and elemental partitioning can be readily calculated if a reliable thermodynamic database is available for the alloy system. Figure 6(a) compares the calculated and experimentally determined aluminum contents in the γ and γ' phases for a number of nickel-base superalloys, while Fig. 6(b–e) are for those of cobalt, chromium, molybdenum, and titanium. All calculations are carried out using the PanNi database (Ref 17), and experimental data from the literature (Ref 18–25) are used for comparison. In all cases, there is good agreement between the calculated and experimental data.

In general, refractory alloying elements, such as molybdenum, niobium, rhenium, and tungsten, are added for solid-solution strengthening of the γ phase and to provide high-temperature creep resistance. A major concern following this is the formation of topologically closed-packed (tcp) phases. The tcp phases are typically rich in refractory alloying elements and are detrimental because they deplete strengthening elements from the matrix phase. Examples of tcp phases include the orthorhombic P phase, the tetragonal σ phase, the rhombohedral R phase, and the rhombohedral μ phase (Ref 26–30). Being able to predict the correlation between the stabilities of the tcp phases and the alloy composition, thereby avoiding the formation of deleterious phases by wisely adjusting the alloy composition, is of critical importance in the design of new alloys and the modification of specification ranges for existing alloys. The CALPHAD approach has been used for such purposes, and one example is shown in Fig. 7. In this example, the effect

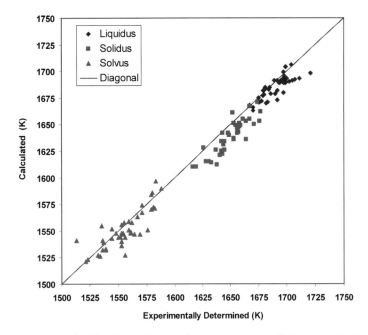

Fig. 4 Comparison of calculated liquidus, solidus, and γ′ solvus temperatures with those determined by experiment. Source: Ref 16

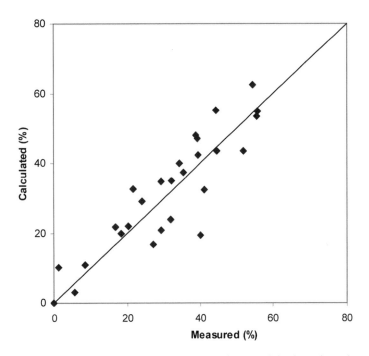

Fig. 5 Comparison of the calculated and experimentally determined percent of γ′ in the γ+γ′ two-phase mixture for a number of Ni-Al-Cr-Mo-Ti-W alloys. Calculations were made by using the PanNi database (Ref 17), and the experimental data are from Ref 18

of rhenium content on the stability of σ is presented for the CMSX-4 alloy. It is seen from this figure that the σ precipitation temperature and its amount strongly depend on the content of rhenium in the alloy, and the higher the rhenium content, the more σ will form. The predicted trend is correct, while the accuracy of the phase transformation temperature and phase amount depends on the quality of the

thermodynamic descriptions for the tcp phases. This point is discussed in detail in the section on the limitations of the approach at the end of this article.

Example 3: Prediction of β-Transus and β-Approach Curves of Commercial Titanium Alloys. Pure titanium has two allotropic forms: the high-temperature body-centered cubic (β) structure and the low-temperature hexagonal

close-packed (α) structure. The β transus, defined as the temperature at which α starts to form from β during solidification, is 883 °C for pure titanium. However, this temperature can vary by several hundred degrees due to the addition of alloying elements. Elements such as aluminum, gallium, germanium, oxygen, carbon, and nitrogen are referred to as α stabilizers because they increase the β transus temperature, while vanadium, iron, molybdenum, chromium, niobium, nickel, copper, tantalum, and hydrogen are β stabilizers because they decrease the β transus. Technical multicomponent titanium alloys are classified as α, β, and α-β alloys, depending on the relative amounts of α and β stabilizers in the alloy. Within the last category are the subclasses near-α and near-β, referring to alloys whose compositions place them near the α/(α+β) or (α+β)/β phase boundaries, respectively. According to different applications, alloy chemistry and heat treatment condition must be carefully optimized to achieve the desired microstructure and mechanical properties. The β-transus temperature is an important reference parameter used by metallurgical engineers in the selection of specified heat treatment conditions for titanium alloys. Current aerospace industrial practice calls for measurement of the β transus on every heat of titanium material, which is costly and time-consuming. These operations and their associate costs can be significantly reduced with the help of a thermodynamic modeling tool, from which the β-transus temperature can be easily calculated, given alloy chemistry.

In the following, examples are presented to apply thermodynamic calculation to two α-β alloys: Ti-6Al-4V (Ti-64) and Ti-6Al-2Sn-4Zr-6Mo (Ti-6246). It is noteworthy to point out that the preceding chemistries are the nominal compositions (wt%). For commercial alloys, the actual composition in each alloy varies. In addition, it is inevitable that commercial alloys always contain minute amounts of impurities, such as iron, silicon, carbon, oxygen, nitrogen, and hydrogen. Ti-64 is the most widely used titanium alloy in the world, with 80% of its usage going toward the aerospace industry. The β-transus temperatures of Ti-64 range from 950 to 1050 °C, depending on the amounts of components and interstitial elements, such as oxygen, carbon, nitrogen, and hydrogen. This temperature can be readily calculated by the CALPHAD approach if the thermodynamic database is available for multicomponent titanium alloys. In this study, PanTi (Ref 31) is used to calculate the β transus for a large number of Ti-64 heats with slightly different chemistries, and Fig. 8 compares the calculated and experimentally measured temperatures (Ref 32). Good agreement between the calculation and measurement is obtained, with an average difference of 5.2 °C. To understand the difference between the predicted values and the experimental data, sensitivity analysis is carried out for Ti-64. It is found that

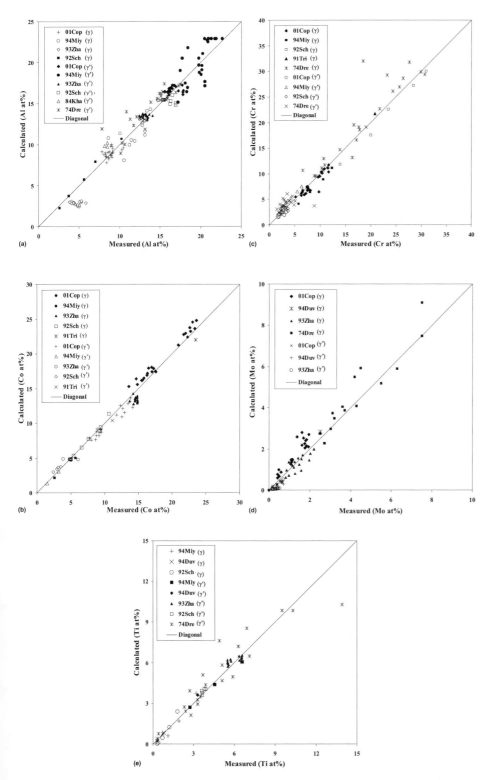

Fig. 6 Comparison between the calculated and measured equilibrium compositions of a variety of elements in γ and γ'. Data sources: 01Cop (γ) and 01Cop (γ'), Ref 25; 94Miy (γ) and 94Miy (γ'), Ref 24; 93Zha (γ) and 93Zha (γ'), Ref 22; 92Sch (γ) and 92Sch (γ'), Ref 21; 84Kha (γ'), Ref 19; 74Dre (γ) and 74Dre (γ'), Ref 18; 91Tri (γ) and 91Tri (γ'), Ref 20; and 94Duv (γ) and 94Duv (γ'), Ref 23

the calculated β transus is very sensitive to the interstitial impurities, such as oxygen, carbon, nitrogen, and hydrogen, as shown in Table 1. In this table, the nominal chemistry of Ti-64 is given as Ti-6Al-4V-0.12O-0.01C-0.01N-0.005H (wt%). It is seen that 10 weight parts per million (wppm) of additional interstitial impurities (oxygen, carbon, nitrogen, hydrogen) will change the β transus of Ti-64 alloy by 0.27, 0.49, 1.03, and −1.08 °C, respectively. The small variation in the composition (ΔC) of these interstitial impurities, measured in wppm, that changes the β transus of Ti-64 by 5 °C is listed in the last column.

As an α-β alloy, the relative amount of α and β in the microstructure plays a key role in the determination of the mechanical properties of the final product. With multicomponent phase equilibrium calculation, the volume fractions of α and β can be calculated as a function of temperature, which provides valuable information for the selection of proper heat treatment conditions to achieve the desired microstructure. Figure 9 shows such an example, in which the α-approach curve, that is, the fraction of α phase as a function of temperature, is calculated and compared with the experimental data for one Ti-64 alloy. In addition to Ti-64, the thermodynamic modeling tool is also applied to other titanium alloys. Figure 10 shows the calculated β-approach curve, that is, the fraction of β phase as a function of temperature, of one Ti-6242 alloy, while Fig. 11 calculates the distribution of aluminum and molybdenum in α and β for the same alloy. Both calculations agree very well with the experimental data from Semiatin (Ref 34). Calculations shown in Fig. 9 to 11 provide valuable guidance in the optimization of processing conditions of titanium alloys, and the good agreement between calculation and experimentation suggests that the PanTi database (Ref 31) can be used for commercial titanium alloys beyond Ti-64.

Example 4: Calculation of Dilation during Phase Transformation of Austenite to Ferrite and Austenite to Martensite. Ferrous alloys such as cast iron and stainless steel that contain many alloy elements have been used worldwide as structural materials. Even though they are considered to be mature materials, there is a continuing demand to further improve their properties, with simultaneous cost reduction. Phase diagrams of this class of multicomponent materials provide the guide for processing optimization, leading ultimately to the improvement of their performance in service. For example, it is known that ferrous alloys undergo dimensional change, referred to as dilation, when subjected to heat treatment. The dilation could cause distortion and lead to cracking of a material component. The origin of this dilation is the result of phase transformation from austenite to ferrite as well as austenite to martensite during heat treatment. Experimental determination of dilation is not only tedious but also very challenging, even for an experienced experimentalist, because the accuracy of the measured data depends strongly on the cooling rate. This is particularly true when a multicomponent alloy is considered, because the phase transformation temperatures are usually unknown. On the contrary, the computational approach is more efficient and provides more consistent results. The following example demonstrates how computational thermodynamics can be used to predict

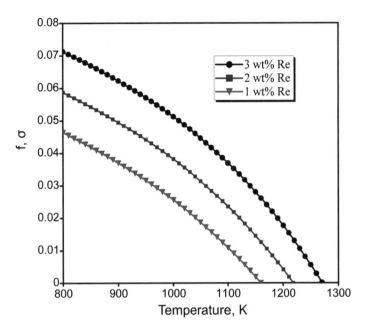

Fig. 7 Effect of rhenium content on the σ-phase precipitation temperature and its amount at different temperatures, as calculated by the PanNi database. Source: Ref 17

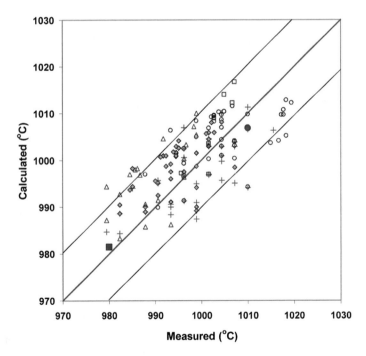

Fig. 8 Comparison of calculated and experimentally determined β transus for a number of Ti-64 alloys with slightly different chemistries. The calculation was carried out using the PanTi database (Ref 31), and the experimental data are from Ref 32

transformation from austenite to ferrite, the fractions of austenite and ferrite at a certain temperature and composition can be readily obtained by thermodynamic calculation if a reliable thermodynamic database is available for ferrous alloys. The PanFe database (Ref 35) is suitable for such a purpose. As shown in Fig. 12, the calculated values of f_V for austenite are in good accordance with known experimental data for a variety of stainless alloys, an indication of the reliability of the PanFe database. These data, when combined with the molar volumes of these two phases, can then be used to calculate the dilation for the phase transformation from austenite to ferrite.

On the other hand, the calculation for the transformation from austenite to martensite is more challenging. Martensite, a supersaturated ferrite with the composition of the parent austenite, is a metastable phase. The total Gibbs energy change for transforming austenite to martensite is the sum of the bulk Gibbs energy difference due to the transformation from austenite to ferrite, the strain energy, the surface energy, and the interfacial work. The first part can be calculated easily because the Gibbs energies of the austenite and ferrite are well developed in the thermodynamic database, while the other three parts must be estimated by empirical equations, such as those proposed by Ghosh and Olson (Ref 43). The martensite transformation temperature, the fraction of martensite, and the retained austenite are usually calculated from semiempirical and empirical equations, such as the one developed by Koistinen and Marburger (Ref 44). However, with the thermodynamic calculation, when the Gibbs energy for the martensite is formulated, the martensite transformation temperature for a given alloy and the fractions of martensite and austenite as a function of temperature can be calculated. These quantities, in combination with the molar volume of each phase, can be used to calculate the dilation due to the phase transformation from austenite to martensite. As an example, the dilation of steel 6-6-2 is calculated and compared with the experimental data from Gordon and Cohen (Ref 45), as shown in Fig. 13. Steel 6-6-2 is a multicomponent alloy with a composition of Fe-0.81C-0.24Mn-0.26Si-5.95W-4.10Cr-1.64V-4.69Mo (wt%). The dashed line and solid squares represent the calculated and experimental determined dilation for the transformation from austenite to ferrite, respectively, while the solid line and the solid circles are those for the transformation from austenite to martensite. By integrating the calculated dilation data with mechanical property prediction models, the strain, distortion, cracking, residual stress, and related properties can be predicted.

Integration with Kinetic and Microstructural Evolution Models

As presented in the last section, knowledge of phase equilibria is needed to understand materials behavior. This section demonstrates how integration of this knowledge with kinetic

the degree of dilation of a chosen ferrous alloy prior to processing this material.

The dilation can be calculated using the following equation:

$$\frac{\Delta l}{l} = \frac{V_i - V_A}{3V_A} \qquad \text{(Eq 2)}$$

where l is the length, V_i is the instantaneous volume, and V_A is the volume of austenite at a

specified temperature. The instantaneous volume, V_i, is calculated from the phase fraction and molar volume of each phase involved in the phase transformation:

$$V_i = \sum_{\varphi} f_V^{\varphi} \cdot V_m^{\varphi} \qquad \text{(Eq 3)}$$

where f_V^{φ} represents the volume fraction, and V_m^{φ} is the molar volume of the φ phase. For the phase

Table 1 Effects of 10wppm interstitial elements on the beta transus of Ti-64

Alloy chemistry (wt%)	Beta transus (°C)	ΔT (°C) (compare to Ti-64)	ΔC (wppm) needs to change ΔT by 5°C
Ti-64: Ti-6Al-4V-0.12O-0.01C-0.01N-0.005H	982.59		
Ti-64+10wppm O	982.86	0.27	185.2
Ti-64+10wppm C	983.08	0.49	102.0
Ti-64+10wppm N	983.62	1.03	48.5
Ti-64+10wppm H	981.51	−1.08	46.3

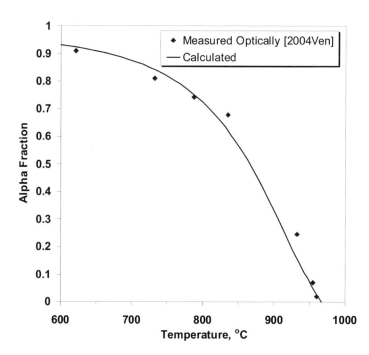

Fig. 9 Plot of the phase fraction of α as a function of temperature for one Ti-64 alloy. The line was calculated by PanTi (Ref 31), and the experimental data are from Ref 33

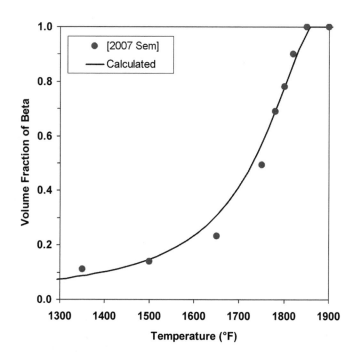

Fig. 10 Plot of the phase fraction of β as a function of temperature for one Ti-6242 alloy. The line was calculated by PanTi (Ref 31), and the experimental data are from Ref 34

and microstructural evolution models greatly enhances the power of the CALPHAD approach in materials design and processing development.

Phase Diagram Calculation Examples

In example 5, thermodynamic calculation coupled with a simple solidification model, that is, the Scheil model (Ref 46), is used to explain the liquation cracking of aluminum welds. In example 6, phase diagram calculation is integrated with kinetic models to study the nucleation, growth, and coarsening of precipitates of nickel-base superalloys. In example 7, thermodynamic calculation is coupled with a phase-field model to predict the microstructural evolution in a Ni-Al-Cr γ+β/γ diffusion couple. Finally, in example 8, thermodynamic modeling is integrated with a microscopic and cellular automaton model to simulate the microstructure and microsegregation of aluminum alloys during solidification.

Example 5: Prediction of Liquation Cracking of Aluminum Welds. In this example, integration of thermodynamic calculation with the multicomponent Scheil model (Ref 46) is used for the study of liquation cracking of aluminum welds. Aluminum welds are susceptible to liquation cracking in the partially melted zone. This is the region immediately outside the fusion zone where liquation occurs due to overheating during welding. Metzger (Ref 47) carried out an extensive investigation on liquation cracking of aluminum welds for alloy 6061 and found that the formation of this type of cracking is closely related to the fillers used. In other words, elimination of liquation cracking in aluminum welds for 6061 can be realized by a wise selection of filler materials. Subsequently, several other researchers had extended his investigation to other aluminum alloys, such as 6063 and 6082, and found similar results (Ref 48–52).

More recently, Huang and Kou (Ref 53, 54) proposed a mechanism, as shown in Fig. 14, to explain the cause of liquation cracking. The top portion of this figure shows a schematic diagram of joining two pieces of aluminum metals, with a weld pool in between. The partially melted zone (PMZ) is a portion of the base metal that experiences partial melting during welding, and the weld pool is a mixture of the base metal and the filler material, with a dilution ratio of 65 to 35. The PMZ is enlarged in the lower part of Fig. 14, in which the interface between the base metal and the weld pool, referred to as the fusion boundary, is clearly shown. Huang and Kou postulated that if less liquid is solidified in the weld pool than that in the PMZ during the later stage of solidification, liquation cracking should not occur. For the reversed case, more liquid is solidified in the weld pool during welding and is pulled away from the PMZ, which will cause cracking. The fraction of solid formed in the PMZ and the weld pool during solidification can be readily calculated using the multicomponent Scheil model (Ref 46), a model for simulating the solidification process. This model assumes

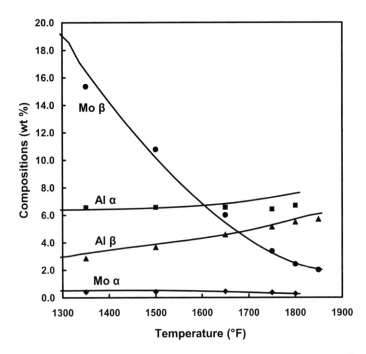

Fig. 11 Distribution of aluminum and molybdenum in the α and β phases for the sameTi-6242 alloy as in Fig. 10. The lines were calculated from PanTi (Ref 31), and the experimental data are from Ref 34

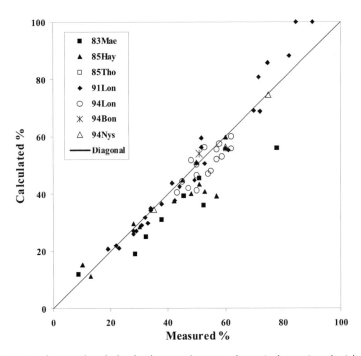

Fig. 12 Comparison between the calculated and measured amounts of austenite for a variety of stainless steels. Data sources: 83Mae, Ref 36; 85Hay, Ref 37; 85Tho, Ref 38; 91Lon, Ref 39; 94 Lon, Ref 40; 94Bon, Ref 41; and 94Nys, Ref 42

phase equilibrium at the liquid-solid interface, no back diffusion in the solid, and complete mixing in the liquid. With these assumptions, only the phase equilibrium information is needed as input for carrying out solidification simulation. By using this model, which has been implemented into the Pandat software

(Ref 6), and the PanAl (Ref 55) thermodynamic database, Huang and Kou (Ref 53, 54) were able to calculate the fraction of solid as a function of temperature for the base metal 6061 in the PMZ and two weld metals. The two welds are made of base metal with two filler materials, 4043 (Al-5Si) and 5356 (Al-5Mg),

respectively, and the weld compositions can be calculated by the equations described by Huang and Kou (Ref 53, 54). The calculated temperature-solid fraction $(T - f_S)$ curves are shown in Fig. 15. As shown in this figure, the calculated f_S for the weld metal made of filler metal 5356 is larger than that of the base metal during the later stage of solidification, which indicates a liquation cracking according to the postulation proposed by Huang and Kou (Ref 53, 54). Contrarily, no cracking will form if filler metal 4043 is used. This is exactly what they have observed, as shown in Fig. 16.

To further extend their research, Kou and his colleagues have applied this approach to other aluminum alloys, such as 7075 and 2024 (Ref 56). The crack susceptibility of welds made with various filler metals was successfully predicted and therefore reduced by adjusting the filler metal composition. In addition to aluminum alloys, Kou and his colleagues have also explored the use of this approach to other metal systems. For example, they have studied the liquation of magnesium alloys in friction-stir spot welding using similar calculations (Ref 57). This example demonstrates that a thermodynamic calculation coupled with a simple kinetic model, such as the multicomponent Scheil model for solidification simulation, can provide the welding industry with an effective approach to understand and therefore reduce liquation cracking during welding.

Example 6: Precipitation Simulation. Precipitation hardening is one of the most important strengthening mechanisms for the nickel-base superalloys. Its success relies on the formation and distribution of the strengthening phases, such as $L1_2$ $γ'$ and DO_{22} $γ''$, which are mostly coherent or semicoherent with the face-centered cubic matrix phase. An optimum size and amount of these strengthening phases is obtained by the deliberate selection of alloy chemistry and heat treatment scheme. Commercial nickel-base superalloys usually contain more than ten alloying elements. These multicomponent and multiphase systems make the microstructures very complicated, and a thorough understanding of the precipitation microstructures by a purely experimental trial-and-error approach is prohibitively expensive. Computational tools, developed by combining reliable experimental evidence with reasonable theoretical basis, thus become an increasingly important substitute for the experiment at a lower cost.

In this example, thermodynamic calculation is integrated with kinetic models to simultaneously treat the three concomitant events: nucleation, growth/dissolution, and coarsening. Thermodynamic calculation, which provides a chemical driving force for each event, is integrated with two kinetic models; one is the model based on the Kampmann-Wagner model in the numerical framework (KWN) (Ref 58), and the other is the fast-acting model (Ref 59). The same set of equations is used to describe the nucleation and growth of these

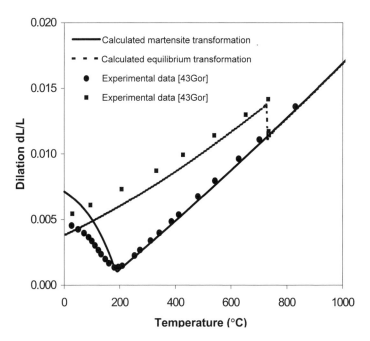

Fig. 13 Comparison between the calculated dilation versus temperature and the experimental data (Ref 45) for steel 6-6-2, a high-speed tool steel

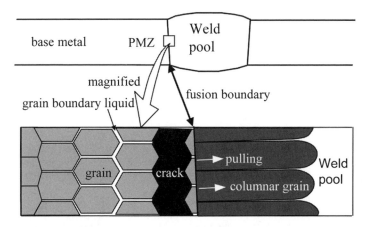

Fig. 14 Schematic diagram to illustrate the mechanism of liquation cracking in full-penetration aluminum welds. PMZ, partially melted zone. Source: Ref 53

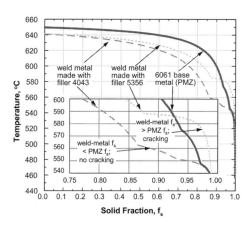

Fig. 15 Calculated solid fraction (f_S) as a function of temperature for the base aluminum alloy (6061) in the partially melted zone (PMZ) and two weld metals composed of 65% base metal and 35% filler 4043 (Al-5Si) or 5356 (Al-5Mg). The calculation is made by the Pandat and PanAl thermodynamic databases

two models. However, in the KWN model, the continuous particle size distribution (PSD) is divided into a large number of size classes, and the program takes a simulation step at every sample time hit. At each simulation step, the number of new particles is first calculated using classical nucleation theory, and then these new particles are allocated to the appropriate size class. From the fast-acting model, only the average size of the particles is calculated.

The kinetic models were first used to simulate the precipitation kinetics of the Ni-14Al (at.%) alloy isothermally annealed at 550 °C, which had been previously studied by Kampmann and Wagner (Ref 58). The simulation results are shown in Fig. 17. While Fig. 17(a) plots the evolution of average γ' particle size with aging time calculated by the KWN and fast-acting models, Fig. 17(b) and (c) plot the

KWN-predicted evolution of particle number density and supersaturation, respectively. The experimental data (Ref 60) are also plotted on these figures for comparison. As can be seen, the predictions and the measurements are in reasonable agreement. The KWN-predicted γ' size distributions for the alloy after aging at 550 °C for 21, 217, and 3720 min are shown in Fig. 17(d). These distributions are plotted in terms of particle frequency, which is computed by normalizing the total number of particles to be one. At the early stages of the precipitation, the nucleation process is dominant, and the number density of γ' particles increases rapidly. Therefore, the PSD is characterized by a sharp peak around the critical nucleus size, as shown in Fig. 17(d) at 21 min. At the late stage, on the other hand, the transformation falls into the coarsening regime, and the PSD is then

flattened, as indicated in the case of 3720 min. It is interesting to point out that the PSD, at an intermediate stage of 217 min, displays a double peak. This phenomenon, however, is consistent with the previous findings from both experimental observation (Ref 61) and model prediction (Ref 62).

Precipitation simulation is then applied to the nickel-base alloy 718. Three major precipitate phases, γ', γ'', and δ, are considered. According to the experimental information (Ref 63–65), γ' particles are treated as spherical shaped, and γ'' particles are treated as lenslike shaped with a constant aspect ratio (thickness/length ratio) of 1/9 to ~1/10. Both phases are considered to nucleate homogeneously within the matrix phase. Two types of δ phase are treated simultaneously; one is for those forming at the grain boundaries, which is treated as lens-shaped with constant aspect ratio, and the other is for those transforming from the metastable γ'', which is treated as plate with time-varying aspect ratio.

The calculated temporal evolutions of the average size for γ'' and γ' have been compared with experimental data (Ref 64) in Fig. 18. The size of γ'' is characterized by the particle length, while that of γ' sphere is characterized by its diameter. The heat treatment is isothermal annealing at three temperatures: 700, 725, and 750 °C. Results for both phases show an excellent agreement at the medium temperature (725 °C). The calculated size change of γ' phase (Fig. 18b) is obviously in a better fit than that of γ'' (Fig. 18a). The reason is partially due to the varying aspect ratios of the γ'' lenses, which are observed to be dependent on alloy composition (Ref 65). Most likely, the particle length also depends on temperature and annealing time, considering the change of the coherency between the γ'' particle and γ matrix. The approximation of constant aspect ratio may therefore cause some artifacts. Despite

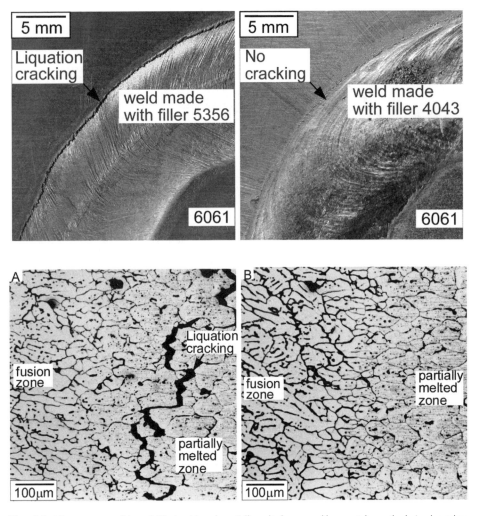

Fig. 16 Microstructures of the solidified weld pool, partially melted zone, and base metal near the fusion boundary. Top figure, low magnification; bottom figure, higher magnification

the discrepancies, however, the overall agreement is still satisfactory.

In addition to the particle size, the model is used to calculate the time-temperature transformation diagram of alloy 718 (Fig. 19). The alloy composition is chosen from that given by Han et al. (Ref 64). Only the starting curves, with a criterion of 1% of volume fraction of the corresponding phases, are plotted in the figure, although the finish curves can be readily calculated.

Example 7: Integration with Phase-Field Model for the Simulation of Microstructural Evolution. The phase-field method is based on the generalized diffusion equation (often referred to as the Cahn-Hilliard equation, Ref 66) that characterizes diffusion with chemical potential gradient and chemical mobility and the time-dependent Ginzburg-Landau equation (Ref 67) that characterizes structural changes accompanying a phase transformation and other materials processes. The phase-field method is well known for its ability to handle arbitrary microstructures consisting of diffusionally and elastically interacting precipitates and

to account self-consistency for topological changes, such as particle coalescence. While quantitative modeling at real length and time scales requires that the fundamental model inputs be linked to multicomponent thermodynamic and mobility databases, special techniques must be developed to overcome the intrinsic length-scale limit. A key step involved in phase-field modeling of microstructural evolution is to construct the total Gibbs energy as a function of concentration, order parameter, and their gradients in a chemically and structurally nonuniform system. Nonequilibrium Gibbs energy is needed by phase-field modeling to simulate the developing process and microstructural evolution from nonequilibrium to equilibrium. Even though it is straightforward to obtain the equilibrium Gibbs energy of each individual phase from the nonequilibrium one by minimizing it with respect to the order parameter, the opposite is not true. There are two approaches to construct the nonequilibrium free energy from the given equilibrium one (Ref 68); one employs a physical order parameter and the other a phenomenological order

parameter. The former has been applied to the study of γ/γ' transformation in Ni-Al-Cr (Ref 69) and the latter to the α/β transformation in Ti-Al-V (Ref 68).

Even though, theoretically, the phase-field method can be applied to multicomponent, multiphase systems, practically it is not an easy task. In the previous studies (Ref 68, 69), the equilibrium Gibbs energy functions are directly implemented in the phase-field modeling code for calculation efficiency. However, this hindered the application of phase-field modeling to multicomponent alloys, because construction of a nonequilibrium Gibbs energy hypersurface in a multidimensional space using those thermodynamic equilibrium Gibbs energies is not trivial. To bypass this obstacle, Zhang et al. (Ref 70) tried to describe multicomponent Ti-64 alloy using a pseudoternary system. Recently, a more robust approach was developed that directly couples the phase-field method with multicomponent thermodynamic calculation. In this approach, the phase-field model obtains the necessary thermodynamic inputs directly from the thermodynamic calculation engine, PanEngine (Ref 71), in each simulation step without the need for tackling with the Gibbs energy functions. It is therefore suitable for the simulation of multicomponent alloys. Figure 20 shows the microstructural evolution in a $\gamma+\beta/\gamma$ diffusion couple during isothermal aging at 1200 °C for 100 h. The top figure is the experimentally determined micrograph, while the bottom one is from phase-field simulation. In this example, the matrix phase of the left half ($\gamma+\beta$) is β (the dark background); therefore, it does not form a continuous matrix phase with the right half (γ). The microstructural changes during the interdiffusion process, including precipitate coalescence and coarsening and especially the motion of the original boundary, are clearly demonstrated in the figure. The ternary system is used in the simulation for the purpose of direct comparison with the available experimental micrograph, although the program is readily applicable to multicomponent systems.

Example 8: Integration with Microscopic and Cellular Automaton Models for Solidification Simulation. The microstructure and microsegregation developed during solidification are of particular importance in the determination of subsequent thermomechanical and heat treatment conditions and the performance of as-cast materials. Solidification simulation of multicomponent alloy systems has become the primary focus of model development, due to the multicomponent nature of commercial alloys. In parallel to the development of rigorous analytical models, various numerical models have been developed to describe the solidification process. For example, one-dimensional (1-D) microscopic models were first developed to analyze the microsegregation and solidification path in multicomponent alloys (Ref 73–75). In the models, the dendrite morphology is usually assumed to be plate, cylinder, or sphere having a length

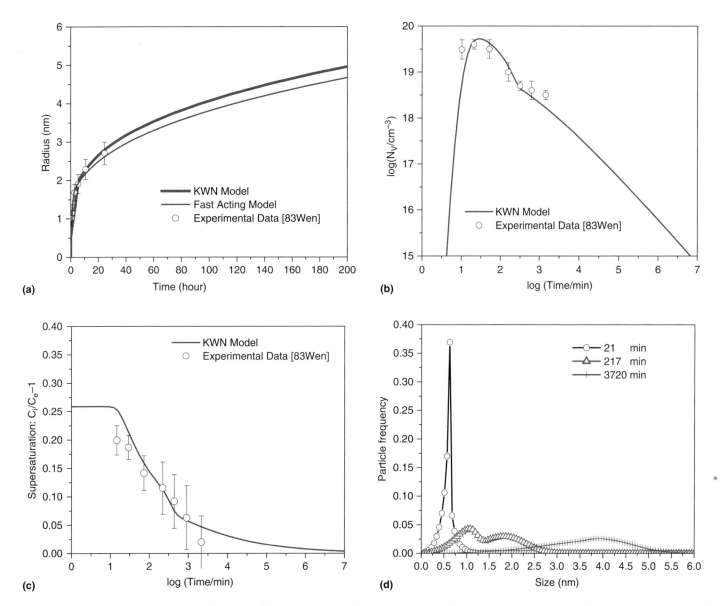

Fig. 17 Comparison of precipitation parameters for a Ni-14Al (at, %) alloy between the predicted values and those determined experimentally. KWN, Kampmann-Wagner model in the numerical framework; 83Wen, Ref 60. (a) Evolution of the average γ′ particle size. (b) Evolution of the γ′ number density. (c) Evolution of supersaturation with time. (d) KWN-predicted particle size distributions for γ′ particles at 21, 217, and 3720 min, respectively

scale of half the secondary dendrite arm spacing. These models are based on a modified Scheil model that assumes complete diffusion in the liquid, while solid back-diffusion, undercooling, and dendrite arm coarsening are also considered. The two key thermodynamic parameters needed for these models are the solute partitioning coefficients and the liquidus slope. In the past, they were usually treated as constants, which are considered to be suitable for simple binary alloy systems. These values, however, may vary significantly with temperature and composition in a multicomponent system. Accordingly, it is necessary to couple the thermodynamic and phase equilibrium calculations with these models for the simulation of microstructure evolution and microsegregation of multicomponent alloys

during solidification. Xie and Yan et al. (Ref 73–75) have coupled their models with the thermodynamic calculation engine PanEngine (Ref 71) and applied these models to the investigation of microsegregation of ternary and higher-order alloys such as aluminum 7050, which contains eleven components and forms six different intermetallic phases in subsequent eutectic reactions during solidification. The predicted solute distributions in the dendrite arms agree very well with the experimentally measured data (Ref 74).

In addition to the 1-D diffusion models, thermodynamic calculation can also be integrated with two-dimensional (2-D) microstructure models, such as the phase-field model discussed in example 7. In this example, a 2-D modified cellular automaton model is coupled with PanEngine

(Ref 71) for the simulation of microstructure evolution and microsegregation of aluminum-rich ternary alloys. This work has been published (Ref 76), and a simulated result is shown in Fig. 21. This figure presents the simulated equiaxed dendrite morphology and solute fields of copper and magnesium for an Al-15Cu-1Mg (wt%) alloy solidified at a cooling rate of 10 K/s. The solute distribution and microsegregation is clearly represented in this figure (produced by the model in color), which is not possible by thermodynamic calculation alone. Yet, the fundamental inputs, such as equilibrium partitioning of elements in phases and the liquidus slope, are provided by phase diagram calculation. Therefore, this example again demonstrates the power of the integrated computational approach.

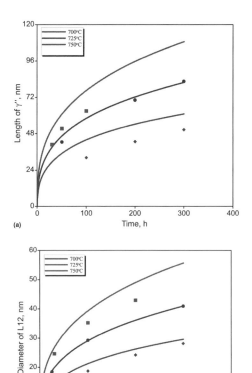

(a)

(b)

Fig. 18 Average size changes of (a) γ'' and (b) γ' phases under temperatures of 700, 725, and 750 °C for nickel alloy 718. The experimental data are taken from Ref 64

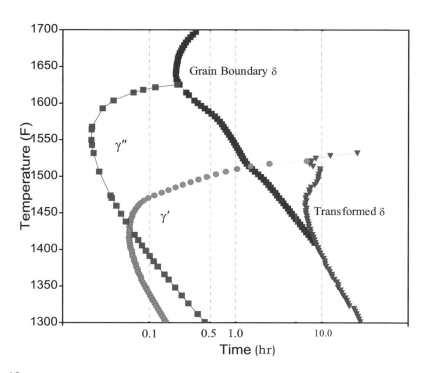

Fig. 19 Calculated time-temperature transformation diagram for alloy 718

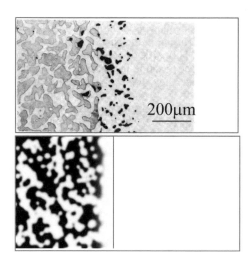

Fig. 20 Microstructural evolution in a $\gamma+\beta$ (left) /γ (right) diffusion couple of Ni-Al-Cr annealed at 1200 °C for 100 h. The top figure is a micrograph (Ref 72), while the bottom one is obtained by phase-field simulation of this work

Limitations of the CALPHAD Approach

The essence of the CALPHAD method is that it provides an efficient approach to convert the experimental information of lower-order systems to self-consistent thermodynamic model parameters, on the basis of which a thermodynamic database of a multicomponent system can be established. As discussed in detail in "Phase Equilibria and Phase Diagram Modeling" in *Fundamentals of Modeling for Metals Processing*, Volume 22A, *ASM Handbook*, 2009, only a limited amount of experimental effort is needed to confirm the reliability of the database thus obtained before it can be used for industrial applications. There is no doubt of the potential power of the CALPHAD approach to save a tremendous amount of experimental work. However, like every other modeling approach, this approach has its limitations, especially when handling new phases that appear in the higher-order systems but not in the constituent lower-order systems. One example is the topologically closed-packed (tcp) phases in the technologically important nickel-base superalloys. These phases, including *P*, σ, μ, and *R*, are typically rich in refractory alloying elements and possess complex crystal structures characterized by close-packed layers of atoms. The tcp phases are detrimental because they deplete strengthening elements from the matrix phase and serve as crack-initiation sites during cyclic loading. If the stabilities of these tcp phases can be accurately predicted as a function of temperature-given alloy chemistry, formation of the deleterious tcp phases can be avoided by deliberately adjusting the alloy chemistry. Thermodynamic descriptions of the tcp phases in the current nickel database are based on the available but limited experimental information in the binaries and some ternaries where they are stable, while their Gibbs energies in other subsystems are estimated. One difficulty in the development of accurate thermodynamic descriptions for the tcp phases is that they may form in the middle of a ternary system similar to a new ternary phase while not stable in any of the three constituent binaries. One typical example is the Ni-Cr-Mo ternary system, in which σ and *P* are stable in the ternary but not any of the constituent binaries, as shown in Fig. 22. In such a situation, the experimentally well-established phase relationship in the ternary is used to optimize the model parameters in the three constituent binaries, as in the case of Ni-Cr-Mo. In reality, however, the phase stabilities of these tcp phases are not known in many subsystems. Their Gibbs energies thus developed for the multicomponent system should therefore be used with caution. First-principles-calculated enthalpy values may be used to improve the Gibbs energy descriptions of these phases, but it is not an easy task to obtain these values in every constituent subsystem due to the complicated structure of the tcp phases. It is also found that the stabilities of these tcp phases are comparable, which means a small difference in the Gibbs energy may lead to a totally different phase relationship. This makes the modeling of these tcp phases even more challenging. The status of the current nickel database is that it exhibits a lack of capability to accurately predict the stabilities of the tcp phases, but it does provide reasonable trends in most cases, such as the example shown in Fig. 7. Although more basic research is needed in this area for additional improvements of the thermodynamic descriptions of the topological phases,

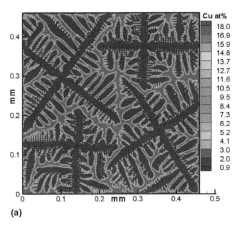
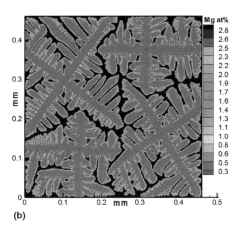

(a)   **(b)**

Fig. 21 Simulated equiaxed dendrite morphology and solute fields of (a) copper and (b) magnesium for an Al-15Cu-1Mg (wt%) alloy solidified at a cooling rate of 10 K/s. The model produces color images. The cruciform dendrites in (a) that appear as darker gray in print indicate a low concentration of copper (at.%) and correspond to the low range of the scale, while the lighter-gray solute corresponds to the upper range of the scale. In (b), the dark solute also corresponds to the upper range of the magnesium at.% scale, while the medium-gray cruciform corresponds to the lower end of the magnesium at.% scale

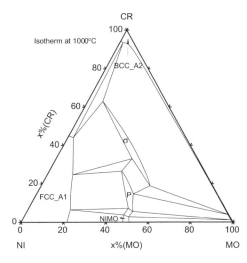

Fig. 22 Isothermal section of the Ni-Cr-Mo ternary system at 1000 °C

computational thermodynamics has been successfully used to obtain phase equilibria for practical applications.

Conclusion

In this article, the powerful features of thermodynamic calculation are demonstrated by several examples, including the design of nickel-base superalloys, calculation of the β-transus and β-approach curves for commercial titanium alloys, prediction of bulk metallic glass formability, and dilation of ferrous alloys during phase transformations from austenite to ferrite and martensite, respectively. Integration of thermodynamic calculation with kinetic and microstructural evolution models makes the computational tools even more powerful. This is demonstrated by integrating thermodynamic

calculation with the multicomponent Scheil model for the prediction of liquation cracking of aluminum welds, with the classic nucleation/growth theory for the simulation of precipitation kinetics of nickel alloys, with the phase-field model for the study of microstructural evolution of a diffusion couple, and with the microscopic and cellular automaton model for simulation of the solidification process. Knowledge of these microstructures will ultimately lead to improved performance of these materials.

ACKNOWLEDGMENT

The authors would like to acknowledge the U.S. Air Force Research Laboratory for financial support through the Small Business Innovation Research and the Metals Affordability Initiative projects.

REFERENCES

1. L. Kaufman, *Computer Calculation of Phase Diagrams,* Academic Press, New York, 1970
2. N. Saunders and A.P. Miodownik, *CALPHAD: A Comprehensive Guide,* R.W. Cahn, Ed., Pergamon Materials Series, 1998
3. K.-C. Chou and Y.A. Chang, A Study of Ternary Geometrical Models, *Ber. Bunsenges. Phys. Chem.,* Vol 93, 1989, p 735
4. F. Zhang et al., Thermodynamic Databases: Useful Tools in the Development of Advanced Materials, *Advances in Materials Technology for Fossil Power Plants: Proceedings from the Fourth International Conference, 2004* (Hilton Head Island, SC), ASM International
5. Y.A. Chang et al., Phase Diagram Calculation: Past, Present and Future, *Prog. Mater. Sci.,* Vol 49, 2004, p 313–345

6. *Pandat—Phase Diagram Calculation Software Package for Multicomponent Systems,* CompuTherm, LLC, Madison, WI, 2000
7. J.F. Loffler, Bulk Metallic Glasses, *Intermetallics,* Vol 11, 2003, p 529
8. R.J. Gottschall, *First Conference on Bulk Metallic Glasses,* United Engineering Foundation, New York, NY, Singapore, 2000
9. X.-Y. Yan et al., A Thermodynamic Approach for Predicting the Tendency of Multicomponent Metallic Alloys for Glass Formation, *Intermetallics,* Vol 9, 2001, p 535–538
10. D. Ma et al., Bulkier Glass Formability Enhanced by Minor Alloying Additions, *Appl. Phys. Lett.,* Vol 87, 2005, p 171914
11. H. Cao et al., Computational Thermodynamics to Identify Zr-Ti-Ni-Cu-Al Alloys with High Glass-Forming Ability, *Acta Mater.,* Vol 54, 2006, p 2975–2982
12. D. Turnbull, Under What Conditions Can a Glass Be Formed?, *Contemp. Phys.,* Vol 10, 1969, p 473
13. X.H. Lin and W.L. Johnson, Formation of Ti-Zr-Cu-Ni Bulk Metallic Glasses, *J. Appl. Phys.,* Vol 78, 1995, p 6514
14. H.B. Cao et al., Synthesis of Copper-Rich Amorphous Alloys by Computational Thermodynamics, *Acta Mater.,* Vol 56, 2008, p 2032–2036
15. T.M. Pollock and S. Tin, Nickel-Based Superalloys for Advanced Turbine Engines: Chemistry, Microstructure, and Properties, *J. Propulsion Power,* Vol 22 (No. 2), 2006, p 361–374
16. F. Ritzert, D. Keller, and V. Vasudevan, "Investigation of the Formation of Topologically Close Packed Phase Instabilities in Nickel-Base Superalloys, Rene N6," NASA/TM, 1999
17. *PanNi—Thermodynamic Database for Multicomponent Nickel Alloys,* CompuTherm, LLC, Madison, WI, 2000
18. R.L. Dreshfield and J.F. Wallace, The Gamma-Gamma Prime Region of the Ni-Al-Cr-Ti-W-Mo System at 850C, *Metall. Trans.,* Vol 5, 1974, p 71–78
19. T. Khan, P. Caron, and C. Duret, The Development and Characterization of a High Performance Experimental Single Crystal Superalloy, Superalloys 1984 (Warrendale, PA), Met. Soc. AIME, p 145–155
20. K. Trinckauf and E. Nembach, Concentration Gradients at the Phase Boundary Between a γ'- Precipitate and the γ -Matrix of a Superalloy, *Acta Metall. Mater.,* Vol 39 (No. 12), 1991, p 3057
21. R. Schmidt and M. Feller-Kniepmeier, Equilibrium Composition and Volume Fraction of the γ'-Phase in a Nickel-Base Superalloy as a Function of Temperature, *Scr. Metal. Mater.,* Vol 26, 1992, p 1919
22. J.S. Zhang et al., Design and Development of Hot Corrosion-Resistant Nickel-Base Single-Crystal Superalloys by the

D-Electrons Alloy Design Theory, Part I: Characterization of the Phase Stability, *Met. Trans. A,* Vol 24, 1993, p 2443

23. S. Duval et al., Phase Composition and Chemical Order in the Single Crystal Nickel Base Superalloy MC2, *Acta Metall. Mater.,* Vol 42 (No. 1), 1994, p 185
24. S. Miyazaki, Y. Murata, and M. Morinaga, Evaluation of Partitioning Ratio of Re Atoms in Ni-Al-Re-X (X: Cr, Mo, W, Ti, Nb, Co) Quaternary Alloys, *Iron Steel (Jpn.),* Vol 80 (No. 2), 1994, p 78–83
25. E.H. Copland, N.S. Jacobson, and F.J. Ritzert, "Computational Thermodynamic Study to Predict Complex Phase Equilibria in the Nickel-Base Superalloy Rene N6," NASA/TM, 2001-210897
26. S.T. Wlodek, The Structure of IN100, *Trans. ASM,* Vol 57, 1964, p 110–119
27. E.W. Ross, Rene 100—A Sigma-Free Turbine Blade Alloy, *J. Met.,* Vol 19 (No. 12), 1967, p 12–14
28. R. Darolia, D.F. Lahrman, and R.D. Field, Formation of Topologically Closed Packed Phases in Nickel Base Single Crystal Superalloys, *Superalloys 1988,* TMS, Warrendale, PA
29. J.D. Nystrom et al., Discontinuous Cellular Precipitation in a High-Refractory Nickel-Base Superalloy, *Metall. Mater. Trans. A,* Vol 28, 1997, p 2443–2452
30. C.M.F. Rae and R.C. Reed, The Precipitation of Topologically Close-Packed Phases in Rhenium-Containing Superalloys, *Acta Mater.,* Vol 49 (No. 10), 2001, p 4113–4125
31. *PanTi—Thermodynamic Database for Multicomponent Titanium Alloys,* CompuTherm, LLC, Madison, WI, 2000
32. D. Furrer, Ladish Co., Inc., Cudahy, WI, beta-transus of titanium alloys, private communication, 2003
33. V. Venkatesh, TIMET, Henderson, NV, titanium alloys, private communication, 2004
34. S.L. Semiatin, T.M. Lehner, J.D. Miller, R.D. Doherty, and D.U. Furrer, Alpha/Beta Heat Treatment of a Titanium Alloy with a Non-Uniform Microstructure, *Metall. Mater. Trans. A,* Vol 38, 2007, p 910
35. *PanFe—Thermodynamic Database for Multicomponent Iron Alloys,* CompuTherm, LLC, Madison, WI, 2000
36. Y. Maehara et al., Effects of Alloying Elements on Sigma-Phase Precipitation in Delta-Gamma Duplex Phase Stainless Steels, *Met. Sci.,* Vol 17, 1983, p 541
37. F.H. Hayes, Phase Equilibria in Duplex Stainless Steels, *J. Less Common Met.,* Vol 14, 1985, p 89
38. T. Thorvaldsson et al., Influence of Microstructure on Mechanical Properties of a Duplex Stainless Steel, *Stainless Steels' 84, Proceedings of the Conference Sponsored and Organized Jointly by Chalmers University of Technology and Jernkontoret (Sweden) with The Metals Society (U.K.),* 1984, p 101–105

39. R.D. Longbottom and F.H. Hayes, *User Aspects of Phase Diagram,* The Institute of Metals, London, 1991, p 32
40. R.D. Longbottom and F.H. Hayes, Effect of Variations in Alloying Element Content and Temperature on the Austenite-Ferrite Phase Balance in Duplex Stainless Steels, *Proceedings of the Conference for Duplex Stainless Steels,* Welding Institute, Cambridge, 1994
41. M. Boniardi, F. Iacoviello, and G.M.L. Vecchia, The Influence of Anisotropy and Stress Ratio on Fatigue Crack Growth of Duplex Stainless Steel, *Proceedings of the Conference for Duplex Stainless Steels,* Welding Institute, Cambridge, 1994
42. M. Nystrom and B. Karlsson, Plastic Deformation of Duplex Stainless Steels with Different Amounts of Ferrite, *Proceedings of the Conference for Duplex Stainless Steels,* Welding Institute, Cambridge, 1994
43. G. Ghosh and G.B. Olson, Kinetics of FCC and BCC Heterogeneous Martensitic Nucleation—The Critical Driving Force for Athermal Nucleation, *Acta Metall. Mater.,* Vol 42 (No. 10), 1994, p 3361–3370
44. D.P. Koistinen and R.E. Marburger, Extent of the Austenite-Martensite Transformation in Pure Iron-Carbon Alloys, *Bull. Am. Phys. Soc.,* Vol 2, 1957, p 122
45. P. Gordon, M. Cohen, and R.S. Rose, The Kinetics of Austenite Decomposition in High Speed Steel, *Trans. ASM,* Vol 3, 1943, p 161–216
46. E. Scheil, Unbroken Series of Solid Solutions in the Binary Systems of the Elements, *Z. Metallkd.,* Vol 34, 1942, p 242
47. G.E. Metzger, Some Mechanical Properties of Welds in 6061 Aluminum Alloy Sheet, *Weld. J.,* Vol 46 (No. 10), 1967, p 457S–469S
48. N.F. Gittos and M.H. Scott, Heat-Affected Zone Cracking of Al-Mg-Si Alloys, *Weld. J.,* Vol 60 (No. 6), 1981, p 95S–103S
49. M. Katoh and H.W. Kerr, Investigation of Heat-Affected Zone Cracking of GTA Welds of Al-Mg-Si Alloys Using the Varestraint Test, *Weld. J.,* Vol 66 (No. 12), 1987, p 60S–368S
50. H.W. Kerr and M. Katoh, Investigation of Heat-Affected Zone Cracking of GTA Welds of Al-Mg-Si Alloys Using the Varestraint Test, *Weld. J.,* Vol 66 (No. 9), 1987, p 251S–259S
51. M. Miyazaki et al., Quantitative Investigation of Heat-Affected Zone Cracking in Aluminum Alloy 6061, *Weld. J.,* Vol 69 (No. 9), 1990, p 362S–371S
52. M.B.D. Ellis, M.F. Gittos, and I. Hadley, Significance of Liquation Cracks in Thick Section Al-Mg-Si Alloy Plate, *Weld. Inst. J. (U.K.),* Vol 6 (No. 2), 1997, p 213–255
53. C. Huang and S. Kou, Liquation Cracking in Partial Penetration Aluminum Welds: Assessing Tendencies to Liquate, Crack and Backfill, *Sci. Technol. Weld. Join.,* Vol 9 (No. 2), 2004, p 149–157

54. C. Huang and S. Kou, Liquation Cracking in Full Penetration Al-Mg-Si Welds, *Weld. J.,* April 2004, p 111S–122S
55. *PanAl—Thermodynamic Database for Multicomponent Aluminum Alloys,* CompuTherm, LLC, Madison, WI, 2000
56. G. Cao and S. Kou, Predicting and Reducing Liquation-Cracking Susceptibility Based on Temperature vs. Fraction Solid, *Weld. J.,* Jan 2006, p 9S–18S
57. Y.K. Yang et al., Liquation of Mg Alloys in Friction-Stir Spot Welding, *Weld. J.,* July 2008, p 167S–178S
58. R. Kampmann and R. Wagner, Kinetics of Precipitation in Metastable Binary Alloys— Theory and Application to Cu-1.9at%Ti and Ni-14at%Al, *Decomposition of Alloys: The Early Stages,* P. Haasen et al., Ed., Pergamon Press, Oxford, 1984, p 91–103
59. K.-S. Wu et al., A Modeling Tool for the Precipitation Simulations of Superalloys during Heat Treatments, *Superalloy 2008* (Seven Springs, PA), TMS, 2008
60. H. Wendt and P. Haasen, Nucleation and Growth of γ'-Precipitates in Ni-14 at.% Al, *Acta Metall. Mater.,* Vol 31 (No. 10), 1983, p 1649–1659
61. S.Q. Xiao and P. Haasen, HREM Investigation of Homogeneous Decomposition in a Ni-12 at.% Al Alloy, *Acta Metall. Mater.,* Vol 39 (No. 4), 1991, p 651–659
62. J.D. Robson, Modelling the Evolution of Particle Size Distribution during Nucleation, Growth and Coarsening, *Mater. Sci. Technol.,* Vol 20, 2004, p 441–448
63. I. Kirman and D.H. Warrington, The Precipitation of Ni₃Nb Phases in a Ni-Fe-Cr-Nb Alloy, *Metall. Trans.,* Vol 1 (No. 10), 1970, p 2667–2675
64. Y.-F. Han, P. Deb, and M.C. Chaturvedi, Coarsening Behavior of γ''- and γ'-Particles in Inconel Alloy 718, *Met. Sci.,* Vol 16 (No. 12), 1982, p 555–561
65. J.P. Collier, A.O. Selius, and J.K. Tien, On Developing a Microstructurally and Thermally Stable Iron-Nickel Base Superalloy, *Superalloys 1988,* TMS, Warrendale, PA, p 43–52
66. J.W. Cahn and J.E. Hilliard, A Microscopic Theory for Antiphase Boundary Motion and Its Application to Antiphase Domain Coarsening, *Acta Met.,* Vol 27, 1979, p 1085
67. J.D. Gunton, M.S. Miguel, and P.S. Sahni, The Dynamics of First-Order Phase Transitions, *Phase Transitions and Critical Phenomena,* Vol 8, C. Domb and J.L. Lebowitz, Ed., Academic Press, New York, 1983
68. Q. Chen et al., Quantitative Phase Field Modeling of Diffusion-Controlled Precipitate Growth and Dissolution in Ti-Al-V, *Scr. Mater.,* Vol 50, 2004, p 471–476
69. K. Wu, Y.A. Chang, and Y. Wang, Simulating Interdiffusion Microstructures in Ni-Al-Cr Diffusion Couples: A Phase Field Approach Coupled with CALPHAD

Database, *Scr. Mater.,* Vol 50, 2004, p 1145–1150

70. F. Zhang et al., Development of Thermodynamic Description of a Pseudo Ternary System for Multicomponent Ti64 Alloy, *J. Phase Equilib. Diff.,* Vol 28 (No. 1), 2007, p 115–120

71. *PanEngine—Thermodynamic Calculation Engine for Multicomponent Phase Equilibrium,* CompuTherm, LLC, Madison, WI, 2000

72. J.A. Nesbitt and R.W. Heckel, Predicting Diffusion Paths and Interface Motion in γ/γ + β, Nickel-Chromium-Aluminum Diffusion Couples, *Metall. Trans. A,* Vol 18 (No. 12), 1987, p 2087

73. F.-Y. Xie et al., Microstructure and Microsegregation in Al-Rich Al-Cu-Mg Alloys, *Acta Mater.,* Vol 47, 1999, p 489–500

74. X. Yan et al., Computational and Experimental Investigation of Microsegregation in an Al Rich Al-Cu-Mg-Si Quaternary

Alloy, *Acta Mater.,* Vol 50, 2002, p 2199–2207

75. F. Xie et al., A Study of Microstructure and Microsegregation of Aluminum 7050 Alloy, *Mater. Sci. Eng. A,* Vol 355, 2003, p 144–153

76. M.-F. Zhu et al., Modeling of Microstructure and Microsegregation in Solidification of Multi-Component Alloys, *J. Phase Equilib. Diff.,* Vol 28 (No. 1), 2007, p 130–138

ASM Handbook, Volume 22B, *Metals Process Simulation*
D.U. Furrer and S.L. Semiatin, editors

Copyright © 2010, ASM International®
All rights reserved.
www.asminternational.org

The Application of Thermodynamic and Material Property Modeling to Process Simulation of Industrial Alloys

N. Saunders, Thermotech/Sente Software Ltd., United Kingdom

MODELING AND SIMULATION of materials processing often requires specific information concerning phase transformations and thermodynamic properties, for example, liquidus and solidus temperatures, the fraction solid as a function of temperature during solidification, Ae_3 and Ae_1 temperatures for steels, β-transus temperatures for titanium alloys, γ' solvus temperatures for nickel-base superalloys, heat evolution during solidification and solid-state phase transformation, specific heats, and so on. The use of thermodynamic modeling tools holds the promise of providing a ready source of such information.

Thermodynamic modeling has been used for many years in the processing of metallic materials. However, it would be true to say that the first major use of such modeling was in extraction and refining, where chemical reactions during processing needed to be understood and predicted. Databases of thermodynamic properties of stoichiometric compounds were developed and made widely available in the literature. Simple rate equations could then be developed and used to model competing chemical reactions. Such methods, while of great value in their own right, were limited when trying to understand the general behavior of metallic alloys, where properties of nonstoichiometric metallic phases with wide ranges of composition, such as austenite and ferrite in steels or the gamma phase in nickel-base alloys, needed to be described.

While thermodynamic models that represent nonstoichiometric solution phases have existed for many decades, it was not until the advent of computer technology that it became possible to contemplate calculation of thermodynamic equilibria between such phases. The seminal works of Kaufman and co-workers (Ref 1, 2) in the development of computer-based methods marked the beginning of what has become known as the CALPHAD approach, after the acronym CALculation of PHAse Diagrams,

and it is interesting to look back and see how many of the early publications centered around the need to describe alloys that were of practical use, such as titanium alloys, nickel alloys, and steels.

As computing power increased, it became possible to use more complex descriptions of thermodynamic properties and move from binary and ternary alloys to the multicomponent alloys used in industrial practice. Although the scientific methodology had been available to do this for some time, practical applications required a minimum computing capability before they became useful.

In the past few decades, groups have been established worldwide that specialize in both the development of software for the calculation of complex multicomponent phase equilibria and the development of thermodynamic databases that describe the properties of the relevant phases in the alloy of interest. It is the linking of software to thermodynamic databases especially developed and validated for multicomponent alloys that has held the key to the practical and everyday use of CALPHAD methods in industry.

As computational speeds further increased, it has become possible to use CALPHAD methods as a general tool that could be linked to material models for phase transformations and subsequently to the modeling of more general material properties, such as physical and mechanical properties.

This article aims to provide background to the CALPHAD method, how it works, and how it can be applied in industrial practice. For more detailed descriptions of the CALPHAD method and its use, the reader is referred to Ref 3 to 5. The extension of CALPHAD methods as a core basis for the modeling of generalized material properties is explored. Such properties include thermophysical and physical properties, temperature- and strain-rate-dependent mechanical properties, properties for use in the modeling of

quench distortion, properties for use in solidification modeling, and so on. Finally, how such modeling may be used in the future is discussed, in particular, focusing on links with finite-element or finite-difference process models.

Calculation of Phase Equilibria in Multicomponent Alloys

There are two main requirements for the calculation of phase equilibria:

- The thermodynamic properties of the phases in an alloy must be represented.
- A Gibbs energy minimization must be performed so that the phase equilibria between the requisite phases can be calculated.

Thermodynamic Models

Thermodynamic modeling of solution phases lies at the very core of the CALPHAD method. In metallic alloys, it is only rarely that calculations involve purely stoichiometric compounds. Solution phases are defined here as any phase in which there is solubility of more than one component. For all solution phases, the Gibbs energy is given by the general formula:

$$G = G^o + G_{mix}^{ideal} + G_{mix}^{xs} \qquad \text{(Eq 1)}$$

where G^o is the contribution of the pure components of the phase to the Gibbs energy, G_{mix}^{ideal} is the ideal mixing contribution, and G_{mix}^{xs} is the contribution due to nonideal interactions between the components, also known as the Gibbs excess energy of mixing.

Random substitutional models are used for phases such as the gas phase or a simple metallic liquid and solid solutions, where components can mix on any spatial position that is available to the phase. For example, in a simple body-centered cubic phase, any of the components could occupy any of the atomic sites that define the cubic structure, as shown in Fig. 1 In a gas or liquid phase, the crystallographic nature of structure is lost, but otherwise, positional occupation of the various components relies on random substitution rather than any preferential occupation of site by a particular component.

Dilute Solution Models. There are a number of areas in materials processing where low levels of alloying are important, for example, in refining and some age-hardening processes. In such cases, it is possible to deal with solution phases by dilute solution models (Ref 6, 7). These have the advantage that there is substantial experimental literature that deals with the thermodynamics of impurity additions, particularly for established materials such as ferrous- and copper-based alloys. However, because of their fundamental limitations in handling concentrated solutions, they are only discussed briefly.

In a highly dilute solution, the solute activity (a_i) is found to closely match a linear function of its concentration (x_i). In its simplest form, the activity is written as:

$$a_i = \gamma_i^o x_i \qquad \text{(Eq 2)}$$

where γ_i^o is the value of the activity coefficient of i at infinite dilution. This is known as Henry's law. Equation 2 may be rewritten in terms of partial Gibbs energies as:

$$\overline{G}_i = \overline{G}_i^{xs} + RT \log_e x_i \qquad \text{(Eq 3)}$$

where $\overline{G}_i^{xs}$ has a constant value in the Henrian concentration range and is obtained directly from γ_i^o, R is the gas constant, and T is the temperature. The expression can also be modified to take into account interactions between the solute elements, and Eq 3 becomes:

$$\overline{G}_i = \overline{G}_i^{xs} + RT \log_e x_i + RT \sum_j \varepsilon_i^j x_i \qquad \text{(Eq 4)}$$

where x_i is the concentration of solute i, and ε_i^j is an interaction parameter taking into account

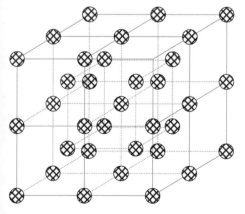

Fig. 1 Simple body-centered cubic structure with random occupation of atoms and all sites consisting of eight-unit cells

the effect of the mixing of component i and j in the solvent.

General solution models include the ideal and nonideal models.

Ideal Solution Model. The simplest solution model is an ideal substitutional solution, which is characterized by the random distribution of components on a lattice with an interchange energy equal to zero. The configurational entropy (S^{config}) is easily calculated and related to the probability of interchange of the components:

$$S^{config} = k \log_e W_p \qquad \text{(Eq 5)}$$

where k is Boltzmann's constant, and W_p is the number of configurations in which the components can be arranged for a given state. From this formula, the well-known equation for the ideal entropy of mixing can be derived as:

$$S_{mix}^{ideal} = R \sum_i x_i \log_e x_i \qquad \text{(Eq 6)}$$

where x_i is the mole fraction of component i, and G_{mix}^{ideal} is then given by:

$$G_{mix}^{ideal} = -TS_{mix}^{ideal} = RT \sum_i x_i \log_e x_i \qquad \text{(Eq 7)}$$

where R is the gas constant. The Gibbs energy of an ideal solution phase will then be:

$$G_m = \sum_i x_i G_i^o + RT \sum_i x_i \log_e x_i \qquad \text{(Eq 8)}$$

with G° defining the Gibbs energy of the phase consisting of pure component i. For the case of gases, ideal mixing is often assumed, and this assumption can often be quite reasonable and used for process models that include the gas phase. However, in condensed phases, there is always some interaction between components.

Nonideal Solutions: Regular and Subregular Solution Models. The regular solution model is the simplest of the nonideal models and basically considers that the magnitude and sign of interactions between the components in a phase are independent of composition. Taking the system A-B and assuming the total energy of the solution (E_o) arises from only nearest-neighbor bond energies, then:

$$E_o = \omega_{AA}E_{AA} + \omega_{BB}E_{BB} + \omega_{AB}E_{AB} \qquad \text{(Eq 9)}$$

where ω_{AA}, ω_{BB}, ω_{AB}, E_{AA}, E_{BB}, and E_{AB} are the number of bonds and energies associated with the formation of different bond types AA, BB, and AB. If the reference states are taken as pure A and B, then Eq 9 can be developed such that the enthalpy of mixing, H_{mix}, is given by:

$$H_{mix} = \frac{Nz}{2} x_A x_B (2E_{AB} - E_{AA} - E_{BB}) \qquad \text{(Eq 10)}$$

If the bond energies are temperature dependent, there will also be an excess entropy of

mixing, leading to the well-known regular solution model for the Gibbs excess energy of mixing:

$$G_{mix}^{xs} = x_A x_B \Omega \qquad \text{(Eq 11)}$$

where Ω is now a temperature-dependent interaction parameter. When generalized and added to Eq 8, this gives:

$$G_m = \sum_i x_i G_i^o + RT \sum_i x_i \log_e x_i + \sum_i \sum_{j>i} x_i x_j \Omega_{ij} \qquad \text{(Eq 12)}$$

However, it has been realized for a long time that the assumption of composition-independent interactions is too simplistic. This led to the development of the subregular solution model, where interaction energies are considered to change linearly with composition. The following expression for G_{mix}^{xs} is then obtained as (Ref 2):

$$G_{mix}^{xs} = x_i x_j \left(\Omega_{ij}^i x_i + \Omega_{ij}^j x_j \right) \qquad \text{(Eq 13)}$$

Taking this process further, more complex composition dependencies of Ω can be considered. It is straightforward to show that a general formula in terms of a power series has the capability to account for most types of composition dependence (Ref 8). The most common method is based on the Redlich-Kister equation, and Eq 13 is expanded to become:

$$G_{mix}^{xs} = x_i x_j \sum_v \Omega_{ij}^v (x_i - x_j)^v \qquad \text{(Eq 14)}$$

where Ω_{ij}^v is a binary interaction parameter dependent on the value of v. Equation 14 is equivalent to the regular solution model when $v = 0$ and subregular when $v = 1$. In practice, the value for v does not usually rise above 2. If it is found necessary for $v > 2$, it is likely that an incorrect model has been chosen to represent the phase.

Most methods of extrapolating the thermodynamic properties of alloys into multicomponent systems are based on the summation of the binary excess parameters. The relevant formulae are based on various geometrical weightings of the mole fractions; the predominant method at the present time uses the Muggianu equation (Ref 9), where, in a ternary system:

$$G_{mix}^{xs} = x_A x_B \sum_v \Omega_{AB}^v (x_A - x_B)^v + x_B x_C$$
$$\sum_v \Omega_{BC}^v (x_B - x_C)^v \qquad \text{(Eq 15)}$$
$$+ x_A x_C \sum_v \Omega_{AC}^v (x_A - x_C)^v$$

Equation 15 can be generalized to:

$$G_{mix}^{xs} = \sum_i \sum_{j>i} x_i x_j \sum_v \Omega_{ij}^v (x_i - x_j)^v \qquad \text{(Eq 16)}$$

and for a multicomponent system, Eq 12 becomes:

$$G_m = \sum_i x_i G_i^o + RT \sum_i x_i \log_e x_i + \sum_i \sum_{j>i} x_i x_j \sum_v \Omega_{ij}^v (x_i - x_j)^v \qquad \text{(Eq 17)}$$

Equation 17 assumes ternary interactions are small in comparison to those that arise from the binary terms. This may not always be the case, and where the need for higher-order interactions is evident, these can be taken into account by a further term of the type $G_{ijk} = x_i x_j x_k L_{ijk}$, where L_{ijk} is an excess ternary interaction parameter. There is little evidence of the need for any higher-order interaction terms, and prediction of the thermodynamic properties of substitutional solution phases in multicomponent alloys is usually based on an assessment of binary and ternary values. Various other polynomial expressions for the excess term in multicomponent systems have been considered (Ref 3, 5, 10, 11). However, these are based on predicting the properties of the higher-order system from the properties of the lower-component systems.

More complex representations of the thermodynamic properties of solution phases exist, for example, the sublattice model, where the phase can be envisioned as composed of interlocking sublattices (Fig. 2) on which the various components can mix. It is usually applied to crystalline phases, but the model can also be extended to consider ionic liquids, where mixing on ionic sublattices is considered. The model is phenomenological in nature and does not define any crystal structure within its general mathematical formulation. It is possible to define internal parameter relationships that reflect structure with respect to different crystal types, but such conditions must be externally formulated and imposed on the model. Equally, special relationships apply if the model is to be used to simulate order-disorder transformations.

Sublattice modeling is now one of the most predominant methods used to describe solution and compound phases. It is very flexible and can account for a variety of different phase types ranging from interstitial phases, such as austenite and ferrite in steels, to intermetallic phases, such as sigma and Laves, which have wide homogeneity ranges. More details of this model and others that are used in CALPHAD calculations can be found in Ref 3 and 5.

Computational Methods

The previous section dealt with thermodynamic models, because these are the basis of the CALPHAD method. However, it is the computational methods and software that allow these models to be applied in practice. In essence, the issues involved in computational methods are less diverse and mainly revolve around Gibbs energy minimization.

It is also worthwhile to make some distinctions between methods of calculating phase equilibrium. For many years, equilibrium constants have been used to express the abundance of certain species in terms of the amounts of other arbitrarily chosen species (Ref 12 to 15). Such calculations have significant disadvantages in that some prior knowledge of potential reactions is often necessary, and it is difficult to analyze the effect of very complex reactions involving many species on a particular equilibrium reaction. Furthermore, unless equilibrium constants are defined for all possible chemical reactions, a true equilibrium calculation cannot be made, and, in the case of a reaction with 50 or 60 substances present, the number of possible reactions is massive.

CALPHAD methods attempt to provide a true equilibrium calculation by considering the Gibbs energy of all phases and minimizing the total Gibbs energy of the system (G). In this circumstance, G can be calculated either from knowledge of the chemical potential ($\overline{G}_i$) of component i by:

$$G = \sum_i n_i \overline{G}_i \qquad \text{(Eq 18)}$$

where n_i is the amount of component i, or alternatively by:

$$G = \sum_\phi N^\phi G_m^\phi \qquad \text{(Eq 19)}$$

where N^ϕ is the amount of phase ϕ, and G_m^ϕ is its Gibbs energy. The number of unknowns is now considerably reduced in comparison to an equilibrium constant approach. Furthermore, the CALPHAD methodology has the significant advantage that, because the total Gibbs energy is calculated, it is possible to derive all of the associated functions and characteristics of phase equilibria, that is, phase diagrams, chemical potential diagrams, and so on.

Calculation of Phase Equilibria. The actual calculation of phase equilibria in a multicomponent, multiphase system is a complex process involving a high level of computer programming. Details of programming aspects are too lengthy to go into detail for this article, but

most of the principles by which Gibbs energy minimization is achieved are conceptually quite simple. This section therefore concentrates on the general principles rather than going into detail concerning the currently available software programs, which, in any case, often contain proprietary code.

Essentially, the calculation must be defined so that the number of degrees of freedom is reduced, the Gibbs energy of the system can be calculated, and some iterative technique can be used to minimize the Gibbs energy. The number of degrees of freedom is reduced by defining a series of constraints, such as the mass balance, electroneutrality in ionic systems, composition range in which each phase exists, and so on.

Most thermodynamic software uses local minimization methods. As such, preliminary estimates for equilibrium must be given so that the process can begin and subsequently proceed smoothly to completion. Such estimates are usually set automatically by the software and do not need to accurately reflect the final equilibrium. However, the possibility that phases may have multiple minima in their Gibbs energy formulations should be recognized and start points automatically set so that the most stable minima are accounted for.

Local minimization tools have the advantage of being rapid in comparison to global minimization methods, which automatically search for multiple Gibbs energy minima. As such, local minimization methods were invariably favored in the early days of CALPHAD. However, with the advent of faster computers, such an advantage becomes less tangible when dealing with relatively simple calculations, and the user will notice little effective difference in speed. Codes such as Thermo-Calc (Ref 16) and PANDAT (Ref 17) now offer global minimization methods as part of their calculation capability. However, if the problem to be solved involves multicomponent alloys with numerous multiple sublattice phases containing a potentially large number of local minima, speed issues will still arise.

Whether to use local or more global methods is a pertinent question if reliability of the final calculation is an issue. For the case of calculation of multicomponent alloys, such as those used by industry and where the composition space is reasonably prescribed, local minimization methods have been used successfully for many years and have proved highly reliable.

One of the earliest examples of Gibbs energy minimization applied to a multicomponent system was by White et al. (Ref 18), who considered the chemical equilibrium in an ideal gas mixture of oxygen, hydrogen, and nitrogen with the species H, H_2, H_2O, N, N_2, NH, NO, O, O_2, and OH being present. The problem here is to find the most stable mixture of species. The Gibbs energy of the mixture was defined using Eq 1 and defining the chemical potential of species i as:

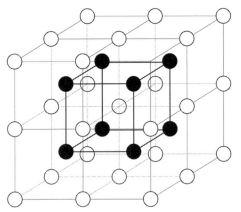

Fig. 2 Simple body-centered cubic structure sites consisting of eight-unit cells with preferential occupation of atoms in the body center and corner positions

$$\overline{G}_i = \overline{G}_i^o + RT \log_e a_i \qquad \text{(Eq 20)}$$

where $\overline{G}_i^o$ is the standard chemical potential of species i. Standard mass balance equations were then made where:

$$\sum_i a_{ij} x_i = n_j \ (j = 1, 2, \ldots m) \qquad \text{(Eq 21)}$$

where a_{ij} represents the number of atoms of element j in the species i, x_i is the number of moles of species i, and n_j is the total number of moles of m different elements in the system. These authors presented two methods of Gibbs energy minimization, one of which used a linear programming method developed by Dantzig et al. (Ref 19), and the other is based on the method of steepest descent using Lagrange's method of undetermined multipliers. The method of steepest descent provides a rapid solution to the minimization problem and was later used by Eriksson (Ref 20, 21) and Eriksson and Rosen (Ref 22) in the software codes SOLGAS and SOLGASMIX. SOLGAS, the earlier code, treated a mixture of stoichiometric condensed substances and an ideal gas mixture, while SOLGASMIX was further able to include nonideal solution phases.

The minimization methods used by later programs such as ChemSage, the successor to SolGAsMix (Ref 23); F*A*C*T (Ref 24); FactSage, the recent combining of ChemSage and F*A*C*T (Ref 25); Thermo-Calc (Ref 16); and MTDATA (Ref 26) are, in the broadest sense, similar in principle to that described previously, although there are clear differences in their actual operation (Ref 27, 28). Thermodynamic models are now more complex, which may make it necessary to consider further degrees of freedom. However, constraints are still made to the system such that the Gibbs energy may be calculated as a function of extensive variables, such as the amount of each phase present in the system. Initial estimates are made for the Gibbs energy as a function of phases present, their amounts and composition, and so on. The Gibbs energy is then calculated and some numerical method is used, whether it be through Lagrangian multipliers (Ref 27) or a Newton-Raphson method (Ref 28), by which new values can be estimated and which will cause the Gibbs energy to be decreased. When the difference in calculated Gibbs energy between the iterative steps reaches some small-enough value, the calculation is taken to have converged.

As mentioned earlier, recent developments have been made to develop more global minimization codes, for example, Thermo-Calc and PANDAT. Such codes use mathematical methods to search out all possible minima and compare Gibbs energies of the local minima so that the most stable equilibrium is automatically calculated. A review of currently available software for calculating phase equilibria as of the year 2002 can be found in a special issue of the CALPHAD journal (Ref 29).

Application of CALPHAD Calculations to Industrial Alloys

Introduction

From the beginning, one of the aims of CALPHAD methods has been to calculate phase equilibria in the complex, multicomponent alloys that are used regularly by industry. Certainly, the necessary mathematical formulations to handle multicomponent systems have existed for some time and have been programmed into the various software packages for calculation of phase equilibria since the middle of the last century. However, it is interesting to note that, until the 1990s (with the exception of steels), there had been very little actual application to the complex systems that exist in technological or industrial practice, other than through calculations using simple stoichiometric substances, ideal gas reactions, and dilute solution models.

Dilute solution modeling has been used for some time, because it is not very intensive in computational terms, and some industrially important materials, although containing many elements, are actually quite low in total alloy or impurity content, for example, high-strength, low-alloy steels. Examples in this area can be found in Ref 30 to 33. The limitations of dilute solution models were discussed earlier. Although useful for certain limited applications, they could not begin to handle, with any accuracy, highly alloyed materials such as stainless steels or nickel-base superalloys. Substance calculations, while containing large numbers of species and condensed phases, are, in many ways, even more limited in their application to alloys, because they do not consider interactions in phases involving substantial mixing of the components.

The main areas of application for more generalized models were, until the 1990s, mainly restricted to binary and ternary systems or limited to "ideal industrial materials," where only major elements were included. The key to the general application of CALPHAD methods in multicomponent systems was the development of sound, validated thermodynamic databases for use in the available computing software. Until then, there had been a dearth of such databases.

Steels were a notable exception to this trend and, in particular, stainless and high-speed steels, where alloy contents can rise to well above 20 wt%. For such alloys, a concentrated solution database (Fe-Base) has existed since 1978, based on work done at the Royal Institute of Technology, Stockholm in Sweden. However, although far more generalized than dilute solution databases, its range of applicability is limited in temperature to between 700 and 1200 °C (1290 and 2190 °F).

The lack of similar databases for other material types presented severe problems for CALPHAD calculations with any of the other commonly used materials and led to a concentration of application to steels. However, during the 1990s, further multicomponent databases were developed for use with aluminum alloys (Ref 34), steels (Ref 3, 35), nickel-base superalloys (Ref 36), and titanium- and TiAl-base alloys (Ref 37, 38). These databases were created mainly for use with industrial, complex alloys, and the accuracy of computed results was validated to an extent not previously attempted. Simple, statistical analysis of average deviation of calculated result from experimental measurement in "real," highly alloyed, multicomponent alloys has demonstrated that CALPHAD methods can provide predictions for phase equilibria whose accuracy lies close to that of experimental measurements.

The importance of validation of computed results cannot be stressed too highly. Computer models, such as those used to simulate materials processing, rely on input data that can be time-consuming to measure but readily predicted via CALPHAD and related methods. For example, it is possible to model the processing of a steel at all stages of manufacture, starting from the initial stages in a blast furnace, through the refinement stages to a casting shop, followed by heat treatment and thermomechanical processing to the final product form. Such a total modeling capability requires that confidence can be placed in the predictions of each of the building blocks, and, in the case of CALPHAD methods, the key to success is the availability of high-quality, validated databases. The following section concentrates on databases.

Databases

Substance Databases. From a simple point of view, substance databases have little complexity because they are assemblages of assessed data for stoichiometric condensed phases and gaseous species. They have none of the difficulties associated with nonideal mixing of substances, which is the case for a "solution" database. However, internal self-consistency still must be maintained. For example, thermodynamic data for $C_{(s)}$, $O_{2<g>}$, and $CO_{2<g>}$ are held as individual entries, which provide their requisite properties as a function of temperature and pressure. However, when put together in a calculation, they must combine to give the correct Gibbs energy change for the reaction $C_{(s)} + O_{2<g>} \leftrightarrow CO_{2<g>}$. This is a simple example, but substance databases can contain more than 10,000 different substances; therefore, it is a major task to ensure internal self-consistency so that all experimentally known Gibbs energy of reaction are well represented. Examples of substance databases of this type can be found in Ref 39.

Solution databases, unlike substance databases, contain thermodynamic descriptions for phases that have potentially very wide ranges of existence, both in terms of temperature and composition. For example, the liquid phase usually extends across the whole of the compositional space encompassed by complete mixing of all of the elements. Unlike an ideal solution, the shape of the Gibbs energy space arising from nonideal interactions can become extremely complex, especially if nonregular terms are used.

Although it may seem an obvious statement, it is important to remember that thermodynamic calculations for multicomponent systems are multidimensional in nature. This means that it becomes impossible to envision the types of Gibbs energy curves illustrated in many teaching texts on thermodynamics and which lead to the easy conceptualization of miscibility gaps, invariant reactions, and so on. In multicomponent space, such things are often very difficult to understand, let alone conceptualize. Miscibility gaps can appear in ternary and higher-order systems, even though no miscibility gap exists in the lower-order systems, and the Gibbs phase rule becomes vitally important in understanding reaction sequences. Partition coefficients that apply in binary systems are usually altered in a multicomponent alloy and may change sign. Also, computer predictions can be surprising at times, with phases appearing in temperature/composition regimes where an inexperienced user may well not expect.

In practice, many scientists and engineers who require results from thermodynamic calculations are not experts in CALPHAD. As such, it is necessary to validate the database for multicomponent systems so that the user can have confidence in the calculated result.

Quantitative Verification of Calculated Equilibria in Industrial Alloys

This section gives examples of how CALPHAD calculations have been validated for materials that are in practical use and are concerned with calculations of critical temperatures and calculations for the amount and composition of phases in duplex and multiphase alloy types. These cases provide an excellent opportunity to compare predicted calculations of phase equilibria against an extensive literature of experimental measurements. This can also be used to show that the CALPHAD route provides results whose accuracy lies very close to what would be expected from experimental measurements. Validation of databases is a key factor in increasing the use of CALPHAD methodology for practical applications.

Calculations of Critical Temperatures. In terms of practical use, one of the most important features of phase equilibria can be the effect of composition on some critical temperature. This can be a liquidus or solidus or a solid-state transformation temperature, such as the β-transus temperature in a titanium alloy. The solidus value can be quite critical, because solution heat treatment windows may be limited by incipient melting. In some materials, a solid-state transformation temperature may be

of prime importance. For example, in titanium alloys, it may be specified that thermomechanical processing is performed at some well-defined temperature below the β-transus temperature. The CALPHAD route provides a method where such temperatures can be quickly and reliably calculated.

Steels. One of the most striking successes of the CALPHAD technique has been in the highly accurate calculation of liquidus and solidus temperatures. Because of their inherent importance in materials processing, there are numerous reported measurements of these values that can be used to judge how well CALPHAD calculations perform in practice. For example, detailed measurements of liquidus and solidus values for steels of all types have been made by Jernkontoret (Ref 40). The values were obtained on cooling at three different cooling rates: 0.5, 1, and 5 °C/s^{-1} (1, 2, and 9 °F/s^{-1}). The effect of cooling rate was not often high on the liquidus but could be quite profound on the solidus, due to the effects of nonequilibrium segregation during the liquid → solid transformation. Calculations for liquidus and solidus were made for these alloys (Ref 3) and compared with results obtained at the lowest cooling rate. Figure 3 shows the results of this comparison, and the accuracy of the predictions is impressive, particularly for the liquidus values, which exhibit an average deviation from

experiment ($\bar{d}$) of only 6 °C (11°F). It is also pleasing to note how well the solidus values are predicted, with an average deviation of just under 10 °C (18 °F). Three solidus values are not matched so well and are highlighted. In these alloys, low-melting eutectics were observed but not predicted, and it is uncertain if the difference is due to an inherent inaccuracy in the prediction or to the persistence of nonequilibrium segregation during solidification.

Titanium Alloys. In titanium alloys, there are numerous measurements of the β transus (the temperature above which the alloy becomes fully β), because this is a critical temperature for these alloys. Figure 4 shows the comparison between predicted (Ref 37) and measured (Ref 41 to 51) values for titanium alloys of all types, ranging from β-type alloys, such as Ti-10V-2Al-3Fe, through to the α-types, such as IMI834. The results exhibit an average deviation from experiment of less than 15 °C (27 °F), which is very good for the measurements of a solid-state transformation such as the β transus.

Nickel-Base Superalloys. In nickel-base superalloys, the temperature window where an alloy can be heat treated in the fully γ state is a critical feature, both in alloy design and practical usage. This heat treatment window is controlled by both the γ′ solvus temperature (γ'_s)

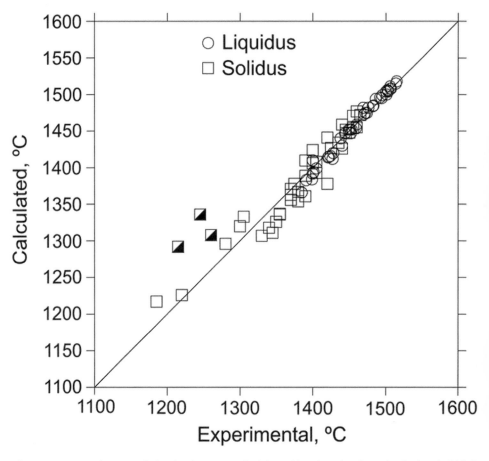

Fig. 3 Comparison between calculated and experimental solidus and liquidus values for steels. The three half-filled symbols are solidus temperatures that do not match as well

and the solidus, and there have been numerous experimental measurements of these properties. A further key feature for cast alloys is the liquidus; thus, numerous measurements have also been made for this temperature. Figure 5 shows a comparison between calculated (Ref 36) and experimental (Ref 52 to 61) values for γ'_s, liquidus (T_l), and solidus (T_s) for a wide variety of nickel-base superalloys. Average deviations from experiment are typically the same as for steels and titanium alloys, with $\bar{d}$ for liquidus and solidus being 6 and 10 °C, respectively, while $\bar{d}$ for the γ'_s is less than 15 °C (27 °F).

Calculations for duplex and multiphase materials are explained in the following.

Duplex stainless steels have been a fruitful area for CALPHAD calculations and have provided an example of the high level of success that has been achieved for practical materials. An early study by Hayes (Ref 62) demonstrated that very reasonable predictions for amounts of austenite could be obtained for a variety of different duplex stainless steels. The later work of Longbottom and Hayes (Ref 63) showed how combining CALPHAD calculation and experiment can provide very accurate formulae for the variation in austenite and ferrite as a function of composition and heat treatment temperature in Zeron 100 stainless steels. These formulae could then be used during production of the material to help define temperatures for thermomechanical processing.

Other calculations (Ref 3) for a wide variety of duplex stainless steels were made for the amounts of austenite as a function of temperature and the partition coefficients of various elements in austenite and ferrite and compared with experimental results (Ref 63 to 77) (Fig. 6, 7). In Fig. 6, the experimental results, given in volume fraction, have been compared with mole percent predictions, which is reasonable because molar volumes of the two phases are very similar. The $\bar{d}$ for the amount of austenite is less than 4%, of the same order as would be expected for experimental accuracy, and the comparison of elemental partition coefficients is extremely good. Carbon and nitrogen levels, which, in practice, are difficult to measure, are automatically calculated at the same time. Where such measurements have been made, the comparison is very good, which further emphasizes the advantage of using a calculation route.

Titanium Alloys. Duplex microstructures are usually formed in titanium alloys, and they are classed using the level of α- or β-titanium in the alloy. There are a few fully α and β alloys; most are duplex in nature. Much work has been done in measuring the β transus, but fewer results are available in the open literature for the variation of volume fraction and composition of α and β. Figures 8(a) to (c) show comparisons between calculated (Ref 37) and experimental (Ref 48, 50, 78 to 80) phase percent versus temperature plots for three types of commercial alloys with varying α levels: Fig. 8(a) Ti-6Al-4V; Fig. 8(b) SP700; and Fig. 8(c) Ti-10V-2Fe-3Al. Because the molar

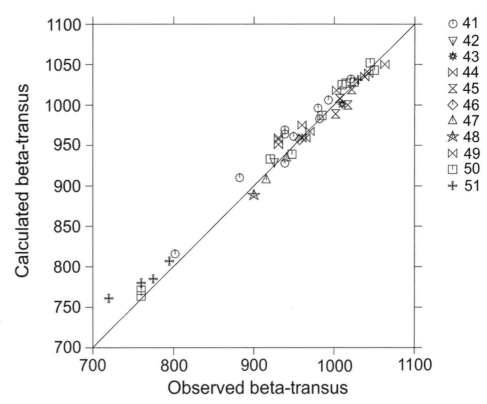

Fig. 4 Comparison between calculated and experimental β-transus temperatures in titanium alloys. Data reference numbers are given in the legend

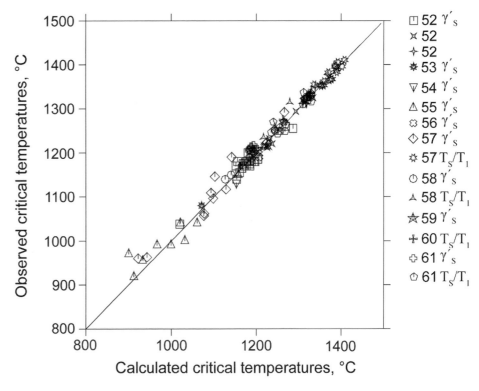

Fig. 5 Comparison between calculated and experimental critical temperatures for nickel-base superalloys. Reference numbers are given in the legend

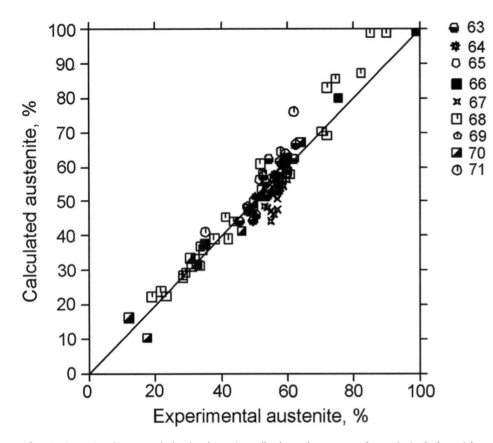

Fig. 6 Comparison between calculated and experimentally observed percentage of austenite in duplex stainless steels. Reference numbers are given in the legend. (Data from Ref 63 represent dual-phase steels.)

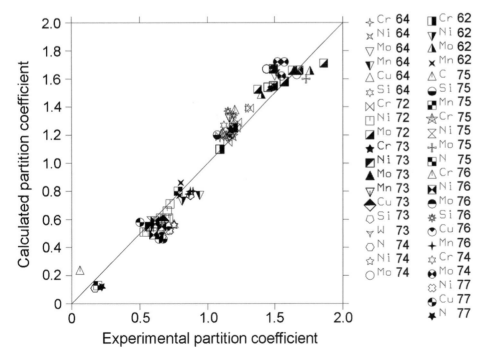

Fig. 7 Comparison between calculated and experimentally observed partition coefficients between austenite and ferrite in duplex stainless steels. Reference numbers are given for elemental partition coefficients

volumes of α and β are very similar, mole percent can be taken as equivalent to volume percent. The agreement is very satisfactory.

Nickel-Base Superalloys. In nickel-base superalloys, considerable work has been done on the determination of γ/γ′ equilibria, and substantial literature exists with which to compare CALPHAD calculations with experimental results. Figure 9 shows a comparison between experiment (Ref 55, 81 to 92) and calculation (Ref 36) for γ′ amounts in a wide variety of superalloys, ranging from low-γ′ types such as Waspaloy through very highly alloyed types such as IN939 to single-crystal alloys such as SRR99. The accuracy is similar to that for the duplex steels, with $\bar{d}$ of the order of 4%. In the comparison, results can be in either weight percent or volume percent. For the latter case, because lattice mismatches are so small, mole percent values give almost identical values to volume percent.

Figures 10(a) to (c) show some of the comparisons for the composition of γ and γ′, where the high standards of comparison between experimental (Ref 81 to 95) and calculated (Ref 36) results are maintained. Where experimental results have been quoted in weight percent, they have been converted to atomic percent to allow for consistency of comparison. The $\bar{d}$ for elements such as aluminum, cobalt, and chromium is close to 1 at.%, while for molybdenum, tantalum, titanium, and tungsten, this value is close to 0.5 at.%. The number of experimental values for hafnium and niobium were found to be too few to be statistically meaningful, but results for average differences appeared to be slightly better than obtained for molybdenum, tantalum, titanium, and tungsten.

Summary

It is clear from the results shown in this section that the CALPHAD route is providing predictions whose accuracy lies close to that expected from experimental measurement. This has significant consequences when considering CALPHAD methods in both alloy design and general everyday usage, because the combination of a high-quality, assessed database and suitable software package can, for a wide range of practical purposes, be considered as an information source that can legitimately replace experimental measurement.

Extending CALPHAD Methods to Model General Material Properties

While thermodynamic modeling provides useful information on specific aspects of materials processing, modern-day simulation packages, often using finite-difference/finite-element methods, rely extensively on more

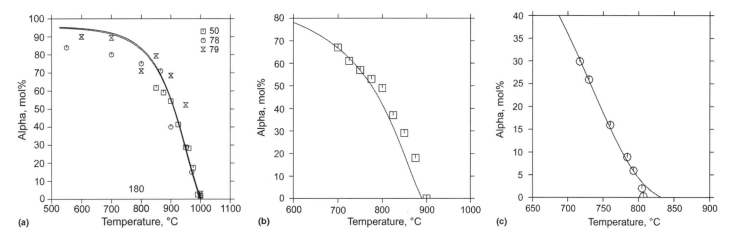

Fig. 8 Comparison of calculated mol% (line Ref 37) and experimental values (vol% considered equivalent to mol%) versus temperature for three duplex titanium alloys: (a) Ti-6Al-4V, (b) SP700, and (c) Ti-10V-2Fe-3Al. Sources of data: (a) given in legend, (b) Ref 48, and (c) Ref 80

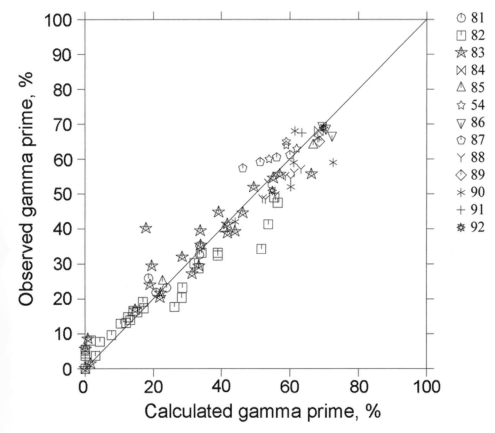

Legend:
- ☽ 81
- ⊡ 82
- ✦ 83
- ⋈ 84
- △ 85
- ☆ 54
- ▽ 86
- ⬡ 87
- Y 88
- ◇ 89
- ✳ 90
- + 91
- ✺ 92

Fig. 9 Comparison between observed and calculated amounts of γ′ in nickel-base superalloys. Reference numbers are given in the legend

energies, and so on could be reliably calculated, rapidly and as required, it became clear that CALPHAD calculations could be used in models already in existence for calculating material properties.

Figure 11 shows a flow chart that encapsulated the status in the late 1990s. Although it was possible to calculate phase formation very accurately and use CALPHAD methods for alloy development (Ref 96 to 100), the quantitative linkage to material properties was lacking. Computer models that directly used thermodynamic calculations for modeling phase transformations already existed, for example, in the calculation of time-temperature transformation (TTT) diagrams in carbon and low-alloy steels (Ref 30, 101 to 103) or the software package DICTRA that could consider more generalized diffusional phase transformations (Ref 104). Hardenability was linked to calculated TTT diagrams in steels (Ref 101), but beyond this, any link to predict material properties was invariably missing.

Over the last decade, the quantitative link to material properties has now been forged, and it is possible to calculate much of the property data required for the modeling and simulation of materials processes, both rapidly and on demand (Ref 105 to 109). The properties that can be calculated are wide ranging, and some are listed as follows:

- Temperature-dependent thermophysical and physical properties
    - a. Specific heat, enthalpy
    - b. Density, thermal expansion coefficient, linear expansion
    - c. γ/γ′ mismatch
    - d. Thermal and electrical conductivity
    - e. Liquid viscosity/diffusivity/surface tension
    - f. Young's/bulk/shear moduli, Poisson's ratio
- Phase transformations
    - a. CCT and TTT diagrams
    - b. γ′ and γ″ coarsening, γ′ microstructure modeling

general material properties, such as thermophysical and physical properties, flow stress data as a function of both temperature and strain rate, continuous cooling transformation (CCT) and CCT diagrams, and so on. The next section discusses how CALPHAD calculations can be directly extended to provide much of the required material property information for the modeling and simulation of

materials processes and to materials behavior in service.

With the increasing adoption of CALPHAD tools in industrial practice during the 1990s, it became clear that there was real potential to extend the method, such that general material properties could be modeled. When it was realized that phase formation, Gibbs energy driving forces for phase transformations, fault

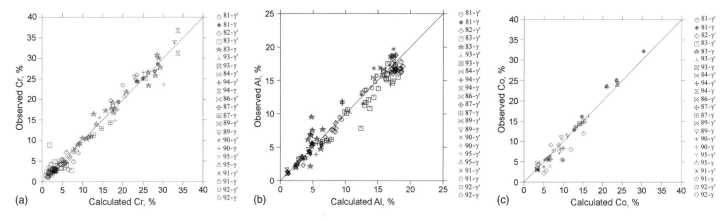

Fig. 10 Comparison of composition of γ and γ' determined experimentally, as given in the legend by reference numbers, with calculated results (Ref 37) in nickel-base superalloys for (a) chromium, (b) aluminum, and (c) cobalt

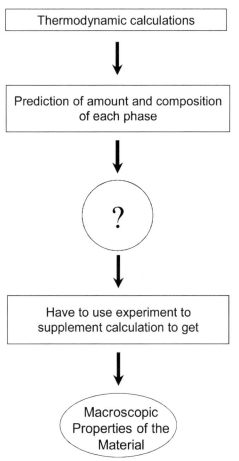

Fig. 11 Flow chart for the status of thermodynamic calculation in relation to predicting material properties

- Temperature- and strain-rate-dependent mechanical properties
  a. Proof stress, tensile stress, and hardness
  b. Stress-strain curves
  c. Creep and rupture life/strength
  d. Mushy zone properties

It is further possible to self-consistently calculate properties such that modeling of quench distortion and heat treatment is enabled by a link between phase transformations and physical and mechanical properties. This section briefly discusses the links that have made such a capability possible while providing a few examples of their use.

Modeling of Physical Properties—Effect of Alloy Composition on Solidification Simulations

Thermophysical and physical properties are an a priori requirement for process simulation of materials, not least of all for casting of alloys. However, while it is possible to calculate phase formation during solidification for a wide range of alloy types using the so-called Scheil-Gulliver (SG) solidification model, none of the requisite physical properties can be provided. The aim of this section is first to describe the SG model, then to show how the link with physical properties can be made, and finally to demonstrate its use for some aluminum casting alloys.

The SG model can be considered as a complementary limiting case to equilibrium solidification, whereby it is assumed that solute diffusion in the solid phase is small enough to be considered negligible and that diffusion in the liquid is extremely fast—fast enough to assume that diffusion is complete. The process that physically occurs can be envisioned as follows.

A liquid of composition C_o is cooled to a small amount below its liquidus. It precipitates out a solid with a composition $C_{S,1}$, and the liquid changes its composition to $C_{L,1}$. However, on further cooling, the initial solid cannot change its composition due to lack of back diffusion, and it is effectively isolated. A local equilibrium is then set up where the liquid of composition $C_{L,1}$ transforms to a liquid of composition $C_{L,2}$, and a solid with composition $C_{S,2}$ is precipitated onto the original solid with composition $C_{S,1}$. This process occurs continuously during cooling and when $k < 1$ leads to the solid phase being lean in solute in the center

of the dendrite and the liquid becoming more and more enriched in solute as solidification proceeds. Eventually, the composition of the liquid will reach the eutectic composition, and final solidification will occur via this reaction.

Any appearance of secondary phases can be easily taken into account in this approach if it assumed that no back diffusion occurs in them. Therefore, all transformations can be accounted for, including the final eutectic solidification. The limit to the SG simulation is that some back diffusion will take place. However, if the degree is small, good results will still be obtained.

For example, Backerud et al. (Ref 110) have experimentally studied almost 40 commercial alloys, and calculated results have been compared to all of these. Results of the comparisons of fraction solid versus temperature for some of these alloys are shown in Fig. 12. The agreement is most striking, and the level of accuracy achieved for these alloys is quite typical of that attained overall in the comparison.

While the CALPHAD route directly supplies important information such as heat evolution and fraction solid as a function of temperature, it does not directly supply any of the other material property requirements such as volume, thermal conductivity, viscosity, various moduli, and so on. To do so requires extensive property databases that can be linked to thermodynamic calculations and that allow the calculation of properties for the individual phases involved.

For individual phases in multicomponent systems, properties such as molar volume, thermal conductivity, Young's modulus, Poisson's ratio, and so on are modeled using pairwise mixture models, similar to those used to model thermodynamic excess functions in multicomponent alloys (see, for example, the section "Thermodynamic Models" in this article). For solidification, properties of the liquid are of prime importance, and Fig. 13 and 14 show comparisons between experimental densities (Ref 111, 112) and thermal conductivities with calculation (Ref 105) for a wide range of alloys in the liquid state. When the property of individual phases is defined, the property of the final alloy can be calculated using mixture

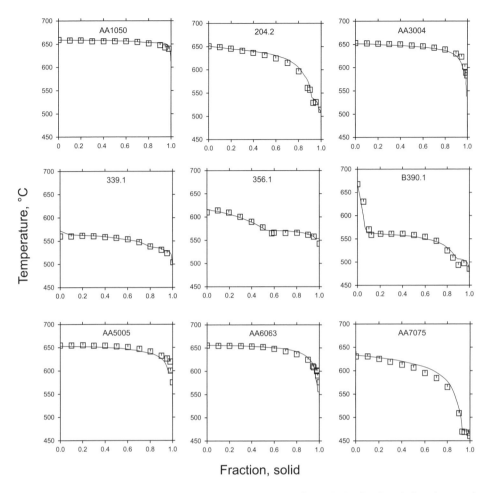

Fig. 12 Fraction solid versus temperature plots for various aluminum alloys calculated under Scheil conditions with experimental results (□) of Backerud et al. (Ref 110) shown for comparison

models that can account for the effect of microstructure on the final property (Ref 113, 114). Such models, which were developed for two-phase systems, have been extended to allow calculations to be made for multiphase structures. As such, physical properties can be calculated for any process where phase formation can be modeled.

Combining phase formation from the Scheil-Gulliver solidification model with physical property models means that almost all of the necessary input data for casting simulation packages can be calculated simultaneously and extremely rapidly. Data can then be readily exported to the software package for simulation to be undertaken.

The use of calculated data has some specific advantages:

- Data for new alloys, or alloys where little or no experimental data exist, can be provided rapidly.
- Data are calculated self-consistently, which is important because solidification is invariably a nonequilibrium process, where it is difficult to measure properties during the solidification process itself. This can mean

that input data for simulation are based on data sheets provided by the alloy supplier, in which case properties may be for heat treated alloys, where properties may be more closely associated with equilibrium rather than the nonequilibrium as-cast state. For example, the solidification range for the nickel-base superalloy 718 is reported in the relevant datasheet as 1260 to 1336 °C (Ref 115). However, the solidus during solidification is much reduced and lies closer to 1150 °C (Ref 116), providing a much extended solidification range. In some cases, it may be that the solidification range during actual casting is experimentally established but the physical properties may not be, in which case a self-consistent set of properties is almost impossible to obtain.

- A further advantage of obtaining data through calculations is that the variation in composition is automatically taken into account, whereas experimentally derived data files of properties held by simulation packages tend to provide data for a generic alloy type. So, for example, solidification simulations for various melts of 356 will use a single set of measured data that do

not take into account the effect of composition variations.

Examples of the use of calculated properties in casting simulations, with particular regard to composition sensitivity, are shown for two aluminum alloys.

Fluidity of Japanese Aluminum Alloy ADC12. Composition specifications provided for alloy types can be wide ranging, and it is known that there is variation in casting properties due to variations in the composition of alloy melts. To this end, a simple casting simulation was set up to model a fluidity spiral for an ADC12 silicon eutectic alloy (alloy similar to AA383) by using two compositions that were appropriate to the lowest and highest elemental levels of the specification.

Figure 15 shows the fraction solid versus temperature plots for the two alloys. The high-specification (HS) alloy is hypereutectic, with primary silicon and intermetallics forming over a significant temperature range, while the low-specification (LS) alloy forms approximately 22% primary aluminum. At the start temperature of eutectic solidification for the HS alloy (565 °C, or 1050 °F), the fraction solid for the LS alloy is ~65%, in comparison to ~5% for the HS alloy. The discrepancies between fraction solid at any temperature remain high for much of the solidification sequence, although both finally solidify via a eutectic involving Al_2Cu. Due to the very different behavior of the two alloys, there is a subsequent effect on all of the properties as a function of temperature. An example is the volume change in the range 450 to 650 °C (840 to 1200 °F), which again is quite different for the two alloys (Fig. 16).

The consequences to castability are very significant; to demonstrate potential effects, a simulation of fluidity using a spiral test was undertaken using ProCAST, with the input material data files created by calculation (Fig. 17). While the high-specification alloy has excellent fluidity, the low-specification alloy behaves, in comparison, quite poorly.

Hotspot Simulation for Casting of A356 into Truck Steps. While the case of the ADC12 alloy shows the effect of large changes in composition, it is also instructive to view the change in a casting simulation when only small changes are made to an alloy. To this end, a simulation for the die casting of the step of a truck, cast from the aluminum alloy 356, is used, and only the minor elements are changed (Ref 117). For example, one composition was Al-0.01Cu-0.2Fe-0.3Mg-0.02Mn-7Si-0.025Zn (wt%), while the other had higher levels of copper (0.25%), manganese (0.3%), and zinc (0.35%).

Although the effect of changes in composition on fraction solid versus temperature behavior is much smaller in comparison to the ADC12 alloy, the effect on the formation of isolated hotspots is quite significant (Fig. 18), leading to potentially important differences in defect formation between the two cases.

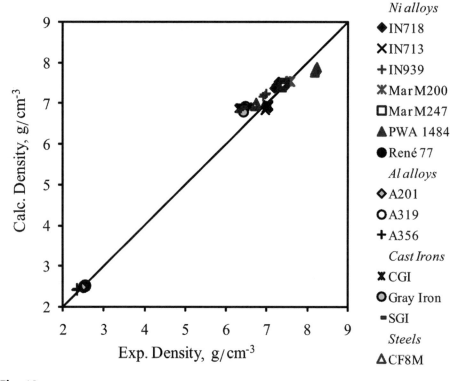

Ni alloys
◆ IN718
✕ IN713
+ IN939
✳ MarM200
☐ MarM247
▲ PWA 1484
● René 77

Al alloys
◇ A201
○ A319
+ A356

Cast Irons
✖ CGI
◎ Gray Iron
▬ SGI

Steels
△ CF8M

Fig. 13 Comparison between calculated and reported (Ref 111) density for various liquid commercial alloys

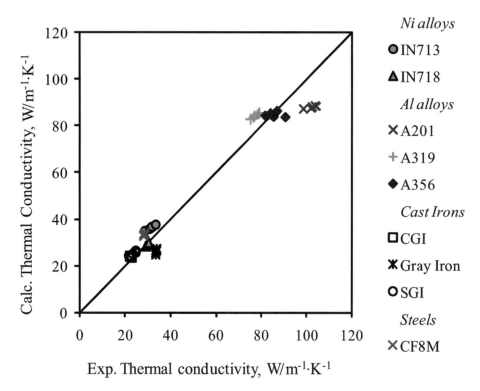

Ni alloys
◉ IN713
▲ IN718

Al alloys
✕ A201
+ A319
◆ A356

Cast Irons
☐ CGI
✖ Gray Iron
○ SGI

Steels
✕ CF8M

Fig. 14 Comparison between calculated and reported (Ref 111, 112) thermal conductivities for various liquid commercial alloys

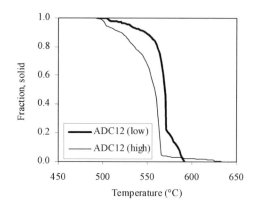

Fig. 15 Calculated fraction solid curves for a low- and high-specification ADC12 aluminum casting alloy

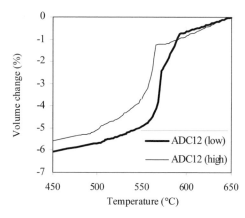

Fig. 16 Calculated volume change versus temperature plots for a low- and high-specification ADC12 aluminum casting alloy

Modeling of Temperature- and Strain-Rate-Dependent Strength

A major success in the extension of thermodynamic modeling to the modeling of more general material properties has been in the area of mechanical properties, particularly with respect to the temperature and strain-rate dependency of high-temperature flow stress (Ref 106, 109). Generally speaking, room-temperature strength decays gradually with increasing temperature up to the point where it enters into a temperature regime where there is a sharp fall in strength, a substantial increase in ductility, and the flow stress becomes much more strongly dependent on strain rate. Figure 19 shows such a behavior for the nickel-base superalloy Nimonic 75. The sharp drop in strength is due to a change from a deformation mechanism dominated by dislocation glide (DG) at low temperatures to one dominated by dislocation climb (DC) at higher temperatures, the controlling mechanism for creep.

To model the strength of an alloy, the first requirement is the strength of the matrix, usually a solid-solution phase such as austenite or

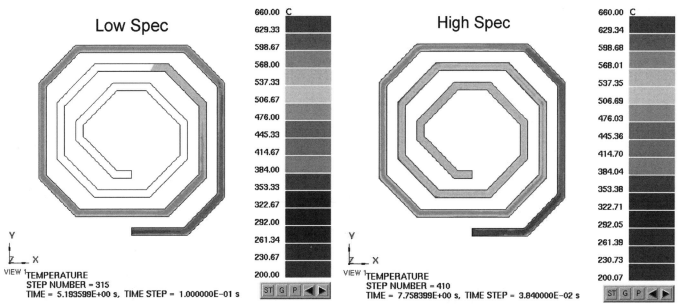

Fig. 17 Comparison of the spiral test result, based on the physical properties calculated from JMatPro for alloy ADC12 of low and high specification. Courtesy of UES Software Asia, generated using ProCast

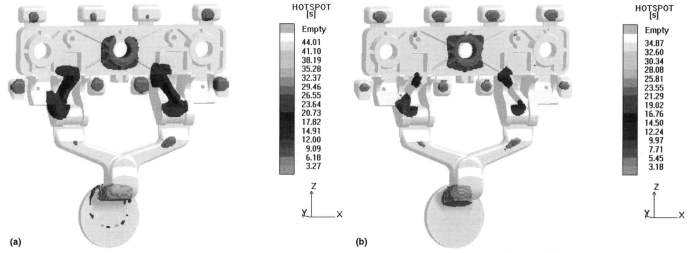

Fig. 18 Comparison of the hotspot result, based on the physical properties calculated from JMatPro for aluminum alloy 356 of (a) low and (b) higher levels of copper, manganese, and zinc. Courtesy of Magma GMBH, generated using MagmaSoft

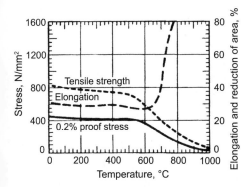

Fig. 19 Tensile properties for a Nimonic 75 nickel-base superalloy. Source: Ref 55

ferrite in steels or the γ phase in nickel-base alloys and so on. Any hardening effects, such as those found with γ' strengthening in nickel-base superalloys, must then be added. Strain-rate and temperature dependency is further required. The next sections describe the development of models to take such effects into account in both the DG and DC regimes.

Solid-Solution Strengthening. In solid-solution alloys, when the phase or phases present in the alloy and their composition are known, it is possible to model strength at lower temperatures in a similar way to physical properties, using pair-wise mixture models. Using a standard Hall-Petch equation, $\sigma_y = \sigma_0 + kd^{1/2}$, strength ($\sigma$) dependence due to grain size (d) can be provided. Figure 20

shows a comparison between experimentally measured and calculated 0.2% proof strength for a variety of solid-solution alloys (Ref 118), which include duplex stainless steels. Using relationships developed by Tabor (Ref 119), it is further possible to interrelate hardness, tensile strength, and proof stress as well as to predict stress/strain curves.

Precipitation Strengthening by γ'. Models for the strengthening of nickel-base superalloys by γ' have existed for many years (Ref 120). However, their use in practice is historically limited, the main reason being that the models require inputs that are extremely difficult to measure and obtain in a self-consistent fashion. Following Brown and Ham (Ref 121), the yield stress of a γ'-hardened alloy can be derived as:

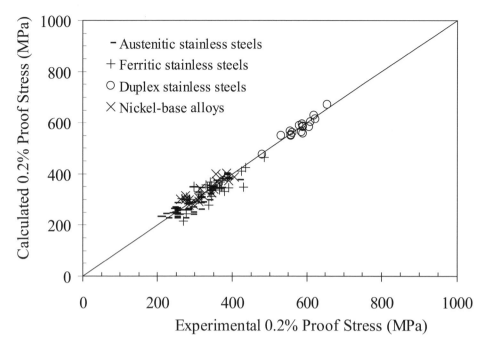

Fig. 20 Comparison between calculated and experimentally measured 0.2% proof stress in iron- and nickel-base solid-solution alloys. Source: Ref 118

$$YS = YS_0 + M \frac{\Gamma}{2\mathbf{b}} \left[A \left(\frac{\Gamma f d}{\tau} \right)^{1/2} - f \right] \quad \text{(Eq 22)}$$

where YS_0 is the yield (proof) stress of the γ matrix, M is the Taylor factor (Ref 122) that relates the proof stress in polycrystalline material and critical shear stress in single-crystal specimens (≈ 3 for face-centered cubic, or fcc, materials, Ref 122), Γ is the antiphase domain boundary (APB) energy of γ' in the {111} plane, $\mathbf{b}$ is the Burgers vector of dislocation, d is the particle diameter, f is the volume fraction of γ' precipitates, τ is the line tension of the dislocation, and A is a numerical factor depending on the morphology of the particles, which, for spherical particles, equals 0.72. Equation 22 is used for small γ' particles, but equations for larger particles exist that use similar input parameters (Ref 123).

While it is conceivable to experimentally determine YS_0 and f, it would be very time-consuming and therefore very restrictive for a predictive model. Other parameters, such as Γ and τ, are, in any case, highly problematical to obtain experimentally. Therefore, it can easily be understood why such equations were rarely used.

However, using thermodynamic calculations, the phase amounts and compositions are rapidly obtained, which provides f. The solid-solution-strength calculations providing YS_0 and τ can be calculated from the shear modulus of the γ solid solution. This leaves Γ as the only remaining input parameter to be obtained, which can be done as follows.

In a perfectly ordered superlattice, such as the $L1_2$, the position of the various unlike and like atoms is prescribed by the ordering of the superlattice. When dislocations pass through this ordered structure, a breakdown of local chemical order ensues, with the subsequent creation of an APB. This boundary has a characteristic energy dependent on the change in the number of like and unlike bonds across the boundary. The number of such bonds across the APB is known from crystallographic considerations and, if the energy of the various bonds can be calculated, the APB energy can also be calculated. Using thermodynamic calculations, the various bond energies can be calculated at will (Ref 124); hence, all of the required input for strength of a γ'-hardened nickel-base superalloy can be obtained a priori. The success of the model has been demonstrated by extensive comparison with experiment (Ref 106, 125).

Temperature- and Strain-Rate Dependency of Strength in the Dislocation-Glide-Controlled Temperature Regime. Examination of the yield/proof stress as a function of temperature, $\sigma(T)$, for many different types of alloys shows a clear correlation between the rate of decrease in $\sigma(T)$ with increasing temperature and the room-temperature 0.2% proof stress (σ_{RT}). The decay is well matched using an exponential form of the following type:

$$\sigma(T) = \alpha + \beta \exp\left(\frac{-Q}{RT} \right) \quad \text{(Eq 23)}$$

where α and β are constants directly related to σ_{RT}, and the value of Q is determined empirically through regression analysis based for each alloy type, for example, whether they are steels,

titanium alloys, austenitic or ferritic stainless steels, and so on.

Well-established approaches to strain-rate sensitivity (Ref 126) can also be applied and take the form:

$$\sigma(T, \dot{\varepsilon}) = \sigma_o(T) \left(\frac{\dot{\varepsilon}}{\dot{\varepsilon}_o} \right)^m \quad \text{(Eq 24)}$$

where $\sigma_o(T)$ is the yield/proof stress at the standard strain rate of $\dot{\varepsilon}_o$, $\dot{\varepsilon}$ is the applied strain rate, and m is the strain-rate-dependency factor.

Strength in the Dislocation-Climb-Controlled Temperature Regime. As mentioned in the introduction to the section "Modeling of Temperature and Strain-Rate-Dependent Strength" in this article, the predominant deformation mechanism at high temperatures is controlled by creep and hence dislocation climb. Therefore, to predictively model high-temperature strength, it is necessary to model creep processes.

In fcc-based alloys, such as nickel-base superalloys, austenitic stainless steels and, in reality, most steels at high temperatures, the secondary creep rate can be calculated using a model that features both a back stress function and takes the stacking fault energy (γ_{SFE}) explicitly into account. The approach has the advantage that it contains parameters that have an identifiable physical basis and can be calculated self-consistently. The ruling equation is taken as (Ref 106, 127):

$$\dot{\varepsilon}_s = AD_{\mathrm{eff}} \left(\frac{\gamma_{SFE}}{G\mathbf{b}} \right)^m \left(\frac{\sigma - \sigma_o}{E} \right)^n \quad \text{(Eq 25)}$$

where $\dot{\varepsilon}_s$ is the secondary creep rate, A is a materials-dependent parameter, D_{eff} is the effective diffusion coefficient, γ_{SFE} is the stacking fault energy of the matrix at the temperature of creep, $\mathbf{b}$ is the Burgers vector, σ is the applied stress, σ_o is the back stress, and G and E are the shear and Young's moduli of γ at the creep temperature, respectively. The back stress, σ_o, is calculated following the treatment of Lagneborg and Bergman (Ref 128), such that $\sigma_o = 0.75\sigma$ when $\sigma < 4\sigma_p/3$ (where σ_p is the critical back stress from strengthening due to precipitates), and $\sigma_o = \sigma_p$ when $\sigma > 4\sigma_p/3$. The exponents m and n can have a range of values in the literature, but self-consistent application in the model shows that fixed values of $m = 3$ and $n = 4$ can account for many alloy types. Models for body-centered cubic materials are similar in general form to Eq 25 but do not use a corresponding fault energy in their formulation.

Application of Eq 25 requires knowledge of the composition of the matrix phase at temperature, so that an effective diffusion coefficient and the shear and Young's moduli can be calculated. Also, the back stress due to precipitation hardening must be estimated. These parameters are now readily calculated as described previously, and the only parameters then required are γ_{SFE} and A. The parameter γ_{SFE} is readily obtained by thermodynamic calculation, because it can be directly related to the Gibbs energy difference between the fcc and close-packed hexagonal phases (Ref 129). This then leaves one parameter, A, which is empirically evaluated and taken as a constant.

The creep model can now be applied directly to high-temperature strength by assuming that the alloy will yield via creep when the strain rate of the mechanical test is equal to or slower than the creep rate at the testing temperature. Figure 21 shows the comparison between experimental and calculated yield stress versus temperature for two alloys (Ref 106), one a solid-solution alloy (Nimonic 75) and the other hardened by γ' precipitates (Nimonic 105). For Nimonic 105, in the creep-controlled region, the alloy is weakened by the gradual removal of γ' to the point that, above its γ'_s of 1025 °C (1877 °F), it becomes fully γ. The method has been applied to a wide variety of nickel-base superalloys, and excellent agreement is found with the temperature dependency of the yield/proof stress (Ref 109).

Calculation of Temperature- and Strain-Rate-Dependent Stress/Strain Diagrams. It is quite straightforward to calculate stress/strain curves in the low-temperature dislocation glide regime using standard formulae. However, no such formulae exist for creep-controlled deformation. To do so requires that creep models include both primary and tertiary creep. This allows the calculation of full creep curves as a function of applied stress and the subsequent construction of a three-dimensional surface that has as its axes stress, strain, and time (Ref 130). By tracking the surface of the stress/strain/time envelope at a given temperature and a constant strain rate, it is possible to calculate the flow stress as a function of time and strain, which directly provides a stress/strain diagram. The model has been tested for a wide range of alloy types, including steels, titanium alloys, and nickel-base superalloys (Ref 109, 130).

Three types of true stress/strain diagrams are produced (Fig. 22):

- *Type 1:* The classic low-temperature type, exhibiting continual work hardening to failure
- *Type 2:* The often-observed form of high-temperature diagram with only a small work hardening in the early strain stages, followed by gradual and increasing flow softening
- *Type 3:* A third region, where there is initially substantial work hardening before flow softening occurs

While the form and type of diagram is well established for the type 1 calculation, it is instructive to more closely compare with experiment the form at high temperatures for the type 2 and 3 regions.

Figure 23 shows calculated stress/strain diagrams for Ti-6Al-4V (extra-low interstitial) at various temperatures between 800 and 1050 °C (1470 and 1920 °F) (Ref 109). While

the agreement with experiment is striking, it is equally noteworthy that Ti-6Al-4V is a two-phase alloy below 1000 °C (1830 °F), so the curves at 850, 900, and 950 °C (1560, 1650, and 1740 °F) are duplex microstructures of α-titanium and β-titanium, with α-titanium becoming predominant at 850 °C (1560 °F), while β-titanium is predominant at 950 °C (1740 °F). Therefore, the success of the calculation not only relies on sound creep models for both phases, but it is necessary to have a sound model for obtaining the amount of each phase. The effect of strain rate is also well matched, with Fig. 24 showing the comparison between calculated and experimental stress/strain curves

for the Ti-6Al-4V alloy at 950 °C (1740 °F) as strain rate is varied between 1×10^{-3} and $100\,s^{-1}$.

Examples of stress/strain curves for type 3 behavior are shown for a carbon steel between 800 and 1200 °C (1470 and 2190 °F) in Fig. 25. There is a clear region of substantial work hardening followed by flow softening, which is again well matched by calculation.

It has been more general to consider that flow softening is a result of initial work hardening, followed by recovery and recrystallization that softens the alloy. However, such models have not achieved significant predictability and are often used only to replicate experimentally observed behavior. By contrast, the model

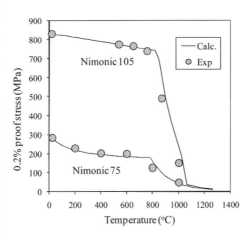

Fig. 21 Comparison between experimental and calculated yield stress for Nimonic 75 and 105 as a function of temperature. Source: Ref 106

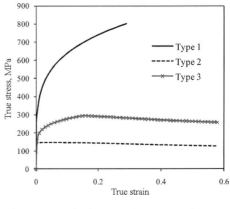

Fig. 22 Calculated true stress/strain curves for a 316 stainless steel showing temperature-dependent types of behavior. Type 1 = low-temperature dislocation glide (DG) controlled; type 2 = high-temperature dislocation climb (DC) controlled; and type 3 = combined DG and DC controlled

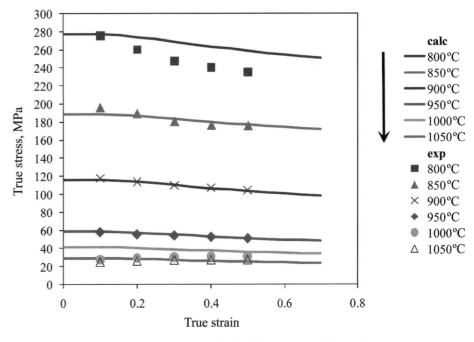

Fig. 23 Comparison between experimental and calculated flow stress curves for Ti-6Al-4V at various temperatures with strain rate 0.1/s. Source: Ref 109

described here is both predictive and reproduces known experimental behavior extremely well. As further examples, Fig. 26 shows the comparison between calculated and experimental flow stress at specific strain rates and temperatures for the nickel-base superalloy 718 (Fig. 26a) and the stainless steel 316 (Fig. 26b).

The reason that the temperature range of flow softening (here, considered due to creep) coincides with recovery and recrystallization may be due to the similarity in some of the fundamental physical processes involved. In both scenarios, substantial diffusion is required, and climb is important for both processes. In addition to accurately predicting high-temperature mechanical properties, there are further reasons to believe that creep-controlled deformation is appropriate. For example, it is noted that flow softening occurs at very high strain rates, >1 s^{-1}, where it is unlikely that there is sufficient time for large-scale subgrain and grain-boundary rearrangements required for recovery and recrystallization to take place. It is further noted that a creep-dominant mechanism also accounts for the frequently observed increase in ductility in the high-temperature range (for example, Fig. 22).

Linking of Thermodynamic, Kinetic, and Material Property Models

Kinetic modeling has been linked with thermodynamic calculations since the inception of the CALPHAD method. There is a natural linkage, in that many kinetic models require driving forces and transition temperatures, for example, the Ae$_3$ or Ae$_1$ temperatures in steels, as well as diffusion coefficients and/or mobilities.

Steels. To this end, particularly in steels, kinetic formalisms have been successfully linked with dilute solution thermodynamic calculations for the calculation of TTT and CCT diagrams (Ref 101–103). The model of Bhadeshia (Ref 102, 103) is of particular interest because the thermodynamic calculations were based on para-equilibrium and include explicit driving forces, rather than parameters based on undercooling below the requisite transition temperatures. The work of Kirkaldy and co-workers (Ref 101) further linked their model to hardenability and presented numerous results of comparison of calculated results with experiment.

The linking of kinetic models with material property models has been applied mostly with steels. This is almost certainly due to austenite decomposition models being well established, and therefore, the requisite model parameters can be empirically derived from experimental TTT or CCT diagrams. Established room-temperature properties of the individual phases present, that is, whether they are strength or thermal conductivity of bainite, pearlite, and so on, can be used in simple mixture models to predict the room-temperature as-quenched property. However, more recent work has attempted to link thermodynamic and kinetic models with material properties over the whole temperature range.

For example, Miettinen and co-workers (Ref 131, 132) have developed a combined thermodynamic and kinetic model for the cooling-rate-dependent solidification of steels and linked this to an austenite decomposition model to provide prediction of phases in low-alloy and stainless steel types from room temperature to the liquid state. Using mainly dilute solution-type models for physical properties, these authors were also able to link the phases present with physical properties and to provide a significant extension to the phase-versus-temperature predictions.

More recently, CALPHAD calculations have been linked to an austenite decomposition model, based on an extended Kirkaldy formalism, to predict TTT and CCT diagrams for a wide range of steels, from carbon and low-alloy types to medium- and high-alloyed steels such as roll steels, tool steels, and various types of stainless steels (Ref 133, 134). When this is linked with both strength and physical property models, it is then possible to calculate a full range of material properties

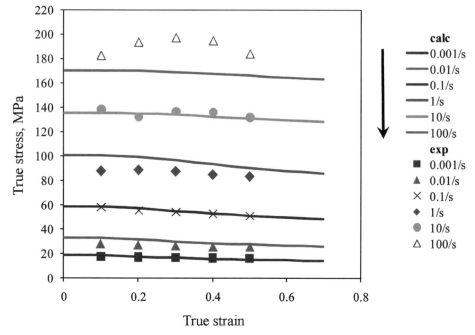

Fig. 24 Comparison between experimental and calculated flow stress curves for Ti-6Al-4V at 950 °C and at various strain rates. Source: Ref 109

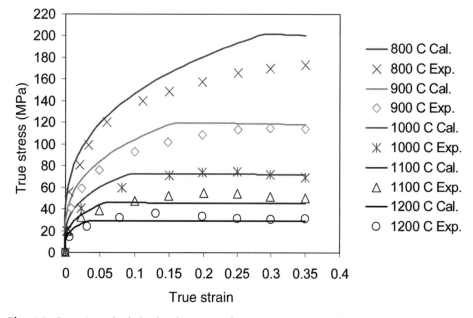

Fig. 25 Comparison of calculated and experimental stress-strain curves at 0.1/s for a carbon steel at various temperatures. Source: Ref 109

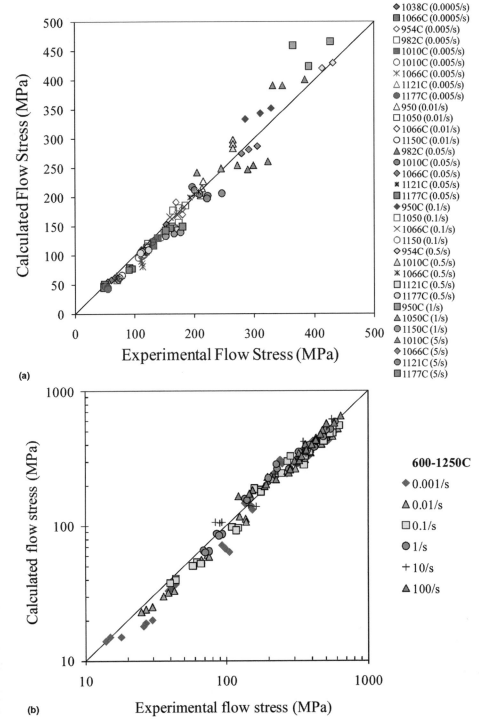

Fig. 26 Comparison between experimental and calculated flow stress for (a) the nickel-base superalloy 718 (Ref 129) and (b) stainless steel 316 at various temperatures and strain rates

to empirically fit model parameters for austenite decomposition so that the subsequently calculated TTT or CCT diagram reasonably matches experiment. Simple kinetic equations, such as a very basic Johnson-Mehl-Avrami type, may be used, for example:

$$X = 1 - \exp(-Kt^n)$$

where X is the fraction transformed at time t, K is a constant, and n is the time exponent, often called the Avrami exponent. Both K and n are then empirically fitted for each transformation product and may be temperature dependent. Knowing the amount of transformation coupled with simple mixture models for individual phase properties thus allows property changes, for example, volume as a function of time/temperature, to be calculated simultaneously with a process simulation that considers both heat transfer and stress evolution.

The models that are successfully used for the predictive modeling of austenite decomposition are invariably more complex and subsequently difficult, if not impossible, to implement interactively. However, it is possible to empirically fit temperature-dependent model parameters for austenite decomposition to the calculated TTT diagram and export calculated material properties for each phase to the requisite data file used by the process simulation software.

Nickel-Base Superalloys. The link between a combined thermodynamic and kinetic model and material property models has been much less explored outside of steels. The lack of general application to other material types has begun to be addressed through a link between γ'-phase evolution models and subsequent material properties of nickel-base superalloys.

The TTT and CCT diagrams have been successfully calculated for nickel-base superalloys (Ref 106, 130), and it has recently proved possible to link a combined thermodynamic and kinetic model for γ' microstructure evolution with strength calculations (Ref 136). This particular case represents the integration of material property modeling with thermodynamic calculation at a high level and provides a very satisfactory last example for this section.

A modified Johnson-Mehl-Avrami model has been developed that can be generally applied to calculate phase transformations for a variety of material types (Ref 137). The model allows the morphology of the precipitate to be considered as well as specifics associated with potential nucleant sites. For γ' precipitation, a spherical particle is assumed, and, for the case of steady-state nucleation, the governing equations can be written as:

$$X = \frac{V}{V_{eq}(T)} = 1 - \exp\left(-fN_rG_r^3t^4\right) \qquad \text{(Eq 26)}$$

where X is the volume fraction of the product phase, V is the volume transformed, $V_{eq}(T)$ is the equilibrium volume amount of the phase at temperature T, f is a shape factor with a value close to unity, N_r is the nucleation rate, G_r is

as a function of processing conditions. For example, Fig. 27 and 28 show calculated (Ref 134) mechanical and physical properties of an 8620 steel as a function of cooling rate. Figure 29 shows the subsequent room-temperature mechanical properties in the form of Jominy hardenability, which is well matched to experiment (Ref 135).

It is possible to export information to process simulation software such that modeling of manufacture of components is possible. For example, such information is of direct relevance to the simulation of quench distortion. A key element for the use of calculated data is that compatibility with the simulation software is required. A trend in heat treatment software is

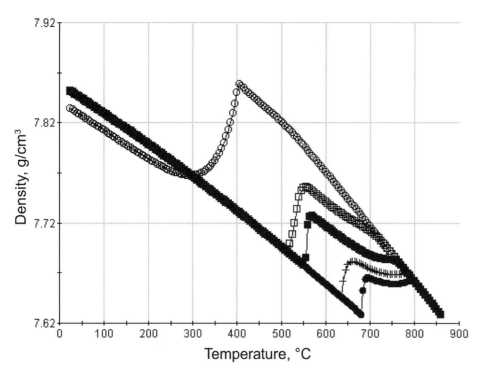

Fig. 27 Calculated density for an 8620 steel during quenching at various cooling rates ranging from 0.01 to 100 °C/s. Source: Ref 134

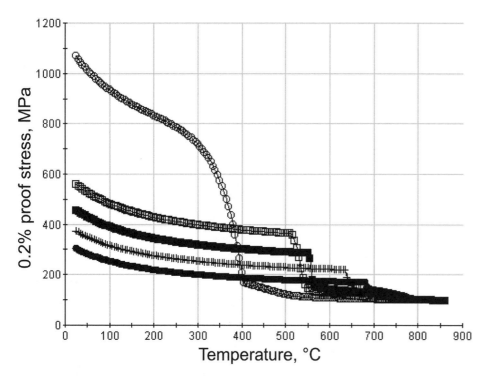

Fig. 28 Calculated 0.2% proof stress for an 8620 steel during quenching at various cooling rates ranging from 0.01 to 100 °C/s. Source: Ref 134

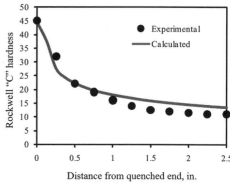

Fig. 29 Comparison between calculated and experimental Jominy hardenability for an 8620 steel. Source: Ref 134

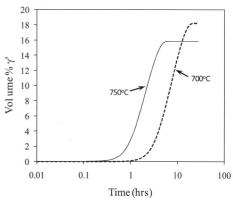

Fig. 30 Calculated volume percent γ' formed at 700 and 750 °C in the nickel-base superalloy Nimonic 80A

where N_o is the total number of active nucleant sites. The methods used for calculating N_r and G_r have been described in detail by Li et al. (Ref 137), and all required information can be obtained from thermodynamic calculation and readily calculable diffusion coefficients. As is known from practice, the formation of γ' is very rapid, with little undercooling below the γ' solvus required before transformation occurs. This rapid transformation also means that site saturation is rapidly achieved, and, for all of the studies attempted so far, it is Eq 27 that appears to be the ruling equation.

Initial testing and validation of the initial kinetic model was performed at usual isothermal, hardening heat treatment temperatures (e.g., <850 °C, or 1560 °F), and Fig. 30 shows the calculated evolution of volume percent γ' versus time at a constant temperature for Nimonic 80A at 750 and 700 °C (1380 and 1290 °F). It is interesting to note that transformation for both cases is virtually complete in the time provided by the heat treatment schedules, 4 and 16 h, respectively, for 750 and 700 °C (1380 and 1290 °F).

At these temperatures, coarsening is not significant within the time scale of heat treatment.

the growth rate, and t is time. For most cases, and especially so for γ', the volume difference between parent and product phases is similar enough that volumes can be interchanged with mole percent values, and V_{eq} in the present case is directly taken from the equilibrium mole percent calculation. For the case where site saturation occurs:

$$X = 1 - \exp\left(-fN_o G_r^3 t^3\right) \qquad \text{(Eq 27)}$$

However, during processing of nickel-base superalloys with medium to high levels of γ', the component is invariably cooled from a high-temperature first-stage treatment, often above the γ' solvus. During the cooling process, γ' is formed, and the amount can be substantial, particularly for alloys with high-γ'-solvus temperatures. In such cases, the temperature range of γ' formation occurs at temperatures and time scales where coarsening occurs. If Eq 27 is the ruling equation, this means that N_o will decrease as particle size increases, and transformation rates will decrease.

Recently, Li et al. (Ref 138) have shown how a combination of thermodynamic calculations and existing theory of Ostwald ripening (Ref 139–141) can be used to calculate coarsening rates of nickel-base superalloys to a high level of accuracy. The thermodynamic calculations provide critical information concerning the composition of γ' and γ'' and allow the calculation of the γ/γ' and γ/γ'' interfacial energy (σ) for use in the relevant kinetic equation:

$$(\bar{r}(t))^3 - (\bar{r}(0))^3 = \left(\frac{8D\sigma N_\alpha(1-N_\alpha)V_m}{9\varepsilon_\alpha(N_\beta - N_\alpha)^2 RT}\right)t$$

(Eq 28)

where $\bar{r}(0)$ is the mean radius at time $t = 0$, D is the diffusion coefficient, N_α and N_β are the mole fractions of solute in the matrix and particle, respectively, V_m is the molar volume of γ', ε_α is the Darken factor, R is the gas constant, and T is the temperature of coarsening. All of the required input can be readily calculated using existing thermodynamic and property models, and agreement with experiment is excellent (Ref 138).

The simplest method to include coarsening, for both isothermal and cooling transformations, is to first calculate X using Eq 26 and 27 and to apply simple additivity rules (Ref 142). Transformation is calculated for discrete time intervals, and a sum of fraction transformed is obtained. For an isothermal case, the transformation occurs at a constant temperature, with constant values for $V_{eq}(T)$, N_r, and G_r. For continuous cooling, small isothermal steps are taken, and $V_{eq}(T)$, N_r, and G_r are calculated for each temperature.

The procedure adopted is to calculate the transformation in the discrete time interval at a constant temperature and to calculate coarsening. During cooling, the time interval corresponds according to the chosen cooling rate. In the first stages of the procedure, the size of γ' is simply calculated using V and either the number of nuclei formed during steady-state nucleation or, when site saturation occurs, by using N_o. However, at some point, the combination of coarsening rates and time will allow γ' to coarsen. As mentioned earlier, because of the rapid transformation kinetics, site saturation occurs rapidly. Equation 27 is invariably the ruling equation at this time, in which case, the effect of coarsening is to reduce the number of γ' particles, effectively reducing the value of N_o. At the next step, a new value for N_o is calculated from V and the coarsened particle diameter.

Figure 31 shows the volume percent γ' versus temperature plot for a U720LI alloy on cooling from a supersolvus heat treatment at 1168 °C (2134 °F). Three cooling rates are shown, 0.5, 1, and 5 °C/s^{-1} (1, 2, and 9 °F/s^{-1}), with associated calculated γ' particle diameters after cooling. The effect of including coarsening during cooling is that there is an initial rapid increase in the γ' particle diameter for the very low volume fraction transformed. This appears to be due as much to the rapid rate of coarsening of very fine γ' particles as to the high coarsening rates existing at temperature. The effect of coarsening on γ' size decreases quite rapidly on further cooling, and the γ' particle diameter then increases almost exclusively by growth of γ' particles due to the increasing volume of γ' that is formed (Fig. 32).

The model is currently implemented for a two-stage heat treatment process, that is, a solution treatment plus an isothermal aging treatment, but can be extended to more complex heat treatment schedules. It can also be used for cast alloys. The model predicts the type of multimodal microstructures that are often formed in nickel-base superalloys, and comparison between calculated and experimentally observed γ' microstructures is in good agreement (Ref 136).

The γ' microstructure information can now be linked directly with strength models, as described in the section "Precipitation Strengthening by γ'" in this article. Figure 33 shows a comparison between the calculated and experimentally determined room-temperature strengths of various nickel-base superalloys (Ref 134). Considering that the only required input to the model is the composition of the alloy and the heat treatment schedule, the agreement is highly satisfactory.

Such modeling has significant use, not least of all in designing heat treatment schedules. Another direct application is in the processing of turbine disks, particularly as disk sizes increase and nickel-base superalloys become more commonly used in industrial gas turbines for power generation. In this case, cooling rates through a large forging may vary far more significantly than

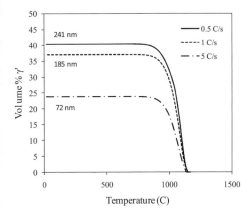

Fig. 31 Calculated volume percent of γ' formed in a U720LI superalloy during cooling at various rates, with calculated γ' particle diameters included

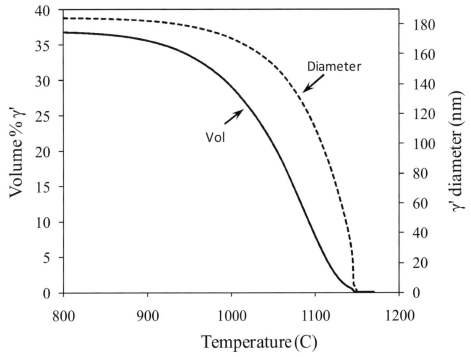

Fig. 32 Calculated volume percent and diameter of γ' formed in U720LI during cooling at 1 °C/s^{-1}

in smaller gas turbines used for aeroengines, producing much larger strength variations within a disk.

Summary and Observations for the Future

This article has aimed to describe the current state-of-the-art with respect to the application of thermodynamic and material property modeling to process simulation of alloys. It has also been the aim to show, as clearly as possible, the key elements that enable the models to be applied to industrial alloys.

Figure 11 summed up the status of CAL-PHAD calculations with respect to predicting more general material properties at the turn of the century. Over the last decade, a direct linkage to material property models has been made, such that the status has been transformed to that shown in Fig. 34. This transformation has been led by the development of the software program JMatPro (Ref 143), which has demonstrated that many of the material properties required by finite-element/finite-difference (FE/FD) process simulation software can be calculated rapidly, on demand, and the results exported such that they can be directly used by the relevant software.

Such a capability has significant consequences because, in practice, it can be very time-consuming and expensive to experimentally determine the full range of material properties required for process simulation modeling. As such, there are very few alloys

where the requisite property data have been fully covered by experiment over the whole temperature range of processing. There are other advantages of using material property modeling, not the least of which is that the property calculations are made on a fully self-consistent basis, and it is possible to calculate properties that are either very difficult or currently impossible to measure, for example, rheological properties of the liquid in the mushy zone.

It has been recognized for some time that material models are key to an integrated approach to computational materials engineering (Fig. 35, Ref 144). However, a further key requirement is that such models are predictive in nature, which is one of the features provided through the linking of thermodynamic and material models. As such, it becomes quite possible to envisage that the virtual simulation of the complete cycle of materials processing, without the requirement for prior experiment, will become achievable within the coming years. Such simulation will additionally include the prediction for the performance of the final manufactured part.

An example of what is currently possible would be the manufacture of a nickel-base turbine disk alloy. The first part of the process route would be solidification processing of an ingot, followed by thermomechanical processing of the ingot into the disk itself, followed by the final heat treatment. Process simulation packages exist that can model each step of the process route, but they all require material properties that can now be calculated. For example, property models exist for general use in solidification

processing, which include rheological properties for the liquid that can be used for prediction of freckles and macrosegregation. All of the requisite properties for thermomechanical processing can be provided, and models now exist for predicting the final γ' microstructure as a function of the heat treatment schedule. When the microstructure is defined, it can be used to predict key properties for use in service, that is, expansion coefficients, thermal conductivity, temperature- and strain-rate-dependent tensile properties, creep-rupture properties, and so on. In addition, γ' coarsening models can be used to follow the potential degradation of properties during prolonged use during service, and potentially deleterious transformations, such as σ formation and $\gamma' \rightarrow \eta$, can be considered.

Other, simpler possibilities exist. For example, in aluminum alloys, it is possible to model a casting as described in the section "Modeling of Physical Properties—Effect of Alloy Composition on Solidification Simulations" in this article and subsequently link with models (Ref 145) that enable the strength throughout the casting to be predicted as a function of local

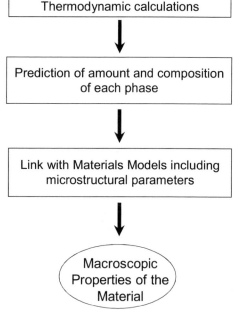

Fig. 34 Current status of thermodynamic calculations and material properties

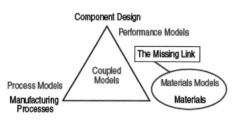

Fig. 35 Integrated computational materials engineering— a new paradigm for the global materials profession. Adapted from Ref 144

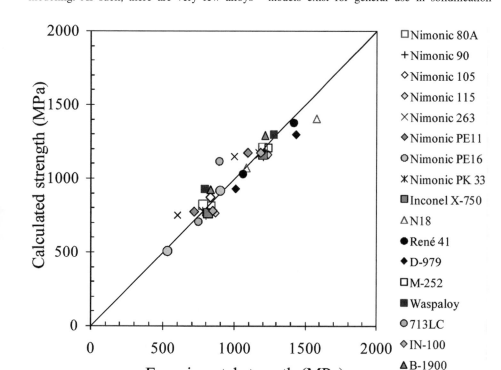

Fig. 33 Comparison between calculated and experimentally reported strengths of various nickel-base superalloys

cooling profile. Another example is that it is possible to simulate the hot press forming of ultrahigh-strength steels a priori (Ref 146).

As the models described here become more used, further possibilities will become apparent. At the present time (2010), property models can be linked to FE/FD simulation packages through the creation of material data input files that are designed for use with the various software packages. The next step will be for the more complex models to be linked to FE/FD packages through a software link, so that property models can be called at will as function dictates and used as required.

In conclusion, thermodynamic and material property models have reached a level whereby it is possible for scientists and engineers who work with process simulation of metallic alloys to call on a wide range of temperature/time-dependent material property data at will. In its own right, such a capability has the potential of providing significant cost-savings. In the longer term, it provides an enabling route by which true virtual simulation of metal processing can be considered as a viable short-term objective. This will enable not only the optimization of existing processing routes but will help fulfil the aims of rapid optimization of new alloys and their subsequent processing conditions.

REFERENCES

1. L. Kaufman, *Phase Stability in Metals and Alloys,* P.S Rudman et al., Ed., McGraw-Hill, 1967, p 125
2. L. Kaufman and H. Bernstein, *Computer Calculations of Phase Diagrams,* Academic Press, New York, 1970
3. N. Saunders and A.P. Miodownik, *CALPHAD—Calculation of Phase Diagrams, Pergamon Materials Series,* Vol 1, R.W. Cahn, Ed., Elsevier Science, 1998
4. M. Hillert, *Phase Equilibria, Phase Diagrams and Phase Transformations,* Cambridge University Press, 1998
5. H.L. Lukas, S.G. Fries, and B. Sundman, *Computational Thermodynamics: The Calphad Method,* Cambridge University Press, 2007
6. C. Wagner, *Thermodynamics of Alloys,* Addison-Wesley, 1951
7. C.H.P. Lupis and J.F. Elliott, *Acta Metall.,* Vol 4, 1966, p 529, 1019
8. J. Tomiska, *CALPHAD,* Vol 4, 1980, p 63
9. Y.M. Muggianu, M. Gambino, and J.P. Bros, *J. Chim. Phys.,* Vol 22, 1975, p 83
10. I. Ansara, *Int. Met. Rev.,* Vol 22, 1979, p 20
11. M. Hillert, *CALPHAD,* Vol 4, 1980, p 1
12. S.R. Brinkley, *J. Phys. Chem.,* Vol 14, 1946, p 563
13. S.R. Brinkley, *J. Phys. Chem.,* Vol 15, 1947, p 107
14. H.J. Kandiner and S.R. Brinkley, *Ind. Eng. Chem.,* Vol 42, 1950, p 850
15. F.G. Krieger and W.B. White, *J. Chem. Phys.,* Vol 16, 1948, p 358
16. J.-O. Anderson, T. Helander, L. Höglund, P. Shi, and B. Sundman, *CALPHAD,* Vol 26, 2002, p 273
17. S.L. Chen, S. Daniel, F. Zhang, Y.A. Chang, F.-Y. Xie, R. Schmid-Fetzer, and W.A. Oates, *CALPHAD,* Vol 26, 2002, p 175
18. W.B. White, S.M. Johnson, and G.B. Dantzig, *J. Chem. Phys.,* Vol 28, 1958, p 751
19. G.B. Dantzig, S.M. Johnson, and W.B. White, "The RAND Corporation, Paper P-1060," April 15, 1957
20. G. Eriksson, *Acta Chem. Scand.,* Vol 25, 1971, p 2561
21. G. Eriksson, *Chem. Scr.,* Vol 8, 1975, p 100
22. G. Eriksson and E. Rosen, *Chem. Scr.,* Vol 4, 1973, p 193
23. G. Eriksson and K. Hack, *Metall. Trans. B,* Vol 21, 1990, p 1013
24. W.T. Thompson, G. Eriksson, A.D. Pelton, and C.W. Bale, *Proc. Met. Soc. CIM,* Vol 11, 1988, p 87
25. C.W. Bale, P. Chartrand, S.A. Degtorov, G. Eriksson, K. Hack, R. Ben Mahfoud, J. Melançon, A.D. Pelton, and S.M. Martin, *CALPHAD,* Vol 26, 2002, p 189
26. R.H. Davies, A.T. Dinsdale, J.A. Gisby, J.A.J. Robinson, and S. Martin, *CALPHAD,* Vol 26, 2002, p 229
27. B. Jansson, "A General Method for Calculating Phase Equilibria Under Different Types of Conditions," TRITA-MAC-0233, Division of Physical Metallurgy, Royal Institute of Technology, Stockholm, Sweden, 1984
28. S.M. Hodson, "MTDATA Handbook: Multiphase Theory," NPL report, Feb 3, 1989
29. *CALPHAD,* Vol 26, 2002, p 141–312
30. J.S. Kirkaldy, B.A. Thomson, and E.A. Baganis, *Hardenability Concepts with Applications to Steel,* J.S. Kirkaldy and D.V. Doane, Ed., TMS, 1978, p 82
31. H.K.D.B. Bhadeshia and H.V. Edmond, *Acta Metall.,* Vol 28, 1980, p 1265
32. K. Hack and P.J. Spencer, *Steel Res.,* Vol 56, 1985, p 1
33. A. Kroupa and J.S. Kirkaldy, *J. Phase Equilibria,* Vol 14, 1993, p 150
34. N. Saunders, *Mater. Sci. Forum,* Vol 217–222, 1996, p 667
35. *Thermo-Calc Database Users Guide,* Thermo-Calc Software AB, Bjornnasvagen 21, SE 113 47 Stockholm, Sweden
36. N. Saunders, *Superalloys 1996,* R.D. Kissinger et al., Ed., TMS, 1996, p 101
37. N. Saunders, *Titanium '95: Science and Technology,* P. Bleckinsop et al., Ed., Inst. Materials, 1996, p 2167
38. N. Saunders, *Gamma Titanium Aluminides 1999,* Y.-W.Kim et al., Ed., TMS, 1999, p 183
39. C.W. Bale and G. Eriksson, *Can. Metall. Q.,* Vol 289, 1990, p 105
40. *A Guide to Solidification of Steels,* Jernkontoret, Stockholm, 1977
41. C.C. Chen and R.B. Sparks, *Titanium Science and Technology,* H. Kimura and O. Izumi, Ed., Met. Soc. AIME, 1980, p 2929
42. Y. Ito, Y. Moriguchi, T. Nishimura, and N. Nagai, *Titanium Science and Technology,* G. Lutjering, U. Zwicker, and W. Bunk, Ed., Deutsche. Gess. fur Metallkunde E.V., 1985, p 1643
43. F.S. Lin, E.A. Starke, Jr., S.B. Chakraborty, and A. Gybor, *Metall. Trans. A,* Vol 15, 1984, p 1229
44. H. Onodera, Y. Ro, T. Yamagata, and M. Yamazaki, *Titanium Science and Technology,* G. Lutjering, U. Zwicker, and W. Bunk, Ed., Deutsche. Gess. fur Metallkunde E.V., 1985, p 1883
45. M. Peters and J.C. Williams, *Titanium Science and Technology,* G. Lutjering, U. Zwicker, and W. Bunk, Ed., Deutsche. Gess. fur Metallkunde E.V., 1985, p 1843
46. T. Sugimoto, K. Kamei, S. Komatsu, and K. Sugimoto, *Titanium Science and Technology,* G. Lutjering, U. Zwicker, and W. Bunk, Ed., Deutsche. Gess. fur Metallkunde E.V., 1985, p 1583
47. G.R. Yoder, F.H. Froes, and D. Eylon, *Metall. Trans. A,* Vol 15, 1984, 183
48. M. Ishikawa, O. Kuboyama, M. Niikura, and C. Ouchi, *Titanium '92 Science and Technology,* F.H. Froes and I.L. Caplan, Ed., TMS, 1993, p 141
49. T. Matsumoto, *Kobe Steel Engineering Reports,* Vol 43, 1993, p 103
50. P. Blenkinsop, "IRC in Materials for High Performance Applications," University of Birmingham, U.K., private communication, 1993
51. S. Lampman, *Properties and Selection: Nonferrous Alloys and Special-Purpose Materials,* Vol 2, *Metals Handbook,* 10th ed., ASM International, 1990, p 592
52. C. Small, Rolls Royce plc, Derby, U.K., private communication, 1993
53. Y. Honnarat, J. Davidson, and F. Duffaut, *Mem. Sci. Rev.,* Vol 68, 1971, p 105
54. E.H. van der Molen, J.M. Oblak, and O.H. Kriege, *Metall. Trans.,* Vol 2, 1971, 1627
55. W. Betteridge and J. Heslop, *The NIMONIC Alloys and Other Ni-Based High Temperature Alloys,* 2nd ed., Edward Arnold, 1974
56. J.R. Brinegar, J.R. Mihalisin, and J. Van der Sluis, *Superalloys 1984,* M. Gell et al., Ed., Met. Soc. AIME, 1984, p 53
57. S.R. Dharwadkar, K. Hilpert, J. Schubert, and V. Venugopal, *Z. Metallkde.,* Vol 83, 1992, p 744
58. S.K. Shaw, "IRC in Materials for High Performance Applications," University of Birmingham, U.K., private communication, 1992
59. S.T. Wlodek, M. Kellu, and D. Alden, *Superalloys 1992,* S.D. Antolovich et al., Ed., TMS, 1992, p 165

60. J. Zou, H.P. Wang, R. Doherty, and E.M. Perry, *Superalloys 1992*, S.D. Antolovich et al., Ed., TMS, 1992, p 165
61. J.S. Zhang, Z.Q. Hu, Y.M. Ata, M. Morinaga, and N. Yukawa, *Metall. Trans. A*, Vol 24, 1993, p 2443
62. F.H. Hayes, *J. Less Common Metals*, Vol 114, 1985, p 89
63. R.D. Longbottom and F.H. Hayes, Paper 124, *Proc. Conf. Duplex Stainless Steels '94*, Welding Institute, Cambridge, 1994
64. X. Li, M.Sc. thesis, University of Birmingham, Edgbaston, U.K., 1995
65. M. Boniardi, F. Iacoviello, and G.M. La Vecchia, Paper 89, *Proc. Conf. Duplex Stainless Steels '94*, Welding Institute, Cambridge, 1994
66. M. Nystrom and B. Karlsson, Paper 104, *Proc. Conf. Duplex Stainless Steels '94*, Welding Institute, Cambridge, 1994
67. W. Gysel and R. Schenk, *Proc. Conf. Duplex Stainless Steels '91*, Les Editions Physique, 1991, p 331
68. R.D. Longbottom and F.H. Hayes, *User Aspects of Phase Diagrams*, Inst. Metals, 1991, p 32
69. T. Thorvaldsson, H. Eriksson, J. Kutka, and A. Salwen, *Proc. Conf. Stainless Steels*, Inst. Metals, 1985, p 101
70. F.H. Hayes, *J. Less Common Metals*, Vol 114, 1985, p 89
71. Y. Maehara, Y. Ohmori, J. Murayama, N. Fujino, and T. Kunitake, *Met. Sci.*, Vol 17, 1983, p 541
72. P. Cortie and J.H. Potgeiter, *Metall. Trans. A*, Vol 22, 1991, p 2173
73. F.H. Hayes, M.G. Hetherington, and R.D. Longbottom, *Mater. Sci. Technol.*, Vol 6, 1990, p 263
74. F. Jomard and M. Perdereau, *Proc. Conf. Duplex Stainless Steels '91*, Les Editions Physique, 1991, p 719
75. P. Merino, X.R. Novoa, G. Pena, E. Porto, and L. Espada, *Proc. Conf. Duplex Stainless Steels '91*, Les Editions Physique, 1991, p 967
76. J. Charles, F. Dupoiron, P. Soulignac, and J.C. Gagnepain, *Proc. Conf. Duplex Stainless Steels '91*, Les Editions Physique, 1991, p 1273
77. E. Hamalainen, A. Laitinen, H. Hanninen, and J. Liimatainen, Paper 122, *Proc. Conf. Duplex Stainless Steels '94*, Welding Institute, Cambridge, 1994
78. A.I. Kahveci and G.E. Welsch, *Scr. Metall.*, Vol 20, 1986, p 1287
79. Y.T. Lee, M. Peters, and G. Welsch, *Metall. Trans. A*, Vol 22, 1991, p 709
80. T.W. Duerig, G.T. Terlinde, and J.C. Williams, *Metall. Trans. A*, Vol 11, 1980, p 1987
81. O.H. Kriege and J.M. Baris, *Trans. ASM*, Vol 62, 1969, p 195
82. W.T. Loomis, J.W. Freeman, and D.L. Sponseller, *Metall. Trans.*, Vol 3, 1972, p 989

83. R.L. Dreshfield and J.F. Wallace, *Metall. Trans.*, Vol 5, 1974, p 71
84. P. Caron and T. Khan, *Mater. Sci. Eng.*, Vol 61, 1983, p 173
85. M. Magrini, B. Badan, and E. Ramous, *Z. Metallkde.*, Vol 74, 1983, p 314
86. T. Khan, P. Caron, and C. Duret, *Superalloys 1984*, M. Gell et al., Ed., Met. Soc. AIME, 1984, p 145
87. Z.-Y. Meng, G.-C. Sun, and M.-L. Li, *Superalloys 1984*, M. Gell et al., Ed., Met. Soc. AIME, 1984, p 563
88. M.V. Nathal and L.J. Ebert, *Superalloys 1984*, M. Gell et al., Ed., Met. Soc. AIME, 1984, p 125
89. D. Blavette, P. Caron, and T. Khan, *Superalloys 1984*, M. Gell et al., Ed., Met. Soc. AIME, 1984, p 305
90. H. Harada, K. Ohno, T. Yamagata, T. Yokokawa, and M. Yamazaki, *Superalloys 1988*, D.N. Duhl et al., Ed., TMS, 1988, p 733
91. R. Schmidt and M. Feller-Kniepmeier, *Scr. Metall. Mater.*, Vol 26, 1992, p 1919
92. S. Duval, S. Chambreland, P. Caron, and D. Blavette, *Acta Metall. Mater.*, Vol 42, 1994, p 185
93. Y. Shimanuki, M. Masui, and H. Doi, *Scr. Metall.*, Vol 10, 1976, p 805
94. K.M. Delargy and G.D.W. Smith, *Metall. Trans. A*, Vol 14, 1983, p 1771
95. K. Trinckhauf and E. Nembach, *Acta Metall. Mater.*, Vol 39, 1991, p 3057
96. K. Ishida and T. Nishizawa, *User Aspects of Phase Diagrams*, F.H. Hayes, Ed., Inst. Materials, 1991, p 185
97. B.J. Lee, *Applications of Thermodynamics in the Synthesis and Processing of Materials*, P. Nash and B. Sundman, Ed., TMS, 1995, p 215
98. G.N. Haidemenopoulos, M. Grujicic, G.B. Olson, and M. Cohen, *J. Alloy. Compd.*, Vol 220, 1995, p 142
99. C.J. Small and N. Saunders, *MRS Bull.*, Vol 24, 1999, p 22
100. M.G. Fahrmann and G.D. Smith, *JOM*, Vol 54 (No. 1), 2002, p 42
101. J.S. Kirkaldy and D. Venugopalan, *Phase Transformations in Ferrous Alloys*, A.R. Marder and J.I. Goldstein, Ed., AIME, 1984, p 125
102. H.K.D.H. Bhadeshia, *Met. Sci.*, Vol 15, 1981, p 175
103. H.K.D.H. Bhadeshia, *Met. Sci.*, Vol 16, 1982, p 159
104. J. Ågren, *ISIJ Int.*, Vol 32, 1992, p 291
105. N. Saunders, X. Li, A.P. Miodownik, and J.-Ph. Schillé, *Modelling of Casting, Welding and Advanced Solidification Processes X*, D. Stefanescu et al., Ed., TMS, 2003, p 669
106. N. Saunders, Z. Guo, X. Li, A.P. Miodownik, and J.-Ph. Schillé, *Superalloys 2004*, K.A. Green et al., Ed., TMS, 2004, p 849
107. N. Saunders, X. Li, A.P. Miodownik, and J.-Ph. Schille, *Ti-2003 Science and*

Technology, G. Luetering, Ed., Wiley-VCH, 2004, p 1397
108. Z. Guo, N. Saunders, A.P. Miodownik, and J.P. Schillé, *Proc. Second International Conference on Heat Treatment and Surface Engineering in Automotive Applications*, June 20–22, 2005 (Riva del Garda, Italy)
109. Z. Guo, N. Saunders, A.P. Miodownik, and J.P. Schillé, *Mater. Sci. Eng. A*, Vol 499, 2009, p 7
110. L. Backerud, E. Krol, and J. Tamminen, *Solidification Characteristics of Aluminium Alloys*, Vol 1 and 2, Tangen Trykk A/S, 1986
111. Solidification Design Center, Auburn University, Alabama
112. P.K. Sung, D.R. Porier, and E. McBride, *Mater. Sci. Eng. A*, Vol 231, 1997, p 189
113. Z. Fan, P. Tsakiropoulos, and A.P. Miodownik, *J. Mater. Sci.*, Vol 29, 1994, p 141
114. Z. Fan, *Philos. Mag. A*, Vol 73, 1996, p 1663
115. Publication SMC-045, Special Metals Corporation, Sept 7, 2007
116. M.G. Burke and M.K. Miller, in *Superalloys 718, 625 and Various Derivatives 1991*, E.A. Loria, Ed., TMS, 1991, p 337
117. Z. Guo, N. Saunders, E. Hepp, and J.-Ph. Schillé, *Solidification Processing 2007*, H. Jones, Ed., University of Sheffield, 2007, p 355
118. N. Saunders, X. Li, A.P. Miodownik, and J.-Ph. Schillé, *Materials Design Approaches and Experiences*, J.-C. Zhao et al., Ed., TMS, 2001, p 185
119. D. Tabor, *The Hardness of Metals*, W. Jackson, H. Frohlich, and N.F. Mott, Ed., Oxford University Press, 1952, p 67
120. N.S. Stoloff, *Superalloys II*, C.T. Sims, N.S. Stoloff, and W.C. Hagel, Ed., J.Wiley & Sons, 1987, p 61
121. L.M. Brown and R.K. Ham, Strengthening Mechanisms in Crystals, *Appl. Sci.*, 1971
122. J.W. Martin, *Precipitation Hardening*, 2nd ed., Butterworth-Heinemann, 1988, p 79
123. W. Hüther and B. Reppich, *Z. Metallkde.*, Vol 69, 1978, p 628
124. A.P. Miodownik and N. Saunders, *Applications of Thermodynamics in the Synthesis and Processing of Materials*, P. Nash and B. Sundman, Ed., TMS, 1995, p 91
125. Z. Guo, N. Saunders, A.P. Miodownik, and J.P. Schillé, *Mater. Sci. Forum*, Vol 546–549, 2007, p 1319
126. G.E. Dieter, *Mechanical Metallurgy*, McGraw-Hill, 1961, p 275
127. A.P. Miodownik, X. Li, N. Saunders, and J.-P. Schillé, *Parsons 2003: Engineering Issues in Turbine Machinery, Power Plant and Renewables*, A. Strang et al., Ed., Inst. MMM, 2003, p 779
128. R. Lagneborg and B. Bergman, *Met. Sci.*, Vol 10, 1976, p 20

129. A.P. Miodownik, *CALPHAD*, Vol 2, 1978, p 207
130. N. Saunders, Z. Guo, A.P. Miodownik, and J.-Ph. Schillé, *Superalloys 718, 625, 706 and Derivatives 2005*, E.A. Loria, Ed., TMS, 2005, p 571
131. J. Miettinen and S. Louhenkilpi, *Metall. Mater. Trans. B*, Vol 25, 1994, p 909
132. J. Miettinen, *Metall. Mater. Trans. B*, Vol 28, 1997, p 281
133. Z. Guo, N. Saunders, A.P. Miodownik, and J.P. Schillé, *Proc. Fifth International Conference on Quenching and Control of Distortion*, AWT, Berlin, 2007, p 183
134. Z. Guo, N. Saunders, and J.P. Schillé, *Proc. of the 17th International Federation for Heat Treatment and Surface Engineering*, Vol 49 (No. 2), Netsu Shori, 2009, p 506
135. *Atlas of Isothermal Transformation and Cooling Transformation Diagrams*, American Society for Metals, 1977
136. N. Saunders, Z. Guo, and J.P. Schillé, "Modelling Microstructural Evolution in Ni-Based Superalloys during Heat Treatment," presented at MRS International Materials Research Conference, June 9–12, 2008 (Chongqing, China)
137. X. Li, A.P. Miodownik, and N. Saunders, *Mater. Sci. Technol.*, Vol 18, 2002, p 861
138. X. Li, N. Saunders, and A.P. Miodownik, *Metall. Mater. Trans. A*, Vol 33, 2002, p 3367
139. I.M. Lifshitz and V.V. Slyozov, *J. Phys. Chem. Solids*, Vol 19, 1961, p 35
140. C. Wagner, *Z. Elektrochem.*, Vol 65, 1961, p 581
141. H.A. Calderon, P.W. Voorhees, J.L. Murray, and G. Gkostorz, *Acta Metall.*, Vol 42, 1994, p 991
142. J.S. Kirkaldy, *Scand. J. Metall.*, Vol 20, 1991, p 50
143. N. Saunders, Z. Guo, X. Li, A.P. Miodownik, and J.-Ph. Schillé, *JOM*, Dec 2003, p 60
144. J. Allison, D. Backman, and L. Christodoulou, *JOM*, Nov 2006, p 25
145. Z. Guo, N. Saunders, A.P. Miodownik, and J.-P. Schillé, *Aluminium Alloys, Their Physical and Mechanical Properties*, J. Hirsch, B. Skrotzki, and G. Gottstein, Ed., Wiley-VCH, 2008, p 1204
146. K. Lee and G.P. Kang, "Numerical Simulation of Hot Press Forming Process," presented at Deform User's Meeting, Nov 12–13, 2008

Simulation of Solidification

ASM Handbook, Volume 22B, *Metals Process Simulation*
D.U. Furrer and S.L. Semiatin, editors

Copyright © 2010, ASM International®
All rights reserved.
www.asminternational.org

Modeling of Transport Phenomena during Solidification Processes

Matthew John M. Krane, Purdue University

IN THE PROCESSING OF MOST METAL PRODUCTS, a critical step is the solidification of an alloy from a melt to make an ingot or a part with a particular shape. Frequently, these products are processed with subsequent heat treatment and/or deformation, and the structure imparted by the solidification will affect the microstructure distribution formed in these downstream processes. The micro- and macro-scale structures and accompanying defects can be traced to the heat and mass transfer and fluid flow during freezing, so these phenomena must be understood to enable prediction of metal quality.

The solid-liquid phase change in casting processes is driven primarily by the extraction of heat from the liquid melt, determined by the temperature difference between the liquid metal and its environment. The rate of heat loss controls the solidification front velocity and the microstructure. Different alloying elements are soluble to varying degrees, and, as the solid forms, there is either a buildup or depletion of the alloying elements in the surrounding liquid. The composition differences that arise between the liquid near the developing solid and in the bulk melt drive a flux of species in the liquid. The temperature and compositional gradients present during the formation of solid contribute to variations in liquid density, which can induce buoyancy-driven flows and advective fluxes of heat and species. Liquid movement around a solidifying alloy also can be caused by motion of the mold, applied electromagnetic fields, and shrinkage of the metal during phase change. Motion caused by these phenomena will affect the developing microstructure and determine the occurrence of defects.

In this article, conservation equations for heat, species, and mass and momentum are discussed, with an eye toward using them to predict transport phenomena during solidification processing. The approach taken here to develop equations governing these transport phenomena is the continuum mixture model (Ref 1–3). This approach treats the entire domain as one phase and predicts mixture quantities for the field variables (e.g., velocity, enthalpy, etc.), combined with different closure relations to calculate the fraction solid and the values of those field variables in the separate phases. (Another approach is a two-phase model, Ref 4 to 7, in which a separate transport equation is solved for each phase present.) Following the presentation of the transport equations, several examples of their application are given to illustrate some of the physics present in alloy solidification. The first two examples demonstrate the utility of scaling analysis to elucidate the fundamental physics in a process and to demonstrate the limitations of simplifying assumptions. Following these approximate analyses, there is a discussion of the solidification behavior of alloys as predicted by full numerical solutions of the transport equations, with some attention paid to experimental validation.

Conservation Equations for Transport Phenomena

Thermal Energy Transport

The primary driving force for solidification is the transfer of heat from an alloy. The temperature difference between the liquid metal and the heat sink and the thermal resistance and capacitance of the heat flow path determine the cooling rates and temperature gradients throughout the process, which, in turn, control the metal microstructure. To describe the thermal response to this heat removal, the conservation equation for energy transport can be written in terms of either the enthalpy of the solid-liquid mixture or the temperature of the system, or a combination of both.

In general, the domain in a solidification process will consist of multiple phases distributed among solid, liquid, and gas fractions. A simplified—but reasonably realistic—system to consider is an alloy in which:

- The transformation from liquid to solid occurs across a temperature range, $\Delta T_{LS} = T_L$ (liquidus temperature) – T_S (solidus temperature).

- All solid phases are lumped into a single solid phase.
- No significant gas voids form; that is, the sum of the volume fractions of the solid (S) and liquid (L) can be taken as unity, $g_S + g_L = 1$.

A conservation equation describing the thermal energy transport in this system is obtained by writing the heat balance in terms of enthalpy. The solid (h_S) and liquid (h_L) phase enthalpies can be defined as:

$$h_S = \int_{T_{ref}}^{T} c_S \, dT \qquad \text{(Eq 1)}$$

and

$$h_L = \int_{T_{ref}}^{T} c_L \, dT + L_f \qquad \text{(Eq 2)}$$

where c is the specific heat at constant pressure, T is temperature, and L_f is the latent heat. A mixture enthalpy ($\bar{h}$) can be defined as a mass fraction (f_i) weighted average of the liquid and solid enthalpies:

$$\bar{h} = f_L \, h_L + f_S \, h_S \qquad \text{(Eq 3)}$$

or, in terms of volume fractions (g_i):

$$\bar{\rho}\bar{h} = g_L \, \rho_L \, h_L + g_S \rho_S \, h_S \qquad \text{(Eq 4)}$$

where the mixture density is defined as:

$$\bar{\rho} = g_L \rho_L + g_S \rho_S \qquad \text{(Eq 5)}$$

and using:

$$f_i = g_i \left(\frac{\rho_i}{\bar{\rho}} \right)$$

With these definitions, an appropriate mixture enthalpy balance for the system under consideration can be written:

$$\frac{\partial \overline{\rho}\overline{h}}{\partial t} + \nabla \cdot (g_S \rho_S h_S\, V_S + g_L \rho_L h_L\, V_L)$$
$$= \nabla \cdot (\overline{k}\nabla T) + S_T \qquad \text{(Eq 6)}$$

where $\overline{k} = g_S k_S + g_L k_L$ is a mixture thermal conductivity, V_L and V_S are the velocities of the liquid and solid phases, respectively, and S_T is a generic heat source term that may include Joule heating or viscous dissipation. The first term on the left side of Eq 6 represents the storage of enthalpy, which is determined by the balance of the other terms, which, respectively, model the advection and conduction of thermal energy. (Advection is defined here as transport due to fluid flow alone, while convection is the combination of advection and diffusion.)

The form of the thermal energy transport equation (Eq 6) is not suitable for immediate implementation in most commercial or general application computer codes. Often, the code requires that the transport equations be rewritten as general advection-diffusion equations:

$$\frac{\partial\,(\rho\,\phi)}{\partial t} + \nabla \cdot (\rho\,\phi\,V) = \nabla \cdot (\Gamma\,\nabla\phi) + S_\phi \qquad \text{(Eq 7)}$$

where ϕ is the transported quantity, Γ is the diffusion coefficient, and S_ϕ is a source/sink term. The advantage of the form in Eq 7 is that it allows for a time-implicit solution using standard algorithms (e.g., Ref 8–11). The manipulation of the energy conservation equation (and the subsequent species equation) into the form in Eq 7 can lead to new terms that typically are lumped into the source S_ϕ.

An example of this formulation using $\phi = \overline{h}$ in Eq 6 is from Ref 1:

$$\frac{\partial\,(\overline{\rho}\,\overline{h})}{\partial t} + \nabla \cdot (\overline{\rho}\,\overline{h}\,\overline{V}) = \nabla \cdot \frac{\overline{k}}{c_S}\nabla\overline{h} -$$
$$\nabla \cdot \left[\overline{\rho}f_S(h_L - h_S)\,(\overline{V} - V_S)\right]$$
$$+ \nabla \cdot \frac{\overline{k}}{c_S}\nabla(h_S - \overline{h}) + S_T$$
$$\text{(Eq 8)}$$

Here, the mixture velocity is $\overline{V} = f_L V_L + f_S V_S$, and the first three terms represent the storage, advection, and diffusion of mixture enthalpy. The algebraic shuffling of terms from Eq 6 to Eq 8 has led to an equation that includes two source terms (the second and third terms on the right side of Eq 8 that arise not from the physics but from the manipulation into the form of Eq 7). Under the assumption of uniform and constant specific heats in each phase and a scaling of temperature to set $T_{\text{ref}} = 0$, the energy equation (Eq 6) can be rewritten in the form of Eq 7 with $\phi = T$:

$$\frac{\partial(\overline{\rho}\,\overline{c}\,T)}{\partial t} + \nabla \cdot (\overline{\rho}\,\overline{c}\,T\,\overline{V}) = \nabla \cdot (\overline{k}\,\nabla T)$$
$$- \frac{\partial(\overline{\rho}f_L\,L_f)}{\partial t} - \nabla \cdot (\overline{\rho}\,\overline{V}\,f_L\,L_f)$$
$$- \nabla \cdot \left[\overline{\rho}f_S\big((c_L - c_S)T + L_f\big)(\overline{V} - V_S)\right] + S_T$$
$$\text{(Eq 9)}$$

Equation 9 has the proper form of Eq 7 if the density in Eq 7 is replaced with $\overline{\rho}\,\overline{c}$. The left-side terms are the storage and advection of the sensible enthalpy, with the latent heat contributions to those effects showing up as "source terms" on the right. The choice of an enthalpy (Eq 8) or temperature (Eq 9) formulation is influenced by the behavior of the system and the ease with which $\overline{h}$ and T can be related to the fraction liquid. During isothermal phase changes, convergence of the solution to the phase change temperature can be difficult because the latent heat is absorbed or released only when the system crosses that discrete point. In this case, tracking mixture enthalpy as the primary variable eliminates the convergence problem because $\overline{h}$, unlike T, varies continuously with heat extraction. On the other hand, the temperature formulation is much easier to relate to the equilibrium phase diagram to obtain the fraction solid, an advantage even more important in multicomponent alloys with complicated relations for phase equilibria. A recent study of the numerical behavior of these two formulations suggests that the enthalpy method generally converges faster than the temperature (Ref 12).

Species Transport

Solutions of Eq 8 and 9 require a relationship between the enthalpy or temperature and the liquid fraction. This function can be derived from a consideration of the phase equilibria of the system in the form of a phase diagram, but the local composition also must be known. To find the composition distribution in the ingot, an equation describing the conservation of solute in the system must be developed and solved. For a given solute species, a mixture concentration can be defined by:

$$\overline{\rho}\overline{C} = \frac{1}{V_{\text{REV}}} \int\limits_{V_{\text{REV}}} [f_S\,\rho_S C_S + (1 - f_S)\rho_L\,C_L]\,dV$$
$$\text{(Eq 10)}$$

where C_i is the composition in phase i, and V_{REV} is a representative elemental volume (REV) in the solidification domain. If the nature of the solid and liquid phases in the REV can be characterized by the volume fraction of liquid, $g_L = 1 - g_S$, and the local solute transport in the liquid phase of the REV is rapid, Eq 10 can be more conveniently written as:

$$\overline{\rho}\,\overline{C} = g_S < \rho_S\,C_S > + g_L \rho_L\,C_L \qquad \text{(Eq 11)}$$

where:

$$g_S < \rho_S\,C_S > = \int_0^{g_S} \rho_S\,C_S\,d\alpha$$

In such a system, the conservation of solute can be written as:

$$\frac{\partial\,\overline{\rho}\,\overline{C}}{\partial t} + \nabla \cdot (g_S < \rho_S C_S > V_S + g_L \rho_L C_L\,V_L)$$
$$= \nabla \cdot (g_L \rho_L D_L\,\nabla C_L)$$
$$\text{(Eq 12)}$$

The first term in Eq 12 is the accumulation of solute, while the other two terms represent advection of solute due to solid and liquid motion and its diffusion in the liquid. The derivation of Eq 12 neglects interdiffusion among various elements in a multicomponent system. Diffusion coefficients, especially those representing the effect of the gradient of one element on the diffusion of another, are not well documented for most liquid metal systems. Fortunately, the assumption to neglect the interdiffusion effect is justified by a scaling analysis (Ref 13) showing that advection dominates species transport on the macroscopic level throughout the process. Equation 12 also neglects macroscopic diffusion in the solid phase because, for most metal alloys (Ref 14), especially if the alloy is a substitutional solution, the diffusion coefficient in the liquid is much larger than that in the solid ($D_L >> D_S$). Equation 12 can also be written in the advection-diffusion form (Eq 7) for mixture composition by setting

$$\phi = \overline{C} = f_S < C_S > + f_L C_L :$$

$$\frac{\partial\,\overline{\rho}\,\overline{C}}{\partial t} + \nabla \cdot (\overline{\rho}\,\overline{C}\,V) = \nabla \cdot (\overline{\rho}f_L D_L \nabla\overline{C}) -$$
$$\nabla \cdot (\overline{\rho}f_S(C_L - < C_S >)(\overline{V} - V_S)) +$$
$$\nabla \cdot (\overline{\rho}f_L\,D_L\,\nabla(C_L - \overline{C}))$$
$$\text{(Eq 13)}$$

As seen in the mixture enthalpy equation (Eq 8), this equation is written in terms of the accumulation, advection, and diffusion of mixture composition ($\overline{C}$), which gives rise to the two "source terms" on the right.

Mass and Momentum Conservation

While attention typically is turned first to the prediction of the transport of heat and species, because these effects directly control the microstructure and final quality of a cast part, the prediction of velocity and pressure in the liquid metal is required also, because they can significantly alter the temperature and composition fields. The basic equations for the flow and pressure are the Navier-Stokes equations that express conservation of mass and linear momentum. Derivation of the general forms of these equations can be found in many standard texts (e.g., Ref 15, 16). For a Newtonian fluid, these equations can be written as:

$$\frac{\partial \rho}{\partial t} + \nabla \cdot \rho V = 0 \qquad \text{(Eq 14)}$$

and

$$\frac{\partial(\rho V)}{\partial t} + \nabla \cdot (\rho V\,V) = \nabla \cdot (\mu\nabla V) - \nabla P + S_M$$
$$\text{(Eq 15)}$$

The left side of Eq 15 represents the local and convective acceleration of the fluid, while the right side contains the forces affecting fluid momentum. The first two terms on the right are the viscous shear and a pressure gradient. The last term accounts for body forces, for example, gravity and electromagnetic forces, and additional viscous terms not expressed by $\nabla \cdot (\mu \nabla \mathbf{V})$(Ref 2, 3).

The central task in using Eq 14 and 15 to predict the fluid velocity and pressure in a casting requires the adaptation of the Navier-Stokes equations to account for physical phenomena specific to two-phase (solid + liquid) solidifying systems. One formal way to achieve this is to use a two-phase volume averaging approach (Ref 7). In this approach, separate solid and liquid equations, expressing mass and momentum conservation at a "microscopic point" in the liquid and solid phases, are developed. These equations are then averaged over a representative elementary volume, containing both the solid and liquid phases, to arrive at macroscopic statements of mass and momentum conservation in each phase. The critical modeling component in this approach is the construction of appropriate interphase transfer terms to describe the momentum and mass exchanges between the phases. The conservation of momentum in the liquid phase can be written in the form of:

$$\frac{\partial (\rho_L \, g_L \, \mathbf{V}_L)}{\partial t} + \nabla \cdot (\rho_L g_L \, \mathbf{V}_L \, \mathbf{V}_L)$$
$$= \nabla \cdot (g_L \mu \, \nabla \, \mathbf{V}_L) - g_L \nabla P + g_L \rho_L \boldsymbol{g} + S_M$$
(Eq 16)

where, in addition to the terms previously noted, the general source (S_M) will also contain extra terms arising from the volume averaging process (e.g., see Table 2 in Ref 5). This statement of the fluid momentum conservation also includes an explicit accounting of the buoyancy force due to gravity, $\boldsymbol{g}$.

A suitable liquid mass conservation that can be used with Eq 16 is:

$$\frac{\partial \overline{\rho}}{\partial t} + \nabla \cdot (\rho_L \mathbf{V}_L) = -\nabla \cdot (\rho_S g_S \, \mathbf{V}_S) \qquad (\text{Eq } 17)$$

where the accumulation and advection of mass in the liquid phase is balanced by the change of mass in the solid phase. When Eq 16 and 17 are combined with similar equations for mass and momentum conservation in the solid, the continuum mixture equations are found (Ref 1):

$$\frac{\partial (\overline{\rho} \overline{\mathbf{V}})}{\partial t} + \nabla \cdot (\overline{\rho} \, \overline{\mathbf{V}} \, \overline{\mathbf{V}}) = \nabla \cdot (\mu \frac{\rho_L}{\overline{\rho}} \nabla \overline{\mathbf{V}}) - \nabla P + S_M$$
(Eq 18)

$$\frac{\partial \overline{\rho}}{\partial t} + \nabla \cdot (\overline{\rho} \, \overline{\mathbf{V}}) = 0 \qquad (\text{Eq } 19)$$

With this system, it is important to note that in the fully liquid domain, the equations revert to the basic form of the Navier-Stokes equations given in Eq 14 and 15.

Arriving at a computationally tractable model for the flow and pressure (P) in a given system requires efforts focused on judicious use of auxiliary relationships and assumptions to construct the source terms in the previous equations. The key tasks that must be undertaken include:

- The appropriate accounting of other forces that drive or restrain the fluid flow, for example, density variations
- The specification of source terms that can model the momentum transfer between the solid and liquid phases
- The accounting of velocity fluctuations arising from turbulence

Elements of these tasks are outlined in more detail next.

Shrinkage-Induced Flow. Most metal alloys shrink by 2 to 10% when they solidify (Ref 14). As the liquid turns to solid, a pressure gradient will develop in the liquid at whatever level is necessary to fill the volume deficit formed. The shrinkage-induced flows arise in Eq 18 and 19, not from the source terms, S_M, but from the temporal variation of mixture density in Eq 19. While the liquid velocities generated can be quite small and have less effect on macrosegregation than buoyancy (see Ref 17 for a scaling analysis of this issue and Ref 18 and 19 for some exceptions), the attendant volume change can cause the casting to pull away from the mold wall, and the pressures induced can cause defects such as hot tears and porosity. The modeling of the evolution of gas porosity requires an additional advection-diffusion equation to track the dissolved gas in the metal, as well as a pore nucleation model. If modeled only at the macroscale, then only the local volume fraction of porosity can be found (e.g., Ref 20–22), while the use of a microscale dendrite morphology model obtains details of pore shape and distribution (e.g., Ref 23). The choice of whether or not to include solidification shrinkage depends on the purpose of the model. If macrosegregation levels are low, and porosity and solid deformation are not an issue, then the advantage of including this effect must be weighed against the increase in computational complexity. Because the volume of the metal is not constant throughout the process, some sort of open boundary, where fluid is allowed to flow in and out based on the local pressure (which will be affected by the shrinkage-induced flow), will be necessary. This approach is the easiest, but it lacks information on the alloy beyond the boundary and so incorporates uncertainties in the behavior near that region (Ref 17). Another option is a more realistic model that simulates the rise and fall of a free surface in a riser (Ref 6). Because of the need to track a moving free surface, this approach can make the calculations more difficult, but it may give better predictions for the fluid flow near the riser.

Buoyancy-Induced Flow. One driving force for flow that occurs in most castings is

buoyancy. Both nonuniform cooling and liquid composition differences due to elemental partitioning during solidification cause density gradients in the liquid. Unless these density variations are stable (cooling from below and interdendritic liquid heavier than the bulk), they will cause natural convection flows in the liquid. In the absence of electromagnetic forces, and when the effects of mold filling die out, buoyancy will be the main driver of flow in most static casting processes. The source term in Eq 18 includes this effect and requires a model for the variation of liquid density with temperature and composition. The most common approach is to use the Boussinesq approximations, which assume that the temperature and composition dependence of density is only in this buoyancy source term and that those density changes are much less than the value of the nominal liquid density ($\Delta\rho_L / \rho_L \ll 1$) (Ref 24). Making these assumptions allows the buoyancy source term to be written as a linear function of temperature and composition changes in the liquid:

$$S_B = \rho_L \boldsymbol{g}$$
$$= \rho_{L,o} \left[\beta_T (T - T_{ref}) + \sum_{i=1}^{N-1} \beta_{S,i} (C_{L,i} - C_{L,i,ref}) \right] \boldsymbol{g}$$
(Eq 20)

where $\rho_{L,o}$ is the liquid density at the reference temperature and composition, and N is the number of species in the alloy. The reference values are usually taken as the initial value of temperature and the nominal alloy composition. The thermal and solutal expansion coefficients are:

$$\beta_T = -\frac{1}{\rho_L} \frac{\partial \rho_L}{\partial T} \qquad \text{and} \qquad \beta_{S,i} = -\frac{1}{\rho_L} \frac{\partial \rho_L}{\partial C_{L,i}}$$
(Eq 21)

In cases of large composition or temperature differences in the liquid metal or other cases in which the Boussinesq approximations break down, the liquid density can be better represented by (Ref 25):

$$\rho_L^{-1} = \left(\sum_{i=1}^{N} \frac{C_{L,i}}{\rho_{L,i} + \Lambda_i (T - T_{ref})} \right) \qquad (\text{Eq } 22)$$

where the Λ_i are related to the thermal expansion coefficient and can be found in Ref 25. After the local liquid composition and temperature are found, Eq 22 can be used to obtain the local liquid density and the buoyancy source term:

$$S_B = \rho_L \boldsymbol{g} \qquad (\text{Eq } 23)$$

Drag in the Mushy Region. A key part of the source term S_M in Eq 18 is a drag force that accounts for the exchange of momentum between the solid and liquid phases in the mushy region. In regions where the solid dendritic matrix is rigid and moves at a prescribed velocity (either zero or a specified casting speed), there is a significant amount of shear transmitted at the solid-liquid interface, and it is worthwhile to consider how this drag will depend on the morphology of this region. At high liquid fractions, where individual primary dendrite arms are largely unconnected to each other, the increased drag on the liquid is due to the no-slip condition on the very high solid surface area per unit volume. While the flow is significantly retarded by this friction, even at very small amounts of solid, there is still enough potential for the flow to cause an exchange of fluid between the mushy zone and the melt pool. Closer to the solidus, where the dendrites are much more interconnected, the surface area/volume ratio is a less important contributor to the frictional drag than the tortous and narrow passages through the solid structure. Here, the drag is so large that typical levels of buoyancy forces cannot induce significant velocities compared to the (albeit very weak) flow controlled by shrinkage.

This frictional drag effect acts on microscopic solid surfaces, but when modeling macroscale transport phenomena, it is treated as a volumetric sink of momentum in the Navier-Stokes equations. Given the slow speed of flows in typical mushy zones, the relation for the drag is usually based on a modified Darcy's law (Ref 26), which gives the drag contribution to the source term in Eq 18:

$$S_D = -\frac{\rho_L}{\rho} \mu \, \mathbf{K}^{-1} (\mathbf{V}_L - \mathbf{V}_S) \qquad \text{(Eq 24)}$$

where $\mathbf{K}$ is a permeability tensor. While this formulation for the interfacial drag looks simple, there is a great complexity lurking in the treatment of the permeability. A ubiquitous approach is to write the permeability model in terms of liquid volume fraction (g_L) and dendritic arm spacing (d). Assuming the mushy region morphology forms an isotropic medium:

$$\mathbf{K} = \frac{d^2}{180} \frac{g_L^3}{(1 - g_L)^2} \mathbf{I} \qquad \text{(Eq 25)}$$

This expression was originally derived for flow perpendicular to a bundle of cylindrical tubes of uniform size and spacing and is only valid at low liquid fractions, which makes it less attractive than it first appears. As noted, with the exception of shrinkage-induced flows, all of the interesting flow effects in the mushy zone occur near the liquidus temperature at high liquid fractions. Also, the geometry for which it is used is much different than a bundle of parallel tubes, because real dendritic microstructures are not evenly spaced and have side branches that vary in size. However, the simplicity of Eq 25, with a very easily determined dependence on microstructural data, has made it a very popular choice for treating mushy zone drag.

Several studies have extended permeability models to high liquid fraction, where Eq 25 gives values of permeability that are much too

high. Because of experimental difficulties, numerical studies have been used to predict flow behavior and relate it to permeability (Ref 27, 28). General correlations are not provided, but the plotted results show that data from a range of Reynolds numbers and simulated dendrite geometry collapse to one curve of a nondimensional, isotropic permeability (K^*) as a function of g_L. To accomplish this simplification, the proper length scale for the permeability was found to be not the dendritic arm spacing, as for Eq 25, but the inverse of the surface area/volume ratio (S_v), so that:

$$K^* = K \, S_v^2 \qquad \text{(Eq 26)}$$

The effect of microstructural anisotropy on the permeability model should also be considered. Flow parallel to the axis of an array of columnar dendrites is obstructed mainly by the secondary dendrite arms, while the liquid moving perpendicular to that axis also must move around the primary arms, giving a very different flow pattern. One example of different permeability functions for flow parallel and perpendicular to the primary dendrite axis for liquid fraction above 0.66 can be found in Ref 29. When g_L is near 1, the ratio of the permeabilities given by these correlations is approximately ½, and both are much smaller than the prediction of the Blake-Kozeny model (Eq 25). The permeability in equiaxed microstructures is much more isotropic.

To solve the mass and momentum equations in the solid + liquid region, the solid velocity must be calculated or prescribed. An obvious assumption is to assume a fixed solid structure, such that $V_S = 0$. In many solidification processes, such as directionally solidified turbine blade casting or electroslag remelting of steels and superalloys, most or all of the solid forms in a rigid structure with little free-floating solid. In such cases, this assumption is sufficiently realistic and results in a much simpler set of equations than if the solid is not rigid. In solving Eq 18 and 19 for the case of a fixed solid structure, a simplification is often made to assume that drag friction S_D is the dominant term in the source S_M; that is, the additional averaging terms present in this source are neglected.

In contrast to the fixed solid, in processes that rely on grain refinement to produce a more uniform, equiaxed microstructure or in cases that have a columnar-to-equiaxed transition, there will be a considerable number of solid particles formed in the melt or detached from the mold walls and the rigid dendritic matrix. These solid particles, moving with the liquid flow or settling toward the bottom of the liquid pool, can substantially alter the flow as well as the macrosegregation and heat extraction patterns, but when these solid particles are present, the difficulties and uncertainties in the model multiply rapidly. Models to predict the motion of this solid during solidification have been developed (Ref 7, 30–33). The predictions of

solid transport depend on the size and shape distributions of the particles as well as the mechanisms by which the dendrites detach from and attach to the rigid structure. Typically, there are too many to track individually, so models usually characterize the free-floating solid particles with a solid volume fraction and submodels for the particle size distribution. Examples of the effects of solid motion are found in Ref 34 to 36.

Turbulence Modeling. While generally the liquid velocities and melt pools in casting processes are small, the flows are not always laminar. In large-scale liquid metal processes, such as vacuum arc or plasma arc remelting, or direct chill or continuous casting, the flows can transition from laminar to turbulent and relaminarize in different regions of the melt pool. Turbulence is also more likely to develop in cases in which the flows are driven by electromagnetic forces, which can produce much higher velocities than buoyancy alone. The hallmark of a turbulent flow is the appearance of a broad spectrum of local velocity fluctuations, resulting in the formation of flow eddies over a large range of length scales. There is a plethora of turbulence models that can be used to simulate the effects of the velocity fluctuations and eddies on the flow, temperature, and composition fields (Ref 37, 38). The basic approach is to time-average the Navier-Stokes equations, removing the details of the fluctuating velocities and replacing them with a time-averaged source term in the momentum equation. This term has the form of the shear stress term in a Newtonian fluid and so is referred to as the Reynolds stress tensor. A common way to represent this term is to use the eddy viscosity concept that models the Reynolds stress via an additional shear viscosity. Typically, the parameters to define this turbulent viscosity are obtained using variations of the k-ε model (Ref 39–41), which includes two additional transport equations for turbulent kinetic energy (k) and dissipation (ε). These two values are also used to modify the heat and species diffusivities (α and D) to account for the effect of turbulence on transport of these quantities.

In contrast to the time-averaged approach, there is also an interest in using turbulence models that can resolve the fluctuating velocity components. Two examples are direct numerical simulation (DNS), which solves the Navier-Stokes equations directly, and large eddy simulation (LES), which uses a filtering equation to explicitly account for the momentum transport due to larger eddies, while the momentum in the smaller eddies is included implicitly via subgrid models. The performance of these more computationally intensive methods has been compared to the k-ε approach in the modeling of the continuous casting of steel (Ref 42). (This process has relatively high momentum compared to other solidification processes and is more likely to undergo transition to turbulence.) In this work, Thomas et al. show that the time-averaged predictions of flow patterns from the k-ε, DNS, and LES methods are very close to each other, although, by its nature, the k-ε model cannot account for transient fluctuations. Proper use of the k-ε model also relies on empirically derived constants that heavily depend on the specific flow conditions. In spite of the fact that use of the incorrect constants in the k-ε model can lead to very wrong results, the DNS and LES are less likely to be used in the near future except in research codes, because they represent a huge increase in computational effort (due largely to the necessary increase in grid resolution).

Summary of Transport Equations

A common method of developing transport equations for mass (Eq 19), momentum (Eq 18), heat (Eq 8 or 9), and species (Eq 13) was outlined previously. The approach is to write conservation equations for each phase and sum all of these advection diffusion equations. Defining mixture density, velocity, enthalpy, and species as volume- or mass-fraction-weighted sums of those quantities for all phases, the mixture transport equations can be written and transformed into a form standard for common numerical methods (Eq 7). The development of these and other types of single-domain models has allowed great progress in the simulation of solidification phenomena since the mid-1980s.

Inspection of these equations shows that there are several more independent variables that must be found than there are equations. Assuming that the four transport equations, for example, Eq 8, 13, 18, and 19, are solved for $\overline{h}$, $\overline{C}$, $\overline{V}$, and P, several variables still remain. Several quantities, h_L, h_S, g_L, and f_L, are defined in Eq 1, 2, and 5 and by $f_L = 1 - f_S$. The solid phase velocity, V_S, is zero in static casting or set in cases with a prescribed casting velocity. As noted previously, configurations with free-floating particles have more complex relationships between the mixture and solid velocities; the reader is referred to Ref 7 and 30 to 36. Finally, the interrelationship among the temperature, mass fraction solid, and liquid and solid compositions is found from the treatment in the model of the equilibrium phase diagram. Several works deal with this subject in detail, for example, Ref 2, 6, and 43 to 45.

Examples of Model Results

The conservation equations listed previously model the transport phenomena that control alloy solidification behavior. Exact solutions for these equations are rare and of limited utility, but other approaches can be used to yield insight into governing physical mechanisms. The results of two methods, scaling analysis and numerical simulations, are shown here to highlight some behaviors of solidifying alloys.

Scaling Analysis

Given the complexity inherent in the solution of the coupled partial differential equations and phase diagrams needed to describe the phenomena occurring during casting, it is useful to begin with an order-of-magnitude, or scaling, analysis (Ref 46, 47). This approach estimates the order of magnitude of each term in the conservation equations, thereby showing which terms represent important effects and which can be neglected; it also can reveal relationships among the different terms in the equations. With practice and some knowledge of the physical phenomena involved, a scaling analysis can give these results with much less time and effort than the exact or a numerical solution requires.

The crucial step in this process is the selection of reference values for all variables. These references must be chosen so that the resulting nondimensional variables vary from 0 to 1 over the space and time of interest in the problem. In many cases, the reference values are not known a priori but are instead revealed by the analysis. Also, the spatial and temporal extent of interest in the scaling analysis is not always the entire domain of the problem (e.g., Ref 48). An example of this analysis applied to the simplest solidification problem is given here, followed by scaling of a much more complicated problem, the fluid flow in the direct chill casting of aluminum.

Heat Transfer and Interface Motion in Sand Casting of Pure Metal. Figure 1 shows the geometry and instantaneous temperature profile of the solidification of a pure substance (at the melting temperature, T_M), where the most significant thermal resistance is in the mold. For simplicity, the liquid has no superheat, the temperature drop across the solid is neglected because the only significant thermal resistance is in the mold, and the mold is treated as semi-infinite and initially at T_o. As the metal is poured, the temperature of the mold at $x = 0$ is very quickly raised to T_M, and heat penetrates the mold to a distance of δ, which increases with time. With no sensible cooling of the solid metal, the heat lost to the mold comes entirely from the latent heat release at

the solid-liquid interface, $M(t)$. This simple problem is chosen as a starting point to illustrate the usefulness of the method and is followed by the analysis of a more complex process.

The analysis of this problem begins with the energy equation in the sand mold:

$$\frac{\partial(\rho_s c_s T)}{\partial t} = \frac{\partial}{\partial x}\left(k_s \frac{\partial T}{\partial x}\right) \qquad \text{(Eq 27)}$$

The maximum temperature range is $\Delta T = T_M - T_o$, over the distance δ, so these quantities are chosen as temperature and spatial reference values. At a given time, the orders of magnitude of the storage and conduction terms in Eq 27 are:

$$\frac{(\rho_s c_s \Delta T)}{t} \sim \frac{1}{\delta}\left(k_s \frac{\Delta T}{\delta}\right) \qquad \text{(Eq 28)}$$

from which an estimate for the time dependence of the penetration depth can be found:

$$\delta \sim \sqrt{\alpha_s\, t}$$

where $\alpha_s = k_s /\rho_s c_s$, the sand thermal diffusivity. This result enables an estimate for the transient heat flux into the mold:

$$q_s'' = -k_s \frac{\partial T}{\partial x}\bigg|_{x=0} \sim k_s \frac{\Delta T}{\delta} \sim k_s \frac{\Delta T}{\sqrt{\alpha_s\, t}} \qquad \text{(Eq 29)}$$

Given the assumption that there is no noticeable temperature drop in the metal during this process, all of the heat entering the mold in Eq 29 must be generated by the latent heat release (LHR) at the moving solid-liquid interface, which can be written as:

$$q_{\text{LHR}} = \frac{\partial(\rho L_f V)}{\partial t}$$

or, with $V = AM$, where A is the area normal to heat flow:

$$q_{\text{LHR}}'' \sim \frac{(\rho L_f M)}{t} \qquad \text{(Eq 30)}$$

Combining Eq 29 and 30, an estimate is obtained for the time dependence of the interface position, M, and for the time for complete solidification, t_f, when $M(t_f) = M_f$:

$$M \sim \frac{(k_s \rho_s c_s)^{1/2}\, \Delta T}{\rho L_f} t^{1/2} \qquad \text{(Eq 31)}$$

and

$$t_f \sim \frac{(\rho L_f M_f)^2}{(k_s \rho_s c_s)\Delta T^2} \qquad \text{(Eq 32)}$$

These estimates are in agreement (to within a factor of order unity) with the exact solutions to this problem found in standard texts (Ref 49, 50).

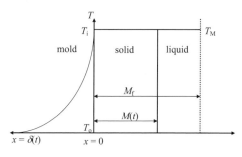

Fig. 1 Schematic of geometry and temperature field for solidification of pure metal in highly resistive mold

Another use for scaling analysis is to test the validity of simplifications, such as the assumption that the temperature drop in the solid metal is small compared to the temperature drop in the mold. For this assumption to be true, the sensible cooling of the solid metal must be much less than the heat flow across it:

$$\frac{\rho c (T_M - T_i) M A}{t} \ll \frac{k(T_M - T_i)A}{M} \qquad \text{(Eq 33)}$$

where T_i is the temperature at the mold-metal interface. This relationship can be rewritten as $M^2/\alpha t \sim (M/\delta)^2 \ll 1$, or, using Eq 31:

$$\frac{(k_s \rho_s c_s)\Delta T^2}{\alpha (\rho L_f)^2} \ll 1 \qquad \text{(Eq 34)}$$

For iron or aluminum in sand or alumina molds, this quantity is found to be on the order of 0.01, so the relationship is true and the assumption is valid.

Fluid Flow in Direct Chill Casting of Aluminum. While the previous example is a very simplified case, scaling analysis can also be used to discern the nature of more complicated casting processes (e.g., Ref 13, 51, 52). One example is direct chill casting, a semicontinuous process in which molten aluminum enters the top of a water-cooled mold, is cooled, and forms a solid ingot drawn out below (Fig. 2). The molten aluminum begins to solidify as a result of the cooling effect of the mold wall. As the billet is withdrawn from the bottom of the mold, water jets impinge on the surface, directly cooling it and forming a solid layer around the molten liquid, commonly referred to as the sump. The flow structure in the sump controls the movement of free-floating solid particles that nucleate there on the grain refiners in the liquid and also influences the fluid flow in the mushy zone. Both of these effects are contributing factors to the growth velocity of the packed solid dendrites and the development of macrosegregation, an ingot-scale inhomogeneity in composition. While one can solve the previous transport equations numerically, it is useful to use a scaling analysis to quantify quickly, to at least an order of magnitude, certain gross parameters in the process. A complete continuum mixture model of the flows in direct chill (DC) casting can be found in Ref 33, and other approaches to single-domain models of this process include Ref 18 and 53.

Figure 3 (Ref 54) shows the expected sump flow structure in DC casting. The fluid enters at the top at temperature $T_M + \Delta T$ (where ΔT is the superheat, and T_M is the liquidus temperature of the alloy) and quickly turns toward the chilled outer radius, forming a boundary layer flow down the interface with the mushy zone (at $T = T_M$ in the figure). The fluid leaves the sump as it is entrained into the mushy zone and frozen. The flow along the cold interface is driven by the downward thermal buoyancy. When it reaches the centerline of the round billet, it wells up into the sump as part of a shear-driven circulation cell. The cell is largely isothermal, with a large temperature gradient in the sump between the cell and the inflow region. There have been many numerical simulations of this process (e.g., Ref 18, 19, 33, 35, 53, 54), but here, the scaling analysis from Ref 54 and 55 is used to determine the order of magnitude of the flow velocity down the interface, the thickness of the boundary layer there, and the spatial extent of the stratified region in the sump.

Beginning in the thermal boundary layer in the liquid next to the mushy zone and assuming all of the pressure gradient is in the buoyancy term, a scaling analysis of the steady-state momentum balance (Eq 18) along the direction of flow gives:

$$u_b\left(\frac{u_b}{L_T}\right), \; v_b\left(\frac{u_b}{\delta}\right) \sim v\left(\frac{u_b}{L_T^2}, \frac{u_b}{\delta^2}\right), \quad g\beta\Delta T \qquad \text{(Eq 35)}$$

where u_b and v_b are velocities parallel and perpendicular to the sump boundary, L_T is the length of the stratified region (in which there is a significant difference in temperature across the boundary layer to drive the flow), δ is the boundary layer thickness, and v (= μ/ρ) is the kinematic viscosity. The first two terms on the left side are inertia, while the right side is composed of two terms for viscous friction, and the last term represents thermal buoyancy. Buoyancy due to compositional gradients is neglected in this analysis. Scaling the continuity equation (Eq 19) in the same region gives:

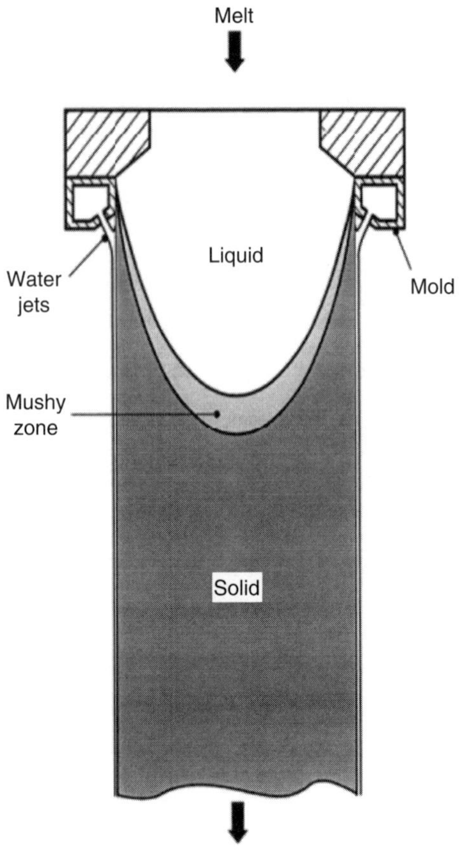

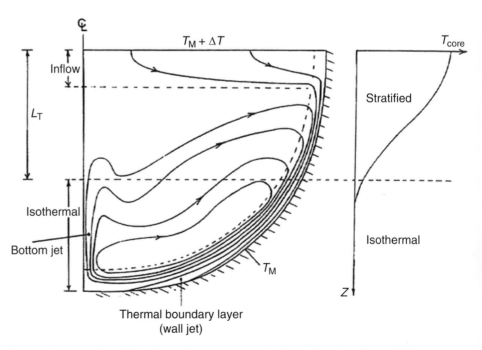

Fig. 2 Schematic of direct chill casting geometry

Fig. 3 Schematic of flow field and centerline temperature in the direct chill casting of a round billet, assuming a hemispherical sump volume. Source: Ref 54

$$\left(\frac{u_b}{L_T}\right) \sim \left(\frac{v_b}{\delta}\right) \quad \text{or} \quad v_b \sim u_b\left(\frac{\delta}{L_T}\right) \qquad \text{(Eq 36)}$$

Using Eq 36 and the assumption that $\delta/L_T \ll 1$, the momentum balance reduces to:

$$\frac{u_b^2}{g\beta\Delta T\, L_T} \sim \frac{\nu u_b}{\delta^2\, g\beta\Delta T} \quad 1 \qquad \text{(Eq 37)}$$

where the remaining three terms, from left to right, represent inertia, friction, and thermal buoyancy. Assuming that the buoyancy force is the effect that accelerates the flow, and friction is not significant in this region:

$$u_b \sim (g\beta\Delta T L_T)^{\frac{1}{2}} \qquad \text{(Eq 38)}$$

Performing a similar analysis on the temperature-based energy Eq 9 in the boundary layer:

$$u_b\left(\frac{\Delta T}{L_T}\right), \quad v_b\left(\frac{\Delta T}{\delta}\right) \sim \alpha\left(\frac{\Delta T}{L_T^2}, \frac{\Delta T}{\delta^2}\right)$$

which, with Eq 36, reduces to:

$$\delta \sim \left(\frac{\alpha\, L_T}{u_b}\right)^{\frac{1}{2}} \qquad \text{(Eq 39)}$$

Davidson and Flood (Ref 54) show that all of the inflow is entrained toward the mold and enters the boundary layer near the corner of the sump bottom and the inlet. If the heat advected into the sump at the top is added to the heat conducted across the stratified region, the sum should be approximately the heat content of the boundary layer:

$$(\rho c\Delta T)(\pi R^2)u_o + k(\pi R^2)\frac{\Delta T}{L_T}$$
$$\sim (\rho c\Delta T)(\pi[R^2 - (R-\delta)^2](R-\delta)^2)u_b \qquad \text{(Eq 40)}$$

where the area of the boundary layer on the right side of Eq 40 is the annulus between $r = R$ and $r = R - \delta$, and the inlet velocity is u_o. Simplifying Eq 40 and assuming $\delta/R \ll 1$ gives:

$$u_o + \frac{\alpha}{L_T} \sim u_b\left(\frac{\delta}{R}\right) \qquad \text{(Eq 41)}$$

Combining Eq 38, 39, and 41, Reese (Ref 55) obtained the following estimates:

$$\frac{L_T}{R} \sim \left(Gr\, Pr^2\right)^{-\frac{1}{7}}\left(1 + u_o^*\right)^{\frac{4}{7}} \qquad \text{(Eq 42)}$$

$$\frac{\delta}{R} \sim \left(Gr\, Pr^2\right)^{-\frac{2}{7}}\left(1 + u_o^*\right)^{\frac{1}{7}} \qquad \text{(Eq 43)}$$

and

$$\frac{u_b R}{\alpha} \sim \left(Gr\, Pr^2\right)^{\frac{3}{7}}\left(1 + u_o^*\right)^{\frac{2}{7}} \qquad \text{(Eq 44)}$$

where:

$$Gr = \frac{g\beta\Delta T R^3}{\nu^2} \quad \text{and} \quad u_o^* = u_o\frac{L_T}{\alpha}$$

The Grashof number, Gr, is a measure of the vigor of the buoyancy-driven flow, and u_o^* is the nondimensional inlet velocity. For typical values of aluminum DC casting, $u_o^* \sim \mathcal{O}(1)$.

These functions indicate that the velocity down the interface, which controls the redistribution of solute by interacting with the flow in the mushy zone and carrying the free-floating solid grains, is only weakly controlled by the ingot size or the superheat (Ref 54):

$$u_b \sim R^{\frac{2}{7}}\Delta T^{\frac{3}{7}} \qquad \text{(Eq 45)}$$

Because u_b is such a weak function of R and ΔT, this result shows that, in the range of those parameters available in practical situations, there is a narrow range of buoyancy-driven velocities. The size of the stratified region, L_T (Eq 42), can be shown to be a very weak function of superheat, so this area will always extend over 10 to 20% of the sump. The scaling analysis shows very quickly that, without the introduction of other forces into this process, the flow structure can only be marginally changed.

In this scaling analysis, Davidson and Flood (Ref 54) and Reese (Ref 55) made assumptions about the momentum balance and the thickness of the boundary layer. Friction was neglected in that momentum balance (Eq 37), which requires:

$$\frac{\nu u_b}{\delta^2\, g\beta\Delta T} \sim Pr \ll 1 \qquad \text{(Eq 46)}$$

having substituted Eq 43 and 44 into Eq 37. Because liquid metals have $Pr \ll 1$, this assumption is generally true. For typical practical values from DC casting of aluminum alloys, $GrPr^2 \gg 1$ and $u_o^* \sim \mathcal{O}(1)$, so it can also be seen from Eq 43 that $\delta/R \ll 1$, confirming the other main assumption in the analysis.

Numerical Analysis

While the previous scaling analyses provide useful estimates of important transport phenomena and elucidate their interrelationships with very little mathematical effort, more detailed information is frequently needed. At that point, numerical solutions to the governing equations for energy (Eq 8 or 9), composition (Eq 13), momentum (Eq 18), and mass (Eq 19) are sought. While this section does not address solution techniques, it does present examples of such simulations of solidification processes, in which coupled predictions of fluid flow and heat and mass transfer are used to predict the origin and extent of defects. (The papers referenced in this section do discuss various numerical techniques to solve the transport equations. These methods are not the subject of this article, so the reader is referred to Ref 8 to 11 for detailed treatments.) The focus of this section

is the prediction of macrosegregation defects. Other choices are possible to illustrate the power of the models to predict physical phenomena in solidifying alloys, but these examples were picked because the primary result (the final composition field) is tightly coupled to the heat transfer and fluid flow during processing and is also of direct interest to the metallurgist.

Macrosegregation in a Binary Alloy. In these studies, the processing defect of interest is macrosegregation, an inhomogeneity in composition that leads to variations in mechanical properties through poor distribution of secondary phases or solid-solution strengthening. The origin of macrosegregation is in the solute redistribution inherent in alloys that freeze over a temperature range. The idealized phase diagram in Fig. 4 shows the equilibrium solid and liquid compositions at a given temperature in a two-phase region. The local liquid composition is, at this point, richer with solute than the nominal composition. If that fluid moves from its initial location and is replaced by fluid with a different composition, then the mixture composition changes locally. Because this process generally occurs much more rapidly than species diffusion (Ref 13), it is the dominant cause of macrosegregation. What causes this fluid flow which redistributes the solute? Both the species gradients arising from the different solubilities of the various alloying elements in the solid and the temperature gradients set up by the cooling process cause variations in the density field of the liquid. The nonuniform density in a gravitational field results in buoyancy-induced flow throughout the interdendritic liquid and the bulk melt. Also, in real systems, the densities of the solid and liquid phases are different (usually with $\rho_S > \rho_L$), and the volume change upon solidification drives fluid motion, which is needed to enforce continuity.

An example of the development of macrosegregation in a static casting is found in Fig. 5. Here, it is assumed that the fluid flow that

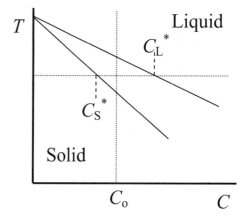

Fig. 4 Idealized binary phase diagram, showing the relationship among the nominal composition (C_o) and the equilibrium liquid (C_L^*) and solid (C_S^*) compositions

causes this local composition change is driven entirely by density changes in the fluid. The liquid density in the mushy zone is increased by the colder temperatures, but here it is assumed that there is a positive solutal buoyancy (i.e., the interdendritic liquid is enriched with a lighter component of the alloy) and that the net effect is an upward buoyancy force there.

When an ingot is cooled from a vertical wall, a buoyancy-driven flow will always arise (usually having much more effect on macrosegregation than the accompanying shrinkage-induced flow) because of the horizontal temperature gradient perpendicular to the gravity vector. The buoyancy force resulting from the composition

gradient can either aid or oppose the downward thermal effect. In Fig. 5, the streamlines and composition fields for opposing solutal and thermal buoyancy are shown for a binary alloy cooled from the left wall. In Fig. 5(a), before solidification begins, there is a counterclockwise flow cell driven entirely by the cooling at the chill wall. When the mushy zone begins to develop (Fig. 5b), the lighter interdendritic fluid rises through the mush (mostly near the liquidus interface, delineated by the white line, where the drag is lowest) and escapes along the top of the mold. The thermal cell, weakened by the cooling of the bulk melt, is forced into the bottom half of the liquid region by the

compositionally driven cell. The beginnings of a cone segregate are seen in the lower part of the mush (darker shading indicates lower solute concentration). A-segregates, indicated by the light and dark bands in the upper mush, have begun to develop. These regions are sites from which much of the lighter fluid may leave the solid-liquid zone, because they have lower solid fractions than the surrounding mushy zone, although Combeau et al. (Ref 36) have shown that they are not necessary to the formation of the overall macrosegregation pattern.

As solidification fronts move forward and the superheat in the melt is all but extinguished, the thermal cell dies out completely, and compositional buoyancy drives the entire flow. The clockwise streamlines in Fig. 5(c) show that the fluid leaves the mush through the A-segregates, the stronger of which feed small recirculation cells in the stratified upper region of the melt. The final macrosegregation pattern, typical of static castings, is seen in Fig. 5(d), which includes a negatively segregated cone, severe A-type segregation, and a solute-rich region along the top and the far vertical wall. This development of macrosegregation in a solidifying binary alloy is typical and has been observed in many single-domain numerical studies, including several early works (e.g., Ref 6, 56, 57).

Although since the 1980s there has been a plethora of numerical studies of the macroscopic heat and mass transfer during alloy solidification, much fewer works are available in which quantitative experimental data are compared to those predictions. Most of the work has been done in metal analog systems (e.g., water-ammonium chloride), which exhibit dendritic growth and are transparent, allowing visualization of flow patterns in the melt and how they interact with the mushy zone. However, there are very few studies comparing composition and temperature data from binary metal alloys to numerical predictions using single-domain models. The first such study was by Shahani et al. (Ref 58), who cast two lead-tin alloys and measured compositions at discrete points. Some agreement was found between predictions and experiments, although the resolution of the measurements was not fine enough to show A-segregates in the casting. Prescott et al. (Ref 59) performed experiments with lead-tin to validate a single-domain mixture model. With the exception of some small nonequilibrium effects, the trends in the predicted transient temperatures matched the simulations well, although the composition measurements only agreed with the trends in the calculated macrosegregation patterns. Krane and Incropera (Ref 60) also performed experiments in lead-tin alloys with similar results. They also measured dendritic arm spacings for use in the Blake-Kozeny permeability model (Eq 25) and showed that the model overpredicts the permeability of the dendritic array near the liquidus temperature (low fraction solid). The evaluation of the agreement between model and

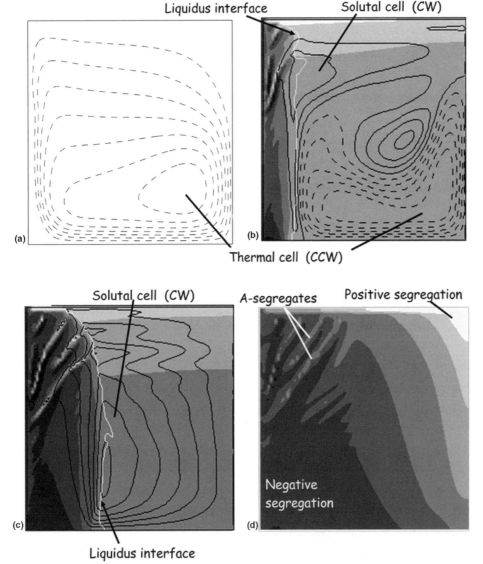

Fig. 5 Development of macrosegregation in an alloy with interdendritic liquid lighter than nominal. (Lighter shade of gray indicates liquid of higher-than-nominal composition; darker regions are depleted in solute. The white lines are the liquidus interface, the solid black lines are clockwise streamlines, and the dashed black lines are counterclockwise streamlines.) The alloy is chilled from the left side and is insulated on other boundaries. (a) Thermally driven flow cell before solidification begins. (b) With some of the ingot solidified, the flow cell driven by solutal buoyancy replaces the enriched interdendritic liquid in the mushy zone with fluid from the melt at the nominal composition. (c) The melt superheat is extinguished, and only a weak solutal cell remains. (d) Segregation pattern in completely solidified ingot

experiment was hampered by the high uncertainty in the composition measurements and in the permeability model. Although composition measurements did not show the presence of A-segregates, these predicted structures were observed in polished sections from the cast ingot. Singh et al. (Ref 61) further investigated the role of permeability in these simulations, evaluating different permeability models (including anisotropic effects) and comparing composition predictions to data from Ref 58 and 60.

Macrosegregation in a Multicomponent Alloy. Most applications of the single-domain models have been simulations of binary alloys, or alloys approximated as binary. These systems have been useful to elucidate the basic transport phenomena, but most commercial alloys contain more than two components, and the interactions of all the elements may have an influence on the convective flows that lead to macrosegregation and other defects. The formation of secondary solid phases can be much more complicated in multicomponent alloys and should be treated using appropriate thermodynamic models.

Mehrabian and Flemings (Ref 62) predicted macrosegregation in an Al-Cu-Ni alloy due only to shrinkage-induced flow, and others have developed models with buoyancy-driven flows limited to the mushy zone (Ref 63) or with only partially coupled transport models (Ref 64). The first fully coupled model that simultaneously solved in all regions of the metal for transport of all the elements, heat, mass, and momentum (and linked to the phase diagram of the alloy) is found in the prediction of solidification of a ten-component, low-alloy steel by Schneider and Beckermann (Ref 44). Their single-domain model, derived using volume averaging methods (Ref 6), predicted developing macrosegregation patterns and fluid flow, assuming that the steel solidified into one solid phase (neglecting secondary phases). For such a lean alloy, the formation of secondary phases will only occur in very small amounts and will not affect the prediction of macrosegregation, because all of the flows redistributing solute are in the vicinity of the liquidus surface; by the time any secondary phases may form, the flow is negligible (Ref 6). The phase equilibrium was modeled by each element independently affecting the liquidus temperature by a constant slope. While small local differences were present, overall macrosegregation patterns of the various elements were found to be similar to each other and, because all of the elements were rejected into the liquid as the iron-rich solid formed, they were redistributed in patterns similar to the binary solidification results mentioned previously. The overall macrosegregation of elements was linearly proportional to their partition coefficient (Fig. 6), and this result did not change if elements with negligible influence on buoyancy were eliminated. Simulation results of the same steel with two different sets of published data for partition coefficient and liquidus slopes showed very different segregation patterns (due to a

calculated reversal in the buoyancy force), indicating the nonlinear sensitivity of the model to proper thermodynamic data. Felicelli et al. (Ref 65) simulated freckle formation in unidirectional solidification of a Ni-Al-Ta-W alloy and showed that the mechanisms for the development of freckles is virtually the same as a binary alloy when the buoyancy force has been determined.

All of the aforementioned studies assumed that the alloys only solidified a primary solid phase, at least until very close to the end of freezing. To evaluate the behavior of a multicomponent alloy in which the liquid composition path changes during freezing, Krane et al. (Ref 45, 66) developed a geometric model for the lead-rich corner of the Pb-Sb-Sn diagram and coupled it to a three-component transport model. In Ref 66, this model was used to simulate various alloys. In some of these alloys, the solidification path had increasing liquid composition for both antimony and tin until solidification ends. In others, the tin concentration increases steadily throughout, while the antimony concentration increases initially but begins to decrease after the start of precipitation of intermetallic phases at approximately $f_s = 0.1$. In the first type of alloy, the basic development of the flow patterns, solidification fronts, and solute redistributions occurs in ways similar to binary alloys. Tin and antimony are rejected throughout the freezing process, and the resulting lighter liquid rises and leaves the mushy zone, creating positive and negative cone segregates with some A-segregates. Alloys of the latter type also began in the primary solidification region but had significant freezing on the twofold saturation curve for the primary lead and SbSn intermetallic. In this case, the tin content of the liquid increased monotonically as the temperature fell, while the liquid antimony concentration first increased (in the primary region) and then decreased (along the binary trough). This reversal of antimony microsegregation led to a unique pattern of macrosegregation.

The convective flows, dominated by the large difference in tin content of the interdendritic fluid and the bulk melt, began to redistribute the antimony as in a two-component alloy. When the binary trough was reached, however, the newly formed solid absorbed, rather than rejected, the antimony, and liquid advected from the mush by the tin-driven flows was depleted in antimony. The resulting pattern (Fig. 7), with negative antimony cone segregates in the lower left and upper right corners, separated by a region of antimony-rich material, is unique to some ternary alloys and is not found in two-component systems or in multicomponent alloys which only experience primary solidification.

Experimental confirmation of multicomponent models is just as scarce as for simulations of binary alloys. Mehrabian and Flemings (Ref 62) found good agreement of their model with experiment, although the thermally and solutally stable configuration only allows shrinkage-induced flows. Vannier et al. (Ref 64) compared the

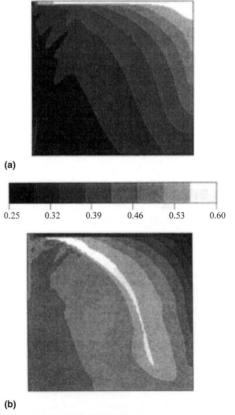

(a)

| 0.25 | 0.32 | 0.39 | 0.46 | 0.53 | 0.60 |

(b)

| 0.025 | 0.035 | 0.043 | 0.055 | 0.065 |

Fig. 7 Solidified composition fields in a Pb-35wt%Sn-5wt%Sb alloy chilled from the left side. The tin pattern (a) shows the standard segregation pattern for a binary alloy (with very weak A-segregates). Because the antimony partitioning had little effect on the buoyancy, the tin segregation dominates the flow. A reversal of antimony microsegregation at low solid fraction and the tin-controlled flow cause the unusual antimony macrosegregation in (b). Source: Ref 68

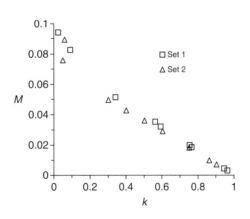

Fig. 6 Dependence of global macrosegregation levels in ten-component steel (all data from two cases in Ref 44). Each point represents the macrosegregation level, M, of an individual component as a function of its partition coefficient, k. The two data sets were predicted under the same conditions and with the same model, except with different sets of partition coefficients and different liquidus surfaces.

predictions from their partially coupled transport model to the carbon concentration profiles in a statically cast six ton steel ingot. Measured compositions near the bottom of the ingot exhibited a negative segregation that was not predicted by the model, which assumed an entirely columnar, rigid mushy zone. In a similar study of a large steel ingot, Gu and Beckermann (Ref 67) made a similar assumption in their fully coupled, single-domain model and produced the same discrepancy. In both studies, the authors concluded that the presence of free-floating, equiaxed dendrites in the ingot (lacking in the models) was responsible for the disagreement. Although their model was capable of simulating the formation of such structures, Gu and Beckermann predicted no A-segregates due to the necessarily coarse grid size. Similar significant variation in the temperature of last solidification was measured and predicted throughout the ingot, owing to the effect of macrosegregation. Du et al. (Ref 53) compared predictions and experiments in DC casting of an Al-Cu-Mg alloy and came to the same conclusions about the need for a free-floating grain model in that process. Combeau et al. (Ref 36) included such a model in predictions of macrosegregation in a large, statically cast steel ingot and found much better agreement with experiment than Ref 64 and 67. They explored the different behaviors of globular and dendritic solid particles. Because the former settles more rapidly than the latter, if globular solid is present, there is much more negative segregation near the ingot bottom. Because the dendritic solid tends to stay flowing with the melt, there is much less segregation with that structure. They also were able to use a much finer grid than in Ref 67 and so were able to predict A-segregates. These two studies showed that the A-segregates are not the primary source of segregated fluid leaving the mushy zone, because they both predicted roughly the same level of segregation near the top of the ingots. A study by Krane et al. (Ref 68) in a Pb-Sb-Sn alloy showed good agreement with temperature measurements but highlighted the sensitivity of composition redistribution to the permeability model.

All of these studies also illustrate the importance of the use of accurate thermodynamic data for the prediction of the relationship among temperature, fraction solid, and phase composition. The temperature and liquid composition in the mushy zone drive the segregating natural convection flow and are very sensitive to models of the liquidus temperature and partition coefficients.

Summary

A set of partial differential equations constituting a common model for transport phenomena in solidification of metal alloys has been reviewed, and the physical phenomena represented by each term has been identified, especially those terms particular to solidification in dendritic alloys. The representations of solidification shrinkage, buoyancy, and drag in the mushy zone are presented, while extensions to account for other effects are possible. These models can include the prediction of electromagnetic effects from Maxwell's equations, gas porosity using a nucleation model and an advection-diffusion equation for the gas, hot tearing when coupled with a stress model, or microstructural development if linked to a method for predicting the movement of phase boundaries. The focus here has been on single-domain mixture models, but multiple-phase approaches are also viable options. (Multiple-domain techniques are used much less often in recent literature.) The need for a closure model relating temperature, solid fraction, and solid and liquid compositions is mentioned, and the reader is referred to Ref 2, 6, and 43 to 45 for more details.

When a set of transport equations has been developed, several examples are given to illustrate their use and elucidate some of the physics present in the solidification of metal alloys. Scaling, or order-of-magnitude, analysis is used to understand the behavior of two processes and to check the modeling assumptions. This simple approach is a good method to obtain behavioral trends quickly. For more detailed knowledge, numerical analysis is necessary, and simulations that show the development of macrosegregation and its interdependence on the flow and temperature fields are presented as examples. Both binary and multicomponent alloys are discussed, and some experimental validation is briefly reviewed.

REFERENCES

1. W.D. Bennon and F.P. Incropera, A Continuum Model for Momentum, Heat and Species Transport in Binary Solid Liquid-Phase Change Systems, Part 1: Model Formulation, *Int. J. Heat Mass Transf.*, Vol 30, 1987, p 2161–2170
2. W.D. Bennon and F.P. Incropera, Numerical Analysis of Binary Solid-Liquid Phase Change Using a Continuum Model, *Num. Heat Transf.*, Vol 13, 1988, p 277–296
3. P.J. Prescott, F.P. Incropera, and W.D. Bennon, *Int. J. Heat Mass Transf.*, Vol 34, 1991, p 2351–2359
4. C. Beckermann and R. Viskanta, Double-Diffusive Convection during Dendritic Solidification of a Binary Mixture, *Physico. Chem. Hydrodyn.*, Vol 10, 1988, p 195–213
5. C. Beckermann and C.Y. Wang, Multi-Phase/-Scale Modeling of Transport Phenomena in Alloy Solidification, *Ann. Rev. Heat Transf.*, Vol 6, 1995, p 115–198
6. M.C. Schneider and C. Beckermann, Numerical Study of the Combined Effects of Microsegregation, Mushy Zone Permeability and Flow, Caused by Volume Contraction and Thermosolutal Convection, on Macrosegregation and Eutectic Formation in Binary Alloy Solidification, *Int. J. Heat Mass Transf.*, Vol 38, 1995, p 3455–3473
7. J. Ni and C. Beckermann, A Volume-Averaged Two-Phase Model for Transport Phenomena during Solidification, *Metall. Trans. B*, Vol 22, 1991, p 349–360
8. S. Patankar, *Numerical Heat Transfer and Fluid Flow*, Hemisphere, 1980
9. C. Hirsch, *Numerical Computation of Internal and External Flows, Volume 1: Fundamentals of Numerical Discretization*, J. Wiley & Sons, 1988
10. H.K. Versteeg and W. Malalasekera, *An Introduction to Computational Fluid Dynamics: The Finite Volume Method*, 2nd ed., Pearson Prentice Hall, 2007
11. J.H. Ferziger and M. Peric, *Computational Methods for Fluid Dynamics*, Springer, 2002
12. V. de Felice, A. Jardy, and H. Combeau, Comparison between Temperature- and Enthalpy-Based Formulations for the Energy Transport Equation in Solidifying Multi-Component Systems, *Int. Symp. on Liquid Metal Processing and Casting*, P.D. Lee, A. Mitchell, and R. Williamson, Ed., TMS, 2009, p 97–105
13. M.J.M. Krane and F.P. Incropera, A Scaling Analysis of the Unidirectional Solidification of a Binary Alloy, *Int. J. Heat Mass Transf.*, Vol 39, 1996, p 3567–3579
14. *Smithell's Metals Reference Handbook*, 7th ed., E.A. Brandes and G.B. Brook, Ed., Butterworth-Heinemann, Ltd., 1992
15. H. Schlichting, *Boundary-Layer Theory*, 7th ed., McGraw-Hill, 1979
16. R.L. Panton, *Incompressible Flow*, 2nd ed., J. Wiley & Sons, 1996
17. M.J.M. Krane and F.P. Incropera, Analysis of the Effect of Shrinkage on Macrosegregation in Alloy Solidification, *Metall. Mater. Trans. A*, Vol 26, 1995, p 2329–2339
18. A.V. Reddy and C. Beckermann, Modeling of Macrosegregation due to Thermosolutal Convection and Contraction-Driven Flow in Direct Chill Continuous Casting of an Al-Cu Round Ingot, *Metall. Mater. Trans. B*, Vol 28, 1997, p 479–489
19. D.G. Eskin, Q. Du, and L. Katgerman, Relationship Between Shrinkage-Induced Macrosegregation and the Sump Profile Upon Direct-Chill Casting, *Scr. Mater.*, Vol 55, 2006, p 715–718
20. K. Kubo and R.D. Pehlke, Mathematical Modeling of Porosity Formation in Solidification, *Metall. Trans. B*, Vol 16, 1985, p 359–366
21. P.K. Sung, D.R. Poirier, S.D. Felicelli, E.J. Poirier, and A. Ahmed, Simulations of Microporosity in IN718 Equiaxed Investment Castings, *J. Cryst. Growth*, Vol 226, 2001, p 363–377
22. A.S. Sabau and S. Viswanathan, Microporosity Prediction in Aluminum Alloy Castings, *Metall. Mater. Trans. B*, Vol 33, 2002, p 243–255
23. R.C. Atwood and P.D. Lee, Simulation of the Three-Dimensional Morphology of Solidification Porosity in an

Aluminium-Silicon Alloy, *Acta Mater.*, Vol 51, 2003, p 5447–5466

24. B. Gebhart, Y. Jaluria, R.L. Mahajan, and B. Sammakia, *Buoyancy-Induced Flows and Transport,* Hemisphere, 1988

25. T. Iida and R. Guthrie, *The Physical Properties of Liquid Metals,* Clarendon Press, Oxford, 1988

26. D.A. Nield and A. Bejan, *Convection in Porous Media,* 2nd ed., Springer, 1999

27. D. Nagelhout, M.S. Bhat, J.C. Heinrich, and D.R. Poirier, Permeability for Flow Normal to a Sparse Array of Fibres, *Mater. Sci. Eng. A,* Vol 191, 1995, p 203–208

28. M.S. Bhat, D.R. Poirier, and J.C. Heinrich, A Permeability Length Scale for Cross Flow Through Model Structures, *Metall. Mater. Trans. B,* Vol 26, 1995, p 1091–1092

29. S.D. Felicelli, J.C. Heinrich, and D.R. Poirier, Simulation of Freckles during Vertical Solidification of Binary Alloys, *Metall. Trans. B,* Vol 22, 1991, p 847–859

30. C.Y. Wang and C. Beckermann, Equiaxed Dendritic Solidification, Part I: Multiscale Modeling, *Metall. Mater. Trans. A,* Vol 27, 1996, p 2754–2764

31. J. Ni and F.P. Incropera, Extension of the Continuum Model for Transport Phenomena Occurring during Metal Alloy Solidification, Part I: The Conservation Equations, *Int. J. Heat Mass Transf.,* Vol 38, 1995, p 1271–1284

32. J. Ni and F.P. Incropera, Extension of the Continuum Model for Transport Phenomena Occurring during Metal Alloy Solidification, Part II: Microscopic Considerations, *Int. J. Heat Mass Transf.,* Vol 38, 1995, p 1285–1296

33. C.J. Vreeman, M.J.M. Krane, and F.P. Incropera, The Effect of Free-Floating Dendrites and Convection on Macrosegregation in Direct Chill Cast Aluminum Alloys, Part I: Model Development, *Int. J. Heat Mass Transf.,* Vol 43, 2000, p 677–686

34. C.Y. Wang and C. Beckermann, Equiaxed Dendritic Solidification, Part II: Numerical Simulations for An Al-4 Wt Pct Cu Alloy, *Metall. Mater. Trans. A,* Vol 27, 1996, p 2765–2783

35. C.J. Vreeman, J.D. Schloz, and M.J.M. Krane, Direct Chill Casting of Aluminum Alloys: Modeling and Experiments on Industrial Scale Ingots, *ASME J. Heat Transf.,* Vol 124, 2002, p 947–953

36. H. Combeau, M. Založnik, S. Hans, and P. E. Richy, Prediction of Macrosegregation in Steel Ingots: Influence of the Motion and the Morphology of Equiaxed Grains, *Metall. Mater. Trans. B,* Vol 40, 2009, p 289–304

37. D.C. Wilcox, *Turbulence Modeling for CFD,* DCW Industries, 2006

38. P.A. Davidson, *Turbulence—An Introduction for Scientists and Engineers,* Oxford University Press, 2004

39. W.P. Jones and B.E. Launder, The Prediction of Laminarization with a Two-Equation Model of Turbulence, *Int. J. Heat Mass Transf.,* Vol 15, 1972, p 301–314

40. W.P. Jones and B.E. Launder, The Calculation of Low-Reynolds-Number Phenomena with a Two-Equation Model of Turbulence, *Int. J. Heat Mass Transf.,* Vol 16, 1973, p 1119–1130

41. B.E. Launder and B.I. Sharma, Application of Energy-Dissipation Model of Turbulence to the Calculation of Flow near a Spinning Disk, *Lett. Heat Mass Transf.,* Vol 1, 1974, p 131–138

42. B.G. Thomas, Q. Yuan, S. Sivaramakrishnan, T. Shi, S.P. Vanka, and M.B. Assar, Comparison of Four Methods to Evaluate Fluid Velocities in a Continuous Slab Casting Mold, *ISIJ Int.,* Vol 41, 2001, p 1262–1271

43. C.Y. Wang and C. Beckermann, *Metall. Mater. Trans. A,* Vol 24, 1993, p 2787–2802

44. M.C. Schneider and C. Beckermann, Formation of Macrosegregation by Multicomponent Thermosolutal Convection during Solidification of Steel, *Metall. Mater. Trans. A,* Vol 26, 1995, p 2373–2388

45. M.J.M. Krane, F.P. Incropera, and D.R. Gaskell, Solution of Ternary Metal Alloys—I. Model Development, *Int J. Heat Mass Transf.,* Vol 40, 1997, p 3827–3835

46. A. Bejan, *Convection Heat Transfer,* 3rd ed., J. Wiley & Sons, 2004

47. J. Dantzig and C. Tucker, *Modeling in Materials Processing,* Cambridge University Press, 2001

48. S. Kimura and A. Bejan, Natural Convection in a Differentially Heated Corner Region, *Phys. Fluids,* Vol 28, 1985, p 2980–2989

49. D.R. Poirier and G.H. Geiger, *Transport Phenomena in Materials Processing,* TMS, 1998

50. J.A. Dantzig and M. Rappaz, *Solidification,* EPFL Press, 2009

51. G. Amberg, Parameter Ranges in Binary Solidification from Vertical Boundaries, *Int. J. Heat Mass Transf.,* Vol 40, 1997, p 2565–2578

52. P.A. Davidson, *An Introduction to Magnetohydrodynamics,* Cambridge University Press, 2001

53. Q. Du, D.G. Eskin, and L. Katgerman, Modeling Macrosegregation during Direct-Chill Casting of Multicomponent Aluminum Alloys, *Metall. Mater. Trans. A,* Vol 38, 2007, p 180–189

54. P.A. Davidson and S.C. Flood, Natural Convection in an Aluminum Ingot: A Mathematical Model, *Metall. Mater. Trans. B,* Vol 25, 1997, p 293–302

55. J.M. Reese, Characterization of the Flow in the Molten Metal Sump during Direct Chill Aluminum Casting, *Metall. Mater. Trans. B,* Vol 28, 1997, p 491–499

56. W.D. Bennon and F.P. Incropera, The Evolution of Macrosegregation in Statically Cast Binary Ingots, *Metall. Trans. B,* Vol 18, 1987, p 611–616

57. V.R. Voller, A.D. Brent and C. Prakash, The Modelling of Heat, Mass and Solute Transport in Solidification Systems, *Int. J. Heat Mass Transf.,* Vol 32, 1989, p 1719–1731

58. H. Shahani, G. Amberg, and H. Fredriksson, On the Formation of Macrosegregations in Unidirectional Solidified Sn-Pb and Pb-Sn Alloys, *Metall. Trans. A,* Vol 23, 1992, p 2301–2311

59. P.J. Prescott, F.P. Incropera, and D. Gaskell, Convective Transport Phenomena and Macrosegregation during Solidification of a Binary Metal Alloy, Part 2: Experiments and Comparisons with Numerical Predictions, *ASME JHT,* Vol 116, 1994, p 742–749

60. M.J.M. Krane and F.P. Incropera, Experimental Validation of Continuum Mixture Model for Binary Alloy Solidification, *ASME JHT,* Vol 119, 1997, p 783–791

61. A.K. Singh, B. Basu, and A. Ghosh, Role of Appropriate Permeability Model on Numerical Prediction of Macrosegregation, *Metall. Mater. Trans. B,* Vol 37, 2006, p 799–809

62. R. Mehrabian and M.C. Flemings, Macrosegregation in Ternary Alloys, *Metall. Trans.,* Vol 1, 1970, p 455–464

63. T. Fujii, D.R. Poirier, and M.C. Flemings, Macrosegregation in a Multicomponent Low-Alloy Steel, *Metall. Trans. B,* Vol 10, 1979, p 331–339

64. I. Vannier, H. Combeau, and G. Lesoult, Numerical Model for Prediction of the Final Segregation Pattern of Bearing Steel Ingots, *Mater. Sci. Eng. A,* Vol l73, 1993, p 317–321

65. S.D. Felicelli, D.R. Poirier, and J.C. Heinrich, Modeling Freckle Formation in Three Dimensions during Solidification of Multicomponent Alloys, *Metall. Mater. Trans. B,* Vol 29, 1998, p 847–855

66. M.J.M. Krane and F.P. Incropera, Solution of Ternary Metal Alloys—II. Predictions of Convective Phenomena and Solidification Behavior in Pb-Sb-Sn Alloys, *Int. J. Heat Mass Transf.,* Vol 40, 1997, p 3837–3847

67. J.P. Gu and C. Beckermann, Simulation of Convection and Macrosegregation in a Large Steel Ingot, *Metall. Mater. Trans. A,* Vol 30, 1999, p 1357–1366

68. M.J.M. Krane, F.P. Incropera, and D.R. Gaskell, Solidification of a Ternary Metal Alloy: A Comparison of Experimental Measurements and Model Predictions in a Pb-Sb-Sn System, *Metall. Mater. Trans. A,* Vol 29, 1998, p 843–853

ASM Handbook, Volume 22B, *Metals Process Simulation*
D.U. Furrer and S.L. Semiatin, editors

Copyright © 2010, ASM International®
All rights reserved.
www.asminternational.org

Modeling of Casting and Solidification Processes

Jianzheng Guo and Mark Samonds, ESI US R&D

CASTING AND SOLIDIFICATION PRO-CESSES are modeled in terms of thermodynamics, heat transfer, fluid flow, stress, defect formation, microstructure evolution, and thermophysical and mechanical properties (Ref 1). Simulation technologies are applied extensively in casting industries to understand the effects of alloy chemistry, heat transfer, fluid transport phenomena, and their relationships to microstructure and the formation of defects. Thanks to the rapid progress of computer calculation capability and numerical modeling technologies, casting simulation can quickly answer some critical questions concerning filling, microstructure, defects, properties, and final shape. Modeling can be used to support quality assurance and can also help to reduce the time for responding to customer inquiries. It shifts the trial-and-error procedure used on the shop floor to the computer, which does it faster, more easily, with greater transparency, and more economically.

Casting simulation software has been available to the foundry industry since the 1980s. Simulation technology has come a long way since the early to mid-1980s, when the design engineer could only work with two-dimensional models. The early days focused on identifying hot spots in the casting. As the computer-aided design and numerical simulation software packages evolved, foundry engineers could increasingly make quick changes to the feeding design to fix these problems with relative ease. Most of the casting simulation packages now on the market can handle fluid flow in the mold and solidification. Today (2010), the foundry industry wants to focus on more advanced predictions, such as stress and deformation, microstructure determination, defect formation, and mechanical properties.

The primary phenomenon controlling casting is the heat transfer from the metal to the mold. The heat-transfer processes are complex. The cooling rates range from an order of one-tenth to thousands of degrees per second, and the corresponding length scales extend from several meters to a few micrometers. These various cooling rates produce different microstructures

and hence a variety of mechanical properties. Solidification kinetics, including nucleation, growth, and coarsening, are now being investigated extensively. The incorporation of these principles into the more traditional thermal, fluid flow, and stress models enable quantitative predictions of microstructure and mechanical properties, such as tensile strength and elongation. The coupling of mechanical analysis with thermal analysis enables the prediction of residual stresses and distortion in the castings and molds. These predictions will enable design engineers to evaluate the effects of nonuniform properties and defects on life-cycle performance of components.

Different casting processes are used to produce different kinds of casting components. Some of the most common ones are sand casting, investment casting, die casting, permanent mold casting, lost foam casting, centrifugal casting, continuous casting, and direct chill casting. Each casting process has its own features. For example, for pressure die casting machines, it is particularly important to optimize not only the casting quality but also the die behavior with respect to thermal stress and strains and life expectancy.

In this article, the topic of computational thermodynamics is first reviewed. The calculation of solidification paths for casting alloys is introduced in which back diffusion is included so that the cooling condition can be accounted for. Then, a brief review of the calculation of thermophysical properties is presented. Fundamentals of the modeling of solidification processes are discussed next. The modeling conservation equations are listed. Several commonly used microstructure simulation methods are presented. Ductile iron casting is chosen as an example to demonstrate the ability of microstructure simulation. Defect prediction is one of the main purposes for casting and solidification simulation. The predictions for the major defects of a casting, such as porosity, hot tearing, and macrosegregation, are highlighted. At the end of this article, several industry applications are presented.

Also see the article "Modeling of Porosity during Solidification" in this Volume.

Computational Thermodynamics

Thermodynamic calculations are the foundation for performing basic materials research on the solidification of metals. It is important to have proper phase information for an accurate prediction of casting solidification (Ref 2).

Calculations for Casting Solidifications

The method of phase diagram calculations was started by Van Laar, who computed a large number of prototype binary phase diagrams with different topological features using ideal and regular solution models. Since then, many researchers began to incorporate phase equilibrium data to evaluate the thermodynamic properties of alloys. It was not until the late 1980s that a number of phase diagram calculation software packages became available. Then, thermodynamic calculations for multicomponent systems became feasible, thanks to the rapid progress in the computer industry.

Solidification proceeds at various rates for castings. Thus, the microstructure and the composition are not homogeneous throughout the casting. The solidification path determines the solidification behavior of an alloy. For complex multicomponent alloys, the solidification path is very complicated. Hence, the equilibrium of each phase at different temperatures must be calculated. The thermodynamic and the kinetic calculations are the base for the prediction of solidification. The diffusion transport in the solid phase must be solved for each element. This requires knowledge of the diffusion coefficient of the element, the length scale, and the cooling conditions. Thermodynamic modeling has recently become increasingly used to predict the equilibrium and phase relationships in multicomponent alloys (Ref 3–5). Currently, several packages are able to simulate solidification using the Scheil model and lever rule, such as Thermo-Calc, Pandat, and JMatPro.

It is critical to have an accurate solidification path for casting simulation (Ref 6–8). Obtaining the solidification path is very important for understanding and controlling the solidification process of the alloy. Normally, there are two ways to predict the solidification path. One is the complete equilibrium approach, which can be calculated by the lever rule. The other is the Scheil model, which assumes that the solute diffusion in the solid phase is small enough to be considered negligible and that diffusion in the liquid is extremely fast—fast enough to assume that diffusion is complete. For almost all practical situations, the solidification occurs under nonequilibrium conditions but does not follow the Scheil model. A modified Scheil model is applied in JMatPro. In this calculation, carbon and nitrogen are treated as completely diffused in the solid, which makes a great improvement, particularly for iron-base alloy solidification. In reality, there is always a finite back diffusion based on the cooling conditions (Ref 9).

There is finite diffusion in the solid, or back diffusion, which is a function of the cooling rate. Back diffusion plays an important role in the calculation of segregation. There are many numerical and analytical models that attempt to handle such phenomena (Ref 10–16). For most of the models, constant partition coefficients are assumed, which is a good approximation for many alloys. Unfortunately, sometimes the partition coefficients can vary dramatically for some commercial alloys. The partition coefficient of an element in an alloy can change from less than one to greater than one, or vice versa, during solidification (Ref 17). For these cases, the analytical or earlier numerical models are no longer valid. The equilibrium of each phase at different temperatures should be calculated. This can be fulfilled by coupling with the thermodynamic calculation. Recently, researchers have started to couple thermodynamic calculations with a modified Scheil model, including back diffusion (Ref 18).

Normally, the liquid is assumed to be completely mixed. The solid concentration is calculated by solving a one-dimensional diffusion equation for each element. The governing equations for conservation of species for multicomponent alloy solidification are (Ref 19):

Liquid species conservation:

$$f_l \frac{\partial C_l^j}{\partial t} = (C_l^j - C_i^j) \frac{\partial f_s}{\partial t} + \frac{SD_j}{L}(C_s^j - C_i^j) \quad \text{(Eq 1)}$$

Solid species conservation:

$$f_s \frac{\partial C_s^j}{\partial t} = (C_i^j - C_s^j)\left(\frac{\partial f_s}{\partial t} + \frac{SD_j}{L}\right) \quad \text{(Eq 2)}$$

where j is the species index, C is concentration, D is a diffusion coefficient, f is the volume fraction of a phase, t is time, L is a diffusion length, and S is the interfacial area concentration. The subscript i refers to the solid-liquid interface, l is the liquid, and s is the solid.

The diffusion length, L, can be determined using a model proposed by Wang and Beckermann (Ref 19) based on the work of Ohnaka (Ref 12) using the one-dimensional platelike dendrite geometry:

$$L = \frac{f_s \lambda}{6}$$

where λ is the secondary dendrite arm spacing, which is a function of cooling rate, $\lambda = a\dot{T}^n$. Here, a and n are constants determined by the alloy composition, and $\dot{T}$ is the cooling rate. The interfacial area concentration, S, is related to the solid volume fraction and the secondary dendrite arm spacing:

$$S = \frac{2}{\lambda}$$

Combining Eq 1 and 2 and then discretizing yields:

$$f_s(C_s^j - C_s^{j^o}) = \left(\Delta f_s + \frac{SD\Delta t}{L}\right)C_i^j - \left(\Delta f_s C_s^{j^o} + \frac{SD\Delta t}{L}C_s^j\right) \quad \text{(Eq 3)}$$

where the superscript "o" refers to the old value of the variable. Hence:

$$C_s^j = \frac{(f_s - \Delta f_s)C_s^{j^o} + \left(\Delta f_s + \frac{SD\Delta t}{L}\right)C_i^j}{f_s + \frac{SD\Delta t}{L}} \quad \text{(Eq 4)}$$

Equation 4 is used for calculating the concentration in the solid. Notice that Eq 4 can automatically turn into the Scheil model or lever rule if the diffusion is zero or infinity:

$$C_s^j = C_i^j$$

when $\frac{SD\Delta t}{L} \to \infty$ (lever rule)

$$C_s^j = C_s^{j^o} + (C_i^j - C_s^o)\Delta f_s / f_s$$

when $\frac{SD\Delta t}{L} = 0$ (Scheil model)

Based on mass conservation, the liquid concentration can be calculated from the solution of the solid concentration profile accordingly.

Example 1: Aluminum Wrought Alloy 2219. Yan (Ref 18) did an experiment for the solidification of a quaternary Al-6.27Cu-0.22Si-0.19Mg alloy (UNS A92219) with a cooling rate of 0.065 K/s. The microstructure of the solidified samples is dendritic. The calculated fraction of solid versus temperature relationship for this quaternary alloy is shown in Fig. 1. According to the Scheil model, the solidification sequence for this alloy is liquid (L) → L + face-centered cubic (fcc) → L + fcc + theta → L + fcc + theta + $Al_5Cu_2Mg_8Si_6$(Q) → L + fcc + theta + $Al_5Cu_2Mg_8Si_6$(Q) + silicon. Experimentally, there were no silicon and $Al_5Cu_2Mg_8Si_6$ phases formed, according to metallographic examination and electron probe microanalysis. The back diffusion model indicates that there are only fcc and theta phases formed during solidification for this cooling condition. The results predicted by the back diffusion model are in agreement with the experimental quantitative image analysis program.

The measured fractions of fcc phase were compared with the calculations from the Scheil model, lever rule, and the current back diffusion model for three different cooling rates. The comparison is shown in Table 1.

The fraction of fcc phase calculated from the Scheil model is less than the measured values, and the fraction of fcc calculated from the lever rule is higher than that from the experiments for all three cooling rates. The back diffusion model, which takes into account the cooling rate, gives good agreement with the experiments.

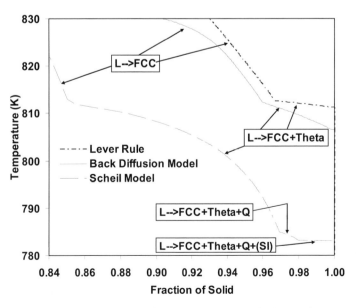

Fig. 1 Solidification paths of a 2219 aluminum alloy, temperature versus solid fraction, from two models and the lever rule. L, liquid; fcc, face-centered cubic

Table 1 Comparison of experimental and calculated fraction of face-centered cubic phase (volume percent)

Cooling rate, K/s	Area scan	Image analysis	Calculation
Lever rule	. . .	. . .	96.7
0.065	96.0	95.4	96.0
0.25	95.8	95.3	95.4
0.75	95.8	94.7	94.3
Scheil model	. . .	. . .	85.4

Thermophysical Properties

The research on thermophysical properties is a very important part of materials science, particularly at the current time, because such data are a critical input for the simulation of metals processing. Lee and co-workers investigated the sensitivity of investment casting simulations to the accuracy of thermophysical properties (Ref 20). They found that the temperature prediction and thermal gradient in the liquid are the most sensitive to the accuracy of the input values used for density and thermal conductivity in the solid. Thermal conductivity in the liquid, specific heat, and density have similar levels of influence on solidification time; increasing their values increases the local solidification time. Thermal conductivity in the solid has the opposite effect compared to all the other properties studied.

Collection and Calculation of Thermophysical Property Data

Accurate thermophysical data are difficult to obtain at high temperatures experimentally, especially for reactive alloys such as titanium- and nickel-base superalloys. An extensive database for the calculation of thermophysical properties has been developed (Ref 7) that uses the phase fraction information predicted with the Gibb's free energy minimization routines developed by Lukas et al. (Ref 3) and extended by Kattner et al. (Ref 4). These properties include density, specific heat, enthalpy, latent heat, electrical conductivity and resistivity, thermal conductivity, liquid viscosity, Young's modulus, and Poisson's ratio. The thermodynamic calculation is based on the thermodynamic databases from CompuTherm LLC. A simple pairwise mixture model, which is similar to that used to model thermodynamic excess functions in multicomponent alloys, can be used to calculate the properties (Ref 5):

$$P = \sum x_i P_i + \sum_i \sum_{j>i} x_i x_j \sum_v \Omega_v (x_i - x_j)^v$$

(Eq 5)

where P is the phase property, P_i is the property of the pure element in the phase, Ω_i is a binary interaction parameter, and x_i and x_j are the mole fractions of elements i and j in that phase.

Thermal Conductivity. The thermal conductivity mainly depends on the chemical composition of an alloy. It also depends to a lesser extent on the precipitates, bulk deformation, microstructures, and other factors (Ref 21, 22). These factors can usually be ignored in the calculation of conductivity for commercial alloys.

The thermal conductivity of alloys is composed of two components: a lattice component and an electronic component. In well-conducting metals, the thermal conductivity is mainly electronic conductivity. The lattice conductivity is usually very small compared to the electronic one. Hence, only the electronic component is considered here. The thermal conductivity, λ, and the electrical resistivity, ρ, are related according to the Wiedeman-Franz-Lorenz law (Ref 23, 24):

$$\lambda = \frac{LT}{\rho}$$

(Eq 6)

where the Lorentz constant is:

$$L = 2.44 \times 10^{-11} W\Omega K^{-2},$$

and T is the temperature.

The thermal conductivity of an element can be calculated, because the electrical resistivity of pure elements can be obtained from experiments or literature. Then, the mixture model (Eq 5) can be applied for multicomponent alloys. Based on this model, an example of the calculated thermal conductivity of an aluminum casting alloy, A356 (UNS A13560), is shown in Fig. 2, with experimental results for comparison. The calculation can accurately predict the thermal conductivity variation with temperatures for this alloy in the liquid, solid, and mushy zone. Figure 3 shows the comparison with experimental results from Auburn University for various alloys at different temperatures. The agreement is good in general.

Liquid Viscosity. Viscosity is an important property to be considered in dealing with fluid flow behavior. The liquid viscosity is a measure of resistance of the fluid to flow when subjected to an external force. There are two approaches to modeling of complex alloy viscosity. One is the fundamental molecular approach, and the other is the semitheoretical procedure. The former one is based mainly on the monatomic nature. There are some models available, but most of them are still under development and do not meet the technological need. The semitheoretical method is applied here to predict the viscosity of alloys. The viscosity, η, of pure liquid metals follows Andrade's relationship (Ref 25):

$$\eta(T) = \eta_o \exp(E/RT)$$

(Eq 7)

where E is the activation energy, and R is the gas constant.

Figure 4 shows an example of the calculated liquid viscosity of a high-temperature nickel-base alloy, IN718, using the mixture model (Eq 5) compared with experimental results. Figure 5 shows the comparison between experimental and calculated results for various alloys at different temperatures.

Density. Currently, the casting simulation models have reached the stage where one of the limiting factors in their applicability is the accuracy of the thermophysical data for the materials to be modeled. Among all the thermophysical data, the temperature-dependent density is one of the most critical ones for the accurate simulation of solidification microstructure and defect formation, such as shrinkage porosity (Ref 7). A database has been developed containing molar volume and thermal volume coefficients of expansion of liquid, solid-solution elements, and intermetallic phases. This is linked to the thermodynamic calculations mentioned previously. Volume calculations are linked to the thermodynamic models such that, when a thermodynamic calculation is made, volume can be directly calculated. The densities of the liquid and solid phases of

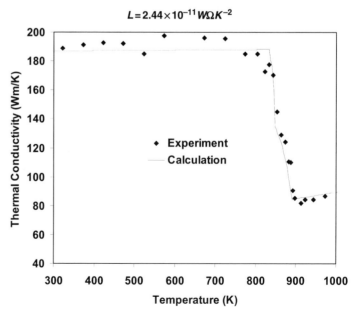

Fig. 2 Comparison between experimental and calculated thermal conductivity versus temperature for an aluminum casting alloy, A356

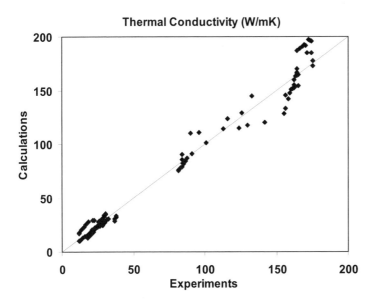

Fig. 3 Comparison between experimental and calculated thermal conductivity for different alloys

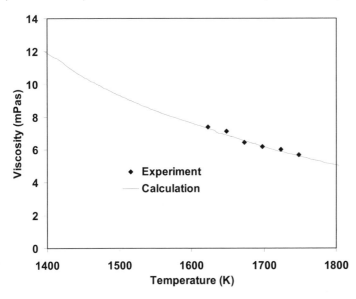

Fig. 4 Comparison between experimental and calculated viscosity for a nickel-base alloy, IN718

multicomponent systems are calculated by the simple mixture model (Ref 26). Figure 6 shows plots comparing experimental values with calculations for the density of different alloys at different temperatures. Figure 7 shows a comparison between the calculated and experimentally reported density for a CF8M stainless steel alloy.

It will increase a modeler's confidence to have reliable calculated thermophysical properties for a casting alloy, especially at high temperatures, where it is very difficult to measure by experiment.

Fundamentals of the Modeling of Solidification Processes

Casting process modeling involves the simulation of mold filling, solidification of the cast metal, microstructure formation, stress analysis on casting and mold, and so on. At the macroscopic scale, these processes are governed by basic equations, which describe the conservation of mass, momentum, energy, and species. Heat transfer is perhaps the most important discipline in casting simulation. The solidification process depends on heat transfer from the part to the mold and from the mold to the environment. The heat can be transferred by conduction, convection, and/or radiation. Conduction refers to the heat transfer that occurs as a result of molecular interaction. Convection refers to the heat transfer that results from the movement of a liquid, such as liquid metal or air. Radiation refers to the heat transfer of electromagnetic energy between surfaces that do not require an intervening medium. Radiation is very important for investment casting processes, for instance.

Mathematical Formulations that Govern the Solidification Process

The following summarize the basic mathematical formulations that govern the solidification process (Ref 27).

Energy Equation. The equation for energy is:

$$\rho \frac{\partial H}{\partial t} + \rho u_i \frac{\partial H}{\partial x_i} - \nabla(k\nabla T) - q(x) = 0 \qquad \text{(Eq 8)}$$

where ρ is density; k is thermal conductivity; and H is enthalpy, a function of temperature that encompasses the effects of specific and latent heat if a fraction-of-solid curve is given as a function of temperature, which is the case assumed here; and:

$$H(T) = \int_0^T C_p dT + L[1 - f_s(T)] \qquad \text{(Eq 9)}$$

where L is latent heat, f_s is fraction of solid, $u_i = f_1 u_{i,l}$ is the component superficial velocity, f_1 is the fraction of liquid, and $u_{i,1}$ is the actual liquid velocity.

Momentum Equations. Assuming that the spatial derivatives of viscosity are small and that the fluid is nearly imcompressible, many terms in the viscous stress tensor can be neglected. The momentum equations can be simplified as:

$$\rho \frac{\partial u_i}{\partial t} + \rho u_j \frac{\partial u_i}{\partial x_j} + \frac{\partial}{\partial x_j}\left(p\delta_{ij}\mu \frac{\partial u_i}{\partial x_j}\right) = \rho g_i - \frac{\mu}{K} u_i$$
$$\text{(Eq 10)}$$

where δ_{ij} is the Kronecker delta, p is pressure, g_i is gravitational acceleration, and K is permeability.

The Continuity Equation. The equation for continuity is:

$$\frac{\partial \rho}{\partial t} + \frac{\partial(\rho u_i)}{\partial x_i} = 0 \qquad \text{(Eq 11)}$$

To solve Eq 11, proper initial conditions and boundary conditions are needed.

The basic initial conditions include temperature, velocity, and pressure:

$$T(x,0) = T_o(x)$$
$$u(x,0) = u_o(x)$$
$$v(x,0) = v_o(x)$$
$$w(x,0) = w_o(x)$$
$$p(x,0) = p_o(x)$$

Boundary Conditions. There are many kinds of boundary conditions. Some of the most common ones are:

Fixed value or Dirichlet boundary condition:

$$Y(x) = Y_d(x)f(t) \text{ on } \Gamma_1 \qquad \text{(Eq 12)}$$

where $Y(x)$ can be temperature, velocity, or pressure; Γ_1 is some subset of the total boundary; $Y_d(x)$ is a specified variable vector; and $f(t)$ is the time function.

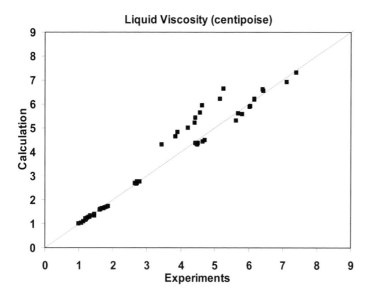

Fig. 5 Comparison between experimental and calculated viscosity for different alloys at different temperatures

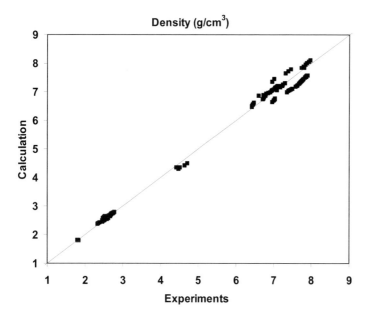

Fig. 6 Comparison between experimental and calculated density for different alloys

Specified heat flux boundary condition:

$$-k\nabla T \cdot \dot{n} = q_n f(t) \text{ on } \Gamma_2 \qquad \text{(Eq 13)}$$

where q_n is the specified heat flux, $\dot{n}$ is a unit vector normal to the surface, and Γ_2 is some subset of the total boundary Γ.

Convective heat flux boundary condition:

$$-k\nabla T \cdot \dot{n} = q_n f(t) h f(t) g(T)[T - T_a] \text{ on } \Gamma_2$$
$$\text{(Eq 14)}$$

where h is the convection coefficient, $g(T)$ is the temperature function, and T_a is the ambient or media temperature, which could be a function of time.

Radiation heat boundary condition:

$$-k\nabla T \cdot \dot{n} = \sigma\varepsilon g(T)[T^4 - T_a^4] \text{ on } \Gamma_2 \qquad \text{(Eq 15)}$$

where σ is the Stefan-Boltzmann constant, and ε is emissivity.

This is a simple version of radiation in which it is assumed that there is only one ambient temperature present. Thus, the view factor is equal to one. The radiation energy exchange between two surfaces changes significantly with their relative direction. The heat exchange is maximum if two surfaces face each other perfectly. The heat exchange will be zero between two surfaces if they cannot see each other. View factor represents the fraction of the radiation directly leaving one surface that is intercepted by another.

For investment casting modeling, the view factor must be calculated carefully. For that case, a view factor radiation model can be used.

A net flux model can be used for more complex view factor radiation. Rather than tracking the reflected radiant energy from surface to surface, an overall energy balance for each participating surface is considered. At a particular surface i, the radiant energy being received is denoted as $q_{in,i}$. The outgoing flux is $q_{out,i}$. The net radiative heat flux is the difference of these two:

$$q_{net,i} = q_{out,i} - q_{in,i} \qquad \text{(Eq 16)}$$

Using the diffuse, gray-body approximation, the outgoing radiant energy can be expressed as:

$$q_{out,i} = \sigma\varepsilon_i T_i^4 + (1 - \varepsilon_i)q_{in,i} \qquad \text{(Eq 17)}$$

The first term in Eq 17 represents the radiant energy, which comes from direct emission. The second term is the portion of the incoming radiant energy, which is being reflected by surface i. The incoming radiant energy is a combination of the outgoing radiant energy from all participating surfaces being intercepted by surface i. Here, the view factor, F_{i-j}, is the fraction of the radiant energy leaving surface j that impinges on surface i. Thus:

$$q_{in,i} = \sum_{j=1}^{N} F_{i-j} q_{out,j} \qquad \text{(Eq 18)}$$

where N is the total number of surfaces participating in the radiation model, and the view factors are calculated from the following integral:

$$F_{i-j} = \frac{1}{A_i} \int_{A_j} \int_{A_i} \frac{\cos\theta_j \cos\theta_i}{\pi r^2} dA_i dA_j \qquad \text{(Eq 19)}$$

where A_i is the area of surface i, θ_i the polar angle between the normal of surface i and the line between i and j, and r is the magnitude of the vector between surface i and j.

Then, the vector of radiosities $q_{out,i}$ can be solved by:

$$\sum_{j=1}^{N} \left[\frac{A_i}{(1 - \varepsilon_i)} \delta_{ij} - A_i F_{i-j} \right] q_{out,j} = \frac{\varepsilon_i A_i}{(1 - \varepsilon_i)} \sigma T_i^4$$
$$\text{(Eq 20)}$$

Hence, the net radiant flux is obtained by:

$$q_{net,i} = \left[\frac{\varepsilon_i}{1 - \varepsilon_i} \right] [\sigma T_i^t - q_{out,i}] \qquad \text{(Eq 21)}$$

This heat flux then appears as a boundary condition for the heat conduction analysis.

Based on the aforementioned equations, the basic heat-transfer and fluid flow problems can be solved with proper initial and boundary conditions.

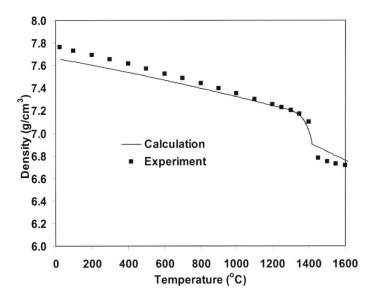

Fig. 7 Comparison between experimental and calculated density for a CF8M cast corrosion-resistant stainless steel (UNS J92900)

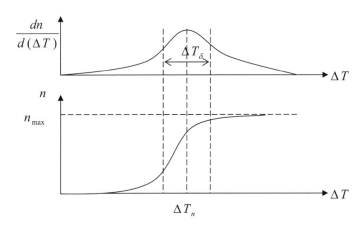

Fig. 8 Nucleation model. The grain density, n, is plotted versus the undercooling, ΔT. The derivative is plotted above.

Microstructure Simulation

The purpose of casting solidification micromodeling is to predict the microstructure of a casting, such as grain size. Understanding the casting solidification process and microstructure formation will greatly facilitate casting design and quality control. Computer modeling also provides the basis for computer-aided manufacturing and product life-cycle management by being able to predict mechanical properties, quality, and useful life (Ref 28–33).

There are several ways to simulate microstructure formation during alloy solidification, such as the deterministic method and the stochastic method. For the deterministic method, the density of grains that have nucleated in the bulk liquid at a given moment during solidification is a deterministic function (e.g., a function of undercooling). The stochastic method is a probabilistic means to predict the nucleation

and growth of the grain, including the stochastic distribution of nucleation locations, the stochastic selection of the grain orientation, and so on. The stochastic method includes cellular automata and the phase-field method.

Deterministic Micromodeling

Modeling of solidification processes and microstructural features has benefited from the introduction of averaged conservation equations and the coupling of these equations with microscopic models of solidification. When conservation equations are averaged over the liquid and solid phases, the interfacial continuity condition automatically vanishes and average entities (e.g. mean temperature or solute concentration) appear. Rappaz et al. (Ref 34, 35) proposed a model using averaging methods to predict the growth of equiaxed grains under isothermal conditions. The nucleation is based on the model

proposed by Thevoz and co-workers (Ref 36), which is illustrated in Fig. 8. At a given undercooling, the grain density, n, is calculated by the integral of the nucleation site distribution from zero undercooling to the current undercooling. Thus, the density of the grain nuclei is:

$$n(\Delta T(t)) =$$

$$\frac{n_{max}}{\sqrt{2\pi} \cdot \Delta T_\delta} \int_0^{\Delta T(t)} \exp\left(-\frac{(\Delta T(t) - \Delta T_n)^2}{2\Delta T_\sigma^2}\right) d(\Delta T(t))$$

(Eq 22)

where n_{max} is the maximum grain nuclei density, ΔT_δ is the standard deviation undercooling, and ΔT_n is the average undercooling.

Rappaz and Boettinger (Ref 28) studied the growth of an equiaxed multicomponent dendrite. In their study, for each element, the supersaturation is:

$$\Omega_j = \frac{c_{l,j}^* - c_{o,j}}{c_{l,j}^*(1 - k_j)} = Iv(Pe_j)$$

(Eq 23)

where $J = 1$, n is the solute element, $c_{l,j}^*$ is the tip liquid concentration, $c_{o,j}$ is the nominal concentration, and k_j is the partition coefficient.

The Peclet number is defined as:

$$Pe_j = \frac{Rv}{2D_j}$$

where D_j is the diffusion coefficient.

The Ivanstsov function is defined as:

$$Iv(Pe) = Pe \cdot \exp(Pe) \cdot E_1(Pe)$$

where $E_1(Pe)$ is the first exponential integral.

Assuming growth at the marginal stability limit, the dendrite radius is calculated by:

$$R = \frac{-2\pi^2 \Gamma}{\sum_{j=1}^n m_j Pe_j \frac{c_{o,j}(1-k_j)}{1-(1-k_j)Iv(Pe_j)}}$$

(Eq 24)

where Γ is the Gibbs-Thomson coefficient.

Hence, the tip velocity is:

$$v = D_1 Pe_1 \frac{2}{R}$$

(Eq 25)

and the tip liquid concentration will be:

$$c_{l,j}^* = \frac{c_{o,j}}{1 - (1 - k_j)Iv(Pe_j)}$$

(Eq 26)

Assume the liquid concentration in the interdendritic region is uniform. The solute profiles in the extradendritic liquid region can be obtained from an approximate model (Ref 28):

$$J_j = D_j \cdot \left(\frac{c_{l,j}^* - c_{o,j}}{\delta_j/2}\right)$$

(Eq 27)

where the solute layer thickness is:

$$\delta_j = \frac{2D_j}{v} \qquad \text{(Eq 28)}$$

So, the solute balance will become:

$$\frac{df_s}{dt}\sum_{j=1}^{n} m_j(k_j-1)c_{l,j}^{*} + (f_g - f_s)\frac{dT}{dt} - \sum_{j=1}^{n} m_j J_j = 0$$

$$\text{(Eq 29)}$$

where f_g is the envelope volume divided by the final grain volume. Please refer to Ref 28 for details about the derivation of the aforementioned equations.

The secondary dendrite arm spacing is calculated by:

$$\lambda_2 = 5.5(Mt_f)^{1/3} \qquad \text{(Eq 30)}$$

where:

$$M = \frac{-\Gamma}{\sum\limits_{j=1}^{n} m_j(1-k_j)(c_{e,j}-c_{o,j})/D_j}$$
$$\cdot \ln\left(\frac{\sum\limits_{j=1}^{n} m_j(1-k_j)c_{e,j}/D_j}{\sum\limits_{j=1}^{n} m_j(1-k_j)c_{o,j}/D_j}\right) \qquad \text{(Eq 31)}$$

Based on the aforementioned equations, the solidification of a multicomponent casting can be predicted. The averaged microstructure, such as grain size and secondary dendrite arm spacing, can be calculated based on the chemistry and cooling conditions. Some examples using such deterministic micromodeling are provided in a later section.

Cellular Automaton Models

Cellular automaton (CA) models are algorithms that describe the discrete spatial and/or temporal evolution of complex systems by applying deterministic or probabilistic transformation rules to the sites of a lattice. In a CA model, the simulated domain is divided into a grid of cells, and each cell works as a small independent automaton. Variables and state indices are attributed to each cell, and a neighborhood configuration is also associated with it. The time is divided into finite steps. At a given time step, each cell automaton checks the variables and state indices of itself and its neighbors at the last time step and then decides the updated results at the present step according to the predefined transition rules. By iterating this operation with each time step, the evolution of the variables and state indices of the whole system is obtained.

The CA model is usually coupled with a finite-element (FE) heat flow solver, such as the CAFÉ model developed by Rappaz and colleagues. The CA algorithm can be used to simulate nucleation and growth of grains. This model can be used to predict columnar-to-equiaxed transition (CET) in alloys. Several

models have been developed over the years for the prediction of microstructure formation in casting (Ref 37–41). One example is shown in Fig. 9.

Cellular automaton/finite-element models are reasonably well suited for tracking the development of a columnar dendritic front in an undercooled liquid at the scale of the shape casting (Ref 39, 40). Although these models do not directly describe the complicated nature of the solid-liquid interface that defines the dendritic microstructure, the crystallographic orientation of the grains as well as the effect of the fluid flow can be accounted for to calculate the undercooling of the mushy zone growth front. Two- and three-dimensional CAFE models were successfully applied to predict features such as the columnar-to-equiaxed transition observed in aluminum-silicon alloys (Ref 39), the selection of a single grain and its crystallographic orientation due to the competition among columnar grains taking place while directionally solidifying a superalloy into a pig-tail shape (Ref 40), as well as the effect of the fluid flow on the fiber texture selected during columnar growth (Ref 41). Coupling with macrosegregation has been developed (Ref 42), thus providing an advanced CAFE model to account for structure formation compared to purely macroscopic models developed previously (Ref 43–45). While both structure and segregation were predicted (Ref 42), comparison with experimental observation concerning structure formation was limited

due to the lack of detailed data (Ref 46). Comparison is thus mainly conducted with the as-cast state.

Wang et al. (Ref 47) investigated the effect of the direction of the temperature gradient on grain growth. As shown in Fig. 10(b), the temperature gradient was inclined at 45° relative to the macroscopic solidification direction, and the magnitude of the gradient was 12 K/mm. It clearly shows that the direction of the temperature gradient can affect both the macro- and microscale dendritic structures as well as the maximum undercooling.

While the CA methods produce realistic-looking dendritic growth patterns and have resulted in much insight into the CET, some questions remain regarding their accuracy. Independence of the results on the numerical grid size is rarely demonstrated. Furthermore, the CA techniques often rely on relatively arbitrary rules for incorporating the effects of crystallographic orientation while propagating the solid-liquid interface. It is now well accepted that dendritic growth of crystalline materials depends very sensitively on the surface energy anisotropy (Ref 48, 49).

Phase-Field Model

An alternative technique for investigating microstructure formation during solidification is the phase-field method. Phase-field models were first developed for simulating equiaxed

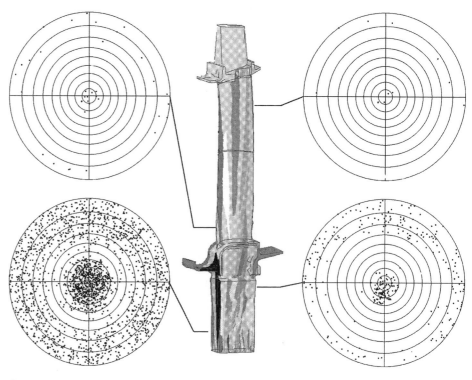

Fig. 9 Three-dimensional view of the final grain structure calculated in the weak coupling mode for a directionally solidified turbine blade. The <100> pole figures are displayed for various cross sections perpendicular to the main blade axis. Source: Ref 40

growth under isothermal conditions (Ref 50, 51). A desirable extension of the model was to study the effect of heat flow due to the release of latent heat. A simplified approach was proposed in which the temperature was assumed to remain spatially uniform at each instant, and a global cooling rate was imposed with consideration of the heat extraction rate and increase of the fraction of solid (Ref 52). The attempt to model nonisothermal dendritic solidification of a binary alloy was made by Loginova et al. by solving both the solute and heat diffusion equations and considering the release of latent heat as well (Ref 53).

Besides equiaxed growth in the supersaturated liquid, the phase-field model was also applied to the simulation of directional solidification, under well-defined thermal conditions (Ref 54, 55). The phase-field model has also been used to simulate the competitive growth between grains with different misorientations with respect to the thermal gradient (Ref 56). Further development in phase-field models includes the extension into three dimensions (Ref 57, 58) and multicomponent systems (Ref 59, 60).

Usually, a regular grid composed of square elements is used in the phase-field models (Ref 51, 52), but an unstructured mesh composed of triangular elements has also been used, which enables the phase-field method to be applicable in a domain with complex geometry shape and also in a large scale. From a physical point of view, the phase-field method requires knowledge of the physical nature of the liquid-solid interface. However, little is known about its true structure. Using Lennard-Jones potentials, molecular dynamics simulations of the transition in atomic positions across an interface have suggested that the interface width extends over several atomic dimensions (Ref 61). At present, it is difficult to obtain usable simulations of dendritic growth with interface thickness in this range due to the limitations of computational resources. Thus, the interface width will be a parameter that affects the results of the phase-field method. It should be realized that in the limit as the interface thickness approaches zero, the phase-field equations converge to the sharp interface formulation (Ref 62, 63). In contrast to CA models that adopt a pseudo-front-tracking technique, phase-field models express the solid-liquid interface as a transitional layer that usually spreads over several cells.

The diffusion equation for heat and solute can be solved without tracking the phase interface using a phase-field variable and a corresponding governing equation to describe the state in a material as a function of position and time. This method has been used extensively to predict dendritic, eutectic, and peritectic growth in alloys and solute trapping during rapid solidification (Ref 64). Figure 11 shows dendrite fragmentation during reheating when an isothermally grown structure was subjected to an instantaneous increase in temperature (Ref 64). The interface between liquid and solid can be described by a smooth but highly localized change of a variable between fixed values such as 0 and 1 to represent solid and liquid phases. The problem of applying boundary conditions at an interface whose location is an unknown can be avoided. Phase-field models have recently become very popular for the simulation of microstructure evolution during solidification processes (Ref 62, 63, 65–67).

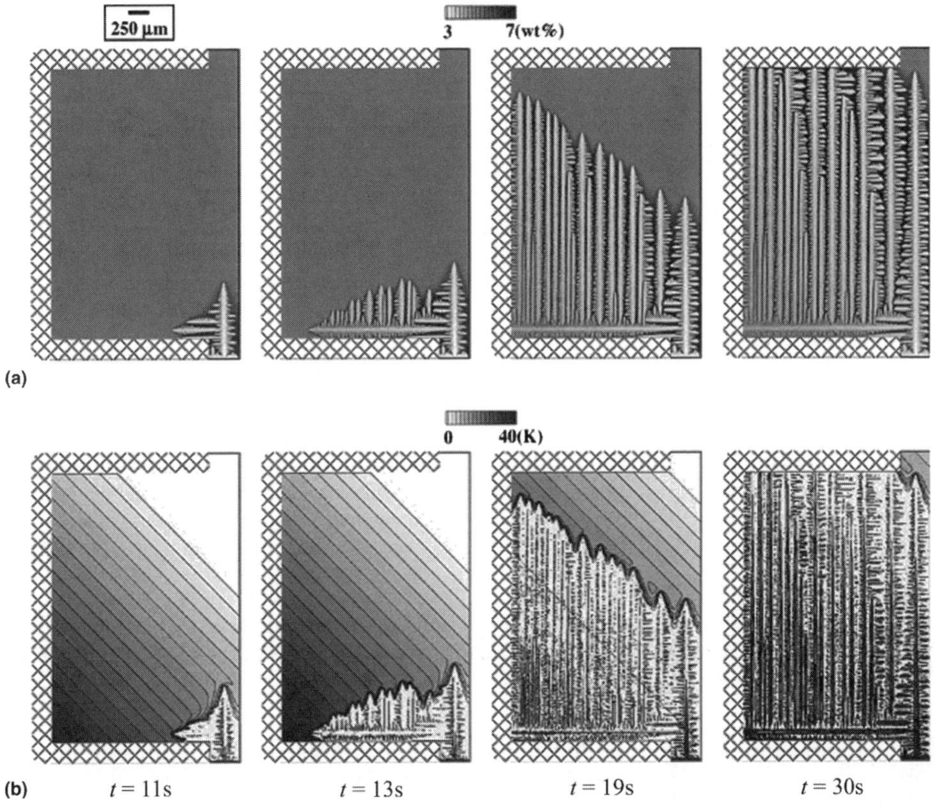

Fig. 10 Cellular automaton model produces realistic dendrite growth. (a) Predicted dendritic structure density. (b) Solutal adjusted undercooling distribution under thermal conditions of 45° inclined isotherms with respect to the growth direction moving at a constant velocity of 150 μm/s. Source: Ref 47

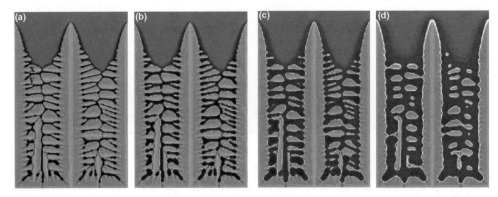

Fig. 11 Melting of dendritic structure and formation of fragments when temperature is increased from the growth temperature of 1574 K shown in (a) to 1589 K shown at later times in (b) through (d). Source: Ref 64

While these models address the evolution of a solid-liquid interface using only one phase-field parameter, interaction of more than two phases or grains, and consequently the occurrence of triple junctions, needed to be included into the multiphase-field approach (Ref 59, 60, 68, 69). One example can be found in Fig. 12, which predicted two-dimensional dendritic solidification of a binary alloy into an undercooled melt with coupled heat and solute diffusion (Ref 69).

The deterministic model is capable of tracking the evolution of the macroscale or average variables, for example, average temperature and the total fraction of solid, but it cannot simulate the structure of grains. The CA models can simulate the macro- and mesoscale grain structures, but it has difficulty in resolving the microstructure. The phase-field method can well reproduce the microstructure of dendritic grains. However, with the current computational power, phase-field models can only work well on a very small scale (up to hundreds of micrometers). The typical scale of laboratory experiments is 1 cm, and the scale of an industry problem can be up to 1 m. Both of them are beyond the capability of the phase-field method. In the industry, for larger castings, deterministic micromodeling is still a main player.

Micromodeling Applications in the Industry

Here, the focus is on deterministic modeling due to its wide application in the casting industry. Thermodynamic calculations are coupled with the macroscale thermal and fluid flow calculations (Ref 70).

Ductile iron is chosen as an example to demonstrate the capability of deterministic micromodeling because of its complex solidification behavior. Ductile irons are still important engineering materials due to their high strength and toughness and relatively low price. In the foundries, ductile irons suffer from shrinkage porosity formation during solidification, which is detrimental to the mechanical properties. To minimize porosity formation, large risers are normally used in the design, which sometimes reduces porosity level but leads to a low yield. Due to the complex solidification behavior of ductile irons and their extreme sensitivity to the process, it is very difficult to optimize the casting design to ensure the soundness of castings. To better understand the shrinkage behavior of ductile iron during solidification, a micromodel was developed to simulate the microstructure formation. The density change during solidification and the room-temperature mechanical properties can be calculated based on the microstructure. The simulation has been compared with the experimental results and found to be in good agreement.

Cast iron remains the most important casting material, with over 70% of the total world tonnage now (Ref 71). Based on the shape of graphite, cast iron can be lamellar (flake) or spheroidal (nodular). In the last 40 years, many papers have been published on the modeling of ductile iron solidification. It started with computational modeling by analytical heat transport and calculation of transformation kinetics (Ref 72–78). The computer model can calculate the cooling curve with an analytical method together with the kinetics calculation of carbon diffusion through the gamma-phase shell. In 1985, Su et al. first coupled heat transfer and a solidification kinetics model using the finite-difference method (Ref 79). Since then, many papers have been published on micromodeling of ductile iron solidification (Ref 80–89). The carbon diffusion-controlled growth through the gamma shell was modeled. In those models, the nodule count, graphite radius, and austenite shell radius were calculated. Onsoien et al. used the internal state variable approach to model the multiple phase changes occurring during solidification and subsequent cooling of near-eutectic ductile cast iron (Ref 90, 91). In their simulation, the effects on the microstructure evolution at various stages of the process by the alloy composition, graphite nucleation potential, and thermal progress were illuminated. The heat flow, fading effect, graphite/austenite eutectic transformation, ledeburite eutectic transformation, graphite growth in austenite regime, and the eutectoid transformation were all modeled. A comprehensive micromodel is developed that can give accurate microstructure information as well as the mechanical properties, such as yield strength, tensile strength, and hardness. The density of austenite, ferrite, pearlite, graphite, liquid, and ledeburite are calculated. The prediction has been compared with the experimental results and found to be in good agreement (Ref 70).

Nucleation Model. Here, Oldfield's nucleation model is applied. In this mode, bulk heterogeneous nucleation occurs at foreign sites that are already present within the melt or intentionally added to the melt by inoculation:

$$N_o = A(\Delta T)^n \qquad \text{(Eq 32)}$$

where A is the nucleation constant, N_o is the nucleation number per unit volume, ΔT is the undercooling, and n is another constant that depends on the effectiveness of inoculation.

Fading Effect. Fading is the phenomenon whereby the effectiveness of inoculation diminishes as the time between inoculation and casting increases. It is believed that the nucleation of graphite occurs on small nonmetallic inclusions that are entrapped in the liquid after inoculation (Ref 88). The small particles will grow with time. The particle diameter can be calculated by:

$$d = (d_o^3 + kt)^{1/3} \qquad \text{(Eq 33)}$$

where d is the particle diameter with time, d_o is the particle diameter at the beginning of the inoculation, and k is a kinetic constant.

Graphite/Austenite Eutectic Transformation. The eutectic growth process in ductile iron is a divorced growth of austenite and graphite, which do not grow concomitantly. At

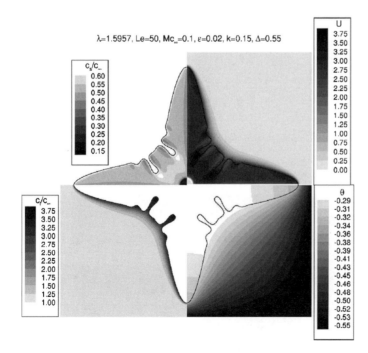

λ=1.5957, Le=50, Mc$_\infty$=0.1, ε=0.02, k=0.15, Δ=0.55

Fig. 12 Predicted results for two-dimensional dendritic solidification of a binary alloy into an undercooled melt with coupled heat and solute diffusion. The upper- and lower-right quadrants show the dimensionless concentration U and temperature fields, respectively; both left quadrants show concentration c/c' fields, with different scales used in the upper and lower quadrants to better visualize the concentration variations in the solid and liquid, respectively. Source: Ref 69

the beginning of the liquid/solid transformation, graphite nodules nucleate in the liquid and grow in the liquid to a small extent. The formation of graphite nodules and their limited growth in liquid depletes the carbon in the melt locally in the vicinity of the nodules. This facilitates the nucleation of austenite around the nodules, forming a shell. Further growth of these nodules is by diffusion of carbon from the melt through the austenite shell. When the austenite shell is formed around each nodule, the diffusion equation for carbon through the austenitic shell is solved in one-dimensional spherical coordinates. The boundary conditions are known from the phase diagram because thermodynamic equilibrium is maintained locally. Conservation of mass and solute is maintained in each grain. Because of the density variation resulting from the growth of austenite and graphite, the expansion/contraction of the grain is taken into account by allowing the final grain size to vary. Toward the end of solidification, the grains impinge on each other. This is taken into consideration by using the Johnson-Mehl approximation.

Using a spherical coordinate system, a mass balance is written as:

$$\rho_G \frac{4}{3}\pi R_G^3 + \rho_\gamma \frac{4}{3}\pi(R_\gamma^3 - R_G^3) + \rho_l \frac{4}{3}\pi(R_l^3 - R_\gamma^3) = m_{av}$$

(Eq 34)

where ρ_G, ρ_γ, ρ_l are the densities of graphite, austenite, and liquid, respectively, and the calculation can be found in the next section; R_G, R_γ, R_l are the radii of graphite, austenite, and the final grain, respectively; and m_{av} is the average mass of the grain.

Assuming complete mixing of the solute in liquid, the overall solute balance is written as:

$$\rho_G \cdot 1 \cdot \frac{4}{3}\pi R_G^3 + \int_{R_c}^{R_a} \rho_\gamma c(r,t) 4\pi r^2 dr + \rho_l c_l \frac{4}{3}\pi(R_l^3 - R_\gamma^3) = c_{av}$$

(Eq 35)

Differentiation of Eq 34 and 35 and the use of Fick's law in spherical coordinates lead to two equations for graphite and austenite growth rates following some manipulation.

Ledeburite Eutectic Transformation. When the temperature reaches below the metastable eutectic temperature, the metastable phase forms. The metastable cementite eutectic is also called ledeburite, in which small islands of austenite are dispersed in the carbide phase. It has both direct and indirect effects on the properties of ductile iron castings. Increasing the volume percent of the hard, brittle carbide results in an increase in the yield strength but a reduction in the tensile strength. Following the assumptions from Onsoien (Ref 90, 91) that the graphite/austenite nodule distribution is approximated by that of a close-packed face-centered space lattice and that the ledeburite

eutectic appears in an intermediate position, the total number of ledeburite nucleation sites would be the same for graphite/austenite nodules. Hence, the grain is assumed to be spherical. The growth of the ledeburite can be calculated as:

$$\frac{dR_{LE}}{dt} = 30.0 \cdot 10^{-6} * (\Delta T)^n$$

(Eq 36)

Thus, the fraction of ledeburite can be written as:

$$f_{LE} = \frac{4}{3}\pi N R_{LE}^3$$

(Eq 37)

Eutectoid Transformation. The eutectoid reaction leads to the decomposition of austenite into ferrite and graphite for the case of the stable eutectoid and to pearlite for the metastable eutectoid transformation. Usually, the metastable eutectoid temperature is lower than the stable eutectoid temperature. Slower cooling rates result in more stable eutectoid structure. Following solidification, the solubility of carbon in austenite decreases with the drop in temperature until the stable eutectoid temperature is reached. The rejected carbon migrates toward graphite nodules, which are the carbon sinks. This results in carbon-depleted regions in austenite around the graphite nodules. This provides favorable sites for ferrites to nucleate, which grow as a shell around the graphite nodules. If the complete transformation of austenite is not achieved when the metastable temperature is reached, pearlite forms and grows in competition with ferrite.

The ultimate goal of process modeling is to predict the final mechanical properties. The mechanical properties (hardness, tensile strength, yield strength, and elongation) of ductile iron castings are a function of composition

and microstructure. The graphite shape, graphite structure, graphite amount, carbide content, and matrix structure (pearlite, ferrite) affect the mechanical properties of ductile iron castings. As for the matrix structure, the increasing of pearlite increases the strength and hardness but reduces the elongation (Ref 92).

To show the capability of this model, a simulated ductile iron casting with a simple geometry was investigated. The dimension of the casting is 10 by 10 by 200 cm. On the left face, it is cooled by contact with a constant-temperature medium (15 °C) at a heat-transfer coefficient of 500 W/m²K. All the other faces are adiabatic. The initial melt temperature is 1400 °C. According to the boundary condition, the left side cools faster than the right side. Figure 13 shows the solidification time for different distances from the cooling end. At the very left end, the solidification time is less than 1 s. On the other hand, the solidification time at 10 cm from the cooling end is more than 100 s. Because of the different cooling, the nodule count varies and is shown in the same figure.

The metastable phase forms when the cooling is too fast. Figure 14 shows the volume fraction of different phases at room temperature. On the very left end, there is approximately 90% volume fraction of ledeburite phase. It reduces gradually from left to right until approximately 3 cm from the chill end. There is no ledeburite phase after 3 cm. At the same time, as cooling decreases, the volume fraction of ferrite increases and that of pearlite decreases. Ledeburite is a very hard, brittle phase. The pearlite phase is harder than ferrite. Hence, the ductility increases as the cooling rate decreases.

From the micromodeling, the calculated grain and graphite radii at different distances from the chill are shown in Fig. 15. Faster cooling results in smaller grain and graphite sizes.

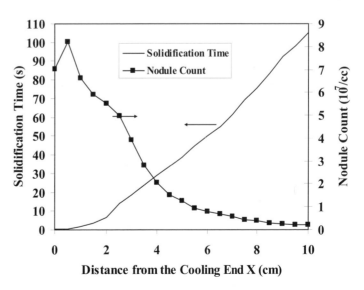

Fig. 13 Solidification time and nodule count at different distances from the chill (cooled end of a ductile iron casting)

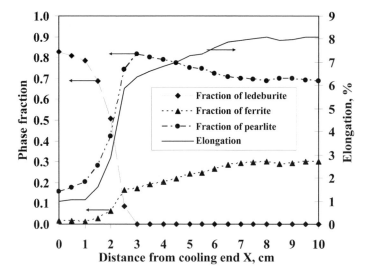

Fig. 14 Phase fractions and elongation of the casting at different distances from the chill. Faster cooling on the left side increases the presence of ledeburite phase in ductile iron. Elongation is increased where the cooling rate is slower.

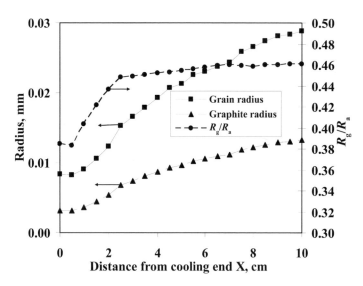

Fig. 15 Grain and graphite size of the casting at different distances from the chill for a ductile iron casting. The ratio of the graphite radius, R_g, to the austenite radius, R_a, is plotted and approaches a constant value. This value is determined by the carbon content of the melt.

The ratio of the radii of graphite and austenite increases as cooling decreases but reaches a constant value of approximately 0.44 even though the radii of graphite and austenite still continue to increase. This constant ratio is determined by the initial carbon content. It can determine the expansion level during solidification.

Based on the microstructure, the mechanical properties can be calculated. As mentioned previously, carbide increases the yield strength but decreases the tensile strength. The yield strength and hardness continuously decrease as the cooling rate decreases. The yield strength is very high on the left part because of the formation of carbide. On the other hand, the carbide decreases the tensile strength. The results are shown in Fig. 16.

Experimental Validations. A series of experiments was performed to validate the micromodel (Ref 93). The three-part cast iron foundry mold containing the gating system is shown in Fig. 17.

The casting is GGG60 ductile iron. The pouring temperature is 1400 °C, the initial die temperature is 165 °C, and the initial sand temperature is 20 °C. To establish the structure of the casting and the morphology of graphite, specimens were taken as shown in Fig. 18. The specimens were then ground, polished, and etched for structure evaluation. It can be seen in the pictures of the microstructure that graphite was segregated in the form of spheroids.

Because of the rapid cooling, a large amount of the metastable phase, ledeburite, was formed

in the corner. The ledeburite phase reduces gradually as the cooling rate decreases. In the center of the casting, no ledeburite phase was found. The radius of the black balls, graphite, increases as cooling decreases. The structure of the metal is formed by pearlite and ferrite. Figure 19 shows the volume fraction of metastable phase (top) and the volume fraction of ferrite (bottom). It is difficult to measure the yield strength of the sample at different locations, because the strength could change dramatically based on the microstructure variation. On the other hand, hardness is an excellent indicator of strength and relatively easy to measure. Figure 20 shows the hardness measurement points on the sample. Table 2 shows the comparison between the measurement and prediction results of the hardness at different locations. It can be concluded that the prediction matches the experiments very well.

Defect Prediction

Defects reduce the performance and increase the cost of castings. It is critical to understand the mechanism of defects and microstructure on the performance so that an effective tool can be developed to prevent defects and control the microstructure. There are many kinds of casting defects. Those defects are dependent on the chemistry of alloys, casting design, and casting processes. Defects can be related to thermodynamics, fluid flow, thermal, and/or stress. For most cases, all of those phenomena are correlated. It is necessary to consider every aspect to prevent the formation of a defect.

Here, some common casting defects in foundries are discussed. They are porosity, hot tearing, and macrosegregation.

Porosity

Porosity formed in castings leads to a decrease in the mechanical properties (Ref 94–99). This porosity may be a combined result of solidification shrinkage and gas evolution. They can occur simultaneously when conditions are such that both may exist in a solidifying casting. One of the most effective ways to minimize porosity defects is to design a feeding system using porosity prediction modeling. In such a way, the model can determine the location of microporosity so that the feeding system can be redesigned. This process is repeated until microporosity is minimized and not likely to appear in the critical areas of the castings.

There are many models that can predict the shrinkage porosity from the pressure drop during interdendritic fluid flow and gas evolution (Ref 97–104). The model of Felicelli et al. can predict the pressure and redistribution of gas and the region of possible formation of porosity by solving the transport of gas solutes (Ref 105). A comprehensive model should

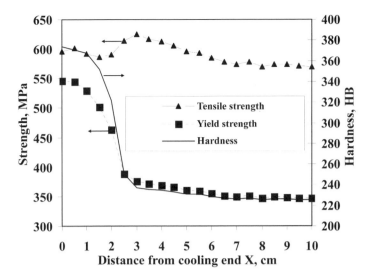

Fig. 16 Mechanical properties of the casting as a function of distance from the chill for the same ductile iron casting

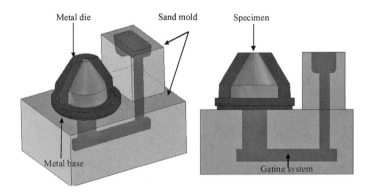

Fig. 17 Experiment setup for casting of a GGG60 ductile iron (German cast iron with nodular graphite)

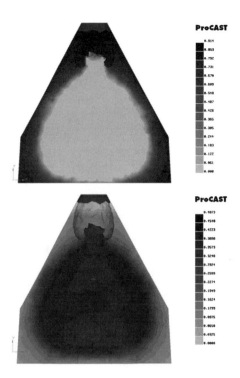

Fig. 19 Simulation results of fraction of metastable phase (top) and fraction of ferrite (bottom) for the same geometry as in Fig. 18

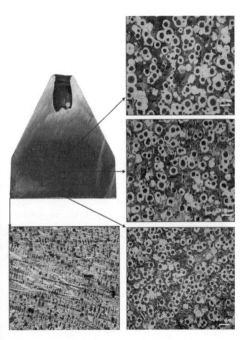

Fig. 18 Microstructure of ductile iron casting from Fig. 17 die at indicated points

calculate the shrinkage porosity, gas porosity, and pore size.

As for many casting defects observed in solidification processes, the mushy zone is the source of microporosity. The basic mechanism of microporosity formation is pressure drop due to shrinkage and gas segregation in the liquid (Ref 94, 95, 97).

The liquid densities of many alloys are lower than that of the solid phase. Hence, solidification shrinkage happens due to the metal contraction during the phase change. The dynamic pressure within the liquid decreases because of the contraction and sometimes cannot be compensated by the metallostatic pressure associated with the height of the liquid metal.

The decrease of pressure lowers the solubility of gas dissolved in the liquid. If the liquid becomes supersaturated, then bubbles can precipitate (Ref 106). Most liquid metals can dissolve some amount of gas. The solubility of gases in the solid phase is usually much smaller than that in the liquid phase. Normally, the rejected gases during solidification do not have enough time to escape from the mushy zone into the ambient air. Being trapped within the interdendritic liquid, the gas can supersaturate the liquid and eventually precipitate under the form of pores if nucleation conditions are met.

The formation of bubbles requires overcoming the surface tension (Ref 107). Homogeneous nucleation is very difficult. In castings, nucleation of pores can be expected to occur primarily on heterogeneous nucleation sites, such as the solid-liquid interface and inclusions (Ref 107, 108).

Generally speaking, there are two ways to predict the level of microporosity in castings. One is a parametric method derived from first principles by using a feeding resistance criterion function combined with macroscopic heat flow calculations (Ref 109–112). Parametric models are easy to apply to shaped castings and have been mainly directed at the prediction of centerline shrinkage. Another approach is a direct simulation method (Ref 97–101, 105, 107). They usually derive governing equations based on a set of simplifying assumptions and solve the resulting equations numerically. By combining the CA technique, some models can not only predict the percentage porosity but also the size, shape, and distribution of the pores (Ref 102–104). There has been some research that attempted to understand the physics of microporosity formation, too (Ref 96).

The earliest work to directly predict microporosity distribution that was general and applicable to shaped castings was done by Kubo and Pehlke (Ref 107). A more accurate fluid flow model was presented by Combeau et al. (Ref 98). In their study, the interdendritic flow for

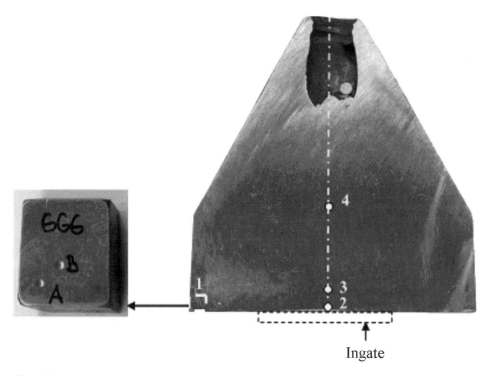

Fig. 20 Location of hardness measurements on sample casting. See Table 2

Table 2 Comparison between measured and predicted hardness

Location	Dimension x, mm	Dimension y, mm	Measurement, HB	Simulated, HB
1 A	4	4	368	371
1 B	10	7	313	320
2	50	4	249	255
3	50	10	236	245
4	50	48	209	203

three-dimensional simulations of mold filling was included but without considering microporosity. Among all these models, no one calculated the diffusion and convection of gas. Felicelli et al. (Ref 105) calculated the redistribution of gas during solidification but did not predict how much porosity forms and the sizes of the pores.

To predict microporosity defects in casting processes accurately, the following factors that contribute to microporosity formation should be considered: macroscopic heat transfer, interdendritic fluid flow, gas redistribution by diffusion and convection, microstructure evolution, and microporosity growth.

The conservation equation beside the equations mentioned early on is gas conservation.

Gas conservation:

$$\frac{\partial \rho \bar{c}}{\partial t} + \rho u_i \nabla c_l = (1-k)c_l \frac{\partial (1-f_l)\rho}{\partial t} + \nabla(f_l \rho \nabla c_l)$$

(Eq 38)

where $\bar{c} = f_l c_l + (1-f_l)c_s$, $c_s = kc_l$, c_l is the gas concentration in the liquid, c_s is the gas

concentration in the solid, and k is the partition coefficient.

In addition to the liquid and solid fraction, which are calculated from the energy equation, the dendrite spacing is needed to estimate pore curvature and permeability in the mushy zone. The pore radius or curvature is taken to be proportional to the dendrite cell spacing through the relationship:

$$r = \frac{1}{2}f_l d$$

where d is the secondary dendrite arm spacing.

Gas Porosity Evolution. Pores will form in a solidifying alloy when the equilibrium partial pressure of gas within the liquid exceeds the local pressure in the mushy zone by an amount necessary to overcome surface tension. Hence, gas porosity develops when:

$$P_g > P_a + P_m + P_d + P_\delta$$

(Eq 39)

where P_g is Sievert pressure, P_a is ambient pressure, P_m is metallostatic pressure, P_d is pressure drop due to the friction within the interdendritic liquid, and $P_\delta = \frac{2\sigma}{r}$ is surface tension, where σ is surface tension, and r is pore radius.

The maximum dissolved gas (hydrogen or nitrogen) in the liquid, g_l, or solid, g_s, and the gas pressure are related through Sievert's law:

$$g_l = \frac{K_g}{f_g}P_g^{1/2} \quad g_s = kg_l$$

(Eq 40)

where K_g and f_g are the equilibrium constant and activity coefficient, respectively, for

hydrogen. For example, the activity coefficient for hydrogen in an aluminum alloy is estimated as (Ref 105, 106):

$$\ln f_H = \sum_{j=1}^{N} a_H^j C^j + \sum_{j=1}^{N} b_H^j (C^j)^2$$

(Eq 41)

where a_H^j and b_H^j are interaction coefficients, and C^j is the concentration of solute element j. The detailed coefficients can be found in Ref 105 and 106.

The equilibrium constant is calculated as:

$$\ln K_H = -3.039 - \frac{6198.47}{T}$$

(Eq 42)

The volume fraction of gas porosity is:

$$c_l f_l + c_s(1-f_l) = g_l f_l + g_s(1-f_l-f_v) + \alpha \frac{P_g f_v}{T}$$

(Eq 43)

where f_v is the volume fraction of gas porosity, and α is the gas conversion factor.

If no pore has formed yet, then:

$$c_l f_l + c_s(1-f_l) = g_l f_l + g_s(1-f_l)$$

(Eq 44)

Shrinkage Porosity. If the pressure drops below the cavitation pressure, it is assumed that liquid feeding ceases and the solidification shrinkage in that computational cell is compensated only by pore growth. In general, cavitation pressure is very small. When the liquid pressure drops below the cavitation pressure, the porosity is determined such that it compensates for the entire solidification shrinkage within the current time step:

$$f_v^{n+1} = f_v^n + \frac{\rho^{n+1} - \rho^n}{\rho^{n+1}/(1-f_v^n)}$$

(Eq 45)

Experimental Validation. In this example, a set of castings with different initial hydrogen content using an iron chill plate was simulated and compared with experimental results for an A319 casting. The geometry and mesh is shown in Fig. 21.

The casting is 132 mm in height, 220 mm in length, and of varying thickness. Wedges are cut horizontally at 35 mm from the bottom end and with a thickness of 12 mm. The initial pouring temperature is 750 °C. Initial hydrogen contents are 0.108, 0.152, and 0.184 ppm. The experimental and simulation results are taken at different distances from the chill end. The comparison of the value of percentage porosity against local solidification time and hydrogen content between simulation and experiment is shown in Fig. 22.

It shows that increasing solidification time and hydrogen content considerably increase the percentage of porosity. Numerical simulation results give excellent agreement with the measurements of percentage of porosity. The results also show that local solidification time

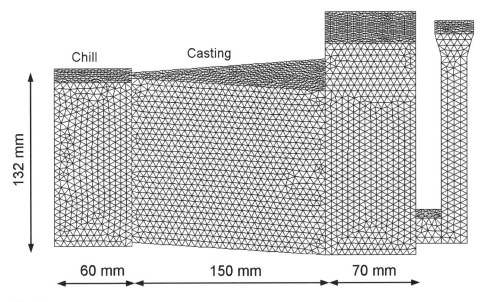

Fig. 21 Finite-element geometry and mesh for a wedge-shaped aluminum A319 casting

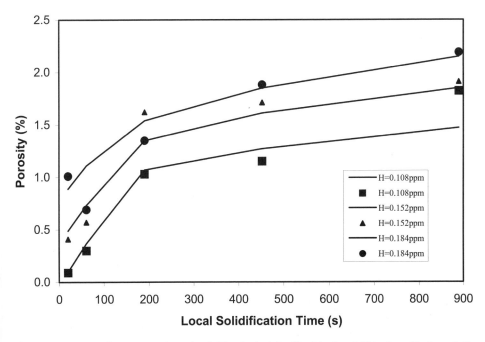

Fig. 22 Comparison between experiment (symbols) and calculation (lines) for three initial values of hydrogen in the A319 casting

and initial hydrogen content are very important factors that influence the formation of porosity.

Macrosegregation

Modeling and simulation of macrosegregation during solidification has experienced explosive growth since the pioneering studies of Flemings and co-workers in the mid-1960s (Ref 113–115).

Beckermann did a comprehensive review of recent macrosegregation models and their application to relevant casting industries (Ref 116). There are numerous factors that can cause

macrosegregation during casting solidification processes. Those include thermal- and solute-induced buoyancy, forced flow, solid movement, and so on.

The application of a multiphase/multiscale macrosegregation model to predict macrosegregation during dendritic alloy solidification, including melt convection and grain movement, is illustrated in an Al-4wt%Cu alloy solidified inside a rectangular cavity cooled from the left sidewall (Ref 38). Figure 23 shows the predicted evolution of the grain density. The effects of grain movement and nucleation rate on macrosegregation are shown in Fig. 24.

Macrosegregation models have been applied extensively in the casting industry, such as steel ingot castings, continuous and direct chill castings, nickel-base superalloy single-crystal castings (freckle simulation, such as in Fig. 25, which shows the prediction of freckling in upward directional solidification of a superalloy from a two-dimensional model), and shape castings as well.

Freckles have been the subject of intense research efforts for approximately 30 years due to their importance as a defect in alloy casting (Ref 117, 118). They represent a major problem in directionally solidified superalloys used in the manufacture of turbine blades (Ref 117–121). Upward directional solidification provides an effective means of producing a columnar microstructure with all the grain boundaries parallel to the longitudinal direction of the casting. In conjunction with a grain selector or a preoriented seed at the bottom of the casting, directional solidification is used to make entire castings that are dendritic single crystals. During such solidification, the melt inside the mushy zone can become gravitationally unstable due to preferential rejection of light alloy elements (for a partition coefficient less than unity) into the melt. Because the mass diffusivity of the liquid is much lower than its heat diffusivity, the segregated melt retains its composition as it flows upward and causes delayed growth and localized remelting of the solid network in the mush. Ultimately, a pencil-shaped vertical channel, devoid of solid, forms in the mushy zone, through which low-density, highly segregated liquid flows upward as a plume or solutal finger into the superheated melt region above the mushy zone. This flow is continually fed by segregated melt flowing inside the mushy zone radially toward the channel. At the lateral boundaries of the channel, dendrite arms can become detached from the main trunk, and those fragments that remain in the channel are later observed as freckle chains.

The complex convection phenomena occurring during freckle formation represent a formidable challenge for casting simulation (Ref 17, 116, 122, 123). In 1991, Felicelli et al. simulated channel formation in directional solidification of lead-tin alloys in two dimensions (Ref 124). Since then, numerous studies have been performed to simulate and predict freckling in upward directional solidification (Ref 17, 125–134). Neilson and Incropera performed the first three-dimensional simulations of channel formation in 1993 (Ref 128). However, the coarseness of the mesh caused a serious lack of resolution and inaccuracies. Three-dimensional simulations have also been performed by Poirier, Felicelli, and co-workers for both binary and multicomponent alloys (Ref 132, 133). Figure 26 illustrates a three-dimensional freckle formation prediction for a binary alloy.

Freckle formation can be simulated with a commercial package, such as ProCAST.

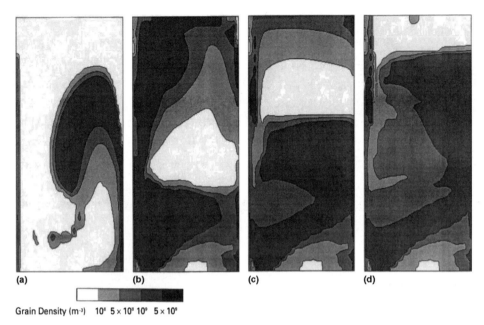

Grain Density (m⁻³)   10⁸ 5 × 10⁸ 10⁹ 5 × 10⁹

Fig. 23 Predicted evolution of grain density during equiaxed dendritic solidification of Al-4Cu (wt%) alloy with grain movement inside 5 × 10 cm rectangular cavity cooled from left sidewall. Source: Ref 38

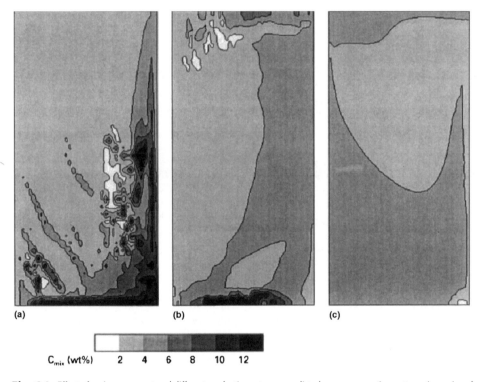

C_mix (wt%)   2   4   6   8   10   12

Fig. 24 Effect of grain movement and different nucleation rates on predicted macrosegregation patterns in equiaxed dendritic solidification of Al-4Cu (wt%) alloy with grain movement inside 5 × 10 cm rectangular cavity cooled from left sidewall. Source: Ref 38

Figure 27 illustrates the freckle formation for a Pb-10%Sn binary alloy directionally solidified in a simple geometry. A temperature gradient is initially imposed to simulate a directional solidification system. Cooling is achieved by lowering the temperatures of the upper and lower walls of the cavity at a constant rate, such that the overall temperature gradient is maintained over the height of the cavity. The lateral walls of the cavity are taken as adiabatic.

Hot Tearing

Hot tearing is one of the most serious defects encountered in castings. Many studies have revealed that this phenomenon occurs in the late stage of solidification, when the fraction of solid is close to one. The formation and propagation of hot tearing have been found to be directly affected by the cooling history, the chemical composition and mechanical properties of the alloy, as well as the geometry of the casting. Various theories have been proposed in the literature on the mechanisms of hot tearing formation. Detailed reviews on the theories and experimental observations of the formation and evolution of hot tearing can be found in Ref 135 and 136 and the references therein.

Most of the existing hot tearing theories are based on the development of strain, strain rate, or stress in the semisolid state of the casting. For the strain-based theory, the premise is that hot tearing will occur when the accumulated strain exceeds the ductility (Ref 137–139). The strain-rate-based theories suggest that hot tearing may form when the strain rate, or strain-rate-related pressure, reaches a critical limit during solidification (Ref 140, 141). The stress-based criterion, on the other hand, assumes that hot tearing will start if the induced stress in the semisolid exceeds some critical value (Ref 138, 142). Although these theories were proposed independently as distinct theories, indeed, they can be considered as somewhat related due to the relationship between strain, strain rate, and stress. It is such a relationship that motivates the development of a hot tearing indicator, which uses the accumulated plastic strain as an indication of the susceptibility of hot tearing. This considers the evolution of strain, strain rate, and stress in the last stage of solidification. A Gurson type of constitutive model, which describes the progressive microrupture in the ductile and porous solid, is adopted to characterize the material behavior in the semisolid state. The proposed hot tearing indicator, while verified specifically for magnesium alloys, has a much wider application.

To reliably predict the formation and evolution of hot tearing in casting by numerical simulations, it is critical to have accurate thermophysical and mechanical properties, especially in the mushy zone. It is also essential that the solidification path of the alloy be accurately described. The prediction of the thermophysical and mechanical properties has recently become possible by using the knowledge of the microstructure, phase fractions, and defects present in a metallic part (Ref 143). The solidification path can be obtained with the help of thermodynamic calculations of phase stability at given temperatures and compositions. A comprehensive multicomponent alloy solidification model, coupled with a Gibbs free-energy minimization engine and thermodynamic databases, has been developed to facilitate such calculations (Ref 7). With the integration of a back-diffusion model in the calculation, solidification conditions, such as cooling rate, can also be taken into account.

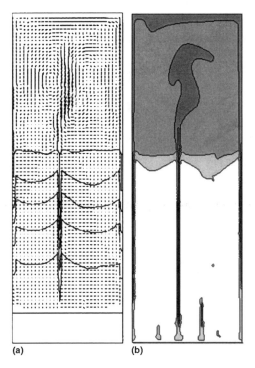

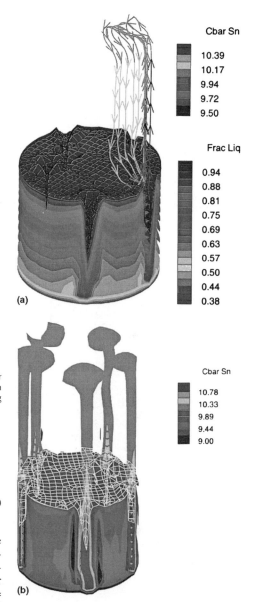

Fig. 25 Solidification of a single-crystal nickel-base superalloy. (a) Predicted velocity vectors (largest vector represents 6.3 mm/s) and solid fraction contours (in 20% increments). (b) Macrosegregation pattern (titanium concentration normalized by initial concentration in equal intervals between 0.87 and 1.34) showing freckle formation during upward directional solidification in 5 × 15 cm rectangular domain. Source: Ref 116

Fig. 26 Solidification of Pb-10Sn (wt%) in a cylinder at 10 min. (a) Isosurfaces of volume fraction of liquid and stream traces emerging from a channel. The color of the stream traces indicates the mixture concentration of tin (Sn), with corresponding levels shown in the upper color bar. The top isosurface φ = 0.98, meshed to show channel penetration, is at z = 9 mm. (b) Isosurfaces of mixture concentration of tin, showing tin enrichment within channels and tin-rich plumes emerging from them. Source: Ref 133

Hot Tearing Indicator. The constitutive model used to describe the material behavior in the semisolid state is the Gurson model (Ref 144–147), which was originally developed for studying the progressive microrupture through nucleation and growth of microvoids in the material of ductile and porous solids.

When the material is considered as elastic-plastic, the yield condition in the Gurson model is of the form:

$$\phi(\boldsymbol{\sigma}, \mathbf{x}, T, \bar{\varepsilon}^p, G_u) = F(\boldsymbol{\sigma}) - G_u(\sigma, \bar{\varepsilon}^p, f_v)\kappa(\bar{\varepsilon}^p, T)$$
$$= 0$$

$$(\text{Eq 46})$$

where $F(\boldsymbol{\sigma}) = (3(\mathbf{s} - \mathbf{x}) : (\mathbf{s} - \mathbf{x})/2)^{1/2}$ is the Mises stress in terms of the deviatoric stress $\mathbf{s} = \boldsymbol{\sigma} - (tr\boldsymbol{\sigma})\mathbf{I}/3$, κ represents the plastic flow stress due to isotropic hardening, and $\mathbf{x}$ denotes back stress due to kinematic hardening. The accumulated effective plastic strain is written as:

$$\bar{\varepsilon}^p = \int_0^t \sqrt{(2/3)\dot{\boldsymbol{\varepsilon}}^p : \dot{\boldsymbol{\varepsilon}}^p} d\tau \qquad (\text{Eq 47})$$

with

$$\dot{\boldsymbol{\varepsilon}}^p = \dot{\gamma}\frac{\partial \phi}{\partial \boldsymbol{\sigma}} \qquad (\text{Eq 48})$$

and $\dot{\gamma}$ being the plastic flow parameter. The Gurson coefficient, G_u, is defined as:

$$G_u = -2f^* q_1 \cosh\left(\frac{tr(\boldsymbol{\sigma})}{2\kappa}\right) + \{1 + (q_1 f^*)^2\} \quad (\text{Eq 49})$$

in which q_1 is a material constant and:

$$f^* = f_v \qquad \qquad \text{for } f_v \le f_c$$
$$f^* = f_c + \frac{f_u - f_c}{f_F - f_c}(f_v - f_c) \quad \text{for } f_v > f_c$$

$$(\text{Eq 50})$$

Here, $f_u = 1/q_1$, f_c is the critical void volume fraction, and f_F is the failure void volume fraction. Their values should be different for different materials. In this calculation for an indicator for hot tearing, constants are used. Here, $q_1 = 1.5$, $f_c = 0.15$, and $f_F = 0.25$, as used in Ref 148. The Gurson coefficient characterizes the rapid loss of material strength due to the growth of void volume fraction, f_v. When $f_v = f_F$, then $f^* = f_u = 1/q_1$, and $G_u = 0$, for zero stress; that is, the stress-carrying capacity of the material vanishes.

The evolution of the void volume fraction is described by the nucleation of the new void and the growth of the existing void:

$$\dot{f}_v = \dot{f}_{\text{nucleation}} + \dot{f}_{\text{growth}} \qquad (\text{Eq 51})$$

with the rate of void growth defined as:

$$\dot{f}_{\text{growth}} = (1 - f^*)tr(\dot{\boldsymbol{\varepsilon}}^p)$$
$$= \dot{\gamma}(1 - f^*)\left(\frac{3f^* q_1}{\kappa}\right)\sinh\left(\frac{tr(\boldsymbol{\sigma})}{2\kappa}\right) \quad (\text{Eq 52})$$

In this study, the nucleation of the void is assumed to be strain controlled and is written as:

$$\dot{f}_{\text{nucleation}} = \dot{e}_{\text{ht}} \qquad (\text{Eq 53})$$

where:

$$e_{\text{ht}} = \int_{t_c}^t \sqrt{(2/3)\dot{\boldsymbol{\varepsilon}}^p : \dot{\boldsymbol{\varepsilon}}^p} d\tau \ t_c \le t \le t_s \quad (\text{Eq 54})$$

is defined as the hot tearing indicator. The symbol t_c represents time at coherency temperature, and t_s denotes time at solidus temperature. It is

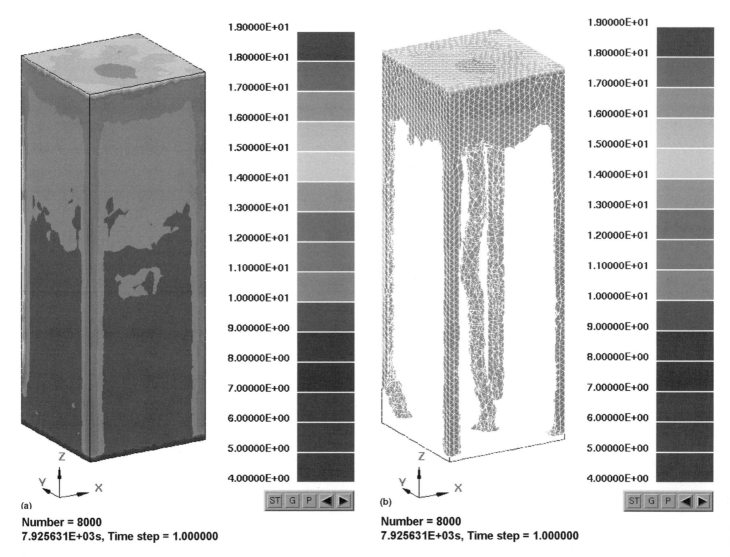

| 1.90000E+01 |
| 1.80000E+01 |
| 1.70000E+01 |
| 1.60000E+01 |
| 1.50000E+01 |
| 1.40000E+01 |
| 1.30000E+01 |
| 1.20000E+01 |
| 1.10000E+01 |
| 1.00000E+01 |
| 9.00000E+00 |
| 8.00000E+00 |
| 7.00000E+00 |
| 6.00000E+00 |
| 5.00000E+00 |
| 4.00000E+00 |

(a)
Number = 8000
7.925631E+03s, Time step = 1.000000

(b)
Number = 8000
7.925631E+03s, Time step = 1.000000

Fig. 27 Macrosegregation for directional solidification of a Pb-10Sn binary alloy. (a) Final tin composition after solidification. (b) Cut-off view of it

observed that the hot tearing indicator is, in fact, the accumulated plastic strain in the semi-solid region, and it corresponds to the void nucleation. Therefore, it should provide a good indication for the susceptibility of hot tearing during solidification. The value of the hot tearing indicator is determined by finite-element analysis (Ref 149). For materials described by the viscoplastic or creep model, yield condition does not exist. The function φ defined in Eq 46 can be used as a potential for the inelastic flow, so that the inelastic part of the strain rate can still be given in the form of Eq 48.

Experiment Validation. Cao et al. performed some experiments to study hot tearing formation during solidification of binary Mg-Al and ternary Mg-Al-Ca alloys in a steel mold (Ref 150, 151), which is shown in Fig. 28. A hot cracking susceptibility was introduced, which is a function of maximum crack width, crack length factor, and the crack location. It

was found that it is easier to have a crack at the sprue end than at the ball end. It is less likely to have a crack in the middle of the rod. Also, the longer rod is easier to crack. Figure 29 shows the simulated results of a hot tearing indicator for a Mg-2%Al alloy casting. The computed hot tearing indicator agrees very well with the experiments.

Figure 30 shows the experimental results of hot tearing at the sprue end of the rods for three different alloys. The calculated hot tearing indicators are shown in Fig. 31 accordingly. It can be seen that hot tearing is less severe as the aluminum content increases from 2 to 4% and then to 8% at the same location for the same casting with the same casting conditions. Again, the simulated hot tearing indicators agree well with the observations. The susceptibility rises sharply from pure magnesium, reaches its maximum at Mg-1%Al, and decreases gradually with further increase in the aluminum content.

The hot tearing indicator is calculated at the end of the sprue for the longest rod with a different alloy composition. For comparison, the hot tearing indicators as well as a crack susceptibility coefficient (CSC), which is defined as the temperature difference between fraction of solid at 0.9 and at the end of solidification, are shown in Fig. 32. Same as the experiment, the susceptibility of hot tearing rises sharply from pure magnesium, reaches its maximum at Mg-1%Al, and decreases gradually with further increase in the aluminum content. Similarly, different ternary magnesium alloys have different hot tearing susceptibility. The calculated CSC by this model for different alloys is shown in Fig. 33. The experimental hot tearing indicator (Ref 150, 151) is included in the same figure for comparison. The addition of calcium to magnesium-aluminum alloys can reduce the temperature range between fraction of solid at 0.9 and end of solidification, which is shown in Fig. 34; hence,

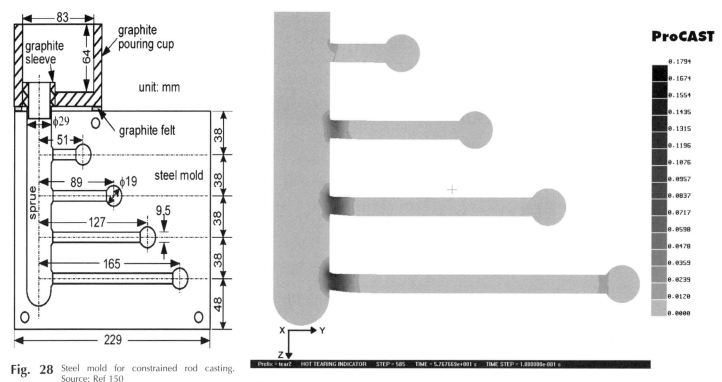

Fig. 28 Steel mold for constrained rod casting. Source: Ref 150

Fig. 29 Hot tearing indicator for a Mg-2Al alloy casting

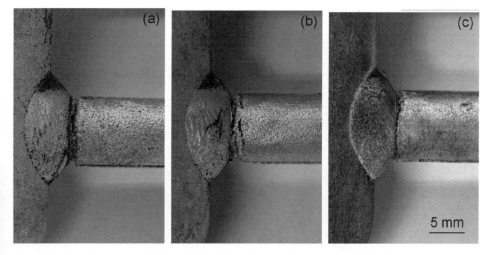

Fig. 30 Close-up views of hot tearing (cracks) in the bottom rods near the sprue. (a) Mg-2Al. (b) Mg-4Al. (c) Mg-8Al

susceptibility to hot tearing formation will decrease as well. It indicates that the current hot tearing indicator can predict hot tearing trends very well. The alloy chemistry, casting geometry, and cooling conditions contribute to the formation of hot tearing, and they are included in this model directly or indirectly.

Examples of Modeling Applied in Casting Industries

To demonstrate the application of modeling in the casting industry, a couple of examples

are presented next. First, a die casting process is modeled while focusing on the stress analysis of the casting and die and the gap formation between mold and part. Next, an example of modeling of an investment casting is presented. As mentioned earlier, radiation is very important for modeling of investment casting. In this example, radiation is discussed extensively.

High- or Low-Pressure Die Casting

For such casting processes, metallic molds are used, occasionally with cooling or heating channels. It is called permanent mold casting

if the filling is by gravity. Otherwise, it is called die casting, for which a pressure is applied to provide rapid filling. These processes are mainly focused on high-volume components such as automobile parts. Fluid flow and stress analysis for both casting and mold are important to other casting processes.

Because the molten metal is introduced into the mold by gravity, permanent mold casting flow analysis is similar to that of sand casting. However, back pressure or trapped gas cannot be ignored, because a metal mold is not permeable, unlike that of a sand mold. The flow analysis can be complicated because of the inclusion of the effect of back pressure in mold filling. If a relatively low pressure is applied to the sealed furnace, it is called a low-pressure casting process. The low pressure pushes the molten metal to fill the mold cavity slowly. For low-pressure casting filling analysis, the boundary conditions are pressure, which is a function of time instead of an inflow velocity. This is to simulate the furnace pressure controlling flow in the low-pressure casting presses.

In permanent mold casting, the molds are used repeatedly. Hence, the molds develop a nonuniform temperature distribution during the initial cycles of the casting process that approaches a periodic quasi-steady-state condition. For a cyclic analysis, all casting parameters, such as the liquid pouring, dwell time, open time, and spraying conditions, must be considered for the calculation before the solidification analysis.

As the modern foundry continues to evolve in implementing new technology, process modeling must also advance to meet the next

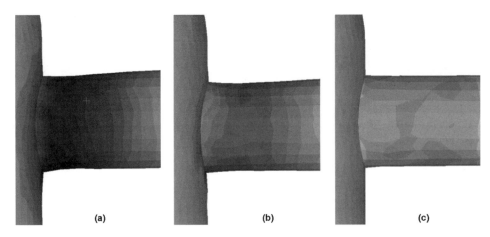

Fig. 31 Hot tearing indicator in the bottom rods near the sprue developed in simulation. (a) Mg-2Al. (b) Mg-4Al. (c) Mg-8Al

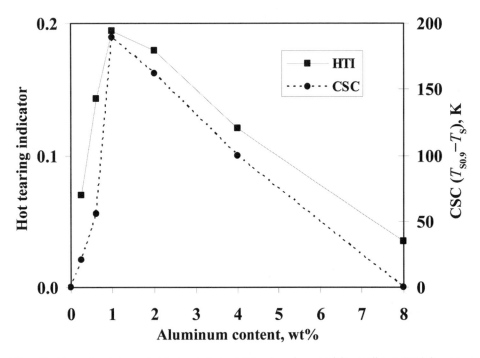

Fig. 32 Comparison between hot tearing indicator (HTI) and crack susceptibility coefficient (CSC) for various aluminum contents in a magnesium alloy

hurdles facing foundry engineers. It is critical that these new hurdles—stress and deformation in the casting and mold, micro- and gas porosity, and as-cast mechanical properties—be accurately predicted and corrected. Beyond simply identifying shrink porosity and fill problems, numerical tools have been developed to predict defects at a microstructural level that can be used effectively by the foundry engineer. Knowing the answers to these questions early in the manufacturing process affords significant time- and cost-savings.

The main goal of the casting process is to approach and achieve net shape. Large deformation or distortion in the part requires more rework, such as hot pressing, even after a heat treatment operation. To aid in the understanding of why parts deform or develop residual stresses, simulation software can now predict thermally induced stresses, which include the effects of the strength or constraint of the mold. Simulating the strength of molding components has been proven as a necessary input in determining the amount of distortion in the casting, because the main prevention against significant warpage is the mold itself.

Coupling the stress analysis with the thermal and fluid calculation gives a more clear understanding of the physical phenomenon. By comparing the mushy zone location and the evolution of stresses while cooling, problems such as hot tearing and cracking can be clearly

indicated. Hot tearing and cracking can occur when the stresses in the local region go beyond the yield stress, thus requiring a tight coupling of the thermal and stress calculation. Another typical phenomenon involves the gap formed between casting and mold as the casting solidifies and shrinks. When this gap develops, there can be a significant reduction in heat transfer from the casting to the mold components where contact is lost (Ref 94). Conversely, there may be locations where, due to the shrinkage of the casting onto the core, an increasing contact pressure will increase the heat-transfer rate from casting to core. By accurately tracking the heat transfer, a better indication of surface shrink can be achieved.

A multibody mechanical contact algorithm is employed to compute the contact and gap formation between the castings and die parts. Contact between different die parts is also considered. An automatic penalty number adjustment technique is implemented in the contact algorithm. Such a technique greatly enhances the stability and robustness of the contact computation algorithm.

The variational form of the equilibrium equation with mechanical contact at any time, t, is written as (Ref 152):

$$\int_\Omega \sigma \cdot \text{grad}(\delta\mathbf{u})d\Omega - \int_\Omega \mathbf{b} \cdot \delta\mathbf{u}d\Omega - \int_{\Gamma_\sigma} \bar{\mathbf{t}} \cdot \delta\mathbf{u}d\Gamma$$
$$+ \int_{\Gamma_c} \xi(\mathbf{u})g(\mathbf{u})\mathbf{n} \cdot \delta\mathbf{u}d\Gamma = 0$$

(Eq 55)

Here, a frictionless contact is considered for simplicity. In Eq 55, Ω represents the geometry of casting and all the mold parts, and Γ represents all the contact interfaces between all parts. The body forces and surface tensions are denoted by $\mathbf{b}$ and $\mathbf{t}$, respectively. The augmented penalty function is given by ξ, while u is the displacement, $\mathbf{n}$ is the surface unit normal, g is the interface gap, and $\bar{\mathbf{t}}$ is the surface traction.

Thermal contact between parts is considered by adjusting the interface heat-transfer coefficient with respect either to the air gap width or the contact pressure, as computed by the mechanical contact algorithm. When the gap width is greater than zero, the adjusted heat-transfer coefficient has the form:

$$h_{\text{eff}} = \frac{1}{\frac{1}{h_0} + \frac{1}{(h_{\text{air}}+h_{\text{rad}})}}$$

(Eq 56)

where h_0 is the initial value of the heat-transfer coefficient, h_{air} is the conductivity of air divided by the gap width (if a vacuum is used, this term equates to zero), and h_{rad} is the radiation heat-transfer coefficient.

If the contact pressure is greater than zero, the effective heat transfer is increased linearly with pressure up to a maximum value.

When the casting is ejected from the die, the mechanical contact is no longer applied to the

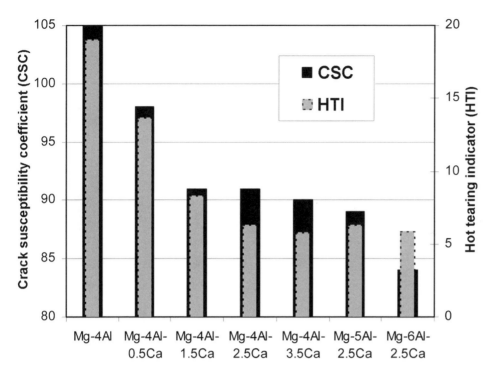

Fig. 33 Comparison between hot tearing indicator (HTI) and crack susceptibility coefficient (CSC) for various magnesium alloy compositions

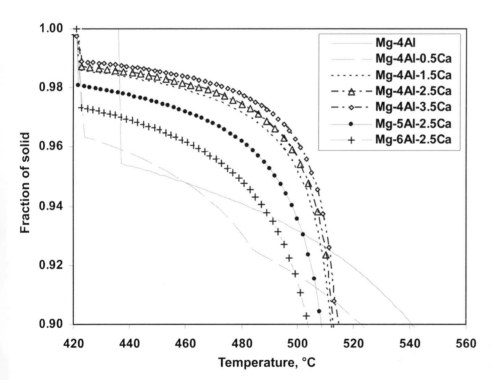

Fig. 34 Solidification paths for some magnesium alloys

casting/die interfaces. Care must then be taken to apply an appropriate displacement constraint to prevent solid body movement.

Example 2: Aluminum Die Casting. In this example (Fig. 35, Ref 153), a simple T-shaped casting of an A356 aluminum alloy in an H13 chromium hot work tool steel mold is simulated. The effective interface heat-transfer coefficient at two different points on the casting is plotted. The top curve is from a point experiencing increasing contact pressure as the casting contracts. The middle curve is from a point where a gap is opening up between casting and mold, assuming the presence of air. The bottom curve is from that same point but assuming a vacuum. The large variation in the coefficient illustrates the importance of accounting for local conditions. In addition, this example illustrates the value of reverse coupling of the mechanical deformations with the energy solution. This effect can be seen in Fig. 35 on the right, where the heat flux contours are plotted. The heat flux is greatest where the contact pressure is highest.

Example 3: Deformation after Ejection. Postejection springback and further relaxation after mold removal can also be tracked, completing the necessary cycle to determine final part shape (Fig. 36). Finally, heat treatment processes may further define the final net shape. By appropriately simulating the thermal loading on the part during heat treatment, any additional shape change may be captured, either independent of the casting process or as a continuance of the deformation and residual-stress evolution formed from casting.

Investment Casting (Ref 154)

The investment casting or lost wax process, as it is commonly called, is one of the oldest known manufacturing processes. The ability to produce near-net shapes minimizing machine cost makes the investment casting process one of the most attractive casting processes, especially for making exotic casting with expensive alloys. This process can be used to make complex shapes, from aircraft jet engine components to small, intricate castings used to make jewelry. The investment casting process does not require elaborate or expensive tooling and has the ability to produce several castings in one pour.

This process is broadly described in the schematic sketch in Fig. 37. The various stages include creating the wax pattern, growing the investment shell, dewaxing, and pouring the casting.

The investment casting process uses ceramic shells as molds. The ceramic shells can be preheated to very high temperatures, up to and above the liquidus temperature. Important and demanding applications of investment castings include aerospace and medical implant applications. Investment castings are commonly made from nickel-base superalloys, titanium alloys, aluminum alloys, cobalt-base alloys, and steels. Nickel-base alloys are used for jet engine structural castings as well as turbine airfoils. Structural castings normally have an equiaxed grain structure, whereas turbine airfoils have equiaxed grain, columnar grain, and single-crystal types of structure.

The unique features of the investment casting process include the use of high-temperature ceramic molds, frequently casting in a vacuum, a furnace withdrawal process for directional solidification, and strict requirements for

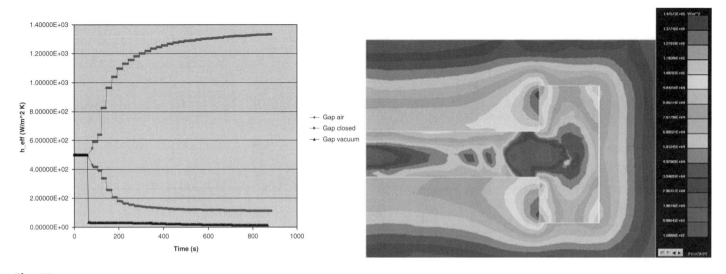

Fig. 35 T-bar example shows the heat flux increasing in locations of higher contact pressure and decreasing where gaps develop

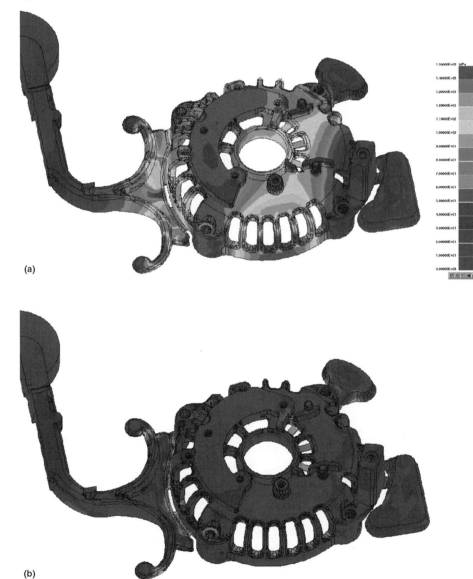

(a)

(b)

Fig. 36 Stresses in the component (a) before and (b) after ejection

microstructure. At times, a centrifugal casting process may be used to improve mold filling. Hence, it is critical to apply some special treatments for modeling of the investment casting process.

The outer contour of a shell mold generally follows the shape of the casting, unlike most other molds. Because the shape of the shell mold is determined by the procedure of repeatedly dipping the wax form into a ceramic slurry, followed by drying, it is somewhat freeform. It is thus difficult to generate a computer-aided design representation of the shell geometry. Hence, it is necessary to use specialized meshing techniques to create a model of the shell. Some commercial packages can generate the shell automatically based on the investment process such that the thickness of shell will be different at different geometrical locations, such as convex sides, concave sides, flat surfaces, or some corners.

Another unique feature for investment casting simulation involves solving the heat-transfer problems. In general, the model should handle heat conduction in the core and the mold, convection and radiation across the metal and mold interface, and radiation and convection at the mold outer interface. If the alloys are cast in a vacuum environment, such as for most nickel- and titanium-base alloy castings, radiation heat transfer is the only method for heat loss from the mold surface to the furnace. For a radiation calculation, proper view factor calculations are very important. Sometimes, the view factors can change, for example, during the withdrawal process for direct chill and single-crystal castings. To change the cooling rate for some investment casting processes, insulation materials, such as kaowool, can be used to wrap specific areas of the shell. For such situations, this can be represented in the model by changing the shell surface emissivity or heat-transfer coefficient.

The most significant mode of heat transfer in an investment casting process is through

COMPUTER SIMULATION

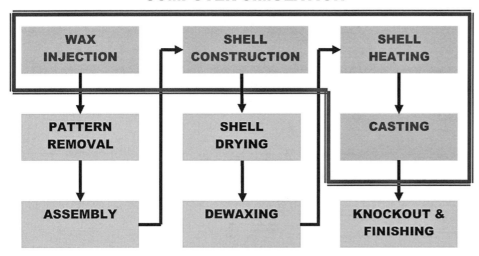

Fig. 37 Flow chart of the investment casting process

radiation. It is critical that this be fully understood and accounted for when planning the process parameters for any casting. One must understand the radiation effects during the metal pouring and cooling cycle, because self-radiation and external casting conditions will affect cooling rates and solidification patterns. Additionally, prepouring soaking determines the shell temperatures before the pour, and if this is not accurately taken into account, it will adversely affect the desired outcome.

This section delves into radiation and the downstream effects in the casting of certain process scenarios:

- Heat loss in the mold before pouring
- Self-radiation effects on solidification
- Cooling of the shell
- Pouring castings in ambient conditions versus in an enclosed chamber (i.e., a can)
- Effects of kaowool and other insulating media

For a simulation program to model the physics of the process, the computer program must accurately mimic the various stages from wax injection to the actual pouring of the molten metal and the solidification process. In addition to the mold filling and solidification analysis, advanced computer simulation programs can also evaluate the thermally induced stress in the casting and underlying grain structure and mechanical properties of the final cast part.

Accurate modeling of the trapped air and shrink porosity is critical to evaluate the existing gating system and feeder locations. Defects are usually observed in the last areas to fill, especially if the permeability of the shell is not appropriate and if the last liquid area during solidification is in the casting and not properly fed. In badly designed gating systems, the liquid metal can prematurely solidify, leading to cold laps and other such defects. It is essential

that the proper heat extraction from the casting to the investment shell is calculated.

Due to the elevated temperature of the mold relative to its surroundings and the nature of the system setup, as has been previously stated, radiative heat transfer is the dominant mode of heat transfer. Because each location on the mold surface sees a different view of its environment, each will have a different rate of heat loss. For example, faces around the outer perimeter of the mold are exposed mostly to the ambient, while the view space of those facing the inside of the mold see mostly other hot regions of the casting. As a result, radiative view factor calculations are required to accurately account for the heat-exchange variations across the geometry. The initial mold temperature and heat loss to the environment prior to casting dictate the thermal profiles that exist at the time of pour. This creates a thermal gradient from the inside to the outside of the shell, depending on the self-radiating effects of the whole assembly.

Most analysis packages on the market today (2010) assume basic radiation heat-transfer calculations making single-body assumptions that do not account for multiple bodies or shapes. Calculating view factors allows for multiple bodies and the shapes of those bodies to be simulated in the radiation heat transfer to give a much more precise and accurate calculation. Under a single-body assumption, each casting would heat or cool exactly the same, regardless of the casting orientation on the tree, even if it was a casting on the end of the runner or a part that was fully surrounded by other hot parts.

Planning setup design for investment casting involves visualizing the "invisible" radiation heat-transfer effects. Therefore, to further understand radiation effects, analysis tools that are able to calculate the view factor radiation and shadowing effects are used to present various common investment casting scenarios.

Example 4: Radiative Cooling. A simple example given in Fig. 38 shows four test bar castings. Figure 38(a), which does not include the effects of self-radiation, shows all four test bars losing the same amount of heat at a given time. Even though the radiative heat effects are considered, there is no reflective heat from the hot surfaces; thus, all the test bars show exactly the same temperature profile, regardless of their location in the model. Figure 38(b) shows the effects of self-radiation on each specimen. As expected, the interior test samples retain heat and take longer to solidify than those on the outside, because the interior bars receive radiated heat from the test bars on either side of them. The outside bars receive heat from only one side.

To prevent freezing of the metal or cold flow during filling, it is desirable to pour the casting as soon as possible when the mold is removed from the furnace. This is especially so for thin-shell molds, conductive shell material, or very thin parts. For small castings, where the mold is typically hand-moved from the furnace to the pouring bed, it is typical to have 10 to 20 s elapse before the metal is poured. Even this small amount of time can cause a significant reduction in temperature on shell faces that are open to the environment. Figures 39(a and b) show the effects of shell cooling between the time it takes the shell to be extracted from the furnace to the beginning of metal entering the mold cavity. In the casting, a 25 to 30 s time delay can cause certain regions of the casting to drop 300 to 400°. Note (Fig. 39a) how the inside walls of the shell on the end have cooled much more rapidly than the walls of central regions.

With large shells or ones that must be moved from a furnace into a vacuum chamber, the elapsed time can be longer, and thus, the difference between internal shell temperatures and external temperatures may be quite large, potentially resulting in unexpected filling issues or patterns.

In many investment castings, when the casting is poured, the shell glows, indicating the large amount of heat inside. During and after mold filling, solidification starts to take place. The energy content of the metal continually decreases by heat conduction through the cooler mold and by radiation from the mold surface. Exterior faces radiate this heat freely and allow for a relatively high rate of cooling. However, internal surfaces or locations where there is mostly part-to-part radiation are, in essence, insulating each other by radiating heat onto itself. It is quite common to have internal parts or parts involved in a high amount of radiation have a solidification pattern that is much different than parts that may be on the end of a row or more exposed to the ambient conditions. Therefore, it may be beneficial to design casting setups such that all of the parts experience the same heat transfer (also know as radiation) effects. By having the same cooling pattern, any changes to the design or rigging of the part

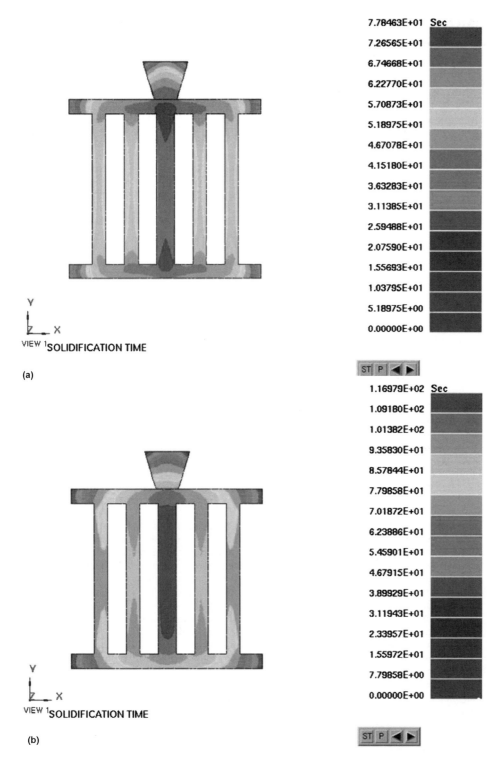

VIEW 1 SOLIDIFICATION TIME

(a)

VIEW 1 SOLIDIFICATION TIME

(b)

Fig. 38 Test bar casting solidification time indicated by coloration. (a) Without self-radiation considered. (b) Effects of self-radiation considered

will apply evenly to each part. Otherwise, the casting engineer may be chasing a defect that occurs in one part position that does not occur in another.

Example 5: Inline Gating. The case study in Fig. 40 shows an in-line gating design. The initial setup shows nine castings stacked next to each other, with a common top-fed runner design. A coupled fluid and solidification analysis was run to understand the problems associated with molten metal flow and porosity. The computer analysis gave a very good indication of the flow pattern of the mold. The molten metal seems to fill most of the central casting

because the sprue is located right above it, then the other castings fill from the center outward. Because the central casting is already filled before the other castings, the temperature in the center casting is initially lower than the others. As the filling progresses to the entire mold cavity, the shell starts to become heated, thus reflecting heat to the outside. The central casting remains much hotter than the "outer" casting region due to the self-radiation effects. A quick look at the solidification plots shows similar results. The outside castings solidify quicker than the inside ones, even though the filling sequence was reverse. Figure 40(b) shows the porosity plot for the in-line gating design. The varying degree and size of porosity validates the fact that even though each part is exactly the same, the location on the tree greatly influences the solidification rate and porosity.

Example 6: Ring Gate. To minimize the erratic flow and solidification pattern, a circular gating design is introduced. A ring gate on the top fed the castings from a central sprue. Due to the symmetric nature of the gating design, the flow pattern for the entire casting is very similar. Also, the fill rate for each part on the tree is the same, so the effects seen in the in-line design, where the central casting filled much earlier than the others, is not observed. Also, due to the circular ring gate design, the radiation view factor that each part "sees" is the same for each part, leading to similar behavior in radiative and reflective heat from each other. An additional benefit of using a ring gate was that this design accommodated 12 castings on the tree instead of 9 in the in-line design. Figure 41 shows the temperature, solidification, and porosity plots of the ring gate design at a critical time in the cooling process. Although the filling pattern showed improvement with the ring gate design, the solidification pattern was similar to the one observed in the in-line design. The porosity magnitude was reduced by approximately 30% compared to the worst porosity observed in some of the in-line design castings.

Example 7: Selective Insulation. The casting engineer has a few variables with which to adjust and optimize when considering the rigging design. The preceding example focused on the configuration of the tree: how to place the various parts in reference to each other. The other controllable parameters are the emissivity of the shell faces, the temperature surrounding the shell (ambient or controlled enclosure temperature), and the emissivity of the casting environment (open-air cooling versus cooling in some chamber). Going to an extreme case, radiation can be eliminated by burying the shell in a sand bed. Figure 42 displays the cooling pattern on the shell for the in-line casting design when local insulation is applied on the shell. By putting a 1.2 to 2.5 cm (0.5 to 1 in.) kaowool (aluminosilicate) insulation on the critical regions to enhance the feeding of the casting, the temperature in

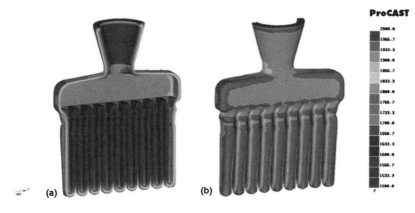

ProCAST

Fig. 39 Test bar casting temperatures (°F). (a) Cooldown inside. (b) Cooldown outside

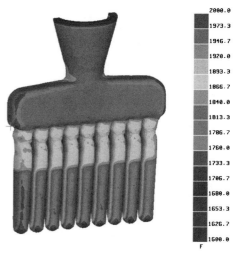

Fig. 42 Test bar temperature plot with local insulation

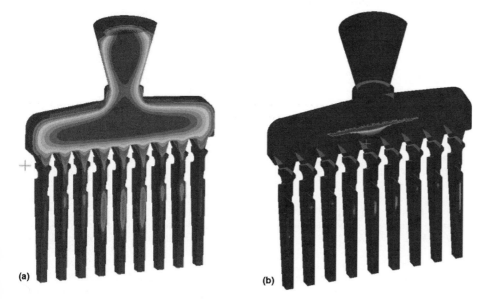

Fig. 40 Test bar. (a) Temperature plot. (b) Porosity plot

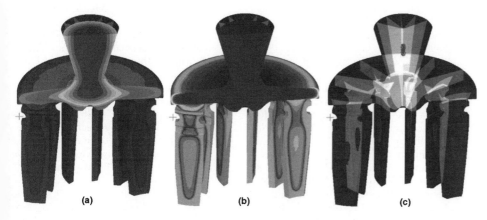

Fig. 41 Ring gate design at a critical time in the cooling process. (a) Temperature. (b) Solidification. (c) Porosity

any area of the shell can be effectively controlled by reducing the radiation heat-transfer loss through that region.

To balance the cooling of the exterior faces of the shell compared to the interior shell locations, a can or other enclosing structure can be used to reflect radiation. Technologically advanced chambers, such as those used in single-crystal casting, are even engineered to force certain solidification behaviors and resultant microstructures by controlling the amount of radiation to and from the casting. With a can, the temperature is not controlled; however, the can itself holds in heat by reflecting some of the heat back onto the part. With a casting chamber, the enclosure emissivity is controlled with specific chamber wall materials, and the temperature can be controlled with heaters or cooling systems.

Other than the examples presented previously, current casting simulation technology has been applied extensively in other casting processes as well, such as sand, shell, semisolid, gravity permanent mold, lost foam, squeeze, continuous, and centrifugal castings.

Conclusions

Modeling of casting and solidification has been used extensively in foundries to solve routine production problems. Casting defects, such as those related to filling, solidification, stress, and microstructure, can be predicted with confidence thanks to comprehensive models and the ability to compute thermophysical and mechanical properties of multicomponent alloys. As always, better understanding and accurate material properties, including mold materials, lead to improved predictive capabilities. Further development efforts should emphasize the enhancement of the accuracy of predicting and eliminating various casting defects. Coupling heat treatment simulation with casting simulation can then predict the final mechanical properties of the part in service.

REFERENCES

1. K.-O. Yu, *Modeling for Casting and Solidification Processing*, Marcel Dekker, Inc., 2002
2. Y.A. Chang, S. Chen, F. Zhang, X. Yan, F. Xie, R. Schmid-Fetzer, and W.A. Oates, Phase Diagram Calculation: Past,

Present and Future, *Prog. Mater. Sci.*, Vol 49, 2004, p 313–345

3. H.L. Lukas, J. Weiss, and E.Th. Henig, Strategies for the Calculation of Phase Diagrams, *CALPHAD*, Vol 6 (No. 3), 1982, p 229–251

4. U.R. Kattner, The Thermodynamic Modeling of Multicomponent Phase Equilibria, *JOM*, Vol 49, 1997, p 14–19

5. N. Saunders and A.P. Miodownik, *CALPHAD: Calculation of Phase Diagrams: A Comprehensive Guide*, Elsevier Science Ltd, New York, NY, 1998, p 299–411

6. J. Guo and M.T. Samonds, Property Prediction with Coupled Macro-Micromodeling and Computational Thermodynamics, *Model. Casting Solid. Proc.*, H. Weng-Sing and Kaohsiung, Ed., Taiwan, 2004, p 157–164

7. J. Guo and M. Samonds, Thermophysical Property Prediction Coupled Computational Thermodynamics with Back Diffusion Consideration, *J. Phase Equilib. Diff.*, Vol 28 (No. 1), Feb 2007, p 58–63

8. J. Guo and M. Samonds, The Thermophysical and Mechanical Property Prediction of Multicomponent Alloys by the Coupling of Macro-Micromodeling and Computational Thermodynamics, *Modeling of Casting, Welding and Advanced Solidification Processes XI* (Opio, France), 2006

9. N. Saunders, Z. Guo, X. Li, A.P. Miodownik, and J.-Ph. Schille, Using JMatPro to Model Materials Properties and Behavior, *JOM*, Dec 2003, p 65

10. H.D. Brody and M.C. Flemings, Solute Redistribution during Dendrite Solidification, *Trans. Met. Soc. AIME*, Vol 236, 1966, p 615

11. T.W. Clyne and W. Kurz, Solute Redistribution during Solidification with Rapid Solid State Diffusion, *Metall. Trans. A*, Vol 12, 1981, p 965

12. I. Ohnaka, Mathematical Analysis of Solute Redistribution during Solidification with Diffusion in Solid Phase, *Trans. ISIJ*, Vol 26, 1986, p 1045–1052

13. C.Y. Wang and C. Beckermann, Unified Solute Diffusion Model for Columnar and Equiaxed Dendritic Alloy Solidification, *Mater. Sci. Eng.*, Vol 171, 1993, p 199

14. L. Nastac and D.M. Stefanescu, An Analytical Model for Solute Redistribution during Solidification of Planar, Columnar and Equiaxed Morphology, *Metall. Trans.*, Vol 24, 1993, p 2107

15. V.R. Voller and C. Beckermann, A Unified Model of Microsegregation and Coarsening, *Metall. Trans.*, Vol 30, 1999, p 2183

16. V.R. Voller, On a General Back-Diffusion Parameter, *J. Cryst. Growth*, Vol 226, 2001, p 562–568

17. M.C. Schneider, J.P. Gu, C. Beckermann, W.J. Boettinger, and U.R. Kattner, Modeling of Micro- and Macrosegregation and Freckle Formation in Single-Crystal Nickel-Base Superalloy Directional Solidification, *Metall. Mater. Trans. A*, Vol 28, 1997, p 1517–1531

18. X. Yan, "Thermodynamic and Solidification Modeling Coupled with Experimental Investigation of the Multicomponent Aluminum Alloys," Ph.D. thesis, University of Wisconsin-Madison, Madison, WI, 2001

19. C.Y. Wang and C. Beckermann, A Multiphase Solute Diffusion Model for Dendritic Alloy Solidification, *Met. Trans. A*, Vol 24, 1993, p 2787–2802

20. X.L. Yang, P.D. Lee, R.F. Brooks, and R. Wunderlich, The Sensitivity of Investment Casting Simulations to the Accuracy of Thermophysical Property Values, *Superalloys*, The Minerals, Metals & Materials Society, 2004

21. B. Alkan, R. Karabulut, and B. Unal, Electrical Resistivity of Liquid Metals and Alloys, *Acta Phys. Pol. A*, Vol 102, 2002, p 385–400

22. P.L. Rossiter, *The Electrical Resistivity of Metals and Alloys*, Cambridge University Press, New York, NY, 1987, p 137–272

23. A. Rudajevova, M. Stanek, and P. Lukac, Determination of Thermal Diffusivity and Thermal Conductivity of Mg-Al Alloys, *Mat. Sci. Eng. A*, Vol 341, 2003, p 152–157

24. P.G. Klemens and R.K. Williams, Thermal Conductivity of Metals and Alloys, *Int. Mat. Rev.*, Vol 31 (No. 5), 1986, p 197–215

25. D. Sichen, J. Bygden, and S. Seetharaman, A Model for Estimation of Viscosities of Complex Metallic and Ionic Melts, *Met. Mat. Trans. B*, Vol 25, Aug 1994, p 519–525

26. N. Saunders, X. Li, A.P. Miodownik, and J.P. Schille, Modeling of the Thermophysical and Physical Properties for Solidification of Al-Alloys, *Light Metals*, 2003, p 999–1004

27. ProCAST user's manual and technical references, ESI, 2008

28. M. Rappaz and W.J. Boettinger, On Dendritic Solidification of Multicomponent Alloys with Unequal Liquid Diffusion Coefficients, *Acta Mater.*, Vol 47 (No. 11), 1999, p 3205–3219

29. W.J. Boettinger, S.R. Coriell, A.L. Greer, A. Karma, W. Kurz, M. Rappaz, and R. Trivedi, Solidification Microstructures: Recent Developments, Future Directions, *Acta Mater.*, Vol 48, 2000, p 43–70

30. J. Guo and M. Samonds, Microstructures and Mechanical Properties Prediction for Ti Based Alloys, *J. Mater. Eng. Perform.*, May 17, 2007

31. J. Guo and M. Samonds, Modeling of Microstructures and Mechanical Properties of α + β Ti Based Alloys, *Mater. Sci. Technol.*, Vol 3, 2005, p 41–50

32. A. Kermanpur, M. Mehrara, N. Varahram, and P. Davami, Improvement of Grain Structure and Mechanical Properties of a Land Based Gas Turbine Blade Directionally Solidified with Liquid Metal Cooling Process, *Mater. Sci. Technol.*, Vol 24 (No. 1), Jan 2008, p 100–106

33. I. Steinbach, C. Beckermann, B. Kauerauf, Q. Li, and J. Guo, Three-Dimensional Modeling of Equiaxed Dendritic Growth on a Mesoscopic Scale, *Acta Mater.*, Vol 47, 1999, p 971–982

34. M. Rappaz and P.H. Thevoz, Solute Diffusion Model for Equiaxed Dendritic Growth, *Acta Metall.*, Vol 35, 1987, p 1487–1497

35. M. Rappaz and P.H. Thevoz, Solute Diffusion Model for Equiaxed Dendritic Growth: Analytical Solution, *Acta Metall.*, Vol 35, 1987, p 2929–2933

36. P. Thevoz, J.L. Desbiolles, and M. Rappaz, Modeling of Equiaxed Microstructure Formation in Casting, *Metall. Trans. A*, Vol 20, 1989, p 311–322

37. M. Rappaz, Modelling of Microstructure Formation in Solidification Processes, *Int. Mater. Rev.*, Vol 34, 1989, p 93

38. C.Y. Wang and C. Beckermann, Equiaxed Dendritic Solidification with Convection: Part I. Multiscale/Multiphase Modeling, *Metall. Mater. Trans. A*, Vol 27, 1996, p 2754–2783

39. Ch.-A. Gandin and M. Rappaz, Coupled Finite Element-Cellular Automation Model for the Prediction of Dendritic Grain Structures in Solidification Processes, *Acta Metall. Mater.*, Vol 42, 1994, p 2233–2246

40. Ch.-A. Gandin, J.-L. Desbiolles, M. Rappaz, and P. Thevoz, Three-Dimensional Cellular Automaton-Finite Element Model for the Prediction of Solidification Grain Structures, *Metall. Mater. Trans. A*, Vol 30, 1999, p 3153–3165

41. H. Takatani, Ch.-A. Gandin, and M. Rappaz, EBSD Characterization and Modelling of Columnar Dendritic Grains Growing in the Presence of Fluid Flow, *Acta Mater.*, Vol 48, 2000, p 675–688

42. G. Guillemot, Ch.-A. Gandin, and H. Combeau, Modeling of Macrosegregation and Solidification Grain Structures with a Coupled Cellular Automaton-Finite Element Model, *Solidification Processes and Microstructures—A Symposium in Honor of Wilfried Kurz*, TMS, Warrendale, PA, 2004, p 157–163

43. N. Ahmad, H. Combeau, J.-L. Desbiolles, T. Jalanti, G. Lesoult, J. Rappaz, M. Rappaz, and C. Stomp, Numerical Simulation of Macrosegregation: A Comparison between Finite Volume Method and Finite Element Method Predictions and a Confrontation with Experiments, *Metall. Mater. Trans. A*, Vol 29, 1998, p 617–630

44. J.-L. Desbiolles, P. Thevoz, and M. Rappaz, Micro-/Macrosegregation Modeling

in Casting: A Fully Coupled 3D Model, *Modeling of Casting, Welding and Advanced Solidification Processes X*, TMS, Warrendale, PA, 2003, p 245–252

45. M. Bellet, V.D. Fachinotti, S. Gouttebroze, W. Liu, and H. Combeau, A 3D-FEM Model Solving Thermomechanics and Macrosegregation in Binary Alloys Solidification, *Solidification Processes and Microstructures—A Symposium in Honor of Wilfried Kurz*, TMS, Warrendale, PA, 2004, p 41–46

46. D.J. Hebditch and J.D. Hunt, Observations of Ingot Macrosegregation on Model Systems, *Metall. Trans.*, Vol 5, 1974, p 1557

47. W. Wang, A. Kermanpur, P.D. Lee, and M. McLean, Simulation of Dendritic Growth in the Platform Region of Single Crystal Superalloy Turbine Blades, *J. Mater. Sci.*, Vol 38, 2003, p 4385–4391

48. J.S. Langer, Existence of Needle Crystals in Local Models of Solidification, *Phys. Rev. A*, Vol 33, 1986, p 435

49. D.A. Kessler and H. Levine, Velocity Selection in Dendritic Growth, *Phys. Rev. B*, Vol 33, 1986, p 7867

50. R.J. Braun, G.B. McFadden, and S.R. Coriell, Morphological Instability in Phase-Field Models of Solidification, *Phys. Rev. E*, Vol 49, 1994, p 4336–4352

51. J.A. Warren and W.J. Boettinger, Prediction of Dendritic Growth and Microsegregation Patterns in a Binary Alloy Using the Phase-Field Method, *Acta Mater.*, Vol 43, 1995, p 689–703

52. W.J. Boettinger and J.A. Warren, The Phase-Field Method: Simulation of Alloy Dendritic Solidification during Recalescence, *Metall. Mater. Trans. A*, Vol 27, 1996, p 657–669

53. I. Loginova, G. Amberg, and J. Agren, Phase-Field Simulation of Nonisothermal Binary Alloy Solidification, *Acta Mater.*, Vol 49, 2001, p 573–581

54. W.J. Boettinger and J.A. Warren, Simulation of the Cell to Plane Front Transition during Directional Solidification at High Velocity, *J. Cryst. Growth*, Vol 200, 1999, p 583–591

55. H.-J. Diepers, D. Ma, and I. Steinbach, History Effects during the Selection of Primary Dendrite Spacing. Comparison of Phase-Field Simulations with Experimental Observations, *J. Cryst. Growth*, Vol 237–239, 2002, p 149–153

56. J. Tiaden and U. Grafe, A Phase-Field Model for Diffusion and Curvature Controlled Phase Transformations in Steels, *Proceedings of the International Conference on Solid-Solid Phase Transformations '99 (JIMIC)*, M. Koiwa, K. Otsuka, and T. Miyazaki, Ed.

57. A. Karma and W.J. Rappel, Phase-Field Method for Computationally Efficient Modeling of Solidification with Arbitrary Interface Kinetics, *Phys. Rev. E*, Vol 53, 1996, p 3017–3020

58. I. Steinbach, B. Kauerauf, C. Beckermann, J. Guo, and Q. Li, Three Dimensional Modeling of Equiaxed Dendritic Growth on a Mesoscopic Scale, *Proceedings of the Eighth International Conference on Modeling of Casting and Welding Processes* (San Diego, CA), B. G. Thomas and C. Beckermann, Ed., 1998, p 565–572

59. I. Steinbach, F. Pezzolla, B. Nestler, M. Seelberg, R. Prieler, G.J. Schmitz, and J. L.L. Rezende, A Phase Field Concept for Multiphase Systems, *Physica D*, Vol 94, 1996, p 135–147

60. J. Tiaden, B. Nestler, H.J. Diepers, and I. Steinbach, The Multiphase-Field Model with an Integrated Concept for Modelling Solute Diffusion, *Physica D*, Vol 115, 1998, p 73–86

61. J.Q. Broughton, A. Bonissent, and F.F. Abraham, The FCC (111) and (100) Crystal-Melt Interfaces: A Comparison by Molecular Dynamics Simulation, *J. Chem. Phys.*, Vol 74, 1981, p 4029

62. A.A. Wheeler, W.J. Boettinger, and G.B. McFadden, Phase-Field Model for Isothermal Phase Transitions in Binary Alloys, *Phys. Rev. A*, Vol 45, 1992, p 7424–9439

63. G. Caginalp and W. Xie, Phase-Field and Sharp-Interface Alloy Models, *Phys. Rev. E*, Vol 48, 1993, p 1897–1910

64. W.J. Boettinger, J.A. Warren, C. Beckermann, and A. Karma, Phase-Field Simulation of Solidification, *Ann. Rev. Mater. Res.*, Vol 32, 2002, p 163–194

65. W.J. Boettinger, A.A. Wheeler, B.T. Murray, G.B. McFadden, and R. Kobayashi, Calculation of Alloy Solidification Morphologies Using the Phase-Field Method, *Modeling of Casting, Welding and Advanced Solidification Processes*, 1993, p 79

66. S.G. Kim, W.T. Kim, and T. Suzuki, Phase-Field Model for Binary Alloys, *Phys. Rev. E*, Vol 60, 1999, p 7186

67. A. Karma, Y.H. Lee, and M. Plapp, Three-Dimensional Dendrite-Tip Morphology at Low Undercooling, *Phys. Rev. E*, Vol 61, 2000, p 3996

68. I. Steinbach and F. Pezzolla, A Generalized Field Method for Multiphase Transformations Using Interface Fields, *Physica D*, Vol 134, 1999, p 385

69. J.C. Ramirez, C. Beckermann, A. Karma, and H.-J. Diepers, Phase-Field Modeling of Binary Alloy Solidification with Coupled Heat and Solute Diffusion, *Phys. Rev. E*, Vol 69 (No. 52), May 2004, p 051607-1 to 051607-16

70. J. Guo, Modeling and Experimental Validation of Cast Iron Solidification, *Fifth Decennial International Conference on Solidification Processing SP07*, July 23–25 2007 (University of Sheffield, U.K.)

71. D.M. Stefanescu, Solidification and Modeling of Cast Iron—A Short History of the Defining Moments, *Mater. Sci. Eng. A*, Vol 413–414, 2005, p 322–333

72. W. Oldfield, A Qualitative Approach to Casting Solidification, *ASM Trans.*, Vol 59, 1966, p 945

73. D.M. Stefanescu and S. Trufinescu, Kinetics of Solidification of Gray Iron, *Z. Metall.*, Vol 9, 1974, p 610

74. H. Fredriksson and L. Svensson, Computer Simulation of Structure Formation and Segregation during the Solidification of Cast Iron, *The Physical Metallurgy of Cast Iron*, H. Fredriksson and M. Hillert, Ed., Elsevier, 1985, p 273–284

75. H. Fredriksson and L. Svensson, Simulation of Grey Cast-Iron Solidification in a Shaped Casting, *Solidification Processing of Eutectic Alloys*, D.M. Stefanescu, G.J. Abbaschian, and R.J. Bayuzick, Ed., The Metallurgical Society, Warrendale, PA, 1988, p 153–162

76. D.M. Stefanescu and C. Kanetkar, Computer Modeling of the Solidification of Eutectic Alloys: The Case of Cast Iron, *Computer Simulation of Micro-Structural Evolution*, D.J. Srolovitz, Ed., The Metallurgical Society, Warrendale, PA, 1985, p 171–188

77. J. Lacaze, M. Castro, C. Selig, and G. Lesoult, Solidification of Spheroidal Graphite Cast Irons, *Modeling of Casting, Welding and Advanced Solidification Processes V*, M. Rappaz, Ed., The Metallurgical Society, Warrendale, PA, 1991, p 473–478

78. E. Fras, W. Kapturkiewicz, and A.A. Burbielko, Micro-Macro Modeling of Casting Solidification Controlled by Transient Diffusion and Undercooling, *Modeling of Casting, Welding and Advanced Solidification Processes VII*, M. Croos and J. Campbell, Ed., The Metallurgical Society, Warrendale, PA, 1995, p 679–686

79. K.C. Su, I. Ohnaka, I. Yamauchhi, and T. Fukusako, Modelling of Solidified Structure of Castings, *The Physical Metallurgy of Cast Iron*, H. Fredriksson and M. Hillert, Ed., Elsevier, 1985, p 181–189

80. S. Chang, D. Shangguan, and D.M. Stefanescu, Prediction of Microstructural Evolution in SG Cast Iron from Solidification to Room Temperature, *Metall. Trans. A*, Vol 22, 1991, p 915

81. D.M. Stefanescu and C.S. Kanetkar, Computer Modeling of the Solidification of Eutectic Alloys: Comparison of Various Models for Eutectic Growth of Cast Iron, *AFS Trans.*, 1987, p 139–144

82. L. Nastac and D.M. Stefanescu, Prediction of Gray-to-White Transition in Cast Iron by Solidification Modeling, *AFS Trans.*, Vol 103, 1995, p 329–337

83. K.M. Pedersen, J.H. Hattel, and N. Tiedje, Numerical Modelling of Thin-Walled Hypereutectic Ductile Cast Iron Parts, *Acta Mater.*, Vol 54 (No. 19), Nov 2006, p 5103–5114

84. G. Lesoult, M. Castro, and J. Lacaze, Solidification of Spheroidal Graphite Cast Irons—I. Physical Modelling, *Acta Mater.*, Vol 46 (No. 3), 1998, p 983–995

85. G. Lesoult, M. Castro, and J. Lacaze, Solidification of Spheroidal Graphite Cast Irons—II. Numerical Simulation, *Acta Mater.*, Vol 46 (No. 3), 1998, p 997–1010

86. F.J. Bradley, A Stereological Formulation for the Source Term in Micromodels of Equiaxed Eutectic Solidification, *Metall. Trans. B*, Vol 24, 1993, p 539

87. R. Vijayaraghavan and F.J. Bradley, Micro-Model for Eutectoid Phase Transformations in As-Cast Ductile Iron, *Scr. Mater.*, Vol 41 (No. 11), 1999, p 1247–1253

88. D. Venugopalan, Prediction of Matrix Microstructure in Ductile Iron, *Trans. Am. Foundrymen's Soc.*, 1990, p 465

89. Q. Chen, E.W. Langer, and P.N. Hansen, Comparative Study on Kinetic Models and Their Effects on Volume Change during Eutectic Reaction of S.G. Cast Iron, *Scand. J. Metall.*, Vol 24, 1995, p 48–62

90. M.I. Onsoien, O. Grong, O. Gundersen, and T. Skaland, Process Model for the Microstructure Evolution in Ductile Cast Iron: Part I. The Model, *Metall. Mater. Trans. A*, Vol 30, 1999, p 1053–1068

91. M.I. Onsoien, O. Grong, O. Gundersen, and T. Skaland, Process Model for the Microstructure Evolution in Ductile Cast Iron: Part II. Applications of the Model, *Metall. Mater. Trans. A*, Vol 30, 1999, p 1069–1079

92. J.R. Davis, *ASM Specialty Handbook: Cast Irons*, ASM International, 1996

93. V. Krutiš and J. Roučka, "Shrinkage of Graphitic Cast Irons," internal report, ESI Group

94. J. Campbell, *Castings*, Butterworth-Heinemann, Oxford, U.K., 1991

95. A.V. Kuznetsov and K. Vafai, Development and Investigation of Three-Phase Model of the Mushy Zone for Analysis of Porosity Formation in Solidifying Castings, *Int. J. Heat Mass Transfer*, Vol 38 (No. 14), Sept 1995, p 2557–2567

96. G.K. Sigworth and C. Wang, Mechanisms of Porosity Formation during Solidification: A Theoretical Analysis, *Metall. Mater. Trans. B*, Vol 24, April 1993, p 349–364

97. D.R. Poirier, K. Yeum, and A.L. Maples, A Thermodynamic Prediction for Microporosity Formation in Aluminum-Rich Al-Cu Alloys, *Metall. Mater. Trans. A*, Vol 18, Nov 1987, p 1979–1987

98. H. Combeau, D. Carpentier, J. Lacaze, and G. Lesoult, Modelling of Microporosity Formation in Aluminum Alloys Castings, *Mater. Sci. Eng. A*, Vol 173, 1993, p 155–159

99. P.K. Sung, D.R. Poirier, S.D. Felicelli, E.J. Poirier, and A. Ahmed, Simulations of Microporosity in IN718 Equiaxed Investment Castings, *J. Cryst. Growth*, Vol 226, 2001, p 363–377

100. J.D. Zhu and I. Ohnaka, Modeling of Microporosity Formation in A356 Aluminum Alloy Casting, *Modeling of Casting, Welding and Advanced Solidification Processes V*, 1991, p 435–442

101. S. Sabau and S. Viswanathan, Microporosity Prediction in Aluminum Alloy Castings, *Metall. Mater. Trans. B*, Vol 33, April 2002, p 243–255

102. D. See, R.C. Atwood, and P.D. Lee, A Comparison of Three Modeling Approaches for the Prediction of Microporosity in Aluminum-Silicon Alloys, *J. Mater. Sci.*, Vol 36, 2001, p 3423–3435

103. R.C. Atwood and P.D. Lee, A Three-Phase Model of Hydrogen Pore Formation during the Equiaxed Dendritic Solidification of Aluminum-Silicon Alloys, *Metall. Mater. Trans. B*, Vol 33, April 2002, p 209–221

104. P.D. Lee, R.C. Atwood, R.J. Dashwood, and H. Nagaumi, Modeling of Porosity Formation in Direct Chill Cast Aluminum-Magnesium Alloys, *Mater. Sci. Eng. A*, Vol 328, 2002, p 213–222

105. S.D. Felicelli, D.R. Poirier, and P.K. Sung, Model for Prediction of Pressure and Redistribution of Gas-Forming Elements in Multicomponent Casting Alloys, *Metall. Mater. Trans. B*, Vol 31, Dec 2000, p 1283–1292

106. P.N. Anyalebechi, Analysis and Thermodynamic Prediction of Hydrogen Solution in Solid and Liquid Multicomponent Aluminum Alloys, *Light Metals*, 1998, p 827–842

107. K. Kubo and R.D. Pehlke, Mathematical Modeling of Porosity Formation in Solidification, *Metall. Trans. B*, Vol 16, June 1985, p 359–366

108. B. Chalmers, *Principles of Solidification*, John Wiley & Sons, Inc., 1964

109. E. Niyama, T. Uchida, M. Morikama, and S. Saito, Predicting Shrinkage in Large Steel Castings from Temperature Gradient Calculations, *AFS Int. Cast Met. J.*, Vol 6, 1981, p 16–22

110. Y.W. Lee, E. Chang, and C.F. Chieu, Modeling of Feeding Behavior of Solidifying Al-7Si-0.3Mg Alloy Plate Casting, *Metall. Trans. B*, Vol 21, 1990, p 715–722

111. V.K. Suri, A.J. Paul, N. El-Kaddah, and J.T. Berry, Determination of Correlation Factors for Prediction of Shrinkage in Castings—Part I: Prediction of Microporosity in Castings; A Generalized Criterion (AFS Research), *AFS Trans.*, Vol 103, 1995, p 861–867

112. R.P. Taylor, H.T. Berry, and R.A. Overfelt, Parallel Derivation and Comparison of Feeding-Resistance Porosity Criteria Functions for Castings, HTD-Vol 323, *National Heat Transfer Conference ASME*, Vol 1, 1996, p 69–77

113. M.C. Flemings and G.E. Nereo, Macrosegregation: Part I, *Trans. Metall. Soc. AIME*, Vol 239, 1967, p 1449–1461

114. M.C. Flemings, R. Mehrabian, and G.E. Nereo, Macrosegregation, Part II, *Trans. Metall. Soc. AIME*, Vol 242, Jan 1968, p 41–49

115. M.C. Flemings and G.E. Nereo, Macrosegregation, Part III, *Trans. Metall. Soc. AIME*, Vol 242, Jan 1968, p 50–56

116. C. Beckermann, Modeling of Macrosegregation: Application and Future Needs, *Int. Mater. Rev.*, Vol 47 (No. 5), 2002, p 243–261

117. A.F. Giamei and B.H. Kear, On the Nature of Freckles in Nickel Base Superalloys, *Metall. Trans.*, Vol 1, 1970, p 2185–2192

118. S.M. Copley, A.F. Giamei, S.M. Johnson, and M.F. Hornbecker, The Origin of Freckles in Unidirectionally Solidified Castings, *Metall. Trans.*, Vol 1, 1970, p 2193–2204

119. A. Hellawell, J.R. Sarazin, and R.S. Steube, Channel Convection in Partly Solidified Systems, *Phil. Trans. R. Soc. (London) A*, Vol 345, 1993, p 507–544

120. M.G. Worster, Convection in Mushy Layers, *Ann. Rev. Fluid Mech.*, Vol 29, 1996, p 91–122

121. T.M. Pollock and W.H. Murphy, The Breakdown of Single-Crystal Solidification in High Refractory Nickel-Base Alloys, *Metall. Mater. Trans. A*, Vol 27, 1996, p 1081–1094

122. J. Guo, "Three-Dimensional Finite Element Simulation of Transport Phenomena in Alloy Solidification," The University of Iowa, Iowa City, Iowa, 2000

123. J. Guo and C. Beckermann, Three-Dimensional Simulation of Freckle Formation during Binary Alloy Solidification: Effect of Mesh Spacing, *Numer. Heat Transf. A*, Vol 44, 2003, p 559–576

124. S.D. Felicelli, J.C. Heinrich, and D.R. Poirier, Simulation of Freckles during Vertical Solidification of Binary Alloys, *Metall. Trans. B*, Vol 22, 1991, p 847–859

125. S.D. Felicelli, D.R. Poirier, A.F. Giamei, and J.C. Heinrich, Simulating Convection and Macrosegregation in Superalloys, *JOM*, Vol 49 (No. 3), March 1997, p 21–25

126. P.J. Prescott and F.P. Incropera, Numerical Simulation of a Solidifying Pb-Sn Alloy: The Effects of Cooling Rate on Thermosolutal Convection and Macrosegregation, *Metall. Trans. B*, Vol 22, 1991, p 529–540

127. C.S. Magirl and F.P. Incropera, Flow and Morphological Conditions Associated with Unidirectional Solidification of Aqueous Ammonium Chloride, *ASME J. Heat Transf.*, Vol 115, 1993, p 1036–1043

128. D.G. Neilson and F.P. Incropera, Effect of Rotation on Fluid Motion and Channel Formation during Unidirectional

Solidification of a Binary Alloy, *Int. J. Heat Mass Transf.*, Vol 36, 1993, p 489–505

129. D.G. Neilson and F.P. Incropera, Three-Dimensional Considerations of Unidirectional Solidification in a Binary Liquid, *Numer. Heat Transf. A*, Vol 23, 1993, p 1–20

130. H.-W. Huang, J.C. Heinrich, and D.R. Poirier, Numerical Anomalies in Simulating Directional Solidification of Binary Alloys, *Numer. Heat Transf. A*, Vol 29, 1993, p 639–644

131. S.D. Felicelli, D.R. Poirier, and J.C. Heinrich, Macrosegregation Patterns in Multicomponent Ni-Base Alloys, *J. Cryst. Growth*, Vol 177, 1997, p 145–161

132. S.D. Felicelli, J.C. Heinrich, and D.R. Poirier, Finite Element Analysis of Directional Solidification of Multicomponent Alloys, *Int. J. Numer. Meth. Fluids*, Vol 27, 1998, p 207–227

133. S.D. Felicelli, J.C. Heinrich, and D.R. Poirier, Three-Dimensional Simulations of Freckles in Binary Alloys, *J. Cryst. Growth*, Vol 191, 1998, p 879–888

134. S.D. Felicelli, D.R. Poirier, and J.C. Heinrich, Modeling Freckle Formation in Three Dimensions during Solidification of Multicomponent Alloys, *Metall. Mater. Trans. B*, Vol 29, 1998, p 847–855

135. J. Guo and J.Z. Zhu, Hot Tearing Prediction for Magnesium Alloy Castings, *Mater. Sci. Forum*

136. J. Guo and J.Z. Zhu, Hot Tearing Indicator for Multi-Component Alloys Solidification, *Fifth Decennial International Conference on Solidification Processing SP07*, July 23–25, 2007 (University of Sheffield, U.K.)

137. A. Stangeland, A. Mo, M. M'Hamdi, D. Viano, and C. Davidson, Thermal Strain in the Mushy Zone Related to Hot Tearing, *Metall. Mater. Trans. A, Phys. Metall. Mater. Sci.*, Vol 37 (No. 3), March 2006, p 705–714

138. D.G. Eskin, Suyitno, and L. Katgerman, Mechanical Properties in the Semi-Solid State and Hot Tearing of Aluminium Alloys, *Prog. Mater. Sci.*, Vol 49 (No. 5), 2004, p 629–711

139. C. Monroe and C. Beckermann, Development of a Hot Tear Indicator for Steel Castings, *Mater. Sci. Eng. A*, Vol 413–414, Dec 15, 2005, p 30–36

140. B. Magnin, L. Maenner, L. Katgermann, and S. Engler, Ductility and Rheology of an Al-4.5% Cu Alloy from Room Temperature to Coherency Temperature, *Mater. Sci. Forum*, Vol 1209, 1996, p 217–222

141. J. Zhang and R.F. Singer, Hot Tearing of Nickel-Based Superalloys during Directional Solidification, *Acta Mater.*, Vol 50 (No. 7), April 19, 2002, p 1869–1879

142. M. Rappaz, J.M. Drezet, and M. Gremaud, New Hot-Tearing Criterion, *Metall. Mater. Trans. A*, Vol 30, 1999, p 449

143. J. Guo and M. Samonds, Microporosity Simulations in Multicomponent Alloy Castings, *Modeling of Casting, Welding and Advanced Solidification Processes X* (Sandestin, FL), D.M. Stefanescu et al., Ed., 2003, p 303–310

144. Z.L. Zhang, C. Thaulow, and J. Odegard, Complete Gurson Model Approach for Ductile Fracture, *Eng. Fract. Mech.*, Vol 67 (No. 2), Sept 2000, p 155–168

145. J. Langlais and J.E. Gruzleski, Novel Approach to Assessing the Hot Tearing Susceptibility of Aluminium Alloys, *Mater. Sci. Forum*, Vol 167, 2000, p 331–337

146. A.L. Gurson, Continuum Theory of Ductile Rupture by Void Nucleation and Growth: Part 1—Yield Criteria and Flow Rules for Porous Ductile Media, *J. Eng. Mater. Technol. (Trans. ASME)*, Vol 99, Ser H (No. 1), Jan 1977, p 2–15

147. V. Tvergaard and A. Needleman, Analysis of the Cup-Cone Fracture in a Round Tensile Bar, *Acta Metall.*, Vol 32, 1984, p 157

148. A. Needleman and V. Tvergaard, Analysis of Ductile Rupture Modes at a Crack Tip, *J. Mech. Phys. Solids*, Vol 35, 1987, p 151

149. O.C. Zienkiewicz, R.L. Taylor, and J.Z. Zhu, *The Finite Element Method: Its Basis and Fundamentals*, Elsevier, 2005

150. G. Cao, S. Kou, and Y.A. Chang, Hot Cracking Susceptibility of Binary Mg-Al Alloys, *Magnesium Technology 2006*, A. A. Luo, Ed., TMS, 2006, p 57–61

151. G. Cao and S. Kou, Hot Tearing of Ternary Mg-Al-Ca Alloy Castings, *Metall. Mater. Trans. A*, Vol 37, Dec 2006, p 3647–3663

152. J.C. Simo, and T.A. Laursen, An Augmented Lagrangian Treatment of Contact Problems Involving Friction, *Comput. Struct.*, Vol 42, 1992, p 97–116

153. M. Aloe, D. Lefebvre, A. Mackenbrock, A. Sholapurwalla, and S. Scott, "Advanced Casting Simulations," internal report, ESI Group, 2006

154. A. Sholapurwalla and S. Scott, "Effects of Radiation Heat Transfer on Part Quality Prediction," internal report, ESI Group, 2006

ASM Handbook, Volume 22B, *Metals Process Simulation*
D.U. Furrer and S.L. Semiatin, editors

Copyright © 2010, ASM International®
All rights reserved.
www.asminternational.org

Computational Analysis of the Vacuum Arc Remelting (VAR) and Electroslag Remelting (ESR) Processes

Kanchan M. Kelkar and Suhas V. Patankar, Innovation Research Inc.
Alec Mitchell, University of British Columbia
Ramesh S. Minisandram, ATI Allvac
Ashish D. Patel, Carpenter Technology Corporation

THE VACUUM ARC REMELTING (VAR) AND ELECTROSLAG REMELTING (ESR) processes are used for the production of high-performance titanium alloys, superalloys, and steels. Due to the complex nature of the process physics and high operating temperatures, a trial-and-error-based approach becomes expensive for process design and optimization. Computational analysis constitutes a scientific approach for evaluating the effects of process parameters on the quality of the final ingot produced and provides a cost-effective method for improving process performance. In this article, an overview of the available studies on computational modeling of the VAR and ESR processes is first described. The most comprehensive models for the analysis of these processes follow a similar approach. Therefore, with a view to describing the state-of-the-art in the modeling of the VAR and ESR processes in a comprehensive manner, this article focuses on a detailed discussion of computational models developed by the authors. These models involve axisymmetric analysis of the electromagnetic, flow, heat-transfer, and phase-change phenomena to predict the pool shape and thermal history of the ingot. Analysis of segregation of alloying elements during solidification that gives rise to macrolevel compositional nonuniformity in titanium alloy ingots is also described. Finally, the calculated thermal history of the ingot is used to analyze the metallurgical structure, and a criterion based on the Rayleigh number of the interdendritic convection is used to predict the probability of formation of freckles. (In unidirectionally solidified ingots, the channel segregation is called a freckle.)

Mathematical formulations of all the aforementioned phenomena are first described for the VAR and ESR processes. Important features of the control-volume-based computational method that address the unique aspects of the VAR and ESR processes for obtaining an efficient and robust solution of the governing equations are then discussed. Applications of the models to practical VAR and ESR processes are presented next, to illustrate the engineering benefits of the models. Finally, future work involving measurement of properties of alloys and slags, analysis of process variants, and extension to three dimensions that is needed for improving the predictive accuracy of the models and expanding their applicability is discussed.

Symbols used in this article are defined in Table 1.

Process Description and Physical Phenomena

Remelting processes involve the melting of previously melted metal into a water-cooled copper crucible, where it undergoes progressive solidification. The most commonly used industrial remelting processes are VAR and ESR. In the VAR process, shown schematically in Fig. 1(a), continuous melting of the consumable electrode is effected by striking a direct current (dc) arc between the electrode and the grounded mold. In the ESR process, shown schematically in Fig. 1(b) for a stationary mold, melting is achieved by immersing the electrode in a bath of molten slag that is electrically heated by passing an alternating current through it. The molten metal falls from the electrode face in the form of metal droplets through the lighter slag and collects in the metal pool in the crucible. Industrial ESR furnaces also use moving or stationary short-collared molds. Often, multiple stages of remelting are performed to refine the alloys and remove the inclusions that may be present in the original material.

The quality of the final ingots produced is determined by the behavior of the underlying physical phenomena, namely electromagnetics, fluid flow, heat transfer, macro/microsegregation, and inclusion motion. The electromagnetic effect common to all processes is the interaction of the current flowing through the ingot (and the slag in the ESR) with the self-induced magnetic field to produce Lorentz forces that influence the motion of the molten metal (and the slag in ESR). Further, in the ESR process, passage of current produces Joule heating in the slag due to its high resistivity. Fluid motion in the metal pool and the slag is induced by the buoyancy and Lorentz forces. High-current VAR processes commonly employ axially oscillating magnetic fields for stabilizing the plasma arc. This magnetic field gives rise to a swirling motion in the circumferential direction that significantly affects mixing in the molten pool. The flow in the molten pool (and the slag in ESR) is turbulent with a very nonuniform mixing, and the motion decays rapidly in the mushy region.

The ingot solidifies progressively due to heat transfer to the mold wall and initially to the base plate. The progressive shrinkage of the solidifying ingot from the mold creates a highly nonuniform heat loss over the mold wall. Segregation of the alloying elements occurs due to the differential partitioning of the solutes between the solid and the liquid phases. Macroscale chemical segregation in the solidified ingot, defined here as the compositional nonuniformity at a scale corresponding to the macroscopic dimensions of the ingot, results from the redistribution of the solute rejected/absorbed by the solid due to the macroscale motion within the pool. The interdendritic

Table 1 Symbols and expressions used in this article

Symbol	Meaning
B	Magnetic flux density, tesla (T)
c	Concentration of an alloying element
C_D	Drag coefficient
$\mathbf{g}$	Gravitational acceleration (m/s^2)
f_{imm}	Immobilization liquid fraction
f_s	Solid fraction
$\mathbf{F}_L$	Lorentz force (N/m^3)
H	Magnetic field intensity (A/m)
h	Sensible enthalpy (J/kg)
J	Current density (A/m^2)
j	In electrical equations $j = \sqrt{-1}$
k	Thermal conductivity (W/(m·K))
K_s	Effective segregation coefficient
l	Length scale (m)
r	Radial direction
Ra	Rayleigh number for interdendritic convection
S	Source term due to mass flux of alloying element (kg/(m^3·s))
Sc	Schmidt number
T	Temperature (K)
t	Time (s)
$\mathbf{u}$	Velocity (m/s)
x	Axial direction
Greek symbols	
ΔH	Latent heat (J/kg)
Γ	Diffusion coefficient (Pa·s)
ε	Turbulence dissipation (m^2/s^3)
φ	Electric potential; an electromagnetic variable (V)
μ	Dynamic viscosity (Pa·s)
μ_0	Permeability of free space (H/m)
μ_{eff}	Effective viscosity (Pa·s)
ρ	Density (kg/m^3)
σ	Electrical conductivity (A/(V·m))
ω	Frequency of current (1/s)
Subscript	
x	Axial direction
r	Radial direction
θ	Angular direction
l	Liquid
conjugate	Complex conjugate
turb	Turbulent value
molecular	Molecular value
i	Alloy element number
s	Solid
freckle	Value for freckle initiation
Superscript	
*	Critical value
1	Liquid
$\wedge$	Complex amplitude
Mathematical operators	
$\nabla\varphi$	Gradient, in rectangular coordinates $\nabla\phi = \left(\frac{\partial\phi}{\partial x}\right)\mathbf{i} + \left(\frac{\partial\phi}{\partial y}\right)\mathbf{j} + \left(\frac{\partial\phi}{\partial z}\right)\mathbf{k}$
$\nabla\cdot$	Divergence, in rectangular coordinates $\nabla\cdot(L\mathbf{i} + M\mathbf{j} + N\mathbf{k}) = \left(\frac{\partial L}{\partial x}\right) + \left(\frac{\partial M}{\partial y}\right) + \left(\frac{\partial N}{\partial z}\right)$
$\nabla\times$	Curl, in rectangular coordinates $\nabla\times(L\mathbf{i} + M\mathbf{j} + N\mathbf{k}) = \begin{vmatrix} \mathbf{i} & \mathbf{j} & \mathbf{k} \\ \frac{\partial}{\partial x} & \frac{\partial}{\partial y} & \frac{\partial}{\partial z} \\ L & M & N \end{vmatrix}$

convection and diffusion processes, which occur due to the prevailing nonuniform thermal and flow conditions, cause the solutes to be redistributed within the mushy region. As a result, microsegregation defects, such as freckles and white spots, can occur. The metallurgical structure of the remelted ingot depends on the temperature gradient and the thermal history of the solidifying metal. The behavior of an inclusion that enters the melt pool is governed by its density, which controls the buoyancy force, and its size and the motion in the molten pool, which govern the drag force. It is important that the inclusions spend sufficient time in the pool so as to be removed by floating and dissolution in the slag (ESR) or by

decomposition (VAR) or dissolution in the metal (titanium VAR) and not enter the solidified ingot.

Computational Modeling of Remelting Processes

Before the advent of computational modeling, the industry had empirically determined that in each process, a melting rate could not be exceeded without deterioration of the ingot macro- and microstructure. Although a qualitative understanding of the underlying mechanisms for such a maximum existed, quantifying or characterizing them in great detail required

expensive ingot cutups and hence was not feasible on a routine basis. Computational modeling offers a scientific approach for examining the effects of operating conditions and ingot sizes on the quality of the ingot produced. When verified, such models can be used to determine optimum processing conditions for various alloys and ingot sizes. Accurate computational models also enable evaluation of the feasibility and advantages of variants of the standard ESR and VAR processes for producing larger ingots and/or obtaining better control over the process performance. In the following section, an overview of the available computational studies and the models developed for the analysis of the ESR and VAR processes is provided.

The initial models for the processes were algebraic (Ref 1–3) and attempted to match predictions to the liquidus profiles observed by cutups in industrial ingots. The models were refined to demonstrate the use of dimensionless numbers in relating the pool volumes among various ingot sizes and melting rates, notably by researchers in the group working in the E.O. Paton Institute (Ref 4–8) not only for steels and nickel-base alloys but also for titanium alloys. These models were useful in estimating initial melt rates for various alloys, particularly when extending the melting program from one ingot diameter to another in the same alloy, but they did not provide any information about the thermal history that is necessary to predict actual metallurgical structure in solidified ingot.

The first models to use finite-difference or finite-element methods for predicting the temperature distribution within the ingots Ref 9–17) were "conduction-only" models that restricted their computational domain to the ingot alone and required specification of thermal conditions on all bounding surfaces. With these restrictions, the formulation of the models becomes essentially the same for both VAR and ESR processes. This approach leads to oversimplified models because, although the ingot/mold and ingot/baseplate heat-transfer coefficients can be measured experimentally with reasonable accuracy, the heat input on the top face is much more complicated and quite distinct for each process. In ESR, slag motion largely determines the energy distribution on the ingot top surface; in VAR, the arc motion has an analogous effect. Further, an added uncertainty in these early models arises because they treat the effect of motion in the pool on heat transfer through an enhanced conductivity of the liquid metal, which is not known and has to be guessed. As a result, these models can predict the liquidus isotherm reasonably well, but the temperature and fluid flow regime in the ingot are essentially unknown. In spite of these drawbacks, the models were used to analyze the segregation behavior of complex alloys (Ref 18, 19) and also to indicate the importance of understanding the flow parameters in both the slag and metal phases of the processes.

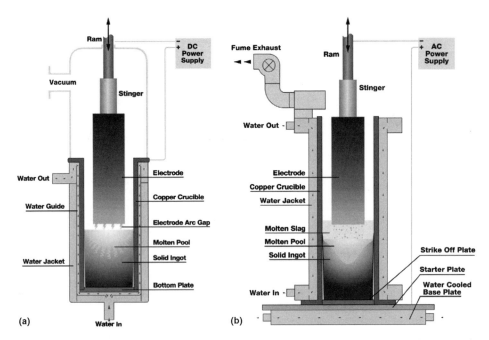

Fig. 1 Schematic diagram of the conventional (a) vacuum arc remelting and (b) electroslag remelting processes. Courtesy of ATI Allvac

The influence of fluid flow on the thermal behavior of the ingot was recognized at a very early stage in the process development. Several investigations (Ref 20–23) into the formation of ingot solidification defects identified uncontrolled electromagnetic stirring as the root cause. Consequently, models using a coupled calculation of electromagnetics based on the solution of Maxwell's equations, the flow field using Navier-Stokes equations, and the temperature field using energy conservation equations were first developed (Ref 24–36) for the ESR process. Early versions of these models focused on the flow in the slag and outlined the effect of changing the electrode/ingot diameter ratio. Later models expanded this concept and suggested that low-temperature analogs of the process (based on liquid mercury) could provide useful information on the process (Ref 37). Such modeling work was also carried out extensively to investigate the formation of liquid metal on the electrode tip as well as the current and voltage distributions in multielectrode ESR configurations (Ref 38–40). Limits in computational capacity restricted the extent to which these models could be applied. However, advances in computing power, along with the development of computational methods for the solution of flow and related transport equations, have spurred the development of computational models for the remelting processes that realistically simulate all physical phenomena in the entire system. A large number of recent studies (Ref 41–65) relate to the development, application, and experimental validation of the models for flow and thermal analysis and the discussion of utility of the corresponding results for understanding the solidification structures in both VAR and ESR, primarily of superalloys and, in some cases, steels.

General-purpose commercial computational fluid dynamics software programs, such as FLUENT (Ref 66) and Calcosoft (Ref 54), have been adapted to simulate the coupled electromagnetic, flow, and thermal phenomena in the ESR and VAR processes. Recent research involving rigorous analyses of the VAR (Ref 42–44, 53, 62, 63) and ESR (Ref 51, 57, 59, 64) processes have resulted in the most advanced computational models. These models have been implemented in specialized software programs, such as Solar (Ref 42), BAR (Ref 43), and MeltFlow-VAR (Ref 63, 67), for the analysis of the VAR process for titanium alloys, and MeltFlow-ESR (Ref 51, 64, 67) for the analysis of the ESR process for superalloys and steels. The results from some of these models have also been used in conjunction with solidification models (largely based on the cellular automata techniques) to predict the growth of the columnar dendritic zone in nickel alloy IN 718 melting, successfully comparing the model predictions with observations in actual ingot macrosections (Ref 26, 57, 64, 68–78). It must be mentioned that the factors relating to structure control and center segregation in ESR ingots contain complicating factors that are presently not defined (Ref 40). Thus, while there is room for improvements in the treatment of the thermal and electrical conditions on the bounding surfaces and in the efficiency of calculation, it is clear that the technology for computational analysis of the flow and thermal behavior of the ingot is adequate for the prediction of the metallurgical structure of the solidified ingot.

It is noteworthy that all the state-of-the-art computational models for the analysis of the VAR and ESR processes use a similar underlying approach. To present all aspects of the computational analysis and its application in a coherent and complete manner, the discussion in this article focuses on the computational models developed by the authors (Ref 51, 63, 64) for the VAR and ESR processes. Note that the model for the VAR process can be readily adapted for the analysis of the electron beam and plasma cold hearth ingot casting processes with appropriate thermal conditions on the ingot top surface, although no explicit discussion is provided here. To facilitate routine use of the models, emphasis is placed during model development on computational efficiency. To this end, physically motivated assumptions are used in the construction of the mathematical formulation, and algorithms grounded in the process physics are developed for efficient computational solution of the governing equations. In the following sections, details of the mathematical formulation, computational method developed for the solution of the governing equations, and application of the resulting models for the analysis of practical VAR and ESR processes are provided.

Analysis of Axisymmetric Behavior and Computational Domain

Most remelting processes involve casting of cylindrical ingots with geometries that are axisymmetric and process conditions that are very nearly invariant in the angular direction. The assumption of axial symmetry enables the analysis to be carried out in two dimensions. Such analysis is an order of magnitude more efficient than a three-dimensional analysis, and it also represents the behavior of the physical system with an accuracy that is sufficient for engineering practice. Hence, this article focuses on the description of two-dimensional axisymmetric models for VAR and ESR.

Remelting Process Models

VAR Process. It is convenient to perform the computational analysis using a frame of reference that is attached to the top surface of the ingot. In this reference frame, the top surface of the ingot appears stationary, and the ingot grows downward with the instantaneous casting velocity. Note that the top surface of the ingot is heated by the plasma arc. It also receives the metal droplets from the melting bottom face of the electrode, and a molten pool forms below the surface. The use of this reference frame allows accurate treatment of the interactions at the top surface, the behavior of the molten pool, and the heat loss to the mold boundary. The model for the VAR process performs an unsteady analysis of the entire process that considers the growth of the ingot and changing melting conditions. The arc behavior is modeled using thermal and electrical boundary conditions based on an overall energy balance.

Therefore, as shown in Fig. 2, the computational domain begins at the top ingot surface and extends to the bottom face of the ingot, which loses heat to the base plate. The computational domain expands in size to accommodate the growth of the ingot.

ESR Process. In this case, the computational domain consists of the ingot with a layer of slag on top. The analysis is performed using a frame of reference that is attached to the top surface of the slag. In this reference frame, the electrode-slag and the slag-ingot interfaces appear stationary. This allows accurate analysis of the interactions at the electrode-slag, slag-ingot, slag-mold, and ingot-mold interfaces as well as analysis of the behavior of the molten slag and the pool. In the present study, the model for the ESR process is developed for steady-state behavior that is attained when the ingot has grown to be sufficiently long and process conditions are steady. Such steady-state analysis enables a computationally efficient evaluation of the effect of operating conditions on the process performance. Further, extension of the formulation to a transient (growing ingot) case is easily incorporated in a similar fashion to the VAR formulation. The electrode is not included in the computational domain. This is because the temperature profile within the electrode shows only a very thin boundary layer of hot metal near the slag-electrode interface. Further, due to the high electrical conductivity of the electrode relative to that of the slag, the distribution of the supplied current is determined primarily by the slag electrical resistivity. Thus, the computational domain begins from the top surface of the slag and includes a sufficiently large ingot length so that the heat transfer at the bottom boundary has little effect on the pool behavior. Finally, the electrode-slag and slag-metal interfaces are assumed to be flat. The resulting computational domain used in the analysis of the ESR process is shown in Fig. 3 for a short-collared mold and involves an exposed length of the solidified ingot.

Analysis of the VAR process involves solution of the unsteady form of governing equations. Further, for constructing a robust computational method, the steady-state behavior for the ESR process is also determined using an unsteady calculation. The resulting equations governing the various physical phenomena and the corresponding boundary conditions for the VAR and ESR processes are described in the next section.

Mathematical Formulation

The physical phenomena in the interior of the ingot are identical for both the VAR and the ESR processes. The primary differences between the two processes occur due to the presence of the slag and the use of alternating current (ac) in the ESR process (as opposed to dc in VAR) and the associated treatments of the conditions on the domain boundaries. Therefore, in this section, the governing equations and boundary

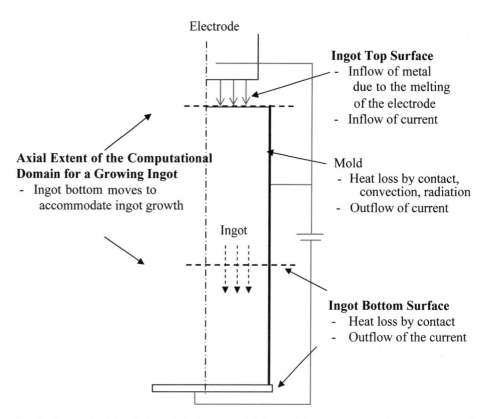

Fig. 2 Computational domain for analysis of the unsteady behavior of the vacuum arc remelting process (frame of reference attached to the top surface of the ingot)

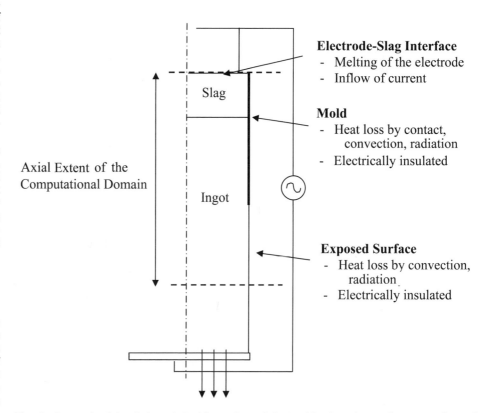

Fig. 3 Computational domain for analysis of the steady-state behavior of the electroslag remelting process (frame of reference attached to the top surface of the slag)

conditions for the each physical phenomenon are first described for the VAR process. This is followed by the discussion of the important extensions necessary for the analysis of the ESR process.

Electromagnetics

The electromagnetic phenomena in remelting processes are governed by the magnetoquasistatic form of Maxwell's equations. Specific formulations of these equations for dc power in VAR and ac power in ESR are discussed as follows.

VAR Process. With dc power, Maxwell's equations reduce to a single equation for governing the distribution of the electric potential, ϕ, in the ingot:

Electric potential: $\nabla \bullet (\sigma \nabla \phi) = 0$ (Eq 1)

The following boundary conditions for current density and electrical resistance are imposed on the ingot surface. They closely follow the practices employed by Bertram et al. (Ref 43):

- *Top surface:* A model for the arc behavior is used for specifying the fraction of the total arc current that enters the top surface of the ingot and its distribution over it.
- *Circumferential and bottom surfaces:* The current entering the ingot flows into the mold through the circumferential and bottom surfaces. The shrinkage of the ingot due to its solidification affects the electrical resistance to current flow from the ingot to the mold. This resistance is calculated from the temperature of the ingot surface, such that there is good contact when the surface temperature exceeds the liquidus temperature of the alloy, and the contact is lost when the ingot surface temperature is below the solidus temperature. Between the liquidus and solidus temperatures, the resistance is linearly ramped down from good contact to poor or no contact.

The distribution of Lorentz force that affects the flow field are deduced from the electric potential. For this purpose, first the distribution of the current density within the ingot is calculated from the gradient of the electric potential.

Current density: $\mathbf{J} = -\sigma \nabla \phi$ (Eq 2)

The self-induced magnetic field in the angular direction is then calculated using Ampere's law:

Magnetic flux density: $B_\theta(r) = \mu_0 \frac{1}{r} \int_0^r J_x dr$ (Eq 3)

Finally, the in-plane (x-r plane) Lorentz force is determined from the current densities and the azimuthal magnetic field in the following manner:

Lorentz force: $\mathbf{F}_L = (-J_x B_\theta \mathbf{e_r} + J_r B_\theta \mathbf{e_x})$ (Eq 4)

ESR Process. With the ac current used in the ESR process, the electromagnetic fields are inherently unsteady. However, the high frequency of the ac power in combination with the density and heat capacity of the slag and the molten metal result in time scales for flow and thermal phenomena that are much larger than the time scales for electromagnetic phenomena. Therefore, analysis of electromagnetics in the ESR process is performed for the periodic steady state, and the cycle-averaged values of the Lorentz force and Joule heating are used in the analysis of the flow and the temperature fields. (Note that the electrochemical reactions taking place at the slag/metal interfaces due to Faradaic reactions are controlled by the ac power cycle and have reaction times on that timescale. However, because these reactions are almost completely reversible and have a negligible impact on the process energy distribution, it is not necessary to include them in the present models.) In the following discussion, the equations governing the analysis of electromagnetics for ac power of a given sinusoidal frequency are first described. The extension of this formulation for the analysis of ac power with a square-wave current is then presented.

In a periodic steady state, an electromagnetic quantity is represented as a product of a spatially-varying complex amplitude and a sinusoidal time variation, as follows:

Periodic steady state: $\varphi = \hat{\phi} \cos(\omega t) + j\hat{\phi} \sin(\omega t)$ (Eq 5)

Further, for axisymmetric coordinates, the magneto-quasistatic form of Maxwell's equations can be reduced to a single equation for the diffusion of the scalar self-induced magnetic field intensity. Thus, the equation for the (complex) amplitude of the magnetic field intensity for the sinusoidal frequency, ω, is as follows:

Magnetic diffusion: $\nabla \bullet \left(\frac{1}{\sigma} \nabla (\hat{H}_\theta \mathbf{e}_\theta)\right) = j\mu_0 \omega \hat{H}_\theta$ (Eq 6)

Magnetic flux density: $\hat{B}_\theta = \mu_0 \hat{H}_\theta$ (Eq 7)

The previous equation is applicable to both the slag and the ingot regions with the use of the appropriate electrical conductivity, σ. Note that the slag-ingot interface is subject to the condition of continuity of the tangential electric field that is inherent in the magnetic diffusion equation, so that the interface is internal to the domain.

The boundary conditions for the magnetic field intensity are deduced from the inflow and outflow of current from the boundaries of the domain in the following manner:

- *Electrode-slag interface:* Because the electrical conductivity of the electrode is much

larger than that for the slag, the electrode is at a uniform potential. This results in a vanishing normal gradient of the magnetic field intensity at this interface.
- *Ingot bottom:* For a sufficiently long ingot, the normal (axial) gradient of the magnetic field intensity is also zero at this interface, analogous to the electrode-slag interface.
- *Exposed slag surface:* Because no current flows out of the exposed section of the slag, the radially-varying magnetic field intensity is specified from the known total current entering the slag from the electrode, using Ampere's law.
- *Circumferential surface of the ingot:* In the present study, the solidified slag skin formed at the slag-mold interface and transferred to the ingot is assumed to be electrically insulating. As a result, the amplitude of the magnetic field intensity is uniform on this boundary, and its value is determined from the total current entering from the electrode, using Ampere's law in a manner analogous to that for the exposed surface of the slag.

The distributions of the current density, Lorentz force, and Joule heating are deduced from the magnetic field intensity. For this purpose, the in-plane (x-r) current density is first calculated using Ampere's law, as follows:

Current density: $\hat{\mathbf{J}} = \nabla \times (\hat{H}_\theta \mathbf{e}_\theta)$ (Eq 8)

The cycle-averaged in-plane (x- and r-directions) Lorentz force is then determined from the following equation:

Cycle-averaged Lorentz force:

$\mathbf{F}_L = Re\left(\frac{1}{2}\hat{\mathbf{J}} \times \hat{\mathbf{B}}_{conjugate}\right)$ (Eq 9)

Finally, the cycle-averaged resistive heating is determined by the following equation:

Cycle-averaged Joule heating:

$S_J = \frac{1}{2\sigma}\hat{\mathbf{J}} \bullet conjugate(\hat{\mathbf{J}})$ (Eq 10)

The Lorentz force and Joule heating affect the flow and the temperature fields in the slag and the ingot.

The aforementioned formulation is easily extended to the analysis of a square-wave alternating current by representing all electromagnetic quantities as corresponding Fourier series sums of sinusoidal waves of frequencies that are harmonics of the basic frequency. Such a representation enables analysis of the electromagnetics in a very convenient manner. This is because the magnetic diffusion equation is linear and the sine functions are orthogonal, so that each sinusoidal frequency acts independent of all the other frequencies. Thus, the magnetic field, current density, Lorentz force, and Joule heating created by the square current are determined by simple linear additions of the corresponding quantities produced by current components of successive harmonic frequencies that constitute the imposed square-wave current.

Fluid Motion

The formulation for analysis of the macro-scale turbulent flow within the molten pool formed in the VAR process and its extension for the coupled analysis of the flow in the molten pool and the slag in the ESR process are discussed as follows.

VAR Process. A pool of molten metal forms under the plasma arc. The size and shape of the pool are determined by the processing conditions and the loss of heat to the mold. The macroscale fluid motion that occurs in the metal pool and the outer mushy region is driven primarily by buoyancy and Lorentz forces and is expected to involve nonuniform turbulent mixing. It is governed by the time-averaged forms of the Navier-Stokes equations that describe the mean velocity:

Continuity: $\dfrac{\partial \rho}{\partial t} + \nabla \bullet \rho \mathbf{u} = 0$ (Eq 11)

Momentum: $\dfrac{\partial(\rho \mathbf{u})}{\partial t} + \nabla \bullet (\rho \mathbf{uu})$
$= -\nabla P + \nabla \bullet \left(\mu_{\text{eff}}(\nabla \mathbf{u} + \nabla \mathbf{u}^T)\right) - \rho \mathbf{g} + \mathbf{F}_{\text{L}}$ (Eq 12)

The third and the fourth terms in Eq 12 represent the buoyancy and Lorentz force, respectively. The effect of turbulent mixing in the pool is felt through the spatially-varying turbulent viscosity that is calculated using the two-equation k-ε turbulent model.

The previous equations are subject to the following boundary conditions:

- *Solidified ingot:* The same set of equations is used to represent the flow in the pool as well as the solidified ingot. The macroscale flow is assumed to cease at the immobilization liquid fraction (f_{imm}), the value of which is typically just under 1. When the liquid fraction is below this value, the metal moves with the instantaneous casting velocity.
- *Ingot top surface:* As part of the arc model, the mass flux profile is used on the top surface to determine the radial distribution of the velocity of the molten metal entering the molten pool. In addition, the molten pool is subject to the shear stress caused by the plasma jet.
- *Circumferential surface:* The no-slip boundary condition is imposed in the region of the ingot where the molten metal contacts the mold surface. When the ingot separates from the mold, the velocity at the outer surface of the ingot is equal to the instantaneous casting velocity.

ESR Process. The single set of equations (Eq 11 and 12) with appropriate values of the fluid density and viscosity represent the turbulent flow in the slag and the molten pool. The motions in the slag and the metal pool interact through the shear force at the slag-metal interface. Analogous to the magnetic diffusion equation, this interface is internal to the domain because the continuity of the interface shear force at the slag-metal interface is inherent in the momentum equation. The special aspects of the boundary conditions that arise due to the presence of the slag are as follows:

- *Melting of the electrode:* The molten metal formed at the electrode-slag interface is assumed to enter the molten pool at a uniform velocity under the slag-metal interface. Further, the effects of this inflow of metal are considered through mass and momentum sources in the continuity and momentum equations, respectively, in the metal pool below the slag-metal interface and in the shadow of the electrode.
- *Slag-mold interface:* A thin layer of solidified slag (slag skin) is formed at the slag-mold interface. Due to its small thickness, the molten slag is assumed to extend nominally to the mold surface, and a no-slip condition is imposed on the flow of the molten slag over this surface.

Magnetic Stirring

Magnetic stirring is used in high-current VAR processes to stabilize the arc. This involves the use of a coil that is wound around the mold, through which a square-wave current (that alternates in direction) is passed. The resulting axial magnetic field interacts with the radial current in the metal to give rise to an oscillating Lorentz force in the angular direction, creating a swirling motion in the molten pool. It should be noted that the angular velocity of the molten metal does not vary in the angular direction. Therefore, the behavior of the process remains axisymmetric.

Analysis of magnetic stirring requires determination of the Lorentz force in the angular direction, its use in the momentum equation for the angular velocity, and the determination of the resulting centrifugal force field for use in the calculation of the in-plane velocity field, using the following equations:

Lorentz force in the angular direction: (Eq 13)
$F_\theta = -J_r B_{\text{axial}}$

Momentum equation in the angular direction:

$\dfrac{\partial(\rho w)}{\partial t} + \nabla \bullet (\rho \bar{u} w) = \nabla \bullet (\mu_{\text{eff}} \nabla w) - \dfrac{\rho u_r w}{r} - \dfrac{\mu_{\text{eff}} w}{r^2}$
$+ F_\theta$ (Eq 14)

Centrifugal force: $F_r = \dfrac{\rho w^2}{r}$ (Eq 15)

The angular velocity is subject to no-slip conditions on the mold wall if the liquid metal touches the mold surface. Further, the angular velocity is zero below the immobilization liquid fraction. Also, there is no shear stress in the angular direction on the top surface of the ingot. The presence of angular motion affects turbulent mixing due to the production of turbulent kinetic energy directly through the gradients of the angular velocity and indirectly through the higher intensity and gradients in the in-plane flow.

Turbulence Model

The two-equation k-ε turbulent model is used for predicting the turbulent mixing in the slag and the molten pool. It involves solution of the transport equations for turbulent kinetic energy, k, and turbulent dissipation, ε. This model is commonly employed in the analyses of conventional flows of gases and liquids. The reader is referred to earlier studies (Ref 79) for details of the governing equations and boundary conditions. The important aspect of this model as used in the VAR and ESR models is that the governing equations are solved only in the molten pool and the slag with standard boundary conditions applied at the mold and pool boundaries. Further, within the molten pool and the slag, the turbulent viscosity and turbulent thermal conductivity are determined from k and ε in the following manner:

Turbulent and effective viscosity: $\mu_{\text{turb}} = c_\mu \dfrac{\rho k^2}{\varepsilon}$
$\mu_{\text{eff}} = \mu_{\text{dynamic}} + \mu_{\text{turb}}$ (Eq 16)

Turbulent and effective conductivity: $\dfrac{k_{\text{turb}}}{C_P} = \dfrac{\mu_{\text{turb}}}{\text{Pr}_{\text{turb}}}$
$k_{\text{eff}} = k_{\text{molecular}} + k_{\text{turb}}$ (Eq 17)

Energy Conservation

This section first describes the formulation of the energy conservation equation for the determination of the temperature distribution in the ingot for the VAR process. Its extension for the prediction of the temperature field in the ESR process involving coupled thermal interactions between the molten slag and the ingot is then described.

VAR Process. The energy conservation equation is formulated using the enthalpy-porosity approach (Ref 80) to enable computation of the phase-change process using a fixed computational grid. In this approach, the total enthalpy is decomposed into a sensible enthalpy, h, and latent heat content, ΔH. The resulting governing equation has the following form:

Energy conservation: $\dfrac{\partial(\rho \mathbf{u} h)}{\partial t} + \nabla \bullet (\rho \mathbf{u} h)$
$= \nabla \bullet (k_{\text{eff}} \nabla T) - \left(\dfrac{\partial(\rho \Delta H)}{\partial t} + \nabla \bullet (\rho \mathbf{u} \Delta H)\right)$ (Eq 18)

Latent heat content: $\Delta H = f(T)L$ for (Eq 19)
$T_{\text{solidus}} \le T \le T_{\text{liquidus}}$

Several points about the previous equation are worth noting. First, a single equation is used for the prediction of the temperature field in the molten pool and the solidified ingot. The pool region itself is not known a priori and is determined from the immobilization liquid fraction that itself is deduced from the temperature field. Second, the energy conservation equation considers the dependence of the thermal conductivity and specific heat of the alloy on temperature. Third, the effective conductivity is calculated from the k-ε model to account for turbulent mixing. Finally, the latent heat content is calculated from alloy-specific variation of the liquid fraction with temperature in the mushy region.

The temperature distribution is determined by the following interactions on the ingot boundaries:

- *Top surface:* The molten metal enters the molten pool in the ingot under the shadow of the electrode with an assumed superheat. Further, the ingot surface exchanges heat by radiation with the electrode temperature in the shadow of the electrode and to the water-cooled mold from the exposed annular region. Finally, the total heat transferred from the arc to the ingot top surface is determined by an overall energy balance for the arc, and its distribution is determined from the arc model (Ref 43).
- *Circumferential surface:* Analogous to the conduction of current, heat loss from the ingot to the mold must account for the shrinkage of the ingot away from the mold as it solidifies. This is accomplished by determining the contact heat-transfer coefficient, which decreases linearly from its maximum value to a value close to zero as the ingot surface temperature decreases from the liquidus temperature to the solidus temperature. The maximum contact heat-transfer coefficient corresponds to combined thermal resistance of the conduction across the mold thickness and the convective heat-transfer coefficient from the mold outer surface to the cooling water. However, unlike current conduction, as the ingot separates from the mold, it also loses heat to the inside surface of the mold by radiation. Additional cooling provided by helium injection in the annular gap formed by the separating ingot is also considered by augmenting the heat-transfer coefficient in the region of separation of the ingot from the mold.
- *Bottom surface:* The heat loss from the bottom surface of the ingot to the mold is considered in a manner similar to that from the circumferential surface.

ESR Process. Analysis of energy conservation in the metal and the slag phases is performed by using the same energy conservation (Eq 18) by using relevant values of the material properties (density, specific heat, and thermal conductivity). For the ESR process, the energy

conservation equation must incorporate the Joule heating produced in the slag as a volumetric source term. (Joule heating in the metal is negligible and hence not considered in VAR.) The continuity of the conduction heat flux at the slag-metal interface is inherent in the energy conservation equation, so that this interface is treated as being internal to the domain.

The ESR process differs from the VAR process in two significant aspects, namely, the melting of the electrode due to the heated slag, and the formation of the slag-skin at the slag-mold interface and its transfer to the continuously forming ingot. Thermal boundary conditions consider these effects in the following manner:

- *Top surface of the slag:* The slag-electrode surface is subject to a uniform heat flux required to melt the electrode at the prescribed melt rate. Alternatively, the temperature at the electrode-slag interface is assumed to be the liquidus temperature of the alloy, and the melt rate is derived from the heat transferred to the electrode. The remaining surface of the slag is assumed to lose heat to the surroundings through radiation and convection.
- *Bottom surface of the ingot:* Under steady-state conditions, the conductive heat loss at the bottom surface of the ingot is calculated from extrapolation within the domain. In addition, the solidified metal carries energy out of this boundary due to its motion at the casting velocity.
- *Slag-mold interface:* The molten slag loses heat to the water-cooled mold to form a thin layer of solidified slag skin. This heat loss is determined from an overall heat-transfer coefficient that is the net result of the thermal conductances of the slag skin and the metal mold and the convective heat-transfer coefficient from the outer mold surface to the cooling water. In turn, this heat loss determines the thickness of the solidified slag skin.
- *Ingot-mold interface:* Analogous to the VAR process, heat loss from the ingot to the cooling water accounts for gradual separation of the ingot from the mold due to the solidification of the alloy. Unlike the VAR process, however, the presence of the slag skin introduces an additional thermal resistance for loss of heat from the ingot. When the ingot separates from the mold, heat loss to the mold surface occurs by thermal radiation and by natural convection in the gap formed between the ingot and the mold. For a short-collared mold, the exposed surface of the ingot loses heat by radiation and convection to the surrounding air.

Redistribution of Alloying Elements due to Macrosegregation in Titanium Alloys

The selective absorption/rejection of individual elements during solidification, combined with imperfect mixing in the molten pool, results in

variation of alloying element concentrations in the radial and axial directions in the solidified ingot. The macroscale flow in the liquid pool penetrates only into the outer mushy region, and any solute rejected (absorbed) during this part of the solidification affects the concentration in the liquid. This, in turn, affects the concentration of that alloying element in the newly formed solid. As the solidification progresses further, the segregated liquid stays between the dendrites, and any microsegregation that occurs in this region is of no interest in titanium alloys because the elements are in solid solution and homogenize during subsequent processing. Thus, the purpose of the analysis of concentrations of alloying elements is to predict the macrolevel composition changes that occur during solidification in the outer mushy region.

The concentration c_i of the ith alloying element within the molten pool and the solid is governed by the convection-diffusion equation:

$$\text{Liquid region}: \frac{\partial(\rho c_i)}{\partial t} + \nabla \bullet (\rho \mathbf{u} c_i)$$
$$= \nabla \bullet (\Gamma_{\text{eff},i} \nabla c_i) + S_{i,\text{solidification-front}} \quad \text{(Eq 20)}$$

$$\Gamma_{\text{eff},i} = \frac{\mu}{\text{Sc}_i} + \frac{\mu_{\text{turb}}}{\text{Sc}_{\text{turb}}} \quad \text{(Eq 21)}$$

$$\text{Solid region}: \frac{\partial(\rho c_i)}{\partial t} + \frac{\partial(\rho u_{\text{cast}} c_i)}{\partial x}$$
$$= -S_{i,\text{solidification-front}} \quad \text{(Eq 22)}$$

Because the velocity is automatically the casting velocity in the solid region, the same equation is used in the solid and the liquid region through the use of local velocity and a relevant diffusion coefficient. In the pool region, the effective diffusion coefficient is a result of molecular diffusion and turbulent diffusion. In the solid region, the diffusion coefficient is zero.

The concentration equation is subject to the following boundary conditions:

- *Ingot top surface:* The concentrations of the alloying elements in the metal flowing into the metal pool through the top surface of the ingot correspond to concentrations at the melt front of the electrode.
- *Pool boundary*: The mass fluxes of an alloying element caused by selective rejection/absorption by the solid give rise to a source/sink in the liquid and a sink/source in the solid regions adjacent to the pool boundary. The segregation mass flux (into the pool) due to the partitioning in the outer mushy region (typically, up to a limiting solid fraction, f_s^*, of ~0.1 to 0.3) for the element i, with an effective segregation coefficient, K_s, close to unity, is given by the following equation:

$$J_i = \rho u_{\text{cast}} c_i^l (1 - K_{s,i}) f_s^* \quad \text{(Eq 23)}$$

In the preceding equation, superscript "l" is for the concentration used to denote its value in the bulk liquid phase. Scheil's equation, which follows and relates the local concentrations of the ith solute in the liquid pool before ($c_{i,0}^l$) and after segregation (c_i^l), can also be employed to calculate the segregation flux for larger values of the limiting solid fraction:

$$c_i^l = c_{i,0}^l \left(1 - f_s^*\right)^{(K_{s,i}-1)} \qquad \text{(Eq 24)}$$

Note that when K_s is less than unity, the solid rejects the solute, and when it is greater than unity, the solid selectively absorbs the solute. Finally, in the present study, the effect of concentration variation on the density is considered to be negligible, thereby decoupling the flow field in the molten pool from the concentration fields of the alloying elements. If, however, the liquid alloy density is strongly dependent on the alloy concentrations, solutal buoyancy forces must be considered in the determination of the flow field in the molten pool.

Metallurgical Structure and Freckle Formation in Superalloys and Steel

For superalloys and steels, microsegregation occurring in the interdendritic region and the associated possibilities of forming freckles are the more important phenomena. However, its prediction from first principles is computationally impractical due to the small scales involved. Instead, following the approach proposed by Auburtin et al. (Ref 81), the thermal history of the solidifying ingot is used to predict the metallurgical structure, and an alloy-dependent criterion based on the Rayleigh number for the thermosolutal convection within the interdendritic region is used to predict the probability of freckle formation. Details of all the equations are available in the study by Auburtin et al. (Ref 81). Hence, the following discussion focuses on the important aspects of this procedure that uses the unsteady thermal history of the ingot in the analysis of the VAR process for superalloys. This is followed by a short description of the simplifications that result in the steady-state analysis for the ESR ingots.

Analysis for Unsteady Conditions. In an unsteady process during which the melt rate is changing continuously, each elemental volume in the solidified ingot experiences a different thermal history. Therefore, the local solidification time (LST) and all quantities related to the metallurgical structure that are used in the calculation of the interdendritic Rayleigh number vary both in the radial and the axial directions within the ingot. Calculation of these quantities is described as follows:

- *Local solidification time and cooling rate:* The LST is defined as the time required for an elemental metal volume to cool from the liquidus temperature to the solidus temperature and is used to calculate carbide sizes. The cooling rate and the LST for each

elemental control volume are computed from the transient variation of the temperature field for each elemental metal volume.
- *Rayleigh number for interdendritic convection:* The Rayleigh number for the thermosolutal convection as the elemental metal volume reaches the alloy-specific freckle-initiation solid fraction uses the following expression:

Rayleigh number for interdendritic flow:

$$\text{Ra} = \frac{g\left(\frac{\partial \rho}{\partial T}\right)\left(\frac{\partial T}{\partial x}\right)}{\mu\left(\frac{k}{\rho C_p}\right)} l^4 \qquad \text{(Eq 25)}$$

The Rayleigh number is dependent on two important parameters: the length scale, l, and the temperature gradient at the freckle-initiation solid fraction. The calculation of the length scale involves several steps. First, the primary and secondary dendrite arm spacings are calculated from the cooling rate experienced by the elemental metal volume at the liquidus temperature. The permeability for the flow of liquid metal parallel to the primary and secondary dendrites is then determined. Next, the angle of the flow (which is caused by thermosolutal forces and occurs in the direction of the gravity vector) relative to the primary dendrites is determined from the slope of the isotherm at the freckle-initiation solid fraction. This angle and the permeabilities determine the effective length scale (or ease of) flow in the interdendritic space. Finally, the local temperature gradient in the gravity direction is determined from the prevailing temperature field as the metal volume crosses the freckle-initiation solid fraction. When this Rayleigh number exceeds the alloy-specific critical Rayleigh number, Ra*, freckle formation is deemed likely.

Analysis for Steady-State Conditions. When the analysis is performed for steady-state behavior, the pool shape and size do not change with time. Therefore, the thermal history experienced by an elemental metal volume as it solidifies depends only on its radial location. Further, these radial variations of the cooling rate can be related to the spacing between the contours of relevant liquid fractions and the casting velocity. When the cooling rate is determined, the calculations of dendrite arm spacings, permeabilities, and finally the Rayleigh number for interdendritic convection for predicting the probability of freckle formation proceed according to the procedure described previously.

Inclusion Motion

In the present study, analysis of the inclusion behavior is performed by introducing inclusions over the top of the molten pool. Because the number of inclusions is expected to be very small, it is assumed that the inclusions do not

affect the motion of the liquid metal in the pool. Further, it is assumed that the residence time of the inclusions is small in comparison to the time scale for ingot growth. Therefore, in an unsteady analysis, the motion of inclusions is calculated based on the motion in the pool at a specific instant during the growth of the ingot. This assumption is made for computational efficiency; the motion of an inclusion can also be tracked within an unsteady flow field.

The motion of an inclusion is governed by Newton's law of motion that describes the interaction of the buoyancy, drag, and inertia forces. The rate of dissolution is expressed in terms of the rate of change of the inclusion diameter. These two equations are given as:

$$\text{Inclusion motion} : m_p \frac{d\bar{u}_p}{dt}$$
$$= C_D A_p \frac{1}{2} \rho_m \left(|\bar{u}_m - \bar{u}_p|\right)\left(\bar{u}_m - \bar{u}_p\right)$$
$$+ m_p \left(1 - \frac{\rho_m}{\rho_p}\right)\bar{g}$$
$$\text{(Eq 26)}$$

$$\text{Inclusion dissolution} : \frac{dD_p}{dt} = -2B \exp\left(-\frac{E}{T}\right)$$
$$\text{(Eq 27)}$$

In the preceding equations, the subscripts p and m denote the inclusion and the liquid metal, respectively. The drag coefficient, C_D, is dependent on the Reynolds number for the inclusion motion and its shape (assumed to be spherical). This Reynolds number is calculated using the velocity of the inclusion relative to the liquid metal. In the presence of swirling motion due to magnetic stirring, the inclusion motion is also subject to centrifugal and Coriolis forces. In addition, the angular velocity of the inclusion must be calculated. Finally, note that, because the inclusions are small, they are assumed to be in thermal equilibrium with the surrounding pool. The rate of dissolution for an inclusion is related to its temperature with inclusion-specific values for the constants B and E.

Computational Solution

The control-volume method of Patankar (Ref 82) is used for the discretization of the equations governing electromagnetics, flow, turbulent mixing, energy conservation, and redistribution of alloy element concentrations. Several enhancements of this method have been carried out to address the unique aspects of the physical phenomena occurring in the VAR and ESR processes. In the following discussion, a brief summary of the basic control volume method is first presented. This is followed by a discussion of the various enhancements incorporated for constructing efficient computational models for the VAR and ESR processes.

Control Volume Computational Method

The control volume computational method involves the use of grid lines to divide the axisymmetric domain into control volumes in the axial and radial directions. Scalar quantities such as pressure, temperature, turbulent kinetic energy, and turbulent dissipation are stored at the main grid points. A staggered grid is used to store the velocity components and current density vectors at the faces of the control volumes. Discretization equations for each variable are constructed by integrating the corresponding transport equations over the respective control volumes. Implicit time-stepping is used for the discretization of the unsteady term in the governing equations. In each time step, the resulting set of discretization equations is solved using the SIMPLER algorithm (Ref 82). The size of the time step is determined automatically from the time scales of the flow and thermal phenomena. Multiple iterations are needed within each time step to address the nonlinearity and coupling between the electromagnetic, flow, and temperature fields.

Treatment of the Growing Ingot in the VAR Process

The growth of the ingot is treated accurately through the motion of the ingot base surface within a fixed grid that spans the entire ingot length. In this treatment, only the base control volume (and not the entire domain) expands in size until the base surface moves to the next control volume, as shown in Fig. 4. This treatment provides several advantages. First, the treatment reduces to the model for steady-state calculation for a long ingot in a natural manner. Second, the time-step size can be controlled independently of the ingot growth. Third, the number of grid points within the ingot grows automatically with the ingot length, and

regridding of the computational domain is not required. Because the base control volume is partially occupied and expands in each time step, a special treatment is required for the discretization of the conservation equations over this volume. This involves formulation of the equations for mass, momentum, energy conservation, turbulence quantities, and concentration transport for an expanding volume (Ref 83) and use of the space conservation law (Ref 84) for their discretization.

Magnetic Stirring in the VAR Process

When magnetic stirring is present, an additional equation for the angular velocity must be solved. This equation for the w-velocity is treated as a scalar transport equation. At each time instant during the time-stepping, the axial magnetic field and the resulting Lorentz force in the angular direction are calculated for use in the w-equation. The in-plane and angular velocity fields are coupled, and this interdependence is handled through multiple iterations within a time step. Further, the time step is chosen to ensure that the stirring cycle is accurately resolved.

Unified Analysis of the Slag and Ingot Regions in the ESR Process

The computational method treats the slag and the metal regions in a unified manner, using a single computational domain.

Interactions across the Slag-Metal Interface. Mathematical formulation of all phenomena requires that the diffusion flux be continuous across the slag-metal interface. A unique feature of the computational method is its use of the harmonic mean of the conductances on the two sides of a control volume face to determine the equivalent conductance at that face. With this treatment, the diffusion flux calculation accounts very accurately for large and

discontinuous changes in the diffusion coefficients across a control volume face. Therefore, continuity of the radial current, the heat flux, and the shear stress at the slag-metal interface in the presence of dissimilar properties of the slag and the metal phases are incorporated in a natural and fully implicit manner, resulting in a robust and efficient computational method.

Treatment of the Melting of the Electrode. The melting of the electrode at the electrode-slag interface gives rise to metal droplets that fall through the slag and enter the metal pool. The interaction of the metal droplets with the slag, as they fall through it, is considered by assuming that the droplets are in thermal equilibrium with the slag. Further, the heat absorbed by the metal droplets from the slag phase is treated as a uniformly distributed volumetric heat sink in the cylindrical volume of the slag under the electrode. Note that the assumption of thermal equilibrium can be readily relaxed by performing a calculation of the motion and energy balance for droplets as they fall through the slag. The inflow of the molten metal into the metal pool is treated through sources of mass, momentum, energy, and turbulence quantities in the control volumes in the metal phase that lie just beneath the slag-metal interface and in the shadow of the electrode. The preceding treatment ensures overall conservation of the energy in the ESR system and results in a convenient and accurate method for handling the melting of the electrode within the unified computational domain.

Analysis of Electromagnetics with ac Power for the ESR Process

The computational method has been extended for the solution of scalar transport equations in which field unknowns are complex variables to enable solution of the magnetic diffusion equation that governs the electromagnetic behavior of the ESR system. The basic framework performs analysis of electromagnetics for a sinusoidal current of a prescribed frequency. Further, a Fourier-series-based procedure involving the analysis for multiple sinusoidal frequencies, which uses the basic single-frequency framework coupled with the linear superposition technique to account for successive harmonics, is also constructed to determine the electromagnetic behavior for the square-wave current pattern.

Redistribution of Alloying Elements due to Macrosegregation in Titanium Alloys

Computational solution of the alloy concentration equation must account for three important aspects of the process of redistribution of the alloying elements. First, the alloying elements do not redistribute within the solidified metal, because diffusion in the solid is absent (Scheil approximation). This behavior is accurately simulated through a special discretization

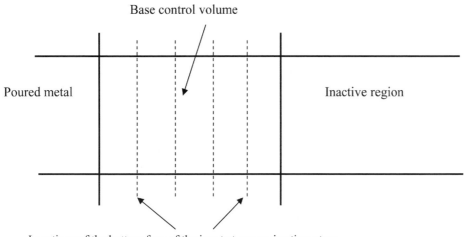

Base control volume

Poured metal

Inactive region

Locations of the bottom face of the ingot at successive time steps

Fig. 4 Subdivision of the base control volume to address the motion of the base plate

method involving explicit time-stepping in the solid region. Secondly, in an unsteady process, the pool volume changes with time. The computational method keeps track of the control volumes containing liquid metal that solidify. The alloying elements that are selectively rejected/absorbed by the newly solidified control volumes give rise to additional segregation source terms at the pool boundary. Finally, the source terms arising from the rejection/absorption of the solute by the solid into the pool are appropriately incorporated in the control volumes in the liquid and the solid regions that are adjacent to the pool boundaries.

Thermal History and Metallurgical Structure

The metallurgical structure of the solidified ingot depends on the thermal history experienced by the metal as it solidifies. In an unsteady process, the thermal history of the elemental volume of metal within the ingot depends on its axial and radial location within the ingot. Due to the use of a Eulerian grid (use of a fixed grid through which the metal moves), each elemental metal volume occupies different positions within the grid as the ingot grows in size. Therefore, a special Lagrangian technique is developed for a precise identification of each elemental metal volume as it moves through the computational grid. Its thermal history is recorded as it cools from the liquidus to the solidus temperature to determine the LST, cooling rate at liquidus temperature, and pool angle. The metallurgical structure parameters are then computed to determine the Rayleigh number for interdendritic convection and to infer the probability of freckle formation.

As discussed in the section on mathematical formulation, the thermal history calculation is somewhat simpler for steady-state conditions, because the thermal history depends only on the radial location of the solidifying elemental metal volume. Therefore, the cooling rate, LST, and pool angle are deduced from the casting velocity and the axial locations of the relevant liquid fractions.

Calculation of the Motion of Inclusions

A Lagrangian approach (Ref 85) is used to solve the equation of motion and dissolution of an inclusion based on the flow field prevailing in the pool at the instant the inclusion is introduced into the pool. The equations of motion and dissolution are solved using a time-marching technique in which the time step is automatically determined by requiring the inclusion to go through multiple time steps for crossing a control volume. Starting with the initial position of the inclusion, the trajectory calculation is continued until the inclusion either dissolves or enters the mushy region, where it is assumed to enter the ingot. Note that the

present method can be extended to calculate the motion of an inclusion for unsteady flow. This will involve the use of the unsteady flow field and the corresponding evolving pool shape along with necessary interpolations to calculate the motion of an inclusion in the transient flow in the pool. However, such a calculation will require a significantly more computational effort.

Application of the Models for the Analysis of Practical Remelting Processes

The computational method described previously has been implemented in two independent computer programs for the analysis of VAR and ESR processes. These computer models have been applied for the analysis of practical VAR and ESR processes for the production of ingots of titanium and superalloys, respectively. The details of the processes modeled and the results obtained are discussed in this section.

Analysis of the VAR Process for the Production of an Ingot of Ti-6Al-4V

The geometry and the important operating and thermal conditions for an industrial VAR process for the casting of ingots of Ti-6Al-4V alloy are listed in Table 2. The corresponding melt schedule is shown in Fig. 5.

Computational Details. The computational model considers the dependence of alloy properties on temperature. For brevity, only the representative values of these properties are listed in Table 2. Transient analysis has been

performed for the entire process, with and without stirring, involving the growth of the ingot. A computational grid of 120 (axial, x) by 50 (radial, r) control volumes is used in the analysis. The analysis of the process with no stirring requires approximately 4 h on a single-processor personal computer with an Intel Pentium4 3.2 GHz processor. Calculations with magnetic stirring require approximately 12

Table 2 Geometry and operating conditions and material properties of alloy Ti-6Al-4V used in the analysis of a practical vacuum arc remelting process

Characteristic	Value
Geometry	
Electrode and ingot diameters	762 and 863.6 mm
Initial and final ingot lengths	0.0 and 2273 mm
Operating conditions	
Inlet superheat	373 K
Strength and oscillation period of the axial magnetic field	50 Gauss and 60 s
Alloy properties (Ti-6Al-4V)	
Density, liquid	3925 kg/m^3
Viscosity, liquid	2.36×10^{-3} Pa·s
Solidus and liquidus temperatures	1868 and 1898 K
Latent heat of fusion	3.5×10^5 J/kg
Specific heat, liquid	852 J/kg·K
Thermal expansion coeff., liquid	6.7×10^{-5} K^{-1}
Thermal conductivity, liquid	29.7 W/(m·K)
Electric conductivity	7.6×10^5 (Ω·m)$^{-1}$
Magnetic permeability	1.257×10^{-6} H/m
Segregation coefficients for Al, V, Fe, O	1.13, 0.95, 0.38, 1.33
Electrode concentrations of Al, V, Fe, O	0.06125, 0.04, 0.004, 0.0016

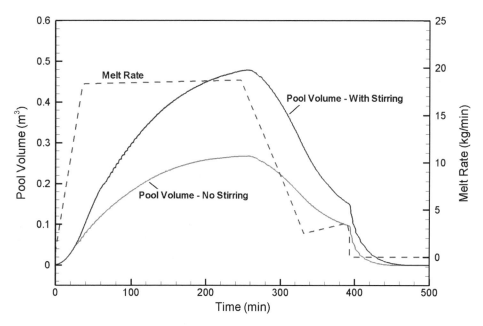

Fig. 5 Process schedule and predicted variation of the melt pool volume for a vacuum arc remelting process for Ti-6Al-4V alloy with and without stirring

h, because a smaller time step is necessary to resolve the behavior of the stirring cycle.

Results and Discussion. The computational model predicts detailed behavior of the electromagnetic, flow, temperature, and alloy concentration fields and the inclusion motion in the ingot during the entire process as well as the thermal history of the final ingot produced in the process. Results of the ingot behavior are presented at various time instants during the initial rampup, steady melting, and hot topping.

Figure 6 shows the variation of the electric current densities and the self-induced magnetic field in the growing ingot in the absence of magnetic stirring. Initially, the current entering the ingot flows out into the mold through both the circumferential and the bottom surfaces. When the ingot becomes long, very little current flows out from the bottom face. The current density interacts with the self-induced magnetic field to produce an in-plane (x-r) Lorentz force field that is directed radially inward and axially downward. In the presence of magnetic stirring, the in-plane behavior of the electromagnetic fields is similar, except that the contact region on the circumferential face of the ingot is longer.

Figures 7(a) and (b) show the evolving liquid fraction, flow, and temperature fields in the ingot without and with magnetic stirring, respectively, while Fig. 5 shows the corresponding evolutions of the pool volume during the casting process. As expected, the size of the molten pool grows during the portion of the process when the melt rate is increasing and then held steady. With the initiation of hot topping, the pool begins to shrink. Further, as the melting stops, the melt pool rapidly solidifies, as seen in Fig. 5. Due to the heating provided by the plasma arc, and in the absence of magnetic stirring, the temperature distribution in the pool shows thermally stable stratification. As a result, the upward flow in the center is

relatively weak, while the metal flows rapidly downward along the pool boundary. The use of magnetic stirring produces a very nonuniform, angular motion in the pool, as shown in Fig. 8. The associated centrifugal forces produce strong in-plane motion in the pool. The resulting pool is both wider and deeper and shows much smaller temperature variation in comparison to the pool produced in the absence of stirring.

For the Ti-6Al-4V alloy, the computational model performs analysis of the macrosegregation phenomenon to predict the variations of the alloying elements in the pool and in the solidified portion of the ingot. Figure 9 shows the variations of the concentrations of oxygen and iron at a time instant near the end of the melting for processes without and with magnetic stirring. Note that the distributions of concentrations in the solidified ingot are a result of the evolution of the pool and mixing within it, selective rejection (absorption) of the alloying element into the pool, and solidification of the solute-enriched (depleted) liquid metal at the melt front during the process. Because oxygen is preferentially absorbed by the solid, as the ingot grows, the average concentration of oxygen in the liquid pool decreases. Therefore, the concentration of oxygen in the upper portion of the ingot (formed later in the process) is lower than that in the lower portion of the ingot (formed earlier in the process). Further, the concentration of oxygen is low in the center portion of the ingot. This is because the flow at the bottom of the pool is weak and does not sufficiently replenish the oxygen absorbed by the newly formed solid in the center region. With magnetic stirring, the increased mixing present significantly reduces the extent of macrosegregation in the radial direction. The variation of the concentration of iron in the pool and the ingot is a mirror image of the corresponding variation of oxygen, because iron is selectively

rejected by the solidifying alloy. The model also predicts macrosegregation of aluminum and vanadium in the solidified ingot. Due to space constraints, the corresponding concentration plots are not included.

Figure 10 shows the thermal history in the solidified portion of the ingot near the end of melting in the form of variation of the time at the start of solidification and the LST in the solidified portion of the ingot. The LST is very small at the bottom and circumferential surfaces of the ingot, because they are closer to the mold and hence cool more rapidly. When magnetic stirring is used, the larger pool, coupled with strong mixing in the pool, increases the cooling rate, and the LST is lower. With the use of proper boundary conditions on the top surface of the ingot after melting has stopped and by performing the calculations until the ingot solidifies completely, the model can predict the thermal history of the entire ingot. As described in the earlier sections, for superalloys and steels, this thermal history is used to predict the metallurgical structure parameters such as dendrite arm spacing, permeability, and, ultimately, the probability of freckle formation based on the Rayleigh number for interdendritic convection throughout the entire ingot. For reasons of conciseness, the results of the analysis of the VAR process for superalloys and steels are not shown here.

The trajectories of light, neutrally buoyant, and heavy inclusions of different sizes (0.1 to 5 mm diameter) are shown in Fig. 11 at a time instant midway through the melting process. All inclusions are introduced at uniformly spaced locations on the top surface of the ingot. Inclusions that are either small in size or neutrally buoyant move with the molten metal. The motion of large inclusions is strongly influenced by their density relative to the liquid metal. Large, heavy inclusions sink quickly to the bottom of the pool, while light inclusions float to the top of the pool, move radially outward, and tend to spend significantly more time in the pool. When stirring is used, inclusions spend a longer time within the pool due to stronger in-plane velocities as well as motion in the angular direction. By performing the analysis of the behavior of many inclusions of different sizes and densities that enter the pool at many different locations over the pool surface, residence times for each type of inclusion, as well as the fraction and distribution of the ending locations for each type of inclusion that survives, are determined.

Analysis of the ESR Process for the Production of an Ingot of IN 718

Table 3 lists the geometry and operating conditions for an industrial ESR process for the production of ingots of IN 718 analyzed in the present study. Note that the process uses ac current at a single sinusoidal frequency of 60 Hz.

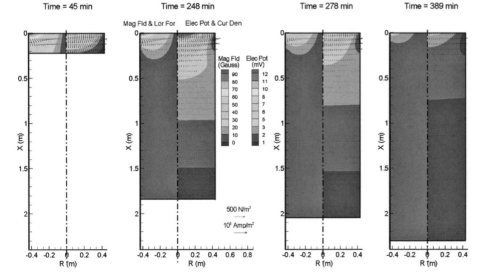

Fig. 6 Distributions of the current density and self-induced magnetic field in the ingot during a vacuum arc remelting process for Ti-6Al-4V alloy without magnetic stirring

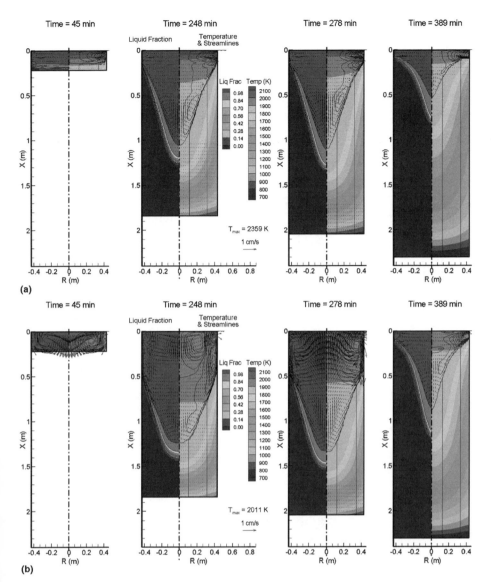

Fig. 7 Liquid fraction, flow, and temperature fields in the ingot during a vacuum arc remelting process for Ti-6Al-4V alloy (a) without and (b) with magnetic stirring

Process Conditions and Computational Details. The model considers the conductivity and specific heat of alloy IN 718 to be temperature dependent. For brevity, material properties of the slag and representative values of the properties of the alloy are also provided in Table 3. Finally, a nonlinear variation of the liquid fraction with temperature in the mushy region, shown in Fig. 12, is used in the analysis. A computational grid of 100 (axial x) by 80 (radial r) control volumes is used in the analysis. The model uses unsteady analysis for the determination of the steady-state conditions in the slag and the ingot. A total of 3 h were required for this analysis on a single-processor personal computer with an Intel Pentium4 3.2 GHz processor.

Results and Discussion. The computational model predicts field variations of the electromagnetic, flow, and temperature fields and the metallurgical structure quantities for steady-state conditions. These variations are presented in the form of various field plots in the ESR system and radial variations in the solidified ingot.

Figure 13(a) shows the instantaneous contours of magnetic flux density and the vectors of current density, and the contours of cycle-averaged Joule heating and vectors of Lorentz force. The current is nonuniformly distributed over the electrode surface, and, after entering the slag, it tends to become more uniform due to the low electrical conductivity of the slag. However, as the current enters the ingot, it migrates radially outward and flows in a small region near the outer radius of the ingot. This corresponds to the skin effect that occurs in ac power. It is noteworthy that, unlike the current density and magnetic field, Lorentz force does not change direction within one ac cycle.

Further, at any spatial location, the local Lorentz force and Joule heating show rapid oscillation (at a frequency twice that of the alternating current) of small amplitudes around the corresponding large cycle-averaged values. Because of the mass and heat capacity, the flow and temperature fields are influenced only by the cycle-averaged values of the Lorentz force and Joule heating. Due to the low electrical conductivity of the slag, the volumetric heating is concentrated in the slag. The Lorentz force is directed radially inward and axially downward in the direction perpendicular to the local current-density vector.

Figure 13(b) shows the contours of the temperature and liquid fraction fields, velocity vectors, and streamlines in the slag and ingot regions. The slag is seen to be hotter than the metal. The heat absorbed by the melting electrode results in an unstable thermal stratification of the slag in the region under the electrode. As a result, there are two recirculating cells in the slag, with a downward motion of the colder slag in the center and over the slag-mold boundary. The metal pool is thermally stably stratified. It involves a single recirculating cell with downward motion along the pool boundary and a slow upward motion in the center region of the pool. The velocities in the slag are an order of magnitude higher than those in the metal. The thermal gradient in the outer mushy region is lower due to the rapid release of latent heat coupled with the convective heat transfer in the pool. The temperature in the solidified ingot gradually decreases as it loses heat to the mold wall. Figure 14 shows the axial variations of the effective heat-transfer coefficient from circumferential surfaces of the slag and ingot to the cooling water and the resulting heat flux. In the slag region, the heat-transfer coefficient is large because the slag skin is in contact with the mold. This behavior also persists in the small contact region in the upper portion of the ingot, where a short liquid metal head is formed. Therefore, the heat flux corresponding to heat loss to the cooling water is very high in the molten slag and contact region of the ingot. As the ingot starts to solidify, it separates from the mold, causing the heat transfer by contact to diminish. As a result, the effective heat-loss coefficient and corresponding heat flux decrease very rapidly. After the ingot separates completely from the mold, heat loss occurs only by radiation and convection in the gap between the ingot and the mold. Therefore, the heat-transfer coefficient and heat flux are significantly lower than that in the contact region. The heat loss occurs by radiation to the chamber walls and convection to the air in the chamber over the portion of the ingot extending beyond the mold. Because the convective heat-loss coefficient and effective emissivity for the exposed ingot are somewhat higher than their values within the mold, the heat-loss quantities show a small discontinuous increase at the location where the ingot becomes exposed.

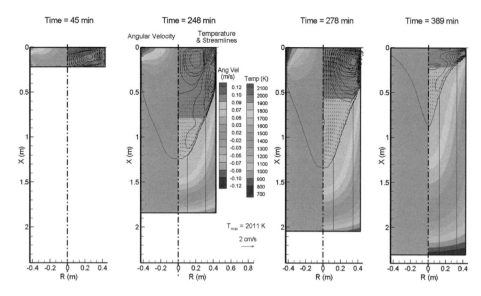

Fig. 8 Angular velocity, in-plane flow, and temperature fields in the ingot during a vacuum arc remelting process for Ti-6Al-4V alloy with magnetic stirring

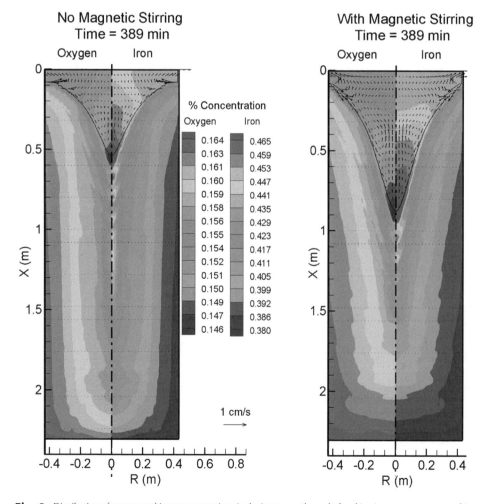

Fig. 9 Distribution of oxygen and iron concentrations in the ingot near the end of melting in a vacuum arc remelting process for Ti-6Al-4V alloy without and with magnetic stirring

Figure 15 shows the radial variations of LST and the primary and secondary dendrite arm spacings in the solidified ingot. The LST is lower at the outer radius than at the center of the ingot because the liquid metal in the outer region of the ingot loses heat to the cooling water more easily. Therefore, the cooling rate increases, and the resulting dendrite arm spacings decrease from the center to the outer radius. Figure 16 shows the radial variations of the Rayleigh number for interdendritic convection that is responsible for freckle formation for two different values of the assumed freckle-initiation solid fractions. It is noteworthy that the Rayleigh number shows a maximum value at midradius location. This is because the ease (length scale) of interdendritic convection is dependent on the angle of the gravity vector relative to the primary dendrites. While the primary dendrite spacing decreases from the center to the outer region, this angle and hence the ratio of the effective permeability to the primary permeability (the relative ease of flow) increases with the radial direction. As a result, the effective length scale, and hence the Rayleigh number, attains its maximum value at a midradius location. Therefore, freckle formation is more likely to occur in the midradius region of the ingot than in the center or circumferential regions. These predictions are consistent with the observed behavior.

Conclusions and Future Work

This article describes comprehensive, robust, and efficient computational models for axisymmetric analysis of electromagnetic, flow, heat-transfer, and phase-change phenomena in the VAR and ESR processes used for the production of ingots of titanium alloys, superalloys, and steels. The computational models also use the transient flow field to analyze macrosegregation phenomena in titanium alloys for the prediction of alloying element distributions in the solidified ingots produced in the VAR process. Similarly, the thermal history of the solidifying metal is used to calculate the local solidification time, metallurgical structure parameters, and Rayleigh number for interdendritic convection to predict the probability of freckle formation throughout the solidified ingots of superalloys and steels. The results of the application for the models for a practical VAR process of Ti-6Al-4V alloy and an ESR process of IN 718 alloy illustrate the ability of the model to provide insights into the physical phenomena that govern the process behavior. Further, quantitative information about the process performance obtained from the models can be used in optimizing process operation to produce ingots with the desired composition and microstructure.

The accuracy of their prediction can be further improved and the scope of their application increased by addressing the following aspects of the computational models:

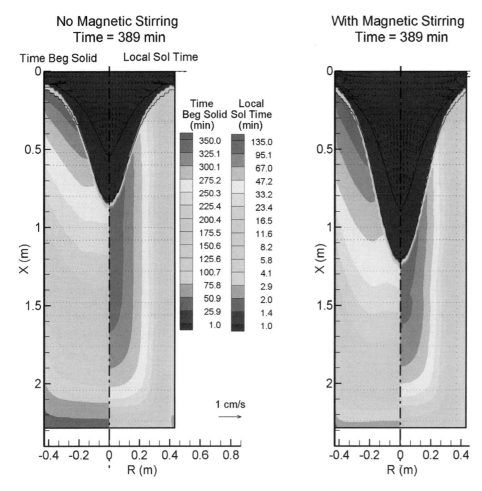

Fig. 10 Variation of local solidification time and time at the beginning of solidification in the ingot near the end of melting in a vacuum arc remelting process for Ti-6Al-4V alloy without and with magnetic stirring

- *Thermophysical properties:* Accurate knowledge of the thermophysical properties of the liquid metal and slag is necessary for an accurate prediction of ingot structure. Although values for the physical properties of both liquid slags (Ref 86) and alloys (Ref 87–90) are available, their accuracy is seldom better than 10% of the value and, in many cases, little better than an estimate. Available software tools such as JMatPro (Ref 91) and Matprop (Ref 92) provide accurate predictions of the properties of solid alloys. However, the accuracy of the predictions of the liquid alloy properties is uncertain. Added uncertainty lies in the radiative heat-transfer aspects of the ESR system. The liquid slags are diathermanous (Ref 93), and the radiative effects in heat transfer should be taken into account, particularly when modeling temperature gradients close to the opaque boundaries of the slag. Also in a parallel manner to the behavior of continuous casting fluxes, it is to be expected that the solid film that deposits from the slag onto the cooled mold surface of the ESR system will have some level of transparency in the operating wavelength range. This factor must be taken into account in defining the bounding heat-transfer mechanisms in future models of the process.

- *ESR process variants:* Variants of the ESR processes, such as electroslag rapid remelting, that employ current-conducting molds to obtain better control over the melt rate are now being considered (Ref 94). Therefore, the models for the standard ESR process must be extended to analyze the advanced versions of the ESR process and to include the possibility of current-path variation.

- *Analysis in three dimensions:* Because most remelted ingots are cylindrical and made from single, cylindrical electrodes, it is reasonable to use axisymmetric analysis for routine engineering use. However, in both ESR and VAR, the current flow to the mold can take place at specific points around the process periphery. Further, in the VAR process, work on the magnetic fields generated by arc movement (Ref 49) has also demonstrated the nonsymmetry of the arc processes. Electroslag remelting practice sometimes employs nonaxisymmetric rectangular slab molds. In both processes, some of the difficulties experienced with the

reproducibility of ingot properties are associated with this three-dimensional behavior (Ref 95). Therefore, the existing axisymmetric models must be extended to three dimensions to analyze these effects. The use of such unsteady three-dimensional analysis will become increasingly feasible as the computing capacity continues to increase in the future.

REFERENCES

1. B.E. Paton, B.I. Medovar, V.F. Demchenko, J.G. Khorungy, V.L. Shevtsoz, J.P. Shtanko, and A.G. Bogachenko, Calculation of Temperature Fields in Plate Ingots and in Ingots of ESR, *Proc. Fifth International Symposium on ESR and Other Technologies,* G.K. Bhat and A. Simkovich, Ed., Mellon Institute, 1974, p 323–345

2. G.F. Ivanova and N.A. Avdonin, Determining the Temperature Field and the Melting Rate of an Electrode in ESR, *Inzh.-Fiz. Zh.,* Vol 20, 1971, p 87–95

3. B.I. Medovar, Chapter III, *Metallurgy of Electroslag Processes,* E.O. Paton Institute, Kiev, 1986, p 47–78

4. V.L. Shevtsov, A.M. Pal'ti, and Y.M. Kamenskii, Temperature Calculation for the Metal Flowing Down from a Melted Consumable-Electrode Tip, *Probl. Spets. Elektrometall.,* Vol 4, 1986, p 16–19

5. I.P. Borodin, V.A. Goryainov, V.S. Koshman, and M.V. Borodina, Solidification of EShP [Electroslag Remelting] Ingots, *Probl. Spets. Elektrometall.,* Vol 4, TsNIITtyazhmash, Sverdlovsk, USSR, 1986, p 19–22

6. Y.G. Emel'yanenko, S.Y. Andrienko, and L.N. Yasnitskii, Study of the Hydrodynamics in Electroslag Remelting by Physical and Mathematical Modeling, *Probl. Spets. Elektrometall.,* Vol 4, 1987, p 5–7

7. F.V. Nedopekin and V.M. Melikhov, Numerical Study of Thermal and Electrical Processes in Electroslag Remelting, *Protsessy Razlivki Stali i Kachestvo Slitka,* Kiev, 1989, p 65–68

8. Y.S. Andrienko, Electroslag Processes under Eddy Current Flow of Melts, *Izv. Akad. Nauk SSSR Met.,* Vol 3, 1991, p 31–36

9. A. Mitchell, J. Szekely, and J.F. Elliot, Heat Transfer Modeling of the ESR Process, *Proc. Int. Symp. Electroslag Refining,* Iron & Steel Soc., United Kingdom, 1973, p 1–14

10. A. Mitchell and S. Joshi, Thermal Fields in ESR Ingots, *Met. Trans.,* Vol 2, 1971, p 449–455

11. B. Hernandez-Morales and A. Mitchell, Review of Mathematical Models of Fluid Flow, Heat Transfer and Mass Transfer in ESR, *Ironmaking Steelmaking,* Vol 26, 1999, p 423–438

12. A.S. Ballantyne and A. Mitchell, Modeling of Ingot Thermal Fields in Consumable Electrode Remelting Processes,

No Magnetic Stirring
Time = 248min

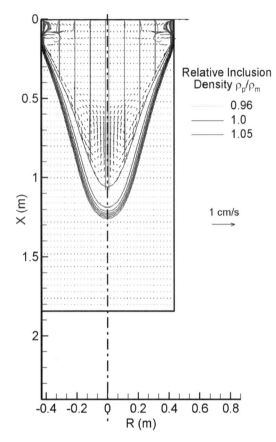

Relative Inclusion
Density ρ_p/ρ_m

— 0.96
— 1.0
— 1.05

1 cm/s
⟶

With Magnetic Stirring
Time = 248min

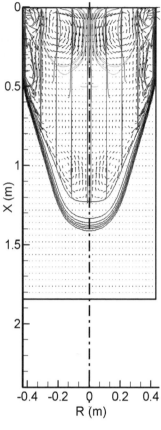

Table 3 Geometry and operating conditions and material properties of alloy IN 718 used in the analysis of a practical electroslag remelting process

Characteristic	Value
Geometry	
Slag height	127 mm
Electrode diameter	431.8 mm
Ingot diameter	355.6 mm
Operating condition	
Electric current	6.5 kA
Total power dissipation	250 kW
Sink temperature, exposed slag surface	400.15 K
Temperature of the cooling water	298.15 K
Alloy properties	
Density, liquid	7500 kg/m^3
Viscosity, liquid	6×10^{-3} kg/m·s
Liquidus temperature	1623 K
Solidus temperature	1473 K
Latent heat of fusion	2.1×10^5 J/kg
Specific heat, liquid	720 J/kgK
Thermal expansion coefficient, liquid	1.5×10^{-4} K^{-1}
Thermal conductivity, liquid	30.52 W/mK
Thermal conductivity, solid	16.72 W/mK
Electric conductivity, liquid	7.1×10^5 (Ω·m)$^{-1}$
Magnetic permeability	1.257×10^{-6} H/m
Slag properties	
Density	2800 kg/m^3
Viscosity	2.5×10^{-3} kg/(m·s)
Specific heat	1255 J/kgK
Thermal expansion coefficient	2.5×10^{-4} K^{-1}
Thermal conductivity	4.0 W/mK

Fig. 11 Trajectories of inclusions of different sizes and densities at a representative time instant during a vacuum arc remelting process for Ti-6Al-4V alloy without and with magnetic stirring

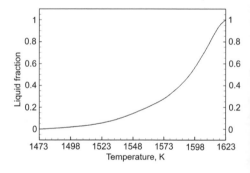

Fig. 12 Variation of the volumetric liquid fraction with temperature in the mushy region for alloy IN 718

Ironmaking Steelmaking, Vol 4, 1977, p 222–234

13. C.L. Jeanfils, J.H. Chen, and H.J. Klein, Temperature Distribution in and ESR Ingot during Transient Conditions, *Proc. Sixth International Vacuum Metallurgy Conf.,* AVS, New York, 1979, p 452–546

14. M.L. Zhadkevich, S.V. Akhonin, and V.N. Boronenkov, Mathematical Simulation of the Refining Processes in Special Electrometallurgy, *Weld. Surf. Rev.,* Vol 6 (No. 2), 1996, p 137–143

15. Z. Jiang and X. Jiang, Mathematical Model of Heat-Generation Distribution in the Electroslag Remelting Bath, *Dongbei Gongxueyuan Xuebao,* Vol 54, 1988, p 63–69

16. J. Kreyenberg, K.H. Tacke, and K. Schwerdtfeger, Prediction of Melting Rate of the Electrode in ESR, *Arch. Eisenhuettenwes.,* Vol 55 (No. 8), 1984, p 369–371

17. T. Umeda, C.P. Hong, and Y. Kimura, Solidification Simulation of ESR Ingots, *Taehan Kumsok Hakhoechi,* Vol 22 (No. 2), 1984, p 104–109

18. B.V. Rao, D.G. Subba, S.J. Emmanual, and J.S.N. Sam, Solidification Modeling of Vacuum Arc Remelted Superalloy 718 Ingot in Superalloys 718, 625, 706 and Various Derivatives, *Proceedings of the International Symposium on Superalloys 718, 625, 706 and Various Derivatives,* E. Loria, Ed., TMS, Warrendale, PA, 1997, p 67–76

19. C.L. Jeanfils, J.H. Chen, and H.J. Klien, Modeling of Macrosegregation in ESR of Superalloys, *Superalloys 1980,* AIME, New York, 1980, p 119–130

20. S.D. Ridder, F.C. Reyes, S. Chakravorty, R. Mehrabian, J.D. Nauman, J.H. Chen, and H.J. Klein, Steady State Segregation and Heat Flow in ESR, *Met. Trans. B,* Vol 9, 1979, p 415–422

21. D. Ablitzer, Transport Phenomena and Modeling in Melting and Refining Processes, *J. Phys.,* Vol 3, 1993, p 873–882

22. J. Wei and Y. Ren, Mathematical Simulation of a Magnetic Field in an Electroslag Remelting System, *Jinshu Xuebao,* Vol 31 (No. 2), 1995, p B52–B59

23. A.S. Ballantyne, A. Mitchell, and R.L. Kennedy, Computer Simulation of Cold Crucible Ingots, *Proc. Fifth International Symposium on ESR and Other Technologies,* G.K. Bhat and A. Simkovich, Ed., Mellon Institute, 1974, p 345–410

24. J. Kreyenberg and K. Schwerdtfeger, Stirring Velocities and Temperature Field in the Slag during Electroslag Remelting, *Arch. Eisenhuettenwes.,* Vol 5, 1979, p 1–6

25. A.H. Dilwari and J. Szekely, Mathematical Model of Slag and Metal Flow in the ESR Process, *Met. Trans. B,* Vol 8, 1977, p 227–236

26. L. Medovar, B. Tsyykulenko, A.V. Chernets, B.B. Fedorovskii, V.E. Shevchenko, I. Lantsman, and V. Petrenko, Investigation of the Effects of an ESR Two-Circuit Diagram on the Size and Shape of the Metal

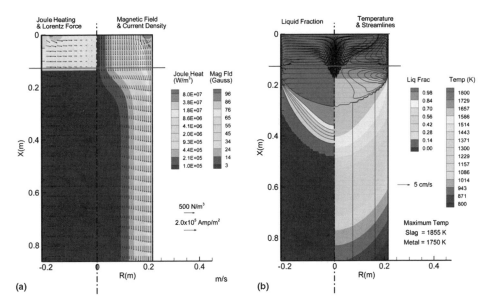

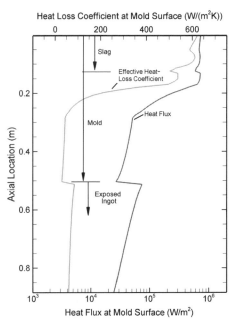

Fig. 13 Electromganetic, flow, temperature, and liquid fraction fields in the slag and ingot during an electroslag remelting process for IN 718

Fig. 14 Axial variations of local heat flux and effective heat-loss coefficient over the circumferential surfaces of the slag and ingot during an electroslag remelting process for IN 718

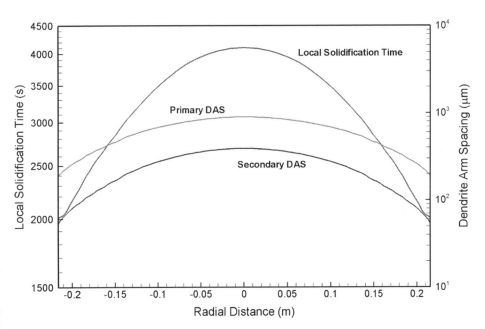

Fig. 15 Predicted radial variations of the local solidification time and dendrite arm spacings (DAS) in the solidified IN 718 ingot formed in an electroslag remelting process under steady-state conditions

Pool, *Probl. Spets. Elekrometall.*, Vol 4, 2000, p 3–7

27. M. Kawakami, K. Nagata, Y. Minoru, S. Naoki, Y. Miyashita, and K. Goto, Profiles of Temperature and Local Heat Generation in ESR, *Tetsu-to-Hagane*, Vol 63, 1977, p 2162–2171

28. M. Choudhary, J. Szekely, B. Medovar, and Y.G. Emel'yanenko, The Velocity Field in the Slag Region of ESR, *Met. Trans. B*, Vol 13, 1982, p 35–43

29. M. Choudhary and J. Szekely, A Comprehensive Representation of Heat and Fluid Flow Phenomena in ESR Systems, *Trans. Iron Steel Soc. AIME*, Vol 3, 1983, p 67–75

30. G. Brueckmann, G. Sick, and K. Schwedtfeger, Slag Movement in ESR of Steel, *Met. Trans. B*, Vol 14, 1983, p 761–764

31. P. Guo, J. Zhang, and Z.-B. Li, Analysis of Factors Influencing Slag Flow in ESR System, *Gangtie Yanjiu Xuebao*, Vol 12, 2000, p 7–10

32. J. Wei and Y. Ren, Mathematical Modeling of Slag Flow Field in ESR System, *Jinshue Xuebao*, Vol 30, 1994, p B481–B490

33. T. Tang and X. Jiang, Mathematical Modeling for the Shape of the Metallic Bath in the Electroslag Remelting Process, *Dongbei Nongxueyuan Xuebao*, Vol 3, 1985, p 78–82

34. B.E. Paton, B.I. Medovar, Y.G. Emel'yanenko, E.V. Shcherbinin, S. Andrienko, Y.A. Chudnovskii, and A.I. Chaikovskii, Magnetohydrodynamic Phenomena in a Slag Pool during Electroslag Melting, *Probl. Spets. Elektrometall.*, Vol 17, 1982, p 3–8

35. L. Willner and F. Varhegyi, Mathematical Model of ESR, Approximating to Service Conditions, *Archiv. Eisenhuttenwes.*, Vol 47, 1976, p 205–209

36. A. Jardy, D. Ablitzer, and J.-F. Wadier, Magnetohydrodynamic and Thermal Behavior of Electroslag Remelting Slags, *Metall. Trans. B, Process. Metall.*, Vol 22 (No. 1), 1991, p 111–120

37. I.V. Chumanov and V.E. Roshchin, Characteristics of Electroslag Remelting Modeling When Using Transparent Models, *Izv. V.U.Z. Chernaya Metall.*, Vol 8, 1998, p 30–35

38. B.R. Mateev, Simulation of the Electroslag Process Using a Low-Temperature Transparent Model, *Zavaryavane (Sofia)*, Vol 12 (No. 3), 1984, p 5–8

39. C. Bratu, F.Stefanescu, L. Sofroni, T. Suzuki, and K. Morita, Prediction of VAR Solidification Conditions by Using Numerical Analysis, *First ISS Tech Conf. Proc.*, April 27–30, 2003 (Indianapolis,

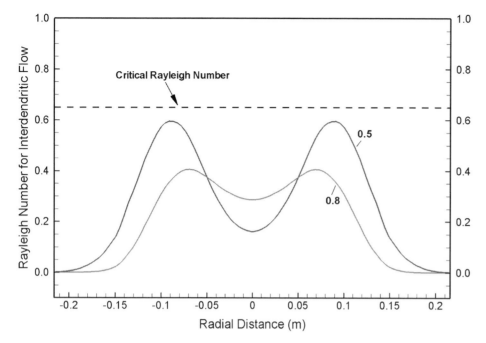

Fig. 16 Predicted radial variations of the interdendritic Rayleigh number for two different freckle-initiation solid fractions in the solidified IN 718 ingot formed in an electroslag remelting process under steady-state conditions

IN), Iron & Steel Society, TMS, Warrendale, PA, 2003, p 1137–1145

40. M.D. Evans and G.E. Kruzynski, Causes and Effects of Centre Segregation in ESR Alloy 718 for Critical Part Applications, *Superalloys 1988,* D.N. Duhl et al., Ed., TMS, Warrendale, PA, 1988, p 91–100

41. Modeling of Electroslag Remelting, *Metal. (Bucharest),* Vol 39 (No. 4), 1987, p 156–159

42. A. Jardy, A.F. Wilson, D. Lasalmonie, and D. Ablitzer, Development of an Enhanced Mathematical Model to Simulate Vacuum Arc Remelting of Titanium and Zirconium Alloys, *Proc. International Symposium on Liquid Metals,* A. Mitchell and J. Vanden Avyle, Ed., AVS, New York, 2001, p 200–211

43. L.A. Bertram, C.B. Adasczik, D.G. Evans, R.S. Minisandram, P.A. Sackinger, D.D. Wegman, and R.L. Williamson, Quantitative Simulation of a Superalloy VAR Ingot at the Macroscale, *Proc. International Symposium on Liquid Metals,* A. Mitchell and P. Auburtin, Ed., AVS, New York, 1997, p 110–133

44. L.A. Bertram, R.S. Minisandram, and K.O. Yu, Vacuum Arc Remeleting and Electroslag Remelting, *Modeling for Casting and Solidification Processing,* K.O. Yu, Ed., Marcel Dekker, 2002, p 565–612

45. R.M. Lothian, P.D. Lee, M. McLean, R.M. Ward, T.P. Johnson, and M.H. Jacobs, Growth Modeling in VAR, *Proc. International Symposium on Liquid Metals,* A. Mitchell and P. Auburtin, Ed., AVS, New York, 1997, p 133–145

46. D. Grandemange, Y. Combres, B. Champin, A. Jardy, S. Hans, and D. Ablitzer, Modeling of Heat, Momentum and Solute Transfer to Optimize the Melting Parameters of VAR, *Proc. International Symposium on Liquid Metals,* A. Mitchell and P. Auburtin, Ed., AVS, New York, 1997, p 204–214

47. R.S. Minisandram, M.J. Arnold, and R.L. Williamson, VAR Pool Depth Measurement and Simulation for a Large Diameter Ti-6Al-4V Ingot, *Proc. International Symposium on Liquid Metals,* P.D. Lee et al., Ed., ASM International, 2005, p 1–7

48. G. Reiter, W. Shutzenhofer, P. Wurtzinger, and S. Zinner, Optimisation and Detailed Validation of a VAR Model, *Proc. International Symposium on Liquid Metals,* P.D. Lee et al., Ed., ASM International, 2005 p 7–13

49. R.M. Ward, B. Daniel, and R.J. Siddall, Ensemble Arc Motion and Solidification during the VAR of a Nickel-Base Superalloy, *Proc. International Symposium on Liquid Metals,* P.D. Lee et al., Ed., ASM International, 2005, p 49–57

50. S. Hans, S. Ryberson, H. Poisson, and P. Heritier, Industrial Applications of VAR Modeling for Special Steels and Nickel-Base Superalloys, *Proc. International Symposium on Liquid Metals,* P.D. Lee, Ed., ASM International, 2005 p 65 – 71

51. K.M. Kelkar, S.V. Patankar, and A. Mitchell, Computational Modeling of the Electroslag Remelting (ESR) Process Used for the Production of Ingots of High-Performance Alloys, *Proc. International*

Symposium on Liquid Metals, P.D. Lee, Ed., ASM International, 2005, p 137–145

52. S. Hans, A. Jardy, and D. Ablitzer, A Numerical Model for the Prediction of Transient Turbulent Flow, Heat Transfer and Solidification during VAR, *Proc. International Symposium on Liquid Metals,* A. Mitchell and J. Fernihough, Ed., AVS, New York, 1994, p 143–155

53. T. Quatravaux, S. Ryberon, S. Hans, A. Jardy, B. Lusson, P.E. Richy, and D. Ablitzer, Transient VAR Ingot Growth Modeling: Application to Nickel-Base and Steel Alloys, *Proc. International Symposium on Liquid Metals,* P.D. Lee et al., Ed., SF2M, Paris, 2003, p 19–29

54. G. Reiter, V. Marronier, C. Sommitsch, M. Gaumann, W. Schutzenhofer, and R. Schneider, Numerical Simulation of VAR with Calcosoft 2D and Its Validation, *Proc. International Symposium on Liquid Metals,* P.D. Lee et al., Ed., SF2M, Paris, 2003, p 19–29

55. A.D. Patel, Analytical Model for Electromagnetic Fields in ESR and VAR Processes, *Proc. International Symposium on Liquid Metals,* P.D. Lee et al., Ed., SF2M, Paris, 2003, p 205–215

56. W. Schutzenhofer, G. Reiter, R. Tanzer, H. Scholz, R. Sorci, F. Arcobello-Varlese, and A. Carosi, Experimental Investigations for the Validation of a Numerical PESR Model, *Proc. International Symposium on Liquid Metals,* P.D. Lee et al., Ed., SF2M, Paris, 2003, p 49–55

57. V. Weber, A. Jardy, B. Dussoubs, D. Ablitzer, S. Ryberon, V. Schmidt, S. Hans, and H. Poisson, A Comprehensive Model of ESR: Description and Validation, *Proc. Liquid Metals Processing and Casting Conference,* P.D. Lee et al., Ed., SF2M, 2007, p 83–89

58. Z. Jiang and Y. Dong, Solidification Model for ESR, *Proc. International Symposium on Liquid Metals,* P.D. Lee et al., Ed., SF2M, Paris, 2003, p 89–95

59. A. Kharicha, W. Schutzenhofer, A. Ludwig, and R. Tanzer, Multiphase Modeling of the Slag Region of ESR, *Proc. International Symposium on Liquid Metals,* P.D. Lee et al., Ed., SF2M, Paris, 2003, p 107–113

60. A. Jardy and D. Ablitzer, Slag Motion and Heat Transfer during Electroslag Remelting, *Mém. Étud. Sci. Rev. Métall.,* Vol 86 (No. 4), 1989, p 225–235

61. A. Jardy and S. Hans, Use of Numerical Modelling to Study the Impact of Operating Parameters on the Quality of a Vacuum Arc Remelted Ingot, *Modeling of Casting, Welding and Advanced Solidification Processes XI, Proceedings from the 11th International Conference on Modeling of Casting, Welding and Advanced Solidification Processes,* C.-A. Gandin and M. Bellet, Ed., May 28–June 2, 2006 (Opio, France), TMS, Warrendale, PA, 2006, p 953–960

62. T. Quatravaux, S. Ryberon, S. Hans, A. Jardy, B. Lusson, P.E. Richy, and D. Ablitzer, Transient VAR Ingot Growth Modelling: Application to Specialty Steels, *J. Mater. Sci.,* Vol 39 (No. 24), 2004, p 7183–7191

63. K.M. Kelkar, S.V. Patankar, A. Mitchell, O. Kanou, N. Fukada, and K. Suzuki, Computational Modeling of the Vacuum Arc Remelting (VAR) Process Used for the Production of Ingots of Titanium Alloys, *Proc. of the Ti-2007 Science and Technology,* M. Ninomi, S. Akiyama, M. Ikeda, M. Hagiwara, and K. Maruyama, Ed., *The Japan Institute of Metals,* Vol II, 2007, p 1279–1282

64. S. Viswanathan, A.D. Patel, K.M. Kelkar, D.G. Evans, D.K. Melgaard, R.S. Minisandram, and S.V. Patankar, 2D Steady-State Analysis of Temperature Profiles, Fluid Flow, and Pool Shape in the ESR Process, *Modeling of Casting, Welding and Advanced Solidification Processes XI, Proceedings from the 11th International Conference on Modeling of Casting, Welding and Advanced Solidification Processes,* May 28–June 2, 2006 (Opio, France), Minerals, Metals & Materials Society, Warrendale, PA, 2006, p 977–984

65. P.A. Davidson, X. He, and A.J. Lowe, Flow Transitions in Vacuum Arc Remelting, *Mater. Sci. Technol.,* Vol 16 (No. 6), 2000, p 699–711

66. ANSYS FLUENT Flow Modeling Software, www.fluent.com, accessed Aug 25, 2009

67. *MeltFlow-VAR and MeltFlow-ESR: Software Tools for the Analysis for Ingot Remelting Processes,* Innovative Research, Inc., Plymouth, MN, www.inres.com, accessed Nov 4, 2009

68. L. Nastic, S. Sundarraj, K.O. Yu, and Y. Pang, Stochastic Modeling of Grain Structure Formation during the Solidification of Superalloy and Ti Alloys Remelt Ingots, *Proc. International Symposium on Liquid Metals,* A. Mitchell and P. Auburtin, Ed., AVS, New York, 1997, p 145–166

69. X. Xu, R.C. Atwood, S. Sridar, P.D. Lee, M. McLean, B. Drummings, R.M. Ward, and M. Jacobs, Grain Size Predictions in VAR: A Critical Comparison of Micromodeling, *Proc. International Symposium on Liquid Metals,* A. Mitchell, L. Ridgway, and M. Baldwin, Ed., AVS, New York, 1999, p 76–90

70. R.C. Atwood, R. Minisandram, R.M. Forbes-Jones, and P.D. Lee, Multiscale Modeling of Microstructure Formation during VAR of Titanium 6-4, *Proc. International Symposium on Liquid Metals,* P.D. Lee et al., Ed., SF2M, Paris, 2003, p 215–223

71. L. Yuan, G. Djambazov, K. Pericleous, and P.D. Lee, Multiscale Modeling of Dendritic Growth during VAR, *Proc. International Symposium on Liquid Metals,* P.D. Lee et al., Ed., SF2M, Paris, 2003, p 43–49

72. X.Q. Wei and L. Zhou, A Cellular Automaton Approach to Simulation of Grain Structure Development in Electroslag Casting, *Acta Metall. Sin. (Engl. Lett.),* Vol 13 (No. 2), 2000, p 794–799

73. L. Nastac, S. Sundarraj, and K.O. Yu, Stochastic Modeling of Solidification Structure in Alloy 718 Remelt Ingots, Superalloys 718, 625, 706 and Various Derivatives, *Proceedings of the International Symposium on Superalloys 718, 625, 706 and Various Derivatives*, E. Loria, Ed., June 15–18, 1997 (Pittsburgh, PA), TMS, Warrendale, PA, 1997, p 55–66

74. X. Xu, W. Zhang, and P.D. Lee, Tree-Ring Formation during Vacuum Arc Remelting of INCONEL 718: Part II, Mathematical Modeling, *Metall. Mater. Trans. A: Phys. Metall. Mater. Sci.,* Vol 33 No. (6), 2002, p 1805–1815

75. W. Zhang, P.D. Lee, and M. McLean, Numerical Simulation of Dendrite White Spot Formation during Vacuum Arc Remelting of Inconel 718, *Metall. Mater. Trans. A: Phys. Metall. Mater. Sci.,* Vol 33 (No. 2), 2002, p 443–454

76. W. Zhang, P.D. Lee, M. McLean, and R.J. Siddall, Simulation of Intrinsic Inclusion Motion and Dissolution during the Vacuum Arc Remelting of Nickel Based Superalloys, Superalloys 2000, *Proceedings of the Ninth International Symposium on Superalloys,* T.M. Pollock, Ed., Sept 17–21, 2000 (Seven Springs, PA), p 29–37

77. W. Zhang, P.D. Lee, and M. McLean, Inclusion Behaviour during Vacuum Arc Remelting of Nickel Based Superalloys, *EUROMAT 99,* Vol 10, D.G. Morris, S. Naka, and P. Caron, Ed., Wiley-VCH Verlag GmbH, Weinheim, Germany, 2000, p 123–128

78. P.D. Lee, B. Lothian, L.J. Hobbs, and M. Mclean, Coupled Macro-Micro Modeling of the Secondary Melting of Turbine Disk Superalloys, *Superalloys 1996,* R.D. Kissinger et al., Ed., TMS, Warrendale, PA, 2006, p 435–442

79. W. Rodi, Turbulence Models and Their Application in Hydraulics, *International Association for Hydraulic Research (IAHR)—Section on Fundamentals of Division II: Experimental and Mathematical Fluid Dynamics,* Rotterdamseweg, 185, 2600 MH Delft, The Netherlands

80. A.D. Brent, V.R. Voller, and K.J. Reid, Enthalpy-Porosity Technique for Modeling Convection-Diffusion Phase Change, *Numer. Heat Transf.,* Vol 13, 1988, p 297–318

81. P. Auburtin, T. Wang, S.L. Cockroft, and A. Mitchell, Freckle Formation and Freckle Criterion in Superalloy Castings, *Metall. Mater. Trans. B,* Vol 31, 2000, p 801–811

82. S.V. Patankar, *Numerical Heat Transfer and Fluid Flow,* Hemisphere, 1980

83. J.C. Slattery, *Momentum, Energy and Mass Transfer in Continua,* Robert E. Krieger Publishing Company, 1981

84. I. Demirdzic and M. Peric, Space-Conservation Law in Finite Volume Calculations of Fluid Flow, *Int. J. Numer. Methods Fluids,* Vol 8, 1988, p 1037–1050

85. C.T. Crowe, M.P. Sharma, and D.E. Stock, The Particle-Source in Cell (PSI-Cell) Model for Gas-Droplet Flows, *Trans. ASME, J. Fluids Eng.,* Vol 30, 1977, p 325–332

86. K.C. Mills and B.J. Keene, Physicochemical Properties of CaF_2-Based Slags, *Int. Metall. Rev.,* Vol 26, 1981, p 21–69

87. A.F. Crawley, Densities of Liquid Metals and Alloys, *Int. Metall. Rev.,* Vol 9, 1974, p 32–51

88. L.D. Lucas, *Physicochemical Measurements in Metals Research,* Vol 1V (Part 2), R.A. Rapp, Ed., Interscience, New York, 1970

89. P.N. Quested, K.C. Mills, R.F. Brooks, A.P. Day, R. Taylor, and H. Szelaogowski, Physical Property Measurements for Simulation Modelling, *Proc. International Symposium on Liquid Metals,* A. Mitchell and P. Auburtin, Ed., AVS, New York, 1997, p 1–18

90. N. Saunders, A.P. Miodownik, and J.-P. Schille, Modeling of the Thermo-Physical and Physical Properties for Solidification of Nickel-Base Superalloys, *Proc. International Conference on Liquid Metals,* P.D. Lee et al., Ed., ASM International, 2005, p 113–129

91. JMatPro, Sentes Software, http://www.sentesoftware.co.uk/, accessed Aug 25, 2009

92. Matprop, Stanford Linear Accelerator Center, Stanford University, www.slac.stanford.edu/comp/physics/matprop.html, accessed Aug 25, 2009

93. A. Mitchell and J.-F. Wadier, Some Observations on the Radiative Properties of ESR Slags, *Can. Metall. Q.,* Vol 20, 1981, p 373–386

94. S.V. Tomilenko, V.I. Us, Yu M. Kuskov, I.A. Lantsman, E.V. Shevtsov, and E.D. Gladkii, Modeling of Magnetohydrodynamic Processes during Electroslag Melting in an Electrically Conductive Mold, *Probl. Spets. Elektrometall.,* Vol 4, 1992, p 19–21

95. F. Zanner, R.L. Williamson, and R. Erdman, On the Origin of Defects in VAR Ingots, *Proc. International Conference on Liquid Metals,* P.D. Lee et al., Ed., ASM International, 2005, p 13–29

ASM Handbook, Volume 22B, Metals Process Simulation
D.U. Furrer and S.L. Semiatin, editors

Copyright © 2010, ASM International®
All rights reserved.
www.asminternational.org

Formation of Microstructures, Grain Textures, and Defects during Solidification

A. Jacot, Ecole Polytechnique Fédérale de Lausanne and CALCOM ESI SA, Switzerland
Ch.-A. Gandin, MINES-ParisTech, France

IN MOST MANUFACTURING PROCES-SES, the structures formed during solidification influence to a large extent the final mechanical properties of the product. (In this article, *macrostructure* refers to the structure of primary dendritic grains only, or the structure of eutectic grains in eutectic alloys, while *microstructure* refers to the internal structure of these grains, such as dendritic, peritectic, and eutectic patterns. The term *structures* covers both the micro- and macrostructures. The terms *micro-* and *macroscopic* are also used to refer to the corresponding length scales.) The influence of the solidification structures is not only important for casting processes but also for wrought alloy materials, because the behavior during subsequent processes, such as homogenization, aging, rolling, extrusion, and so on, is largely dependent on the as-cast state. Strict specifications of the as-cast state are therefore generally imposed regarding the grain size, type, volume fraction, and morphology of the various microstructure constituents. As part of the process of controlling and optimizing the as-cast structures, numerical simulation is becoming increasingly important.

Macroscopic models of solidification, which describe thermal exchange and fluid flow on the scale of casting processes, are now well advanced and have been used in practice for many years (Ref 1, 2). A general tendency is to incorporate into such simulations a description of the solidification structures. Various methods are used, depending on the aspects to be addressed. Some of them can be applied to the entire casting, by performing a calculation at every node of the grid used for the resolution of the heat- and mass-transfer problems, while other methods focus on a selected region but lead to a much more detailed view of the structure. Despite recent efforts, coupling between these scales largely remains to be done.

Several modeling approaches of solidification structures are presented in the article "Modeling of Dendritic Grain Solidification" in this Volume. One of them is the indirect modeling approach of grain structures, which is based on volume averaging concepts. The method can be integrated in macroscopic simulations to describe, typically, the average grain size, the average dendrite arm or eutectic spacing, and the volume fraction of phases as a function of the local conditions in terms of cooling rate, fluid flow, and solid transport (Ref 3 to 11). These methods allow for a description of the structure in the entire ingot and for more accurate calculations of segregations.

At a much smaller scale, numerical models have been developed with the objective of performing a direct simulation of the microstructure. This category of models, which is also presented in the article "Modeling of Dendritic Grain Solidification" as direct models of the dendritic structure, encompasses the phase-field (Ref 12 to 18), the level-set (Ref 19, 20), and the front-tracking methods (Ref 21, 22). These models consider a small element of volume, which is representative of the microstructure. The evolution of the solidifying phase(s) in this volume is described by a numerical resolution of the heat and/or solute diffusion equations, taking into account the local conditions at the solid/liquid interface, and the influence of anisotropic solid/liquid interfacial energies and mobilities. Direct models of the microstructure have been used to calculate solidification patterns, giving rise to spectacular images of dendrites and eutectic structures. They have reached a point where they can be applied to industrially relevant alloy compositions and solidification conditions (Ref 23, 24), although the computation time remains an important limitation.

At an intermediate scale, cellular automata (CA) have been developed with the goal of achieving a direct representation of the grain structure only, while still using an indirect approach for the internal structure of the grains (Ref 25 to 28). The CA models are described in the article "Modeling of Dendritic Grain Solidification" in this Volume, where they are referred to as direct models of the grain structure. They provide a detailed description of the grain structure and, in particular, textures in cast parts, including statistical aspects due to random nucleation events. Unlike direct models of the microstructure, which are very computationally intensive and therefore limited to a few grains, CA calculations can be performed on the scale of the entire cast component and can be coupled with macroscopic calculations of heat and solute flow for the prediction of macrosegregation (Ref 28).

This article reviews various aspects of the simulation of solidification microstructures and grain textures, focusing on applications of the approaches that are presented in the article "Modeling of Dendritic Grain Solidification." It is organized according to physical topics, beginning with grain structures (first section) and finishing with the morphology of dendrites or eutectics that compose the internal structure of the grains (second section). A particular emphasis has been put on the simulation of defects related to grain textures and microstructures, which is the topic of a third section. For each topic, the application of the most important simulation approaches used so far is reviewed, and a short status of numerical simulation in the field is reported.

Simulation of the Grain Structure

Structures of Equiaxed Eutectic Grains

Structure formation during equiaxed solidification of eutectic alloys was one of the first solidification problems to be addressed by numerical simulation. The first models were based on an indirect approach in which the eutectic grain structure was represented by an average number density of grains instead of being directly described. (This approach has sometimes been referred to as deterministic, Ref 29.) The early work of Oldfield already outlined most of the concepts of the indirect approach of equiaxed eutectic solidification (Ref 30). This model was developed to calculate the cooling curves during a casting process of a lamellar cast iron. It was based on a combination of analytical expressions giving the rates of nucleation and growth of equiaxed eutectic grains as a function of undercooling. The approach was later formalized by Rappaz (Ref 3), who showed how to couple the heat flow equation with the nucleation and growth laws.

The amount of latent heat released by solidification can be incorporated into the heat flow equation through the calculation of the evolution of the volume fraction of solid provided by the microstructure model. Assuming spherical eutectic grains:

$$\frac{dg_s}{dt} = 4\pi n \langle r^2 \rangle v \psi \qquad \text{(Eq 1)}$$

where g_s is the volume fraction of solid, n is the current number density of grains, r is the grain radius, v is the growth velocity, and ψ is the grain impingement factor, which expresses as $1 - g_s$ if the grains are randomly located in the volume (Ref 31). The second moment of the grain size distribution at time t, $n \langle r^2 \rangle (t)$, is obtained by numerical integration of the nucleation and growth laws:

$$n \langle r^2 \rangle (t) = \int_0^t \left[\int_\tau^t v(\Delta T(t')) dt' \right]^2 \frac{dn(\Delta T(\tau))}{d(\Delta T)} \frac{dT(\tau)}{dt} d\tau \qquad \text{(Eq 2)}$$

where $\Delta T(t) = T_{eutectic} - T(t)$ is the undercooling, and $dn/d(\Delta T)$ is the distribution of nucleation undercoolings, which is often represented by a Gaussian function (Ref 32). In Eq 2, the inner integral expresses the radius of the grains that nucleated at time τ and grew until time t. The growth velocity in Eq 1 and 2 is calculated as a function of undercooling, using generally the form $v(\Delta T) = A\Delta T^2$ (where A is a constant), which comes from classical eutectic growth theory (Ref 33) and is briefly summarized in the section "Eutectics and Other Multiphase Microstructures" in this article.

By coupling these equations with a numerical calculation of heat diffusion in the casting, the cooling curve and the average grain size can be calculated at every point of the grid. The indirect approach was used to address different problems associated with equiaxed eutectic solidification, in particular, competition between gray and white iron (Ref 34 to 36). The approach was also adapted to address equiaxed solidification of nodular cast iron, taking into account diffusion of carbon through the austenite shell toward the graphite nodule at the center of the grain (Ref 37 to 39). In addition to the grain size, the ductile iron models also

provide the proportion of graphite and the carbon content of austenite.

The indirect approach provides the average grain size as a function of the location in the cast part, which is directly linked to the number density, n. The grain size distribution, however, is not described or only limited to a calculation of the first, second, and third moment of the distribution, following the numerical method proposed in Ref 40. Another shortcoming is the assumption of uniform temperature on the scale of the grain, which is not always satisfied. Elongated grain shapes resulting from growth in high-temperature gradients cannot be described with this approach.

To have a more detailed description of the grain structure resulting from equiaxed eutectic solidification, direct stochastic approaches were proposed (Ref 41 to 43). In these models, the nucleation centers are chosen randomly in a representative simulation domain and are assigned nucleation undercoolings according to the Gaussian nucleation model described in Ref 32. The interfaces of the grains are mapped with a large number of facets. Each grain is described by a series of variables that include the position of the center, the radius, and facet states indicating whether a given facet is still in contact with the liquid. The growth rate is calculated as a function of the undercooling, using classical eutectic growth theory (Ref 33).

The direct stochastic approach has been used to evaluate the effect of grain movement (Ref 41). It showed that grain sedimentation substantially modifies the impingement factor and thereby the cooling curve, as compared with a perfectly random spatial distribution. The same approach was also used to investigate the shape of eutectic grains solidifying in a temperature gradient (Ref 42). It was also coupled with the calculation of heat flow on the scale of an entire ingot (Ref 43). Figure 1 illustrates the stochastic approach applied to cast iron solidification. It directly shows the influence of the local cooling conditions on the grain structure.

Structures of Equiaxed Dendritic Grains

The average grain size resulting from equiaxed dendritic solidification can also be described with indirect approaches that are similar to those used for eutectic solidification. As

compared with equiaxed eutectic solidification, an additional difficulty arises from the fact that the dendritic grains are not fully solid. A first solution to this problem was given by Rappaz and Thévoz (Ref 4, 44). The central idea of this model is the division of the volume associated with a growing equiaxed dendritic microstructure into two regions. The first region is the mushy zone made of a mixture of solid and interdendritic liquid, while the second region is an external liquid with a composition different from the interdendritic liquid. By combining a solute balance over the two regions, a dendrite tip kinetics model based on marginal stability theory (Ref 45) and a local heat balance, the evolution of the grain envelope, and its internal fraction of solid can be calculated as a function of the local variation of enthalpy. Associated with an appropriate nucleation law, the model can predict realistic cooling curves and recalescence. The model was coupled with a finite-element (FE) method for the calculation of heat transfer on the process scale. The model can thus predict the grain size at every FE node as a function of the local cooling conditions.

In a later but similar approach, M'Hamdi et al. (Ref 8) developed a combined Eulerian/Lagrangian method to deal with solid transport in continuous casting of steel, which permits the description of both equiaxed and columnar grain morphologies. Beckermann and colleagues (Ref 5, 6) further developed the modeling approach of Rappaz and Thévoz based on a procedure of formal averaging of the conservation equations. (See the article "Modeling of Dendritic Grain Solidification" in this Volume for more details on this approach.) Their model describes the effect of the transport of free solid on the final structure in the casting (Ref 6). It was used to show the influence of nucleation and solid transport on the final compositions and grain sizes (Ref 46).

The size of equiaxed dendritic grains can also be calculated with the CA method. (Refer to the article "Modeling of Dendritic Grain Solidification" in this Volume as well as Ref 25 to 27 for a description of this method.) As compared with indirect approaches, the CA method has the advantage of directly representing each individual grain, including the shape. This is particularly important when temperature inhomogeneities are no longer negligible on the

Fig. 1 Stochastic simulation of equiaxed eutectic solidification applied to a gray cast iron ingot. A constant heat flux is applied on the left border of the 0.03 by 0.11 m domain, while the other borders are assumed to be perfectly insulated. The cooling rate decreases globally for an increasing distance from the left border, which leads to slightly larger equiaxed eutectic grains. The dashed line corresponds to the eutectic temperature isotherm. Source: Ref 43

scale of the grains. The CA method permits an introduction of different nucleation behaviors at the surface of the casting and in the bulk, thereby allowing for a more realistic distribution of grain sizes near the mold. The method also describes the growth of columnar grains and the columnar-to-equiaxed transition (Ref 47, 48).

Structures of Columnar Dendritic Grains

The simulation of columnar grains is somewhat easier to perform than for equiaxed grains because growth is constrained by the thermal field. The growth velocity of the columnar front is directly linked to the velocity of the liquidus isotherm, v_T. It is generally assumed that growth takes place under steady-state conditions with respect to diffusion of solute species. The velocities of the front and of the isotherms can further be assumed to be equal if the temperature gradient is maintained.

As dendrites are constrained to grow along certain crystallographic orientations, the growth velocity of the dendrite tips, v^*, depends on their angle of inclination with respect to the temperature gradient, θ, according to the expression $v^* = v_T/\cos\theta$. This implies that misaligned dendrites must grow faster than those that are well aligned in order to reach the same isotherm speed. Misaligned dendrites therefore require a higher undercooling, and their tips progress slightly behind those of well-aligned tips. Thus, they tend to be eliminated by

well-aligned dendrites that block them, either directly, in the case of converging trunks, or by the emission of secondary arms in the gap that is formed between the diverging trunks.

The CA method is well adapted to the simulation of grain selection during columnar solidification (Ref 26). It takes into account the random location and orientation of the nuclei. The CA growth algorithm also accounts for preferential <100> growth directions of the dendrites and the fact that misaligned tips grow behind well-oriented ones. Coupled with the FE prediction of the heat flow, it does not require any assumption on the temperature gradient.

The coupled CAFE method was applied successfully to the description of the grain texture evolution in directionally solidified superalloys. Figure 2 shows the evolution of the calculated <100> pole figures as a function of the distance from the chill (Ref 26). Near the surface of the chill, the distribution of the angle θ is close to the theoretical curve for randomly oriented grains, which is a direct consequence of a perfectly random selection of the orientation of the nuclei. As the distance increases, the distribution becomes narrower and is displaced toward small values of θ. Grain texture evolutions calculated with the CA method have been compared with electron backscatter diffraction measurements and showed good agreement (Ref 49, 50).

Recently, the selection of grains during columnar solidification was also addressed with the phase-field method (Ref 51, 52). As opposed to the CA method, which is limited

to the prediction of the grain envelope, the phase field has the advantage of a direct description of the secondary arms and of the solute diffusion fields around the growing tips. This allows for a more detailed description of the blocking mechanisms. In particular, the roles of interacting solutal fields for converging dendrite trunks and secondary arms for diverging trunks can be directly analyzed. Figure 3 shows a phase-field simulation of columnar growth in an AZ31 alloy (Ref 51). It illustrates the elimination of a highly misoriented grain (second grain from the left) by better oriented dendrites (first and third grains from the left), as well as the lower position of the tips of the weakly misoriented grain as compared to the perfectly aligned dendrites. This study has recently been extended to three dimensions, which is a necessity for quantitative investigations of texture evolutions (Ref 52).

Columnar-to-Equiaxed Transition

In many castings, transitions from columnar to equiaxed grain structures can occur at some distance from the walls. Such transitions occur if solid can nucleate (or detach from columnar dendrites) and grow in an equiaxed manner in the undercooled region ahead of the columnar front (Fig. 4). If this equiaxed free solid occupies a high volume fraction or has sufficiently enriched the liquid in front of the columnar dendrites, it can block the growth of the columnar structure, and the transition to an equiaxed grain structure can occur. Predicting the occurrence and the position of the columnar-to-equiaxed transition (CET) in a cast component is of great technical importance because the properties of the casting can be substantially influenced by the grain structure.

A criterion for the CET was proposed by Hunt (Ref 53). This model considers unidirectional steady-state solidification. Equiaxed grains are assumed to nucleate at a prescribed undercooling, ΔT_n, and grow at a velocity given

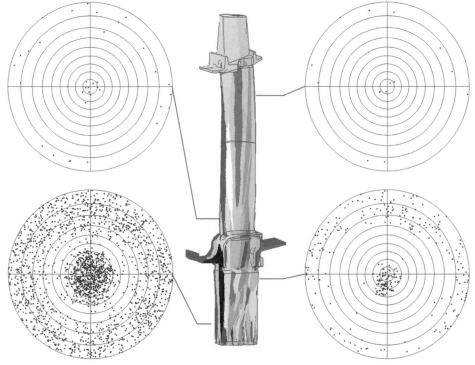

Fig. 2 Grain structure in a directionally solidified superalloy turbine blade simulated with the cellular automaton method. The <100> pole figures are displayed for various cross sections perpendicular to the main blade axis. Source: Ref 26

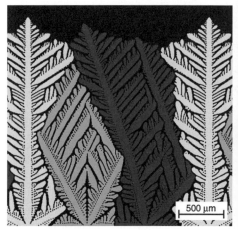

Fig. 3 Phase-field simulation of grain selection during directional dendritic solidification. Source: Ref 51

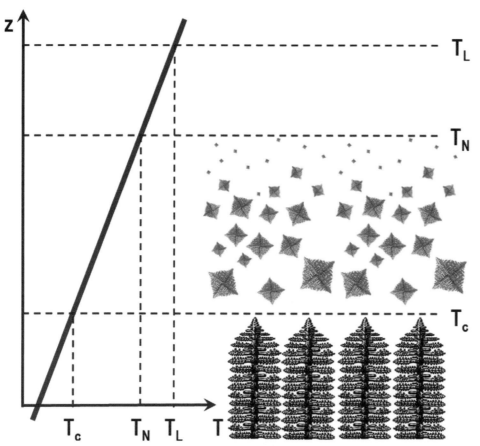

Fig. 4 Schematic illustration of the columnar-to-equiaxed transition. Grains nucleating in the undercooled region ahead of the columnar front, with a nucleation temperature (T_N) that is comprised between the columnar tip temperature (T_C) and the liquidus temperature (T_L), can block the growth of the columnar structure if they are sufficiently developed

by the parabolic expression $v = A \, \Delta T^2$. By integrating this expression over the time elapsed between the nucleation event and the capture of the nucleation center by the columnar front, the volume fraction occupied by the equiaxed grains when they are reached by the front can be calculated. To determine the CET, Hunt considers that the front is blocked mechanically by the equiaxed grains if their volume fraction is larger than 0.49 (Ref 53). Although this value was derived from geometrical considerations, it should be considered as rather arbitrary. This criterion leads to the following condition for the CET:

$$G < \left(\frac{4\pi n_g}{81 \times 0.49}\right)^{1/3} \left(\frac{v_T}{A}\right)^{1/2} \left[1 - \left(\frac{\Delta T_n}{(v_T/A)^{1/2}}\right)^3\right]$$

(Eq 3)

where G is the temperature gradient, n_g is the number of equiaxed grains per unit volume, and v_T is the velocity of the liquidus isotherm.

The CET criteria such as Eq 3 are often represented in so-called CET diagrams, where v_T is plotted against G. From such diagrams, one can determine the combinations of v_T and G that give a fully columnar, fully equiaxed, or mixed grain structure. The criterion can be

easily implemented into a heat flow model to predict the location of the CET in a cast component.

Several modifications of Hunt's criterion have been proposed. One can mention the use of a more recent dendrite growth model, which takes into account nonequilibrium effects (Ref 54), or the incorporation of the solutal interaction between dendrites (Ref 55), which avoids the arbitrary definition of the critical volume fraction of equiaxed grains to block the columnar front. The influence of a distribution of nucleation undercoolings (Ref 25, 56) and the influence of the misalignment of the columnar dendrites with respect to the temperature gradient (Ref 48) can also be considered.

Another CET criterion was proposed by Gandin, who used a comprehensive one-dimensional micro-macro model to track the position of the columnar dendritic front in a directionally solidified ingot (Ref 57). It was observed that the velocity of the columnar front goes through a maximum, which is associated with the full dissipation of the superheat in the remaining liquid, and that the temperature gradient in the liquid ahead of the columnar front vanishes shortly after this maximum. This maximum could be used as a CET criterion. The results were compared with casting experiments

also used in Ref 55. The adjustment of the nucleation undercooling for equiaxed grains in Ref 55 led to the same conclusions as in Ref 57.

Browne (Ref 58) developed another criterion based on an equiaxed index that can be interpreted as the integral of the undercooling over the undercooled liquid regions in the domain. It was proposed that the peak equiaxed index can be used as an indicator to give the tendency to form an equiaxed zone in a casting of fixed dimension. Again, comparisons with Ref 57 led to the same conclusions (Ref 59).

The CET can also be predicted with volume averaging approaches (Ref 55, 60). In a first model, the nucleation undercooling of the equiaxed grains was assumed to be zero (Ref 60). This assumption was relaxed in a later model, which also includes the effect of solutal interaction, eliminating the need of the mechanical blocking criterion (Ref 55). In a similar approach, columnar and equiaxed solids were considered as distinct phases, which allows for better description of the influence of the melt convection on the CET (Ref 61).

The CA method provides a direct representation of the grain structure and can be used to predict the position of the CET. In the CA model, no specific distinction is made between equiaxed and columnar grains. The CET is determined by visual inspection of the simulated grain structure such as on a macrograph (Ref 25, 48, 56).

Figure 5 shows a series of longitudinal sections in CA calculations of directional solidification (Ref 48). A standard set of conditions was defined, which correspond to a temperature gradient of $G = 3 \times 10^4$ K/m, a solidification velocity of $v_T = 10$ mm/s, and no misalignment of the columnar grain with respect to the temperature gradient ($\psi = 0$). Then, v_T, G, and ψ were varied independently, as can be seen in Fig. 5(a) to (c). These calculations clearly illustrate that nucleation ahead of the columnar front is favored by high velocities, low-temperature gradients, and highly misoriented columnar dendrites. In the latter case, one notes a grain orientation selection among the "equiaxed" grains, which leads to the development of a secondary columnar grain structure that is better aligned with the temperature gradient than the original one.

As can be seen in Fig. 5, the CA method provides a realistic description of the CET. The method does not require any blocking criterion, because mechanical blocking is automatically taken into account by the capture algorithm of the cells. In its most recent development, the method also takes into account the solutal interaction. This is the case of the two-dimensional simulations of the grain structure presented in the article "Modeling of Dendritic Grain Solidification" in this Volume. As compared with indirect approaches, such as the previously mentioned volume averaging models (Ref 6, 46), coupling the CA method with convection in the melt and transport of the grains is more difficult. However, such a coupling has already been achieved in two dimensions (Ref 62).

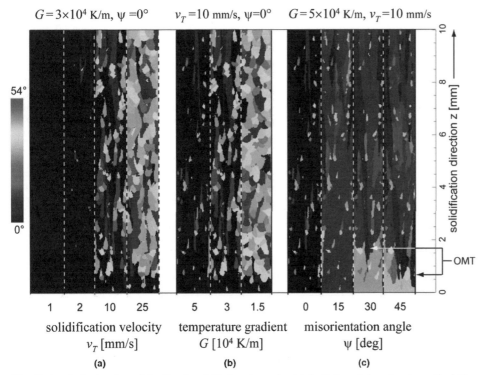

Fig. 5 Longitudinal sections of directionally solidified specimens simulated with the cellular automaton method. The grain structure is shown as a function of (a) the solidification velocity (G, ψ constant), (b) the temperature gradient (v_T, ψ constant), and (c) the seed orientation (G, v_T constant). Black regions have the same orientation as the seed, while misoriented regions are marked with a lighter shade of gray (or color online). OMT, oriented-to-misoriented transition. Adapted from Ref 48

The CET was also studied using the phase-field method (Ref 63). This investigation provided evidence of some CET mechanisms, such as the deactivation of seeds by solutal interactions and the sensitivity of the CET with respect to the crystal anisotropy. Until now, such simulations have only been achieved in two dimensions because of high computation times.

Some attempts have also been made to simulate the CET at an intermediate scale between the CA and phase-field methods (Ref 64). The approach is based on a resolution of the diffusion equation on CA grids, with a typical size of a few micrometers. By having a direct representation of the dendritic structures, they can capture most of the CET mechanisms. However, such methods suffer from shortcomings associated with the difficulty to resolve the solute diffusion boundary layer with sufficient spatial resolution while being limited to relatively small calculation domains, as compared to the CA methods.

Simulation of the Internal Grain Structure

Introduction

In addition to being influenced by the grain structure, the mechanical properties of a solidified material strongly depend on the type, morphology, chemistry, and spatial distribution of the different phases in the interior of the grains.

A large variety of mathematical models aimed at predicting these various aspects of the internal grain structure can be found in the literature and sometimes also in commercial simulation packages. Depending on the objectives, the internal grain structure can be approached with either simple analytical models or volume averaging approaches, cellular automata, phase field, level set, and so on. This section is organized according to the objective of the models in terms of the characteristic being simulated: phase proportions, phase chemistry, dendritic morphologies, eutectic structures, and so on. Each aspect is illustrated by examples of simulations with various modeling approaches.

Phase Proportions and Chemical Composition

Analytical and One-Dimensional Microsegregation Models. The type, volume fraction, and composition of the microstructural constituents are important quantities characterizing the internal structure of a grain after solidification. These quantities are directly related to the solute partitioning taking place at the solid/liquid interfaces and to the solidification path taken in the phase diagram. They can be obtained by microsegregation models, which have been the subject of research and development for several decades (Ref 65 to 74). Commercial software can provide a description of the microsegregation and solidification paths for the limiting cases of equilibrium (Lever rule), no diffusion in the solid (Gulliver-Scheil approximation), and their derivation based on partial equilibrium (Ref 68, 71, 72). These calculations are based on extensive thermodynamic databases for complex thermodynamic systems and the CALPHAD approach of phase diagram calculation (Ref 75 to 77). More advanced microsegregation models take into account diffusion in the growing solid based on a modified Gulliver-Scheil equation or a numerical solution of the diffusion problem (Ref 5, 10, 11, 65 to 69). Microsegregation models sometimes incorporate a coarsening law to account for the effect of the variation of the dendrite arm or cell spacing on the solute distribution (Ref 68, 69). A more detailed review of this topic can be found in Ref 74.

Volume Averaging Approach. The indirect models of the grain structure based on volume averaging usually provide only limited information on the volume fractions and compositions of the phases (Ref 4, 5). An exception is the equiaxed dendritic model of Ref 10 and 11, which has been extended to incorporate the formation of the peritectic and eutectic structures during the second stage of solidification. In this model, the solidification sequence is decomposed into several stages, an illustration of which is given in Fig. 6 (Ref 10). During the first stage (S1), a mushy zone, which is

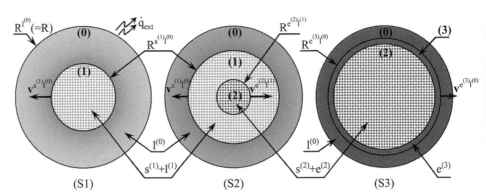

Fig. 6 Schematized solidification sequence of a binary alloy in three steps, denoted S1, S2, and S3, which correspond respectively to: (S1) growth of a mushy zone (region 1) into the undercooled extradendritic liquid (region 0), (S2) growth of an interdendritic eutectic (region 2) into the interdendritic liquid of the mushy zone, and (S3) growth of the extradendritic eutectic structure (region 3). Source: Ref 10

composed of the dendritic solid and interdendritic liquid (region 1 in Fig. 6), grows into the undercooled extradendritic liquid (region 0). During the second stage (S2), an interdendritic eutectic structure develops from the center (region 2). Finally, in a third stage (S3), an extradendritic eutectic structure (region 3) develops and surrounds the dendritic grain. The model predicts the evolution of the volume fraction and average composition of the extradendritic liquid, the interdendritic liquid, the dendritic solid, the interdendritic eutectic structure, and the extradendritic structure. Compared with experimental data for hypoeutectic aluminum-copper alloys, it shows good predictability, provided that the nucleation undercoolings of the microstructures are known (Ref 10, 11).

Although they do not explicitly calculate the redistribution and diffusion of the solute elements, the indirect models of eutectic solidification in cast iron can also provide information on phases in the sense that they can predict the type of eutectics (white or gray) that will form (Ref 34 to 36). In this case, the type and volume fractions of the phases are largely determined by the nucleation phenomena. Metastable carbides will appear if the nucleation conditions for the stable graphite phase are not met.

Two- and three-dimensional microsegregation models, such as the phase-field (Ref 12 to 18), level-set (Ref 19, 20), and pseudo-front tracking methods (Ref 21, 22), provide a detailed description of the solute distribution in the solid and in the liquid during solidification. They can therefore be used to calculate the proportion of eutectic phases as a function of the solidification conditions, taking into account the effect of the morphology of the primary phase. An important difficulty in these approaches is associated with the spatial resolution. Simulations addressing the morphology of the primary phase are performed in a computational domain that corresponds approximately to the grain size. Therefore, the calculation grid is generally too coarse to allow for a direct description of the interdendritic particles. Based on this observation, mixed approaches have been proposed.

The model in Ref 78 combines, on one hand, a direct description of the primary phase formation using a pseudo-front tracking method (Ref 22) to describe the formation of the primary solid phase and, on the other hand, a volume average method for the formation of the secondary phases in the interdendritic regions. The model is coupled to a CALPHAD method to obtain the phase equilibrium in the interdendritic mixture and the concentrations to be prescribed at the primary phase/mixture interface. This model was used to predict the volume fractions of Mg_2Si and iron-bearing particles in an AA5182 direct chill casting slab (Ref 23). The model was also employed to explain the experimentally observed relationship between the cooling rate and the nonequilibrium eutectic fraction obtained in an inoculated Al-2.5wt%Cu binary alloy (Ref 79). It was observed that the eutectic

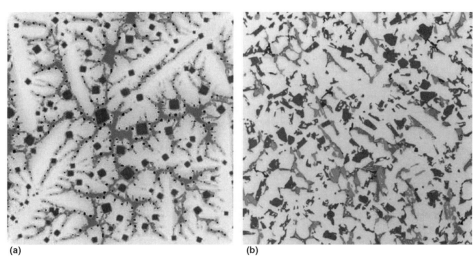

(a)   (b)

Fig. 7 Comparison between (a) simulation and (b) experiment in an Al-Cu-Si-Mg alloy (KS1295). The area is 400 by 400 μm for both images. Source: Ref 24

fraction exhibits a maximum for intermediate cooling rates of approximately 1 K/s. While both the classic Brody-Flemings model and one-dimensional numerical calculations failed to explain this behavior, the two-dimensional model could successfully reproduce it. The difference is explained by the capacity of the two-dimensional model to capture the variation of the primary phase morphology, which varies from globular to dendritic for increasing cooling rate, and the influence of coarsening, which strongly affects the dendrite arm spacing during solidification.

With constantly increasing computer power, first calculations combining a direct description of both the primary phase and secondary particles recently became possible. A multiphase field model coupled with a thermodynamic database was used to address the formation of microstructure in a multicomponent Al-Cu-Si-Mg alloy (Ref 23). The simulation provides realistic images of the dendritic grains and of the square-shaped silicon particles forming in the interdendritic region (Fig. 7). However, such calculations are limited to microstructures that are characterized by relatively coarse dendrites and large secondary particles.

Dendritic Structures

Average Dendrite Arm Spacing. One of the most important aspects of solidification microstructures is dendrite arm spacing. A fine dendritic microstructure is recognized as yielding superior strength and ductility. It also reduces the homogenization time during subsequent heat treatment processes, because microsegregation (coring) is less pronounced in fine structures. For these reasons, dendrite arm spacing measurements are often found to be fully part of quality-control programs in foundries producing high-strength castings. The columnar structures are characterized in terms of the spacing between primary trunks, λ_1, and between secondary

arms, λ_2. Prediction of the secondary dendrite arm spacing is also required for the calculation of microsegregation and the solidification path, because it determines the influence of back diffusion and thus the composition of the remaining liquid. Fluid flow calculations in solidification processes also require an estimation of the secondary arm spacing, because this quantity strongly influences the permeability of the mushy zone.

Predictions of the average primary and secondary dendrite arm spacings of a dendritic microstructure can be obtained with analytical expressions relating these quantities with the solidification time or the local temperature gradient and the cooling rate. Different theoretical formulas for λ_1 and λ_2 have been proposed in the literature and are reviewed and compared in Ref 80. The following expression, which comes from the coarsening theory, is often used for λ_2 predictions:

$$\lambda_2 = M t_f^{1/3} \qquad \text{(Eq 4)}$$

where t_f is the solidification time, and M is a parameter that depends on the interfacial energy, the diffusion coefficients in the liquid, and the phase diagram. In directional solidification, the solidification time can be estimated as $t_f = \Delta T_0/(G\, v_T)$, where ΔT_0 is the solidification interval, v_T is the isotherm velocity, and G is the thermal gradient.

These expressions can easily be incorporated a posteriori into heat and fluid flow simulations of solidification processes, simply by recording the local solidification time. Using this approach, the distribution of the secondary dendrite arm spacing can be obtained in a whole cast component. Based on that, a prediction of the local mechanical properties can be made. Indirect approaches of dendritic grain structures, as well as CA models, do not provide a direct description of the morphology of the solidifying phases. Therefore, they generally also rely on analytical expressions such as Eq 4 to calculate an average secondary dendrite arm spacing.

Direct simulation of dendritic structures has been used to address different aspects of dendritic solidification: growth directions, tip kinetics, and dendritic morphology in the presence of fluid flow.

Growth Directions. The production of calculated images of dendritic structures has been one of the most spectacular applications of the direct simulation methods of solidification microstructures. The front-tracking (Ref 21, 22), level-set (Ref 19, 20), and, above all, phase-field methods (Ref 12 to 18) have been widely used to simulate the growth of dendrites and their complicated morphologies. An important aspect of the problem is the introduction of appropriate functions to describe the anisotropy of the interfacial energy, which determines the growth directions of the dendrite arms and also plays a crucial role in the tip kinetics.

Fourfold symmetry dendrites can be obtained in two-dimensional simulations by making the interfacial energy depend on the orientation of the solid/liquid interface. The mathematical formalism to incorporate this dependency into a phase-field model was developed in Ref 81 and 82. The first numerical application was shown in Ref 82, based on the following anisotropy function, which is representative of a cubic system in two dimensions:

$$\sigma = \sigma_0[1 + \varepsilon_4 \cos(4\theta)] \qquad \text{(Eq 5)}$$

where σ_0 is the nominal interfacial energy, ε_4 is a relative anisotropy parameter, and θ is the angle between the normal vector to the solid/liquid interface and the <100> crystallographic orientation.

A more general method is to describe the interfacial energy with spherical harmonics, which also has the great advantage of being applicable to three-dimensional simulations (Ref 83). Dendritic growth in cubic metals can be described by two anisotropy parameters associated with the four- and sixfold symmetries:

$$\sigma = \sigma_0[1 + \varepsilon_1 K_1(\mathbf{n}) + \varepsilon_2 K_2(\mathbf{n})] \qquad \text{(Eq 6)}$$

where $K_1(\mathbf{n})$ and $K_2(\mathbf{n})$ are sixfold anisotropy cubic-harmonic functions that can be expressed as functions of the components n_i of the normal unit vector to the solid/liquid interface, $\mathbf{n}$, as follows:

$$K_1(\mathbf{n}) = n_1^4 + n_2^4 + n_3^4 - 3/5$$
$$K_2(\mathbf{n}) = 3\left(n_1^4 + n_2^4 + n_3^4\right) + 66n_1^2 n_2^2 n_3^2 - 17/7 \qquad \text{(Eq 7)}$$

This approach has been used to simulate the transition of growth direction from <100> to <110> that has been observed in aluminum-zinc for an increasing zinc content (Ref 84). The transition of the growth direction in aluminum-zinc was attributed to a dependency of the anisotropy of the interfacial energy with the zinc content. Figure 8 shows a series of

dendritic morphologies calculated with different parameters for the anisotropy of the interfacial energy. A transition of growth directions from <100> to <110> can be obtained by varying the parameter ε_1. Atomistic simulation can be used as a complementary approach to calculate the interfacial energy and its anisotropy parameter, which are quantities difficult to determine experimentally (Ref 85, 86). However, this approach is currently limited to pure metals and model alloys.

Tip Kinetics. The growth kinetics of a free dendrite tip have been investigated with various direct modeling approaches, such as the level-set (Ref 19), pseudo-front tracking (Ref 22), and phase-field methods (Ref 87 to 89). The main objective of such studies is to determine the operating point of the dendrite tip, that is, the tip velocity and tip radius for a given undercooling. The most significant studies were performed using a three-dimensional phase-field model, which has been applied for both pure substances (Ref 87) and binary alloys (Ref 88, 89). Comparisons were made with experimental data and with classic dendrite tip models based on the Ivantsov solution and marginal stability theory (Ref 90 to 92), or with microscopic solvability (Ref 93 to 96), which has the advantage of also including the influence of the anisotropy of the interfacial energy on the operating point. Quantitative three-dimensional calculations, at both low

and high undercooling, showed that the dendrite tip kinetics predicted with the phase-field method are in good quantitative agreement with the sharp interface solvability theory, showing even better agreement with experimental results because it is exempt from certain symmetry assumptions (Ref 87).

Dendritic Morphology in the Presence of Fluid Flow. Direct simulations of solidification in the presence of fluid flow have been performed using the phase-field (Ref 97 to 100) and other methods (Ref 101 to 103). Figure 9 shows a three-dimensional phase-field simulation of free dendritic growth in a forced flow (Ref 99). It illustrates how the liquid flow modifies the convective heat transport away from the dendrite tip and favors growth in the upstream directions, while dendrite arms pointing downstream grow very slowly. The computational capabilities currently remain a limitation for direct simulation of the convection effects on dendritic growth. This type of calculation currently is used to make comparisons with analytical theories. These investigations are of particular interest for the calculation of grain structures and macrosegregation with indirect approaches based on volume averaging and CA methods, which require analytical expressions for the dendrite tip kinetics. The effect of fluid flow can be taken into account in these models by using modified expressions coming from such investigations.

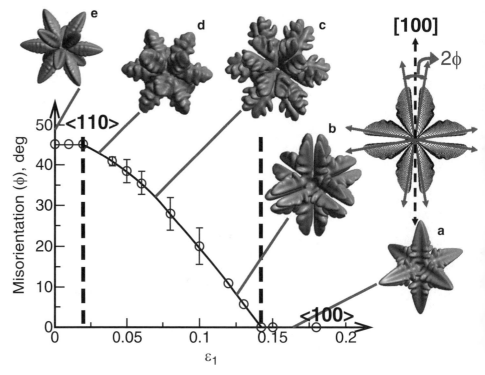

Fig. 8 Phase-field simulations showing the equiaxed growth morphologies obtained for different coefficients of the first two cubic harmonics that characterize the anisotropy of the interfacial free energy (ε_1 varied, fixed $\varepsilon_2 = -0.02$). The growth directions of the dendrite arms are contained in the {100} planes. However, within a certain transition interval delimited by the dashed lines, the misorientation (defined as the angle, ϕ, between the growth directions in these planes and the principal crystal axes) varies continuously from 0 (<100> dendrites) to 45° (<110> dendrites) when ε_1 is decreased. Source: Ref 84

Eutectics and Other Multiphase Microstructures

Average Eutectic Spacing. The morphology and spacing of eutectic microstructures has a great influence on the mechanical properties of eutectic alloys. As for dendritic structures, analytical models can be used to calculate the average spacing of eutectics. The classic Jackson and Hunt's theory is based on the steady-state solution of the diffusion equation for the solutal field ahead of a growing periodic arrangement of lamellae (Ref 33). The model relates the lamellar spacing, λ, and the undercooling, ΔT, with the growth velocity of the eutectic structure, v:

$$\lambda^2 v = C_1 \qquad (Eq\ 8)$$

$$\Delta T = C_2 v^{1/2} \qquad (Eq\ 9)$$

where C_1 and C_2 are parameters that depend on the phase diagram, the interfacial energy, and the diffusion coefficient in the liquid.

Jackson and Hunt's theory was later extended by Magnin and Kurz to describe the growth of irregular eutectics, that is, eutectic structures such as iron-carbon or aluminum-silicon, which are characterized by the presence of a facetted phase associated with a high-melting entropy (Ref 104). A compilation of the constants C_1 in Eq 8 was provided by Guzik and Kopycinsky, who also proposed a modified version of the Magnin-Kurz theory (Ref 105). A theoretical model for the growth of rod eutectics in the form of a two-dimensional array was recently proposed in Ref 106.

The volume averaging approaches and the stochastic models of equiaxed eutectic solidification are generally based on Jackson and Hunt's theory to calculate the growth rate and lamellar spacing as a function of the local undercooling. Therefore, these models can provide the spatial distribution of the average lamellar spacing in the casting as a function of the local solidification conditions. This approach can be used to estimate the mechanical properties in a cast part, taking into account the influence of the eutectic spacing.

Direct Simulation of Multiphase Microstructures. Except for some early work based on random-walker analysis (Ref 107), direct descriptions of the formation of multiphase microstructures such as eutectics, peritectics, or monotectics have been achieved using the phase-field method. Formulating a phase-field model for eutectic or peritectic structures is more challenging than for dendritic structures because of the presence of three (or more) phases. In one of the first phase-field models of eutectic solidification, the problem was addressed with only one phase-field variable, by introducing a maximum in the free energy curve of the solid for intermediate compositions, which can yield a eutectic phase diagram if an appropriate choice of parameters is made (Ref 108). However, the idea of introducing a phase-field variable for each phase of the

system rapidly emerged as a better choice for addressing real systems, which led to the so-called multiphase-field formulation (Ref 109). In this approach, each phase is attributed a phase-field variable that ranges between 0 and 1, where 1 corresponds to the bulk of that phase and 0 to regions where it is absent. The sum of the phase-field variables at a given location must always be 1. The multiphase-field method allows for the combination of the free energy function of each phase into a single free energy expression entering into the functional. The free energy functions provided by the CALPHAD-type approaches (Ref 75–77) can be used for that purpose (Ref 110, 111). The method can be coupled with a resolution of the solute diffusion equations (Ref 112). It can also account for the anisotropy of the surface energy and interface kinetics (Ref 113). Some numerical difficulties of this approach, which were associated with the undesired systematic presence of the third phase in the interfaces and excessive solute trapping, have been solved by using another form of the free energy functional and antitrapping currents (Ref 114).

The multiphase-field method has found several applications in the formation of multiphase microstructure patterns. Figure 10 shows an example of a three-dimensional phase-field simulation of a zigzag instability developing in a lamellar array during directional eutectic solidification (Ref 115). Other specific examples related to eutectic solidification are the selection process of the eutectic lamellar spacing (Ref 116), cells of ternary eutectics (Ref 117), and stability of the lamellae at the triple junction (Ref 118). The phase-field approach was also applied to peritectic solidification in iron alloys (Ref 119, 120) and monotectic solidification (Ref 121). More examples can be found in recent review articles on this topic (Ref 16 to 18).

Direct simulation of multiphase microstructures is seen as instrumental in gaining an

understanding of the process of pattern selection in multiphase solidification microstructures. In particular, the role of external forces such as convection, stresses, and off-axis temperature gradients can be investigated in a systematic manner with simulation (Ref 122). The influence of the crystalline anisotropy, in particular for irregular eutectic structures, is another topic where simulation can be of great value. However, a quantitative phase-field model for multiphase systems containing both facetted and nonfacetted solid phases is still lacking to carry out such studies.

Simulation of Texture and Microstructure Defects

Different types of texture defects can appear in solidification processes. A common texture defect in casting processes is the development of regions of columnar grains, which usually exhibit poor mechanical properties as compared with equiaxed regions, due to a larger grain size. Columnar grains are also usually undesirable because of their larger tendency to develop hot tears. As seen in the first section of this article, numerical simulation is a powerful tool to describe the development of columnar grains and the CET. Therefore, it can be used to anticipate this type of defect by providing a detailed description of the grain structure in a cast component. Because this topic has already been presented in the first section, it is not discussed further here.

Because the definition of a texture or microstructure defect varies considerably with the process, the alloy, and the objectives in terms of mechanical properties, defects are addressed through a series of specific examples coming from existing industrial problems. In this respect, texture and microstructure defects in nickel-base superalloys, aluminum casting, and galvanized coatings are addressed successively.

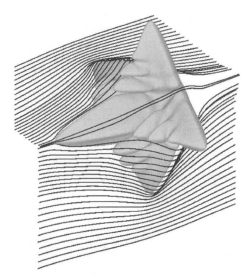

Fig. 9 Phase-field simulation of dendritic growth in the presence of fluid flow. Source: Ref 99

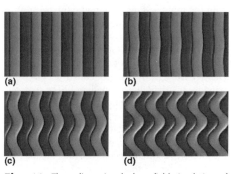

Fig. 10 Three-dimensional phase-field simulation of the development of a zigzag instability in a binary eutectic alloy (viewed from the top). The instability is activated by an excessive lamellar spacing. The calculation was initiated with a reduced lamellar spacing of $\Lambda = \lambda/\lambda_{JH} = 1.3$, where λ_{JH} is the stable spacing obtained with Jackson and Hunt's theory. The four images correspond to frames taken at increasing times. Steady state has been reached in frame (d). Source: Ref 115

The role of numerical simulation in the process of understanding and ultimately avoiding texture defects is pointed out.

Stray Grain and Intragranular Misorientations in Single-Crystal Nickel-Base Superalloys

Nickel-base superalloy components produced by investment casting often are solidified with a low-temperature gradient, which leads to a fine, equiaxed grain structure showing superior mechanical properties. To improve the service lifetime of the most exposed parts of gas turbines, the grain texture is optimized through directional solidification (DS) in a temperature gradient. The objective of the DS technology is to grow a structure of columnar grains along the direction of the main load axis of the component in service. If columnar grains nucleate at the base of the blade and reach the other end without being stopped by equiaxed grains, the grain boundaries remain approximately aligned with the load axis, and better creep properties are obtained. As a result of the growth competition, [001] crystal orientations are selected by the DS process, which is favorable for thermal fatigue resistance because this direction corresponds to the minimum of Young's modulus of the crystal. The grain structure can be further optimized by growing single crystals, which is the current practice in the field of investment casting of superalloy components for gas turbines.

Stray Grains. One of the major problems in manufacturing single-crystal cast components is avoiding nucleation and growth of so-called stray grains ahead of the columnar front. Because stray grains are likely to be highly misoriented with respect to the main crystal, they have a detrimental impact on the mechanical properties. Stray grains are most frequently observed in single-crystal turbine blades at the platform end (Ref 123). In this region of the blade, the melt can experience large undercoolings caused by higher heat exchanges with the mold and, most importantly, because of the time needed by the dendritic structure to expand at the outlet of the platform through multiple branching events. A criterion for the formation of stray grains was proposed by de Bussac and Gandin (Ref 123). They obtained a processing window that has three boundaries. The first one is given by a critical value of the ratio v_T/G^2, above which stray grains will form ahead of the columnar front by the same mechanisms as in the CET. At very high cooling rates, stray grains will form at the platform end. The second boundary of the processing window is thus given by the critical cooling rate, $\dot{T}_{crit} = G v_T$. The third boundary is given by a minimum value for v_T to avoid a long holding time of the melt, which favors chemical reactions with the mold and shell creep, while reducing productivity of the solidification process.

The simulation of stray grains is a difficult task because it requires an accurate description of phenomena occurring on very different scales. On a macroscopic level, it requires a detailed calculation of heat flow to correctly predict the undercooling reached in the melt, in particular at the platform ends. The microstructure model should provide an accurate description of the kinetics of the columnar dendrites, including the effect of the crystallographic orientation. The solute interactions and the branching mechanisms are also expected to play an important role.

Because CA address most of these phenomena, they are obviously the method of choice for the prediction of stray grains. In many CA simulations, the thermal field is simply prescribed, ignoring any feedback from the CA method into the heat flow calculation. Such "weak" coupling schemes are not adequate for the prediction of stray grains, because the influence of the latent heat released by the solidification front must be described accurately. The CA methods based on "strong" coupling schemes have been used for the prediction of stray grains (Ref 47, 124). These simulations proved to reproduce qualitatively the formation of stray grains. However, being based on a Scheil model for the calculation of the solid fractions inside the grains, they suffer from too rudimentary a coupling method for the incorporation of the latent heat. The solute interaction and branching mechanisms now can be taken into account in CA models (Ref 125). Progress toward the goal of preventing stray grain formation is therefore expected from these developments, in particular, if simulation can be combined with dedicated experiments aimed at determining the nucleation conditions.

Misorientations. Even if the thermal conditions are perfectly controlled and the formation of stray grains avoided, texture defects can arise in the form of intragranular misorientations. In cast single-crystal superalloys, variations of the crystallographic orientation by 5 to 10° are commonly observed between the bottom and the top of the blade, which has a negative effect on the creep properties.

The origin of intragranular misorientations in various systems has been the object of several studies, which were recently reviewed in Ref 126. In single-crystal superalloys, several factors have been found to promote the development of misorientations. The first factor is the large undercoolings that can be reached in certain regions of the blade, such as the platform (Ref 127). In these regions where the dendritic structure grows at large supersaturation, dendrite arms are very thin and are more prone to undergo plastic deformation from thermal stresses (Ref 128 to 130). A second factor is the combination of numerous branching events and complex heat transfer, which can cause the solidification front to take complex shapes and possibly split into several parts before forming a convergence fault where they join higher in the casting (Ref 131 to 133). A numerical model has been proposed by Napolitano and Schaefer to track the convergence fault in the casting (Ref 133). The model is based on a schematic description of the dendritic array composed of a series of dendrite tips growing and branching along <100> directions. The dendrite tip kinetics are described with a power-law approximation. Although very innovative and useful, the model does not describe the various forces that lead to dendrite misorientations. Instead, the misorientation at a convergence fault is calculated as a simple function of the lengths of the shortest paths to follow backward until a common point is found in the dendritic arrays. Thus, an integrated simulation approach encompassing the mechanical forces responsible for the misorientations is currently still missing.

Feathery Grains

The growth of feathery grains, also sometimes referred to as twinned dendrites, is a texture defect of aluminum alloys that has been studied for several decades (Ref 134–141). This defect is typically found during semicontinuous casting of aluminum alloys, when the alloy is not inoculated and when critical thermal conditions (temperature gradient of approximately 100 K/cm, solidification rate of 1 mm/s) and a slight convection in the melt are present. Figure 11(a) shows an example of feathery grains formed in a directionally solidified Al-10wt% Zn alloy (Ref 141). The feathery grains are made of an ensemble of twinned columnar dendrites that nucleate at some point in the ingot (small circles in Fig. 11a). After nucleation, twinned dendrites grow and expand laterally by a branching mechanism and rapidly invade the ingot, overgrowing ordinary columnar grains. Feathery grains exhibit a characteristic fan-shaped morphology due to systematic misorientations between each twinned dendrite trunk (Ref 141). The twinned dendrites are made of <110> primary trunks that are split by a coherent (111) twin plane. While the primary growth direction is <110>, a complex structure of secondary arms develops by growth along <110> but also sometimes along <100> directions (Ref 141).

Several attempts have been made to explain the kinetic advantage of twinned dendrites as compared to a regular morphology (Ref 136, 137, 139, 141). It was suggested that the mechanical equilibrium involving the solid/liquid interfacial energy and the twin energy at the triple junction stabilizes a grooved tip (Ref 136). An alternative explanation is that the anisotropy of the solid/liquid interfacial energy could stabilize a sharp pointed tip due to torque terms (Ref 137). However, this mechanism can be ruled out in aluminum alloys because of the weak anisotropy of the solid/liquid interfacial energy in this system (Ref 139). Finally, it was proposed that the primary trunks of twinned dendrites would grow as doublons, as illustrated in Fig. 11(b) (Ref 139, 140). However, recent results obtained in more concentrated alloys do not support this mechanism

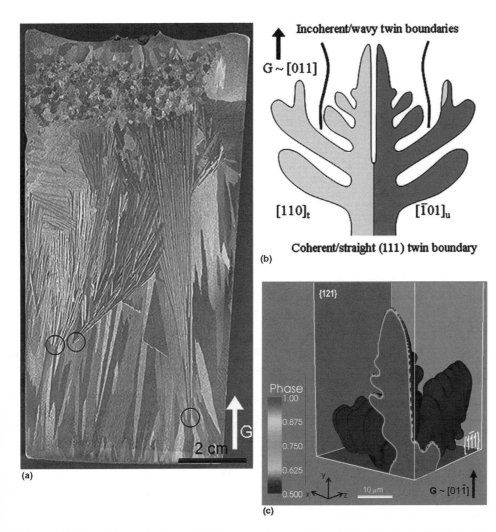

Fig. 11 (a) Longitudinal section of a typical feathery grain structure observed in a directionally solidified Al-10wt% Zn specimen (Ref 141). (b) Schematics of the twinned dendrite doublon proposed by Henry (Ref 139). (c) Phase-field simulation of a Al-9at.%Zn twinned dendrite growing under a temperature gradient of 10^5 K/m and a cooling rate of –70 K/s. The yz-plane corresponds to the {111} coherent twin plane. A wetting condition has been imposed at this boundary to account for the twin energy. After some time, the solid detaches from the twin boundary, leaving a narrow liquid channel in between. A secondary arm grows parallel to the twin plane, at 60° from the primary arm. Courtesy of M.A. Salgado-Ordorica, EPFL

materials. Hot dip coatings usually present large grains with diameters on the order of millimeters, often referred to as spangles. The formation of spangles has been examined in some detail for zinc coatings (Ref 144 to 146) and also to some extent for Al-Zn-Si coatings (Ref 147, 148). In Zn-0.2wt%Al coatings, a strong basal texture ((0001) parallel to the sheet surface) can develop. This texture can be caused by several factors, such as the mechanisms of dendritic growth in the hexagonal close-packed zinc system, which exhibits higher velocities for dendrite tips growing in the basal plane. Another factor is the influence of the substrate, which can favor the nucleation of certain crystallographic orientations. Preferred crystallographic orientation in relation with the substrate orientation cannot be excluded, but a direct link between the grain textures in the steel and in the coating is difficult to make because of the presence of thin layers of intermetallic compounds at the interface.

Another difference between bulk and coating microstructures is the development of large intragranular variations of the crystallographic orientation. Continuous variations of the crystallographic orientation as large as 35° were observed within individual grains in 55Al-43.4Zn-1.6Si (wt%) hot dip galvanized coatings (Ref 126, 147). This phenomenon has been attributed to plastic deformation of the dendritic network during solidification as a result of important capillary forces at the free surface (Ref 126).

Several studies of microstructure and texture in coatings have been based on simulation (Ref 149 to 152). Before presenting some examples, dendritic growth in a confined space is briefly discussed. This is a fundamental aspect of coatings structure, which can lead to the formation of a complex surface appearance and plays an important role in the formation of textures.

Phase-Field Simulation of Dendritic Growth in Confined Spaces. The velocity of a dendrite tip growing in a confined space can differ substantially from growth in the bulk, because the dendrite can be forced to stop at the boundary or allowed to continue to grow along unusual crystallographic directions. The growth velocity of a free dendrite developing in an infinite surrounding melt can be calculated with the marginal stability (Ref 90 to 92) or solvability theory (Ref 93 to 96). However, this is not the case for a dendrite tip whose diffusion field is interacting with the boundaries of the melt. Therefore, numerical simulation, based mainly on phase field, has been used to describe the growth of dendrites in confined spaces. Analyses were carried out for dendrites that are aligned with the confining boundaries (Ref 153) or for an arbitrary incidence angle (Ref 149, 150).

Figure 12 shows a phase-field simulation of dendrite tips interacting either weakly (Fig. 12a) or strongly (Fig. 12b) with the domain boundaries (Ref 150). The figure shows

(Ref 141). They indicate that twinned dendrite tips probably have a groove to equilibrate the surface tensions. This groove would deepen to form a doublon only for dilute alloys (Ref 141).

The morphology of twinned dendrites has been investigated by phase-field simulations (Ref 142, 143). In a first attempt, two-dimensional calculations were used to show that a grooved tip can propagate in a stable manner if the anisotropy of the interfacial energy is described with second-order harmonics (Ref 142). Recent three-dimensional calculations, which are illustrated in Fig. 11(c), showed the important influence of the wetting angle at the twin boundary, which can promote grooved tips or doublons (Ref 143). However, the question of the kinetic advantage of a doublon over a regular dendritic morphology remains to be addressed. The prediction of twinned dendrites by numerical simulation remains a challenge,

because this texture defect seems to be due to a complex selection mechanism associated with the weak anisotropy of the solid/liquid interfacial energy in aluminum alloys and convection in the melt.

Grain Texture and Microstructures in Galvanized Coatings

Galvanized coatings are applied on steel sheets for corrosion protection. In the hot dipping process, the steel sheets are coated continuously by immersion in a bath of molten zinc or aluminum alloys. At the exit of the bath, some liquid is taken away with the steel sheet and solidifies to form a coating with a thickness of approximately 10 to 30 µm, which is controlled by air knives.

The macrostructures of hot dip coatings can differ considerably from those of bulk

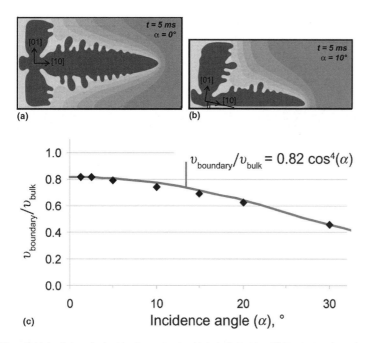

Fig. 12 Phase-field simulation of a dendrite tip growing into (a) the bulk liquid and (b) impinging a boundary shortly after nucleation (Al-24at.%Zn; supersaturation of 0.2; domain dimensions of (a) 16 by 30 μm^2 and (b)12 by 30 μm^2; the nucleation centers are indicated by the origin of the set of axes). (c) Steady-state velocity of a dendrite tip growing along a boundary reported to the growth rate in the bulk as a function of the incidence angle, α. Source: Ref 150

substantial differences in the dendrite shape between the two cases. If the dendrite tip approaches the boundary with an incidence angle, α, the dendrite cannot reach the boundary, because a solute-rich liquid film remains along the boundary and prevents any contact. By running a series of similar calculations, an expression for the tip velocity along the boundary, $v_{boundary}$, as a function of α could be determined (Fig. 12c). According to the simulations, even for very small incidence angles, the tip velocity along the interface reaches only 82% of the velocity of a free dendrite (Fig. 12c). This can be understood by looking, in Fig. 12 (b), at the diffusion fields ahead of the dendrite tip growing along the boundary. The case of a dendrite tip following an interface without wetting it corresponds to a double tip with a wider boundary layer. For an incidence angle, α, close to 0°, this situation would correspond to a doublon. The low tip velocity observed even for very small incidence angles indicates that the doublon is not the most favorable growth morphology under these conditions.

Orientation Texture. The formation of grain textures in hot dip galvanized coatings has been studied with the means of CA simulations (Ref 146, 152). The objectives of such simulation were to determine the texture and shape of the grain envelopes as a function of the nucleation conditions (Ref 146) or, using an inverse modeling approach, to determine the nucleation undercoolings based on comparisons of calculated and measured grain densities for various cooling rates (Ref 152).

The model developed in Ref 152 combines a CA method for the description of the grain structure and a finite-volume method for the resolution of heat diffusion in the coated sheet. The heterogeneous nucleation of zinc grains in the coating was simulated using two families of nucleation sites that were attributed different distributions of nucleation undercoolings and crystallographic orientations. The first family has random crystallographic orientations, while the other one exhibits a basal texture; that is, (0001) tends to be parallel to the sheet surface. Grain growth is described using cell-capture algorithms similar to the CA models of bulk materials (Ref 25 to 28). The dendrite tip velocity depends on the local undercooling and on the incidence angle with respect to the free surface, using a dependency that is similar to the one shown in Fig. 12(c). In addition, it depends on the growth direction to distinguish between growth along $< 10\bar{1}0 >$ and $< 0001 >$.

The model was used to determine the nucleation parameters of zinc in the coating using an inverse modeling approach. The method consisted of adjusting the nucleation parameters to recover the grain orientation distributions and the number densities of grains determined by electron backscatter diffraction. Figure 13 shows a typical calculation and a comparison between experimental and calculated number densities of grains. It was shown that basal grains nucleate at a minimal undercooling of approximately 0.55 K, which is approximately 0.15 K lower than for grains having a random orientation.

Surface Appearance. The grains in hot dip galvanized coatings generally exhibit a macrostructure of alternating shiny and dull sectors, which have a large impact on the visual appearance of the coating. An example of the visual aspect in an Al-Zn-Si coating (55Al-43.4Zn-1.6Si, wt%) is shown in Fig. 14(a). Such a surface appearance can be harmful for certain products, in particular for the building industry, where the material is used in the unpainted condition.

The formation of microstructure in Al-Zn-Si is considerably different from zinc coatings, namely because of the cubic crystallographic structure of the primary aluminum phase and the absence of systematic texture. A particular feature of the Al-Zn-Si coating alloy is that dendrite tips grow along 24 crystallographic directions, very close to <320> rather than along the usual six <100> directions commonly observed in face-centered cubic alloys (Ref 84, 147, 151). This makes interpretation of the solidification microstructure much more difficult. Similar to zinc coatings, the growth of dendrites in a 20 to 30 μm thick layer is strongly influenced by the boundaries that confine the melt.

The problem of surface appearance in galvanized coatings has been addressed with a geometrical approach (Ref 150). The method consists of describing the trajectories of the dendrite tips based on a series of criteria for arm branching and for the interaction with the melt boundaries (Ref 150). The tip velocity is calculated based on the current undercooling of the melt and the relationship of Fig. 12(c) for the quantification of the effect of the incidence angle at the boundary.

The model was used to predict the development of the dendrite skeleton within the confined volume of the coating as a function of the crystallographic orientation of the nucleus (Ref 150). Good agreement was found between the shiny sectors observed on the experimental micrographs and the surface regions predicted by the simulation to solidify through dendritic growth along the free surface (Fig. 14). Hence, it was proposed that areas exhibiting a shiny surface appearance are caused by growth along the free surface, whereas dull areas result from growth along the substrate/coating interface. This hypothesis was also in agreement with surface characterization by optical microscopy, laser profilometry, and wavelength-dispersive spectroscopy chemical analysis (Ref 150).

ACKNOWLEDGMENT

The authors would like to thank Mr. Mario Salgado-Ordorica for the helpful discussions on twinned dendrites and for providing new results of phase-field simulation of this morphology.

REFERENCES

1. *J. Met.,* Vol 54 (No. 1), 2002
2. M. Aloe and M. Gremaud, *Cast Met. Diecast. Times,* April/May 2007, p 35-3

(a)

Number of grains

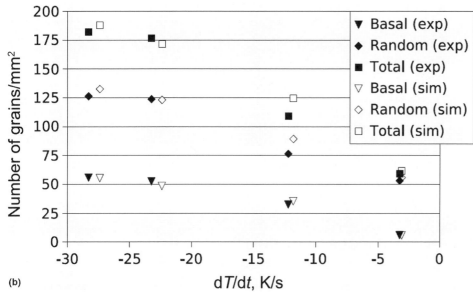

▼ Basal (exp)
◆ Random (exp)
■ Total (exp)
▽ Basal (sim)
◇ Random (sim)
□ Total (sim)

(b)

Fig. 13 (a) Image of a cellular automaton simulation of solidification in a Zn-0.2wt%Al galvanized coating. The 5 by 3 mm domain is cooled at –12 K/s with a positive temperature gradient from left to right. (b) Experimental and calculated number densities of grains as a function of the cooling rates obtained after inverse modeling. By minimizing the differences between experimental and simulation results, the nucleation conditions of basal and randomly oriented grains could be determined. Source: Ref 152

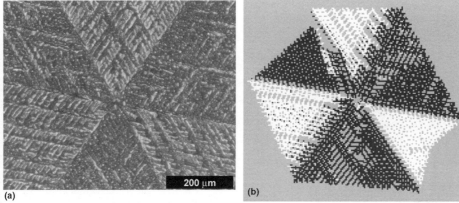
(a)   (b)

Fig. 14 (a) Optical micrograph showing shiny and dull sectors in an Al-Zn-Si coating. (b) Numerical simulation of the surface appearance with a geometrical approach for a grain having the same crystallographic orientation with respect to the coating in (a). Source: Ref 150

8. M. M'Hamdi, M. Bobadilla, H. Combeau, and G. Lesoult, *Modeling of Casting, Welding and Advanced Solidification Processes VIII,* B.G. Thomas and C. Beckermann, Ed., TMS, 1998, p 375–382

9. O. Nielsen, B. Appolaire, H. Combeau, and A. Mo, *Metall. Mater. Trans. A,* Vol 32, 2001, p 2049–2060

10. Ch.-A. Gandin, S. Mosbah, T. Volkmann, and D.M. Herlach, *Acta Mater.,* Vol 56, 2008, p 3023–3035

11. D. Tourret and Ch.-A. Gandin, *Acta Mater.,* Vol 57, 2009, p 2066–2079

12. L.Q. Chen, *Ann. Rev. Mater. Res.,* Vol 32, 2002, p 113–140

13. W.J. Boettinger, J.A Warren, C. Beckermann, and A. Karma, *Ann. Rev. Mater. Res.,* Vol 32, 2002, p 163–194

14. D. Lewis, J. Warren, W. Boettinger, T. Pusztai, and L. Granasy, *J. Met.,* Vol 56, 2004, p 34–39

15. H. Emmerich, *Adv. Phys.,* Vol 57, 2008, p 1–87

16. M. Plapp, *J. Cryst. Growth,* Vol 303, 2007, p 49–57

17. I. Steinbach, *Model. Simul. Mater. Sci. Eng.,* Vol 17, 2009, p 073001

18. I. Singer-Loginova and H.M. Singer, *Rep. Prog. Phys.,* Vol 71, 2008, p 106501

19. Y.T. Kim, N. Goldenfeld, and J. Dantzig, *Phys. Rev. E,* Vol 6, 2000, p 2471–2474

20. L.J. Tan and N. Zabaras, *J. Comp. Phys.,* Vol 221, 2007, p 9–40

21. D. Juric and G. Tryggvason, *J. Comp. Phys.,* Vol 123, 1996, p 127–148

22. A. Jacot and M. Rappaz, *Acta Mater.,* Vol 50, 2002, p 1909–1926

23. A. Jacot and Q. Du, *A Symposium in Honor of Wilfried Kurz, TMS 2004,* M. Rappaz, C. Beckermann, and R. Trivedi, Ed., TMS, 2004, p 504–511

24. B. Böttger, J. Eiken, and I. Steinbach, *Acta. Mater.,* Vol 54, 2006, p 2697–2704

25. Ch.-A. Gandin and M. Rappaz, *Acta Mater.,* Vol 42, 1994, p 2233–2246

26. Ch.-A. Gandin, J.-L. Desbiolles, M. Rappaz, and Ph. Thévoz, *Metall. Mater. Trans. A,* Vol 30, 1999, p 3153–3165

27. G. Guillemot, Ch.-A. Gandin, H. Combeau, and R. Heringer, *Model. Simul. Mater. Sci. Eng.,* Vol 12, 2004, p 545–556

28. G. Guillemot, Ch.-A. Gandin, and M. Bellet, *J. Cryst. Growth,* Vol 303, 2007, p 58–68

29. M. Rappaz and Ch.-A. Gandin, *Acta Metall.,* Vol 41, 1993, p 345–360

30. W. Oldfield, *Trans. ASM,* Vol 59, 1966, p 945–961

31. M. Avrami, *J. Chem. Phys.,* Vol 8, 1940, p 212

32. Ph. Thévoz, J.L. Desbiolles, and M. Rappaz, *Metall. Trans. A,* Vol 20, 1989, p 311–322

33. K.A. Jackson and J.D. Hunt, *Trans. Met. Soc. AIME,* Vol 236, 1966, p 1129–1142

34. D. Goettsch and J. Dantzig, *Metall. Mater. Trans. A,* Vol 25, 1994, p 1063–1079

3. M. Rappaz, *Inter. Mater. Rev.,* Vol 34, 1989, p 93–123

4. M. Rappaz and Ph. Thévoz, *Acta Metall.,* Vol 35, 1987, p 1487–1497

5. C.Y. Wang and C. Beckermann, *Metall. Trans.,* Vol 24, 1993, p 2787–2802

6. C.Y. Wang and C. Beckermann, *Metall. Mater. Trans. A,* Vol 27, 1996, p 2754–2764

7. O. Grong, A.K. Dahle, M.I. Onsoien, and L. Arnberg, *Acta Mater.,* Vol 46, 1998, p 5045–5052

35. E. Fras and H.F. López, *Acta Metall. Mater.,* Vol 41, 1993, p 3575–3583

36. A. Jacot, D. Maijer, and S. Cockcroft, *Metall. Trans. A,* Vol 31, 2000, p 2059–2068

37. K.C. Su, I. Ohnaka, I. Yamusuchi, and T. Fukusado, *The Physical Metallurgy of Cast Iron,* H. Fredriksson and M. Hillert, Ed., North-Holland, 1985, p 181–189

38. D.M. Stefanescu and C.S. Kanetkar, *AFS Trans.,* Vol 68, 1987, p 139–144

39. G. Lesoult, M. Castro, and J. Lacaze, *Acta Mater.,* Vol 46, 1998, p 983–995

40. J.D. Hunt, *Mater. Sci. Eng.,* Vol 65, 1984, p 75–83

41. C. Charbon, A. Jacot, and M. Rappaz, *Acta Metall. Mater.,* Vol 42, 1994, p 3953–3966

42. M. Rappaz, C. Charbon, and R. Sasikumar, *Acta Metall. Mater.,* Vol 42, 1994, p 2365–2374

43. C. Charbon and R. LeSar, *Model. Simul. Mater. Sci. Eng.,* Vol 5, 1997, p 53–65

44. Ph. Thévoz and M. Rappaz, *Acta Metall.,* Vol 35, 1987, p 2929–2933

45. J. Lipton, M.E. Glicksman, and W. Kurz, *Metall. Trans. A,* Vol 18, 1987, p 341–345

46. C.Y. Wang and C. Beckermann, *Metall. Mater. Trans. A,* Vol 27, 1996, p 2765–2783

47. M. Rappaz, Ch.-A. Gandin, J.-L. Desbiolles, and Ph. Thévoz, *Metall. Mater. Trans. A,* Vol 27, 1996, p 695–705

48. S. Mokadem, C. Bezençon, A. Hauert, A. Jacot, and W. Kurz, *Metall. Trans. A,* Vol 38, 2007, p 1500–1510

49. Ch.-A. Gandin, M. Rappaz, D. West, and B.L. Adams, *Metall. Mater. Trans. A,* Vol 26, 1995, p 1543–1551

50. P. Carter, D.C. Cox, Ch.-A. Gandin, and R.C. Reed, *Mater. Sci. Eng. A,* Vol 280, 2000, p 233–246

51. B. Böttger, J. Eiken, M. Ohno, G. Klaus, M. Fehlbier, R. Schmid-Fetzer, I. Steinbach, and A. Bührig-Polaczek, *Adv. Eng. Mater.,* Vol 8, 2006, p 241–247

52. J. Eiken, *Int. J. Cast Met. Res.,* 2009, in press

53. J.D. Hunt, *Mater. Sci. Eng.,* Vol 65, 1984, p 75–83

54. M. Gäumann, R. Trivedi, and W. Kurz, *Mater. Sci. Eng. A,* Vol 226–228, 1997, p 763–769

55. M.A. Martorano, C. Beckermann, and Ch.-A. Gandin, *Metall. Trans. A,* Vol 34, 2003, p 1657–1674

56. M.A. Martorano and V.B. Biscuola, *Acta Mater.,* Vol 57, 2009, p 607–615

57. Ch.-A. Gandin, *Acta Mater.,* Vol 48, 2000, p 2483–2501

58. D.J. Browne, *ISIJ Int.,* Vol 45, 2005, p 37–44

59. S. McFadden, D.J. Browne, and Ch.-A. Gandin, *Metall. Mater. Trans. A,* Vol 40, 2009, p 662–672

60. C.Y. Wang and Ch. Beckermann, *Metall. Trans. A,* Vol 25, 1994, p 1081–1093

61. A. Ludwig and M. Wu, *Mater. Sci. Eng. A,* Vol 413–414, 2005, p 109–114

62. G. Guillemot, Ch.-A. Gandin, and H. Combeau, *ISIJ Int.,* Vol 46, 2006, p 880–895

63. A. Badillo and C. Beckermann, *Acta Mater.,* Vol 54, 2006, p 2015–2026

64. H.B. Dong and P.D. Lee, *Acta Mater.,* Vol 53, 2006, p 659–668

65. H.D. Brody and M.C. Flemings, *Trans. AIME,* Vol 236, 1966, p 615–624

66. T.W. Clyne and W. Kurz, *Metall. Trans. A,* Vol 12, 1981, p 965–971

67. S. Kobayashi, *J. Cryst. Growth,* Vol 88, 1988, p 87–96

68. A. Roósz and H.E. Exner, *Modeling of Casting, Welding and Advanced Solidification Processes VI,* T.S. Piwonka, V. Voller, and L. Katgerman, Ed., TMS, 1993, p 243–250

69. S. Sundaraj and V.R. Voller, *Int. J. Heat Mass Transf.,* Vol 36, 1993, p 713–723

70. B. Sundman and I. Ansara, *The SGTE Casebook: Thermodynamics at Work,* K. Hack, Ed., The Institute of Materials, London, 1996, p 94–98

71. M. Hillert and B. Sundman, *Calphad,* Vol 14, 1990, p 111–114

72. Q. Chen and B. Sundman, *Mater. Trans.,* Vol 43, 2002, p 551–559

73. T. Kraft, M. Rettenmayr, and H.E. Exner, *Model. Simul. Mater. Sci. Eng.,* Vol 4, 1996, p 161–177

74. T. Kraft and Y.A. Chang, *J. Met.,* Vol 47, 1997, p 20–28

75. L. Kaufman and H. Bernstein, *Computer Calculation of Phase Diagrams,* Academic Press, New York, 1970

76. N. Saunders and P. Miodownik, *Calphad,* Vol 1, *Pergamon Materials Series,* R.W. Cahn, Ed., 1998

77. L. Lukas, S.G. Fries, and B. Sundman, *Computational Thermodynamics, The Calphad Method,* Cambridge University Press, 2007

78. Q. Du and A. Jacot, *Acta Mater.,* Vol 53, 2005, p 3479–3493

79. Q. Du, D.G. Eskin, A. Jacot, and L. Katgerman, *Acta Mater.,* Vol 55, 2007, p 1523–1532

80. D. Bouchard and J. Kirkaldy, *Metall. Mater. Trans. B,* Vol 28, 1997, p 651–663

81. G.B. McFadden, A.A. Wheeler, R.J. Braun, and S.R. Coriell, *Phys. Rev. E,* Vol 48, 1993, p 2016–2024

82. R. Kobayashi, *Phys. D,* Vol 63, 1993, p 410–423

83. A. Karma and W.J. Rappel, *Phys. Rev. Lett.,* Vol 77, 1996, p 4050–4053

84. T. Haxhimali, A. Karma, F. Gonzales, and M. Rappaz, *Nature Mater.,* Vol 5, 2006, p 660–664

85. J.J. Hoyt, M. Asta, and A. Karma, *Mater. Sci. Eng. R,* Vol 41, 2003, p 121–163

86. C.A. Becker, D. Olmsted, M. Asta, J.J. Hoyt, and S.M. Foiles, *Phys. Rev. Lett.,* Vol 98, 2007, p 125701–4

87. A. Karma and W.-J. Rappel, *J. Cryst. Growth,* Vol 174, 1997, p 54–64

88. W.J. Boettinger and J.A. Warren, *Acta Mater.,* Vol 43, 1995, p 689–703

89. J.C. Ramirez and C. Beckermann, *Acta Mater.,* Vol 53, 2005, p 1721–1736

90. J.S. Langer and H. Müller-Krumbhaar, *Acta Metall.,* Vol 26, 1978, p 1681, 1689, 1697

91. J. Lipton, M.E. Glicksman, and W. Kurz, *Mater. Sci. Eng.,* Vol 65, 1984, p 57–63

92. J. Lipton, M.E. Glicksman, and W. Kurz, *Metall. Trans. A,* Vol 18, 1987, p 341–345

93. P. Pelcé, *Dynamics of Curved Fronts,* Academic Press, 1988

94. A. Karma and B. Kotliar, *Phys. Rev. A,* Vol 30, 1984, p 3147

95. A. Barbieri and J.S. Langer, *Phys. Rev. A,* Vol 39, 1989, p 5314–5325

96. M. Ben Amar and E.A. Brener, *Phys. Rev. Lett.,* Vol 71, 1993, p 589–592

97. C. Beckermann, H.J. Diepers, I. Steinbach, A.Karma, and X. Tong, *J. Comp. Phys.,* Vol 154, 1999, p 468–496

98. R. Tönhardt and G. Amberg, *J. Cryst. Growth,* Vol 194, 1998, p 406–425

99. J.-H. Jeong, N. Goldenfeld, and J.A. Dantzig, *Phys. Rev. E,* Vol 64, 2001, p 041602

100. Y. Lu, C. Beckermann, and J.C. Ramirez, *J. Cryst. Growth,* Vol 280, 2005, p 320–334

101. N. Al-Rawahi and G. Tryggvason, *J. Comp. Phys.,* Vol 194, 2004, p 677–696

102. S. Chakraborty and D. Chatterjee, *Phys. Lett. A,* Vol 351, 2006, p 359–367

103. D. Li, R. Li, and P. Zhang, *Appl. Math. Model.,* Vol 31, 2007, p 971–982

104. P. Magnin and W. Kurz, *Acta Metall.,* Vol 35, 1987, p 1119–1128

105. E. Guzik and D. Kopycinski, *Mater. Trans. A,* Vol 37, 2006, p 3057–3067

106. J. Teng, S. Liu, and R. Trivedi, *Acta Mater.,* Vol 56, 2008, p 2819–2833

107. A. Karma, *Phys. Rev. Lett.,* Vol 59, 1987, p 71–74

108. A. Karma, *Phys. Rev. E,* Vol 49, 1994, p 2245

109. I. Steinbach, F. Pezzolla, B. Nestler, M. Seesselberg, R. Prieler, G.J. Schmitz, and J.L.L. Rezende, *Phys. D,* Vol 94, 1996, p 135–147

110. U. Grafe, B. Bottger, J. Tiaden, and S.G. Fries, *Scr. Mater.,* Vol 42, 2000, p 1179–1186

111. H. Kobayashi, M. Ode, S. Gyoon Kim, W. Tae Kim, and T. Suzuki, *Scr. Mater.,* Vol 48, 2003, p 689–694

112. J. Tiaden, B. Nestler, H.J. Diepers, and I. Steinbach, *Phys. D,* Vol 115, 1998, p 73–86

113. B. Nestler and A. Wheeler, *Phys. Rev. E,* Vol 57, 1998, p 2602–2609

114. R. Folch and M. Plapp, *Phys. Rev. E,* Vol 72, 2005, p 011602

115. A. Parisi and M. Plapp, *Acta Mater.,* Vol 56, 2008, p 1348–1357

116. B. Nestler and A.A. Wheeler, *Phys. D,* Vol 138, 2000, p 114–133

117. M. Plapp and A. Karma, *Phys. Rev. E,* Vol 66, 2002, p 061608
118. S. Akamatsu, M. Plapp, G. Faivre, and A. Karma, *Phys. Rev. E,* Vol 66, 2002, p 030501
119. J. Tiaden, *J. Cryst. Growth,* Vol 198–199, 1999, p 1275–1280
120. T.S. Lo, S. Dobler, M. Plapp, A. Karma, and W. Kurz, *Acta Mater.,* Vol 51, 2003, p 599–611
121. B. Nestler, A.A. Wheeler, L. Ratke, and C. Stöcker, *Phys. D,* Vol 141, 2000, p 133–154
122. M. Asta, C. Beckermann, A. Karma, W. Kurz, R. Napolitano, M. Plapp, G. Purdy, M. Rappaz, and R. Trivedi, *Acta Mater.,* Vol 57, 2009, p 941–971
123. A. de Bussac and Ch.-A. Gandin, *Mater. Sci. Eng. A,* Vol 237, 1997, p 35–42
124. X. Yang, D. Ness, P.D. Lee, and N. D'Souza, *Mater. Sci. Eng.,* Vol 413–414, 2005, p 571–577
125. S. Mosbah, M. Bellet, and Ch.-A. Gandin, *Modeling of Casting, Welding and Advanced Solidification Processes XII,* S.L. Cockcroft and D.M. Maijer, Ed., TMS, 2009, p 485–493
126. C. Niederberger, J. Michler, and A. Jacot, *Acta Mater.,* Vol 56, 2008, p 4002–4011
127. U. Paul, P.R. Sahm, and D. Goldschmidt, *Mater. Sci. Eng. A,* Vol 173, 1993, p 49–54
128. M. Newell, K. Devendra, P.A. Jennings and N. D'Souza, *Mater. Sci. Eng. A,* Vol 412, 2005, p 307–315
129. N. D'Souza, M. Newell, K. Devendra, P.A. Jennings, M.G. Ardakani, and B.A. Shollock, *Mater. Sci. Eng. A,* Vol 413–414, 2005, p 567–570
130. A. Wagner, B.A. Shollock, and M. McLean, *Mater. Sci. Eng. A,* Vol 374, 2004, p 270–279
131. M. Rappaz and E. Blank, *J. Cryst. Growth,* Vol 74, 1986, p 67–76
132. E.V. Agapova, G.N. Pankin, V.V. Ponomarev, V.N. Larinov, and A.Y. Denisov, *Russ. Metall.,* (No. 2), 1989, p 101–105
133. R.E. Napolitano and R.J. Schaefer, *J. Mater. Sci.,* Vol 35, 2000, p 1641–1659
134. J. Hérenguel, *J. Met.,* Vol 4, 1952, p 385–386
135. H. Fredriksson and M. Hillert, *J. Mater. Sci.,* Vol 6, 1971, p 1350–1354
136. J.A. Eady and L.M. Hogan, *J. Cryst. Growth,* Vol 23, 1974, p 129–136
137. H.J. Wood, J.D. Hunt, and P.V. Evans, *Acta Mater.,* Vol 45, 1997, p 569–574
138. K.I. Dragnevsi, R.F. Cochrane, and A.M. Mullis, *Metall. Mater. Trans. A,* Vol 35, 2004, p 3211–3220
139. S. Henry, P. Jarry, and M. Rappaz, *Metall. Trans. A,* Vol 29, 1998, p 2807–2817
140. S. Henry, Ph.D thesis, No. 1943, EPFL, 1998
141. M.A. Salgado-Ordorica and M. Rappaz, *Acta Mater.,* Vol 56, 2008, p 5708–5718
142. A.M. Mullis, *Modeling of Casting, Welding and Advanced Solidification Processes XI,* Ch.-A. Gandin, M. Bellet, and J. Allison, Ed., TMS, 2006, p 497–504
143. M.A. Salgado-Ordorica and M. Rappaz, *Modeling of Casting, Welding and Advanced Solidification Processes XII,* S.L. Cockcroft and D.M. Maijer, Ed., TMS, 2009, p 545–552
144. F.A. Fasoyinu and F. Weinberg, *Metall. Mater. Trans. B,* Vol 21, 1990, p 549–558
145. J. Strutzenberger and J. Faderl, *Metall. Mater. Trans. A,* Vol 29, 1998, p 631–646
146. A. Sémoroz, L. Strezov, and M. Rappaz, *Metall. Mater. Trans. A,* Vol 33, 2002, p 2695–2701
147. A. Sémoroz, Y. Durandet, and M. Rappaz, *Acta Mater.,* Vol 49, 2001, p 529–541
148. R.Y. Chen and D.J. Willis, *Metall. Mater. Trans. A,* Vol 36, 2005, p 117–128
149. A. Sémoroz, S. Henry, and M. Rappaz, *Metall. Mater. Trans. A,* Vol 31, 2000, p 487–495
150. C. Niederberger and A. Jacot, *Modeling of Casting, Welding and Advanced Solidification Processes XI,* Ch.-A. Gandin, M. Bellet, and J. Allison, Ed., TMS, 2006, p 481–488
151. C. Niederberger, J. Michler, and A. Jacot, *Phys. Rev. E,* Vol 74, 2006, p 021604
152. A. Mariaux, T.V. De Putte, and M. Rappaz, *Modeling of Casting, Welding and Advanced Solidification Processes XII,* S.L. Cockcroft and D.M. Maijer, Ed., TMS, 2009, p 667–674
153. B.P. Athreya and J.A. Dantzig, *Solidification Processes and Microstructures: A Symposium in Honor of Wilfried Kurz,* M. Rappaz, C. Beckermann, and R. Trivedi, Ed., TMS, 2004, p 357–368

ASM Handbook, Volume 22B, *Metals Process Simulation*
D.U. Furrer and S.L. Semiatin, editors

Copyright © 2010, ASM International®
All rights reserved.
www.asminternational.org

Modeling of Dendritic Grain Solidification

Ch.-A. Gandin, MINES ParisTech, France
A. Jacot, Ecole Polytechnique Fédérale de Lausanne and CALCOM ESI SA, Switzerland

THREE DIFFERENT MODELING approaches for grain structures formed during solidification of metallic alloys are presented in this article. They are direct modeling of dendritic structure, direct modeling of grain structure, and indirect modeling of grain structure. For each method, the main construction bases, the scale at which it applies, and the mathematical background are presented. For the purpose of comparison, simulations are conducted for a single equiaxed dendritic grain. More illustrative applications are gathered in the article "Formation of Microstructures, Grain Textures, and Defects during Solidification" in this Volume.

Approaches to Modeling Dendritic Solidification

Dendritic crystals are by far the most common structural feature observed in solidification of metallic alloys. They form as a result of the destabilization of the solid-liquid interface. In a sufficiently strong temperature gradient, and in the case where a limited number of crystals nucleate into the liquid, the dendritic crystals tend to grow in the direction opposite to the heat flow. As a result, the grains adopt an elongated shape referred to as columnar grains. In a shallower temperature gradient, or in the case of intense nucleation, the crystals are more isotropic and form a grain structure known as equiaxed. Equiaxed grains are of technical relevance for achieving isotropic mechanical properties of the cast parts. Their internal structure, characterized by a dendrite arm spacing and its associated distribution of chemical species (known as segregation), also contributes to the control of properties.

If the dendritic structure first forms on cooling from the liquid state, it is generally followed by several others structures. After primary dendritic solidification, the liquid that remains between the grains and the dendrite arms can successively transform into peritectic and/or eutectic structures. While modeling of primary dendritic solidification is very richly

illustrated in the literature, this is not as true for peritectic and eutectic solidification.

Several types of models have been developed to simulate the development of a solid-liquid interface (Ref 1 to 5). A typical size of the simulation domain and of the representative elementary volume (REV) is associated with each modeling approach. Differences are mainly due to the scale at which the interfacial heat and mass transfers are applied and the corresponding geometrical approximations. In this article, they are distinguished as three main approaches schematized in Fig. 1.

The first approach is referred to as direct modeling of the dendritic structure. It starts with a thorough physical representation and, as an output, directly simulates the dynamics of the development of the solid phase from the liquid phase, the resulting destabilization of the interface and, eventually, the dendritic structure (Fig. 1a). By assigning the interface between the solid and liquid being considered with an a priori thickness, an approximation for the spatial distribution of the phases within the interface is introduced. The drawback of such modeling is the limited domain size that can be simulated. It is typically limited to one or only a few crystals. No direct modeling of the entire volume of a casting is permitted. Even the solidification structure of a spherical particle produced by gas atomization of a liquid alloy, as small as 100 μm in diameter, can hardly be quantitatively simulated with such modeling approaches.

For this reason, a method of averaged modeling of the grain structure has been developed (Ref 6 to 10) and applied to ingot solidification (Ref 11 to 15). This second approach is referred to as indirect modeling of the grain structure. The goal is to predict the average grain size and volume fraction of the phases, while the solid phase is not topologically described (Fig. 1c). As a consequence, and unlike the previous type of model, no direct visualization of the dendritic structure is accessible with this approach, because the grains are not tracked

individually. Coupling with modeling of the casting process becomes feasible, and one can produce maps of the grain density, n, in cast parts as a result of local cooling rate and solid and liquid flows, as well as the variation of composition. As such, this indirect modeling of the grain structure can be seen as an extension of classical segregation models based on mass balances on an isolated representative volume of the structure found in text books (Ref 16). Indeed, both the dendrite arm spacing and the grain size are described concomitantly. However, approximations concerning the grain shape are rough (e.g., spherical grain for equiaxed solidification, cylindrical primary dendrite as constitutive geometrical element for columnar grain solidification). The latter is assumed a priori, and no possibilities for its evolution as a function of the local time evolution of the heat and solute flows have been implemented so far. As a consequence, competition between columnar and equiaxed grains, and the possible texture evolution that accompanies dendritic solidification, are difficult to model.

Thus, an intermediate type of model was introduced to achieve a direct representation of the grain structure only, while still simplifying the description of the topology of the solid phase (Fig. 1b). This third approach is referred to as direct modeling of the grain structure. Such models describe the development of the boundary between the envelope of the dendritic grains and the liquid melt. This envelope contains the mushy zone made of the mixture of the dendritic structure and the interdendritic liquid and is delimited by the tips of the dendrites growing into the undercooled melt. Using a growth model that mimics the preferential growth directions of the dendrite trunks and arms, the development of the grain envelope can be computed, resulting in columnar or equiaxed, textured or isotropic, dendritic grain structures. Similar segregation models as those developed by the indirect averaging modeling approaches are coupled for the prediction of the space distribution of the fraction of solid within the envelope of the grains.

The three modeling approaches are hereafter named direct modeling of the dendritic structure, based on tracking of the solid-liquid interface (Fig. 1a); direct modeling of the grain structure, based on tracking of the mushy zone-liquid boundary (Fig. 1b); and indirect modeling of the grain structure, based on averaged grain structure (Fig. 1c). Their description is successively given in this article. Applications are proposed in the article "Formation of Microstructures, Grain Textures, and Defects during Solidification" in this Volume.

Modeling

The mathematical frameworks of heat and mass conservation for a multiphase system domain are presented in Ref 16 and 17. The derivations yield a set of conservation equations averaged over a REV made of a mixture of solid and liquid phases. The general approximation of static phases and a binary system is made. Additionally, the densities of the solid and liquid phases are assumed constant and equal. The system is only made of a binary mixture of constituents. The objective is to remain as simple as possible for the sake of comparing the several modeling approaches. It is also assumed that neither heat production nor mass production takes place in the domain. As mentioned in the introduction, the scale of the structure at which information is desired defines the REV to be used in order to apply the general formulation of the average equations. In the following, three REVs are described with the goal of tracking the solid-liquid interface (Fig. 1a), tracking the mushy zone-liquid boundary (Fig. 1b), and modeling the grain structure by means of average quantities (Fig. 1c).

Direct Modeling of the Dendritic Structure

In this modeling approach, the objective is to perform a direct simulation of the solidification microstructure in a small region of the casting (typically 1 mm in size). If the temperature at the solid-liquid interface is known, the growth of the primary phase from the liquid can be described by formulating a solute balance at the solid-liquid interface (which must include the departure from thermodynamic equilibrium due to curvature) and solving the solute diffusion equations in both phases. A direct solution of this difficult moving boundary problem can be obtained using the so-called front-tracking methods (Ref 18). To avoid the difficulty of evolving meshes, it is generally preferred to use fixed grid methods such as the pseudo-front tracking (Ref 4), the level set (Ref 2, 19 to 22), and finally the phase field (Ref 3, 23 to 31), which is currently the most widely used

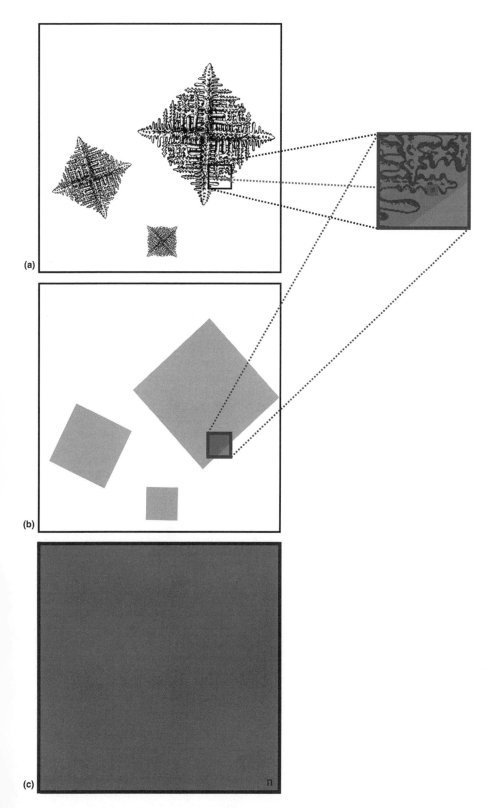

Fig. 1 Schematics of the approximations used to represent the structure of a dendritic alloy solidifying in a given volume of a casting. The direct modeling approaches are based on tracking methods of (a) the solid-liquid interface or (b) the mushy zone-liquid boundary, while (c) indirect averaged modeling does not provide a direct representation of the structure and thus requires the grain density, n, as an additional variable. The squares (shown in color in the Online Edition) schematize the typical length scale of the representative elementary volume (REV) used by the three methodologies, with (a) the very small box straddling the dendrite tip, shown within the offset box (red in the Online Edition), (b) the offset box and its relation to the dendritic structure of Fig. 1(a) (green in the Online Edition), and (c) the gray-scale region (blue in the Online Edition). Note that the scale of the REV in Fig. 1(a) (red) for tracking the solid-liquid interface is overestimated in this representation; a better estimation is shown in Fig. 2.

technique to address this problem. To illustrate the direct modeling approach of the dendritic structure, it has been chosen to present here a level-set method, which has the advantage of a relatively simple derivation as compared to the phase field.

Model Formulation. Figure 2(a) introduces the geometrical description chosen for the distribution of the solid and liquid phases in order to reach a direct simulation of the development of the dendritic structure. The model is based on the level-set method presented in Ref 19. It is based on tracking the position with time of an isopleth of a chosen function. In the present case, the function is the signed geometric distance to the solid-liquid interface, and the "0" isopleth of this function is tracked. For that purpose, an expression of the velocity of the selected isopleth must be known.

A level-set function, ϕ, is introduced at each position x of domain Ω. It is defined by the signed distance, $d(x, \gamma)$, from the solid-liquid interface, γ:

$$\phi = d(x, \gamma) \text{ for } \{x \in \Omega\} \qquad \text{(Eq 1)}$$

where $\gamma = \{x \in \Omega; d(x) = 0\}$. Isopleth "0" of function ϕ thus defines the solid-liquid interface, γ. The time evolution of the interface position can be deduced by applying a pure advection of the level-set function at the velocity of the solid-liquid interface, $v^{s/l}$, that is, solving the equation:

$$\frac{\partial \phi}{\partial t} + v^{s/l} \cdot \text{grad } \phi = 0 \qquad \text{(Eq 2)}$$

The solution of Eq 2 deviates from the distance function given by Eq 1. Initialization of the level-set function thus must be conducted after each advection step, the description of which is given elsewhere (Ref 32). The solid-liquid interface satisfies the following condition between its temperature, $T^{s/l}$, liquid composition, $w^{l/s}$, curvature, K, and normal velocity, and $v_n^{s/l} = v^{s/l} \cdot n^{s/l}$:

$$T^{s/l} = T_M + m_L w^{l/s} - \Gamma K - \frac{1}{\mu_k} v_n^{s/l} \qquad \text{(Eq 3)}$$

where T_M and m_L are properties of a linear binary phase diagram, that is, the melting temperature of the solvent and the slope of the liquidus line, respectively. Material properties are the attachment kinetics coefficient, μ_k, and the Gibbs-Thomson coefficient, Γ. While the kinetics coefficient is kept constant in the following, the Gibbs-Thomson coefficient, $\Gamma = T_M (\sigma^{s/l}(\theta) + d^2\sigma^{s/l}/d\theta^2)/(\rho \Delta H_f)$, is the product of the free energy of the solid-liquid interface, $\sigma^{s/l}(\theta)$, and the melting temperature, T_M, divided by the volumetric enthalpy of fusion ($\rho \Delta H_f$). The interface free energy depends on the orientation of the normal of the solid-liquid interface with respect to the crystallographic orientation, $\theta - \theta_0$, with θ_0 being the orientation of the [10] axis of the crystal, and $\theta = \arctan(n_y^{s/l}/n_x^{s/l})$ being the angle between the normal and the horizontal axis. Here, a simple fourfold anisotropy is considered, $\sigma^{s/l}(\theta) = \sigma_0 [1 + \varepsilon_4 \cos(4(\theta - \theta_0))]$, where σ_0 is an average of the free energy over all directions, and ε_4 is the strength of the anisotropy. The curvature of the interface can be directly computed from the divergence of the normal unit vector to the solid-liquid interface and thus with the level-set function:

$$K = \text{div} \cdot n^{s/l} \text{ with } n^{s/l} = -\frac{\text{grad } \phi}{|\text{grad } \phi|} \qquad \text{(Eq 4)}$$

To compute the velocity of the solid-liquid interface with Eq 3, it is necessary to calculate the interface composition, $w^{l/s}$, and temperature, $T^{s/l}$. To do so, the distribution of the solid phase, g^s, and hence of the liquid phase, $g^l = 1 - g^s$, are arbitrarily defined as a sinusoidal function within an interval of thickness $2W$ centered on the interface (Fig. 2a):

$$g^s = \begin{cases} 1 & \text{for } \{x \in \Omega; \phi > W\} \\ \frac{1}{2}\left[1 + \sin\left(\frac{\pi\phi}{2W}\right)\right] & \text{for } \{x \in \Omega; -W < \phi > W\} \\ 0 & \text{for } \{x \in \Omega; \phi < -W\} \end{cases}$$
$$\text{(Eq 5)}$$

Further assuming a constant segregation ratio between the solid and liquid average compositions everywhere within the interface (i.e., where a mixture of the solid and liquid phases coexist), $k = \langle w^s \rangle / \langle w^l \rangle$, mathematical developments of the average conservation equations lead to:

$$\frac{\partial \langle w \rangle}{\partial t} - \text{div}\left[\tilde{D} \text{ grad}\langle w \rangle\right] - div\left[\tilde{D} \frac{(1-k)\langle w \rangle}{kg^s + (1-g^s)} \text{ grad } g^s\right] = 0 \qquad \text{(Eq 6)}$$

where $\tilde{D} = (k g^s D^s + (1 - g^s)D^l)/(k g^s + (1 - g^s))$ is an average diffusion coefficient for the solute based on the diffusion coefficients for the solid and liquid phases, respectively D^s and D^l.

Similarly, assuming a constant enthalpy of fusion, ΔH_f, and constant heat capacity in the solid and liquid phases, C_p^s and C_p^l, the average enthalpy of the solid phase, $H^s = C_p^s T$, and of the liquid phase, $H^l = C_p^l T + \Delta H_f$, can be introduced in the average conservation of energy, leading to:

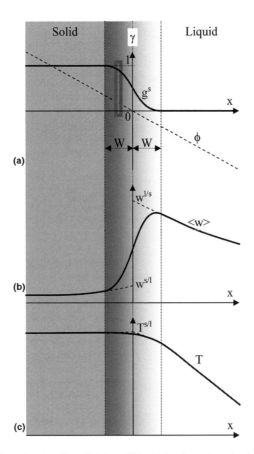

Fig. 2 Schematic one-dimensional profiles of (a) the solid-liquid interface using a level-set approach (Fig. 1a). The level-set function, ϕ, is defined by the signed distance from the interface, γ, chosen here at $x = 0$. Within the interface thickness, $2W$, the fraction of solid, g^s, continuously increases from 0 in the liquid to 1 in the solid. This is made possible by the introduction of an arbitrary trigonometric function of the distance to the interface. The half-interface thickness, W, becomes a parameter of the model that must be studied in order to retrieve correct (b) composition and (c) temperature profiles in the liquid and solid phases. The width of the narrow vertical rectangle (red in the Online Edition) in (a) informs on the typical size of a representative elementary volume required for direct modeling of the solid-liquid interface, that is, smaller than the parameter W. It is added here because the size provided by the small (red) square in Fig. 1(a) does not give an adequate representation.

$$\langle C_{\mathrm{p}} \rangle \frac{\partial T}{\partial t} - \Delta H_{\mathrm{f}} \frac{\partial g^{\mathrm{s}}}{\partial t} - \mathrm{div}[\langle \kappa \rangle \mathbf{grad}\, T] = 0 \quad (\text{Eq 7})$$

with $\langle C_{\mathrm{p}} \rangle = g^{\mathrm{s}} \langle C_{\mathrm{p}}^{\mathrm{s}} \rangle^{\mathrm{s}} + g^{\mathrm{l}} \langle C_{\mathrm{p}}^{\mathrm{l}} \rangle^{\mathrm{l}}$, $\langle \kappa \rangle = g^{\mathrm{s}} \langle \kappa^{\mathrm{s}} \rangle^{\mathrm{s}} + g^{\mathrm{l}} \langle \kappa^{\mathrm{l}} \rangle^{\mathrm{l}}$, and where κ^{s} and κ^{l} are the thermal conductivities in the solid and liquid phases, respectively. Finally, the composition and temperature fields are obtained from the solution of Eq 6 and 7. The temperature at the interface, $T^{\mathrm{s/l}}$, is taken as the temperature at the location of the interface, $T(\gamma)$, and the liquid composition, $w^{\mathrm{l/s}}$, is chosen as the average composition of the liquid at the interface, $\langle w^{\mathrm{l}} \rangle^{\mathrm{l}}(\gamma)$. Figures 2(b) and (c) show schematic representations of the composition and temperature fields on each side of the interface as well as within the interface. In practice, normalizations are useful prior to solving the aforementioned equations, an example of which is available in Ref 29. The normalization exercise is of little interest and consequently is omitted in this article. Similarly, various numerical methods are used in the literature to solve the set of conservation equations derived by the models. In the following, the finite-element method is used, together with mesh adaptation techniques, which are not explained hereafter.

In the present model, the level-set method is designed to define and track the interface dynamics by solving Eq 1 to 4, while volume averaging over a thickness centered on the interface is used in Eq 5 to 7. The later concept of an interface thickness is the main feature of phase-field modeling (Ref 29). The present model could thus be considered as a mixture of the classical level-set method for tracking the interface during dendritic growth (Ref 30) combined with the phase-field methodology, for which the evolution of the fraction of solid within the interface thickness is prescribed, and mesh adaptation (Ref 31).

Application. Figure 3 presents the application of the level-set model to simulate the development of an equiaxed dendrite (Ref 22). The material properties are those of a succinonitrile/1.3 wt% acetone alloy. Nucleation takes place at the bottom left corner of the square shown in Fig. 3(a). The nucleus chosen is a quarter solid circle of size $8.65 \cdot 10^{-6}$ m. Its initial temperature and composition are equal to 291.82 K (18.67 °C, or 65.61 °F) and 0.15 wt% and 268.52 K (−4.63 °C, or 23.67 °F) and 5.146 wt%, respectively, in the solid and in the liquid. Symmetry conditions are applied at the left and bottom boundaries of the simulation domain in Fig. 3(a), explaining the arrangement of the four representations (a) to (d) in Fig. 3. At the time chosen for the representation, 0.094 s, the equiaxed dendrite has only propagated to a small portion of the domain. In fact, Fig. 3(a) only displays approximately 1/10 of the $435 \cdot 10^{-3} \times 435 \cdot 10^{-3}$ m² simulation domain. This magnification permits clear observation of the simulated structure. The distribution of the solid phase, $g^{\mathrm{s}} = 1$, and the liquid phase, represented by $g^{\mathrm{s}} = 1 - g^{\mathrm{l}} = 0$, are shown in Fig. 3(a), together with the finite-element mesh. As can be seen, the solid phase develops with four main trunks and numerous dendrite arms, that is, forming the dendritic

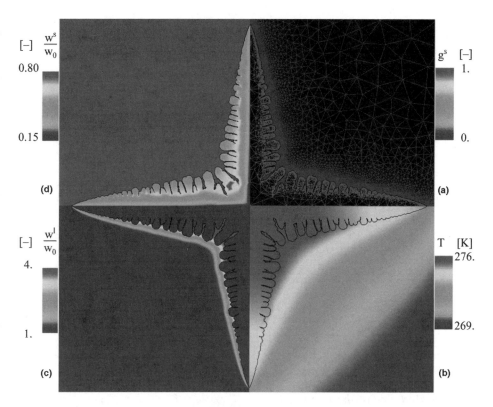

Fig. 3 Direct modeling of solidification dendritic structure is characterized by full access to the topology of the phases. This is possible by using a representative elementary volume smaller than the interface thickness (Fig. 2). As a result of simulation, the detailed distributions of (a) the solid and liquid phases, (b) the temperature, and (c) the solute composition can be computed. Among other advantages, it is also possible to extract the dendrite tip growth kinetics, the characteristic dendrite arm spacing, the shape of the grain envelope, as well as the solute flow inside and outside the grain envelope—information that are inputs to more approximate modeling approaches. The main drawback is the excessively high computational resources required, which prevents utilization for practical applications to casting technologies. Current research in the field deals with (a) automatic remeshing of the domain in which the solution is computed. Source: Ref 22

morphology. In addition to the phase distributions, the composition and temperature are fully determined as a result of the simulation.

Figure 3(b) displays the temperature field inside and outside the structure. Because the latent heat is released upon solidification, the development of the structure is accompanied by local heating. This typical thermal signature observed for equiaxed growth is a result of nucleation and growth of the solid phase in an initially undercooled melt with no or shallow temperature gradient. It is known as recalescence.

The composition field in the liquid phase is shown in Fig. 3(c). As can be seen, the liquid composition (i.e., solute concentration) between the arms of the dendrite is at its highest and is almost uniform. The highest composition gradient in the liquid is found outside the dendritic structure, that is, outside a fictitious envelope that would be formed by linking all leading tips of the dendrite arms and trunks. The inside of such an envelope would thus contain a mixture of solid plus liquid phases, known as the mushy zone, while the outside is fully liquid. Almost the same observation can be made with the temperature field in Fig. 3(b), although the gradient is not as steep. The reason for this difference is the ratio existing between the

thermal and chemical diffusivities, $(\kappa^{\mathrm{l}}/C_{\mathrm{p}}^{\mathrm{l}})/D^{\mathrm{l}}$, known as the Lewis number. For the present alloy, its value is equal to 50. Consequently, the diffusion length that characterizes heat diffusion is much larger than for solute diffusion in the liquid. This is the reason why the size of the mesh shown in Fig. 3(a) has been adapted with respect to the solute field rather than the temperature field.

However, it should be noted that in the case of a pure alloy, the solute field does not exist anymore, and the mesh size should follow the temperature gradient only. Hence, a criterion for mesh adaptation based on a combination of the composition and temperature fields is required in order to track correctly the composition and temperature gradients. It can also be shown that a correct composition profile in the solid and liquid phases is only reached when the interface thickness, W, is small enough and when the number of meshes within the interface is sufficient. Thus, the smallest mesh size is defined as a fraction of the interface thickness. In this way, segregation of solute between solid and liquid can be computed as shown in Fig. 3(d). The ratio between the solute diffusion in the liquid and in the solid, $D^{\mathrm{l}}/D^{\mathrm{s}}$, is of the order of 10^3. Because diffusion in the solid is thus much smaller than in the liquid and because the size of the dendrite arms is also small, correct description of the solid phase requires keeping track of its history in a precise manner. For this reason, a fine mesh must be kept in the solid during the entire simulation. This is also required in order to compute the interface curvature via Eq 4 and hence the variation of composition in the liquid that dictates coarsening between the dendrite arms.

Despite the powerful methods developed for direct modeling of the dendritic structure, one must realize that no direct application to industrial solidification processes has been achieved yet. Indeed, the typical simulation time for the dendritic structure shown in Fig. 3 is several hours, while the simulated domain is only $435 \cdot 10^{-3} \times 435 \cdot 10^{-3}$ m^2, even though two-dimensional and for a simple binary alloy. Computation of the dendritic structure with as much detail as in Fig. 3 for an industrial casting will remain unrealistic for several decades. In fact, this level of detail is normally not necessary for the design and optimization of casting processes. For these reasons, other modeling methodologies have been proposed, two of them being presented hereafter.

Indirect Modeling of the Grain Structure

Model Formulation. As explained previously from the observations of the simulation results shown in Fig. 3, the composition field in the liquid is almost uniform in between the dendrite arms. Oppositely, liquid composition is not uniform outside the envelope of the grain. Also recall that the slowest diffusion process takes place in the solid phase. In the case of metals, the Lewis number is typically higher than 300, while only a ratio equal to 50 was used in the simulation shown in Fig. 3. The temperature is thus approximately uniform at the scale of the entire grain and its surroundings. For these reasons, simplifications have been proposed to describe the dendritic grain structure, which are presented in Fig. 4. The nucleation event of the solid phase takes place at the center of the domain at a defined undercooling, ΔT_N, below the equilibrium temperature of the alloy, T_L. The solid structure propagates from the center of the domain in the radial direction. After nucleation, the grain envelope is assumed to develop with the velocity of the tips of the dendritic structure, v_n^s, with a predefined morphology that can be as simple as a sphere in the case of an equiaxed grain. This is schematized in Fig. 4(a). Inside the grain envelope, the internal dendritic structure is not modeled directly. It is replaced by a uniform volume fraction of the solid phase, g^s, and an average dendrite arm spacing, λ_2. The purpose of these approximations is to permit one-dimensional geometrical approximations for the diffusion processes. This is shown in Fig. 4(b) by sketching the diffusion profile in

the extradendritic liquid, l, from positions R^s and up to the limit of the grain, R^l, the latter being linked to the final grain density, n, using the relationship $R^l = [(4/3)\pi n]^{-1/3}$.

Similarly, Fig. 4(c) shows the approximation of a diffusion profile in the solid phase, s, within a fraction of half the dendrite arm spacing, $[g^s/(g^s + g^d)] \cdot [\lambda_2/2]$. The composition of the liquid located within the grain envelope, or interdendritic liquid, d, is uniform, as shown in Fig. 4(b) and (c). The ratio $[g^s/(g^s + g^d)]$ corresponds to the internal fraction of solid in the mushy zone, that is, the actual volume fraction of solid in the domain delimited by R^l divided by the volume fraction of the mushy zone, $(g^s + g^d)$. With such approximations, only the diffusion fluxes in the solid and the extradendritic liquid, both connected with the interdendritic liquid, are considered.

This modeling approach started with the work of Rappaz and Thévoz (Ref 6). It was enhanced to account for diffusion in the solid (Ref 7, 8) and peritectic and eutectic transformations (Ref 9, 10). It consists of solving the following set of equations to determine the time evolution of the volume fraction of the phases, g, and their average composition, $<w>$:

$$\frac{\partial}{\partial t}(g^s <w^s>^s) = S^{s/d}\overline{w^{s/d}}\;\overline{v_n^{s/d}} + S^{s/d}\frac{D^s}{l^{s/d}}(\overline{w^{s/d}} - <w^s>^s)$$

(Eq 8)

$$\frac{\partial}{\partial t}(g^d <w^d>^d) = -S^{s/d}\overline{w^{s/d}}\;\overline{v_n^{s/d}} - S^{l/d}\overline{w^{l/d}}\;\overline{v_n^{l/d}}$$
$$- S^{s/d}\frac{D^s}{l^{s/d}}(\overline{w^{s/d}} - <w^s>^s)$$
$$- S^{l/d}\frac{D^l}{l^{l/d}}(\overline{w^{l/d}} - <w^l>^l)$$

(Eq 9)

$$\frac{\partial}{\partial t}(g^l <w^l>^l) = S^{l/d}\overline{w^{l/d}}\;\overline{v_n^{l/d}} + S^{l/d}\frac{D^l}{l^{l/d}}(\overline{w^{l/d}} - <w^l>^l)$$

(Eq 10)

where $\overline{w^{s/d}}$ is the average composition of the solid phase at the s/d interface, and $\overline{w^{l/d}}$ is the average composition of the liquid phase at the l/d interface. Mass exchanges are considered between the solid phase and the interdendritic liquid phase through the interfacial area concentration, $S^{s/d}$, as well as between the extradendritic liquid phase and the interdendritic liquid phase through the interfacial area concentration, $S^{l/d}$, while the mass exchange between the solid phase and the extradendritic liquid phase is neglected. Solute profiles are assumed in the solid phase and the extradendritic liquid phase, respectively, characterized by the diffusion lengths $l^{s/d}$ and $l^{l/d}$. Such lengths are schematized in Fig. 4(b) in the extradendritic liquid from the interdendritic liquid/extradendritic liquid boundary, $l^{l/d}$, and in Fig. 4(c) in the solid from the solid/interdendritic liquid interface, $l^{s/d}$. The expressions for the interfacial area concentrations and the diffusion

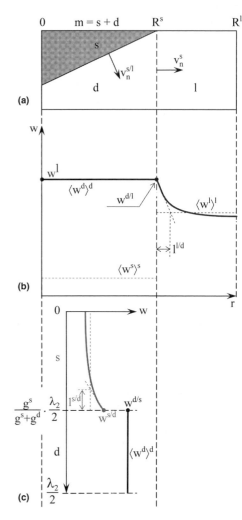

Fig. 4 Schematic one-dimensional representation of a dendritic grain envelope used for indirect averaged modeling (Fig. 1c). The complex geometry of the dendritic structure is not directly modeled. Instead, the dendritic grain envelope is described as (a) a mushy zone, m, made of a mixture of the solid phase, s, plus an interdendritic liquid phase, d. Its position, R^s, is deduced by time integration of the dendrite tip growth kinetics, v_n^s. The mushy zone develops in an extradendritic liquid phase, l, that extends up to R^l. The grain density, n, being proportional to $[(4/3)p(R^l)^3]^{-1}$, evaluation of R^l can be deduced from time integration of a nucleation kinetics of new grains. Composition profiles (plain lines) and average compositions, $<w>$ (dashed lines), in the solid and the liquid phases are shown (b) along the direction of the propagation of the structure limited by the final grain size and (c) in a representative elementary volume pertinent to model interdendritic segregation, that is, with a typical length scale proportional to the secondary dendrite arm spacing, l_2. Mass balances are developed that link the velocity of the solid-liquid interface, $v_n^{s/l}$, to the other variable introduced. The outputs of the present model are the volume fractions of each phase and their average composition. The lengths $l^{s/d}$ and $l^{l/d}$ sketch the diffusion fluxes in the solid and the extradendritic liquid, respectively.

lengths, derived with the same assumptions as in the appendices of Ref 7 and 33, are provided in Ref 34. Complete mixing of the interdendritic liquid composition and continuity of the composition at interface l/d are assumed,

$\overline{w^{l/d}} = <w^d>^d$, together with equilibrium at the s/d interface. Thus, at temperature T, readings of the liquidus and solidus curves of the equilibrium phase diagram, respectively, give $<w^d>^d$ and $w^{s/d}$. With k the partition ratio, one can also write $w^{s/d} = kw^{l/d}$.

The time derivative of the volume fraction of the solid phase s, $\partial g^s/\partial t$, can be written as a function of its interfacial area concentration, $S^{s/d}$, and the normal velocity of the s/d interface, $\overline{v_n^{s/d}}$, as $S^{s/d}\overline{v_n^{s/d}} = \partial g^s/\partial t = -\partial g^d/\partial t$. Similarly, one can write $S^{l/d}\overline{v_n^{l/d}} = \partial g^l/\partial t = -\partial g^m/\partial t$. The volume fraction of the interdendritic liquid phase and the external liquid phase are respectively defined and computed as $g^d = g^m - g^s$ and $g^l = 1 - g^m$. The growth rate of the mushy zone, $\partial g^m/\partial t$, is calculated with a dendrite tip growth kinetics model (Ref 35):

$$v_n^{s/l} = \frac{4\sigma^* D^l m_L (k-1) w^{l/s}}{\Gamma} \left(\mathrm{Iv}^{-1}(\Omega) \right)^2 \qquad \text{(Eq 11)}$$

$$\Omega = \frac{w^{l/s} - w^{l,\infty}}{w^{l/s}(1-k)} \qquad \text{(Eq 12)}$$

where Γ is the Gibbs-Thomson coefficient, Iv^{-1} is the inverse of the Ivantsov function (Ref 36), σ^* is a stability constant taken equal to $1/(4\pi^2)$ (Ref 35), m_L is the liquidus slope of the phase diagram, and Ω is the local supersaturation defined at the tip of a growing dendrite located at the growth front, that is, between the mushy zone and the extradendritic liquid. The local supersaturation in Eq 12 is defined as the deviation of the liquid composition at the dendrite tip, $w^{l/s}$, from the composition far away from the dendrite tip, $w^{l,\infty}$, that is, in the extradendritic liquid, normalized by the composition jump between the liquid phase and the solid phase, $w^{l/s}(1-k)$. The curvature undercooling is taken into account by adding its contribution to the solutal undercooling and assuming local equilibrium at the dendrite tip for the calculation of the liquid composition, $w^{l/s}$ (Ref 37, 38).

Dendrite tip models assume steady-state growth of the structure in an undercooled liquid with an initial uniform composition taken equal to the nominal alloy composition, $w^{l,\infty} = w_0$. However, to account for the solutal interactions between grain boundaries, Wang and Beckermann (Ref 7) and Martorano et al. (Ref 33) choose to use the value of the liquid composition averaged over the extradendritic domain that remains in the predefined grain envelop, $\langle w \rangle^l$ (Ref 7, 33). While the two strategies have been used in the literature, the choice was made to use $w^{l,\infty} = w_0$ for comparison with other numerical results made available in the literature (Ref 39).

Globular growth may occur if solidification leads to a fraction of solid inside the grain envelope that is greater than unity. In such a case, the approach proposed in Ref 33 and 40 applied. The growth velocity of the mushy zone

is adapted to find a solution while constraining the system to reach an internal solid fraction equal to one in the mushy zone. The velocity of the expanding mushy zone is then directly derived from the formation of the internal solid phase, and Eq 11 and 12 are no longer used.

The enthalpy of fusion per unit volume, ΔH_f, and the heat capacity per unit volume, C_p, being assumed constant, the uniform temperature condition leads to a global heat balance:

$$C_p \frac{\partial T}{\partial t} - \Delta H_f \frac{\partial g^s}{\partial t} = \frac{3}{R^l} h_{ext}(T - T_{ext}) \qquad \text{(Eq 13)}$$

where T is the time-dependent temperature of the system, and g^s is the fraction of solid. The external temperature, T_{ext}, and the heat-transfer coefficient, h_{ext}, are chosen constant over time. Heat exchange occurs at the outer boundary of the domain with a heat extraction rate, $\dot{q}_{ext} = h_{ext}(T - T_{ext})$. Equations 8 to 13 form a system of partial differential equations solved using an iterative method. When the prescribed growth temperature of the eutectic structure is reached, a simple isothermal transformation is assumed in order to transform the remaining liquid phase, $(1 - g^s)$, into a volume fraction of eutectic, g^E. During the formation of the eutectic, only Eq 13 is solved, considering no temperature variation over time, and calculating the total fraction of solid from the variation of enthalpy until completion of solidification (Ref 37, 39).

Application. Figure 5 presents the application of indirect modeling of the grain structure to simulate the development of a single equiaxed dendritic grain with a given final radius, $R = 125 \cdot 10^{-6}$ m. This situation corresponds to the simulation of the formation of a single crystal in a droplet. It can also represent a typical grain at a given location in a casting. The material properties are those of the Al-10wt%Cu alloy: $\Gamma = 2.41 \cdot 10^{-7}$ K m, $C_p = 3 \cdot 10^6$ J m^{-3} K^{-1}, $\Delta H_f = 9.5 \cdot 10^8$ J m^{-3}, $D^l = 4.37 \cdot 10^{-9}$ m^2 s^{-1}, $D^s = 0$ m^2 s^{-1}. Nucleation takes place at a given value of the nucleation undercooling, $\Delta T_N = 30$ °C (54 °F). A Fourier boundary condition is applied on the spherical boundary of the droplet using a constant heat-transfer coefficient, $h_{ext} = 490$ W m^{-2} °C^{-1}, and a constant external temperature, $T_{ext} = 373$ K (100 °C, or 212 °F). The phase diagram is simplified by linear monovariant lines defined by a liquidus slope, $m_L = -3.37$ wt% K^{-1}; a segregation coefficient, $k = 0.17$; a eutectic temperature, $T_E = 817.74$ K (544.59 °C, or 1012.26 °F); a eutectic composition, $w_E = 34.38$ wt%; and the liquidus temperature of the alloys, $T_L = 899.9$ K (626.8 °C, or 1160.2 °F), for the alloy composition, $w_0 = 10$ wt%. The initial temperature of the system is uniform and equal to 900.9 K (627.8 °C, or 1162.0 °F). Note that these approximations follow the study by Heringer et al. (Ref 39), while a reasonable value for the diffusion of copper in the primary aluminum phase would be $D^s = 5 \cdot 10^{-13}$ m^2 s^{-1} (Ref 40).

Typical results of such indirect modeling of the grain structure are the temperature and phase volume fraction histories shown by the thick gray curves in Fig. 5(a) and (b), respectively. The initial temperature is above the liquidus temperature, T_L (i.e., the equilibrium temperature at which the solid and liquid phases can start to coexist). Upon heat extraction, the system first cools down while staying in the liquid state. When the nucleation temperature is reached, that is, $T_L - \Delta T_N$, the solid starts to grow, thus increasing the size of the grain envelope and hence the volume fraction of the mushy zone, g^m, as well the fraction of solid, g^s. The ratio $g^{sm} = g^s/g^m$, known as the internal fraction of solid, is at first very high, that is, close to unity. It progressively decreases proportionate to the development of the mushy zone. This corresponds to an increased growth rate of the solid fraction large enough to reheat the system, as is shown by the recalescence in Fig. 5(a). It is thus the same phenomenon described in Fig. 3.

After the recalescence, the grain envelope becomes fully developed, g^m reaches unity, and no more extradendritic liquid remains. Then, the latent heat is released only by the solidification of the interdendritic liquid, which is not sufficient to maintain the recalescence. As a consequence, solidification is accompanied by a temperature reduction. Such progressions are confirmed but described in better detail by Heringer et al. (Ref 39). These authors have proposed a one-dimensional mushy zone front-tracking (1-D MZFT) model that can also predict the local solidification evolution inside the grain envelope. It is based on a one-dimensional numerical solution of the averaged conservation equations for heat and solute mass. The finite volume method and fixed grid is used; tracking of the boundary between the mushy zone and the liquid melt is achieved by a time integration of Eq 11 and 12.

Comparison of this numerical solution with the prediction of the indirect grain structure simulation is also possible by considering a volume averaging over the one-dimensional spherical domain of the temperature and volume fractions with time. The results of such averaging are also shown in Fig. 5. Very good agreement is reached, thus validating the approximations and solutions made by indirect grain structure modeling. However, the 1-D MZFT model also shows that, on recalescence, the internal structure remelts, thus explaining the decrease of g^{sm} predicted in Fig. 5(b). The reason why the total fraction of solid, g^s, still increases in Fig. 5(b) is because of the development of the mushy zone that transforms the extradendritic liquid into solid by growing the grain envelope.

At this stage, it is interesting to point out that only time evolutions of the average compositions in the solid and liquid phases normally are predicted by indirect modeling of the grain structure. As a consequence, a composition profile inside the envelope of the grain is not

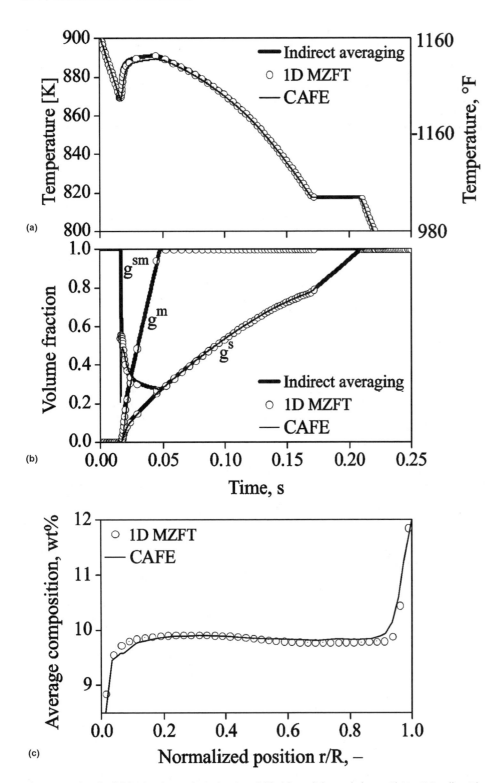

Fig. 5 Predicted solidification for a spherical grain solidified by radial growth for an Al-10wt%Cu alloy. The prediction by indirect modeling of the grain structure (thick gray curves) (Ref 9) is compared with a one-dimensional mushy zone front-tracking (1-D MZFT) solution (open symbols) for the solidification of a single equiaxed grain (Ref 39). The curves show the time evolution of (a) the temperature and (b) the fraction of solid, g^s, the fraction of mushy zone, g^m, and the internal fraction of solid, $g^{sm} = g^s/g^m$. Note that (c) the composition profile within the grain is not directly predicted by indirect modeling of the grain structure, and comparison with the prediction of the 1-D MZFT model only is possible with the CAFE simulation presented later in Fig. 7. The composition profile within the grain is not yet available with direct modeling for the dendritic structure of Fig. 3.

available. This is yet a direct output of the 1-D MZFT model, as shown in Fig. 5(c), thus identifying a main difference compared to direct modeling. However, coupling of indirect modeling of grain structure for simulation of solidification processes has been demonstrated in the past, examples of which are available in Ref 11 to 15. The average composition profile can then be predicted at a scale that is larger than the grain size. It must be underlined that the set of equations presented for indirect modeling have also been extended to model columnar structures (Ref 7). Geometrical approximation must be changed to an a priori unit columnar dendrite shape for the description of mass exchange, for example, a cylinder of radius given by the primary dendrite arm spacing instead of the sphere for the equiaxed grain. The main variables to be adapted are then the interfacial area concentrations and the diffusion lengths. Coupling of the geometric descriptions is also possible for the prediction of the columnar-to-equiaxed transition, a well-known feature of casting that is illustrated later in this article.

As illustrated and explained, numerous limitations are found when using indirect grain structure modeling. Yet, this approach remains the only possible way to achieve a practical simulation of the structure at the scale of a large casting, such as forge ingots corresponding to several tons of steels. For smaller castings, however, an alternative way has been developed, which is described next. It permits direct modeling of the grain structure.

Direct Modeling of the Grain Structure

Model Formulation. Direct modeling of the grain structure presented in this section is based on coupling a cellular automaton (CA) method and a finite-element (FE) method. Other variants of a CA method have been proposed (Ref 5). However, their objective was to solve both the heat and solute diffusion fields while tracking the solid-liquid interface, which has not yet been achieved realistically compared to phase-field or level-set tracking methods. Instead, the present CA method only tracks the boundary between the mushy zone and the liquid, as explained in the introduction. This boundary represents the envelope of the dendritic grains. Two other methods have also been proposed, based on a mesoscopic phase-field method (Ref 41) and a front-tracking method (Ref 42).

Figure 6(a) gives a schematic presentation of the tessellations using the FE and CA methods. The continuous domain is divided into an FE mesh using coarse triangles "F" defined by nodes n_i^F (i = [1, 3]). A regular lattice of fine squares defining the cells of the CA grid is superimposed onto the FE mesh. Each cell ν located in an element "F" is defined by the coordinates of its center, C_ν. Linear interpolation coefficients are defined between a node n_i^F (i = [1, 3]) and the cell ν. A variable defined at the FE nodes can thus be

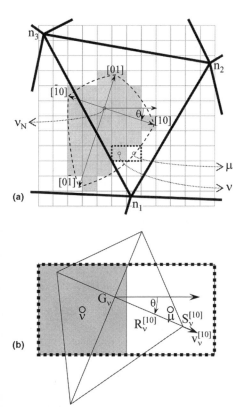

Fig. 6 Schematic two-dimensional geometric description of a dendritic grain using direct modeling of the grain structure by means of the cellular automaton (CA) method coupled with the finite-element (FE) method (Fig. 1b). A representation is given of (a) a unit triangular mesh used by the FE tessellation, defined by nodes n_1 to n_3, on which is superimposed a regular CA grid of square cells. The cell color corresponds to the value of the state index: White cells are in the liquid state, and gray cells are in a mushy state, that is, made of a mixture between the primary solid dendritic phase and the liquid phase. The shape of a single grain nucleated at the center of cell n_N is given by the assembly of contiguous gray cells. It may be compared with the desired grain envelope provided by the dashed line contour. When nucleated with selected <10> crystallographic directions (see text), its development from the center of cell n_N is modeled by the use of (b) a growth algorithm applied to each mushy cell. For a given mushy cell, ν, having at least one neighboring liquid cell, μ, it consists of an integration over time of a dendrite tip growth kinetics in all of its four <10> directions, $\nu_n^{<10>}$, thus defining four apexes $S_n^{<10>}$ that delimit a growing shape from its growing center, G_n (only $\nu_n^{[10]}$ and $S_n^{[10]}$ are shown for clarity). When the center of a neighboring liquid cell, μ, is entrapped in the growing shape, its state index is switched to the mushy state, and it starts to develop its own growing shape with the same <10> directions in order to continue propagating the dendritic grain envelope.

interpolated at a given CA cell. Similarly, information computed onto the CA grid can be summed up and projected onto the FE mesh (Ref 37, 43 to 46). The averaged equations derived for a mixture of the solid and liquid phases are combined in the following way:

$$\frac{\partial <H>}{\partial t} - \mathrm{div}(<\kappa> \mathbf{grad}\, T) = 0 \qquad \text{(Eq 14)}$$

The average enthalpy $<H>$ is chosen as the primary unknown in Eq 14. The solution of this

nonlinear equation is obtained by using a Newton-Raphson procedure, which requires calculating the derivative $\partial <H>/\partial T$ at each node of the FE mesh (Ref 47). This is done by summing this quantity over the cells of the CA grid (Ref 46). Similarly, the averaged equation becomes:

$$\frac{\partial <w>}{\partial t} - \mathrm{div}(g^l D^l \mathbf{grad} <w^l>^l) = 0 \qquad \text{(Eq 15)}$$

where the term $\mathrm{div}(g^s\, D^s\, \mathbf{grad}\langle \rho^s w^s \rangle^s)$ has been neglected. Indeed, the ratio of the diffusion coefficient in the primary solid phase, D^s, over the diffusion coefficient in the liquid phase, D^l, is typically of the order of 10^3. This shows that diffusion in the solid phase can be neglected at the scale of the casting. The primary unknown considered in Eq 15 is the average composition of solute, $<w>$. The average composition of the liquid phase is eliminated following the work by Prakash and Voller, who introduced a split operator technique with a Euler backward scheme (Ref 48). Mesh adaptation is also used and explained elsewhere (Ref 49 to 52).

Nucleation and growth algorithms previously designed to track the development of the grain envelopes are briefly explained hereafter (Ref 37, 46). Each cell, ν, has a state index, I_ν, which is initialized to zero and corresponds to the liquid state. Upon cooling, when the temperature of a chosen nucleation cell ν_N, T_{ν_N}, is lower than a prescribed nucleation temperature, T_N, the cell state, I_{ν_N}, is changed from liquid, $I_{\nu_N} = 0$, to mushy, $I_{\nu_N} = 1$, that is, a mixture of the solid and liquid phases. Propagation of the solid is realized by the growth algorithm. An equilateral quadrangular surface is defined with its center, G_{ν_N}, located at cell center C_{ν_N} (identified by an open circle centered in cell ν_N in Fig. 6a) and four perpendicular <10> lengths, $R_{\nu_N}^{<10>}$. The directions of the <10> lengths are defined by an angle, θ, with respect to the (x, y) frame of coordinates, as illustrated in Fig. 6(a). For cubic metals, they correspond to the main growth directions of the dendritic structure, that is, four of the six <100> crystallographic directions located in a common (100) plane. In other words, the two other <100> crystallographic directions are assumed perpendicular to the simulation domain and are not simulated in a two-dimensional approximation. This limiting assumption is removed with a three-dimensional approach (Ref 44, 45). The lengths of the four <10> directions associated with the equilateral quadrangular surface centered in G_{ν_N} are updated by integration over time of a dendrite tip growth kinetics, $\nu_{\nu_N}^{<10>} = \partial R_{\nu_N}^{<10>}/\partial t$, as long as the neighboring cells are in a liquid state, that is, not yet captured by the growing cell ν_N. When the growth shape of cell ν_N encompasses the center of one of the neighboring liquid cells, the capture criterion is verified. The state of the captured neighboring liquid cell is then changed from liquid to mushy, and its growth shape is initialized.

An illustration of the growth shape is given in Fig. 6(b) for a cell ν located at the mushy zone/liquid boundary. At the time chosen for the representation, the neighboring liquid cell μ is captured. Variables I_μ, G_μ, and $R_\mu^{<10>}$ are initialized or updated. Note that the growth center G does not necessarily correspond to the cell center C; the crystallographic orientation of the nucleation cell is preserved upon propagation of the mushy zone, thus propagating the orientation of the grain, and the capture is only permitted for the first neighbors of a growing cell. If all its neighboring cells are in a mushy state, the state index is updated as $I_\nu = -1$. The latter permits simple identification of the growing cells located at the mushy zone/liquid boundary by the state index value +1.

A mushy cell ν is made of a mixture of the solid and liquid phases. As mentioned previously, the liquid entrapped in between the dendrite arms and more generally inside a fictitious envelope of the grain, which is defined by the leading dendrite tips and arms, has an almost uniform composition. Consequently, the liquid phase is subdivided into an interdendritic liquid, d, plus an extradendritic liquid phase, l. A total of three phases is thus considered for mass balances, as in indirect modeling. A mushy zone volume fraction assigned to each cell ν, g_ν^m, is defined as the volume fraction of the solid phase s, g_ν^s, plus the interdendritic liquid phase d, g_ν^d: $g_\nu^m = g_\nu^s + g_\nu^d$. It is estimated by an average of the four lengths $R_\nu^{<10>}$ as $g_\nu^m = (1/4) \sum_{<10>} (R_\nu^{<10>}/R_\nu^f)^2$. The final radius associated with cell ν, R_ν^f, is defined by the spatial limit for the growth of the equilateral quadrangle, which is of the order of several secondary arm spacings. In the case of a dendritic structure, this limit is chosen proportional to the primary dendrite arm spacing, $R_\nu^f = \lambda_1/2$.

To determine the average enthalpy, $<H_n>$, and solute composition, $<w_n>$, at the FE node n being deduced from the solution of Eq 14 and 15, conversions are required into a temperature, T_n, and a fraction of solid, g_n^s. Instead of applying a solidification path at the FE nodes (Ref 47, 53), the conversions are first carried out for each CA cell ν to compute the temperature, T_ν, and fraction of solid, g_ν^s, from the interpolated enthalpy, $<H_\nu>$, and average composition, $<w_\nu>$. The fields at the CA cells are finally projected back to the FE nodes (Ref 37, 46). Assuming equal and constant densities in all phases, one can write $g_\nu^s + g_\nu^d + g_\nu^l = 1$ and $<w_\nu> = g_\nu^s <w_\nu^s>^s + g_\nu^d <w_\nu^d>^d + g_\nu^l <w_\nu^l>^l$. A segregation model is required to model the time evolution of the average volume fraction and composition of the solid phase s, $g_\nu^s <w_\nu^s>^s$, the interdendritic liquid phase d, $g_\nu^d <w_\nu^d>^d$, and the extradendritic liquid phase l, $g_\nu^l <w_\nu^l>^l$. This is done using a modified version of Eq 8 to 10.

Source terms are added in Eq 9 and 10 to account for the solute mass exchange of the cell ν with its surroundings, respectively $\dot{\varphi}_\nu^d$ and $\dot{\varphi}_\nu^l$. Solute exchange between cells is only based on diffusion in the liquid, that is, through the interdendritic liquid phase, d, and through the extradendritic liquid phase, l. By summing up $g_v^d \dot\varphi_v^d$ and $g_v^l \dot\varphi_v^l$, the equivalent terms are obtained at the scale of the CA model of the solute diffusion term in Eq 15, $\mathrm{div}(g^l D^l \, \mathbf{grad}\langle w^l\rangle)^l$, computed by the FE model and interpolated at cell v. The relative portions, $\dot\varphi_v^l$ and $\dot\varphi_v^d$, can be quantified by introducing a partition ratio for diffusion in the liquid, $\varepsilon_{D^l} = \dot\varphi_v^l/\dot\varphi_v^d$. The following correlation is proposed as a function of the volume fraction of the interdendritic liquid phase and the extradendritic liquid phase: $\varepsilon_{D^l} = g_v^l/(g_v^l + g_v^d)$. Hence, terms $g_v^d \dot\varphi_v^d$ and $g_v^l \dot\varphi_v^l$ can be evaluated from the solution of Eq 15. Finally, with Eq 8 to 12 and a local heat balance for cell v, $\partial\langle H_v\rangle/\partial t = C_p \, \partial T_v/\partial t - \Delta H_f \partial g_v^s/\partial t$, a complete system of differential equations is obtained. A splitting scheme is applied to the differential equations, together with a first-order Taylor series. An iterative algorithm is implemented to calculate the solution. More details are available in Ref 34.

Application. Figure 7 presents the application of direct modeling of the grain structure to simulate the development of a single equiaxed dendritic grain. It corresponds to the same problem as the one presented in Fig. 5. For symmetry reasons, the simulation domain is limited to a quarter-disk geometry with axisymmetrical conditions with respect to its two perpendicular rectilinear edges. A single grain nucleation event is imposed at the corner of

the simulation domain. The grain is further restricted to grow with a spherical shape, also for comparison purposes. Thus, the value of all the parameters already has been listed, except for those added in the caption of Fig. 7. For instance, the primary and secondary dendrite arm spacings are chosen equal to $\lambda_1 = \lambda_2 = 20 \cdot 10^{-6}$ m but have little influence in the present set of the parameters because no solid diffusion is assumed.

Figure 7 shows the model predictions when the volume fraction of the mushy zone in the droplet has reached 0.64. The triangular elements are displayed in Fig. 7(a), while the temperature, the volume fraction of solid, and the average solute composition are displayed in Fig. 7(b) to (d), respectively. The location where the fraction of solid drops to zero is observable in Fig. 7(c). It compares favorably with the position of the thick black line drawn on top of the FE mesh in Fig. 7(a), the latter being deduced from the CA simulation by drawing the boundary between the growing mushy cells and the liquid cells. Figure 7(d) reveals the sudden increase of the average composition in the vicinity of the grain envelope, due to the solute pileup in the liquid ahead of the growth front. Comparison of Fig. 7(a) and (d) thus gives an illustration of the method used to adapt the FE mesh size. Figure 7(b) also shows the temperature field inside the droplet. The maximum temperature variation only

reaches a few degrees during the propagation of the mushy zone and is localized at the growing interface. This is due to the release of the latent heat at the grain envelope. The mushy zone is actually remelting due to the recalescence taking place at its boundary (Ref 39). Finally, as shown in Fig. 5(c), the present model retrieves the final segregation profile predicted by Heringer et al. (Ref 39) as well as time evolutions shown in Fig. 5(a) and (b) for indirect modeling.

An advanced application of direct modeling of the grain structure is illustrated in the following example for the solidification of a Sn-10wt%Pb alloy. The experimental setup is designed for the study of segregation induced by thermosolutal buoyancy forces. It is largely inspired from the previous work proposed by Hebditch and Hunt (Ref 54). A parallelepiped geometry is used with dimensions $10 \times 6 \times 1$ cm^3 and is filled with a Sn-10wt%Pb alloy. The smallest surfaces of the casting, that is, the two opposite faces with dimension 6×1 cm^2, are in contact with temperature-controlled heat exchangers. Their temperature changes are thus time dependent by imposing either constant heating/cooling rates or holding temperatures. All other surfaces are made adiabatic, as explained in detail in Ref 55.

The experiment proceeds by first melting the ingot and maintaining it in a liquid state, with sufficient time for temperature homogenization at 250 °C (480 °F), which is monitored by an array of thermocouples. A temperature gradient of 200 K m^{-1} is then imposed through the liquid by prescribing a heating rate and a temperature plateau at 270 °C (518 °F) with the right-side heat exchanger. After a holding period, an identical cooling rate of 0.03 K s^{-1} (0.05 °F s^{-1}) is imposed on both heat exchangers while maintaining the initial 20 K (36 °F) temperature difference between the smallest faces of the parallelepiped geometry, thus leading to solidification of the alloy. An array of 50 thermocouples is used to record the temperature during the experiment. It consists of five rows with ten thermocouples, each thermocouple being distant from its neighbor by 1 cm in both the vertical and horizontal directions. This grid of thermocouples is used to measure the evolution of experimental temperature maps at the surface of one of the 10×6 cm^2 faces. Careful analysis of the flow within the thickness of the ingot (1 cm) shows almost no temperature gradient in the transverse direction. The temperature is thus assumed uniform in the thickness. This assumption is verified using a three-dimensional simulation of the heat flow. Comparison of the grain structure on the opposite large face of the ingot also reveals minor differences, as shown in Fig. 8(a). In addition to the in situ measurement of the temperature maps, the average concentration of lead has been measured in the as-solidified sample using the same grid (Ref 55). The 50 values collected are then used to plot the final average lead concentration, as shown in Fig. 8(b).

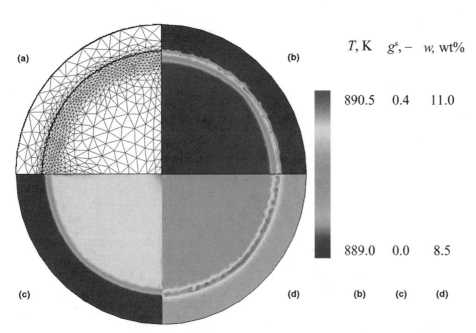

$T, \mathrm{K} \quad g^s, - \quad w, \mathrm{wt\%}$

890.5   0.4   11.0

889.0   0.0   8.5

Fig. 7 Direct modeling of solidification of a single equiaxed grain using the cellular automaton (CA) method coupled with the finite-element (FE) method is a refinement of the indirect modeling approach (Fig. 5). Integration over time on the geometrical CA grid of kinetics laws for nucleation and growth permits topological description of the development of the grain envelope. The latter is thus made of an assembly of interconnected CA cells in a mushy state. Advantages with respect to indirect modeling are direct access to the grain structure, a numerically computed composition field inside and outside the grain envelope, and no requirement for an a priori shape of the grain (Fig. 8), possibly accounting for the crystallographic orientation of the grains. The same system as in Fig. 5 is used. Simulation is carried out using an axisymmetrical coordinate representation of a spherical domain. The present drawing shows (a) the FE mesh and the CA growth front (thick black line), (b) the temperature field, (c) the volume fraction of solid phase, and (d) the average composition field. CA cell size: $10 \cdot 10^{-6}$ m; minimum FE mesh: $30 \cdot 10^{-6}$ m; maximum FE mesh: $200 \cdot 10^{-6}$ m; objective relative error on $\langle w\rangle$: 10^{-4}

A two-dimensional Cartesian CAFE simulation has been performed, the results of which are presented in Fig. 8. Values for the parameters and materials properties are available in Ref 56. Unlike previous simulations presented in this article, segregation is caused mainly by thermosolutal convection. This means that advection terms are added in the conservation Eq 14 and 15 (Ref 37, 46). Upon solidification, the liquid is cooled down and enriched with lead, thus generating buoyancy forces caused by the density variation with temperature and composition. Note that the calculation still considers diffusion in both the solid and liquid phases, although these effects are not the main mechanisms to explain the final segregation map shown in Fig. 8. Similarities exist with the simulated segregation in other tin-lead alloys (Ref 57) and gallium-indium alloys (Ref 58) also using rectangular castings. Unlike previous aluminum-silicon alloys, properties are not well known for tin-lead alloys. Comparison between measurement and simulation could only be reached by adjusting the thermal expansion coefficient.

The simulated grain structure shown in Fig. 8 (a) appears much coarser than the experimental grain structure. The main reason for this difference is the absence of grain sedimentation in the present simulation. In the simulated grain structure, after nucleation in the undercooled liquid, grains remain fixed in space and can only grow. This was already shown to be a primary limitation if one wishes to compare grain structure and segregation (Ref 57).

An interesting feature is the peculiar shape of the grains that are seen at the bottom part of the ingot. These grains were not nucleated at the mold wall but grew from the inside of the casting toward the mold wall. The reason is understood easily when considering the average composition map shown in Fig. 8(b). A large zone enriched in lead forms at the bottom part of the ingot, which is seen in both the experimental and simulation results. This zone is the last to solidify because the local liquidus temperature is lower. Consequently, this pocket of liquid is solidified by growth of existing grains from the inside of the ingot toward the bottom, explaining the elongated shape of the grains in this area.

Model Comparison and Summary

Table 1 presents a partial summary of the main inputs and outputs of the three modeling approaches presented in this article, so that the methods can be compared concisely. As immediately seen in the table, the fewer the inputs, the larger the output, and vice versa. Direct modeling of the dendritic structure is thus, in principle, preferred. However, applying it to casting is totally out of reach because of the required computational resources, thus fully justifying the other modeling approaches.

Direct modeling of the dendritic structure based on the level-set method is shown to

Experimental

(a)

Simulated

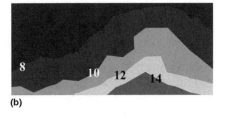
(b)

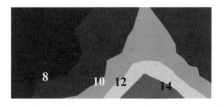

Fig. 8 CAFE predictions (Ref 56) versus measurements (Ref 20) of the final distributions of (a) the grain structure and (b) the lead composition (labeled in wt%) on solidification of a Sn-10wt%Pb alloy in a 10 cm wide × 6 cm high × 1 cm thick rectangular cavity. Heat extraction only takes place from the left-side, vertical limit of the domain; all other boundaries are insulated. Upon cooling and solidification, buoyancy forces act on the melt due to density variation of the liquid phase with temperature and lead concentration. As a result, the mushy zone forming from the left side of the casting is washed away by a solute-driven, counterclockwise, natural convection flow, thus forming the final segregation pattern shown. The size of the representation in (b) corresponds to that of the dashed boxes displayed in (a).

Table 1 Comparison of the main inputs/outputs, approximations, numerical methods, kinetics laws, and applications for the three approaches to modeling of dendritic grain solidification presented in this article

Modeling approach	Dendritic structure	Direct grain structure	Indirect grain structure
Domain size, m	10^{-6} to 10^{-3}	10^{-4} to 10^{-1}	10^{-3} to 1
Solidification time, s	10^{-3} to 10^{2}	10^{-1} to 10^{4}	1 to 10^{6}
Methods(a)	Level set	Cellular automaton	Global averaging
Inputs	Material properties	Material properties	Material properties
	Nucleation kinetics	Nucleation kinetics	Nucleation kinetics
	Crystal orientation	Crystal orientation	Growth kinetics for the mushy zone/liquid boundary
	Distribution of phases within the interface thickness	Dendrite tip growth kinetics for the mushy zone/liquid boundary	Diffusion length and area concentration for mass exchange at the phases interface
		Diffusion length and area concentration for mass exchanges at interfaces and boundaries	Grain envelope morphology
Outputs	Space distribution (morphology) of phases	Space distribution of the mushy zone and of its internal fraction of phases	Space distribution of the mushy zone
	Space distribution of crystal orientation and possible texture for a few grains	Space distribution of crystal orientation and possible texture for a large population of grains	Internal average fraction of phases within the mushy zone
	Space distribution of species in phases	Space distribution of species in the liquid and in the mushy zone	Space distribution of species in the liquid
	Space distribution of temperature	Space distribution of temperature	Internal average composition of species within the mushy zone
	Kinetics of the phase transformation (including dendrite tip growth kinetics)		

(a) Presented in this article (see text for references to other methods)

capture the development of a solid-liquid interface. As a consequence, it provides information, as illustrated in Fig. 3—composition and temperature fields in each phase, and kinetics everywhere at the solid-liquid interface. The latter means that not only is the dendrite tip growth computed but also the coarsening of the dendritic structure. The texture of a grain structure when the development of several grains is considered can also be predicted, as described in the article "Formation of Microstructures, Grain Textures, and Defects during Solidification" in this Volume. With such a detailed approach, the dynamic of a mushy zone could, in principle, also be extracted. However, it usually is not done for computational time reasons. The main drawback of such an approach is the difficulty of dealing with the scale of cast parts. For this reason, alternative methods have been developed.

Direct modeling of the grain structure is presented using a cellular automaton approach. The objective is not to model the dendritic structure anymore but rather the development of the grain envelope, that is, of the mushy zone/liquid boundary. The solid-liquid interface is not directly modeled; thus, another model is required to compute the growth velocity of the grain envelopes. This is possible by using approximate dendrite tip kinetics models, which has been well developed in the literature. Similarly, because the topological description of the solid phase in the mushy zone is not made, an average fraction of the phases must be determined. This is done by considering segregation models based on mass balances at interfaces, requiring approximations with respect to the diffusion profile in each phase. Similarly, only the average compositions in each phase of the mushy zone are predicted. Typical outputs have been presented in Fig. 7 and 8. As is further developed in the article "Formation of Microstructures, Grain Textures, and Defects during Solidification" in this Volume, application to an entire cast part is possible. It is still limited to small-volume castings.

Averaging can be applied at the scale of a grain. This indirect modeling approach requires assuming an a priori grain morphology. While the outputs of indirect approaches do not directly give access to a grain structure, maps of average grain size can be shown. The grain structure can be quantitatively calculated with indirect modeling and compared with direct modeling of the grain structure, as shown in Fig. 5. Comparison with direct modeling of the dendritic structure remains qualitative, however, with the latter still requiring too much computational resources to reach that scale in a quantitative way.

Reading the article "Formation of Microstructures, Grain Textures, and Defects during Solidification" in this Volume and "Direct Modeling of Structure Formation" in *Casting*, Volume 15 of *ASM Handbook* (Ref 59), is recommended for further examples and comparison of modeling methods for the prediction of solidification structures.

REFERENCES

1. A. Karma and W.-J. Rappel, *J. Cryst. Growth*, Vol 54, 1997, p 174
2. L. Tan and N. Zabaras, *J. Comput. Phys.*, Vol 211, 2006, p 36
3. I. Steinbach, *Model. Simul. Mater. Sci. Eng.*, Vol 17, 2009, p 1
4. A. Jacot and M. Rappaz, *Acta Mater.*, Vol 50, 2002, p 1909
5. W. Wang and P.D. Lee, *Acta Mater.*, Vol 51, 2003, p 2971
6. M. Rappaz and Ph. Thévoz, *Acta Metall.*, Vol 35, 1987, p 2929
7. C.Y. Wang and C. Beckermann, *Metall. Trans. A*, Vol 24, 1993, p 2787
8. B. Appolaire, H. Combeau, and G. Lesoult, *Mater. Sci. Eng. A*, Vol 487, 2008, p 33
9. S. Mosbah, M. Bellet, and Ch.-A. Gandin, *Solidification and Gravity 2008*, Materials Science Forum, A. Roósz et al., Ed., Trans Tech Publications, in press
10. D. Tourret and Ch.-A. Gandin, *Acta Mater.*, Vol 57, 2009, p 2066
11. Ph. Thévoz, M. Rappaz, and J.-L. Desbiolles, *Light Metals*, Ch.M. Bickert, Ed., TMS, Warrendale, PA, 1990, p 975
12. C.Y. Wang and C. Beckermann, *Metall. Mater. Trans. A*, Vol 27, 1996, p 2754
13. M.C. Schneider and C. Beckermann, *Metall. Mater. Trans. A*, Vol 26, 1995, p 2373
14. M. Wu and A. Ludwig, *Metall. Mater. Trans. A*, Vol 38, 2007, p 1465
15. S. Gouttebroze, M. Bellet, and H. Combeau, *C.R. Mécan.*, Vol 335, 2007, p 269
16. J.A. Dantzig and M. Rappaz, *Solidification*, EPFL Press, Lausanne, 2009
17. *Numerical Modeling in Materials Science and Engineering*, Soft Cover Edition, Springer-Verlag, Berlin, Heidelberg, New York, 2010
18. Y. Saito, G. Goldbeck-Wood, and H. Müller-Krumbhaar, *Phys. Rev. A*, Vol 38, 1988, p 2148
19. J.A. Sethian and P. Smereka, *Ann. Rev. Fluid Mech.*, Vol 35, 2003, p 341
20. K. Wang, A. Chang, L.V. Kale, and J.A. Dantzig, *J. Parallel Distrib. Comput.*, Vol 66, 2006, p 1379
21. A. Chang, J.A. Dantzig, B.T. Darr, and A. Hubel, *J. Comput. Phys.*, Vol 226, 2007, p 1808
22. J.-F. Zaragoci, L. Silva, Ch.-A. Gandin, and Th. Coupez, in preparation
23. L.Q. Chen, *Ann. Rev. Mater. Res.*, Vol 32, 2002, p 113
24. W.J. Boettinger, J.A. Warren, C. Beckermann, and A. Karma, *Ann. Rev. Mater. Res.*, Vol 32, 2002, p 163
25. D. Lewis, J. Warren, W. Boettinger, T. Pusztai, and L. Granasy, *J. Met.*, Vol 56, 2004, p 34
26. H. Emmerich, *Adv. Phys.*, Vol 57, 2008, p 1
27. M. Plapp, *J. Cryst. Growth*, Vol 303, 2007, p 49
28. I. Singer-Loginova and H.M. Singer, *Rep. Prog. Phys.*, Vol 71, 2008, p 106501
29. J.-C. Ramirez, C. Beckermann, A. Karma, and H.-J. Diepers, *Phys. Rev. E*, Vol 69, 051607, 2004, p 1
30. S. Chen, B. Merriman, S. Osher, and P. Smereka, *J. Comp. Phys.*, Vol 135, 1997, p 8
31. N. Provatas, N. Goldenfeld, and J.A. Dantzig, *Phys. Rev. Lett.*, Vol 80, 1998, p 3308
32. Th. Coupez, *Journées Activités Universitaires de Mécanique*, La Rochelle, 2006
33. M.A. Martorano, C. Beckermann, and Ch.-A. Gandin, *Metall. Mater. Trans. A*, Vol 34, 2003, p 1657
34. S. Mosbah, Ch.-A. Gandin, and M. Bellet, *Metall. Mater. Trans. A*, Vol 41, 2010, p 651
35. W. Kurz and D.J. Fisher, *Fundamentals of Solidification*, Trans Tech Publications, 1992
36. G.P. Ivantsov, *Dokl. Akad. Nauk SSSR*, Vol 58, 1947, p 567
37. G. Guillemot, Ch.-A. Gandin, and M. Bellet, *J. Cryst. Growth*, Vol 303, 2007, p 58
38. Ch.-A. Gandin, G. Guillemot, B. Appolaire, and N.T. Niane, *Mater. Sci. Eng. A*, Vol 342, 2003, p 44
39. R. Heringer, Ch.-A. Gandin, G. Lesoult, and H. Henein, *Acta Mater.*, Vol 54, 2006, p 4427
40. Ch.-A. Gandin, S. Mosbah, Th. Volkmann, and D. Herlach, *Acta Mater.*, Vol 56, 2008, p 3023
41. D.J. Browne, *ISIJ*, Vol 45, 2005, p 37
42. I. Steinbach and C. Beckermann, *Acta Mater.*, Vol 47, 1999, p 971
43. Ch.-A. Gandin and M. Rappaz, *Acta Metall.*, Vol 42, 1994, p 2233
44. Ch.-A. Gandin, J.-L. Desbiolles, M. Rappaz, and Ph. Thévoz, *Metall. Mater. Trans. A*, Vol 30, 1999, p 3153
45. H. Takatani, Ch.-A. Gandin, and M. Rappaz, *Acta Mater.*, Vol 48, 2000, p 675
46. G. Guillemot, Ch.-A. Gandin, and H. Combeau, *ISIJ Int.*, Vol 46, 2006, p 880
47. W. Liu, "Finite Element Modelling of Segregation and Thermomechanical Phenomena in Solidification Processes," Ph. D. Thesis, Ecole Nationale Supérieure des Mines de Paris, Paris, France, 2005
48. C. Prakash and V. Voller, *Numer. Heat Transf. B*, Vol 15, 1989, p 171
49. M. Fortin, Estimation d'Erreur a Posteriori et Adaptation de Maillages, *Revue Européenne des Eléments Finis*, Vol 9, 2000, p 467
50. F. Alauzet and P.J. Frey, INRIA report, 2003, p 4759
51. M. Hamide, E. Massoni, and M. Bellet, *Int. J. Num. Meth. Eng.*, Vol 73, 2008, p 624
52. C. Gruau and T. Coupez, *Comput. Meth. Appl. Mech. Eng.*, Vol 194, 2005, p 4951
53. M. Bellet, V.D. Fachinotti, S. Gouttebroze, W. Liu, and H. Combeau, *Solidification Processes and Microstructures: A Symposium in Honor of Prof. W. Kurz*, M. Rappaz, C. Beckermann, and R. Trivedi, Ed., TMS, Warrendale, PA, 2004, p 15

54. D.J. Hebditch and J.D. Hunt, *Metall. Trans.*, Vol 5, 1974, p 1557
55. X.D. Wang, P. Petitpas, C. Garnier, J.P. Paulin, and Y. Fautrelle, *Modeling of Casting, Welding and Advanced Solidification Processes XI*, TMS, Warrendale, PA, 2006
56. Ch.-A. Gandin, J. Blaizot, S. Mosbah, M. Bellet, G. Zimmermann, L. Sturz, D.J. Browne, S. McFadden, H. Jung, B. Billia, N. Mangelinck, H. Nguyen-Thi, Y. Fautrelle, and W. Wang, *Solidification and Gravity 2008*, Materials Science Forum, A. Roósz et al., Ed., Trans Tech Publications
57. G. Guillemot, Ch.-A. Gandin, and H. Combeau, *ISIJ Int.*, Vol 46, 2006, p 880
58. G. Guillemot, Ch.-A. Gandin, and M.Bellet, *J. Cryst. Growth*, Vol 303, 2007, p 58
59. Ch.-A. Gandin and I. Steinback, Direct Modeling of Structure Formation, *Casting*, Vol 15, *ASM Handbook*, ASM International, 2008

ASM Handbook, Volume 22B, *Metals Process Simulation*
D.U. Furrer and S.L. Semiatin, editors

Copyright © 2010, ASM International®
All rights reserved.
www.asminternational.org

Modeling of Laser-Additive Manufacturing Processes

Anil Chaudhary, Applied Optimization, Inc.

ADDITIVE MANUFACTURING produces change in the shape of a substrate by adding material progressively. This addition of material can be as a solid, liquid, or a mixture thereof. For example, in laser and electron beam deposition, the incoming material can be powder or wire, which is melted by the energy of the beam. The material enters into a melt pool, which solidifies to fuse with the substrate when the beam moves away. In spray forming, a thin stream of molten metal is broken into droplets by gas jets, which impinge on the substrate in a semisolid state, solidify fully, and bond. In ultrasonic-additive manufacturing, the material is added as solid. The force and vibration caused by the ultrasonic energy causes the new material to fuse with the substrate. A common feature of these processes is that, at any time, the joining of new material occurs over a region that is small compared to the substrate dimensions. This region moves in tandem with the input energy source. Within this region, the substrate and the material deposit experience intensely nonlinear thermal and mechanical behavior. Control of this behavior to within a desired thermomechanical window is fundamental to defect-free additive manufacturing. Online closed-loop control is increasingly being used for this purpose. It tunes the process parameters in real-time on the basis of thermal imaging data. However, there is a physical limit to which online control can correct the process, because the material temperatures cannot be changed instantaneously, and timely initiation of corrective action to avoid incipient flaws may not always be possible. Process simulation is fundamental to overcome this limitation. It is useful to discover a feasible process path a priori, which can then be monitored and controlled in situ. Also, the process may be designed to be robust with respect to perturbations in material and process parameters, such that the 100th deposit is the same as the first one. This is the basic theme for the process simulation procedures in this article. The concepts are described by considering laser deposition as an example. Modeling and simulation of other additive manufacturing processes (e.g., electron beam and spray forming processes) are briefly reviewed relative to modeling of laser-additive processes.

Laser Deposition

Laser deposition manifests a collection of thermomechanical phenomena that are driven by the interaction between the laser energy, additive material, and the substrate. Their occurrence is intense in a small region in the proximity of the laser, and it is milder elsewhere. Some of the phenomena are unique to laser deposition (e.g., powder flow driven by the gas, convective flow in the melt pool), while others are similar to traditional manufacturing processes. Thus, to simulate laser deposition is to orchestrate multiple simulation components automatically such that each component emulates a single aspect of the laser deposition physics. There is a significant body of literature on such individual simulation methods and components. There are detailed, intricate methods reported in the literature that delve deep into the physics and can require input data on material constitutive properties that are not readily available. There also are methods that make simplifying assumptions on the material and/or process behavior and provide a quicker, rougher result. A rule of thumb is to choose a simulation method for which the required input data are either available, can be generated, or calibrated at the desired level of accuracy. Datum that is essential to the simulation yet not easily available is the temperature dependence of laser absorptance. Indeed, the error in the simulation is proportional to the error in these data. Thus, the following description includes a method to calibrate laser absorptance and a process simulation that requires thermophysical property data that are commonly available. These methods aid in problem solving in a way that the time and expense needed for the simulation and the fidelity of its answers is on par with the level of accuracy that is commonly attainable in the process-monitoring sensors, and its return-on-investment calculation is simple.

The addition (or deposition) of material on a substrate is performed in accordance with a path specified in computer-aided design. At the start of deposit, the substrate is at ambient temperature. The substrate temperature rises as material continues to be deposited. This increase in temperature depends on the thickness and heat capacity of the substrate material. For a thin substrate, if the beam energy is kept constant, the melt pool size can increase progressively. This is undesirable because it changes the solidification conditions significantly in the deposit from one location to another and can increase variability in the microstructure and mechanical properties. A larger melt pool also causes deeper remelting in the preceding layer and higher temperature oscillations in the lower layers. This affects the distribution and character of the residual stresses and the solid-state microstructure transformations in the lower layers. Occurrence of such variable melt pool conditions gives rise to several hard-to-answer questions. How and where are the test samples chosen to characterize the deposit material? How much of the testing must be repeated for a different deposit and substrate? How are confidence intervals established? How is the material qualified? The use of simulation minimizes such questions by creating the ability to predict the process parameters that result in consistent solidification conditions. These conditions are a function of numerous factors, such as the deposit path and geometry, substrate geometry, material thermophysical properties, additive process parameters, and so on. Process simulation can unravel the confounding of these factors and illuminate a feasible solution.

Fundamentals of Process Modeling

The principal thermomechanical phenomena during the laser deposition process are:

- Absorption of laser radiation
- Heat conduction, convection, and phase change
- Elastic-plastic deformation

The absorption of laser radiation drives the heat transfer, which in turn results in residual stress and deformations. The governing equations of thermal and mechanical phenomena are universal; only their application is different in ways that are manifested by the initial and boundary conditions during laser deposition. These boundary conditions are governed by the process parameters (Fig. 1). In general, they change with time no matter if the model was developed in a fixed or moving reference frame (i.e., if the model was Lagrangian or Eulerian), because the laser moves and mass is added on a path that traces the desired deposition geometry on a substrate that can be of any shape. This makes it difficult to obtain exact analytical solutions for thermomechanical phenomena in laser deposition.

Modeling of Laser Energy Absorption

Fundamentals of modeling of laser energy absorption are at the heart of process modeling. It governs how fast a given laser can travel without compromising the balance between the competing thermal and mechanical phenomena in a way that results in relatively steady-state melt pool conditions, irrespective of the location in the deposit. A thermal balance is attained when the absorbed laser energy is apportioned between the additive material and the wetting depth on the substrate, such that both reach the desired melt pool temperature simultaneously. A certain percentage of energy is conducted away in the substrate when the wetting depth is generated. This energy blends with the energy from the previously solidified deposit. Upon combining both, a mechanical balance is attained when the thermal gradients

in the solidifying deposit and around it in the substrate are such that the material plastic deformation and the resulting residual stresses are maintained below acceptable limits.

Absorptance is the complement of reflectance. It is a measure of interaction between the laser beam photons and the free and bound electrons in the irradiated surface materials. This interaction results in heat generation in a very thin surface layer. Typically, absorptance increases with the surface material temperature and its roughness. It decreases with the laser wavelength. Consequently, absorptance is difficult to measure accurately. Four examples of the difficulty are as follows:

- When a laser is incident on a rough surface, a part of the reflected energy is trapped by the surface asperities, which results in a higher value of absorptance.
- When a laser is incident upon a cluster of powder particles, the energy incident on each particle is partly absorbed, and the balance is reflected or scattered. This energy is intercepted by particles in the vicinity, and the process repeats until such time that the scattered energy escapes away from the substrate and the additive material. This results in a higher value of absorptance.
- The re-reflection also occurs between the substrate surface and the powder particles. The laser spot is occluded by the particles in the power jet prior to reaching the substrate surface. A part of the laser energy is reflected by the substrate, which is intercepted by the powder jet particles. The process repeats until the reflected energy escapes the deposit region.

- When material absorbs laser energy, its temperature rises, which increases absorptance. Thus, absorption measurement data need careful deconvolution to solve for absorptance.

The controlling parameter for heat flow is the net rate of energy input, which equals the product of laser power and net absorptance. The power input per unit area, or the power density, is the ratio of this net rate of energy input to the area swept by the laser spot per second (i.e., the product of laser spot diameter and travel speed). The power density value must be tuned to match the process parameters for powder and substrate (Fig. 1). This is because the timing of temperature rise for the substrate and powder material must match each other, such that the molten material can blend together consistently. This is attained when the heating of the powder material prior to reaching the substrate complements the substrate wetting depth, its area, and superheat. This involves a chain of synchronized events. Specifically, the powder particles plunge into a superheated molten metal wetting pool. The mass flux of the plunging powder is the feed rate. Upon plunge, the particles melt and blend into the pool. This lowers the superheat of the molten metal just as it creates a melt pool bead that rises above the substrate. The time it takes to create a melt pool bead of desired size, thickness, and a (lower) superheat temperature is less than or equal to the time it takes for the laser to travel a distance equal to its spot diameter. All of these are nonlinear thermal phenomena for which no analytical solutions are available that take into account all of the nonlinearities. Nevertheless, the classical Rosenthal solution has proven to be highly useful for preliminary modeling of laser deposition.

Rosenthal Solution

This solution expresses the temperature distribution in the substrate due to a moving point heat source. In laser deposition, the heat source is the net rate of energy input by the laser. The solution assumes that the material properties are independent of temperature and there is no phase change. Because there is phase change, it is used by making two substitutions. First, the material properties are chosen at the temperature of primary interest. For melt pool analysis, it is the melting temperature. For determination of temperatures that would be used for residual-stress calculations, it is a fraction of the melting temperature. Second, the solution is used in its nondimensional form (Ref 1–6). The use of such a nondimensional solution is of fundamental importance because it allows reuse of experimental observations data from one set of process parameters to another set under the conditions of dynamic similarity. In other words, it becomes feasible to perform subscale testing of a process. The subscale process is chosen such that the ratio of heat flux values at the corresponding locations in the two processes is constant. Three variations on the fundamental solution by

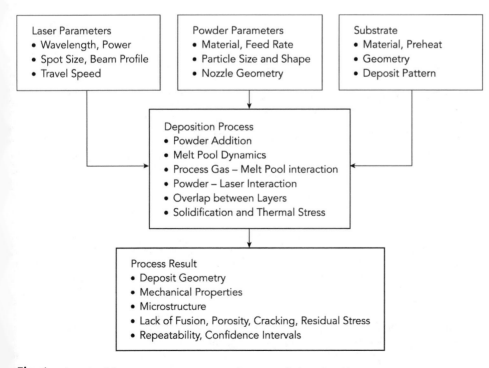

Fig. 1 Schematic of the components, parameters, and outcome of a laser deposition process

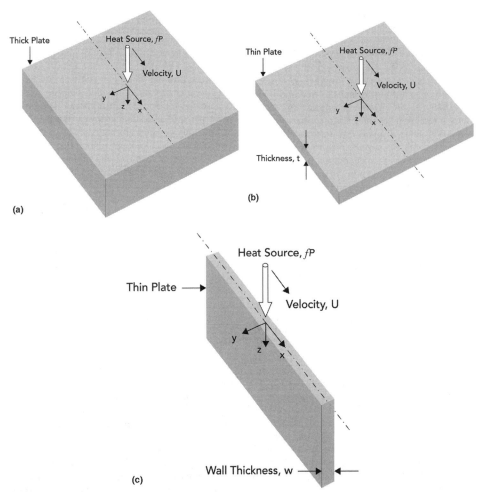

Fig. 2 Substrate geometry in the three variations of the Rosenthal solution. (a) Thick plate. (b) Thin plate. (c) Thin wall

Rosenthal are given in the following sections (Fig. 2). The first is for an infinitely thick and wide substrate (i.e., a half-space). The second is for a thin substrate of infinite width (i.e., a thin plate). The third is for an infinitely thick substrate of a small width (i.e., a thin wall). All three solutions can be computed in a standard spreadsheet program and are very useful for a first-cut simulation.

Thick-Plate Solution. The nondimensional temperature distribution in an infinitely thick plate subjected to a moving heat source of constant velocity is given by (Ref 2, 6):

$$2\pi T^* = \frac{1}{R^*}\exp[-(x^* + R^*)/2]$$
$$T^* = \frac{\alpha k}{fPU}(T - T_0)$$
$$x^* = \frac{xU}{\alpha}\ y^* = \frac{yU}{\alpha}\ z^* = \frac{zU}{\alpha}$$
$$R^* = \left(x^{*2} + y^{*2} + z^{*2}\right)^{1/2}$$

where T_0 is the ambient temperature of the substrate, T is the substrate temperature at position (x, y, z), P is the laser power, U is the laser velocity, f is the fraction of the laser energy absorbed by the substrate, α is the thermal diffusivity, k is the thermal conductivity, T^* is the dimensionless temperature, (x^*, y^*, z^*) are the dimensionless coordinates, and R^* is the dimensionless distance. The laser position is at the origin, and it travels along the +x-direction.

Thin-Plate Solution. The nondimensional temperature distribution in a thin plate subjected to a moving heat source of constant velocity is given as follows (Ref 2):

$$2\pi T^* = \exp(-x^*/2)K_0(R^*/2)$$
$$T^* = \frac{tk}{fP}(T - T_0)$$
$$R^* = \left(x^{*2} + y^{*2}\right)^{1/2}$$

where t is the plate thickness, and K_0 is the Bessel function of the second kind and zero order. Note that the expressions for the dimensionless temperature and distance are different from the case of a thick plate. The laser position is at the origin, and it travels along the +x-direction.

Thin-Wall Solution. The nondimensional temperature distribution in a thin wall subjected to a moving heat source of constant velocity is given as follows (Ref 6):

$$2\pi T^* = \exp(-x^*/2)K_0(R^*/2)$$
$$T^* = \frac{wk}{fP}(T - T_0)$$
$$R^* = \left(x^{*2} + z^{*2}\right)^{1/2}$$

where w is the wall thickness. The laser position is at the origin, and it travels along the +x-direction. Note that the dimensionless temperature and distance are defined differently than the previous two cases. Three ways in which these Rosenthal solutions can be used are described next.

A first way is to set the value of parameter f equal to the net absorptance. The solution then provides the temperature distribution as a function of position. Because it does not include the latent heat effect, the region where the temperature is below the melting point is closer to the real answer. The locus of all points where the temperature equals the melting temperature is postulated as the boundary of the melting pool (Ref 6). This is a useful approximation because it provides a quick calculation of melt pool size as a function of the process parameters, using a spreadsheet program and making a graph of temperature versus distance. The values of temperature inside the melt pool will be higher than the melting temperature. An estimate of the melt pool temperature can be obtained by correcting the mean temperature of the melt pool given by the Rosenthal solution to account for latent heat.

A second way is useful in order to choose a feed rate for the additive material. An adiabatic estimate for the energy required to melt the additive material equals the energy needed to raise its temperature to the melting temperature plus the latent heat. The rate of energy input required to melt the additive material is subtracted from the net energy that will be input by the laser (i.e., the product of the laser power and net absorptance). The value of fP is then set equal to the energy that creates a melt pool within the substrate material. Once again, the size and depth of this melt pool can be computed using the Rosenthal solution. This is the wetting melt pool. The feed rate of the additive material can be adjusted such that a desired depth of wetting is attained.

A third way is to make use of the nondimensional form of this solution. For example, in case of thick plates, for any two processes that have the same value of the product fPU, the dimensionless temperature is the same. This allows transformation of the temperature contours from one process to another, using the formulae for dimensionless coordinates. Also, the dimensionless temperature is proportional to the temperature change above the ambient substrate temperature. This is relevant for the situation when there are repeated passes of laser deposition, which cause the ambient substrate temperature to rise progressively. If this rise is measured with a pyrometer, the Rosenthal solution can be used to determine the reduced values of laser power needed to maintain consistent melt pool size through the multiple layers of the deposit.

Due to a variety of nonlinearities that are present in the laser deposition process, the computational mechanics-based methods are a mainstay of its process modeling. These methods can be based on either the finite-element method or the finite-volume method. They may be formulated in a fixed or moving coordinate frame. The fundamental formulation of these methods remains unchanged from that in the modeling of more traditional processes, except that they account for the continuous change in the boundary conditions and the corresponding interweaving of thermal and mechanical phenomena. They use automated updating of the analysis model as the additive material continues to be deposited. Indeed, if any of these automated techniques were absent, the modeling of laser deposition would be impractical. The computational methods provide detailed information on the evolution of temperature, residual stress, and distortion at every location in the part. These data serve as input for the modeling of microstructure and defect generation, as described in the following sections.

Fundamentals of Modeling Microstructure

Consistency of microstructure between the substrate and the deposit is fundamental in laser deposition. In this regard, the temperature gradient, G, and solidification velocity, R, at the onset of solidification are the two most important parameters. Their product equals the solidification cooling rate. The G and R parameters govern the character of grain morphology, grain size, and texture upon solidification (Ref 7). The postsolidification cooling rate controls the character of fine-scale microstructure. A more detailed evolution of microstructure during and after solidification can be modeled using methods such as Monte Carlo simulations, cellular automata, and phase-field simulation. The effect of G and R on the process outcome is frequently represented as a processing map, and the simulation results are superimposed on this map to predict the microstructure characteristics. In this regard, the nondimensional Rosenthal solution is used for initial modeling of microstructure. The G and R parameters are computed using the Rosenthal solution in accordance with the process parameters, and their values are interpreted using the processing maps (Ref 5, 6, 8). This is beneficial for narrowing the choice of process parameters that can result in a desired microstructure.

Microstructure modeling in laser deposition is similar to predicting microstructure upon multiple-pass welding (Ref 9). The welding joins two pieces using multiple passes by a heat source that also adds (i.e., fills) material between the two pieces. Each pass causes a thermal cycle in the adjoining material, and its cumulative effect is manifested in the microstructure. In laser deposition, multiple layers are deposited atop the substrate, and each layer causes a thermal cycle in the adjoining layers below. In the initial layers, the melt pool thermal energy can diffuse into the substrate, which acts as an efficient heat sink due to its size. As the deposit is built up, the new layer is further removed from the substrate, and the diffusion of thermal energy is slower. This reduces the solidification cooling rate. Process simulation is used to predict the thermal cycling, and these data are used for microstructure prediction.

Fundamentals of Modeling Defect Generation

Modeling defect generation in laser deposition is similar to prediction of defects in casting. The primary defects in laser deposition are lack of fusion, shrinkage, porosity, and cracking. They can be compared with weld-line defect, shrinkage, porosity, and cracking defects in casting, respectively, as described next. Consequently, modeling defect generation in laser deposition draws upon the corresponding techniques in casting simulation.

A weld-line defect in casting occurs when two free surfaces meet during mold filling, and the liquid metal temperature of the two surfaces is inadequate to allow formation of a complete bond between them. Similarly, the lack-of-fusion defect occurs when the laser energy is insufficient to melt the incoming material, melt a small thickness of the substrate, and create sufficient superheat in the resulting melt pool to allow a liquid flow front as the new material joins with the previous deposit and/or the substrate. This can be a transient occurrence that results from local geometry and temperature conditions, such as in the case of a complex multilayer deposit whose seams blend into each other. The propensity for a lack-of-fusion defect can be predicted by tracking the superheat of the melt pool, depth of melting of the substrate, and the time it takes for the melt pool material to reach the solidus temperature as the laser moves away.

Shrinkage defects in casting are a result of competing phenomena, such as progress of the solidification front, feeding of liquid metal, and deformation of the solidified material. In general, shrinkage of the liquid phase must be compensated, and thus, the last region to solidify is the probable location for occurrence of shrinkage defects. Gas porosity defects occur due to entrapped and dissolved gases. In laser deposition, solidification occurs rapidly, and an isolated internal region in the melt pool can be enclosed such that it solidifies last and results in shrinkage porosity. It also can occur in cases of deposits with overlapping layers. Porosity can occur when the material in the new layer must blend and bond fully with the previous layers. Gas porosity can occur when the laser deposit is performed in the presence of process gases, which are used to control the solidification rate of the new deposit. The process gases can be entrapped due to melt pool dynamics and deposit geometry. The propensity for occurrence of these defects can be predicted from postprocessing of temperature data from process simulation.

Cracking in casting originates from nonuniform cooling and the resulting development of a thermo-elastic-plastic stress state in the material because it cannot freely contract upon solidification. Cracking occurs if the stress level exceeds the material ultimate tensile strength. Hot tearing are cracks that occur during solidification. Cold cracks occur after the completion of solidification. The occurrence of cracking in laser deposition has the same origin. There is nonuniform, constrained cooling because the melt pool is a small region and the adjoining substrate is solid. The propensity of occurrence of cracking can be predicted by postprocessing the thermal and stress data from process simulation.

Input Data for Modeling and Simulation

Description of Required Data

Four sets of data are needed, namely, material constitutive data, solid model, initial and boundary conditions, and laser deposition process parameters. These are described in the following.

Material Constitutive Data. Three types of data are needed:

- Laser absorptance as a function of temperature
- Thermophysical properties of the additive material and substrate from room temperature to the melt pool superheat temperature
- Constitutive model for the elastic-plastic deformation behavior of the additive material and substrate from room temperature to the mushy zone temperature

Thermophysical data and constitutive models are available in the technical literature. The absorptance data, however, are rarely available as a function of temperature, and it must be generated. Thus, a sample procedure is described in the following.

This procedure uses an insulated cylinder with an axial hole (Fig. 3). The cylinder is made of the substrate material, and the axial hole is for powder placement. The axial hole has a hemispherical bottom. Figure 3(a) shows the geometry, which is expressed in terms of the radius of the axial hole. The cylinder is instrumented with three thermocouples, which are located midway through the wall thickness. Figure 3(b) shows the experimental setup during irradiation. The cylinder is wrapped in insulation, except at the opening of the axial hole. A measured mass of powder is placed in the axial hole, and a graphite mask with a same-sized hole in the center is placed on top of the

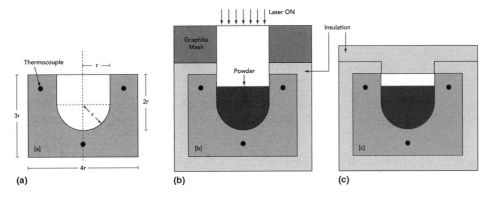

Fig. 3 (a) Cylinder with an axial hole. (b) Setup during irradiation. (c) Insulation on all sides to trap absorbed heat

cylinder. The powder is then irradiated by the laser for a measured duration of time, and the temperature data at the three thermocouples are logged. All thermocouples register the same temperature at the beginning, but upon irradiation, the temperatures become different. When the irradiation is completed, the graphite mask is replaced with insulation (Fig. 3c). The collection of thermocouple data is continued until such time that all thermocouples register the same temperature. This occurs because the trapped energy diffuses into the cylinder by conduction to equalize the temperature at all thermocouples. The total energy absorbed is then equal to the energy needed to raise the temperature of the cylinder and the powder from its initial value to its final value.

This procedure seeks to create an adiabatic system that traps the absorbed energy. The diameter of the axial hole is a few percent larger than the laser spot size. The graphite mask is to allow the laser spot to irradiate only the cross section of the axial hole. When no powder is placed, the absorptance value is for the substrate material. The temperature dependence of this absorptance can be determined by progressively increasing the duration of irradiation. A lower-power laser may be used for this purpose to avoid vaporization of the surface material for longer durations of irradiation. Similar experiments can be performed with powder placed in the hole.

The temperature dependence of absorptance is determined using a small number of such experiments. The duration of irradiation is progressively increased so that a higher temperature is reached in the powder or substrate material surface in each experiment compared to the previous one. The data are used to determine the absorptance in three steps:

1. The total energy absorbed in each experiment is known from the thermocouple data. Create a graph of values for the total energy absorbed versus the duration of irradiation, and fit a polynomial curve through the points. This may be done using a standard spreadsheet program. The slope of this polynomial curve is equal to the absorption heat flux, which is a function of time. The ratio of absorption heat flux to laser power is absorptance, which now becomes known as a function of time.

2. It is necessary to determine the temperature reached upon irradiation for each experiment. To this end, build a computer model of the cylinder and the additive material. Use the absorption heat flux versus time as the thermal loading, and perform a heat-transfer simulation. Postprocess the results to generate the values of temperature at the irradiated surface versus time.

3. In the table of data for absorptance versus time, use the computed value of irradiated surface temperature in place of time for each data point. This creates data for the temperature dependence of absorptance.

Solid Model. The solid model for laser deposition is different from the traditional manufacturing processes in that it must have the ability to represent the incremental addition of the deposit material. Its definition depends on whether it is to be used for a simulation that makes use of a fixed mesh or one that makes use of a moving mesh. For a fixed-mesh simulation, the solid model includes the substrate and the completed deposit as input. The actual, accumulated deposit shape is computed during simulation, and the geometry "fills up" with material as the process continues. In the case of a moving-mesh simulation, the solid model begins as a substrate model, and it is progressively augmented as deposit is continued. In either case, the solid model definition is supplemented by information on the deposit sequence.

Initial and Boundary Conditions. Accurate specification of initial and boundary conditions is necessary in order to obtain a correct solution. This is an involved, intricate process that is typically automated in the modern software so that it is transparent to the user. The intricacy occurs because the process parameters manifest themselves as additional time-dependent boundary conditions.

Typically, the initial conditions consist of the specification of the ambient temperatures of the environment, substrate, and powder. Thermal boundary conditions include the convection and radiation heat loss to the environment and the moving heat source due to the laser. However, as the deposit is continued, the old boundary conditions become annihilated on the substrate surface that gets covered by the new deposit. Specifically, three salient changes take effect:

- A new, internal heat flux condition is created on this covered substrate surface due to the energy delivered by the laser to the substrate.
- New boundary conditions come into existence on the boundary of the new deposit material for the heat loss to the environment.
- The initial temperature of the incoming powder is computed based on the absorbed energy that is apportioned to the powder. These changes occur continuously and in tandem with the laser path.

The setting of mechanical boundary conditions can also be quite challenging. This is because the substrate is typically placed onto a table and not clamped. Ideally, this can be specified as a contact boundary condition between the bottom of the substrate and the table. A contact boundary condition is where the contact forces can only be compressive. This means that if the substrate warps upon deposition, a part of the substrate resting on the table can lift off. However, it is unknown if or where the substrate could lift off. Thus, the mechanical boundary conditions are commonly specified as a clamping condition on the edges, which means that the contact force can be compressive or tensile. Care must be taken so that the mechanical boundary conditions do not impede substrate distortion due to an artificial clamping constraint.

Process Parameters. The selection of process parameter values that merit further evaluation with a detailed simulation is performed using processing maps. This is an invaluable step because it minimizes the simulation effort. This is described in the next section.

Simulation of Additive Manufacturing

The purpose of simulation is to reduce trial and error and thereby reduce cost. Any error that occurs late in the additive manufacturing process is difficult to debug by trial and error. This is because its occurrence is a culmination of the sequence of thermomechanical events that have occurred prior to reaching that state. To fix the error, it is necessary to roll back the process in time as far as needed and to make an early change in the process parameters. The merit of simulation is that it allows an engineer to see this future in the computer. In this regard, there are three aspects of simulation described in the following:

- *Simulation for initial selection of process parameter setup:* This is the use of processing maps, which express the historical data on process success in terms of dimensionless parameters, such as the dimensionless beam power and travel speed (Ref 10). The maps show regions in the non-dimensional space where the process will result in heating, melting, or vaporization of the substrate or additive material. The maps also superimpose on these regions the contours of relevant parameters, such as the constant energy density, wetting depth, and so on, and delineate zones that are suited for hardening, cladding, cutting, and so forth. Maps are reported for a variety of materials, and they are a trove of well-organized historical knowledge. They can be used to select the process parameters by superimposing on them the parameter window that is feasible with the equipment to be used for additive manufacturing. Process parameters within that window that are suitable for deposition represent the available choice. However, it is important to note that the parameters change during the deposit, and there must be sufficient flexibility such that their values remain in the feasible window even as they are changed during the process.

- *Simulation for in situ process control:* This is the use of data from a thermal camera, pyrometer, or thermocouple to adjust the process parameters from one layer of deposit to the next. Each layer of deposit changes the ambient temperature of the substrate, T_0. This can be measured and used as input to the nondimensional Rosenthal solution to solve for new values of process parameters that maintain the consistent size and temperature of the melt pool. It is also useful to determine if dwell time is necessary between passes to attain a desired ambient temperature in the substrate prior to commencement of the next pass.
- *Simulation for ex situ process optimization:* This blends computational mechanics with multidisciplinary optimization to emulate the essence of in situ process control in the computer. It is ex situ because it must be performed in advance of the deposition due to its computational needs. The "measurements" are data in the thermomechanics solution, which represents how the process wants to evolve under its present parameter setup (Ref 11–15). The "feedback" is the correction computed by the optimization solution, which represents how much the parameter setup must change to conform to a performance requirement specified by the user (Ref 16, 17).

Computational mechanics simulation of laser deposition consists of three principal thermo-mechanical phenomena described in the fundamentals of process modeling. Among the three, the mathematics for simulation of phase change, heat conduction, and convection are akin to simulation of solidification processes,

while that for elastic-plastic deformation is akin to simulation of bulk deformation processes. The mathematics for laser absorption comprise calculations that are based on the analytical geometry of occlusion and re-reflection of the laser by the powder jet and the substrate (Ref 18). A challenge in the simulation is the representation of time-evolving boundary conditions and the accurate modeling of build height and cross section in complex deposits. This is laborious. Newly available simulations automate boundary condition generation so that it is transparent to the user. On the other hand, the procedures for automated modeling of build profile and build height in complex deposits are current areas of active research. This is because it is intricately tied to melt pool dynamics and high-temperature material properties, both of which are difficult to characterize. The reported works on the subject to date are limited to studies on simple, linear geometry of deposit.

Simulation of laser deposition is performed step-by-step (Fig. 4). Each step represents a small increment in time, Δt. To begin, the initial conditions are given, and so the solution is "known" at time zero. The first time step computes the solution at time Δt. The time value is updated to Δt, and the second step solves for the solution at $2\Delta t$, and so on. Within each time increment, temperature and deformation solutions are obtained, commonly using a staggered approach. This has three basic parts:

- The known values of deformations from the preceding time increment are used to update the substrate geometry. The laser position is updated in accordance with its travel velocity and direction.
- A new increment of deposit is considered to occur on the updated geometry. The mass of

this deposit is governed by the powder feed rate. The deposit geometry is automatically generated and merged with the substrate. A first estimate of interaction between the laser and the additive material increment is computed using an analytical model. This provides a first-principles calculation for the additive material temperature and the heat flux apportioned to the substrate. Thermal and mechanical boundary conditions are updated to reflect the addition of deposit.
- The thermal solution is computed. The temperature solution is used as initial conditions for the deformation solution. For each solution, the governing equations are linearized about the last known state of the process and material, and the results are obtained using a nonlinear iterative procedure. This procedure repeats for every time increment.

A simulation may consist of hundreds of time increments, depending on the type and extent of the deposit. It has now become a practical, worthwhile endeavor due to automation of the deposit geometry generation and transfer of boundary conditions from one time increment to the next.

Consider a situation when a computational mechanics solution determines that certain desired performance criteria are not fulfilled by the chosen set of process parameters. Examples of the performance criteria are permissible range of wetting depth, melt pool size, melt pool temperature, and so on. They represent conditions that minimize the propensity of defect generation while minimizing cost. In analogy to defects, which are typically not pervasive but occur in isolated locations, a loss of performance occurs locally in a few time increments. Knowing how to manually correct the process parameters to mitigate the loss of

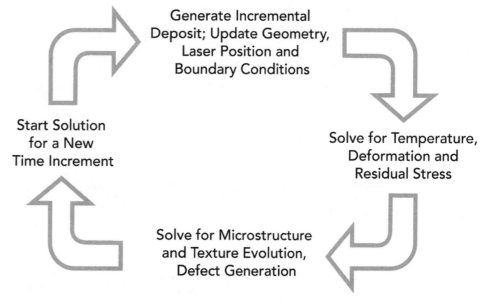

Fig. 4 Schematic of computational mechanics solution

performance locally can be, in the least, a tough, time-consuming, and error-prone activity. This is mitigated by the optimization component of the solution, which monitors the performance criteria in each time increment just as they evolve and alters the process to mitigate the loss conditions automatically. It seeks to emulate the in situ process control in the computer. It can be driven by a more complete set of performance criteria than what is feasible in an in situ process control. This is because an optimization solution has access to the full flora of results of the computed solution. On the contrary, in situ process control depends on the sensor data, which, at best, can provide information on the state of the material on the top surface of the deposit.

Figure 5 illustrates the optimization procedure. Central to this procedure is the specification of a performance criterion. In Fig. 5, the criterion is that the melt pool size and superheat be within a specific range of values. This range is defined based on a compilation of collective knowledge and correlation of how melt pool size and superheat will manifest themselves in terms of material properties, defects, and so on. This is at the heart of optimization, and indeed, it is reported that developing a meaningful performance criterion can take up to half the total effort needed to optimize a process (Ref 17). Clearly, while seeking to optimize one aspect of the process, it is imperative that any other desirable attributes of the current or known set of process parameters not be adversely affected. This may be constrained by issues such as scatter in the material properties, equipment capability, production throughput, cost of operations, and so forth. Thus, due diligence in quantifying the pluses and minuses of the chosen performance criterion is essential. The criterion itself can be specified in ways such as "the lower the better," "the higher the better," or "nominal is best" (Ref 16). For example, optimization of melt pool size and superheat may be started as "nominal is best." The first principal stress component caused by residual stresses is an example of "the lower the better," and so on.

The process simulation begins with user-specified process parameters and computes a solution for the first time increment. Acceptance of this solution is conditional on it fulfilling the performance criterion. If this were not the case, a best guess for change in the process parameters is determined using a variety of methods. For example, it may use analytical solutions to project the solution forward or backward. Or, it may use a search-based technique, such as golden section, steepest descent, orthogonal projection, and so on. The time increment is then resolved until the performance criterion can be fulfilled. Such internal embedding of the optimization loop is central to minimizing the hours in a day that an engineer must expend to resolve process challenges. Optimization is attained at the expense of computer time, which can be included into the process planning, such that it remains a cost driver no longer but becomes a daemon that pays for itself by contributing unforeseeable and actionable information for process development instead.

The optimization solution has a particularly salient role during periods of increased transience of thermomechanical phenomena. Consider, for example, a long line of deposit on a thick substrate. At the beginning of deposit, the substrate is cold, and the entire dynamics of the melt pool must be initiated. The quasi-steady state of deposit must be attained as quickly as possible, so that the length of the end tab, or the region of transient solidification conditions, is minimized. Toward the end of the line, the melt pool would be in a quasi-steady state. Upon reaching the end, if the laser reverses the travel direction, it will act upon a hotter substrate than what it encountered during its way forward. The laser power must be reduced to maintain the consistency of the melt pool conditions. This is archetypical optimization. Similar examples abound in resolving the ever-present challenge of control of distortions caused by residual stresses. Candidate solutions include alterations in the deposit path and generating a self-balancing state of residual stress by making deposits on the top and bottom sides of the substrate. These are a metaphor of the classic problem of traveling salesman route optimization. The desired end result is always the same: consistency and robustness of deposit and substrate materials in a way that the part can be certified for service. In this regard, the process simulation results serve as input to the simulation of microstructure and texture, which is addressed next.

Simulation of Microstructure and Texture Evolution

The procedures for simulation of microstructure and texture evolution apply to laser deposition. Among these methods, the R- and G-based microstructure prediction is of common interest in many situations. The R and G values are obtained by postprocessing the thermal solution, which can be an analytical solution or a computational mechanics solution. The two parameters are related to microstructure by using processing maps. For example, see Ref 8 for a Ti-6Al-4V map. Because R and G are a function of the process parameters, the processing-microstructure relationships can be derived from these maps. For example, see Ref 19 for such a map, which was developed for single-crystal laser deposition of superalloys.

Integration of Modeling and Simulation with Design

A unique opportunity for integration of modeling and simulation (M&S) with design lies in the ability of M&S to streamline the component certification process, which relies on the material strength specification and its confidence level. Depending on the application, the A-, B-, or S-basis material properties may be required. The M&S can ensure consistency of deposit material and point to locations where test samples may be taken in order to span the complete range of conditions that occur during additive manufacturing. Indeed, ASTM International and the Society of Manufacturing Engineers have partnered to form the ASTM Committee F-42 on Additive Manufacturing Technologies to develop additive manufacturing technologies standards. One of its goals is to allow manufacturers to better compare and contrast the performance of different additive processes. At a single-component level, the M&S role is to quantify the similarities and differences between the various choices for process parameters and help the components produced with the chosen process attain certification.

Computational Mechanics and Analytical Solutions

Examples of Multiple Concepts

Each of the following examples is presented to illustrate one or more concepts. All computational mechanics solutions were obtained using the SAMP software (Ref 20). It automates the various components of additive manufacturing simulation and solves for optimal process parameter values that can result in consistent solidification conditions for the melt pool material. The analytical solutions were obtained

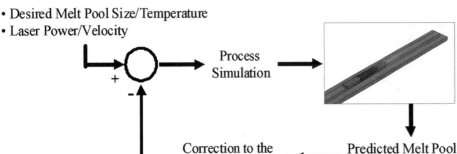

- Desired Melt Pool Size/Temperature
- Laser Power/Velocity

Fig. 5 Ex situ control loop in the optimization solution

using a Rosenthal solver that implements the fundamental solution in a generalized manner, such that it can account for the finite size of the laser spot, heat sinks, and so on (Ref 21).

Single-Line, Multilayer Deposit on a Thick Plate. The deposition conditions in this example are similar to those reported in Ref 9, which is a doctoral dissertation. One objective in the dissertation work was to calibrate a thermal model for certain existing Ti-6Al-4V deposits for which only the microstructure data were available. The question was to identify thermal conditions that would result in the observed microstructure. A solution was found using the finite-difference method and by establishing a correlation between the time-temperature history and the end microstructure. Because of the detail to which the dissertation describes this work, it is a good benchmark about which process alterations may be explored, and their effect on the results can be reconciled to create a better understanding of the process.

The deposit comprises a single pass of laser glazing followed by eight layers of deposit on a thick substrate. In each pass, the laser travels in a straight line from the aft end of the substrate to its forward end. Upon reaching the forward end, the laser reverses to the aft end. During the reverse, its power is off. The resulting interpass time is 20 s. The initial temperature of the substrate and the environment temperature are 300 K. The process parameters are shown in Fig. 6. The interface heat-transfer coefficient for free convection to the environment is 25 W/m^{-2} · K^{-1} and for forced convection to the process gases is 100 W/m^{-2} · K^{-1}. The laser absorptance is 0.19. Or, the net power input by the laser for this deposit is 19% of 13 kW, or 2.47 kW. This value of absorptance is too low; it is used to be consistent with Ref 9. A more typical value is 0.39. Thus, the following results may be interpreted as the material behavior for a process for which the net power input is 2.47 kW. The emissivity for solid- and liquid-phase materials is 0.63. The powder efficiency is 100%.

Figure 7(a) shows the temperature distribution midway through each pass. In the glazing pass, the temperatures in the wake of the laser drop significantly because the substrate is a thick plate. The temperature, however, rises steadily with each pass due to the accumulated thermal energy, and the high temperature penetrates deeper in the substrate. The substrate temperature is observed to rise by approximately 300 K due to the glazing and the eight

layers of deposit (Fig. 7b). In the creation of the melt pool, approximately 7% of the absorbed laser energy is consumed by the powder jet particles before they plunge in the melt pool. The laser power, however, is specified to be constant throughout the process, which causes the melt pool size to increase, as shown in Fig. 7(c). This increase in the melt pool size is predicted to be 23%.

If the Rosenthal solution was to be used for in situ process control for this process, one may assume that a thermal imager would observe the temperature rise in the substrate and use its value as parameter T_0. In this regard, the solution becomes simpler to use if it is applied to the locations on the top surface and along the centerline of the deposit. These locations are where $y* = z* = 0$ and $x* = R*$. For the present case, the Rosenthal solution predicts the increase in the melt pool size to be 16%. This is a useful result, given that it predicts the correct trend and provides a wealth of insight at virtually no computational effort. It is useful in the development of new processes, such that sensor data may be acquired at the right locations. It may even be feasible to establish a correlation between the Rosenthal solution and the computational solution in a way that the analytical result, when scaled by a calibration factor, provides an even better insight in real-time.

Figure 7(d) shows the time-temperature history at six locations in the deposit. Material is deposited at location A before location B, and so on. Thus, the location A thermal history spans a longer duration than the history at other locations. It has nine peaks, which correspond to the glazing and the eight deposit layers. The temperature at location A exceeds 1923 K (i.e., the liquidus temperature for Ti-6Al-4V) during the glazing pass and in the first layer of deposit. In other words, melting occurs twice. As a result, the solidification microstructure at location A is governed by the G and R parameter values that occur during the last time solidification occurs at location A. This is the starting state, which evolves due to the solid-state microstructure transformations that occur during the rest of the process. Figure 7(e) shows a processing-microstructure correlation map (Ref 8), which shows the relationship between the solidification microstructures and the R and G parameter values. The shaded region is the window where the R and G values predicted by the simulation reside. These values are when the material solidifies for the last time.

Single-Line Deposit on a Thick Plate with Optimization. The process parameters in this example are adapted and extended from those reported in Ref 22 to 25, all of which are doctoral dissertations that provide a wealth of knowledge in methodical detail on the laser deposition process. This example considers a notional cloverleaf-like deposit on a thick substrate. The deposit is continuous until eight layers are deposited, except for an interpass time of 3 s. The cloverleaf contour meanders on the substrate surface such that, at times, it is close to the outside boundary of the substrate, and at other times, it is adjacent to a recently deposited material. This creates increased transience in the process. The purpose of this simulation is to illustrate how laser power must be reduced progressively in a way that the melt pool size can remain consistent throughout the deposit. The initial temperature of the substrate and the environment temperature are 300 K. Just as in the previous example, the substrate becomes progressively warmer with continued deposit. The process parameters are given in Fig. 8. The interface heat-transfer coefficient for free convection to the environment is 13.5 W/m^{-2} · K^{-1} and for forced convection to the process gases is 135 W m^{-2} · K^{-1}. The laser absorptance is 0.39. Or, the net power input by the laser for this deposit is 39% of 500 W, or 195 W. The apportionment of this energy to the powder and substrate, including the re-reflection considerations, is succinctly illustrated in Ref 22.

The simulation was set up to control the melt pool size to be between 4 and 4.5 mm. Figure 9 (a) shows the temperature distribution in each layer. The first six distributions are for the first layer. The next three are for layers two, five, and eight, respectively. The simulation is automatically tuning the laser power such that the melt pool size remains consistent. It accounts for the simultaneous occurrence of various thermomechanical phenomena and computes the required change in the process parameters. Figures 9(b and c) show how the substrate temperature increases with layer number and the required reduction in laser power to counterbalance the transience. The laser power reduces by over 50% in eight layers. This is ex situ control, which provides information a priori so that the requirements for any in situ control can be less demanding. Once again, the Rosenthal solution can be used to support in situ process control.

As the layers are built up, the transience arising from the proximity of the deposit to the substrate boundary subsides. The power requirement levels off. The process simulation can be continued for additional layers only at the expense of additional computational time. This is a salient benefit made feasible by the geometry and boundary condition automation features in a computational process simulation. However, with increasing number of layers, even though the thermal conditions settle, issues such as the accumulation of residual stresses and the propensity for distortion or cracking take on increasing importance. Because the residual stresses cannot be measured in situ, the

Laser Parameters	
Type	Carbon Dioxide
Wavelength	10.6 micron
Power	13 kW
Spot Size	15 mm
Travel Speed	2.54 mm/sec

Powder Parameters	
Material	Ti-6Al-4V
Feed Rate	190.5 cubic mm/sec
Powder Velocity	2.5 m/sec
Particle Size	130 micron
Jet Diameter	13 mm
Jet Angle	60 degrees

Substrate Parameters	
Material	Ti-6Al-4V
Length	300 mm
Width	100 mm
Thickness	32 mm

Deposit Parameters	
Length	250 mm
Width	15 mm
Thickness	5 mm

Fig. 6 Process parameters for the single-line, multilayer deposit

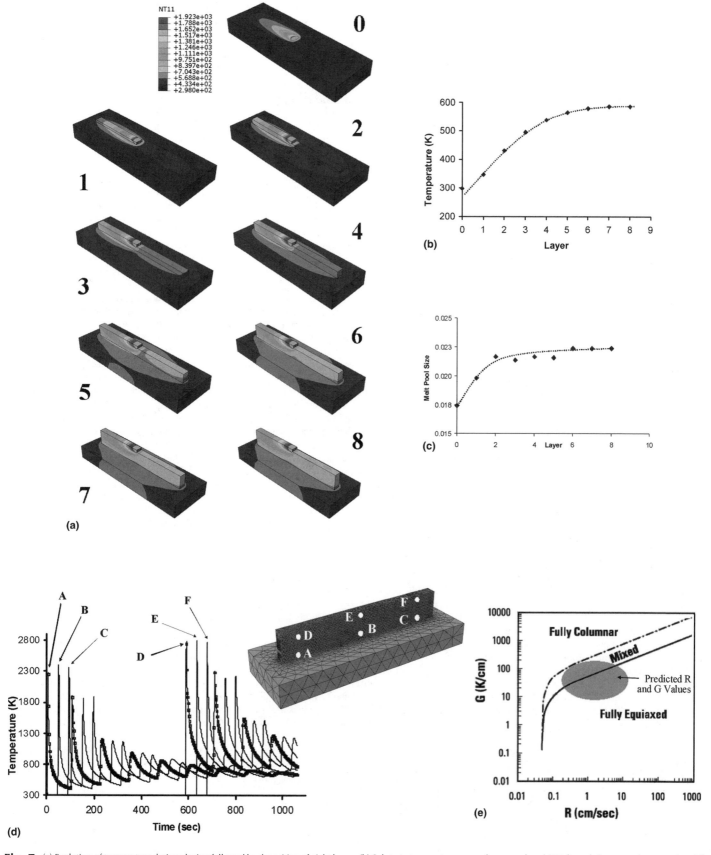

Fig. 7 (a) Evolution of temperature during glazing followed by deposition of eight layers. (b) Substrate temperature versus layer number. (c) Melt pool size versus layer number. (d) Peaks in temperature-time history at various locations in the deposit. (e) Correlation of microstructure to solidification rate and thermal gradient during laser deposition

simulation becomes increasingly useful. In this regard, the through-thickness thermal gradient has been observed to correlate with the magnitude of residual stress (Ref 5).

Inadequacy in the melt pool thermal conditions has the propensity to initiate a lack-of-fusion defect, and this is addressed by the ex situ control simulation. The last region to solidify is of concern due to its propensity to originate shrinkage defects. In this regard, the temperature distribution at the end of layer

8 is of interest, because it can be viewed as the last region to solidify.

Multiline, Multilayer Deposit on a Thin Plate. The geometry and process parameters for this process are created to represent a notional repair of surface scour on a thin plate. This situation is different from the preceding two examples in that the substrate has only a limited heat capacity locally, because of which it cannot function as an effective heat sink. This can be represented by the thin-plate Rosenthal

solution. Due to its smaller thickness, the substrate temperature can rise even with a smaller deposit. This example, like the previous one, is to control the melt pool size between 10 and 11 mm. The area of repair is rectangular. Its dimensions are such that it requires multiple layers of multiple side-by-side passes of deposit to repair the scour area. There are two layers of deposit. Upon the second layer of deposit, there is overbuild. In practice, such overbuild will be machined off. Process parameters are given in Fig. 10. The deposit sequence consists of a laser that travels back and forth to cover the rectangular area with two layers. There are eight lines of deposit in each layer, for a total of sixteen lines of deposit.

Figure 11 shows the temperature distributions for a few lines. Because the substrate is thin, the role of process simulation is to determine a workable level of superheat for the melt pool. The simulation shows that a melt pool size of 10 to 11 mm and a melt pool temperature of 2100 K result in acceptable wetting depth. Figure 12 shows the laser power versus line number. Within the two layers, the laser power drops by approximately 40%.

Single-Line, Multilayer Deposit on a Thin Wall. This is a corollary of the previous example. It is edge-replenishment repair instead of surface-scour repair. The substrate thickness is the same as in the previous example, but because the deposit is on its edge, the substrate has the ability to conduct the heat away from the deposit. Thus, the temperature rise in the substrate is smaller. The analytical solution for this case is the thin-wall Rosenthal solution. Process parameters are given in Fig. 13. The temperature distributions upon each line of deposit are shown in Fig. 14. They can be contrasted with the results of the previous example to gain insight on the effect of substrate orientation. Laser power versus time is given in Fig. 15.

Modeling and Simulation of Other Additive Processes

Additive manufacturing is feasible with a variety of approaches. Although one approach may manifest itself differently compared to others, many approaches have much in common regarding the thermomechanics of the substrate and deposit and their process simulation. The fundamental components of the simulation persist between approaches, but the boundary conditions in each approach are different. This may leave an engineer in a situation where process simulation is available for one approach but not quite for another. In such a case, nondimensional analysis can be used to draw a metaphor between the two approaches, and based on this, the simulations performed for one approach can be transposed to represent the behavior for another approach. This may first be applied for analytical solutions to confirm the chosen metaphor. In this regard, the three variants of the Rosenthal solution presented

Laser Parameters	
Type	Carbon Dioxide
Wavelength	10.6 micron
Power	1200 W
Spot Size	2.54 mm
Travel Speed	2.54 mm/sec

Powder Parameters	
Material	Ti-6Al-4V
Feed Rate	3.275 cubic mm/sec
Powder Velocity	2.5 m/sec
Particle Size	42.5 micron
Jet Diameter	2.2 mm
Jet Angle	65 degrees

Substrate Parameters	
Material	Ti-6Al-4V
Diameter	76.2 mm
Thickness	23.8 mm

Deposit Parameters	
Width	2.54 mm
Thickness	0.5 mm

Fig. 8 Process parameters for the cloverleaf-shaped deposit on a cylindrical substrate

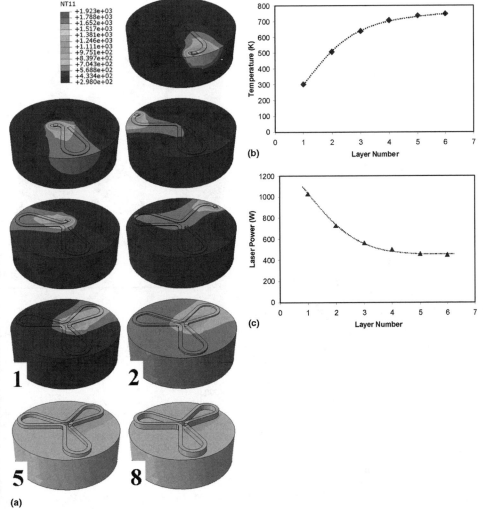

Fig. 9 (a) Evolution of temperature during eight layers of a cloverleaf-shaped deposit. (b) Increase in substrate temperature with layer number. (c) Decrease in laser power with layer number

Laser Parameters	
Type	Carbon Dioxide
Wavelength	10.6 micron
Power	2500 W
Spot Size	6.35 mm
Travel Speed	2.54 mm/sec

Powder Parameters	
Material	Ti-6Al-4V
Feed Rate	20.48 cubic mm/sec
Powder Velocity	2.5 m/sec
Particle Size	42.5 micron
Nozzle Diameter	5.5 mm
Jet Angle	65 degrees

Substrate Parameters	
Material	Ti-6Al-4V
Length	76.2 mm
Width	63.5 mm
Thickness	6.35 mm

Deposit Parameters	
Width	6.35 mm
Thickness	1.27 mm

Fig. 10 Process parameters and deposit sequence for scour repair on a thin plate

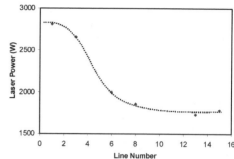

Fig. 12 Laser power versus line number

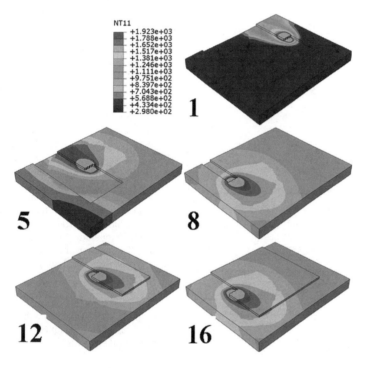

Fig. 11 Temperature distributions for selected lines in the deposit

Fig. 14 Temperature distributions for selected lines in the deposit

Laser Parameters	
Type	CO2
Wavelength	10.6 micron
Power	1300 W
Spot Size	6.35mm
Travel Speed	2.54 mm/sec

Powder Parameters	
Material	Ti-6Al-4V
Feed Rate	20.4 mm³/sec
Powder Velocity	2.5 m/sec
Particle Size	42.5 micron
Nozzle Diameter	5.5 mm
Jet Angle	65 degrees

Substrate Parameters	
Material	Ti-6Al-4V
Length	63.5 mm
Width	6.35 mm
Thickness	76.2 mm

Deposit Parameters	
Width	6.35 mm
Thickness	1.27 mm

Fig. 13 Process parameters for a thin-plate edge repair

Fig. 15 Laser power versus line number

earlier in this article play a vital, baseline role. Several additional analytical solutions and corollaries thereof are given in Ref 26, which are all useful for the same purpose.

For example, in additive manufacturing using the electron beam, the value of absorptance is significantly larger than in laser deposition. It is ~95%, which is more than two times higher than in laser deposition. Its thermomechanics are similar to the laser deposition process, but the process control has additional constraint for temperature control due to the occurrence of selective vaporization of the substrate and deposit materials. For example, in electron beam deposition of Ti-6Al-4V, selective evaporation of 8 to 10% of the aluminum has been reported (Ref 27). Thus, the control of the melt pool temperature is fundamental.

In electron beam deposition, the additive material can be a wire. The wire melts and joins with the substrate. The cross section of the deposit bears a similarity to the circular cross section of the wire. It can be more round than in the case of the cross section obtained with powder as the additive material. This change in shape affects the local temperature and residual-stress values. However, further away, the thermal and mechanical behavior is governed by the overall shape and extent of the deposition process. Thus, the result obtained using a simplification for the deposit cross section is generally adequate, and a rectangular cross section for the additive material deposit is commonly used in process simulation.

A two-part simulation procedure for additive manufacturing using spray forming is described in Ref 28. These two parts are described separately in Ref 29 and 30 by the same authors. The first part is for simulation of droplet spray formation, droplet mass and enthalpy evolution, droplet deposition and redeposition, and calculation of porosity distribution within the deposit. The second part is modeling of heat flow and solidification, which conforms to the boundary conditions that result from the droplet deposition. In a situation when the first part of the simulation is unavailable, the process simulation of spray forming may be performed by using the boundary conditions that result from a mean character of droplet deposition.

Summary

The foregoing discussion was organized as a précis of this Handbook in that it covered various aspects of simulation, such as the fundamentals of process modeling, microstructure and defect generation, input data requirements, analytical and computational modeling procedures, integration of modeling and simulation with design, and examples that each illustrated one or more salient merits of simulation. The information was presented from the viewpoint of an engineer in an industrial environment who seeks to develop or enhance additive manufacturing processes while operating in a tight envelope of constraints in schedule, costs, and productivity. The information is organized with a central theme that control of the additive manufacturing process is fundamental to its success and consequently for component certification. The simulation schemes are those that have matured to a point that they can be used reliably by an engineer in industry with minimal effort. Clearly, the state-of-the-art for simulation of additive manufacturing processes has advanced beyond these methods in academic or research laboratory environments, where the principal focus is science. There are detailed models in the literature that address the various individual physical phenomena at great detail, and the technology as a whole is progressing in leaps and bounds. The use of electronic databases can locate these works easily, and the developments therein can be

adapted as applicable. In the years to come, there will be several additions to the capability of simulations that can be performed in the industrial environment. In the coming years, this will include melt pool fluid flow simulation and melt pool free-surface tracking and its interaction with the previously deposited material. This will allow the simulation to attain process control with greater precision and will render it ever more irreplaceable for obtaining insight into the inner workings of additive manufacturing.

REFERENCES

1. N. Christensen, V. Davies, and K. Gjermundsen, The Distribution of Temperature in Arc Welding, *Br. Weld. J.*, Vol 12 (No. 2), 1965, p 54–65
2. S. Kou, Welding, Glazing, and Heat Treating—A Dimensional Analysis of Heat Flow, *Metall. Trans. A*, Vol 13, 1982, p 363–371
3. A. Vasinonta, J.L. Beuth, and M.L. Griffith, A Process Map for Consistent Build Conditions in the Solid Freeform Fabrication of Thin-Walled Structures, *J. Manuf. Sci. Eng.*, Vol 123, Nov 2001, p 615–622
4. S. Bontha, N.W. Klingbleil, P.A. Kobryn, and H.L. Fraser, Thermal Process Maps for Prediction of Solidification Microstructure in Laser Fabrication of Thin-Wall Structures, *J. Mater. Proc. Technol.*, Vol 178, 2006, p 135–142
5. A. Vasinonta, J.L. Beuth, and M.L. Griffith, Process Maps for Predicting Residual Stress and Melt Pool Size in the Laser-Based Fabrication of Thin-Walled Structures, *J. Manuf. Sci. Eng.*, Vol 129, Feb 2007, p 101–109
6. S. Bontha, N.W. Klingbeil, P.A. Kobryn, and H.L. Fraser, Effects of Process Variables and Size-Scale on Solidification Microstructure in Beam-Based Fabrication of Bulky 3D Structures, *Mater. Sci. Eng. A*, Vol 513–514, July 15, 2009, p 311–318
7. K.-O. Yu, *Modeling for Casting and Solidification Processing*, CRC Press, 2001
8. P.A. Kobryn and S.L. Semiatin, Microstructure and Texture Evolution during Solidification Processing of Ti-6Al-4V, *J. Mater. Proc. Technol.*, Vol 135, 2003, p 330–339
9. S.M. Kelly, "Thermal and Microstructure Modeling of Metal Deposition Processes with Application to Ti-6Al-4V," Ph.D. thesis, Virginia Polytechnic Institute, 2004
10. J.C. Ion, H.R. Shercliff, and M.F. Ashby, Diagrams for Laser Materials Processing, *Acta Metall. Mater.*, Vol 40 (No. 7), 1992, p 1539–1551
11. M. Alimardani, E. Toyserkani, and J.P. Huissoon, A 3D Dynamic Numerical Approach for Temperature and Thermal Stress Distributions in Multilayer Laser Solid Freeform Fabrication Process, *Opt. Lasers Eng.*, Vol 45, 2007, p 1115–1130
12. X. He and J. Mazumdar, Modeling of Geometry and Temperature during Direct Metal Deposition, *Laser Materials Processing Conference, ICALEO 2006*, p 1022–1029
13. A.M. Deus and J. Mazymdar, Three-Dimensional Finite Element Models for the Calculation of Temperature and Residual Stress Fields in Laser Cladding, *Laser Materials Processing Conference, ICALEO 2006*, p 496–505
14. S. Ghosh and J. Choi, Fully Coupled Temperature-Stress Finite Element Analysis for Thermal Stresses in Laser Aided DMD Process, *Laser Materials Processing Conference, ICALEO 2006*, p 999–1008
15. J. Cao and J. Choi, "A Multi-Scale Modeling of Laser Cladding Process," Technical Report AFRL-ML-WP-TP-2006-429, Air Force Research Laboratory
16. A. Chaudhary and S. Vaze, Design Optimization for Dies and Preforms, *Forming and Forging*, Vol 14, *Metals Handbook*, ASM International, 1988
17. J.S. Arora, *Introduction to Optimum Design*, McGraw Hill, 1989
18. M. Picasso, C.F. Marsden, J.D. Wagniere, A. Frenk, and M. Rappaz, A Simple but Realistic Model for Laser Cladding, *Metall. Mater. Trans. B*, Vol 25, p 281–291
19. M. Gaumann, C. Bezencon, P. Canalis, and W. Kurz, Single-Crystal Laser Deposition of Superalloys: Processing-Microstructure Maps, *Acta Mater.*, Vol 49, 2001, p 1051–1062
20. "SAMP: An Automated 3-D Software for Simulation of Additive Manufacturing Processes," Applied Optimization, Inc., www.appliedO.com
21. "Rosenthal Solver: An Implementation of Non-Dimensional Rosenthal Analytical Solution for Preliminary Design of Additive Manufacturing Processes," Applied Optimization, Inc., www.appliedO.com
22. J. Kummailil, "Process Models for Laser Engineered Net Shaping," Ph.D. thesis, Worcester Polytechnic Institute, 2004
23. P.C. Collins, "A Combinatorial Approach to the Development of Composition-Microstructure-Property Relationships in Titanium Alloys Using Directed Laser Deposition," Ph.D. thesis, The Ohio State University, 2004
24. D.S. Salehi, "Sensing and Control of Nd: YAG Laser Cladding Process," Ph.D. thesis, Swinburne University of Technology, 2005
25. C. Semetay, "Laser Engineered Net Shaping (LENS) Modeling Using Welding Simulation Concepts," Ph.D. thesis, Lehigh University, 2007
26. J.F. Ready, Ed., *LIA Handbook of Laser Materials Processing*, Laser Institute of America, 2001
27. K.M.B. Taminger, "Electron Beam Freeform Fabrication: A Fabrication Process That Revolutionizes Aircraft Structural

Designs and Spacecraft Supportability," ARMD Technical Seminar, May 22, 2008

28. J. Mi, P.S. Grant, U. Fritsching, O. Belkessam, I. Garmendia, and A. Landaberea, Multiphysics Modelling of the Spray Forming Process, *Mater. Sci. Eng. A*, Vol 477, 2008, p 2–8

29. J. Mi and P.S. Grant, Modelling the Shape and Thermal Dynamics of Ni Superalloy Rings during Spray Forming, Part I: Shape Modelling—Droplet Deposition, Splashing and Redeposition, *Acta Mater.*, Vol 56, 2008, p 1588–1596

30. J. Mi and P.S. Grant, Modelling the Shape and Thermal Dynamics of Ni Superalloy Rings during Spray Forming, Part II: Thermal Modelling—Heat Flow and Solidification, *Acta Mater.*, Vol 56, 2008, p 1597–1608

ASM Handbook, Volume 22B, *Metals Process Simulation*
D.U. Furrer and S.L. Semiatin, editors

Copyright © 2010, ASM International®
All rights reserved.
www.asminternational.org

Modeling of Porosity Formation during Solidification

Peter D. Lee and Junsheng Wang, Imperial College London

THE SOLIDIFICATION OF CASTING is normally thought of as a transition from liquid to solid; however, in most instances a third gas phase forms, termed porosity. Pores form due to inadequate feeding of the volumetric change from liquid to solid and the partitioning of solutes such as hydrogen, nitrogen, and oxygen. The combination of these two driving forces means that porosity, both as macroshrinkage and microporosity, can be found in most castings. The presence of porosity can be highly detrimental to the final mechanical properties of components ranging from continuous casting of steel alloys to aluminum alloy sand castings (Ref 1, 2). Figure 1 shows the strong influence of pore size on fatigue life, which decreases by a factor of 8 as the pore size increases from 100 to 600 μm for a secondary dendrite arm spacing of 48 ± 5 μm (Ref 3). Several studies have concluded that any pores larger than the secondary dendrite arm spacing may act as the initiation sites for fatigue failure (Ref 3–5), as shown by the fractograph inset in Fig. 1. In summary, the failure of cast components, such as automotive wheels and engines, may be dominated by the level of porosity, especially when they experience cyclic loading (either mechanically or thermally induced) (Ref 4–7). Therefore, there is a need for models that predict the percentage and size of porosity formed during solidification in order to effectively predict mechanical properties.

Simulating porosity first requires the identification of the mechanisms governing pore nucleation and growth. Over 50 years ago, Whittenberger and Rhines classified microporosity according to the two main driving forces: gas or shrinkage (Ref 8). Although this article focuses on these two driving forces, as detailed by Campbell, there are many other possible reasons for pores to form, ranging from entrapped air (common in high-pressure die casting) to thermal stresses opening up tears (Ref 2). Successive investigations and advances in the scientific understanding of pore formation have led to the development of many different modeling approaches, ranging from easy-to-implement criteria functions (Ref 9–11),

then analytical solutions (Ref 12), flow simulations (Ref 13), and finally to complex direct simulations of the nucleation of multiple solid and gas phases coupled to macroscopic shrinkage models (Ref 14).

The first class of models, criteria functions, was initiated in 1953 by Pellini (Ref 11), who presented one of the first predictive criteria relating the percentage of porosity to the thermal gradient and geometric criteria. Walther et al.

(Ref 12) presented one of the earliest of the second class of models, analytic solutions, in 1956 by analytically solving for a relationship between the formation of centerline shrinkage and the feeding of liquid down a simple cylinder. This simple idea formed the basis of almost all of the current shrinkage-driven models when it was extended to a bundle of cylinders by Piwonka and Flemings in 1966 (Ref 15). In 1985, Kubo and Pehlke (Ref 13) initiated the

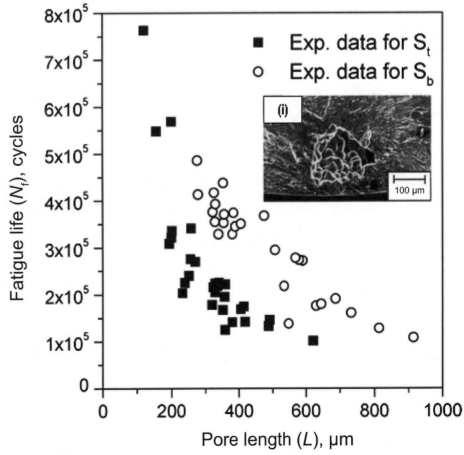

Fig. 1 Experimentally observed fatigue life as a function of initiating pore length for a secondary arm spacing of 48 ± 5 μm (S_t) and 24 ± 5 μm (S_b). Inset (i): SEM image of an initiating pore. Adapted from Ref 3

third class of models, thermal/fluid flow simulations, by using Darcy's law (Ref 16) rather than analytic solutions for the flow-down cylinders, such as the Hagen Poiseuille equation. Kubo and Pehlke (Ref 13) wrote a computer program to solve a simplified set of transport equations for interdendritic flow using Darcy's equation and then calculated the pressure drop and related that to conditions under which shrinkage pores were likely to form.

This third class of models, thermal/fluid flow models, is perhaps the most popular, with many showing excellent correlation to experiment in terms of the predicted percentage porosity (Ref 17–24). However, the kinetics of porosity nucleation is usually ignored in these models, which is very important when predicting not only the percentage porosity but also pore size. The size distribution of porosity is essential for the accurate prediction of fatigue life (Ref 3–5); therefore, the next logical step was the inclusion of the kinetics governing pore nucleation and growth, forming the fourth class of models, kinetic/microstructural-based models. This class of models was initiated in 1995 by Lee and Hunt (Ref 25), who presented a model that solved the diffusion of the gas-forming elements causing pore growth and included the nucleation kinetics obtained from in situ x-ray radiography experiments as well as the restriction imposed by the dendrites on pore growth. Because of the direct coupling of this model with experiments, it predicted not only percentage porosity but also pore size distribution, although only in two dimensions and for the limited case of columnar growth.

This article first provides an overview of the equations governing pore formation and then reviews the four classes of models, highlighting both the benefits and drawbacks of each class. The accurate use of such porosity models has recently received increasing attention because of the drive to produce ever higher-performance components with reduced weight to minimize energy consumption and hence environmental concerns (Ref 26). Extensive experimental studies on pore formation in a range of alloys have led to ever more complex models that simulate both the diffusion of gaseous elements and the feeding of solidification shrinkage in both simplified and commercial alloys (Ref 14, 27–31). These models are now becoming more commonly incorporated into commercial software packages to predict the location, percentage, and even size distribution of pores in industrial castings (Ref 27, 32). Table 1 provides a list of symbols used in the mathematical expressions.

Governing Mechanisms

The formation of porosity results from a combination of inadequate feeding of the volumetric change and segregation of the dissolved gas-forming elements (e.g., hydrogen in

aluminum-base alloys, hydrogen and oxygen in copper-base alloys, and hydrogen, nitrogen, and oxygen in steel). Inadequate feeding, combined with solidification shrinkage, leads to a local reduction in pressure, causing shrinkage pores to form, normally at a high solid fraction (∼0.9) (Ref 33). However, this reduction in pressure also causes the solubility of any dissolved gases in the liquid to decrease, increasing the supersaturation of that species and hence serving as the driving force for forming gas pores. In addition to these two driving forces, two other mechanisms must also be considered: pore nucleation and the interfacial energy between the pore and the surrounding liquid or solid phases. These various factors are shown schematically in Fig. 2. Physical entrapment of gas, such as during mold filling, is a significant source of gas for porosity formation but is beyond the scope of this article.

One of the underlying physics principles, the flow of interdendritic liquid to feed shrinkage, is governed by Navier-Stokes and continuity equations, which are detailed in fluid dynamics textbooks, such as Bird et al. (Ref 34), or metallurgically focused texts, such as Szekely and Thermelis (Ref 35) or Geiger and Poirier (Ref 36). However, in the semisolid region, most models use the simplified solution postulated by Darcy that assumes Stoke's flow, termed Darcy's law (Ref 16):

$$\bar{u} = -\frac{K}{\mu}(\nabla P - \rho g) \qquad \text{(Eq 1)}$$

where $\bar{u}$ is the superficial velocity (i.e., the average velocity over both the liquid and solid), K is the permeability of the porous medium, μ is the viscosity, ∇P is the pressure drop, ρ is the density of the molten metal, and g is the acceleration due to gravity. From this simplified convection equation, the pressure drop due to volumetric shrinkage, termed P_s, can be evaluated.

The pressure a pore experiences, P_l, is a combination of the shrinkage pressure together with the external pressure, P_{ex} (either atmospheric or applied in the case of low- or high-pressure casting), together with any metallostatic pressure, P_m:

$$P_l = P_{ex} + P_m + P_s \qquad \text{(Eq 2)}$$

If there are no dissolved gases in the liquid, this is the full story. However, Whittenberger and Rhines experimentally demonstrated in a magnesium-aluminum alloy system that even at extremely high negative pressures in the liquid, no pores nucleated unless dissolved gases were present (Ref 8). Therefore, this treatment must be extended to calculate the influence of any gas species present. Assuming that a pore already exists and is in equilibrium with both the local liquid pressure and temperature (i.e., that the ideal gas law holds, and hence, the volume of the pore, V, is equal to nRT/P_g), the pressure of gas within needed for pore growth, P_g, is given by combining Eq 2 with the Young-Laplace equation (Ref 37):

Table 1 List of symbols

Symbol	Definition
$\%P$	Percentage porosity
C	Concentration
D	Diffusion coefficient
f	Fraction of a phase (f_l, fraction liquid; f_s, fraction solid; etc.)
f_H	Hydrogen interaction coefficient
f_r	Friction coefficient
g	Gravitational acceleration
G	Temperature gradient
K	Permeability
L	Length of the mushy zone
m	Liquidus slope
P	Pressure
r	Spherical and cylindrical coordinate
R	Cooling rate
R_{eq}	Equivalent pore radius, determined by calculating the radius of a circle/sphere of equivalent area/volume to that measured
R_H	Hydrogen consumption or generation by pores in the liquid
S	Solubility limit
SSn	Supersaturation needed for pore nucleation
S_t	Time-dependent source term, representing the rejection of hydrogen into the liquid phase
t	Time
T	Temperature
t_s	Solidification time (liquidus temperature to the solidus temperature)
V	Solidification velocity
α	Ideal gas constant
β	Volume shrinkage upon solidification
γ	Surface tension
κ	Gas partition coefficient
λ_1	Primary dendrite arm spacing
λ_2	Secondary dendrite arm spacing
μ	Viscosity
ν	Flow velocity
ρ	Density
τ	Tortuousity factor

Subscript/superscript

a	Ambient
avg	Average
c	Critical
e	Eutectic or effective
g	Gas phase
H	Hydrogen
l	Liquid state
m	Metallostatic
max	Maximum
o	Initial condition
P	Pore
s	Solid state
S	Shrinkage
V	Vapor

$$P_g \geq P_l + P_\gamma \qquad \text{(Eq 3)}$$

where P_γ is the pressure due to the formation of the interface between the gas and liquid phases (or gas/solid if the pore contacts a dendrite or other solid). For a simple spherical bubble surrounded by liquid, this is given by:

$$P_\gamma = \frac{2\gamma}{R} \qquad \text{(Eq 4)}$$

where γ is the gas-liquid interfacial energy, and R is the radius of the bubble.

Clearly, pore pressure (P_g) varies during solidification due to the transport of liquid (P_l) and through the exchange of gaseous species

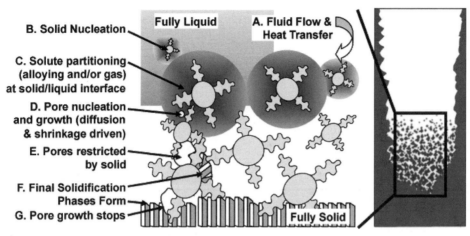

Fully Liquid

A. Fluid Flow & Heat Transfer

B. Solid Nucleation

C. Solute partitioning (alloying and/or gas) at solid/liquid interface

D. Pore nucleation and growth (diffusion & shrinkage driven)

E. Pores restricted by solid

F. Final Solidification Phases Form

G. Pore growth stops

Fully Solid

Fig. 2 Schematic diagram of the various physical processes involved in the formation of microporosity

atoms with the surrounding liquid. The atoms will enter or leave the pore, depending on the local supersaturation level. The supersaturation will depend on the diffusion of solute atoms partitioned out of the solidifying phases and the solubility of that species in the liquid. Therefore, the solubility is a function of many factors, including the temperature, pressure, and other solute concentrations (Ref 38, 39).

Although the total number of moles of gaseous species in the system is conserved to a first approximation, the concentration of gas-forming elements steadily increases due to partitioning from the solid, where its solubility is an order of magnitude lower (Ref 38). The rejection of gaseous species from the solid phase increases the concentration in the interdendritic liquid, producing a concentration gradient, and eventually, a significant supersaturation can occur.

When there is sufficient supersaturation, a pore should nucleate. However, the supersaturations predicted by classical homogeneous or heterogeneous nucleation theory are huge. Jones et al. (Ref 40) reviewed the theory of nucleation of gas bubbles in a supersaturated solution, classifying the nucleation mechanisms into four types: I, classical homogeneous nucleation; II, classical heterogeneous nucleation; III, pseudoclassical nucleation; and IV, nonclassical nucleation. Details are in Ref 40, but, in summary, type III and IV assume pre-existing sites and hence have much reduced critical supersaturations. As mentioned previously, Whittenberger and Rhines (Ref 8) experimentally demonstrated that the actual supersaturation reached upon nucleation of pores during solidification is orders of magnitude less than that predicted by types I or II nucleation theories. Therefore, either type III or IV nucleation is most likely happening, as postulated by many authors, occurring on inclusions or oxide films (Ref 41, 42). Type IV nucleation of pores is also analogous to the free growth barrier mechanism proposed by Greer et al. (Ref 43, 44) for the nucleation of solid phases.

When nucleated, the pores act as sinks for gas-forming elements, which, together with

partitioning at the solid-liquid interface, can further increase the concentration gradient. Diffusion must therefore also be simulated. Diffusion is governed by Fick's second law (Ref 34, 45):

$$\frac{\partial}{\partial t} \cdot (\rho C_g) + \nabla \cdot j = R_g \qquad \text{(Eq 5)}$$

where C_g is the concentration of gas-forming elements (e.g., hydrogen in liquid aluminum), R_g is the source/sink of that solute per unit volume, and j is the diffusion or Fickian flux, given by:

$$j = -j = \rho \nabla C_g \qquad \text{(Eq 6)}$$

Dropping the subscript g and replacing it with an s or l to denote whether C is in the solid or liquid, Eq 5 and 6 can be combined to obtain (Ref 46):

$$\frac{\partial}{\partial t} \cdot (\rho_l C_l f_l + \rho_s C_s f_s) = \nabla \cdot (\rho_l D_e \nabla C_l) + R_g \qquad \text{(Eq 7)}$$

where D_e is the effective diffusivity given as a function of T and f_s. Equation 7 must be solved together with the appropriate sinking of gaseous species into nucleated pores to simulate the driving force generated by dissolved gas-forming elements on pore formation.

It is useful at this stage to illustrate the relative importance of these different mechanisms using a few simplified calculations in a binary Al-10wt%Cu casting. Assuming that solidification follows the Scheil-Gulliver equation and that the initial hydrogen level is 0.25 mL standard temperature and pressure (STP)/100 g and partitions following the Lever rule, the hydrogen concentration in the interdendritic liquid will be over 10 times the equilibrium concentration by the eutectic temperature (Ref 47), more than sufficient for the type III or IV nucleation of pores.

Looking now at the influence of shrinkage in this Al-10wt%Cu alloy, the pressure drop can be approximated by an upper-bound calculation (Ref 47):

$$\Delta P = -\frac{\mu \beta V}{K} \int_0^L \partial x = -\frac{\mu \beta V L}{K}$$

$$= -\frac{1.5 \times 10^{-3} \times 0.05\, VL}{2.0 \times 10^{-12}} = -3.8 \times 10^7\, VL \qquad \text{(Eq 8)}$$

where L is the length of the mushy zone (assumed to be 20 mm), V is the velocity of the eutectic front (0.5 mm/s), μ is the viscosity (1.5×10^{-3} Pa·s), β is the solidification shrinkage (5%), and K is the permeability. Using Eq 8 for Al-10wt%Cu, where the fraction liquid is approximately 0.23 when the eutectic forms, the worst-case bound on the pressure drop is 7500 Pa (using Poirier et al.'s K value of $\sim 10^{-13}$m²) (Ref 48), or less than 8% of an atmosphere, an insufficient pressure drop for even type IV nucleation to occur.

In Al-10wt%Cu, hydrogen is most likely to be the most important driving force for pore formation. However, in an Al-4wt%Cu alloy, where the eutectic forms at a liquid fraction of ~ 0.05, the permeability is $\sim 10^{-15}$, giving a pressure drop of over 7 atm (for the same conditions as mentioned previously)—more than enough to nucleate and grow pores.

It can also be argued that diffusion is important by using a simple characteristic diffusion length (l) calculation:

$$l = \sqrt{Dt}$$

where D is the diffusivity of hydrogen in liquid aluminum, 3.18×10^{-7} m²/s at 660 °C (1220 °F), and t is time (Ref 49). For a time scale of 1 s, l is ~ 0.5 mm. Therefore, in 1 s, hydrogen is able to diffuse into pores at the length scale of the typical grain size in a casting.

These calculations, which roughly match experimental observations in aluminum-copper systems (Ref 32), illustrate that both mechanisms can be important; further, in most cases, pore formation is governed by a combination of these mechanisms.

Porosity Model Types

Having provided an overview of the governing mechanisms and illustrated that there are many mechanisms influencing pore formation, the different types of porosity models are now grouped into four classes to facilitate discussing their benefits and drawbacks. These classifications build on two prior reviews of shrinkage (Ref 22) and other models (Ref 50):

• Criteria functions
• Analytic models
• Continuum models
• Kinetic models

The evolution of these four categories of models, together with a selection of representative papers (the list is far from complete but is instead meant as a starting guide), is shown schematically in Fig. 3.

Criteria Functions

Arguably, Pellini (Ref 11) developed the first model of pore formation in 1953 when he related the propensity for pore formation to the thermal gradient and geometric features. His model can be classified as a criteria function; that is, it is a simple quantitative rule based on the local solidification conditions, such as thermal gradient and cooling velocities. Criteria functions became popular when Niyama et al. (Ref 9) developed a simple correlation between the thermal gradient, G, the cooling rate, R, and the propensity for shrinkage porosity to form. They suggested that when:

$$G/\sqrt{R}$$

is greater than an alloy-specific constant, shrinkage porosity would form. Although the Niyama criterion is widely used, many studies have illustrated that it is most applicable to shrinkage cavities, rather than microporosity (Ref 10).

Most other criteria functions have also been based on thermal parameters and hence allow immediate prediction of the regions of a casting where pores are likely to occur. Newer criteria have been developed using extensive statistical correlations of experimentally measured porosity to thermal parameters, where the latter are measured with thermocouples or determined from heat-transfer models (Ref 9, 11, 51–58). Taylor et al. (Ref 59) and Viswanathan et al. (Ref 60) have reviewed criteria functions in detail, and a summary of the key ones is given in Table 2.

Although these criteria functions benefit from being easy to implement within heat-transfer models of castings, they suffer from a number of significant limitations:

- Most are based on shrinkage mechanisms.
- They are experimentally fit to a particular alloy and cannot be safely extrapolated to include even minor alloying changes, for example, strontium additions in an aluminum-silicon alloy (Ref 50).
- Most do not include any influence of processing conditions beyond thermal control, such as hydrogen content, grain refining additions, and so on.

Thus, extensive casting tests are required to generate enough data to fit constants within a selected criteria function, and it makes them expensive to develop and difficult to apply to new casting techniques, alloys, or even, in some cases, geometries. Criteria functions still are being further refined; recently, Carlson and Beckermann (Ref 61) developed a dimensionless Niyama criterion that incorporates alloy properties (e.g., viscosity and volumetric shrinkage) and microstructural features (e.g., secondary dendrite arm spacing), significantly improving the predictions of percentage shrinkage porosity.

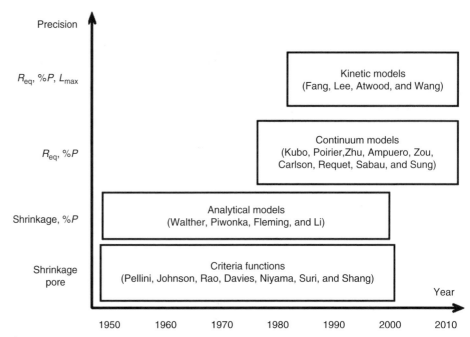

Fig. 3 History of the development of model classes and the values they can predict

Table 2 Criteria functions for porosity prediction

Authors	Criteria function	Alloy	Conditions	C_{H0}
Pellini 1953 (Ref 11)	Thermal gradient: G	Al-Si	Sand	Unknown
Rao et al. 1973 (Ref 52)	Feeding index: G/t_s	Al-Cu-Si (LM4)	Sand	Unknown
Niyama et al. 1982 (Ref 9)	Thermal gradient and cooling rate: $G/R^{1/2}$	Steel	Sand	Unknown
Kao et al. 1995 (Ref 56) and Li et al. 2001 (Ref 57)	Feeding resistance: $G^{-0.38}/V_s^{1.62}$	A356	Sand	0.15 ± 0.005 mL/100 g STP(a)
Shang et al. 2004 (Ref 58)	Solidification time and solidus velocity: $t_s^{1.18}/V_s^{1.13}$	A319, A356, and A332	Low P_0	0.20–0.30 mL/100 g STP(a)

(a) STP, standard temperature and pressure

Analytical Models

Many investigators have shown that by making a number of assumptions, the complex interactions of shrinkage and gas can be simplified sufficiently to allow analytic solutions to be developed, termed analytic models in this article. Most of these solutions have based the theoretical analysis on feeding behaviors. The two primary assumptions frequently made are that both the thermal gradient and solidification velocity are constant. Walther et al. (Ref 12) presented the first such analysis by solving liquid feeding down a long tube solidifying inward in the radial direction, deriving the pressure drop, ΔP, as:

$$\Delta P = \frac{f_r}{g} \frac{64 B^4 \beta L^2}{r^4} \left(\frac{1}{2} + \frac{\beta \phi_r L}{3r} \right) \quad \text{(Eq 9)}$$

where β is the volumetric change upon solidification, B is a constant, L is the length, r is the radius of the liquid central cylinder, and f_r is

the friction factor. Results from this approach showed a good match with their experiments. Many other authors (Ref 18, 21, 62–65) have derived different formulations for the calculation of pressure drop in the mushy zone. These models are inherently limited by their simplifying assumptions; however, they form the basis of the more recent Darcy's law models, as described subsequently.

Continuum Models

Darcy Flow Models. Kubo and Pehlke (Ref 13) presented the first comprehensive Darcy flow continuum model by coupling Darcy's law (Eq 1) to the continuity equations, the Fourier equation for heat transport, and by calculating the solid fraction using the Scheil-Gulliver equation. They also included the partitioning of hydrogen between the solid and liquid phases, although not its diffusion-limited transport. The resulting model showed good qualitative agreement to experiment for percentage porosity. An interesting feature of their model is that the majority of the pore growth occurred during the last 20% of solidification as the permeability decreased by several orders of magnitude, as shown in Fig. 4. This is a common feature of all the Darcy flow-based models, but it is in conflict with in situ experimental observations made by Lee et al. (Ref 47) in both aluminum-copper and aluminum-silicon alloys, where the pores were observed to grow over a much wider range of solid fraction, as shown in Fig. 5, noting that a solid fraction of 80% is not reached until a temperature of 550 °C (1020 °F) for the Al-10Cu alloy shown. Further, in Fig. 5, most of the growth was experimentally observed to occur above a temperature of 600 °C, when the fraction of solid is less than 0.6 (Ref 47). Although many investigators (Ref 18, 21, 28, 29, 62–65) have developed more complex models of flow through the semisolid region, removing many aspects of the assumptions made by Kubo and Pehlke (Ref 13), the lack of experimental correlation in terms of kinetics has led to the development of models that combine the influence of gas diffusion and segregation together with shrinkage effects.

Continuum Flow Models Including the Influence of Gas. The continuum flow models of Kubo and Pehlke (Ref 13) and many later authors typically solve the energy and momentum equations (i.e., the Fourier and Navier-Stokes equations; see standard fluid dynamics texts, e.g., Ref 34). These solutions were extended to include mass transport for solute, with the formulations of Poirier and coauthors (e.g., Ref 18, 21, 62) typifying the methodology. In 1987, Poirier et al. (Ref 18) published one of the first continuum flow models to incorporate gas concentration, although not its diffusion. They tracked the partitioning of hydrogen between the liquid and solid, calculating its supersaturation and using nucleation criteria based on the secondary arm spacing to determine when microporosity would form. As shown in Fig. 6, this model illustrated the importance of initial gas concentration as well as shrinkage.

More recently, Poirier and coauthors (Ref 19, 21, 62) have significantly expanded this model to calculate the pressure and redistribution of gas-forming elements in multicomponent systems (ranging from aluminum alloys to steels and nickel superalloys) during solidification.

Their methodology is based on a volume averaging of the liquid, solid, and gas within each element, allowing the momentum equation to be reformulated for the mushy zone (assuming it is a porous medium), as:

$$f_l \frac{\partial}{\partial t}\left(\frac{u}{f_l}\right) + u \cdot \nabla\left(\frac{u}{f_l}\right) = -\frac{f_l}{\rho_l}\nabla P + \frac{\mu}{\rho_l}\nabla^2 u - \frac{\mu}{\rho_l}\frac{f_l}{K}u + \frac{\mu\beta}{3\rho_l}\nabla\left(\frac{\partial f_l}{\partial t}\right) + \frac{\rho f_l}{\rho_l}g \qquad \text{(Eq 10)}$$

where β is the solidification shrinkage, which is defined as $\beta = (\rho_s - \rho_l)/\rho_l$, and K is the permeability in the equiaxed structure. Solute conservation is determined using:

$$\frac{\partial \overline{C}^j}{\partial t} + u \cdot \nabla C_1^j = -\nabla \cdot j^j - \beta\frac{\partial f_l}{\partial t}C_1^j \qquad \text{(Eq 11)}$$

with

$$j = -\rho_l f_l D_l \nabla C_l - \rho_s f_s D_s \nabla C_s \qquad \text{(Eq 12)}$$

Their most recent application of the model was for AISI 8620 steel castings, where they used Eq 10 to 12 combined with a model of the thermodynamics of gas-forming elements (both nitrogen and hydrogen as well as nitrogen interaction with titanium in the melt). This allowed them to predict the gas supersaturation and, via Sievert's law, estimate the pressure. This was then compared with the local interdendritic pressure, assuming shrinkage, to predict the pressure difference, or potential for pore formation, as shown in Fig. 7(a). By altering the initial nitrogen and hydrogen concentrations, a process map can be produced for each individual casting shape/condition, to predict how well the initial gas concentration must be controlled to prevent pore formation (Fig. 7b). Unfortunately, this model still does not predict the size of the pores. Further, the initial gas concentrations (predicted in Fig. 7a) are impractically low. Note that in steels, Poirier et al. illustrated that the effect of nitrogen is effectively mitigated by the common practice of adding titanium, the dashed line in Fig. 7(b).

Several other groups have also published similar models, each with different extensions/additions. Sabau and Viswanathan (Ref 67) used a drag coefficient to account for the momentum loss due to the flow around and through the dendrite structures instead of a direct source term for the Darcy flow. This allowed them to add in higher flow velocity terms in addition to the Darcy flow resistance for situations such as squeeze casting. However, this was found not to be important under normal casting conditions. They improved upon the solution method by applying a variable projection method to give more stable solutions for larger time steps, particularly when incorporating the effect of pore growth on reducing liquid flow. They are one of the few investigators to present the pressure calculations including pore growth, which significantly reduces the

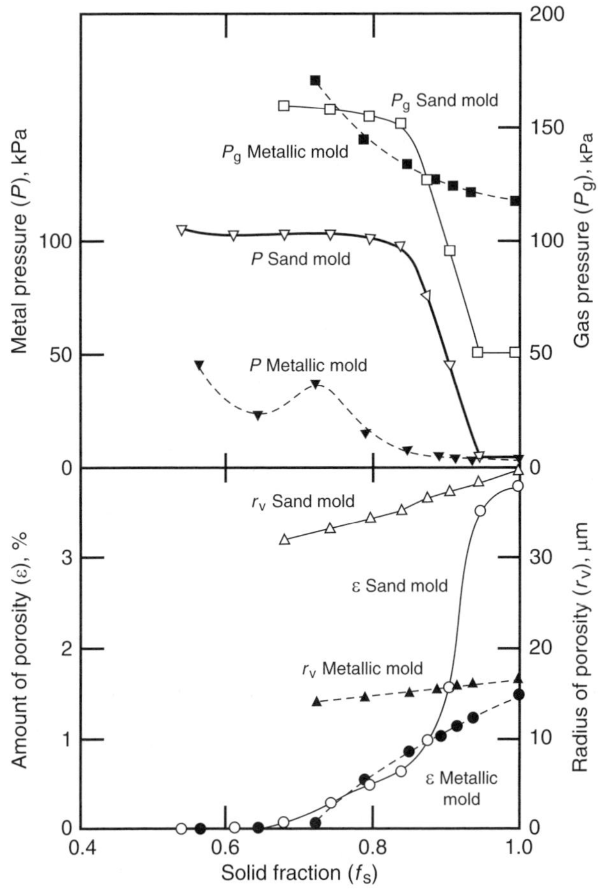

Fig. 4 Percentage porosity predictions by Kubo et al. illustrating how the sudden drop in metallostatic pressure when the solid fraction exceeds 0.8 causes the prediction of rapid growth of microporosity in an Al-4.5Cu plate. Adapted from Ref 13

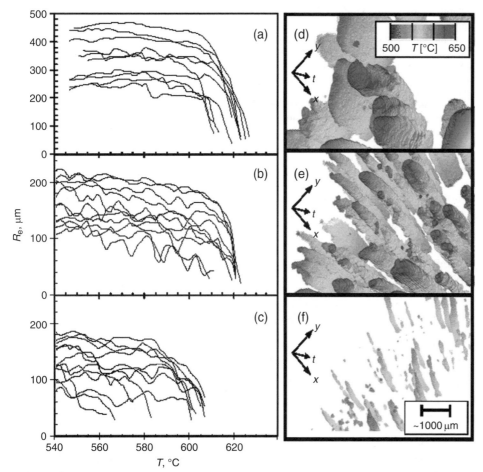

Fig. 5 Experimentally measured pore growth rates as a function of temperature for three different local solidification times: (a,d) 100 s, (b,e) 50 s, and (c,f) 28 s. In (a–c), the equivalent radius is plotted, while in (d) and (e), the growth of the pores with time is rendered in three dimensions, showing how they nucleate as spheres (rounded in two-dimensional cross section) and become irregular in shape as they are restricted by the solid around them. Adapted from Ref 66

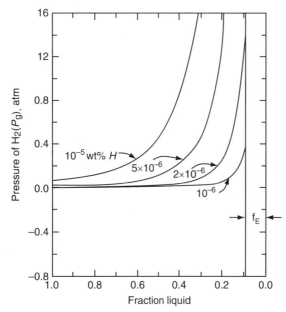

Fig. 6 Predicted hydrogen gas pressure in the interdendritic liquid for four initial hydrogen concentrations. Adapted from Ref 18

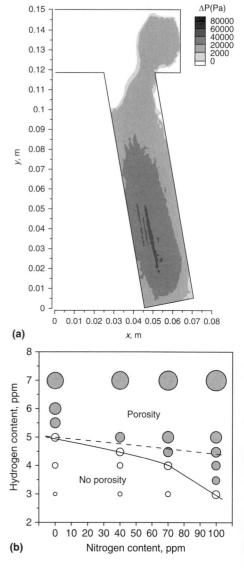

Fig. 7 (a) Predicted difference between gas and local pressure in a steel casting, with the greatest difference indicating the highest chance of microporosity formation. (b) For the same casting, a process map shows how increasing hydrogen or nitrogen content increases the chance of pore formation (filled circles). Adapted from Ref 19, 21

pressure drop in the mushy zone at high solid fractions, as shown in Fig. 8. When no pores form (solid line in Fig. 8), the pressure drops quickly as the fraction liquid approaches zero. However, when there is significant hydrogen present (>0.2 mL/100 g STP), the pressure only drops to approximately 0.2 atm. The authors explain that microporosity partially compensates for the solidification shrinkage, reducing the feeding demand and leading to a lower pressure drop. Very few other models incorporate the feedback effect of microporosity on the fluid flow, despite this work demonstrating its need when developing an accurate model of porosity formation.

Zhu, Cockcroft, and Maijer (Ref 28, 29) have also developed a continuum heat-transfer and

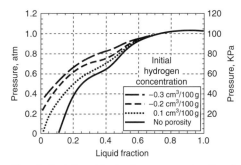

Fig. 8 Evolution of the local pressure as a function of fraction liquid for different initial hydrogen contents using the model of Sabau and Viswanathan. Adapted from Ref 67

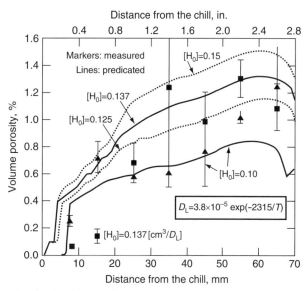

Fig. 9 Comparison of predicted and measured microporosity in A356 castings using the model of Zhu et al. Adapted from Ref 29

Darcy flow model. They first solve the thermal problem in a commercial code and then use the same mesh to solve the pressure drop and hydrogen partitioning in the mushy zone, using their own code. Using this framework, they tested the influence of the nucleation supersaturation on the percentage porosity, concluding that the final amount of porosity was independent of the nucleation supersaturation. However, altering the nucleation supersaturation alters the average distance between pores. In their first publication (Ref 28), they used a criterion-based correlation between local pressure and pore formation, which they improved upon in their later publication (Ref 29), where they show an excellent correlation between predicted and measured percentage porosity for directionally solidified plates of A356 alloy (Fig. 9).

Most of the models reviewed in the previous three sections are based on the assumption that the main driving force for pore formation is the inadequate feeding of volumetric shrinkage, inducing a local pressure drop. Although some of the studies reviewed solved for gas partitioning and hence incorporate the effect of initial gas content on the predicted percentage porosity, only a few used a microstructural feature criterion to relate the local pressure and hence gas supersaturation to pore size. The next section examines those models that have concentrated on gas evolution and those that try to combine multiple driving forces via direct simulation at a microstructural level.

Kinetic Models

In the previous sections, most models were formulated with the premise that shrinkage is the main driving force for microporosity; however, there is a separate class of models based on the premise that the diffusion-limited transport of gas is equally important. Because the diffusivity of the soluble gas atom (e.g., hydrogen in aluminum) is very dependent on the state of the alloy (i.e., solid or liquid), these gas diffusion models are frequently combined with grain nucleation and growth models. By combining microstructure, gas, and shrinkage, such

models predict not only the percentage porosity but also the pore size distribution and morphology (Ref 25, 66, 68–70), which can be critical for fatigue predictions (Ref 3, 71).

Gas-Controlled Growth Models. Fang and Granger (Ref 72, 73) developed one of the earliest pore growth models based on gas concentration. They separated pore growth into three stages:

1. Between the liquidus and eutectic temperature, the pore was assumed to grow spherically, based on the amount of excess hydrogen rejected as solidification progresses, assuming isolated spherical pockets of alloy.
2. When the eutectic temperature is reached, the pores are allowed to grow isothermally, assuming partitioned hydrogen contributes to pore growth.
3. After a set eutectic fraction has evolved, it is assumed that feeding is completely cut off, and the pore volume is increased by the fraction of shrinkage of the residual eutectic phase.

This methodology gave a good correlation in final average pore size to that measured experimentally in A356 but was very dependent on the ad hoc size of isolated spherical pockets selected, which effectively controlled the size. Physically, each pocket represents a region in the melt where the hydrogen diffusion fields interact and, in effect, acts as a fitting parameter.

Stochastic Nucleation and Diffusion-Controlled Growth Models. Based on in situ experimental observations of the kinetic pore nucleation and growth in aluminum-copper alloys using an x-ray temperature gradient stage, Lee and Hunt (Ref 47) concluded that,

for a small mushy zone, microporosity formation is controlled primarily by gas diffusion. Accordingly, they developed the first stochastic model to simulate the porosity formation in directional solidification of aluminum-copper alloys, which includes the nucleation kinetics by stochastic functions and incorporates growth using a finite-difference solution of gas diffusion in two dimensions (Ref 66), calculating the hydrogen concentration in the liquid, C_l, from Fick's second law:

$$\frac{\partial}{\partial t}[C_l(\rho_l f_l + k_{PH}\rho_s f_s)] = \nabla \cdot (\rho_l D_e \nabla C_l) + Q_H$$

(Eq 13)

where k_{PH} is the hydrogen partition coefficient, Q_H is a source term representing the generation or consumption of hydrogen by the pores from the metal, and D_e is the effective diffusion coefficient of hydrogen in the mushy zone. It was the first model to implement a stochastic pore nucleation model via assigning each nucleus with a potential (or activity), based on experimentally measured values.

Based on experimental observations during columnar dendritic growth, the pore morphology was simulated as spherical until impinging on the solid, when they became elongated in the solidification direction, growing as hemispherically capped segmented cones, as shown schematically in Fig. 10(a). Although this model was the first to predict a pore size distribution and showed good correlation to experimentally measured values (Ref 66), the growth morphology could not be applied to the more commercially important case of equiaxed-dendritic grains. A later work by Atwood et al. (Ref 74) developed a very computationally efficient one-dimensional (in spherical coordinates) diffusion-limited growth model for pores in

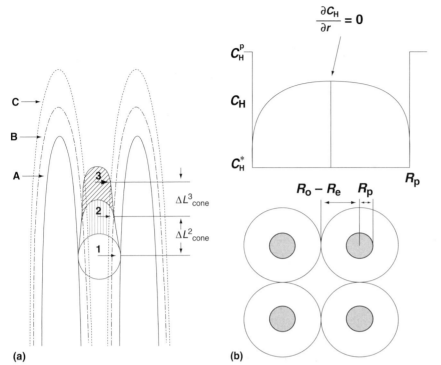

Fig. 10 Schematic of the pore growth morphologies implemented by (a) Lee and Hunt (Ref 66) during columnar dendritic solidification and (b) Atwood et al. (Ref 74) for equiaxed solidification

equiaxed grains, assuming the impingement and domain shown in Fig. 10(b), but this model suffered the same problem as Fang and Granger's (Ref 72, 73): It required the ad hoc selection of a liquid pocket size.

Huang et al. (Ref 75) also allowed a random distribution of the location of nuclei for both pores and grains. They modeled the nucleation and growth of both grains and pores during the solidification of A356 alloys using a two-dimensional cellular automata (CA) model. Their mechanism for grain nucleation and growth was similar to that used in the CA models published by Rappaz and Gandin (Ref 76, 77) but was extended by allowing some cells to become gaseous in addition to liquid and solid. The formation of the gaseous cells (pores) was based on the evolution of hydrogen gas as it was partitioned and rejected from the newly formed solid phase. The volume of porosity was then calculated using the ideal gas law and across the entire domain, and this additional volume was randomly added to the existing pores. The diffusion of hydrogen was not considered.

Following the methodology of combining microstructural predictions with pore evolution, Atwood and Lee (Ref 69) simulated the combined diffusion-controlled growth of both the primary phase and hydrogen porosity in an aluminum-silicon alloy using a CA method. This model extended the finite-difference solution of hydrogen diffusion-controlled growth from Lee and Hunt (Ref 66) to simulate the diffusion-controlled growth of the primary phase, as well as the interaction between phases, all

in three dimensions. The model did not directly predict shrinkage but instead used a metallostatic pressure input as a function of time. Both the α-aluminum grains and pores were randomly nucleated, with the potentials for the pores obtained from the experiments of Lee and Hunt (Ref 47). This allowed predictions of not only the percentage porosity and average pore size but also the distribution in pore sizes. The model showed a good qualitative correlation to experimental observations in an Al-7wt %Si alloy, but quantitative correlations were dependent on the ad hoc entry of local metallostatic pressure as a function of time (or temperature). The model has also been applied to simulate iron-rich intermetallic formation and investigate its influence on pore nucleation and growth (Ref 31) as well as multicomponent (Al-Si-Cu) effects on pore morphology (Ref 68), as shown in Fig. 11.

Multiscale Models. Many separate shrinkage and gas-dominated models have been developed and have shown that, in most cases, both factors control the pore size distribution and percentage porosity. Experimentally, it has been shown that both factors must be considered (e.g., Ref 8). However, hydrogen diffusion occurs at a relatively small scale (i.e., hundreds of micrometers), while the mushy zone over which shrinkage is being fed can be very large (tens to hundreds of millimeters). To solve across these two scales, several authors have implemented multiscale models, where shrinkage is solved using a macroscopic computational fluid dynamics (CFD) and heat-transfer code, while hydrogen diffusion is solved on a

local scale, either using analytic or one-dimensional, spherical coordinate models (Ref 27, 28, 78) or microstructurally explicit models (Ref 14, 31, 70).

In 2002, Lee et al. (Ref 70) published the first multiscale, microstructure-explicit model of microporosity formation during solidification, although it was applied to direct chill casting rather than shape casting. This model coupled the thermal and pressure profiles from a macroscopic CFD code (EKK, MI) into the local diffusion model of Atwood and Lee (Ref 30). The results showed that for aluminum-magnesium alloys, it was critical to incorporate the growth restriction of the pores impinging on the solid. However, the local hydrogen diffusion and microstructure model was so computationally costly, it could only be run at a few locations within the casting.

Later, Lee et al. (Ref 14) further developed this model, successfully applying it to a complex W319 casting, predicting the maximum pore size and distribution in an engine block cast via the Cosworth process, and producing a good correlation between predictions and experimental measurements (Fig. 12). However, this was done by running the micromodel independently from the macromodel over a wide range of parameters and regression fitting a model-based constitutive equation for maximum pore size prediction, which was a function of thermal, pressure, and alloy properties. This methodology had very little computational cost, but the constitutive equation cannot be used to extrapolate beyond the parameter space in which it was derived. Maijer et al. (Ref 7) illustrated the potential of coupling such a model-based pore size constitutive equation into a through-process model to predict final in-service fatigue life based on pores formed during the solidification of an automotive wheel casting.

In 2000, Hamilton et al. (Ref 78) developed a one-dimensional spherical coordinate, microstructural-level model of the local diffusion around a pore and coupled it into a macroscopic CFD code (CAPFLOW, EKK, MI), producing a multiscale model that could predict pore formation everywhere in a large casting. Unfortunately, this model had restrictions similar to Fang and Granger's (Ref 73) and required the number density of pores to be input. Further, the coupling was limited to temperature and did not include pressure, a major limitation.

Carlson et al. (Ref 27) developed an approximate one-dimensional spherical solution of the diffusion of hydrogen and implemented this using a volume-averaged technique within a macroscopic code. This allows solution of the local hydrogen diffusion around an average pore to be tracked (and this technique has been shown to be extensible to tracking bins of sizes) at a microstructural scale with great computational efficiency within a full macroscopic code for solving heat, mass, and momentum transfer. The implementation had some restrictions requiring fitting to experiment, the main ones

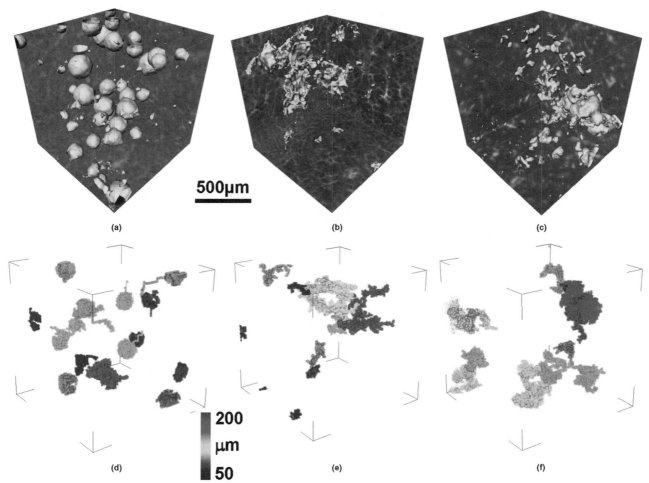

Fig. 11 Comparisons of multiscale simulations of pore morphology with three wedge casting experiments. (a), (b), and (c) are x-ray tomography images of pores in Al-4Cu, Al-7Si, and Al-7.5Si-3.5Cu, respectively. (d), (e), and (f) are simulated pores in these three alloys. Source: Ref 68

being that the number (and size, although this was shown not to be sensitive) of active nuclei must be specified, and the model does not directly simulate the pore-dendrite interaction; hence, highly tortuous pores are not well

described. However, this technique has great promise in providing computationally efficient coupling across the scales, demonstrating that even more accurate models will be developed in the future.

Conclusions

The modeling of porosity formation during the solidification of aluminum alloys has changed tremendously over the past five decades, ranging from analytic solutions to highly complex simulations of evolving the kinetics of porosity and microstructure with stochastic nucleation and growth.

Each of the four types of models reviewed has limitations, such as:

- Analytic solutions are applicable only to directional solidification.
- Criteria functions cannot be extrapolated to new alloys or processes.
- Thermal/fluid flow models only show a good correlation to experiment for percentage porosity, not pore morphology.
- Continuum kinetic models (which predict the distribution of porosity and maximum pore size) are computationally very expensive.

An ideal model would correct these limitations; however, because of the vast number of material properties and boundary conditions required, an "ideal" model may be too complex to be industrially viable. Several techniques are

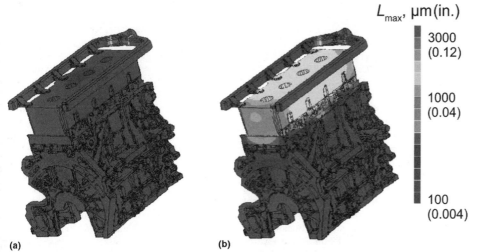

Fig. 12 Application of multiscale model to an industrial automotive components application, predicting the percentage porosity and maximum pore length in a V2.3L engine block cast via the Cosworth process. (a) $C_H = 0.1$ mL/100 g. (b) $C_H = 0.2$ mL/100 g. Source: Ref 14

being developed to span the scales from microstructure to macroscopic heat transfer and fluid flow, and the next generation of models will most likely use these techniques. Coupled with parallelized solvers, the limitations in computational speed may well be overcome.

Together with the development of new models, a greater understanding of the physical processes that govern the formation of porosity and its interaction with the developing microstructure is required. This insight will only be obtained by parallel development of careful experimental investigations and simulations.

REFERENCES

1. M.C. Flemings, *Solidification Processing*, McGraw-Hill, Inc., New York, 1974
2. J. Campbell, *Castings*, 2nd ed., Butterworth-Heinemann, London, 2003, p 337
3. Y.X. Gao, J.Z. Yi, P.D. Lee, and T.C. Lindley, A Micro-Cell Model of the Effect of Microstructure and Defects on Fatigue Resistance in Cast Aluminum Alloys, *Acta Mater.*, Vol 52 (No. 19), 2004, p 5435–5449
4. J.F. Major, Porosity Control and Fatigue Behavior in A356-T61 Aluminum Alloy, *AFS Trans.*, Vol 105, 1998, p 901–906
5. K. Tynelius, J.F. Major, and D. Apelian, A Parametric Study of Microporosity in the A356 Casting Alloy System, *Trans. Am. Foundrymen's Soc.*, Vol 101, 1994, p 401–413
6. M.J. Couper, A.E. Neeson, and J.R. Griffiths, Casting Defects and the Fatigue Behaviour of an Aluminium Casting Alloy, *Fatigue Fract. Eng. Mater. Struct.*, Vol 13 (No. 3), 1990, p 213–227
7. D.M. Maijer, Y.X. Gao, P.D. Lee, T.C. Lindley, and T. Fukui, A Through-Process Model of an A356 Brake Caliper for Fatigue Life Prediction, *Metall. Mater. Trans. A*, Vol 35 (No. 10), 2004, p 3275–3288
8. E.J. Whittenberger and F.N. Rhines, Origin of Porosity in Castings of Magnesium Aluminum and Other Alloys, *J. Met.*, 1952, p 409–420
9. E. Niyama, T. Uchida, M. Morikawa, and S. Saito, A Method of Shrinkage Prediction and Its Application to Steel Casting Practice, *AFS Int. Cast Met. J.*, Vol 9, 1982, p 52–63
10. J.A. Spittle, M. Almeshhedani, and S.G.R. Brown, The Niyama Function and Its Proposed Application to Microporosity Prediction, *Cast Met.*, Vol 7 (No. 1), 1994, p 51–56
11. W.S. Pellini, Factors Which Determine Riser Adequacy and Feeding Range, *AFS Trans.*, Vol 61 (No. 67), 1953, p 61–80
12. W.D. Walther, C.M. Adams, and H.F. Taylor, Mechanism for Pore Formation in Solidifying Metals, *AFS Trans.*, Vol 64, 1956, p 658–664

13. K. Kubo and R.D. Pehlke, Mathematical Modeling of Porosity Formation in Solidification, *Metall. Trans. B*, Vol 16, 1985, p 359–366
14. P.D. Lee, A. Chirazi, R.C. Atwood, and W. Wang, Multiscale Modelling of Solidification Microstructures, Including Microsegregation and Microporosity, in an Al-Si-Cu Alloy, *Mater. Sci. Eng. A*, Vol 365 (No. 1–2), 2004, p 57–65
15. T.S. Piwonka and M.C. Flemings, Pore Formation in Solidification, *AIME Metall. Soc. Trans.*, Vol 236 (No. 8), 1966, p 1157–1165
16. H. Darcy, *Les Fontaines Publiques de la Ville de Dijon*, Delmont, Paris, 1856
17. K. Kubo and R.D. Pehlke, Porosity Formation in Solidifying Castings, *Trans. Am. Foundrymen's Soc.*, Vol 94, 1986, p 753–756
18. D.R. Poirier, K. Yeum, and A.L. Maples, A Thermodynamic Prediction for Microporosity Formation in Aluminum-Rich Al-Cu Alloys, *Metall. Trans. A*, Vol 18 (No. 11), 1987, p 1979–1987
19. S.D. Felicelli, D.R. Poirier, and P.K. Sung, A Model for Prediction of Pressure and Redistribution of Gas-Forming Elements in Multicomponent Casting Alloy, *Metall. Mater. Trans. B*, Vol 31 (No. 6), 2000, p 1283–1292
20. D.R. Poirier, *A Continuum Model of Microporosity in an Aluminum Casting Alloy*, Vol 1, 2001
21. P.K. Sung, D.R. Poirier, and S.D. Felicelli, Continuum Model for Predicting Microporosity in Steel Casting, *Modell. Sim. Mater. Sci. Eng.*, Vol 5, 2002, p 551
22. D.M. Stefanescu, Computer Simulation of Shrinkage Related Defects in Metal Castings—A Review, *Int. J. Cast. Met. Res.*, Vol 18, 2005, p 129–143
23. D.R. Poirier, Phenomena Related to the Formation of Microporosity in Castings, *Modeling of Casting, Welding and Advanced Solidification Processes VIII*, TMS, 1998
24. K.D. Carlson, Z.P. Lin, C. Beckermann, G. Mazurkevich, and M. Schneider, Modeling of Porosity Formation in Aluminum Alloys, *Aluminium Alloys 2006*, Pt 1 and 2, 2006, p 1699–1706
25. P.D. Lee and J.D. Hunt, A Model of the Interaction of Porosity and the Developing Microstructure, *Conference Modeling of Casting*, Welding and Advanced Solidification Processes VII, Sept 10–15, 1995 (London, U.K.), Minerals, Metals and Materials Society/AIME, Warrendale, PA, 1995
26. J. Allison, M. Li, C. Wolverton, and X. Su, Virtual Aluminum Castings: An Industrial Application of ICME, *JOM*, 2006, p 28–35
27. K. Carlson, Z. Lin, and C. Beckermann, Modeling the Effect of Finite-Rate Hydrogen Diffusion on Porosity Formation in Aluminum Alloys, *Metall. Mater. Trans. B*, Vol 38 (No. 4), 2007, p 541–555

28. J.D. Zhu, S.L. Cockcroft, D.M. Maijer, and R. Ding, Simulation of Microporosity in A356 Aluminium Alloy Castings, *Int. J. Cast Met. Res.*, Vol 18, 2005, p 229–235
29. J. Zhu, S. Cockcroft, and D. Maijer, Modeling of Microporosity Formation in A356 Aluminum Alloy Casting, *Metall. Mater. Trans. A*, Vol 37 (No. 12), 2006, p 1075–1085
30. R.C. Atwood and P.D. Lee, A Three-Phase Model of Hydrogen Pore Formation during the Equiaxed Dendritic Solidification of Aluminum-Silicon Alloys, *Metall. Mater. Trans. B*, Vol 33 (No. 2), 2002, p 209–221
31. J. Wang, M. Li, J. Allison, and P.D. Lee, Multiscale Modeling of the Influence of Fe Content in a Al-Si-Cu Alloy on the Size Distribution of Intermetallic Phases and Micropores, *J. Appl. Phys.*, Vol 107 (No. 6), 2010, p 061804-1
32. P.D. Lee, J. Wang, and R.C. Atwood, Microporosity Formation during the Solidification of Aluminum-Copper Alloys, *JOM-e*: Visualization: Defects in Casting Processes (USA), 2006, www.tms.org/pubs/journals/JOM/0612/Lee/Lee-0612.html
33. J. Dantzig and M. Rappaz, *Solidification: Methods, Microstructure and Modeling*, EFPL Press, Lausanne, Switzerland, 2009
34. R.B. Bird, W.E. Stewart, and E.N. Lightfoot, *Transport Phenomena*, John Wiley & Sons, 1960
35. J. Szekely and N.J. Thermelis, *Rate Phenomena in Process Metallurgy*, John Wiley & Sons Inc., New York, 1971
36. G.H. Geiger and D.R. Poirier, and M. Cohen, *Transport Phenomena in Metallurgy*, Addison-Wesley Series in Metallurgy and Materials, Addison-Wesley Publishing Company, Reading, MA, 1973
37. T. Young, An Essay on the Cohesion of Fluids, *Philos. Trans.*, 1805, p 65
38. C.E. Ransley and H. Neufeld, The Solubility of Hydrogen in Liquid and Solid Aluminium, *J. Inst. Met.*, Vol 74, 1948, p 599–620
39. D. Doutre, Internal report, Alcan Int. Ltd., 1991
40. S.F. Jones, G.M. Evans, and K.P. Galvin, Bubble Nucleation from Gas Cavities—A Review, *Adv. Colloid Interface Sci.*, Vol 80 (No. 1), 1999, p 27–50
41. S. Fox and J. Campbell, Visualisation of Oxide Film Defects during Solidification of Aluminium Alloys, *Scr. Mater.*, Vol 43 (No. 10), 2000, p 881–886
42. J. Campbell, An Overview of the Effects of Bifilms on the Structure and Properties of Cast Alloys, *Metall. Mater. Trans. B*, Vol 37, 2006, p 857–863
43. A.L. Greer, Grain Refinement of Alloys by Inoculation of Melts, *Philos. Trans. R. Soc. A*, Vol 361 (No. 1804), 2003, p 479–495
44. A.L. Greer, A.M. Bunn, A. Tronche, P.V. Evans, and D.J. Bristow, Modelling of Inoculation of Metallic Melts: Application

to Grain Refinement of Aluminium by Al-Ti-B, *Acta Mater.*, Vol 48 (No. 11), 2000, p 2823–2835

45. A. Fick, *Philos. Mag.*, Vol 10, 1855, p 30
46. D.R. Poirier, P.J. Nandapurkar, and S. Ganesan, The Energy and Solute Conservation Equations for Dendritic Solidification, *Metall. Mater. Trans. B*, Vol 22 (No. 6), 1991, p 889–900
47. P.D. Lee and J.D. Hunt, Hydrogen Porosity in Directional Solidified Aluminium-Copper Alloys: In Situ Observation, *Acta Mater.*, Vol 45 (No. 10), 1997, p 4155–4169
48. M.S. Bhat, D.R. Poirier, and J.C. Heinrich, Permeability for Cross Flow through Columnar-Dendritic Alloys, *Metall. Mater. Trans. B*, Vol 26 (No. 5), 1995, p 1049–1056
49. W. Eichenauer and J. Markopoulos, Messung des Diffusionskoeffizienten von Wasserstoff in Flussigem Aluminium, *Metallkde*, Vol 65 (No. 10), 1974, p 649–652
50. P.D. Lee, A. Chirazi, and D. See, Modeling Microporosity in Aluminum-Silicon Alloys: A Review, *J. Light Met.*, Vol 1 (No. 1), 2000, p 15–30
51. W.H. Johnson and J.K. Kura, Some Principles for Producing Sound Al-7Mg Alloy Casting, *AFS Trans.*, Vol 67, 1959
52. G.V. Kutmba Rao and V. Panchanathan, Chill Action on LM-4 Alloy Castings, *Aluminium*, Vol 49 (No. 7), 1973
53. V.d.L. Davies, Feeding Range Determined by Numerically Computed Heat Distribution, *AFS Cast Met. Res. J.*, Vol 11, 1975, p 33
54. Y.W. Lee, E. Chang, and C.F. Chieu, Modeling of Feeding Behavior of Solidifying Al-Si-0.3Mg Alloy Plate Casting, *Met. Trans. B*, Vol 21, 1990, p 715–722
55. V.K. Suri, A.J. Paul, N. El-Kaddah, and J.T. Berry, Determination of Correlation Factors for Prediction of Shrinkage in Castings, Part I: Prediction of Microporosity in Castings; A Generalized Criterion, *Conference Ninety-Eighth Annual Meeting of the American Foundrymen's Society*, May 1–4, 1994 (Hamilton, Ontario, Canada), American Foundrymen's Society, Inc., Des Plaines, IL, 1994
56. S.T. Kao and E. Chang, The Role of the Pressure Index in Porosity Formation in A356 Alloy Castings, *Cast Met.*, Vol 7, 1995, p 219
57. K.-D. Li, M.-C. Cheng, and E. Chang, Effect of Pressure on the Feeding Behavior

of A356 Alloy in Low-Pressure Casting, *AFS Trans.*, Vol 01-026, 2001, p 1–9
58. L.H. Shang, F. Paray, J. Gruzleski, S. Bergerson, C. Mercadante, and C.A. Loong, Prediction of Microporosity in Al-Si Castings in Low Pressure Permanent Mould Casting Using Criteria Functions, *Int. J. Cast. Met. Res.*, Vol 17 (No. 4), 2004, p 193–200
59. R.P. Taylor, J.T. Berry, and R.A. Overfelt, Parallel Derivation and Comparison of Feeding-Resistance Porosity Criteria Functions for Castings, *The 1996 31st ASME National Heat Transfer Conference* (Part 1 of 8), Aug 3–6, 1996 (Houston, TX), ASME, New York, NY, 1996
60. S. Viswanathan, V.K. Sikka, and H.D. Brody, The Application of Quality Criteria for the Prediction of Porosity in the Design of Casting Processes, *Modell. Cast. Weld. Adv. Solid. Proc. VI*, 1993
61. K.D. Carlson, S.Z. Ou, and C. Beckermann, Feeding of High-Nickel Alloy Castings, *Metall. Mater. Trans. B*, Vol 36 (No. 6), 2005, p 843–856
62. P.K. Sung, D.R. Poirier, S.D. Felicelli, E.J. Poirier, and A. Ahmed, Simulations of Microporosity in IN718 Equiaxed Investment Casting, *J. Cryst. Growth*, Vol 226 (No. 2–3), 2001, p 363–377
63. G. Couturier, Effect of Volatile Elements on Porosity Formation in Solidifying Alloys, *Modell. Sim. Mater. Sci. Eng.*, Vol 14 (No. 2), 2006, p 253–271
64. P. Rousset, M. Rappaz, and B. Hannart, Modeling of Inverse Segregation and Porosity Formation in Directionally Solidified Aluminum Alloys, *Metall. Mater. Trans. A*, Vol 26 (No. 9), 1995, p 2349–2358
65. C.C. Pequet, M. Gremaud, and M. Rappaz, Modeling of Microporosity, Macroporosity, and Pipe-Shrinkage Formation during the Solidification of Alloys Using a Mushy-Zone Refinement Method: Applications to Aluminum Alloys, *Metall. Mater. Trans. A*, Vol 7, 2002, p 2095
66. P.D. Lee and J.D. Hunt, Hydrogen Porosity in Directionally Solidified Aluminium-Copper Alloys: A Mathematical Model, *Acta Mater.*, Vol 49 (No. 8), 2001, p 1383–1398
67. A.S. Sabau and S. Viswanathan, Microporosity Prediction in Aluminum Alloy Castings, *Metall. Mater. Trans. B.*, Vol 33 (No. 2), 2002, p 243–255

68. J.S. Wang and P.D. Lee, Simulating Tortuous 3D Morphology of Microporosity Formed during Solidification of AlSiCu Alloys, *Int. J. Cast. Met. Res.*, Vol 20, 2007, p 151–158
69. R.C. Atwood and P.D. Lee, Simulation of the Three-Dimensional Morphology of Solidification Porosity in an Aluminium-Silicon Alloy, *Acta Mater.*, Vol 51 (No. 18), 2003, p 5447–5466
70. P.D. Lee, R.C. Atwood, R.J. Dashwood, and H. Nagaumi, Modeling of Porosity Formation in Direct Chill Cast Aluminum-Magnesium Alloys, *Mater. Sci. Eng. A*, Vol 328 (No. 1–2), 2002, p 213–222
71. J.Z. Yi, Y.X. Gao, P.D. Lee, H.M. Flower, and T.C. Lindley, Scatter in Fatigue Life due to Effects of Porosity in Cast A356-T6 Aluminum-Silicon Alloys, *Metall. Mater. Trans. A*, Vol 34 (No. 9), 2003, p 1879–1890
72. Q.T. Fang and D.A. Granger, Porosity Formation in Modified and Unmodified A356 Alloy Castings, *AFS Trans.*, Vol 209, 1989, p 927–935
73. Q.T. Fang and D.A. Granger, Prediction of Pore Size due to Rejection of Hydrogen during Solidification of Aluminum Alloys, *Light Met.*, Feb 27–March 3, 1989 (Las Vegas, NV), TMS/AIME, Warrendale, PA, 1989
74. R.C. Atwood, S. Sridhar, W. Zhang, and P.D. Lee, Diffusion-Controlled Growth of Hydrogen Pores in Aluminium-Silicon Castings: In Situ Observation and Modelling, *Acta Mater.*, Vol 48 (No. 2), 2000, p 405–417
75. J. Huang, J.G. Conley, and T. Mori, Simulation of Microporosity Formation in Modified and Unmodified A356 Alloy Casting, *Metall. Trans. B*, Vol 29 (No. 6), 1998, p 1249–1260
76. Ch.-A. Gandin, C. Charbon, and M. Rappaz, Probabilistic Modeling of Grain Information in Solidification Processes, *Conf. Model. Cast. Weld. Adv. Solid. Proc. VI*, March 21–26, 1993 (Palm Coast, FL), TMS, 1993
77. Ch.-A. Gandin and M. Rappaz, A 3D Cellular Automaton Algorithm for the Prediction of Dendritic Grain Growth, *Acta Mater.*, Vol 45 (No. 5), 1997, p 2187–2195
78. R.W. Hamilton, D. See, S. Butler, and P.D. Lee, Multiscale Modelling for the Prediction of Casting Defects in Investment Cast Aluminium Alloys, *IBF Castcon 2000*, June 30, 2000, Stratford, IBF

Simulation of Metal Forming Processes

ASM Handbook, Volume 22B, *Metals Process Simulation*
D.U. Furrer and S.L. Semiatin, editors

Copyright © 2010, ASM International®
All rights reserved.
www.asminternational.org

Finite Element Method Applications in Bulk Forming*

Soo-Ik Oh, John Walters, and Wei-Tsu Wu

METALWORKING, with its thousands of years of history, is one of the oldest and most important materials processing technologies. During the last 30 years, with the continuous improvement of computing technology and the finite element method (FEM) as well as the competition for a lower-cost and better-quality product, metalworking has evolved rapidly. This article gives a summary of overall development of the FEM and its contribution to the materials forming industry. Because significant efforts were carried out with great success by many universities and research institutes with a similar objective and application, this article is focused on the overall philosophy and evolution of the FEM for solving bulk forming issues. The program used to demonstrate this success is the commercial code named DEFORM (Scientific Forming Technologies Corp.). A number of examples of the application of FEM to various bulk forming processes are also summarized.

This article provides an overview of FEM applications. In this section, a number of applications of FEM are presented in the order they would be used in a typical manufacturing process sequence: primary materials processing, hot forging and cold forming, and product assembly. Material fracture and die stress analysis are covered, and optimization of the design of forming processes is also reviewed.

Historical Overview

Lee and Kobayashi first introduced the rigid-plastic formulation in the 1970s (Ref 1). This formulation neglects the elastic response of deformation calculations. In the late 1970s and early 1980s, a processing science program (Ref 2) funded by the United States Air Force was performed at the Battelle Memorial Institute Columbus Laboratories to develop a process model for the forging of dual-property titanium engine disks. These disks are required to have excellent creep and high stress-rupture properties in the rim and high fatigue strength in the bore region. A FEM-based code, ALPID (Ref 3), was developed under this program. Thermo-viscoplastic FEM analyses (Ref 4) were also performed to investigate the temperature variation during hot-die disk-forging processes. The flow stress of thermo-rigid-viscoplastic material is a function of temperature, strain, and strain rate. Approximately five aerospace manufacturers pioneered the use of the code. Based on the same foundation, DEFORM was developed for two-dimensional applications in 1986. Due to the large deformation in the metal-forming application, the updated Lagrangian method always suffers from mesh distortion and consequently requires many remeshings to complete one simulation. Two-dimensional metal-forming procedures became practical for industrial use when automated remeshing became available in 1990 (Ref 5). In the beginning of the 1980s, the PDP11 and the CDC/IBM mainframe computers were used. In the mid-1980s, the VAX workstation became the dominant machine for running the simulations. In the late 1980s, UNIX workstations became the primary computing facility.

Unfortunately, the majority of the metalworking processes are three-dimensional (3-D), where a two-dimensional (2-D) approach cannot approximate reality satisfactorily. The initial 3-D code development began in the mid-1980s (Ref 6). One simulation with backward extrusion in a square container was reported to take 152 central processing unit (CPU) hours on a VAX-11/750. In addition to the need for remeshing, a more complicated process was estimated to take several weeks. Due to the lack of computing speed in the 1980s for 3-D applications, the actual development was delayed until the 1990s (Ref 7). Since then, many ideas to develop a practical 3-D numerical tool were evaluated and tested. The successful ones were finally implemented. After the mid-1990s, significant computing speed improvement was seen in personal computer (PC) technology, coupled with a lower price as compared to UNIX-based machines. For this reason, the PC has become the dominant computing platform. Due to the competition for better product quality at a lower production cost, process modeling gradually became a necessity rather than a research and development tool in the production environment.

Although FEM programs were initially developed for metalworking processes, it was soon realized that metalworking is just one of the many operations before the part is finally installed. Prior to forging, the billet is made by primary forming processes, such as cogging or bar rolling from a cast ingot. After forging, the part is heat treated, rough machined, and finish machined. The microstructure of the part continuously evolves together with the shape. The residual stress within the part and the associated distortion are also changing with time. To really understand product behavior during the service, it is essential to connect all the missing links, not only the metalworking. In the mid-1990s, a small business innovative research program was awarded by the U.S. Air Force and the U.S. Navy to develop a capability for heat treatment and machining (Ref 8). To track the residual-stress distribution, elastoplastic and elastoviscoplastic formulations were used. Microstructural evolution, including phase transformation and grain-size evolution, was implemented. Distortion during heat treatment and material removal during machining processes can thus be predicted.

During the 1990s, most efforts were focused on the development of the FEM for computer-aided engineering applications. However, the engineer's experience still plays a major role in achieving a solution to either solving a production problem or reaching a better process design. The FEM solution-convergence speed depends highly on the engineer's experience, and the interpretation of the results requires complete understanding of the process. As the computing power continues to improve, optimization using systematic search becomes more and more attractive (Ref 9).

In the following sections, a brief overview of the methodologies and some selected representative applications focusing on the bulk forming process are given.

*Reprinted from *ASM Handbook*, Volume 14A, *Metalworking: Bulk Forming*, p 617–639

Methodologies

To account for the complicated thermal-mechanical responses to the manufacturing process, four FEM modules, as shown in Fig. 1, are loosely coupled. They are the deformation model, the heat-transfer model, the microstructural model, and, in the case of steel, a carbon diffusion model.

Thermalmechanical Models

Deformation Model. For metalworking applications, the formulation must take into account the large plastic deformation, incompressibility, material-tool contact, and (when necessary) temperature coupling. To avoid deformation locking under material incompressibility, the penalty method and selective integration method are usually used for the 2-D quadrilateral element and 3-D brick element, while the mixed formulation for the 3-D tetrahedral element is employed. It is generally agreed that the quadrilateral element and brick element are preferred in FEM applications. Due to the difficulty in both remeshing and (frequently) the initial meshing with a brick mesh in most forming applications, a tetrahedral mesh is generally used.

Due to its simplicity and fast convergence, the rigid-plastic and rigid-viscoplastic formulations are used primarily for processes when residual stress is negligible. The elastoplastic and elasto-viscoplastic formulations are important for calculating residual stress, such as in heat treatment and machining applications. However, it is very difficult to accurately characterize residual-stress evolution for forming at an elevated temperature, especially when there is significant microstructural changes, including phase transformation, precipitation, recrystallization, texture changes, and so on.

Because metalforming processes are transient, the updated Lagrangian method has been the primary FEM method for metalforming

applications. Using this method for certain steady-state processes such as extrusion, shape rolling, and rotary tube piercing, however, may not be computationally efficient. In these special applications, the arbitrary Lagrangian Eulerian (ALE) method recently has been used with great success.

Heat-Transfer Model. The heat-transfer model solves the energy balance equation. The three major modes of heat transfer are conduction, convection, and radiation. Conduction is the transfer of heat through a solid material or from one material to another by direct contact. Generally speaking, below 540 °C (1000 °F), convection has a much more pronounced effect than radiation. Above 1090 °C (2000 °F), however, radiation becomes the dominant mode of heat transfer, and convection can essentially be considered a second-order effect. Between these temperatures, both convection and radiation play an important role.

In order to predict the temperature evolution accurately during metalworking processes, several important thermal boundary conditions must be considered:

- Radiation heat with view factor to the surrounding environment
- Convection heat to/from the surrounding environment, including the tool contact, free air, fan cool, water or oil quench
- Friction heat between two contacting bodies. It is also noted that friction heating is the primary heat source in the friction-stir welding process.
- Deformation, latent heat, and eddy current are the primary volume heat sources. Deformation heat is important for large, localized deformation and fast processes, because the adiabatic heat will increase the local temperature quickly, and material is likely to behave differently at elevated temperatures. It plays an important role in metalworking, inertial welding, translational friction welding, and the cutting process. The latent heat comes from the phase transformation or

phase change, and eddy-current heat is generated by electromagnetic fields.

Microstructural Model. Grain size is an important microstructural feature that affects mechanical properties. For example, a fine grain size is desirable to resist crack initiation, while a larger grain size is preferred for creep resistance. To obtain optimal mechanical properties, precise control of the grain size is crucial. In order to achieve a desirable microstructural distribution, as-cast materials usually undergo multiple stages of forming, such as billet conversion and closed-die forging, and multiple heat treatment steps, such as solution heat treating and aging.

During thermomechanical processing, a dislocation substructure is developed as deformation is imposed. The stored energy can provide the driving force for various restorative processes, such as dynamic recovery or recrystallization. On the completion of recrystallization, the energy can be further reduced by grain growth, in which grain-boundary area is reduced. The kinetics of recrystallization and grain-growth processes are complex. In order to predict the grain-size distribution in finished components, a basic understanding of the evolution of microstructural evolution during complex manufacturing sequences, including the primary working processes (ingot breakdown, rolling, or extrusion), final forging, and heat treatment, must be obtained. Hence, the development of microstructural evolution models has received considerable attention in recent years. Recrystallization behavior can be classified into three broad categories: static, metadynamic, and dynamic recrystallization (Ref 10).

The description of each recrystallization mode as well as static grain growth is well documented (Ref 11). Sellars' model has been used for static and metadynamic recrystallization, and the Yamada model has been used for dynamic recrystallization. Microstructural evolution in superalloys is complicated by the precipitation of γ', γ'', and δ phases (Ref 8). However, the present phenomenological approach neglects the specific effect that such phases have on the mechanisms of microstructural evolution.

Phase transformation is also another important aspect for material modeling (Ref 12). It is not only critical to achieve desirable mechanical properties but also to better understand the residual stress and the associated distortion. Phase transformation can be classified into two categories: diffusional and martensitic. Using carbon steel as an example, the austenite-ferrite and austenite-pearlite structure transformations are governed by diffusional-type transformations. The transformation is driven by a diffusion process depending on the temperature, stress history, and carbon content and is often represented by the Johnson-Mehl equation:

$$\Phi = 1 - \exp(-bt'')$$

where Φ is the fraction transformed as a function of time, t, and b and n are material

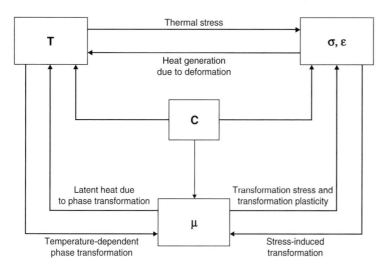

Fig. 1 View of the various coupled phenomena within metalforming

coefficients. The diffusionless transformation from austenite to martensite usually depends on temperature, stress, and carbon content.

Primary Materials Processing Applications

Cogging

Ingot conversion, also known as cogging, is one of the most common processes used to break down the coarse, cast micro-structure of superalloy ingots. As shown in Fig. 2, the ingot is held by a pair of manipulators at one of the two ends and is forged between two dies during the conversion process. The primary objective of the conversion process is to produce a fine grain structure for subsequent secondary forging operations. In essence, the process consists of multiple open-die forging (and reheating) operations in which the ingot diameter is reduced and its length is increased. Excessive furnace heating may promote undesirable grain growth. On the other hand, insufficient heating or excessive forging time may result in cracking. Control of the forging temperature, the amount of deformation, the forging time, and the precipitation of second phases is especially important for producing a desirable grain structure. Modeling the microstructural evolution of the ingot during the cogging process has been of great interest in recent years.

In the following example (Ref 13), the billet material was assumed to be nickel alloy 718 with an initial grain size of 250 μm (10 mils) (ASTM 1). The workpiece was taken to be octagonal in cross section (with a breadth of 380 mm, or 15 in., across the flat faces) and 2 m (7 ft) in length. Typical industrial processing conditions were applied. One deformation sequence comprising four passes without reheating was simulated.

Figure 3(a) shows the average grain size at the end of the fourth pass, as predicted by FEM. Predicted microstructures at approximately one-quarter of the workpiece length are shown in Fig. 3(b) to (e). After four passes, the simulation predicted that recrystallization would be rather inhomogeneous, and a number of dead zones would have developed near the

surface. These trends are consistent with industrial observations.

Rotary Tube Piercing

Tube piercing modeling (Ref 14) illustrates the use of the ALE technique. The rotary piercing of a solid bar into a seamless tube, also known as the Mannesmann process, is a very fast rolling process. In the process, the preheated billet is cross rolled between two barrel-shaped rolls at a high speed, as shown in Fig. 4. The updated Lagrangian approach was first used in the investigation. Due to the dominantly rotational velocity field, the time-step size is limited to a small value, and the whole part must be modeled for better solution accuracy. It therefore increases the computing effort. To reduce the CPU time, a new method with the Eulerian approach was developed. Geometry updating is carried out in the feeding direction, while the nodal coordinates in the hoop direction remain unchanged. With this approach and the rotational symmetry treatment,

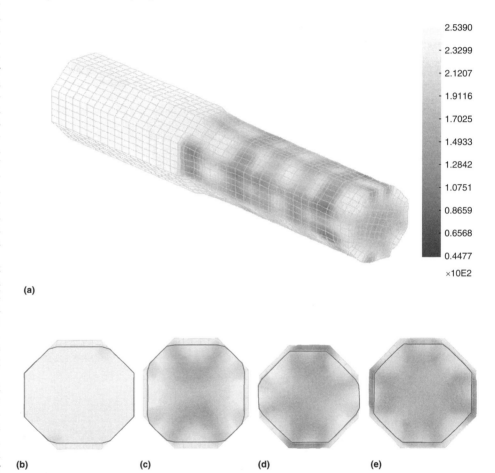

2.5390
2.3299
2.1207
1.9116
1.7025
1.4933
1.2842
1.0751
0.8659
0.6568
0.4477
×10E2

(a)

(b)　　　　(c)　　　　(d)　　　　(e)

Fig. 3 Finite element method-predicted average grain size. (a) On the free surface after the final pass or within the workpiece after pass number (b) 1, (c) 2, (d) 3, or (e) 4

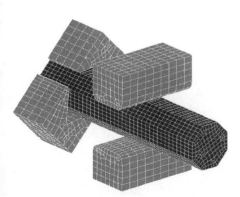

Fig. 2 Meshes used for finite element method simulation of the cogging process

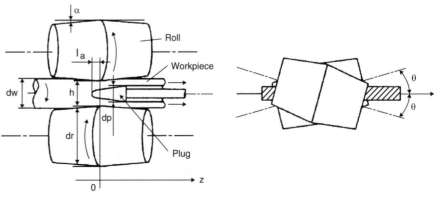

Fig. 4 Schematic illustration of rotary tube piercing process

only half of the model is simulated, due to the symmetry condition.

During the process, tensile stress is created within the workpiece near the plug tip, and fracture continuously takes place to make the hole as the solid cylinder/tube is pulled through the rollers. The relative plug position with respect to the rollers is an import process design variable that will affect the occurrence of the rear-end defect. Figures 5(a) and (b) show the backend defect of a tube from experiment and simulation, respectively.

Rolling

The following are application examples in shape rolling and FEM evaluation of roll deflection.

Tram Rail Shape Rolling. Voestalpine Schienen GmbH simulated the multipass rolling of a rail section (Ref 15). This type of rail was used for the public tramways in many European cities. There were several passes making up the processing route of this rail section, but the first few were not considered critical. The final four roll passes were simulated.

A 76 cm (30 in.) length of the rail was modeled and initially contained 60,000 elements. After numerous automatic remeshings over the course of the simulations, the mesh had increased to approximately 75,000 elements. All simulations were carried out in nonisothermal mode to allow accurate modeling of any roll chilling and deformation heating effects on the predicted material flow.

Snapshots of the final four passes are shown in Fig. 6 to 9. The actual guide vanes for maintaining rail straightness were included in the simulations as rigid bodies. Without these guide vanes, the rail section could distort quite significantly. In addition, a pusher was applied to obtain the initial feeding of the rail into the roll gap. In all of these figures, the rolls are shown as semitransparent, and the guide vanes and pusher were omitted for clarity. The final two passes included side rolls and can be seen in Fig. 8 and 9. The predicted rail geometry after all rolling operations is shown exiting the final pass rolls in Fig. 9.

The purpose of the side roll in the second-from-last pass was to form the groove in the head of the rail. This was the critical rolling pass. The material flow had to be optimized to give approximately the same pressure or load on the upper and lower faces of the side roll. If this was not achieved, cracking would result in the side roll after a very short service life. Figure 10 shows an end-on view of the groove being formed in the head of the rail.

Elastic Roll Deflection. During flat rolling operations, it is not uncommon to obtain rolled sheet or plate having greater thickness in the center as compared to the edges. This is due to the problem of roll deformation. The material being rolled exerts a reaction force on the rolls. The reaction force bends the rolls, which are supported by bearings at their ends (Fig. 11), and flattens the roll locally due to the contact pressure. The rolls are elastically deformed, and there is less plastic deformation being imparted to the workpiece, resulting in a rolled stock of greater thickness than intended. In order to compensate for the roll deformation and obtain the desired workpiece dimensions, crowned rolls, as shown in Fig. 11, are often used to reduce this effect.

An elastic roll analysis was carried out in a FEM simulation. The analysis was of the ALE type (Ref 14), with a rigid-plastic, aluminum 1100-series alloy rolling stock. The analysis accounted for thermal effects also. The roll was set to a temperature of 425 °C (800 °F), and the roll stock was set to 540 °C (1000 °F). The roll speed was 15 rpm. Half-symmetry was applied, and the rolling configuration is shown in Fig. 12. Figure 12 also shows the predicted roll deformation deflection along its length. The deflection was determined from the brick element nodal coordinates.

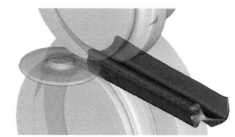

Fig. 8 Rail section exiting the second-from-last pass

Fig. 9 Rail section existing the final pass

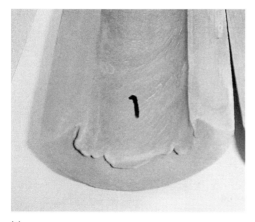

(a)

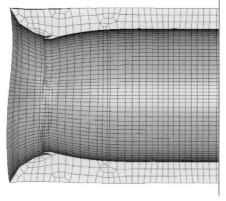

(b)

Fig. 5 Backend defect of 26.7 cm (10.5 in.) diameter billet. (a) Experimental. (b) Predicted. Source: Ref 14

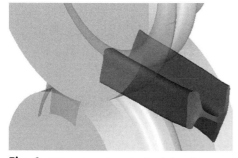

Fig. 6 Rail section existing the fourth-from-last pass

Fig. 7 Rail section exiting the third-from-last pass

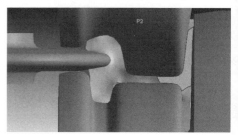

Fig. 10 End-on view of the rail section being rolled in the second-from-last pass

The predicted elastic roll stress is shown in Fig. 13. The roll is shown sectioned, having been sliced at half-length with contours of *y*-component stress. The stress at point P1 (Fig. 13) evolves and converges to a steady state, as illustrated in this figure.

Shape Drawing

While material being formed always follows the path of least resistance, that path is not always intuitive. Process simulation is a powerful tool in the prediction of material flow, especially in 3-D

processes. One such process is shape drawing. When drawing a shape, there are several potential defects. These include die underfill, bending, ductile fracture, peeling at the die entry, and necking after the die exit. To illustrate the capability, a drawing process was analyzed using three input shapes into a shaped-draw die. The process is performed at room temperature. The goal of this process is to draw a shape that matches the exit cross section of the die. The first simulation used a round input material. The result was an underfill on the outside features (Fig. 14). This did not satisfy the final shape requirements of this case. The second simulation used a larger-diameter round input. The result was unstable flow, resulting in peeling (Fig. 15). Peeling is an undesirable effect where material is scraped off the wire into long slivers before entering the input port of the draw die. Because material follows the path of least resistance, it was clear that the round input stock was less than optimal. Finally, a shaped input was simulated with good results (Fig. 16). The hex-shaped initial shape placed more material where it was required to fill the exit cross section. In a process such as drawing, it is difficult to determine whether the input shape will yield a successful output, because the material has several competing directions to flow. Several process parameters, such as friction and temperature, can affect this result. Simulation can give insight into such a process before prototyping a die set.

Hot Forging Applications

Billet-Heating Processes

Billet heating is an important process in hot forging and heat treatment. The heating time and heat rate are the typical control process parameters. Cracking can occur when an excessive heating rate is used. Long heating time wastes energy and may result in poor microstructural properties. Insufficient heating time can result in high forming load, poor material

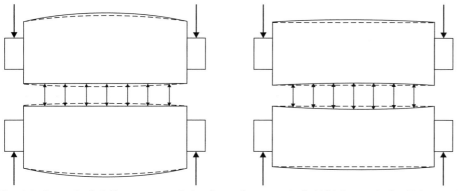

Fig. 11 Crowned rolls (left) to compensate for bending, and uncrowned rolls (right) that may lead to thickness variation in rolled stock

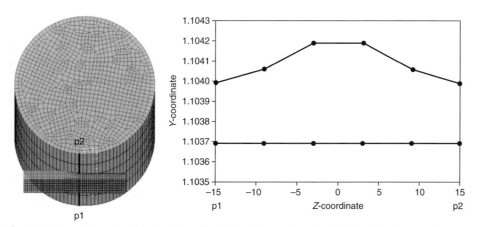

Fig. 12 Arbitrary Lagrangian Eulerian elastic roll analysis. Rolling configuration (left) shows the half-symmetry setup. Nodal coordinates of the brick meshed roll along the line p1–p2 represent roll bending deflection (right)

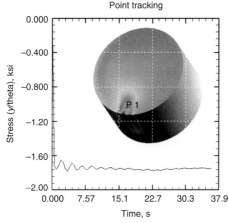

Fig. 13 Roll sectioned at its midlength showing the initial *y* stress component. After a short duration, the arbitrary Lagrangian Eulerian stress is predicted (from the point-tracking curve)

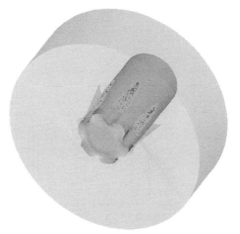
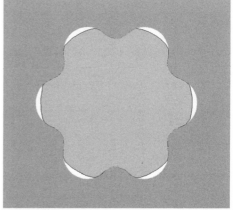

Fig. 14 Drawing the initial round input shape resulted in an underfill on the outside features. The die contact can be seen through the transparent die by dark areas representing die contact (left). A cross-sectional slice also clearly shows the underfill (right)

flow, and fracture. It is therefore important to understand the temperature evolution within the workpiece during the heating process. The most frequently used heating methods are induction heating and furnace heating.

In addition to thermal, mechanical, and microstructural models, an electromagnetic model is needed to analyze the induction heating process. The electromagnetic model is first conducted to compute the magnetic field intensity and the eddy-current density. The heat generation based on the ohmic loss is then used to compute the temperature field. Microstructural and deformation information can be computed if necessary.

This method has been successfully applied to heating titanium billets and induction hardening of steel bearings (Ref 16). A scanning induction process (Ref 17, 18) is given to illustrate the methodology. In this process, approximately 300 mm (12 in.) of a 440 mm (17 in.) long, 23 mm (0.9 in.) diameter SAE 1055 steel shaft is induction hardened by moving it through a system comprising a two-turn 20 kHz induction heating coil and a water quench ring.

The FEM model used in the simulation is shown in Fig. 17. Induction heating was applied from the start of the simulation, but there was no cooling or relative movement between the induction unit and the workpiece for the first 2 s. Subsequently, the heating/cooling assembly moved a distance of 300 mm (12 in.) upward with a constant speed of 10 mm/s (0.4 in./s). The workpiece was then allowed to cool to room temperature, specified as 20 °C (70 °F). In practice, the shaft moves, and the induction unit remains stationary; however, in the simulation, it remained fixed, and the coil and quench ring moved upward.

A power of 15 kW was assumed for heating the workpiece, operating at a frequency of 20 kHz in order to concentrate the heating at the surface. A heat-transfer cooling window, representing the quench ring, was specified a temperature of 20 °C (70 °F) and a convection coefficient of 20 kW/m^2 · K, which are representative of a water quench. The predicted temperature field is shown in Fig. 18. It is noted that the shaft surface temperature is highest (represented by the square symbol) near the coils. Downward along the shaft surface, the temperature reaches its lowest point due to the water quench. Further down, the temperature increased again due to heat conduction from the hotter interior.

Full details of the overall methodology for induction heating/hardening are contained in Ref 16.

For furnace heating, radiation heat dominates the temperature distribution of the workpiece. Radiative heat energy emitted by a body depends on the emissivity of the body surface. This emissivity depends on the material type and the surface condition, and its value ranges from 0 to 1. For example, the emissivity of a black body (absorbs all energy incident on it) is 1, while the emissivity of aluminum and carbon steel is 0.1 and 0.4, respectively. If multiple bodies are involved, the net radiant exchange between the bodies depends on the geometry and orientation of the parts as well as the relative distance between the bodies. This net radiant exchange between the multiple bodies is represented by the view factor, F_{1-2}. Modeling radiation with view factor is imperative to achieve accurate simulation results for high-temperature heating and cooling processes.

As an example of this dependence on part geometry and orientation, the heating of nine billets (15 cm diameter by 30 cm high, or 6 in. diameter by 12 in. high) in a 1095 °C (2000 °F) furnace was modeled (Fig. 19). The billets were loaded in three rows and spaced 8 cm (3 in.) apart from each other. From the predicted temperature distribution in the billets, the effect of radiation shadowing can easily be seen. The influence of shadowing can be analyzed through the use of a radiation view factor in the Stephan-Boltzmann equation. The slower heating rate of the center billet can be seen by comparing a plot

Fig. 15 A larger-diameter wire resulted in unstable flow and subsequent peeling

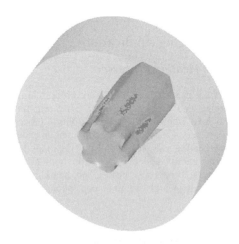

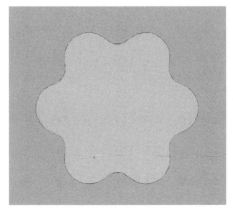

Fig. 16 A hex-shaped input stock resulted in good die contact (top) and no underfill (bottom)

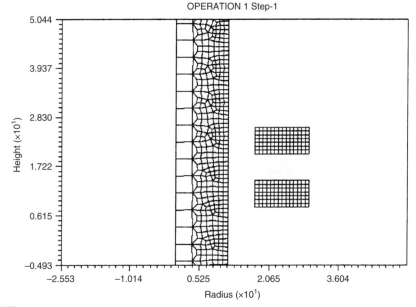

Fig. 17 Mesh illustration of workpiece and heating coils

of temperature versus time for a point sampled with this effect on and off (Fig. 20). If view factor radiation had not been considered in the simulation, all of the billets would have heated the same. Because view factor was incorporated, the difference in temperature between the billets in the corner and in the center of the loading pattern was realistically predicted.

Axle-Beam Forging

A major commercial vehicle part manufacturer discovered a problem with an axle-beam

forging. A forming lap or fold defect was evident on the finish-forged product, as seen in Fig. 21. History and experience guided the designers to concentrate their efforts on the blocker and finisher stages of manufacture. However, changes to these did not eliminate the lap.

The manufacturing process was simulated. Four discrete operations were involved: roll former, bender, blocker, and finisher operations, as seen in Fig. 22. Because the material behavior can be highly temperature sensitive in a hot forming process, all stages of the process were modeled with full thermal coupling. Also

included were intermediate operations, such as transfer times from the furnace to the press and times when the forging was resting on relatively cool dies. In this approach, surface chilling of the workpiece was accounted for.

The simulation results highlighted the fold occurring during the bender operation, as seen in Fig. 23. The defect was carried through to the finish-forged axle beam. After reviewing the simulation result, the designers were able to locate the defect on the actual part, as seen in Fig. 24. The designers modified the pads on the bottom die to revise material flow and eliminate the lap.

In addition to overcoming the forming defect, the bender die-pad modifications resulted in a reduced forging load in the bender, blocker, and finisher operations. The production trials correlated very well with the simulations, which also predicted lower forging loads with the modified pads.

After forging is completed and the flash is removed, the axle beam is heat treated to provide mechanical properties required for its service life. The axle beams are heat treated in batches and are supported on their pads; that is, they are heat treated upside down relative to the orientation on the truck. Distortion during the quenching operation is undesirable. In any case, distortion does occur due to the volume increase associated with the austenite-to-martensite phase transformation. Figure 25 illustrates the comparison between the as-forged and asquenched axle-beam predictions, both at room temperature. The as-forged part is shown in the foreground, and the same part after quenching is shown in the background. Both are at room temperature. Note how the heat treated beam is noticeably longer than the original forging.

Knee-Joint Forging

The forging industry continues to expand with the rest of innovation and often finds new and interesting applications. One such application is the medical implant industry. In this case, an artificial knee implant is considered. The orthopedic surgeon can remove the patella (kneecap), shave the heads of the femur and tibia, and implant the prosthesis. Special bone cement is used for suitable adhesion, and the implants can be seen in their locations in Fig. 26.

An analysis was carried out on the hot forging operations of the tibial part of the Ti-6A1-4V knee-joint prosthetic device. In this case, there were three operations: blocker, finisher, and restrike operation. Each of the three operations consisted of a furnace heat, forming operation, and flash trim. Different friction conditions were applied for the extruded part and the coined portion of the prosthesis. This was important because, in practice, only the extruded part of the dies is lubricated; the coined section is formed dry. Figure 27 shows the tibial part at the end of the blocker, finisher, and restrike simulations.

Furnace temperatures were specified as 940 °C (1725 °F) for the blocker and 925 °C (1700 °F) for the finisher and restrike operations.

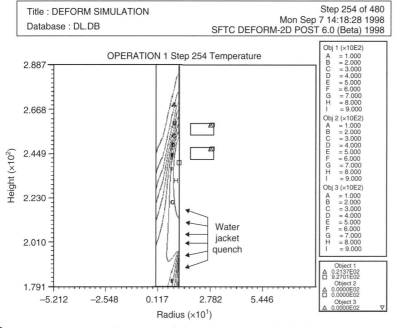

Fig. 18 Steady-state temperature distribution in workpiece 25.4 s into the simulation

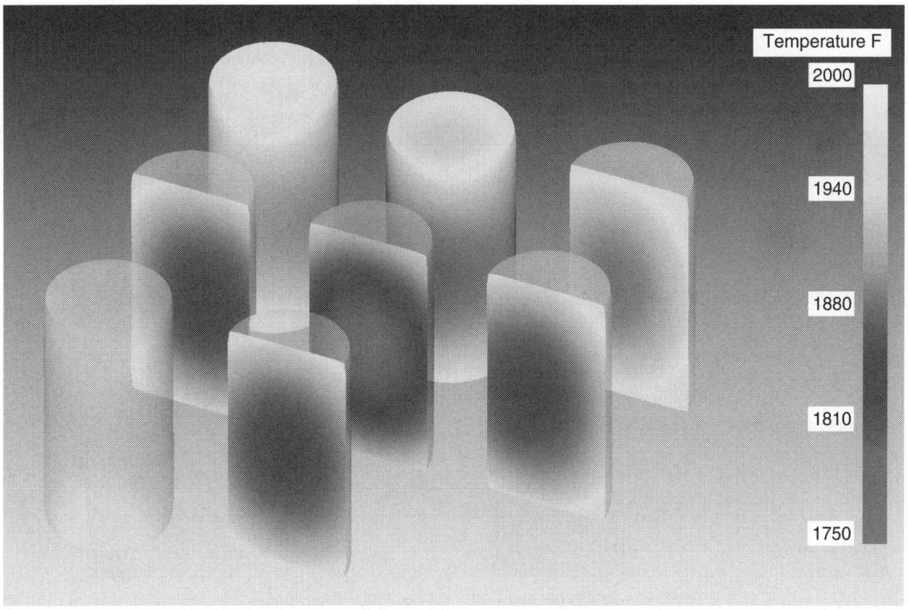

Fig. 19 Temperature of nine billets (eight are shown) in a 1095 °C (2000 °F) furnance. Note that the proximity between the parts affects the temperature

A rigid-plastic workpiece and rigid dies were used in this analysis. After the blocker-operation simulation, the workpiece was trimmed, as shown in Fig. 28.

The surface curvature weighting was set high in this analysis. As a consequence, the tighter radii of the webs and ribs received a finer element size, whereas the larger, flatter surfaces were

assigned coarser elements. This is illustrated clearly in Fig. 29.

Cold Forming Applications

Cold-Formed Copper Welding Tip

Multiple folds were observed during the production of a copper welding electrode, as seen in Fig. 30. The cause of the defects was not entirely understood. The entire forming process was simulated to gain a better understanding of why the folds were developing (Ref 19).

The actual part underwent a total of five operations to form the finished electrode:

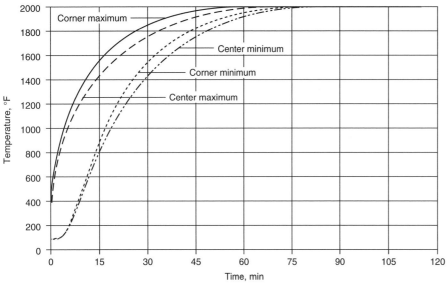

Fig. 20 Temperature plot comparing a point on a billet considering and not considering view factor radiation effects

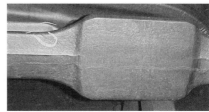

Fig. 21 Noticeable fold on a finished axle-beam forging

Fig. 22 The four operations in the process are initial perform in the forming rolls, after bending (shown in bottom die), after blocker operation (shown in bottom die), and the finished axle beam after flash removal

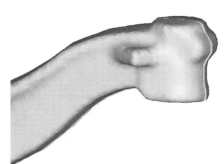

Fig. 25 The as-forged shape is shown in the front (lighter color), with the heat treated shape behind (sliced and darker color). The phase transformation was the key reason for this distortion

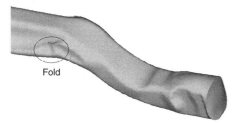

Fig. 23 Simulation predicted the fold occurring during the bender operation, as shown

Fig. 24 Once the fold location was established from simulation, the defect was identified on the actual axle beam after bending (location shown by chalk)

Fig. 26 After surgery, the femoral and tibial prosthetic parts are in position, securely adhered with bone cement

shearing, squareup, preform, backward extrusion, and final forming (Fig. 31). The sheared rectangular slug was assumed to be the starting material for the simulations, and therefore, the shear and squareup were not simulated. The preform and backward extrusion operations were axisymmetric in nature and were therefore simulated in two dimensions. When the 2-D

backward extrusion was finished, the final operation was simulated in three dimensions.

During the preform operation, a stepped die was used to distribute the volume. This sharp step then gets pushed into the tapered die during the backward extrusion, creating a fold on the exterior of the part. The fold develops midstroke and moves upward as the extrusion progresses. The location of this external lap predicted in the simulation is identical to that seen on the actual formed electrode.

The final operation involves piercing the extruded slug with a splined punch. At this stage, the outside of the part is close to finish shape, and the punch is used to form the intricate cooling fins on the inside of the electrode.

Midway through the finish operation, it is seen that quite a few defects are being formed by the punch (Fig. 32). Simultaneously, folds are created from material smearing onto the interior wall, peeling and eventually smearing onto the bottom internal surface (Fig. 33, 34), and lapping on the inside tip and faces of the cooling fins.

All of the defects observed in the simulations correlated very well with those seen on the actual formed electrodes. The results of the simulations proved invaluable in determining why the various defects occurred.

Pipe-Type Defects in Aluminum Components

A 6061 aluminum suspension component was simulated as an impact extrusion. The part was formed in one operation on a mechanical press. The simulation was performed to test the feasibility of producing a defect-free part in one forming operation. The simulation predicted a pipe-type defect prior to the dies being manufactured. The actual part produced can be seen in

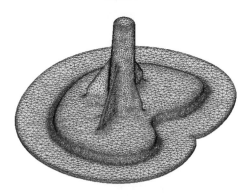

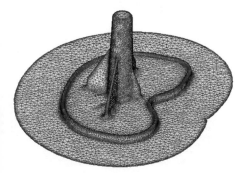

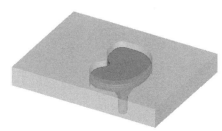

Fig. 27 Shape and finite element method mesh from the tibial knee-joint forming simulations. From top to bottom: at the end of the blocker, finisher, and restrike operations

Fig. 28 Simulated tibial knee forging implant after trimming in between the blocker and finisher operations

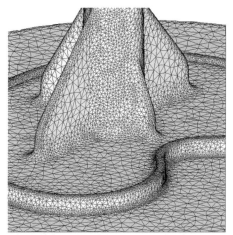

Fig. 29 Finite element method mesh after the final operation. Note how the finer elements are concentrated in locations of smaller features

Fig. 30 Arrows highlight the defects observed after forming the electrode

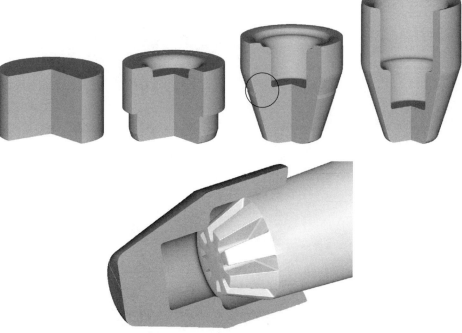

Fig. 31 The progression is shown on the top from left to right. In the bottom center, the punch is shown positioned prior to the final forming operation. The fold created during extrusion is circled

Fig. 35. This piping or "suck" defect occurs due to volume deficiency. As the section between the punch and die becomes thin, an inadequate volume of material is available to feed the extrusion. When this occurs, longitudinal tensile stresses form under the nose of the punch. At that time, surface material is pulled into a cavity as it forms. This defect is shown in Fig. 36.

Another case where a pipe-type defect occurs is in the forward extrusion of a pressure valve (Fig. 37). Although the process is 3-D, the problem was successfully simulated assuming plane-strain deformation (Ref 20) during the late 1980s. In the following discussion, a true 3-D model was used.

The material for the pressure valve is aluminum alloy 6062. Because of symmetry, only one-fourth of the part was simulated. For better resolution of the defect, more elements were placed near the center of the part. The predicted part geometry at different stages of the extrusion process is shown in Fig. 38. From these figures, the defect starts at the center of the part and propagates in the transverse direction where the part is being extruded. This behavior is seen in the flow lines of the extrusion process (Fig. 39). A striking similarity between the predicted flow lines (Fig. 39) and the actual part (Fig. 37) can also be observed.

Fastener Forming

In the development of metalforming processes, designers balance many complex parameters to accomplish a workable progression design. These parameters include the number of intended operations, required volumetric displacements, final part geometry, starting material size, available forming equipment, and the material behavior of the workpiece. Frequently, variations have existed between the designer's concept of the progression and the actual shop trial. When unexpected metal flow occurs, a part with underfill, excessive loads, die breakage, laps, or other production problems can result.

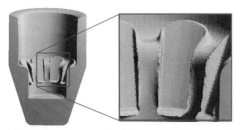

Fig. 32 Early stage of tooth forming. At this stage, the initial defect formations are clear

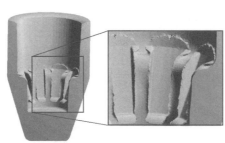

Fig. 33 As the teeth continue to form, the smearing of the material above the teeth is visible

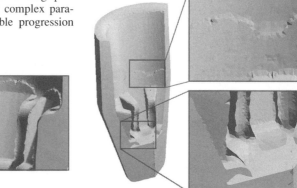

Fig. 34 Material peeling down as the teeth are formed

Fig. 35 Aluminum impact extrusion

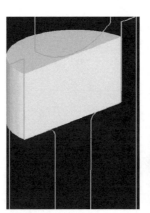
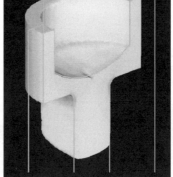

Fig. 36 Predicted geometry at different stages of the extrusion process

Fig. 37 Pressure valve extrusion showing pipe-type defect in the center

Fig. 38 Predicted geometry at different stages of the extrusion process

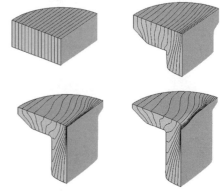

Fig. 39 Flow lines of the pressure valve extrusion

In one case (Ref 21), a fastener manufacturer noted a small defect during the shop trial of an automotive part (Fig. 40). The forming process was simulated, and the simulation reproduced this superficial defect and helped the manufacturer to understand the root cause. Additionally, the simulation (Fig. 41) revealed a severe lap that had originally been overlooked. When the trial parts were cut up, this defect was present as predicted. In this case, the cause of the lap was apparent from the simulation. Each station of a redesigned process was analyzed prior to shop trials. This approach resulted in a lap-free part (Fig. 41).

Bevel-Gear Forging

Bevel gears are important components in the automotive industry, such as in transmission differentials. Many of these components are forged at a hot temperature to minimize the amount of load required to form the part. This creates a part as seen in Fig. 42.

There has been some interest in forming these parts at room temperature for a better net shape and an improved surface finish. To study this as a 2-D process, an axisymmetric assumption is used. One tooth is isolated, and the circumferential flow is neglected. The radius of the tooth was specified to consider the volume of the actual part. The flow is seen in Fig. 43 in the case where the top and bottom die move together at the same speed. The flownet result can be compared to different movement conditions, such as when only the top die moves downward, as seen in Fig. 44, or in the case where only the bottom die moves, as seen in Fig. 45. Note that the filling of the material occurs well in the cases where the dies moved together and in the case where the bottom moves upward, but there was folding predicted in the case where only the top die moved downward. In each case, there is a marked difference in the grain orientation after forging, which can be seen from the flow lines. A similar study was performed as a comparison of simulation to plasticine deformation, and it was shown that the simulation was very accurate in predicting the material flow (Ref 22).

Fracture Prediction

Chevron Cracks

Forward extrusion is a process used extensively in the automotive manufacturing industry. Certain extruded components, such as axle shafts, are considered critical for safe operation of the vehicle and must be free of defects. During the mid-1960s, automotive companies encountered severe axleshaft breakage problems. In addition to the obvious visible external

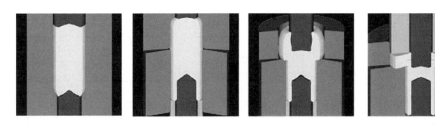

Fig. 41 The top four images show a lap formation on the inside diameter of the head region on a cold-formed automotive part. The bottom four images show the redesign with no lap

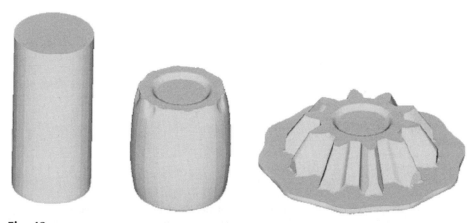

Fig. 42 Hot forging of a bevel gear from a round billet to the final shape

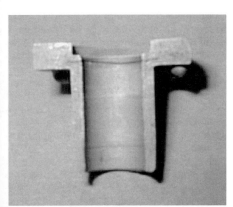

Fig. 40 Fastener showing a lap

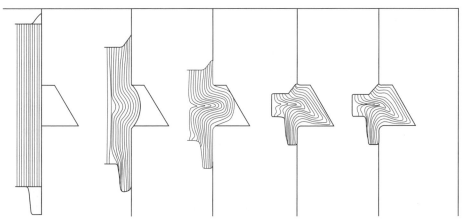

Fig. 43 Two-dimensional flownet of a single tooth in a bevel-gear forming process. The top and bottom dies move together

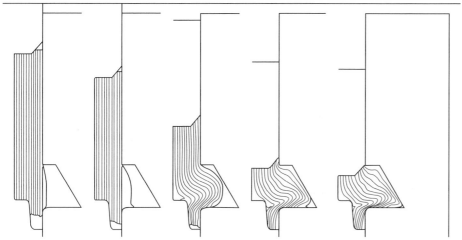

Fig. 44 Two-dimensional flownet of a single tooth in a bevel-gear forming process. Only the top die moves in a downward direction

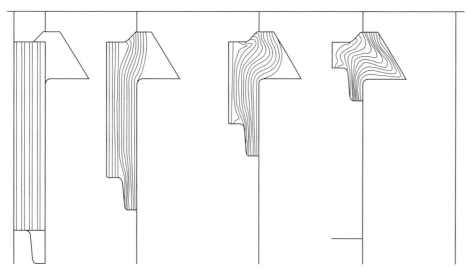

Fig. 45 Two-dimensional flownet of a single tooth in a bevel-gear forming process. The bottom die moves in an upward direction

defects, internal chevron cracks were also present (Ref 23). The problem was so serious that a number of manufacturers adopted 100% ultrasonic testing procedures, with automatic rejection of suspect shafts.

To avoid 100% inspection, in the early 1970s, the Chrysler Corporation developed conservative guidelines for forward extrusion in conical dies, guaranteeing chevron-free parts (Ref 23). Upper-bound methodologies were used to determine the conditions under which chevrons would form. Based on die-cone angle, process reduction, and friction conditions, Avitzur derived mathematical expressions to describe the central bursting phenomenon during the wire drawing or extrusion of a non-strain-hardening material (Ref 24). To validate this work, experiments were carried out at Lehigh University on AISI 1024 plain carbon steel bars. Drawing was carried out in dies having an 8° semicone angle, with greater than 22% reduction, that is, in the safe region of Avitzur's curves. No central bursts were reported (Ref 24).

Avitzur derived similar criteria for central bursting in strain-hardening materials (Ref 25) that were later validated by Zimerman and Avitzur (Ref 26). Experimental results are shown (Fig. 46) where it is clear that no central bursting occurred in the safe zone. This work also illustrates the die angles and drawing forces where central bursting occurs. This is overlaid on a schematic of the drawing conditions providing sound flow, dead-zone formation, and shaving (Fig. 47) (Ref 26). DaimlerChrysler implemented Avitzur's curves in the early 1970s, and the chevron cracking problems were no longer troublesome.

More recently, work has been carried out using FEM in conjunction with ductile fracture criteria to determine the occurrence of central bursting. The parameter damage is a cumulative measure of the deformation under tensile stress and has been associated with chevron cracking. Researchers have evaluated seven different damage or ductile fracture criteria (Ref 27). The various criteria express ductile fracture as

a function of the plastic deformation of the material, taking into account the geometry, damage value, stresses, and strain within the workpiece. When the maximum damage value (MDV) of the material exceeds the critical damage value (CDV), crack formation is expected.

A specific damage model proposed by Cockcroft and Latham states that fracture occurs when the cumulative energy density due to the maximum tensile stress exceeds a certain value. This criterion has provided good agreement at predicting the location of the MDV. The Cockcroft and Latham criterion is shown as follows in both dimensional and nondimensional forms:

$$\int\limits^{\bar{\varepsilon}_f} \sigma^* d\bar{\varepsilon} = C_a$$

$$\int\limits^{\bar{\varepsilon}} \frac{\sigma^*}{\bar{\sigma}} d\bar{\varepsilon} = C_b$$

where σ^* is the principal (maximum tensile) stress, $\bar{\sigma}$ is the effective stress, $d\bar{\varepsilon}$ is the increment of effective strain, and C_a and C_b are constant values.

Under ideal drawing or extrusion conditions, the strain distribution across the component cross section would be uniform. However, the occurrence of subsurface redundant deformation causes the strain distribution to become nonuniform. The amount of redundant deformation increases with increasing die angle and can cause extremely high tensile stresses. These internal tensile stresses can in turn lead to microvoiding and ultimately to cracking (Ref 28). If the die angle/draw reduction combination cannot maintain compressive axial stresses in the drawn component, the center portion will be stretched, and the tensile stress may increase to a level where bursting occurs (Fig. 48). In the first forming operation for a component, the stresses in the component may be high, but the damage will still be quite low because the damage accumulates with deformation. For this reason, fracture does not generally occur until the second or third draw or extrusion operation. A triple extrusion (last image in Fig. 48) was simulated to demonstrate this phenomenon. It is clear that in this case, chevrons do not form until the third reduction.

One method to determine the CDV of a material is to perform poorly lubricated compression and notched tension tests until cracking is detected. After testing, simulations can be performed, matching the geometry and process conditions of the experimental tests, to calculate damage values. The predicted MDV at the instant of fracture is a good representation of the CDV of the material. Because the crack is not visible until after it has formed in the tensile test, a higher estimation of the CDV is likely. Averaging the CDVs calculated from the compression and notched tensile tests is a reasonable approach.

A comparison was made between two automotive shaft designs manufactured using a double extrusion. The only parameter changed was the die semicone angle for the minor diameter. Five-hundred steel shafts (AISI 1024) were produced with a nominal 22.5° extrusion die angle, and 500 were produced with a 5° die angle. Chevron cracks were observed on 1.2% of the shafts made from the 22.5° die, but none were observed in the product produced using the 5° angle.

Process simulation was used to analyze both processes. The simulation indicated a higher damage value for the product extruded with the 22.5° die angle than for the parts produced with the 5° angle. The high damage value correlated well with the location of the chevron cracks (Fig. 49).

When damage levels are very high, fracture will occur consistently. When processes are well below the ductility limit, fractures are not expected to occur. A narrow range exists in between these two regions where the chance of cracking is probabilistic and a higher damage prediction can be interpreted as a greater chance of fracture. Because the damage is cumulative in nature, the prior working history of the wire or workpiece is an important factor to be considered for the likelihood of fracture occurrence.

Fracture during Cold Forming

During an initial trial of a cylindrical cold-formed part (Fig. 50), the manufacturer observed a severe fracture originating in the inside diameter of the part after the second operation. There was also a die underfill in the area of the fracture. The defect was observed after the end of the forming operation, as shown in Fig. 51. As seen in Fig. 51, the flowlines during the two operations show the grain orientation of the part after forming. Also seen in Fig. 51 are the underfill at both the inner and outer diameters. A plot of damage versus time is shown in Fig. 52, where the separation and contact of the sampled points have been highlighted in the plot. It can be noted that the damage at point 2 remains 0 during the first operation and starts to increase in the second operation. As shown in Fig. 51, in the second operation, point 2 originally resides underneath the punch. The point contacts the punch and moves outward as the punch moves downward. As the material is backward extruded, this point moves upward and loses contact with the punch. The separation between the material and the punch forms an underfill on the part inside diameter surface. The predicted damage value continues to increase, as shown in Fig. 52, when point 2 passed the punch corner, backward extruded, and separated from the punch. At point 1, the damage increases when the material is backward extruded and separated from the bottom die. The damage value at point 1 is significantly lower than point 2.

In order to manufacture this part, subsequent analysis revealed that the damage factor could be reduced. Analysis was used in conjunction with shop trials to eliminate the fracture and die underfill in this part (Fig. 53).

Die Stress Analysis

Four Common Modes of Die Failure

The four common modes of die failure are catastrophic fracture, plastic deformation, low-cycle fatigue, and wear.

Catastrophic failure occurs when a die is loaded to stress levels that exceed the ultimate strength at temperature of the die material. This can occur due to gross overloading of the die structure, inadequate support to transmit the load to an adjoining component, or a stress concentration at a sharp feature. Dies that fail in this mode can release enormous amounts of energy, resulting in a serious safety hazard.

Plastic deformation occurs when the stress on a die exceeds the yield strength at the operating temperature. Plastic deformation can occur as a localized effect or a widespread condition. Examples of a localized effect are rolling the corner of a punch or the initial yielding in a stress concentration prior to a fatigue failure. In these cases, the overall die dimensions do not change. Large-scale yielding occurs when a bolster plate is dished or the entire inside diameter of a shrink ring becomes oversized.

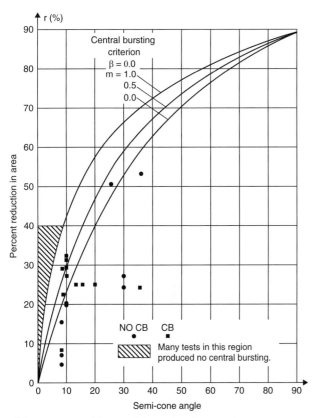

Fig. 46 Comparison of theoretical data and results

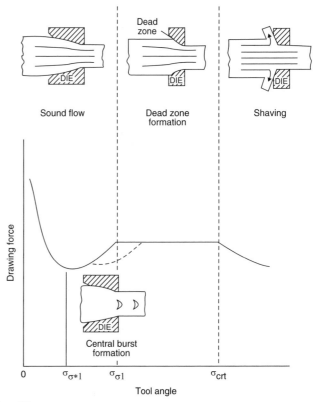

Fig. 47 Effect of tool angle and mode of flow on drawing force

Low-cycle fatigue is a process that can occur when a mechanical or thermal tensile stress is cyclically applied to a die as each part is produced. If the stress intensity or number of cycles is sufficient, a very small crack can initiate. After crack initiation, the crack propagates with each additional cycle. As the defect grows, the die structure weakens, which results in higher stress concentrations and an increased crack growth rate. Eventually, the die fails due to a catastrophic fracture, although the energy released during final fracture is generally small relative to a single-cycle catastrophic failure. Low-cycle fatigue is well understood by the manufacturers of automobiles, aircraft engines, and other critical service systems. Materials used in these applications can be characterized by stress-number of cycles curves that relate the fatigue life to cycles to failure. Typically, these data are very expensive to obtain. Thus, it is extremely rare to find fatigue data for forging or cold heading dies in the literature.

Die stress analysis is used to determine the stress level on a die during service. More than one stress state may be significant. Initially, the effective stress is used to understand the magnitude of stress on the tool. Ideally, the stress should be well below the yield strength at the local service temperature of the die. Additionally, maximum principal stress is used to determine the tensile component of this stress. This can be used to predict the likelihood of a fatigue failure. Stress components are used to quantify the direction of stress. A clear understanding of the stress direction is required to ensure that the correct solution is developed.

Die wear is also a frequent topic of discussion related to coatings, die materials, and lubricants. Considerable research is being performed on the topic of wear relative to a wide range of manufacturing processes.

Die-Insert Fracture

A multistation cold heading process was used to produce a high-volume automotive part from AISI 4037 steel wire. The process involved extruding the body, upsetting, and back extruding the flange area of the part. Piercing out the center finished the part. While this part was not completely axisymmetric, the deviations from this were subtle; thus, using a 2-D analysis was reasonable and very fast (Ref 29).

The first station was primarily a squareup of the cutoff, with a relatively small amount of forward extrusion. The second station was a finished forward extrusion coupled with a small amount of backward extrusion around the punch pin. The third (final) station was a backward extrusion around the punch, combined with upsetting in the flange area. With the deformation analysis results completed, the interface pressure values were interpolated along the tooling surfaces to perform a decoupled-die stress analysis.

This part was run several times in production prior to the investigation. History showed that the die insert was breaking prematurely in the

first and third stations, as shown in Fig. 54. Additionally, the punch was failing in the second station, and other tooling components were failing in the piercing station. While all stations were analyzed, the first station is summarized here.

It has been learned that to accurately analyze die stresses, complete tooling geometry must be used, including shrink rings and mounting dies. Analyzing a single die in isolation does not provide meaningful results in cases such as the die insert, where the

interaction with the shrink ring is critical. In the present example, the dies were analyzed as elastic bodies during the analysis. In the case of the steel tools, the effective stress was used as a yield criterion, but in the carbide components, maximum and minimum principal stresses were used, along with their components. These low-cycle fatigue failures in the carbide inserts indicated tensile stresses in the crack-initiation site, even though the maximum principal stresses were less than 690 MPa (100 ksi).

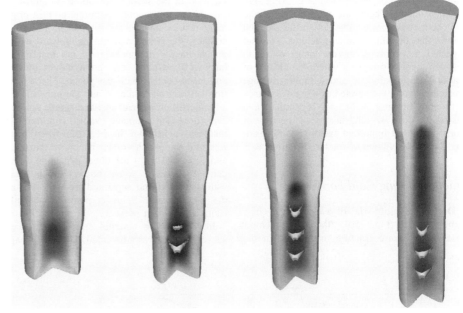

Fig. 48 A computer simulation demonstrates the formation of chevron cracks. The dark colors represent higher values of damage

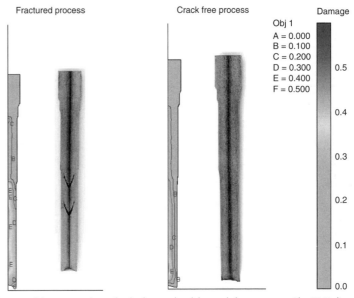

Fig. 49 Contours of damage are shown for the fractured and the crack-free processes. The 22.5° die angle (left) exhibits chevron cracking due to high damage. The 5° die angle (right) has lower damage, and no central bursting was observed

Results for the first station tooling assembly can be seen in Fig. 55 and 56. This assembly shows the punch (D40), the insert (D70), and the shrink ring/sleeve assembly (S7 and M300). The ring/sleeve assembly did not indicate any overstress condition and performed well in field service. The insert did indicate a positive maximum principal stress in the region where the fracture was observed. The principal stress direction was predominantly axial. After carefully studying the simulations, it was decided to run a simulation with a two-piece insert, as shown in Fig. 57. The analysis indicated a significantly lower stress in the region of the fracture.

The tooling was modified based on the simulation results. The insert in the first station was split into two components. This resulted in a 550% life improvement. Other tooling modifications resulted in similar improvements in other stations of the process.

Secondary effects influenced tool life on the other components as a result of the aforementioned modifications. For instance, tooling components that were not modified in each station also showed substantial improvements, most of which were above a 150% increase. The net result, based on historical data, showed a 43% reduction in overall tooling costs and a 54% reduction in downtime associated with tooling problems on this product.

Turbine Spool Die Failure

A company was experiencing a dimensional problem with a hot-forged turbine disk (Ref 30, 31), as shown in Fig. 58. The forging was undersized on a number of features on an inside diameter. After an investigation of potential causes, the dies were thoroughly inspected. The inside diameter of a die liner and the outside container were oversized. This was somewhat surprising, because the outside diameter of the forging was in tolerance.

A stress analysis indicated that the effective stress exceeded the yield strength at temperature in the region where the die had yielded (Fig. 59). A range of redesign options was developed and analyzed. The redesign that was selected involved a thicker die wall (Fig. 60). Criteria for the redesign included cost, ease of assembly, and structural integrity. It was critical that the die wall possessed sufficient strength to avoid plastically deforming during the forging operation.

This design was successful in that the dimensional deviations were eliminated and there was a significant reduction in die maintenance cost. The root cause of this problem was large-scale plastic deformation. This case is interesting because the original symptom of a problem was a dimensional deviation in the forging. Because this die material was quite ductile, no fracture was observed. In fact, the inside features were actually forged within tolerance, but the outside was forged oversized. During the ejection process, the part was essentially extruded, moving the dimensional deviation from the outside of the forging to the inside, as shown in Fig. 61.

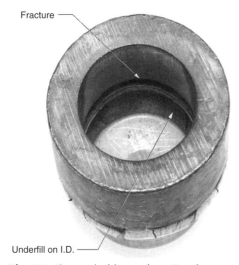

Fracture

Underfill on I.D.

Fig. 50 Photograph of the actual part. Note the severe fracture (dark area) and die underfill (below dark area) on the inside diameter (ID)

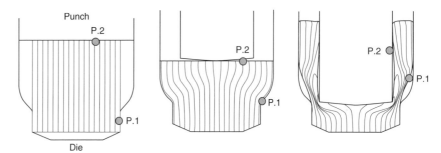

Punch

P.2

P.1

Die

P.2

P.1

P.2

P.1

Fig. 51 Flowlines for the forming of the component. Note that the free surfaces are very clear on both the inner and outer diameters

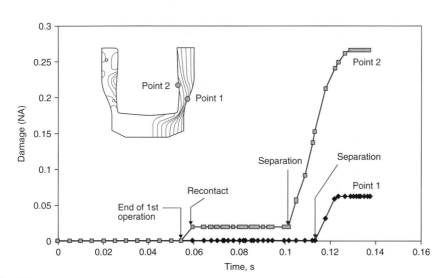

Point 2

Point 1

Separation

Separation

Point 2

Recontact

End of 1st operation

Point 1

Time, s

Damage (NA)

Fig. 52 Point tracking the damage factor of two points on the surface of the component

Fig. 53 After the redesign, the inside diameter is free of defects

Product Assembly

Staked Fastener Installation

The development of complex forming processes, such as self-penetrating fasteners, staked studs, and rivets, involves complexities over and above traditional forming operations due to the interactions of multiple plastic deforming bodies. In such cases, fastener installation is influenced by plastic strain (work hardening) induced during prior cold forming operations. Therefore, it is not practical to perform installation trials with machined blanks. The Fabristeel Corporation used computer simulation to develop their self-piercing mechanically staked fasteners for sheet metal parts (Fig. 62). The patented drawform stud was fully developed using simulation. The development process included forming the stud, the installation process itself (Fig. 63), and a pullout test (Fig. 64). Based on damage values in the sheet, the original design was modified to prevent fracture in the panel.

Breakaway Lock Development

Recently, process simulation has been used to develop an aluminum breakaway padlock (Fig. 65). Pull strength was a critical requirement for the intended application. Unlike a typical keyed padlock, this lock was developed to be a one-time-use item.

The lock was developed so that during installation, the bolt is torqued down until the head shears off at an undersized (recessed) diameter. At the installation, the end of the bolt should plastically deform the shackle, resulting in a permanent installation. The grade of aluminum, heat treatment, and geometry of the lock were the primary design variables. Common aluminum alloys were analyzed, including 6061 and 6062, as were different heat treatments for the shackle, bolt, and body. Shackle diameters of 5 and 6 mm (0.20 and 0.24 in.) were modeled to determine the effect of diameter on the pull strength.

The lock shackle was modeled as a rigid-plastic material, and for this portion of the analysis, the bolt was considered rigid. The bolt was twisted and subsequently pushed into the shackle (rotated and translated inward) until the tip of the bolt was fully engaged. The

localized deformation can be seen in Fig. 66, where effective strain is displayed in the shackle.

The first question for this lock design was the mode of failure. Localized plastic deformation at the tip of the bolt could occur, as seen in Fig. 67, allowing the shackle to become dislodged. Alternatively, the shackle could unbend or neck, allowing the free end to be pulled out of the lock body. Because the deformation in both the bolt and the shackle was needed to accurately predict the failure, both were analyzed as plastic objects.

Pull tests were performed by assigning a constant (upward) velocity to a rigid object placed inside of the shackle. The bottom of the lock body was fixed using boundary conditions. The contact between the bolt and the shackle was considered, as was the interaction between the shackle and the lock body.

It was shown from the simulations that both deformation modes (tip deformation and shackle unbending) occurred during this process. At the start of the pull test, the tip of the bolt plastically deformed, as seen in Fig. 67. After this initial deformation, however, the bolt remained structurally sound, and the shackle started to unbend. The unbending continued until the free end of the shackle pulled free from the lock base, as seen in Fig. 68. This bending was the primary mode of failure observed in all of the simulated pull tests, and it matched prototype tests conducted.

The failure mode and pull strength were studied for each design, as seen in Fig. 69. When using the same-diameter shackle, the lock made from 6061-T6 material demonstrated a higher pull strength than the lock made from 6062. Likewise, for the same material, the simulations showed that the larger the shackle diameter, the higher the pull strength. After this study, 6061-T6 material and a 6 mm (0.24 in.) diameter shackle were selected for the final lock design.

The trend in pull strength that was observed in the pull tests was quite intuitive. Simulation was not only able to confirm the trend, but it was also able to determine the amount of load that each lock could withstand.

The ability to determine how the lock would fail was the real strength of the FEM simulations in this example. Depending on the shackle material and diameter, it was conceivable that some locks would fail due to bolt tip deformation, while others would fail due to shackle unbending. Simulation was able to determine that for the materials and diameters studied, all locks would fail the same way.

Solid-State Welding Process

Inertia welding is a solid-state welding process in which the energy required for welding is obtained from a rotating flywheel. The frictional heat developed between the two joining surfaces rubbing against each other under axial load produces the joint (Ref 32, 33). One of the two components is attached to a rotating flywheel, while the other component remains stationary. During the process, the rotating component is pushed against the fixed one, causing heat generation through contact friction and consequent rise in the interface temperature. As the temperature increases, the material starts to soften and deform. At the final stage,

Fig. 54 The failure mode of the die insert in the first station was a low-cycle fatigue fracture, as shown

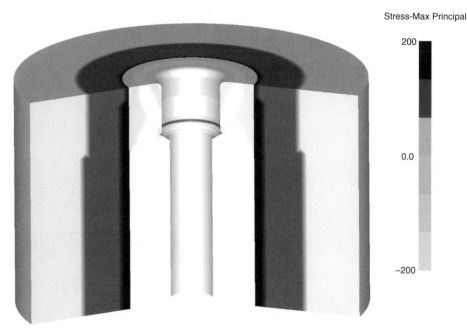

Stress-Max Principal

200

0.0

−200

Fig. 55 The contours of maximum principal stress on the original design are shown, with the dark colors representing tension and the light colors representing compression. The tensile stress on the inside of the shrink ring is expected. The tensile stress at the inside corner of the die insert was problematic. This is coincident with the location of the low-cycle fatigue fracture

the flywheel stops as the inertial energy decays to 0, and both sides of the material near the interface get squeezed out of the original contact position and form a complex-shaped flash, as shown in Fig. 70. The quality of the welded joint depends on many process parameters, including applied pressure, initial rotational speed and energy, interface temperature, the amount of upset and flash expelled, and the residual stresses in the joint. Being able to accurately model the process is essential to understand, control, and optimize the process. After forging and machining, several aircraft engine disks can be joined together using this process. Figure 71 shows a cutaway view, with a close-up view of the weld, for a typical aircraft engine disk welding.

A typical comparison of predicted and measured upset versus time is shown in Fig. 72. From this figure, it can be seen that the characteristics of this process include three distinct stages: the upset rate is 0 at the initial stage; the upset rate is increasing first and then decreasing in the middle of the process; and the

upset rate approaches 0 at the end of the process. In the first stage, the energy is used to raise the temperature on the contacting surface. The material starts to deform in the second stage. As the flash forms, the contacting surface area increases, the rate of upset starts to slow down, and the inertial energy is eventually consumed in the third stage.

Figure 73 shows a comparison of the predicted temperature at the end of the weld with a macrograph of the weld. There is very good agreement between the predicted temperature field and the observed heat-affected zone and also the flash shape and thickness.

It is an added challenge to weld two dissimilar materials, due to the differences in thermal as well as deformation behavior over a wide range of temperature. Proper process welding conditions are crucial to achieve the desired weld geometry and properties. As shown in Fig. 74, a compressor spool of two adjacent stages made of different materials is joined. Good agreement between the predicted and observed weld shapes is observed.

Optimization of Forging Simulations

In the past, FEM has been primarily used as a numerical tool to analyze forming processes. Decision-making to optimize a process is strongly dependent on the designer's experience in an actual process and on numerical modeling. Ultimately, it would be very desirable to develop an optimization technique to achieve an acceptable die shape automatically.

In forging operations, for example, intermediate shapes are often used to ensure proper metal distribution and flow. The design of intermediate shapes, also called blockers and preforms, are of critical importance for the success of forging processes. In the following sections, optimization in 2-D and 3-D preform die design is presented.

Two-Dimensional Case

An axisymmetric example to demonstrate the potential application of design optimization is shown in Fig. 75.

In this example, the primary objective was to completely fill the finishing tools during the last operation. A numerical measure of this criterion was obtained by comparing the outline of the desired part and that of the actual forged part. It was also important to obtain homogeneous materials properties within the component; a homogeneous distribution of effective strain provided a satisfactory approximate criterion to achieve this goal.

The shape of a preform die was represented by B-spline curves. The design variables were the control points of B-spline or a piecewise linear curve.

An isothermal forging condition was used in the simulation. The flow stress of the workpiece was assumed to be in the form of $\overline{\sigma} = 100\overline{\varepsilon}^{0.2}\dot{\overline{\varepsilon}}^{0.03}$ (ksi). A constant shear friction factor of 0.5 was assumed, as well as a billet radius of 2.5 cm (1.0 in.) and a length of 3.8 cm (1.5 in.). The velocity of both the preform and the finish dies was 2.5 mm/s (0.1 in./s).

Figure 75(a) shows the forging process without using a preform die, and underfill was

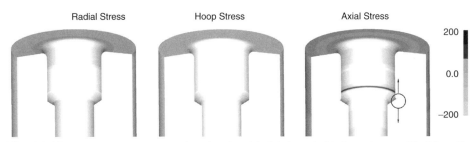

Fig. 56 The component stresses were evaluated on the original design to isolate the root cause of the failure. The dark colors represent tension and the light colors compression. Hoop and radial stresses (light color) are essentially compressive. The circle highlights a tensile stress in the axial direction, as shown with the arrows

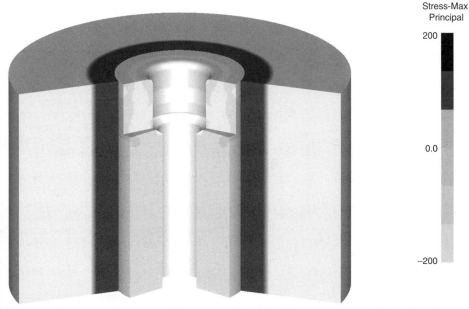

Fig. 57 The contours of maximum principal stress on the redesigned insert are shown, with the dark colors representing tension and the light colors representing compression. This design resulted in an extended die life due to the carbide insert remaining in a compressive stress state throughout the process

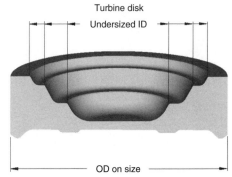

Fig. 58 The inside features of a turbine disk forging were undersized. The outside diameter (OD) was in tolerance. ID, inside diameter

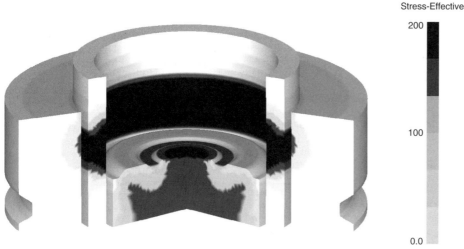

Fig. 59 Analysis of original design. Contours of effective stress depict a very high stress level (darkest contour) through the die wall in the contact zone. This stress in the die wall exceeded the yield strength of the die material at the operating temperature, resulting in plastic deformation

Fig. 60 Analysis of new design using the same scale. The effective stress levels are much lower throughout the die wall. While high stresses were calculated in the workpiece contact zone, this design possessed sufficient strength to avoid large-scale plastic deformation

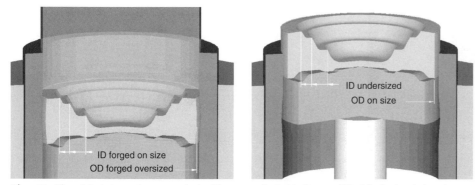

Fig. 61 The original observation was undersized features on the inside diameter (ID) of the forging. In fact, the outside diameter (OD) was forged oversized, as shown on the left. A swaging operation was inadvertently performed while ejecting the part, as shown on the right. This resulted in the initial investigation focusing on the top die (punch)

predicted, as indicated by the arrows. The initial guess of the preform die shape prior to the final forging is shown in Fig. 75(b). As indicated in the same figure, an underfill problem remained. Through the optimization iterations, the preform die shape evolved, as shown in Fig. 75(c). It took approximately seven iterations to resolve the underfill problem and to obtain the homogenized strain distribution, as shown in Fig. 75(d) Figures 75(e) to (g) show the distribution of effective plastic strain without preform die, with initial preform die, and with optimized preform die, respectively. It can be seen that the distribution of effective strain became more uniform through the iterative optimization process.

Three-Dimensional Case

Most forming processes are, in fact, 3-D. Tooling geometry is most likely prepared using commercially available computer-aided design (CAD) systems. In the CAD systems, nonuniform rational B-spline surface is a very popular way of representing the complex tooling geometry. In the FEM model, the geometry is generally represented by a set of polygons. At the current stage, the stereolithography (STL) representation is used as a vehicle between CAD and FEM to transfer the geometry definition. Parametric design and feature-based representation, which may be available in the CAD system (dependent on the methodology of the CAD system) during the designing process, are unfortunately lost at the end of the STL transformation. The FEM can only be used as a numerical tool to evaluate the given tooling design. Although the CAD and FEM can conceptually be used as a black box and integrated by a closed-loop optimization procedure, numerous 3-D FEM simulations will be needed to evaluate the sensitivity of design parameters. The procedure can be extremely computational, demanding, and impractical for daily use. Unlike the 2-D approach, being able to generate a reasonable initial design in order to facilitate the design process has been the near-term objective.

In a recent approach, preform design based on the so-called filtering method was used to determine an initial preform shape. For a given final part configuration, there are three steps to generate an initial preform shape (Ref 34–37):

1. *Digitizing process:* The final part geometry is imported into the system as a triangulated surface, in STL or other format. The digitizing procedure is first performed to convert the triangulated surface into a point cloud array.
2. *Filtering process:* The resulting surface will be passed through a filtering procedure to produce a smoothed shape. In the filtering procedure, the geometric domain will be converted into a frequency domain by using a Fourier transformation (Ref 38–40). By using the filtering function, high-frequency regions (sharp corners, edges, or small

geometric details of a surface) will be removed. An inverse Fourier transformation will then be applied to obtain a smoothed surface.

3. *Trimming process:* The trimming procedure is used to control the boundary shape of a smoothed shape, so as to obtain a realistic initial preform shape.

An aluminum structural part, as shown in Fig. 76(a), is used to illustrate the aforemen-

tioned procedure. The overall dimensions of the part are 61.5 by 16.5 by 9.1 cm (24.2 by 6.5 by 3.6 in.). The filtered smoothed shape is shown in Fig. 76(b). After trimming, the designed initial preform is shown in Fig. 76(c).

In order to improve and eventually automate the design modification procedure, a systematic method to evaluate the material flow based on the FEM result must be developed. Because the material flow history is known from the FEM simulation, the defect location can be

traced, and modifications can be made accordingly. To illustrate the overall procedure, the forging of the example part was simulated, as shown in Fig. 77.

In this simulation, only the upper part was simulated due to symmetry. Aluminum 6061 was selected as a workpiece material. The initial temperature was assumed to be 480 °C (900 °F) for the workpiece and 205 °C (400 °F) for the die. The total die stroke from the initial contact to the final position was 3.612 cm (1.422 in.). A flash thickness of 2.5 mm (0.1 in.) was used. The estimated maximum load was 6300 klbf. As shown in Fig. 77, the flash shape was not uniform in the areas indicated by arrows, and underfill was predicted, indicated by dashed lines.

To identify the region where modification was necessary, an ideal forged component with flash was defined first in order to compare with the simulation result. At the present stage, the ideal final forged part is defined in such a way that it has uniform flash width and does not have underfill condition. By comparing the ideal forged shape and the predicted shape at the reference step, the location of the underfill defect and the amount of additional volume can be identified. The material flow history information from a FEM simulation can be used to modify the preform shape by adding volume (if material is not sufficient) or by removing volume (if excessive flash was

Fig. 62 Front (left) and back (right) view of an installed stud

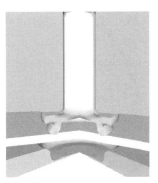
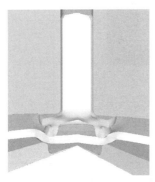

Fig. 63 Installation of the drawform stud. The plastic strain is shown as the shaded color

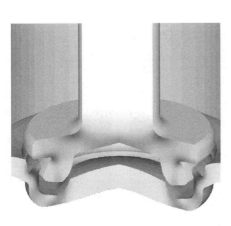

Fig. 64 After the forming and installation are completed, the pull test is simulated. This figure shows the process as the stud is breaking out of the plate. Experimental values matched the simulation results within 5%

Fig. 65 Aluminum breakaway lock. Courtesy of Hercules Industries

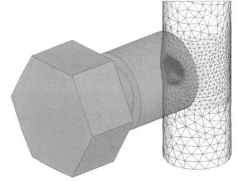

Fig. 66 Plastic deformation (dark area is higher strain) after the installation

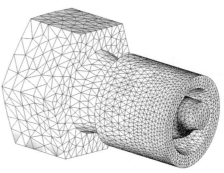

Fig. 67 Plastic deformation of the bolt at the start of the pull test

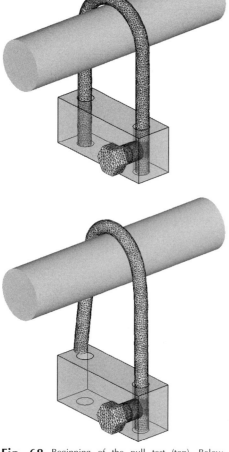

Fig. 68 Beginning of the pull test (top). Below, unbending of the shackle (failure mode)

found). Another filtering procedure will be applied to obtain a smooth geometry. An example to illustrate the procedure is shown in Fig. 78. Figure 78(a) is the ideal forged component with uniform flash. The flash width is 25.4 mm (1 in.), and the flash thickness is 1.3 mm (0.05 in.) (upper part only). The predicted shape from the simulation at a stroke of 35.8 mm (1.41 in.) is shown in Fig. 78(b). The areas that need to be modified are shown in Fig. 78(c). Finally, the modified preform shape is shown in Fig. 78(d). In this example, an additional 0.6% volume was added to make the modified preform shape.

To validate the modified preform, another FEM simulation was performed. The new simulation result is shown in Fig. 79(a). From this figure, it is clear that the underfill defect is removed and the flash is more uniform

compared to the initial design. The outer boundaries of the final product without flash, the predicted forging using the initial preform, and the predicted forging using the modified preform are shown in Fig. 79(b).

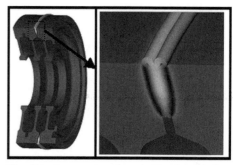

Fig. 71 Cutaway view of a typical inertia weld of aircraft engine disks and a close-up view of the welded region. Source: Ref 33

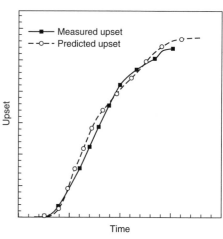

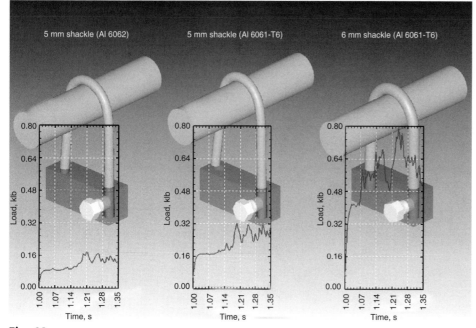

Fig. 70 Inertia welding of two disk-shaped objects showing the initial geometry and the temperatures at the completion of weld. Note the flash expulsion at both the inside and outside diameters. This is a two-dimensional model shown in a three-dimensional view for easy visualization. Source: Ref 33

Fig. 72 Measured and predicted upsets as a function of time for a typical inertia welding process. Note the good agreement between the predicted and the measured upset rates. Source: Ref 33

Fig. 69 Load results for various pull tests

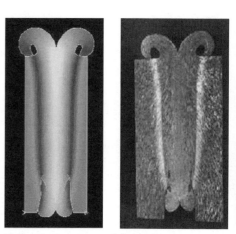

Fig. 73 Comparison of the predicted temperature (left) at the completion of the weld with a macrograph (right) showing the heat-affected zone. Source: Ref 33

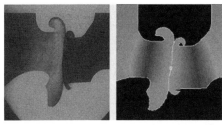

Fig. 74 Comparison of actual (left) with predicted (right) weld shape during the inertia welding of two dissimilar materials, as would occur in a compressor spool with two adjacent stages made of different materials. Due to differing material characteristics on the two sides of the weld interface, modeling is crucial in determining the right weld geometry and parameters. Source: Ref 33

Conclusion

Growing together with computing power, FEM has evolved rapidly and is making great contributions to conventional metalworking technologies. The thousand-years-old technology has quickly evolved in the last 20 years in such a way that the FEM method is now routinely and successfully used in a wide range of bulk forming applications, including material flow, defect prediction, microstructure/property predictions, heat transfer, forming equipment response to the workpiece, die stress and deflection, and so on. With continuous improvement in computing resources and demand from industry, it is expected that future developments will likely be in the following areas:

- *Optimization:* Currently, FEM users play an important role in solving a specific process design by performing sensitivity analysis of multiple FEM models. Convergence speed is highly dependent on the user's understanding of the FEM model and the actual process. It is believed that optimization techniques will be extremely helpful in finding an optimal solution systematically and quickly on the computer. Currently, optimization has been used with success in conjunction with FEM to determine not only the heat-transfer coefficient but also the material parameters and friction factor, the interface heat-transfer coefficient between two contact bodies (Ref 41) and in the preform design to remove the underfill flow defect in 2-D application (Ref 9). In order to carry out the 3-D optimal shape design, a closer integration of computer-aided design/computer-aided engineering (FEM) is necessary and is expected in the near future. For a true optimal product and process design, the costs that are associated with the production process, such as tooling, material, equipment, heating (when necessary), inspection, and inventory, should also be characterized and considered in the model.

- *Computer modeling of microstructural features:* Much work has been carried out to predict microstructural features, such as grain size and phase transformation, in the past by using phenomenological approaches. Texture is another important microstructural feature that will affect mechanical properties. Crystal-plasticity modeling has made excellent progress in predicting texture (Ref 42). As the computing environment continues to

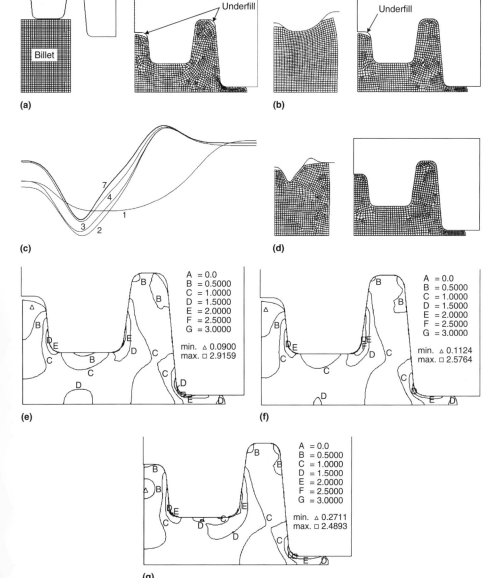

Fig. 75 An example of optimal preform design in two-dimensional application. (a) Final forging shape without preform die. (b) Initial preform and final forging shape. (c) Evolved preform die shapes during optimization iteration. (d) Optimal preform and final forging shape. (e) Effective plastic strain distribution without preform die. (f) Effective plastic strain distribution with initial preform die. (g) Effective plastic strain distribution with optimized preform die

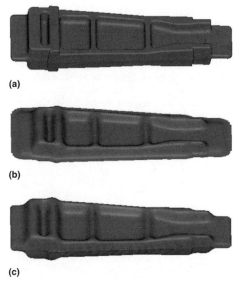

Fig. 76 Example of the design of initial preform. (a) Final product shape. (b) Filtered smoothed shape. (c) Designed initial preform shape

improve, the crystal-plasticity method coupled with the FEM code (CPFEM) to describe the texture/anisotropy evolution during the forming process will become realistic and practical. The texture model can be further integrated with the microstructural model so that transformation, grain-size evolution, and grain-size effects can be taken into account throughout the various stages of thermal-mechanical processes. The framework established for CPFEM and the predicted crystallographic texture will also facilitate further investigations of mechanical properties, fracture, and life prediction.

With the continuously improving understanding of material response to thermal-mechanical changes and proven success in metalforming applications, it is believed that FEM will make significant contributions to the postforming operations, such as heat treatment and machining. In due time, it should be possible to analyze the entire manufacturing process, from the cast ingot, open-die forging, closed-die forging, heat treatment, and machining. This will allow designers to include residual stress and grain flow in their product design and application analysis. The benefits from this should include reduced product life-cycle cost and increased safety margins on critical service components.

ACKNOWLEDGMENTS

In the past 20 years, many of our colleagues have worked on this subject and made process simulation successful not only for academic research but also for practical industrial use. It is not possible to properly thank everyone who has made significant contributions to the FEM methodology for metalforming applications. The authors would like to thank and acknowledge Professor T. Altan at The Ohio State University and Drs. S.L. Semiatin at the U.S. Air Force Research Laboratory, S. Srivatsa at General Electric Aircraft Engine, and K. Sawamiphakdi at The Timken Company. We thank them for their work and continuous guidance, support, and encouragement. The authors would also like to thank the entire staff from Scientific Forming Technologies Corporation for their enthusiasm and diligent work to meet all the technical challenges throughout the years. The authors would also like to thank Mr. Jeffrey Fluhrer for editing this manuscript.

REFERENCES

1. C.H. Lee and S. Kobayashi, New Solutions to Rigid-Plastic Deformation Problem Using a Matrix Method, *J. Eng. Ind. (Trans. ASME)*, Vol 95, 1973, p 865
2. G.D. Lahoti and T. Altan, "Research to Develop Process Models for Producing a Dual Property Titanium Alloy Compressor Disk," Interim Annual Report, AFML-TR-79-4156, Battelle Columbus Laboratories, Dec 1979
3. S.I. Oh, Finite Element Analysis of Metal Forming Problems with Arbitrarily Shaped Dies, *Int. J. Mech. Sci.*, Vol 24, p 479
4. N. Rebelo and S. Kobayashi, A Coupled Analysis of Viscoplastic Deformation and Heat Transfer, Parts I and II, *Int. J. Mech. Sci.*, Vol 22, 1980, p 699, 707
5. W.T. Wu, S.I. Oh, T. Altan, and R. Miller, "Automated Mesh Generation for Forming Simulation—I," ASME International Computers in Engineering Conference, 5–9 Aug 1990
6. J.J. Park and S.I. Oh, "Three Dimensional Finite Element Analysis of Metal Forming Processes," NAMRC XV, 27–29 May 1987 (Bethlehem, PA)
7. G. Li, W.T. Wu, and J.P. Tang, "DEFORM-3D—A General Purpose 3-D Finite Element Code for the Analysis of Metal Forming Processes," presented at the Metal Forming Process Simulation in Industry International Conference and Workshop, 28–30 Sept 1994 (Baden-Baden, Germany)
8. W.T. Wu, G. Li, J.P. Tang, S. Srivatsa, R. Shankar, R. Wallis, P. Ramasundaram, and J. Gayda, "A Process Modeling System for Heat Treatment of High Temperature Structural Materials," AFRL-ML-WP-TR-2001-4105, June 2001
9. J. Oh, J. Yang, W.T. Wu, and H. Delgado, "Finite Element Method Applied to 2-D and 3-D Forging Design Optimization," NumiForm 2004, 13–17 June 2004 (Columbus, OH)
10. C.M. Sellars and J.A. Whiteman, Recrystallization and Grain Growth in Hot Rolling, *Met. Sci.*, Vol 13, 1979, p 187–194
11. G. Shen, S.L. Semiatin, and R. Shivpuri, Modeling Microstructure Development during the Forging of Waspaloy, *Metall. Mater. Trans. A*, Vol 26, 1995, p 1795–1803
12. K. Arimoto, G. Li, A. Arvind, and W.T. Wu, "The Modeling of Heat Treating Process," ASM 18th Heat Treating Conference, 12–15 Oct 1998 (Chicago, IL), ASM International
13. D. Huang, W.T. Wu, D. Lambert, and S.L. Semiatin, Computer Simulation of Microstructure Evolution During Hot Forging of Waspaloy and Nickel Alloy 718, *Microstructure Modeling and Prediction During Thermomechanical Processing*, TMS, 2001, p 137–147
14. J. Yang, G. Li, W.T. Wu, K. Sawamiphakdi, and D. Jin, "Process Modeling for Rotary Tube Piercing Application," Materials Science and Technology 2004, 26–29 Sept 2004 (New Orleans, LA)
15. D. Lambert, P.R. Jepson, and H. Pihlainen, Process Simulation Development for Industrial Rolling Applications, *Modeling, Control, and Optimization in Ferrous and Nonferrous Industry*, Materials Science and Technology 2003, 9–12 Nov 2003 (Chicago, IL), p 529–543
16. K. Sawamiphakdi, C. Ramos, T.J. Favenyesi, R.H. Klundt, and G.D. Lahoti, "Finite Element Modeling of Induction Hardening Process," The First International Conference on Numerical Modeling and Computer Applications on Thermal Process of Automobile Components, 27–29 Jan 2003 (Bangkok, Thailand)
17. D. Lambert, W.T. Wu, K. Arimoto, and J. Ni, "Computer Simulation of Induction Hardening Process Using Coupled Finite Element and Boundary Element Methods," ASM 18th Heat Treating Conference and

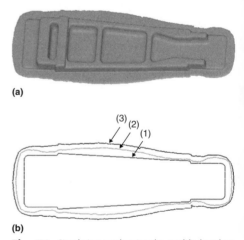

(a)

(b)

Fig. 79 Simulation result using the modified preform design. (a) Deformed shape. (b) Outer boundaries of (1) final product without flash, (2) predicted forging using the initial design, and (3) predicted forging using the modified design

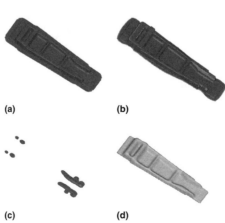

(a) (b)

(c) (d)

Fig. 78 Example of the preform modification scheme. (a) Ideal forged component with uniform flash. (b) Predicted shape from the simulation result. (c) Areas that need to be modified. (d) Modified preform shape

Fig. 77 Simulation result using the initial preform design/deformed shape

Exposition, 12–15 Oct 1998 (Rosemont, IL), ASM International

18. F. Ikuta, K. Arimoto, and T. Inoue, in Proc. Conf. Quenching and the Control of Distortion, 4–7 Nov 1996 (Cleveland, OH)

19. W.T. Wu, J.T. Jinn, J.B. Yang, J.Y. Oh, and G.J. Li, "The Finite Element Method and Manufacturing Processes," EAS-FORM, Feb 2003

20. W.T. Wu, S.I. Oh, T. Altan, and R.A. Miller, "Optimal Mesh Density Determination for the FEM Simulation of Forming Process," NUMIFORM '92, 14–18 Sept 1992 (Valbonne, France)

21. J. Walters, W.T. Wu, A. Arvind, G. Li, D. Lambert, and J.P. Tang, "Recent Development of Process Simulation for Industrial Applications," presented at the Fourth International Conference on Precision Forging Technology, 12–14 Oct 1998 (Columbus, OH); *J. Mater. Process. Technol.*, Vol 98, 2000, p 205–211

22. M. Knoerr, J. Lee, and T. Altan, Application of the 2-D Finite Element Method to Simulation of Various Forming Processes, *J. Mater. Process. Technol.*, Vol 33, 1992, p 31–55

23. J. Hoffmann, C. Santiago-Vega, and V. Vazquez, Prevention of Ductile Fracture in Forward Extrusion with Spherical Dies, *Trans. North Am. Res. Inst. SME*, Vol XXVIII, 2000, p 155

24. B. Avitzur, Analysis of Central Bursting Defects in Extrusion and Wire Drawing, *J. Eng. Ind. (Trans. ASME)*, Feb 1968, p 79–91

25. B. Avitzur, Strain-Hardening and Strain-Rate Effects in Plastic Flow through Conical Converging Dies, *J. Eng. Ind. (Trans. ASME)*, Aug 1967, p 556–562

26. Z. Zimerman and B. Avitzur, Analysis of the Effect of Strain Hardening on Central Bursting Defects in Drawing and Extrusion, *J. Eng. Ind. (Trans. ASME)*, Feb 1970, p 135–145

27. H. Kim, M. Yamanaka, and T. Altan, "Prediction and Elimination of Ductile Fracture in Cold Forgings Using FEM Simulations," Engineering Research Center for the Net Shape Manufacturing Report ERC/NSM-94–42, Aug 1994

28. J. Walters, C.E. Fischer, and S. Tkach, Using a Computer Process Model to Analyze Drawing Operations, *Wire Cable Technol. Int.*, Jan 1999

29. J. Walters, S. Kurtz, J.P. Tang, and W.T. Wu, The 'State of the Art' in Cold Forming Simulation, *J. Mater. Process. Technol.*, Vol 71, 1997, p 64–70

30. A. Lau, "Finite Element Methods Applied to Forging Die Stress Analysis," presented at the North American Forging Technology Conference, 4 Dec 1990 (Orlando, FL)

31. J. Walters, Application of Finite Element Method in Forging: An Industry Perspective, *J. Mater. Process. Technol.*, Vol 27, (No. 1–3), p 43–51; presented at An International Symposium in Honor of Professor Shiro Kobayashi, 15–17 Aug 1991 (Palo Alto, CA)

32. K.K. Wang, "Friction Welding," WRC Bulletin 204, 1975

33. K. Lee, A. Samant, W.T. Wu, and S. Srivatsa, "Finite Element Modeling of Inertia Welding Processes," NumiForm 2001

34. S.I. Oh and S.M. Yoon, A New Method to Design Blockers, *Ann. CIRP*, Vol 43 (No. 1), 1994, p 245–248

35. S.M. Yoon, "Automatic 3-D Blocker Design for Closed Die Hot Forging with Low Pass Filter," Ph.D. dissertation, Seoul National University, 1996

36. J.Y. Oh, "A Study on the Blocker Design and Model Experiment of Closed Die Forging," M.S. dissertation, Seoul National University, 1995

37. S.M. Yoon, J.Y. Oh, and S.I. Oh, A New Method to Design Three-Dimensional Blocker, *J. Mater. Process. Technol.*, submitted in 2002

38. E.O. Brigham, *The Fast Fourier Transform and Its Applications*, Prentice-Hall, 1988, p 240

39. J.S. Walker, *Fast Fourier Transforms*, CPS Press, 1991

40. G.D. Bergland, A Guided Tour of the Fast Fourier Transform, *IEEE Spectrum*, Vol 6, July 1969, p 41

41. J. Yang, DEFORM User's Group Meeting, Fall 2002

42. T.J. Turner, M.P. Miller, and N.P. Barton, The Influence of Crystallographic Texture and Slip System Strength on Deformation Induced Shape Changes in AA 7075 Thick Plate, *Mech. Mater.*, Vol 34, 2002, p 605–625

ASM Handbook, Volume 22B, *Metals Process Simulation*
D.U. Furrer and S.L. Semiatin, editors

Copyright © 2010, ASM International®
All rights reserved.
www.asminternational.org

Sheet Metal Forming Simulation

SOFTWARE PROGRAMS continue to provide increasingly important tools in computer-aided engineering of metal stamping operations. With continuing improvements in the speed and memory of computers, simulation programs are becoming ever more effective tools in reducing the need for physical testing and the avoidance of costly downstream problems by solving the problems upfront in the early development stage.

This article provides an overview on metal stamping simulation based on the finite-element methods or model (FEM). Although the term *computer-aided engineering* may refer to any number of computer protocols, FEM technology is increasingly powerful and accurate with improvements in computing power and supporting technology, such as automatic mesh generation. Hence, emphasis is given to software tools using three-dimensional FEM technology after a brief review of history and applied analysis of simple forming operations.

Simulation of sheet metal forming processes normally also covers a wide range of various processes, the most common one being stamping, but other processes such as tube hydroforming, stretch forming, superplastic forming, and so on are also included in the general topic of sheet metal forming simulation. This article concentrates on the most common techniques used, keeping in mind that special variation of processes requires special adjustments to the input, process setup, and sometimes even the code used.

Overview on Sheet-Forming Simulation

Over the years, successful stamping simulation has helped to considerably reduce the costs and lead time for the automotive industry, partially due to the technology growth. The industry applications of sheet metal forming simulation have grown enormously in recent years, as process simulations eliminate or reduce the need for extensive shop trials.

In the fourth quarter of the 20th century, the use of the finite-element analysis (FEA) for sheet metal forming applications created strong interest for researchers and industry users. Throughout the 1970s and early 1980s, most studies were limited to either axisymmetric or plane-strain problems or to three-dimensional (3-D) problems with simple geometry, such as cups. However, these studies only provided general information on a variety of important issues on sheet metal forming and were of little use for the complicated 3-D sheet metal forming problems. Except for can making, the technology did not offer practical industry applications (Ref 1).

The technology continued to grow in the ensuing 10 years, with an emphasis on real applications and research on studying various issues related to computer-aided engineering (CAE) methodology for sheet metal forming. For example, more practical contact treatments were developed, and better material models, such as Barlat's 1989 and Hill's 1990 yield loci, were published (Ref 2, 3). In the 1990s, more vigorous research and development was being carried out to improve the reliability and accuracy of sheet metal forming simulation. Since 1993, the technology has made significant and rapid progress in many different areas. Considerable efforts were made in improving the challenges for dealing with complex industrial parts so that more process details and bigger model sizes can be analyzed. Several critical technology breakthroughs occurred in the mid-1990s. Most notable were the successful development of automatic meshing for dies and the adaptivity for deformable sheet metal. The main emphasis during that period was focused on formability evaluation from blanking through trimming operations. Users were looking for answers to straining, thinning, wrinkling, and failure predictions.

There were two major breakthroughs in the next period, beginning in 1999. One was the improvement of springback prediction, as a result of several global multiyear consortia. The most important of these was the Springback Predictability Project, funded by the National Institute of Standards and Technology Advanced Technology Program (NIST-ATP). Due to the work of this consortium, the industry finally had a code that could provide decent springback prediction for some parts. Another major development was the emphasis of a user-friendly metal-forming system that integrated all needed functions, including some computer-aided design (CAD) work that used to be done outside of the system.

The rapid developments of software technology and lower-cost computer hardware have enabled many metal-forming operations to be modeled cost-effectively. Simulation lead time has been decreasing significantly since 1990. Advances in the technology also have moved beyond discussions on the validity of forming simulation methods, such as the debate about implicit versus explicit solution schemes and comparison or validation of results. The current focus has been on the drive to simulate the entire forming process and to provide tolerance assessments with respect to springback, gage fit, trimline accuracy, and also surface quality (Ref 4).

In a typical stamping operation, the entire line of dies has five basic stages (Ref 1):

1. Blanking
2. Forming
3. Trimming
4. Flanging
5. Trimming, flanging, and restriking

However, a complete stamping simulation requires more steps:

1. Blanking
2. Gravity loading
3. Binder closure
4. Forming
5. Springback
6. Trimming
7. Springback
8. Flanging
9. Springback
10. Trimming, flanging, and restriking
11. Springback

Simulation and the integrated numerical analysis of all these steps are beyond the scope of this article. The focus here is just on steps 4 and 5 of forming and its associated springback.

There are many different approaches to analyzing formability. Before 3-D FEM became practical, for example, one common method was length-of-line analysis, which simply calculates the ratio between the final length of a cut section and its original length. This unbelievably crude method (by current standards) was even used well into the 1990s before any computer-aided methods became truly useful.

Until the mid-1990s, 3-D FEM was not very practical, and there were two major reasons to use the two-dimensional (2-D) approach. First, the 2-D approach was much faster and more stable. Secondly, developing the die face on the computer was a very slow process in those days. The 2-D solution provided some guidelines to the initial die-face design. Although the technology was relatively crude compared to a 3-D analysis, it was better than experience-based die-design methodology. Two-dimensional solutions could take care of the challenging sliding contact problems in an ideal condition. A more advanced approach was to use shell elements with additional boundary conditions to force a 3-D model to behave like a 2-D problem. The first approach is much faster but has very limited applications. The shell-model approach offers more realistic contact, especially for a more complex section line, but is much slower compared to other 2-D methods.

With current analysis tools, the need for 2-D FEM simulation has diminished tremendously. There are many technical reasons why 3-D FEM simulation for metal forming did not become immediately practical for production use, compared to other application areas of computer simulation. To obtain reasonably reliable results for real production parts, the following formulation issues for FEM technology needed to be resolved:

- Large displacement
- Large deformation
- Large rotation
- Large sliding with nonlinear contact
- Efficient and reliable contact treatment
- Element formulation to handle through-thickness variation in order to model combined stretching and multibending and unbending
- Advanced material model to handle anisotropy and cyclic loading
- Advanced friction model to capture the surface traction
- Intelligent adaptivity algorithm to refine the mesh as needed

As noted, significant advancement in this area began in the early 1990s. The majority of the issues listed have been overcome with new technology that was developed and enhanced between 1990 and 2000. Although most of the early work was done by academia, codes developed by some commercial software companies have been the primary workhorses in real-world applications. The current market is primarily dominated by AutoForm of AutoForm Engineering, LS-DYNA of Livermore Software Technology Corporation, and PAM-STAMP of ESI Group. They successfully built their own codes based on early codes from academia. Over the years, they continuously enhanced their codes with in-house development or implementation of published work. Few people

in the industry use codes other than these three codes.

FEM Simulation of Sheet Forming

Finite-element analysis is a very general tool, and Ref 5 to 10 are presented in approximate order of increasing difficulty. In metals forming analysis, FEA consists of the following steps (Ref 11):

1. Establish the governing equations: equilibrium (or momentum for dynamic cases), elasticity and plasticity rules, and so on.
2. Discretize the spatially continuous structure by choosing a mesh and element type.
3. Convert the partial differential equations representing the continuum motion into sequential sets of linear equations representing nodal displacements.
4. Solve the sets of linear equations sequentially, step forward and repeat.

Items 1 and 2 of this list are of particular importance for springback analysis. For item 1, many choices of material model may be used, but most forming simulations rely on two basic governing equations: either static equilibrium is imposed in a discrete sense (i.e., at nodes rather than continuous material points), or else a momentum equation in the form of $F = MA$ is satisfied for dynamic approaches. For nearly all commercial codes used for metal forming, the static equilibrium solutions are obtained with implicit methods that solve for equilibrium at each time step by iteration, starting from a trial solution. Thus, such programs are often referred to as static implicit. Examples of static implicit include ABAQUS Standard and ANSYS. Forming programs that solve a momentum equation typically use explicit methods that convert unbalanced forces at each time increment into accelerations but do not iterate to find an assured solution. Examples include ABAQUS Explicit and LS-DYNA.

The essence of the FEM, as opposed to other discrete treatments such as the finite-difference method, lies in an equivalent work principle. Continuous displacements within an element are represented by a small number of nodal displacements (and possibly other variables) for that element. Similarly, the work done by the deformation throughout the element is equated to the work done by the displacements of the nodes and thus are defined by equivalent internal forces at these nodes.

The internal work is computed by integrating the stress-strain relation over the volume of the element. This frequently cannot be accomplished in closed form, so certain sampling points, or integration points, inside the element are chosen to simplify this integration. The number and location of integration points may be selected to provide the desired balance

between efficiency and accuracy, or between locking and hourglassing, as mentioned subsequently.

For forming and springback analysis, the procedure consists of applying boundary conditions (i.e., the motion of a punch or die, the action of draw beads, frictional constraints, and so on), stopping at the end of the forming operation, replacing the various contact forces by fixed external forces (without changing the shape of the workpiece), and then relaxing the external forces until they disappear. The last step (or steps) produces the springback shape.

Because the choices of program, element, and procedure usually apply to both the forming and springback steps, it is difficult to separate discussions of accuracy between the two stages. The deformation history established in the forming operation is used in the springback simulation via the final shape, loads, internal stresses, and material properties.

Two choices of particular importance in FEA of sheet forming and springback analysis are:

- Type of solution algorithm/governing equation
- Type of element

These two aspects are discussed as follows, followed by an example of FEM calculation code.

Explicit and Implicit Solution Algorithms

One of the most important technologies in a 3-D FEM code is the time-integration scheme. Generally speaking, this can be divided into two major families: explicit time integration and implicit time integration. Table 1 lists general advantages and disadvantages of the two methods.

Both algorithms can be applied to either static or dynamic governing equations. Hence, there are four possible fundamental formulations used by the FEM codes:

- Static implicit
- Static explicit
- Dynamic implicit
- Dynamic explicit

Among them, the static implicit formulation is widely used for static or quasi-static problems such as structural analyses. The dynamic explicit formulation is widely used for dynamic problems such as drop tests and vehicle crash simulations. However, the choice of integration algorithm for metal-forming applications was not that obvious in the early years as FEM started to be used in this field. The majority of researchers thought static implicit formulation was more rigorous and suitable for accurate results for metal-forming simulations.

By 1990, almost all successful sheet metal forming simulations published in major conferences and technical journals used the static implicit approach, with a few exceptions using the static explicit formulation. As papers using dynamic explicit formulations began to increase throughout the 1990s, the debate about this issue continued for many years. One major reason fueling the debate was that the majority, if not all, of the codes used a single time integration method. Most debates seemed to focus on justifying the superiority of their own codes rather than a fair comparison of the advantages and disadvantages of each method. With the realization that both methods have usefulness, this debate is now moot. Most software vendors now use both.

Most applied sheet-forming analysis in industry currently uses dynamic explicit methods. The complicated die shapes and contact conditions that occur in complex industrial forming are more easily handled by the very small steps required by the dynamic explicit methods. The stress solutions are of little importance. Often, even simulations that are inaccurate in an absolute sense can be used by experienced die designers to guide sequential modifications leading to improved dies. On the other hand, if a certain set of tools cannot be simulated successfully, that is, with an implicit method that does not converge, the die improvement process is stymied completely.

As discussed in the section "Springback Analysis" in this article, springback simulation is much more sensitive to numerical procedure than forming analysis. The reason is simple: Springback simulation relies on accurate knowledge of stresses throughout the part at the end of the forming operation. Conversely, forming analysis is primarily concerned with the distribution of strain within the shape of the part. The shape of the final part is largely determined by the shape of the dies because, near the end of the forming operation, contact occurs over a large fraction of the workpiece. Therefore, the oscillatory nature of the stresses obtained at the end of a dynamic explicit analysis may be unsuited for accurate springback analysis. Poor and uncertain results have been reported (Ref 12).

Developments are currently proceeding in attempts to artificially smooth or damp dynamic explicit forming solutions in the hope of providing a stable base for springback calculations. It is too early to be confident that these approaches will be successful. Certain isolated results can appear promising; however, as is shown later in this section, it is not unusual to obtain fortuitously accurate results in springback analysis. For this reason, great caution should be used in drawing conclusions from a small number of apparently accurate predictions.

The foregoing refers to the drawbacks of a dynamic explicit simulation of a forming operation prior to a springback analysis. The springback simulation itself is also much better suited to implicit methods because the operation is

dominated by quasi-static elastic deformation that is computed very inefficiently by dynamic explicit methods. For this reason, implicit springback analysis is often favored even after explicit forming analysis. It is for these two reasons that static implicit methods are better suited to forming analysis where accurate springback predictions are required. The obvious drawback is the uncertain convergence of current versions of such methods.

Table 2 gives some well-known codes representing each method. Note that the codes listed in Table 2 were those popular during that period. Some of them may not be available today (2010). SHEET-x represents several codes from The Ohio State University. The dynamic implicit method was available as an option in some dynamic-explicit-based codes.

The static implicit formulation used for metal-forming simulation is very similar to the one used for structural analysis. It transforms all externally applied force (such as the moving punch) into internal energy, due to the deformation of the sheet metal. The equilibrium state is maintained at every state. This method has been demonstrated to be far superior to other methods simulating small-scale lab-type parts such as cups. The drawbacks of the static implicit method are its storage and central processing unit (CPU) requirements that grow nearly as a cubic function of the number of elements. As the size or the complexity of the geometry grows, the static implicit approach becomes less practical. As computer random access memory price decreases and CPU speed increases as Moore's law predicts, this may

become more possible but still less efficient compared to dynamic explicit codes.

There are other issues that limit the widespread use of static implicit codes in a production environment. Sheet metal forming is not really a "pseudostatic" problem. In a typical stamping operation, the actual forming completes the total travel of 100 to 150 mm (4 to 6 in.) in 2 to 3 s, with the peak speed close to 500 mm/s, (20 in./s). The sheet metal flows in various directions and changes in the process of forming. Most areas (nodes in FEA) of the sheet metal have large displacement and large rotation in addition to large deformation (in elements). This limits the step size of static implicit codes in order to capture the physical behavior of the metal.

The other challenge is that the contact between the metal and the dies is highly nonlinear. Correctly capturing the contact and calculating the friction force is not an easy task for static implicit codes. Capturing the onset of wrinkling and correctly predicting the actual formation of wrinkles are also difficult challenges for static implicit codes. One fundamental characteristic of the static implicit algorithm is the need to invert the stiffness matrix. As the complexity and the size of the problem grow, convergence slows down tremendously or cannot be accomplished at all.

AutoForm, from AutoForm Engineering, is a unique code. Some people classify it as a fast implicit code. It uses the same governing equation as the static implicit codes. The major reasons for its fast performance are that it uses membrane element and some decoupled proprietary techniques. Among them, the matrix is

Table 1 Advantages and disadvantages of static implicit and dynamic explicit finite-element programs

Program type	Advantages	Disadvantages
Static implicit	• Known accuracy • Equilibrium satisfied • Smooth stress variation • Elastic solutions are possible • Unconditionally stable	• Solution not always assured • Complex contact difficult to enforce • Long computer processing times for complex contact
Dynamic explicit	• Solution always obtained • Simple contact • Short computer processing times with mass scaling	• Uncertain accuracy • Equilibrium not satisfied in general • Mass scaling introduces error in static problems • Oscillatory stress variation • Elastic solutions are difficult and slow • Conditionally stable

Table 2 Summary of metal stamping codes by integration algorithm

Method	Code names
Static implicit	ABAQUS (standard), ADINA, MARC, INDEED, NIKE, Metalform
Dynamic explicit	LS-DYNA, PAM-STAMP, OPTRIS, RADIOSS, LLNL-DYNA, ABAQUS/Explicit, DYTRAN, DYNAMIC
Static explicit	ITAS, ROBUST, SHEET-x, Panelform

Note: ABAQUS, ABAQUS, Inc.; ADINA, ADINA R&D, Inc.; NIKE, Lawrence Livermore National Laboratory; LS-DYNA, Livermore Software Technology Corp.; PAM-STAMP, ESI Group; OPTRIS, Dynamic Software; RADIOSS, Mecalog Group; LLNL-DYNA, Lawrence Livermore National Laboratory; DYTRAN, MSC Software; SHEET-x, The Ohio State University

decoupled and simplified. This allows the matrix to remain small with a high convergence rate. Although many experts doubt the technical soundness of some of the techniques used by AutoForm, the code seems to generate relatively good results in most cases.

Dynamic Explicit Formulation. The basic governing equation of dynamic explicit codes has an inertia term similar to the following equation. Different codes may use different formulations, but the concept of the momentum equation is the same:

$$F_{int}(\dot{u}) + c + m\ddot{u} = F_{ext} \qquad \text{(Eq 1)}$$

where F_{int} is the internal force, F_{ext} is the external force, c is damping, m is mass, u is displacement, $\dot{u}$ is velocity, and $\ddot{u}$ is acceleration.

Critics of dynamic explicit codes argue that equilibrium is never maintained. This statement is a little misleading. In dynamic explicit codes, equilibrium is dynamically balanced. It is true that the inertia force may be exaggerated, mostly due to user error. In that case, the quality of the solution will deteriorate. If a user chooses a proper setting so that the inertia is no more than 5%, the solution quality should be good.

Static explicit codes were presented as taking the advantages of both static implicit and dynamic explicit codes. It was actively promoted in the 1990s. The results seem to indicate that static explicit codes have the weaknesses of both codes. The popularity of this group greatly diminished in the late 1990s after the third NUMISHEET conference in 1996.

Elements and Mesh Generation

For the simulation of sheet metal forming processes, normally FEM is used. All the components of a calculation (metal sheet or tube, tools) are shown as meshes, or a discrete representation of the geometry. For nondeformable tools, the mesh is only a representation of the geometry, and the finite elements are only facets to be used for contact description. In contrast, for the blank, the tube, or a deformable tool, the finite elements in the mesh represent small pieces of the material that deforms with a prescribed behavior.

The choice of element refers to the number of nodes per element, the number of degrees of freedom at each node, and the relationship with the assumed interior configuration (among other things) that define the element type. The nodal displacements are the primary variables to be solved for. A fixed relationship between the displacements of points within the finite element to the displacements of the nodes is assumed. In this way, the continuous nature of the deformation within the element is related to a small number of variables. Of course, the distribution in the element may be quite different from the continuum solution, but this difference can often be progressively reduced by refining the mesh, that is, by choosing finer and finer elements. By comparing the solutions, an adequate mesh size can be determined.

Different element types are used for different calculation purposes. The three main types of elements are:

- Membrane elements
- Shell elements
- Volume elements

Each of these elements can have a different number of nodes and different behavior and element formulations. Some examples of commonly used element types are described subsequently. More details on element types and use can be found in the section "The Basis of Finite Elements" in *Materials Selection and Design,* Volume 20 of *ASM Handbook* (Ref 13).

Elasto-Plastic Shell Elements with three or four nodes (Fig. 1) have the following characteristics:

- Midlin-Reissner's theory: The transverse shear strain, constant in the thickness, is taken into account.
- Each node has three degrees of freedom in translation and three degrees of freedom in rotation.
- Subintegration: There is only one integration point on the surface of the element and in its thickness.

- No transverse shear locking
- Excellent CPU time

Elasto-plastic volume elements (Fig. 2) have the following characteristics:

- Four to eight nodes per element
- Element C^0: trilinear interpolation polynomials
- Each node has three degrees of freedom in translation and no degree of freedom in rotation.
- Subintegration: There is only one integration point in the element.

Bar elements can be employed to discretize cables, springs, and trusslike members that are unable to transmit bending and torsion moments. In the case of cables, these elements cannot transmit compression forces. Bar elements have the following characteristics:

- Each node has three degrees of freedom in translation.
- No moment is applied to the nodes by the element.

Bar elements can also be used to model gas springs within a die. To do this, the compressive behavior of the spring is defined using a force-versus-displacement curve.

Non-elasto-plastic shell elements with three or four nodes have the following characteristics:

- Each node has three degrees of freedom in translation.
- No stiffness

Tools are normally defined as rigid bodies.

Adaptivity. Proper mesh definition is always important for all FEM simulation. However, it is not practical to use fine mesh for the entire blank from the beginning of a forming operation due to the high CPU hours needed. Adaptivity can intelligently refine the blank mesh only in the areas needed. Users do not have to figure out a "smart" mesh pattern to capture the deformation gradient correctly. It is especially essential for the deep drawing operation, when the sheet metal often has large displacement. Adaptivity has been around for many years but was not popular until 1995, due to some limitations of earlier adaptive algorithms.

Since 1997, it has become standard practice among the majority of CAE analysts to use adaptivity regardless of the software used, because the technology provides fast and reliable results. By the late 1990s, all

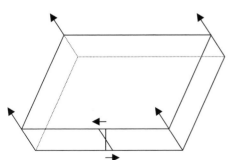

Element C°: bilinear interpolation polynomials

Fig. 1 Elasto-plastic shell elements

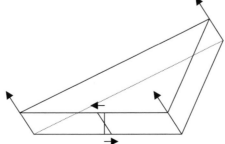

Element C°: degree one interpolation polynomials

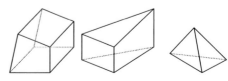

Fig. 2 Elasto-plastic volume elements

metal-forming software also developed intelligent algorithms that can foresee contact problems and refine the mesh before it is needed. Although the details of the algorithms may be different, they are all based on the element normal differences between adjacent elements in the approaching tooling mesh.

The algorithms work well in most stamping cases. However, a different criterion may be needed in some special cases. It does not work at all in other types of simulation, such as crashworthiness. It is desirable to modify the algorithm to use a different index other than angle differences. Those indexes may include thinning, effective stress, or effective strain.

The purpose of adaptivity is to refine the mesh. There are three fundamentally different approaches: *h*-adaptivity, *p*-adaptivity, and *r*-adaptivity. The most common method used by sheet metal forming software is *h*-adaptivity. This method splits the elements when the criterion is met; hence, it is dubbed as "fission." Although there are some variations in *h*-adaptivity, each element is often split into four elements. Adaptive constraints are applied to middle nodes to take care of connectivity requirements.

FEM Calculation Code

The mechanical phenomena that occur in a blank or in a tube are faithfully reproduced using a large number of these elements. Within reason, the finer the mesh to be generated, the better the quality of the results, whereas the higher the number of elements, the longer the calculation time. In any simulation, if the detail size is smaller than that of the elements, it cannot be represented. The size of the elements defines the precision of the simulation.

A finite element can be a two-node element (bar), a three-node element (triangle), a four-node element (quadrangle), or a six- or eight-node volume element (hexahedron), and it is constructed from nodes that are defined in its corners. Each node has two types of degrees of freedom: translation and rotation. The translation degree of freedom of a node represents its ability to move in translation along a direction, whereas a rotation degree of freedom of a node represents its ability to rotate about an axis. A node with three degrees of freedom in translation and three degrees of freedom in rotation can move along three axes—*X*, *Y*, and *Z*—and can rotate about these three axes.

Depending on the calculation type (implicit or explicit), the calculation is subdivided into increments or time-steps. Generally, implicit increments are large with respect to the explicit time-steps. Positions, velocities, accelerations, and forces are permanently calculated at the nodes (Fig. 3), which are points linked to the material. Within the elements, strains are calculated from positions.

Corresponding stresses are then obtained, which result in forces on the nodes. This calculation is repeated over all the elements for the entire duration of the calculation. Boundary conditions are used to remove degrees of freedom (locking), while velocities and forces further define the kinematic behavior of the FEM. To describe the actual deformation process, material properties and thickness must be assigned to an element.

Explicit, Implicit, and Advanced Implicit Algorithms

Algorithms used by the solver of numerical simulation work step-by-step to find dynamic equilibrium at each step. Various types of algorithms can be used: explicit, implicit, and advanced implicit. The main differences are highlighted through this section. The principle of the explicit and the implicit time integration of a one-dimensional (1-D) system with one degree of freedom can be represented by a linear spring system (Fig. 4). The spring system is a linear damped spring system with an equilibrium equation of:

$$m \cdot a_n + c \cdot v_n + k \cdot x_n = f_n$$

where *n* means the time increment.

Explicit Calculation Type. In the explicit method, the nodal velocities are written down at times $t_{n-1/2}$, $t_{n+1/2}$ and nodal displacements and accelerations at times t_{n-1}, t_n, t_{n+1}. At time t_n, the nodal displacement x_n is known, and the acceleration a_n is computed from the internal and external forces. Nodal velocity, $v_{n-1/2}$, is known at time $t_{n-1/2}$. The algorithm searches for the nodal velocity $v_{n+1/2}$ at time $t_{n+1/2}$ and the nodal displacement x_{n+1} at time t_{n+1}.

The application of the central-difference method gives nodal velocity at time $t_{n+1/2}$ and the nodal displacement at time t_{n+1} (assuming that $\triangle T_n$ is small):

$$a_n = m^{-1} \cdot (f_n(t) - k \cdot x_n)$$

in case of no damping applied.

$$a_n = \frac{V_{n+1/2} - V_{n-1/2}}{(\triangle T_{n-1} + \triangle T_n)/2}$$

$$V_{n+1/2} = \frac{X_{n+1} - X_n}{\triangle T_n}$$

For complex processes (other than a 1-D system), *m* is a matrix; it is diagonal and can be immediately calculated without any matrix inversion. Unfortunately, this method is stable only if a small time step $\triangle T_n$ is used

Implicit Calculation Type. When using the implicit method, stamping simulations are considered as static, using an incremental method (based on loading or tool kinematics). The dynamic effects are neglected, and the velocity and the acceleration are set to zero.

Within one increment, the solver automatically tries to find the solution of a set of nonlinear equations, using linear iterations, also known as Newton iterations, with convergence criteria. An example of Newton iterations for a force (*F*) as a function of the displacement variable, *u*, is illustrated in (Fig. 5). The two iterations in Fig. 5 are calculated as follows:

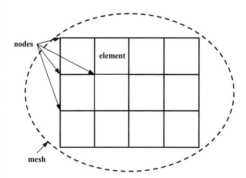

Fig. 3 Nodes in a mesh

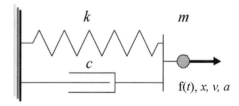

Fig. 4 Linear spring constant (*k*) with damping constant (*c*) for a mass (*m*) with position (*x*), velocity (*v*), and acceleration (*a*) as a function of time, f(*t*)

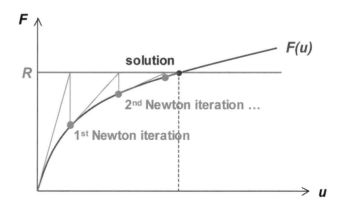

Fig. 5 Example of Newton iteration. *R* is the applied external force (F_{ext}), and F(*u*) is a function of displacement (*u*)

First iteration

$$F(u) = F(0) + \delta F/\delta u(0)\delta u, \text{ with}$$
$$\to \delta u_1 = K^{-1}(0)(F_{\text{ext}} - F(0))$$
$$\to u_1 = \delta u_1$$

where K^{-1} is a matrix inversion.
Second iteration

$$F(u) = F(u_1) + \delta F/\delta u(u_1)\delta u, \text{ with}$$
$$\to \delta u_2 = K^{-1}(u_1)(F_{\text{ext}} - F(u_1))$$
$$\to u_2 = \delta u_1 + \delta u_2$$

In the case of a forming example, the force function $F(u)$ could represent an external applied force (F_{ext}), where $F_{\text{ext}} = 0$ in the case of springback. Total displacement (u_n) is:

$$u_n = \delta u_1 + \delta u_2 + \ldots + \delta u_n$$

and the displacement convergence is reached when:

$$|\delta u_n|/\max(|\delta u_i|) < \text{Tolerance}$$

In Fig. 5, the solution is $R = F_{\text{ext}}$.

Convergence Criteria. Different convergence criteria can be used. Two typical criteria that are used to check the convergence of the solution are:
Displacement convergence tolerance

$$|\delta u_n|/\max(|\delta u_i|) < \text{Tol}_d$$

Energy convergence tolerance

$$|\delta E_n|/\max(|\delta E_i|) < \text{Tol}_E$$

where $\delta E_n = \delta u_n \cdot (F_{\text{ext}} - F(u_{n-1}))$
The energy convergence tolerance is normally used for springback simulation.

Matrix Inversion. At each Newton iteration, there is a matrix inversion, K^{-1}, which can be done by several methods. Some typical methods are described as follows.

Direct. The matrix is directly inverted using the decomposition method of K into two triangular matrices ($K = LU$).

The preconditioned conjugate gradient method solves $Ku = R$ with an iterative technique. With this method, the matrix is not really inverted. Iterations are performed on vectors u_i, which converge to the solution $u_n = K^{-1}R$.

Iterative. The linear system $Ku = R$ is also solved with an iterative technique. With this method, the matrix is not really inverted. Iterations are performed on vectors u_i, which converge to the solution $u_n = K^{-1}R$.

Time-Steps and Increments. Implicit and explicit integration schemes are the two types of solvers commonly used. Implicit and explicit algorithms work step by step to find dynamic or static equilibrium at each step. While the explicit integration scheme makes use of time-steps, the implicit solver uses increments to discretize the calculation.

Implicit Increments. For implicit calculation, the equation is not solved using the whole imposed displacement or load; displacement and load conditions are broken up into several steps, called increments. Each step is called a load increment (Fig. 6). For each increment, the solution of the sets of nonlinear equations requires iterations, called Newton iterations. A convergence criterion is needed to stop the iteration loop.

Explicit Time-Steps. For explicit calculation, the state of the simulation is not continuously calculated; the time is broken up into a large number of steps called cycles, and the state of the simulation is calculated for each cycle. The interval between two cycles (Fig. 7) is called a time-step.

A time-step must be small enough to satisfy the stability condition of the explicit integration scheme and to assume small displacement approximations. A local time-step is associated with each element. This element time-step, ΔT_{el}, is equal to the time taken by an elastic wave to pass through the element. Hence, it depends on the size, density, and elastic modulus of the element. A global time-step, ΔT, used for the calculation is computed from these element time-steps. Only the time-step of deformable elements is used for the calculation of the global time-step.

Mass Scaling. The purpose of mass scaling is to reduce CPU time by increasing the time-step. This increase of the time-step leads to a reduction of the number of cycles and thus to a reduction of the CPU time, because the latter is proportional to the number of cycles. To do this, the mass of some elements is artificially increased; thus, the element time-steps are increased too.

Kinematics and Loading. The kinematics conditions are attributes applied to an object to determine its movement conditions (fixing, imposed velocity, or displacement). When kinematics or loadings are applied to an object, they are applied to all the entities of lower hierarchical level included in the initial entities. Therefore, the user should adopt a "physical" approach rather than a numerical one when choosing the objects to which kinematics will be applied. A metal sheet side, for example, is a set of edges and not a set of nodes. A tool is a set of elements and not a set of nodes.

Each tool object normally gets one of these kinematic attributes:

- *Cartesian kinematics:* Given a coordinate system, in each direction, a locking condition, displacement, or velocity can be defined, also as a curve over time.
- *Rotational kinematics:* The rotational kinematics attribute enables angular velocities to be imposed for an object. For example, it is used to model the rotation of the bend-die object in a tube-forming process.
- *Kinematic path:* This is a special type of kinematics used, for example, in roll-hemming simulation to move a robot along a 3-D curve.

A tool can also be controlled by loading, which is a force attribute. Normally, a force is added in a given direction; this can depend on time. A force along a curve can also be defined. Through force special processes, such as hydroforming, pressure can be applied directly on the element normals.

Normally, the blank is deformed and/or kept in place by tool objects. In addition, boundary conditions can be added directly on the blank; for example, the nodes on one side can be locked.

Material Yield Criteria

There are different yield criteria used as models of material deformation, depending on the type of material that is simulated. The most commonly used for standard steel is still a planar-isotropic model developed by Hill in 1948 (Ref 14, 15). However, the different behaviors of other materials, such as dual-phase steels or transformation-induced plasticity steels, aluminum, titanium, and so on, require special material laws adapted to the behavior of that specific material. While the thinning calculation in dynamic explicit codes is less dependent on the stress and material model, it is crucial to have a good material model and stress calculation for springback analysis.

There are several important criteria for choosing a proper model. First, the parameters needed for the model can be easily acquired through relatively simple tests. Second, the model should not create too great a cost penalty in using it. Third, it should not cause too many numerical complications that compromise the accuracy due to approximation or round-off error. Fourth, it should not involve parameters that require adjustment from case to case.

Many more advanced models and yield criteria have been proposed since 1989. However, these advanced material models, although theoretically sound, often do not give better springback prediction. The confusing results may be due, in part, to the stress noise commonly seen in analyses with dynamic explicit codes. The stress noise is the result of stress-wave propagation and penalty-method-based contact algorithm.

The general consensus is that isotropic hardening does not accurately describe the sheet metal behavior and always notably underpredicts springback. This helped encourage a substantial amount of work on the kinematic

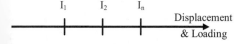

Fig. 6 Increment used in implicit type of finite-element analysis calculation

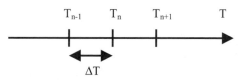

Fig. 7 Time-step in explicit types of finite-element analysis calculation

hardening model. However, the results do not support the theory that kinematic hardening models are generally superior to isotropic hardening models. Springback calculations using a kinematic hardening model often (but not always) overpredict the springback amounts. The Chaboche model was implemented by some metal-forming codes. However, the results in terms of springback prediction are inconclusive for various reasons. A combination of isotropic and kinematic hardening models that account for anisotropic behavior may be a potential material model to use.

Models Used for Yield Criteria

Hill Yield Criteria. The Hill 1948 yield criterion is an improvement over the von Mises yield criterion by accounting for normal anisotropy. This planar-isotropic yield criterion for plane stress is:

$$\bar{\sigma}^2 = \sigma_{11}^2 - \frac{2R}{(R+1)}\sigma_{11}\sigma_{22} + \sigma_{22}^2 + \frac{2(2R+1)}{(R+1)}\sigma_{12}^2 \tag{Eq 2}$$

where σ is stress, and R is the anisotropy parameter.

Considering only the principal stresses, this equation can be simplified as:

$$\bar{\sigma}^2 = \sigma_1^2 + \sigma_2^2 + R(\sigma_1 - \sigma_2)^2 \tag{Eq 3}$$

This yield surface is an improvement over the Tresca yield surface (an irregular hexagon) and the von Mises yield surface (an ellipse). However, it seems to exaggerate the influence of the anisotropy, especially in the biaxial stretching quadrant. Hosford proposed a yield criterion in 1979 that is mathematically similar but with a high-order exponent (Ref 16):

$$\bar{\sigma} = \sigma_1^m + \sigma_2^m + R(\sigma_1 - \sigma_2)^m \tag{Eq 4}$$

where m is a mathematical exponent.

Exponents of $m = 6$ and $m = 8$ were proposed for body-centered cubic materials such as ferritic steel and face-centered cubic metals such as aluminum, respectively. This locus lies between the Tresca criterion and Hill's quadratic form and matches experiments better, compared to the other three criteria. However, this minimizes the influence of anisotropy due

to its high exponent, and it does not address the planar-anisotropy issue. Deep-drawn cups such as beverage cans made of certain metals have a strong earing phenomenon. This is due to the directional anisotropy in the plane of certain metals, such as the extra-deep-draw-quality steel. The following equation is a general form of the aforementioned yield function that considers the anisotropy in both the rolling and transverse directions:

$$2\bar{\sigma}^m = \frac{2}{(1+r_0)}[\sigma_1^m + (r_0/r_{90})\sigma_2^m + r_0(\sigma_1 - \sigma_2)^m] \tag{Eq 5}$$

where r is the anisotropy parameter, also known as the Lankford anisotropy coefficients r_0, r_{45}, and r_{90} along directions $0°$, $45°$, and $90°$ with respect to the rolling direction, such that:

$$r_\alpha = \frac{\varepsilon_w(\alpha)}{\varepsilon_t(\alpha)}$$

where α is the angle relative to the rolling direction, ε_w is the width strain, and ε_t is the thickness strain. When $m = 2$, Eq 5 is the same as the von Mises yield criterion if $r_0 = r_{90} = 1$, and it represents Hill's 1948 quadratic yield criterion if $r_0 = r_{90} = R$. It coincides with Hosford's 1979 yield criterion given in Eq 4 when $r_0 = r_{90} = R$. Lankford anisotropy coefficients must be defined. The need for Lankford's coefficients depends on the anisotropic type used:

- *Orthotropic:* The anisotropy in three directions is taken into account, so three values, r_0, r_{45}, and r_{90}, are needed.
- *Normal:* Only the anisotropy through the thickness is taken into account, so only one average value r, independent of the direction, is needed.
- *Isotropic:* There is no anisotropy, so there is no need for Lankford's coefficient.

A traction test (Fig. 8) is performed on the test sample, then $\varepsilon_w(\alpha)$ and $\varepsilon_t(\alpha)$ are measured. If Hill's coefficients are known and not Lankford's coefficients, it is possible to calculate them into Lankford's coefficients.

Barlat-Lian Model. Barlat and Lian proposed a model in 1989 that was developed with aluminum alloys in mind (Ref 2). This model

offers improved agreement compared to models using Hill's 1948 model with planar isotropy. It is less complex compared to Barlat's later models (see the article "Modeling and Simulation of the Forming of Aluminum Sheet Alloys" in *Metalworking: Sheet Forming,* Volume 14B of *ASM Handbook,* Ref 17); hence, it is easier for implementation:

$$2\bar{\sigma}^m = a\{|K_1 + K_2|^m + |K_1 - K_2|^m\} + c \cdot |2K_2|^m \tag{Eq 6}$$

where K_1, K_2, a, c, and h are defined as:

$$K_1 = 1/2(\sigma_1 + h\sigma_2)$$
$$K_2 = \sqrt{\left(\frac{\sigma_1 - h\sigma_2}{2}\right)^2 + p^2\sigma_{12}^2}$$
$$a = 2\left(1 - \sqrt{\frac{r_0}{1+r_0}\frac{r_{90}}{1+r_{90}}}\right) \tag{Eq 7}$$
$$c = 2\sqrt{\frac{r_0}{1+r_0}\frac{r_{90}}{1+r_{90}}}$$
$$h = \sqrt{\frac{r_0}{1+r_0}\frac{1+r_{90}}{r_{90}}}$$

The variable p does not have a close form solution, and it is found by an iterative search to solve $g(p)$:

$$g(p) = [2m\bar{\sigma}^m/(\partial\Phi/\partial\sigma_1 + \partial\Phi/\partial\sigma_2) \cdot \sigma_{45}] - 1 - r_{45} \tag{Eq 8}$$

The Corus-Vegter model is a refined case of the original Vegter model, and it is essentially a discrete yield description, derived from several experimental measurements. This yield model has the potential to be more accurate in representing the yield surface and effective in improving the results of simulations, in both formability and springback predictions.

The Corus-Vegter model may have some advantages over the commonly used yield models. A comparison between Hill 1948, Hill 1990, and Vegter yield loci is shown in Fig. 9. The Vegter model is constructed through four points on a quarter of the yield ellipse; both

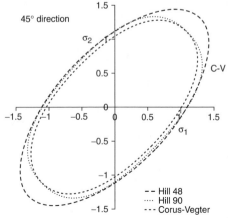

Fig. 9 Comparison of Hill 1948, Hill 1990, and Vegter yield ellipses in the 45° direction

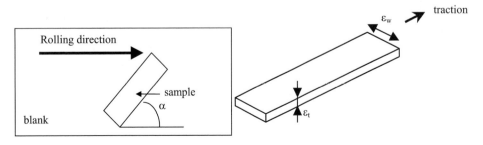

Fig. 8 Traction test

the point and the slope (or tangent) are either measured or are known. Bezier interpolation between the points and two symmetry planes are then used to construct the planar ellipse. Cosine interpolation between the 0°, 45°, and 90° measurement sets is used to construct the entire yield surface description.

Contact and Friction (Adapted from Ref 18)

One of the least understood and most significant factors in sheet metal forming is the role of friction between the sheet metal and the die as the metal slides across the die face as the part is being formed. The advent of computer simulation and the future direct-engineering methods have provided the ability to put the effects of friction into the part and die designs mathematically. The problem (circa 2006) is that the effects are not well quantified. The practical approach has been to adjust a single friction coefficient used in the analysis until the analysis matches well a known condition, and then use that coefficient for future analyses. As with material properties, analysts have tended to err on the side of conservative friction values.

There are three separate frictional conditions acting in a draw die. These are:

- Metal passing through a draw bead
- Metal clamped in the binder
- Metal sliding across a die radius while it is simultaneously changing the wrap angle on the radius and is being increasingly plastically strained

The contact and friction are normally defined simultaneously for each pair of objects by declaring that one of the objects (the master) is impenetrable by the other (the slave). The contact then permanently prohibits the nodes of the slave object, generally the blank, from penetrating master object elements, generally the tool.

The most commonly used friction law is the Coulomb friction:

$$F_f = \mu F_n$$

where μ is the friction coefficient that must be defined, depending on the lubricant used. Examples of Coulomb's coefficient are:

−0.05 to 0.1	Excellent sliding
−0.1 to 0.15	Conventional values
−0.2	Rough surface

Even though a constant value for Coulomb friction is the most commonly used, variations where the friction depends on, for example, pressure and/or velocity are also used.

Contact Types

Various contact types are used, depending on the project type and on the kinematics and loading conditions, with typical contact algorithms being penalty, nonlinear penalty, Lagrangian,

or self-contact. As a sample of contact type, the penalty contact is described here.

Penalty Contact. This contact type allows small penetrations between slave and master objects. The tool geometry is less respected than in more accurate contact types. The penalty forces are generated in response to existing or potential penetrations between two objects. The forces are applied on the nodes of slave objects penetrating the master.

When the sheet is compressed between two rigid tools, the normal elasticity provided by the penalty spring ensures that contact pressure is distributed over a finite area rather than being concentrated at a single slave node. The effective stiffness of contact is a function of the original slave mesh size, increasing with the fineness of the slave mesh. Best modeling results are obtained when this stiffness is equivalent to the normal elasticity of the physical tools.

Normals of the tool elements must be oriented consistently, such that they define the contact surfaces of the tools. They must point toward the blank. This contact type is simple to use, but the results are not so accurate. On the other hand, when the quality of the tool CAD data is not so good, then penalty contact will still provide a usable contact.

Frictional Conditions Acting in a Draw Die

The draw bead, illustrated in Fig. 10, has been the object of the greatest investigative effort. Several physical draw bead simulators exist, and many data have been extracted from them and many papers written on those investigations. However, the draw bead performance of specific materials with specific lubricants has not been published as any type of accepted, and greatly needed, industry standard.

The amount of restraining force generated by a draw bead is a function of the sheet material being pulled through it, the lubricant, the die condition (e.g., material roughness, etc.), and the dimensions, depth (D) and width (W), of the bead. For any specific set of material, die condition, dimension W, and lubricant, the restraining force as a function of dimension D can be established in the laboratory by experimentation with a draw bead simulator (DBS). The results are illustrated in Fig. 11.

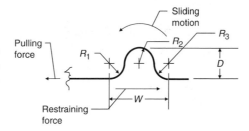

Fig. 10 Draw bead. The draw bead creates a restraining force on the metal as it slides through the binder into the die cavity by bending, unbending, and friction as it is pulled through three (or four) radii (R). D, depth; W, width

The bead restraining force is not used in the direct engineering of the die design. This analytical technique inputs (or calculates) the restraining force. The bead force is used after the design has been finalized to determine the bead geometry that will create the calculated required restraining force. That geometry is then built into the CAD data files for the die face.

There are several draw bead computer models in use today (2010). Some of these are commercially available, and some are proprietary. These models allow interpolation between, and extrapolation beyond, actual test data to provide bead performance data on a wide range of conditions. For these models to work correctly, they must be supplied with the bead geometry, the material performance data (as described previously), and the friction for the die, material, and lubricant combination. The friction values are calculated from the DBS test data. These friction values represent a complex and low-pressure condition, and it can be argued that the values calculated by the normal methods are not true friction coefficients but rather some combined friction coefficient and error correction factor. The use of the friction coefficients is valid as long as the coefficients are used for the friction models and are not used for other conditions (described subsequently) that exist in the die.

The binder is that section of the die outside the part that first clamps the sheet metal before the punch enters and stretches the sheet metal into the part shape. The binder serves several functions. The first is to preform the sheet into a developable shape that places, to the best possible extent, the right amounts of material into the right areas for subsequent forming. The second function is to keep the sheet metal within the binder from wrinkling. The third function is to create the necessary backforce for properly forming the part as the punch stretches it into the desired shape. A part of that third function is to be the mounting platform for the draw bead.

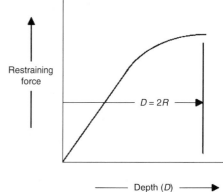

Fig. 11 Effect of depth on draw bead restraining force. The draw bead restraining force is nearly proportional to the depth until the maximum depth of twice the bead radius (R) is approached. As the bead depth exceeds twice the radius, no additional restraining force is generated

As the sheet metal is stretched in the generally horizontal part area to set the part shape into the sheet metal, some of the extra metal (from stretching) will move (displace) into the more generally vertical stretch or draw wall between the part area and the binder, as indicated by the diverging arrows in the part area shown in Fig. 12. Similarly, any metal being pulled (drawn) from the binder into the stretch or draw wall will result in hoop compression, as indicated by the converging arrows in the binder area shown in Fig. 12. Two elements of sheet metal in the binder area located at points A and B, respectively, will move to points A′ and B′, respectively, causing them to become closer together. The material between the two elements must do some combination of:

- Be compressed and thicken
- Become elongated in the direction of the arrows by having sufficient pulling force in the direction of the arrows to plastically deform the material to the uniaxial tension condition or greater
- Wrinkle

One purpose of the binder is to control the wrinkling to within some allowable amount (what is allowable can vary from application to application) and thereby force some combination of the other two possibilities into the sheet metal.

The binder should not actually clamp the sheet metal. Instead, the binder should be closed onto blocks outside of the sheet metal and those blocks hold the binder open from the sheet metal by approximately one-fourth of the metal thickness. The binder should be held onto the blocks with sufficient force that the metal cannot push it off the blocks as the metal starts to wrinkle. Although the metal will wrinkle slightly, it cannot wrinkle as much as the displacement arrows imply. Hence, the material must plastically deform.

If that deformation can be represented by a circle that is deformed into an ellipse, the largest diameter of the ellipse (the major strain axis) is aligned in the direction of the displacement arrow. The plastic deformation of the material (one can think of this as rearranging the surface area) requires a pulling force on the metal in the direction of the displacement arrow that is equal to a force in the opposite direction restraining the sliding motion. The slight wrinkling of the metal creates a normal force on the binder that is a function of the material internal resistance to deformation, the material thickness, and the height of the wrinkle. The friction between the metal and the binder is small but can still generate a significant force acting to restrain (but not stop) the sliding motion of the material through the binder and toward the part being formed. The friction-induced restraining force is additive to the plastic deformation restraining force. Figure 13 is a view of the section from Fig. 12 and illustrates the wrinkling that occurs as a result of the converging material displacements in the binder. When more force is needed to adequately restrain the sliding motion of the material, draw beads are added.

The friction coefficients representative of what is happening in the binder can be measured by pulling strips of material through flat plates. Such friction tests are available and should be used to generate the data. These conditions are very low-pressure situations, and the data should not be used for the internal areas of the die where the pressures and the metal straining conditions are much more severe.

Internal Die Radii. The third and most complex friction condition is where material is sliding across the curved punch and/or die faces and simultaneously being stretched in plastic deformation. The condition is illustrated in Fig. 14. The pressure of the sheet metal against the die face is a function of the radius and the stress in the metal and is calculated with the pressure vessel equation:

Pressure = Instantaneous stress in the metal

× (Metal thickness/Die radius)

The equation shows frictional restraining force to be small when the die radius-to-metal-thickness ratio is large (e.g., >15) but can be quite large when the radius-to-thickness ratio is small (e.g., <5).

The actual value of the restraining force is a function of:

- Die radius
- Included arc angle
- Amount of straining in the sheet metal
- Stress in the sheet metal in the direction of motion
- Velocity of the sheet metal relative to the die
- Displacement distance (separate from velocity)
- Die material and coating (e.g., chrome plating)
- Die surface finish
- Lubricant type
- Lubricant amount

The value also depends on interactions between these factors. In the late 1990s, a team at the General Motors Metal Fabricating Division created a friction test that captured the global effects of the conditions by having the die radius on a separate block resting on load cells. This work showed the variability of friction and the ability to accurately measure it at actual press speeds and with large specimens. The test equipment, procedure, and results of that work are included in the final report of the National Institute of Standards and Technology (NIST) Springback Predictability Project (Ref 19). Such a test measures the combined effects of friction and the forces necessary to accomplish the bending and unbending.

To reduce the data to just friction, several data-processing steps are required. First, the test must be modeled by the simulation program in which the test data are going to be used. That simulation modeling of the test is done with the friction value set to zero and the simulation postprocessed to output the same data curves as derived from the actual test. These curves

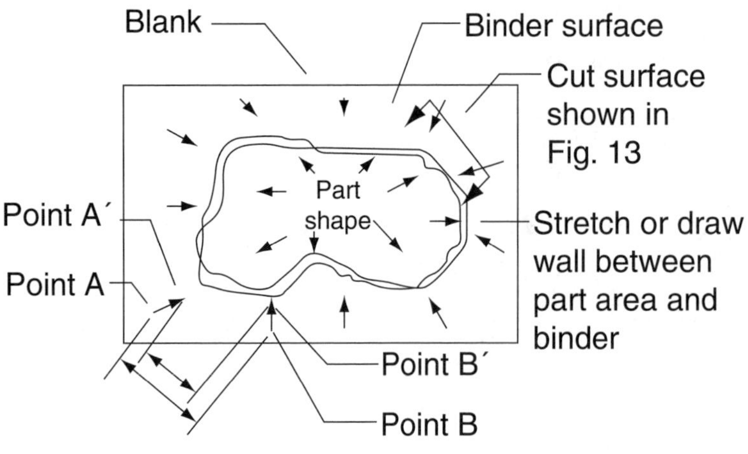

Fig. 12 Effect of a binder on sheet metal behavior. As the sheet metal blank is drawn into a part shape, the material that forms the part shape is generally stretched outward, and the material within the binder slides inward, as indicated by the arrows. The converging pattern of the inwardly sliding material in the blank implies that the material will try to wrinkle

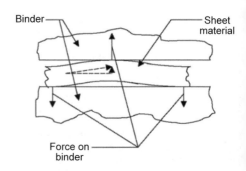

Fig. 13 Detail of binder effect. As the material wrinkles, it bumps against the binder in this view of the section cut from Fig. 12. The binder limits the height of the wrinkle and withstands the force of the material pushing against it. The force vector diagram within the material illustrates the force component pushing on the die

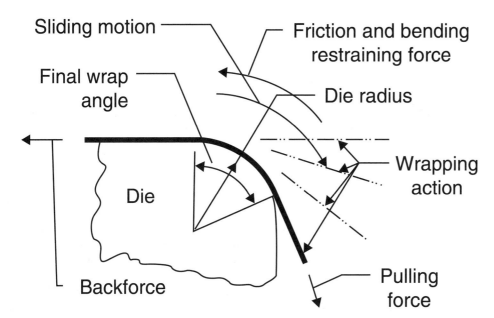

Fig. 14 Force acting to restrain the sliding motion generated by friction and bending as sheet metal slides across a die face. The magnitude of that force is strongly affected by the radius and wrap angle

simulate the zero friction condition, so they represent what the test data would be if there was no friction. Thus, these simulations show what the simulation code would predict for only the effects of bending and unbending. Then, the difference between the simulation data curves and the actual data collected can be mathematically analyzed to deduce the friction values. The difference between the simulation and the test data is a value that includes the friction and is also a correction for any errors in how the simulation software computes the bending contributions.

Because the various simulation software programs have different methods of computing bending and different interpretations of how friction is applied, an adaptation of the test data must be done for each simulation package. The results of the test data must be reduced by regression analysis to coefficients of a basic equation. The form of the basic equation can be standardized, and a set of coefficients can be created for each sheet material, die material, die coating, and lubricant combination. Such data should be created and made available as an industry standard.

The NIST Springback Predictability Project, completed in 2000, also enhanced the LS-DYNA (Livermore Software Technology Corp.) software package to enable the program to work with the user's friction models at every node and time-step, to ensure the use of the correct friction factor at all times and at all locations during the simulation.

Sheet-Forming FEA Results

From FEA calculations, a wide range of results can be obtained. Examples include

contours of variation quantities. Other important results in sheet-forming analysis are forming-limit diagrams and forming-limit curves.

Contour Results

A normal and useful option is to display results as contours onto the part. The most common contour values are described as follows.

Thickness and thinning for shell elements of the deformable part (blank or tube) are the basic result for sheet forming.

Strain contours include:

- The principal major and minor strains are observed to determine the zones with maximum traction or compression strains. Visualizing the associated vectors can be very useful to determine the principal directions.
- The strain mode makes it possible to know the type of stress (simple traction, biaxial traction, restrained) to which a zone is submitted.
- The plastic strains enable the sheet hardening—hence the consolidation of the blank—to be assessed.

Stresses are essential to analyze the stretching of the part as well as the risk of wave formation. They usually indicate what will happen with the springback, especially when analyzing them on fibers.

Kinematics and Position. These contours give information on the velocity, displacement, or positions of the nodes. These contours must be analyzed on nodes, with smooth display. It can be used for tools (such as the blank holder) or for analyzing the springback of the blank.

Mesh Quality and Undercut. All the contours about mesh quality are useful for

analyzing the quality of the tool mesh, especially for the use of advanced contact or for doing an offset.

Draw Depth. This contour is used for checking the die design and optimizing it.

Distance. This contour is calculated on the node and is used to check the distance between the blank and tools or between two blanks. The second blank can be imported or displayed with the multistate mode. The distance is calculated from the node of the first set to the element of the second set. The signed distance option enables positive and negative values to be given by the scalar product between the vector (node/element) to the oriented normal of the element.

Contact contours help to analyze the contact between the blank and tools or a specific zone of tools. This is used for external parts that should not be marked by the radii of the tool, to avoid shape defect. It can also be interesting information for tool wear.

Pressure gives numerical information on the normal and tangential nodal contact between the blank and tool. The pressure is a nodal pressure, calculated by dividing the contact force (applied by all the elements that are in contact) at each node by the average surface of all the elements to which the node belongs. Using the mark contour (see the following), the user can know with which tool the blank is in contact.

Marks give instantaneous or cumulated information on blank nodes in contact with a tool (binary information).

Detachment can be used for analyzing surface defects, by checking the loss of contact between the blank and tools during simulation.

Energy. This group of contours gives the various energies of the blank (internal, hourglass, membrane, etc.). It can be used for very specific processes.

Forming-Limit Curve

During sheet forming, the material experiences stretching in two directions, as the material is formed in the die by forces that can pull or push the material in all three directions. The forming-limit diagram (FLD) (Fig. 15) is a very efficient way of describing the biaxial stretching that can occur during sheet forming. The diagram is a plot of strains along the major and minor axes, and the forming limit curve (FLC) on the FLD is the demarcation of when cracking or excessive thinning occurs. The "Stretch" circle in Fig. 15 is the region where the material experiences stretching in both directions. Stretching in only one direction is illustrated by the "Plane strain" circle, while the "Draw" circle represents stretching in one direction with a reducing strain in the other direction. In contrast, the strain path from a tensile test represents a single mode of deformation of pulling in only one direction while the thickness and width of the pulled specimen are unrestrained.

By comparing the FLC/FLD material property to FEA-predicted strains, the propensity

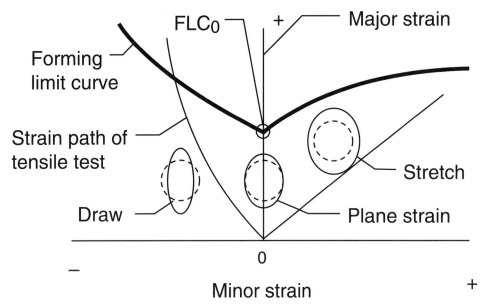

Fig. 15 The forming-limit diagram and forming-limit curve

for failure during sheet forming can be assessed. The FLD enables the user to evaluate the risks of rupture or necking by comparing the position of these points to one or several limit curves (which correspond to necking, rupture, or other curves, taking into account a safety coefficient). The forming limit shows the conditions at which the material will start to neck, after which it will soon fracture. The FLC shown in Fig. 15 captures the necking failure limit of the material through all the common deformation modes.

For most low-carbon sheet steels, the FLC_0 point (the lowest point on the FLC) can be calculated from the sheet thickness and n-value. The shape of the FLC remains the same. It is moved up or down on the graph to fit to the FLC_0 point. However, that simple relationship does not hold true for other materials, such as aluminum and stainless steel, and it may not hold true for all steels in the emerging range of new products, such as transformation-induced plasticity steels.

As part of FEA results, a point in a bi-dimensional space is associated to each element. The abscissa is the minor strain of the element; the ordinate is the major strain of the element. All these points constitute the strain FLD. The FLD is displayed in a 2-D window and is juxtaposed to the 3-D window, where the same results are displayed onto the structure (Fig. 16). The various zones in Fig. 16 are as follows:

- *Zone 1:* Cracks. Points located above the FLC
- *Zone 2:* Marginal zone, excessive thinning. Points located between the FLC and the same curve decreased by 10% of the curve value at $x = 0$. The percentage can be modified by the user.

- *Zone 3:* Safe zone
- *Zone 4:* Insufficient stretching. Points located inside the circle, with its center as the origin and with a radius value of 0.002. It can be modified by the user.
- *Zone 5:* Wrinkling trend. Points located above the straight line of major strain = minor strain ($y = x$) and under the line $y = x [(-1 - R_m)/R_m]$, where R_m is the average Lankford coefficient
- *Zone 6:* Strong wrinkling trend. Points located under the $y = -x$ straight line

The FLC is usually an experimental curve defined for each material and thickness range. It is possible to use an approximate FLC calculated with different models, but in that case, the results are less precise and must be cautiously analyzed.

The FLD can be displayed with strains or with stresses. Usually, strains are used because it is easier to experimentally define the FLC for strain than for stress. However, the FLD by stresses is useful because the analysis depends less on the path. The FLD is based on true strains (not engineering strains). The FLD can be analyzed for each fiber (membrane, lower and upper) that can be displayed separately.

Springback Analysis (Adapted from Ref 20)

As noted, one topic at the forefront of stamping analysis simulation is springback prediction. Springback is the elastically driven change-of-shape response that occurs when the external forming forces are removed. Springback is usually undesirable and can be

compensated for by overbending or by a secondary operation such as restriking. However, the design and simulation of proper compensation requires reliable prediction of springback, typically expressed as a change of curvature (that is, the original curvature minus the final curvature, which is generally greater than 1).

It is crucial to have a good material model (yield criterion model) and stress calculation for springback analysis. As the use of aluminum and advanced high-strength steel increases, springback becomes a pronounced concern. The springback of some ultrahigh-strength steels can be even greater than the thickness of the casting in the die face. Accordingly, the accuracy of springback prediction is vital before it can be minimized. The first step in solving the problems mathematically is to obtain accurate stress distribution. To this end, the choice of the right material model and the yield criterion is very important (see the section "Material Yield Criteria" in this article).

In addition, springback is a greater concern with the use of advanced high-strength steels in automotive panels. Until the advent of these advanced or ultrahigh-strength steels, many companies did not really make use of springback prediction capabilities, preferring to solve the springback concerns during physical tryout, by process or die geometry adjustments. This approach is not effective, because this makes the problem a mandatory concern to predict springback.

Important aspects for accuracy in springback prediction for typical sheet-forming operations (R/t assumed to be in the range of 5 to 25) involve:

- Sheet tension (most critical)
- Strain-hardening law
- Presence of anticlastic curvature

The analysis of springback even in pure bending also involves a number of underlying assumptions that include (Ref 5):

- Plane sections remain planar
- No change in sheet thickness
- Two-dimensional geometry, either plane strain or plane stress in width direction
- Constant curvature (i.e., no instability of shape)
- No stress in the radial, or through-thickness, direction
- The neutral (stress-free) axis is the center fiber and is the zero-extension fiber.
- No distinction between engineering and true strain
- Isotropic, homogeneous material behavior
- Elastic straining only during springback

Although fairly small errors are associated with each of these assumptions, these errors can grow when elastic-plastic laws are considered and when bending and unbending occur. The validity of these assumptions is discussed in the section "Approximations in Classical

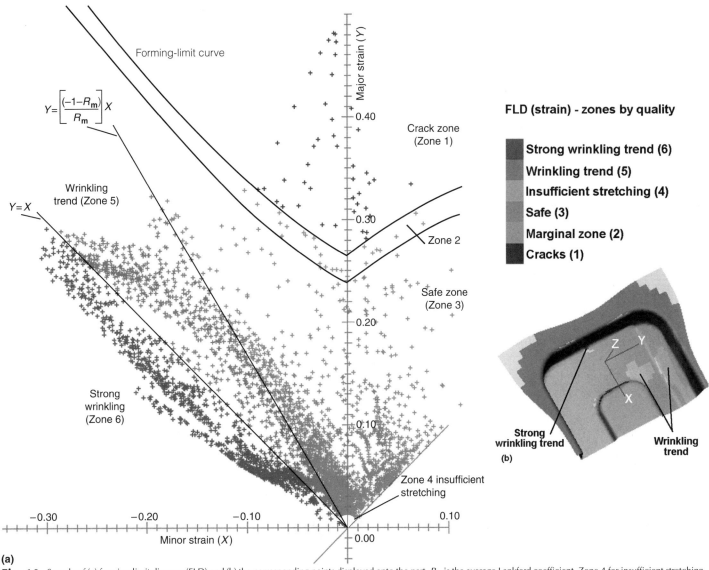

Fig. 16 Sample of (a) forming-limit diagram (FLD) and (b) the corresponding points displayed onto the part. R_m is the average Lankford coefficient. Zone 4 for insufficient stretching is selected here as points within a radius of 0.002 from the origin of the FLD. Courtesy of ESI Group Inc.

Bending Theory" in the article "Springback" in *Metalworking: Sheet Forming*, Volume 14B of *ASM Handbook* (Ref 20).

For a typical industrial sheet-forming operation, the sheet workpiece is pressed between nearly rigid tools with draw-in constraints enforced, usually via draw beads. The general operation may be arbitrary in three dimensions, and conformance to the tools is by no means assured (Ref 21), thus making it impossible to know, a priori, the bend radius of the sheet. Many material elements undergo bending and unbending with superimposed tension, whereas the closed-form analyses usually assume a flat starting configuration in both in-plane directions. Determining sheet tension, which was shown in the last section to be critically important in springback, is complex, depending on friction, bending and unbending, and boundary constraints. All of

these variables may change throughout the part and the forming stroke, over small distances and times.

For arbitrary 3-D forming operations, FEM simulation (or a similar numerical method) is required throughout the forming operation to obtain a final, as-formed state. This configuration (with tool contact forces) may then be used as a basis for a general springback analysis using the same, or a different, FEA model. However, application of 2-D closed-form methods may be possible and profitable for some classes of forming operations. In spite of the difficulties of applied springback analysis, certain applied problems have sufficiently restrictive characteristics to provide a basis for closed-form or empirical analysis.

For example, pure bending by dies in two dimensions may be analyzed using closed-form bending solutions if the workpiece is assumed

to conform to the punch surface. Results have been presented for U-bending and V-bending (Fig. 17) using such analysis (Ref 22–27) and empirical approaches (Ref 28–31). A closely related application in sheet-metal forming is flanging, for which analysis (Ref 32) and empirical approaches (Ref 33) have been presented.

Closely related operations involving significant tension are often called stretch-bend or draw-bend problems. These operations involve the bending and unbending of sheet as it is progressively drawn over a die. The typical application is often referred to as a top-hat section (Fig. 18c, 19) and may be called channel forming, among other common names. For large draw-in, the principal springback typically occurs in the form of sidewall curl, which is the curling of the material that was drawn over the die radius (and which was flat while the workpiece was held in the dies during the operation itself). Much of the experimental work appearing in the literature for draw bending must be examined critically, because the tensile stress or load is often not carefully controlled or measured. In a few exceptional cases (Ref 34–36), direct control was imposed to obtain draw-bend results. For other work, experiments rely on indirect control of tension via friction, draw beads, or die clearances to establish the essentially unknown value of sheet tension. As shown in the last section, the tensile stress has a dominant effect on springback, particularly for values approaching the yield tension, thus leading to large uncertainties in measured results unless the tensile stress is known accurately.

A wide range of experimental data for a draw-bend problem from various sources appears as part of the NUMISHEET '93 *U-Channel Benchmark* (Ref 37). Geometry, material, lubrication, and forming parameters were fully specified by the conference organizers, and numerous laboratories were asked to carry out independent measurements and simulations. The results, shown in Fig. 15(b), illustrate the wide scatter that was obtained. In general, the experimental scatter was greater than or equal to the simulation scatter, illustrating the difficulty in carrying out such experiments with normal industrial forming machinery. It appears

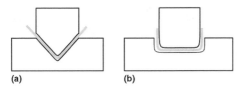

Fig. 17 Schematics of (a) V-bend and (b) U-bend forming operations

that the scatter of experimental results is typical for experiments employing indirect control of sheet tension. The sources of error in FEM simulations are considered in more detail subsequently, along with a summary of draw-bend results for which the sheet tension is carefully controlled.

Draw-Bend Testing

The advantages and pitfalls of FEA of springback for many sheet-forming operations are revealed by the draw-bend test (Fig. 20), which closely represents the situation in channel forming (Fig. 17b, 19a) and many other press-forming operations. The advantage of the test is that sheet tension may be closely controlled in parallel with finite-element simulations for evaluation (Ref 36, 38).

The material in the test is drawn over a round tool under the action of a pulling displacement (and corresponding front force) and resisting force (back force). The workpiece may or may not conform completely to the tool surface when under load; it undergoes bending and then unbending under tension, then rebending under the final unloading when it is released from the fixtures. When released, the drawn length of the strip specimen adopts a final radius of curvature (r'). This is precisely analogous to the channel-forming operation, where the pulling displacement is provided by the punch displacement, the back force is provided by a draw bead or frictional resistance over a binder surface, and the final radius of curvature of the drawn section is referred to as sidewall curl. When the drawn distance is sufficiently large, the final springback changes are dominated by the sidewall curl (radius r') rather than changes in the small region in contact with the tool at the end of the test.

Finite-element sensitivity studies (Ref 39–41) of the draw-bend test revealed that accurate springback prediction requires much tighter tolerances and closer attention to numerical parameters than does forming analysis. Furthermore, the tighter tolerances must be maintained throughout the forming operation; that is, it is not sufficient to do a coarse forming simulation followed by a precise springback simulation.

Using meshes, tolerances, and numerical parameters typical for forming analysis to analyze the draw-bend test gave nonphysical predictions, including, under some conditions, simulated springback opposite to the direction observed. Figure 21(a) (Ref 34) shows the initial simulations and the final ones (i.e., with appropriate choices of model parameters). A mesh size four times finer along the draw direction was required, combined with a number of integration points ten times larger than normal.

The finite-element sensitivity results based on the draw-bend test may be summarized as:

- The finite elements in contact with tooling should be limited in size to approximately 5 to 10° of the turning angle. This is approximately 2 to 4 times the refinement typically recommended for simulation of forming operations.
- The convergence tolerance and contact tolerance must typically be set tighter than for forming analysis. There are a variety of ways to define such measures, depending on the programs used, so again, the best policy is to refine the measure until the differences become insignificant.
- While most applied sheet-forming analyses use shell elements with three to seven integration points through the thickness, up to 51 integration points are required to assure simulated springback results within 1% of the "converged" solutions. (Converged

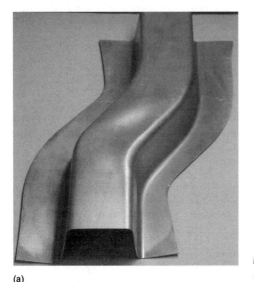

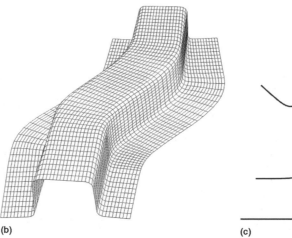

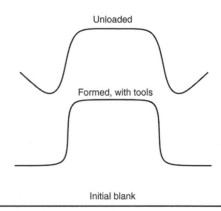

(a)   **(b)**   **(c)**

Fig. 18 Typical automotive sheet-formed part, the S-rail. (a) Formed part. (b) Finite-element representation, as formed. (c) Cross-sectional schematics at three forming stages

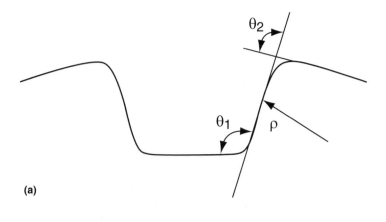

(a)

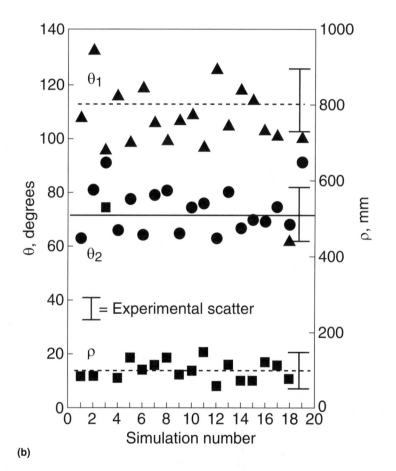

(b)

Fig. 19 U-channel forming and benchmark results. (a) Cross-sectional geometry, after springback. (b) Simulation results (points) and superimposed estimated experimental scatter (bars) for results reported by various laboratories. θ_1 and θ_2, angles characterizing springback in top-hat samples; ρ, radius of curvature of sidewall curl

solutions were obtained by using very large numbers of integration points, until no appreciable change in springback was observed.) More typically, 25 integration points were found to be sufficient for many simulations.

The last of these conclusions represents a dramatic divergence from current practice and remains surprisingly controversial, with researchers continuing to recommend using numbers of integration points ranging from five to nine. For this reason, non-FEA numerical studies were undertaken to explore the errors associated with numerical integration for finite numbers of integration points. Results from the finite-element sensitivity studies and from non-FEA numerical studies are described further in the section "Draw-Bend Experience" in the article "Springback"

in *Metalworking: Sheet Forming,* Volume 14B of *ASM Handbook* (Ref 20).

ACKNOWLEDGMENT

Editors thank Martin Skrikerud of ESI group for contribution to this article.

REFERENCES

1. C.Y. Sa, Computer-Aided Engineering in Sheet Metal Forming, *Metalworking: Sheet Forming,* Vol 14B, *ASM Handbook,* ASM International, 2006, p 766–791
2. F. Barlat and J. Lian, Plastic Behaviour and Stretch Ability of Sheet Metals, Part 1: A Yield Function for Orthotropic Sheets under Plane Stress Conditions, *Int. J. Plast.,* Vol 5, 1989, p 51–66
3. R. Hill, *J. Mech. Phys. Solids,* 1990, p 405–417
4. D. Ling, J.-L. Babeau, M. Skrikerud, Y. Dammak, and F. El Khaldi, Integrated Forming Simulation Using State of the Art Methodologies, *NUMIFORM,* 2007
5. R.D. Cook, D.R. Malkus, M.E. Plesha, and R.J. Witt, *Concepts and Applications of Finite Element Analysis,* 4th ed., Wiley, 2002
6. J.N. Reddy, *Introduction to the Finite Element Method,* 2nd ed., McGraw-Hill, 1993
7. O.C. Zienkiewicz and R.L. Taylor, *The Finite Element Method: Volume 1, The Basis,* 5th ed., Butterworth-Heinemann, Oxford, U.K., 2000
8. O.C. Zienkiewicz and R.L. Taylor, *The Finite Element Method: Volume 2, Solid Mechanics,* 5th ed., Butterworth-Heinemann, Oxford, U.K., 2000
9. T.J.R. Hughes, *The Finite Element Method: Linear Static and Dynamic Analysis,* Prentice-Hall, 1987
10. K.-J. Bathe, *Finite Element Procedures,* Prentice-Hall, 1995
11. R.H. Wagoner and J.-L. Chenot, *Metal Forming Analysis,* Cambridge University Press, Cambridge, U.K., 2001
12. N. He and R.H. Wagoner, Springback Simulation in Sheet Metal Forming, *Proceedings of NUMISHEET '96* (Columbus, OH), 1996, p 308–315
13. D.L. Dewhirst, Finite-Element Method, *Materials Selection and Design,* Vol 20, *ASM Handbook,* ASM International, 1997
14. R. Hill, Theory of Yielding and Plastic Flow of Anisotropic Metals, *Proc. R. Soc. (London),* 1948, p 193–281
15. R. Hill, *The Mathematical Theory of Plasticity,* Clarendon Press, Oxford, 1950
16. W.R. Hosford, *Proc. Seventh North American Metalworking Conf.* (Dearborn, MI), Society of Manufacturing Engineers, 1979, p 1912–1916
17. J.W. Yoon and F. Barlat, Modeling and Simulation of the Forming of Aluminum Sheet Alloys, *Metalworking: Sheet*

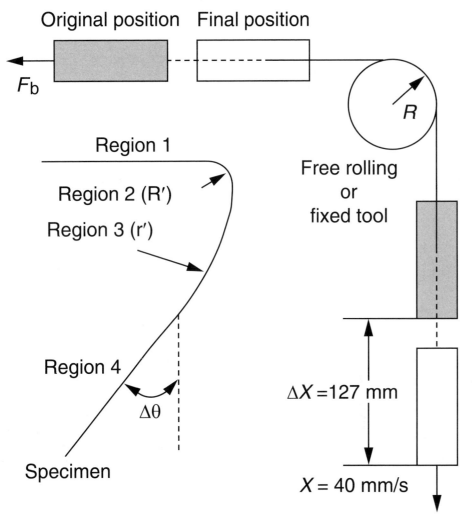

Forming, Vol 14B, *ASM Handbook,* ASM International, 2006, p 792–826

18. E. Herman, D.J. Schaeffler, and E.J. Vineberg, CAD/CAM and Die Face Design in Sheet Metal Forming, *Metalworking: Sheet Forming,* Vol 14B, *ASM Handbook,* ASM International, 2006, p 756–765

19. "Springback Predictability Project (SPP)," NIST-ATP Final Technical Report, NIST Cooperative Agreement 70NANB5H1149, Dec 2000

20. R.H. Wagoner, J.F. Wang, and M. Li, Springback, *Metalworking: Sheet Forming,* Vol 14B, *ASM Handbook,* ASM International, 2006, p 733–755

21. W. Johnson and T.X. Yu, *Int. J. Mech. Sci.,* Vol 23, 1981, p 687–696

22. C.T. Wang, G. Kinzel, and T. Altan, *J. Mater. Process. Technol.,* Vol 39, 1993, p 279–304

23. C. Sudo, M. Kojima, and T. Matsuoka, Some Investigations on Elastic Recovery of Press Formed Parts of High Strength Steel Sheets, *Proceedings of the Eighth Biennial Congress of IDDRG,* Vol 2, Sept 1974 (Gothenburg, Sweden), The International Deep Drawing Research Group, p 192–202

24. M.L. Chakhari and J.N. Jalinier, Springback of Complex Parts, *Proceedings of the 13th Biennial Congress of IDDRG,* Feb 1984 (Melbourne, Australia), The International Deep Drawing Research Group, p 148–159

25. L.C. Zhang and Z. Lin, *J. Mater. Process. Technol.,* Vol 63, 1997, p 49–54

26. A. Focellese, L. Fratini, F. Gabrielli, and F. Micari, *J. Mater. Process. Technol.,* Vol 80–81, 1998, p 108–112

27. H. Ogawa and A. Makinouchi, Small Radius Bending of Sheet Metal by Indentation with V-Shape Punch, *Advanced Technology of Plasticity, Proceedings of the*

Fig. 20 Schematics of the draw-bend test and final configuration of the unloaded specimen. F_b, normalized back force; R, tool radius; R', radius of curvature of region in contact with tool, after unloading; r', radius of curvature in curl region, after springback; $\Delta\theta$, springback angle; ΔX, displacement, the distance between the original and final positions; $\dot{X}$, displacement rate

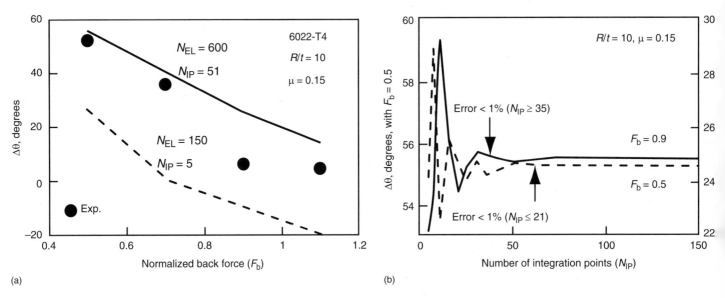

(a)

(b)

Fig. 21 Sensitivity of simulated draw-bend springback to mesh size (N_{EL}) and number of through-thickness integration points (N_{IP}). (a) Nonphysical springback predictions obtained using typical sheet-forming simulation parameters. (b) Accuracy of selected springback solutions depending on the number of through-thickness integration points. $\Delta\theta$, springback angle; R/t, ratio of bending radius to thickness; μ, friction coefficient

Sixth ICTP, Vol 2, Sept 19–24, 1999 (Nuremberg, Germany), p 1059–1064

28. T.X. Yu and W. Johnson, *J. Mech. Work. Technol.,* Vol 6, 1982, p 5–21

29. B.S. Levy, *J. Appl. Metalwork.,* Vol 3 (No. 2), 1984, p 135–141

30. S.S. Han and K.C. Park, An Investigation of the Factors Influencing Springback by Empirical and Simulative Techniques, *Proceedings of NUMISHEET '99* (Besancon, France), 1999, p 53–57

31. L.R. Sanchez, D. Robertson, and J.C. Gerdeen, "Springback of Sheet Metal Bent to Small Radius/Thickness Ratios," Paper 960595, Society of Automotive Engineers, 1996, p 650–656

32. C. Wang, An Industrial Outlook for Springback Predictability, Measurement Reliability, and Compensation Technology, *Proceedings of NUMISHEET 2002* (Jeju Island, Korea), 2002, p 597–604

33. R.G. Davies, *J. Appl. Metalwork.,* Vol 1 (No. 4), 1981, p 45–52

34. S. Takahashi, T. Kuwabara, and K. Ito, Springback Analysis of Sheet Metal Subjected to Bending-Unbending under Tension II, *Advanced Technology of Plasticity, Proceedings of the Fifth ICTP,* Vol 2 (Columbus, OH), 1996, p 747–750

35. Y.C. Liu, *J. Eng. Mater. Technol. (Trans. ASME),* Vol 110, 1988, p 389–394

36. W.D. Carden, L.M. Geng, D.K. Matlock, and R.H. Wagoner, *Int. J. Mech. Sci.,* Vol 44, 2002, p 79–101

37. A. Makinouchi, E. Nakamachi, E. Onate, and R.H. Wagoner, Ed., *Proceedings of NUMISHEET '93* (Isehara, Japan), The Institute of Physical and Chemical Research, 1993

38. K.P. Li, W.P. Carden, and R.H. Wagoner, *Int. J. Mech. Sci.,* Vol 44, 2002, p 103–122

39. K.P. Li, L.M. Geng, and R.H. Wagoner, Simulation of Springback: Choice of Element, *Advanced Technology of Plasticity, Proceedings of the Sixth ICTP* Vol III (Nuremberg, Germany), 1999a, p 2091–2098

40. K.P. Li, L.M. Geng, and R.H. Wagoner, Simulation of Springback with the Draw/Bend Test, *Proceedings of IPMM '99* (Vancouver, B.C., Canada), 1999b, p 1

41. K.P. Li and R.H. Wagoner, Simulation of Deep Drawing with Various Elements, *Proceedings of NUMISHEET '99* (Besancon, France), 1999, p 151–156

Simulation of Powder Metallurgy Processes

ASM Handbook, Volume 22B, *Metals Process Simulation*
D.U. Furrer and S.L. Semiatin, editors

Copyright © 2010, ASM International®
All rights reserved.
www.asminternational.org

Modeling of Powder Metallurgy Processes

Howard Kuhn, University of Pittsburgh

POWDER METALLURGY (PM) is a process of shaping metal powders into near-net or net shape parts combined with densification or consolidation processes (e.g., sintering) for the development of final material and design properties. Powder metallurgy parts are used in a wide variety of industries, with three broad categories of applications:

- Lower-cost components (typically with lower mechanical properties than wrought product)
- High-performance parts with the advantages of PM in achieving uniform alloying and microstructures
- Enhanced-property parts that are difficult to produce by wrought processing

This article introduces general considerations, models, and applications in the modeling of PM processes. Additional articles on PM modeling in this Volume include:

- "Modeling and Simulation of Press and Sinter Powder Metallurgy"
- "Modeling of Hot Isostatic Pressing"
- "Modeling and Simulation of Metal Powder Injection Molding"

General Considerations of Process Modeling

Processing is one of three major design considerations that contribute to the functional and economic success of a final product. Processes, materials, and geometric design interact to form a three-legged foundation (Fig. 1) supporting product development. Geometric features (shape, dimensions, tolerances, etc.) in conjunction with the selected material properties (physical, mechanical, thermal, electromagnetic, etc.) are specified to meet the functional and usability requirements of the product. The material and process interactions are controlled to deliver the required metallurgical structure and therefore the properties of the product. Finally, the geometric design must be feasible within the constraints of the process yet take advantage of the geometric flexibility offered by the process.

A variety of computer models serve as useful supplements to experience in the selection of materials, specification of geometric features, and definition of process parameters, as indicated in Fig. 1. For example, structural and thermal finite-element analysis (FEA), thermal finite-difference modeling (FDM), computational fluid dynamics (CFD), and electromagnetic field analysis are used to develop virtual prototypes of the product for the purpose of optimizing its performance. These same computational methods, generally in more complicated and nonlinear form, are used to model the various processes in an attempt to minimize or eliminate trial-and-error efforts to reach a practical set of processing parameters. Material representation for both product design and process analysis presently relies on a myriad of tests to generate data in the proper form for the analysis codes, but databases of material data are growing and being made available for specific applications. At some point in the future, the rapidly advancing field of material analysis using microscopic, nanometric, and first-principles models will enable the prediction of material properties from composition and material structural features. Conversely, such models will permit design of a material meeting a specific set of property requirements.

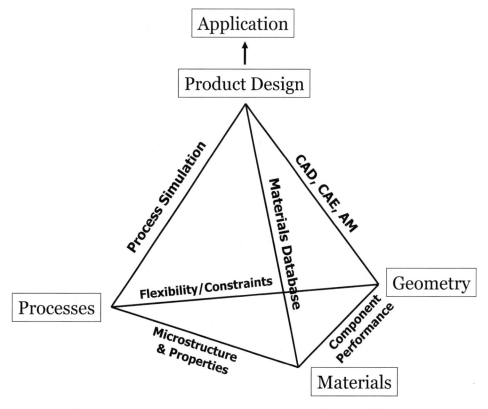

Fig. 1 Geometric design, process parameters, and materials selection contribute to product development. Computer-aided tools are useful as a complement to experience and history in determining the material, geometry, and process parameters for product design. CAD, computer-aided design; CAE, computer-aided engineering; AM, analytical modeling

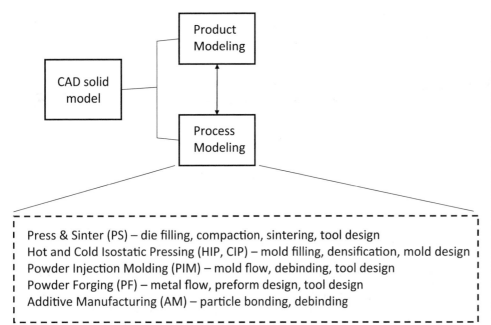

Fig. 2 Information flow from the computer-aided design (CAD) solid model to product and process modeling activities. For each process, various subprocesses are amenable to modeling

In casting and forming industries, process modeling is increasingly being used to produce the correct first article. These process models combine continuum field calculations with defect criteria to display the material behavior during processing. This presents an analytical platform for predictive avoidance of defective parts. More advanced codes even provide predictions of microstructure and microstructure distribution in the finished part. Appropriate material databases and simulation codes for casting and forming, combined with ever-decreasing computing cost and ever-increasing computing power, now make process modeling capabilities available to even small foundries and mills, similar to the widespread implementation of computer numerical control simulation codes in machine shops in recent decades.

Powder Metallurgy Process Modeling

Extrapolating this success in process modeling to PM methods of shapemaking is complicated by the larger number of process steps involved in PM, the wide variety of physical processes involved in those steps, and the stochastic nature of particulate behavior. Powder particles have high surface area per unit volume, and surface interactions between particles, whether by mechanical motion, localized deformation, or atomic diffusion, lead to an abundance of opportunities for aberrant behavior. The size and shape of particles in a particular batch can vary widely, resulting in localized variation of density throughout the powder mass. These pockets of density variation may be of significant size relative to the feature size and tolerances required in the final product.

From a modeling perspective, the assumption of uniform density entering each process step may be significantly incorrect and lead to unexpected deviations from reality. Similarly, micromodels of powder behavior based on uniform particle size and shape may lead to erroneous results compared to the behavior of actual powder masses.

Nevertheless, the attractiveness of PM processes for net shape forming of commercial and defense components inspires the use of process models to facilitate product realization. Powder metallurgy modeling (Fig. 2) represents the interaction and data transfer between various computer applications. The starting point in all cases is a three-dimensional (3-D) computer-aided design (CAD) solid model of the part geometry, generally derived from a sketch of the design concept and embodying the design intent. This electronic representation of the part is used in every subsequent product and process design step and may be altered as necessary to meet the constraints of the various steps. The CAD file facilitates concurrent engineering by permitting rapid communication between product and process design management teams.

Referring to Fig. 2, first the CAD model can be used in conjunction with FEA or CFD codes to evaluate product performance and enable virtual geometry or material changes to meet performance requirements. For example, for parts subjected to cyclic loading, design geometry features can be derived from life-prediction analysis—the combination of FEA for stress-intensity factors and crack propagation measurements for the material of interest—to avoid premature failure. As another example, impeller vane geometry for increased pump efficiency

can be determined from CFD analysis of fluid flow patterns and velocities.

Second, the CAD model can be used in conjunction with other physics-based codes for simulation of the shapemaking operations and their various subprocesses. These processes include press and sinter (PS), cold isostatic pressing (CIP), hot isostatic pressing (HIP), powder injection molding (PIM), bulk deformation processes (particularly powder forging, or PF), and the emerging additive manufacturing (AM) processes (now evolving from rapid prototyping technologies to production applications). All of these cases except AM involve specification of tool and mold parameters, so the CAD data are also useful for definition of a CAD model of the tool geometry, generally by generating an inverse image of the product. Finishing operations, such as heat treating, also use the CAD file in conjunction with tailored FEA codes to determine distortion, residual stress, and final microstructures.

Finally, the resulting collection of computational data in Fig. 2 can be used upstream for customer communications and marketing and downstream for product testing, process planning, and process equipment control models.

Powder Metallurgy Process Descriptions

Press and Sinter

Powder compaction is the most widely used powder shapemaking process. The process involves compression of a powder mass confined in a die cavity and acted upon by upper and lower punches, as shown in Fig. 3. First, powder flows into the die cavity from a feed shoe. Upper and lower punches then move toward each other to compress the powder. The pressure first rearranges the powder particles and then deforms each particle, leading to an increase in overall density by decreasing the pore space between particles. Because of friction along the die walls and between each particle, the density is not uniform but is greatest in the regions of contact with the punches (Fig. 4). Lubrication is added in the form of admixed powder lubricants or as coatings on the particles to reduce friction. The particle size of powder used in die compaction is comparatively large (>60 μm) to reduce the interparticle surface area and the resulting nonuniformities in density during filling of the die and during compaction.

The pressure during die compaction is not isostatic but is greatest in the pressing direction, while the constraining lateral pressure of the die cavity may be as low as one-half of the axial pressure. This ratio of constraining pressure-to-axial applied pressure increases as the compacted density increases. Typically, the initial density of the loose powder in the die is 50 to 60% of full density, and the compacted densities are 80 to 95%.

Parts having multiple levels must be compacted with upper and lower punches that are segmented. The vertical motions of these punch

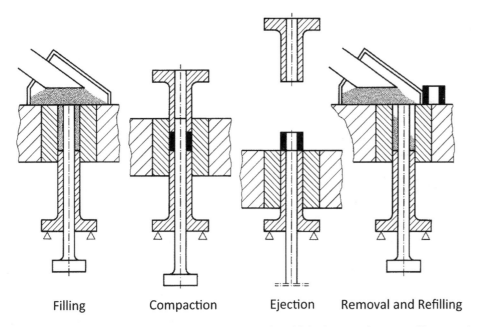

Fig. 3 Steps in powder compaction. A feed shoe provides powder to fill the die cavity, the upper and lower punch move toward each other to compact the powder, the lower and upper punches move upward to eject the part from the die, and the fill shoe removes the previous part and refills the die cavity

Filling   Compaction   Ejection   Removal and Refilling

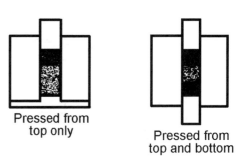

Pressed from top only

Pressed from top and bottom

Fig. 4 Density in compaction is greatest at the surfaces of the moving punches

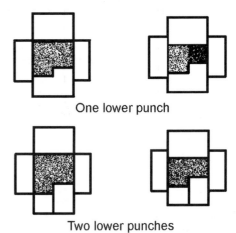

One lower punch

Two lower punches

Fig. 5 Multilevel parts require independent motions of the punches so that the compaction ratio is the same in all levels

segments are controlled independently so that the compaction ratio (ratio of filled height to compressed height) is the same in each level of the part. Figure 5 shows a simple two-level part that is pressed between a solid-step bottom punch (left) and the same part pressed with a split bottom punch (right). The right side of the part undergoes a greater compaction ratio and greater densification if the bottom punch is solid, while more uniform density is achieved if the right punch segment of the bottom punch moves upward with less displacement than the left punch segment. In complex, multilevel parts, careful control of the punch motions is necessary to assure uniform compacted density and to avoid shear cracks in the powder by excessive movement of powder from one region to another.

When the compacted part is ejected from the die, it expands because the pressures acting on the powder during compaction are relieved and elastic recovery, or springback, occurs. During ejection, cracks may form as the part emerges from the die, and tensile stresses occur due to the upward force of the ejection punch and the downward force of friction along the die walls. In addition to large cracks, microdelaminations may occur perpendicular to the pressing direction.

Modeling opportunities abound in powder pressing. Simulation of the flow of powder into the die from the feed shoe (Fig. 3), particularly in complex, multilevel parts, would aid the development of powder-feed procedures for improved uniformity of density during die filling. They could also provide a map of density distribution after die filling that would be the starting point for subsequent compaction modeling. Modeling the compaction process itself is particularly useful for determining the punch and die design and tool motions for increased density uniformity in the compacted part. In addition, models of the interaction between the tooling (die and punches) and the compacted powder part during ejection are useful for making tooling or lubricant modifications to prevent cracking during ejection. Models of the tooling itself are also used to estimate stresses and improve the fatigue life of the dies and punches.

Collectively, these modeling efforts are used to establish the setup conditions for a pressing operation. However, because of the variability of powder lots and the subtle influence of environmental conditions, adjustments are made to the press parameters in subsequent trials and part measurements. In progressive plants, these adjustments are guided by the compaction models.

Sintering follows the pressing operation to develop metallurgical bonds between the powder particles, increase the density, and impart strength to the finished part. The process is also used to provide strength to parts produced by PIM and AM. Pressing and sintering is also used to manufacture preforms for PF.

Solid-state sintering involves several micromechanisms that bond particles and lead to densification. These include surface, grain-boundary, and bulk diffusion; evaporation-condensation; and plastic flow. All of these mechanisms depend on thermal activation, so temperature and time are the primary variables affecting the degree of sintering, as depicted in Fig. 6. Bonding of the particles begins at the junction points between particles, where small necks form and grow to close up porosity between the particles (Fig. 7) (Ref 1). Sintering is driven by reduction in surface energy, so sintering is more highly activated in powder masses of small particle size. Localized stresses at interparticle junctions increase as particle size decreases, which enhances localized surface diffusion and evaporation/condensation mechanisms of sintering. Some nanosized particles sinter at near-ambient temperatures.

Some materials, such as tungsten, require very high sintering temperatures, while others, such as aluminum, have an adherent oxide layer on the particle surfaces that interferes with the sintering mechanisms. For these cases, liquid-phase sintering is used because diffusion is greatly enhanced by the liquid phase. The liquid-phase source may be the addition of a lower-melting-point material, such as copper powder added to tungsten powder. In other cases, the powder may be an alloy, and sintering is carried out in a two-phase region where the alloy is partially liquid, as depicted in Fig. 8.

An alternative to sintering to high density is infiltration of a partially sintered part. The sintered part must have continuous porosity to allow infiltration of a lower-melting-point alloy into the pores of the sintered part. Infiltration occurs by surface tension between the liquid alloy and the powder particle surfaces.

Because most metals oxidize at high temperatures, the sintering process must be carried

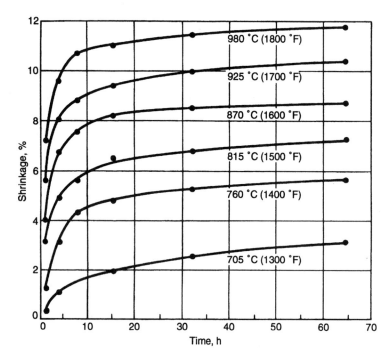

Fig. 6 Increase in density due to shrinkage during sintering of copper powders. Increasing temperature and time increases the effects of sintering

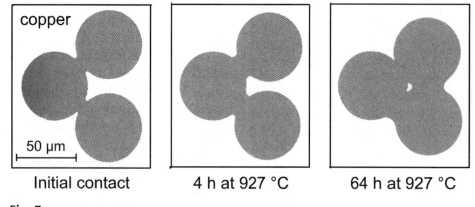

Fig. 7 Sintering begins by forming necks at the contact points between particles. Source: Ref 1

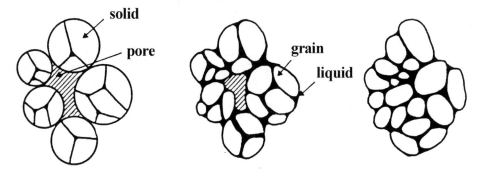

Fig. 8 Illustration of liquid-phase sintering of an alloy. When heated above its solidus, a liquid phase forms in the alloy powders and fills the pores between powder particles

out in a furnace that contains an inert or reducing atmosphere. Some ceramics, glass, and gold powders do not oxidize and do not require a protective atmosphere during sintering.

Sintering models are used to determine dimensional changes, which then are used to modify tool dimensions. Advanced applications of sintering models include the starting density distribution of pressed parts as determined by powder compaction models.

Isostatic Pressing (CIP and HIP)

Isostatic pressing is an important powder shapemaking process for large parts, limited-production numbers, and highly reactive powder materials. The process is carried out at ambient temperature (CIP) or at elevated temperature (HIP). The CIP parts are not formed to full density and must be sintered to develop strength. The HIP parts, however, reach full density because the pressure and temperature environment collapses all pores in the mold. In one sense, HIP is the simultaneous application of compaction and sintering.

In either CIP or HIP, loose powder is filled into a mold, gas is removed from the powder, and the mold is sealed (Fig. 9). The mold containing powder is then placed in a pressure chamber, and pressure is applied by a fluid—a liquid such as kerosene for CIP or an inert gas such as argon for HIP. Thus, the applied pressure is isostatic, or equal in all directions. In either case, this pressure acts on the mold, and deformation of the mold causes transmission of the fluid pressure onto the powder within the mold (Fig. 9). However, at corners and other regions of sharp dimensional change (Fig. 10), the transmission of pressure through the mold may not be exactly uniform, leading to small regions of low density in the compacted powder.

Isostatic pressing is a near-net shape process because the flexible mold is subject to slight variations in pressure transmission from part to part, or pressure shielding at transition zones in the mold, or slight variations in density of powder distribution in the mold. However, net surfaces can be formed in isostatically pressed parts by using hard inserts, such as a core rod to form an internal hole to precise diameter (Fig. 9).

In addition to forming discrete parts, HIP is used to complete the densification in parts produced by other processes, such as PIM, PS, and AM, when such parts require increased strength. Hot isostatic pressing is also used to close up the porosity in cast parts and to repair turbine blades by closing the cavities caused by creep during their high-temperature, high-stress use.

The primary objective of CIP and HIP modeling is to determine the mold shape and dimensions that will produce as close as possible the required part shape and dimensions. However, most modeling approaches require the mold shape as a starting point for the

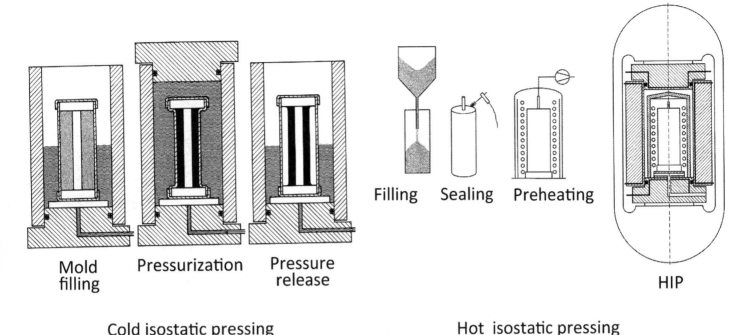

Fig. 9 Schematic of a cold isostatic pressing (CIP) mold (left) and a hot isostatic pressing (HIP) mold (right)

simulation. By iterative use of the model, the mold shape and dimensions required to achieve the part shape and dimensions can be determined.

Powder Injection Molding

Powder injection molding is based on the conventional injection molding process commonly used for production of plastic parts. Pellets of the plastic are fed into the injection molding machine, which conveys them along a screw turning inside a long barrel. Heating bands around the barrel as well as heat generated by intense shear deformation in the screw melt the pellets (Fig. 11). The melted plastic is then pushed by a plunger into the mold cavity, where it is packed under pressures up to 70 MPa (10 ksi) and then solidified into the part shapes. The mold material is a hard alloy, such as beryllium-copper, that has high conductivity for rapid heat removal.

In PIM, metal or ceramic particles are mixed into a polymer binder that is then extruded into pellets, similar to those used in plastic injection molding. The polymer/powder feedstock pellets are formed into the required part shapes using conventional injection molding machines (Fig. 11). The polymer binder is partially removed from the parts by a chemical process, followed by a two-stage furnace operation that removes the remaining polymer binder and then sinters the powder particles together. Because the powders are not compacted together by this operation, after debinding the density of the remaining powder is on the order of 50 to 60% of full density.

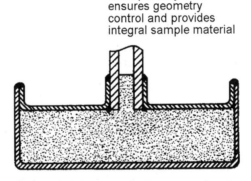

Tube partially filled ensures geometry control and provides integral sample material

Fig. 10 Example of a hot isostatic pressing mold produced by welding steel sheet. The stiffness of the mold corners may shield the powder from the applied pressure

Powders used for PIM are generally much smaller in size (<20 μm) than those used in conventional PS or isostatically pressed powders. Their smaller size promotes sintering, which is necessary to fill the large gap from its original density of 50 to 60% to nearly full density.

One of the key processes in PIM is removal of the polymer binder so that interparticle contact can be made for sintering to near full density. While the parameters for effective debinding by chemical and thermal means can be determined experimentally, models of debinding can be used to reduce or eliminate experimental determinations. Such models are particularly useful for determining the debinding parameters for large parts.

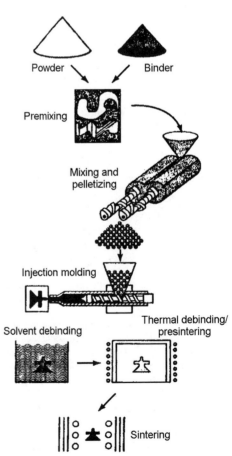

Fig. 11 The steps in powder injection molding

Another objective of PIM modeling is prediction of the flow of polymer/powder mix during mold filling so that complete filling occurs with no unbonded weld lines. In addition, flow conditions are monitored to detect segregation of powder from the polymer, resulting in non-uniform powder distribution. To simulate PIM, combined fluid-flow and heat-transfer models are used, incorporating criteria for powder segregation and satisfactory flow-front welding. Refinement of PIM models is particularly important in the ongoing development of processes for micro-PIM.

Bulk Deformation Processing—Powder Forging

Full-density PM parts require high pressures at high temperatures to completely eliminate the porosity between particles. Hot isostatic pressing, as described previously, is one approach to reaching full density, but the flexible molds used in HIP do not lead to a net shape part, and the process is not suitable to high production rates. Both of these limitations are overcome through the process of PF, which has found its niche in production of connecting rods and automotive drivetrain parts. Hot rolling and hot extrusion of powders are other bulk deformation processes applied to powders, but neither is used to produce large quantities of powder metal parts.

Powder forging involves large plastic deformation of pressed-and-sintered preforms to finished net shapes in a trap die, that is, a die set consisting of one upper and one lower punch within a die cavity, similar to conventional compaction tooling (Fig. 3). In this case, the preform shape is considerably simpler and smaller than the finished part shape and dimensions (Fig. 12). This allows for significant plastic flow during the forging stroke. During plastic deformation, large shear stresses accompany the forging pressure, which is effective in not only eliminating the porosity in the pressed-and-sintered preform but also assuring structural integrity across the collapsed pore interfaces (Fig. 13). While increasing the amount of plastic deformation increases the efficiency of densification and pore interface bonding, it also increases the risk of cracking. The forging preforms consist of pressed-and-sintered material with some porosity, which can act as initiation points for ductile crack formation during plastic deformation.

Modeling of PF is used to determine the preform shape that will lead to full densification without cracking. In addition, modeling is used to determine tooling design to prolong fatigue life and minimize die wear.

Additive Manufacturing

Rapid prototyping has evolved during the past twenty years from a process for production of form-and-fit models to a suite of technologies that are capable of producing metal parts

Fig. 12 Pressed-and-sintered preform (foreground) and powder-forged connecting rod

having properties equivalent to cast and wrought materials. These processes are also known as direct digital manufacturing processes or rapid manufacturing processes. The AM processes are distinctly different from other PM shapemaking processes in that no tooling is involved; the parts are built up one layer at a time by selectively bonding particles within each layer. While they are still undergoing extensive development, the processes are gaining favor for part production because of their limitless geometric flexibility and ability to make one-of-a-kind parts without the expense of hard tooling. One major area of application of AM is in thermal and fluid management systems, such as heat exchangers, because the processes can produce parts with internal channels that are nonround and nonstraight, a distinct advantage over conventional hole drilling or trepanning of internal passages.

All AM processes begin with powders; in some cases, the powders are coated with polymer to aid bonding. In most processes, the powder is spread in a thin layer, the powder is selectively bonded by one of several means, and a new layer of powder is spread, and so on. The powders are bonded by localized melting due to heat from a laser or electron beam (Fig. 14), or they are bonded by binder droplets from a printhead, such as in 3-D printing (Fig. 15). In any case, the beam path or the droplet pattern is guided by instructions derived from the electronic format of the CAD model of the part. Another approach to AM uses a moving laser head to melt a powder stream and deposit it, one layer at a time, on a substrate or previously deposited material to form a part shape (Fig. 16).

Powders bonded by thermal means (laser or electron beam) produce the required part shape in one step, followed by removal of support structures, while those produced by 3-D printing require subsequent sintering or sintering and infiltration to achieve structural integrity but do not require support structures. Thermally bonded powders often have properties that exceed cast and wrought materials of the same composition, because the localized region of heating is rapidly cooled by surrounding

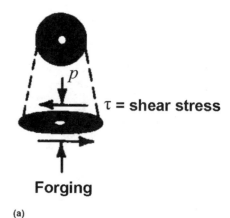

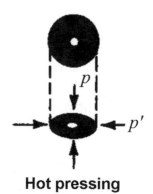

Fig. 13 Comparison of pore collapse by hydrostatic pressure and by forging. Forging with lateral flow causes shear in the material, which efficiently collapses the pores and ensures a sound metallurgical bond across the collapsed pore surfaces

material, leading to finer microstructures than would be obtained by castings that cool at a much slower rate.

In addition to modeling of the part performance, modeling of thermal AM processes focuses on the localized melting occurring at the point of intersection between the laser spot and the metal powder layer. In the case of 3-D printing, modeling is used to optimize the debinding and sintering process, which is similar to that of PIM. In addition, models of infiltration are used to optimize this final densification process.

Models and Applications

Various models for PM processes are available and have been applied extensively in attempts to optimize process efficiency and product quality. Each model has been devised to treat one specific process or another, but there is a great deal of commonality in many of the models. It is convenient to categorize these models according to their common characteristics and then illustrate their adaptations to the various processes. The three groups of models are discrete-element models, linear

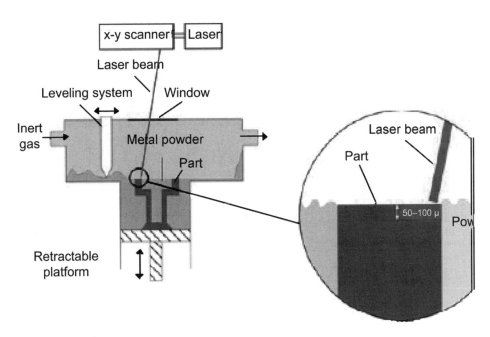

Fig. 14 Laser or electron beam methods for selectively bonding metal particles

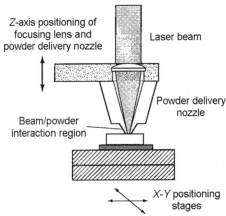

Fig. 16 Laser system for melting a stream of powder particles and placing them in a selected pattern

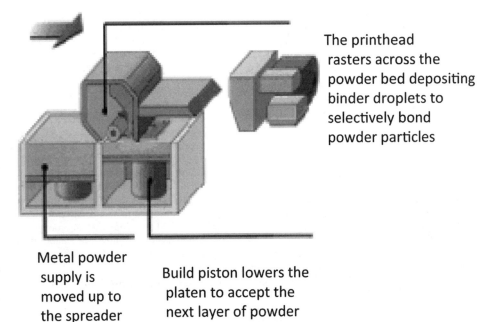

The printhead rasters across the powder bed depositing binder droplets to selectively bond powder particles

Metal powder supply is moved up to the spreader

Build piston lowers the platen to accept the next layer of powder

Fig. 15 Three-dimensional printing using droplets from a printhead to selectively bond metal particles

continuum models, and nonlinear continuum models. Discrete-element modeling treats individual particles and their interactions in a powder mass, which can be used to simulate the behavior of powder during free flow in die and mold filling and the early stages of sintering. Linear continuum models include linear elastic finite element, heat transfer, and fluid mechanics, which can be applied to product design, tool design, PIM simulation, and continuum approaches to sintering. Nonlinear continuum models involve FEA of nonlinear

materials, which are used for simulations of powder die compaction, cold and hot isostatic compaction, and powder forging.

In all process modeling applications, high-fidelity simulations require the input of detailed material representations in a form that is suitable for the appropriate modeling tools. In addition, defect criteria are required if the model is to simulate defect formation during the process. For example, in linear elastic FEA, the material parameters are simply the elastic modulus and Poisson's ratio for the material. If the material is highly

anisotropic, elastic parameters must be input for each of the principal directions in the material. A failure or defect criterion for linearly elastic material would simply be the yield strength. In more extensive applications, stress analysis from FEA can be used in conjunction with crack propagation data for the material to estimate fatigue life of a part or tooling component.

Examples of the use of modeling to solve specific problems in PM processing are given in the following sections.

Powder Metallurgy Process Models

Discrete-element modeling (DEM) was devised to address problems of individual particle interactions in geological applications (Ref 2, 3). Each particle in an assembly of particles is treated as a discrete body interacting through point contacts with surrounding particles. Newton's second law for translation and rotation (six degrees of freedom in 3-D systems) is integrated for each individual particle by taking into account gravity and the normal and tangential tractions occurring at each point of contact between adjacent particles. The resulting displacements describe the individual particle motions and, collectively, the powder mass movement characteristics. With present-day computing power, upward of one million particles can be represented, and the particles can have general, nonspherical shapes. While this does not represent the total number of particles in a typical part, it is sufficient to describe the macroscopic effect of various process parameters on the powder mass response and does provide insight into its general behavior.

Input parameters for DEM include particle shape, elastic properties, and surface roughness, which are translated into spring constants, friction, and damping coefficients at each point contact. The particle interactions can also include adhesive and hydrodynamic forces. Adhesive forces could include interparticle bonds due to binders, so the DEM method could be used to simulate the behavior of particles during 3-D printing or during liquid-phase sintering. Adhesive forces could also include the localized forces at the necks between particle contact points, which opens the possibility of simulating the early stages of solid-state sintering. Including hydrodynamic forces on individual particles permits the use of DEM to simulate the dynamics of individual particles in a fluid stream, such as the particle flow during PIM.

Linear Continuum Models. Finite-element and finite-difference methods are well established and are widely used to represent the response of solid and fluid materials to forces and heat input. In all cases, the material is represented as an assembly of discrete elements, and the conservation laws of mass, momentum, and energy are applied to each element. Constitutive laws representing the material behavior permit application of the methods to specific materials, but they require careful measurement of material properties, often in unconventional ways. Defect criteria are also included in the material representation so that limiting conditions can be determined in the process and parameter changes made in the models to prevent defects.

For example, for linear elastic stress analysis, the material parameters are elastic modulus and Poisson's ratio, while the primary limit is the yield strength of the material. For linear elastic simulation of porous materials, the material parameters will depend on the level of porosity and must be measured on the actual material rather than taken from handbook data for the bulk material. The most convenient and accurate method for determination of the elastic properties is the resonant frequency test. From the resonant frequency of a test bar excited in axial and shear modes, the elastic constants—elastic modulus, E, and shear modulus, G—can be determined for various porosity levels. Then, the Poisson's ratio can be calculated from the relationship $\nu = (E/2G) - 1$.

The output of linear elastic stress analysis includes stresses and strains in all principal directions as well as the displacement of the boundaries. Stress combinations such as the deviatoric stress component can also be output for comparison with the material yield strength for evaluation of failure.

For thermal analysis, the primary material inputs to the simulation codes are conductivity and specific heat. If the material to be simulated is a porous solid, these parameters will depend on the level of porosity and, again, must be measured for the actual material rather than taken from handbook data for the bulk material. In applications to sintering or HIP, the porosity is changing throughout space and time and must be included to represent the changing values of the material thermal properties during the process. In applications to product analysis where thermal analysis is carried out in conjunction with stress analysis to determine thermal stresses, an additional material input is the coefficient of thermal expansion.

Computational fluid mechanics codes require the viscosity of the fluid as well as its thermal properties. The major application of CFD in PM processes is the simulation of polymer flow during PIM. The viscosity of the fluid is highly dependent on the solid particle content and must be measured for the powder/binder combination of interest over the expected range of temperatures occurring in the process.

Nonlinear Continuum Models. All mechanical means of powder consolidation—die compaction, isostatic compaction, and powder forging—involve large deformations of the powder beyond their elastic limit and therefore invoke considerable nonlinear complications in their representation of material behavior. Fully dense materials in isotropic form behave plastically according to the well-known yield criterion of von Mises:

$$S_o^2 = [(S_1 - S_2)^2 + (S_2 - S_3)^2 + (S_3 - S_1)^2]/2 \tag{Eq 1}$$

where the S_i are principal stress components, and S_o is the yield strength of the material. When the combination of stresses acting on a material satisfies this criterion, yielding occurs and plastic deformation commences as long as the applied stresses continue to increase. For example, if a single uniaxial stress is applied, yielding will occur when that stress equals S_o. If a state of pure shear is applied, $S_1 = -S_2$, and yielding will occur when the shear stress $S_1 = S_o/3^{1/2}$. However, if a hydrostatic state of stress is applied, $S_1 = S_2 = S_3$, yielding will not occur regardless of the magnitude of the stress. For this reason, it is common to represent the total stress state in a material in terms of the hydrostatic component, or mean stress:

$$S_m = (S_1 + S_2 + S_3)/3 \tag{Eq 2}$$

plus the deviatoric component represented by Eq 1, which leads to plastic deformation.

While the hydrostatic stress does not cause yielding in a full-density material, porous materials are known to yield and densify under the action of hydrostatic pressure (see the section "Constitutive Behavior" in the article "Powder Metallurgy Process Modeling and Design" in *Powder Metal Technologies and Applications*, Volume 7 of *ASM Handbook*). To accommodate this observation, the yield criterion can be modified to incorporate a term that reflects yielding if hydrostatic stress is applied:

$$S_o^2(D) = [(S_1 - S_2)^2 + (S_2 - S_3)^2 + (S_3 - S_1)^2 - A(D)S_m^2]/2 \tag{Eq 3}$$

where the yield strength, S_o, and the coefficient A are functions of density, D, expressed as a fraction of full density. As the density approaches full density, D = 1, the coefficient A becomes zero, and the hydrostatic component of stress has no effect on yielding. An elegant and detailed development of the constitutive equations for porous materials is given in Ref 4.

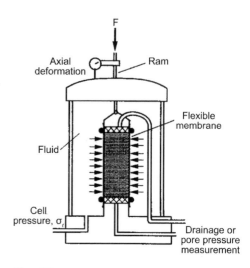

Fig. 17 Schematic of a triaxial compaction cell

Whereas the material parameter S_o in the yield criterion for fully dense material, Eq 1, can be determined through a simple compression for tension test, the parameters defining the yield criterion for porous material must be determined from combined stress tests that simultaneously apply a hydrostatic stress and a shear, or deviatoric, stress. One instrument for this purpose is a triaxial compaction cell, illustrated in Fig. 17 (Ref 5). Through various combinations of triaxial and axial pressure, yielding can be determined and represented as a failure envelope in coordinates of deviatoric stress versus hydrostatic pressure.

Another method for obtaining the required material coefficients involves a series of compression, compaction, and shear tests, which is much less costly than triaxial compaction testing (Fig. 18). A shear test cell (Fig. 19) borrowed from soil mechanics can also be used to test powder metals under combined normal and shear stresses (Ref 6). An extensive series of material tests using these test methods was used to construct the yield envelope for an iron powder (Fig. 20). The results show the change in the yield envelope with increasing density (Ref 6).

Press and Sinter Modeling

A comprehensive approach to modeling of the press and sinter operations is given in the article "Modeling and Simulation of Press and Sinter Powder Metallurgy" in this Volume. The article describes the history, spectrum of models, and integration of compaction and sintering models to optimize the overall process. In the following, examples are given of modeling applied to practical problems in PM outside of those covered in the other article.

Die Filling. As shown in Fig. 3, powder pressing consists of die filling, compaction, and ejection of the part. The first stage of die compaction involves filling the die cavity with powder flowing freely from a feed shoe. Modifying the die filling operation to improve uniformity of density is a major objective. A

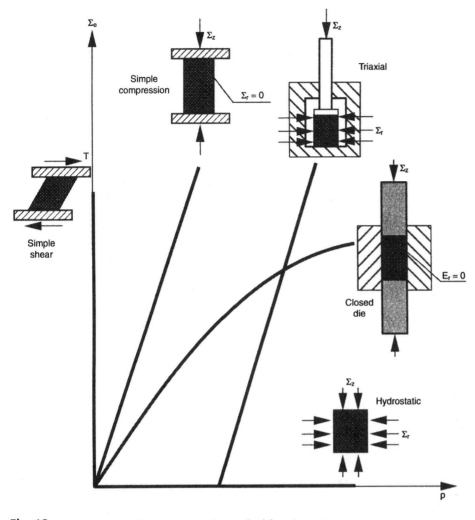

Fig. 18 Various stress states imposed on a powder mass by deformation tests

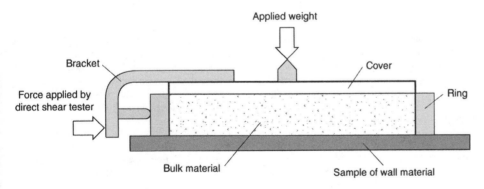

Fig. 19 Schematic of shear cell for testing powder materials under various combinations of normal pressure and shear

the die. Another application of DEM to simulate die filling was performed by Bierswich et al. (Ref 8). In this work, the effects of shoe vibration and die vibration on loose powder die filling were determined for a cylindrical two-level part. The density distribution in vibrated and nonvibrated dies is more effective than die shoe vibration. Rotational vibrations of 1 mm amplitude at 200 Hz were shown to be most effective. Similar advice applies to mold filling for isostatic pressing operations.

Compaction. The second stage, die compaction, can be simulated using the nonlinear finite-element method with appropriate material constitutive equations. Many contributions to powder compaction models have been made since the late 1980s, when large-strain modeling methods and efficient solution algorithms were developed. One practical example of die compaction modeling follows, taken from Ref 4. The model was applied to an axisymmetric two-level part for which extensive experimental data were available through a study by Kergadallan et al. (Ref 9). This case study illustrates the iterative use of modeling and experimentation to refine the process parameters to achieve a defect-free part with a minimum of trial-and-error effort. It also illustrates the use of a model to guide the downstream correction of production process parameters.

The part geometry, shown in Fig. 22, consists of a thin outer rim, a hub, and a bore, which is representative of many common powder metal parts (e.g., engine camshaft timing pulleys and one-way mechanical diode clutch plates). The powder blend that was used is based on the diffusion-alloyed iron powder, Hoeganaes Distalloy AE.

Figure 23 shows a schematic of the press and tooling. The tooling consists of four moving components: an outer die and an inner core that move together, a top punch, a lower inner punch, and a stationary lower outer punch. The press was instrumented with strain gages to measure loads on tooling members and potentiometer displacement transducers to measure tooling displacements.

Fill positions of the punches are given by R_1 and R_2, as shown in Fig. 24. The initial density in the die cavity was estimated by assuming uniform density within the rim section and the hub section of the part. Displacement histories for each of the moving tooling members are shown in Fig. 24.

Two specific samples of compacted parts, part 30 and part 34, are used for description of the model application. Compaction of part 30 resulted in low density in the rim section. The hub and rim densities were approximately 7.06 and 6.90 g/cm^3, respectively. In addition, a distinct crack appeared around the rim inside diameter, 2 mm below the hub. Very fine cracks were also present on the outer surface of the rim, close to the bottom end. Although it is not possible to state exactly when the cracks were formed, it is very likely that they appeared during load removal/ejection. For part 30, the

similar objective applies to filling molds for cold or hot isostatic pressing, an important step toward greater dimensional control in the finished parts. As described previously, the DEM approach is appropriate for simulation of free-flowing powder as it fills a die or mold cavity. Although the method is currently too computationally intense to be used as an everyday analysis tool, focused applications can be

used to determine some overall guidelines for die and mold filling.

One example of the use of DEM for die filling (Ref 7) is shown in Fig. 21. Note that the powder flow is similar in appearance to fluid flow around a corner, leaving a cavity next to the inner die wall. For a given cavity morphology, die shoe configurations can be evaluated for uniform flow and density distribution in

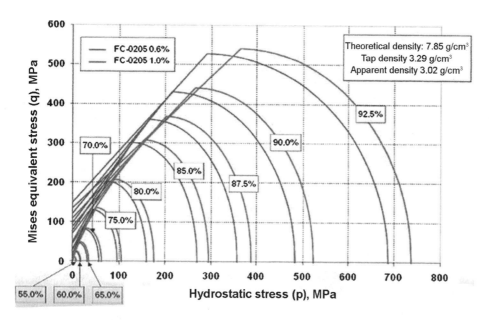

Fig. 20 Evolution of the yield curve with increasing density

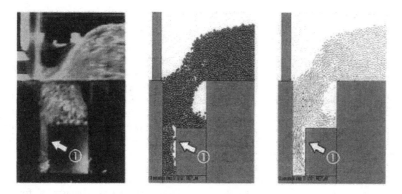

Fig. 21 Results of discrete-element modeling applied to filling a die cavity with free-flowing powder particles

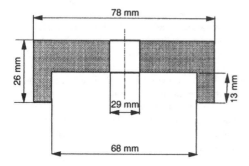

Fig. 22 Part geometry for case study of powder compaction modeling

maximum loads on the inner and outer lower punches were 2.1 and 0.59 MN, respectively. The high load on the lower inner punch resulted in high punch deflection. During ejection, elastic recovery of the lower inner punch resulted in tension on the rim, causing cracking.

Compaction modeling was then applied to part 34 in an effort to reduce the density

imbalance between the hub and rim sections and to eliminate the cracks. The powder material constitutive equation was determined by a series of hydrostatic and triaxial compaction tests on the powder and was implemented in the finite-element code Abaqus/Standard through a user-defined material subroutine. An axisymmetric model of the compact was set up using four-noded axisymmetric elements. Figure 25 shows a 3-D representation of the finite-element mesh obtained by revolving the part about the axis of symmetry.

To eliminate the cracks that appeared in part 30, part 34 was pressed with higher fill in the rim section and reduced fill in the hub section. The tool motions applied were modified to accommodate the different fill positions; otherwise they were very similar to the motions used with part 30. Figure 26 illustrates the compaction process as predicted by the numerical model for part 34. The deformed mesh shows that there is very limited transfer of powder between the hub and rim sections of the part. This behavior is expected because the tooling

motions were designed to minimize powder transfer to avoid the formation of a defect at the corner between the two sections. The measured peak load values are given in Table 1 and are shown to be very close to the predicted loads. For part 34, the lower inner punch load is reduced to 1.61 MN, while the lower outer punch force increases slightly to 0.61 MN. The hub and rim densities are approximately 6.97 and 6.94 g/cm³, much closer than in part 30. The reduced load in the lower inner punch results in lower deflection and eliminates tension during load removal/part ejection. Part 34 was defect free.

Ejection. The third stage of die compaction is ejection of the part from the die. Under the pressure applied by the punches, the powder tends to expand outward but is constrained by the dies. Depending on the powder characteristics and lubricant, the radial pressure resulting from this constraint is greater than 50% of the axial applied pressure. After the punch pressure is released, this radial pressure partially remains, which maintains elastic expansion of the die and radial compression on the compacted powder. Then, as the compact is ejected from the die, it expands elastically (Fig. 27). Prediction of this elastic recovery is important for tool design of parts that must maintain tight tolerances. An approach to modeling of the springback and stresses occurring during powder compact ejection has been presented by Gubanick et al. (Ref 10) to determine the die geometry modifications necessary for improved yield of WC-Co compacts. Good agreement was obtained between measured and model results for both lateral and vertical springback.

Die Design. Modeling of the elastic behavior of dies during compaction can be used to address some particularly perplexing design problems. For example, shelf dies of the type shown in Fig. 28 are common in compaction of axisymmetrical parts. The bottom support for the flange section can be provided by the inside core of the die set or by a more expensive separate lower outer punch with separation at the outer diameter of the flange. When a separate punch is not used, cracking of the die can occur at the location of the inside radius, indicated by "r" in Fig. 28.

To address this problem, Amantani et al. (Ref 11) performed extensive linear finite-element modeling of the shelf die configuration (Fig. 29). A wide variety of diameter ratios (d_b/d_s), height ratios (h_2/h_1), corner radius (r), and interference fits on the die insert were considered. Figure 29 shows the mesh pattern and boundary conditions used for the die body and insert. Diameter ratios $d_b/d_s < 1.6$ were found to be successful with one-piece shelf dies. Also, the die inside radius, r, must be greater than 1.0 mm to ensure sufficiently low stress concentration to prevent die cracking for production runs.

Surface Densification. Gear teeth are one of the most severely stressed sections produced by PM processes. The roots of the teeth should be densified beyond that accomplished by

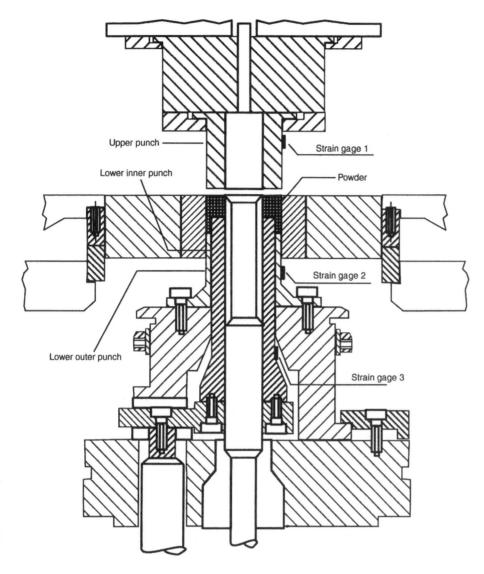

Fig. 23 Instrumented press used for the case study of powder compaction modeling

Labels in figure: Upper punch; Strain gage 1; Lower inner punch; Powder; Strain gage 2; Lower outer punch; Strain gage 3

and temperature-dependent phase-change kinetics and porosity- and temperature-dependent mechanical, physical, and thermal properties. The data were obtained for an F-4065 heat treatable PM steel and applied to a PM component. Predictions and measurements of distortion, dimensional change, and phase changes were carried out. Validation was determined by measuring the dimensional change of a hole in the component and by measuring the amount of retained austenite. Comparison of the calculated and measured results showed excellent agreement.

REFERENCES

1. R. German, *Powder Metallurgy and Particulate Materials Processing,* Metal Powder Industries Federation, 2005, p 235
2. P.A. Cundill and O.D.L. Strack, Discrete Numerical Model for Granular Assemblies, *Geotechnique,* Vol 29, 1979, p 47–65
3. R.M. O'Connor et al., Discrete Element Modeling of Sand Production, *Int. J. Rock Mech. Min. Sci.,* Vol 34, 1997, p 3–4
4. J.R.L. Trasorras, R. Parameswaran, and A.C.F. Cocks, Mechanical Behavior of Metal Powders and Powder Compaction Modeling, *Powder Metal Technologies and Applications,* Vol 7, *ASM Handbook,* ASM International, 1998, p 326–342
5. D.M. Wood, *Soil Behavior and Critical State Soil Mechanics,* Cambridge University Press, 1990
6. Y. Hammi et al., Modeling for PM Component Design and Performance Prediction, *Proceedings of the 2008 World Congress on Powder Metallurgy and Particulate Materials,* Metal Powder Industries Federation, 2008, p 1–99
7. C.Y. Wu et al., Die Filling and Powder Transfer, *Int. J. Powder Metall.,* Vol 39 (No. 4),2003, p 51–64
8. C. Bierswich, Simulation of Die Filling in 3D with Special Emphasis on Vibration Supported Filling, *Adv. Powder Metall.,* 2008, p 130
9. Y. Kergadallan, G. Puente, P. Doremus, and E. Pavier, Compression of an Axisymmetric Part, *Proc. of the Int. Workshop on Modelling of Metal Powder Forming Processes* (Grenoble, France), 1997, p 277–285
10. R.J. Gubanick et al., Modeling of Springback during Ejection of Hardmetal PM Parts Following Compaction, *Proceedings of the 2008 World Congress on Powder Metallurgy and Particulate Materials,* Metal Powder Industries Federation, 2008, p 1–35
11. E. Amentani et al., FEM Analysis of Stress and Deformation States of Shelf Dies for Metal Powder Compaction, *Proceedings of the 2008 World Congress on Powder Metallurgy and Particulate Materials,*

conventional compaction and sintering to enhance their cantilever bend-load capability. Gear rolling and a gear cogging operation can be used for densification, but the design of tool and preform profiles that maximize the tooth root density is left to trial and error and speculation.

An approach to tool and preform design for gear tooth densification by modeling has been presented by Planitzer et al. (Ref 12). Opposing hammers on opposite ends of a diameter across the gear are pressed onto the gear teeth individually as the gear is indexed. By finite-element modeling (FEM), the tooth preform shape was determined that increased the density distribution at the root of the tooth. The possible preforms are shown in Fig. 30. The basic preform conforms to the required tooth profile but allows considerable deformation at the tooth root. Preform versions 01, 02, and 03 allow for compression of the tooth height as well as direct deformation of the tooth root. Results

for densification of the various preform profiles, as predicted by FEM, are shown in Fig. 31, which indicate that the greatest densification area in the critical zone of the tooth root is obtained from preform versions 02 and 03. Although the cogging loads are higher for these preform profiles, the resulting gear life is shown to be significantly longer.

Heat Treatment. Distortion and even cracking can occur in PM parts as they undergo heat treatment. Recently, a modeling approach known as DANTE was developed for practical prediction of the heat treatment response of wrought alloys (Ref 13). DANTE contains a mechanics module coupled to a stress/displacement solver, a phase transformation module coupled to a thermal solver, and a diffusion module coupled to a diffusion solver, as illustrated in Fig. 32. To include the effects of porosity, the method was modified for application to heat treatment of PM parts (Ref 14). Extensive data were generated for porosity-

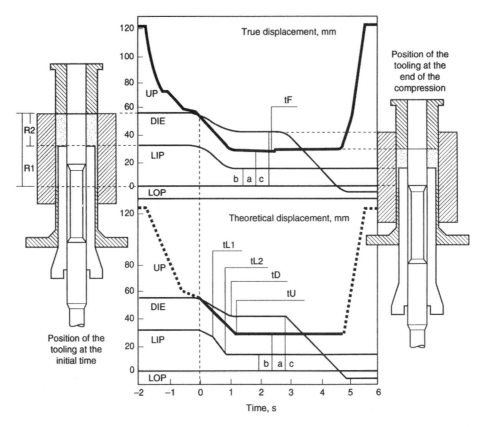

Fig. 24 Fill positions and strokes of the punches for compacting the test part. UP, upper punch; LIP, lower inner punch; LOP, lower outer punch

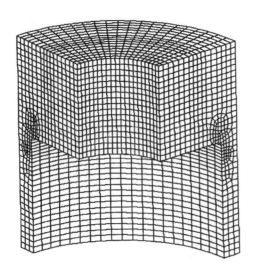

Fig. 25 Finite-element model used to simulate the compaction of the part shown in Fig. 22

Metal Powder Industries Federation, 2008, p 1–52

12. F. Planitzer et al., Investigation of a Cogging Process for Surface Densification of P/M Gears by FEM Simulation, *Proceedings of the 2009 World Congress on Powder Metallurgy and Particulate Materials,* Metal Powder Industries Federation, 2009, p 1–67

13. Deformation Control Technology, Inc., www.deformationcontrol.com/dct_products.htm

14. V. Warke et al., Predicting the Effects of Heat Treatment on Pressed and Sintered Steel Parts, *Proceedings of the 2008 World Congress on Powder Metallurgy and Particulate Materials,* Metal Powder Industries Federation, 2008, p 1–42

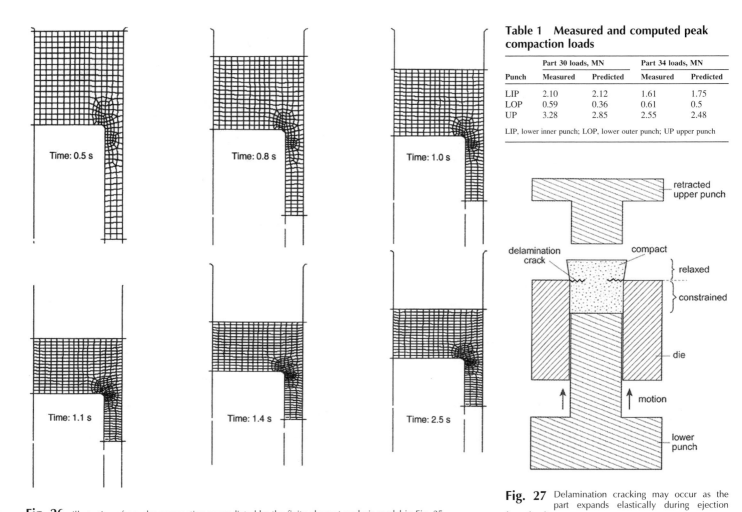

Fig. 26 Illustration of powder compaction as predicted by the finite-element analysis model in Fig. 25

Table 1 Measured and computed peak compaction loads

	Part 30 loads, MN		Part 34 loads, MN	
Punch	Measured	Predicted	Measured	Predicted
LIP	2.10	2.12	1.61	1.75
LOP	0.59	0.36	0.61	0.5
UP	3.28	2.85	2.55	2.48

LIP, lower inner punch; LOP, lower outer punch; UP upper punch

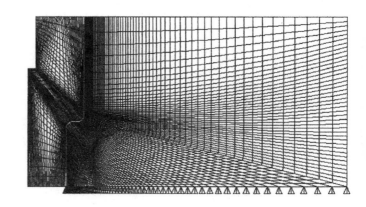

Fig. 27 Delamination cracking may occur as the part expands elastically during ejection from the die

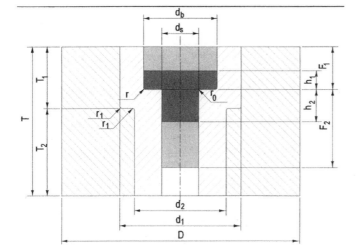

Fig. 28 Shelf die for compaction of an axisymmetrical part

Fig. 29 Finite-element model of the shelf die in Fig. 28

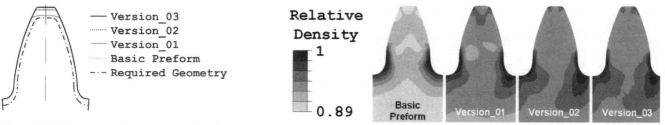

Fig. 30 Various proposed preform profiles for densification of a powder metallurgy gear

Fig. 31 Densification due to cogging of powder metallurgy gear teeth as predicted by finite-element modeling for the profiles shown in Fig. 30

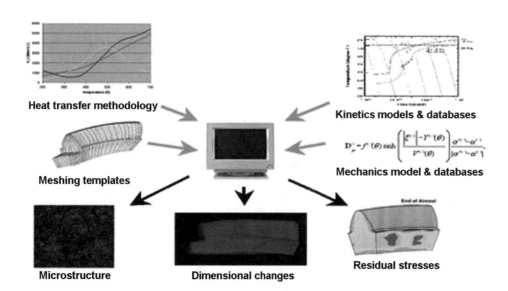

Fig. 32 Illustration of the computational modules and their interaction in DANTE heat treatment simulation software

ASM Handbook, Volume 22B, *Metals Process Simulation*
D.U. Furrer and S.L. Semiatin, editors

Copyright © 2010, ASM International®
All rights reserved.
www.asminternational.org

Modeling and Simulation of Press and Sinter Powder Metallurgy

Suk Hwan Chung, Hyundai Steel Company
Young-Sam Kwon, CetaTech, Inc.
Seong Jin Park, Pohang University of Science and Technology

EFFECTIVE COMPUTER SIMULATIONS of metal powder compaction and sintering are at the top of the powder metallurgy industry's wish list. There is much anticipated advantage to such efforts, yet there are problems that will inhibit widespread implementation. Press-sinter powder metallurgy computer simulations currently focus on the use of minimal input data to help with process setup. Although the simulations are reasonably accurate, a large data array is required to hone in on current industrial practice. For example, final dimensions for automotive transmission gears are required to be held within 10 μm, but the simulations are not capable of such accuracy. Simple factors such as frictional tool heating are missing from the simulations. Additionally, powders vary in particle size distribution between production lots, but the simulations assume a nominally uniform powder. Because it is expensive to test each powder lot, the logic is to assume a nominal set of characteristics. In production, such process and powder variations are handled by constant adaptive control techniques. As an example, when an outside door is opened on the press room, it is common that press adjustments will be required to hold sintered dimensions. The press-sinter powder metallurgy simulations have not advanced to such levels of sophistication. Instead, the press-sinter powder metallurgy simulations are used to help set up production operations, with heavy reliance on experienced operators to make final trial-and-error adjustments.

In practice, the variations in powder, press, tooling, and other process variables are handled through skilled technicians, quality charts, and adaptive process control that relies on frequent sampling and periodic equipment adjustments. The gap between press-sinter practice and modeling may close if more rapid data-generation routes were developed. For example, a study on modeling the press-sinter production of a main bearing cap required 10,000 measures to isolate the behavior. It is not economically

feasible to repeat this testing for each 20 ton lot of powder. Even so, a great benefit comes from the fact that the computer simulations have forced the technical community to organize our knowledge and determine where there are problems.

Computer simulations of press-sinter operations trace to the 1960s (Ref 1 to 20). The early simulations were generally unstable and two-dimensional (for example, the sintering of aligned wires). By 1975, a variety of two-dimensional sintering approaches existed. With the expansion in computer power, the implementation of three-dimensional simulations arose to provide realistic outputs. In more recent times, the simulations have provided valuable three-dimensional treatments to predict the final component size and shape after sintering. Because the pressed green body is not homogeneous, backward solutions are desired to select the powder, compaction, and sintering attributes required to deliver the target properties with different tool designs, compaction presses, and sintering furnaces. In building toward this goal, various simulation types have been evaluated: Monte Carlo, finite difference, discrete element, finite element, fluid mechanics, continuum mechanics, neural network, and adaptive learning. Unfortunately, the input data and some of the basic relations are not well developed; accurate data are missing for most materials under the relevant conditions. For example, rarely is the strength measured for a steel alloy at the typical 1120 °C sintering temperature. Further, constitutive models do not exist for the conditions relevant to sintering; for example, friction in die compaction changes during the split-second pressure stroke, because lubricant (polymer) particles deform and undergo viscous flow to the die wall, effectively changing friction constantly during compaction. Thus, the simulations are approximations using extrapolated data and simplified relations. For this reason, computer simulations of press-

sinter routes work best in the setup mode. The simulations help define the processing window and set initial operating parameters. Presented here is information relevant to computer simulation of first-article production, what is best termed setup calculations, realizing that practice relies heavily on adaptive process control to keep the product in specification after the initial setup is accomplished.

Brief History

The first major publication on computer simulation of sintering came out in 1965 (Ref 1). Early simulations were two-dimensional (sintering two wires) with a single diffusion mechanism. These simulations were slow, requiring ten times more computer time than the actual physical sintering time. Most damaging, these early models were unstable, because they lost volume and increased energy. However, within 20 years the concept was extended to include multiple transport mechanisms, multiple sintering stages, and even pressure-assisted sintering (Ref 6, 9 to 11). These simulations predicted density versus compaction pressure, sintering time, peak temperature, heating rate, green density, and particle size.

One of the first realizations was the limitations arising from the assumed isothermal conditions and simplistic microstructure coarsening. Dilatometry experiments show that most sintering occurs on the way to the peak temperature, so isothermal models poorly reflect actual behavior (Ref 20). Indeed, production powder metallurgy often simply "kisses" the peak temperature, a situation far from what is assumed in the simulations. Also, the assumed homogeneous and ideal microstructure unrealistically limits the models. Today (2009), the sintering body first treated with a compaction or shaping simulation to predict the green microstructure gradients, and subsequent sintering simulations, use those density gradients, via finite-element analysis, to predict the final size, shape, and properties (Ref 16 to 19).

Theoretical Background and Governing Equations

The methodologies used to model the press and sinter powder metallurgy include continuum, micromechanical, multiparticle, and molecular dynamics approaches. These differ in length scales. Among the methodologies, continuum models have the benefit of shortest computing time, with an ability to predict relevant attributes such as the component density, grain size, and shape.

Mass, volume, and momentum conservation are evoked in the continuum approach. Although such assumptions may seem obvious, powder metallurgy processes are ill-behaved and difficult to properly simulate. For example, polymers are added to the powder for tool lubrication, but the polymers are pyrolyzed during sintering, resulting in 0.5 to 1.5 wt% mass loss. Likewise, pore space is not conserved during compaction and sintering, so bulk volume is not conserved. Even so, mass conservation equations are invoked to track densification, while momentum conservation is used to follow force equilibrium, including the distortion effect from gravity. Energy conservation is also essential in the continuum approach. However, it is typical to assume temperature is uniform in the compact—set to room temperature during compaction and following an idealized thermal cycle (often isothermal) during sintering. Both are incorrect, because tool heating occurs with repeated compaction strokes, and compact position in the sintering furnace gives a lagging thermal history that depends on location. Indeed, because dimensional precision is the key to powder metallurgy, statistical audits have repeatedly found that subtle factors such as position in the furnace are root causes of dimensional scatter. For example, fluid flow and heat-transport calculations show considerable temperature differences associated with atmosphere flow and component shadowing within a furnace. Because such details are not embraced by the models, the typical assumption is to ignore temperature distribution within the component, yet such factors are known to cause part distortion during production.

Additionally, constitutive relations are required to describe the response of the compact to mechanical force during compaction and sintering. Many powder metallurgy materials are formed by mixing powders that melt, react, diffuse, and alloy during sintering. This requires sophistication in the models to add phase transformations, alloying, and other factors, many of which depend on particle size and other variations (Ref 15). From conservation laws and constitutive relations, a system of partial differential equations is created that includes the initial and boundary conditions. These must be integrated with microstructure and property models so that final compact properties can be predicted. Because the constitutive relations for compaction and sintering are completely different, they are described here in two separate sections.

Constitutive Relation during Compaction

Continuum plasticity models are frequently used to describe the mechanical response of metal powders during compaction. These phenomenological models, originally developed in soil mechanics, are characterized by a yield criterion, a hardening function, and a flow rule. Representative models include those known as the Cam-Clay (Ref 21), Drucker-Prager-Cap (Ref 22), and Shima-Oyane (Ref 23) models. Of these, the most successful for metal powders has been the Shima-Oyane model, although for ceramics, soils, and minerals, other relations are generally more successful.

The typical initial and boundary conditions during compaction are as follows:

- *Initial condition for the powder:* Tap density
- *Boundary conditions:* Velocity prescribed in upper and bottom punches and friction condition in the tooling side wall; usually assumed the same for all tool surfaces independent of wear and independent of lubricant flow during the compaction stroke

During compaction and ejection, a damage model, such as the Drucker-Prager failure surface (Ref 22) and failure separation length (FSL) idea (Ref 24), is required. To predict crack formation, the FSL assumes there is an accumulated separation length from the Drucker-Prager failure surface, which provides the possibility of crack formation, as shown in Fig. 1. The equation for the FSL is expressed as:

$$F_S = q + p \tan \beta - d \qquad \text{(Eq 1)}$$

where q and p are the effective stress and hydrostatic pressure, respectively. Note that d and β are the offset stress and slope, respectively, for the Drucker-Prager failure surface shown in Fig. 1. Because the models predict green density versus location, defect sensitivity is possible. For example, elastic relaxation occurs on ejection, and if the stress exceeds the green strength, then green cracking occurs. It is in this area that the compaction models are most effective.

Constitutive Relation during Sintering

Continuum modeling is the most relevant approach to modeling grain growth, densification, and deformation during sintering. Key contributions were by Ashby (Ref 6, 9), McMeeking and Kuhn (Ref 25), Olevsky et al. (Ref 17, 19), Riedel et al. (Ref 13, 26, 27), Bouvard and Meister (28), Cocks (29), Kwon et al. (Ref 30, 31), and Bordia and Scherer (Ref 32 to 34) based on a sintering mechanism such as surface diffusion, grain-boundary diffusion, volume diffusion, viscous flow (for amorphous materials), plastic flow (for crystalline materials), evaporation condensation, and rearrangement. For industrial application, the phenomenological models are used for sintering simulations with the following key physical parameters:

- Sintering stress (Ref 20) is a driving force of sintering due to interfacial energy of pores and grain boundaries. Sintering stress depends on the material surface energy, density, and geometric parameters such as grain size when all pores are closed in the final stage.

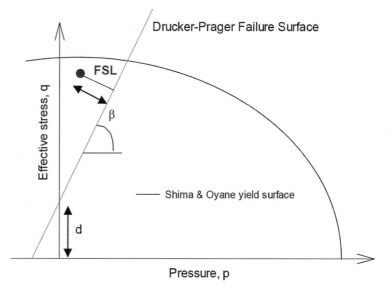

Fig. 1 Definition of failure separation length (FSL) based on Shima-Oyane yield model and Drucker-Prager failure surface. *d*, offset stress; β, slope

- Effective bulk viscosity is a resistance to densification during sintering and is a function of the material, porosity, grain size, and temperature. The model of the effective bulk viscosity has various forms according to the assumed dominant sintering mechanism.
- Effective shear viscosity is a resistance to deformation during sintering and is also a function of the material, porosity, grain size, and temperature. Several rheological models for the effective bulk viscosity are available.

The preceding parameters are a function of grain size. Therefore, a grain-growth model is needed for accurate prediction of densification and deformation during sintering.

Typical initial and boundary conditions for the sintering simulations include:

- *Initial condition:* Mean particle size and grain size of the green compact for grain growth and initial green density distribution for densification obtained from compaction simulations
- *Boundary conditions:* Surface energy condition imposed on the free surface and friction condition of the component depending on its size, shape, and contact with the support substrate

The initial green density distribution within the pressed body raises the necessity of starting the sintering simulation with the output from an accurate compaction simulation, because die compaction induces green density gradients that depend on the material, pressure, rate of pressurization, tool motions, and lubrication. The initial and boundary conditions help determine the shape distortion during sintering from gravity, nonuniform heating, and the green body density gradients.

Numerical Simulation. Even though many numerical methods have been developed, the finite-element method (FEM) is most popular for continuum models of the press and sinter process. The FEM approach is a numerical computational method for solving a system of differential equations through approximation functions applied to each element, called domain-wise approximation. This method is very powerful for the typical complex geometries encountered in powder metallurgy. This is one of the earliest techniques applied to materials modeling and is used throughout industry today (2009). Many powerful commercial software packages are available for calculating two- and three-dimensional thermomechanical processes such as found in press and sinter powder metallurgy.

To increase the accuracy and convergence speed for the press and sinter simulations, developers of the simulation tools have selected explicit and implicit algorithms for time advancement, numerical contact algorithms for problems such as surface separation, and remeshing algorithms as required for large

deformations such as seen in some sintered materials, where up to 25% dimensional contraction is possible.

Figure 2 shows the typical procedure for computer simulation for the press-sinter process, which consists of five components: simulation tool, pre- and postprocessors, optimization algorithm, and experimental capability. Pre- and postprocessors are important for using the simulation tools efficiently. The preprocessor is a software tool to prepare input data for the simulation tool, including computational domain preparation such as geometry modeling and mesh generator. Figure 3 is an example of the component, compaction, and sintering models, in this case for an oxygen sensor housing. Executing this model requires considerable input, including a material property database (including strain effects during compaction and temperature effects during sintering) and a processing condition database (loading schedule of punches and dies for compaction simulation and heating cycle for sintering simulation). A postprocessor is a software tool to visualize and analyze the simulation results, which enhances the usefulness of the simulations. From the standpoint of process setup calculations, the optimization algorithm is essential to maximize computer simulation capability providing the optimum part, die, and process condition design. Experimental capability is very important in computer simulation, providing a means to evaluate changes in materials, powders, compaction schedules, heating cycles, and generally to provide verification of the simulation results.

Experimental Determination of Material Properties and Simulation Verification

Material Properties and Verification for Compaction

One of the first needs is to measure the powder density as a function of applied pressure to generate the material parameters in the constitutive model for compaction, including the Coulomb friction coefficient between the powder and die. Note that these factors vary with the powder lot, lubricant, tool material, and even tool temperature. The procedure to obtain the material properties based on the generalized Shima-Oyane model is as follows (Ref 24):

- Measure the pycnometer and tap densities of the powder.
- Conduct a series of uniaxial compression tests with die wall lubrication to minimize the die wall friction effect. The tap density is considered the starting point (after particle rearrangement) corresponding to zero compaction pressure. By curve fitting, six material parameters (α, γ, m, a, b, and n) are determined for the yield surface, Φ:

$$\Phi = \left(\frac{q}{\sigma_m}\right)^2 + \alpha(1-D)^\gamma \left(\frac{p}{\sigma_m}\right) - D^m \quad \text{(Eq 2)}$$

where q and p are the effective stress and hydrostatic stress or pressure, respectively; D is the relative or fractional density; and

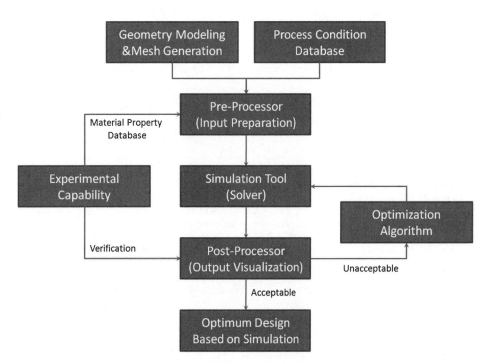

Fig. 2 Typical procedure for computer simulation for press and sinter process

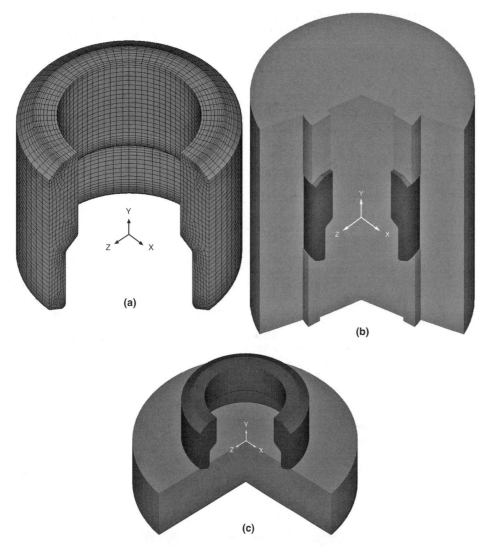

Fig. 3 Modeling and mesh-generation example showing the simulation of compaction and sintering for an oxygen sensor housing. (a) Mesh generation for the compact. (b) Modeling of punches and dies during compaction for the press simulation. (c) Modeling of compact in contact with the substrate during sintering simulation

σ_m is the flow stress of the matrix material, which is expressed as:

$$\sigma_m = a + b\bar{\varepsilon}_m^n \qquad (Eq\ 3)$$

where $\bar{\varepsilon}_m$ is the effective strain of the matrix material.

- A series of uniaxial compression tests are performed without wall lubricant, then the Coulomb frictional coefficient is obtained by FEM simulation.

Figure 4 is an illustration of the compaction curve for an iron-base powder (Distaloy AE, Höganäs) in a simple cylindrical geometry. Uniaxial compression tests are provided for two samples: the smaller sample of 4.5 g in a 12 mm diameter die with die wall lubrication, and the larger sample of 9.3 g without wall lubrication. By curve fitting, the six material parameters are (Ref 24):

$\alpha = 6.20, \gamma = 1.03, \ m = 7.40, a = 184$ MPa, $b = 200$ MPa, and $n = 0.240$

By FEM simulation, the Coulomb frictional coefficient was obtained as 0.1. Table 1 shows the example of a complete set of material properties of an iron-base powder as input data for the compaction simulation.

Verification of the predicted density gradients in the green compact has been approached by many techniques. The most reliable, direct, and sensitive comes from taking hardness or microhardness traces on a polished cross section. Thus, to verify the compaction simulation results, the relationship between hardness and green density is conducted according to the following procedure:

- Use the same samples as used for obtaining the material parameters.

- Presinter the compacts at a temperature sufficient to bond the particles but below the temperature range where dimensional change or chemical reactions occur.
- Carefully prepare a metallographic cross section of the presintered samples and treat with a vacuum annealing cycle to minimize any hardness change induced by the cutting process.
- Measure the hardness of each sample with a known green density, and from that develop a correlation between density and hardness.
- Apply the same procedure and hardness traces to real components, and from precise measurements of hardness and location develop a contour plot of the green density distribution for comparison with the computer simulation.

As an example, Fig. 5 is a plot of the correlation between green density and hardness for the WC-Co system. For this plot, the presintering cycle of the WC-Co system was at 790 °C for 30 min, and the annealing cycle was at 520 °C for 60 min; in this case, a Rockwell 15T hardness scale was used. The obtained correlation is (Ref 35):

$$D = 0.638 + 1.67 \times 10^{-3}H - 5.44 \times 10^{-7}H^2 \qquad (Eq\ 4)$$

where H is the 15T Rockwell hardness number, and D is the fractional density. Figure 6 compares the simulation results taken from a commercial software package (PMsolver) with the experimental results for a cutting tool geometry formed from a cemented carbide powder, based on Eq 4.

Material Properties and Verification for Sintering

In the development of a constitutive model for sintering simulation, a wide variety of tests are required, including data on grain growth, densification (or swelling), and distortion. These are approached as follows:

- *Grain growth:* Quenching tests are conducted from various points in the heating cycle, and the mounted cross sections are analyzed to obtain grain-size data to implement grain-growth models. A vertical quench furnace is used to sinter the compacts to various points in the sintering cycle and then to quench those compacts in water. This gives density, chemical dissolution (for example, diffusion of one constituent into another), and grain size as instantaneous functions of temperature and time. The quenched samples are sectioned, mounted, and polished prior to optical or scanning electron microscopy (SEM). Today (2009), automated quantitative image analysis provides rapid determination of density, grain size, and phase content versus location in the compact. Usually during sintering, the mean grain size, G, varies from the starting mean grain size, G_0 (determined on the

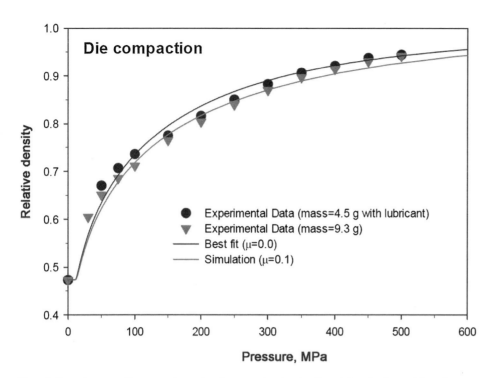

Fig. 4 Example compaction curve for an iron-base powder showing the relative density model results versus compaction pressure with no friction and with a friction coefficient of 0.1 for two compact samples.
Source: Ref 24

Table 1 Complete set of material properties for die compaction of an iron-base powder

Densities	Pycnometer density	7.8 g/cm³
	Fractional tap density	0.45
Yield function (Shima-Oyane)	α	6.2
	γ	1.03
	m	7.4
Flow stress of matrix materials (work hardening)	a	184 MPa
	b	200 MPa
	n	0.24
Friction coefficient		0.1
Failure surface	d	0.01
	$\tan \beta$	3.41

Source: Ref 24

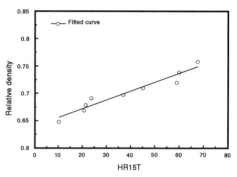

Fig. 5 Plot of relative density and Rockwell 15T hardness scale for the die compaction of a WC-Co powder. Source: Ref 35

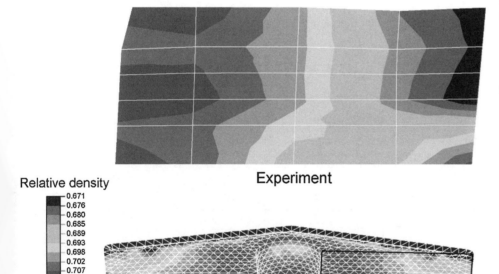

Fig. 6 Comparison between the computer-simulated green density gradients (PMsolver) and the experimental results taken from hardness tests on a cross-sectioned green compact of a cutting tool formed from WC-Co powder.
Source: Ref 35

green compact). A new master sintering curve concept is applied to fit the experimental grain-size data to an integral work of sintering (Ref 36), because actual cycles are a complex combination of heats and holds. The resulting material parameters trace to an apparent activation energy as the only adjustable parameter. Figure 7 shows SEM micrographs after a quenching test and grain-growth modeling for a W-8.4wt%Ni-3.6wt%Fe mixed powder compact during liquid-phase sintering.

• *Densification:* To obtain material parameters for densification, constant heating rate dilatometry is used for in situ measurement of shrinkage, shrinkage rate, and temperature. By fitting the experimental data to models that include the sintering stress, σ_s, and bulk viscosity, K, as functions of density and grain size, again relying on the master sintering curve concept (Ref 37), the few unknown material parameters are extracted. Figure 8 shows the dilatometry data and model curve-fitting results used to obtain the missing material parameters during sintering of a 316L stainless steel (Ref 38).

• *Distortion:* Powder metallurgy compacts reach very low strength levels during sintering. Accordingly, weak forces such as gravity, substrate friction, and nonuniform heating will induce distortion and even cracking. To obtain the material parameters related to distortion, three-point bending or sinter forging experiments are used for in situ measurement of distortion (Ref 39). By fitting the experimental data with FEM simulations for shear viscosity, μ, with grain growth, the parameters such as apparent activation energy and reference shear viscosity are extracted. Figure 9 shows an in situ bending test and FEM results for obtaining material parameters in shear viscosity for a

316L stainless steel powder doped with 0.2% B to induce improved sintering.

Such data-extraction techniques have been allied to several materials, including tungsten alloys, molybdenum, zirconia, cemented carbides, niobium, steel, stainless steel, and alumina. Table 2 is an example set of material properties for W-8.4 wt%Ni-3.6wt%Fe as used as input data for the sintering simulation (Ref 40). The preceding experiment techniques can be used for verification of sintering simulation results.

Demonstration of System Use

Time Advance Algorithm and Compaction Simulation Accuracy

Time advance in the simulation models is a concern with respect to the balance between accuracy and computational speed. The explicit method is fast but sometimes exhibits convergence and accuracy problems, while the implicit method is accurate but slow. Figure 10 illustrates this case for a simple cylindrical geometry in the WC-Co system (Ref 35). As shown in Fig. 10 (b), the implicit method is more accurate for this case.

Gravitational Distorting in Sintering

The rheological data for the sintering system allow the system to respond to the internal sintering stress that drives densification and any external stress, such as gravity, that drives distortion. When a compact is sintered to high density, it is also necessary to induce a low strength. (The material is thermally softened to a point where the internal sintering stress can induce densification.) Figure 11 shows

sintering simulation results for a tungsten heavy alloy, relying on test data taken on Earth and under microgravity conditions, to then predict the expected shapes for various gravitational conditions: Earth, Moon, Mars, and in space. The results show that gravity affects shape distortion during sintering (Ref 40). Accordingly, the computer simulations can be used to reverse engineer the green component geometry to anticipate the distortion to achieve the desired sintered part design.

Compaction Optimization

There are two different simulation approaches used to optimize die compaction. One is based on the concept of design of experiment, and the other is a derivative-based optimization scheme. The process designer must first select a reasonable initial guess, an

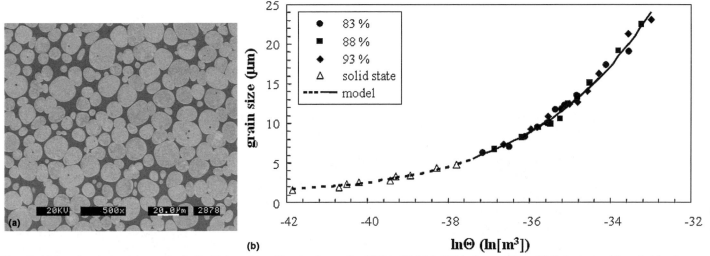

Fig. 7 (a) Scanning electron micrograph of a liquid-phase sintered tungsten heavy alloy (W-8.4wt%Ni-3.6wt%Fe) after quenching. (b) Grain-size model results taken from an integral work of sintering concept that includes only the thermal cycle (time-temperature path) to predict grain size for any point in a heating path for three different tungsten contents. Source: Ref 36

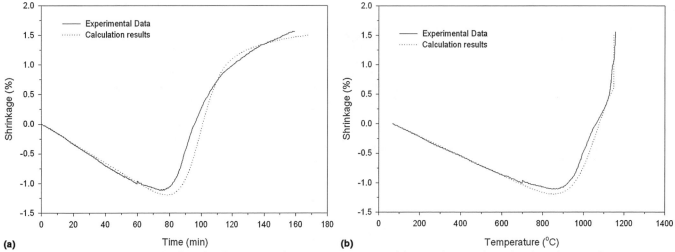

Fig. 8 Dilatometry data showing in situ shrinkage data during constant heating rate experiments and the curve-fitting results used to obtain the material parameters to predict densification of a 316L stainless steel powder. (a) Shrinkage with time. (b) Shrinkage with temperature. Source: Ref 38

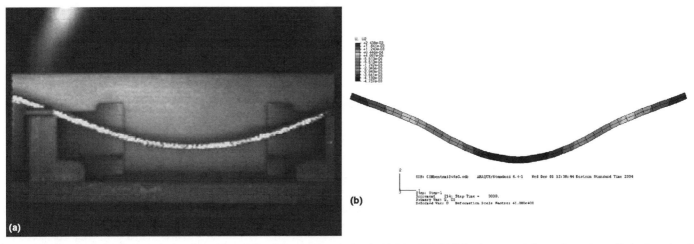

Fig. 9 (a) Video image taken during in situ bending test for a 316L stainless steel sample doped with 0.2 wt% B. (b) Finite-element modeling results used to verify the shear viscosity property as a function of time, temperature, grain size, and density during heating. Source: Ref 39

Table 2 Complete set of material properties for simulating the sintering of W-8.4wt%Ni-3.6wt%Fe

Pycnometer density		16.75 g/cm^3
Initial density distribution, ρ_0		Results from press simulation
Initial mean grain size		0.30 μm
Surface energy, γ		2.5 J/m^2
Transition temperature from solid state to liquid phase		1460 °C
Friction coefficient		0.3

Sintering stress	$\sigma_s = \dfrac{6\gamma}{G} \dfrac{\rho^2(2\rho - \rho_0)}{\theta_0}$ for $\rho < 0.85$	
	$\sigma_s = \dfrac{2\gamma}{G} \left(\dfrac{6\rho}{\theta}\right)^{1/3}$ for $\rho_W > 0.95$	
	$\sigma_s = \dfrac{(\rho_2 - \rho)}{(\rho_2 - \rho_1)} \sigma_{si} + \dfrac{(\rho - \rho_1)}{(\rho_2 - \rho_1)} \sigma_{sf}$ for $0.85 \leq \rho_W \leq 0.95$	

State			**Solid**	**Liquid**
Grain growth	$\dfrac{dG}{dt} = \dfrac{k_0 \exp(-Q_G/RT)}{G^l}$	k_0 (m^{n+1}/s)	2.8×10^{-13}	1.1×10^{-15}
		Q_G (kJ/mol)	241	105
		l	2.0	2.0
Bulk viscosity	$K_i = \dfrac{\rho(\rho - \rho_0)^2}{8\theta_0^2} \dfrac{TG^3}{\alpha_i \exp(-Q_D/RT)}$ for $\rho \leq 0.92$	Q_D (kJ/mol)	250	250
	$K_f = \dfrac{\rho}{8\theta_0^{1/2}} \dfrac{TG^3}{\alpha_f \exp(-Q_D/RT)}$ for $\rho > 0.92$	α_i (m$^6 \times$K/s)	1.3×10^{-17}	5.0×10^{-17}
	with $\alpha_f = \dfrac{\theta_0^2}{\sqrt{0.08}(0.92 - \rho_0)^2} \alpha_i$			
Shear viscosity	$\mu_i = \dfrac{\rho^2(\rho - \rho_0)}{8\theta_0} \dfrac{TG^3}{\beta_i \exp(-Q_D/RT)}$ for $\rho \leq 0.92$			
	$K_f = \dfrac{\rho}{8} \dfrac{TG^3}{\beta_f \exp(-Q_D/RT)}$ for $\rho > 0.92$	β_i (m$^6 \times$K/s)	1.3×10^{-17}	1.3×10^{-12}
	with $\beta_f = \dfrac{\theta_0}{0.92(0.92 - \rho_0)} \beta_i$			

Note that R is the universal gas constant, ρ is the density, θ ($= 1 - \rho$) is the porosity, θ_0 ($= 1 - \rho_0$) is the initial porosity, and ρ_W is the density of the tungsten skeleton. Source: Ref 40

objective function that needs to be minimized, and design variables for both approaches.

Figure 12 shows the first approach used to optimize the loading schedule to generate a uniform green density during die compaction (Ref 35). In this case, the target is a cutting tool formed from WC-Co. The displacement of the upper punch was set as the design variable, and the lower punch displacement was automatically calculated because the final dimension was fixed. The objective function was set to the standard deviation in green density, which is called the nonuniformity, and this was to be minimized. Figure 12(a) plots the density histograms for five different processing conditions. The first one (black) is the initial compaction process design, and the fourth one (yellow) shows the optimum design for maximum uniformity with only a change in the upper punch motion. The density distributions of those two cases are shown in Fig. 12 (b). For optimization in more complicated systems, the Taguchi method with an orthogonal array can be used to create simulation experiments to efficiently isolate solutions.

In using a derivative-based optimization scheme, it is important to define the searching

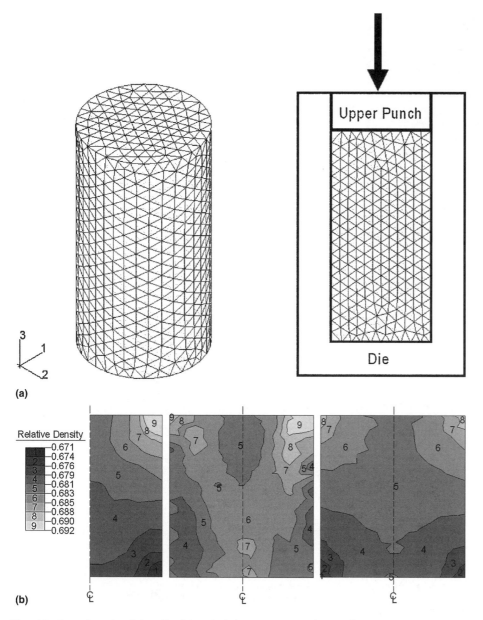

(a)

(b)

Relative Density
1 — 0.671
2 — 0.674
3 — 0.676
4 — 0.679
5 — 0.681
6 — 0.683
7 — 0.685
8 — 0.688
9 — 0.690
— 0.692

Fig. 10 Comparison of explicit and implicit methods for compaction simulation in the WC-Co system. (a) Meshed geometry. (b) Left, half-axis experimental result; center, full-axis explicit method result; and right, implicit method result. Source: Ref 35

direction and stepping size. The searching direction is decided by the direct differentiation or adjoint variable methods, and the stepping size is usually selected by polynomial curve fitting. Using the results of a finite-element simulation with a mesh system generated by the given design variables, design sensitivities are calculated by the algorithm of the searching direction. The searching direction is selected by the conjugate gradient method, and the proper stepping size is selected by the polynomial curve fitting, with the objective functions obtained by additional finite-element simulations. The design parameters are iteratively updated until the convergence criteria are satisfied. Figure 13 shows the procedure for optimizing the loading schedule to have uniform density distribution during die compaction for a hub part formed using a steel alloy powder (Ref 24). The design variables considered in this example were the loading schedules of upper, inner lower, core rod, and die, and the objective function is the nonuniformity after die compaction. The goal was to minimize the nonuniformity. Figure 13(a) shows the compaction tool set and analysis domain. Figure 13(b) plots the objective function during optimization, and Fig. 13(c) plots the density distributions of the initial and optimum designs.

Sintering Optimization

Usually, a small grain size is desired to improve properties for a given sinter density. In this illustration, the design variable is the sintering cycle. To obtain maximum density and minimum grain size, the following objective function, F, is proposed (Ref 41):

$$F = \alpha \left[\frac{\Delta \rho}{\rho} \right] + (1 - \alpha) \left[\frac{\Delta G}{G} \right] \qquad \text{(Eq 5)}$$

where α is an adjustable parameter. Figure 14(a) shows an example for maximum density and minimum grain size for a 17-4 PH stainless steel powder. For example, the minimum grain size will be 21.9 μm if the specified sintered density is 95% or theoretical. Figure 14(b) shows the corresponding

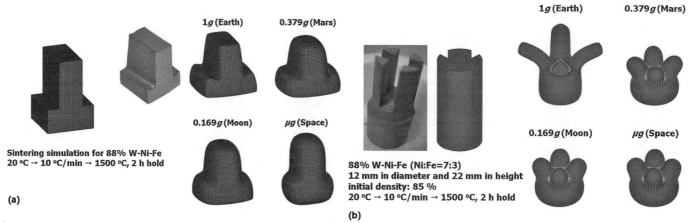

(a) Sintering simulation for 88% W-Ni-Fe
20 °C → 10 °C/min → 1500 °C, 2 h hold

1g (Earth)   0.379g (Mars)
0.169g (Moon)   μg (Space)

88% W-Ni-Fe (Ni:Fe=7:3)
12 mm in diameter and 22 mm in height
initial density: 85 %
20 °C → 10 °C/min → 1500 °C, 2 h hold

1g (Earth)   0.379g (Mars)
0.169g (Moon)   μg (Space)

(b)

Fig. 11 Final distorted shape by sintering under various gravitational environments for complicated test geometries. (a) T-shape. (b) Joint part. Source: Ref 40

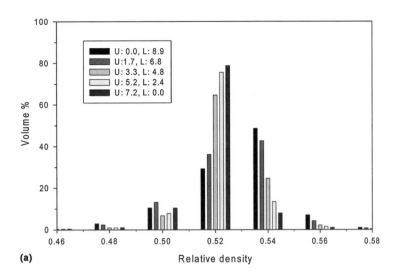

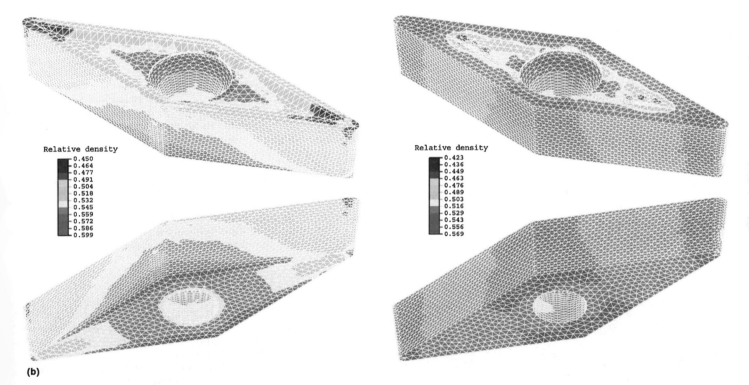

Fig. 12 Optimization for uniform density distribution during die compaction for a cutting tool fabricated from WC-Co. (a) Histogram for various processing conditions. (b) Green density distributions in the initial and optimum designs. Source: Ref 35

sintering cycle by matching the value of the adjustable parameter α.

Coupled Press and Sinter Optimization

Coupled simulations are necessary to predict the quality of the final component after the combined press and sinter operations. Figure 15 illustrates this for the case for a shaped cutting tool formed from WC-Co. In practice, the final component exhibited cracking along the corners, but there was no sign of cracking after die compaction, as shown in Fig. 15(b). From the simulation results for the initial design, the density distribution is found to vary from a fractional density of 0.583 to 0.814. After optimization of the tool loading schedule with the objection function to make the green density as uniform as possible, the green density distribution ranged from 0.638 to 0.675. Implementation of this loading cycle resulted in elimination of cracking in the sintered component.

Figure 16 illustrates the distortion in final shape for oxygen sensor components. After optimization of the tool loading schedule and the sintering cycle, the final distortion from the target shape is significantly reduced for both the holder and sleeve.

Conclusion

Computer simulations of the press-sinter cycle in powder metallurgy have advanced considerably and, in combination with standard finite-element techniques, show a tremendous

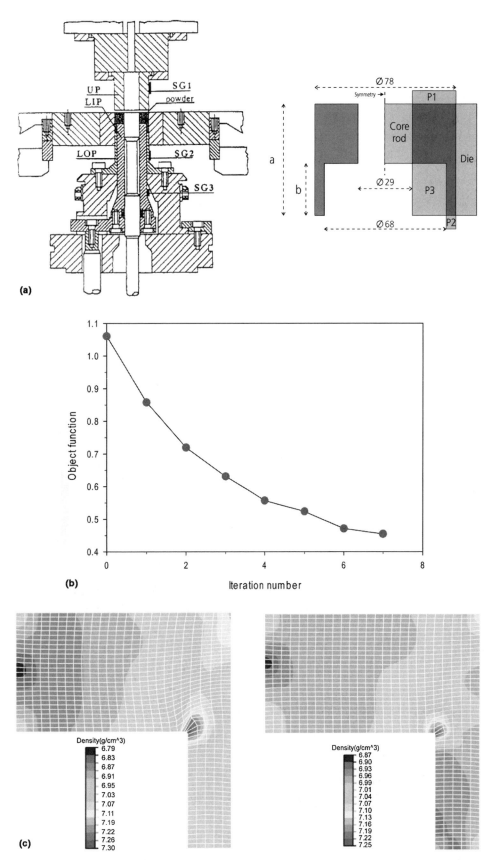

Fig. 13 Optimization to minimize the green density gradients during die compaction of a steel hub component. (a) Compaction tool set and analysis domain. (b) Variation of objective function during optimization iteration. (c) Green density distributions in the initial and optimum designs. Source: Ref 24

ability to guide process setup. Illustrated here are the compaction and sintering concepts required to perform process optimization, typically via selection of appropriate compaction tool motion. Although the models are only approximations to reality, they are still of value in forcing a careful inspection of what is understood about the press-sinter process. In this regard, the greatest value of modeling is in the forced organization of process knowledge.

There remain several barriers to widespread implementation. The largest is that traditional powder metallurgy is largely dependent on adaptive process control, because many of the important factors responsible for dimensional or quality variations are not measured. The variations in particle size, composition, tool wear, furnace location, and other factors, such as reactions between particles during heating, impact the important dimensional-control aspects of press-sinter powder metallurgy. However, nominal properties, such as strength, hardness, or fatigue life, are dominated by the average component density. In that regard, especially with respect to the initial process setup, the computer simulations are of great value. Nevertheless, important attributes such as dimensional tolerances and internal cracks or other defects are outside the cost-benefit capabilities of existing simulations. Further, the very large number of materials, processes, tool materials, sintering furnaces, and process cycles makes it difficult to generalize; significant data collection is required to reach the tipping point where the simulations are off-the-shelf. Thus, much more research and training is required to move the simulations into a mode where they are widely applied in practice. Even so, commercial software is available and shows great value in the initial process definition to set up a new component.

REFERENCES

1. F.A. Nichols and W.W. Mullins, Morphological Changes of a Surface of Revolution due to Capillarity-Induced Surface Diffusion, *J. Appl. Phys.,* Vol 36, 1965, p 1826–1835
2. F.A. Nichols, Theory of Sintering of Wires by Surface Diffusion, *Acta Metall.,* Vol 16, 1968, p 103–113
3. K.E. Easterling and A.R. Tholen, Computer-Simulated Models of the Sintering of Metal Powders, *Z. Metallkd.,* Vol 61, 1970, p 928–934
4. A.J. Markworth and W. Oldfield, Computer Simulation Studies of Pore Behavior in Solids, *Sintering and Related Phenomena,* G.C. Kuczynski, Ed., Plenum Press, New York, NY, 1973, p 209–216
5. K. Breitkreutz and D. Amthor, Monte-Carlo-Simulation des Sinterns durch Volumen-und Oberflachendiffusion, *Metallurgie,* Vol 29, 1975, p 990–993

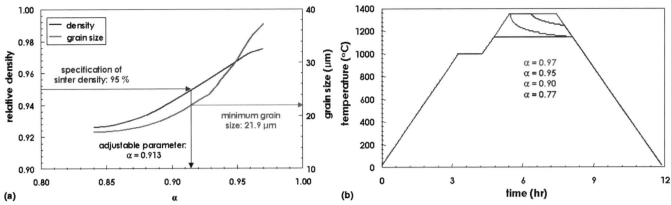

Fig. 14 (a) Minimum grain size for a given final sinter density and (b) the corresponding sintering cycle for achieving this goal in a 17-4 PH stainless steel. Source: Ref 41

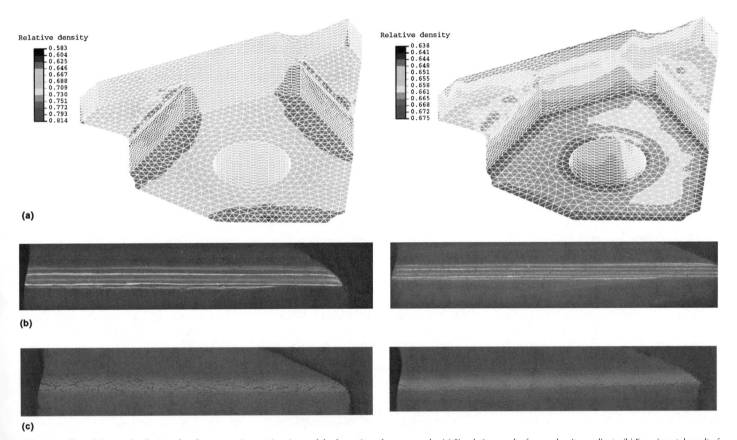

Fig. 15 Effect of density distribution after die compaction on sintering and the formation of corner cracks. (a) Simulation result of green density gradients. (b) Experimental result of green compact. (c) Experimental result of sintered compact.

6. M.F. Ashby, A First Report on Sintering Diagrams, *Acta Metall.*, Vol 22, 1974, p 275–289
7. R.M. German and J.F. Lathrop, Simulation of Spherical Powder Sintering by Surface Diffusion, *J. Mater. Sci.*, Vol 13, 1978, p 921–929
8. P. Bross and H.E. Exner, Computer Simulation of Sintering Processes, *Acta Metall.*, Vol 27, 1979, p 1013–1020
9. F.B. Swinkels and M.F. Ashby, A Second Report on Sintering Diagrams, *Acta Metall.*, Vol 29, 1981, p 259–281

10. C.M. Sierra and D. Lee, Modeling of Shrinkage During Sintering of Injection Molded Powder Metal Compacts, *Powder Metall. Int.*, Vol 20 (No. 5), 1988, p 28–33
11. K.S. Hwang, R.M. German, and F.V. Lenel, Analysis of Initial Stage Sintering Through Computer Simulation, *Powder Metall. Int.*, Vol 23 (No. 2), 1991, p 86–91
12. R.M. German, Overview of Key Directions and Problems in Computational and Numerical Techniques in Powder Metallurgy, *Computational and Numerical Techniques in Powder Metallurgy*, D.S. Madan,

I.E. Anderson, W.E. Frazier, P. Kumar, and M.G. McKimpson, Ed., The Minerals, Metals, Materials Society, Warrendale, PA, 1993, p 1–15
13. H. Riedel, D. Meyer, J. Svoboda, and H. Zipse, Numerical Simulation of Die Pressing and Sintering—Development of Constitutive Equations, *Int. J. Refract. Met. Hard Mater.*, Vol 12, 1994, p 55–60
14. K.Y. Sanlituurk, I. Aydin, and B.J. Briscoe, A Finite Element Approach for the Shape Prediction of Ceramic Compacts During

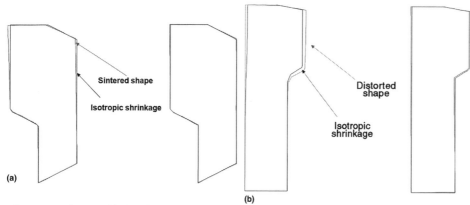

Fig. 16 Final part profiles based on the combined press-sinter simulation for the optimized design of a holder and sleeve for an oxygen sensor. (a) Initial and optimum holder designs. (b) Initial and optimum sleeve designs

Sintering, *J. Am. Ceram. Soc.,* Vol. 82, 1999, p 1748–1756

15. R. Raman, T.F. Zahrah, T.J. Weaver, and R.M. German, Predicting Dimensional Change During Sintering of FC-0208 Parts, *Advances in Powder Metallurgy and Particulate Materials—1999,* Vol 1, Metal Powder Industries Federation, Princeton, NJ, 1999, p 3.115–3.122

16. O. Gillia and D. Bouvard, Phenomenological Analysis and Numerical Simulation of Sintering Application to WC-Co System, *Proceedings of the 2000 Powder Metallurgy World Congress,* Part 1, K. Kosuge and H. Nagai, Ed., Japan Society of Powder and Powder Metallurgy, Kyoto, Japan, 2000, p 82–87

17. V. Tikare, M.V. Braginsky, E.A. Olevsky, and R.T. Dehoff, A Combined Statistical-Microstructural Model for Simulation of Sintering, *Sintering Science and Technology,* R.M. German, G.L. Messing, and R.G. Cornwall, Ed., Pennsylvania State University, State College, PA, 2000, p 405–409

18. A. Zavaliangos and D. Bouvard, Numerical Simulation of Anisotropy in Sintering due to Prior Compaction, *Int. J. Powder Metall.,* Vol 36 (No. 7), 2000, p 58–65

19. V. Tikare, E.A. Olevsky, and M.V. Braginsky, Combined Macro-Meso Scale Modeling of Sintering, Part II: Mesoscale Simulations, *Recent Developments in Computer Modeling of Powder Metallurgy Processes,* A. Zavaliangos and A. Laptev, Ed., ISO Press, Ohmsha, Sweden, 2001, p 94–104

20. R.M. German, *Sintering Theory Practice,* John Wiley & Sons, Inc., New York, NY, 1996

21. A.N. Schofield, Original Cam-Clay, *Proceedings of the International Conference on Soft Soil Engineering,* Nov 8–11, 1993 (Guangzhou)

22. D.C. Drucker and W. Prager, Soil Mechanics and Plastic Analysis or Limit Design, *Q. Appl. Math,* Vol 10, 1952, p 157–164

23. S. Shima and M. Oyane, Plasticity Theory for Porous Metals, *Int. J. Mech. Sci.,* Vol 18, 1976, p 285–292

24. Y.-S. Kwon, S.-H. Chung, K.T. Kim, R.M. German, and H.I. Sanderow, Numerical Analysis and Optimization of Die Compaction Process, *Advances in Powder Metallurgy and Particulate Materials,* Part 4, 2003, p 4-37 to 4-50

25. R.M. McMeeking and L. Kuhn, A Diffusional Creep Law for Powder Compacts, *Acta Metall. Mater.,* Vol 40, 1992, p 961–969

26. T. Kraft and H. Riedel, Numerical Simulation of Die Compaction and Sintering, *Powder Metall.,* Vol 45, 2002, p 227–231

27. P.E. McHugh and H. Riedel, A Liquid Phase Sintering Model: Application to Si_3N_4 and WC-Co, *Acta Metall. Mater.,* Vol 45, 1997, p 2995–3003

28. D. Bouvard and T. Meister, Modelling Bulk Viscosity of Powder Aggregate During Sintering, *Model. Simul. Mater. Sci. Eng.,* Vol 8, 2000, p 377–388

29. A.C.F. Cocks, The Structure of Constitutive Laws for the Sintering of Fine Grained Materials, *Acta Metall.,* Vol 45, 1994, p 2191–2210

30. Y.S. Kwon and K.T. Kim, High Temperature Densification Forming of Alumina Powder—Constitutive Model and Experiments, *J. Eng. Mater. Technol.,* Vol 118, 1996, p 448–455

31. Y.S. Kwon, Y. Wu, P. Suri, and R.M. German, Simulation of the Sintering Densification and Shrinkage Behavior of Powder-Injection-Molded 17-4 PH Stainless Steel, *Metall. Mater. Trans. A,* Vol 35, 2004, p 257–263

32. R.K. Bordia and G.W. Scherer, On Constrained Sintering—I. Constitutive Model for a Sintering Body, *Acta Metall.,* Vol 36, 1998, p 2393–2397

33. R.K. Bordia and G.W. Scherer, On Constrained Sintering—II. Comparison of Constitutive Models, *Acta Metall.,* Vol 36, 1998, p 2399–2409

34. R.K. Bordia and G.W. Scherer, On Constrained Sintering—III. Rigid Inclusions, *Acta Metall.,* Vol 36, 1998, p 2411–2416

35. S.H. Chung, Y.S. Kwon, C.M. Hyun, K.T. Kim, M.J. Kim, and R.M. German, Analysis and Design of a Press and Sinter Process for Fabrication of Precise Tungsten Carbide Cutting Tools, *Advances in Powder Metallurgy and Particulate Materials,* Part 8, 2004, p 8-26 to 8-39

36. S.J. Park, J.M. Martin, J.F. Guo, J.L. Johnson, and R.M. German, Grain Growth Behavior of Tungsten Heavy Alloys Based on Master Sintering Curve Concept, *Metall. Mater. Trans. A,* Vol 37, 2006, p 3337–3343

37. S.J. Park, J.M. Martin, J.F. Guo, J.L. Johnson, and R.M. German, Densification Behavior of Tungsten Heavy Alloy Based on Master Sintering Curve Concept, *Metall. Mater. Trans. A,* Vol 37, 2006, p 2837–2848

38. S.H. Chung, Y.S. Kwon, C. Binet, R. Zhang, R.S. Engel, N.J. Salamon, and R.M. German, Application of Optimization Technique in the Powder Compaction and Sintering Processes, *Advances in Powder Metallurgy and Particulate Materials,* Part 9, 2002, p 9–131 to 9–146

39. D. Blaine, R. Bollina, S.J. Park, and R.M. German, Critical Use of Video-Imaging to Rationalize Computer Sintering Simulation Models, *Comput. Ind.,* Vol 56, 2005, p 867–875

40. S.J. Park, S.H. Chung, J.L. Johnson, and R.M. German, Finite Element Simulation of Liquid Phase Sintering with Tungsten Heavy Alloys, *Mater. Trans.,* Vol 47, 2006, p 2745–2752

41. S.J. Park, R.M. German, P. Suri, D. Blaine, and S.H. Chung, Master Sintering Curve Construction Software and Its Application, *Advances in Powder Metallurgy and Particulate Materials,* Part 1, 2004, p 1-13 to 1-24

ASM Handbook, Volume 22B, *Metals Process Simulation*
D.U. Furrer and S.L. Semiatin, editors

Copyright © 2010, ASM International®
All rights reserved.
www.asminternational.org

Modeling of Hot Isostatic Pressing

Victor Samarov, Synertech P/M Inc.
Vassily Goloveshkin, LNT

THE MAJOR GOALS FOR HOT ISO-STATIC PRESSING (HIP) MODELING for engineers involved in process development are to build certain technological algorithms to establish the main process parameters for a new material and/or shape and to provide the necessary specifications without (or with the minimum of) laborious trial-and-error experiments.

Introduction to the HIP Process

Hot isostatic pressing has two major applications:

- Healing of inherent internal defects (mainly porosity) in castings and welds
- Consolidation of powder materials into solid, 100% dense blanks and structures

HIP of Casting and Welds

The first process is actually a special heat treatment under high pressure and is controlled mainly by the trajectory of the HIP cycle (combinations of pressure, temperature, and time). These are generally established for the main processed materials. During the HIP process, high-pressure gas (usually argon, although combinations of other gasses are used) is applied to the dense surface skin of the casting at a temperature where internal porosity can be healed by plastic deformation of the pores. The level of porosity in most castings is low, so dimensional changes in the casting are typically insignificant.

Technological models of HIP densification are based on the well-known properties of heat transfer in solid materials and phase transformations during heat treatment (Ref 1, 2). Time, temperature, and pressure conditions for the HIP cycle are selected to assure permanent removal of porosity. It is important to mention, however, that because certain phase transformations are associated with the volume changes of the appropriate phases, the applied HIP pressure can alter the value of these previously established temperatures (Ref 3).

Consolidation of Powders

Hot isostatic pressing of powder materials is a much more complicated process than HIP of a monolithic solid, because the formation of a part and consolidation of powder during HIP is provided by a capsule that gives the initial shape and dimensions for the powder bulk. The capsule transfers the external isostatic pressure of the gas medium with partial shielding onto the powder.

Figures 1 and 2 present the views of a scanning electron microscopy photo of titanium powder and an optical micrograph obtained after HIP of this powder.

The deformation and densification of the powder and shaping capsules is controlled by various physical mechanisms (Ref 4, 5):

- Plastic deformation of a metal capsule (mold) acting as a deforming shaping tool for the contained powder
- Viscous-plastic flow of powder material during densification
- Diffusion bonding of powder particles
- Pressure-activated diffusion sintering of the micropores at the final stages of HIP to form a 100% dense material typical for HIP

As a result, the technological models of the HIP process for powder materials must account for:

- Influence of the complex loading trajectory (pressure, time, temperature) during HIP on the densification, material microstructure, and properties
- Influence of the aforementioned process parameters on the deformation pattern (shape change)
- Statistical dispersion of geometrical and quality parameters of HIPed parts as a result of a normal scatter in the process parameters and the starting properties of the processed materials
- Influence of the rheological properties of the materials involved during HIP of a capsule with powder on the geometrical and material quality parameters
- Dimensional scale factors

The models must also provide:

- Accurate prediction of the capsule deformation during powder consolidation
- Necessary dimensional and metallurgical quality of as-HIPed surfaces
- Full densification of powder under the given HIP trajectory
- Process parameters for bonding of dissimilar materials during HIP of multimaterial structures

Formation of tailored, engineered, and functionally gradient structures and multimaterial parts is another new task in the technological

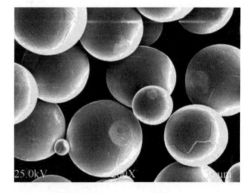

Fig. 1 Scanning electron microscopy photo of PREP Ti-6Al-4V (ASTM grade 5) powder

Fig. 2 Microstructure of Ti-6Al-4V powder after hot isostatic processing consolidation. Original magnification: 100×

modeling of HIP. This requires the successful completion of the three following modeling tasks:

- Optimal geometry and metallurgical quality of the interface
- Mechanical and performance properties of the interface
- Mechanical properties of the base materials resulting from a compromise heat treatment

Bimetallic Parts. For any bimetal part manufactured by HIP, the interface between the two bonded materials is always a net shape surface; therefore, special modeling and design is needed to control deformations there. In addition, the geometry of the interface itself needs optimization after its mechanical and physical properties are well defined.

The chemical, structural, and physical compatibility of the materials forming a bimetallic or a functionally gradient part is of major importance. Experimental modeling must precede the development of a general mathematical model for new combinations of materials. The final structural properties of the heterogeneous material made by HIP are defined as a result of a compromise heat treatment, which is another technological tool enabling control by modeling the microstructure of the interface as well as that of both base materials.

HIP Constituent Processes and Modeling Tools. The entire technological process of HIP for powder materials involves the following manufacturing steps. These may require additional technological algorithms and models, including cost models:

- Manufacturing of shaped containers (capsules) for powder (with the outer and inner shape to provide the necessary flow pattern during HIP)
- Preparation of these containers for filling (welding, cleaning, inspection)
- Filling containers with powder to provide the necessary tap density and its uniformity, to avoid segregation of powder particles
- Evacuation and outgassing of containers with powder to remove the absorbed gases and moisture
- Sealing and inspection of the containers to ensure that they can successfully sustain the pressure and temperature of HIP
- HIP consolidation to provide the desired density, mechanical properties, shape, and dimensions
- Decanning of the HIPed products and providing the necessary surface quality

Although all process models are important to ensure the success of the development and manufacturing, the applications, tasks, and modeling techniques involved in the optimization of these processes are numerous. Therefore, this article concentrates on the main process of HIP consolidation and forming of powder materials as the most important in this chain. The others, however, are also needed to optimize the entire process. A major source of

information describing these algorithms can be found in the *Proceedings of the International HIP Conferences* that have been held regularly since 1988 (Ref 6–15).

Any modeling for such a complex technological process as HIP is based either on the empirical knowledge and process design rules developed through elaborate (and often proprietary) experience or on the virtual mathematical/physical models supported by databases with the properties of the corresponding materials, or both.

It must be mentioned that virtual mathematical models for the processes involving mass and heat transfer (high-temperature deformation) of materials, such as HIP, always contain a certain degree of approximation caused by the physics of the models themselves, and their further complications must be considered carefully for the practical needs.

Two major technological tasks must be solved during HIP:

- Densification of powder material to (usually) 100% density
- Shaping to the desired geometry as a result of this densification

From the physical and engineering point of view, these two tasks are different.

In the early decades of HIP development (approximately 1955 to 1975), a lot of efforts were directed toward increasing the precision of the description of densification during HIP in order to optimize the main process parameters: temperature, time, and pressure. The major modeling approach was based on the micromechanical models describing the behavior of powder at a particle scale. This involved imitating the real contacts between particles and using the rheological properties of the powder particles obtained by special experiments (Ref 16, 17). The increased complexity of the models accounting for the real particle distribution and friction between particles allowed greater precision in describing the densification process up to the final diffusion healing of the fine pores. Micromechanical modeling was acknowledged as a reliable and available modeling technique (Ref 18, 19).

However, the ability of the HIP process to achieve the full density by increasing any of the main process parameters (temperature, pressure, and time) as well as further progress in HIP technology and instrumentation enabled the improvement of the microstructure and properties of components. This made refinement of these fundamental modeling results less urgent for practical applications.

The various tasks of shaping during HIP, especially when meeting net shape geometry and surface requirements, cannot be solved by improving the equipment or process parameters. In addition, these tasks require individual solutions for every new part.

During HIP, there are no rigid shaping tools typical for other processes of material forming, such as dies, molds, stamps, and rolls. Densification and shaping are done by the isostatic

pressure of high-temperature gas acting on the wall of a capsule, providing the initial shape for the powder bulk.

The final dimensions of a component after HIP consolidation are provided by the appropriate design of the HIP tooling and predicted by modeling of its deformation during HIP.

To develop a given complex-shaped part from powder material by way of HIP, the problem of shape control during HIP densification must be solved. Although the HIP pressure is isostatic, temperature uniformity and the properties of the powder bulk are presumably isotropic. The final geometry of a part is formed as a result of nonuniform shrinkage of the complex-shaped metal can with powder inside (Ref 20, 21). The deformation pattern during HIP becomes even more complicated if inserts are placed inside a capsule to provide for internal cavities in the part.

Figure 3 is an example of capsules used to HIP a relatively simple-shaped blank for a turbine disk. A more complex-shaped capsule and inserts for an impeller with an internal flow passage are shown in Fig. 4. Capsules are usually removed by machining after HIP. The complex-shaped inserts are leached out using a chemical agent that will not affect the consolidated material.

It was realized in the early 1980s (Ref 20) that the major factor defining deviations from isotropic deformation during HIP is the stiffness of the capsule. The thicker the capsule wall and the stronger the capsule material, the more control it has of the deformation pattern.

Fig. 3 Capsule elements for hot isostatic pressing of a blank for a turbine disk

Fig. 4 Capsule and insert for hot isostatic pressing of a net shape impeller

Mechanically, control relates to the relative amount of work needed to deform a capsule in a given direction versus the work needed to densify the powder. For identical capsules, it has been shown that the deformation is more uniform for the stronger high-temperature superalloys than for titanium alloys or steels; conversely, a stronger capsule material (stainless steel versus mild low-carbon steel) leads to a more nonuniform deformation for the same shape and powder (Ref 22). It has also been demonstrated that the stiffness of the capsule considerably modifies the stresses within the powder material inside it (Ref 23).

A mathematical description of deformations for complex shapes under such complex loading requires modeling with multiaxial stresses, and this type of modeling is essentially limited to the numerical finite-element techniques (Ref 6). A HIP process model that is adequate to design the capsules and inserts to provide the necessary configuration of a component after densification is based on the following developments.

New Engineering Models for Powder Material Behavior during HIP. To be efficient, these models must account for both plastic deformation of the capsule and nonuniform shrinkage of powder during HIP. Full models of the powder behavior involve various mechanisms of mass and heat transfer from the elementary structural adjustments of rigid particles to the final diffusion-activated sintering of micropores and "healing" the final porosity under isostatic external stresses. At the very last stages of HIP, density inside the powder becomes uniform, and the stresses are isotropic. If the stress inside the capsules with powder was uniform as well, the task of modeling would be trivial, controlled only by the initial density of the powder. However, to describe the peculiar deformation pattern, it is necessary to account for the nonuniform stresses in the powder caused at the initial stages of HIP by the capsule and temperature field and to account for the resulting nonuniform densification. The deviatoric (anisotropic) components of strain and stress tensors are therefore responsible for the deviations from the uniform deformation field. This is why efficient process models concentrate on the adequate description of plastic deformations initiated by shear caused by the nonuniform stiffness of the HIP tooling, which is the main cause of nonisotropic shrinkage of the powder during HIP.

Adequate Databases for the Rheological Properties of the Powder and HIP Tooling Materials. Systematic experiments with HIPed powder materials have proved that after reaching 80 to 85% density, further densification and shrinkage become practically uniform for most materials. The major nonuniformity of deformation for the capsules with powder is introduced at the initial stages of the process, when material is transformed from a loose powder into a continuous porous medium under the influence of a relatively stiff capsule.

Therefore, the main rheological parameters of the powder material must correspond to the interval of 60 to 85% of the apparent density. The techniques of producing such samples with the interrupted HIP cycles in conjunction with the accepted plastic yield criteria for the powder (Ref 24–26) provide the main material parameters and an adequate description of deformations during HIP (Ref 27–29).

Techniques for the Solution of the Reverse Problem of HIP. This means determining the HIP capsule geometry from the results of modeling. Although mathematical modeling rather precisely predicts the shape and dimensions of a given capsule after HIP (actually limited only by the precision of the physical model itself), the definition of the initial geometry of the HIP tooling for the given part by the results of modeling (the reverse problem) presents a complicated task.

Because the deformation field for a complex-shaped part is far from uniform, this type of reconstruction of the capsule by the results of modeling and experiments requires special numerical techniques and procedures for interpretation of the calculations and experimental data (Ref 30).

Special Designs for HIP Tooling for Complex-Shaped Parts. The HIP capsules must be designed with specific geometric and strength anisotropy to provide the necessary flow pattern during HIP deformation. These designs must also account for the stochastic factors of the HIP process, such as deviations of density, material properties, and process parameters (Ref 23).

Adaptation of the Developed Engineering Models to the Modern Computer Software Commercial Packages. Initially, the packages that describe the plastic flow of materials were not intended to solve the problems of HIP. Special program modules were developed for ANSYS, ABAQUS, and other commercial packages to describe the powder densification and deformation in plastic capsules (Ref 28). It is also important that the results of numerical modeling and reverse engineering design for HIP tooling be presented in formats acceptable for the computer-aided design (CAD) and computer-aided manufacturing software for computer numerical control machining of this tooling and coordinate-measuring machining of the HIPed parts.

Within these five developments, there are at least four main aspects of modeling that enable production of complex net shape HIP components and provide reliable and reproducible manufacturing:

- Rheological and numerical models for the HIP process and powder material for adequate description and optimization of the flow pattern
- Optimal design of HIP tooling to eliminate distortions and to provide dimensional stability

- Experimental step through an informative demonstrator revealing additional details of the deformation field
- Iterative information loop based on the computer analysis of numerically predicted and experimental dimensional data and the reverse engineering technique

Consistent application of these tools provides the possibility of modeling, developing, and efficient manufacturing of complex-shaped parts by HIP.

Evolution of Approaches to HIP Modeling

HIP modeling must provide adequate complex mathematical modeling of HIP shrinkage accounting for actual two- and three-dimensional geometry, boundary conditions at the powder-capsule interface, and special rheology of the processed medium consisting of compressible (powder) and noncompressible insert materials. The models must include techniques for the design of capsules and forming inserts. These solutions leave far behind the traditional concept of a capsule as just a can transferring isostatic pressure to the powder and providing its 100% density. The capsule for a complex-shaped part appears to be a plastically deformed, special technological shaping tool.

It is also important to keep in mind that, because of the stochastic character of most technological parameters involved during HIP, it becomes impossible to ensure net shape geometry over all the part, including its internal surfaces. This analysis and statement, expressed in Ref 31, is generally true, but the current level of computer modeling and CAD, together with the design principles for HIP tooling, open the possibilities for efficient control of shrinkage and the stable manufacturing by HIP of the selectively net shape parts with the most important and critical surfaces that do not require further machining.

Design of the HIP Tooling

The design procedure of the tooling for net shape HIP must account for:

- Objective inadequacy, even in the case of the most complete simulation of the process and actual phenomenon
- All technological factors during filling, handling, and HIP itself that lead to irregularities in density, properties, and temperature fields
- Objective lack of information on the material properties and dimensions prior to and after HIP
- Deformation flow pattern (most important) and all macroscopic effects revealed from the results of modeling

The design goals include:

- Stable and reproducible shrinkage, with minimized scatterband of the final dimensions

- No macrowarping or other distortions caused by irregular plastic stiffness of the capsule elements
- Well-controlled flow pattern during shrinkage and densification, so that all small distortions, such as ellipticity, meniscus, and local bends, are placed out to the less important and easily machined areas of the part
- Net precise and reproducible shape in the most critical areas by sacrificing dimensional precision in less important areas
- Efficient recalculation procedure for transfer from demonstrators to prototypes, based on the results of experiments, remodeling, and material flow pattern revealed from the experiments

The problem of mathematical modeling for HIP of near-net or net shape parts from powder materials can be split into the following five subproblems:

- Mathematical description of outgassing of powder materials
- Adequate mathematical model to describe the mutual deformation of porous material in a metal capsule also containing solid inserts
- Description of mechanical and physical processes at the interface between the powder and the surface of the capsule and inserts
- Developing a mathematical method of designing a HIP capsule by the specified final geometry of the part (the "first reverse problem" of HIP modeling)
- Developing methods to specify the initial geometry of the HIP capsule using the results of modeling and experimental data on the first part (the "second reverse problem" of HIP modeling)

These subproblems are considered in detail.

Mathematical Modeling of Outgassing of Powder Materials. This problem includes the description of the following two processes:

- Evacuation of atmospheric air or the cover gas from the space between the powder particles
- Outgassing of adsorbed gas and moisture from the surface of the powder particles

The major challenge for mathematical description and technological control of this process is caused by the molecular regime of the gas flow in the channels between the powder particles at low pressures. Under the molecular flow, when the mean free path of gas molecules exceeds the size of the pores in the powder material, the intensity of outgassing is no longer controlled by the pressure or the capacity of a vacuum pump but only by the temperature of the gas. The problem is obviously complicated by the very low thermal conductivity of powder materials in vacuum.

The regime of molecular flow is reached rather quickly during outgassing, at the pressure of

approximately 4 to 5 torr for powder particles of approximately 100 µm (0.004 in.) size. In this case, the evolution of vacuum pressure (P) is described by the following differential equation:

$$\frac{\partial P}{\partial t} = \nabla \cdot (D \nabla P) + I$$

where D is the coefficient of diffusion, and I is an additive describing desorption of gases from the surface of powder particles. The coefficient of diffusion, D, can be estimated as:

$$D = \frac{2}{3} \bar{v} R_p$$

where R_p is the typical size of the pores, and the root mean square velocity of the gas particles is:

$$\bar{v} = \sqrt{\frac{RT}{\pi\mu}}$$

where R is the universal gas constant, μ is the molecular weight, and T is the absolute temperature.

To obtain a theoretical assessment for the diffusion coefficient, the Monte Carlo method can be used to analyze the chaotic movement of gas molecules inside the system of packed spheres when their average diameter and normal size distribution are known. Analysis of the process has shown that the characteristic dimension of the pores constitutes ⅛ the average size of the spherical powder particles (Ref 32).

Another important additive is the volume of desorption from the surface of the powder particles. The literature provides varying estimates for this value, from the amount equal to the volume of gas in the interparticle space to a value 10 times greater. Regardless, removal of the absorbed air and moisture is technologically very important; it can be achieved only by heating the powder particles. This is difficult in the vacuum die due to the absence of convection heating and low thermal conductivity.

The quantity Q (the pressure created if all gas is desorbed into the interparticle space) can be assessed using the following differential equation:

$$\frac{\partial Q}{\partial t} = -I$$

where I can be calculated as follows:

$$I = Q_d \beta_k \frac{6(1-\theta)}{\theta} \frac{RT_0}{P_0} \frac{1}{R} \frac{(P_u - P)}{\sqrt{2\pi\mu RT}} H(P_u - P)$$

where $P_u(T)$ is the pressure of the saturated vapor as a function of temperature. T_0 and P_0 are the initial temperature and pressure, R is the universal gas constant, $H(x)$ is the Heaviside function, and β_k is an experimental coefficient related to the porosity of powder material and the average radii of the powder particles. Keeping in mind the strong temperature dependence of $P_u(T)$, different amounts of absorbed gases can be removed by a simple evacuation.

In addition, due to the very low thermal conductivity of the powder medium when powder particles have only point contacts, the temperature field in an outgassed powder may be extremely nonuniform. Therefore, to fully analyze, model, and control the powder outgassing process, it is necessary to resolve jointly the thermal conductivity equation for powder and the diffusion equation for gas pressure (Ref 33).

Mathematical Modeling of HIP Densification and Shrinkage of Powder Materials. In general, the task of mathematical modeling for HIP is to adequately describe the densification and shrinkage of the given powder in the specified HIP cycle to be able to design the HIP capsule so that after HIP, it produces the desired shape and dimensions of the part.

The tolerance requirements for the part may require high precision. Parts must be manufactured with tolerances of less than 100 µm (several thousandths of an inch), and this demands corresponding precision in the model and stability of the process itself.

There are two basic challenges. The first is that the HIP process is characterized by comparatively large deformations, approximately 15 to 20% linear. The initial density is approximately 60 to 65% of the theoretical density, which is usually reached by the end of the process. From a mathematical point of view, it means that the constitutive equations describing the material flow will be nonlinear, and the boundary conditions will be preset on the moving boundary.

The second problem is more daunting. The constitutive equations defining relations between the components of the stress tensor and the parameters of material flow cannot be precisely defined. This problem is typical for all modeling attempts to describe the material behavior beyond its elasticity level. Each calculation will have a degree of approximation.

This leads to a very important conclusion: Each HIP development effort based on modeling must be an iterative process to become really efficient.

The essence of this iterative process is as follows. At the first step, a mathematical model of the HIP process is built, modeling is done, and the parameters of the HIP tooling are defined. After HIP, the geometry of the part is compared to the specification, and corresponding corrections are introduced into the model. A converging iterative process can be built by applying the negative values of the deformations obtained as a result of the first modeling and by correcting the initial geometry while applying the law of mass conservation (Ref 30). This approach may be considered analogous to the one suggested much earlier for the problems of plasticity for solid materials (Ref 34). A suitable mathematical model of HIP:

- Provides a close first approximation for the geometry of the part (1 to 2% geometrical precision)

- Accounts for the main deformation effects and distortions and the influence of the process parameters (mainly the functions of density in the yield criterion) based on the results of the experiments and, if necessary, introduces additional parameters into the model

There exist several different approaches to the mathematical description of powder densification and shrinkage. Most of them consider powder material as a continuous medium, because in the modeling of the technological process, the main interest is in the kinematic aspects of HIP densification that are similar to the behavior of a standard continuous medium.

The constituent equations used for mathematical description for the deformation of powder materials have a significant difference from the equations used in the classical theories of plasticity. The latter are based on small or even zero volume deformations. The volume deformation or its equivalent, such as densification or porosity, is the important parameter characterizing the densification of the powder material during HIP.

However, for technological control and modeling of HIP, the most important are the shear components of the deformation tensor, because they define important deviations from the isotropic densification.

Usually, HIP parameters target densification of the powder materials to 100% density; as far as the initial density can be defined precisely, the value of the volume deformation can also be considered well known in every point of the part after the process is completed. Therefore, it is important for modeling to be reasonably precise in the prediction of each dimension of a part; an error in one dimension leads to errors in the others. If a modeling technique or design of the HIP tooling enables the precise definition of one set of dimensions (for example, radial deformations) after HIP, the other dimensions will be quite precise due to the mass conservation law.

It is important to develop models that can adequately provide the necessary precision of predictions, fully accounting for the necessary effects of deformation without being too mathematically complicated. It is important that the models describe the behavior for the complete system consisting of capsules, powder, and inserts.

Green's plasticity theory (Ref 24) is often used for mathematical modeling of HIP. A complete mathematical definition of the HIP model includes the following equation:

$$\frac{\partial \sigma_{ij}}{\partial x_j} = 0$$

where σ_{ij} is the stress tensor.

To describe the behavior of powder material at the yield limit, the following elliptic Green's yield criterion (Ref 24) is used:

$$\frac{\sigma^2}{f_2^2} + \frac{s^2}{f_1^2} = T^2$$

where $\sigma = 1/3\sigma_{ii}$ is the first invariant of the stress tensor (the value of hydrostatic pressure) and:

$$s_{ij} = \sigma_{ij} - \sigma\delta_{ij} \text{ and } s^2 = \frac{3}{2}s_{ij}s_{ij}$$

is the second invariant of the stress tensor deviator, where:

$$s_{ij} = \sigma_{ij} - \sigma\delta_{ij}$$

$f_1(\rho)$ and $f_2(\rho)$ are functions of material density, ρ, defined experimentally using interrupted HIP cycles and other techniques (Ref 21, 35, 36), and T is the yield limit of the powder material after complete densification ($\rho = 1$), that is, a known function of temperature.

Relations between the components of the deformation tensor and deformation rate tensor are defined by the associated flow law. Mathematically, this law means that the increments of plastic deformation occur in a direction orthogonal to the yield surface, and it can be written in the following form:

$$d\varepsilon_{ij} = \frac{\partial\Phi}{\partial\sigma_{ij}}dA$$

where $\Phi(\sigma_{ij}) = 0$ is the equation describing the yield surface, $d\varepsilon_{ij}$ is infinitesimal increments of deformations, and dA is an infinitesimal increment for the work of internal stresses.

To describe the mechanical behavior of the solid materials of capsule and inserts, the ideal plasticity criterion is used:

$$s^2 = T_1^2$$

together with the noncompressibility equation, $\nabla \cdot \bar{u} = 0$, where $\bar{u}$ is the displacement rate. Also, the value of the yield limit for the solid material, T_1, is considered to be the known function of temperature. To define density, the continuity equation is used:

$$\frac{d\rho}{dt} + \rho\nabla\bar{u} = 0$$

Hot isostatic pressing is usually performed under a rather uniform external temperature field; however, the conductivity of the powder medium is very low at the initial stages of HIP, and substantial temperature gradients can occur inside the capsule with powder, especially for the parts with large cross sections. This may also affect the deformation pattern.

For temperature, the conductivity equation is used:

$$\rho c_m \frac{\partial T}{\partial t} = \nabla \cdot (\lambda\nabla T)$$

where c_m is the specific conductivity, and λ is the thermal conductivity coefficient.

There is a significant dependence of the conductivity coefficient, λ, on the powder density; during densification, it can change more than two degrees. The boundary conditions on the surface of the capsule are the derivatives of the HIP cycle; isostatic pressure and temperature are the known functions of time.

It is important to note that Green's theory considers the powder medium under deformation as a medium with initially isotropic properties. However, in the vicinity of the stiff capsule elements and inserts, deformation may be practically unidirectional. As shown in Ref 37 under these conditions, the powder continuum may obtain a kind of deformational anisotropy, so that pores in powder acquire a preferable direction for densification. One of the possible ways to account for such deformational anisotropy is to build new appropriate functions for Green's criterion. There are three of these functions of deformation tensor components, as well as density (Ref 38). This will also correspondingly modify Green's basic criterion.

Shielding of Stresses. One more aspect is important in the theoretical study and practical modeling of HIP. When complex-shaped parts are HIPed, the capsule and inserts may substantially shield the outside isostatic stresses. This often happens at the initial stages of HIP, when the main distortions and deviations from the uniform deformation pattern (which are the most important for shape control) occur. Shielding of stresses inside the HIP tooling may lead to formation of some rigid zones in the powder and inserts. Because traditional numerical modeling usually does not take this phenomenon into account, it is necessary to develop a numerical modeling technique to account for the possibility of the rigid zones in the powder and the effects caused by them.

Such an accounting is especially important for large parts when the temperature field inside the part is substantially nonuniform. Because the coefficient of thermal conductivity has a substantial increase with temperature for the materials of interest, and the value of the yield strength decreases with temperature, a kind of densification front can form (Ref 39, 40). This front will move from the surface of the capsule inward. Modeling (Ref 40) shows that there is cold and practically nondensified powder before the front and hot and substantially densified material behind the front. The latter, due to its higher strength, can cause additional shielding of stresses, especially for the axisymmetric parts.

Within Green's approach and experimental determination of the functions $f_1(\rho)$ and $f_2(\rho)$ of density (based on the interrupted HIP cycles), the influence of temperature is present in the equations only in the value of the yield strength of solid materials. It is necessary to understand whether these important functions of the yield criterion depend on density only or whether they also depend on the temperature or, more generally, on the profile of the HIP cycle and temperature field inside the capsule.

Consider the definition for the function $f_2(\rho)$. It is defined from the experiments on isostatic pressure (Ref 21) and actually represents the limiting pressure that causes the yield of the powder particles. This pressure is definitely a function of the temperature itself as well as the yield strength of the solid material. The following analysis shows, however, that this function can be presented as a simple product of the function of density and temperature and is therefore irrelevant to the specific HIP cycle.

If the powder material is described by the ideal plasticity law, the limiting pressure $f_2(\rho)$ and deformation rate field are defined from the condition of minimum power of the external forces. This corresponds to the minimum of the following functional:

$$\min \frac{\iiint\limits_{V} \sigma_s \varepsilon\, dV}{\iint\limits_{S} \left(\vec{u}\ \vec{\nu}\right) ds}$$

where V is the volume of material, S is the external surface of the capsule with powder, ε is the intensity of the deviatoric deformation rates, $\vec{u}$ is the displacements of the surface, $\vec{\nu}$ is the normal vector to the surface, and σ_s is the yield strength.

It can be seen that the value of the yield strength as a function of temperature is a linear constituent of the $f_2(\rho)$ function. The same analysis can be done for the $f_1(\rho)$ function.

Modeling the Interaction of Powder with the Capsule Surface. Despite the fact that powder material can be considered a continuous medium during modeling, the powder behavior at the interface of inserts within a capsule requires special consideration. As a rule, powder particles indent the surface of the tooling, which is usually a low-grade steel. This causes certain problems with the surface finish for the net shape parts and also intensifies the diffusion of iron and carbon into the powder materials. However, for the case of diffusion bonding of powder and solid materials during HIP, the influence of this indentation and the resulting deformation of the surface can be favorable for the quality of the bond. While modeling this interaction between powder and solid materials, the powder can no longer be considered as a continuous medium; a micromechanical approach must be used to analyze the effects of this mutual deformation.

The influence of trajectory of temperature and pressure versus time is quite important here, because the ratio between yield stresses for the different materials may vary substantially during the HIP cycle.

The First Reverse Problem of HIP Modeling. The design of the initial shape of the HIP tooling, capsule, and insert cannot be solved by direct recalculation from the final shape of the part, because this problem is mathematically incorrect when the plastic flow law is considered. This is because the function $f_2(\rho)$ infinitely increases near the value of $\rho = 1$. (Mathematically infinite pressure is needed to reduce the porosity to zero using just the plastic flow law.)

Therefore, the final density must be preset close to but below 1. As a result, these small deviations of density accounting for the behavior of $f_2(\rho)$ may lead to large deviations of the calculated initial shape of the capsule.

The following approach to CAD of the capsules is mathematically correct and technically efficient. It consists of the solution of a set of direct problems of HIP modeling. At the first step, the initial shape of the HIP tooling is defined as the final shape of the part, and HIP modeling is carried out. After this, using the deformation map obtained by mathematical modeling, the new shape is defined by "bloating" the geometry of the part. Then, modeling is repeated for this new geometry and compared with the part definition. The new deformation map is then built, accounting for the residuals. The process is continued until it converges (the convergence has been proved mathematically). Some practical criteria must be applied to optimize the external shape of the capsule from a technological point of view.

The Second Reverse Problem of Hip Modeling. This problem is generated by the following issues. Due to the approximations introduced in the constituent equations of the mathematical models, the first prototype built according to the results of modeling and the solution of the first reverse problem are still different from the geometrical specification. The problem is how to introduce the small changes into the initial geometry, accounting for the results of the first experimental iteration.

The essence of the approach is as follows. The deformation field is built based on the numerical modeling of the shrinkage for the HIP tooling. However, it is very important that the results come close to the experiment. After the experimental analysis, corrections to the deformation field are built (in the most critical points of special interest of the part). Considering that these deviations are generated by the approximations and errors of the constitutive equations of the mathematical models, the integral model parameters are altered. For example, small changes are introduced into the functions of the Green criterion so that those effects missed during the first modeling are revealed. After that, the second tooling is corrected by the results of the new modeling, and the second experimental iteration is produced. Usually, it answers all dimensional requirements, because the changes are small and the mass conservation law is quite precise.

Example of the Modeling Process

Setup and Evaluation

Finite-Element Method Formulations. Within this model, a finite-element method (FEM) was used for the numerical solution of the aforementioned nonlinear equations. Standard three-noded linear triangular elements with two degrees of freedom were used to discrete the field variables.

Numerical Solutions. The constitutive relations were described in an incremental format. Accordingly, the numerical integrations of the plastic constitutive equations must be performed incrementally over a sequence of loading steps. The load was divided into a number of increments, and for each increment, the successive approximation method was employed. The calculations were continued over the whole load increment in the same manner, and the total stress and strain are calculated by accumulating each incremental solution.

Computer-Aided Three-Dimensional Design. The aforementioned model has been developed into software (Ref 21, 41) that uses Visual C++ computer codes and is used as an analytical tool for predicting the deformation and shape changes that occur during HIP. This FEM computing model was integrated with an FEM meshing tool and a CAD module for the analysis and tooling design. The FEM meshing tool automatically generated and adjusted the meshes that were needed for the FEM calculation. The predicted deformation or shape change of the HIPed component was visualized in CAD, through which deviations between the predicted shape and the target shape were calibrated and used as a reference for subsequent capsule and tooling design.

Numerical Modeling and Tooling Design of a Casing Component Demonstration (Ref 41)

A casing component demonstrator was modeled, designed, and manufactured into a near-net shape using commercial Ti-6Al-4V powder. The casing, as shown in Fig. 5, consists of an axial-symmetric cylindrical body with two bosses, two edge flanges, two body rings, and two reinforced ribs. The whole design and manufacturing process involved the initial prediction of the shape changes during HIP (FEM model), subsequent tooling design (CAD drawing), and final realization of the target shape through HIP. In manufacturing this component, the Ti-6Al-4V powder with a tap density of 67% was canned in a predesigned tooling capsule and HIP consolidated. The two different types of material and property databases were assumed in the model, that is, for the capsule of low-carbon A1018 steel and for the powder material of Ti-6Al-4V. Firm bonding at the interface between the two materials was assumed. Before the final HIP tooling design, the deformation profile of the metal powders and the associated distortion of the capsule and tooling during HIP were predicted by the FEM model. Figure 6 shows the results of modeling for the different cross sections of the casing.

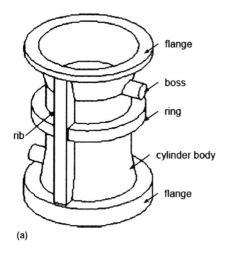

(a)

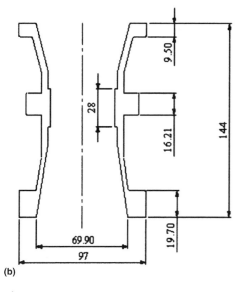

(b)

Fig. 5 Ti-6Al-4V casing

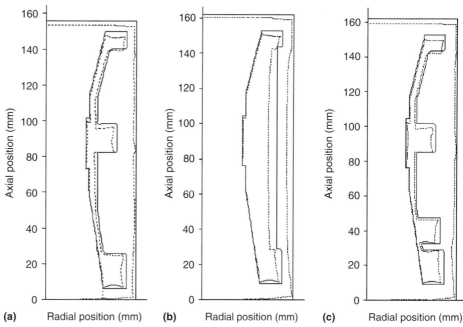

Fig. 6 Results of modeling for the different cross sections of the casing

The calculated deformation map was recorded and reconstructed in the CAD module, through which the discrepancies between the deformed and the target geometries were precisely calibrated using standard CAD measuring tools. The magnitudes of these discrepancies were then used to adjust the cavity geometry in the next iteration of the calculation. A computer-aided automatic iterative remeshing scheme (Ref 28) was used for local regions, where necessary. For a general geometry as complex as this, three to four computing iterations are usually adequate to precisely define the necessary capsule dimensions. The results of the last iteration are shown by the dotted lines in Fig. 6.

Dimensional Analysis of the Casing Demo and Validation of Model Predictions. The tooling set was made according to the design and was assembled to encapsulate the powder during HIP. The powder was consolidated from the initial relative density, $\rho_0 = 67\%$, to the fully dense state. Figure 7 presents the HIPed

casing demo after removal of the capsule by pickling in nitric acid.

The dimensions of the pickled component after HIP were measured using a coordinate-measuring machine through high-energy x-ray tomography body scanner technology. The digitized surface data were imported into the CAD module and compared with the model predictions. The predictions and the measurements for the internal and external contours of the casing as well as the positions for each of the ring and bosses were in good agreement, allowing the analysis to proceed with the solution of the second reverse problem of HIP modeling, as described previously.

REFERENCES

1. M. Koizumi and M. Nishihara, *Isostatic Pressing: Technology and Applications*, Elsevier, 1991, p 377
2. H. Atkinson and B. Rickinson, *Hot Isostatic Processing*, United Kingdom, 1991
3. N. Wain, J. Mei, M. Loretto, and X. Wu, Effect of HIP Pressure on Equilibrium Microstructure of Ti Alloys, *HIP'08, Proceedings of the 2008 International Conference on Hot Isostatic Pressing*, 2008
4. H.J. Frost and M.F. Ashby, *Deformation Mechanism Maps*, Pergamon Press, United Kingdom, 1982
5. A. Laptev, V. Samarov, and S. Podlesny, Parameters of Hot Isostatic Pressing of Porous Materials, *News of the USSR Academy of Sciences, Metals*, (No. 2), 1989
6. HIP—Theories and Applications, *Proceedings of the International Conference on Hot Isostatic Pressing*, CENTEC, Sweden 1988, p 568

7. Applications and Developments, *Proceedings of the International Conference on Hot Isostatic Pressing of Materials*, Flemish Society of Engineers, 1988, p 455
8. Hot Isostatic Pressing. Theory and Applications, *Proceedings of the International Conference*, ASM International, 1989, p 359

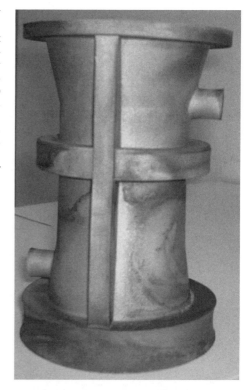

Fig. 7 Hot isostatic pressed Ti-6Al-4V casing demo

9. Hot Isostatic Pressing. Theory and Applications, *Proceedings of the International Conference*, Elsevier, 1991, p 589

10. Hot Isostatic Pressing '93, *Proceedings of the International Conference on Hot Isostatic Pressing—HIP'93*, Elsevier, 1993, p 601

11. Hot Isostatic Pressing, *Proceedings of the International Conference on Hot Isostatic Pressing*, ASM International, 1996, p 297

12. Hot Isostatic Pressing, *Proceedings of the International Conference on Hot Isostatic Pressing*, International Academic Press, China, 1999, p 396

13. *Proceedings of the International Conference on Hot Isostatic Pressing—HIP'02*, VILS, Russia, 2003, p 250

14. *HIP'05, Proceedings of the 2005 International Conference on Hot Isostatic Pressing*, SF2M, France, 2005, p 287

15. *HIP'08, Proceedings of the 2008 International Conference on Hot Isostatic Pressing*, p 318

16. W. Kaysser, Present State of Modeling of Hot Isostatic Pressing. Theory and Applications, *Proceedings of the International Conference, HIP-1989*, ASM International

17. D. Bovuard and M. Lafer, New Developments in Modeling of Hot Isostatic Pressing, *Proceedings of the International Conference, HIP-1989*, ASM International

18. M.F. Ashby, A First Report of Sintering Diagrams, *Acta Metall.*, Vol 22, 1974, p 275–284

19. E. Arzt, M.F. Ashby, and K.E. Easterling, Practical Application of Hot Isostatic Pressing Diagrams: Four Case Studies, *Metall. Trans. A*, Vol 14, 1983, p 211–221

20. G. Garibov, V. Samarov, and L. Buslavskii, Capsules Deformation during the Hot Isostatic Pressing of Disks from Powders, *Powder Metall. Met. Ceram.*, Vol 9 (No. 9), 1980

21. D. Seliverstov, V. Samarov, V. Goloveshkin, and P. Extrom, Capsule Design for HIP of Complex Shape Parts, *Proceedings of the International Conference on Hot Isostatic Pressing—HIP'93*, Elsevier, 1993, p 555–561

22. V. Samarov, Industrial Application of HIP for Near Net Shape Critical Parts and Components, *Proceedings of the International Conference on Hot Isostatic Pressing—HIP'93*, Elsevier, 1994, p 171–185

23. W. Kaysser, M. Aslan, and E. Arzt, Microstructural Development and Densification during HIP of Ceramics and Metals, *Powder Metall.*, Vol 31, 1988, p 63–39

24. R.J. Green, *Int. Mech. Sci.*, Vol 14 (No. 4), 1972, p 215–224

25. S. Shima and M. Oyane, Plasticity Theory for Porous Materials, *Int. J. Mech. Sci.*, Vol 18, 1976, p 285–291

26. M. Abouaf, J. Chenot, and G. Raisson, Finite Element Simulation of the Hot Isostatic Pressing of Metal Powders, *J. Numer. Meth. Eng.*, Vol 25, 1988, p 191–212

27. A. Vlasov and D. Seliverstov, New Phenomenological Model of HIP Simulation Based on the Development of Plasticity Theory, *Proceedings of the International Conference on Hot Isostatic Pressing HIP'02*, VILS, Russia, 2003

28. K. Zadeh, Finite Element Simulation of Near Net Shape Parts Produced by Hot Isostatic Pressing, *Proceedings of the International Conference on Hot Isostatic Pressing*, ASM International, 1996, p 297

29. D. Seliverstov and V. Samarov, "HIP Modeling of Complex Shape Parts: Experience Trends and Perspectives," 1994 Powder Metallurgy World Congress (Paris), 1994

30. G. Raisson and V. Samarov, "Design Rules for Net Shape or Near Net Shape Components Produced by Hot Isostatic Pressing of Superalloy Powders," 1994 Powder Metallurgy World Congress (Paris), 1994

31. R. Widmer, Future Prospects for Hot Isostatic Pressing, *Proceedings of the International Conference*, Elsevier, 1991, p 589

32. A. Bochkov, V. Samarov, J. Mei, and X. Wu, Modeling and Experimental Study of Out-Gassing of Ti5Al4V Powder for HIP, *Proceedings of the 2005 International Conference on Hot Isostatic Pressing*, SF2M, France, 2005

33. E. Khomyakov, V. Nepomnyatchy, and V. Samarov, Analysis of Efficiency of Different Out-Gassing Techniques for Capsules with Powder Prior to HIP, *Proceedings of Powder Metallurgy World Congress 2004*, Vienne

34. A.A. Ilyushin, *Plasticity*, Academy of Sciences, USSR, 1963

35. V. Seetharaman, S.M. Doraivelu, and H.L. Gegel, Plastic Deformation Behavior of Compressible Solids, *J. Mater. Shap. Technol.*, Vol 8, 1990, p 239–248

36. H.A. Kuhn and C.L. Downey, Deformation Characteristics and Plasticity Theory of Sintered Powder Material, *Int. J. Powder Metall.*, (No. 1), 1971, p 15–25

37. R. Dutton and S. Semiatin, The Effect of Density Anisotropy on the Yielding and Flow Behavior of Partically Consolidated Powder Compacts, *Metall. Mater. Trans. A*, Vol 29, May 1998, p 1471–1475

38. V. Samarov, V. Goloveshkin, and R. Dutton, HIP Modeling Methodology Based on the Inherent Process of Anisotropy, *Proceedings of the 2005 International Conference on Hot Isostatic Pressing*, SF2M, France, 2005

39. W. Li, M. Ashby, and K. Easterling, On Densification and Shape Change during Hot Isostatic Pressing, *Acta Metall.*, Vol 35, 1987, p 2832–2842

40. B. Dryanov and V. Samarov, Isostatic Densification of the Powder Material in a Non-Uniform Temperature Field, *Poroshk. Metall.*, (No. 3), 1989, p 25029

41. W.X. Yuan, J. Mei, V. Samarov, D. Seliverstov, and X. Wu, Computer Modelling and Tooling Design for Near Net Shaped Components Using Hot Isostatic Pressing, *J. Mater. Proc. Technol.*, Vol 182 (Issues 1–3, No. 2), 2007, p 39–49

ASM Handbook, Volume 22B, *Metals Process Simulation*
D.U. Furrer and S.L. Semiatin, editors

Copyright © 2010, ASM International®
All rights reserved.
www.asminternational.org

Modeling and Simulation of Metal Powder Injection Molding

Seokyoung Ahn, The University of Texas-Pan American
Seong-Taek Chung, CetaTech, Inc.
Seong Jin Park, Pohang University of Science and Technology
Randall M. German, San Diego State University

INJECTION MOLDING is used extensively in the forming of plastics, because of the low overall cost in a high level of shape complexity. Powder injection molding (PIM) builds on the long-recognized success of plastic molding by using a high-particle-content thermoplastic as feedstock. The steps in PIM involve first mixing selected small powders (usually smaller than 20 μm) and polymer binders. The particles are small to aid in sintering densification and often have near-spherical shapes to improve flow and packing. The thermoplastic binders are mixtures of waxes, polymers, oils, lubricants, and surfactants. When molten, the binder imparts viscous flow characteristics to the mixture to allow filling of complex tool geometries. A favorite binder system relies on a mixture of paraffin wax and polypropylene, with a small quantity of stearic acid. The combination of powder and binder that works best gives a paste with approximately the same consistency as toothpaste, with no voids; this often leads to a formulation near 60 vol% powder and 40 vol% binders. This mixture is heated in the molding machine, rammed into a cold mold, and when the binder freezes in the mold, the component is ejected. Next, the binder is removed by heat and solvents (some of the binders are water soluble), and the remaining 60% dense powder structure is then sintered to near-full density. The product may be further densified, heat treated, machined, or plated. The sintered compact has the shape and precision of an injection-molded plastic but is capable of performance levels unattainable with polymers. Almost half of all PIM is applied to stainless steels, but a wide variety of compositions are in production (Ref 1).

The equipment used for shaping the compact is the same as used for plastic injection molding, so software for molding machine control is the same as found in plastics. Most molding machines fill a die through a gate from a pressurized and heated barrel. (The barrel and gate are connected by a nozzle and runner.) A plunger or reciprocating screw generates the pressure needed to fill the die. Computer simulations used for mold filling in plastics are not directly useful for powder injection molding, because inertia, thermal conductivity, and powder-binder separation are new concerns with PIM feedstock. The feedstock enters the barrel as cold granules, and during compression to remove trapped air, it is heated above the binder melting temperature. Because the feedstock is hot and the die is cold, filling must be accomplished in a split second to avoid premature freezing. After filling the die, packing pressure is maintained on the feedstock during cooling to eliminate shrinkage voids. After sufficient cooling, the hardened compact is ejected and the cycle repeated. It is common to have multiple cavities (four or more) and cycle times of 15 s, so production rates of 16 parts/min are often observed.

The PIM process is practiced for a very wide range of materials, including most common metals, many ceramics, and cemented carbides. The largest uses are in metal powder injection molding. Other variants are ceramic injection molding and cemented carbide injection molding.

Simulations used for plastics have been applied to PIM, but the high solid content often makes for differences that are ignored in the plastic simulations. Several situations demonstrate the problems, such as powder-binder separation at weld lines, high inertial effects such as in molding tungsten alloys, and rapid heat loss such as in molding copper and aluminum nitride. Also, powder-binder mixtures are very shear-rate sensitive. Thus, the computer simulations to support molding build from the success demonstrated in plastics but adapt those concepts in new, customized PIM simulations for filling, packing, and cooling.

Theoretical Background and Governing Equations

A typical injection-molded component has a thickness much smaller than the overall largest dimension. A typical wall thickness is in the 1 to 3 mm range, while the longest dimension may range near 25 mm with an overall mass near 10 g. There is much variation, but these values offer a glimpse at the typical components (Ref 2). In molding such components, the molten powder-binder feedstock mixture is highly viscous. As a result, the Reynolds number (a dimensionless number characterizing a ratio of inertia force to viscous force) is low, and the flow is modeled as a creeping flow with lubrication, as treated with the Hele-Shaw formulation. With the Hele-Shaw model, the continuity and momentum equations for the melt flow in the injection molding cavity are merged into a single Poisson equation in terms of the pressure and fluidity. Computer simulation is usually based on a 2.5-dimensional approach because of the thin wall and axial symmetry. However, the Hele-Shaw model has its limitations and cannot accurately describe three-dimensional (3-D) flow behavior in the melt front, which is called fountain flow, and special problems arise with thick parts with sudden thickness changes, which cause racetrack flow.

Nowadays, several 3-D computer-aided engineering simulations exist that successfully predict conventional plastic advancement and pressure variation with changes in component design and forming parameters (Ref 3). For PIM 3-D simulation, Hwang and Kwon (Ref 4) developed a filling simulation with slip using an adaptive mesh refinement technique to capture the large deformation of the free surfaces, but this is computationally intensive (Ref 4–7), so further research is moving toward simplified solution routes (Ref 3). This section focuses on the axisymmetric 2.5-dimensional approach rather than a full 3-D approach, because the 2.5-dimensional approach is more robust and better accepted by industry. The postmolding sintering simulation is described in the earlier press-sinter simulation.

Filling Stage

Powder injection molding involves a cycle that repeats every few seconds. At the start of the cycle, the molding machine screw rotates in the barrel and moves backward to prepare molten feedstock for the next injection cycle while the mold closes. The mold cavity fills as the reciprocating screw moves forward, acting as a plunger, which is called the filling stage. During the filling stage, a continuum approach is used to establish the system of governing equations:

- *Mass and momentum conservation:* With the assumption of incompressible flow, the mass conservation, also called continuity equation, is expressed as:

$$\frac{\partial u}{\partial x} + \frac{\partial v}{\partial y} + \frac{\partial w}{\partial z} = 0 \qquad \text{(Eq 1)}$$

where x, y, and z are Cartesian coordinates, and u, v, and, w are corresponding orthogonal velocity components. As for the momentum conservation, with lubrication and the Hele-Shaw approximation, the Navier-Stokes equation is modified for molten feedstock during the filling stage (Ref 8, 9):

$$\frac{\partial P}{\partial x} = \frac{\partial}{\partial z}\left(\eta \frac{\partial u}{\partial z}\right)$$
$$\frac{\partial P}{\partial y} = \frac{\partial}{\partial z}\left(\eta \frac{\partial v}{\partial z}\right) \qquad \text{(Eq 2)}$$
$$\frac{\partial P}{\partial z} = 0$$

where P is the pressure, z is the thickness, and η is the viscosity of the PIM feedstock. Combining Eq 1 and 2 with integration in the z-direction (thickness direction) gives:

$$\frac{\partial}{\partial x}\left(S \frac{\partial P}{\partial x}\right) + \frac{\partial}{\partial y}\left(S \frac{\partial P}{\partial y}\right) = 0 \qquad \text{(Eq 3)}$$

where

$$S \equiv \int_{-b}^{b} \frac{z^2}{\eta}\,dz \qquad \text{(Eq 4)}$$

Equation 3 is the flow-governing equation for the filling stage. This is exactly the same form of steady-state heat conduction equation obtained by substituting temperature T into P and thermal conductivity k into S. In this analogy, S is the flow conductivity or fluidity. As a simple interpretation of this flow-governing equation, molten PIM feedstock flows from the high-pressure region to the low-pressure region, and the speed of flow depends on the fluidity, S. During the calculation, the fluidity increases as the thickness of the component increases and the viscosity of PIM feedstock decreases for feedstock cooling. After obtaining the pressure field, the velocity components u and v are obtained by integrating Eq 2 in the z-direction (thickness direction).

- *Energy equation:* In accordance with the lubrication and Hele-Shaw approximations during the filling stage, the energy equation is simplified as:

$$\rho C_p\left(\frac{\partial T}{\partial t} + u\frac{\partial T}{\partial x} + v\frac{\partial T}{\partial y}\right) = k\frac{\partial^2 T}{\partial z^2} + \eta\dot{\gamma}^2 \qquad \text{(Eq 5)}$$

where ρ is the molten PIM feedstock density, C_p is the molten PIM feedstock specific heat, and:

$$\dot{\gamma} = \sqrt{(\partial/\partial z)^2 + (\partial v/\partial z)^2}$$

is the generalized shear rate, and k is the thermal conductivity of the feedstock.

In addition, a constitutive relationship is necessary to describe the molten PIM feedstock response to its flow environment during cavity filling, which requires a viscosity model. Several viscosity models for polymers containing high concentrations of particles are available. Generally, they include temperature, pressure, solids loading, and shear strain rate; selected models are introduced later in this article. The selection of a viscosity model depends on the desired simulation accuracy over the range of processing conditions, such as temperature and shear rate, as well as access to the experimental procedures used to obtain the material parameters.

After acquiring the system of differential equations from continuum-based conservation laws and the constitutive relations for analysis of the filling stage, then boundary conditions are necessary. Typical boundary conditions during the filling stage are as follows:

- *Boundary conditions for flow equation:* Flow rate at injection point, free surface at melt front, and no slip condition at cavity wall
- *Boundary conditions for energy equation:* Injection temperature at injection point, free surface at melt front, and mold-wall temperature condition at cavity wall

Note that the only required initial condition is the flow rate and injection temperature at the injection node, which is one of the required boundary conditions.

For a more rigorous approach during the filling stage, a few efforts have invoked a full 3-D model and have included fountain flow, viscoelastic constitutive models, slip phenomena, yield phenomena, and inertia effects in governing equations and interface (Ref 3, 5, 6, 10).

Packing Stage

When mold filling is nearly completed, the packing stage starts. This precipitates a change in the ram control strategy for the injection molding machine, from velocity control to pressure control, which is called the switchover point. As the cavity nears filling, the pressure control ensures full filling and pressurization of the filled cavity prior to freezing of the gate. It is important to realize the packing pressure is used to compensate for the anticipated shrinkage in the following cooling stage. Feedstock volume shrinkage results from the high thermal expansion coefficient of the binder, so on cooling there is a measurable contraction. By appropriate pressurization prior to cooling, the component shrinks sufficiently after the gate

freezes that there are no sink marks (too low a packing pressure) and no difficulty with ejection (too high a packing pressure).

For the analysis of the packing stage, it is essential to include the effect of melt compressibility. Consideration is given to the melt compressibility using a dependency of the specific volume on pressure and temperature, leading to a feedstock specific pressure-volume-temperature (pVT) relationship, or the equation of state. Several models are available to describe the pVT relation of PIM feedstock, such as the two-domain modified Tait model and the IKV model. These models predict an abrupt volumetric change for both semicrystalline polymers used in the binder and the less abrupt volume change for amorphous polymers used in the binder.

With the proper viscosity and pVT models, the system of governing equation for the packing stage based on the continuum approach is as follows:

- *Mass conservation:* The continuity equation of compressible PIM feedstock is expressed as:

$$\frac{\partial \rho}{\partial t} + \frac{\partial(\rho u)}{\partial x} + \frac{\partial(\rho v)}{\partial y} + \frac{\partial(\rho w)}{\partial z} = 0 \qquad \text{(Eq 6)}$$

With the assumption that pressure convection terms may be ignored in the packing stage, this becomes:

$$\kappa\frac{\partial p}{\partial t} - \beta\left(\frac{\partial T}{\partial t} + u\frac{\partial T}{\partial x} + v\frac{\partial T}{\partial y}\right) + \left(\frac{\partial u}{\partial x} + \frac{\partial v}{\partial y}\right) = 0 \qquad \text{(Eq 7)}$$

where κ is the isothermal compressibility coefficient of the material $\partial\rho/\rho\partial p)$, and β measures the volumetric expansivity of the material $(\partial\rho/\rho\partial T)$. Those are easily calculated from the equation of state. Note that the same momentum conservation is used as in Eq 2, regardless whether the material to be considered is compressible or not.

- *Energy equation:* The energy equation is derived as:

$$\rho C_p\left(\frac{\partial T}{\partial t} + u\frac{\partial T}{\partial x} + v\frac{\partial T}{\partial y}\right) = k\frac{\partial^2 T}{\partial z^2} + \eta\dot{\gamma}^2 + \beta T\frac{\partial p}{\partial t} \qquad \text{(Eq 8)}$$

That is, the shear rate for the compressible case in packing is, for practical purposes, the same as for the filling phase.

Typical initial and boundary conditions during the packing stage are as follows:

- *Initial conditions:* Pressure, velocity, temperature, and density from the results of the filling stage analysis
- *Boundary conditions for equations of mass and momentum conservation:* Prescribed pressure at injection point, free surface at melt front, and no slip condition at cavity wall
- *Boundary conditions for energy equation:* Injection temperature at injection point, free

surface at melt front, and mold-wall temperature condition at the cavity wall, which is interfaced with the cooling-stage analysis

Cooling Stage

Of the three stages in the injection molding process, the cooling stage is of greatest importance because it significantly affects the productivity and the quality of the final component. Cooling starts immediately upon injection of the feedstock melt, but formally, the cooling time is referred to as the time after the gate freezes and no more feedstock melt enters the cavity. It lasts up to the point of component ejection, when the temperature is low enough to withstand the ejection stress. In the cooling stage, the feedstock volumetric shrinkage is counteracted by the pressure decay until the local pressure drops to atmospheric pressure. Thereafter, the material shrinks with any further cooling, possibly resulting in residual stresses due to nonuniform shrinkage or mold constraints (which may not be detected until sintering). In this stage, the convection and dissipation terms in the energy equation are neglected, because the velocity of a feedstock melt in the cooling stage is almost zero (Ref 11–13). Therefore, the objective of the mold-cooling analysis is to solve only the temperature profile at the cavity surface to be used as boundary conditions of feedstock melt during the filling and packing analysis.

When the injection molding process is in steady state, the mold temperature will fluctuate periodically over time during the process, due to the interaction between the hot melt and the cold mold and circulating coolant. To reduce the computation time for this transient process, a 3-D cycle-average approach is adopted for the thermal analysis to determine the cycle-averaged temperature field and its effects on the PIM component. Although the mold temperature is assumed invariant over time, there is still a transient for the PIM feedstock (Ref 14), leading to the following features:

- *Mold cooling analysis:* Under this cycle-average concept, the governing equation of heat transfer for the injection mold cooling system is written as:

$$\nabla^2 \bar{T} = 0 \tag{Eq 9}$$

where $\bar{T}$ is the cycle-average temperature of the mold.

- *PIM component cooling analysis:* Without invoking a flow field, the energy equation is simplified as:

$$\rho C_p \frac{\partial \bar{T}}{\partial t} = k \frac{\partial^2 \bar{T}}{\partial z^2} \tag{Eq 10}$$

Typical initial and boundary conditions applied during the cooling stage are as follows:

- *Initial conditions:* Temperature as calculated from the packing-stage analysis
- *Boundary conditions—mold:* Interface input from the PIM feedstock cooling analysis,

convection heat transfer associated with the coolant, natural convection heat transfer with air, and thermal resistance condition from the mold platen
- *Boundary conditions—component:* Interface input from the mold cooling analysis

Note that the boundary conditions for the mold and PIM feedstock cooling are coupled to each other. More details on this are given in the section, "Numerical Simulation," in this article.

For a more rigorous approach during the cooling stage, some researchers have included more than two different mold materials with flow-analysis that includes the cooling channel details. This enables corresponding heat-transfer analysis with any special cooling elements, such as baffles, fountains, thermal pins, or heat pipes (Ref 12).

Numerical Simulation

As far as the numerical analysis of injection molding is concerned, several numerical packages are already available for conventional thermoplastics. One may try to apply the same numerical analysis techniques to PIM. However, the rheological behavior of a powder-binder feedstock mixture is significantly different from that of a thermoplastic. Hence, the direct application of methods developed for thermoplastics to PIM requires caution (Ref 4, 10). Commercial software packages, including Moldflow (Moldflow Corp., Framingham, MA), Moldex3D (CoreTech System Co., Ltd., Chupei City, Taiwan), PIMsolver (Ceta-Tech, Sacheon, Korea), and SIMUFLOW (C-Solutions, Inc., Boulder, CO), are available for PIM simulation. Further, several research groups have written customized codes, but these are generally not released for public use.

It is well known that powder-binder feedstock mixtures used in PIM exhibit a peculiar rheological feature known as wall slip (Ref 10, 15). Therefore, a proper numerical simulation of the PIM process essentially requires a proper constitutive equation representing the slip phenomena of powder-binder feedstock mixtures (Ref 10, 14).

Filling and Packing Analysis

For numerical analysis of the filling and packing stages of PIM, both the pressure and energy equations must be solved during the entire filling and packing cycle. This is achieved using the finite-element method for Eq 2, while a finite-difference method is used in the z-direction (thickness), making use of the same finite elements in the x-y plane for solving Eq 3.

The finite-difference method (FDM) is a relatively efficient and simple numerical method for solving differential equations. In this method, the physical domain is discretized in the form of finite-difference grids. A set of algebraic equations is generated as the

derivatives of the partial differential equations and expressed by finite differences of the variable values at the grid points. The resulting algebraic equation array, which usually forms a banned matrix, is solved numerically. Generally, the solution accuracy is improved by reducing the grid spacing. However, because the FDM is difficult to apply to a highly irregular boundary or a complicated domain typical of injection molding, the use of this method must be restricted to regular and simple domains or used with the finite-element method (FEM) as an FDM-FEM hybrid scheme (Ref 3).

The FEM has excellent flexibility in treating complex geometries and irregular boundaries, which is a key advantage of this method. It requires discretizing the physical domain into several finite elements. The field variables are represented with shape functions and nodal values over each finite element. Using residual minimization techniques (or, equivalently, variational techniques) such as the Galerkin method, the governing equations are transformed into discretized forms (Ref 8, 9). For 3-D simulation of injection molding, the resulting global matrix system from the algebraic equations is typically large and sparse, which requires large memory space and processing time. The central processor use time may be estimated based on the number of elements in the meshes and the degrees of freedom per node. In transient problems, a finite-difference expression for the time derivatives is typically used in conjunction with the finite-element discretization.

For the numerical analysis of the filling process of PIM, one must solve both the pressure equation and the energy equation during the entire filling cycle until the injection mold cavity is filled. An FEM method is employed to solve Eq 2 and 3, while FDM is used thickness-wise (z-direction), making use of the same finite elements in the x-y plane (Ref 9).

Cooling Analysis

For numerical analysis of the cooling process in PIM, the boundary-element method (BEM) is widely used due to its advantage in reduction of the dimensionality of the solution. The BEM discretizes the domain boundary rather than the interior of the physical domain. As a result, the volume integrals become surface integrals, significantly reducing the number of unknowns, computation effort, and mesh generation (Ref 3).

A standard BEM formulation for Eq 9 based on Green's second identity leads to the following:

$$\alpha T(x) = \int_S \left[\frac{1}{r} \left(\frac{\partial T(\zeta)}{\partial n} \right) - T(\zeta) \left(\frac{1}{r} \right) \right] dS(\zeta) \tag{Eq 11}$$

Here, **x** and **ζ** relate to the positional vector in the mold, $r = |\zeta - x|$, and α denotes a solid angle formed by the boundary surface. Equation 11 for two closed surfaces, such as defined by the component shape, leads to a redundancy in the final system of linear algebraic equations, so a modified procedure is used (Ref 16). For a circular hole, a special formulation is created based on the line-sink approximation. This approach avoids discretization of the circular channels along the circumference and significantly saves computer memory and time.

For the thermal analysis of a PIM component, the FDM is used with the Crack-Nicholson algorithm for time advancement. The mold and PIM component analyses are coupled with each other in boundary conditions, so iteration is required until the solution converges.

Coupled Analysis between Filling, Packing, and Cooling Stages

The filling, packing, and cooling analyses are coupled to each other. When the filling and packing stages are analyzed, the cavity-wall temperature is a boundary condition for the energy equation. This cavity-wall temperature is obtained from the cooling analysis. On the other hand, when the cooling stage is analyzed, the temperature distribution of the powder-binder feedstock mixture in the thickness direction at the end of filling and packing is an initial condition for the heat transfer of powder-binder feedstock. This initial temperature distribution is obtained from the filling and packing analysis. Therefore, the coupled analysis among the filling, packing, and cooling stages may be made for accurate numerical simulation results (Ref 14).

Figure 1 shows a typical procedure for computer simulation of the PIM process, which consists of three components: input data, analysis, and output data. The quality of the input data is essential to success. The preprocessor is a software tool used to prepare a geometric model and mesh for the component and mold; it includes:

- Material data for feedstock, mold, and coolant
- Processing conditions for filling, packing, and cooling processes

Figure 2 shows one example of geometry modeling and mesh generation for a U-shaped component, including the delivery system and cooling channels.

Experimental Material Properties and Verification

Material Properties for Filling Stage

Successful simulation of the filling stage during PIM depends on measuring the material properties, including density, viscosity, and thermal behavior. Among these, the viscosity of the PIM feedstock and its variation with temperature, shear rate, and solid volume fraction are special concerns (Ref 17–19). The following procedure is an example of the method used to obtain these material properties. For this

illustration, assume a spherical stainless steel powder in a wax-polymer binder:

1. Melt densities, heat capacities, and thermal conductivities of the binder and feedstock are required. This is attained using a helium pycnometer, differential scanning calorimeter, and laser flash thermal conductivity device.
2. The transition temperature for the feedstock is measured, again using differential scanning calorimetry.
3. The binder viscosity is measured and fit to a model. For the characterization of the binder viscosity, a rotational-type rheometer is widely used due to relatively low viscosity of the wax. From the measured slow strain-rate viscosity data, binder viscosity is obtained by curve fitting to a Newtonian binder viscosity model (η_b) with temperature (T) dependency:

$$\eta_b = B_b \exp\left(\frac{T_{b,b}}{T}\right) \qquad \text{(Eq 12)}$$

where B_b is the constant amplitude, and $T_{b,b}$ is an Arrhenius-type coefficient, also called the reference temperature.

4. The feedstock viscosity is measured versus key parameters. The rheological behavior of the feedstock is measured by capillary rheometry. The slip characteristics of the PIM feedstock are determined using three different capillary dies with different length-to-diameter ratios (Ref 20). By using high length-to-diameter-ratio capillaries, the pressure loss correction, called Bagley's correction, is avoided. Rabinowitch's correction is extracted to obtain the true shear rate from the apparent shear rate for a non-Newtonian fluid, characteristic of the feedstock (Ref 10). The variation of viscosity with temperature is determined by testing the feedstock at different temperatures above the transition temperature. Such capillary measurements are typically carried out three times to confirm the repeatability of the data, giving a total of 27 tests—3 different diameter dies, 3 different temperatures, and 3 replications.
5. The feedstock viscosity data are modeled using standard loaded polymer concepts. A concentrated powder-binder feedstock mixture has a yield strength with shear-

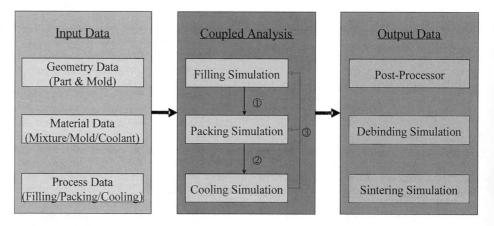

Fig. 1 Overall structure of computer-aided engineering analysis for powder injection molding parts. 1, pressure, temperature, and slip layer distribution at the end of filling; 2, temperature distribution and slip layer at the end of packing; 3, temperature distribution on the cavity surface. Source: Ref 14

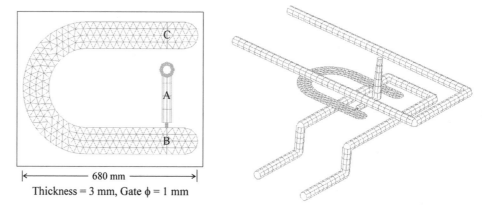

Fig. 2 Geometry modeling and mesh generation for U-shaped part, including delivery system and cooling channels. Pressure measurements at A, B, and C. Source: Ref 14

thinning viscous behavior (Ref 15). The modified-Cross model for viscosity as a function of the effective shear rate and temperature is the most useful treatment:

$$\eta_m(\dot{\gamma},T)=\frac{\eta_0}{1+(\eta_0\dot{\gamma}/\tau^*)^{1-n}}+\frac{\tau_y}{\dot{\gamma}} \quad \text{(Eq 13)}$$

where temperature enters the same as in Eq 12:

$$\eta_0 = B\exp(T_b/T) \quad \text{(Eq 14)}$$

In Eq 13 and 14, the subscript m indicates the powder-binder feedstock mixtures, while η_0, n, τ^*, B, and T_b denote the zero shear-rate viscosity, the power-law exponent, the transition shear stress, the constant amplitude, and the reference temperature (absolute scale) for the Arrhenius temperature dependence,

respectively. A term τ_y is added to the modified-Cross model for the yield stress (Ref 14, 21). To introduce the slip phenomenon, two models can be used: slip velocity and slip layer. The slip velocity model is expressed as:

$$V_S = \alpha_1\exp(\beta_1 T)\tau_w^{m_1} \quad \text{(Eq 15)}$$

The slip layer model is expressed as:

$$\delta = \alpha_2\exp(\beta_2 T)\tau_w^{m_2} \quad \text{(Eq 16)}$$

In Eq 15 and 16, the subscripts 1 and 2 indicate the slip velocity and slip layer models, respectively, and α, β, and m are the material constants. Figure 3 illustrates the concept of the slip layer and slip velocity. All the material parameters for feedstock viscosity based on Eq 13 to 16 are obtained by curve fitting from the measured viscosity data.

Table 1 gives an example of the material properties used in the PIM filling stage simulation for a 316L stainless steel powder with a median particle size of 8.0 μm in a standard wax-polymer binder system at a solid volume of 53%. Figure 4 demonstrates the importance of introducing slip phenomena to a viscosity model for the aforementioned feedstock at 100 °C. This plot shows the raw viscosity versus shear strain rate using three capillary diameters and the collapse of those results into a single curve with the slip correction.

Material Properties for the Packing Stage

To simulate the packing stage, the Hele-Shaw flow of a compressible viscous melt of PIM feedstock under nonisothermal conditions is assumed. For this, the two-domain modified Tait model is adopted to describe the phase behavior of the feedstock (Ref 8). A dilatometer is used to measure dimensional changes as a function of temperature and other variables, and the results are extracted by curve fitting.

Table 2 and Fig. 5 give one example of pvT material properties based on the following two-domain modified Tait model for the packing-stage simulation for the same stainless steel feedstock. For the solid-liquid phase:

$$v(p,T)=v_0(T)\left[1-0.0894\ln\left(1+\frac{p}{B(T)}\right)\right]+v_t(p,T)$$
$$v_0(T)=b_1+b_2\bar{T}$$
$$B(T)=b_3\exp(-b_4\bar{T})$$
$$\bar{T}=T-b_5$$
$$v_t(p,T)=b_7\exp(b_8\bar{T}-b_9 P)$$

$$\text{(Eq 17)}$$

with the transition temperature, T_g, which is calculated as $T_g(p)=b_5+b_6 p$.

Material Properties for the Cooling Stage

For the cooling-stage simulation, material properties of the mold material and coolant must be measured. Table 3 gives one example of the material properties for a typical H13 tool steel as the mold material, with water as the coolant.

Verification of the simulation is a critical step prior to any effort to optimize a design based on simulations. This verification usually includes the validation of the model used in developing the software. To demonstrate the verification of the simulation tool through experiment, the U-shaped test mold shown in Fig. 2 is selected with the stainless steel PIM feedstock reported previously and an H13 mold. Three pressure transducers were used to compare the simulation results with the experimental data. The cavity thickness is 3 mm, and the gate diameter is 1 mm. The coolant inlet temperature is 20 °C, the inlet flow rate is 50

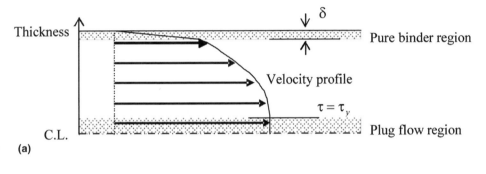

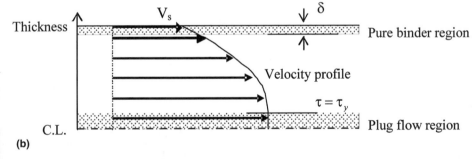

Fig. 3 Schematic diagram of mold cavity filling flow with (a) the slip layer model and (b) the slip velocity model with yield stress. Source: Ref 10

Table 1 Complete set of material properties of powder injection molding (PIM) feedstock with stainless steel powder 316L stainless steel (PF-15F, median particle size: 8.0 μm, Atmix Corp., Japan) combined with a wax-polymer-based binder system at a solid volume of 53%

Property		PIM feedstock		Binder system
Melt density	ρ_m	3.98×10^3 kg/m³	ρ_b	7.49×10^2 kg/m³
Heat capacity	C_m	6.85×10^2 J/kg · K	C_b	2.29×10^3 J/kg · K
Thermal conductivity	k_m	1.84 W/m · K	K_b	0.178 W/m · K
Transient temperature	T_g	52.8 °C	...	...
Viscosity	B	5.19×10^{-3} Pa · s	B_b	5.72×10^{-4} Pa · s
	T_b	5370 K	$T_{b,b}$	3650 K
	n	0.180	...	...
	τ^*	6.37×10^4 Pa	...	...
	τ_y	100 Pa	...	...
Slip phenomena		**Velocity model**		**Layer model**
	α_1	5.42×10^{-14} m/s	α_2	2.73×10^{-9} m
	β_1	2.75×10^{-2} /K	β_2	4.23×10^{-3} /K
	m_1	1.50	m_2	0.513

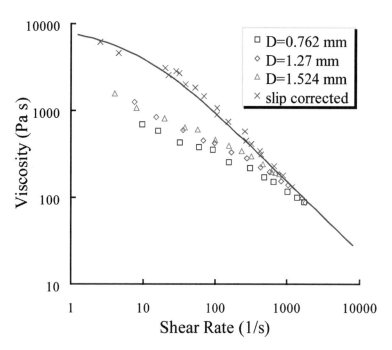

Fig. 4 Slip-corrected viscosity of 316L stainless steel feedstock using three different capillary diameters. Source: Ref 14

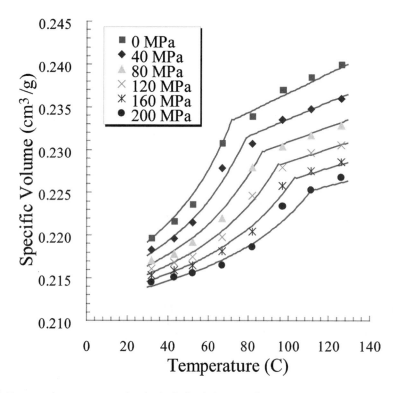

Fig. 5 Pressure-volume-temperature data for the feedstock. Source: Ref 14

Table 2 Pressure-volume-temperature material properties of powder injection molding feedstock with stainless steel powder 316L stainless steel (PF-15F, median particle size: 8.0 μm, Atmix Corp., Japan) combined with a wax-polymer-based binder system at a solid volume of 53%

Parameter	Solid	Liquid
b_1	2.190×10^{-4} m³/kg	2.334×10^{-4} m³/kg
b_2	7.716×10^{-8} m³/kg · K	1.156×10^{-7} m³/kg · K
b_3	1.000×10^{9} Pa	2.940×10^{8} Pa
b_4	1.000×10^{-2} /K	4.689×10^{-2} /K
b_5	3.450×10^{2} K	
b_6	1.990×10^{-7} K/Pa	
b_7	1.446×10^{-5} m³/kg	
b_8	3.388×10^{-2} /K	
b_9	9.328×10^{-9} /Pa	

Table 3 Material properties for HP-13 tool steel as the water-cooled mold material

Material	Property	Value
HP-13 for mold	Thermal conductivity	1.05×10^{2} W/m · K
Water for coolant	Density	974 kg/m³
	Heat capacity	4.20×10^{3} J/kg · K
	Thermal conductivity	0.688 W/m · K
	Viscosity	3.65×10^{-4} Pa · s

Table 4 Processing conditions for 316L stainless steel U-shaped specimen

Filling time	1 s
Filling/packing switchover	98%
Packing time	5 s
Packing pressure	13 MPa
Injection temperature	120 °C
Cooling time	10 s
Coolant inlet temperature	20 °C
Coolant inlet flow rate	50 cm³/s

between maximum and minimum values is 18 °C. This variation is large enough to cause a significant difference in the solidification layer development during the packing stage. Therefore, one may expect the simulation to have a significant error in pressure prediction during the packing stage without consideration of the cooling effect. Figure 6(c) shows the distribution of the slip layer thickness at the end of filling. The predicted slip layer thickness is from 0.8 to 6.2 μm, which is less than the 8 μm median particle size. A high slip layer thickness becomes an insulator during cooling, resulting in nonuniform cooling and increased cooling time.

To examine the validation and importance of the slip phenomena and the coupled analysis between the filling/packing and cooling stages, the pressure traces were compared between simulation and experiment, as shown in Fig. 7 using the three positions indicated in Fig. 2. Figure 7 gives the pressure-time plot obtained from the experiment and simulation. The simulation results were obtained from the filling and

cm³/s, and the cooling time is 10 s, as summarized in Table 4.

Figure 6 describes some of the simulation results obtained using PIMsolver. Figure 6(a) shows the filling pattern, indicating how the mold cavity fills as a function of time with the slip layer model. The filling time was 1.28 s.

Figure 6(b) shows the average mold-temperature distributions on the upper and lower surfaces of the cavity from the cooling analysis results. The highest average temperature is 51°C and occurs at the base of the "U," and the lowest temperature is 33 °C at the runner inlet. The mold-wall temperature is not uniform, and the difference

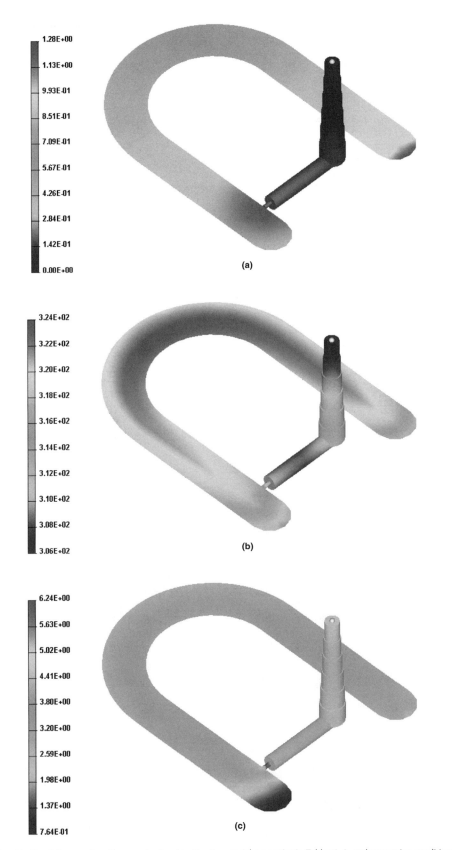

Fig. 6 Simulation results with geometry input in Fig. 2, material properties in Tables 1–3, and processing conditions in Table 4. (a) Filling pattern (filling time 1.28 s). (b) Average mold cavity-wall temperature distributions (K) between the upper and lower surfaces obtained from the cooling analysis (maximum temperature = 51 °C; minimum temperature = 33 °C). (c) Slip layer thickness distributions at the end of filling obtained from the filling analysis (maximum thickness = 6.24 µm; minimum thickness = 0.76 µm). Source: Ref 14

packing analyses at a constant cavity-wall temperature of 30 °C without and with consideration of slip layer. As shown in Fig. 7(a) and (b), the simulation results obtained with the slip layer provide a better fit to the experimental data. Without the slip correction, there is an underestimation of the pressure. However, the simulation results with the slip correction continue to show significant deviations from experimental results. The results with the distributed mold-wall temperature interfaced with the results from the cooling analysis, as shown in Fig. 7(c), explain this deviation. If one considers the cavity-wall temperature distributions from the cooling analysis (coupled analysis), then that temperature enables the best agreement to the experimental results. This is because the constant mold-wall temperature is lower than obtained from the cooling analysis, and the mold-wall temperature is not uniform. In addition, the accuracy of pressure prediction at the end of the packing stage (approximately 4 s) is improved due to the difference in the development of the solidification layer, as mentioned previously. The cooling analysis is very important in PIM, because the thermal conductivity of PIM feedstock is much higher than for a common plastic, so the slip layer plays a significant role as an insulator.

Demonstration of Usefulness and Optimization

This section presents simulation results from some of the 2.5-dimensional examples to demonstrate the usefulness of the computer-aided engineering (CAE) analysis and optimization capability of the PIM process. It has been shown that the developed CAE tool for the PIM process is capable of predicting the filling pattern, temperature distribution, clamping force, and other important variables. This section demonstrates how to use this basic information from the simulation tool to predict injection-molding-related defects and also presents a systematic way of using the CAE tool to develop an optimal injection molding process.

Basic Capability—Short Shot, Flash, Weld Line, Air Vent, and Other Features

This section demonstrates the use of simulation results to predict typical molding defects. There are many types of molding defects, which can be identified mainly as basic defects, dimensional defects, and other defects. The basic defects are traced to the molding parameters.

For the basic defects, simulations use the pressure field analysis to predict short shots and flashing as well as the filling pattern to identify trapped air and weld line location. A short shot occurs when the molded part is filled incompletely because insufficient material was injected into the mold. Several factors cause the short shot defect, such as an insufficient-sized or restrictive flow area, low melt temperature, low mold-wall temperature, lack of vents, low injection pressure, and premature solidification. Through computer simulations, short shot defects are predicted and minimized by analyzing the mold-filling pattern from the pressure, velocity, and temperature results. Flash is a defect where excessive material is found at locations where the mold separates, notably the parting surface, moveable core, vents, or venting ejector pins. The causes of flash are low clamping force, gap within the mold, molding conditions, and improper venting. Flash is avoided by using the clamping force calculation results and the simulation results. Figure 8 shows examples of short shot and flash of PIM components.

Weld line and the resulting mark or knit line is another flaw that is also a potential weakness in a molded plastic part. Weld lines are formed by the union of two or more streams of feedstock flowing together, such as when flow passes around a hole, insert, or in the case of multiple gates or variable thicknesses in the

component. Consequently, a weld line reduces the strength of the green component and leaves an undesirable surface appearance. It should be avoided when possible. The results from the computer analysis are used to predict the weld line location. Air trap or air vent is a defect caused by air that is caught inside the mold cavity. The air trap locations are usually in areas that fill last. The air trap is predicted from the filling pattern analysis and can be avoided by reducing the injection speed and enlarging or properly placing vents. Figure 9 shows the predicted locations of short shot, air trap, and weld line for the injection-molded PIM components.

Other defects, such as burn marks, flow marks, meld lines, jetting, surface ripples, sink marks, and so on, are also accessible using computer simulation tools in the design process stage. Especially important in production are control of factors related to dimensional uniformity. These are analyzed by checking all three main stages of the injection molding. Good component quality with uniform mass and uniform green density is important to hold final dimensional control. Table 5 is an example of the solution windows used to drive PIM toward dimensional stability during each stage.

Imbalanced Filling of Multiple-Cavity Tooling

By increasing the number of quality components produced during a given molding cycle (multicavity molds), the cost of tooling increases, yet the cost of production is reduced. This assumes that each component produced in each cavity of the multicavity mold will be identical. However, despite almost identical cavities, flow paths, cooling, and other control parameters, variations often exist between molded components. These variations significantly limit the benefit of a multicavity mold. The use of computer simulations allows balancing of the multiple-cavity filling event. First, from the cooling analysis, the cooling channel is configured to obtain a uniform mold-wall temperature around each component. Second, from the 3-D delivery system, analysis of the conduction effect of the flow in each branch allows the delivery system to be appropriately configured. Combining these results with a viscous heating analysis leads to optimization of the processing variables as well as the cavity, delivery system, and cooling channel designs.

Figure 10(a) shows the FEM meshing for an eight-cavity delivery system. Figure 10(b)

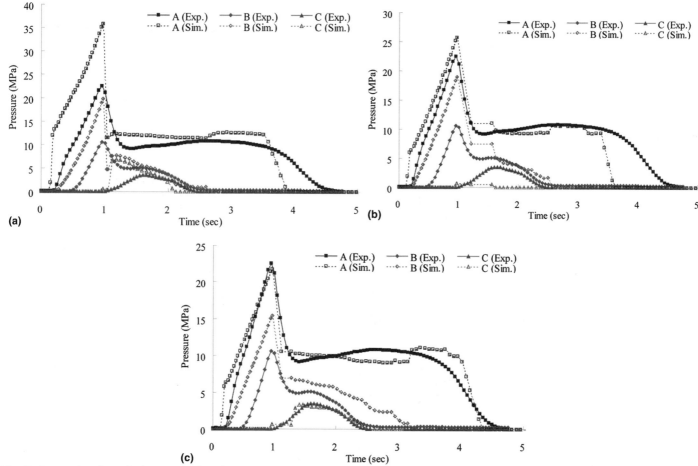

Fig. 7 Pressure-time plots at the three points indicated in Fig. 2. Simulation results are obtained from filling, packing, and cooling analyses with a constant cavity-wall temperature of 30 °C. (a) Without slip modeling and without interface with cooling analysis. (b) With slip modeling and without interface with cooling analysis. (c) With slip modeling and with interface with cooling analysis. Source: Ref 14

(a)

(b)

Fig. 8 (a) Short shot is predicted by checking the filling pattern. (b) Flash is avoided by calculating clamping force from the simulation results. Source: Ref 22

visually shows the filling difference associated with a shift in the mold-filling time. In this case, slightly slower filling was beneficial.

Figure 11(a) shows the cooling analysis and temperature difference in two neighboring cavities. The difference in temperature distributions in the two cavities results from the combination of the cooling channel configuration and the viscous heating effect in the delivery channel.

Balanced Filling

Flow balancing during filling in a multiple-cavity tool set requires understanding of the melt-front velocity (MFV) and melt-front area (MFA) profiles. As the name suggests, MFV is the melt-front advancement speed, and MFA is the cross-sectional area advancement; they are either the length of the melt front multiplied by the thickness of the component, the cross-sectional area of the runner, or a sum of both if the melt is flowing in both places. At any time, the product of local MFV and MFA along all moving fronts is equal to the volumetric flow rate (Ref 27, 28):

$$MFV \equiv \frac{Q}{MFA} \qquad \text{(Eq 18)}$$

where Q is the volumetric flow rate. A high feedstock velocity at the melt front gives a higher

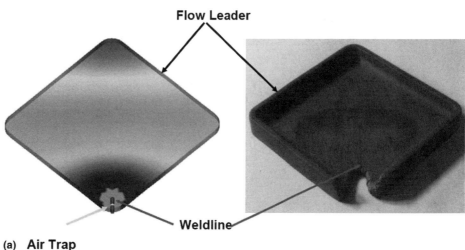

(a) Air Trap

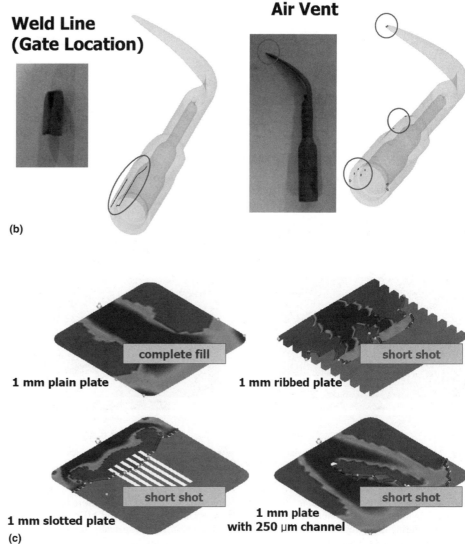

Fig. 9 Weld lines and air trap/vent are predicted by using a computer-aided engineering tool for the powder injection molding process. (a) Plate. Source: Ref 22. (b) Dental scaler tip. Source: Ref 23. (c) Geometry evaluation of microfeatures. Source: Ref 24

Table 5   Relationship of variables to improve quality at each molding stage

Objective functions	Filling stage	Packing stage	Cooling stage
	Design variables		
Minimum injection pressure	Optimum filling time	Optimum packing time to avoid backflow	...
Maximum productivity	...	...	Optimum cooling time
Minimum deformation	Optimum ram-speed profile for uniform velocity at melt-front	Optimum packing pressure for profile uniform shrinkage	Optimum cooling system design for uniform and even cooling and eliminating hot spot

Source: Ref 25

Short-Shot Test with Various t_f

(b) $t_f = 4.0$ s $t_f = 0.5$ s

Fig. 10 (a) Finite-element modeling mesh for a multicavity and delivery system. (b) Filling experiment showing the different filling pattern by changing the filling time (t_f) from 4.0 to 0.5 s. Source: Ref 26

surface stress with more molecular orientation and particle migration. A variable velocity of the melt front during filling results in differential sintering shrinkage and component warpage. Therefore, it is desirable to maintain a constant velocity at the melt front to generate uniform molecular orientation and minimized particle migration in the flow direction (not in thickness direction), which results in uniform material properties. Therefore, MFV and MFA are important design parameters, especially for balancing the flow during cavity filling. For example, MFA is used to quantitatively compare the degree of flow balance.

Figure 12 illustrates MFV and MFA variations in a plate component that has four gates. Due to the variable gate locations and filling pattern, a constant MFV is not guaranteed, even with a constant volumetric flow rate (or equivalently, a constant ram speed). Portions of the melt front reach the end of the cavity while other portions are still moving. Optimizing the ram-speed profile or relocation of the gate location by minimizing the MFA in the cavity is discussed later in the section on optimization in this article.

Sensitivity Analysis

Simulating and optimizing the PIM process is difficult because several material, component, and process parameters are linked. It is difficult to identify critical parameters in the computer design because multiple objective functions must be considered. An important question arises about how variations in material property influence the errors arising in predictions of various simulated parameters. During the design of a component, adjustment is made to its size to improve the functional and aesthetic attributes of the component. Additionally, changes are made to the location and dimensions of the melt-delivery system to improve the manufacturability of the component. These issues also raise questions about how small changes in dimensions influence process variations. Finally, several process settings are controlled by the operator on the injection molding machine during the production stage. It is important to understand how a computerized design tool captures the influence of such process variations in its prediction (Ref 27).

Process, design, and material parameters are optimized using sensitivity analysis. It is a valuable tool for the design engineer who determines the critical input parameters as well as for the production engineer who must optimize production. For the sensitivity analysis, the input parameters are varied over a fixed range (for example, $\pm 5\%$), and the response of the output parameters is monitored. The sensitivity is calculated as the slope of the dimensionless dependent variable with respect to the dimensionless independent variable, according to:

$$\text{Sensitivity} \equiv \frac{\text{Percentage change in output}}{\text{Percentage change in input}}$$
$$\equiv \frac{\text{Increment in output/Initial value of output}}{\text{Increment in input/Initial value of input}}$$

(Eq 19)

This definition of sensitivity was used to compare input and output parameters having different dimensions. For example, a sensitivity value of -1.5 means that the percentage decrement of output is 1.5% if the percentage increment of input is 1.0%. It is independent of input and output parameter units. The result from Atre et al. (Ref 27) is shown in Fig. 13, showing all normalized sensitivity values between the input and output parameters, which means that both the pressure- and temperature-related output parameters are sensitive to the process and geometry conditions and feedstock

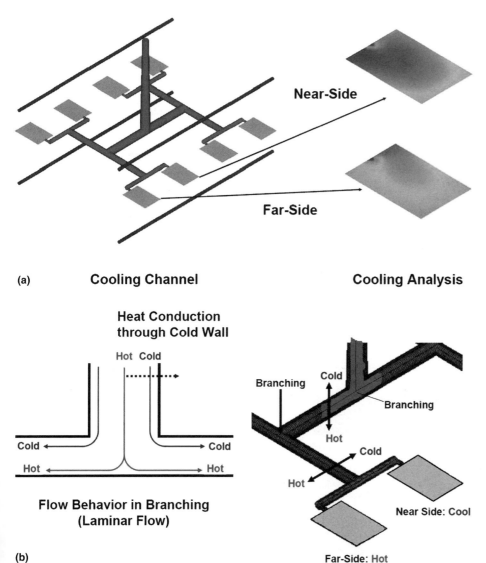

(a) Cooling Channel

Cooling Analysis

Heat Conduction through Cold Wall

Hot Cold

Cold ← → Cold

Hot ← → Hot

Flow Behavior in Branching (Laminar Flow)

(b)

Branching — Cold — Branching

Cold

Hot

Hot

Near Side: Cool

Far-Side: Hot

Fig. 11 (a) Multiple-cavity cooling analysis showing the temperature difference in neighboring cavities in what would appear to be a balanced flow system. (b) Illustration of how the viscous heating effect leads to this temperature imbalance. Source: Ref 26

Optimization 1—Filling Time

Filling time is an important variable that is optimized to reduce the required injection pressure. From the CAE filling analysis, the injection pressure is calculated by varying the filling time. When plotting the required injection pressure versus various filling times, the optimal filling-time ranges are determined for the lowest injection pressure. The curve is U-shaped because, on one hand, a short fill time involves a high melt velocity and thus requires a higher injection pressure to fill the mold. On the other hand, the injected feedstock cools more with a prolonged fill time. This results in a higher melt viscosity and thus requires a higher injection pressure to fill the mold. The shape of the curve of injection pressure versus fill time depends very much on the material used as well as on the cavity geometry and mold design. If the required injection pressure exceeds the maximum machine capacity, the process conditions or runner system must be modified. Figure 14(a) and (b) show the optimum filling-time selection for the selected multicavity mold.

Optimization 2—Gate Location and Number of Gates

For a given component geometry, a filling analysis is generated based on the filling patterns in Fig. 15. The injection molding pressure can be reduced by trying other locations for the gate, from a single end to a single center location. Based on this new gate location, the injection pressure is calculated. Compared to the previous design, the injection pressure is reduced. Further reduction of the injection pressure can be obtained by using multiple gates. With three gates evenly distributed along the diagonal direction, the injection pressure is reduced from 24.2 MPa for the single end gate to 6.29 MPa.

Optimization 3—Delivery System and Ram-Speed Profile

Figure 16(a) shows the meshed geometry and solid layer fraction distribution of a cellular phone housing with a center gate. Figure 16(b) shows the corresponding optimal ram-speed profile that minimizes the MFV based on the current gate location. Table 6 shows that all other variables have also been improved due to this delivery and ram-speed profile optimization.

Optimization 4—Design of Experiments

There are many injection molding parameters that affect green component properties. Therefore, a design of experiment (DOE) approach is used where there are many inputs. The most frequently used methods are the partial or full fractional design and the Taguchi approach

properties, while the flow-related output parameters are sensitive to the process and geometry conditions. The abbreviations used in Fig. 13 include the following.

For classification of input parameters:

- *PC (processing conditions):* Filling time (t_f), switchover point (SO), injection temperature (T_i), and mold-wall temperature (T_w)
- *GC (geometry conditions):* Part thickness (PT), gate diameter (GD), runner diameter (RD), and sprue diameter (SD)
- *FP (feedstock material properties):* Density (ρ), specific heat capacity (C_p), thermal conductivity (k), transition temperature (T_g), eject temperature (T_e), viscosity parameters (n, B, C, T_b, τ_y)

- *BP (binder material properties):* Density (ρ_b), specific heat capacity ($C_{p,b}$), thermal conductivity (k_b), viscosity parameters (B_b, $T_{b,b}$, α, β, m)

For classification of output parameters:

- *Pressure related:* Injection pressure (P_i), clamping force (f_c), and maximum shear stress (τ_{max})
- *Temperature related:* Melt-front temperature difference ($\triangle$MFT), packing time (t_p), and cooling time (t_c)
- *Velocity related:* Maximum shear rate ($\dot{\gamma}_{max}$), average MFV (μMFV), standard deviation of MFV (σMFV), average MFA (μMFA), and standard deviation of MFA (σMFA).

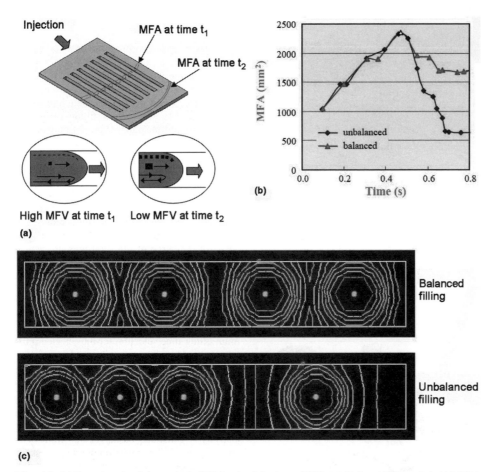

Fig. 12 (a) Illustration of melt-front velocity (MFV) and melt-front area (MFA) at t_1. Reduced MFA results in high MFV and must be adjusted by balancing the gates or optimizing the ram-speed profile. (b) Reduced MFA variation after gate balancing for the cavity shown in (c). (c) Variation in MFA and its corresponding filling pattern with balanced and unbalanced filling. Source: Ref 25

(Ref 18, 29). If an appropriate DOE method is used, one can easily establish whether the inputs have an effect on the outputs of the system. An optimization study is demonstrated for the multicavity shown in Fig. 14. The Taguchi method follows this procedure:

- Determine the objective function to be optimized.
  - Minimize the weight of delivery and injection pressure.
- Identify the control factors and their levels.
  - Injection temperature (A)
  - Mold-wall temperature (B)
  - Diameter of main runner (C)
  - Diameter of branched runner (D)
- Identify constraints.
  - Capability of injection molding machine is the constraint.
- Design the matrix experiment and define the data analysis procedure.
  - L_9 (3^4) orthogonal array
- Predict the performance at these levels.
  - Simulation or experiment based on L_9 array
- Analyze the data and determine optimum levels for control factors.
  - The larger the signal-to-noise ratio, the better.

- Conduct the matrix experiment.

The four input parameters (injection temperature, mold-wall temperature, diameter of main runner, and diameter of branched runner) are the control factors, and the optimization procedure involves the determination of the "best" levels of control factors. The best levels of control factors are those that maximize the signal-to-noise (S/N) ratio. Maximizing the S/N ratio results in minimizing any property that is sensitive to noise. The "larger-the-better" characteristic of the S/N ratio is chosen because a high value of S/N implies that the signal is higher than the uncontrollable noise factors. The S/N ratio was calculated using an equation that is measured in unit decibels:

$$S/N = -\log_{10}[\text{mean sum of square of the measured data}]$$
(Eq 20)

The objective function to be minimized is:

$$F = \alpha(\text{Normalized weight of delivery system}) + (1 - \alpha)(\text{Normalized injection pressure})$$
(Eq 21)

where α is an adjustable parameter selected by the user.

Table 7 depicts the Taguchi L_9 orthogonal array and the results from the experiment. Figure 17 shows the S/N ratio analysis of the experiment, where 1, 2, and 3 represent low, medium, and high settings, respectively. From the analysis, the total contribution of the injection temperature to the objective function is 47.22%, the mold-wall temperature is 34.3%, the diameter of the main runner is 12.42%, and the diameter of the branch is 6.06%.

The conditions corresponding to No. 9 prove optimal, and the combination minimizes the objective function when the injection temperature is 140 °C, the mold-wall temperature is 45 °C, the diameter of the main runner is 5 mm, and the diameter of the branch runner is 4 mm. This combination provides a scrap savings of 24% and a reduction in the injection pressure of 0.1% from the initial design.

Summary

This article shows the unique attributes for PIM filling simulations. After molding, assuming the component is rigid, the components are subjected to a heating cycle where both binder burnout and sintering take place. The constitutive models for sintering are covered in the article "Modeling and Simulation of Press and Sinter Powder Metallurgy" in this Handbook. However, for PIM components, the sintering shrinkage is large, because the compact starts near 60% dense and shrinks approximately 15% in attaining a sintered density near 98% of theoretical. To date, simulations for the final size and shape are proving accurate. The current efforts are aimed at adding the debinding cycle onto the integrated molding and sintering simulations. This is obviously a topic for more research.

ACKNOWLEDGMENT

Considerable support in this simulation development was provided by Young-Sam Kwon of CetaTech and Sundar Atre of Oregon State University.

REFERENCES

1. R.M. German and A. Bose, *Injection Molding of Metals and Ceramics,* Metal Powder Industries Federation, Princeton, NJ, 1997
2. R.M. German, *PIM Design and Applications User's Guide,* Innovative Material Solutions, State College, PA, 2003
3. S. Kim and L. Turng, Developments of Three-Dimensional Computer-Aided Engineering Simulation for Injection Molding, *Model. Simul. Mater. Sci. Eng.,* Vol 12, 2004, p 151–173
4. C.J. Hwang and T.H. Kwon, A Full 3D Finite Element Analysis of the Powder Injection Molding Filling Process Including Slip Phenomena, *Polym. Eng. Sci.,* Vol 42 (No. 1), 2002, p 33–50

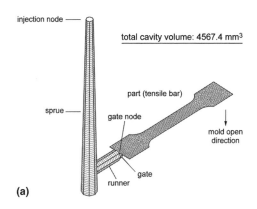

total cavity volume: 4567.4 mm³

(a)

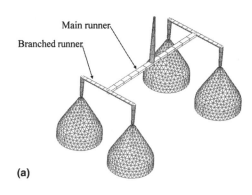

(a)

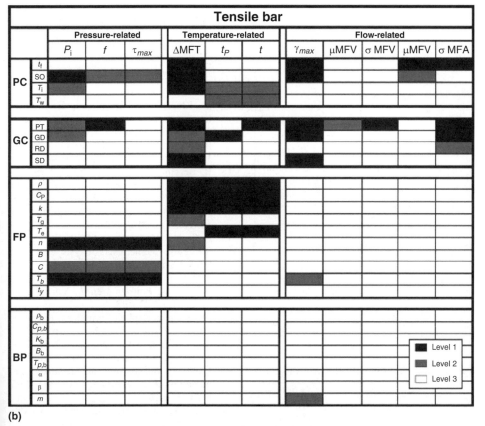

Fig. 13 (a) Description of tensile components used for analysis. (b) Sensitivity analysis of output parameters toward input parameters for a tensile bar. See text for explanation of abbreviations. Black = strong level; gray = medium level; white = weak level. Source: Ref 27

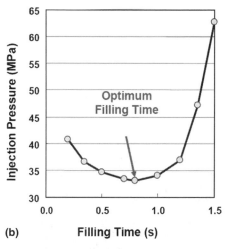

Fig. 14 Filling-time optimization by using a computer-aided engineering tool for powder injection molding. (a) Mesh used for optimization study. (b) Optimum filling-time finding that minimizes injection pressure. Source: Ref 25

5. K. Mori, K. Osakada, and S. Takaoka, Simplified Three-Dimensional Simulation of Non-Isothermal Filling in Metal Injection Moulding by the Finite Element Method, *Eng. Comput.,* Vol 13 (No. 2), 1996, p 111–121

6. V.V. Bilovol, L. Kowalski, J. Duszczyk, and L. Katgerman, Comparison of Numerical Codes for Simulation of Powder Injection Moulding, *Powder Metall.,* Vol 46 (No. 1), 2003, p 55–60

7. C. Binet, D.F. Heaney, R. Spina, and L. Tricario, Experimental and Numerical Analysis of Metal Injection Molded Products, *J. Mater. Process. Technol.,* Vol 164–165, 2005, p 1160–1166

8. H.H. Chiang, C.A. Hieber, and K.K. Wang, A Unified Simulation of the Filling and Post Filling Stages in Injection Molding, Part I: Formulation, *Polym. Eng. Sci.,* Vol 31 (No. 2), 1991, p 116–124

9. C.A. Hieber and S.F. Shen, A Finite-Element/Finite-Difference Simulation of the Injection Molding Filling Process, *J. Non-Newton. Fluid Mech.,* Vol 56, 1995, p 361

10. T.H. Kwon and S.Y. Ahn, Slip Characterization of Powder-Binder Mixtures and Its Significance in the Filling Process Analysis of Powder Injection Molding, *Powder Technol.,* Vol 85, 1995, p 45–55

11. S.J. Park and T.H. Kwon, Sensitivity Analysis Formulation for Three-Dimensional Conduction Heat Transfer with Complex Geometries Using a Boundary Element Method, *Int. J. Numer. Methods Eng.,* Vol 39, 1996, p 2837–2862

12. S.J. Park and T.H. Kwon, Optimal Cooling System Design for the Injection Molding Process, *Polym. Eng. Sci.,* Vol 38, 1998, p 1450–1462

13. S.J. Park and T.H. Kwon, Thermal and Design Sensitivity Analyses for Cooling System of Injection Mold, Part 1: Thermal Analysis, *ASME J. Manuf. Sci. Eng.,* Vol 120, 1998, p 287–295

14. S. Ahn, S.T. Chung, S.V. Atre, S.J. Park, and R.M. German, Integrated Filling, Packing, and Cooling CAE Analysis of Powder Injection Molding Parts, *Powder Metall.,* Vol 51 (No. 4), 2008, p 318–326

15. D.M. Kalyon, Apparent Slip and Viscoplasticity of Concentrated Suspensions, *J. Rheol.,* Vol 49, 2005, p 621–640

16. M. Rezayat and T. Burton, A Boundary-Integral Formulation for Complex Three-Dimensional Geometries, *Int. J. Numer. Methods Eng.,* Vol 29, 1990, p 263–273

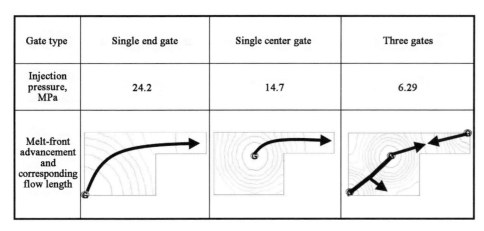

Gate type	Single end gate	Single center gate	Three gates
Injection pressure, MPa	24.2	14.7	6.29
Melt-front advancement and corresponding flow length			

Fig. 15 Optimization of gate location and number of gates. Source: Ref 25

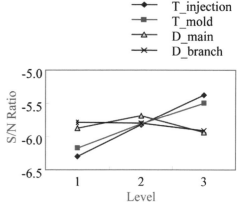

Fig. 17 Signal-to-noise (S/N) analysis of the multicavity showing that injection temperature is the most sensitive factor for achieving the objective function. Source: Ref 25

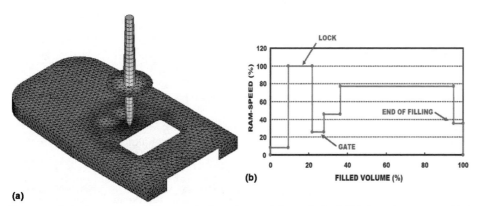

Fig. 16 Optimal ram-speed profile to maintain uniform melt-front velocity (MFV) from the computer-aided engineering simulation of a cell phone shell application. (a) Meshed geometry. (b) Optimal ram-speed profile that minimizes variation of the MFV. Source: Ref 22

Table 6 Optimum filling-stage scenario after optimization (multicavity shown in Fig. 14)

Design parameters	Initial	Optimum
Injection pressure, MPa	62.4	54.0 (−13.4%)
Clamping force, ton	15.3	13.2 (−13.7%)
Maximum shear stress, MPa	0.456	0.393 (−13.8%)
Difference in melt-front temperature (ΔMFA), °C	41.6	9.8 (−76.4%)
Deviation of melt-front velocity (σMFV), mm/s	1411	120 (−91.5%)
Solidification fraction, %	8.0	4.7(−41.4%)

Table 7 The Taguchi L_9 orthogonal array used for the experiment and its results

No.	$T_{injection}$, °C	T_{mold}, °C	D_{main}, mm	D_{branch}, mm	Objective function	Scrap, g	Pressure, MPa
1	120	35	4.0	4.0	2.15	36.67	50.05
2	120	40	5.0	4.5	2.02	46.18	40.24
3	120	45	6.0	5.0	2.03	57.48	34.12
4	130	35	5.0	5.0	2.02	49.56	38.33
5	130	40	6.0	4.0	1.96	51.06	35.60
6	130	45	4.0	4.5	1.88	39.70	39.53
7	140	35	6.0	4.5	1.95	54.09	33.31
8	140	40	4.0	5.0	1.88	43.09	37.30
9	140	45	5.0	4.0	1.75	43.15	33.07
Initial design	130	40	6.0	5.0	2.00	57.48	33.11

17. M. Khakbiz, A. Simchi, and R. Bagheri, Investigation of Rheological Behaviour of 316L Stainless Steel-2 Wt-% TiC Powder Injection Moulding Feedstock, *Powder Metall.,* Vol 48 (No.2), 2005, p 144–150

18. B. Berginc, Z. Kampus, and B. Sustarsic, Influence of Feedstock Characteristics and Process Parameters on Properties of MIM Parts Made of 316L, *Powder Metall.,* Vol 50 (No. 2), 2007, p 72–183

19. P. Suri, R.M. German, J.P. de Souza, and S.J. Park, Numerical Analysis of Filling Stage during Powder Injection Moulding: Effects of Feedstock Rheology and Mixing Conditions, *Powder Metall.,* Vol 47 (No. 2), 2004, p 137–143

20. M. Mooney, Explicit Formulas for Slip and Fluidity, *J. Rheol.,* Vol 2, 1931, p 210–222

21. M.L. Foong, K.C. Tam, and N.H. Loh, Yield Stress Behaviour of Metal Injection Moulding Suspensions at Elevated Temperatures, *J. Mater. Sci.,* Vol 30, 1995, p 3625–3632

22. S.J. Park, S.T. Chung, A. Schenck, S.V. Atre, and R.M. German, Avoiding Molding Defects during the PIM of 316L Stainless Steel, *Proceedings PIM 2003,* Innovative Material Solutions, State College, PA, 2003

23. C.J. Hwang, Y.B. Ko, H.P. Park, S T. Chung, and B.O. Rhee, Computer Aided Engineering Design of Powder Injection Molding Process for a Dental Scaler Tip Mold Design, *Mater. Sci. Forum,* Vol 534–536, 2007, p 341–344

24. C. Wu, S.V. Atre, S. Laddha, S. Lee, K. Simmons, S.J. Park, R.M. German, and D.T. Whychell, Sr., Material Homogeneity in Powder Injection Molded Ceramic Microchannel Arrays, *PIM Int.,* Vol 2 (No. 2), 2008, p 68–73

25. *PIMsolver User's Guide,* Ver. 2.0, Cetatech, 2005

26. S.V. Atre, S.J. Park, A. Schenck, T.G. Kang, and R.M. German, Imbalance Filling

of Multi-Cavity Tooling: A Comparison of Plastics and Metals, *Proceedings PIM 2003,* Innovative Material Solutions, State College, PA, 2003

27. S.V. Atre, S.J. Park, R. Zauner, and R.M. German, Process Simulation of Powder Injection Moulding: Identification of Significant Parameters during Mould Filling Phase, *Powder Metall.,* Vol 50 (No.1), 2007, p 76–85

28. L.S. Turng, *C-Mold Design Guide—A Resource for Plastics Engineers,* 3rd ed., Advanced CAE Technology Inc., Troy, NY, 1998

29. R. Urval, S. Lee, S.V. Atre, S.J. Park, and R.M. German, Optimization of Process Conditions in Micro Powder Injection Molding Using Taguchi Robust Design Method, Part I: Primary Design Parameters, *Powder Metall.,* Vol 51 (No. 2), 2008, p 133–142

Simulation of Machining Processes

ASM Handbook, Volume 22B, *Metals Process Simulation*
D.U. Furrer and S.L. Semiatin, editors

Copyright © 2010, ASM International®
All rights reserved.
www.asminternational.org

Modeling and Simulation of Machining

Christian E. Fischer, Scientific Forming Technologies Corporation

CONVENTIONAL MACHINING describes a family of processes in which metal is removed locally by the shearing action of a cutting tool. Cutting typically involves a single- or multiple-point cutting tool. Processes can broadly be grouped into:

- *Turning:* A slowly translating tool is moved into a rapidly rotating part, typically on a lathe.
- *Milling:* A rapidly rotating tool is used to remove material from the surface of a stationary or slowly translating and/or rotating workpiece.
- *Drilling:* A rotating tool is used to form a hole in a workpiece.
- *Shaping or broaching:* Material is removed by using a multiple-edge tool with progressively evolving shapes.
- *Thread cutting:* A tap or die or used.

These processes are used to precisely shape a part. They are selected over other shaping processes because of their potential for geometric accuracy and their ability to produce shapes that cannot be produced by other processes. They also offer an economic advantage for one-of-a-kind or many low-to-moderate production-run products (Ref 1). Nonetheless, metal cutting is frequently an expensive and time-consuming process. The economics of machining are governed largely by the cost of owning and maintaining equipment and by labor costs. Reducing the production cost per part frequently means minimizing the time that part is being processed on a machine while still maintaining acceptable part quality.

The factors that limit production rate can be grouped into three categories: workpiece quality, machine capabilities, and chip control. The elements influencing workpiece quality include:

- *Dimensional accuracy:* The accuracy and repeatability of the tool path and the distortion of the workpiece govern the dimensional accuracy.
- *Surface integrity:* Excessive heat or distortion can adversely affect workpiece fatigue performance.
- *Residual stress:* Surface and bulk stresses in the part can influence fatigue life and product performance.

- *Surface finish:* Requirements for roughness and other surface characteristics must be met.

The elements influencing machine capability include:

- *Horsepower:* The machine must provide adequate horsepower to maintain the desired material-removal rate.
- *Stiffness:* Deflection of the machine and workholding system should be small enough to allow repeatable processing.
- *Dynamic stability:* Many cutting processes introduce cyclic loading, which can lead to chatter and extremely adverse effects on tool life and surface finish. Resistance to vibration is a key factor in avoiding this effect.
- *Control response:* In numerical-controlled machining, the time required to speed up, slow down, and change directions can influence processing time, particularly in high-speed machining.
- *Cutting tool life/changeover time/frequency:* The time required for a tool change operation is nonproductive time on a machine. In general, faster material-removal rates lead to higher tool wear rates and more frequent tool changes. A balance must be struck between maximizing material-removal rates and minimizing tool change time.

The elements influencing chip control include:

- *Chip breakage and removal:* Chips that remain in or around the cutting zone can adversely influence surface finish and tool life. A critical consideration in cutting tool design is achieving a chip that is easily removed from the cutting zone.

Analytical and computer modeling offers the potential for an improved understanding of the process, which can assist in selecting optimum or near-optimum processing conditions. There are two primary motivations for modeling any process or product, regardless of the application:

- Reduce or eliminate the need for experiments or prototypes and thereby reduce the cost and/or time required to develop a new product or process

- Gain a better understanding of phenomena that cannot be easily measured or studied in experiments, and use that information to guide future design decisions

Fundamentals and General Considerations

Machining Fundamentals

The Chip Formation Process. Fundamentally, all chip forming processes rely on the same shearing action at the tool edge. The process can be idealized by a two-dimensional model, as illustrated in Fig. 1 and 2. This so-called orthogonal model assumes that a rigid tool engages a workpiece at a depth t_0. The metal shears along a plane oriented at an angle ϕ between the tool edge and the cutting plane. This shearing action forms a chip of thickness t in front of the tool. The orientation of the tool face relative to the cutting plane is known as α, the rake angle of the tool (Ref 2).

Simple geometric analysis gives the relationship between the uncut thickness, D, the chip thickness, t, the rake angle, α, and the shear plane angle, ϕ (as shown in Fig. 3), by the equation:

$$\tan \phi = \frac{r \cos \alpha}{\cos(\phi - \alpha)}$$

where $r = D/t$. There is a relatively minor amount of deformation at the cut surface of the workpiece caused by contact with the side or flank edge of the tool.

The direct forces on a cutting tool are a result of pressure exerted on the tool by the chip, pressure on the flank edges, and friction on the tool, as shown in Fig. 4. Friction acts between the chip and the tool and between the newly cut workpiece surface and the flank edge of the tool.

During the cutting process, heat is generated by plastic deformation in the chip and by friction between the tool, chip, and workpiece. The temperature of the chip and cutting tool increases, as does the newly cut surface of the workpiece.

General Considerations Modeling and Simulation of Machining Processes

In computer models and simulations, it is common practice to use discretization to describe the physical behavior of a process and product. For example, finite-element simulation is widely used because of its ability to describe arbitrary shapes and field-variable distributions. However, any sort of discretization is challenged by extreme ranges of length scale for the process being modeled. Machining processes present exactly this sort of length-scale challenge, where submillimeter-scale phenomena such as surface microstructure and residual stress can influence the performance of meter-scale products.

Current computer modeling and simulation technology does not permit the concomitant simulation of small-scale effects over the entirety of a large part. For this reason, metal cutting models are currently grouped into large-scale and small-scale models. Large-scale models capture overall part and machine tool behavior. Small-scale models focus on the interaction of the cutting tool and the workpiece in a localized region of the workpiece.

Perhaps the most straightforward large-scale models are used to verify numerical control (NC) toolpaths; check final part shape; check for interference between the cutting tool, spindle, fixture, and workpiece; and perform basic material-removal rate and volume calculation. The most basic models do not include analytical equations, only volume and geometric interaction. More sophisticated models rely on basic analytical equations to predict cutting force and power requirements. Many NC toolpath planning codes, such as those from Catia, Unigraphics, Pro Engineer, Delcam, Esprit, Gibbs, Mastercam, and others, include basic verification. Vericut, by CG Tech, is a stand-alone verification program that also includes optimization functions based on geometric chip-load calculations. The AdvantEdge Production Module from Third Wave Systems uses finite-element-based power calculations to optimize feeds and speeds.

Using finite-element methods, distortion due to relief of prior residual stresses and cutting-induced forces can be studied. As of 2010, state-of-the-art simulations consider material removal as a geometric Boolean difference operation, without consideration of the removal method. However, several ongoing projects seek to consider not only prior stresses but the forces and stresses introduced during the metal-removal process.

Dynamic stiffness, vibration, and chatter can adversely affect part quality and tool life. There are several approaches to modeling chatter that characterize the dynamic response of the machine/workpiece system and seek to find stable operating conditions.

Chatter is due primarily to forced vibrations where the forcing frequency is generally due to spindle rotation speed and periodic engagement and release of cutting edges. Koenigsberger and Tlusty (Ref 3) developed an analytic model of the forcing function as a function of the undeformed chip thickness. The simplest approach is to model the machine tool as a single degree of freedom system, but it can also be modeled as a multiple degree of freedom system. In all cases, the chip load is taken stepwise, considering the effect of vibration-induced scallops from the prior cut. These vibration models are used to identify cutting speeds and chip loads that will create dynamically stable and unstable cuts. Altintas (Ref 4) extended this method of analysis to develop an analytical prediction of three-dimensional chatter stability in milling.

Much of the implementation of these models involves measurement and characterization of the dynamic stiffness of the system. A detailed description of the models and their application is given in Ref 5.

The primary focus of this article is smaller-scale models, which do not seek to characterize the full system but rather only the workpiece/tool/chip interface and behaviors closely associated with that.

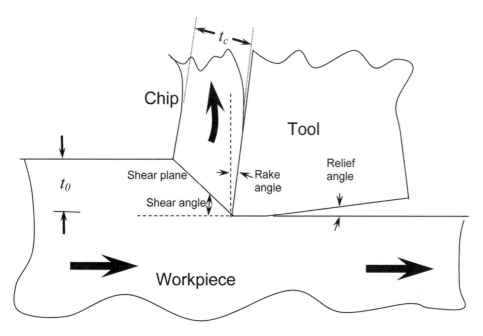

Fig. 1 Schematic illustration of two-dimensional approximation of a cutting process. Adapted from Ref 2

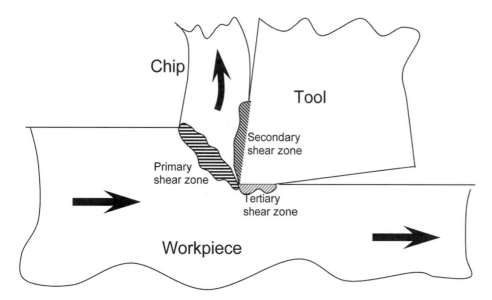

Fig. 2 Primary, secondary, and tertiary shear contact zones

Analytical Models

The earliest and simplest models apply empirical observations and basic relationships between force, velocity, and power to predict cutting forces. While simplistic, they are often adequate for sizing equipment or designing fixtures. Analytical chip-forming models apply plasticity theory to characterize the interaction between tool edge, workpiece, and chip. Ernst (Ref 6) and Merchant (Ref 7) were the first to apply this theory to predict cutting forces in 1938 and 1945, respectively. They assumed that shearing occurs along a single plane and that the shear stress on the plane is equal to the shear yield stress of the material. From this assumption, a force balance was derived, and cutting and thrust forces are predicted. Cutting power is the product of cutting speed and cutting force.

Merchant theorized that the shear plane angle that minimized the cutting power was the best-fit shear plane angle. Solution of the equation requires identifying a so-called friction angle. The friction angle, β, is defined from the ratio between the forces parallel (P) and perpendicular (N) to the rake face:

$$\beta = \arctan\left(\frac{P}{N}\right)$$

A geometric analysis relates the rake angle, α, friction angle, β, shear plane angle, ϕ, uncut chip thickness, a, and out-of-plane thickness, b, to the cutting force, F_c:

$$F_c = \frac{kab\,\cos(\beta - \alpha)}{\sin\phi\,\cos(\phi + \beta - \alpha)}$$

and the thrust force, F_t:

$$F_t = \frac{kab\,\sin(\beta - \alpha)}{\sin\phi\,\cos(\phi + \beta - \alpha)}$$

where k is the shear flow stress, assumed to be a constant.

Merchant assumed that ϕ would assume a value that minimized power consumption, given by $V_c F_c$, where V_c is the cutting speed.

Differentiating the cutting force equation with respect to ϕ yields:

$$\phi = \frac{\pi}{4} - \beta + \alpha$$

In many cases, there is poor agreement between Ernst and Merchant's models and experimental data, but their approach established a basic procedure for analysis. Lee and Shaffer (Ref 8) developed a slip line field assuming multiple shear planes, which also fell short in terms of accuracy but provided further insight into the mechanics of chip formation.

Semiatin and Rao (Ref 9) extended Merchant's theory and the temperature predictions of Loewen and Shaw to develop a criterion for transition from stable to unstable (serrated) chip formation.

Loewen and Shaw (Ref 10) calculated the temperature rise in an infinitesimally thin shear zone as:

$$T - T_{amb} = \left(0.95\int_0^{\Gamma}\tau d\Gamma/\rho C J\right)(1 - \beta)$$

where ρ is the density, C is the specific heat, J is the mechanical equivalent of heat, β is the fraction of heat lost to conduction and transport in the chip, and Γ is the length of the shear zone.

Semiatin and Rao referenced prior theoretical and experimental work that defined a stability criterion, α, where the softening rate sufficiently exceeds hardening:

$$\alpha \equiv \frac{\sqrt{3}}{\Gamma}\frac{d\dot{\Gamma}}{d\Gamma} = \frac{1}{\dot{\varepsilon}}\frac{d\dot{\varepsilon}}{d\varepsilon} > 5$$

When α exceeds 5, flow localization is imminent.

Along with the equation for temperature rise, they found that for materials with power-law strain hardening:

$$\frac{1}{\tau}\frac{dT}{d\Gamma} = \frac{0.95}{\rho C J}\left\{Z - \frac{0.664}{1 + n}\left(\frac{\Delta\Gamma}{vd}\right)^{1/2}Z^2\right\}$$

where v is the cutting speed, d is the uncut chip thickness, and Z is given by:

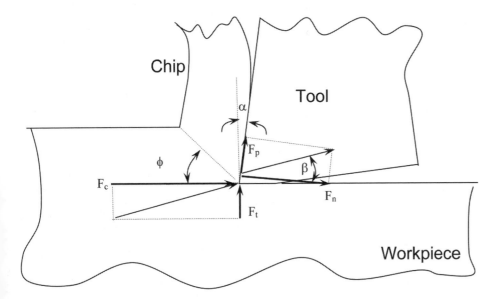

Fig. 3 Force and angle measurements used in Ernst and Merchant's cutting analysis

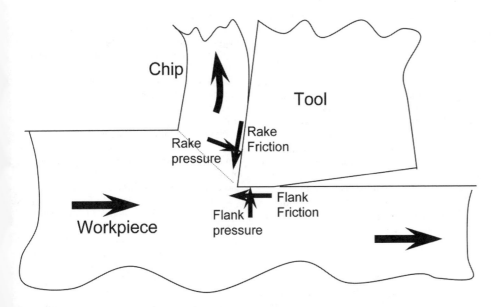

Fig. 4 Forces acting on the rake and flank surfaces of a cutting tool

$$Z = \left[1 + 1.328 \left(\frac{\Delta \Gamma}{vd} \right)^{1/2} \right]^{-1}$$

While the approach showed some promise, it still uses a shear plane rather than a shear zone.

Rather than shear planes, Oxley (Ref 11) proposed a method of analysis using shear zones. He assumed a parallel-sided primary shear zone and a secondary shear zone along the rake face of the cutting tool. Material properties vary through the zone as a function of strain, temperature, and cutting velocity. The average shear strain rate in the primary shear zone is:

$$\dot{\gamma} = C_0 \frac{V_s}{l}$$

where V_s is the shearing velocity along the primary shear plane, given by:

$$V_s = \frac{V_c \cos \alpha}{\cos(\phi - \alpha)}$$

and l is the length of the plane. C_0 is an empirically determined constant, which is approximately 5.9 for mild steels.

The average strain in the primary deformation zone is:

$$\gamma = \frac{\cos \alpha}{2 \sin \phi \cos(\phi - \alpha)}$$

The shear flow stress in the primary shear zone is given by k, which is a function of the strain, strain rate, and temperature.

So, the shear force per unit thickness is:

$$\frac{F_s}{b} = \frac{k a}{\sin \phi}$$

where a is the uncut chip thickness, and b is the out-of-plane thickness.

The cutting and thrust forces, F_c and F_t, are related to the shear force by:

$$F_s = F_c \cos \phi - F_t \sin \phi$$

This method has generally been shown to yield accurate results.

Other researchers have extended Oxley's shear zone model, using improved estimates of the state variable variation through the zone. Tounsi and his coworkers (Ref 12) assumed an arbitrary one-dimensional velocity variation across the primary shear zone. From this variation, an incompressibility assumption, boundary conditions of the shear zone, and a distribution of strain rate, pressure, and stress can be derived.

These analytical models have the advantage that they can calculate values very quickly. However, because they rely on assumed velocity fields and tool-edge representation, there are inherent limits to how accurately they can capture actual chip behavior.

Finite-Element Modeling and Simulation

Finite-element models that incorporate large deformation plasticity theory represent a more detailed alternative to the analytical models. They can accurately predict three-dimensional chip geometry, cutting forces, and temperature effects in the tool, workpiece, and chip. Models can use solid models of actual tool geometry rather than simplified shapes. The drawback to finite-element models is calculation time, which can extend into hours or days, even on very fast computers.

Where analytic models rely on assumptions about material flow in a specific process, finite-element models are typically developed for more general conditions. Fundamental physics is incorporated into the model via a simple element shape rather than a complex shear zone behavior. The process geometry is discretized. Deformation calculations in a single element are relatively simple. Using a computer to assemble thousands of simple elements allows a very detailed description of flow behavior. Significantly fewer a priori assumptions about the process are required.

In general, finite-element models fall into two categories:

- Steady-state models represent a snapshot in time. An assumed chip shape is used. The shape may be modified as part of the solution calculation.
- Transient models track the evolution of chip geometry with time. Tool entry and exit behavior as well as evolution of chip shape and contact are modeled.

Each modeling approach has its advantages and disadvantages. Transient models capture more details about the process, including unsteady chip behavior such as chip serration. However, these models require special treatment for chip separation, either through a predefined separation surface or through frequent remeshing to manage severe deformation at the tool edge. Calculation times can also be significant, and reaching steady-state tool temperatures is not practical without special treatment.

Steady-state models inherently capture steady temperature and chip behavior. Remeshing is generally not required, because any distortion of elements is small. No special treatment is required for separation of the chip, because the tool position does not change.

For all models, calculation time tends to increase with increasing resolution in models. Element size establishes discretization detail. Smaller elements provide better resolution of detail but require proportionally more computation effort.

Various Modeling Approaches

Transient Models. In a transient, or Lagrangian, finite-element model, the workpiece geometry and associated field variables evolve with time. In general, the finite-element mesh will track specific material points, such that the mesh shape evolves with the deforming material. Because of the severe deformation associated with chip formation, some sort of remeshing, rezoning, or mesh smoothing is generally required in close association with the deformation calculations. Chip separation at the tool edge can be modeled by a predefined line, by distortion and remeshing, or by a fracture criterion.

The approach is well suited to modeling unsteady behavior such as varying chip morphology, exit burr formation, or inherently unsteady processes such as milling. However, it is difficult to predict steady-state behavior using a transient model. In particular, several seconds of cutting time and several meters of cut are required for a tool to reach a stable temperature. Because runtimes for most Lagrangian analyses extend into hours or even days for a several centimeter cut lasting milliseconds, running a transient analysis to reach steady temperature is impractical.

Continuous Cutting Model. Klocke and coworkers (Ref 13) modified a commercial finite-element package to create a simulation method that adds material in advance of the cutting edge and deletes material from the workpiece beyond the cutting edge.

The simulation method features the advantage of Lagrangian models in that they can model unstable chip-forming behavior. With a smaller workpiece, the total volume of material and thus the degrees of freedom required for solution are smaller.

Steady-State Model. In a steady-state, or Eulerian, model, the workpiece geometry boundary is fixed in space. Material convects through the mesh, but the shape of the mesh does not change. In this approach, cutting force, temperature, and state variables such as strain can be calculated along stream lines in the workpiece.

Because there is no evolution in the shape of the part, a single-step analysis can be performed significantly more quickly than a Lagrangian model. However, it is not possible to capture unsteady behavior such as chip serration, chip breaking, or burr formation. Furthermore, it is necessary to assume the shape of the chip a priori.

Hybrid Model. Chigurupati (Ref 14) reported on the development of a hybrid model. An initial transient simulation is performed to obtain a preliminary chip shape. Then, using the chip shape as a starting condition, an arbitrary Lagrangian Eulerian (ALE) simulation is performed. In the ALE model, the material flow source and sink are assumed on the workpiece, and a second sink is assumed at the free end of the chip. For a fixed point in time, the workpiece geometry is updated until flow lines are tangent to the surface. Temperature is updated until steady-state conditions are reached. When both of these conditions are met, a steady-state solution is reached.

When steady-state tool temperature is established, the transient analysis is completed to analyze exit behavior.

Two-Dimensional versus Three-Dimensional Analysis. Two-dimensional (2-D) plane-strain or plane-stress finite-element models are appropriate for orthogonal cutting simulation. Compared to three-dimensional models, they have significantly shorter calculation time for a given level of resolution. The 2-D models are appropriate for applications requiring a great deal of resolution, such as studies of surface residual stress or microstructure, or for fundamental studies of physical behavior, development, and verification of material data, and so on.

Three-dimensional (3-D) models are appropriate for studying geometric details of actual cutting tools, which, in most cases, cannot be adequately represented by a 2-D model. However, 3-D simulations with small elements require extremely long calculation times, limiting the practicality for applications where a very fine level of detail is required.

Pioneering Developments in Finite-Element Analysis. Today (2010), research in finite-element analysis is focused primarily on developing applications, improving input data, and refining numerical techniques. The CIRP (Ref 15) has sponsored a series of workshops on the modeling of machining processes, starting in Atlanta in 1996 and meeting annually since then (Ref 16). Together, the proceedings of these conferences provide an excellent overview of the evolution of the state-of-the-art of machining process modeling. The current growing industrial application of finite-element simulation of machining owes its existence to fundamental research and development beginning in the late 1980s and continuing through the 1990s.

In 1985, Strenkowski and Carroll described a finite-element model of orthogonal metal cutting (Ref 17) that used a Lagrangian approach to finite-element analysis. The model used a predefined separation line, with element separation determined by a strain criterion. The accuracy of this method suffered from a strong dependence on the selection of the separation criterion.

Through the mid-1990s, several research groups introduced Lagrangian simulations using automated remeshing techniques rather than a predefined separation plane. In 1992, Sekhon and Chenot (Ref 18) simulated orthogonal cutting with automated remeshing using the code Forge2, which had been developed at the French research institute CEMEF. In 1993, Camacho, Marusich, and Ortiz introduced a continuous remeshing technique (Ref 19), which Marusich used to develop a finite-element system for high-speed machining (Ref 20). In 1996, Cerretti, Fallboehmer, Wu, and Altan adopted the remeshing procedure in the commercial software DEFORM-2D to model orthogonal cutting (Ref 21).

In 1997, Strenkowski and Athavale refined Eulerian simulation to the point where it was suitable for simulating tools with chip control (Ref 22).

Because of computational complexity, early research was entirely conducted using 2-D plane-strain models. In 2000, Cerretti (Ref 23) modeled the turning process in 3-D, which was perhaps the first 3-D simulation result using a commercial code with automated remeshing.

To varying degrees, all of these researchers demonstrated the capability of finite-element simulation to give reasonable predictions of cutting forces, chip shape and morphology, and cutting zone temperature. Subsequent work has focused primarily on improving and better understanding input data and numerical techniques.

Input Data for Modeling and Simulation

In 1998, Childs addressed material property needs for simulation of metal machining (Ref 24). The primary and secondary shear zones exhibit extreme deformation rates, temperature rises, and pressures against cutting tools. Traditional material models and testing methods frequently do not accurately capture the complete physics of the process. In particular, there are no clear standards for determination of material flow stress behavior or friction between the workpiece and tool. There is, however, substantial research into the topic.

Flow Stress Measurements

Flow stress is a measure of the stress required for an increment of plastic deformation of a material at a given strain, strain rate, and temperature. It generally is also a function of prior processing, including grain size, phase, and starting hardness.

Traditional testing methods involve measurement of the force required for deformation of a material sample under tension, compression, or shearing at a temperature and deformation rate similar to the process being simulated. However, strain rates in machining can exceed 10^5/s or even 10^6/s in the primary shear zone. Standard compression tests are suitable only for strain rates under 10^2/s. Even high-speed compression tests can only produce usable results at strain rates up to 450/s (Ref 25).

The Kolsky or split Hopkinson bar test measures dynamic stress-strain response via a series of bars that transmit a pressure pulse through the test sample. Strain gages are used to measure incident and reflected pulse speed and intensity. However, even this impact approach is limited to strain rates of little more than 10^3/s.

This limitation has led several researchers to use cutting tests on a lathe or milling machine to measure deformation force, then use inverse analysis techniques to fit coefficients to an equation that will result in analytical prediction of similar forces under similar cutting conditions.

Constitutive Models. The typical approach is to adjust the coefficients of a constitutive equation until predicted forces match the experimental results.

Johnson and Cook (Ref 26) suggested a simple description of flow stress as a function of strain, strain rate, and temperature:

$$\bar{\sigma} = (A + B\varepsilon^n)(1 + C\ln(\dot{\bar{\varepsilon}}/\dot{\bar{\varepsilon}}_0))(1 - [(T - T_{amb})/(T_{melt} - T_{amb})]^m)$$

In the Johnson-Cook equation and its variants, strain hardening, strain-rate sensitivity, and thermal softening are treated in separate, independent terms. A represents the basic strength, and B and n the strain-hardening behavior. C represents rate sensitivity, and m the thermal softening behavior. Various authors have proposed variations of this basic equation to account for phenomena such as dynamic strain hardening (i.e., blue brittleness) in steels.

Zerilli and Armstrong (Ref 27) suggested an alternate formulation, based on dislocation dynamics:

$$\bar{\sigma} = C_1 + C_2\varepsilon^{0.5}\exp[(-C_3 + C_4\ln\dot{\bar{\varepsilon}})T]$$

for face-centered cubic materials

or

$$\bar{\sigma} = C_1 + C_5\varepsilon^n + C_2\exp[(-C_3 + C_4\ln\dot{\bar{\varepsilon}})T]$$

for body-centered cubic metals

Marusich and Ortiz observed that there is a transition in fundamental behavior from low-strain-rate to high-strain-rate behavior. This led to the development of a two-stepped constitutive model (Ref 20):

$$\left(1 + \frac{\dot{\varepsilon}^p}{\dot{\varepsilon}_0^p}\right) = \left(\frac{\bar{\sigma}}{g(\varepsilon^p)}\right)^{m_1}, \text{ if } \dot{\varepsilon}^p \le \dot{\varepsilon}_t$$
$$\left(1 + \frac{\dot{\varepsilon}^p}{\dot{\varepsilon}_0^p}\right)\left(1 + \frac{\dot{\varepsilon}_t}{\dot{\varepsilon}_0^p}\right)^{m_2/m_1 - 1} = \left(\frac{\bar{\sigma}}{g(\varepsilon^p)}\right)^{m_2}, \text{ if } \dot{\varepsilon}^p > \dot{\varepsilon}_t$$

where the g term accounts for strain hardening and thermal softening:

$$g = [1 - \alpha(T - T_0)]\sigma_0\left(1 + \frac{\varepsilon^p}{\varepsilon_0^p}\right)^{\frac{1}{n}}$$

Umbrello and his coworkers adopted the fundamentals of the Johnson-Cook model, with independent terms for strain hardening, thermal softening, and rate sensitivity (Ref 28):

$$\sigma(\varepsilon, \dot{\varepsilon}, T, HRC) = B(T)(C\varepsilon^n + F + G\varepsilon)[1 + (\ln(\dot{\varepsilon})^m - A)]$$

where HRC is hardness on the Rockwell C scale, $B(T)$ is a thermal softening term, F and

G are functions of the hardness of the material, and *A* and *m* are related to rate sensitivity. The authors used Brozzo's (Ref 29) damage criteria to predict fracture, which results in chip segmentation.

Inverse Testing Methods for Flow Stress/ Constitutive Behavior. Researchers at The Ohio State University Engineering Research Center for Net Shape Manufacturing (ERC/ NSM) (Ref 25, 30) applied Oxley's slip line field description of the primary shear zone as a basis for inverse analysis. The researchers modified Oxley's method such that flow stress could be defined as any arbitrary function of strain, strain rate, and temperature. They then selected the Johnson-Cook flow stress model, or a variant thereof, to define the flow stress.

In a cutting test, researchers measure cutting force, thrust force, and, using a quick-stop test, shear angle and the thickness of the secondary shear zone. An optimization scheme is then used to vary the coefficients of the Johnson-Cook equation and the tool-chip friction to find the best-fit values.

Experimental testing achieves strain rates of 10^5 to 10^6/s. Testing has been performed using both a lathe and an end mill (Ref 25). The end mill has the advantage of continuously varying chip load. Multiple data points can be taken from a single experiment, and thereby, the total number of experiments can be reduced.

Interface Friction

In metal cutting, the primary friction force is tangent to the rake face. There is a small secondary friction force acting on the flank wear land. The vast majority of research has been devoted to understanding rake-face friction behavior, because it makes a significant contribution to the total thrust force on the tool.

Most authors suggest that friction behavior is nonlinear (Ref 31). In the tool-chip interface, there is generally to be a low-pressure region of sliding contact and a high-pressure region of sticking contact, where the friction stress rises to the shear flow stress of the chip material.

Zorev (Ref 32) proposed a simple stepwise friction model:

$$\tau = \mu P \quad \text{where} \quad \tau < k$$

and

$$\tau = k \quad \text{where} \quad \mu P > k$$

where μ is the coulomb friction coefficient, τ is the friction stress, k is the shear flow stress, and P is the normal stress on the rake face.

Shirakashi and Usui (Ref 33) proposed an equation that relates friction stress to the normal stress at the tool-chip interface and the shear flow stress of the chip:

$$\tau = k\left[1 - \exp\left(-c\frac{P}{k}\right)\right]$$

where c is a constant that is dependent on the workpiece and tool materials.

A large number of other authors have assumed a wide range of linear shear or coulomb friction coefficients based as much for convenience in the commercial simulation software used as for any scientific principle.

A major issue with the determination of a friction coefficient is that many researchers assume that most or all normal forces arise from tool-chip friction. A significant problem with a large number of finite-element simulation approaches is the underprediction of thrust forces for material data and friction conditions that give reasonable results. While many authors have commented on this effect, none seems to have uncovered a single cause.

Unpublished studies suggest that there may be a significant contribution of normal cutting force from flank-wear effects. The tool flank-wear land is frequently disregarded in finite-element simulations, or mesh discretization is too coarse to consistently capture forces on a micrometer-scale wear land. However, unpublished studies on heavily worn tools suggest that proper simulation of this land with adequate mesh discretization at the flank gives improved thrust force prediction.

Tool Design

Application Examples

Tool Design for Chip Removal. As previously noted, chips that are not effectively removed from the cutting zone can adversely affect workpiece surface finish and cutting tool life. Cutting tool geometry plays a significant role in chip control. Development of new cutting tool geometry generally involves testing a number of prototype tools before suitable performance is realized. For pressed and sintered cobalt carbide inserts, the cost of a single cutting tool may be $5,000 to $10,000, and lead times for production may be several weeks.

Fischer (Ref 34) and Kammermeier (Ref 35) reported on the redesign of a polycrystalline diamond insert for the Kennametal Fix Perfect face mill (Fig. 5, 6). The mill is used for finishing flat aluminum surfaces. Feedback from field application indicated that chips were not clearing the cutting zone but rather were becoming trapped under the mill body and marring the finished surface. The original insert design featured a 0° axial rake angle and a 22° effective radial rake angle.

Kennametal proposed two possible redesigns of this insert. Both involved adding an axial rake to pull the chip away from the workpiece and improve evacuation. The standard insert as well as the two redesigns were simulated using DEFORM-3D. To validate the simulation results, prototype inserts were manufactured and tested at Kennametal facilities.

Two prototype geometries were developed. Design 1 featured a 20° radial rake and 8° axial rake. Design 2 featured a 9° radial and 7° axial rake. Figures 7 and 8 show the proposed redesigns of the insert, with a 20° radial rake angle and an 8° axial rake angle. This combination of radial plus axial rake lifts the chip away from the workpiece face and tends to coil it. The lifting can be observed clearly in Fig. 8 (a), and the tight coiling of the chip can be seen in Fig. 8(b).

Simulations of the second redesign also showed suitable performance. Kennametal engineers selected that design to provide a balance of performance with a stronger cutting edge. The prototype testing results compared

Fig. 5 Kennametal Fix Perfect milling cutter with inserts. Source: Ref 34

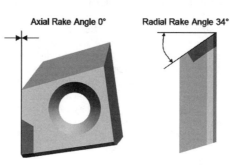

Fig. 6 Geometry of baseline standard product. Source: Ref 34

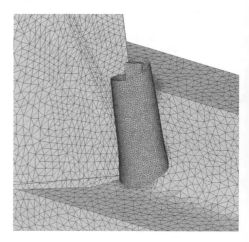

Fig. 7 Chip form for first proposed redesign: 20° radial rake angle and 8° axial rake angle. Source: Ref 34

extremely well with simulation results, as shown in Fig. 9.

Effect of Tool Geometry on Burr Formation. Simulation is useful for studying chip geometry but can also be used to study burr formation. Researchers at The Ohio State University, working with a major automotive original equipment manufacturer, used finite-element analysis to study the effect of tool wear on burr formation in face milling high-silicon aluminums with polycrystalline diamond (PCD) tooling (Ref 36). Because of the extremely long life of PCD, it is impractical to conduct experiments to wear out a tool. This makes simulation an interesting alternative.

Figure 10 shows a finite-element model of a workpiece at the end of a face mill cut. Simulations were run using zero rake and high-positive (20°) rake tools, with both sharp and worn geometry. Burrs for sharp and worn zero rake tools are shown in Fig. 11.

Figure 12 shows a comparison in burr size between the four tool/wear combinations. As can be seen from the graph, the high-positive rake tool produces a smaller burr when sharp,

and the size of the burr is less sensitive to wear than it is for the zero rake tool. This gives guidance to the manufacturer in selecting a tool.

Burrs on actual parts tend to be larger than those simulated, because the physical burrs are accumulations of multiple passes. However, the trends show good agreement between simulation and physical samples.

Tool Temperature Calculation. Cutting tool geometry is an important factor in initial performance. However, tool temperature is frequently a limiting factor in material-removal rates because increasing temperature tends to shorten tool life.

To validate finite-element temperature predictions, experimental results on insert temperatures were taken from the literature (Ref 37). Here, the workpiece material was 1045 steel, and the insert was uncoated tungsten carbide. Cutting speed was 222 m/min, and depth of cut was 1.5 mm. Simulations were performed at feed rates of 23, 31, 40, and 48 μm, respectively. Experimental temperature measurements on the rake face were made using infrared microscopy. Accuracy was reported to

be ±40 °C at 700 °C. As can be seen from Fig. 13, good comparisons were seen between the model predictions and measured temperature values on the rake surface from the experiments.

Effect of Coatings on Tool Temperatures. One of the many roles of coatings on cutting tools is to provide a thermal barrier to protect the substrate. Yen and his coworkers (Ref 38) have demonstrated two methods by which coating layers can be simulated and evaluated using finite-element analysis.

The first method is to model individual layers, with discrete material properties assigned to discrete thin layers of "coating" mesh elements (Fig. 14). The second method is to assign composite properties based on the thermodynamic principle of one-dimensional heat transfer through a composite wall, as expressed by (Ref 39):

$$\frac{\Delta x}{K_{eq}} = \frac{\Delta x_1}{K_1} + \frac{\Delta x_2}{K_2} + \frac{\Delta x_3}{K_3}$$

where Δx_i is the thickness value of individual coating layers, K_i is the thermal conductivity of individual coating layers, Δx is the total thickness of the composite layer ($= \sum \Delta x_i$), and K_{eq} is the equivalent thermal conductivity of the composite layer.

Both approaches showed good correlation between experimentally measured and simulated temperatures (Fig. 15).

Tool Wear

Application Examples

Tool wear is an unavoidable effect of metal cutting. It is generally influenced by temperature, pressure, sliding, properties of the workpiece material, tool materials, and coatings.

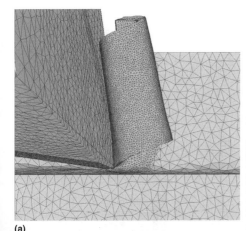
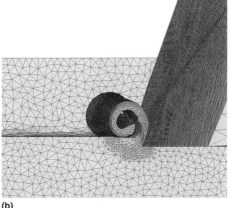

(a)   **(b)**

Fig. 8 First proposed redesign: (a) radial view and (b) axial view. The chip is lifted away from the workpiece, which promotes better evacuation. Source: Ref 34

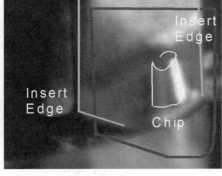

Fig. 9 Comparison of computer-aided design solid model of cutter (image from high-speed video) and finite-element modeling simulation result, all from the same orientation. Insert with 7° axial rake and 9° effective radial rake. Source: Ref 34

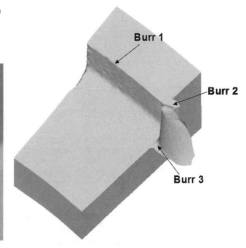

Fig. 10 Workpiece after face milling. Burr formation was observed at the cutting edge and the side and face of the exit. Source: Ref 36

Fig. 11 Strain-rate distribution at exit burr for (a) sharp tool and (b) worn tool (v = 0.4 mm) of the Clapp-Dico cutter. Source: Ref 36

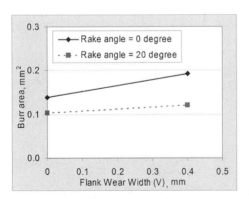

Fig. 12 Comparison of burr area in relation to flank wear width (v) between Clapp-Dico tool (axial rake angle = 0°) and Kennametal high-shear tool (axial rake angle = 20°). The points on the left indicate sharp tools, and the points on the right indicate worn tools. Source: Ref 36

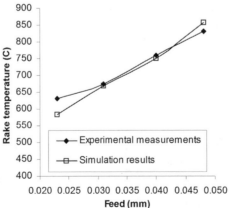

Fig. 13 Comparison of experimentally measured and simulated steady-state temperatures for four different feed rates. Source: Ref 14

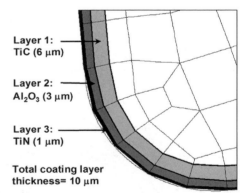

Fig. 14 Mesh model for a TiC/Al$_2$O$_3$/TiN-coated tool as used in experiments, based on the individual layer model. Source: Ref 38

Yen, Soehner, and their associates (Ref 40) have implemented a tool wear model based on work by Usui (Ref 41). The model predicts wear rate as a function of stress and temperature in the insert and sliding velocity of the chip at the insert interface. The wear rate is given by:

$$dW/dt = A\sigma_t V_s \exp\left(-\frac{B}{T}\right)$$

where σ_t is the effective stress in the tool, V_s is the relative sliding velocity, and T is the absolute temperature. The constants A and B are evaluated by curve fitting experimental measurements on flank wear.

The wear rates are used to update node positions on the tool geometry to simulate the actual wear process (Fig. 16). Wear is a nonlinear process characterized by primary, steady-state, and tertiary zones. An insert/workpiece material combination will exhibit a family of curves based on cutting speed. For a sample insert, the simulated result yielded a nonlinear curve that fit well with the family of curves for that insert (Fig. 16).

Workpiece Surface Integrity. Depending on the end application of the product being machined, maintaining good workpiece surface integrity can be a critical consideration in process definition. Worn tools have a significant impact on residual stress and burr formation. In some cases, workpiece surface integrity may be the limiting factor in the utilization of worn tools.

White layer is a thin layer of altered microstructure that can develop under abusive machining conditions. It can have an extreme adverse effect on fatigue life and must be avoided in fatigue-loaded parts.

Fischer et al. (Ref 42) have used simulation software to predict white-layer formation in 52100 steel, quenched and tempered to a hardness of 53.5. With a worn tool, simulation results compare extremely well to experimentally measured white layer at 100 and 300 m/min. Simulation results show that, with a sharp tool, white layer is avoided, even at the faster 300 m/min cutting speed.

Figure 17 shows simulation results with microstructure modeling enabled. A detail of Fig. 17 is shown in Fig. 18, with a high-

martensite layer at the surface of the workpiece. Figure 19 shows a comparison between experimental and simulated martensite depth for two different cutting speeds.

Conclusions

Computer simulation and modeling is becoming a valuable tool for the evaluation and improvement of cutting processes. The capability of simulation to predict chip shape, cutting zone temperature, tool wear, and cutting forces has been well established. While there has been some significant success in modeling larger-scale workpiece effects, there is still a need for more fundamental development in this area.

In many cases, the major impediment to wider industrial application of simulation technology is simulation time. In many situations, the time for a simulation may be substantially longer than the time required for production trials. Because of this, simulation has proven its value primarily in applications where there is a substantial cost for prototype process or product trials. These include the development of new cutting tools, the development of cutting parameters for expensive materials or costly parts, and the development of processes that will see their first testing on production equipment.

As computer speed and software technology continue to improve, the bar for return on investments in simulation technology will continue to fall.

REFERENCES

1. D.A. Stephenson and J.S. Agapiou, *Metal Cutting Theory and Practice,* 2nd ed., Taylor & Francis, 2006
2. S. Kalpaakjian, Manufacturing Processes for Engineering Materials, 2nd ed., Addison-Wesley, 1996
3. F. Koenigsberger and J. Tlusty, Machine Tool Structures, Vol 1: Stability Against Chatter, Pergamon Press, 1967
4. Y. Altintas, Analytical Prediction of Three Dimensional Chatter Stability in Milling, *JSME Int. J. I, Ser. C,* Vol 44 (No. 3), 2001

Fig. 15 Comparison of predicted average tool-chip interface temperatures with the experimental data. AISI 1045, coated tool, cutting speed (V_c) = 220 m/min, feed = 0.16 mm/rev. FEM, finite-element model. Source: Ref 38

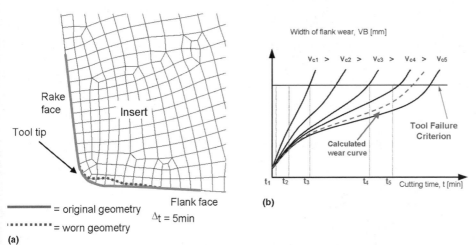

Fig. 16 (a) Original and updated mesh geometry in cutting tool. (b) Tool wear curve family, with simulated tool curve superimposed. Courtesy of The Ohio State University ERC/NSM. Source: Ref 40

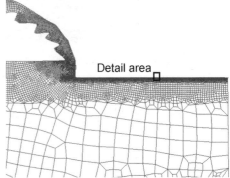

Fig. 17 Overview of the simulation result in quenched and tempered 52100 showing an area of defined mesh refinement and the detail area, which is enlarged in Fig. 18. Source: Ref 42

300 m/min

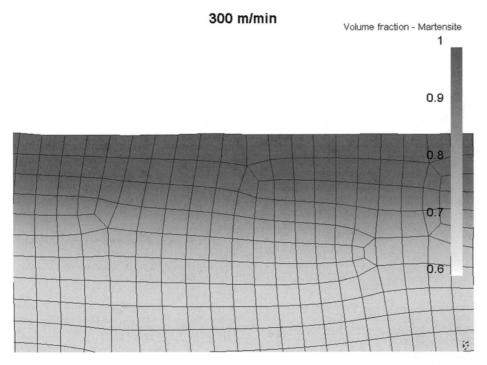

Volume fraction - Martensite

Fig. 18 Detail of surface martensite (Fig. 17) for 300 m/min simulation in 52100. Darker gray indicates a higher volume fraction of martensite. Source: Ref 42

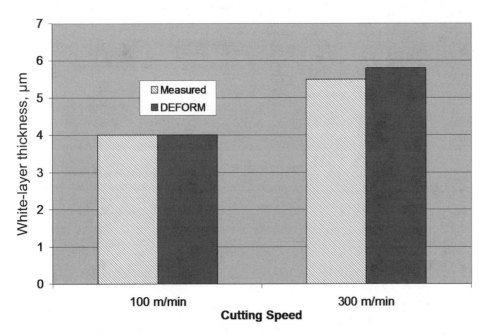

Fig. 19 Comparison of measured and simulated white-layer thickness in 52100 for 100 and 300 m/min cutting speeds. Source: Ref 42

5. Y. Altintas, Manufacturing Automation: Metal Cutting Mechanics, Machine Tool Vibrations, and CNC Design, Cambridge University Press, 2000
6. H. Ernst, Physics of Metal Cutting, *Machining of Metals,* American Society for Metals, Metals Park, OH, 1938, p 24

7. M.E. Merchant, Mechanics of the Metal Cutting Process I. Orthogonal Cutting of a Type 2 Chip, *J. Appl. Phys.,* Vol 16, 1945, p 267–275
8. E.H. Lee and B.W. Shaffer, The Theory of Plasticity Applied to a Problem of Machining, *ASME J. Appl. Mech.,* Vol 18, 1951, p 405–412

9. S.L. Semiatin and S.B. Rao, Shear Localization during Metal Cutting, *Mater. Sci. Eng.,* Vol 61, 1983, p 185–192
10. E.G. Loewen and M.C. Shaw, On the Analysis of Cutting Tool Temperature, *Trans. ASME,* Vol 76, 1954, p 217
11. P.L.B. Oxley, *Mechanics of Machining, An Analytical Approach to Assessing Machinability,* Halsted Press, A Division of John Wiley & Sons Limited, New York, 1989
12. N. Tounsi, J. Vincenti, A. Otho, and M.A. Elbestawi, From the Basic Mechanics of Orthogonal Metal Cutting Toward the Identification of the Constitutive Equation, *Int. J. Mach. Tools Manuf.,* Vol 42 (No. 12), Sept 2002, p 1373–1383
13. F. Klocke, H.W. Raedt, and S. Hoppe, 2D FEM Simulation of the Orthogonal High Speed Cutting Process, *Mach. Sci. Technol.,* Vol 5 (No. 3), Nov 2001, p 323–340
14. P. Chigurupati, J.T. Jinn, J.Y. Oh, Y. Yin, H. Zhang, and W.T. Wu, Advances in Machining Process Modeling, *Numiform 2004: International Conference on Numerical Methods in Industrial Forming Processes* (Columbus, OH), 2004
15. *Proceedings of the CIRP International Workshop on Modeling of Machining Operations,* I.S. Jawahir, A.K. Balaji, R. Stevenson, Ed., May, 1998, Atlanta, GA, USA
16. *Proceedings of the 10th CIRP International Workshop on Modeling of Machining Operations,* F. Micari and L. Filice, Ed., August, 2007, Reggio Calabria, Italy
17. J.S. Strenkowski and J.T. Carroll, A Finite Element Model of Orthogonal Metal Cutting, *ASME J. Eng. Ind.,* Vol 107, 1985, p 346–354
18. G.S. Sekhon and J.L. Chenot, Some Simulation Experiments in Orthogonal Cutting, *Numerical Methods in Industrial Forming Processes,* 1992, p 901–906
19. G. Camacho, T. Marusich, and M. Ortiz, Adaptive Meshing Methods for the Analysis of Unconstrained Plastic Flow, *Advanced Computational Methods for Material Modeling,* D.J. Benson and R.J. Asaro, Ed., ASME AMD, Vol 180/PVP Vol 268, 1993, p 71–83
20. T.D. Marusich and M. Ortiz, Modeling and Simulation of High Speed Machining, *Int. J. Num. Meth. Eng.,* Vol 38, 1995
21. E. Ceretti, P. Fallbohmer, W.T. Wu, and T. Altan, Application of 2D FEM to Chip Formation in Orthogonal Cutting, *Int. J. Mater. Proc. Technol.,* Vol 59, 1996
22. J.S. Strenkowski and S.M. Athavale, A Partially Constrained Eulerian Orthogonal Cutting Model for Chip Control Tools, *J. Manuf. Sci. Eng.,* Vol 119, Nov 1997, p 681
23. E. Ceretti, C. Lazzaroni, L. Menegardo, and T. Altan, Turning Simulations Using a Three Dimensional FEM Code, *J. Mater. Proc. Technol.,* Vol 8, 2000, p 99–103

24. T.H.C. Childs, Material Property Needs in Modeling Metal Machining, *Mach. Sci. Technol.,* Vol 2 (No. 2), 1998, p 303–316

25. M.N. Shatla, "Prediction of Forces, Stresses, Temperatures, and Tool Wear in Metal Cutting," Ph.D. dissertation, The Ohio State University, 1999

26. G.R. Johnson and W.H. Cook, A Constitutive Model and Data for Metals Subjected to Large Strains, High Strain Rates, and High Temperatures, *Proceedings of the Seventh International Symposium on Ballistics* (The Hague, The Netherlands), April 1983

27. F.J. Zerilli and R.W. Armstrong, Dislocation-Mechanics-Based Constitutive Relations for Material Dynamics Calculations, *J. App. Phys.,* Vol 61, 1987, p 1816

28. D. Umbrello, J. Hua, and R. Shivpuri, Hardness Based Flow Stress and Fracture Model for Numerical Simulation of Hard Machining AISI 52100 Bearing Steel, *Mater. Sci. Eng. A,* 2004

29. P. Brozzo, B. De Luca, and R. Rendina, A New Method for the Prediction of Formability Limit of Metal Sheet, *Proceedings of the Seventh Biennial Conference of the International Deep Drawing Research Group,* 1972

30. T. Ozel and T. Altan, Determination of Workpiece Flow Stress and Friction at the Chip Tool Contact for High Speed Cutting, *Int. J. Mach. Tools Manuf.,* Vol 40 (No. 1), Jan 2000, p 133

31. E. Ceretti, L. Filice, and F. Micari, Basic Aspects and Modelling of Friction in Cutting, *Friction and Flow Stress in Forming and Cutting,* P. Boisse, T. Altan, and K. Van Luttervelt, Ed., 2001

32. N.N. Zorev, Inter-Relationship between Shear Processes Occurring Along Tool Face and Shear Plane in Metal Cutting, *Int. Res. Prod. Eng.,* ASME, New York, 1963, p 42–49

33. T. Shirakashi and E. Usui, Friction Characteristics on Tool Face in Metal Machining, *J. JSPE,* Vol 39 (No. 9), 1973, p 966–971

34. C.E. Fischer and N.K.R. Mylavaram, Design and Simulation of Cutting Tools, *CIRP International Conference on High Performance Cutting,* Oct 19–20, 2004 (Aachen, Germany)

35. D. Kammermeier and N.K.R. Mylavaram, Simulation of the Micro Tool Geometry Opens New Horizons in Chip Forming, *Eighth CIRP International Workshop on Modeling of Machining Operations* (Chemniz, Germany), 2005

36. T. Altan and P. Sartkulvanich, OSU/ERC, private communication

37. M.A. Davies, H. Yoon, T.L. Schmitz, T.J. Burns, and M.D. Kennedy, High Resolution Measurement of the Temperature Distribution at the Tool Chip Interface in AISI 1045 Steel and Comparison to Predictions, *Proceedings of the Fourth CIRP International Workshop on Modeling of Machining Operations,* C.A. van Luttervelt, Ed., 2001, p 19–25

38. Y.C. Yen, A. Jain, P. Chigurupati, W.T. Wu, and T. Altan, Computer Simulation of Orthogonal Cutting Using a Tool with Multiple Coatings, *Proceedings of the Sixth CIRP International Workshop on Modeling of Machining Operations,* 2003

39. J.P. Holman, *Heat Transfer,* McGraw-Hill Book Co., New York, 1986

40. Y.C. Yen, P. Sartkulvanich, I. Al-Zkeri, J. Soehner, and T. Altan, Prediction of Cutting Forces, Tool Stresses, Temperatures, and Tool Wear in High Speed Machining—A Progress Report, *Proceedings of the 2002 NSF Design, Service, and Manufacturing Grantees and Research Conference,* Jan 2002 (San Juan, Puerto Rico)

41. E. Usui, A. Hirota, and M. Masuko, *Trans. ASME,* Vol 100, 1978, p 222–228

42. C.E. Fischer and A.R. Bandar, Finite Element Simulation of Surface Microstructure Effects in Metal Cutting, *Third CIRP Workshop on High Performance Cutting,* June 2008 (Dublin, Ireland)

ASM Handbook, Volume 22B, *Metals Process Simulation*
D.U. Furrer and S.L. Semiatin, editors

Copyright © 2010, ASM International®
All rights reserved.
www.asminternational.org

Modeling Sheet Shearing Processes for Process Design

Somnath Ghosh, The Ohio State University
Ming Li, Alcoa Technical Center

SHEARING PROCESSES are among those most frequently used in sheet metal manufacturing and forming operations. These processes are widely used in the primary metal, automotive, aerospace, rigid packaging, electronics, household appliance, building and construction industries, to name just a few. Sheet shearing processes may be categorized into two broad groups, depending on the tool geometry and applications. The first group comprises two top and bottom blades (cutting dies), which can be either straight or curved, as shown in the schematic of Fig. 1(a). The actions of the two blades are discontinuous and in the form of strokes. The blade cutting edge is in the plane of the sheet or parallel to it. Processes belonging to this category include blanking, trimming, piercing, guillotining, and cropping. The second category includes shear slitting and side- or edge-trimming processes. These processes use two rotary top and bottom knives, which can be of block or disk type. Schematic diagrams of the blade setup and various views of the blades are shown in Fig. 2. In these shearing processes, the actions of the top and bottom knives are continuous. They are used in primary metal industry from hot and cold mill side- or edge-trimming

to slitting and side- or edge-trimming at finishing for manufacturing sheet (or foil) metal coils. They are also used in the paper, plastic film, and even fabric manufacturing industries. From a mechanics point of view, in-plane shear or mode II fracture dominates processing in the first category, while the second group is predominantly governed by antiplane shear or mode III fracture.

Process Parameters

A number of important process parameters govern the performance of these shearing processes. For example, for blanking and trimming processes in the first group, blade geometry and sharpness, blanking line or trim line curvature (required by the part shape), and the blade shear angle are dominant parameters. This blade shear angle is normally very small, its purpose being to reduce the overall cutting force and the press power (or tonnage) requirement. Another critical parameter established in Ref 1 is the angle of blade travel direction with respect to the plane of the sheet for trimming processes. This blade cutting angle is shown in Fig. 1(a).

The focus of industry, in terms of research and development issues, has been on punch force-displacement relations for reducing energy consumption as well as blade penetration. However, critical concerns with industrial applications of such processes as blanking, trimming, and piercing are burr formation, cut-edge quality, sliver or fine generation, and tool life. An optical micrograph of the cut surface from these processes is shown in Fig. 1 (b), while the sheet cut-edge profile is schematically shown in Fig. 1(c). In general, a few characteristic zones can be distinctly identified in these surfaces. They are:

- *Rollover zone:* Corresponds to the part of the edge that is drawn into the sheet
- *Burnish or sheared zone:* Smooth surface region that is formed by the cutting tool before the onset of ductile fracture
- *Fracture zone:* With a rough surface formed by the mechanism of ductile fracture as the tool progresses through the sheet
- *Burr region:* Due to the specific location of fracture initiation. The extent of the burr region depends on the crack path in the final stages of shearing (Ref 2).

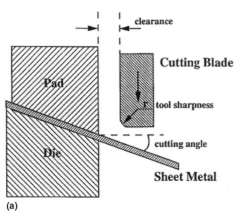

(a)

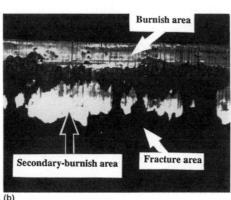

(b)

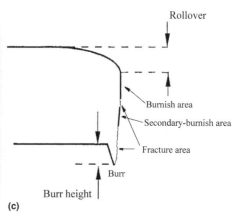

(c)

Fig. 1 The shearing process. (a) Schematic diagram of a shearing process. (b) Optical micrograph of the cut surface. (c) Schematic of the sheet cut-edge profile

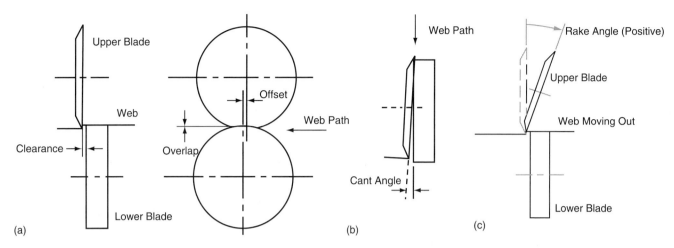

Fig. 2 Schematic diagram of a shear-slitting process. (a) Side, (b) top, and (c) front views

Burrs formed in blanking processes often scratch the part surface and also impose safety hazards during handling. Rough cut edges initiate cracks in the subsequent forming processes, such as stretch flanging. Burr problems are encountered in the computer and appliance industries, which require specific deburr operations. Slivers are generated during trimming of aluminum automotive body sheets. These slivers are carried through downstream processes and cause damage to the surface of formed parts, resulting in costly manual repair, as discussed in Ref 3.

For the second group of shearing, for example, shear slitting and trimming, as shown in Fig. 2, important process parameters include clearance, overlap, cant and rake angles, knife type and tip geometry (sharpness), knife run-out and diameter, sheet thickness, line speed, and tension (Ref 4, 5). Significant problems can occur with cut-edge quality, burrs, knife life, hairs/fines and debris, and so on in these processes. The hairs or fines are small aluminum metal debris generated during the shearing processes. When their shapes are long and slender, they are called hairs, and when their shapes are dustlike, they are called fines. In general, they are the major resources of contamination during the shearing processes.

The cut-edge quality is measured by edge waviness, coil wall appearance, and percentage of burnish area (nick) versus fracture area (break). Burrs can seriously impair subsequent processes by cutting into rolls. Hairs, fines, and debris can be carried to accumulators, levelers, work rollers, and backup bearing rollers and can cause dents on coil surfaces. These can build up on the side wall of coils and cause contamination in manufacturing processes. Hence, it is a major reason for customer rejection of products in the metals industry.

Conventional sheet shearing processes may be partitioned into three main stages with respect to deformation and damage mechanisms. The first stage of elastic deformation yields very little bending deformation of the sheet. The second stage is realized as stresses exceed the material yield strength. A plastic zone is formed in the

clearance region between the blade and the die. Small cracks initiate in this region with continuing blade travel. Finally, in the third stage, cracks grow in size and coalesce to cause final material separation by the mechanism of ductile fracture. Cracking by ductile fracture of metals and alloys is a complex phenomenon. It depends on macroscopic process parameters as well as on the sheet microstructure and its material properties.

Brief Literature Review on Modeling Shearing Processes

While research on metal shearing dates back to the early 1900s, Chang and Swift (Ref 6) were among the first to provide detailed experimental information on the effects of clearance and tool sharpness for industrial metals. The basic characteristics of shearing processes include plastic flow and crack propagation. These were described in Ref 6 for soft metals such as lead and tin as well as for relatively hard metals such as mild steels. Mechanisms of the double-shear phenomenon, which results in rough cut surface quality with zero or very small clearances, were identified. Double shear refers to a phenomenon where cracks initiating from the top and bottom knives do not meet naturally in the sheet metal being sheared. This work also demonstrated that the shearing load decreases with increasing clearance due to the development of a tensile stress component in the shear zone at the late stages of shearing processes. The phenomenon can be explained better in light of the current understanding of the dependence of hydrostatic tensile stress on void growth rate (Ref 7) and the mechanisms of crack initiation and propagation in the shear zone. Johnson and Slater (Ref 8) conducted a comprehensive survey on the effects of punch speed (strain rate) and temperature on blanking. Johnson et al. (Ref 9) also surveyed the literature until the early 1980s and summarized the effects of punch semiangle (describing the shape of the punch) and in-plane anisotropy on piercing and hole flanging. The normal punch semiangle is 90°, or a right angle.

A series of papers by Atkins et al. (Ref 10–12) have significantly advanced the understanding of shearing mechanisms and their applications. They proposed a critical punch travel distance criterion for the onset of cracking in Ref 10 and used rigid-plastic fracture mechanics to analyze the guillotining process in Ref 11. They were the first to link fracture toughness to plastic flow and crack propagation in the shear zone. The papers also introduced the concept of plastic instability leading to shear band formation and propagation in the shear zone. In Ref 12, many interesting questions were raised on cut surfaces produced by guillotining, and significant insight was provided into the process. They concluded that the main mechanism of fracture during shearing is void growth and coalescence. The paper also emphasized the importance of combined shear and tensile stresses in the shear zone. Zhou and Wierzbicki developed a tension zone model in Ref 13, based on the assumption that large material rotations occur in the shear zone, which gives rise to large tensile deformation.

More recently, the experimental work of Li et al. (Ref 1–5) focused on the micromechanics of deformation and fracture to investigate burr formation in shearing processes. Both groups of shearing processes, continuous (rotary) and discontinuous (straight or curved blade), were discussed in this work. The first group was motivated by the rapid rise in the use of aluminum in automotive applications to meet fuel-efficiency requirements. Investigations on the effect of clearances, blade sharpness, and blade travel direction or cutting angle on the sheared surface quality and burr height variations were reported in Ref 1. This work revealed that, contrary to conventional wisdom, blade travel perpendicular to the sheet (0° cutting angle) at appropriate cutting angles can make the surface quality insensitive to the blade sharpness. Negligible burrs are produced for large clearances and extremely dull blades. A plot of burr height as a function of the cutting angle for various blade sharpnesses is shown in Fig. 3. The cutting angle introduces a tensile stress component even at the initial stage of the shearing process, which accelerates void growth in the shear zone. The findings suggest that the robustness of current shearing practices for aluminum sheets can be improved considerably, while requiring less frequent tool sharpening and less restrictions on clearance control. A new technology that can significantly improve the burr height and cut-edge quality was developed in Ref 14 and successfully implemented in many production lines.

A different critical challenge is encountered in press forming or stamping of aluminum alloy sheets to form body panels. Appreciable amounts of slivers, which consist of debris and small aluminum pieces, are produced during aluminum trimming operations using dies that are conventionally designed for steel sheets. High production volume causes slivers generated in each stroke to spread all over the trimming dies. The slivers are subsequently carried into downstream processes and may cause damage to the surface of formed parts. A systematic experimental study in Ref 3 concluded that slivers can be reduced or even eliminated by modifying trimming tools and dies. The largest amount of slivers was produced for aluminum alloy sheets with 0° cutting conditions.

The second group of shearing processes includes the continuous shear slitting and side-trimming processes for materials in the form of long slab, sheet, or coil (web). In the primary metal or textile industries, coils can be miles long and are called webs. The term *web-handling* describes the finishing processes such as winding, leveling, coiling, and so on. The work studied in Ref 4 focused on the effect of three major slitting parameters, namely clearance, overlap, and cant angle on the burr height. Clearances higher than a critical value for shear slitting and side trimming of aluminum webs were found to yield unacceptable burr heights. This is evidenced in Fig. 4 for aluminum alloys 1050-H18, 3004-H19, and 5182-H19. In this work, the critical clearance, c_{cr}, was related to the material Young's modulus, E, yield strength, σ_y, and maximum fracture strain, ε_f, as:

$$c_{cr} - 0.004 \frac{Et}{\sigma_y} \frac{1}{\sqrt{\varepsilon_f}} (\alpha_{cant} + d) + 0.18t \quad \text{(Eq 1)}$$

where α_{cant} is the cant angle, t is the sheet thickness, and d is a constant.

The effects of rake angle were also investigated in Ref 2. For shear slitting and side trimming, this angle is the equivalent of the cutting angle in die or press cutting operations (Ref 1, 3, 4). This new slitting configuration induces a local tension in the shear zone that facilitates early separation. Side trimming of aluminum sheets over a wide range of slitting conditions indicates that the new configuration is insensitive to clearance and overlap. This results in very low edge burr height, even when the clearance is considerably larger than the sheet thickness. As shown in Fig. 5, even a large clearance (330% of the sheet thickness, $t = 0.15$ mm) yielded negligible burr height.

Numerical Modeling of Shearing Processes

Empirical models are generally not capable of accurately predicting the size of the fracture zone and burr in shearing processes due to the complex phenomenon of ductile fracture. The ductile fracture surface is generally a complex path of coalesced voids in the material microstructure that can develop quickly and localize in very narrow bands. The evolution of ductile failure depends on various factors, such as stress triaxiality (ratio of mean stress to equivalent stress), local morphology, void volume fraction, effective plastic strain, and void initiation, growth, and coalescence.

A host of numerical studies have been performed to resolve challenging questions related to the prediction of material damage and

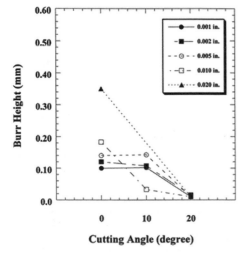

Fig. 3 Burr height as a function of the cutting angle, showing insensitivity to blade sharpness at 20° cutting angle. The legend indicates five blade sharpnesses (r).

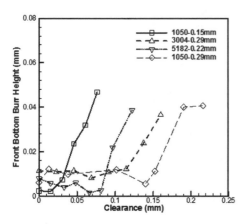

Fig. 4 Burr height dependence with clearance, establishing a critical clearance for four alloy thickness combinations

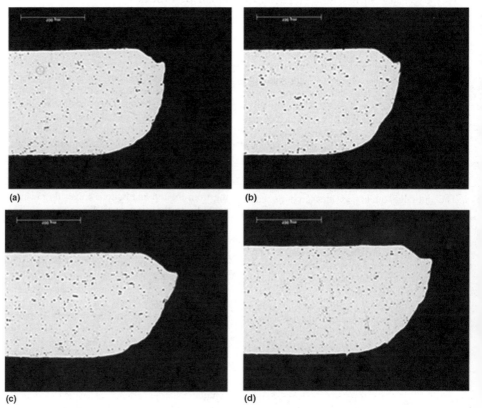

Fig. 5 Cross sections of side-trimmed edge sheet of thickness 0.147 mm for 15° rake angle. Unit shown is 400 μm. Clearance: (a) 0, (b) 0.30, (c) 0.43, and (d) 0.50 mm

failure. In many of these studies, ductile fracture causing sheet shearing was modeled with damage laws expressed in terms of the plastic strain and stress tensor. Taupin et al. (Ref 15) used a rigid plastic material model with a failure criterion proposed by McClintock (Ref 16) in their finite-element model for fracture prediction. Failure in the McClintock model is represented as the void growth rate $d\mathbf{R}/\mathbf{R}$, which is expressed in terms of the rate of change of the radii of circular cylindrical voids. This growth rate is written in terms of the transverse stresses (σ_{xx}, σ_{yy}), strains (ε_{xx}, ε_{yy}), equivalent stresses and strains (σ_{eq}, ε_{eq}), and the strain-hardening exponent in a power-law model, n, as:

$$\frac{d\mathbf{R}}{\mathbf{R}} = \frac{\sqrt{3}}{2}\frac{n}{n-1}d\varepsilon_{eq}\sinh\left(\frac{\sqrt{3}(\sigma_{xx}+\sigma_{yy})}{2\sigma_{eq}}\frac{n}{n-1}\right) + \frac{d\varepsilon_{xx}+d\varepsilon_{yy}}{2}$$

(Eq 2)

This model does not incorporate the effect of void interactions. The LeMaitre damage model (Ref 17) was used to predict crack initiation and propagation in the simulation of blanking processes by Hambli (Ref 18, 19) and Miguel et al. (Ref 20). The LeMaitre damage model follows the framework of continuum damage mechanics models that assume a homogeneous distribution of microvoids and is based on the hypothesis of strain equivalence. Evolution equations for internal variables are derived by assuming the existence of a dissipation potential, Ψ, which is a scalar convex function of the state variables that may be decomposed into plastic (Ψ^p) and damage (Ψ^d) components. For isotropic hardening and damage, the function is expressed as:

$$\Psi = \Psi^p + \Psi^d = \Phi + \frac{r}{(1-\mathbf{D})(s+1)}\left(\frac{-\mathbf{Y}}{r}\right)^{s+1}$$

(Eq 3)

Here, r and s are material- and temperature-dependent properties, and Φ and $\mathbf{Y}$ are the yield function and damage strain energy, respectively. In Ref 19, the author combined predictive finite-element modeling, including damage and fracture models, with neural network modeling to obtain the optimum punch-die clearance for sheet metal blanking processes.

Fang et al. (Ref 21) conducted finite-element modeling of the blanking process for optimizing punch-die clearance values for fixed thickness of aluminum sheets. Fracture associated with the shearing process was modeled using the Cockroft and Latham (CL) fracture criterion (Ref 22) discussed in the section "The Cockroft-Latham Model" in this article. The ratio of maximum principal tensile stress to the equivalent stress drives damage evolution in this model. MacCormack et al. (Ref 23) simulated a trim die forging process using the DEFORM finite-element package together with the CL fracture criterion to model induced damaged.

All of the aforementioned simulations incorporated fracture through an element-deletion technique, in which an element is removed when the damage criterion is achieved at an integration point. Although a simple and effective method, the solution with this method is inherently mesh dependent. Brokken et al. (Ref 24) averted this shortcoming by introducing a discrete crack propagation approach for simulating fracture. Their strategy incorporated an operator split arbitrary Lagrangian-Eulerian formulation in conjunction with remeshing to resolve extremely localized deformation in the blanking processes. The initiation and propagation of discrete cracks was controlled by a fracture potential field that was calculated using Rice and Tracey's void growth model (Ref 7). In this model, the rate of size change of a spherical void in a linear strain-hardening plastic matrix, subjected to a tensile deformation with a remote hydrostatic stress, is expressed as:

$$\frac{d\mathbf{R}}{\mathbf{R}} = (1+\mathbf{G})\frac{d\varepsilon_{kk}}{2} + \left[\mathbf{D}\sqrt{\frac{2}{3}d\varepsilon_{kk}d\varepsilon_{kk}}\right]\mathbf{R}_0$$

(Eq 4)

where $\mathbf{R}_0$ is the initial radius of the sphere, and $\mathbf{D}$ and $\mathbf{G}$ are material constants that depend on the stress state and the material strain hardening. A modified form of Rice and Tracey's model was used by Brokken et al. (Ref 25) with different factors of stress triaxiality, as proposed by Goijaerts et al. (Ref 26). The fracture potential function was calculated and extrapolated to nodes of the finite-element assembly in every load increment. When the potential function at a node reaches a critical value, a crack is initiated or, in the case of a crack tip, propagated in the direction of maximum fracture potential. The direction was determined by sampling a finite number of radially spaced points around the node. Continuous remeshing is necessary for accommodating new and moving cracks with this method.

Klingenberg et al. (Ref 27) investigated the behavior of blank material in punching and blanking processes by finite-element simulations and experiments to develop a system for on-line characterization of the blank material properties. The finite-element simulation was conducted with the ABAQUS code (Ref 28), incorporating the Gurson-Tvergaard-Needleman (GTN) models (Ref 29, 30) for modeling ductile fracture by void nucleation, growth, and coalescence. The GTN models are quite effective in modeling ductile failure and are discussed in the section "Continuum Damage Models for Failure and Their Calibration" in this article. In a comprehensive treatise on the analysis of guillotining and slitting processes, Wisselink (Ref 31) developed arbitrary Lagrangian-Eulerian finite-element models for plastic deformation and ductile fracture using the Oyane model (Ref 32, 33). Both two-dimensional transient models of

orthogonal shearing and three-dimensional stationary models of guillotining and slitting were developed in this work. The steady-state conditions were estimated from the initial geometry, including a crack front. The position of the crack front was fixed during the simulations, and the assumed position was verified by a fracture criterion. A problem with this method is that the position of the initial crack front is not known in most cases. Simulations performed by Kubli et al. (Ref 34) predicted the rollover zone in a shearing operation that was simulated by continuous remeshing. In a majority of the work discussed previously, emphasis is on the analysis and optimization of the punch force in the blanking process. Relatively little work has been reported on the prediction of crack profile explicitly to minimize the formation of surface defects such as burrs. Ad hoc methods of design by trial and error can become extremely tedious and expensive, as the number of design variables and design space corresponding to each design variable become large. Various techniques of metal forming process design have been proposed in the literature. These methods use gradient-based optimization techniques based on sensitivity analysis, genetic algorithms, neural networks, and so on. A cogent review of some of the currently used optimization techniques has been presented by Nicholson (Ref 35).

Preview of the Following Sections— Experimental-Numerical Studies of Shearing and Burr Formation

The remainder of this article discusses the development of experimentally validated finite-element models for analyzing two classes of shearing processes mentioned in the beginning of this article. The discussions are based on experimental and numerical studies that were conducted by Ghosh et al. in Ref 36 and 37 to facilitate process parameter design. The simulations use validated fracture models to examine the influence of process parameters on burr formation. Three different models of ductile fracture, namely the GTN models (Ref 29, 30), the shear failure model (Ref 28), and the CL model (Ref 22), are examined for effective modeling of shearing processes. Results of the numerical simulations are used for sensitivity analysis with respect to various process and tool parameters. An overview of experimental analysis of some shearing processes for understanding the role of tool parameters on the material deformation is presented in the next section. Continuum models of ductile failure are then assessed for their effectiveness in representing the actual failure in shearing processes. Microstructural characterization with stereology is used to render three-dimensional volumetric parameters in this section. This is followed by the numerical simulation of an edge-shearing process, along with sensitivity studies with respect to process and tool parameters. Finally, the shear slitting problem is studied, and concluding remarks are provided.

Experimental Studies for Material and Process Characterization

This section discusses results of some experiment studies with two aluminum alloys, 6022-T4 and 5182-H19, that were conducted at Alcoa Technical Center using a double-action laboratory die. The aluminum sheet was clamped between a bottom die and a clamping pad (Ref 1). The die was placed in a small commercial press with a punch travel speed of 0.05 m/s. This is typical of automotive production trimming operations. As the punch travels down, the first action was to clamp the sheet and press the bottom die down against the die springs. The blade then proceeds to cut off the sheet metal in subsequent action. At the cutting position, the pad pressure was designed to be 1 MPa applied on the sheet uniformly. This closely simulates the trimming operation under actual production conditions.

Various geometric features of the processing tools affect the overall surface quality of the product. The quality is represented by the extent of primary and secondary burnish areas, the fracture area, as well as the process deformities such as the burr and rollover. The cutting angle, clearance, and the blade sharpness (represented by the blade-edge radius) were identified as the most critical tool geometric parameters from extensive experimental studies in Ref 1. While in most trimming operations it is conventional to keep the cutting angle as close to 0° as possible, blade travel at an angle to the sheet plane is more practical. This additional parameter takes advantage of the microstructural characteristics of aluminum alloys (Ref 1). The die is capable of cutting the sheet metal at various angles, and 0, 10, and 20° were selected. The clearance was controlled at 5, 10, and 20% of the sheet thickness by using very thin metal shims. Three blade sharpnesses with blade edge radii of 0.025, 0.125, and 0.250 mm were machined and measured by an optical comparator. The results of tests with several combinations of these three parameters are shown in Table 1. While the sheet thickness for the experiments was 1 mm, the general results apply to other thicknesses as well, because the clearances are measured as percentages of sheet thickness.

The initial deformation in shearing processes is plastic indentation with the formation and propagation of strain localization zones (Ref 2). Figures 6(a) and (b) show micrographs of the cross section of an interrupted test sample that was stopped prior to complete separation. The plastic strain localization zones propagate much faster from the moving (top) blade than from the bottom stationary blade (or die). The deformation zone is nonsymmetric with respect to the top and bottom blades. The nonsymmetrical behavior of the localization zone is due

to stress components in the shear zone induced by bending effects. Most theoretical models and numerical simulations of the shearing processes prescribe symmetric boundary conditions, which would certainly result in a symmetrical deformation pattern.

Continuum Damage Models for Failure and Their Calibration

Macroscopic damage models are necessary to predict the onset and propagation of ductile fracture that causes material separation. A major

Table 1 Burr heights measured from experiments with various process parameters

Cutting angle, degrees	Clearance thickness, %	Burr height for various edge radii (R), mm		
		$R = 0.025$	$R = 0.125$	$R = 0.250$
0	5	0.05	0.05	0.06
	10	0.1	0.14	0.183
	20	0.15	0.183	0.233
10	5	0.021	0.024	0.022
	10	0.1	0.14	0.03
	20	0.13	0.2	0.22
20	5	0.005	0.005	0.006
	10	0.006	0.007	0.007
	20	0.07	0.18	0.3

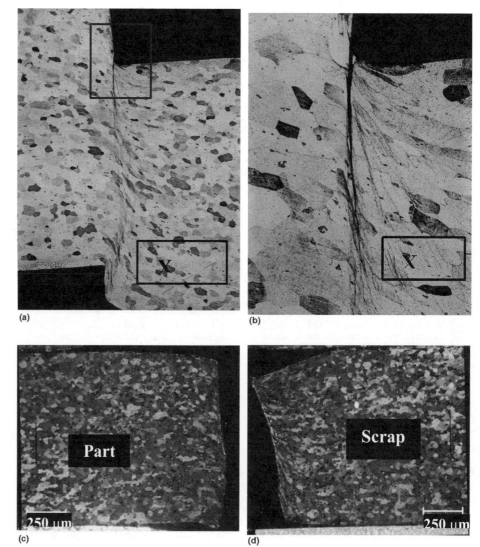

Fig. 6 Micrographs at stages of shearing. (a) Initiation and propagation of the plastic localization zone and shear bands produced with a sharp moving (top) blade at 0° cutting angle and 5% clearance. (b) Zoom-in of the boxed region showing void sheets in front of the crack tip. (c) and (d) Micrographs of the sheared part and scrap, respectively

mechanism of ductile fracture initiation in metals is void nucleation caused by dislocation pileups, second-phase particle cracking, or other imperfections. Voids grow with increasing plastic deformation and subsequently coalesce to form dominant evolving cracks (Ref 21). Various damage laws have been proposed in the literature to predict the onset of failure as functions of stresses and strain history. Of these, those commonly used for ductile fracture are by Clift et al. (Ref 38), Gurson-Tvergaard-Needleman (Ref 29, 30), Rice and Tracey (Ref 7), McClintock (Ref 16), Oyane et al. (Ref 32, 33), and Cockroft and Latham (Ref 22). In many of these models, a material point is assumed to have failed when a local damage parameter reaches a critical value at this location. In the numerical simulation of shearing processes, this damage parameter is often a function of the equivalent plastic strain and invariants of the stress tensor, for example, stress triaxiality, hydrostatic stress, principal stress, and so on. An accumulated equivalent plastic-strain-based shear failure model, featured in ABAQUS (Ref 28), is also taken as a viable macroscopic model of ductile fracture. Ghosh et al. (Ref 36, 39) calibrated three different damage models by experiments and examined for effectiveness in the shearing process simulations. These are the GTN model, the CL model, and the shear failure model. The element-deletion algorithm was used for representing progressive material separation and crack propagation with these continuum damage models. In this algorithm, the element stress-carrying capacity at an integration point is reduced to zero when a critical damage condition is attained. The three damage models are briefly described next.

The Gurson-Tvergaard-Needleman model (Ref 29, 30, 40) is a widely used phenomenological continuum damage model for progressively cavitating plastic solids. It accounts for the loss of load-carrying capacity through void nucleation, growth, and coalescence in the microstructure with increasing plastic strain. Void coalescence shifts a relatively homogeneous deformation state to a highly localized one in the microstructure. Plastic deformation is modeled using an associated flow rule. In this model, the evolving pressure-dependent yield surface is parameterized as a function, Φ, of the homogenized von Mises stress tensor, $\overline{\sigma}$, the tensile flow stress, σ_y, in the matrix material, and the void volume fraction, f, as:

$$\Phi = \left(\frac{\overline{\sigma}}{\sigma_y}\right)^2 + 2q_1 f^* \cosh\left(-q_2 \frac{3p}{2\sigma_y}\right) - (1 + q_3 f^{*2})$$
$$= 0 \qquad (Eq\ 5)$$

where p is the hydrostatic pressure, and q_1, q_2, and q_3 are experimentally evaluated constant coefficients. An acceleration function, f^*, has been introduced to model the complete loss of material stress-carrying capacity due to void coalescence as:

$$f^* = \begin{cases} f & \forall f \leq f_c \\ f_c + \frac{f_u^* - f_c}{f_f - f_c}(f - f_c) & \forall f > f_c \end{cases} \qquad (Eq\ 6)$$

Here, f_c is the critical void volume fraction at which void coalescence first occurs with accelerated void growth. In this phase, the material starts to soften until the void volume fraction reaches an ultimate value, f_f, at final failure. As the void volume fraction $f \to f_f$, the acceleration function becomes:

$$f \to f_u^* = \frac{1}{q_1}$$

At this value, the yield surface shrinks to a point, manifesting a complete loss of material stress-carrying capacity corresponding to ductile failure. Elements for which all integration points have failed are removed from the analysis. The rate of change of void volume fraction is governed by the equations:

$$\dot{f} = \dot{f}_{growth} + \dot{f}_{nucleation} \qquad (Eq\ 7)$$

where the change in volume fraction due to growth of existing voids is expressed as:

$$\dot{f}_{growth} = (1 - f)\dot{\varepsilon}_{\sim kk}^{\ pl} \qquad (Eq\ 8)$$

The matrix void growth rate is due to the plastic dilatation strain rate, $\dot{\varepsilon}_{\sim kk}^{\ pl}$, and Eq 8 accounts for plastic incompressibility in the underlying matrix. The change in volume fraction due to nucleation of new voids is governed by a strain-controlled nucleation criterion that has been developed in Ref 40 as:

$$\dot{f}_{nucleation} = A\dot{\overline{\varepsilon}}^{pl} \text{ where}$$
$$A = \frac{f_N}{S_N\sqrt{2\pi}} \exp\left[-\frac{1}{2}\left(\frac{\overline{\varepsilon}^{pl} - \varepsilon_N}{S_N}\right)\right] \qquad (Eq\ 9)$$

Here, ε_N and S_N are the mean and standard deviation of the nucleating strain distribution, and f_N is the volume fraction of nucleating voids.

The Cockroft-Latham model (Ref 22), also discussed in Ref 32 and 33, determines the likelihood of tensile fracture during plastic deformation from the consideration of plastic work. The material is assumed to undergo ductile fracture when a damage parameter, D, expressed as a function of the stress and plastic strain components, reaches a critical value, D_c:

$$D = \int_0^{\overline{\varepsilon}^{pl}} \left(\frac{\sigma^*}{\overline{\sigma}}\right) d\,\overline{\varepsilon}^{pl} \to D_c \qquad (Eq\ 10)$$

Here, $\overline{\varepsilon}^{pl}$ is the equivalent plastic strain, $\overline{\sigma}$ is the effective von Mises stress, and σ^* is the maximum positive principal stress. The material loses its stress-carrying capacity at the fracture plastic strain, $\overline{\varepsilon}_{frac}^{pl}$. This model is implemented in ABAQUS-Explicit (Ref 28) using the user material subroutine VUMAT and is called from all integration points in the finite-element model.

The shear failure model in Ref 28 uses the accumulated equivalent plastic strain as a damage indicator. Failure occurs when a plastic-

strain-dependent damage parameter, w, satisfies the condition:

$$w = \sum_n \left(\frac{\Delta\overline{\varepsilon}^{pl}}{\overline{\varepsilon}_f^{pl}}\right) \geq 1 \qquad (Eq\ 11)$$

where $\Delta\overline{\varepsilon}^{pl}$ is the increment of equivalent plastic strain, $\overline{\varepsilon}_f^{pl}$ is the plastic strain at failure, and n corresponds to the number of increments in the finite-element analysis. While $\overline{\varepsilon}_f^{pl}$ can be a function of the plastic strain rate, stress triaxiality, and temperature, it is assumed to depend on stresses only in the present analysis.

Mechanical Testing and Damage Characterization

A set of experiments is designed to characterize the parameters in the damage models discussed in the section "Continuum Damage Models for Failure and Their Calibration" in this article. The dogbone specimens of Fig. 7(a) are used to measure the evolving void volume fraction at regions of interest. Samples are strained to failure, and the average failure strain is recorded from extensometer readings to be approximately 11.3%. In addition, interrupted tests are conducted with specimens loaded to various levels of prefailure strains. The strains at interruption are concentrated near the fracture strain to capture the rapid growth of the local void volume fraction in the necked regions. Each specimen is sectioned after load interruption to extract samples from the notched regions, thus focusing on studies around each notch near the center of each specimen. These samples are polished using standard metallographic techniques, and micrographs of each sample are taken with a Philips XL-30 environmental scanning electron microscope. Figure 8(a) shows the micrograph near the edge at 11.3% strain level. The actual microstructural geometry of the voids is generally quite complex. Equivalent elliptical voids that closely approximate the actual morphology are generated following techniques developed in Ref 41. In this work, the volume of each void in the micrograph is retained through equivalent major and minor axes of the ellipse.

Three-Dimensional Volume Fraction from Two-Dimensional Data Using Stereology. It is necessary to map two-dimensional (2-D) microstructural section data onto three-dimensional (3-D) microstructural representations for comparison with computational models. Such data can include size and shape distributions of elliptical inclusions in the microstructure. Methods of stereology enable the mapping of certain 3-D characteristics by projecting 2-D data (Ref 42). For example, distribution of 3-D spheroids in the microstructure can be estimated from the size and shape distribution of 2-D elliptical cross-sectional information. Following developments in Ref 41, the Saltykov method is used to generate 3-D void volume fraction data. A modified version of the Saltykov method incorporating shape effects has

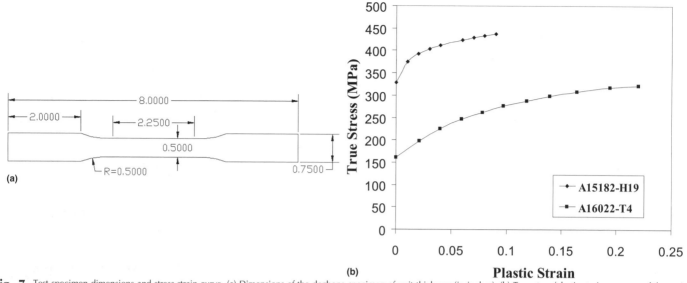

Fig. 7 Test specimen dimensions and stress-strain curve. (a) Dimensions of the dogbone specimen of unit thickness (in inches). (b) True stress/plastic strain response of the various aluminum alloys

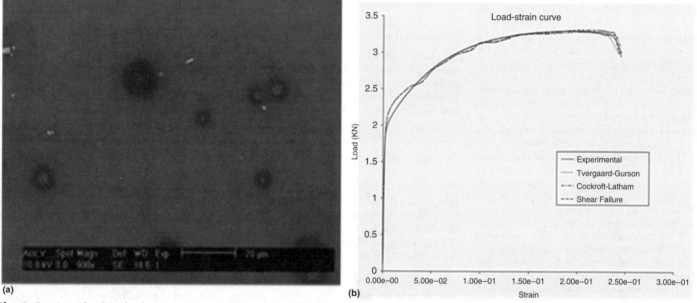

Fig. 8 Experimental and simulated stress-strain curve with micrograph. (a) Sample micrograph near the notch at 11.3% strain. (b) Experimental load-strain curve for the dogbone specimen showing failure, together with simulated results of three damage models

been suggested in Ref 42 for generating prolate or oblate spheroid geometries.

To construct spheroids, the major axis length ($a \leq a_{\max}$) and the minor axis length ($b \leq b_{\max}$) for each ellipse in the section are first recorded. Here, $a_{\max}$ and $b_{\max}$ are the maximum possible lengths of major and minor axes. The range of 2-D ellipses and the 3-D spheroids is divided into m intervals or classes. For prolate spheroids, the size of these intervals depends on the largest minor axis length, $b_{\max}$, while the largest major axis length, $a_{\max}$, is needed for oblate spheroids. The size of the interval, Δ, for prolate spheroids is:

$$\Delta = \frac{b_{\max}}{m}$$

and for oblate spheroids is:

$$\Delta = \frac{a_{\max}}{m}$$

In addition, a shape parameter is defined as:

$$y^2 = 1 - \left(\frac{b}{a}\right)^2, \forall \, y^2 \in (0,1)$$

The range of y^2 is divided into k intervals of equal size $1/k$. The 2-D and 3-D objects with size in the range $[(i\text{-}1)\Delta, i\Delta]$ and shape parameter y^2 in the range $[(j\text{-}1)k, jk]$ belong to the $(ij)^{\text{th}}$ size-shape interval. The 3-D spheroid size-shape distribution is obtained

from the 2-D size-shape distribution (Ref 41, 42) as:

$$N_v(i,j) = \frac{1}{\Delta} \sum_{\alpha=i}^{m} \sum_{\beta=j}^{k} p^{i\alpha} \, q^{j\beta} \, N_A(\alpha,\beta) \qquad \text{(Eq 12)}$$

where $N_v(i,j)$ is the number of prolate or oblate spheroids in the ij^{th} size-shape interval, $N_A(\alpha,\beta)$ is the number of ellipses in the size-shape interval $\alpha\beta$, and the coefficients $p^{i\alpha}$ and $q^{\beta j}$ for the prolate and oblate categories are taken from tables in Ref 42. Void volumes for each interval ij are obtained from the mean size $(i\text{-}0.5)\Delta$ and mean shape parameter $\frac{i-0.5}{k}$. The total volume of voids near the notch is calculated for specimens loaded to various strain levels by adding the volumes from all intervals. Micrographs are taken at three different locations near the edge, and the average values of void volume fraction are tabulated in Table 2. The void volume fraction is observed to increase with overall strain level in an exponential manner.

Evaluating Damage Parameters from Experiments and Simulations. Damage parameters in the various models of ductile fracture include f_c, f_f, ε_N, S_N, and f_N in the GTN model, D_c in the CL model, and the failure strain $\bar{\varepsilon}_f^{pl}$ in the shear failure model. These are evaluated using an inverse method involving numerical simulations and experiments. The experimental load-displacement curve showing failure for the first set of experiments is depicted in Fig. 8(b). The experimental specimen is simulated by a finite-element model in ABAQUS-Explicit (Ref 28). The specimen model contains a small geometric imperfection at the center to initiate necking. The material plastic constitutive behavior follows the stress-strain curves in Fig. 7(b). The damage parameters in each model of ductile fracture are adjusted in the finite-element simulations such that the results of the simulation match the experimental results. Evaluation of the nucleation parameters ε_N, S_N, and f_N and in the GTN model (Ref 29, 40) requires the microscopic void volume fraction data as a function of overall strain. The

nucleation parameters are determined by a microgenetic algorithm (Ref 37), which minimizes the least-square difference between simulated and experimental void volume fraction.

Figure 9 shows the void volume fraction, f, as a function of the strain from experiments as well as simulations. The experimental points correspond to each of the interrupted load tests. These void volume fraction data are found to be best represented by an exponential function of the strain, which matches the simulation results quite well. The damage parameters for the various models are tabulated as follows:

Damage parameter	Value
f_c	0.08
f_f	0.1
ε_N	0.5
S_N	0.1
f_N	0.04
D_c	0.6
$\bar{\varepsilon}_f^{pl}$	0.35

The macroscopic load-displacement curves and failure behavior resulting from these simulations are shown in Fig. 8(b). All the simulated results match the experimental results rather well. The corresponding crack profiles with the damage models are compared with the experimental crack in the dogbone sample in Fig. 10. Following the material and damage parameter calibration, the model is ready for shearing process simulations.

Edge-Shearing Process Simulation and Parametric Studies

From the second category of shearing, edge trimming is now examined.

Finite-Element Model and Simulations

A plane-strain finite-element model is set up in ABAQUS-Explicit for modeling shearing of thin aluminum sheets. The plain-strain constraint restricts the edge of the cutting tool to be parallel to the width of the sheet, avoiding skewness in cuts. The clamping pad, die, and blade, shown in the schematic of Fig. 1(a), are modeled as rigid surfaces. The metal sheet of dimensions 1 by 2 mm is meshed using approximately 7000 plane-strain QUAD4 elements with reduced integration and hourglass control (CPE4R). Because the deformations and stresses are localized in a very narrow region near the cutting tool, a graded mesh that facilitates high gradients is used. As shown in Fig. 11, a very fine mesh is used in the region of strain localization, with a coarse mesh representation for the remaining area. Excessive element distortion can occur in these simulations involving large plastic deformation in the presence of sharp contact zones and localized material failure. Loss of element accuracy and/or misrepresented boundary conditions can lead to high degrees of inaccuracy, sometimes causing termination of analysis. To avert this, the arbitrary Lagrangian-Eulerian (ALE) option is used in ABQUS-Explicit simulations. The ALE algorithm can effectively avoid excessive element distortion and improper contact surface penetration for the shearing problems, as discussed in Ref 43 and 44. In this analysis, the rigid die and clamping pad are assumed to be totally constrained, while translation in the horizontal direction as well as rotation are constrained for the blade. A vertical velocity boundary condition of 0.05 m/s is specified for the blade. Contact between the rigid surfaces and the metal sheet is modeled using surface-to-surface contact with a penalty formulation in ABAQUS-Explicit. Master surfaces are associated with the rigid bodies, while the top and the bottom

Table 2 Void volume fraction at the notch edge as a function of overall strain

Overall strain, %	Void volume fraction
8	3.52×10^{-4}
9	9.62×10^{-5}
10	1.75×10^{-3}
11	8.23×10^{-4}
11.2	3.40×10^{-3}
11.25	1.91×10^{-2}

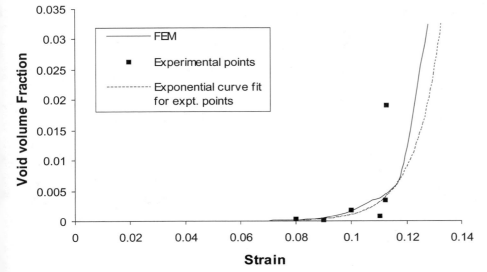

Fig. 9 Void volume fraction as a function of strain for experiments and simulations with Gurson-Tvergaard-Needleman model. FEM, finite-element model

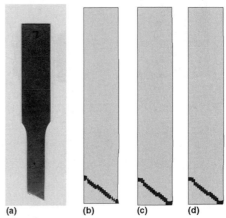

Fig. 10 Crack profiles in the dogbone sample for (a) experiment and simulations with (b) the Gurson-Tvergaard-Needleman, (c) the Cockroft-Latham, and (d) the shear failure models

surfaces of the sheet are defined as slave surfaces in the contact algorithm. A Coulomb friction model is used between the contacting surfaces. A friction coefficient of $\mu = 0.05$ is arrived at by comparing results of simulations with experiments. The low value of μ is attributed to well-lubricated contact surfaces.

Numerical simulations of shearing processes with the GTN model, the CL model, and the shear failure model are conducted and compared with experiments for establishing their suitability in process simulations. A contour plot of the void volume fraction at high strains with the GTN model is shown in Fig. 12(a). The material does not adequately separate for this model, because the void volume fraction does not evolve to the critical void volume fraction due to a small increase in the hydrostatic stress. Consequently, crack propagation becomes arrested in the thickness direction. From Eq 4, it is known that the void growth, $\dot{f}_{growth}$, depends only on the dilatation part of the plastic strain tensor and does not account for the deviatoric parts. However, ductile fracture in the shearing processes is predominantly shear dominated, and hence, material separation in problems with low triaxiality cannot be modeled accurately with the GTN model. In recent years, extensions to the GTN models have been made (Ref 45) to incorporate damage growth under low triaxiality straining for shear-dominated states. In Ref 45, an additional term that scales with plastic shear strain increment is added to the expression for $\dot{f}_{growth}$ to represent void softening in shear.

The CL model, however, develops complete material separation due to crack propagation, as shown in Fig. 12(b). When the simulated crack profile is compared with the experimental profile in Fig. 6, a significant difference is observed. Also, the burr heights obtained from the simulations using the CL model do not match the experimental results for various values of clearance, tool radius, and cutting angle. For all cases, the crack propagation ends very near the corner of the bottom die. This is not observed experimentally. This model is based on the macroscopic value of plastic work as a damage parameter, and it predicts an earlier failure than experiments. This prediction is consistent with continuum damage laws based on damage work, where homogenization of local variables often overpredicts failure. This CL model is considered inappropriate with respect to the objectives of this study. Finally, the shear failure model is able to successfully model material separation and crack propagation. As shown in Fig. 13 and 14, the simulated crack profile with this model agrees with the experimental results. The appropriate representation of the burr heights with the shear failure model confirms its suitability in shearing process simulations. Three-dimensional simulation of the shearing process with 3-D brick elements is conducted, as shown in Fig. 13(d). The clearance is set to 5% of the sheet thickness, and the

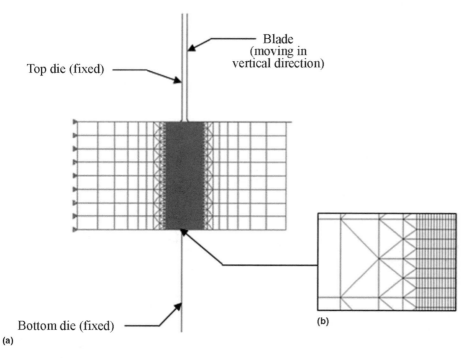

Fig. 11 Finite-element model setup of the edge-shearing problem. (a) Full cross section. (b) Zoomed-in view in the deformation zone showing the transition between coarse and fine mesh

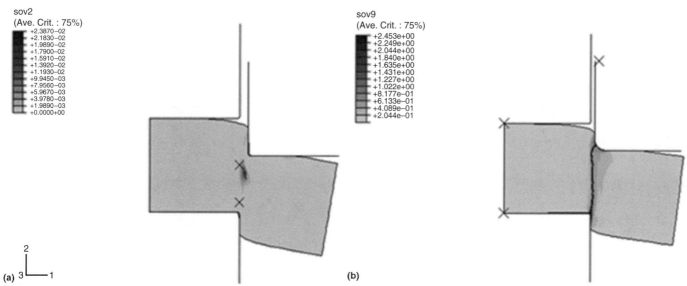

Fig. 12 Simulation of edge-shearing process. (a) Contour plot of void volume fraction with the Gurson-Tvergaard-Needleman model showing arrested cracking. (b) Contour plot of equivalent plastic strain with the Cockroft-Latham model

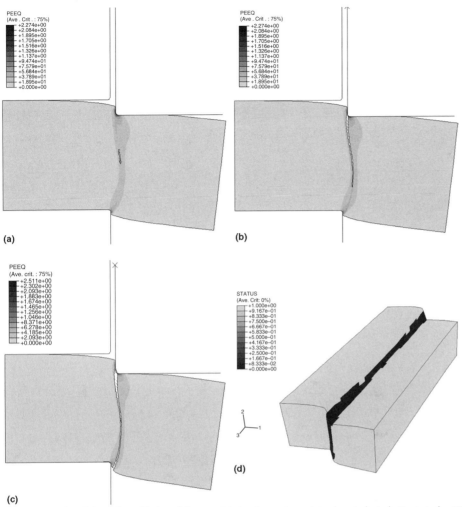

Fig. 13 Results of simulation with shear failure model showing contour plots of equivalent plastic strain for 5% clearance, 0.025 mm blade radius, and 0° cutting angle. At (a) 0.14, (b) 0.15, and (c) 0.18 mm blade travel. (d) Three-dimensional model showing the cut surface

tool radius is 0.025 mm. The cut-surface profile for the 3-D model is similar to that for the 2-D plane-strain model.

Parametric Studies on the Edge-Shearing Process and Design

The validated finite-element model is now used for parametric design of edge-shearing process parameters to reduce burr formation. The clearance, blade radius, and cutting angle are identified as the most critical design parameters. Experimental evidence suggests that there is no significant effect of the punch velocity on burr formation. Figures 14 and 15 depict the sensitivity of burr height to the clearance, blade radius, and cutting angle obtained from simulations and experimental results. The predicted burr heights from simulations are smaller than the experimental values, especially at larger clearances. The plot of burr height versus clearance for ∼0° cutting angle shows a clear trend that burr height increases with increasing clearance and large blade radius (duller blade). When the clearance is very small (≤5% of sheet thickness), the burr height does not vary significantly with blade sharpness. This suggests that, in conventional shearing, very small clearances may provide an advantage of less tool sharpening. To understand the effects of cutting angle, burr heights are plotted as a function of cutting angle in Fig. 15(a) for 5% clearance. The burr height decreases as the cutting angle increases, with the heights almost zero at 20° cutting angle for all blade radii. At large cutting angles, burr heights are insensitive to the blade sharpness at small clearance (∼5%). The values and trends predicted by simulation match the experimental values for small clearance at all cutting angles. The crack is found

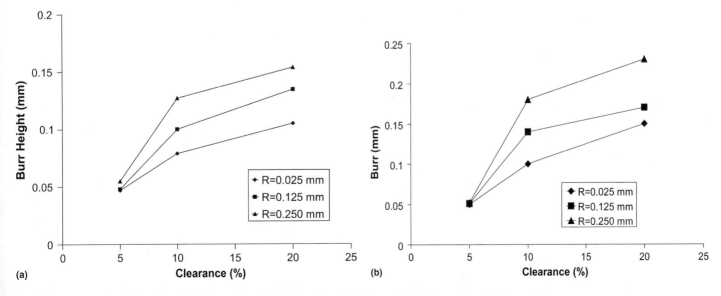

Fig. 14 (a) Simulated and (b) experimentally observed burr height as functions of the clearance for various radii for 0° cutting angle

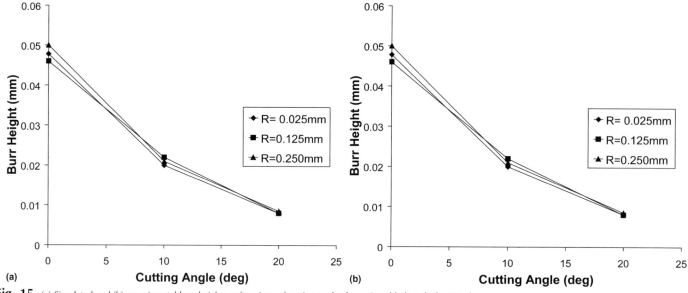

Fig. 15 (a) Simulated and (b) experimental burr heights as functions of cutting angles for various blade radii for 5% clearance

to propagate faster as the cutting angle increases. This explains the reduction in the extent of burr formation with increasing cutting angle.

A genetic algorithm (GA) program is implemented in conjunction with the finite-element model simulations to generate an optimal combination of process variables that will yield the minimum burr height. The resulting design process variables are:

Clearance, %	Radius, mm	Cutting angle, degrees
5.34	0.08	18.4

The results converge to a value very close to the minimum clearance of 5%. The blade radius from GA is found to be 0.08 mm instead of the minimum radius 0.025 mm, because at very small clearances, burr heights are almost insensitive to the blade radius. The cutting angle of 18.4° is close to the 20° angle found in parametric studies. It is observed that beyond 17°, burr heights are insensitive to the cutting angle. Burr heights are negligible for cutting angles in the range of 17 to 24° for the 5% clearance, as shown in Fig. 16. Thus, burr heights are at a minimum for small clearances close to 5% and for cutting angles greater than 17°. Because the burr height is insensitive to the blade radius at these clearances and cutting angles, even a dull blade can result in a negligible burr formation.

Shear-Slitting Process Simulation and Parametric Studies

Current practices of shear slitting of aluminum sheets and foils rely heavily on operator experience and a trial-and-error method. This second type of continuous shearing is discussed, and numerical simulations of the shear-slitting process develop relationships between the slit-edge quality and process parameters. Various process parameters such as clearance, blade overlap, cant angle, rake angle, web speed, and web tension have been considered in numerical studies in Ref 36. As shown in Fig. 2, clearance is the distance between the blade and the lower die cutting edges in the horizontal direction. The cant angle is the angle between two blade planes viewed in the direction perpendicular to the web plane. Rake angle is between the blade and the bottom blade, viewed in the direction parallel to the web plane. Experimental work in Ref 1 has inferred that cutting angle or rake angle can have a significant effect on the burr height. Technology to improve burr height and cut-edge quality for nonzero rake angles was developed (Ref 14). This section discusses numerical simulations focusing on the effects of rake angle and cant angle on burr formation, conducted in Ref 36. The computational model is validated with laboratory and field experiments discussed in the section "Experimental Studies for Material and Process Characterization" in this article. Parametric studies are done to study the effect of various parameters on the extent of burr formation.

The material considered is the aluminum alloy 5182-H19, commonly used in the high-volume-usage beverage can end and tab stock. The sheet materials are normally supplied in narrow-width coil form. Therefore, it requires the high-speed, high-efficiency, and high-quality slitting process to cut multistrip coils (up to 40 cuts) from a master coil. A 3-D finite-element model for the shear-slitting process is developed in ABAQUS-Explicit (Ref 28), as shown in Fig. 17. The nature of

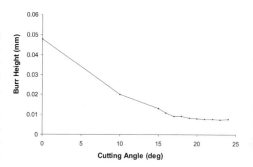

Fig. 16 Simulated burr height as a function of the cutting angle at 5% clearance and 0.025 mm blade radius

boundary conditions and the geometry of the setup, for example, cant and rake angles, necessitate a 3-D computational model. A metal sheet of dimensions $0.2159 \times 2 \times 6$ mm is meshed using approximately 16,000 eight-noded brick elements (C3D8). A graded mesh that facilitates high gradients in a very narrow region near the cutting tool is used to account for localized deformation and stresses. The ALE option is again used in ABAQUS-Explicit. Based on observations made in the 2-D shearing-process simulations, the shear failure model with a calibrated critical shear strain of $\bar{\varepsilon}_f^{pl} = 0.29$ is used in these simulations.

The clamping pad and blades are modeled as rigid surfaces and are assumed to be totally constrained. The effects of friction and wear of the tooling material are much more significant than its elastic deformation. It is assumed that rigid blades (both top and bottom) have little effect on the cutting edge quality. For the blade, translation in the horizontal direction as well as rotation is constrained. A vertical boundary velocity of 0.5 m/s is specified for the blade. Contact between the rigid surfaces and the metal sheet is modeled using surface-to-surface contact with a penalty formulation in ABAQUS. The Coulomb friction model with a friction coefficient of μ = 0.05 is used between contacting surfaces. The deformed configuration for the process simulation with 0° rake and 0° cant angles is shown in Fig. 17(b).

Parametric Studies for Burr Formation in Shear Slitting

The effect of variations in the rake and cant angles on the extent of burr formation is studied in this section. Simulations are conducted for three different rake angles of 0, 10, and 20° and for three cant angles of 0, 0.25, and 0.5°. The simulated burr height is measured at various locations over the cut edge in the z-direction, and the mean of all these values is used as the final burr height that is compared with experiments. Figure 18(a) shows the plot of burr heights for various rake and cant angles. The burr height decreases with increasing rake angle and is a minimum near a rake angle of 20°. This effect of the rake angle

on burr height is similar to that of cutting angle in the edge-shearing processes. The rake angle introduces a normal component of cutting force, which accelerates crack initiation and growth. Consequently, the crack propagates faster through the thickness of the sheet, resulting in a reduction of burr height. From Fig. 18(a), it is also seen that the burr height decreases with an increase in cant angle. Values of the simulated burr heights for various cant angles are compared with those obtained from experiments with cant angles of 0, 0.25, and 0.5°. A comparison of the simulated and experimental results is shown in Fig. 18(b). Simulated values of the burr height are approximately 15 to 18% lower than the experimental values. This difference may be attributed to errors in measurements, lack of proper lubrication, and temperature effects in the model. Additionally, the fact that slitting is simulated using continuum damage laws will also lead to some errors in the predictions. However, both sets of results show similar trends in burr height reduction as a function of the cant angle.

The effect of material properties on the burr height is studied with additional simulations for

Al6022-T4 alloy, for which the material and damage properties are discussed in the section "Experimental Studies for Material and Process Characterization" in this article. The simulations are performed for 0° rake angle and various cant angles. The comparison of the burr heights for the two materials at various cant angles is shown in Fig. 19. The burr heights for the Al5182-H19 alloy are slightly less than the burr heights obtained using the Al6022-T4 alloy. This can be explained from the fact that the equivalent plastic strain at failure for Al5182-H19 is lower than that of Al6022-T4, implying lower ductility for the latter material. The extent of burr formation is related to the ductility of the material and is lower for materials with lower strain-to-failure values.

Discussions and Summary

This article discusses a set of experimental and computational studies aimed at understanding the effect of various processing parameters on the extent of burr and other defect formation during sheet edge-shearing and slitting

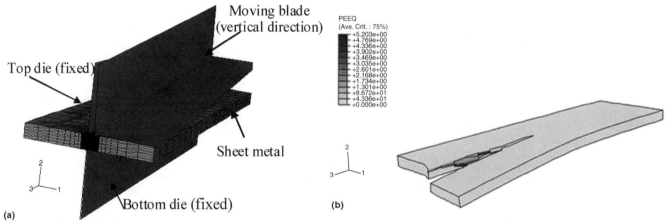

Fig. 17 Geometry of shear-slitting simulation. (a) Finite-element model setup for the shear-slitting process. (b) Deformed configuration for 0° rake and 0° cant angles

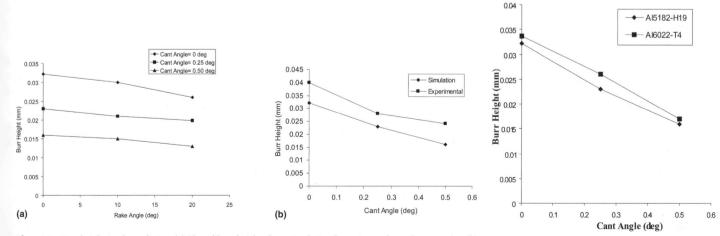

Fig. 18 Burr height in shear slitting. (a) Plot of burr heights from simulation for various rake and cant angles. (b) Comparison of burr heights obtained from simulation with experimentally observed burr heights for various cant angles

Fig. 19 Comparison of burr heights for 5182-H19 and 6022-T4 aluminum alloys for 0° rake angle

processes. The computational simulations use macroscopic constitutive and damage models that are calibrated with experimental results and microscopic image analysis. Ductile fracture, observed to be the primary mechanism governing the shearing process, may be modeled using various damage models. While macroscopic damage models have limitations with respect to representing precise microscopic crack initiation and plastic flow localization, as discussed in Ref 46 and 47, experimental verifications demonstrate that the constitutive and shear-based damage models can describe macroscopic responses and microscopic features, such as surface profiles and burr heights, with reasonable accuracy.

Sensitivity analyses exhibit the same computational and experimental trends in the dependence of burr height on process parameters such as blade-die clearances, blade radii, and rake and cant angles. Material ductility is also found to affect the development of burrs. A critical stage in shearing of ductile materials is the formation and propagation of plastic localization zones. Burr heights are predominantly influenced by the shape of the localization zones. These are not necessarily straight under an inhomogeneous state of stresses. The results can be used to estimate process parameters needed for minimizing burr formation. On the other hand, burnish depths are primarily influenced by blade sharpness and moderately by clearances and the cutting angle. Important conclusions have been reached through these simulations in Ref 36 and 39 as well as in the experimental studies of Ref 1 to 5. The best cutting conditions for aluminum alloys can be achieved through a range of cutting angles, irrespective of large clearances and dull blades. This lays the foundation for the design of robust processes that would require much less frequent tool sharpening and fewer restrictions on clearance control than the current practices.

Even though research on shearing ductile metals and alloys began in the early 1900s, and progress has been made, much more can be achieved through modern experimental and evolving computational methods. It is difficult to conceptualize the entire complexity of the deformation process to failure from a purely macroscopic mechanics perspective without taking the material microstructure into consideration. For example, in Ref 1 to 5 it has been observed that at certain cutting angles with intermediate clearances, the burr height first increases and then unexpectedly decreases with increasing blade-edge radii. Another observation is the existence of instability conditions in the parameter space, at which the cut surface quality and burr heights are extremely sensitive to cutting parameters. Also, there is a dramatic change from a shear mode to a shear-and-tear mode of the cut surface under certain cutting conditions. Macroscopic continuum mechanics considerations suggest that the deformation and fracture in the shear zone should be 2-D because of the 2-D applied loads (Ref 1). However, conspicuous nonuniformity of the cut surface implies that material inhomogeneity plays a key role.

Considerations at both the micro- and macroscale are necessary to understand some of the interesting phenomena discussed previously. Coupling multiscale experiments, microscopy, and image analysis with powerful multiscale computational analysis (Ref 46–48) can significantly improve the precision of predictive capabilities and provide important solutions to problems encountered in these industrial processes. Insightful microscale information can be fed into higher-scale simulations, as discussed in Ref 46, 47, and 49 in these analyses. Multiscale, multidisciplinary approaches can be extremely beneficial to integrate microstructure into macroscopic material performance and processing to provide solutions for important industrial problems.

REFERENCES

1. M. Li, An Experimental Investigation on Cut Surface and Burr in Trimming Aluminum Autobody Sheet, *Int. J. Mech. Sci.,* Vol 42 (No. 5), 2000, p 889–906
2. M. Li, Micromechanisms of Deformation and Fracture in Shearing Aluminum Alloy Sheet, *Int. J. Mech. Sci.,* Vol 42 (No. 5), 2000, p 907–923
3. M. Li, Sliver Generation Reduction in Trimming of Aluminum Autobody Sheet, *ASME J. Manuf. Sci. Eng.,* Vol 125, 2003, p 128–137
4. J. Ma, H. Lu, M. Li, and B. Wang, Burr Height in Shear Slitting of Aluminum Webs, *ASME J. Manuf. Sci. Eng.,* Vol 128, 2006, p 46–55
5. H. Lu, J. Ma, and M. Li, Edge Trimming of Aluminum Sheets Using Shear Slitting at a Rake Angle, *ASME J. Manuf. Sci. Eng.,* Vol 128, 2006, p 866–873
6. T.M. Chang and H.W. Swift, Shearing of Metal Bars, *J. Inst. Metall.,* Vol 78, 1950, p 119–146
7. J.R. Rice and D.M. Tracey, On the Ductile Enlargement of Voids in Triaxial Stress Fields, *J. Mech. Phys. Solids,* Vol 17, 1969, p 201–217
8. W. Johnson and R.A. Slater, Survey of Slow and Fast Blanking of Metals at Ambient and High Temperatures, *Proc. Int. Conf. Manuf. Technol., CIRP-ASTME,* 1967, p 825–851
9. W. Johnson, S.K. Ghosh, and S.R. Reid, Piercing and Hole-Flanging of Sheet Metals: A Survey, *Mem. Sci. Rev. Metall.,* Vol 77, 1980, p 585–606
10. A.G. Atkins, On Cropping and Related Process, *Int. J. Mech. Sci.,* Vol 22 (No. 4), 1980, p 215–231
11. A.G. Atkins, On the Mechanics of Guillotining Ductile Metals, *J. Mater. Proc. Technol.,* Vol 24, 1990, p 245–257
12. A.G. Atkins, Surfaces Produced by Guillotining, *Philos. Mag. A,* Vol 43, 1981, p 627–641
13. Q. Zhou and T. Wierzbicki, A Tension Zone Model of Blanking and Tearing of Ductile Metal Plate, *Int. J. Mech. Sci.,* Vol 38 (No. 3), 1995, p 303–324
14. M. Li and G. Fata, Trimmed Aluminum Sheet, U.S. Patent 5,820,999, 1998
15. E. Taupin, J. Breitling, W. Wu, and T. Altan, Material Fracture and Burr Formation in Blanking Results of FEM Simulations and Comparison with Experiments, *J. Mater. Proc. Technol.,* Vol 59, 1996, p 68–78
16. F.A. McClintock, Criteria for Ductile Fracture by the Growth of Holes Subjected to Multi-Axial Stress States, *Trans. ASME J. Appl. Mech.,* Vol 35, 1968, p 363–371
17. J. Lemaitre, A Continuous Damage Mechanics Model for Ductile Fracture, *J. Eng. Mater. Technol.,* Vol 107, 1985, p 83–89
18. R. Hambli, Numerical Fracture Prediction during Sheet Metal Blanking Processes, *Eng. Fract. Mech.,* Vol 68 (No. 3), 2000, p 365–378
19. R. Hambli, Prediction of Burr Height Formation in Blanking Using Neural Networks, *Int. J. Mech. Sci.,* Vol 44, 2002, p 2089–2102
20. V. Miguel and J.D. Bressan, A Computational Approach to Blanking Processes, *J. Mater. Proc. Technol.,* Vol 125–126, 2002, p 206–212
21. G. Fang, P. Zeng, and L. Lou, Finite Element Simulation of the Effect of Clearance on the Forming Quality in the Blanking Process, *J. Mater. Proc. Technol.,* Vol 122, 2002, p 249–254
22. M.G. Cockroft and D.J. Latham, Ductility and Workability of Metals, *J. Inst. Met.,* Vol 96, 1968, p 33–39
23. C. MacCormack and J. Monaghan, Failure Analysis of Cold Forging Dies Using FEA, *J. Mater. Proc. Technol.,* Vol 117, 2001, p 209–215
24. D. Brokken, W.A.M. Brekelmans, and F.P.T. Baaijens, Predicting the Shape of Blanked Products: A FEM Approach, *J. Mater. Proc. Technol.,* Vol 103, 2000, p 51–56
25. D. Brokken, W.A.M. Brekelmans, and F.P.T. Baaijens, Discrete Ductile Fracture Modeling for the Metal Blanking Process, *Comput. Mech.,* Vol 26, 2000, p 104–114
26. A.M. Goijaerts, L.E. Govaert, and F.P.T. Baaijens, Evaluation of Ductile Fracture Models for Different Metals in Blanking, *J. Mater. Proc. Technol.,* Vol 110, 2001, p 312–323
27. W. Klingenberg and U.P. Singh, FE Simulation of the Punching Process Using In-Process Characterization of Mild Steel, *J. Mater. Proc. Technol.,* Vol 134, 2003, p 296–302
28. ABAQUS Version 6.2, User's Manual, Hibbitt Karlsson and Sorensen Inc.

29. V. Tvergaard, Material Failure by Void Growth, *Adv. Appl. Mech.,* Vol 27, 1990, p 83–147

30. N. Aravas, On the Numerical Integration of Class of Pressure-Dependent Plasticity Models, *Int. J. Numer. Meth. Eng.,* Vol 24, 1987, p 1395–1416

31. H.H. Wisselink, "Analysis of Guillotining and Slitting: Finite Element Simulations," Ph.D. thesis, University of Twente, The Netherlands, Jan 2000

32. M. Oyane, T. Sato, K. Okimoto, and S. Shima, Criteria for Ductile Fracture and Their Applications, *J. Mater. Proc. Technol.,* Vol 4, 1980, p 65–81

33. M. Oyane, Criteria of Ductile Fracture Strain, *Jpn. Soc. Mech. Eng.,* Vol 15 (No. 90), 1972, p 1507–1513

34. W. Kubli, M. Maurer, and J. Reissner, The Use of FE Process Simulation in Tool and Machine Optimization of the Fine Blanking, *Adv. Met. Form. Mach.,* 1989, p 369–377

35. T.A.J. Nicholson, *Optimization in Industry,* Vol 1 and 2, Longman, London, 1971

36. S. Ghosh, M. Li, and A. Khadke, Three Dimensional Simulations of Rotary Shear-Slitting Process for Aluminum Alloys, *J. Mater. Proc. Technol.,* Vol 162 (No. 1), 2005, p 91–102

37. S. Roy, S. Ghosh, and R. Shivpuri, A New Approach to Optimal Design of Multi-Stage Metal Forming Processes by Micro Genetic Algorithms, *Int. J. Mach. Tools Manuf.,* Vol 37 (No. 1), 1997, p 29–44

38. S.E. Clift, P. Hartley, E.N. Sturgess, and G. W. Rowe, Fracture Prediction in Plastic Deformation Processes, *Int. J. Mech. Sci.,* Vol 32 (No. 1), 1990, p 1–17

39. A. Khadke, S. Ghosh, and M. Li, Numerical Simulations and Design of Shearing Process for Aluminum Alloys, *ASME J. Manuf. Sci. Eng.,* Vol 127 (No. 3), 2005, p 612–621

40. C. Chu and A. Needleman, Void Nucleation Effects in Biaxially Stretched Sheets, *J. Eng. Mater. Technol.,* Vol 102, 1980, p 249–256

41. M.S. Li, S. Ghosh, O. Richmond, H. Weiland, and T.N. Rouns, Three Dimensional Characterization and Modeling of Particle Reinforced Metal Matrix Composites, Part 1: Quantitative Description of Microstructural Morphology, *Mater. Sci. Eng. A,* Vol 265, 1999, p 153–173

42. E.R. Weibel, *Stereological Methods: Theoretical Foundation,* Vol 2, Academic Press, London, 1980

43. S. Ghosh and S. Raju, R-S Adapted Arbitrary Lagrangian Eulerian Finite Element Method for Metal Forming Analysis with Strain Localization, *Int. J. Numer. Meth. Eng.,* Vol 39, 1996, p 3247–3272

44. S. Ghosh, Arbitrary Lagrangian-Eulerian Finite Element Analysis of Large Deformation in Contacting Deformable Bodies, *Int. J. Numer. Meth. Eng.,* Vol 33, 1992, p 1891–1925

45. K. Nahshon and J. Hutchinson, Modification of the Gurson Model for Shear Failure, *Eur. J. Mech.,* Vol 27, 2008, p 1–17

46. C. Hu, J. Bai, and S. Ghosh, Micromechanical and Macroscopic Models of Ductile Fracture in Particle Reinforced Metallic Materials, *Mod. Simul. Mater. Sci. Eng.,* Vol 15, 2007, p S377–S392

47. S. Ghosh, K. Lee, and P. Raghavan, A Multi-Level Computational Model for Multiscale Damage Analysis in Composite and Porous Materials, *Int. J. Solids Struct.,* Vol 38 (No. 14), 2001, p 2335–2385

48. S. Ghosh, Adaptive Concurrent Multi-Level Model for Multi-Scale Analysis of Composite Materials Including Damage, *Multiscale Modeling and Simulation of Composite Materials and Structures,* Y. Kwon, D.H. Allen, and R. Talreja, Ed., Springer, 2008, p 83–164

49. S. Ghosh, J. Bai, and D. Paquet, A Homogenization Based Continuum Plasticity-Damage Model for Ductile Failure of Materials Containing Heterogeneities, *J. Mech. Phys. Solids,* Vol 57, 2009, p 1017–1044

ASM Handbook, Volume 22B, *Metals Process Simulation*
D.U. Furrer and S.L. Semiatin, editors

Copyright © 2010, ASM International®
All rights reserved.
www.asminternational.org

Modeling of Residual Stress and Machining Distortion in Aerospace Components

Kong Ma and Robert Goetz, Rolls-Royce
Shesh K. Srivatsa, GE Aviation

THE INSERTION OF NEW MATERIALS into aircraft systems takes several years and many millions of dollars. Experimental trials to define the manufacturing process to meet the specifications can add significant time and cost. Many military programs have small lot production in either initial engine development programs or specialized production, providing additional criticality to improving the first-time yield of manufacturing processes and quickly resolving production issues. Additionally, the impact of an unintended process change is unknown without evaluating the component, again adding time and cost to issue resolution. Therefore, new approaches are required to facilitate the rapid certification of materials and processes technologies. Significant improvements in manufacturing processes have been realized by process modeling tools such as DEFORM (Scientific Forming Technologies Corp.) and FORGE (Transvalor) for metal forming and Pro-CAST (ESI Group) for casting, which are now in routine industrial use. Modeling and simulation are critical for increasing the affordability of current and future aerospace materials and products and in developing and certifying materials in a shorter timeframe that more closely matches the product design cycle.

Introduction—Residual Stress, Distortion, and Modeling

Technical Need

Aircraft engine and airframe structural components that are machined from forgings or plate stock represent a significant cost for both military and commercial aircraft. Typical component applications, as shown in Fig. 1, are rotating disks in aircraft engines and structural components in airframes. The buy-to-fly weight ratio, which is the ratio of the forged material

weight to the finished component weight, is typically between 4 and 10 for such components. The excess material is removed by various machining operations, which are a major contributor to the cost of forged components.

Metallic components undergo various forming processes, such as casting, forging, rolling, and so on, in which the material is heated to high temperatures. A typical wrought component of a titanium- or nickel-base alloy begins as an ingot. The cast structure is broken down into billet form, which is then forged into the rough shape of the component, with positive stock surrounding the finished shape. Deformation occurring during the forming process causes residual stresses that can be compounded by thermal gradients. After forming, the components are subjected to a series of heat treatment processes to improve the microstructure and material properties (e.g., toughness, strength, creep, fatigue). Most heat treatment processes involve heating the material to a high temperature to produce a change in microstructure (e.g., phase transformation, recrystallization).

Nickel-base superalloy disks used in aircraft engines typically undergo a two-step heat treatment process: solutionizing and aging. The first step is a solution heat treatment, and the solution temperatures are often high enough (~1100 °C, or 2000 °F, depending on the alloy) to almost completely relax any preexisting residual stress induced during the forming process. When the components are removed from the heat treat furnace at the end of the heating and soaking process, they are transported to the quenching station, during which the components lose heat to the ambient atmosphere due to radiation to the surrounding surfaces, natural convection through air, and conduction to the handling mechanism through the direct contact areas. This period is called the quench delay or transfer time and usually is very short (15 to 60 s).

This period is followed by rapid quenching. When the component is subjected to the much cooler quench medium (typically oil, water, polymer, salt, or forced air), the outer surface of the component cools down rapidly, contracts, and metallurgically stabilizes when it reaches a relatively low temperature (e.g., below 480 °C, or 900 °F, for nickel alloys), while the interior is still at a high temperature. At this point, the outside of the component is under tensile stress and yields, while the interior material is under compression because the outer volume cannot contract (inward) against the yet-to-contract hot interior. Gradually, the heat from the inside dissipates outward. The interior material then tries to contract, but now the outer volume is already relatively fixed because it is at a much

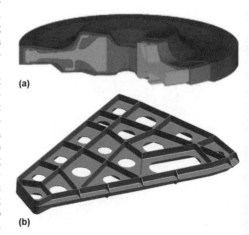

Fig. 1 Aircraft engine and airframe components with large buy-to-fly ratios and high machining costs. (a) Typical aircraft engine forging. Blue (dark outer region): forging shape; red (central region): intermediate shape; green (bright core region): finish machined shape; large volume of material machined away. (b) Typical airframe structural forging. Intricate geometrical features result in a large volume of material being machined away

lower temperature. Therefore, the interior is under tension and the outer region is under compression, because it is being pulled inward by the inner material. Plastic deformation subsequent to yielding induces bulk residual stresses in all directions.

Temperature gradients during quenching cause thermal stresses, which drive localized plastic deformation and residual stress buildup. Upon cooling to room temperature, residual stresses can exceed half of the alloy tensile strength. Often, to obtain favorable material strength and microstructure, fast cooling (quenching) is applied. Higher cooling rates result in higher residual stresses. Residual stresses resulting from thermomechanical processing can cause in-process cracking, machining distortion, in-service distortion, and/or lowered life.

The stress profile within a component depends on the local geometry features and the temperature difference between the near surface area and the quench medium (which determines the rate of heat loss). An area with a thinner cross section usually has lower stress than an area with a thicker section. Variations in residual stress occur due to variability in manufacturing process conditions, for example, the loading pattern of components in the furnace and quench medium, the agitation level in quench tanks, and the nature of the heat treat fixtures.

During the second heat treatment step, known as aging, the component is reheated to a temperature much lower than the solution temperature (typically 650 to 820 °C, or 1200 to 1500 °F) to form secondary and tertiary gamma prime in nickel-base superalloys. This step completes the transformation to the desired microstructure and properties, with the added benefit of stress relaxation through creep and recovery processes. The amount of stress relaxation depends on the time and temperature of the age cycle and the magnitude of the initial residual stress. Higher temperatures result in a greater degree of relaxation. Stress relaxation is related to the creep behavior of the material, and therefore, the microstructure (grain size and gamma prime) also affects stress relaxation. If the level of residual stress is below the steady-state relaxation stress, further relaxation will not occur unless higher temperatures are used.

Thus, residual stress cannot be eliminated, only reduced during final aging, and the component still has enough residual stress to affect its behavior during machining and in service.

Component distortion can be caused by material bulk stresses resulting from heat treating operations and/or by local near-surface machining-induced stresses. When the component cross section is thick, bulk residual stresses dictate component distortion. As the cross section is machined thinner (~3 mm, or 0.125 in.), surface residual stress begins to play a more significant role in component distortion.

The prediction of residual stresses at quantified levels of uncertainty can improve processing methods, component design, robustness, performance, and quality as well as achieve more efficient material utilization and aircraft system efficiency, which result in lower environmental impact. Distortion of machined titanium and nickel alloys contributes significantly to the cost of these components. Heat treatment and machining are the two critical operations in the manufacture of engine and airframe components that influence residual stress. Residual stresses and associated distortion have a significant effect on manufacturing cost in four distinct ways.

First, the forging and intermediate heat treat shapes contain additional material to account for expected distortion (Fig. 2). This material, added to ensure a positive material envelope over the finished component shape, represents a raw material cost and increases the machining cost. This additional material also imposes a limit on the benefit of near-net shape forging, which is being vigorously pursued by the industry.

Second, component distortion during machining requires that the machining process engineer plan machining operations and fixtures so that distortion does not compromise the finished component shape. Distortion during and after machining can result in added operations, such as lineup and straightening, rework, or scrap. Typically, additional machining operations and setups are added in a time-consuming and costly trial-and-error approach to minimize the effects of component distortion. For example, components such as disks are machined alternately on either side in an attempt to

stepwise balance the distortion. The time spent "flipping" components erodes productivity for thick, stiff components; for thin components, the strategy may be inadequate.

Third, residual stresses and associated distortion add complexity to machining process development and shop operations. Distortion affects the details of the machining plan and the way the component interfaces with machining fixtures. These effects generally vary between material suppliers, from lot to lot, and from one machining process to another. Distortion thereby not only influences the effort incurred during initial development of machining plans but may require adjustments after the initial plan has been set.

Finally, distortion results in preload of aircraft structures and fasteners and can cause assembly problems. Manufacturing residual stresses can adversely impact the behavior of the components during service. Distorted and prestressed components can result in fatigue capability degradation by increasing the local mean stresses. Residual stresses affect the dimensional stability of rotating components in aircraft engines. These components are exposed to high temperatures for long times, and distortions can affect system tolerances, clearances, and efficiency. For an accurate analysis of component behavior during service, the manufacturing residual stresses must be included as initial conditions.

The physics and mathematics of residual-stress redistribution within a component during machining are well understood. Determining residual stresses and subsequent distortion requires modeling using finite-element methods. This method has been used to evaluate the effect of processing conditions on residual-stress development and the effect of residual stresses on distortion during machining. The buildup of residual stresses during heat treatment and machining are difficult to assess using intuition, engineering judgment, or empirical methods. The physical interplay of quench heat transfer, elevated-temperature mechanical behavior, and localized plastic deformation is complex. Subtle changes in processing conditions and component geometry can significantly affect the magnitude and pattern of residual stresses.

For routine use, a fast-acting, validated, physics-based model with sufficient fidelity and robustness is needed to accurately predict the effects of thermomechanical processing and reduce scatter in residual stress, microstructure, mechanical properties, and their measurement. Residual-stress modeling technology must be standardized to meet an industry requirement for accuracy and capability in manufacturing (distortions), service (dimensional stability), service-life estimation (fatigue life, crack initiation, crack growth and propagation), and material testing (measurement scatter and sampling effects).

There is a need to understand the effects of heat treating and machining on distortion and

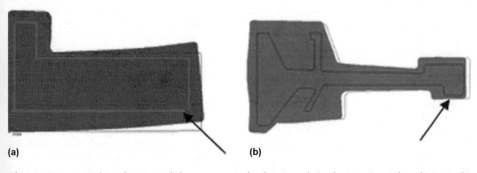

Fig. 2 Extra material envelope is needed to compensate for distortions during heat treating and machining to the inside finished component shape

(a)   (b)

to predict, minimize, and control these distortion-related processes to achieve robust six-sigma quality. There is a need to develop heat treatment and machining processes for minimizing distortions, realizing that this is not always the same as minimizing residual stresses. In addition, there is a need to accurately predict residual stress and consider its impact on component life and behavior in service. The industrial drive is toward stronger and longer-lasting components with higher-temperature capability. As new materials with better properties to meet more exacting requirements are introduced, they will be more difficult to machine. While materials scientists are developing higher-temperature materials, it is also possible to further improve existing designs and materials. One way to improve component design and increase life is by understanding the distribution of residual stress from the manufacturing process and linking it with the product life cycle. Modeling will help reduce machining problems and thereby enable more rapid introduction of high-performance materials and components.

Objectives of Residual Stress and Machining Distortion Modeling

The overall objective is to develop and validate a high-productivity modeling method that accurately predicts the magnitude and pattern of distortion during machining of forgings used in aircraft engines and airframe structures and to establish an approach for using machining modeling to generate machining plans that yield less component distortion and reduce the cost of machining.

Metals Affordability Initiative Programs

For almost a decade, the United States Air Force (USAF) Metals Affordability Initiative (MAI) has devoted significant resources to understanding the impact of residual stresses on component variability and machining distortion. The MAI team consists of materials developers, forgers, software developers, universities, aircraft engine makers, and airframers to bring the real-world perspectives of the entire supply chain to the project. The methods developed represent a sound engineering practice for predicting machining distortion and are available for licensing in commercial codes. Many aerospace original equipment manufacturers and their suppliers now have established in-house analysis methods. The MAI projects have reduced the time to implementation of the process technology by permitting a focused, larger-scale, complete effort across engine manufacturers, airframers, and material and machining suppliers than would be possible if efforts were conducted by individual companies alone.

Significant progress was made in the development and validation of two-dimensional (2-D) modeling tools for predicting machining

distortions in the USAF MAI Dual-Use Science and Technology (DUST-7) Program, Cooperative Agreement F33615-99-2-5216. This program advanced the state of the art in going from the previous state of a time-consuming, manual, partially validated, not-production-ready procedure to an automated, high-productivity, user-friendly, fast-acting, validated, commercially supported, and production-ready analysis tool that can be used to achieve significant cost-savings. It was shown that 2-D distortions can be predicted to within the typical process variability of ± 20 % or ± 5 mils (0.125 mm).

Aircraft engine rotating components are 2-D axisymmetric up until the final machining operations, when three-dimensional (3-D) features such as dovetail slots, cooling holes, and so on are machined. For this reason and because of the simpler nature of 2-D models as compared to 3-D, the development of 2-D tools was addressed first. The 2-D model was rigorously validated first on simple-shaped forgings and later for complex shapes in a production environment. Two-dimensional rotating disks account for the majority of aircraft engine forgings. The machining model developed realistically captures the process boundary conditions (tooling constraints) for any user-specified sequence of machining operations. The method was rigorously validated first on simple shapes in a well-controlled situation and then extended to complex shapes in a production environment. The material chosen for this program was cast and wrought U720, but the model/method is pervasive and can be employed for other materials.

The 2-D model has been implemented at various aircraft engine makers, and it has been used successfully for many production components. This analytical tool guides machining operation sequence and tooling design for rotor hardware to minimize component distortion, which was previously predominately an experience-based trial-and-error process.

Following the successful completion of the 2-D program, the MAI team has developed 3-D distortion modeling tools. The results presented in this article are largely based on the MAI programs.

Modeling of Heat-Treat-Induced Residual Stress

Finite-Element Residual-Stress Analysis

Residual-stress analysis involves:

- Determination of heat-transfer coefficients during quenching
- Measurement of material constitutive behavior (elastic-plastic creep) at processing conditions
- Finite-element analysis to calculate thermal and stress fields
- Finite-element analysis for machining distortion

- Finite-element analysis for in-service distortion and for strength and service-life estimation

Commercially available process modeling tools, such as DEFORM and FORGE, are finite-element-based analysis tools. They employ an elastic-plastic formulation, which is the necessary basis for the model formulation. The detailed steps of setting up a model for the heat treatment process vary depending on the software used. However, they all involve several general steps:

1. Construct the heat treat geometrical model from computer-aided design (CAD) tools.
2. Obtain the thermophysical material properties for the component material. This includes thermal conductivity, heat capacity, density, thermal expansion coefficient, Young's modulus, and flow stress describing plastic behavior. All properties are temperature dependent and also depend on the microstructure and previous processing history of the material. The use of realistic material data is critical to the success of any modeling effort.
3. Obtain the detailed process information. This includes the heat treat solution temperature, soak time, quench delay time, quench medium type, medium temperature, component loading configuration, and so on. Similar processing data are needed for any subsequent stress relief (or aging) processes.
4. Determine the heat-transfer coefficients (HTCs) for the interface between the quenching medium and the component. Using the analytical approach, one can use the sophisticated two-phase (gaseous and liquid) flow computational fluid dynamic methods to simulate the quenching agitation interacting with the specific component geometry. This requires very detailed characterization of the physical quenching configuration and is time-consuming. A much simpler approach is to experimentally determine the effective HTCs by instrumenting an experimental component with thermocouples. The HTCs are then input into the modeling software as time- and temperature-dependent boundary conditions.
5. Mesh the component geometry. All the general rules and guidelines used for standard finite-element analysis apply to the process model; for example, quad elements are better than triangular elements, a higher-density mesh yields more accurate results, and so on. One point worth noting is that the elastic-plastic formulation is used to perform the thermomechanical analysis of components. Therefore, particular attention should be given to the mesh density in the areas subjected to steep thermal and mechanical changes during the heat treat cycle.
6. Run the heat treat model. Because of the complexity of the algorithm and the time and temperature base of the process model, it usually takes from a few hours for a

1. "One-step" from heat treat shape to finished part shape

- Method: Material removal in one pass
- Pros: Simple, predicts trends
- Cons: Path-dependence and clamping ignored; inaccurate for large distortions

2. "Multi-step" procedure with predetermined material removal

PRE-HT   POST-HT   PASS 1   PASS 2   PASS 3   FINAL

- Method: Remove material in a pre-determined multi-pass sequence
- Pros: In-process distortion prediction
- Cons: Path-dependence and clamping ignored; inaccurate for large distortions

3. "Multi-step" procedure with path-dependent material removal

- Method: Multi-pass material removal; complete remeshing at each step
- Pros: More realistic; workpiece/tooling interactions considered
- Cons: Remeshing interpolation errors; more involved to set up model

Fig. 3 Comparison of various machining distortion prediction methods. The multistep procedure with path-dependent material removal most accurately represents what is happening in practice

mid-sized 2-D model to several days for a complex 3-D model. Like any finite-element analysis, if the solution fails to converge, the use of smaller simulation time-steps and/or altering the mesh can help eliminate the problem.

Distortion from previous steps carried over

(a)

Distortion from previous steps _not_ carried over

(b)

Fig. 4 (a) Boolean and (b) remeshing procedures

Modeling Procedures (2-D)

Generally, three procedures have been commonly used to model the machining distortion (Fig. 3). All of these techniques use different methods for how the material is removed during machining and the subsequent re-equilibration of residual stresses. All of the techniques described as follows neglect the surface residual stress induced by the interaction between the machine tool and the component, and therefore, the effect of cutting conditions is also ignored. The effect of machining-induced stresses is addressed in a later section in this article.

Method 1. In the one-step procedure, plastic strains from the heat treat shape are mapped onto the machined shape, and the strains and stresses are re-equilibrated to obtain the resulting distortion. In essence, this method means that all the material is machined off instantaneously in one machining pass. This method is straightforward and easy to implement; it avoids remeshing associated with modeling each machining operation, and it predicts bulk distortions and trends correctly. However, it ignores the influence of the machining path and the effect of in-process shape change on the workpiece/fixture interface, and its accuracy decreases with increasing distortion.

Method 2. The multistep procedure with predetermined material removal is similar to the one step procedure, but material is removed in multiple passes based on a predetermined machining sequence. The workpiece is meshed up front to follow the machining sequence. This method also avoids remeshing associated with modeling each machining operation, and it predicts bulk distortions and trends correctly. However, it ignores the influence of the machining path and the effect of in-process shape change on the workpiece/fixture interface, and its accuracy decreases with increasing distortion. The initial meshing is more involved than the one-step method. If changes in machining sequences are to be evaluated, the heat treat analysis must be completely redone.

Method 3. In the multistep procedure with path-dependent material removal, a complete remeshing is performed at each machining operation, and the material removal follows the actual machining sequences. This is a more realistic representation of the machining process, and it accounts for in-process distortions and workpiece/tooling interactions. It is most involved to set up the model. The material

removal can be accomplished in two ways: Boolean or remeshing.

Boolean Procedure. The material inside the machining path is removed to obtain the new geometry by a Boolean operation of the current geometry and the machining path (Fig. 4a). The new geometry follows the distorted geometry from the preceding step everywhere except where the machining cut is taken. This new geometry, which represents the workpiece shape after machining, is remeshed and the stresses/strains re-equilibrated to obtain the unrestrained distortion. When the Boolean cut is very thin (e.g., on the last pass) and the workpiece is not constrained during cutting, there will be very little additional distortion, and the cut face will follow the cutting tool path.

Remeshing Procedure. The plastic strains from the current geometry are mapped onto the machined geometry and the stresses re-equilibrated to obtain the resulting distortion. The new geometry follows what the user has predefined and not the distorted geometry from the previous step. The new geometry is indicated by the solid line. Remeshing to this new

Fig. 5 Distorted airframe component. This component was machined flat. The material stress and machining-induced stress are causing it to distort

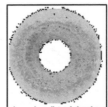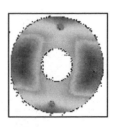

Fig. 6 Optical scan pictures showing axisymmetric and nonaxisymmetric distortion (due to nonaxisymmetric fixturing) following heat treatment. Colors indicate axial distortion.

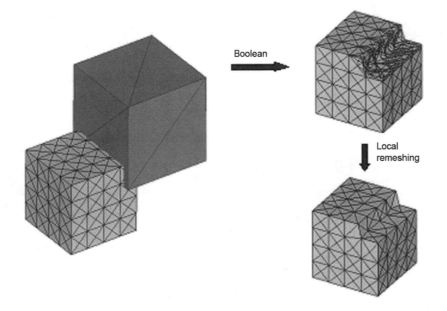

Fig. 7 Local remeshing of elements affected by the machining process

geometry causes the distortion from the preceding steps not to be carried through, as shown on the uncut faces in Fig. 4(b).

Of the three methods shown in Fig. 3, method 1 is straightforward, and methods 2 and 3 require a large amount of user time to set up the problem for a general multistep machining process. As a result, such analyses are not performed routinely. An automated version of method 3 that minimizes model set-up time, streamlines the overall procedure, and minimizes the interpolation error was developed in the MAI programs.

Modeling Procedures (3-D)

Three-dimensional distortion modeling is needed to address airframe structures and complex engine components. Figure 5 shows a distorted airframe component with a 3-D geometry. Figure 6 shows both axisymmetric and nonaxisymmetric distortions of a nominally 2-D axisymmetric-shaped engine disk forging.

In the modeling of the material-removal process during machining, a new finite-element mesh must be generated on the as-machined shape. Residual stresses and strains must be interpolated from the premachined shape (mesh) to the postmachined shape (mesh). This process of interpolation introduces errors in the simulation that can be significant if the component geometry has thin walls adjacent to thicker sections. This problem is more acute in 3-D modeling than in 2-D modeling because of the increased geometrical complexity of 3-D shapes (thin/thick sections) and because of the limitations on the fineness of the mesh that can be employed in 3-D to keep the computations manageable. The solution accuracy was improved with a combination of controlling the local mesh density to have finer elements in thin geometrical features and high-stress gradient regions, local remeshing, and improved interpolation schemes.

Local Remeshing. During global remeshing, a completely new mesh is generated over the entire volume of the workpiece. Therefore, every element and node in the model is changed, and the state variables are interpolated from the old mesh to the new mesh. This introduces large interpolation errors. During material removal, generally only a small volume of the workpiece geometry is altered. Figure 7 shows a schematic of the material-removal process. In the local remeshing methodology, only the elements along the machined surface and their neighboring elements are remeshed. As compared with the original mesh, 86.7% nodes and 79.6% elements remain unchanged in this example. During data interpolation, only the modified nodes/elements are affected. Therefore, interpolation error can be avoided for the major part of the mesh that remains unchanged.

Improved Interpolation. Two new interpolation schemes were developed. In the first, the interpolation is performed based on a local polynomial fit. In the second, the element variables are also stored at the nodes and the nodal values used during interpolation, which avoids the error during transfer of element data to nodes. The interpolation error was reduced significantly with the new interpolation schemes. When interpolating onto the same mesh, the error in radial distortion was reduced to 2.5% as compared to 50% with the old method.

The method of combining local remeshing, improved interpolation, and mesh windows to control element size helps reduce errors. In general, this method results in peak values of stress and strain being retained more accurately than the previous methods; that is, there is less smoothing error with the new method. Simulation results were found to be in good agreement with experimental data. All predictions were within 20% of the measurements with the new method.

To easily set up the model for multiple machining operations and passes, a template (Fig. 8) was set up that can position the workpiece, the fixtures and loads, and the material removal in a user-friendly setup. A preview of all the machining steps ensures error-free setup before the simulations are commenced.

To facilitate the material-removal process for machining distortion modeling, the machining path information (described by G-code) was converted to a geometry that can be used to generate the machined workpiece configuration (Fig. 9). A G-code interpreter was developed and tested with several examples (Fig. 10).

During the simulation of residual stresses and machining distortions, it is necessary to constrain the six degrees of freedom in the workpiece to eliminate rigid body motion. Guidelines were developed on the selection of the nodes and the manner in which they must be constrained. A preprocessing function was developed to automate/guide the definition of these boundary conditions. The method is:

1. Fix one node in x-, y-, z-directions. This removes the three degrees of freedom in translation.
2. Find a point at the same x and z but different y. Fix this in the z-direction; it removes x-rotation.
3. Find a point at the same y and x but different z. Fix this in the x-direction; it removes y-rotation.
4. Find a point at the same z and y but different x. Fix this in the y-direction; it removes z-rotation.

The machining distortion solution is dependent on the chosen reference boundary conditions to constrain rigid body motion. A facility was developed to allow the user to easily select reference points/planes/axes to represent the predicted distortion in the selected frame of reference. This feature enables the display of the distortion solution in any frame of reference and enables easy comparison with measured data. In the free-state distortion simulations, six degrees of freedom are removed by assigning boundary condition constrains to the model. Depending on the locations where the boundary condition constraints are applied, the distortion results may appear to be different. For the 2-D example shown in Fig. 11, the distortion results appear to be different with respect to where the constraints are applied. However, if the distorted models are rotated and translated appropriately, the distortion results would be the same. This means that results using different boundary condition constraints can be converted to the same results using a suitable reference frame definition. Various other improvements were made to facilitate the display of distortions in an easily usable format, for example, axial runout display, and so on.

Modeling Data Requirements

Material Characterization

The MAI programs have focused on three materials: Ti-64, U720, and alloy 718. However, the model/method is pervasive and can be applied to other materials. The data needed to do this are listed as follows. All data should cover the range from room temperature up to the heat treat temperature:

- Constitutive behavior (stress-strain in plastic region)
- Young's modulus
- Creep and stress-relaxation data
- Poisson's ratio
- Thermal expansion coefficient
- Heat capacity
- Thermal conductivity
- Heat-transfer coefficients during the entire quench process

On-Cooling Tensile Tests

On-cooling tests are used to generate data describing the constitutive behavior of the material: stress as a function of strain, strain rate, and temperature. Data should be generated at a minimum of two different strain rates. Details of the testing procedure are as follows:

- Heat the tensile specimen to the heat treatment solution temperature and hold at this temperature for 20 min.
- Cool the tensile specimen at a specified cooling rate (representative of the cooling rate during the actual quenching process) from the solution temperature to the test temperature, then hold at the test temperature for approximately 10 min for temperature stabilization.

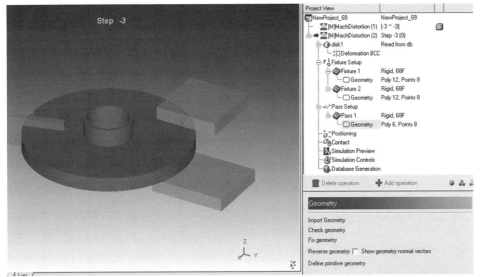

Fig. 8 Machining distortion template

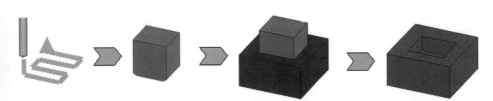

Fig. 9 Boolean geometry creation from machining G-code, material removal, and machined component

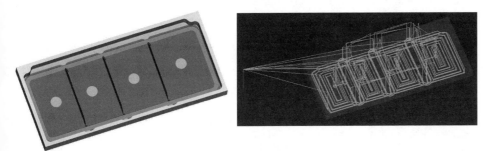

Fig. 10 G-code converter: machining path (G-code) converted to material-removal geometry

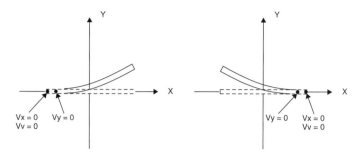

Fig. 11 Two-dimensional illustration showing different distortion results using different constraints

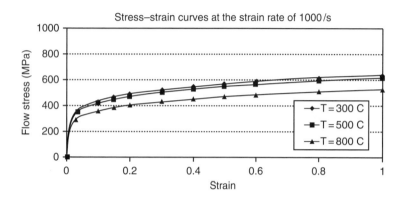

Fig. 13 Typical high-strain-rate flow-stress data

Fig. 12 Heat-transfer coefficient measurement using a disk outfitted with thermocouples

- Conduct tensile testing at this test temperature at a strain rate of 0.005 in./in./min to yield, then at a strain rate of 0.05 in./in./min to fracture.
- This thermal cycling procedure follows the thermal history of the forging during heat treatment, and so, the data generated are representative of the heat treated material.
- Conduct the tests over the temperature range from room temperature to the heat treat temperature.
- These tests give both the Young's modulus and the plastic behavior of the material.

Stress-Relaxation/Creep Tests

There are two methods for generating the data needed for modeling the stress-relaxation behavior during aging. One is to use stress-relaxation curves for the appropriate temperatures and heat treatment condition of the aging cycle. The second is to use data from creep tests. The stress-relaxation test consists of prestraining a tensile specimen to a high elastic strain or just over the yield limit. The displacement is then fixed, and the stress-relaxation curve is recorded for the entire aging time, if possible. Typically for superalloys, stress decreases linearly with log(time). In a creep test, the strain is measured under a given applied stress that is held fixed over time. The stress-relaxation technique requires fewer tests than the traditional creep technique to cover the entire stress/temperature/time behavior of the material. Stress-relaxation curves can be converted to creep strain rate versus stress for use in finite-element models to analyze stress relaxation. It is critical to generate these data with the appropriate microstructure material.

Thermophysical Property Tests

There are ASTM International standard tests for measuring the various thermophysical properties—thermal expansion coefficient, thermal conductivity, and heat capacity—so these tests are not described here (see the Section "Input Data for Simulations" in this Handbook for several articles on thermophysical properties).

Heat Treat Thermocouple Tests

Heat-transfer coefficient data should include transfer from the furnace to the cooling station in addition to the main quench itself (fan, water, polymer, salt, or oil). Typically, the HTCs are a function of temperature and the position on the workpiece. These data are specific to the quench facility used. Accurate HTC data are critical for correctly predicting residual stresses and subsequent machining distortions. The prevalent method of determining HTCs for furnace heatup, transfer, and quench (various media) uses thermal data from a quenching experiment (Fig. 12). This method involves a number of subjective decisions that can significantly impact the accuracy of the results. Inverse methods (2-D or 3-D) for obtaining HTCs are prone to instability and nonunique solutions. Problems exist on the validity of transferring a set of HTCs obtained on one shape to a different shape and in capturing localized distributions at critical geometrical features. An alternative method is to use computational fluid dynamics to predict coolant flow and obtain HTCs using well-established correlations to fluid flow. Computational fluid dynamics has only been used occasionally for this purpose, due to its complexity and lack of accuracy for boiling heat transfer in oil or water quench.

High-Strain-Rate Flow Stress for Machining

For realistic modeling of the machining process, accurate material property data are needed. Flow-stress data are needed over the range of strain, strain rate, and temperature that exist in machining operations. Obtaining the flow stress for use in metal-cutting simulation is difficult because of the high values of strain and strain rate that are involved. Conventional tests (e.g., compression and tensile tests) cannot be used to obtain reliable flow-stress data under cutting conditions. Flow-stress data were measured for mill-annealed Ti-6Al-4V and for alloy 718 by the Engineering Research Center for Net Shape Manufacturing at The Ohio State University. Cutting forces were measured for slot milling tests on plate samples. Flow stress was calculated from the experimental forces and plastic zone thicknesses. The flow-stress data were then validated through finite-element method simulations of orthogonal turning. The advantages of this approach are reduced experimental effort and cost compared to conventional material testing (e.g., compression and tensile tests). Typical high-strain-rate flow-stress data are shown in Fig. 13.

This method is limited by Oxley's assumptions (Ref 1):

- Tool edge is assumed to be sharp.
- Chip formation is of the continuous type (no serrations).
- The width of the cut must be more than 10 times the feed rate to satisfy the plane-strain assumption.

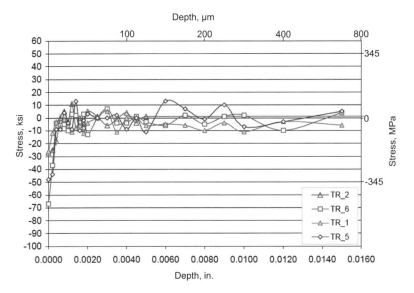

Fig. 14 Residual stresses in the tool axis direction. X-ray diffraction measurements; Ti 6-4 sample 90° orientation. Values less than zero indicate compressive residual stress and values greater than zero indicate tensile residual stress.

- Stress and temperature on the shear plane and the tool-chip interface are averaged.
- No built-up edge appears on the tool.

A methodology to obtain flow-stress data suitable for machining simulation using inverse numerical analysis was developed. In the inverse calculation, the error between the measured cutting forces and the forces predicted by DEFORM are minimized using an optimization approach. DEFORM is capable of modeling the machining process in either a transient or a steady-state mode. The steady-state approach is significantly faster (~15 min) than the transient approach (~20 h), with year 2010 state-of-the-art personal computers. The inverse analysis requires multiple simulations to reach the minimum of the objective function (least error between the measured and predicted cutting forces). To shorten the simulation time, the steady-state approach was used to perform the inverse analysis. The 2-D steady-state method was validated by comparing it with the transient approach.

The Johnson-Cook flow-stress equation was used:

$$\sigma_{eq} = (A + B\varepsilon^n)$$
$$\left(1 + C \ln\left(\frac{\dot{\varepsilon}}{\dot{\varepsilon}_0}\right)\right)\left(1 - \left(\frac{T - T_{room}}{T_m - T_{room}}\right)^m\right)$$
$$\text{(Eq 1)}$$

where σ_{eq} is the material flow stress; A, B, n, C, and m are five material constants; T is the absolute temperature; T_{room} is the room temperature; and T_m is the melting temperature. (See the article "Evaluation of Workability for Bulk Forming Processes" in *Metalworking: Bulk Forming,* Volume 14A of *ASM Handbook*.)

The procedure was validated using experimental cutting force data from the Aerospace Manufacturing Technology Center. The process was longitudinal turning of a tube made of alloy 718. The inverse analysis was carried out on an Itanium machine, and it took approximately 31 h and 70 iterations to find the minimum of the objective function.

Residual-Stress and Distortion Measurement Techniques

Residual-Stress Measurement

All residual-stress measurement methods are indirect and rely on converting a measured strain (e.g., slotting, hole-drilling, ring-core) to a stress. The inverse procedure leads to high measurement scatter (>100% between different sources and methods) for complicated 3-D stress states. In addition, some destructive measurement techniques can change the residual-stress state as part of the measurement itself. There is an ASTM International standard that outlines the limitations of various measurement methods. Measurement techniques differ in respect of the stress components measured, depth (near surface versus through thickness), mapping dimensionality (1-D, 2-D, 3-D), spatial resolution, sensitivity at low stress levels, destructive versus nondestructive, and near-surface resolution. No single measurement technique is applicable in all cases, and validation requires a combination of measurements depending on component geometry and surface-versus-internal measurements. Validation requirements are dependent on the application requirements (manufacturing distortion, in-service distortion, service-life estimation).

In x-ray diffraction (XRD), the stresses are obtained from the measurement of the crystal lattice strain. The stress obtained is an average over the x-ray beam volume. The accuracy of XRD depends on grain size and therefore the material type. To measure the stress profiles inside or near the surface, some material must be removed to expose the target area. This will affect the stress equilibrium in the component. Therefore, correction methods are needed to obtain the original stress before the material removal. In spite of these limitations, the measurement technique that is most robust for determining the machining stress profile is XRD. However, even this method has the difficulty of collecting data at a sufficient number of points due to the small depth of machining-induced stresses. In addition, it is expensive to gather a large amount of x-ray data.

To test the applicability of the XRD method in measuring machining-induced residual stresses, measurements were performed on four specimens. Measurements were made at 0.0002 in. (5 μm) intervals up to 0.002 in. (50 μm) into the workpiece. Afterward, measurements were made at 0.0005 in. (13 μm) intervals until 0.005 in. (125 μm). Beyond 0.005 in., measurements were taken at 0.001 in. (25 μm) intervals. These results were corrected for material removal at various depths. The XRD measurements indicate high surface stresses but with shallow depth. The stresses do not propagate greater than 0.001 in. below the surface (Fig. 14). With measurements taken at 0.0002 in. depth intervals, only three to five meaningful data points were obtained. The repeat measurement at another test laboratory indicated the same magnitude of stresses but twice the depth of penetration. This shows the uncertainties and variability inherent in all stress measurement techniques. All measurement techniques are indirect and convert a measured strain to stress.

Other less-common techniques can also be employed, such as microslitting, synchrotron, contour method, and so on.

Feasibility Demonstration of Microslot Milling and Distortion Measurement

Tests were performed at Microlution, a designer and builder of micromilling machines, to determine the feasibility of using microslot milling to remove very fine layers of material from the machined surface and to measure the resulting distortion to investigate machining-induced stresses present in the sample. Typically, machining-induced stresses are within the first few hundred micrometers from the surface in titanium. Figure 15 shows a schematic of this process. The sample is clamped near one edge, cantilevering the remainder. First, a measurement device (e.g., confocal laser) is used to measure the initial contour along the path shown by the dotted lines. Next, a small strip of material is removed by micromilling in the form of a rectangular slot the full length of the sample with width w and depth d. The measurement along the dotted lines is then repeated to determine any distortion caused by the layer removal. This process is repeated multiple times to measure the change in distortion caused by each layer removal. The measured change in distortion is related to the removal of machining-induced stresses present in the layer that was last removed and the stiffness properties of the sample.

For aluminum samples, the distortions are approximately 0.005 in. in magnitude. The measurement noise using a laser triangulation measurement system is approximately 0.0001 to 0.0002 in. This provides a signal-to-noise ratio of 50 to 1, and the method is effective. However, the method was not feasible for titanium samples, due to low signal-to-noise ratio at the surface where the larger stresses are located. The sample is stiffest during the removal of the first few layers, when nearly all of the material is intact, and most of the stresses are present in the first few layers. Thus, the distortion measurement is least sensitive in the most critical regime.

Distortion Measurement

When validating the modeling predictions of machining distortions with experimental measurements, the data should be gathered on the face opposite to the one machined in the current operation, before and after this operation. The difference between the before/after measurements gives the distortion induced during this machining operation. This quantity should be compared with modeling predictions. Distinction should also be made between distortion when the component is clamped in the machine fixture versus the free state when all external loads are removed.

If an attempt is made to correlate the modeling predictions of machining distortions with experimental measurements on a face just machined, this can potentially involve large errors. After the workpiece is removed from the fixtures, a free-state dimensional inspection is made. The difference between the nominal undistorted shape and the free-state dimension is the machining distortion induced during this operation. However, in practice, the nominal undistorted state may be slightly offset. Such offsets can occur because the workpiece is not perfectly flat or axisymmetric and cannot be exactly positioned in the fixture on the cutting machine. So, machining distortions induced during the current operation on the cut face are confounded with positioning errors. Axial drops on the just-machined face measured relative to the reference point will be sensitive to the exact amount of material removed and the distortion that occurs as the material is being removed. Therefore, a realistic comparison of the predicted and measured distortions on the just-machined face is difficult.

Dial indicators are the most commonly used for in-process measurement. When the component is still clamped in the fixture, one can use the machine turret to carry a dial indicator to scan the prismatic surfaces and note the distortion contour.

Coordinate measurement machines (CMMs) are also very common for measuring multiple features on more complex parts (Fig. 16). When compared with the modeling results, the model must be in a free-standing state unless the part stays in the same fixture during machining when presented to the CMM.

Optical scanning techniques (laser or fringe projection) have become more mature and more accurate in recent years (Fig. 16). Optical scanning has a unique advantage because it is capable of providing a large amount of digital data of the component profile in a very short time, which none of the other techniques can offer. Because of the size of the $X/Y/Z$ point cloud data file (can be up to millions of data points), special software such as Geomagic, Polyworks, RapidForm, or Surfacers is needed to interrogate the data and compare with CAD models.

Model Validation on Engine-Disk-Type Components

2-D Residual-Stress Validation on Engine-Disk-Type Components

Model validation was conducted first on simple pancake shapes and then on complex production shapes. The experimental heat treat conditions were selected to maximize residual stresses and subsequent machining distortions. The intent was to intentionally generate large residual stresses and machining distortions in order to measure them accurately and to avoid large errors in experimental measurements, which can prevent meaningful model validation. The heat treat cycle consisted of heating the U720 forgings from room temperature to the solution temperature of ~1100 °C (2000 °F), holding at temperature for 2 h, followed by a 30 s transfer time from the furnace to the fan cooling station, then fan cooling for 10

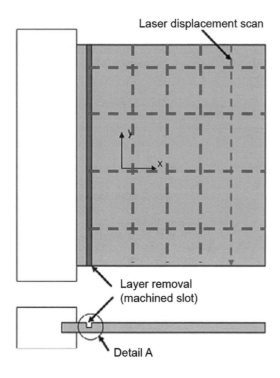

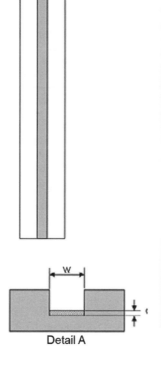

Fig. 15 Layer removal and distortion measurement schematic

Fig. 16 Coordinate measurement machine inspection and optical scanning setup for measuring distortions

min, after which the forgings were cooled to room temperature in still air.

The heat treatment of a Rolls-Royce production disk (Fig. 17) was performed for two cases: the current oil quench process and a proposed fan quench process as an improvement of the current process. The current production oil quench process has high cooling rates. The proposed fan cooling process results in a close-to-uniform cooling rate in a large volume of the forging of a magnitude that would meet the mechanical property requirements for the disk. This uniformity in cooling rate reduces residual stresses, heat treat distortions, and subsequent machining distortions as compared to the nonuniform cooling rates achieved by the oil quench process. The tensile residual stresses in the middle of the disk are reduced by more than 50% in the fan cooling process as compared to the oil quench process. The heat treat distortion is reduced by approximately 70% in the fan cooling process as compared to the oil quench process.

Residual stresses at the end of heat treatment were predicted. For simplicity, it was assumed that any residual stresses from prior forging operations were not significant and were relieved during the heatup-and-hold stage of solution heat treatment. This assumption is reasonable because the yield stress and creep strength of the material are small at the solution temperature, and therefore, any prior manufacturing residual stresses would be relaxed. The residual stresses in the forgings were primarily induced during quenching. Sensitivity studies were performed to establish that the results were only slightly affected (~5%) with respect to finite-element mesh size and variations in the HTCs. An uncertainty of ±10% in the HTCs is typical of production conditions.

Given an accurate residual-stress profile, well-defined constraints imposed during machining, accurate material properties, and a well-characterized metal-removal plan, prediction of component distortion should agree reasonably well with measured dimensional changes. However, prior attempts to match measured distortion values against prediction have shown only qualitative agreement. The validation of complex models can easily be frustrated by experimental and analysis inaccuracies as well as by confounding of multiple effects. Therefore, a three-step statistically designed procedure was conducted to validate all the submodules and the overall model:

• Validate the thermal models by conducting thermocouple tests.
• Validate residual stresses and distortions by conducting stress and CMM measurements.
• Validate machining distortions by conducting CMM measurements.

For each step, validation was done in a systematic step-wise manner by testing each feature in the model one at a time and then all together. This helped isolate the shortcomings of the model and remove them before proceeding to an overall validation. Validation was

performed on both simple and complex 2-D and later 3-D shapes and on both airframe and engine materials.

First, the thermal model was validated by conducting experiments using a pancake instrumented with thermocouples to measure the thermal response during quench. Heat-transfer coefficients were calculated from the measured temperature-time data. Good correlation was established between simulation and experiment, thus validating the thermal model. The same procedure was repeated on a production shape for both oil and fan quench. The accurate prediction of thermal response is a prerequisite for the accurate prediction of residual stresses and subsequent machining distortions.

Radial and hoop residual stresses were measured along three sections and two clock positions (2 and 10 o'clock) in one pancake forging using XRD. The measurements were conducted up to half the forging thickness. Selected stress measurements were repeated at another test laboratory to evaluate the reproducibility of the measurements and assess the

accuracy of the data. The two sets of results differ by approximately 30 to 150%. The accurate measurement of stresses is difficult. Any stress measurement technique is indirect and relies on the measurement of a strain (either by strain gages, hole drilling, chemical milling, x-ray, or neutron diffraction) and converting the strain to a stress measurement. This can lead to large errors in the measured stresses when the state of stress is triaxial with a complicated distribution, as in these forgings. The large differences between the measurements from the two testing sources confirm the inaccuracies involved in the measurement of residual stresses. The validation of the model itself was based on measured distortion data.

A significant amount of material is machined out as the residual-stress measurements are made at increasing depth, as shown in Fig. 18. This material removal will influence the state of stress in the forging. The predicted residual stresses were corrected to account for the material removal. A 3-D 90° model of the forging was created, and the 2-D residual stresses were

 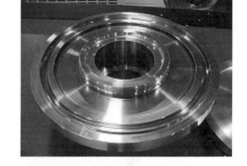

Fig. 17 Machined production disks

Fig. 18 Machining of slots during measurement of residual stresses

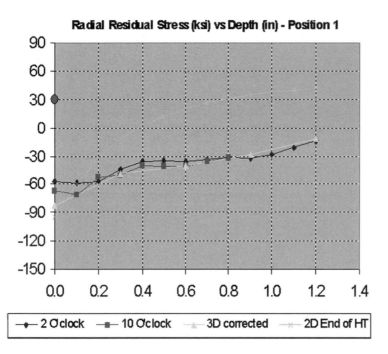

Fig. 19 Comparison of predicted and measured residual radial stress

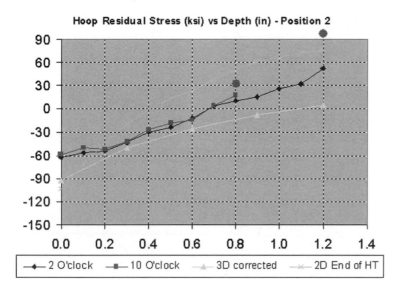

Fig. 20 Comparison of predicted and measured residual hoop stress

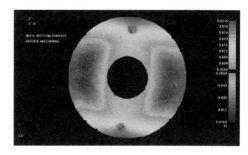

Fig. 21 Optical scan data showing distortions after heat treatment

where the stress gradients are steep. Repeat measurement(s) are shown by closed circles and show the variability between two different measurement laboratories.

2-D Machining Distortion Validation on Engine-Disk-Type Components

Pre- and post-heat treat CMM inspection of the forgings consisted of taking measurements at various radial locations at 45° intervals to obtain the distortion induced during heat treatment (Fig. 21). Forgings that were heat treated identically and also had the same support during heat treat show similar distortions. This demonstrates that the measured heat treat distortions are reproducible. All the forgings had a 3-D warpage as a result of the heat treat process. The measurements for the fan-cooled forgings were more tightly bunched together, showing less 3-D warpage with fan quench as compared with oil quench. The 3-D effect was averaged to allow a comparison with the 2-D cross section results, which are based on the assumption that the component is perfectly axisymmetric (i.e., no warpage). The amount of nonaxisymmetry decreases as the machining progresses. Note that the distortions are almost axisymmetric after machining. The nonaxisymmetry introduced during heat treatment has been removed during machining.

The measured distortions are the result of deformations occurring during heatup from room temperature to the solution temperature, holding at solution temperature, and subsequent quenching back to room temperature. The meaningful validation of predicted heat treat distortions is confounded by the interplay between several factors and by the fact that the distortions are small (~0.25 mm, or 10 mils, generally). The modeling predictions show the distortions induced only during the quenching part of the process. The measured and predicted heat treat distortions do not show good agreement, because the distortions occurring due to creep and sagging during heatup and holding have been ignored in the model. The modeling of these distortions requires creep material property data at high temperatures and the inclusion of gravity-induced sagging. This influences the distortions strongly. However, because the internal residual stresses

mapped onto the 3-D model. In the 3-D model, the machining of the slots was carried out in depths of 7.5 mm (0.3 in.). At each depth, after the material was removed, the stresses and strains were allowed to re-equilibrate. This corrects for the state of stress due to material removal. The predicted stress after successive material-removal passes was compared with measured values to provide a better assessment of the modeling predictions.

Figures 19 and 20 show a comparison of the predicted and measured residual radial and hoop stresses at the three measurement positions. The measurements at the two o'clock and ten o'clock positions are compared with

the 2-D predictions at the end of heat treatment and the 3-D predictions corrected for material removal. The 2-D predictions not corrected for material removal do not agree well with the measurements, especially at increasing depth as more and more material is removed. On the other hand, there is good agreement between the measurements and the 3-D corrected predictions. The discrepancy between measurements and predictions is largest at the surface. Possible causes of this discrepancy are residual surface hardening from machining not removed by etching, extrapolation of stresses from the finite-element centroids to the surface, and larger experimental errors near the surface are relieved during holding at solution temperature and regenerated during the cooling process, this assumption has a negligible effect on the prediction of residual stresses and subsequent machining distortions.

For the pancake forgings, the finished shape shown in Fig. 22 was chosen for the purpose of achieving large distortion (for easy measurement). The material was removed in the four quadrants (top/bottom, inside diameter/outside diameter). Several alternate machining shapes were investigated, and this one was chosen to obtain distortions in the 10 to 20 mils range.

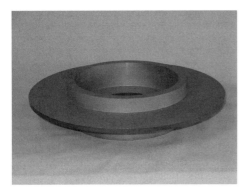

Fig. 22 Forging of U720 after all four quadrants have been machined

Distortions in this range are required in order to measure them accurately and to avoid large errors in experimental measurements, which can prevent meaningful model validation.

Initial predictions of machining distortions showed poor agreement with the measured data. For some cases, the predicted distortion was in a direction opposite to that measured. Measurements showed that the distortions caused by clamping forces while the forging was machined were negligible. All modeling inputs and procedures were examined carefully, and five improvements were made to obtain better agreement between the measured and predicted residual stresses and machining distortions:

- Exact stress-strain behavior instead of a simplified bilinear representation
- Strain-rate dependency of stress-strain data
- Material removal in layers versus single pass to predict the correct distortion direction
- Kinematic versus isotropic hardening
- Temperature-dependent Poisson's ratio

Of these five changes, the first three had the most significant effect on modeling predictions. The last two had a smaller effect. The plots shown in Fig. 23 to 25 are a small sample of all the results and show the general behavior. These figures show, in general, a good

agreement between the predicted and measured machining distortions, considering the extent of nonaxisymmetric deformation at some operations. In most of the cases, the agreement is within ±20%. When the distortions are very small (<5 mils, 125 µm), the noise in the measurements is large relative to the measurement. This can show up as a large percentage error but small absolute error. Process improvements by changing the machining sequence have been demonstrated using the model and were implemented successfully, resulting in cost-savings.

The conclusions of the distortion validation study are:

- Distortion measurements are more reliable and were used for model validation.
- Measurements and predictions show the same trend for all cases.
- Predictions agree better with measurements for smaller depths of cut.
- Predictions agree better with measurements for oil quench than for fan quench.
- Moving the finished shape axially changes the distortion approximately the same as the 3-D variation.
- The machining distortions are ~50% less with fan quench than with oil quench due to reduced residual stresses. This is a potential process improvement.
- Possible reasons for the discrepancy include:
  a. 3-D heat-transfer coefficient variation not exactly captured in the 2-D axisymmetric model
  b. Inaccuracy in extrapolated low-strain-rate stress-strain data
  c. Sag in the furnace: effect of heat treat fixtures

In practice, material is removed on one side of the component, the component is flipped over, and material is then removed on the other side. This process is repeated until one gradually approaches the finished component shape by successively removing smaller amounts of material on each side. This requires a number of machining operations, especially for distortion-prone geometries and/or materials. A possible machining strategy is to model the material removal to increasing depths on one side, up to the point where there is positive material left over the finished component shape. At this point, the forging would need to be flipped over and the process repeated on the other side.

2-D Machining Distortion Validation using National Aeronautics and Space Administration Data

The National Aeronautics and Space Administration's (NASA) Integrated Design and Processing Analysis Technology and Advanced Subsonics Technologies programs studied residual stress and machining distortions in advanced disk alloys. This work was extended to predict the effect of heat treatment on residual stress and

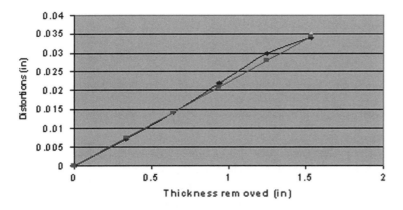

Fig. 23 Comparison of measured and predicted distortions for pancake forgings

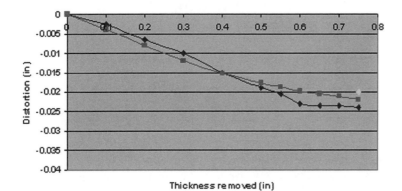

Fig. 24 Comparison of measured and predicted distortions for pancake forgings

Oil quench operation 20B distortion

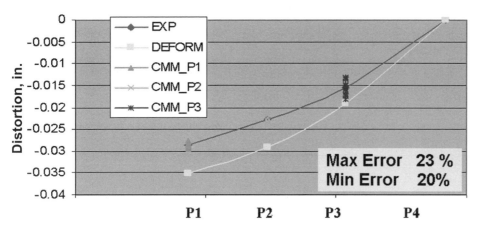

Fan cool operation 20B distortion

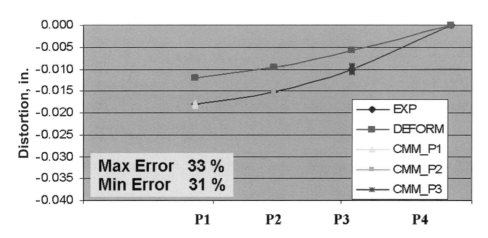

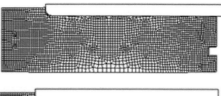

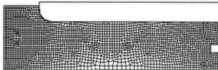

Fig. 26 Two machining operations

Fig. 25 Machining distortions: fan vs. oil quench. Distortion data represent the average of the eight experimental measurements at 45° intervals. The data points (coordinate measurement machine, CMM) show all eight measurements and the extent of nonaxisymmetric distortion. DEFORM represents the modeling predictions

agreement was obtained for the distortion of the other disks, also.

3-D Model Validation on Engine-Disk-Type Components

The machined U720 forging shown in Fig. 22 was selected for broaching distortion validation. The heat treatment and prior machining of this forging had been well characterized. Several simulations were carried out to define the machined geometry that would result in measurable distortions. Distortions should be large enough so that they can be measured accurately and used meaningfully for model validation. Small distortions are likely to have noise in the data, making such data unsuitable for model validation.

Two slots, each 5 cm (2 in.) deep, were broached in the U720 disk (Fig. 28). These slots simulate dovetail slots for blades in aircraft engine rotating disks. This was a well-controlled experiment to generate meaningful data for model validation. The finite-element mesh was fine in the vicinity of the slot to accurately capture the stress and distortions in this region. Radial, axial, and hoop distortion measurements were taken in the slot region after the machining of each slot.

Two tapered pockets with a wall thickness of 0.5 cm (0.2 in.) were milled in another U720 forging (Fig. 29). These pockets simulate features in airframe structural components. The pocket wall thickness was large enough to avoid distortions induced by cutting forces and surface residual-stress effects. The finite-element mesh was fine in the vicinity of the pockets to accurately capture the stress and distortions in this region. Radial, axial, and hoop distortion measurements were taken in the slot region after the machining of each pocket.

Model validation was completed on experimental 3-D shaped components similar to production forgings. Alloy 718 pancake forgings were made from 20 cm (8 in.) billet weighing 55 kg (120 lb) and forged to ~35 cm (14 in.) in diameter and 5 to 7.5 cm (2 to 3 in.) thick. One forging was used for gathering temperature data during quench for obtaining HTCs. Figure 30 shows the comparison between measured and predicted temperatures at two

subsequent machining distortions of simple forgings made of an advanced disk alloy (Ref 2, 3). Four pancake-shaped disks, weighing approximately 45 kg (100 lb) each, were isothermally forged to a pancake shape 35 cm (14 in.) in diameter by 4.8 cm (1.9 in.) thick. The four forgings were given different heat treatments. Heat treatments 2, 3, and 4 produced a fine-grained microstructure as a result of subsolvus solution temperature (1135 °C, or 2075 °F) and were designed to yield progressively lower residual stress. The first heat treatment produced a coarse-grained microstructure as a result of the supersolvus temperature (1182 °C, or 2160 °F) and was included to provide a direct comparison with the subsolvus, stabilized heat treatment. The dimensions of the four forgings were measured to obtain the initial distortion/warpage resulting from heat treatment.

DEFORM was used to simulate the four heat treatments to predict the initial residual-stress distribution prior to machining. Following this, two machining operations were performed (Fig. 26), which consisted of two face cuts on the top surface of each forging. The first cut went to a depth of 0.24 in. (6 mm), and the second cut went an additional 0.24 in. for a total depth of 0.48 in. After each cut, the disk was unclamped, and warpage and thickness measurements were made. These data were gathered under controlled conditions for multipass machining operations and are therefore very suitable for model validation. Figure 27 shows a comparison of the axial distortion data measured by NASA (dotted lines) and the simulation data from the DEFORM (solid lines) machining distortion model. The measurements show that the disks are not perfectly axisymmetric. The measured distortion is an average of the eight sampling points around the circumference. The agreement between measurements and predictions is very good. Similar good

thermocouples representing the best and worst matches. This figure also shows the layout of the thermocouples. A total of 13 thermocouples were used to capture the HTC variations around the forging.

Production disklike features were machined in four forgings, modified to accentuate the machining distortions: dovetail slots in the rim, holes in the web, and stem slots (Fig. 31).

Prior to machining the forgings, the process was modeled to define the machined geometry and machining sequence. The objective was to define conditions that would result in measurable distortions. Distortions were measured at each machining step. The measured distortions of each disk were compared to the corresponding numerical prediction. Much data were gathered at all steps of machining. Here,

only the distortions introduced during the 3-D machining steps are shown for the four disks.

A comparison of the measured and predicted distortions at the stem (disk 1) and at the outside diameter (OD) (disk 2) is shown in Fig. 32. A comparison of the measured and predicted distortions at the OD (disk 3) and at the stem (disk 4) is shown in Fig. 33. In all cases, the measured and predicted machining distortions matched within +30% on average.

Thermocoupled trials, residual stress, and machining distortion analyses have been completed on various production aircraft engine disks at the various original equipment manufacturers. The modeling results were generally in good agreement with the measurements.

Machining-Induced Residual Stresses and Distortions

For airframe-type components, machining-induced surface residual stresses are generally the main cause of distortions. The 3-D process model predicts component dimensional changes as a function of the initial residual-stress state, cutting tool forces, machining-induced surface stresses, machining plan design, and machine fixtures. Measurement and modeling of machining-induced residual stresses and distortions in subscale rib/web geometries were performed. Machining-induced residual stresses were obtained from one of four methods:

- *Detailed finite-element analysis of the cutting process:* Slow, expensive to run, reasonable accuracy
- *Simple fast-acting mechanistic model:* Fast, cheap to run, reasonable accuracy after calibration
- *Semiempirical linear stress model:* Fast, cheap to run, good accuracy after calibration
- *X-ray diffraction measurements:* Empirical, slow, expensive

The first three methods are described in the following sections. X-ray diffraction measurements have already been described in the preceding sections. Stresses from these models

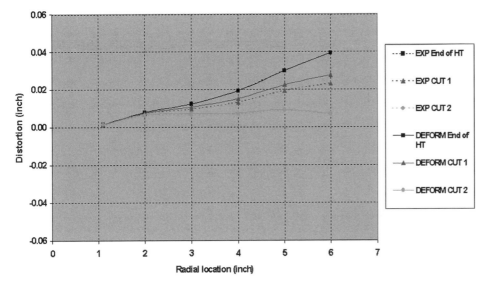

Fig. 27 Good agreement between DEFORM predictions (solid lines) and NASA's measurements (dotted lines) of the axial distortions of disk 1 after heat treatment and after two machining cuts

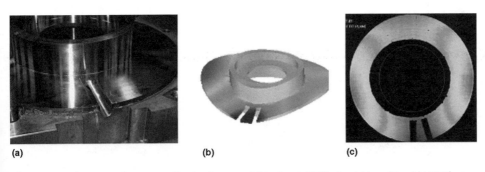

Fig. 28 Good agreement between predicted and measured distortions in U720 after slot broaching. (a) U720 forging being machined. (b) Predicted axial distortion. (c) Measured distortion

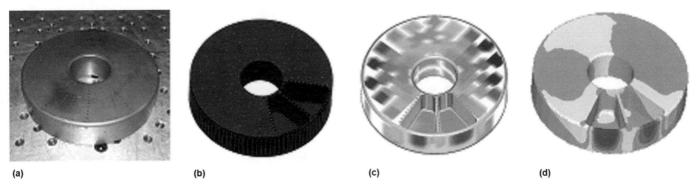

Fig. 29 Good agreement between predicted and measured distortions after pocket milling in U720. (a) U720 disk. (b) Measurement holes. (c) Measured distortion. (d) Predicted distortion

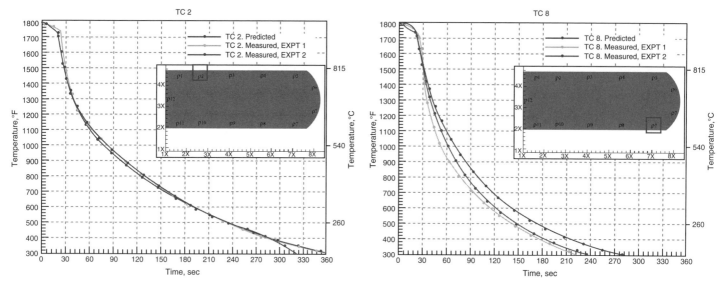

Fig. 30 Comparison between measured and predicted temperature during water quench: thermocouples (TC) 2 and 8

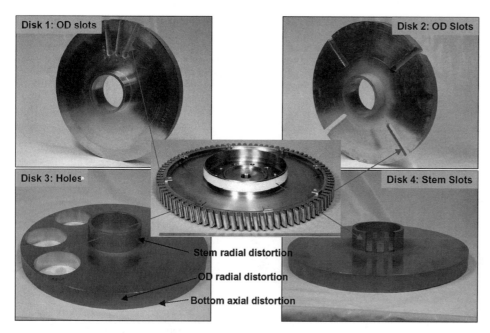

Fig. 31 Machining of production-like features in engine-disk-type forgings. OD, outside diameter

and/or measurements were input into a 3-D distortion finite-element model to predict component distortion. Distortion data were gathered after machining and compared with the modeling predictions.

Finite-Element Prediction of Machining-Induced Stress

Detailed finite-element modeling of the machining process can be performed using commercial software such as DEFORM or AdvantEdge (Third Wave Systems). Here, results from AdvantEdge are reported.

Simulations were performed using AdvantEdge for selected conditions of cutting speed, feed, radial and axial depths of cut, cutter geometry (including edge preparation, axial and radial rake angles, number of flutes), and material grade. The simulations predict temperatures, forces, and machining-induced residual stress. The tool used had a 35° helix angle, 8 flutes, a 19.05 mm (750 mils) diameter, with a 3.048 mm (120 mils) corner radius and an edge sharpness of 0.0508 mm (2 mils).

Hole-drilling measurements were conducted at Los Alamos National Laboratory (LANL). The error for each measurement was estimated

as 5% by LANL, based on historical evidence, with the exception of the first measurement (0.05 mm, or 2 mil, depth), which was estimated to have a 10% error due to the dish angle of the drill. Due to the nature of the hole-drilling experiments, axial stresses could not be obtained, and the first point measured was at 0.05 mm depth. Figure 34 shows a comparison of the predicted and measured tangential and radial stresses for a cutting speed of 121 surface meters per minute (smm) and a feed of 0.0508 mm/tooth. Both exhibit maximum compressive stress values at approximately 0.05 mm; however, the simulation results underpredict the magnitude compared to the measurements.

AdvantEdge 3-D predictions satisfactorily captured the effects of variations in chip loads and cutting speeds on the workpiece residual stresses. Trends of peak stress as a function of feed and speed were similar between the simulations and hole-drilling measurements. Cutting speeds were observed to have a significant effect on surface stresses in the simulations. With increased cutting speed (and correspondingly higher temperatures on the tool and workpiece), the surface residual stresses were observed to increase and become more tensile. Increased chip loads (feeds) were observed to have a pronounced effect on subsurface stresses. With increased chip loads, subsurface stresses (below peak compressive zone of stresses) were observed to become less compressive in nature. Mesh refinements did not result in a substantial change in the predicted results.

Detailed finite-element models of the chip formation process are time-consuming to run and are not yet fully validated. Meaningful results can be obtained if the cutting process can be approximated as 2-D (e.g., turning), with computational times of 4 to 8 h. For 3-D cutting processes, several days of computational time are required. Therefore, these models are not yet production-ready to be used in the industry on a routine basis.

Mechanistic Machining Model

Mechanistic machining models have been developed for quickly predicting (in seconds as opposed to several hours or days for finite-element methods) cutting forces, temperature, and machining-induced residual stresses for broaching and milling processes. The speed with which these models generate results provides the potential for analyzing a wide range of conditions in a short period of time to establish a set of conditions for use in a production environment. The overall procedure consists of obtaining cutting forces from the mechanics of cutting, computing stresses from the applied cutting loads, and relaxing the stresses to obtain the residual stresses in the workpiece (Ref 4). Mechanistic models must be calibrated with experimental data and are good over a limited range of cutting conditions close to the calibration data set. The models predict the cutting forces reasonably well for both broaching and milling operations. The residual stresses are captured with respect to trends and depth of penetration.

Linear Stress Model

Samples of ribs and webs that are representative of large airframe structural components were used to evaluate machining-induced residual stresses and distortions. The principal stresses for rib coupons are aligned with the helix angle of the cutter. For the web coupons, the principal stresses are aligned tangential and normal to the cutter radius. Process parameters used as control variables included spindle speed, feed rate, cutting tool material, cutting tool geometry, and edge sharpness, which were defined using Taguchi methods. Ribs were made by finishing with the side of a cutter, and webs were made with the bottom of a cutter. The geometry was chosen to allow a 5 by 5 cm (2 by 2 in.) sample for stress and distortion measurement. Industry-standard milling cutters were selected to machine the samples (Fig. 35).

A thin sample distorts after machining, thus relieving some of the machining-induced stresses. Therefore, the residual stress measured in a thin sample is not the same as the machining stresses. To accurately measure the machining-induced stresses, samples much thicker than typical thicknesses were used. This eliminates the postmachining distortion and partial relief of stresses and accurately captures the stresses induced by machining.

Based on the rib and web distortion experimental data, a linear stress model was developed for the mapping of residual stresses on an airframe-type component and for obtaining its distortion due to machining-induced residual stresses. Based on experimental and numerical observations, the following assumptions were made:

- Machining-induced effects are concentrated in a thin surface layer.
- Machining-induced effects from previous cuts are removed, and a new surface stress layer is created during each pass of the tool. Therefore, only the machining parameters in the last pass are needed to determine the machining-induced effects.
- Machining stresses depend on the thickness direction only and can be averaged over the machined surface.
- Machining-induced plastic strains do not depend on the shape of the component for a given set of tools, material, and machining parameters.
- Machining-induced effects at joints (e.g., filleted regions) are not significant and therefore are ignored for the determination of distortions and residual stresses.

The distortion of the rib and web samples was measured using laser interferometry. The measured distortion was fitted with polynomial functions (linear coefficients for the x-, z-, and xz-directions). The three coefficients represent bending in the two directions and the twist, respectively. Figure 36 shows the contribution to the distortion caused by each one of the linear terms (x, z, and xz). The ability to obtain a good fit of the distortion using linear terms indicates that the linearity assumption is valid.

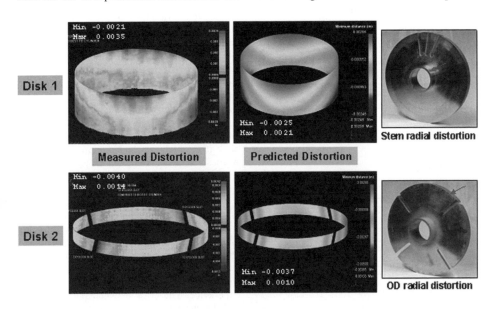

Fig. 32 Comparison of measured and predicted distortions for disks 1 and 2. OD, outside diameter

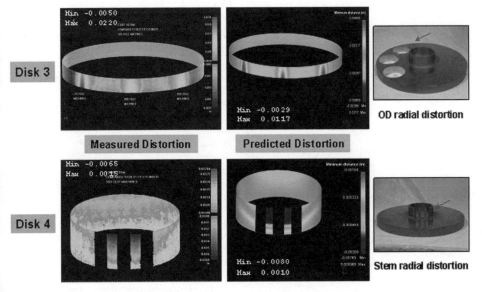

Fig. 33 Comparison of measured and predicted distortions for disks 3 and 4. OD, outside diameter

Coupons with a worse fit had small distortions with a small signal-to-noise ratio.

For a component that is a collection of ribs and webs of relatively uniform thickness joined by fillets, the distortion can be predicted by using as input linear terms determined through the experiments described here. Based on risk, cost, and schedule feasibility, for production use, an empirical combination of XRD with calibration by linear stress modeling was selected.

This approach combines the best measurement of the shape of the machining stress profile (x-ray) with the best measurement of the magnitude of the machining stress profile (linear stress model). The x-ray data defined the shape of the stress gradient as starting negative (compressive) and quickly decaying to zero. The coefficients in the linear stress model were obtained by matching the area under the stress profile (weighted by the distance normal to the surface). Figure 37 is one example of the stress input.

Four rib and four web coupons were modeled (Fig. 38, 39). The dimension of the rib and web coupons was 5 by 5 cm (2 by 2 in.). The thickness of the coupon was assumed to be uniform. Eight-noded linear brick elements were used. Surface meshes were generated to capture the initial stress variation through the thickness direction. Six nodal points are enough to capture this input curve. The coupon was then allowed to re-equilibrate under the applied stress field. The resulting distortion was compared with the measurements. Numerical tests were conducted to evaluate the effect of mesh size on the distortion results. Increasing the number of thickness layers had minimal impact on the results. However, increasing the number of in-plane elements had a significant impact. A mesh size of 96 by 96 in-plane elements with 12 thickness layers provided a mesh-independent converged solution.

The model was validated on a selected subset of rib/web samples using residual stresses from a mechanistic model and from x-ray measurements. Figure 40 shows a comparison between the measured and prediction distortions for typical rib and web samples. The ribs twist and the webs bow out, which is consistent with prior experience. The error between the predictions and the measurements ranges from 2 to 29%. Similar agreement was obtained on production components that cannot be shown here due to proprietary reasons.

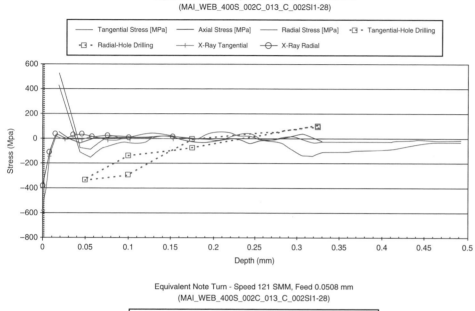

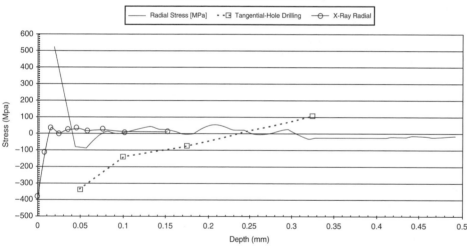

Fig. 34 Turning/hole-drilling comparison for a cutting speed of 121 surface meters per minute (smm) and a feed of 0.0508 mm/tooth

Fig. 35 (a) Tungsten carbide and AISI M-42 cutting tools. (b) Machining of subscale webs. (c) Machining of subscale ribs

Integration of Machining Stresses into DEFORM

Inclusion of surface stresses and cutting tool forces is important for components with thin section sizes. Figure 41 shows a flow chart of the production distortion model. The machining stresses are imported into DEFORM using a graphical user interface (GUI), taking into account the cutter direction, path, and type. The GUI enables easy, error-free import of data. Bulk residual stresses, if significant, can be superposed on the machining stresses. The overall stress field is then equilibrated to obtain the component distortion. If the distortion is outside prescribed limits, the process is repeated with a different machining process until the distortions fall within the prescribed limits.

The simulation procedure consists of these steps:

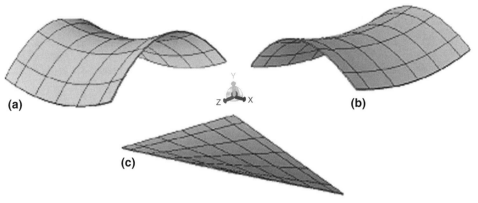

Fig. 36 Displacement computed for linear shape functions. (a) x-component. (b) z-component. (c) xz-component

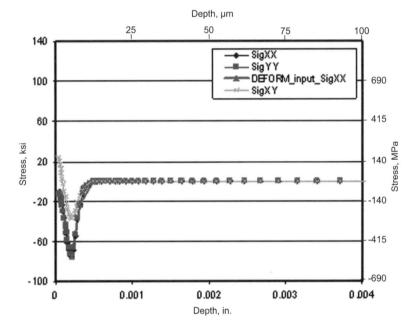

Fig. 37 Example of stress input curve

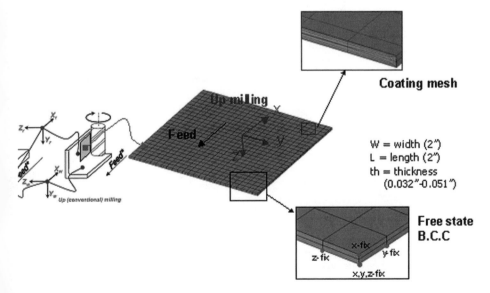

Fig. 38 Finite-element model for rib coupons. bcc, body-centered cubic

1. Generate a brick mesh for the final machined component.
2. Generate multiple layers of near-surface mesh to capture the machining stresses.
3. Interpolate bulk residual stress from heat treatment to this mesh, if needed.
4. Import machining-induced residual stresses to the near-surface mesh (details follow).
5. Carry out a stress equilibrium simulation.
6. Repeat the process for subsequent machining operations, if needed.

The following procedure was developed for importing cutting-induced residual stress:

1. Generate a fine surface mesh.
2. Pick surface nodes in the region where the stresses are to be imported.
3. Input cutting direction for the region.
4. Define machining-induced residual stress as a function of depth or a constant value.
5. Interpolate imported stress components to the mesh nodal locations. Rotate the stress components to the model coordinate system. The rib region is cut by the flutes on the cutter, and the principal residual stresses in the cutting and the transverse directions are oriented with respect to the helix angle. The web region is cut by the bottom of the cutter, and the principal residual stresses in the cutting and the transverse directions change depending on the tool path direction.

Aircraft structural components typically consist of multiple thin walls, as shown in Fig. 42. To predict the distortion of thin ribs and/or webs, meshing of thin walls is important for accurate results. Because thin walls can be easily modeled by a structured mesh system, a brick mesh is often used for thin-walled aircraft components. The advantages of tetrahedral meshes are that it is possible to automate initial mesh generation, remeshing, and near-surface mesh generation, which makes it possible to automate the modeling of multiple machining operations. Because automatic brick mesh generators are not available, it is not possible to do this with brick meshes. However, brick meshes provide greater accuracy, and a much smaller number of brick elements is needed to define large thin-walled airframe geometries, which reduces the computing required. A large number of tetrahedral elements are required for thin-walled airframe geometries, thereby significantly increasing the computational effort. The selection of the approach must be evaluated on a case-by-case basis depending on the component geometry, machining operations, and the distortion information required from the model.

Scientific Forming Technologies Corporation has developed a procedure to realistically model the machining process and streamline the analysis of multistep machining with the commercially available software DEFORM. A custom machining template was developed for a user to perform all the simulation steps in an automated sequence. A series of

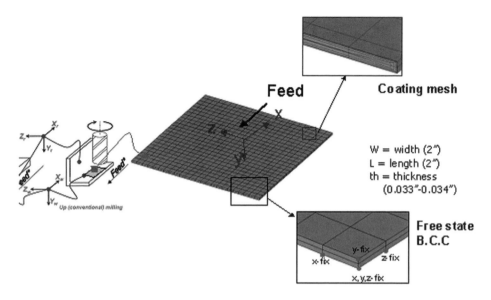

Fig. 39 Finite-element model for web coupons. bcc, body-centered cubic

Fig. 40 Rib/web distortion. Predictions and measurements match 2 to 29%.

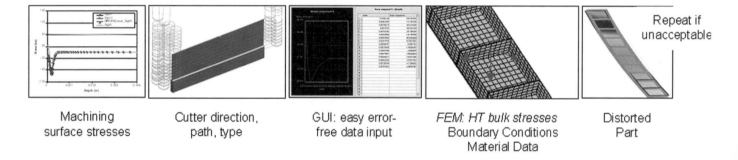

Machining surface stresses

Cutter direction, path, type

GUI: easy error-free data input

FEM: HT bulk stresses Boundary Conditions Material Data

Distorted Part

Repeat if unacceptable

Fig. 41 Production model flow chart. GUI, graphical user interface; FEM, finite-element model; HT, heat treatment

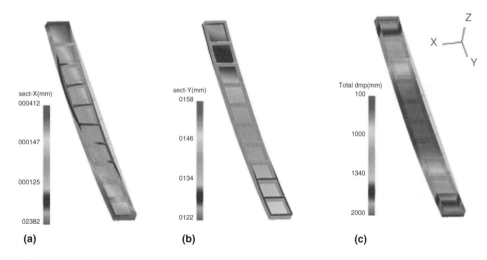

Fig. 42 Machining distortion example. (a) x-displacement. (b) y-displacement. (c) Total displacement

machining distortion simulations can be triggered from the procedure to simulate the complete process, from the heat treat shape to the final machined shape. Because machining involves the complexity of multiple passes and multiple operations, it is essential to graphically preview the relative location of the fixtures, workpiece, and machining paths prior to the analysis. This feature enables an upfront review of the entire material-removal process and ensures that all data have been input correctly. An important objective of the improved machining simulation method is to bring the modeling analysis and methods into closer alignment with the physical machining process as understood by a machining process engineer. The key challenge is to achieve the appropriate balance between improved functionality and ease of use for the resulting simulation method.

Modeling steps include Boolean operation for material removal, stress re-equilibrium under clamping condition (after material is removed), and free-state distortion (after clamps are removed). The approach starts with a residual-stress pattern and distorted heat treat shape generated using DEFORM for heat treatment process modeling. These results are then mapped onto the mesh used for machining simulation along with the geometry of the machining plan generated using CAD software and numerical control machining information. The geometry, representing the machined shape, is then meshed, clamping loads are added, and machining is carried out through element removal. A subsequent analysis is required to verify that the tooling has sufficient stiffness to withstand the rebalanced loads following machining. The machined, distorted shape is calculated following removal of clamping loads, and the entire cycle is repeated for each operation of metal removal until the finished component shape is reached. The results are presented in a format that is directly comparable to dimensional measurements. Typical results are shown in Fig. 42. For a realistic model size of a typical aircraft component, a total of approximately 150,000 elements are expected. Depending on the computer, solution method, and boundary conditions used, this model has a runtime of approximately 15 min to 3 h. Therefore, it is concluded that using DEFORM with the brick approach can be practical from the perspective of computational requirements.

Modeling Benefits

Although machining is a mature manufacturing process, the drive toward affordability continues to press established machining operations to increase metal-removal rates, increase machine utilization, and eliminate machining steps. These efforts are worthwhile because machining costs are a significant fraction of the total cost of manufacturing for aerospace forged components.

In the near-term, savings will accrue from reduced machining costs, reduced scrap, improved manufacturing lead and cycle times, reduced time to first article, and improved component performance and life during service, resulting in reduced operating costs. A more significant additional cost-savings is the ability to go to nearer-net shape forgings after this technology has been more extensively validated. Accurate prediction of distortions will enable a reduction of the material envelope needed to compensate for distortions, especially for the high-cost powder metal alloys used in rotating disks. The technology developed here is applicable to all military (United States Air Force and United States Navy) and commercial aircraft and engines.

The program is well aligned with the philosophy to achieve affordable metallic materials and processes with accelerated implementation for aerospace systems. Benefits are a reduction in acquisition costs of metallic components. Additional benefits also include potential for design of more robust components that have reduced tendency to distort during engine operation, which may affect engine clearances, efficiency, and performance.

Modeling provides a data-driven understanding of residual stress, validated commercially supported tools, and standardized modeling and measurement procedures. The MAI programs represent a major technology advance for the industry and have advanced the state of the art to a user-friendly, validated, commercially supported, and production-ready analysis tool for 3-D machining problems, which can be used to achieve significant cost-savings. Because process modeling can be used to improve both the fabrication processes and the component performance during service, it should be incorporated into the integrated design environment in the organization to achieve design for manufacturability and design for process excellence. Various design disciplines can take advantage of process models, such as service-life estimation, inspection, supplier/original equipment manufacturer (OEM) collaborations, repair, and overhaul.

The supply chain, consisting of manufacturers of aerospace components in addition to the OEMs, stands to benefit from the use of modeling. The OEMs will see a reduction in machining costs, and the forging suppliers will benefit by being able to better control the heat treatment process. Distortion problems pose the biggest challenge to new components and/or new suppliers. Modeling technology will help shorten that learning curve. Current components with distortion problems will benefit during a change of suppliers. New components will benefit right from the start. Although modeling has been demonstrated here for only selected engine and airframe materials, the model/method is pervasive and can be applied to other materials, adding to the total savings.

The methodology of this program will include the capability to evaluate the full range of process conditions for production hardware and to define process sensitivities relative to material and process variations early in the production process. This information will better define the process window. In addition, these tools could be used for evaluations when it was determined that the process window was breached.

Modeling Implementation in a Production Environment

Successful completion of the various MAI programs has permitted technology implementation on a wide variety of components. Implementation has occurred initially on new components in which process(es) could be integrated into the original design, thus reducing or eliminating additional certification costs. Subsequent production implementation to address distortion problems on existing components is based on the cost benefit balanced against any additional certification costs. Specific applications with noted cost-reduction potential include superalloy rotating components and titanium structural components. Implementation of the 2-D model is more widespread, and it has been used successfully for several production components at several OEMs. As the models become more accurate with more validation, the use and benefits will grow.

The mode and extent of use of the machining model will be somewhat user-specific, depending on the extent of validation carried out, the problem the user is trying to solve or avoid, the certainty with which the various boundary conditions and material property data during heat treatment and machining can be quantified, and so on. Here, only some general guidelines can be provided. The general implementation approach is shown in Fig. 43. The details of implementation will differ for large/small suppliers and airframe/engine components. The OEMs, forge/heat treat suppliers, and machining suppliers are involved at various stages.

This article demonstrates that finite-element modeling can be a powerful tool to predict the residual stresses developed during heat treatment processes and the distortion during machining operations. The use of commercially available software minimizes maintenance and enhancement risks. The machining template in DEFORM also provides an easy way to model the distortions developed during multioperation machining sequences. These models have been integrated with standard engineering tools and implemented within the modeling organizations at the OEMs and at their forging and/or machining suppliers.

Future Work

Future work should focus on establishing standard material characterization, measurement, and modeling methods to ensure accurate and repeatable residual-stress predictions. Additional model validation on more materials and different types of components is also needed. Suggested future work includes the following.

Roadmap. A roadmap is needed to formalize plans to address the various issues relative to residual-stress modeling, development, and rapid implementation of modeling tools that link various materials and process models and provide a known level of accuracy and uncertainty. The roadmap should identify risks and a risk mitigation plan, balancing risk, cost, payoff, and maturity. Lessons learned from engine programs should be leveraged to airframe components, recognizing the tremendous scaleup in computational requirements from 2-D engine disks to large 3-D airframe components.

Modeling and Measurement Accuracy. For the modeling results to be useful, different levels of accuracy are needed, depending on the application. The bulk residual-stress modeling and measurement accuracy required for a range of applications should be established, including manufacturing (heat treat and machining distortions), service (dimensional stability), service-life estimation (fatigue life, crack initiation and propagation), and material characterization. Various residual-stress measurement methods should be compared to develop standardized procedures and recommendations. An assessment of the accuracy and variability of the predicted and measured residual-stress profiles and their impact on manufacturing, service, and service-life estimation should be determined. Model accuracy, capability, and user-friendliness should be addressed to obtain an industrially usable tool.

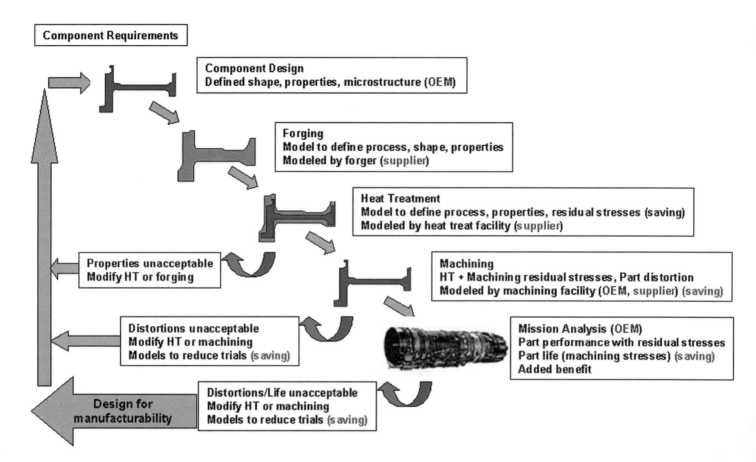

Fig. 43 General implementation approach. OEM, original equipment manufacturer; HT, heat treatment

Fig. 44 Typical aircraft engine and airframe applications

Material Data. Material constitutive properties (tensile and creep) are needed as inputs to the residual-stress models. Development of standard material test methods (on-cooling tensile and creep/stress relaxation) and an industry-wide data set for commonly used alloys will reduce uncertainty and improve modeling accuracy. These data sets could more thoroughly cover the full range of temperatures, strain rates, and microstructural conditions than would be economically feasible for a single company. Modeling enhancements are also needed to incorporate these data into the model in a standardized way and to develop a physics-based model, which includes microstructure evolution and deformation mechanisms to describe material behavior during heat treatment. Effects of evolving microstructural features and crystallographic texture on elevated-temperature mechanical properties should be evaluated. If these are significant, a testing plan to capture these effects should be developed. The methodology should also include aluminum airframe and nickel-base engine disk materials.

Validation. Residual-stress predictions require further validation to support their quantitative application to various applications. Validation is needed on subscale and full-scale components in a production environment, streamlining and integration of commercial codes for user-friendly industrial implementation, and developing industry guidelines for model usage. Use cases that codify the methodology and describe the problem-solving steps have been used successfully in prior programs to demonstrate the modeling framework. Standard benchmark use cases should be defined to design a heat treatment and machining process to better balance properties and distortions and to identify optimal parameters. This involves the generation of experimental data under controlled and production conditions and extensive model validation followed by implementation on production hardware (Fig. 44). This effort will allow comparison and transfer of residual-stress predictions seamlessly through the supply chain, including mills, forge suppliers, OEMs, and machining suppliers.

Modeling Sensitivity Studies. Traditionally, engineering analysis is performed for nominal conditions. The design must account for various sources of uncertainty inherent in materials behavior, manufacturing processes, models, and so on to arrive at a robust control strategy to ensure minimal variability in the component characteristics. The error in residual-stress predictions can be estimated by a Monte Carlo analysis driven by probability density functions that describe the uncertainty in inputs (e.g., heat-transfer coefficients and material properties). An error propagation analysis should be used to quantify the compounding of errors as the analysis progresses through various steps. This will establish confidence limits on the modeling predictions and experimental measurements.

A sensitivity study is recommended to establish which inputs most strongly impact the modeling outputs of interest. Variations in the critical inputs should be quantified to assess the accuracy of the modeling outputs. Efforts can then be focused on reducing the variability in component distortion by studying the most critical steps. The sensitivity analysis can also potentially define the resolution needed in the input material property data.

Qualitative analysis is the capability to predict the trend under different processing conditions. Engineers can use this to carry out many "what-if" studies without the need to rely on expensive experiments. Quantitative analysis is the capability to accurately predict the component behavior. This requires an accurate modeling algorithm and input parameters/data, including both the boundary conditions and material properties.

Industry Standards. Residual-stress modeling and measurement techniques and the procedures to generate the various modeling inputs lack a standardized approach. An industry standard must be established that can be used throughout the supply chain (mills, component producers, and OEMs) to enable integrated design, material, and processing technology efforts. As a "best practice," the analysis and experimental methods should include metrics, red flags, and/or guidelines to permit a quantitative assessment of the adequacy of each analysis and measurement. It should also include instructions about the range of applicability of the associated methods. Standards for modeling and measurement procedures, material data, and boundary condition inputs should be prepared. The goal would be to develop standard methods in the form of an Aerospace Material Specification. Developed best practices (input data, simulation, postprocessing) should be aimed at producing consistent results, independent of the user, with acceptable accuracy.

ACKNOWLEDGMENTS

The authors would like to acknowledge the contributions of the team members in the various Metals Affordability Initiative Programs: Keith Young and James Castle (Boeing); B.K. Chun, W.T. Wu, M. Knezevic, and J.Y. Oh (Scientific Forming Technologies Corp.); Michael Glavicic and Robert Ress (Rolls-Royce); and Jeff Simmons (Air Force Research Laboratory).

REFERENCES

1. P.L.B. Oxley, *The Mechanics of Machining: An Analytical Approach to Assessing Machinability,* E. Horwood, Halsted Press, New York, 1989, p 242
2. J. Gayda, "The Effect of Heat Treatment on Residual Stress and Machining Distortions in Advanced Nickel-Based Disk Alloys," NASA/TM-2001-210717
3. W.-T. Wu, G. Li, J. Tang, S. Srivatsa, R. Shankar, R. Wallis, P. Ramasundaram, and J. Gayda, "A Process Modeling System for Heat Treatment of High Temperature Structural Materials," Final report, USAF Contract F33615-95-C-5238, June 2001
4. J.-C. Su, "Residual Stress Modeling in Machining Processes," Ph.D. thesis, Georgia Institute of Technology, 2006

SELECTED REFERENCES

- T.G. Byrer, Ed., *Forging Handbook,* Forging Industry Association, American Society for Metals, 1985
- D. Dye, K.T. Conlon, and R.C. Reed, Characterization and Modeling of Quenching-Induced Residual Stresses in the Nickel-Based Superalloy IN718, *Metall. Mater. Trans. A,* Vol 35, June 2004, p 1703
- T.P. Gabb, J. Telesman, P.T. Kantzos, and K. O'Connor, "Characterization of the Temperature Capabilities of Advanced Disk Alloy ME3," NASA/TM 2002-211796
- D.L. McDowell and G.J. Moyar, A More Realistic Model of Nonlinear Material Response: Application to Elastic-Plastic Rolling Contact, *Proceedings of the Second International Symposium on Contact Mechanics and Wear of Rail/Wheel Systems* (Kingston, RI), 1986
- T. Reti, Z. Fried, and I. Felde, Computer Simulation of Steel Quenching Process Using a Multi-Phase Transformation Model, *Comput. Mater. Sci.,* Vol 22, 2001, p 261–278
- M.A. Rist, S. Tin, B.A. Roder, J.A. James, and M.R. Daymond, Residual Stresses in a Quenched Superalloy Turbine Disc: Measurements and Modeling, *Metall. Mater. Trans. A,* Vol 37, Feb 2006, p 459
- D. Rondeau, "The Effects of Part Orientation and Fluid Flow on Heat Transfer Around a Cylinder," M.S. thesis, Worcester Polytechnic Institute, 2004
- G. Shen and D. Furrer, Manufacturing of Aerospace Forgings, *J. Mater. Proc. Technol.,* Vol 98, 2000, p 189–195
- "Standard Test Method for Determining Residual Stresses by the Hole-Drilling Strain-Gage Method," E 837-08, ASTM
- "Standard Test Methods for Stress Relaxation for Materials and Structures," E 328-02 (Reapproved 2008), ASTM
- P.J. Withers and H.K.D.H. Bhadeshia, Overview: Residual Stress Part 1—Measurement Techniques, *Mater. Sci. Technol.,* Vol 17, April 2001, p 355
- P.J. Withers and H.K.D.H. Bhadeshia, Overview: Residual Stress Part 2—Nature and Origins, *Mater. Sci. Technol.,* Vol 17, April 2001, p 366

Simulation of Joining Operations

ASM Handbook, Volume 22B, *Metals Process Simulation*
D.U. Furrer and S.L. Semiatin, editors

Copyright © 2010, ASM International®
All rights reserved.
www.asminternational.org

Introduction to Integrated Weld Modeling

Sudarsanam Suresh Babu, The Ohio State University

THE SCOPE of this article is to provide an overview of integrated weld modeling. It is not intended to provide a comprehensive review of integrated weld modeling activities in the literature but introduces methodology and relevant resources that can be accessed by the reader for further development, evaluation, and deployment.

Integrated weld modeling is an important activity that crosscuts many industries. In early 2000, the American Welding Society, Department of Energy, Edison Welding Institute, and industrial members from the heavy industry, aerospace, petroleum/energy, and automotive industries developed a research roadmap for the welding industry (Ref 1). The strategic goals for the welding industry by 2020 were identified to be the following:

- Increase the uses of welding by 25%, decrease the cost, and increase the productivity
- Enhance the process technology that allows for the use of welding across all manufacturing sectors
- Develop new welding technology along with new materials so that it can be used for all applications
- Assure that welding can be part of the six-sigma quality environment
- Increase the knowledge base of people employed at all levels of the welding industry
- Reduce energy use by 50% through productivity improvements

Although welding itself does not consume extensive energy, welding does play a critical role in the development and deployment of the materials for energy exploration, transfer, conversion, efficiency, and storage. The roadmap also identified that the engineering solutions for joining materials are not unique and do differ depending on the geometry, materials, and applications. Due to this complexity of the problem, the development of joining technology for a given material is associated with extensive experimental trial-and-error optimization. To minimize this experimental

approach, an integrated computational modeling was suggested as a solution (Fig. 1).

This leads to a fundamental question: Is it possible to develop physics-based computational models to describe the behavior of existing and emerging materials subjected to joining processes? A review of the literature shows that the development of physics-based models is indeed challenging due to the complex interaction between physical processes during welding. Some of these physical processes include heat and mass transfer, phase transformations, electromagnetic phenomena, plastic strain, and reactions with the environment during welding/joining. Researchers, including Ashby

(Ref 2), Bhadeshia (Ref 3), Cerjak (Ref 4), David (Ref 5), DebRoy (Ref 6), Eagar (Ref 7), Easterling (Ref 8), Goldak (Ref 9), Grong (Ref 10), Kirkaldy (Ref 11), Koseki (Ref 12), Kou (Ref 13), Leblond (Ref 14), Matsuda (Ref 15), Rappaz (Ref 16), Szekely (Ref 17), Vitek (Ref 18), Yurioka (Ref 19), and Zacharia (Ref 20), developed a framework for linking thermomechanical histories to microstructure development and mechanical heterogeneity in welds. These developments can be summarized in the form of a schematic diagram (Ref 11) published by Kirkaldy (Fig. 2). According to this diagram, by integrating individual submodels for heat, mechanical, and material models, one can

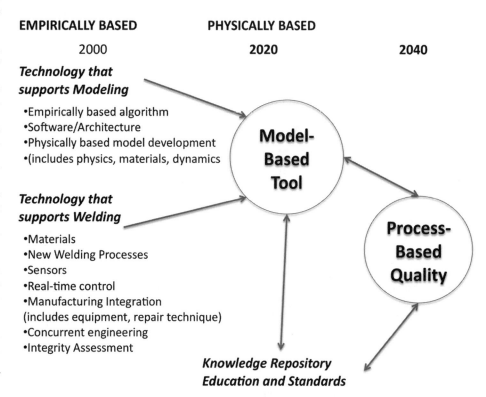

EMPIRICALLY BASED　　　　**PHYSICALLY BASED**

2000　　　　　　　2020　　　　　　　2040

Technology that supports Modeling

- Empirically based algorithm
- Software/Architecture
- Physically based model development
- (includes physics, materials, dynamics

Technology that supports Welding

- Materials
- New Welding Processes
- Sensors
- Real-time control
- Manufacturing Integration
(includes equipment, repair technique)
- Concurrent engineering
- Integrity Assessment

Model-Based Tool

Process-Based Quality

Knowledge Repository Education and Standards

Fig. 1　Suggested roadmap for the development of a model-based tool in the year 2020 as a way of ensuring process-based quality by the year 2040

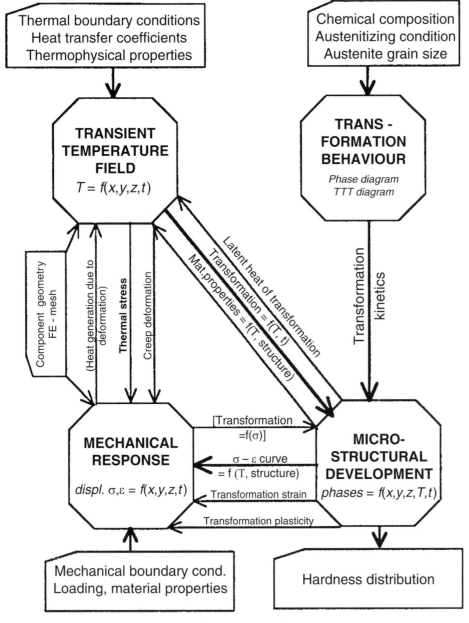

Fig. 2 Overview of integrated welding modeling methodology proposed by Kirkaldy (Ref 11) shows the importance of predicting transformation behavior and microstructural development as well as thermal and mechanical behavior

manufacturing process. The reader is also referred to some of the series of conference proceedings titled *Mathematical Modeling of Weld Phenomena* (Ref 4, 28–34) as well as *Trends in Welding Research* conference proceedings published periodically by ASM International (Ref 35, 36). These conference proceedings provide a detailed progression of integrated weld modeling and its capability for a wide range of joining processes. In addition, the reader is also referred to three classic textbooks related to modeling of welding processes (Ref 37–39). This article discusses some of the salient features of each subprocess model that is indicated as part of the integrated weld modeling shown in Fig. 2. Most of the materials presented in this review have been adopted from Ref 40.

Process Modeling

The goal of process modeling is to predict weld pool shape, thermal cycles, and fluid flow conditions under transient (coordinate system fixed to the part) or steady-state (coordinate system moving with the welding heat source) conditions. This section discusses the fundamentals of the underlying physics and the methodologies to solve the same.

Underlying Physics and Methodologies

Modeling of Heat Transfer during Welding. Heat transfer in welding can be represented by equations of heat conduction in solids. For example, the spatial variation of the heating or cooling rate in a Cartesian coordinate can be related to the second derivative of the temperature gradients in *x*-, *y*-, and *z*-directions (Ref 39):

$$\frac{\partial T}{\partial t} = \frac{\lambda}{\rho c}\left[\frac{\partial^2 T}{\partial x^2} + \frac{\partial^2 T}{\partial y^2} + \frac{\partial^2 T}{\partial z^2}\right]$$
$$= a\left[\frac{\partial^2 T}{\partial x^2} + \frac{\partial^2 T}{\partial y^2} + \frac{\partial^2 T}{\partial z^2}\right] \quad \text{(Eq 1)}$$

where T is temperature, t is time, λ is thermal conductivity, ρ is density, c is specific heat capacity, and a is the thermal diffusivity of the material being welded. With boundary conditions imposed by the welding, Eq 1 can be solved for both transient and steady-state conditions of welding. A famous solution of the generic equation for steady-state distribution of temperature in a plate during arc welding was given by Rosenthal (Ref 41) as:

$$T\{x,R\} = T_0 + \frac{\eta VI}{2\pi\lambda}\left(\frac{1}{R}\right)\exp\left\{-\frac{v}{2a}(R+x)\right\}; R$$
$$= \sqrt{x^2 + y^2 + z^2}$$

(Eq 2)

where $T\{x,R\}$ is the temperature as a function of the radial distance (R) and distance (x) along the welding centerline, T_0 is the preheat or

predict the overall performance of welded structures. The approach starts with a heat-transfer model that will simulate temperature distributions in three dimensions [$T = f\{x, y, z, \text{time}\}$] as a function of process parameters and time. Thermal cycle data will be used by material models to predict the microstructure evolution and its impact on transient mechanical (σ-ε relations) properties. The transient changes in temperature and mechanical properties will be fed into a finite-element structural model to predict plastic strain distribution. This information allows for the prediction of final properties, residual stress, and distortion in a complex welded geometry. This interdisciplinary approach may appear simple; however, it

requires collaboration between experts in metallurgy, finite-element analysis, welding process, and computer science (Ref 21, 22). To a limited extent, this vision has become a reality by pioneering work in many organizations and commercial software companies (Ref 23–27).

It is noteworthy that the scope of this article is not to provide a comprehensive review of integrated weld modeling activities in the literature. It is designed to provide an overview of the methodology and relevant resources that can be accessed by the reader for further development, evaluation, and deployment. The reader is also referred to other articles within this Volume that describe modeling of inertia welding, diffusion bonding, and the additive

interpass temperature, V is the arc voltage, I is the welding current, v is the welding speed, a is the thermal diffusivity, and η is the arc efficiency. This solution is often referred as the Rosenthal equation and has been extensively used for obtaining approximate temperature isotherms during welding of plate as a function of process parameters. For example, Eq 2 was used to calculate the steady-state temperature distributions on a steel plate surface for two different welding speeds. The calculations (Fig. 3) show the formation of a teardrop-shaped weld pool with an increase in welding speed. The aforementioned distribution also can be used to calculate the heating and cooling rate. Equation 2 assumes a point heat source, which is not strictly applicable to arc welding heat sources. Modification of the previous equations for distributed heat sources has been attempted by many researchers. A review of the same can be seen in the book by Grong (Ref 39). In addition, this book also provides analytical solutions for other steady-state and transient welding cases.

Although analytical solutions provide a faster estimation, these solutions often ignore boundary conditions imposed by the geometries of a realistic welded structure. To address this limitation, finite-difference and finite-element formulations (Ref 42) have been developed. These formulations allow for both steady-state and transient solutions (Ref 37). However, one of the biggest challenges in these formulations is to describe the heat flux into the weld pool. The extent and the distributions of heat flux determine the shape of the weld pool, peak temperature, and heating and cooling rates (Ref 43, 44). This challenge has been elegantly addressed by researchers from Goldak's group to distribute the power density within front and rear quadrants of the weld area. This methodology is schematically explained in Fig. 4. The power density distribution in the front quadrant is given by Eq 3:

$$q\{x,y,z,t\} = \frac{6\sqrt{3}f_f Q}{abc\pi\sqrt{\pi}}\exp\left\{\frac{-3x^2}{a^2}\right\}$$
$$\exp\left\{\frac{-3y^2}{b^2}\right\}$$
$$\exp\left\{\frac{-3[z+v(\tau-t)]^2}{c^2}\right\}; \text{front} \qquad \text{(Eq 3)}$$

$$q\{x,y,z,t\} = \frac{6\sqrt{3}f_r Q}{abc\pi\sqrt{\pi}}\exp\left\{\frac{-3x^2}{a^2}\right\}$$
$$\exp\left\{\frac{-3y^2}{b^2}\right\}$$
$$\exp\left\{\frac{-3[z+v(\tau-t)]^2}{c^2}\right\}; \text{rear} \qquad \text{(Eq 4)}$$

where Q is the energy input rate (in watts); the parameters a, b, and c are the dimensions that describe the ellipsoidal shape in x-, y-, and z-coordinates, respectively; τ is the lag factor; t is time, and f_f and f_r are the fractions of heat source deposited in the front and rear quadrant and are related to each other by Eq 5:

$$f_f + f_r = 1 \qquad \text{(Eq 5)}$$

By changing the dimensions of the a, b, c, f_f, and f_r values, it is possible to match most of the complex weld pool shapes that are observed in fusion welds. It is important to note that these parameters must be obtained by the optimization procedure. For example, Kelly (Ref 45) must consider more than 2000 sets of such parameters to match the experimental weld pool size. In such calibration studies, it is often impossible to match all the experimental parameters, such as weld pool width, depth, curvature, and spatial variation of thermal cycles. Improvements to the aforementioned double ellipsoidal model are being pursued by many researchers (Ref 45). In this regard, it is important to understand fundamental reasons for such variations of the weld pool shape by using coupled heat- and mass-transfer models.

When the weld pool flux is calibrated using the double-ellipsoidal models, it is easier to describe steady and transient thermal cycles for complex welding conditions. An example of transient thermal simulation of laser cladding to build a nickel-base superalloy structure on a substrate is shown in Fig. 5 (Ref 46). The simulations were performed with commercial finite-element analysis software (ABAQUS) and user-defined subroutines. (Reference to any commercial software in this article does not imply endorsement of this software by the author or ASM International. Readers are requested to independently evaluate software for their own applications.) During the early stages of cladding, thermal simulations showed a small molten pool (Fig. 5a) due to the three-dimensional heat conduction mode. With the progress of the buildup, the heat conduction mode changes to two dimensional. As a result, the molten pool size increases, and there is remelting of the previous layers. The spatial and temporal variations of thermal cycles also change. These simulations demonstrated that for efficient laser

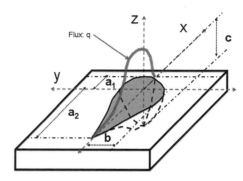

Fig. 4 Schematic of parameters used in describing the heat flux during welding. The reader is referred to Eq 3 and 4 in the text

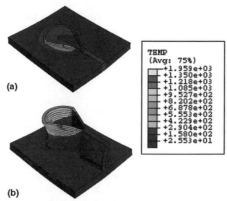

Fig. 5 Simulation of temperature transients while building a complex shape through the laser cladding process. (a) Thermal gradients are large at the early stage of cladding due to three-dimensional heat-transfer conditions. (b) With the buildup of sufficient layers, the heat-transfer condition changes to two dimensional. As a result, the thermal gradient is shallower. Such changes will affect the ensuing solidification and solid-state microstructure

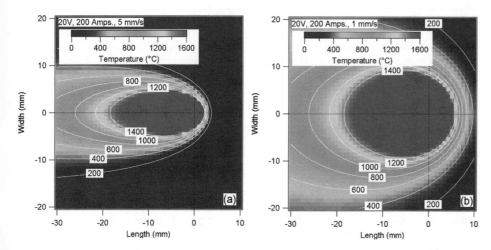

Fig. 3 Calculated temperature isotherms during welding of low-alloy steels are shown in image and contour plots. The plots demonstrate the change of weld pool shape with an increase in welding speed from 1 to 5 mm/s

cladding of the complex shapes shown in Fig. 5, the heat input must be changed based on local heat-transfer conditions. By coupling these thermal models with optimization software, it is possible to a priori design the path and heat input during laser cladding. The reader is referred to the article "Modeling of Laser-Additive Manufacturing Processes" in this Volume for more details.

Fluid Flow Modeling. As mentioned earlier, a complex weld pool shape develops due to heat transfer through conduction in the liquid and solid as well as convective transfer by way of electromagnetic force, buoyancy forces, and shear stresses due to the spatial variation of surface tension with temperature. The effect of fluid flow on weld pool shape was initiated by the pioneering work of Kou (Ref 13), Zacharia (Ref 20), and DebRoy (Ref 6) and their collaborators. The underlying methodology relies on solving equations of energy and momentum conservation as well as continuity conditions. The first governing equation is the conservation of momentum:

$$\rho \nabla(V.V) = -\nabla P + \mu \nabla.(\nabla V) + (S_V - \rho \nabla.(U_s V))$$
(Eq 6)

where ρ is density; V is the fluid flow velocity vector in the x-, y-, and z-directions, which is represented by u, v, and w; respectively; U_s is the welding velocity; P is the effective pressure on the welding pool; and S_V is the source term that takes into account the buoyancy factor (S_b), Marangoni shear stress (τ_{mar}), and electromagnetic force (Ref 13).

The buoyancy term (S_b) is described by $\rho g \beta(T - T_{ref})$, in which β is the thermal expansion coefficient at a temperature, T, with reference to a reference temperature (T_{ref}). The shear stress (τ) is given by $f_1(d\gamma/dT)\nabla T$. In this relation, the parameter $d\gamma/dT$ is the variation of surface tension with temperature. This can take either positive or negative value and is shown to have large influence in the weld pool fluid flow conditions. The other parameter, f_1, takes care of the reduction in shear stress near the mushy (mixture of liquid and solid) regions.

The second governing equation deals with the conservation of energy:

$$\rho \nabla(V.h) = \nabla.\left(\frac{k}{C_p}\nabla h\right) + S_l + S_h - \rho \nabla.(U_s h)$$
(Eq 7)

where the total enthalpy of the material, h, is related to the sum of sensible heat and latent heat content, which takes into account the phase change (e.g., solid to liquid and liquid to solid); C_p is the specific heat; k is the thermal conductivity; S_h is a source term that takes into account the heat input from the welding source and convective and radiation heat loss; and S_l is a source term that accounts for the latent heat of melting and the convective transport of latent heat. This term also takes care of the phase change during solidification. The aforementioned governing equations can be solved by

numerical methods to calculate the spatial variation of fluid flow velocity (u, v, w) and temperature.

In the early 1990s, the previously mentioned methodologies were used on simple butt welds. In addition, the sensitivity of these calculations to the magnitudes of thermophysical properties was documented (Ref 47). Recently, the aforementioned methodology has been extended to fillet welds using coordinate transformation methods (Ref 48). The methodology to transform the fillet weld geometry in a real x, y, and z Cartesian coordinate system to a computational domain is shown in Fig. 6(a). After this coordinate transformation, the governing Eq 6 and 7 are solved in this domain. A typical result from such calculations is shown in Fig. 6(b). The plot shows the ability of fluid flow calculations to very effectively simulate the curvature of the welds as well as the penetration of the weld pool into the base materials. Comparisons of the calculated weld pool shapes to experimental weld shapes (Ref 49) are shown in Fig. 6(c). This example shows the predictive power of computational heat- and mass-transfer models. It is important to note that the change in weld pool shape shown in Fig. 6(c) also changes the spatial variation of cooling rates. As a result, the predictions of cooling rates are expected to be more accurate with the use of heat and fluid flow models. The fidelity of such calculations on the sensitivity of steel weld metal microstructure was demonstrated by DebRoy and his collaborators (Ref 50, 51).

Microstructure Modeling

The goal of microstructure modeling is to predict phase fractions as well as grain size in the heat-affected zone (HAZ) and weld metal region as a function of alloy composition and thermal cycles. Because thermal cycles are predicted by the process models described earlier, in this section, methods to predict the microstructures in the HAZ and weld metal are presented. The reader is also referred to a classic textbook by Grong for in-depth treatment of this subject (Ref 10). In this section, the pioneering work done to predict the HAZ and weld metal microstructure in the early 1980s and 1990s is presented first. Later, the application of computational thermodynamic and kinetic models is also presented.

Prediction of HAZ Microstructure

Methods for predicting HAZ properties range from simple equations using the base material or weld metal chemistry to complex models that take many factors into account, including chemistry, initial grain size, heat input, and cooling rate. The overall goals and approaches are summarized schematically in Fig. 7. The approaches can be broadly classified into two main themes. In the first approach, the microstructure is not predicted explicitly; rather, the

input parameters, such as base metal composition, microstructure, and welding process parameters, are related to one or more HAZ properties. In the second approach, the input parameters are used to calculate the HAZ microstructure and then are correlated with different properties. The historical development of these approaches is briefly discussed and is adopted from Ref 52.

First-generation approaches involve simple correlations between the steel composition and various properties, such as hardness, tensile strength, and cracking tendency. All of the formulas have a form of carbon equivalence that ignores the kinetics and does not consider the microstructure evolution explicitly (Ref 53). The second-generation methods (developed between the 1980s and the 1990s) build on the carbon equivalence formulas to consider the intricacies due to the peak temperature, initial precipitate type, cooling rates, and microstructure evolution. Yurioka et al. (Ref 54) have developed a carbon equivalent number (CEN):

$$CEN = C + A(C)*\left(\frac{Si}{24} + \frac{Mn}{6} + \frac{Cu}{15} + \frac{Ni}{20} + \frac{Cr + Mo + Nb + V}{5} + 5B\right)$$
(Eq 8)

where $A(C) = 0.75 + 0.25 \tan h\{20(C - 0.12)\}$.

The third-generation methods (developed between 1990 and 2000) started to focus on evaluating the HAZ microstructure with detailed thermodynamics and kinetics, because thermodynamic and kinetic models are generic and allowed for extrapolation to a wide range of steels. However, relating the microstructure to properties still relied on empirical correlations. Ion et al. (Ref 55) used a kinetic approach to predict the hardness and microstructure of weld HAZs. Their approach uses a modification of the Rosenthal equations (Ref 41) to provide a circular disc heat source rather than a point source. The hardness prediction is based on estimating the volume fractions of microstructural constituents (martensite, bainite, ferrite, and pearlite) and using the rule of mixtures to approximate the hardness. The model takes into account many variables, including heat flow, austenite grain growth, precipitate dissolution and/or coarsening, and chemistry to predict the volume fraction of the constituents. An example calculation is shown in Fig. 8. Using the model developed by Ion et al. (Ref 55), the microstructure and hardness of the HAZ for a given cooling rate (10 Ks^{-1}) and austenite grain size (50 m) for two different steels with 0.1 and 0.2 wt% C is shown. These calculations show that a change in carbon content changes the HAZ microstructure from ferrite to a microstructure with predominantly martensite. In addition, the predictions also show an increase in hardness. Such calculations can be used to provide guidance for weld cooling rate control in the HAZ. An online version of these calculations is available (Ref 56).

Recently, prediction of HAZ microstructure and/or properties has relied on artificial neural

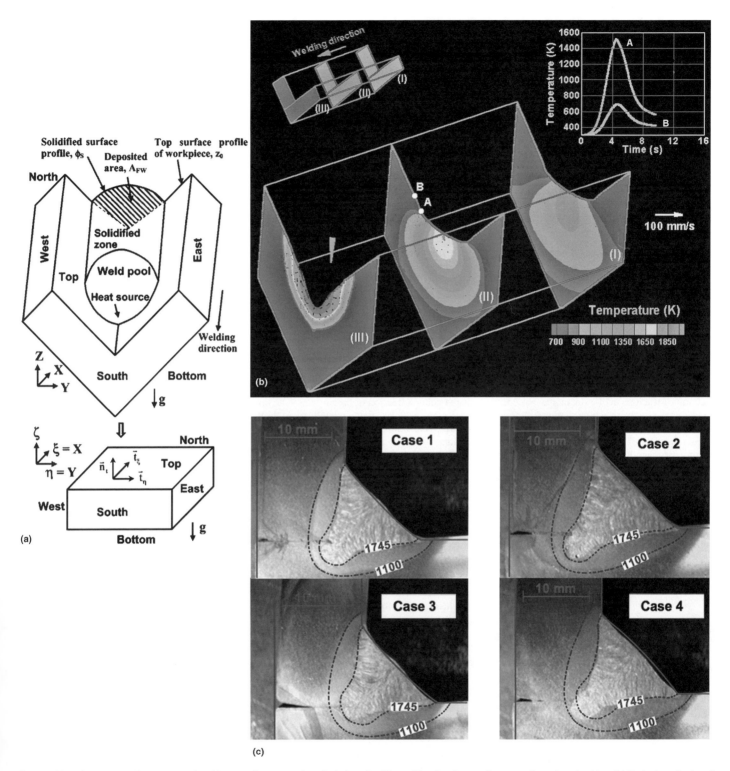

Fig. 6 (a) Technique to perform computational heat- and mass-transfer calculations for fillet welds using the coordinate transformation algorithm. (b) Typical result of such simulation shows the weld pool curvature as well as transients of temperature distributions. (c) Comparisons of predicted shape and size of the weld pool and the heat-affected zone for various welding process conditions show good agreement. Note the ability of these calculations to predict a change in curvature of the weld pool as one moves from top cap to the root of the weld. Such curvatures cannot be predicted without considering the fluid flow effects

networks (Ref 57) and detailed computational materials models coupled with either thermomechanical or fluid flow calculations. There has been much published literature that focuses on calculating the nucleation and growth of ferrite from austenite as a function of composition and cooling rate. These are essentially based on the calculation of time-temperature transformation diagrams and converting them to continuous cooling transformation diagrams (Ref 58). Advanced models that consider simultaneous formation of grain-boundary ferrite, Widmanstätten ferrite, pearlite, and bainite are also available (Ref 59). Some of these models are currently available in the form of public

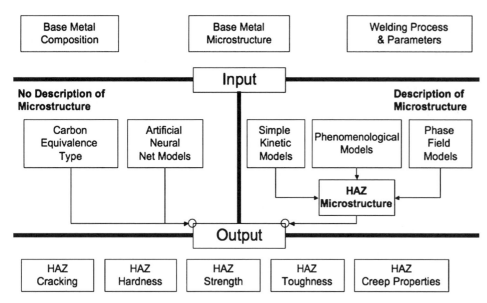

Fig. 7 Schematic illustration of various modeling methodologies to describe the heat-affected zone (HAZ) of steel welds

domain software (Ref 60). The models to predict microstructure evolution in the HAZ and weld metal region of age-hardening aluminum alloys can be found in the classic work of Myhr and Grong (Ref 61, 62). In this work, the dissolution of age-hardening precipitates and subsequent precipitation was described by isokinetic equations and assumed the additivity law.

In the previous treatments of microstructure evolution, in addition to phase transformation, an estimation of grain growth is important. For example, the steel microstructure predictions by equations given by Ion et al. (Ref 55) are sensitive to the prior austenite grain size. The austenite grain growth can be described by the following classic relationship:

$$\frac{dD}{dt} = nK\left[\frac{1}{D} - \frac{1}{\overline{D}_{\text{lim}}}\right]^{(1/n-1)} \qquad \text{(Eq 9)}$$

where D is the grain size (in micrometers); D_{lim} is the limiting grain size (in micrometers), which is determined by the stability of the precipitates; and K and n are the rate constant and time exponent of grain growth, which depend on the materials and temperature, respectively. Therefore, if the composition and thermal cycle can be acquired from the steels and the welding parameters, the only thing needed for grain growth is the limiting grain size. In this article, the limiting grain size is determined by the pinning effect of precipitates (Ref 63):

$$\overline{D}_{\text{lim}} = k\frac{r}{f} \qquad \text{(Eq 10)}$$

where $\overline{D}_{\text{lim}}$ is the limiting grain size; k is the Zener coefficient, which was first derived by Zener to be 0.75; r is the average radius of

precipitates, with the unit of micrometers; and f is the volume fraction of precipitates. The limiting grain size is then related to stability (growth, coarsening, and dissolution) of grain-boundary pinning precipitates. Although, the aforementioned methodology has been extensively used, it does not provide morphological changes in grain shape and size. In this regard, pioneering work was done by Radhakrishnan et al. (Ref 64, 65) using Monte Carlo simulations. Using this methodology, thermal pinning of grain growth was also successfully simulated. This work has been pursued by other researchers to describe the HAZ grain growth in other alloy systems (Ref 66).

Prediction of Weld Metal Microstructure

The earliest models to calculate weld metal microstructure can be tracked to a classic paper by Bhadeshia et al. (Ref 3) for steel welds. In this model, the sequential decomposition of the austenite phase into allotriomorphic ferrite (also known as grain-boundary ferrite), Widmanstätten ferrite, acicular ferrite, and martensite-austenite constituents is calculated. The basis of this model emanates from an ability to calculate the time-temperature transformation (TTT) diagram as a function of steel composition. The TTT diagrams are then converted into continuous cooling transformation (CCT) diagrams. Given an expression for the weld cooling curve and austenite grain size, the weld metal microstructure can be calculated using the flow chart shown in Fig. 9(a). Comparison of predicted microstructure and measured microstructure is shown in Fig. 9(b). The aforementioned model has been coupled with

detailed heat- and mass-transfer models by DebRoy and his collaborators (Ref 50, 51). Microstructure development in the weld metal region of aluminum alloys (Ref 67) and nickel alloys (Ref 68) has been addressed through computational thermodynamics and kinetics, which are introduced in the next section (Ref 40).

Application of Computational Thermodynamics and Kinetics Tools

There are standard mathematical models to calculate multicomponent thermodynamic and diffusion-controlled growth of the product phase into a parent phase (Ref 69). A typical calculation of multicomponent equilibrium is demonstrated with the Fe-Cr-Ni system with liquid, δ-ferrite (body-centered cubic), and γ-austenite (face-centered cubic) phases at 1700 K (Fig. 10). Thin lines that go across the phase boundaries in this diagram correspond to tie lines. For further details on this subject, the readers are referred to other sections in this Handbook. In this section, application of these tools to weld metal microstructure prediction is described (Ref 40) for various reactions that happen as a function of high-to-low temperature during fusion welding.

Liquid-Gas Reactions. The prediction of weld metal composition in gas-shielded processes, including gas metal arc welding (GMAW) (Ref 70), gas tungsten arc welding (GTAW), laser beam welding, and low-pressure electron beam welding, has always remained a challenge due to competing phenomena with the arc, plasma, shielding gas, atmosphere, and consumables. For example, stainless steels with high nitrogen levels (Ref 71, 72) have been developed to reduce nickel concentration but still maintain an austenitic microstructure. However, predicting the amount of nitrogen that remains after GMAW or GTAW in these stainless steels as a function of welding parameters is indeed a challenge. Hertzman et al. (Ref 73, 74) concluded that the final nitrogen concentration is decided by the equilibrium between nitrogen activity and composition of the liquid metal after a critical time.

It is important to consider the effect of plasma environment, which may lead to enhanced nitrogen dissolution in steel welds. This phenomenon may play an important role at the early stages of nitrogen balance, as discussed by Hertzman et al. (Ref 73). Mundra and DebRoy (Ref 75), followed by Palmer and DebRoy (Ref 76), have developed models to describe nitrogen dissolution (mono- and diatomic fashion) into liquid metal from the plasma environment. Enhanced nitrogen dissolution from laser plasma was leveraged by Babu et al. (Ref 77) to induce fine-scale carbonitrides during surface alloying of iron alloys. Similar to the kinetics of nitrogen dissolution (Ref 78–80), it is possible to describe the dissolution of oxygen (Ref 81) and hydrogen (Ref

Fe-0.1C-0.2Si-1Mn-0.5Ni-0.5Cr-0.5Mo 0.1V-0.1Co-0.01N-0.001B (Austenite Grain Size = 50μm)

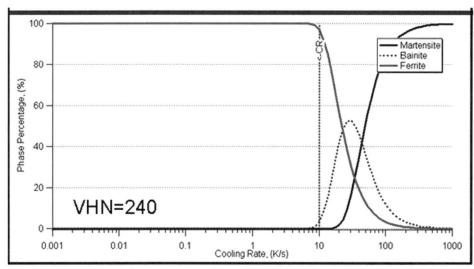

(a)

Fe-0.2C-0.2Si-1Mn-0.5Ni-0.5Cr-0.5Mo 0.1V-0.1Co-0.01N-0.001B (Austenite Grain Size = 50μm)

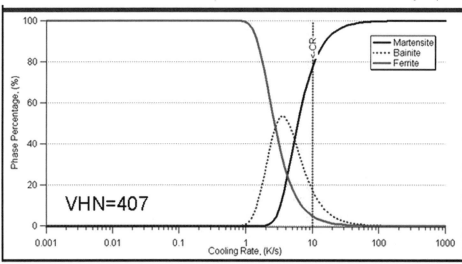

(b)

Fig. 8 Typical calculations of heat-affected zone microstructural constituent and hardness as a function of cooling rate. For a given low-carbon steel composition, (a) a slow cooling rate of 10 °C/s leads to a soft microstructure and (b) a faster cooling rate leads to a hard martensitic microstructure

82–84). Dissolved oxygen and sulfur (Ref 85) have a major influence on the surface tension of liquid ($\gamma_{surface}$) and its variation with temperature ($d\gamma_{surface}/dT$). The sign and magnitude ($d\gamma_{surface}/dT$) has a large influence on the fluid flow characteristics in the weld pool (Ref 86, 87). In addition, evaporation from the weld pool has also been addressed by various researchers using thermodynamic and kinetic descriptions (Ref 88–90) and comprehensive heat- and mass-transfer models. Some preliminary efforts are also being made to predict the composition of the fume particles using the liquid-gas equilibrium (Ref 91).

Liquid-Slag Reactions. Most fusion welding processes encounter liquid and slag equilibrium, for example, shielded metal arc welding, submerged arc welding (SAW), and flux cored arc welding. There is a close similarity between these processes and steelmaking. As a result, ladle thermodynamics have been leveraged to describe liquid-slag equilibrium in welds. In the classic works of Mitra and Eager (Ref 92–94), thermodynamic theories were used to predict the final weld metal composition of SAW welds as a function of process, flux, and consumable characteristics. Many researchers considered interactions between flux/slag chemistry and consumables to evaluate final weld metal compositions (Ref 95–98) as well as the inclusion formation. As expected, these phenomenological theories were based on ladle thermodynamics (Ref 99).

Inclusion Formation. Inclusions that form in welds are either oxides, nitrides, carbides, sulfides, or a combination thereof. Certain oxide inclusions, with special characteristics, promote the formation of acicular ferrite (Ref 100, 101) during solid-state decomposition of austenite. Acicular ferrite is known to promote the toughness of welds. This beneficial effect of inclusion outweighs the deleterious effects on fracture initiation. Early work on predicting inclusion formation using ladle thermodynamics originated from many researchers, notably Olson, Liu, and Edwards (Ref 102), Thewlis (Ref 103), as well as Kluken and Grong (Ref 99). These researchers predicted that complex inclusions form due to sequential formation of oxides, including Al_2O_3, SiO_2, Ti_xO_y, MnO, and complex spinels. Even with the aforementioned classic work, the prediction of inclusion composition, size, number density, and type of surface oxide remained a challenge for some time. By coupling ladle thermodynamic theory, computational thermodynamics (Ref 104–106), and overall transformation kinetic theories (Ref 107–110), it is now possible to comprehensively predict inclusion characteristics. These models have also been extended to laser-surface alloying to predict dissolution of hard particles (Ref 111). These models have also been integrated with computational heat- and mass-transfer models (Ref 112–114). For example, Hong et al. evaluated the cyclic growth and dissolution of inclusions in a weld pool. The trajectory of the inclusion within a weld pool is shown in Fig. 11(a). This movement was calculated by using the fluid flow velocities (u, v, and w) calculated in x-, y-, and z-coordinates and by assuming no slip between the inclusion surface and liquid steel flow. The corresponding variation of temperature and the associated change in radius of the inclusion is shown in Fig. 11(b). Using such calculations and microstructure observations, the rapid rate of inclusion growth was rationalized based on a collision and coalescence mechanism (Ref 115).

Solidification. To comprehensively understand the weld metal microstructure, the following must be predicted:

- First phase to form from the liquid
- Solidification temperature range ($\Delta T_{equilibrium} = T_L - T_s$)
- Extent of alloying element segregation
- Morphology of the solidification grains
- Cracking tendency

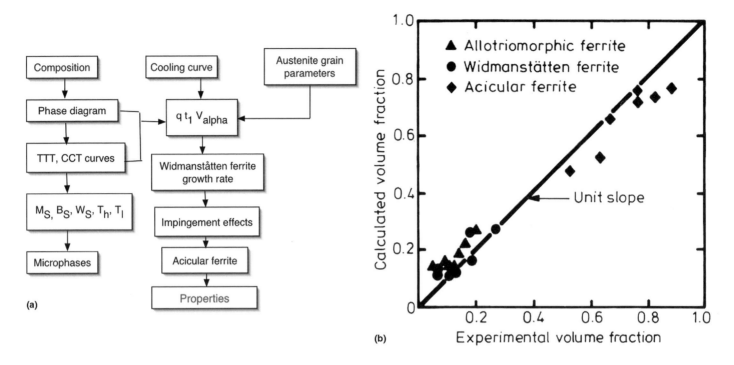

Fig. 9 (a) Overview of methodology developed by Bhadeshia et al. (Ref 3) to predict the microstructure constituent in the as-welded region of low-alloy steels. In contrast to the carbon equivalence formula, this model is based on quasi-chemical thermodynamics and kinetic equations. TTT, time-temperature transformation; CCT, continuous cooling transformation. (b) Comparison of predicted and measured microstructure in the as-welded region shows good agreement

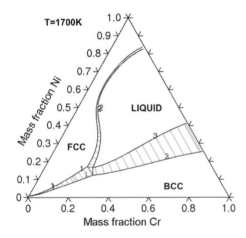

Fig. 10 Example of a calculated ternary phase diagram with tie lines and invariant triangles (three-phase equilibrium) in an Fe-Cr-Ni system at 1700 K. fcc, face-centered cubic; bcc, body-centered cubic

- Nonequilibrium solidification as a function of composition and cooling rates

A multicomponent thermodynamic phase diagram can be calculated using computational thermodynamics tools, and this information can be used to predict the first phase to form from liquid during welding (Ref 116–118). Using Scheil-Gulliver calculations, the nonequilibrium solidification range as well as the maximum extent of segregation can be predicted (Ref 119). It is important to note that equilibrium and Scheil-Gulliver calculations

provide two extreme conditions of weld solidification. In nominal weld cooling rates, some extent of diffusion is expected in the solid phase. This must be considered in the Scheil-Gulliver calculations. Classic papers from Dupont's group (Ref 120–122) focus on using a modified form of the computational thermodynamics tools to evaluate the solidification path and also the microsegregation effect in similar and dissimilar metal welding conditions for a wide range of nickel-base and stainless steel alloys.

To track the solid-liquid interface, diffusion in both liquid and solid phases as a function of cooling rate must be considered. Early work on modeling the weld solidification rate was done by Koseki et al. (Ref 123, 124). With the introduction of sidewise dendrite growth models in DictTra software (Ref 69), weld solidification for a wide range of alloy systems and cooling rates can be done easily. With an increase in cooling rate, an increase in the liquid-solid interface velocity is expected. This may affect solute partitioning between the liquid and solid and induce morphological changes in the solidification microstructure. Often, this increase in the liquid-solid interface may also trigger a nonequilibrium phase selection during solidification. For example, in steels, a transition from ferrite (body-centered cubic) solidification and austenite (face-centered cubic) solidification may occur (Ref 125, 126) with an increase in the liquid-solid interface velocity. Using levitation melting methods, Koseki and Flemings (Ref 127) developed phase stability and growth velocity

criteria to describe these transformations. Fukumoto and Kurz (Ref 128) used growth velocity considerations for the phase selection of austenite instead of ferrite during rapid cooling conditions, using interface-response function models. This is illustrated in Fig. 12. The selection of the ferrite or austenite phase is governed by its dendrite tip temperature. This temperature can be calculated using the interface-response function theories. The calculations show that for slow interface velocities, the dendrite tip temperature of ferrite is always higher than that of austenite. Under this condition, only ferrite solidification is expected. However, above a critical velocity, the dendrite tip temperature of austenite is higher than ferrite. This condition will lead to the austenite mode of solidification. Using these theories, the transition from ferrite to austenite solidification modes during laser welding was rationalized.

Solid-State Transformation. The role of various microalloying elements (Ref 129) on the HAZ microstructure in thermomechanical-controlled processing in low-alloy steels has been evaluated with computational thermodynamics (CT) and computational kinetics (CK) tools. The ferrite-to-austenite transformation in stainless steel welds (Ref 130) was simulated for a condition that is close to limited partitioning of the substitutional condition, which is in between the paraequilibrium and local equilibrium conditions. The effect of retained austenite in the steel microstructure on the reaustenitization kinetics during heating was rationalized using a thermodynamic criterion (Ref 131). To design new steels with improved properties

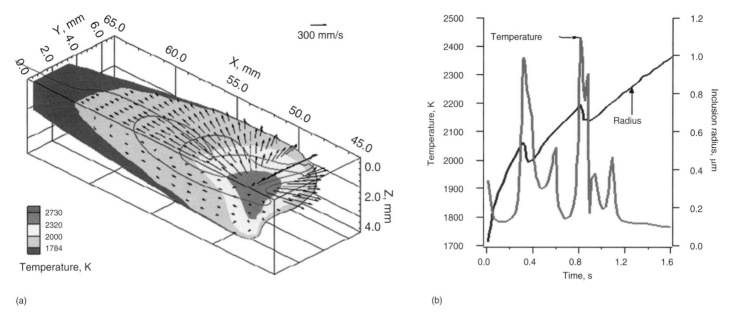

(a)   (b)

Fig. 11 Prediction of inclusion size distribution in low-alloy steel welds by tracking the growth and dissolution of thousands of inclusions as they gyrate through the weld pool. (a) Trace of an inclusion as it traverses through the weld pool with different temperature regions, before being trapped by the advancing liquid-solid boundary. (b) Corresponding variation of temperature and the associated growth and dissolution of an inclusion. The calculations show that the observed inclusion diameters could be interpreted without invoking Ostwald ripening, which is typical of a stagnant molten pool

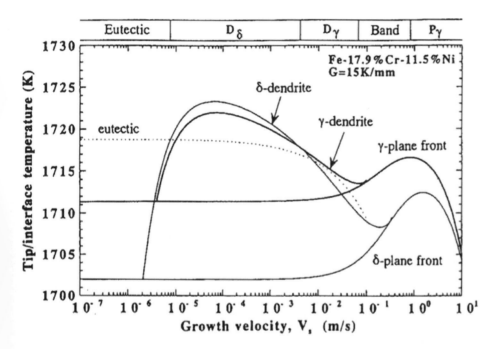

Fig. 12 Transition from body-centered cubic (bcc) to face-centered cubic (fcc) mode of solidification with an increase in the liquid-solid interface velocity. The calculations show that in an Fe-Cr-Ni weld, the bcc mode of solidification is preferred below 2×10^{-3} m/s, and the fcc mode of solidification is preferred at higher velocities. These plots are used to rationalize the transition from δ-ferrite to the austenite mode of solidification in laser welds. Source: Ref 127

and weldability, CT and CK tools are very effective. For example, a transformation-induced plasticity (TRIP) steel with good welding and galvanizing performance was designed with CT and CK tools. In this work, carbon, silicon, and phosphorous concentrations were optimized, and the microstructure evolution during steel processing was modified (Ref 132, 133). Kaputska et al. used CT tools to describe the stabilization of δ-ferrite in the HAZ of aluminum-base TRIP steels (Ref 134). Using CT tools, blast-resistant steels with good weldability (Ref 135) have been developed. The HAZ microstructure evolution in duplex stainless steel weld metal was modeled using overall transformation kinetic models based on CT tools. This work stressed the importance of nitrogen concentration on the stabilization of austenite (Ref 136). Similar work on duplex steels has been performed by other researchers (Ref 137). In addition, the effect of minor variations in boron was rationalized based on the CCT diagrams predicted by CT and CK tools (Ref 138). Barabash et al. (Ref 139) showed the dissolution of gamma-prime precipitate and its effect on dislocation activity in HAZ regions. These interactions were rationalized based on transmission electron microscopy, synchrotron radiation, and computational thermodynamic calculations.

Kelly (Ref 45) used CT and CK tools to obtain the parameters for the Johnson-Mehl-Avrami-Kologoromov model, which is then coupled with a thermal model for laser near-net shaping of Ti-6Al-4V alloys. Using this thermal-microstructure model, Kelly described the microstructure evolution as a function of the heating and cooling cycle. This model was used to describe the formation of various morphologies of α+β, including the grain-boundary, colony, basketweave, massive, and martensitic microstructures. Similar approaches to describe microstructure gradients in multipass welds in steels were performed by Reed and Bhadeshia (Ref 140). In these analyses, the austenite formation and its effect on increasing the fraction of the reheated region was described by coupling heat-transfer models with thermodynamic and kinetic calculations (Ref 141). Extensive work has been done by Keehan et al. (Ref 142–146) to develop high-strength and high-toughness weld metal

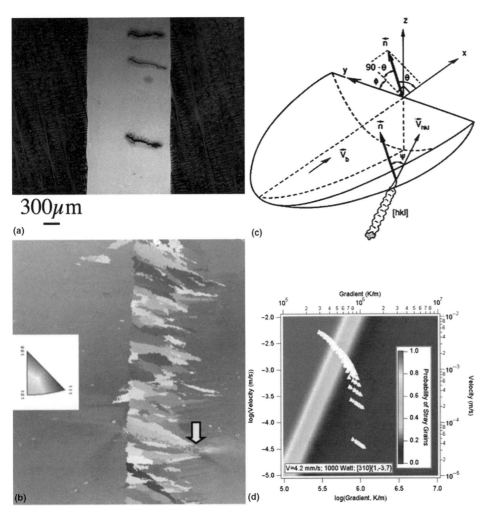

Fig. 13 (a) Optical micrograph of an autogenous gas tungsten arc weld made on a single-crystal nickel-base superalloy shows the transverse crack on only one side of the weld. (b) Orientation imaging microscopy showing the close correlation between the crack and the presence of high-angle stray grain boundaries. (c) Schematic illustration of a geometric model of the weld pool used for estimating the dendrite growth velocity as a function of welding speed. (d) Predicted tendency to form stray grains as a function of dendrite growth velocity and temperature gradients

compositions that are based on bainite microstructure. These developments relied on CT and CK tools (Ref 147). The reliability of weld microstructure models and their relation to properties have been reviewed by Bhadeshia (Ref 148).

To address the stability of weldment structures during postweld heat treatment and service, CT and CK tools have been used. In certain 9Cr steels, the addition of boron leads to a fine distribution of $M_{23}C_6$ and an increase in creep resistance. Thermodynamic calculations showed that boron will segregate to martensite lath boundaries and increase the driving force of boride precipitation. These precursor phases will act as heterogeneous nucleation sites for boron containing $M_{23}C_6$ (Ref 149). Simultaneous transformation kinetic theories of carbide precipitation (Ref 150) and neural network models (Ref 151) have been used to develop creep-resistant weld metals. The nucleation and kinetics of chi, sigma, and

secondary gamma (face-centered cubic) phases during aging of a super duplex stainless steel weld metal were investigated with experimental characterization and thermodynamic driving force calculations (Ref 152). Similar work has been performed by Sieurin et al. (Ref 153, 154) to describe the sigma-phase precipitation and reforming of austenite during isothermal and continuous cooling. Ramirez et al. (Ref 155) used analytical electron microscopy and thermodynamic modeling to evaluate secondary austenite formation and its interaction with Cr_2N precipitation. Another form of degradation in high-chromium stainless steel welds is the tendency for low-temperature embrittlement. This occurs due to the phase separation of chromium-rich and chromium-lean regions through a spinodal decomposition (Ref 156) and the subsequent precipitation of gamma phase. Many researchers have characterized this decomposition using the atom probe technique and compared it with CT and CK tools (Ref

157). The CT and CK models also have been used for evaluating the stability of dissimilar material welds. One successful example is the prediction of carbon migration in dissimilar material welds (Ref 158, 159).

Performance Modeling

The goal of performance modeling relates to the prediction of weldability, geometrical distortion, and/or locked-in residual stresses as a function of material, restraint, process, and process parameters as well as service temperature.

Weldability Prediction

One of the main uses of numerical modeling is to predict the cracking tendency in welds. Weld solidification cracking is governed by severe restraints (stress/strain state) and the composition of the base metal (affects epitaxial growth) and the filler metal (affects the solidification range). To predict the solidification-cracking tendency in welds, there is a need to couple the CT and CK calculations with finite-element analysis that predicts the spatial variation of stress, strain, and temperature (Ref 160–164). This is explained with an example from the published literature.

There is a growing need to develop joining processes and process parameters for single-crystal nickel-base superalloys (Ref 165). During such a development cycle, autogenous welding experiments showed cracking (Fig. 13a) on only one side of the weld. Orientation imaging microscopy showed that the cracking was associated with the formation of stray grains on only one side of the weld (Fig. 13b). The reason for such asymmetrical cracking and grain structure evolution was rationalized with integrated process-microstructure models. The first step was to rationalize the dendrite growth velocities on the left and right side of the weld. The weld pool shape was modeled with a simple analytical heat-transfer equation (Ref 166). Using a geometric model (Fig. 13c) developed by Rappaz et al. (Ref 16), the spatial variation of the temperature gradient and dendrite growth velocity was calculated as a function of weld pool shape and crystallographic orientation of the single crystal. Next, these spatial variations of gradients and velocities were plotted (shown as triangular markers) on a processing map (Fig. 13d) that predicts the probability of stray grain formation. This map is developed using the constitutional supercooling theory for nucleation of new grains ahead of the growing liquid-solid interface. The comparison of these values showed that the tendency for the formation of stray grains is more on the right side of the weld. The formation of high-angle grain boundaries has been attributed as the cause for weld metal cracking (Ref 167). Solidification cracking also requires mechanical driving force. Park et al. (Ref 162) integrated these material models with thermomechanical models to develop processing maps for cracking tendency due to stray grain formation. The processing map developed by Park et al. focused only on weld cracking for single crystals; however, in other polycrystalline alloys, the weldability evaluations must consider weld penetration, porosity formation, HAZ liquation, centerline grain formation, and solidification cracking.

Dye et al. (Ref 163) developed such weldability maps (Fig. 14) for nickel alloys using a different functional relationship between power and velocity. This map shows microstructural evolutions that impact weld penetration, tendency for porosity formation, HAZ liquation, solidification cracking, and center grain formation as a function of welding power and welding speed. For a given welding power, an increase in welding speed will lead to a reduction in weld penetration. An increase in welding power, more than needed for weld penetration, will lead to the formation of porosity. Similarly, with a large increase in power at the same velocity, the thermal gradient in the middle of the weld will be reduced. Under these conditions, the formation of centerline grains will be promoted. The overlaying of all these vulnerabilities leads to a small processing window where the welds can be made without undesirable defects. It is important to note that this optimum process parameter region will shift depending on the materials and thermomechanical conditions.

Residual Stress and Distortion Modeling

The reader is referred to Ref 168 for a comprehensive review of various methodologies and applications for processes and mechanisms of welding residual stress and distortion. In this section, some examples are presented to show

the capability of numerical models to predict residual stress and distortion during welding.

Residual-Stress Prediction. The origin of residual stresses during materials processing is discussed in depth in a review by Withers (Ref 169). For in-depth understanding of the mechanics, the reader is also referred to a classic textbook by Masubuchi (Ref 170). Macroscopic residual stresses develop in welding due to localized plastic strain induced by a severe temperature gradient brought about by localized heat sources. The predictions of residual stresses are usually done by finite-element analyses in either two- or three-dimensional conditions. Most of these computational simulations are also performed in a sequential fashion, as explained subsequently. First, for a given set of geometry and welding process parameters, the thermal distribution is predicted. In the next step, using thermal distribution and the associated variation of mechanical properties, the localized plastic strain distribution is predicted. Due to the mechanical equilibrium constraint, this localized plastic strain accumulation at high temperature leads to a distribution of locked-in elastic residual stresses. It is important to note that these predictions are sensitive to the boundary conditions used in the mechanical equilibrium calculations. For example, local constraints (clamps and fixtures) may affect the evolution of residual stress as well as postweld heat treatments such as temper-bead techniques (Ref 171, 172) and surface peening (Ref 173, 174). It has been demonstrated that the initial distribution of residual stress before the welding operation is important in the estimation of residual stress due to welding. For example, in automotive industries, the sheet metal forming process may induce a pattern of residual-stress distribution due to its inhomogeneous

distribution of plastic strain (Ref 175). In addition, the postweld heat treatment affects the distribution of residual stress due to relaxation.

Residual stresses in welds can be measured by hole-drilling, x-ray diffraction, and neutron diffraction methods (Ref 169). The presence of residual stress has a large effect on fatigue life and stress-corrosion cracking. An example calculation of residual-stress prediction and experimental measurements using x-ray diffraction is shown in a steel weld (Fig. 15). The results (Ref 49) show that it is indeed possible to predict the distribution of residual stresses in welds with good accuracy. There has been a growing impetus to predict the effect of phase transformation on the magnitude of residual stresses in welds (Ref 176). This is partly driven

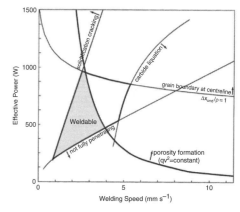

Fig. 14 Calculated weldability map showing the tendency for various weld-defect formations as a function of welding power and speed. Many of these phenomena, including liquation, are predicted using computational thermodynamics and computational kinetics tools. Source: Ref 162

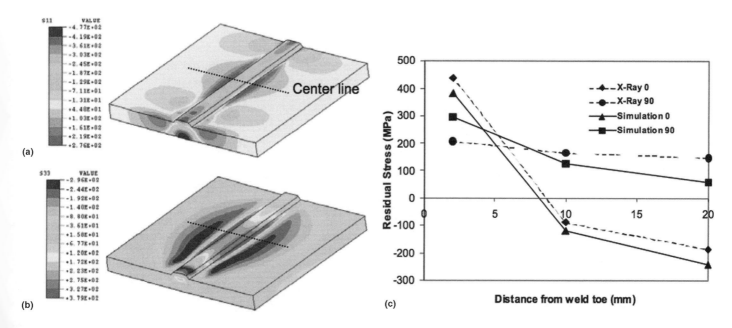

Fig. 15 Simulated residual stress in a butt weld in two directions: (a) perpendicular and (b) parallel to the welding direction. (c) The calculations are compared with the measurements by x-ray diffraction and show good agreement. Source: Ref 49

by the need for the use of low-temperature-transformation wires to induce compressive residual stress near the toe of fillet welds to improve the fatigue life of welded components (Ref 49, 177).

Distortion Prediction. The distortion of the welded structure is essentially due to dimensional changes that occur during or after welding due to the imbalance of locked-in stresses. The distortion simulation can be performed by finite-element analysis in three dimensions (Ref 23, 168, 178), with consideration of elastic and plastic strain accumulation during welding. However, the application of finite-element simulation to large-scale structures (e.g., ship superstructure) with consideration of detailed local welding process effects and constraints may become computationally expensive and time-consuming. In this regard, a simplified procedure that is based on applied plastic strain has been developed (Ref 179 to 181). A comparison of three-dimensional elastic-plastic analysis and the simplified plasticity strain method is shown in Fig. 16. The predicted distortions are comparable and prove the applicability of this technique to large-scale structures. This technique is very useful for consideration of buckling distortion in long welds on thin sheet structures. The technique was used to derive the process parameters for the transient thermal tensioning technique to minimize buckling distortion in ship panel structures (Ref 182).

Property Predictions

After the microstructural gradients, residual stress, and distortion conditions in a weld have been described, the next step is to predict the low- and high-temperature properties of the weld. Some of the methodologies are presented as follows.

Low-Temperature Properties. Relevant low-temperature properties for welds are hardness distributions, tensile properties, toughness, and fatigue properties. Hardness distributions in welds are often predicted using empirical equations, which relate the composition of the alloys and the microstructure. For steel welds, this is discussed in detail by Ion et al. (Ref 8) and Bhadeshia (Ref 148). Using these hardness models, Santella et al. (Ref 183) and Yuping et al. (Ref 184) predicted the performance of steel spot welds. Similarly, the tensile properties of the welds are often predicted using additive law (Ref 185) and neural networks (Ref 151). An example of toughness prediction in steel weld metals, using artificial neural networks, is shown in Fig. 17. Using such maps, new filler metal compositions have been derived for low-temperature service (Ref 146). In welds, fatigue properties are often controlled by the crack propagation conditions. This is partly because most of the welds often have

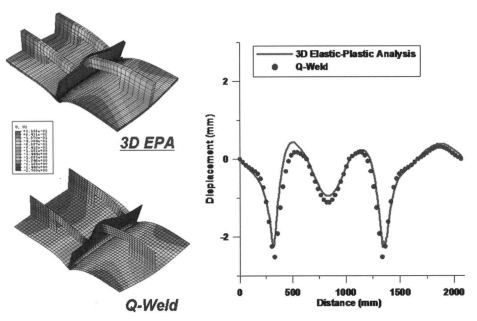

Fig. 16 Results from distortion modeling based on plasticity strain (Q-Weld) compared with fully three-dimensional coupled thermal elastic-plastic analysis (3D-EPA) model. The overall shape distortion (exaggerated) is comparable. In addition, the predicted displacement of the edges from the bottom plane is compared for both methods, demonstrating the applicability of the strain-based method for predicting the distortion of large-scale structures

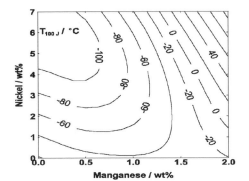

Fig. 17 Contour plot of Charpy toughness values, at a given temperature, as a function of manganese and nickel concentration constructed using an artificial neural network model. Such plots can be used to design welding consumables

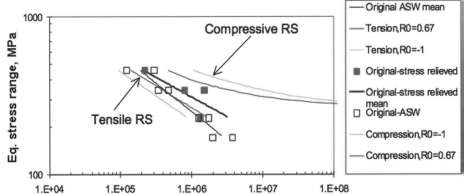

Fig. 18 Calculated increase in fatigue life predicted by the modification of residual stresses (RS) in the weld. Note that an order of magnitude or more increase in fatigue life is predicted due to the presence of compressive stress near the root of the weld. Source: Ref 49

submicroscopic features that provide an easy site for fatigue crack initiation. As a result, most of the research on prediction of fatigue-life estimation has relied on crack propagation conditions. Crack propagation is strongly affected by geometry (Ref 186) and the presence of compressive residual stresses (Ref 49,

171, 174, 175). An example calculation on the effect of tensile and residual stress is shown in Fig. 18. The calculations indeed show an improvement of fatigue life near the toe of a fillet weld due to the presence of compressive residual stress. This has been experimentally evaluated by measurements (Ref 49).

High-Temperature Properties. The inherent creep properties of weld regions without the consideration of geometry can be predicted by using neural network models (Ref 148, 151, 187) or with hybrid models that consider the phase stability (Ref 188 to 190). It is also possible to develop detailed creep property calculations using detailed computational models, as described by To et al. (Ref 191). When the inherent properties are known, there is a need to map them to predict the creep life of structural components (Ref 192). Such calculations have become critical to evaluate the performance of welds made on new-generation high-chromium (P91) steels. These welds often show a soft zone in the HAZ. These regions will lead to localized reduction in creep strength. An example calculation of creep failure in the HAZ of a girth-welded pipe is shown in Fig. 19. In this model, different creep models (constitutive relations) were applied to the base metal and HAZ regions. In the next step, the damage accumulation in these regions was predicted for a given operating condition. These calculations have also been validated with experimental measurements.

Access and Delivery of Integrated Weld Process Models

The aforementioned examples clearly show that with an integrated weld-modeling framework, it is possible to evaluate the welded structure, as envisioned by Kirkaldy (Ref 11). Some of the commercially available integrated weld-modeling software providers are listed in Table 1.

In addition, there are public domain software tools (Ref 193) that can be used for integrated modeling of welds. The listing of these resources is not an endorsement of the products, and readers are requested to make independent testing and evaluation for their own applications. It is important to note that this article did not include any reference to other forms of fusion welding (resistance, laser and electron beam) solid-state modeling techniques. There has been extensive work in this area, which is covered in other resources.

The discussions and references demonstrate that it is indeed possible to describe the weld properties in a wide range of alloy systems as a function of alloy composition and welding process parameters using integrated models. However, these models should be able to consider different processes (submerged, gas metal, gas tungsten, laser, etc.), process parameters, geometrical conditions (restraints, distortion), and mechanical integrity (residual stress, low- and high-temperature properties) in a simpler methodology. In addition, these integrated models should be seamlessly accessible to academic and industrial users. Although commercially available weld-modeling software exists (Table 1), it requires expertise in finite-element analysis as well as a computational and welding

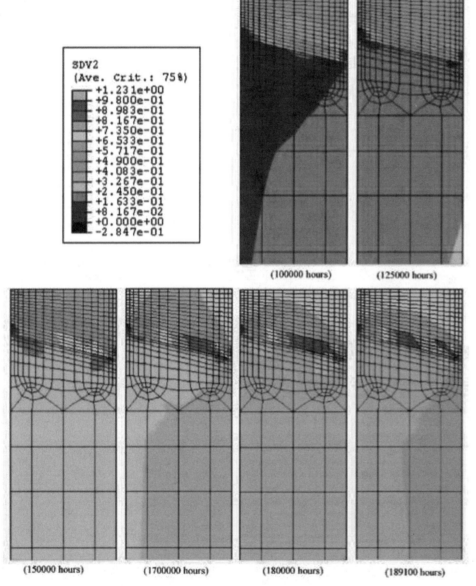

Fig. 19 Predicted accumulation of creep damage in the heat-affected zone of a chromium-molybdenum steel using constitutive equations as a function of service lifetime. Source: Ref 192

Table 1 Software tools for integrated weld modeling

Software	Internet link	Platform
SYSWELD	http://www.esi-group.com/products/welding	Desktop
VrWeld	http://goldaktec.com/vrweld.html	Desktop
VFT	http://www.battelle.org/	Desktop
WELDSIM	http://www.aws.org/wj/2008/05/wj200805/wj0508-36.pdf	Desktop
SORPAS	http://www.swantec.com/sorpas.htm	Desktop
E-WeldPredictor	http://calculations.ewi.org/VJP/	Internet

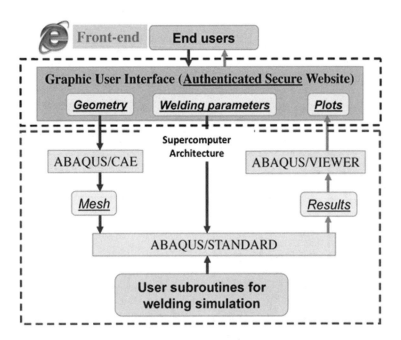

Fig. 20 Architecture for deploying computational weld mechanics models through internet technology and supercomputing architecture. This architecture allows for calculation of thermal cycle, microstructure, residual stress, and distortion within a single computer model

background. However, the presence of such expertise is often limited in small- and medium-scale industries. As a result, the use of computational weld mechanics has been widespread only in large-scale industries. To provide access to these models for small- and medium-scale industries, there is a need to develop methodologies for rapid deployment and easy-to-use applications.

To meet this crucial need, the E-WeldPredictor approach has been developed by researchers from the Edison Welding Institute (Ref 194). In this methodology, all of the expertise for developing and running a finite-element analysis to calculate thermomechanical-metallurgical responses was ported into a supercomputing architecture. Simple, accessible, internet-based applications have been developed in collaboration with supercomputing expertise and have been deployed through a client-server computational architecture (Fig. 20). The steps are explained as follows.

First, the user provides the information for the welding processes. This allows the model to set the effective heat-transfer efficiency. In the next step, the user provides the overall geometry of the welding (pipe or plate) and the boundary conditions (thickness, diameter, and length). This information is used to set the overall simulation geometry within the finite-element software. In the next step, the user provides the detailed joint geometry (e.g., V- or J-groove). This information is used to calculate the extent of weld metal volume to be used for filling the void within the calculation framework. In the next step, the user provides the materials used for the base and filler metals. This information is used to calculate the microstructure and hardness. Currently, these material models do not consider the initial

microstructure of the base metal, and a ferrite microstructure is assumed. In the next step, the user provides the information for heat input and geometry of the individual weld bead that will be used for filling up the joint. These parameters are then used to set the heat-flux conditions for simulating the weld pool (Ref 42). Finally, the model allows the user to check whether the intended weld geometry is indeed the one the user would like to consider. As soon as the user accepts and presses the submit button, the overall thermal-microstructure-mechanical model is evaluated in a sequential manner within a supercomputer architecture at the Ohio Supercomputing Center. The results of the calculations are formatted into a PDF and delivered to the user in an average time frame of 15 min. The calculation time does increase with the number of beads. Currently, the E-WeldPredictor approach has been deployed for predicting microstructure, distortion, and residual stresses in pipe and plate welds for a wide range of steels. The reader is referred to the internet site listed in Ref 195 for more details and usage scenarios. In this section, a typical application of the tool for a case study to evaluate the use of X-65 steels instead of 2.25Cr-1Mo steels for a welded pipeline construction with the same filler and process parameters is discussed.

Case Study Evaluating Use of X-65 Steels

Case Study Involving Typical Use of the E-WeldPredictor Tool. While welding a 2.25Cr1Mo steel pipeline, a hard zone was observed in the HAZ. An ER70S6 filler wire was used in this situation. The focus of the case

study was to substitute the 2.25Cr1Mo steels with traditional X-65 steels for the same process and filler wire conditions, without adverse effect in residual stress and distortion. The individual steps in running the simulations and input parameters are shown in Fig. 21. The results obtained based on these inputs are summarized in Fig. 22. The calculations indicated that there are no significant differences in the HAZ width, residual stress, or distortions by substitution of the 2.25Cr1Mo with the X-65 steel (Fig. 22). The most important difference was the formation of martensite in the HAZ of the 2.25Cr1Mo steels compared to a bainitic microstructure in the X-65 steels. This microstructural distribution is also reflected in the reduction of peak hardness in the HAZs of the X-65 steels. With this tool, the range of other process parameter and material combinations was evaluated, and some key conditions are being considered for detailed experimental evaluations. It is important to note that this tool now can be used for seamlessly evaluating the process-material effects and satisfies the need identified earlier.

Currently, this tool is not comprehensive for a wide range of geometries (fillet welds), boundary conditions (restraints), processes (resistance, laser, friction stir welding, etc.), alloy systems (aluminum, titanium, etc.), and performances (toughness, creep, fatigue, etc.). However, the framework can be modified for these needs with the development of submodels for these processes, materials, and performances (Fig. 21).

Use of Optimization Methodologies

With the development of these integrated models, it is now possible to use the models as a virtual process design tool. Similar to experimental trial-and-error optimization, these models can be run over a wide range of process and material parameters. However, there are two challenges in the implementation of this. These are the inability of the models to do an exhaustive search and the lack of robust material parameters that describe the physical processes. In this regard, optimization methodologies have proved to be useful. Murugananth et al. (Ref 196) have used commercial optimization software so that optimum weld metal composition can be designed for maximizing the weld metal toughness. In this work, the exhaustive search of an artificial neural network model for weld metal toughness was avoided by using generic and stochastic optimization algorithms. The optimization exercise led to a weld metal composition that was in agreement with experimental measurements (Fig. 23). Although, in this exercise, the time for one set of calculations is not more than 10 s, the previously mentioned methodology proved the utility of the optimization tools. Recently, the aforementioned methodology was used to arrive at weld metal compositions that will maximize acicular ferrite in pipeline steel welds (Ref 197).

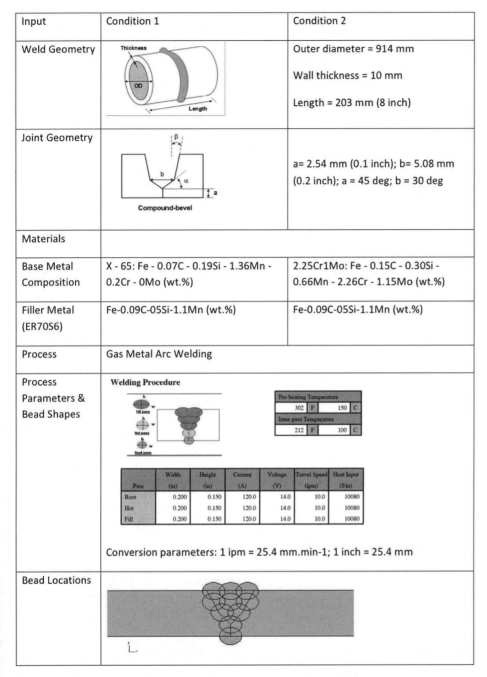

Input	Condition 1	Condition 2
Weld Geometry		Outer diameter = 914 mm Wall thickness = 10 mm Length = 203 mm (8 inch)
Joint Geometry	Compound-bevel	a= 2.54 mm (0.1 inch); b= 5.08 mm (0.2 inch); a = 45 deg; b = 30 deg
Materials		
Base Metal Composition	X - 65: Fe - 0.07C - 0.19Si - 1.36Mn - 0.2Cr - 0Mo (wt.%)	2.25Cr1Mo: Fe - 0.15C - 0.30Si - 0.66Mn - 2.26Cr - 1.15Mo (wt.%)
Filler Metal (ER70S6)	Fe-0.09C-05Si-1.1Mn (wt.%)	Fe-0.09C-05Si-1.1Mn (wt.%)
Process	Gas Metal Arc Welding	
Process Parameters & Bead Shapes	Welding Procedure Pre-heating Temperature: 302 F / 150 C Inter-pass Temperature: 212 F / 100 C | Pass	Width (in)	Height (in)	Current (A)	Voltage (V)	Travel Speed (ipm)	Heat Input (J/in)
---	---	---	---	---	---	---
Root	0.200	0.150	120.0	14.0	10.0	10080
Hot	0.200	0.150	120.0	14.0	10.0	10080
Fill	0.200	0.150	120.0	14.0	10.0	10080 | Conversion parameters: 1 ipm = 25.4 mm.min-1; 1 inch = 25.4 mm	
Bead Locations		

Fig. 21 Table showing an overview of input parameters that can be considered by a computational weld mechanics model within the E-WeldPredictor architecture. Similar simulations can be performed by other models (Table 1); however, they require a dedicated workstation

The previous example is based on a condition that a model for Charpy toughness is verified and validated. However, in certain conditions, the uncertainties of the modeling results are related to the parameters and constants used in the models. For example, in fluid flow simulation of arc welding, there are uncertainties in the arc efficiency (η), arc radius ($r*$), arc energy distribution parameter (d), effective thermal conductivity ($k*$), and effective viscosity (μ) of the liquid metal. To predict the weld pool shape with greater accuracy, it is necessary to obtain these parameters by calibration with a limited number of experimental data. Similar to the earlier example, an exhaustive search of parameter space for these variables (η, $r*$, d, $k*$, and μ) is computationally impractical. To address this challenge, De and DebRoy (Ref 198) developed an elegant optimization methodology to derive these parameters using optimization algorithms. The algorithm evaluates various combinations of parameter space and tries to minimize an objective function that is a function of difference between the experimentally observed weld pool size and the predicted weld pool size. The progress of the optimization is shown in Fig. 24. As soon as, η, $r*$, d, $k*$, and μ are calibrated using the aforementioned method, the heat-transfer and fluid flow model can be used for a wide range of process parameters. The previously mentioned methodology has become a standard technique among weld-modeling researchers.

Concluding Remarks

The integrated weld process modeling methodology presented in this article is in alignment with the needs identified by the National Materials Advisory Board. In a report, integrated computational materials engineering is considered as a transformational discipline for improved competitiveness and national security (Ref 199). The integrated weld-modeling activity will be crucial for developing and deploying advanced materials in practical applications. Moreover, these tools will also accelerate the critical development and deployment of hybrid materials (Ref 200), which rely on a wide range of materials joined in a specific geometric shape and are expected to fill the holes in the material property space (Ref 201).

ACKNOWLEDGMENTS

The author would like to thank the collaboration with world-class researchers in the area of integrated computational weld modeling from The Pennsylvania State University, Oak Ridge National Laboratory, Caterpillar, Edison Welding Institute, Lincoln Electric Company, Engineering Mechanics Corporation of Columbus, Lawrence Livermore National Laboratory, BAM (Germany), and The Ohio State University. Many of the plots and graphs used in the article are derived from these collaborative works.

The author also acknowledges the permission to use various figures derived from the published work of international researchers in this area.

REFERENCES

1. Welding Technology Roadmap, U.S. Department of Energy, Sept 2000, http://files.aws.org/research/roadmap.pdf
2. M.F. Ashby and K.E. Easterling, A First Report on Diagrams for Grain Growth in Welds, *Acta Metall.*, Vol 30, 1982, p 1969–1978
3. H.K.D.H. Bhadeshia, L.-E. Svensson, and B. Gretoft, A Model for the Development of Microstructure in Low-Alloy Steel (Fe-Mn-Si-C) Weld Deposits, *Acta Metall.*, Vol 33, 1985, p 1271–1283
4. H. Cerjak and K.E. Easterling, Ed., *Mathematical Modeling of Weld Phenomena,* The Institute of Materials, London, 1993

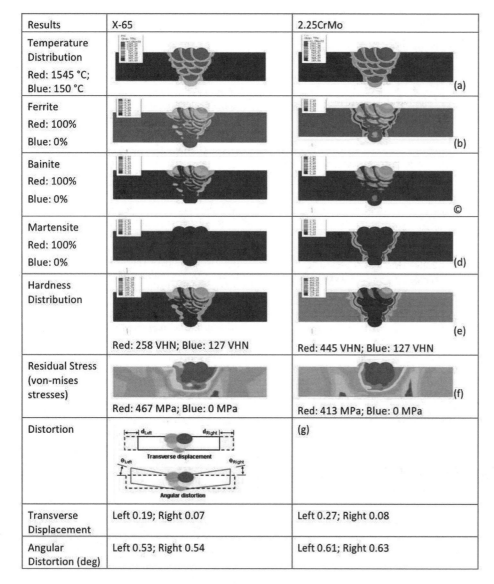

Results	X-65	2.25CrMo
Temperature Distribution Red: 1545 °C; Blue: 150 °C		(a)
Ferrite Red: 100% Blue: 0%		(b)
Bainite Red: 100% Blue: 0%		(c)
Martensite Red: 100% Blue: 0%		(d)
Hardness Distribution	Red: 258 VHN; Blue: 127 VHN	Red: 445 VHN; Blue: 127 VHN (e)
Residual Stress (von-mises stresses)	Red: 467 MPa; Blue: 0 MPa	Red: 413 MPa; Blue: 0 MPa (f)
Distortion		(g)
Transverse Displacement	Left 0.19; Right 0.07	Left 0.27; Right 0.08
Angular Distortion (deg)	Left 0.53; Right 0.54	Left 0.61; Right 0.63

Fig. 22 Results from E-WeldPredictor software for routine comparison of thermal cycle, microstructure, residual stress, and distortion as a function of two steel compositions

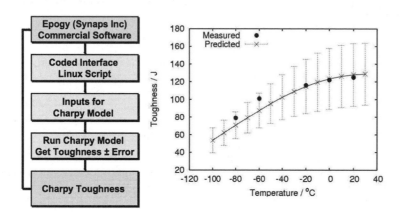

Fig. 23 Methodology used to optimize welding consumable composition. The methodology integrates commercial optimization software with a neural net model for weld metal toughness. The model predicted the optimized weld metal composition for maximizing toughness as a function of temperature. The model predictions are also compared with experimental joining measurements. The agreement shows the feasibility of optimization methodologies.

5. S.A. David and J.M. Vitek, Correlation between Solidification Parameters and Weld Microstructures, *Int. Mater. Rev.*, Vol 34, 1989, p 213–245

6. T. DebRoy and S.A. David, Physical Processing in Fusion Welding, *Rev. Mod. Phys.*, Vol 67, 1995, p 85–112

7. A.H. Dilawari, T.W. Eager, and J. Szekely, Analysis of Heat and Fluid Flow Phenomena in Electro Slag Welding, *Weld. J.*, Vol 57, 1978, p s24–s30

8. J. Ion and K.E. Easterling, Computer Modeling of Weld-Implant Testing, *Mater. Sci. Technol.*, Vol 1, 1985, p 405–411

9. D.F. Watt et al., An Algorithm for Modeling Microstructural Development in Weld Heat-Affected-Zones, A. Reaction—Kinetics, *Acta Metall.*, Vol 36, 1988, p 3029–3035

10. O. Grong, *Metallurgical Modeling of Welding, Materials Modeling Series,* The Institute of Materials, London, 1994

11. J.S. Kirkaldy, Diffusion-Controlled Phase Transformations in Steels—Theory and Applications, *Scand. J. Metall.*, Vol 20, 1991, p 50–61

12. T. Koseki et al., Numerical Modeling of Solidification and Subsequent Transformation of Fe-Cr-Ni Alloys, *Metall. Mater. Trans. A.*, Vol 25, 1994, p 1309–1321

13. S. Kou and Y.H. Wang, Computer Simulation of Convection in Moving Arc Weld Pools, *Metall. Trans. A—Phys. Metall. Mater. Sci.*, Vol 17, 1986, p 2271–2277

14. J.B. Leblond and J. Devaux, A New Kinetic Model for Anisothermal Metallurgical Transformations in Steels Including Effect of Austenite Grain Size, *Acta Metall.*, Vol 32, 1984, p 137–146

15. F. Matsuda, H. Nakagawa, and J. Lee, Numerical Analysis of Micro-Segregation during Welding, *Q. J. Jpn. Weld. Soc.*, Vol 9, 1991, p 85–92

16. M. Rappaz et al., Development of Microstructures in Fe-15Ni-15Cr Single Crystal Electron Beam Welds, *Metall. Trans. A.*, Vol 20, 1989, p 1125–1138

17. J. Szekely and G. Oreper, Transient Heat and Fluid Flow Phenomena in Arc Welding, *J. Met.*, Vol 35, 1983, p 49

18. J.M. Vitek, S.A. Vitek, and S.A. David, Numerical Modeling of Diffusion Controlled Phase Transformations in Ternary Systems and Application to the Ferrite to Austenite Transformation in the Fe-Cr-Ni System, *Metall. Mater. Trans. A.*, Vol 26, 1995, p 2007–2025

19. N. Yurioka et al., Determination of Necessary Preheating Temperature in Steel Welding, *Weld. J.*, Vol 62, 1983, p s147–s153

20. T. Zacharia et al., Weld Pool Development during GTA and Laser-Beam Welding of Type 304 Stainless Steel, 1. Theoretical Analysis, *Weld. J.*, Vol 68, 1989, p s499–s509

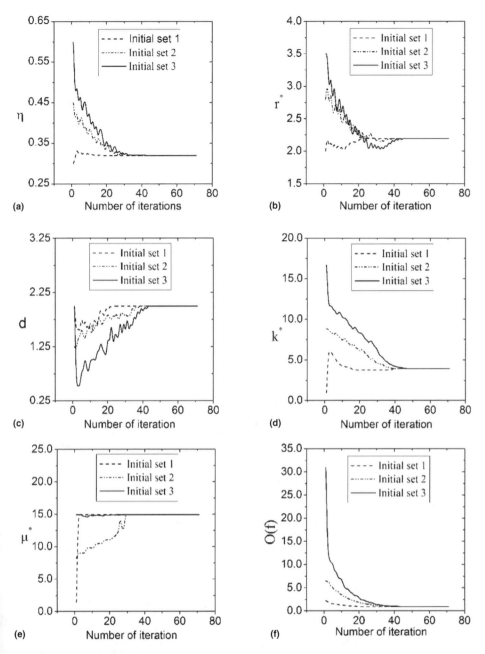

Fig. 24 Optimization methodology for arriving at parameters that are not possible or are difficult to measure. In this demonstration, uncertainty in the arc efficiency, arc radius, energy distribution parameter, effective thermal conductivity, and effective viscosity is reduced through optimization schemes.

21. J.H. Hattel, Integrated Modeling in Materials and Process Technology, *Mater. Sci. Technol.*, Vol 24, 2008, p 137–148

22. J. Ronda and G.J. Oliver, Consistent Thermo-Mechano-Metallurgical Model of Welded Steel with Unified Approach to Derivation of Phase Evolution Laws and Transformation Induced Plasticity, *Comput. Methods Appl. Mech. Eng.*, Vol 189, 2000, p 361–417

23. SYSWELD, ESI Group, http://www.esi-group.com/products/welding-old/sysweld/sysweld_brochure.pdf

24. X.L. Chen et al., Achieving Manufacturing Quality and Reliability Using Thru-Process Simulation, *J. Phys. IV (France)*, Vol 120, 2004, p 793–800

25. VrWeld, Goldak Technologies Inc., http://goldaktec.com/newsandevents.html

26. O.R. Myhr and O. Grong, Utilizing a Predictive Tool for Designing Welded Aluminum Components, *Weld. J.*, May 2008

27. E-WeldPredictor, EWI, http://calculations.ewi.org/vjp/EWeldPredictor.html

28. H. Cerjak and H.K.D.H. Bhadeshia, Ed., *Mathematical Modeling of Weld Phenomena 2*, The Institute of Materials, United Kingdom, 1995

29. H. Cerjak and H.K.D.H. Bhadeshia, Ed., *Mathematical Modeling of Weld Phenomena 3*, The Institute of Materials, United Kingdom, 1997

30. H. Cerjak and H.K.D.H. Bhadeshia, Ed., *Mathematical Modeling of Weld Phenomena 4*, The Institute of Materials, United Kingdom, 1998

31. H. Cerjak and H.K.D.H. Bhadeshia, Ed., *Mathematical Modeling of Weld Phenomena 5*, The Institute of Materials, United Kingdom, 2001

32. H. Cerjak and H.K.D.H. Bhadeshia, Ed., *Mathematical Modeling of Weld Phenomena 6*, The Institute of Materials, United Kingdom, 2002

33. H. Cerjak, H.K.D.H. Bhadeshia, and E. Kozeschnik, Ed., *Mathematical Modeling of Weld Phenomena 7*, Verlag der Technischen Universität Graz, 2005

34. H. Cerjak, H.K.D.H. Bhadeshia, and E. Kozeschnik, Ed., *Mathematical Modeling of Weld Phenomena 8*, The Institute of Materials, United Kingdom, 2007

35. S.A. David, Ed., *Trends in Welding Research in the United States*, American Society for Metals, 1982

36. S.A. David, T. DebRoy, J.C. Lippold, H.B. Smartt, and J.M. Vitek, Ed., *Trends in Welding Research, Seventh International Conference Proceedings*, ASM International, 2006

37. J.A. Goldak and M. Akhlagi, *Computational Welding Mechanics*, Springer, 2005

38. D. Radjaj, *Welding Residual Stresses and Distortion: Calculation and Measurement*, DVS Verlag, Germany, 2003

39. O. Grong, *Metallurgical Modeling of Welding*, 2nd ed., The Institute of Materials, United Kingdom, 1997

40. S.S. Babu, Thermodynamic and Kinetic Models for Describing Microstructure Evolution during Joining of Metals and Alloys, *Int. Mater. Rev.*, Vol 54 (No. 6), 2009, p 333–367

41. D. Rosenthal, Mathematical Theory of Heat Distribution during Welding and Cutting, *Weld. J.*, Vol 20, 1941, p 220s–234s

42. J. Goldak, A. Chakravarti, and M. Bibby, A New Finite Element Model for Welding Heat Sources, *Metall. Trans. B*, Vol 15, 1984, p 299–305

43. S.M. Kelly and S.L. Kampe, Microstructural Evolution in Laser-Deposited Multilayer Ti-6Al-4V Builds: Part I. Microstructural Characterization, *Metall. Mater. Trans. A*, Vol 35, 2004, p 1861–1867

44. S.M. Kelly and S.L. Kampe, Microstructural Evolution in Laser-Deposited Multilayer Ti-6Al-4V Builds: Part II. Thermal Modeling, *Metall. Mater. Trans. A*, Vol 35, 2004, p 1869–1879

45. S.M. Kelly, "Thermal and Microstructure Modeling of Metal Deposition Processes with Application to Ti-6Al-4V," Ph.D. thesis, Virginia Polytechnic Institute and University, Blacksburg, VA, 2004

46. Y. Yang, Edison Welding Institute, Columbus, OH, unpublished work

47. K. Mundra et al., Role of Thermophysical Properties in Weld Pool Modeling, *Weld. J.*, Vol 71, 1992, p s313–s320

48. W. Zhang, C.-H. Kim, and T. DebRoy, Heat and Fluid Flow in Complex Joints during Gas Metal Arc Welding, Part I: Numerical Model of Fillet Welding, *J. Appl. Phys.*, Vol 95, 2004, p 5210–5219

49. Z. Yang and H.W. Ludewig, "Virtual Welded-Joint Design Integrating Advanced Materials and Processing Techniques," Final technical report, DF-FC36-01ID14242, http://www.osti.gov/bridge/purl.cover.jsp?purl=/862362-8Vmn0R/, accessed on May 1, 2009

50. K. Mundra, T. DebRoy, S.S. Babu and S.A. David, Weld Metal Microstructure Calculations Form Fundamentals of Transport Phenomena in the Arc Welding of Low-Alloy Steels, *Weld. J.*, Vol 76, 1997, p S163–S171

51. Z. Yang and T. DebRoy, Modeling of Macro- and Microstructures of Gas Metal Arc Welded HSLA-100 Steel, *Metall. Mater. Trans. A*, Vol 30, 1999, p 483–493

52. S. Fiore, S.S. Babu, and B. Glodowski, "Factors Controlling Microstructure and Properties of Steel Heat-Affected Zones—An Overview," unpublished research, 2009

53. J. Dearden and H. O'Neill, A Guide to the Selection and Welding of Low Alloy Structural Steel, *Trans. Inst. Weld.*, Vol 3,1940, p 203–214

54. N. Yurioka, H. Suzuki, S. Ohshita, and S. Saito, Determination of Necessary Preheating Temperature in Steel Welding, *Weld. J. Res. Suppl.*, Vol 62 (No. 6), June 1983, p 147s–153s

55. J.C. Ion, K.E. Easterling, and M.F. Ashby, A Second Report of Diagrams of Microstructure and Hardness for Heat Affected Zones in Welds, *Acta Metall.*, Vol 32 (No. 11), 1984, p 1949–1962

56. Edison Welding Institute, http://calculations.ewi.org

57. H.K.D.H. Bhadeshia, Neural Networks in Materials Science, *ISIJ Int.*, Vol 39, 1999, p 966–979

58. S.S. Babu and H.K.D.H. Bhadeshia, The Transition from Bainite to Acicular Ferrite in Reheated Fe-Cr-C Weld Deposits, *Mater. Sci. Technol.*, Vol 6, 1990, p 1005–1020

59. S.J. Jones and H.K.D.H. Bhadeshia, Kinetics of the Simultaneous Decomposition of Austenite into Several Transformation Kinetics, *Acta Mater.*, Vol 45 (No. 7), 1997, p 2911–2920

60. Materials Algorithm Project, University of Cambridge, United Kingdom, http://www.msm.cam.ac.uk/map/mapmain.html

61. O.R. Myhr and O. Grong, Process Modeling Applied to 6082-T6 Aluminum Weldments 1: Reaction Kinetics, *Acta Metall. Mater.*, Vol 39, 1992, p 2693–2702

62. O.R. Myhr and O. Grong, Process Modeling Applied to 6082-T6 Aluminum Weldments 2: Application of Model, *Acta Metall. Mater.*, Vol 39, 1992, p 2703–2708

63. C. Zener (quoted in C.S. Smith), Grains, Phases and Interphases; An Interpretation of Microstructure, *Trans. Metall. Soc.*, Vol 175, 1948, p 15–51

64. B. Radhakrishnan and T. Zacharia, Monte-Carlo Simulation of Grain-Boundary Pinning in the Weld Heat-Affected-Zone, *Metall. Mater. Trans. A*, Vol 26, 1995, p 2123–2130

65. B. Radhakrishnan and R.G. Thompson, Kinetics of Grain-Growth in the Weld Heat-Affected-Zone of Alloy-718, *Metall. Mater. Trans. A*, Vol 24, 1993, p 2773–2785

66. S. Mishra and T. DebRoy, Non-Isothermal Grain Growth in Metals and Alloys, *Mater. Sci. Technol.*, Vol 22, 2006, p 253–278

67. S.S. Babu et al., Solidification and Microstructure Modeling of Welds in Aluminum Alloys 5754 and 6111, *Sci. Technol. Weld. Join.*, Vol 6, 2001, p 31–40

68. S.S. Babu et al., Microstructural Development in PWA1480 Electron Beam Welds—An Atom Probe Field Ion Microscopy Study, *Appl. Surf. Sci.*, Vol 94/95, 1996, p 280–287

69. J.O. Andersson et al., Thermo-Calc and DicTra Computational Tools for Materials Science, *Calphad*, Vol 26, 2002, p 273–312

70. O. Grong and N. Christensen, Factors Controlling MIG Weld Metal Chemistry, *Scand. J. Metall.*, Vol 12, 1983, p 155–165

71. M. Liljas and J.O. Nilsson, Development of Commercial Nitrogen-Rich Stainless Steels, High Nitrogen Steels '98, *Mater. Sci. Forum*, Vol 318 (No. 3), 1999, p 189–200

72. V. Muthupandi et al., Effect of Nickel and Nitrogen Addition on the Microstructure and Mechanical Properties of Power Beam Processed Duplex Stainless Steel (UNS 31803) Weld Metals, *Mater. Lett.*, 2005, p 2305–2309

73. S. Hertzman et al., Influence of Shielding Gas Composition and Welding Parameters on the N-Content and Corrosion Properties of Welds in N-Alloyed Stainless Steel Grades, *ISIJ Int.*, Vol 36, 1996, p 968–976

74. S. Hertzman, The Influence of Nitrogen on Microstructure and Properties of Highly Alloyed Stainless Steels, *ISIJ Int.*, Vol 41, 2001, p 580–589

75. K. Mundra and T. DebRoy, A General Model for Partitioning of Gases between a Metal and Its Plasma Environment, *Metall. Mater. Trans. A*, 1995, p 149–157

76. T.A. Palmer and T. DebRoy, Physical Modeling of Nitrogen Partition between the Weld Metal and Its Plasma Environment, *Weld. J.*, Vol 75, 1996, p s197–s207

77. S.S. Babu et al., Reactive Gas Shielding during Laser Surface Alloying for Production of Hard Coatings, *Surf. Coat. Technol.*, Vol 200, 2006, p 2663–2671

78. M. Du Toit and P.C. Pistorious, Nitrogen Control during Arc Welding of Stainless Steel—Part 1: Experimental Observations, *Weld. J.*, Vol 82, 2003, p 219s–224s

79. M. Du Toit and P.C. Pistorious, Nitrogen Control during Arc Welding of Stainless Steel—Part 2: A Kinetic Model for Nitrogen Absorption and Desorption, *Weld. J.*, Vol 82, 2003, p 231s–237s

80. A. Gruszczyk, The Kinetics of Nitrogen Absorption by Arc-Melted Fe-C-Mn-Type Filler Metals, *Weld. J.*, Vol 83, 2004, p 94s–101s

81. R.E. Francis, J.E. Jones, and D.L. Olson, Effect of Shielding Gas Oxygen Activity on Weld Metal Microstructure of GMA Welded Microalloyed HSLA Steel, *Weld. J.*, Vol 69, 1990, p s408–s415

82. S.A. Gedeon and T.W. Eager, Thermochemical Analysis of Hydrogen Absorption in Welding, *Weld. J.*, Vol 69, 1990, p s264–s271

83. J.H. Kiefer, Effect of Moisture Contamination and Welding Parameters on Diffusible Hydrogen, *Weld. J.*, Vol 75, 1996, p s155–s161

84. K. Mundra, J.M. Blackburn, and T. DebRoy, Absorption and Transport of Hydrogen during Gas Meta Arc Welding of Low Alloy Steel, *Sci. Technol. Weld. Join.*, Vol 2, 1997, p 174–184

85. M.J. McNallan and T. DebRoy, Effect of Temperature and Composition on Surface Tension in Fe-Ni-Cr Alloys Containing Sulfur, *Metall. Trans. B*, Vol 4, 1991, p 557–560

86. C.R. Heiple and J.R. Roper, Mechanism for Minor Element Effect on GTA Fusion Zone Geometry, *Weld. J.*, Vol 61, 1982, p S97–S102

87. P. Sahoo, T. DebRoy, and M.J. McNallan, Surface Tension of Binary Metal—Surface Active Solute Systems under Conditions Relevant to Welding Metallurgy, *Metall. Mater. Trans. B*, Vol 19, 1988, p 483–491

88. A. Powell, Mathematical Modeling of Vapor-Plume Focusing in Electron-Beam Evaporation, *Metall. Mater. Trans. A*, Vol 32, 2001, p 1959–1966

89. K. Mundra and T. DebRoy, Calculation of Weld Metal Composition Change in High-Power Conduction Mode Carbon-Dioxide Laser Welded Stainless Steels, *Metall. Mater. Trans. B*, Vol 24, 1993, p 145–155

90. X. He, T. DebRoy, and P.W. Fuersch-bach, Alloying Element Vaporization during Laser Spot Welding of Stainless Steel, *J. Phys. D—Appl. Phys.*, Vol 36, 2003, p 3079–3088

91. J.W. Sowards, "Development of a Chromium Free Consumable for Joining Stainless Steels," Ph.D. thesis, The Ohio State University, 2009

92. U. Mitra and T.W. Eagar, Slag-Metal Reactions during Welding 1. Evaluation and Reassessment of Existing Theories, *Metall. Trans. B*, Vol 22, 1991, p 65–71

93. U. Mitra and T.W. Eagar, Slag-Metal Reactions during Welding 2. Theory, *Metall. Trans. B*, Vol 22, 1991, p 73–81

94. U. Mitra and T.W. Eagar, Slag-Metal Reactions during Welding 3. Verification of the Theory, *Metall. Trans. B*, Vol 22, 1991, p 83–100

95. O. Grong, T.A. Siewert, G.P. Martins, and D.L. Olson, A Model for the Silicon-Manganese Deoxidation of Steel Weld Metals, *Metall. Trans. A*, Vol 17, 1986, p 1797–1807

96. C.S. Chai and T.W. Eager, Slag-Metal Equilibrium during Submerged Arc-Welding, *Metall. Trans. B*, Vol 12, 1981, p 539–547

97. T. Lau, G.C. Weatherly, and A. McLean, The Sources of Oxygen and Nitrogen Contamination in Submerged-Arc Welding Using CaO-Al$_2$O$_3$ Based Fluxes, *Weld. J.*, Vol 64, 1985, p s343–s347

98. M. Zinigrad, Computational Methods for Development of New Welding Materials, *Comput. Mater. Sci.*, Vol 37, 2006, p 417–424

99. A.O. Kluken and O. Grong, Mechanisms of Inclusion Formation in Al-Ti-Si-Mn Deoxidized Steel Weld Metals, *Metall. Trans. A*, Vol 20, 1989, p 1335–1349

100. H. Terashima and P.H.M. Hart, Effect of Aluminum on C-Mn-Nb Steel Submerged-Arc Weld Metal Properties, *Weld. J.*, Vol 63, 1984, p S173–S183

101. H. Homma et al., Improvement of HAZ Toughness in HSLA Steel by Introducing Finely Dispersed Ti-Oxide, *Weld. J.*, Vol 66, 1987, p S301–S309

102. F.C. Liao and S. Liu, Effect of Deoxidation Sequence on Carbon Manganese Steel Weld Metal Microstructures, *Weld. J.*, Vol 71, 1992, p S94–S103

103. G. Thewlis and D.R. Milner, Inclusion Formation in Arc Welding, *Weld. J.*, Vol 56, 1977, p S281–S288

104. K.C. Hsieh, S.S. Babu, and J.M. Vitek, Calculation of Inclusion Formation in Low-Alloy Steel Welds, *Mater. Sci. Eng. A*, Vol 215, 1996, p 84–91

105. T. Koseki, S. Ohkita, and N. Yurioka, Thermodynamic Study of Inclusion Formation in Low Alloy Steel Weld Metals, *Sci. Technol. Weld. Join.*, Vol 2, 1997, p 65–69

106. K. Ichikawa, T. Koseki, and M. Fuji, Thermodynamic Estimation of Inclusion

Characteristics in Low Alloy Steel Weld Metals, *Sci. Technol. Weld. Join.*, Vol 2, 1997, p 231–235

107. S.S. Babu et al., Model for Inclusion Formation in Low Alloy Steel Welds, *Sci. Technol. Weld. Join.*, Vol 4, 1999, p 276–284

108. J. Lehmann, P. Rocabois, and H. Gaye, Kinetic Model of Non-Metallic Inclusions Precipitation during Steel Solidification, *J. Non-Cryst. Solids*, Vol 282, 2001, p 61–71

109. T. Hong and T. DebRoy, Time-Temperature-Transformation Diagrams for the Growth and Dissolution of Inclusions in Liquid Steels, *Scr. Mater.*, Vol 44, 2001, p 847–852

110. T. Hong and T. DebRoy, Nonisothermal Growth and Dissolution of Inclusions in Liquid Steels, *Metall. Mater. Trans. B*, Vol 34, 2003, p 267–269

111. S.S. Babu et al., Toward Prediction of Microstructural Evolution during Laser Surface Alloying, *Metall. Mater. Trans. A*, Vol 33, 2002, p 1189–1200

112. K. Mundra et al., Weld Metal Microstructure Calculations from Fundamentals of Transport Phenomena in the Arc Welding of Low-Alloy Steels, *Weld. J.*, Vol 76, 1997, p S163–S171

113. T. Hong, W. Pitscheneder, and T. DebRoy, Quantitative Modeling of Motion, Temperature Gyrations, and Growth of Inclusions in Weld Pool, *Sci. Technol. Weld. Join.*, Vol 3, 1998, p 33–41

114. T. Hong et al., Modeling of Inclusion Growth and Dissolution in the Weld Pool, *Metall. Mater. Trans. B*, Vol 31, 2000, p 161–169

115. S.S. Babu et al., Coarsening of Oxide Inclusions in Low Alloy Steel Welds, *Sci. Technol. Weld. Join.*, Vol 1, 1996, p 17–27

116. J.M. Vitck et al., Welding of Single-Crystal Nickel-Based Superalloys, *Mathematical Modeling of Weld Phenomena 7*, H. Cerjak, H.K.D.H. Bhadeshia, and E. Kozeschnik, Ed., Technical University of Graz, 2005, p 235–250

117. J.M. Vitek et al., Analysis of Stray Grain Formation in Single-Crystal Nickel-Based Superalloy Welds, *Superalloys 2004*, K.A. Green, T.M. Pollock, H. Harada, T.E. Howson, R.C. Reed, J.J. Schira, and S. Walston, Ed., The Minerals, Metals & Materials Society, Warrendale, PA, 2004, p 459–466

118. S.N. Banovic, J.N. Dupont, and A.R. Marder, Dilution Control in Gas Tungsten Arc Welding Involving Super Austenitic Stainless Steels and Nickel Based Alloys, *Metall. Mater. Trans. B*, Vol 32, 2001, p 1171–1176

119. E.P. George, S.S. Babu, S.A. David, and B.B. Seth, IN-939 Based Superalloys with Improved Weldability, *Proceedings of BALTICA V* (Helsinki, Finland), 2001

120. D.F. Susan et al., A Solidification Diagram from Ni-Cr-Mo-Gd Alloys Estimated by Quantitative Microstructural Characterization and Thermal Analysis, *Metall. Mater. Trans. A*, Vol 37, 2006, p 2817–2825

121. J.N. Dupont, Mathematical Modeling of Solidification Paths in Ternary Alloys: Limiting Cases of Solute Redistribution, *Metall. Mater. Trans. A*, Vol 37, 2006, p 1937–1947

122. T.D. Anderson et al., Phase Transformations and Microstructure Evolution of Mo Bearing Stainless Steels, *Metall. Mater. Trans. A*, Vol 38, 2007, p 671–685

123. T. Koseki et al., Numerical Modeling of Weld Solidification of Austenitic Stainless Steels, *Proceedings of the Seventh International Symposium on Physical Simulations*, 1997, p 75–80

124. T. Koseki, H. Inoue, and A. Nogami, Prediction and Control of Weld Solidification in Steels and Ni-Base Alloys, *Trends in Welding Research*, J.M. Vitek, S.A. David, J.A. Johnson, H.B. Smart, and T. DebRoy, Ed., ASM International, 1999, p 751–760

125. J.M. Vitek and S.A. David, Prediction of Non-Equilibrium Solidification Modes in Austenitic Stainless Steel Laser Welds, *Laser Materials Processing IV*, J. Mazumder, K. Mukerjee, and B.L. Modrikes, Ed., TMS, Warrendale, PA, 1994, p 153–167

126. W. Löser and D.M. Herlach, Theoretical Treatment of the Solidification Undercooled Fe-Cr-Ni Melts, *Metall. Trans. A*, Vol 23, 1992, p 1585–1591

127. T. Koseki and M.C. Flemings, Solidification of Undercooled Fe-Cr-Ni Alloys: 3. Phase Stability in Chill Castings, *Metall. Mater. Trans. A*, Vol 28, 1997, p 2385–2395

128. S. Fukumoto and W. Kurz, Solidification Phase and Microstructure Selection Maps for Fe-Cr-Ni Alloys, *ISIJ Int.*, Vol 39, 1999, p 1270–1279

129. R. Lagneborg et al., The Role of Vanadium in Microalloyed Steels, *Scand. J. Metall.*, Vol 28, 1999, p 186–241

130. J.M. Vitek, E. Kozeschnik, and S.A. David, Simulating the Ferrite to Austenite Transformation in Stainless Steel Welds, *Calphad*, Vol 25, 2001, p 217–230

131. J.R. Yang and H.K.D.H. Bhadeshia, Re-Austenitization Experiments on Some High Strength Steel Weld Deposits, *Mater. Sci. Eng. A*, Vol 118, 1989, p 155–170

132. L. Li et al., Design of TRIP Steel with High Welding and Galvanizing Performance in Light of Thermodynamics and Kinetics, *J. Iron Steel Res. Int.*, Vol 14, 2007, p 37–41

133. L. Li et al., Effects of Alloying Element on the Concentration Profile of Equilibrium Phases in Transformation Induced

Plasticity Steel, *J. Mater. Sci. Technol.*, Vol 19, 2003, p 273–277

134. N. Kaputska et al., Effect of GMAW Process and Material Conditions on DP 780 and TRIP 780 Welds, *Weld. J.*, Vol 87, 2008, p 135–149

135. A. Saha and G.B. Olson, Computer Aided Design of Transformation Toughened Blast Resistant Naval Hull Steels: Part I, *J. Comput.-Aided Mater. Des.*, Vol 14, 2007, p 177–200

136. S. Hertzman et al., An Experimental and Theoretical Study of Heat Affected Zone Austenite Formation in Three Duplex Stainless Steels, *Metall. Mater. Trans. A*, Vol 28, 1997, p 277–285

137. H. Lee et al., Effect of Tungsten Addition on Simulated Heat Affected Zone Toughness in 25%Cr Base Super Duplex Stainless Steels, *Mater. Sci. Technol.*, Vol 14, 1998, p 54–60

138. S.S. Babu et al., Effect of Boron on the Microstructure of Low-Carbon Steel Resistance Seam Welds, *Weld. J.*, Vol 77, 1998, p 2495–2535

139. O.M. Barabash et al., Evolution of Dislocation Structure in the Heat-Affected-Zone of a Nickel Based Single Crystal, *J. Appl. Phys.*, Vol 96, 2004, p 3673–3679

140. R.C. Reed and H.K.D.H. Bhadeshia, A Model for Multipass Welds, *Acta. Metall. Mater.*, Vol 42, 1994, p 3663–3678

141. R.C. Reed, "The Characterization and Modeling of Multipass Weld Heat-Affected Zones," Ph.D. thesis, University of Cambridge, United Kingdom, 1990

142. E. Keehan et al., New Developments with C-Mn-Ni High Strength Steel Weld Metals, Part A—Microstructure, *Weld. J.*, Vol 85, 2006, p 200s–210s

143. E. Keehan et al., New Developments with C-Mn-Ni High Strength Steel Weld Metals, Part B—Mechanical Properties, *Weld. J.*, Vol 85, 2006, p 218s–224s

144. E. Keehan et al., Influence of Carbon, Manganese and Nickel on Microstructure and Properties of Strong Steel Weld Metals Part 1—Effect of Nickel Content, *Sci. Technol. Weld. Join.*, Vol 11, 2006, p 1–8

145. E. Keehan et al., Influence of Carbon, Manganese and Nickel on Microstructure and Properties of Strong Steel Weld Metals Part 2—Impact Toughness Gain Resulting from Manganese Reductions, *Sci. Technol. Weld. Join.*, Vol 11, 2006, p 9–18

146. E. Keehan et al., Influence of Carbon, Manganese and Nickel on Microstructure and Properties of Strong Steel Weld Metals Part 3—Increased Strength Resulting from Carbon Additions, *Sci. Technol. Weld. Join.*, Vol 11, 2006, p 19–24

147. H.K.D.H. Bhadeshia et al., Coalesced Bainite, *Trans. Indian Inst. Met.*, Vol 15, 2006, p 689–694

148. H.K.D.H. Bhadeshia, Reliability of Weld Microstructure and Property Calculations, *Weld. J.*, Vol 83, 2004, p 237s–243s

149. P. Hofer et al., Atom Probe Field Ion Microscopy Investigation of Boron Containing Martensitic 9% Cr Steel, *Metall. Mater. Trans. A*, Vol 31, 2000, p 975–984

150. D. Cole and H.K.D.H. Bhadeshia, Design of Creep-Resistance Steel Welds, *Proc. Mathematical Modeling of Weld Phenomena—V*, H. Cerjak and H.K.D.H. Bhadeshia, Ed., Institute of Materials, 2001, p 431–448

151. H.K.D.H. Bhadeshia, Neural Networks in Materials Science, *ISIJ Int.*, Vol 39, 1999, p 966–979

152. J.O. Nilsson et al., Structural Stability of Super Duplex Stainless Weld Metals and Its Dependence on Tungsten and Copper, *Metall. Mater. Trans. A*, Vol 27, 1996, p 2196–2208

153. H. Sieurin et al., Sigma Phase Precipitation in Duplex Stainless Steel 2205, *Mater. Sci. Eng.*, Vol 444, 2007, p 271–276

154. H. Sieurin and R. Sandstorm, Austenite Reformation in the Heat-Affected Zone of Duplex Stainless Steels, *Mater. Sci. Eng. A*, Vol 418, 2006, p 250–256

155. A.J. Ramirez, S.D. Brandi, and J.C. Lippold, Secondary Austenite and Chromium Nitride Precipitation in Simulated Heat-Affected Zones of Duplex Stainless Steels, *Sci. Technol. Weld. Join.*, Vol 9, 2004, p 301–313

156. J.M. Vitek et al., Low-Temperature Aging Behavior of Type-308 Stainless Steel Weld Metal, *Acta Metall. Mater.*, Vol 39, 1991, p 503–516

157. F. Danoix and P. Auger, Atom Probe Studies of the Fe-Cr System and Stainless Steels Aged at Intermediate Temperature: A Review, *Mater. Charact.*, Vol 44, 2000, p 177–201

158. V. Jan et al., Weld Joint Simulations of Heat-Resistant Steels, *Arch. Metall. Mater.*, Vol 49, 2004, p 469–480

159. T. Helander et al., Structural Changes in 12-2.25 Cr Weldments—An Experimental and Theoretical Approach, *Mater. High Temp.*, Vol 17, 2000, p 389–396

160. Z. Feng, T. Zacharia, and S.A. David, Thermal Stress Development in a Nickel Based Superalloy during Weldability Test, *Weld. J.*, Vol 76, 1997, p s470–s483

161. Z. Feng et al., Quantification of Thermomechanical Conditions for Weld Solidification Cracking, *Sci. Technol. Weld. Join.*, Vol 2, 1997, p 11–19

162. J.W. Park et al., Stay Grain Formation, Thermomechanical Stress and Solidification Cracking in Single Crystal Nickel Base Superalloy Welds, *Sci. Technol. Weld. Join.*, Vol 9, 2004, p 472–482

163. D. Dye, O. Hunziker, and R.C. Reed, Numerical Analysis of the Weldability of Superalloys, *Acta Mater.*, 2001, p 683–697

164. T. Bollinghaus and H. Herold, Ed., Section III, *Hot Cracking Phenomena in Welds*, Springer, p 185–245

165. S.S. Babu et al., Joining of Nickel Base Superalloy Single Crystals, *Sci. Technol. Weld. Join.*, Vol 9, 2004, p 1–12

166. J.M. Vitek, The Effect of Welding Conditions on Strain Grain Formation in Single Crystal Welds—Theoretical Analysis, *Acta Mater.*, Vol 53, 2005, p 53–67

167. S. Mokadem, Laser Repair of Superalloy Single Crystals with Varying Substrate Orientations, *Metall. Mater. Trans. A*, Vol 38, 2007, p 1500–1510

168. Z. Feng, Ed., *Processes and Mechanisms of Welding Residual Stress and Distortion*, CRC Press, 2005

169. P.J. Withers, Residual Stress and Its Role in Failure, *Rep. Prog. Phys.*, Vol 70, 2007, p 2211–2264

170. K. Masubuchi, *Analyses of Welded Structures*, Pergamon Press, 1980

171. A. Saxena, Role of Nonlinear Fracture Mechanics in Assessing Fracture and Crack Growth in Welds, *Eng. Fract. Mech.*, Vol 74, 2007, p 821–838

172. N. Yurioka and Y. Horii, Recent Developments in Repair Welding Technologies in Japan, *Sci. Technol. Weld. Join.*, Vol 11, 2006, p 255–264

173. W. Sagawa et al., Stress Corrosion Cracking Countermeasure Observed on Ni-Based Alloy Welds of BWR Core Support Structure, *Nucl. Eng. Des.*, Vol 239, 2009, p 655–664

174. O. Hatamleh, A Comprehensive Investigation on the Effects of Laser and Shot Peening on Fatigue Crack Growth in Friction Stir Welded AA 2195 Joints, *Int. J. Fatigue*, Vol 31, 2009, p 974–988

175. W.J. Kang and G.H. Kim, Analyses of Manufacturing Effects on Fatigue Failure of an Automotive Component Using Finite Element Methods, *Fatigue Fract. Eng. Mater. Struct.*, Vol 32, 2009, p 619–630

176. M. Mochizuki and M. Toyoda, Strategy of Considering Microstructure Effect on Weld Residual Stress Analysis, *J. Press. Vessel Technol.*, Vol 129, 2007, p 619–629

177. A. Ohta et al., Superior Fatigue Crack Growth Properties in Newly Developed Weld Metal, *Int. J. Fatigue*, Vol 21, 1999, p s113–s118

178. D. Deng et al., Determination of Welding Deformation in Fillet-Welded Joint by Means of Numerical Simulation and Comparison with Experimental Measurements, *J. Mater. Proc. Technol.*, Vol 183, 2007, p 219–225

179. L. Zhang et al., Evaluation of Applied Plastic Strain Methods for Welding Distortion Prediction, *Trans. ASME*, Vol 129, 2007, p 1000–1010

180. G.H. Jung, A Shell-Element Based Elastic Analysis Predicting Welding-Induced Distortions for Ship Panels, *J. Ship Res.,* Vol 51, 2007, p 128–136

181. G.H. Jung and C.L. Tsai, Fundamental Studies on the Effect of Distortion Control Plans on Angular Distortion in Fillet Welded T-Joints, *Weld. J.,* Vol 83, 2004, p 213s–223s

182. J. Song et al., Sensitivity Analysis and Optimization of Thermo-Elastic-Plastic Processes with Applications to Welding Side Heater Design, *Comput. Meth. Appl. Mech. Eng.,* Vol 193, 2004, p 4541–4566

183. S.S. Babu et al., Modeling of Resistance Spot Welds: Process and Performance, *Weld. World,* Vol 45, 2001, p 18–24

184. Y.P. Yang et al., Integrated Computational Model to Predict Mechanical Behavior of Spot Weld, *Sci. Technol. Weld. Join.,* Vol 13, 2008, p 232–239

185. C.H. Young and H.K.D.H. Bhadeshia, Strength of Mixtures of Bainite and Martensite, *Mater. Sci. Technol.,* Vol 10, 1994, p 209–214

186. P. Dong, A Robust Structural Stress Method for Fatigue Analysis of Offshore/Marine Structures, *J. Offshore Mech. Arctic Eng.—Trans. ASME,* Vol 127, 2005, p 68–74

187. H. Fujii et al., Prediction of Creep Rupture Life in Nickel Base Superalloys Using Bayesian Neural Network, *J. Jpn. Inst. Met.,* Vol 63, 1999, p 905–911

188. F. Brun et al., Theoretical Design of Ferritic Creep Resistant Steels Using Neutral Network, Kinetics and Thermodynamic Models, *Mater. Sci. Technol.,* Vol 15, 1999, p 547–554

189. H.K.D.H. Bhadeshia, Performance of Neural Networks in Materials Science, *Mater. Sci. Technol.,* Vol 25, 2009, p 504–510

190. H.K.D.H. Bhadeshia, Mathematical Models in Materials Science, *Mater. Sci. Technol.,* Vol 24, 2008, p 128–136

191. A.C. To et al., Materials Integrity in Microsystems: A Framework for a Peta-Scale Predictive-Science-Based Multi-Scale Modeling and Simulation System, *Comput. Mech.,* Vol 42, 2008, p 485–510

192. J. Storesund et al., Creep Behavior and Life Time of Large Welds in 20 CrMoV 12 1—Results Based on Simulation and Testing, *Int. J. Pressure Vessel Piping,* Vol 83, 2006, p 875–883

193. Materials Algorithm Project, University of Cambridge, United Kingdom, http://www.msm.cam.ac.uk/map/mapmain.html

194. W. Zhang et al., Automatic Weld Modeling Based on Finite Element Analysis and High Performance Computing, *AWS Annual Meeting* (Chicago, IL), 2007

195. Edison Welding Institute, http://calculations.ewi.org

196. M. Murugananth et al., Optimization of Shielded Metal Arc Weldmetal for Charpy Toughness, *Weld. J.,* Vol 83, 2004, p 267s–276s

197. S.S. Babu and M. Murugananth, Computational Optimization of Weld Metal Composition for Maximizing Acicular Ferrite, *Proceedings of Eighth International Conference on Trends in Welding Research,* S.A. David, T. DebRoy, J.N. Dupont, T. Koseki, and H.B. Smartt, Ed., 2008 (Pine Mountain, GA), ASM International, 2009, p 568–574

198. A. De and T. DebRoy, Probing Unknown Welding Parameters from Convective Heat Transfer Calculation and Multivariable Optimization, *J. Phy. D,* Vol 37, 2004, p 140–150

199. *Integrated Computational Materials Engineering: A Transformational Discipline for Improved Competitiveness and National Security,* Committee on Integrated Computational Materials Engineering, National Research Council, 2008, http://www.nap.edu/catalog/12199.html

200. M.F. Ashby and Y.J.M. Brechet, Designing of Hybrid Materials, *Acta Mater.,* Vol 51, 2003, p 5801–5821

201. M.F. Ashby, Hybrids to Fill Holes in Material Property Space, *Philos. Mag.,* Vol 85, 2005, p 3235–3257

ASM Handbook, Volume 22B, *Metals Process Simulation*
D.U. Furrer and S.L. Semiatin, editors

Copyright © 2010, ASM International®
All rights reserved.
www.asminternational.org

Simulation of Rotational Welding Operations

Philip J. Withers and Michael Preuss, Manchester University, United Kingdom

THE PRINCIPLE OF RUBBING TWO OBJECTS TOGETHER, thereby causing frictional heating, is one dating back many centuries. It now forms the basis of many friction-joining, surfacing, and processing techniques. When joining objects by friction, the workpieces are rubbed together under high pressure to generate the required frictional energy such that the parts form a solid-state joint without melting.

For axisymmetric components (e.g., cylinders, tubes, disks, etc.), rotational friction welding methods are appropriate (Fig. 1a, b). The two most common variants are direct-drive rotational friction welding (DD-RFW) and inertia friction welding (IFW). In rotational welding, the heat at the interface is generated by rotation of one part maintained in sliding contact with a stationary part (Fig. 1). The main difference between the two welding techniques is that during DD-RFW, the energy supplied to the rotating part comes from a large-capacity motor, while for IFW, the rotating part is connected to a flywheel. The drive motor is disconnected from the flywheel before the rotating and stationary parts are pushed together, so that energy is supplied to the joint through loss of kinetic rotational energy. In both cases, friction/plasticity at the interface heats up the material very rapidly. As a consequence, the material at the interface becomes softened and is ejected to form a flash (Fig. 1a), while cold material is pushed toward the weld interface, resulting in a loss of length (often termed the upset). Accordingly, the American Welding Society has categorized friction welding as a special form of forge welding (Ref 1).

Historical Development

Bevington was probably the first to exploit friction welding. In 1891, he obtained a patent in which the concept of using frictional heat for extrusion and welding processes was applied (Ref 2). However, it was not until the 1950s that this concept was more widely considered and reached commercial viability. Friction welding was first put to commercial use in Russia in approximately 1956 to 1957 due to the efforts of Chudikov, who successfully demonstrated the possibility of achieving high-quality butt welds between metal rods (Ref 3). In 1959, Vill (Ref 4) and other researchers (Ref 5–7) started to define welding parameters more systematically by analyzing the energy distribution in friction-welded steel bars. The American Machine and Foundry Company introduced the process to the western world in the 1960s, reporting the first experimental work and a discussion of the basic characteristics of the process (Ref 8, 9). Cheng later investigated analytically the temperature distribution during the welding of similar (Ref 10), as well as for the first time dissimilar (Ref 11), metal tubes. Hazlett and Gupta first reported mechanical properties of friction welds with several metal combinations, showing excellent weld properties (Ref 12–14). In the same period, research on friction welding in Western Europe was driven mainly by the welding institutes based in the United Kingdom and West Germany. In Japan, the burgeoning interest in friction welding among the industrial and academic circles led to the formation of the Japan Friction Welding Association in 1964.

Until 1964, all friction-welding processes relied on a direct-drive process. In 1965, the Caterpillar Tractor Company introduced the concept of using a flywheel attached to the rotating spindle (Ref 15). In this way, all the energy required for heat generation at the interface can be provided by the kinetic energy associated with the inertia of the rotating flywheel. Inertia friction welding has since become a popular joining technique for the transport industry. In recent years, IFW has become particularly important for joining difficult-to-weld aeroengine materials such as nickel-base superalloys (Ref 16, 17) and titanium alloys (Ref 18).

Basic Principles

Friction welding is based on the rapid introduction of heat, causing the temperature at the interface to rise sharply and leading to local softening. During rotational friction welding, the two hot mating surfaces are then immediately forged together to form a sound metallurgical bond. Figure 2 compares the characteristics of DD-RFW (Fig. 2a) and IFW (Fig. 2b). In both cases, the rotating part is first brought up to an initial rotational speed (N), and the welding cycle starts when the nonrotating part is brought into contact with the rotating part, resulting in very rapid heating at the interface. In the case of DD-RFW, the rotational speed, N, is kept approximately constant while the initial axial pressure is relatively modest; the aim of this stage is to heat the material rather than to forge it. When the weld region has attained the required forging temperature, the drive from the motor is disconnected, and the rotation soon arrests (Ref 19). In Fig. 2(a), this braking phase is associated with the rotational speed curve falling to zero. Usually, the axial pressure is increased during the braking phase to achieve the required loss of length (upset) of the material by squeezing material out to form the flash, thereby forging the two parts together. This is then identified as the forging phase, which lasts beyond the point at which the rotational speed has dropped to zero (Ref 1). An appreciable shortening of the workpiece is generally already seen during the heating phase after the torque has peaked for the first time (Fig. 2a). In some cases, a more gradual increase of axial pressure is employed to avoid overloading the drive motor (Ref 19). Inertia friction welding is slightly different in that the entire energy needed to form the joint is stored as kinetic energy in the flywheel spinning at the initial rotational speed, N, before the driving power is cut off and the nonrotating part is pushed against the rotating part. This results in a reduction of speed over the entire weld cycle, but because of the large inertia associated with the flywheels, the braking takes place over a longer time period compared to DD-RFW. Consequently, traditional IFW is carried out under a constant axial pressure but at variable rotational speed, as compared with

Fig. 1 Principle of rotational friction welding. (a) Schemati. (b) Jaws of a commercial inertia friction welding machine designed for joining aeroengine turbine disks

conventional DD-RFW which uses stepped pressure and constant rotational speed. In other words, IFW blends the heating and forging actions. It should be noted that, in principle, it is also possible to have a two-stage pressure approach for IFW (Ref 1).

One of the advantages of IFW is the relatively small number of process parameters involved, namely the initial rotational speed (N), the axial pressure (P) (Fig. 2b), and the moment of inertia (I) of the attached flywheel and rotating part. The kinetic energy available for the welding process is determined by N and I, although it is important to remember that not all the kinetic energy will be transformed to heat at the interface due to the internal friction of the IFW machine. The axial pressure largely affects the welding time (Ref 1), although the level of inertia from the flywheel also plays an important role. Essentially, both low pressure and high inertia extend the welding time, which will result in an increased heat-affected zone and a more gradual peak temperature profile across the weld line. The simplicity of only having to optimize three welding parameters when using IFW is often considered to be a great advantage because it requires less effort to develop optimum welding parameters than the DD-RFW process. However, in some cases, the limited number of welding parameters can also be a downside, because this may not provide sufficient flexibility when trying to join a difficult-to-weld material. By contrast, DD-RFW has approximately seven parameters that have an effect on weld characteristics (Ref 20). These parameters are the rotational speed (N), axial pressure (P_1), frictional heating time (t_1),

deceleration time (t_2), delay time (t_3), forging time (t_4), and the forging pressure (P_2) (Fig. 2a).

The capability to store energy in the flywheel makes the motor power requirements of IFW inherently lower than the DD-RFW process. This renders it more appropriate when joining large cross sections or materials requiring high levels of energy input (Ref 19).

Despite some differences, both friction-welding processes display similar characteristics of frictional torque at the interface, that is, two peaks in the recorded torque: one during the initial stage and the other near the end of the welding cycle (Fig. 2). The initial stage of friction welding is dominated by dry friction and wear behavior under severe load. As the rubbing interface changes from partial to complete contact, fluctuations in the rising torque curves can commonly be observed (Ref 21). In the case of DD-RFW, the control of the peak and equilibrium torque levels are considered to be crucial for the quality of the weld (Ref 22, 23). When developing welding parameters for IFW, the initial rotational speed and the size of the flywheel must first be considered in order to supply enough kinetic energy to the process. Oberle proposed that the minimum required rotational speed for successfully joining components depends on the yield strength of the material (Ref 24), although the high-temperature yield strength may be a more appropriate parameter when joining some gas turbine engine materials. The axial pressure applied during friction welding will affect the rate of shortening/upset (or the burn-off rate), which is usually constant during the equilibrium

torque period (Ref 1). It has been reported for DD-RFW that the burn-off rate can be linearly proportional to the axial pressure (Ref 25).

In the case of IFW, burn off occurs at the end of stage 2 and beginning of stage 3 (Fig. 2b), when the torque starts to increase and reaches its second, and usually much higher, peak. It is at this moment that "fresh" subsurface material is brought into intimate contact to form a metallurgical bond, while the original and possibly contaminated surface layer has been forged into the flash. The high torque is attributed to the remaining angular momentum carried by the flywheel. At this stage, the torsional forging force becomes crucial, because the rotational speed during stage 3 is very low. Thus, when welding difficult-to-forge material such as nickel-base superalloys, large flywheels are required to prolong the forging period and to raise the level of the peak torsion (Ref 1). Because the flywheel maintains a significant torsional force toward the end of stage 3, the material is forced out along an essentially spiral path compared with the more radial path associated with DD-RFW, due to the predominance of the axial force in the forging stage of the process (Ref 26).

As solid-state welding techniques, both DD-RFW and IFW are suitable for joining dissimilar materials. Nevertheless, metals that have significantly different high-temperature flow properties or that open up the possibility of intermetallic phases forming at the weld line present challenges when developing such processes. Ways to overcome these issues are to join components with different cross sections and to keep the welding time as short as possible (Ref 27–29).

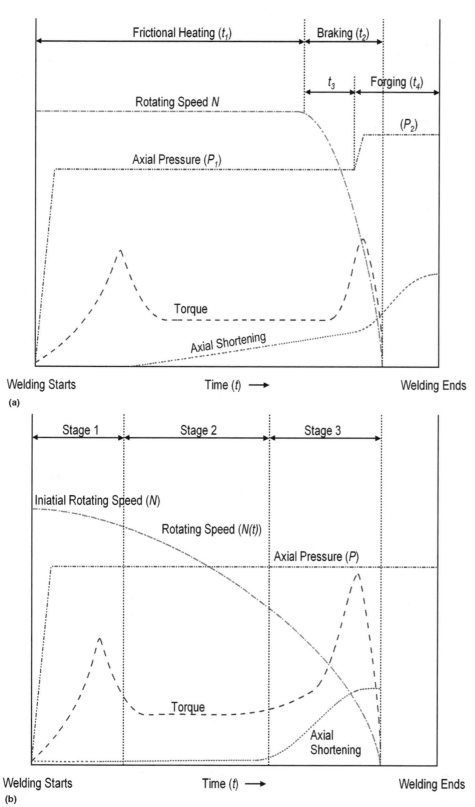

Fig. 2 Process characteristics of typical (a) direct-drive rotational friction-welding and (b) inertia friction-welding processes

One of the important aspects of joint quality is whether DD-RFW or IFW can cause melting of the material at the interface. Generally, postmortem metallurgical evidence and the observation of essentially continuous torque curves suggest that DD-RFW and IFW are truly solid-state welding processes (Ref 1). Because heating rates are predicted to be approximately 10 to 60 times faster than for DD-RFW, IFW is probably more prone to melting at the interface, particularly when joining materials that have a low thermal conductivity, such as titanium alloys (Ref 1, 24).

Weld Microstructure

Both friction-welding processes result in a narrow (typically 0.5 to 5mm) weld region of severely deformed material but with a highly refined grain structure usually termed the thermomechanically affected zone (TMAZ). The TMAZ is surrounded by a wider heat-affected zone (HAZ), that is, a region that was exposed to relatively high temperature during welding but was not significantly deformed. In certain engineering alloys (e.g., ferritic steels, titanium alloys, nickel-base superalloys), the temperature profile across the weld region (TMAZ + HAZ) can result in substantial phase transformation in addition to the severe deformation in the TMAZ. As a result, a highly complex and often very dramatic microstructural variation can be observed across the weld region, which requires advanced characterization tools to fully describe the changes (Ref 16, 30, 31). Figures 3(a) and (b) show a low-magnification macroimage displaying the microstructural variations across the weld line joining two different steels (SCMV, 0.3C-3Cr-1.5Mo steel, against Aermet 100, 0.2C-2.5Cr-10Ni-13Co-1.5Mo-3.5Nb steel). By using a color etch, the TMAZ and, to a lesser degree, the HAZ have been revealed.

When joining precipitation-strengthened alloys, such as nickel-base superalloys, aluminum alloys, and so on, the high temperature in the TMAZ and HAZ will result in dissolution of strengthening precipitates. When the welding process has completed, these may remain in solid solution due to the very high cooling rates. In the case of precipitation-strengthened nickel-base superalloys, the softening of the HAZ can be seen in alloys that have a relatively low volume fraction of precipitates, while in other alloys richer in precipitate-stabilizing alloying elements, precipitation will occur even under severe cooling rates (Fig. 4). In the latter case, the HAZ tends to display a mechanical strength at least equal to that of the base material (Ref 32).

It should be noted that direct-drive friction welding is often preferred for welding steels because the thermal conductivity of steel is relatively high compared to nickel and titanium. This means that the duration of an inertia welding process may not be long enough to reach a sufficiently high temperature (Ref 34). The heating time for direct-drive welding is more easily controlled.

The development of more advanced high-temperature materials for aeroengine application in the last two decades has provided an

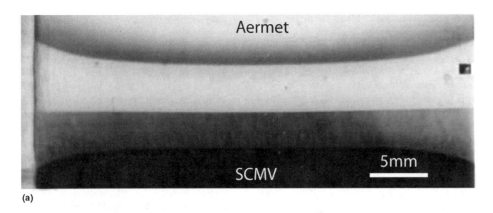

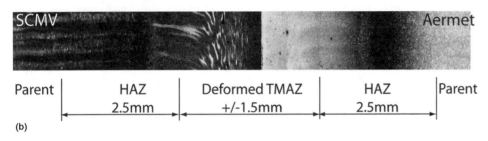

Fig. 3 Dissimilar steel inertia friction weld of SCMV against Aermet 100. (a) Macrograph. (b) Micrograph showing the microstructural variations across the weld revealed by using a color etch. HAZ, heat-affected zone; TMAZ, thermomechanically affected zone. Source: Ref 31

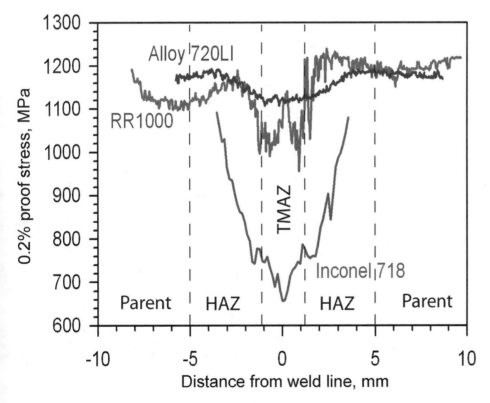

Fig. 4 Variation in 0.2% proof stress across the thermomechanically affected zone (TMAZ) and heat-affected zone (HAZ) of three inertia-friction-welded nickel-base superalloys in the as-welded condition. The measurements were made on cross-weld samples using electron speckle pattern interferometry to map the strain response across the weld line during tensile testing. Inconel 718 is an alloy with approximately 25 vol% precipitates in the standard condition, while RR1000 and alloy 720LI contain approximately 45 vol% precipitates when fully aged. Source: Ref 32. The approximate extents of the TMAZ regions are inferred from Ref 33 and the HAZs from Ref 32

important area for further developing friction welding, because some of these alloys cannot be joined by fusion welding. In addition to the complex microstructure and mechanical property variations across the weld line, the issue of residual stresses generated during joining has become of great importance. To assess the residual stresses deep inside the welded component, advanced diffraction techniques, such as neutron diffraction, can be used that enable a detailed quantification of the residual-stress field of the weld region (Ref 16, 32, 35). The so-called contour method, a destructive method of determining residual stresses, can also be used to assess the residual stresses in inertia friction welds (Ref 36). Generally, such work has demonstrated that the residual stresses generated during IFW are largely dependent on the high-temperature capability of the material to be joined. In addition, phase transformation as a result of the joining process can have a very pronounced effect on the stresses generated in the hot region (Ref 32, 35). It is also important to note that when joining high-performance alloys, a post weld heat treatment is generally required to relieve or at least reduce the welding stresses. One important aspect of this heat treatment is to reduce the residual stresses sufficiently without compromising the microstructure of the material and the mechanical properties of the welded component. This can be particularly challenging when undertaking a post weld heat treatment of a dissimilar joint, because one material may restrict the temperature range to an extent that the residual-stress-relief heat treatment becomes ineffective for the other material, leaving high residual stresses in the weld (Ref 37). In conclusion, much of the metallurgical and residual-stress analysis carried out on friction joints today (2010) is to develop an improved understanding of the welding process and to provide data that can be used to validate process models and thereby optimize the process window in an efficient manner.

Modeling the Welding Process

Models are important in the drive to move away from a trial-and-error approach to optimizing weld process parameters, whether that be to obtain a given upset, optimize the microstructure, or to control residual stresses. Weld process modelers have generally shied away from explicitly modeling the complete rotational welding process due to the computational demands associated with such a task. This is because of the very severe circumferential strains that are introduced in the narrow weld region. Instead, researchers have tended to neglect the rotational motion itself, preferring to capture indirectly the effects of the high rate of straining and the extensive shearing that it causes at, or near, the contact surfaces. As a result, most models regard rotational welding largely as a forging together of two axisymmetric materials where the heating and the resulting local plastic properties are determined by the rate at which rotational energy or work is transferred to the workpieces. In other words, the rate at which heat is generated has been calculated in terms of the work done by the motor in DD-RFW or the rate of change of the kinetic energy of the flywheel in the case of IFW.

During IFW, the operator has relatively little direct control over the upset, thermal evolution, and welding time compared to direct-drive processes. As a result, to achieve a specified upset, weld microstructure, and residual stresses, the welding engineer either must rely on a matrix of welding trials—varying the initial angular velocity, flywheel momentum, and pressure to arrive at the desired result by trial and error—or have access to a weld process model that provides the link between them. In the inertia welding of aero-engine assemblies (Ref 16, 38), for example, dimensional tolerances of 100 μm are required so that the assembled engine is within specification, while a high level of control over microstructure and residual stress is also desirable. This places tremendous requirements on the control and repeatability of the inertia welding process. Furthermore, a trial-and-error approach would simply be too expensive on full-scale assemblies, while the extrapolation of process parameters from subscale prototype testing would not be feasible. Consequently, inertia friction weld models are of great importance commercially.

Obtaining a Realistic Friction Law

Modeling the effective "friction" response of the materials is central to simulating the welding process. Friction has been treated in modeling strategies in a number of ways. Early work involved the assumption that the friction coefficient, μ_{fr}, is either constant or varies radially (Ref 8, 39). However, it is typically sensitive to pressure, P, temperature, T, and sliding speed, v (Fig. 5a). When modeling IFW

nickel-base superalloys, Moal et al. (Ref 41, 42) did not consider the effect of temperature on friction behavior explicitly, as in Fig. 5(a), but rather simply plotted the change in shear stress/pressure, where the shear stress is a function of angular velocity. Balasubramanian et al. (Ref 34) undertook a combined experimental and theoretical analysis of steel DD-RFW to try and derive an appropriate "friction" law by fitting numerical models to measured thermal profiles. Using a regression analysis for 1045 steel, they obtained the relation:

$$\mu_{fr} = 35.7T^{-0.395}P^{-0.601}v^{-0.022} \qquad \text{(Eq 1)}$$

valid from $T = 20$ to 1200 °C, $P = 4$ to 125 MPa, and $v = 2$ to 3.4 m/s^{-1}. Although empirically based, it shows an inverse relation to pressure, temperature, and sliding velocity in decreasing order of importance. For GH4169 steel, Du et al. (Ref 43) have arrived at a similar equation but a different dependency on temperature:

$$\mu_{fr} = 0.12(T - 273)^{0.471}P^{-0.233}e^{-0.739v} \qquad \text{(Eq 2)}$$

Duffin and Bahrani (Ref 44) identified a series of distinct "frictional" stages during continuous-drive friction welding, with "frictional" force being heavily dependent on the sliding speed:

- *Stage I:* Initially, sliding takes place between unlubricated surfaces, forming isolated, strongly adhering points of real contact. The adhesion and seizure between the rubbing surfaces increase the frictional force and the resisting torque and raise the temperature at the interface.
- *Stage II:* This is a transition stage during which the point contacts change into a layer of plasticized material. The plasticized layer offers less resistance to rubbing, and the resisting torque falls.
- *Stage III:* This is the equilibrium stage, during which the temperature at the interface and the rate of upset remain steady. The plasticized

layer will therefore be relatively thick at low rubbing speed, and its thickness will decrease as the speed is increased. However, because the plasticized material is soft, it tends to be forged out when the axial force is high. The plasticized layer is considered to behave similar to a layer of liquid of high viscosity being sheared between two solid cylinders.

- *Stage IV:* As the speed decreases during the deceleration stage, the thickness of the plasticized layer increases, and the frictional behavior becomes similar to that at low rubbing speed. Consequently, the resisting torque rises to a terminal peak value and then falls to zero when the relative motion ceases. The increase in torque is accompanied by an increase in the rate of upset.

In response to this, various researchers (Ref 45, 46) have used a two-stage friction law. When the temperature is low, a Coulomb relation is used:

$$\text{Shear stress} = -\mu_{fr}\sigma_N \qquad \text{(Eq 3)}$$

where σ_N is the normal stress, and μ_{fr} has the form given by Eq 2. When the temperature is high, the shear stress is taken to be the shear yield stress.

Thermal Modeling and the Input of Heat

Rotational friction welding is characterized by very rapid heating and cooling cycles, the details of which are important from both a hot forming viewpoint as well as a microstructural development viewpoint. In addition, the severe local thermal excursion also plays a pivotal role in determining the residual stresses. As a consequence, modeling of the evolution of the thermal field has been an important objective since the early days of rotational friction welding.

In some respects, the modeling of rotational friction welding is simpler than fusion welding because of the straightforward manner in which heat is transferred to the workpiece. In contrast to difficulties associated with estimating the efficiency of welding torches, weld pool dynamics, and so on, the heat is generated at, or close to, the contacting surfaces, either by conventional sliding friction or by plastic flow. In this regard, DD-RFW is perhaps the simplest case because the rotational velocity is maintained by the drive. Consequently, the energy is provided at a rate, q, determined by applying a force pair or torque, τ, to the rotating tool ($q = \tau\omega$, where ω is the relative angular velocity). In IFW, the energy is provided by the loss of rotational kinetic energy ($\frac{1}{2} I\omega^2$, where I is the moment of inertia) as the tool slows. In essence, thermal models examine the competition between heat input and heat dissipation. The rate of heat input, q, is often described simply by:

$$q = \int_0^{r_0} P\mu_{fr}\varpi r2\pi r\,dr = 2/3P\mu_{fr}\varpi 2\pi r_0^3 \qquad \text{(Eq 4)}$$

where P is the pressure, μ_{fr} is the friction coefficient, ω is the rotational velocity, and r_0 is the

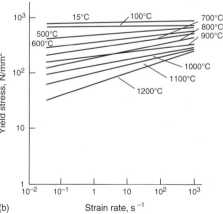

Fig. 5 Effect of temperature on weld parameters. (a) Friction coefficient as a function of temperature at two pressures (curve a: 40N/mm^2; curve b: 60N/mm^2). The continuous lines denote the relation used for the numerical simulation in the section "Analytical Thermal Models" in this article. Source: Ref 40. (b) Temperature effect on yield stress of 20G steel as a function of strain rate

radius of the workpiece. Alternatively, Fu et al. (Ref 47) have used the internal rate of plastic work:

$$q = 0.9\sigma_{eq}\dot{\varepsilon}_{eq} \qquad \text{(Eq 5)}$$

where σ_{eq} is the equivalent stress:

$$\sigma_{eq} = \left[\frac{1}{2} \cdot \left\{ (\sigma_{11} - \sigma_{22})^2 + (\sigma_{22} - \sigma_{33})^2 \right. \right.$$
$$\left. \left. + (\sigma_{11} - \sigma_{33})^2 + 6(\sigma_{12}^2 + \sigma_{22}^2 + \sigma_{31}^2) \right\} \right]^{\frac{1}{2}}$$
$$\text{(Eq 6)}$$

and $\dot{\varepsilon}_{eq}$ is the equivalent strain rate that is similarly defined, and the constant, 0.9, represents the thermal efficiency of plastic deformation.

Dissipation is related to the transfer of heat away from the weld, primarily by conduction but also by radiation (Ch 3 in Ref 48). In this context, heat transfer either across free surfaces into the environment or into the jigging must be considered. Some heat is stored within the plastically deformed body as stored dislocations or elastic strains, but these terms are normally neglected. For certain systems, there is the potential for energy changes arising from phase transformations, but this is not normally a significant term in the energy balance equation.

Analytical Thermal Models

In practice, the welding process is very complex. However, by incorporating some basic assumptions about heat flow in one direction (axial) and neglecting the variation in material properties with temperature, it is possible to develop simple analytical solutions for the heat flow (Ref 5, 26, 49, 50), at least for direct-drive welding. Sluzalec and Sluzalec (Ref 49) wrote the one-dimensional (z) heat flow equation as:

$$\frac{\partial T}{\partial t} = q\frac{\partial^2 T}{\partial z^2} - \frac{2h}{cpr_0}T + \frac{q_0\exp(mt)\delta(z)}{c\rho} \qquad \text{(Eq 7)}$$

where z is the direction along the rotation axis, h is the surface film conductance, c is the heat capacity, ρ is the density, m is a parameter of the process, and $\delta(z)$ is the Dirac delta function. This equation can be solved using a Laplace transform to give an error function solution.

Predictions for the DD-RFW of 50mm diameter cylinders of 20G steel (0.25C, 0.35Si, 1Mn, 0.05P, 0.05S) at 20 MPa pressure for a heating time of 20s, where $q = 84\times10^5\text{W/m}^{-2}$, are shown in Fig. 6(a). For the distances from the weld line plotted, the temperatures increase steadily until welding is complete. Midling and Grong (Ref 51) also obtained similar results for welding at low power. For high-power direct-drive welding, the near-weld region is predicted to reach a maximum temperature when a dynamic balance is achieved between heat conduction and heat generation at the interface (Fig. 6b). They used Rykalin's one-dimensional infinite rod solution (Ref 5),

breaking the welding process down into three stages.

Stage I ($t < t_h$) is the heating stage for which a continuous, planar heat source warms a long rod at a constant rate, q:

$$T - T_0 = \frac{q_0\sqrt{t}}{A\rho c\sqrt{\pi a}}\left[\exp\left(\frac{-z^2}{4at}\right) - \left(\sqrt{\frac{\pi}{4at}}z\right)\text{erfc}\left(\frac{z}{\sqrt{4at}}\right)\right] \qquad \text{(Eq 8)}$$

where q_0 is the net power.

In steady-state stage II ($t_h < t < t_s$), the contact temperature remains constant (at T_{max}), while the temperature everywhere else slowly approaches this value:

$$T - T_0 = \frac{q_0\sqrt{t_h}}{A\rho c\sqrt{\pi a}}\left[\exp\left(\frac{-z^2}{4at}\right) - \left(\sqrt{\frac{\pi}{4at}}z\right)\text{erfc}\left(\frac{z}{\sqrt{4at}}\right)\right] \qquad \text{(Eq 9)}$$

In the cooling stage III ($t_s < t$), a negative heat source is added at $z=0$ to stop the input of further heat.

Fu and Duan (Ref 52) and others (Ref 34, 53) extended this approach using a two-dimensional (z, r) heat-diffusion equation:

$$\frac{\rho C_p}{k}\frac{\partial T}{\partial t} = \frac{\partial^2 T}{\partial^2 r} + \frac{\partial^2 T}{\partial^2 z} + \frac{1}{r}\frac{\partial T}{\partial r} + \frac{Q}{k} \qquad \text{(Eq 10)}$$

where k is the thermal conductivity, C_P is the specific heat capacity, and Q is the rate of volumetric heat generation.

Few analytical inertial weld process models have been described in the open literature. Dave et al. (Ref 54) produced a thermal model to predict the transient thermal profiles when inertia welding tubes to aid parametric optimization. Many assumptions were made, such as averaged temperature and material properties. Reasonable agreement was found with thermocouple results.

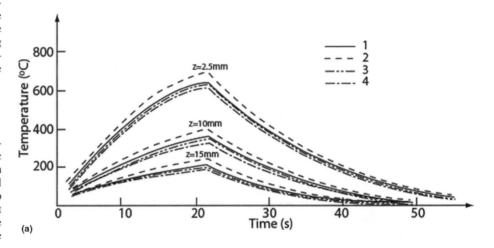

(a)

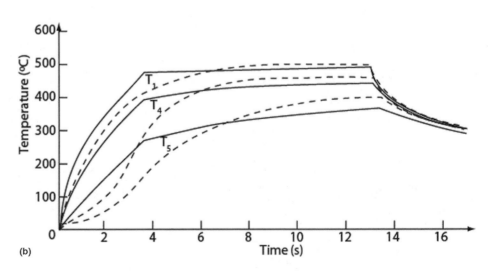

(b)

Fig. 6 Characteristic temperature curves for direct-drive welding. (a) Low-power welding of 20G steel cylinders at different distances from the weld line ($z=0$). Line 1: analytical solution; line 2: experiment; line 3: numerical solution with temperature-independent material properties; line 4: with temperature-dependent properties. Source: Ref 49. (b) High-power welding of AlCuMg$_2$ alloys welded under different conditions showing measured (dashed) and predicted (continuous) profiles. Source: Ref 51

Numerical Thermal Models

In reality, many properties vary in a complex manner as a function of temperature and strain rate. Storm (Ref 55), Plunkett (Ref 56), and Yang and Szewczyk (Ref 57) were among the first to come up with heat-transfer solutions where the thermal conductivity and specific heat are temperature dependent. Furthermore, as illustrated in Fig. 5(b), both the yield stress and the friction behavior tend to be strain-rate sensitive. As a result, realistic modeling requires the use of the finite-difference method (Ref 10, 11, 58, 59) or, more commonly, the finite-element method (Ref 40, 60).

When modeling rotational welding, one of two approaches can be used. These have recently been termed (Ref 61) the representative and the predictive modeling techniques:

- In the representative modeling technique (also called the energy balance method) (Ref 62), the need for a constitutive law describing the "friction" between the faying surfaces is circumvented by using data on the slowing of the flywheel (for inertia welding) or the torque applied (for direct-drive welding), taken directly from the welding process measurements, to calculate the rate at which heat is input into the workpiece. This allows the welding conditions to be simulated, providing information about the temperature profiles, material flow, and residual stresses (Ref 33, 47, 62).
- In the predictive modeling technique, a friction law, which is a function of temperature, pressure, rotational velocity, and material properties, is employed. In this way, the model is able to calculate the instantaneous rotational velocity or torque from a knowledge of the initial conditions and the "friction" law, taking into account some loss of energy due to the mechanics of the machine (Ref 41, assumed 10% loss). In this way, the complete welding process can be modeled (Ref 51, 53, 63).

In the predictive scheme, one must calculate the heating rate via the mechanical work being introduced into the weld by the forces acting where the faying surfaces are in contact; if a region moves out of contact, as may occur in the flash, then the force locally is zero. On the other side of the energy balance sheet, heat is transported from the weld region by a combination of conduction and convection (less important). Another important contribution can be the heat-transfer coefficient between the workpiece and the jig constraining it (Ref 33).

For the friction welding of two 50mm diameter 20G steel cylinders, considered analytically in the previous section, a predictive model has been developed using the friction response shown in Fig. 5(a). It can be seen from Fig. 6 that in simple cases such as this, the use of thermally dependent material properties (lines

numbered 4) is not especially important and that the analytical model (lines numbered 1) does essentially as well as the finite-element method. It has been argued that such systems are, to some extent, self-regulating, thereby inhibiting melting. Indeed, it has been proposed that the maximum temperature at the joint cannot be greater than the temperature at which the yield stress of the workpiece is equal to the pressure applied (Ch 10 in Ref 48).

Probably, the first model of the inertia welding process was a finite-difference model constructed by Wang and Nagappan (Ref 59) to predict solely the thermal evolution during welding. More recently, Fu et al. (Ref 47) adopted a representative modeling approach that imposes a rotational speed-time trajectory, taken directly from their measurements, onto their model. The thermal predictions were correlated with infrared measurements of the surface temperature field. Likewise, Balasubramanian et al. (Ref 62) also took the representative modeling approach for modeling inertia welding of alloy IN718 (UNS N07718). The measured temperature profiles are shown in Fig. 7 for locations originally 4.67 and 8.48mm from the interface, from which it is evident that the temperature rise is initially very rapid before falling away. This decrease is caused by a fall in the angular velocity of the workpiece and has also been observed by others (Ref 53, 59).

A representative model for the IFW of RR1000 nickel-base superalloy tube geometries has been developed by Grant et al. (Ref 64). In this study, a series of run-down tests was used to determine the power loss. A run-down test involves taking the IFW machine up to welding speed, N, and subsequently allowing it to decelerate under its own internal resistance while recording $\omega(t)$. The tests were performed for a number of different flywheel inertias and starting speeds. This study indicates that increasing the welding pressure slightly increases the maximum temperature at the weld line and increases the rate at which the temperature

drops with distance from the weld (Fig. 8). Bennett et al. (Ref 61) compared the performance of predictive and representative models for inertia welding alloy IN718 using DEFORM-2D by initially training the heating model using data from previously completed welds. The thermal predictions for the two approaches were very similar, with a good match to thermocouple data near the weld (1 mm) but underpredicting the temperature further from the weld line (9 mm).

At a practical level, one of the difficulties associated with rotational welding is the determination of the temperature variation close to the weld, due to the difficulties encountered when placing thermocouples very close to the weld (Ref 34). Pyrometers have been used (Ref 63), but their accuracy can be unsatisfactory, and the flash formation will interfere with such measurements. In certain cases, it is possible to validate the thermal predictions by comparing observed and predicted microstructures (Ref 63). For some materials, it is possible to use certain microstructural features, which change with the peak temperature the material has been exposed to during joining. An example is the γ'-strengthened nickel-base superalloys, where the γ' precipitates start dissolving at approximately 800 °C until they are fully dissolved, generally between 1100 and 1200 °C. By experimentally simulating the temperature history of the material near the weld line and comparing the γ' distribution with microstructural observations across the weld line, it was possible in Ref 64 to validate temperature predictions very close to the weld interface (Fig. 9).

Modeling Mechanical Aspects of Welding

The thermal model is normally coupled to a mechanical model. Consequently, an important aspect is the adoption of suitable constitutive

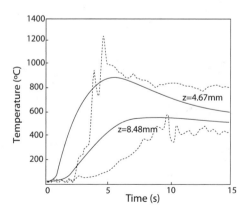

Fig. 7 Comparison of experimental measured (continuous) and numerically simulated (dashed) temperature profiles for an IN718 inertia weld for two axial distances (z) from the weld line. Source: Ref 62

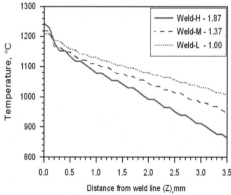

Fig. 8 Peak temperature as a function of distance from the weld line attained in an RR1000 weld inertia-welded tube wall for low (L), medium (M ~ 37% higher welding pressure), and high pressure (H ~ 87% higher welding pressure). Source: Ref 64

laws by which materials behavior can be represented. DD-RFW is characterized by strain rates reaching $1000 \ s^{-1}$ locally (Ref 51). As a result, the strain-rate dependency of the stress-strain response (Fig. 5b) must be captured by reliable constitutive equations.

Rich and Roberts (Ref 26) used a plasticity model to describe the softened region. Equations developed for hot working are often used to describe the deformation behavior, for example, the Norton-Hoff constitutive law:

$$s = 2K\left(\sqrt{3\dot{\varepsilon}_{eq}}\right)^{m-1}\dot{e} \qquad \text{(Eq 11)}$$

where s is the deviatoric stress tensor, $\dot{e}$ is the deviatoric strain-rate tensor, K is a constant, m is the strain-rate sensitivity, and $\dot{\varepsilon}_{eq}$ is the equivalent strain rate. The deformation of the softened material has also been modeled as a viscous Newtonian fluid (Ref 50). Because large deformations can occur in the weld zone as well as the flash, it is often the case that remeshing is necessary (Ref 33, 41, 45).

When mechanically modeling the DD-RFW process, it is usual to derive the heat input directly from the torque offered by the workpieces. As described previously, this may be predicted in terms of the frictional response of the workpiece or measured by experiment and subsequently imposed on the model as a heating term. Zhang et al. (Ref 53) calculated the heat introduced into the weld region for solid GH 4139 steel cylinders welded by DD-RFW from a knowledge of the rotational speed, torque, and pressure. The model was validated by thermocouple measurements of the temperature at key locations and the shape of the flash ejected. The amount of flash ejected is often deliberately engineered to be extensive in order to sweep away oxide and dust on the original joint surfaces into the flash. This work identified that the largest deformation rate is found in the center, near the weld centerline.

Moal and co-workers (Ref 41, 42) took the predictive scheme to derive the slowing of the workpiece and the upset (Fig. 10a, b). It appears that while the onset of the upset is well

predicted, the final moments as the workpiece comes to a halt are not so well predicted. Somewhat better agreement is achieved by Lee et al. (Ref 65), who extended the functionality of DEFORM-2D to include torsional effects by introducing a special axisymmetric element that included radial, axial, and circumferential velocity components to model the IFW of an aeroengine disk component. Plots of the variation in angular velocity and upset with time for the training phase of a representative model for alloy 718 (Ref 61) were similar to those in Fig. 10, with the velocity measurements in better agreement than the upset (10% underprediction).

Modeling of Residual Stresses

A number of researchers have exploited their coupled thermal and mechanical finite-element models to predict the residual stress introduced by rotational welding (Ref 33, 45, 61, 64, 66). Figure 11 shows the measured and predicted residual hoop, axial, and radial stresses within 5mm of an RR1000 nickel-base superalloy inertia-friction-welded tube (Ref 64). The maximum tensile hoop stress was found at the weld line close to the inner wall for both predictions and measurements. The finite-element model predicts a maximum tensile hoop stress of 1420 MPa compared to 1565 ± 70 MPa measured by neutron diffraction, which should be considered an excellent agreement. As well as having similar peak stresses, the predicted and the measured hoop stress fields (Fig. 11a) have very similar forms. Both demonstrate little axial variance near the weld line before exhibiting a knee at 3mm, after which the tensile hoop stresses drop quickly over a relatively short distance (a few millimeters).

The trends in the axial stresses are also well captured by the model, varying approximately linearly from positive to negative from the inner to the outer wall, indicative of a bending stress (Fig. 11b), as seen elsewhere (Ref 67). However, the peak axial stresses predicted by the model are significantly smaller than the

measured results, varying at the weld line from 470 MPa near the inner wall to −470 MPa near the outer. This compares with measured values of 900 and −740 MPa, respectively. The axial stresses arise as a combination of the tourniquet-like contraction of the weld metal as it cools and the effect of the clamping arrangement. In the modeling, it was found that a higher tooling pressure increases the axial bending stress. Because the wall thickness of the weld is small (8 mm), the radial stresses were found to be negligible (<150 MPa) for both the measurements and the predictions (Fig. 11c).

It should be remembered that for dissimilar welds, there is also the capacity for residual stresses to arise from the mismatch in coefficient of thermal expansion and other properties between the two materials being joined (Ref 36, 61). Kim et al. (Ref 66) found that for titanium-stainless steel welds, the lower thermal conductivity of titanium meant that higher peak temperatures were achieved in the steel near the bond line compared to joining steel against itself. Close to the bond line, the radial residual-stress components were tensile in the stainless steel and compressive in the titanium. At the surface, the axial residual-stress components were compressive in the steel near the bond line and tensile further away. In titanium, the axial component was tensile, except very close to the bond line. The titanium close to the bond line was predicted to be severely plastically strained.

In steel rotational welds, there is also the opportunity for martensitic phase transformations (Ref 31). On cooling, the phase transformations from austenite to martensite or bainite can occur displacively, bringing about both a shear and a volume change that can reduce tensile stresses in the near-weld zone (Ref 68). In this context, it is important to remember that austenitization occurs at a higher temperature on reheating than on cooling, so that only material in the HAZ that has exceeded this temperature will transform. Bennett et al. (Ref 61) have calculated that for the CMV steel under study, the martensite start temperature is approximately 300 °C

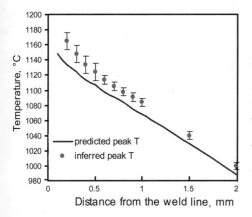

Fig. 9 Comparison of predicted peak temperatures and those inferred from a microstructural study as a function of distance from the weld line. Source: Ref 64

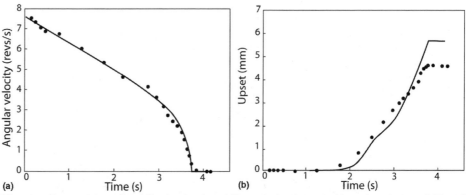

Fig. 10 Characteristic predicted (lines) and measured (data points) variations in (a) angular velocity and (b) axial upset through the inertia welding process for nickel alloy cylinders ($\omega_0 = 7.667$ rev/s; $I_0 = 102 \ kg/m^2$; internal radius = 42 mm; external radius = 48 mm; $P = 360$ MPa). Source: Ref 41

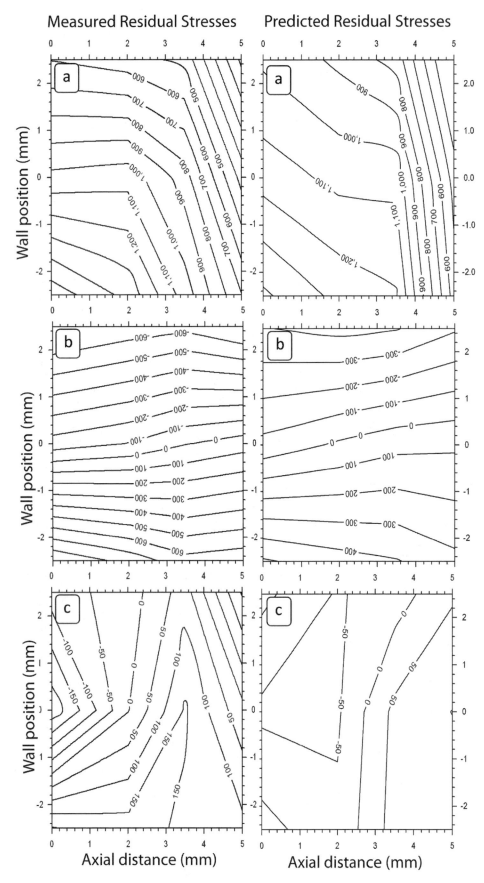

Measured Residual Stresses   **Predicted Residual Stresses**

Fig. 11 Contour plots of the measured (left side) and modeled (right side) stresses in the (a) hoop, (b) axial, and (c) radial directions. All stresses are plotted in MPa. The typical accuracy of the stress characterization by neutron diffraction was ±70 MPa. Source: Ref 64

(Fig. 12a). Taking this into account within DEFORM, the predicted residual stresses are shown in Fig. 12(b). It can be seen that the transformation has a significant effect, reducing the stress by approximately 50% of the yield stress, not just within the HAZ but beyond. This is very beneficial because it moves the peak stresses from the HAZ region into the parent metal.

Finally, an important aspect for the modeling of residual stress is to predict the effect of subsequent postweld heat treatments (PWHTs). Postweld heat treatments are often necessary to reduce the residual stresses introduced by the process. Currently, successful PWHT procedures are found by a combination of trial and error coupled with detailed residual-stress measurements, as shown in Fig. 13. In principle, it is possible to incorporate annealing formulations into finite-element models to optimize PWHT procedures, thereby accelerating the process development cycle. This is especially important for the joining of high-γ' nickel-base superalloys, because they have better high-temperature strength than conventional superalloys and thus are more resistant to PWHT. Annealing models must take into account both short-timescale stress relief brought about by a lowering of the yield stress at elevated temperature (Fig. 13c) and creep models to capture longer-term relaxation phenomena (Fig. 13d). This remains an area for future work.

Modeling of Microstructure and Validation

To date (2010), very little work is available in the public domain describing any modeling effort to predict microstrutcure and the performance of rotation-friction-welded materials/components. Soucail and Bienvenu (Ref 70) produced one of the first attempts to fully predict the microstructural variation across an inertia friction weld for Astroloy, based on the thermal evolution. Validation of the temperature predictions was performed using a pyrometer and thermocouple measurements as well as metallographic techniques. Good agreement in the thermal profiles was found close to the weld line; however, at distances greater than 300 µm from the weld line, the model diverged considerably from the experimental measurements.

REFERENCES

1. K.K. Wang, Friction Welding, *Weld. Res. Counc. Bull.*, Vol 204, 1975, p 1–22
2. J.H. Bevington, Ed., Spinning Tubes Mode of Welding the Ends of Wire, Rods, etc. and Mode of Making Tubes, Patent 1208, U.P.N. British Patent Office, United Kingdom, 1891
3. A.I. Chudikov, Ed., Friction Welding, R.P.R. 106270, United Kingdom, 1956

4. V.I. Vill, Energy Distribution in Friction Welding of Steel Bars, *Weld. Prod.*, Vol 10, 1959, p 31–41

5. N.N. Rykalin, A.I. Pugin, and V.A. Vasireva, The Heating and Cooling of Rods Butt Welded by the Friction Process, *Svar. Proizvod. (Weld. Prod.)*, Vol 6, 1959, p 42–52

6. R.I. Zakson and V.D. Voznesenskii, Power and Heat Parameters of Friction Welding, *Weld. Prod.*, Vol 10, 1959, p 63–70

7. A.E. Geldman and M.P. Sanders, Power and Heating in the Friction Welding of

8. Thick-Walled Steel Pipes, *Weld. Prod.*, Vol 10, 1959, p 53–62

9. M.B. Hollander, Development in Friction Welding, *Metall. Eng. Q.*, Vol 2, 1962, p 14–24

10. M.B. Hollander, C.J. Cheng, and J.C. Wyma, Friction Welding Parameter Analysis, *Weld. J., Res. Suppl.*, Vol 42, 1963, p 495s–501s

10. C.J. Cheng, Transient Temperature Distribution during Friction Welding of Two Similar Materials in Tubular Form, *Weld. J., Res. Suppl.*, Vol 41 (No. 12), 1962, p 542–550

11. C.J. Cheng, Transient Temperature Distribution during Friction Welding of Two Dissimilar Materials in Tubular Form, *Weld. J., Res. Suppl.*, Vol 41 (No. 5), 1963, p 233–240

12. T.H. Hazlett, Properties of Friction Welded Plain Carbon and Low Alloy Steels, *Weld. J., Res. Suppl.*, Vol 41 (No. 2), 1962, p 49s–52s

13. T.H. Hazlett, Properties of Friction Welds between Dissimilar Metals, *Weld. J., Res. Suppl.*, Vol 41 (No. 10), 1962, p 448s–450s

14. K.K. Gupta and T.H. Hazlett, Friction Welding of High Strength Structural Aluminium Alloys, *Weld. J., Res. Suppl.*, Vol 42 (No. 11), 1963, p 490s–494s

15. R.R. Irvine, Welders Are Stirred "Inertia," *Iron Age*, Vol 195 (No. 18), 1965, p 29

16. M. Preuss, J.W.L. Pang, P.J. Withers, and G.J. Baxter, Inertia Welding Nickel-Based Superalloy: Part I. Metallurgical Characterization, *Metall. Mater. Trans. A*, Vol 33 (No. 10), 2002, p 3215–3225

17. M. Preuss, J.W.L. Pang, P.J. Withers, and G.J. Baxter, Inertia Welding Nickel-Based Superalloy: Part II. Residual Stress Characterization, *Metall. Mater. Trans. A*, Vol 33 (No. 10), 2002, p 3227–3234

18. A. Barussaud and A. Prieur, Structure and Properties of Inertia Welded Assemblies of Ti Based Alloy Disks, *Proc. of Eighth World Conference on Titanium* (Birmingham, U.K.), Institute of Materials, 1995

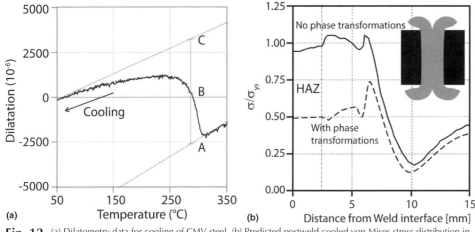

Fig. 12 (a) Dilatometry data for cooling of CMV steel. (b) Predicted postweld cooled von Mises stress distribution in a CMV steel after inertia welding with and without phase transformation. The austenitized zone is shown highlighted (inset). HAZ, heat-affected zone. Source: Ref 61

19. T.K. Johnson and T.W. Daines, "A Comparison of the Capabilities of Continuous Drive Friction and Inertia Welding," Technical paper, AD73, Society of Manufacturing Engineers, 1973, p 221

20. F. Eichhorn and R. Schaefe, Investigation of the Process in the Welding Zone during Conventional Friction Welding of Steel, *Weld. Res. Abroad*, 1969, p 2–20

21. J.F. Squires Thermal and Mechanical Characterisation of Friction Welding Mild Steel, *Br. Weld. J.*, Vol 13 (No. 11), 1966, p 652–657

22. J.C. Needham and C.R.G. Ellis, Automation and Quality Control in Friction Welding: Part I, *Weld. Res. Inst. Res. Bull.*, Vol 12, 1971, p 333–339

23. J.C. Needham and C.R.G. Ellis, Automation and Quality Control in Friction Welding: Part II, *Weld. Res. Inst. Res. Bull.*, Vol 13, 1972, p 47–51

24. T.L. Oberle, "Inertia Welding," International Automotive Eng. Congress (Detroit, MI), SAE, 1973

25. R.G. Ellis, Continuous Drive Friction Welding of Mild Steel, *Weld. J. Res. Suppl.*, Vol 51, 1972, p 183s–197s

26. T. Rich and R. Roberts, Thermal Analysis for Basic Friction Welding, *Met. Constr. Br. Weld. J.*, Vol 3, 1971, p 93–98

27. V.V. Satyanarayana, G.M. Reddy, and T. Mohandas, Dissimilar Metal Friction Welding of Austenitic-Ferritic Stainless Steels, *J. Mater. Proc. Technol.*, Vol 160 (No. 2), 2005, p 128–137

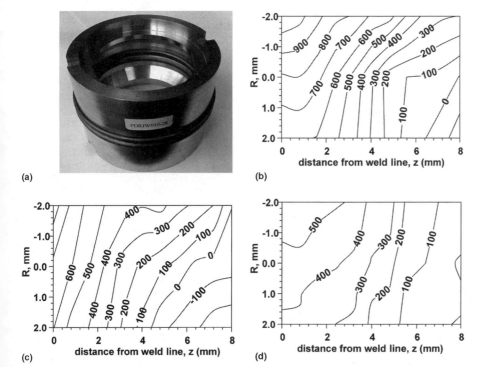

Fig. 13 (a) 143 mm diameter Udimet 720LiNi superalloy inertia-welded testpiece. Hoop stresses measured by neutron diffraction (inner wall at R = –4; outer wall at R = 4) for (b) the as-welded condition and after the following heat treatments: (c) 760 °C (no dwell) and (d) 8 h at 760 °C. Source: Ref 69

28. M. Sahin, Joining of Stainless-Steel and Aluminium Materials by Friction Welding, *Int. J. Adv. Manuf. Technol.*, 2008, p 0268–3768

29. P.L. Threadgill, The Welding Institute, Abington, U.K., 2008

30. M. Preuss, J.Q. da Fonseca, I. Kyriakoglou, P.J. Withers, and G.J. Baxter, Characterisation of Gamma' Across Inertia Friction Welded Alloy 720Li, *Superalloys 2004*, K.A. Green et al., Ed., 2004, p 477–484

31. R. Moat, M. Karadge, M. Preuss, S. Bray, and M. Rawson, Phase Transformations Across High Strength Dissimilar Steel Inertia Friction Weld, *J. Mater. Proc. Technol.*, Vol 204 (No. 1–3), 2008, p 48–58

32. M. Preuss, P.J. Withers, and G.J. Baxter, A Comparison of Inertia Friction Welds in Three Nickel Base Superalloys, *Mater. Sci. Eng. A—Struct. Mater. Prop. Microstruct. Proc.*, Vol 437 (No. 1), 2006, p 38–45

33. L. Wang, M. Preuss, P.J. Withers, G. Baxter, and P. Wilson, Energy-Input-Based Finite-Element Process Modeling of Inertia Welding, *Metall. Mater. Trans. B—Proc. Metall. Mater. Proc. Sci.*, Vol 36 (No. 4), 2005, p 513–523

34. V. Balasubramanian, Y.L. Li, T. Stotler, J. Crompton, A. Soboyejo, N. Katsube, and W. Soboyejo, A New Friction Law for the Modelling of Continuous Drive Friction Welding: Applications to 1045 Steel Welds, *Mater. Manuf. Proc.*, Vol 14 (No. 6), 1999, p 845–860

35. R.J. Moat, D.J. Hughes, A. Steuwer, N. Iqbal, M. Preuss, S.E. Bray, and M. Rawson, Residual Stresses in Inertia Friction Welded Dissimilar High Strength Steels, *Metall. Mater. Trans. A*, Vol 40 (No. 9), 2009, p 2098–2108

36. M.B. Prime, M.R. Hill, A.T. DeWald, R.J. Sebring, V.R. Dave, and M.J. Cola, Residual Stress Mapping in Welds Using the Contour Method, *Trends in Welding Research, Proceedings*, S.A. David et al., Ed., 2003, p 891–896

37. M. Preuss, A. Steuwer, P.J. Withers, G.J. Baxter, and S. Bray, The Development of Residual Stresses in Inertia Welding of Dissimilar Nickel-Based Superalloys, *Sixth International Trends in Welding Research Conference* (Pine Mountain, GA), ASM International, 2003

38. M. Karadge, B. Grant, P.J. Withers, G. Baxter and M. Preuss, Thermal Relaxation of Residual Stresses in Nickel Based Superalloy Inertia Friction Welds, to be submitted, 2010

39. V.I. Vill, *Friction Welding of Metals*, Van Nostrand Reinhold, 1962

40. A. Sluzalec, Thermal Effects in Friction Welding, *Int. J. Mech. Sci.*, Vol 32 (No. 6), 1990, p 467–478

41. A. Moal and E. Massoni, Finite-Element Simulation of the Inertia Welding of 2 Similar Parts, *Eng. Computat.*, Vol 12 (No. 6), 1995, p 497–512

42. A. Moal, E. Massoni, and J.-L. Chenot, A Finite Element Modelling for the Inertia Welding Process, *Proc. Int. Conf. on Computational Plasticity* (Barcelona), Pineridge Press, 1992

43. S.G. Du, L.Y. Duan, S.C. Wu, and G.S. Cheng, *Mech. Sci. Technol.*, Vol 16 (No. 4) (Suppl.), 1997, p 703s (in Chinese)

44. F.D. Duffin and A.S. Bahrani, Frictional Behaviour of Mild Steel in Friction Welding, *Wear*, Vol 26, 1973, p 53–74

45. L. D'Alvise, E. Massoni, and S.J. Walloe, Finite Element Modelling of the Inertia Friction Welding Process between Dissimilar Materials, *J. Mater. Proc. Technol.*, Vol 125, 2002, p 387–391

46. Q.Z. Zhang, L.W. Zhang, W.W. Liu, X.G. Zhang, W.H. Zhu, and S. Qu, 3D Rigid Viscoplastic FE Modelling of Continuous Drive Friction Welding Process, *Sci. Technol. Weld. Join.*, Vol 11 (No. 6), 2006, p 737–743

47. L. Fu, L.Y. Duan, and S.G. Du, Numerical Simulation of Inertia Friction Welding Process by Finite Element Method, *Weld. J.*, Vol 82 (No. 3), 2003, p 65S–70S

48. A. Sluzalec, *Theory of Thermomechanical Processes in Welding*, Springer, 2005

49. A. Sluzalec and A. Sluzalec, Solutions of Thermal Problems in Friction Welding—Comparative Study, *Int. J. Heat Mass Trans.*, Vol 36 (No. 6), 1993, p 1583–1587

50. A. Francis and R.E. Craine, On a Model for Frictioning Stage in Friction Welding of Thin Tubes, *Int. J. Heat Mass Trans.*, Vol 28 (No. 9), 1985, p 1747–1755

51. O.T. Midling and Ø. Grong, A Process Model for the Friction Welding of Al-Mg-Si Alloys and Al-SiC Metal-Matrix Composites, 1. HAZ Temperature and Strain-Rate Distribution, *Acta Metall. Mater.*, Vol 42 (No. 5), 1994, p 1595–1609

52. L. Fu and L.Y. Duan, The Coupled Deformation and Heat Flow Analysis by Finite Element Method during Friction Welding, *Weld. J.*, Vol 77 (No. 5), 1998, p S202–S207

53. L.W. Zhang, J.B. Pei, Q.Z. Zhang, C.D. Liu, W.H. Zhu, S. Qu, and J.H. Wang, The Coupled FEM Analysis of the Transient Temperature Field during Inertia Friction Welding of GH4169 Alloy, *Acta Metall. Sin. (Engl. Lett.)*, Vol 20 (No. 4), 2007, p 301–306

54. V.R. Dave, M.J. Cola, and G.N.A. Hussen, Heat Generation in the Inertia Welding of Dissimilar Tubes, *Weld. J.*, Vol 80 (No. 10), 2001, p 246S–252S

55. M.L. Storm, Heat Conduction in Simple Metals, *J. Appl. Phys.*, Vol 22 (No. 7), 1951, p 940–951

56. R. Plunkett, A Method for the Calculation of Heat Transfer in Solids with Temperature Dependent Properties, *Trans. ASME*, Vol 73, 1951, p 605–608

57. K.T. Yang and A. Szewczyk, An Approximate Treatment of Unsteady Heat Conduction in Semi Infinite Solids with Variable Thermal Properties, *ASME Trans. J. Eng. Mater. Tech.*, Vol 81, 1959, p 215–252

58. K.K. Wang and W. Lin, Flywheel Friction Welding Research, *Weld. J.*, Vol 53 (No. 6), 1974, p S233–S241

59. K.K. Wang and P. Nagappan, Transient Temperature Distribution in Inertia Welding of Steels, *Weld. Res. Suppl.*, Vol 49 (No. 9), 1970, p S419–426

60. M. Kleiber and A. Sluzalec, Finite Element Analysis of Heat Flow in Friction Welding, *Eng. Trans.*, Vol 32, 1984, p 107–113

61. C.J. Bennett, T.H. Hyde, and E.J. Williams, Modelling and Simulation of the Inertia Friction Welding of Shafts, *Proc. Inst. Mech. Eng. L, J. Mater.—Des. Applic.*, Vol 221 (No. L4), 2007, p 275–284

62. V. Balasubramanian, Y.L. Li, T. Stotler, J. Crompton, N. Katsube, and W. Soboyejo, An Energy Balance Method for the Numerical Simulation of Inertia Welding, *Mater. Manuf. Proc.*, Vol 14 (No. 5), 1999, p 755–773

63. M. Soucail, A. Moal, L. Naze, E. Massoni, C. Levaillant, and Y. Bienvenu, Microstructural Study and Numerical-Simulation of Inertia Friction Welding of Astroloy, *Superalloys 1992*, S.D. Antolovich et al., Ed., 1992, p 847–856

64. B. Grant, M. Preuss, P.J. Withers, G. Baxter, and M. Rowlson, Finite Element Process Modelling of Inertia Friction Welding Advanced Nickel-Based Superalloy, *Preparation*, 2008

65. K. Lee, A. Samant, W.T. Wu, and S. Srivatsa, Finite Element Modeling of Inertia Welding Processes, *Simulation of Materials Processing: Theory, Methods and Applications*, Proc. of NUMIFORM Conf., 2001, p 1095–1100

66. Y.C. Kim, A. Fuji, and T.H. North, Residual Stress and Plastic Strain in AISI 304L Stainless-Steel Titanium Welds, *Mater. Sci. Technol.*, Vol 11 (No. 4), 1995, p 383–388

67. J.W.L. Pang, M. Preuss, P.J. Withers, G.J. Baxter, and C. Small, Effects of Tooling on the Residual Stress Distribution in an Inertia Weld, *Mater. Sci. Eng. A—Struct. Mater. Prop. Microstruct. Proc.*, Vol 356 (No. 1–2), 2003, p 405–413

68. H. Dai, J.A. Francis, H.J. Stone, H.K.D.H. Bhadeshia, and P.J. Withers, Characterizing Phase Transformations and Their Effects on Ferritic Weld Residual Stresses with X-Rays and Neutrons, *Metall. Mater. Trans. A*, 2008

69. P.J. Withers, Mapping Residual and Internal Stress in Materials by Neutron Diffraction, *C.R. Physique*, Vol 8, 2007, p 806–820

70. M. Soucail and Y. Bienvenu, Dissolution of the Gamma' Phase in a Nickel Base Superalloy at Equilibrium and Under Rapid Heating, *Mater. Sci. Eng. A—Struct. Mater. Prop. Microstruct. Proc.*, Vol 220 (No. 1–2), 1996, p 215–222

ASM Handbook, Volume 22B, *Metals Process Simulation*
D.U. Furrer and S.L. Semiatin, editors

Copyright © 2010, ASM International®
All rights reserved.
www.asminternational.org

Simulation of Friction Stir Welding

Junde Xu and Jeff J. Bernath, Edison Welding Institute Incorporated

FRICTION STIR WELDING (FSW) is a relatively new solid-state joining technique (Ref 1, 2). In FSW, material joining is facilitated by a rotating and traveling tool that penetrates into the workpiece material. The interaction between the tool and the workpiece material plasticizes and heats the material. As the welding tool advances along the joint line, welds are made by displacing and mixing the softened material from ahead of the tool to behind the tool. Compared with the traditional gas metal arc welding, workpiece material experiences a much lower temperature in FSW. Therefore, weld defects such as porosity, slag, and cracking commonly associated with solid-liquid transformation can be essentially eliminated. All these advantages have made FSW a rapidly developing technique in the past two decades, with its applications extending from the joining of aluminum alloys to titanium alloys and steels. As part of the efforts, various methodologies and approaches have been developed to model the FSW process, evaluate the impact of tool design details and process variables, and guide the process optimization (Ref 3 to 29).

Fundamentals of Friction Stir Welding

Figure 1 shows a typical FSW operation. As shown, an FSW tool consists of a shoulder and a pin. In Fig. 1, the workpiece is firmly fixed on the anvil to preclude any significant movement during the welding process. The length of the protruding pin is slightly shorter than the desired depth of the weld nugget (thickness of the workpiece in a butt welding) so that the annular end surface of the shoulder can have full contact with the workpiece but not leave a considerable groove on the top surface after the weld is made. Due to the rotation of the welding tool, FSW is not a symmetric process. The side of the workpiece material that observes the same tool travel direction and rotation direction is customarily called the advancing side, whereas the opposite side is called the retreating side (Fig. 1). Figures 2 and 3 show the various designs of the shoulder and pin, respectively. The shoulder can be flat, concave, or with scrolls, the latter two being adopted to force the heated surface material to move toward the pin and minimize the depth of the groove on the top surface. A cylindrical pin with helical threads was first used in the FSW of aluminum alloys. Since then, various features, such as the frustum-shaped pin and the threaded pin with flutes or flats, have been introduced in the design of pin geometry to obtain more desirable material flow, enhance heat generation, and reduce process load.

A typical FSW process starts with the plunging of the rotating pin into the workpiece. After an intimate contact between the tool shoulder and workpiece is reached, the welding tool begins to move along the joint line while maintaining the aforementioned contact. In this process, the interaction between the rotating tool shoulder and the top surface of the workpiece is twofold. On the one hand, the relative motion between the two causes friction on the interface, heating both the shoulder and the surface material. On the other hand, the rotating shoulder tends to carry the heated surface material to move with it. This results in large plastic deformation in a region that is usually limited to the material near the surface layer and is distinctive

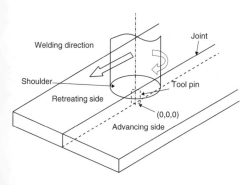

Fig. 1 Schematic illustration of friction stir welding process

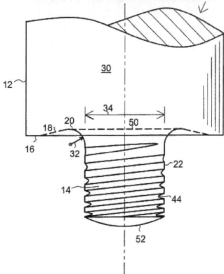

Fig. 2 Typical friction stir welding tool

Fig. 3 Different tool designs for friction stir welding

in the micrograph shown in Fig. 4. Heat is generated by both friction and plastic deformation in this region. For a threaded pin, the interaction between the pin and the workpiece is very complex. It has been observed that a three-dimensional material flow field exists in FSW (Ref 30 to 32). The pin, whether threaded or not, relocates the base material when it advances along the joint line. Again, large plastic deformation is involved, which also generates heat. Due to the rotation of the pin, the removed base material is transported to behind the pin mainly through the retreating side. Therefore, heat transfer is contributed by both conduction and convection in FSW.

Compared with other material joining techniques, FSW yields unique microstructures (Ref 33). Figure 4 is a micrograph of a friction-stir-welded aluminum alloy. In addition to the base material, three distinct regions can be identified. The material in the weld nugget experiences the most severe deformation and the generated high temperature, the latter facilitating a recrystallization process that forms and grows new grains. This region is often referred to as the dynamic recrystallization zone or stir zone. Outside the weld nugget, the material also undergoes certain plastic deformation and high temperature but no recrystallization. This forms a thermomechanically affected zone (TMAZ) in which the microstructure is impacted by both the deformation and the thermal history. Between the TMAZ and the base material is the heat-affected zone,

where no plastic deformation occurs, but the heat generated in the process causes some microstructural change. It should be pointed out that the previous description is qualitative because the actual size and shape of the microstructural zones depend on welding tool design, workpiece material, and welding parameters. Even for the same tool design and workpiece material, variations in process conditions, such as travel speed or cooling conditions, can result in different microstructures.

Governing Equations for Heat Transfer

Because material joining is enabled by the heat generated in FSW, heat generation and transfer are important issues in the analysis of FSW. Based on the descriptions in the previous section, both conduction and convection contribute to heat transfer. For a fixed Cartesian coordinate system, as shown in Fig. 5, the governing partial differential equation of the combined heat conduction and convection is as follows:

$$\rho c \left(\frac{\partial \theta}{\partial t} + \dot{u}_x \frac{\partial \theta}{\partial x} + \dot{u}_y \frac{\partial \theta}{\partial y} + \dot{u}_z \frac{\partial \theta}{\partial z} \right)$$
$$= \frac{\partial}{\partial x} \left(K \frac{\partial \theta}{\partial x} \right) + \frac{\partial}{\partial y} \left(K \frac{\partial \theta}{\partial y} \right) + \frac{\partial}{\partial z} \left(K \frac{\partial \theta}{\partial z} \right)$$
$$+ \eta \left(\sigma'_{xx} \dot{\varepsilon}^p_{xx} + \sigma'_{yy} \dot{\varepsilon}^p_{yy} + \sigma'_{zz} \dot{\varepsilon}^p_{zz} + 2 \sigma'_{xy} \dot{\varepsilon}^p_{xy} \right.$$
$$\left. + 2 \sigma'_{yz} \dot{\varepsilon}^p_{yz} + 2 \sigma'_{zx} \dot{\varepsilon}^p_{zx} \right)$$

(Eq 1)

The symbols and nomenclature are summarized in Table 1.

In Eq 1, θ is temperature, t is time, ρ is the density, c is the specific heat, K is the thermal conductivity, and η is the fraction of plastic dissipation. In Eq 1, ρ varies with temperature because of thermal expansion, whereas c and K are, in general, temperature dependent. Let i, j, and k denote the unit vectors in the Cartesian coordinate system shown in Fig. 5; the displacement vector of the workpiece material is:

$$\underset{\sim}{u} = u_x \underset{\sim}{i} + u_y \underset{\sim}{j} + u_z \underset{\sim}{k}$$

(Eq 2)

Then, the velocity vector of the material is:

$$\underset{\sim}{\dot{u}} = \dot{u}_x \underset{\sim}{i} + \dot{u}_y \underset{\sim}{j} + \dot{u}_z \underset{\sim}{k}$$

(Eq 3)

The relation of the deviatoric stress, σ', with the stress, σ, is:

$$\underset{\sim}{\sigma'} = \begin{pmatrix} \sigma'_{xx} & \sigma'_{xy} & \sigma'_{xz} \\ \sigma'_{yx} & \sigma'_{yy} & \sigma'_{yz} \\ \sigma'_{zx} & \sigma'_{zy} & \sigma'_{zz} \end{pmatrix}$$
$$= \begin{pmatrix} \sigma_{xx} + p & \sigma_{xy} & \sigma_{xz} \\ \sigma_{yx} & \sigma_{yy} + p & \sigma_{yz} \\ \sigma_{zx} & \sigma_{zy} & \sigma_{zz} + p \end{pmatrix}$$

(Eq 4)

where

$$p = -\frac{\sigma_{xx} + \sigma_{yy} + \sigma_{zz}}{3}$$

(Eq 5)

is the pressure or the negative of the mean stress. The plastic strain rate, $\dot{\varepsilon}^p$, is:

$$\underset{\sim}{\dot{\varepsilon}^p} = \begin{pmatrix} \dot{\varepsilon}^p_{xx} & \dot{\varepsilon}^p_{xy} & \dot{\varepsilon}^p_{xz} \\ \dot{\varepsilon}^p_{yx} & \dot{\varepsilon}^p_{yy} & \dot{\varepsilon}^p_{yz} \\ \dot{\varepsilon}^p_{zx} & \dot{\varepsilon}^p_{zy} & \dot{\varepsilon}^p_{zz} \end{pmatrix}$$

(Eq 6)

Neglecting the elastic deformation in the workpiece, the components of the plastic strain rate are:

Table 1 Symbols and nomenclature used in this article

c	Specific heat (J/m^2 s K^4)
e	Emissivity
E	Young's modulus
g	Gravitational acceleration (m/s^2)
h	Thickness of workpiece (m)
H	Heat-transfer coefficient (J/m^2 s K)
k	Stefan-Boltzmann constant (J/m^2 s K^4)
K	Thermal conductivity (J/m s K)
n	Outer normal direction
p	Pressure or negative of mean stress (N/m^2)
$\bar{p}$	Pressure between shoulder and workpiece (N/m^2)
q	Heat input power per unit thickness of the plate (J/s m)
Q	Heat input power (J/s)
$\bar{Q}$	Activation energy (J/mol)
r_0	Outer radius of cylindrical shoulder (m)
r_r	Radius at pin root (m)
r_t	Radius at pin tip (m)
R	Gas constant (J/mol K)
t	Time (s)
u	Displacement (m)
V	Tool advancing speed (m/s)
X	Welding direction
Y	Direction perpendicular to welding direction and thickness direction
Z	Thickness direction
α	Thermal diffusivity (m^2/s)
α_L	Thermal expansion coefficient (1/K)
δ	Percentage slip
ε	Strain
$\dot{\varepsilon}^e$	Elastic strain rate (1/s)
$\dot{\varepsilon}^p$	Plastic strain rate (1/s)
$\dot{\varepsilon}^{th}$	Thermal strain rate (1/s)
λ	Plastic multiplier
μ	Viscosity
η	Fraction of plastic dissipation
μ_f	Frictional coefficient
ν	Poisson's ratio
ρ	Density (g/m^3)
θ	Temperature (K)
σ	Stress (N/m^2)
σ'	Deviatoric stress (N/m^2)
ω	Pin rotating speed (1/s)
Γ_c	Portion of boundary where convective heat loss is taken into account
Γ_r	Portion of boundary where radiative heat loss is taken into account
ψ	Helix angle of thread

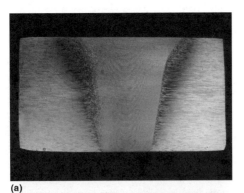

(a)

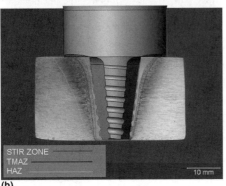

STIR ZONE
TMAZ
HAZ

10 mm

(b)

Fig. 4 Micrograph of friction-stir-welded 25 mm thickness 2000-series aluminum. (a) As-welded macro section. (b) Overlay including friction stir welding tool and weld region map. TMAZ, thermomechanically affected zone; HAZ, heat-affected zone

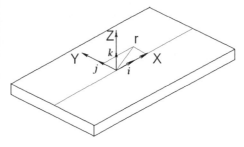

Fig. 5 Cartesian coordinate system for formulating the governing equations for friction stir welding

$$\dot{\varepsilon}^p_{xx} = \frac{\partial \dot{u}_x}{\partial x}, \dot{\varepsilon}^p_{yy} = \frac{\partial \dot{u}_y}{\partial y}, \dot{\varepsilon}^p_{zz} = \frac{\partial \dot{u}_z}{\partial z}$$

$$\dot{\varepsilon}^p_{xy} = \dot{\varepsilon}^p_{yx} = \frac{1}{2}\left(\frac{\partial \dot{u}_x}{\partial y} + \frac{\partial \dot{u}_y}{\partial x}\right)$$

$$\dot{\varepsilon}^p_{yz} = \dot{\varepsilon}^p_{zy} = \frac{1}{2}\left(\frac{\partial \dot{u}_y}{\partial z} + \frac{\partial \dot{u}_z}{\partial y}\right) \qquad \text{(Eq 7)}$$

$$\dot{\varepsilon}^p_{zx} = \dot{\varepsilon}^p_{xz} = \frac{1}{2}\left(\frac{\partial \dot{u}_z}{\partial x} + \frac{\partial \dot{u}_x}{\partial z}\right)$$

The heat in the FSW workpiece transfers to the surroundings through radiation and convection. The radiative heat loss can be calculated from:

$$-K\frac{\partial \theta}{\partial n}\bigg|_{\Gamma_r} = k\,e\left(\theta^4 - \theta_0^4\right) \qquad \text{(Eq 8)}$$

where k is the Stefan-Boltzmann constant, e is the emissivity, Γ_r is the portion of boundary where radiative heat loss is taken into account, and $\frac{\partial \theta}{\partial n}$ is the temperature gradient on the boundary in the outer normal direction. As examples:

$$\frac{\partial \theta}{\partial n} = \frac{\partial \theta}{\partial z} \qquad \text{(Eq 9)}$$

on the top surface of the plate shown in Fig. 5, whereas:

$$\frac{\partial \theta}{\partial n} = -\frac{\partial \theta}{\partial z} \qquad \text{(Eq 10)}$$

on the bottom surface of the plate.

The convective heat loss can be calculated from:

$$-K\frac{\partial \theta}{\partial n}\bigg|_{\Gamma_c} = H\left(\theta - \theta_0\right) \qquad \text{(Eq 11)}$$

where H is the heat-transfer coefficient or film coefficient, and Γ_c is the portion of boundary where convective heat loss is taken into account.

For workpiece material outside the weld nugget and TMAZ, no plastic deformation is involved in FSW. Therefore:

$$u_x = u_y = u_y = 0 \qquad \text{(Eq 12)}$$

and Eq 1 reduces to:

$$\rho\,c\,\frac{\partial \theta}{\partial t} = \frac{\partial}{\partial x}\left(K\frac{\partial \theta}{\partial x}\right) + \frac{\partial}{\partial y}\left(K\frac{\partial \theta}{\partial y}\right)$$
$$+ \frac{\partial}{\partial z}\left(K\frac{\partial \theta}{\partial z}\right) \qquad \text{(Eq 13)}$$

Although the weld nugget and TMAZ occupy only a small region of the workpiece, the transported material carries most of the generated heat in the process.

Analytical Solution. One of the early attempts to study the heat transfer in FSW was to neglect the temperature gradient in the plate thickness direction and the heat convection in the workpiece. Under these conditions, Eq 13 further reduces to:

$$\rho\,c\,\frac{\partial \theta}{\partial t} = \frac{\partial}{\partial x}\left(K\frac{\partial \theta}{\partial x}\right) + \frac{\partial}{\partial y}\left(K\frac{\partial \theta}{\partial y}\right) \qquad \text{(Eq 14)}$$

Consider an infinitely large plate with constant thermal properties (ρ, c, K) and a uniform initial temperature of $\bar{\theta}$ (Fig. 5). When a point heat source travels at a speed of V in the X-direction and no heat is lost to the surrounding environment, the solution of Eq 14 (Rosenthal's solution, Ref 34) is:

$$\Delta\theta = \theta - \bar{\theta} = \frac{q}{2\pi K}\exp\left[-\frac{Vx}{2\alpha}\right]K_0\left[\frac{Vr}{2\alpha}\right] \qquad \text{(Eq 15)}$$

where

$$\alpha = \frac{K}{\rho\,c} \qquad \text{(Eq 16)}$$

is the thermal diffusivity, q is the heat input power per unit thickness of the plate, K_0 is the modified Bessel function of the second kind of order zero,

$$r = \sqrt{x^2 + y^2} \qquad \text{(Eq 17)}$$

and x and y are the X and Y coordinates of the location of interest, respectively (Fig. 5).

The two-dimensional Rosenthal's solution is capable of calculating the temperature everywhere in the plate except for the location $x = 0$ and $y = 0$. This is because the function K_0 approaches to infinity when r approaches to zero. As described previously, in FSW the heat is generated in a certain volume of the workpiece. Neglecting this detail not only causes the solution to fail to predict the temperature at $x = 0$ and $y = 0$ but also provides invalid solution for the adjacent locations. Besides, temperature-dependent thermal properties should be considered in the solution. A complete analysis model should also take into account the temperature gradient through the plate thickness and the heat loss to the backing material and surrounding air. Including all these complexities in the analysis model makes it impossible to derive analytical solutions like Rosenthal's. Therefore, various numerical techniques have been adopted to solve the equations and simulate the FSW process.

Modeling Neglecting Convective Heat Transfer in the Workpiece

Solutions for Heat Transfer in the Workpiece

A series of analysis work has been conducted to study the thermal process in FSW by neglecting the convective heat transfer in the workpiece. As described previously, convective heat transfer occurs in the weld nugget and TMAZ, where the material transportation and deformation change the location of some workpiece material. The analysis work in this class assumes the workpiece material does not experience any displacement while introducing heat source models to simulate the heat generated in FSW.

A commonly used heat source model for the shoulder-workpiece interface is as follows. Let ω denote the rotating speed of the welding tool, $\bar{p}$ the pressure between the shoulder and workpiece, and μ_f the frictional coefficient. The frictional force per unit interface area, f, is:

$$f = \mu_f\,\bar{p} \qquad \text{(Eq 18)}$$

For a location having a radius of r on the interface (Fig. 6), the relative speed between the shoulder and the workpiece, v_r, is:

$$v_r = 2\pi\omega r \qquad \text{(Eq 19)}$$

Assuming that the work done by the frictional force is completely transformed into heat on the interface, the heat generation per unit time (power) on a unit area is $v_r f$. The total heat input power, Q, is:

$$Q = \int_{r_r}^{r_0}(v_r f)\,2\pi r\,dr = \int_{r_r}^{r_0}2\pi\omega r^2\mu p\,dr \qquad \text{(Eq 20)}$$

where r_0 is the outer radius of the cylindrical shoulder, and r_r is the radius of the pin root (Fig. 6). In Eq 20, μ and p are, in general, functions of location and temperature; μ may also depend on p. Therefore, a simple expression for Q is not available. For the heat-source model for the pin-workpiece interface, some analysis work has either neglected it or replaced the integration interval in Eq 20 from $[r_r, R]$ to $[0, R]$ to include the contributions from the pin. Although these approximations could be reasonable when the workpiece thickness is very small compared with the shoulder diameter, attempts were also made to model the pin-workpiece interaction.

The heat generated due to pin-workpiece interaction consists of:

- The heat generated by shearing the base material

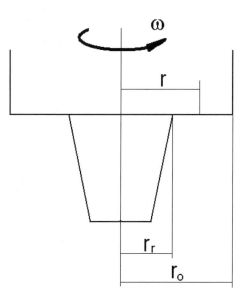

Fig. 6 Geometry of a friction stir welding tool

- The heat generated by the friction on the threaded surface of the pin
- The heat generated by friction on the cylindrical pin surface

One of the approximations suggests the total generated heat is (Ref 4):

$$Q_{\text{pin}} = \frac{2\pi r_r h K \tau V_m}{\sqrt{3}} + \frac{2\mu \pi r_r h K \tau V_{rp}}{\sqrt{3(1+\mu^2)}}$$
$$+ \frac{4\mu F_p V_m \cos \phi}{\pi} \qquad \text{(Eq 21)}$$

where

$$V_m = \frac{\sin \lambda}{\sin(\psi + \phi)} r_r \omega \qquad \text{(Eq 22)}$$

$$V_{rp} = \frac{\sin \phi}{\sin(\psi + \phi)} r_r \omega \qquad \text{(Eq 23)}$$

$$\phi = 90° - \psi - \tan^{-1}\mu \qquad \text{(Eq 24)}$$

where h is the thickness of the workpiece, τ is the average shear stress on the pin-workpiece interface, F_p is the translational force on the tool, and ψ is the helix angle of the thread.

Another similar approximation assumes a uniform shear stress, τ, between the workpiece, the shoulder, and the pin (Ref 7). Therefore, Eq 18 becomes:

$$\tau = \mu \bar{p} \qquad \text{(Eq 25)}$$

and Eq 20 can be calculated as:

$$Q_{\text{shoulder}} = \frac{2}{3}\pi \omega \tau R^3 \qquad \text{(Eq 26)}$$

where the integration interval is set to $[0, R]$. For a cylindrical pin surface, the heat-input power is modeled as:

$$Q_{\text{pin}} = (2\pi r_r h)\tau \omega r_r = 2\pi \omega \tau r_r^2 h \qquad \text{(Eq 27)}$$

Heat-input models, as described previously, can be easily implemented in finite-element analysis. Because material transportation is not taken into account, the problem becomes symmetric with respect to the welding direction. Therefore, the finite-element model only analyzes half of the plate. An incremental heat conduction analysis with appropriate radiative and convective boundary conditions will be required to solve the evolving temperature field in the workpiece. Figure 7 shows the temperature distribution of an aluminum alloy plate being friction stir welded when the pin tool travels to the midpoint of the welding line.

Solutions for Structural Distortion in FSW

Similar to traditional arc welding, the nonuniform heating and subsequent cooling cause nonuniform volume change of the workpiece

Fig. 7 Temperature distribution of an aluminum plate in friction stir welding when the tool travels to the midpoint of the welding line

material. Residual stress develops in this process, which deforms the workpiece and results in structural distortion. Based on the solved temperature history, the development of the residual stress accompanying the heating-cooling process in FSW and the resulting distortion can be simulated using the methods well established in solid mechanics.

Consider the equilibrium equations of the workpiece, as shown in Fig. 5:

$$\frac{\partial \sigma_{xx}}{\partial x} + \frac{\partial \sigma_{xy}}{\partial y} + \frac{\partial \sigma_{xz}}{\partial z} = 0$$
$$\frac{\partial \sigma_{yx}}{\partial x} + \frac{\partial \sigma_{yy}}{\partial y} + \frac{\partial \sigma_{yz}}{\partial z} = 0 \qquad \text{(Eq 28)}$$
$$\frac{\partial \sigma_{zx}}{\partial x} + \frac{\partial \sigma_{zy}}{\partial y} + \frac{\partial \sigma_{zz}}{\partial z} + \rho g = 0$$

where g is the gravitational acceleration. The total strain rate is:

$$\dot{\underset{\sim}{\varepsilon}} = \begin{pmatrix} \dot{\varepsilon}_{xy}^e & \dot{\varepsilon}_{xy}^e & \dot{\varepsilon}_{xz}^e \\ \dot{\varepsilon}_{yx}^e & \dot{\varepsilon}_{yy}^e & \dot{\varepsilon}_{yz}^e \\ \dot{\varepsilon}_{zx}^e & \dot{\varepsilon}_{zy}^e & \dot{\varepsilon}_{zz}^e \end{pmatrix} + \begin{pmatrix} \dot{\varepsilon}_{xx}^p & \dot{\varepsilon}_{xy}^p & \dot{\varepsilon}_{xz}^p \\ \dot{\varepsilon}_{yx}^p & \dot{\varepsilon}_{yy}^p & \dot{\varepsilon}_{yz}^p \\ \dot{\varepsilon}_{zx}^p & \dot{\varepsilon}_{zy}^p & \dot{\varepsilon}_{zz}^p \end{pmatrix}$$
$$+ \begin{pmatrix} \dot{\varepsilon}_{xx}^{th} & 0 & 0 \\ 0 & \dot{\varepsilon}_{yy}^{th} & 0 \\ 0 & 0 & \dot{\varepsilon}_{zz}^{th} \end{pmatrix} \qquad \text{(Eq 29)}$$

where the superscripts "e," "p," and "th" denote the contribution from elastic deformation, plastic deformation, and thermal expansion, respectively. The thermal strain rate can be calculated from the solved temperature:

$$\dot{\varepsilon}_{xx}^{th} = \dot{\varepsilon}_{yy}^{th} = \dot{\varepsilon}_{zz}^{th} = \alpha_L \dot{\theta} \qquad \text{(Eq 30)}$$

where α_L is the thermal expansion coefficient, and $\dot{\theta}$ is the temperature changing rate. The elastic strain rate can be calculated from Hooke's law:

$$\dot{\varepsilon}_{xx}^e = \frac{1}{E}\left[\dot{\sigma}_{xx} - \nu(\dot{\sigma}_{yy} + \dot{\sigma}_{zz})\right]$$
$$\dot{\varepsilon}_{yy}^e = \frac{1}{E}\left[\dot{\sigma}_{yy} - \nu(\dot{\sigma}_{zz} + \dot{\sigma}_{xx})\right]$$
$$\dot{\varepsilon}_{zz}^e = \frac{1}{E}\left[\dot{\sigma}_{zz} - \nu(\dot{\sigma}_{xx} + \dot{\sigma}_{yy})\right]$$
$$\dot{\varepsilon}_{xy}^e = \frac{1+\nu}{E}\dot{\sigma}_{xy} \quad \dot{\varepsilon}_{yz}^e = \frac{1+\nu}{E}\dot{\sigma}_{yz} \quad \dot{\varepsilon}_{zx}^e = \frac{1+\nu}{E}\dot{\sigma}_{zx}$$
$$\text{(Eq 31)}$$

where, again, an overdot denotes the changing rate. The plastic strain rate can be calculated based on the theory of plasticity. For most metals, the von Mises yield criterion and the associated flow rule have been widely adopted in the analysis. Based on this:

$$\dot{\varepsilon}_{xx}^p = \dot{\lambda}\frac{3\sigma'_{xx}}{2\sigma_e} \quad \dot{\varepsilon}_{yy}^p = \dot{\lambda}\frac{3\sigma'_{yy}}{2\sigma_e} \quad \dot{\varepsilon}_{zz}^p = \dot{\lambda}\frac{3\sigma'_{zz}}{2\sigma_e}$$
$$\dot{\varepsilon}_{xy}^p = \dot{\lambda}\frac{3\sigma'_{xy}}{2\sigma_e} \quad \dot{\varepsilon}_{yz}^p = \dot{\lambda}\frac{3\sigma'_{yz}}{2\sigma_e} \quad \dot{\varepsilon}_{zx}^p = \dot{\lambda}\frac{3\sigma'_{zx}}{2\sigma_e}$$
$$\text{(Eq 32)}$$

where $\dot{\lambda}$ is the plastic multiplier, and σ_e is the effective stress:

$$\sigma_e = \left[\frac{3}{2}\left((\sigma'_{xx})^2 + (\sigma'_{yy})^2 + (\sigma'_{zz})^2 + 2(\sigma'_{xy})^2\right.\right.$$
$$\left.\left. + 2(\sigma'_{yz})^2 + 2(\sigma'_{zx})^2\right)\right]^{\frac{1}{2}} \qquad \text{(Eq 33)}$$

As an example, the temperature history of the plate shown in Fig. 7 is used to simulate the resulting residual stress and distortion. Figures 8 and 9 show the distribution of residual longitudinal and transverse stress (S33 and S11 in the figures), respectively. Figures 8 and 9 correspond to a state when the plate has fully cooled to environmental temperature but is still fixed on the anvil. When the constraints are removed, the residual stress will cause the plate to distort. Figure 10 shows the distorted shape and the out-of-plane (Y-direction) displacement of the plate.

The aforementioned modeling approaches can be readily implemented in solid mechanics-based engineering software packages such as finite-element packages. While they are particularly useful in dealing with residual stress and structural distortion issues, they do not aim to investigate the intricate interaction between tool and workpiece. As a matter of fact, it is the complex material transportation, its dependence on the tool geometry and welding parameters, and its consequences in weld quality that have driven another class of analysis, which is detailed in the following section.

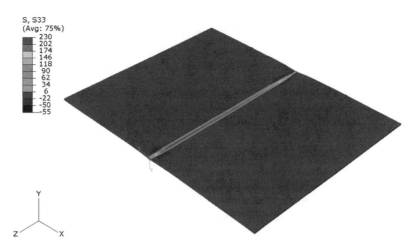

Fig. 8 Distribution of residual longitudinal stress in an aluminum plate after it is friction stir welded but is fixed on the anvil

Modeling Considering Convective Heat Transfer in the Workpiece

Basic Equations

The analysis work in this class tries to include the material deformation and displacement in the analysis model. This requires the solution of Eq 1, which represents the conservation of energy, to obtain both the temperature and velocity of the workpiece material. To focus on the large deformation and displacement of the workpiece material such as the weld nugget and TMAZ, the elastic deformation is neglected because it is comparatively small (Eq 7). Under this assumption, mass conservation guarantees the incompressibility of the workpiece material, which can be expressed as:

$$\frac{\partial \dot{u}_x}{\partial x} + \frac{\partial \dot{u}_y}{\partial y} + \frac{\partial \dot{u}_z}{\partial z} = 0 \qquad \text{(Eq 34)}$$

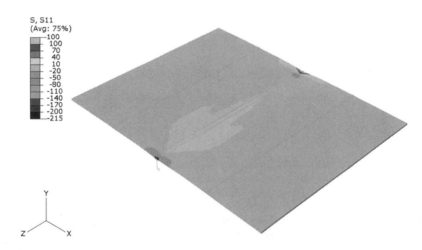

Fig. 9 Distribution of residual transverse stress in an aluminum plate after it is friction stir welded but is fixed on the anvil

Finally, in this class of analysis work, the steady-state movement of the workpiece material is governed by the conservation of momentum, which yields the following Navier-Stokes equations:

$$\rho \left(\frac{\partial \dot{u}_x \dot{u}_x}{\partial x} + \frac{\partial \dot{u}_x \dot{u}_y}{\partial y} + \frac{\partial \dot{u}_x \dot{u}_z}{\partial z} \right)$$
$$= -\frac{\partial p}{\partial x} + \frac{\partial \sigma_{xx}}{\partial x} + \frac{\partial \sigma_{xy}}{\partial y} + \frac{\partial \sigma_{xz}}{\partial z}$$
$$\rho \left(\frac{\partial \dot{u}_y \dot{u}_x}{\partial x} + \frac{\partial \dot{u}_y \dot{u}_y}{\partial y} + \frac{\partial \dot{u}_y \dot{u}_z}{\partial z} \right)$$
$$= -\frac{\partial p}{\partial y} + \frac{\partial \sigma_{yx}}{\partial x} + \frac{\partial \sigma_{yy}}{\partial y} + \frac{\partial \sigma_{yz}}{\partial z} \qquad \text{(Eq 35)}$$
$$\rho \left(\frac{\partial \dot{u}_z \dot{u}_x}{\partial x} + \frac{\partial \dot{u}_z \dot{u}_y}{\partial y} + \frac{\partial \dot{u}_z \dot{u}_z}{\partial z} \right)$$
$$= -\frac{\partial p}{\partial z} + \frac{\partial \sigma_{zx}}{\partial x} + \frac{\partial \sigma_{zy}}{\partial y} + \frac{\partial \sigma_{zz}}{\partial z}$$

In the aforementioned equations, the pressure, p, is defined by Eq 5.

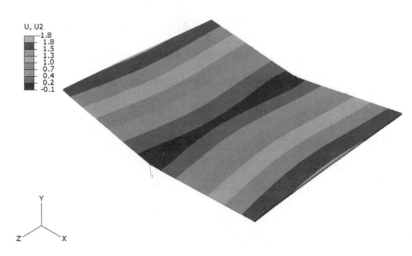

Fig. 10 Distorted shape of an aluminum plate after it is friction stir welded and the clamping removed

Material Properties

Equations 1, 34, and 35 constitute a set of basic equations that have been used to analyze the steady-state flow of incompressible fluid. In other words, the workpiece material has been idealized as a fluid. The material property that correlates the flow stress and the strain rate is viscosity, which is defined as:

$$\mu = \frac{\sigma_e}{3\,\dot{\varepsilon}_e} \qquad \text{(Eq 36)}$$

where σ_e is the effective stress or the flow stress as defined by Eq 33, and the effective strain rate, $\dot{\varepsilon}_e$, is defined as:

$$\dot{\varepsilon}_e = \sqrt{\frac{2}{3}\left((\dot{\varepsilon}_{xx})^2+(\dot{\varepsilon}_{yy})^2+(\dot{\varepsilon}_{zz})^2+2\,(\dot{\varepsilon}_{xy})^2+2\,(\dot{\varepsilon}_{yz})^2+2\,(\dot{\varepsilon}_{zx})^2\right)}$$

$$\text{(Eq 37)}$$

respectively. Experimental data show that viscosity depends on material, strain rate, and temperature. Based on the measurement results, various functional forms have been proposed for the relation between the flow stress and the strain rate. Many relations have used the following Zener-Hollomon parameter to include the influence of temperature and to suggest a temperature-compensated strain rate (Ref 35):

$$Z = \dot{\varepsilon}_e\,\exp\!\left(\frac{\overline{Q}}{R\,\theta}\right) \qquad \text{(Eq 38)}$$

In Eq 38, $\overline{Q}$ is the temperature-independent activation energy, R is the gas constant, and θ is the absolute temperature. It was found for a given material that the measured flow stress can be satisfactorily expressed as simple functions of the aforementioned Zener-Hollomon parameter. One of the widely used relations is the following hyperbolic-sine relation between the Zener-Hollomon parameter and the flow stress (Ref 36):

$$\sigma_e = \frac{1}{a}\sinh^{-1}\!\left[\left(\frac{Z}{A}\right)^{\frac{1}{n}}\right] \qquad \text{(Eq 39)}$$

or

$$\dot{\varepsilon}_e = A\,(\sinh(a\,\sigma_e))^n\,\exp\!\left(-\frac{\overline{Q}}{R\,\theta}\right) \qquad \text{(Eq 40)}$$

where A, a, n, and $\overline{Q}$ are to be determined by curve-fitting the test data.

Solutions of the Steady-State Temperature and Velocity Fields

The aforementioned basic equations and material properties make the problem ready to be solved using the techniques developed in computational fluid dynamics (CFD). In the numerical implementation of the solution, it is found convenient to fix the axis of the rotating pin while having the workpiece move at the pin advancing speed but in an opposite direction. Figure 11 is a two-dimensional description

of the problem, in which the rectangle represents the workpiece, and the circular hole in the rectangle models the pin tool. Because the material is modeled as a fluid, the size of the rectangular plate is chosen to be large enough such that the pin advancing speed can be imposed as the far-field boundary conditions, that is, the velocity at the inlet, outlet, and the moving wall (Fig. 11). The rotating speed of the pin surface is imposed on the boundary of the circular hole (no-slip condition). To adequately simulate the large velocity gradient near the pin, the mesh used in the CFD analysis is usually very fine near the circular hole but becomes coarser as the distance to the hole increases (Fig. 12). Figure 13 shows a two-dimensional solution of the streamlines in the FSW of aluminum alloy 6061. The results

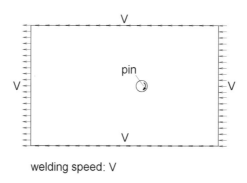

Fig. 11 Schematic of flow domain and boundary conditions in the computational fluid dynamics solution

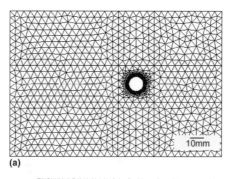

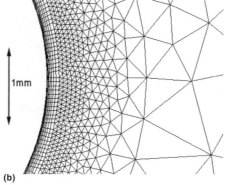

Fig. 12 (a) Two-dimensional mesh for computational fluid dynamics analysis for friction stir welding. (b) Mesh details near pin tool

indicate material transport occurs mainly around the pin in the retreating side. Based on the solved material flow field, the final position of material points can be predicted. Figure 14 shows a comparison between the experimentally observed final positions of marker inserts and the predictions. In the experiments, small marker particles were lined up in the transverse direction in the flow field. The final positions were determined in a postweld procedure involving cutting, etching, and imaging. As shown, the two-dimensional simulation is capable of matching the experimental results.

The fluid flow problem has also been solved three dimensionally. Compared with the two-dimensional solutions, both the pin and the shoulder are taken into account in imposing the boundary conditions for the analysis of

Fig. 13 Plot of simulated streamlines at pin. Source: Ref 21

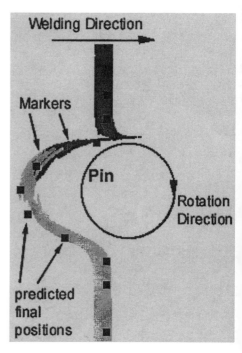

Fig. 14 Comparison between predicted final position and marker inserts. Source: Ref 21

material flow and heat generation/transfer. The corresponding velocity boundary conditions can be expressed as:

$$\dot{u}_x = \omega r \sin\phi - V$$
$$\dot{u}_y = \omega r \cos\phi \qquad \text{(Eq 41)}$$

where

$$r = r_r = r_t \qquad \text{(Eq 42)}$$

for the surface contacting a cylindrical pin and:

$$r_t \leq r \leq r_0 \qquad \text{(Eq 43)}$$

for the top surface contacting the flat shoulder.

The aforementioned velocity boundary conditions assume no slipping between the rotating tool and the workpiece. To account for the relative motion at the tool-workpiece interface, various models have been developed (Ref 14, 15, 17–19). A simple approach is to introduce a parameter, percentage slip, δ, to quantify the slipping. δ is a number between 0 and 1. $\delta = 0$ denotes a sticking boundary condition, whereas $\delta = 1$ indicates slipping on the tool-workpiece interface. Using δ, the velocity boundary conditions are expressed as:

$$\dot{u}_x = (1-\delta)(\omega r \sin\phi - V)$$
$$\dot{u}_y = (1-\delta)\omega r \cos\phi \qquad \text{(Eq 44)}$$

A different approach to model the slipping between the tool and workpiece is to prescribe a threshold value for the shear stress on the interface (Ref 14). Using this approach, two different types of boundary conditions can exist simultaneously on different parts of the interface:

- *Sticking condition:* For the part of the interface where the shear stress is lower than the threshold value, the tool velocity is imposed on the contacting workpiece material.
- *Slipping condition:* For the part of the interface where the shear stress is higher than the threshold value, the shear stress on the interface is truncated to the threshold value. As a result, slipping occurs.

In the aforementioned modeling approaches, both the percentage slip and the threshold shear stress must be determined experimentally.

Figure 15 shows the streamlines at different horizontal planes in the FSW of a plate. Comparing Fig. 15 against Fig. 13, it can be seen that the flow fields from the three-dimensional solution display similar features as their two-dimensional counterparts, although the influence of the shoulder diminishes quickly in the plate thickness direction. The streamlines shown in Fig. 15(a) depict the material flow pattern only 0.35 mm from the top surface. They are similar to the results in Fig. 16, which show the parallel incoming SiC markers being swept around the retreating side, and the flow

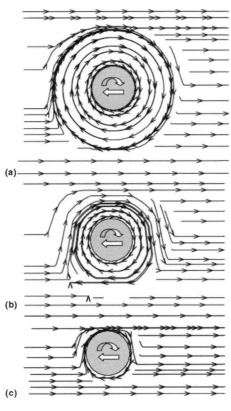

Fig. 15 Stream traces on horizontal planes (a) 0.35, (b) 1.59, and (c) 2.28 mm below the top surface of a 304 stainless steel of 3.18 mm thickness that was friction stir welded at a welding speed of 4 mm/s and a tool rotation at 300 rpm. Source: Ref 17

separation near the advancing side. The temperature distributions at various horizontal planes are plotted in Fig. 17. Also similar to the two-dimensional case are the temperature distributions, which display slight asymmetry, with the material on the advancing side showing marginally higher temperature.

Analysis work adopting this class of modeling approaches has been used to study and improve tool design (Ref 14, 15). Attempts have also been made to evaluate weld quality based on the solved strain rate and temperature (Ref 14, 15). Typically, the computation for this class of models is time-consuming. With the rapid development of computer technology, it is believed that, in the future, this class of analysis will be employed routinely for tool design.

Active Research Topics in the Simulation of Friction Stir Welding

Due to the need to improve tool design and optimize the process of FSW, the simulation of FSW has been a very active research area. To date, tool design and the selection of the accompanying FSW process parameters still adopt a trial-and-error methodology, which is

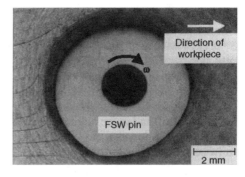

Fig. 16 Macrograph showing the parallel incoming SiC markers being swept around the retreating side and the flow separation near the advancing side. Source: Ref 37

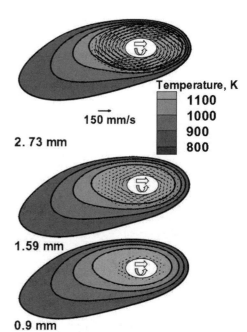

Fig. 17 Plots of temperature and velocity fields on horizontal planes (a) 0.35, (b) 1.59, and (c) 2.28 mm below the top surface of a 304 stainless steel of 3.18 mm thickness that was friction stir welded at a welding speed of 1.693 mm/s and a tool rotation at 300 rpm. Source: Ref 17

costly and time-consuming. The complex physics involved in FSW has posed various challenges for simulation and has motivated a series of attempts to include them in the analysis model.

The fact that tool design directly impacts the heat generation, material flow, and quality of the joints made in FSW has long been recognized. It is hoped that a full-scale three-dimensional analysis taking into account the tool geometry features will help investigate the aforementioned issues. This is particularly true for tools having complex geometry, which results in three-dimensional material flow patterns, such as the threaded frustum-shaped pin widely used in the FSW of aluminum alloys. Challenges in building up the analysis model come mainly from modeling the interaction between the tool and the workpiece. For a non-axisymmetric pin (a pin having a varying shape of axisymmetric cross section, such as pins with flats), the contact between the pin and the workpiece changes periodically, which implies there is no mathematically strict steady state of material flow. Experiments have shown that the flats are not fully in contact with the workpiece material (Ref 32). For a cylindrical pin with three flats, two-dimensional CFD analysis has been performed by assuming that the whole outer surface of the pin is in contact with the workpiece material. The calculation results indicate that the tool orientation has only small effects (Ref 14). The same approach has been used in three-dimensional analysis (Ref 15). It is expected that further verification of the three-dimensional analysis results with the experimental results will help establish a model that is capable of simulating the three-dimensional material flow pattern, hence ultimately assisting the design of FSW tools and the selection of process parameters.

The localized intense deformation in FSW has also motivated various numerical approaches to accommodate the deformation feature. Among these approaches, the arbitrary Lagrangian-Eulerian method has attracted many researchers. In continuum mechanics, there are two basic descriptions of the motion of material points: the Lagrangian description and the Eulerian description. The Lagrangian description observes the motion of material points while moving with the material, but the Eulerian description observes the motion at a fixed location. Because the Lagrangian description tracks the history of material motion, it makes it very convenient to find the deformation of materials. The Lagrangian description has been widely used in solid mechanics, although it does not work properly when shear deformation is large. The Eulerian description, on the other hand, provides the velocity field for given locations and given time. The solutions of most fluid dynamics problems have adopted the Eulerian description. Because it does not directly deal with the deformation history, large shear deformation can be handled with relative ease. Based on the aforementioned, it can be seen that the Lagrangian description is appropriate for material away from the weld joint, whereas the Eulerian description is suitable for material in the vicinity of the stir zone. A numerical technique known as the arbitrary Lagrangian-Eulerian method has been developed to combine the strength of the two descriptions of material motion. This method has also been used for the simulation of FSW (Ref 25 to 28).

While research is being conducted to include contributing physics in the FSW model and to develop efficient numerical schemes for the solution, attempts have been made to use the existing FSW models to find guidelines in selecting welding parameters, predicting weld quality, and improving tool design. Parametric studies have been conducted to simulate the material flow pattern, traversing force, torque, and temperature in FSW. The parameters investigated include the tool profile and welding process variables, such as travel speed and rotation speed. Guidelines have been provided to evaluate tool designs. The predicted flow field and temperature field are used to estimate the quality of the joints. On the other hand, experimental data are also collected to verify the simulation results. The synergy between the continuous improvement of the FSW process and the development of comprehensive FSW models has made the simulation of FSW a very *challenging*, although *attractive*, research area.

REFERENCES

1. W.M. Thomas, Friction Stir Butt Welding, International Patent Application PCT/GB92/02203, 1991
2. C.J. Dawes and W.M. Thomas, Friction Stir Welding Process Welds Aluminum Alloys, *Weld. J.,* Vol 75 (No. 3), 1996, p 41–45
3. J.E. Gould and Z. Feng, Heat Flow Model for Friction Stir Welding of Aluminum Alloys, *J. Mater. Process. Manuf. Sci.,* Vol 7, 1998, p 185–194
4. P. Colegrove, M. Painter, D. Graham, and T. Miller, "3-Dimensional Flow and Thermal Modelling of the Friction Stir Welding Process," Session 2, Paper 2, Second Int. Symposium on Friction Stir Welding (Gothenburg, Sweden), The Welding Institute, 2000
5. P. Ulysse, Three-Dimensional Modeling of Friction Stir-Welding Process, *Int. J. Mach. Tools Manuf.,* Vol 42, 2002, p 1549–1557
6. M.Z.H. Khandkar, J.A. Khan, A.P. Reynolds, and M.A. Sutton, Predicting Residual Thermal Stresses in Friction Stir Welded Metals, *J. Mater. Process. Technol.,* Vol 174, 2006, p 195–203
7. Z.H. Khandkar, J.A. Khan, and A.P. Reynolds, Prediction of Temperature Distribution and Thermal History during Friction Stir Welding: An Input Torque Based Model, *Sci. Technol. Weld. Join.,* Vol 8, 2003, p 165–174
8. Z.H. Khandkar and J.A. Khan, Thermal Modeling of Overlap Friction Stir Welding for Al-Alloys, *J. Mater. Process. Technol.,* Vol 10, 2001, p 91–105
9. V. Soundararajan, S. Zekovic, and R. Kovacevic, Thermo-Mechanical Model with Adaptive Boundary Conditions for Friction Stir Welding of Al 6061, *Int. J. Mach. Tools Manuf.,* Vol 45, 2005, p 1577–1587
10. X.K. Zhu and Y.J. Chao, Numerical Simulation of Transient Temperature and Residual Stress in Friction Stir Welding of 304L Stainless Steel, *J. Mater. Process. Technol.,* Vol 146, 2004, p 263–272
11. C.M. Chen and R. Kovacevic, Finite Element Modeling of Friction Stir Welding—Thermal and Thermomechanical Analysis, *Int. J. Mach. Tools Manuf.,* Vol 43, 2003, p 1319–1326
12. R.L. Goetz and K.V. Jata, Modeling Friction Stir Welding of Titanium and Aluminum Alloys, *Friction Stir Welding and Processing,* K.V. Jata, M.W. Mahoney, R.S. Mishra, S.L. Semiatin, and D.P. Field, Ed., TMS (The Minerals, Metals & Materials Society), 2001
13. W.D. Lockwood, B. Tomaz, and A.P. Reynolds, Mechanical Response of Friction Stir Welded AA2024: Experiment and Modeling, *Mater. Sci. Eng. A,* Vol 323, 2002, p 348–353
14. P.A. Colegrove and H.R. Shercliff, Development of Trivex Friction Stir Welding Tool: Part 1—Two-Dimensional Flow Modeling and Experimental Validation, *Sci. Technol. Weld. Join.,* Vol 9, 2004, p 352–361
15. P.A. Colegrove and H.R. Shercliff, Development of Trivex Friction Stir Welding Tool: Part 2—Three Dimensional Flow Modeling, *Sci. Technol. Weld. Join.,* Vol 9, 2004, p 352–361
16. P.A. Colegrove and H.R. Shercliff, 3-Dimensional CFD Modeling for Flow Round a Threaded Friction Stir Welding Tool Profile, *J. Mater. Process. Technol.,* Vol 169, 2005, p 320–327
17. R. Nandan, G.G. Roy, and T. DebRoy, Numerical Simulation of Three-Dimensional Heat Transfer and Plastic Flow during Friction Stir Welding, *Metall. Mater. Trans. A,* Vol 37, 2006, p 1247–1259
18. R. Nandan, G.G. Roy, T.J. Lienert, and T. DebRoy, Numerical Modeling of 3D Plastic Flow and Heat Transfer during Friction Stir Welding of Stainless Steel, *Sci. Technol. Weld. Join.,* Vol 11, 2006, p 526–537
19. R. Nandan, G.G. Roy, T.J. Lienert, and T. DebRoy, Three-Dimensional Heat and Material Flow during Friction Stir Welding of Mild Steel, *Acta Mater.,* Vol 55, 2007, p 883–895
20. T. Long and A.P. Reynolds, Parametric Studies of Friction Stir Welding by Commercial Fluid Dynamics Simulation, *Sci. Technol. Weld. Join.,* Vol 11, 2006, p 200–208
21. T.U. Seidel and A.P. Reynolds, Two-Dimensional Friction Stir Welding Process Model Based on Fluid Mechanics, *Sci. Technol. Weld. Join.,* Vol 8, 2003, p 175–183
22. J.H. Cho, D.E. Boyce, and P.R. Dawson, Modeling Strain Hardening and Texture Evolution in Friction Stir Welding of Stainless Steel, *Mater. Sci. Eng. A,* Vol 398, 2005, p 146–163
23. M. Song and R. Kovacevic, Thermal Modeling of Friction Stir Welding in a Moving Coordinate System and Its Validation, *Int. J. Mach. Tools Manuf.,* Vol 43, 2003, p 605–615
24. G.J. Bendzsak, T.H. North, and Z. Li, Numerical Model for Steady-State Flow in

Friction Welding, *Acta Mater.,* Vol 45, 1997, p 1735–1745

25. H. Schmidt and J. Hattel, A Local Model for the Thermomechanical Conditions in Friction Stir Welding, *Model. Simul. Mater. Sci. Eng.,* Vol 13, 2005, p 77–93

26. H.W. Zhang, Z. Zhang, and J.T. Chen, 3D Modeling of Material Flow in Friction Stir Welding under Different Process Parameters, *J. Mater. Process. Technol.,* Vol 183, 2007, p 62–70

27. Z. Zhang and H.W. Zhang, Numerical Studies on Controlling of Process Parameters in Friction Stir Welding, *J. Mater. Process. Technol.,* Vol 209, 2009, p 241–270

28. S. Guerdoux and L. Fourment, ALE Formulation for the Numerical Simulation of Friction Stir Welding, *Int. Conf. Comput. Plasticity,* E. Onate and D.R.J. Owen, Ed. (Barcelona), 2005

29. A. Askari, S. Silling, B. London, and M. Mahoney, Modeling and Analysis of Friction Stir Welding Processes, *Friction Stir Welding and Processing,* K.V. Jata, N.W. Mahoney, R.S. Mishra, S.L. Semiatin, and D.P. Field, Ed., TMS (The Minerals, Metals & Materials Society), 2001, p 43–54

30. T.U. Seidel and A.P. Reynolds, Visualization of the Material Flow in AA2195 Friction-Stir Welds Using a Marker Insert Technique, *Metall. Mater. Trans. A,* Vol 32, 2001, p 2879–2884

31. K.J. Colligan, Material Flow Behavior during Friction Stir Welding of Aluminum, *Weld. J.,* Research Supplement 78, 1999, p 229s–237s

32. K.J. Colligan and S.K. Chopra, Examination of Material Flow in Thick Section Friction Stir Welding of Aluminum Using a Stop-Action Technique, *Proceedings of the Fifth International Symposium on Friction Stir Welding* (Metz, France), TWI, 2004

33. M.W. Mahoney, C.G. Rhodes, J.G. Flintoff, R.A. Spurling, and W.H. Bingel, Properties of Friction-Stir-Welded 7075 T651 Aluminum, *Metall. Mater. Trans. A,* Vol 29, 1998, p 1955–1964

34. H.S. Carslaw and J.C. Jaeger, *Conduction of Heat in Solids,* 2nd ed., Clarendon Press, Oxford, 1959

35. C. Zener and J.H. Hollomon, Effect of Strain Rate upon Plastic Flow of Stress, *J. Appl. Phys.,* Vol 15, 1944, p 22–32

36. C.M. Sellars and W.J. McTegart, On the Mechanism of Hot Deformation, *Acta Metall.,* Vol 14, 1966, p 1136–1138

37. B. London, M. Mahoney, W. Bingel, M. Calabrese, and D. Walden, *Proceedings of the Fifth International Symposium on Friction Stir Welding* (Kobe, Japan), TWI, 2001

ASM Handbook, Volume 22B, *Metals Process Simulation*
D.U. Furrer and S.L. Semiatin, editors

Copyright © 2010, ASM International®
All rights reserved.
www.asminternational.org

Modeling of Diffusion Bonding

C.C. Bampton, Pratt & Whitney Rocketdyne

THE PROCESS OF DIFFUSION BONDING may be described as one in which no macro-plastic deformation (not more than a few percent) of the components being joined takes place. Joining is achieved mainly by atomic transport at the mating interfaces to remove bond line voids. To ensure sufficient atomic mobility while minimizing macroplastic deformation, an appropriate balance between the process variables of temperature, pressure, and time must be established. Typically, the process variables are up to several hours at moderate temperatures and pressures. However, the balance also depends on the properties (yield stress, creep behavior, diffusion constants) of the materials being joined and on the condition of the two faying surfaces.

Typically, the two surfaces to be bonded are far from smooth on the atomic scale, and the initial contact area between the two surfaces constitutes a very small fraction of the total surface area. Nonetheless, where the atoms on each surface are within an atomic distance of one another, there is the opportunity for a metallic bond to form, physically joining the two surfaces together (Ref 1). However, on application of pressure, the contact areas will expand to support that pressure. Again, it is assumed that, over the contact area, atoms on the surface of each component are within an atomic distance of one another and that a metal-to-metal bond forms across the interface. The fraction of bonded area has thus increased. The aspect ratio of the voids is governed by the roughness profile of the original surfaces. The majority of the diffusion bonding process is spent in bringing the surfaces of these voids to within an atomic spacing so that a metallic bond may form over the whole of the surface to be bonded.

The goals of modeling diffusion bonding can be regarded as twofold: to optimize the selection of the process variables for a given material and to provide an understanding of the mechanisms by which bonding is achieved. With regard to the latter, diffusion bonding takes place as a consequence of one or more competing mechanisms, the contributions from each changing not only from bond to bond as process variables and/or materials are changed

but also during the time an individual bond is fabricated. Given this complexity, existing models of diffusion bonding described in the following section tend to assume that the surfaces to be joined are free of contaminants and oxide, that bonding occurs between similar materials, and that the materials are single-phase metals. These points are discussed further in the final section on the limitations of existing models.

Models for Metallic Alloys

The majority of models simplify the geometries (profiles) of the contacting surfaces and assume that they can be regarded as a series of identical asperities touching at their tips to create any array of identical voids. The manner by which one (or part) of such a representative void is removed is then studied in the various models.

Cline (Ref 2) proposed that diffusion bonding could occur in two ways. In the first, the applied load was sufficient to cause considerable plastic deformation of the asperities. In the second, bond development was dominated by diffusion-controlled processes, more akin to the original recrystallization model of Parks (Ref 3). The combination of these two approaches, albeit refined considerably, provides the basis for all subsequent models; most later quantitative models assume that, initially, plastic deformation occurs until the contacting area of deforming asperities increases sufficiently that the stress on these areas falls below the yield stress of the material, and that subsequently, bonding continues by various diffusional processes (including creep).

The next qualitative model was proposed by King and Owczarksi (Ref 4), who suggested that bonding takes place in three stages. The first two stages were as described previously, but migration of the interface away from the voids was assumed to occur during the second stage, and so, volume diffusion was required in the third stage to ensure removal of the remaining isolated voids. In fact, interface migration does not necessarily occur in all materials.

Hamilton (Ref 5) developed a simple but remarkably accurate quantitative model based on careful microstructural observations of the faying surfaces evolution and measurements of creep-type behavior. To illustrate the sequence of surface asperity deformation, a diffusion bond was produced with an equiaxed, two-phase Ti-6Al-4V alloy under a pressure gradient. The pressure gradient was achieved by machining a thickness taper into one Ti-6Al-4V sheet and bonding a second as-rolled sheet of uniform thickness to it. The processing conditions were 1700 °F, 5 h, and nominal 2000 psi over the specimen plan area. Due to the taper, local bonding pressure actually varied from 0 to well over 2000 psi. The resulting sequence of bonding is shown in Fig. 1, where the effect of pressure gradient on the removal of asperities is clearly illustrated. It is apparent that the surface deformation across the bond line is directly related to the pressure applied. Also, as the interface voids shrink, they become spherical, indicating that surface tension and diffusional processes are becoming effective. The apparent bridging that is seen to occur under very light contact suggests that vapor-phase transport may also occur at this bonding temperature. This sequence illustrates the importance of applying sufficient pressure during initial stages of bonding to overcome workpiece separations due to the surface asperities.

Hamilton (Ref 5) used this microstructural evolution and materials behavior to develop the first quantitative model to predict the time required to attain full interfacial contact between two rough surfaces. Hamilton assumed that the surfaces to be joined consisted of triangular-section asperities in point-to-point contact (Fig. 2). The asperities were thought to collapse by time-dependent plasticity (in this case, by super-plastic flow) under plane-strain conditions.

The asperities were treated as deforming in plane strain. Hamilton calculated the deformation of the entire asperity using a mean stress value. He derived the required equations to determine the relationship between applied bond pressure and time to overcome asperities, provided that the required material properties, relating to stress and strain rate, are available.

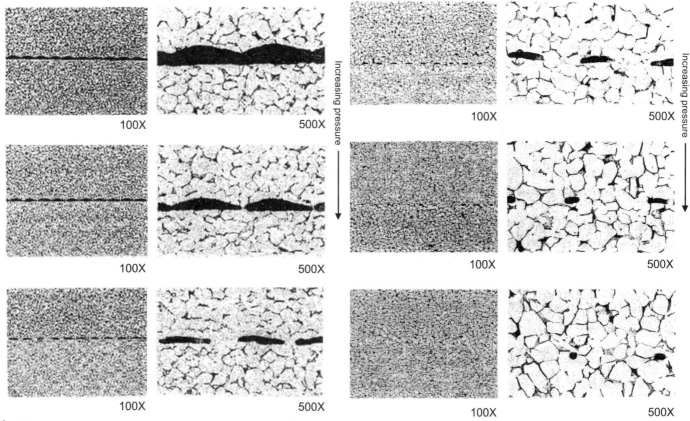

Fig. 1 Sequence of diffusion bonding in Ti-6Al-4V under a pressure gradient. Source: Ref 5

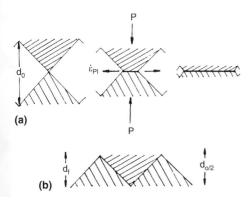

Fig. 2 Idealized geometry of Hamilton's model. (a) Plastic collapse of asperities. (b) Effective final thickness of bond zone. Source: Ref 6

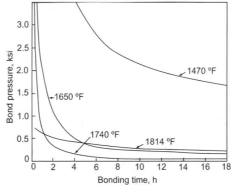

Fig. 3 Bond pressure requirements as a function of bond time as determined analytically. Source: Ref 5

To determine the pressure requirement, it is necessary to calculate the average effective strain rate, establish the corresponding flow stress for the material and temperature under consideration (from experimentally determined data), and then compute the required bond pressure. The average stress in the asperity was related to the applied pressure and the calculated mean strain rate. The bonding time was then calculated from both the displacement required to fully overlap the asperities and the axial strain rate.

The Hamilton model was based on plastic deformation and creep, and favorable agreement between experiment and prediction was found for the titanium alloy Ti-6Al-4V. Figure 3 shows the calculated bond pressure requirements evaluated as a function of bonding time. However, in such an alloy and with the assumed surface roughness (large wavelength of 100 μm), creep would be expected to be the dominant mechanism under normal diffusion bonding conditions. Thus, a model based on just this one diffusion-controlled

mechanism, creep, would be expected to give good results; the agreement may be poorer for other materials or if bonding conditions were chosen such that creep was no longer the dominant mechanism. Interestingly, the highest diffusion bonding temperature, as shown in Fig. 3, did not correspond with the shortest bonding time. This reflects the change in microstructure, due to traversing the alpha-phase solvus, creating a single phase (beta), which is then free to grain grow, hence reducing creep rate.

Garmong et al. (Ref 7) refined Hamilton's model both by treating each ridge as a series of horizontal slices and then summing the response of each slice to the applied stress and also by including diffusion mechanisms. In addition, the surface was modeled as two superimposed wavelengths (Fig. 4), a feature that has not appeared in some later models, even though it more accurately represents a real surface in terms of both waviness (large wavelength) and roughness (short wavelength equivalent to the dimensions of a ridge in later models).

The collapse of the two wavelengths was treated sequentially by plastic deformation and creep. The removal of the final small voids was modeled as the shrinkage of a thick-walled hollow plastic sphere under hydrostatic pressure; the various diffusion mechanisms were analyzed using sintering equations (Ref 8). The model by Garmong et al. embodied not

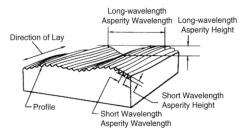

Fig. 4 Schematic of bimodal topography of mechanical surfaces. Source: Ref 7

only the important features of earlier models but also stands out for several reasons:

- It includes the use of two wavelengths to describe surface roughness, which perhaps should be adopted again in future models, given that long-range waviness may control the ease of bonding.
- It was the first to use sintering theory to model void closure, an approach that has been developed considerably in later models.
- It was potentially applicable to materials other than the titanium alloy for which experimental data were given.

The use of sintering theory was adopted more extensively in the model by Derby and Wallach (Ref 9, 10). One aim of their model was to be able to relate the extent of bonding for any single-phase metallic alloy to the various process variables and thus examine the mechanisms by which bonding was achieved in different alloys. The original interface was regarded as a series of straight-sided ridges contacting peak-to-peak; this simplified modeling because plane-strain conditions could be assumed in the subsequent analysis. Because diffusion bonding does not take place by one mechanism alone, seven possible mechanisms were considered.

- Plastic yielding deforming an original contacting asperity
- Surface diffusion from a surface source to a neck
- Volume diffusion from a surface source to a neck
- Evaporation from a surface source to condensation at a neck
- Grain-boundary diffusion from an interfacial source to a neck
- Volume diffusion from an interfacial source to a neck
- Power-law creep

Given an initial geometry (i.e., wavelength and height of voids, based on measurement of the original surface roughness), rate equations for each of the seven mechanisms operating independently are summed to give an overall void shrinkage rate and hence to predict the small extent of bonding for a correspondingly small increment of time. The void geometry is naturally changed by this bonding in a manner

that depends on the extent of the contribution from each mechanism. Using the new geometry, the extent of bonding for the next time increment can again be predicted from the rate equations. This iterative process continues until bonding is complete. The results are displayed in a manner similar to that adopted by Ashby (Ref 11) and Wilkinson and Ashby (Ref 12) for pressure sintering, that is, plots of the extent of bonding against one of the major process variables (temperature, pressure, void wavelength, or void height) and with time contours superimposed on the plots. In this model, three stages of void closure were employed:

- Instantaneous plastic flow
- Straight-sided void geometry until the void aspect ratio reached unity
- Circular cross-sectional void. (Hence, mechanisms for which the driving force arises from variations in surface curvature, e.g., surface diffusion, can no longer contribute to bonding.)

Good agreement with experiment was observed for a number of materials (in particular, iron and copper), although the change in geometry from second to final stage resulted in a discontinuity in bonding rate. This discontinuity was reduced, but not eliminated, by remodeling the creep mechanism (Ref 10).

In commenting on an early version of the previous model, Allen and White (Ref 13) pointed out that the original assumption of long cylinders in the model by Derby and Wallach (Ref 10) would lead to an overestimate of bonding time. In practice, a long cylinder would be expected to break up into a series of isolated spherical voids, and the shrinkage rate of these is considerably greater (by 5 to 20 times) than that of the corresponding cylinder. Additional points by Allen and White (Ref 13) related to the effects of contaminants and, in a qualitative way, to grain size.

Pilling et al. (Ref 14) proposed a diffusive creep cavitation model, based on original work by Chen and Argon (Ref 15), rather than using sintering as an analog. An initial circular void geometry was adopted, instead of the elliptical shape generally observed in most micrographs of void cross sections, because it was assumed that rates of surface diffusion mechanisms were sufficient to maintain circular voids. Thus, the contributions from mechanisms whose driving force relies on differences in surface curvature are precluded, and so, the model is not as universal as those in which surface curvature mechanisms are considered. Because surface diffusion mechanisms are important when bonding many materials (e.g., steels under most conditions, copper at low temperatures), this model seems to be intended mainly for predicting bonding of fine-grained superplastic alloys in which creep mechanism tends to dominate. Nonetheless, an innovation of the model by Pilling et al. (Ref 14) was the inclusion of an analysis of grain size; the material surrounding each

void is divided into horizontal slices, each of thickness equal to the grain size, and the contribution of each boundary intersecting the void is added to the grain-boundary contribution.

Pilling (Ref 6) further developed his modeling approach to predict the diffusion bonding kinetics for Ti-6A1-4V under a variety of bonding conditions and material parameters. For a given surface finish, the predicted bonding time was found to be insensitive to the bonding temperature as long as the temperature remains within the superplastic regime (from 900 to 940 °C). Increasing the bonding pressure by a factor of 10 reduced the bonding time by a factor of 3. Measurements of the roughness of as-received and pickled surfaces, ground surfaces, and grit-blasted surfaces showed widely different asperity geometries. The variation in bonding time with pressure predicted by Pilling's isostatic model showed that the surface finish exerts the greatest influence on the time required to attain full interfacial contact. The reasons postulated are twofold. First, as the voids decrease in size, the surface tension becomes progressively larger with respect to the applied pressure and contributes increasingly toward the effective stress in the wall of the cylinder. Second, for a fixed amount of diffusion, the fractional change in the volume of the interfacial void increases as the void dimensions decrease—diffusion is more effective at closing smaller voids. It was also evident that the bonding time tends asymptotically toward an upper limiting value as the applied pressure is decreased—the self-sintering limit.

Comparison of the predicted bonding times with those measured experimentally shows that Pilling's isostatic model underestimates the time required to attain full interfacial contact. Consideration of only the short-wavelength asperities does not allow an adequate representation of the surface geometry, and it would appear that the long-wavelength roughness, or waviness, of the surfaces to be joined should also be considered. Because the bond zone is usually relatively narrow with respect to the thickness of the pieces being bonded, the applied pressure will probably be unable to force the two surfaces to make point-to-point contact over the entire interface. The attainment of contact over much larger dimensions than the short-wavelength asperities is therefore perhaps the more important step in ensuring the integrity of diffusion bonds. As the planar dimensions of the interfacial voids increase, the time taken to attain full interfacial contact becomes decreasingly dependent on the actual void dimensions, the surface tension, and the diffusive mass transfer. The kinetics of void closure tend toward those given by the size-independent form and toward that predicted by the plane-strain model of Hamilton (Ref 5).

A subsequent model, developed by Hill and Wallach (Ref 16), aimed to overcome some of the approximations and limitations inherent in previous models. First, by using an ellipse to represent the initial void shape, the need for discrete stages with their own geometries is removed, because an elliptical cross-sectional void can develop into a round void by successive incremental changes to the two axes of the ellipse, that is, without a discontinuity. In fact, a further advantage of describing a void as an ellipse is that surface source mechanisms can be reactivated naturally if a void that has become circular during bonding then changes its shape back to an ellipse as a consequence of contributions to bonding from interface sources and creep mechanism. Second, the elliptical geometry permits more rigorous analyses of plastic deformation and creep than those used earlier. Third, a simple statistical analysis enables the contributions from grain-boundary sources in fine-grained materials to be more accurately included, and so, the effects of grain size can be incorporated. In other ways, the same philosophy as proposed by Derby and Wallach (Ref 10) was continued; the contributions from the seven possible bonding mechanisms were summed iteratively and the results displayed in a similar fashion by a mapping technique. Good agreement with experiment is reported (Ref 16).

Most recently, Li et al. (Ref 17) have developed a promising approach for a probabilistic model for the prediction of diffusion bonding time, based on the identification and characterization of deterministic variables and random variables. The probabilistic distribution of surface roughness is introduced into a deterministic model of diffusion bonding by taking the parameters of void radius and height as random variables. The cumulative distribution function of the bonding time was calculated using Monte Carlo simulation and a numerical integration method. After comparing the predicted bonding time of the probabilistic model with those of the existing deterministic models and with experimental results from previous work on the (superplastic) Ti-6Al-4V alloy, they concluded that the bonding time predicted by the probabilistic model is in best agreement with the test results. An evaluation of the effect of surface roughness on bonding time could be shortened greatly by improving the surface quality.

Current Status of Modeling

For single-phase metallic alloys, there is a reasonable understanding of the dominant bonding mechanisms and of the extent to which each mechanism contributes for different materials and bonding conditions. Therefore, the use of a model can lead to optimization of process variables in a much faster manner than can be achieved by experiment alone.

To achieve further improvements in kinetic modeling, there is the problem of evaluating the predictions of any model, that is, the experimental difficulty of accurately measuring the bonded area (Ref 18). Also, there are difficulties in obtaining, with sufficient accuracy, the necessary input data to a model for a specific material; for example, diffusion and creep data generally are obtained from tabulated data and not measured exactly for the sample being bonded. (The thermomechanical history and precise surface condition will differ for each nominally identical batch of the same alloy.) With only slight adjustments to, say, process parameters and input data, model predictions can change considerably from poor to exact agreement with experimental data from bonding. This has been considered by Hill (Ref 19), who has shown that for many materials, the values of shear modulus, creep, and surface diffusion data (and their temperature dependence) are critical. The latter two values are particularly important because, for most materials, creep and/or surface diffusion mechanisms dominate and contribute significantly to bonding. Unfortunately, it is precisely these two properties that are difficult to measure accurately.

Future of Modeling

Future challenges include:

- Improved surface profile descriptions (e.g., returning to the two-wavelength model of Garmong et al., Ref 7)
- Recognition of the roles of surface contaminants, surface oxides, and surface coatings applied to enhance diffusion bonding
- Modeling of two-phase materials
- Modeling of bonding between dissimilar materials
- Modeling of bonding in nonmetals

REFERENCES

1. K.E. Easterling and A.R. Tholen, *Acta Metall.,* Vol 20, 1972, p 1001–1008
2. C.L. Cline, *Weld. J.,* Vol 45, 1966, p 481–489
3. J.M. Parks, *Weld. J.,* Vol 32, 1953, p 209–222
4. W.H. King and W.A. Owczarski, *Weld. J.,* Vol 46, 1967, p 289–290
5. C.H. Hamilton, *Titanium Science and Technology,* Vol 1, R.I. Jaffee and H.M. Burte, Ed., Plenum, 1973, p 625–648
6. J. Pilling, *Mater. Sci. Eng.,* Vol 100, 1988, p 137–144
7. G. Garmong, N.E. Paton, and A.S. Argon, *Metall. Trans. A,* Vol 6, 1975, p 1269–1279
8. R.L. Coble, *J. Appl. Phys.,* Vol 41, 1970, p 4798–4807
9. B. Derby and E.R. Wallach, *Met. Sci.,* Vol 16, 1982, p 49–56
10. B. Derby and E.R. Wallach, *Met. Sci.,* Vol 18, 1984, p 427–431
11. M.F. Ashby, *Acta Metall.,* Vol 22, 1974, p 275–289
12. D.S. Wilkinson and M.F. Ashby, *Acta Metall.,* Vol 23, 1975, p 1277–1285
13. D.J. Allen and A.A.L. White, *Proc. Conf. on the Joining of Metal Practice and Performance,* Institution of Metallurgists, Warwick, 1981
14. J. Pilling, D.W. Livesey, J.B. Hawkyard, and N. Ridley, *Met. Sci.,* Vol 18, 1984, p 117–122
15. I.-W. Chen and A.S. Argon, *Acta Metall.,* Vol 29, 1981, p 1759–1768
16. A.D. Hill and E.R. Wallach, *Acta Metall.,* Vol 37 (No. 9), 1987, p 2425–2437
17. S.-X. Li, S.-T. Tu, and F.-Z. Xuan, *Mater. Sci. Eng. A,* Vol 407, 2005, p 250–255
18. B. Derby, G.A.D. Briggs, and E.R. Wallach, *J. Mater. Sci.,* Vol 18, 1983, p 2345–2353
19. A.D. Hill, Ph.D. thesis, University of Cambridge, 1984

Simulation of Heat Treatment Processes

ASM Handbook, Volume 22B, *Metals Process Simulation*
D.U. Furrer and S.L. Semiatin, editors

Copyright © 2010, ASM International®
All rights reserved.
www.asminternational.org

Heating and Heat-Flow Simulation

Gang Wang, Yiming Rong, and Richard D. Sisson, Jr.,
Center for Heat Treating Excellence, Worcester Polytechnic Institute

THE FIRST STEP IN EVERY HEAT TREATING PROCESS is heating the parts to the desired temperature. The heating process takes valuable time, and this time at elevated temperatures may have a significant influence on the microstructure of the parts. The heating time also adds to the total cycle time. The heating rate for each part in a furnace load is a function of the heating method, furnace design parameters, part size, and part loading patterns (i.e., racking). The temperature distribution within the furnace and temperature variation with time will also influence the response of the parts to the heat treatment.

The temperature of a part at a specific location in the furnace as well as the temperature on and within the part may vary with time, as shown in Fig. 1. The important segments of the temperature profile (Eq 1) are total cycle time (t_{cycle}), heating time ($t_{heating}$), soaking time (t_{soak}), and cooling time (t_{cool}). For a particular furnace, these times may vary with the racking as well as part size:

$$t_{cycle} = t_{heating} + t_{soak} + t_{cool} \tag{Eq 1}$$

Heat Transfer during Furnace Heating

The heating process in a furnace is complicated in real industrial settings. Behavior depends upon the furnace configuration, part loading patterns, material properties, and the heat-transfer physics. It is almost impossible to model all the details of the heating process because of the complex geometry of furnaces and the large number of parts. Simplifications are needed to build an acceptable model.

The General Model

The fundamental principle of the model is the concept of energy conservation, which means the energy into the furnace system is equal to the heat absorbed by the furnace atmosphere, the parts, and the accessories, and the losses through the wall and in emitted air. Also, the relationship among the components

in the system, such as heating elements, parts, fixtures, and furnace, is presented in the model. A series of mathematical equations have been used to represent the performance and behavior of the heat source, furnace, and parts based on the radiation, conduction, and convection heat transfers. The following mathematical models can be established:

- Heat-source model
- Conduction heat-transfer model of parts and fixtures
- Radiation heat-transfer model in the furnace
- Convection heat-transfer model in the furnace

Energy Conservation Concept. The entire furnace, including the accessories, parts, and the medium, such as air, is regarded as an independent system. Based on the energy conservation principle, the heat input is equal to the heat loss plus the storage energy.

The general governing equation for the heat source is:

$$Q_{HS} = Q_{Medium} + Q_{partwp} + Q_{furnace} + Q_{loss} \tag{Eq 2}$$

where Q_{HS} is the heat generated by some heat source; Q_{Medium} is the heat absorbed by the medium in the furnace, such as gases; Q_{wp} is the heat absorbed by workpieces; $Q_{Furnace}$ is the heat

absorbed by parts (trays/fixtures) and furnace walls; and Q_{loss} is the loss taken by the cooling source and exhausted out of the furnace.

Heat-Transfer Analysis. From the point of view of physics, all modes of heat transfer must be considered during the heating process, that is, conduction within parts or accessories, convection of the gas medium, and the radiation among the parts.

The general governing equation for the temperature of a constituent is:

$$\rho c V \frac{dT}{dt} = A(q_{in} - q_{out}) \tag{Eq 3}$$

where ρ, c, and V are the density, specific heat, and volume of the heat-transfer medium, respectively; A is the surface area of the heat-transfer medium; q_{in} is the input heat flux; and q_{out} is the heat flux lost at the surface.

The treatment of the components in furnaces is very important for the prediction of their thermal behavior:

- *Parts:* The parts on the tray/fixture are fully explored. The overlapping of parts is not considered.
- *Wall:* The furnace wall is assumed to be covered with fully heat-insulating materials.
- *Heating elements:* The radiant-tube-heated or electrically heated heating elements have

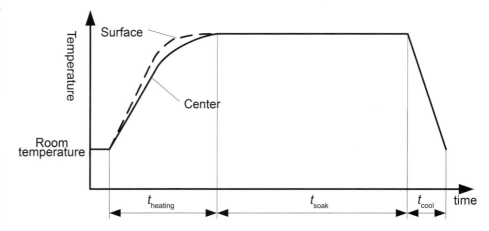

Fig. 1 Temperature versus time for a furnace cycle

been modeled with different configurations (e.g., either two- or three-sided heating configurations in box-type furnaces).

- *Medium:* The atmosphere in the furnace is either protective gas or other heat treating gas (e.g., endothermic gas for carburization). There is a fan to circulate the gas in the furnace where the gas flow is assumed to be steady.
- *Fixture or tray:* The parts are loaded into a batch furnace with either box-type or cylinder geometric shape.

Conduction

Heat conduction occurs inside parts and fixtures in intimate contact. Fourier's law is used to describe the temperature evolution within the parts (Ref 1):

$$\vec{q_c} = -k\vec{\nabla}T \qquad \text{(Eq 4)}$$

where $\vec{q_c}$ is the heat flux, that is, the heat flow per unit surface area along the normal direction; k is the thermal conductivity of the material; and T is the temperature.

The governing equation of transient heat conduction in isotropic metal materials is expressed in Cartesian coordinates as (Ref 2):

$$\frac{\partial}{\partial x}\left(k\frac{\partial T}{\partial x}\right) + \frac{\partial}{\partial y}\left(k\frac{\partial T}{\partial y}\right) + \frac{\partial}{\partial z}\left(k\frac{\partial T}{\partial z}\right) + \dot{S}$$
$$= \rho \cdot c \cdot \frac{\partial T}{\partial t} \qquad \text{(Eq 5)}$$

where $\dot{S}$ is the volumetric heating rate, ρ is the density, and c is the specific heat of the part/fixture material.

Coupling with certain boundary conditions, Eq 5 represents the conduction within the parts.

Radiation Calculation

Because of the high furnace temperature, heat transfer by radiation is a significant factor affecting the heat distribution.

Radiant Heat. Thermal radiation happens between any two objects above absolute zero. The emitted heat from an ideal blackbody can be expressed by the Stefan-Boltzmann law (Ref 3):

$$q_r = \sigma T^4 \qquad \text{(Eq 6)}$$

where q_c is the total emissive flux of a blackbody over all wavelengths (in W/m^2), T_s is the surface temperature measured in absolute degrees (in Kelvin), and σ is the Stefan-Boltzmann constant with the value of 5.669×10^{-8} W/m^2/K^4.

Actually, the real surfaces of the parts in the furnace do not radiate as much energy as an ideal blackbody; however, the total radiation emitted by these bodies, which are called "gray," still generally follows the T^4 proportionality. At the same time, the fact that not all the radiation leaving surface 1 will reach surface 2, because radiation travels in straight lines, must be taken into account. The net

radiant heat transfer from one object, surface 1, to another, surface 2, under vacuum condition can be expressed as (Ref 4):

$$Q_{12} = \varepsilon_1 \cdot F_{12} \cdot \sigma \cdot A_1 \cdot \left(T_1^4 - T_2^4\right) \qquad \text{(Eq 7)}$$

where Q_{12} is the net rate of energy exchange by radiation between two surfaces; ε_1 is the emissivity of surface 1; A_1 is the area of the heating surface; T_1 is the absolute temperature of the heating surface; T_2 is the absolute temperature of the target surface; and F_{12} is a view factor, the fraction of energy leaving surface 1 and intercepted by surface 2.

When the gas medium must be taken into account, the net radiant heat transfer from one object to another with participating media is shown as follows:

- Black surface surrounded by gas (Ref 5):

$$Q_{g-s} = \sigma \cdot A_s \cdot \left(\varepsilon_g T_g^4 - \alpha_g T_s^4\right) \qquad \text{(Eq 8)}$$

where Q_{g-s} is the net rate of heat transfer from the gas to the black enclosure, A_s is the surface area of the enclosure; T_s is the surface temperature of the enclosure, T_g is the temperature of the gas, ε_g is the overall gas emissivity, and α_g is the absorptivity of the gas.

- Two gray surfaces surrounded by gas (Ref 6):

$$Q_{1-2} = \frac{\sigma\left(T_1^4 - T_2^4\right)}{\frac{1-\varepsilon_1}{\varepsilon_1 A_1} + R_\Delta + \frac{1-\varepsilon_2}{\varepsilon_2 A_{12}}} \qquad \text{(Eq 9)}$$

$$R_\Delta = \frac{(A_1 F_{12}\tau_g)^{-1}\left[(A_1 F_{1g}\varepsilon_g)^{-1} + (A_2 F_{2g}\varepsilon_g)^{-1}\right]}{(A_1 F_{12}\tau_g)^{-1} + (A_1 F_{1g}\varepsilon_g)^{-1} + (A_2 F_{2g}\varepsilon_g)^{-1}}$$
$$\text{(Eq 10)}$$

where R_Δ is the equivalent series resistance of the loop of the network, and τ_g is the total transmissivity of the gas.

View factor (Ref 4) is the fraction of energy emitted by surface 1, dA_1, that reaches object 2, dA_2, as shown in Fig. 2. The temperatures of dA_1 and dA_2 are T_1 and T_2, respectively. Then, the view factor F_{1-2} can be defined as:

$$F_{12} = \frac{1}{A_1}\int_{A_1}\int_{A_2}\frac{\cos\beta_1 \cdot \cos\beta_2}{\pi s^2}dA_1 dA_2 \qquad \text{(Eq 11)}$$

where A_1 and A_2 are the areas of surface 1 and surface 2, respectively; β_1 and β_2 are the angles between the line$_{A1-A2}$ and the normal of the surfaces, A_1 and A_2, of the two objects, respectively; and s is the distance the heat travels between dA_1 and dA_2.

Convection Calculation

A comprehensive convection model of the heating process should involve the flow details for the gas medium that are affected by the

furnace geometry, the arrangement of the work load, the nozzle configurations, and process parameters. However, this is too complex and time-consuming for industrial applications. Instead, the convection heat-transfer coefficient has been used to simplify the effect of gas flow on the heat exchange between the parts and the environment.

Relevant assumptions and restrictions (Ref 6) are as follows:

- Newtonian fluid and incompressible
- Laminar flow
- The effects of radiant transport and viscous dissipation are omitted.
- Constant-property fluid in nonflow directions is assumed.

The heat energy that enters a part by means of convection heat transfer can be calculated using:

$$Q_c = h \cdot A_S \cdot (T_{flow} - T_S) \qquad \text{(Eq 12)}$$

where h is the convection heat-transfer coefficient, A_S is the surface area of the part, and T_{flow} and T_S are the temperature of the fluid and part surface, respectively. The T_{flow} is assumed to be approximately equal to the furnace temperature. The average convection heat-transfer coefficient, h, is generally calculated by (Ref 1):

$$h = \frac{k_g}{L^*} \cdot Nu_{L^*} \qquad \text{(Eq 13)}$$

where k_g is the thermal conductivity of the gas (in W/m·K), L^* is the equivalent length of the part related to the part geometry and size, and Nu_{L^*} is the Nusselt number:

$$Nu_{L^*} = f(Ra, Pr, \text{ geometric shape, boundary conditions})$$

In the calculation of h, the key effort is calculating the Nusselt number, Nu_{L^*}, which, as shown, is a function of four variables, including the Rayleigh number (Ra) and the Prandtl number (Pr). The problem can be classified into two types, natural convection and forced convection, according to the furnace conditions.

In actual industrial operations, the Nu is too complicated to be obtained with a high degree of precision. Based on some empirical functions, the expression of Nu can be elicited by considering

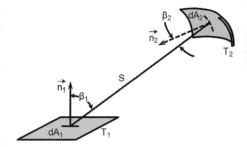

Fig. 2 Radiant heat exchange between two objects

the part-loading patterns and gas flow rate. This requires knowledge of the equipment itself before proceeding with example calculations.

Industrial Furnace Types

To better model the heating processes, furnace types, heating methods, and control systems must be studied.

Furnace Types

According to the loading pattern, the furnaces used for heat treating generally are classified into two categories: batch furnaces and continuous furnaces. In a batch furnace, parts normally are loaded and unloaded into and out of the furnace chamber in batches. A continuous furnace has an automatic conveying system that provides a constant part load through the unit.

Batch furnaces consist of an insulated chamber with an external reinforced steel shell, a heating system for the chamber, and one or more access doors to the heated chamber. Two kinds of shapes of the furnace body are commonly used in industry: box type and cylinder type. A box-type-shaped furnace is shown in Fig. 3. The cylinder-type furnace can be set into horizontal or vertical arrangement, as shown in Fig. 4 (Ref 7) and Fig. 5 (Ref 8), respectively.

Vacuum furnaces are considered a special kind of batch furnace that accommodates heat treated metals in a sealed enclosure. The enclosure is evacuated to certain partial pressures according to the metals and processes. Vacuum is substituted for the usual protective gas atmospheres used during part or all of the heat treatment. Vacuum furnaces are usually cylindrically shaped, with a horizontal or vertical arrangement. The part loading can be bottom-loading or top-loading. Figure 4 is a typical bottom-loading vacuum furnace.

Continuous furnaces consist of the same basic components as batch furnaces but are operated in uninterrupted cycles as the parts move through.

Pusher furnaces include skid-rail furnaces and roller-rail furnaces. A pusher furnace uses the tray-on-tray concept to move parts through the furnace. The pusher mechanism pushes a solid row of trays from the charge end until the first tray is properly located in position at the discharge end for removal. At timed intervals, the trays are moved successively through the furnace. The cycle time through the furnace is varied by changing the push intervals. Figure 6 shows a schematic view of tray movement in a pusher furnace. Figure 7 shows another type of pusher furnace with a rotary hearth in it instead of the usual linear motion. Figure 8 shows a typical pusher furnace for continuous carburizing.

Conveyor-type furnaces include roller-hearth furnaces and continuous-belt furnaces. Roller-hearth furnaces move the parts through a heating zone with powered, shaft-mounted rollers that contact the parts or trays. Continuous-belt furnaces move the parts on mesh or cast-link belts. Conveyors used include woven belts of suitable material, chains with projecting lugs, or pans or trays connected to roller chains. A gas atmosphere seal is used to maintain atmosphere integrity in the furnace chamber, and fans are used for recirculating the atmosphere.

Heating Elements

The heating elements can be classified according to their heating methods: direct fired, radiant tube, and electrical.

Direct-Fired Heating. Parts are exposed directly to the products of combustion, typically gas fired. Generally, the gaseous fuel is natural gas, propane, a propane-air mix, or a relatively low-energy manufactured gas. The flue gases can be controlled or varied by adjusting the fuel-air ratio of the combustion system. Figure 9 shows a gas-fired pit furnace (Ref 7), where Fig. 9(a) is the exterior appearance of the furnace, Fig. 9(b) is the flow with burner on, and Fig. 9(c) is the flow with burner off.

Radiant-Tube Heating. The work chamber is protected from the products of combustion using tubes, usually in either "U-" or trident shape, as shown in Fig. 10 and 11 (Ref 9). The radiant tubes are heated by gas-fired or electrical resistance heating elements. The work chamber normally contains a controlled atmosphere, as dictated by the process.

Fig. 3 Car-bottom batch furnace for homogenizing large cylindrical parts. Courtesy of Despatch Industries, Inc.

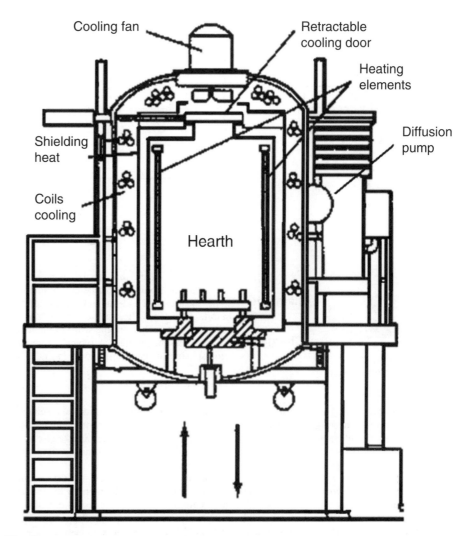

Fig. 4 Vertical cylinder-type batch furnace. Source: Ref 7

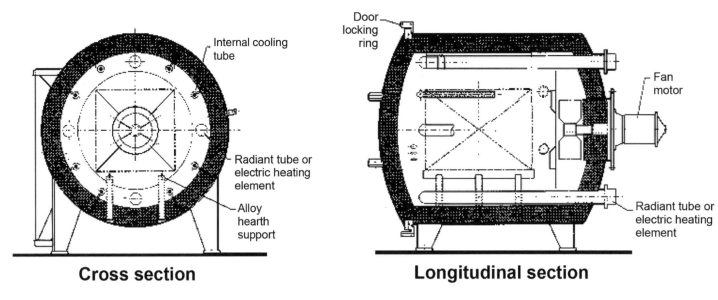

Fig. 5 Horizontal cylinder-type batch furnace. Source: Ref 8

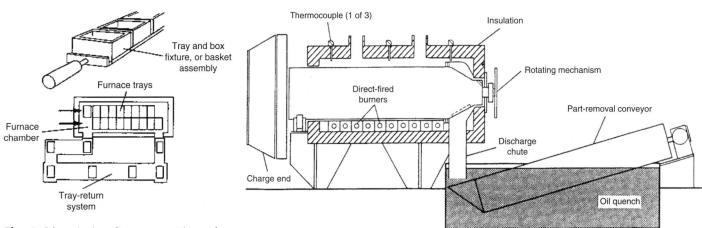

Fig. 6 Schematic view of tray movement in a pusher furnace

Fig. 8 Pusher furnace

Fig. 7 Rotary-hearth furnace

Electrical Heating. The basic consideration in selecting the type of heating elements is to determine whether the elements are to be the open type or the indirect type. The open type means the elements are exposed to the furnace environment, as shown in Fig. 12, where Fig. 12(a) shows a typical layout of the heating and cooling elements along the axis of a cylinder furnace, and Fig. 12(b) shows the typical layout of heating and cooling elements around the circumference of the cylinder body. The

indirect type means that the heating elements are protected from the furnace environment by some means, such as a radiant tube, muffle, or retort. Resistance heating elements can be metallic (e.g., nickel chrome strip) or nonmetallic (e.g., silicon carbide). A typical example of atmosphere flow pattern is shown in Fig. 12(c).

Furnace Control

Furnace temperature control is essential to ensure quality heat treatment. Understanding the furnace control operation is likewise essential for modeling the furnace temperature distribution.

Furnace Performance Test under Unloaded Condition. To monitor the temperature uniformity in a furnace space, nine thermocouples are placed at specified positions in the furnace. When the furnace is heated without a parts load, the temperature is measured at these

locations. The temperature range among these thermocouples should be within a specified range, say, ± 2.5, 5.5, or 8 °C (4.5, 10, or 14 °F), typically. This test is conducted after furnace installation and periodically thereafter as a quality measure and to calibrate the furnace.

Heating Control. To control the temperature in the furnace, the furnace temperature is measured under the loaded condition. A thermocouple is placed at a designated position in the chamber of the furnace and monitors the air in the furnace, while avoiding any contact with parts. When the furnace is heated up, the temperature is read automatically. According to the temperature obtained from this thermocouple, the heating control (and/or fan) is turned on or off to match the desired heating cycle. In a multizone furnace, one thermocouple is set at a specified position in each zone of the furnace. The temperature is controlled separately in different zones. It can be seen that the air temperature in the furnace is actually controlled. Thermocouples can be placed nearer the part location to monitor the part temperature. Usually, there is a delay in the temperature rise of the part compared to the air temperature. The delay (especially on the interior of large parts with low thermal conductivity) could be as long as several hours. To accommodate the delay, additional heating time is needed before the soaking time (Fig. 1) starts. The additional time contributes to the cycle time/productivity and may cause variations of material properties, mainly due to the variation of temperature distribution.

Furnace Temperature Control Procedure. Automatic temperature controllers are used in most furnace operations. The operating procedure is:

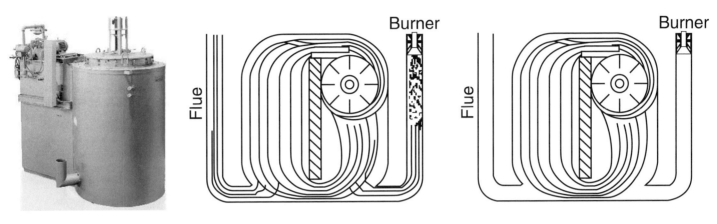

Fig. 9 Gas-fired furnace and its flow pattern. Source: Ref 7

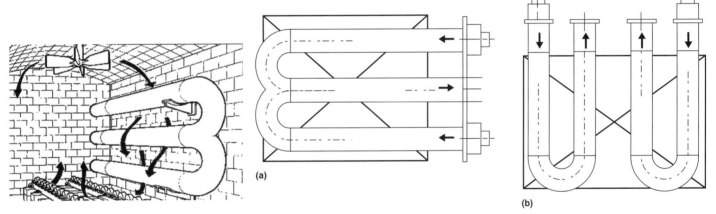

Fig. 10 Wind flow pattern of the trident radiant tube. Source: Ref 9

Fig. 11 Work zone of radiant-tube heating

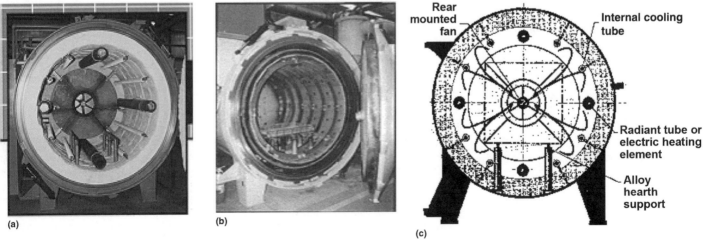

Fig. 12 Arrangement and flow pattern of electrically heated furnace

- Set the thermal schedule temperature as a function of time for a specific heat treatment case.
- Measure the temperature at the set point periodically.
- Compare with the set point temperature in the thermal schedule specified in advance.
- Heat up continuously if the measured temperature is lower than the setting temperature in the thermal schedule; otherwise, turn the heating off.
- Check the thermal schedule time at a predetermined interval for another measurement.

There is no direct control of the part temperature, and the quality of heat treatment is ensured according to the furnace performance, knowledge of the thermal behavior of the parts, and thermal schedule setting.

Some furnaces may use thermistors or infrared devices to monitor and control the temperature. The more complicated control methods, such as the proportional-integral method or the proportional-integral-derivative method, have been applied to the real heating control procedure.

Part Types and Load Pattern

When the heat transfer in the furnace is analyzed, both parts and fixtures should be included in the temperature model. In the heating process, various kinds of parts are heat treated, such as gears, bolts, shafts, flanges, blades, and even steel rod coils. To ensure a good heat treat quality, the parts are arranged into a specified load pattern in a tray and/or a fixture. Several typical trays/fixtures are shown in Fig. 13. These trays/fixtures must be designed into different sizes to meet the requirements of holding the parts.

Figure 13(a) shows a bar frame-type basket, and Fig. 13(b) is a pit furnace basket. They are used in situations where the parts size does not permit them to be loaded directly on a furnace hearth, tray, or grid. Figure 13(c) is a typical carburizing furnace tray. Generally, a tray or grid should be of sufficient section size to provide reasonable part loading. When a part requires special positioning, a fixture such as in Fig. 13(d or e) can be used. The parts may range from simple shapes, such as round, square rectangular, or fluted bars, to extremely intricate shapes. Figure 13(d) is a fixture designed for heat treating shafts, and Fig. 13 (e) is a fixture designed for heat treating lawn mower blades. There are other types of fixtures to be found in shops.

Simulation of Heating

The model to calculate the thermal history of the parts is very important to understanding the heating process and to controlling the process parameters. Computational methods have been developed to predict the part temperature and the furnace temperature, which are two of the most significant indexes.

Flowchart for Temperature Calculation

Figure 14 shows a flow chart of temperature calculation (Ref 10). When the parts and furnace condition are specified, the part load design and thermal schedule determination can be conducted through the user interface. Then, the furnace temperature and part temperature can be estimated. The system includes a part information database (Workpiece DB), a furnace information database (Furnace DB), a knowledge base representing the relationships of heat treatment requirements (Knowledge base/H.T. Spec DB), and part material properties (Material DB). It should be noted that when parts are in different locations of the furnace, the radiation view factors are different. Therefore, the part thermal history is different for different parts.

Part Temperature Calculation

There are four main factors affecting part temperature in heat treatment: part properties, part loading, part materials, and furnace condition and control, as shown in Fig. 15. The part thermal history in the heat treating furnace is a three-dimensional conduction heat-transfer process with hybrid convection/radiation boundary conditions. It depends on the furnace temperature, working conditions, and part properties. Part load patterns and the thermal profile design are the main tasks in the heat treatment process planning to guarantee the heat treatment quality. To simplify the temperature calculation, three basic assumptions are made:

- The austenitization of parts contributes only to the change of specific heat.
- Air temperature in the furnace space is uniform.
- Convection heat transfer on the surface of a single part is uniform.

Boundary Conditions for Heat Transfer to the Parts. In a heat treating process, parts are subjected to both convection and radiation conditions. The energy balance equation of the part is:

$$E_{storage} = E_{convection} + E_{radiation} \qquad (Eq\ 14)$$

where $E_{storage}$ is the heat stored in the part, and $E_{convection}$ and $E_{radiation}$ are the heat obtained from convection and radiation heat transfer, respectively.

Let the volume and surface area of the part be V and A, respectively. The energy terms in Eq 14 can be calculated using the following equations:

- Radiation at heat-source temperature, T_{heater}:

$$E_{radiation} = \varepsilon \cdot \sigma \cdot F \cdot A \cdot \left(T_{fce}^4 - T^4\right) \qquad (Eq\ 15)$$

where ε is the emissivity, σ is the Stefan-Boltzmann constant, F is the view factor between the furnace wall and the part, T_{fce} is the furnace temperature in Kelvin, and T is the temperature of the part.

- Convection under the ambient temperature, T_{fce}:

$$E_{convection} = h \cdot A \cdot (T_{fce} - T) \qquad (Eq\ 16)$$

where h is the convection heat-transfer coefficient, and T is the part temperature.

Natural Convection. If there is no recirculating fan in the furnace, the problem is one of natural convection. There is no specific effect of part arrangement on the natural convection heat-transfer calculation. In the next sections, the calculation of the equivalent length of the part, L^*, and the Nusselt number are presented first, and then some effect factors of the convection coefficient are discussed.

Calculations of L^* and Nu_{L*} with Specified Part Shapes. Figure 16 shows various part shapes with their arrangements. The equivalent length can be calculated by (Ref 5):

$$L^* = A^{1/2} \qquad (Eq\ 17)$$

where A is the surface area of the part (in meters squared).

The Nusselt number is calculated by (Ref 5):

$$Nu_{L*} = Nu_{L*}^0 + \frac{0.67 \cdot G_{L*} \cdot Ra_{L*}^{1/4}}{\left[1 + \left(\frac{0.492}{Pr}\right)^{9/16}\right]^{4/9}} \quad (0 < Ra_{L*} < 10^8) \qquad (Eq\ 18)$$

where Nu_{L*}^0 and G_{L*} are the conduction limit Nusselt number and the geometric parameter, respectively. Table 1 lists the values of these two constants for the parts with shapes and orientations shown in Fig. 16.

The Prandtl number, Pr, is expressed as:

$$Pr = \frac{\nu}{\alpha} \qquad (Eq\ 19)$$

where ν is the kinematic viscosity of the furnace gas (in m^2/s); α is the thermal diffusivity of the furnace gas (in m^2/s) and $\alpha = k/(\rho \cdot c_p)$, where k is the conductivity (in W/m·K), ρ is the density (in kg/m^3), and c_p is the specific heat of the gas (in J/kg·K). Table 1 lists a set of typical values of these numbers.

Ra_{L*} is the Rayleigh number:

$$Ra_{L*} = \frac{g \cdot \beta \cdot (T_w - T_\infty) \cdot L^{*3}}{\alpha \cdot \nu} \quad (0 < Ra_{L*} < 10^8) \qquad (Eq\ 20)$$

where g is the gravitational acceleration (in m/s^2); β is the coefficient of volumetric thermal expansion for the ideal gas (in K^{-1}), where $\beta = 1/T$ (Charles' law of volume at constant pressure. T is in Kelvin); T_w is the sphere surface temperature (in Kelvin); T_∞ is the free-stream temperature (in Kelvin); and L^* is the equivalent length (in meters).

General Equation for Any Shape of Parts. Because the values in Table 1 do not vary appreciably, a general expression based on the

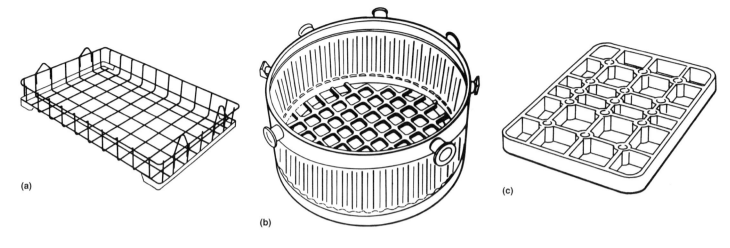

Fig. 13 Tray/fixture types for loading workpieces

average values of $Nu_{L^*}^0$ and G_{L^*} and valid for $Pr \geq 0.7$ is:

$$Nu_{L^*} = 3.47 + 0.51 \cdot Ra_{L^*}^{1/4} \qquad \text{(Eq 21)}$$

where Ra_{L^*} can be calculated using Eq 20.

Forced Convection. There are two situations in forced convection heat transfer: single parts with specified shapes and an array of parts. In both cases, the calculation of equivalent length, L^*, can use Eq 17. However, there are various methods for calculating the Nusselt number that depend not only on the medium properties but also the parts arrangement, that is, part-load pattern.

Calculation of Nu_{L^*} for Specified Part Shapes. For a single part with the specified shapes in Fig. 16, the following experimental equation can be used to calculate the Nusselt number, Nu_{L^*} (Ref 5):

$$Nu_{L^*} = Nu_{L^*}^0$$
$$+ \left[0.15 \cdot \left(\frac{p}{L^*} \right)^{1/2} \cdot Re_{L^*}^{1/2} + 0.35 \cdot Re_{L^*}^{0.566} \right] \cdot Pr^{1/3}$$
$$\text{(Eq 22)}$$

where p is the maximum perimeter of the part perpendicular to the flow direction, U_∞ (U_∞ is the velocity of flow, in m/s); and $Nu_{L^*}^0$ is the overall Nusselt number in the no-flow (pure conduction) limit; representative values of this constant are listed in Table 1. Pr is the Prandtl number, and Re_{L^*} is the Reynolds number that is equal to:

$$Re_{L^*} = \frac{U_\infty \cdot L^*}{\nu} \qquad \text{(Eq 23)}$$

Note that Eq 22 is recommended under the following conditions:

$$0 < Re_{L^*} < 2 \times 10^5, Pr > 0.7, 0 < C/B$$
$$< 5 \text{(when spheroid)} \qquad \text{(Eq 24)}$$

where C and B are geometrical parameters for spheroids, as shown in Fig. 16.

For the case of multiple parts, either aligned or staggered, as in Fig. 17, the following calculation of the Nusselt number applies.

Assume that the air fluid velocity is U_∞ and the temperature of flow is T_∞, the transverse

and longitudinal pitches are S_T and S_L, respectively, and the number of rows of parts transverse to the flow is N. For convenience, the equivalent length is called the equivalent diameter, D^*. The calculation of D^* is the same as L^*.

For the calculation of Nu_{D^*}, there are two different equations according to the number of rows of parts (Ref 11):

$$\begin{cases} Nu_{D^*}^N = \frac{1+(N-1)\cdot\Phi}{N} \cdot Nu_{D^*}^1 & (N < 10) \\ Nu_{D^*}^N = \Phi \cdot Nu_{D^*}^1 & (N \geq 10) \end{cases} \qquad \text{(Eq 25)}$$

where N is the number of rows of parts transverse to the flow, and Φ is an arrangement factor expressed as follows:

$$\begin{cases} \phi_{\text{aligned}} = 1 + \frac{0.7}{\Psi^{1.5}} \cdot \frac{S_L/S_T - 0.3}{(S_L/S_T + 0.7)^2} & \text{for aligned parts} \\ \phi_{\text{staggered}} = 1 + \frac{2}{3P_L} & \text{for staggered parts} \end{cases}$$
$$\text{(Eq 26)}$$

where S_L and S_T are the on-center distances between parts, as shown in Fig. 17; and ψ is a factor defined as:

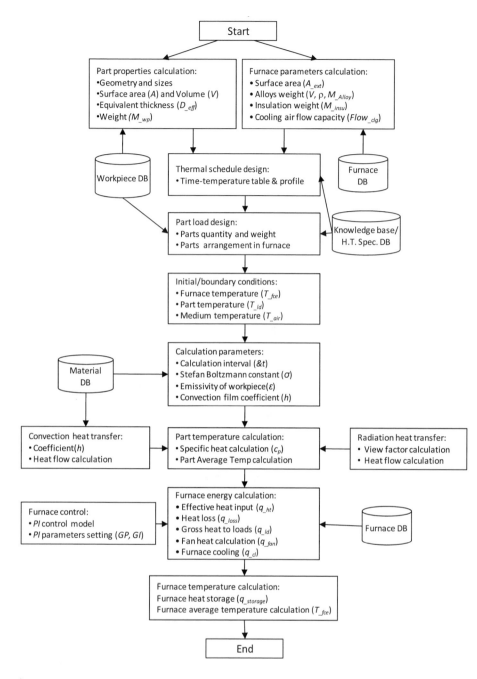

Fig. 14 Flowchart for temperature calculation of loaded furnace. Source: Ref 10

Uniform Temperature within the Part with Lumped-Mass Assumption. Heat storage inside the part is expressed as:

$$E_{\text{storage}} = \rho \cdot c_p \cdot V \cdot \frac{dT}{d\tau} \tag{Eq 30}$$

where ρ is the density of the part, c_p is the specific heat, and τ is the time.

For the part with a small Biot number, a stronger lumped-mass assumption can be made that the temperature distribution in the part is uniform. The Biot number, Bi, is a dimensionless number used in non-steady-state heat-transfer calculations and is defined as:

$$\text{Bi} = \frac{hL_c}{k} \tag{Eq 31}$$

where h is the heat-transfer coefficient, L_c is the characteristic length of the part, and k is the thermal conductivity of the part.

This number determines whether or not the temperature inside a body varies significantly in space. The assumption means that the conduction inside the part is much faster than the convection outside. Because of the good thermal conductivity of most metals, this is reasonable when the size of the part is small and the surface-to-volume ratio is large.

Combine Eq 15, 16, and 30 into Eq 14, then:

$$\rho \cdot c \cdot V \cdot \frac{dT}{d\tau} = h \cdot A \cdot (T_{\text{fce}} - T) + \varepsilon \cdot \sigma \cdot F \cdot A \\ \cdot (T_{\text{fce}}^4 - T^4) \tag{Eq 32}$$

Because the part is taken as the lumped-mass body, an equivalent part or "visual part" (i.e., the radiation from one surface can reach the other surfaces along a straight line) is applied to represent the real part, as shown in Fig. 18. In this way, the calculation can be further simplified.

Let $V = A \cdot D_{\text{equ}}$, where D_{equ} is the equivalent thickness of the part. Applying the finite-difference method to the left side of Eq 32, then:

$$\frac{\partial T}{\partial \tau} = \frac{T^{k+1} - T^k}{\delta \tau} + O(\delta \tau) \tag{Eq 33}$$

where k is the time step. By combining Eq 32 and 33, at time step k, the equation becomes:

$$T^{k+1} = T^k + \frac{\delta \tau}{\rho \cdot c \cdot t_{\text{equ}}} \left[h \cdot (T_{\text{fce}}^k - T^k) \right. \\ \left. + \varepsilon \cdot \sigma \cdot F \cdot \left((T_{\text{fce}}^k)^4 - (T^k)^4 \right) \right] \tag{Eq 34}$$

Equation 34 can be used to calculate the part temperature with uniform temperature distributions at any time in a heat treating cycle.

Nonuniform Temperature within the Part. The partial differential conduction equation has been built and solved for the temperature distribution on other parts where the geometries must be considered.

$$\begin{cases} \Psi = 1 - \frac{\pi}{4P_T} & P_L \geq 1 \\ \Psi = 1 - \frac{\pi}{4P_T \cdot P_L} & P_L < 1 \end{cases} \tag{Eq 27}$$

where P_T is the transverse pitch, and $P_T = S_T / D^*$; P_L is the longitudinal pitch, and $P_L = S_L / D^*$; and $Nu_{D^*}^1$ is the Nusselt number for the first row. Use Eq 22 to calculate $Nu_{D^*}^1$. In calculating Re_{D^*}, the maximum velocity of fluid in the space between parts is used:

$$Re_{L^*} = \frac{U_{\text{max}} \cdot D^*}{\nu} \tag{Eq 28}$$

where $D^* = A^{1/2}$ is the equivalent diameter.

The maximum fluid velocity, U_{max}, is calculated as follows (Ref 12):

$$\begin{cases} U_{\text{max}} = \frac{U_\infty \cdot S_T}{S_T - D^*} & \text{for aligned parts} \\ U_{\text{max}} = \frac{U_\infty \cdot S_T}{S_T - D^*} & \text{when } S_{D^*} > \frac{1}{2}(S_T + D^*) \quad \text{for staggered parts} \\ U_{\text{max}} = \frac{U_\infty \cdot S_T}{2(S_D - D^*)} & \text{when } S_{D^*} < \frac{1}{2}(S_T + D^*) \\ S_{D^*} = \left[S_L^2 + \left(\frac{S_T}{2} \right)^2 \right]^{1/2} \end{cases} \tag{Eq 29}$$

where U_∞ is the average fluid velocity in the furnace working space, and S_L, S_T, and D^* are as shown in Fig. 17.

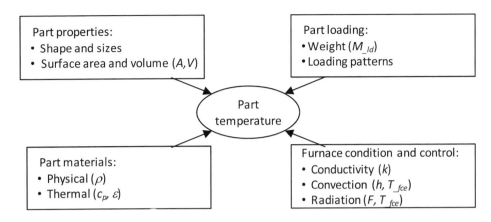

Fig. 15 Factors affecting part temperature

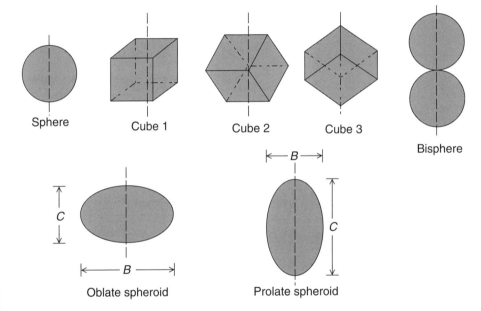

Fig. 16 Natural convection for various shapes of parts

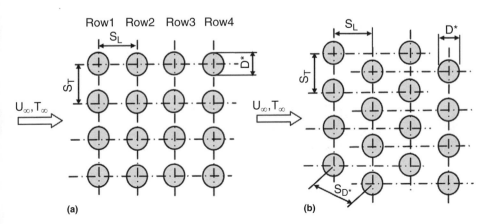

Fig. 17 Configurations of arrays of parts

In Cartesian coordination, the heat-conduction equation is shown as follows:

$$\rho c_p \frac{\partial T}{\partial t} = \boxed{?} \cdot (\lambda \nabla T) + \dot{Q} \qquad \text{(Eq 35)}$$

where ∇ is the Laplace operator.

Analytical Solution. In some cases with simple geometries and temperature-independent physical properties, the analytical solution can be obtained. For example, when one-dimensional unsteady heat conduction happens with no source and constant thermal properties, the equation is as follows:

$$\rho c_p \frac{\partial T}{\partial t} = \lambda \cdot \frac{\partial^2 T}{\partial x^2} \qquad \text{(Eq 36)}$$

If a certain temperature is set as the boundary condition, the solution of the temperature inside for a one-dimensional infinite plane is expressed as (Ref 1):

$$T = \frac{T_0 \; \text{erfc} \; x}{2\sqrt{\lambda t}} \qquad \text{(Eq 37)}$$

where T_0 is the initial temperature through the part, and erfc is the complementary error function.

This analytical method has strong limitations, as previously mentioned. The numerical method is used instead to simulate the temperature field inside the part for more realistic and more complex cases.

Numerical Solution. Equation 35 can be solved by using numerical computation based on classic finite-difference or finite-element methods (Ref 13), with the certain discretization and boundary conditions that come from

Table 1 Typical values of Nu_{L*}^0 and G_{L*}

Body shape	Nu_{L*}^0	G_{L*}
Sphere	3.545	1.023
Bisphere	3.475	0.928
Cube 1	3.388	0.951
Cube 2	3.388	0.990
Cube 3	3.888	1.014
Vertical cylinder	3.444	0.967
Horizontal cylinder	3.444	1.019
Cylinder at 45 °C ($H = D$)	3.444	1.004
Prolate spheroid ($C/B = 1.93$)	3.566	1.012
Oblate spheroid ($C/B = 0.5$)	3.529	0.973
Oblate spheroid ($C/B = 0.1$)	3.342	0.768

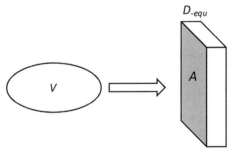

Fig. 18 Equivalent part

the discussion in the section "Boundary Conditions for Heat Transfer to the Parts" in this article.

Furnace Temperature Calculation

The last step in the calculation flowchart (Fig. 14) is the furnace temperature calculation.

Mathematical Model. The temperature history of the loaded furnace in the heat treating cycle is a hybrid convection/radiation heat-transfer process. According to the heat-transfer theory and energy-balance equation, the heat exchange of the loaded furnace can be mathematically presented in following model:

$$q_{\_storage} = \sum_{i=1}^{N} \left(\rho_i \cdot c_{pi} \cdot V \cdot \frac{\partial T_i}{\partial \tau} \right) \quad \text{(Eq 38)}$$

where N is the quantity of accessories in the furnace.

All the accessories are assumed to be at the same temperature, and the temperature distribution inside is assumed to be uniform. Combining Eq 33 and 38 and letting the time step be k and the time calculation interval be $\delta\tau$, the equation can be rearranged in the time step $(k + 1)$ for the average furnace temperature:

$$T_{\_fce}^{k+1} = T_{\_fce}^{k} + \frac{q_{\_storage}^{k} \cdot \delta\tau}{\sum_{i=1}^{N} (\rho_i \cdot c_{pi} \cdot V)} \quad \text{(Eq 39)}$$

where $q_{\_storage}$ is the average storage of heat flux in the loaded furnace; $\delta\tau$ is the time interval; and ρ_i, c_{pi}, and V_i are the density, specific heat, and volume, respectively, of accessory i in the furnace.

These accessories include all the components involved in the heat treating process except for the parts. Equation 39 can be used to calculate the furnace average temperature at time step $(k + 1)$.

Factors Affecting Furnace Temperature Calculation. From Eq 39, the main factors affecting the furnace temperature are material properties of loads and heat storage. Loads include parts and all the furnace accessories, and heat storage is related to the furnace heat input, furnace control method, and furnace heat loss, as shown in Fig. 19.

Loads Model for Temperature Calculation. In a loaded furnace, many components are involved in the heat treating cycle besides parts, including furnace component alloys, insulation, and heating elements, as shown in Fig. 20. In a broad sense, they are called loads. Generally, the nonpart loads are divided into three types:

- Furnace accessories (predominantly metal alloy), including grate, firing ring, U-tube, fixture, roller, fan and diffuser
- Heating elements
- Insulation

The total heat flux, $q_{\_acc}$ is:

$$\Sigma(\rho_i \cdot c_{pi} \cdot V_i) = c_{p\_alloy} \cdot m_{\_alloy} + c_{p\_htr} \cdot m_{\_htr} + c_{p\_insul} \cdot m_{\_insul}$$

$$\text{(Eq 40)}$$

where $m_{\_alloy}$, $m_{\_htr}$, and $m_{\_insul}$ are the weight of the furnace alloy, heater, and insulation; and c_{p_alloy}, c_{p_htr}, c_{p_insul} are their specific heat, respectively.

Heat Storage in Furnace. Heat storage is the summation of all the heat energy obtained by the furnace:

$$q_{\_storage} = q_{\_ht} - q_{\_loss} - q_{\_ld} - q_{\_cl} + q_{\_fan} \quad \text{(Eq 41)}$$

where $q_{\_ht}$ is the effective heat input, $q_{\_loss}$ is the heat loss; $q_{\_ld}$ is the heat to the workpieces, $q_{\_cl}$ is the cooling energy, and $q_{\_fan}$ is the heat input by fan.

These heat energy terms can be calculated by using the following ways:

- The heat to parts, $q_{\_ld}$, is mainly subject to the hybrid convection and radiation heat transfer. It can be calculated as:

$$q_{\_ld} = k_{\_ld} \cdot \frac{(dT_{\_ld})}{dt}$$
$$= h \cdot (T_{\_fce}) - T_{\_ld} + \sigma \cdot \varepsilon \cdot (T_{\_fce})^4 - T_{\_ld}^4 \quad \text{(Eq 42)}$$

- The heat loss, $q_{\_loss}$, depends on the design factor for the heat loss of the furnace and the surface area of the furnace. The following is an empirical equation:

$$q_{\_loss} \, DF \cdot (a_{\_loss} + b_{\_loss} \cdot T_{\_fce}) \cdot A_{\_ext} \quad \text{(Eq 43)}$$

where DF is the design factor, $a_{\_loss}$ and $b_{\_loss}$ are the empirical constants, and $A_{\_ext}$ is the surface area of the furnace.

- The fan heat input, $q_{\_fan}$, is calculated based on an empirical equation:

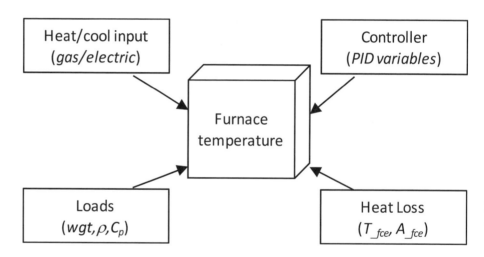

Fig. 19 Factors that affect the furnace temperature

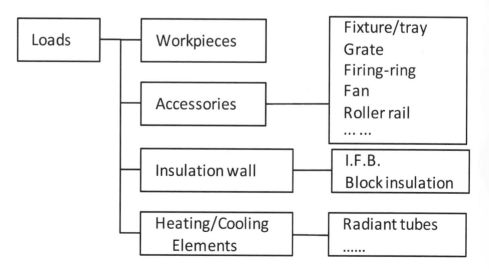

Fig. 20 Loads in a heat treating furnace

$$q_{\_fan} = P_{\_fan} \cdot \left(\frac{a_{\_fan}}{b_{\_fan} + T_{\_fce}} \right) \qquad \text{(Eq 44)}$$

where $P_{\_fan}$ is the power of the recirculating fan (in Watts), $a_{\_fan}$ and $b_{\_fan}$ are the empirical constants, and $T_{\_fce}$ is the furnace temperature (in °C).

- The effective heat input, $q_{\_ht}$, is:

$$q_{\_ht} = Htg \cdot AHC \cdot q_{\_conn} \qquad \text{(Eq 45)}$$

where Htg is a coefficient adjusted by using a certain control strategy; AHC is the corrected coefficient to the gross heat input, which is furnace-specific; and $q_{\_conn}$ is the nominal power from the heating elements.

Cooling Air Input. There are two empirical equations used in heating and cooling processes:

- At heating process:

$$q_{\_cl} = Clg \cdot Flow_{\_clg} \cdot k10.019 \cdot (T_{\_fce}) - T_{\_clg} \qquad \text{(Eq 46)}$$

where $Clg = Htg$ is the adjustment for cooling air input, and $Flow_{\_clg}$ is the cooling air flow capacity:

$$\frac{Flow_{\_clg} = q_{\_conn}}{1000} \cdot k29.44 \cdot (1 + xs_{\_air}) \qquad \text{(Eq 47)}$$

- At cooling process:

$$q_{\_cl} = \frac{Tubes}{k3\ 6.718e - 6 + \frac{k47.582}{(T_{\_fce})^{\frac{5}{2}}}} \qquad \text{(Eq 48)}$$

where $k1$, $k2$, $k3$, and $k4$ are constants, and $Tubes$ is the number of effective cooling tubes.

Furnace Temperature Control Model. The typical temperature control method in the furnace is a proportional, integral, and derivative (PID) controller. Some research work has been done on the furnace temperature control (Ref 14, 15). When the measured temperature is different from the set temperature, the heating or cooling input is controlled to minimize the error. It is a typical feedback control system, as shown in Fig. 21.

Generally, the PID method uses the following equation to control the loop:

$$u = P \cdot \left[(T_s - T_0) + D \cdot \frac{d}{dx}(T_s - T_0) \right.$$
$$\left. + I \cdot \int (T_s - T_0)dt \right] \qquad \text{(Eq 49)}$$

where T_s is the set point temperature, T_0 is the furnace temperature, the error value $e = (T_s - T_0)$; and P, D, and I are the proportional gain, damping, and integral gain, respectively.

In a PID control process, it is important to adjust the P, D, and I components, tuning these constants so that the weighted sum of the proportional, integral, and derivative terms produces a controller output that steadily drives the process variable in the direction required to minimize the error.

There are several methods that can be used to determine the P, D, and I values. One of them is the Ziegler and Nichols approach (Ref 16), which is a practical method of estimating the values of K, T, and d experimentally. K is the process gain used to represent the magnitude of the controller effect on the process variable, T is the process time constant used to represent the severity of the process lag, and d is the dead time used to represent another kind of delay present in many processes, where the sensor used to measure the process variable is located some distance from the actuator used to implement the controller corrective efforts. With the controller in manual mode (no feedback), a step change is included in the controller output. Then, the process reaction is analyzed graphically (Fig. 22). The process gain, P, can be approximated by dividing the net change of the process variable by the size of the step change generated by the controller. The dead time is estimated from the interval between the controller step change and the beginning of a line drawn tangent to the reaction curve at its steepest point. Ziegler and Nichols also used the inverse slope of that line to estimate the time constant, T:

$$\begin{cases} P = \frac{1.2T}{K \cdot d} \\ I = \frac{0.6T}{K \cdot d^2} \\ D = \frac{0.6T}{K} \end{cases} \qquad \text{(Eq 50)}$$

Sometimes, the sensor measuring the oven temperature is susceptible to other electrical interference, a derivative action that can cause the heater power to fluctuate wildly. In this case, $D = 0$, and a *PI* controller is often used instead of a *PID* controller. The application of *PD* control in a furnace is to control heat/cool gases input as well as energy input.

The adjustable coefficient of heat input, Htg, depends on the control strategy of heating processes according to furnaces. The typical expression of Htg is shown in the following equation:

$$Htg = \frac{PG}{Span} \cdot (T_{\_sp}) - T_{\_fce}$$
$$+ \frac{PG \cdot IG}{Span} \cdot \int_0^\tau (T_{\_sp}) - T_{\_fce}dt \qquad \text{(Eq 51)}$$

The proportional gain, PG, and the integral gain, IG, can be preset based on previous experience.

Model Verification and Case Studies

Heating simulations can be validated by comparison with measured results in full-scale furnaces. Several case studies have been conducted and are presented to illustrate the use of simulations.

Alumina Rod in Random Load Pattern

A large number of alumina rods were loaded randomly in a container. The effective thermal conductivity was first calculated based on measured values, and then the model was used to predict the heating of new loads.

Experimental and Simulated Calculation of the Effective Thermal Conductivity. The test was run in a small, electrically heated Allcase (Surface Combustion, Inc.) furnace. The material was loaded in a round stainless steel drum measuring 267 mm in diameter by 298 mm high. Thermocouples were buried at various points in the load, and the temperatures were recorded against time. The locations of the thermocouples are shown in Fig. 23. The material in test 1 contained Al_2O_3 in the form of small cylinders, as shown in Fig. 24. The furnace was heated to 900 °C (1650 °F), and then

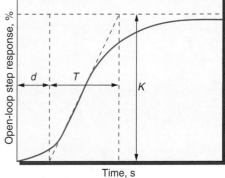

Time, s

Fig. 22 Ziegler-Nichols reaction curve. Adapted from Ref 16

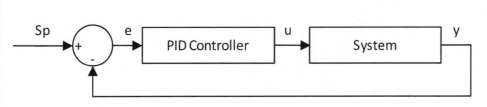

Sp + e → PID Controller → u → System → y

Fig. 21 Control feedback loop of proportional, integral, and derivative (PID) control method. *Sp*, set point; *e*, error; *u*, output; *y*, measured temperature

the load was put inside. The profile charts of the recorded temperature data are shown in Fig. 25.

To calculate the effective thermal conductivity, a two-dimensional finite-difference enmeshment was used because of symmetry in the circumferential direction of the whole load, as shown in Fig. 26. The measured data of the surface elements (thermocouples 1, 2, 4, 5, and 6) define the boundary conditions. The temperatures of other surface elements are interpolated from the previously measured surface temperatures. An initial value of effective

thermal conductivity is assumed. Then, thermal diffusivity can be assumed as constant.

Next, the central element temperature is calculated based on the virtual conductivity and measured surface temperatures. This calculated temperature is compared to the measured value (TC3) of the center element; then, the effective thermal conductivity is adjusted by an iteration equation until the error comes to an acceptable level. In the calculation, the initial temperature of all elements is 25 °C (77 °F), and the surface element temperature is known from measured results.

For alumina rods with thermal diffusivity α = 1.7 cm^2/min, the error between the calculated temperature and the measured one of the center element is smallest, as shown in Fig. 27.

For α = 1.7 cm^2/min, if the effective density ρ_{eff} = 1500 kg/m^3 and the specific heat c = 850 J/kg·K, then the effective thermal conductivity is λ_{eff} = 3.6 W/m-K. (The thermal conductivity of alumina is 26.9 to 27.6 W/m-K.)

Prediction of Heating Time for New Loads. The operating condition is defined for the new load. The furnace temperature is 900 °C (1650 °F), and the initial load temperature is 20 °C (68 °F). The container is 320 mm in diameter and 402 mm tall, with a wall of 3 mm. The heating interval was chosen to be when the final core temperature reached 550 °C (1020 °F).

The effective thermal diffusivity is used to calculate the heating process of the new load. Assuming emissivity = 0.8, the heating curves are calculated and plotted in Fig. 28. From the curve of the core, it can be seen that the core temperature reaches 550 °C (1020 °F) after 80 min. Thus, t_3 = 550 °C = 80 min.

Steel Parts in Arranged Load Pattern

The furnace is a vacuum furnace, and the soaking temperature is approximately 980 °C (1800 °F).

Part and Process Conditions. The workpieces in this case study are shown in Fig. 29. The material is Kovar (Fe-29Ni-17Co), and the weight of each part is 0.5 kg (1.03 lb). The furnace used is a vacuum furnace, shown in Fig. 30. The furnace parameters are listed in Table 2.

The parts are arranged in the aligned manner, as shown in Fig. 31. The details of the parts,

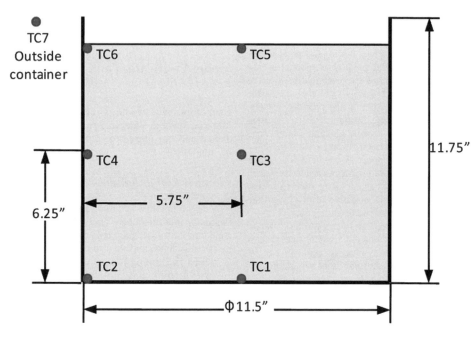

Fig. 23 Positions of thermocouples

Fig. 24 Alumina rods and the load

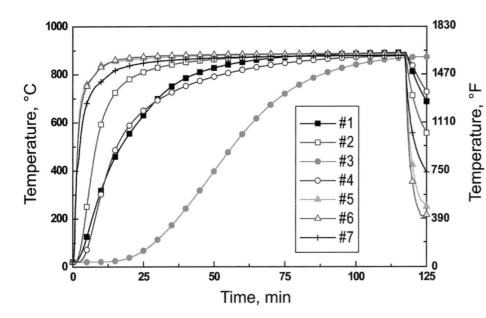

Fig. 25 Measured thermal profile for alumina rods

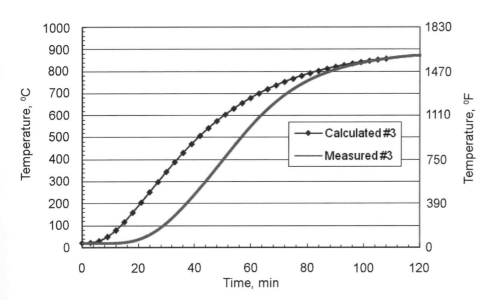

Fig. 27 Comparison of calculated and measured temperatures of alumina rods during heating (thermal diffusivity, α = 1.7 cm²/min)

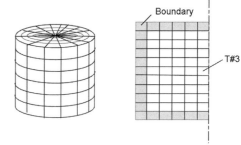

Fig. 26 Finite-difference enmeshment of cylindrical load

Part and Process Conditions. The parts are flat rings (Fig. 33) of AISI 1008 steel. The weight for each part is 0.525 lb, with a 2 in. internal diameter, 3.5 in. external diameter, and 0.3 in. height. The tempering study was carried out at 177 °C (350 °F) in the furnace shown in Fig. 34. The furnace parameters are listed in Table 4. The parts were heated in the presence of natural gas combustion products. The rings were mounted on the rod, as shown in Fig. 35. The details of the load pattern are listed in Table 5. In the tempering process, the parts were heated to 175 °C (350 °F) and soaked for 60 to 90 min.

Calculation and Discussion. The furnace is a direct gas-fired furnace. The combustion chamber is at the bottom of the furnace, and the hot combustion gas is blown into the load chamber through the side wall of the furnace by a fan. The flow propelled by the fan can be calculated from the fan information for an unrestricted case, but the flow in the chamber is not the same. Therefore, a coefficient, a, is used to calculate the fraction of gas flow in the load chamber from the combustion chamber:

$$a = \frac{U_{\text{flow}_{\text{load}_{\text{chamber}}}}}{U_{\text{flow}_{\text{combustion}_{\text{chamber}}}}} \qquad \text{(Eq 52)}$$

The calculated results are given for a equal to 0.05, 0.1, and 0.2, as shown in Fig. 36. It can be seen that there is almost no difference between parts that are fastest and slowest to heat. As a increases, the heating rate of the parts increased. However, the increase is not significant. Therefore, the soaking time of approximately 80 min for the parts is reasonable in industrial production.

Summary

A comprehensive model of the heating process in the furnace involves accounting for complicated furnace control, gas convection, radiation coupling, and heat conduction inside parts, as well as the part-loading patterns. Assumptions and simplifications have been applied to calculate the heating process in a furnace probably loaded with many small parts. The temperature distribution and evolution in

fixtures, and load pattern are listed in Table 3. As seen (Fig. 31a), the parts are arranged in such a manner so that they can be properly heat treated and do not merge into one another after heating. Thermocouples are located as shown in Fig. 31(b), with one in the load center and one at the corner of the load.

Calculation and Discussion. The comparison of the calculated and measured values at the load center and corner points is given in Fig. 32. The curves show that there is not much difference between the measured temperature values of the slowest and fastest parts to heat and the calculated temperature values of the slowest and fastest parts to heat in the system.

There is some overlap between parts because of their rims, but the view factor calculation is based on the assumption of no overlap.

Tempering Process

Tempering is a common heat treatment activity. The tempering process time is determined based on experience, so it is important to examine a case study on tempering. Usually, tempering is conducted at relatively low temperatures, where convection plays a more dominant role than radiation. In this section, a tempering case study is considered to validate the convection model.

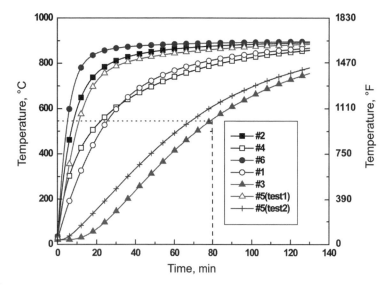

Fig. 28 Calculated temperature profiles of surfaces and core ($t_3 = 550\ ^\circ C = 80$ min)

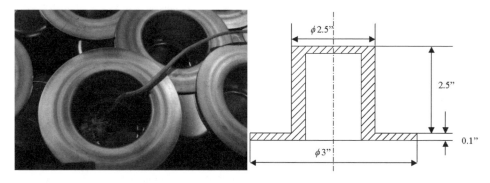

Fig. 29 Part shape and size of steel parts, case study 2

Fig. 30 Furnace used in case study 2 of steel parts

Table 2 Furnace data

Layout	Horizontal
Total size	ϕ 1.2 × 1.5 m (4 × 5 in.)
Part	61 × 61 × 122 cm (24 × 24 × 48 in.)
Heat input	280 kW
Heating elements (total weight)	80 kg (176 lb)
Supports	18 kg (40 lb)
Roller rails	68 kg (150 lb)
Insulation	Front, side, and back walls: 1.3 cm (0.5 in.) graphite, 1.3 cm (1.5 in.) Kaowool, 6.4 cm (2.5 in.) graphite

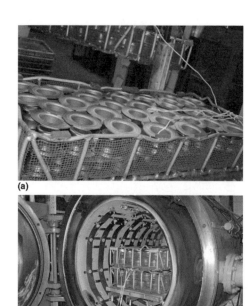

Fig. 31 Load pattern and arrangement of thermocouples

Table 3 Load pattern

Each fixture weight	14 kg (30 lb), rectangular (netlike)
Each fixture size	109 × 56 × 18 cm (43 × 22 × 7 in.)
Fixture configuration	1 row, 1 column, 2 layers
Parts configuration in each fixture	12 rows, 6 columns, 2 layers
Total number of parts in each fixture	144
Total number of parts	288
Total weight of parts in each fixture	67 kg (148.32 lb)
Total weight of parts	135 kg (296.64 lb)

the furnace and in the parts has been predicted and compared to measured values.

Based on the temperature simulation, the recipe for the heating process can be optimized to shorten the heating time, increase the efficiency, and save energy without reducing the quality of the parts.

This article concentrates on the heating process of a loaded furnace with more than many parts. The simplification has been made statistically to avoid the details of part geometry and exact positions involved, which could make the calculation impossible. In the case of large parts, geometry and positioning play an important role in the thermal response, and a three-dimensional model must be built using finite-element modeling or computational fluid dynamics software packages, such as Fluent or Ansys.

The heating process modeling and control are based on a furnace-specific design that must consider the subtleties of furnaces and practical control effectiveness. This article tries to summarize common heat treating practices in terms of the functionality for general furnaces, not for a specific furnace or process. If the specific information on a particular furnace and process is involved in the modeling and calculation, better prediction results can be expected but may not be useful for other processes.

ACKNOWLEDGMENT

Special thanks to the member companies of the Center of Heat Treating Excellence at Worcester Polytechnic Institute for the great support and the constructive suggestions. In particular, thanks go to Surface Combustion for many valuable inputs during the study of furnace modeling.

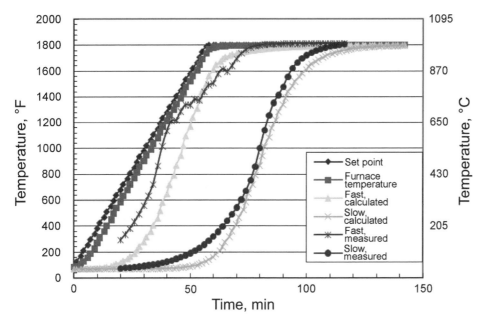

Fig. 32 Comparison of the measured and calculated temperature values for parts that heat slowly and parts that heat quickly

Fig. 35 Arrangement of load in tempering furnace

Fig. 33 Flat rings for tempering, case study 3

Table 4 Tempering furnace specifications

Type	Horizontal (gas fired)
Total size	142 × 94 × 97 cm (56 × 37 × 38 in.)
Workspace	61 × 122 × 61 cm (24 × 48 × 24 in.)
Heat input	900,000 Btu/h
Roller rails	20 kg (43 lb)
Insulation	Top fiber: 23 cm (9 in.)
	Side fiber: 23 cm (9 in.)
	Bottom: 9 cm (3.5 in.) tile, 8 cm (3 in.) brick, and 20 cm (8 in.) castable

Table 5 Load pattern

Each fixture weight	45 kg (100 lb), rectangular (netlike)
Each fixture size	61 × 122 × 15 cm (24 × 48 × 6 in.)
Fixture configuration	1 row, 1 column, 5 layers
Parts configuration in each fixture	44 rows, 8 columns, 1 layer
Total number of parts in a single fixture	352
Total number of parts	1760
Total weight of parts in a single fixture	83.8 kg (184.8 lb)
Total weight of the workpiece	419 kg (924 lb)

Fig. 34 Tempering furnace for case study 3

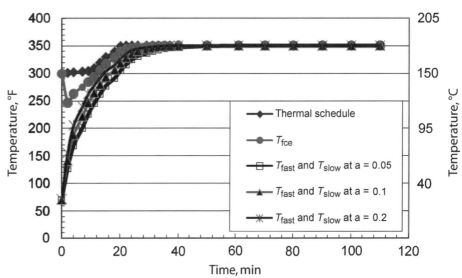

Fig. 36 Calculated results of tempering the ring

REFERENCES

1. J.P. Holman, *Heat Transfer,* 9th ed., The McGraw-Hill Companies, Inc., 2002
2. T.-M. Shih, *Numerical Heat Transfer,* Hemisphere Publishing Corporation, 1984
3. L.M. Jiji, *Heat Transfer Essentials—A Textbook,* Begell House, Inc., 1998
4. J.H. Lienhard, *A Heat Transfer Textbook,* Prentice-Hall, Inc.
5. A. Bejan, *Heat Transfer,* John Wiley & Sons, Inc., 1993
6. R.W. Lewis et al., *The Finite Element Method in Heat Transfer Analysis,* John Wiley & Sons, Inc., 1996
7. *Furnace Specifications,* Lindberg, Riverside
8. *Furnace Specifications, Surface Combustion,* Maumee
9. W.S. Janna, *Engineering Heat Transfer,* PWS Engineering, Boston
10. Y. Rong, Q. Lu, and J. Kang, "Development of an Analytical Tool for Part Load Design and Temperature Control within Loaded Furnace and Parts," CHTE Project Report, WPI, Worcester, MA, May 2001
11. A.F. Mills, *Heat and Mass Transfer,* Richard D. Irwin, Inc., 1995
12. F. Kreith, *The CRC Handbook of Thermal Engineering,* CRC Press, 2000
13. S.V. Patankar, *Numerical Heat Transfer and Fluid Flow,* Hemisphere Pub. Corp., Washington; McGraw-Hill, New York, 1980
14. W.D. Disler, Using On-Line Predictive Computer Modeling to Optimize Heat Treat Processing, *Proceedings of the Second International Conference on Carburizing and Nitriding with Atmospheres,* Dec 1995 (Cleveland, OH), p 29–34
15. W. Sobol et al., Fuzzy Logic Control of Industrial Heat Treatment Furnaces, *18th International Conference of the North American Fuzzy Information Processing Society,* 1999, p 839–843
16. A. O'Dwyer, *Handbook of PI and PID Controller Tuning Rules,* 2nd ed., Imperial College Press, 2006

ASM Handbook, Volume 22B, *Metals Process Simulation*
D.U. Furrer and S.L. Semiatin, editors

Copyright © 2010, ASM International®
All rights reserved.
www.asminternational.org

Simulation of Induction Heating Prior to Hot Working and Coating

Valery Rudnev, Inductoheat, Inc.

TEMPERATURE GREATLY AFFECTS THE FORMABILITY OF METALS. Heating of a component to temperatures that correspond to the plastic deformation range creates a favorable condition for metal to be subsequently forced by various means into a desired shape. Prior to hot forming (i.e., forging, upsetting, rolling, and extrusion), there are many ways to heat workpieces, including the use of induction heaters, gas-fired furnaces, infrared heaters, electric, and other fuel-fired furnaces. In recent decades, induction heating has become an increasingly popular choice among producers for heating metals prior to hot working. This tendency continues to grow at an increasing pace due to an ability of induction heating to create high heat intensity quickly, not just at the surface of a workpiece but internally as well. This leads to low process cycle time (high productivity) with repeatable high quality while using a minimum of shop floor space. Induction heating is more energy efficient and inherently environmentally friendlier than most other heat sources. (It is free from CO_2 emission.) A considerable reduction of heat exposure also contributes to the environmental friendliness. Induction heating offers other attractive features, such as:

- A measurable reduction of scale
- Short start-up and shut-down times
- Readiness for automatization with lower labor cost
- Ability to heat in a protective atmosphere if required

Other advantages of heating by induction are discussed in (Ref 1).

Workpieces

Steel components represent by far the majority of hot-worked workpieces, although other metals, including titanium, aluminum, copper, brass, bronze, tungsten, and nickel alloys, are also inductively heated and hot worked for a variety of commercial applications. The most popular metal hot working processes for which induction heating is applied are (Ref 1):

- *Forging:* Billets or bars are heated either fully (Fig. 1) or partially in cut lengths or continuously and are forged in presses, hammers (repeated blows), or upsetters (which gather and form the metal).
- *Forming:* Hot forming includes a variety of metalworking operations generally encompassing bending, expanding, spinning, and so on. The versatility of induction heating is that it can selectively heat through specific areas of the workpiece or can heat areas to different temperatures, providing desirable temperature gradients and heating profiles.
- *Extrusion:* Extruding is the process of forcing or squeezing metal through a die. Both ferrous and nonferrous metals are heated by induction prior to direct or indirect extrusion.
- *Rolling:* Bars, billets, rods, slabs, blooms, strips, blanks, and sheets are processed in rolling mills. These components are made from ingots or continuous cast metals and their alloys.

As expected, the goal of using induction heating in the aforementioned applications is to provide the metal workpiece at the hot working stage with the desired (typically uniform) temperature across its diameter/thickness as well as along its length and across the width or perimeter.

Required heating temperature depends on alloy and specifics of the metalworking process. Commonly required temperatures for selected ferrous and nonferrous alloys prior to hot working are shown in Tables 1 and 2, respectively (Ref 1–9). It is also understandable that some specific applications may require heating temperatures outside of those ranges.

Size and Type of Induction Heaters

The power ratings of induction heating machines range from less than 100 kW up to dozens of megawatts typically using low and medium frequency (60 Hz to 30 kHz). Cylindrical and rectangular solenoid (helical) multiturn induction coils are most often used in induction heating prior to metal hot working. Usually, the initial temperature of the workpiece prior to induction heating is uniform and corresponds to an ambient temperature. However, there are cases when an initial temperature is not uniform. Induction heaters installed between the continuous casting operation and rolling operation can serve as a typical example. Due to uneven cooling of different areas of the continuously cast workpiece (i.e., slab, transfer bar, or bloom) as it progresses during and after continuous casting toward the rolling mill, the surface layers and particularly the edge areas become appreciably cooler than the central and internal regions.

Fig. 1 Billets or bars are fully or partially heated in cut lengths or continuously and are forged in presses, hammers, or upsetters

Heating Modes

There are four basic heating modes in induction heating prior to hot working (Fig. 2). They are described as follows (Ref 1).

Static Heating. In the static heating mode, the workpiece, such as a billet or slab, is placed into the horizontal or vertical inductor for a given period of time, while a set amount of power is applied until the component reaches the desired heating conditions. Upon reaching the required conditions, the heated component is extracted from the induction heater and delivered to the metal hot working station. The next cold workpiece is reloaded into the coil, and the process repeats.

Progressive Multistage Heating. This heating mode occurs when two or more heated workpieces (i.e., billets, blanks, bars) are moved (via pusher, indexing mechanism, walking beam, etc.) through a single coil or multicoil induction heater. Therefore, components or their different regions are sequentially heated (progressive manner) at certain predetermined heating stages.

Continuous Heating. With the continuous heating mode, the workpiece is moved in a continuous motion through one or more in-line induction heating coils. This heating mode is commonly used when it is required to heat long components, such as bars, slabs, tubes, wires, and strips.

Oscillating Heating. In this heating mode, a workpiece moves back and forth (oscillates) during the process of heating inside of a single coil or multicoil induction heater with an oscillating stroke.

When designing induction heating systems, an ability to achieve desired temperature uniformity is only one of the goals. Additional design criteria include maximum production rate, minimum metal losses (scale reduction), and the ability to provide flexible and compact systems that maximize system overall efficiency (Ref 1, 9, 10). Other important factors include quality assurance, process repeatability, automation capability, environmental friendliness, lean manufacturing, reliability, controllability, and maintainability of the equipment. The last criterion, but not the least, is the competitive cost of an induction heating system.

The main components of an induction heating system are an inductor or induction coil, power supply, cooling system, control and monitoring, and the heated workpiece itself. Inductors or induction coils are usually designed for specific applications and therefore are found in a wide variety of shapes and sizes. Due to space limitation, a detailed description of a particular technological induction heating process is not provided here. The interested reader can find such information in Ref 1. This article focuses on estimation techniques to determine basic induction heating process parameters, including but not limited to required coil power, length of heating line, and frequency selection.

Experience gained on previous jobs and the ability to provide accurate mathematical modeling of the process serve as a comfort factor when designing new in-line induction heating systems. By combining advanced software with a sophisticated engineering background, induction heating specialists possess the unique ability to analyze, in a few hours, complex technological problems that could take days or even weeks to solve by trying to set experiments or through physical modeling using the pilot models. Numerical computational methods allow manufacturers of induction equipment to determine comprehensive details of the process that would be extremely difficult, if not impossible, to determine experimentally. Prior to this discussion, it would be beneficial to briefly review several physical phenomena related to induction heating.

Basic Electromagnetic Phenomena in Induction Heating

Induction heating is a complex combination of electromagnetic, heat transfer, and metallurgical phenomena involving many factors. Heat transfer and electromagnetics are nonlinear and tightly interrelated, because the physical properties of heated materials depend strongly on both

Table 1 Commonly required temperatures when heating selected ferrous alloys

Alloys, AISI	Typical forging temperatures	
	°C	°F
Nickel	880–1230	1616–2246
Low-carbon steel	1100–1315	2012–2400
Medium- and high-carbon steel	1050–1220	1922–2228
Plain carbon steels		
1010, 1015	1315	2400
1020, 1030	1288	2350
1040, 1050	1260	2300
1060	1182	2160
1070	1150	2100
1080	1204	2200
1095	1177	2150
Alloy steels		
4130	1204	2200
4140	1232	2250
4320	1232	2250
4340	1288	2350
4615	1204	2200
5160	1204	2200
6150	1204	2200
8620	1232	2250
9310	1232	2250
Stainless steels		
2xx, 3xx	1100–1250	2012–2282

Source: Ref 1

Table 2 Commonly required temperatures when heating selected nonferrous alloys

Alloys	Working temperatures	
	°C	°F
Aluminum	360–560	680–1040
Copper	680–960	1256–1760
Titanium	830–1150	1526–2100
Magnesium	320–380	608–716
Tungsten	~1350	~2462

Source: Ref 1

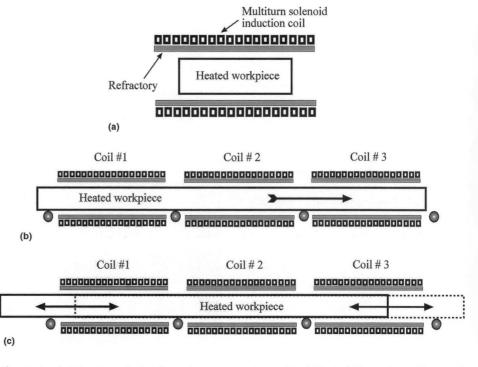

(a)

(b)

(c)

Fig. 2 Four basic heating modes in induction heating prior to hot working. (a) Static. (b) Progressive multistage and continuous. (c) Oscillation. Source: Ref 1

temperature and magnetic field intensity. The metallurgical phenomenon is also a nonlinear function of temperature, heating intensity, cooling severity, alloy chemical composition, prior microstructure, and other factors.

The basic electromagnetic phenomenon of induction heating is quite simple; it is discussed in the literature (Ref 1–6) and can be found in college physics textbooks. A conventional induction heating system that consists of a cylindrical workpiece surrounded by a multiturn induction coil is shown in Fig. 3. An alternating voltage applied to the induction coil terminals results in an alternating electrical current flow in the coil circuit. Coil current produces in its surroundings a time-variable magnetic field that has the same frequency. This magnetic field induces eddy currents in the workpiece located inside the coil or in its close proximity. It is imperative to remember that eddy currents could also be induced in other electrically conductive objects that are located near the coil. Induced currents have the same frequency; however, their direction is opposite to the coil current. Eddy currents produce heat by the Joule effect (I^2R).

Due to several electromagnetic phenomena, the eddy current (heat source) distribution within the workpiece is not uniform, potentially causing a nonuniform temperature profile. A nonuniform current distribution can be caused by several electromagnetic phenomena, including but not limited to skin, proximity, ring, and electromagnetic end and edge effects. These effects play an important role in understanding the induction heating principles (Ref 1, 3, 6). Before exploring the distribution of the magnetic field and eddy current, it is imperative to understand the nature of electromagnetic physical properties of heated metals.

Electromagnetic Properties of Metals and Alloys

Unlike fuel-fired and infrared furnaces, the performance of induction heaters first and foremost is affected by the electromagnetic properties of the heated metal. Electromagnetic properties of materials encompasses a variety of characteristics including magnetic permeability, electrical resistivity (electrical conductivity),

Fig. 3 Conventional induction heating system that consists of a cylindrical workpiece surrounded by a multiturn induction coil

saturation flux density, coercive force, and many others. While recognizing the importance of all electromagnetic properties, two of them—electrical resistivity (electrical conductivity) and magnetic permeability—have the most pronounced effect on the performance of an induction heating system, coil efficiency, and selection of the main design and process parameters (Ref 1).

Two Properties with the Most Pronounced Effect

Electrical Resistivity (Electrical Conductivity). The ability of material to conduct electric current is specified by electrical conductivity, σ. The reciprocal of the conductivity, σ, is electrical resistivity, ρ. The units for ρ are ohm-meter (Ω ·m) and for σ, siemens per meter (S/m or formerly mho/m). Both characteristics can be used in engineering practice; however, the majority of data books consist of data for electrical resistivity (Ref 11–15, 17). Therefore, the value of electrical resistivity is primarily used here. Electrical resistivity is an imperative physical property. It affects practically all important parameters of an induction heating system, including depth of heating, heat uniformity, coil electrical efficiency, power and frequency selection, coil impedance, load-matching capability, and others (Ref 1). Electrical resistivity of a particular metal varies with temperature, chemical composition, microstructure, and grain size. For most metals and alloys, ρ increases with temperature.

The resistivity of pure metals can often be represented as a linear function of the temperature (unless there is a change in the crystal structure of metal):

$$\rho(T) = \rho_0[1 + \alpha(T - T_0)] \qquad \text{(Eq 1)}$$

where ρ_0 is the resistivity at ambient temperature, T_0; $\rho(T)$ is the resistivity at temperature T; and α is the temperature coefficient of the electrical resistivity. The metric unit for α is 1/°C. For most metals and alloys (including carbon steels, alloyed steels, copper, aluminum, titanium, and tungsten), electrical resistivity increases with temperature; thus, α is a positive number. Table 3 consists of the values of α for selected metals.

Table 3 Temperature coefficient for some metals

Metals (at room temperature)	Temperature coefficient of resistance (α), 1/°C
Aluminum	0.0043
Cobalt	0.0053
Copper	0.004
Gold	0.0035
Iron	0.005
Lead	0.0037
Nichrome	0.0004
Nickel	0.0069
Silver	0.004
Titanium	0.0035
Tungsten	0.0045
Zinc	0.0042

Source: Ref 17

Figure 4 shows electrical resistivities of some commonly used metals as a function of temperature (Ref 1). For some electrically conductive materials, the electrical resistivity decreases with temperature, and therefore, the value of α can be negative. For other materials, resistivity is a nonlinear function of temperature. At the melting point, the electrical resistivity of metals is sharply increased. Grain size has a pronounced effect on electrical resistivity (i.e., higher ρ typically corresponds to finer grains) as well as plastic deformation, heat treatment, and some other factors. At the same time, the effect of temperature and chemical composition are two of the most pronounced factors (Fig. 5).

It is important to remember that impurities observed in metals and alloying elements distort the crystal structure and can affect the behavior of ρ to a considerable extent. This is particularly true for alloys (Ref 12–15). For some binary alloys, the behavior of ρ versus the concentration of alloying elements is represented by a bell-shaped curve. This curve often has the maximum of electrical resistivity at the concentration of alloying elements equal to approximately 50% of the atomic weight (Ref 12–15). Figures 5 to 7 illustrate this phenomenon. In most cases, however, instead of the bell-shaped curve, electrical resistivity continuously increases with concentration of alloys.

In contrast to bulk metals, some materials (i.e., composites, powder metallurgy parts, laminated materials) introduce additional challenges in computer modeling due to the anisotropic nature of physical properties manifesting itself in the directionally dependent nature of thermal and electromagnetic properties, including electrical resistivity. Material anisotropy can appreciably change the field generated by the induction coil and heating pattern.

One should not confuse electrical resistivity, ρ (Ω ·m), with electrical resistance, R (Ω). The relationship between these parameters can be expressed as:

$$R = \frac{\rho l}{a} \qquad \text{(Eq 2)}$$

where l is the length of the current-carrying conductor, and a is the area of the conductor cross section where the electric current is flowing.

Magnetic Permeability. Relative magnetic permeability, μ_r, indicates the ability of a material to conduct the magnetic flux better than vacuum or air, and μ_r is a nondimensional parameter. Relative magnetic permeability has a pronounced effect on coil calculations, computation of electromagnetic field distribution, and selection of process parameters.

The constant $\mu_0 = 4\pi \times 10^{-7}$ H/m or Wb/(A · m), is called the permeability of free space (the vacuum). The product of relative magnetic permeability and permeability of the free space is called magnetic permeability, μ, and

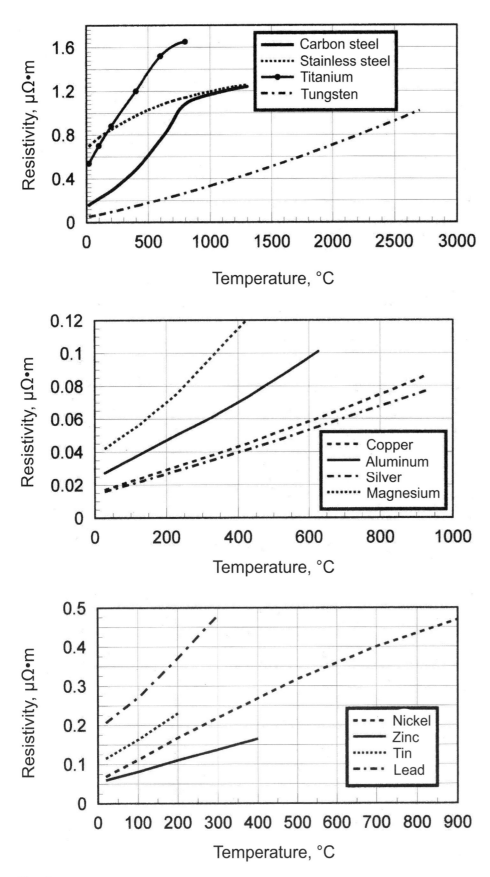

Fig. 4 Electrical resistivity of some commonly used metals as a function of temperature. Source: Ref 1

corresponds to the ratio of the magnetic flux density (B) to the magnetic field intensity (H):

$$\frac{B}{H} = \mu_r \mu_0 = \mu \text{ or } B = \mu_r \mu_0 H = \mu H \qquad \text{(Eq 3)}$$

The relative magnetic permeability is closely related to magnetic susceptibility by the following expression:

$$\mu_r = \chi + 1 \text{ or } \chi = \mu_r - 1 \qquad \text{(Eq 4)}$$

In other words, magnetic susceptibility, χ, shows the amount by which μ_r differs from unity.

Based on their magnetization ability, materials can be divided into paramagnetic, diamagnetic, and ferromagnetic (Ref 1, 12, 13). Relative magnetic permeability of paramagnetic materials is slightly greater than 1 ($\mu_r > 1$). The value of μ_r for diamagnetic materials is slightly less than 1 ($\mu_r < 1$). Due to insignificant differences of μ_r for both paramagnetic and diamagnetic materials, those materials are simply called nonmagnetic materials in induction heating practice. The relative magnetic permeability, μ_r, of nonmagnetic materials is assumed to be equivalent to that of air and is assigned a value of 1. Typical nonmagnetic metals include aluminum, copper, titanium, and tungsten. In contrast to paramagnetic and diamagnetic materials, ferromagnetic materials exhibit an appreciable value of relative magnetic permeability ($\mu_r \gg 1$), including iron, cobalt, nickel, and some rare earth metals, and are simply called magnetic materials.

A material ferromagnetic property is also a complex function of grain structure, chemical composition, frequency, magnetic field intensity, and temperature. As shown in Fig. 8(a), the same kind of carbon steel at the same temperature and frequency can have a different value of μ_r due to differences in the intensity of the magnetic field. For example, μ_r of magnetic steels commonly used in induction heating can vary from small values of $\mu_r = 3$ or 4 to very high values exceeding 500, depending on the magnetic field intensity (H) and temperature (T). The temperature at which a ferromagnetic body becomes nonmagnetic is called the Curie temperature (Curie point). Figure 9 shows a dramatic illustration of the complex relationship among μ_r, temperature, and magnetic field intensity for medium-carbon steel. Figure 8(a) implies that magnetic permeability always decreases with increasing temperature. This is indeed the case in the majority of induction heating applications prior to hot working. However, in a relatively weak magnetic field, μ_r may first increase with temperature, and only near the Curie point would it begin to drastically decline. Discussion of this phenomenon is provided in Ref 1.

Chemical composition is another factor that has a marked effect on the Curie point. The Curie point of plain low-carbon steel corresponds to the A_2 critical temperature on the

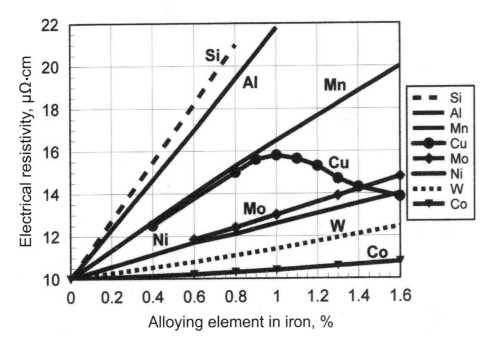

Fig. 5 Dependence of electrical resistivity on the addition of small amounts of various alloying elements to iron. Source: Ref 12

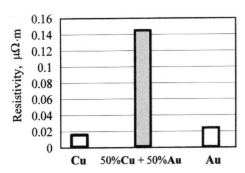

Fig. 6 Electrical resistivity vs. percentage of alloying elements in a copper-gold binary alloy. Source: Ref 12

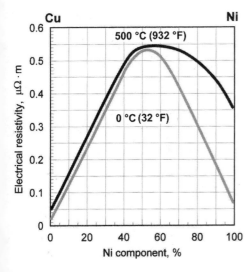

Fig. 7 Electrical resistivity of copper-nickel alloys at different temperatures. Source: Ref 15

$$\delta = 503 \sqrt{\frac{\rho}{\mu_r F}} \qquad \text{(Eq 5)}$$

where ρ is the electrical resistivity of the metal, measured in $\Omega \cdot m$; μ_r is the relative magnetic permeability; F is the frequency in hertz (cycles per second); and δ is measured in meters, or:

$$\delta = 3160 \sqrt{\frac{\rho}{\mu_r F}} \qquad \text{(Eq 6)}$$

where δ is measured in inches, and ρ is electrical resistivity, measured in $\Omega \cdot in$.

Mathematically speaking, the penetration depth, δ, is the distance from the surface of the conductor toward its core at which the current decreases exponentially to 1/exp its value at the surface. The power density at this distance will decrease to $1/\exp^2$ its value at the surface. Figure 10 illustrates the skin effect appearance by showing the distribution of current density and power density from the surface of the cylinder workpiece toward the core (Ref 1, 11). As one can see from Fig. 10(b), at a distance equal to one penetration depth from the surface ($y = \delta$), the current will equal 37% of its surface value. However, the power density will equal 14% of its surface value. From this, it can be concluded that approximately 63% of the current and 86% of the power in the workpiece will be concentrated within a surface layer of thickness δ.

Analysis of Eq 5 and 6 shows that the penetration depth has different values for different materials and is also a function of frequency and temperature. During the heating cycle, ρ can increase to four to six times its initial value. Therefore, the penetration depth increases correspondingly during the process cycle, even when heating nonmagnetic metals. Table 4 shows some penetration depths of metals that are most commonly used with induction heating versus temperature and frequency.

Equations 5 and 6 represent the definition of current penetration depth in its classical form and do not have a fully determined meaning for magnetic materials below the Curie point, because of the nonconstant distribution of μ_r within the magnetic workpiece (Ref 1). In engineering practice, when heating magnetic materials, the value of relative magnetic permeability at the surface of the workpiece, μ_r^{surf}, is typically used to provide a determination of those expressions in definite form. Table 5 shows the values of the penetration depth in medium-carbon steel at an ambient temperature (21 °C) as a function of frequency and magnetic field intensity, H, at the workpiece surface.

Figure 11 provides a general illustration of penetration depth as a function of temperature when heating ferromagnetic metals (Ref 1, 20). At the beginning of the heating cycle, the current penetration depth into the carbon steel slightly increases because of the increase in electrical resistivity with temperature. With a further rise of temperature (at approximately 550 °C, or 1022 °F), μ_r starts to decrease at an accelerated

iron-iron carbide phase transformation diagram. Therefore, even among plain carbon steels, the Curie temperature may be noticeably different due to the carbon content, as shown in Fig. 8 (b). For example, the Curie temperature of plain carbon steel AISI 1008 is clearly different from steel AISI 1060 (768 versus 732 °C, or 1414 versus 1350 °F). Depending on the heat intensity (°C/s or °F/s), there can be some shifting of the Curie temperature due to thermal hysteresis.

Electromagnetic Skin Effect and Current Penetration Depth

One of the challenges in induction heating arises from the necessity to provide the required surface-to-core temperature uniformity. The workpiece core tends to be heated slower than its surface. The main reason for the heat deficit in the core of the heated component is the skin effect (Ref 1–6). The skin effect is considered to be a fundamental property of induction heating representing a nonuniform distribution of an alternating current within the electrical conductor cross section. This effect will also be observed in any electrically conductive body (workpiece) located inside an induction coil or in its close proximity. Eddy currents induced within the heated workpiece primarily flow in the surface layer, or within the skin, where 86% of all induced power will be concentrated. The thickness of this layer is called the reference depth or current penetration depth, δ. The value of the penetration depth varies with the square root of the electrical resistivity and inversely with the square root of frequency and the relative magnetic permeability according to Eq 5 and 6:

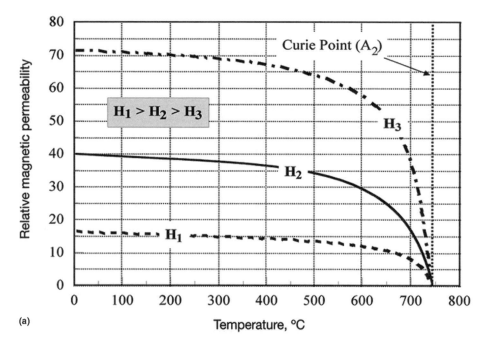

(a)

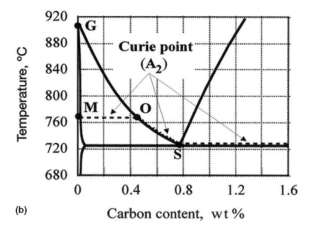

(b)

pace. Near a critical temperature, T_c, known as the Curie point, permeability drastically drops to unity because the metal becomes nonmagnetic. After heating above the Curie temperature, the penetration depth will continue to rise due to the increase in electrical resistivity of the metal (Fig. 11). However, the rate of change is not as significant as it was during the transition through the Curie temperature. The variation of δ during induction heating drastically changes the degree of skin effect. Figure 12 shows typical variation (in folds) of the current penetration depth of some metals during induction heating prior to hot working.

An assumption of a distribution of current density along the workpiece thickness/radius as always exponentially decreasing from the workpiece surface toward its core is a substantially simplified assumption. It is important to remember that this classical assumption is appropriate only for a solid workpiece having constant electrical resistivity and magnetic permeability. Therefore, realistically speaking, this assumption can be made for only some unique cases of induction heating. For the great majority of induction heating applications, the temperature distribution during different heating stages is not uniform, and appreciable thermal gradients within the heated workpiece are present. These temperature gradients result in nonuniform distributions of electrical resistivity and relative magnetic permeability within the workpiece, meaning that the classical definition of current penetration depth that assumes an exponential distribution of current often does not fully apply and can lead to a deceiving result. A classical definition of δ can only be used for rough estimates for induction heating of nonmagnetic materials (i.e., aluminum, copper, and some stainless steels) and through heating of magnetic steels at temperatures appreciably exceeding the Curie point.

Magnetic-Wave Phenomenon

In such applications as surface hardening and heating carbon steels to working temperatures just above the Curie point, the power density

Fig. 8 Magnetic properties of steel. (a) Effect of temperature field intensity on relative magnetic permeability of medium-carbon steel. (b) Effect of carbon content on Curie temperature of plain carbon steel at a heating rate less than 70 °C/s. Source: Ref 1

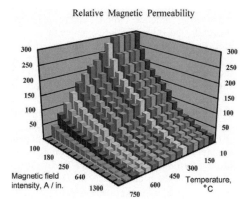

Fig. 9 Relative magnetic permeability as a function of magnetic field intensity (range 100 to 1500 A/in., or 39 to 590 A/cm) and temperature (range 10 to 750 °C, or 50 to 1382 °F). Source: Ref 55

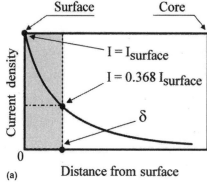

(a)

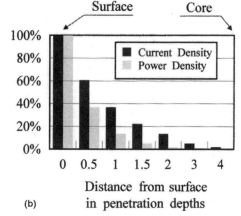

(b)

Fig. 10 Skin effect. (a) Current density versus distance from surface. (b) Current density and power density distributions as percentage versus penetration depth due to skin effect. Source: Ref 1

Table 4 Penetration depth of nonmagnetic metals measured in millimeters

Metal	T °C	T °F	ρ μΩ·m	ρ μΩ·in.	0.06	0.50	1	2.5	4	8	10	30	70	200	500
Aluminum	20	68	0.027	1.06	10.7	3.70	2.61	1.65	1.30	0.92	0.83	0.48	0.31	0.18	0.12
	250	482	0.053	2.09	15.0	5.18	3.66	2.32	1.83	1.29	1.16	0.67	0.44	0.26	0.16
	500	932	0.087	3.43	19.2	6.64	4.69	2.97	2.35	1.66	1.48	0.86	0.56	0.33	0.21
Copper	20	68	0.018	0.71	8.81	3.05	2.16	1.36	1.08	0.76	0.68	0.39	0.26	0.15	0.10
	500	932	0.050	1.97	14.5	5.03	3.56	2.25	1.78	1.26	1.12	0.65	0.43	0.25	0.16
	900	1652	0.085	3.35	19.3	6.67	4.72	2.98	2.36	1.67	1.49	0.86	0.56	0.33	0.21
Brass	20	68	0.065	2.56	16.6	5.74	4.06	2.56	2.03	1.43	1.28	0.74	0.48	0.29	0.18
	400	752	0.114	4.49	21.9	7.60	5.37	3.40	2.69	1.90	1.70	0.98	0.64	0.38	0.24
	900	1632	0.203	7.99	29.3	10.1	7.17	4.53	3.58	2.53	2.27	1.31	0.86	0.51	0.32
Stainless steel	20	68	0.690	27.2	53.9	18.7	13.2	8.36	6.61	4.67	4.18	2.41	1.58	0.93	0.59
	800	1472	1.150	45.3	69.6	24.1	17.1	10.8	8.53	6.03	5.39	3.11	2.04	1.21	0.76
	1200	2192	1.240	48.8	72.3	25.1	17.7	11.2	8.86	6.26	5.60	3.23	2.12	1.25	0.79
Silver	20	68	0.017	0.67	8.34	2.89	2.04	1.29	1.02	0.72	0.65	0.37	0.24	0.14	0.09
	300	572	0.038	1.50	12.7	4.39	3.10	1.96	1.55	1.10	0.98	0.57	0.37	0.22	0.14
	800	1472	0.070	2.76	17.2	5.95	4.21	2.66	2.10	1.49	1.33	0.77	0.50	0.30	0.19
Tungsten	20	68	0.050	1.97	14.5	5.03	3.56	2.25	1.78	1.26	1.12	0.65	0.43	0.25	0.46
	1500	2732	0.550	21.7	48.2	16.7	11.8	7.46	5.90	4.17	3.73	2.15	1.41	0.83	0.53
	2800	5072	1.040	40.9	66.2	22.9	16.2	10.3	8.11	5.74	5.13	2.96	1.94	1.15	0.73
Titanium	20	68	0.500	19.7	45.9	15.9	11.3	7.11	5.62	3.98	3.56	2.05	1.34	0.80	0.50
	600	1112	1.400	55.1	76.8	26.6	18.8	11.9	9.41	6.65	5.95	3.44	2.25	1.33	0.84
	1200	2192	1.800	70.9	87.1	30.2	21.3	13.5	10.7	7.54	6.75	3.90	2.55	1.51	0.95

Table 5 Penetration depth of carbon steel 1040 at ambient temperature of 21 °C (70 °F)

Magnetic field intensity A/mm	A/in.	60 mm	60 in.	500 mm	500 in.	3000 mm	3000 in.	10,000 mm	10,000 in.	30,000 mm	30,000 in.	100,000 mm	100,000 in.
10	250	2.50	0.100	0.88	0.034	0.36	0.014	0.2	0.008	0.11	0.004	0.06	0.002
40	1000	4.70	0.185	1.63	0.064	0.67	0.026	0.36	0.014	0.21	0.008	0.12	0.005
80	2000	6.30	0.249	2.20	0.086	0.9	0.035	0.49	0.019	0.28	0.011	0.16	0.006
120	3050	7.76	0.306	2.69	0.106	1.1	0.043	0.6	0.024	0.35	0.014	0.19	0.007
160	4050	8.76	0.345	3.03	0.119	1.24	0.049	0.68	0.027	0.39	0.015	0.21	0.008
200	5100	9.63	0.379	3.33	0.131	1.36	0.054	0.75	0.029	0.43	0.017	0.24	0.009
280	7100	11.20	0.442	3.89	0.153	1.59	0.062	0.87	0.034	0.50	0.020	0.27	0.011

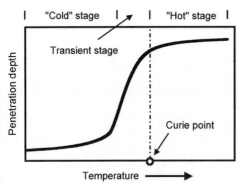

Fig. 11 Typical variations of current penetration depth during induction heating of a carbon steel workpiece. Source: Ref 20

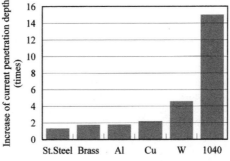

Fig. 12 Possible variation of current penetration depth of various metals during induction heating. Source: Ref 55

tightly coupled electrothermal processes taking place in induction heating. Of course, it also was not possible to measure the power/current density distribution inside the solid body. Research studies that applied advanced numerical computer modeling, conducted in the mid-1990s (Ref 20), enabled a quantitative estimation of this magnetic-wave phenomenon (also known as the dual-properties phenomenon).

The magnetic-wave phenomenon is always present during the transition from an initial heating stage (below the Curie point) to a final heating stage (appreciably above the Curie temperature). Figures 13 and 14 show the results of numerical computer modeling: temperature distribution and power density profiles at different heating stages during in-line multicoil (Fig. 15) induction heating of 75 mm (3 in.) diameter carbon steel bars. The magnetic-wave phenomenon takes place during an interim heating stage, when the workpiece temperature rises from below the Curie temperature to above it. In applications such as surface hardening, the magnetic-wave phenomenon exists during the majority of the heat cycle and plays a very important role in the prediction of the final temperature profile and hardened case depth. On the other hand, in applications such as through hardening, normalizing, or induction heating of steels prior to hot working, the duration of the transition (interim) stage is much shorter compared with the hot heating stage (above the Curie temperature). For example, the hot stage when heating steels prior to forging is usually 60 to 70% of the total heating time. Here and in other similar applications, the magnetic-wave phenomenon has an insignificant effect on the final temperature distribution. More discussion on the magnetic-wave phenomenon is provided subsequently and in the article "Simulation of Induction Heat Treatment" in this Volume.

Note that the magnetic-wave phenomenon can also play an important role in some low-temperature induction heating applications, such as metal coating and plating. For example, if a nonmagnetic, electrically conductive coating is applied to a carbon steel part, this phenomenon can be quite pronounced, depending on the applied frequency and thickness of the nonmagnetic but electrically conductive surface deposit.

Mathematical Modeling

Mathematical modeling is one of the major factors in successful design of induction heating systems. The choice of a particular theoretical model depends on several factors, including the complexity of the engineering problem, required accuracy, time limitations, and cost (Ref 1). Many mathematical modeling methods and software programs exist or are under development. Work in this field is done at universities, including Michigan State University (United States), Hannover University

distribution along the radius/thickness has a unique wave shape, which differs significantly from the commonly assumed, classical exponential distribution. Here, the maximum power density is located at the surface and decreases toward the core. Then, at a certain distance from the surface, the power density increases, reaching a maximum value before again decreasing. This magnetic-wave phenomenon

was introduced by Simpson (Ref 4) and Losinskii (Ref 3). They intuitively felt that there should be conditions where the power density (heat source) distribution would differ from that of the commonly accepted classical exponential form. They provided a qualitative description of a magnetic-wave phenomenon based on intuition. At that time, it was difficult to obtain a quantitative estimation of this phenomenon due to limited computer modeling power and the lack of software that could simulate the

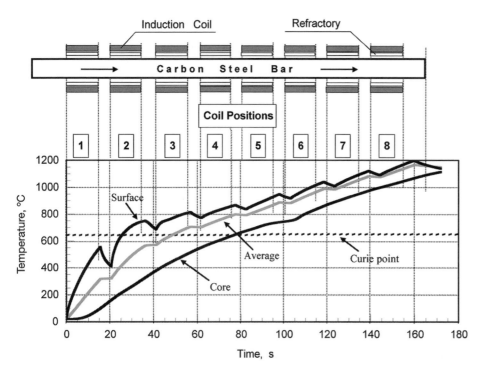

Fig. 13 In-line induction billet heating. Time history of surface, core, and average temperatures. Source: Ref 21

(Germany), and Padua University (Italy); research labs; inside large manufacturers of induction heating machinery, such as Inductoheat, Inc.; and by specialized software-developing companies, such as Integrated Engineering Software Inc., Infolytica Inc., MagSoft Corporation, COMSOL Group, Vector Fields Inc., ANSYS Corporation, Ansoft Corporation, ESI Group, and others (Ref 19, 22–30).

A great majority of the commercial codes used for computer modeling of induction heating processes are all-purpose programs that were developed to be used in other areas of engineering (for example, design of transformers, alternating current motors, antennas, and nondestructive testing). It is important to be aware that, unfortunately, many generalized programs could experience difficulties in simulating certain subtleties of the induction heating process, including but not limited to:

- The presence of thermal refractory can have steady-state or transient thermal conditions.
- Motion of heated workpiece. The heated bar or billet can continuously move through several in-line coils, or it can oscillate inside of a multicoil induction line.
- The induction heating system may consist of several power sources using different frequencies and power levels.
- Multilayer and multiphase inductors can be used.

Modern induction heating specialists routinely use a variety of numerical computer modeling techniques that are based on finite-difference, finite-element, mutual impedance, and boundary-element methods. These approaches are used in the simulation of both electromagnetic and heat-transfer problems. Each of these methods has certain advantages and has been used alone or in combination with others. In recent years, the finite-element method became a dominant numerical simulation tool for a variety of engineering applications, including mechanical engineering, fluid flow, heat transfer, acoustics, and so on. Although the finite-element method is a very effective modeling technique for induction hardening, stress relieving, and tempering applications, it cannot be considered an ultimate computational tool for all induction heating applications. In some induction applications, a combination of finite-difference and boundary-element techniques or a combination of finite-difference and mutual impedance techniques is more effective (Ref 1, 30, 31). The appreciable amount of time required for mesh generation and process computation is an obvious drawback of using the finite-element method for real-time control systems. Experience shows that there is not a single universal computational method that optimally fits all induction heating applications. For each problem or family of similar problems, certain numerical methods or software programs are preferred. Taking into consideration the specifics of a particular application, it is advantageous to have a number of subject-oriented and highly efficient programs rather than searching for a single universal program. The article "Simulation of Induction Heat Treatment" in this Volume is devoted to a discussion of different numerical computation approaches for coupled electromagnetic and heat-transfer processes taking place in induction heating and induction heat treating.

In the fast-paced world economy, the ability of induction heating manufacturers to minimize the time between a customer request for a quote and the quote itself is critical for a company's success. In contrast to academia, the fast pace of industry often does not allow the luxury of waiting several weeks or even days to obtain the results of numerical modeling but instead demands a quick estimate regarding the main process parameters of an induction system. Recognizing the significant benefits of using numerical simulations, it is also important to be able, within a few hours or even minutes, to develop a general understanding regarding the most critical parameters of a particular induction heating system. This article focuses on such techniques.

Rough Estimation of the Required Power for Induction Heating

The selection of power, frequency, and coil length in induction heating prior to metal hot working is highly subjective and is affected by designer past experience, the type of heated metal, required temperature uniformity and cycle time, application specifics, and so on. One quick way to estimate a rough order of magnitude of the required workpiece power (P_w) is based on an average value of specific heat, c, of the heated metal (Ref 1, 32, 33). The specific heat, c, represents the amount of required energy to be absorbed by a unit mass of the workpiece to achieve a unit temperature increase, measured in J/(kg · °C) or Btu/(lb · °F). A high value of specific heat corresponds to a high required power to heat a unit mass to a unit temperature. The values of the specific heat of some commonly used metals are shown in Fig. 16 (Ref 8, 35–37).

Equation 7 allows the estimation of the kilowatts required to be induced by induction coil within the heated workpiece (i.e., billet or bar) in order to provide the required rise of an average temperature at the required production rate:

$$P_w = mc\frac{T_f - T_{in}}{t} \qquad \text{(Eq 7)}$$

where m is the mass of the heated workpiece in kilograms; c is the average value of the specific heat, measured in J/(kg · °C); T_{in} and T_f are the average values of the initial and final temperatures, respectively, in °C; and t is the required heat time in seconds.

Methods for Heating Copper Billet

Example 1: Copper Billet. To heat a copper cylinder (0.12 m outside diameter, 0.4 m long) from ambient temperature (20 °C, or 68 °F) to an average temperature of 620 °C (1148 °F) in 120 s, the power required to be induced within the workpiece can be determined according to Eq 7. In this case, the mass of the heated metal can be calculated as:

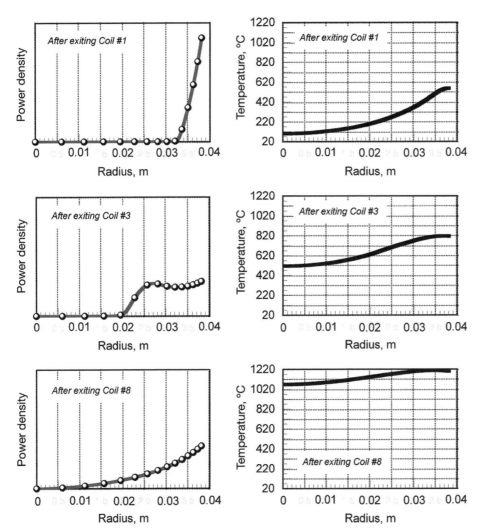

Fig. 14 Power density and temperature profiles at different bar positions (refer to Fig. 13) inside an in-line induction heater. Source: Ref 21

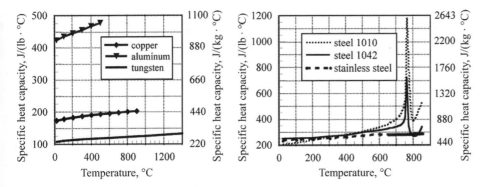

Fig. 16 Specific heat capacity of some commonly used metals vs. temperature. Source: Ref 1

Fig. 15 In-line multicoil induction bar/billet heater. Courtesy of Inductoheat, Inc.

Example 2: Alternate Method for Copper Billet. Some practitioners prefer using the value of the heat content of the material instead of a specific heat for determining the value of P_w. Heat content (HC) is measured in kW × h/t. In this case, Eq 7 can be rewritten as:

$$P_w = HC \times \text{Production} \qquad (\text{Eq 8})$$

where production is measured in t/h.

Figure 17 shows the heat content values for commonly used metals. As an example, the required power (P_w) is calculated for the previous example using the heat content value. From Fig. 17, the required heat content value would be approximately equal to 70 kW × h/t:

$$\text{Production} = \frac{40.3\,\text{kg}}{120\,\text{s}} = \frac{0.0403\,t}{0.033\,\text{h}} = 1.22\,t/\text{h}$$

$$P_w = HC \times \text{Production} = 70 \times 1.22 = 85.4\,\text{kW}$$

Such simplified formulas as Eq 7 and 8 are convenient to use in applications such as induction heating of the classical-shaped workpieces (i.e., billets, bars, slabs, blooms, tubes, pipes, etc.) where relatively uniform through heating is required. These formulas have the advantage of providing a quick estimate of the value of the required workpiece power (P_w). It is important to remember that power induced in the workpiece (P_w) does not have a constant value during the heating cycle and varies depending on the change in electrical resistivity and relative magnetic permeability of the heated material. This is why, instead of using P_w, the value of P_w^{av} (meaning an average power per heating cycle) is often used for rough estimation.

Total Efficiency of the Coil

Power (P_w) does not represent the required power at the coil terminals (the so-called coil power). Equation 9 provides a correlation between the required average coil power (P_c^{av}) and the required average workpiece power (P_w^{av}):

$$P_c^{av} = \frac{P_w^{av}}{\eta} \qquad (\text{Eq 9})$$

where η is the coil total efficiency.

According to Eq 10, the total efficiency of the induction coil (η) is a combination of both

$$m = \frac{\pi D^2}{4}l\gamma = \frac{3.14 \times 0.12^2}{4} \times 0.4 \times 8.91 \times 10^3$$
$$= 40.3\,\text{kg}$$

where γ is the density in kg/m³ (for copper, $\gamma = 8.91 \times 10^3$ kg/m³); D is the outside diameter in meters; and l is the billet length in meters. $c = 420$ J/(kg · °C) can be used as an average value

of the specific heat of copper in the temperature range of 20 to 620 °C. Therefore, applying Eq 7, the power required to be induced within the workpiece (P_w) should be as follows:

$$P_w = mc\frac{T_f - T_{in}}{t} = 40.3 \times 420 \times \frac{620 - 20}{120}$$
$$= 84{,}630\,\text{W} = 84.6\,\text{kW}$$

coil electrical efficiency (η_{el}) and coil thermal efficiency (η_{th}):

$$\eta = \eta_{el}\,\eta_{th} \qquad \text{(Eq 10)}$$

where both η_{el} and η_{th} are in the range of 0 to 1.

Defining Electrical and Thermal Efficiencies

Coil Electrical Efficiency. The value of η_{el} represents the amount of electrical losses (P_{loss}^{el}) compared to the power induced in the workpiece (P_c^{av}) according to:

$$\eta_{el} = \frac{P_w^{av}}{P_w^{av} + P_{loss}^{el}} \qquad \text{(Eq 11)}$$

where P_{loss}^{el} includes power loss in actual coil copper turns (P_{loss}^{turns}) and power loss in electrically conductive bodies located in the inductor surrounding (P_{loss}^{sur}) and can be determined as:

$$P_{loss}^{el} = P_{loss}^{turns} + P_{loss}^{sur} \qquad \text{(Eq 12)}$$

The value of P_{loss}^{sur} represents an undesirable heating of tools, liners, flux concentrators, magnetic shunts, fixtures, enclosures, support beams, and other electrically conductive structures that are located near the induction coil and where appreciable eddy currents can be induced.

As shown in Ref 33, when heating a solid cylinder in a long solenoidal coil, the value of η_{el} can be calculated according to:

$$\eta_{el} = \frac{1}{1 + \frac{D_1'}{D_2'}\frac{\rho_1}{\rho_2}\frac{\delta_1}{\delta_2}} = \frac{1}{1 + \frac{D_1'}{D_2'}\sqrt{\frac{\rho_1}{\mu_r \rho_2}}} \qquad \text{(Eq 13)}$$

where D_1' is an equivalent coil inside diameter, $D_1' = D_1 + \delta_1$; D_2' is an equivalent outside diameter of a cylinder, $D_2' = D_2 - \delta_2$; δ_1 and δ_2 are current penetration depths in the coil copper and cylinder (workpiece), correspondingly; ρ_1 and ρ_2 are electrical resistivities of the coil and workpiece, respectively; and μ_r is the relative magnetic permeability of the cylinder.

Equation 13 has been obtained under the following assumptions:

- The skin effect is pronounced.
- The coil is a stand-alone electrical device, and there are no electrically conductive structures located in its proximity.
- The inductor is a single-layer, infinitely long, and tightly wound solenoid producing a homogeneous magnetic field without any disturbances and flux leakage.
- Electrically thick copper tubing is used for coil fabrication.

The ratio:

$$\frac{D_1'}{D_2'}\sqrt{\frac{\rho_1}{\mu_r \rho_2}}$$

is called the coil electrical efficiency factor (Ref 33). High coil electrical efficiency corresponds to the low value of the electrical efficiency factor. Therefore, high coil electrical efficiency takes place when heating magnetic workpieces that have high electrical resistivity (assuming an absence of eddy current cancellation within the heated workpiece), using the smallest possible coil-to-workpiece gap ($D_1/D_2 \to 1$). For example, a coil electrical efficiency, η_{el}, when heating carbon steel cylinders below

the Curie temperature is usually in the range of 0.8 to 0.95. In contrast, when heating billets made from silver or copper to low temperatures using conventional solenoid inductors, η_{el} is typically in the range of 0.35 to 0.45.

When heating rectangular bodies, including slabs, round-cornered square bars, and plates, Eq 13 instead of Eq 12 should be applied (Ref 33):

$$\eta_{el} = \frac{1}{1 + \frac{F_1}{F_2}\sqrt{\frac{\rho_1}{\mu_r \rho_2}}} \qquad \text{(Eq 14)}$$

where F_1 and F_2 are equivalent perimeters of the rectangular coil openings and heated slab, correspondingly.

The coil efficiency of induction heaters is a complex function of several design parameters, including but not limited to the coil-to-workpiece radial gap, physical properties of the heated metal, presence of magnetic flux concentrators and Faraday rings, length of the coil, and applied frequency. As an illustration, Fig. 18 shows that there will be high coil electrical efficiency when the frequency is greater than F1, which corresponds to a ratio of cylinder outside diameter (OD) to current penetration depth (δ) greater than three (OD/δ > 3). The use of a frequency that results in a ratio of OD/δ > 8 will only slightly increase the coil efficiency. The use of very high frequencies (frequency > F2) tends to decrease the total efficiency due to higher transmission losses and decreased thermal efficiency, because it will require a long heat time to provide the sufficient surface-to-core temperature uniformity. Coil electrical efficiency will dramatically decrease if the chosen frequency results in a ratio of OD/δ < 3 (frequency less than F1, Fig.18). This takes place due to the cancellation of induced eddy currents circulating in the opposite sides of the solid cylinder (Ref 1).

Coil Thermal Efficiency. The value of η_{th} designates coil thermal efficiency and represents the amount of heat losses (P_{loss}^{th}) during the whole heating cycle compared to actual heating power and can be determined as:

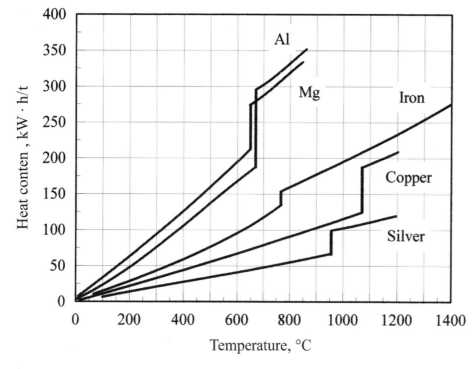

Fig. 17 Heat content of metals when heating to various temperatures. Source: Ref 1

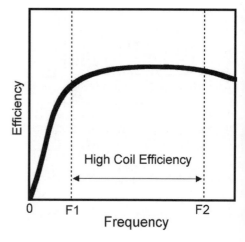

Fig. 18 Coil electrical efficiency (arbitrary units) vs. frequency (arbitrary units). Source: Ref 1

$$\eta_{th} = \frac{P_w^{av}}{P_w^{av} + P_{loss}^{th}} \qquad \text{(Eq 15)}$$

P_{loss}^{th} includes heat losses from the workpiece surface due to radiation and convection as well as heat loss due to thermal conduction (for example, heat conduction losses from the billet to water-cooled liners, guides, or support structures).

Thermal insulation (thermal refractory) of the induction coil significantly decreases heat losses from the workpiece surface. It is fabricated from poor thermal-conductive materials. Accurate estimation of the value of P_{loss}^{th} can be determined after numerical computer modeling, which is discussed in the article "Simulation of Induction Heat Treatment" in this Volume. At the same time, there are several empirical formulas that allow one to obtain a rough estimate of those losses. Some of those recommendations are provided in Ref 34 for cylindrical coils with concrete blocks used as a refractory. According to those recommendations, the value of surface thermal losses can be determined as:

$$P_{loss}^{th} = 3.74 \times 10^{-2} \frac{l}{\lg\left(\frac{D1}{D3}\right)} \qquad \text{(Eq 16)}$$

where P_{loss}^{th} is the heat loss from the workpiece surface in kilowatts; l is the coil length in centimeters; $D1$ is the inside diameter of the induction coil in centimeters; and $D3$ is the inside diameter of the refractory in centimeters.

At this point, it would be beneficial to review three modes of heat transfer that are related to P_{loss}^{th}.

Three Modes of Heat Transfer—Thermal Conduction, Thermal Convection, and Thermal Radiation

In induction heating, all three modes of heat transfer—conduction, convection, and radiation—are present (Ref 1, 38–46).

Thermal Conduction

Heat is transferred by thermal conduction from the high-temperature regions of the workpiece toward the low-temperature regions. Fourier's law is the basic law that describes heat transfer by conduction:

$$q_{cond} = -k\,grad(T) \qquad \text{(Eq 17)}$$

where q_{cond} is the heat flux by thermal conduction, k is a coefficient of thermal conductivity, and T is temperature.

As can be seen from Eq 17, according to Fourier's law, the rate of heat transfer in a workpiece is proportional to the temperature gradient (temperature difference) and the value of thermal conductivity. In other words, a substantial surface-to-core temperature differential and a high value of thermal conductivity of the metal result in intensive heat transfer from the hot surface of the workpiece toward its colder core. Conversely, the rate of heat transfer by thermal conduction is inversely proportional to the distance between regions with different temperatures.

Thermal conductivity, k, designates the rate at which heat travels across the workpiece. A material with a high k-value conducts the heat faster than a material with a low k. In choosing a material for an inductor refractory, a lower value of k is preferable, corresponding to high thermal efficiency (η_{th}) and lower heat losses that leak through a refractory. On the other hand, when the thermal conductivity of the heated metal is high, it is easier to obtain a uniform temperature distribution within the workpiece. This would be advantageous in through-heating applications such as heating prior to hot working, through hardening, normalizing, annealing, and so on. However, in selective heating applications (for example, bar-end heating, Fig. 19), a high value of k is quite often a disadvantage because of its tendency to provide an intense heat transfer in the axial direction, equalizing (soaking) the temperature distribution. This leads to heating not only in the workpiece area that is required to be heated but also in a much greater area, resulting in an appreciable heating of the bar shank due to thermal conduction (the so-called cold sink effect). Heat flow that occurs due to the heat sink effect not only leads to a redistribution of the temperature pattern but also affects the amount of total mass of the metal being heated, which directly affects the total amount of required energy, for example. The values of thermal conductivity of some commonly used metals are shown in Fig. 20 (Ref 35–41). As may be noted, thermal conductivity is a nonlinear function of temperature.

Thermal Convection

Heat transfer by convection is carried out by fluid, gas, or air (i.e., from the surface of the heated workpiece to the surrounding area). The well-known Newton's law can describe the convection heat transfer (Ref 37–42). It states that the heat-transfer rate is directly proportional to the temperature difference between the workpiece surface and the ambient area:

$$q_{conv} = \alpha(T_s - T_a) \qquad \text{(Eq 18)}$$

where q_{conv} is the heat flux density by convection, measured in W/m² or W/in.²; α is the convection surface heat-transfer coefficient, measured in W/(m² · °C) or W/(in.² · °F); T_s is the surface temperature in °C or °F; and T_a is the ambient temperature in °C or °F.

The convection heat-transfer coefficient is primarily a function of the thermal properties and geometry of the workpiece; the thermal properties of the surrounding air, gas, or fluids; and their viscosity or the velocity of the heated workpiece if it moves or rotates at an appreciable speed (e.g., induction heating of a fast-moving strip or spinning shaft). It is particularly important to take this mode of heat transfer into account when designing low-temperature induction heating applications where convection losses often exceed surface heat losses due to thermal radiation.

In different applications, a magnitude of convection losses can vary dramatically depending on the surface temperatures of the workpiece and the outside temperature as well as workpiece geometry, its surface conditions, the presence of refractory, and whether it is free or forced convection. The heat losses due to forced convection can be substantially higher (i.e., often five- to tenfold greater) compared to free convection losses of the stationary heated workpiece.

There are several empirical formulas that can provide a rough engineering estimation of the free convection losses, q_{conv}, including Eq 19 (Ref 47) and Eq 20 (Ref 48):

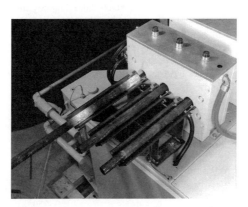

Fig. 19 Induction bar-end heater. Courtesy of Inductoheat, Inc.

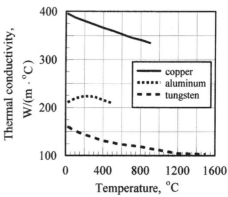

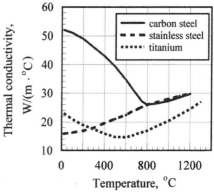

Fig. 20 Thermal conductivity of some commonly used metals vs. temperature. Source: Ref 1

$$q_{conv} = 1.86(T_s - T_a)^{1.3} \, (\text{W/m}^2) \qquad \text{(Eq 19)}$$

$$q_{conv} = 1.54(T_s - T_a)^{1.33} \, (\text{W/m}^2) \qquad \text{(Eq 20)}$$

where T_s and T_a are the surface temperature and ambient temperature, correspondingly, in °C.

Example 3: Free Convection Heat Loss. The free convection heat loss density from the workpiece surface heated to 640 °C into the surrounding ambient atmosphere ($T_a = 20$ °C) is calculated using each formula. According to Eq 19:

$$q_{conv} = 1.86 \times 620^{1.3}$$
$$q_{conv} = 7.94 \, \text{kW/m}^2$$

According to Eq 20:

$$q_{conv} = 1.54 \times 620^{1.33}$$
$$q_{conv} = 7.97 \, \text{kW/m}^2$$

Convection heat-transfer mode plays a particularly important role in the quenching process where the surface heat-transfer coefficient is associated with the cooling intensity during quenching.

Thermal Radiation

Radiation Mode of Heat Transfer. In the third mode of heat transfer, which is thermal radiation, the heat may be transferred from the hot workpiece into a nonmaterial region (vacuum). The effect of heat transfer by thermal radiation can be introduced as a phenomenon of the electromagnetic energy propagating due to a temperature difference (Ref 36–46). This phenomenon is governed by the Stefan-Boltzmann law of thermal radiation, which states that the heat-transfer rate by radiation is proportional to a radiation-loss coefficient, C_s, and the value of $T_s^4 - T_a^4$.

Because radiation losses are proportional to the fourth power of temperature, these losses are the major part of the total surface heat losses in high-temperature applications (for example, induction heating prior to forging). The radiation heat-loss coefficient includes emissivity, radiation shape factor (the view factor), and surface conditions. For example, the value of emissivity increases with increased surface oxidation. At the same time, polished metal radiates noticeably less heat to the surroundings than will nonpolished metal. A comparison of emissivities of some common polished metals versus nonpolished metals is shown in Table 6.

The radiation heat-loss coefficient can be determined approximately as $C_s = \sigma_s \varepsilon$, where

ε is the emissivity of the metal, and σ_s is the Stefan-Boltzmann constant ($\sigma_s = 5.67 \times 10^{-8}$ W/(m$^2 \cdot$ K^4)). Figure 21 shows radiation heat-loss density as a function of temperature and ε. Similar to convection losses, there is a formula that provides a rough engineering estimation of the free radiation losses, q_{rad} (Ref 34):

$$q_{rad} = 5.67 \\ \times 10^{-8} \varepsilon [(T_s + 273)^4 - (T_a + 273)^4] (\text{W/m}^2) \qquad \text{(Eq 21)}$$

For example, freely radiated carbon steel slab ($\varepsilon = 0.8$) at a temperature of 1200 °C (2192 °F) loses into the surrounding atmosphere ($T_a = 20$ °C) the amount of heat that corresponds to the value of radiation loss density of:

$$q_{conv} = 5.67 \times 10^{-8} \times 0.8 \times (1473^4 - 293^4) \\ = 213 \times 10^3 \, \text{W/m}^2 = 213 \, \text{kW/m}^2$$

The previously described estimation of radiation heat losses is a valid assumption for classical workpiece geometry when there is free heat radiation from a heated body into the surroundings. However, there are some applications where the radiation heat-transfer phenomenon can be complicated and such a simple approach would not be valid. Details regarding subtleties of three modes of heat transfer can be found in Ref 36 to 46.

As discussed previously, in conventional induction heating, the heat transfer by convection and thermal radiation reflects the value of surface heat losses. Its high value reduces the total efficiency of the induction heater. As expected, these heat losses are highly nonlinear. Analysis shows (Fig. 22) that convection losses are the major part of the surface heat losses in low-temperature induction heating applications (i.e., aluminum, lead, zinc, tin, and magnesium; steel at a temperature lower than 350 °C, or 662 °F). In the majority of hot working applications (i.e., induction heating of steels, titanium, tungsten), thermal radiation losses are much greater than convection losses, representing the major portion of total surface heat losses.

It is wise to remember that according to Eq 10, the total efficiency of the induction coil is a product of both coil electrical efficiency and coil thermal efficiency. Nearly all coils used in induction heating prior to hot working use thermal insulation (i.e., refractory or liners) positioned between the coil and the heated workpiece. Besides reducing the heat losses from the workpiece surface and increasing thermal efficiency, the thermal insulation protects

the coil windings from heat exposure by acting as a thermal barrier. At the same time, the use of refractory results in the need for extra space for its installation, resulting in a larger coil diameter and consequently a larger coil-to-workpiece electromagnetic gap. This deteriorates coil-to-workpiece electromagnetic coupling and, in turn, decreases coil electrical efficiency (Fig. 23).

Therefore, on one hand, thermal refractory (heat insulation) allows improvement of coil thermal efficiency. On the other hand, it reduces coil electrical efficiency. A decision to use or not to use the refractory is always a reasonable compromise. In some cases, it is more energy- and cost-efficient not to use any refractory and minimize the coil-to-workpiece gap, thus maximizing coil electrical efficiency as is typically done in induction hardening, brazing, soldering, and other applications that apply relatively short heat cycles and high frequencies.

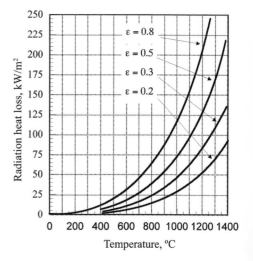

Fig. 21 Variation of the thermal radiation-loss density vs. temperature and emissivity (ε). Source: Ref 1

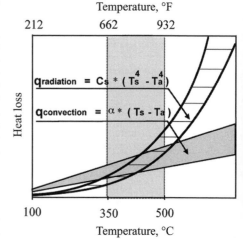

Fig. 22 Convection and thermal radiation heat losses versus temperature in typical induction heating applications. Source: Ref 32

Table 6 Comparison of emissivities, ε, of some commonly used polished metals vs. nonpolished metals

Surface condition	Aluminum	Carbon steel	Copper	Brass and zinc
Polished	0.042–0.053	0.062	0.026–0.042	0.03–0.039
Nonpolished, oxidized	0.082–0.40	0.71–0.8	0.24–0.65	0.21–0.50

Source: Ref 1

In other cases, it is advantageous to use refractory, thereby minimizing surface heat losses and overpowering the reduction of electrical efficiency with substantial improvement in thermal and total efficiency. This usually takes place in induction bar and billet heating prior to forging, rolling, and extrusion. Numerical computer modeling helps make an appropriate decision whether it is advantageous to use or not to use a thermal refractory.

Surface-to-Core Temperature Uniformity

It is typically much easier to provide surface-to-core temperature uniformity for metals with high thermal conductivity, such as aluminum, silver, or copper. Metals with poor thermal conductivity, including stainless steel, titanium, and carbon steel, require extra care to obtain the desired temperature uniformity. This extra care includes proper selection of heating mode, frequency choice, process time, and other parameters.

Figure 24 shows the time-temperature curve that is typical for the static mode (Fig. 2a) of heating a nonmagnetic solid cylinder. Immediately after the beginning of heating, the surface temperature and average temperature begin to rise. In contrast, the core temperature starts to grow after some delay. If the material properties are constant and surface heat losses are absent, then at a certain time, a steady state will take place (a linear region) where all three temperatures (surface, average, and core) are represented by three parallel lines (solid lines in Fig. 24). As soon as the power is cut off, the surface temperature decays rapidly due to heat conduction toward a cooler core; similarly, the core temperature would continue to rise although no power is applied to the induction coil.

In reality, the surface-to-core temperature differential starts to decline before the soaking stage begins (dotted curve in Fig. 24 represents a realistic surface temperature rise). This takes place due to the increasing surface heat losses with temperature and an increase of current

penetration depth, δ. It should be mentioned that the soaking stage can take place when the heated workpiece is inside the induction coil and/or during transfer of the heated workpiece to the next metalworking operation. The latter approach allows the total process time to be minimized (Ref 1, 10).

To reduce the heating time while providing required surface-to-core temperature uniformity, power pulsing can be applied (Fig. 25). Power pulsing refers to a technique that applies short bursts of power of greater magnitude than in the previously discussed case (Fig. 24), thus permitting a maximum allowable temperature to be quickly reached, which, in turn, leads to more intense heat flow from the surface toward the core. Pulse heating consists of a series of "Heat ON" and "Heat OFF" cycles, allowing the maintenance of maximum permissible surface temperature until the desired average temperature and uniformity are obtained. Depending on the application specifics, the process time reduction with pulse heating can exceed 40% compared to the conventional heating mode (Ref 1).

Although Fig. 25 illustrates typical time-temperature profiles during static pulse heating, similar curves can also represent progressive multistage heating and continuous heating modes as well. In these cases, the time axis represents the length of the induction heating line (Fig. 13 and 15). Bursts of power would represent the power of a respected in-line coil. Each coil may be of different lengths, having different windings, and/or can be individually fed from different power supplies with the ability to adjust the output power and frequency, or certain coils can also have series/parallel electrical connections being fed by a single

inverter. The previously described rough-estimation techniques can be used to obtain the powers of each in-line coil. However, it is important to remember that simplified formulas typically fall short in revealing process details such as the prediction of transit and final temperature distributions. Numerical modeling is typically required in order to obtain accurate surface-to-core temperature profiles (Fig. 14) within the heated workpiece at different heating stages.

Length of Induction Line

The determination of the length of the coil line is another important step in specifying an induction heating system. In making the decision of how long the induction line needs to be, it is actually determining the time required for heating the workpiece to an acceptable temperature condition. The minimum required length of induction line is a complex function of various factors including the workpiece geometry, applied frequency, heating mode, power density, maximum permissible temperature, material properties, the required heat uniformity, and production rate.

Numerical computation (for example, using the software ADVANCE and some other numerical codes) can be helpful in making a decision regarding the required coil length as well as other important coil design parameters. A rough estimate for determining the minimum heat time and coil length for progressive or continuous heating when heating low and medium plain carbon steel cylinders from ambient temperature to the typical forging temperatures

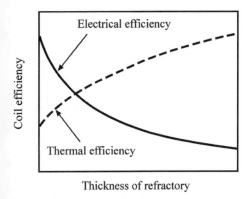

Fig. 23 Electrical and thermal efficiencies of induction coil vs. thermal refractory thickness. All arbitrary units. Source: Ref 1

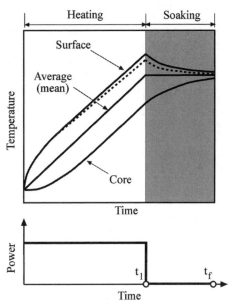

Fig. 24 Time-temperature profile during static heating of a nonmagnetic solid cylinder. Broken line is surface temperature with thermal losses considered

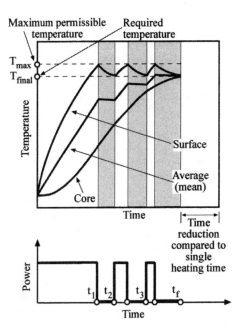

Fig. 25 Power pulsing can reduce induction heating stabilization time. Source: Ref 1

can be obtained using the empirical Eq 22 and 23 (Ref 48, 49):

$$\text{Minimum heat time} = K_{\text{time}} \times D^2 \qquad \text{(Eq 22)}$$

$$\text{Minimum coil length} = K_{\text{length}} \times PR \qquad \text{(Eq 23)}$$

where time is measured in seconds, D is the diameter of the solid cylinder, K_{time} is a coefficient that depends on the units of diameter D ($K_{\text{time}} = 25$ if D is measured in inches; $K_{\text{time}} = 38{,}750$ if D is measured in meters), and PR is production rate. If PR is measured in pounds/hour, then the length is measured in inches, and $K_{\text{length}} = 0.03$. If PR is measured in kilograms/hour, then length is measured in meters, and $K_{\text{length}} = 0.00168$. These expressions assume a conventional coil design (profiled or accelerated coil design approaches are excluded). Numerical computer modeling provides better accuracy in estimating required coil length (see the article "Simulation of Induction Heat Treatment" in this Volume).

Selection of Coil Copper Tubing

The great majority of induction coils are water cooled. Rectangular or round copper tubing is typically used for fabrication of coils for metal hot working. Copper tubing is naturally profiled for the water passage. It is important to make sure that the water passage is sufficient in extracting the heat generated within the coil turns due to Joule losses and heat transfer from the heated workpiece. The amount of heat (kilowatts) required to be dissipated by water cooling can be estimated based on power losses, $P_{\text{loss}}^{\text{turns}}$, in coil copper turns. This is determined based on total coil efficiency, η, and coil power, P_{c}, according to the expression $P_{\text{loss}}^{\text{turns}} = P_{\text{c}} \cdot (1 - \eta) \cdot K_{\text{R}}$, where K_{R} is a dimensionless coefficient representing an amount of heat being absorbed by the induction coil due to heat leakage through the refractory (if used) or the amount of heat transferred from the surface of the heated workpiece to the coil turns due to the heat convection and thermal radiation (if refractory is not used). For induction forging applications, K_{R} is usually within the range of 1.05 to 1.15.

Tube wall thickness should be chosen based on the operating frequency that directly affects the current penetration depth in the coil copper (δ_{Cu}). The minimum copper tubing wall thickness should increase as the frequency decreases. For example, a system with low operating frequency requires a thicker wall tube than a high-frequency system. To minimize coil electrical losses, the wall thickness of copper tubing that faces the heated workpiece should be greater than $1.6\delta_{\text{Cu}}$. An excessive coil copper loss will take place if the tubing wall is smaller than $1.6\delta_{\text{Cu}}$, resulting in a reduction in coil efficiency. The copper tubing wall may be thicker than calculated according to the suggested expression, because it may not be mechanically reliable to use too-thin wall tubing due to the mechanical flexing caused by electromagnetic forces. As frequency is lowered, more attention must be paid to coil support because there is more vibration at lower frequencies, especially at the turns near both ends of multiturn solenoid coils (Ref 1, 50).

Electromagnetic Forces

A current-carrying conductor placed in a magnetic field experiences a force that is proportional to the current and magnetic flux density (Ref 1, 50, 51). Thanks to Ampere and Biot-Savart, this force is quantified. If a current-carrying element (dl), carrying a current (I), is placed in an external magnetic field (B), it will experience a force (dF), according to:

$$dF = I \times B \, dl = I \, B \, dl \sin\phi \qquad \text{(Eq 24)}$$

where F, I, and B are vectors, and ϕ is the angle between the direction of the current, I, and the magnetic flux density, B.

In SI units, the force is measured in Newtons (1 N = 0.102 kg-force = 0.225 lb-force). Figure 26 shows that the direction of the magnetic force experienced by the element dl of the current-carrying conductor placed in an external magnetic field B can be determined based on the left-hand rule. According to the rule, if the middle finger of the left hand follows the direction of current flow and the pointer finger follows the direction of the magnetic flux of the external field (magnetic field lines head into the palm), then the thumb will show the direction of the force.

It is important to remember from Eq 24 that if the angle (ϕ) between the direction of current (I) and magnetic field (B) is equal to zero, then $\sin\phi = 0$, and therefore, no force will be experienced by the current-carrying conductor. In other words, if the current-carrying conductor is parallel to an external magnetic field, it will not experience any force from that field.

If two current-carrying conductors (such as bus bars or coil copper turns) having currents flowing in opposite directions are located near each other, then each conductor will experience forces oriented in the opposite direction (Fig. 27a) that attempt to separate the conductors, $F_{12} = -F_{21}$. In contrast, if two conductors are carrying currents oriented in the same direction (Fig. 27b), the resultant forces will act as attractive forces, compressing conductors toward each other, $F_{12} = F_{21}$. What follows are simplified calculations of attractive magnetic forces occurring between two thin wires, each carrying a current of 200 A and separated by a distance of 20 mm (0.8 in.). According to basic electromagnetics, each of the parallel current-carrying wires produces a magnetic field according to the equation:

$$B = \mu_0 I / 2\pi R \qquad \text{(Eq 25)}$$

where R is the radial distance between the wires (Fig. 28). Therefore, the magnetic force experienced by the second wire, according to Eq 24, will be:

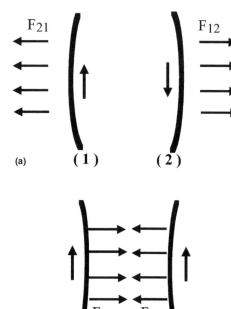

Fig. 27 Magnetic force in current-carrying conductors 1 and 2. (a) Current flow in opposite direction. (b) Current flow in same direction. Source: Ref 1

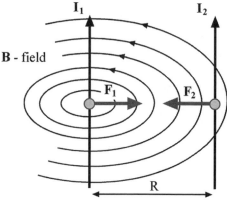

Fig. 28 Magnetic interaction between two thin wires. Source: Ref 1

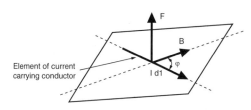

Fig. 26 Left-hand rule of magnetic force (F) on current element (Idl) caused by magnetic flux (B)

$$F = I_2(\mu_0 I_1/2\pi R)l$$

and the force per unit length will be:

$$F/l = I_2(\mu_0 I_1/2\pi R)$$

In this case, the force per unit length will be:

$$F/l = [4\pi \times 10^{-7}\ \text{Wb}/(\text{A} \times \text{m})(200\ \text{A})^2]/2\pi(0.02\ \text{m})$$
$$= 0.4\ \text{N/m}$$

These phenomena can also be applied to a multiturn solenoid coil (Fig. 29). Alternating voltage applied to a multiturn solenoid results in a current flow within it, producing electromagnetic forces. Because the currents flowing in each turn of the multiturn solenoid are oriented in the same direction, the turns will experience longitudinal compressive stresses. Assuming an infinitely long solenoid and a homogeneous magnetic field, it can be shown that the longitudinal magnetic pressure, f_l (density of the magnetic force in N/m^2), inside the infinitely long and homogeneous solenoid can be expressed as:

$$f_l = F_l/\text{Area} = \mu_0 H_t^2/2 = B_t^2/(2\mu_0) \quad \text{(Eq 26)}$$

where H_t is the root mean square tangential component of the magnetic field intensity, H:

$$H_t = NI/l \quad \text{(Eq 27)}$$

where N represents a number of turns of the long solenoid, l is its length, and I is its current. At the same time, the turns of the solenoid experience tensile forces in the radial direction, because the current flowing on the opposite side of each turn is oriented in the opposite direction. The radial tensile magnetic pressure, f_R, can be described as:

$$f_R = \mu_0 H_t^2/2 = B_t^2/(2\mu_0) \quad \text{(Eq 28)}$$

Another assumption used when deriving Eq 26 and 28 is that the solenoid is empty or consists of an infinitely long nonmagnetic load with a constant electrical resistivity. It must be emphasized that because eddy currents induced by the induction coil within the heated workpiece are oriented in a direction opposite to that of the coil current, the coil turns experience tensile magnetic pressure, whereas the workpiece is under compressive pressure. To provide a rigid and reliable coil design, this magnetic pressure should be taken into consideration.

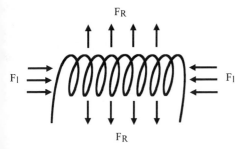

Fig. 29 Magnetic forces, axial (F_l) and radial (F_R), in a solenoid multiturn coil. Source: Ref 1

Coil End Effects

The discussion so far has considered only an infinitely long solenoid. However, when the workpiece and, in particular, induction coil are of finite length (which is the realistic case), the electromagnetic end and edge effects have a pronounced impact on the orientation, magnitude, and distribution of the magnetic forces. Electromagnetic end and edge effects are discussed in detail in Ref 1 and 51 to 54.

Two typical examples are shown in Fig. 30. If a nonmagnetic bar is partially placed inside a multiturn inductor (for bar-end heating, for example), the magnetic force will be trying to eject the bar from the coil (Fig. 30a). Stronger forces appear when heating bars of low-electrical-resistivity metals (such as aluminum, copper, or brass). However, the situation is quite different when a magnetic bar (carbon steel, nickel, or cast iron, for instance) is partially placed inside a multiturn inductor (Fig. 30b). The resulting force is a combination of two forces: one resulting from the demagnetization effect, which attempts to remove the bar from the inductor, and the other resulting from the magnetization effect, which attempts to pull the bar toward the center of the coil (Ref 1, 51, 54). The force due to magnetization is typically the stronger of the two, resulting in the net force, which attempts to pull the bar inside the coil.

In most induction heating applications, the electromagnetic (EM) force has a complex three-dimensional distribution. Depending on application specifics, one of three force components—longitudinal (axial), radial, or hoop (tangential)—may be significantly greater than the others. It is important to remember that three-dimensional distribution of forces is also a function of the geometry of the system, and

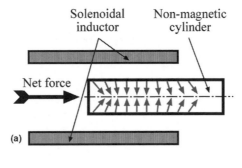

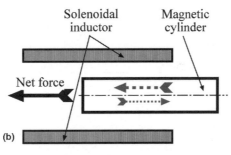

Fig. 30 Magnetic forces in bar-end heating of magnetic and nonmagnetic bars. Source: Ref 1

its magnitude and orientation can vary appreciably during the heating cycle.

It is obvious that the basic formulas discussed previously can be applied only in simplified cases. For the majority of induction heating applications, only numerical computer modeling study can help a designer to accurately predict EM forces experienced by the inductor, heated workpiece, and other components of the induction system and what actions should be taken to fabricate robust and reliable coil designs. For example, experienced users of induction forging equipment who are involved in maintaining multiturn forge coils are likely to have noticed that failure of such inductors is often associated with the failure of turns located at its ends (Ref 1, 51, 54). Arcing, copper overheating, excessive vibration, development of a short circuit between turns, and liner and/or refractory disintegration are only a few of the failure modes observed at end turns that are related to a distortion of the electromagnetic field in the coil end areas (Fig. 31) and the appearance of excessive forces there (Ref 1, 51, 54).

In coil areas located away from its ends (the so-called regular area of the inductor), the magnetic field distribution can often be considered relatively uniform and homogeneous, particularly for electromagnetically long coils having large values of the turn space factor, K_{space}. As discussed in Ref 1, the coil turn space factor indicates how tightly the coil turns were wound. It can be calculated as follows:

$$K_{space} = \text{Turn width}/(\text{Turn width} + \text{Axial gap between turns})$$

For example, a single-turn coil has a space factor of 1. For the majority of multiturn inductors used in annealing, stress relieving, and heating for hot working, $K_{space} = 0.7$ to 0.9. This means that coil turns are tightly wound, and slight disturbances in the magnetic field between turns will not have an appreciable effect on the magnetic field distribution in the regular area of the coil (Fig. 31). In contrast, a distortion of the magnetic field at the coil end is quite dramatic and affects several factors that can be directly related to specific modes of premature coil failure in this area. Examples of these factors include:

- Distortion of current-density distribution within turns located in the coil end area
- Redistribution of the power losses and voltage drops per turn
- Complex distribution of magnetic field force or magnetic field pressure experienced by turns

Figure 32 shows the distribution of three magnetic force components (axial, hoop, and radial) experienced by the turns of the coil design shown in Fig. 31 (frequency = 3 kHz). In this particular case, the radial component of the magnetic force is minor and can be neglected in a coil design. The greatest force experienced by turn 1 is related to its axial component. This component gradually decreases, becoming as low as the radial component in the regular area of the coil, away from the end. Note, however, that the hoop component of the force gradually increases, reaching its maximum value in the regular area. Knowledge of the magnitude and orientation of magnetic forces is essential to prevent excessive vibration of coil turns and to ensure the fabrication of a durable and sound inductor.

To conclude this discussion regarding magnetic forces, it should be mentioned that the magnetic field and force distribution in the coil end area is a complex function of several often-interrelated factors. These include but are not limited to:

- Electromagnetic properties of the heated metal (i.e., magnetic versus nonmagnetic; good electrical conductor versus poor conductor)
- Frequency and magnetic field intensity
- Workpiece geometry
- Heating mode
- Specifics of coil design, including coil overhang, coil-to-workpiece coupling, space factor of turns (K_{space}), coil copper tubing, presence of magnetic flux concentrators and shunts, presence of Faraday rings, flux extenders, and other electromagnetic devices (Ref 1, 51, 54)

It is important to remember that, depending on the application, these factors can have different impacts. Although each application should be considered individually, there are some general recommendations or tendencies. For example, power losses within the coil end turns are typically greater when heating nonmagnetic metals that have low electrical resistivities than when heating magnetic materials that have high resistivity values. Even such frequently overlooked parameters as the profile of the coil turns (their height and width) could, in some cases, have a noticeable effect on field force and power loss distribution along the coil turns. Simplified formulas typically fall short in the prediction of magnetic forces experienced by different coil components and turns. Numerical computer modeling provides the reliable capability to predict how numerous interrelated factors will affect magnetic forces.

Frequency Selection for Various Induction Heating Systems

Frequency Selection for Heating Solid Cylinders. Power supply frequency is one of the most critical parameters of the induction heating system. If the frequency is too low, an eddy current cancellation within the heated body takes place, resulting in poor coil electrical efficiency (Fig. 18). On the other hand, when the frequency is too high, the skin effect will be highly pronounced, resulting in a current concentration in a very fine surface layer compared to the diameter/thickness of the heated component. To provide sufficient heating of the core by thermal conduction, a long heating time (thus a long induction heating line) is needed. Prolonged heat time is associated with an increase of the radiation and convection heat losses that, in turn, reduce the total efficiency of the induction heater. The choice of frequency is based on a reasonable compromise. Table 7 shows minimum bar/billet diameters as a function of frequency and temperature for efficient induction heating solid cylinders made from selected metals.

Frequency Selection for Tube and Pipe Heating. Figure 33 shows that there is a difference in the frequency selection for induction heating of tubular workpieces as compared to solid cylinders. In induction tube and pipe heating, the optimal frequency, which corresponds to maximum coil efficiency, is shifted toward lower frequencies (frequencies are between F_1 and F_2 for tubes instead of F_2 to F_3 for solid cylinders). The most desirable frequency for heating hollow cylinders typically provides a current penetration depth larger than the tube wall thickness (except for heating of relatively small tube sizes). This condition can result in measurable increase in coil electrical efficiency. In some instances, an improvement in electrical efficiency can exceed 10 to 16%. In others, the increase in coil efficiency is less pronounced and may not even be noticeable (Ref 1, 57, 67, 68).

Generally speaking, the value of an appropriate frequency when heating tubular workpieces is a complex function of several parameters, including the electromagnetic properties of

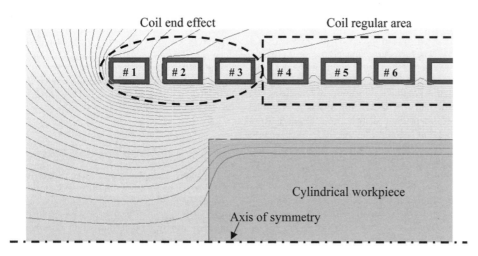

Fig. 31 Magnetic field distribution in a multiturn induction coil showing the coil end effect. Source: Ref 54

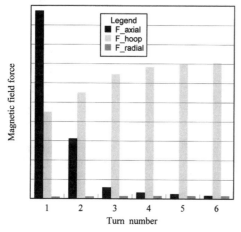

Fig. 32 Distribution of the three components of magnetic field force along the turns of a multiturn induction coil. Source: Ref 54

Table 7 Minimum bar/billet diameters (mm) for efficient induction heating

Material	Temperature °C	°F	Frequency, kHz 0.06	0.2	0.5	1	2.5	10	30
Copper	900	1652	68	35	23	17	11	5	3
Aluminum	500	932	68	35	23	17	11	5	3
Brass	900	1652	102	56	35	26	16	8	5
Titanium	1200	2192	304	168	105	74	47	23	13
Tungsten	1500	2732	168	92	58	43	27	14	8
Steel	1200	2192	253	140	94	65	41	19	12

Source: Ref 1

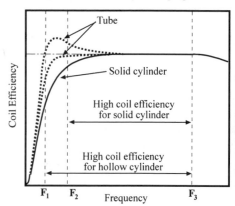

Fig. 33 Coil electrical efficiency vs. frequency when heating hollow and solid cylinders. Source: Ref 67

heated metal, coil length, the ratio of coil inside diameter to tube outside diameter, and the ratio of tube wall thickness to current penetration depth, with the last two being the most critical factors. Numerical computer modeling allows the selection of the optimal frequency for a particular application. At the same time, several simplified formulas are in use in the tube/pipe industry. For electromagnetically long solenoid-type inductors, a quick rough estimate of the appropriate frequency can be done using Eq 29 and 30. Table 8 consists of frequencies obtained by using the formula recommended in Ref 36 when heating different metals:

$$Frequency = 34.6 \frac{\rho}{A_m h} (Hz) \qquad (Eq\ 29)$$

where ρ is the electrical resistivity of the heated metal ($\mu\Omega \cdot in.$); A_m is the average diameter, A_m = (Tube outside diameter – h) in inches; and h is the wall thickness, in inches or:

$$F_{optimal} = 8.65 \frac{\rho 10^5}{A_m h} (Hz) \qquad (Eq\ 30)$$

where the units are ρ ($\Omega \cdot m$), and A_m and h are measured in meters (Ref 33).

It should be taken into consideration that in cases when induction heaters cannot be considered to be electromagnetically long but rather electromagnetically short inductors, the value of the optimum frequency will be higher than the values shown in Table 8.

It is also necessary to mention that in tube and pipe heating, an audible noise can become a critical factor that greatly affects the selection of frequency. An undesirable combination of tube geometry, frequency, and power density could lead to a measurable emitting of resonant sound waves. A frequency selection different than the previously suggested values may be used to help reduce an undesirable audible noise.

Frequency Selection for Heating Rectangular Workpieces. Induction heating is often the most popular choice for heating of rectangular-shaped workpieces, which include slabs, blooms, round-cornered-square (RCS) billets, plates, and strips. Coil arrangements for heating rectangular workpieces include solenoid coils, transverse flux inductors, traveling-wave inductors, and "C"-core inductors (Ref 1, 21, 22, 52, 53, 58). The great majority of rectangular workpieces are heated using solenoid multiturn inductors (Fig. 34, 35). The complex nature of various electrothermal phenomena (for example, electromagnetic edge and end effects, thermal edge effects) that takes place when heating rectangular workpieces is discussed in Ref 1, 58.

The choice of frequency has a major effect on coil electrical parameters and the temperature profile within the slab body, including heat uniformity along the length, thickness, and across the width. When heating a rectangular-shaped body, it is convenient to quantify the

Table 8 Desirable frequencies for some tubes/pipes using solenoid-type inductors under the assumption that these coils can be considered to be electromagnetically long

Tube outside diameter		Wall thickness, mm	Optimal frequency, kHz		
mm	in.				
Nonmagnetic stainless steel			20 °C (68 °F)	800 °C (1472 °F)	1200 °C (2192 °F)
12.7	0.5	1	51	85	92
		2	28	47	50
		3	21	34	37
25.4	1.0	1	25	41	44
		2	13	21	23
		3	8.9	15	16
		5	5.9	9.8	11
50.8	2.0	1	12	20	22
		2	6.1	10.1	11
		3	4.2	6.9	7.5
		5	2.6	4.3	4.7
76.2	3.0	1	7.9	13.2	14.3
		2	4	6.7	7.2
		3	2.7	4.5	4.9
		5	1.7	2.8	3
102	4.0	1	5.9	9.9	10.6
		2	3	5	5.4
		3	2	3.4	3.6
		5	1.2	2.1	2.2
Brass			20 °C (68 °F)	400 °C (752 °F)	900 °C (1632 °F)
12.7	0.5	1	4.8	8.4	15
		2	2.6	4.6	8.2
		3	1.9	3.4	6
25.4	1.0	1	2.3	4	7.2
		2	1.2	2.1	3.8
		3	0.84	1.5	2.6
		5	0.6	1	1.7
50.8	2.0	1	1.1	2	3.5
		2	0.6	1	1.8
		3	0.4	0.7	1.2
		5	0.25	0.43	0.8
Copper			20 °C (68 °F)	500 °C (932 °F)	900 °C (1632 °F)
12.7	0.5	1	1.33	3.7	6.3
		2	0.73	2	3.4
		3	0.54	1.5	2.5
25.4	1.0	1	0.64	1.8	3
		2	0.33	0.92	1.6
		3	0.23	0.64	1.1
50.8	2.0	1	0.31	0.87	1.5
		2	0.16	0.44	0.8
		3	0.11	0.3	0.51

Source: Ref 1

Fig. 34 Inductor for heating metal plates and strips. Courtesy of Inductoheat, Inc.

skin effect using the ratio of d/δ, where d is the thickness of the slab, and δ is the current penetration depth calculated according to Eq 5 or 6. More uniform temperature profiles along the slab thickness correspond to a lower ratio of d/δ. If d/δ is appreciably greater than eight, the temperature distribution along the slab thickness and, in particular, across the width will be noticeably nonuniform. An increase of cycle time in combination with power density reduction leads to more uniform heating because thermal conductivity helps to equalize the thermal gradients. Keep in mind, though, that an increase in cycle time results in a corresponding increase in heat losses and a respective reduction of thermal efficiency, η_{th}.

Similar to induction heating of cylinders, the choice of frequency affects not only the required temperature profile within the slab but also the coil electrical efficiency (Ref 1, 22, 50, 52, 53, 58). There is an optimal frequency ($F_{el.eff}$) that corresponds to the maximum value of coil electrical efficiency, η_{el}. The use of a frequency higher than the optimal will only slightly change the efficiency. However, a selection of frequency significantly higher than optimal tends to substantially overheat corners, requiring longer heat time and reducing overall efficiency. If the chosen frequency is noticeably lower than the optimal value, the electrical efficiency can dramatically decrease due to cancellation of the induced currents circulating in the opposite sides of the slab cross section. Table 9 shows the minimum slab thickness for efficient slab heating of a variety of metals.

High coil electrical efficiency will be obtained if the ratio of slab thickness, d, to penetration depth, δ, is 2.8 or more. Frequency that corresponds to the maximum coil electrical efficiency while heating infinitely wide slabs can be determined as follows (Ref 1, 22, 50, 52, 53, 58):

For nonmagnetic slab or a magnetic slab heated above the Curie temperature:

$$\frac{d}{\delta_{nonmagn.}} \cong 3 - 3.5 \qquad \text{(Eq 31)}$$

For magnetic steel heated below the Curie temperature:

$$\frac{d}{\delta_{magn.}} \cong 2.8 - 3.1 \qquad \text{(Eq 32)}$$

where $\delta_{nonmagn.}$ is the current penetration depth in nonmagnetic slab, and $\delta_{magn.}$ is the current penetration depth in magnetic slab.

It is important to be aware that the presence of surface heat losses and the need to provide heat uniformity in two-dimensional edges and three-dimensional corners increase desirable frequencies compared to those recommended in accordance with Eq 31 and 32.

While discussing the efficiency of heating, it is imperative to mention that smaller coil-to-work-piece radial air gaps and tighter windings of coil turns improve the coil efficiency. The higher ratio of b/d (where b is the width of the slab) also corresponds to the higher coil electrical efficiency (assuming the same slab-to-coil gaps).

In the past, coil calculations for heating blooms and RCS bars/billets with a square cross section were conducted using formulas for equivalent cylinders (cylinders with equivalent diameters). An error in such calculations is usually within 5 to 10%. A calculation error quickly increases with the higher ratio of b/d if $b/d > 1.5$. Therefore, a calculation approach based on equivalent diameters should not be used. In cases such as this, numerical simulations offer better results (Ref 1, 22, 58). For illustration, the companion article, "Simulation of Induction Heat Treatment," in this Volume provides several case studies using numerical computer simulation of RCS carbon steel billets using frequencies lower and higher than the optimal one.

Fig. 35 Unique 6000 kW/110 Hz induction coil provides heating of the world's largest steel slab. Slab geometry: 3.2 m wide and 0.22 m thick. Courtesy of Inductotherm Corp.

Table 9 Minimum thickness (mm) for efficient heating of nonmagnetic wide slabs and plates using longitudinal flux inductors

Material	Temperature		Frequency, kHz					
	°C	°F	0.06	0.5	1	2.5	4	10
Aluminum	100	212	31	10.7	7.6	4.8	3.8	2.4
	250	482	37	12.9	9	5.8	4.6	2.9
	500	932	48	16.6	11	7.4	5.9	3.7
Copper	100	212	24	8	6	3.7	2.9	1.9
	500	932	36	12.6	8.9	5.6	4.4	2.8
	900	1652	48	16.7	11.8	7.5	5.9	3.7
Brass	100	212	44	15	10.9	6.9	5.4	3.4
	500	932	59	20	14	9.2	7.3	4.6
	900	1652	73	25	17.9	11	9	5.7
Silver	100	212	24	8.5	6	3.7	3	1.9
	300	572	31	11	7.8	4.9	3.9	2.4
	800	1472	43	14.9	10	6.6	5.2	3.3
Nonmagnetic stainless steel	100	212	143	49	35	22	17.5	11
	800	1472	174	60	42	27	21.3	13
	1200	2192	180	63	44	28	22	14
Tungsten	100	212	38	13	9	5.8	4.6	2.9
	900	1652	90	31	22	13.9	11	7
	1500	2732	120	42	30	18.6	14.7	9
Titanium	100	212	131	45	32	20	16	10
	600	1112	192	66	47	30	23	14.9
	1500	2732	218	75	53	33	26	17

Source: Ref 1

Conclusion and Example Calculations

The previously described rough-estimation techniques in combination with an experience gained on previous jobs allow a basic idea to be quickly obtained regarding the critical parameters of the induction system for heating of

Fig. 36 Multicoil induction in-line heater. Courtesy of Inductoheat, Inc.

a particular component. This knowledge is imperative for developing a general understanding of the process and for developing gut-feelings regarding what would it take to heat a particular workpiece according to basic process requirements. Unfortunately, the great majority of simplified formulas have many restrictions, making it difficult to get design details. This is true not only for applications related to induction heating of metals prior to hot working but, to an even greater degree, for induction heat treating. Advanced numerical simulation software based on tightly coupled electromagnetic and thermal phenomena enable induction heating designers to determine details of the process that could be costly, time-consuming, and, in some cases, extremely difficult or impossible to determine experimentally.

Recent advances in developing fast-performing computers with a substantial memory, hard-drive capacity, and speed of computations make it increasingly critical to develop subject-oriented, highly efficient numerical simulation codes to be used as a part of the process control system (including the programmable logic controller, the human machine interface, and other controls), assuring the achievement of optimal process parameters and maximizing system performance.

The subject of numerical computer simulation of induction heat treating and induction heating

processes is discussed in the companion article, "Simulation of Induction Heat Treatment," in this Volume. It is imperative, however, at this point to provide a few case studies to illustrate the capabilities of numerical simulation in comparison to rough-estimation techniques.

Case Studies of Numerical Simulation

Example 4: Electromagnetic and Thermal Subtleties of In-Line Induction Heating. Depending on the applications, some induction systems (for example, billet, rod, or bar heating systems) may consist of several in-line induction coils (Fig. 36). The challenge with in-line induction heating arises from the fact that the surface-to-core temperature profile continues to change as the bar passes through the line of induction coils. As discussed earlier, due to the physics of induction heating, the bar core tends to be heated slower than its surface. At the same time, the leading and trailing ends have a tendency be heated differently than the central (regular) region of the bar due to transient end effect, which is responsible for the appearance of a nose-to-tail temperature nonuniformity (Ref 1). Figure 37 shows the results of numerical simulation of the temperature distribution resulting from a transient end effect during induction heating of titanium alloy

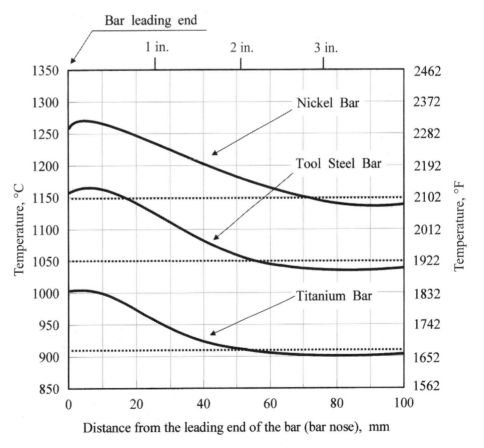

Fig. 37 Temperature profile of in-line induction heating of titanium bar with a diameter of 25.4 mm using a frequency of 30 kHz at 50.8 mm/s. Source: Ref 1

Fig. 38 Results of numerical simulation of transient end effect in the bar leading end zone (axial temperature distribution) when heating different materials. Source: Ref 1

(Ti-6Al-4V) cylinder bars of 25.4 mm (1 in.) diameter traveling at a speed of 50.8 mm/s (2 in./s) using a frequency of 30 kHz. As an example, Fig. 38 shows the results of numerical simulation of the transient end effect in the bar leading end area (axial temperature distribution) when heating different materials. Keep in mind that leading and trailing ends of the same bar are typically heated differently.

It is important to remember that the transient end effect can manifest itself not only in overheating of the bar ends but also in their underheating. Bar ends typically have a heat surplus when heating nonmagnetic metals. However, nose-to-tail temperature distribution can be quite different when heating ferromagnetic bars. The appearance of transient end effect depends on a number of factors, including physical properties of the heated materials, design features of the induction system, and the process recipe.

Figure 13 shows the results of the simulation of induction heating of a 76 mm (3 in.) diameter carbon steel bar and its surface-to-core temperature profile along the induction line in the steady-state condition using the numerical modeling software ADVANCE. The coil parameters include the following:

- Inside diameter: 152 mm (6 in.)
- Refractory thickness: 12 mm (0.5 in.)
- Coil length: 1 m (40 in.)
- Number of coils: 8
- Gap between coils: 0.3 m (12 in.)
- Frequency: 1 kHz
- Production rate: 65 mm/s (2.56 in./s)

The results of the simulation reveal comprehensive details of the process that would be extremely difficult, if not impossible, to determine experimentally.

At the initial heating stage, the entire bar is magnetic, and the skin effect is pronounced. Power induced in the magnetic bar concentrates in the fine surface layer, which typically does not exceed 6 mm (0.25 in.) for frequencies greater than 500 Hz. In addition, surface heat losses due to thermal radiation and convection are relatively low at this stage. Both factors lead to a rapid increase in temperature at the bar surface, with practically no change at its core. An appreciable surface-to-core temperature gradient could be present due to an intensive heating of bar surface layers.

Figure 14 (top) shows a surface-to-core temperature profile and radial power density (heat source) distribution after the bar exits coil 1 (scales of power density profiles are different for various coil positions). The temperature profile does not match the heat source profile because of thermal conductivity, which spreads the heat from the surface toward its core.

During the initial heating stage, the coil efficiency is quite high (typically 80%) and continuously increases due to an increase in steel electrical resistivity with temperature (Fig. 4). Because the surface temperature is below the Curie point, the magnetic permeability remains relatively high. After a short time, the coil efficiency reaches its maximum value, and the efficiency starts to decline because the temperature of the surface layers starts to approach a Curie temperature (Fig. 8a). The interim heating stage takes place when the bar surface loses its magnetic properties (i.e., after exiting coil 3, Fig. 14). The intensity of heating started to decrease at this stage due to the following factors:

- Because the surface of the carbon steel bar loses its magnetic properties and the relative magnetic permeability drops to 1, the coil electrical efficiency declines, and the power density induced within the bar surface is decreased.
- Specific heat has its maximum value (a peak) near the Curie point (Fig. 16, right). The value of the specific heat denotes the amount of energy that must be absorbed by the material to achieve the required temperature rise.
- Latent heat is present.

At the end of an initial heating stage, the electrical resistivity of the carbon steel increases approximately two to three times compared to its value at ambient temperature. At the same time, the decrease in magnetic permeability is much more pronounced (thirtyfold or more). Both factors cause an increase in current penetration depth of 6 to 12 times. A significant portion of the power is now induced in the internal layers of the bar. Although the bar surface becomes nonmagnetic, its internal layers retain their magnetic properties. This is the reason why the next heating stage is called the interim heating stage, showing a strong presence of the magnetic-wave phenomenon.

When the thickness of the surface layer that is heated above the Curie point exceeds the penetration depth in hot steel, the magnetic-wave phenomenon disappears, and the interim heating stage ends. The induced power density will be very similar to a classical exponential distribution (Fig. 10a), and the final heating stage begins.

In most instances, a reasonable compromise must be reached when specifying a frequency for an in-line bar heater because bar processing companies can rarely heat a dedicated bar size using a particular induction heating system. Therefore, it is often necessary to heat a variety of bars. For some of them, the frequency of the supplied inverter must be higher than optimal, but for others, it may be lower than optimal. Some of the solid-state inverters lend themselves to dual- or multifrequency configurations for the load circuit, which, to a large extent, can overcome this problem.

Because an induction heating line often must process bars of several different diameters, the specified cross-sectional range to be processed in a given set of induction coils and the required number of coil sets involve the consideration of a number of factors affecting the total energy efficiency of the system. Besides frequency and the other factors discussed previously, coil heating efficiency is largely a matter of the coil fill factor (area of the workpiece to be heated compared to the inside diameter of the coil windings). As the coil fill factor decreases, the coil efficiency decreases as well, requiring more power. Energy costs rise as the heating efficiency decreases. On the other hand, the savings in energy cost by using a number of coil sets having different inside diameters that are oriented on processing certain bar sizes maximizing coil fill factor are diminished by the capital cost of investing in several sets of coils. There is also a production loss due to the time required to change coil sets, although advanced quick-change or shuttle design features can minimize this downtime. A careful analysis of the product mix is necessary to determine how often the bar size may change and what the duration of each product run will be.

It is often cost-effective for the customer using the coil set designed for heating "big runners" to heat the small-sized bars as well, which, under normal conditions, would require separate coil sets. Computer simulation helps the user to make an intelligent decision regarding the use of the second and possibly third coil set as well as the bar thermal conditions of the entire product range. Note that below the Curie point, heating efficiency is not as greatly affected by bar size variation; thus, it may be worthwhile to change only the so-called above-Curie coils of an in-line multicoil induction system.

Figure 39 shows an InductoForge modular billet heater, representing a novel technology that was specifically developed for improving the performance of in-line induction systems used for heating a wide range of bar/billet sizes.

Fig. 39 InductoForge modular billet heater. Courtesy of Inductoheat, Inc.

The frequency of each module can vary from 500 Hz up to 6 kHz. The flexibility of this feature alone can be illustrated by the following practical example. A frequency of 500 Hz provides in-depth and highly efficient heating when processing 4 in. diameter steel billets to forging temperatures. However, if the future market situation changes and a shop needs to run 1 in. diameter bars, then 500 Hz or even 1 kHz frequency will not be able to efficiently heat those parts to forging temperatures due to eddy current cancellation, resulting in dramatic coil efficiency reduction. Low frequency can still be used in the initial heating stage, when the bar retains its magnetic properties. Further down the heating line, when the bar becomes nonmagnetic, it is more efficient to use a higher frequency to avoid current cancellation and undesirable reduction of coil efficiency. Therefore, by switching the output frequency of inductor sections that correspond to the final heating stage from 500 Hz to 6 kHz, the InductoForge modular billet heater will not only be able to process smaller-diameter parts but also will be able to maintain maximum electrical efficiency and required temperature uniformity.

If required, InductoForge modules can easily be combined in-line to form a heater that provides the required high production rate. It is also easy to add or remove modules to the heating line to match changes in production, or modules can be redistributed to different production areas (shops), making it possible to run different parts at a lower production rate. The multifrequency modular design concept of the induction bar heater also allows profiling of the power distribution along an induction heating line, providing true optimization of the heating process.

When choosing a numerical simulation code, it is imperative that it be capable of taking into consideration the aforementioned features of modern in-line induction heaters, allowing the simulation of the effect of different process recipes and system design subtleties, including steady-state and transient heating conditions (such as those shown in Fig. 13, 14, 37, 38) and discovering information that helps to optimize the performance of the induction heating line.

Example 5: Subsurface Overheating Phenomenon. One of the challenges in induction heating of billets and bars arises from the need to ensure particular surface-to-core temperature uniformity. Some practitioners incorrectly assume that with induction heating, the coldest temperature is always located at the core of the heated billet, and the maximum temperature is always located at the billet surface. It is often assumed that overheating does not occur if the surface temperature, measured by a pyrometer, does not exceed the maximum permissible level.

It is imperative to recognize that the presence of heat loss from the billet surface could shift the temperature maximum an appreciable distance further away from its surface, marking

its location somewhere beneath the billet surface (Ref 9, 58–61). Positioning and magnitude of the subsurface heat surplus is a complex function of five major factors: type of heated metal, frequency, refractory, final temperature, and distribution of heating power along the induction heating line.

Effect of Frequency. Lower frequency increases current penetration depth, resulting in more in-depth heating and providing faster heating of the billet core. This shortens the induction line; however, depending on the specifics of an application, it can also aggravate subsurface overheating by increasing its magnitude and shifting the location of the maximum temperature further away from the surface.

Effect of Refractory. The use of an appreciably thick refractory with improved thermal insulation properties does just the opposite. It increases the coil thermal efficiency, reduces subsurface overheating, and shifts the billet maximum temperature toward its surface.

Effect of Heating Temperature. An increase in the final required temperatures leads to an effect similar to lowering the frequency in regards to the location and severity of subsurface overheating.

Effect of Power Distribution. The effect of power distribution along the induction heater (if a single inductor is used) or along the heating line (if multiple coils are used) on the final temperature uniformity is quite complex and seldom discussed in the literature. In most publications devoted to in-line induction heating of billets and bars, it is strongly suggested to have a graded (profiled) power distribution along the induction line by putting more power into the coils located at the beginning of a multicoil heating line. Putting more power upfront may seem to be a universal rule of thumb. It forces more energy into the billet at the front of the heating line, giving the heat more time to soak into the core of the billet and resulting in better surface-to-core uniformity at the end of heating, because the temperature in the center of the billet can reach the forging temperature in a shorter period of time. This also may reduce the length of the coil line. Traditionally, such an approach uses a single inverter that powers several coils with graded (profiled) or conventional uniform coil winding and/or series/parallel coil circuit connections.

Industry has accumulated different recommendations regarding a preferable distribution of power along an induction heating line. This includes the following rules of thumb: 60/40 rule, 70/30 rule, 80/20 rule, and so on. The first two digits represent a percentage of the total power that should be applied to the first half of the induction heating line. The last two digits correspond to a percentage of the total power that should be applied to the second half of the induction heating line.

However, the problem with this approach is that the power distribution along the heating line cannot be easily modified if the production rate, type of heated metal, or billet size

changes. For example, with a conventional induction design, if the production rate is reduced, the subsurface overheating typically worsens, negatively affecting the billet subsurface microstructure (Ref 9, 61). Because the system puts more energy into the billet in the coils positioned at the beginning of the induction line, too much energy may soak down into the billet subsurface area and its core when the line runs slowly. The presence of surface heat loss can reverse an expected (traditional) radial temperature profile, leading to the temperature inside the billet being appreciably higher than the surface temperature. It is also very common for billet-sticking problems to be more pronounced with graded power distribution along the induction line when the system runs at a rate slower than the nominal for which it was designed. This is because the subsurface temperature may be hot enough to cause the billets to fuse together.

The effect of subsurface overheating is particularly pronounced when heating smaller-sized billets at a lower rate, using an induction line designed for heating larger billets at a nominal rate. As an example, Fig. 40 shows a surface-to-core temperature profile when heating 50.8 mm (2 in.) diameter billets at a substantially slower rate, using a conventional induction heating line designed for processing 63.5 mm (2.5 in.) billets at a nominal rate (Fig. 41) (Ref 61). Note that the billet surface temperature, which would normally be measured by pyrometers in both cases, is the same. Further reduction in the diameters of the heated billets could lead to more severe subsurface overheating.

Besides the potential danger of premature die wear of hammers and presses, improperly heated billets can raise some safety issues when forging billets that have appreciably higher subsurface temperatures and can also cause problems related to altering the quality of forged parts (Ref 9, 61). Subsurface overheating

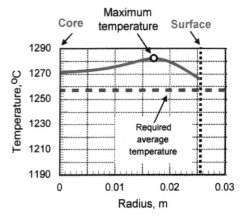

Fig. 40 Surface-to-core temperature profile when heating 50.8 mm diameter steel billets at a slower rate, using a conventional induction heating line designed for processing 63.5 mm billets at a nominal rate. Source: Ref 61

weakens the grain boundaries, which, in the presence of a tensile stress, causes them to break, resulting in a defect known as hot shortness. Burning, intergranular melting, or grain-boundary liquation are related to localized melting at austenite grain boundaries. During the forging process, metal undergoes deformational and frictional heating. If this heating, in combination with the billet predeformation temperature, is high enough to allow intergranular liquation, then failure may occur by intergranular cracking. Subsurface overheating could be further aggravated by the presence of excessive residuals and possible variations in the chemistry of a given steel (Ref 9). Taking into consideration that pyrometers can only reliably measure the billet surface temperature, there is always the danger of missing the appearance of severe subsurface overheating.

It should be mentioned here that the moderate and well-controlled subsurface temperature surplus could be advantageous in some applications. It helps to reduce billet surface cooling during its transportation from the induction heater to the next metal hot working operation.

A precise process control recipe based on an accurate prediction of the billet inside temperatures, using numerical computer simulation techniques, is essential for avoiding the previously described problems related to subsurface overheating. Unfortunately, rough-estimation techniques do not provide any information regarding the potential appearance of subsurface overheating. Only numerical computation can reliably predict temperature distribution within the inductively heated workpiece and the appearance, location, and magnitude of subsurface overheating and ways to control or avoid it. Numerical simulation can also be used as one of the main components of the process control system, assuring the proper subsurface thermal conditions of heated workpieces.

Example 6: Bar-End Heating. Although many of the bars, billets, or rods being manufactured today (2010) lend themselves to processes in which entire workpieces are heated and fed into a roll former or other type of hot working machine, in some cases it is necessary to hot form only a certain portion of the workpiece, for example, its end (Ref 1, 62). Examples of these types of parts are sucker rods for oil-country goods or various structural linkages in which an eye or a thread may be added to one or both ends of the bar. Placing the end of the bar into a multiturn coil and heating it for a specified amount of time generally accomplishes induction bar-end heating (Fig. 19). The choice of design parameters (i.e., selection of frequency and power) when heating ends of bars, billets, and rods is similar to cases where a whole body is heated. At the same time, there are some specifics.

As for all through-heating applications, a certain minimum time is necessary to obtain a required surface-to-core temperature uniformity, and the choice of frequency affects not only temperature uniformity but the overall efficiency of the system as well. Some bar heating applications require a specific temperature profile (uniform or nonuniform) along the heated length of the workpiece, including sharp or gradual cutoff of the heat pattern and/or a certain length of the longitudinal transition zone. Although there are a variety of coil designs for bar-end heating (i.e., oval coil, channel inductor, solenoid coil, split-return inductor, etc.), the basic principles for obtaining the required temperature profile within the bar end are quite similar. In its simplest form, a single-shot heating mode can be used to heat the end of the bar by placing a workpiece into the proper position within a coil and allowing the appropriate power at a given frequency to do its work, providing a desired heating pattern and required production rate.

In the case of the induction bar-end heater, the temperature distribution within the bar is affected, among other factors, by the electromagnetic end effect, which manifests itself by distortion of the electromagnetic field at the end regions of the induction coil (Fig. 42). Figure 43 shows a normalized surface power density distribution along the length of the heated bar. The electromagnetic end effect in the extreme end of the cylindrical bar (Fig. 43, A-B region) is defined primarily by four variables: the skin effect, R/δ; the coil overhang, σ; the ratio R_i/R; and the coil turn space factor, K_{space}, where R is the radius of the heated bar, R_i is the inside radius of the coil copper, and δ is the current penetration depth in the heated bar. An incorrect combination of these factors can lead to either underheating or overheating the extreme end of the bar (Ref 1, 62). Studies show that when heating carbon steel bars, the electromagnetic end effect area may extend toward the central region of the bar no further than 1.5 times the bar diameter ($l_{AB} < 3R$, where l_{AB} is the length of the end effect at the extreme end of the heated bar). Higher frequency and large coil overhang lead to a power surplus in the extreme end of the bar. As a result, noticeable overheating may take place in that area. Low frequencies, large coil-to-bar radial gaps, and small coil overhang could cause a substantial power deficit at the extreme end of the bar, which therefore will be underheated.

It should be pointed out that a uniform power distribution along the extreme end of the bar will not correspond to a uniform temperature profile because of the additional heat losses (due to thermal radiation and convection) at the bar extreme end area compared to its central part. By the proper choice of design parameters, it is possible to obtain a condition where the additional heat losses at the end of the bar are compensated for by the additional induced power (power surplus) due to the electromagnetic end effect. This allows a reasonably uniform temperature distribution to be obtained within the required heated area of the bar.

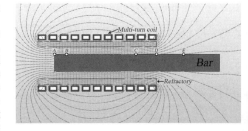

Fig. 42 Electromagnetic coil end effects appeared in bar-end heating applications

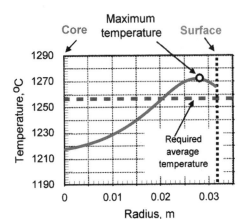

Fig. 41 Surface-to-core temperature profile when heating 63.5 mm diameter steel billets processed at a nominal rate. Source: Ref 61

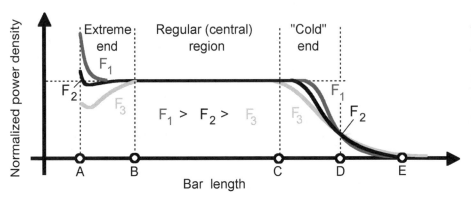

Fig. 43 Effect of frequency on a normalized surface power density distribution along the length of the heated bar

Figure 43 (zone C-D-E) shows the appearance of electromagnetic end effect and coil field distribution in the bar near the right end of the induction coil (the so-called cold end). Distribution of heat sources within the transition zone primarily depends on the radii ratio, R_i/R; the coil design; and the skin effect, R/δ. Due to the physics of the electromagnetic end effect in that area, there is always a power deficit under the coil end at any frequency, unless graded (profiled) coils with tightly wound turns near the cold end are used. The analytical solution in a simplified case, assuming the coil is an empty solenoid, helps to better appreciate the electromagnetic end effect near the right end of the induction coil (Fig. 42, 43).

Axial distribution of the magnetic field in the end area of the empty ideal solenoid (helical) coil can be obtained using an expression that describes the magnetic field distribution in a single loop of wire. The assumption of an ideal solenoid presumes the following conditions:

- The solenoid turns are tightly wound using thin wire.
- The coil current is uniformly distributed within each turn, and the skin effect is absent.
- There is no leakage magnetic flux.
- There are no electrically conductive bodies located within close proximity to the solenoid inductor.

Figure 44 shows a sketch of such an ideal solenoid of length l and radius R that has N tightly wound turns. A current-carrying empty loop produces a B-field (B is magnetic flux density) along the loop axis with a z-component according to the following expression (Ref 1, 63–66):

$$B_z = \frac{\mu_0 R^2 I}{2(R^2 + Z^2)^{3/2}} \qquad \text{(Eq 33)}$$

where Z is the axial distance from the loop to the area of interest, I is the loop current, and

μ_0 is the permeability of free space (a vacuum), $\mu_0 = 4\pi\ 10^{-7}$ H/m (or Wb/(A·m)).

The magnetic field at the center of the empty loop can be obtained by assuming $Z = 0$ in Eq 33. This results in:

$$B_z = \frac{\mu_0 I}{2R} \qquad \text{(Eq 34)}$$

The magnetic field distribution along the axis of an empty solenoid can be obtained by expansion of B_z of a single wire loop on a multiturn coil. Taking into consideration the assumption of tightly wound solenoid turns, the contribution of a small current-carrying section, dZ, on the total field in the center of the solenoid will be:

$$dB_z = \frac{\mu_0 R^2}{2(R^2 + Z^2)^{3/2}} \frac{NI}{l} dZ$$
$$= \frac{\mu_0 R^2 NI}{2l} \left(\frac{dZ}{(R^2 + Z^2)^{3/2}} \right) \qquad \text{(Eq 35)}$$

The total magnetic field in the center of the coil can be obtained by taking into account the contributions of all current-carrying sections. Therefore, after integrating dB_z along the coil length, the magnetic field along the coil center can be written as:

$$B_z = \frac{\mu_0 R^2 NI}{2l} \int_{-L/2}^{L/2} \frac{dZ}{(R^2 + Z^2)^{3/2}} \qquad \text{(Eq 36)}$$

After simple mathematical operations, the total axial field in the center of the solenoid will be:

$$B_z = \frac{\mu_0 NI}{\sqrt{(4R^2 + l^2)}} \qquad \text{(Eq 37)}$$

If the length of the solenoid is much greater then its radius, $l \gg R$ (electromagnetically long coil), then it is possible to neglect R with respect to l, and Eq 37 can be rewritten as:

$$B_z = \frac{\mu_0 NI}{l} \qquad \text{(Eq 38)}$$

This is the well-known expression for the axial component of the B-field at the center of an electromagnetically long solenoid. It is possible to show that by allowing the corresponding limits in Eq 36 that are appropriate for the end of the empty coil, Eq 33 and 36 can be transformed into:

$$B_z = \frac{\mu_0 NI}{2\sqrt{(R^2 + l^2)}} \qquad \text{(Eq 39)}$$

For an electromagnetically long coil, Eq 39 can be approximated as:

$$B_z = \frac{\mu_0 NI}{2l} \qquad \text{(Eq 40)}$$

Therefore, a comparison of Eq 38 and 40 shows that at the ends of the empty coil, the magnetic flux density, B_z, drops to one-half its value at the center. It is appropriate to extend this conclusion for illustration of the end effect for an electromagnetically long multiturn coil with an infinitely long homogeneous nonmagnetic workpiece.

After comparing Eq 38 and 40, one can conclude that the density of the induced current under the coil end is two times less than in its center. It means that the power density under the coil end is equal to a quarter of that in the center ($P_{\text{end}} = 0.25 \cdot P_{\text{center}}$). The length of zone "C-D-E" primarily depends on the skin effect in the heated workpiece, the ratio of coil inside radius to workpiece outside radius, and the coil turn space factor, K_{space}. Depending on application, this zone may be as long as five times the coil radius (if the skin effect is not pronounced and the ratio of coil inside radius to workpiece radius is large) or twelvefold an equivalent radial gap between the coil and the surface of the bar (assuming pronounced skin effect and small air gaps).

Typically, the length of zone "C-D-E" (Fig. 43) is equal to 1.5 to 4.5 times the coil radius. High frequency, short cycle time, pronounced skin effect, and small coil copper to bar radial air gaps lead to a shorter end-effect zone. Application of the external magnetic flux concentrators and magnetic shunts having a "U-" or "L"-shape also shortens an end-effect zone.

Another important feature that defines the coil length is the fact that in zone "C-D-E," which is often defined as an axial transition zone, there is a significant longitudinal temperature gradient that results in axial heat flow due to thermal conduction from the high-temperature region of the bar toward its cold area, manifesting itself as the heat sink phenomenon. This phenomenon is more pronounced when heating metals having high thermal conductivity (such as aluminum, copper, silver, and gold). In some applications, there is a limitation on the permissible (typically maximum) length of the transition zone that could be related to bar-handling issues when using robots. The proper choice of process parameters and coil

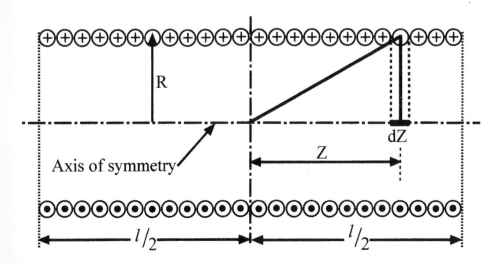

Fig. 44 Sketch of a tightly wound multiturn solenoid for end-effect simulation. Source: Ref 1

R

Axis of symmetry

Z

dZ

$l/2$ $l/2$

design features helps to minimize this cooling effect and obtain the required uniform temperature profile for the bar with a minimum transition zone.

In the great majority of cases, induction bar-end heaters cannot be considered infinitely long ideal solenoids. From another perspective, the extreme end of the workpiece (Fig. 31) offers no thermal path for heat conduction, while its cold end (Fig. 43, zone "C-D-E") provides a ready heat sink and easy conduction path. Heat flow that occurred due to the heat sink effect not only leads to a redistribution of temperature pattern but also affects the amount of mass being heated, which directly corresponds to the amount of total required energy.

The accurate prediction of an electromagnetic field and temperature distribution in bar-end heating applications, as well as required coil design parameters and process recipe, can be obtained using numerical simulations. Figure 45 shows a sketch of an induction bar-end heater. The material is plain carbon steel AISI 1035. The outside diameter of the bar is 25.4 mm (1 in.). Figure 46 shows the dynamics of the induction carbon steel bar-end heating obtained after numerical simulation.

At the initial heating stage, the whole bar is magnetic (cycle time < 3 s), leading to a pronounced skin effect ($R/\delta > 20$). Because of the pronounced skin effect, the induced power appears in the fine surface layer of the bar. The core of the bar is heated from the surface and subsurface layers by thermal conduction. Consequently, the surface of the bar is heated much faster than the core (Fig. 46, top). Due to relatively low temperatures, the surface heat losses are relatively low. An appreciable radial temperature gradient appears at this stage.

With time (cycle time approximately 5 s), an interim heating stage takes place. The temperature of the bar surface approaches the Curie temperature, which is accompanied by a correspondent loss in magnetic properties. Relative magnetic permeability, μ_r, in the surface layer drops to 1. In addition, due to the temperature rise, the electrical resistivity of the carbon steel increases approximately two to three times compared to its values at the initial heating

stage. Both factors—a reduction of relative magnetic permeability and an increase of electrical resistivity—lead to substantial increase in the current penetration depth, δ, providing a deeper heating effect. Because the bar surface becomes nonmagnetic but the internal area of the bar and its core retain magnetic properties, a magnetic-wave phenomenon appears at this stage.

Finally, the total bar end located under the coil, including its core, becomes nonmagnetic. The penetration depth becomes quite large, and the skin effect becomes noticeably less pronounced ($R/\delta < 2.5$). As a result, a significant amount of energy is induced within the internal area of the bar, and the core starts to heat more intensely, equalizing the radial temperature distribution. The combination of substantial current penetration depth, increased heat losses from the surface of the bar, and radial heat flow due to thermal conduction results in achieving the required heat uniformity at the desired final temperature (cycle time of 22 s).

Additional case studies using different tightly coupled electromagnetic-thermal numerical computer simulation software for the needs of induction heating prior to metal forging, rolling, and extrusion are provided in the article "Simulation of Induction Heat Treatment" in this Volume.

REFERENCES

1. V. Rudnev, D. Loveless, R. Cook, and M. Black, *Handbook of Induction Heating*, Marcel Dekker, 2003, p 800
2. C.A. Tudbury, *Basics of Induction Heating*, Rider, New York, 1960
3. M.G. Lozinskii, *Industrial Applications of Induction Heating*, Pergamon, London, 1969
4. P.G. Simpson, *Induction Heating: Coil and System Design*, McGraw-Hill, New York, 1960
5. S.L. Semiatin and D.E. Stutz, *Induction Heat Treatment of Steel*, ASM International, 1986
6. "Induction Heating," Course 60, ASM International, 1986
7. T. Byrer, *Forging Handbook*, Forging Industry Association, 1985
8. J.R. Davis, Ed., *Metals Handbook Desk Edition*, 2nd ed., ASM International, 1998
9. V. Rudnev, D. Brown, C. Van Tyne, and K. Clarke, Intricacies for the Successful Induction Heating of Steels in Modern Forge Shops, *Proc. 19th International Forging Congress* (Chicago, IL), 2008
10. E. Rapoport and Y. Pleshivtseva, *Optimal Control of Induction Heating Processes*, CRC Press, 2007
11. V. Rudnev, Systematic Analyses of Induction Coil Failures, Part 11b: Frequency Selection, *Heat Treat. Prog.*, ASM International, Sept/Oct 2007, p 23–25
12. R.M. Bozorth *Ferromagnetism*, IEEE Press, New York, 1993, p 968
13. P. Neelakanta, *Handbook of Electromagnetic Materials*, CRC Press, 1995, p 591
14. M. Hansen and K. Anderko, *Constitution of Binary Alloys*, McGraw-Hill, New York, 1958, p 1305
15. K. Schroder, *CRC Handbook of Electrical Resistivities of Binary Metallic Alloys*, CRC Press, 1983, p 442
16. J.R. Howell, *A Catalog of Radiation Configuration Factors*, McGraw-Hill, New York, 1982
17. I. Grigor'ev and E. Melikov, *Physical Values*, Energoatomizdat, Moscow, 1991
18. V. Rudnev, A Common Misassumption in Induction Hardening, *Heat Treat. Prog.*, ASM International, Sept/Oct 2004, p 23–25
19. ESI Group, www.esi-group.com
20. V. Rudnev et al., Progress in Study of Induction Surface Hardening of Carbon Steels, Gray Irons and Ductile (Nodular) Irons, *Ind. Heat.*, March 1996
21. V. Rudnev et al., Efficiency and Temperature Considerations in Induction Re-Heating of Bar, Rod and Slab, *Ind. Heat.*, June 2000
22. V. Rudnev, "Mathematical Simulation and Optimal Control of Induction Heating of Large Cylinders and Slabs," Ph.D. thesis, St. Petersburg El. Eng. Univ., Russia, 1986
23. Integrated Engineering Software, Inc., www.integratedsoft.com
24. Infolytica Corp., www.infolytica.com
25. MagSoft Computer and Software, www.magsoft.com
26. Multiphysics Modeling and Simulation, COMSOL, Inc., www.comsol.com
27. Vector Fields—Software for Electromagnetic Design, Cobham Technical Services, www.vectorfields.com
28. ANSYS, Inc., www.ansys.com
29. Ansoft Electronic Design Products, ANSYS, Inc., www.ansoft.com
30. S. Gurevich, A. Vasiliev, and B. Polevodov, General Tendencies of Mathematical Modeling of Induction and Dielectric Heating, *Electrotechnica*, Vol 8, 1982
31. V. Rudnev, Subject-Oriented Assessment of Numerical Simulation Techniques for

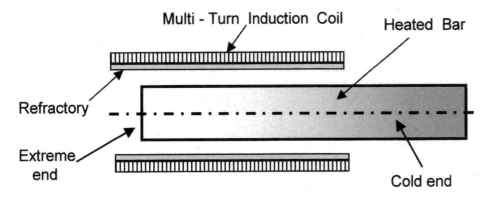

Fig. 45 Sketch of induction bar-end heater

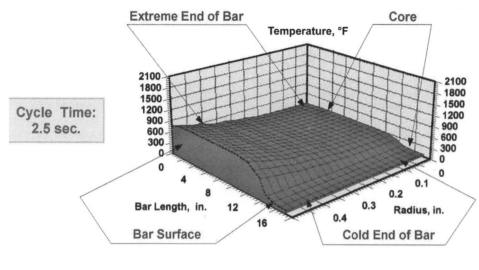

Cycle Time: 2.5 sec.

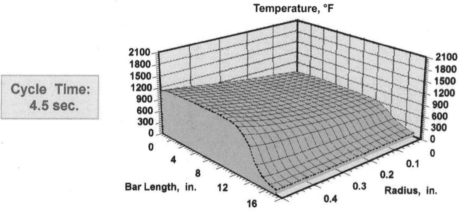

Cycle Time: 4.5 sec.

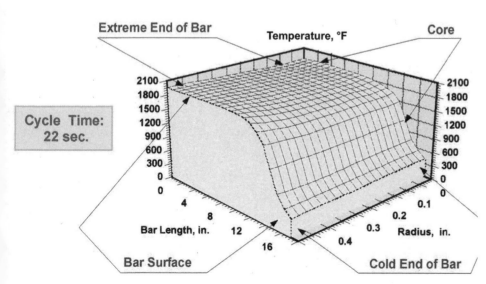

Cycle Time: 22 sec.

Fig. 46 Dynamics of induction carbon steel bar-end heating. Source: Ref 62

Induction Heating Applications, *Int. J. Mater. Prod. Technol.*, Vol 29 (No. 1–4), 2007, p 43–51

32. V. Rudnev, D. Loveless, R. Cook, and M. Black, Chapter 11A, *Induction Heat Treatment, Steel Heat Treatment Handbook*, Marcel Dekker, 1997, p 810

33. A.E. Slukhotskii and S.E. Ryskin, *Inductors for Induction Heating*, Energy Publ., St. Petersburg, Russia, 1974

34. N. Pavlov, *Engineering Calculations of Induction Heaters*, Energy, Moscow, Russia, 1978

35. *Properties and Selection: Nonferrous Alloys and Pure Metals*, Vol 2, *Metals Handbook*, 9th ed., American Society for Metals, 1979

36. J. Vaughan and J. Williamson, Design of Induction Heating Coils for Cylinder Non-Magnetic Loads, *Trans. AIEE*, Vol 64, Aug 1945

37. *High-Temperature Property Data: Ferrous Alloys*, ASM International, 1988

38. B. Gebhart, *Heat Transfer*, McGraw-Hill, New York, 1970

39. S. Patankar, *Numerical Heat Transfer and Fluid Flow*, Hemisphere, New York, 1980

40. F.P. Incropera and D.P. Dewitt, *Fundamentals of Heat Transfer*, Wiley, New York, 1981

41. R.F. Myers, *Conduction Heat Transfer*, McGraw-Hill, New York, 1972

42. J.A. Adams and D.F. Rogers, *Computer Aided Analysis in Heat Transfer*, McGraw-Hill, New York, 1973

43. W.M. Rohsenow and J.P. Hartnett, *Handbook of Heat Transfer*, McGraw-Hill, New York, 1973

44. R. Siegel and J.R. Howell, *Thermal Radiation Heat Transfer*, 2nd ed., McGraw-Hill, New York, 1980

45. J. Wiebelt, *Engineering Radiation Heat Transfer*, Holt, Rinehart and Winston, New York, 1966

46. E. Sparrow and R. Cess, *Radiation Heat Transfer*, Wadsworth, Englewood Cliffs, NJ, 1966

47. A. Shamov and V. Bodazhkov, *Design and Maintenance of High Frequency Systems*, Mashgiz, St. Petersburg, Russia, 1963

48. M. Orfueil, *Electric Process Heating*, Battelle Press, 1987

49. K. Schweigert, Induction Heating Prior to Hot Working, *Proc. of the Sixth Int. Induction Heating Seminar*, Vol 2, 1995

50. V. Rudnev and D. Loveless, Longitudinal Flux Induction Heating of Slabs, Bars, and Strips Is No Longer "Black Magic," Part 2, *Ind. Heat.*, March 1995

51. V. Rudnev, Electromagnetic Forces in Induction Heating, *Heat Treat. Prog.*, July 2005

52. V. Rudnev and D. Loveless, Induction Slab, Plate and Bar Edge Heating for Continuous Casting Lines, *Met. Produc.*, Vol 33, Oct 1994

53. V. Rudnev and D. Loveless, Longitudinal Flux Induction Heating of Slabs, Bars, and Strips Is No Longer "Black Magic," Part 1, *Ind. Heat.*, Jan 1995

54. V. Rudnev, Systematic Analyses of Induction Coil Failures, Part 6: Coil End Effect, *Heat Treat. Prog.*, ASM International, May/June 2006, p 19–20

55. V. Rudnev, V. Demidovich, Advanced Computation Software for Everyday Use in Modern Induction Heat Treating, *Proc.*

of 17th ASM Heat Treating Conference, Indianapolis, 15–18 Sept 1997, p.551–555

56. V. Rudnev, Systematic Analysis of Induction Coil Failures, Part 13: Electromagnetic Proximity Effect, *Heat Treat. Prog.,* ASM International, Oct 2008, p 23–26

57. P. Ross and V. Rudnev, Innovative Induction Heating of Oil Country Tubular Goods, *Ind. Heat.,* May 2008

58. V. Rudnev, Successful Induction Heating of RCS Billets, *Forge Mag.,* July 2008

59. V. Rudnev, How Do I Select Inductors for Billet Heating, *Heat Treat. Prog.,* May 2008

60. V. Rudnev, Question and Answer Regarding Induction Heating Large Billets, *Heat Treat. Prog.,* Jan 2009

61. D. Brown, V. Rudnev, J. Lin, and T. Nakagawa, Superior Induction Heating Technologies for Modern Forge Shops, *Proc. of Japanese Forging Association Conference (JFA),* July 2009, p 54–59

62. V. Rudnev and R. Cook, Bar End Heating, *Forging,* Winter 1995

63. P. Hammond, *Electromagnetism for Engineers,* Pergamon, New York, 1978

64. M.A. Plonus, *Applied Electromagnetics,* McGraw-Hill, New York, 1978

65. I.E. Tamm, *Fundamentals of the Theory of Electricity,* Moscow, Russia, 1981

66. W.H. Hayt, *Engineering Electromagnetics,* McGraw-Hill, New York, 1981

67. V. Rudnev, A Fresh Look at Induction Heating of Tubular Products: Part 1, *Heat Treat. Prog.,* May 2004, p.17–19

68. V. Rudnev, A Fresh Look at Induction Heating of Tubular Products: Part 2, *Heat Treat. Prog.,* July 2004, p.23–25

ASM Handbook, Volume 22B, *Metals Process Simulation*
D.U. Furrer and S.L. Semiatin, editors

Copyright © 2010, ASM International®
All rights reserved.
www.asminternational.org

Simulation of Induction Heat Treating

Valery Rudnev, Inductoheat, Inc.

MATHEMATICAL MODELING IS ONE OF THE MAJOR FACTORS in successful design of induction heating systems. Computer modeling provides the ability to predict how different factors may impact the transitional and final heat treating conditions of the workpiece and what must be accomplished in the design of the induction heat treating system to optimize it, improve the effectiveness of the process, and guarantee the desired results.

Theoretical models may vary from a simple hand-calculated formula to a very complicated numerical analysis, which can require a few days of model preparation and several hours of actual computer runtime, even when using modern computers. The choice of a particular theoretical model depends on several factors, including the complexity of the engineering problem, required accuracy, emergency of the project (time limitation for obtaining results), and cost.

Before an engineer starts to provide a mathematical simulation of any process, he/she should have a sound understanding of the nature and physics of the process (Ref 1). (Also see the article "Simulation of Induction Heating Prior to Hot Working and Coating" in this Volume.) The user should also be aware of the limitations of applied mathematical models, assumptions, and sensitivity of the chosen model to sometimes poorly defined parameters. Such parameters are boundary conditions, material properties, and nonuniform initial temperature distribution. Underestimation of application specifics or overly simplified assumptions can lead to an incorrect mathematical model (including improperly chosen governing equations) that will not be able to provide the required accuracy of modeling.

In contrast, correctly applied mathematical modeling provides an assurance for designing novel induction heating systems by determining details of the process that could be costly, time-consuming, and, in some cases, extremely difficult, if not impossible, to determine experimentally.

Metal Heat Treating by Induction

Induction heat treatment of metals represents a wide range of applications (Ref 1 to 11). This includes but is not limited to annealing, normalizing, surface (case) hardening, through hardening, tempering, and stress relieving of a variety of workpieces. Steel and cast iron by far represent the majority of induction heat treated components, although other materials, including light metals, aerospace materials, superalloys, and composites, are also inductively heat treated or preheated prior to coating and hot and warm forming.

Descriptions of Various Applications

Normalizing can be done by heating the steel to a temperature of approximately 100 °C above the upper transformation temperature and then allowing it to air cool to room temperature. Normalizing is done to improve homogenization of cold-worked and cast parts, refine grain structure, obtain relatively low hardness, reduce residual stresses, and provide desirable conditions of a material prior to the next technological operation (i.e., hardening).

Annealing is a broad term that is used by heat treat practitioners to describe a variety of processes and properties related to microstructure, machinability, formability, and internal stresses. There are two basic types of annealing:

- Full annealing
- Process annealing

The purpose of full annealing is much like that of normalizing. Annealing results in hardness reduction and an improvement of the material ductility, machinability, and homogenization (Ref 3 to 9). After heating to the annealing temperature range, which depends on metal chemical composition, the component is slowly cooled, producing essentially stress-free structures. The cooling rate is noticeably slower with full annealing compared to normalizing. Full annealing temperatures of hypoeutectoid steels are just above the A_3 critical temperature, but hypereutectoid steels are just above the A_1 critical line.

A process that often is associated with full annealing is homogenization. At the same time, the major purpose of homogenization is to obtain a homogeneous structure by eliminating alloy segregation. It is usually performed at higher temperatures than full annealing to create favorable conditions for diffusion processes required for homogenization. Homogenization of steels usually occupies the temperature range of 1000 to 1150 °C. Typically, full annealing and homogenization is done in gas-fired or electric furnaces. Very seldom is induction heating used for these processes.

In applications that do not require achieving fully annealed structures, a process annealing may be applied instead of full annealing. Process annealing represents an intermittent stage between a fully annealed condition and a cold-worked condition. The main purpose of process annealing is to improve the ductility of cold-worked workpieces.

For process annealing, a workpiece is typically heated to temperatures 20 to 120 °C below the low critical temperature A_1, held for some time at temperature, and then air cooled. Induction is often a preferable choice for process annealing applications.

Hardening. The most popular application of induction heat treating is the hardening of steels and cast irons. A typical heat treatment procedure for induction hardening involves heating the component to the austenitizing temperature, holding it at temperature for a period long enough to complete the formation of austenite (if required), then rapidly cooling the alloy until it is below the martensite start (M_s) temperature (Fig. 1). Rapid cooling allows replacement of the diffusion-dependent transformation process by a shear-type transformation of the initial structure of steel or cast iron into a much harder constituent called martensite. Hardening may be done either on the surface of the workpiece or throughout the entire cross section.

Because of the physics of the induction phenomenon, heating can be localized to areas where the metallurgical or mechanical changes are desired. The goal of induction surface hardening is to provide a martensitic layer on specific areas of the component to increase the hardness, strength, and wear resistance while allowing the remainder of the part to be unaffected (Ref 1 to 3, 15 to 21). Figure 2 shows a small portion of the virtually endless variety of steel and cast iron components that can be induction hardened. This includes, but is not limited to, various transmission and engine components (e.g., axles, constant velocity joints, camshafts, crankshafts, gears, cylinders, connecting rods, and rocker arms); fasteners (e.g., bolts, screws, and studs); off-road, farm, and mining

Fig. 1 Induction scan hardening requires rapid cooling. Courtesy of Inductoheat, Inc.

Table 1 Hardness levels recommended by SAE to determine an effective case depth in induction-hardened steels

Carbon content, %	Effective case depth hardness, HRC
0.28–0.32	35
0.33–0.42	40
0.43–0.52	45
>0.53	50

Source: Ref 22

Fig. 2 Selection of the variety of steel and cast iron components that can be induction hardened. Courtesy of Inductoheat, Inc.

Fig. 3 Various applications and part geometries that require specific hardness profiles. Courtesy of Inductoheat, Inc.

equipment (i.e., shafts, clutch plates, pins, track links, valve-spring wire, drill bits, and plows); cutting tools; bearings; inner/outer race sprockets; spindles; skid plates; and fixtures (Ref 1).

Obtaining a certain hardness at a particular depth often represents customer hardening specifications. The definition of case depth (hardness depth) is quite subjective and often defined as the surface area where the microstructure is at least 50% martensite. Below the case depth, the hardness begins to decrease drastically. The distance below the part surface where the hardness drops to 10 HRC lower than the surface hardness is often called the effective case depth by practitioners.

At the same time, SAE standard J423 (Ref 22) defines an effective case depth as, "The perpendicular distance from the surface of a hardened case to the furthest point where a specified level of hardness is maintained...For determination of effective case depth, the 50 HRC criterion is generally used. The sample or part is considered to be through hardened when the hardness level does not drop below the effective case depth hardness value." It is important to be aware that for induction surface (case) hardening applications, the SAE standard makes an exemption, suggesting the use of a lower-hardness criterion. Recommended hardness levels for various nominal carbon levels are tabulated in Table 1.

Various applications and part geometries require certain hardness profiles (Fig. 3). The obtained hardness pattern primarily depends on the following factors: temperature distribution, microstructure of the metal, chemical composition, quenching conditions, grain size, and hardenability of the steel. Temperature distribution in induction surface hardening is controlled by selection of frequency, power density, time of heating, and coil geometry.

Steel selection depends on specifics of the component working conditions, required hardness, and cost (Ref 4 to 14). Plain carbon steels and low-alloy steels are the least expensive steels used successfully for a variety of hardening applications. It is important to remember that the carbon content of steel plays a critical role in the determination of the maximum achievable hardness and also affects the amount of retained austenite and the steel hardenability.

Medium-carbon steels (i.e., ASTM/SAE 1035 to 1060) are the most common steels used in industry. Low-carbon steels are used where toughness rather than high hardness is required, such as in clutch plates or pins for farm equipment. A wide application range of high-carbon steels in industry is limited by their low ductility, poor machinability, and higher cost compared to medium-carbon steels. At the same time, there is a variety of applications, including valve-spring wire, drill bits, grinding balls, cutting tools, and others, where high-carbon steels (such as AISI 1060 to 1090) are specified and provide a noticeable advantage over medium- and low-carbon steels.

Although plain carbon steels, being the least expensive steels, are widely used in industry, there are many engineering applications where the properties of plain carbon steels and low-alloy steels are not suitable for meeting a particular engineering requirement or combination of requirements, such as the depth of hardness (hardenability), corrosion resistance, impact and fatigue strength, reduced oxidation, and susceptibility of heat treated parts to cracking and shape distortion. Alloy steels help to enhance particular properties or its combination. Examples of alloy steels used in induction hardening are 4140, 4150, 4340, 5150, 51100, and 52100.

Generally speaking, shallow case depths require higher frequencies, low energy, and high power densities (Ref 1). In some cases of surface hardening of massive parts with shallow case depths (typically less than 1 mm), it is possible to use self–quenching techniques (also called mass quenching). If the heated surface layer is sufficiently thin and its mass is appreciably small compared to the mass of the unheated core, it is possible to obtain a sufficiently rapid cooling of the component surface due to thermal conduction of the heat toward the cold core. In this case, the mass of the cold core acts as a large heat sink, and the rate of cooling may be severe enough to form a martensitic structure. Therefore, self-quenching can make the use of fluid quenchants unnecessary, except for rinsing heat treated parts with the purpose of cooling for safe handling.

The specification of hardness pattern, including the case depth, is a complex function of the application specifics and required component performance parameters. The cross-sectional areas of the parts and the magnitudes of the loads they must handle dictate the case depth. Typically, very shallow case depths of 0.3 to 0.75 mm (0.01 to 0.03 in.) are solely for wear resistance. These case depths are obtained by using frequencies between 600 and 100 kHz, correspondently. Parts that require both wear resistance and moderate loading, such as camshafts and crankshafts, are usually induction hardened to case depths of 1 to 4 mm (0.04 to 0.16 in.). Deeper case depths strengthen the part dramatically, because load stresses drop exponentially from the surface. To obtain these case depths, the frequency selection is usually

between 100 and 8 kHz. Parts that should withstand heavy loads, such as axle shafts, wheel spindles, and large heavy-duty gears, demand even greater case depths. Components for heavy machinery (i.e., off-road machines, ships, mill rolls, earth-moving machines) could require hardened case depths anywhere from 4 to 16 mm (0.16 to 0.625 in.) and even greater, which call for a frequency between 10 kHz and 500 Hz and even lower. For example, track links for earth-moving machines require a case depth of 12 to 16 mm (0.5 to 0.625 in.).

Complex-shaped parts can present some challenges in obtaining the required hardness patterns, demanding complex geometries of heat treating inductors (Fig. 4). The presence of various holes, grooves, sharp corners, voids, notches, and other surface and subsurface discontinuities and stress raisers noticeably affects an inductor design and process recipe. Proper selection of process parameters, including frequency, power density, time settings, scan rate, and specifics of spray quenching, is critical for assuring the required case hardening conditions and hardness pattern. Computer modeling is imperative for determining optimal inductor geometry and process recipes.

Spray quenching is typically used in induction hardening applications. Spray quenching

works best if the component is rotated during the quenching operation, which ensures cooling uniformity. By rotating parts, the workpiece essentially experiences a constant impingement rather than many small impingements. The intensity of spray quenching (intensity of heat removal) depends on the quenchant flow rate; the impingement angle at which the quenchant strikes the workpiece; the temperature, purity, and type of quenchant; as well as the part temperature. Quenchants used include water, aqueous polymer solutions, and, to lesser extent, oil, water mist, and forced air (Ref 8). Water and aqueous polymer solutions are the most popular.

Through hardening may be needed for parts requiring high strength, such as snowplow blades, springs, chain links, truck bed frames, certain fasteners (including nails and screws), and so on. In these cases, the entire component is raised above the A_{c3} transformation temperature and then quenched. Through hardening applications require uniform heating. Selection of the correct induction heating frequency is very important for achieving a sufficient surface-to-core temperature uniformity in the shortest time with the highest heating efficiency. Typically, the quenching intensity of the surface-hardened component is greater than

Fig. 4 Variety of inductors for various induction heat treating applications. Courtesy of Inductoheat, Inc.

the cooling intensity of a through-hardened workpiece. This is so because in surface hardening, additional cooling of the surface layers takes place due to the cold core, which plays the role of a heat sink (Ref 1, 18).

Tempering and Stress Relieving. The tempering process takes place after the steel is hardened but is no less important. The transformation to martensite through quenching creates a very hard and brittle structure. Untempered martensite retains a high level of residual stresses and typically is too brittle for many commercial applications, except components that primarily require wear resistance. Reheating of as-quenched steels and cast irons for tempering produces a tempered martensite microstructure. A variety of microstructures and mechanical properties of steels and cast irons can be produced by tempering, including increasing the steel toughness, yield strength, and ductility, relieving internal stresses, eliminating brittleness, and, in some cases, improving the shape stability of the as-quenched component.

A conventional way of tempering is to run the parts through a tempering furnace (gas fired or infrared), which is typically located in a separate production area and therefore requires extra space, labor, and time for part transportation. Tempering in the furnace is a time-consuming process that may take 2 to 3 h. Short-time induction tempering was developed to overcome these drawbacks.

Tempering temperatures are always below the lower transformation temperature (A_1) and usually in the range of 120 to 650 °C (248 to 1200 °F). There are four overlapping tempering stages on reheating as-quenched martensite of plain carbon steels (Ref 9):

1. Precipitation of ε-carbide and partial loss of tetragonality in martensite (temperature below 250 °C)
2. Decomposition of retained austenite (200 to 300 °C)
3. Replacement of ε-carbides by cementite; martensite loses tetragonality (200 to 350 °C)
4. Coarsening and spheroidization of cementite, accompanied by recrystallization of ferrite (above 350 °C)

Depending on the stage of tempering, various amounts of iron carbides (Fe_3C) are formed, affecting the properties of hardened and tempered structures. Low-temperature tempering of carbon steels occurs typically at 120 to 300 °C (248 to 572 °F). Depending on the tempering time, the hardness reduction may not exceed 1 to 3 points HRC. If plain carbon steel is tempered above 600 °C (1112 °F), practically all of the residual stresses existing in the part will be removed, and changes in microstructure may lead to a significant loss in hardness (Fig. 5, 6) (Ref 2 to 9, 23). The hardness reduction may exceed 15 points HRC. Note that induction tempering of alloyed steel, even at a

temperature level of 600 °C (1112 °F), may not result in significant hardness loss due to the relatively short process time of induction tempering (Fig. 7) (Ref 2, 3). It is important to recognize that when tempering alloy steel, the hardness does not always gradually decrease with temperature increase, and some alloys exhibit the phenomenon called secondary hardening (Fig. 8) (Ref 2, 3).

Time and temperature are two of the most critical parameters of tempering. To provide a similar effect in short-time induction tempering as in long-time oven tempering, it is necessary to use higher temperatures (Ref 1, 2 to 7). There are several ways to determine the time-temperature correlation between conventional long-time low-temperature oven tempering and short-time higher-temperature induction tempering, including the Hollomon-Jaffe equation, the Grange-Baughman tempering correlation, and so on.

Heat Treating Inductors

The induction coil is the most critical component of any induction heating system. Basic physical principles of the induction heating phenomenon are discussed in Ref 1 to 3 as well as in the article "Simulation of Induction Heating Prior to Hot Working and Coating" in this Volume. It is assumed that readers are familiar with physics of induction heating, which can be summarized as follows. An inductor or induction coil is an electrical device positioned in close proximity to the electrically conductive workpiece required to be heat treated. Alternating current

flowing in the inductor generates a time-varying magnetic field that provides an electromagnetic link between the inductor and workpiece, resulting in contactless heating of part or all of the workpiece due to the Joule effect.

A particular inductor configuration depends on specifics of the application that include, but are not limited to, geometry of the workpiece, heating mode, process recipe, production rate, required heating profile, available power/frequency source, and the material handling to be used for production (i.e., how the part is moved into the inductor or the inductor indexes into the part, whether rotation of the part is required, or how the part is transferred after heat treatment). Figure 4 shows a small sample of possible coil designs. Inductors for induction metal heat treating applications can be divided into four major categories: scanning, progressive, single-shot, and special inductors. Coils for hardening are typically computerized numerical control machined from a solid copper bar, which makes them very rigid, durable, and repeatable. In other cases, a copper tube (rectangular or round cross section) may be used (Fig. 9). In some low-coil-current/low-workpiece-temperature applications, Litz wire can be successfully used for fabrication inductors (Ref 1, 26). Litz wire derives its name from the German word *litzendraht*, meaning woven wire. It consists of a number of individually insulated wires twisted or braided into a uniform pattern, so that each strand tends to take all possible positions in the cross section of the entire multistrand conductor. Litz wire eliminates the skin effect of alternating current that flows through it, leading to an increase in current-carrying area, reduced

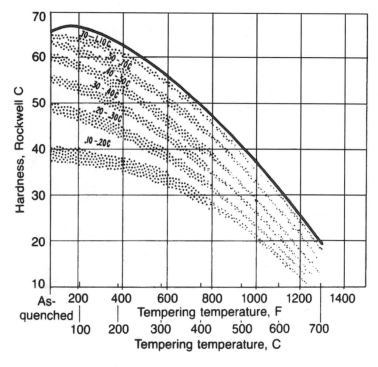

Fig. 5 Rockwell C hardness of tempered martensite in carbon and low-alloy steels versus tempering temperature. Source: Ref 5, 23

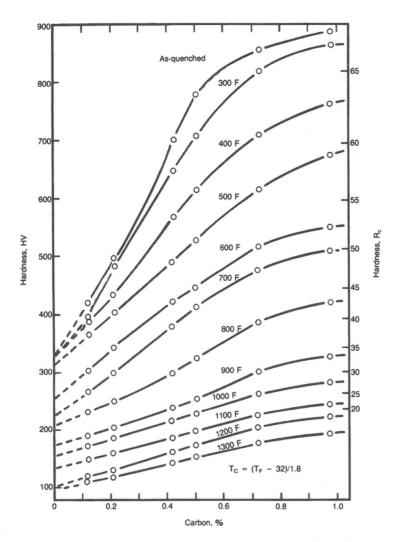

Fig. 6 Vickers and Rockwell C hardnesses of tempered martensite in carbon and low-alloy steels as a function of carbon content at various tempering temperatures. Source: Ref 6

$T_C = (T_F - 32)/1.8$

Descriptions of Various Hardness Patterns

Through Hardening. Guidelines for selection of frequency and power for induction through hardening, normalizing, and full annealing are similar to other induction applications that require through heating of workpieces, such as heating prior to metal hot working. This is discussed in the companion article "Simulation of Induction Heating Prior to Hot Working and Coating" in this Volume (Tables 7 to 9 and Fig. 32 and 34). A system designed to operate in the 400 kHz range would have a very shallow depth of heating, whereas a 1 kHz system will produce relatively deep heating. As in any application that requires through heating, the main concern in choosing frequency is an ability to avoid an eddy current cancellation that takes place if the selected frequency is too low (Ref 1, 27).

For example, when through hardening small parts, it would not be practical to try to harden to a 6.3 mm (0.25 in.) diameter part or a thin-wall part using a 1 kHz system. The current penetration depth in hot steel (above the Curie point) at 1 kHz (approximately 17 mm, or 0.66 in.) is simply too great for this small diameter. Eddy current cancellation occurs upon reaching the Curie temperature, resulting in a dramatic reduction of coil electrical efficiency and making the heated part practically transparent to the electromagnetic field generated by the induction coil.

Over the years, industry has accumulated numerous "rules of thumb" regarding the calculation of process parameters that focus on hardening steel cylinders. Such recommendations are shown in Tables 2 and 3. Table 2 shows typical operating conditions for progressive through hardening of steel parts.

Comments and clarifications for Table 2 include (Ref 3):

- Note the use of dual frequencies for round cylinders.
- Power is transmitted by the inductor at the operating frequency indicated. This power is approximately 25% less than the power input to the machine, because of losses within the machine.
- Inductor input is at the operating frequency of the inductor.

Table 3 shows approximate power densities required for through heating of steel for hardening and tempering. Comments and clarifications for Table 3 include (Ref 3):

- The values in this table are based on the use of proper frequency and normal overall operating efficiency of the equipment.
- In general, these power densities are for section sizes of 13 to 50 mm (½ to 2 in.). Higher inputs can be used for smaller sizes, and lower inputs may be used for larger section sizes.

coil resistance and coil losses, and increased overall electrical efficiency.

As discussed in Ref 1 to 3 as well as the article "Simulation of Induction Heating Prior to Hot Working and Coating" in this Volume, induction heating is a complex combination of electromagnetic, heat-transfer, and metallurgical phenomena that are tightly interrelated, because the physical properties of heat treated materials depend strongly on temperature, magnetic field intensity, as well as chemical composition and microstructure. The metallurgical aspects of metal heat treating have been discussed in Ref 1 to 6. This article concentrates on simulation techniques of the electromagnetic and thermal processes that occur during induction heat treating.

Estimation Techniques for Frequency and Power

The companion article "Simulation of Induction Heating Prior to Hot Working and Coating" in this Volume discusses various estimation techniques to determine basic process parameters of induction heating prior to metal hot forming, the limitations of those techniques, and their subjective nature. The selection of frequency and power in induction heat treating applications is subjective to an even greater degree, being affected by not only the process specifics but also by the type of alloy, its initial microstructure, specifics of inductor design, and the required hardness pattern. Steel hardening is conventionally described as heating the entire component or a part of the component that must be hardened to the austenitizing temperature, holding it, if necessary, for a period long enough to obtain a complete transformation to homogeneous austenite, and then rapidly cooling it to below the M_s temperature, where the martensitic transformation begins. The first step in designing an induction hardening machine is to specify the required hardness pattern, including surface hardness, case depth, and transition zone. The hardness pattern is directly related to temperature distribution and is controlled by selection of frequency, time, power, and workpiece/coil geometry.

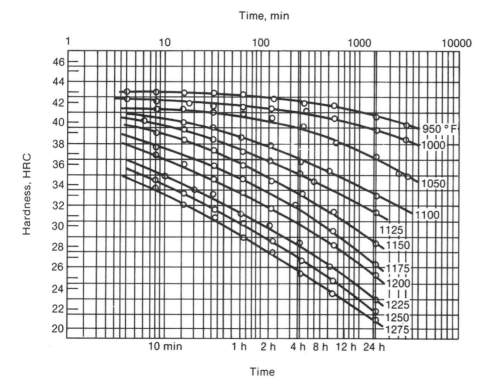

Fig. 7 Tempering behavior (Rockwell C hardness) of ASTM/SAE 4340 steel at various temperatures as a function of time. Source: Ref 2, 24

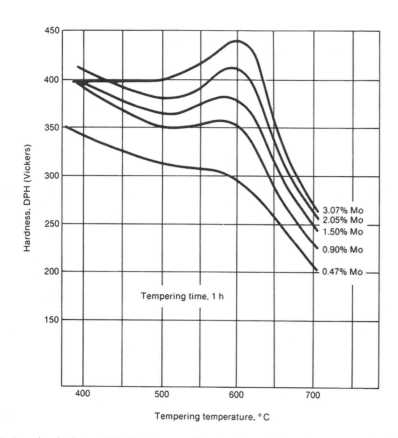

Fig. 8 Secondary hardening (Vickers) during tempering of steels containing various amounts of molybdenum. Source: Ref 25

It is important to remember that application of any rule of thumb is very subjective and limited to certain process conditions and prior microstructure of the steel (Ref 15, 16, 18 to 21).

Surface Hardening (Case Hardening). In surface hardening (case hardening) applications, a certain hardness layer at a particular surface (for example, inside diameter, outside diameter, or butt-end face) of the heat treated component is desired. Frequency selection is appreciably more subtle here than in through heating applications.

According to a classical definition of the skin effect, approximately 86% of all power induced by the induction coil will be concentrated in the surface layer, which is called the current penetration depth. Frequency and temperature have a dominant effect on the magnitude of current penetration depth in steels and cast irons. By controlling the depth of current penetration, it is possible to austenitize selective areas of the component that require hardening without affecting the rest of the component. Depending upon a required hardness case depth, the frequency selection for surface hardening applications ranges from 60 Hz (hardening of large mill rolls) to greater than 450 kHz (hardening small pins).

The frequency selection can also be affected by an economical factor as well. A higher frequency rating usually reflects a higher capital cost for the equipment.

Induction heat treating practitioners typically divide frequencies into three categories: low frequency (below 10 kHz), medium frequency (from 10 to 70 kHz), and high frequency (above 70 kHz). With the frequency increase, the depth of heating is decreased according to Eq 5 and 6 in the companion article "Simulation of Induction Heating Prior to Hot Working and Coating" in this Volume.

Various case depths and hardness patterns can be achieved with the same inductor by changing the frequency, power density, and duration of heating. Figure 10 shows an impressive example of a diversity of achievable induction hardening patterns when hardening gearlike components (Ref 1, 28, 106). The left side of Fig. 10 shows various hardness patterns that were obtained on the same carbon steel shaft with variations in heat time, frequency, and power density. The right side of Fig. 10 shows a similar effect when surface hardening a spur gear. As a rule, when it is necessary to harden only the gear tooth tips, a higher frequency and high power density are applied; to harden the tooth roots, use a lower frequency. A high power density in combination with the relatively short heat time generally results in a shallow pattern, while a low power density and extended heat time produces a deep pattern with wide transition zones.

For illustration purposes, Fig. 11 shows sketches of three cases related to frequency selection when surface hardening a solid shaft. In all three cases, it was possible to achieve the same case depth (indicated as a dotted line) while using three appreciably different frequencies, designated as too high, too low, and the optimal frequency. If the frequency is too high for the specified case depth, resulting in too small a current penetration depth (Fig. 11, left), additional heating time is needed to allow the heat to be conducted to the desired depth to obtain the required hardness depth. Not only does this add unnecessary cycle time, but it may also lead to a noticeable overheating of the surface, resulting in excessive grain growth, decarburization, scale formation, incipient melting, grain-boundary liquation, and other undesirable metallurgical phenomena.

Conversely, if the chosen frequency is appreciably lower than the optimal frequency, it produces a depth of heating that is significantly greater than necessary. The result is a large heat-affected zone, additional shape distortion of the heat treated component, and unnecessary waste of energy for heating a greater-than-required mass of metal (Fig. 11, middle). In some cases, the penetration depth can be so large compared to the required hardened case depth that it will not be possible to meet a pattern specification, because the maximum permissible case depth is exceeded or the pattern repeatability may be altered.

Generally speaking, the optimal frequency results in a current penetration depth at a temperature above the Curie point within the range of 1.2 to 2.4 times the required hardness case depth. In this case, the surface of the component will not be overheated, and additional heat that is generated below the case depth will be just sufficient to compensate for the cooling/soaking effect of the cold core.

Keeping in mind the previously mentioned drawbacks of using other than optimal frequencies, in many cases it is possible to achieve the various hardness patterns by using an appropriate combination of power density, frequency, and heat time (as illustrated in Fig. 11). The basic rule is that when a particular case depth is required, it may be possible to achieve the same case depth with a frequency that is lower than recommended by using a higher power density and a shorter heat time (faster scan rate). Conversely, if a higher-than-optimal frequency is available, then a lower power

Fig. 9 Crankshaft hardening inductors made of copper bar or tube. Comparison of U-shaped inductors (top) versus C-inductors (bottom)

Table 2 Typical operating conditions for progressive through-hardening of steel parts by induction

Section size		Material	Frequency(a), Hz	Power(b), kW	Total heating time, s	Scan time		Work temperature				Production rate		Inductor input(c)	
mm	in.					s/cm	s/in.	Entering coil		Leaving coil		kg/h	lb/h	kW/cm²	kW/in.²
								°C	°F	°C	°F				
Rounds															
13	½	4130	180	20	38	0.39	1	75	165	510	950	92	202	0.067	0.43
			9600	21	17	0.39	1	510	950	925	1700	92	202	0.122	0.79
19	¾	1035 mod	180	28.5	68.4	0.71	1.8	75	165	620	1150	113	250	0.062	0.40
			9600	20.6	28.8	0.71	1.8	620	1150	955	1750	113	250	0.085	0.55
25	1	1041	180	33	98.8	1.02	2.6	70	160	620	1150	141	311	0.054	0.35
			9600	19.5	44.2	1.02	2.6	620	1150	955	1750	141	311	0.057	0.37
29	1⅛	1041	180	36	114	1.18	3.0	75	165	620	1150	153	338	0.053	0.34
			9600	19.1	51	1.18	3.0	620	1150	955	1750	153	338	0.050	0.32
49	1¹⁵⁄₁₆	14B35H	180	35	260	2.76	7.0	75	165	635	1175	195	429	0.029	0.19
			9600	32	119	2.76	7.0	635	1175	955	1750	195	429	0.048	0.31
Flats															
16	⅝	1038	3000	300	11.3	0.59	1.5	20	70	870	1600	1449	3194	0.361	2.33
19	¾	1038	3000	332	15	0.79	2.0	20	70	870	1600	1576	3474	0.319	2.06
22	⅞	1043	3000	336	28.5	1.50	3.8	20	70	870	1600	1609	3548	0.206	1.33
25	1	1036	3000	304	26.3	1.38	3.5	20	70	870	1600	1595	3517	0.225	1.45
29	1⅛	1036	3000	344	36.0	1.89	4.8	20	70	870	1600	1678	3701	0.208	1.34
Irregular shapes															
17.5–33	¹¹⁄₁₆–1⁵⁄₁₆	1037 mod	3000	580	254	0.94	2.4	20	70	885	1625	2211	4875	0.040	0.26

(a) Note use of dual frequencies for round sections. (b) Power transmitted by the inductor at the operating frequency indicated. This power is approximately 25% less than the power input to the machine, because of losses within the machine. (c) At the operating frequency of the inductor. Source: Ref 3

Table 3 Approximate power densities required for through heating of steel for hardening, tempering, or forming operations

Frequency(a), Hz	150–425 °C (300–800 °F)		425–760 °C (800–1400 °F)		760–980 °C (1400–1800 °F)		980–1095 °C (1800–2000 °F)		1095–1205 °C (2000–2200 °F)	
	kW/cm²	kW/in.²	kW/cm²	kW/in.²	kW/cm²	kW/in.²	kW/cm²	kW/in.²	kW/cm²	kW/in.²
60	0.009	0.06	0.023	0.15	(c)	(c)	(c)	(c)	(c)	(c)
180	0.008	0.05	0.022	0.14	(c)	(c)	(c)	(c)	(c)	(c)
1000	0.006	0.04	0.019	0.12	0.08	0.5	0.155	1.0	0.22	1.4
3000	0.005	0.03	0.016	0.10	0.06	0.4	0.085	0.55	0.11	0.7
10,000	0.003	0.02	0.012	0.08	0.05	0.3	0.070	0.45	0.085	0.55

(a) The values in this table are based on use of proper frequency and normal overall operating efficiency of equipment. (b) In general, these power densities are for section sizes of 13 to 50 mm (½ to 2 in.). Higher inputs can be used for smaller section sizes, and lower inputs may be required for larger section sizes. (c) Not recommended for these temperatures. Source: Ref 3

Fig. 10 Variety of induction hardening patterns obtained by using variations in frequency, heat time, and power density. Source: Ref 28

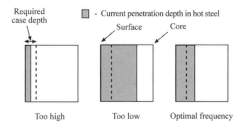

Fig. 11 Comparison of required case depth versus current penetration depths in hot steel at various frequencies. Source: Ref 1

density should be used in combination with a longer heat time.

In the majority of cases, the frequencies used in surface hardening of regular-shaped components result in a pronounced skin effect, which eliminates the concern regarding an eddy current cancellation. However, if the workpiece has an irregular shape (C-shaped tubes, odd-shaped parts, or slotted cylinders), the eddy currents may be forced to flow on the outside and inside areas of the part, with the potential of canceling each other (Ref 1). The basic rule of thumb that avoids a current cancellation in such cases is that the chosen frequency should result in a value of current penetration depth that would be no more than one-fourth the thickness of the current-conducting path (Ref 1).

Over the years, industry has accumulated numerous standard practices to calculate process parameters for induction surface hardening. Most of these practices are extremely subjective and are related only to the selection of frequency and power when single-shot hardening or scan hardening regular-shaped cylinders (such as solid shafts) made from plain carbon steels. Some of these recommendations are shown in Tables 2 to 4.

According to another common practice, the required frequency for surface hardening of solid cylinders with a case depth ranging from 1.6 to 5 mm can be determined based on the conditions in Eq 1 or 2:

$$\left(\frac{6.5}{X_{CD}}\right)^2 < \text{Frequency (kHz)} < \left(\frac{16.6}{X_{CD}}\right)^2 \quad (\text{Eq 1})$$

$$\text{Frequency (kHz)} \cong \left(\frac{9.8}{X_{CD}}\right)^2 \quad (\text{Eq 2})$$

where X_{CD} is a required case depth (in millimeters).

Calculating Frequency Range for Case Depth

Example 1: Case Depth. A 3 mm case depth is required; the calculation for the frequency range is:

$$\left(\frac{6.5}{3}\right)^2 < \text{Frequency (kHz)} < \left(\frac{16.6}{3}\right)^2$$

According Eq 1, an appropriate frequency range is 4.7 to 30.6 kHz. In comparison, according to Eq 2, the frequency is 10.7 kHz. Components of smaller sizes (smaller outside diameter) shift Eq 1 toward using higher frequencies. In contrast, lower frequencies could be used when induction case hardening larger-sized parts.

Heat Duration

The final parameter is the duration of heat. Many induction-hardened components are heated to hardening temperatures within several seconds or even during a fraction of a second (i.e., gear contour hardening using a dual-frequency source) (Ref 28). Upon choosing an appropriate frequency and power density based on one of the aforementioned criteria, the duration of heat can also be determined after several heat trials and evaluation of the obtained hardness patterns.

It is imperative to keep in mind that most rules of thumb assume hardening of fine-grained, normalized, homogeneous ferritic-pearlitic prior structures with a carbon content of approximately 0.4 to 0.5%. The microstructure of steel prior to heat treatment (also referred to as the initial structure, the structure of the parent material, or the structure of the "green" part) has a pronounced effect on the results of the heat treatment and the required process parameters, such as the austenitizing temperature and the amount of time the part is required to be held at that temperature. As shown in Ref 1, 2, and 30, even for steel of the same grade (1042), the required induction hardening temperature range depends on the heat intensity as well as the prior microstructure of the steel:

- 880 to 1095 °C (1620 to 2000 °F) for annealed prior microstructures
- 840 to 1000 °C (1550 to 1830 °F) for normalized prior microstructures
- 820 to 930 °C (1510 to 1710 °F) for quenched and tempered prior microstructures

The effect of prior structure can be explained as follows (Ref 1, 4, 5, 15 to 21). A quenched and tempered prior structure is the most favorable. It consists of tempered martensite and ensures rapid transformation, which lowers the temperature required for austenite formation. This results in a fast, consistent response of the steel to induction hardening, with minimum amounts of grain growth, shape/size distortion, and surface oxidation; a minimum required heating energy; and a well-defined, "crisp" hardness pattern having a narrow transition zone. This type of initial structure can even result in slightly higher hardness and a deeper hardened case depth as well as lower coil power compared with other prior structures. Normalized initial microstructures consisting of homogeneous fine pearlitic-ferritic structures are generally considered to be the next-favorable initial structures.

Unfavorable Structures. If the initial microstructure of a steel component has a significant amount of coarse pearlite and, most importantly, coarse ferrites or clusters or bands of ferrites, then the structure cannot be considered favorable. Ferrite is practically a pure iron and does not contain the sufficient amount of the carbon required for martensitic transformation. Pure ferrite consists of less than 0.025% C. Large areas (clusters or bands) of ferrite require a long time for carbon to diffuse into carbon-poor areas for formation of homogeneous austenite. Those ferrite clusters or bands could act as one very large grain of ferrite and are often retained in the austenite upon rapid heating (Ref 1). After quenching a heterogeneous austenite or mixed austenite, a complex microstructure can form that contains martensite and upper transformation products. Scattered soft and hard spots and poor mechanical properties characterize this structure. Appreciably higher temperatures and longer heating times are required to austenitize steels having these structures. It is strongly recommended to avoid severely segregated and banded initial microstructures. Steels with large stable carbides (spheroidized microstructures) also have poor response to induction hardening and also result in the need for prolonged heating and significantly higher temperatures to complete austenitization. Longer heating times lead to grain growth, the formation of coarse martensite, a larger transition zone, surface oxidation/decarburization, and increased shape distortion. Coarse martensite has a negative effect on such important properties as toughness, impact strength, and bending fatigue strength and is susceptible to cracking (Ref 1 to 3; also see the article "Simulation of Induction Heating Prior to Hot Working and Coating" in this Volume).

Summarizing the effect of prior structures, one can conclude that components having quenched and tempered, normalized, annealed, or spheroidized structures have measurably different required austenitizing temperatures and holding times at those temperatures, dramatically affecting the selection of heat time and power. In particular, prior microstructures have a pronounced effect when shorter heat times are used.

Bear in mind that rapid heating reduces the effect of thermal conduction and tends to austenitize primarily the area where eddy currents were induced (i.e., the outer surface layer). This results in a very short transition zone. With the

Table 4 Power densities required for surface hardening of steel

Frequency, kHz	Depth of hardening(a) mm	Depth of hardening(a) in.	Low(d) kW/cm²	Low(d) kW/in.²	Optimum(e) kW/cm²	Optimum(e) kW/in.²	High(f) kW/cm²	High(f) kW/in.²
500	0.381–1.143	0.015–0.045	1.08	7	1.55	10	1.86	12
	1.143–2.286	0.045–0.090	0.46	3	0.78	5	1.24	8
10	1.524–2.286	0.060–0.090	1.24	8	1.55	10	2.48	16
	2.286–3.048	0.090–0.120	0.78	5	1.55	10	2.33	15
	3.048–4.064	0.120–0.160	0.78	5	1.55	10	2.17	14
3	2.286–3.048	0.090–0.120	1.55	10	2.33	15	2.64	17
	3.048–4.064	0.120–0.160	0.78	5	2.17	14	2.48	16
	4.064–5.080	0.160–0.200	0.78	5	1.55	10	2.17	14
1	5.080–7.112	0.200–0.280	0.78	5	1.55	10	1.86	12
	7.112–8.890	0.280–0.350	0.78	5	1.55	10	1.86	12

(a) For greater depths of hardening, lower kilowatt inputs are used. (b) These values are based on use of proper frequency and normal overall operating efficiency of equipment. These values may be used for both static and progressive methods of heating; however, for some applications, higher inputs can be used for progressive hardening. (c) Kilowattage is read as maximum during heat cycle. (d) Low kilowatt input may be used when generator capacity is limited. These kilowatt values may be used to calculate largest part hardened (single-shot method) with a given generator. (e) For best metallurgical results. (f) For higher production when generator capacity is available. Source: Ref 3

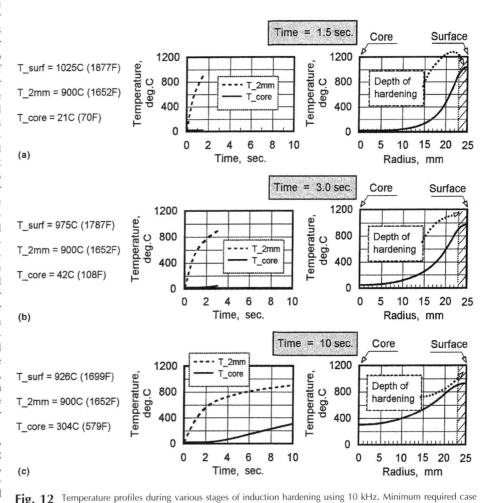

Fig. 12 Temperature profiles during various stages of induction hardening using 10 kHz. Minimum required case depth of 2 mm will be achieved in all three cases. Source: Ref 1

increase in heat time, thermal conductivity starts to play an increasingly dominant role, allowing the heat to soak from higher-temperature surface areas toward lower-temperature internal regions, leading to a fuzzy transition area of the hardened pattern and requiring a greater amount of energy. Figure 12 shows the results of a computer

simulation illustrating this effect (Ref 1) when surface hardening 50 mm diameter solid shafts.

As seen in Fig. 12(a), a heating time of 1.5 s shows no rise in core temperature and a slight increase in temperature in the internal areas of the workpiece. A heating time of 3.0 s (Fig. 12b) leads to a small rise in core temperature and a moderate increase in temperature of the internal areas, while a heating time of 10.0 s (Fig. 12c) results in a significant rise in temperature of the internal areas and the core. In all three cases, the temperature at the hardened depth (2 mm from the surface) is approximately the same. Below the hardened depth is where the difference is primarily seen, because the temperature of subsurface layers begins to increase with increasing duration of the heat.

Physics indicates that if a greater amount of heat is induced in the workpiece, then a greater mass of steel will be heated, leading to a greater workpiece expansion and thus often resulting in a greater shape distortion. Therefore, to decrease the distortion of hardened components (for example, gears), it is desirable to keep the heating time as short as possible. However, there are some limitations. First, the material must reach the minimum required transformation temperature that is sufficient to create a homogeneous austenite at the depth to be hardened. A combination of unreasonably high frequency and/or surface power density can lead to surface overheating, resulting in known undesirable microstructures. Second, such a combination can produce substantial temperature gradients during heating. Thermal stresses can reach their critical values, and cracking can develop when hardening brittle materials (i.e., high-carbon steels and gray cast irons). In such cases, it may be desirable to use longer heating times with lower power density. Third, hot and cold spots can become pronounced if geometrical irregularities are present (for example, sharp corners, holes, keyways, edges, etc.). In these cases, a less intensive heat and longer heat times are also desired, including the use of some dwells before quenching to allow the thermal conduction to reduce the temperature of hot spots.

Figure 13 consists of an alternative chart developed by industry for choosing frequency, power density, and heat time for surface hardening of solid, regular-shaped components without irregularities (Ref 3). Table 5 shows an estimation of power requirements for induction hardening of spur gears (Ref 3).

It is imperative to remember that the procedure for determining an optimal combination of process parameters using rules of thumb could easily be misleading. Even a cursory look at Table 4 and Fig. 13 shows an appreciable difference for the recommended power densities for seemingly similar application. Each component to be induction hardened has its own "personality" with respect to material specifics, prior structure, geometry, and functionality.

Numerical Computer Simulation

As outlined previously, an estimation of the most appropriate process parameters based on rules of thumb as well as the analytical methods and equivalent circuit coil design methods popular in the 1960s and 1970s is very subjective, with inherent appreciable restrictions (Ref 1, 31, 32; also see the article "Simulation of Induction Heating Prior to Hot Working and Coating" in this Volume). Those techniques should be used for quick estimation of the approximate ballpark parameters of induction heat treating systems and only in the cases of regular-shaped workpieces. It is imperative to be aware that in many applications (especially in the cases of heat treating irregular-shaped workpieces), erroneous and inadequate results can be obtained when such calculation techniques are used.

Advances in modern computers, the increasing complexity of induction heating applications, and increasing demands to manufacture higher-quality parts in combination with the necessity of improving cost-effectiveness by shortening the learning curve and reducing development time that is naturally related to reducing the delivery time of equipment have significantly restricted the usefulness of simplified formulas, as well as analytical and seminumerical methods.

Rather than using computational techniques with many restrictions and disputable accuracy, modern induction heating specialists turned to highly effective numerical methods, such as finite-difference, finite-element, mutual impedance, and boundary-element methods. These numerical techniques are widely and successfully used in the computation of coupled electromagnetic, heat-transfer, and metallurgical phenomena. Each of these simulation techniques has certain pros and cons and has been used alone or in a combination with others.

Any computational analysis can produce, at best, only the results that are derived from the properly determined physical properties of heated materials and correctly defined governing equations. Therefore, the first important step in any computer simulation is to obtain physical properties of heated materials from properly conducted experiments or trusted sources. The well-known but rude saying "garbage in, garbage out" indicates the necessity of having accurate physical properties. The second but as important step in any mathematical simulation is to choose an appropriate theoretical model that correctly represents the technological process or phenomenon.

Generally speaking, to conduct a computer modeling of an induction heat treating system, it is necessary to model several process stages, including but not limited to heating, soaking (dwelling), cooling during quenching, as well as reheating for tempering or stress relieving (Ref 1).

Governing Electromagnetic Equations

The technique of calculating electromagnetic fields depends on the ability to solve Maxwell's equations, which are a set of equations that describe the interrelated nature of electric and magnetic fields. For general time-varying electromagnetic fields, when there are no substantially fast-moving components, Maxwell's equations in differential form are (Ref 1, 33 to 37):

$$\nabla \times \mathbf{H} = \mathbf{J} + \frac{\partial \mathbf{D}}{\partial t} \quad \text{(from Ampere's law)} \quad \text{(Eq 3)}$$

$$\nabla \times \mathbf{E} = -\frac{\partial \mathbf{B}}{\partial t} \quad \text{(from Faraday's law)} \quad \text{(Eq 4)}$$

$$\nabla \bullet \mathbf{B} = 0 \quad \text{(from Gauss's law)} \quad \text{(Eq 5)}$$

$$\nabla \bullet \mathbf{D} = \rho^{\text{charge}} \quad \text{(from Gauss's law)} \quad \text{(Eq 6)}$$

where $\mathbf{E}$ is the electric field intensity, $\mathbf{D}$ is the electric flux density, $\mathbf{H}$ is the magnetic field intensity, $\mathbf{B}$ is the magnetic flux density, $\mathbf{J}$ is the conduction current density, and ρ^{charge} is the electric charge density. Equations 3 to 6 include the vector algebra ∇, $\nabla \bullet$, and $\nabla \times$, These symbols express the differential operators of gradient (grad), divergence (div), and curl (curl), respectively.

The same fundamental laws governing the general time-varying electromagnetic field can also be written in integral form (Ref 33 to 37):

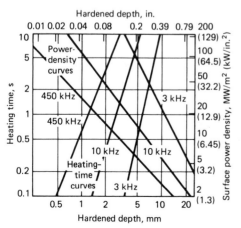

Fig. 13 Interrelationship among heating time, surface power density, and hardened depth for various frequencies. Source: Ref 3

Table 5 Power requirements for induction hardening of gear teeth

Tooth	Diametral pitch	Approximate length of tooth profile		Surface area per tooth(a)		Power required per tooth(b), kW	Total power required(c), kW
		mm	in.²	cm²	in.²		
A	3	50	2.0	12.9	2.0	20	800
B	4	38	1.5	9.7	1.5	15	600
C	5	33	1.3	8.4	1.3	13	520
D	6	25	1.0	6.5	1.0	10	400
E	7	23	0.9	5.8	0.9	9	360
F	8	19	0.75	4.8	0.75	7.5	300

(a) For a face width of 25 mm (1 in.). (b) At a power density of 1.55 kW/cm² (10 kW/in.²). (c) For a gear having 40 teeth. Source: Ref 3

$$\oint_C \mathbf{H} \cdot d\mathbf{l} = \int_S \mathbf{J} \cdot d\mathbf{A} + \int_S \frac{\partial \mathbf{D}}{\partial t} \cdot d\mathbf{A} \text{ (from Ampere's law)}$$

$$\text{(Eq 7)}$$

$$\oint_C \mathbf{E} \, d\mathbf{l} - \int_C \mathbf{B} \times v \cdot d\mathbf{l} = -\frac{d}{dt} \int_S \mathbf{B} \cdot d\mathbf{A} \quad \text{(Eq 8)}$$
$$\text{(from Faraday's law)}$$

$$\oint_S \mathbf{B} \cdot d\mathbf{A} = 0 \qquad \text{(from Gauss's law)} \qquad \text{(Eq 9)}$$

$$\oint_S \mathbf{D} \, d\mathbf{A} = \int_V \rho^{charge} \, dV \qquad \text{(from Gauss's law)}$$

$$\text{(Eq 10)}$$

where dV represents the differential of volume, V, surrounded by surface, S; $d\mathbf{A}$ denotes the differential vector element of surface region "A" with small magnitude and direction normal to surface S; $d\mathbf{l}$ represents the vector element of the path length that is tangential to contour C surrounding surface c; and v is the instantaneous velocity of element $d\mathbf{l}$.

Physical interpretation of Maxwell's equations is provided in Ref 1 and 33 to 37.

Finite-element and finite-difference methods are numerical modeling techniques that typically use the differential form of Maxwell's equations, while the mutual impedance and boundary-element methods usually apply its integral form. Transformation of the differential form of Maxwell's equations representing the electromagnetic field is discussed here.

Maxwell's equations (Eq 3 to 6) are in indefinite form, because the number of equations is less than the number of unknowns. These equations become definite when the relations between the field quantities are specified. The following constitutive relations are additional and hold true for a linear, isotropic medium:

$$\mathbf{D} = \varepsilon \, \varepsilon_o \, \mathbf{E} \qquad \text{(Eq 11)}$$

$$\mathbf{B} = \mu_r \, \mu_o \, \mathbf{H} \qquad \text{(Eq 12)}$$

$$\mathbf{J} = \sigma \, \mathbf{E} \quad \text{(Ohm's law)} \qquad \text{(Eq 13)}$$

where the parameters ε, μ_r, and σ denote, respectively, the relative permittivity, relative magnetic permeability, and electrical conductivity of the material; $\sigma = 1/\rho$, where ρ is the electrical resistivity.

Selected basic units in the electromagnetic field (SI international) are shown in Table 6.

The constant $\mu_o = 4\pi \times 10^{-7}$ H/m [or Wb/(A $\times$ m)] is called the permeability of free space (the vacuum), and similarly, the constant $\varepsilon_o = 8.854 \times 10^{-12}$ F/m is called the permittivity of free space. Both relative magnetic permeability, μ_r, and relative permittivity, ε, are nondimensional parameters and have very similar meanings. Relative magnetic permeability indicates the ability of a material to conduct the magnetic flux better than vacuum or air. The article "Simulation of Induction Heating Prior

to Hot Working and Coating" in this Volume provides detailed discussion regarding the nonlinear nature of relative magnetic permeability. Similar to μ_r, relative permittivity (or the dielectric constant) indicates the ability of a material to conduct the electric field better than vacuum or air. By taking Eq 11 and 13 into account, Eq 3 can be rewritten as:

$$\nabla \times \mathbf{H} = \sigma \mathbf{E} + \frac{\partial(\varepsilon_0 \, \varepsilon_r \, \mathbf{E})}{\partial t} \qquad \text{(Eq 14)}$$

For the great majority of applications involving induction heating of metals, the impact of induced conduction current density, $\mathbf{J}$, on the total heat generation is much greater than the impact of displacement current density:

$$\frac{\partial \mathbf{D}}{\partial t}$$

Therefore, the last term on the right side of Eq 14 can be neglected, and it can be simplified as:

$$\nabla \times \mathbf{H} = \sigma \mathbf{E} \qquad \text{(Eq 15)}$$

It is important to realize that, depending on the magnetic and electric properties, certain quantities of the electromagnetic field can be continuous or discontinuous on different sides of an interface divided by two different media. Designating two different media with a common interface by subscripts "1" and "2," the basic field vectors must satisfy the following boundary conditions (Ref 33 to 38):

$$(\mathbf{B}_2 - \mathbf{B}_1) \cdot n = 0 \qquad \text{(Eq 16)}$$

$$(\mathbf{H}_2 - \mathbf{H}_1) \times n = \mathbf{J}^{surface} \qquad \text{(Eq 17)}$$

$$(\mathbf{E}_2 - \mathbf{E}_1) \times n = 0 \qquad \text{(Eq 18)}$$

Boundary conditions (Eq 16 to 18) can be interpreted as follows. According to Eq 16, the normal component of the magnetic flux density, $\mathbf{B}$, is continuous across the interface of media 1 and 2. In the presence of the surface current, $\mathbf{J}^{surface}$, as follows from Eq 17, the tangential components of the magnetic field strength, $\mathbf{H}$, are discontinuous across the two media with a magnitude of $\mathbf{J}^{surface}$. If the surface current is absent (as in the case of electrically conductive materials), then the tangential component is continuous, and the right side of Eq 17 will be 0. According to Eq 18, the tangential components of the electric field intensity, $\mathbf{E}$, are continuous across the interface. Equations 16 to 18 can be very helpful in providing a simple test of whether the results of numerical computer modeling are physically correct.

After some vector algebra, Eq 3, 4, and 12 can be written:

$$\nabla \times \left(\frac{1}{\sigma} \nabla \times \mathbf{H}\right) = -\mu_r \mu_o \frac{\partial \mathbf{H}}{\partial t} \qquad \text{(Eq 19)}$$

$$\nabla \times \left(\frac{1}{\mu_r} \nabla \times \mathbf{E}\right) = -\sigma \mu_o \frac{\partial \mathbf{E}}{\partial t} \qquad \text{(Eq 20)}$$

Because the magnetic flux density, $\mathbf{B}$, satisfies a zero divergence condition (Eq 5), meaning that magnetic field lines (B-lines) always form continuous loops (without having any points of origin or termination), it can be expressed in terms of a magnetic vector potential, $\mathbf{A}$, as:

$$\mathbf{B} = \nabla \times \mathbf{A} \qquad \text{(Eq 21)}$$

After substituting $\mathbf{B}$ in Eq 4 and using Eq 21, it is possible to derive the following:

$$\nabla \times \mathbf{E} = -\nabla \times \frac{\partial \mathbf{A}}{\partial t} \qquad \text{(Eq 22)}$$

Therefore, after integration, one can obtain:

$$\mathbf{E} = -\frac{\partial \mathbf{A}}{\partial t} - \nabla \, \varphi \qquad \text{(Eq 23)}$$

where φ is the electric scalar potential. Equation 13 can be written as:

$$\mathbf{J} = -\sigma \frac{\partial \mathbf{A}}{\partial t} + \mathbf{J}_s \qquad \text{(Eq 24)}$$

where $\mathbf{J}_s = -\sigma \nabla \varphi$ is the induction coil current density (source current density).

Nonlinearity. One of the major difficulties in electromagnetic field and heat-transfer simulation is related to the nonlinear nature of material properties (Ref 1, 41, 72; see also the article "Simulation of Induction Heating Prior to Hot Working and Coating" in this Volume). Electromagnetic properties of materials encompass a variety of characteristics, including magnetic permeability, electrical resistivity (electrical conductivity), saturation flux density, and coercive force. While recognizing the importance of all electromagnetic properties, two of them—electrical conductivity, σ (electrical resistivity, ρ), and relative magnetic permeability, μ_r—have the most pronounced effect on the process of induction heating of metals. Electrical conductivity, σ, and its reciprocal, electrical resistivity, ρ, vary with temperature, chemical composition, microstructure, grain size, and so on. For most metals and alloys (including carbon steels, alloyed steels, copper, aluminum, and titanium), electrical resistivity, ρ, increases with temperature (Fig. 4 in the article "Simulation of Induction Heating Prior to Hot Working and Coating" in this Volume).

Relative magnetic permeability, μ_r, is not only a complex function of grain structure,

Table 6 Basic electromagnetic field properties

Quantity	Symbol	SI units
Electric field density	$\mathbf{E}$	V/m
Electric flux density	$\mathbf{D}$	C/m^2
Conduction current density	$\mathbf{J}$	A/m^2
Electrical resistivity	ρ	Ω m
Electrical conductivity	σ	S/m
Magnetic field intensity	$\mathbf{H}$	A/m
Magnetic flux density	$\mathbf{B}$	T

chemical composition, and temperature but also frequency and magnetic field intensity. As shown in Fig. 8(a) of the article "Simulation of Induction Heating Prior to Hot Working and Coating," the same kind of carbon steel at the same temperature and frequency can have a substantially different value of μ_r due to differences in the intensity of the magnetic field. The temperature at which a ferromagnetic body becomes nonmagnetic is called the Curie temperature (Curie point). Figure 9 of the companion article "Simulation of Induction Heating Prior to Hot Working and Coating" shows a dramatic illustration of the complex relationship among μ_r, temperature, and magnetic field intensity for medium-carbon steel. The piecewise continuous nature of material properties is postulated in the majority of mathematical modeling approaches. After neglecting the hysteresis and magnetic saturation, it can be shown that:

$$\frac{1}{\mu_r \mu_o}(\nabla \times \nabla \times \mathbf{A}) = \mathbf{J}_s - \sigma \frac{\partial \mathbf{A}}{\partial t} \qquad \text{(Eq 25)}$$

For the great majority of induction metal heat treating applications (such as surface hardening, through hardening, annealing, and normalizing), the heating effect due to hysteresis loss does not typically exceed 7 to 8% compared to the heat effect due to Joule heat generated by eddy current, because during the majority of the heat cycle, the surface temperature of the component is well above the Curie point. This makes valid the assumption of neglecting hysteresis losses. However, in some low-temperature applications, where heated metal retains its magnetic properties during the entire heating cycle (for example, induction tempering, stress relieving, heating prior to galvanizing and coating of ferrous strips and wires, etc.), hysteresis heat generation can be appreciable compared to Joule heat generated by eddy current losses, and the assumption of neglecting magnetic hysteresis will not be valid.

In some applications, certain components of the induction heating system move with substantial speed in respect to the induction heater. This includes such induction heat treating applications as high-speed strip and wire heating prior to metallic and nonmetallic coating (such as galvanizing) and heating of fast-rotated discs. In such applications, the mathematical model (Eq 26) includes the velocity of workpiece movement:

$$\nabla \times \frac{1}{\mu \mu_o} \nabla \times \mathbf{A} + \sigma \frac{\partial \mathbf{A}}{\partial t} - \sigma(v \times \nabla \times \mathbf{A}) = \mathbf{J}_s$$
$$\text{(Eq 26)}$$

where v represents the velocity of component movement.

It is possible to simplify the governing equations (Eq 19, 20, 25, and 26) by assuming a time-harmonic electromagnetic field (a quasi-stationary assumption). Under this assumption, it is possible to conclude that the electromagnetic field quantities in Maxwell's equations are harmonically oscillating functions with a single frequency. Therefore, for appreciably fast-moving components, Eq 26 can be rewritten in the following form:

$$\nabla \times \nabla \times \mathbf{A} + j\,\omega\,\sigma\,\mu_r\,\mu_o\,\mathbf{A}$$
$$- \sigma\,\mu_r\,\mu_o(v \times \nabla \times \mathbf{A}) = \mu_r\,\mu_o\,\mathbf{J}_s \qquad \text{(Eq 27)}$$

For the great majority of induction metal heat treating applications (including hardening, tempering, and bar and billet heating), the velocity of component movement is appreciably small compared to the applied frequency, and Eq 19, 20, or 25 is used instead of Eq 26 and 27.

The following expressions can be written after applying some vector algebra to Eq 19, 20, and 25 and assuming the quasi-stationary (time-harmonic) nature of the electromagnetic field with an angular frequency of ω:

$$\frac{1}{\sigma}\nabla^2 \mathbf{H} = j\,\omega\,\mu_r\,\mu_o\,\mathbf{H} \qquad \text{(Eq 28)}$$

$$\frac{1}{\mu_r}\nabla^2 \mathbf{E} = j\,\omega\,\sigma\,\mu_o\,\mathbf{E} \qquad \text{(Eq 29)}$$

$$\frac{1}{\mu_r \mu_o}\nabla^2 \mathbf{A} = -\mathbf{J}_s + j\,\omega\,\sigma\,\mathbf{A} \qquad \text{(Eq 30)}$$

where ∇^2 denotes the Laplace operator, which has different forms in Cartesian and cylindrical coordinates. In Cartesian coordinates:

$$\nabla^2 \mathbf{A} = \frac{\partial^2 \mathbf{A}}{\partial X^2} + \frac{\partial^2 \mathbf{A}}{\partial Y^2} + \frac{\partial^2 \mathbf{A}}{\partial Z^2} \qquad \text{(Eq 31)}$$

In cylindrical coordinates (axisymmetric case):

$$\nabla^2 \mathbf{A} = \frac{1}{R}\frac{\partial}{\partial R}\left(R\frac{\partial \mathbf{A}}{\partial R}\right) + \frac{\partial^2 \mathbf{A}}{\partial Z^2} \qquad \text{(Eq 32)}$$

An assumption of a single-frequency oscillating current means that harmonics are absent in both the impressed and induced currents. The governing equations (Eq 28 to 30) with the appropriate boundary condition can be solved with respect to $\mathbf{H}$, $\mathbf{E}$, or $\mathbf{A}$ (Ref 1, 33 to 40). Equations 28 to 30 are valid for general three-dimensional fields and allow the determination of all the required induction system design parameters, such as coil current, voltage, power, impedance, coil power factor, and eddy current distribution.

Two-Dimensional Approximation. Although there is considerable theoretical and practical interest in solving three-dimensional problems, an appreciable amount of induction metal heat treating applications can be effectively modeled using two-dimensional assumptions. For many induction heat treating applications, certain quantities of the magnetic field (such as magnetic vector potential, electric field intensity, and magnetic field intensity) may be assumed to be entirely directed. For example, in the longitudinal cross section of a multiturn solenoid coil (Fig. 14, top), both $\mathbf{A}$ and $\mathbf{E}$ vectors have only one component, which is entirely Z-directed. In contrast, in the case of a transverse section of a solenoid coil (Fig. 14, bottom), $\mathbf{H}$ and $\mathbf{B}$ vectors have only one component. This allows the three-dimensional field to be reduced to an evaluation of behavior of a combination of two-dimensional forms. For example, in the case of the magnetic vector potential, Eq 30 can be expressed as follows.

For a two-dimensional Cartesian system:

$$\frac{1}{\mu_r \mu_0}\left(\frac{\partial^2 \mathbf{A}}{\partial X^2} + \frac{\partial^2 \mathbf{A}}{\partial Y^2}\right) = -\mathbf{J}_s + j\,\omega\,\sigma\,\mathbf{A} \qquad \text{(Eq 33)}$$

For an axisymmetric cylindrical system:

$$\frac{1}{\mu_r \mu_0}\left(\frac{\partial^2 \mathbf{A}}{\partial R^2} + \frac{1}{R}\frac{\partial \mathbf{A}}{\partial R} + \frac{\partial^2 \mathbf{A}}{\partial Z^2} - \frac{\mathbf{A}}{R^2}\right) = -\mathbf{J}_s + j\,\omega\,\sigma\,\mathbf{A}$$
$$\text{(Eq 34)}$$

If the geometry of an induction system has symmetries, then the corresponding boundary conditions should be defined on the symmetry axes. A typical boundary condition in electromagnetism imposes a zero field at infinite by assuming that boundaries are moved sufficiently far away from the inductor, making the magnetic vector potential $\mathbf{A}$ have zero value along the boundary (Dirichlet condition) or have a Neumann boundary condition when its gradient is negligibly small along the boundary compared to its value elsewhere in the region:

$$\left(\frac{\partial \mathbf{A}}{\partial n} = 0\right)$$

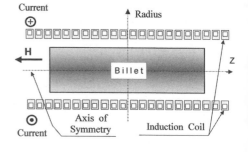

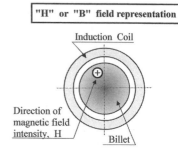

Fig. 14 E or A and H or B field representations in a cylindrical induction system. Source: Ref 41

By using simple vector algebra manipulations, it is possible to obtain governing equations similar to Eq 33 and 34 that can be formulated with respect to **E**, **B**, or **H**. Therefore, by choosing a proper field representation, a general problem can be solved in respect to either **A**, **E**, **B**, or **H**. For example, governing equations that are formulated with respect to **A** or **E** are very convenient for describing the electromagnetic field in a longitudinal cross section of the induction heating system using cylinder multiturn solenoid inductors (Fig. 14, top). In contrast, the electromagnetic field distribution in a transverse cross section of the workpiece (for example, a heat source distribution within the transverse cross section of noncylindrical components, such as slabs, blooms, strips, gears, etc.) can be more conveniently described by governing equations formulated with respect to **B** or **H** (Fig. 14, bottom).

Examples of Electromagnetic Field Formulations

Example 2: Split-Return Inductor. Two-dimensional electromagnetic field formulation in respect to magnetic vector potential, **A**, is used for computer modeling and optimization of electromagnetic parameters of split-return inductors.

Although induction heat treating has an interrelated nature, where electromagnetic and thermal phenomena are coupled together, a noticeable amount of important process subtleties and design information can be obtained by conducting just an electromagnetic computation and by reviewing electrical parameters of the system. This includes the distribution of the electromagnetic field as well as the distribution of the inductor current, induced eddy current density, and magnetic field intensity. Such information allows a better appreciation of process subtleties and develops know-how in determining conditions for the "sound" design of an induction system.

Split-return inductors are used in various applications, including hardening, annealing, tempering, stress relieving, bonding, brazing, and soldering. For example, Fig. 15 shows split-return inductors for use in in-line continuous pipe seam annealing (top) and selective hardening (bottom) applications. A split-return inductor has a main (center) leg that splits into two equal return legs, producing a unique eddy current distribution within the workpiece. Figure 16 shows an electrical circuit of a split-return inductor. If the copper width of the main and return legs are the same, the current density in the main leg of the inductor is double that in the return legs. Also, the power density (heat source) under the main leg is four times higher than the power density under its return legs. This ratio can be even greater if the width of the copper tubing of the return legs is greater than the width of the copper tubing for the main leg, or if a flux concentrator is applied for the main leg, which results in further increase of power density in a narrow band of the workpiece located under the main leg. Unless it is a transverse scanning application, it is typically considered that kilowatt losses induced within the workpiece by the return legs are a waste of energy, and steps should be taken to minimize these losses. In contrast, kilowatt losses induced in the workpiece by the main leg represent useful Joule heat, and therefore, steps should be taken to maximize it to obtain maximum electrical efficiency.

Figure 17 shows magnetic field distribution with (right) and without (left) a U-shaped magnetic flux concentrator located around the central leg of a split-return inductor. Computations were conducted by Inductoheat's personnel using the two-dimensional electromagnetic boundary-element software Oersted (Integrated Engineering Software Corp.). Oersted performs an electromagnetic field simulation using the integral form of Maxwell's equations formulated in respect to a magnetic vector potential, **A**. Computer modeling reveals that without a

Fig. 15 Split-return inductors for use in pipe seam annealing (top) and selective hardening (bottom) applications. Source: Ref 42

Fig. 16 Electrical circuit of a split-return inductor. The current in the main coil is double that in the return legs. Source: Ref 42

Fig. 17 Magnetic field distribution without (left) and with (right) a U-shaped magnetic flux concentrator located around the central leg of a split-return inductor. Source: Ref 42

concentrator, the magnetic flux spreads around the surroundings of the inductor legs (Fig. 17, left). Because current flow in the return legs is in the opposite direction of that in the main leg, the electromagnetic proximity effect shifts the currents to the sides of the coil facing each other. This dramatically reduces overall electrical efficiency (requiring much higher coil current to provide the needed heating) and increases coil copper losses (wasting energy). The situation worsens if the distance between the main and return legs is reduced or the inductor-to-part air gap is increased. Computer simulation shows that coil electrical efficiency in such systems can drop to 20 to 30% or lower (particularly when heating nonmagnetic metals or heating magnetic metals above the Curie temperature).

Analysis of the magnetic field distribution reveals that the U-shaped magnetic concentrator positioned around the main leg forms a magnetic path to channel the main magnetic flux in a well-defined area and magnetically separates fields produced by the main and return legs (Fig. 17, right). Available materials for magnetic flux concentrator fabrication include laminations, pure ferrites, and materials that consist of iron-base and ferrite-base compressed powder particles (i.e., Fluxtrol, AlphaFlux, AlphaForm, Ferrotron) (Ref 1, 43, 44). Analysis of pros and cons of different concentrator materials are provided in Ref 1 and 43 to 45. Thanks to the magnetic flux concentrator, the inductor current density is shifted toward the workpiece surfaces, improving heating efficiency, reducing required coil current and magnetic forces, and improving electromagnetic coupling and overall conditions to increase inductor life. This is the reason why it is strongly recommended to use magnetic flux concentrators in the split-return inductor design (Ref 42).

Computer simulation also reveals that even with frequency as low as 1 kHz, edges of copper corners that face the workpiece experience the highest current density and the possibility of being overheated, particularly when U-shaped concentrators are used (Fig. 17, right). This is the reason why sharp copper edges should be avoided. Instead, it is beneficial if the inductor legs have small radii.

It is important to remember that all concentrators degrade in service. Even under normal working conditions, their ability to concentrate magnetic fields begins to slowly decline due to, for example, degradation of magnetic particles and/or the binder (i.e., epoxy) used to hold magnetic powder particles together being exposed to high temperatures, quenchants, and rusting (Ref 1, 44 to 46). It is important to recognize this fact when using electromagnetic properties of flux concentrators as input data for the computer modelling. As one can conclude, in split-return inductors, the flux concentrator is positioned in the highest flux density region (Fig. 17) and is exposed to thermal radiation from the heated workpiece surface. The effectiveness and life of concentrators may be jeopardized due to a magnetic saturation and localized overheating. Figure 18 shows such

an improperly designed inductor, with laminations that were degraded due to excessive magnetic saturation of corners and subsequent overheating.

The computer modeling capability of the induction coil manufacturer helps to optimize inductor design and provides an appreciable comfort factor for building long-lasting split-return inductors where magnetic saturation of flux concentrators will be avoided and inductor performance is optimized.

Example 3: Rectangular Workpieces. Field formulation in two dimensions with respect to magnetic field intensity, **H**, is used for computation of induction heating of rectangular workpieces using the finite-element method.

Induction heating is a popular approach to heating noncylindrical workpieces, such as rectangular-shaped parts including round-cornered-square (RCS) billets, bars, blooms, plates, slabs, and so on. There are three basic induction approaches to heating RCS billets: static, progressive, and oscillating heating (Ref 1, 47). With progressive multistage horizontal heating (Fig. 19) being the most popular approach, billets or bars are moved through a single coil or multicoil horizontal induction heater. As a result, the billet or bar is sequentially (progressively) heated at predetermined locations inside the induction heater. Depending on the application, different coils positioned in-line can have various power levels and frequencies.

The bulk temperature of the billet must be raised to a specified level with a certain degree of heat uniformity. The uniformity requirement may include maximum tolerable thermal gradients: surface to core, end to end, and side to side. In most cases, a billet that is heated with an appreciable nonuniformity can negatively affect the quality of heated products and the life of critical components of metalworking machines.

Due to the noncylindrical geometry of RCS billets, in addition to the surface-to-core temperature uniformity, customers often specify temperature uniformity in its transversal cross section, including maximum allowable central part-to-corner temperature nonuniformity. Depending on the specifics of the process parameters, edge areas of RCS billets can be underheated, overheated, and heated uniformly. Transversal electromagnetic edge effect and thermal edge effect are primarily responsible for temperature distribution within the transverse cross section of RCS billets including edge regions (Ref 1, 40, 47 to 49).

Electromagnetic edge effect represents a distortion of the electromagnetic field and induced heat sources in the corner areas of RCS billets. The maximum value of the eddy current density is located on the surface of the central part of the RCS billet. However, it does not mean that the maximum temperature is always located there. Smaller penetration depths and higher frequencies make the skin effect more pronounced. In this case, the path of the induced current closely matches the contour of the

heated RCS billets. If the skin effect is pronounced, then the eddy current and power density are approximately the same along the billet perimeter, except its edge areas, where the distortion of induced power takes place. Even though heat losses at the edge (corner) area are higher than heat losses at the central part, the edge areas can be overheated compared to the central part. The phenomenon of edge overheating is more likely to occur in the induction heating of magnetic steels, aluminum, silver, or copper slabs, where the skin effect is typically pronounced.

If the skin effect is not pronounced, then underheating of the edge areas may occur. In this case, the path of eddy currents in the transverse cross section does not match the contour of the billet, and most of the induced currents close their loops earlier without reaching the corners and edge areas. As a result, there will be a deficit of power densities and heat sources in the edge areas compared to corresponding values in the central part, resulting in colder corners and requiring the use of dual- or multiple-frequency designs (Ref 1, 47).

As an example, Fig. 19 shows induction heating of RCS billets (top left) and RCS bars (bottom right). Figure 19 (top right) shows the results of computer simulation using a differential form of Maxwell's equations solved in respect to magnetic field intensity, **H**, for two polar cases of the temperature distribution within one-quarter of the transverse cross section of the RCS billet. For that particular billet size, a frequency of 30 kHz resulted in a pronounced skin effect, acting as a good example of when the chosen frequency is too high. This may lead to dramatically overheated corners in the RCS billet. In contrast, 500 Hz resulted in a penetration depth that was too large, leading to a clearly underheated corner. It is important to note that although a frequency of 30 kHz appeared to be too high in this case, in other applications when the sizes of billets or slabs would be much smaller, the frequency could act as too low. The terms *too high* or *too low* for the frequency are relative to a particular geometry of the workpiece and its material properties.

Fig. 18 Degraded laminations resulting from excessive magnetic saturation and overheating due to improper design. Source: Ref 42

Figure 19 (bottom left) shows two-dimensional temperature profiles of ¼ of the RCS bar transverse cross section, representing the dynamics of induction heating of a rectangular round-cornered-square carbon steel bar with 0.1 m (4 in.) cross section and using a frequency of 500 Hz. Appreciable temperature gradients occur within the bar cross section. It is important to have a clear understanding of the magnitude of these gradients, not only at the end of heating but during the intermediate and particularly the initial heating stage. With intensive heating, longitudinal and transverse cracks can occur as a result of substantial thermal stresses (thermal shocks) that are caused by different magnitudes of temperature and temperature gradients. As can be seen in Fig. 19, at the end of heating, the temperature distribution within the ¼ cross section of the RCS bar is quite uniform.

Mathematical Modeling of Thermal Processes

As discussed previously, the steady-state assumption of the electromagnetic field quasi-stationary (time-harmonic) nature with an angular frequency of ω is very popular. It is effectively used to simplify electromagnetic simulations of the great majority of induction heat treating applications without appreciably compromising the accuracy of computation while minimizing computation time. Unfortunately, the steady-state approach cannot be used for modeling thermal processes of induction heat treating, because, according to the steady-state approach, the temperature would remain constant at a given point of the component at all times. Because the component temperature distribution varies with time during induction heat treating (Fig.19), only the transient (time-dependent) approach should be used to model thermal processes that take place in induction heat treating. With a transient heat-transfer problem, the temperature is a function of not only the space coordinates but also of time.

The transient heat-transfer process in a metal workpiece can be described by the Fourier equation (Ref 1, 50 to 54):

$$c\gamma \frac{\partial T}{\partial t} + \nabla \cdot (-k \nabla T) = Q \qquad \text{(Eq 35)}$$

where T is temperature; γ is the density of the metal, c is the specific heat, k is the thermal conductivity of the metal, and Q is the heat-source density produced by eddy currents per unit time in a unit volume (the so-called heat generation). This heat-source density is obtained as a result of solving the electromagnetic problem. Selected basic units in heat-transfer analysis (SI international) are shown in Table 7.

Both k and c are substantially nonlinear functions of temperature (see Fig. 16 and 18 in the article "Simulation of Induction Heating Prior to Hot Working and Coating" in this Volume). An assumption of the constant value of thermal conductivity could result in appreciable errors in predicting transient temperature distribution and the final heat treating pattern. Rough approximation of specific heat, c, that postulates its constant value could lead to a significant error in obtaining the required power and can also deform the expected temperature profile. Therefore, both physical properties must be treated as nonlinear functions of temperature.

Equation 35, with suitable boundary and initial conditions, represents the three-dimensional temperature distribution at any time and at any point in the workpiece. The initial temperature condition refers to the temperature profile within the workpiece at time $t = 0$. The initial temperature distribution is usually uniform and corresponds to the ambient temperature. However, in some cases, the initial temperature distribution is appreciably nonuniform due to the residual heat that remains after a previous technological process (i.e., preheating, interrupted

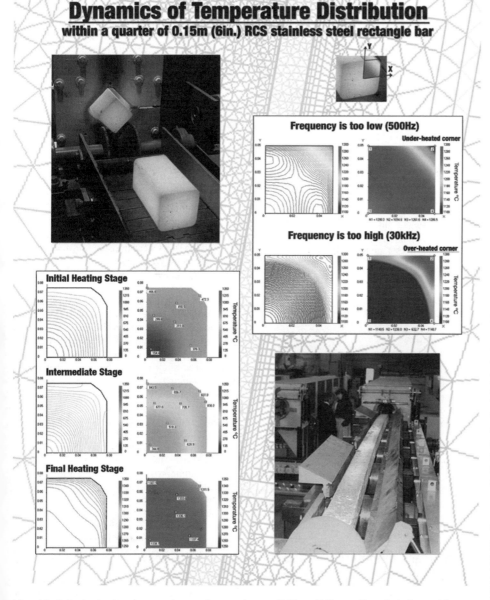

Fig. 19 Induction heating of rectangular, round-cornered-square (RCS) steel billets and bars. Only the top-right quarter of the bar is simulated because of symmetry. Source: Ref 47

Table 7   Basic properties of heat-transfer analysis

Quantity	Symbol	SI units
Temperature	T	K
Thermal conductivity	k	W/(m K)
Density	γ	kg/m³
Specific heat	c	J/(kg K)

quenching, or continuous casting). As an example, Fig. 20 shows the world's largest oscillating induction heater, which was engineered and manufactured by Inductotherm, Corp. After installing the world's largest continuous caster, Geneva Steel, Utah, was looking for a method to reheat large slabs that weighed 800 to 1000 lb/in. width. The goal for the reheating system was the lowest capital cost possible while adding capacity on line in the shortest time possible. Single slabs 1.8 to 3.2 m (71 in. to 126 in.) wide or twin-cast slabs 1.07 to 1.6 m (42 to 63 in.) wide are produced in the continuous caster. It was required that the induction reheater be located in line between the existing continuous caster and the rolling mill.

This task was accomplished by a 42,000 kW induction oscillating system with the capacity to reheat 500 short tons per hour from a bulk input (initial) temperature of 1090 °C (1994 °F) to a bulk output temperature of 1260 °C (2300 °F). Distribution of the slab initial temperature prior to induction reheating was substantially nonuniform. Transverse corners were the coldest areas (approximately 800 °C), while the core temperature was substantially higher (approximately 1150 °C). The induction reheater heats two slabs side by side, four slabs side by side and end to end, three slabs end to end, two slabs end to end, or one large single slab. The overall length of the reheater is 14 m, while the overall width is 4 m. Seven solenoid coils are placed in line with one another at a distance of 1.71 m center to center. Each coil can deliver to the slab up to 6000 kW at 110 Hz of power. Slabs go through an oscillation stroke of 1.71 m and continue to oscillate back and forth.

Computer modeling was a decisive factor in the success of this installation. Mathematical simulation was the only valid option in determining the proper process parameters and coil design features, because the size and required heating power of such a system make it impractical to run lab tests.

For most engineering induction heating problems, thermal boundary conditions represent the combined effect of surface conduction and the heat losses due to heat radiation and convection (Ref 1):

$$-k \frac{\partial T}{\partial n} = \alpha \left(T_s - T_a\right) + c_s \left(T_s^4 - T_a^4\right) + Q_s$$

$$(\text{Eq 36})$$

where $\partial T / \partial n$ is the temperature gradient in a direction normal to the surface at the point under consideration, α is the convection surface heat-transfer coefficient, c_s is the radiation heat-loss coefficient, Q_s is the surface conduction loss (such as during quenching or as a result of workpiece contact with cold rolls, water-cooled guides, or liners), and n denotes the normal to the boundary surface.

As one may see from Eq 36, the heat losses at the workpiece surface are highly nonlinear. If the heated body is geometrically symmetrical along the axis of symmetry, the Neumann boundary condition can be formulated as:

$$\frac{\partial T}{\partial n} = 0 \qquad (\text{Eq 37})$$

The Nuemann boundary condition implies that the temperature gradient in a direction normal to the axis of symmetry is zero. In other words, there is no heat exchange at the axis of symmetry. In the case of heating a cylindrical workpiece, Eq 35 can be rewritten as:

$$c \gamma \frac{\partial T}{\partial t} = \frac{\partial T}{\partial Z} \left(k \frac{\partial T}{\partial Z}\right) + \frac{1}{R} \frac{\partial}{\partial R} \left(k R \frac{\partial T}{\partial R}\right) + Q$$

$$(\text{Eq 38})$$

However, when heating rectangular workpieces (i.e., heat transfer in slab, RCS bars, or plate), Eq 35 can be written in Cartesian coordinates as:

$$c \gamma \frac{\partial T}{\partial t} = \frac{\partial T}{\partial X} \left(k \frac{\partial T}{\partial X}\right) + \frac{\partial}{\partial Y} \left(k \frac{\partial T}{\partial Y}\right) + \frac{\partial}{\partial Z} \left(k \frac{\partial T}{\partial Z}\right) + Q$$

$$(\text{Eq 39})$$

Equations 38 and 39 with boundary conditions in Eq 36 and 37 are the most popular equations for mathematical modeling of the heat-transfer processes in induction heating and heat treating applications.

Similar to mathematical modeling of electromagnetic processes, an appreciable amount of thermal simulations can also be effectively modeled using two-dimensional assumptions, with a correspondent reduction of time required for computer modeling and data preparation.

Numerical Computation of the Induction Heat Treating Processes

Many mathematical modeling methods and programs exist or are under development (Ref 1, 10, 55 to 60). Work in this field is done in universities, including Leibniz University (Hannover, Germany), Padua University (Padua, Italy), Michigan State University, University of Latvia (Riga, Latvia), and others; research labs; inside large companies such as Inductoheat, Inc.; and by specialized software companies such as Integrated Engineering Software, Inc., Infolytica Corp., MagSoft Corp., ANSYS, Inc., Comsol Group, ESI Group, Ansoft Corp., Vector Fields Inc., and others (Ref 1, 10, 55 to 72). For each problem or family of similar problems, certain numerical methods or software are preferred. There is not a single universal computational method that is optimal for solving all induction heating problems. The right choice of numerical computational technique depends on the application (Ref 1).

Because of space limitations, this article does not provide an exhaustive review of the methods available for electromagnetic field and heat-transfer simulations. There are many publications that describe the features and applications of various mathematical modeling techniques. An interested reader can study the description of the most popular computational techniques used for simulation of heat-transfer and electromagnetic processes in Ref 1, 38, 39, 40, 45, 51, 53, 69, and 72 to 99. Only a brief review is provided here.

Finite-Difference Method

The finite-difference method (FDM) was the earliest numerical technique (Ref 41, 73 to 76) used for mathematical simulation of various processes. The FDM has been used extensively for solving both heat-transfer and electromagnetic problems. It is particularly easy to apply when the modeling area has classical geometries: cylindrical or rectangular. Finite-difference mesh typically represents a rectangular grid (orthogonal mesh) consisting of numerous increments, nodes, and cells. Depending on a particular FDM approach, nodes can be placed at the cell element corners or in the cell element center. The choice of positioning nodes affects the formation of the global matrix, the treatment of boundary equations, and some other factors related to the computing process. To simplify the introduction to FDM, an approach placing the nodes at the cell corners is discussed here. Because of the rectangular grid (Fig. 21), the discretization algorithm is quite simple. An approximate solution of the governing equation is found at the mesh points defined by the intersections of the lines.

The computation procedure consists of replacing each partial derivative of the governing equation (Eq 33, 34, 38, or 39) by a finite-difference "stencil" that couples the value of the unknown variable (temperature, magnetic vector potential, or magnetic field intensity) at a node of approximation with its value in the surrounding area. This method provides a pointwise approximation of the partial

Fig. 20 Induction reheating of the world's largest carbon steel slab. Maximum slab width: 3.2 m; thickness: 0.22 m at 540 tons/h; total power: 42,000 kW. Courtesy of Inductotherm Corp.

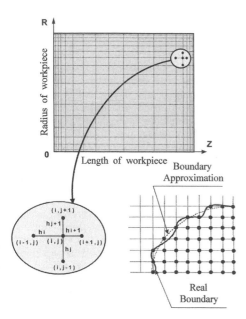

Fig. 21 Finite-difference approximation and rectangular grid. Source: Ref 41

differential equation and is quite universal because of its generality and relative simplicity to apply (Ref 74). By Taylor's theorem for two variables, the value of a variable at a node on the mesh can be expressed in terms of its neighboring values and separation distance (called a space step), h, as in the following expressions (stencils):

$$\frac{\partial T}{\partial X} \Rightarrow \frac{T_{i+1} - T_i}{h} + O(h) \qquad \text{(Forward difference)}$$

(Eq 40)

$$\frac{\partial T}{\partial X} \Rightarrow \frac{T_i - T_{i-1}}{h} + O(h) \qquad \text{(Backward difference)}$$

(Eq 41)

$$\frac{\partial T}{\partial X} \Rightarrow \frac{T_{i+1} - T_{i-1}}{2h} + O(h) \qquad \text{(Central difference)}$$

(Eq 42)

$$\frac{\partial^2 T}{\partial X^2} \Rightarrow \frac{T_{i+1} - 2T_i + T_{i-1}}{h^2} + O(h^2)$$

(Eq 43)

Here, the notation $O(h)$ is used to show that the error involved in the approximation is on the order of h. Similarly, $O(h^2)$ is for the approximation error on the order of h^2, which is more accurate than one on the order of h. Substitution of the finite-difference stencils into the electromagnetic and heat-transfer partial differential equations provides the local approximation. By assembling all local approximations and taking into account the proper initial and boundary conditions, one can obtain a set of simultaneous algebraic equations that can be solved with respect to unknown variables of electromagnetic and heat-transfer problems (i.e., T, $\mathbf{A}$, $\mathbf{E}$, $\mathbf{H}$, or $\mathbf{B}$) at each node of the mesh. The solution can be obtained either by iterative

techniques or by direct matrix inversion methods. The matrices are sparsely occupied, having nonzero elements in the neighborhood of the diagonal only. This allows the application of special simplification in the computational procedure.

As an example, the FDM for modeling heat-transfer processes for heating cylindrical billet (Fourier equation) is demonstrated. To mathematically describe a heat-transfer process in a cylindrical billet, the governing Eq 35 can be rewritten as:

$$c\gamma \frac{\partial T}{\partial t} = \frac{\partial}{\partial Z}\left(k\frac{\partial T}{\partial Z}\right) + \frac{1}{R}\frac{\partial}{\partial R}\left(kR\frac{\partial T}{\partial R}\right) + Q(Z, R)$$

(Eq 44)

To describe two-dimensional heat-transfer processes in rectangular bodies (slab, plate, RCS bar and bloom), Eq 35 can be rewritten as:

$$c\gamma \frac{\partial T}{\partial t} = \frac{\partial}{\partial X}\left(k\frac{\partial T}{\partial X}\right) + \frac{\partial}{\partial Y}\left(k\frac{\partial T}{\partial Y}\right) + Q(X, Y)$$

(Eq 45)

Parameters c and k are nonlinear functions of the temperature. The partial differential equation (Eq 44) may be expressed in a more concise form by introducing the finite-difference operators:

$$\frac{1}{c\gamma}\frac{\partial}{\partial Z}\left(k\frac{\partial T}{\partial Z}\right) \Rightarrow \Lambda_Z T$$

(Eq 46)

$$\frac{1}{c\gamma}\frac{1}{R}\frac{\partial}{\partial R}\left(kR\frac{\partial T}{\partial R}\right) \Rightarrow \Lambda_R T$$

(Eq 47)

Substitution of Eq 46 and 47 into Eq 44 results in the following finite-difference format:

$$\frac{\partial T}{\partial t} = \Lambda_Z T + \Lambda_R T + \frac{1}{c\gamma}Q(Z, R)$$

(Eq 48)

To take into consideration a nonlinear nature, the material properties are considered to be piecewise constants. Therefore, the coefficients of Eq 44 and 45 vary at different mesh nodes. The finite-difference stencil with respect to the Z-coordinate can be written as:

$$\frac{1}{c\gamma}\frac{\partial}{\partial Z}\left(k\frac{\partial T}{\partial Z}\right) \Rightarrow \Lambda_Z T^\tau =$$

$$= \frac{2}{c(T)\gamma(T)(h_i + h_{i+1})}\left(k_{i+1,j}\frac{T_{i+1,j}^\tau - T_{i,j}^\tau}{h_{i+1}} - k_{i,j}\frac{T_{i,j}^\tau - T_{i-1,j}^\tau}{h_i}\right)$$

(Eq 49)

In FDM, it is important that the boundaries of the mesh region coincide with the boundaries of the appropriate regions of the induction heating system. Experience in using FDM in induction heating computations has shown that noncoincidence of the boundaries has a strong negative effect on the accuracy of the calculation. Approximation of the boundary conditions by $Z = 0$ and $Z = ZZ$ is shown as:

$$Z = 0, \Rightarrow k_{1,j}\frac{T_{1,j}^\tau - T_{0,j}^\tau}{h_{i+1}} = P_{z=0}$$

$$Z = ZZ, \Rightarrow -k_{ZZ,j}\frac{T_{N,j}^\tau - T_{N-1,j}^\tau}{h_N} = P_{z=NN}$$

(Eq 50)

where the i, j, and τ indexes correspond to the Z-axis, the R-axis, and the time, respectively. The finite-difference expressions for differential operators with respect to radius will be similar to Eq 49 and 50 (Ref 74). When boundaries of the mesh do not coincide with boundaries of the components of the modeled system, then the values corresponding to the temperature at the boundary nodes are the values they have at the neighboring nodes of the real boundary (Fig. 21, bottom right). The accuracy of the numerical computation depends on both the errors in the governing equation approximation and the error from approximating the boundary conditions. Therefore, care should be taken in approximating not only the governing equations but the boundary conditions as well.

Another factor that emphasizes the importance of a good approximation of the boundary condition and approximation of the subsurface area is the fact that, because of the skin effect and some other electromagnetic phenomena discussed in Ref 1, the heat sources penetrate from the workpiece surface toward the core. Subsequently, the most significant amount of heat sources is located at the surface and immediate subsurface areas. Therefore, a rough approximation in these areas can have a detrimental effect on the overall accuracy of the calculations.

An important feature of heat-transfer simulation is the fact that induction heating is a nonlinear time-dependent (transient) process. There are several formats available to address these features of nonlinearity and time-dependency (Ref 1, 38, 39, 74). Each algorithm has its own advantages and disadvantages. The choice of a particular numerical procedure depends on several factors, including the specifics of the application, computer capabilities, and individual experience using numerical methods. Finite-difference formats for the heat-transfer transient problem range from explicit to implicit forms (Ref 74). Implicit forms require solving a set of algebraic equations at each time step.

Explicit Formulation. The explicit approximation is the simplest technique. In explicit forms, the temperature distribution is obtained directly in a step-by-step manner. A forward-difference approximation with respect to time leads to the explicit finite-difference formulation:

$$\frac{T_{i,j}^{\tau+1} - T_{i,j}^\tau}{h_\tau} = \Lambda_Z T_{i,j}^\tau + \Lambda_R T_{i,j}^\tau + \frac{1}{c\gamma}Q_{i,j}^\tau$$

(Eq 51)

From Eq 51, the unknown temperatures corresponding to the $(\tau + 1)$ time step are obtained as functions of the known material properties, heat sources, and temperatures at time τ (Fig. 22a). The temperature distribution within the workpiece is achieved after the first

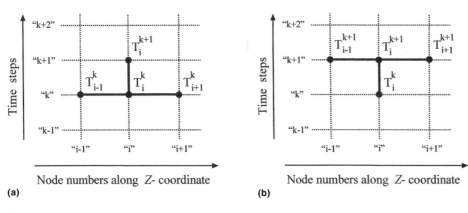

Fig. 22 Examples of (a) simplified explicit and (b) implicit forms for one-dimensional approximation. Source: Ref 1

time step h_τ, which is found by the given initial condition (initial temperature condition is assumed to be ambient in the majority of induction heat treating applications; however, in some cases, there may be pre-existing nonuniform temperature profiles) and the appropriate boundary conditions. Therefore, the unknown temperatures are obtained explicitly from their initially known or previously calculated values. No time-extensive procedures are required to execute a computer code.

The ability to provide a stable and accurate numerical solution is primarily a concern when using an explicit finite-difference format. Accuracy is a measure of the closeness of the numerical approximation to the exact solution (Ref 1, 73 to 76). The finite-difference format is said to be numerically stable if, at sufficiently small time steps, τ, and space steps, h_τ, the equation has a unique solution and that solution does not increase its magnitude with small variations of material properties, τ and/or h_τ.

The stability condition depends on the properties of the finite-difference format and is, in many cases, independent of the governing partial differential equation or physical phenomena. Unfortunately, explicit methods are accurate and stable only for certain relations between the time, space steps, and values of material properties. Sometimes, those relations can contradict each other. The stability condition usually leads to extremely small time steps.

With explicit formats, it is not unusual to have the situation where decreasing time steps and space steps will not improve the solution but rather worsen it. This is a typical case of unstable or ill-conditioned systems. In such cases, the use of different stencils may help. For example, instead of a central-difference stencil, a forward- or backward-difference approximation can be used and visa versa. Thus, regardless of the simplicity and convenience of the explicit algorithms, the concern for obtaining an accurate and stable solution (particularly taking into consideration essentially nonlinear material properties) leads to the limited use of these algorithms for simulating processes of induction heating.

Implicit methods are more popular, due to their ability to provide more stable solutions compared to explicit algorithms and to have a relatively independent choice of mesh parameters (Fig. 22b). The use of any implicit method requires the calculation of a system of algebraic equations. Several implicit methods were developed to reduce computational efforts (Ref 73 to 76).

Generally speaking, when using implicit methods for modeling of heat-transfer problems, the finite-difference format can be written as (Ref 76):

$$\frac{T_{i,j}^{\tau+1} - T_{i,j}^{\tau}}{h_\tau} = \xi\left(\Lambda_Z T_{i,j}^{\tau+1} + \Lambda_R T_{i,j}^{\tau+1}\right) + (1-\xi)\left(\Lambda_Z T_{i,j}^{\tau} + \Lambda_R T_{i,j}^{\tau}\right) + \frac{1}{c\gamma} Q_{i,j}^{\tau}$$

(Eq 52)

The choice of the parameter ξ is a balance between accuracy and stability. The value of this parameter varies between 0 and 1. For $\xi = 0.5$, the well-known Crank-Nicolson format represents an intermediate approximation of the partial derivatives (half-way between two levels of time, τ and $\tau + 1$). The complete implicit format is obtained when $\xi = 1$.

The implicit method is said to be unconditionally stable; however, certain computational oscillations could still appear when coarse mesh and large time steps are used. The time step is restricted by the desired accuracy. The following finite-difference implicit formats are commonly used for solving the transient heat-transfer problem.

A locally one-dimensional format (proposed by Samarskii, Ref 74) is also known as the alternating directional implicit (ADI) method. According to the ADI method, a two-dimensional set of equations is replaced by two sets of single-dimension equations alternating the solution in correspondent directions:

$$\frac{T_{i,j}^{\tau+0.5} - T_{i,j}^{\tau}}{h_\tau} = \Lambda_Z T_{i,j}^{\tau+0.5} + \frac{1}{2c\gamma} Q_{i,j}^{\tau}$$

(Eq 53)

$$\frac{T_{i,j}^{\tau+1} - T_{i,j}^{\tau+0.5}}{h_\tau} = \Lambda_R T_{i,j}^{\tau+1} + \frac{1}{2c\gamma} Q_{i,j}^{\tau}$$

(Eq 54)

The set of Eq 53 and 54 is said to be stable for all sizes of time step h_τ. The main restriction for choosing a large h_τ is avoiding significant truncation errors. Physically, Eq 53 and 54 can be interpreted as a complex combination of two heat-transfer processes: first, along the Z-axis, and second, along the R-axis.

The transition from time level τ to time level $\tau + 1$ is assumed to be made in two stages, using intermittent time step 0.5 h_τ. This means that the transition from a known temperature field distribution of $T_{i,j}^{\tau}$ to an unknown temperature of $T_{i,j}^{\tau+1}$ is made through the intermediate temperature distribution of $T_{i,j}^{\tau+0.5}$. In each direction, the Fourier equation is approximated implicitly, with the necessity of solving two sets of simultaneous algebraic equations. After substituting the respective finite-difference stencils into Eq 53 and 54 and after some simple algebraic operations, Eq 53 can be rewritten as:

$$\xi_i T_{i-1,j}^{\tau+0.5} - \psi_i T_{i,j}^{\tau+0.5} + \upsilon_i T_{i+1,j}^{\tau+0.5} = -F_{i,j}^{\tau}$$

(Eq 55)

and, respectively, Eq 54 will be written as:

$$\xi_i T_{i,j-1}^{\tau+1} - \psi_i T_{i,j}^{\tau+1} + \upsilon_i T_{i,j+1}^{\tau+1} = -F_{i,j}^{\tau}$$

(Eq 56)

where ξ, ψ, and υ are coefficients.

The matrices of the algebraic Eq 55 and 56 are sparsely occupied and have a tridiagonal matrix structure, meaning that nonzeros occupy only the main diagonal and its neighborhood. Thanks to this feature, several computational procedures can be effectively used to solve Eq 55 and 56.

The Peaceman-Rachford format (Ref 74) is written as:

$$\frac{T_{i,j}^{\tau+0.5} - T_{i,j}^{\tau}}{0.5\,h_\tau} = \Lambda_Z T_{i,j}^{\tau+0.5} + \Lambda_R T_{i,j}^{\tau} + \frac{1}{c\gamma} Q_{i,k}^{\tau}$$

(Eq 57)

$$\frac{T_{i,j}^{\tau+1} - T_{i,j}^{\tau+0.5}}{0.5\,h_\tau} = \Lambda_Z T_{i,j}^{\tau+0.5} + \Lambda_R T_{i,j}^{\tau+1} + \frac{1}{c\gamma} Q_{i,k}^{\tau}$$

(Eq 58)

Equation 57 is implicit in direction Z and explicit in R. On the contrary, Eq 58 is explicit in direction Z and implicit in R. A set of algebraic equations that corresponds to the Peaceman-Rachford format is similar to Eq 53 and 54.

There have been two general techniques developed for solving algebraic equations obtained after substitution of the finite-difference stencils into the partial differential equations. Direct methods represent one large group of computational methods, and iterative algorithms (such as the Jacobi method, Gauss-Seidel method, overrelaxation techniques, etc.) are the second (Ref 73 to 76). One of the most widely used methods for solving a tridiagonal matrix is the Gaussian two-step elimination method. This algorithm is relatively simple and effective, taking advantage of the tridiagonal matrix and requiring minimum computer memory and short execution time.

The optimal choice of mesh generation and time steps has a pronounced effect on the accuracy and stability of a calculation using any of the numerical modeling techniques; however, it becomes particularly critical when using FDM. In FDM, as in any of the numerical techniques, smaller space and time steps are recommended for greater accuracy. At the same time, it is quite clear that the large number of nodes results in a more cumbersome and time-consuming solution. Therefore, there should be a reasonable compromise between mesh size, time steps, computation time, and accuracy of modeling. Naturally, it is recommended to select a finer mesh for regions of intense variations of electromagnetic or thermal fields, and a coarser mesh for areas where there is insignificant variation of variables. The optimal combination of mesh parameters and time steps is usually determined by computational experiments. The calculations are provided for the various mesh sizes and time steps, and the results are compared. If the comparison shows a large difference, then it is necessary to repeat the calculation for a finer mesh and/or smaller time steps until the difference between the calculations is insignificant. The rule of thumb is that if the computation is done correctly, the values of the unknown variables (i.e., temperature) should converge as the space mesh becomes finer and the time steps become smaller.

The reduction of the space steps leads to the reduction of the truncation error. At the same time, a finer mesh has a larger number of nodes and smaller space steps. The number of algebraic equations can grow tremendously; however, a computer deals with only a limited number of arithmetic units. This can lead to a crucial level of round-off errors. Therefore, the accuracy of the computations can be improved by refining the space mesh and reducing the time steps, unless the round-off errors become excessive (Fig. 23). Developers and users of numerical computation software have developed a certain know-how that allows them to relatively quickly overcome the problems resulting from computation failure due to round-off errors. For example, one can successfully rerun the program by only slightly changing some insignificant material properties or by a slight remeshing. The other way to avoid computation failures caused by round-off error is to use double-precision arithmetic. Of course, it will lead to an increase in software execution time and the necessity of using 64-bit processors.

The aforementioned remarks regarding the various aspects of mesh generation and computation errors are valid not only for FDM but for the majority of other numerical techniques as well (including finite and boundary elements). When modeling coupled problems that represent multiphysical phenomena (i.e., electrothermal, heat-transfer-phase transformation, and electromechanical), it is very attractive to use a single universal mesh. This may seem like a time-saving approach, allowing one to save the time for meshing. However, if the physical phenomena are inherently different (i.e., heat-transfer phenomena and eddy current), it is often more efficient to use different, subject-optimized meshes.

The FDM has been illustrated based on the most commonly used first- and second-order finite-difference approximations and placing nodes at cell corners (Ref 73 to 76). The accuracy of the numerical calculations may be improved by employing higher-order finite-difference approximations. Such approximations will allow one to reduce truncation error, but, at the same time, this approach increases the number of nodes involved in the local approximation. Therefore, a matrix of algebraic equations will no longer be tridiagonal but five- or seven-diagonal. This makes the program code more complex, with a noticeable increase in execution time and memory use. However, taking into consideration the recent achievements in developing high-performance multicore processors, the increase in execution time and memory utilization become less critical.

Finite-Element Method

The finite-element method (FEM) is another group of numerical techniques devoted to obtaining an approximate solution for various technical problems, including those encountered in induction heating. This numerical technique was originally applied in mechanical engineering. Later, applications of FEM expanded to other areas of engineering. It has become the most popular numerical tool for a variety of scientific and engineering applications. The tremendous improvement in computer capabilities (particularly within the past three decades) has boosted the development of several variations of the FEM (Ref 77 to 86). Some of these are:

- Weighted residual method (weak form of the governing equations)
- Different types of the Ritz method
- Different types of the Galerkin method
- Pseudovariational methods
- Methods based on minimization of the energy functional

As described previously, the FDM provides a pointwise approximation; however, the FEM provides an element-wise approximation of the governing equations. Different finite-element approaches may be better suited for certain problems. For example, the weighted residuals formulation has been very effectively used for computation of heat-transfer problems. An interested reader can find a description of various finite-element techniques in Ref 77 to 86.

Induction heating is a complex combination of electromagnetic and heat-transfer phenomena. In the previous section, the use of FDM was illustrated for modeling a heat-transfer problem; in this section, the use of FEM for solving electromagnetic problems is discussed.

Electromagnetic Processes. Several different FEM codes have been developed for modeling electromagnetic processes taking place in electric generators and motors, circuit breakers, transformers, nondestructive testing equipment, and induction heating systems. Many worthwhile texts, conference proceedings, and articles have been written on the subject of finite-element modeling (Ref 69, 70, 72, 77 to 86) as well as techniques based on infinite and edge elements. The large number of papers on the subject of FEM applications for electromagnetic field computation makes it impossible to mention all of the contributions. At the same time, some of the proposed finite-element models are similar in form. However, it should be mentioned here that Silvester and Chari (Ref 77, 78) presented the first general nonlinear variational formulation of magnetic field analysis using FEM. Essential input into the development of FEM was provided by Lord, Trowbridge, Sabonnadiere, Udpa, Konrad, Salon, Brauer, Bossavit, and many others (Ref 69, 79 to 86). The following is a short description of one form of FEM.

Due to the general postulate of the variational principle, the solution of electromagnetic field computation is typically obtained by minimizing the energy functional that corresponds to the governing equation (e.g., Eq 33, 34, or 35) instead of solving that equation directly. The energy functional is minimized for the integral over the total area of simulation, which includes the workpiece, coil, flux concentrators, tooling, and surrounding area.

The principle of minimum energy (Ref 79 to 86) requires that the vector potential distribution correspond to the minimum of the stored field energy per unit length. As a result of that assumption, it is necessary to solve the global set of simultaneous algebraic equations with respect to the unknown, for example, magnetic vector potential at each node. The

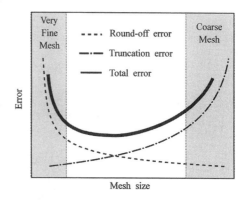

Fig. 23 Correlation among the round-off error, truncation error, and mesh size. Source: Ref 1

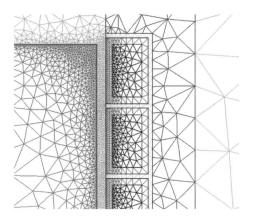

Fig. 24 Example of finite-element discretization

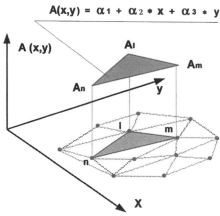

Fig. 25 First-order triangular element. Source: Ref 41

formulation of the energy functional, its minimization to obtain a set of finite-element equations, and the solution techniques (the solver) were created for both two-dimensional (Cartesian system) and axisymmetric (cylindrical system) problems. Magnetic vector potential, **A**, in the two-dimensional case (longitudinal cross section) acts in the direction of the current density, **J**, and is described by a two-dimensional partial differential equation (Eq 33). The boundary of the region can be selected in such a way that the magnetic vector potential, **A**, is zero along the boundary (Dirichlet condition) or Neumann condition, $(\partial \mathbf{A}/\partial n = 0)$, meaning that its gradient is negligibly small along the boundary compared to the value elsewhere in the region. The energy functional corresponding to the two-dimensional governing equation (Eq 33) can be written in the following form (Ref 77, 78):

$$F = \int_V \left(\frac{1}{2\mu_r \mu_0} \left(\left| \frac{\partial \mathbf{A}}{\partial X} \right|^2 + \left| \frac{\partial \mathbf{A}}{\partial Y} \right|^2 \right) + j \frac{\omega \sigma}{2} |\mathbf{A}|^2 - \mathbf{J}_S \mathbf{A} \right) dV$$

(Eq 59)

where V is the total area of modeling, and $\mathbf{J}_S$ is the source current density. The first, second, and third terms inside of the integrand represent the energy of the magnetic field, the eddy currents, and the source current, correspondingly. The minimization of the functional (Eq 59) corresponds to the solution of the two-dimensional eddy current field problem, taking into consideration the corresponding boundary conditions.

The area of study is divided into nonoverlapping numerous finite elements (mesh), as shown in Fig. 24; therefore, the minimization of this functional provides the minimization of energy at every node of each element.

There is a variety of geometric arrangements and shapes of finite elements. In fact, the flexibility of their shapes allows them to satisfy regions of practically any geometry. The simplest two-dimensional finite element is the first-order triangle (Fig. 25). In the axisymmetric cylindrical case, such a finite-element mesh may be represented as a set of rings. Each ring

revolves around the axis of symmetry and has a triangular cross section (so-called triangular torus element). The use of high-order iso-parametric elements allows the required total number of elements to be reduced at the expense of an increase in the computation time and algorithm complexity.

Similar to FDM, space discretization is a very important aspect of FEM analysis. The following are some general remarks concerning finite-element discretization (mesh generation), which has some similarities with FDM:

- The area of study is subdivided into non-overlapping finite elements (finite-element mesh, Fig. 24). The sides of the finite elements intersect at nodes. The number and location of these elements depend on personal judgment or specially developed mesh-generation algorithms. However, to obtain reasonable accuracy of the numerical solution, the finite-element mesh must be relatively fine (sizes of finite elements must be smaller) in the regions where high gradients of the unknown are present. Special effort should be made to generate sufficiently fine mesh within three current penetration depths in the workpiece surface where the eddy current is induced. This means that using higher frequencies dictates generating finer meshes.
- All the finite elements should have the same unit depth in the Z-direction.
- The current density, flux density, electrical conductivity, magnetic permeability, and other material properties are postulated to be constant within each element. At the same time, they can be different from element to element.
- The designer should take advantage of the symmetry involved in the system geometry, for example, disappearing normal derivative values of **A** along the symmetry.

Assuming that local behavior of the electromagnetic field is linear across each finite element and can be approximated by a linear

law, and supposing that the chosen finite elements are first-order parametric triangulars, then the magnetic potential behavior within a triangular can be defined as:

$$A(X, Y) = \alpha_1 + \alpha_2 X + \alpha_3 Y \qquad \text{(Eq 60)}$$

Based on the laws of a two-dimensional linear approximation, the coefficients α_1, α_2, and α_3 are constant and can be calculated from the three independent simultaneous equations by assuming vertex values of A_l, A_m, and A_n of a magnetic vector potential, **A**, at the three nodes of a triangular. Therefore, the local set of equations can be rewritten as:

$$A_l = \alpha_1 + \alpha_2 X_l + \alpha_3 Y_l$$
$$A_m = \alpha_1 + \alpha_2 X_m + \alpha_3 Y_m \qquad \text{(Eq 61)}$$
$$A_n = \alpha_1 + \alpha_2 X_n + \alpha_3 Y_n$$

The matrix notation of Eq 61 can be written as:

$$\begin{bmatrix} A_l \\ A_m \\ A_n \end{bmatrix} = \begin{bmatrix} 1 & X_l & Y_l \\ 1 & X_m & Y_m \\ 1 & X_n & Y_n \end{bmatrix} \begin{bmatrix} \alpha_1 \\ \alpha_2 \\ \alpha_3 \end{bmatrix} \qquad \text{(Eq 62)}$$

The determinant of the square matrix in Eq 62 can be introduced as a value of twice the triangular area. Knowing the geometry of the elements and the magnetic vector potential at each node in every element, it is possible to obtain the value of **A** at any point inside the element. By extending a local approximation to all the elements that represent the total area of interest, it is possible to obtain an approximation for the magnetic vector potential throughout the area of modeling (Fig. 24). Energy balance within the area of modeling is determined by minimizing the energy functional at every node. This can be arranged by setting equal to zero the first partial derivative of the functional with respect to each node. Instead of performing the minimization of the functional node by node, it is reasonable to perform it element by element.

The total (global) energy associated with a whole area being modeled equals the sum of the energies of all elements. As a result, a set of the simultaneous algebraic equations with respect to the unknown values of the magnetic vector potential at each node can be obtained. After some algebraic operations, the local matrix equation, which represents the minimization of the energy functional within any triangular element, can be written as:

$$[[V]_e + j[W]_e] \, [\mathbf{A}] = [Q]_e \qquad \text{(Eq 63)}$$

where

$$[V]_e = \frac{1}{4\mu_r \mu_0 \Delta}$$

$$\begin{bmatrix} (b_l b_l + c_l c_l) & (b_l b_m + c_l c_m) & (b_l b_n + c_l c_n) \\ (b_m b_l + c_m c_l) & (b_m b_m + c_m c_m) & (b_m b_n + c_m c_n) \\ (b_n b_l + c_n c_l) & (b_n b_m + c_n c_m) & (b_n b_n + c_n c_n) \end{bmatrix}$$

(Eq 64)

$$\begin{bmatrix} a_l & a_m & a_n \\ b_l & b_m & b_n \\ c_l & c_m & c_n \end{bmatrix} =$$

$$\begin{bmatrix} (X_m Y_n - X_n Y_m) & (X_n Y_l - X_l Y_n) & (X_l Y_m - X_m Y_l) \\ (Y_m - Y_l) & (Y_n - Y_l) & (Y_l - Y_m) \\ (X_n - X_m) & (X_l - X_n) & (X_m - X_l) \end{bmatrix}$$

$$\text{(Eq 65)}$$

$$[W]_e = \frac{\omega \sigma \Delta}{12} \begin{bmatrix} 2 & 1 & 1 \\ 1 & 2 & 1 \\ 1 & 1 & 2 \end{bmatrix} \quad \text{(Eq 66)}$$

$$[Q]_e = \frac{J_s \Delta}{3} \begin{bmatrix} 1 \\ 1 \\ 1 \end{bmatrix} \quad \text{(Eq 67)}$$

$$[A]_e = \begin{bmatrix} A_l \\ A_m \\ A_n \end{bmatrix} \quad \text{(Eq 68)}$$

where Δ is the cross-sectional area of a particular triangular, and j is:

$$\sqrt{-1}$$

After assembling all local matrices of finite elements and specifying the corresponding boundary conditions, a global matrix equation can be obtained:

$$[G][A] = [Q] \quad \text{(Eq 69)}$$

It is necessary to mention here that there are several commonly used ways to specify the boundary conditions in Eq 69. One of the most popular techniques is called blasting the diagonal. This technique requires multiplying the diagonal terms of the equations representing the nodes where the value of the magnetic vector potential is known by a significantly large number (i.e., 10^{30}). At the same time, the corresponding right sides of those equations are replaced by known values of boundary conditions times the new diagonal. Such an artificial approach is very effective and easy to apply.

For the axisymmetric case, the local and global matrix will be similar to Eq 63 to 69 (Ref 80, 81). Parameters of the local matrix of Eq 63 for the axisymmetric problem (i.e., cylindrical system) are:

$$[V]_e = \frac{R_c}{4 \mu_r \mu_0 \Delta}$$

$$\begin{bmatrix} (\beta_l \beta_l + c_l c_l) & (\beta_l \beta_m + c_l c_m) & (\beta_l \beta_n + c_l c_n) \\ (\beta_m \beta_l + c_m c_l) & (\beta_m \beta_m + c_m c_m) & (\beta_m \beta_n + c_m c_n) \\ (\beta_n \beta_l + c_n c_l) & (\beta_n \beta_m + c_n c_m) & (\beta_n \beta_n + c_n c_n) \end{bmatrix}$$

$$\text{(Eq 70)}$$

where R_c is the radius of the finite-element centroid, and:

$$\beta_i = b_i + \frac{2\Delta}{3R_c}, \quad i = 1, m, n$$

$$\begin{bmatrix} a_l & a_m & a_n \\ b_l & b_m & b_n \\ c_l & c_m & c_n \end{bmatrix} =$$

$$\begin{bmatrix} (R_m Z_n - R_n Z_m) & (R_n Z_l - R_l Z_n) & (R_l Z_m - R_m Z_l) \\ (Z_m - Z_n) & (Z_n - Z_l) & (Z_l - Z_m) \\ (R_n - R_m) & (R_l - R_n) & (R_m - R_l) \end{bmatrix}$$

$$\text{(Eq 71)}$$

$$[W]_e = \frac{\omega \sigma R_c \Delta}{12} \begin{bmatrix} 2 & 1 & 1 \\ 1 & 2 & 1 \\ 1 & 1 & 2 \end{bmatrix} \quad \text{(Eq 72)}$$

$$[Q]_e = \frac{J_s R_c \Delta}{3} \begin{bmatrix} 1 \\ 1 \\ 1 \end{bmatrix} \quad \text{(Eq 73)}$$

$$[A]_e = \begin{bmatrix} A_l \\ A_m \\ A_n \end{bmatrix} \quad \text{(Eq 74)}$$

A significant portion of the computation work for FEM consists of solving the large system of matrix equations. The global matrix in Eq 63 can be solved using either iterative methods or direct matrix inversion techniques, while taking into consideration the sparse nature and banded symmetry of the matrix.

As mentioned previously, the accuracy of the numerical approximation of the governing partial differential equations improves with a finer mesh. In other words, the more finite elements used in the simulation, the better the approximation will be and the closer a numerical solution gets to the exact solution of the governing equations. The optimal number of elements depends on the problem of modeling. In the developing and testing stage of a finite-element code, a developer can judge the obtained accuracy of the finite-element approximation based on its comparison with reliable experiments, available analytical solution, and satisfaction of conditions (Eq 16 to 18). After solving the system of algebraic equations and obtaining the distributions of the magnetic vector potential in the region of modeling, it is possible to find all of the required output parameters of the electromagnetic field. The induced current density in conductors is:

$$J_e = -j\omega \sigma A \quad \text{(Eq 75)}$$

The total current density in the conductor is:

$$J = J_s - j\omega \sigma A \quad \text{(Eq 76)}$$

The magnetic flux density components, B_x and B_y, can be calculated from Eq 21 as follows (Ref 80, 81):

$$\frac{\partial A}{\partial Y} = -B_x; \quad \frac{\partial A}{\partial X} = B_y \quad \text{(Eq 77)}$$

From Eq 77, the flux density can be obtained as:

$$B = \left[B_x^2 + B_y^2\right]^{1/2} \quad \text{(Eq 78)}$$

For the axisymmetric case of a cylindrical workpiece, the magnetic flux density components, B_R and B_Z, can be calculated as:

$$B_R = -\frac{\partial A}{\partial Z}; \quad B_Z = \frac{\partial A}{\partial R} + \frac{A}{R} \quad \text{(Eq 79)}$$

For the magnetic field intensity:

$$H = \frac{B}{\mu_r \mu_0} \quad \text{(Eq 80)}$$

For the electric field intensity:

$$E = -j\omega A \quad \text{(Eq 81)}$$

The electromagnetic force density in current-carrying conductors and the workpiece can be computed from the cross product of the vector of total current density and the vector of magnetic flux density:

$$F_x = J \times B_y; \quad F_y = -J \times B_x \quad \text{(Eq 82)}$$

From a vector potential solution, it is possible to compute the other important quantities of the process, such as stored energy, flux leakage, total power, and coil impedance.

It is important to mention here that, in contrast to metals, some materials (i.e., composites, powder metallurgy parts, laminated materials, etc.) introduce additional challenges in computer modeling due to the anisotropic nature of physical properties manifesting itself in the directionally dependent nature of electromagnetic and thermal properties. Material anisotropy can appreciably change the electromagnetic field distribution and heating pattern. For example, in electromagnetic computations, Ohm's law for an anisotropic electrically conductive medium (Eq 13), should be written according to:

$$J_x = \sigma_x E_x$$
$$J_y = \sigma_y E_y \quad \text{(Eq 83)}$$
$$J_z = \sigma_z E_z$$

Similar to electromagnetics, anisotropy can also have a marked effect on heat transfer; for example, thermal conductivity, k, in the matrix is directionally dependent, taking different values in different directions, k_x, k_y, and k_z. Fortunately, in the great majority of induction heating and heat treating applications, isotropic workpieces are encountered, and the assumption of heated metals being isotropic is applicable from an engineering perspective. At the same time, there are cases when heated materials are appreciably anisotropic, such as electrically conductive composites or flux concentrators (i.e., laminations or powder metallurgy concentrators), and the effect of anisotropy must be taken into account, requiring correspondent treatment of governing equations.

While discussing both FDM and FEM as well as any numerical simulation techniques, it is imperative to mention the necessity of having user-friendly pre- and postprocessing procedures. It is particularly true for using three-dimensional software, because the most critical (from having an error-free solution) and often most time-consuming part of the solution process is related to input data preparation and describing the area of interest in terms of lines, arcs, segment, bends, and so on.

Superficially, the FDM and FEM appear to be different; however, they are closely related.

As outlined previously, FDM starts with a differential statement of the problem of interest and requires that the partial derivative of the governing equation be replaced by a finite-difference stencil to provide a pointwise approximation. The FEM starts with a variational statement and provides element-wise approximation. Both methods discretize a continuous function (e.g., magnetic vector potential or temperature) and result in a set of simultaneous algebraic equations to be solved with respect to its nodal values. Therefore, the two methods are actually quite similar.

Finite-difference stencils overlap one another and, in the case of complex workpiece geometry, could have nodes outside the boundary of the workpiece, coil, or other components of the induction heating system. Finite elements do not overlap one another, do not have nodes outside the boundaries, and fit the complicated shape boundary perfectly. In electromagnetic field computation, finite elements are usually introduced as a way to minimize a functional. In fact, FDM can also be described as a form of a functional minimization (so-called finite-difference energy method). Therefore, FDM and FEM are different only in the choice of mesh generation and the way in which the global set of the algebraic equation is obtained. They have approximately the same accuracy; however, the required computer time and memory are often less when FDM is used for modeling of classical-shaped bodies. For example, the computer time needed to form global matrices is usually four to nine times greater with FEM than with FDM. As one would expect, a comparison of the efficiency of the two methods depends on the type of problem and program organization.

The FDM is usually not as well suited as FEM for a simulation of induction heat treatment systems with complicated shaped boundaries or in the case of a mixture of materials and forms (e.g., heat treating of camshafts, crankshafts, gears, and other critical components). In this case, FEM has a distinct advantage over FDM.

Example 4: Shaft Hardening. It is sometimes challenging to properly induction harden the fillet area of a shaft (Fig. 26, bottom row) without overheating certain neighboring regions. Mainly, there are two critical factors that make fillets unique regions, from an induction hardening perspective. Those factors are:

- There is a large mass of metal located near the fillet area, acting as a substantial cold sink and requiring a greater amount of heat to be induced within the fillet.
- The geometry makes it challenging to induce a sufficient amount of current in the fillet area due to an unfavorable combination of electromagnetic phenomena, including skin, proximity, end, and ring effects (Ref 1).

Instead of using a cut-and-try method for obtaining coil design specifics and a process

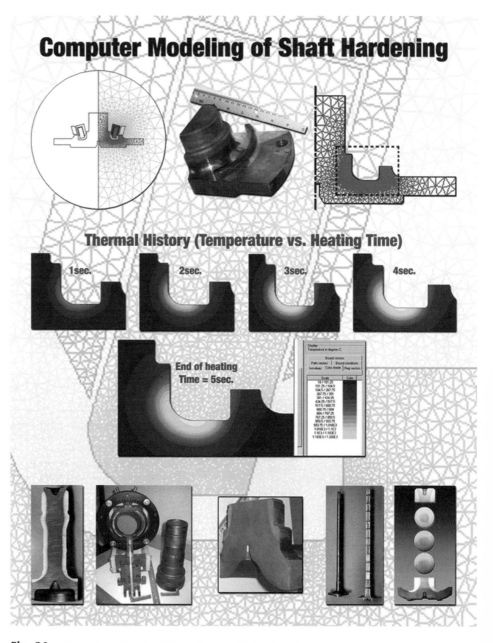

Fig. 26 Induction hardening of shaft fillets. Courtesy of Inductoheat, Inc.

recipe that would provide the required heat pattern, it is more effective to simulate the induction heating process by using a finite-element code. Figure 26 shows the results of a computer simulation of induction heating a shaft fillet area using a single-turn inductor. Coil copper was profiled, providing optimized distribution of the heat sources. The addition of a magnetic flux concentrator and a Faraday ring added complexity to the inductor design but increased the efficiency. The top of Fig. 26 shows a section of the shaft, its fillet area, and the finite-element mesh. Variation of the temperature distribution at various stages of the heat cycle is shown in the middle row of Fig. 26. Flux-2D software was used to conduct this study.

As shown previously, both FDM and FEM require a network mesh of the area of modeling. That network includes induction coil(s), the workpiece, flux concentrators, and electrically conductive structures positioned in close proximity to the inductor. Unfortunately, to suit the condition of smoothness criteria and continuity of the governing differential equation, it is also necessary to generate a mesh within electrically nonconductive areas, such as the air space regions. In most inductor design cases, electromagnetic field distribution in the air can be considered useless information. Such information may be of interest only during the final design stage when evaluating electromagnetic field exposure from the induction heater (Ref 101). The need to always carry out a computation of the electromagnetic field in the air can be considered a noticeable disadvantage of both FDM and FEM.

Another difficulty that appears when using FDM or FEM for electromagnetic field computation is the treatment that extends to infinity. This relates to the infinite nature of electromagnetic wave propagation. Several methods have been used, taking into account the phenomenon of an infinite exterior region. Some of those methods are the ballooning method, mapping technique, and a combination of finite and infinite elements. However, each of the aforementioned methods has certain shortcomings.

Mutual Impedance Method

The inductors involved in induction heating of rods, billets, bars, or slabs prior to the metal hot forming processes, including forging, upsetting, rolling, and extrusion, are quite different compared to inductors for surface hardening. Induction billet/bar heaters are typically designed as multiturn solenoidal coils of cylindrical or rectangular shape. Such induction heaters can consist of one or several in-line coils. The total length of a system sometimes exceeds 10 m. The inside diameter of some coils can be as large as 1 m, and the length of a single coil can exceed 2 m. Depending on the specifics of the application, coils can be fabricated as single- or multilayer solenoids connected to a single- and/or multiphase power source with normal and/or complex drive circuit connections (i.e., autotransformer connection).

Two of the well-known disadvantages of both FDM and FEM relate to the difficulty of simulating appreciably long systems, because of the necessity of generating an enormously large finite-element or finite-difference mesh.

As an alternative to FDM and FEM, the mutual impedance method (MIM) can be used to solve a circuit analysis problem combined with an eddy current problem of induction heating for cylindrical-shaped systems (i.e., induction billet, tube, pipe, or bar heating). In some rare cases, MIM can be used for computer modeling of rectangular systems (heating of slabs and plates by induction). As an alternative to FDM and FEM, the MIM applies an integral form of electromagnetic equations instead of using its differential formulation. The integral equation approach typically requires less computer memory and execution time. It is not required to make an artificial assumption for external boundary conditions at infinitely propagating regions. The integral equation is complete as it is, thanks to the explicit appearance of the boundary values in the integrals.

Another advantage of using integral equations deals with the fact that the area of integration (computation) is limited only to surfaces of electrically conductive bodies. In other words, the electrically conductive bodies of the induction heating system limit the mesh of discretization. Therefore, the areas requiring discretization include induction coils, workpieces, magnetic shunts, concentrators, and electrically conductive tooling. Unlike FEM and FDM, integral formulations do not include the free space areas (such as air) into the general consideration.

As with FDM and FEM, there are several different formulations of MIM devoted to simulation of the induction heating process. One of the earliest texts describing this technique was reported by Kolbe and Reiss in 1962 (Ref 96). Further development of this technique was done by Tozoni (Ref 97), Dudley and Burke (Ref 95), and several other researchers (Ref 38, 39). A brief introduction to MIM is given here based on the concepts discussed in Ref 38 and 39. First, consider two axisymmetric, multiturn, coaxial coils (Fig. 27) connected in series and driven by a sinusoidal voltage source. A quasi-stationary (time-harmonic) field is also assumed. Both coils are placed around an axisymmetric nonmagnetic workpiece (such as copper, aluminum, titanium, or nonmagnetic austenitic stainless steel billets). The electromagnetic field distribution in such a system can be described with respect to the current densities occurring in the electrically conductive parts of the induction system by the Fredholm integral equation of the second kind:

$$2\pi R_Q \rho_Q J + j\omega \int\limits_{P \in H, W} M_{QP} J_P \, dS_P = V_Q \quad \text{(Eq 84)}$$

where $\rho_Q = 1/\sigma_Q$ is the resistivity of element Q; R_Q is the average radius of element Q; J_Q and J_P are the current densities in elements Q and P, respectively; M_{QP} is the mutual inductance between elements Q and P, representing a mutual electromagnetic interaction of the elements (current-carrying rings); S_P is the computation areas ($P \in H, W$, where H represents the induction heater, and W represents the heated workpiece); and V_Q is the source voltage of the element. The value of V_Q is zero for all elements of the workpiece.

The method of solving the integral equation in its most general case has been described in Ref 97. The solution of Eq 84 in its simplified form (Ref 38, 39) is presented here. The electrically conductive regions of the induction heating system, including the induction heater and workpiece (Fig. 27, areas H and W, correspondingly), are subdivided into appropriate elements. As with FEM, eddy current densities and material properties are assumed to be constant within each element. If the skin effect in the coil is pronounced, then the multiturn induction coils can be considered as acting similar to multiturn solenoids. The integral Eq 84 can be rewritten as:

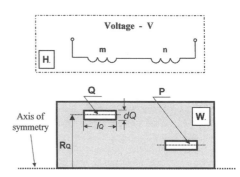

Fig. 27 Representation of the induction system for the mutual impedance method. Multiturn coils have m and n turns. Workpiece has differing current densities at P and Q. W, cylinder workpiece; H, multiturn induction heater. Source: Ref 1

$$r_Q I_Q + j\omega \sum_{P \in H, W} M_{QP} I_P = V_Q \quad \text{(Eq 85)}$$

where r_Q is the resistance of element Q.

As seen from Eq 84 and 85, the Fredholm integral equation of the second kind is converted into an impedance equation representing the well-known Kirchhoff's law. After assembling equations that correspond to all electrically conductive elements of the induction system, the global set of impedance equations can be obtained. To illustrate the previously described procedure, a set of global equations representing the induction system shown in Fig. 27 is obtained as follows. According to the sketch shown in Fig. 27, the induction heating system consists of two elements of the workpiece (P and Q) and two induction coils (m and n) connected in series. The global set of the impedance equations for this case is shown as:

$$
\begin{aligned}
(r_Q + j\omega M_{QQ})I_Q &+ & j\omega M_{QP}I_P &+ & j\omega(M_{Qn} + M_{Qm})I_{mn} &= 0 \\
j\omega M_{PQ}I_Q &+ & (r_P + j\omega M_{PP})I_P &+ & j\omega(M_{Pn} + M_{Pm})I_{mn} &= 0 \\
j\omega(M_{nQ} + M_{mQ})I_Q &+ & j\omega(M_{nP} + M_{mP})I_P &+ & (r_n + r_m + j\omega(M_{mn} + M_{nm}))I_{mn} &= V
\end{aligned}
\quad \text{(Eq 86)}
$$

where M_{QQ}, M_{PP}, M_{nn}, and M_{mm} are the self-inductances of the elements and coils, respectively; and M_{QP}, M_{Qn}, M_{Qm}, M_{PQ}, M_{Pn}, M_{Pm}, M_{nQ}, M_{nP}, M_{nm}, M_{mQ}, M_{mP}, and M_{mn} are the mutual inductances representing the interaction of all the current-carrying elements.

The formulas for calculation of the various self-inductances and mutual inductances with their range of applicability are given in Ref 98 to 100. The resistances of the rings (r_Q and r_P) and the resistances of the coils (r_n and r_m) can be calculated as:

$$r_Q = \frac{2\pi \rho_Q R_Q}{d_Q l_Q} \quad \text{(Eq 87)}$$

$$r_n = \frac{2\pi \rho_n R_n}{l_n \delta_n g} N \quad \text{(Eq 88)}$$

where g is the space factor of the coil, and N is the number of turns of coil n. After some simple algebraic operations (Ref 1, 38, 39), the resulting matrix equation can be rewritten as:

$$\begin{bmatrix} a_W & a_{WH} \\ a_{HW} & a_H \end{bmatrix} \begin{bmatrix} I_W \\ I_H \end{bmatrix} = \begin{bmatrix} 0 \\ V_H \end{bmatrix} \qquad \text{(Eq 89)}$$

where a_W and a_H are the matrices of the self-impedances of the workpiece and induction coils, and a_{WH} and a_{HW} are the matrices of the mutual inductances.

By evaluation of the equations for the mutual inductances between elements, it is obvious that $M_{PQ} = M_{QP}$ and $a_{WH} = a_{HW}$. Therefore, the matrix of:

$$\begin{bmatrix} a_W & a_{WH} \\ a_{HW} & a_H \end{bmatrix}$$

is symmetric, and the set of Eq 89 can be rewritten as:

$$[S_r + j\,S_x]\ \ [I_r + j\,I_x] = [V_r + j\,V_x] \qquad \text{(Eq 90)}$$

where S_r is the diagonal matrix, consisting of the resistivities of elements of the coils and the workpiece; S_x is the square matrix of the self-inductances and mutual inductances; I_r and I_x are the column matrices, representing that the currents have both a real and an imaginary part of the components ($I = I_r + j\,I_x$); and V_r and V_x are the column matrices of the voltages ($V = V_r + j\,V_x$), which are similar to the column matrices of the currents.

The set of Eq 90 can be rewritten as:

$$\begin{bmatrix} S_r & -S_x \\ S_x & S_r \end{bmatrix} \begin{bmatrix} I_r \\ I_x \end{bmatrix} = \begin{bmatrix} V_r \\ V_x \end{bmatrix} \qquad \text{(Eq 91)}$$

The execution time and computer memory required for storage of all matrices can be reduced by applying a special computational procedure for solving the set of Eq 91 that takes into consideration the fact that the matrix S_r is a diagonal matrix and the matrix S_x is a symmetrical square matrix (Ref 38, 39):

$$[S_r + S_x\,S^{-1}{}_r\,S_x]\ \ [I_x] = [V_r - S_x\,S^{-1}{}_r\,V_r] \qquad \text{(Eq 92)}$$

$$[I_r] = [S_r]^{-1}\,[V_r + S_x\,I_x] \qquad \text{(Eq 93)}$$

After solving the set of Eq 92 and 93, one can obtain the coil currents and eddy currents as well as power densities, heat-source distribution in the workpiece, and electrically conductive components located in close proximity to the induction coil, as well as other important output parameters of the induction system, including coil power, power induced within the workpiece, coil electrical efficiency, and the power factor.

As in any of the numerical techniques, a proper space discretization is critical when applying MIM. One of the obvious advantages of MIM is its ability to relatively easily incorporate circuit connection features into a general consideration. Some of the typical

circuits that have been studied using this technique are shown in Fig. 28. The MIM was extended for the computation of induction heating of magnetic workpieces (Ref 39).

Unfortunately, MIM does not appear to be an effective computational technique for complex-shaped bodies, due to the known limitations of calculating self-inductances and mutual inductances of arbitrarily oriented elements. However, in the case of classical geometries, it does allow one to easily incorporate circuit connection features into the general consideration when modeling regular-shaped workpieces. This is an important feature for calculating systems such as line frequency induction heating of large aluminum cylinder billets using multiturn and multilayer coil arrangements (typically using four to six coil layers) connected to a multiphase power source.

Boundary-Element Method

The fourth family of numerical techniques devoted to induction heating computation is called the boundary-element method (BEM). This method started to be widely used for modeling of the processes related to induction heating in the late 1980s and early 1990s. The mathematics required to discuss BEM are more advanced than that needed for FDM, FEM, or MIM. The interested reader will find several texts, conference proceedings, and journal articles (Ref 87 to 89, 91 to 94) that describe various modifications of BEM.

In contrast to FDM and FEM, when applying BEM, an integral form of Maxwell's equation is used as a governing equation for the electromagnetic problem. This allows taking into consideration only conductive bodies in the computation. In this respect, BEM is similar to MIM.

With BEM, unknown characteristics of the electromagnetic field (i.e., magnetic vector potential) are expressed in terms of an integral over the boundary of the area of interest (Fig. 29). In this case, the problem of mathematical modeling of induction processes may be divided into two tasks: external and internal electromagnetic problems. Using an iterative procedure, both tasks can be solved. The internal problem describes the electromagnetic field distribution within the body of the workpiece. The external problem describes the field distribution in external regions. In some cases of substantially nonlinear material properties, an error can be accumulated in solving an internal problem using BEM. In cases such as this, BEM is used for solving the external problem, and FEM or FDM is used to solve the internal problem. Such an approach uses the best of both numerical methods. As an example, Fig. 13 and 14 in the article "Simulation of Induction Heating Prior to Hot Working and Coating" in this Volume show temperature profiles when heating carbon steel bars using an in-line multicoil induction heater. A combination of

BEM and FDM was the best approach to simulate such a system.

In contrast to FDM and FEM, the use of Green's function eliminates the need for meshing the whole domain of modeling. Because with BEM, unknowns are only located on the boundaries or interfaces, it requires discretization only of the boundaries of the electrically conductive components of the induction system (Fig. 29). This substantially simplifies one of the most time-consuming parts of numerical model preparation when FDM and FEM are used (compare meshes shown in Fig. 24 and 29). A computational procedure establishes the unknown surface qualities (i.e., equivalent current densities along surfaces), which would satisfy the global solution.

According to one of many forms of BEM (Ref 87), it is assumed that the surface impedance is initially known and could be determined as:

$$Z_0 = \frac{E_t}{H_t} \qquad \text{(Eq 94)}$$

This assumption has been proposed by Leontovich for problems exhibiting pronounced skin effect. It is obvious that Z_0 is not constant along the workpiece surface and is a function of the electromagnetic field. For many induction heating applications, the surface impedance of the workpiece can be defined at a particular node, ξ, similar to the surface impedance of an infinite conducting half-plane as (Ref 45):

$$Z_0^{\xi} = \frac{\rho_{\xi}\,(U + jV)}{\delta_{\xi}} \qquad \text{(Eq 95)}$$

where U and V are coefficients, and U and $V \leq 1$, depending on the application; ρ_{ξ} is the electrical resistivity; and δ_Q is the current penetration depth at node Q.

As shown in Ref 45, if the skin effect in a nonmagnetic load is pronounced, then

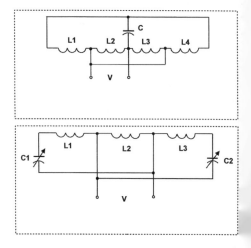

Fig. 28 Examples of circuit connections of induction heating coils (L) with fixed or variable capacitors (C). V, coil voltage. Source: Ref 1

$U = V = 1$ (Leontovich condition). In the case of a magnetic body placed in a relatively strong magnetic field, those coefficients will be $U =1.32 - 1.37$ and $V = 0.97$. In the edge region and corner areas of the surface, Eq 94 is not valid, and different approaches should be used (Ref 39, 45).

As mentioned previously, a thorough discussion of BEM is beyond the scope of this article. Details about BEM can be found in Ref 87 to 89 and 91 to 94. Such advantages as the reduction of computation time, simplicity, user-friendliness of the mesh generation, and good accuracy (particularly when working with nonmagnetic materials or materials with linear properties) make this technique quite attractive in certain applications compared to other methods.

Example 5: Camshaft Hardening. BEM was used to evaluate design approaches that would allow undesirable temper-back effect to be avoided when induction hardening camshafts (Fig. 30).

One of the major challenges in induction hardening of steel camshafts is to avoid undesirable heating of adjacent areas that have previously been hardened (the so-called temper-back or annealing effect of adjacent regions). The complexity of this problem arises from the fact that, due to electromagnetic field propagation, eddy currents are induced not only in the workpiece that is located under the inductor and meant to be heated but in adjacent areas as well. Computer modeling was conducted using the software Oersted, which uses BEM to simulate an electromagnetic field distribution around a single-turn inductor. Computer simulation helped to better understand the physics of this quite complex process, and it would be beneficial to briefly describe it here (Ref 1, 43, 44, 102).

Without a concentrator (Fig. 30, top left), the magnetic flux would spread around the coil and link with electrically conductive surroundings, which include neighboring areas of the part (i. e., cam lobes and journals) and possibly certain areas of tooling and the fixture. As a result of induced eddy currents, the heat will be produced in those electrically conductive regions. This heat can cause undesirable metallurgical changes, resulting, for example, in strength reduction of the tooling or a decrease in hardness of lobes that were hardened during a previous operation. At the initial stage of the heating cycle, the entire workpiece is magnetic. Because of better electromagnetic coupling, surface areas of the camshaft located under the coil will have substantially more intense heating than other areas in the surrounding coil. An intermediate heating stage begins after the surface of the heated lobe reaches the Curie temperature and its relative magnetic permeability, μ_r, drops to 1. Therefore, the surface layer of the heated lobe becomes nonmagnetic, and its heating intensity drastically decreases. At this intermediate heating stage, the electromagnetic coupling between the coil and the

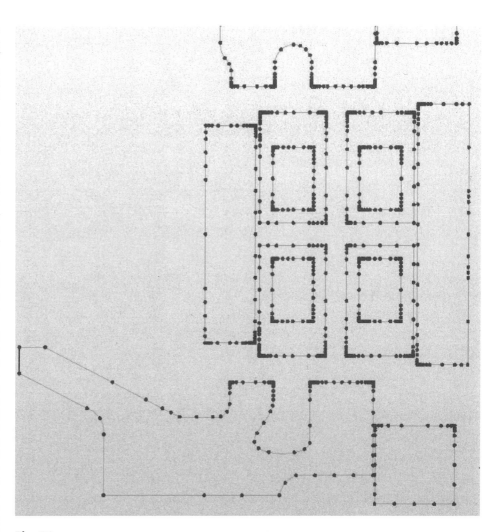

Fig. 29 Boundary-element method requires discretization only on boundaries and interfaces

lobe will not be as strong as it had been during the initial heating stage (when the whole camshaft was magnetic). Although the surface of the heated lobe has lost its magnetic properties, the adjacent areas retain theirs. Consequently, the electromagnetic coupling of the inductor with those areas will not decrease. Actually, the greater portion of the electromagnetic field will start to link with the adjacent ferromagnetic regions located in close proximity to the inductor. This may include certain areas of previously hardened lobes. In addition, in order to have a short cycle time and to keep the heat intensity of the surface area located under the coil at the same heating rate as during the initial stage, the control system could automatically increase the coil current when the intermediate stage begins. This will further increase the heating intensity of the magnetic regions of already hardened lobes located in proximity to the inductor, manifesting itself in potential undesirable tempering back of those areas.

After installation of a U-shaped magnetic flux concentrator around the inductor (Fig. 30, left bottom), a much smaller portion of the

inductor electromagnetic field will link with adjacent lobes located near the lobe that is to be hardened. Therefore, a magnetic flux concentrator decouples the induction coil and the areas of adjacent lobes. Mathematical simulation shows that an appropriate use of magnetic flux concentrators allows a four- to twelvefold reduction in the power density induced in adjacent lobes compared to using a bare coil (Ref 1, 102).

Depending on the application specifics and camshaft geometry, some lobes may be positioned too close to each other, and minor tempering back may still be present. The application of a Faraday ring (also called a robber ring) can compliment the use of magnetic flux concentrators, further reducing an external magnetic field (Fig. 30, left middle).

Example 6: Heating Large Aluminum Billets. Control of electromagnetic field exposure around a large multiturn, multiphase inductor for heating large aluminum billets using BEM is examined.

Any electrical device must comply with certain standards, regulations, and recommendations (different in various countries) regarding

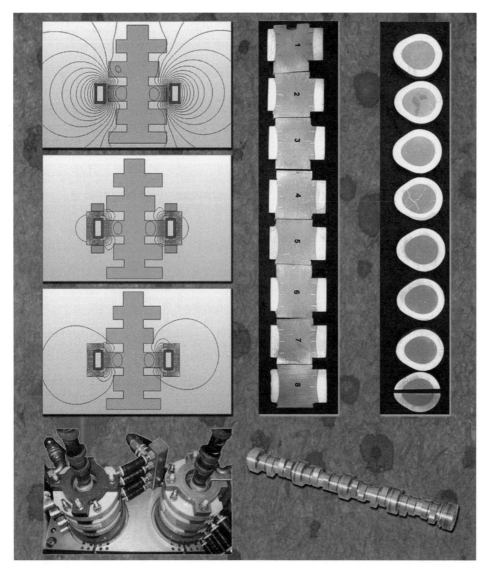

Fig. 30 Electromagnetic control of undesirable temper back of previously hardened camshaft lobes

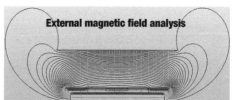

Fig. 31 Distribution of the magnetic field around a large multilayer, multiturn inductor for heating aluminum billets prior to direct extrusion. (Due to symmetry, only the top half of the system is shown.)

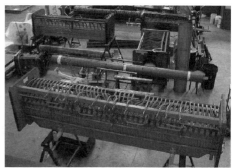

Fig. 32 Multilayer induction heating coils. Courtesy of IHWT, England

safe levels of magnetic field exposure (Ref 101). It is imperative to be able to predict the strength of the electromagnetic field in the induction coil surroundings, defining safety areas where humans may be present and evaluating means of reducing an external magnetic field (for example, installing magnetic shunts, shields, etc.). The BEM is a particularly effective modeling tool for simulation of external electromagnetic field exposure. Figure 31 shows the distribution of a magnetic field around a large multilayer, multiturn inductor for heating aluminum billets prior to direct extrusion. (Due to symmetry, only the top half of the system is shown.) The size of the aluminum billets typically ranges from 100 to 600 mm in diameter and up to 1.8 m in length. An example of such inductors is shown in Fig. 32. Depending on application specifics, such induction coils can consist of up to six layers fabricated using numerous disc-pair copper windings (Ref 1). Single- or multiphase arrangements of electrical connection of multilayer inductor sections can be used. All these design features complicate the use of FDM and FEM, requiring an appreciable amount of time to address all critical design features under consideration and to generate an appropriate mesh. At the same time, BEM simulates such applications quickly and effectively.

Coupling of Electromagnetic and Thermal Problems

As stated previously, one of the major features of induction heat treating computation deals with the fact that both the electromagnetic and heat-transfer phenomena are tightly coupled due to the interrelated nonlinear nature of the material properties. This feature dictates the necessity of developing special computational algorithms that are able to deal with these interrelated effects. It is important to note that the time scales (time constants) of the electromagnetic and heat-transfer processes are quite different. Electromagnetic processes used in induction heating are very fast, with time constants significantly less than 0.02 s (depending on the frequency). At the same time, the heat-transfer processes are much longer. For example, heat cycles even for medium-sized induction billet heaters can easily exceed 300 s, depending on frequency, geometry, and material properties of the billet. Heat times for hardening typically range from 3 to 12 s using medium and radio frequencies. Even for contour gear hardening, which is known to be one of the fastest induction heat treating processes, the heat time could be as short as 1 s with an applied frequency of approximately 400 kHz. This frequency results in an electromagnetic process time constant of approximately 0.0000025 s. As can be seen, the difference in the time constants of electromagnetic field and thermal processes in induction heating and heat treating is in orders of magnitude.

Two-Step Coupling

There are several ways to couple the electromagnetic and heat-transfer problems. The simplest method is called the two-step coupling approach (Fig. 33). The electromagnetic problem is solved during the first computational step. Obtained current densities and heat-source distribution are used to determine parameter Q in Eq 35, 38, and 39 for solving the thermal problem, assuming that the electromagnetic field distribution and electromagnetic physical properties (electrical resistivity and magnetic permeability) do not change during heating. The two-step approach is known for its short

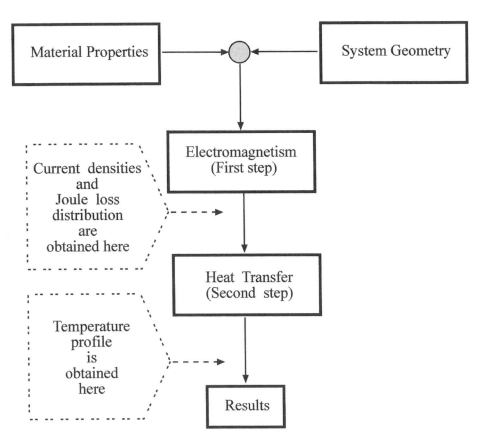

Fig. 33 Schematic of the two-step approach for coupling in induction heating simulation. Source: Ref 1

execution time and moderate computer memory requirements.

Even a cursory look at the behavior of the material properties discussed in the companion article, "Simulation of Induction Heating Prior to Hot Working and Coating" in this Volume, reveals the danger in using the two-step approach. The electrical resistivity of some metals can vary more than eight times during the process of heating. At the same time, the variation of relative magnetic permeability typically ranges from 20 to 80 times during a typical heat cycle. Therefore, an assumption that material properties are constant during the entire heating time is a very rough postulation and can result in significant calculation errors. Therefore, the two-step coupling approach can be used in a very limited number of induction heating applications. Low-temperature heating (i.e., heating from 100 to 200 °C) of aluminum or copper strips, rod, or wires serves as an example of when the two-step coupling approach can be used quite effectively. (In cases such as this, the average values of the metal properties are used.)

Indirect Coupling. The most common approach to coupled electromagnetic and heat-transfer problems is called the indirect coupling method (Ref 1, 10). This method calls for an iteration process (Fig. 34), which consists of an electromagnetic computation and then recalculation of heat sources to provide a heat-

transfer computation. This coupling approach assumes that temperature variations are not significant during certain heat-time intervals predetermined by the user. This means that the electromagnetic material properties remain approximately the same, and the heat-transfer process continues to be simulated without correcting the heat sources. The temperature distribution within the workpiece obtained from such time-stepped heat-transfer computations is used to update the values of specific heat and thermal conductivity at each time step. As soon as the heat-source variations become significant (due to the variations of electrical conductivity and magnetic permeability), the convergence condition will no longer be satisfied, and recalculation of the electromagnetic field and heat sources takes place.

Depending on the application, at least five to eight iteration times are typically required to make corrections in the electromagnetic field distribution with recalculation of the heat sources. For the great majority of induction heating applications, the indirect coupling approach is valid and very effective. However, there are rear cases where this approach could possibly lead to noticeable errors. In these cases, the direct coupling method should be applied (Ref 1, 10).

Direct Coupling. To provide a direct coupling of the electromagnetic and heat-transfer problems, it is necessary to formalize a set of

governing equations in such a way that the unknown parameters of the electromagnetic field (for example, a magnetic vector potential or magnetic field intensity) and the unknown parameters of the thermal problem (i.e., temperature) will be part of one global matrix, which will be solved simultaneously (Fig. 35). Direct coupling results in an extremely intensive execution time of computer simulations, and it should be used only in cases where it is absolutely needed.

The ability to simulate coupled electromagnetic and heat-transfer phenomena helps to reveal previously unknown phenomena or clarify some common misassumptions related to induction heat treating.

Example 7: Uncovering a Common Misassumption in Induction Hardening with Computer Modeling. When discussing induction heating, reference is often made to the phenomenon of skin effect. The skin effect is considered to be a fundamental property of induction heating that represents a nonuniform distribution of an alternating current within the conductor cross section. This effect is found in any electrically conductive workpiece located inside an induction coil or in its close proximity. Eddy currents induced within the workpiece will primarily flow in the surface layer (the skin), where 86% of all induced power will be concentrated. This layer is called the reference depth or current penetration depth, δ. The degree of skin effect depends on the frequency and electromagnetic properties (electrical resistivity, ρ, and relative magnetic permeability, μ_r) of the conductor. It is often recommended to calculate the distribution of the current density along the workpiece thickness (radius) by using Bessel functions. For electromagnetically thick workpieces, the following simplified expression is frequently used:

$$I = I_0 \cdot e^{-y/\delta} \qquad \text{(Eq 96)}$$

where I is the current density (A/m^2) at distance y (m) from the workpiece surface toward the core, I_0 is the current density at the surface (A/m^2), and δ is the current penetration depth (m). According to this equation, an eddy current density induced within an inductively heated workpiece has its maximum value at the surface and falls off exponentially.

Current penetration depth, δ, is described (in meters) as:

$$\delta = 503 \times (\rho/\mu_r F)^{1/2} \qquad \text{(Eq 97)}$$

where ρ is the electrical resistivity of the metal ($\Omega \cdot$m), μ_r is the relative magnetic permeability, and F is the frequency (Hz), or (in inches):

$$\delta = 3160 \times (\rho/\mu_r F)^{1/2} \qquad \text{(Eq 98)}$$

where electrical resistivity, ρ, is in units of $\Omega \cdot$in.

Thus, the value of penetration depth varies with the square root of electrical resistivity

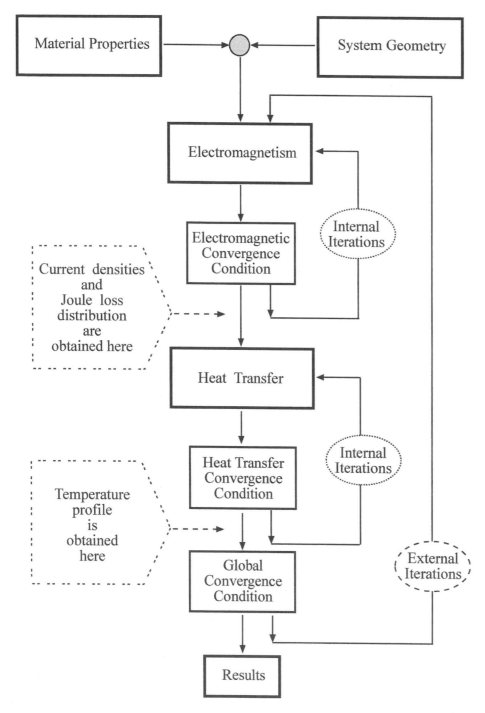

Fig. 34 Schematic of the indirect coupling approach to solving electromagnetic and heat-transfer problems. Source: Ref 1

distributions of current density and power density (heat-source distributions) along the thickness/radius of the workpiece are simplified and, as described previously, assumed to exponentially decrease from the surface toward its core (Eq 96 and Fig. 37). It is important to remember that this assumption is appropriate only for a solid body (workpiece) having both electrical resistivity and magnetic permeability as constants. Therefore, realistically speaking, this assumption can be made only for some unique cases of induction heating and heat treating. For the great majority of induction heat treating applications, the current density (heat-source) distribution is not uniform, and there are always thermal gradients within the heated workpiece. These thermal gradients result in nonuniform distributions of electrical resistivity and magnetic permeability within the workpiece. This nonlinearity means that the classical definition of current penetration depth does not fit its principal assumption.

The assumption of exponential current density distribution can be used to some extent for rough engineering estimates for induction heating of nonmagnetic materials (i.e., aluminum, copper, titanium, etc.) and through heating of stainless steels to forging temperature. However, in induction surface hardening, the power density distribution along the radius/thickness has a unique wave shape, which differs significantly from the commonly assumed exponential distribution. Here, the power density has its maximum value at the surface and decreases toward the core. Then, at a certain distance from the surface, the power density suddenly starts to increase again, reaching a maximum value before it starts a final decline.

Originally, a hypothesis regarding the magnetic-wave phenomenon was introduced by Losinskii (Ref 46) and Simpson (Ref 105). They intuitively felt that there should be situations where the power density (heat-source) distribution would differ from that of the traditionally accepted exponential form. They provided a qualitative description of this phenomenon based on their intuition and understanding of the physics of the process. At the time, a quantitative evaluation of this phenomenon could not be developed due to the limitation in computer modeling capabilities and the lack of software that could simulate the tightly coupled electrothermal phenomena of induction heat treating processes. Of course, it also was not possible to measure the power/current density distribution inside the solid workpiece without appreciably disturbing an eddy current flow. Later, a nonexponential skin effect distribution was also briefly mentioned (Ref 39). To the author's knowledge, the first publication that provides a quantitative assessment of the magnetic-wave phenomenon was published in Ref 107, with further research provided in Ref 1, 31, 41, and 104.

Modern coupled electromagnetic-thermal numerical software, such as ADVANCE, enables a quantitative estimation of the

and inversely with the square root of frequency and relative magnetic permeability. Mathematically speaking, the penetration depth, δ, in Eq 96 is the distance from the surface of the conductor toward its core, at which the current decreases exponentially to $1/\exp$, its value at the surface. The power density at this distance will decrease to $1/\exp^2$, its value at the surface.

Figure 36 illustrates the skin effect, showing the distribution of current density from the

workpiece surface toward the core. At one penetration depth from the surface ($y = \delta$), the current will equal 37% of its surface value. However, the power density will only correspond to 14% of its surface value. From this, one can conclude that approximately 63% of the current and 86% of the induced power in the workpiece will be concentrated within a surface layer of thickness δ.

In the great majority of publications devoted to induction heating and induction heat treating,

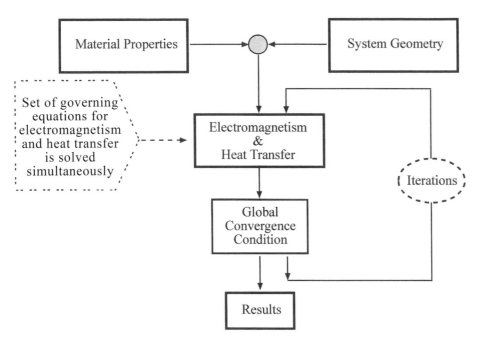

Fig. 35 Schematic of the direct coupling approach to solving electromagnetic and heat-transfer problems. Source: Ref 1

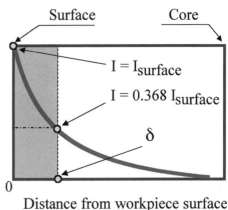

Fig. 37 Current distribution and current versus distance from surface of workpiece due to a classical definition of the skin effect. Source: Ref 42

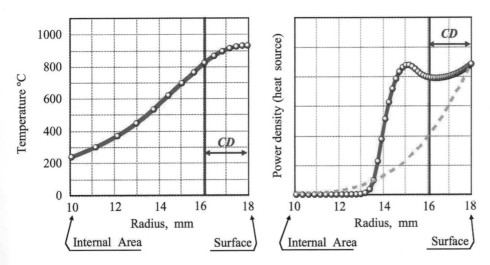

Fig. 36 Actual temperature profile and power density distribution for induction surface hardening of carbon steel shafts using 10 kHz frequency. Case depth (CD) is 2 mm. The dashed line in the graph at right is the commonly assumed power density distribution. Source: Ref 1, 104

magnetic-wave phenomenon (also known as the dual-properties phenomenon) based on its ability to properly simulate interrelated electromagnetic and thermal problems. An example is given in Fig. 36, which shows the temperature profile (left) and power density distribution (right) along the radius of a 36 mm (1.42 in.) diameter carbon steel shaft at the final stage of heating using a 10 kHz frequency. For comparison, the dotted curve corresponding to the power density distribution represents the commonly assumed exponential curve, and the solid curve shows the actual magnetic-wave distribution. The cause of this phenomenon relates to

the phenomenon that steel retains its magnetic properties in the subsurface region. Note that in some unique applications, due to this phenomenon, the maximum value of heat sources can be located in an internal layer of the workpiece and not at its surface. Detailed discussion regarding this phenomenon can be found in Ref 1, 31, 41, 104, 106, and 107.

Consideration of the wavelike distribution of power density (heat source) makes a significant impact on the optimal frequency choice for induction surface hardening. The following statement can be made from the preceding discussion: Frequency selection is not as obvious

and easy a task as it may appear by using rules of thumb that require a comprehensive computer simulation.

Example 8: Case Study of a Superior Induction Heating Design with Computer Modeling. In an attempt to further advance piston manufacturing technology for both traditional and high-performance engines, Federal-Mogul specialists have combined their efforts with experts from Spinduction Weld Inc., who developed a novel welding process called Spinduction, and Inductoheat, Inc. (Ref 29). The Monosteel piston (Federal-Mogul) starts life as two separate forgings (Fig. 38): a top or crown forging and a lower or skirt forging. Before joining, the forged top and bottom halves are prepared with two concentric circular end-face lands, which are simultaneously welded.

Spinduction welding provides relative motion between the joining parts, as in friction welding, to minimize impurity entrapment in the weld zone and to produce fast, high-integrity welds with consistent quality. The important advantages of Spinduction are that it produces consistently flawless welds at very low rotational velocities, well below the minimum forging velocity for friction welding, with minimal or no flash projection, which is appreciably different compared to the traditionally used friction welding processes (compare Fig. 39 and 40). The Spinduction process uses induction heating for over 95% of the weld energy input, so when considering very large workpieces, the Spinduction process is more advantageous than inertia or friction welders. It is easier and more cost-efficient to build a high-power induction system than a mechanical drive line with comparable kinetic energy.

Induction preheating is a critical part of this technology, providing the required heat input quickly, efficiently, and with the required uniformity within four circular end-face lands. Immediately after completion of the heating stage, the inductor is retracted within a fraction of a second, and the two piston halves are rapidly pressed together. Rotation begins just

Fig. 38 The Monosteel piston starts out as two separate forgings. Source: Ref 29

Fig. 39 Cross section of a friction-welded piston. Source: Ref 29

Fig. 40 Spinduction produces welds at very low rotational velocities with minimal flash projection. Source: Ref 29

before the surfaces of both halves come into contact. Rotational displacement and the axial forging force are simultaneously applied and controlled until the required tangential displacement is achieved and rotation stops.

Some of the innovations and know-how of this technology are related to the ability to accurately control heating and cooling stages to produce sound welds. There are several phenomena that contribute to the challenge of satisfying the required temperature distribution of joining surfaces as well as closely controlling an optimal heat-affected transition zone beyond those faces. These include:

- Interaction of magnetic fields generated by different inductor segments takes place due to the close proximity of four circular faces that must be heated. Besides that, electrically conductive bodies, such as fixtures and tooling (i.e., grippers), also distort those fields, acting as Faraday rings.
- Induced eddy currents have a tendency to take the shortest path of the smallest impedance, resulting in an electromagnetic ring effect that has a tendency to force the eddy current to flow nonuniformly across heating faces.
- Transient time between the end of heating and beginning of joining should be as short as possible and preferably less than 1 s. Longer inductor retraction time results in a dramatic reduction of the temperature of joining faces due to thermal conduction toward colder areas as well as heat-surface losses (thermal radiation and convection). For example, simulations show that if the inductor retraction time is greater than 1.5 s, the temperature drop can easily exceed 450 °C (810 °F).

- The existence of varied background masses behind the heating faces (Fig. 40) leads to appreciably nonuniform heat sinks that, depending on the piston cross section, provide substantially different cooling effects during heating and inductor retraction. To complicate matters further, all of these factors are nonlinear and multidimensional in nature.
- The thermal profile in the transition zone has a marked effect on the cooling rate of the welded joint, which determines the final microstructure and also affects the steel self-tempering capabilities.

The nature of the Spinduction process made it difficult to see or measure the temperature of the heated surface, making the old cut-and-try method of inductor development impractical. Inductoheat's superior computer modeling capability and design experience were decisive factors in the successful development of a novel inductor design (Ref 29). Using the finite-element code Flux-2D to model helped to determine not only the subtleties of the inductor geometry but also the process recipe that would guarantee the optimal temperature profiles and superior quality of the pistons. Figures 41 and 42 show the finite-element mesh and variation of temperature profiles during heating of the

top and bottom halves at 0° and 90° cross sections, respectively. Total heat time was approximately 6 s. The advanced inductor concept made possible a retraction time of less than 1 s.

Example 9: Computer Simulation of Heating and Quenching Stages in Nonrotational Crankshaft Hardening. Crankshafts are used in internal combustion engines, pumps, and compressors. In automobiles, they typically weigh 13 to 39 kg (30 to 85 lb), depending on the engine. Some crankshafts exceed 1000 kg (2000 lb) in marine or stationary engines for power generators. A crankshaft is cast or forged and comprises a series of crankpins (pins) and main journals (mains) interconnected by webs and counterweights (Fig. 43). Steel forgings, nodular iron castings, microalloy forgings, and austempered ductile iron castings are among the materials most frequently used for crankshafts. High strength and elasticity, good wear resistance, light weight, small vibration, geometrical accuracy, short length, and low cost are some of the most important crankshaft requirements. Most of these attributes are augmented by the induction hardening process. Because the diameters of crank journals (mains and pins) are much smaller compared to the external dimensions of the counterweights (webs), the conventional encircling-type coils could not freely pass from one heat treated feature to another. This feature dictates having a specific inductor design. To use the required hardening while not having to rotate the crankshaft, a patented stationary hardening process for crankshafts (SHarP-C technology) was developed (Ref 1, 108).

According to the patented nonrotational hardening process, an inductor consists of two coils: a top (passive) coil and a bottom (active) coil. The bottom coil, being active, is connected to a medium- or high-frequency power supply, while the top coil (passive) represents a short circuit (a loop). The bottom coil is a stationary coil, while a top coil can be opened and closed. Each coil has two semicircular areas where the crankshaft features are located. A robot loads a crankshaft into the heating position, the top inductor pivots into the closed position, and power is applied from the power supply to the bottom (active) inductor. The current starts to flow in the bottom inductor. Being electromagnetically coupled to the top coil, the current flowing in the bottom coil will induce the currents in the top coil in the opposite direction, similar to a transformer effect. Any heated feature of the crankshaft "sees" the SHarP-C inductor as a classical and highly efficient fully encircling coil (Ref 1, 108).

Intensive theoretical and computer simulation studies have been conducted by Inductoheat's personnel in cooperation with leading world experts, such as Dr. Lynn Ferguson and his team from Deformation Control Technology, Inc., to determine the favorite distribution of transitional and residual stresses of heat treated crank components (Ref 108). Figure 43 shows the temperature variation during various stages of heating and quenching of the

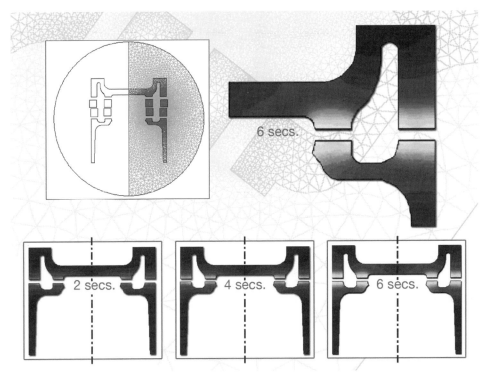

Fig. 41 Inductoheat's computer modeling capability was a decisive factor in predicting temperature profiles during heating of top and bottom complex-shaped halves (0° cross section). Source: Ref 29

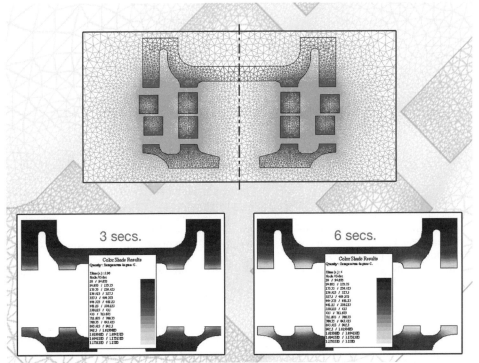

Fig. 42 Finite-element analysis computer modeling helps to determine the transient and final thermal conditions (90° cross section) of the piston halves as well as an optimal inductor design and process recipe. Source: Ref 29

crankshaft journal (Ref 1). Note the presence of residual heat accumulated during an interrupted quenching. This well-controlled residual heat complements the induction tempering of as-

hardened journals. Computer modeling of temperature profiles during spray quenching of crankshaft journals and the prediction of austenite formation are shown in Fig. 44(b).

Challenges Faced by Developers of Modern Simulation Software

In the fast-paced world economy, the ability of induction heating manufacturers to minimize the time between a customer request for quotation and delivery of the quotation through efficient computer modeling is critical for a company's success. Also, contrary to academia, the rapid pace of industry does not allow the luxury of waiting several days to obtain the computer modeling results but demands those results within a couple of hours.

Modern computer simulation techniques are capable of effectively simulating electromagnetic and thermal phenomena for many processes that involve electromagnetic induction. The great majority of commercial codes used for computer modeling of induction heating processes are all-purpose programs. Regardless of the well-recognized impressive capabilities of modern commercial simulation tools, most of the generalized programs experience certain difficulties in effectively handling particular features of the induction heating process, including but not limited to the presence of thermal refractory, scanning operations that combine heating and quenching, simultaneous movement or oscillation of a heated workpiece in respect to the induction coil, simultaneous use of two frequencies for contour gear hardening, and so on. There is an appreciable family of induction heat treating applications where the ability to simulate just-coupled electromagnetic and heat-transfer phenomena is not sufficient. It is also imperative to be able to simulate processes of a metallurgical nature, such as phase transformations, prediction of transitional and residual stresses, shape/size distortion of heat treated components, and probability of crack development. There are a number of induction thermal processes that also exhibit certain process subtleties that cannot be easily modeled using the majority of commercially available software.

The necessity to sell products to as many customers as possible forces developers of coupled electromagnetic-thermal software to produce a universal simulation tool that can be used by various industries (i.e., motors, transformers, magnetic recording, medical, nondestructive testing, and others, including induction heating users). As a result, certain process subtleties of induction heat treating were overlooked by developers or substantially simplified, posing noticeable limitations for using universal software. Some of those applications that create appreciable challenges and limitations for a majority of presently available commercial software are discussed as follows.

How to Treat the Realities of Prior Microstructures and How to Predict Final Structure, Taking into Consideration the Subtleties of Inductive Heat Treating. When iron is alloyed with various percentages of carbon, the critical temperatures are often determined by

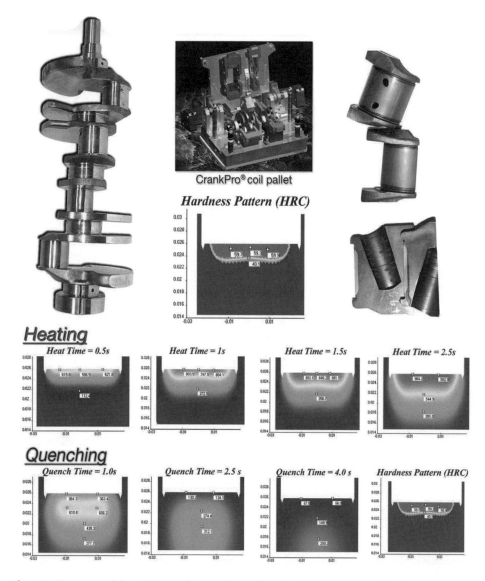

Fig. 43 Computer modeling of induction heating and quenching when hardening a V-8 crankshaft journal. Courtesy of Inductoheat, Inc.

the iron-iron carbide phase-transformation diagram (Fe-Fe$_3$C diagram). The lower left portion of this diagram represents heat treating of steels and is shown in Fig. 45. This widely used diagram is a graph of temperature versus carbon content of steel and shows how heating to elevated temperatures or cooling from an elevated temperature can cause a transformation in the steel crystal structure. However, it is important to be aware that this phase diagram may be misleading in the majority of induction hardening applications, because it is valid only for the equilibrium condition of plain carbon steels. Nonequilibrium conditions, appreciable amounts of alloying elements, pressure, and certain prior treatments can noticeably shift the critical temperatures (Ref 1, 19).

One of the major requirements of an equilibrium condition is enough time at temperature. Ideally, in the case of a sufficiently slow heating/cooling, transformation temperatures should be approximately the same for both heating and

cooling; in other words, there should be no appreciable difference between A$_c$ and A$_r$ critical temperatures. Any observed difference between A$_c$ and A$_r$ represents thermal hysteresis, which is a function of several factors, including the metal chemical composition and the heating/cooling rate. The greater the rate of heating/cooling, the greater the difference between those temperatures. Practically speaking, the equilibrium condition simply does not exist in induction hardening. This eliminates using those classical diagrams in an attempt to predict critical temperatures.

Induction hardening is a very fast process. The intensity of heating or the heating rate often exceeds 100 °C/s (180 °F/s) and, in some cases, 1000 °C/s (1800 °F/s). Therefore, the phase transformation cannot by any means be considered as equilibrium, and the phenomenon of thermal hysteresis is always pronounced. Rapid heating drastically affects the kinetics

of austenite formation, shifting it toward higher temperatures to create conditions conducive to the required diffusion-based processes and resulting in a homogeneous austenitic structure with a uniform distribution of carbon.

The presence of heterogeneous austenite can result in an as-quenched part having an unacceptable microstructure. The degree of heterogeneity in the microstructure of the as-quenched part can be reduced by increasing the hardening temperature. Observation of ghost pearlite or an excessive amount of free ferrite during a metallographic evaluation of as-hardened specimens can also indicate the presence of heterogeneous austenite. Figure 46 shows the effect of heating rate on the A$_3$ critical temperature of medium carbon steel (Ref 2, 30). The inability of the classical Fe-Fe$_3$C diagram to take into account heating intensity limits its use for predicting the temperatures required for induction hardening applications.

Probably the most comprehensive study regarding the correlation of heat intensity with the ability to obtain homogeneous austenite was conducted by Orlich, Rose, and colleagues at the Max-Planck-Institut für Eisenforschung GmbH in Düsseldorf, Germany. They developed atlases that consist of more than 500 pages of nonequilibrium time-temperature-austenitizing diagrams for a variety of steels after induction heating with heat intensities ranging from 0.05 to 2400 °C/s (0.09 to 4320 °F/s). Those diagrams are much better suited for predicting the required final temperatures for austenitization than the conventional Fe-Fe$_3$C diagram.

Another critical factor that is missed on practically all known phase-transformation diagrams is related to the presence of irregularities in prior microstructures (also referred to as the initial structure, structure of the parent material, or structure of the green part). Examples of such irregularities include microstructural and chemical segregation (Fig. 47), banded structures (Fig. 48), and the presence of inclusions, heterogeneous grains, clusters (Fig. 49), and so on. Upon fast austenitization and quenching of severely banded steel structures, the as-quenched structure often exhibits traces of banding, pearlitic-ferritic networking, and ghost structures. Alloy segregation and a microstructure consisting of large clusters are considered to be undesirable structures as well and should be avoided. Large graphite flakes of cast irons or clusters having a preferable orientation of flakes and located near the casting surface serve as appreciable stress raisers (Fig. 49) and make gray iron castings more sensitive to cracking during rapid heating and fast quenching. In addition, upon hardening of gray irons, soft spots may occur in areas of large cluster locations. It has been reported that eutectoid carbides can encourage cracking in virtually all types of cast irons suitable for surface hardening (Ref 3 to 12), due to insufficient holding time at the austenitizing temperature.

At this time (2010), the author is not aware of any software that can simulate the effect of

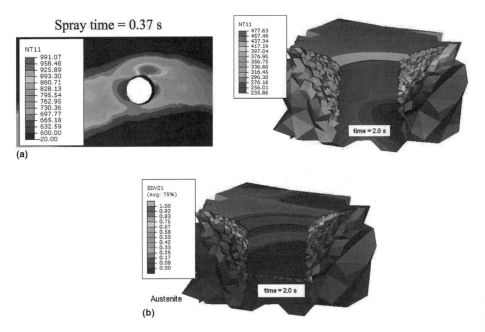

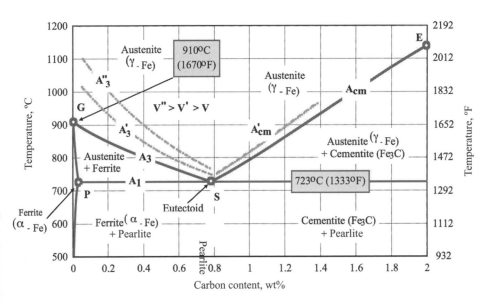

Fig. 44 Computer simulation of (a) temperature profiles during spray quenching of a crankshaft journal and (b) prediction of austenite transformation. Courtesy of Deformation Control Technology, Inc.

Fig. 45 Lower portion of the Fe-Fe₃C equilibrium phase transformation diagram and its deformation when an intensive heating rate is used. Note: A_3'', A_3', and A_{cm}' and A_3 and A_{cm} at heating rates (°C/s) V'', V', and V, respectively (V'' > V' > V). Source: Ref 19

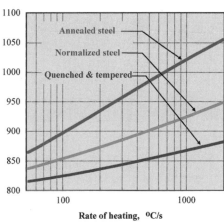

Fig. 46 Effect of initial microstructure and heating rate on A_3 critical temperature for ASTM SAE 1042 steel. Source: Ref 2, 30

Fig. 47 Micrograph showing microstructural segregation on low-carbon steel. Original width of image is 0.75 mm.

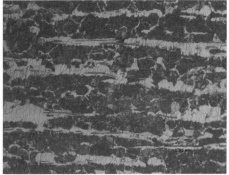

Fig. 48 Micrograph showing banded structure of medium-carbon steel and presence of inclusions. Original width of image is 0.3 mm.

these irregularities. It would be very beneficial to be able to simulate the phase-transformation mechanism of the aforementioned structural irregularities and ways to modify the induction heat treating process recipes obtained based on the assumption that irregularities are absent.

How to Properly Model Spray Quenching and Scan Hardening. Spray quenching (Fig. 50) is the most popular method used in induction hardening (Ref 1, 2, 18, 70, 90). Spray quenching works best if the component (axle shafts, spindles, rods, and gears, for example)

is rotated during the quench, ensuring cooling uniformity. Uneven quenching could have a detrimental effect on the microstructure of the heat treated part and could make distortion and cracking pronounced.

The intensity of spray quenching depends on the quenchant flow rate, the angle at which the quenchant strikes the workpiece, as well as the temperature, purity, and type of quenchant, depth of heating, and the temperature of the heat treated component. Quenchants used include water, aqueous polymer solutions, and,

to a lesser extent, oil, water mist, and forced air (Ref 8). Water and aqueous polymer solutions are the most popular. Spray quenching typically leads to a greater quench severity

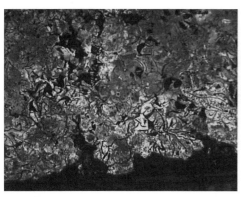
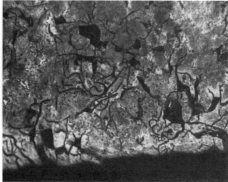

Fig. 49 Micrograph of large graphite flakes of gray cast irons or clusters having a preferable orientation of flakes and located near the casting surface. These serve as an appreciable stress raiser. Source: Ref 1

Fig. 50 Induction dual-scan hardening uses spray quenching. Courtesy of Inductoheat, Inc.

compared to immersion quenching, resulting in higher hardness and greater compressive surface stresses.

There were several attempts to use continuous cooling transformation (CCT) diagrams and classical cooling intensity curves to simulate the effect of cooling severity during the quenching stage of induction hardening. However, it is important to remember that classical CCT diagrams have inherent and appreciable limitations when applied to induction hardening (Ref 1, 13, 14). The CCT diagrams were developed assuming homogeneous austenite, which is not always the case in induction hardening. Rapid induction heating noticeably affects the kinetics of austenite formation and the carbon distribution within it. If heterogeneous austenite results, it means, among other factors, that there is a nonuniform distribution of carbon. Upon quenching, decomposition of heterogeneous austenite first begins in regions of lower carbon concentration. This results in a shift of the CCT curve to the left and an increase in the M_s temperature. The CCT curve for regions having excessive amounts of carbon will be shifted in the opposite direction with a corresponding decrease in the M_s temperature. Therefore, cooling of high- and low-carbon-concentration areas of heterogeneous austenite would be accompanied by different critical cooling curves.

The cooling curves of CCT diagrams assume a particular (typically constant) cooling rate, which is often not a valid assumption in spray quenching, particularly in induction scan hardening applications. In addition, the samples used to develop these diagrams typically have small cross sections. Therefore, there will be inherent errors in trying to apply classical cooling curves to larger components and parts having complex shapes (gears and shafts, for example). Note also that components having different surface-area-to-volume ratios have different cooling characteristics.

In induction hardening, the heat-exchange process between the surface of the heated component and the quenchant is a function, among other factors, of the surface temperature, which is not constant. Besides that, the workpiece temperature prior to quenching (the austenitizing temperature) is typically not identical to that in the material CCT diagram. In addition, the temperature distribution prior to quenching is also nonuniform.

There is a common misunderstanding regarding the ability to apply the widely published, classical cooling curves for immersion quenching to induction hardening applications. Classical cooling curves representing three stages of quenching in a liquid—vapor blanket (stage A), nucleate boiling (stage B), and convective cooling (stage C)—cannot be applied directly to simulate spray quenching. The differences are both quantitative and qualitative and include but are not limited to specifics of film formation and heat transfer through the vapor blanket during the initial stage (A), as well as the kinetics of formation, growth, and removal of bubbles from the surface of the heated component during nucleate boiling (stage B) (Ref 8). Due to the nature of spray quenching, stages A and B are greatly suppressed in time, while cooling during the convection stage (C) is noticeably more intense, compared with the process represented by classical cooling curves (Ref 18). Also, the thickness of the vapor blanket film during stage A is typically much thinner during spray quenching than when the part is submerged in a quench tank and depends on impingement angle, flow rate, part rotation, and other characteristics of the quenching system. This vapor film is unstable and could be frequently ruptured. In addition, the transition between stages A and B is smoother with spray quenching than that shown by classical cooling curves for immersion quenching. During nucleate boiling (stage B), bubbles are smaller because they have less time to grow. Appreciably larger number of bubbles form during spray quenching, and the intensity with which they remove heat from the surface of the component is substantially greater compared with immersion quenching.

Another factor that has a considerable effect on quench severity in surface hardening by induction is the thermal sink effect provided by the component cold regions (i.e., core). In the majority of induction surface hardening applications, the core temperature does not rise significantly, primarily due to a pronounced skin effect, high heat intensity, and short heating time. As a result, heat transfer from the surface of the workpiece to its core during the heating stage is not sufficient to significantly raise core temperature. A cold core complements spray quenching by further increasing the cooling intensity at the surface and in subsurface regions of the part. Note that in some induction surface hardening applications requiring shallower case depths (0.5 to 2.5 mm, or 0.02 to 0.1 in.), self-quenching can be used. Here, the effect of thermal conduction away from the surface by a sufficiently cold core provides a cooling intensity that is high enough to miss the nose of the continuous cooling curve. This technique (also called mass quenching or autoquenching) allows a shallow case to be obtained without the use of a liquid quenchant (Ref 1). When higher frequencies and shorter heating times are used for larger parts (greater diameters or thicknesses), the heat sink or cold core effect is more pronounced, which increases the overall cooling severity.

To complicate matters further, in induction scan hardening applications, there is both radial cooling and axial (longitudinal) heat transfer. It is particularly important to take axial thermal conduction into consideration when low and moderate scan rates are applied. Additionally, a certain combination of coil geometry, applied frequency, and impingement angle of the quenchant can result in the presence of appreciable eddy current heating due to a residual electromagnetic field (electromagnetic end effect of the inductor) while the quenchant is being applied. In addition, latent heat, which takes place due to phase transformation during heating and quenching, should be taken into consideration in computer modeling.

Generally speaking, the process of part cooling during quenching can be easily simulated using a governing equation according to Eq 35 and boundary conditions according to Eq 36. The well-known saying "good things come in small packages" can be related to the ability to model the spray quenching process. The greatest difficulty in simulating spray quenching is the ability to properly quantify the surface conduction parameter, Q_s, in Eq 36, which has a substantially nonlinear nature and depends on a number of factors, including:

- Quenchant-related factors:
  a. Type of quenchants (water, polymer-based quenchants, forced air, oils, etc.)
  b. Pressure
  c. Temperature
  d. Concentration
  e. Cleanliness / contamination
  f. Flow
- Real-life process and design-related features:
  a. Integral quenching versus quench follower
  b. Number of quench holes (orifices) and their size
  c. Impingement angle
  d. Distance between quench block and workpiece surface
  e. Quenching time
  f. Part geometry

Over the years, manufacturers of induction heat treating machinery have accumulated proprietary knowledge in determining appropriate correlations that allow the determination of proper boundary conditions for induction hardening using spray quenching. Computer modeling can help to analyze the dynamics of induction scan hardening as a sequence of different process stages. As an example, Fig. 51 shows the results of computer modeling the sequential dynamics of induction scan hardening of a hollow shaft using a two-turn machined integral quench inductor with an L-shaped flux concentrator ring (frequency = 9 kHz) (Ref 109). At the beginning (Fig. 51a, b), a 2.6 s power dwell is applied to properly pre-heat the shaft fillet area. During this stage, an inductor is energized but does not move, and quenching is not applied. Upon completing the dwell stage, the shaft fillet is sufficiently preheated, and scanning begins. Scan rate and coil power are varied during scanning to allow proper accommodation of changes in shaft geometry. Computer modeling (FEM) reveals several important process subtleties:

- During scanning, appreciable heating of the shaft begins at a distance a good deal above the top copper turn, creating a preheating effect. Factors responsible for preheating are heat flow in the axial direction due to thermal conduction, propagation of the external magnetic field, and generation of heat sources outside of the induction coil.
- The presence of an external magnetic field outside the induction coil is also responsible for the postheating of shaft areas located immediately below the bottom turn and, in some cases, even in regions where the quenchant impinges the shaft surface. With insufficient quenching, the latter can dramatically reduce quenching severity and potentially create conditions for crossing the nose of the CCT curve, resulting in the formation of mixed structures with the presence of upper transformation products (e.g., bainitic/pearlitic structures or ghost networking).

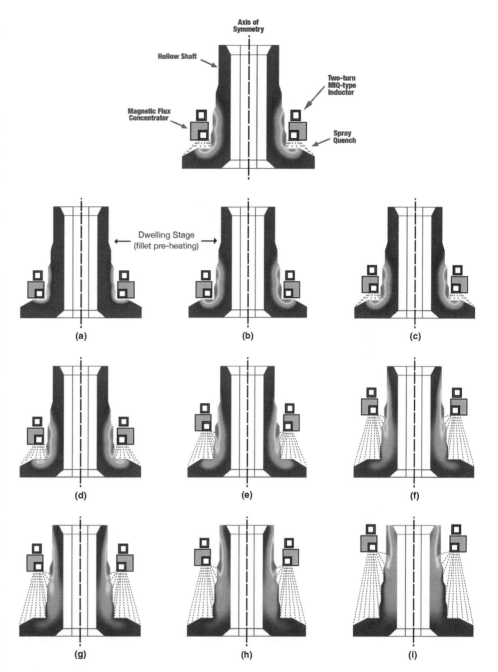

Fig. 51 Computer simulation of the sequential dynamics of induction scan hardening a hollow shaft using a two-turn machined integral quench (MIQ) inductor with an L-shaped magnetic flux concentrator ring (frequency = 9 kHz). See Fig. 52 and 53. Source: Ref 109

Such microstructures are notorious for scattering and lower hardness readings.
- Both the electromagnetic proximity effect and coil end effect (Ref 1) cause hot spots to appear on a shoulder near a shaft diameter change. During scanning, the magnetic field preferably couples to the shoulders, leading to a power density surplus at those locations. The presence of hot spots produced by power surplus necessitates having prolonged cooling to remove excessive heat, ensuring martensitic formation and obtaining

sufficient hardness at these locations. At the same time, a heat-source deficit could occur in the undercut region and transition area near the shaft smaller diameter.

Comet-Tail Effect. It is imperative to take into consideration the comet-tail effect when developing a scan hardening process recipe. Figures 52 and 53 show the magnified temperature pattern of an intermediate process stage (Fig. 51e, f). The comet-tail effect manifests itself as a heat accumulation in shaft subsurface

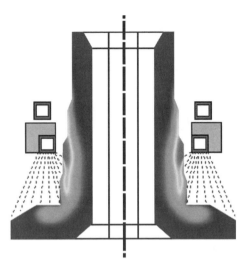

Fig. 52 Magnified temperature pattern of an intermediate process stage (Fig. 51e) of scan hardening a shaft, showing the comet-tail effect. Source: Ref 109

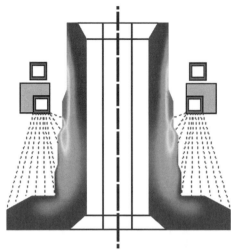

Fig. 53 The comet-tail effect manifests itself as a heat accumulation in shaft subsurface regions below the scan inductor (Fig. 51f). Source: Ref 109

Fig. 54 Single-shot induction hardening of shafts using a channel-type inductor. Courtesy of Inductoheat, Inc.

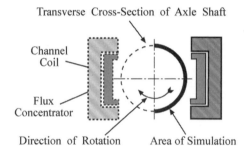

Fig. 55 Sketch of single-shot induction hardening of an axle shaft. Note: The right half of this induction system is computer modeled in Fig. 56

regions below the scan inductor, being pronounced in the areas of a diameter change. Upon quenching, the temperature of the shaft surface can be cooled sufficiently below the M_s temperature. At the same time, the heat accumulated in the shaft subsurface may be sufficient for tempering back of as-quenched surface regions, leading to soft spots.

Critical Issue with Computer Modeling of Scan Hardening. A limitation of the great majority of commercially available induction heating software is that they are not capable of taking into consideration a comet-tail effect when trying to model induction scan hardening. In addition, some software cannot properly handle pre- and postheating effects due to external magnetic field propagation and axial thermal conduction. These restrictions dramatically limit their use. When designing inductors and developing optimal process recipes, it is imperative to properly model both heating and spray quenching stages. Otherwise, crucial aspects of the process may be missed, having a negative impact on modeling accuracy and usefulness.

How to Implement a Part Rotation in a Simulation of Single-Shot Induction Hardening. Channel-type single-shot inductors are made of tubing or machined from solid copper, but unlike scanning inductors, they produce an axial rather than a circumferential current path (Fig. 54). Because the predominant current flow is axial, sufficient rotation must be used to produce a uniform case depth through the circumference of the part. Typically, the inductor legs have two horseshoe-shaped loops that join them, making a loop for current flow. Coil copper is often profiled. Laminations or other types of magnetic flux concentrators are often used along with profiling of the copper to provide the required pattern by accommodating the geometry of the heated component.

Single-shot inductors are the preferable choice when the workpiece has an irregular shape or varying diameters, shoulders, or fillets. Typically, quench blocks with a staggered hole pattern are directed at the part, with the intent to completely cover the part with quench fluid. Single-shot hardening is also the preferable choice when shorter heat times are desired. Normally, the heat time for single-shot hardening ranges from 2 to 20 s (with 3 to 8 s being the most typical).

Single-shot hardening systems introduce several challenges to computer modeling; some of them are related to inherent nonsymmetry of the heat-source distribution. As can be seen from Fig. 54 and 55, eddy currents are induced only in part areas located under the inductor legs. Heating of the rest of the part is delayed until the shaft rotates, bringing unheated areas in proximity to the inductor legs. Therefore, each elementary volume of the shaft is going through a repeatable heating, cooling, heating, cooling cycle. Insufficient part rotation (particularly when short heat times are used) may lead to an appreciably nonuniform temperature distribution along the shaft perimeter. Upon quenching, this could result in a nonuniform martensite formation and hardness distribution, leading to excessive shaft distortion and potentially not meeting the heat treating specification. To properly simulate such systems, in addition to the necessity for coupling electromagnetic and heat-transfer phenomena, it is also necessary to incorporate the part rotation into consideration. Accommodation of the shaft rotation may be challenging due to several factors. It requires developing a special procedure for dynamic mesh adaptation and repositioning finite elements that carry the effect of the shaft rotation. It is also necessary to constantly modify the changing initial temperature distribution when correspondent regions of the shaft are positioned in proximity to the respective leg

of a single-shot inductor. When simulating a heat-transfer phenomenon in such systems, it is necessary to take into account a forced thermal convection that occurs due to the shaft rotation. As an example, Fig. 55 shows the sketch of a single-shot induction hardening system. Taking advantage of symmetry, only the right side of such a system was modeled using finite-element analysis. Figure 56 shows the result of computer simulation of initial, interim, and final heating stages, taking into consideration the shaft rotation. Insufficient part rotation resulted in a nonuniform temperature distribution along the shaft perimeter (Fig. 56, left). Proper shaft rotation results in a sufficiently uniform temperature pattern (Fig. 56, right).

How to Properly Model Transient Billet Heating Processes. Progressive multistage horizontal heating is the most popular approach to heating metals prior to hot forming by forging, upsetting, rolling, extrusion, and other methods. Billets or bars are moved (via pusher, indexing mechanism, or walking beam, for example) through a multicoil horizontal

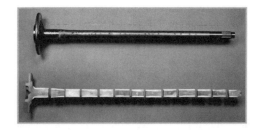

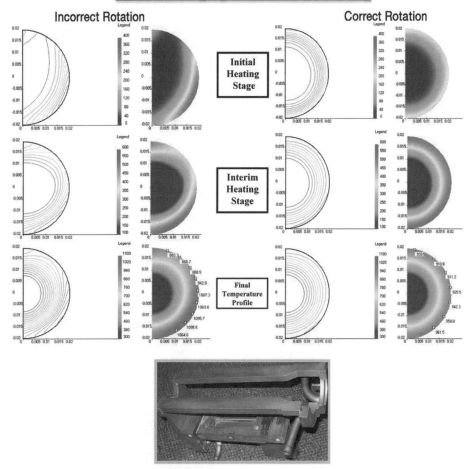

Fig. 56 Results of numerical simulation of heating an axle shaft by using a single-shot inductor. Courtesy of Inductoheat, Inc.

Fig. 57 In-line induction bar end heater. Courtesy of Inductoheat, Inc.

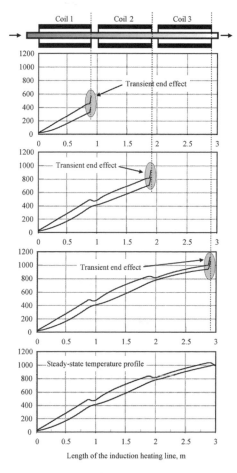

Fig. 58 Computer modeling of in-line induction heating of a 25.4 mm diameter titanium bar using 30 kHz at 50.8 mm/s demonstrates transient end effects. Source: Ref 1

induction heater (Fig. 57). As a result, the billet is sequentially (progressively) heated at predetermined positions inside of the induction heater. Progressive multistage horizontal heating is a popular choice for small- and medium-sized billets (usually less than 0.2 m in diameter). Several commercially available software allow simulating the steady-state condition of such systems. Most software predict surface-to-core temperature distribution at various stages but only take into consideration a steady-state condition. At the same time, a customer often requires a guarantee of surface-to-core as well as nose-to-tail temperature uniformity. Severe temperature nonuniformity may appear during transient stages that include

holding, start-up or shut-down stages. Figure 58 shows the results of computer modeling to predict the temperature profiles that appeared due to a transient end effect during in-line induction heating of 25.4 mm (1 in.) diameter titanium alloy bars traveling at a speed of 50.8 mm/s (2 in./s) and using a frequency of 30 kHz (Ref 1).

With increasing frequency or power density, the surplus of heat sources induced within the billet end area will increase as well. It should be mentioned here that the axial heat-source distribution and the temperature distribution along the bar-end-effect area are not the same. Due to the additional heat losses from the bar end and the soaking action of thermal conductivity,

the longitudinal temperature nonuniformity will not be as pronounced as the nonuniformity of the power density. As one may expect, due to the transient end effect and depending on the metal material properties and process parameters, the end of the bar can be not only overheated but underheated as well. For example, when heating magnetic bars to temperatures below the Curie point, the maximum temperature may occur at a distance of 40 to 75 mm (1.5 to 3 in.) from the butt end of the bar. The transient end effect in the bar-end zone is similar to the end effect during static heating (Ref 1); however, there are several features that make this process unique and must be taken into account when simulating such systems. One of the main features deals with the variation of coil current (in the case of constant coil voltage) or coil voltage (in the case of constant coil current) or both (in the case of constant or regulated coil power) when the leading end of the first bar or trailing end of the last bar moves through the induction coil.

In addition to the transient electromagnetic end effect, computer modeling should take into account the transient thermal effect that takes place when the leading end of the first bar moves through the inductor. During the start-up process, the transient thermal effect occurs due to the nonconstant temperature of the refractory. The start-up process begins with a cold refractory, or, in the case of an intermediate start, the refractory is partially heated. This results in increased heat losses from the bar surface due to increased thermal radiation and convection and requires a special control algorithm that provides an extra coil power to compensate for these losses. It is imperative to recognize the fact that besides the temperature

nonuniformity that occurs in the areas of the leading end of the first bar and trailing end of the last bar, there may be other situations that would result in the appearance of nose-to-tail temperature nonuniformity (Ref 1).

When bars, rods, or billets travel end-to-end through an induction heating line, the nose-to-tail temperature uniformity is typically not a problem. However, sometimes there is a 0.1 to 0.25 m (4 to 10 in.) axial air gap between the leading and following bars. The existence of such air gaps between bars or billets could result in noticeable temperature nonuniformity along the length of the bar as well. Obtaining the required temperature distribution along the length of long products that travel one after another with a certain axial air gap requires the ability to manage the electromagnetic end effects (Ref 1, 39 to 41, 48, 49, 110 to 114). This phenomenon is pronounced when heating large billets using horizontal coil arrangements (Fig. 59). Figure 60 shows deformation of the electromagnetic field when the billet is retracted from its final heating position inside of the induction coil. An improperly designed inductor and/or incorrectly chosen process recipe could lead to appreciable nose-to-tail temperature gradients. The complexity of transient processes worsens

due to variation of the coil electrical parameters (power, voltage, current, impedance) while it is empty, partially loaded, or fully loaded. Computer modeling helps to simulate such cumbersome phenomena, assisting in the proper design of induction heating systems and avoiding unpleasant surprises by providing not only radial but longitudinal temperature uniformity as well.

How to Properly Simulate Static Induction Heating of Large Billets Using the Vertical Coil Design Approach. In static heating, a billet or slag is placed into an induction coil having a vertical or horizontal arrangement for a given period of time, while a set amount of power is applied until it reaches the desired heating conditions (temperature and degree of uniformity). The heated billet is then extracted from the inductor and delivered to the metal-forming station. Another cold billet is then loaded into the coil, and the process repeats. Either approach (vertical or horizontal) can use a protective atmosphere if required. As discussed previously, progressive multistage horizontal heating is popular for small- and medium-sized billets (usually less than 0.2 m in diameter). When heating large-diameter steel or titanium billets, it is often advantageous to use static heating with a vertical coil arrangement or a combination of the progressive multistage horizontal method for preheating and the static vertical method for final heating (Ref 110). The four photos in Fig. 61 show a billet being discharged after static heating in a vertical inductor (Fig. 62). Billet transfer, tipping,

Fig. 59 Progressive multistage horizontal heating of large steel billets. Courtesy of Inductoheat, Inc.

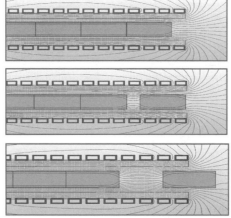

Fig. 60 Computer-modeled electromagnetic field distribution at exiting end of induction coil (transient electromagnetic end effect). Source: Ref 115

Fig. 62 Vertical inductor for static heating of large billets. Courtesy of IHWT, United Kingdom

Fig. 61 Discharge sequence for a steel billet after being heated in a static vertical inductor. Source: Ref 110

and charging mechanisms are located below the platform (Ref 1, 110).

When the heating cycle is completed, the control system checks whether the press is ready to accept the billet. If it is not, the inductor changes its mode from heating to holding, using lower coil power to compensate for heat losses from the workpiece surface. The advantages and disadvantages of the progressive multistage horizontal and static vertical billet heating approaches are given in Ref 110.

Several software are available on the market for modeling progressive multistage horizontal induction heaters in the steady-state condition. It is more difficult to find software appropriate for computer modeling of static vertical billet heaters. At first glance, the design of an induction vertical system for heating large billets may seem to be relatively straightforward. However, it is deceiving because, in reality, the vertical coil design may incorporate an even greater amount of subtleties and know-how than horizontal designs. To decide whether software is suitable for modeling vertical heaters, its capabilities must be thoroughly understood. The following is a checklist to be considered when evaluating potential software for modeling vertical induction billet heating systems:

- Electromagnetic end effect
- Thermal edge effects (Lambert's law and view factors, for example)
- Tight coupling of both electromagnetic and thermal processes. (Two-step coupling should be avoided because it does not provide the accuracy required for most applications; step-by-step coupling should be used instead.)
- Presence of thermal insulation (thermal refractory) and the ability to model heat exchange between the refractory and heated billet
- Nonlinear interrelated nature of physical properties
- Specifics of coil windings and copper tubing
- Surface heat losses due to thermal radiation and convection
- Presence of magnetic flux concentrators, shunts, diverters, and/or flux extenders
- Heat transfer between heated billet and pedestal
- Transient processes (cold and warm startup and the ability to hold the heated billet inside the induction coil if the forming machine is not ready to accept it)
- Possibility of the billet having a nonuniform initial temperature distribution prior to induction heating (for example, existence of radial and longitudinal temperature gradients after piercing or continuous casting)
- Cooling of the billet in air during its transfer from inductor to forming machine

As an example, Fig. 63 shows the computer-modeled dynamics of induction heating a Ti-6Al-4V titanium alloy billet in a static vertical inductor using a line frequency (60 Hz). The

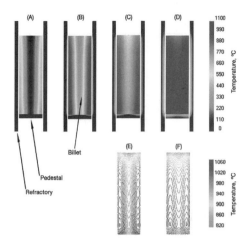

Fig. 63 Computer modeling of sequential vertical induction heating of a Ti-6Al-4V billet to provide a temperature profile having the bottom warmer than the top. Sequential heating of the thermal refractory is shown. The flux concentrator (located below the pedestal) and multiturn vertical induction coil are not shown. (a) Initial heating stage. (b, c) Intermediate stages. (d, e) End of heating. (f) After air cooling for 60 s during transfer from inductor to forming machine. Courtesy of Inductoheat, Inc.

study was conducted by Inductoheat personnel. The billet was 200 mm (7.8 in.) in diameter and 665 mm (26.2 in.) long. A cold startup was assumed, where the refractory insulation was initially at ambient temperature. The billet was positioned on a nonelectrically conductive pedestal. The stack of laminations acting as a magnetic flux concentrator was located below the pedestal. A longitudinal temperature gradient (heat profile) was desired, with the top of the billet being cooler. Distortion of the temperature profile upon billet transportation on air to the extrusion press is also shown. By adjusting the coil overhangs at top and bottom and the position of the flux concentrator, it is possible to obtain either a uniform temperature distribution or a reverse heat pattern, having the top of the billet warmer, which emphasizes flexibility as one of the advantages of static vertical billet heaters while heating various-sized billets.

How to Properly Simulate Induction Heat Treating of Model Gearlike Components. In contrast to carburizing and nitriding, induction hardening does not require heating the entire gear. With induction, heating can be localized to only those areas in which metallurgical changes are required (Ref 1, 28, 103, 106, 107, 116, 117). For example, the flanks, roots, and tips of gear teeth can be selectively hardened (Fig. 10). Gear performance characteristics (including load condition and operating environment) dictate the required surface hardness, core hardness, hardness profile, residual-stress distribution, grade of steel, and prior microstructure of the steel. External spur and helical gears, worm gears, and internal gears, racks, and sprockets are among those that typically are induction hardened (Fig. 64). A major goal of induction gear hardening is to provide a

Fig. 64 Induction-hardened gears and gearlike components. Courtesy of Inductoheat, Inc.

fine-grained martensitic layer on specific areas of the gear teeth.

Depending on the required hardness pattern and tooth geometry, gears are induction hardened by encircling the gear with a coil (so-called spin hardening) or, for larger gears, heating them tooth-by-tooth. Spin hardening is particularly appropriate for gears having fine- and medium-sized teeth. Gears are rotated during heating to ensure an even distribution of energy. When applying encircling coils, there are five parameters that play important roles in obtaining the required hardness pattern: frequency, power, cycle time, coil geometry, and quenching conditions. There are four popular heating modes used for the induction spin hardening of gears that employ encircling-type coils: conventional single-frequency, pulsing single-frequency, conventional dual frequency, and various simultaneous dual-frequency concepts (Ref 1, 106, 116). All four can be applied in either a single-shot or scanning heat treating approach. The choice of heating mode depends on the application and equipment cost.

Remember that hardening is a two-step process: heating and quenching. Both are important. There are three ways to quench gears in spin hardening applications:

- Submerge the gear in a quench tank. This technique is particularly applicable for large gears.
- Small- and medium-sized gears are usually quenched in-place, using an integrated spray quench.
- Use a separate, concentric spray quench block (quench ring) located below the inductor.

Complex shapes or gears and gearlike components present obvious challenges in modeling both heating and quenching stages. Most simulations were applied for modeling spur gears, which have the simplest geometry among gearlike components. For example, Fig. 65 to 68

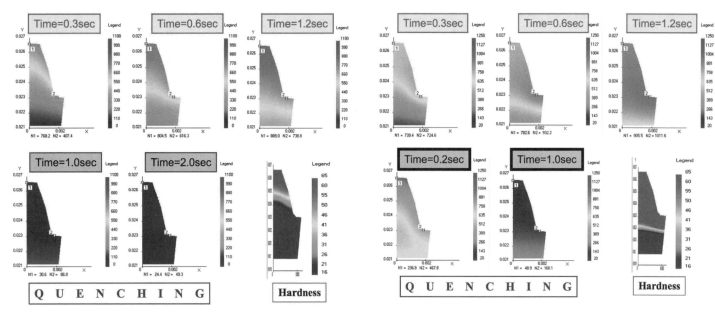

Fig. 65 Dynamics of induction hardening fine-pitched gears using a frequency of 300 kHz. Heat time is 1.2 s. Computer modeling data. Source: Ref 106

Fig. 66 Dynamics of induction hardening fine-pitched gears using a frequency of 30 kHz. Heat time is 1.2 s. Computer modeling data. Source: Ref 106

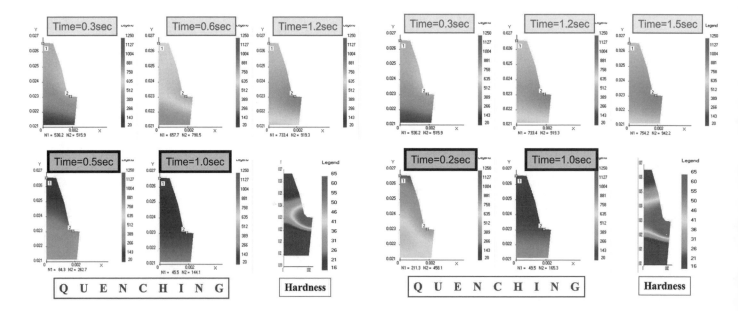

Fig. 67 Dynamics of induction hardening fine-pitched gears using a frequency of 10 kHz. Heat time is 1.2 s. Computer modeling data. Source: Ref 106

Fig. 68 Dynamics of induction hardening fine-pitched gears using a frequency of 10 kHz. Heat time is 1.5 s. Computer modeling data. Source: Ref 106

show the dynamics of temperature distribution during heating and quenching of a fine-pitch gear using various frequencies: radio frequency (300 kHz), moderate frequency (30 kHz), and low frequency (10 kHz), respectively. The FEM has been used to simulate this process. Maxwell's equations were solved in respect to magnetic field intensity, **H**. The heat-transfer phenomenon was also solved using finite elements applied to the two-dimensional form of Fourier Eq 35 with boundary conditions of Eq 36 and 37. Advantage was taken in respect to

symmetry of the gear tooth, allowing the modeling of only ½ of the tooth.

As expected, when a radio frequency of 300 kHz is applied, an eddy current induced in the gear follows the contour of the gear (Fig. 65). Because the highest concentration of current density will be in the tip of the tooth, there will be a power surplus in the tip compared to the root. Also, taking into account that the tip of the tooth has the minimum amount of metal to be heated, compared with the root, the tip will experience the most intensive

temperature rise during the heating cycle. In addition, from the thermal perspective, the amount of metal beneath the gear tooth root represents a much larger heat sink compared with that beneath the tooth tip. Another factor that contributes to more intensive heating of the tooth tip is a better electromagnetic proximity effect between the inductor and tooth tip versus its root. Higher frequency has the tendency to make a proximity effect more pronounced (Ref 1, 31, 116 to 119). These factors provide rapid austenitization of the tooth tip, which, upon quenching, produces a martensitic layer there. Note that the root area has not been hardened, because insufficient heat was generated for austenitization.

The next simulation was conducted using a moderate frequency, being tenfold lower (using 30 kHz). At the end of heating, the whole tooth was heated quite uniformly, achieving the austenitizing temperature (Fig. 66) and resulting in a through-hardened tooth. The study was continued by further reduction of frequency, using 10 kHz for heating these fine-toothed gears. With 10 kHz, the eddy current flow and temperature distribution in the gear tooth are quite different (Fig. 67). Keep in mind that the heat time in all three cases was the same: 1.2 s. A frequency reduction from 300 to 10 kHz noticeably increases the eddy current penetration depth in the hot steel—from 1 to 5.4 mm. (0.04 to 0.21 in.). In a fine-toothed gear, such an increase in current penetration depth results in a current cancellation phenomenon in the tooth tip and pitch line area when applying a frequency of 10 kHz. This makes it much easier for induced current to take a short path, following the base circle or root line of the gear instead of following along the tooth contour. The result is more intensive heating of the tooth root area compared with its tip as well as the development there of martensite upon quenching. In contrast to using 300 kHz, tips were not hardened

at all with 10 kHz. Notice the effect of a slight increase in heat time when using 10 kHz (compare Fig. 67 and 68).

The results of modeling support the experimentally obtained hardening patterns and confirm the previous explanation of the physics of the electrothermal processes that occur during induction spin hardening of gears using different frequencies (Fig. 10). It is important to remember that the terms *high frequency* and *low frequency* are not absolute. For example, depending on gear geometry, a frequency of 10 kHz may be considered low when heating fine-toothed gears but would be considered high when hardening large gears having coarse teeth. Similarly, a frequency of 300 kHz could act as a very low frequency for certain gear geometries (i.e., tall and skinny teeth).

As stated previously, because of the relative simplicity of the process, most computer simulations have been done to simulate induction hardening of spur gears using a single frequency. However, there are other types of gears, for example, bevel gears (Fig. 69), that present extra complexity in simulation and require three-dimensional modeling. There are a few software-developing companies on the market that provide three-dimensional codes. This includes Infolytica Corp., Integrated Engineering Software, Inc., ANSYS, Comsol, and MagSoft Corp. Figures 70 and 71 show

examples of three-dimensional simulations. Figure 70 shows a current distribution when heating the inside diameter of gears. Figure 71 shows the results of heating selected areas of a plate using a hairpin inductor. However, wide application of 3-D software in induction hardening is quite limited. Hopefully, in the future, three-dimensional software will find a wider use in the heat treating industry.

Quite often, to prevent problems such as pitting, spalling, tooth fatigue, and endurance and impact limitations, it is necessary to harden the contour of the gear or to have a so-called gear contour hardening (Fig. 72). This often maximizes beneficial compressive stresses within the case depth and minimizes distortion of as-hardened gears. Many times, obtaining a true contour-hardened pattern using a single frequency can be a difficult task due to the difference in current density (heat-source) distribution and heat-transfer conditions within a gear tooth. As discussed previously, there are two main factors complicating the task of obtaining the contour hardness profile. First, with encircling-type coils, the coupling at the gear tip has a better electromagnetic proximity with the inductor compared with that at the root area. Therefore, it is often more difficult to induce energy into the gear root. Second, there is a significant heat sink located under the gear root (below the base circle).

The pulsing single-frequency concept was developed as the first step in overcoming difficulties in obtaining contourlike hardness patterns (Ref 1, 28, 106, 116, 117). Power pulsing provides the desirable heat flow toward the root of the gear without appreciably overheating the tooth tip, enabling the attainment of the desired metallurgical structures and helping to provide a gear contourlike hardness pattern. Preheat times are typically from several seconds to a minute, depending on the size and shape of the gear teeth and its prior

Fig. 69 Bevel gear shape increases complexity of model

Fig. 70 Three-dimensional computer model of current density distribution when heating the inside diameter of a gear. Courtesy of Integrated Engineering Software, Inc.

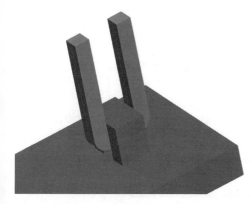

Fig. 71 Three-dimensional computer model of heating selected areas of a plate by using a hairpin inductor. Courtesy of Infolytica Corp.

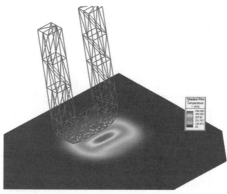

Fig. 72 Gear contour hardening. Courtesy of Inductoheat, Inc.

microstructure. Obviously, preheating reduces the amount of energy required in the final heat.

After preheating, there may be a soak time ranging from 2 to 10 s to achieve a more uniform temperature distribution across the teeth of the gear. Depending on the application, preheating can consist of several stages (preheat power pulses). Final heat times can range from less than 1 s to several seconds. In simulating induction heating using power pulsing, the final condition after soaking serves as a nonuniform initial thermal condition for the next power pulsing.

The pulsing dual-frequency concept (Ref 1, 28, 116, 117) is a further improvement in the attempt to obtain a contour hardness pattern and to minimize gear distortion. Two different power supplies are required. The idea of using two different frequencies to produce the desired contourlike pattern has been around since the late 1950s. Since then, several companies have pursued this idea, and various names and abbreviations have been used to describe it. The gear is induction preheated to a temperature determined by the process features, typically being 350 to 100 °C below the critical temperature A_{c1}. The preheat temperature depends on the type and size of the gear, tooth shape, prior microstructure, required hardness pattern, acceptable distortion, and the available power source. Usually, preheating is accomplished by using a medium frequency (3 to 10 kHz). A lower frequency results in greater eddy current penetration depths, which lead to a more in-depth preheating effect. A high frequency (30 to 450 kHz) and high power density are applied during the final heating stage. Selected frequency allows the eddy current to penetrate only to the desired depth.

Depending on the application, a single-coil design or two-coil arrangement (Fig. 73, top and bottom, respectively) can be used. A single-coil design has many limitations. Some of them are related to low reliability and maintainability of the induction coil. With the second-coil arrangement, one coil provides preheating, and another coil provides final heating. Both coils work simultaneously if the scanning mode is applied. In the case of a single-shot mode, a two-step index-type approach is used (Fig. 73, bottom).

In some cases, dual-frequency machines produce parts having lower distortion and more favorable distribution of residual stresses compared with single-frequency techniques. However, depending on tooth geometry, the time delay between low-frequency preheating and high-frequency final heating could have a detrimental negative effect on obtaining truly contour-hardened patterns or minimization of shape distortion.

Some induction practitioners have heard about simultaneous dual-frequency gear hardening, which uses two single-frequency inverters working on the same coil at the same time. Low frequency helps to austenitize the roots of the teeth, and high frequency helps to

austenitize the teeth flanks and tips. However, it is not advantageous to have two different frequencies working simultaneously all the time. Many times, depending on the gear geometry, it is preferable to apply a lower frequency at the beginning of the heating cycle, and, after achieving a desirable root heating, the higher frequency can complement the initially applied lower frequency, thereby completing a job by working together. Figure 74 shows a sketch of the circuitry of Inductoheat's single-coil dual-frequency gear hardening system. Figure 75 shows waveforms of coil current and coil

voltage when two appreciably different frequencies are applied at the same time to a single inductor. This machine was designed specifically for induction hardening of an internal wide-face, gearlike component having a minor gear diameter of 176 mm (6.9 in.) and a major gear diameter of 186 mm (7.3 in.) using a single-shot dual-frequency heating mode.

The total power exceeds 1200 kW, comprising medium-frequency (10 kHz) and high-frequency (120 to 400 kHz) modules working not just simultaneously but in any sequence

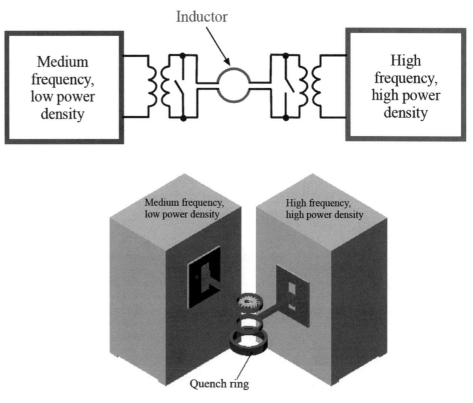

Fig. 73 Conventional dual-frequency induction hardening using a single inductor (top) and two inductors (bottom). Source: Ref 28

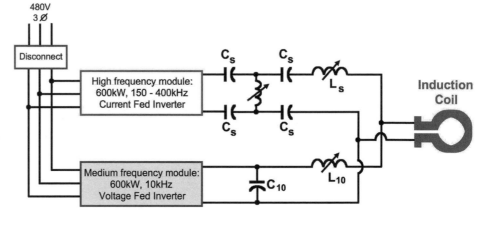

Fig. 74 Sketch of the circuitry of a single-coil dual-frequency inverter. Source: Ref 28

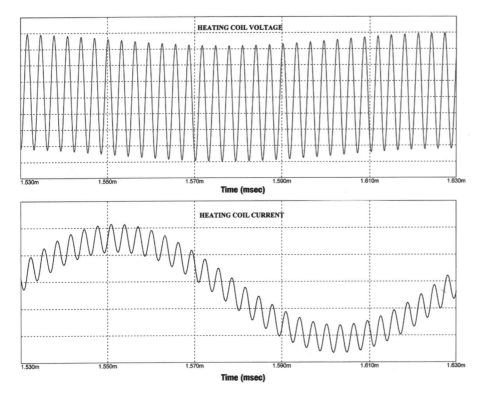

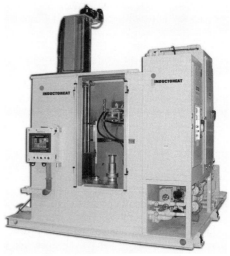

Fig. 76 Inductoheat's single-coil dual-frequency system comprises medium-frequency (10 kHz) and high-frequency (120 to 400 kHz) modules working simultaneously or in any sequence desirable to optimize properties of induction heat treated gearlike components. Total power exceeds 1200 kW. Source: Ref 28

Fig. 75 Waveforms of coil current and coil voltage of a single-coil dual-frequency induction hardening technology. Source: Ref 28

Fig. 77 Snakeskin hardness pattern. Source: Ref 1

desirable to optimize properties of the gears heat treated using this technology. Total heat time was minimized to approximately 1.5 s. At the beginning of the heating cycle, a medium frequency is applied for 0.8 s, providing the required root heating. For the remainder of the heat cycle, two frequencies were working together, complementing each other. Figure 76 shows Inductoheat's two-frequency induction gear hardening system.

At this point, computer simulation of gears and gearlike components using the dual-frequency approach represents one of the most difficult tasks in developing computer modeling software. It is hoped that such software will be available in the near future.

Computer Modeling of Barber-Pole Phenomenon. Heat treat practitioners sometimes observe unusual effects in induction hardening, such as the striping phenomenon, barber-pole effect, fish-tail effect, soft spotting, snakeskin effect, and others (Ref 1, 41, 46, 120, 121). As an example, Fig. 77 and 78 show the appearance of the snakeskin and striping phenomena, correspondingly. These phenomena have never been obtained by computer modeling and have been viewed only in practical applications or in laboratory experiments. Some of them are considered to be mysterious phenomena. For example, certain forms of striping phenomena suddenly occur and then disappear. There is no seemingly single explanation of this phenomenon.

Experience shows that striping can appear in several different ways. However, in the majority of cases, very narrow bright stripes (rings) appear at the beginning of the heating cycle (Fig. 78). Over time, the narrow stripes widen. At this stage, the maximum temperature moves from the center of each ring toward the edges of each bright-hot ring. During the heating process, the stripes sometimes move back and forth along the workpiece surface area under the coil. With longer heating cycles, the striping phenomenon usually disappears.

It has been observed that the appearance of stripes depends on a complex function of the frequency, magnetic field intensity, and thermal, electrical, and magnetic properties and structure of the steel. It seemingly occurs only when relatively high power densities are applied. If the power density is relatively low, the temperature will equalize between the neighboring bright (high-temperature) and dark (low-temperature) rings because of the thermal conductivity of the steel. Regardless of several attempts (Ref 1, 41, 46, 120, 121) to provide postulated explanations of these phenomena, it would be beneficial if some of those mysterious phenomena could be computer modeled.

Conclusion

It is important to recognize that the use of modern numerical software (including finite elements, finite differences, boundary elements, etc.) does not guarantee obtaining correct computational results. It must be used in conjunction with

Fig. 78 Multiple stripes appear due to a striping phenomenon. (Note that the induction coil has only four turns.) Source: Ref 121

experience in numerical computations and engineering knowledge to achieve the required accuracy of mathematical simulation. This is especially so because even in modern commercial software, regardless of the amount of testing and verification, a computation program may never have all of its possible errors detected. The engineer must consequently be on guard against various types of possible errors. The more powerful the software, the more complex it is, with potentially having the greater probability of errors. Common sense and an engineering "gut feeling" are always the analyst's helpful assistants. Computer modeling provides the ability to predict how various factors may impact the transient and final heat treating conditions of the workpiece, load matching, and what must be accomplished in the design of the induction heating system to improve the effectiveness of the process and guarantee the desired heating results and its load matching capability (Ref 1, 122).

In the fast-paced global economy, the ability of induction heating manufacturers to minimize the development time and shorten the learning curve through efficient computer modeling is critical for a company's success. By combining the most advanced software with a computational and engineering background, modern induction heating specialists possess the unique ability to analyze in a few hours complex induction heating problems that could take days or even weeks to solve using other methods, including physical experiment. This leads to the saving of prototyping dollars and facilitates the building of reliable, competitive products in a short design cycle.

Recent advances in developing fast-performing computers with a substantial memory, speed of computation, and hard-drive capacity make it increasingly critical to develop effective numerical simulation software to be used as a part of the process control and quality-assurance systems, thereby assuring the achievement of optimal process parameters and system performance.

Experience gained on previous jobs and the ability to computer model induction processes provide a comfort zone when designing new induction heating systems. This combination of advanced software and a sophisticated engineering background enables manufacturers of induction heating equipment to quickly determine details of the process that could be costly, time-consuming, and, in some cases, extremely difficult, if not impossible, to determine experimentally.

REFERENCES

1. V. Rudnev, D. Loveless, R. Cook, and M. Black, *Handbook of Induction Heating,* Marcel Dekker, 2003, p 800
2. S.L. Semiatin and D.E. Stutz, *Induction Heat Treatment of Steel,* American Society for Metals, 1986
3. *Heat Treating,* Vol 4, *ASM Handbook,* ASM International, 1991
4. Ch. Brooks, *Principles of the Heat Treatment of Plain Carbon and Low Alloy Steels,* ASM International, 1996
5. G. Krauss, *Steels: Heat Treatment and Processing Principles,* ASM International, 2005
6. M.A. Grossman and E.C. Bain, American Society for Metals, 1964
7. J.R. Davis, Ed., *Metals Handbook Desk Edition,* 2nd ed., ASM International, 1998
8. G. Totten and M. Howes, *Steel Heat Treatment Handbook,* Marcel Dekker, 1997
9. R.W.K. Honeycombe and H.K.D.H. Bhadeshia, *Steels: Microstructure and Properties,* Arnold, Great Britain, 1995
10. V. Rudnev, Subject-Oriented Assessment of Numerical Simulation Techniques for Induction Heating Applications, *Int. J. Mater. Prod. Technol.,* Vol 29 (No. 1–4), 2007, p 43–51
11. *Heat Treater's Guide: Practices and Procedures of Irons and Steels,* ASM International, 1999
12. *Properties and Selection: Nonferrous Alloys and Pure Metals*, Vol 2, *Metals Handbook*, 9th ed., American Society for Metals, 1979
13. *Atlas of Isothermal Transformation and Cooling Transformation Diagrams,* American Society for Metals, 1977
14. G. Vander Voort, *Atlas of Time-Temperature Diagrams for Irons and Steels,* ASM International, 1991
15. V. Rudnev, Metallurgical Insights for Induction Heat Treaters, Part 5: Super-Hardening Phenomenon, Professor Induction Series, *Heat Treat. Prog.,* Sept 2008
16. V. Rudnev, Metallurgical Insights for Induction Heat Treaters, Part 4: Obtaining Fully Martensitic Structures Using Water Spray Quenching, Professor Induction Series, *Heat Treat. Prog.,* March/April 2008
17. V. Rudnev, Metallurgical Insights for Induction Heat Treaters, Part 3: Limitations of TTT and CCT Diagrams, Professor Induction Series, *Heat Treat. Prog.,* Nov/Dec 2007
18. V. Rudnev, Metallurgical Insights for Induction Heat Treaters, Part 2: Spray Quench Subtleties, Professor Induction Series, *Heat Treat. Prog.,* Aug 2007
19. V. Rudnev, Metallurgical Insights for Induction Heat Treaters, Part 1: Induction Hardening Temperatures, Professor Induction Series, *Heat Treat. Prog.,* May/June 2007
20. V. Rudnev, Metallurgical "Fine Points" of Induction Heating, Part 1, *Ind. Heat.,* March 2005
21. V. Rudnev, Metallurgical "Fine Points" of Induction Heating, Part 2, *Ind. Heat.,* May 2005
22. "Methods of Measuring Case Depth," J423, Society of Automotive Engineers, Feb 1998
23. R.A. Grange, C.R. Hribal, and L.F. Porter, Hardness of Tempered Martensite in Carbon and Low-Alloy Steels, *Metall. Trans. A,* Vol 8, 1977
24. D.K. Bullens, *Steel and Its Heat Treatment,* Wiley, New York, 1948
25. E.C. Rollason, *Fundamental Aspects of Molybdenum in Transformation of Steel,* Climax Molybdenum Co., London
26. V. Rudnev, Systematic Analysis of Induction Coil Failures, Part 3: Coil Copper Selection, Professor Induction Series, *Heat Treat. Prog.,* Nov/Dec 2005, p 19–20
27. V. Rudnev, Systematic Analysis of Induction Coil Failures, Part 11b: Frequency Selection, Professor Induction Series, *Heat Treat. Prog.,* Sept/Oct 2007, p 23–25
28. V. Rudnev, Single-Coil Dual-Frequency Induction Hardening of Gears, Professor Induction Series, *Heat Treat. Prog.,* Oct 2009
29. V. Rudnev, D. Loveless, C. Ribeiro, and J. Boomis, Unleashing a Superior Induction Heating Design with Computer Modeling, *Ind. Heat.,* Aug 2009
30. W. Feuerstein and W. Smith, *Trans. ASM,* Vol 46, 1954, p 1270
31. V. Rudnev, R. Cook, D. Loveless, and M. Black, Induction Heat Treatment: Basic Principles, Computation, Coil Construction and Design, Ch 4, *Steel Heat Treatment: Equipment and Process Design,* G. Totten, Ed., CRC, 2007
32. A.E. Slukhotskii and S.E. Ryskin, *Inductors for Induction Heating,* Energy Publ., St. Petersburg, Russia, 1974
33. L. Landau and E. Lifshitz, *Theoretical Physics,* Vol 8, *The Electrodynamics of Continuous Media,* Nauka, Moscow, 1992
34. R. Feynman, R. Leighton, and M. Sands, *The Feynman Lectures on Physics,* Vol 2, Addison-Wesley Publ., 1964
35. K.J. Binns, P.J. Lawrenson, and C.W. Trowbridge, *The Analytical and Numerical Solution of Electric and Magnetic Fields,* John Wiley & Sons, Ltd., 1992, p 470
36. P. Hammond, *Electromagnetism for Engineers,* Pergamon, New York, 1978
37. M.A. Plonus, *Applied Electromagnetics,* McGraw-Hill, New York, 1978
38. S. Gurevich, A. Vasiliev, and B. Polevodov, General Tendencies of Mathematical Modeling of Induction and Dielectric Heating, *Electrotechnica,* Vol 8, 1982, p 37–40
39. V. Nemkov and V. Demidovich, *Theory of Induction Heating,* Energoatomizdat, St. Petersburg, Russian, 1988
40. V. Rudnev, "Mathematical Simulation and Optimal Control of Induction Heating of Large-Dimensional Cylinders and Slabs," Ph.D. thesis, St. Petersburg El. Eng. Univ., Russia, 1986
41. V. Rudnev, D. Loveless, R. Cook, and M. Black, Induction Heat Treatment, Ch 11A, *Steel Heat Treatment Handbook,* Marcel Dekker, 1997, p 810
42. V. Rudnev, Systematic Analysis of Induction Coil Failures, Part 14: Split-Return Inductors and Butterfly Inductors,

Professor Induction Series, *Heat Treat. Prog.,* March/April 2009, p 17–19

43. V. Rudnev, An Objective Assessment of Magnetic Flux Concentrators, Professor Induction Series, *Heat Treat. Prog.,* Nov/Dec 2004, p 19–23

44. V. Rudnev, Magnetic Flux Concentrators: Myths, Realities and Profits, *Metal Heat Treat.,* March/April 1995

45. L. Neiman, *Skin Effect in Ferromagnetic Bodies,* Gosenergoizdat, St. Petersburg, Russia, 1949

46. M.G. Lozinskii, Industrial Applications of Induction Heating, Pergamon Press, London, 1969

47. V. Rudnev, Successful Induction Heating of RCS Billets, *Forge Mag.,* July 2008

48. V. Rudnev and D. Loveless, Longitudinal Flux Induction Heating of Slabs, Bars, and Strips Is No Longer "Black Magic," Part 2, *Ind. Heat.,* March 1995

49. V. Rudnev, Question and Answer Regarding Induction Heating Large Billets, Professor Induction Series, *Heat Treat. Prog.,* Jan 2009

50. B. Gebhart, *Heat Transfer,* McGraw-Hill, New York, 1970

51. S. Patankar, *Numerical Heat Transfer and Fluid Flow,* Hemisphere, New York, 1980

52. F.P. Incropera and D.P. Dewitt, *Fundamentals of Heat Transfer,* Wiley, New York, 1981

53. J.A. Adams and D.F. Rogers, *Computer Aided Analysis in Heat Transfer,* McGraw-Hill, New York, 1973

54. W.M. Rohsenow and J.P. Hartnett, *Handbook of Heat Transfer,* McGraw-Hill, New York, 1973

55. E. Rapoport and Y. Pleshivtseva, *Optimal Control of Induction Heating Processes,* CRC Press, 2007, p 349

56. F. Dughiero and M. Forzan, Electromagnetic Thermal Metallurgical Models for Induction Hardening, *Proc. Int. Scientific Colloquium Modeling for Electromagnetic Processing,* Hannover, 2008, p 117–118

57. A. Muhlbauer, Short Historical Overview of Induction Heating and Melting, *Proc. Int. Symposium on Heating by Electromagnetic Sources (HES-04)* (Padua, Italy), June 2004

58. B. Nacke, Recent Development in Induction Heating and Future Prospects in Industrial Applications, *Proc. Int. Symposium on Heating by Electromagnetic Sources (HES-07)* (Padua, Italy), June 2007

59. E. Baake, H. Schulbe, and A. Nikanorov, Alternative Processing for Laminar Compound Materials Based on Induction Heating, *Proc. Int. Symposium on Heating by Electromagnetic Sources (HES-07)* (Padua, Italy), June 2007

60. S. Lupi, M. Fozran, and A. Aliferov, Characteristics of Installations for the Direct Resistance Heating of Ferromagnetic Bars

of Square Cross-Section, *Proc. Int. Scientific Colloquium Modeling for Electromagnetic Processing (MEP-08),* Hannover, 2008, p 43–51

61. Integrated Engineering Software, Inc., www.integratedsoft.com

62. Infolytica Corp., www.infolytica.com

63. MagSoft Computer and Software, www.magsoft.com

64. Multiphysics Modeling and Simulation, COMSOL, Inc., www.comsol.com

65. Vector Fields—Software for Electromagnetic Design, Cobham Technical Services, www.vectorfields.com

66. ANSYS, Inc., www.ansys.com

67. Ansoft Electronic Design Products, ANSYS, Inc., www.ansoft.com

68. ESI Group, www.esi-group.com

69. S.-H. Kang and Y.-T. Im, Three-Dimensional Finite Element Analysis of the Quenching Process of Plain Carbon Steel with Phase Transformation, *Metall. Mater. Trans. A,* Vol 36, Sept 2005, p 2315–2325

70. R. Thomas et al., Analytical/Finite Element Modeling and Experimental Verification of Spray-Cooling Process in Steel, *Metall. Mater. Trans. A,* Vol 29, May 1998, p 1485–1498

71. J. Grum, Overview of Residual Stresses after Induction Surface Hardening, *Int. J. Mater. Prod. Technol.,* Vol 29 (No. 1–4), 2007, p 9–42

72. J. Grum and V. Rudnev, Induction Heating & Hardening & Welding, *Int. J. Mater. Prod. Technol.,* Interscience Publishers, Geneva, Switzerland, 2006

73. G. Smith, *Numerical Solution of Partial Differential Equations: Finite Difference Methods,* 3rd ed., Oxford University Press, Oxford, U.K., 1985

74. A.A. Samarskii, *Theory of Finite Difference Schemes,* Moscow, Russia, 1977

75. S.V. Patankar and B.R. Baliga, A New Finite-Difference Scheme for Parabolic Differential Equations, *Num. Heat Transf.,* Vol 1, 1978, p 27–30

76. J.C. Strikwerda, *Finite Difference Schemes and Partial Differential Equations,* Wadsworth & Brooks, Belmont, CA, 1989

77. M.V.K. Chari, "Finite Element Analysis of Nonlinear Magnetic Fields in Electric Machines," Ph.D. dissertation, McGill University, Montreal, PQ, Canada, 1970

78. M.V.K. Chari and P.P. Silvester, Finite Element Analysis of Magnetically Saturated DC Machines, *IEEE Trans. PAS,* Vol 90, 1971, p 2362

79. J. Donea, S. Giuliani, and A. Philippe, Finite Elements in Solution of Electromagnetic Induction Problems, *Int. J. Num. Methods Eng.,* Vol 8, 1974, p 359–367

80. W. Lord, Application of Numerical Field Modeling to Electromagnetic Methods of

Nondestructive Testing, *IEEE Trans. Mag.,* Vol 19 (No. 6), 1983, p 2437–2442

81. W. Lord, "Development of a Finite Element Model for Eddy Current NDT Phenomena," Electrical Engineering Dept., Colorado State University, 1983

82. P.P. Silvester and R.L. Ferrari, *Finite Elements for Electrical Engineers,* Cambridge University Press, New York, 1983

83. D.A. Lowther and P.P. Silvester, *Computer Aided Design in Magnetics,* Springer, Berlin, 1986

84. W. Lord, Y.S. Sun, S.S. Udpa, and S. Nath, A Finite Element Study of the Remote Field Eddy Current Phenomenon, *IEEE Trans. Mag.,* Vol 24 (No. 1), 1988, p 435–438

85. A. Muhlbauer, S. Udpa, V. Rudnev, and A. Sutjagin, Software for Modeling Induction Heating Equipment by Using Finite Elements Method, *Proc. Tenth All-Union Conf. High-Frequency Application* (St. Petersburg, Russia), Part 1, 1991, p 36–37

86. O.C. Zienkiewicz and R.L. Taylor, *The Finite Element Method: Basic Formulation and Linear Problems,* Vol 1, 4th ed., McGraw-Hill, New York, 1989

87. C.A. Brebbia, *The Boundary Element Method for Engineers,* Pentech Press, London, 1978

88. Y.B. Yildir, "A Boundary Element Method for the Solution of Laplace's Equation in Three-Dimensional Space," Ph.D. thesis, University of Manitoba, Winnipeg, Canada, 1985

89. T. Inuki and S. Wakao, Novel Boundary Element Analysis for 3-D Eddy Current Problems, *IEEE Trans. Mag.,* Vol 29 (No. 2), 1993, p 1520–1523

90. V. Rudnev, Spray Quenching in Induction Hardening Applications, *J. ASTM Int.,* Vol 6 (No. 2), 2009

91. Y. Yildir, B. Klimpke, and D. Zheng, *A Computer Program for 2D/RS Eddy Current Problem,* Integrated Engineering Software, Inc., 1993

92. D. Zheng and K. Davey, Pushing the Limits of 2-D Boundary Element Eddy Current Codes—Connectivity, *Int. J. Num. Model.: Electronic Networks, Devices and Fields,* Vol 9, 1996, p 115–124

93. B. Klimpke, "A Hybrid Magnetic Field Solver Using a Combined Finite Element/Boundary Element Field Solver," The U.K. Magnetic Conference Advanced Electromagnetic Modeling and CAD for Industrial Applications, Feb 19, 2003

94. *User Manual,* Integrated Engineering Software, Inc., 2003

95. R. Dudley and P. Burke, The Prediction of Current Distribution in Induction Heating Installations, *IEEE Trans. Ind. App.,* Vol 9 (No. 5), 1972

96. E. Kolbe and W. Reiss, Eine Methode zur Numerischen Bestimmung der Stromichteverteilung, *Wiss. Z.. Hochshule*

Elektrotechnik, Ilmenau, Germany, Vol 9 (No. 3), 1963

97. O. Tozoni, *Calculation of the Electromagnetic Fields Using Computers,* Kiev, Ukraine, 1967

98. F. Grover, *Inductance Calculations Working Formulas and Tables,* Dover, New York, 1946

99. V. Demidovich and V. Nemkov, Computation of Induction Heating of Non-Magnetic Cylinders, *Proceedings of VNIITVCh: Industrial Applications of High Frequency Currents,* Vol 15, 1975

100. P. Kalantarov and L. Tscheitlin, *Inductance Calculations,* Energoatoizdat, St. Petersburg, Russia, 1986

101. "IEEE Standard for Safety Levels with Respect to Human Exposure to Radio Frequency Electromagnetic Fields," C95.1-2005

102. V. Rudnev, R. Cook, and D. Loveless, Keeping Your Temper with Flux Concentrators, *Mod. Appl. News,* Nov 1995

103. G. Doyon, D. Brown, and V. Rudnev, Induction Heating Helps Put Wind Turbines in High Gear, *Heat Treat. Prog.,* Sept 2009, p 55–58

104. V. Rudnev, A Common Misassumption in Induction Hardening, Professor Induction Series, *Heat Treat. Prog.,* Sept/Oct 2004, p 23–25

105. P.G. Simpson, *Induction Heating: Coil and System Design,* McGraw-Hill, New York, 1960

106. V. Rudnev, Spin Hardening of Gears Revisited, Professor Induction Series, *Heat Treat. Prog.,* March/April 2004, p 17–20

107. V. Rudnev et al., Progress in Study of Induction Surface Hardening of Carbon Steels, Gray Irons and Ductile (Nodular) Irons, *Ind. Heat.,* March 1996

108. G. Doyon, D. Brown, and V. Rudnev, Taking Crank Out Crankshaft Hardening, *Ind. Heat.,* Dec 2008

109. V. Rudnev, Q & A Regarding Quenching and Case Hardening of Solid Shafts vs. Hollow Shafts, Professor Induction Series, *Heat Treat. Prog.,* Sept 2009, p 29–32

110. V. Rudnev, How Do I Select Inductors for Billet Heating, Professor Induction Series, *Heat Treat. Prog.,* May 2008

111. P. Ross and V. Rudnev, Innovative Induction Heating of Oil Country Tubular Goods, *Ind. Heat.,* May 2008

112. V. Rudnev and D. Loveless, Induction Slab, Plate and Bar Edge Heating for Continuous Casting Lines, *33 Metal Producing,* Oct 1994

113. V. Rudnev and D. Loveless, Longitudinal Flux Induction Heating of Slabs, Bars, and Strips Is No Longer "Black Magic," Part 1, *Ind. Heat.,* Jan 1995

114. V. Rudnev et al., Intricacies of Successful Induction Heating of Steels for in Modern Forge Shops, *Proc. 19th Int. Forging Congress* (Chicago, IL), 2008

115. G. Doyon, D. Brown, V. Rudnev, and Ch. Van Tyne, Ensuring Superior Quality of Inductively Heated Billets, *Forge,* April, 2010

116. V. Rudnev, Induction Hardening of Gears and Critical Components, Part 1, *Gear Technol.,* Sept/Oct 2008

117. V. Rudnev, Induction Hardening of Gears and Critical Components, Part 2, *Gear Technol.,* Nov/Dec 2008

118. V. Rudnev, Systematic Analysis of Induction Coil Failures, Part 2: Effect of Current Flow on Crack Propagation, Professor Induction Series, *Heat Treat. Prog.,* Sept/Oct 2005, p 33–35

119. V. Rudnev, Systematic Analysis of Induction Coil Failures, Part 13: Electromagnetic Proximity Effect, Professor Induction Series, *Heat Treat. Prog.,* Oct 2008, p 23–26

120. V. Rudnev, Metallurgical Insights for Induction Heat Treaters, Part 7: Barber-Pole, Snake-Skin, and Fish-Tail Phenomena, Professor Induction Series, *Heat Treat. Prog.,* May 2009

121. V. Rudnev, Metallurgical Insights for Induction Heat Treaters, Part 6: Striping Phenomenon, Professor Induction Series, *Heat Treat. Prog.,* Nov/Dec 2008, p 21–22

122. D. Loveless, R. Cook, and V. Rudnev, Induction Heat Treatment: Modern Power Supplies, Load Matching, Process Control, and Monitoring, Ch 5, *Steel Heat Treatment: Equipment and Process Design,* G. Totten, Ed., CRC, 2007

ASM Handbook, Volume 22B, *Metals Process Simulation*
D.U. Furrer and S.L. Semiatin, editors

Copyright © 2010, ASM International®
All rights reserved.
www.asminternational.org

Modeling of Quenching, Residual-Stress Formation, and Quench Cracking

Ronald A. Wallis, Wyman Gordon Forgings

OVER THE PAST TWO DECADES, the use of computer modeling to predict the mechanical properties and stresses in heat treated components has increased significantly (Ref 1). The advent of low-cost, relatively high-speed computers helped in this progress, but the developments occurred primarily as a result of the need for manufacturers to become more competitive and because customers demanded lower-cost components. This is particularly true when dealing with alloys containing high levels of expensive elements, such as nickel and titanium. The driving force behind this work is therefore the need to produce a component that meets the customers' specification the first time. Avoiding multiple trials to develop a heat treatment practice that produces the properties needed, avoids quench cracking, and limits the distortion during quenching or machining saves both time and money.

To be a useful tool for industry, the models must be accurate. Several commercial finite-element computer codes are now available that incorporate the necessary theoretical formulations and material models to calculate transient temperatures and stresses in a part during heat treatment. Although there are still improvements being made to the computer codes, the shortcoming at present is in generating data for the models and confirming that the model/material data combination is producing accurate results. It is obvious that the accuracy obtained from such models depends to a large extent on the data input. Some of these data are relatively easy to obtain (from literature and commercial testing laboratories), but other data are difficult to obtain and have required the development of specialized testing methods. As one would expect, the sequence in terms of application followed the availability or ease of developing the data for the models, and this first led to the calculation of transient temperature distributions in the components being heat treated. This enabled the determination of cooling rates within a part, which, in some alloy systems, correlates to mechanical properties, such as tensile and creep strength. Progress in phase transformation models also allowed the prediction of hardness in steel components.

Thermal models were quickly followed by the calculation of stresses in the part, which was used to help design processes to reduce distortion during heat treatment. Such models also led to a better understanding of the residual stresses in the final component, allowing their use to assist in developing processes with reduced distortion during subsequent machining. Thermomechanical models have also evolved to the point where computer modeling can be used to develop techniques to help solve quench cracking problems.

Prediction of Transient Temperatures in a Part

Heat-Transfer Boundary Conditions

To model the heat transfer in a component being cooled, accurate boundary conditions must be applied to the surface of the part. The heat extraction from the part is usually quantified as a heat-transfer coefficient (HTC), which gives the heat-removal rate from the surface as a function of the surface temperature. Heat-transfer data for one type of quenching oil is shown in Fig. 1 (Ref 2) and follows the typical trend for boiling heat transfer. At high temperatures (above approximately 850 °C, or 1560 °F), film boiling occurs, and the vapor blanket formed on the surface of the component limits heat transfer. As the surface temperature falls, transitional boiling heat transfer occurs, and the vapor film becomes unstable. The heat extraction rate in this region is significantly higher. Nucleate boiling, characterized by the rapid formation of bubbles, appears to take place between approximately 480 and 400 °C (895 and 750 °F). The rate of bubble generation decreases with a decrease in surface temperature, and hence, the heat extraction rate over this temperature range decreases rapidly. At lower temperatures (below approximately 400 °C, or 750 °F), the heat transfer is mainly by convection.

To be able to predict accurate temperatures in the component, it is critically important to

quantify the way that the heat is extracted from the part, whether from quenching in a liquid bath, by liquid sprays, or from forced air/gas cooling. In the case of quenching in a liquid bath, the type of liquid and agitation has an effect on the cooling rate and hence the HTCs (Ref 3). The same quench oil may exhibit different characteristics from tank to tank when different agitation methods are employed. Polymer quenchants have also been shown to exhibit significant variation in cooling rates depending on the level of agitation (Ref 3).

Generally, the HTCs for a specific medium and mechanical cooling configuration (tanks, sprays, fans, etc.) are obtained from tests on parts outfitted with thermocouples. The time-temperature data obtained at locations within a part are used in a computer program to determine the heat extraction rate needed on the surface to get that cooling rate profile. Correct placement of the thermocouples in such an experiment is critical to obtaining accurate data. If only the heat transfer at a small area in the center of a disk is being considered (Fig. 2a), the system may be treated as one dimensional, and the instrumentation is relatively easy, with only a line of thermocouples needed. The size of the disk must be considered, however, to prevent the heat transfer from the edges becoming a source of error. If it is

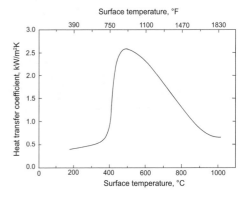

Fig. 1 Typical heat-transfer coefficients for oil quenching. Source: Ref 2

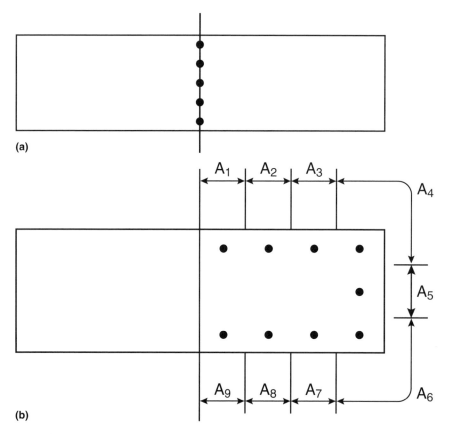

Fig. 2 Typical thermocouple arrangement in disks for determination of heat-transfer coefficients when heat transfer is considered (a) one dimensional and (b) two dimensional. ● = thermocouple location; A_n = surface area typically associated with thermocouple and for which the heat flux/heat-transfer coefficient is obtained

required to obtain the variation of the HTCs around a part, the experimental system quickly becomes more complicated. In the simple case of a disk, a series of thermocouples may be placed to cover the perimeter (Fig. 2b), but the number of measurement points and their distance from the surface and the corners must be carefully considered to obtain accurate data. This is an area of ongoing research, and to date, there are no guidelines for the design of such experiments.

The positioning of thermocouples in more complex shapes, such as turbine disks, has been done, but it requires considerable care and is relatively expensive. In some cases, researchers have applied thermocouples on the surface of a part (Fig. 3a) (Ref 4). However, in general, these thermocouples are not as robust and are prone to failure during testing, particularly in a manufacturing environment. There is also a concern that they may not be measuring the true surface temperature but some temperature intermediate between the part surface and the cooling medium (Ref 5). It is more common to instrument the part with subsurface thermocouples, as shown in Fig. 3(b) (Ref 6), but some studies have shown that this technique also has sources of temperature measurement errors, although they are considered to be much smaller (Ref 7, 8). It should be noted that the variation of the HTC around the part was not reported in either of the last two examples.

In industry, the calculation of HTCs has generally been obtained either by trial and error or by inverse engineering techniques. In either case, the surface of a part will be divided into zones, and normally, each zone will have a

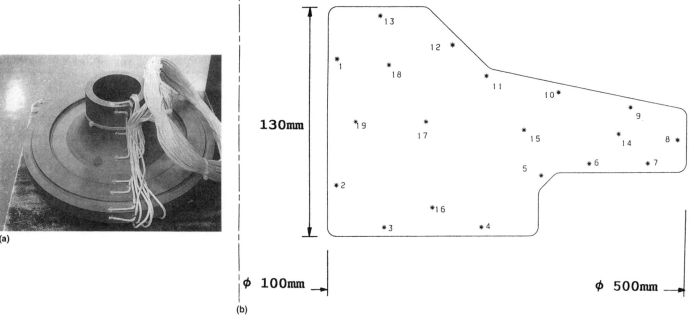

Fig. 3 (a) Surface thermocouples used on an instrumented high-pressure turbine disk to establish heat-transfer coefficients. Source: Ref 4. (b) Internal thermocouples used on an instrumented turbine disk to establish heat-transfer coefficients. Source: Ref 6

thermocouple associated with it that is beneath the surface and somewhere midzone (as illustrated in Fig. 2b). In the trial-and-error method, estimated HTCs, which may vary with temperature, are applied to each surface zone in the computer model. The temperatures predicted by the model are then compared to the experimental data, the HTCs are subsequently modified based on the errors, and the model is rerun. This procedure is repeated until the experimental and model results have converged. Needless to say, this can be a tedious process. An alternate method is to use inverse engineering techniques (Ref 9–13), where, as with the iterative method, the part is divided into surface areas. However, in the inverse case, the measured temperature data are input into a nonlinear finite-difference or finite-element computer code. This program calculates the temperature distribution in the part and, through mathematical techniques, determines the time-dependent surface flux conditions for each area that results in temperatures that best match the test data over the entire test period. These surface fluxes may then be converted into HTCs. Clearly, this is a more complex method, but it is less tedious because it does not involve repetitive, manually interactive iterations. For either of the two methods, the analysis is somewhat sensitive to how the surface area is divided into subareas and the location of points where temperature-time data are obtained in the experiments. Poor choices of either will result in an inadequate description of the true boundary condition and thus lead to errors in the predicted temperatures.

Optimizing codes have also been applied to the problem (Ref 14–18). Such techniques generally rely on the user to apply an upper and lower value to the surface HTCs to limit the

number of iterations required and, in most cases, have been applied to the relatively simple case of forced-air cooling. The application of the technique to quenching that involves boiling heat transfer around complex shapes still requires some development, although steady progress is being made. For example, in a study by Kim et al. (Ref 18), optimizing techniques were used to obtain HTCs on a cylindrical probe during polymer quenching. Temperature-dependent HTCs were obtained, although even for this relatively simple geometry, some of the reported cases required between 1000 and 2000 iterations to converge to a solution.

A few researchers have also applied neural network models to the inverse heat conduction problem, although, to date, the applications have been limited to relatively simple cases (Ref 19–21).

Historical Perspective. Turning attention to the actual reporting of experiments and HTCs, some excellent work has been carried out, some of it going back over 50 years. In the 1950s, Paschkis and Stolz (Ref 22) used cylinders, spheres, and slabs to determine the HTCs for various quench oils and water. Their findings for three quench oils are shown in Fig. 4. In the 1970s, Mitsutsuka and Fukuda (Ref 23) used low-carbon steel plates to determine HTCs in still water and examined the effect of water temperature. In the early 1980s, Price and Fletcher (Ref 24) carried out some well-reported work where thin plates were used to determine the HTCs for water, oil, and polymer quenching (Fig. 5 to 7). Also in the 1980s, Bamberger and Prinz (Ref 25) used copper, aluminum, nickel, and steel billets to determine HTCs for water and water-organic mixtures as well as water sprays. They also compared the

experimental data with an equation developed to predict the HTC in water (Fig. 8). Additionally, in the 1980s, Trujillo and Wallis (Ref 26), Wallis (Ref 27, 28), and Segerberg and Bodin (Ref 29) determined HTCs or fluxes for oil quenching in two dimensions around disks, the latter including cases for disks with a hole in the center. Examples of their results are presented in Fig. 9 and 10 (Ref 26, 29). In the 1990s, work in this area continued, and the determination of HTCs around flat disks and a contour shape were reported by Cross et al. (Ref 30). In addition, Stringfellow (Ref 31) published data for HTCs during oil quenching in which the oil was either static, flowing parallel, or flowing perpendicular to the surface of the testpiece. The approach taken to determine the HTCs was similar to that used by Price

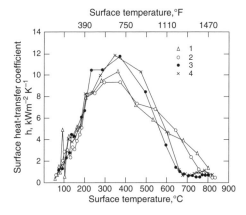

Fig. 5 Effect of temperature on surface heat-transfer coefficient obtained during quenching of 3.3 mm strip in water. Numbers refer to order in which quenches were carried out on single specimens. Source: Ref 24

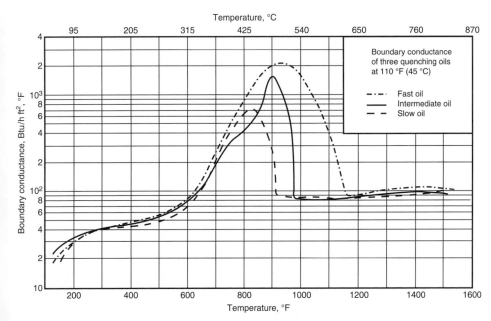

Fig. 4 Boundary conductance of three quenching oils at 45 °C (110 °F). Source: Ref 22

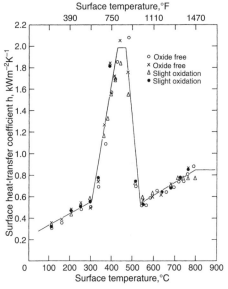

Fig. 6 Effect of temperature on surface heat-transfer coefficient obtained during quenching of 15 mm plate in oil. Source: Ref 24

and Fletcher (Ref 24); a modified version of one of Stringfellow's graphs is presented in Fig. 11. In the 1990s and 2000s, the work progressed to the point where HTCs were being obtained around complex shapes such as full-sized gas turbine engine disks. This work continues, but unfortunately, to date, most of the data remain proprietary and hence unpublished.

Data have also been reported that show the effect of oil viscosity on heat transfer (Fig. 12) (Ref 32, 33), and although HTCs were not calculated, the effect of water temperature on the cooling rate may be seen in Fig. 13 (Ref 3, 34). Quenchant flow has also been shown to affect the heat-transfer rate (Ref 3, 32, 33, 35 to 37), with work reported by Mason and Capewell on the cooling rate of an Inconel

600 probe when quenched in oil, polymer, and water shown in Fig. 14 (Ref 37) and the variation of the HTC for an aqueous solution given in Fig. 15 (Ref 32, 33). Other factors, such as the age of the oil and contaminants, have also been shown to affect the cooling characteristics of a quenchant (Ref 31, 38), an example of which is shown in Fig. 16. The surface finish on the component is also considered to affect the HTC (Ref 3, 24, 39 to 41); however, quantitative data are hard to come by. Results from one study reported by Allen and Fletcher (Ref 42) are presented in Fig. 17. The aforementioned

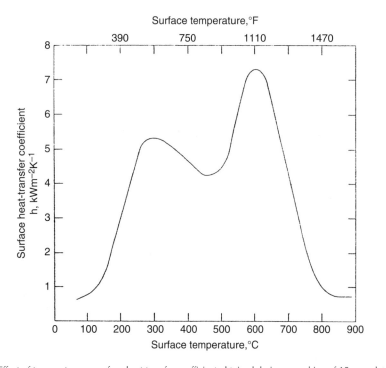

Fig. 7 Effect of temperature on surface heat-transfer coefficient obtained during quenching of 15 mm plate in 25% polymer solution; average of six quenches. Source: Ref 24

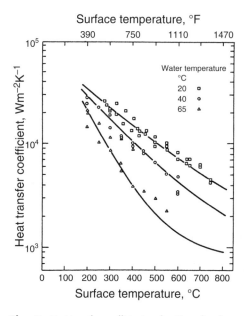

Fig. 8 Heat-transfer coefficient as function of surface temperature during immersion in water at 20, 40, and 65 °C (70, 105, and 150 °F); lines are calculated values. Source: Ref 25

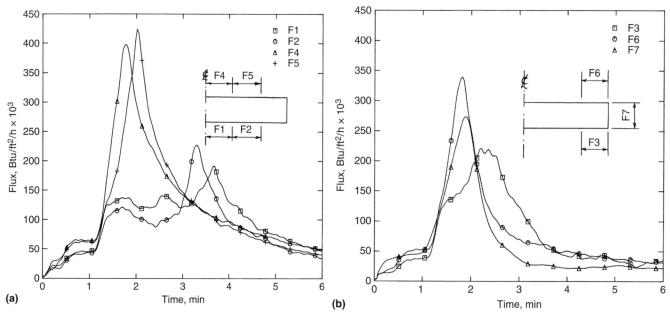

Fig. 9 Variation of heat flux with time for (a) the center area of a disk and (b) the outer area of a disk. Source: Ref 26

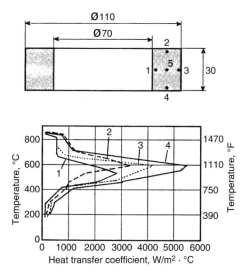

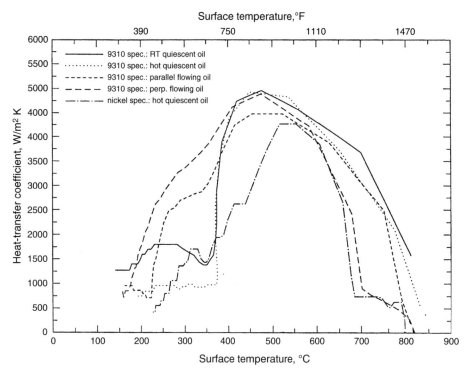

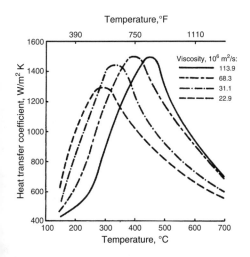

Fig. 10 Heat-transfer coefficients around a ring. Cooling was carried out in stationary oil at a temperature of 70 °C (160 °F) with the ring in a horizontal position. Source: Ref 29

Fig. 11 Calculated variation of the surface heat-transfer coefficient with temperature for various oil quenching conditions (oil movement). RT, room temperature. Source: Ref 31

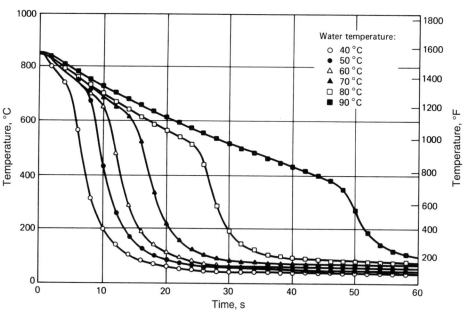

Fig. 12 Heat-transfer coefficients for hardening oil of various viscosities. Source: Ref 32

Fig. 13 Effect of bath temperature on cooling curves measured in the center of an Inconel 600 probe (12.5 mm diameter × 60 mm) quenched into water flowing at 0.25 m/s. Source: Ref 34

citations cover only a small portion of the work reported in the open literature. They do, however, show the complexity of obtaining reliable data to put into a thermal model. Figures 12 through 17 indicate that in addition to the experimental and numerical issues associated with obtaining HTCs during quenching, the variables that may be encountered in a production environment over time may also contribute to differences found between model results and actual measurements of cooling rates in a component.

The foregoing implies that caution should be used before applying HTCs directly out of the literature in a model, because the data really only apply to the particular system and quench media tested. It must be realized that using these data may not give accurate results when they are used for predicting temperatures in a

part that is not heat treated in the same tank and using the same media that was tested.

Tests to determine the variation of the heat transfer that occurs around the periphery of a three-dimensional object can be relatively complex. Even in two dimensions, it may still be a challenge, for example, a simple cylinder. If this is quenched in oil, in a horizontal

orientation, the heat extraction from the bottom surface may be significantly lower than that from the top surface (Fig. 18) (Ref 2). This is because the vapor that is formed at high temperature has difficulty escaping from the bottom surface, and which acts as a thermal barrier and limits the heat transfer. This effect may become even more marked if fixtures are used to support the component, because these may make the path for vapor to escape even more difficult. This phenomenon is important because the cooling rate in the bottom portion of a component may be different from that in the top. When heat treating alloys that are cooling-rate sensitive, this may translate into different properties being obtained in the top and bottom parts of a component. In most industrial quench tanks, agitation is not tremendously powerful in the working area. Ducts or impellers are used to direct flow into the quench zone, but they generally must be some distance away from the work area for their protection. Thus, the flow of liquid around the bottom surfaces may be low and easily impeded. In the case of air or spray quenching, the geometry of the part and the manner in which the air/liquid is directed can have a significant effect on the local heat transfer around the part.

Generating HTCs around complex shapes may be carried out using techniques similar to those described previously, but clearly the testing is more involved because more thermocouples are needed to cover the surface area of the part. Examples where the variation of the HTC around a simple shape such as a pancake were described previously and examples shown in Fig. 9 and 10 (Ref 26 to 29). For many years, the same technique has been used in industry for more complex, contoured shapes, but unfortunately, to date the results remain mostly proprietary. Examples of such work were presented in Fig. 3(a) (Ref 4) and Fig. 3(b) (Ref 6), but, as stated, no HTCs were reported in either of these publications.

One detailed study aimed at the determination of HTCs around disks of various shapes was carried out by Bass et al. (Ref 43 to 46) and also reported later by Cross (Ref 47). These researchers started with disks approximately 20 cm (8 in.) in diameter and up to 5 cm (2 in.) thick and later moved on to look at subscale versions of turbine disks approximately 23 cm (9 in.) in diameter, with a degree of contour to the shape, some of which had a central hole bored in them. The disks were fitted with insulated sheathed thermocouples, but, in this case, the thermocouples that were spaced around the periphery of the disk were installed such that the tip of the thermocouple was flush with the surface (Fig. 19). The disks were quenched in oil in a purpose-built experimental tank in

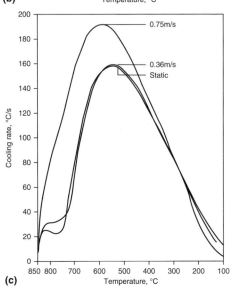

Fig. 14 Effect of quenchant flow rate on the cooling characteristics of (a) reference oil, (b) 20% polyalkylen glycol polymer quenchant, and (c) water at 40 °C (105 °F). Source: Ref 37

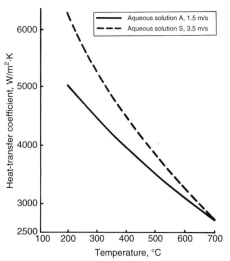

Fig. 15 Heat-transfer coefficients for aqueous solutions at various flow rates. Source: Ref 32

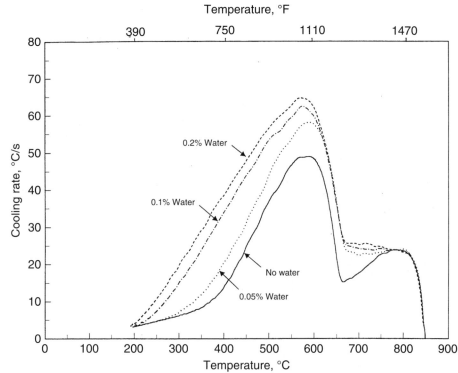

Fig. 16 Effect of water contamination in oil on the cooling rate (laboratory test). Source: Ref 38

which the flow rate of the oil could be varied. The data were reduced using the commercial Ansys code (Ref 48), and the software optimizing capabilities were employed to obtain the HTCs. One contour disk is shown in Fig. 20, with the location of the thermocouples given in Fig. 21. The variation of the HTC obtained for this disk is presented in Fig. 22 (Ref 47).

From the foregoing, it becomes apparent that, ideally, the HTCs applied in a model should be obtained from the facility in which the part is going to be heat treated and should include the effects of any fixtures that are used to support the part. However, in most cases, this is impractical because it implies developing the data for the specific part being heat treated, and this involves significant time and expense. This approach may be deemed appropriate when the cost of a component is high and the accuracy of the model is critical to prevent quench cracking, or the model is being used to predict residual stresses in the component and to develop a process to reduce distortion during subsequent machining operations.

In the case of lower-cost components, however, at this point in time, the increased accuracy obtained from a model does not justify the expense of carrying out individual experiments for a particular part, and the heat-transfer data from general experiments are applied to various shapes. Therefore, one concern that arises in thermal modeling is how to generate general HTC curves that may be subsequently used in models for various shapes. Heat-transfer coefficients generated from tests on a flat pancake will have limited accuracy when applied to a complex shape. In general, as shown later, it has been found that HTCs generated in one dimension during liquid quenching may be applied to relatively flat components (but with some degree of contour) with reasonable accuracy, because boiling heat transfer tends to dominate the process, particularly on free open surfaces, and this reduces the effect of shape on

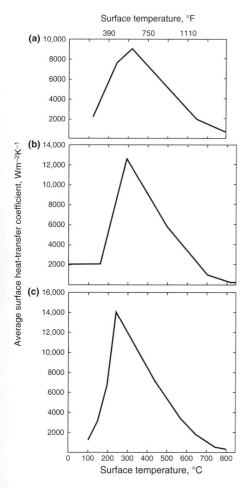

Fig. 17 Effect of surface finish on surface heat-transfer coefficient during water quenching. (a) 120, (b) 400, and (c) 600 grade. Note: 600 grade is the smoothest finish, and 120 grade is the roughest. Source: Ref 42

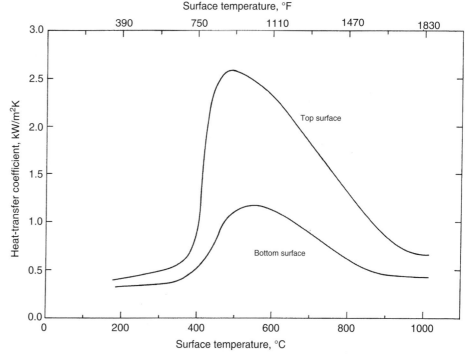

Fig. 18 Typical heat-transfer coefficients for oil quenching showing variation between top and bottom surfaces of a disk when quenched horizontally. Source: Ref 2

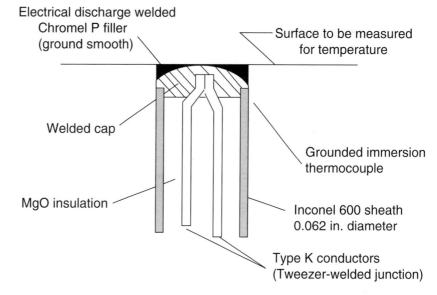

Fig. 19 Installed grounded immersion thermocouple used to measure temperatures around surface of disk. Source: Ref 47

the surface HTC. Similarly, tests with parts having a hole in the center have enabled reasonable estimates to be made of HTCs in holes. However, in the cases of liquid spray or forced-air cooling, the HTCs may vary significantly around a contoured part, and obtaining detailed HTC data around the specific part through testing may be the only way to obtain accurate results from the model.

Computational fluid dynamics (CFD) codes have also been used in limited cases to predict the HTCs around a part. In particular, significant work has been done in vacuum furnaces, where the flow of gases in the chamber and around parts has been modeled (Ref 49 to 55). However, while CFD has been used to model flows in liquid quench tanks (Ref 56 to 62) and around parts (Ref 61 to 66), a model to simulate boiling heat transfer still requires some development work before becoming a main industrial tool for HTC determination. This is an area of active research (Ref 17, 67 to 69), but it is difficult to estimate how long it will take to become an option in commercial CFD codes.

Work has been published describing the use of CFD to determine the HTCs around a gear blank quenched in a molten salt bath (Fig. 23, 24) (Ref 70). One limiting factor cited in this study was the lack of temperature-dependent physical property data for the salt that covered the entire temperature range encountered during quenching. This led to extrapolation of data for use in the CFD model.

In another study (Ref 71), CFD was used to calculate the HTCs around a gear during the intensive quenching process (Ref 3, 72). Intensive quenching involves flooding the part with high-speed water so that heat is extracted quickly and uniformly, with the goal of virtually eliminating the vapor and nucleate boiling phase. With complex boiling phenomena ignored, the CFD model was used to predict the HTCs around a section of the gear teeth, as shown in Fig. 25 (Ref 71). The computational domain and grid used for the steady-state three-dimensional CFD modeling is shown in Fig. 26. The estimated HTCs at the start of the quench process, with the part hot, and later, when the part has cooled, are shown in Fig. 27.

In a recent investigation (Ref 73), equations describing the heat transfer during the boiling regions encountered during quenching (i.e., film boiling, transition boiling, and nucleate boiling) were incorporated in user-defined functions for use with the commercial CFD code Fluent (Ref 74). Fluent was used to predict the HTCs around a turbine disk during oil quenching as well as to calculate the transient temperature distribution in the disk. Abaqus (Ref 75) was used subsequently to predict the development of residual stresses. Because equations were used for the separate boiling regions, the starting temperature for each region was required. These transition points were obtained from data from an experiment based on ISO 9950 (Ref 76), with an Inconel probe quenched in a sample of the quenching oil.

The disk and its surrounding area were meshed using quad elements, and the rest of the tank was meshed using triangular elements, as shown in Fig. 28. In practice, the disk is moved up and down in the quench tank, and this movement was modeled via another user-defined function describing the linear velocity with time. All the quad elements were moved up and down, while the remaining triangular elements were adapted using smoothing and local remeshing methods. Unfortunately, no HTCs were given in the paper, but the work represents an interesting development in modeling boiling heat transfer during quenching. The problem of defining the onset of the various boiling regimes for a particular quenchant and the difficulty in predicting when vapor locking occurs (where vapor becomes trapped in pockets formed by the shape of the component) represent a significant challenge to CFD modeling.

It is interesting to note that a similar approach (using boiling heat-transfer equations) was used in another study described by Vorster et al. (Ref 77), where residual stresses were calculated in a water-quenched cylinder, 100 mm long by 16 mm in diameter, made of AISI 317L material. In this case, however, the equations were used directly in a finite-element code to predict the temperature distribution in the cylinder rather than in a CFD analysis. The residual-stress portion of this work is described later in this article, but it is worth pointing out here that the authors state that care must be taken when choosing boiling heat-transfer correlations from literature, because most apply to steady-state conditions rather than transient cooling operations (as found in quenching).

The determination of HTCs during quenching by CFD modeling is an area that holds a great deal of promise. It is a subject receiving considerable attention, and, with continued effort, this technique will eventually become viable for generating data where boiling heat transfer occurs. As with any modeling work, validation is an important step that should be carried out in CFD work.

Material Property Data

The material properties used in a model also have a significant impact on the accuracy of the results obtained. Generally, the temperature-dependent thermal properties of the material, such as thermal conductivity, specific heat capacity, and density, are relatively easy to obtain. (Note: These thermal properties are also needed to carry out the conduction calculations, described previously, to determine the HTCs during cooling.) Data for many materials have been published (Ref 78, 79), and several laboratories specialize in generating these types of data. Thus, when the heat-transfer boundary conditions have been obtained, it is relatively easy to model the transient temperatures in the component. Therefore, one of the earliest uses of modeling was to determine the cooling rates in a part.

Predicting Cooling Rates in a Component

Many commercially available finite-element codes are capable of handling the heat-transfer analysis to obtain the transient temperature distribution on components during cooling. However, validation is an important part of any

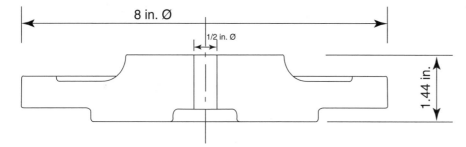

Fig. 20 One disk used in the test showing overall dimensions. Source: Ref 47

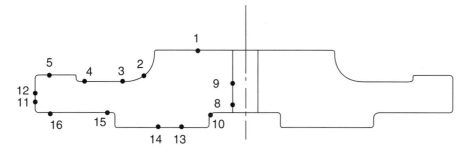

Fig. 21 Diagram of disk showing surface thermocouples (constructed from information in Ref 46)

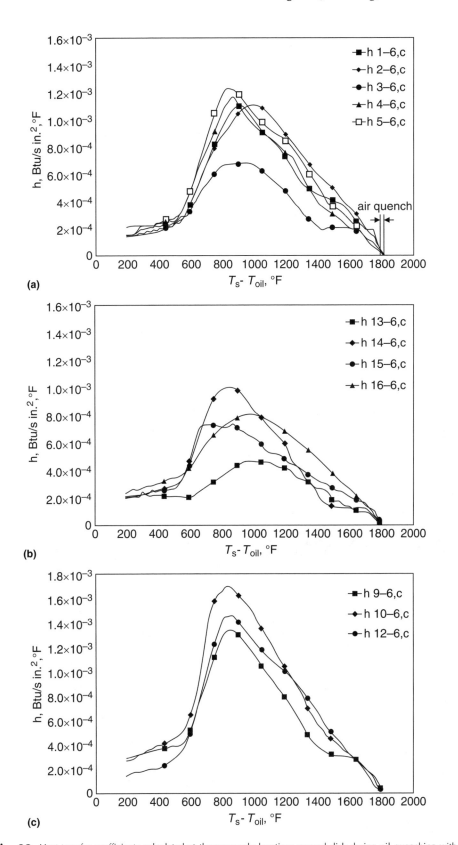

(a)

(b)

(c)

Fig. 22 Heat-transfer coefficients calculated at thermocouple locations around disk during oil quenching with an oil velocity of 100 ft/min. (a) Top surface, (b) bottom surface, and (c) side heat-transfer coefficients. Source: Ref 47

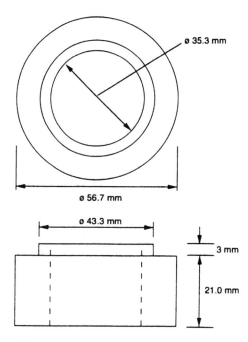

Fig. 23 Geometry of a thick ring. Source: Ref 70

modeling study, and Fig. 29 shows the comparison between temperatures predicted by a computer model and those measured in a test (Ref 2, 27, 64, 80). The example shown is a full-sized turbine disk approximately 61 cm (24 in.) in diameter and up to approximately 7 cm (3 in.) thick in the bore area and made from a nickel-base material. The disk was heated to 1175 °C (2145 °F) and then transferred from the furnace to a forced-air cooling system in 0.83 min, forced-air cooled for 2 min, transferred to an oil tank in 0.75 min, and finally oil quenched. In this example, the HTCs for both the forced-air and oil quenching steps were obtained from tests using simple-shaped, thermocouple-equipped disks and employing the inverse conduction techniques described previously. The correlation between the measured and calculated temperatures may be seen to be good even in a relatively complicated, multiple-step heat treatment process.

In alloys where the final mechanical properties are a function of the cooling rate obtained during quenching, computer models have been used to accurately predict the properties obtained and also to design processes to achieve the desired properties and reduce distortion (Ref 2, 64, 80). Figure 30 shows a case for a nickel-base superalloy disk where the residual stresses generated during oil quenching caused severe distortion problems during machining. The cooling-rate property relationship for the alloy was obtained for room- and elevated-temperature yield strength, as indicated by the two sets of solid lines. Modeling showed that the oil quench provided a very fast quench rate, but the data also indicated that the desired mechanical properties could be obtained by forced-air cooling. Forced-air cooling rates (90 to 135 °C/min, or 160 to 245 °F/min) resulted in lower cooling rates than oil quenching (235 to

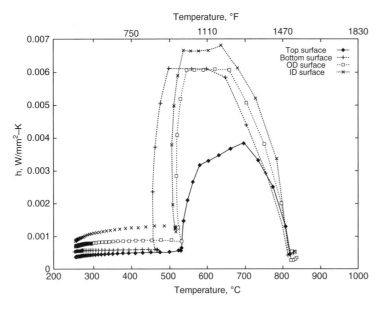

Fig. 24 Heat-transfer coefficients derived from the calculated surface heat flux. OD, outside diameter; ID, inside diameter. Source: Ref 70

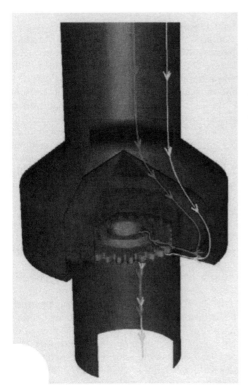

Fig. 25 Cutaway view of the quench fixture showing the gear, upstream deflector, and representative water path lines passing over and under the gear. Source: Ref 71

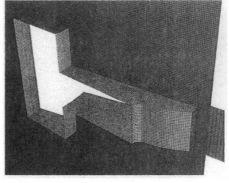

Fig. 26 Computational domain and grid used for three-dimensional gear model. Source: Ref 71

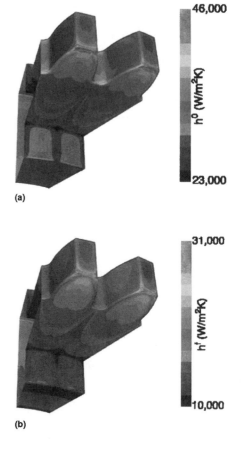

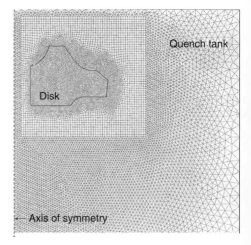

Fig. 27 Heat-transfer coefficients when (a) the part is hot and (b) the part has cooled. Source: Ref 71

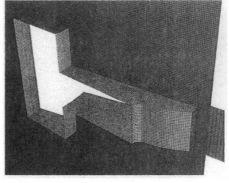

Fig. 28 Mesh used for disk and quench tank in computational fluid dynamics study. Source: Ref 73

300 °C/min, or 425 to 540 °F/min), with resultant lower residual stresses. Thus, the process was changed and the part successfully heat treated, meeting the property requirements while eliminating the distortion problem.

In another study (Ref 4), a cooling-rate model also was used to predict final properties in a nickel-base Merl76 disk. In this case, a microstructural prediction model was developed and combined with material strength predictions based on regression analysis of tensile properties. Figure 31 shows a limited comparison between predicted and measured yield strength at two subsurface points in a part quenched in oil. In this example, the HTCs used in the model were obtained using data gathered from tests with the thermocouple-equipped disk shown in Fig. 3(a).

Prediction of Residual Stresses

Material Property Data Required

When a model has been developed to calculate the transient temperature distribution in a component during quenching, it is possible to extend the analysis to the prediction of transient stresses in the material. This ultimately leads to the determination of the stresses that are present in the part at the end of the process: the residual stresses. This analysis, however, requires an additional set of mechanical property data, some of which are relatively complex and not generally available in literature. Properties such

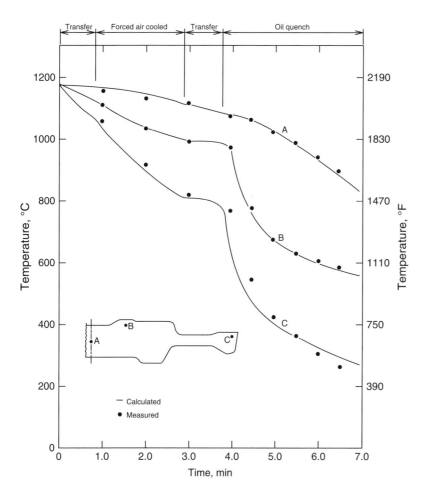

Fig. 29 Comparison between calculated and measured temperatures during heat treatment of a full-sized turbine disk forging. Source: Ref 64

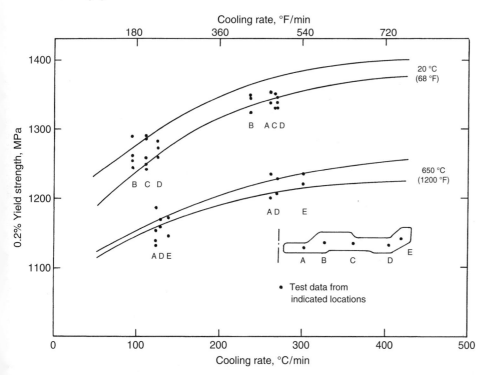

Fig. 30 Yield strength as a function of cooling rate together with test data for forced-air cooled and oil-quenched parts. Source: Ref 64

as the thermal expansion coefficient, Poisson's ratio, and the modulus of elasticity are required from ambient temperature to the highest temperature encountered during heat treatment. Stress-strain data are also needed as a function of temperature and possibly strain rate and cooling rate. The lack of suitable data for models is one of the major factors that has limited the use of models to predict residual stresses.

The detailed theory behind the various stress models used in industry is beyond the scope of this article. The works of Boley and Weiner (Ref 81) and Fletcher (Ref 82) are two excellent starting sources for those who wish to delve into the theory behind the development of thermal stresses.

Generally, the part being modeled is considered to be in a stress-free state and at a uniform temperature at the start of the quenching analysis. If considered important, however, the nonuniform temperatures and stresses may be carried over from a prior operation, such as forging or heating.

In the case of steels, the accurate prediction of residual stresses must also include material data that take into account the effect of phase transformations. During some phase transformations, relatively large volume changes take place in the material, and these must be taken into account in the model. These effects and the methods used to incorporate them into models are briefly described later in this article, and the theory behind the microstructure models is explained in more detail in sections of *Fundamentals of Modeling for Metals Processing*, Volume 22A, of the *ASM Handbook* (Ref 83 to 85). The phase changes that occur in some alloy systems are very small and, to date, have been neglected, thus making the calculation of residual stresses less complicated. For convenience, modeling related to these latter alloys systems is dealt with first.

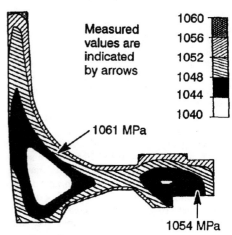

Fig. 31 Predicted versus measured yield strength for Mer176 disk forging quenched from 1079 °C (1975 °F). Source: Ref 4

Models Neglecting Material Transformation Effects

In general, to date, most work involving the calculation of stresses generated in nickel- or titanium-base parts during processing has ignored any volumetric changes that take place due to phase changes that occur during heat treatment. The volumetric changes in these alloy systems are generally small enough to be ignored. Some of the data needed for the model, such as thermal expansion coefficient, Poisson's ratio, and the modulus of elasticity as a function temperature, may be obtained from tests that are carried out routinely by commercial laboratories. However, tests to obtain the stress-strain relationship of the material as a function not only of temperature but also of strain rate and cooling rate are more specialized and thus, at this time, are carried out by relatively few establishments.

It is clearly important that the input data capture the true behavior of the material during quenching. Over the years, various material models have been used by researchers in this field, and because these can affect the accuracy of the residual stresses predicted, the different features are described in more detail.

A schematic representation of a typical stress-strain curve for a material, obtained from a tensile test, is shown in Fig. 32. In a relatively simple form, this curve may be represented by a bilinear material model where one straight line (the slope representing the elastic modulus) describes the material behavior up to the yield point, σ_y, and another straight line represents the plastic behavior beyond this point (sometimes called the plastic hardening modulus). Clearly, the drawback of this model is that if the material exhibits nonlinear flow hardening or flow softening, errors result (note the difficulty in fitting a straight line to the curve in Fig. 32). In practice, tests are carried out to obtain data to cover the range of temperatures encountered in heat treatment (Fig. 33). Each curve may be described independently, having its own elastic modulus, yield point, and plastic modulus. The interval between the test temperatures should be such that the interpolation carried out in the model will not result in significant errors. A more accurate model is one in which the plastic curve may be input in tabular form, which enables any nonlinear effects to be taken into account.

In many materials, the strength is also a function of strain rate, as indicated in Fig. 34. Models that ignore this effect and consider only the effect of temperature (as in Fig. 33) are sometimes referred to as rate-independent models. Clearly, adding the effect of strain rate enables the material behavior to be better described in the model. However, the increased accuracy obtained must be weighed against the additional material testing required. If strain-rate effects are small, it may not be worth the extra expense, particularly if the other possible

sources of errors in the model are considered. In the strain-rate model, tests must be carried out at various strain rates for each temperature tested. Thus, if it is determined that tests are required at 12 temperatures, a rate-independent model requires 12 tests. If three strain rates are determined to be necessary for a strain-rate model, then 36 tests are required.

In some alloys, the strength is also a function of the cooling rate that the material undergoes from the heat treatment solution temperature (Fig. 35), and this leads to further complications in the testing. In such cases, the testing is carried out by using a specialized machine, where the material is held at the solution temperature and then cooled at a specified rate before being tensile tested at the specified test temperature and at a specified strain rate to obtain the stress-strain curve (Ref 86). This is sometimes referred to as an on-cooling tensile test. Where the cooling rate is important, the data described previously for the strain-rate model (Fig. 32 to 34) may be obtained using the test method described previously but using an appropriate cooling rate. While cooling-rate effects have been reported (Ref 86), the effect on strength has generally been relatively small (Fig. 36). While it is important to obtain a certain minimum cooling rate in most nickel- and titanium-base turbine engine disks, the strength is generally not very sensitive to the range of cooling rates found in a part during quenching. In such cases, a single cooling rate representative of the cross section being considered is selected. It should be noted that in Ref 86, the

cooling rates of 220 and 1110 °C/min (400 and 2000 °F/min) are relatively high, representing what would be expected near the surface of the part during the initial period of quenching. In practice, lower cooling rates would be present in the material away from the surface and also on the surface later in the quenching period. Models incorporating the cooling-rate effect have not been described in the literature, and, at present, commercial codes can only handle this by linking with special subroutines written by the user. Clearly, adding cooling rate to the mechanical test array increases the number of tests significantly. Simply adding two cooling rates to the example given previously brings the total number of tests to 72.

For many alloys, it is important that the material used for the tests be in the condition it would be in prior to heat treatment (i.e., in the state being modeled). Thus, if a forged component is being considered, the starting material for the tests would be in the as-forged state, not in the final forged and fully heat treated state.

It may be seen from the foregoing that a complete set of data to describe material behavior during quenching can be quite extensive. Because of this, results from early residual-stress models presented in the literature generally included only the effect of temperature in the material databases. Generally, tests were carried out to determine the sensitivity of the data to changes in strain rate and cooling rate, and a single value of each of these parameters was selected to generate a dataset for the model. However, this practice is rapidly

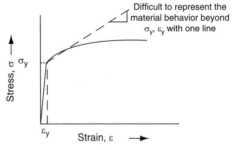

Fig. 32 Typical stress-strain curve

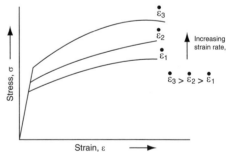

Fig. 34 Stress-strain curves showing effect of test strain rate

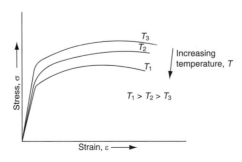

Fig. 33 Stress-strain curves showing effect of test temperature

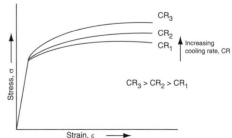

Fig. 35 Stress-strain curves showing effect of test cooling rate (see text for description of test)

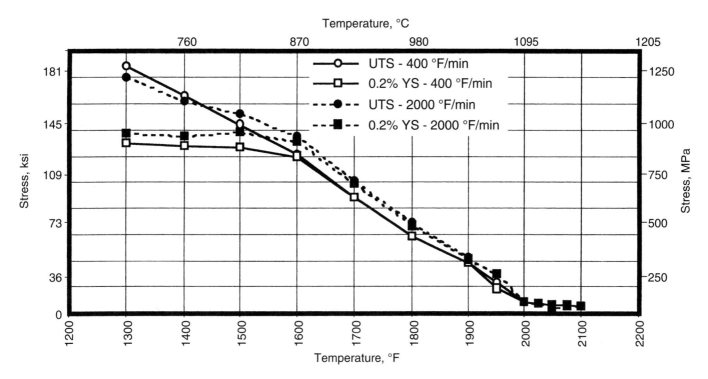

Fig. 36 On-cooling tensile test data showing the effect of cooling rate on ultimate tensile strength (UTS) and yield strength (YS) of a nickel-base superalloy. Source: Ref 86

changing, and over the past few years, the drive for improved accuracy from the models has led to more variables being included in analyses. Current state-of-the-art modeling of residual stress of nickel- or titanium-base alloys for aircraft engine components is for the data to take into account both the temperature and strain-rate effects. The tests are generally carried out at a single cooling rate from the solution temperature, but the effect of cooling rate is examined when the databases are being generated. In the future, one would expect residual-stress models to also incorporate the cooling-rate effect into the material property databases for some alloys, where it is found to have a marked effect on strength. It is also probable that metallurgical effects will be incorporated in the models of the future. Due to the complexity of the subject, this is an area that has not been explored in any depth, and it is considered ripe for future development and studies.

Validation of Residual Stress Models

Many commercial computer codes are available for the calculation of residual stresses; the main codes being used for this type of work are, alphabetically, Abaqus, Ansys, Dante, Deform, Hearts, LS-Dyna, and Sysweld (Ref 48, 75, 87 to 91). Published data showing validation of residual-stress models have been confined mainly to simple shapes such as cylinders and plates. Examples for more complex manufactured shapes are still not common, probably due in part to the expense of making

the residual-stress measurements. Such studies have certainly been carried out more than indicated in the literature, but unfortunately, most are still considered proprietary.

One relatively early published study in which computed and measured residual stresses were compared was carried out on a nickel-base turbine disk (Ref 64, 92). This work is particularly interesting in that two different computer models were used independently. A full-sized (450 kg) nickel-base superalloy forging was solution heat treated and oil quenched, and then, residual-stress measurements were made by employing the strain gage/hole drilling technique. Measurements of circumferential, radial, and axial residual stresses were made at six locations (Fig. 37) through the thickness of the disk, and these six measurement positions were repeated at three locations (0, 120, and 240°) around the disk to check for symmetry.

The HTCs used in the model were obtained from the tank employed to quench the disk, using the inverse conduction techniques described earlier in this article. The model results (Fig. 38) show that upon entering the quench tank, the temperatures in the surface layers of the disk fall relatively rapidly. The heat extraction from the center of the disk is limited by the thermal characteristics of the material, and hence, the temperature in the bulk of the disk falls slower. It is this surface-to-center temperature difference that generates thermal stresses within the part. If the stresses generated in the disk were always below the yield point for the material, the disk would elastically deform during quenching, but upon

cooling down completely, the disk would revert to its original shape and be in a stress-free state at the end of quenching. However, the disk is quenched from high temperature, and, at elevated temperatures, the strength of the material is significantly lower than it is at room temperature (i.e., a yield strength of 30 MPa, or 4 ksi, at 1100 °C, or 2010 °F, compared with 990 MPa, or 144 ksi, at room temperature). Shortly after being immersed in the quench tank, the surface area of the disk is cooled to a temperature well below that of the center. The surface therefore contracts but is prevented from contracting the full amount desired by the hotter center region. This places the surface in tension, and because the material is relatively weak at high temperatures, plastic deformation occurs. Thus, the disk takes on a modified shape as cooling proceeds, and residual stresses are present in the part when it has cooled down to ambient temperature.

Figure 39 shows the circumferential and radial stress time plot for six points in the disk. It may be seen that during the first 5 min, the surface is placed in tension, and the subsurface points are in compression. As time progresses, the center cools more rapidly than the surface, but because the surface is cooler, it is stronger and prevents the center from contracting the desired amount. Hence, a stress reversal takes place, and after 8 min, the surface is placed in compression, and tensile stresses are induced in the center region. These stresses increase significantly over the next 10 min, but as the disk cools down, the rate of stress change decreases so that by approximately 50 min, equilibrium is

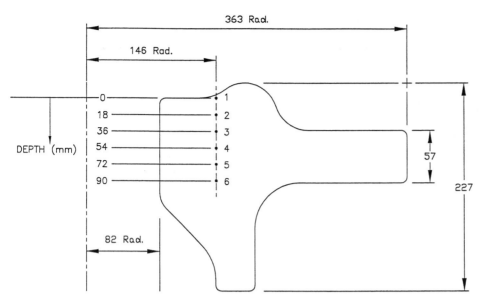

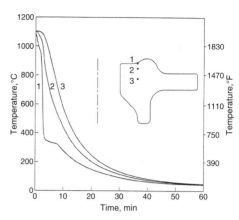

Fig. 38 Cooling curves for several points in forging during quenching. Source: Ref 64

Fig. 37 Layout of disk showing the position of strain-gage measurement of residual stress. Source: Ref 64

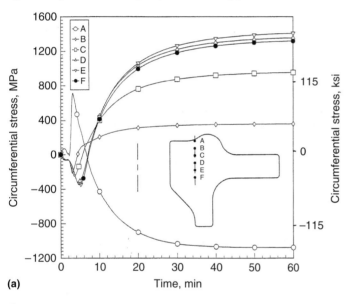

(a)

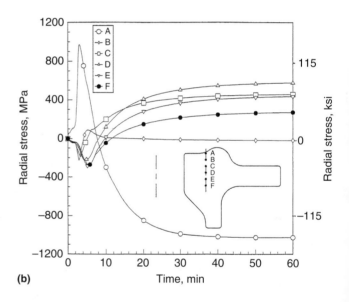

(b)

Fig. 39 Variations of (a) circumferential and (b) radial stress with time. Source: Ref 64

attained. The stresses at a time of 60 min, when the disk is at ambient temperature, represent the locked-in or residual stresses.

The comparison between the predicted and measured circumferential and radial stresses is presented in Fig. 40. The two finite-element models used were Abacus (Ref 75) (a commercial code) and Topaz/NIKE (Ref 93, 94) (codes available at the time of the study from Lawrence Livermore National Laboratory). Two curves are shown for the measured values, and these represent the maximum and minimum values found at any of the three circumferential measurements (i.e., the measured stresses were not exactly symmetrical). It may be seen that good agreement exists between the measured and calculated values.

Clearly, work of this nature has inherent errors. In the case of measurements, the

proprietary techniques employed resulted in an estimated error of +5% compared with the generally accepted +10% (Ref 95). In the model, errors exist due to inaccuracies in the boundary HTC data obtained from a pancake being applied to a disk, as discussed in the previous section. However, the largest source of error was attributed to the inadequate input of the plastic behavior of the material. Mechanical testing was carried out on material that was in the as-forged condition, as it would be in practice at the start of the quench. In practice, the plastic behavior of the material in the disk is different on the surface than in the center region (i.e., the material deforms at different strain rates). In the study, however, mechanical testing was only carried out at a single strain rate, and this inevitably led to an error in the

model. A strain rate representative of the mid-thickness was chosen for testing. In the case of NIKE, errors were also present due to the inadequate material model used in the code. At the time, the plastic behavior could only be input as a function of temperature by specifying a yield point and a plastic hardening modulus. It is difficult to properly describe the behavior of material using this approach, and this could be the source of the difference between the two computer code results. Abaqus, on the other hand, allowed the stress-strain curve to be input as a series of points, and this allows a better description of the test data. This highlights the need for care in applying data in the models as well as the need for validation to determine some estimate of the errors present. Also, conventional rules concerning

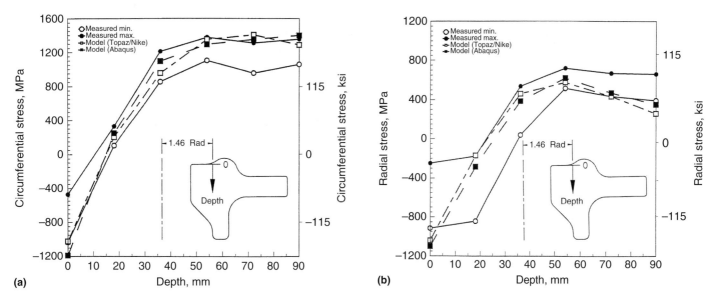

Fig. 40 Measured and calculated (a) circumferential and (b) radial stresses in the disk. Source: Ref 92

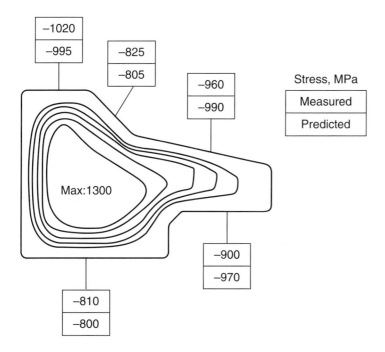

Fig. 41 Comparison between measured and predicted tangential surface stress. Source: Ref 6

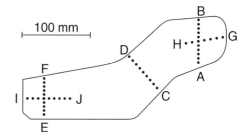

Fig. 42 Positions of strain measurements made in the disk, shown on the plane of symmetry. Source: Ref 96

finite-element mesh size and time step used in the analysis must be followed to reduce the effect of these potential sources of errors, which, with some care, can be minimized.

Other published data on a nickel-base disk also have shown good agreement between model and experimental measurements. Figure 41 shows a comparison of measured and predicted tangential residual stresses at five points around the surface of an Astroloy disk that was quenched in oil (Ref 6). To obtain the heat-transfer boundary conditions for this investigation, a full-sized part was instrumented with 19 thermocouples (as shown in Fig. 3b),

mostly around the perimeter, and quenched. Inverse techniques similar to those described previously were used to back out the HTCs. The variation of the HTC around the part was not published, but the calculated and measured stresses around the disk agree to within 10%. This study also concluded that the transformation plasticity was negligible compared to the viscoplastic flow occurring in the disk during quenching.

A detailed validation study on an IN718 disk forging quenched in water has been reported, where neutron diffraction measurements of residual stresses were made through the

thickness of the disk. Measurements were made along several planes in the disk, as shown in Fig. 42 (Ref 96). In one validation study by Dye et al. (Ref 97), the commercial code Abaqus (Ref 75) was used, and in work published later by Rist et al. (Ref 96), Deform 2D (Ref 88) was employed. In both cases, the HTCs used in the model were obtained from immersion quench tests with an IN718 bar. Both workers used similar material models, with Rist et al. (Ref 96) using a rate-independent elastic-plastic model, where plastic flow was described by a temperature-dependent initial yield stress, σ_y, and a linear hardening modulus, H. Thus, the effective flow stress was given by $\sigma_{ff} = \sigma_y + H\,\varepsilon_{ff}$, where ε_{ff} is the effective strain. A constant value of $H = 100$ MPa was assumed, and it was stated that the results were found to be quite insensitive to changes in this parameter. The temperature-dependent yield strength of the material was determined by taking the room-temperature yield stress of solution-treated material and assuming that the stress-temperature behavior would take the same form as for fully heat treated material (for which data were available). The yield and modulus used in the investigation are shown in Fig. 43. The effect of any microstructural phase changes on the stress-strain behavior of the material was not taken into account.

Figure 44 shows the comparison between measured and predicted residual stresses through the thickness of the disk at two locations (one region near the bore and one near the rim). In both regions, the trend and magnitude of the predicted stresses agree reasonably well with the measurements, but differences were observed. In general, the model predicted stresses up to approximately 100 MPa (15 ksi) higher than measured in the center thickness regions. Overall, both of the investigators concluded that the agreement between the measured stresses and the model were satisfactory but identified further work aimed at improving the accuracy. Among these were validated thermal boundary conditions for the disk and further development of material property data to enable the use of an improved material model.

Another validation study involving neutron diffraction measurements on a relatively flat 320 mm diameter IN718 gas turbine compressor disk with a thickness up to 26 mm was carried out by Cihak et al. (Ref 98, 99). The commercial code Deform (Ref 88) was used for the simulations, and the material model took into account temperature and strain rate. In this case, a thermocouple-equipped disk was water quenched in a vertical orientation, and simulations were carried out with different boundary conditions in an attempt to match the test data (a true inverse heat-condition analysis was not carried out). Comparisons between measured and residual stresses were presented for a range of cases where either a constant HTC was used or where the coefficient varied with time or temperature. Table 1 gives the time dependent HTCs used, and Fig. 45(a) shows the temperature-dependent HTC. The comparison between the predicted and measured stresses is shown in Fig. 45(b) (Ref 100), which indicates that reasonable results were obtained in all three cases presented.

The aforementioned validation examples indicate that, given reasonable input data, the residual stresses predicted by models of quenched aircraft engine disks are sufficiently accurate to be useful to engineers. This has led to the use of such models to investigate changes in heat treatment processes to aid in the ease of manufacturing. One example is presented in Fig. 46 (Ref 101), in which the residual stresses were modeled in a nickel-base superalloy disk for three different quenching media: oil, fan plus water, and air mist. This clearly shows that the stresses in all three directions are much lower for air-mist quenching than for oil quenching. The stress gradients in the oil-quenched part may give rise to distortion during machining as, say, a layer on the top part of the disk is machined off. As the material is removed, the part will move and take on a slightly different shape to maintain an equilibrium stress state. This movement results in machining issues, possibly resulting in added operations (having to machine a little from one side at a time). The lower stresses in the spray quench make this less likely. (This may be determined by machining modeling, which is described in the article "Modeling of Residual Stress and Machining Distortion in Aerospace Components" in this Volume.) The possibility of being able to vary the cooling rate

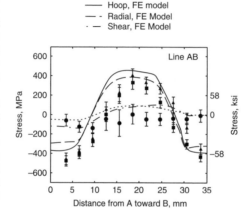

Axial, measured ●
Hoop, measured ▲
Radial, measured ■
Axial, FE model ·······
Hoop, FE model ———
Radial, FE Model – – –
Shear, FE Model – · –

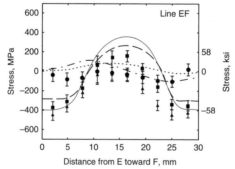

Fig. 44 Residual-stress components derived from strain measurements and from finite-element (FE) model predictions. Source: Ref 96

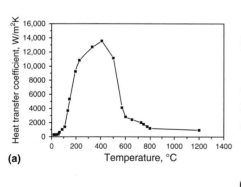

Fig. 43 Flow stress and elastic modulus data used in the finite-element model of the quenching process. Source: Ref 96

Table 1 Heat-transfer coefficients assumed in the model for cooling in water (c2 in Fig. 45b)

Time, s	Heat-transfer coefficient, W/m^{2}K
0–1	8000
1–4	60
4–40	8000
40–70	3000
70–120	2000

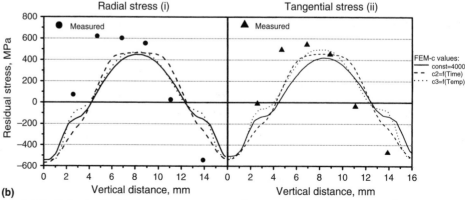

Fig. 45 (a) Variation of the heat-transfer coefficient. Source: Ref 99. (b) Simulated residual stresses using a constant heat-transfer coefficient of 4000 W/m^2 K, a heat-transfer coefficient varying with time (c2, Table 1) or with temperature (a). Radial (i) and tangential (ii) stresses along the cross section of the disk (b). The symbols indicate the stress values determined by neutron diffraction (measurement error approximately ±50 MPa). FEM, finite-element method. Source: Ref 99, 100

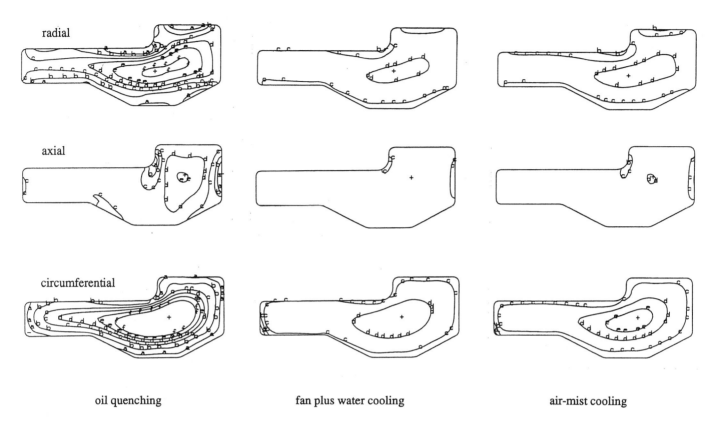

Fig. 46 Comparison of residual stresses in a disk for oil quenching, fan-plus-water cooling, and air-mist cooling. Stress contour levels (MPa): a = −600, b = −360, c = −120, d = 120, e = 360, and f = 600. Source: Ref 101

both around the part and with time by the position of nozzles and the use of computers makes air-water sprays attractive when trying to balance cooling rate with stresses (Ref 102 to 107).

Optimization of Residual Stress Models

In an enhancement of straight forced-air cooling, the air flow, and hence the heat transfer, may be directed to different areas around the circumference of the part and also be controlled by a computer. This allows the HTC to be varied both around the circumference of a part, by the positioning of different ducts/jets, and with time by controlling the air flow. Generating heat-transfer data for a model for this type of system is time-consuming, but the technique has great potential in optimizing cooling rates while limiting distortion problems.

Optimizing techniques have also been employed at the component design stage to reduce residual stresses and ease machining-related distortion. The applications described in the open literature have been for forced-air cooling, where, between combinations of ducts and air jets, a relatively wide range of HTCs is possible. The optimizing process starts by specifying a target range for the minimum cooling rate in the various sections of the part, so that the required mechanical properties are met. The surface of the disk is then divided into zones, and, to reduce the computational time,

the HTC for the zones may be bound between the upper and lower limits (based on prior experience with the equipment used to cool the piece). While this technique, or manual variations of it, has found niche applications in the aerospace industry, it is not used routinely in general industry.

One example of applying an optimizing technique to a generic turbine disk (material not specified) was presented by Rohl et al. (Ref 108, 109). The aim of the exercise was to reduce the high residual-stress levels found in an oil-quenched disk by changing the process to fan cooling, while still maintaining the minimum cooling rate necessary to obtain the required mechanical properties in the disk. To reduce computing time, the problem may be decoupled, with the heat-transfer and stress analyses being carried out separately. It is recognized that uniform cooling of a part reduces residual stresses, and thus, the aim is to formulate an objective function that penalizes nonuniform cooling while, at the same time, ensuring that the target minimum cooling rate is met. In this example, it was assumed that the fan-cooling equipment had the capability of controlling the airflow (and hence the HTC) to individual sections of the part. The heat-transfer boundary of the disk was divided into nine zones, as shown in Fig. 47. The temperatures and stresses in the disk were calculated using Deform (Ref 88), and the iSIGHT (Ref 110)

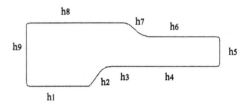

Fig. 47 Turbine disk geometry and fan-cooling variables for model routine optimizing study. Source: Ref 108. Reprinted with permission of the American Institute of Aeronautics and Astronautics

code was used for the optimizing routine. The optimizing was started by specifying the initial estimated HTCs around the disk for the fan-cooling case. This resulted in the cooling-rate distribution shown in Fig. 48 (normalized with respect to the target cooling rate). After optimizing, the cooling-rate distribution shown in Fig. 49 was obtained, which clearly shows how much more uniform the cooling is even when compared to the initial starting point for fan cooling. In this example, 13 iterations of the model were needed for the problem to converge to a solution.

The difference between the predicted hoop stresses in the original oil-quenched disk and in the final fan-cooled disk may be seen by comparing Fig. 50 and 51 (which give the stresses normalized with respect to the maximum tensile stress in the oil-quenched part). The residual stresses for the optimized disk are significantly lower, with the tensile stresses being reduced by almost 1 order of magnitude and the compressive stresses by a factor of approximately 6 to 7. The reductions of the maximum tensile and compressive hoop and radial stresses are shown in Table 2.

Furrer et al. (Ref 111) reported a heat treat optimization study on a U720LI turbine disk, again using the combination of the Deform and iSIGHT computer codes. The variations of the HTCs for forced-air cooling used in the optimization are shown in Fig. 52, and the resultant reduction in residual stresses from the original oil quenching to the optimized heat treatment are shown in Fig. 53.

In practice, most disks used in aircraft engines undergo at least one further heat treatment cycle (aging), where the disk is raised to a temperature (lower than that for the prior solution cycle) and held for a given period. The relaxation of the residual stresses induced from the solution/quench cycle during this subsequent heat treatment step may also be modeled, but this requires the generation of another set of property data: creep relaxation data. These data describe how the stresses relax at temperature as a function of time. Such modeling is described in the article "Stress-Relief Simulation" in this Volume. Furthermore, the final residual-stress state in the component (after stress relieving has taken place during the aging cycle) may be used as input into a model to predict the distortion during the machining operation. Such modeling is described in the article "Modeling of Residual Stress and Machining Distortion in Aerospace Components" in this Volume. It is being carried out more and more as time goes on and leads to modeling being used to compare residual stresses and machining-related distortion problems with various quench methods and part design. The ability to model the entire heat treating process and the subsequent machining operation has resulted in the models being used to help develop the next generation of heat treat equipment.

Models Incorporating Material Transformation Effects

As mentioned earlier, in the case of steels, the volumetric changes during phase transformations may be relatively large and must be considered for a model to accurately predict the stresses that are generated in the part during quenching. Figure 54 (Ref 112) shows the basic relationship between factors that are important to the generation of residual stresses and is similar to many diagrams presented in the literature. This can lead to the impression that the interactions are relatively simple to implement into a computer model. However, this is not necessarily the case, as may be seen from Fig. 55, which shows a schematic representation of the interactions included in a finite-element program claiming to predict residual stresses in a component including metallurgical effects (Ref 113, 114). As may be seen, the interaction between thermal, metallurgical, and mechanical effects is extremely complex, and current models used in industry may or may not include all the factors involved.

The development of equations that describe the thermal/metallurgical/mechanical interactions and the generation of data for the models add a level of complexity so far not needed when working with nickel- and titanium-base alloys. (Although Denis et al., Ref 115, does refer to some initial research that was carried out on a titanium alloy where phase changes were considered.) Details of the phase-change mechanisms occurring in steel and the methods used to formulate these into computer codes are described very well by Ericsson in the article titled "Principles of Heat Treating of Steels," by Kirkaldy in the article "Quantitative Prediction of Transformation Hardening in Steels," and by Gergely et al. in the section "Computerized Properties Prediction and Technology Planning in Heat Treatment of Steels" in *Heat Treating*, Volume 4, of the *ASM Handbook* (Ref 114, 116, 117). The reviews by Denis et al. (Ref 115) and Rohde and Jeppsson (Ref 118) also describe how phase transformations are incorporated into simulation tools.

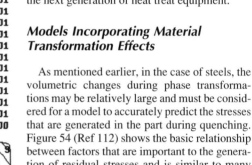

Color Index

—	B	0.833E+01
—	A	0.767E+01
—	0	0.700E+01
—	9	0.633E+01
—	8	0.567E+01
—	7	0.500E+01
—	6	0.433E+01
—	5	0.367E+01
—	4	0.300E+01
—	3	0.233E+01
—	2	0.167E+01
—	1	0.999E+00

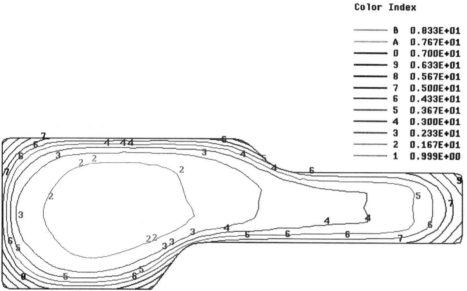

Fig. 48 Initial cooling-rate distribution. Source: 108. Reprinted with permission of the American Institute of Aeronautics and Astronautics

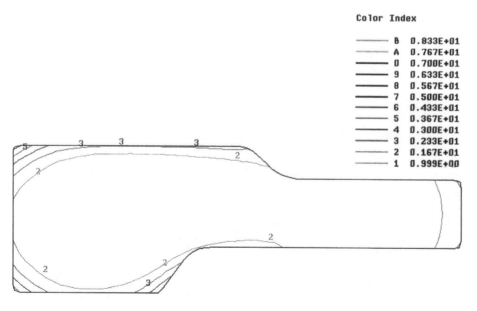

Fig. 49 Final cooling-rate distribution. Source: 108. Reprinted with permission of the American Institute of Aeronautics and Astronautics

Two excellent examples that clearly indicate the differences between the stresses predicted by a model with and without the inclusion of the austenite-to-martensite transformation are cited in Ref 116 and are worth describing again. In the first example, only pure thermal effects are considered. Figure 56 (Ref 116, 119) shows the thermal stresses developed in a 100 mm (4 in.) diameter bar that is water quenched from an austenitizing temperature of 850 °C (1560 °F). The top graph shows the temperature of the surface (S) and the core (C) of the bar during the quench. As expected, the temperature at the surface falls more rapidly than at the core, and at time w, the temperature difference between the surface and core attains a maximum of approximately 550 °C (1020 °F). The result of this cooling is that the specific volume in the hotter core portion of the bar is greater than that at the surface. The volume contraction of the cooling surface region is prevented by the hotter, higher-specific-volume core region, and this results in the development of thermal stresses, as shown in the lower portion of Fig. 56. Curve "a" represents the theoretical thermal stress at the surface, assuming only elastic material behavior. Curve "b" represents the thermal stresses taking into account the yield strength of the material and showing the reduction of the stress predicted because the yield strength cannot be exceeded, and thus, plastic flow occurs. The thermal stress developed is approximately proportional to the temperature difference, so at time w, the stress is tensile in the surface region and compressive in the center. As cooling continues, the temperature difference decreases. Eventually, a point is reached where the cooler surface resists the contraction of the hotter, but more rapidly cooling core, and a stress reversal takes place. At the end of cooling, the final or residual-stress pattern is compressive in the surface and tensile in the core (right side of Fig. 56).

This example leads to several general observations regarding the magnitude of thermally induced residual stresses:

- For a given shape, the thermal stress is higher in a material having a low thermal conductivity, high heat capacity, or high thermal expansion coefficient.
- Larger section size or increased surface cooling intensity of the cooling medium increases the temperature difference and hence the thermal stresses in the component.
- Materials with a higher yield stress at elevated temperatures will undergo less plastic flow, and thus, the residual stress and the yield stress at ambient temperature put an upper limit on the residual stress.

Color Index

—	B	0.833E+00
—	A	0.665E+00
—	0	0.498E+00
—	9	0.331E+00
—	8	0.164E+00
—	7	-0.359E-02
—	6	-0.171E+00
—	5	-0.338E+00
—	4	-0.505E+00
—	3	-0.673E+00
—	2	-0.840E+00
—	1	-0.101E+01

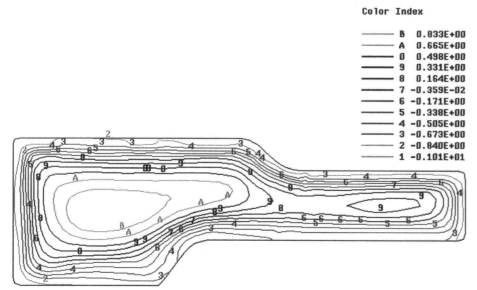

Fig. 50 Hoop stress for oil quenching process. Source: 108. Reprinted with permission of the American Institute of Aeronautics and Astronautics

Color Index

—	B	0.896E-01
—	A	0.671E-01
—	0	0.445E-01
—	9	0.220E-01
—	8	-0.577E-03
—	7	-0.231E-01
—	6	-0.457E-01
—	5	-0.682E-01
—	4	-0.907E-01
—	3	-0.113E+00
—	2	-0.136E+00
—	1	-0.158E+00

Fig. 51 Hoop stress optimized process. Source: 108. Reprinted with permission of the American Institute of Aeronautics and Astronautics

Table 2 Stress reduction in a disk obtained by changing from oil quenching to fan cooling and using optimization:

Stresses are normalized with respect to the maximum tensile stress of the oil-quenched part.

	Oil quenched	Fan cooled (after optimization)
Hoop stress, MPa (ksi)		
Tension	6.895 (1.000)	0.841 (0.122)
Compression	6.943 (1.007)	1.096 (0.159)
Radial stress, MPa (ksi)		
Tension	6.9 (1.0)	0.641 (0.093)
Compression	8.646 (1.254)	1.227 (0.178)

Source: Ref 108

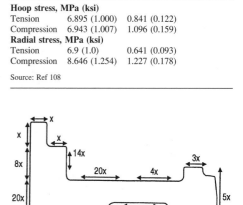

Fig. 52 Initial heat-transfer coefficient profiles for optimizing study on a generic nickel-base superalloy disk configuration. Source: Ref 111

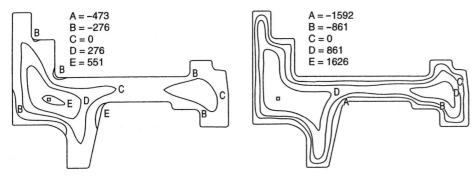

Fig. 53 Finite-element analysis model prediction of residual stresses for a generic turbine disk made from U720LI heat treated with (a) the supercooler method, resulting in a maximum and minimum residual stress of 551 and −473 MPa, respectively, and (b) the oil quenching method, resulting in a maximum and minimum residual stress of 1626 and −1592 MPa, respectively. Source: Ref 111

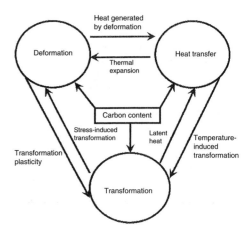

Fig. 54 Schematic showing interaction between mechanical, thermal, and metallurgical processes that must be taken into account in models to predict residual stresses when phase transformation is important. Source: Ref 112

causing a volumetric expansion that the surface resists, leading to another stress reversal. Finally, after cooling, transformation-induced tensile stresses at the surface dominate over the thermally induced compressive stresses. The residual-stress pattern in this case is the reverse of that in Fig. 56, where only thermal effects were considered.

The foregoing two examples clearly illustrate why models for the prediction of residual stresses in a steel component must include metallurgical transformation effects if accurate results are to be obtained. The analysis is further complicated by other factors that may need to be considered:

- The martensitic transformation temperature, M_s, is decreased by the presence of compressive stress and increased when the material is under tensile stress (Ref 116, 121).
- Transformation plasticity (the permanent strain that occurs during an ongoing phase transformation under applied stress lower than the yield stress) should also be considered. Examples showing the importance of this factor are given later.
- The incubation time of nonmartensitic transformations has been shown to be prolonged in a material under compressive stress (Ref 116, 122) and shortened when tensile stress is present (Ref 116, 123).

It has already been shown that when calculating the residual stresses that are developed in a steel component during heat treatment, it is imperative to incorporate the effects of phase transformations if accurate results are to be obtained. Among other factors, transformation affects the flow stress of the material, which is a critical parameter used in the calculations. Clearly, the first step in modeling transformation is to define the volume fraction of each possible phase in each subvolume of a meshed component at any time during the simulation. The phase-transformation models currently being used by most researchers developing residual-stress models are based on curve-fitting of the

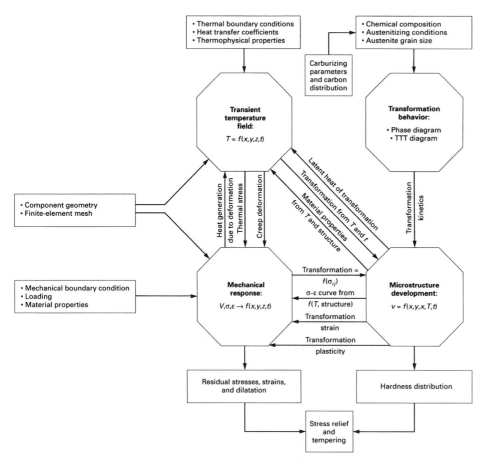

Fig. 55 General system overview of a program to predict the thermomechanical behavior of low-alloy steels. Source: Ref 113, 114

Now consider the case where the effect of austenite-to-martensite transformation is included in the analysis. The upper portion of Fig. 57 (Ref 116, 120) shows the surface and core temperature history of the piece, and superimposed on this figure is also the transformation curve. (Note: The time axis scale changes from seconds to minutes at 60 s). At time t_1, the surface temperature falls below the M_s (martensite start) temperature, and the

surface starts to transform. Because the specific volume of martensite is higher than that of austenite, the transformation results in a volumetric expansion such that the surface attempts to expand. As indicated in the lower portion of Fig. 57, this counteracts the pure thermal tensile stresses, and a stress reversal starts. This stress reversal takes place earlier than when transformation effects are not included. At time t_2, the core reaches the transformation temperature, experimentally derived transformation diagrams (e.g., the time-temperature transformation, or TTT, diagram, also referred to as the isothermal transformation diagram). Research is taking place to develop models that will determine the phase transformations without the need for so much experimental data (Ref 124, 125), but these are not advanced to the point of being useful to engineers at this time.

A TTT diagram, such as Fig. 58 (Ref 126), can be used to determine the volume fraction of each phase present by tracking the path of the material with respect to time and temperature. The diagram gives the start (subscript "s") and finish (subscript "f") points for transformation. (The "s" and "f" usually designate the time for 1 and 99% of the transformation, respectively, to occur, but other percentages may also be used.) Computational derivation of the TTT diagram from first principles and using the steel composition is the subject of ongoing research. However, the TTT curves for many steels have been published (e.g., Ref 126 to 128), and the classic "C-" shaped curve may be approximated in a computer model (Ref 114, 116, 117, 128 to 131). There are also commercial software packages available that are able to calculate the TTT and continuous cooling transformation

diagrams for general steels (Fig. 59) (Ref 132 to 135). Using these approaches, the time-temperature path of individual subvolumes, or elements, of the model may be tracked through the TTT diagram and the volume fractions of the transformed phases determined. Several different techniques to determine the volume of the phases present at each time step in the analysis are employed by researchers and code developers (Ref 136 to 145).

When calculating the transformation kinetics of austenite to ferrite, pearlite, or bainite, the Johnson-Mehl-Avrami equation is generally used:

$$V = 1 - \exp(-kt^n) \qquad \text{(Eq 1)}$$

where V is the transformed volume fraction, t is time, and k and n are temperature-dependent constants evaluated from the TTT diagram.

The transformation to martensite may also be determined using one of several different techniques. That given by Ericsson (Ref 116, 144) is:

$$V_M = (1 - V_F - V_P - V_B - V_C)$$
$$(1 - \exp(-0.011 \, (M_s - T))) \qquad \text{(Eq 2)}$$

where subscripts F, P, B, and C represent ferrite, pearlite, bainite, and cementite,

respectively; M_s is the martensite start temperature; and T is the temperature at a given time in the analysis.

With a TTT diagram available, some computer codes will allow the user to describe the curve by inputting individual corresponding time and temperature points. This circumvents the need for curve-fitting, such as described previously.

The kinetics of individual phase transformations have also been curve-fitted to various equations and the resulting empirical equations used to predict the phase volume fractions (Ref 112). Lusk et al. (Ref 146, 147) reported using an internal state variable approach to modeling the phase transformation that is based on a set of differential equations that describe the rates of formation of the various phases. Because this is a very active research field, the methods are sure to improve over the coming years.

When the phases present are determined, the parameters such as volume change, latent heat of transformation, and the appropriate flow stress to use for the time step are selected. These are generally found by using the rules of mixtures.

The thermomechanical model for calculating residual stresses in materials that undergo phase transformation requires a relatively large material database. The data must be capable of adequately describing all possible material behavior during the quench cycle; different parts of the component will cool at different rates, giving rise to possible different phase transformations. Flow stress data (stress-strain curves) for each phase are needed that cover the range of strain rates and temperatures that

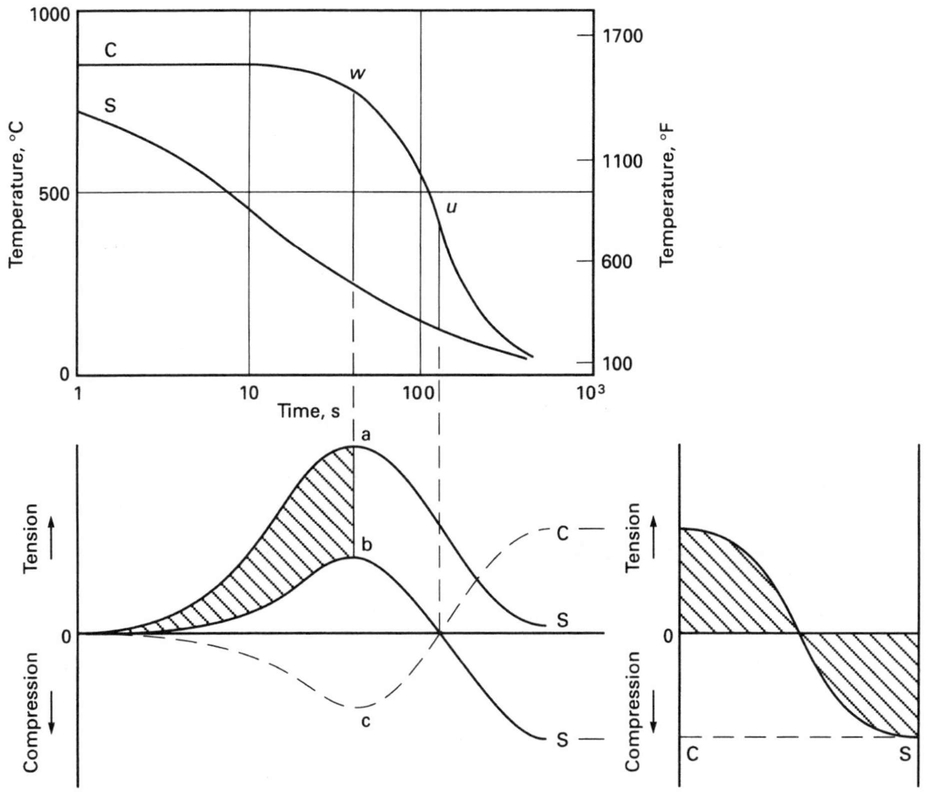

Fig. 56 Formation of thermal stresses on cooling in a 100 mm (4 in.) steel specimen. C designates the core, S the surface, u the instant of stress reversal, and w the time instant of maximum temperature difference. The top graph shows the temperature variation with time at the surface and at the core; the graph below shows the hypothetical thermal stress, a, which is proportional to the temperature difference between the surface and the core, the actual stress at the surface, b, which can never exceed the yield stress, and the actual stress in the core, c. To the right is shown the residual-stress distribution after completion of cooling as a function of the specimen radius. Source: Ref 116, 119

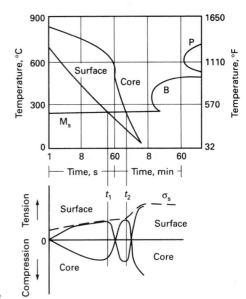

Fig. 57 Formation of residual stress on cooling, considering thermal expansion and the austenite-to-martensite transformation. The dashed line is the yield stress, σ_s, at the surface. See text for details. Source: Ref 116, 120

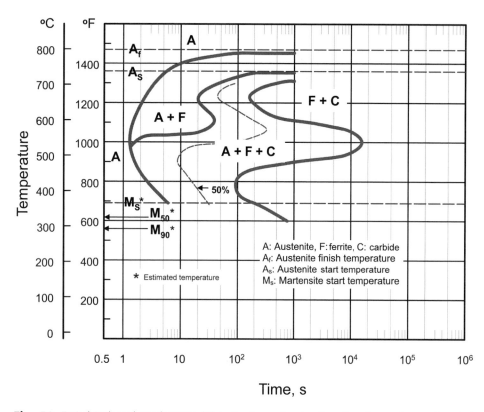

Fig. 58 Typical isothermal transformation (IT) or time-temperature transformation diagram for a carbon steel. Source: Ref 126

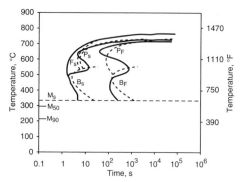

Fig. 59 Comparison between experimental (bold lines) and calculated time-temperature transformation diagram (dashed lines) for a 5140 steel, composition Fe-0.42C-0.68Mn-0.16Si-0.93Cr (wt%). M, martensite; P, pearlite; F, ferrite; B, bainite, subscript S, start; subscript F, finish. Source: Ref 131

(Ref 149, 150). Testing and evaluation of the material data needs for the models is also described by Eriksson et al. (Ref 151).

A number of thermomechanical models have been developed to describe the behavior of material that undergoes a phase transformation. The references cited here (Ref 82, 115, 143, 147, 152 to 159), although not all inclusive, cover descriptions of several of these models and are good starting points for those interested in learning more about the detailed formulations. However, in models of this type, the material behavior is generally described by assuming that the total strain rate, $\dot{E}_T$, is derived from five different sources:

$$\dot{E}_T = \dot{E}_E + \dot{E}_{TH} + \dot{E}_p + \dot{E}_{TR} + \dot{E}_{TP} \qquad (Eq~3)$$

where $\dot{E}_E$ is the elastic strain rate, $\dot{E}_{TH}$ is the thermal strain rate, $\dot{E}_p$ is the plastic strain rate, $\dot{E}_{TR}$ is the strain rate arising from the volume change associated with phase transformation, and $\dot{E}_{TP}$ is the strain rate associated with transformation plasticity.

Clearly, modeling phase changes represents a significant complication, and the accuracy of a model is not easily determined. Validation of the models may be broken down into two parts. The first part is the verification of the thermal and phase transformation model, where the cooling rates and hardness predictions are matched to experiments. The second part is a comparison of measured and predicted residual stresses.

Model Validation. Ferguson et al. (Ref 160) reported on the development of a software tool, Dante (Ref 87), to simulate hardening of alloy steel. Hardness may be used to support the accuracy of the cooling-rate and phase-prediction portion of the model. Figure 60 shows the results for a Jominy simulation for 5120 and 5140 steel bars. Agreement between the model-prediction and measured values is good.

A comparison between the measured and predicted hardness on a gear tooth was reported by Cai et al. (Ref 161). The commercial

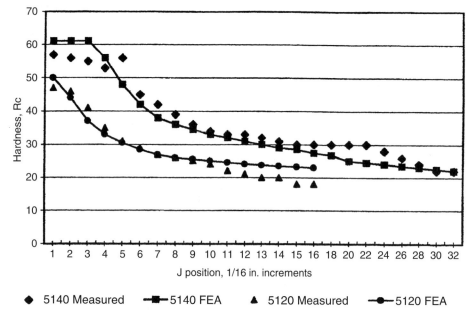

● 5140 Measured   ■—5140 FEA   ▲ 5120 Measured   ●—5120 FEA

Fig. 60 Jominy hardness data for 5120 and 5140 steel predicted by finite-element analysis (FEA) and measured. Source: Ref 160

are encountered in the heat treatment process. Additional data, also as functions of temperature, for each phase are needed, such as specific heat capacity, thermal conductivity, coefficient of thermal expansion, density, elastic modulus, and Poisson's ratio. Few of these data exist in the open literature, and it is a significant task to generate a dataset for a single material. Li et al. presented a method to derive the phase-transformation kinetics from dilatometry experiments (Ref 148), and recently, an ASTM International standard was issued on this topic

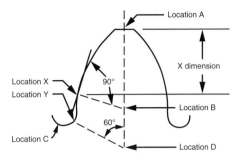

Fig. 61 Gear location label. Source: Ref 161

finite-element code Abaqus was coupled with a Caterpillar-developed heat treat simulation package, QSIM. Microhardness measurements were made along the depth of the tooth at the locations indicated by "A," "X," and "C" in Fig. 61. The predicted hardness was found to agree well with the measurements, as shown in Fig. 62. Good agreement between predicted and measured hardness values around a carburized and quenched low-alloy steel (JIS SCr420) gear blank was also reported in a detailed study by Shichino (Ref 162), as shown in Fig. 63.

Liščič et al. (Ref 163) also recently reported on the development of a two-dimensional model to predict the hardness in components.

The example presented was an axisymmetrical workpiece quenched in oil (Fig. 64). The comparison between predicted and measured hardness (Fig. 65) shows very good correlation, with differences within 2 HRC.

Incorporating transformation plasticity effects has been shown to be an important feature of models to predict distortion and residual stresses. Yamanaka et al. (Ref 164) used the code Hearts (Ref 89) to examine the effect of transformation plasticity on the distortion of a chromium-molybdenum steel ring (Fig. 66) undergoing a carburizing and oil quench process. The transformation plasticity in the model took a form based on the theory put forward by Greenwood and Johnson (Ref 165), with the transformation strain rate being proportional to the martensitic transformation rate, $\dot{\xi}$, and the deviatoric stress, s_{ij}:

$$\dot{\varepsilon}_{ij}^{tp} = \frac{3}{2} K\, h(\xi)\dot{\xi}s_{ij}$$
$$h(\xi) = 2(1-\xi)$$

(Eq 4)

where K, the transformation coefficient, was assumed to be constant and to be characterized by experiment.

Figure 67 shows the predicted distortion of a cross section of the ring at the end of the two processes simulated (the distortion is magnified 100 times). Normal quenching refers to the case where the ring is simply oil quenched and was used to isolate the effect of carburizing. The initial shape is given by the dotted line, and the shape predicted with the model that included transformation plasticity is shown by

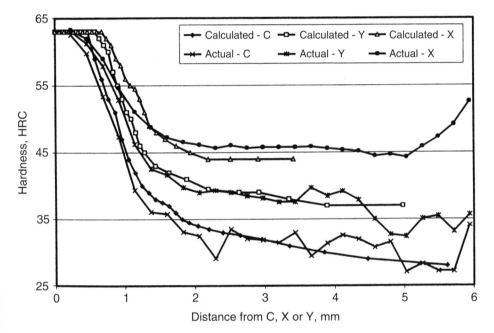

Fig. 62 Microhardness measurements and QSIM prediction comparison at various locations of the gear shown in Fig. 61. Source: Ref 161

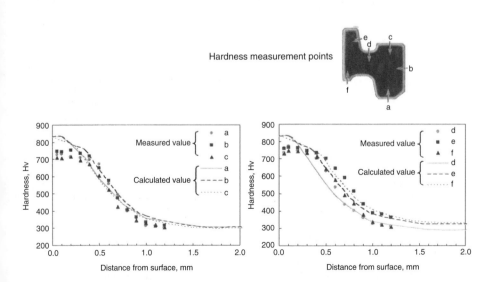

Fig. 63 Comparison between measured and predicted hardness distribution on a gear blank after carburizing and quenching. Source: Ref 162

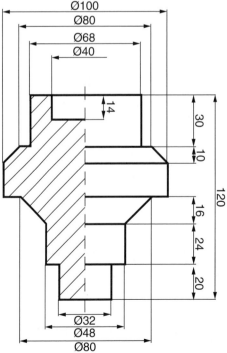

Fig. 64 Example of axially symmetric workpiece. Source: Ref 163

Hardness HRC

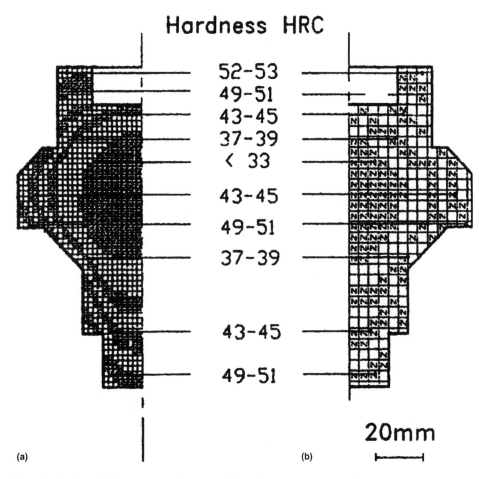

52-53
49-51
43-45
37-39
< 33
43-45
49-51
37-39
43-45
49-51

(a)   (b)

20mm

Fig. 65 Hardness distribution across the section of the workpiece shown in Fig. 64 after quenching in oil. (a) By mathematical modeling. (b) By experiment. Source: Ref 163

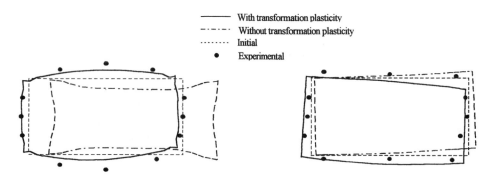

——— With transformation plasticity
—·—· Without transformation plasticity
·········· Initial
● Experimental

Fig. 67 Simulated distortion after (a) carburized and (b) normal quenching with measured data. Enlarged 100 times and central axis on the left. Source: Ref 164

Chemical composition of SCM420, wt%

C	Si	Mn	P	S	Cr	Mo
0.20	0.21	0.80	0.009	0.016	1.06	0.15

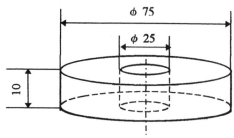

φ 75

φ 25

10

Fig. 66 Dimension and chemical composition of chromium-molybdenum steel specimen. Source: Ref 164

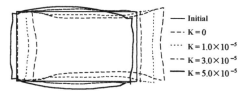

——— Initial
– – – K = 0
·–·–· K = 1.0×10⁻⁵
·· ·· K = 3.0×10⁻⁵
——— K = 5.0×10⁻⁵

Fig. 68 Effect of intensity of transformation plasticity on simulated distortion pattern. Enlarged 100 times and central axis on the left. Source: Ref 164

a solid line. It may be seen that in the carburized case, the effect of adding transformation plasticity is dramatic. Without transformation plasticity included, the predicted shape has concave surfaces compared to convex surfaces when the effect is taken into account. The distortion predicted by the model agrees well with measurements taken for both carburized and normal quenching.

It is interesting to note the effect of the transformation coefficient, K, on the predicted shapes. Figure 68 shows how the shape changes as the intensity of transformation plasticity is increased. The "tuning" of this parameter plays an important part in the accuracy of these models.

The importance of the transformation coefficient, K, was also shown by Franz et al. (Ref 166) for three steels, with German designations 20NiCrMo4-3-5, 50NiCrMo4-3-5, and 80NiCrMo4-3-5, having identical alloying element contents but with carbon contents of 0.2, 0.5, and 0.8 wt%. The cylinders were 20 mm (0.8 in.) in diameter and 60 mm (2.4 in.) long and quenched in helium at 20 °C (68 °F), which

was expanded from 10 bar to 1 bar (10 atm to 1 atm). The residual-stress distribution along the surface of the cylinders was obtained by x-ray diffraction measurements. The influence of the transformation plasticity constant, K, may be seen in Fig. 69, which shows the measured and predicted residual-stress distribution for all three steels. This clearly shows that in two of the steels, neglecting transformation plasticity results in the prediction of much higher residual-stress levels and a larger difference between the maximum and minimum values. The difference was not as significant for the third steel tested.

Work carried out by Allen and Fletcher (Ref 42) expanded on earlier studies (Ref 82, 167, 168) and used the HTCs generated experimentally to predict the residual stresses in a steel plate during water quenching. They compared the results with measurements made via a strain gage/material-removal technique. The effect of surface roughness of the plates was examined (as observed by the differences in the HTCs obtained during quenching, Fig. 17), and the model they developed incorporated phase transformation and transformation plasticity effects. Comparison between measured and experimental residual stresses in a 20 mm (0.8 in.) thick plate quenched in water was found to be very good (Fig. 70). It was concluded that the changes in the heat transfer resulting from surface condition had relatively little effect on the calculated stress distribution in the plate.

In a study by Vorster et al. (Ref 77), boiling heat-transfer equations were used to describe the boundary conditions on a 100 mm (4 in.)

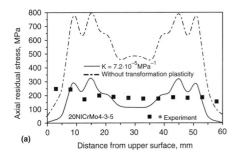

(a)

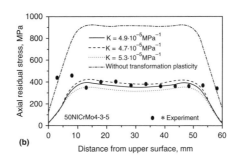

(b)

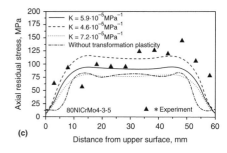

(c)

Fig. 69 Influence of the transformation plasticity constant, K, on the axial residual-stress distribution along the surface of the cylinders. Source: Ref 166

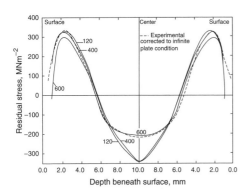

Fig. 70 Comparison between measured and calculated residual-stress distribution in water-quenched plates accounting for surface finish. Note: 600 grade is the smoothest finish, and 120 grade is the roughest. Source: Ref 42

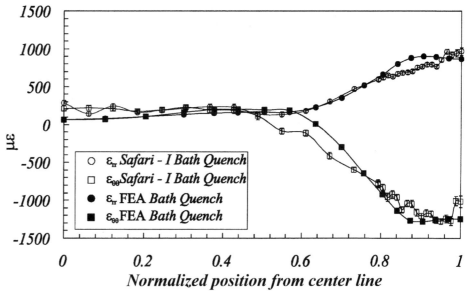

Fig. 71 Comparison of measured and predicted residual elastic radial and hoop strain. Measurements were made at Safari-I experimental nuclear facility, and the predictions were computed using the finite-element analysis (FEA) code Abaqus. Source: Ref 77

long by 16 mm (0.6 in.) diameter cylinder of AISI 317L steel. The commercial finite-element code Abaqus (Ref 75) was used to predict the residual strains in the cylinder, and these were compared with strain measurements made by neutron diffraction. A comparison between the measured and predicted residual radial and hoop strains is shown in Fig. 71. This shows good agreement, and the authors concluded that it is possible to accurately predict residual stress in quenched components with negligible error using boiling heat-transfer correlations found in the literature. They did state, however, that care must be taken when choosing heat-transfer correlations, because most apply to steady-state conditions rather than transient cooling operations (as found in quenching). Modeling transitional boiling with a linear relationship instead of a power law was found to significantly improve the numerical results.

Optimization Models. Optimizing has also been applied to the heat treatment of steel components. Li and Grandhi (Ref 169) presented a case where optimizing was used on a steel gear blank to reduce distortion during gas quenching. Deform (Ref 88) was used to predict the temperatures and stresses in the part during quenching, with phase-transformation effects included in the model. As in the optimizing models described earlier, the surface of the part was divided into regions in

which the HTC could be varied with time. The objective function was to minimize distortion, and constraints were placed on the surface hardness to meet the minimum required. Uniform residual stress was considered preferable in the example, and the standard deviation of the residual stress was used as a constraint to improve results. Figure 72 shows the quarter-section of the axisymmetric component together with the division of the surface into three regions. The hardness distribution and maximum principal residual-stress distribution for a reference design and the optimum design of the HTCs are shown in Fig. 73 and 74. In both cases, the distribution is more uniform after the optimization (which took 22 iterations).

One study reported by Li et al. (Ref 170) used the finite-element-based software Dante (Ref 87) to optimize the heat treatment of an induction-hardened, thin-walled spur gear made from AISI 5120 or AISI 5130. The heat generated by the induced eddy currents was used as input to the heating model, and the optimization was carried out to improve the

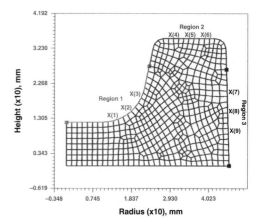

Fig. 72 Finite-element mesh of component showing the three heat-transfer regions. Source: Ref 169

water-spray quenching step of the hardening process. The objective was to minimize the distortion of the gear tooth while satisfying the residual-stress and microstructural requirements and placing a constraint on the maximum bore surface temperature. The geometry of the spur gear is shown in Fig. 75. A simplified, two-dimensional plane-strain model of a single gear tooth was employed for the optimization (Fig. 76). The range of design variables is listed in Table 3.

The ability to spray both the outside and inside diameters of the ring gear was assumed, and the HTCs were assumed to be constant with time. A delay time between the end of high-frequency induction heating and spray quenching was also a design variable. To reduce the number of iterations required, upper and lower bounds were placed on the HTCs. The evolution of the design variables is shown in Fig. 77, and the evolution of the critical parameters in the gear is shown in Fig. 78. The dashed line on the radial displacement plot is the expected displacement due to phase transformation. The optimizing process took 16 iterations, and the optimized results are shown in Fig. 79. The residual-stress distribution obtained was subsequently used as the starting condition to model the tooth loading during operation, with a view to assessing the gear fatigue life.

Modeling to Prevent Cracking during Heating or Quenching

Models to predict, and thereby eliminate, the risk of a part cracking during the heating or quenching process are less advanced than those used for predicting cooling rates or residual stresses. This is partly due to the lack of the necessary material property data to put into a model and/or the detailed criteria needed to interpret the resulting output from the model and predict cracking. The main reason, however, is that cracking during heating and quenching is not a major problem in industry, and hence, the subject has not seen the attention that has been devoted to other areas of process modeling.

The open literature in this area is sparse. Often, the reported analyses have been carried out after a part has cracked, and the models have subsequently been tuned to predict the cracking. Only after this tuning, and perhaps after further detailed material testing, are criteria developed that allow a technique to be established to predict and prevent cracking on other parts made of the same material.

There are two quite different aspects to this work: cracking upon heating and cracking upon cooling.

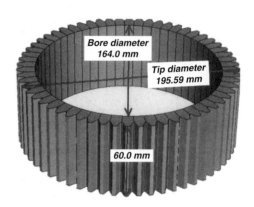

Fig. 75 Spur gear geometry with 60 teeth. Source: Ref 170

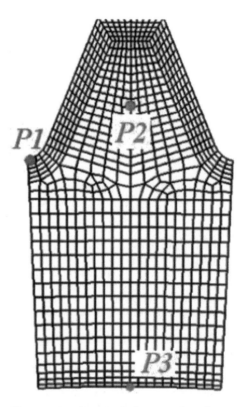

Fig. 76 Two-dimensional plane-strain finite-element mesh. Source: Ref 170

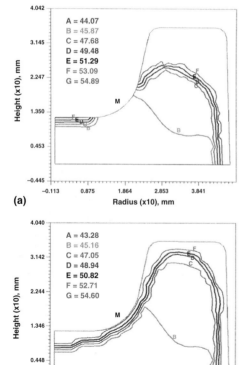

Fig. 73 Hardness distribution (HRC). (a) Reference design. (b) Optimum design. Source: Ref 169

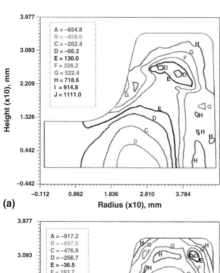

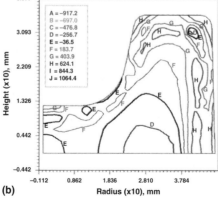

Fig. 74 Maximum principal residual stress (MPa). (a) Reference design. (b) Optimum design. Source: Ref 169

Table 3 Lower and upper bounds of design variables (DV)

	DV 1 (H on tooth surface), W/mm² K	DV 2 (H on bore surface), W/mm² K	DV 3 (delay time), s
Lower bound	0.005	0.005	0.0
Upper bound	0.02	0.02	2.0

Source: Ref 170

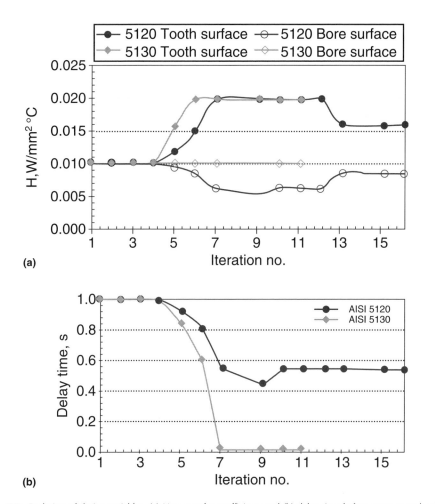

Fig. 77 Evolution of design variables. (a) Heat-transfer coefficients and (b) delay time before spray quench with optimization iterations. Source: Ref 170

Cracking upon Heating

A model to predict the temperatures and stresses during heating is virtually the same as one described previously for quenching. The same thermal property data for the material are needed, and the mechanical property data required are similar. Generally, the stress-strain data used are temperature dependent but ignore strain-rate and heating-rate effects. Heating models are described in the article "Heating and Heat-Flow Simulation" in this Volume, but briefly, if the part is being heated in a furnace chamber by gas or electric, the heat transfer to the part is generally by a combination of radiation and convection (with radiation dominating at temperatures higher than approximately 500 °C, or 930 °F). When multiple parts are in the furnace, the view that a particular part has of the walls, roof, and hearth of the furnace is partially obstructed by adjacent parts in the furnace, and this reduces the heat-transfer rate. This restriction is expressed by the term *view factor,* which has a value of between 0 and 1.0 and represents the portion of radiation emitted by one surface that falls onto the second surface. The analysis carried out may or

may not include radiation shadowing by adjacent pieces in the furnace; some commercially available codes are capable of calculating the view factors between the part and furnace components, taking into account the obstruction by other parts in the furnace. The second term affecting the rate of heat transfer from the furnace to the part is the emissivity of the materials comprising the part, the refractory materials, and, in the case of a gas/oil-fired furnace, the emissivity of the combustion products. The emissivity of a material may vary with temperature and surface condition, making the situation even more complex. Thus, the calculation of heat transfer in a furnace can be extremely complicated, particularly if the radiation interchanges between gases and the surfaces in the combustion chamber are accounted for (Ref 171, 172).

Fortunately, for most practical purposes, many variables can be ignored and the model simplified to where the heating model considers only a single part (the view factor from the furnace to the part either being set at 1.0 or varied around the surface, but shadowing by adjacent pieces is generally ignored). Similarly, the emissivity is generally fixed at a single value.

Generally, cracking during the heating cycle has only been expressed as a concern during processing of large ingots. Alam et al. (Ref 173, 174) describe a case where a finite-element model was used to compare the stresses generated in a large alpha-2 titanium aluminide ingot, Ti-24Al-11Nb, with 0.76 m (2.5 ft) diameter and 7.6 m (25 ft) length during two different heating cycles. The model used was a combination of Alpid (a forerunner of Deform, Ref 88) and Nike (Ref 94). Two cases were modeled to compare with real experience where ingots placed directly in a furnace at 1093 °C (2000 °F) cracked, whereas those charged into a furnace initially at 649 °C (1200 °F) for 10 h and then raised at 140 °C/h (250 °F/h) did not. Comparing the stresses generated to the yield stress and ultimate tensile strength (UTS) for the material (Fig. 80, Ref 174), it may be seen that the model indicates the probability of failure in the first case, where the stresses calculated are well above the yield stress and UTS of the material. Because a simple elastic analysis was used in the model, whereas, in reality, at higher temperatures some plastic and creep behavior would be expected, the model did not have the capability to directly predict failure. However, it may be seen that the calculated stresses remain below the yield point throughout the cycle in the case where the ingot is held at a lower temperature for the first 10 h (Fig. 81). Thus, the model correlates with practical experience where the ingot did not crack.

An interesting piece of work reported by Hara in 1963 (Ref 175 to 179) describes an investigation into the cause of cracking in large, high-carbon-chromium steel ingots when being heated to rolling temperature in a reheating furnace. At certain heating rates, cracks developed in the ingots, and modeling was used to examine the thermal stresses developed in the ingots under various furnace operating conditions. Circular ingots 400 mm (16 in.) in diameter by 1.5 m (5 ft) long and square ingots 400 mm (16 in.) on the sides by 1.5 m (5 ft) long were examined. As was explained previously in the description of models to predict cooling rate and residual stresses in steels, models to predict cracking in steels must also take into account the volumetric changes that result from the material undergoing phase transformations. Hara used a finite-difference model to calculate the temperature distribution through the thickness of the ingot and from this calculated the stresses and strains in the ingot, accounting for a phase change in the material. The calculations for three heating conditions (i.e., different heating rates before and after the start of phase transformation) are given in Fig. 82. From a series of calculations (Ref 176) and corresponding heating tests with ingots, it was concluded that cracking occurred only in ingots whose core had completed transformation with a surface-center temperature difference of more than 240 °C (465 °F). From this study, a practice was developed to reheat the ingots in a

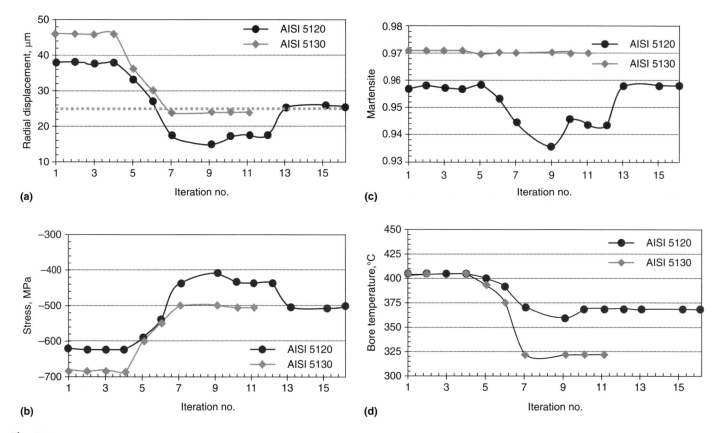

Fig. 78 Evolution of (a) distortion, (b) residual stress, (c) martensite volume fraction, and (d) bore temperature with optimization iterations. Source: Ref 170

reasonable time without cracking them Fig. 83 (Ref 178, 179).

In both of the aforementioned examples, it may be noted that modeling was carried out after cracking had occurred in production. The model was then used to explain the cause of cracking and to develop a practice to overcome the problem.

Cracking during Cooling

As one would expect, the prediction of quench cracking requires a relatively high level of accuracy from the model. The input data required for the model include all that described previously for cooling and residual-stress modeling but also require special material properties and theories of failure. Again, there are differences in the data required depending on the composition of the material.

In the case of nickel-base superalloys, early published work indicated success when modeling alloys that have little ductility over a certain temperature range. Stresses calculated at the surface of the part during quenching were simply compared to the yield point of the material. An example of a material having little ductility is shown in Fig. 84 (Ref 86), where the data plotted for on-cooling tensile tests indicate virtually zero ductility at a test temperature of approximately 870 °C (1600 °F) when cooled at a rate of 1100 °C/min (2000 °F/min) between the solution temperature and the test temperature.

The practical example shown in Fig. 85 (Ref 2) is for a disk quenched in one of two orientations: one upside down compared to the other. In one orientation, the disks were successfully quenched, but when quenched in the other orientation, the disks cracked. In this early quench-crack model, material property data were only available to allow an elastic analysis to be performed. However, this clearly shows the difference between the two cases, with stresses in one orientation staying below the yield point over the critical temperature range, while, in the other orientation, the yield point of the material is exceeded. Of interest in this case is that on-cooling tensile testing, described previously for residual-stress modeling, was carried out to obtain the yield stress, UTS, and percent reduction of area (a parameter indicating ductility) of the material. In this test, the material is taken to the solution heat treat temperature, cooled at a specific rate to the test temperature, and then pulled in tension at a certain strain rate until failure. Because the strength of the material being considered in the analysis depended to some extent on the cooling rate and strain rate, these parameters were selected based on the path that the material close to the surface of the disk would experience during quenching, as predicted by thermal models. The model used at the time was not capable of incorporating cooling-rate effects into the material model, so, a single, relatively high cooling rate was chosen for the

testing, corresponding to what the material at the surface of the part would experience. Strictly, a database for this type of model would cover a range of cooling rates and strain rates. However, creating such a database would require extensive testing and a model capable of handling such data. In general, to date, it has been found adequate to work with data generated at one cooling rate and strain rate but to carry out testing at a few other selected parameters to determine the sensitivity.

In the example detailed previously, the cracking criterion was very simple; that is, the calculated effective stress was simply compared to the yield point of the material. If the predicted stresses exceeded or came close to the yield point over a temperature range where the material had a very low ductility, then it was assumed that the part was at risk for cracking. This technique has been used to design the shape of parts to reduce the risk of cracking (for example, modifying the radius at cross-sectional changes and modifying the general shapes in areas where large cross-sectional changes occur). Some research has developed other criteria for failure, but, to date, these remain proprietary to individual companies.

Other researchers have attempted a fracture mechanics approach to the problem of cracking on cooling. Mao et al. (Ref 180, 181) used a specimen (Fig. 86) that was precracked at room temperature prior to conducting an on-cooling test. The precracks were in the range of 0.45

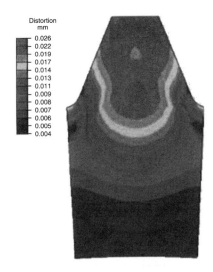

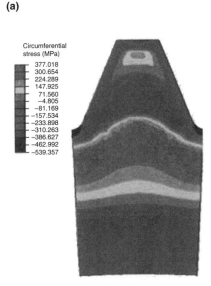

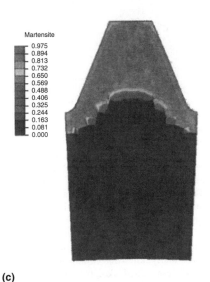

Fig. 79 Optimized result for gear made of AISI 5130. (a) Radial distortion. (b) Circumferential stress. (c) Martensite distribution. Source: Ref 170

$< a/w < 0.55$, where a is the crack length, and w is the width of the specimen. The specimen was heated to temperature, and, when stabilized, the system employed a constant displacement rate while the specimen was cooled at a specified rate. The decrease in temperature causes the quenching stress (load) to increase due to thermal contraction. The stress-intensity factor, K, was converted from Tada's empirical equation according to the on-cooling load and the precrack length (Ref 182):

$$K = \frac{P}{B\sqrt{W}} \frac{\sqrt{2\tan\frac{\pi a}{2w}}}{\cos\frac{\pi a}{2w}} \left[0.752 + 2.02\left(\frac{a}{w}\right) + 0.37\left(1 - \sin\frac{\pi a}{2w}\right)^3 \right] \quad \text{(Eq 5)}$$

where a is the precrack length at the onset of quenching, measured from the failed specimen; W is the width of the specimen; B is the thickness of the specimen; and P is the load recorded during quenching.

In this analysis, quench cracking is assumed to occur when K attains a certain value, at which either cracking occurs or a maximum load is reached. This point was defined as the cracking toughness, K_Q, and the temperature at which it occurred was called the failing temperature. However, no examples were cited where this method has been used to either predict cracking in a component or design shapes to prevent cracking.

As was explained in the discussion on cooling rate and residual-stress models, models to predict cracking in steels must take into account the volumetric changes that result from the material undergoing a transition from austenite to other phases.

Arimoto et al. (Ref 183) reported on a model developed in Japan by Inoue et al. (Ref 184) to investigate the cracking of a 12% Cr, 0.16% C martensitic stainless steel during

water quenching. The quenching process produced a circumferential crack, as shown in Fig. 87. The thermal properties used in the model were taken to be the same as AISI 1045, for which data were available, and the quenching HTC was assumed to vary with temperature. Most of the mechanical properties and flow stress data were also assumed to be the same as AISI 1045, including transformation strain and transformation plasticity. As expected, the simulation showed that the transformation to martensite started at the corners of the part and that the material at the fracture point transformed at an early stage in the quench. The radial stress history for a number of points in the part is presented in Fig. 88. This shows that the material at the fracture point is initially placed in compression, but after 154 s moves into tension. A little later in the quench, the maximum stress moves from point 4 to point 1, deeper in the part, where the austenite has not yet transformed to martensite. Because austenite is known to have better ductility than martensite, it was concluded that this stress level was more likely to result in plastic deformation rather than fracture. As in much of the early work reported on this topic, the criterion was simply a comparison of stresses and was carried out after a part had cracked.

A study to predict cracking in a purposely designed shape was reported by Arimoto et al. (Ref 185). The code Deform (Ref 88) was used to predict the three-dimensional stresses in a specimen designed by Owaku (Ref 186) to investigate quench cracking (Fig. 89). Two materials were used (Japanese standard S45C and SK4); quench tests were carried out in plain water and a 10% polyalkylen glycol (PAG) solution. Three test conditions were examined: nonagitated water, water with a velocity of 0.7 m/s, and nonagitated PAG. Microstructural changes in the material were considered in the

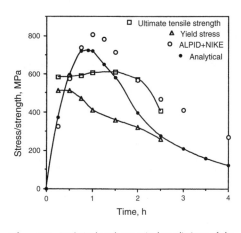

Fig. 80 Analytical and numerical predictions of the maximum axial stress developed at the center of a 0.76 m (2.5 ft) ingot placed directly into a furnace at 1093 °C (2000 °F). The model stress predictions are compared to the material strength data corresponding to the predicted instantaneous temperature at this location. Source: Ref 174

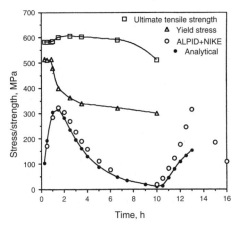

Fig. 81 Analytical and numerical predictions of the maximum axial stress developed when a 0.76 m (2.5 ft) ingot is charged into a 649 °C (1200 °F) furnace for 10 h, after which the furnace temperature is raised at a rate of 140 °C/h (250 °F/h). The stresses are compared with material strength data evaluated at the predicted temperature. Source: Ref 174

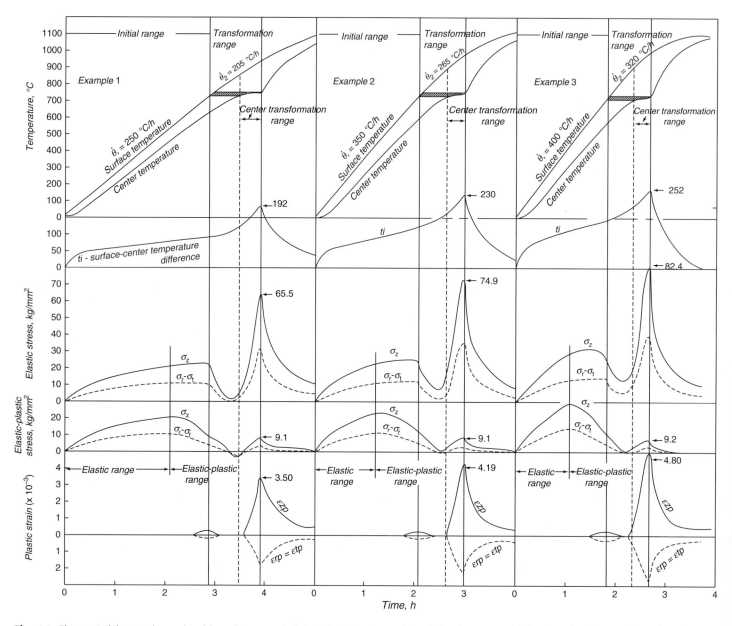

Fig. 82 Three typical theoretical examples of thermal stresses and plastic strains in ingot cores during the heating process of high-carbon-chromium steel ingots during heating. (Subscripts r, t, and z are radial, tangential, and axial stresses and strains, respectively.) Source: Ref 178

model, and tests were carried out with instrumented specimens to confirm that the temperature-dependent HTCs being used were reasonable. Figure 90 shows a typical cracked specimen after being quenched in still water, together with the predicted maximum principal stresses after 3.76 s into the quench. Point 3 on the surface of the hole, as shown in Fig. 91, corresponds very closely to one of the crack locations shown in Fig. 90. The maximum principal stress and the martensitic transformation history are presented in Fig. 92 for all three quench conditions and both steel types. The highest predicted stresses occurred in the cases of agitated water. Polymer quenching resulted in no apparent cracks in specimens of either of the steel compositions. When quenched in agitated water, the S45C alloy did not crack.

In still water, specimens made from both alloys cracked. The conclusion of the study was that quench cracking remains a complex phenomenon that is time dependent and subject to variables such as vapor blankets and localized boiling of the quenchant.

Cai et al. published work on the development of a model for an investigation into quench-cracking problems at the root of keyways machined in shafts (Ref 187). The study covered two steel types, SAE 1040 and SAE 4140, and compared the stresses developed during quenching in either water or oil. The shaft was 50 mm (2 in.) in diameter and 200 mm (8 in.) long, with a 10 mm (0.4 in.) wide and 10 mm (0.4 in.) deep keyway slot, as shown in Fig. 93. One corner at the bottom of the slot (the "A" location) had a sharp corner, and the

other corner (the "B" location) had a radius to reduce the stress-concentration effect. The model used the simple criterion for quench cracking as "stresses exceeding the strength of the material at temperature," that is, comparing the stresses predicted to the UTS and yield strength of the material. However, detailed tests were carried out to obtain the yield strength and UTS as functions of temperature. The tests comprised heating up the specimen to austenitizing temperature, cooling at an appropriate rate, and isothermally holding to form the desired microstructure at the desired temperatures. Finally, after this equalization, the tensile test was conducted. A dataset for SAE 4140 is shown in Fig. 94.

The model was internally developed at Caterpillar (Ref 188), with database routines built around the commercial code Abaqus (Ref 75). It includes thermodynamics data, transient behavior, phase transformation, and related transformation strain and plasticity data. It was determined that the most critical period was during the first few seconds of the quench, when the surface was cooling rapidly from the austenitizing temperature and before it reached the martensitic start temperature, M_s. When the temperature fell below the M_s and transformation began, the stress reversed rapidly from tension to compression at the near-surface region. The results of the simulations are summarized in Fig. 95. In the case where the SAE 1040 steel shaft was oil quenched (Fig. 95a), the low hardenability of the material coupled with the relatively mild cooling did not give

rise to large tensile stresses at point "A" on the keyway. In the simulation of water quenching (Fig. 95b), however, the tensile stresses predicted at "A" were significantly higher, a peak of close to 412 MPa and above the UTS in the early stages of the quench. The model for the SAE 4140 material showed higher stresses due

to its higher hardenablility. Quenching in oil forms high tensile stresses at the sharp corner (Fig. 95c), and a tensile stress of 470 MPa was predicted during water quenching (Fig. 95d). The results shown in Fig. 95 indicated that stresses developed at the sharp corner of the keyway during the early part of quenching

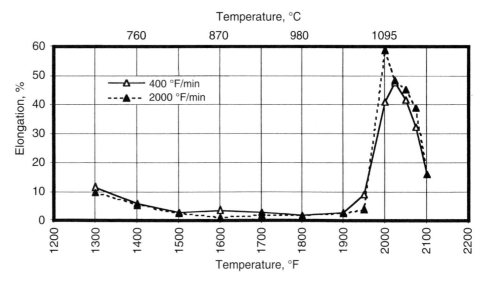

Fig. 84 On-cooling tensile test data showing effect of cooling rate on percent elongation of a nickel-base superalloy. Source: Ref 86

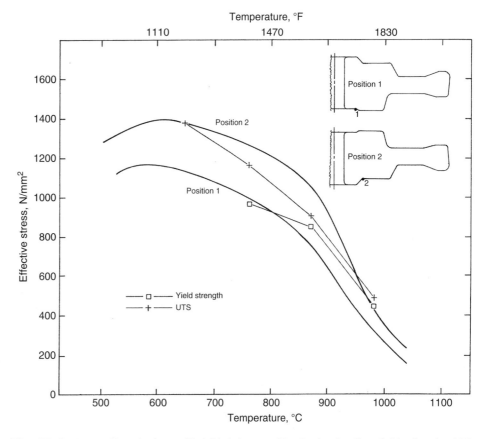

Fig. 83 Method of heating high-carbon-chromium steel ingots and the calculated corresponding values of thermal stresses and plastic strain in the core. (Subscripts r, t, and z are radial, tangential, and axial stresses and strains, respectively.) Source: Ref 178

Fig. 85 Stress-temperature plot for a turbine disk during quenching showing the effect of disk orientation. UTS, ultimate tensile strength. Source: Ref 2

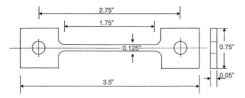

Fig. 86 Schematic of quench-cracking specimen. Source: Ref 180

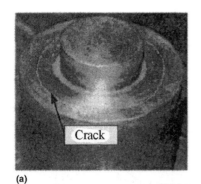

(a)

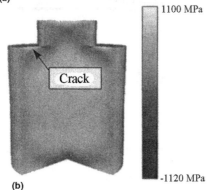

(b)

Fig. 87 (a) Quench crack and (b) simulation model (radial stress). Source: Ref 183

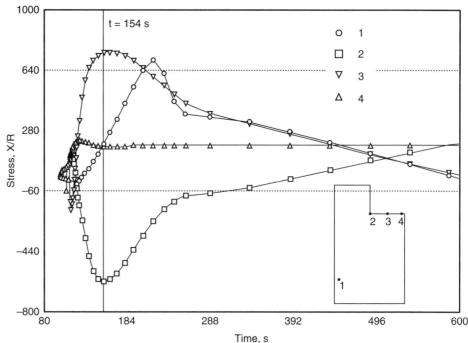

Fig. 88 Radial stress as a function of time for four workpiece locations. Source: Ref 183

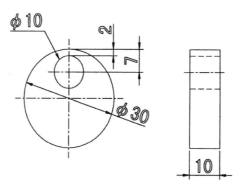

Fig. 89 Crack test specimen geometry and dimensions. Source: Ref 185

would exceed the UTS for the SAE 4140 material for both oil and water quenching and for the SAE 1040 material during water quenching. The only case where the stresses remained below the UTS during quenching with a sharp corner on the keyway was when an SAE 1040 shaft was quenched in oil.

Eliminating the sharp corner reduced the stresses significantly. In the case of SAE 1040, placing a radius at the root of the keyway, point "B," resulted in stresses below the UTS of the material for either quench process. For SAE 4140, a radius at the root kept the stresses below the UTS during oil quenching, while, during water quenching, the UTS was exceeded very briefly at approximately 0.5 s into the quench.

The model was validated empirically by carrying out quench tests with actual shafts of both materials, all having keyways with sharp corners. Tests were conducted where the bars were either quenched for 2 or 8 s before being removed and slowly cooled. It was reported that the experiments showed that quench cracks

initiate very early in the process and that the shafts quenched in water cracked, while those quenched in oil did not. Although not perfect (i.e., the oil-quenched SAE 4140 shafts did not crack), the model was concluded to be a valuable tool.

Clearly, developing accurate criteria for preventing quench cracking is difficult and thus far remains somewhat elusive. Arimoto et al. (Ref 189) proposed a methodology for a combination of tests and simulations in an attempt to develop a validated model. As far as is known by the author, no one has taken up this challenge.

Summary

Computer modeling of the heat treatment process has come a long way in the past couple

of decades. In some industries, it is common practice to use models to predict cooling rates (and hence the expected mechanical properties) and residual stresses in components that are heated and quenched during processing. In many cases, the shortcoming is not in the computer models but in the material properties and thermal boundary conditions needed as input data. Inverse methods are well established to determine the HTCs during quenching but require relatively time-consuming and expensive testing with thermocouple-equipped parts. Computational fluid dynamics codes are improving at a rapid pace and are likely to be used increasingly over the coming years to reduce or eliminate the need for testing.

Extensive work has been carried out to characterize the response of materials during phase changes and to develop models that incorporate the thermal, material, and mechanical interactions during heat treatment. This area of modeling is also likely to advance significantly over the next few years.

The number of cases where models predicting residual stresses have been validated with measurements has increased over the past few years. The results to date have been encouraging and show that, given good input data, the models are capable of predicting residual-stress levels of sufficient accuracy to be useful to engineers who develop processes and design components.

Predicting cracking during thermal processing is an area that pushes modeling to the current limits. The problem lies not so much in the computer codes as in the development of the criteria to apply to predict whether a material

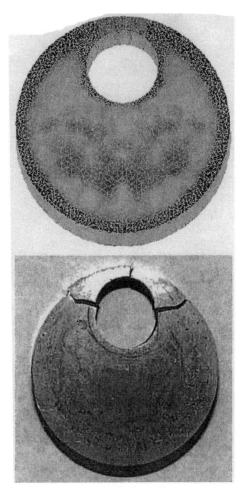

Fig. 90 Comparison for SK4 steel after still-water quench. Predicted maximum principal stress after 3.76 s (top) and cracked specimen (bottom). Source: Ref 185

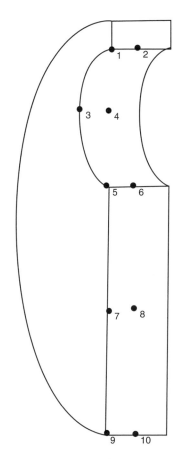

Fig. 91 Location of point-tracking analysis. Source: Ref 185

will fail or not. However, as materials are pushed to their limits, this area of heat treat modeling may rise in importance and receive much more attention. This is particularly true in the aerospace industry, where some of the newer materials that have been developed are more prone to quench cracking than the previous materials used.

ACKNOWLEDGMENTS

The author would like to thank the many authors of papers cited in this article who were contacted for their help to clarify points in their papers; there are too many to list individually here. However, there is one individual that deserves a special mention, and that is Dr. George Totten. He has been a friend for many years, and the author is very grateful for his generosity in pointing out a number of relevant references that have undoubtedly improved this article and, hopefully, made it better for students and researchers to find sources to assist them to continue their investigations beyond the introduction provided in these pages.

REFERENCES

1. J. Mackerle, Finite Element Analysis and Simulation of Quenching and Other Heat Treatment Processes: A Bibliography (1976–2001), *Comput. Mater. Sci.,* Vol 27, 2003, p 313–332
2. R.A. Wallis, N.M. Bhathena, P.R. Bhowal, and E.L. Raymond, "The Application of Process Modeling to Heat Treatment of Superalloys," Paper presented to the Specialists Meeting on Aerospace Materials Process Modelling, 65th Panel Meeting (Turkey), AGARD, Fall 1987, and *Ind. Heat.,* Vol LV (No. 1), Jan 1988, p 30–33
3. G.E. Totten, Ed., *Steel Heat Treatment: Metallurgy and Technologies,* CRC Press, Taylor & Francis Group, Boca Raton, FL, p 583–589, 606–609
4. J.J. Schirra and S. Goetschius, Development of an Analytical Model Predicting Microstructure and Properties Resulting from the Thermal Processing of a Wrought Powder Nickel-Base Superalloy Component, *Superalloys 1992,* The Metallurgical Society, AIME, p 437–446
5. T.C. Tszeng and V. Sarat, A Study of Fin Effects in the Measurement of Temperature Using Surface-Mounted Thermocouples, *Trans. ASME, J. Heat Transf.,* Vol 125, Oct 2003 p 926–935
6. J.M. Franchet, F. Devy, P.E. Mosser, Y. Honnorat, and A. Benallal, Residual Stress Modeling during the Oil Quenching of an Astroloy Turbine Disk, *Superalloys 1992,* The Metallurgical Society, AIME, p 73–82
7. M.A. Wells and K.J. Daun, Accurate Determination of Surface Heat Fluxes during Quenching Characterized by Boiling Water Heat Transfer, *J. ASTM Int.,* Vol 6 (No. 1), Paper ID JAI101818, Dec 2008
8. D. Li and M.A. Wells, Effect of Subsurface Thermocouple Installation on the Discrepancy of the Measured Thermal History and Predicted Surface Heat Flux during a Quench Operation, *Metall. Trans. B,* Vol 36, June 2005, p 343–354
9. G. Stoltz, Numerical Solutions to an Inverse Problem of Heat Conduction for Simple Shapes, *Trans. ASME, J. Heat Transf.,* Feb 1960, p 20–26
10. J.V. Beck, B. Blackwell, and C.A. St. Clair, *Inverse Heat Conduction,* 1st ed., John Wiley, 1985
11. J.V. Beck, B. Blackwell, and A. Haji-Sheikh, Comparison of Some Inverse Heat Conduction Methods Using Experimental Data, *J. Heat Mass Transf.,* Vol 39 (No. 19), 1996, p 3649–3657
12. D.M. Trujillo and H.R. Busby, *Practical Inverse Analysis in Engineering,* 1st ed., CRC Press, 1997
13. N. Zabaras, Inverse Problems in Heat Transfer, Chapter 17, *Handbook of Numerical Heat Transfer,* 2nd ed., W.J. Minkowycz, E.M. Sparrow, and J.Y. Murthy, Ed., John Wiley & Sons, 2006
14. D.E. Smith, Optimization-Based Inverse Heat Transfer Analysis for Salt Quenching of Automotive Components, *Int. J. Veh. Des.,* Vol 25 (No. 1–2), Special Issue, 2001, p 23–39
15. D.E. Smith, Computing Heat Transfer Coefficients for Industrial Processes, *19th ASM Heat Treating Society Conference Proceedings, Including Steel Heat Treating in the New Millennium,* ASM International, Nov 1999, p 325–333
16. V. Sahai, "Analysis of Heat Transfer during Quenching of a Gear Blank," Lawrence Livermore National Laboratory Rpt UCRL-JC-133520, March 1999
17. M. Li and J.E. Allison, Determination of Thermal Boundary Conditions for the Casting and Quenching Process with the Optimization Tool OptCast, *Metall. Trans. B,* Vol 38 (No. 4), Aug 2007, p 567–574
18. J.T. Kim, C.H. Lim, J.K. Chio, and Y.K. Lee, A Method for the Evaluation of Heat Transfer Coefficient by Optimization Algorithm, *Solid State Phenom.,* Vol 124–126, 2007, p 1637–1640

19. V. Dumek, M. Druckmuller, M. Raudensky, and K.A. Woodbury, Novel Approaches to the ICHP: Neural Networks and Expert Systems, Inverse Problems in Engineering: Theory and Practice, *Proceedings of the First International Conference on Inverse Problems in Engineering: Theory and Practice,* June 13–18, 1993 (Palm Coast, FL), ASME Book No. I00357, p 275–282

20. S.S. Sablani, A. Kacimov, J. Perret, A.S. Mujumdar, and A. Campo, Non-Iterative Estimation of Heat Transfer Coefficients Using Artificial Neural Network Models, *Int. J. Heat Mass Transf.,* Vol 48, 2005, p 665–679

21. E.H. Shiguemori, F.P. Harter, H.F. Campos Veldo, and J.D.S. da Silva, Estimation of Boundary Conditions in Conduction Heat Transfer by Neural Networks, *Tendências em Matemática Aplicada e Computacional,* No. 2, 2002, p 189–195

22. V. Paschkis and G. Stolz, *J. Met.,* Aug 1956, p 1074–1075

23. M. Mitsutsuka and K. Fukuda, The Transition Boiling and Characteristic Temperature in Cooling Curve during Water Quenching of Heated Metal, *Trans. ISIJ,* Vol 16, 1976, p 46–50

24. R.F. Price and A.J. Fletcher, Determination of Surface Heat Transfer Coefficients during Quenching of Steel Plates, *Met. Technol.,* May 1980, p 203–211

25. M. Bamberger and B. Prinz, Determination of Heat Transfer Coefficients during Water Cooling of Metals, *Mater. Sci. Technol.,* Vol 2, April 1986, p 410–415

26. D.M. Trujillo and R.A. Wallis, Determination of Heat Transfer from Components during Quenching, *Ind. Heat.,* Vol LVI (No. 7), July 1989, p 22–24

27. R.A. Wallis, Using Computer Programs to Calculate Heat Transfer, *Heat Treat.,* Vol XXI (No. 112), Dec 1989, p 26, 27, 31

28. R.A. Wallis, The Estimation of Surface Heat Fluxes and Heat Transfer Coefficients during Quenching, *Proceedings of Third Annual Inverse Problems in Engineering Seminar,* J.V. Beck, Ed., Michigan State University, June 1990

29. S. Segerberg and J. Bodin, Variation in the Heat Transfer Coefficient around Components of Different Shapes during Quenching, *Proceedings of the First International Conference on Quenching and Control of Distortion,* Sept 1992 (Chicago, IL), p 165–170

30. M.F. Cross, J.C. Bennett, Jr., and R.W. Bass, Developing Empirical Equations for Heat Transfer Coefficients on Metallic Disks, *Proc. 19th ASM Heat Treating Society Conference,* Nov 1–4, 1999 (Cincinnati, OH), ASM International, p 335–342

31. R. Stringfellow, "Prediction and Control of Heat Treat Distortion of Helicopter Gears," Aviation Applied Technology Directorate, U.S. Army Aviation and Troop Command Report USATCOM TR 95-D-5, Arthur D. Little Inv., Cambridge, MA, 1995

32. G.E. Totten, C.E. Bates, and N.A. Clinton, *Handbook of Quenchants and Quenching Technology,* ASM International, 1993, p 106–107

33. H. Zieger, *Neue Hutte,* Vol 9, 1986, p 339–343

34. C.E. Bates, G.E. Totten, and R.L. Brennan, Quenching of Steels, *Heat Treating,* Vol 4, *ASM Handbook,* ASM International, 1991, p 67–120

35. H.M. Tensi, H.J. Spies, A. Spengler, and A. Stich, "Wechselwirkung zwischen Benetzungskinematik und Stahlhärtung beim Tauchkühlen," DFG Rep., Contracts Te 65/35-1,2 and Sp 367/1-1,2, Deutsche Forschungsgemeinschaft, Bonn, 1994

36. M. Sedighi and C.A. McMahon, The Influence of Quenchant Agitation on the Heat Transfer Coefficient and Residual

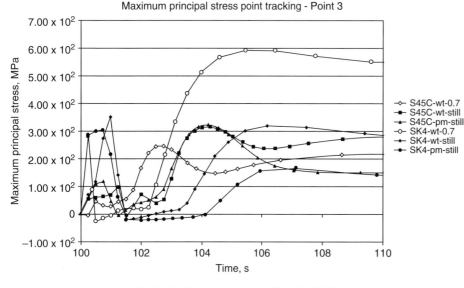

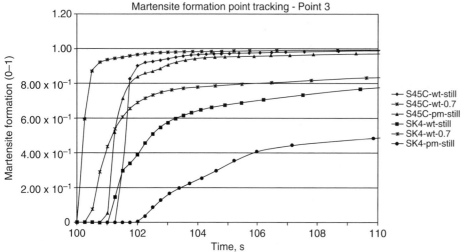

Fig. 92 Point 3 (see Fig. 91) tracking analysis for maximum principal stress (top) and martensite transformation (bottom). Quenching starts at a time of 100s. Source: Ref 185

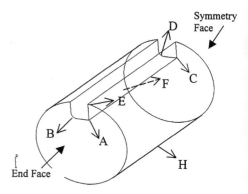

Fig. 93 Keyway shaft and characteristic locations. Source: Ref 187

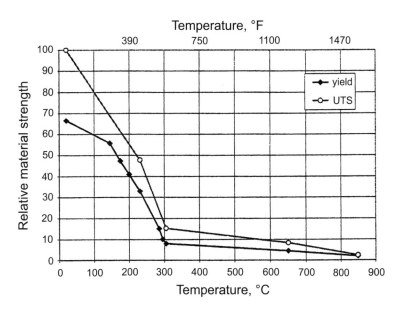

Fig. 94 Relative ultimate tensile strength (UTS) and yield strength for SAE 4140 as a function of temperature in a fully martensitic transformation case. Source: Ref 187

Stress Development in the Quenching of Steels, *Proc. Inst. Mech. Eng.,* Vol 214 Part B, 2000, p 555–567

37. K.J. Mason and I. Capewell, The Effect of Agitation on the Quenching Characteristics of Oil and Polymer Quenchants, *Heat Treat. Met.,* Vol 4, 1986, p 99–103

38. R.A. Wallis, R.I. San Pedro, and S. Owens, Four Years Experience Using an Inconel Probe to Test the Cooling Characteristics of a Production Quench Oil, *17th ASM Heat Treating Society Conference Proceedings,* Sept 1997, ASM International, 1998, p 457–466

39. M. Narazaki, Effects of Work Properties on Cooling Characteristics during Quenching, *Netsu Shori (J. Jpn. Soc. Heat Treat.),* Vol 35, 1995, p 221–226

40. M. Narazaki, G.E. Totten, and G.M. Webster, Hardening by Reheating and Quenching, *Handbook of Residual Stress and Deformation of Steels,* G.E. Totten, T. Inoue, and M. Howes, Ed., ASM International, 2002, p 248–295

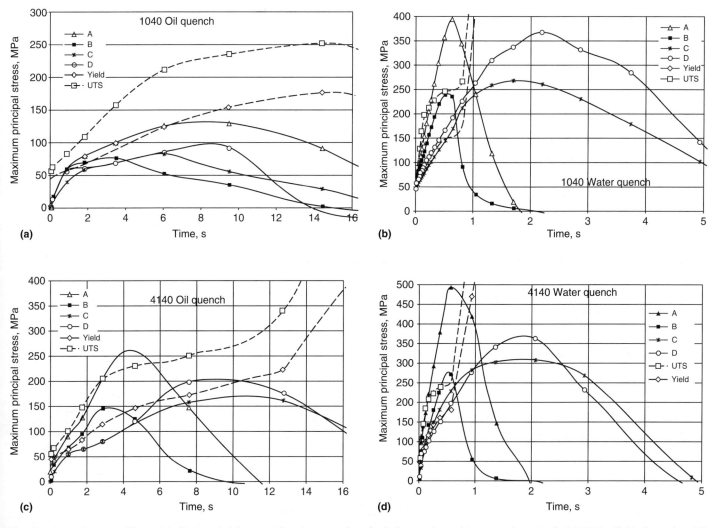

Fig. 95 Results from quenching model of keyway shaft for two steels and two quench media. Early stage of quenching stress at key points for SAE 1040 with (a) oil quench and (b) water quench. Early stage of quenching stress at key points for SAE 4140 with (c) oil quench and (d) water quench. UTS, ultimate tensile strength. Source: Ref 187

41. K. Narayan Prabhu and P. Fernandes, Effect of Surface Roughness on Metal/Quenchant Interfacial Heat Transfer and Evolution of Microstructure, *Mater. Des.,* Vol 28, 2007, p 544–550

42. F.S. Allen and A.J. Fletcher, Effect of Surface Conditions on Generation of Stress and Strain in Quenched Steel Plates, *Mater. Sci. Technol.,* Vol 3, April 1987, p 291–298

43. R. Bass, D. Leonard, M. Allen, J.C. Bennett, M. Cross, and K. Brown, Heat Transfer of Turbine Disks in a Liquid Quench: Part I—Experimental Setup, *Proceedings of the 17th ASM Heat Treating Society Conference,* Sept 1997 (Indianapolis, IN), ASM International, 1998, p 341–345

44. R. Bass, D. Leonard, J. Morral, J.C. Bennett, M. Cross, M. Allen, and K. Brown, Heat Transfer of Turbine Disks in a Liquid Quench: Part II—Experimental Results for a Solid Disk, *Proceedings from First International Automotive Heat Treating Conference,* July 1998 (Puerto Vallarta, Mexico), p 339–347

45. R. Bass, D. Leonard, M. Allen, J. Bennett, M. Cross, J. Morral, and K. Brown, Heat Transfer of Turbine Disks in a Liquid Quench: Part III—Experimental Results for a Disk with a Bore, *Proceedings of the 18th ASM Heat Treating Society Conference,* Oct 1998, ASM International, 1999, p 552–556

46. R. Bass, M. Allen, D. Leonard, M. Cross, K. Brown, J. Bennett, and J. Morral, "Heat Transfer of Turbine Disks in a Liquid Quench, Program Summary," Nov 17, 1998 (University of Connecticut)

47. M. Cross, "Applications in Quenching Heat Transfer: Liquid Quenching of Turbine Disks and Chill Down in Cryogenic Fluid Transfer Lines," Ph.D. Thesis, University of Connecticut, 2003

48. Ansys Inc., Canonsburg, PA

49. J. Ferrari, N. Ipek, N. Lior, and T. Holm, Gas Quench Vessel Efficiency: Experimental and Computational Analysis, *Proceedings of ASM 18th Heat Treating Society Conference,* Oct 1998, ASM International, 1999, p 500–505

50. X. Chen, D. Zang, L. Meekisho, J. Pan, and M. Hu, Temperature and Cycle, Optimization in Conventional and Vacuum Heat Treatment Processes, *Proceedings of the 16th ASM Heat Treating Society Conference and Exposition,* March 1996 (Cincinnati, OH), ASM International, p 197–203

51. G. Pellegrino, F. Chaffotte, J.F. Douce, S. Denis, J.P. Bellot, and P. Lamesle, Efficient Numerical Simulation Techniques for High Pressure Gas Quenching, *Proceedings of the 23rd ASM Heat Treating Society Conference,* Sept 2005 (Pittsburgh, PA), ASM International, p 320–328

52. Y. Luo, J. Kang, B. Liu, and Y. Rong, Numerical Simulation of Gas Quenching Process of Workpieces, *Proceedings of 24th ASM Heat Treating Society Conference,* Sept 2007 (Detroit, MI), ASM International, p 385–389

53. M. Hunkel, Th. Lübben, O. Belkessam, U. Fritsching, F. Hoffmann, and P. Mayr, Modeling and Simulation of Coupled Gas and Material Behavior during Gas Quenching, *Proceedings of the 21st ASM Heat Treating Society Conference,* Nov 2001 (Indianapolis, IN), ASM International, p 32–40

54. S. Schuettenberg, M. Hunkel, U. Fritsching, and H.-W. Zoch, Distortion Compensation by Means of Quenching in Flexible Jet Fields, *Proceedings of the 23rd ASM Heat Treating Society Conference* (Pittsburgh, PA), ASM International, 2005, p 211–215

55. O. Macchion, "CFD in the Design of a Gas Quenching Furnace," Thesis, Department of Mechanics, Royal Institute of Technology, Stockholm, Sweden, June 2005

56. D.R. Garwood, J.D. Lucas, R.A. Wallis, and J. Ward, "Computer Simulation of the Quenching Process: Prediction of Cooling Rates and Fluid Flow," Paper presented at First International Conference on Quenching and Control of Distortion, Sept 1992 (Chicago, IL); also published as Modeling of the Flow Distribution in an Oil Quench Tank, *JMEPEG,* Vol 1, Dec 6 1993, p 781–787

57. N. Bogh, Quench Tank Agitation Design Using Flow Modeling, *Proceedings of the International Heat Treating Conference: Equipment and Processes,* ASM International, 1994, p 51–54

58. W.W. Bower, A.B. Cain, and T.D. Smith, Computational Simulations of Quench Tank Flow Patterns, *Proceedings of the 17th ASM Heat Treating Society Conference,* ASM International, Sept 1997, p 389–383

59. D.S. MacKenzie, A. Kumar, and H. Metwally, Optimizing Agitation and Quench Uniformity Using CFD, *Proceedings of the 23rd ASM Heat Treating Society Conference,* ASM International, 2005, p 271–278

60. A.J. Baker and N.S. Winowich, Issues Critical to CFD Simulation of Heat Treating and Quenching Processes, *Proceedings of the Third International Conference on Quenching and Control of Distortion* (Prague, Czech Republic), ASM International, 1999, p 390–402

61. S. Kernazhitskiy, "Numerical Modeling of a Flow in a Quench Tank," Master of Science Thesis, Mechanical Engineering Dept., Portland State University, Portland, OR, 2003

62. N. Chen, B. Liao, J. Pan, Q. Li, and C. Gao, Improvement of the Flow Rate Distribution in Quench Tank by Measurement and Computer Simulation, *Mater. Lett.,* Vol 60, 2006, p 1659–1664

63. S. Kernazhitskiy and G. Recktenwald, Numerical Modeling of Flow in a Large Quench Tank, *Proceedings of HT-FED2004, 2004 ASME Heat Transfer/Fluids Engineering Summer Conference,* Paper HT-FED2004-56419, July 2004 (Charlotte, NC)

64. R.A. Wallis, D.R. Garwood, and J. Ward, The Use of Modeling Techniques to Improve the Quenching of Components, *Proceedings of the International Heat Treating Conference: Equipment and Processes,* April 1994 (Schaumburg, IL), ASM International, p 105–116

65. D. Rondeau, "The Effects of Part Orientation and Fluid Flow on Heat Transfer around a Cylinder," Masters Thesis, Materials Science and Engineering, Worcester Polytechnic Institute, Worcester, MA, May 2004

66. A. Kumar, H. Metwally, S. Paingankar, and D.S. MacKenzie, Evaluation of Flow Uniformity Around Automotive Pinion Gears during Quenching, *IFHTSE Fifth International Conference on Quenching and Control of Distortion,* April 2007 (Berlin), p 69–76

67. I. Golobič, E. Pavolvič, J. Von Hardenberg, M. Berry, R.A. Nelson, D.B.R. Kenning, and L.A. Smith, Comparison of a Mechanisitic Model for Nucleate Boiling with Experimental Spatio-Temporal Data, *Trans. IChemE,* Part A, April 2004; *Chem. Eng. Res. Des.,* Vol 82 (A4), p 435–444

68. M. Maniruzzaman and R. Sisson, Bubble Dynamics during Quenching of Steel, *Proceedings of the 21st ASM Heat Treating Society Conference,* Nov 2001 (Indianapolis, IN), ASM International, 2001, p 104–111

69. P. Krukovskyi, N. Kobasoc, and A. Polubinskiy, CFD—Analysis of a Part Under Quenching as a Heat Transfer Conjugate Problem, *Int. ASME Trans.,* Vol 2 (No. 9), Nov 2005, p1723–1728

70. D. Schick, D. Chenoweth, N. Palle, C. Mack, W. Copple, W.-T. Lee, W. Elliot, J. Park, R. Lenarduzze, H. Walton, and M. Howes, Development of a Carburizing and Quenching Simulation Tool: Determination of Heat Transfer Boundary Conditions in Salt, *Proceedings of the Second International Conference on Quenching and the Control of Distortion* (Cleveland, OH), ASM International, 1996, p 357–366

71. A. Banka, J. Franklin, L. Zhichao, B.L. Ferguson, and M. Aronov, Applying CFD to Characterize Gear Response during Intensive Quenching Process, *Proceedings of the 24th ASM Heat Treating Society Conference,* Sept 2007 (Detroit, MI), ASM International, p 147–155

72. N.I. Kobasco, B.K. Ushakov, and W.S. Morhunink, Design of Steel Intensive Quench Process, *Handbook of Metallurgical Process Design,* G.E. Totten, K.

Funatani, and L. Xie, Ed., Marcel Dekker, New York, NY, 2004, p 733–764

73. M. Springmann and A. Kühhorn, Coupled Thermal-Multiphase Flow Analysis in Quenching Processes for Residual Stress Optimization in Compressor and Turbine Disks, *Proceedings of PVP2008, 2008 ASME Pressure Vessels and Piping Division Conference,* Paper PVP2008-61126, July 27–31, 2008 (Chicago, IL)

74. Fluent/Ansys, Lebanon, NH

75. *Abaqus, Simulia—Dassault Systèmes,* Rising Sun Mills, Providence, RI

76. *Industrial Quenching Oils—Determination of Cooling Characteristics—Nickel-Alloy Probe Test Method,* 1st ed., ISO 9950, 1995

77. W.J.J. Vorster, M.W. Van Der Watt, A.M. Venter, E.C. Oliver, D.G.L. Prakash, and A.M. Korsunsky, Influence of Quenchant Hydrodynamics and Boiling Phase Incipient Temperature Shifts on Residual Stress Formation, *Heat Transf. Eng.,* Vol 30 (No. 7), 2009, p 564–573

78. *Aerospace Structural Metals Handbook,* Belfour Stulen, Inc., Syracuse University Press, 1977

79. *Alloy Digest,* ASM International

80. R. Wallis and P.R. Bhowal, Property Optimization in Superalloys through the Use of Heat Treat Process Models, *Superalloys 1988,* The Metallurgical Society, AIME, p 525–534

81. B. Boley and J.H. Weiner, *Theory of Thermal Stresses,* Reprint edition 1985 with corrections, Robert Krieger Publishers, 1985

82. A.J. Fletcher, *Thermal Stress and Strain Generation in Heat Treatment,* 1st ed., Elsevier Science Publishers Ltd., 1989

83. Fundamentals of the Modeling of Microstructure and Texture Evolution, *Fundamentals of Modeling for Metals Processing,* Vol 22A, *ASM Handbook,* ASM International, 2009, p 153–321

84. Materials Fundamentals, *Fundamentals of Modeling for Metals Processing,* Vol 22A, *ASM Handbook,* ASM International, 2009, p 441–488

85. Modeling of Microstructures, *Fundamentals of Modeling for Metals Processing,* Vol 22A, *ASM Handbook,* ASM International, 2009, p 489–582

86. R.D. Kissinger, Cooling Path Dependent Behavior of a Supersolvus Heat Treated Nickel Base Superalloy, *Superalloys 1996,* The Metallurgical Society, AIME, p 687–695

87. Dante, Deformation Control Technologies Inc., Cleveland, OH

88. Deform, Scientific Forming Technologies Corporation, Columbus, OH

89. T. Inoue and K. Arimoto, Development and Implementation of CAE System "HEARTS" for Heat Treatment Simulation Based on Metallo-Thermo Mechanics, *J. Mater. Eng. Perform.,* Vol 6, 1997, p 51–60

90. LS-Dyna, Livermore Software Technologies Corporation, Livermore, CA

91. Sysweld, ESI Group, Paris, France

92. R. Wallis and I. Craighead, Prediction of Residual Stresses in Gas Turbine Components, *JOM,* Vol 47 (No. 10), Oct 1998, p 69–71

93. A.B. Shapiro and A.L. Edwards, "TOPAZ Heat Transfer Code Users Manual and Thermal Property Data Base," Lawrence Livermore National Laboratories Report UCRL-ID104558, May 1990

94. J.O. Hallquist, "LS-NIKE2D, A Vectorized Implicit Finite Deformation Code for Analyzing the Static and Dynamic Response of 2-D Solids with Interactive Rezoning and Graphics," Livermore Software Technology Report 1006, July 1990

95. G.S. Schajer, Measurement of Non-Uniform Residual Stresses Using the Hole Drilling Method, Part II—Practical Application of the Integral Method, *ASME. J. Eng. Mater. Technol.,* Vol 110, 1988, p 334–349

96. M.A. Rist, S. Tin, B.A. Roder, J.A. James, and M.R. Daymond, Residual Stresses in a Quenched Superalloy Turbine Disc: Measurements and Modeling, *Metall. Trans. A,* Vol 37, Feb 2006, p 459–467

97. D. Dye, B.A. Roder, S. Tin, M.A. Rist, J.A. James, and M.R. Daymond, Modeling and Measurement of Residual Stresses in a Forged IN718 Superalloy Disk, *Superalloys 2004,* The Metallurgical Society, AIME, p 315–322

98. U. Cihak, P. Staron, W. Marketz, H. Leitner, J. Tockner, and H. Clemens, Residual Stresses in Forged IN718 Turbine Discs, *Z. Metalkd.,* Vol 95, 2004, p 663–667

99. U. Cihak, M. Stockinger, P. Staron, J. Tockner, and H. Clemens, Characterization of Residual Stresses in Compressor Discs for Aeroengines: Neutron Diffraction and Finite Element Simulations, *Proceedings of the Sixth International Symposium on Superalloys 718, 625, 706 and Derivatives,* E.A. Loria, Ed., Oct 2005, TMS, Warrendale, PA, 2005, p 517–526

100. U. Cihak, Correction to Fig. 7 in Ref 99, private communication, July 2009

101. R.A. Wallis, R. Garwood, J. Ward, and Q. Xia, Quenching Using Air-Water Mixtures, *Proceedings of the Second International Conference on Quenching and the Control of Distortion,* Nov 4–7, 1996 (Cleveland, OH), ASM International, p 463–472

102. D.V. Budrin and V.M. Kondratov, Quenching in Air-Water Mixtures, *Metalloved. Term. Obrab. Met.,* No. 6, June 1965, p 367–370 (in Russian)

103. P. Archambault, G. Didier, F. Moreaux, and G. Beck, Computer Controlled Spray Quenching, *Met. Prog.,* Oct 1984, p 67–72

104. S. Segerberg, Spray Quenching—A Method with Great Flexibility, *Proceedings of the Sixth International Congress on Heat Treatment of Materials,* G. Krauss, Ed., Sept 1988 (Chicago, IL), p 177–181

105. N.V. Antonishin, M.A. Geller, and M.S. Zheludkevich, Quenching Articles with a Computer-Controlled Gas-Liquid Stream, Translated from *Metalloved. Term. Obrab. Met.,* No. 3, March 1989, p 7–9 (in Russian)

106. M. de Oliveira, J. Ward, D.R. Garwood, and R.A. Wallis, Quenching of Aerospace Forgings from High Temperatures Using Air-Assisted, Atomized Water Sprays, *JMEPEG,* Vol 11, 2002, p 80–85

107. S. Schuettenberg, F. Krause, M. Hunkel, H.-W. Zoch, and U. Fritsching, Quenching with Fluid Jets, *Mat.-wiss. Werkst. tech.,* Vol 40 (No. 5–6), 2009, p 408–413

108. P.J. Röhl and S.K. Srivatsa, "A Comprehensive Approach to Engine Disk IPPD," American Institute of Aeronautics and Astronautics Paper AIAA-1997-113, 38th AIAA/ASME/ASCE/AHS/ASC Structures, Structural Dynamics and Materials Conference and Exhibit and AIAA/ASME/AHS Adaptive Structures Forum, April 1997 (Kissimmee, FL), Collection of Technical Papers, Pt. 2 (A97-24112 05-39)

109. P.J. Röhl, B. He, and P. Finningan, "A Collaborative Optimization Environment for Turbine Engine Development," American Institute of Aeronautics and Astronautics Paper AIAA-1998-4734, Seventh AIAA/USAF/NASA/ISOO Symposium on Multidisciplinary Analysis and Optimization, Sept 1998 (St. Louis, MO), Collection of Technical Papers, Pt. 1 (A98-39701 10-31)

110. iSight, Engineous Software, Inc., Raleigh, NC

111. D.U. Furrer, R. Shankar, and C. White, Optimizing the Heat Treatment of Ni-Based Superalloy Turbine Discs, *JOM,* March 2003, p 32–34

112. *Deform-2D Users Manual,* Scientific Forming Technologies, Columbus, OH

113. B. Buchmayr and J.S. Kirkaldy, Modelling of the Temperature Field, Transformation Behavior, Hardness and Mechanical Response of Low Alloy Steels during Cooling from the Austenite Region, *J. Heat Treat.,* Vol 8, 1990, p127–136

114. J.S. Kirkaldy, Quantitative Prediction of Transformation Hardening in Steels, *Heat Treating,* Vol 4, *ASM Handbook,* ASM International, 1991, p 20–32

115. S. Denis, P. Archambault, E. Gautier, A. Simon, and G. Beck, Prediction of Residual Stress and Distortion of Ferrous and Non-Ferrous Metals: Current Status and Future Developments, *JMEP,* Vol 11 (No. 1), Feb 2002, p 92–102

116. T. Ericsson, Principles of Heat Treating Steels, *Heat Treating,* Vol 4, *ASM Handbook,* ASM International, 1991 p 3–19

117. M. Gergely, S. Somogyi, T. Réti, B. Donát, and T. Konkoly, Computerized Properties Prediction and Technology Planning in Heat Treatment of Steels, *Heat Treating*, Vol 4, *ASM Handbook*, ASM International, 1991, p 638–656

118. J. Rohde and A. Jeppsson, Literature Review of Heat Treatment Simulations with Respect to Phase Transformation, Residual Stress and Distortion, *Scand. J. Metall.*, Vol 29, 2009, p 47–62

119. A. Rose and H.R Hougardy, Transformation Characteristics and Hardenability of Carburizing Steels, *Symposium on the Transformation and Hardenability in Steels*, Climax Molybdenum Corp., 1967

120. R. Chatterjee-Fischer, Beispiele für durch Wärmebehandlung bedingte Eigenspannungen und ihre Auswirkungen, *Härt.-Tech. Mitt.*, Vol 28, 1973, p 276–288

121. S. Denis, E. Gautier, A. Simon, and G. Beck, Stress-Phase Transformation Interactions—Basic Principles, Modeling and Calculation of Internal Stresses, *Mater. Sci. Technol.*, Vol 1, 1985, p 805–814

122. S. Denis, E. Gautier, S. Sjöström, and A. Simon, Influence of Stresses on the Kinetics of Pearlite Transformation during Continuous Cooling, *Acta Metall.*, Vol 35, 1987, p 1621–1632

123. E. Guatier, A. Simon, and G. Beck, Plasticité de Transformation Durant la Transformation Perlitique d'un Acier Eutectoide, *Acta Metall.*, Vol 35, 1987, p 1367–1375

124. M.R. Varma, R. Sasikumar, S.G.K. Pillai, and P.K. Nair, Cellular Automation Simulation of Microstructure Evolution during Austenite Decomposition under Continuous Cooling Conditions, *Bull. Mater. Sci.*, Vol 25 (No. 3), June 2001, p 305–312

125. Y.J. Lan, N.M. Xiao, D.Z. Li, and Y.Y. Li, Mesoscale Simulation of Deformed Austenite Decomposition into Ferrite by Coupling a Cellular Automation Method with a Crystal Plasticity Finite Element Model, *Acta Mater.*, Vol 53, 2005, p 991–1003

126. *Atlas of Isothermal Transformation and Cooling Transformation Diagrams*, American Society for Metals, 1977

127. "Atlas of Isothermal Transformation Diagrams of BS EN Steels," Special Report 56, The Iron and Steel Institute, London, 1956

128. M. Atkins, *Atlas of Continuous Cooling Transformation Diagrams for Engineering Steels*, British Steel Corporation, Sheffield, 1977

129. J.-L. Lee and H. Bhadesia, A Methodology for the Prediction of Time-Temperature-Transformation Diagrams, *Mater. Sci. Eng. A*, Vol 171, 1993, p 223–230

130. N. Saunders, Z. Guuo, X. Li, A.P. Miodownik, and J.-P. Schillè, "The Calculation of TTT and CCT Diagrams for General Steels," Internal report, Sente Software Ltd., 2004

131. Z. Guo, J.P. Schillè, N. Saunders, and A.P. Miodownik, Modelling Material Properties Leading to the Prediction of the Distortion during Heat Treatment of Steels for Automotive Applications, *Proc. Second International Conference on Heat Treatment and Surface Engineering in Automotive Applications*, June 20–22 2005 (Riva del Garda, Italy)

132. JMatPro, Sente Software Ltd., United Kingdom

133. N. Saunders, Z. Guo, X. Li, A.P. Miodownik, and J.-Ph. Schillé, Using JMatPro to Model Materials Properties and Behavior, *JOM*, Dec 2003, p 60–65

134. Thermocalc, Thermo-Calc Software, Stockholm, Sweden

135. Pandat, CompuTherm LLC, Madison, WI

136. S. Denis, D. Farias, and A. Simon, Mathematical Model Coupling Phase Transformations and Temperature Evolutions in Steels, *ISIJ Int.*, Vol 32 (No. 3), 1992, p 316–325

137. Y. Nagasaka, J.K. Brimacombe, E.B. Hawbolt, I.V. Samarasekera, B. Hernandez-Morales, and S.E. Chidiac, Mathematical Model of Phase Transformations and Elasto-Plastic Stress in the Water Spray Quenching of Steel Bars, *Metall. Trans. A*, Vol 24, April 1993, p 795–808

138. P.R. Woodard, S. Chandrasekar, and H.T. Y. Yang, Analysis of Temperature and Microstructure in the Quenching of Steel Cylinders, *Metall. Trans. B*, Vol 30, Aug 1999, p 815–822

139. Y. Ghanimi and B. Buchmayr, A Procedure to Implement Microstructural Evolution in a Commercial FE Program, *Proceedings of the ASM Heat Treating Society Third International Conference on Quenching and Control of Distortion*, March 24–26, 1999 (Prague, Czech Republic), ASM International, p173–181

140. J.-M. Bergheau, F. Boitout, V. Toynet, S. Denis, and A. Simon, Finite Element Simulation of Coupled Carbon Diffusion, Metallurgical Transformation and Heat Transfer with Applications in the Automobile Industry, *Proceedings of the ASM Heat Treating Society Third International Conference on Quenching and Control of Distortion*, March 24–26, 1999 (Prague, Czech Republic), ASM International, p 145–156

141. T. Reti, Z. Fried, and I. Felde, Multi-Phase Modeling of Austenite Transformation Processes during Quenching, *Proceedings of the ASM Heat Treating Society Third International Conference on Quenching and Control of Distortion*, March 24–26, 1999 (Prague, Czech Republic), ASM International, p 157–167

142. S.-H. Kang and Y.-T. Im, Three Dimensional Finite Element Analysis of the Quenching Process of Plain-Carbon Steel

143. C. Şimşar and C.H Gür, A Review on Modeling and Simulation of Quenching, *J. ASTM Int.*, Vol 6 (No. 2), Paper ID JAI101766, ASTM International, 2009

144. D.P. Koistinen and R.E. Marburger, A General Equation Prescribing the Extent of the Austenite-Martensite Transformation in Pure Iron-Carbon Alloys and Plain Carbon Steel, *Acta Metall.*, Vol 7, 1959, p 59–60

145. J. Rhode, A. Thuvander, and A. Melander, Using Thermodynamic Information in Numerical Simulation of Distortion due to Heat Treatment, *Proceedings of the Fifth ASM Heat Treatment and Surface Engineering Conference in Europe Incorporating the Third International Conference on Heat Treatment with Atmospheres*, E.J. Mittemeijer and J. Grosch, Ed., ASM International, 2000, p 21–29

146. M. Lusk, G. Krauss, and H.-J. Jou, A Balance Principle Approach to Modeling Phase Transformation, *J. Phys.(France) IV*, Colloque 8, 1994, p 279–284

147. D. Bammann, V. Prantil, A. Kumar, J. Lathrop, D. Mosher, M. Callabresi, H.-J. Jou, M. Lusk, G. Krauss, B. Elliot, G. Ludtka, T. Lowe, B. Dowling, D. Shick, and D. Nikkel, Development of a Carburizing and Quenching Simulation Tool: A Material Model for Carburizing Steels Undergoing Phase Transformations, *Proceedings of the Second International Conference on Quenching and the Control of Distortion*, Nov 1996 (Cleveland, OH), p 367–375

148. Z. Li, B.L. Ferguson, and A.M Freborg, Data Needs for Modeling Heat Treatment of Steel Parts, *Proc. MS&T '04, Continuous Casting Fundamentals, Engineered Steel Surfaces, and Modeling and Computer Applications in Metal Casting, Shaping and Forming Processes*, Vol 2, TMS, 2004, p 219–226

149. M. Mehta and T. Oakwood, "Development of a Standard Methodology for the Quantitative Measurement of Steel Phase Transformation Kinetics and Dilation Strains Using Dilatometric Methods," Final report, AISI/DOE Technology Roadmap Program Office, Pittsburgh, PA, Cooperative Agreement DE-FC36-97ID13554

150. "Standard Practice for Quantitative Measurement and Reporting of Hypoeutectoid Carbon and Low-Alloy Steel Phase Transformations," A 1033-04, ASTM International

151. M. Eriksson, M. Oldenburg, M.C. Somani, and L.P. Karjalainen, Testing and Evaluation of Material Data for Analysis of Forming and Hardening of Boron Steel Components, *Model. Simul. Mater. Sci. Eng.*, Vol 10, 2002, p 277–294

152. T. Inoue and K. Tanaka, An Elastic-Plastic Stress Analysis of Quenching When Considering a Transformation, *Int. J. Mech. Sci.*, Vol 17, 1975, p 361–367

153. Y.Y. Li and Y. Chen, Modeling Quenching to Predict Residual Stress and Microstructure Distribution, *Trans. ASME, J. Eng. Mater. Technol.*, Vol 110, Oct 1988, p 372–388

154. M.T. Todinov, Mechanism for Formation of the Residual Stresses from Quenching, *Model. Simul. Mater. Sci. Eng.*, Vol 6, 1998, p 273–291

155. M. Narazaki, D.-Y. Ju, K. Osawa, and S. Fuchizawa, Influence of Transformation Plasticity on Quenching Distortion of Carbon Steel, *Proceedings of the ASM Heat Treating Society Third International Conference on Quenching and Control of Distortion*, March 24–26, 1999 (Prague, Czech Republic), p 405–415

156. C. Liu, D.-Y. Ju, and T. Inoue, A Numerical Modeling of Metallo-Thermo-Mechanical Behavior in Both Carburized and Carbonitrided Quenching Processes, *ISIJ Int.*, Vol 42 (No. 10), 2002, p 1125–1134

157. C.C. Liu, X.J. Xu, and Z. Lui, A FEM Modeling of Quenching and Tempering and Its Application in Industrial Engineering, *Finite Elem. Anal. Des.*, Vol 39, 2003, p 1053–1070

158. H. Alberg, "Material Modelling for Simulation of Heat Treatment," Licentiate Thesis, Department of Applied Physics and Mechanical Engineering, Luleå University of Technology, 2003

159. C. Şimşar and C.H. Gür, A FEM Based Framework for Simulation of Thermal Treatments: Application to Steel Quenching, *Comput. Mater. Sci.*, Vol 44, 2008, p 588–600

160. B.L. Ferguson, G.J. Petrus, and T. Pattok, A Software Tool to Simulate Quenching of Alloy Steels, *Proceedings of the ASM Heat Treating Society Third International Conference on Quenching and Control of Distortion*, March 24–26, 1999 (Prague, Czech Republic), p 188–200

161. J. Cai, L. Chuzhoy, C. Berndt, G. Keil, K. Burris, and M. Taft, Application of Heat Treatment Simulation in Continuous Improvement of Product Reliability, *19th ASM Heat Treating Society Conference Proceeding Including Steel Heat Treating, The New Millennium*, Nov 1999, ASM International, 2000, p 299–306

162. H. Shichino, Construction of Heat Treatment Database and Enhancement of Simulation Technique, *Komatsu Tech. Rep.*, Vol 51 (No. 155), 2005

163. B. Liščič, S. Singer, and B. Smoljan, Prediction of Quench-Hardness within the Whole Volume of Axially-Symmetric Workpieces of Any Shape, *J. ASTM Int.*, Vol 7 (No. 2), Dec 2009, Paper ID JAI102647

164. S. Yamanaka, T. Sakanoue, T. Yoshii, T. Kozuka, and T. Inoue, Influence of Transformation Plasticity on the Distortion of Carburized Quenching Process of Cr-Mo Steel Ring, *Proceedings of ASM 18th Heat Treat Society Conference*, Oct 12–15, 1998, ASM International, p 657–664

165. G.W. Greenwood and H.R. Johnson, *Proc. R. Soc. (London) A*, Vol 283, 1965, p 403–422

166. C. Franz, G. Besserdich, V. Schulze, H. Müller, and D. Löhe, Influence of Transformation Plasticity on Residual Stresses and Distortions due to the Heat Treatment of Steels with Different Carbon Contents, *J. Phys. (France) IV*, Vol 120, 2004, p 481–488

167. A.J. Fletcher, Prediction of Thermal Stresses on Quenched Plates of EN 30B (E35 M30) Steel, *Met. Technol.*, June 1977, p 307–316

168. A.J. Fletcher and R.F. Price, Generation of Thermal Stresses and Strain during Quenching of Low-Alloy Steel Plates, *Met. Technol.*, Nov 1981, p 427–445

169. Z. Li and R.V. Grandhi, Optimum Design of Heat Transfer Coefficient during Gas Quenching Process, *Proceedings of the 21st ASM Heat Treating Society Conference*, Nov 2001, ASM International

170. Z. Li, A. Freborg, and B.L. Ferguson, Effective Design of Heat Treat Processes Using Computer Simulations, *Proceedings of the 24th ASM Heat Treating Society Conference*, Sept 2007 (Detroit, MI), ASM International, p 205–213

171. H.C. Hottel and A.F. Sarofim, *Radiative Heat Transfer*, McGraw-Hill, 1967

172. J.M. Rhine and R.J. Tucker, *Modeling of Gas-Fired Furnaces and Boilers*, British Gas, McGraw-Hill, 1991

173. M.K. Alam and S.L. Semiatin, Thermal Stress Development during Processing of Ingots, *Processing and Fabrication of Advanced Materials for High Temperature Applications—II*, V.A. Ravi and T. S. Srivatsan, Ed., TMS, 1993

174. M.K. Alam, R.L. Goetz, and S.L. Semiatin, Modeling of Thermal Stresses and Thermal Cracking during Heating of Large Ingots, *Trans. ASME, J. Manuf. Sci. Eng.*, Vol 118, May 1996, p 235–243

175. T. Hara, On the Interior Temperature Distribution of Ingots during Heating Process, *Tetsu-to-Hagané*, Vol 49 (No. 11), Oct 1963, p 1669–1675 (in Japanese)

176. T. Hara, On the Thermal Stress of Ingots during Heating Process, *Tetsu-to-Hagané*, Vol 49 (No. 12), Nov 1963, p 1765–1772 (in Japanese)

177. T. Hara, The Measurement of Modulus of Elasticity at High Temperature and Coefficient of Thermal Expansion for the Comparison of the Rate of Frequency of Thermal Stress Cracking in Various Steels, *Tetsu-to-Hagané*, Vol 49 (No. 13), Dec 1963, p 1885–1891 (in Japanese)

178. T. Hara, On the Formation of Internal Cracks in Heating Process of High Carbon Steels Ingots, *Tetsu-to-Hagané*, Vol 50 (No. 1), Jan 1964, p 29–37 (in Japanese)

179. T. Hara, Study of Thermal Stress Cracks in Steel Ingots, *Tetsu-to-Hagané* (overseas), Vol 4 (No. 3), Sept 1964, p 296–303

180. J. Mao, V.L. Keefer, K.-M. Chang, and D. Furrer, An Investigation on Quench Cracking Behavior of Superalloy Udimet 720LI Using a Fracture Mechanics Approach, *JMEPEG*, Vol 9, 2000, p 204–212

181. J. Mao, K.-M. Chang, and D. Furrer, Quench Cracking Characterization of Superalloys Using Fracture Mechanics Approach, *Superalloys 2000*, The Metallurgical Society, AIME, p 109–116

182. H. Tada, P. Paris, and G. Irvin, *The Stress Analysis of Cracks Handbook*, Del Research Corp., Hellertown, PA, 1973, p 2.10–2.12

183. K. Arimoto, G. Li, A. Arvind, and W.T. Wu, The Modeling of Heat Treating Process, *Proceedings of 18th ASM Heat Treating Conference*, Oct 1998, ASM International, p 23–30

184. T. Inoue, K. Haraguchi, and S. Kimura, Stress Analysis during Quenching and Tempering, *J. Soc. Mater. Sci., Jpn.*, Vol 25 (No. 273), 1976, p 521 (in Japanese)

185. K. Arimoto, D. Lambert, K. Lee, W.T. Wu, and M. Narazaki, Prediction of Quench Cracking by Computer Simulation, *Proceedings of the 19th ASM Heat Treating Society*, Nov 1999, ASM International, 2000, p 435–440

186. S. Owaku, *Netsu Shori*, Vol 30 (No. 2), 1990, p 63–67 (in Japanese)

187. J. Cai, L. Chuzhot, and K.W. Burris, Numerical Simulation of Heat Treatment and Its Use in Prevention of Quench Cracks, *Proceedings of the 20th ASM Heat Treating Society Conference*, Oct 9–12, 2000 (St. Louis, MO), ASM International, p 701–707

188. L. Chuzhoy, T.E. Clemens, J.E. McVicker, and K.W. Burris, Numerical Simulation of Quenching Process at Caterpillar, *SAE Trans.*, Document 931172, 1993

189. K. Arimoto, F. Ikuta, T. Horino, S. Tamura, M. Narazaki, and Y. Mikitam, Preliminary Study to Identify Criterion for Quench Crack Prevention by Computer Simulation, *14th Congress of IFHTSE*, Oct 2004 (Shanghai, China); *Trans. Mater. Heat Treat.*, Vol 25 (No. 5), 2004, p 486–493

ASM Handbook, Volume 22B, *Metals Process Simulation*
D.U. Furrer and S.L. Semiatin, editors

Copyright © 2010, ASM International®
All rights reserved.
www.asminternational.org

Simulation of Diffusion in Surface and Interface Reactions

Paul Mason, Thermo-Calc Software Inc., McMurray, Pennsylvania
Anders Engström, Thermo-Calc Software AB, Stockholm, Sweden
John Ågren, Royal Institute of Technology, Stockholm, Sweden
Samuel Hallström, Thermo-Calc Software AB, Stockholm, Sweden

HISTORICALLY, THE FIELD OF MATERIALS SCIENCE AND ENGINEERING has focused on establishing the processing-structure-property relationships of materials through experimental trial and error with an understanding of the influence of microstructure—the microstructure, processing, and chemistry all being directly related to the crystallography, kinetics, and thermodynamics of the materials (Ref 1). In comparatively recent years, as computational methods have evolved, it has been possible to predict some such properties and relationships through numerical simulation.

For example, in the area of computational thermodynamics, the computer calculation of phase diagrams (CALPHAD) approach (Ref 2) has been successfully employed to predict the thermodynamic properties of complex multicomponent, multiphase systems based on mathematical models that describe the Gibbs energy as a function of temperature, pressure, and composition for each individual phase in a system. Adjustable parameters in the expressions of the numerical models capture the composition dependence in binary and ternary systems and are adjusted to best correspond to the experimental data available. This enables both the experimental data of the binary and ternary systems to be reproduced through calculations and also, more importantly, predictions to be made for higher-order systems.

As in the physical world, the interrelationships between crystallography, kinetics, and thermodynamics are being increasingly connected in the world of simulation, also. For example, in the CALPHAD approach, the naming and models that describe the individual phases are inherently linked to an understanding of the crystallographic structure of those phases to be able to describe mixing in solid solutions. Ab initio techniques are now also being employed more routinely to calculate thermodynamic data that are very difficult to measure experimentally. The CALPHAD approach has also been extended to modeling

other properties, including, for example, volume data for individual phases that enable the prediction of density, and also atomic mobilities that enable diffusion coefficients as a function of temperature and local compositions to be derived.

Chen et al. (Ref 3) have described the interconnection between some of the different computational approaches used to model the thermodynamics and kinetics of multicomponent systems within the CALPHAD framework, and this is shown in Fig. 1. Moving from the atomistic scale to the mesoscale, ab initio calculations provide data that can supplement experimental thermodynamic and kinetic data upon which most of the existing CALPHAD-type predictions are based.

However, thermodynamic calculations, while useful, do not consider the changes over time. As mentioned elsewhere (see the article "Modeling Diffusion in Binary and Multicomponent Alloys" in *Fundamentals of Modeling for Metals Processing*, Volume 22A of *ASM Handbook*), diffusion can play an important role in both materials processing and materials degradation during the service life of a part. For example, most heat treatments undertaken to modify the microstructure and improve the performance of the material involve diffusion. With respect to materials degradation, the kinetics of the degradation process can determine the service life of the part. In recent years, the sharp interface approach, the phase-field method, and the mean-field Langer-Schwartz-type approach have all been employed to consider the dimension of time and simulate the evolution of microstructures that are diffusion controlled.

For a more detailed review of the methods used to model the evolution of phase

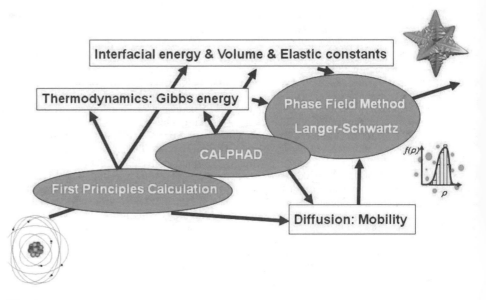

Fig. 1   Atoms to microstructure. Relationship between scales of different simulation methods. Source: Ref 3

boundaries at the meso- and nanoscales, readers are directed to the paper by Thornton et al. (Ref 4). This article focuses on the modeling and simulation of diffusion-controlled processes related to both materials processing, such as heat treatments, and materials degradation from a practical perspective by using the one-dimensional (1-D) sharp interface approach. The theoretical background of this approach and the modeling of diffusivity in multicomponent systems are initially considered. Some examples for both heat treatment and materials degradation are then described, with an emphasis on the approaches used and the lessons learned from performing such simulations.

1-D Finite-Difference Method

With the sharp interface approach, the interface is represented as a dividing surface where the properties are assumed to change discontinuously from the bulk properties of one phase to those of another phase. By approximating the interface as the dividing surface and examining the excess thermodynamic quantities over the bulk values that are assumed to change as one proceeds from one phase to the other, Gibbs showed that interfaces possess an excess grand canonical energy, the so-called interfacial energy (Ref 4, 5). For two phases separated by a curved interface, the presence of an interfacial energy results in a jump of pressure across the interface. This jump in pressure gives rise to a dependence of the interfacial temperature or composition of the phases at the interface on interfacial curvature, which is approximately represented by the so-called Gibbs-Thomson equation.

The sharp interface approach becomes inadequate when the length scale of the structure of interest becomes of the order of the interface thickness. Although the microstructural scale is much larger than that of the interfacial scale in many physical systems, such assumptions can be violated in specific situations. Nucleation and spinodal decomposition are two important examples. In nucleation, the nucleus can be quite small, and thus, its size may be of the same order of magnitude as the interfacial thickness. During spinodal decomposition, however, the structure grows by the amplification of composition perturbations, and thus, the sharp interface description is clearly inadequate. Nevertheless, such processes can be described by a diffuse interface approach, for example, such as the phase-field method. DICTRA (Ref 6) is a commercial software package that has been developed to simulate diffusion-controlled phase transformations on the basis of the following assumptions:

- The movement of a phase interface is controlled by the mass balance obtained from the fluxes of the diffusing elements across the interface.
- Local equilibrium is assumed at the moving interface. In multicomponent systems, the

operating tie-line is determined by the condition that the mass balance of every diffusing element gives the same interface velocity.

- Diffusion is treated in terms of mobilities and true thermodynamic driving forces, that is, chemical potential gradients. DICTRA calculates diffusion coefficients according to the scheme shown in Fig. 2, where so-called mobilities are combined with the second derivative of the Gibbs energy obtained from a thermodynamic equilibrium code such as Thermo-Calc (Ref 7). This is discussed further in the section "Diffusivity and Mobility in Multicomponent Systems" in this article.
- Diffusion-controlled phase transformations are treated in geometries with one space variable, that is, in planar, cylindrical, or spherical cells with one or more different phases present.

DICTRA is a general code and can treat not only phase transformations, that is, moving boundary problems, but also diffusion in one-phase systems and reactions in dispersed systems. DICTRA has been used to make the calculations referenced in this article unless noted otherwise.

Diffusion Simulation Models

One-Phase Simulations. In one-phase diffusion simulations, the multicomponent diffusion equations are solved by a numerical procedure developed by Ågren (Ref 8). Combining the continuity equation with Fick's first equation provides the fundamental differential equation of diffusion, sometimes referred to as Fick's second equation, which, in the 1-D planar case reads:

$$\frac{\partial [c]}{\partial t} = \frac{\partial}{\partial z}[D]\frac{\partial [c]}{\partial z} \qquad \text{(Eq 1)}$$

where [c] is a vector containing the concentration of diffusing species, t is a time variable, z is distance, and [D] is a matrix that contains the interdiffusion coefficients. Assuming that [D] is concentration dependent gives a system of coupled parabolic partial differential equations, and

these equations must be solved numerically so that the evolution of the concentration profiles is calculated as a function of time.

Some example applications of one-phase simulations are homogenization and the interdiffusion between a coating and a substrate if they are both one-phase materials containing the same phase.

Moving Phase-Boundary Simulations. In moving phase-boundary simulations, the movement of a phase boundary due to diffusion is simulated. The necessary boundary conditions at the phase boundary are calculated in Thermo-Calc (Ref 7), assuming that local equilibrium holds at the phase boundary. The diffusion equations in each phase are solved as described previously, and the displacement of the phase boundary due to the diffusion flux is calculated as shown in Fig. 3.

Dispersed System Simulations. In dispersed system simulations, long-range diffusion through multiphase structures is simulated, *long-range* meaning that the diffusion distances in the multiphase structure are long compared to the distances between the precipitates. A model for this was developed by Engström et al. (Ref 9), and it is based on the assumptions that local equilibrium is established in each volume element at each time-step in the calculation and that diffusion occurs in a continuous matrix phase. The model calculates chemical diffusion and phase transformations in two steps. First, the diffusion equations for the matrix phase are solved for a given time-step at each grid point. Second, at the end of each time-step, a calculation of the equilibrium between the matrix and the precipitate phases is made by automatically calling Thermo-Calc, and the new phase compositions are obtained. Thus, the matrix composition changes, and DICTRA solves the diffusion equations for this new matrix composition for the next time-step, and so on.

The model has been successfully used to simulate processes such as carburization of nickel-chromium alloys (Ref 9), carbon

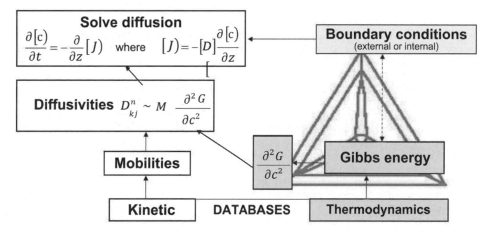

Fig. 2 Relationship between thermodynamic and kinetic data within DICTRA

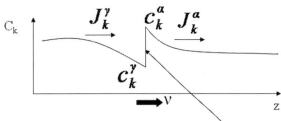

C_k

J_k^γ c_k^α J_k^α

c_k^γ

v

z

Sharp interface with assumption of local equilibrium

n-1 unknowns:

n-2 chemical potentials.

Velocity of phase boundary, v

n-1 Flux Balance Equations:

$$v(c_k^\alpha - c_k^\gamma) = J_k^\alpha - J_k^\gamma$$

F-B Equations solved as:

$$\sum_k^{n-1} \left(v(c_k^\alpha - c_k^\gamma) - (J_k^\alpha - J_k^\gamma) \right)^2 < \varepsilon$$

Fig. 3 Representation of the sharp interface method used within DICTRA

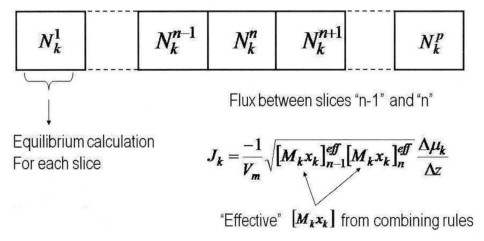

N_k^1 N_k^{n-1} N_k^n N_k^{n+1} N_k^p

Equilibrium calculation
For each slice

Flux between slices "n-1" and "n"

$$J_k = \frac{-1}{V_m} \sqrt{[M_k x_k]_{n-1}^{eff} [M_k x_k]_n^{eff}} \frac{\Delta \mu_k}{\Delta z}$$

"Effective" $[M_k x_k]$ from combining rules

Fig. 4 Representation of the homogenization model within DICTRA

diffusion in welds between dissimilar steels (Ref 9), and gradient sintering of cemented carbide cutting tools (Ref 10). Several examples where the dispersed-systems model has been employed are given in the section "Case Studies" in this article.

There are, however, some drawbacks with this model. First, the microstructure of the problem should be such that the assumption that diffusion occurs in a continuous matrix phase is reasonable. Thus, one continuous phase should be present in the diffusion zone. Another drawback is the assumption that local equilibrium is established in each volume element at each time-step, which neglects the effect of a finite rate of growth and dissolution of individual precipitates. This may be important when considering, for example, nonisothermal conditions where the fraction of precipitates in a volume element may change

more rapidly in the calculation than the rate of growth or the dissolution of precipitates may allow for in reality.

Homogenization Model. To overcome some of the limitations in the dispersed model, and specifically the requirement for a continuous matrix phase, the so-called homogenization model developed by Larsson and Engström (Ref 11) was recently introduced into the software. Using this approach, fluxes are calculated in a lattice-fixed frame of reference using locally averaged kinetic properties or so-called effective mobilities. The resulting concentration fields are mapped to a number-fixed frame of reference, assuming that the number-fixed frame of reference is defined by the condition that the number of atoms on each side of a reference plane stays constant. By assuming locally minimized Gibbs energy and locally averaged kinetic properties, this approach can

be applied to a range of different problems, including diffusion in multiphase mixtures. Locally minimized Gibbs energy means that the local phase fractions, phase compositions, and so on correspond to the equilibrium values given by the local composition. As such, a simulation can be represented, as in Fig. 4.

Diffusivity and Mobility in Multicomponent Systems

Computer software such as DICTRA, some phase-field packages such as MICRESS (Ref 12 to 14), and other simulation software such as PrecipiCalc (Ref 15) use kinetic data in the form of atomic mobilities that are assessed in an approach similar to the CALPHAD method described by Andersson and Ågren (Ref 16). This section briefly describes how such data are derived, although for more detail, readers should go to the cited references.

Knowing the atomic mobilities and assuming a vacancy exchange diffusion mechanism in a crystalline phase, the diffusion coefficient in a lattice-fixed frame of reference, D_{kj}^L, can be calculated as the product of a mobility term and a thermodynamic factor:

$$D_{kj}^L = \sum_{i=1}^n \delta_{ik} x_i M_i \frac{\partial \mu_i}{\partial x_j} \quad \text{(Eq 2)}$$

where δ represents the Kronecker delta, and $\delta_{ik} = 1$ when $i = k$ and $\delta_{ik} = 0$ when $i \neq k$; M_i is the atomic mobility; μ_i is the chemical potential; and x_i is the mole fraction for element i. If the partial molar volumes are assumed to be constant, then the diffusivities in Eq 2 can be transformed to the volume-fixed frame of reference:

$$D_{kj}^V = \sum_{i=1}^n (\delta_{ik} - x_k) x_i M_i \frac{\partial \mu_i}{\partial x_j} \quad \text{(Eq 3)}$$

The species k is the diffusing specie, and j is the gradient specie. The partial derivative of the chemical potential, μ_i, with respect to the mole fraction, x_i, defines the so-called thermodynamic factors, which may be easily calculated from thermodynamic data.

Finally, the chemical diffusivity (interdiffusion coefficient) is obtained from Eq 4 after selecting a solvent or reference specie, n. Thus, Eq 2 to 4 establish a relation between the atomic mobility, M_i, and the chemical diffusivity, $\tilde{D}_{kj}^n$:

$$\tilde{D}_{kj}^n = D_{kj}^V - D_{kn}^V \quad \text{(Eq 4)}$$

From absolute-reaction rate theory arguments, Andersson and Ågren (Ref 16) suggested that the atomic mobility that enters into Eq 1 and 2 is divided into a frequency factor, M_i^o, and an activation enthalpy, Q_i:

$$M_i = M_i^o \exp\left(\frac{-Q_i}{RT}\right) \frac{1}{RT} \quad \text{(Eq 5)}$$

where R is the gas constant, and T is the absolute temperature. Jönsson (Ref 17) modified the aforementioned equation after finding it superior to model the logarithm of the atomic mobility rather than the value itself:

$$RT \ln(RTM_i) = RTM_i^o - Q_i \qquad \text{(Eq 6)}$$

In a CALPHAD-type approach, the composition and temperature dependence of the atomic mobility has been successfully represented by a Redlish-Kister polynomial:

$$\phi - \sum_j x_j \phi_i^j \sum_p \sum_{j>p} x_p x_j \left[\sum_{r=0}^{m} {}^r\phi_i^{p,j}(x_p - x_j)^r \right] \qquad \text{(Eq 7)}$$

where ϕ_i in the equation may represent either RTM_i^0 or Q_i, present in Eq 6.

The unknown parameters in Eq 7 (i.e., ϕ_i^j and ${}^r\phi_i^{p,j}$) are determined following a CALPHAD-type assessment procedure, which typically will include a least-squares fit to available experimental information.

Engström and Ågren (Ref 18) reported an assessment of diffusional mobilities in face-centered cubic Ni-Cr-Al alloys, and this describes more details of the approach used to assess these data.

Case Studies

The following case studies illustrate some examples where diffusion simulations have been applied to industrial-based problems.

Gas Carburization of Highly Alloyed Steels

Gas carburization is the addition of carbon to the surface of low-carbon steels at temperatures generally between 850 and 950 °C (1560 and 1740 °F), at which austenite, with its high solubility for carbon, is the stable crystal structure. Carburizing is usually performed to increase the superficial hardness and the overall mechanical characteristics of the surface to obtain good wear and fatigue resistance.

The case hardness of carburized steels is primarily a function of carbon content, and the case depth of carburized steel is a function of carburizing time and the available carbon potential at the surface. When prolonged carburizing times are used for deep case depths, a high carbon potential produces a high surface-carbon content, which can result in excessive

retained austenite or the precipitation of carbides near the surface of the alloy. Both of these microstructural characteristics have adverse effects on the distribution of residual stress in the case-hardened part.

For highly alloyed steels, such as stainless steels, gas carburizing results in the formation of chromium-rich carbides during carbon transfer into the steel, which causes precipitation hardening in the surface. However, the precipitation of these chromium-rich carbides results in a decrease of chromium in the matrix phase, which has a detrimental effect on the corrosion resistance of the alloy. Therefore, a compromise must be made to maintain a good corrosion resistance in the carburized layer. Traditional methods to balance these objectives have been by experiment only, but computational tools now offer alternatives to this approach, which enables the assessment of a number of parameters involved in the carburizing process (such as alloying elements, temperature, carbon flux, etc.).

From a theoretical perspective, gas carburization is a complex process that involves a carbon enrichment step, a diffusion step, and an austenitization step, followed by a quench and a temper. Knowing the necessary final carbon profile, the metallurgist must take into account all the treatments that are made at different temperatures. From a mechanistic perspective, to present a reasonable model of the overall process, at least the following subprocesses must be properly dealt with, each of which are interconnected:

- Carbon diffusion in the steel surface
- Reactions on the steel surface
- Changes of furnace atmosphere caused by reactions with the steel and by the gas flow

Turpin et al. (Ref 19) have reported a study combining both experimental work and theoretical simulations, using both thermodynamic and diffusion simulations, to investigate the carbon diffusion and phase transformation during gas carburization of high-alloyed stainless steels and thus provide a useful case study. The study was focused on martensitic stainless steels, which are widely used in applications requiring both high mechanical strength and corrosion resistance. Two alloys of industrial importance were considered, and the compositions of these alloys are shown in Table 1.

Prior to making kinetic simulations, it is typical that some thermodynamic calculations are made first to determine the equilibrium state of the system and hence the direction that the

system is driving toward in terms of the phases formed, should the system be given enough time for equilibrium to be attained. The authors used such calculations to predict the temperature and carbon content dependence of the phase composition, and these simulations helped to determine the optimal composition of the initial alloys. The main conclusions arising from these calculations, which were made using Thermo-Calc and the TCFe2000 database (Ref 20), were:

- At 955 °C (1750 °F), which corresponds to the austenitization temperature, calculations show that as the amount of carbon content in the alloy is increased, $M_{23}C_6$ and then M_7C_3 will be the first carbides to precipitate in the austenite phase.
- If the amount of carbon exceeds 3.8 wt%, then M_3C carbides are predicted to be stable. M_3C carbides have a structure similar to cementite and preferentially precipitate at the grain boundaries, which weakens the microstructure. To avoid these phases, the overall content of carbon in the steel must be below this amount at the end of the carburizing process.
- Above 1.7 wt% C, the mole fraction (an indicator of the volume fraction) of M_7C_3 carbides exceeds 20%, and the chromium content of the alloy associated with these carbides is 65 wt%. Therefore, there is a correspondingly strong depletion of chromium from the matrix.
- To balance the desire for adding carbon into the matrix phase to obtain hardness while depleting the matrix of carbon, it was determined that the optimal amount of carbon in the matrix phase should not exceed 1 wt%. That is, the thermodynamic calculations were used to establish a limit, without yet any consideration of the kinetics or time.

The second step of the simulation was then performed to consider the diffusional reactions in the multicomponent system during the carbon enrichment step and the diffusion step of the gas carburizing process, and to determine how the composition and the amount of each phase vary with time and distance from the gas/solid interface and the carbon profile of the alloy as a function of time and distance. These calculations were made using the dispersed model, which was previously described.

Carbon Enrichment Step. Several boundary conditions can be used for such a simulation, for example, fixing the carbon activity or carbon concentration at the surface or the carbon flux. In this work, the carbon flux was determined experimentally using thermogravimetric measurements. By knowing the surface area of the sample, Turpin et al. were able to use the following expression, where $n_c(t)$ is the number of moles of carbon transferred through the gas/solid interface as a function of time, and S represents the surface area of the austenitic phase:

Table 1 Chemical composition of two industrial stainless steels: Fe-13Cr-5Co-3Ni-2Mo-0.07C (alloy A) and Fe-12Cr-2Ni-2Mo-0.12C (alloy B)

Alloy	Composition, wt%								
	C	Cr	Co	Ni	Mo	Mn	V	Si	Nb
A	0.07	13.15	5.34	2.57	1.78	0.52	0.62	0.36	0.07
B	0.115	11.63	0	2.45	1.72	0.6	0	0.29	0

Source: Ref 19

$$J_c(t) = \frac{1}{S}\delta n_c(t)/\delta t \qquad (Eq\ 8)$$

Because the value of S will reduce as carbide precipitates form at the surface, Turpin et al. used image analysis with an optical microscope to measure the variation of S as a function of the overall carbon content at the surface. For a carbon content of 4 mass%, the austenitic surface area represents no more than half of the sample surface area if the temperature is 955 °C (1750 °F).

Diffusion Step. During the diffusion step, the N_2-CH_4 mixture is replaced with pure N_2, and the carbon flux at the surface of the samples is zero. To simulate this step, a zero carbon flux was applied as the boundary condition for 2 h.

Figure 5 shows the simulated carbon profiles compared with experimental data obtained for one of the alloys, Fe-13Cr-5Co-3Ni-2Mo-0.07C. As can be observed, there is generally good agreement between the calculated and experimental values, and the authors conclude that the carbon profile can be calculated and followed at any time if the boundary condition evolution at the gas-solid interface is known during the carburizing treatment.

Gas Nitriding and Nitrocarburizing

Nitriding and nitrocarburizing are thermochemical surface treatments by which nitrogen or carbon and nitrogen are introduced into the steel workpiece to produce hard and wear-resistant surfaces. The case produced can be subdivided into a compound layer consisting predominantly of ε and/or γ′ (Fe_4N) phases, which is responsible for good tribological and anticorrosion properties of the surface and a diffusion zone where nitrogen or carbon and nitrogen are dissolved interstitially in the ferritic matrix, leading to improved fatigue resistance.

Nitriding. As with gas carburizing, the more high-alloy steels, for example, tool steels, show a more complex behavior with internal nitriding. A more rigorous modeling approach must be adopted to simulate these complex materials. Nitriding of iron has been modeled successfully by several research groups (Ref 21 to 23). More recently, Larsson and Ågren (Ref 24) reported simulations made using the dispersed model and a comparison with experimental work for high-vanadium steels. The compositions of the alloys studied are given in Table 2. However, to simplify the calculations, some minor elements that were present in the alloys were excluded from the simulations. The elements iron, carbon, chromium, molybdenum, tungsten, vanadium, and nitrogen were considered in the calculations, as were those phases that were identified experimentally, that is, ferrite, the M(C,N) carbonitride, and the M_6C carbide.

As with the work of Turpin et al., described in the previous section on gas carburizing, Larsson and Ågren also emphasized the importance

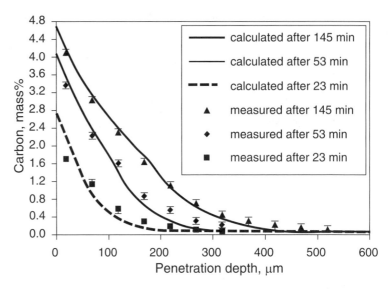

Fig. 5 Evolution of the experimental and calculated carbon profiles of three samples of the Fe-13Cr-5Co-3Ni-2Mo-0.07C grade, carburized in the same conditions during 23, 53, and 145 min at 955 °C (1750 °F). Source: Ref 19

Table 2 Composition of investigated alloys

Alloy	Composition, mass%									
	C	Si	Mn	Cr	Mo	W	Co	V	N	Fe
1	2.4	0.4	0.3	4.2	2.9	4.0	12.1	9.8	0.1	bal
2	2.3	0.5	0.3	4.0	3.0	4.0	0.3	9.7	0.1	bal
3	2.1	0.5	0.3	3.9	3.0	3.9	0.3	9.8	0.1	bal
4	2.0	0.5	0.3	4.0	3.0	3.9	0.3	10.1	0.1	bal

Source: Ref 24

of choosing good boundary conditions for their simulations. However, the authors note that this is a very complex issue for two reasons: The dissociation rates of ammonia vary widely (Ref 25), and there is usually an incubation time before nitriding proceeds at a steady rate (Ref 26). A constant nitrogen activity of 100 (reference state is N_2 at 1 atm and ambient temperature) was therefore chosen as the boundary condition, and it was argued that this was a reasonable assumption, because a constant surface-activity boundary condition for nitrogen should be a good description when the incubation time has been reached. As for carbon, it was acknowledged that some decarburization may have occurred, but in the absence of oxygen, the only possible reaction path is via the formation of methane, which is a very slow process (Ref 25, 27). Thus, a zero flux-boundary condition was chosen for carbon.

Two different kinds of simulations were made. The first kind, designated "Sim 1" in the figures, was made using the dispersed model, according to Engström et al. (Ref 9). A second set of simulations was then made where the MC initially present was assumed to be inert. The reason why Larsson and Ågren decided to run the second set of simulations is illustrated in Fig. 6, which shows the calculated carbon and nitrogen profiles as a function of distance from the nitride surface compared with

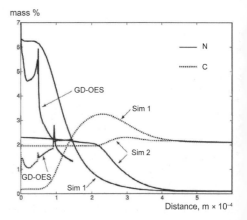

Fig. 6 Results from the simulation of nitriding and comparison with experimental data obtained by glow discharge optical emission spectroscopy (GD-OES) analysis. Carbon and nitrogen profiles in mass percent. Nitrided surface at origin of x-axis. Source: Ref 24

experimental glow discharge optical emission spectroscopy (GD-OES) data. It can be seen in the profile from the first simulation that the carbon content in the nitrided zone is far too low compared with the GD-OES profiles. However, scanning electron microscopy (SEM)-backscattered electron images also showed that the M_6C carbide had practically disappeared in the near-surface zone. Therefore, to derive bounds for the carbon and nitrogen profiles, it was decided to perform simulations where the MC initially present was assumed to be inert. Results from this simulation are designated "Sim 2." The inert MC was taken into account with regard to the labyrinth factor, which is one of the parameters of the dispersed model and the resulting profiles. This approach was successful insofar as the GD-OES profiles are bounded by the simulation profiles. However, the nitriding depth—the case depth—came out somewhat too large.

While the simulations did not show a full quantitative agreement with the experimental results due to equilibrium not being fully established locally, the results obtained from the simulations did agree qualitatively with the experimental results, and an increased understanding of the process was gained. Larsson and Ågren concluded by stating that an improved fit with experimental data could have been obtained by additional assumptions on the state of the material, the labyrinth factor, and so on, but at the time, this was deemed too ad hoc to be of any value. Together with the work of Turpin et al., described in the previous section, these two examples illustrate the importance of selecting boundary conditions and initial conditions to represent the state of the real problem being modeled.

Nitrocarburizing. Modeling the growth kinetics of the compound layer during nitrocarburizing of iron is more complex than modeling nitriding or carburizing alone, due to the complexity of the microstructural evolution. During nitrocarburizing of iron, the compound layer formed is quite inhomogeneous in microstructure, for example, particles of cementite and/or γ' phases are present in the ε layer (Ref 28). However, the ε/γ' bilayer could also be expected during nitrocarburizing if the applied carbon activity is relatively low and the nitrogen activity is relatively high (Ref 29, 30). Additionally, local equilibrium at the compound-surface layer during gaseous nitrocarburizing does not apply, while local equilibrium at the interfaces in the compound layer appears to be maintained, but the interface compositions of phases in local equilibrium are time dependent. Even so, Du et al. (Ref 31) assert that modeling of nitrocarburization can be based on solving multicomponent, multiphase diffuse equations, provided that the surface conditions are known and local equilibrium prevails at the phase boundary.

For example, Du and Ågren (Ref 32) investigated modeling the nitrocarburizing of iron for the growth of ε/γ' bilayers, comparing both an analytical approach and a numerical approach. For the numerical DICTRA simulations, the thermodynamic data for Fe-N and Fe-C-N were based on Du (Ref 33). Diffusion data of nitrogen in the ε and γ' phases were based on an assessment by Du and Ågren (Ref 34), and the diffusion data of carbon and nitrogen in the α phase were based on an evaluation by Jönsson (Ref 35).

As part of the work, some simulations were made to study the phase constitution of the compound layer under different nitrogen and carbon activities at the surface during the nitrocarburizing process. Three simulations were reported for $a_N = 800$: $a_C = 0.2$, 1.5, and 4.0, respectively (reference state: C, graphite; N, 1 atm N_2 gas) and assuming initially that the compound layer was comprised only of ε. The diffusion paths calculated at 2 h are shown in Fig. 7 for each surface condition.

The dashed line corresponding to $a_C = 0.2$ cuts through the $\varepsilon+\gamma'$ two-phase region, which means that γ' should actually be stable between the ε and α phases. The line for $a_C = 1.5$ only passes through the $\alpha+\varepsilon$ two-phase region, indicating that there are no solutions to the flux balance equations assuming ε/γ'-bilayer growth. With $a_C = 4.0$, the line cuts through the cementite + ε and α + cementite + ε regions, and therefore, the compound layer is predicted to contain the cementite phase.

Du et al. (Ref 31) identified that further development of DICTRA was required to handle phase transformations at a moving interface across which two matrix phases are different. This work predated the implementation of the homogenization model into the software, which addresses this point.

Gradient Sintering

Cemented carbides consist of hard WC grains embedded in a more ductile binder phase, usually rich in cobalt. Often, other hard-phase grains are also present in the form of carbides or carbonitrides with a cubic structure. A common application area for cemented carbides, often in the form of tool inserts, is the cutting of metal. To increase the lifetime of the cemented carbide inserts during metal working, the wear surfaces are usually coated with a thin layer of a hard material, such as TiC, TiN, or Al_2O_3, or with multilayers of such hard materials. In the industrial production of tool materials, these layers are usually grown by chemical vapor deposition (CVD) at temperatures of approximately 1000 °C (1830 °F). Due to differences in thermal expansion coefficients between coating and insert, cracks are formed in the coating during cooling after the CVD. To prevent these cracks from propagating into the bulk and causing failure, a tough surface zone is created prior to coating. This zone is enriched in binder phase and depleted in cubic carbides. One of the important issues to understand is identifying the rate-controlling step for the formation of this zone.

Ekroth et al. (Ref 36) performed some experimental studies and computer simulations using DICTRA to resolve a question related to identifying the rate-controlling step for the formation of this zone and also to begin development of a model for the predictions of the influence on the gradient zone, which would take into account parameters such as the carbon and nitrogen content in the material, sintering time,

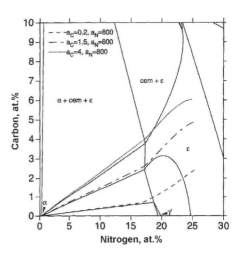

Fig. 7 Calculated diffusion path at 2 h under a fixed nitrogen activity but different carbon activities at the iron surface for 575 °C (1070 °F). Source: Ref 32

temperature, atmosphere, grain size, type of gamma phase, and insert geometry.

Prior to this study, Suzuki et al. (Ref 37) had reported that if nitrogen-containing alloys are sintered under denitriding conditions, the zone thickness obeys a parabolic growth law, which led to the hypothesis that outward nitrogen diffusion in the liquid binder was the rate-controlling step for zone formation. However, cubic nitrides are thermodynamically very stable, and questions were asked as to whether the denitriding effects alone were strong enough to cause the formation of a nitride-free zone.

Schwarzkopf et al. (Ref 38) subsequently suggested that the process is controlled by inward titanium diffusion driven by the outward gradient in nitrogen content. That is, as nitrogen diffuses out from the material, the strong thermodynamic attraction between nitrogen and titanium causes titanium to diffuse in the opposite direction to regions of high nitrogen content, and thus, a surface layer depleted of both titanium and nitrogen forms.

To investigate these hypotheses further, Ekroth prepared some samples based on a mixture of WC, (Ti,W)C, and Ti(C,N). The chemical composition of the sintered material was 6.85Co-5.8Ti-0.38N-6.35C-balW (amounts in weight percent). The samples were heated in two stages. The first heat treatment was a presintering with a dewaxing step, followed by deoxidation before the actual sintering. The controlled atmosphere was introduced at 50 mbar (0.7 psi) when the temperature reached 1350 °C (2460 °F). After reaching 1390 °C (2530 °F) and a 15 min hold time, the furnace was turned off. After presintering, the samples had no visible binder gradient. After this, the samples were reheated to 1450 °C (2640 °F) for 2 h in a nitrogen-free atmosphere (containing mostly carbon monoxide and argon) to develop a controlled binder-phase gradient. The microstructure and phase volume fractions were then studied using SEM and image

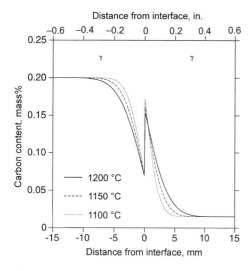

Fig. 8 Calculated concentration profile for carbon after 10 h at 1200, 1150, and 1100 °C (2190, 2100, and 2010 °F). Source: Ref 45

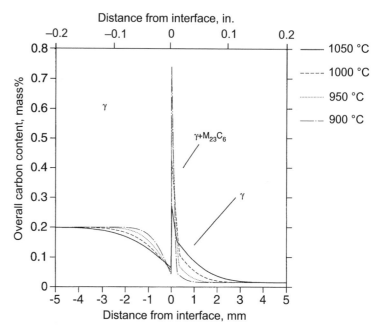

Fig. 9 Calculated concentration profile obtained for carbon after 10 h at 1050, 1000, 950, and 900 °C (1920, 1830, 1740, and 1650 °F). Source: Ref 45

analysis, and elemental profiles were obtained using electron probe microanalysis (EPMA).

Simulations were performed using the dispersed model described earlier. Thermodynamic data were obtained from a database for nitrogen-containing cemented carbides (Ref 39), with modifications for short-range order in the face-centered cubic phase (Ref 36). In the absence of other data, it was assumed that the elements cobalt, titanium, tungsten, carbon, and nitrogen had the same mobility in the liquid phase. The activation energy, Q, was assumed to be 65,000 J·mol^{-1} (Ref 36), and the frequency factor was chosen to be equal to 9.24×10^{-7} m^{2}s^{-1}.

A comparison between the experimental determination of volume fractions and the simulation of mole fractions was made for the different phases. The fractions of the phases present after sintering as a function of distance were also compared with simulations, and these were also found to be in good agreement with the experimental data. Ekroth concluded that the experimental and simulated data were in sufficiently good agreement to indicate that diffusion and the thermodynamic properties are the two major factors that control the gradient structure formation, as predicted by the diffusion-local equilibrium hypothesis. Ekroth also noted that the formation can be reasonably well predicted without considering convective flow and capillary forces. However, one noticeable deviation of the experimental data from the predictions was that the experimental profile for the solid binder was lower than the simulated binder. This was attributed to the possible formation of (Ti,W)(C,N) and WC during the solidification of the liquid cobalt, resulting in a lower amount of solid cobalt binder.

Welds

Diffusion in joints of dissimilar steels has been the topic of a number of experimental

(Ref 40 to 42) and theoretical studies (Ref 43, 44) due to the numerous practical applications in welding and joining. The microstructure and alloy composition in the steels may change as a result of the interdiffusion and influence the mechanical properties, such as toughness and creep strength of the joint. However, the ability to predict weld microstructure is a complex problem related to the interaction of phase stability, multicomponent diffusion, steep temperature gradients, and morphological instabilities during rapid heating and cooling. In the past two decades, computational thermodynamic and kinetic models have been developed to predict these interactions in a wide range of alloys. For example, Helander and Ågren (Ref 45) simulated the diffusion of a typical joint of a stainless steel and a low-alloy steel in a temperature range between 600 and 1200 °C (1110 and 2190 °F) and compared the calculated carbon concentration profiles with experimental data at 650 °C (1200 °F).

Diffusion simulations were performed in the temperature range between 600 and 1200 °C (1110 and 2190 °F) for 10 h for two alloys comprising 18Cr-10Ni-0.015C-balFe (stainless steel) and 0.1Cr-0.1Ni-0.2C-balFe (low-alloy steel). For the lower temperature range (600 to 850 °C, or 1110 to 1560 °F), a ferritic layer was predicted to form in the low-alloy steel at the interface between the steels, while the stainless steel was predicted to carburize with carbide precipitates forming in an austenite matrix; that is, there is a change of the matrix phase where the diffusion will occur at the joint. Because the homogenization model was not included in the software when these simulations were made, a continuous matrix phase was required to perform the calculations. For

this reason, the cell model was employed whereby each "steel" was placed in a separate cell where the cell interface is fixed and cannot move. DICTRA then iterates until it finds chemical potentials at the cell boundary for which the flux balances are obeyed, and the chemical potentials are then treated as ordinary boundary conditions for the diffusion calculation in each cell. In presenting the results, Helander and Ågren noted four distinct temperature ranges that exhibited markedly different microstructures.

The calculated carbon concentration profiles for 1200, 1150, and 1100 °C (2190, 2100, and 2010 °F), where both steels are fully austenitic, are given in Fig. 8. The discontinuity observed at the original joint (where distance is given as zero) is due to the difference in alloy content of the two steels and the much slower diffusion of substitutional alloy elements.

The calculated overall carbon profiles at 1050, 1000, 950, and 900 °C (1920, 1830, 1740, and 1650 °F), where carbides form in the stainless steel close to the joint, are given in Fig. 9. The fraction of $M_{23}C_6$ in the stainless steel is plotted as a function of distance for these temperatures in Fig. 10. The maximum fraction of carbides is displaced further away from the interface for the higher simulation temperatures, and this is due to the increased diffusion distances of the substitutional elements at the higher temperatures.

At 850 and 835 °C (1560 and 1535 °F), the ferritic layer grows into the austenitic matrix of the low-alloy steel due to decarburization. In the stainless steel, the M_7C_3 carbide is formed at the interface. Further away, where the carbon content in the matrix is lower, the $M_{23}C_6$ would precipitate from the austenitic phase. Figure 11 shows the mole fraction as a function of distance of M_7C_3 and $M_{23}C_6$ in the stainless steel for these two temperatures after 10 hours.

A similar plot is given in Fig. 12 for 725, 700, 650, and 600 °C (1340, 1290, 1200, and 1110 °F). In this temperature range, the low-alloy steel is a mixture of cementite and ferrite, while the stainless steel is predicted to contain ferrite, austenite, $M_{23}C_6$, and sigma in the lower part of the temperature range. To simplify the calculations at these temperatures, Helander and Ågren decided to ignore the sigma phase, because it was just a few percent, based on equilibrium calculations, and the ferrite phase was omitted from the calculations on the stainless steel side because the formation of this phase would be very sluggish at these low temperatures. As shown in Fig. 12, the high diffusivity of carbon in ferrite compared to austenite results in very high carbon concentrations in the stainless steel at the interface. Also, the M_7C_3 carbide forms at the interface, with the $M_{23}C_6$ carbide forming further away. Comparison of the experimental data to the simulations shows that the measured width of the ferritic layer for a diffusion couple produced in this temperature range is 530 μm compared with a predicted value of 560 μm. In the stainless steel, the carburized layer had a measured width of 100 μm compared with a calculated value of 70 μm. This discrepancy corresponds to an uncertainty in the carbon diffusivity of a factor of 2, which is within the accuracy reported by Jönsson (Ref 46) in his assessment.

The authors concluded that the results of the simulations were in general agreement with practical experience and noted that such simulations are used by Sandvik Steel AB for optimizing the heat treatment of composite tubes.

Coatings

Commercial nickel-base superalloys have been developed with compositions and heat treatments to optimize high-temperature strength as well as resistance against creep, fatigue, oxidation, and hot corrosion. However, although nickel-base superalloys have excellent strength and creep resistance, they normally require protective coatings (i.e., oxidation resistance or thermal barrier coatings with bond coats) to meet the high-temperature demand of higher-performance turbine engines (Ref 47). The MCrAlY-type coatings are one such example.

The lifetime of the MCrAlY-type coating depends on the availability of aluminum and chromium that form protective oxides at the outer surface. In both MCrAlY and NiAl-base coatings, the aluminum-rich β phase serves as an aluminum reservoir for the formation of continuous, stable, and protective Al_2O_3 scale (Ref 47–53). During high-temperature exposure, the aluminum-rich β phase dissolves through depletion of aluminum from the coating, which occurs by two mechanisms:

- Loss of aluminum by interdiffusion toward the surface of the coating to form Al_2O_3
- Compositional differences between the coating and the superalloy, leading to significant loss of the active elements through interdiffusion

Several studies have shown that the interdiffusion between the coating and the superalloy substrate may contribute more to the overall aluminum depletion (Ref 54–65) than the aluminum depletion caused by oxidation does, although repeated spallation of the Al_2O_3 scale may accelerate the loss caused by oxidation. In general, when the aluminum concentration in the coating falls below approximately 10 at.%, after complete dissolution

of the phase, these coatings can no longer maintain the continuity of the scales and are considered no longer effective.

It has been postulated that the lifetime of the coatings, defined by depletion of aluminum, may be enhanced by controlling the interdiffusion fluxes of individual components, particularly for aluminum. Therefore, while a coating may be selected/designed for environmental degradation resistance, and a superalloy can be selected/developed based on strength and creep and fatigue resistances, a system selection/development should be optimized to control the interdiffusion fluxes of diffusing components, so that aluminum interdiffusion flux across the coating/superalloy interface is minimized.

With this aim in mind, Perez et al. (Ref 66) experimentally studied diffusion couples consisting of single-phase B2 β-NiAl with various commercial superalloys and then determined the concentration profiles in the single-phase β-NiAl side of the couple using EPMA.

Diffusion couples were prepared from hot extruded β-NiAl and several commercial superalloys. The average chemical compositions of the superalloys were measured using EPMA, and values for the trace elements (which fall below the detection limits for EPMA) were based on values from the literature. The β-NiAl used for all the diffusion couples was of near-stoichiometric composition. The diffusion couples were annealed isothermally at 1050 °C (1920 °F) for 96 h and then cooled by quenching with water. Microstructural analysis of the diffusion couples was then carried out by backscatter electron imaging, and the concentration profiles within the diffusion couples were determined using EPMA.

Engström et al. (Ref 67) subsequently simulated some of these diffusion couples using the homogenization model. Thermodynamic data were taken from the TCNi1 (Ref 68–70) thermodynamic database for nickel-base

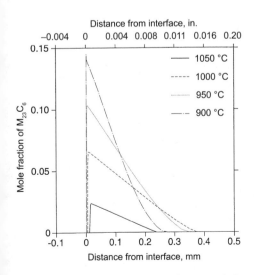

Fig. 10 Calculated mole fractions of $M_{23}C_6$ carbide formed after 10 h at 1050, 1000, 950, and 900 °C (1920, 1830, 1740, and 1650 °F). Source: Ref 45

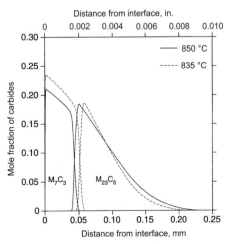

Fig. 11 Calculated mole fractions of carbides formed after 10 h at 850 and 835 °C (1560 and 1535 °F). Source: Ref 45

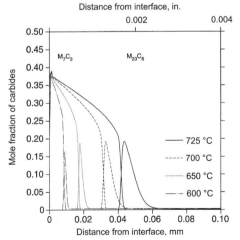

Fig. 12 Calculated mole fractions of carbides formed after 10 h at 725, 700, 650, and 600 °C (1340, 1290, 1200, and 1110 °F). The first peak at each temperature represents the mole fraction of M_7C_3. The second peak shows the fraction of $M_{23}C_6$. Source: Ref 45

superalloys. Although this database contains only seven elements (nickel, aluminum, chromium, cobalt, rhenium, titanium, tungsten), and larger databases were available for use with Thermo-Calc and DICTRA, it was necessary to use this database because it was the only one that used an order-disorder description to model both the body-centered cubic (bcc)_B2 (NiAl, B2) and face-centered cubic (fcc)_L12 (Ni₃Al, γ′) phases. (See the phase diagram for nickel-aluminum in Fig. 13.) Vacancies are included in the description of the bcc_B2 phase, and it is important to simulate the concentration dependence of diffusion of vacancies, because chemical ordering creates structural vacancies.

Mobility data were taken from MOBNi1 (Ref 72), a mobility database for nickel-base superalloys, which contains assessed mobility data for the disordered fcc_A1 (γ) and liquid phases. This was extended to add some descriptions for the mobilities in the bcc_B2 and fcc_L12 ordered phases.

Chemical ordering is handled in the mobility databases by using a phenomenological model suggested by Helander and Ågren (Ref 73). Data for the bcc_B2 and fcc_L12 phases for Ni-Al-Cr were taken from Campbell (Ref 74), and preliminary descriptions were used for the remaining elements, that is, cobalt, titanium, and tungsten. It was assumed that cobalt behaved as nickel, and that titanium and tungsten have the same contribution for ordering as chromium.

Simulations were made for two alloys: GTD111 and IN939. Compositions are given in Table 3. Carbon, molybdenum, and tantalum were excluded from the calculations because these are not included in the TCNi1 database, and the amounts of these elements were left as balance of nickel.

Simulated concentration profiles are shown for aluminum and nickel compared with experimental data for GTD111 in Fig. 14 and IN939 in Fig. 15. As can be seen from the figures, even with the simplifications of treating just a few of the major elements in the nickel superalloy, and making some simplified assumptions regarding the mobilities of some of these elements in the ordered phases, it is possible to reproduce reasonably well the trends in the experimental data. Work is continuing to improve the data for these ordered phases.

Although not compared with the experimental data of Perez et al., it was predicted that phase fractions can be plotted as a function of distance (relative to the interface of the coating and the nickel superalloy being zero), and these are shown in Fig. 16 for GTD111 and Fig. 17 for IN939. These figures show the growth of the bcc_B2 phase into the alloy and the formation of the zone enriched in γ′ (fcc_L12).

Similar comparisons reported previously by Dahl et al. (Ref 75) using the dispersed model indicated that such predictions can qualitatively predict the interface behavior in terms of correct trends in both the composition profiles and phase fraction diagrams. Dahl et al. found the fit to

experimentally measured layer widths to be quite good, even in this case without any diffusion data for the fcc_L12 phase. It was postulated that the reason for the surprisingly good agreement is that diffusion in the ordered phase is very slow compared with the disordered fcc_Al (γ) phase and

therefore works as an effective block for diffusion in both the actual microstructure and in simulations where it was treated as a "diffusion = none" phase.

Both Engström and Dahl used DICTRA for their simulations, which restricts the

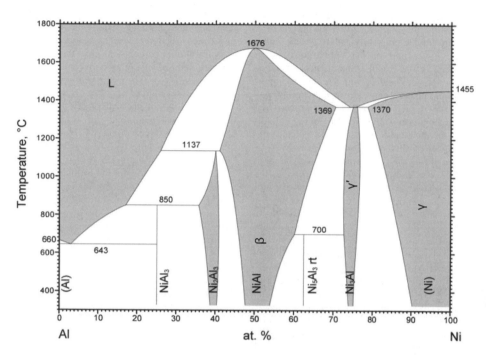

Fig. 13 Calculated phase diagram for aluminum-nickel using Thermo-Calc and TCNi1. Source: Ref 71

Table 3 Composition of GTD111 and IN939

Alloy	Ni	Al	C	Co	Cr	Mo	Ta	Ti	W
				Composition, at.%					
GTD111	bal	6.9	0.48	9.5	16.5	0.97	0.89	6.24	0.97
IN939	bal	4.45	0.71	18.6	25.3	0.43	0.37	4.07	0.51

Source: Ref 66, 67

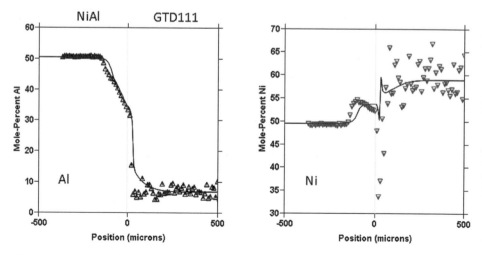

Fig. 14 Calculated mole fractions of aluminum and nickel for alloy GTD111 compared with experimental data of Perez et al. (Ref 66) as a function of distance. Position zero is the interface between the coating and substrate. Source: Ref 67

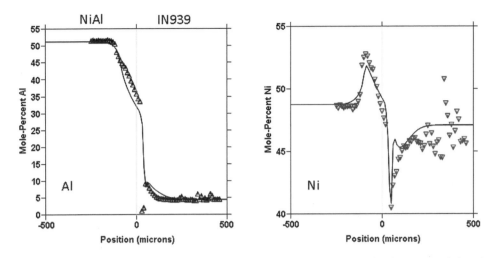

Fig. 15 Calculated mole fractions of aluminum and nickel for alloy IN939 compared with experimental data of Perez et al. (Ref 66) as a function of distance. Position zero is the interface between the coating and substrate. Source: Ref 67

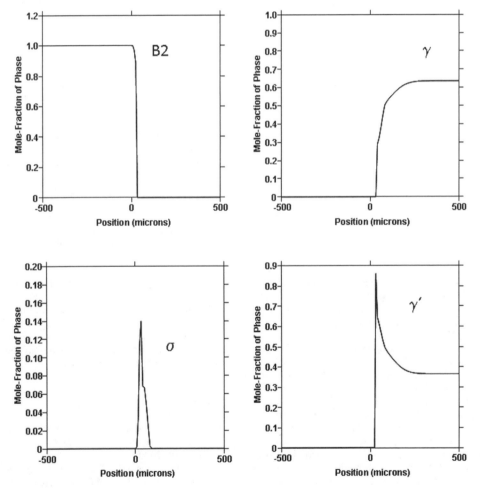

Fig. 16 Calculated mole fractions of B2, γ, σ, and γ' phase for alloy GTD111. Position zero is the interface between the coating and substrate. Source: Ref 67

precipitate morphology, evolution, dissolution of the γ+β regions, and diffusion paths to compare with experimental data.

Thermodynamic data were taken from the reference of Huang and Chang (Ref 77), and kinetic mobility data for the γ phase was obtained from Engström and Ågren (Ref 18). Data were evaluated for the β phase based on data by Hopfe et al. (Ref 78). The phase-field equations were solved with an explicit numerical method for two dimensions using periodic boundary conditions and a mesh size of 256 × 1024.

Simulations for six γ+β/γ diffusion couples were made, which corresponded to experimental studies performed by Nesbitt and Heckel (Ref 79). The initial average composition of the γ+β alloy was the same for all simulations where the mole fraction of chromium was 0.126 and the mole fraction for aluminum was 0.242, which is close to those of the experimental study.

The evolution of precipitate morphology during interdiffusion of γ+β/γ diffusion couples was simulated. β was calculated to have a volume fraction of 0.53, and γ, initially modeled in the two-phase region as randomly distributed circular particles, was calculated to have a diameter of 60 μm (compared with experimental values of 64 μm) and a Gaussian size distribution. The calculations suggest that the microstructure quickly evolves into a highly connected structure that somewhat resembles the experimental microstructures. However, the microstructure coarsens at an unrealistic rate due to the large surface tension for the γ/β interface required for the current phase-field method at this length scale. Also, the simulations were found to overpredict the recession in every case. This was explained by Wu et al. as being due to the thermodynamic data of Huang and Chang, which predicts the phase boundary between the γ+β and β regions to be several percent richer in aluminum than that predicted by other databases and experimental data. As a result, the fraction of β phase in the simulation was less than that measured experimentally.

Oxidation

A very recent addition to the functionality of the DICTRA software package has been the introduction of models to support the simulation of diffusion in oxides and other ionic-type systems. The motivation to develop such an approach is to enable, for example, predicting the oxidation of steels and the influence of alloying elements on the oxidation rate and also to investigate the degradation of coatings on nickel-base superalloys within the CALPHAD-type framework.

Oxidation occurs by diffusion of cations and anions through one or more oxide scales. This is a complex process to model at the level of the basic thermodynamics and kinetic simulations, because both mass conservation and charge balance must be considered during the

simulations to those of one dimension. Wu et al. (Ref 76) have also investigated the formation of microstructures in γ+β/γ diffusion couples made of Ni-Cr-Al alloys in two

dimensions using a phase-field model. In this work, the authors systematically looked at the effect of γ-alloy composition on the interdiffusion microstructure and simulated the

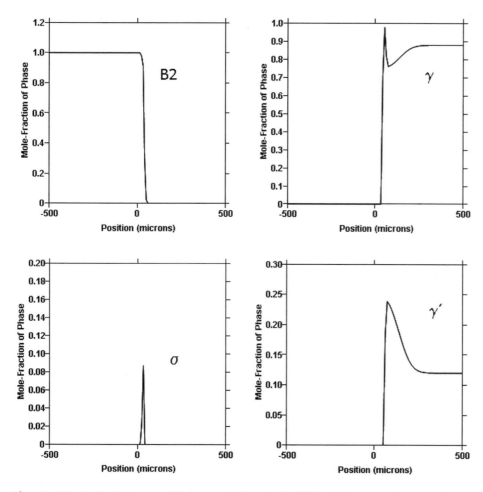

Fig. 17 Calculated mole fractions of B2, γ, σ, and γ′ phase for alloy IN939. Position zero is the interface between the coating and substrate. Source: Ref 67

numerical simulation. New models introduced into DICTRA Version 25 and onward can now treat diffusion in basically any oxide, provided that there are diffusion data available and that the necessary thermodynamic data and mobility data have been assessed. The first system assessed has been the important iron-oxygen system, and this work was performed by Hallström et al. (Ref 80).

The iron-oxygen system was chosen for its industrial importance and also because the necessary thermodynamic data for modeling the metal oxide system are available and well established. The iron-oxygen system contains three oxides: wustite, hematite, and magnetite.

The thermodynamic data are based on an evaluation by Sundman (Ref 81) for the iron-oxygen system. Mobility data for the wustite phase are based on tracer diffusion data from Chen and Peterson (Ref 82); the derived chemical diffusion coefficient agrees within a factor of 2 compared with measured values reported by Millot and Berthon (Ref 83). Mobility data for magnetite were based on an assessment of tracer diffusion data measured by Dieckmann and Schmalzried (Ref 84), Peterson et al. (Ref 85), Aggarwal and Dieckmann (Ref 86), and Becker et al. (Ref 87).

Mobility data for hematite were based on the data of Atkinson and Taylor (Ref 88), Hoshino and Peterson (Ref 89), and Amami et al. (Ref 90). Figure 18 shows a comparison of the calculated diffusivity for magnetite compared with experimentally measured values based on the optimized mobilities for this phase.

The mobility database for iron oxides contains fully assessed kinetic data for cation diffusion in all three oxides. However, the oxidation of iron does not only involve diffusion of cations. Looking at cross sections of oxidized iron, it is apparent that diffusion of oxygen must also play an important role since the oxides also grow inward. This can only be accomplished by oxygen diffusion through the oxides. At the time this work was reported however, oxygen diffusion through the oxides (and subsequent inward growth of the oxides) had not been considered. Also, at the temperatures of interest for these materials, grain boundaries and other fast diffusion paths may also play important roles in the oxidation process and must be taken into account. At the time of writing this article, such a model is under development but has not yet been implemented into the programming code.

However, it is possible to account for these effects in a simplified manner by multiplying the diffusivities in the various phases by arbitrary user-defined functions. For example, one way to take grain-boundary diffusion into account is by assigning the grain boundaries a certain width and a fraction of the total volume (e.g., from the approximation $f^{gb} = \delta/D$, where δ is the grain-boundary width, and D is the average grain size). From this, the effective diffusion coefficient can be calculated as a weighted sum of the various contributions, provided that the grain-boundary mobility is known. The grain-boundary width can be approximated to 5 Å, and experimental results from Pujilaksono (Ref 91) have been used to estimate the necessary parameters. If the average grain size of magnetite is approximately 3 μm and that of hematite is 0.1 μm, the grain-boundary fractions become 1.7×10^{-4} and 5×10^{-3}, respectively. By making some estimate of the activation energy of grain-boundary diffusion relative to the activation energy for bulk diffusion, it is possible to relate some effective mobility to the evaluated bulk mobility and use that effective value to make some simulations.

As an example, simulation conditions comparable to the experiments reported by Pujilaksono for an oxygen partial pressure of 0.05 atm (0.7 psi) at 600 °C (1110 °F) for 24 h were used as the basis for a prediction of oxide growth. Using an activation energy for grain-boundary diffusion of half the value for bulk diffusion resulted in predicted thicknesses of hematite, magnetite, and wustite that are approximately 0.2, 2.7, and 46 μm, respectively. The experimental thicknesses vary with location on the sample but are typically 2, 11, and 21 μm, respectively. While these predictions are of approximately the same order of magnitude, changing the activation energy for grain-boundary diffusion to one-third that of bulk diffusion resulted in the growth curves shown in Fig. 19 and predicted thicknesses of 2, 11, and 45 μm, respectively, at the end of the simulation, that is, much closer to the experimental values. Further work is needed to develop this model, but this is one example of a relatively new development in the area of diffusion-based simulations that could lead to future improvements in alloy design and understanding of the oxidation process.

REFERENCES

1. Z.-K. Liu, A Materials Research Paradigm Driven by Computation, *JOM,* Vol 61 (No. 10), 2009
2. L. Kaufman and H. Bernstein, *Computer Calculation of Phase Diagram,* Academic Press Inc., New York, 1970
3. Z. Chen et al., "Precipitation Simulation in Multicomponent Ni-Based Alloys," AEROMAT 2009 Conference (Dayton, OH)
4. K. Thornton, J. Agren, and P.W. Voorhees, Modelling the Evolution of Phase Boundaries in Solids at the Meso- and Nano- Scales, *Acta Mater.,* Vol 51, 2003, p 5675–5710

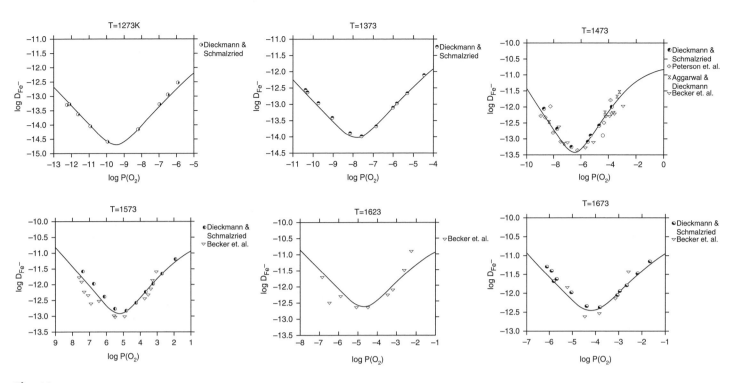

Fig. 18 Comparison of the calculated diffusivity for magnetite compared with experimentally measured values based on the optimized mobilities for this phase. Source: Ref 80

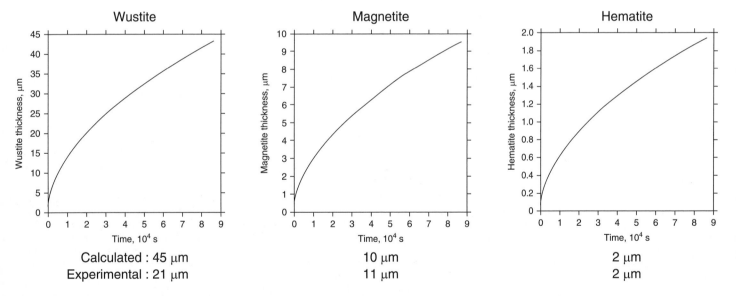

	Wustite	Magnetite	Hematite
Calculated :	45 μm	10 μm	2 μm
Experimental :	21 μm	11 μm	2 μm

Fig. 19 Prediction of oxide growth for iron compared with experimental data. Source: Ref 91

5. J.W. Gibbs, *The Collected Works of J. Willard Gibbs,* Oxford University Press, Oxford, 1948

6. A. Borgenstam et al., DICTRA, A Tool for Simulation of Diffusional Transformations in Alloys, *J. Phase Equilibria,* Vol 21, 2000, p 269–280

7. J.O. Andersson et al., Thermo-Calc and DIC-TRA, Computational Tools for Materials Science, *CALPHAD,* Vol 26, 2002, p 273–312

8. J. Ågren, Numerical Treatment of Diffusional Reactions in Multicomponent Alloys, *J. Phys. Chem. Solids,* Vol 43, 1982, p 385–391

9. A. Engström, L. Höglund, and J. Ågren, Computer Simulation of Diffusion in Multiphase Systems, *Metall. Mater. Trans. A,* Vol 25, 1994, p 1127–1134

10. A. Engström and J. Ågren, *Defect and Diffusion Forum,* Vol 143–147, 1997, p 677–682

11. H. Larsson and A. Engström, A Homogenization Approach to Diffusion Simulations Applied to α and γ Fe-Cr-Ni Diffusion Couples, *Acta Mater.,* Vol 54, 2006, p 2431–2439

12. I. Steinbach et al., A Phase Field Concept for Multiphase Systems, *Phys. D,* Vol 94, 1996, p 135–147

13. S.G. Fries et al., Upgrading CALPHAD to Microstructure Simulation: The Phase-Field Method, *Int. J. Mater. Res.,* Vol 100, 2009, p 2

14. J. Eiken, B. Böttger, and I. Steinbach, Multiphase-Field Approach for Multicomponent Alloys with Extrapolation Scheme

for Numerical Application, *Phys. Rev. E,* Vol 73, 2006, p 066122

15. H.-J. Jou, P. Voorhees, and G.B. Olson, PrecipiCalc Simulations for the Prediction of Microstructure/Property Variation in Aeroturbine Disks, *Superallloys 2004,* K.A. Green et al., Ed., TMS, 2004, p 877–886

16. J.O. Andersson and J. Ågren, Models for Numerical Treatment of Multicomponent Diffusion in Simple Phases, *J. App. Phys.,* Vol 72, 1992, p 1350–1355

17. B. Jönsson, Ferromagnetic Ordering and Diffusion of Carbon and Nitrogen in bcc Cr-Fe-Ni Alloys, *Z. Metallkd.,* Vol 85, 1994, p 498–501

18. A. Engström and J. Ågren, Assessment of Diffusional Mobilities in Face Centered Cubic Ni-Cr-Al Alloys, *Z. Metallkd.,* Vol 87, 1996, p 92–97

19. T. Turpin, J. Dulcy, and M. Gantois, Carbon Diffusion and Phase Transformations during Gas Carburizing of High-Alloyed Stainless Steels: Experimental Study and Theoretical Modeling, *Metall. Trans. A,* Vol 36, 2005, p 2751–2760

20. "Thermodynamic Database for Fe-Based Alloys and Steels," TCFe2000, Thermo-Calc Software AB, 2000

21. M.A.J. Somers and E.J. Mittemeijer, *Metall. Trans. A,* Vol 26, 1995, p 57–74

22. H. Du and J. Ågren, *Z. Metallkd.,* Vol 86, 1995, p 522–529

23. L. Torchane et al., *Metall. Trans. A,* Vol 27, 1996, p 1823–1835

24. H. Larsson and J. Ågren, Gas Nitriding of High Vanadium Steels—Experiments and Simulations, *Metall. Trans. A,* Vol 35, 2004, p 2799–2802

25. *Heat Treating,* Vol 4, *Metals Handbook,* 9th ed., American Society for Metals, 1981

26. H. Du, Doctoral thesis, Royal Institute of Stockholm, Stockholm, Sweden, 1994

27. C.R. Brooks, *Principles of the Surface Treatment of Steels,* Technomic, Lancaster, PA, 1992

28. H. Du, M.A. Somers, and J. Ågren, *Trita-Mac 0565,* Royal Institute of Technology, Stockholm, Sweden, 1994

29. A. Wells, Ph.D. thesis, The University of Liverpool, Liverpool, U.K., 1982

30. M.A. Somers and E.J. Mittemeijer, *Surf. Eng.,* Vol 3, 1987, p 123–137

31. H. Du, M.A. Somers, and J. Ågren, Microstructural and Compositional Evolution of Compound Layers during Gaseous Nitrocarburizing, *Metall. Trans. A,* Vol 31, 2000, p 195–211

32. H. Du and J. Ågren, Theoretical Treatment of Nitriding and Nitrocarburizing of Iron, *Metall. Trans. A,* Vol 27, 1996, p 1073–1080

33. H. Du, *J. Phase Equilibria,* Vol 14, 1993, p 682–693

34. H. Du and J. Ågren, *Trita-Mac-0565,* Royal Institute of Technology, Stockholm, 1994

35. B. Jönsson, *Trita-Mac-0514,* Royal Institute of Technology, Stockholm, 1993

36. M. Ekroth et al., Gradient Zones in WC-Ti (C,N)-Co-Based Cemented Carbides—Experimental Study and Computer Simulations, *Acta Mater.,* Vol 48, 2000, p 2177–2185

37. H. Suzuki, K. Hayashi, and Y. Taniguchi, *Trans. Jpn. Institut. Met.,* Vol 22, 1981, p 758

38. M. Schwarzkopf et al., *Mater. Sci. Eng.,* Vol A105/106, 1988, p 225

39. CAMPADA, Thermodynamic Carbo-Nitride Database, CN1b, Royal Institute of Stockholm, Stockholm, 1999

40. R. Ayer et al., *Metall. Trans. A,* Vol 20, 1989, p 665–681

41. B. Buchmayr et al., *Second Int. Conf. Trends in Welding Research* (Gatlinburg, TN), ASM International, 1989, p 237–242

42. B. Lopez, I. Gutierrez, and J.J. Urcola, *Mater. Charact.,* Vol 28, 1992, p 49–59

43. J. Ågren, *Scand. J. Metall.,* Vol 10, 1981, p 134–140

44. B. Buchmayr and J.S. Kirkaldy, *Fundamentals and Applications of Ternary Diffusion,* G.R. Purdy, Ed., Pergamon Press, New York, 1990, p 164–172

45. T. Helander and J. Ågren, Computer Simulation of Multicomponent Diffusion in Joints of Dissimilar Steels, *Metall. Trans. A,* Vol 28, 1997, p 303–308

46. B. Jönsson, *Z. Metallkd.,* Vol 85, 1994, p 502–509

47. K. Hunecke, *Jet Engines,* Motorbooks Int., 1997, 6th print, 2003

48. C.T. Simms, N.S. Stoloff, and W.C. Hagel, *Superalloys II,* John Wiley and Sons, 1987

49. M.J. Donachie and S.J. Donachie, *Superalloys: A Technical Guide,* 2nd ed., ASM International, 2002

50. J.A. Nesbitt and R.W. Heckel, Modeling Degradation and Failure of Ni-Cr-Al Overlay Coatings, *Thin Solid Films,* Vol 119, 1984, p 281–290

51. J.A. Nesbitt et al., Diffusional Transport and Predicting Oxidation Failure during Cyclic Oxidation of β-NiAl Alloys, *Mater. Sci. Eng. A,* Vol 153, 1992, p 561–566

52. J.A. Nesbitt and R.W. Heckel, Diffusional Transport during the Cyclic Oxidation of γ + β, Ni-Cr-Al (Y,Zr) Alloys, *Oxid. Met.,* Vol 29, 1988, p 75–102

53. N.S. Cheruvu, K.S. Chan, and G.R. Leverant, Coating Life Prediction for Combustion Turbine Blades, *International Gas Turbine and Aeroengine Congress and Exhibition,* June 2–5, 1998 (Stockholm, Sweden), Vol 98-GT-478

54. E.Y. Lee et al., Modeling the Microstructual Evolution and Degradation of M-Cr-Al-Y Coatings during High Temperature Oxidation, *Surf. Coat. Technol.,* Vol 32, 1987, p 19–39

55. M.S. Thompson and J.E. Morral, Kinetics of Coatings/Substrate Interdiffusion in Multicomponent System, *Proc. High Temperature Coatings,* M. Khobaib and R.C. Krutenat, Ed., Metallurgical Soc., 1986, p 55–66

56. K.L. Luthra and M.R. Jackson, Coating/Substrate Interactions at High Temperature, *Proc. Symp. High Temperature Coatings,* M. Khobaib and R.C. Krutenat, Ed., Metallurgical Soc., 1986, p 85–100

57. J.E. Morral, Interdiffusion and Coating Design, *Surf. Coat. Technol.,* Vol 43/44, 1990, p 371–380

58. M.J. Fleetwood, Influence of Nickel-Base Alloy Composition on the Behavior of Protective Aluminide Coatings, *J. Inst. Met.,* Vol 98, 1970, p 1–7

59. M.A. Dayananda, Multicomponent Diffusion Studies in Selected High Temperature Alloy Systems, *Mater. Sci. Eng. A,* Vol 121, 1989, p 351–359

60. C.W. Yeung et al., Interdiffusion in High Temperature Two Phase Ni-Cr-Al Coating Alloys, *Mater. Sci. Forum,* Vol 163–165, 1994, p 189–194

61. J.E. Morral and R.H. Barkalow, Analysis of Coatings/Substrate Interdiffusion with Normalized Distance and Time, *Scr. Metall.,* Vol 16, 1982, p 593–594

62. K.A. Ellison, J.A. Daleo, and D.H. Boone, Interdiffusion Behavior in NiCoCrAlYRe-Coated IN-738 at 940 C and 1050 C, *Superalloys 2000, Proc. Ninth Int. Symp. Superalloys,* T.M. Pollock et al., Ed., Sept 17–21, 2000 (Seven Springs, PA), p 649

63. B. Wang et al., Interdiffusion Behavior of Ni-Cr-Al-Y Coatings Deposited by Arc-Ion Plating, *Oxid. Met.,* Vol 56, 2001, p 1–13

64. B. Wang et al., The Behavior of MCrAlY Coatings on Ni₃Al Base Superalloy, *Mater. Sci. Eng. A,* Vol 357, 2003, p 39–44

65. B. Gleeson, E. Basuki, and A. Crosky, Interdiffusion Behavior of an Aluminide Coated Nickel-Base Alloy at 1150 C, *Elevated High Temperature Coatings: Science and Technology IV,* N.B. Dahotre, J.M. Hampikian, and J.E. Morral, Ed., TMS, 2001, p 119–132

66. E. Perez, T. Patterson, and Y. Sohn, Interdiffusion Analysis for NiAl versus Superalloys Diffusion Couples, *J. Phase Equilibria Diff.,* Vol 27, 2006, p 659–664

67. A. Engström et al., "Simulation of Interdiffusion Kinetics in Ni-Base Superalloy/NiAl-Coating Systems," AEROMAT 2009 (Dayton, OH), 2009

68. Thermodynamic Database for Ni-Based Superalloys, TCNi1, Thermo-Calc Software AB

69. I. Ansara, N. Dupin, and H.L. Lukas, *J. Alloy Compd.,* Vol 247, 1997, p 20–30

70. N. Dupin and B. Sundman, *Scand. J. Metall., Proceedings of the Discussion Meeting on Thermodynamics of Alloys* (Stockholm), Vol TOFA2000

71. I. Ansara, "Al-Ni Phase Diagram," ASM Alloy Phase Diagrams Center, ASM International, 1997

72. Mobility Database for Ni-Based Superalloys, MOBNi1, Thermo-Calc Software AB, Stockholm, Sweden

73. T. Helander and J. Ågren, A Phenomenological Treatment of Diffusion in Al-Fe and Al-Ni Alloys Having B2-B.C.C. Ordered Structure, *Acta Mater.,* Vol 47, 1999, p 1141–1152

74. C.E. Campbell, Assessment of the Diffusion Mobilites in the Γ' and B2 Phases in the Ni-Al-Cr System, *Acta Mater.,* Vol 56, 2008, p 4277–4290

75. K.V. Dahl, J. Hald, and A. Horsewell, Interdiffusion between Ni-Based Superalloy and MCrAlY Coating, *Defect Diff. Forum,* Vol 258–260, Trans Tech Publications, Switzerland, 2006, p 73–78

76. K. Wu et al., Multiphase Ni-Cr-Al Diffusion Couples: A Comparison of Phase Field Simulations with Experimental Data, *Acta Mater.,* Vol 56, 2008, p 3854–3861

77. W. Huang and Y.A. Chang, *Intermetallics,* Vol 7, 1999, p 863

78. W.D. Hopfe et al., *Diffusion in Ordered Alloys,* TMS, Warrendale, PA, 1993, p 69

79. J.A. Nesbitt and R.W. Heckel, *Metall. Trans. A,* Vol 18, 1987, p 2061

80. S. Hallstrom, L. Höglund, and J. Ågren, "Modeling of Diffusion in Wustite and Simulation of Oxidation of Iron At 600°," The Sixth European Stainless Steel Conference, June 10–13 2008 (Helsinki), Science and Market

81. B. Sundman, *J. Phase Equilibria,* Vol 12, 1991, p 127–140

82. W.K. Chen and N.L. Peterson, Effect of the Deviation from Stoichiometry on Cation Self-Diffusion and Isotope Effect in Wüstite Fe, *J. Phys. Chem. Solids,* Vol 36, 1975, p 1097–1103

83. F. Millot and J. Berthon, Isothermal Transport Properties of $Fe_{1-\delta}O$ from a Polarization Method, *J. Phys. Chem. Solids,* Vol 47, 1986, p 1–10

84. R. Dieckmann, H. Schmalzried, and B. Bunsenges, *Phys. Chem.,* Vol 81, 1977, p 344

85. N.L. Peterson, W.K. Chen, and D. Wolf, *J. Phys. Chem. Solids,* Vol 41, 1980, p 709–719

86. S. Aggarwal and R. Dieckmann, *Phys. Chem. Minerals,* Vol 22, 2002, p 707–718

87. K.D. Becker, V. von Wurmb, and F.J. Litterst, *J. Phys. Chem. Solids,* Vol 54, 1993, p 923–935

88. A. Atkinson and R.I. Taylor, *J. Phys. Solids,* Vol 46, 1985, p 469–475

89. K. Hoshino and N.L. Peterson, *J. Phys. Solids,* Vol 46, 1985, p 375–382

90. B. Amami et al., *Ionics,* Vol 5, 1999, p 358–370

91. T. Jonsson et al., Seventh International Symposium on High Temperature Corrosion and Protection of Materials, May 18–23, 2008 (Les Embiez, France)

Integration of Modeling and Simulation in Design

ASM Handbook, Volume 22B, *Metals Process Simulation*
D.U. Furrer and S.L. Semiatin, editors

Copyright © 2010, ASM International®
All rights reserved.
www.asminternational.org

Solid Modeling

Stephen M. Samuel, Design Visionaries Inc.

ALL DESIGN STARTS IN THE IMAGINA-TION of a person with a need. If water needs to be transported from one place to another, a bridge needs to be constructed, or portable music is desired, there is a great design waiting to be discovered. The design process begins as soon as a capable person says there should be a product. Should that product have any type of physical manifestation, such as sheet metal panels, a molded composite housing, or even a wooden frame, the design becomes complex enough that a modern computer-aided design (CAD) system can help. The more complex the design, the more advantageous a CAD system becomes. For the design of a large jet engine with hundreds of thousands of separate components, a CAD system is an absolute necessity. Although it is possible to create a fantastic jet engine using little more than slide rules and drafting tables, high-end CAD is essential to be competitive in an extremely sophisticated market place. For the design of a wooden chair, although not as important, the CAD system still delivers huge benefits to the designer or engineer who is proficient with a high-end CAD tool. *Solid modeling* is the term that describes three-dimensional (3-D) geometric models that are produced with a CAD tool. *Product definition data* is the term used to describe not only the 3-D geometry but everything else that goes along with a product definition, such as material properties, color, and manufacturing processes.

In the most ideal case, perhaps sometime in the future, there will be a CAD system that somehow works on brain waves and renders a design to fit the desired application. The design will be perfect in every conceivable way and will automatically translate into the instructions to run the most sophisticated manufacturing tools. The proverbial button will be pressed, and the fully functioning device or product will emerge. The evolution of CAD systems is a consistent march to that lofty goal. However, as of this writing (2010), CAD systems and solid modeling must be driven using computer keyboards and mice, valuator devices, and myriads of commands and menus. Several sources of solid modeling systems are given in Table 1. The most powerful CAD systems must be studied for quite some time for users to attain an expert level.

Those that use CAD to a professional level and are involved in the design process across many industries have converged on certain agreed-upon terms. Mechanical CAD is MCAD; the term is necessary to differentiate it from electrical CAD, or ECAD, which is used to lay out circuit boards. Solid modeling is the act of creating 3-D models of various components and systems, as opposed to two-dimensional (2-D) layout work. An assembly is a collection of those components. A component is built up of certain entities whose terminology varies a bit among the various CAD systems that populate the marketplace. Most CAD systems offer some way of creating basic geometric building blocks in 3-D space, called primitives. These are the cube, sphere, cone, cylinder, and so on. Another category of entities is 3-D but volumeless. These are called sheets or surfaces. A user can create flat surfaces, cylindrical surfaces, airfoil-like surfaces, and many others that are discussed later. Other important entities are curves, such as lines, arcs, and splines. In addition, there are conics, text entities, planes, axes, and coordinate systems. A user can also create helixes, parabolas, hyperbolas, and any other function curve that can be described mathematically.

When a good solid model is created to represent a part, there are a huge number of functions that it will support. One can determine the weight, verify its fit into an assembly, create 2-D drawings for it, or machine cutter paths. One can animate the solid model as it is driven kinematically along a predetermined path, and the various views of it, with shading and reflections in perspective, are illustrated.

Solid Modeling

One of the most important principles of solid modeling is the concept of the closed and unified volume. A solid model is only a solid model when there is a fully closed volume of surface entities that have the extra information of which side of the surface is inside or outside, those surfaces are unified, and they have an assigned density. For example, one may place six square surfaces together, such that the shape formed is a perfect cube, but if the surfaces are not unified or associated with each other, they do not form a solid. Figure 1 shows two sets of surfaces arranged in cubes. When a hole is drilled in the one that is not a solid, it opens up to the interior; when the same hole is drilled into the solid, new surfaces are created to represent the material inside.

The modern designer must be able to quickly operate on these entities—trimming, scaling, copying, and moving them, if desired. When necessary, solids can be added and subtracted from each other to form more complex solid models. For example, when one subtracts a small cylinder from a larger cylinder, a ring results, as shown in Fig. 2. Similarly, if the two shapes are added, a wheel with an axle is produced.

Fundamental Approaches

The essence of solid modeling can be captured by two basic analogies. One may start with a block of material and keep whittling away at it, similar to a marble sculptor, or one may begin with a small solid and keep adding to it, similar to a clay sculptor. A combination of both may also be used. For example, one

Table 1 Typical solid modeling software

Software	Provider	Website(a)
NX/Unigraphics	Siemens PLM Software	http://www.plm.automation.siemens.com/en_us/products/nx/
Pro/Engineer	Parametric Technology Corporation	http://www.ptc.com/products/proengineer/
SolidWorks	Dassault Systems	http://www.solidworks.com/
Catia	Dassault Systems	http://www.3ds.com/products/catia
Inventor	Autodesk, Inc.	http://usa.autodesk.com/
Rhino	McNeel North America	http://www.rhino3d.com/
Alias	Autodesk, Inc.	http://usa.autodesk.com/adsk/servlet/pc/index?id=14437167&siteID=123112

(a) Accessed April 2010

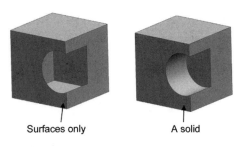

Fig. 1 Solids as opposed to surfaced models. The distinction between a set of surfaces and a unified solid is shown

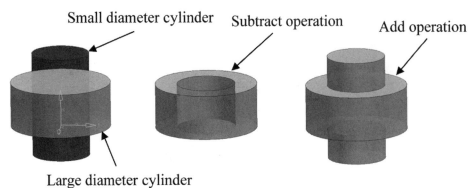

Fig. 2 Boolean operations are used to create complex shapes by subtracting or adding simple shapes

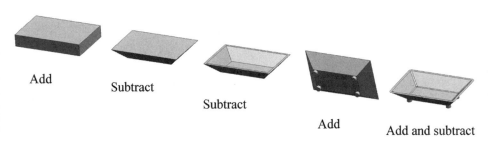

Fig. 3 A final model is created through a sequence of add-and-subtract operations

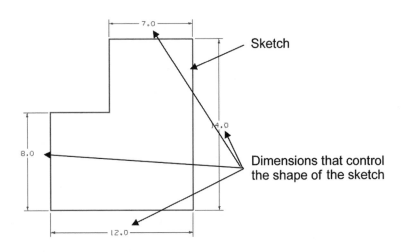

Fig. 4 Two-dimensional sketch consisting of a dimensioned closed loop

may start with a block, as shown in Fig. 3, make angled cuts on all the sides, scoop out the center, add small cylinders at each corner, and smooth out the corners to make a serving tray.

On a higher level, solids can be created using sketch entities. The sketch entity is usually a collection of lines, arcs, splines, points, dimensions, and constraints that are strung together to create a profile. In many cases, it is a closed loop (Fig. 4). After a sketch is created, it can be extruded along some axis, revolved about an axis, or driven along some other curve or set of curves. In the case of an extrusion, this usually forms a solid with capped ends; in the case of a revolve, this usually results in a solid that emulates a turning operation; and in the case of a sketch driven along some path, this usually results in a solid that is the equivalent of a piece of material that has been drawn and then formed (Fig. 5, 6).

Simulating Manufacturing Operations. It is easy to understand solid modeling if one thinks about a host of operations that can be done in the manufacturing field. For example, to smooth out the sharp edges on a flat, square piece of wood, one could use a router or a piece of sandpaper. In many CAD systems, this is called an edge blend, as shown in Fig. 7. In most cases, the same entity covers exterior blending as well as interior, as shown in Fig. 8.

A similar way to envision the fillet is rolling a ball against two surfaces and creating a surface of circular cross section everywhere the ball touches, as shown in Fig. 9. An entity similar to the fillet is the chamfer (Fig. 10).

Parametric Approach. Many solid modeling CAD programs follow a parametric, features-based paradigm. That is, models are created by adding many features together in a sequential format that are driven by parameters. For example, one can create a model of a beer mug, starting with a cylinder that is 8 in. high and 4 in. in diameter, as shown in Fig. 11. The cylinder is the feature, and the dimensions are the parameters. Next, one performs a hollow or shell operation so that the cylinder has room for the beer. The shell command specifies the wall thickness at 0.5 in. Then, a handle is added, using a sketch and a sweep command. The list of operations is then cylinder, hollow, sketch, sweep, and add. The list of operations is called the constructive solid geometry

(CSG) tree. It is the backbone of feature-based parametric modeling. Along with the CSG tree, there is a database of every numerical value that was entered as the geometry is created. The beauty and power of the feature-based parametric paradigm is that any time a design change is required, all one has to do is change one of the underlying parameters in the database, and the entire model is recreated. In the case of the beer mug, when one wants a larger mug, the parameter that captures the original 8 in. height is changed form 8 to 12 in. All the other features then update. Figure 11 illustrates every step of the example.

In many CAD systems, modeling is greatly facilitated by using constrained sketches. They

provide the user with the ability to truly capture design intent from a 2-D perspective at first, then 3-D later. The entity is created on a default flat plane, or datum plane, and used in subsequent operations to create 3-D geometry. A sketch captures the design intent by allowing a user to dimension and or constrain the various entities that make up the sketch. Each dimension of each sketch shows up in the database as a parameter that controls the 2-D shape in the same way parameters control solid features. When the shape of the sketch must be changed, one may access the dimensions, and the entire shape updates. The true power of the sketch is its capability to capture more than just the dimensions. For example, to create a model of

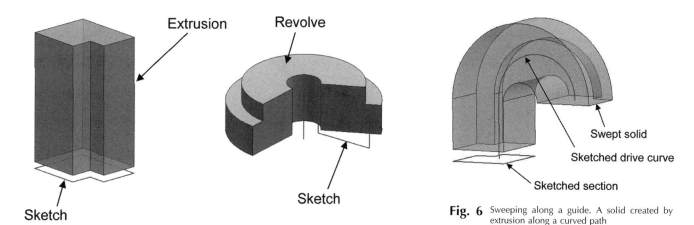

Extrusion

Revolve

Sketch

Sketch

Swept solid

Sketched drive curve

Sketched section

Fig. 6 Sweeping along a guide. A solid created by extrusion along a curved path

Fig. 5 Revolve and extrude operations. Extruding a two-dimensioned closed loop along an axis or rotating it about an axis defines very different solids

a window with the height 1.618 times the width (the golden rectangle relationship), one may easily enter that mathematical relationship into the database. The beauty of this capability is that it allows one to iterate on the width of the window and have the height automatically update for any width. It is the automatic nature of constraints and dimensions that provides the greatest ease when creating a sketch and subsequent model. To illustrate the point further, consider the two simple shapes in Fig. 12. The one on the left is dimensioned as shown, with the inner profile dimensioned from the left side. Due to the numerical values, the overall width being 60, the width of the window being 30, and the distance from the left side being 15, the widow is centered. The contour on the right has a centering rule for the spacing of the window. When a design change is made where the overall width is made larger, the shape on the left does not update properly, as shown in Fig. 13. Therefore, the sketch on the right captures the true design intent. The sketch on the left does not.

In addition to sketches that capture design intent, modern solid modeling is augmented by smart features such as holes, bosses, pads, pockets, ribs, and many more. What makes a smart feature smart is its capability to capture not just the geometry that was intended when it was first created but the true design intent of its creation. For example, the hole in the solid shown in Fig. 14 is intended to have a depth of half the thickness of the block that it is drilled into. When the block thickness is changed, the hole depth automatically updates to half the new thickness. Other smart features are the boss, pad, keyhole, threaded hole, and groove, to name a few (Fig. 15).

Reference Entities. The creation of complex solid models with challenging geometry often requires the use of reference entities, such as datum coordinate systems, planes, axes, and datum points. These help to position other entities, such as sketches, curves, and so on. Modern CAD systems allow the creation of these entities in almost every conceivable way.

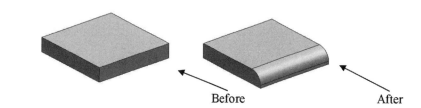

Before

After

Fig. 7 A blend exterior. The edge blend in a computer-aided design system is analogous to sanding or routing a slab. In most cases, the same entity covers exterior blending as well as interior, as shown in Fig. 8

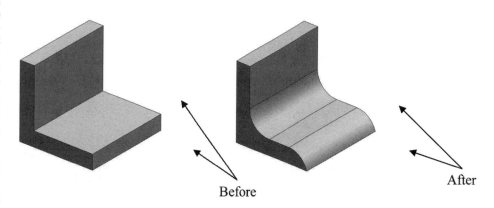

Before

After

Fig. 8 A blend interior. A simple angular piece is evolved into a more realistic structural angle

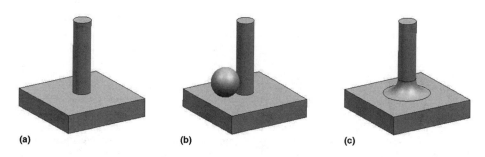

(a)   (b)   (c)

Fig. 9 Rolling ball fillet. A method of creating fillets. (a) A cylindrical solid is attached to a block. (b) A fillet is created by using a rolling ball. (c) Finished form; lines indicate points of contact between ball and surfaces

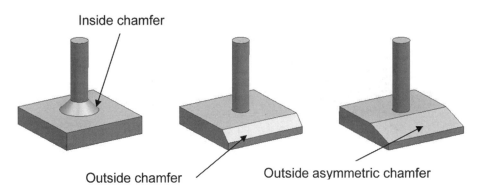

Fig. 10 Inside and outside chamfers

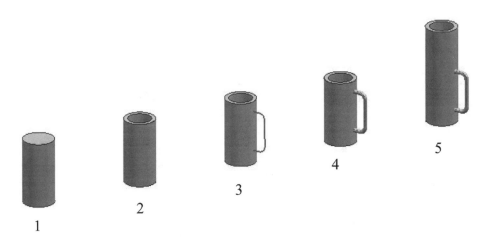

Fig. 11 Parametric model. Mug design based on design parameters

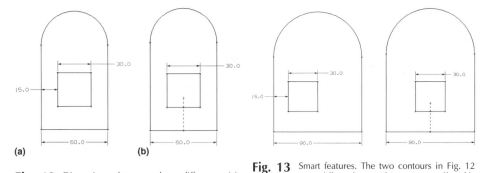

Fig. 12 Dimension schemes make a difference. (a) Two dimensions reference the left side of the figure. (b) A centering rule is applied

Fig. 13 Smart features. The two contours in Fig. 12 react differently to a change in overall width

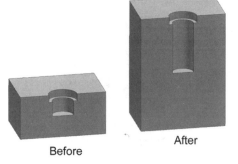

Fig. 14 Form features. The hole depth is set at half the block thickness

For example, a datum plane can be created in the following ways: offset from a face, through three points, through two nonparallel axes, through one point and an axis, tangent to a cylindrical face, and many others. Parametric datum planes update automatically and serve as the backbone for many complex models. For example, a propeller model is created by creating a series of datum planes through a cylinder, then using the last datum plane in the series to create a propeller blade. The pitch of the blade is controlled by the angle of the datum plane. Each entity is necessary for the creation of the next (Fig. 16).

Datum curves are created in 3-D space to serve as the skeleton for all sorts of geometry. Figure 17 shows the creation of a curved mesh surface that is created from a series of curves.

Datum points can be placed anywhere in 3-D space, in almost any conceivable way. The points generally support the creation of other 3-D geometry. For example, points can be distributed on a surface to serve as the locations to make blister features (Fig. 18).

Along with datum axes, planes, points, and curves, most good solid modeling systems provide the capability to create datum coordinate systems. A datum coordinate system is essential when positioning solid models in 3-D space. The datum coordinate system is an entity that allows one to position it by keying in offsets from the global coordinates system. When geometry is created on datum coordinate systems, the geometry is moved when the parameters of the datum coordinate system are varied.

Expressions and Variables

Most modern solid modeling systems allow the creation of expressions or variables. A numerical value is assigned to a variable name, such as thickness (thk) = 0.25, in a consistent unit system. Then, as various features continue to be created, the expression is used. The geometry that is created is then associated to that expression. For example, a simple enclosure is created with a wall thickness of thk = 0.25. When the ribs are created to strengthen the enclosure, the rib thickness is related to the variable (0.65 × thk). The design intent that is captured is the desire for the rib to always be smaller than the overall thickness of the enclosure, thereby avoiding the heat-sink problems that can occur during an injection molding process. The usefulness of this capability is evident when a design change is needed. When it is determined that the overall wall thickness must be greater, and thk is reset to 0.35, all the ribs update automatically. In this way, expressions serve the purpose of capturing the design intent and the numerical relationships between various features (Fig. 19).

Expressions are also great for driving geometry that follows a mathematical function. For example, the set of parametric equations in Fig. 20 creates the crazy double-helical shape shown.

Surfacing

Surfacing is extremely important whenever a model is to be created that has the shape of an airfoil, human form, consumer product,

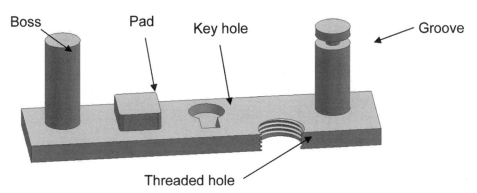

Boss　Pad　Key hole　Groove

Threaded hole

Fig. 15 Design features: boss, pad, keyhole, threaded hole, and groove

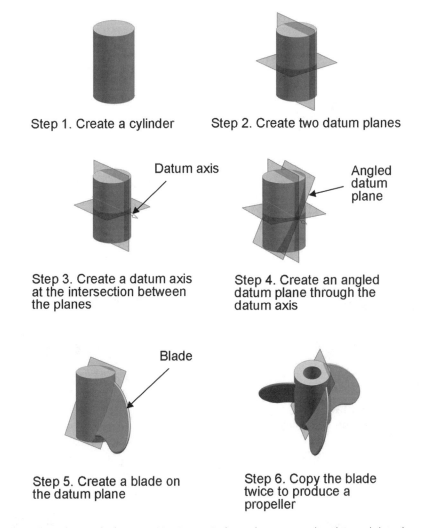

Step 1. Create a cylinder

Step 2. Create two datum planes

Datum axis

Step 3. Create a datum axis at the intersection between the planes

Angled datum plane

Step 4. Create an angled datum plane through the datum axis

Blade

Step 5. Create a blade on the datum plane

Step 6. Copy the blade twice to produce a propeller

Fig. 16 Datum planes and reference entities. Parametric datum planes serve as the reference skeleton for complex designs

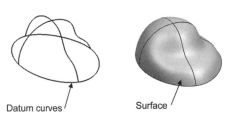

Datum curves　Surface

Fig. 17 Datum curves and surfaces used to generate a curved mesh to evolve a complex irregular surface

connect the section curves, it is usually called a mesh surface (Fig. 22).

Creating Various Surfaces

Extrusion of a Changing Section. Many forms in engineering are constructed from various sections swept along a path. Whether one is creating a diverging fuel nozzle or a sound-attenuating device, there must be a way of creating surfaces along some central path or drive curve. In Fig. 23, two semicircular cross sections of different radii are swept along a spiral. The result is the surface shown.

Transitions. In most solid modeling systems, there is a surface that can be used to make a transition between two existing surfaces of different shapes. This is used, for example, to make a model of the ductwork behind the dashboard of a car or perhaps a model of the latest hip implant. The procedure involves selecting two edges of two different surfaces and instructing the transition surface operator to create a smooth and continuous surface that flows from one edge of one surface to the other edge of the other surface while maintaining tangency at the take-off edges (Fig. 24).

Modeling Real Objects. In many cases, surfaces are created from real objects that are already in existence. It is usually a multistep process where an object is scanned with a laser or a coordinate measuring system to produce a point cloud. The point cloud is most useful when it follows certain rules, such as the points being arranged in horizontal rows with equal numbers of points in each row (Fig. 25).

A surface differs from a solid model in that it has no mass, no inside or outside, and is infinitely thin. Surface modeling is a useful tool during solid modeling because there are many ways to convert a surface or set of surfaces into a solid. For example, one can add thickness to a surface, automatically creating a solid (Fig. 26).

Quilt or Sew. One can also take a number of surfaces and stitch them together on the edges. Some programs refer to this operation as a quilt or a sew. The collection of surfaces must be airtight (all edges must be aligned) to become a solid (Fig. 27).

Surface Subtraction. A surface can be used to carve away at an already established solid body. In an operation that can be described as a cut, trim,

sporting good, toy, or many other applications. Perhaps the most common way of creating a surface is to create a number of sections that vary in shape and to "stretch" a skin over them, such as an airplane wing (Fig. 21).

When a surface must be controlled to a greater degree than planer sections can afford, it may be necessary to supply curves in between the sections. When a surface is created with a set of section curves and a set of curves that

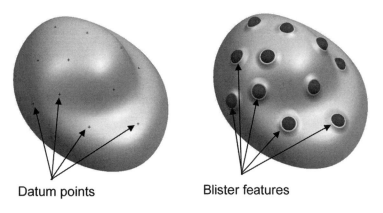

Datum points   Blister features

Fig. 18 Datum points on a complex surface anchor the locations of blister features

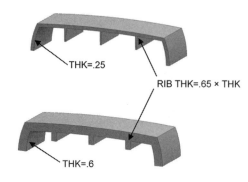

THK=.25

RIB THK=.65 × THK

THK=.6

Fig. 19 Parametric rib structure and wall thickness. The automatic updating of a design where one dimension is a function of another. In this case, the rib thickness is dependent on the wall thickness (THK).

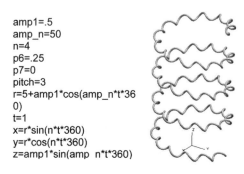

```
amp1=.5
amp_n=50
n=4
p6=.25
p7=0
pitch=3
r=5+amp1*cos(amp_n*t*36
0)
t=1
x=r*sin(n*t*360)
y=r*cos(n*t*360)
z=amp1*sin(amp_n*t*360)
```

Fig. 20 Parametric equations on the left define the complex double-helical curve

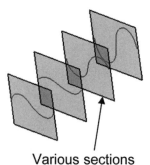

Various sections

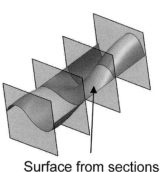

Surface from sections

Fig. 21 Surface through section curves. Surface is created by defining the profile at various sections and stretching a skin over them

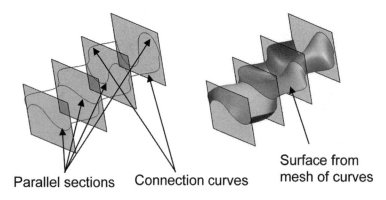

Parallel sections   Connection curves

Surface from mesh of curves

Fig. 22 Surface through mesh of curves. A more complicated surface uses connection curves in addition to section curves to create the surface

Rectangular surface

Circular surface   Transition surface

Fig. 24 Surface transition between fixed-end surfaces

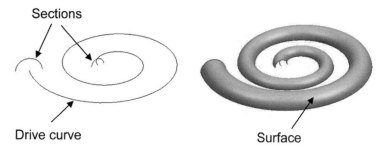

Sections

Drive curve

Surface

Fig. 23 Surface swept from sections

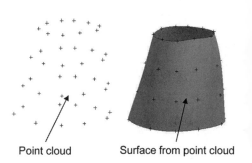

Point cloud   Surface from point cloud

Fig. 25 Surface generated from a point cloud

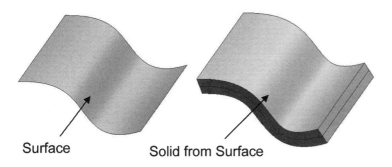

Surface   Solid from Surface

Fig. 26 Thickened surface creates a solid

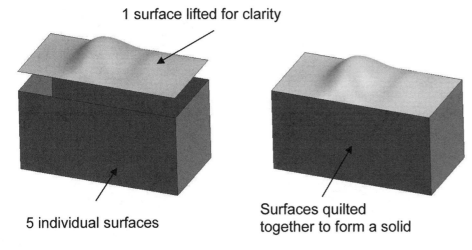

1 surface lifted for clarity

5 individual surfaces

Surfaces quilted together to form a solid

Fig. 27 A set of surfaces stitched together (all edges aligned) makes a solid

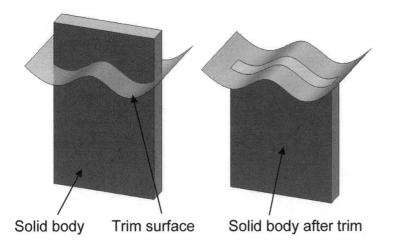

Solid body   Trim surface   Solid body after trim

Fig. 28 Surface used to trim a solid

or carve, a surface is placed in a strategic location relative to a solid body. The operation is performed, and one side of the solid is cut away (Fig. 28). This is a very useful technique for certain consumer products that have a number of highly sculpted regions that perform various functions, such as speaker locations on a radio, the front face of a fax machine, or the input card orifices on a laptop computer.

Sheet Metal

Modern high-end CAD systems afford a number of sheet metal functions. The user normally starts with a flat panel with a uniform thickness, and various sheet metal features are added on until the component is done. The main advantage to using sheet metal features is that they carry the extra information required to produce an accurate flat pattern when the design is done. Figure 29 shows a piece of sheet metal that has been created very simply, then flattened.

The sheet metal options also include a number of other common sheet metal contours, such as louvers, dimples, bends, and cutouts (Fig. 30).

Assemblies

When the aforementioned techniques are used to create individual components, they can be assembled. Typically, a simple assembly is a number of separate part files with separate models in each that are being "collected" by one assembly part file. For example, the assembly in Fig. 31 of a simple drawer unit shows each and every separate component in its own part file.

In this case, the main part file that calls all the other part files and automatically opens them is called the drawer assembly (Fig. 32).

Location Method. The assembly file contains all the information that associates each and every component. There are two main ways an assembly defines the geometric relationships between components: 3-D location and geometric constraint. When components are related by 3-D location, they are positioned relative to a central or global coordinate system. It is a lightweight method, but it does not automatically adjust the locations of the components when one in the group changes shape. For example, a three-stage rocket has a first stage that is 10 ft long, a center stage that is 5 ft long, and a final stage that is also 10 ft long. If each component is stacked in the right location, one on top of the other, the entire rocket is 25 ft high. With the 3-D location method, if the center stage grows to 10 ft, the final stage does not necessarily relocate to the correct new height of 30 ft. The stages do not "feel" each other (Fig. 33). There will be an overlap, either between stages 1 and 2, or 2 and 3, or both. In this example, it is very easy to see, but in more complex assemblies, it may not be.

Geometric Constraints. When assemblies are put together using geometric constraints, surfaces, edges, centerlines, planes, and other 3-D entities are related to each other. For example, the top of stage 1 of the example rocket assembly would have a relationship with the bottom of stage 2, and the top of stage 2 would touch the bottom of stage 3 (Fig. 34).

At first glance, it would seem that using the constraints method is the way to go, but there are many other considerations. Using the constraints method can get very difficult when working on an assembly with many components. When many different groups of people are working with a large assembly, it adds another level of complexity that makes the global coordinate or 3-D location method much easier. Each contributor knows where their component must end up in 3-D space, which enables much work to be done without the overhead of having all the other components loaded.

Off-the-Shelf Items. Whether assemblies are performed with the global coordinate method or the constraints method, users still

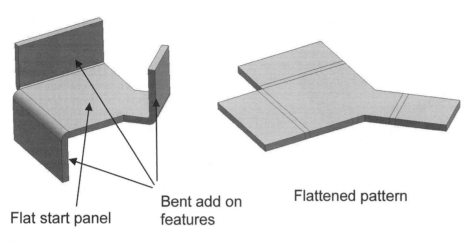

Flat start panel

Bent add on features

Flattened pattern

Fig. 29 Sheet metal bend and flat pattern

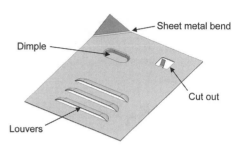

Sheet metal bend

Dimple

Cut out

Louvers

Fig. 30 Common sheet metal features: dimple, bend, cutout, and louvers

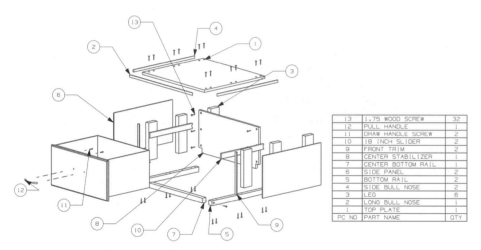

13	1.75 WOOD SCREW	32
12	PULL HANDLE	1
11	DRAW HANDLE SCREW	2
10	18 INCH SLIDER	2
9	FRONT TRIM	2
8	CENTER STABILIZER	1
7	CENTER BOTTOM RAIL	1
6	SIDE PANEL	2
5	BOTTOM RAIL	2
4	SIDE BULL NOSE	2
3	LEG	6
2	LONG BULL NOSE	1
1	TOP PLATE	1
PC NO	PART NAME	QTY

DRAWER ASSEMBLY

Fig. 31 Exploded view and bill of materials for a drawer assembly

Fig. 32 Visualization of the drawer assembly created from the individual parts files

get the benefit of collaborative work and part reuse. It is very common for designers who are working on systems such as consumer products or machinery to save huge amounts of time by downloading models of various purchased components right from the internet. Websites make available models of many of the components, which they provide for free download (Ref 1, 2). The models come in a variety of formats, and all major CAD modeling programs are able to use one or more of the formats for each component. The assemblies paradigm makes possible rapid design because entire subassemblies are used from previous projects. When the situation allows, a good designer can employ various previously used components and can benefit greatly from the fact that the drawing is probably there already, along with all sorts of specification work as well as test data.

Advanced Modeling

There are many other modeling techniques that do not fall into any particular category but must be present for any high-end CAD system to be competitive. To name a few, these include:

- Driving geometry with a spreadsheet or program
- Linking the parameters of one part file to another
- Borrowing features and/or groups of features from one part file and transferring to another

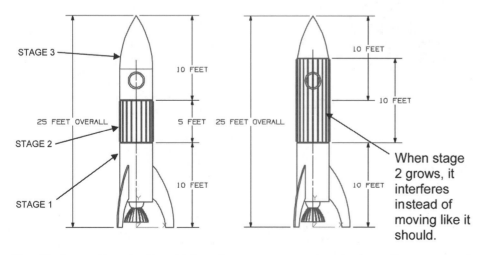

Fig. 33 An assembly made with a global coordinate system may experience interference if one component is changed

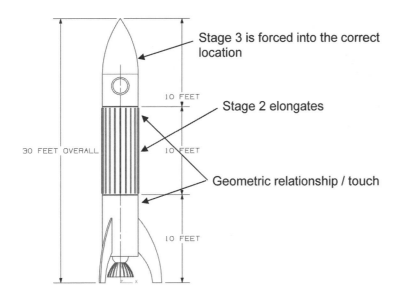

Fig. 34 Assembly made by using geometric constraints where the part characteristics are mated. This allows for the modification of one part

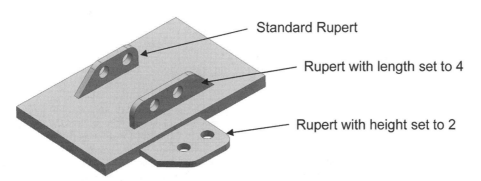

Fig. 35 User-defined features. Example of adapting a standard part for the program library

- Performing advanced logical functions on geometry, such as defining limits for values
- New features turn on based on expressions.

Much of the advanced functionality in solid modeling enhances the ability to manipulate difficult surfaces. An incomplete list includes:

- Matching edges of various surfaces
- Extending surfaces
- Changing the shape of surfaces by manipulating the underlying poles
- Trimming and untrimming surfaces
- Performing nonuniform scale on surfaces

A powerful solid modeling program gives the user an ability to create libraries of their features. For example, an aerospace company designs a product with a certain standard flange with two standard holes through it. Hypothetically, it is called a rupert flange. A user builds a parametric rupert flange and puts it into the library for all those in the company that use a similar design with slight variations. Now, anyone who needs a rupert flange accesses the library, inputs their dimensional requirements, and then places their special copy of the rupert flange directly on their geometry (Fig. 35).

Explicit-Parametric Modeling

When solid modeling was new, everything was modeled explicitly, meaning that the dimensions used to create geometry were no longer accessible when the geometry was created. For example, when the creation operation of a 3 in. cube was over, the only way one could change the size of the cube was to perform some type of transform on it. One could not access the original data, make a change, and have that change cascade through the solid and subsequent solid features. Explicit solid modeling was the state of the art for years until sometime in the late 1980s and early 1990s, when the parametric paradigm swept over and dominated the CAD world establishment. When the parametric paradigm caught on, many of the most modern programs abandoned explicit modeling all together. One of the latest and most exciting developments in solid modeling is explicit-parametric modeling, which is an incredibly

powerful combination of the two techniques. With explicit-parametric modeling, a user is allowed to employ a powerful set of parametric entities that, in essence, overwrites the previous parametric entities. The user is allowed to change the design intent of a particular solid model without in-depth and detailed consideration of what had been done previously. For example, a model of a speaker housing on an alarm clock has a flat, horizontal face on the top where the sleepy user is meant to pound his fist in the morning. In a desire to make the experience a better one, the designer decides to have a little fun with the top surface and tilts it a bit toward the user. If the software that the designer is using does not have explicit-parametric capability, the job can become complex and lengthy. The designer will have to go back in the database and find the original revolve that was responsible for the top, then somehow reorient it and hope that all the little hole features and the subsequent chamfers update. However, by using the explicit-parametric operation, the designer simply highlights all the surfaces to be changed, then instructs them to tilt about the axis (Fig. 36). The change is explicit in that it partially overwrites the design intent and features that were created before. If the hole pattern or any of the other features that preceded the tilt are changed, it will cascade through and update. In addition, the change is parametric, in that if one goes back in the database and changes the numerical value of the tilt angle, all the surfaces will retilt to the new angle.

Model Verification

All of the great solid modeling systems offer some type of model verification. As a user creates product definition data, there is a cadre of data verifications that are required. These range from measuring the distance between two points to finding out the full mass properties of a complex model. For those who are working with geometry that is to be injection molded, there is a way to check all the surfaces to ensure that they have at least the minimum draft on them so they are easily ejected. There are ways to find out the optimal parting line of a model that is to be cast and ways of finding how a surface will reflect light when it is finally a real product. There are even ways of finding out how and if components in an assembly are interfering. For example, if a peg that was 2.5 in. in diameter was inserted into a hole that is too small, say 2 in. in diameter, the modeler will highlight the surfaces that are interfering and even create a separate interference solid upon request (Fig. 37).

Associativity and Concurrent Engineering

When a CAD model is created, it typically captures the 3-D geometry of a component, otherwise known as the shape data. It usually does not capture the nonshape data, such as surface

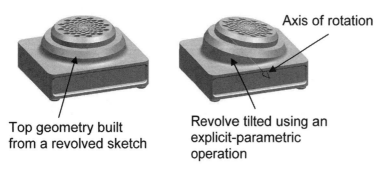

Fig. 36 Explicit-parametric operations. Design change in an alarm clock

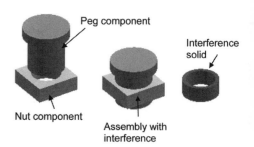

Fig. 37 Model verification of interference

finishes, geometric tolerance, heat treatment, or any other special manufacturing instructions. These nonshape data are usually captured on a 2-D drawing. The 2-D drawing is linked or associated to the 3-D part. Every powerful CAD system affords the user the ability to create a separate part file for the 2-D drawing that, in essence, has the 3-D geometry part file as a one-level associated component. This way, the 2-D views for the drawing, along with all the dimensions, can be placed on the drawing. This capability allows two different users to work on a job at virtually the same time. The 3-D designer begins the job, and shortly afterward, the 2-D draftsperson begins the drawing. As the 3-D designer improves the design, the 2-D drafting model receives updates. The associative nature of modern CAD systems affords concurrent engineering and collaboration to those who are responsible for all the other product design functions, such as stress analysis, manufacturing, test, kinematic analysis, mold flow, and anything else. Most competitive CAD packages are sold with options to do all of the aforementioned in one seamless, integrated package. In 2010, these systems can be purchased for prices from $1200 to ten times that price and beyond, depending on the options.

Product Lifecycle Management

Product lifecycle management (PLM) includes everything that is necessary to streamline the innovation and management of product definition data (PDD), old and new. The PDD can include a vast array of CAD part files, specifications in the form of portable document formats, Word documents, manufacturing and inspection instructions, user manuals, shipping and storage requirements, and so on. Because these data are created by users employing CAD systems, they may be augmented by other data found in libraries, other folders, and many other places. It all must be organized so that the right data are given to and/or created by the right person at the right time. Whether it is a simple toy or a complex jet fighter, it is critical that the PDD be organized, never misplaced, and never accidentally substituted or altered by the wrong procedure or person. As a product

is defined in its lifecycle, good PLM is able to not only organize the data but keep track of all the procedures that were put in place to create the data. A good PLM system indicates very quickly the current definition of a product and its history from cradle to grave. A good PLM system gives one the power to do these things and, in doing so, provides a huge advantage to those who are innovating in any competitive market. Most CAD modelers have a PLM system that can be purchased with the CAD software. Most PLM systems can manage the files that come from a CAD package that is not in the same product family, but, at the time of this writing (2010), there is usually smoothing lost in translation.

Collaborative Engineering

Making progress on a project involving engineers of varied disciplines, designers, subtier manufacturers, marketing considerations, and so on can be a very difficult and challenging process. There is a "Holy Grail" of product design that most in the business are always trying to achieve. Products are defined in such a way that they optimize the input of all functional groups. In essence, it is the widget that looks exactly the way marketing would like, is as functional as engineering would like, is as easy to manufacture as manufacturing personnel would like, is all set up to be recycled as easily as any product can be, and sells like hotcakes to make everyone wealthy. To achieve this Holy Grail, each product must have as much consideration from each of the contributing groups as early and as often as possible. The essence of collaborative engineering is to manage the input of all these various people with various focuses and ideas.

For this, engineers and product designers all over the globe use PLM systems. A good PLM system has an incredible capability to allow access to PDD, no matter where the team members are in the world. The premise is that "if all the members of the cross-functional groups have access to all the data all the time, the different phases of product development will be in series instead of parallel." For example, if the manufacturing engineer, who usually does not see the design until the drafting

department throws it "over the wall," can gain access to the data and helps define a better shape for a certain casting that is to be produced, this will mean greater gains in manufacturability. The Holy Grail is in sight as long as everyone communicates as upfront and effectively as possible.

The Future of CAD

Speculating about the future is always risky. The quantum leaps that occur over time, especially in any business that has to do with software and computers, make predictions extremely difficult. However, it is apparent what CAD users want, and hopefully, what CAD users want is what the industry will provide, sooner or later. All over the globe, CAD users have difficulty in finding the commands they are looking for. It is extremely important for solid modeling systems to become better at organizing and streamlining the user interface so that various commands are easier to find. To some degree, the most powerful systems are at a disadvantage because of the sheer volume of commands and abilities they have compared to lesser systems. It can be confusing, especially for new users.

Another general thrust in CAD systems comes from the fact that most design engineers would love to find some way to dispense with the entire drafting process. Indeed, high-end systems already have the capability to capture features such as geometric dimensioning and tolerancing, surface finish callouts, and nonshape entities right in the 3-D model. In the future, this will undoubtedly be made easier and more common.

Another controversial improvement in CAD is the total abandonment of parametrics. There are a large number of engineers who never felt that the move to parametric modeling was a benefit. Commensurately, there are a small yet growing number of CAD software designers that are determined to make a nonparametric paradigm that is as productive and easy to perform design iterations with as in parametric modeling. The future may yield a solid modeling package that gives anyone at any time the choice to be parametric, nonparametric, and lightweight. By and large, all CAD modelers look at all the features and functions of all the others and make sure that, somehow, they have everything that everyone else has. In recent years, CAD software companies have bought each other out and made new programs that incorporate everything that the other program had, plus what they had originally. It is also a great way to increase the installed base of users.

It is the opinion of the author that CAD systems of the future will be fewer, far more powerful, less expensive, better known to newly graduated users and engineers, and easier to employ by an expert or a novice user. It is also the opinion of the author that the more powerful the systems become over time, the more exciting and fun they will be to use. More tasks will be easier to do, with more computing power and capabilities, and designers will be asking them to do increasingly difficult geometry. To some degree, designers limit the shape of their designs based on the limitations of the CAD systems that they use. The future will bring more freedom and better designs.

REFERENCES

1. McMaster-Carr Supply Company, www.McMaster.com, accessed May 2010
2. Digi-Key Corporation, www.Digikey.com, accessed May 2010

SELECTED REFERENCES

- J. Bird and C. Ross, *Mechanical Engineering Principles,* Reed Elsevier, 2002
- G. Qi, *Engineering Design Communication and Modeling,* Thompson, 2006
- S.M. Samuel, A. Pragada, B. Stevenson, and E.D. Weeks, *Basic to Advanced NX6 Modeling, Drafting and Assemblies,* San Jose, CA, 2008
- S.M. Samuel, E.D. Weeks, and M.A. Kelley, *Teamcenter Engineering and Product Lifecycle Management Basics,* San Jose, CA, 2006
- S. Samuel, E.D. Weeks, and B. Stevenson, *Advanced Simulation Using Nastran NX5/NX6,* San Jose, CA, 2009
- S. Tickoo, *Catia V5R16 for Designers,* CAD-CIM Technologies

ASM Handbook, Volume 22B, *Metals Process Simulation*
D.U. Furrer and S.L. Semiatin, editors

Copyright © 2010, ASM International®
All rights reserved.
www.asminternational.org

Design Optimization Methodologies

Alex Van der Velden, Patrick Koch, and Santosh Tiwari, SIMULIA

OPTIMIZATION FINDS APPLICATION in every branch of engineering and science. The process of optimization involves choosing the best solution from a pool of potential candidate solutions such that the chosen solution exceeds the rest in certain aspects. Design optimization is the process whereby a selected set of input design variables is varied automatically by an algorithm to achieve more desired outputs. These outputs typically represent the variation from a target, minimal cost, and/or maximal performance.

For design variables to be varied automatically, algorithms that search the design domain are necessary. For this purpose, it is necessary to be able to compute the outputs of interest automatically. Even though the word *optimization* is used, only trying out all relevant combinations can guarantee that, indeed, the best design parameters for an arbitrary complex space have been found. In practice, this takes too much time. Consider, for instance, that a simple problem with ten possible discrete values for five parameters and a 5 min analysis time would require a year to analyze all the combinations.

As such, the practical value of a particular optimization algorithm is its ability to find a better solution—within a given clock time—than a solution obtained by a manual search method or another algorithm. This clock time includes the effort it takes to set up the simulation process and to configure the optimization methods. To minimize the set-up time, commercial software such as Isight can be used (Ref 1). The setup of the simulation process varies from problem to problem; this article focuses on the optimization methodologies.

No-Free-Lunch Theorem

The no-free-lunch theorem of Wolpert and Macready (Ref 2) states that "...for any [optimization] algorithm, any elevated performance over one class of problems is exactly paid for in performance over another class."

This concept is shown in Fig. 1. On the diagonal is plotted a set of arbitrary problems, f_i, ordered by the minimum number of iterations, n_{min}, and required by a set of methods, A, B, ..., Z, to solve this particular problem.

For instance, consider the problem of finding x_i for Min[f_1 (x_i)] = constant. There exists a method A that finds the minimum value of f_1 (x_i) with a single-function evaluation. Method A, called the "lucky guess" method, simply tries a random number with a fixed seed. Its first guess happens to be the optimal value of this problem. Obviously, this method is not very efficient for any other problem. The efficient performance for problem 1 goes at the expense of the efficiency to solve all other problems 2, ..., n.

The minimum value of f_1 is also found by method B, but this method is not as efficient as the "lucky guess" method. Method B is a genetic algorithm. In its first iteration, method B first computes a number of random samples of x_i with respect to f_1 before deciding the next set of samples (second iteration) based on proximity to the lowest function values in the first iteration. Because method B already requires several samples before the first iteration, method B is not as efficient as method A for problem 1. However, it does pretty well on a variety of problems, including 1, 2, 4, 5, 7, and 8. It is the most efficient method for problem 8.

Method C is a gradient method and may need to evaluate the gradients of x_i with respect to f_1 before completing the first iteration step. Because of that, it is obviously not as efficient as the "lucky guess" method for problem 1. Even though it is the most efficient method for problem 4 (which happens to be a linear function), for most problems, method B is more robust. The gradient method often gets stuck in local minima. Method C is not as robust as method B because it only gets the best answer two times versus method B's nine times for the set of methods and problems being considered.

Method D was also tried. Method D samples the space with a design of experiment techniques and shrinks the search space around the best point in the array for each iteration. It is able to solve quite a few problems, but it is inefficient and would therefore be considered dominated by other methods over the set of problems $f_1, f_2 \cdots f_n$.

This meant that it was necessary to develop an environment that allowed the introduction of many algorithms specifically suited to solve

certain classes of customer problems. The open-component architecture of Isight (Ref 1, 3) allows the development of these design drivers independently from the product-release cycle. However, in many cases, customers do not have such specialized algorithms available and are looking for a commercial product to improve their designs.

For that purpose, a set of best-of-class general- and specialized-purpose algorithms has been provided that works out of the box. These optimizers solve both deterministic and nondeterministic single- and multiple-objective functions. A deterministic function always returns the same result when called with a specific set of input values, while a nondeterministic (stochastic) function may return different results when they are called with a specific set of input values. The following sections give a description of all of these classes of problems and the optimization methods that solve them.

Deterministic Single-Objective Problem

In the case of a single-objective problem, a single output is being maximized or minimized and/or a set of outputs is being constrained to stay within a certain range. Minimize $f(\mathbf{x})$ where:

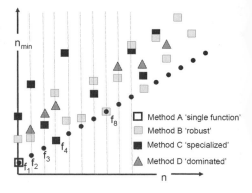

Fig. 1 Illustration of the no-free-lunch theorem for different types of optimizers

$$f(\mathbf{x}) = w_1 f_1(\mathbf{x}) + w_2 f_2(\mathbf{x}) \ldots + w_m f_m(\mathbf{x})$$

Subject to:

$$g_j(\mathbf{x}) \le 0, j = 1, 2, \ldots, J$$
$$h_k(\mathbf{x}) = 0, k = 1, 2, \ldots, K$$
$$x_i^{(L)} \le x_i \le x_i^{(U)}, i = 1, 2, \ldots, N$$

A good example of a single-objective material-processing application is data matching (also know as model fitting or parameter estimation), shown in Fig. 2. Here, the objective is to minimize an error function between a parametric model and experimental data.

Improving the Optimization Process

The selection of the right objective is the most critical aspect of optimization. In the case of Fig. 2, the objective is a straightforward error minimization between model and experiment. The only question here is whether the selected parametric form does not overfit the data. To make a convincing argument that the model is valid in general, the same model should be fit to several sets of experimental data. The single-objective error function could be averaged over the set.

Design Variables. Apart from the selection of objectives, the second most important thing the user can do to improve the optimization process is to select an appropriate set of design variables. Often, variables can be coupled in such a way that the volume fraction of good solutions in the design space is maximized. This can be done according to the Buckingham PI theorem (Ref 5) or other scaling methods.

Reducing the design space complexity will make it easier for the algorithm to find improvements. However, to find an improvement, it is necessary to have at least as many active degrees of freedom to improve the design as there are active constraints. If not enough degrees of freedom are available, the design is effectively frozen in its current state.

Example 1: Deep-Drawing Molybdenum. A good example of systematic selection of variables was given by Kim et al. (Ref 6) for the multistage deep-drawing process of molybdenum sheet (Fig. 3). Because molybdenum has a low drawing ratio, it requires many drawing stages to be transformed into a cup shape. For each part of the drawing process, the authors investigated the proper set of design parameters considering process continuity (clearance, die corner radius, intake angle, etc.). The purpose of this study is to find out the proper eight-stage drawing process that minimizes the maximum strain in the resulting cup shape.

Prior to executing the optimization, the authors did a thorough study of how the design variables interacted with each other. For instance, the authors discovered that the intake angle, θ, and the drawing radius, R_d, had an impact on the maximum stroke, L, for every stage. The maximum stroke is defined as the value of L at which the maximum strain in the cup exceeds the limiting

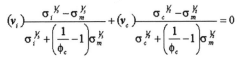
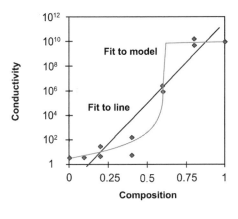

$$(v_i)\frac{\sigma_i^{1/t} - \sigma_m^{1/t}}{\sigma_i^{1/t} + \left(\frac{1}{\phi_c} - 1\right)\sigma_m^{1/t}} + (v_c)\frac{\sigma_c^{1/t} - \sigma_m^{1/t}}{\sigma_c^{1/t} + \left(\frac{1}{\phi_c} - 1\right)\sigma_m^{1/t}} = 0$$

Fig. 2 Data-matching application. Composite conductivity should behave according to the McLachlan equation. Fitting the parameters (design variables) of this equation gives a better fit than a linear function. Source: Ref 4

material strain value. The reason for the importance of the intake angle was that the frictional force between the flange and the die hindered the material flow into the punch-die gap, and the flange-die contact area decreases as the intake angle increases. Therefore, if the intake angle were not a design variable, it would significantly influence the outcome of any optimization effort.

Even after the critical design variable interactions have been found, it is nontrivial to find the optimal combination of the 24-design-variable problem. Most of the effects are coupled due to the nonlinearity of the geometry and the material response. The authors chose the adaptive simulated annealing optimizer (Ref 7) to find a cup design with a 22% lower strain compared to the initial process.

Single-Objective Optimization Methodologies

Complementary Approaches

This section describes optimization algorithms that provide a good set of complementary approaches to solve a wide variety of single-objective mechanical engineering applications.

Gradient Methods. For differentiable functions, gradient methods can be used. The constraints are handled directly without being converted into penalty functions. The gradient methods are very suitable for parallel execution because the gradients can be computed independently from each other.

Example 2: Rosenbrock Function. The process of gradient optimization can be easily illustrated with a Rosenbrock function with a local and a global minimum (Fig. 4):

$$Z = 100(Y - X^2)^2 + (1 - X)^2$$

Local minimum [0.71, 0.51]; $Z = 0.0876$

Global minimum: [1,1] ; $Z = 0$

For this purpose, the large-scale generalized reduced gradient (LSGRG) (Ref 4) optimizer,

described later in this section, was used. The first attempt starts from the left part of the design space $[-1,1]$, and the optimizer finds a solution $z = 0.033$ close to the local optimum following the path of steepest descend [0.81, 0.66]. If the optimizer is started from the top right corner of the design space [2,3], it does find a value of the objective $z = 1.2 \times 10^{-7}$ extremely close to the global minimum [0.99, 0.99], even though it is initially "sidetracked" in its search. Gradient methods, including very good ones like LSGRG, only find minima in the path of the steepest descend. The optimization is stopped when a convergence condition, such as the Kuhn-Tucker criterion, is satisfied:

- The vector of first derivatives of the objective function f (projected toward the feasible region if the problem is constrained) at the point x^* is zero.
- The matrix of second derivatives of the objective function f (projected toward the feasible region G in the constrained case) at the point x^* is positive definite.

For most engineering problems (with multiple minima), the Kuhn-Tucker criterion can be satisfied without having found the global minimum. The implication is that the solution found by the gradient optimizer is now dependent upon the starting point, and this is not very desirable. For this reason, other techniques that may be less efficient but more reliable must also be considered.

Nonlinear Programming by Quadratic Lagrangian (NLPQL)-Sequential Quadratic Programming (SQP) (Ref 8). This method builds a quadratic approximation to the Lagrange function and linear approximations to all output constraints at each iteration, starting with the identity matrix for the Hessian of the Lagrangian, and gradually updating it using the Broydon-Fletcher-Goldfarb-Shanno method. On each iteration, a quadratic programming problem is solved to find an improved design until the final convergence to the optimum design.

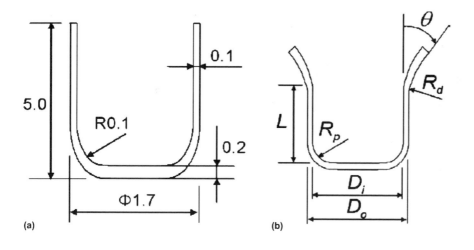

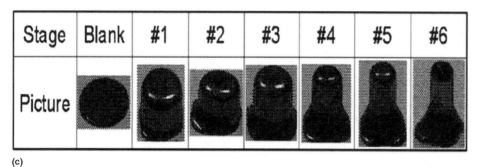

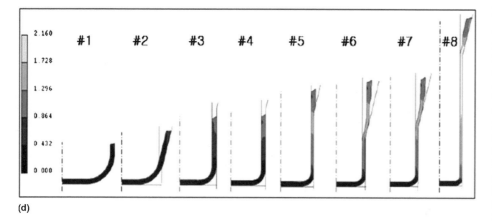

Predicted maximum effective strains ($\bar{\varepsilon}_{max}$) after each stage for the initial and the optimum design

Stage	Initial design (A)	Optimum design (B)	Strain ratio (B/A)
#2	1.11	0.77	0.69
#3	1.59	1.32	0.83
#4	1.86	1.43	0.77
#5	2.12	1.54	0.73
#6	2.56	1.53	0.60
#7	2.66	1.55	0.58
#8	2.78	2.16	0.78

(e)

Fig. 3 Simulated drawing process to produce a molybdenum cup. Design variables for (a) the final target shape and (b) the intermediate stages. (c) Representative presentation of the draw process for the initial design. (d) Cross sections of the optimal design for each of the drawing stages. (e) Reduction in maximum effective strain in the cup during the optimization process. Source: Ref 6

Modified Method of Feasible Directions (MMFD) (Ref 9). This method exploits the local area around the initial design point, handles inequality and equality constraints directly, and rapidly obtains a local optimum design. The MMFD is best used when starting from a feasible design point. It usually requires multiple iterations consisting of a search direction calculation (using gradients of each variable) and a one-dimensional search. The MMFD follows the active constraints during the search until no further improvement can be made. It is well suited for highly nonlinear design spaces but is not suited for discontinuous design spaces. The method operates on the set of real numbers, and the gradient evaluation can be executed in parallel.

Large-Scale Generalized Reduced Gradient (LSGRG) (Ref 10). This method uses the generalized reduced gradient algorithm for solving constrained nonlinear optimization problems. The algorithm uses a search direction such that any active constraints remain precisely active for some small move in that direction. The generalized reduced gradient method is an extension of an earlier reduced gradient method that solved equality-constrained problems only.

The next group of optimization methods does not require gradient information and can be used on nondifferentiable functions. The search direction relies on the information obtained by sampling the design space. Constraints violations are added as penalties to the objectives.

Hooke-Jeeves Direct Search (Ref 11). The Hooke-Jeeves algorithm examines points near the current point by perturbing design variables, one axis at a time, until an improved point is found. It then follows the favorable direction until no more design improvement is possible. The size of variable perturbations is determined by the relative step size. It is gradually reduced by applying the step size reduction factor until convergence is detected. It is not easily possible to parallelize this method, but it can be very efficient on moderately coupled problems. Hooke-Jeeves is a pattern-search method and not a gradient method. During the search, it covers a wide range of the design space. The idea behind this is that the nature of the function is not known a priori, so one must do a wide exploration and not just go down the path of steepest descend. This means that if Hooke-Jeeves is used on a quadratic function similar to the one shown in Fig. 5, it is obviously less efficient than gradient methods. However, with some tweaking of the tuning parameters, such as step sizes and number of iterations, it does find the optimum and does so even if multiple local minima are present. The Hooke-Jeeves method does not have a convergence criterion and stops whenever a preset maximum number of runs is reached.

Nelder and Mead Downhill Simplex (Ref 12). This method samples the space across a subregion and moves from the worst point in the direction of the opposite face of the simplex toward better solutions. The simplex is a geometrical body with $N + 1$ vertices represented by a triangle in two dimensions and a tetrahedron in three

Attempt 1 starting point

X = -1.0; Y = 1.0

Best design point:

X = 0.81 ; Y = 0.66

Objective = 0.033

Close to local optimum

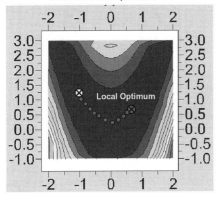

Attempt 2 starting point

X = 2.0; Y = 3.0

Best design point:

X = 0.99; Y = 0.99

Objective = 1.2E-7

Close to Global Optimum

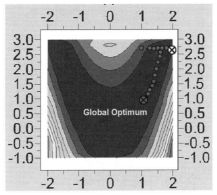

Fig. 4 Gradient optimization exercise with the Rosenbrock function from two starting points

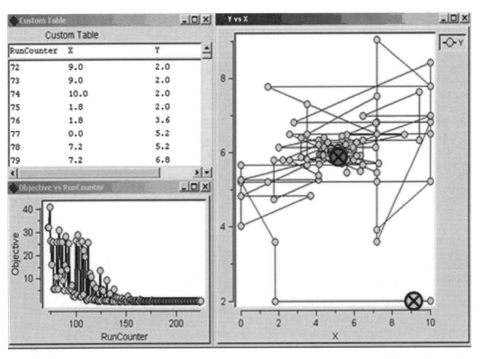

Fig. 5 Hooke-Jeeves optimization exercise with a quadratic function $y = (x1 - 5)^2 + (x1 - 6)^2$ starting from point [9, 2]

dimensions. The method calculates and compares the objective function at the vertices of a simplex in the variable space, selects the worst one, and moves this point through the opposite face of the simplex to a lower point. If this new vertex is better, the old one is deleted. If there is no improvement after a number of steps, the method "shrinks" the simplex by reducing the length of each side, thus trapping the optimum solution. It is not easily possible to parallelize this method, but it can be very efficient on moderately coupled problems.

Adaptive Simulated Annealing (ASA) (Ref 7). This algorithm is very well suited for solving highly nonlinear problems with short-running analysis codes, when finding the global optimum is more important than a quick improvement of the design. Each iteration in the simulated annealing perturbates the current solution by a random step size that is chosen based on a probability that depends on the difference between corresponding function values and a global parameter, T. The algorithm is inspired by the annealing process, and

temperature (T) starts out large and is reduced to very small values as the process advances. The parameter T is automatically adjusted.

Multi-Island Genetic Algorithm (MIGA) (Ref 13). Genetic algorithms work well because they incorporate randomness in their search. It gives the algorithm the ability to correct deterministic search bottlenecks that are caused by the reasoning in the previous two "space-sampling" methods and the gradient methods. The MIGA algorithm divides the population into several islands, performs traditional genetic operations on each island separately, and then migrates individuals between the islands. It searches many designs and multiple locations of the design space. Genetic algorithms such as MIGA tend to be less efficient than the two previous methods in this class, but they have the advantage that function evaluations can be executed in parallel.

Hybrid algorithms combine the benefits of several algorithms, usually at some computational expense.

Multifunction Optimization System Tool (MOST) (Ref 14). The MOST can be efficiently used for both continuous optimization problems and integer or discrete design space optimization, where one or more design variables are restricted to an integer domain. The MOST initially executes an SQP algorithm to obtain a continuous solution to the problem. At this stage, all integer variables are treated as continuous variables with a minimum step size of 1.0. If there are any integer (or discrete) variables, MOST will use the continuous solution as the starting point for its modified branch-and-bound algorithm. During this stage, integer variables are dropped one at a time. The reduced continuous optimization problem is solved for each of the dropped variables, fixing their values at integer levels above and below their previously found optimum values. Again, all remaining integer variables are treated as continuous variables with a minimum step size of 1.0.

Pointer Automatic Optimizer (Ref 15). Pointer is an automatic optimization engine that controls a set of standard optimization techniques. It currently controls four optimization methods: an evolutionary algorithm (Ref 16), the Nelder and Mead downhill simplex method, sequential quadratic programming (NPQL), a linear solver, and a Tabu method (Ref 17). This complementary set of algorithms was selected because each succeeds and fails for different topography features. It has been found that a hybrid combination of these methods solves a broad range of design optimization problems. The Pointer automatic optimizer can control one algorithm at a time or all four at once. As the optimization proceeds, the technique determines which algorithms are most successful as well as optimal internal control parameter settings (step sizes, numbers of iterations, number of restarts, etc.). This procedure is hidden from the user. The goal is to enable the nonoptimization expert to successfully use these methods.

The Pointer algorithm performance can be seen as an example of a robust method, as illustrated in Fig. 6. All of these optimizers were compared against a standard benchmark test created by Dr. Sandgren (Ref 18) to present a wide variety of single-objective optimization problems in many fields of mathematics and engineering. No attempt was made to use parallelization of hardware.

Figure 6 shows the results of the test. In each case, there was no expert intervention, and all algorithms were used in one setting considered suitable for the benchmark problems considered (ASA, NLPQL, MOST, LSGRG, and indirect optimization on the basis of self-organization). Although in a few cases, Pointer was also the most efficient method, on average, Pointer was more than two times as expensive as the most efficient algorithm for the individual benchmark problem. In the case of problem 13, Pointer was the only algorithm to find a solution. This shows, again, that there is no free lunch.

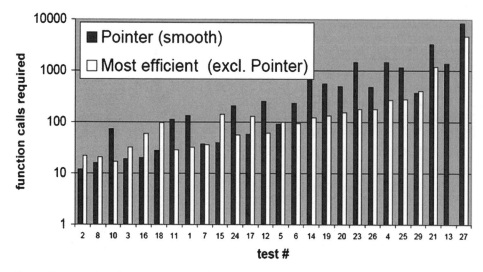

Fig. 6 The Pointer broadband optimizer using the smooth topology setting on differentiable functions is compared to the most efficient method. Test problems are sorted by the minimum number of function calls required by any method using the benchmark set.

The Deterministic Multiobjective Problem

While the single-objective design formulation with an efficient algorithm is computationally the least expensive solution for a particular problem, most real-world problems often involve multiple conflicting objectives. Therefore, a significant amount of research has been performed toward the design of multiobjective optimization algorithms. The multiobjective optimization problem that such algorithms attempt to solve is formally stated as:

$$\text{Minimize } (f_1(\mathbf{x}), f_2(\mathbf{x}), \dots, f_M(\mathbf{x}))$$
$$\text{Subject to } g_j(\mathbf{x}) \leq 0, j = 1, 2, \dots, J$$
$$h_k(\mathbf{x}) = 0, k = 1, 2, \dots, K$$
$$x_i^{(L)} \leq x_i \leq x_i^{(U)}, i = 1, 2, \dots, N$$

In most scenarios, the outcome of a multiobjective optimization process is a set of nondominated Pareto solutions (Ref 19). The usual definition of Pareto domination that is used in the present context is: A feasible solution "a" dominates another feasible solution "b" for an M-objective minimization problem if the following conditions are met:

- $f_i^a \leq f_i^b$ for all $i = 1, 2, \dots, M$
- $f_i^a < f_i^b$ for at least one $i \in \{1, M\}$

The identified Pareto solutions define a Pareto front or M-dimensional Pareto surface. Plotting and visualizing the Pareto front is key to understanding the solution space and evaluating trade-offs between the M objectives. The Pareto front can be a simple smooth curve or a complex discontinuous set of curves/surfaces. Two such examples of Pareto fronts are shown in Fig. 7.

Multiobjective problems can be solved by single-objective methods using a single weighted-sum-type objective. The minimum

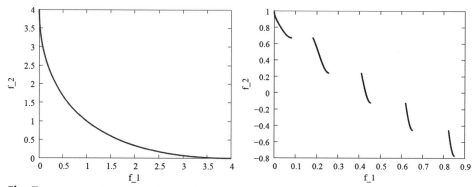

Fig. 7 Two example Pareto fronts depicting the trade-off between objectives f_1 and f_2

summed objective for each set of weights ($w_1 f_1(\mathbf{x}) + w_2 f_2(\mathbf{x}) \dots + w_m f_m(\mathbf{x})$) represents one particular optimal solution on the Pareto front. Even though this is the most efficient way of finding a multiobjective trade-off, there are a number of drawbacks to this approach. First, the weighted sum creates a convex combination of objectives, and optimal solutions in nonconvex regions are not detected (Fig. 8). Second, the proper weighting between objectives and constraints is not always clear up front.

The method by Kim and de Weck (Ref 20) addresses the problem of finding solutions in nonconvex regions using weighted sums, but other methods are more widespread. These true multiobjective methods are easy to use and do well at capturing the Pareto front at the expense of more function calls.

Multiobjective Optimization Methodologies

Multiobjective optimization has become mainstream in recent years, and many algorithms to solve multiobjective optimization problems have

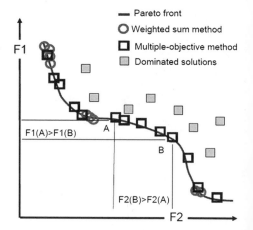

Fig. 8 Using the weighted-sum approach and true multiobjective algorithms to find the Pareto front

been suggested. The use of multiobjective optimization in industry has been accelerated by the availability of faster processing units and the computational analysis models for various engineering problems and disciplines. Multiobjective optimization algorithms, especially those based on evolutionary principles, have seen wide acceptability because, for most engineering problems, a quick computation of approximate solutions is often desirable. Evolutionary algorithms (EAs) are adaptive search techniques inspired by nature, and their working principle is based on Darwin's theory of survival of the fittest (Ref 21, 22). The adaptive nature of EAs can be exploited to design optimization algorithms by designing suitable variation operators and an appropriate fitness function. The genetic algorithm (GA) (Ref 23) is one of the evolutionary techniques that have been successfully used as an optimization tool. Typically, a GA works with a population (a set of solutions) instead of a single solution (individual). This property of a GA makes it an ideal candidate for solving multiobjective optimization problems where the outcome (in most cases) is a set of solutions rather than a single solution. The population approach of a GA also makes it resilient to premature convergence, thereby making it a powerful tool for handling highly nonlinear and multimodal functions.

We found that the following genetic algorithms provide a good set of complimentary approaches to solve multiobjective problems in mechanical engineering applications.

Complementary Approaches

Nondominated Sorting Genetic Algorithm (NSGA-II) (Ref 24). In the NSGA-II, each objective parameter is treated separately. Standard genetic operation of mutation and crossover are performed on the designs. The selection process is based on two main mechanisms: nondominated sorting and crowding distance sorting. By the end of the optimization run, a Pareto set is constructed where each design has the best combination of objective values, and improving one objective is impossible without sacrificing one or more of the other objectives. The NSGA-II is widely used and has become a de facto standard against which the performance of other algorithms is compared.

Neighborhood Cultivation Genetic Algorithm (NCGA) (Ref 25). In NCGA, each objective parameter is treated separately. Standard genetic operation of mutation and crossover are performed on the designs. The crossover process is based on the neighborhood cultivation mechanism, where the crossover is performed mostly between individuals with values close to one of the objectives. By the end of the optimization run, a Pareto set is constructed where each design has the best combination of objective values, and improving one objective is impossible without sacrificing one or more of the other objectives.

Archive-Based Microgenetic Algorithm (AMGA) (Ref 26). The AMGA is an evolutionary optimization algorithm and relies on genetic variation operators for creating new solutions. The generation scheme deployed in this algorithm can be classified as generational because, during a particular iteration (generation), only solutions created before that iteration take part

in the selection process. The algorithm, however, generates a small number of new solutions (recommended value is two solutions per iteration) at every iteration, and therefore, it can also be classified as an almost steady-state genetic algorithm. The algorithm works with a small population size (recommended value is four) and maintains an external archive of good solutions obtained. At every iteration, a small number of solutions are created using genetic variation operators. The algorithm is referred to as archive-based microgenetic algorithm because it works with a very small population size and uses an archive to maintain its search history. It is recommended to use a large size for the archive, and the best results are obtained if the size of the archive is the same as the number of function evaluations allowed (i.e., the algorithm stores its complete search history). The size of the archive determines the computational complexity of the algorithm; however, for computationally expensive optimization problems, the actual time taken by the algorithm is negligible. The parent population is updated using the archive, and binary tournament selection is performed on the parent population (for creating the mating population).

Multiobjective Optimization Study

Evaluating and comparing multiobjective optimization algorithms is more involved than evaluating and comparing single-objective algorithms. Rather than simply comparing the objective value at a single solution point and the number of system evaluations required to achieve the solution objective value, evaluating the solution set provided by a multiobjective optimization algorithm requires comparison of Pareto sets for accuracy and completeness.

Pareto Fronts

In recent studies, four unary performance indicators are commonly used to compare the ability of multiobjective optimization algorithms to characterize the Pareto front (i.e., they compare a single nondominated set to a Pareto-

optimal frontier). The performance indicators are delineation, distance, diversity, and hypervolume (Ref 27). A brief description of each performance indicator follows:

1. *Delineation metric:* It measures how much of the true Pareto-optimal front is represented by the obtained solution set.
2. *Distance metric:* It measures the average Euclidean distance between the true Pareto-optimal front and the obtained solution set.
3. *Diversity Metric:* It measures the uniformity of distribution and the spread of the obtained solution set.
4. *Hypervolume metric:* It measures the fraction of search space not dominated by the obtained solution set in comparison to the true Pareto-optimal set.

To use these performance indicators, the true Pareto-optimal front must be known. A smaller value for a performance indicator means a better solution set. Ideally, if the original Pareto-optimal front is used as the solution set, all the performance indicators should evaluate to zero. Because a finite number of points (1000 points for the problems presented here) are used to represent the true Pareto-optimal frontier, a value of 0.01 or less for a performance indicator implies that the obtained solution set is virtually indistinguishable from the Pareto-optimal front. If the value of the performance indicator is 0.5 or more, it implies that an acceptable solution set was not obtained. All the objectives are normalized (the Pareto-optimal set is mapped to the range [0, 1]) before the performance indicators are computed. Only the nondominated solutions belonging to rank 1 are considered for computing the performance indicators.

Results from NSGA-II, NCGA, and AMGA are presented in Table 1 for five common multiobjective optimization test problems, taken from Ref 28: ZDT1, ZDT2, ZDT3, ZDT4, and ZDT6. Each problem has two objectives. Minimization for both of the objectives for all of the test problems is assumed. ZDT1-3 has 30 design variables, and ZDT4 and ZDT6 have 10 design variables.

Table 1   Median value for all multiobjective optimization performance metrics

Problem	Algorithm	Delineation	Distance	Diversity	Hypervolume
ZDT1	NSGA-II	0.055573	0.050177	0.130880	0.109433
	NCGA	0.068537	0.063598	0.109231	0.107068
	AMGA	**0.033043**	**0.023070**	**0.071707**	**0.069749**
ZDT2	NSGA-II	0.080988	0.070241	0.326157	0.257817
	NCGA	0.149708	0.136145	0.502602	0.415069
	AMGA	**0.037432**	**0.029006**	**0.095623**	**0.135636**
ZDT3	NSGA-II	0.038556	0.029505	0.143943	0.100310
	NCGA	0.043921	0.028587	0.156448	0.127858
	AMGA	**0.020222**	**0.009450**	**0.088270**	**0.055998**
ZDT4	NSGA-II	0.029920	0.014183	**0.085797**	0.059543
	NCGA	0.958743	1.011684	1.884524	0.993992
	AMGA	**0.026019**	**0.010017**	0.157624	**0.047144**
ZDT6	NSGA-II	0.136153	0.132502	0.388023	0.349802
	NCGA	1.367150	1.249455	3.120534	1.000000
	AMGA	**0.099351**	**0.097869**	**0.279079**	**0.288113**

NSGA, nondominated sorting genetic algorithm; NCGA, neighborhood cultivation genetic algorithm; AMGA, archive-based microgenetic algorithm

Even though these algorithms are quite robust, a search from a given starting point still produces a slightly different answer because of the randomness in the search procedure. To account for this in the comparison, 15 simulation runs were executed, starting with different random seeds for each algorithm-problem pair (a total of $15 \times 3 \times 5 = 225$ simulations performed). The size of the initial population used for all of the algorithms is 100. The number of generations used is 40 for all except ZDT4, which used 100 generations, giving a total number of function evaluations of 10,000 for ZDT4 and 4000 for all other problems.

Table 1 presents the median value for each performance indicator for each algorithm, across the 15 executions of each algorithm, for each test problem. The best (lowest) metric values for each problem are highlighted in bold. It is evident from the simulation results that AMGA has the overall best performance, obtaining the best metric values for all problems and all metrics except for diversity for ZDT4, for which NSGA-II obtained the best value. The AMGA is capable of reporting a large number of nondominated solutions for the same number of function evaluations. The development of AMGA can also be perceived as an exercise in combining the best features of different algorithms and best practices into a unified optimization framework. Although the computational complexity of the algorithm is more than either NSGA-II or NCGA, the execution time is not affected drastically. Almost the entire execution time is consumed by the analysis routines, and therefore, the algorithm can afford to perform expensive operations if such operations can result in a reduced number of function evaluations. The two guiding principles that have shaped the design of the AMGA algorithm are focused on reducing the number of function evaluations for the same degree of convergence and making the algorithm immune to changes in sizing or tuning parameters. Limited success has been achieved by AMGA in fulfilling these goals.

The results for problem ZDT3 are shown graphically in Fig. 9. The Pareto frontier for ZDT3 is the second, discontinuous front shown in Fig. 7. In this problem, AMGA gives performance metric values that are 40 to 80% better than the next-best value. Note, however, that NCGA is designed to work with bit strings, whereas NSGA-II and AMGA are designed to work with real variables. All of the problems considered in this study involve real variables that may have impacted the performance of NCGA. For problems involving discrete variables, NCGA can produce better results. Also, the use of bit strings gives NCGA the capability to produce a more uniform distribution across the Pareto-optimal frontier. Again, the no-free-lunch theorem (Ref 2) applies.

It should also be noted that the number of Pareto points output by AMGA has been deliberately limited to 20 for the purposes of a more fair comparison. If the number of points output by AMGA is not restricted, to exploit one of the advantages of AMGA, the value of performance parameters would be significantly better.

Nondeterministic, Stochastic Optimization Problem

Real-world engineered products and processes do not behave in a deterministic manner. Most systems behave stochastically, involving chance or probability. Variation is inherent in material characteristics, loading conditions, simulation model accuracy, geometric properties, manufacturing precision, actual product usage, and so on. Application of deterministic optimization strategies, without incorporating uncertainty and measuring its effects, leads to designs that cannot be called optimal but instead are potentially high-risk solutions that can have a high probability of failing in use. Optimization algorithms tend to push a design toward one or more constraints until the constraints are active. With a design sitting on one or more constraint boundary, even slight uncertainties in the problem formulation or changes in the operating environment could produce failed, unsafe designs and/or result in substantial performance degradation.

Traditionally, many uncertainties are removed through assumptions, and others are handled through crude safety methods, which often lead to overdesigned products and do not offer insight into the effects of individual uncertainties and the actual margin of safety of a design.

More recently, stochastic optimization methods—often called probabilistic optimization or robust optimization—have been developed to address uncertainty and randomness through statistical modeling and probabilistic analysis (Ref 29, 30). These probabilistic approaches have been developed to convert deterministic problem formulations into stochastic formulations to model and assess the known uncertainties of the effects. Until recently, however, the computational expense of stochastic methods, in terms of the number of function evaluations necessary to accurately capture performance variation, has made the application of these methods impractical for all but academic investigations or very critical cases. With the steady increases in computing power, large-scale parallel processing capabilities, and availability of probabilistic analysis and optimization tools and systems, however, the combination of these technologies can facilitate effective stochastic analysis and optimization for complex design problems, allowing the identification of designs that qualify as not only feasible but as consistently feasible in the face of uncertainty.

A stochastic optimization problem can be formally stated as follows:

$$\text{Minimize } f_m(\mu_{y_m}(\mathbf{x}), \sigma_{y_m}(\mathbf{x})), m = 1, 2, \ldots, M$$
$$\text{Subject to } g_j(\mu_{y_j}(\mathbf{x}), \sigma_{y_j}(\mathbf{x})) \leq 0, j = 1, 2, \ldots, J$$
$$h_k(\mathbf{x}) = 0, k = 1, 2, \ldots, K$$
$$x_i^{(L)} + n\sigma_{x_i} \leq x_i \leq x_i^{(U)} - n\sigma_{x_i},$$
$$i = 1, 2, \ldots, N$$

The stochastic optimization problem models both nominal (or mean) performance and performance variation through statistics and/or probabilities. For example, the constraints can

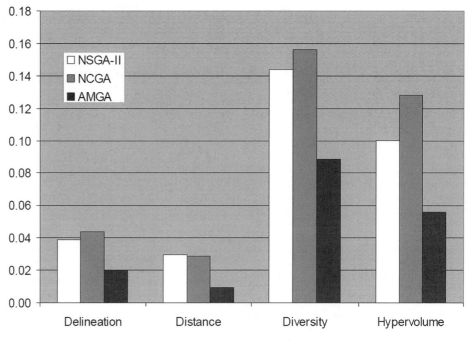

Fig. 9 ZDT3 performance results for all three algorithms and all performance metrics. NSGA, nondominated sorting genetic algorithm; NCGA, neighborhood cultivation genetic algorithm; AMGA, archive-based microgenetic algorithm

be modeled as a mean value adjusted by a specified number of standard deviations:

$$\mu_y - n\sigma_y \geq \text{Lower limit}$$
$$\mu_y + n\sigma_y \leq \text{Upper limit}$$

Or as a probability of violating the specified limit:

$$g(\mathbf{x}) \leq 0 \text{ becomes } P_f = P(g(\mathbf{x}) \leq 0) \leq P^U$$

The stochastic optimization problem is inherently a multiobjective problem: Each performance measure has two objective components corresponding to nominal/mean performance and performance variation. Multiobjective optimization strategies can be used to effectively assess the trade-offs between the performance measures and between the nominal performance and measured variation of performance. The stochastic optimization problem can also be formulated as a single objective problem, using the weighted-sum approach as follows:

$$F = \sum_{m=1}^{M}\left[\frac{w_{1_m}}{s_{1_m}}\left(\mu_{y_m} - T_m\right)^2 + \frac{w_{2_m}}{s_{2_m}}\sigma_{y_m}^2\right]$$

where w_{1m} and w_{2m} are the weights, and s_{1m} and s_{2m} are the scale factors for the mean on target and minimize variation objective components, respectively, for performance response m; T_m is the target for performance response m; and M is the number of performance responses included in the objective. For the case in which the mean performance is to be minimized or maximized rather than directed toward a target, the objective formulation can be modified as follows, where the first term is positive when the response mean is to be minimized and negative when the response mean is to be maximized:

$$F = \sum_{m=1}^{M}\left[(+/-)\frac{w_{1_m}}{s_{1_m}}\mu_{y_m} + \frac{w_{2_m}}{s_{2_m}}\sigma_{y_m}^2\right]$$

Calculation of the statistics and probabilities necessary to implement a stochastic optimization strategy requires the identification and characterization of random variables and the incorporation of a sampling strategy within the stochastic optimization search. Random variables are inputs to a simulation with known variation. They are described through probability distributions and associated properties. By sampling from these distributions following their prescribed properties, the effects of this input variation on performance can be assessed. Through this sampling, stochastic analysis is used to measure design quality (reliability and robustness). Many methods have been developed for stochastic sampling, including Monte Carlo methods (Ref 31, 32), structural reliability analysis methods (Ref 33–35), sensitivity-based methods, based on Taylor's expansion (Ref 36, 37), and design of experiments (Ref 38). Examples of computed-quality

attributes are mean, sigma level, defects per million, probability of success, and probability of failure.

Stochastic optimization is then used to improve design quality. Any of the authors' multi- and single-objective optimization methods (Ref 1) mentioned earlier in this article can be used to optimize the attributes of design quality as measured by the stochastic analysis.

The concept of stochastic, robust optimization is illustrated in Fig. 10. If the function in Fig. 10 is to be minimized, the solution given by point 1 would be chosen if uncertainty and performance variation are not considered, as with deterministic optimization. Given uncertainty in the design parameter x, defined as a variation of $\pm\Delta x$ around the chosen value, the solution at point 1 leads to a large level of variation, Δf_1, of the performance function $f(x)$. To the right of point 1 in the figure, there exists a more flat region of $f(x)$, which can be shown to be more robust or less sensitive to variation in the design parameter x. If point 2 is chosen, for the same design parameter variation, $\pm\Delta x$, the variation of the performance function, Δf_2, is significantly less than that at point 1. The sacrifice in choosing point 2 is the increase in the nominal value of $f(x)$, which is higher at point 2 than at point 1. This is the trade-off that must be evaluated in searching for a robust solution as opposed to a solution with optimal mean performance. It can be seen in the figure that an even flatter region than that at point 2 exists further to the right of point 2 (direction of increasing x). Although the performance variation may be even less in this region, the mean performance may not be acceptable. Both elements of desired mean performance and reduced performance variation must be incorporated in a robust optimization formulation.

A very common way to express the quality of a design is the number of sigmas (standard deviations) a design is away from failing to meet the specifications. Six sigmas correspond to three defects per million, a widely quoted (Ref 39) quality goal in manufacturing processes.

Stochastic Optimization Studies

Metal Production and Product Design

Examples of the stochastic approach are given for metal production and product design.

Example 3: Steel Mill. In Fig. 11, the robust optimization of a steel mill is presented. In this case, the operation of the mill under varying conditions (times, temperature profiles, cooling air velocities, etc.) is of interest, so the deviation from the specification (material characteristics, dimensions, etc.) is minimized with constraints on equipment operation. In this particular case, robust optimization reduced the mean specification score by 1%, but the standard deviation of the specification score was reduced by an impressive 95%. This provided significant savings in scrap cost.

Example 4: Automobile Crashworthiness. One quality engineering design application currently of high visibility in the automotive industry is vehicle structural design for crashworthiness. These design problems are not only particularly complex, in terms of understanding the problem and defining design requirements and design alternatives, but also involve a very high degree of uncertainty and variability. Velocity of impact, mass of vehicle, angle of impact, and mass/stiffness of barrier are just a few of many uncertain parameters. A balance must be struck in designing for crashworthiness between designing the vehicle structure to absorb/manage the crash energy (through

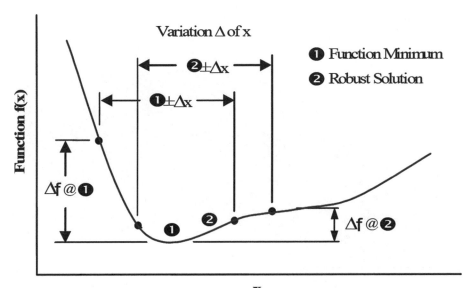

Fig. 10 Robust optimization concept

structure deformation) and maintaining passenger compartment integrity, all in the face of uncertainty and variability in materials, structural configuration, and crash scenario.

One specific crash scenario is investigated in Ref 29 (side impact) using the stochastic analysis and optimization tools available in Isight. A typical vehicle side-impact model is shown in Fig. 12. Including a finite-element dummy model, the total number of elements in this model is approximately 90,000, and the total number of nodes is approximately 96,000. The initial lateral velocity of the side-deformable barrier is 31 miles per hour. The central processing unit time for a RADIOSS simulation of the model is approximately 20 h on an SGI Origin 2000.

For side-impact protection, the vehicle design should meet the requirements for the National Highway Traffic Safety Administration side-impact procedure specified in the Federal Motor Vehicle Safety Standards or the European Enhanced Vehicle-Safety Committee (EEVC) side-impact procedure. In this study, the EEVC side-impact test configuration was used. The dummy performance is the main concern in side impact, which includes head injury criterion, abdomen load, pubic symphysis force (pelvic load), viscous criterion, and rib deflections (upper, middle, and lower). These dummy responses must at least meet EEVC requirements. Other concerns in side-impact design are the velocity of the B-pillar at middle point and the velocity of the front door at B-pillar.

For side impact, increase of gage design variables tends to produce better dummy performance. However, it also increases vehicle weight, which is undesirable. Therefore, a balance must be sought between weight reduction and safety concerns. The objective is to reduce the weight with imposed constraints on the dummy safety. Here, eleven design parameters are used for side-impact optimization. Nine design variables include two materials of critical parts and seven thickness parameters. The material design variables are discrete, either mild steel or high-strength steel. All thickness and material design variables are also random variables, normally distributed with standard deviation of 3% of the mean for the thickness variables and 0.6% for the material properties. The final two parameters, barrier height and hitting position, are pure random variables continuously varying from −30 to 30 mm according to the physical test.

A deterministic optimization was initially applied for this problem using the NLPQL-SQP algorithm. Starting from an infeasible baseline design, this optimization solution results in a feasible design with a weight reduction of nearly 20% but also results in three active constraints associated with rib deflection, pubic symphysis force, and door velocity. When a stochastic analysis is performed at this design solution, using Monte Carlo descriptive sampling with 2000 points, reliability values of 40% are obtained (60% probability of failure). After applying stochastic optimization, the reliability is increased, but at the expense of the vehicle weight, as shown in Fig. 13. For this problem, when the number of standard deviations of performance maintained within the required limits (original optimization constraints) reaches approximately 3σ, the weight is nearly equal to that of the baseline design; no weight savings is achieved, but the quality level is increased. As the quality level is increased further, the weight is increased over the baseline.

Closing

Optimization is a useful and effective simulation-based design tool for identifying one or more designs that best achieves a set of requirements or for improving an existing design. Optimization has been used to solve a wide range of industrial problems and is being used more and more. One conclusion of these industrial applications of optimization is that no one optimization algorithm—or even class of optimization algorithms—is appropriate or

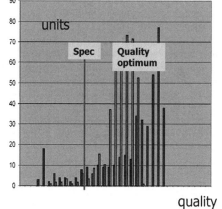

- ◆ **Vary tool set points (recipes)**
- ◆ **Minimize deviation from spec with constraints on equipment operation**
- ◆ **Key issue is uncertainty – manufacturing conditions are not known precisely but vary according to certain probability distributions**
- ◆ **Goal is an optimum manufacturing setup that is insensitive to this uncertainty**

Fig. 11 Tool-set point optimization of a seamless steel tube mill. Note: Results are for illustration only. Source: Ref 40

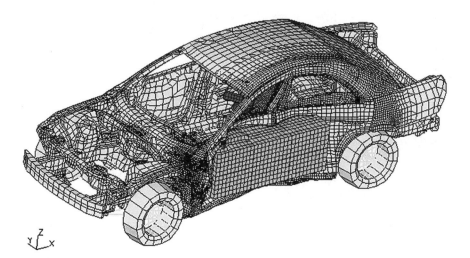

Fig. 12 Automotive crashworthiness robust optimization, side-impact model. Source: Ref 29

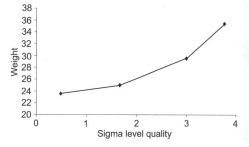

Fig. 13 Performance versus quality trade-off, side-impact problem. Source: Ref 29

even capable for solving a wide range of problems well. Again, there is no free lunch. Many optimization strategies have been developed to address different types of problems. These optimization strategies can be combined and applied to create new strategies, allowing both global and local search, continuous and discontinuous spaces, continuous and discrete variables, smooth and noisy topologies, and so on, to become more robust toward solving a wider range of problems.

There is also always a trade-off between efficiency and accuracy. Formulating and solving a single-objective problem is usually the most efficient approach but may not give the best trade-off solution or allow the problem to be explored sufficiently. Extending a problem to multiple objectives and truly evaluating the trade-offs between the objectives requires increased computational effort. Incorporating uncertainty and searching for designs of higher quality, requiring additional sampling during a search strategy, add even significantly more computational expense.

For complex simulation models requiring minutes or hours per analysis, optimization—and, especially, stochastic optimization—can quickly become impractical. However, additional simulation-based design-enabling technologies can be combined to support the implementation of optimization to even highly complex problems. Two such current technologies are parallel processing and approximation methods. Many analyses during optimization and sampling for stochastic analysis are independent analyses known in advance: gradient runs for gradient-based methods, a population in a genetic algorithm, Monte Carlo or design of experiments samples, and so on. These analyses can be executed in parallel on multiprocessor machines and/or a network of machines. For the most computationally expensive analyses, requiring days for a single analysis, a set of sampled design points can be executed in parallel and used to construct surrogate models, or approximation models, to replace the high-fidelity code during optimization. Polynomial response surfaces and, more recently, radial basis functions or Kriging surrogate models have been employed for this purpose.

REFERENCES

1. A. Van der Velden, CAD to CAE Process Automation Through iSIGHT-FD, Engineous Software, GT 2007-27555, *Proceedings of the ASME Turbo Expo* (Montreal), 2007
2. D.H. Wolpert et al., No Free Lunch Theorems for Optimization, *Evolutionary Computation, IEEE Transactions on Evolutionary Computation,* 1996
3. B. Wujek, P.N. Koch, M. McMillan, and W.-S. Chiang, "A Distributed, Component-Based Integration Environment for Multidisciplinary Optimal and Quality Design," AIAA-2002-5476, Ninth AIAA/ISSMO Symposium on Multidisciplinary Analysis and Optimization (Atlanta, GA)
4. J. Runyan, R.A. Gerhardt, and R. Ruh, Electrical Properties of BN Matrix Composites—Part I: Analysis of the McLachlan Equation in Modeling the Conductivity of BN-B4C and BN-SiC Composites, *J. Am. Ceram. Soc.,* Vol 84 (No. 7), 2001, p 1490–1496
5. E. Buckingham, The Principle of Similitude, *Nature,* 1915, p 96
6. H.-K. Kim, S.K. Hong, J.-K. Lee, B.-H. Jeon, J.J. Kang, and Y.-M. Heo, "Finite Element Analysis and Optimization for the Multi-Stage Deep Drawing of Molybdenum Sheet," CP778 Vol A, Numisheet, American Institute of Physics, 0-7354-0265-5/05, 2005
7. L. Ingber, Simulated Annealing: Practice versus Theory, *Math. Comput. Model.,* Vol 18, 1993, p 29–57
8. K. Schittkowski, NLPQL: A Fortran Subroutine for Solving Constrained Nonlinear Programming Problems, *Ann. Oper. Res.,* Vol 5, 1985, p 485–500
9. G.N. Vanderplaats, An Efficient Feasible Directions Algorithm for Design Synthesis, *AIAA J.,* Vol 22 (No. 11), Oct 1984, p 1633–1640
10. L.S. Lasdon, A.D. Waren, A. Jain, and M. Ratner, Design and Testing of a Generalized Reduced Gradient Code for Nonlinear Programming, *ACM Transactions on Mathematical Software,* Vol 4 (No. 1), March 1978, p 34–50
11. R. Hooke and T.A. Jeeves, Direct Search Solution of Numerical and Statistical Problems, *J. ACM,* Vol 8, April 1961, p 212–229
12. J.A. Nelder and R. Mead, Downhill Simplex Method in Multidimensions, *Comput. J.,* Vol 7, 1965, p 308–313
13. M. Miki, T. Hiroyasu, M. Kaneko, and K. Hatanaka, A Parallel Genetic Algorithm with Distributed Environment Scheme, *IEEE International Conference on Systems, Man, and Cybernetics,* Vol 1, 1999, p 695–700
14. C.H. Tseng, W.C. Liao, and T.C. Yang, "Most 1.1 User's Manual," Technical Report AODL-96-01, Dept. of ME, National Chiao Tung University, Taiwan ROC, 1996
15. A. Van der Velden and D. Kokan, The Synaps Pointer Optimization Engine, *Proceedings of DETC/CIE, ASME Computers and Information in Engineering Conference* (Montreal, Canada), 2002
16. H.P. Schwefel, "Evolutionsstrategie und Numerische Optimierung," Ph.D. thesis, Verfarenstechnik TU Berlin, 1975
17. F. Glover and D. de Werra, *Tabu Search,* Baltzer Science Publishers, 1997
18. E. Sandgren, "The Utility of Nonlinear Programming Algorithms," Ph.D. thesis, Purdue University, Dec 1977
19. Y. Sensor, Pareto Optimality in Multi-Objective Problems, *App. Math. Optimiz.,* Vol 4 (No. 1), March 1977, p 41–59
20. I.Y. Kim and O. de Weck, Adaptive Weighted Sum Method of Multiobjective Optimization, *Tenth AIAA/ISSMO MDAO Conference 2004,* p 2004–4322
21. R. Dawkins, *The Selfish Gene,* Oxford University Press, New York, 1976
22. N. Eldredge, *Macro-Evolutionary Dynamics: Species, Niches and Adaptive Peaks,* McGraw-Hill, New York, 1989
23. D.E. Goldberg, *Genetic Algorithms for Search, Optimization, and Machine Learning,* Addison-Wesley, Reading, MA, 1989
24. K. Deb, S. Agrawal, A. Pratap, and T. Meyarivan, A Fast and Elitist Multi-Objective Genetic Algorithm: NSGA-II, *IEEE Trans. Evolution. Comput.,* Vol 6 (No. 2), 2002, p 182–197
25. S. Watanabe, T. Hiroyasu, and M. Miki, NCGA: Neighborhood Cultivation Genetic Algorithm for Multi-Objective Optimization Problems, *Proceedings of the Genetic and Evolutionary Computation Conference GECCO 2002,* p 458–465
26. S. Tiwari, P. Koch, G. Fadel, and K. Deb, "AMGA: An Archive-Based Micro Genetic Algorithm for Fast and Reliable Convergence," Genetic and Evolutionary Computation Conference GECCO 2008 (Atlanta, GA), July 12–16, 2008
27. E. Zitzler, L. Thiele, M. Laumanns, C.M. Fonseca, and V.G. Da Fonseca, Performance Assessment of Multiobjective Optimizers: An Analysis and Review, *IEEE Trans. Evolution. Comput.,* Vol 7 (No. 2), April 2003, p 117–132
28. K. Deb, *Multi-Objective Optimization Using Evolutionary Algorithms,* Wiley, Chichester, U.K., 2001
29. P.N. Koch, R.-J. Yang, and L. Gu, Design for Six Sigma through Robust Optimization, *J. Struct. Multidiscip. Optimiz.,* Vol 26 (No. 3–4), 2004, p 235–248
30. A.D. Belegundu, Probabilistic Optimal Design Using Second Moment Criteria, *J. Mech. Transm. Automat. Des.,* Vol 110 (No. 3), 1988, p 324–329
31. J.M. Hammersley and D.C. Handscomb, *Monte Carlo Methods,* Chapman and Hall, London, 1964
32. E. Saliby, Descriptive Sampling: A Better Approach to Monte Carlo Simulation, *J. Oper. Res. Soc.,* Vol 41 (No. 12), 1990, p 1133–1142
33. A.M. Hasofer and N.C. Lind, Exact and Invariant Second Moment Code Format, *J. Eng. Mech., ASCE,* Vol 100 (No. EM1), Feb 1974, p 111–121
34. R. Rackwitz and B. Fiessler, Structural Stability under Combined Random Load Sequences, *Comput. Struct.,* Vol 9, 1978, p 489–494
35. R.E. Melchers, *Structural Reliability: Analysis and Prediction,* 2nd ed., Ellis Horwood Series in Civil Engineering, John Wiley & Sons, New York, 1999
36. W. Chen, J.K. Allen, K.-L. Tsui, and F. Mistree, A Procedure for Robust Design:

Minimizing Variations Caused by Noise Factors and Control Factors, *ASME J. Mech. Des.,* Vol 118 (No. 4), 1996, p 478–485

37. C.-C. Hsieh and K.P. Oh, MARS: A Computer-Based Method for Achieving Robust Systems, *FISITA Conference, The Integration of Design and Manufacture,* Vol 1, 1992, p 115–120

38. D.C. Montgomery, *Design and Analysis of Experiments,* John Wiley & Sons, New York, 1996

39. M. Harry, "Six Sigma," Currency book, 2000

40. R.V. Kolarik and I. Chan, "U.S. Department of Energy Technology Success Story: Enhanced Spheroidized Annealing Cycle for Tube and Pipe Manufacturing," Timken Company, DOE

SELECTED REFERENCES

• H. Ishibuchi and T. Yoshida, Hybrid Evolutionary Multi-Objective Optimization Algorithms, *Soft Computing Systems: Design, Management and Applications (Frontiers in Artificial Intelligence and Applications),* Vol 87, A. Abraham, J. Ruiz-del-Solar, and M. Köppen, Ed., IOS Press, 2002, p 163–172

• C. Wooffindin, "Corus Smart Weld Optimiser," Engineous North America User Meeting (Orlando, FL), 2007

ASM Handbook, Volume 22B, *Metals Process Simulation*
D.U. Furrer and S.L. Semiatin, editors

Copyright © 2010, ASM International®
All rights reserved.
www.asminternational.org

Stress-Relief Simulation

Dennis J. Buchanan, University of Dayton Research Institute

THE ADVANCES IN PROCESSING OF METALLIC MATERIALS for improved physical, thermal, and mechanical properties are numerous. Cast, wrought, and powder metallurgy processes have been optimized to achieve a balance of material properties. The temperature and deformation processing histories are carefully controlled during production to increase uniformity of material properties throughout the body of the component. However, during the cooldown stage, after thermomechanical processing, the outer surfaces cool faster than the interior, resulting in a temperature gradient. This produces thermally generated bulk residual stresses that remain in the body even after the entire part achieves room temperature. In most cases, these manufacturing residual stresses are undesirable. In fact, the magnitude of bulk residual stresses in turbine engine disks may be a large fraction of the material yield strength (Ref 1). The drawback of these undesirable residual stresses is when they are superimposed with the applied mechanical and thermal stresses associated with the operating conditions of the component. The consequence of not incorporating the generation of residual stresses and residual stress relaxation into model simulations may result in excessive scrap rates, manufacturing costs during production, and reduced service life of the finished component.

Bulk residual stresses must be quantified and modeled for relaxation or relief as a function of thermal and mechanical loading history. Numerous studies document the existence of residual stresses through measurements and have described the effects of residual stresses for a variety of materials and components (Ref 2–9). This article summarizes many of the existing approaches used to simulate relaxation of bulk residual stresses in components. The capability to simulate the evolution of residual stresses as a function of thermal and mechanical loading history is critical to design, manufacture, and service-life prediction. If residual stresses are neglected or ignored in the design or manufacture of the component, a premature or catastrophic failure may result in a costly shutdown or loss of life.

The remainder of this article on stress-relief simulation describes a progression of different creep models and solution procedures, starting with a simple analytic model with a closed-form solution to a complex displacement-based finite-element numerical solution that is capable of incorporating material and geometric nonlinearities. This simple-to-complex progression of analyses serves two useful purposes and suggests that this approach should also be followed by the analyst in developing solutions to real engineering problems. First, many problems can be solved easily by imposing simple approximations or assumptions in either the material behavior or the solution procedure. A quick, simple analysis should be performed first to evaluate the time, expense, and potential accuracy of available modeling solutions. Furthermore, a simple analysis approach can quickly demonstrate the need for more sophisticated methods that include temperature-dependent material properties, rate dependence, and material and geometric nonlinearities. For example, the need for temperature-dependent material properties can be evaluated with a simple model by executing the analysis with different material properties that are representative of expected temperature ranges. If the results of these analyses differ greatly over the temperature range, then temperature-dependent material properties should be considered for a more accurate solution. When a more accurate analysis is required, the simple analysis will serve as a benchmark comparison for more detailed analyses. Second, not all analysts will have the same knowledge about the critical factors that are important to the solution or the same level of problem-solving experience. In either case, the analyst can gain experience and build knowledge by starting with simple material models and solution methods before progressing to the complex solution methods.

Examples of Complicated Residual Stress States in Simple Bodies

In many instances, residual stresses are intentionally engineered into the design to extend the life of the component. For example, residual stresses can be introduced by monotonic overloading, or what is often called presetting a component under a prescribed loading state such that yielding occurs at weak points. The yielding accomplishes several tasks, including hardening the material, increasing the elastic working range, establishing compressive stresses around notches and existing subcritical crack tips, and ultimately prooftesting the component prior to service. Unfortunately, both manufacturing-related and engineered residual stress states are often multiaxial, with complicated distributions throughout the entire body of the component. In fact, the unintended consequence of engineering residual stresses into specific locations of components is the resulting balancing stresses that develop throughout the body that are required to satisfy equilibrium.

Analyzing Results

Pure Bending of a Rectangular Beam. A case in point is the simple example of a rectangular beam subjected to pure bending, with cross-sectional dimensions shown in Fig. 1. In this example, elastic perfectly plastic material behavior is used in the deformation analysis along with the enforcement of load equilibrium and compatibility conditions. Figure 2 shows the one-dimensional residual stress and strain

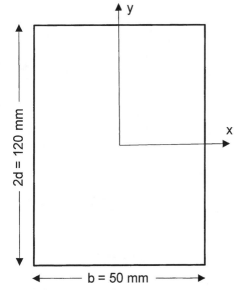

Fig. 1 Cross-sectional dimensions of a rectangular beam under pure bending

distributions that remain in the beam after being subject to plastic deformation during the loading, followed by the elastic unloading. The resulting distributions exhibit some interesting characteristics. Inside the elastic region ($y <$ ±40 mm), where stress and strain are still linearly related, the residual stress and residual strain have the same sign. In contrast, the top and bottom surfaces of the beam have stresses and strains of opposite sign. This is where yielding occurs first and deformation from plastic behavior is dominant. Additional details of this analysis are described elsewhere (Ref 10).

Thermal Stress in Annular Disk. A second example is the thermal stresses that develop in an annular disk of uniform thickness due to a radial temperature gradient, as shown in Fig. 3. This type of temperature profile, shown on the left side of the figure, is often observed in the cooling of castings and forgings and in the service operating conditions of rotating turbine engine components. Here, one can observe the additional complexity of a two-dimensional stress state compared to the previous example. The stress profiles are a function of radial position only and are shown on the right side of the figure. The radial residual stress (σ_{rr}) is totally compressive in the interior and zero at the inner and outer radii, while the circumferential stress ($\sigma_{\theta\theta}$) is compressive at the inner radius and transitions to tension at the outer radius. Additional details of this thermal stress analysis can be found in the literature (Ref 11, 12).

These analytical examples highlight the complexity of the residual stress and strain distributions observed in simple geometries, with ideal material behavior, and trivial loading and boundary conditions. Furthermore, these examples provide a baseline or comparison solution for more advanced numerical solutions that are typically used to solve real engineering problems. Real components with a complicated thermomechanical loading history, which contains fillets, holes, and other geometric features, will require nonlinear material and geometric solution procedures and will result in complex multiaxial residual stress and residual strain distributions. Many experienced stress analysts will agree that the best course of action in the solution of these problems is to start with the simple solution first, gain confidence in the results, and then move on to more advanced solutions that incorporate temperature-dependent properties, nonlinear material and geometric behavior, coupling, and so on. The remainder of this article discusses approximate and advanced solution techniques that can be employed in practice for simulation of residual stress relief.

Stress relaxation and creep deformation are essentially the same material phenomenon; the only difference is what boundary condition is prescribed and what response is observed or measured. In a stress-relaxation test, an applied displacement is prescribed, and a decreasing load (or stress) is observed with time. In a creep test, an applied load (or stress) is prescribed, and an increasing displacement is observed with time. Both approaches characterize the same time-dependent material behavior required to develop a constitutive material relating stress and strain. However, time-dependent material behavior is usually shown as creep data and modeled as creep deformation mechanisms. Therefore, this article discusses implementation of creep data and models for stress-relief simulation.

Approximate Solution Technique— Reference Stress Method with Steady Creep

The approach of the reference stress method (RSM) is to relate the creep deformation response at a specific point in a structure to that of a laboratory test specimen. The relationship typically takes the form of:

$$\dot{\varepsilon}^c_{structure} = \beta \dot{\varepsilon}^c_{specimen} \qquad \text{(Eq 1)}$$

where, β is a geometric scaling factor between the structure and the specimen. Maturation of the technique occurred predominantly in the United Kingdom during the 1960s. However, the first documented application of the RSM was made by Soderberg (Ref 13) in 1941 in the United States. Detailed discussions of the RSM, with additional examples, may be found in Ref 14–16.

A simple example is the stress distribution in a rectangular beam under pure bending, with the cross-sectional dimensions (height = $2d$; width = b) of Fig. 1. The steady-state creep analysis is accomplished using the three field equations of solid mechanics as applied to elementary beam theory. The first equation is the compatibility equation, which relates the strain in the beam (ε) to the beam curvature (κ) times the distance from the neutral axis (y):

$$\varepsilon = \kappa y \qquad \text{(Eq 2)}$$

The second equation is the equilibrium equation, which relates the applied bending moment (M) to the stresses (σ) integrated over the cross-sectional area of the beam:

$$M = \int_A \sigma y \, dA \qquad \text{(Eq 3)}$$

Finally, the third equation is the constitutive equation, which takes the form of a power-law creep model relating creep strain rate ($\dot{\varepsilon}^c$) to stress through:

$$\dot{\varepsilon}^c = B\sigma^n \qquad \text{(Eq 4)}$$

where B and n are fitting parameters describing the creep deformation behavior of the material. These equations may be combined and normalized by the maximum stress (σ^e_{max}) in an elastic beam into a nondimensional form for stresses in the beam:

$$\frac{\sigma}{\sigma^e_{max}} = \frac{2n+1}{3n}\left(\frac{y}{d}\right)^{1/n} \qquad \text{(Eq 5)}$$

This equation relates the normalized stress in the beam to the normalized distance from the neutral axis (y/d) and the exponent of the power-law equation. Figure 4 is a plot of this equation for different values of the exponent. When $n = 1$, this reduces to linear elastic material behavior, shown as a straight line in the plot. For larger values of the exponent ($n = 3$, 5), the stress profile is nonlinear. The necessary material properties for this analysis are the elastic modulus and secondary creep rate as a function of applied stress, as shown by Eq 4. A detailed analysis of this example is described in Boyle and Spence (Ref 15).

The solution described previously assumes that elastic strains and elastic strain rates are negligible in size compared to creep strains and creep strain rates. Because the elastic strain rate is negligible, the stress rate is also

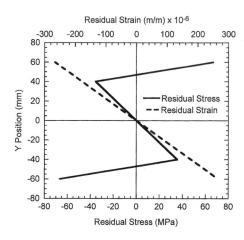

Fig. 2 Residual stress and residual strain distributions in a rectangular beam after being subjected to plastic deformation under pure bending and then unloaded

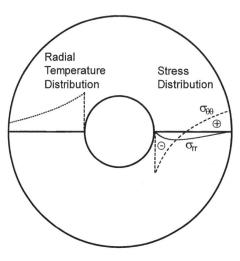

Fig. 3 Multiaxial stress distribution in uniform-thickness disk for a radial temperature profile typically associated with cooling. σ_{rr}, radial residual stress; $\sigma_{\theta\theta}$, circumferential stress

negligible, by way of Hooke's law, and the resulting stress does not change with time. Despite this simplifying assumption, this technique provides a good first approximation to the limiting solution. Additional examples with approximate solutions may be found in a number of references (Ref 14 to 18).

Advanced Solution Techniques

This section describes more advanced techniques that are applicable to transient creep, advanced constitutive models, and complicated stress and temperature loading histories. In addition, the next example highlights the difference between a steady-state and transient creep analysis solution using a spinning disk that creeps at elevated temperature.

Finite-Difference Method for Solution of Idealized Problems

In this section, another simplification is removed from the previous analyses to allow for transient creep response, sometimes referred to in the literature as including elastic strains. Furthermore, this section provides a comparison of the resulting stress state using another example of steady creep and transient creep analysis from the literature.

This example uses the uniform-thickness (h) disk geometry shown in Fig. 3, with an inner and outer radius (r) of 31.8 and 152.4 mm, respectively. The disk was spun at a constant angular velocity (ω) of 15,000 revolutions per minute and an elevated temperature of 538 °C. The analysis highlights the relaxation and redistribution of circumferential stresses in a spinning disk that creeps at elevated temperature. This same approach can be used to model relaxation of residual stresses in a stationary disk with a temperature gradient as shown in Fig. 3. Details of the experimental work are found in Ref 19, while the description of the analysis herein follows Ref 14 and 17. In this particular example, a cylindrical coordinate system is adopted. The two strain compatibility equations can be combined into a single equation:

$$\varepsilon_r - \varepsilon_\theta = r\frac{d\varepsilon_\theta}{dr} \qquad \text{(Eq 6)}$$

relating the radial (ε_r) and circumferential (ε_θ) strains in a differential form. Furthermore, the equilibrium equation also takes a differential form:

$$\frac{d}{dr}(hr\sigma_r) - h\sigma_\theta + \rho\omega^2 r^2 h = 0 \qquad \text{(Eq 7)}$$

where radial (σ_r) and circumferential (σ_θ) stresses are a function of the density (ρ). The stress-strain constitutive equations are cast as:

$$\varepsilon_r = \frac{1}{E}(\sigma_r - \nu\sigma_\theta) + \alpha T + \varepsilon_r^C + \Delta\varepsilon_r^C \qquad \text{(Eq 8)}$$

$$\varepsilon_\theta = \frac{1}{E}(\sigma_\theta - \nu\sigma_r) + \alpha T + \varepsilon_\theta^C + \Delta\varepsilon_\theta^C \qquad \text{(Eq 9)}$$

$$\varepsilon_z = \frac{\nu}{E}(\sigma_r + \sigma_\theta) + \alpha T - \varepsilon_r^C - \varepsilon_\theta^C - \Delta\varepsilon_r^C - \Delta\varepsilon_\theta^C \qquad \text{(Eq 10)}$$

where the accumulated radial (ε_r^C), circumferential (ε_θ^C), and axial (ε_z^C) creep strains and their respective increments ($\Delta\varepsilon_r^C$, $\Delta\varepsilon_\theta^C$) are summed with the elastic and thermal strains (αT). The rotating disk has a multiaxial stress state; hence, a von Mises effective stress (σ_e) and an effective creep strain increment ($\Delta\varepsilon_e^C$), which are respectively:

$$\sigma_e = \sqrt{\sigma_r^2 + \sigma_\theta^2 - \sigma_r\sigma_\theta} \qquad \text{(Eq 11)}$$

$$\Delta\varepsilon_e^C = \frac{2}{\sqrt{3}}\sqrt{(\Delta\varepsilon_r^C)^2 + (\Delta\varepsilon_\theta^C)^2 + \Delta\varepsilon_r^C\Delta\varepsilon_\theta^C} \qquad \text{(Eq 12)}$$

are used in the creep rate calculations, which typically take the form of a power-law model:

$$\frac{\Delta\varepsilon_e^C}{\Delta t} = B\sigma_e^n \qquad \text{(Eq 13)}$$

Furthermore, it is assumed that the Prandtl-Reuss plastic flow relation is applicable to creep flow, so that individual increments in creep strain may be computed as:

$$\Delta\varepsilon_r^C = \frac{\Delta\varepsilon_e^C}{2\sigma_e}(2\sigma_r - \sigma_\theta) \qquad \text{(Eq 14)}$$

$$\Delta\varepsilon_\theta^C = \frac{\Delta\varepsilon_e^C}{2\sigma_e}(2\sigma_\theta - \sigma_r) \qquad \text{(Eq 15)}$$

Finally, hydrostatic stresses are assumed to have negligible effect on yield behavior, and during plastic and creep flow, the material is incompressible, which results in the equation for the axial strain increment:

$$\Delta\varepsilon_z^C = -(\Delta\varepsilon_r^C + \Delta\varepsilon_\theta^C) \qquad \text{(Eq 16)}$$

The solution of these equations can be solved using either the integral equation method or the finite-difference method. A detailed description of both solution procedures for this example problem may be found in Ref 17 and therefore is not be repeated here. The solution to this problem has been solved using a power-law

creep model with transient creep analysis and steady creep, as shown in Fig. 5. The necessary material properties for this analysis are the elastic modulus, Poisson's ratio, and secondary creep rate as a function of applied stress that follows the power-law relation of Eq 13. The plot displays the circumferential stress at the inner radius (bore) and outer radius (rim) versus time. The solid lines are the solution for a transient creep analysis, while the dotted lines correspond to the steady creep analysis. The bore exhibits a significant decrease in stress starting from a peak value of 370 MPa to a lower value of 200 MPa. In contrast, the rim exhibits an increase in stress from the lowest value of 100 MPa up to 150 MPa. The steady creep analysis, shown as dotted lines, displays bore and rim stresses of 200 and 150 MPa, respectively. In this particular example, the steady creep solution significantly underpredicts the initial tensile stress at the bore and overpredicts the initial tensile stress at the rim. After approximately 10 h, the two solution methods display agreement. After 100 h of creep deformation, with the disk radius increasing and the thickness decreasing, the transient creep solution displays an increase in the circumferential stress at both the bore and the rim. Experimental strain results from Ref 19 show good agreement with the transient creep analysis.

The problem described previously has been solved by both the integral equation method and the finite-difference method (Ref 17). The integral equation method is somewhat less preferred because it is difficult to incorporate variations in material properties and dimensions as a function of coordinate position in the disk. The finite-difference method works well for this simple geometry. However, for complicated geometries, the finite-element method is preferred over the finite-difference method. Furthermore, for complicated material behavior, geometric nonlinearity, and coupled physics problems, commercial finite-element software packages are the best choice.

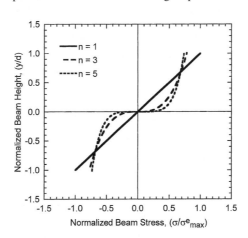

Fig. 4 Stress distributions in a rectangular beam for various exponents (n) of the power-law creep model obtained using the reference stress method

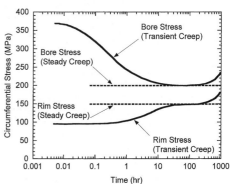

Fig. 5 Stress relaxation predictions in a spinning disk of uniform thickness at elevated temperature. Source: Data from Ref 17

Finite-Element Method for Solution of Complex Problems

Ultimately, the complexity of the physical shape of engineering components and the material models employed in stress-relief simulation require advanced analysis tools such as finite-element (FE) modeling. Most of the commercially available FE applications have some type of aforementioned constitutive model available for complex thermal and mechanical loading history analysis. This makes FE well suited to simulate complex mechanical processing steps that result in material and geometric nonlinearity with temperature- and strain-rate-dependent material properties. Unfortunately, these capabilities require a significant amount of material data, including temperature-, rate-, and time-dependent properties. Furthermore, many of the microstructural models require specialized tests to determine model parameters. In general, material input data that support a microstructural physics-based model must be representative of the temperatures, strain rates, and microstructural conditions of the stress-relief simulation.

One example of the use of FE is the simulation of stress evolution during quenching and the resulting retained residual stresses after cooling of a superalloy forging (Ref 7). The objective of this research was to develop numerical techniques that could reliably predict the retained residual stresses in components following heat treatment. These residual stresses are the result of trapped thermal stresses that developed upon cooling. The outside surface typically cools faster than the interior during the quenching process. Shown in Fig. 6 is the temperature-versus-time history at a point on the top surface and a point in the interior of the forging. Prior to quenching, the forging is at a uniform temperature of 1104 °C. Following the start of the quenching, the surface temperature drops sharply, while the interior temperature decreases at a much slower rate. After the surface temperature has dropped below 400 °C, the surface cooling rate decreases. At this point, the temperature difference between the surface and the interior is greater than 600 °C. Furthermore, the cooling rate on the surface is now less than the cooling rate in the interior. After approximately 50 min, the surface and interior have reached ambient temperature. However, it is the temperature gradients that exist during the cooling process that produce the thermal stresses in the forging.

Shown in Fig. 7 are the circumferential and radial stresses on the top surface and in the interior of the disk versus time. During the first minutes of rapid surface cooling, tensile circumferential and radial stresses develop on the surface as the forging cools and tries to contract around a hotter interior. In contrast, the interior develops compressive circumferential and radial stresses as it is squeezed by the contracting exterior. After approximately 5 min, when the surface cooling rate has decreased, the stresses start to decrease in magnitude and begin to change sign. The resulting surface residual stresses in both the circumferential

and radial directions are compressive, with a magnitude of approximately 1000 MPa. The interior residual stresses are both tensile, with magnitudes of 560 and 1350 MPa in the radial and circumferential directions, respectively. Experimental measurements using the hole-drilling technique with strain gages show good agreement with the model predictions (Ref 7). Additional heat treatment processes generally follow to reduce the magnitude of these residual stresses to lower levels.

The FE method is probably the most powerful and advanced tool available for stress-relief simulation. Most commercial FE packages provide the capability to perform stress analysis, heat transfer, and fluid flow for both static and dynamic solutions with material and geometric nonlinearities. Many of these codes include multiphysics analysis, coupled modeling capabilities, and can be augmented with user-written subroutines for custom constitutive material models. There are several textbooks available that describe the fundamentals and specific applications of the FE technique (Ref 20 to 26). Unfortunately, stress-relief simulation of metallic materials is a complex subject and is not usually adequately addressed in the literature. However, two sources (Ref 23, 26) provide a good starting point on the theoretical and numerical background for implementation of constitutive models in FE analysis. The challenge is selecting or developing an appropriate constitutive model that is applicable over the range of processing conditions. For example, the constitutive model should incorporate temperature-dependent material properties, a plasticity model with mixed nonlinear isotropic-kinematic hardening, and a creep/relaxation model that represents the active physical deformation mechanisms. The next section highlights many of the important material behavior aspects of a physics-based constitutive model that should be considered in a stress-relief simulation.

Transient Creep with Variable Stress History and Physics-Based Material Models

Although the topic of this article is focused on stress relief through relaxation of stresses, the experiments, physical understanding of

deformation mechanisms, and theoretical models are usually cast as creep models instead of relaxation models. Hence, the following discussion is focused on creep deformation mechanisms and their associated models.

The primary variables associated with creep deformation and creep rate are stress, temperature, and time. Much of the early work characterizing creep behavior was aimed at fitting empirical equations as a function of stress, temperature, and time. Furthermore, it is typically assumed that these equations are products separable into functions for stress and temperature. For variable stress and temperature loading conditions, a rate formulation must be adopted. The creep strain rate equation is often cast in either a time-hardening formulation:

$$\dot{\varepsilon}_c = F_1(\sigma)F_2(T)F_3(t) \qquad \text{(Eq 17)}$$

or a strain-hardening formulation:

$$\dot{\varepsilon}_c = F_1(\sigma)F_2(T)F_4(\varepsilon_c) \qquad \text{(Eq 18)}$$

Both formulations are state variable approaches that may be applied to complex thermal and mechanical loading histories. Figure 8 shows the creep strain versus time history under a simple stepped stress loading history using both the time- and strain-hardening approaches. The plots display the creep-strain-versus-time curves for the three constant stress (σ_1, σ_2, σ_3) histories along with the creep strain history for the variable stress history. These figures of time and strain hardening show how the creep strain history is constructed using the different constant stress histories for the two formulations. In most creep analyses, the strain-hardening approach is implemented.

A major element missing from the aforementioned creep models is the evolution of material microstructure with time or deformation history. It is often assumed that the microstructure, and hence the material properties, remain unchanged throughout the deformation history. Aspects of the material microstructure, such as grain size, dislocation structure, inclusions, vacancies, and so on, have an impact on the deformation rate. Both the time- and

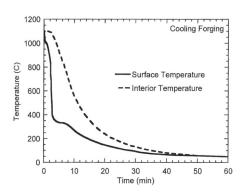

Fig. 6 Temperature history of a superalloy forging during quenching at points on the surface and in the interior. Source: Data from Ref 7

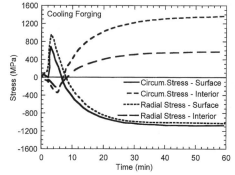

Fig. 7 Stress history simulation of a superalloy forging during quenching at points on the surface and in the interior. Source: Data from Ref 7

strain-hardening approaches are suitable if the creep rate is dominated by a single deformation mechanism. If multiple deformation mechanisms are active, or the dominant mechanism changes with thermal and mechanical loading history, simple time- and strain-hardening approaches fail to capture the loading response. Schoeck (Ref 27) presents a more general formulation for creep rate that accounts for multiple independent creep mechanisms in the form:

$$\dot{\varepsilon}_c = \sum_i f_i(\sigma, T, st) e^{-U_i(\sigma, T, st)/kT} \qquad \text{(Eq 19)}$$

where f_i is the creep rate function (i.e., creep mechanism), U_i is the activation energy for the creep mechanism, σ is the applied stress, T is the temperature, k is the Boltzmann constant, and st is the structure of the material.

Numerous models have been developed for the dominant creep mechanisms, such as glide and climb of dislocations and diffusion through grains and along grain boundaries. Nowick and Machlin (Ref 28) and Weertman (Ref 29) developed the early dislocation creep models to describe climb and glide deformation mechanisms, which gave rise to many of the commonly used exponential and hyperbolic sine formulations for creep rate. Interaction or competition between deformation mechanisms may become complex. The Bailey-Orowan equation (Ref 30) formulates competing strain-hardening and recovery mechanisms with steady-state creep as a balance between mechanisms. The θ-projection (Ref 30) approach describes primary and tertiary creep using exponential forms, with secondary or minimum creep rate as an inflection point in between. Initial approaches to represent material degradation under creep loading include

the continuum damage mechanics (CDM) approaches of Kachanov (Ref 31) and Rabotnov (Ref 32) that incorporate a single damage parameter and associated evolution equation. The simple damage parameters in the CDM approach have been replaced by specific degradation models representing mechanisms such as cavity nucleation and growth, subgrain coarsening, multiplication of mobile dislocations, and thermally and environmentally driven mechanisms (Ref 33, 34). A number of modeling approaches (Ref 35 to 41) have been developed to account for the combined contributions of plasticity and creep deformation. The trend has been to incorporate plasticity and creep into a single, unified, inelastic model. These unified models have evolved to include complex nonlinear hardening rules to capture the Bauschinger effect, cyclic hardening or softening, and residual stress relaxation. The selection of a constitutive model for stress-relief simulation is dependent on a number of factors. However, the critical requirements are that the correct physical deformation mechanisms be represented in the model and that material data or model constants required by the model are available.

Stress-Relief Simulation

The previous sections describe the different modeling approaches in order of increasing modeling sophistication. As the sophistication of the analysis increases, the analyst is required to have a greater knowledge of materials behavior, more experience with modeling tools, and generally more material data for the model. Furthermore, it becomes more challenging to interpret and analyze the validity of the

simulation results. The late Professor Richard W. Hamming (Ref 42) stated the following as the motto of his book on numerical methods: "The purpose of computing is insight, not numbers." He went on to emphasize that knowing what to do with the answer is more important than the answer itself. In regard to stress-relief simulation, the purpose of computing is to understand which material parameters, boundary conditions, and model variables drive stress relaxation and ultimately result in minimizing the retained residual stresses.

A stress-relief simulation will most likely incorporate a complicated geometric body, temperature-dependent material properties, physics-based models of deformation mechanisms, a variable thermomechanical loading history, and a nonlinear solution. Many of the commercial FE codes provide design optimization tools to evaluate the sensitivity of inputs (dimensions, material properties, boundary conditions, applied loads, etc.) on analysis results. These design tools provide automated methods to execute parametric studies that can be used to explore potential changes in processing parameters to minimize residual stresses. Furthermore, these design tools can identify strengths, weaknesses, and limitations of candidate materials models and guide potential experimental research for new processing approaches to minimize retained residual stresses.

Discussion/Summary

Stress-relief simulation can be as simple as applying empirical material models with analytical solutions to as complex as nonlinear time-dependent FE analysis with a physics-based microstructural model written into a user subroutine. The best approach is to start with a simple model and build experience and confidence in a solution before moving to more sophisticated methods. In most cases, development of modeling and simulation capabilities provides positive benefits to design, manufacture, and reliability of costly components.

ACKNOWLEDGMENT

The author gratefully acknowledges the support of the Air Force Research Laboratory, Materials and Manufacturing Directorate (AFRL/RXLMN), Wright-Patterson Air Force Base, Ohio, under on-site contract FA8650-04-C-5200.

REFERENCES

1. D. Furrer and H. Fecht, Ni-Based Superalloys for Turbine Disks, *J. Mater.*, Jan 1999, p 14–17
2. D.R. Mack, Measurement of Residual Stress in Disks from Turbine-Rotor Forgings, *Exp. Mech.*, Vol 2 (No. 5), May 1962, p 155–158

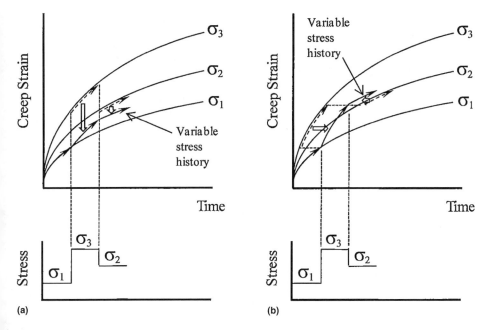

Fig. 8 Strain history paths for (a) time-hardening and (b) strain-hardening approaches for variable stress histories

3. P.I. Rusin and V.M. Shapkin, Residual Stresses in Ferritic Wrought Iron Quenched after High-Frequency Heating, *Met. Sci. Heat Treat.*, Vol 6 (No. 7), July 1964, p 454–456

4. N.G. Ivanova and N.A. Moskaleva, Residual Stresses in Centrifugally-Cast Iron Pipe, *Met. Sci. Heat Treat.*, Vol 14 (No. 8), Aug 1972, p 712–714

5. M.W. Joerms, Calculation of Residual Stresses in Railroad Rails and Wheels from Saw Cut Displacement, *Proceedings of ASM's Conference on Residual Stress—In Design, Process and Materials Selection,* W.B. Young, Ed., April 1987 (Cincinnati, OH), p 205–209

6. A.V. Clark, R.E. Schramm, H. Fukuoka, and D.V. Mitrakovic, Ultrasonic Characterization of Residual Stress and Flaws in Cast Steel Railroad Wheels, *IEEE Ultrasonics Symposium,* 1987, p 1079–1082

7. R.A. Wallis and I.W. Craighead, Predicting Residual Stresses in Gas Turbine Components, *J. Mater.,* Oct 1995, p 69–71

8. M.P. Mungi, S.D. Rasane, and P.M. Dixit, Residual Stresses in Cold Axisymmetric Forging, *J. Mater. Proc. Technol.,* Vol 142 (No. 1), Nov 2003, p 256–266

9. D. Greving, M. Gorelick, and H. Kington, Manufacturing Related Residual Stresses and Turbine Disk Life Prediction, *American Institute of Physics, Conf. Proceedings,* Vol 760, April 2005, p 1339–1346

10. F.P. Beer and E.R. Johnston, Jr., *Mechanics of Materials,* McGraw-Hill Book Company, 1981, p 187–189

11. R.D. Cook and W.C. Young, *Advanced Mechanics of Materials,* Macmillan Publishing Company, 1985, p 112–114

12. J. Chakrabarty, *Theory of Plasticity,* McGraw-Hill Book Company, 1987, p 359–363

13. C.R. Soderberg, Interpretation of Creep Tests on Tubes, *Trans. ASME,* Vol 63, 1941, p 737–748

14. H. Kraus, *Creep Analysis,* John Wiley & Sons, Inc., 1980

15. J.T. Boyle and J. Spence, *Stress Analysis for Creep,* Butterworth & Co., 1983

16. R.K. Penny and D.L. Marriot, *Design for Creep,* 2nd ed., Chapman & Hall, 1995

17. A. Mendelson, *Plasticity: Theory and Applications,* The Macmillan Company, 1968

18. J. Skrzypek, *Plasticity and Creep Theory, Examples and Problems,* CRC Press, 1993

19. A.M. Wahl, G.A. Sankey, M.J. Manjoine, and E. Shoemaker, Creep Tests of Rotating Disks at Elevated Temperature and Comparison with Theory, *J. Appl. Mech.,* Vol 21, 1954, p 222–235

20. K.J. Bathe and E.L. Wilson, *Numerical Methods in Finite Element Analysis,* Prentice-Hall, 1976

21. O.C. Zienkiewicz, *The Finite Element Method,* McGraw-Hill, 1977

22. E. Hinton and D.R.J. Owen, *Finite Element Programming,* Academic Press, 1977

23. D.R.J. Owen and E. Hinton, *Finite Elements in Plasticity: Theory and Practice,* Pineridge Press, Ltd., 1980

24. J.E. Akin, *Application and Implementation of Finite Element Methods,* Academic Press, 1982

25. R. Cook, D. Malkus, M. Plesha, and R. Witt, *Concepts and Applications of Finite Element Analysis,* 4th ed., John Wiley & Sons, 2001

26. F. Dunne and N. Petrinic, *Introduction to Computational Plasticity,* Oxford University Press, New York, NY, 2005

27. G. Schoeck, Theories of Creep, *Mechanical Behavior of Materials at Elevated Temperatures,* J.E. Dorn, Ed., McGraw-Hill Book Company, Inc., 1961, p 79–107

28. A.S. Nowick and E.S. Machlin, "Quantitative Treatment of the Creep of Metals by Dislocation and Rate Process Theories," Technical Note 1039, National Advisory Committee for Aeronautics (NACA), 1946

29. J. Weertman, Theory of Steady-State Creep Based on Dislocation Climb, *J. App. Phys.,* Vol 26 (No. 10), Oct 1955, p 1213–1217

30. R.W. Evans and B. Wilshire, *Creep of Metals and Alloys,* The Institute of Metals, 1985

31. L.M. Kachanov, Time of the Rupture Process under Creep Conditions, *Izv. Akad. Nauk. SSR, Otd Tekh., Nauk,* No. 8, 1958, p 26–31

32. Y.N. Rabotnov, *Creep Problems in Structural Members,* North Holland, Amsterdam, 1969

33. M.F. Ashby, A First Report on Deformation-Mechanism Maps, *Acta Metall.,* Vol 20, 1972, p 887–897

34. M. McLean and B.F. Dyson, Modeling the Effects of Damage and Microstructural Evolution on the Creep Behavior of Engineering Alloys, *J. Eng. Mater. Technol.,* Vol 122, 2000, p 273–278

35. S.R. Bodner, Constitutive Equations for Elastic-Viscoplastic Strain-Hardening Materials, *J. App. Mech., Trans. ASME,* June 1975, p 385–389

36. A.K. Miller, An Inelastic Constitutive Model for Monotonic, Cyclic, Creep Deformation: Part I—Equations Development and Analytical Procedures; Part II—Applications to Type 304 Stainless Steel, *J. Eng. Mater. Technol., Trans. ASME,* Vol 96, 1976, p 97–113

37. D.N. Robinson, "A Unified Creep-Plasticity Model for Structural Metals at High Temperature," ORNL TM-5969, Oct 1978

38. K. Walker, "Research and Development Program for Nonlinear Structural Modeling with Advanced Time-Temperature Dependent Constitutive Relationships," NASA Contractor Report, NASA-CR-165533, 1981

39. D.R. Sanders, "A Comparison of Several Creep Constitutive Theories for the Prediction of Elastic-Plastic-Creep Response and Their Application to Finite Element Analysis," Ph.D. thesis, Texas A&M University, 1986

40. V.G. Ramaswamy, "A Constitutive Model for the Inelastic Multiaxial Cyclic Response of a Nickel Base Superalloy René 80," NASA Contractor Report 3998, 1986

41. J.L. Chaboche, Continuum Damage Mechanics: Part I—General Concepts; Part II—Damage Growth, Crack Initiation, and Crack Growth, *J. App. Mech.,* Vol 55, 1988, p 59–72

42. R. Hamming, *Numerical Methods for Scientists and Engineers,* McGraw-Hill, 1962, p 702–710

ASM Handbook, Volume 22B, *Metals Process Simulation*
D.U. Furrer and S.L. Semiatin, editors

Copyright © 2010, ASM International®
All rights reserved.
www.asminternational.org

Uncertainty Management in Materials Design and Analysis

Janet K. Allen, University of Oklahoma

COMPUTATIONAL MATERIALS DESIGN is in its infancy. There is tremendous scope for the multiscale computational design of materials. However, in spite of ever-increasing computational power, many approximations and simplifications must be made to compute results. In this article, an approach to managing the uncertainty present in materials design is presented. First, it is necessary to understand the sources of uncertainty—the region of space considered, the number and independence of input variables, and the way they are sampled—followed by creating surrogate models and studying error propagation among several computations. Then, the opportunities that a designer has to control or manage this uncertainty are presented.

Frame of Reference

The pace of industry has led to a demand for more rapid materials development, and this has led to replacing physical experimentation with computer experimentation. Advances over the last few decades in modeling materials, combined with the rapid increase in computational power, make computer experimentation increasingly feasible. However, uncertainty—and the accompanying lack of accuracy—is one of the greatest problems in the computational development of materials, that is, in the design of materials.

Uncertainty is defined as unpredictability, indeterminacy, or indefiniteness of the estimated amount or percentage by which an observed or calculated value may differ from the true value (Ref 1). There are a myriad of sources of uncertainty in materials themselves. The most obvious material heterogeneity is the variation in microstructure, but there are other sources of uncertainty. Further, although the resolution of the various forms of physical observation of material properties is improving, measurement variability introduces an additional degree of uncertainty. Modeling and simulation of physical behavior introduces more uncertainty, and often, in an effort to reduce computation time, the models themselves are simplified, either by creating surrogate models or by using simplifying assumptions. Again, this introduces uncertainty.

Approaches to Dealing with Uncertainty

Mitigating Uncertainty. Because uncertainty is inherent in materials, conceptually there are two ways of introducing it into computations. The first is to refine computations indefinitely in the costly effort to generate models that precisely represent the material. This is termed mitigating uncertainty.

In complex designs, this requires that all simulations contributing to the design be refined simultaneously. In practice, if several groups are working on different aspects of the design, there is the temptation for each group to continue to refine one portion of the design. However, when the overall problem is analyzed these refinements may be swamped by the uncertainty in a few calculations. Refining models can be a very expensive process, both in terms of modeling efforts and in terms of computations, so it is important not to waste resources on parts of the problem that do not contribute significantly to an improved solution.

Managing Uncertainty. The second approach to dealing with uncertainty is to manage uncertainty, that is, to recognize and tolerate uncertainty in specific simulations that contribute to the overall design and to evolve and refine models where precision is important. There can be uncertainty in the modeling methods, in the knowledge of constraints, or in the knowledge of the goals or objectives. Each must be dealt with individually and appropriately but not necessarily by removing uncertainty completely. A part of uncertainty management is to mitigate uncertainty selectively where there is the greatest benefit in terms of the solution proposed. In addition, it is wise to seek solutions that are robust: solutions that are insensitive to changing conditions and therefore are unaffected by uncertainty and thus reduce the required accuracy of the solution. In all cases, in the context of the specific design problem, a designer balances the accuracy (lack of uncertainty) in the results against the resources required to obtain that result. Thus, to successfully manage uncertainty, it is important to consider carefully the context and the time and money available for generating results. Of course, this is usually most successful if the designer is able to design a comprehensive plan for obtaining results.

Inductive and Deductive Approaches. So far, only uncertainty produced by a single experiment, model, or analysis has been considered. This is much more complex when moving away from traditional homogenization procedures and considering multiple analyses. Olson has developed a hierarchical model of interactions that produce material performance (Fig. 1). This starts with materials processing methods, the (micro)structure that they develop, the material properties that the microstructure then yields, and the performance that the properties dictate. Taking this one step further, knowing materials performance is a step toward product design, which then involves a whole hierarchy of other activities. This is a bottom-up or deductive approach to creating materials. Alternatively, a designer may start from the top of the hierarchy and determine a desired performance, followed by identifying the properties that will yield this performance, the microstructure to yield properties, and the processing that creates the microstructure. This is an inductive approach. For either inductive or deductive

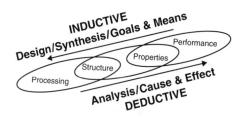

Fig. 1 Hierarchical diagram of materials design based on the work of Olson. Source: Ref 2

approaches, it is important to be able to manage the uncertainty both at single levels and as uncertainty propagates from level to level.

Whether considered from the top down inductively or the bottom up deductively, design problems such as these can either be solved sequentially or concurrently in an all-in-one approach. Much of the existing work has been done using all-in-one approaches. This has the disadvantage that any modification requires completely recomputing the entire design problem, which can be extremely costly when the uncertainty inherent in the material is modeled. In this article, a perspective on the capabilities and limitations of the various methods is offered, not the detailed mathematical foundations of the different methods. Ample references are provided to guide the interested reader. This is an extensively studied topic, and the references presented here are representative and not intended as an exhaustive collection of all work on the subject.

Whether materials development is viewed from an inductive or deductive perspective, uncertainty in the starting point is magnified as it passes through multiple computations. This leads to a situation in which minor variations (uncertainty) in input propagate through a series of stages with an ever-expanding amount of uncertainty. The situation is complicated by the fact that several models or simulations may be required to describe the material at any one of these levels.

Although it may be possible to take comprehensive measurements for every value in a physical sample, this is generally both expensive and time-consuming. Accordingly, a subset of possible measurements is evaluated or sampled, and a distribution of values for that property is computed. The basic approach is to construct approximations of the analysis codes that are efficient to run and yield insight into the functional relationship between x and y. If the true nature of the system behavior or the computer code is:

$$y = f(\boldsymbol{x})$$

then a model of the model or metamodel or surrogate model, $\hat{y}$, of the analysis code is:

$$\hat{y} = g(x)$$

These approximations then can replace the existing analysis code.

In what follows, a designer is assumed to have a comprehensive understanding of the desired accuracy of results and to be aware of the resources available; therefore, the focus in this article is to provide an understanding of opportunities for managing uncertainty and the decisions that influence the accuracy of the results. Because mitigating uncertainty is a part of uncertainty management, the early part of this article is on uncertainty mitigation. The starting place is to select the size of the region under consideration and to identify input variables that have the greatest impact on the results. Because many simulations are either more accurate or simpler

if the variables are independent, methods for identifying independent input variables are discussed. Next, methods for sampling and experimental design are presented. This section on uncertainty mitigation is followed by a discussion of uncertainty management, starting with an understanding of uncertainty propagation and multiscale robust design.

Input Data for Surrogate Modeling

Managing uncertainty starts with determining the context of the design problem and identifying options for uncertainty mitigation. The first important decision is to determine the size of the region that will be considered. Clearly, a large region will provide more scope to reach a good solution but also demands increased numbers of samples to be studied, and evaluating each sample can be costly.

The number of simulations required to create a surrogate model increases exponentially with the number of input variables (Ref 3). Some variables may have minimal effect on the output, and these may be set to constant values in further calculations, to the extent possible. Ideally, this is accomplished while also removing or reducing correlations between input variables. This procedure is called variable screening.

Variable Screening

Reducing the Number of Input Variables. One approach to variable reduction is to use a k-level factorial or fractional factorial design, where k is usually a low number, for example, 2 or 3. This is followed by an analysis of variance that indicates which variables account for most of the variance in output variables (Ref 4). Alternatively, fractional factorial designs may be computed iteratively to identify factors that significantly affect the output (Ref 5). Full and fractional factorial designs are discussed later in this article.

Several other approaches have been proposed. For example, in Ref 6, Daniel proposes a one-at-a-time procedure in which the effects of varying one variable at a time are monitored. Although Daniel's method alone is not effective for eliminating interactions among variables, Morris proposes a modification of this method that accounts for these interactions (Ref 7).

Sobel has developed the global sensitivity analysis for reducing the numbers of input variables by determining the total effect—the main effect and the interaction effects—of each variable and therefore determining those that can be ignored (Ref 8). For these calculations, the variances are usually determined by Monte Carlo simulations using a crude model that is usually assumed to be adequate for this purpose but not for an accurate description of system behavior.

A different approach is to group the variables by nondimensionalizing them, thus reducing the number of variables. This can be done systematically by using the Vaschy-Buckingham theorem

and guarantees that the minimum number of parameters is identified, which provides an exact representation of the output for a given design problem (Ref 9, 10). Because this method is a different approach from the screening procedures discussed earlier in this section, these methods may be used together to further enhance the reduction in the number of input variables.

Ensuring the Independence of Input Variables. Some methods of analyzing materials, for example, finite-element analyses, generate a large amount of data that is not necessarily independent. However, most of the methods for error propagation are based on the assumption that the input variables are independent. Proper orthogonal decomposition (POD), also known as Karhunen-Loeve expansion, may be used both to reduce the dimensionality of the input variables and to provide a set of input variables that are uncorrelated. The POD can reduce the number of input variables by several orders of magnitude (Ref 11).

Given a set of N simulation samples, POD identifies a low dimensional basis. If U is a field within the sampling domain, K is the number of basis vectors, and α_k is the POD coefficients on the K basis vectors, then:

$$U = \sum_{k=1}^{K} \alpha_k \Phi_k = \sum_{k=1}^{K} \langle U_k \Phi_k \rangle \; \Phi_k \; k = 1 \ldots K$$

where U is the vector representation of the field, and $\{\Phi_k\}_{k=1..K}$ are the basis vectors for the reduced-dimension, orthogonal basis (Ref 12).

Note that many of the techniques presented in this section may also be used to assess output dimensionality and create independent variables if the output of a simulation is to be used in further analyses. When the appropriate input variables have been identified, then samples of the input variables are selected to cover the possible spectrum of the input variables; these samples are produced by various experimental designs.

Experimental Designs

The various methods for design of experiments (DOE) have been developed to identify an efficient set of simulations and use these simulations to create a representation of the output of the simulation based on the distribution of input variables. Experimental designs are discussed in this section. Metamodeling approaches for using the results are discussed in the next section.

A Survey of Experimental Designs

A DOE represents a sequence of experiments to be performed, expressed in terms of factors (design variables) set at specified levels (predefined values). An experimental design is represented by a matrix X where the rows denote experiment runs, and the columns represent particular factor settings.

Full Factorial Design. The most basic DOE is a full factorial design in which two or more factors (variables) are considered. Each of these factors is assumed to cover a range of values, and those ranges are divided into levels. For each complete trial of the experiment, all possible combinations of the levels of the factors are considered. The most common designs are 2^k (for evaluating main effects and interactions) and 3^k (for evaluating main and quadratic effects and interactions) for k factors at 2 and 3 levels, respectively. The advantages of factorial designs are their flexibility and orthogonality. A flexible design is one that can be used for multiple models, thus permitting the possibility of sequential experimentation. An orthogonal design is one in which the input variables are uncorrelated, and the regression coefficients of the fitted linear model are also uncorrelated (Ref 13). More information about orthogonality and flexibility follows.

Fractional Factorial Design. The size of a full factorial experiment increases exponentially with the number of factors; this leads to an unmanageable number of experiments. Fractional factorial designs are used when experiments are costly and many factors are required. A fractional factorial design is a fraction of a full factorial design; the most common are $2^{(k-p)}$ designs in which the fraction is $1/2^{(p)}$. The reduction of the number of design points in a fractional factorial design is not without a price, however. The 2^3 full factorial design allows estimation of all main effects (x_1, x_2, x_3), all two-factor interactions (x_1x_2, x_1x_3, and x_2x_3), as well as the three-factor interaction ($x_1x_2x_3$). For a 2^{3-1} fractional factorial design, the main effects are aliased (or biased) with the two-factor interactions. Aliased effects cannot be estimated independently.

Often, 2^k and $2^{(k-p)}$ designs are used to identify or screen for important factors. Thus, the less important factors can be removed from consideration. When there are many factors, the sparsity-of-effects principle (Ref 2) can be invoked, and the system is assumed to be dominated by main effects and low-order interactions. In this case, two-level fractional factorial designs are used to screen factors to identify those with the greatest effects. The sparsity-of-effects principle is not always valid, however; Hunter (Ref 14) notes that every design provides aliased estimates. Quadratic and cubic effects bias the estimates of the mean and main effects when a two-level fractional factorial design is used.

One specific family of fractional factorial designs frequently used for screening is the two-level Plackett-Burman designs (Ref 15). These are used to study $k = n - 1$ factors in $n = 4m$ design points, where m is an integer. If interactions are negligible, these designs allow unbiased estimation of all main effects and require only one more design point than the number of factors; they also give the smallest possible variance (Ref 16). Myers et al. (Ref 4) present a more complete discussion of factorial designs and aliasing of effects. To estimate quadratic effects, 3^k or $3^{(k-p)}$ designs can be used, but those often require an unmanageable number of design points.

Often, it is desirable to use the smallest number of factor levels in a DOE. One common class of these designs is the Box-Behnken designs (Ref 17). They are formed by combining 2^k factorials with incomplete block designs. More information about central composite designs and Box-Behnken designs can be found in Ref 18.

Measures of Merit for Evaluating Experimental Designs

Selecting the appropriate design is essential for effective experimentation; the desire to gain as much information as possible about the response-factor relationships must be balanced against the cost of experimentation. Several measures of merit are available and useful for evaluating and comparing experimental designs based on the assumption of classical experimental design (not experimental designs for computer experiments).

Desirable Characteristics. To facilitate the efficient estimation of parameters, there are four desirable characteristics for an experimental design:

- Orthogonality
- Rotatability
- Robustness
- Blocking

A design is orthogonal if, for every pair of factors x_i and x_j, the sum of the cross products of the N design points is zero:

$$\sum_{u=1}^{N} x_{iu}x_{ju} = 0$$

In the late 1950s, Box and his coworkers demonstrated that rotatability and the minimization of bias from higher-order terms are the essential criteria for good forecasting. A design is rotatable if:

$$N \bullet \mathrm{Var}\left[\hat{y}(x)/\sigma^2\right]$$

has the same value at any two locations that are the same distance from the design center. Where σ is the standard deviation of the predicted value, $\hat{y}(x)$ is the surrogate function, Var is the variance, and N is the number of design points.

For a first-order model, the estimates of all coefficients will have minimum variance if the design is configured so that:

$$\sum_{u=1}^{N} x_{iu}^2 = N$$

and the variance of predictions, $\hat{y}$, also has constant variance at a fixed distance from the center of the design; then the design is also rotatable.

If the design still performs well when the assumptions are violated, then the experimental design is said to be robust; it can either be model robust (Ref 19, 20), distribution robust (Ref 21, 22), or both (Ref 13). Note that this is a different use of the term *robust* than robust design of the entire system.

In blocking, the experiment is designed so that the experiments are run in small groups or blocks so that the error within the blocks can be reduced.

Qian et al. (Ref 23) recognize that often a designer has available both approximate experiments and more accurate and costly experiments, and they propose an experimental design that can combine both types of data effectively.

A saturated design is one in which the number of design points is equal to one more than the number of factor effects to be estimated. Saturated fractional factorial designs allow unbiased estimation of all main effects with the smallest possible variance and size (Ref 16). The most common examples of saturated designs are the Plackett-Burman two-level design and Taguchi's orthogonal arrays (Ref 24, 25). In many cases, the primary concern in the design of an experiment is its size.

Unsaturated Design. Most designs are unsaturated in that they contain at least two more design points than the number of factors. For estimating second-order effects, small composite designs have been developed to reduce the number of required design points. A small composite design is saturated if the number of design points is $2k + k(k - 1)/2 + 1$ (the number of coefficients to be estimated for a full quadratic model). Myers et al. (Ref 4) note that these designs may not always be good; additional comments on small composite designs can be found in Ref 26 and 27.

Supersaturated Design. In supersaturated designs, the number of design points is less than or equal to the number of factors (Ref 28). It is desirable to use unsaturated designs for predictive models, unless running the necessary experiments is prohibitively expensive.

When comparing experiments based on the number of design points and the information obtained, the D-optimal and D-efficiency statistics are often used, but these statistics depend on the assumption that the underlying model is stochastic. Thus, the use of D-optimality or D-efficiency for computer experiments is questionable (Ref 29).

Sampling. Initially, a designer may not be able to determine a region of interest; therefore, experiments may be performed sequentially. The preliminary experiments are used to either limit the size of the region explored or the number of variables considered by focusing on the most significant variables, using a preliminary screening technique (Ref 18). Sequential sampling can reduce computational costs by allowing the designer to focus on those factors that are most influential for the final design.

Another way of reducing the number of samples required is to reduce the size of the space explored. Several heuristic approaches to reducing the design space have been proposed (Ref 30–33). Wujek and

Renaud (Ref 34, 35) developed and compared several strategies for focusing the function approximation in a more desirable portion of the design space. Trust regions are used to focus surrogate models on successively smaller regions of design space (Ref 36).

The methods for the design of experiments discussed so far are suitable for physical experiments, and they have been used occasionally for computer experiments; however, some methods of experimental design have been developed specifically for computer experiments, as discussed in the following section.

Computer Experimentation

The methods described in the section "Experimental Designs" are founded on the assumption that physical experiments are being performed and that there is measurement error when each value is determined. However, computational experiments are typically deterministic, in that a specific input leads to a single output each and every time the computation is performed, and errors are not due to random effects. Therefore, it is preferable to use experimental designs that are not based on these assumptions about the underlying model(s) (Ref 37), although the classical design methods described previously can be effective for computer experiments in which there is a large sample size (Ref 38). Additional recommendations for experimental designs for use in computer-based experimentation are given here.

The Monte Carlo sampling method is popular in industry. Because it involves repeated sampling of the output based on randomly generated input, it is suitable for computer experimentation, but it is very inefficient. However, it can be useful for situations in which there is a high degree of correlation among the variables. Several variations of Monte Carlo methods have been proposed (Ref 39). Monte Carlo procedures with large numbers of samples have been used to determine the exact shape of functions, but again, this is a very costly procedure.

Space-Filling Designs. For sampling deterministic computer experiments, many researchers recommend the use of space-filling designs that, in some sense, treat all regions of the design space nearly uniformly (Ref 18). This is desirable if the designer does not know where the regions of interest lie. Further, most space-filling designs do not replicate runs. Fang et al. (Ref 40) present various algorithms for constructing uniform designs and several methods for assessing uniformity. The most frequently recommended space-filling designs are orthogonal arrays (Ref 24, 41, 42), Latin hypercube designs (Ref 43–45), Hammersley sequences (Ref 46, 47), and uniform designs (Ref 48).

Park (Ref 44) discusses optimal Latin hypercube designs for computer experiments that either minimize the integrated mean square error or maximize entropy and are used to spread the points out over the design region. Shewry and Wynn (Ref 49) and Currin et al.

(Ref 50) also use the maximum entropy principle to develop designs for computer experiments. Morris and Mitchell (Ref 51) propose maximin distance designs found within the class of Latin hypercube arrangements, because they "offer a compromise between the entropy/maximin criterion, and good projective properties in each dimension." Simpson et al. (Ref 52) compare several space-filling design methods; they conclude that uniform designs provide good sampling for accurate approximations for different sample sizes.

Bayesian Approach. If the designer has some knowledge about the system under consideration, Chaloner and Verdinelli (Ref 53) recommend a Bayesian approach to experimental design, in which the designer's knowledge is used to influence the choice of experiment, experimental design, and sample size.

Model Fitting and Model Choices

After selecting an appropriate experimental design and performing the necessary simulations, the next step is to choose an appropriate model-fitting method. There have been several reviews of model-fitting methods. A brief description is given here; for more information, see Ref 39 and 54 to 56.

Although some surrogate modeling methods have specific ways of fitting the data, the standard method for fitting a model to experimental data is based on the assumption that the expectation of the error is zero and therefore that the fit is unbiased. Usually, a root mean square fit of the data is performed, but the partial least-squares method for fitting the data has also been proposed (Ref 57).

Using any fitting method, however, there is always room for error at any specific point, so it can happen that a solution found using the surrogate is infeasible when compared with an accurate assessment at the proposed point. This can be a serious difficulty, but it can be mitigated by artificially demanding a higher level of performance from the surrogate, either by adding a constant to the required prediction response to create a safety margin or by multiplying the prediction response by a constant to create a safety factor (Ref 58, 59). Alternatively, the surrogate model may be biased by requiring the model to fit the data conservatively by demanding that the fit attain a high level of significance (Ref 60, 61). Other methods have been proposed; however, Picheny (Ref 62) finds that none of the proposed approaches are clearly superior and therefore recommends the simplest method, the safety margin approach.

Methods for Modeling

Response-Surface Method. There are many methods for surrogate modeling; probably the best known is the response-surface method (Ref 4). Given a response, y, and a vector of independent factors, x, influencing y, the relationship between y and x is:

$$y = f(x) + \varepsilon$$

where ε, the random error, is assumed to be normally distributed with mean zero and standard deviation, σ. Predicted values are obtained by fitting the curve $\hat{y} = f(x)$. Usually, the functions $f(x)$ are low-order polynomials; however, higher-order polynomials can be used to fit surfaces with higher curvature. Note that this approach is based on the assumption that the data are noisy and that the fitted model is exact.

Kriging, an interpolation method, starts with the assumption that the data are exact and may be fit with a Gaussian process:

$$y(x) = f(x) + Z(x)$$

where $y(x)$ is the unknown function, $f(x)$ is a known polynomial function of x, and $Z(x)$ is the realization of a normally distributed Gaussian random process with mean zero, variance σ^2, and nonzero covariance (Ref 63–65). The designer is offered a choice of Gaussian process models and, depending on the choices made, kriging can either honor the data and give an exact interpolation of the data or smooth the data with an inexact interpolation (Ref 66). Especially for nonlinear functions, kriging can often provide a better model of the data than response-surface models. Local regression is also a method for fitting local regions of the data. It uses a moving weighted least-squares approach (Ref 67).

Kernels. Instead of fitting the data to a single function, several methods have been proposed that rely on kernels or bases containing multiple functions. If a multiquadratic kernel is used, the radial basis function scheme can interpolate sample points and is also fairly easy to compute (Ref 68, 69).

Neural Networks. Some methods rely on "training" the model to fit a set of data. A neural network, for example, is composed of computational neurons that are multiple linear regression models with a nonlinear (usually sigmoidal) transformation on y. If the inputs to each neuron are $\{x_1, x_2, \ldots, x_n\}$, and if the regression coefficients have the weights w_i, then the output, y, becomes:

$$y = \frac{1}{1 + e^{-\eta/T}}$$

where $\eta = \Sigma w_i x_i + \beta$ gives β, the bias value of a neuron; and T is the slope parameter of the sigmoidal function defined by the user. The neural network is then created by assembling the neurons to create an architecture. If the architecture is large enough, a neural network can approximate almost any set of training data, although the computational costs can be exorbitant (Ref 70). For more details of neural-network modeling, see Ref 71 and 72.

The support vector regression method is used to identify feasible and infeasible regions (Ref 73). It also requires training on a data set and hence can be computationally intensive, especially if several variables are considered. It has been compared to other methods for creating surrogate models for a series of simulations and is claimed to have greater accuracy than kriging, multivariate regression splines (Ref 74), or radial basis functions (Ref 70).

Multiple Surrogates. Regardless of how good a fit a single model is for the data, one way of obtaining a better understanding of the function that is being modeled is by fitting several surrogate models to the same data. Multiple surrogates add information, because the calculations are based on different assumptions, and the surrogates can vary considerably in their accuracy in different regions of the design space. Viana et al. (Ref 75) argue that using multiple surrogates acts as an insurance policy against poorly fitted data without substantially adding to the computational costs.

When choosing a method for developing a surrogate model, it is important to consider the way it will be used. Wang and Shan (Ref 39) note that some of the areas where surrogate models can play an important role include:

- Model approximation for computationally expensive models
- Design space exploration
- Problem formulation
- Support for optimization. There are several different types of optimization schemes—global, multiobjective, stochastic, multidisciplinary design, and so on—and surrogate models can be introduced into them in various ways to reduce computational costs.

Surrogate models can also provide a good approximation of noisy data. However, if the surrogates are to be used for robust design when variance is an important consideration, low-order response surfaces should not be used (Ref 76) because they are often inaccurate for the computation of variance. Here, the surrogate models are used for error propagation among a series of computations, and, at this point, the focus shifts from opportunities for mitigating uncertainty to uncertainty management.

Management of Uncertainty— Uncertainty Propagation

When a model or an analysis has an uncertain input, this error is then propagated through to the solution. Of particular interest for materials are the analyses relating the process-microstructure-properties-performance, as shown in Fig. 1; however, this difficulty is not unique to the study of materials. First, a general method for error propagation is presented, and then a method that is especially useful for multilevel hierarchical systems, the inductive design exploration method, is presented.

When selecting a procedure for propagating uncertainty, several factors enter into the decision:

- The degree of accuracy of knowledge about the distributions of input variables as well as the anticipated use of the knowledge about the output and the degree of accuracy required are important factors. For example, if the results are to be used in the early stages of design, accuracy is typically relaxed in favor of the rapid exploration of possible outputs. As regions of interest are identified, then more accurate simulations are required.
- If the input variables are correlated, then either they must be transformed into uncorrelated variables, or simulation-based methods must be used.
- Some methods are preferred for computations involving highly nonlinear performance functions.
- The intended use for the information being calculated. If error propagation is required for the robustness computations, it may only be necessary to compute the first and second moments of the output distributions (Ref 77). However, if knowledge about reliability is required, then more comprehensive knowledge of the output distribution is needed (Ref 78, 79).
- Computational costs also affects the choice of propagation methods.

Of course, regardless of input distributions, the output distribution may always be determined similarly to the way the shape of the input is determined.

Lee and Chen (Ref 80) identify five categories of the more popular types of uncertainty propagation methods:

- *Simulation-based methods:* For example, Monte Carlo sampling (Ref 81–83), importance sampling (Ref 84, 85), and adaptive sampling (Ref 86). To generalize, these methods require extensive computations and can be costly.
- *Local expansion methods:* For example, perturbation methods and Taylor series expansions (Ref 81–83, 87). In general, these methods are of limited value when there are multiple uncertain inputs and the outputs are highly nonlinear.
- *Most probable point methods:* For example, the first-order reliability and second-order reliability methods (Ref 88, 89).
- *Functional expansion-based methods:* Polynomial chaos expansions are classified here, along with Newman expansions (Ref 87, 90, 91).
- *Numerical integration-based methods:* This group includes dimensional reduction, by which a multidimensional moment integral is approximated by reduced dimensional integrals based on the additive decomposition of the performance function (Ref 92–95).

While it is clear that the simulation-based methods, such as Monte Carlo simulation, most accurately reflect performance, they are also extremely costly.

Lee and Chen (Ref 80) evaluate some of these methods based on their performance on selected design problems. Although their results are not comprehensive, they provide some insight into the selection of methods. They note that for situations in which there are no interactions among variables, univariate dimensional reduction is both rapid and reasonably accurate. However, if there are interactions among the variables, then the higher-order moments (skewness and kurtosis) of the variables are susceptible to error, but the lower-order moments (mean and variance) retain their accuracy. Univariate dimensional reduction also performs well for non-normal inputs. Polynomial chaos expansions also perform adequately for finding the moments, but accuracy depends on the method used to find the coefficients.

If probability density functions and probability estimations are needed, the entire spectrum of performance must be computed. This is especially important for error propagation in a chain of calculations, and therefore, Lee and Chen recommend polynomial chaos expansion for design problems of this sort (Ref 80).

The univariate dimensional reduction method is very efficient compared with the other methods. As the number of random variables considered increases, the computational costs increase linearly. However, this is accomplished by ignoring interaction effects among the variables, which, if they are present, reduces the accuracy of the third and fourth moment calculations. When used in design, these methods for error propagation are of limited value. Design and materials design, in particular, are usually multilevel/multiscale problems in which a goal is specified, and the process is inductive. In other words, the performance output is specified, and values of input variables are determined that can attain this performance.

Computational frameworks to accomplish these analyses are not the focus of this article, but further information about them is available in Ref 96 and in the article "Design Optimization Methodologies" in this Volume.

An entirely different approach to managing uncertainty is to use robust computations. That is, instead of searching for optimal solutions, search for solutions that are insensitive to variations (Ref 46, 97, 98), in other words, solutions that occur in a "flat" region of the design space.

Choi et al. (Ref 99) incorporate considerations of robustness into the inductive design exploration method. They focus on chains of computations and have developed a modular approach to the computations. This is especially important in materials design because, repeatedly, large regions of design space must be explored, and, as a design evolves, successively more accurate computations may be inserted in the computation scheme without the need for reconfiguring the software. Consider a process that proceeds deductively: $x \rightarrow y \rightarrow z$. First, identify the rough regions of interest, the design spaces. To compute the inductive process, $z \rightarrow y \rightarrow x$, the designer then discretizes the design spaces x and y. In each of these design spaces, the discretized points are mapped onto the higher-level design space; x is mapped onto y, and y is mapped onto z, being sure to include all the known amounts of uncertainty. This mapping can be run in parallel. Store this information in a database. Using this information, the designer is now able to perform an inductive design procedure, starting at the highest level, level z, to identify a region of interest and then using the database to retrieve information about the mapping from z to y and then to level x. This method has been demonstrated for the design of a multifunctional structural material (Ref 96, 99).

The focus of this article is on the management of uncertainty in the computational design of materials. This is complex and closely tied to the specific research under consideration. There are multiple decision points at which a designer's choices determine the accuracy of the results and the expense of obtaining them (Table 1). As with other decisions, the considerations must be trade-offs between risk and the cost of obtaining the results (both computational/programming time and the costs of research/development of the underlying physical models). These trade-offs are particularly difficult to make because the decisions are made under considerable uncertainty.

In general, the earlier the uncertainty is introduced, the greater its impact will be on the final result. Therefore, it is especially important to focus on choosing a tolerable amount of uncertainty in the early stages of experimentation, and it is advisable to plan the course of experimentation in advance. Panchal et al. (Ref 100–102) recommend controlling the process of design using a value-of-information metric based on the probable impact that model refinement will have on the design decisions being made. This has been demonstrated for the design of a blast-resistant panel (Ref 100).

Appendix 1—Glossary of Terms

Blocking. A schedule for conducting treatment combinations in an experimental study such that any effects on the experimental results due to a known change in raw materials, processes, operators, machines, and so on become concentrated in the levels of the blocking variable. The reason for blocking is to isolate a systematic effect and prevent it from obscuring the main effects. Adapted from Ref 103

Center points. Points at the center value of all factor ranges. Adapted from Ref 103

Deduction. The process of starting at the lowest level and reasoning toward a conclusion. In the context of materials design, starting with the chemical composition and processing conditions and moving toward a desired final performance through the microstructure and properties developed. Compare this process with *induction*. Adapted from Ref 1

Deterministic experiment. An experiment in which, when the experiment is repeated with identical input, the same output is obtained. There is no randomness associated in the development of the answer.

Induction. The process of finding a desired property by assuming that that property is true and then finding the necessary conditions under which it must be true. As presented in this article, induction is the process of assuming the desired material performance is true and then finding the properties, microstructure, and processing conditions that are required. Compare this process with *deduction*. Adapted from Ref 1

Kriging. An interpolation technique first used in geostatistics to predict the location of ore veins. It is named after D.G. Krige, a mining engineer from South Africa.

Kurtosis. A measure of whether a data distribution is peaked or flat relative to a normal distribution. Data sets with high kurtosis tend to have a distinct peak near the mean and decline rather rapidly. Data sets with low kurtosis tend to have a flat top near the mean, rather than a sharp peak. Adapted from Ref 103

Latin hypercube design. An experimental design consisting of n trials and for which each factor has n distinct levels. Usually, the factor levels are equally spaced. There is only one sample in each row and each column. Latin hypercube designs are especially useful for computer experiments. Adapted from Ref 103

Managing uncertainty. The process of balancing the needed accuracy of the solution against the cost of obtaining that solution.

Mitigating uncertainty. The process of removing uncertainty by refining the model(s), the simulation(s), or the input variables.

Monte Carlo methods. A class of computational methods that rely on repeated random sampling to compute results. This approach is most useful when it is impossible or infeasible to compute an exact result when studying systems with a large number of coupled parameters. Adapted from Ref 104

Orthogonality. An experimental design is orthogonal if the effects of any factor balance out (sum to zero) across the effects of the other factors. Adapted from Ref 103

Robust design. Designing products or processes so that they are minimally affected by uncertainty.

Rotatability. An experimental design is rotatable if the variance of the predicted response at any point x depends only on the distance of x from the design center point. A design with this property can be rotated around its center point without changing the prediction variance at x. Adapted from Ref 103

Skewness. A measure of symmetry or the lack of symmetry. A distribution is symmetric if it looks the same on the left and right of the center point. Adapted from Ref 103

Uncertainty. Uncertainty refers to unpredictability, indeterminacy, indefiniteness, or the estimated percentage by which an observed or calculated value may differ from the true value.

Variance. A measure of the squared dispersion of observed values or measurements expressed as a function of the sum of the squared deviations from the population mean or sample average. Adapted from Ref 105

ACKNOWLEDGMENT

The author gratefully acknowledges the suggestions made by Professor Farrokh Mistree on the development of this article.

REFERENCES

1. Dictionary.com, www.dictionary.com, accessed Feb 20 and March 31, 2010
2. G.B. Olson, Computational Design of Hierarchically Structured Materials, *Science,* Vol 288 (No. 5468), 1997, p 993–998
3. P.N. Koch, T.W. Simpson, J.K. Allen, and F. Mistree, Statistical Approximations for Multidisciplinary Optimization: The Problem of Size, *J. Aircraft,* Vol 36 (No. 1), 1999, p 275–286
4. R.H. Myers, D.C. Montgomery, and C.M. Anderson-Cook, *Response Surface Methodology: Process and Product Optimization Using Designed Experiments,* 2nd ed., John Wiley & Sons, New York, 2002

Table 1   Decision points in computational materials design where a designer can manage uncertainty

Source of uncertainty	Comments
Selection of size of region explored	As the size of the considered region increases, the number of datapoints needed to provide adequate coverage increase.
Choice of input variables Number and independence	A larger number of input variables requires more computation time; variables that are not independent introduce biases when surrogate models are created.
Sampling input variables (design of experiments)	The type of experimental designs used and the number of samples influence the accuracy of the computed surrogate models.
Analysis code Assumptions, simplifications	Consider robust methods.
Sampling output variables	The type of experimental designs used and the number of samples influence the accuracy of the results.
Surrogate model creation	Both the nature of the selected surrogate model and the level of required accuracy determine the error.
Error propagation	When multiple computations are required, error passes through each computation, generally increasing with the move from model to model. Introduce robustness.

5. A. Saltelli, T. Andres, and T. Homma, Sensitivity Analysis of Model Output. Performance of the Iterated Fractional Factorial Design, *Comput. Stat. Data Anal.,* Vol 20, 1995, p 387–407

6. C. Daniel, One-at-a-Time Plans, *J. Am. Stat. Assn.,* Vol 68, 1973, p 353–360

7. M. Morris, Factorial Sampling Plans for Preliminary Computational Experiments, *Technometrics,* Vol 33, 1991, p 161–174

8. I. Sobel, Sensitivity Estimates for Non-Linear Mathematical Models, *Math. Modeling Comput. Exp.,* Vol 4, 1993, p 407–414

9. A.A. Sonin, *The Physical Basis of Dimensional Analysis,* 2nd ed., Massachusetts Institute of Technology, Cambridge, MA, 2001

10. E. Buckingham, On Physically Similar Systems: Illustrations on the Use of Dimensional Equations, *Phys. Rev.,* Vol 4, 1914, p 345–376

11. N. Rolander, J. Rambo, Y. Joshi, J.K. Allen, and F. Mistree, An Approach to Robust Design of Turbulent Convective Systems, *J. Mech. Design,* Vol 128, 2006, p 844–855

12. I.T. Jolliffe, *Principal Component Analysis,* Springer, New York, NY, 2002

13. V.C.P. Chen, K.-L. Tsui, R.B. Barton, and J.K. Allen, A Review of Design and Modeling in Computer Experiments, *Statistics in Industry,* R. Khattree and C.R. Rao, Ed., Elsevier Science, 2003, p 231–262

14. J.S. Hunter, Statistical Design Applied to Product Design, *J. Qual. Technol.,* Vol 17 (No. 4), 1985, p 210–221

15. R.L. Plackett and J.P. Burman, The Design of Optimum Multifactorial Experiments, *Biometrika,* Vol 33, 1946, p 305–325

16. G.E.P. Box, J.S. Hunter, and W.G. Hunter, *Statistics for Experimenters: An Introduction to Design, Data Analysis, and Model Building,* Wiley, Inc., New York, 1978

17. G.E.P. Box and D.W. Behnken, Some New Three Level Designs for the Study of Quantitative Variables, *Technometrics,* Vol 2 (No. 4), 1960, p 455–475; Errata, Vol 3 (No. 4), p 576

18. D.C. Montgomery, *Design and Analysis of Experiments,* 7th ed., John Wiley and Sons, Inc., 2009

19. G.E.P. Box and N.R. Draper, A Basis for the Selection of Response Surface Designs, *J. Amer. Stat. Assoc.,* Vol 54, 1959, p 622–654

20. G.E.P. Box and N.R. Draper, The Choice of a Second Order Rotatable Design, *Biometrika,* Vol 50, 1963, p 335–352

21. G.E.P. Box and N.R. Draper, Robust Designs, *Biometrika,* Vol 62, 1975, p 347–352

22. P.J. Huber, Robustness and Designs, *A Survey of Statistical Designs and Linear Models,* J.N. Srivastava, Ed., North-Holland, New York, 1975

23. Z. Qian, C.C. Seepersad, V.R. Joseph, and J.K. Allen, Building Surrogate Models Based on Approximate and Detailed Simulations, *J. Mech. Des.,* Vol 128 (No. 4), 2006, p 668–677

24. G. Taguchi, Y. Yokoyama, and Y. Wu, *Taguchi Methods: Design of Experiments,* American Supplier Institute, Allen Park, MI, 1993

25. G. Taguchi, *System of Experimental Design: Engineering Methods to Optimize Quality and Minimize Costs,* UNI-PUB/Krauss International Publications, 1987

26. G.E.P. Box and N.R. Draper, *Empirical Model Building and Response Surfaces,* John Wiley and Sons, New York, 1987

27. J.M. Lucas, Optimum Composite Designs, *Technometrics,* Vol 16 (No. 4), 1974, p 561–567

28. N.R. Draper and D.K.J. Lin, Connections between Two-Level Designs of Resolutions III and IV, *Technometrics,* Vol 32 (No. 3), 1990, p 283–288

29. J. Kiefer, Optimal Experimental Designs, *J. Roy. Stat. Soc. Ser. B,* 1959 (21), p 272–316

30. W. Chen, J.K. Allen, D.P. Schrage, and F. Mistree, Statistical Experimentation Methods for Achieving Affordable Concurrent Systems Design, *AIAA J.,* Vol 35 (No. 5), 1997, p 893–900

31. G.G. Wang and S. Shan, Design Space Reduction for Multiobjective Optimization and Robust Design Optimization Problems, *SAE Trans. J. Mater. Manuf.,* 2004, p. 101–110

32. S. Shan and G.G. Wang, Space Exploration and Global Optimization for Computationally Intensive Design Problems: A Rough Set Based Approach, *Struct. Multidiscip. Optim.,* Vol 28 (No. 6), 2004, p 427–441

33. G.G. Wang and T.W. Simpson, Fuzzy Metamodeling for Space Reduction and Design Optimization, *Eng. Optimiz.,* Vol 36 (No. 3), 2004, p 313–335

34. B.A. Wujek and J.E. Renaud, New Adaptive Move-Limit Management Strategy for Approximate Optimization, Part I, *AIAA J.,* Vol 36 (No. 10), 1998, p 1911–1921

35. B.A. Wujek and J.E. Renaud, New Adaptive Move-Limit Management Strategy, Part 2, *AIAA J.,* Vol 36 (No. 10), 1998, p 1922–1937

36. J.F. Rodriguez, J.E. Renaud, and L.T. Watson, Trust Region Augmented Lagrangian Methods for Sequential Response Surface Approximation and Optimization, *J. Mech. Des.,* Vol 120, 1998, p 58–66

37. R. Jin and T.W. Simpson, Comparative Studies of Metamodeling Techniques under Multiple Modeling Criteria, *J. Struct. Optimiz.,* Vol 23 (No. 1), 2001, p 1–13

38. J. Sacks, W.J. Welch, T.J. Mitchell, and H.P. Wynn, Design and Analysis of Computer Experiments, *Stat. Sci.,* Vol 4 (No. 4), 1989, p 409–435

39. G.G. Wang and S. Shan, Review of Metamodeling Techniques in Support of Engineering Design Optimization, *J. Mech. Des.,* Vol 129, 2007, p 370–380

40. K.T. Fang, R. Li, and A. Sudjianto, *Design and Modeling for Computer Experiments,* Taylor & Francis Group, Boca Raton, FL, 2006

41. A.B. Owen, Orthogonal Arrays for Computer Experiments, Integration and Visualization, *Stat. Sinica,* Vol 2, 1992, p 439–452

42. A.S. Hedayat, N.J.A. Sloane, and J. Stufken, *Orthogonal Arrays: Theory and Applications,* Springer, New York, 1999

43. B. Tang, Orthogonal Array Based Latin Hypercubes, *J. Am. Stat. Assn.,* Vol 88 (No. 424), 1993, p 1392–1397

44. J.-S. Park, Optimal Latin Hypercube Designs for Computer Experiments, *J. Stat. Plan. Infer.,* Vol 39 (No. 1), 1994, p 95–111

45. K.Q. Ye, W. Li, and A. Sudjianto, Algorithmic Construction of Optimal Symmetric Latin Hypercube Designs, *J. Stat. Plan. Infer.,* Vol 90, 2000, p 145–159

46. M. Meckesheimer, R.R. Barton, T.W. Simpson, F. Limagem, and B. Yannou, Metamodeling of Combined Discrete/Continuous Responses, *AIAA J.,* Vol 40 (No. 10), 2002, p 2053–2060

47. E.W. Weisstein, "Hammersley Point Set," MathWorld—A Wolfram Web Resource, http://mathworld.wolfram.com/HammersleyPointSet.html, accessed May 19, 2010

48. K.T. Fang, D.K.J. Lin, P. Winkler, and Y. Zhang, Uniform Design Theory and Application, *Technometrics,* Vol 39 (No. 3), 2000, p 237–248

49. M.C. Shewry and H.P. Wynn, Maximum Entropy Sampling, Vol 14 (No. 2), *J. Appl. Stat.,* p 165–170

50. C. Currin, T. Mitchell, M. Morris, and D. Ylvisaker, Bayesian Predication of Deterministic Functions, with Applications to the Design and Analysis of Computer Experiments, *J. Am. Stat. Assn.,* Vol 86, 1991, p 953–963

51. M.D. Morris and T.J. Mitchell, Exploratory Designs for Computational Experiments, *J. Stat. Plan. Inf.,* Vol 43 (No. 3), 1995, p 381–402

52. T.W. Simpson, D.K.J. Lin, and W. Chen, Sampling Strategies for Computer Experiments, *Int. J. Reliab. Appl.,* Vol 2 (No. 3), 2001, p 209–240

53. K. Chaloner and I. Verdinelli, Bayesian Experimental Design: A Review, *Stat. Sci.,* Vol 10 (No. 3), 1995, p 273–304

54. J. Sobieszczanski-Sobieski and R.T. Haftka, Multidisciplinary Aerospace Design Optimization: Survey of Recent Developments, *34th Aerospace Sciences*

Meeting and Exhibit, Jan 15–18, 1996 (Reno, NV)

55. R.T. Haftka, E.P. Scott, and J.R. Cruz, Optimization and Experiments: A Survey, *Appl. Mech. Rev.,* Vol 51 (No. 7), 1998, p 435–448

56. J.F.M. Barthelemy and R.T. Haftka, Approximation Concepts for Optimal Structural Design—A Review, *Struct. Optim.,* Vol 5, 1993, p 129–144

57. A.-L. Boulesteix and K. Strimmer, Partial Least Squares: A Versatile Tool for the Analysis of High-Dimensional Genomic Data, *Brief. Bioinfomatics,* Vol 7, 2006, p 32–44

58. E. Acer, A. Kale, and R.T. Haftka, Comparing Effectiveness of Measures That Improve Aircraft Structural Safety, *J. Aerospace Eng.,* Vol 20 (No. 3), 2007, p 186–199

59. I. Elishakoff, *Safety Factors and Reliability: Friends or Foes?* Kluwer Academic Publishers, 2004

60. J. Lee, H. Jeong, D.H. Choi, V. Volovoi, and D. Mavris, An Enhancement of Constraint Feasibility in BPN Based Approximate Optimization, *Comp. Meth. Appl. Mech. Eng.,* Vol 196, 2007, p 2147–2160

61. J. Lee, H. Jeong, and S. Kang, Derivative and GA-Based Methods in Metamodeling of Back-Propagation Networks for Constrained Approximate Optimization, *Struct. Multidisc. Optimiz.,* Vol 35 (No. 1), 2008, p 29–40

62. V. Picheny, "Improving Accuracy and Compensating for Uncertainty in Surrogate Modeling," Ph.D. dissertation, University of Florida, 2009

63. M.L. Stein, *Interpolation of Spatial Data: Some Theories for Kriging,* Springer, 1999

64. T.W. Simpson, T. Mauery, J.J. Korte, and F. Mistree, Kriging Models for Global Approximation in Simulation-Based Multidisciplinary Design Optimization, *AIAA J.,* Vol 39 (No. 2), 2001, p 2233–2241

65. J.P.C. Kleijnen, Kriging Metamodeling in Simulation: A Review, *Euro. J. Op. Res.,* Vol 192 (No. 3), 2009, p 707–716

66. N. Cressie, Spatial Prediction and Ordinary Kriging, *Math. Geol.,* Vol 20 (No. 4), 1998, p 405–421

67. W.S. Cleveland and E. Gross, Computational Methods for Local Regression, *Stat. Comput.,* Vol 1, 1991, p 47–62

68. N. Dyn, D. Levin, and S. Rippa, Numerical Procedures for Surface Fitting of Scattered Data by Radial Basis Function, *SIAM J. Sci. Stat. Comput.,* Vol 7 (No. 2), 1986, p 41–47

69. H. Fang and M.F. Horstmeyer, Global Response Approximations with Radial Basis Functions, *Eng. Optimiz.,* Vol 38 (No. 4), 2006, p 407–424

70. B. Cheng and D.M. Titterington, Neural Networks: A Review from a Statistical Perspective, *Stat. Sci.,* Vol 9 (No. 1), 1994, p 2–54

71. H.K.D.H. Bhadaeshia and H.J. Stone, Neural-Network Modeling, *Fundamentals of Modeling for Metals Processing,* Vol 22A, *ASM Handbook,* ASM International, p 435–439, 2009

72. W. Sha and S. Malinov, Application of Neural-Network Models, *Fundamentals of Modeling for Metals Processing,* Vol 22A, *ASM Handbook,* ASM International, p 553–565, 2009

73. S.M. Clarke, J.H. Griebsch, and T.W. Simpson, Analysis of Support Vector Regression for Approximation of Complex Engineering Analyses, *J. Mech. Des.,* Vol 127, 2005, p 1077–1087

74. J.H. Friedman, Multivariate Adaptive Regressive Splines, *Ann. Stat.,* Vol 19 (No. 1), 1991, p 639–659

75. F.A.C. Viana, R.T. Haftka, and V. Steffen, Jr., Multiple Surrogates: How Cross-Validation Errors Can Help Us to Obtain the Best Predictor, *Struct. Multidisc. Optim.,* Vol 39, 2009, p 439–457

76. M. Rippel, "Improved Robustness Formulations and a Simulation-Based Robust Concept Exploration Method," M.S. thesis, The George W. Woodruff School of Mechanical Engineering, Georgia Institute of Technology, Atlanta, GA, 2009

77. J.K. Allen, C.C. Seepersad, H.-J. Choi, and F. Mistree, Robust Design for Multiscale and Multidisciplinary Applications, *J. Mech. Des.,* Vol 128 (No. 4), 2006, p 832–843

78. B.D. Youn, K.K. Choi, and Y.H. Park, Hybrid Analysis Method for Reliability-Based Design Optimization, *J. Mech. Des.,* Vol 125 (No. 2), 2003, p 221–232

79. S.-K. Choi, R.V. Grandhi, and R.A. Canfield, *Reliability-Based Structural Design,* Springer, 2006

80. S.H. Lee and W.A. Chen, Comparative Study of Uncertainty Propagation Methods for Black-Box Type Problems, *Struct. Multidisc. Optim.,* Vol 37, 2009, p 239–253

81. H.O. Madsen, S. Krenk, and N.C. Lind, *Methods of Structural Safety,* Dover Publications, Mineola, NY, 2006

82. A. Der Kiueghian, Structural Reliability Methods for Seismic Safety Assessment: A Review, *Eng. Struct.,* Vol 18 (No. 6), 1996, p 412–424

83. P. Christensen and M.J. Baker, *Structural Reliability Theory and Its Applications,* Springer, New York, NY, 1982

84. R.E. Melchers, Importance Sampling in Structural Systems, *Struct. Safety,* Vol 6 (No. 1), 1989, p 3–10

85. S. Engelund and R. Rackwitz, A Benchmark Study on Importance Sampling Techniques in Structural Reliability, *Struct. Safety,* Vol 12, 1993, p 255–276

86. C.G. Bucher, Adaptive Sampling—An Iterative Fast Monte Carlo Procedure, *Struct. Safety,* Vol 5 (No. 2), 1988, p 119–126

87. R.G. Ghanem and P.D. Spanos, *Stochastic-Finite Elements: A Spectral Approach,* Springer-Verlag, New York, NY, 1991

88. A.M. Hasofer and N.C. Lind, Exact and Invariant Second Order Code Format, *J. Eng. Mech. Div.—ASCE,* Vol 100 (NEMI), 1974, p 111–121

89. B. Fiessler, R. Rackwitz, and H. Nuemann, Quadratic Limit States in Structural Reliability, *J. Eng. Mech. Div.—ASCE,* Vol 105 (No. 4), 1979, p 661–676

90. D. Xiu and G.E. Karniadakias, Modeling Uncertainty in Flow Simulations via Generalized Polynomial Chaos, *J. Comp. Phys.,* Vol 187 (No. 1), 2003, p 137–167

91. S.K. Choi, R.V. Grandhi, and R.A. Canfield, Structural Reliability Under Non-Gaussian Stochastic Behavior, *Comp. Struct.,* Vol 82 (No. 13–14), 2004, p 1113–1121

92. D.H. Evans, An Application of Numerical Integration Techniques to Statistical Tolerancing III: General Distributions, *Technometrics,* Vol 14 (No. 1), 1972, p 23–35

93. H.S. Seo and B.M. Kwak, Efficient Statistical Tolerance Analysis for General Distributions Using Three-Point Information, *Int. J. Prod. Res.,* Vol 40, 2002, p 931–944

94. T.W. Lee and B.M. Kwak, Response Surface Augmented Moment Method for Efficient Reliability Analysis, *Struct. Safety,* Vol 28 (No. 3), 2006, p 261–272

95. S. Rahman and H. Xu, A Univariate Dimension-Reduction Method for Multi-Dimensional Integration in Stochastic Mechanics, *Prob. Eng. Mech.,* Vol 19 (No. 4), 2004, p 393–408

96. D.L. McDowell, J.P. Panchal, H.-J. Choi, C.C. Seepersad, J.K. Allen, and F. Mistree, *Integrated Design of Multiscale, Multifunctional Materials and Products,* Elsevier, 2010

97. G.-J. Park, T.H. Lee, K.H. Lee, and K.H. Hwang, Robust Design: An Overview, *AIAA J.,* Vol 44 (No. 1), 2006, p 181–191

98. H.G. Beyer and B. Sendhoff, Robust Optimization—A Comprehensive Survey, *Comp. Meth. Appl. Mech. Eng.,* Vol 196 (No. 33–34), 2007, p 3190–3218

99. H.J. Choi, D.L. McDowell, D.W. Rosen, J.K. Allen, and F. Mistree, An Inductive Design Exploration Method for Robust Multiscale Materials Design, *J. Mech. Des.,* Vol 130, 2008, p 031402

100. J.H. Panchal, C.J.J. Paredis, J.K. Allen, and F. Mistree, A Value-of-Information Based Approach to Simulation Model

Refinement, *Eng. Optimiz.,* Vol 40 (No. 3), 2008, p 223–251
101. J.H. Panchal, C.J.J. Paredis, J.K. Allen, and F. Mistree, Design Process Simplification via Scale and Decision Decoupling—A Vale of Information Based Approach, Vol 9 (No. 2), 2009, p 1–12
102. M. Messer, J.H. Panchal, J.K. Allen, F. Mistree, V. Krishnamurthy, B. Klein, and D.P. Yoder, Designing Embodiment Design Processes Using a Value-of-Information Based Approach with Applications for Integrated Product and Materials Design, *J. Mech. Des.,* (in press)
103. *NIST/SEMATECH e-Handbook of Statistical Methods,* www.itl.nist.gov/div898/handbook, accessed March 2010
104. Wikipedia, www.wikipedia.org
105. J.R. Davis, Ed., *ASM Materials Engineering Dictionary,* ASM International, 1992

ASM Handbook, Volume 22B, *Metals Process Simulation*
D.U. Furrer and S.L. Semiatin, editors

Copyright © 2010, ASM International®
All rights reserved.
www.asminternational.org

Manufacturing Cost Estimating*

David P. Hoult and C. Lawrence Meador

COST ESTIMATION is an essential part in the design, development, and use of products. In the development and design of a manufactured product, phases include concept assessment, demonstrations of key features, and detailed design and production. The next phase is the operation and maintenance of the product, and finally, its disposal. Cost estimation arises in each of these phases, but the cost impacts are greater in the development and design phases.

Anecdotes, experience, and some data (Ref 1, 2) support the common observation that by the time 20% of a product is specified, 70 to 80% of the costs are committed, even if those costs are unknown! Another common perception is that the cost of correcting a design error (cost overrun) rises very steeply as product development proceeds through its phases. What might cost one unit of effort to fix in the concept assessment phase might cost a thousand units in detailed engineering. These experiences of engineers and designers force the cost estimator to think carefully about the context, timing, and use of cost estimates.

General Concepts

In this article, the focus is on products defined by dimensions and tolerances, made from solid materials and, fabricated by some manufacturing process. Two issues should be apparent: first, accurate cost estimates of a product in its early stages of design is difficult; second, there are a very large number of manufacturing processes, so one must somehow restrict the discussion to achieve sensible results.

In dealing with the first issue a series of cost estimates corresponding to the phases in the product-development program should be considered. As more details of the product are specified, the cost estimates should become more accurate. Thinking this way, it is plausible that different tools for cost estimation may be employed in different phases of a program. In this review examples are given of methods of cost estimation that may be used in the contexts of the different phases of a development program.

Managing Data and Costs

Domain Limitation. The second issue gives rise to the important principle of *domain limitations* (Ref 3, 4), meaning data that form the basis for a cost estimate must be specific to the manufacturing process, the materials used, and so on. Cost estimates only apply within specific domains. This leads directly to another difficulty: most products, even if they are only moderately complex, like a dishwasher with 200 unique parts, have a least three or five domains in which cost estimates apply (i.e., sheet metal assembly, injection-molded plastic parts, formed sheet metal parts, etc.). More complex products, such as an aircraft jet engine with 25,000 unique parts, might have 200 different manufacturing processes, each of which defines one or more domains of cost estimation. Clearly, as one considers still more complex products, such as a modern military radar system with 10,000 to 20,000 unique parts, or a large commercial airliner with perhaps 5 million unique parts, the domains of cost estimation expand dramatically. So, although domain limitation is necessary for cost-estimates accuracy, it is not a panacea.

Database Commonality. Estimating the costs of a complex product through various phases of development and production requires organization of large amounts of data. If the data for design, manufacturing, and cost are linked, there is *database commonality*. It has been found (Ref 3) that having database commonality results in dramatic reductions in cost and schedule overruns in military programs. In the same study, domain limitation was found to be essential in achieving database commonality.

Having database commonality with domain limitation implies that the links between the design and specific manufacturing processes, with their associated costs, are understood and delineated. Focusing on specific manufacturing processes allows one to collect and organize data on where and how costs arise in specific processes. With this focus, the accuracy of cost estimates can be determined, provided that uniform methods of estimation are used, and provided that, over time, the cost estimates are compared with the actual costs as they arise in production. In this manner, the accuracy of complex cost estimates may be established and improved.

In present engineering and design practice, many organizations do not have adequate database commonality, and the accuracy of cost estimates is not well known. Database commonality requires an enterprise-wide description of cost-dominant manufacturing processes, a way of tracking actual costs for each part, and a way of giving this information—in an appropriate format—to designers and cost estimators. Most "empirical methods" of cost estimation, which are based on industrywide studies of statistical correlation of cost, may or may not apply to the experience of a specific firm (see the discussion in the sections that follow).

Costs are "rolled up" for a product when all elements of the cost of a product are accounted for. Criteria for cost estimation using database commonality is simple: speed (how long does it take to roll up a cost estimate on a new design), accuracy (what is the standard deviation of the estimate, based on comparison with actual costs) and risk (what is the probability distribution of the cost estimate; what fraction of the time will the estimate be more than 30% too low, for example). One excellent indicator of database commonality is the roll-up time criteria. World-class cost-estimation roll-up times are minutes to fractions of days. Organizations that have such rapid roll-up times have significantly less cost and schedule overruns on military projects (Ref 3).

Cost allocation is another general issue. Cost allocation refers to the process by which the components of a design are assigned target costs. The need for cost allocation is clear: how else would an engineer, working on a large project, know how much the part being designed should cost? And, if the cost is unknown and the target cost is not met, there will be time delays, and hence costs incurred due to unnecessary design iteration. It is generally recognized that having integrated product teams (IPTs) is a good industrial practice. Integrated product teams should allocate costs at the earliest stages of a development program. Cost estimates should be performed concurrently with the design effort throughout the development process. Clearly, estimating costs at early stages in a development program, for example, when the concept of the product is being assessed, requires quite different tools

---

* Reprinted from the article, "Manufacturing Cost Estimating," by David P. Hoult and C. Lawrence Meador in *ASM Handbook*, Volume 20, *Materials Selection and Design*, 1997, p 716–722.

than when most or all the details of the design are specified. Various tools that can be used to estimate cost at different stages of the development process are described later in this section.

Elements of Cost. There are many elements of cost. The simplest to understand is the cost of material. For example, if a part is made of aluminum and is fabricated from 10 lb of the material, if the grade of aluminum costs $2/lb, the material cost is $20. The estimate gets only a bit more complex if, as in the case of some aerospace components, some 90% of the materials will be machined away; then the sale on scrap material is deducted from the material cost.

Tooling and fixtures are the next easiest items to understand. If tools are used for only one product, and the lifetime of the tool is known or can be estimated, then only the design and fabrication cost of the tool is needed. Estimates of the fabrication costs for tooling are of the same form as those for the fabricated parts. The design cost estimate raises a difficult and general problem: cost capture (Ref 4). For example, tooling design costs are often classified as overhead, even though the cost of tools relates to design features. In many accounting systems, manufacturing costs are assigned "standard values," and variances from the standard values are tabulated. This accounting methodology does not, in general, allow the cost engineer to determine the actual costs of various design features of a part. In the ledger entries of many accounting systems, there is no allocation of costs to specific activities or no activity-based accounting (ABC) (Ref 5). In such cases there are no data to support design cost estimates.

Direct labor for products or parts that have a high yield in manufacturing normally have straightforward cost estimates, based on statistical correlation to direct labor for past parts of a similar kind. However, for parts that have a large amount of rework the consideration is more complex, and the issues of cost capture and the lack of ABC arise again. Rework may be an indication of uncontrolled variation of the manufacturing process. The problem is that rework and its supervision may be classified all, or in part, as manufacturing overhead. For these reasons, the true cost of rework may not be well known, and so the data to support cost estimates for rework may be lacking.

The cost estimates of those parts of overheads that are associated with the design and production of a product are particularly difficult to estimate, due to the lack of ABC and the problem of cost capture. For products built in large volumes, of simple or moderate complexity, cost estimates of overheads are commonly done in the simplest possible way: the duration of the project and the level of effort are used to estimate the overhead. This practice does not lead to major errors because the overhead is a small fraction of the unit cost of the product.

For highly engineered, complex products built in low volume, cost estimation is very difficult. In such cases the problem of cost capture is also very serious (Ref 4).

Machining costs are normally related to the machine time required and a capital asset model for the machine, including depreciation, training, and maintenance. With a capital asset model, the focus of the cost estimate is the time to manufacture. A similar discussion holds for assembly costs: with a suitable capital asset model, the focus of the cost estimate is the time to assemble the product (Ref 1).

Methods of Cost Estimations. There are three methods of cost estimation discussed in the following sections of this article. The first is parametric cost estimation. Starting from the simplest description of the product, an estimate of its overall cost is developed. One might think that such estimates would be hopelessly inaccurate because so little is specified about the product, but this is not the case. The key to this method is a careful limitation of the *domain* of the estimate (see the previous section). This example deals with the estimate of the weight of an aircraft. The cost of the aircraft would then be calculated using dollars/pound typical of the aircraft type. Parametric cost estimation is the generally accepted method of cost estimation in the concept assessment phases of a development program. The accuracy is surprisingly good—about 30% (provided that recent product-design evolution has not been extensive).

The second method of cost estimation is empirically based: one identifies specific design features and then uses statistical correlation of costs of past designs to estimate the cost of the new design. This empirical method is by far the most common in use. For the empirical method to work well, the features of the product for which the estimate is made should be unambiguously related to features of prior designs, and the costs of prior designs unambiguously related to design features. Common practice is to account for only the major features of a design and to ignore details. Empirical methods are very useful in generating a rough ranking of the costs of different designs and are commonly used for that purpose (Ref 1, 6, 7). However, there are deficiencies inherent in empirical methods commonly used.

The mapping of design features to manufacturing processes to costs is not one-to-one. Rather, the same design feature may be made in many different ways. This difficulty, the feature mapping problem, discussed in Ref 4, limits the accuracy of empirical methods and makes the assessment of risk very difficult. The problem is implicit in all empirical methods. The problem is that the data upon which the cost correlation is based may assume the use of manufacturing methods to generate the features of the design that do not apply to the new design. It is extraordinarily difficult to determine the implicit assumptions made about manufacturing processes used in a prior empirical correlation. A commonly stated accuracy goal of empirical cost estimates is 15 to 25%, but there is very little data published on the actual accuracy of the cost estimate when it is applied to new data.

The final method discussed in this article is based on the recent development called complexity theory. A mathematically rigorous definition of complexity in design has been formulated (Ref 8). In brief, complexity theory offers some improvement over traditional empirical methods: there is a rational way to a assess the risk in a design, and there are ways of making the feature mapping explicit rather than implicit. Perhaps the most significant improvement is the capability to capture the cost impact of essentially all the design detail in a cost estimate. This allows designers and cost estimators to explore, in a new way, methods to achieve cost savings in complex parts and assemblies.

Parametric Methods

An example for illustrating parametric cost estimation is that of aircraft. In Ref 9, Roskam—a widely recognized researcher in this field—describes a method to determine the size (weight) of an aircraft. Such a calculation is typical of parametric methods. To determine cost from weight, one would typically correlate costs (inflation adjusted) of past aircraft of similar complexity with their weight. Thus weight is surrogate for cost for a given level of complexity.

Most parametric methods are based on such surrogates. For another simple example, consider that large coal-fired power plants, based on a steam cycle, cost about $1500/kW to be built. So, if the year the plant is to be built (for inflation adjustment) and its kW output is known, parametric cost estimate can be readily obtained.

Parametric cost estimates have the advantage that little needs to be known about the product to produce the estimate. Thus, parametric methods are often the only ones available in the initial (concept assessment) stages of product development.

The first step in a parametric cost estimation is to limit the domain of application. Roskam correlates statistical data for a dozen types of aircraft and fifteen sub types. The example he uses to explain the method is that of a twin-engine, propeller-driven airplane. The mission profile of this machine is given in Fig. 1 (Ref 9).

Inspection of the mission specifications and Fig. 1 shows that only a modest amount of information about the airplane is given. In particular, nothing is specified about the detailed design of the machine! The task is to estimate the total weight, W_{TO} or the empty weight, W_E, of the airplane. Roskam argues that the total weight is equal to the sum of the empty weight, fuel

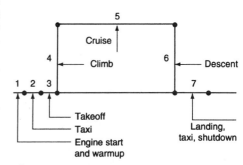

Fig. 1 Mission profile

weight, W_F, payload and crew weight, $W_{PL} + W_{crew}$, and the trapped fuel and oil, which is modeled as a fraction, M_{tfo}, to the total weight. M_{tfo} is to be a small (constant) number, typically 0.001 to 0.005. Thus the fundamental equation for aircraft weight is:

$$W_{TO} = W_E + W_F + W_{PL} + W_{crew} + M_{tfo}W_{TO} \quad \text{(Eq 1)}$$

The basic idea of Roskam is that there is an empirical relationship between aircraft empty and total weights, which he finds to be:

$$\log_{10} W_E = \{(\log_{10} W_{TO}) - A\}/B \quad \text{(Eq 2)}$$

The coefficients, A and B, depend on which of the dozen types and fifteen subtypes of aircraft fit the description in Table 1 and Fig. 1. It is at this point that the principle of domain limitation first enters. For the example used by Roskam, the correlation used to determine $A = 0.0966$ and $B = 1.0298$ for the twin-engine, propeller-driven aircraft spans a range of empty weights from 1000 to 7000 lb.

The method proceeds as follows to determine the weight of fuel required in the following way. The mission fuel, W_F, can be broken down into the weight of the fuel used and the reserve fuel:

$$W_F = W_{Fres} + W_{Fused} \quad \text{(Eq 3)}$$

Roskam models the reserve fuel as a fraction of the fuel used (see Table 1). The fuel used is modeled as a fraction of the total weight, and depends on the phase of the mission, as described in Fig. 1. For mission phases that are not fuel intensive, a fixed ratio of the weight at the end of the phase to that at the beginning of the phase is given. Again, these ratios are specific to the type of aircraft. For fuel-intensive phases, in this example the cruise phase, there is a relationship between the lift/drag ratio of the aircraft, the engine fuel efficiency, and the propeller efficiency. Again, these three parameters are specific to the type of aircraft.

When the fuel fraction of the total weight is determined by either a cruise calculation, or by the ratio of weight at the end of a mission

Table 1 Mission specification for a twin-engine, propeller-driven airplane

1. Payload	Six passengers at 175 lb each (including the pilot) and 200 lb total baggage
2. Range	1000 statute miles with maximum payload
3. Reserves	25% of mission fuel
4. Cruise speed	250 knots at 75% power at 10,000 ft and at takeoff weight
5. Climb	10 min to 10,000 ft at takeoff weight
6. Takeoff and landing	1500 ft ground run at sea level, standard day. Landing at 0.95 of takeoff weight
7. Powerplants	Piston/propeller
8. Certification base	FAR23

phase to the beginning of a mission phase, the empty weight can be written in terms of the total weight. Then Eq 2 is used to find the total weight of the aircraft.

For the problem posed, Roskam obtains an estimated total weight of 7900 lb. The accuracy can be estimated from the scatter in the correlation used to determine the coefficients A and B, and is about 30%. For details of the method Roskam uses for obtaining the solution, refer to Ref 9.

Some limitations of the parametric estimating method are of general interest. For example, if the proposed aircraft does not fit any of the domains of the estimating model, the approach is of little use. Such an example might be the V-22, a tilt wing aircraft (Ref 10), which flies like a fixed-wing machine, but tilts its wings and propellers, allowing the craft to hover like a helicopter during take-off and landing. Such a machine might be considered outside the domain of Roskam's estimating model. The point is not that the model is inadequate (the V-22 is more recent than Roskam's 1986 article), but the limited product knowledge in the early stages of development makes it difficult to determine if a cost estimate for the V-22 fits in a well-established domain.

Conversely, even complex machines, such as aircraft, are amenable to parametric cost estimates with fairly good accuracy, provided they are within the domain of the cost model. In the same article, Roskam presents data for transport jets, such as those used by airlines. It should be emphasized that the weight (and hence cost) of such machines, with more than one million unique parts, can be roughly estimated by parametric methods.

Of course, cost is not the same as weight or, for that matter, any other engineering parameter. The details of the manufacturing process, inventory control, design change management, and so forth, all play a role in the relationship between weight and cost. The more complex the machine, the more difficult it is to understand if the domain of the parametric cost-estimating model is the same as that of the product being estimated.

Empirical Methods of Cost Estimation

Examples for Sheet Metal and Riveted Parts

Almost all the cost-estimating methods published in the literature are based on correlation of some feature or property of the part to be manufactured. Two examples are presented. The first is from the book by Boothroyd, Dewhurst, and Knight (Ref 1), hereafter referred to as BDK. Chapter 9 of this book is devoted to "Design for Sheet Metalworking." The first part of this chapter is devoted to estimates of the costs of the dies used for sheet metal fabrication. This example was chosen because the work of these authors is well recognized. (Boothroyd and Dewhurst Inc. sells widely used software for design for manufacture

and design assembly.) In this chapter of the book, the concept of "complexity" of stamped sheet metal parts arises. The complexity of mechanical parts is discussed in the section "Complexity Theory" in this article.

Example 1: Cost Estimates for Sheet Metal Parts. Sheet metal comes in some 15 standard gages, ranging in thickness from 0.38 to 5.08 mm. It is commonly available in steel, aluminum, copper, and titanium. Typical prices for these materials are 0.80–0.90$/lb for low-carbon steel, $6.00-$7.00/lb for stainless steel, $3.00/lb for aluminum, $10.00/lb for copper, and $20.00/lb for titanium. It is typically shipped in large coils or large sheets.

Automobiles and appliances use large amounts of steel sheet metal. Aluminum sheet metal is used in commercial aircraft manufacture, but in lesser amounts due to the smaller number of units produced.

Sheet metal is fabricated by shearing and forming operations, carried out by dies mounted in presses. Presses have beds, which range in size from 50 by 30 cm to 210 by 140 cm (20 by 12 in. to 82 by 55 in.). The press force ranges from 200 to 4500 kN (45 to 1000 lbf). The speed ranges from 100 strokes/min to 15 strokes/min in larger sizes.

Dies typically have four components: a basic die set; a punch, held by the die set, which shears or forms the metal; a die plate through which or on which the punch acts; and a stripper plate, which removes the scrap at the end of the fabrication process.

BDK estimate the basic die set cost (C_{ds}, in U.S. dollars) as basically scaling with usable area (A_u, in cm^2):

$$C_{ds} = 120 + 0.36A_u \quad \text{(Eq 4)}$$

The coefficients (Eq 4) arise from correlating about 50 data points of die set cost with useable area. The tooling elements (the punch, die plate, and stripper plate) are estimated with a point system as follows: let the complexity of the part to be fabricated be X_p. Suppose that the profile has a perimeter P (cm), and that the part has an over width and length of W (cm) and L (cm) of the smallest dimensions which surround the punch. The complexity of the part is taken to be:

$$X_p = (P/L)(P/W) \quad \text{(Eq 5)}$$

The assessment of how part complexity affects cost arises repeatedly in cost estimating. The subject is discussed at length in the next section "Complexity Theory." From the data of BDK, the basic time to manufacture the die set (M, in hours) can be estimated by the following steps: Define the basic manufacturing points (M_{po}) as

$$M_{po} = 30 + 0.56X_p^{2/3} \quad \text{(Eq 6)}$$

Note that the manufacturing time increases a bit less than linearly with part complexity. This is consistent with the section "Complexity Theory." BDK goes on to add two correction factors to M_{po}. The first is a correction factor due to plate size and part complexity, f_{LW}. From BDK data it is found:

$$f_{LW} = 1 + 0.0276 \, LW \, X_p^{0.093} \qquad \text{(Eq 7)}$$

The second correction factor is to account for the die plate thickness. BDK cites Nordquist (Ref 11), who gives a recommended die thickness, h_d, as:

$$h_d = 9 + 2.5 \log_e(U/U_{ms})Vh^2 \qquad \text{(Eq 8)}$$

where U is the ultimate tensile stress of the sheet metal, U_{ms} is the ultimate stress of mild steel, a reference value, V, is the required production volume, and h is the thickness (in mm) of the metal to be stamped. BDK recommends the second correction factor to be:

$$f_d = 0.5 + 0.02h_d \text{ or } f_d = 0.75 \qquad \text{(Eq 9)}$$

whichever is greater.

The corrected labor hours, M_p, are then estimated as:

$$M_p = f_d f_{LW} M_{po} \qquad \text{(Eq 10)}$$

The cost of the die is the sum of the corrected labor hours times the labor rate of the die fabricator plus the basic die set cost, from Eq 4.

As a typical example of the empirical cost estimating methods, the BDK method takes into account several factors such as the production volume, the strength of the material (relating to how durable the die needs to be), the die size, and complexity of the part. These factors clearly influence die cost. However, the specific form of the equations are chosen as convenient representations of the data at hand. (As, indeed, are Eq 6 and 7, derived by fitting BDK data.)

The die cost risk (i.e., uncertainty of the resulting estimate of die cost) is unknown, because it is not known how the model equations would change with different manufacturing processes or different die design methods.

It is worth noting carefully that only some features of the design of the part enter the cost estimate: the length and width of the punch area, the perimeter of the part to be made, the material, and the production volume. Thus, the product and die designers do not need to be complete in all details to make a cost estimate. Hence, the estimate can be made earlier in the product-development process. Cost trades between different designs can be made at an early stage in the product-development cycle with empirical methods.

Example 2: Assembly Estimate for Riveted Parts. The American Machinist Cost Estimator (Ref 7) is a very widely used tool for empirical cost estimation. It contains data on 126 different manufacturing processes. A spreadsheet format is used throughout for the cost analysis. One example is an assembly process. It is proposed to rivet the aluminum frame used on a

powerboat. The members of the frame are made from 16-gage aluminum. The buttonhead rivets, which are sized according to recommendations in Ref 12, are 5/16 in. in diameter and conform to ANSI standards. Figure 2 shows the part.

There are 20 rivets in the assembly, five large members of the frame, and five small brackets. Chapter 21 in Ref 7 includes six tables for setup, handling, pressing in the rivets, and riveting. A simple spreadsheet (for the first unit) might look like Table 2. The pieces are placed in a frame, the rivets are inserted, and riveted. The total cycle time for the first unit is 18.6 min. There are several points to mention here. First, the thickness of the material and the size of the rivets play no direct part in this simple calculation. The methods of Ref 7 do not include such details.

Yet common sense suggests that some of the details must count. For example, if the rivet holes are sized to have a very small clearance, then the "press-in-hardware" task, where the rivets are placed in the rivet holes, would increase. In a like manner, if the rivets fit looser in the rivet holes, the cycle time for this task might decrease. The point of this elementary discussion is that there is some implied tolerance with each of the steps in the assembly process.

In fact, one can deduce the tolerance from the standard specification of the rivets. From Ref 12, in the tolerance on 5/16 in. diameter buttonhead rivets is 0.010 in. So the tolerance of the hole would be about the same size.

The second point is that there are 30 parts in this assembly. How the parts are stored and how they are placed in the riveting jig or fixture determines how fast the process is done. With experience, the process gets faster. There is a well-understood empirical model for process learning. The observation, often repeated in many different industries, is that inputs decrease by a fixed percentage each time the number of units produced doubles. So, for example, L_i is the labor in minutes of the ith unit produced, and L_0 is the labor of the first unit, then:

$$L_i = L_0 i^{\log \phi / \log 2} \qquad \text{(Eq 11)}$$

The parameter ϕ measures the slope of the learning curve. The learning curve effects were first observed and documented in the aircraft industry, where a typical rate of

improvement might be 20% between doubled quantities. This establishes an 80% learning function, that is, $\phi = 0.80$. Because this example is fabricated from aluminum, with rivets typical of aircraft construction, it is easy to work out that the 32nd unit will require 32.7% of the time (6.1 min) compared to the first unit (18.6 min).

Learning occurs in any well-managed manual assembly process. With automated assembly, "learning" occurs only when improvements are made to the robot used. In either case, there is evidence that, over substantial production runs and considerable periods of time, the improvement is a fixed percentage between doubled quantities. That is, if there is a 20% improvement between the tenth and twentieth unit, there will likewise be a 20% improvement between the hundredth and two hundredth unit.

The cost engineer should remember that, according to this rule, the percentage improvement from one unit to the next is a steeply falling function. After all, at the hundredth unit, it takes another hundred units to achieve the same improvement as arose between the 10th and 20th units (Ref 13).

Complexity Theory

Up to now this article has dealt with the cost-estimation tools that do not require a complete description of the part or assembly to make the desired estimates. What can be said if the design is fully detailed? Of course, one could build a prototype to get an idea of the costs, and this is often done, particularly if there is little experience with the manufacturing methods to be used. For example, suppose there is a complex wave feed guide to be fabricated out of aluminum for a modern radar system. The part has some 600 dimensions. One could get a cost estimate by programming a numerically controlled milling machine to make the part, but is there a simpler way to get a statistically meaningful estimate of cost, while incorporating all of the design details? The method that fulfills this task is complexity theory.

There has been a long search for the "best" metric to measure how complex a given part or assembly is. The idea of using dimensions and tolerances as a metric comes from Wilson

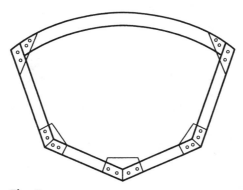

Fig. 2 Powerboat frame assembly

Table 2 Spreadsheet example for assembly of frame (Fig. 2)

Source (a)	Process description	Table time, min	Setup, min
21.2-S	Setup		15
21.2-1	Get 5 frame members from skid	1.05	
21.2-1	Get 5 brackets from bench	0.21	
21.2-2	Press in hardware (20 rivets)	1.41	
21.2-3	Set 20 rivets	0.93	
Total cycle time (minutes)		**3.60**	**15**

(a) Tables in Ref 7, Chapter 21

(Ref 14). The idea presented here is that the metric is a sum of log (d_i/t_i), where d_i is the ith dimension and t_i is its associated tolerance (i ranges over all the dimensions needed to describe the part). According to complexity theory, how complex a part is, I, is measured by:

$$I = \sum_i \log(d_i/t_i) \qquad \text{(Eq 12)}$$

Originally, the log function was chosen from an imperfect analogy with information theory. It is now understood that the log function arises from a limit process in which tolerance goes to zero while a given dimension remains fixed. In this limit, if good engineering practice is followed, that is, if the accuracy of the machine making the part is not greatly different than the accuracy required of the part, and if the "machine" can be modeled like a first-order damped system, then it can be shown that the log function is the correct metric. Because of historical reasons, the log is taken to the base 2, and I is measured in bits. Thus Eq 12a is written:

$$I = \sum_i \log_2(d_i/t_i) \qquad \text{(Eq 12a)}$$

There are two main attractions of the complexity theory. First, I will include all of the dimensions required to describe the part. Hence, the metric captures all of the information of the original design. For assemblies, the dimensions and tolerances refer to the placement of each part in the assembly, and second, the capability of making rigorous statements of how I effects costs. In Ref 8 it is *proven* that if the part is made by a single manufacturing process, the average time (T) to fabricate the part is:

$$T = A \cdot I, \quad A = \text{const} \qquad \text{(Eq 13)}$$

Again, in many cases, the coefficient A must be determined empirically from past manufacturing data. The same formula applies to assemblies made with a single process, such as manual labor. The extension to multiple processes is given in Ref 8.

A final aspect of complexity theory worth mentioning is risk. Suppose a part with hundreds of dimensions is to be made on a milling machine. The exact sequence in which each feature of the part is cut out will determine the manufacturing time. But there are a large number of such sequences, each corresponding to some value of A. Hence there is a collection of As, which have a mean that corresponds to the average time to fabricate the part. That is the meaning of Eq 13.

It can be shown that the standard deviation of manufacturing time is:

$$s_T = \sigma_A I \qquad \text{(Eq 14)}$$

where σ_T is the standard deviation of the manufacturing time, and σ_A is the standard deviation of the coefficient A. σ_A can be determined from past data. These results have a simple interpretation: Parts or assemblies with tighter (smaller) tolerances take longer to make or assemble because with dimensions fixed, the *log* functions increase as the tolerances decrease. More complex parts, larger I, take longer to make (Eq 13), and more complex parts have more cost risk (Eq 14). These trends are well known to experienced engineers.

In Eq 8, a large number of parts from three types of manufacturing processes were correlated according to Eq 13. The results of the following manual lathe process are typical of all the processes studied in Eq 8. Figure 3 shows the correlation of time with I, the dimension information, measured in bits. An interesting fact, shown in Fig. 4 is that the accuracy of the estimate is no different than that of an experienced estimator.

In Eq 13, the coefficient, A, is shown to depend on the machine properties such as speed, operation range, and time to reach steady-state speed. Can one estimate their value from first principals? It turns out that for manual processes one can make rough estimates of the coefficient.

The idea is based on the basic properties of human performance, known as Fitts' law. Fitts and Posner reported the maximum human information capacity for discrete, one-dimensional positioning tasks at about 12 bits/s (Ref 15). Other experiments have reported from 8 to 15 bits/s for assembly tasks (Ref 16).

The rivet insertion process discussed previously in this article is an example. The tolerance of the holes for the rivets is estimated to be 0.010 in., that is, the same as the handbook value of the tolerance of the barrel of the rivet (Ref 12). Then it is found that $d/t \approx 0.312/0.010 = 31.2$ and $\log_2 = 4.815$ bits for each insertion. The initial rate of insertion (Ref 7) was 20 units in 1.41 min. That corresponds to $A = 1.14$ bits/s. Clearly, there is some considerable improvement available if the maximum values quoted (Ref 15, 16) can be achieved for rivet insertion.

Estimating Assembly Time

Example 3: Manual Assembly of a Pneumatic Piston. In Ref 1 there is an extensive and helpful section on manual assembly. The method BDK used categorizes the difficulty of assembling parts by a number of parameters, such as the need to use one or two hands, the need to use mechanical tools, part symmetry, and so on. Figure 5 (reproduced from an example in Ref 17) shows the assembly of a small pneumatic piston. Table 3 lists assembly times.

Consider the entries for the two screws. The handling code, 68, describes a part with 360° symmetry that can be handled with standard tools. The insertion code, 39, describes a part not easy to align or position. The time for assembly of the screws is nominally 32 s, less an allowance of 31 s for repetitive operations.

Now consider a simplified design (Fig. 6 and Table 4). The same tables from Chapter 21 in Ref 7 are used as in Table 2. Software is available to automate the table look-up process. For the same problem using complexity theory, there is only one coefficient, $A = 1.5$ bits/s for

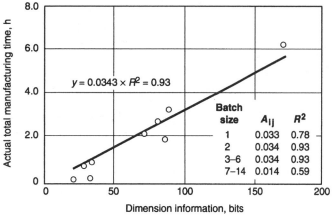

Fig. 3 Manufacturing time and dimension information for the lathe process (batch size 3 to 6 units)

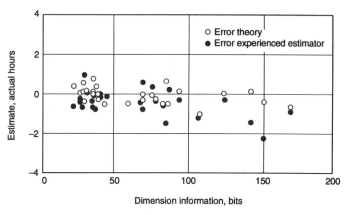

Fig. 4 Accuracy comparison for the lathe process

the small manual assembly. This value is found by calculating the bits of information in the initial design and using the time found in Ref 17 to determine A.

The tolerances were obtained in the following way. For the screws, the size was chosen to be M3X0.5—a coarse thread metric size consistent with insertion into molded nylon. As reported in Ref 12, the tolerance is ANSI B1.13M-1979.

The spring tolerance is derived using a standard wire size, 0.076 in. (0.193 mm), which gives a spring index $D/d = 12.95$, (D = spring diameter, d = wire diameter) well within the Spring Manufacturers Institute recommended range. The tolerance quoted is the standard tolerance for this wire diameter and spring index.

The plastic parts are presumed to be injection molded, with a typical tolerance of 0.1 mm. The screws are assumed to tighten to a tolerance of 60/360 = 1/6th of a turn.

These data, which can be easily verified in practice, give essentially the same results as the other empirical methods. Calculations for the original design (Fig. 5) and the simplified design (Fig. 6) are compared in Tables 5 and 6, respectively. This compares well with the results of Ref 17 (73% reduction) versus 76% reduction here.

There are two comments to make:

- This method requires only one coefficient for hand assembly of small parts. $A \approx 1.6$ bits/s and no look-up tables.
- One can make small changes in design, for example, change the screw size, and get an indication of the change in assembly time.

If one started with a preliminary design, the assembly time estimate would grow more accurate as more details of the design itself and the manufacturing process to build the design become known. However, the methods of Boothroyd, and others (Ref 1, 6, 17), require less data than does the complexity theory.

In the simplified manufacturing process, the coefficient $A = 1.6$ bits/s is substantially less than the Fitts' law (discussed earlier in this section on complexity theory) value of 8 bits/s. The discrepancy may lie in the time it takes an assembler to pick up and orient each part before the part is assembled. Jigs and trays, and so forth, that reduce this pick-and-place orientation effort would save assembly time. As before, there is some considerable improvement available if the maximum values quoted (Ref 15, 16) can be achieved for manual assembly. The value obtained here ($A = 1.16$) is close to that deduced from Ref 7 for the hand insertion of rivets.

Using complexity theory and a single assembly process, the ratio of the assembly times can be calculated without any knowledge of the coefficient, A. Thus complexity theory offers advantages when a single process is used, even if little or nothing is known about the performance of the process.

Cost Estimation Recommendations

Which type of cost estimate one uses depends on how much is know about the design. In the early stages of concept assessment of a new part or product, parametric methods, based upon past experience, are preferred. Risk is hard to quantify for these methods, because it can be very difficult to determine whether the new product really is similar to those used to establish the parametric cost model.

If there is some detailed information about the part or product, and the method of manufacturing is well known, then the empirical methods should be used. They can indicate the relative cost between different designs and give estimates of actual costs.

If detailed designs are specified, and a single manufacturing process is to be used, complexity theory should be used to compare the relative costs and cost risks of the different designs, even if the manufacturing process is poorly understood.

If there are detailed designs available, and well-known manufacturing methods are used, either complexity theory or empirical methods can be used to generate cost estimates. If a rigorous risk assessment is needed, complexity theory should be used.

REFERENCES

1. G. Boothroyd, P. Dewhurst, and W. Knight, *Product Design for Manufacture and Assembly*, Marcel Dekker, 1994, Chapt. 1
2. K.T. Ulrich and S.A. Pearson, "Does Product Design Really Determine 80% of Manufacturing Cost?," working paper 3601-93-MSA, MIT Sloan School, Nov 1994
3. D.P. Hoult and C.L. Meador, "Cost Awareness in Design: the Role of Database Commonality," SAE 96008, Society of Automotive Engineers, 1996
4. D.P. Hoult and C.L. Meador, "Methods of Integrating Design and Cost Information to Achieve Enhanced Manufacturing Cost/Performance Trade-Offs," Save International Conference Proceedings, Society for American Value Engineers, 1996, p 95–99
5. H.T. Johnson and R.S. Kaplan, *Relevance Lost, the Rise and Fall of Management Accounting*, Harvard Business School Press, 1991
6. G. Boothroyd, *Assembly Automation and Product Design*, Marcel Dekker, 1992
7. P.F. Ostwald, "American Machinist Cost Estimator," Penton Educational Division, Penton Publishing, 1988
8. D.P. Hoult and C.L. Meador, "Predicting Product Manufacturing Costs from Design Attributes: A Complexity Theory Approach," No. 960003, Society of Automotive Engineers, 1996
9. J. Roskam, Rapid Sizing Method for Airplanes, *J. Aircraft*, Vol 23 (No.7), July 1986, p 554–560
10. The Bell-Boeing V-22 Osprey entered Low Rate Initial Production with the MV-22

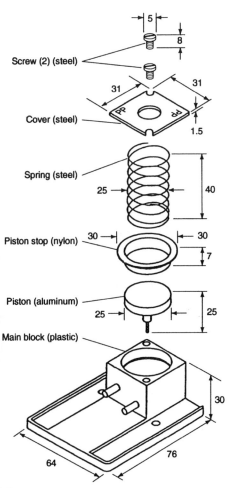

Fig. 5 Assembly of pneumatic piston. Dimensions in millimeters

Table 3 Assembly times for piston example (Fig. 5)

Part ID No.	No. times operation carried out	Handling code	Time, s	Insertion code	Insertion time, s	Total time, s	Pneumatic piston
7	1	30	1.95	00	1.5	3.45	Block
6	1	15	2.25	22	6.5	8.75	Piston
5	1	10	1.50	00	1.5	3.0	Piston stop
4	1	80	4.10	00	1.5	5.6	Spring
3	1	28	3.18	09	7.5	10.7	Cover
2	2	68	8.00	39	8.0	29.0	Screw
						Total time: 60.5	

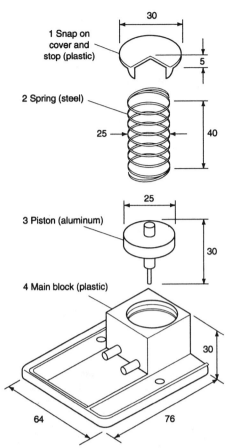

Fig. 6 Simplified assembly of pneumatic piston. Dimensions in millimeters

Table 4 Assembly times for simplified piston design (Fig. 6)

Part ID No.	No, times operation carried out	Handling code	Times, s	Insertion code	Insertion time, s	Total time, s	Pneumatic piston
4	1	30	1.95	00	1.5	3.45	Block
3	1	15	2.25	00	1.5	3.75	Piston
2	1	80	4.10	00	1.5	5.6	Spring
1	1	10	1.50	30	2.0	3.5	Cover/stop
					Total time: 16.3		

Table 5 Original manual assembly design

Feature (No.)	Dimensions, mm	Tolerance	Bits	Notes
2	3	0.063	11.14693	Diameter dimension
2	30	0.1	16.45764	Horizontal location
2	30	0.1	16.45764	Horizontal location
2	2160	60	10.33985	6 turns to install screw
Subtotal			**54.40206**	
1	31	0.1	8.276124	Horizontal location of plate
1	31	0.1	8.276124	Horizontal location of plate
Subtotal			**70.95431**	
1	25	0.43	5.861448	Spring location
Subtotal			**76.81575**	
1	25	0.1	7.965784	Piston location
Subtotal			**84.78154**	
1	25	0.1	7.965784	Piston stop
Total:			**92.74732**	**1.5 bits/s**

contract signed June 7, 1996, *Tiltrotor Times*, Vol 1 (No. 5), Aug 1996
11. W.N. Nordquist, *Die Designing and Estimating*, 4th ed., Huebner Publishing, 1955
12. E. Oberg, F.D. Jones, and H.L. Horton, *Machinery's Handbook*, 22nd ed., Industrial Press, 1987, p 1188–1205
13. G.J. Thuesen, and W.J. Fabrycky, *Engineering Economy*, Prentice Hall, 1989, p 472–474
14. D.R. Wilson, "An Exploratory Study of Complexity in Axiomatic Design," Doctoral Thesis, Massachusetts Institute of Technology, 1980
15. P.M. Fitts, and M.I. Posner, Human Performance, Brooks/Cole Publishing, Basic Concepts in Psychology Series, 1967
16. J. Annett, C.W. Golby, and H. Kay, The Measurement of Elements in an Assembly Task — The Information Output of the Human Motor System, *Quart. J. Experimental Psychology*, Vol 10, 1958
17. G. Boothroyd and P. Dewhurst, *Product Design for Assembly*, Boothroyd Dewhurst, 1989

Table 6 Simplified manual assembly design

Feature (No.)	Dimensions, mm	Tolerance	Bits	Notes
1	25	0.1	7.965784	Piston location
1	25	0.43	5.861448	Spring location
1	25	0.1	7.965784	Piston cap
Total:			**21.79302**	**%reduction 76.50281**

Reference Information

ASM Handbook, Volume 22B, *Metals Process Simulation*
D.U. Furrer and S.L. Semiatin, editors

Copyright © 2010, ASM International®
All rights reserved.
www.asminternational.org

Software for Computational Materials Modeling and Simulation

This compilation is intended to demonstrate the depth and breadth of commercial and third-party software packages available to simulate metals processes. It also represents the spectrum of applications, from simulation of atomic-level effects to manufacturing optimization. This is a first attempt to present commercial software together in one place and is recognized as being incomplete. Indeed, most vendors listed in these tables offer many other software packages tailored to specific simulation applications.

Because software is continually updated, no guarantee is made regarding the accuracy of the software packages or availability, vendor, or website information. Readers are invited to recommend additions or changes to this compilation. Please send the information to MemberServiceCenter@asminternational.org, with the subject line "Attn Handbook Editor, V22B."

Table 1   Electronic structure methods (density functional theory, quantum chemistry)

Inputs: atomic number, mass, valence electrons, crystal structure and lattice spacing, Wyckoff positions, atomic arrangement. Outputs: electronic properties, elastic constants, free energy vs. structure and other parameters, activation energies, reaction pathways, defect energies and interactions.

Software name	Function or process applications	Vendor or developer	URL
a.chem	a.chem is short for a.chem.upenn.edu, a multiuser, UNIX-based minisupercomputer at the Chemistry Computer Facility of the Department of Chemistry at the University of Pennsylvania. It offers molecular modeling, simulation, mathematical software packages, and the Cambridge Structural Database.	University of Pennsylvania	http://help.chem.upenn.edu/
CASTEP	CASTEP is a software package that uses density functional theory to provide an atomic-level description of materials and molecules. Information can be provided about total energies, forces, and stresses on an atomic system as well as calculation of optimal geometries, band structures, optical spectra, and phonon spectra. It can perform molecular dynamics simulations.	Accelrys	www.castep.org www.accelrys.com
DACAPO	DACAPO is a total-energy program based on density functional theory. It uses a plane-wave basis for the valence electronic states and describes the core-electron interactions with Vanderbilt ultrasoft pseudopotentials.	Technical University of Denmark, Denmark	https://wiki.fysik.dtu.dk/dacapo/Dacapo
GAMESS	GAMESS (General Atomic and Molecular Electronic Structure System) is a general program for ab initio molecular quantum chemistry calculations and functional theory approximation.	Ames Laboratory/Iowa State University	http://www.msg.ameslab.gov/gamess/index.html
Gaussian	Gaussian is an electronic structure program that predicts the energies, molecular structures, vibrational frequencies, and molecular properties from the fundamental laws of quantum mechanics. Gaussian models can be applied to both stable species and short-lived intermediate and transition structures.	Gaussian, Inc.	http://www.gaussian.com/index.htm
SIESTA	SIESTA (Spanish Initiative for Electronic Simulations with Thousands of Atoms) performs electronic structure calculations and ab initio molecular dynamics simulations of molecules and solids. It uses a density functional theory code that predicts the physical properties of a collection of atoms. Properties that can be predicted using the code include Kohn-Sham band structures, electron density, and Mulliken populations.	Nanotec Electronica	www.nanotec.es http://www.icmab.es/siesta/
VAMP	VAMP is a semiempirical molecular orbital package for molecular organic and inorganic systems. It is an intermediate module between force-field and first-principles methods and is capable of calculating many physical and chemical molecular properties.	Accelrys	www.accelrys.com http://accelrys.com/products/datasheets/vamp.pdf
VAMP/VASP	VAMP/VASP performs ab initio quantum-mechanical molecular dynamics using pseudopotentials and a plane-wave basis set.	University of Vienna, Austria	http://cms.mpi.univie.ac.at/vasp/Welcome.html
WIEN2k	WIEN2k performs electronic structure calculations of solids using density functional theory. It is based on the full-potential (linearized) augmented plane-wave and local orbitals method.	Vienna University of Technology, Austria	http://www.wien2k.at/index.html

Table 2 Atomistic simulations (molecular dynamics, Monte Carlo)

Inputs: interaction scheme, potentials, methodologies, benchmarks. Outputs: thermodynamics, reaction pathways, structures, point defect and dislocation mobility, grain-boundary energy and mobility, precipitate dimensions.

Software name	Function or process applications	Vendor or developer	URL
Cerius2	Cerius2 offers a wide range of scientific application modules for various molecular environments, including materials science and life science.	Accelrys	www.accelrys.com http://accelrys.com/resource-center/case-studies/archive/studies/catshape_two_ligands.html
DL_POLY	DL_POLY is a parallel molecular dynamics simulation package designed to facilitate molecular dynamics simulations of macromolecules, polymers, ionic systems, and solutions.	Science and Technology Facilities Council, Daresbury Laboratory, U.K.	www.stfc.ac.uk http://www.cse.scitech.ac.uk/ccg/software/DL_POLY/
DMol^3	DMol^3 uses density functional theory to simulate chemical processes and predict properties of materials. It investigates the nature and origin of the chemical, electronic, and structural properties of a system without the need for experimental input.	Accelrys	www.accelrys.com http://accelrys.com/products/datasheets/dmol3.pdf
LAMMPS	LAMMPS (Large-Scale Atomic/Molecular Massively Parallel Simulator) is a classical molecular dynamics code created for simulating molecular and atomic systems such as proteins in solution, liquid crystals, polymers, zeolites, or simple Lenard-Jonesium. LAMMPS features parallelism via a spatial decomposition algorithm; short-range pairwise Lennard-Jones and Coulombic interactions; long-range Coulombic interactions via Ewald or particle-mesh Ewald; harmonic molecular potentials; class II molecular potentials; NVE, NVT, and NPT dynamics; constraints on atoms or groups of atoms, including SHAKE; rRESPA long-timescale integrator; and energy minimizer (Hessian-free truncated Newton method).	Sandia National Laboratory	http://lammps.sandia.gov
ParaDyn	ParaDyn is a parallel F77 code that implements all the features of the popular serial EAM code for metals and metal alloys known as DYNAMO. Nearly all of ParaDyn's functionality is now in LAMMPS.	Sandia National Laboratory	http://www.sandia.gov/~sjplimp/download.html#pd

Table 3 Dislocation dynamics

Inputs: crystal structure and lattice spacing, elastic constants, boundary conditions, mobility laws. Outputs: stress-strain behavior, hardening behavior, effect of size scale.

Software name	Function or process applications	Vendor or developer	URL
microMegas (mM)	mM is an open-source dislocation dynamics program intended for the modeling and visualization of dislocation-based plastic deformation in crystals. The output of the simulation yields information about the microstructure, local quantities of interest (internal stresses, dislocation densities, slip systems activity), and the global mechanical response.	Laboratoire d'Etude des Microstructures	http://zig.onera.fr/mm_home_page/index.html#Introduction
ParaDiS	ParaDiS (Parallel Dislocation Simulator) is a large-scale dislocation dynamics simulation code to study the fundamental mechanisms of plasticity.	Lawrence Livermore National Laboratory	https://www.llnl.gov/str/November05/Bulatov.html
PARANOID	PARANOID simulates dislocation behavior in a wide variety of situations. The code is based on elastic theory in the continuum limit and should be applicable to situations in which the dislocation cores are more than a few nanometers apart. The stress tensor that moves the dislocations is calculated at every point by evaluating the full Peach-Koehler expression over all of the dislocations present.	IBM Research	http://www.research.ibm.com/dislocationdynamics

Table 4 Thermodynamic methods (CALPHAD)

Inputs: free-energy data from electronic structure, calorimetry data, free-energy functions fit to materials databases. Outputs: phase predominance diagrams, phase fractions, multicomponent phase diagrams, free energies.

Software name	Function or process applications	Vendor or developer	URL
FactSage	FactSage calculates the conditions for multiphase, multicomponent equilibria.	Thermfact and GTT-Technologies	http://www.factsage.com
Gemini	Gibbs Energy MINImizer. Thermodynamic software	Thermodata	www.gmini.org
JMat Pro	JMatPro is aimed at multicomponent alloys used in industrial practice. It makes calculations for stable and metastable phase equilibrium, solidification behavior and properties, mechanical properties, thermophysical and physical properties, phase transformations, and chemical properties.	Sente Software	www.sentesoftware.co.uk
MeltCalc	Thermodynamic modeling spreadsheet for Excel	Corbett Research	www.meltcalc.com
MTDATA	MTDATA calculates phase equilibria and thermodynamic properties in multicomponent, multiphase systems using critically assessed thermodynamic data. It has applications in the fields of metallurgy, chemistry, materials science, and geochemistry, depending on the data available. Problems of mixed character can be handled, for example, equilibria involving the interaction between liquid and solid alloys and matte, slag, and gas phases.	National Physical Laboratory, U.K.	www.npl.co.uk/advanced-materials/measurement-techniques/modelling/mtdata
Pandat	Pandat software is an integrated computational environment for phase diagram calculation and materials property simulation of multicomponent systems based on the CALPHAD (CALculation of PHAse Diagram) approach.	CompuTherm LLC	http://www.computherm.com/home.html
Thermo-Calc	Thermo-Calc performs thermodynamic and phase diagram calculations for multicomponent systems of practical importance. Databases are available for steels; Ti, Al, Mg, and Ni alloys; multicomponent oxides; and many other materials.	Thermo-Calc Software	www.thermocalc.com

Table 5   Microstructural evolution methods (phase field, front-tracking methods, Potts models)

Inputs: free-energy and kinetic databases (atom mobilities), interface and grain-boundary energies, (anisotropic) interface mobilities, elastic constants. Outputs: solidification and dendritic structure, microstructure during processing, deployment, evolution in service.

Software name	Function or process applications	Vendor or developer	URL
DICTRA	DICTRA simulates diffusion in multicomponent alloy systems. Processes that can be simulated include homogenization of alloys, carburizing and decarburizing of steels, microsegregation during solidification, carbide dissolution during austenitization of steels, and coarsening of precipitates.	Thermo-Calc Software	http://www.thermocalc.com
JMat Pro	JMatPro is aimed at multicomponent alloys used in industrial practice. It makes calculations for stable and metastable phase equilibria, solidification behavior and properties, mechanical properties, thermophysical and physical properties, phase transformations, and chemical properties.	Sente Software Ltd.	http://www.sentesoftware.co.uk/jmatpro.aspx
MICRESS	MICRESS (MICRostructure Evolution Simulation Software) applies a multiphase-field method for multicomponent alloys and enables the treatment of multiphase, multigrain, and multicomponent problems in the fields of solidification, grain growth, recrystallization, or solid-state phase transformations. It covers phase evolution, solutal and thermal diffusion, and transformation strain in the solid state.	ACCESS e.V., Aachen University of Technology (RWTH), Germany	http://web.access.rwth-aachen.de/MICRESS
μMatIC	μMatIC three-dimensional code predicts microstructural features of solidification, including dendritic morphology, spacing, grain structure, concentration profile, pores, casting defects, etc.	Imperial College London, U.K.	http://www3.imperial.ac.uk/advancedalloys/software
PrecipiCalc	PrecipiCalc calculates three-dimensional multiparticle diffusive precipitation kinetics of multiple phases. PrecipiCalc adopts multicomponent thermodynamics and mobility in its precipitation models to allow realistic and mechanistic modeling on nucleation, growth, and coarsening.	QuesTek	http://www.questek.com

Table 6   Micromechanical and mesoscale property models (solid mechanics and finite-element analysis)

Inputs: microstructural characteristics, properties of phases and constituents. Outputs: properties of materials, for example, modulus, strength, toughness, strain tolerance, thermal/electrical conductivity, and permeability; possibly creep and fatigue behavior.

Software name	Function or process applications	Vendor or developer	URL
DARWIN	DARWIN integrates finite-element stress analysis results, fracture mechanics-based life assessment for low-cycle fatigue material anomaly data, probability of anomaly detection, and inspection schedules to predict the probability of fracture as a function of applied operating cycles.	Southwest Research Institute (SwRI)	http://www.swri.org/4org/d18/mateng/probfrac/default.htm
FRANC3D	FRANC3D provides a mechanism for representing the geometry and topology of three-dimensional structures with arbitrary nonplanar cracks, along with functionality for discretizing or meshing the structure, attaching boundary conditions at the geometry level and allowing the mesh to inherit these values, and modifying the geometry to allow crack growth, but with only local remeshing required during crack growth.	Cornell Fracture Group	http://www.cfg.cornell.edu/software/franc3d_osm.htm
JMat Pro	JMatPro is aimed at multicomponent alloys used in industrial practice. It makes calculations for stable and metastable phase equilibria, solidification behavior and properties, mechanical properties, thermophysical and physical properties, phase transformations, and chemical properties.	Sente Software Ltd.	http://www.sentesoftware.co.uk/jmatpro.aspx
OOF	OOF Software is designed to help materials scientists calculate macroscopic properties from images of real or simulated microstructures. It reads an image, assigns material properties to features in the image, and conducts virtual experiments to determine the macroscopic properties of the microstructure.	National Institute of Standards and Technology	http://www.ctcms.nist.gov/oof/oof2
Voro++	Voro++ is an open-source software library for the computation of Voronoi diagrams (a special kind of decomposition of a metric space determined by distances to a specified set of objects in the space).	University of California-Berkeley, Lawrence Berkeley Laboratory	http://math.lbl.gov/voro++/about.html
Zencrack	Zencrack is a state-of-the-art software tool for three-dimensional fracture mechanics simulation. The program uses finite-element analysis to allow calculation of fracture mechanics parameters, such as energy-release rate and stress-intensity factors.	Zentech International Limited	http://www.zentech.co.uk/zencrack.htm

Table 7   Microstructural imaging software

Inputs: images from optical microscopy, electron microscopes, x-rays, etc. Outputs: image quantification and digital representations.

Software name	Function or process applications	Vendor or developer	URL
3D-DOCTOR	3D-DOCTOR is an advanced three-dimensional modeling, image processing, and measurement software for magnetic resonance imaging, computed tomography, positron emission tomography, microscopy, and scientific and industrial imaging applications. It supports both grayscale and color images stored in DICOM, TIFF, Interfile, GIF, JPEG, PNG, BMP, PGM, MRC, RAW, or other image file formats. 3D-DOCTOR creates three-dimensional surface models and volume rendering from two-dimensional cross-sectional images in real time.	Able Software Corp.	http://www.ablesw.com/3d-doctor
Amira	Amira is a multifaceted software platform for visualizing, manipulating, and understanding life science and biomedical data coming from all types of sources and modalities, such as clinical or preclinical image data, nuclear data, optical or electron microscopy imagery, molecular models, vector and flow data, simulation data on finite-element models, and all types of multidimensional image, vector, tensor, and geometry data.	Visage Imaging, Inc.	www.amira.com
IDL	IDL provides a computing environment for data analysis, visualization, and application development.	ITT Visual Information Solutions	http://www.ittvis.com/ProductServices/IDL.aspx
ImageJ	ImageJ is an open-source program for image processing and analysis in JAVA. It can display, edit, analyze, process, save, and print 8-, 16-, and 32-bit images. It can read image formats including TIFF, GIF, JPEG, BMP, DICOM, FITS, and RAW.	National Institutes of Health	www.nih.gov http://rsb.info.nih.gov/ij

(continued)

Table 7 (continued)

Software name	Function or process applications	Vendor or developer	URL
	It is multithreaded, so time-consuming operations such as image file reading can be performed in parallel with other operations.		
ITK	ITK (Insight Segmentation and Registration Toolkit) is an open-source, cross-platform system that provides developers with a suite of software tools for image analysis. Developed through extreme programming methodologies, ITK's algorithms allow for registering and segmenting multidimensional data.	National Library of Medicine, ITK Project	http://www.itk.org
Mimics	Mimics software processes and edits two-dimensional image data, such as computed tomography, microcomputed tomography, or magnetic resonance imaging. From these two-dimensional data, the software constructs three-dimensional models. Mimics bridges the gap between image data and a wide variety of engineering applications, such as finite-element analysis, design, surgical simulation, additive manufacturing, etc.	Materialise	http://www.materialise.com/mimics
ParaView	ParaView is an open-source, multiplatform data analysis and visualization application. ParaView builds visualizations to analyze data using qualitative and quantitative techniques. The data exploration can be done interactively in three dimensions or programmatically using ParaView's batch processing capabilities. The software was developed to analyze extremely large datasets using distributed memory computing resources. It can be run on supercomputers to analyze datasets of terascale as well as on laptops for smaller data.	Kitware Inc.	www.paraview.org
VGStudio MAX	VGStudio MAX is an industrial computer tomography data analysis software package. It allows two- and three-dimensional dimensioning, wall-thickness analysis, and extraction of STL surface models and point clouds. It can be used for defect/porosity/inclusion detection, crack segmentation, material distribution in composite materials, and component assembly analysis.	Volume Graphics GmbH	http://www.volumegraphics.com/products/vgstudiomax
VTK	VTK (Visualization Toolkit) is an open-source software system for three-dimensional computer graphics, image processing, and visualization. VTK supports a wide variety of visualization algorithms, including scalar, vector, tensor, texture, and volumetric methods and advanced modeling techniques, such as implicit modeling, polygon reduction, mesh smoothing, cutting, contouring, and Delaunay triangulation.	Kitware Inc.	www.vtk.org

Table 8 Mesoscale structure models (mesoscale material-behavior models)

Inputs: processing thermal and strain history. Outputs: microstructural characteristics (for example, grain size, texture, precipitate dimensions).

Software name	Function or process applications	Vendor or developer	URL
JMat Pro	JMatPro is aimed at multicomponent alloys used in industrial practice. It makes calculations for stable and metastable phase equilibria, solidification behavior and properties, mechanical properties, thermophysical and physical properties, phase transformations, and chemical properties.	Sente Software Ltd.	http://www.sentesoftware.co.uk/jmatpro.aspx
PrecipiCalc	PrecipiCalc calculates three-dimensional multiparticle diffusive precipitation kinetics of multiple phases. PrecipiCalc adopts multicomponent thermodynamics and mobility in its precipitation models to allow realistic and mechanistic modeling on nucleation, growth, and coarsening.	QuesTek	http://www.questek.com

Table 9 Part-level finite-element analysis, finite difference, and other continuum models

Inputs: part/tooling geometry, manufacturing processing parameters, component loads, material properties. Outputs: distribution of temperatures, stresses, and deformation; electrical currents; magnetic and optical behavior; etc. Can be linked with other models for prediction of microstructure and mechanical properties.

Software name	Function or process applications	Vendor or developer	URL
Abaqus/Standard	Abaqus/Standard is designed to model static and low-speed dynamic events where highly accurate stress solutions are critically important. Examples include sealing pressure in a gasket joint, steady-state rolling of a tire, or crack propagation in a composite airplane fuselage. Within a single simulation, it is possible to analyze a model in both the time and frequency domain. Following the mounting analysis, the prestressed natural frequencies of the cover can be extracted, or the frequency domain mechanical and acoustic response of the prestressed cover-to-engine induced vibrations can be examined.	Dassault Systèmes	http://www.simulia.com
ANSYS 12	ANSYS 12 offers dramatic simulation for every major physics discipline, including structural analysis, explicit analysis, thermal analysis, fluid dynamics, electromagnetics-low frequency, electromagnetics-high frequency, advanced materials characteristics, and coupled physics.	ANSYS, Inc.	www.ansys.com
ANSYS FLUENT	ANSYS FLUENT software models flow, turbulence, heat transfer, and reactions for industrial applications ranging from air flow over an aircraft wing to combustion in a furnace, from bubble columns to oil platforms, from blood flow to semiconductor manufacturing, and from clean-room design to wastewater treatment plants. It has special models for modeling in-cylinder combustion, aeroacoustics, turbomachinery, and multiphase systems.	ANSYS, Inc.	www.fluent.com
ANSYS ICEM CFD	ANSYS ICEM CFD meshing software starts with advanced computer-aided design/geometry readers and repair tools that allow the user to quickly move to a variety of geometry-tolerant meshers and produce high-quality volume or surface meshes. The software provides advanced mesh diagnostics, interactive and automated mesh editing, output to a	Ansys Inc.	www.ansys.com

(continued)

Table 9 (continued)

Software name	Function or process applications	Vendor or developer	URL
	wide variety of computational fluid dynamics and finite-element analysis solvers, and multiphysics postprocessing tools.		
AnyCasting	AnyCasting is applicable to all casting processes, including sand mold casting, permanent casting, low- and high-pressure die casting, tilt casting, and investment casting.	AnyCasting Co., Inc.	www.anycasting.com www.anycasting.com/en/ software/anycastingtm. php
CAPCAST	CAPCAST is a suite of finite-element-method-based software for solidification modeling, including thermal, solidification, flow, porosity, and stress analysis tools.	EKK Inc.	http://www.ekkinc.com/ tools.html
COMSOL	COMSOL is a modular suite of software tools for multiphysics modeling and simulation. Available modules include alternating/direct current, microelectromechanical systems, radio frequency, Optimization Lab, acoustics, heat transfer, structural mechanics, computer-aided design import, chemical engineering, COMSOL Reaction Engineering Lab, and Materials Library.	COMSOL AB	www.comsol.com
DANTE	DANTE is a finite-element-method-based heat treating simulation software that provides an accurate description of heat transfer during heating and cooling processes. It addresses immersion quenching, intensive quenching methods, gas quenching, press quenching, and low-temperature tempering processes.	Deformation Control Technology, Inc.	www.deformationcontrol. com
DEFORM	DEFORM is an engineering software that enables designers to analyze metal forming, heat treatment, machining, and mechanical joining processes. It simulates distribution of temperatures during heat treatment, stresses, strains, and residual stresses.	Scientific Forming Technologies Corporation	www.deform.com
DYNA3D	DYNA3D is an explicit finite-element program for structural/continuum mechanics problems. It uses small time steps to integrate the equations of motion and is efficient in solving transient dynamic problems. The material library includes isotropic elastic, orthotropic elastic, elastic-plastic, orthotropic elastic-plastic, rate-dependent elastic-plastic, temperature-dependent elastic-plastic, concrete, and rubberlike materials. Its element library includes solid, shell, beam, bar, spring, and damper elements. DYNA3D also has various contact surface options for interaction effects between two bodies.	Lawrence Livermore National Laboratory	https://www-eng.llnl.gov/ mdg/mdg_codes_dyna3d. html
EDEM	EDEM is a general-purpose computer-aided engineering tool using discrete element modeling technology for the simulation and analysis of particle handling and manufacturing operations and creates a parameterized model of granular solids systems. Computer-aided design models of real particles can be imported to obtain an accurate representation of their shape. EDEM manages information about each individual particle (mass, temperature, velocity, particle shape, etc.) and the forces acting on it.	DEM Solutions Ltd.	http://www.dem-solutions. com/edem.php
FARADAY	FARADAY is a three-dimensional, time-harmonic eddy current field solver that uses boundary-element method (BEM), finite-element method (FEM), or hybrid method (BEM and FEM combination) technologies. It is capable of large open-region analysis, exact modeling of boundaries, and solving problems where dealing with small skin depths is critical.	Integrated Engineering Software Sales, Inc.	www.integratedsoft.com http://www.integratedsoft. com/products/faraday/ default.aspx
FLUX	FLUX is a general-purpose computer-aided engineering software that models quantities such as flux density or heat distribution. It can use variation curves versus time or position, or make spatial spectrum computations. It can analyze global quantities such as inductance, energy, force, and torque.	Magsoft Corporation	www.magsoft-flux.com http://www.magsoft-flux. com/Products/Flux
FORGE	FORGE simulates hot, warm, and cold forging of both three-dimensional parts (e.g., steering knuckles, crankshafts, twin connection rods, lower arms, constant-velocity joints, bevel gears, aircraft landing gears, fan blades, engine mountings, and wing components) and two-dimensional geometry parts (axisymmetric parts and parts with high length-to-width ratios, such as cylinders, impacts, extrusions, axles, shafts, gear blanks, rings, fasteners and wire drawing, aircraft disks, blades and wheels, bearing cages, and railway wheels).	Transvalor Inc. (Europe) Technology Management Services (U.S. and Australia)	www.transvalor.com http://www.transvalor.com/ forge_gb.php
INDUCTO	INDUCTO is a coupled two-dimensional/RS magnetic and thermal analysis program for applications involving inductive heating. It uses the boundary-element method (BEM), finite-element method (FEM), or hybrid method (BEM and FEM combination) solver. The software simulates static and phasor analysis modes and nonlinear, permanent magnet and lossy magnetic materials based on boundary conditions such as temperature, heat flux, temperature gradient, and convective and radiative heat exchange. Heat sources are assigned in the form of volume heat and surface heat. A material table stores thermal conductivity, specific heat, mass density or absorption, and scattering coefficients.	Integrated Engineering Software, Inc.	www.integratedsoft.com http://www.integratedsoft. com/products/inducto/ default.aspx
JSCast	JSCast is a three-dimensional integrated computer-aided engineering system for casting optimization and includes mold-filling, solidification, and deformation simulations.	QUALICA, Inc.	http://www.jscast.com.cn/ english.htm
KIVA	KIVA is a three-dimensional, coupled mechanics analysis software package for use in understanding and designing heat treatment processes. The focus is on thermal convection heat treatment.	Deformation Control Technology, Inc.	http://deformationcontrol. com/dct_products.htm
LS-DYNA	LS-DYNA is a general-purpose transient dynamic finite-element program capable of simulating complex real-world problems. It is optimized for shared and distributed memory Unix, Linux, and Windows-based platforms.	Livermore Software Technology Corp.	http://www.lstc.com/ lsdyna.htm
MAGMASOFT	MAGMASOFT is a simulation tool for the technological and quality-focused production of castings. Simulation capabilities provide an understanding of mold filling, solidification, mechanical properties, thermal stresses and distortions, etc. It addresses gating and feeding problems, predicts casting quality, aids permanent mold design, and reduces fettling costs.	MAGMA GmbH	http://www.magmasoft.de/ms/ products_en_magmasoft/ index.php
MagNet	MagNet is a two-/three-dimensional simulation software for electromagnetic fields to model and predict the performance of electromagnetic or electromechanical devices, such as electric motors/generators, magnetic levitation, transformers, actuators, sensors/ nondestructive testing, induction heating, loudspeakers, magnetic recording heads, magnetic resonance imaging, and transcranial magnetic simulations.	Infolytica Corp.	www.infolytica.com/ http://www.infolytica.com/ en/products/magnet
MAVIS-FLOW	MAVIS-FLOW simulates mold filling and solidification of cast alloys. A rapid solidification model is employed to simulate solidification and internal shrinkage defects in sand, investment, gravity die, and high-pressure die castings. The 2000 update added a Navier-	Alphacast Software Ltd.	www.alphacast-software. co.uk/index4.htm

(continued)

Table 9 (continued)

Software name	Function or process applications	Vendor or developer	URL
	Stokes fluid flow simulator coupled with heat transfer and solidification, which improved in-gate, overflow, and turbulence simulation.		
Maxwell	Maxwell is an electromagnetic field simulation software used for the design and analysis of two-/three-dimensional structures, such as motors, actuators, transformers, and other electric and electromechanical devices common to automotive, military/aerospace, and industrial systems. Based on the finite-element method, Maxwell solves static, frequency-domain and time-varying electromagnetic, and electric fields.	ANSYS, Inc.	www.ansys.com http://www.ansoft.com/ products/em/maxwell
MeltFlow	MeltFlow-VAR and MeltFlow-ESR use advanced computational fluid dynamics techniques that have been developed specifically for a detailed analysis of the electromagnetic (direct current in vacuum arc remelting, or VAR; alternating current in electroslag remelting, or ESR), fluid flow, heat transfer, and alloy element redistribution phenomena occurring in the VAR and ESR processes. The tools enable prediction of pool evolution during the entire process and the thermal history, distributions of alloy element concentrations, dendrite arm spacings, and freckle formation probabilities in the solidified ingot.	Innovative Research, Inc.	http://inres.com/products/ meltflow/Overview.html
Moldflow	Moldflow provides part-level finite-element analysis for filling, cooling, and packing analysis for plastic injection molding analysis. The simulation software, part of the Autodesk solution for digital prototyping, provides tools to optimize the design of plastic parts and injection molds and study the injection molding process. It simulates how changes to wall thickness, gate location, material, and geometry affect manufacturability.	Autodesk, Inc.	http://www.moldflow.com/ stp/
OERSTED	OERSTED is a two-dimensional/RS time-harmonic eddy current field solver for solving magnetic design problems that require large open-field analysis, exact modeling of boundaries, and for applications where dealing with small skin depths is critical. It calculates force, torque, displacement current, flux linkage, induced voltage, power, and impedance and is used for applications such as magnetic resonance imaging, nondestructive testing systems, bus bars, charging fixtures, induction heating coils, magnetic recording heads, magnetic shielding, coils and transformers, and induction motors.	Integrated Engineering Software, Inc.	www.integratedsoft.com
Opera	Opera provides tools for electromagnetic design, simulation, and analysis of results. It consists of a preprocessing environment for creating design models (or importing them from computer-aided design programs) plus a finite-element analysis solver. Three generic solvers are optionally available: static electromagnetic fields, low-frequency time-varying electromagnetic fields, and high-frequency time-varying electromagnetic fields. It is available in several forms optimized for specific design problems, including linear and rotating machinery design, superconducting magnet quenching, space charge effects from particle beams, permanent magnet magnetization/demagnetization, thermal and stress analysis (stand-alone or coupled), and electric field analysis in conducting-dielectric media.	Vector Fields Ltd.	www.vectorfields.com http://www.vectorfields. com/opera.php
PAM-STAMP 2G	PAM-STAMP 2G is an integrated, scalable, and streamlined stamping simulation package. It covers the entire tooling process from quotation and die design through formability and validation, including springback prediction and simulation. It provides modeling and simulation for automotive, aerospace, and general stamping processes.	ESI Group	www.esi-group.com http://www.esi-group.com/ products/metal-forming/ pam-stamp-2g
PHYSICA	PHYSICA models the complex interactions of physical and chemical phenomena that involve a number of complex interactions that are at the foundation of manufacturing process industries. Examples include fluid-structure interaction, aeroacoustics, magneto-hydrodynamics with heat transfer and solidification/melting, and thermally driven material defects. Manufacturing processes include casting and solidification processes, electronic manufacturing processes, granular flow, welding, forming processes (e.g., extrusion), primary metals processing (e.g., direct smelting, heap leaching), etc.	Physica Ltd.	www.physica.co.uk
PIMsolver	PIMsolver is computer-aided engineering analysis software for the powder injection molding (PIM) process. The software simulates PIM part design, mold design, and machine processing conditions for optimization during the design stage.	Ceta Tech	www.cetatech.com
PMsolver	PMsolver is computer-aided engineering software for simulation of conventional powder metallurgy, die compaction, and sintering. It is capable of predicting the density distribution after compaction, crack formation, and the deformed shape after sintering caused by heterogeneous density distributions.	Ceta Tech	www.cetatech.com
ProCAST	ProCAST provides predictive evaluations of the entire casting process, including mold filling, solidification, microstructure, and thermomechanical simulations. It covers a wide range of casting processes and alloy systems, including high- and low-pressure die casting, sand casting, gravity die casting and tilt pouring, investment casting, shell casting, lost foam, and centrifugal casting. The software can simulate steady-state conditions as well as the initial and final stages of continuous and semicontinuous casting processes, including horizontal and vertical continuous and semicontinuous casting, direct chill casting, strip casting, twin-roll casting, and the Hazlett process.	ESI Group	www.esi-group.com http://www.esi-group.com/ products/casting/procast
SimLAM	SimLAM is a software simulation package for laser-additive manufacturing. SimLAM simulates a laser deposition process to perform heat-transfer and residual-stress analyses and optimizes laser deposition processes based on user-defined criteria to minimize residual stress and maximize efficiency.	Applied Optimization, Inc.	www.appliedo.com http://www.appliedo.com/ products.html
SIMTEC	SIMTEC's finite-element method simulation enables computer-aided calculation of mold filling and solidification for a full range of casting types, including common casting processes such as high-pressure die casting, low-pressure die casting, sand casting, lost foam, permanent mold (conventional/tilt pour), investment casting, and shell casting. It also simulates special casting processes such as continuous casting, centrifugal casting, squeeze casting, and semisolid casting.	SIMTEC Inc.	www.simtec-inc.com
Simufact	Simufact is a modular simulation system for a wide variety of different production areas, including solid forming, sheet metal forming, mechanical joining, and welding. Simulation data and a project management system are integrated components of the software.	Simufact Engineering GmbH	http://www.simufact.de/en/ index.html

(continued)

Table 9   (continued)

Software name	Function or process applications	Vendor or developer	URL
SINDA/FLUINT	SINDA/FLUINT is a generalized tool for simulating complex thermal/fluid systems such as those found in the electronics, automotive, petrochemical, turbomachine, and aerospace industries. The software is a comprehensive finite-difference, lumped parameter (circuit or network analogy) tool for heat-transfer design analysis and fluid flow analysis in complex systems and uses text-based input files.	Cullimore & Ring Technologies, Inc.	www.crtech.com
SOLIDCast	SOLIDCast can be used to simulate castings poured in gray iron, ductile iron, steel, aluminum, copper-base, magnesium, nickel-base, and almost any other alloy. A database of several hundred alloys, with their properties, is included. It will simulate other foundry activities such as green sand, chemically bonded sand, investment and permanent mold, chills, hot topping, cooling channels, heating elements, riser and gating design, etc.	Finite Solutions Inc.	www.finitesolutions.com
STAR-CD	STAR-CD bridges the gap between fluid dynamics and structural mechanics by performing structural analysis calculations using computational fluid dynamics. It solves for flow, thermal, and stress simulation in a single, general-purpose, commercial finite-volume code.	CD-adapco	www.cd-adapco.com
ThermNet	ThermNet simulates the steady-state and transient temperature distribution of specified heat sources. It simulates temperature distributions caused by specified heat sources in the presence of thermally conducting materials.	Infolytica Corp.	www.infolytica.com/ http://www.infolytica.com/ en/products/thermnet/
Welding Simulation Solution	Welding Simulation Solution is used to determine whether a weld will perform as well as the parent material without compromising the parent material specified properties. Choose the best possible material, determine the best possible microstructure after welding, evaluate material exposure, control residual stresses, and minimize tensile stresses to avoid buckling.	ESI Group	www.esi-group.com http://www.esi-group.com/ products/welding

Table 10   Code and systems integration

Inputs: format of input and output of modules and the logical structure of integration of initial input. Outputs: parameters for optimized design, sensitivity to variations in inputs or individual modules.

Software name	Function or process applications	Vendor or developer	URL
Isight (formerly Fiper)	Isight is a suite of visual tools for creating simulation process flows consisting of a variety of applications, including commercial computer-aided design/engineering software, internally developed programs, and Excel spreadsheets, to automate the exploration of design alternatives and identification of optimal performance parameters.Isight enables users to automate simulation process flows and leverage advanced techniques such as design of experiments, optimization, approximations, and design for Six Sigma.	Dassault Systèmes	http://www.simulia.com/ products/isight2.html
QMD-FLAPW	QMD-FLAPW provides functionalities and applications that include precise total energies and atomic forces with full structure optimization; phonon dynamical matrix and vibrational frequencies; spin-orbit-induced magnetic phenomena: magneto-crystalline anisotropy, magnetostriction, magnetic circular dichroism, surface magneto-optic Kerr effect, beyond local density approximation (LDA) with screened exchange-LDA, and model GW for excited-state properties (band gaps, band offsets, etc.); optical properties totally from first principles with full exponential form for matrix elements; and magnetic and electric quadrupole hyperfine interactions. Typical outputs are band structure, wave function, Fermi surface, density of states, total energy, atomic forces, charge density, spin density, optimized geometry, elastic stiffness constants, dielectric functions, spin moment, magneto-crystalline anisotropy, hyperfine field, orbital moment, magnetic circular dichroism, magneto-optic Kerr effect, and magnetostriction.	Quantum Materials Design, Inc.	http://flapw.com/products. html

Table 11   Statistical tools and design optimization (neural nets, principal component analysis)

Inputs: composition, process conditions, properties. Outputs: correlations between inputs and outputs; mechanistic insights.

Software name	Function or process applications	Vendor or developer	URL
Isight (formerly Fiper)	Isight is a suite of visual tools for creating simulation process flows consisting of a variety of applications, including commercial computer-aided design/engineering software, internally developed programs, and Excel spreadsheets, to automate the exploration of design alternatives and identification of optimal performance parameters.Isight enables users to automate simulation process flows and leverage advanced techniques such as design of experiments, optimization, approximations, and design for Six Sigma.	Dassault Systèmes	http://www.simulia.com/products/isight2.html
MATLAB	MATLAB is a high-level technical computing language and interactive environment for algorithm development, data visualization, data analysis, and numeric computation. Applications of MATLAB include signal and image processing, communications, control design, test and measurement, financial modeling and analysis, and computational biology.	The MathWorks, Inc.	http://www.mathworks.com/products/matlab
Minitab	Minitab provides a collection of statistical tools for quality improvement and quality assurance, including descriptive statistics, hypothesis tests, confidence intervals, and normality tests.	Minitab Inc.	http://www.minitab.com/en-US/default.aspx
PatternMaster	PatternMaster is a neural-network trainer capable of dealing with extremely large problems. Factor and schema analysis is achieved in a three-dimensional virtual reality environment. Training sessions are governed and archived via an XML scheme called ANNML (Artificial Neural Network Markup Language).	Imagination Engines, Inc.	http://www.imagination-engines.com

(continued)

Table 11 (continued)

Software name	Function or process applications	Vendor or developer	URL
PHX ModelCenter	PHX ModelCenter provides a graphical environment for process integration and design automation that enables engineers to perform design iterations during product development and verification to meet targets on quality, cost, and delivery schedules. It works with computer-aided engineering software tools.	Phoenix Integration, Inc.	www.phoenix-int.com
Pro/ENGINEER	Pro/ENGINEER is a three-dimensional product design package that uses integrated, parametric, three-dimensional computer-aided design/manufacturing/engineering solutions. It automatically propagates design changes to all downstream deliverables, has virtual simulation capability, and can generate associative tooling design and manufacturing deliverables.	Parametric Technology Corp.	www.ptc.com
SAS/STAT	SAS/STAT software provides statistical analysis from traditional statistical analysis of variance and predictive modeling to exact methods and statistical visualization techniques. It is capable of handling large data sets from disparate sources.	SAS Institute Inc.	http://www.sas.com/software
Spotfire S+	Spotfire S+ is a statistical analysis, graphics, and programming package for Windows, UNIX, and Linux based on S-programming language	TIBCO Software Inc.	http://spotfire.tibco.com/Products/S-Plus-Overview.aspx
SYSTAT 13	SYSTAT 13 is a statistical analysis and graphics software that provides common procedures, such as linear regression, analysis of variance, and nonparametric tests, and advanced methods, such as mixed-model analysis, advanced regression (e.g., robust, nonlinear, partial least squares, etc.), and response-surface methods. Also included are new statistical methods such as ARCH and GARCH models in time series, best subsets regression, confirmatory factor analysis, environment variables in best statistics, and polynomial regression.	Cranes Software International Ltd.	http://www.systat.com/Products.aspx

Table 12   Material property and selection databases

Inputs: composition, process conditions, properties. Outputs: comparative analysis of material specifications or grades.

Software name	Function or process applications	Vendor or developer	URL
Alloy Finder	Alloy Finder features information on thousands of alloys from around the world. Included is key alloy information, including composition, producer, tensile properties, and similar alloys.	ASM International	www.asminternational.org http://products.asminternational.org/alloyfinder/index.jsp
AMPTIAC	AMPTIAC supports researchers, designers, and decision makers in their efforts to enhance the performance of systems in any environment through proper materials selection, processing, and use. It consists of five separate material components: ceramic and ceramic-matrix composites; organic structural materials and organic-matrix composites; monolithic metals, alloys, and metal-matrix composites; electronic, optical, and photonic materials; and environmental-protection and special-function materials.	Advanced Materials and Processes Technology Information Analysis Center	http://iac.dtic.mil/iac_dir/AMPTIAC.html
CES Selector	CES Selector, based on the Cambridge Engineering Selector (CES) technology, is a tool for the rational selection of engineering materials (metals, ceramics, polymers, composites, woods) and manufacturing processes (shaping, finishing, joining, and surface treatment). The CES Selector can be implemented with a proprietary database supplied by the user or with data products from Granta Design.	Granta Design	www.grantadesign.com
CINDAS	CINDAS provides material properties databases for thermal, mechanical, electrical, physical, and other properties of various materials including aerospace alloys. The web-based application contains over 5000 materials and approximately 50,000 data curves in the Thermophysical Properties of Matter Database, over 750 materials and over 18,000 data curves in the Microelectronics Packaging Materials Database, and over 80,000 data curves in the Aerospace Structural Metals Database.	Cindas LLC	https://cindasdata.com
KEY to METALS	KEY to METALS contains over 4 million property records for steel, aluminum, copper, titanium, and other metals from more than 40 countries and standards, including chemical compositions, international cross-reference tables, mechanical properties, physical properties, high-temperature properties, heat treatment data and diagrams, fatigue data, and applications guidelines.	Key to Metals AG	www.key-to-metals.com
MatWeb	MatWeb is a searchable database of material properties and includes data sheets of thermoplastic and thermoset polymers such as acrylonitrile-butadiene-styrene, nylon, polycarbonate, polyester, polyethylene, and polypropylene; metals such as aluminum, cobalt, copper, lead, magnesium, nickel, steel, superalloys, titanium, and zinc alloys; ceramics; plus semiconductors, fibers, and other engineering materials.	MatWeb	www.matweb.com

Copyright © 2010, ASM International®
All rights reserved.
www.asminternational.org

Metric Conversion Guide

This Section is intended as a guide for expressing weights and measures in the Système International d'Unités (SI). The purpose of SI units, developed and maintained by the General Conference of Weights and Measures, is to provide a basis for worldwide standardization of units and measure. For more information on metric conversions, the reader should consult the following references:

- *The International System of Units*, SP 330, 1991, National Institute of Standards and Technology. Order from Superintendent of Documents, U.S. Government Printing Office, Washington, DC 20402-9325
- *Metric Editorial Guide,* 5th ed. (revised), 1993, American National Metric Council, 4340 East West Highway, Suite 401, Bethesda, MD 20814-4411

- "Standard for Use of the International System of Units (SI): The Modern Metric System," IEEE/ASTM SI 10-1997 Institute of Electrical and Electronics Engineers, 345 East 47th Street, New York, NY 10017, USA
- *Guide for the Use of the International System of Units (SI)*, SP811, 1995, National Institute of Standards and Technology, U.S. Government Printing Office, Washington, DC 20402

Base, supplementary, and derived SI units

Measure	Unit	Symbol	Measure	Unit	Symbol
Base units			Entropy	joule per kelvin	J/K
			Force	newton	N
Amount of substance	mole	mol	Frequency	hertz	Hz
Electric current	ampere	A	Heat capacity	joule per kelvin	J/K
Length	meter	m	Heat flux density	watt per square meter	W/m^2
Luminous intensity	candela	cd	Illuminance	lux	lx
Mass	kilogram	kg	Inductance	henry	H
Thermodynamic temperature	kelvin	K	Irradiance	watt per square meter	W/m^2
Time	second	s	Luminance	candela per square meter	cd/m^2
			Luminous flux	lumen	lm
Supplementary units			Magnetic field strength	ampere per meter	A/m
Plane angle	radian	rad	Magnetic flux	weber	Wb
Solid angle	steradian	sr	Magnetic flux density	tesla	T
			Molar energy	joule per mole	J/mol
			Molar entropy	joule per mole kelvin	$J/mol \cdot K$
Derived units			Molar heat capacity	joule per mole kelvin	$J/mol \cdot K$
Absorbed does	gray	Gy	Moment of force	newton meter	$N \cdot m$
Acceleration	meter per second squared	m/s^2	Permeability	henry per meter	H/m
Activity (of radionuclides)	becquerel	Bq	Permittivity	farad per meter	F/m
Angular acceleration	radian per second squared	rad/s^2	Power, radiant flux	watt	W
Angular velocity	radian per second	rad/s	Pressure, stress	pascal	Pa
Area	square meter	m^2	Quantity of electricity, electric charge	coulomb	C
Capacitance	farad	F	Radiance	watt per square meter steradian	$W/m^2 \cdot sr$
Concentration (of amount of substance)	mole per cubic meter	mol/m^3	Radiant intensity	watt per steradian	W/sr
Current density	ampere per square meter	A/m^2	Specific heat capacity	joule per kilogram kelvin	$J/kg \cdot K$
Density, mass	kilogram per cubic meter	kg/m^3	Specific energy	joule per kilogram	J/kg
Does equivalent, dose equivalent index	sievert	Sv	Specific entropy	joule per kilogram kelvin	$J/kg \cdot K$
Electric change density	coulomb per cubic meter	C/m^3	Specific volume	cubic meter per kilogram	m^3/kg
Electric conductance	siemens	S	Specific volume	cubic meter per kilogram	m^3/kg
Electric field strength	volt per cubic meter	V/m	Surface tension	newton per meter	N/m
Electric flux density	coulomb per square meter	C/m^2	Thermal conductivity	watt per meter kelvin	$W/m \cdot K$
Electric potential, potential difference, electromotive force	volt	V	Velocity	meter per second	m/s
			Viscosity, dynamic	pascal second	$Pa \cdot s$
Electric resistance	ohm	Ω	Viscosity, kinematic	square meter per second	m^2/s
Energy, work, quantity of heat	joule	J	Volume	cubic meter	m^3
Energy density	joule per cubic meter	J/m^3	Wavenumber	1 per meter	1/m

Conversion factors

To convert from	to	multiply by
Angle		
degree	rad	1.745 329 E − 02
Area		
in.^{2}	mm^2	6.451 600 E + 02
in.^{2}	cm^2	6.451 600 E + 00
in.^{2}	m^2	6.451 600 E − 04
ft^2	m^2	9.290 304 E − 02
Bending moment or torque		
lbf · in.	N · m	1.129 848 E − 01
lbf · ft	N · m	1.355 818 E + 00
kgf · m	N · m	9.806 650 E + 00
ozf · in.	N · m	7.061 552 E − 03
Bending moment or torque per unit length		
lbf · in./in.	N · m/m	4.448 222 E + 00
lbf · ft/in.	N · m/m	5.337 866 E + 01
Current density		
A/in.^{2}	A/cm^2	1.550 003 E − 01
A/in.^{2}	A/mm^2	1.550 003 E − 03
A/ft^2	A/m^2	1.076 400 E + 01
Electricity and magnetism		
gauss	T	1.000 000 E − 04
maxwell	μWb	1.000 000 E − 02
mho	S	1.000 000 E + 00
Oersted	A/m	7.957 700 E + 01
Ω · cm	Ω · m	1.000 000 E − 02
Ω circular−mil/ft	μΩ · m	1.662 426 E − 03
Energy (impact, other)		
ft · lbf	J	1.355 818 E + 00
Btu (thermochemical)	J	1.054 350 E + 03
cal (thermochemical)	J	4.184 000 E + 00
kW · h	J	3.600 000 E + 06
W · h	J	3.600 000 E + 03
Flow rate		
ft^3/h	L/min	4.719 475 E − 01
ft^3/min	L/min	2.831 000 E + 01
gal/h	L/min	6.309 020 E − 02
gal/min	L/min	3.785 412 E + 00
Force		
lbf	N	4.448 222 E + 00
kip (1000 lbf)	N	4.448 222 E + 03
tonf	kN	8.896 443 E + 00
kgf	N	9.806 650 E + 00
Force per unit length		
lbf/ft	N/m	1.459 390 E + 01
lbf/in.	N/m	1.751 268 E + 02
Fracture toughness		
ksi $\sqrt{in.}$	MPa $\sqrt{m}$	1.098 800 E + 00
Heat content		
Btu/lb	kJ/kg	2.326 000 E + 00
cal/g	kJ/kg	4.186 800 E + 00
Heat input		
J/in.	J/m	3.937 008 E + 01
kJ/in.	kJ/m	3.937 008 E + 01

To convert from	to	multiply by
Impact energy per unit area		
ft.lbf/ft^2	J/m^2	1.459 002 E + 01
Length		
Å	nm	1.000 000 E − 01
μin.	μm	2.540 000 E − 02
mil	μm	2.540 000 E + 01
in.	mm	2.540 000 E + 01
in.	cm	2.540 000 E + 00
ft	m	3.048 000 E − 01
yd	m	9.144 000 E − 01
mile, international	km	1.609 344 E + 00
mile, nautical	km	1.852 000 E + 00
mile, U. S. statute	km	1.609 347 E + 00
Mass		
oz	kg	2.834 952 E − 02
lb	kg	4.535 924 E − 01
ton (short, 2000 lb)	kg	9.071 847 E + 02
ton (short, 2000 lb)	kg × 10^3(a)	9.071 847 E − 01
ton (long, 2240 lb)	kg	1.016 047 E + 03
Mass per unit area		
oz/in.^{2}	kg/m^2	4.395 000 E + 01
oz/ft^2	kg/m^2	3.051 517 E − 01
oz/yd^2	kg/m^2	3.390 575 E − 02
lb/ft^2	kg/m^2	4.882 428 E + 00
Mass per unit length		
lb/ft	kg/m	1.488 164 E + 00
lb/in.	kg/m	1.785 797 E + 01
Mass per unit time		
lb/h	kg/s	1.259 979 E − 04
lb/min	kg/s	7.559 873 E − 03
lb/s	kg/s	4.535 924 E − 01
Mass per unit volume (includes density)		
g/cm^3	kg/m^3	1.000 000 E + 03
lb/ft^3	g/cm^3	1.601 846 E − 02
lb/ft^3	kg/m^3	1.601 846 E + 01
lb/in.^{3}	g/cm^3	2.767 990 E + 01
lb/in.^{3}	kg/m^3	2.767 990 E + 04
Power		
Btu/s	kW	1.055 056 E + 00
Btu/min	kW	1.758 426 E − 02
Btu/h	W	2.928 751 E − 01
erg/s	W	1.000 000 E − 07
ft · lbf/s	W	1.355 818 E + 00
ft · lbf/min	W	2.259 697 E − 02
ft · lbf/h	W	3.766 161 E − 04
hp (550 ft · lbf/s)	kW	7.456 999 E − 01
hp (electric)	kW	7.460 000 E − 01
Power density		
W/in.^{2}	W/m^2	1.550 003 E + 03
Pressure (fluid)		
atm (standard)	Pa	1.013 250 E + 05
bar	Pa	1.000 000 E + 05
in. Hg (32 °F)	Pa	3.386 380 E + 03
in. Hg (60 °F)	Pa	3.376 850 E + 03
lbf/in.^{2} (psi)	Pa	6.894 757 E + 03
torr (mm Hg, 0 °C)	Pa	1.333 220 E + 03

To convert from	to	multiply by
Specific heat		
Btu/lb · °F	J/kg · K	4.186 800 E + 03
cal/g · °C	J/kg · K	4.186 800 E + 03
Stress (force per unit area)		
tonf/in.^{2} (tsi)	MPa	1.378 951 E + 01
kgf/mm^2	MPa	9.806 650 E + 00
ksi	MPa	6.894 757 E + 00
lbf/in.^{2} (psi)	MPa	6.894 757 E − 03
MN/m^2	MPa	1.000 000 E + 00
Temperature		
°F	°C	5/9 · (°F − 32)
°R	°K	5/9
K	°C	K−273.15
Temperature interval		
°F	°C	5/9
Thermal conductivity		
Btu · in./s · ft^2 · °F	W/m · K	5.192 204 E + 02
Btu/ft · h · °F	W/m · K	1.730 735 E + 00
Btu · in./h · ft^2 · °F	W/m · K	1.442 279 E − 01
cal/cm · s · °C	W/m · K	4.184 000 E + 02
Thermal expansion(b)		
in./in. · °C	m/m · K	1.000 000 E + 00
in./in. · °F	m/m · K	1.800 000 E + 00
Velocity		
ft/h	m/s	8.466 667 E − 05
ft/min	m/s	5.080 000 E − 03
ft/s	m/s	3.048 000 E − 01
in./s	m/s	2.540 000 E − 02
km/h	m/s	2.777 778 E − 01
mph	km/h	1.609 344 E + 00
Velocity of rotation		
rev/min (rpm)	rad/s	1.047 164 E − 01
rev/s	rad/s	6.283 185 E + 00
Viscosity		
poise	Pa · s	1.000 000 E − 01
stokes	m^2/s	1.000 000 E − 04
ft^2/s	m^2/s	9.290 304 E − 02
in.^{2}/s	mm^2/s	6.451 600 E + 02
Volume		
in.^{3}	m^3	1.638 706 E − 05
ft^3	m^3	2.831 685 E − 02
fluid oz	m^3	2.957 353 E − 05
gal (U.S. liquid)	m^3	3.785 412 E − 03
Volume per unit time		
ft^3/min	m^3/s	4.719 474 E − 04
ft^3/s	m^3/s	2.831 685 E − 02
in.^{3}/min	m^3/s	2.731 177 E − 07

(a) kg × 10^3 = 1 metric ton or 1 megagram (Mg). (b) Preferred expression is 10^{-6}/K or 10^{-6}/F as length units are unnecessary.

ASM Handbook, Volume 22B, *Metals Process Simulation*
D.U. Furrer and S.L. Semiatin, editors

Copyright © 2010, ASM International®
All rights reserved.
www.asminternational.org

Useful Formulas for Metals Processing

Table 1 Casting and solidification

Law or rule	Equation	Importance	Ref
Partition coefficient	$k_i = \dfrac{x_i^S}{x_i^L}$ k_i = partition coefficient x_i^S = concentration of specie i at the solidus x_i^L = concentration of specie i at the liquidus	Describes the difference in concentration for solute i that would be at equilibrium at a temperature within the freezing range as given by the equilibrium tie line on the phase diagram.	1
Scheil-Gulliver	$C_S^* = (1 - f_S)^{k-1}$ C_S^* = composition of solid f_S = volume fraction of solid k = partition coefficient	Composition of the primary phase forming at the solid-liquid interface. Assumes little or no diffusivity of solute in the solid (microsegregation).	2
Lever rule	$f_S = \dfrac{C_L^* - C_O}{C_L^* - C_S^*}$ C_L^* = composition of liquid C_O = initial melt composition	Assumes complete solute diffusivity through the solid.	2
Transport equations Mass balance	 $\dfrac{\partial \rho}{\partial t} + \nabla \cdot (\rho \mathbf{v}) = 0$ ρ = density $\mathbf{v}$ = velocity t = time	Balance equations during solidification	3
Momentum balance	$\dfrac{\partial \rho \mathbf{v}}{\partial t} + \nabla \cdot (\rho \mathbf{v}\mathbf{v}) = \nabla \cdot \boldsymbol{\sigma} + \rho \mathbf{b}$ $\mathbf{v}\mathbf{v}$ = tensor with components $\mathbf{v}_i \mathbf{v}_j$ $\boldsymbol{\sigma}$ = stress tensor $\mathbf{b}$ = body force, e.g., gravity		
Energy balance	$\dfrac{\partial \rho H}{\partial t} + \nabla \cdot (\rho H \mathbf{v})$ $\qquad = \nabla \cdot (k \nabla T) - p \nabla \cdot \mathbf{v} + \boldsymbol{\tau} : \mathbf{D} + \dot{Q}$ H = enthalpy density p = pressure $\mathbf{D}$ = rate of deformation tensor $\boldsymbol{\tau} : \mathbf{D}$ = work done by the shear stress $\dot{Q}$ = rate of heat transfer		
Solute balance	$\dfrac{\partial \rho C}{\partial t} + \nabla \cdot (\rho C \mathbf{v}) = \nabla \cdot (D \nabla \rho C)$ C = overall alloy composition D = diffusivity		
Local solute redistribution	$\dfrac{df_S}{dC_L} = \dfrac{(1 - f_S)}{C_L} \dfrac{(1 - \beta)}{(1 - k)} \left(1 - \dfrac{u_n}{v_T} \right)$ f_S = volume fraction solid β = solidification shrinkage u_n = liquid flow velocity normal to isotherms v_T = isothermal velocity	Solute balance in the mushy zone (macrosegregation).	4
v'ant Hoff	$k = 1 - \dfrac{m_L \Delta H_f^A}{R (T_f^A)^2}$ k = partition coefficient m_L = liquidus slope ΔH_f^A = latent heat for pure solvent A T_f^A = melting point of pure solvent A R = ideal gas constant	Relationship for liquid-solid equilibrium	5
Heat flow	$\dfrac{\partial T}{\partial t} = \alpha_m \dfrac{\partial^2 T}{\partial x^2}$ α_m = thermal diffusivity	Describes transient, one-dimensional heat flow	6

(continued)

Table 1 (continued)

Law or rule	Equation	Importance	Ref
Solidified metal thickness	T = temperature t = time x = distance $$S = \frac{2}{\sqrt{\pi}} \underbrace{\left(\frac{T_M - T_0}{\rho_s H}\right)}_{\text{Metal}} \underbrace{\sqrt{K_m \rho_m c_m}}_{\text{Mold}} \sqrt{t}$$ S = thickness solidified T_M = temperature of mold surface T_0 = ambient temperature ρ_S = density of solid H = heat of fusion K_m = thermal conductivity of mold ρ_m = density of mold c_m = specific heat of mold material t = time	Relates thermal properties of the metal and the mold to the freezing rate of a metal cast into a relatively insulating mold.	6
Stokes-Einstein	$$D = \frac{k_B T}{6 \pi \eta r}$$ D = diffusion coefficient of sphere k_B = Boltzman's constant η = viscosity of the diffusion medium r = radius of sphere	Relationship between viscosity and diffusion of spherical particles through a liquid with low Reynolds number.	
Grueneisen equation	$$\beta = \frac{\gamma C_p}{V B_a}$$ β = thermal expansion γ = Grueneisen parameter C_p = specific heat at constant pressure V = volume of the solid B_a = adiabatic bulk modulus	Temperature dependence of thermal expansion for solids	7
Arrhenius equation	$k = A \, e^{\left(\frac{-E}{RT}\right)}$ k = reaction rate constant E = activation energy A = constant R = ideal gas constant	Relates temperature to chemical reaction rate.	
Bragg's law	$n\lambda = 2d \sin \theta$ n = an integer λ = x-ray wavelength d = interplanar spacing of crystalline phase θ = scattering angle	Relates the intensity of a diffracted beam depends to crystalline structure	8
Fick's 1st law	$$J = -D \frac{dC}{dx}$$ J = diffusion flux D = diffusion coefficient $\frac{dC}{dx}$ = concentration gradient	Diffusion flux with respect to the concentration gradient of the diffusing element	
Fick's 2nd law	$$\frac{\partial C}{\partial t} = \frac{\partial}{\partial x}\left(D \frac{\partial C}{\partial x}\right)$$	Relates diffusive flux to a concentration field that changes with time	
Griffith equation for fracture toughness	$$\sigma_G = \left[\frac{2E\gamma_s}{\pi a}\right]^{1/2}$$ σ_G = stress to fracture E = Young's Modulus γ = specific surface energy $2a$ = crack size	Relates crack geometry to fracture stress.	9
Sievert's law	$[\%X] = K \, (p_X)^{1/2}$ K = equilibrium constant p_X = partial pressure of gas over melt	Relates atomic fraction of gas, X, dissolved in melt to partial pressure of gas over the melt.	10

Table 2 Dimensionless groups in fluid mechanics

Name of dimensionless number	Definition	Physical interpretation	Importance	Ref
Archimedes number	$$Ar = \frac{gL^3(\rho_p - \rho)\rho}{\mu^2}$$	$$\frac{\text{Inertial forces} \times \text{Buoyancy forces}}{(\text{Viscous forces})^2}$$	Particle settling	11
Bingham number	$$Bm = \frac{\tau_y L}{\mu_\infty V}$$	$$\frac{\text{Yield stress}}{\text{Viscous stress}}$$	Flow of Bingham plastics = yield number, Y	11
Bingham Reynolds number	$$Re_B = \frac{LV\rho}{\mu_\infty}$$	$$\frac{\text{Inertial force}}{\text{Viscous force}}$$	Flow of Bingham plastics	11
Biot number	$$Bi = \frac{hL}{k}$$	$$\frac{\text{Internal resistance to heat conduction}}{\text{External resistance to heat conduction}}$$		11, 12
Bond number	$$Bo = \frac{(\rho_L - \rho_G)L^2 g}{\sigma}$$	$$\frac{\text{Gravitational force}}{\text{Surface tension force}}$$	Atomization = Eotvos number, Eo	11
Capillary number	$$Ca = \frac{\mu V}{\sigma}$$	$$\frac{\text{Viscous force}}{\text{Surface tension force}}$$	Two-phase flows, free surface flows	11

(continued)

Table 2 (continued)

Name of dimensionless number	Definition	Physical interpretation	Importance	Ref
Eckert number	$E = \dfrac{V^2}{c_p \Delta T}$	$\dfrac{\text{Kinetic energy}}{\text{Thermal energy}}$	Dissipation	12
Euler number	$Eu = \dfrac{\Delta p}{\rho V^2}$	$\dfrac{\text{Frictional pressure loss}}{2 \times \text{velocity head}}$	Fluid friction in conduits	11
Fanning fraction factor	$f = \dfrac{D' \Delta p}{2\rho V^2 L} = \dfrac{2\tau_w}{\rho V^2}$	$\dfrac{\text{Wall shear stress}}{\text{Velocity head}}$	Fluid friction in conduits; Darcy friction factor $= 4f$	11
Fourier number	$Fo = \dfrac{k\theta}{\rho c_p L^2}$	$\dfrac{\text{Dimensionless time for transient}}{\text{Conduction}}$		11, 12
Froude number	$Fr = \dfrac{V^2}{gL}$	$\dfrac{\text{Inertial force}}{\text{Gravity force}}$	Often defined as $Fr = V/\sqrt{gL}$	11
Graetz number	$Gz = \dfrac{mc_p}{kL}$		Relates thermal capacity of fluid to convection heat transfer	11
Grashof number	$Gr = \dfrac{L^3 \rho^2 \beta g \Delta T}{\mu^2}$	$\dfrac{(\text{Buoyancy forces})(\text{Inertia forces})}{(\text{Viscous forces})^2}$		11, 12
Knudsen number	$Kn = \dfrac{\lambda}{L}$	$\dfrac{\text{Molecular mean free path}}{\text{Length}}$	Continuum approximation in fluids	12
Lewis number	$Le = \dfrac{D\rho}{\alpha}$	$\dfrac{\text{Mass diffusivity}}{\text{Thermal diffusivity}}$		12
Mach number	$M = \dfrac{V}{c}$	$\dfrac{\text{Fluid velocity}}{\text{Sonic velocity}}$	Compressible flow	11
Nusselt number	$Nu = \dfrac{h_c L}{k}$	$\dfrac{\text{Convective heat transfer}}{\text{Conductive heat transfer}}$		11, 12
Peclet number	$Pe = \dfrac{LV}{D}$	$\dfrac{\text{Convective transport}}{\text{Diffusive transport}}$	Heat, mass transfer, mixing	11
Prandtl number	$Pr = \dfrac{c_p \mu}{\alpha}$	$\dfrac{\text{Momentum diffusivity}}{\text{Thermal diffusivity}}$		11, 12
Reynolds number	$Re = \dfrac{LV\rho}{\mu}$	$\dfrac{\text{Inertial force}}{\text{Viscous force}}$		11
Schmidt number	$Sc = \dfrac{\mu}{\rho D}$	$\dfrac{\text{Momentum diffusivity}}{\text{Mass diffusivity}}$		11, 12
Sherwood number	$Sh = \dfrac{KL}{D}$		Ratio of convective to diffusive mass transport	12
Stanton number	$St = \dfrac{h}{c_p V \rho}$	$\dfrac{\text{Heat transfer at wall}}{\text{Convective heat transfer}}$		11, 12
Weber number	$We = \dfrac{\rho V^2 L}{\sigma}$	$\dfrac{\text{Inertial force}}{\text{Surface tension force}}$	Bubble, drop formation	11

Symbol		Symbol	
c	sonic velocity	V	characteristic of average fluid velocity
D	diffusivity	α	thermal diffusivity
D'	diameter of pipe	β	volumetric thermal expansion
g	acceleration of gravity	λ	mean free path
L	characteristic length	μ	fluid viscosity
p	pressure	μ_∞	infinite shear viscosity (Bingham plastics)
Δp	frictional pressure drop	ρ	fluid density
K	mass transfer coefficient	ρ_G, ρ_L	gas, liquid densities
k	thermal conductivity	σ	surface tension
h	interface heat transfer coefficient	c_p	specific heat
h_c	convective heat transfer coefficient	ΔT	temperature change
		θ	time
		τ_y	yield strength

Table 3 Effective stress, strain, and strain rate (isotropic material) in arbitrary coordinates

Variable or quantity	Symbol or equation
Stress tensor components	σ_{ij}
Strain increment components	$d\varepsilon_{ij}$
Strain-rate components	$\dot\varepsilon_{ij}$
Von Mises effective stress ($\bar\sigma$)	$\bar\sigma = \sqrt{\dfrac{1}{2}\{(\sigma_{xx}-\sigma_{yy})^2 + (\sigma_{yy}-\sigma_{zz})^2 + (\sigma_{zz}-\sigma_{xx})^2\} + 3\sigma_{xy}^2 + 3\sigma_{yz}^2 + 3\sigma_{zx}^2}$ where σ_{xy}, σ_{yz}, and σ_{zx} are generalized tensor notation for shear stresses τ_{xy}, τ_{yz}, τ_{zx}, respectively
Von Mises effective strain increment ($d\bar\varepsilon$)	$d\bar\varepsilon = \sqrt{\dfrac{2}{9}\{(d\varepsilon_{xx}-d\varepsilon_{yy})^2 + (d\varepsilon_{yy}-d\varepsilon_{zz})^2 + (d\varepsilon_{zz}-d\varepsilon_{xx})^2\} + \dfrac{4}{3}d\varepsilon_{xy}^2 + \dfrac{4}{3}d\varepsilon_{yz}^2 + \dfrac{4}{3}d\varepsilon_{zx}^2}$
Von Mises effective strain rate ($\dot\varepsilon = d\bar\varepsilon/dt$)	$\dot\varepsilon = \sqrt{\dfrac{2}{9}\{(\dot\varepsilon_{xx}-\dot\varepsilon_{yy})^2 + (\dot\varepsilon_{yy}-\dot\varepsilon_{zz})^2 + (\dot\varepsilon_{zz}-\dot\varepsilon_{xx})^2\} + \dfrac{4}{3}\dot\varepsilon_{xy}^2 + \dfrac{4}{3}\dot\varepsilon_{yz}^2 + \dfrac{4}{3}\dot\varepsilon_{zx}^2}$

Table 4 Effective stress, strain, and strain rate (isotropic material) in principal coordinates

Variable or quantity	Symbol or equation
Principal stress components	$\sigma_1, \sigma_2, \sigma_3$
Principal strain-increment components	$d\varepsilon_1, d\varepsilon_2, d\varepsilon_3$
Principal strain-rate components	$\dot{\varepsilon}_1, \dot{\varepsilon}_2, \dot{\varepsilon}_3$
Von Mises effective stress ($\bar{\sigma}$)	$\bar{\sigma} = \sqrt{\dfrac{1}{2}\{(\sigma_1 - \sigma_2)^2 + (\sigma_2 - \sigma_3)^2 + (\sigma_3 - \sigma_1)^2\}}$
Von Mises effective strain increment ($d\bar{\varepsilon}$)	$d\bar{\varepsilon} = \sqrt{\dfrac{2}{9}\{(d\varepsilon_1 - d\varepsilon_2)^2 + (d\varepsilon_2 - d\varepsilon_3)^2 + (d\varepsilon_3 - d\varepsilon_1)^2\}}$
Von Mises effective strain rate ($\dot{\bar{\varepsilon}} = d\bar{\varepsilon}/dt$)	$\dot{\bar{\varepsilon}} = \sqrt{\dfrac{2}{9}\{(\dot{\varepsilon}_1 - \dot{\varepsilon}_2)^2 + (\dot{\varepsilon}_2 - \dot{\varepsilon}_3)^2 + (\dot{\varepsilon}_3 - \dot{\varepsilon}_1)^2\}}$

Table 5 Formulas for compression testing of isotropic material

Variable or quantity	Symbol or relation
Uniaxial compression under uniform deformation conditions	
Initial sample dimensions	Height (h_0)
	Diameter (d_0)
	Area (A_0), $A_0 = \pi d_0^2/4$
Instantaneous (final) sample dimensions	Height (h)
	Diameter (d)
	Area (A), $A = \pi d^2/4$
Crosshead speed	v
Applied load	P
Constant-volume assumption of plastic flow	$A_0 h_0 = Ah \rightarrow d_0^2 h_0 = d^2 h$
Height reduction (R), %	$R = \left[\dfrac{(h_0 - h)}{h_0}\right] \times 100$
Nominal (engineering) axial strain (e), %	$e = \left[\dfrac{(h - h_0)}{h_0}\right] \times 100$
True axial strain (ε)	$\varepsilon = \ln\left[\dfrac{h}{h_0}\right]$
True axial strain rate ($\dot{\varepsilon}$)	$\dot{\varepsilon} = -\text{v}/h$
Nominal (engineering) axial stress (S)	$S = -\dfrac{P}{A_0}$
True axial stress (σ)	$\sigma = -\dfrac{P}{A} = -\dfrac{Ph}{A_0 h_0} = \dfrac{Sh}{h_0}$
Effective stress ($\bar{\sigma}$)	$\bar{\sigma} = -\sigma$
Effective strain ($\bar{\varepsilon}$)	$\bar{\varepsilon} = -\varepsilon$
Uniaxial compression with friction correction	
Friction shear factor (m_s)	$m_s \approx \sqrt{3}\mu$ where $\mu \equiv$ Coulomb coefficient of friction
Friction correction for flow stress	$\dfrac{\bar{\sigma}}{p_{av}} = \left(1 + \dfrac{m_s d}{(3\sqrt{3})h}\right)^{-1}$
	where $\bar{\sigma}$ denotes flow stress (under homogeneous frictionless conditions), and p_{av} denotes the average pressure
Homogeneous plane-strain compression	
Through-thickness true strain	ε_3
Effective strain ($\bar{\varepsilon}$)	$\bar{\varepsilon} = \dfrac{-2\varepsilon_3}{\sqrt{3}}$
True stress	σ_3
Effective stress ($\bar{\sigma}$)	$\bar{\sigma} = \dfrac{-\sigma_3\sqrt{3}}{2}$

Table 6 Formulas for tension testing of isotropic material

Variable or quantity	Symbol or relation
Uniaxial tension under uniform deformation conditions, constant crosshead speed	
Initial gage (reduced section) dimensions	Length (L_0)
	Diameter (d_0)
	Area (A_0), $A_0 = \pi d_0^2/4$
Instantaneous reduced section dimensions	Length (L)
	Diameter (d)
	Area (A), $A = \pi d^2/4$
Crosshead speed	v
Applied load	P
Constant-volume assumption of plastic flow	$A_0 h_0 = Ah \rightarrow d_0^2 h_0 = d^2 h$
Nominal (engineering) axial strain (e), %	$e = \dfrac{(L - L_0)100}{L_0}$
Nominal (engineering) axial strain rate $(\dot{e})$	$\dot{e} = \dfrac{v}{L_0}$
True axial strain (ε)	$\varepsilon = \ln\left[\dfrac{L}{L_0}\right] = \ln(1 + e)$
	where e is expressed as a decimal fraction
True axial strain rate $(\dot{\varepsilon})$	$\dot{\varepsilon} = \dfrac{v}{L}$
Nominal (engineering) axial stress (S)	$S = \dfrac{P}{A_0}$
True axial stress (σ)	$\sigma = \dfrac{P}{A} = \dfrac{PL}{A_0 L_0} = \dfrac{SL}{L_0} = S(1 + e)$
	where e is expressed as a decimal fraction
Effective stress $(\overline{\sigma})$	$\overline{\sigma} = \sigma$
Effective strain $(\overline{\varepsilon})$	$\overline{\varepsilon} = \varepsilon$
Strain-hardening exponent (n)	$n = \dfrac{\partial \ln \sigma}{\partial \ln \varepsilon}$
	evaluated at a fixed strain rate and temperature
Strain-rate sensitivity exponent (m)	$m = \dfrac{\partial \ln \sigma}{\partial \ln \dot{\varepsilon}}$
	evaluated at fixed strain and temperature
Postuniform deformation in uniaxial tension of round-bar samples	
Initial gage (reduced section) dimensions	Length (L_0)
	Diameter (d_0)
	Area (A_0), $A_0 = \pi d_0^2/4$
Length of gage section at failure	L_f
Gage-section diameter at failure (minimum) section	d_f
Gage-section area at failure (minimum) section (A_f)	$A_f = \pi d_f^2/4$
Total elongation (e_t), %	$e_t = \dfrac{(L_f - L_0)100}{L_0}$
Reduction in area (RA), %	$RA = \dfrac{(A_0 - A_f)100}{A_0}$
True fracture strain (ε_f)	$\varepsilon_f = \ln\left[\dfrac{A_0}{A_f}\right] = 2 \ln\left[\dfrac{d_0}{d_f}\right]$
Necking during tension testing of round-bar samples	
Sample radius at symmetry plane of neck	a
Profile radius of neck	R
Bridgman correction for necking during tension testing of round-bar samples	$\overline{\sigma} = \dfrac{\sigma_x}{\left[1 + \frac{2R}{a}\right] \cdot \left[\ln\left(1 + \frac{a}{2R}\right)\right]}$
	where:
	• $\overline{\sigma}$ denotes the flow stress
	• σ_X denotes the average axial stress (which is equal to applied load $\div$ sample cross-sectional area at neck symmetry plane)

Table 7 Formulas for torsion testing of isotropic material (solid round-bar sample)

Variable or quantity	Symbol or relation
Reduced section dimensions	Length (L) Outer radius (R)
Radial coordinate	r
Twist (θ), in radians	θ (in radians) = twist in degrees $\times \dfrac{\pi}{180}$
Twist rate (in radians per second)	$\dot{\theta}$
Shear strain (γ)	$\gamma = \dfrac{r\theta}{L}$
Shear strain rate ($\dot{\gamma}$)	$\dot{\gamma} = \dfrac{r\dot{\theta}}{L}$
Effective strain ($\bar{\varepsilon}$)	$\bar{\varepsilon} = \gamma/\sqrt{3}$
Effective strain rate $\dot{\bar{\varepsilon}}$	$\dot{\bar{\varepsilon}} = \dot{\gamma}/\sqrt{3}$
Torque	M
Shear stress (τ) corresponding to shear strain/shear strain rate at $r = R$	$\tau = \dfrac{(3 + n^* + m^*) \cdot M}{2\pi R^3}$ where: n^* = slope of ln M vs. ln θ plot $\approx$ strain-hardening exponent (n) m^* = slope of ln M vs. ln $\dot{\theta}$ plot $\approx$ strain-rate sensitivity exponent (m)
Effective stress ($\bar{\sigma}$)	$\bar{\sigma} = \sqrt{3} \cdot \tau$

Table 8 Formulas related to flat (sheet) rolling

Variable or quantity	Symbol or relation
Underformed roll radius	R
Rolling speed (roll surface velocity) (v_R), m/s	$v_R = (2\pi R) \times$ (angular velocity, in revolutions per second)
Initial sheet thickness	h_0
Final sheet thickness	h_f
Draft (Δh)	$\Delta h = h_0 - h_f$
Thickness reduction, %	$\dfrac{(h_0 - h_f)100}{h_0}$
True thickness strain (ε_3)	$\varepsilon_3 = \ln\left[\dfrac{h_f}{h_0}\right]$
Rolling load (roll separating force), P_L	$P_L = \dfrac{2\bar{\sigma}}{\sqrt{3}}\left[\dfrac{1}{Q}(\varepsilon^Q - 1)b\sqrt{R'\Delta h}\right]$ where $\bar{\sigma}$ denotes the flow stress under homogeneous uniaxial stress conditions $Q = \mu L_p/\bar{h}$ $\mu \equiv$ Coulomb coefficient of friction $L_p \equiv$ projected contact length $\bar{h} \equiv (h_0 + h_f)/2$ $b \equiv$ sheet width $R' \equiv$ flattened roll radius
Hitchcock equation for flattened roll radius, R':	$R' = R\left\{1 + \dfrac{16(1 - v^2)P_L}{b\pi E(\Delta h)}\right\}$ where: $v \equiv$ Poisson's ratio of roll material $E \equiv$ Young's modulus of roll material
Projected contact length (L_p)	$L_p = \sqrt{R'\Delta h}$
Average effective strain rate ($\dot{\bar{\varepsilon}}$)	$\dot{\bar{\varepsilon}} = \dfrac{v_R}{h_0}\sqrt{\dfrac{2(h_0 - h_f)}{R'}}$

Table 9 Formulas related to conical-die extrusion

Variable or quantity	Symbol or relation
Initial billet dimensions	Diameter (d_0) Area (A_0), $A_0 = \pi d_0{}^2/4$
Final billet dimensions	Diameter d_f, Area (A_f), $A_f = \pi d_f^2/4$
Die semicone angle	α
Ram speed	v
Reduction ratio	$R = A_0/A_f$
Reduction (r), %	$r = \dfrac{(A_0 - A_f) \cdot 100}{A_0}$
Extrusion (ram) pressure, p_{av}	$\dfrac{p_{av}}{\bar{\sigma}} = a + b \cdot \ln R$ in which $\bar{\sigma}$ denotes material flow stress, and a and b denote material-dependent constants
Average effective strain rate in deformation zone, $\dot{\bar{\varepsilon}}$	$\dot{\bar{\varepsilon}} = \dfrac{6v d_0^2 (\tan\alpha)\ln R}{d_0^3 - d_f^3}$

Table 10 Formulas related to wire drawing

Variable or quantity	Symbol or relation
Initial wire diameter	d_0
Final wire diameter	d_f
Die semicone angle	α
Reduction ratio (R)	$R = A_0/A_f$
Reduction (r), %	$r = \dfrac{(A_0 - A_f)100}{A_0}$
Effective strain	$\ln\left[\dfrac{A_0}{A_f}\right]$
Average drawing stress (σ_{dwg})	$\dfrac{\sigma_{dwg}}{\bar\sigma} = \left(\dfrac{3.2}{\Delta + 0.9}\right) \cdot (\alpha + \mu)$ where: $\bar\sigma$ denotes material flow stress $\mu \equiv$ Coulomb coefficient of friction Δ denotes the deformation-zone geometry parameter (see below)
Deformation-zone geometry parameter (Δ)	$\Delta \equiv \dfrac{\alpha}{r}(1 + \sqrt{1 - r})^2$ where the reduction (r) is expressed as a decimal fraction, not as a percentage

Table 11 Formulas related to bending

Variable or quantity	Symbol or relation
Bending of sheet	
Bend radius	R
Sheet thickness	h
Minimum bend radius, $(2R/h)_{min}$	$(2R/h)_{min} = \dfrac{1}{\varepsilon_f} - 1$ where ε_f denotes the true fracture strain in uniaxial tension
Bending of bars	
Bar diameter	d
Mandrel diameter	D
Strain imposed on outer fiber	$\varepsilon = \dfrac{1}{(1 + D/d)}$

Table 12 Formulas related to deep drawing of cups from sheet metal

Variable or quantity	Symbol or relation
Blank diameter	D
Cup diameter	d
Reduction (R), %	$R = \dfrac{(D - d)100}{D}$
Limiting drawing ratio (LDR)	$LDR \equiv D_{max}/d$ where D_{max} denotes the maximum blank diameter that can be drawn without cup failure

Table 13 Formulas for anisotropic sheet materials (See also Table 14)

Variable or quantity	Symbol or relation
Basic definitions	
Normal plastic anisotropy (R)	$R = d\varepsilon_w/d\varepsilon_t \approx \varepsilon_w/\varepsilon_t$ where ε_w, ε_t denote the true width and thickness strains during uniform, uniaxial tension of a sheet specimen
Average normal plastic anisotropy ($\bar R$)	$\bar R = (R_{0°} + 2R_{45°} + R_{90°})/4$
Planar plastic anisotropy (ΔR)	$\Delta R = (R_{0°} - 2R_{45°} + R_{90°})/2$
Effective stress ($\bar\sigma$) and effective strain increment ($d\bar\varepsilon$) assuming *plane-stress* conditions in principal coordinates and planar plastic isotropy ($R_{0°} = R_{90°} = R_{45°} = R$)	$\bar\sigma = \sqrt{\dfrac{3(1+R)}{2(2+R)}} \cdot \sqrt{(\sigma_1)^2 + (\sigma_2)^2 - \dfrac{2R}{(1+R)}(\sigma_1\sigma_2)}$ $d\bar\varepsilon = \sqrt{\dfrac{2(2+R)}{3(1+2R)^2}} \cdot \sqrt{(d\varepsilon_1 - Rd\varepsilon_3)^2 + (d\varepsilon_2 - Rd\varepsilon_3)^2 + R(d\varepsilon_1 - d\varepsilon_2)^2}$
Hill quadratic yield function (orthotropic texture)	$F(\sigma_{yy} - \sigma_{zz})^2 + G(\sigma_{zz} - \sigma_{xx})^2 + H(\sigma_{xx} - \sigma_{yy})^2 + 2L\sigma_{yz}^2 + 2M\sigma_{zx}^2 + 2N\sigma_{xy}^2 = 1$
Uniaxial sheet tension test assuming uniform deformation and planar isotropy ($R_{0°} = R_{90°} = R_{45°} = R$):	
Axial true stress	σ_1
Axial true strain	ε_1
Effective stress ($\bar\sigma$)	$\bar\sigma = \sqrt{\dfrac{3(1+R)}{2(2+R)}}\,\sigma_1$

(continued)

Table 13 (continued)

Variable or quantity	Symbol or relation
Effective strain ($\bar{\varepsilon}$)	$\bar{\varepsilon} = \sqrt{\dfrac{2(2+R)}{3(1+R)}}\, \varepsilon_1$

Plane-strain compression of sheet assuming uniform deformation and planar isotropy ($R_{0°} = R_{90°} = R_{45°} = R$):

Through-thickness true stress	σ_3
Through-thickness true strain	ε_3
Effective stress ($\bar{\sigma}$)	$\bar{\sigma} = -\sqrt{\dfrac{3(1+2R)}{2(1+R)(2+R)}}\, \sigma_3$
Effective strain $\bar{\varepsilon}$	$\bar{\varepsilon} = -\sqrt{\dfrac{2(2+R)(1+3R+2R^2)}{3(1+2R)^2}}\, \varepsilon_3$

Ratio of plane-strain flow stress (σ_{ps}) to uniaxial flow stress (σ_{uni}) (at equal levels of plastic work)

Ratio of plane-strain flow stress (σ_{ps}) to uniaxial flow stress (σ_{uni})	$\dfrac{\sigma_{ps}}{\sigma_{uni}} = \dfrac{(1+R)}{\sqrt{1+2R}}$

Table 14 Barlat's anisotropic yield function Yld2000-2d for plane-stress deformation of sheet material (Ref 13, 14)

Variable or quantity	Symbol or relation(a)
Basic definitions	
Yield function (ϕ) and associated effective stress ($\bar{\sigma}$)	$\phi = \lvert \tilde{S}_1' - \tilde{S}_2' \rvert^a + \lvert 2\tilde{S}_2'' + \tilde{S}_1'' \rvert^a + \lvert 2\tilde{S}_1'' + \tilde{S}_2'' \rvert^a = 2\bar{\sigma}^a$ $(\tilde{S}_1', \tilde{S}_2')$ and $(\tilde{S}_1'', \tilde{S}_2'')$ are the principal values of two linearly transformed stress deviators (see below). Recommended values for the exponent a are 6 and 8 for body-centered cubic and face-centered cubic metals, respectively.
Yield condition	$\bar{\sigma} = \left\{ \dfrac{\phi}{2} \right\}^{1/a} = h(\bar{\varepsilon})$
Work-equivalent effective strain ($\bar{\varepsilon}$), defined incrementally for plane stress	$d\bar{\varepsilon} = \dfrac{\sigma_{xx}}{\bar{\sigma}} d\varepsilon_{xx} + \dfrac{\sigma_{yy}}{\bar{\sigma}} d\varepsilon_{yy} + 2\dfrac{\sigma_{xy}}{\bar{\sigma}} d\varepsilon_{xy}$
Hardening function [$h(\bar{\varepsilon})$]	Most prevalent choices: • Flow stress in uniaxial tension in the rolling direction: $d\bar{\varepsilon} = d\varepsilon_{xx}$ • Flow stress in balanced biaxial tension ($\sigma_{yy} = \sigma_{xx}$), $d\bar{\varepsilon} = -d\varepsilon_{zz}$
First linear transformation	
Components of the transformed stress deviator	$\tilde{s}_{xx}' = \alpha_1 (2\sigma_{xx} - \sigma_{yy})/3$ $\tilde{s}_{yy}' = \alpha_2 (2\sigma_{yy} - \sigma_{xx})/3$ $\tilde{s}_{xy}' = \alpha_7 \sigma_{xy}$
Principal values of the transformed stress deviator	$\tilde{S}_1' = \left\{ \tilde{s}_{xx}' + \tilde{s}_{yy}' + \sqrt{\left(\tilde{s}_{xx}' - \tilde{s}_{yy}'\right)^2 + 4\tilde{s}_{xy}'^2} \right\} \Big/ 2$ $\tilde{S}_2' = \left\{ \tilde{s}_{xx}' + \tilde{s}_{yy}' - \sqrt{\left(\tilde{s}_{xx}' + \tilde{s}_{yy}'\right)^2 + 4\tilde{s}_{xy}'^2} \right\} \Big/ 2$
Second linear transformation	
Components of the transformed stress deviator	$\tilde{s}_{xx}'' = \{(8\alpha_5 - 2\alpha_6 - 2\alpha_3 + 2\alpha_4)\sigma_{xx} + (4\alpha_6 + \alpha_3 - 4\alpha_4 - 4\alpha_5)\sigma_{yy}\}/9$ $\tilde{s}_{yy}'' = \{(4\alpha_3 - 4\alpha_4 - 4\alpha_5 + \alpha_6)\sigma_{xx} + (8\alpha_4 + 2\alpha_5 - 2\alpha_6 - 2\alpha_3)\sigma_{yy}\}/9$ $\tilde{s}_{xy}'' = \alpha_8 \sigma_{xy}$
Principal values of the transformed stress deviator	$\tilde{S}_1'' = \left\{ \tilde{s}_{xx}'' + \tilde{s}_{yy}'' + \sqrt{(\tilde{s}_{xx}'' - \tilde{s}_{yy}'')^2 + 4\tilde{s}_{xy}''^2} \right\} \Big/ 2$ $\tilde{S}_2'' + \left\{ \tilde{s}_{xx}'' + \tilde{s}_{yy}'' - \sqrt{\left(\tilde{s}_{xx}'' - \tilde{s}_{yy}''\right)^2 + 4\tilde{s}_{xy}''^2} \right\} \Big/ 2$
Coefficients and input data(c)	
Anisotropy coefficients(b)	α_1 to α_8
Flow stresses in uniaxial tension for directions at 0, 45, and 90° from rolling direction	$\sigma_0, \sigma_{45}, \sigma_{90}$
Bulge test flow stress (assumed to be the balanced biaxial flow stress)	σ_b
r values in uniaxial tension for directions at 0, 45, and 90° from rolling direction	r_0, r_{45}, r_{90}
Biaxial r values measured with the disk compression test	$r_b = d\varepsilon_{yy}/d\varepsilon_{xx}$

(a) The axes (x, y, z) denote the rolling direction, transverse direction, and normal direction of a sheet, respectively. (b) The eight anisotropy coefficients α_1 to α_8 can be calculated using a numerical solver (e.g., Newton-Raphson) using the eight physical parameters (flow stresses and r values) as input. (c) Note that σ_b and r_b are important experimental data to define the shape of the yield locus. However, if not available, they can be estimated using another yield function or a crystal-plasticity model.

REFERENCES

1. U.R. Kattner, Thermodynamics and Phase Diagrams, *Casting*, Vol 15, *ASM Handbook*, ASM International, 2008, p 269–275
2. H.D. Brody, Microsegregation, *Casting*, Vol 15, *ASM Handbook*, ASM International, 2008, p 338–347
3. J.A. Dantzig, Transport Phenomena During Solidification, *Casting*, Vol 15, *ASM Handbook*, ASM International, 2008, p 288–292
4. C. Beckermann, Macrosegregation, *Casting*, Vol 15, *ASM Handbook*, ASM International, 2008, p 348–352
5. J.H. Perepezko, Nucleation Kinetics and Grain Refinement, *Casting*, Vol 15, *ASM Handbook*, ASM International, 2008, p 276–287
6. M.C. Flemings, *Solidification Processing*, McGraw-Hill College, 1974
7. J.J. Valencia and K-O. Yu, Thermophysical Properties, *Modeling for Casting and Solidification Processing*, K-O. Yu, ed.; Marcel Dekker, 2002, pp 189–237
8. R.P. Goehner and M.C. Nichols, X-Ray Powder Diffraction, *Materials Characterization*, Vol. 10, *ASM Handbook*, ASM International, 1986, p 333–343
9. S.D. Antolovich, B.F. Antolovich, An Introduction to Fracture Mechanics, *Fatigue and Fracture*, Vol. 19, *ASM Handbook*, ASM International, 1996, p 371–380
10. B. Mishra, Steelmaking Practices and Their Influence on Properties, *Metals Handbook Desk Edition, 2nd ed.*, J.R. Davis, ed., ASM International, 1998
11. R.H. Perry and D.W. Green, eds., *Perry's Chemical Engineers' Handbook*, 7th ed., McGraw Hill, New Yourk, NY, 1997
12. B.T.F. Chung, "Heat Transfer," p. 477–584, *ASM Handbook of Engineering Mathematics*, ASM, Metals Park, OH, 1983
13. F. Barlat, J.C. Brem, J. W. Yoon, K. Chung, R.E. Dick, D.J. Lege, F. Pourboghrat, S.-H. Choi, and E. Chu, Plane Stress Yield Function for Aluminum Alloy Sheets—Part I: Theory, *Int. J. Plast.*, Vol 19, 2003, p 1297–1319
14. J.W. Yoon, F. Barlat, R.E. Dick, K. Chung, and T.J. Kang, Plane Stress Yield Function for Aluminum Alloy Sheets—Part II: FE Formulation and Its Implementation, *Int. J. Plast.*, Vol 20, 2004, p 495–522

ASM Handbook, Volume 22B, *Metals Process Simulation*
D.U. Furrer and S.L. Semiatin, editors

Copyright © 2010, ASM International®
All rights reserved.
www.asminternational.org

Glossary of Terms

A

ab initio. From the beginning (Latin); often used to refer to first-principles modeling approaches.

abnormal grain growth. Rapid, nonuniform, and usually undesirable growth of one or a small fraction of grains in a polycrystalline material during annealing. The phenomenon is most frequent in fine-grained materials in which a larger-than-average grain (or grains) consumes surrounding small grains whose growth is limited by particle pinning. Also known as *secondary recrystallization.*

acicular alpha. A product of nucleation and growth from β to the lower-temperature allotrope α-phase. It may have a needlelike appearance in a micrograph and may have needle, lenticular, or flattened bar morphology in three dimensions. This type of transformation product is commonly observed in titanium and zirconium alloys.

aging. A change in material property or properties with time. See also *quench aging* and *strain aging.*

alligatoring. The longitudinal splitting of flat slabs in a plane parallel to the rolled surface.

allotriomorphic crystal. A crystal having a normal lattice structure but an outward shape that is imperfect, because it is determined to some extent by the surroundings. The grains in a metallic aggregate are allotriomorphic crystals.

allotropy. The property by which certain elements may exist in more than one crystal structure.

allotropic transformation. The ability of a material to transform from one crystal structure to another. Closely synonymous with polymorphism.

alloying element. An element added to and remaining in a metal that changes structure and properties.

amorphous material. A material that lacks the long-range three-dimensional atomic periodicity that is characteristic of a crystalline solid.

angle of bite. In the rolling of metals, the location where all of the force is transmitted through the rolls; the maximum attainable angle between the roll radius at the first contact and the line of roll centers. Operating angles less than the angle of bite are termed contact angles or rolling angles.

angstrom (Å). A unit of linear measure equal to 10^{-10} m, or 0.1 nm. Although not an accepted SI unit, it is occasionally used for small distances, such as interatomic distances, and some wavelengths.

angularity. The conformity to, or deviation from, specified angular dimensions in the cross section of a shape or bar.

anion. A negatively charged ion that migrates toward the anode (positive electrode) under the influence of a potential gradient.

anisotropy. Variations in one or more physical or mechanical properties with direction with respect to a fixed reference system in the material.

annealing. A generic term denoting a treatment—heating to and holding at a suitable temperature followed by cooling at a suitable rate—used primarily to soften metallic materials but also to produce desired changes simultaneously in other properties or in microstructure. When applied only for the relief of stress, the process is called stress relieving or stress-relief annealing. In ferrous alloys, annealing is carried out above the upper critical temperature, but the time-temperature cycles vary widely in maximum temperature attained and cooling rate used, depending on composition, material condition, and desired results. In nonferrous alloys, annealing cycles are designed to remove part or all of the effects of cold working (recrystallization may or may not be involved), cause coalescence of precipitates from the solid solution in relatively coarse form, or both, depending on composition and material condition.

aspect ratio. The ratio of the length of one axis to that of another, for example, *c/a*, or the continued ratio of three axes, such as *a:b:c.*

asperities. Protrusions rising above the general surface contours that constitute the actual contact areas between touching surfaces.

austempering. A heat treatment for ferrous alloys in which a part is quenched from the austenitizing temperature at a rate fast enough to avoid formation of ferrite or pearlite and then held at a temperature just above the martensite start temperature until transformation to bainite is complete. Although designated as bainite in both austempered steel and austempered ductile iron, austempered steel

consists of two-phase mixtures containing ferrite and carbide, while austempered ductile iron consists of two-phase mixtures containing ferrite and austenite.

austenite. A high-temperature form of iron. In steel heat treating, the steel is heated into the austenite region before rapidly cooling (quenching).

austenite stabilizer. An alloying element that, when added to iron, increases the region of the phase diagram in which austenite (face-centered cubic iron) is stable. Strong austenite stabilizers are carbon, nickel, and manganese.

average grain diameter. The mean diameter of an equiaxed grain section whose size represents all the grain sections in the aggregate being measured. See also *grain size.*

Avrami plot/(Avrami equation). Plot describing the kinetics of phase transformations in terms of the dependence of fraction X of microstructure that has transformed (e.g., recrystallized, decomposed, etc.) as a function of time (t) or strain (ε). Avrami plots usually consist of a graph of log $[\ln(1/(1 - X))]$ versus log t (or log ε) and are used to determine the so-called Avrami exponent n in the relation $X = 1 - \exp(-Bt^{n})$.

axis (crystal). The edge of the unit cell of a space lattice. Any one axis of any one lattice is defined in length and direction relative to other axes of that lattice.

B

bar. A section hot rolled from a *billet* to a form, such as round, hexagonal, octagonal, square, or rectangular, with sharp or rounded corners or edges and a cross-sectional area of less than 105 cm² (16 in.²). A solid section that is long in relationship to its cross-sectional dimensions, having a completely symmetrical cross section and a width or greatest distance between parallel faces of 9.5 mm (⅜ in.) or more.

barreling. Convexity of the surfaces of cylindrical or conical bodies, often produced unintentionally during upsetting or as a natural consequence during compression testing.

basal plane. That plane of a hexagonal or tetragonal crystal perpendicular to the axis of highest symmetry. Its Miller indices are (0001).

Bauschinger effect. A reduction in yield strength on straining a material in the opposite direction to the initial testing.

bendability. The ability of a material to be bent around a specified radius without fracture.

bending. The straining of material, usually flat sheet or strip metal, by moving it around a straight axis lying in the neutral plane. Metal flow takes place within the plastic range of the metal, so that the bent part retains a *permanent set* after removal of the applied stress. The cross section of the bend inward from the neutral plane is in compression; the rest of the bend is in tension.

bending and forming. The processes of bending, flanging, folding, twisting, offsetting, or otherwise shaping a portion of a blank or a whole blank, usually without materially changing the thickness of the metal.

bending moment. The moments (force times distance) that tend to bend a beam in the plane of the loads.

bending stress. A stress involving tensile and compressive forces, which are not uniformly distributed. Its maximum value depends on the amount of flexure that a given application can accommodate. Resistance to bending can be termed stiffness.

bending under tension. A forming operation in which a sheet is bent with the simultaneous application of a tensile stress perpendicular to the bend axis.

bend (longitudinal). A forming operation in which the axis is perpendicular to the rolling direction of the sheet.

bend or twist (defect). Distortion similar to warpage generally caused during forging or trimming operations. When the distortion is along the length of the part, it is termed bend; when across the width, it is termed twist. When bend or twist exceeds tolerances, it is considered a defect. Corrective action consists of hand straightening, machine straightening, or cold restriking.

bend radius. The radius measured on the inside of a bend that corresponds to the curvature of a bent specimen or the bent area in a formed part.

bend test. Evaluation of a sheet metal response to a bending operation, such as around a fixed radius tool.

bend (transverse). A forming operation in which the bend axis is parallel to the rolling direction of the sheet.

beta structure. Structurally analogous body-centered cubic phases (similar to β-brass) or electron compounds that have ratios of three valence electrons to two atoms.

beta transus. The minimum temperature above which equilibrium α does not exist. For β eutectoid additions, the β transus ordinarily is applied to hypoeutectoid compositions or those that lie to the left of the eutectoid composition. This temperature pertains primarily to the transformation of titanium or zirconium alloys.

biaxial stretchability. The ability of sheet material to undergo deformation by loading in tension in two directions in the plane of the sheet.

billet. A semifinished section that is hot rolled from a metal ingot, with a rectangular cross section usually ranging from 105 to 230 cm^2 (16 to 36 $in.^2$), the width being less than twice the thickness. Where the cross section exceeds 230 cm^2 (36 $in.^2$), the term *bloom* is properly but not universally used. Sizes smaller than 105 cm^2 (16 $in.^2$) are usually termed bars. A solid semifinished round or square product that has been hot worked by forging, rolling, or extrusion. See also *bar*.

bite. Advance of material normal to the plane of deformation and relative to the dies prior to each deformation step.

blank. In forming, a piece of sheet material, produced in cutting dies, that is usually subjected to further press operations. A piece of stock from which a forging is made.

blanking. The operation of punching, cutting, or shearing a piece out of stock to a predetermined shape.

blister. A local protuberance in the surface of the sheet, often elongated, resulting from an internal separation due to the expansion of entrapped gas. The gas may be entrapped during casting, pickling, annealing, or electroplating in a previously existing subsurface defect.

block and finish. The forging operation in which a part to be forged is blocked and finished in one heat through the use of tooling having both a block impression and a finish impression in the same die block.

blocking. (1) A forging operation often used to impart an intermediate shape to a forging, preparatory to forging of the final shape in the finishing impression of the dies. Blocking can ensure proper working of the material and can increase die life. (2) A schedule for conducting treatment combinations in an experimental study such that any effects on the experimental results due to a known change in raw materials, processes, operators, machines, and so on become concentrated in the levels of the blocking variable. The reason for blocking is to isolate a systematic effect and prevent it from obscuring the main effects. Adapted from Ref 1

bloom. A semifinished hot rolled product, rectangular in cross section, produced on a blooming mill. See also *billet*. For steel, the width of a bloom is not more than twice the thickness, and the cross-sectional area is usually not less than approximately 230 cm^2 (36 $in.^2$). Steel blooms are sometimes made by forging.

board hammer. A type of forging hammer in which the upper die and ram are attached to "boards" that are raised to the striking position by power-driven rollers and let fall by gravity. See also *drop hammer*.

boss. A relatively short, often cylindrical protrusion or projection on the surface of a forging.

bottom draft. Slope or taper in the bottom of a forge depression that tends to assist metal flow toward the sides of depressed areas.

boundary condition. A requirement to be met by a solution to a set of differential equations on a specified set of values of the independent variables.

bow. The tendency of material to curl downward during shearing, particularly when shearing long, narrow strips.

Bravais lattices. The 14 possible three-dimensional arrays of atoms in crystals (see *space lattice*).

brick element. The element for three-dimensional finite-element modeling that is brick shaped (six faces) and has eight nodes.

Bridgman correction. Factor used to obtain the flow stress from the measured axial stress during tension testing of metals in which necking has occurred.

Brillouin zones. Energy states for the free electrons in a metal, as described by the use of the band theory (zone theory) of electron structure. Also called electron bands.

brittle fracture. A fracture that occurs without appreciable plastic deformation.

brittleness. A tendency to fracture without appreciable plastic deformation.

buckling. A bulge, bend, kink, or other wavy condition of the workpiece caused by compressive stresses.

bulge test. A test wherein the blank is clamped securely around the periphery and, by means of hydrostatic pressure, the blank is expanded. The blank is usually gridded so that the resulting strains can be measured. This test is usually performed on large blanks of 20 to 30 cm (8 to 12 in.) in diameter.

bulging. The process of increasing the diameter of a cylindrical shell (usually to a spherical shape) or of expanding the outer walls of any shell or box shape whose walls were previously straight.

bulk forming. Forming processes, such as extrusion, forging, rolling, and drawing, in which the input material is in billet, rod, or slab form and a considerable increase in surface-to-volume ratio in the formed part occurs under the action of largely compressive loading. Compare with *sheet forming*.

Burgers vector. The crystallographic direction along which a dislocation moves and the unit displacement of dislocations; the magnitude of the Burgers vector is the smallest unit distance of slip in the direction of shear due to the movement of one dislocation.

burnishing. The smoothing of one surface through frictional contact with another surface.

burr. A thin ridge or roughness left on forgings or sheet metal blanks by cutting operations such as slitting, shearing, trimming, blanking, or sawing.

C

CAD/CAM. An abbreviation for computer-aided design/computer-aided manufacturing.

camber. The tendency of material being sheared from sheet to bend away from the sheet in the same plane.

canned extrusion. A coextrusion process in which the billet consists of an outer material, or can, that is relatively ductile and nonreactive, and the core is a reactive, brittle powder or other material.

canning. A dished distortion in a flat or nearly flat sheet metal surface, sometimes referred to as oil canning. Distortion of a flat or nearly flat metal surface that can be deflected by finger pressure but will return to its original position when the pressure is removed. Enclosing a highly reactive metal within a relatively inert material for the purpose of hot working without undue oxidation of the active metal.

casting. (1) Metal object cast to the required shape by pouring of injecting liquid metal into a mold, as distinct from one shaped by a mechanical process. (2) Pouring molten metal into a mold to produce an object of desired shape. (3) Ceramic forming process in which a body slip is introduced into a porous mold, which absorbs sufficient water from the slip to produce a semirigid circle.

casting modulus. A simplified approach to determining solidification time. The time is proportional to the square of the section modulus (the ratio of volume to surface area), known as Chvorinov's rule.

cation. A positively charged ion that migrates through the electrolyte toward a cathode (negative electrode) under the influence of a potential gradient.

Cauchy distribution. The general formula for the probability density function of the Cauchy distribution is:

$$f(x) = \frac{1}{s\pi(1 + ((x - t)/s)^2)}$$

where t is the location parameter, and s is the scale parameter. The case where $t = 0$ and $s = 1$ is called the standard Cauchy distribution. The equation for the standard Cauchy distribution reduces to:

$$f(x) = \frac{1}{\pi(1 + x^2)}$$

Cauchy distributions look similar to a normal distribution; however, they have much heavier tails. When studying hypothesis tests that assume normality, the manner in which the tests perform on data from a Cauchy distribution is a good indicator of how sensitive the tests are to heavy-tail departures from normality. The mean and standard deviation of the Cauchy distribution are undefined. The practical meaning of this is that collecting 1000 data points gives an estimate of the mean and standard deviation that is no more accurate than from a single point. Adapted from Ref 1

cavitation. The formation of microscopic cavities during the cold or hot deformation of metals, generally involving a component of tensile stress. Cavities may nucleate at second-phase particles lying within grains or at grain boundaries (with or without particles) as a result of slip intersection or grain-boundary sliding. Under severe conditions, cavities may grow and coalesce to give rise to fracture. In liquids, cavitation is the formation and instantaneous collapse of cavities or bubbles caused by rapid and intense pressure changes. Cavitation caused by untrasonic radiation is sometimes used to effect violent local agitation. Cavitation caused by severe turbulent flow often leads to damage of adjacent material surfaces.

cell. Micron-sized volume bounded by low-misorientation walls comprised of dislocation tangles.

center bursting. Internal cracking due to tensile stresses along the central axis of products being extruded or drawn.

center points. Points at the center value of all factor ranges. Adapted from Ref 1

chamfer. A beveled surface to eliminate an otherwise sharp corner. A relieved angular cutting edge at a tooth corner.

check. A crack in a die impression corner, generally due to forging strains or pressure, localized at some relatively sharp corner. Die blocks too hard for the depth of the die impression have a tendency to check or develop cracks in impression corners. One of a series of small cracks resulting from thermal fatigue of hot forging dies (often called a heat check or heat checking).

chord modulus. The slope of the chord drawn between any two specific points on a stress-strain curve. See also *modulus of elasticity*.

circle grid. A regular pattern of circles, often 2.5 mm (0.1 in.) in diameter, marked on a sheet metal blank.

circle-grid analysis. The analysis of deformed circles to determine the severity with which a sheet metal blank has been deformed.

clad. Outer layer of a coextruded or codrawn product. See also *sleeve*.

clamping pressure. Pressure applied to a limited area of the sheet surface, usually at the periphery, to control or limit metal flow during forming.

clearance. In punching and shearing dies, the gap between the die and the punch. In forming and drawing dies, the difference between this gap and metal thickness.

closed-die forging. The shaping of hot metal completely within the walls or cavities of two dies that come together to enclose the workpiece on all sides. The impression for the forging can be entirely in either die or divided between the top and bottom dies. Impression die forging, often used interchangeably with the term closed-die forging, refers to a closed-die operation in which the dies contain a provision for controlling the flow of excess material, or *flash*, that is generated. By contrast, in flashless forging, the material is deformed in a cavity that allows little or no escape of excess material.

closed dies. Forging or forming impression dies designed to restrict the flow of metal to the cavity within the die set, as opposed to open dies, in which there is little or no restriction to lateral flow.

closed pass. A pass of metal through rolls where the bottom roll has a groove deeper than the bar being rolled and the top roll has a collar fitting into the groove, thus producing the desired shape free from *flash* or *fin*.

close-tolerance forging. A forging held to unusually close dimensional tolerances so that little or no machining is required after forging. See also *precision forging*.

cluster mill. A rolling mill in which each of two small-diameter work rolls is supported by two or more backup rolls.

coarsening. The increase in the average size of second-phase particles, accompanied by the reduction in their number, during annealing, deformation, or high-temperature service exposure. Coarsening thus leads to a decrease in the total surface energy associated with the matrix-particle interfaces.

codrawing. The simultaneous drawing of two or more materials to form an integral product.

coefficient of friction. A measure of the ease with which one body will slide over another. It is obtained by dividing the tangential force resisting motion between the two bodies by the normal force pressing the two bodies together.

coefficient of thermal expansion (CTE). Change in unit of length (or volume) accompanying a unit change of temperature, at a specified temperature or temperature range.

coextrusion. The simultaneous extrusion of two or more materials to form an integral product.

cogging. The reducing operation in working an ingot into a billet with a forging hammer or a forging press.

coil breaks. Creases or ridges that appear as parallel lines transverse to the direction of rolling and extend across the width of the sheet. Coil breaks are caused by the deformation of local areas during coiling or uncoiling of annealed or insufficiently temper-rolled steel sheets.

coining. A closed-die squeezing operation in which all surfaces of a workpiece are confined or restrained, resulting in a well-defined imprint of the die on the work. A *restriking* operation used to sharpen or change an existing radius or profile. Coining can be done while forgings are hot or cold and is usually performed on surfaces parallel to the parting line of the forging.

cold forming. See *cold working*.

cold heading. Working metal at room temperature such that the cross-sectional area of a portion or all of the stock is increased. See also *heading* and *upsetting*.

cold lap. A flaw that results when a workpiece fails to fill the die cavity during the first forging. A seam is formed as subsequent dies force metal over this gap to leave a seam on the workpiece surface. See also *cold shut*.

cold rolled sheet. A mill product produced from a hot rolled pickled coil that has been given substantial cold reduction at room temperature. After annealing, the usual end product is characterized by improved surface, greater uniformity in thickness, increased tensile strength, and improved mechanical properties as compared with hot rolled sheet.

cold shut. A fissure or lap on a forging surface that has been closed without fusion during the forging operation. A folding back of metal onto its own surface during flow in the die cavity; a forging defect.

cold trimming. The removal of flash or excess metal from a forging at room temperature in a trimming press.

cold-worked structure. A microstructure resulting from plastic deformation of a metal or alloy below its recrystallization temperature.

cold working. The plastic deformation of metal under conditions of temperature and strain rate that induce *strain hardening*. Usually, but not necessarily, conducted at room temperature. Also referred to as cold forming or cold forging. Contrast with *hot working*.

columnar structure. A coarse structure of parallel, elongated grains formed by unidirectional growth that is most often observed in castings but sometimes seen in structures. This results from diffusional growth accompanied by a solid-state transformation.

compact (noun). The object produced by the compression of metal powder, generally while confined in a die.

compact (verb). The operation or process of producing a compact; sometimes called pressing.

compression test. A method for assessing the ability of a material to withstand compressive loads.

compressive strength. The maximum compressive stress a material is capable of developing. With a brittle material that fails in compression by fracturing, the compressive strength has a definite value. In the case of ductile, malleable, or semiviscous materials (which do not fail in compression by a shattering fracture), the value obtained for compressive strength is an arbitrary value dependent on the degree of distortion that is regarded as effective failure of the material.

compressive stress. A stress that causes an elastic or plastic body to deform (shorten) in the direction of the applied load. Contrast with *tensile stress*.

computational fluid dynamics (CFD). An area of computer-aided engineering devoted to the numerical solution and visualization of fluid-flow problems.

computer-aided design (CAD). Any design activity that involves the effective use of the computer to create or modify an engineering design. Often used synonymously with the more general term computer-aided engineering (CAE).

computer-aided materials selection system. A computerized database of materials properties operated on by an appropriate knowledge base of decision rules through an *expert system* to select the most appropriate materials for an application.

concurrent engineering. A style of product design and development that is done by concurrently using all of the relevant information in making each decision. It replaces a sequential approach to product development in which one type of information was predominant in making each sequential decision. Concurrent engineering is carried out by a multifunctional team that integrates the specialties or functions needed to solve the problem. Sometimes called simultaneous engineering.

constitutive equation. Equation expressing the relation between stress, strain, strain rate, and microstructural features (e.g., grain size). Constitutive equations are generally phenomenological (curve fits based on measured data) or mechanism-based (based on mechanistic model of deformation and appropriate measurements). Phenomenological constitutive equations are usually valid only within the processing regime in which they were measured, while mechanism-based relations can be extrapolated outside the regime of measurement, provided the deformation mechanism is unchanged. A mathematical relationship that describes the flow stress of a material in terms of the plastic strain, the strain rate, and temperature.

constraint modeling. A form of computer modeling in which constraints are used to create a set of rules that control how changes are made to a group of geometric elements (lines, arcs, form features, etc.). These rules are typically embodied in a set of equations. Constraint models are defined as either parametric or variational.

continuum mechanics. The science of mathematically describing the behavior of continuous media. The same basic approach can apply to descriptions of stress, heat, mass, and momentum transfer.

controlled rolling. Multistand plate or bar rolling process, typically for ferrous alloys, in which the reduction per pass, rolling speed, time between passes, and so on are carefully chosen to control recrystallization, precipitation, and phase transformation in order to develop a desired microstructure and set of properties.

conventional forging. Forging process in which the work material is hot and the dies are at room temperature or slightly elevated temperature. To minimize the effects of die chilling on metal flow and microstructure, conventional forging usually involves strain rates of the order of 0.05 s^{-1} or greater. Also known as nonisothermal forging.

core. (1) Inner material in a coextruded or codrawn product. (2) In casting, a specially formed material inserted in a mold to shape the interior part of the casting that cannot be shaped as easily by the pattern. (3) In ferrous alloys prepared for case hardening, that portion of the alloy that is not part of the case.

coring. (1) A central cavity at the butt end of a rod extrusion; sometimes called *extrusion pipe*. (2) A condition of variable composition between the center and surface of a unit of microstructure (such as a dendrite, grain, or carbide particle); results from nonequilibrium solidification, which occurs over a range of temperature.

corrugating. The forming of sheet metal into a series of straight, parallel alternate ridges and grooves with a rolling mill equipped with matched roller dies or a press brake equipped with specially shaped punch and die.

corrugations. Transverse ripples caused by a variation in strip shape during hot or cold reduction.

Coulomb friction. Interface friction condition for which the interface shear stress is proportional to the pressure normal to the interface. The proportionality constant is called the Coulomb coefficient of friction, μ, and takes on values between 0 (perfect lubrication) and $1/\sqrt{3}$ (sticking friction) during metalworking. See also *friction shear factor*.

cracked edge. A series of tears at the edge of the sheet resulting from prior processing.

crank. Forging shape generally in the form of a "U" with projections at more or less right angles to the upper terminals. Crank shapes are designated by the number of throws (for example, two-throw crank).

creep. Time-dependent strain occurring under stress.

creep forming. Forming, usually at elevated temperatures, where the material is deformed over time with a preload, usually weights placed on the parts during a stress-relief cycle.

crimping. The forming of relatively small *corrugations* in order to set down and lock a seam, to create an arc in a strip of metal, or to reduce an existing arc or diameter. See also *corrugating*.

cross breaks. Visually apparent line-type discontinuities more or less transverse to the coil rolling direction, resulting from bending the coil over too sharp a radius and thus kinking the metal.

crystal. A solid composed of atoms, ions, or molecules arranged in a pattern that is periodic in three dimensions.

crystal lattice. A regular array of points about which the atoms or ions of a crystal are centered.

crystalline. That form of a substance comprised predominantly of (one or more) crystals, as opposed to glassy or amphorous.

crystalline defects. The deviations from a perfect three-dimensional atomic packing that are responsible for much of the structure-sensitive properties of materials. Crystal defects can be point defects (vacancies), line defects (dislocations), or surface defects (grain boundaries).

crystal-plasticity modeling. Physics-based modeling techniques that treat the phenomena of deformation by way of slip and twinning in order to predict strength and the evolution of crystallographic texture during the deformation processing of polycrystalline materials. See also *deformation texture, slip, Schmid's law, Taylor factor*, and *twinning.*

crystal system. One of seven groups into which all crystals may be divided: triclinic, monoclinic, orthorhombic, hexagonal, rhombohedral, tetragonal, and cubic.

cube texture. A texture found in wrought metals in the cubic system in which nearly all the crystal grains have a plane of the type (100) parallel or nearly parallel to the plane of working and a direction of the type [001] parallel or nearly parallel to the direction of elongation.

cumulative distribution function (CDF). A frequency distribution arranged to give the number of observations that are less than given values. 100% of the observations will be less than the largest class interval of the observations.

cup. (1) A sheet metal part; the product of the first drawing operation. (2) Any cylindrical part or shell closed at one end.

cup fracture (cup-and-cone fracture). A mixed mode fracture, often seen in tensile test specimens of a ductile material, in which the central portion undergoes plane-strain fracture and the surrounding region undergoes plane-stress fracture. One of the mating fracture surfaces looks like a miniature cup; it has a central depressed flat-face region surrounded by a shear lip. The other fracture surface looks like a miniature truncated cone.

cupping. (1) The first step in *deep drawing.* (2) Fracture of severely worked rods or wire in which one end looks like a cup and the other a cone.

cupping test. A mechanical test used to determine the ductility and stretching properties of sheet metal. It consists of measuring the maximum part depth that can be formed before fracture. The test is typically carried out by stretching the testpiece clamped at its edges into a circular die using a punch with a hemispherical end. See also *cup fracture, Erichsen test*, and *Olsen ductility test.*

curling. Forming of an edge of circular cross section along a sheet or along the end of a shell or tube, either to the inside or outside, for example, the pinholes in sheet metal hinges and the curled edges on cans, pots, and pans.

D

damage. General term used to describe the development of defects such as cavities, cracks, shear bands, and so on that may culminate in gross fracture in severe cases. The evolution of damage is strongly dependent on material, microstructure, and processing conditions (strain, strain rate, temperature, and stress state).

DBTT. See *ductile-to-brittle transition temperature (DBTT).*

dead-metal zone. Region of metal undergoing limited or no deformation during bulk forming of a workpiece, generally developed adjacent to the workpiece-tooling interface as a result of friction, die chilling, or deformation-zone geometry.

deduction. The process of starting at the lowest level and reasoning toward a conclusion. In the context of materials design, starting with the chemical composition and processing conditions and moving toward a desired final performance through the microstructure and properties developed. Compare this process with *induction.* Adapted from Ref 2

deep drawing. Forming operation characterized by the production of a parallel-wall cup from a flat blank of sheet metal. The blank may be circular, rectangular, or a more complex shape. The blank is drawn into the die cavity by the action of a punch. Deformation is restricted to the flange areas of the blank. No deformation occurs under the bottom of the punch—the area of the blank that was originally within the die opening. As the punch forms the cup, the amount of material in the flange decreases. Also called cup drawing or radial drawing.

deflection. The amount of deviation from a straight line or plane when a force is applied to a press member. Generally used to specify the allowable bending of the bed, slide, or frame at rated capacity with a load of predetermined distribution.

deformation (adiabatic) heating. Temperature increase that occurs in a workpiece due to the conversion of strain energy, imparted during metalworking, into heat.

deformation energy method. A metalforming analysis technique that takes into account only the energy required to deform the workpiece.

deformation limit. In *drawing*, the limit of deformation is reached when the load required to deform the flange becomes greater than the load-carrying capacity of the cup wall. The deformation limit (limiting drawing ratio) is defined as the ratio of the maximum blank diameter that can be drawn into a cup without failure, to the diameter of the punch.

deformation-mechanism map. Strain rate/temperature map that describes forming or service regimes under which deformation is controlled by micromechanical processes such as dislocation glide, dislocation climb, and diffusional flow limited by bulk or boundary diffusion.

deformation processing. A class of manufacturing operation that involves changing the shape of a workpiece by plastic deformation through the application of a compressive or tensile force. Often carried out at ambient or elevated temperature.

deformation texture. Preferred orientation of the crystals/grains comprising a polycrystalline aggregate that is developed during deformation processing as a result of slip and rotation within each crystal that comprises the aggregate.

Demarest process. A *fluid forming* process in which cylindrical and conical sheet metal parts are formed by a modified rubber bulging punch. The punch, equipped with a hydraulic cell, is placed inside the workpiece, which in turn is placed inside the die. Hydraulic pressure expands the punch.

density. Weight per unit volume.

design of experiments. Methodology for choosing a small number of screening experiments to establish the important material and process variables in a complex manufacturing process.

deterministic experiment. An experiment in which, when the experiment is repeated with identical input, the same output is obtained. There is no randomness associated in the development of the answer.

developed blank. A sheet metal blank that yields a finished part without trimming or with the least amount of trimming.

deviatoric. The nonhydrostatic component of the state of stress on a body. It is the deviatoric component that causes shape change (plastic deformation).

die. (1) A tool, usually containing a cavity, that imparts shape to solid, molten, or powdered metal primarily because of the shape of the tool itself. Used in many press operations (including blanking, drawing, forging, and forming), in die casting, and in forming green powder metallurgy compacts. Die-casting and powder metallurgy dies are sometimes referred to as molds. See also *forging dies.* (2) A complete tool used in a press for any operation or series of operations, such as forming, impressing, piercing, and cutting. The upper member or members are attached to the slide (or slides) of the press, and the lower member is clamped or bolted to the bed or bolster, with the die members being so shaped as to cut or form the material placed between them when the press makes a stroke. (3) The female part of a complete die assembly as described previously.

die assembly. The parts of a die stamp or press that hold the die and locate it for the punches.

die block. A block, often made of heat treated steel, into which desired impressions are machined or sunk and from which closed-die forgings or sheet metal stampings are produced using hammers or presses. In forging, die blocks are usually used in pairs, with part of the impression in one of the blocks and the rest of the impression in the other. In sheet metal forming, the female die is used in conjunction with a male punch. See also *closed-die forging*.

die button. An insert in a die that matches the punch and is used for punching and piercing operations. The die button is readily removable for sharpening or replacement as an individual part of a die.

die cavity. The machined recess that gives a forging or stamping its shape.

die check. A crack in a die impression due to forging and thermal strains at relatively sharp corners. Upon forging, these cracks become filled with metal, producing sharp, ragged edges on the part. Usual die wear is the gradual enlarging of the die impression due to erosion of the die material, generally occurring in areas subject to repeated high pressures during forging.

die chill. The temperature loss experienced by a billet or preform when it contacts dies that are maintained at a lower temperature.

die clearance. Clearance between a mated punch and die; commonly expressed as clearance per side. Also called clearance or punch-to-die clearance.

die coating. Hard metal incorporated into the working surface of a die to protect the working surface or to separate the sheet metal surface from direct contact with the basic die material. Hard-chromium plating is an example.

die forging. A forging that is formed to the required shape and size through working in machined impressions in specially prepared dies.

die forming. The shaping of solid or powdered metal by forcing it into or through the *die cavity*.

die impression. The portion of the die surface that shapes a forging or sheet metal part.

die insert. A relatively small die that contains part or all of the impression of a forging or sheet metal part and is fastened to the master *die block*.

die line. A line or scratch resulting from the use of a roughened tool or the drag of a foreign particle between tool and product.

die lock. A phenomenon in which the deformation is limited in a forging near the die face due to chilling of the workpiece and/or friction at the workpiece-die interface.

die lubricant. In forging or forming, a compound that is sprayed, swabbed, or otherwise applied on die surfaces or the workpiece during the forging or forming process to reduce friction. Lubricants also facilitate release of the part from the dies and provide thermal insulation. See also *lubricant*.

die match. The alignment of the upper (moving) and lower (stationary) dies in a hammer or press. An allowance for misalignment (or mismatch) is included in forging tolerances.

die radius. The radius on the exposed edge of a deep-drawing die, over which the sheet flows in forming drawn shells.

die section. A section of a cutting, forming, or flanging die that is fastened to other sections to make up the complete working surface. Also referred to as cutting section.

die set. (1) The assembly of the upper and lower die shoes (punch and die holders), usually including the guide pins, guide pin bushings, and heel blocks. This assembly takes many forms, shapes, and sizes and is frequently purchased as a commercially available unit. (2) Two (or, for a mechanical upsetter, three) machined dies used together during the production of a *die forging*.

die shift. The condition that occurs after the dies have been set up in a forging unit in which a portion of the impression of one die is not in perfect alignment with the corresponding portion of the other die. This results in a mismatch in the forging, a condition that must be held within the specified tolerance.

die stamping. The general term for a sheet metal part that is formed, shaped, or cut by a die in a press in one or more operations.

dimpling. (1) The stretching of a relatively small, shallow indentation into sheet metal. (2) In aircraft, the stretching of metal into a conical flange for a countersunk head rivet.

direct (forward) extrusion. See *extrusion*.

directional solidification. Controlled solidification of molten metal in a casting so as to provide feed metal to the solidifying front of the casting. Usually, this results in the metal solidifying in a preferred direction. In the limit, the solidification can be controlled to grow as a single grain (single-crystal casting).

discontinuous yielding. The nonuniform plastic flow of a metal exhibiting a yield point in which plastic deformation is inhomogeneously distributed along the gage length. Under some circumstances, it may occur in metals not exhibiting a distinct yield point, either at the onset of or during plastic flow.

dishing. The formation of a shallow concave surface in which the projected area is very large compared with the depth of the impression.

dislocation. A linear imperfection in a crystalline array of atoms. Two basic types include: (1) an edge dislocation corresponds to the row of mismatched atoms along the edge formed by an extra, partial plane of atoms within the body of a crystal; and (2) a screw dislocation corresponds to the axis of a spiral structure in a crystal, characterized by a distortion that joins normally parallel planes together to form a continuous helical ramp (with a pitch of one interplanar distance) winding about the dislocation. Most prevalent is the so-called mixed dislocation, which is any combination of an edge dislocation and a screw dislocation.

dislocation density. The total length of dislocation lines per unit volume, or the number of dislocation lines that cut through a unit cross-sectional area.

dispersion strengthening. The strengthening of a metal or alloy by incorporating chemically stable submicron-sized particles of a nonmetallic phase that impede dislocation movement at elevated temperature.

double-cone test. Simulative bulk forming test consisting of the compression of a sample shaped like a flying saucer between flat dies. The variation of strain and stress state developed across the sample is used to obtain a large quantity of data on microstructure evolution and failure in a single experiment.

draft. The amount of taper on the sides of the forging and on projections to facilitate removal from the dies; also, the corresponding taper on the sidewalls of the die impressions. In *open-die forging*, draft is the amount of relative movement of the dies toward each other through the metal in one application of power.

drawability. A measure of the *formability* of a sheet metal subject to a drawing process. The term is usually used to indicate the ability of a metal to be deep drawn. See also *drawing* and *deep drawing*.

draw bead. An insert or riblike projection on the draw ring or hold-down surfaces that aids in controlling the rate of metal flow during deep-drawing operations. Draw beads are especially useful in controlling the rate of metal flow in irregularly shaped stampings.

draw forming. A method of curving bars, tubes, or rolled or extruded sections in which the stock is bent around a rotating form block. Stock is bent by clamping it to the form block, then rotating the form block while the stock is pressed between the form block and a pressure die held against the periphery of the form block.

drawing. A term used for a variety of forming operations, such as *deep drawing* a sheet metal blank; *redrawing* a tubular part; and drawing rod, wire, and tube. The usual drawing process with regard to sheet metal working in a press is a method for producing a cuplike form from a sheet metal disk by holding it firmly between blankholding surfaces to prevent the formation of wrinkles while the punch travel produces the required shape.

drawing die. A type of die designed to produce nonflat parts such as boxes, pans, and so on. Whenever practical, the die should be designed and built to finish the part in one stroke of the press, but if the part is deep in proportion to its diameter, redrawing operations are necessary.

drawing ratio. The ratio of the blank diameter to the punch diameter.

draw radius. The radius at the edge of a die or punch over which sheet metal is drawn.

draw stock. The forging operation in which the length of a metal mass (stock) is increased at the expense of its cross section; no *upset* is involved. The operation includes converting ingot to pressed bar using "V," round, or flat dies.

drop forging. The forging obtained by hammering metal in a pair of closed dies to produce the form in the finishing impression under a *drop hammer*; forging method requiring special dies for each shape.

drop hammer. A term generally applied to forging hammers in which energy for forging is provided by gravity, steam, or compressed air.

drop hammer forming. A process for producing shapes by the progressive deformation of sheet metal in matched dies under the repetitive blows of a gravity-drop or power-drop hammer. The process is restricted to relatively shallow parts and thin sheet from approximately 0.6 to 1.6 mm (0.024 to 0.064 in.).

drop through. The type of ejection where the part or scrap drops through an opening in the die.

dry-film lubricant. A type of lubricant applied by spraying or painting on coils or sheets prior to blanking, drawing, or stamping. The lubricant can have a wax base and be sprayed hot onto the sheet surface and solidify on cooling, or be a water-based polymer and be roll coated onto the surface (one or both sides) and be heated to cure and dry. Such lubricants have uniform thickness, low coefficients of friction, and offer protection from corrosion in transit and storage.

ductile fracture. Failure of metals as a result of cavity nucleation, growth, and coalescence. Ductile fracture may occur during metal forming at both cold and hot working temperatures.

ductile-to-brittle transition temperature (DBTT). A temperature or range of temperatures over which a material reaction to impact (high strain rate) loads changes from ductile, high-energy-absorbing to brittle, low-energy-absorbing behavior. The DBTT determinations are often done with Charpy or Izod test specimens measuring absorbed energy at various temperatures.

ductility. A measure of the amount of deformation that a material can withstand without breaking.

dynamic friction. The friction forces between two surfaces in relative motion. See also *static friction.*

dynamic material modeling. A methodology by which macroscopic measurements of flow stress as a function of temperature and strain rate are used with continuum criteria of instability to identify regions of temperature and strain rate in which voids, cracks, shear bands, and flow localization are likely to occur.

dynamic recovery. Recovery process that occurs during cold or hot working of metals, typically resulting in the formation of low-energy dislocation substructures/subgrains within the deformed original grains. Dynamic recovery reduces the observed level of strain hardening due to dislocation multiplication during deformation.

dynamic recrystallization. The formation of strain-free recrystallized grains during hot working. It results in a decrease in flow stress and formation of equiaxed grains, as opposed to dynamic recovery in which the elongated grains remain.

dynamic strain aging. A behavior in metals in which solute atoms are sufficiently mobile to move toward and interact with dislocations during deformation. This results in strengthening over a specific range of elevated temperature and strain rate.

E

earing. The formation of ears or scalloped edges around the top of a drawn shell, resulting from directional differences in the plastic-working properties of rolled metal with, across, and at angles to the direction of rolling.

edge dislocation. A line imperfection that corresponds to the row of mismatched atoms along the edge formed by an extra, partial plane of atoms within the body of a crystal.

edger (edging impression). The portion of a die impression that distributes metal during forging into areas where it is most needed in order to facilitate filling the cavities of subsequent impressions to be used in the forging sequence.

edge strain. Repetitive areas of local deformation extending inwardly from the edge of temper-rolled sheet.

edging. (1) In sheet metal forming, reducing the flange radius by retracting the forming punch a small amount after the stroke but before release of the pressure. (2) In rolling, the working of metal in which the axis of the roll is parallel to the thickness dimension. Also called edge rolling. The result is changing a rounded edge to a square edge. (3) The forging operation of working a bar between contoured dies while turning it 90° between blows to produce a varying rectangular cross section.

effective draw. The maximum limits of forming depth that can be achieved with a multiple-action press; sometimes called maximum draw or maximum depth of draw.

effective strain. The (scalar) strain conjugate to effective stress defined in such a manner that the product of the effective stress and the effective strain increment is equal to the increment in imposed work during a deformation process.

effective stress. A mathematical way to express a two- or three-dimensional stress state by a single number.

elastic deformation. A change in dimensions that is directly proportional to and in phase with an increase or decrease in applied force; deformation that is recoverable when the applied force is removed.

elasticity. The property of a material by which the deformation caused by stress disappears upon removal of the stress. A perfectly elastic body completely recovers its original shape and dimensions after the release of stress.

elastic limit. The maximum stress a material can sustain without any permanent strain (deformation) remaining upon complete release of the stress.

elastic modulus. See *Young's modulus.*

elastohydrodynamic lubrication. A condition of lubrication in which the friction and film thickness between two bodies in relative motion is determined by the elastic properties of the bodies in combination with the viscous properties of the lubricant.

electrical resistivity. The electrical resistance offered by a material to the flow of current, times the cross-sectional area of current flow and per unit length of current path; the reciprocal of the conductivity. Also called resistivity or specific resistance.

electric-discharge machining (EDM). Metal-removal (machining) process based on the electric discharge/spark erosion resulting from current flowing between an electrode and workpiece placed in close proximity to each other in a dielectric fluid. The electrode may be a wire (as in wire EDM) or a contoured shape (so-called plunge EDM); the latter technique is used for making metalworking dies.

electromagnetic forming. A process for forming metal by the direct application of an intense, transient magnetic field. The workpiece is formed without mechanical contact by the passage of a pulse of electric current through a forming coil. Also known as magnetic pulse forming.

electron backscatter diffraction (EBSD). Materials characterization technique conducted in a scanning electron microscope (and sometimes a transmission electron microscope) used to establish the crystallographic orientation of individual (micronsized) regions of material through analysis of Kikuchi patterns formed by backscattered electrons. Automated EBSD systems can thus be used to determine the texture over small-to-moderate-sized total volumes of material.

elongation. (1) A term used in mechanical testing to describe the amount of extension of a testpiece when stressed. (2) In tensile testing, the increase in the gage length, measured after fracture of the specimen within the gage length, usually expressed as a percentage of the original gage length.

elongation, percent. The extension of a uniform section of a specimen expressed as a percentage of the original gage length:

$$\text{Elongation}, \% = \frac{(L_x - L_0)}{L_0} \times 100$$

where L_0 is the original gage length, and L_x is the final gage length.

embossing. A process for producing raised or sunken designs in sheet material by means of male and female dies, theoretically with minimal change in metal thickness. Examples are letters, ornamental pictures, and ribs for stiffening. Heavy embossing and *coining* are similar operations.

engineering strain. A term sometimes used for average linear strain or nominal strain in order to distinguish it from true strain. In tensile testing, it is calculated by dividing the change in the gage length by the original gage length.

engineering stress. A term sometimes used for average linear stress or nominal stress in order to differentiate it from true stess. In tension testing, it is calculated by dividing the load applied to the specimen by its original cross-sectional area.

Erichsen test. A *cupping test* used to assess the ductility of sheet metal. The method consists of forcing a conical or hemispherical-ended plunger into the specimen and measuring the depth of the impression at fracture.

error function. The function that often results as a solution to a partial differential equation. The error function is defined as:

$$\mathrm{erf}(x) = \frac{2}{\pi} \int_0^x e^{-x^2} dx$$

The error function is also called the probability integral.

erosion resistance. The ability of a die material to resist sliding wear and thus maintain its original dimension.

etching. Production of designs, including grids, on a metal surface by a corrosive reagent or electrolytic action.

Euler angles. Set of three angular rotations used to specify unambiguously the spatial orientation of crystallites relative to a fixed reference frame.

expert system. A computer-based system that captures the knowledge of experts through the integration of databases and knowledge bases using search and logic deduction algorithms.

extruded hole. A hole formed by a punch that first cleanly cuts a hole and then is pushed farther through to form a flange with an enlargement of the original hole. This may be a two-step operation.

extrusion. The conversion of an ingot or billet into lengths of uniform cross section by forcing metal to flow plastically through a die orifice. In forward (direct) extrusion, the die and ram are at opposite ends of the extrusion stock, and the product and ram travel in the same direction. Also, there is relative motion between the extrusion stock and the die. In backward (indirect) extrusion, the die is at the ram end of the stock, and the product travels in the direction opposite that of the ram, either around the ram (as in the impact extrusion of cylinders, such as cases for dry cell batteries) or up through the center of a hollow ram.

extrusion billet. A metal slug used as *extrusion stock.*

extrusion forging. Forcing metal into or through a die opening by restricting flow in other directions. A part made by the operation.

extrusion pipe. A central oxide-lined discontinuity that occasionally occurs in the last 10 to 20% of an extruded bar. It is caused by the oxidized outer surface of the billet flowing around the end of the billet and into the center of the bar during the final stages of extrusion. Also called *coring.*

extrusion stock. A rod, bar, or other section used to make extrusions.

eyeleting. The displacing of material about an opening in sheet or plate so that a lip protruding above the surface is formed.

F

fatigue. The phenomenon leading to fracture under repeated or fluctuating stresses having a maximum value less than the ultimate tensile strength of the material. Fatigue failure generally occurs at loads that, applied statically, would produce little perceptible effect. Fatigue fractures are progressive, beginning as minute cracks that grow under the action of the fluctuating stress.

fatigue limit. The maximum cyclic stress that a material can withstand for an infinitely large number of stress cycles. Also called endurance limit.

fatigue-strength reduction factor. The ratio of the fatigue strength of a member or specimen with no stress concentration to the fatigue strength with stress concentration. This factor has no meaning unless the stress range and the shape, size, and material of the member or specimen are stated.

fiber texture. Crystallographic texture in which all or a large fraction of the crystals in a polycrystalline aggregate are oriented such that a specific direction in each crystal is parallel to a specific sample direction, such as the axis of symmetry of a cylindrical object. Often found in wrought products such as wire and round extrusions that have been subjected to large axisymmetric deformation.

fillet. The concave intersection of two surfaces. In forging, the desired radius at the concave intersection of two surfaces is usually specified.

film strength. The ability of a surface film to resist rupture by the penetration of asperities during sliding or rolling of two surfaces over each other.

fin. The thin projection formed on a forging by trimming or when metal is forced under pressure into hairline cracks or die interfaces.

finishing dies. The die set used in the last forging step.

finish trim. Flash removal from a forging; usually performed by trimming but sometimes by band sawing or similar techniques.

finite-element analysis (FEA). A computer-based technique used to solve simultaneous equations that is used to predict the response of a workpiece or structure to applied loads and temperature. The FEA is a tool used to model deformation and heat treating processes.

finite-element modeling (FEM). A numerical technique in which the analysis of a complex part is represented by a mesh of elements interconnected at node points. The coordinates of the nodes are combined with the elastic-plastic properties of the material to produce a stiffness matrix, and this matrix is combined with the applied loads to determine the deflections at the nodes and hence the stresses. All of the aforementioned is done with special FEM software. The FEM approach also may be used to solve other field problems in heat transfer, fluid flow, acoustics, and so on. Also known as finite-element analysis (FEA).

fixture. A tool or device for holding and accurately positioning a piece or part on a machine tool or other processing machine.

flame hardening. A heat treating method for surface hardening steel of the proper specifications in which an oxyacetylene flame heats the surface to a temperature at which subsequent cooling, usually with water or air, will give the required surface hardness.

flame straightening. The correction of distortion in metal structures by localized heating with a gas flame.

flange. A projecting rim or edge of a part; usually narrow and of approximately constant width for stiffening or fastening.

flanging. A bending operation in which a narrow strip at the edge of a sheet is bent down (up) along a straight or curved line. It is used for edge strengthening, appearance, rigidity, and the removal of sheared edges. A flange is often used as a fastening surface.

flaring. The forming of an outward acute-angle *flange* on a tubular part.

flash. Metal in excess of that required to completely fill the blocking or finishing forging impression of a set of dies. Flash extends out from the body of the forging as a thin plate at the line where the dies meet and is subsequently removed by trimming. Because it cools faster than the body of the component during forging, flash can serve to restrict metal flow at the line where dies meet, thus ensuring complete filling of the impression.

flash extension. That portion of flash remaining on a forged part after trimming; usually included in the normal forging tolerances.

flash line. The line left on a forging after the flash has been trimmed off.

flash pan. The machined-out portion of a forging die that permits the flow through of excess metal.

flattening. (1) A preliminary operation performed on forging stock to position the metal for a subsequent forging operation. (2) The removal of irregularities or distortion in sheets or plates by a method such as *roller leveling or stretcher leveling.* (3) For wire, rolling round wire to a flattened condition.

flattening dies. Dies used to flatten sheet metal hems; that is, dies that can flatten a bend by closing it. These dies consist of a top and bottom die with a flat surface that can close one section (flange) to another (hem, seam).

fleck scale. A fine pattern of scale marks on the sheet surface that can be either dark or light. The dark pattern originates from scale and other impurities embedded in the strip during hot rolling that are not removed during pickling. The light pattern originates from a scale pattern imprinted on the work rolls in the finishing stands in the hot mill being printed onto the strip.

flex roll. A movable roll designed to push up against a sheet as it passes through a roller leveler. The flex roll can be adjusted to deflect the sheet any amount up to the roll diameter.

flex rolling. Passing sheets through a *flex roll* unit to minimize yield point elongation in order to reduce the tendency for *stretcher* strains to appear during forming.

flexural strength. A property of solid material that indicates its ability to withstand a flexural or transverse load.

flow curve. A curve of true stress versus true strain that shows the stress required to produce plastic deformation. A graphical representation of the relationship between load and deformation during plastic deformation.

flow lines. (1) Texture showing the direction of metal flow during hot or cold working. Flow lines can often be revealed by etching the surface or a section of a metal part. (2) In mechanical metallurgy, paths followed by minute volumes of metal during deformation.

flow localization. A situation where material deformation is localized to a narrow zone. Such zones often are sites of failure. Flow localization results from poor lubrication, temperature gradients, or flow softening resulting from adiabatic heating, generation of softer crystallographic texture, grain coarsening, or spheroidization of second phases.

flow softening. Stress-strain behavior observed under constant strain-rate conditions characterized by decreasing flow stress with increasing strain. Flow softening may result from deformation heating as well as a number of microstructural sources, such as the generation of a softer crystallographic texture and the spheroidization of a lamellar phase.

flow stress. The uniaxial true stress required to cause plastic deformation at a particular value of strain, strain rate, and temperature.

flow through. A forging defect caused by metal flow past the base of a rib with resulting rupture of the grain structure.

fluid-cell process. A modification of the Guerin process for forming sheet metal, the fluid-cell process uses higher pressure and is primarily designed for forming slightly deeper parts, using a rubber pad as either the die or punch. A flexible hydraulic fluid cell forces an auxiliary rubber pad to follow the contour of the form block and exert a nearly uniform pressure at all points on the workpiece. See also *fluid forming* and *rubber-pad forming*.

fluid forming. A modification of the Guerin process, fluid forming differs from the fluid-cell process in that the die cavity, called a pressure dome, is not completely filled with rubber but with hydraulic fluid retained by a cup-shaped rubber diaphragm.

fluting. A series of sharp parallel kinks or creases that can occur when sheet steel is formed clylindrically. Fluting is caused by inhomogeneous yielding of these sheets.

foil. Metal in sheet form less than 0.15 mm (0.006 in.) thick.

fold. A forging defect caused by folding metal back onto its own surface during its flow in the die cavity.

forgeability. Term used to describe the relative ability of material to deform without fracture. Also describes the resistance to flow from deformation.

forging. The process of working metal to a desired shape by impact or pressure in hammers, forging machines (upsetters), presses, rolls, and related forming equipment. Forging hammers, counterblow equipment, and high-energy-rate forging machines apply impact to the workpiece, while most other types of forging equipment apply squeeze pressure in shaping the stock. Some metals can be forged at room temperature, but most are made more plastic for forging by heating.

forging dies. Forms for making forgings; they generally consist of a top and bottom die. The simplest will form a completed forging in a single impression; the most complex, consisting of several die inserts, may have a number of impressions for the progressive working of complicated shapes. Forging dies are usually in pairs, with part of the impression in one of the blocks and the rest of the impression in the other block.

forging plane. In forging, the plane that includes the principal die face and is perpendicular to the direction of ram travel. When the parting surfaces of the dies are flat, the forging plane coincides with the parting line. Contrast with *parting plane*.

forging quality. Term used to describe stock of sufficient quality to make it suitable for commercially satisfactory forgings.

forging stock. A wrought rod, bar, or other section suitable for subsequent change in cross section by forging.

formability. The ease with which a metal can be shaped through plastic deformation. Evaluation of the formability of a metal involves measurement of strength, ductility, and the amount of deformation required to cause fracture. The term *workability* is used interchangeably with formability; however, formability refers to the shaping of sheet metal, while workability refers to shaping materials by *bulk forming*. See also *forgeability*.

formability-limit diagram. An empirical curve showing the biaxial strain levels beyond which failure may occur in sheet metal forming. The strains are given in terms of major and minor strains measured from deformed circles, previously printed onto the sheet.

formability parameters. Material parameters that can be used to predict the ability of sheet metal to be formed into a useful shape.

forming. The plastic deformation of a billet or a blanked sheet between tools (dies) to obtain the final configuration. Metalforming processes are typically classified as *bulk forming* and *sheet forming*. Also referred to as metalworking.

forming-limit diagram (FLD) or forming-limit curve (FLC). An empirical curve in which the major strains at the onset of necking in sheet metal are plotted vertically and the corresponding minor strains are plotted horizontally. The onset-of-failure line divides all possible strain combinations into two zones: the safe zone (in which failure during forming is not expected) and the failure zone (in which failure during forming is expected).

form rolling. Hot rolling to produce bars having contoured cross sections; not to be confused with the *roll forming* of sheet metal or with *roll forging*.

forward extrusion. Same as direct extrusion. See also *extrusion*.

fracture. The irregular surface produced when a piece of metal is broken.

fracture criterion. A mathematical relationship among stresses, strains, or a combination of stresses and strains that predicts the occurrence of ductile fracture. Should not be confused with fracture mechanics equations, which deal with more brittle types of fracture.

fracture-limit line. An experimental method for predicting surface fracture in plastically deformed solids. Is related to the forming-limit diagram used to predict failures in sheet forming.

fracture load. The load at which splitting occurs.

fracture-mechanism map. Strain rate/temperature map that describes regimes under which different damage and failure mechanisms are operative under either forming or service conditions.

fracture strain (ε_f). The true strain at fracture defined by the relationship:

$$\varepsilon_f = \ln \left[\frac{\text{Initial cross} - \text{sectional area}}{\text{Final cross} - \text{sectional area}} \right]$$

fracture strength. The engineering stress at fracture, defined as the load at fracture divided by the original cross-sectional area. The fracture strength is synonymous with the breaking strength.

fracture stress. The true stress at fracture, which is the load for fracture divided by the final cross-sectional area.

fracture toughness. A generic term for measures of resistance to extension of a crack. The term is sometimes restricted to results of fracture mechanics tests, which are directly applicable in fracture control. However, the term commonly includes results from simple tests of notched or precracked specimens not based on fracture mechanics analysis. Results from tests of the latter type are often useful for fracture control, based on either service experience or empirical correlations with fracture mechanics tests. See also *stress-intensity factor*.

free bending. A bending operation in which the sheet metal is clamped at one end and wrapped around a radius pin. No tensile force is exerted on the ends of the sheet.

friction. The resisting force tangential to the common boundary between two bodies when, under the action of an external force, one body moves or tends to move relative to the surface of the other.

friction coextrusion. A solid core along with a tube made of a cladding material is friction extruded.

friction extrusion. A rotating round bar is pressed against a die to produce sufficient frictional heating to allow softened material to extrude through the die.

friction hill. Shape of the normal pressure-position plot that pertains to the axisymmetric and plane-strain forging of simple and complex shapes. The pressure is approximately equal to the flow stress at the edge of the forging and increases toward the center, thus producing the characteristic hilllike shape. The exact magnitude of the increase in pressure is a function of interface friction and the diameter-to-thickness or width-to-thickness ratio of the forging.

friction shear factor. Interface friction coefficient for which the interface shear stress is taken to be proportional to the flow stress divided by $\sqrt{3}$. The proportionality constant is called the friction shear factor (or interface friction factor) and is usually denoted as m. The friction shear factor takes on values between 0 (perfect lubrication) and 1 (sticking friction) during metalworking. See also *Coulomb friction*.

Fukui cup test. A cupping test combining stretchability and drawability in which a round-nosed punch draws a circular blank into a conical-shaped die until fracture occurs at the nose. Various parameters from the test are used as the criterion of formability.

fuzzy logic. The use of fuzzy sets in the representation and manipulation of vague information for the purpose of making decisions or taking actions. Fuzzy logic enables computers to make decisions based on information that is not clearly defined.

G

gage. (1) The thickness of sheet or the diameter of wire. The various standards are arbitrary and differ with regard to ferrous and nonferrous products as well as sheet and wire. (2) An aid for visual inspection that enables an inspector to determine more reliably whether the size or contour of a formed part meets dimensional requirements.

gage length. The original length of that part of a test specimen over which strain or other characteristics are measured.

galling. A condition whereby excessive friction between high spots results in localized welding with subsequent spalling and further roughening of the rubbing surface(s) of one or both of two mating parts.

Gaussian distribution. See *normal distribution*.

grain. An individual crystal in a polycrystalline metal or alloy.

grain boundary. A narrow zone in a metal or ceramic corresponding to the transition from one crystallographic orientation to another, thus separating one grain from another; the atoms in each grain are arranged in an orderly pattern.

grain-boundary sliding. The sliding of grains past each other that occurs at high temperature. Grain-boundary sliding is common under creep conditions in service, thus leading to internal damage (e.g., cavities) or total failure, and during superplastic forming, in which undesirable cavitation may also occur if diffusional or deformation processes cannot accommodate the sliding at a sufficient rate.

grain growth. The increase in the average size of grains in a crystalline aggregate during annealing (static conditions) or deformation (dynamic conditions). The driving force for grain growth is the reduction in total grain-boundary area and its associated surface energy.

grain size. A measure of the area or volume of grains in a polycrystalline material, usually expressed as an average when the individual sizes are fairly uniform.

gridding. Imprinting an array of repetitive geometrical patterns on a sheet prior to forming for subsequent determination of deformation. Imprinting techniques include: (1) Electrochemical marking (also called electrochemical or electrolytic etching)—a grid-imprinting technique using electrical current, an electrolyte, and an electrical stencil to etch the grid pattern into the blank surface. A contrasting oxide usually is redeposited simultaneously into the grid. (2) Photoprint—a technique in which a photosensitive emulsion is applied to the blank surface, a negative of the grid pattern is placed in contact with the blank, and the pattern is transferred to the sheet by a standard photographic printing practice. (3) Ink stamping. (4) Lithographing.

gripper dies. The lateral or clamping dies used in a forging machine or mechanical upsetter.

Guinier-Preston (G-P) zone. A small precipitation domain in a supersaturated metallic solid solution. A G-P zone has no well-defined crystalline structure of its own and contains an abnormally high concentration of solute atoms. The formation of G-P zones constitutes the first stage of precipitation and is usually accompanied by a change in properties of the solid solution in which they occur.

H

Hall-Petch dependence. A reflection of the effect of grain size on the yield strength of a metal. It states that the yield strength is inversely proportional to the square root of the grain size.

Hall-Petch relationship. A general relationship for metals that shows that the yield strength is linearly related to the reciprocal of the square root of the grain diameter.

hammering. The working of metal sheet into a desired shape over a form or on a high-speed hammer and a similar anvil to produce the required dishing or thinning.

hand forge (smith forge). A forging operation in which forming is accomplished on dies that are generally flat. The piece is shaped roughly to the required contour with little or no lateral confinement; operations involving mandrels are included. The term hand forge refers to the operation performed, while hand forging applies to the part produced.

hardness test. A test to measure the resistance to indentation of a material. Tests for sheet metal include Rockwell, Rockwell Superficial, Tukon, and Vickers.

Hartmann lines. See *Lüders lines*.

heading. The *upsetting* of wire, rod, or bar stock in dies to form parts that usually contain portions that are greater in cross-sectional area than the original wire, rod, or bar.

healed-over scratch. A scratch that occurred during previous processing and was partially obliterated in subsequent rolling.

hemming. A bend of 180° made in two steps. First, a sharp-angle bend is made; next, the bend is closed using a flat punch and a die.

high-angle boundary. Boundary separating adjacent grains whose misorientation is at least 15°.

high-energy-rate forging. The production of forgings at extremely high ram velocities resulting from the sudden release of a compressed gas against a free piston. Forging is usually completed in one blow. Also known as HERF processing, high-velocity forging, and high-speed forging.

high-energy-rate forming. A group of forming processes that applies a high rate of strain to the material being formed through the application of high rates of energy transfer. See also *high-energy-rate forging* and *electromagnetic forming*.

hodograph. A curve traced in the course of time by the tip of a vector representing some physical quantity. In particular, the path traced by the velocity vector of a given particle.

hole expansion test. A formability test in which a tapered punch is forced through a punched or a drilled and reamed hole, forcing the metal in the periphery of the hole to expand in a stretching mode until fracture occurs.

hole flanging. The forming of an integral collar around the periphery of a previously formed hole in a sheet metal part.

homogenization. Heat treatment used to reduce or eliminate nonuniform chemical composition that develops on a microscopic scale (microsegregation) during the solidification processing of ingots and castings. Homogenization is commonly used for aluminum alloys and nickel-base superalloys.

Hooke's law. A generalization applicable to all solid material, which states that stress is directly proportional to strain and is expressed as:

$$\text{Stress/strain} = \sigma/\varepsilon = \text{Constant} = E$$

where E is the modulus of elasticity (Young's modulus). The constant relationship between stress and strain applies only below the proportional limit.

hot forming. See *hot working.* Similar to hot sizing, however, the forming is done at temperatures above the annealing temperature, and deformation is usually larger.

hot isostatic pressing (HIP). A process for simultaneously heating and forming a powder metallurgy compact in which metal powder, contained in a sealed flexible mold, is subjected to equal pressure from all directions at a temperature high enough for full consolidation to take place. Hot isostatic pressing is also frequently used to seal residual porosity in castings and to consolidate metal-matrix composites. A process that subjects a component (casting, powder forgings, etc.) to both elevated temperature and isostatic gas pressure in an autoclave. The most widely used pressurizing gas is argon. When castings are hot isostatically pressed, the simultaneous application of heat and pressure virtually eliminates internal voids and microporosity through a combination of plastic deformation, creep, and diffusion.

hot rolled sheet. Steel sheet reduced to required thickness at a temperature above the point of scaling and therefore carrying hot mill oxide. The sheet may be flattened by cold rolling without appreciable reduction in thickness or by roller leveling, or both. Depending on the requirements, hot rolled sheet can be pickled to remove hot mill oxide and is so produced when specified.

hot shortness. A tendency for some alloys to separate along grain boundaries when stressed or deformed at temperatures near the melting point. Hot shortness is caused by a low-melting constituent, often present only in minute amounts, that is segregated at grain boundaries.

hot size. A process where a preformed part is placed into a hot die above the annealing temperature to set the shape and remove springback tendencies.

hot strip or pickle line scratch. Scratches that are superficially similar to slivers or skin laminations but originate from mechanical scoring of the strip in the hot mill, pickle line, or slitter.

hot upset forging. A *bulk forming* process for enlarging and reshaping some of the cross-sectional area of a bar, tube, or other product form of uniform (usually round) section. It is accomplished by holding the heated forging stock between grooved dies and applying pressure to the end of the stock, in the direction of its axis, by the use of a heading tool, which spreads (upsets) the end by metal displacement. Also called hot heading or hot upsetting. See also *heading* and *upsetting.*

hot working. The plastic deformation of metal at such a temperature and strain rate that recrystallization or a high degree of recovery takes place simultaneously with the deformation, thus avoiding any *strain hardening.* Also referred to as hot forging and hot forming. Contrast with *cold working.*

hub. A *boss* that is in the center of a forging and forms a part of the body of the forging.

hubbing. The production of die cavities by pressing a male master plug, known as a *hub,* into a block of metal.

hydrodynamic lubrication. A system of lubrication in which the shape and relative motion of the sliding surfaces causes the formation of a liquid film having sufficient pressure to separate the surfaces. See also *elastohydro-dynamic lubrication.*

hydrostatic extrusion. A method of extruding a *billet* through a die by pressurized fluid instead of the ram used in conventional *extrusion.*

hydrostatic stress. The average value of the three normal stresses. The hydrostatic stress is a quantity that is invariant relative to the orientation of the coordinate system in which the stress state is defined.

I

IACS. See *percent IACS (%IACS).*

impact extrusion. The process (or resultant product) in which a punch strikes a slug (usually unheated) in a confining die. The metal flow may be either between punch and die or through another opening. The impact extrusion of unheated slugs is often called cold extrusion.

impact line. A blemish on a drawn sheet metal part caused by a slight change in metal thickness. The mark is called an impact line when it results from the impact of the punch on the blank; it is called a recoil line when it results from transfer of the blank from the die to the punch during forming, or from a reaction to the blank being pulled sharply through the draw ring.

impact strength. A measure of the resiliency or toughness of a solid. The maximum force or energy of a blow (given by a fixed procedure) that can be withstood without fracture, as opposed to fracture strength under a steady applied force.

impression. A cavity machined into a forging die to produce a desired configuration in the workpiece during forging.

inclusion. A physical and mechanical discontinuity occurring within a material or part, usually consisting of solid, encapsulated foreign material. Inclusions are often capable of transmitting some structural stresses and energy fields, but to a noticeably different degree than from the parent material.

increase in area. An indicator of sheet metal forming severity based on percentage increase in surface area measured after forming.

induction. The process of finding a desired property by assuming that that property is true and then finding the necessary conditions under which it must be true. Induction is the process of assuming the desired material performance is true and then finding the properties, microstructure, and processing conditions that are required. Compare with *deduction.* Adapted from Ref 2

ingot. A casting intended for subsequent rolling, forging, or extrusion.

ingot conversion. A primary metalworking process that transforms a cast ingot into a wrought mill product.

ingot metallurgy. A processing route consisting of casting an ingot that is subsequently converted into mill products via deformation processes.

inoculant. Materials that, when added to molten metal, modify the structure and thus change the physical and mechanical properties to a degree not explained on the basis of the change in composition resulting from their use. Ferrosilicon-base alloys are commonly used to inoculate gray irons and ductile irons.

intellectual property. Knowledge-based property, usually represented by patents, copyrights, trademarks, or trade secrets.

interface heat-transfer coefficient (IHTC). Coefficient defined as the ratio of the heat flux across an interface to the difference in temperature of material points lying on either side of the interface. In bulk forming, the IHTC is usually a function of the die and workpiece surface conditions, lubrication, interface pressure, amount of relative sliding, and so on.

intermetallic alloy. A metallic alloy usually based on an ordered, stoichiometric compound (e.g., Fe_3Al, Ni_3Al, $TiAl$) and often possessing exceptional strength and environmental resistance at high temperatures, unlike conventional (less highly alloyed) disordered metallic materials.

interstitial-free steels. Steels where carbon and nitrogen are removed in the steelmaking process to very low levels, and any remaining interstitial carbon and nitrogen is tied up with small amounts of alloying elements that form carbides and nitrides, that is, titanium and niobium. Although these steels have low strength, they exhibit exceptional formability.

ionic bond. (1) A type of chemical bonding in which one or more electrons are transferred completely from one atom to another, thus converting the neutral atoms into electrically charged ions. These ions are approximately spherical and attract each other because of their opposite charges. (2) A primary bond arising from the electrostatic attraction between two oppositely charged ions.

ironing. An operation used to increase the length of a tube or cup through reduction of wall thickness and outside diameter, the inner diameter remaining unchanged.

isostatic pressing. A process for forming a powder metallurgy compact/metal-matrix composite or for sealing casting porosity by applying pressure equally from all directions. See also *hot isostatic pressing (HIP)*.

isothermal forging. A hot forging process in which a constant and uniform temperature is maintained in the workpiece during forging by heating the dies to the same temperature as the workpiece.

isothermal transformation (IT) diagram. A diagram that shows the isothermal time required for transformation of austenite to begin and to finish as a function of temperature.

isotropy. The condition in which the properties are independent of the direction in which they are measured.

J

J-integral. A mathematical expression involving a line or surface integral that encloses the crack front from one crack surface to the other, used to characterize the *fracture toughness* of a material having appreciable plasticity before fracture. The *J*-integral eliminates the need to describe the behavior of the material near the crack tip by considering the local stress-strain field around the crack front; J_{Ic} is the critical value of the *J*-integral required to initiate growth of a pre-existing crack.

Joffe effect. Change in mechanical properties, especially the fracture strength, resulting from testing in an environment that modifies the surface characteristics of the material.

K

Keeler-Goodwin diagram. The *forming-limit diagram* for low-carbon steel commonly used for sheet metal forming.

kinetics. Term describing the rate at which a metallurgical process (e.g., recovery, recrystallization, grain growth, phase transformation) occurs as a function of time or, if during deformation, of strain.

kinks. Sharp bends or buckles caused by localized plastic deformation of a sheet.

klink. An internal crack caused by too rapid heating of a large workpiece.

knockout. A mechanism for releasing workpieces from a die.

knockout mark. A small protrusion, such as a button or ring of flash, resulting from depression of the *knockout pin* from the forging pressure or the entrance of metal between the knockout pin and the die.

knockout pin. A power-operated plunger installed in a die to aid removal of the finished forging.

Kriging. An interpolation technique first used in geostatistics to predict the location of ore veins. It is named after D.G. Krige, a mining engineer from South Africa.

Kronecker symbol. A second-order tensor, δ_{ij}. $\delta_{ij} = 1$ for $i = j$; $\delta_{ij} = 0$ for $i \neq j$.

kurtosis. A measure of whether a data distribution is peaked or flat relative to a normal distribution. Data sets with high kurtosis tend to have a distinct peak near the mean and decline rather rapidly. Data sets with low kurtosis tend to have a flat top near the mean rather than a sharp peak. Adapted from Ref 1

L

lancing. Cutting along a line in the workpiece without detaching a slug from the blank.

laser cutting. A cutting process that severs material with the heat obtained by directing a laser beam against a metal surface. The process can be used with or without an externally supplied shielding gas.

lateral extrusion. An operation in which the product is extruded sideways through an orifice in the container wall.

Latin hypercube design. An experimental design consisting of *n* trials and for which each factor has *n* distinct levels. Usually, the factor levels are equally spaced. There is only one sample in each row and each column. Latin hypercube designs are especially useful for computer experiments. Adapted from Ref 1

lattice. A regular geometrical arrangement of points in space.

lattice constants. See *lattice parameter*.

lattice parameter. The length of any side of a unit cell of a given crystal structure. The term is also used for the fractional coordinates x, y, z of lattice points when these are variable.

leveling. The flattening of rolled sheet, strip, or plate by reducing or eliminating distortions.

limiting dome height (LDH) test. A mechanical test, usually performed unlubricated on sheet metal, that simulates the fracture conditions in a practical press-forming operation. The results are dependent on the sheet thickness.

limiting drawing ratio (LDR). See *deformation limit*.

linear elastic fracture mechanics. A method of fracture analysis that can determine the stress (or load) required to induce fracture instability in a structure containing a crack-like flaw of known size and shape. See also *stress-intensity factor*.

liners. Thin strips of metal inserted between the dies and the units into which the dies are fastened.

location and scale parameters. A probability distribution is characterized by location and scale parameters, which are typically used in modeling applications. The effect of the location parameter is to translate the graph of the probability density function, relative to the standard normal distribution; that is, a location parameter simply shifts the graph left or right on the horizontal axis. The effect of the scale parameter is to stretch out the graph. A scale parameter greater than 1 stretches the probability density function. A scale parameter less than 1 compresses the probability density function. The compression approaches a spike as the scale parameter goes to 0. A scale parameter of 1 leaves the probability density function unchanged (if the scale parameter is 1 to begin with). Nonpositive scale parameters are not allowed. The standard form of any distribution has location parameter 0 and scale parameter 1. For the normal distribution, the location and scale parameters correspond to the mean and standard deviation, respectively. However, this is not true for most other distributions. Adapted from Ref 1

loose metal. A defect in an area of a stamping where very little contour is present. The metal in the area has not been stretched, resulting in a shape with no stiffness. The area may have waves in it or may sag so that there is a dish in an area that is intended to be flat or nearly flat. This defect differs from oil canning in that the metal cannot be snapped back into the desired shape when a load is removed or reversed on the area.

low-angle boundary. Boundary separating adjacent grains whose misorientation is less than 15°. See also *subgrain*.

lubricant. A material applied to dies, molds, plungers, or workpieces that promotes the flow of metal, reduces friction and wear, and aids in the release of the finished part.

lubricant residue. The carbonaceous residue resulting from lubricant that is burned onto the surface of a hot forged part.

Lüders lines. Elongated surface markings or depressions, often visible with the unaided eye, that form along the length of a round or sheet metal tension specimen at an angle of approximately 55° to the loading axis. Caused by localized plastic deformation, they result from discontinuous (inhomogeneous) yielding. Also known as Lüders bands, Hartmann lines, Piobert lines, or stretcher strains.

M

lumped-parameter model. A mathematical model in which the distributed properties of physical quantities are replaced with their lumped equivalents. When a problem can be analyzed in terms of a finite number of discrete elements, it can be expressed by ordinary differential equations. To describe the more realistic case of distributed parameters having many values spread over a field in space requires the use of partial differential equations.

machinability. The relative ease with which material is removed from a solid by controlled chip-forming in a machining process.

major strain. Largest principal strain in the sheet surface. Often measured from the major axis of the ellipse resulting from deformation of a circular grid.

malleability. The characteristic of metals that permits plastic deformation in compression without fracture.

managing uncertainty. The process of balancing the needed accuracy of solution against the cost of obtaining that solution.

mandrel. (1) A blunt-ended tool or rod used to retain the cavity in a hollow metal product during working. (2) A metal bar around which other metal can be cast, bent, formed, or shaped. (3) A shaft or bar for holding work to be machined.

mandrel forging. The process of rolling or forging a hollow blank over a mandrel to produce a weldless, seamless ring or tube.

Mannesmann process. A process for piercing tube billets in making seamless tubing. The billet is rotated between two heavy rolls mounted at an angle and is forced over a fixed mandrel.

Marforming process. A *rubber-pad forming* process developed to form wrinkle-free shrink flanges and deep-drawn shells. It differs from the Guerin process in that the sheet metal blank is clamped between the rubber pad and the blankholder before forming begins.

martensite. A generic term for microstructures formed by diffusionless phase transformation in which the parent and product phases have a specific crystallographic relationship. Martensite is characterized by an acicular pattern in the microstructure in both ferrous and non-ferrous alloys. In alloys where the solute atoms occupy interstitial positions in the martensitic lattice (such as carbon in iron), the structure is hard and highly strained; but where the solute atoms occupy substitutional positions (such as nickel in iron), the martensite is soft and ductile. The amount of high-temperature phase that transforms to martensite on cooling depends to a large extent on the lower temperature attained, there being a rather distinct beginning temperature (M_s) and a temperature at which the transformation is essentially complete (M_f).

mass-conserving process. A manufacturing process in which the mass of the starting material is approximately equal to the mass of the final product or part. Examples are casting, precision forming, and powder processes.

match. A condition in which a point in one die half is aligned properly with the corresponding point in the opposite die half, within specified tolerance.

matched edges (match lines). Two edges of the die face that are machined exactly at 90° to each other and from which all dimensions are taken in laying out the die impression and aligning the dies in the forging equipment.

material heat. The pedigree of the starting stock or billet used to make a forging.

matrix phase. The continuous (interconnected) phase in an alloy with two or more phases. In cast or wrought materials, the matrix phase is often comprised of the first phase to solidify.

mechanical texture. Directionality in the shape and orientation of microstructural features such as inclusions, grains, and so on.

mechanical working. The subjecting of material to pressure exerted by rolls, hammers, or presses in order to change the shape or physical properties of the material.

mechanistic modeling. An approach that requires a fundamental understanding of the physics and chemistry governing the process. These laws and principles are used to describe the process and its parameters. The results are then validated against controlled test results.

mesh. (1) The number of screen openings per linear inch of screen; also called mesh size. (2) The screen number on the finest screen of a specified standard screen scale through which almost all of the particles of a powder sample will pass.

metadata. Descriptive data about the material for which data are reported. Metadata include a complete description of the material (producer, heat number, grade, temper, etc.), a complete description of the test method, and information about the test plan.

metal. An opaque lustrous elemental chemical substance that is a good conductor of heat and electricity and, when polished, a good reflector of light. Most elemental metals are malleable and ductile and are, in general, denser than other elemental substances. As to structure, metals may be distinguished from nonmetals by their atomic binding and electron availability. Metallic atoms tend to lose electrons from the outer shells, the positive ions thus formed being held together by the electron gas produced by the separation. The ability of these "free electrons" to carry an electric current, and the fact that this ability decreases as temperature increases, establish the prime distinctions of a metallic solid.

microalloyed steel. A low-to-medium-carbon steel usually containing small alloying additions of niobium, vanadium, nitrogen, and so on whose thermomechanical processing is controlled to obtain a specific microstructure and thus a suite of properties.

microhardness test. An indentation test using diamond indentors at very low loads, usually in the range of 1 to 1000 g.

microstructure. The structure of polished and etched metals as revealed by a microscope.

mill finish. A nonstandard (and typically non-uniform) surface finish on mill products that are delivered without being subjected to a special surface treatment (other than a corrosion-preventive treatment) after the final working or heat treating step.

mill product. Any commercial product of a mill.

mill scale. The heavy oxide layer that forms during the hot fabrication or heat treatment of metals.

minimum bend radius. The smallest radius about which a metal can be bent without exhibiting fracture. It is often described in terms of multiples of sheet thickness.

minor strain. The principal strain in the sheet surface in a direction perpendicular to the major strain. Often measured from the minor axis of the ellipse resulting from deformation of a circular grid.

mischmetal. From the German *mischmetall*, with roots mischen (to mix) and metall (metal), it is a natural mixture of rare earth metals containing approximately 50 wt% Ce, 25% La, 15% Nd, and 10% other rare earth metals, iron, and silicon. It is commonly used to make rare earth additions to alloys (e.g., magnesium alloys), rather than using more expensive pure forms of the rare earth metals.

mismatch. The misalignment or error in register of a pair of forging dies; also applied to the condition of the resulting forging. The acceptable amount of this displacement is governed by blueprint or specification tolerances. Within tolerances, mismatch is a condition; in excess of tolerance, it is a serious defect. Defective forgings can be salvaged by hot reforging operations.

misorientation. Angular difference between the orientations of two grains adjacent to a grain boundary, between a twin and its parent matrix, and so on.

mitigating uncertainty. The process of removing uncertainty by refining the model(s), the simulation(s), or the input variables.

mixed dislocation. Any combination of a *screw dislocation* and an *edge dislocation*.

model. A physical, mathematical, or otherwise logical representation of a system, entity, phenomenon, or process.

modeling. Application of a standard, rigorous, structured methodology to create and validate a physical, mathematical, or otherwise logical representation of a system, entity, phenomenon, or process.

modulus of elasticity (E). The measure of rigidity or stiffness of a metal; the ratio of stress, below the proportional limit, to the corresponding strain. In terms of the stress-strain diagram, the modulus of elasticity is the slope of the *stress-strain curve* in the range of linear proportionality of stress to strain. Also known as *Young's modulus*. For materials that do not conform to *Hooke's law* throughout the elastic range, the slope of either the tangent to the stress-strain curve at the origin or at low stress, the secant drawn from the origin to any specified point on the stress-strain curve, or the chord connecting any two specific points on the stress-strain curve is usually taken to be the modulus of elasticity. In these cases, the modulus is referred to as the *tangent modulus, secant modulus, or chord modulus*, respectively.

modulus of rigidity. See *shear modulus*.

Monte Carlo methods. A class of computational methods that rely on repeated random sampling to compute results. This approach is most useful when it is impossible or infeasible to compute an exact result when studying systems with a large number of coupled parameters. Adapted from Ref 3

Monte Carlo modeling. Numerical modeling technique, based on statistical mechanics, that can be used to describe the migration of grain boundaries in polycrystalline aggregates during annealing or deformation processes and thus is applied to describe recrystallization, grain growth, and the accompanying evolution of texture. Also referred to as the Potts technique.

m-value. See *strain-rate sensitivity*.

N

near-net shape forging. A forging produced with a very small finish allowance over the final part dimensions and requiring some machining prior to use.

necking. (1) The reduction of the cross-sectional area of metal in a localized area by uniaxial tension or by stretching. (2) The reduction of the diameter of a portion of the length of a cylindrical shell or tube.

necklace recrystallization. Partial static or dynamic recrystallization that nucleates heterogeneously on grain boundaries in various steels, nickel-base superalloys, and so on. A microstructure of fine (necklace-like) grains lying on the original grain boundaries is thus produced.

net shape forging. A forging produced to finished part dimensions that requires little or no further machining prior to use.

neural network. Nonlinear regression-type methodology for establishing the correlation between input and output variables in a physical system. For example, neural networks can be used to correlate processing variables to microstructural features or microstructural features to mechanical properties.

neuron. A node in a neural-network system that can be considered as an internal variable and whose value is a function of the neurons in the previous layer.

Newtonian fluid. A fluid exhibiting Newtonian viscosity wherein the shear stress is proportional to the rate of shear.

no-draft (draftless) forging. A forging with extremely close tolerances and little or no draft that requires minimal machining to produce the final part. Mechanical properties can be enhanced by closer control of grain flow and by retention of surface material in the final component.

nominal strain. The unit elongation given by the change in length divided by the original length. Also called *engineering strain*.

nominal stress. The unit force obtained when the applied load is divided by the original cross-sectional area. Also called *engineering stress*.

nonfill (underfill). A forging condition that occurs when the die impression is not completely filled with metal.

normal anisotropy. A condition in which a property or properties in the sheet thickness direction differ in magnitude from the same property or properties in the plane of the sheet.

normal distribution. (1) The *probability density function* used to describe the various properties of materials and the distribution of most random variables encountered in engineering design. (2) The general formula for the probability density function of the normal distribution is:

$$f(x) = \frac{e^{-(x-\mu)^2/(2\pi^2)}}{\sigma\sqrt{2\pi}}$$

where μ is the location parameter, and σ is the scale parameter. The case where $\mu = 0$ and $\sigma = 1$ is called the standard normal distribution. The equation for the standard normal distribution is:

$$f(x) = \frac{e^{-x^2/2}}{\sqrt{2\pi}}$$

The normal distribution is probably the most important distribution in statistics. In modeling applications, such as linear and nonlinear regression, the error term is often assumed to follow a normal distribution with fixed location and scale. Adapted from Ref 1

notching. An unbalanced shearing or blanking operation in which cutting is done around only three sides (usually) of a punch.

n-value. A term commonly referred to as the strain-hardening exponent derived from the relationship between true stress, σ, with true strain, ε, given by $\sigma = K\varepsilon^n$. Preferably called the *strain-hardening exponent*.

O

objective function. Mathematical function describing a desired material or process characteristic whose optimization is the goal of process design. In bulk forming, typical objective functions may include forging weight (minimum usually is best), die fill (minimum underfill is best), and uniformity of strain or strain rate (maximum uniformity is best). In design optimization, it is the grouping of design parameters that is attempted to be maximized or minimized, subject to the problem constraints. Also known as criterion function.

offal. Sheet metal section trimmed or removed from the sheet during the production of shaped blanks or the formed part. Offal is frequently used as stock for the production of small parts.

offset. The distance along the strain coordinate between the initial portion of a stress-strain curve and a parallel line that intersects the stress-strain curve at a value of stress (commonly 0.2%) that is used as a measure of the *yield strength*. Used for materials that have no obvious *yield point*.

offset yield strength. The stress at which the strain exceeds by a specified amount (the *offset*) an extension of the initial proportional portion of the stress-strain curve; expressed in force per unit area.

Olsen ductility test. A *cupping test* in which a piece of sheet metal, restrained except at the center, is deformed by a standard steel ball until fracture occurs. The height of the cup at the time of fracture is a measure of the ductility.

open-die forging. The hot mechanical forming of metals between flat or shaped dies in which metal flow is not completely restricted. Also known as hand or smith forging. See also *hand forge (smith forge)*.

optimization. The process of searching for the best combination of design parameters. Design optimization suggests that, for a given set of possible designs and design criteria, there exists a single design that is best or optimal.

orange peel. In metals, a surface roughening in the form of a pebble-grained pattern that occurs when a metal of unusually coarse grain size is stressed beyond its elastic limit.

orientation-distribution function (ODF). Mathematical function describing the normalized probability of finding grains of given crystallographic orientations/Euler angles. Because crystallographic orientations are in terms of Euler angles, the description of texture using ODFs is unambiguous, unlike pole figures. See also *texture, preferred orientation*, and *pole figure*.

orthogonality. An experimental design is orthogonal if the effects of any factor balance out (sum to zero) across the effects of the other factors. Adapted from Ref 1

Ostwald ripening. The increase in the average size of second-phase particles, accompanied by the reduction in their number, during annealing, deformation, or high-temperature service exposure. Ostwald ripening leads to a decrease in the total surface energy associated with matrix-particle interfaces. Also known as *coarsening*.

outliers. Observed values much lower or higher than most other observations in a data set. Values identified as outliers should be investigated. Data should be screened for outliers, because they have substantial influence on statistical analysis.

oxidation. A reaction where there is an increase in valence resulting from a loss of electrons. Such a reaction occurs when most metals or alloys are exposed to atmosphere and the reaction rate increases as temperature increases.

oxide-dispersion-strengthened (ODS) alloys. A class of materials in which fine oxide particles are incorporated in metal powders, compacted, and then fabricated into finished forms by deformation processing. The resulting material has improved thermal softening resistance with excellent thermal and electrical conductivity. Examples are ODS copper alloys and sintered aluminum powder.

oxidized surface. A tightly adhering oxide surface layer that results in modified surface color and reduced reflectivity. It is often accompanied by surface penetration of oxide that causes brittleness.

P

pack rolling. Hot, flat rolling process in which the workpiece (or a stack of workpieces) in the form of plate, sheet, or foil is encased in a sacrificial can to reduce/eliminate contamination (e.g., oxygen pickup) or poor workability due to roll chill.

pancake forging. A rough forged shape, usually flat, that can be obtained quickly with minimal tooling. Usually made by upsetting a cylindrical billet to a large height reduction in flat dies. Considerable machining is usually required to attain the finish size.

parting. A shearing operation used to produce two or more parts from a stamping.

parting line. The line along the surface of a forging where the dies meet, usually at the largest cross section of the part. *Flash* is formed at the parting line.

parting plane. The plane that includes the principal die face and is perpendicular to the direction of ram travel. When parting surfaces of the dies are flat, the parting plane coincides with the parting line. Also referred to as the forging plane.

pass. (1) A single transfer of metal through a stand of rolls. (2) The open space between two grooved rolls through which metal is processed.

peak count. In surface measurements, the number of asperities above a given (defined) height cut-off level and within a given width cut-off. Frequency is taken at 50 μin./in.

peak density. The average number of peaks within the specified cut-off levels.

peak height. Peak-to-valley magnitude as measured by a suitable stylus instrument. Peak height is related to roughness height, depending on uniformity of surface irregularities.

peen forming. A dieless, flexible-manufacturing technique used primarily in the aerospace industry for forming sheet metals by way of the deformation imparted by the controlled-velocity impact of balls.

percent IACS (%IACS). In 1913, values of electrical conductivity were established and expressed as a percent of a standard. The standard chosen was an annealed copper wire with a density of 8.89 g/cm^3, a length of 1 m, a weight of 12 g, with a resistance of 0.1532 Ω at 20 °C (70 °F). The 100% IACS (International Annealed Copper Standard) value was assigned with a corresponding resistivity of 0.017241 $\Omega mm^2/m$. The percent IACS for any material can be calculated by %IACS = 0.017241 $\Omega mm^2/m$ × 100/volume resistivity.

perforating. The punching of many holes, usually identical and arranged in a regular pattern, in a sheet, workpiece blank, or previously formed part. The holes are usually round but may be any shape. The operation is also called multiple punching. See also *piercing*.

permanent set. The deformation or strain remaining in a previously stressed body after release of the load.

permeability (magnetic). A general term used to express various relationships between magnetic induction and magnetizing force. These relationships are either "absolute permeability," which is a change in magnetic induction divided by the corresponding change in magnetizing force, or "specific (relative) permeability," the ratio of the absolute permeability to the permeability of free space.

phenomenological model. Empirical or data-based modeling that relies on collecting data from observations, specifying the correlation structure between variables, using numerical techniques to find parameters for the structure such that the correlation between the data is maximized, and then validating the model against a test data set.

phonon (wave). An organized movement of atoms or molecules, such as a sound wave.

pickup. Small particles of oxidized metal adhering to the surface of a *mill* product.

piercing. The general term for cutting (shearing or punching) openings, such as holes and slots, in sheet material, plate, or parts. This operation is similar to blanking; the difference is that the slug or piece produced by piercing is scrap, while the blank produced by *blanking* is the useful part.

pilgering. Also called rocking or tube reducing. The most common machines employ a set of reciprocating grooved rolls and a tapered mandrel. As the larger tube is fed into this device, it is rolled in small increments to the smaller size. Tube reducing is of commercial importance for two reasons: Very heavy reductions (up to 85%) can be applied to mill length tubes, and the process can be applied to the refractory alloys that are difficult to cold draw because of high power requirements. Adapted from Ref 1

pinchers. Surface defects having the appearance of ripples or of elongated areas of variable surface texture extending at an acute angle to the sheet rolling direction and often branching. Pinchers are usually caused by poor hot rolled strip shape or improper drafting or incorrect crown on the cold reduction mill. They are sometimes referred to as feather pattern.

pinning. The retardation or complete cessation of grain growth during annealing or deformation by second-phase particles acting on grain boundaries.

Piobert lines. See *Lüders lines*.

pit. A small, clean depression in a sheet surface caused by the rolling-in of foreign particles such as sand, steel, and so on that subsequently fall out.

planar anisotropy. A term indicating variation in one or more physical or mechanical properties with direction in the plane of the sheet. The planar variation in plastic strain ratio is commonly designated as Δr, given by $\Delta r = (r_0 + r_{90} - 2r_{45})/2$. The earing tendency of a sheet is related to Δr. As δr increases, so does the tendency to form ears.

plane strain. Deformation in which the normal and shear components associated with one of the three coordinate directions are equal to zero. Bulk forming operations that approximate plane-strain conditions include sheet rolling and sheet drawing.

plane stress. Stress state in which the normal and shear components of stress associated with one of the three coordinate directions are equal to zero. Most sheetforming operations are performed under conditions approximating plane stress.

planishing. Smoothing a metal surface by rolling, forging, or hammering; usually the last pass or passes of a shaping operation.

plastic deformation. The permanent (inelastic) distortion of metals under applied stresses that strain the material beyond its *elastic limit*. The ability of metals to flow in a plastic manner without fracture is the fundamental basis for all metalforming processes.

plastic flow. The phenomenon that takes place when metals or other substances are stretched or compressed permanently without rupture.

plastic instability. The deformation stage during which plastic flow is nonuniform and necking occurs.

plasticity. The property of a material that allows it to be repeatedly deformed without rupture when acted upon by a force sufficient to cause deformation and that allows it to retain its shape after the applied force has been removed.

plastic-strain ratio (*r*-value). A measure of normal plastic anisotropy is defined by the ratio of the true width strain to the true thickness strain in a tensile test. The average plastic strain ratio, r_m, is determined from tensile samples taken in at least three directions from the sheet rolling direction, usually at 0, 45, and 90°. The r_m is calculated as $r_m = (r_0 + 2r_{45} + r_{90})/4$. The ratio of the true width strain to the true thickness strain in a sheet tensile test is $r = \varepsilon_w/\varepsilon_t$. A formability parameter that relates to drawing, it is also known as the anisotropy factor. A high r-value indicates a material with good drawing properties.

ploughing. Plastic deformation of the surface of the softer component of a friction pair.

point lattice. A set of points in space located so that each point has identical surroundings. There are 14 ways of so arranging points in space, corresponding to the 14 *Bravais lattices*.

Poisson distribution. The Poisson distribution is used to model the number of events occurring within a given time interval. The formula for the Poisson probability mass function is:

$$p(x, \lambda) = \frac{e^{-\lambda}\lambda^x}{x!} \quad \text{for } x = 0, 1, 2, \cdots$$

where λ is the shape parameter, which indicates the average number of events in the given time interval. Adapted from Ref 1

Poisson's ratio (ν). The absolute value of the ratio of transverse (lateral) strain to the corresponding axial strain resulting from uniformly distributed axial stress below the *proportional limit* of the material in a tensile test.

pole figure. Description of crystallographic texture based on a stereographic-projection representation of the times-random probability of finding a specific crystallographic pole with a specific orientation relative to sample reference directions. For axisymmetric components, the sample reference directions are usually the axis and two radial directions; for a sheet material, the rolling, transverse, and sheet-normal directions are used. Because pole figures provide information only with regard to the orientation of one crystallographic pole, several pole figures or an orientation-distribution function (derivable from pole-figure measurements) are needed to fully describe crystallographic texture. See also *orientation-distribution function*.

polycrystalline aggregate. The collection of grains/crystals that form a metallic material.

polygonization. A recovery-type process during the annealing of a worked material in which excess dislocations of a given sign rearrange themselves into low-energy, low-angle tilt boundaries.

population. A statistical concept describing the total set of objects or observations under consideration.

porosity. Fine holes or voids within a solid; the amount of these pores is expressed as a percentage of the total volume of the solid.

postforming. Any treatment after the part has been formed, such as annealing, trimming, finishing, and so on.

powder forging. The plastic deformation of a powder metallurgy *compact or preform* into a fully dense finished shape by using compressive force; usually done hot and within closed dies.

powder metallurgy (PM). The technology and art of producing metal powders and using metal powders for production of mass materials and shaped objects.

precipitation hardening. Hardening in metals caused by the precipitation of a constituent from a supersaturated solid solution.

precision. In testing, a measure of the variability that can be expected among test results. The precision of an instrument indicates its ability to reproduce a certain reading. Precision is the inverse of standard deviation. A decrease in the scatter of test results is represented by a smaller standard deviation, leading directly to an increase in precision.

precision forging. A forging produced to closer tolerances than normally considered standard by the industry.

preferred orientation. Nonrandom distribution of the crystallographic orientations of the grains comprising a polycrystalline aggregate.

preform. (1) The forging operation in which stock is preformed or shaped to a predetermined size and contour prior to subsequent die forging operations. When a preform operation is required, it will precede a forging operation and will be performed in conjunction with the forging operation and in the same heat. (2) The initially pressed powder metallurgy *compact* to be subjected to *repressing*.

prelubed sheet. A sheet or coil that has had a lubricant applied during mill processing to serve as a forming lubricant in the fabrication plant.

press forging. The forging of metal between dies by mechanical or hydraulic pressure; usually accomplished with a single work stroke of the press for each die station.

press forming. Any sheet metal forming operation performed with tooling by means of a mechanical or hydraulic press.

pressing. The product or process of shallow drawing sheet or plate.

press load. The amount of force exerted in a given forging or forming operation.

principal strain. The normal strain on any of three mutually perpendicular planes on which no shear strains are present.

principal strain direction. The direction of action of the normal strains.

principal stress. One of the three normal stresses in the coordinate system in which all of the shear stresses are equal to zero.

prior particle boundary (PPB). An apparent boundary between the pre-existing powder metal particles that is still evident within the microstructure of consolidated powder metallurgy products because of the presence of carbide or other phases that form at these boundaries.

probability density function (PDF). (1) A mathematical function that, when integrated between two limits, gives the probability that a random variable assumes a value between these limits. (2) For a continuous function, the probability density function is the probability that the variate has the value x. For continuous distributions, the probability at a single point is zero; this is often expressed in terms of an integral between two points:

$$\int_a^b f(x)dx = Pr[a \leq X \leq b]$$

For a discrete distribution, the probability density function is the probability that the variate takes the value x:

$$f(x) = Pr[X = x]$$

Probability distributions are typically defined in terms of the probability density function. Adapted from Ref 1

processing map. A map of strain rate versus temperature that delineates the regions that should be avoided in processing to prevent the formation of poor microstructures or voids or cracks. These maps are generally created by the dynamic material modeling method or by mapping extensive results of processing experience.

process model. A mathematical description of the physical behavior underlying a manufacturing process that is used to predict performance of the process in terms of operating parameters. Most often, process models are reduced to software and are manipulated with computers.

process modeling. Computer simulation of deformation, heat treating, and machining processes for the purpose of improving process yield and material properties.

profile (contour) rolling. In *ring rolling*, a process used to produce seamless rolled rings with a predesigned shape on the outside or the inside diameter, requiring less volume of material and less machining to produce finished parts.

progressive die. A die planned to accomplish a sequence of operations as a strip or sheet of material is advanced from station to station, manually or mechanically.

progressive forming. Sequential forming at consecutive stations with a single die or separate dies.

projection welding. Electrical resistance welding in which the welds are localized at embossments or other raised portions of the sheet surface.

proof load. A predetermined load, generally some multiple of the service load, to which a specimen or structure is submitted before acceptance for use.

proof stress. The stress that will cause a specified small permanent set in a material. A specified stress to be applied to a member or structure to indicate its ability to withstand service loads.

proportional limit. The greatest stress a material is capable of developing without a deviation from straight-line proportionality between stress and strain. See also *elastic limit* and *Hooke's law.*

punch. (1) The male part of a die—as distinguished from the female part, which is called the die. The punch is usually the upper member of the complete die assembly and is mounted on the *slide* or in a *die set* for alignment (except in the inverted die). (2) In double-action draw dies, the punch is the inner portion of the upper die, which is mounted on the plunger (inner slide) and does the drawing. (3) The act of piercing or *punching* a hole. Also referred to as punching.

punching. The die shearing of a closed contour in which the sheared-out sheet metal part is scrap.

punch nose radius. The shape of the punch end, contacting the material being formed to allow proper material flow or movement.

punch press. (1) In general, any mechanical press. (2) In particular, an endwheel gap-frame press with a fixed bed, used in piercing.

Q

quarter hard. A temper of nonferrous alloys and some ferrous alloys characterized by tensile strength approximately midway between that of dead soft and half-hard tempers.

quench aging. Hardening by precipitation that results after the rapid cooling from solid solution to a temperature below which the elements of a second phase become supersaturated. Precipitation occurs after the application of higher temperatures and/or times and causes increases in yield strength, tensile strength, and hardness.

quenching. Rapid cooling of metals from a suitable elevated temperature, generally accomplished by immersion in water, oil, polymer solution, or salt, although forced air is sometimes used.

R

rabbit ear. Recess in the corner of a metal-forming die to allow for wrinkling or folding of the blank.

radial draw forming. The forming of sheet metals by the simultaneous application of tangential stretch and radial compression forces. The operation is done gradually by tangential contact with the die member. This type of forming is characterized by very close dimensional control.

radial forging. A process using two or more moving anvils or dies for producing shafts with constant or varying diameters along their length or tubes with internal or external variations in diameter. Often incorrectly referred to as *rotary forging.*

radial roll (main roll, king roll). The primary driven roll of the rolling mill for rolling rings in the radial pass. The roll is supported at both ends.

ram. The moving or falling part of a drop hammer or press to which one of the dies is attached; sometimes applied to the upper flat die of a steam hammer. Also referred to as the *slide.*

R-curve. In fracture mechanics, a plot of crack-extension resistance as a function of stable crack extension, which is the difference between either the physical crack size or the effective crack size and the original crack size. *R*-curves normally depend on specimen thickness and, for some materials, on temperature and strain rate.

recovery. Process occurring during annealing following cold or hot working of metals in which defects such as dislocations are eliminated or rearranged by way of mechanisms such as dipole annihilation, the formation of subgrains, and subgrain growth. Recovery usually leads to a reduction in stored energy, softening, reduction or elimination of residual stresses, and, in some instances, changes in physical properties. Recovery may also serve as a precursor to static recrystallization at sufficient levels of prior cold or hot work. See also dynamic *recovery.*

recrystallization. A process of nucleation and growth of new strain-free grains or crystals in a material. This process occurs upon heating above the recrystallization temperature (approximately 40% of the metal absolute melting temperature) during/after hot working or during annealing after cold working. Recrystallization can be dynamic (occurring during straining), static (occurring following deformation, typically during heat treatment), or meta-dynamic (occurring immediately after deformation due to the presence of recrystallization nuclei formed during deformation).

recrystallization texture. Crystallographic texture formed during static or dynamic recrystallization. The specific texture components that are formed are dependent on the nature of the stored work driving recrystallization and the nucleation and growth mechanisms that underlie recrystallization.

redrawing. The second and successive deep-drawing operations in which cuplike shells are deepened and reduced in cross-sectional dimensions.

reduction. (1) In cupping and deep drawing, a measure of the percentage of decrease from blank diameter to cup diameter, or of the diameter reduction in redrawing. (2) In forging, extrusion, rolling, and drawing, either the ratio of the original to the final cross-sectional area or the percentage of decrease in cross-sectional area.

reduction in area. The difference between the original cross-sectional area and the smallest area at the point of rupture in a tensile test; usually stated as a percentage of the original area.

redundant work. Energy in addition to that required for uniform flow expended during processing due to inhomogeneous deformation.

relative density. Ratio of density to pore-free density.

relief. Clearance obtained by removing material, either behind or beyond the cutting edge of a punch or die.

repressing. The application of pressure to a sintered compact; usually done to improve a physical or mechanical property or for dimensional accuracy.

rerolling quality. Rolled billets from which the surface defects have not been removed or completely removed.

residual stress. An elastic stress that exists in a solid body without an imposed external force. Residual stresses often result from forming or thermal processing and are caused by such factors as cold working, phase changes, temperature gradients, or rapid cooling.

response surface modeling. A statistical, mathematical, or graphical model that describes the variation of the response variable in terms of the parameters of the problem.

restriking. (1) The striking of a trimmed but slightly misaligned or otherwise faulty forging with one or more blows to improve alignment, improve surface condition, maintain close tolerances, increase hardness, or effect other improvements. (2) A sizing operation in which coining or stretching is used to correct or alter profiles and to counteract distortion. (3) A salvage operation following a primary forging operation in which the parts involved are rehit in the same forging die in which the pieces were last forged.

retained austenite. An amount of the high-temperature face-centered cubic phase of iron (austenite) that does not transform to martensite (is retained) when quenched to room temperature.

reverse drawing. *Redrawing* of a sheet metal part in a direction opposite to that of the original drawing.

reverse flange. A sheet metal flange made by shrinking, as opposed to one formed by stretching.

reverse redrawing. An operation after the first drawing operation in which the part is turned inside out by inverting and redrawing, usually in another die, to a smaller diameter.

rheology. The science of deformation and the flow of matter.

ring compression test. A workability test that uses the expansion or contraction of the hole in a thin compressed ring to measure the frictional conditions. The test can also be used to determine the flow stress.

ring rolling. The process of shaping weldless rings from pierced disks or shaping thick-walled ring-shaped blanks between rolls that control wall thickness, ring diameter, height, and contour.

robust design. Designing products or processes so that they are minimally affected by uncertainty.

rod. A solid round section 9.5 mm (⅜ in.) or less in diameter whose length is greater in relation to its diameter.

roll. Tooling used in the rolling process to deform material stock.

roll bending. The curving of sheets, bars, and sections by means of rolls.

rolled-in scale. Localized areas of heavy oxide not removed by the hot mill descaling sprays and rolled out to elongated streaks during further processing.

roller leveling. *Leveling* by passing flat sheet metal stock through a machine having a series of small-diameter staggered rolls that are adjusted to produce repeated reverse bending.

roll feed. A mechanism for feeding strip or sheet stock to a press or other machine. The stock passes between two revolving rolls mounted one above the other, which feed it under the dies a predetermined length at each stroke of the press. Two common types of drive are the oscillating-lever type and the rack-and-pinion type. The single-roll feed may be used to either push or pull the stock to or from the press. The double-roll feed is commonly used with wider presses (left to right) or in other cases where a single-roll feed is impractical.

roll flattening. The flattening of sheets that have been rolled in packs by passing them separately through a two-high cold mill with virtually no deformation. Not to be confused with roller leveling.

roll forging. A process of shaping stock between two driven rolls that rotate in opposite directions and have one or more matching sets of grooves in the rolls; used to produce finished parts of preforms for subsequent forging operations.

roll former. A device with three or more rolls positioned to progressively plastically form sheet or strip metal into curved or linear shapes.

roll forming. A process in which coil sheet or strip metal is formed by a series of shaped rolls into the desired configuration. Metal-forming through the use of power-driven rolls whose contour determines the shape of the product; sometimes used to denote power *spinning*.

rolling. The reduction of the cross-sectional area of metal stock, or the general shaping of metal products, through the use of rotating rolls.

rolling mandrel. In ring rolling, a vertical roll of sufficient diameter to accept various sizes of ring blanks and to exert rolling force on an axis parallel to the main roll.

rolling mills. Machines used to decrease the cross-sectional area of metal stock and to

produce certain desired shapes as the metal passes between rotating rolls mounted in a framework comprising a basic unit called a stand. Cylindrical rolls produce flat shapes; grooved rolls produce rounds, squares, and structural shapes.

roll mark. A mark in light relief on the sheet surface produced by an indentation in the cold reduction mill work roll surface.

roll straightening. The straightening of metal stock of various shapes by passing it through a series of staggered rolls (the rolls usually being in horizontal and vertical planes) or by reeling in two-roll straightening machines.

roll threading. The production of threads by rolling the piece between two grooved die plates, one of which is in motion, or between rotating grooved circular rolls.

roping. A surface defect consisting of a series of generally parallel markings or ripples on areas of rolled formed sheet parts that have undergone substantial strain. The ripples are always parallel to rolling direction.

rotary forging. A process in which the workpiece is pressed between a flat anvil and a swiveling (rocking) die with a conical working face; the platens move toward each other during forging. Also called orbital forging. Compare with *radial forging*.

rotary shear. A sheet metal cutting machine with two rotating-disk cutters mounted on parallel shafts driven in unison.

rotary swager. A swaging machine consisting of a power-driven ring that revolves at high speed, causing rollers to engage cam surfaces and force the dies to deliver hammerlike blows on the work at high frequency. Both straight and tapered sections can be produced.

rotary swaging. A *bulk forming* process for reducing the cross-sectional area or otherwise changing the shape of bars, tubes, or wires by repeated radial blows with one or more pairs of opposed dies.

rotatability. An experimental design is rotatable if the variance of the predicted response at any point x depends only on the distance of x from the design center point. A design with this property can be rotated around its center point without changing the prediction variance at x. Adapted from Ref 1

rough blank. A blank for a forming or drawing operation, usually of irregular outline, with necessary stock allowance for process metal, which is trimmed after forming or drawing to the desired size.

roughing stand. The first stand (or several stands) of rolls through which a reheated billet or slab passes in front of the finishing stands. See also *rolling mills*.

roughness cut-off level. Terms used in the measurement of surface roughness. (a) Width cut-off: the greatest spacing of repetitive surface irregularities used in the measurement of roughness, usually 0.030 in. (b) Height cut-off: the minimum surface irregularity in peak count determinations, usually 50 μin.

roughness height. The average height of surface irregularities with reference to a mean or nominal surface as determined by height and width cut-offs. It may be expressed as the deviation from the nominal surface, as arithmetic average, or as root mean square.

rubber forming. A sheet metal forming process in which rubber is used as a functional die part.

rubber-pad forming. A sheet metal forming operation for shallow parts in which a confined, pliable rubber pad attached to the press slide (ram) is forced by hydraulic pressure to become a mating die for a punch or group of punches placed on the press bed or baseplate. Developed in the aircraft industry for the limited production of a large number of diversified parts, the process is limited to the forming of relatively shallow parts, normally not exceeding 40 mm (1.5 in.) deep. Also known as the Guerin process. Variations of the Guerin process include the *Marforming process*, the *fluid-cell process*, and *fluid forming*.

run. The quantity produced in one setup.

r-value. The ratio of true width strain to true thickness strain. Often called *plastic-strain ratio*.

S

saddening. The process of lightly working an ingot in the initial forging operation to break up and refine the coarse, as-cast structure at the surface.

scale parameter. See *location and scale parameters*.

scale pattern. A transverse surface pattern on cold rolled sheet caused by intermittent removal of the scale in the scale-breaker operation prior to pickling. The result is a pattern of overpickled areas that are not eliminated in cold reduction.

Schmid factor. In a uniaxial tension test, the geometric factor that corresponds to the product of the cosine of the angle between the tension axis and the slip-plane normal and the cosine of the angle between the tension axis and the slip direction. Often denoted as m.

Schmid's law. Criterion that slip in metallic crystals is controlled by a critical resolved shear stress that depends on specific material, strain rate, and test temperature but is independent of the stress normal to the slip plane.

scoring. (1) The marring or scratching of any formed part by metal pickup on the punch or die. (2) The reduction in thickness of a material along a line to weaken it intentionally along that line.

scratches. Lines generally caused by sliding of the sheet surface over sharp edges of processing equipment or over other sheets.

scratch resistance. The ability of a material to resist scratching. It is a function of the material hardness, although the lubricity of the surface will also play a part.

screw dislocation. A line imperfection that corresponds to the axis of a spiral structure in a crystal and is characterized by a distortion joining normally parallel lines together to form a continuous helical ramp (with a pitch of one interplanar distance) winding about the dislocation.

screw press. A high-speed press in which the ram is activated by a large screw assembly powered by a drive mechanism.

scuffing. Localized damage caused by the occurrence of solid-phase welding between sliding surfaces. No local surface melting occurs.

seam. A surface defect appearing as thin lines in the rolling direction of sheet metals due to voids elongated during rolling.

seaming. The process of joining sheet metal parts by interlocking bends.

secant modulus. The slope of the secant drawn from the origin to any specified point on the stress-strain curve. See also *modulus of elasticity*.

secondary recrystallization. See *abnormal grain growth*.

secondary sheet. A shearing action that occurs between soft work metal and a cutting edge as a result of insufficient clearance.

secondary tensile stress. Tensile stress that develops during a bulk deformation process conducted under nominally compressive loading due to nonuniform metal flow resulting from geometry, friction, or die-chilling effects. Secondary tensile stresses are most prevalent in open-die forging operations.

segregation. A nonuniform distribution of alloying elements, impurities, or microphases.

seizure. The stopping of relative motion between two bodies as the result of severe interfacial friction. Seizure may be accompanied by gross surface welding.

semifinisher. An impression in a series of forging dies that only approximates the finish dimensions of the forging. Semifinishers are often used to extend die life or the finishing impression, to ensure proper control of grain flow during forging, and to assist in obtaining desired tolerances.

seminotching. A process similar to notching except that the cutting operation is a partial one only, permitting the cut shape to remain with the blank or part.

set. The shape remaining in a stamped or press-formed part after the punch force is removed. See also *permanent set*.

severe plastic deformation. Processes of plastic deformation with accumulated natural logarithmic strains more than 4 that are usually used to change material structure and properties.

shank. The portion of a die or tool by which it is held in position in a forging unit or press.

shape distortion. A dimensional change due to warping or bending resulting mainly from thermal treatment.

shape fixability. The ability of a material to retain the shape given to it by a forming operation.

shaving. Backflow of the clad or sleeve material during hydrostatic coextrusion.

shear. (1) A machine or tool for cutting metal and other material by the closing motion of two sharp, closely adjoining edges, for example, squaring shear and circular shear. (2) An inclination between two cutting edges, such as between two straight knife blades or between the punch cutting edge and the die cutting edge, so that a reduced area will be cut each time. This lessens the necessary force but increases the required length of the working stroke. This method is referred to as angular shear. (3) The act of cutting by shearing dies or blades, as in a squaring shear. (4) The type of force that causes or tends to cause two contiguous parts of the same body to slide relative to each other in a direction parallel to their plane of contact.

shear band. Region of highly localized shear deformation developed during bulk forming (and sometimes during sheet forming) as a result of material properties (such as a high flow-softening rate and low rate sensitivity of the flow stress), metal flow geometry, friction, chilling, and so on.

shear burr. A raised edge resulting from metal flow induced by blanking, cutting, or punching.

shearing. A cutting operation in which the work metal is placed between a stationary lower blade and movable upper blade and severed by bringing the blades together. Cutting occurs by a combination of metal shearing and actual fracture of the metal.

shear modulus (G). The ratio of shear stress to the corresponding shear strain for shear stresses below the proportional limit of the material. Values of shear modulus are usually determined by torsion testing. Also known as modulus of rigidity.

shear strength. The maximum shear stress a material can sustain. Shear strength is calculated from the maximum load during a shear or torsion test and is based on the original dimensions of the cross section of the specimen.

shear stress. (1) A stress that exists when parallel planes in metal crystals slide across each other. (2) The stress component tangential to the plane on which the forces act.

sheet. Any material or piece of uniform thickness and of considerable length and width as compared to its thickness. With regard to metal, such pieces under 6.5 mm (¼ in.) thick are called sheets, and those 6.5 mm (¼ in.) thick and over are called plates. Occasionally, the limiting thickness for steel to be designated as sheet steel is No. 10 Manufacturer's Standard Gage for sheet steel, which is 3.42 mm (0.1345 in.) thick.

sheet forming. The plastic deformation of a piece of sheet metal by tensile loads into a three-dimensional shape, often without significant changes in sheet thickness or surface characteristics. Compare with *bulk forming*.

shell four-ball test. A lubricant test in which three balls are clamped in contact, as in an equilateral triangle. The fourth ball is held in a rotating chuck and touches each of the stationary balls. Load is applied through a lever arm system that pushes the stationary balls upward against the rotating ball.

shim. A thin piece of material used between two surfaces to obtain a proper fit, adjustment, or alignment.

shot peening. A method of cold working metals in which compressive stresses are induced in the exposed surface layers of parts by impingement of a stream of shot (small spherical particles), directed at the metal surface at high velocity under controlled conditions. It differs from blast cleaning in primary purpose and in the extent to which it is controlled to yield accurate and reproducible results. Although shot peening cleans the surface being peened, this function is incidental. The major purpose of shot peening is to increase fatigue strength. Shot for peening is made of iron, steel, or glass.

shrinkage. The contraction of metal during cooling after hot forging. Die impressions are made oversized according to precise shrinkage scales to allow the forgings to shrink to design dimensions and tolerances.

shrink flanging. The reduction of the length of the free edge after the flanging process.

shuttle die. A multiple-station die in which the separated workpieces are fed from station to station by bars that are positioned in the die proper. Also known as a transfer die.

sidepressing. A deformation process in which a cylinder is laid on its side and deformed in compression. It is a good test to evaluate the tendency for fracture at the center of a billet, or for evaluating the tendency to form shear bands.

side thrust. The lateral force exerted between the dies by reaction of a forged piece on the die impressions.

single-stand mill. A rolling mill designed such that the product contacts only two rolls at a given moment.

sinking. The operation of machining the impression of a desired forging into die blocks.

sintering. The densification and bonding of adjacent particles in a powder mass or compact by heating to a temperature below the melting point of the main constituent.

size effect. The behavior in which the dimensions of the test specimen affect the value of the mechanical property measured. Most prominent for fatigue properties and strength of brittle materials, where strength is lower for large section size.

sizing. (1) Secondary forming or squeezing operations needed to square up, set down, flatten, or otherwise correct surfaces to produce specified dimensions and tolerances. (2) Some burnishing, broaching, drawing, and shaving operations are also called sizing.

(3) A finishing operation for correcting ovality in tubing. (4) Final pressing of a sintered powder metallurgy part.

skewness. A measure of symmetry or the lack of symmetry. A distribution is symmetric if it looks the same on the left and right of the center point. Adapted from Ref 1

skin lamination. A subsurface separation that can result in surface rupture during forming.

slab. A flat-shaped semifinished rolled metal ingot with a width not less than 250 mm (10 in.) and a cross-sectional area not less than 105 cm^2 (16 in.^{2}).

slabbing. The hot working of an ingot to a flat rectangular shape.

sleeve. Outer layer of a coextruded or codrawn product. See also *clad*.

slide. The main reciprocating member of a press, guided in the press frame, to which the punch or upper die is fastened; sometimes called the *ram*. The inner slide of a double-action press is called the plunger or punch-holder slide; the outer slide is called the blankholder slide. The third slide of a triple-action press is called the lower slide, and the slide of a hydraulic press is often called the platen.

slide adjustment. The distance that a press slide position can be altered to change the shut height of the die space. The adjustment can be made by hand or by power mechanism.

sliding friction test—flat dies. A test in which a sheet steel sample is placed between two flat, hardened die faces and pulled through the dies under conditions that permit recording of the load applied to the dies and the force required to pull the strip.

sliding friction test—wedge dies. Similar to the flat die test assembly except that one or both of the flat dies has a wedge configuration to confine the edges of the specimen. This permits development of unit loadings in excess of compressive strength of the specimen, and cold reduction of drawn strip is readily accomplished.

slip. Crystallographic shear process associated with dislocation glide that underlies the large plastic deformation of crystalline metals and alloys. Slip is usually observed on close-packed planes along close-packed directions, in which case it is referred to as restricted slip. In body-centered cubic materials, such as alpha iron, slip occurs along several different planes containing a close-packed direction and is referred to as pencil glide.

slip-line field. Graphical technique used to estimate the deformation and stresses involved in plane-strain metalforming processes.

slitting. Cutting or shearing along single lines to cut strips from a sheet or to cut along lines of a given length or contour in a sheet or workpiece.

sliver. A surface defect consisting of an elongated thin layer of partially attached metal.

slotting. A stamping operation in which elongated or rectangular holes are cut in a blank or part.

slug. (1) The metal removed when punching a hole in a forging; also termed punchout. (2) The forging stock for one workpiece cut to length.

smith forging. See *hand forge (smith forge)*.

smudge. A dark-appearing surface contamination on annealed sheet generally resulting from cold-reduction oil residues or carbon deposited from annealing gas with an unfavorable CO/CO^2 ratio. Smudge may adversely affect painting or plating but may be beneficial in the prevention of galling.

smut. A contaminant consisting of fine, dark-colored particles on the surface of pickled sheet products. This usually results from heavy oxidation of the steel surface during hot rolling.

snap through. Shock in a die due to the sudden beginning and completion of fractures in cutting dies, causing the compressed punch to elastically snap into tension.

solid modeling. A form of computer modeling in which the three-dimensional features of the part or object are represented. With solid modeling, a cut through the model reveals interior details. The method also permits accurate calculation of mass properties (e.g., mass and moment of inertia), and, with full color, shading, and shadowing, it creates realistic displays. Solid models may be integrated with motion analysis software to create realistic simulations. Solid models may also be linked with finite-element models.

sow block. A block of heat treated steel placed between the anvil of the hammer and the forging die to prevent undue wear to the anvil. Sow blocks are occasionally used to hold insert dies. Also called anvil cap.

space lattice. A set of equal and adjoining parallelepipeds formed by dividing space by three sets of parallel planes, the planes in any one set being equally spaced. There are seven ways of so dividing space, corresponding to the seven *crystal system* structures. The unit parallelepiped is usually chosen as the unit cell of the system. Due to geometrical considerations, atoms can only have one of 14 possible arrangements, known as *Bravais lattices*.

spalling. The removal of small pieces of metal from the working face of the die, usually as a result of severe heat checking. Spalling is most likely in hard materials with low ductility.

spank. A press operation used to reform parts that have already had their major contour formed or drawn in the conventional manner. The spank operation is often used where it is not possible to produce the final contour, such as sharp creases or corners, in a single forming operation. It is also used at the end of a production line where large sheet metal parts have become distorted due to previous operations, such as trimming, punching, forming, and flanging. Spanking is used to

bring the panels back to the desired contour. See also *restriking*.

special boundary. A grain boundary between two grains whose crystallographic lattices have a certain fraction ($1/N$, in which N is an integer) of coincident lattice points. Such boundaries, denoted using the notation ΣN, may have low mobility and surface energy.

specific heat. Amount of heat required to change the temperature of one unit weight of a material by one degree.

specific modulus. The material elastic modulus divided by the material density.

specific strength. The material strength divided by the material density.

spheroidization. Process of converting a lamellar, basketweave, or acicular second phase into an equiaxed morphology via deformation, annealing, or a combination of deformation followed by annealing.

spinning. The forming of a seamless hollow metal part by forcing a rotating blank to conform to a shaped mandrel that rotates concentrically with the blank. In the typical application, a flat-rolled metal blank is forced against the mandrel by a blunt, rounded tool; however, other stock (notably, welded or seamless tubing) can be formed. A roller is sometimes used as the working end of the tool.

spinoidal hardening. Strengthening caused by the formation of a periodic array of coherent face-centered cubic solid-solution phases on a submicrostructural size level.

springback. The elastic recovery of metal after stressing. The extent to which metal tends to return to its original shape or contour after undergoing a forming operation. This is compensated for by overbending or by a secondary operation of restriking.

stacking-fault energy (SFE). The energy associated with the planar fault formed by dissociated dislocations in crystalline materials. Low-SFE materials typically have wide stacking faults, and high-SFE materials very narrow or no stacking faults. The SFE affects a number of material properties, such as work-hardening rate and recrystallization. Materials with low SFE undergo rapid dislocation multiplication and hence show high work-hardening rates and relative ease of dynamic recrystallization because of the difficulty of dynamic recovery. Materials with high SFE energies usually exhibit low work-hardening rates because of the ease of dynamic recovery and are difficult to recrystallize.

stamping. A general term to denote all press-working. In a more specific sense, stamping is used to imprint letters, numerals, and trademarks in sheet metal, machined parts, forgings, and castings. A tool called a stamp, with the letter or number raised on its surface, is hammered or forced into the metal, leaving a depression on the surface in the form of the letter or number.

standard deviation. A measure of the dispersion of observed values or results from the average, expressed as the positive square root of the variance.

static friction. The force tangential to the interface that is just sufficient to initiate relative motion between two bodies under load.

sticker breaks. Repetitive transverse lines, often curved, caused by localized welding of coil wraps during annealing and subsequent separation of these welded areas during an uncoiling operation. Also referred to as sticker marks.

stiffness. Resistance to elastic deformation.

stochastic search methods. A large group of optimization techniques that uses probabilistic methods. Two common methods are genetic algorithms and simulated annealing.

stock. A general term used to refer to a supply of metal in any form or shape and also to an individual piece of metal that is formed, forged, or machined to make parts.

stoichiometric. Having the precise weight relation of the elements in a chemical compound, or quantities of reacting elements or compounds being in the same weight relation as the theoretical combining weight of the elements involved.

straightening. A finishing operation for correcting misalignment in a forging or between various sections of a forging.

strain. The unit of change in the size or shape of a body due to force, in reference to its original size or shape.

strain aging. The changes in ductility, hardness, yield point, and tensile strength that occur when a metal or alloy that has been cold worked is stored for some time. In steel, strain aging is characterized by a loss of ductility and a corresponding increase in hardness, yield point, and tensile strength.

strain hardening. An increase in hardness and strength caused by plastic deformation at temperatures below the recrystallization range. Also known as work hardening.

strain-hardening coefficient or exponent. The value n in the relationship $\sigma = K\varepsilon^n$, where σ is the true stress; ε is the true strain; and K, which is called the strength coefficient, is equal to the true stress at a true strain of 1.0. The strain-hardening exponent, also called n-value, is equal to the slope of the true stress/true strain curve up to maximum load, when plotted on log-log coordinates. The n-value relates to the ability of a sheet material to be stretched in metalworking operations. The higher the n-value, the better the formability (stretchability). Also called work-hardening exponent.

strain lines. Surface defects in the form of shallow line-type depressions appearing in sheet metals after stretching the surface a few percent of unit area or length. See also *Lüders lines*.

strain rate. The time rate of deformation (strain) during a metalforming process.

strain-rate sensitivity. The degree to which mechanical properties are affected by changes in deformation rate. Quantified by the slope of a log-log plot of flow stress (at fixed strain and temperature) versus strain rate. Also known as the m-value.

strength. The ability of a material to withstand an applied force.

strength coefficient (K). A constant related to the tensile strength used in the power-law equation $\sigma = K\varepsilon^n$. In mechanical engineering nomenclature, it is called so, and the power-law equation is given as $\sigma = \sigma_o\varepsilon^n$. See also n-value.

stress. The intensity of the internally distributed forces or components of forces that resist a change in the volume or shape of a material that is or has been subjected to external forces. Stress is expressed in force per unit area. Stress can be normal (tension or compression) or shear.

stress concentration. On a macromechanical level, the magnification of the level of an applied stress in the region of a notch, void, hole, or inclusion.

stress-concentration factor (K_t). A multiplying factor for applied stress that allows for the presence of a structural discontinuity such as a notch or hole; K_t equals the ratio of the greatest stress in the region of the discontinuity to the nominal stress for the entire section. Also called theoretical stress-concentration factor.

stress-intensity factor. A scaling factor, usually denoted by the symbol K, used in linear elastic fracture mechanics to describe the intensification of applied stress at the tip of a crack of known size and shape. At the onset of rapid crack propagation in any structure containing a crack, the factor is called the critical stress-intensity factor, or the fracture toughness. Various subscripts are used to denote different loading conditions or fracture toughnesses.

stress raisers. Design features (such as sharp corners) or mechanical defects (such as notches) that act to intensify the stress at these locations.

stress relaxation. Drop in stress with time when material is maintained at a constant strain. The drop in stress is a result of plastic accommodation processes.

stress relief. The removal or reduction of residual stress by thermal treatment, mechanical treatment (shot peening, surface rolling, stretching, bending, and straightening), or vibratory stress relief.

stress-strain curve. A graph in which corresponding values of stress and strain from a tension, compression, or torsion test are plotted against each other. Values of stress are usually plotted vertically (ordinates or y-axis) and values of strain horizontally (abscissas or x-axis). Also known as stress-strain diagram.

stretchability. The ability of a material to undergo stretch-type deformation.

stretcher leveling. The leveling of a piece of sheet metal (that is, removing warp and distortion) by gripping it at both ends and subjecting it to a stress higher than its yield strength.

stretcher straightening. A process for straightening rod, tubing, and shapes by the application of tension at the ends of the stock. The products are elongated a definite amount to remove warpage.

stretcher strains. Elongated markings that appear on the surface of some sheet materials when deformed just past the yield point. These markings lie approximately parallel to the direction of maximum shear stress and are the result of localized yielding. See also *Lüders lines*.

stretch flanging. The stretching of the length of the free edge after the flanging process.

stretch forming. The shaping of a sheet or part, usually of uniform cross section, by first applying suitable tension or stretch and then wrapping it around a die of the desired shape.

stretching. The mode of deformation in which a positive strain is generated on the sheet surface by the application of a tensile stress. In stretching, the flange of the flat blank is securely clamped. Deformation is restricted to the area initially within the die. The stretching limit is the onset of metal failure.

striking surface. Those areas on the faces of a set of dies that are designed to meet when the upper die and lower die are brought together. The striking surface helps protect impressions from impact shock and aids in maintaining longer die life.

strip. A flat-rolled metal product of some maximum thickness and width, arbitrarily dependent on the type of metal; narrower than *sheet*.

stripping. The removal of the metal strip from the punch after a cutting operation. Also a term referring to the removal of a part adhering to the punch on the upstroke after forming.

subcritical crack growth (SCG). A failure process in which a crack initiates at a preexisting flaw and grows until it attains a critical length. At that point, the crack grows in an unstable fashion, leading to catastrophic failure. Typical examples of SCG processes are fatigue failure and stress corrosion.

subgrain. Micron-sized volume bounded by well-defined dislocation walls. The misorientations across the walls are low angle in nature, that is, $<15°$.

subpress die. A die that is closed by the press ram but opened by springs or other means because the upper shoe is not attached to the ram.

subsow block (die holder). A block used as an adapter in order to permit the use of forging dies that otherwise would not have sufficient height to be used in the particular unit or to permit the use of dies in a unit with different *shank* sizes.

superplastic forming. Forming using the superplasticity properties of material at elevated temperatures.

superplastic forming and diffusion bonding. The process of combining the diffusion bonding cycle into the superplastic forming.

superplasticity. The ability of certain metals to develop extremely high tensile elongations at elevated temperatures and under controlled rates of deformation. Materials that show high strain-rate sensitivity (≥ 0.5) at deformation temperatures often exhibit superplasticity. The phenomenon is often developed through a mechanism of grain-boundary sliding in very fine-grained, two-phase alloys.

support plate. A plate that supports a draw ring or draw plate. It also serves as a spacer.

surface finish. The classification of a surface in terms of roughness and peak density.

surface hardening. A localized heat treating process that produces a hard-quenched surface in steel without introducing additional alloying elements. Surface hardening can be produced by flame, induction, or laser or electron beam thermal treatments.

surface hardness. The hardness of that portion of the material very near the surface, as measured by microhardness or superficial hardness testers.

surface oxidation. Development of an oxide film or layer on the surface of metals in oxidizing environments. Oxidation at high temperatures is occasionally referred to as scaling.

surface roughness. The fine irregularities in the surface texture that result from the production process. Considered as vertical deviations from the nominal or average plane of the surface.

surface texture. Repetitive or random deviations from the nominal surface that form the pattern of the surface. Includes roughness, waviness, and flaws.

surface topography. The fine-scale features of a surface as defined by the size and distribution of asperities. Surface topography is measured by surface roughness and the direction of surface features (lay).

swage. (1) The operation of reducing or changing the cross-sectional area of stock by the fast impact of revolving dies. (2) The tapering of bar, rod, wire, or tubing by forging, hammering, or squeezing; reducing a section by progressively tapering lengthwise until the entire section attains the smaller dimension of the taper.

sweep device. A single or double arm (rod) attached to the upper die or slide of a press and intended to move the operator's hands to a safe position as the dies close, if the operator's hands are inadvertently within the point of operation.

Swift cup test. A simulative test in which circular blanks of various diameter are clamped in a die ring and deep drawn into a cup by a flat-bottomed cylindrical punch. The ratio of the largest blank diameter that can be drawn successfully to the cup diameter is known as the *limiting drawing ratio or deformation limit*.

T

Taguchi method. A technique for designing and performing experiments to investigate processes in which the output depends on many factors (e.g., material properties, process parameters) without having to tediously and uneconomically run the process using all possible combinations of values of those variables. By systematically choosing certain combinations of variables, it is possible to separate their individual effects.

tailor-welded blank. Blank for sheet forming typically consisting of steels of different thickness, grades/strengths, and sometimes coatings that are welded together prior to forming. Tailor-welded blanks are used to make finished parts with a desirable variation in properties such as strength, corrosion resistance, and so on.

tangent bending. The forming of one or more identical bends having parallel axes by wiping sheet metal around one or more radius dies in a single operation. The sheet, which may have side flanges, is clamped against the radius die and then made to conform to the radius die by pressure from a rocker-plate die that moves along the periphery of the radius die. See also *wiper forming (wiping)*.

tangent modulus. The slope of the stress-strain curve at any specified stress or strain. See also *modulus of elasticity*.

Taylor factor. The ratio of the required stress for deformation under a specified strain state to the critical resolved shear stress for slip (or twinning) within the crystals comprising a *polycrystalline aggregate*. The determination of the Taylor factor assumes uniform and identical strain within each crystal in the aggregate and provides an upper bound on the required stresses. The Taylor factor averaged over all crystals in a polycrystalline aggregate ($= \overline{M}$) provides an estimate of the effect of texture on strength.

tearing. Failure and localized separation of a sheet metal.

temper. In nonferrous alloys and in some ferrous alloys (steels that cannot be hardened by heat treatment), the hardness and strength produced by mechanical or thermal treatment, or both, and characterized by a certain structure, mechanical properties, or reduction in area during cold working.

temperature-compensated strain rate. Parameter used to describe the interdependence of temperature and strain rate in the description of thermally activated (diffusion-like) deformation processes. It is defined as $\dot{\varepsilon}$ $\exp(Q/RT)$, in which $\dot{\varepsilon}$ denotes the strain rate, Q is an apparent activation energy characterizing the micromechanism of deformation, R is the gas constant, and T is the absolute temperature. Flow stress, dynamic recrystallization, and so on at various strain rates and temperatures are frequently interpreted in terms of the temperature-compensated strain rate. Also known as the *Zener-Hollomon parameter (Z)*.

tempering. (1) In heat treatment, reheating hardened steel to some temperature below the eutectoid temperature to decrease hardness and/or increase toughness. (2) The process of rapidly cooling glass from near its softening point to induce compressive stresses on the surface balanced by interior tension, thereby imparting increased strength.

template (templet). A gage or pattern made in a die department, usually from sheet steel; used to check dimensions on forgings and as an aid in sinking die impressions in order to correct dimensions.

tensile ratio. The ratio of the tensile strength to yield strength. It is the inverse of the yield ratio.

tensile strength. In tensile testing, the ratio of maximum load to original cross-sectional area. Also known as ultimate strength. Compare with *yield strength*.

tensile stress. A stress that causes two parts of a body to pull apart. Contrast with *compressive stress*.

tension. The force or load that produces elongation.

tensor order. A measure of the number of directional dimensions associated with a quantity. A scalar, for example, has a tensor order of 0, indicating that it has no directionality associated with it. A vector, having a single direction, is a quantity with a tensor order of 1. The stress tensor has an order of 2. Higher-order tensors exist with a number of directions equal to their tensor order.

tetrahedral element. The element for three-dimensional finite-element modeling that is tetrahedron shaped (four faces) and has four nodes; also called the tet element.

texture. The description of the relative probability of finding the crystals comprising a polycrystalline aggregate in various orientations.

thermal conductivity. Ability of a material to conduct heat. The rate of heat flow, under steady conditions, through unit area, per unit temperature gradient in the direction perpendicular to the area. It is given in SI units as watts per meter Kelvin (W/m K); in customary units as (Btu/ft^2 °F).

thermal fatigue. Fracture resulting from the presence of temperature gradients that vary with time in such a manner as to produce cyclic stresses in a structure.

thermocouple. A device for measuring temperature, consisting of two dissimilar metals that produce an electromotive force roughly proportional to the temperature difference between their hot and cold junction ends.

thermomechanical processing (TMP). A general term covering a variety of processes combining controlled thermal and deformation treatments to obtain synergistic effects, such as improvement in strength without loss of toughness. Same as thermal-mechanical treatment.

thick-film lubrication. A condition of lubrication in which the film thickness of the lubricant is appreciably greater than that required to cover the surface asperities when subjected to the operating load. See also *thin-film lubrication*.

thin-film lubrication. A condition of lubrication in which the film thickness of the lubricant is such that the friction and wear between the surfaces is determined by the properties of the surfaces as well as the characteristics of the lubricant.

three-point bending. The bending of a piece of metal or a structural member in which the object is placed across two supports and force is applied between and in opposition to them.

tilt boundary. Grain boundary for which the crystal lattices of the grains on either side of the boundary are related by a rotation about an axis that lies in the plane of the boundary.

torsion. A twisting deformation of a solid or tubular body about an axis in which lines that were initially parallel to the axis become helices.

torsional stress. *The shear stress* on a transverse cross section resulting from a twisting action.

total elongation. The total amount of permanent extension of a testpiece broken in a tensile test; usually expressed as a percentage over a fixed gage length. See also *elongation, percent*.

toughness. The ability of a material to resist an impact load (high strain rate) or to deform under such a load in a ductile manner, absorbing a large amount of the impact energy and deforming plastically before fracturing. Such impact toughness is frequently evaluated with Charpy or Izod notched impact specimens. Impact toughness is measured in terms of the energy absorbed during fracture. Fracture toughness is a measure of the ability of a material to withstand fracture in the presence of flaws under static or dynamic loading of various types (tensile, shear, etc.). An indicator of damage tolerance, fracture toughness is measured in terms of Mpa$\sqrt{m}$ or ksi$\sqrt{in.}$

transformation-induced plasticity (TRIP). A phenomenon occurring chiefly in certain highly alloyed steels that have been heat treated to produce metastable austenite or metastable austenite plus martensite, whereby, on subsequent deformation, part of the austenite undergoes strain-induced transformation to martensite. Steels capable of transforming in this manner, commonly referred to as TRIP steels, are highly plastic after heat treatment but exhibit a very high rate of strain hardening and thus have high tensile and yield strengths after plastic deformation at temperatures between 20 and 500 °C (70 and 930 °F). Cooling to –195 °C (–320 °F) may or may not be required to complete the transformation to martensite. Tempering usually is done following transformation.

transformation temperature. The temperature at which a change in phase occurs. This term is sometimes used to denote the limiting temperature of a transformation range.

transition temperature. (1) An arbitrarily defined temperature that lies within the temperature range in which metal fracture characteristics (as usually determined by tests of notched specimens) change rapidly, such as the ductile-to-brittle transition temperature (DBTT). The DBTT can be assessed in several ways, the most common being the temperature for which the structure is 50% ductile. The DBTT is commonly associated with temper embrittlement and radiation damage (neutron irradiation) of low-alloy steels. (2) Sometimes used to denote an arbitrarily defined temperature within a range in which the ductility changes rapidly with temperature.

Tresca yield criterion. Prediction of yielding in ductile materials when the maximum shear stress on any plane reaches a critical value, $\tau = \tau_c$.

triaxiality. The ratio of the hydrostatic (mean) stress to the flow (effective) stress. Triaxiality provides a measure of the tendency for cavities to grow during deformation processing.

trimming. The mechanical shearing of flash or excess material from a forging with a trimmer in a trim press; can be done hot or cold.

triple junction/triple point. Point at which three grains meet in a polycrystalline aggregate. Also, region in which high stress concentrations may develop during hot working or elevated-temperature service, thus nucleating wedge cracking.

true strain. (1) The ratio of the change in dimension, resulting from a given load increment, to the magnitude of the dimension immediately prior to applying the load increment. (2) In a body subjected to axial force, the natural logarithm of the ratio of the gage length at the moment of observation to the original gage length. Also known as natural strain.

true stress. The value obtained by dividing the load applied to a member at a given instant by the cross-sectional area over which it acts.

tube stock. A semifinished tube suitable for subsequent reduction and finishing.

twinning. Also called deformation or mechanical twinning, it is a deformation mechanism, similar to dislocation slip, in which small (often plate- or lens-shaped) regions of a crystal or grain reorient crystallographically to adopt a twin relationship to the parent crystal. It is particularly common in noncubic metals (e.g., alpha-titanium and tetragonal tin) and in many body-centered cubic metals deformed at high rates and/or low temperatures. Twinning is often accompanied by an audible crackling sound, from which "crying tin" gets its name.

twist boundary. Grain boundary for which the crystal lattices of the grains on either side of the boundary are related by a rotation about an axis that lies perpendicular to the plane of the boundary.

TZM. A high-creep-strength titanium, zirconium, and molybdenum alloy used to make dies for the isothermal forging process.

U

ultimate strength. The maximum stress (tensile, compressive, or shear) a material can sustain without fracture; determined by dividing maximum load by the original cross-sectional area of the specimen. Also known as nominal strength or maximum strength.

ultrasonic inspection. The use of high-frequency acoustical signals for the purpose of nondestructively locating flaws within raw material or finished parts.

uncertainty. Uncertainty refers to unpredictability, indeterminacy, indefiniteness, or the estimated percentage by which an observed or calculated value may differ from the true value.

underfill. A portion of a forging that has insufficient metal to give it the true shape of the impression.

uniform elongation (e$_u$). The elongation that occurs at maximum load and immediately preceding the onset of necking in a tensile test.

upset. The localized increase in cross-sectional area of a workpiece or weldment resulting from the application of pressure during mechanical fabrication or welding.

upset forging. A forging obtained by *upset* of a suitable length of bar, billet, or bloom.

upsetter. A horizontal mechanical press used to make parts from bar stock or tubing by *upset forging*, piercing, bending, or otherwise forming in dies. Also known as a header.

upsetting. The working of metal so that the cross-sectional area of a portion or all of the stock is increased. See also *heading*.

V

vacuum forming. Sheetforming process most commonly used for titanium in which a blank is placed into a chamber that has a heated die, and a vacuum is applied to creep form the part onto the die. The part is usually covered with an insulating material, and the bag is outside this material.

validation. The process of substantiating that material property test data have been generated according to standard methods and practices, or other indices of quality, reliability, and precision. The validation process is the first step toward ratification or confirmation of the data, making them legally effective and binding in some specified application.

variability of data. The degree to which random variables deviate from a central value or mean. In statistical terms, this is measured by the sample standard deviation or sample variance.

variance. A measure of the squared dispersion of observed values or measurements expressed as a function of the sum of the squared deviations from the population mean or sample average. Adapted from Ref 4

vent. A small hole in a punch or die for admitting air to avoid suction holding or for relieving pockets of trapped air that would prevent die closure or action.

vent mark. A small protrusion resulting from the entrance of metal into die vent holes.

viscoelasticity. A property involving a combination of elastic and viscous behavior that makes deformation dependent on both temperature and strain rate. A material having this property is considered to combine the features of a perfectly elastic solid and a perfect fluid.

viscosity. Bulk property of a fluid or semifluid that causes it to resist flow.

visioplasticity. A physical-modeling technique in which an inexpensive, easy-to-deform material (e.g., clay, wax, lead) is gridded and deformed in subscale tooling to establish the effects of die design, lubrication, and so forth on metal flow and defect formation by way of postdeformation examination of grid distortions.

W

warm working. Deformation at elevated temperatures below the recrystallization temperature. The flow stress and rate of strain hardening are reduced with increasing temperature; therefore, lower forces are required than in cold working. See also *cold working and hot working.*

wear plates. Replaceable elements used to face wearing surfaces on a hydraulic press.

wear resistance. Resistance of a sheet metal to surface abrasion. See also *erosion resistance.*

web. A relatively flat, thin portion of a forging that effects an interconnection between ribs and bosses; a panel or wall that is generally parallel to the forging plane.

wedge compression test. A simple workability test in which a wedge-shaped specimen is compressed to a certain thickness. This gives a gradient specimen in which material has been subjected to a range of plastic strains.

Weibull distribution. The formula for the probability density function of the general Weibull distribution is:

$$f(x) = \frac{\gamma}{\alpha}\left(\frac{x-\mu}{\alpha}\right)^{(\gamma-1)} \exp(-((x-\mu)/\alpha)^\gamma)$$
$$x \geq \mu; \gamma, \alpha > 0$$

where γ is the shape parameter, μ is the location parameter, and α is the scale parameter. The case where $\mu = 0$ and $\alpha = 1$ is called the standard Weibull distribution. The case where $\mu = 0$ is called the two-parameter Weibull distribution. The equation for the standard Weibull distribution reduces to:

$$f(x) = \gamma x^{(\gamma-1)} \exp(-(x^\gamma)) \quad x \geq 0; \gamma > 0$$

The Weibull distribution is used extensively in reliability applications to model failure times. Adapted from Ref 1

Widmanstätten structure. Characteristic structure produced when preferred planes and directions in the parent phase are favored for growth of a second phase, resulting in the precipitated second phase appearing as plates, needles, or rods within a matrix.

wiper forming (wiping). Method of curving sheet metal sections or tubing over a form block or die in which this form block is rotated relative to a wiper block or slide block.

wire. A thin, flexible, continuous length of metal, usually of circular cross section and usually produced by drawing through a die.

wire drawing. Reducing the cross section of wire by pulling it through a die.

wire drawing test. A test in which a cylindrical draw die is used to reduce the diameter of wire. The drawing force is measured and reflects lubricant effectiveness.

wire rod. Hot rolled coiled stock that is to be cold drawn into wire.

workability. See also *formability*, which is a term more often applied to sheet materials. The ease with which a material can be shaped through plastic deformation in bulk forming processes. It involves both the measurement of the resistance to deformation (the flow properties) and the extent of possible plastic deformation before fracture occurs (ductility).

work hardening. See strain hardening.

work-hardening exponent. See strain-hardening exponent.

workpiece. General term for the work material in a metalforming operation.

wrap forming. See *stretch forming.*

wrinkling. A wavy condition obtained in deep drawing of sheet metal in the area of the metal between the edge of the flange and the draw radius. Wrinkling may also occur in other forming operations when unbalanced compressive forces are set up.

wrought material. Material that is processed by plastic deformation, typically to produce a recrystallized microstructure. Cast and wrought materials are produced by ingot casting and deformation processes to produce final mill products.

Y

yield. Evidence of plastic deformation in structural materials. Also known as plastic flow or creep.

yield point. The first stress in a material, usually less than the maximum attainable stress, at which an increase in strain occurs without an increase in stress. Only certain metals—those that exhibit a localized, heterogeneous type of transition from elastic to plastic deformation—produce a yield point. If there is a decrease in stress after yielding, a distinction can be made between upper and lower yield points. The load at which a sudden drop in the flow curve occurs is called the upper yield point. The constant load shown on the flow curve is the lower yield point.

yield point elongation. The extension associated with discontinuous yielding that occurs at approximately constant load following the onset of plastic flow. It is associated with the propagation of Lüder's lines or bands.

yield ratio. The ratio of the yield strength to the tensile strength. It is the inverse of the tensile ratio.

yield strength. The stress at which a material exhibits a specified deviation from proportionality of stress and strain. An offset of 0.2% is used for many metals. Compare with *tensile strength.*

yield stress. The stress at which a material exhibits the first measurable permanent plastic deformation.

Young's modulus. A measure of the rigidity of a metal. It is the ratio of stress, within the proportional limit, to corresponding strain. Young's modulus specifically is the modulus obtained in tension or compression.

Z

Zener-Hollomon parameter (Z). See *temperature-compensated strain rate.*

REFERENCES

1. *NIST/SEMATECH e-Handbook of Statistical Methods,* http://www.itl.nist.gov/div898/handbook/, accessed March 2010
2. Dictionary.com, www.dictionary.com, accessed March 31, 2010
3. Wikipedia, www.wikipedia.org

4. J.R. Davis, Ed., *ASM Materials Engineering Dictionary,* ASM International, 1992

SELECTED REFERENCES

- J.R. Davis, Ed., A*SM Materials Engineering Dictionary*, ASM International, 1992
- *DoD Modeling and Simulation (M&S) Glossary*, U.S. Department of Defense, (DoD 5000.59-M), Jan 1998
- *Introduction to Finite-Difference Methods for Numerical Fluid Dynamics,* Los Alamos National Laboratories, released to OSTI, 1996
- *Dictionary of Scientific and Technical Terms,* 5th ed., McGraw-Hill
- *Metallogragphy and Microstructures,* Vol 9, *ASM Handbook,* ASM International, 2004
- *Metalworking: Sheet Forming,* Vol 14B, ASM Handbook, ASM International, 2006
- *Materials Selection and Design*, Vol 20, *ASM Handbook,* ASM International, 1997

Index